Lecture Notes in Computer Science 2130

Edited by G. Goos, J. Hartmanis and J. van Leeuwen

Springer
Berlin
Heidelberg
New York
Barcelona
Hong Kong
London
Milan
Paris
Tokyo

Georg Dorffner Horst Bischof
Kurt Hornik (Eds.)

Artificial Neural Networks – ICANN 2001

International Conference
Vienna, Austria, August 21-25, 2001
Proceedings

Springer

Series Editors

Gerhard Goos, Karlsruhe University, Germany
Juris Hartmanis, Cornell University, NY, USA
Jan van Leeuwen, Utrecht University, The Netherlands

Volume Editors

Georg Dorffner
University of Vienna, Dept. of Mecidal Cybernetics and Artificial Intelligence
Freyung 6/2, 1010 Vienna, Austria
E-mail: georg@ai.univie.ac.at

Horst Bischof
Technical University of Vienna, Institute for Computer Aided Automation
Pattern Recognition and Image Processing Group
Favoritenstr. 9/1832, 1040 Vienna, Austria
E-mail: bis@prip.tuwien.ac.at

Kurt Hornik
Wirtschaftsuniversität Wien, Institut für Statistik
Augasse 2-6, 1090 Wien, Austria
E-mail: Kurt.Hornik@ci.tuwien.ac.at

Cataloging-in-Publication Data applied for

Die Deutsche Bibliothek - CIP-Einheitsaufnahme

Artificial neural networks : international conference ; proceedings / ICANN 2001, Vienna, Austria, August 21 - 25, 2001. Georg Dorffner ... (ed.). - Berlin ; Heidelberg ; New York ; Barcelona ; Hong Kong ; London ; Milan ; Paris ; Singapore ; Tokyo : Springer, 2001
(Lecture notes in computer science ; Vol. 2130)
ISBN 3-540-42486-5

CR Subject Classification (1998): F.1, I.2, I.5, I.4, G.3, J.3, C.2.1, C.1.3

ISSN 0302-9743
ISBN 3-540-42486-5 Springer-Verlag Berlin Heidelberg New York

Springer-Verlag Berlin Heidelberg New York
a member of BertelsmannSpringer Science+Business Media GmbH

http://www.springer.de

Typesetting: Camera-ready by author, data conversion by Olgun Computergrafik
Printed on acid-free paper SPIN 10840062 06/3142 5 4 3 2 1 0

Preface

This book is based on the papers presented at the *International Conference on Artificial Neural Networks*, ICANN 2001, from August 21–25, 2001 at the Vienna University of Technology, Austria. The conference is organized by the Austrian Research Institute for Artifical Intelligence in cooperation with the Pattern Recognition and Image Processing Group and the Center for Computational Intelligence at the Vienna University of Technology. The ICANN conferences were initiated in 1991 and have become the major European meeting in the field of neural networks.

From about 300 submitted papers, the program committee selected 171 for publication. Each paper has been reviewed by three program committee members/reviewers. We would like to thank all the members of the program committee and the reviewers for their great effort in the reviewing process and helping us to set up a scientific program of high quality. In addition, we have invited eight speakers; three of their papers are also included in the proceedings.

We would like to thank the European Neural Network Society (ENNS) for their support. We acknowledge the financial support of Austrian Airlines, Austrian Science Foundation (FWF) under the contract SFB 010, Austrian Society for Artificial Intelligence (ÖGAI), Bank Austria, and the Vienna Convention Bureau. We would like to express our sincere thanks to A. Flexer, W. Horn, K. Hraby, F. Leisch, C. Schittenkopf, and A. Weingessel. The conference and the proceedings would not have been possible without their enormous contribution.

Vienna, June 2001

Georg Dorffner
Horst Bischof
Kurt Hornik
Program Co-Chairs
ICANN 2001

Organization

ICANN 2001 is organized by the Austrian Research Institute for Artificial Intelligence in cooperation with the Pattern Recognition and Image Processing Group and the Center for Computational Intelligence at the Vienna University of Technology.

Executive Committee

General Chair:	Georg Dorffner, Austria
Program Co-Chairs:	Horst Bischof, Austria
	Kurt Hornik, Austria
Organizing Committee:	Arthur Flexer, Austria
	Werner Horn, Austria
	Karin Hraby, Austria
	Friedrich Leisch, Austria
	Andreas Weingessel, Austria
Workshop Chair:	Christian Schittenkopf, Austria

Program Committee

Shun-Ichi Amari, Japan
Peter Auer, Austria
Pierre Baldi, USA
Peter Bartlett, Australia
Shai Ben David, Israel
Matthias Budil, Austria
Joachim Buhmann, Germany
Rodney Cotterill, Denmark
Gary Cottrell, USA
Kostas Diamantaras, Greece
Wlodek Duch, Poland
Peter Erdi, Hungary
Patrick Gallinari, France
Wulfram Gerstner, Switzerland
Stan Gielen, The Netherlands
Stefanos Kollias, Greece
Vera Kurkova, Czech Republic
Anders Lansner, Sweden
Aleš Leonardis, Slovenia
Hans-Peter Mallot, Germany
Christoph von der Malsburg, Germany
Klaus-Robert Müller, Germany
Thomas Natschläger, Austria
Lars Niklasson, Sweden
Stefano Nolfi, Italy
Erkki Oja, Finland
Gert Pfurtscheller, Austria
Ulrich Ramacher, Germany
Stephen Roberts, UK
Mariagiovanna Sami, Italy
Jürgen Schmidhuber, Switzerland
Olli Simula, Finland
Peter Sincak, Slovakia
John Taylor, UK
Carme Torras, Spain
Volker Tresp, Germany
Michel Verleysen, Belgium

Additional Referees

Panagiotis Adamidis
Esa Alhoniemi
Gabriela Andrejkova
Matthias Bethge
Leon Bobrowski
Mikael Bodén
Sander Bohte
Roman Borisyuk
Ronald Bormann
Leon Bottou
Raffaele Calabretta
Angelo Cangelosi
Cristiano Cervellera
Eleni Charou
Rizwan Choudrey
Paolo Coletta
Marie Cottrell
Aaron D'Souza
Eric de Bodt
Daniele Denaro
Andrea Di Ferdinando
Markus Diesmann
Hubert Dinse
Sara Dolnicar
José Dorronsoro
Douglas Eck
Julian Eggert
Örjan Ekeberg
Udo Ernst
Christian Eurich
Richard Everson
Sergio Exel
Attila Fazekas
Arthur Flexer
Mikel L. Forcada
Felipe M. G. Franca
Paolo Frasconi
Frederick Garcia
Philippe Gaussier
Apostolos Georgakis
Felix Gers
Marc-Oliver Gewaltig
Zoubin Ghahramani
Rémi Gilleron
Michele Giugliano
Christian Goerick
Mirta Gordon
Bernhard Graimann
Ying Guo
John Hallam
Inman Harvey
Robert Haschke
Rolf Henkel
Tom Heskes
Sepp Hochreiter
Jaakko Hollmen
Anders Holst
Dirk Husmeier
Marcus Hutter
Ari Hämäläinen
Robert Jacobs
Thomas M. Jorgensen
Christian Jutten
Paul Kainen
Samuel Kaski
Yakov Kazanovich
Richard Kempter
Werner M. Kistler
Jens Kohlmorgen
Joost N. Kok
Jeanette H. Kotaleski
Jerzy Korczak
Constantine Kotropoulos
Stanislav Kovacic
Stefan Kozak
Brigitte Krenn
Malgorzata Kretowska
Norbert Krüger
Ben Kröse
Vlado Kvasnicka
Ivo Kwee
Peter König
Jorma Laaksonen
Krista Lagus
Wee Sun Lee
Charles-Albert Lehalle
Fritz Leisch
Uros Lotric
Dominique Martinez
Peter Meinicke
Thomas Melzer
Risto Miikkulainen
Sebastian Mika
José del R. Millán
Igor Mokris
Noboru Murata
Jean-Pierre Nadal
Hiroyuki Nakahara
Ralph Neuneier
Athanasios Nikolaidis
Nikos Nikolaidis
Klaus Obermayer
Vladimir Olej
Luigi Pagliarini
Domenico Parisi
Olivier Teytaud
Helene Paugam-Moisy
Stavros Perantonis
Conrad Perez
Markus Peura
Frederic Piat
Fernando Pineda
Gianluca Pollastri
Gunnar Rätsch
Kimmo Raivio
Erhard Rank
Carl Edward Rasmussen
Iead Rezek
Carlos Ribeiro
Helge Ritter
Tobias Rodemann
Raul Rojas
Agostinho Rosa
Volker Roth
Ulrich Rückert
Vicente Ruiz de Angulo
Pal Rujan
David Saad
Mariagiovanna Sami
Marcello Sanguineti
Petr Savicky
Christian Schittenkopf

Michael Schmitt
Bernhard Schölkopf
Nicol Schraudolph
René Schüffny
Walter Senn
Terezie Sidlofova
Jiři Sima
Zenon Sosnowski
Andreas Stafylopatis
Arnost Stedry
Branko Ster
Volker Steuber
Petr Stepan
Piotr Suffczynski
Johan Suykens
Peter Sykacek
Anastasios Tefas
Andreas Thiel
Peter Tino
Marc Tommasi
Mira Trebar
Edmondo Trentin
Sofia Tsekeridou
Koji Tsuda
Panagiotis Tzionas
Shiro Usui
Nikolaos Vassilas
Juha Vesanto
Ricardo Vigario
Sethu Vijayakumar
Alessandro Villa
Nikos Vlassis
Alpo Värri
Andreas Weingessel
Heiko Wersing
Nicholas Wickström
Wim Wiegerinck
Stefan Wilke
Laurenz Wiskott
Rolf P. Würtz
Tony Zador
Tom Ziemke

Sponsoring Institutions

- Austrian Airlines
- Austrian Research Institute for Artificial Intelligence
- Austrian Science Foundation (FWF) under the contract SFB 010
- Austrian Society for Artificial Intelligence (ÖGAI)
- Bank Austria
- Vienna Convention Bureau
- Vienna University of Technology

Table of Contents

Invited Papers

Data Analysis and Pattern Recognition

Theory

Kernel Methods

Topographic Mapping

Independent Component Analysis

Signal Processing

Time Series Processing

Special Session: Agent-Based Economic Modeling

Selforganization and Dynamical Systems

Robotics and Control

Vision and Image Processing

Computational Neuroscience

Part I

Invited Papers

The Complementary Brain (Abstract)

Stephen Grossberg

Center for Adaptive Systems and Department of Cognitive and Neural Systems
Boston University
677 Beacon Street, Boston, MA 02215, USA
steve@bu.edu

How are our brains functionally organized to achieve adaptive behavior in a changing world? This talk will survey evidence supporting a computational paradigm that radically departs from the computer metaphor suggesting that brains are organized into independent modules. Evidence is reviewed that brains are organized into parallel processing streams with complementary properties. This perspective clarifies, for example, how parallel processing in the brain leads to visual percepts and object recognition, and how perceptual mechanisms differ from those of movement control. Multiple modeling studies predict how the neocortex is organized into parallel processing streams such that pairs of streams obey complementary computational rules (as when two puzzle pieces fit together). Hierarchical interactions within each stream and parallel interactions between streams create coherent behavioral representations that overcome the complementary deficiencies of each stream and support unitary conscious experiences. For example, visual boundaries and surfaces seem to obey complementary rules in the Interblob and Blob streams from area V1 to V4. They interact to generate visible representations of 3D surfaces in which partially occluded objects are mutually separated and completed. Visual boundaries and motion in the Interblob and Magnocellular cortical processing streams seem to obey complementary computational rules in cortical areas V2 and MT. They interactively form representations of object motion in depth. Predictive target tracking (e.g., targets moving relative to a stationary observer) and optic flow navigation (e.g., an observer moving relative to its world) seem to obey complementary computational rules in ventral and dorsal MST. These regions interact to track moving targets, and to determine an observer's heading and time-to-contact. Spatially-invariant object recognition seems to obey complementary computational rules as compared to spatial representation and the control of action in Inferotemporal and Parietal cortex, where they, respectively, learn to stably recognize objects in a changing world, and rapidly relearn spatial and action parameters when motor parameters change. Their interaction enables spatially-invariant object recognition categories of valued objects (in the What stream) to direct spatial attention and actions (in the Where stream) towards these objects in space. These results help to quantitatively simulate many data and to make surprising predictions. They provide a more global perspective on how the brain controls

G. Dorffner, H. Bischof, and K. Hornik (Eds.): ICANN 2001, LNCS 2130, pp. 3–4, 2001.

behavior, and what are the brain's computational units that have behavioral significance, including conscious perceptual representations. In particular, the role of synchronous processing in binding complementary properties together, and how it may persist, collapse, and be reset in temporal cycles of coherent and incoherent processing, will be noted. These insights also provide new designs for technological applications wherein adaptive autonomous control in uncertain and unpredictable environments is needed.

Acknowledgement

Supported in part by AFOSR, DARPA, NSF, and ONR.

References

[Grossberg 2000] Grossberg S., The Complementary Brain: Unifying Brain Dynamics and Modularity, *Trends in Cognitive Sciences*, 4, 233-245, 2000.

Neural Networks for Adaptive Processing of Structured Data

Alessandro Sperduti

Dip. di Informatica, Università di Pisa, Corso Italia, 40, 56125 Pisa, Italy

Abstract. Structured domains are characterized by complex patterns which are usually represented as lists, trees, and graphs of variable sizes and complexity. The ability to recognize and classify these patterns is fundamental for several applications that use, generate or manipulate structures. In this paper I review some of the concepts underpinning Recursive Neural Networks, i.e. neural network models able to deal with data represented as directed acyclic graphs.

1 Introduction

The processing of structured data is usually confined to the domain of symbolic systems. Recently, however, there has been some effort in trying to extend the computational capabilities of neural networks to structured domains. While earlier neural approaches were able to deal with some aspects of processing of structured information, none of them established a practical and efficient way of dealing with structured information. A more powerful approach, at least for classification and prediction tasks, was proposed in [15] and further extended in [7]. In these works, Recursive Neural Networks, a generalization of recurrent neural networks for processing sequences to the case of directed graphs, were defined. These models are able to learn a mapping from a domain of ordered (or positional) directed acyclic graphs, with labels attached to each node, to the set of real numbers. The basic idea behind the models is the extension of the concept of *unfolding* from the domain of sequences to the domain of directed ordered graphs (DOGs).

In this paper, I briefly present some of the basic concepts underpinning Recursive Neural Networks. Some supervised and unsupervised models are presented, together with an outlook of the main computational, complexity, and leaning results obtained up to now.

The possibility of processing structured information using neural networks is appealing for several reasons. First of all, neural networks are universal approximators; in addition, they are able to learn from a set of examples and very often, by using the correct methodology for training, they are able to reach a quite high generalization performance. Finally, they are able to deal with noise and incomplete, or even ambiguous, data. All these capabilities are particularly useful when dealing with prediction tasks where data are usually gathered experimentally, and thus are partial, noisy, and incomplete. A typical example of

G. Dorffner, H. Bischof, and K. Hornik (Eds.): ICANN 2001, LNCS 2130, pp. 5–12, 2001.

such a domain is chemistry, where compounds can naturally be represented as labeled graphs.

2 Data Structures and Notation

In this paper we assume that instances in the learning domain are DOAGs (directed ordered acyclic graphs) or DPAGs (directed positional acyclic graphs). A DOAG is a DAG $\mathcal{D}$ with vertex set $\mathrm{vert}(\mathcal{D})$ and edge set $\mathrm{egd}(\mathcal{D})$, where for each vertex $v \in \mathrm{vert}(D)$ a total order on the edges leaving from v is defined. DPAGs are a superclass of DOAGs in which it is assumed that for each vertex v, a bijection $P : \mathrm{egd}(\mathcal{D}) \to I\!N$ is defined on the edges leaving from v. The *indegree* of node v is the number of incoming edges to v, whereas the *outdegree* of v is the number of outgoing edges from v.

We shall require the DAG (either DOAG or DPAG) to possess a supersource[1], i.e. a vertex $s \in \mathrm{vert}(\mathcal{D})$ such that every vertex in $\mathrm{vert}(\mathcal{D})$ can be reached by a directed path starting from s. Given a DAG $\mathcal{D}$ and $v \in \mathrm{vert}(\mathcal{D})$, we denote by $\mathrm{ch}[v]$ the set of children of v, and by $\mathrm{ch}_k[v]$ the k-th child of v.

We shall use lowercase bold letters to denote vectors, uppercase bold letters to denote matrices, and calligraphic letters for representing graphs. A data structure $\mathcal{Y}$ is a DAG whose vertices are labeled by vectors of real-valued numbers which either represent numerical or categorical variables. Subscript notation will be used when referencing the labels attached to vertices in a data structure. Hence $\boldsymbol{y}_v$ denotes the vector of variables labeling vertex $v \in \mathrm{vert}(\mathcal{Y})$.

In the following, we shall denote by $\#^{(i,c)}$ the class of DAGs with maximum indegree i and maximum outdegree c. A generic class of DAGs with bounded (but unspecified) indegree and outdegree, will simply be denoted by $\#$. The class of all data structures defined over the label universe domain $\boldsymbol{\mathcal{Y}}$ and skeleton in $\#^{(i,c)}$ will be denoted as $\boldsymbol{\mathcal{Y}}^{\#^{(i,c)}}$. The void DAG will be denoted by the special symbol ξ.

3 Recursive Neural Networks

Recursive Neural Networks are neural networks able to perform mappings from a set of labeled graphs to the set of real vectors. Specifically, the class of functions which can be realized by a recursive neural network can be characterized as the class of functional graph transductions $\mathcal{T} : \mathcal{I}^{\#} \to I\!R^k$, where $\mathcal{I} = I\!R^m$, which can be represented in the following form $\mathcal{T} = g \circ \widehat{\tau}$, where $\widehat{\tau} : \mathcal{I}^{\#} \to I\!R^n$ is the *encoding* (or *state transition*) function and $g : I\!R^n \to I\!R^k$ is the *output* function. Specifically, given a DOAG $\mathcal{D}$, $\widehat{\tau}$ is defined recursively as

$$\widehat{\tau}(\mathcal{D}) = \begin{cases} \mathbf{0} \text{ (the null vector in } I\!R^n) & \text{if } \mathcal{D} = \xi \\ \tau(\boldsymbol{y}_s, \widehat{\tau}(\mathcal{D}^{(1)}), \ldots, \widehat{\tau}(\mathcal{D}^{(c)})) & \text{otherwise} \end{cases} \tag{1}$$

[1] If no supersource is present, a new node connected with all the nodes of the graph with null *indegree* can be added.

where τ is defined as $\tau : I\!R^m \times \underbrace{I\!R^n \times \cdots \times I\!R^n}_{\text{c times}} \to I\!R^n$ where $I\!R^m$ denotes the label space, while the remaining domains represent the encoded subgraphs spaces up to the maximum out-degree of the input domain $\mathcal{I}^\#$, c is the maximum out-degree of DOAGs in $\mathcal{I}^\#$, $s = source(\mathcal{D})$, $\boldsymbol{y}_s$ is the label attached to the super-source of $\mathcal{D}$, and $\mathcal{D}^{(1)}, \ldots, \mathcal{D}^{(c)}$ are the subgraphs pointed by s. A typical neural realization for τ is

$$\tau(\boldsymbol{u}_v, \boldsymbol{x}^{(1)}, \ldots, \boldsymbol{x}^{(c)}) = \boldsymbol{F}(\boldsymbol{B}\boldsymbol{u}_v + \sum_{j=1}^{c} \boldsymbol{A}_j \boldsymbol{x}^{(j)} + \boldsymbol{\theta}), \tag{2}$$

where $\boldsymbol{F}_i(\boldsymbol{v}) = f(v_i)$ (sigmoidal function), $\boldsymbol{u}_v \in I\!R^m$ is a label, $\boldsymbol{\theta} \in I\!R^n$ is the bias vector, $\boldsymbol{B} \in I\!R^{m \times n}$ is the weight matrix associated with the label space, $\boldsymbol{x}^{(j)} \in I\!R^m$ are the vectorial codes obtained by the application of the encoding function $\hat{\tau}$ to the subgraphs $\mathcal{D}^{(j)}$ (i.e., $\boldsymbol{x}^{(j)} = \hat{\tau}(\mathcal{D}^{(j)})$), and $\boldsymbol{A}_j \in I\!R^{m \times m}$ is the weight matrix associated with the jth subgraph space. Concerning the output function g, it can be defined as a map $g : I\!R^n \to I\!R^k$, and in general, it is realized by a feed-forward network.

The encoding process of an input graph can be represented graphically by unfolding equation 1 through the input graph, and using equation 2, obtaining in this way the so called *encoding network*. An example of encoding network obtained by using two recursive neurons an a single output neuron is shown in Figure 1. The output of the encoding network depends on the values of the weights and it will used as a numerical vectorial code to represent the input graph. This code will then be further processed by the parametric function $g()$ (in the example, the single output neuron) so to obtain the desired regression (or classification) value for the specific input graph.

Given a cost function E and a training set $T = \{(\mathcal{D}_i, t_i)\}_{i=1,..,L}$, where for each data structure $\mathcal{D}_i$ a desired target value t_i is associated, using eq. 1, eq. 2, and the neural network implementing the output function g, for each $\mathcal{D}_i$ a feed-forward network can be generated and trained so to match the corresponding desired target value t_i. Since the weights are shared among all the generated feed-forward networks, the training converges to a set of weight values which reproduces the desired target value for each data structure in the training set.

For each $\mathcal{D}_l$, its vertexes are enumerated according to a chosen inverse topological order as $\tilde{v}_1, \tilde{v}_2, \ldots, \tilde{v}_{P_l}$. Moreover, let $\hat{v}_1, \hat{v}_2, \ldots, \hat{v}_h$ be the set of vertexes, belonging to any DAG in the training set, for which a target value is defined. For simplicity, here we assume that only the supersources have a target defined (thus $h = L$). $\boldsymbol{U} \doteq [\boldsymbol{U}_1, \cdots, \boldsymbol{U}_L] \in R^{m+1, P^*}$ collects all the labels of the data structures (including the bias components always to 1), where $P^* \doteq \sum_{l=1}^{L} P_l$. Similarly, for each graph $\mathcal{D}_l$, and each $k \in [1, \ldots, c]$, the matrices $\boldsymbol{X}_l^{(k)} \in I\!R^{n, P_l}$ and defined as $\boldsymbol{X}_l^{(k)} \doteq [\boldsymbol{x}_{l,ch_k[\tilde{v}_1]}, \cdots, \boldsymbol{x}_{l,ch_k[\tilde{v}_{P_l}]}]$, where $\boldsymbol{x}_{l,ch_k[\tilde{v}_i]} = \boldsymbol{x}_0$ if $ch_k[\tilde{v}_i] = \emptyset$, collect the status information for each pointer k. All the information concerning a pointer k is stored into matrices $\boldsymbol{X}^{(k)} \doteq \left[\boldsymbol{X}_1^{(k)}, \ldots, \boldsymbol{X}_L^{(k)}\right] \in I\!R^{n, P^*}$. Moreover, we define the matrix collecting the information needed to compute the output

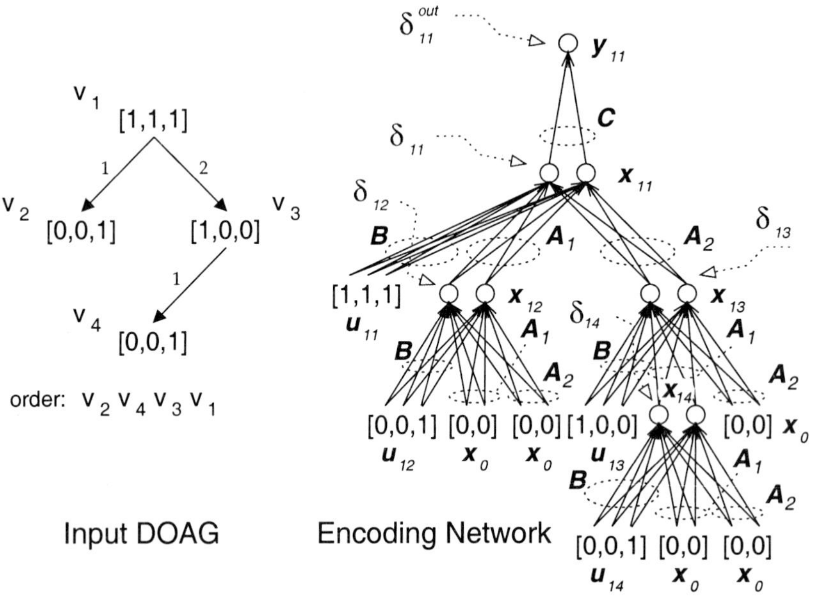

Fig. 1. Example of DOAG with corresponding encoding network.

associated to the training graphs as $\boldsymbol{X}^{target} = \begin{bmatrix} \boldsymbol{x}_{\hat{v}_1} & \boldsymbol{x}_{\hat{v}_2} & \dots & \boldsymbol{x}_{\hat{v}_L} \\ 1 & 1 & & 1 \end{bmatrix}$. Let us define $\delta_{ilv} \doteq \partial E_l / \partial net_{ilv}$. For node v of a given DOAG $\mathcal{D}_l$, the corresponding delta error δ_{ilv} for the state variables can be collected in vector $\boldsymbol{\delta}_{lv} \doteq [\delta_{1lv}, \dots, \delta_{nlv}]^t$ and the contributions from all the graph's nodes can be collected in matrix $\boldsymbol{\Delta}_l \doteq [\boldsymbol{\delta}_{l,1}, \dots, \boldsymbol{\delta}_{l,P_l}] \in I\!R^{n,P_l}$, where the order of the columns follows the inverse topological order chosen for the graph. Finally, $\boldsymbol{\Delta} \doteq [\boldsymbol{\Delta}_1, \dots, \boldsymbol{\Delta}_L] \in I\!R^{n,P^*}$ contains the delta errors for all the graphs of the learning environment, whereas the delta error corresponding to the output unit is denoted by $\delta^{out}_{ls} \doteq \frac{\partial E_l}{\partial net_{ls}}$ and collected into the matrix $\hat{\boldsymbol{\Delta}}$. The gradient of the cost can be calculated by using Backpropagation in each encoding network, that is by propagating the error through the given structure, similarly to what happens in recurrent networks when processing sequences. The gradient of the cost can be written in a compact form by using the vectorial notation $\boldsymbol{G}_B \doteq \left[\frac{\partial E}{\partial b_{ij}}\right] \in I\!R^{m+1,n}$ and $\mathbf{G}_{A_k} \doteq \left[\frac{\partial E}{\partial a_{ijk}}\right] \in I\!R^{n,n}$. Based on these definitions, $\boldsymbol{G}_B$ and $\boldsymbol{G}_{A_k}$ can be computed as follows

$$\boldsymbol{G}_B = \sum_{l=1}^{L} \boldsymbol{G}_{B,l} = \sum_{l=1}^{L} \sum_{v \in \mathrm{vert}(\mathcal{D}_l)} \boldsymbol{u}_{lv} \boldsymbol{\delta}^t_{lv} = \boldsymbol{U}\boldsymbol{\Delta}^t,$$

$$\boldsymbol{G}_{A_k} = \sum_{l=1}^{L} \boldsymbol{G}_{A_k,l} = \sum_{l=1}^{L} \sum_{v \in \mathrm{vert}(\mathcal{D}_l)} \boldsymbol{x}_{l,ch_k[v]}\, \boldsymbol{\delta}^t_{lv} = \boldsymbol{X}^{(k)} \boldsymbol{\Delta}^t, \tag{3}$$

where $\mathbf{G}_{B,l}$, $\mathbf{G}_{A_k,l}$ are the gradient contributions corresponding to E_l, that is to DOAG $\mathcal{D}_l$. Specifically, let $Q_k(v) \doteq \{u \mid ch_k[u] = v\}$. The *delta-error* δ_{ilv} can be computed recursively according to the following equation

$$\delta_{ilv} = \sigma'(net_{ilv}) \sum_{k=1}^{o} \sum_{j=1}^{n} a_{kji} \left(\sum_{z \in Q_k(v)} \delta_{jlz} \right) \tag{4}$$

where if $Q_k(v) = \emptyset$ then $\sum_{z \in Q_k(v)} \delta_{jlz} = 0$.

The above equation can be rewritten in compact form as

$$\boldsymbol{\delta}_{lv} = \boldsymbol{J}_{lv} \sum_{k=1}^{o} \boldsymbol{A}_k \left(\sum_{z \in Q_k(v)} \boldsymbol{\delta}_{lz} \right), \tag{5}$$

where $\boldsymbol{J}_{lv}$ is a diagonal matrix with elements $[\boldsymbol{J}_{lv}]_{ii} = \sigma'(net_{ilv})$. Moreover, by applying recursively equation (5), we obtain

$$\boldsymbol{\delta}_{lv} = \left(\sum_{p \in Paths_l(s,v)} \prod_{(u,ch_k[u]) \in p}^{(s \rightarrow v)_{left}} \boldsymbol{J}_{l,ch_k[u]} \boldsymbol{A}_k^t \right) \boldsymbol{\delta}_{ls} \tag{6}$$

where $Paths_l(s, v)$ is the set of paths in $\mathcal{D}_l$ from the supersource s to node v, and the product is left-hand starting from the supersource s and ending to node v. This equation gives rise to the *Back-Propagation Through Structure* ($BPTS$) gradient computational scheme [9, 15].

Following the same approach, it is not difficult to generalize any supervised learning algorithm, such as RTRL, to the treatment of structured data. Constructive algorithms, such as Recurrent Cascade-Correlation, can be generalized as well [15].

A common features of all the supervised models is that, assuming stationarity, causality, and discrete labels, the training set can be optimized so to reduce the computational complexity of training. The basic idea is to represent each subgraph in the training set only once: if two graphs share the same subgraph, this subgraph only needs to be represented once, as well as only once the state associated to it must be computed. Applying this basic idea, the training set can be collapsed into a single minimal DOAG (see Figure 2 for an example) in time which is $O(P^* \log P^*)$.

3.1 Unsupervised Learning: Self-Organizing Maps for Structured Data

A Self-Organizing Map model for processing structured data has been recently proposed [11]. This model hinges on the recursive encoding idea described in eq. 1. The aim of the SOM learning algorithm is to learn a *feature map* $\mathcal{M} : \boldsymbol{\mathcal{I}} \rightarrow \boldsymbol{\mathcal{A}}$ which given a vector in the spatially continuous input space $\boldsymbol{\mathcal{I}}$ returns a point in the spatially *discrete* output display space $\boldsymbol{\mathcal{A}}$. In fact, SOM is performing

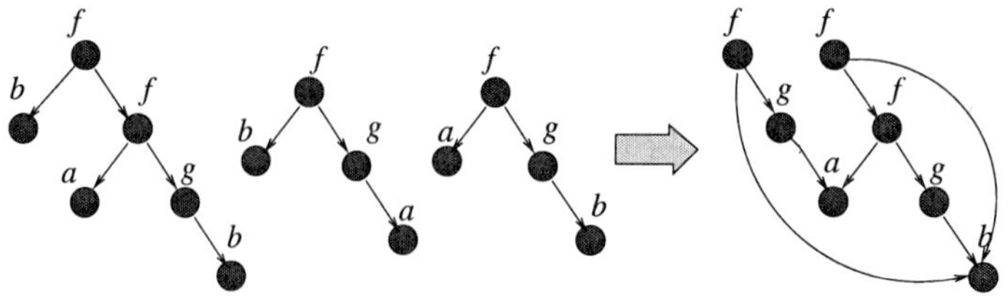

Fig. 2. Optimization of the training set.

data reduction via a vector quantization approach. When the input space is a structured domain with labels in $\mathcal{Y}$, i.e., $\mathcal{I} \equiv \mathcal{Y}^{\#^{(i,c)}}$, the map $\widehat{\mathcal{M}} : \mathcal{Y}^{\#^{(i,c)}} \to \mathcal{A}$ can be realized by the following recursive definition:

$$\widehat{\mathcal{M}}(\mathcal{D}) = \begin{cases} nil_{\mathcal{A}} & \text{if } \mathcal{D} = \xi \\ \mathring{\mathcal{M}}\left(\boldsymbol{y}_s, \widehat{\mathcal{M}}(\mathcal{D}^{(1)}), \ldots, \widehat{\mathcal{M}}(\mathcal{D}^{(c)})\right) & \text{otherwise} \end{cases}$$

where $nil_{\mathcal{A}}$ is a special coordinate (the void coordinate) into the discrete output space $\mathcal{A}$, and $\mathring{\mathcal{M}}: \mathcal{Y} \times \mathcal{A}^c \to \mathcal{A}$ is a SOM, defined on a generic node, which takes in input the label of the node and the "encoding" of the subgraphs $\mathcal{D}^{(1)}, \ldots, \mathcal{D}^{(c)}$ according to the $\widehat{\mathcal{M}}$ map. By "unfolding" the recursive definition, it turns out that $\widehat{\mathcal{M}}(\mathcal{D})$ can be computed by starting to apply $\mathring{\mathcal{M}}$ to leaf nodes (i.e., nodes with null outdegree), and proceeding with the application of $\mathring{\mathcal{M}}$ bottom-up from the frontier to the supersource of the graph $\mathcal{D}$.

Assuming that each label in $\mathcal{Y}$ is encoded in $\mathcal{U} \subset I\!R^m$, for each $v \in \text{vert}(\mathcal{D})$, we have a vector $\boldsymbol{u}_v$ of dimension m. Moreover, the display output space $\mathcal{A}$, where $\mathcal{A} \equiv [1..n_1] \times [1..n_2] \times \cdots \times [1..n_q]$, is realized through a q dimensional lattice of neurons. So, the winning neuron is represented by the coordinates $(i_1, .., i_q)$.

With the above assumptions, we have that $\mathring{\mathcal{M}}: I\!R^m \times ([1..n_1] \times \cdots \times [1..n_q])^c \to [1..n_1] \times \cdots \times [1..n_q]$, and the $m+cq$ dimensional input vector $\boldsymbol{v}$ to $\mathring{\mathcal{M}}$, representing the information about a generic node v, is defined as $\boldsymbol{v} = \left[\boldsymbol{u}_v \ \mathcal{D}_{ch_1[v]} \ \mathcal{D}_{ch_2[v]} \ \cdots \ \mathcal{D}_{ch_c[v]}\right]$, where $\mathcal{D}_{ch_i[v]}$ is the coordinates vector[2] of the winning neuron for the subgraph pointed by the i-th pointer of v.

Of course, each neuron with coordinates vector $\boldsymbol{c}_j$ in the q dimensional lattice has an associated vector weight $\boldsymbol{w}_{\boldsymbol{c}_j} \in I\!R^{m+cq}$. The weights associated with each neuron in the q dimensional lattice $\mathring{\mathcal{M}}$ can be trained using the standard SOM's learning process, while the training algorithm for $\widehat{\mathcal{M}}$ is shown in Figure 3.

The coordinates for the (sub)graphs are stored in $\mathcal{D}_v$, once for each processing of graph $\mathcal{D}$, and then used when needed[3] for the training of $\mathring{\mathcal{M}}$. Of course, the

[2] The null pointer $nil_{\mathcal{A}}$ can be defined, for example, by the vector with all components at -1.

[3] Notice that the use of an inverted topological order guarantees that the updating of the coordinate vectors $\mathcal{D}_v$ is done before the use of $\mathcal{D}_v$ for training.

Unsupervised Stochastic Training Algorithm for $\widehat{\mathcal{M}}$

input: $\{\mathcal{D}_i\}_{i=1,\ldots,N}$, c, $\mathring{\mathcal{M}}$;
begin
  randomly set the weights for $\mathring{\mathcal{M}}$;
  repeat
    randomly select $\mathcal{D} \in T$ with uniform distribution;
    `List(`$\mathcal{D}$`)` $\leftarrow$ an inverted topological order for vert($\mathcal{D}$);
    for v $\leftarrow$`first(List(`$\mathcal{D}$`))` **to** `last(List(`$\mathcal{D}$`))`**do**
      train($\mathring{\mathcal{M}}\left(\left[\boldsymbol{u}_v\ \mathcal{D}_{ch_1[v]}\ \mathcal{D}_{ch_2[v]}\ \cdots\ \mathcal{D}_{ch_c[v]}\right]\right)$);
      $\mathcal{D}_v \leftarrow \mathring{\mathcal{M}}\left(\left[\boldsymbol{u}_v\ \mathcal{D}_{ch_1[v]}\ \mathcal{D}_{ch_2[v]}\ \cdots\ \mathcal{D}_{ch_c[v]}\right]\right)$;
end

Fig. 3. The unsupervised SOM for structured data.

stored vector is an approximation of the true coordinate vector for the graph rooted in v, however, if the learning rate is small, this approximation may be negligible. The call *train*() refers to a single step of standard SOM training.

Preliminary experimental results showed that the model was actually able to cluster similar (sub)structures and to organize the input (sub)graphs according to both information encoded in the labels and in the structures.

Computational, Complexity, and Learnability Issues The computational power of Recursive Neural Networks has been studied in [14] by using hard threshold units and frontier-to-root tree automata. In [3], several strategies for encoding finite-state tree automata in high-order and first-order sigmoidal recursive neural networks have been proposed.

Complexity results on the amount of resources needed to implement frontier-to-root tree automata in recursive neural networks are presented in [10].

Results on function approximation and theoretical analysis of learnability and generalization of recursive neural networks (referred as folding networks) can be found in [12, 13]. Finally, an analysis of sufficient conditions that guarantee the absence of local minima when training recursive neural networks can be found in [6].

4 Conclusion

Neural networks can deal with structured data. Here I have discussed some of the basic concepts underpinning Recursive Neural Networks. Applications of Recursive Neural Networks are starting to emerge. They include learning search-control heuristics for automated deduction systems [8], logo recognition [5], chemical applications (QSPR/QSAR) [1, 2], incremental parsing of natural language [4]. More work is needed to progress in this field. For example, no accepted proposal for expanding the processing of neural networks to cyclic graphs has yet been established.

References

1. A.M. Bianucci, A. Micheli, A. Sperduti, and A. Starita. Application of cascade correlation networks for structures to chemistry. *Applied Intelligence*, 12:115–145, 2000.
2. A.M. Bianucci, A. Micheli, A. Sperduti, and A. Starita. Analysis of the internal representations developed by neural networks for structures applied to quantitative structure-activity relationship studies of benzodiazepines. *J. Chem. Inf. Comput. Sci.*, 41(1):202–218, 2001.
3. R.C. Carrasco and M.L. Forcada. Simple strategies to encode tree automata in sigmoid recursive neural networks. *IEEE TKDE*, 13(2):148–156, 2001.
4. F. Costa, P. Frasconi, V. Lombardo, and G. Soda. Towards incremental parsing of natural language using recursive neural networks. *Applied Intelligence*, To appear.
5. E. Francesconi, P. Frasconi, M. Gori, S. Marinai, J. Q. Sheng, G. Soda, and A. Sperduti. Logo recognition by recursive neural networks. In R. Kasturi and K. Tombre, editors, *Graphics Recognition – Algorithms and Systems*, pages 104–117. LNCS, Vol. 1389, Springer Verlag, 1997.
6. P. Frasconi, M. Gori, and A. Sperduti. On the efficient classification of data structures by neural networks. In *IJCAI*, pages 1066–1071, 1997.
7. P. Frasconi, M. Gori, and A. Sperduti. A general framework for adaptive processing of data structures. *IEEE TNN*, 9(5):768, 1998.
8. C. Goller. *A Connectionist Approach for Learning Search-Control Heuristics for Automated Deduction Systems.* PhD thesis, Technical University Munich, Computer Science, 1997.
9. C. Goller and A. Küchler. Learning task-dependent distributed structure-representations by backpropagation through structure. In *IEEE International Conference on Neural Networks*, pages 347–352, 1996.
10. M. Gori, A. Kuchler, and A. Sperduti. On the implementation of frontier-to-root tree automata in recursive neural networks. *IEEE Transactions on Neural Networks*, 10(6):1305 –1314, 1999.
11. M. Hagenbuchner, A. Sperduti, and A. C. Tsoi. A supervised self-organising map for structured data. In *WSOM'01*, 2001.
12. B. Hammer. On the learnability of recursive data. *Mathematics of Control Signals and Systems*, 12:62–79, 1999.
13. B. Hammer. Generalization ability of folding networks. *IEEE TKDE*, 13(2):196–206, 2001.
14. A. Sperduti. On the computational power of recurrent neural networks for structures. *Neural Networks*, 10(3):395–400, 1997.
15. A. Sperduti and A. Starita. Supervised neural networks for the classification of structures. *IEEE Transactions on Neural Networks*, 8(3):714–735, 1997.

Bad Design and Good Performance: Strategies of the Visual System for Enhanced Scene Analysis

Florentin Wörgötter

University of Stirling, Department of Psychology, Stirling FK9 4LA, Scotland
`http://www.cn.stir.ac.uk`

Abstract. The visual system of vertebrates is a highly efficient, dynamic scene-analysis machine even though many aspects of its own design are at a first glance rather inconvenient from the viewpoint of an neural network- or computer vision engineer. For several of these apparently imperfect design principles, it seems, however, that the system is able to turn things around and instead make a virtue out of them.

I will concentrate on three examples where the design and the actual performace of the visual system seem to mismatch. In particular I will try to show how visual perception in creatures and computers can actually be improved when treating such "faulty" signals in the "right way". Starting with older studies about visual signal transmission delays and their possible use in image segmentation, I will then present novel ideas about the positive effects of noise in visual signals. Finally, I will present data about changing the spatio-temporal resolution of cortical cell responses. For the first time, this was measured in an open-loop paradigm, achieved by specifically eliminating the cortico-thalamic feedback loop from within an otherwise intact cortex.

1) The visual world at the level of a single cortical cell is anything but constant. Receptive fields of cortical cells very often encounter new situations due to fast (saccadic) eye movements, which occur at a rate of up to 5 Hz, and/or due to object motion in the viewed scene. This process becomes more complicated by the fact that the visual activity reaches higher visual areas only after a certain delay, the visual latency, which is heavily contrast dependent: activity from dark objects arrive much later than from bright objects. Interestingly, normally we do not see a delay between the perception of bright versus dark objects. In a series of older studies, we had suggested that, instead of interfering with perception, visual latency might actually be used to segment images into their dark and bright parts and thereby help in process of object recognition. The naturally arising delay between bright-elicited and dark-elicited activity is enough to drive such a process. In fact, image segmentation can be accelerated and improved rather dramatically in technical systems when latencies are combined with a spin-relaxation model for image segmentation.

2) Latency differences may play a role especially after a saccadic eye movement, when the eye stabilizes on a new image. However, even during fixation or

G. Dorffner, H. Bischof, and K. Hornik (Eds.): ICANN 2001, LNCS 2130, pp. 13–15, 2001.

smooth pursuit, retinal positioning errors and the effects of eye-tremor induce target shifts that lead to a rapid change in cortical cell stimulation. In spite of this, we do not notice motion blurring. This effect is so hard to correct in technical systems (cameras) when they have to operate under such adverse conditions. Recording from the visual cortex, we recently found that (simulated) eye-tremor superimposed on a moving stimulus drives cortical cells harder compared with a smoothly moving stimulus. We attribute this effect to stochastic resonance and it may well be that this effect leads to a contrast enhancement at edges. Thus, tremor seems to enhance cortical signal amplitudes. In addition to this, we found that tremor can - quite paradoxically - also substantially increase visual resolution. Hyper-acuity describes the fact that our visual resolution is better than predicted from the distance between two photoreceptors. One expects that motion noise should lower visual resolution. Interestingly, we shown in a model that hyper-acuity can actually be assisted by eye-tremor. This is based on the fact that many photoreceptors are randomly moved across the stimulus. The temporal integration properties of the retina and the divergence/convergence structure of the primary visual pathway are also instrumental in this process. Technical systems (camera chips) can in a similar way benefit from tremor - as has been pointed out by Mitros and his co-workers from CALTECH (in press).

3) All these effects arise mainly from the processing properties of our "visual front-end", the eye or the retinal network, respectively. However, along the ascending pathways, additional sources, which shape and modify the signals, come into play. At the level of the retina, visual signals (especially from X-cells) still bear a high degree of linearity, but more and more non-linear disturbances occur higher up in the visual hierarchy. This is mostly due to the recursive action of feedback loops which interfere with ascending signals. This happens for the first time in the visual thalamus through the action of the cortico-thalamic feedback. Already, early models have suggested that such non-linear disturbances could actually enhance the signal analysis properties of a system. A classic example is the proposal that a shift in the visual attention's spotlight could be introduced by this (or another) feedback loop. Later on, data became available which showed that the cortico-thalamic feedback loop also changes the spatio-temporal resolution of thalamic cell responses in a non-linear way. In a model, we predicted that also in the cortex, this feedback loop is actively involved in the process of locally enhancing the visual resolution. All this has been known or suggested for some time. But how can one measure the effects of a closed loop in the cortex itself? From an engineering perspective, it would be ideal to compare the normal (closed-loop) situation with the so called the "open-loop" condition. This has been so far impossible to investigate in the cortico-thalamic system. It would require disentangling cells and fibers and eliminating specifically those from which the feedback arises. By means of a novel, complicated, 3-step experimental protocol we are now able to do this for the first time. These experiments indeed support the prediction that the closed-loop cortico-thalamic system enhances cortical cell responses without loosing the spatial precision of their receptive fields.

The results presented here are mainly intended to provide a proof of concept for the underlying ideas. Obviously, it is very hard to try to find unequivocal experimental support for this. Studies concerning the action of the cortico-thalamic feedback loop performed by many groups are probably gradually reaching a state where - despite a lack of many details - the conclusions start to converge. In addition, these studies show that it is sometimes possible to trace seemingly unaddressable model predictions by designing dedicated experiments. Thus, adopting the pessimist's view, it does not seem to be entirely hopeless to use such proof of concept models as a step in trying to understand brain function. Those who feel that this statement is still too frustrating may find consolation in the fact that ideas such as those put forward here have already often been successfully implemented in technical systems.

Acknowledgements

The Support of the DFG and the HFSP are gratefully acknowledged.

Part II

Data Analysis and Pattern Recognition

Fast Curvature Matrix-Vector Products

Nicol N. Schraudolph

Institute of Computational Sciences, Eidgenössische
Technische Hochschule, CH-8092 Zürich, Switzerland
nic@inf.ethz.ch

Abstract. The Gauss-Newton approximation of the Hessian guarantees positive semi-definiteness while retaining more second-order information than the Fisher information. We extend it from nonlinear least squares to all differentiable objectives such that positive semi-definiteness is maintained for the standard loss functions in neural network regression and classification. We give efficient algorithms for computing the product of extended Gauss-Newton and Fisher information matrices with arbitrary vectors, using techniques similar to but even cheaper than the fast Hessian-vector product [1]. The stability of SMD [2–5], a learning rate adaptation method that uses curvature matrix-vector products, improves when the extended Gauss-Newton matrix is substituted for the Hessian.

1 Definitions and Notation

Network. A neural network with m inputs, n weights, and o linear outputs is usually regarded as a mapping $\mathbb{R}^m \to \mathbb{R}^o$ from an input pattern $\boldsymbol{x}$ to the corresponding output $\boldsymbol{y}$, for a given vector $\boldsymbol{w}$ of weights. Here we formalize such a network instead as a mapping $\mathcal{N} : \mathbb{R}^n \to \mathbb{R}^o$ from weights to outputs (for given inputs), and write $\boldsymbol{y} = \mathcal{N}(\boldsymbol{w})$. To extend this formalism to networks with nonlinear outputs, we define the output nonlinearity $\mathcal{M} : \mathbb{R}^o \to \mathbb{R}^o$ and write $\boldsymbol{z} = \mathcal{M}(\boldsymbol{y}) = \mathcal{M}(\mathcal{N}(\boldsymbol{w}))$. For networks with linear outputs, $\mathcal{M}$ is the identity.

Loss Function. We consider neural network learning as the minimization of a scalar loss function $\mathcal{L} : \mathbb{R}^o \to \mathbb{R}$ defined as the log-likelihood $\mathcal{L}(\boldsymbol{z}) \equiv -\log \Pr(\boldsymbol{z})$ of the network output $\boldsymbol{z}$ under a suitable statistical model [6]. For supervised learning, $\mathcal{L}$ may also implicitly depend on given targets $\boldsymbol{z}^*$ for the network outputs. Formally, the loss can now be regarded as a function $\mathcal{L}(\mathcal{M}(\mathcal{N}(\boldsymbol{w})))$ of the weights, for a given set of inputs and (if supervised) targets.

Jacobian. The Jacobian $J_\mathcal{F}$ of a function $\mathcal{F} : \mathbb{R}^m \to \mathbb{R}^n$ is the $n \times m$ matrix of partial derivatives of the outputs of $\mathcal{F}$ with respect to its inputs. For a neural network defined as above, the gradient of the loss with respect to the weights is given by

$$\frac{\partial}{\partial \boldsymbol{w}} \mathcal{L}(\mathcal{M}(\mathcal{N}(\boldsymbol{w}))) \;=\; J'_{\mathcal{L}\circ\mathcal{M}\circ\mathcal{N}} \;=\; J'_\mathcal{N} J'_\mathcal{M} J'_\mathcal{L}\,, \tag{1}$$

where $\circ$ denotes function composition, and $'$ the matrix transpose. We use J as an abbreviation for $J_{\mathcal{L}\circ\mathcal{M}\circ\mathcal{N}}$.

G. Dorffner, H. Bischof, and K. Hornik (Eds.): ICANN 2001, LNCS 2130, pp. 19–26, 2001.

Matching Loss Functions. We say that the loss function $\mathcal{L}$ *matches* the output nonlinearity $\mathcal{M}$ iff $J'_{\mathcal{L}\circ\mathcal{M}} = A\boldsymbol{z} + \boldsymbol{b}$, for some A and $\boldsymbol{b}$ not dependent on $\boldsymbol{w}$. The standard loss functions used in neural network regression and classification — sum-squared error for linear outputs, and cross-entropy error for softmax or logistic outputs — are all matching loss functions with $A = I$ and $\boldsymbol{b} = -\boldsymbol{z}^*$, so that $J'_{\mathcal{L}\circ\mathcal{M}} = \boldsymbol{z} - \boldsymbol{z}^*$ [6, chapter 6]. This will simplify some of the calculations described in Section 3 below.

Hessian. The instantaneous Hessian $H_{\mathcal{F}}$ of a scalar function $\mathcal{F} : \mathbb{R}^n \to \mathbb{R}$ is the $n \times n$ matrix of second derivatives of $\mathcal{F}(\boldsymbol{w})$ with respect to its inputs $\boldsymbol{w}$:

$$H_{\mathcal{F}} \equiv \frac{\partial J_{\mathcal{F}}}{\partial \boldsymbol{w}'}, \quad \textit{i.e.,} \quad (H_{\mathcal{F}})_{ij} = \frac{\partial^2 \mathcal{F}(\boldsymbol{w})}{\partial w_i \, \partial w_j}. \tag{2}$$

For a neural network as defined above, we abbreviate $H \equiv H_{\mathcal{L}\circ\mathcal{M}\circ\mathcal{N}}$. The Hessian proper, which we denote $\bar{H}$, is obtained by taking the expectation of H over inputs: $\bar{H} \equiv \langle H \rangle_{\boldsymbol{x}}$. For matching loss functions, $H_{\mathcal{L}\circ\mathcal{M}} = AJ_{\mathcal{M}} = J'_{\mathcal{M}}A'$.

Fisher Information. The instantaneous Fisher information matrix $F_{\mathcal{F}}$ of a scalar log-likelihood function $\mathcal{F} : \mathbb{R}^n \to \mathbb{R}$ is the $n \times n$ matrix formed by the outer product of its first derivatives:

$$F_{\mathcal{F}} \equiv J'_{\mathcal{F}} J_{\mathcal{F}}, \quad \textit{i.e.,} \quad (F_{\mathcal{F}})_{ij} = \frac{\partial \mathcal{F}(\boldsymbol{w})}{\partial w_i} \frac{\partial \mathcal{F}(\boldsymbol{w})}{\partial w_j}. \tag{3}$$

Note that $F_{\mathcal{F}}$ always has rank one. As before, we abbreviate $F \equiv F_{\mathcal{L}\circ\mathcal{M}\circ\mathcal{N}}$. The Fisher information matrix proper, $\bar{F} \equiv \langle F \rangle_{\boldsymbol{x}}$, describes the geometric structure of weight space [7] and is used in the *natural gradient* descent approach [8].

2 Extended Gauss-Newton Approximation

Problems with the Hessian. The use of the Hessian in second-order gradient descent for neural networks is problematic: for nonlinear systems, $\bar{H}$ is not necessarily positive definite, so Newton's method may diverge, or even take steps in uphill directions. Practical second-order gradient methods should therefore use approximations or modifications of the Hessian that are known to be reasonably well-behaved, with positive semi-definiteness as a minimum requirement.

Fisher Information. One alternative that has been proposed is the Fisher information matrix $\bar{F}$ [8], which — being a quadratic form — is positive semi-definite by definition. On the other hand, $\bar{F}$ ignores *all* second-order interactions, thus throwing away a lot of potentially useful information. By contrast, we shall derive an approximation of the Hessian that is positive semi-definite even though it does retain certain second-order terms.

Gauss-Newton. An entire class of popular optimization techniques for nonlinear least squares problems — as implemented by neural networks with linear outputs and sum-squared loss function — is based on the well-known Gauss-Newton (*aka* "linearized", "outer product", or "squared Jacobian") approximation of the Hessian. Here we extend the Gauss-Newton approach to other standard loss functions — in particular, the cross-entropy loss used in neural network classification — in such a way that even though some second-order information is retained, positive semi-definiteness can still be proven. Using the product rule, the instantaneous Hessian of our neural network can be written as

$$H \;=\; \frac{\partial}{\partial \boldsymbol{w}'}(J_{\mathcal{L}\circ\mathcal{M}}\, J_{\mathcal{N}}) \;=\; J'_{\mathcal{N}}\, H_{\mathcal{L}\circ\mathcal{M}}\, J_{\mathcal{N}} + \sum_{i=1}^{o} (J_{\mathcal{L}\circ\mathcal{M}})_i\, H_{\mathcal{N}_i}\,, \tag{4}$$

where i ranges over the o outputs of $\mathcal{N}$ (the network proper), with $\mathcal{N}_i$ denoting the subnetwork that produces the ith output. Ignoring the second term above, we define the extended, instantaneous Gauss-Newton matrix

$$G \equiv J'_{\mathcal{N}}\, H_{\mathcal{L}\circ\mathcal{M}}\, J_{\mathcal{N}}\,. \tag{5}$$

Note that G has rank $\leq o$ (the number of network outputs), and is positive semi-definite, regardless of the choice of network $\mathcal{N}$, provided that $H_{\mathcal{L}\circ\mathcal{M}}$ is.

G models the second-order interactions among the network outputs (via $H_{\mathcal{L}\circ\mathcal{M}}$) while ignoring those arising within the network itself ($H_{\mathcal{N}_i}$). This constitutes a compromise between the Hessian (which models all second-order interactions) and the Fisher information (which ignores them all). For systems with a single, linear output and sum-squared error, G reduces to F; in all other cases it provides a richer source of curvature information.

Standard Loss Functions. For the standard loss functions used in neural network regression and classification, G has additional interesting properties:

Firstly, the residual $J'_{\mathcal{L}\circ\mathcal{M}} = \boldsymbol{z} - \boldsymbol{z}^*$ vanishes at the optimum for realizable problems, so that the Gauss-Newton approximation (5) of the Hessian (4) becomes exact in this case. For unrealizable problems, the residuals at the optimum have zero mean; this will tend to make the last term in (4) vanish in expectation, so that we can still assume $\bar{G} \approx \bar{H}$ near the optimum.

Secondly, in each case we can show that $H_{\mathcal{L}\circ\mathcal{M}}$ (and hence G, and hence $\bar{G}$) is positive semi-definite: for linear outputs with sum-squared loss — *i.e.*, conventional Gauss-Newton — $H_{\mathcal{L}\circ\mathcal{M}} = J_{\mathcal{M}}$ is just the identity I; for independent logistic outputs with cross-entropy loss it is $\mathrm{diag}[\boldsymbol{z}\,(1-\boldsymbol{z})]$, positive semi-definite because $(\forall i)\; 0 < z_i < 1$. For softmax output with cross-entropy loss we have $H_{\mathcal{L}\circ\mathcal{M}} = \mathrm{diag}(\boldsymbol{z}) - \boldsymbol{z}\boldsymbol{z}'$, which is also positive semi-definite since $(\forall i)\; z_i > 0$ and $\sum_i z_i = 1$, and thus

$$\begin{aligned}(\forall \boldsymbol{v} \in \mathbb{R}^o)\; \boldsymbol{v}'[\mathrm{diag}(\boldsymbol{z}) - \boldsymbol{z}\boldsymbol{z}']\,\boldsymbol{v} \;&=\; \sum_i z_i v_i^2 \;-\; (\sum_i z_i v_i)^2 \\ &=\; \sum_i z_i v_i^2 \;-\; 2\,(\sum_i z_i v_i)(\sum_j z_j v_j) \;+\; (\sum_j z_j v_j)^2 \\ &=\; \sum_i z_i (v_i - \sum_j z_j v_j)^2 \;\geq\; 0\,. \end{aligned} \tag{6}$$

3 Fast Curvature Matrix-Vector Products

3.1 The Passes

We now describe algorithms that compute the product of F, G, or H with an arbitrary n-dimensional vector $\boldsymbol{v}$ in $O(n)$. They are all constructed from the same set of passes in which certain quantities are propagated through the network in either forward or reverse direction. For implementation purposes it should be noted that *automatic differentiation* software tools[1] can automatically produce these passes from a program implementing the basic forward pass f_0.

$\boldsymbol{f_0}$. This is the ordinary forward pass of a neural network, evaluating the function $\mathcal{F}(\boldsymbol{w})$ it implements by propagating activity forward through $\mathcal{F}$.

$\boldsymbol{r_1}$. The ordinary backward pass of a neural network, calculating $J'_{\mathcal{F}}\boldsymbol{u}$ by propagating $\boldsymbol{u}$ backwards through $\mathcal{F}$. Uses intermediate results from the f_0 pass.

$\boldsymbol{f_1}$. Following Pearlmutter [1], we define the *Gateaux* derivative

$$\mathcal{R}_{\boldsymbol{v}}(\mathcal{F}(\boldsymbol{w})) \equiv \left.\frac{\partial \mathcal{F}(\boldsymbol{w} + r\boldsymbol{v})}{\partial r}\right|_{r=0} = J_{\mathcal{F}}\,\boldsymbol{v} \tag{7}$$

which describes the effect on a function $\mathcal{F}(\boldsymbol{w})$ of a weight perturbation in the direction of $\boldsymbol{v}$. By pushing $\mathcal{R}_{\boldsymbol{v}}$ — which obeys the usual rules for differential operators — down into the equations of the forward pass f_0, one obtains an efficient procedure which calculates $J_{\mathcal{F}}\boldsymbol{v}$ from $\boldsymbol{v}$; see [1] for details and examples. This f_1 pass uses intermediate results from the f_0 pass.

$\boldsymbol{r_2}$. When the $\mathcal{R}_{\boldsymbol{v}}$ operator is applied to the r_1 pass for a scalar function $\mathcal{F}$, one obtains an efficient procedure for calculating the Hessian-vector product $H_{\mathcal{F}}\,\boldsymbol{v} = \mathcal{R}_{\boldsymbol{v}}(J'_{\mathcal{F}})$. Again, see [1] for details and examples. This r_2 pass uses intermediate results from the f_0, f_1, and r_1 passes.

3.2 The Algorithms

The first step in all three matrix-vector products is the computation of the gradient J' of our neural network model by standard backpropagation:

Gradient. J' is computed by an f_0 pass through the entire network ($\mathcal{N}$, $\mathcal{M}$, and $\mathcal{L}$), followed by an r_1 pass propagating $\boldsymbol{u} = 1$ back through the entire network ($\mathcal{L}$, $\mathcal{M}$, then $\mathcal{N}$). For matching loss functions there is a shortcut: since $J'_{\mathcal{L}\circ\mathcal{M}} = A\boldsymbol{z} + \boldsymbol{b}$, we can limit the forward pass to $\mathcal{N}$ and $\mathcal{M}$ (to compute $\boldsymbol{z}$), then r_1-propagate $\boldsymbol{u} = A\boldsymbol{z} + \boldsymbol{b}$ back through just $\mathcal{N}$.

Fisher Information. To compute $F\boldsymbol{v} = J'J\boldsymbol{v}$, simply multiply the gradient J' by the inner product between J' and $\boldsymbol{v}$. If there is no random access to J' or $\boldsymbol{v}$ — *i.e.*, its elements can be accessed only through passes like the above — the

[1] See `http://www-unix.mcs.anl.gov/autodiff/`

scalar $J\boldsymbol{v}$ can instead be calculated by f_1-propagating $\boldsymbol{v}$ forward through the network. This step is also necessary for the other two matrix-vector products.

Hessian. After f_1-propagating $\boldsymbol{v}$ forward, r_2-propagate $\mathcal{R}_{\boldsymbol{v}}(1)=0$ back through the entire network to obtain $H\boldsymbol{v} = \mathcal{R}_{\boldsymbol{v}}(J')$ [1]. For matching loss functions, the shortcut is to f_1-propagate $\boldsymbol{v}$ through just $\mathcal{N}$ and $\mathcal{M}$ to obtain $\mathcal{R}_{\boldsymbol{v}}(\boldsymbol{z})$, then r_2-propagate $\mathcal{R}_{\boldsymbol{v}}(J'_{\mathcal{L}\circ\mathcal{M}}) = A\,\mathcal{R}_{\boldsymbol{v}}(\boldsymbol{z})$ back through $\mathcal{N}$.

Gauss-Newton. Following the f_1 pass, r_2-propagate $\mathcal{R}_{\boldsymbol{v}}(1) = 0$ back through $\mathcal{L}$ and $\mathcal{M}$ to obtain $\mathcal{R}_{\boldsymbol{v}}(J'_{\mathcal{L}\circ\mathcal{M}}) = H_{\mathcal{L}\circ\mathcal{M}} J_{\mathcal{N}}\boldsymbol{v}$, then r_1-propagate that back through $\mathcal{N}$, giving $G\boldsymbol{v}$. For matching loss functions we do not require an r_2 pass: since

$$G \;=\; J'_{\mathcal{N}} H_{\mathcal{L}\circ\mathcal{M}} J_{\mathcal{N}} \;=\; J'_{\mathcal{N}} J'_{\mathcal{M}} A' J_{\mathcal{N}}\,, \tag{8}$$

we can limit the f_1 pass to $\mathcal{N}$, multiply the result with A', then r_1-propagate it back through $\mathcal{M}$ and $\mathcal{N}$. Alternatively, one may compute the equivalent $G\boldsymbol{v} = J'_{\mathcal{N}} A J_{\mathcal{M}} J_{\mathcal{N}} \boldsymbol{v}$ by continuing the f_1 pass through $\mathcal{M}$, multiplying with A, and r_1-propagating back through $\mathcal{N}$.

Batch Average. To calculate the product of a curvature matrix $\bar{C} \equiv \langle C\rangle_{\boldsymbol{x}}$ — where C is one of F, G, or H — with vector $\boldsymbol{v}$, average the instantaneous product $C\boldsymbol{v}$ over all input patterns $\boldsymbol{x}$ (and associated targets $\boldsymbol{z}^*$, if applicable) while holding $\boldsymbol{v}$ constant. For large training sets, or non-stationary streams of data, it is often preferable to estimate $\bar{C}\boldsymbol{v}$ by averaging over "mini-batches" of (typically) just 5–50 patterns.

3.3 Computational Cost

Table 1 summarizes the curvature matrix C corresponding to various gradient methods, the passes needed (for a matching loss function) to calculate both the gradient $\bar{J}' \equiv \langle J'\rangle_{\boldsymbol{x}}$ and the fast matrix-vector product $\bar{C}\boldsymbol{v}$, and the associated computational cost in terms of floating-point operations (flops) per weight and pattern in a multi-layer perceptron. These figures ignore certain optimizations — *e.g.*, not propagating gradients back to the inputs — and assume that any computation at the network's nodes is dwarfed by that required for the weights.

Table 1. Passes needed to compute gradient $\bar{J}'$ and fast matrix-vector product $\bar{C}\boldsymbol{v}$, and associated cost (for a multi-layer perceptron) in flops per weight and pattern, for various choices of curvature matrix C.

Method	Pass	f_0	r_1	f_1	r_2	Cost
	result:	$\mathcal{L}$	$J'\boldsymbol{u}$	$J\boldsymbol{v}$	$H\boldsymbol{v}$	(for $\bar{J}'$
$C =$ name	cost:	2	3	4	7	& $\bar{C}\boldsymbol{v}$)
I	steepest descent	√	√			6
F	natural gradient	√	√			10
G	Gauss-Newton	√	√√	√		14
H	Newton's method	√	√	√	√	18

4 Application to Stochastic Meta-Descent (SMD)

Algorithm. SMD [2–5] is a new, highly effective online algorithm for local learning rate adaptation. It updates the weights $\boldsymbol{w}$ by the simple gradient descent

$$\boldsymbol{w}_{t+1} \;=\; \boldsymbol{w}_t \;-\; \boldsymbol{p}_t \cdot J', \tag{9}$$

where $\cdot$ denotes element-wise multiplication, and J' the stochastic gradient. The vector $\boldsymbol{p}$ of local learning rates is adapted multiplicatively:

$$\boldsymbol{p}_t \;=\; \boldsymbol{p}_{t-1} \cdot \max(\tfrac{1}{2},\; 1 + \mu\, \boldsymbol{v}_t \cdot J'), \tag{10}$$

using a scalar meta-learning rate μ. This update minimizes the network's loss with respect to $\boldsymbol{p}$ by *exponentiated* gradient descent [9], but has been re-linearized so as to avoid the computationally expensive exponentiation operation [10]. The auxiliary vector $\boldsymbol{v}$ used in (10) is itself updated iteratively via

$$\boldsymbol{v}_{t+1} \;=\; \lambda\, \boldsymbol{v}_t \;+\; \boldsymbol{p}_t \cdot (J' \;-\; \lambda\, C \boldsymbol{v}_t), \tag{11}$$

where C is the curvature matrix, and $0 \leq \lambda \leq 1$ a forgetting factor for nonstationary tasks. $C\boldsymbol{v}_t$ is computed via the fast algorithms described above.

Benchmark Setup. We illustrate the behavior of SMD on the "four regions" benchmark [11]: a fully connected feedforward network $\mathcal{N}$ with two hidden layers of 10 tanh units each (Fig. 1, right) is to classify two continuous inputs in the range [-1,1] into four disjoint, non-convex regions (Fig. 1, left). We use the standard softmax output nonlinearity $\mathcal{M}$ with matching cross-entropy loss $\mathcal{L}$, meta-learning rate $\mu = 0.05$, initial learning rates $\boldsymbol{p}_0 = 0.1$, and uniformly random initial weights in the range [-0.3,0.3]. Training patterns are generated online by drawing independent, uniformly random input samples; they are presented in mini-batches of 10 patterns each. Since each pattern is seen only once, the empirical loss provides an unbiased estimate of generalization ability.

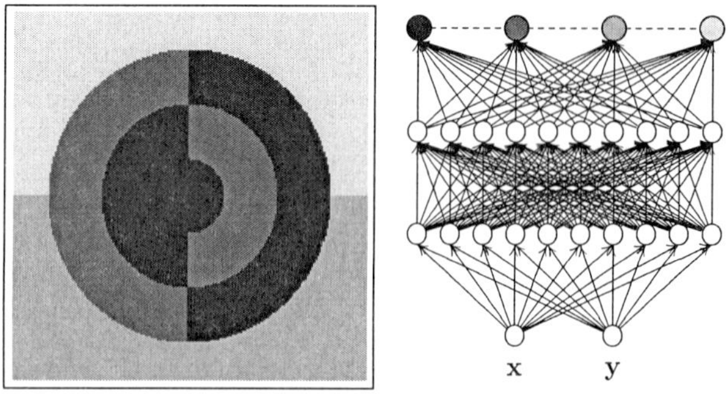

Fig. 1. The four regions task (left), and the network we trained on it (right).

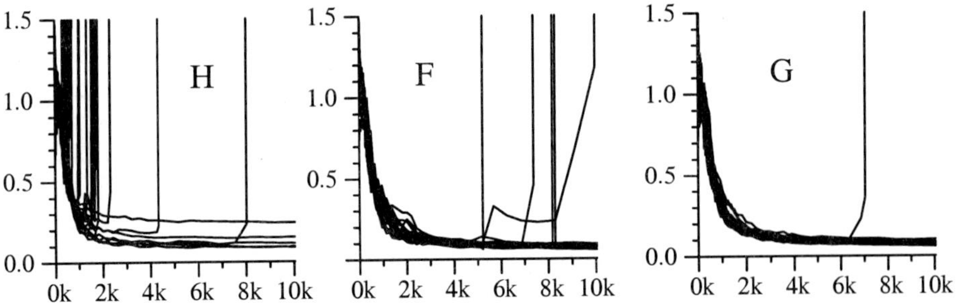

Fig. 2. Loss curves for 25 runs of SMD with $\lambda = 1$, when using the Hessian (left), the Fisher information (center), or the extended Gauss-Newton matrix (right) for C in Equation (11). Vertical spikes indicate divergence.

Curvature Matrix. Fig. 2 shows loss curves for SMD with $\lambda = 1$ on the four regions problem, starting from 25 different random initial states, using the Hessian, Fisher information, and extended Gauss-Newton matrix, respectively, for C in Equation (11). With the Hessian (left), 80% of the runs diverge — most of them early on, when the risk that H is not positive definite is greatest. When we guarantee positive semi-definiteness by switching to the Fisher information matrix (center), the proportion of diverged runs drops to 20%; those runs that still diverge do so only relatively late. Finally, for our extended Gauss-Newton approximation (right) only a single run diverges, illustrating the benefit of retaining certain second-order terms while preserving positive semi-definiteness.

Stability. The residual tendency of SMD to occasionally diverge can be suppressed further by slightly lowering the λ parameter. By curtailing the memory of iteration (11), however, this can compromise the rapid convergence of SMD, resulting in a stability/performance tradeoff (Fig. 3):

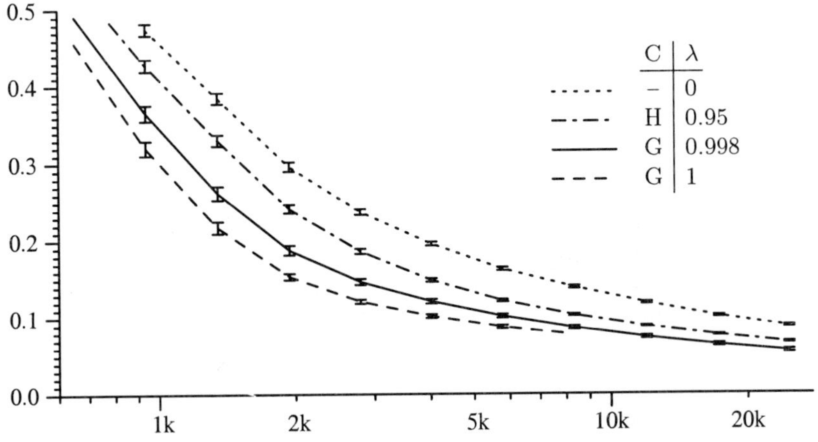

Fig. 3. Average loss over 25 runs of SMD for various combinations of curvature matrix C and forgetting factor λ. Memory ($\lambda \to 1$) is key to rapid convergence.

With the extended Gauss-Newton approximation, a small reduction of λ to 0.998 (solid line) is sufficient to prevent divergence, at a moderate cost in performance relative to $\lambda=1$ (dashed). When the Hessian is used, by contrast, λ must be set as low as 0.95 to maintain stability, and convergence is slowed much further (dash-dotted). Even so, this is still significantly faster than the degenerate case of $\lambda=0$ (dotted), which in effect implements IDD [12], the to our knowledge best competing online method for local learning rate adaptation.

From these experiments it appears that memory (*i.e.*, λ close to 1) is key to achieving the rapid convergence characteristic of SMD. We are now investigating other, more direct ways to keep iteration (11) under control, aiming to ensure the stability of SMD while maintaining its excellent performance at $\lambda = 1$.

Acknowledgment.. We would like to thank Jenny Orr and Barak Pearlmutter for many helpful discussions, and the Swiss National Science Foundation for the financial support provided under grant number 2000–052678.97/1.

References

1. B. A. Pearlmutter, "Fast exact multiplication by the Hessian," *Neural Computation*, vol. 6, no. 1, pp. 147–160, 1994.
2. N. N. Schraudolph, "Local gain adaptation in stochastic gradient descent," in *Proc. 9th Int. Conf. Artificial Neural Networks*, pp. 569–574, IEE, London, 1999.
3. N. N. Schraudolph, "Online learning with adaptive local step sizes," in *Neural Nets – WIRN Vietri-99: Proc. 11th Italian Workshop on Neural Networks* (M. Marinaro and R. Tagliaferri, eds.), Perspectives in Neural Computing, (Vietri sul Mare, Salerno, Italy), pp. 151–156, Springer Verlag, Berlin, 1999.
4. N. N. Schraudolph, "Fast second-order gradient descent via O(n) curvature matrix-vector products," Tech. Rep. IDSIA-12-00, IDSIA, Galleria 2, CH-6928 Manno, Switzerland, 2000. Submitted to *Neural Computation.*
5. N. N. Schraudolph and X. Giannakopoulos, "Online independent component analysis with local learning rate adaptation," in *Adv. Neural Info. Proc. Systems* (S. A. Solla, T. K. Leen, and K.-R. Müller, eds.), vol. 12, pp. 789–795, The MIT Press, Cambridge, MA, 2000.
6. C. M. Bishop, *Neural Networks for Pattern Recognition.* Oxford: Clarendon, 1995.
7. S.-i. Amari, *Differential-Geometrical Methods in Statistics*, vol. 28 of *Lecture Notes in Statistics.* New York: Springer Verlag, 1985.
8. S.-i. Amari, "Natural gradient works efficiently in learning," *Neural Computation*, vol. 10, no. 2, pp. 251–276, 1998.
9. J. Kivinen and M. K. Warmuth, "Additive versus exponentiated gradient updates for linear prediction," in *Proc. 27th Annual ACM Symp. Theory of Computing*, (New York, NY), pp. 209–218, Association for Computing Machinery, 1995.
10. N. N. Schraudolph, "A fast, compact approximation of the exponential function," *Neural Computation*, vol. 11, no. 4, pp. 853–862, 1999.
11. S. Singhal and L. Wu, "Training multilayer perceptrons with the extended Kalman filter," in *Adv. Neural Info. Proc. Systems: Proc. 1988 Conf.* (D. S. Touretzky, ed.), pp. 133–140, Morgan Kaufmann, 1989.
12. M. E. Harmon and L. C. Baird III, "Multi-player residual advantage learning with general function approximation," Tech. Rep. WL-TR-1065, Wright Laboratory, WL/AACF, 2241 Avionics Circle, Wright-Patterson AFB, OH 45433-7308, 1996.

Architecture Selection in NLDA Networks

José R. Dorronsoro, Ana M. González, and Carlos Santa Cruz*

Dept. of Computer Engineering and Instituto de Ingeniería del Conocimiento
Universidad Autónoma de Madrid, 28049 Madrid, Spain

Abstract. In Non Linear Discriminant Analysis (NLDA) an MLP like architecture is used to minimize a Fisher's discriminant analysis criterion function. In this work we study the architecture selection problem for NLDA networks. We shall derive asymptotic distribution results for NLDA weights, from which Wald like tests can be derived. We also discuss how to use them to make decisions on unit relevance based on the acceptance or rejection of a certain null hypothesis.

1 Introduction

Non Linear Discriminant Analysis (NLDA) is a novel feature extraction procedure proposed by the authors in which a Multilayer Perceptron (MLP) architecture is used but where optimal weights are selected by minimizing a Fisher discriminant criterion function. More precisely, in this work we shall consider two class problems and the simplest possible such architecture, having D input units, a single hidden layer with H units and a single, linear output unit ($C-1$ outputs are used in Fisher's linear discriminants for a C class problem). We denote D dimensional network inputs as $X = (x_1, \ldots, x_D)^t$ (A^t denotes the transpose of A), the weights connecting the D inputs with the hidden unit h as $W_h^H = (w_{1h}^H, \ldots, w_{Dh}^H)^t$ and the weights connecting the hidden layer to the single output by W^O. We assume that $x_D \equiv 1$ for bias effects. The network transfer $F(X, W)$ function is thus given by $y = F(X, W) = (W^O)^t O$, with $O = O(X, W) = (o_1, \ldots, o_H)^t$ the outputs of the hidden layer, that is, $o_h = f((W_h^H)^t \cdot X)$, and f denoting the sigmoidal function. In other words, $y = F(X, W)$ performs a standard MLP transformation. The difference lies in the selection of the optimal weights $W_* = (W_*^O, W_*^H)$, for which the criterion function $J(W) = s_T/s_B$ is to be minimized. Here $s_T = E[(y - \overline{y})^2]$ is the total covariance of the output y and $s_B = \pi_1(\overline{y}_1 - \overline{y})^2 + \pi_2(\overline{y}_2 - \overline{y})^2$ the between class covariance. By $\overline{y} = E[y]$ we denote the overall output mean, $\overline{y}_c = E[y|c]$ denotes the output class conditional means and π_c the class prior probabilities. NLDA networks allow a general and flexible use of any other of the criterion functions customarily employed in linear Fisher analysis. In [1, 7] other properties of NLDA networks are studied, and it is shown that the features they provide can be more robust than those obtained through MLPs in problems where there is a large class overlapping and where class sample sizes are markedly unequal.

* With partial support from Spain's CICyT, grant TIC 98–247.

G. Dorffner, H. Bischof, and K. Hornik (Eds.): ICANN 2001, LNCS 2130, pp. 27–32, 2001.

As with MLPs, optimal NLDA network architectures have to be somehow decided upon. We shall analyze here how this can be done. In our setting we have to decide on the number of hidden units. If the relevance of the original features is also to be decided, the number of input units cannot be taken to be fixed in advance and may have to be lowered. Architecture selection is a long studied question for MLPs, and many approaches, such as bootstrapping, pruning, regularization, stopped training or information criteria have been proposed (see [3, 6, 8] for an overview of these approaches). A common starting point for several of these methods can be found in the asymptotic theory of network weights. We shall derive asymptotic distribution results for the w_{dh} weights connecting the input and hidden layers, from which we will obtain a statistic for weight relevance. For the linear weights w_h between the hidden units and the output we shall adapt some statistics used in classical Fisher analysis to determine the significance of discrimination features. These results are given in the next section, while the third one contains an illustration of these procedures on a synthetic problem.

2 Unit Relevance in NLDA Networks

We shall study unit relevance, working with sets of weights related to a concrete unit. To do so we could consider either input or output weights. However, and as it happens with MLPs, NLDA networks may present identifiability problems. For instance, if all the weights w_{kl} starting at a given hidden unit k are zero, the weights w_{hk} leading to that unit can take any value without affecting network outputs. To avoid these effects we shall consider what we could term "output" unit relevance, that is, the joint relevance of the weights leaving a unit. For a two class NLDA network and a $D \times H \times 1$ architecture, the relevance of the hidden unit h is thus given in terms of the weight w_h^O, $1 \leq h \leq H$, while that of an input unit d, $1 \leq d \leq D$, is jointly determined by the weight set w_{dh}^H, $1 \leq h \leq H$. Notice that the non relevance of an input unit has to be interpreted as the lack of relevance of the associated feature, that can thus be deleted without any loss of discriminating information. Each relevance has to be dealt with separately and we begin with the input units, for which we shall derive a Wald like test. This requires to obtain the asymptotic distribution of these weights. The starting point is the following result [4].

Theorem 1. *Let $\Psi(X, W)$ be a vector valued function depending on a random variable X and a parameter vector W such that $\nabla_W \Psi(X, W)$ and $\nabla_W^2 \Psi(X, W)$ exist for all X and are bounded by integrable functions. If $E[\Psi(X, W_*)] = 0$ at an isolated point W_* and the matrices*

$$\mathcal{H}^* = \mathcal{H}(W_*) = E[\nabla_W \Psi(X, W_*)], \quad \mathcal{I}^* = \mathcal{I}(W_*) = E[\Psi(X, W_*)\Psi(X, W_*)^t]$$

are respectively definite positive and non singular. Then, if X_n is a sequence of i.i.d. random vectors distributed as X, the equation $\sum_1^N \Psi(X_n, W) = 0$ has a solution sequence $\widetilde{W}_n$ converging in probability to W_* and such that $\sqrt{N}(\widetilde{W}_n -$

$W_*)$ *converges in distribution to a multivariate normal with 0 mean and variance* $\mathcal{C}^* = \mathcal{C}(W_*) = (\mathcal{H}^*)^{-1}\mathcal{I}^*(\mathcal{H}^*)^{-1}$.

If S is now a 0–1 selection vector with a given number K of 1's, then, under the null hypothesis $H_0 : SW_* = 0$, it follows that $\sqrt{N}S\widetilde{W}_N \to N(0, S^t\mathcal{C}^*S)$ in distribution. Therefore, the random variable $\zeta_N = \sqrt{N}(S^t\mathcal{C}^*S)^{-1/2}S\widetilde{W}_N$ converges to a K dimensional $N(0, I)$ and

$$||\zeta_N||^2 = N(S\widetilde{W}_N)^t(S^t\mathcal{C}S)^{-1}(S\widetilde{W}_N)$$

converges in distribution to a chi–square χ^2_K with K degrees of freedom.

We shall apply this result to NLDA network architecture selection. Optimal weights are obtained iterating a two step procedure. Assuming that we have the values (W_t^O, W_t^H) at step t, we first obtain W_{t+1}^O applying standard linear Fisher analysis to the outputs $O = O(X, W_t^H)$ of the hidden layers. Next, with W_{t+1}^O fixed, W_{t+1}^H is obtaining by numerically minimizing of J as a function $J(W^H) = J(W_{t+1}^O, W^H)$ of the W^H. We derive first asymptotic results for the hidden weights W^H, obtaining Ψ from the gradient with respect to these weights of the network criterion function J, which we compute next. Since the output weights W^O are computed as in Fisher's linear discriminant analysis, they are therefore invariant with respect to translations in the last hidden layer. In particular, we can center around zero the last hidden layer outputs without affecting the W^O or the value of J. We will thus assume that at the last hidden layer, $\overline{O} = 0$, which implies that the network outputs y will have zero mean as well, for $\overline{y} = (W^O)^t\overline{O}$. Under these simplifications, we therefore have $s_T = E[y^2]$ and $s_B = \pi_1(\overline{y}_1)^2 + \pi_2(\overline{y}_2)^2$. It now follows that

$$\frac{\partial s_T}{\partial w_{kl}} = 2E\left[y\frac{\partial y}{\partial w_{kl}}\right] = 2w_l E\left[yf'(a_l)x_k\right];$$

$$\frac{\partial s_B}{\partial w_{kl}} = 2\sum_c \pi_c\overline{y}_c E_c\left[\frac{\partial y}{\partial w_{kl}}\right] = 2w_l\sum_c \pi_c\overline{y}_c E_c\left[f'(a_l)x_k\right],$$

where we recall that $E_c[z] = E[z|c]$ denotes class conditional expectations. Therefore, considering now the W^O fixed and viewing therefore J as a function $J(W^H)$ of only the W^H weights, it can be shown that

$$\frac{\partial J}{\partial w_{kl}} = \frac{2w_l}{s_B}\left(E\left[yf'(a_l)x_k\right] - J\sum \pi_c\overline{y}_c E_c\left[f'(a_l)x_k\right]\right). \tag{1}$$

Let us now define $\Psi = (\Psi_{kl})$ as $\Psi_{kl}(X, W) = w_l y f'(a_l)x_k - \lambda_{kl}(W) = z_{kl}(X, W) - \lambda_{kl}(W)$, where the nonrandom term $\lambda_{kl}(W)$ is given by $\lambda_{kl}(W) = w_l J(W)\sum_c \pi_c\overline{y}_c E_c[f'(a_l)x_k]$. At an isolated minimum W_* of J, (1) shows that

$$E[\Psi_{kl}(X, W_*)] = \frac{s_B^*}{2}\frac{\partial J}{\partial w_{kl}}(W_*) = 0,$$

with $s_B^* = s_B(W_*)$; in particular, $E[z_{kl}(\cdot, W_*)] = \lambda_{kl}(W_*)$. Moreover, at W_*

$$\mathcal{H}^* = E\left[\frac{\partial \Psi_{kl}}{\partial w_{mn}}(X, W_*)\right] = \frac{\partial}{\partial w_{mn}}\left(\frac{s_B}{2}\frac{\partial J}{\partial w_{kl}}\right)(W_*) = \frac{s_B^*}{2}\frac{\partial^2 J}{\partial w_{mn}\partial w_{kl}}(W_*),$$

and, hence, $\mathcal{H}^* = E[\nabla_W \Psi(X, W_*)]$ is definite positive. The second partials at a minimum W_* of J can be computed using (1) and their value is

$$\begin{aligned}
&\frac{\partial^2 J}{\partial w_{mn} \partial w_{kl}}(W_*) \\
&= \frac{2w_l}{s_B^*} \left\{ E[f'(a_n)x_m f'(a_l)x_k] + \delta_{ln} E[y f''(a_l) x_m x_k] \right\} \\
&- \frac{2w_l}{s_B^*} J^* \left\{ \sum \pi_c \left(E_c[f'(a_n)x_m] E_c[f'(a_l)x_k] + \delta_{ln} \overline{y}_c E_c[f''(a_l)x_m x_k] \right) \right\}.
\end{aligned}$$

Here $J^* = J(W_*)$ and δ_{ln} denotes Kronecker's δ. We also have

$$\begin{aligned}
(\mathcal{I}^*)_{(kl)(mn)} &= E[\Psi_{kl}(X, W_*) \Psi_{mn}(X, W_*)^t] \\
&= E[z_{kl}(X, W_*) z_{mn}(X, W_*)] - \lambda_{kl}(W_*) \lambda_{mn}(W_*).
\end{aligned}$$

Therefore $(\mathcal{I}^*)_{(kl)(mn)} = cov(z_{kl} z_{mn})$, and if it is non singular, it follows that for N random variables $X_1, \ldots, X_N$, i.i.d. as X, there is a solution sequence $\widetilde{W}_N$ of

$$\frac{1}{N} \sum_{j=1}^{N} \left(z_{kl}^j(X, W) - \lambda_{kl}(W) \right) = \frac{1}{N} \sum_{j=1}^{N} \left(y^j d_l^j x_k^j - \lambda_{kl}(W) \right) = 0, \tag{2}$$

that converges in probability to W_* and such that $\sqrt{N}(\widetilde{W}_N - W_*)$ converges in distribution to a multivariate normal $N(0, \mathcal{C}^*)$, where $\mathcal{C}^* = (\mathcal{H}^*)^{-1} \mathcal{I}^* (\mathcal{H}^*)^{-1}$.

We define the relevance R_d of the d input unit by choosing now a selection vector $S = (s_{(pq)})$, $1 \leq p \leq D$, $1 \leq q \leq H$, where $s_{(dq)} = 1$ and all the others are zero, and setting

$$R_d = N(S\widehat{W}_N)^t (S\hat{\mathcal{C}}S)^{-1} (S\widehat{W}_N) \simeq N \sum_h \frac{(\hat{w}_{dh}^*)^2}{((\mathcal{H}^*)^{-1} \mathcal{I}^* (\mathcal{H}^*)^{-1})_{(dh)(dh)}}. \tag{3}$$

We can thus decide to keep or remove the d–th feature according to the chosen confidence level of a χ_H^2 distribution with H degrees of freedom when computed on the value R_d.

Gradient and hessian computations can also be derived for the linear output weights. In fact, since $s_T = (W^O)^t S_T^O W^O$, $s_B = (W^O)^t S_B^O W^O$, considering now J as a function $J(W^O)$, it is easy to see that $\nabla_W J = 2(S_T^O - J S_B^O) W^O / s_B$, and it also follows that $\nabla_W^2 J(W_*) = 2(S_T^O - J S_B^O)/s_B^*$, with again $s_B^* = s_B(W_*)$. However, it is clear that $\nabla_W^2 J(W_*)$ is singular, for the optimal W_*^O is an eigenvalue of $S_T^O - J^* S_B^O$ and the preceding arguments leading to asymptotic weight distributions will not hold for the W^O. As an alternative we can consider input relevance tests used in classical Fisher analysis. Let $W^O = (S^O)^{-1}(\overline{O}_1 - \overline{O}_2)$ be the optimal Fisher weights, $D^2 = (\overline{O}_1 - \overline{O}_2)^t (S_W^O)^{-1} (\overline{O}_1 - \overline{O}_2)$ and $T = (S_T^O)^{-1}$ be the inverse of the total covariance matrix of the hidden layer outputs. Then it can be shown for two class problems [5] that if the inputs come from two d–dimensional multinormal distributions with a common covariance, the statistic

$$R_h = \frac{(N - d - 1) c^2 w_h^2}{(N - 2) T_{hh} (N - 2 + c^2 D^2)}, \tag{4}$$

with $c^2 = N_1 N_2 / N$, N_i the number of sample patterns in class i and $N = N_1 + N_2$, follows an $F_{1,N-2}$ distribution under the null hypothesis $w_h = 0$. In our context, the values of the last hidden outputs would be taken as the inputs of a classical Fisher transformation. Of course, the preceding common multinormal distribution assumption may not be true for these hidden layer outputs and, therefore, tests for significance at a given level may not be strictly applicable. However, a value of the statistic that would warrant the rejection of the null hypothesis could be taken as an indicator of that weight's relevance. On the other hand, a value of the statistic compatible with the acceptance of the null hypothesis should indicate the convenience of that weight removal.

3 A Numerical Illustration

We will illustrate the preceding techniques on a toy synthetic problem with 2 unidimensional classes. The first one, C_0, follows a $N(0, 0.5)$ distribution and the other one, C_1, is given by a mixture of two gaussians, $N(-2, 0.25)$ and $N(2, 0.25)$. The prior probabilities of the two classes are 0.5. As a classification problem, it is an "easy" one, for the mean error probability (MEP) of the optimal Bayes classifier has a rather low value of about 0.48 %. On the other hand, the class distributions are not unimodal, and neither linearly separable. NLDA networks will thus realize feature enhancing rather than feature extraction. We shall work with the sample versions $\widehat{\mathcal{I}}$, $\widehat{\mathcal{H}}$ of $\mathcal{I}$, $\mathcal{H}$ and we denote by $\hat{w}_h$, $\hat{w}_{dh}$ the sample derived weights. We start with a minimal network with all units being relevant and shall "grow" the network's hidden layers. That is, we will add new units in a step by step fashion and stopping if the last added unit fails the corresponding test. Once we arrive at the optimal number of hidden units, we shall prune input units also in a step by step fashion, removing each time the feature found to be less relevant.

It is easily seen that the optimal architecture for this problem just needs 2 hidden units: one hidden unit is not enough for neither MLPs nor NLDA networks (although it could be if, for instance, $f(x) = x^2$ is used instead of the sigmoidal) while with 2 units the MEP of the δ_{NLDA} classifier coincides with the optimal Bayes value of 0.48. To illustrate input feature selection we shall consider 3–dimensional features, where we add two noisy inputs. The first one is given by a uniform distribution on $[-8, -12]$, and the second by a $N(10, 1)$ normal.

As just explained, we start with a minimal $3 \times 1 \times 1$ network and grow it using the statistic R_h to detetect when an irrelevant hidden unit has been added. The first half of table 1 shows the values of $R(h)$ for networks with 2 and 3 hidden units and what would be the probabilities of obtaining the statistic's value assuming the null hypothesis to be true. It is clear from that table that all hidden units are relevant in networks with 2 such units. A third unit is not, however, relevant: the corresponding value of the statistic R_h is 0.631 and the probability of getting that value or higher under H_0 is about 0.85. This shows the optimal number of hidden units to be 2. Once we have arrived at the

Table 1. Typical hidden and input unit relevance values and probabilities of getting such a value or higher under the null hypothesis.

Hidden unit relevance						
No. Hidden U.	R_h			H_0 Prob.		
2	9.14 10^4	9.07 10^4		0	0	
3	7.41 10^4	8.90 10^4	0.63	0	0	0.85
Input unit relevance						
No. Inputs	R_d			H_0 Prob.		
3	5.84 10^3	10.2	4.20	0	0.06	0.13
2	6.01 10^3	1.07		0	0.58	

$3 \times 2 \times 1$ architecture optimal at this point, we will study input feature relevance, removing step by step those features found not to be relevant. This is done using the statistic (3) which now approximately follows a χ^2_2 distribution. The second half of table 1 shows the values of R_d for networks with 3 and 2 input units and the corresponding probability levels under the null hypothesis. The 3–input row shows that the null hypothesis can be very safely rejected for the first input, while it would be rejected at the 6 % level for the second input and at a relatively high 13 % level for the third input. Observe also that the first input relevance is about 1000 times bigger than those of the other inputs. If the third feature (gaussian inputs) is rejected and its input unit removed, the 2–input row shows that H_0 for the second feature (uniform inputs) would have to be rejected at a clearly much too high 58 % level, while the null hypothesis can be rejected for the first input: under H_0, the probability of getting such a value for the statistic is again 0. This yields the final optimal $1 \times 2 \times 1$ architecture.

References

1. J. Dorronsoro, F. Ginel, C. Sánchez, C. Santa Cruz, "Neural Fraud Detection in Credit Card Operations", IEEE Trans. in Neural Networks 8 (1997), 827–834.
2. K. Fukunaga, **Introduction to Statistical Pattern Recognition**, Academic Press, 1972.
3. R. Golden, **Mathematical Models for Neural Network Analysis and Design**, MIT Press, 1996.
4. E.B. Manoukian, **Modern Concepts and Theorems of Mathematical Statistics**, Springer, 1986.
5. K.V. Mardia, J.T. Kent, J.M. Bibby, **Multivariate Analysis**, Academic Press, 1979.
6. B.D. Ripley, **Pattern Recognition and Neural Networks**, Cambridge U. Press, 1996.
7. C. Santa Cruz, J.R. Dorronsoro, "A nonlinear discriminant algorithm for feature extraction and data classification", IEEE Transactions in Neural Networks 9 (1998), 1370–1376.
8. H. White, "Learning in artificial networks: a statistical perspective", Neural Computation 1 (1989), 425–464.

Neural Learning Invariant to Network Size Changes

Vicente Ruiz de Angulo and Carme Torras

Institut de Robòtica i Informàtica Industrial (CSIC-UPC),
Parc Tecnològic de Barcelona-Edifici U, Llorens i Artigas 4-6, 08028-Barcelona, Spain
ruiz@iri.upc.es, torras@iri.upc.es

Abstract. This paper investigates the functional invariance of neural network learning methods. By functional invariance we mean the property of producing functionally equivalent minima as the size of the network grows, when the smoothing parameters are fixed. We study three different principles on which functional invariance can be based, and try to delimit the conditions under which each of them acts. We find out that, surprisingly, some of the most popular neural learning methods, such as weight-decay and input noise addition, exhibit this interesting property.

1 Introduction

This work stems from an observation we made in analyzing the behaviour of a deterministic algorithm to emulate neural learning with random weights. We found that, for a fixed variance greater than zero, there is a number of hidden units above which the learned function does not change, or the change is slight and tends to zero as the size the network grows [7]. Here we study the conditions a neural learning algorithm should satisfy in order to lead to the same function, irrespective of network size.

Methods for complexity reduction [1] usually include one parameter (and sometimes more than one) to regulate the simplicity or smoothness imposed on the function implemented by the network. Each method simplifies the network in a way that is supposed to be optimal for the class of functions that is being approximated. Thus, ideally, the optimal level of smoothing should be obtained only by manipulating the above-mentioned parameter. Variability caused by other sources must be considered spurious uncertainty. For example, functionally different minima obtained by the same algorithm when running on different architectures or when departing from different initial points are embarrassing for the practitioner, who would desire to be freed from having to optimize the algorithm also along these lines. In particular, the selection of the number of hidden units of the architecture can influence decisively the result and is computationally cumbersome.

This motivates the interest in complexity minimization methods that show dependence only on the explicit complexity parameter and not on the size of the chosen architecture. However, there are no claims about functional invariance for the known methods, although Neal [4] devised a prior such that the complete bayesian procedure using it can be considered functionally invariant (see Section 3.1). In what

G. Dorffner, H. Bischof, and K. Hornik (Eds.): ICANN 2001, LNCS 2130, pp. 33–40, 2001.

follows we put forth some theoretical arguments and present some experimental results indicating that functional invariance may be a rather common phenomenon, even in well-known methods used for a long time by the connectionist community.

We also try to delimit the conditions that a complexity reduction method must satisfy in order to yield functional invariance. The paper focuses on the regularization methods for complexity reduction [1] and those that can be made equivalent to them. Regularization consists in adding a penalty function to the error function that regulates the complexity of the implemented network via a multiplicative factor called regularizer coefficient.

2 The phenomenon: learned-function invariance

We shall first define clearly the phenomenon under study, namely learned-function invariance. Let $F(X,W)$ and $G(X,W')$ denote the input-output functions implemented by two feedforward networks having equal number of input and output units, but different number of hidden units, and with weight vectors W and W', respectively. The functional distance between $F(\bullet,W)$ and $G(\bullet,W')$ is defined as

$$dist(F(\bullet,W),G(\bullet,W')) = \frac{1}{Vol(\Omega(s))} \int_{\Omega(s)} \left(F(X,W) - G(X,W')\right)^2 dX$$

when s tends to infinity, $\Omega(s)$ being a cube of side s in the input space.

Now, let $M(F,\lambda)$ be the optimum weight vector obtained with network F by a learning method M involving some complexity reduction regulated by the parameter λ. The name "method" is used here to denote an idealized algorithm, usually characterized by the minimization of an objective function, that always find global optima. If M is a regularization method then

$$M(F,\lambda) = \underset{W}{\operatorname{argmin}}\; C(W) = \underset{W}{\operatorname{argmin}}\; E(W) + \lambda R(W),$$

where $E(W)$ is the standard error function and $R(W) > 0$ is the regularization term.

Finally, let $\{F_n\}$ be a family of one hidden-layer architectures differing only in the number n of hidden units. We say that the algorithm M yields functional invariance for the network family $\{F_n\}$ if $dist(F_i(\bullet,M(F_i,\lambda)),F_{i+1}(\bullet,M(F_{i+1},\lambda)))$ tends to zero when i tends to infinity for every $\lambda > 0$.

It is necessary to make some remarks about these definitions. First, we only consider global minima of $C(W)$ in the definition of $M(F,\lambda)$, and not local minima, saddle points or other points resulting from a numerical optimization of $C(W)$. Second, all the global minima of $C(W)$ must be functionally equivalent, or the distance $dist(F_i(\bullet,M(F_i,\lambda)),F_{i+1}(\bullet,M(F_{i+1},\lambda)))$ would not be well defined. Obviously it is impossible to fulfill this condition for $\lambda = 0$, but not for $\lambda > 0$. This is related to the explicit exclusion of local minima from the functional invariance definition, since global and local minima, having different values of E, produce different outputs for the training patterns, which implies that the two minima cannot be functionally equivalent. It is possible to extend this definition using families of architectures that are not limited to one hidden layer of units. The only condition is that the elements of the family can be indexed in such a way that, given an arbitrary precision, for any given functional continuous mapping (from the input to the output space), there exists an index value such that architectures with higher indices can approximate the mapping with that precision. The last remark is that the definitions are independent from the training set, the only implicit requirement being that any non-empty training set would bring the functional distance limit to zero.

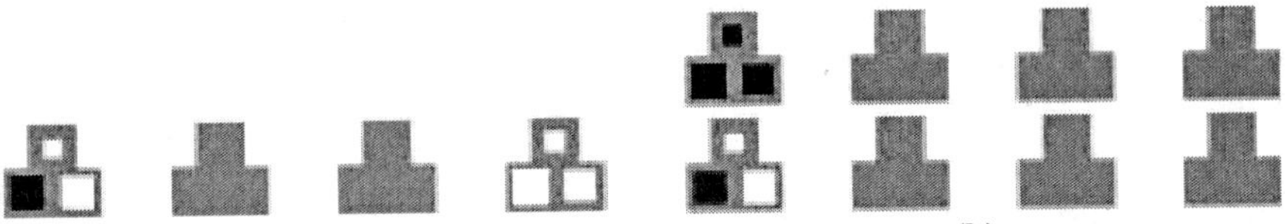

(a) **(b)**

Fig. 1. The results of training a 4-HU network (a), and an 8-HU network (b) are shown. Each of the blocks in the figure contains all the weights associated to a hidden unit. Weights are represented by squares, white if their value is positive, and black if their value is negative. The size of the square is proportional to the magnitude of the weight. The top weight in the block is the weight from the hidden unit to the output unit. The weights in the bottom row correspond to the weights incoming to the hidden unit.

Figure 1 shows two networks trained with the same 20 training points, randomly drawn from the function $.3x^3 + .3x^2 + 10/\ 3(x+3)^2$ in the interval [-1.5 , 1.5], using a deterministic algorithm to emulate learning with random weights [5, 6]: the mean of the weight distribution is adapted to minimize the average error over the distribution. The complexity of the function implemented by that mean is regulated via the variance of the distribution of the weights. Both the four hidden-units (HU's) network and the eight HU's network were trained using the same variance. It can be seen that the distance between the two resulting networks is null, i.e., they are functionally equivalent. Clearly, the weights of the first unit in Figure 1(a) are the same as those of the fifth unit in Figure 1(b). Moreover, the fourth unit in Figure 1(a) shows a direct correspondence with the first unit in Figure 1(b): the weights have pairwise the same magnitude, and since the signs of both the input weights and the output weights are inverted, the two units have the same functionality. It can be concluded that the two networks implement the same approximation of the desired function.

Applying the algorithm to any network with a large number of hidden units, we obtain the same units in different positions and combinations of sign inversions that produce always the same input-output function. Testing the algorithm with other variances produces other configurations that clearly converge to some function in the limit of infinite HU's. However, the closer is the variance to zero, the more difficult is the optimization, as the configurations found become more and more complex, and convergence is attained much more slowly as the number of HU's grows. This is a feature common to all the algorithms that we have explored: it is hard to check functional invariance when the complexity reduction constraints are loose.

3 Conditions for the Appearance of Invariance

Initially, when we found functional invariance while experimenting with the random weights learning algorithm, we thought it was a rather unique phenomenon, in the sense that it was particular to the kind of weight configurations that our algorithm created, or at least, that any algorithm exhibiting functional invariance should produce weights sharing the same essential properties. However, a deeper reflection revealed that functional invariance can appear when using different algorithms, and due to very diverse reasons. Up to now, we have identified three types of algorithms corresponding to three principles in which functional invariance can be based.

3.1 Neal's Type Priors

Regularization methods can be considered under a bayesian perspective by viewing $E(W)$ and $\lambda R(W)$ as the negative log probability (disregarding some constants) of the output distribution of the function being approximated and the weight distribution, respectively. Then, $C(W)$ can be shown to be equivalent to the negative log of the posterior weight probability, and its minimization corresponds to a "maximum a posteriori" procedure.

However, the use of the same prior on two different architectures does not imply a direct relationship between the functions they implement. In fact, a prior over the weights in different architectures can induce different priors over the output functions.

Neal [4] devised a prior over the weights that, although inducing also a different prior over functions for each network, converges to a unique one as the number of hidden units tends to infinity.

Convergence of the input-output functions posterior implies convergence of its mode. Thus, a procedure optimizing this posterior may be used to obtain functional invariance. Despite this, a prior over functions in the infinite number of HU's limit is not enough to directly imply the functional invariance of the minimization of $C(W)$. In fact, this minimization finds the most probable weight vector, which *does not correspond* to the most probable input-output function, because there is a Jacobian determinant factor mediating the two probability densities [8], and the minimization of $C(W)$ optimizes the posterior of the weights, not that of the input-output functions.

The true bayesian procedure, however, does not consist simply in finding the mode of the weight posterior, as carried out by the usual regularization method. Instead, it takes into account the complete probability distribution to generate an answer. For example, the bayesian answer to the question "what is the *best* output for a given X?" under a quadratic loss function would be guessing the average of the values for that point of the posterior of the input-output functions. This involves an integration over the probability space that cancels out the Jacobian determinant, so that it is the same to integrate over the weight posterior as to integrate over the input-output function posterior. Thus, this type of answer, considered as the output of the learning algorithm, makes the complete bayesian procedure functionally invariant as we have defined it.

3.2 Regularizers Implying a Target Mean Function

Input noise addition during training is equivalent to Tikhonov regularization when the number of patterns of the training set is infinite [1]. Even with finite training sets, it is equivalent to the addition of a penalization term [5, 6], although this is not a classical regularizer because it involves the output training patterns and depends on $E(W)$. However, the function invariance property of noise addition is better understood by taking a wider perspective. The minimization of a quadratic $E(W)$ can be viewed as an attempt to estimate with the network the mean values taken by the output patterns for a given input. Usually, there are none or very few desired values for each point in the input space. However, with input noise addition, potentially infinite patterns are available, and the expected output pattern for a given input point is [3]:

$$m(X) = \sum_{s=1}^{np} Y_s \, p(X - X_s) \Big/ \sum_{k=1}^{np} p(X - X_k)$$

where $\{X_i, Y_i\}_{i=1..np}$ are the original (not noisy) training patterns and p is the probability density function of the noise. We require $p(X)$ to be non zero for every X in the input space, so that the randomly generated patterns cover it completely. This assures that $m(X)$ is always well-defined, since otherwise the denominator could be zero somewhere. We need $m(X)$ to be well-defined over the entire input space, because the zones of the input space that do not have desired values leave the network free to interpolate arbitrarily. Instead, if a target function is defined over all the input space, any family of networks with enough approximation power has to approximate the same function without degrees of freedom left. Thus, the fundamental condition for noise addition to be a function invariant method is the use of infinite domain probability density functions.

An important remark is that the existence and uniqueness of $m(X)$ independently of the architecture used, makes input noise addition functionally invariant in the more general sense given in Section 2, not limited to one-hidden layer architectures.

3.3 Decomposable Regularizers

Let us now talk about regularizers for one-hidden layer networks that are additively decomposable, $R(W) = \sum_u r(\mathbf{w}_u)$, where $\mathbf{w}_u$ is the vector of all incoming and outcoming weights of hidden unit u. We require also that $r(\mathbf{w}_u)$ has a minimum value of zero attained when $\mathbf{w}_u = \mathbf{0}$[1]. These regularizers exhibit functional invariance if there exists a threshold for $r(\mathbf{w}_u)$, such that when the number of units tends to infinite, all the units under this threshold tend to the minimum of $r(\mathbf{w}_u)$, i.e., their associated weights tend to zero.

This can be argued as follows: suppose we have a network with n hidden units that has been brought to a global minimum of $C(W)$. In the limit of $n = \infty$, if we order the values of $r(\mathbf{w}_u)$, this sequence must tend to zero or, otherwise, $R(W) = \infty$ in the

[1] As a matter of fact, it would suffice that the weights from the input layer (excluding the bias unit) to u were zero in the minimum of $r(\mathbf{w}_u)$.

minimum. Thus, there exists a finite number N of units whose $r(\mathbf{w}_u)$ is above the threshold and, therefore, are significantly non linear. The remaining $n-N$ units contribute globally with a purely linear mapping A to the input of the output layer.

For a network with $n-1$ hidden units, a configuration with the same N nonlinear units and all but one of the same $n-N$ linear units, reproduces the same function implemented by the n HU's with a difference that tends to zero as n grows. With an appropriate scaling of the linear units, the cost $C(W)$ would be the same because, on the one hand, since the $n-N-1$ units are linear, the mapping A can be recovered perfectly (therefore keeping $E(W)$) and, on the other hand, since they are on the minimum of $r(\mathbf{w}_u)$, the infinitesimal scaling does not affect the regularization cost, because $\partial r/\partial \mathbf{w}_u = 0$.

This is a global minimum of the $n-1$ HU's network. If there was a lower minimum with different nonlinear units or a different linear mapping, it would be easy to build a weight configuration in the n HU's having the same $E(W)$ and the same or lower $R(W)$, which would contradict the hypothesis that the original minimum of $C(W)$ for the n HU's network was global. Thus, functional invariance is guaranteed.

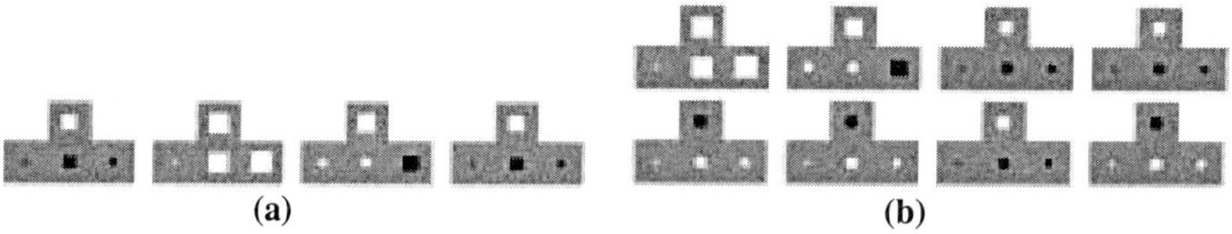

Fig. 2. Results of training a 4-HU network and a 8-HU network with 25 samples of the function $\sin(x_1+x_2)$, using a weight-decay coefficient value of .4. The resulting configurations contain replicated units that fill all the spare units available.

The deterministic algorithm to emulate learning with random weights, which triggered this work, belongs to this category. It relies on the equivalence of weight noise addition with the addition of a decomposable regularizer, which for networks with linear function activations at the linear units takes the form [6,7]:

$$r(\mathbf{w}_u) = a\, y_u^2 + b\textstyle\sum_m w_{mu}^2 y_u'^2$$

where a and b are constants, w_{mu} is the weight from the hidden unit u to the output unit m, and y_u and y_u' are the activation function of u and its derivative, respectively. It can be observed that this regularizer satisfies the condition of having minimum at zero only when symmetric activation functions are used.

But the most surprising example of this type of regularizer is the old good weight-decay. Careful experiments indicate that it satisfies the non-obvious condition mentioned in the first paragraph of this section. The theoretical demonstration is work still in progress. As an example, see Figure 2, where the weights resulting from training a 4-HU network and a 8-HU network with 25 samples of the function $\sin(x_1+x_2)$ are shown. Linear activation functions were used at the output units. One can see that in Figure 2(a) the first and the fourth units are replicated. In Figure 2(b), the same non-replicated units as in (a) appear, and the two units that were replicated in (a) appear six times in (b) but with smaller weight magnitudes. This is the scaling we were talking about before. When the number of hidden units is very large, the weight magnitudes of the replicated units become practically null, but they keep globally forming the same linear mapping.

We approximate experimentally the functional distance as the mean square distance between the outputs of two networks in a grid of 10,000 points regularly distributed in the input domain. Figure 3 shows the functional distances between architectures with different number of hidden units, minimized for several weight-decay coefficient values α. Since for some α's, some of the networks fell in local minima, in these cases we used the best minimum selected from three different random trials. It is evident that as α grows, all the architectures tend to produce the same results. But the most interesting observation from this graph is that, for any $\alpha > 0$, the distances between architectures decrease very quickly as the number of hidden units grows, and are indistinguishable from zero above 50 units. Notice that the comparisons involve networks that differ the more in the number of hidden units, the larger are the architectures. The above observation agrees with the expectation of a tendency to closer similarity for larger nets. Of course all the architectures exhibit almost the same generalization error for any positive α, especially those above 50 HU's.

Discussion

In this paper we have put forth a definition of functional invariance, which basically states that a learning method is functionally invariant if, when applied to increasingly large networks, the output for every possible input tends to a limit, and have examined what kinds of methods possess this property. We have identified three mechanisms that can originate functional invariance:

- bayesian treatment of neural networks with weight priors that converge to a prior over functions,
- implicit definition of a mean target for the complete input space, and
- additively decomposable regularizers that produce minima with a finite number of nonlinear units in the limit of infinite units.

Examples of the first two types of mechanisms are bayesian learning with Neal's type weight priors, and input noise addition using probability density functions taking non-zero values in all their domain, respectively. Examples of the third type are the regularizer that emulates learning with random weights and classical weight-decay. There are relations between these mechanisms, but it is difficult to see a unifying principle. For example, the second type can be viewed as defining a prior over input-output functions, as the first type does, namely one that always concentrates the probability on a single function. However, this prior is the same for all networks using the second type of mechanism, while in the first mechanism the prior over functions is approximated only for large networks. In addition, since the target function is completely defined in the second type of mechanism, the distance to that function is directly minimized, whereas in the first type, averaging over the probability distributions is required to guarantee functional invariance.

There are also differences in the type of units that these mechanisms generate. The third type produces only a finite number of nonlinear units in the infinite limit, while the second gives rise to an infinite number of them. Take into account that the implicit target function can be anyone and, therefore, an infinite number of nonlinear units is required to approximate it [2].

It could seem strange that no one (up to our knowledge) had observed the functional invariance of, for example, weight-decay. However, careful optimizations are required to observe regular patterns in the weight configurations and thus see this property with sharpness. This does not mean that these results are not of practical relevance; as two different but large networks are brought moderately close to a global minimum, the functional distance between them becomes very low. The problem of falling into local minima can be significant, but their frequency using weight-decay is apparently not high. For example, in the experiment of Figure 3 comparing the functional distances of several nets with different weight-decay coefficients, among the numerous optimizations required, only three times we got a local minimum in a single trial.

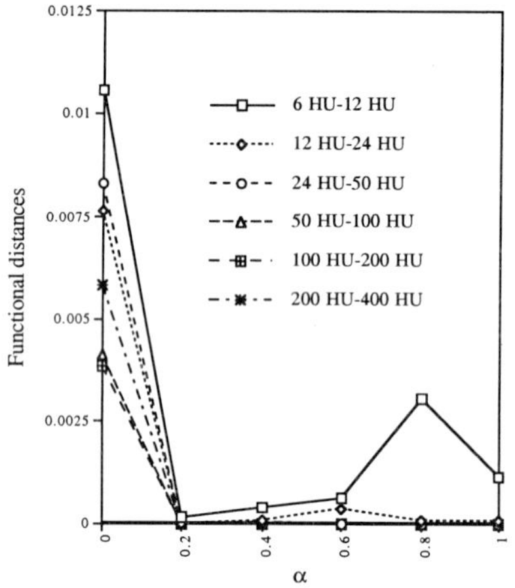

Fig. 3. Evaluation of the similarity between the functions implemented by architectures with different numbers of hidden units. Values spanning a wide range of the weight-decay coefficient α are tested.

References

1. Bishop, C.M.: "Neural networks for pattern recognition". Oxford University Press, 1995.
2. Hornik, K.: "Approximation capabilities of multilayer feedforward networks". *Neural Networks*, Vol 4 (2), pp. 251-257, 1991.
3. Koistinen, P. and Holmstrom, L.: "Kernel regression and backpropagation training with noise", Advances in Neural Information Processing Systems 4, Morgan-Kauffman, 1992.
4. Neal, R.M.: "Bayesian learning for neural networks". Springer-Verlag, New York, 1996.
5. Ruiz de Angulo, V. and Torras, C.: "Random weights and regularization", ICANN´94, Sorrento, pp. 1456-1459, 1994.
6. Ruiz de Angulo, V. and Torras, C.: "A deterministic algorithm that emulates learning with random weights", *Neurocomputing* (to appear).
7. Ruiz de Angulo, V. and Torras, C.: "Architecture-independent approximation of functions", *Neural Computation*, Vol. 13, No. 5, pp. 1119-1135, May 2001.
8. Wolpert, D.H. (1994): "Bayesian backpropagation over I-O function rather than weights", Advances in Neural Information Processing Systems 6, Morgan Kauffman, 1999.

Boosting Mixture Models for Semi-supervised Learning

Yves Grandvalet[1], Florence d'Alché-Buc[2], and Christophe Ambroise[1]

[1] Heudiasyc, UMR CNRS 6599,
Université de Technologie de Compiègne,
BP 20.529, 60205 Compiègne cedex, France
{Yves.Grandvalet,Christophe.Ambroise}@hds.utc.fr
[2] LIP6, UMR CNRS 7606,
Université Pierre et Marie Curie,
4, place Jussieu, 75252 Paris Cedex, France
florence.dalche@lip6.fr

Abstract. This paper introduces MixtBoost, a variant of AdaBoost dedicated to solve problems in which both labeled and unlabeled data are available. We propose several definitions of loss for unlabeled data, from which margins are defined. The resulting boosting schemes implement mixture models as base classifiers. Preliminary experiments are analyzed and the relevance of loss choices is discussed. MixtBoost improves on both mixture models and AdaBoost provided classes are structured, and is otherwise similar to AdaBoost.

1 Introduction

In many discrimination problems, unlabeled examples are massively available while labeled ones are scarce. Such situations arise in applications where class labeling cannot be automated and requires an expert. The concerned domains range from ecology to industrial process diagnosis but the two archetypal ones are web and text mining. Nowadays the latter receives much attention, and recent papers show the considerable improvements provided by taking into account a large pool of unclassified documents [8].

This paper investigates whether boosting can be adapted to semi-supervised learning. Boosting is so successful in supervised classification, that Breiman, quoted by Friedman et al. [5], called AdaBoost with trees the "best off-the-shelf classifier in the world". Basically, boosting concentrates on difficult examples, also qualified as near-miss examples. It is not obvious that it may be adapted to unclassified examples, for which misclassification cannot be assessed, and still less that it can take advantage of them.

From the brief review of AdaBoost given in section 2, ways of extending boosting to semi-supervised problems are discussed in section 3. Section 4 then motivates the use of mixture models as base classifier and details an original implementation. The resulting algorithm for boosting mixture models, MixtBoost, is tested on preliminary experiments in section 5, while section 6 summarizes the main conclusions and presents perspectives of this work.

G. Dorffner, H. Bischof, and K. Hornik (Eds.): ICANN 2001, LNCS 2130, pp. 41–48, 2001.

2 AdaBoost

AdaBoost is an acronym standing for adaptive boosting. Boosting refers to the problem of improving the performance of any rough "base classifier" by combining. The adaptivity of AdaBoost refers to its capacity to boost without the prior knowledge of the accuracy of the base classifiers.

There are several versions of boosting, and among them several forms of AdaBoost. For clarity sake, we only consider here the special case of binary classification, but multi-class and regression tasks can also be tackled by this algorithm [3–5].

AdaBoost is an iterative learning method which produces a combination of simple classifiers. Provided the latter have error rates slightly better than random guessing, the final combination is ensured to be arbitrarily accurate. The final aggregated classifier is defined as $\text{sign}(g_T)$, with

$$g_T(\mathbf{x}) = \sum_{t=1}^{t=T} \alpha_t h_t(\mathbf{x}) \ , \tag{1}$$

where α_t denotes the (positive) weight of base classifier h_t in the combination.

The procedure for generating h_t and α_t from a learning set of $\{\mathbf{x}_i, y_i\}_{i=1}^n$ is summarized in Fig. 1. It only requires to define either a loss function l or a consistent margin ρ. Throughout this paper, we use the original proposition of Freund and Schapire [4], with $h(\mathbf{x}) \in [-1, 1]$ and $y \in \{-1, 1\}$. The loss is defined as

$$l(h(\mathbf{x}), y) = \frac{1}{2}|h(\mathbf{x}) - y| \ , \tag{2}$$

with the corresponding margin

$$\rho(h(\mathbf{x}), y) = 1 - 2l(h(\mathbf{x}), y) \ . \tag{3}$$

The latter relationship between loss and margin provides the usual definition of margin $\rho(h(\mathbf{x}), y) = h(\mathbf{x})y$.

1. Start with weights $w_1(i) = 1/n$, $i = 1, \dots, n$, and $g_0 = 0$
2. Repeat for $t = 1 \dots T$:
   (a) Fit the base classifier $h_t(\mathbf{x}) \in [-1, 1]$, providing it with the weighted empirical distribution defined by $\mathbf{w}_t = (w_t(1), \dots, w_t(n))$.
   (b) Compute the weighted error: $\epsilon_t = \sum_{i=1}^n w_t(i) l(h_t(\mathbf{x}_i), y_i)$.
   (c) Compute $\alpha_t = \log(1 - \epsilon_t) - \log(\epsilon_t)$, and $g_t = g_{t-1} + \alpha_t h_t$
   (d) Set the weights $w_{t+1}(i) = \exp(-\rho(g_t(\mathbf{x}_i), y_i))$ and normalize so that $\sum_{i=1}^n w_{t+1}(i) = 1$.
3. Output the final classifier $\text{sign}(g_T(\mathbf{x}))$

Fig. 1. AdaBoost algorithm

Many empirical studies have shown that AdaBoost is an efficient scheme for reducing the error rates of classifiers such as trees or neural networks. In the experiments of Schapire et al. [10], boosting almost systematically improves upon a single classifier. This success is explained by the beneficial effect of boosting on margin distributions, the latter being related to bounds on the prediction error. The relevance of this connection is challenged by Breiman [3], who furthermore indicates that boosting may be viewed as an iterative optimization technique minimizing some decreasing function of the margin. With this point of view, the number of boosting steps T (1) is similar to the number of iterations in other optimization algorithms. Overfitting can be avoided by an early stopping procedure (such as in [10]), by regularizing to get "soft margins" [9], or by replacing $\exp(-\rho)$ at step 2(d) of Fig. 1, by other margin functionals such as $1 - \tanh(\rho)$ [6].

3 Generalizing AdaBoost to Semi-supervised Data

Once AdaBoost is provided with a base classifier, its main components are the loss l and the margin ρ (at steps 2(b) and 2(d) in Fig. 1). The simplest way to generalize to semi-supervised problems consists thus in defining a loss/margin for unlabeled data. The generalization to unlabeled data should not affect the loss for labeled examples, and should penalize inconsistencies between the classifier output $h(\mathbf{x})$ and available information.

We first assume that a true label exists for each example. Furthermore, missing labels are not due to censoring, but simply reflect the absence of class information. In particular, this excludes the case where labels are missing depending on the true label. In probabilistic terms, they are supposed to be missing at random. Missing labels are thus interpreted as "the example belongs to one class, but this class is unknown".

On the one hand, as a missing label indicates that any label is possible, no output is inconsistent with this information. Our first "basic" generalization, denoted l_0, consists in giving a null loss to unlabeled examples.

On the other hand, even if the label is missing, the truth is that only one of them is correct. The loss should thus penalize inconsistencies with plausible labels. A first candidate solution is the loss associated with the most probable label

$$l_{\mathrm{P}}(h(\mathbf{x}), m) = l(h(\mathbf{x}), y_{\max}) \ \text{, with } y_{\max} = \operatorname*{Argmax}_{y} P(Y = y | X = \mathbf{x}) \ , \tag{4}$$

where m denotes a missing label. Our second proposal is the expected loss with respect to possible labels

$$l_{\overline{\mathrm{P}}}(h(\mathbf{x}), m) = \mathbb{E}(l(h(\mathbf{x}), Y)) \ . \tag{5}$$

Posterior probabilities $\mathrm{P}(Y = y | X = \mathbf{x})$ being unknown, some estimates have to be used. They can be derived from the classifier h or from the combined classifier g.[1] Considering that $(h(\mathbf{x})+1)/2$ or $(g(\mathbf{x})/\|\boldsymbol{\alpha}\|_1+1)/2$ estimates the posterior probability $\mathrm{P}(Y=1|X=\mathbf{x})$, two approximations of losses (4) (respectively denoted l_h and l_g) and (5) (respectively denoted $l_{\overline{h}}$ and $l_{\overline{g}}$) are defined.

With these losses, it should be kept in mind that the posterior probabilities are estimated. Paradoxical situations may arise when the latter are not provided by the classifier at hand h. In particular, an unlabeled example may be reported to be badly classified while its true label is unknown. This situation is prohibited by restricting the loss function to be in $[0,1/2]$ for l_g and $l_{\overline{g}}$.

Table 1 summarizes the losses and margins for missing labels which are considered throughout our experiments at step 2(b) of Fig. 1. Margins for h are obtained from losses by the linear transformation (3). Margins for g, required at step 2(d), are computed by replacing h by the normalized classifier $g/\|\boldsymbol{\alpha}\|_1$ in the second column, and unnormalizing $(\rho(g(\mathbf{x}),m)=\|\boldsymbol{\alpha}\|_1\rho(g(\mathbf{x})/\|\boldsymbol{\alpha}\|_1,m))$.

Table 1. Losses and margins for a missing label m

Loss	$l(h(\mathbf{x}),m)$	$\rho(h(\mathbf{x}),m)$	$\rho(g(\mathbf{x}),m)$
l_0	0	1	$\|\boldsymbol{\alpha}\|_1$
l_h	$\frac{1}{2}\lvert h(\mathbf{x})-\mathrm{sign}(h(\mathbf{x}))\rvert$	$\lvert h(\mathbf{x})\rvert$	$\lvert g(\mathbf{x})\rvert$
$l_{\overline{h}}$	$\frac{1}{2}(1-h(\mathbf{x})^2)$	$h(\mathbf{x})^2$	$g(\mathbf{x})^2 / \|\boldsymbol{\alpha}\|_1$
l_g	$\frac{1}{2}\min(\lvert h(\mathbf{x})-\mathrm{sign}(g(\mathbf{x}))\rvert,1)$	$\max(h(\mathbf{x})\mathrm{sign}(g(\mathbf{x})),0)$	$\lvert g(\mathbf{x})\rvert$
$l_{\overline{g}}$	$\frac{1}{2}\min(1-h(\mathbf{x})g(\mathbf{x}) / \|\boldsymbol{\alpha}\|_1,1)$	$\max(h(\mathbf{x})g(\mathbf{x}) / \|\boldsymbol{\alpha}\|_1,0)$	$g(\mathbf{x})^2 / \|\boldsymbol{\alpha}\|_1$

All loss options amount to assume that any classifier output is to some extent correct for unlabeled examples. Their error measure is systematically lower than $1/2$, and their corresponding margin is positive. For l_0, all outputs are equally correct, for l_h and $l_{\overline{h}}$, self-confident outputs are privileged, and for l_g and $l_{\overline{g}}$, outputs consistent with the aggregated classifier are preferred.

4 Base Classifier

The base classifier should be able to make use of the unlabeled data provided by the boosting algorithm. Mixture models are well suited for this purpose, as shown by their extensive use in clustering for representing different subpopulations.

Hierarchical mixtures provide flexible discrimination tools, where each conditional distribution $f(\mathbf{x}|y=k)$ is modelled by a mixture of components [2]. At the high level, the distribution is described by

$$f(\mathbf{x}|\Phi)=\sum_{k=1}^{K}p_k f_k(\mathbf{x};\theta_k) \ , \tag{6}$$

[1] The outputs of the normalized combined classifier $g(\mathbf{x})/\|\boldsymbol{\alpha}\|_1$, where $\|\boldsymbol{\alpha}\|_1=\sum_{t=1}^{T}\alpha_t$ is the ℓ^1 norm of $\boldsymbol{\alpha}$, span $[-1,1]$.

where K is the number of classes, p_k are the mixing proportions, θ_k the conditional distribution parameters, and Φ denotes all parameters $\{p_k; \theta_k\}_{k=1}^K$. The high-level description can also be expressed as a low-level mixture of components, as shown here for binary classification:

$$f(\mathbf{x}|\Phi) = \sum_{k_1=1}^{K_1} p_{k_1} f_{k_1}(\mathbf{x}; \theta_{k_1}) + \sum_{k_2=1}^{K_2} p_{k_2} f_{k_2}(\mathbf{x}; \theta_{k_2}) \ . \quad (7)$$

In this setting, the EM algorithm can be used to maximize the log-likelihood with respect to Φ considering that the incomplete data is $\{\mathbf{x}_i, y_i\}_{i=1}^n$ and the missing data is the component label c_{ik}, $k = 1, \ldots, K_1 + K_2$ [7].

We consider an original implementation of EM based on the concept of possible labels [1]. It is well adapted to hierarchical mixtures, where the class label y provides a subset of possible components. When $y = 1$ the firsts K_1 modes are possible, when $y = -1$ the lasts K_2 modes are possible, and when an example is unlabeled, all modes are possible.

A binary vector $\mathbf{z}_i \in \{0, 1\}^{(K_1+K_2)}$ indicates the components from which feature vector $\mathbf{x}_i$ may have been generated, in agreement with the assumed mixture model and the (absence of) label y_i. Assuming that the training sample $\{\mathbf{x}_i, \mathbf{z}_i\}_{i=1}^n$ is i.i.d, the weighted log-likelihood is given by

$$L(\Phi; \{\mathbf{x}_i, \mathbf{z}_i\}_{i=1}^n) = \sum_{i=1}^n w_t(i) \log\left(f(\mathbf{x}_i, \mathbf{z}_i; \Phi)\right) \ , \quad (8)$$

where $w_t(i)$ are the weights provided by boosting at step t. L is maximized using the following EM algorithm:

E-Step Computation of the expectation of $L(\Phi; \{\mathbf{x}_i, \mathbf{z}_i\}_{i=1}^n)$ conditionally to $\{\mathbf{x}_i, \mathbf{z}_i\}_{i=1}^n$ and the current value of Φ (denoted Φ^q):

$$Q(\Phi|\Phi^q) = \sum_{i=1}^n \sum_{k=1}^{K_1+K_2} w_t(i) u_{ik} \log\left(p_k f_k(\mathbf{x}_i; \theta_k)\right) \ , \quad (9)$$

$$\text{with } u_{ik} = \frac{z_{ik} p_k f_k(\mathbf{x}_i; \theta_k)}{\sum_\ell z_{i\ell} p_\ell f_\ell(\mathbf{x}_i; \theta_\ell)} \ .$$

M-Step Maximization of $Q(\Phi|\Phi^q)$ with respect to Φ.
Assuming that each mode k follows a Gaussian distribution with mean $\boldsymbol{\mu}_k$, and covariance $\boldsymbol{\Sigma}_k$, $\Phi^{q+1} = \{p_k^{q+1}; \boldsymbol{\mu}_k^{q+1}; \boldsymbol{\Sigma}_k^{q+1}\}_{k=1}^{K_1+K_2}$ is given by:

$$p_k^{q+1} = \frac{\sum_i w_t(i) u_{ik}}{\sum_i w_t(i)} \ ; \ \boldsymbol{\mu}_k^{q+1} = \sum_i \frac{w_t(i) u_{ik} \mathbf{x}_i}{\sum_i w_t(i) u_{ik}} \ ; \quad (10)$$

$$\boldsymbol{\Sigma}_k^{q+1} = \frac{1}{\sum_i w_t(i) u_{ik}} \sum_{i=1}^n w_t(i) u_{ik} (\mathbf{x}_i - \boldsymbol{\mu}_k^{q+1})(\mathbf{x}_i - \boldsymbol{\mu}_k^{q+1})^t \ . \quad (11)$$

The parameters of the hierarchical model are thus estimated altogether.

5 Preliminary Experiments

5.1 Experimental Setup

Preliminary tests of the algorithm are performed for three benchmarks of the boosting literature: banana [9], twonorm and ringnorm [3]. Banana is a two-dimensional two classes benchmark with 400 training patterns and 4900 test patterns. The class shapes are complex and the best reported test error rate is 10.7% [9]. Twonorm and ringnorm are 20-dimensional two classes benchmarks with 300 training patterns and 4000 test patterns. Classes are Gaussian clusters, Bayes error rates are low (respectively 1.5% and 2.3%), and Rätsch et al. [9] report error rates of 1.6% and 2.7%.

For all benchmarks, training is performed on 10 different samples and in five different settings where 0%, 50%, 75%, 90% and 95% of labels are missing. MixtBoost is tested for all margins defined in section 3, and is compared to mixture models and AdaBoost. MixtBoost and AdaBoost are trained identically, the only difference being that AdaBoost is not fed with missing labels. Both algorithms are run for $T = 100$ boosting steps, without protection against overfitting. The base classifier is a hierarchical mixture model with an arbitrary choice of 4 modes per class. Mixture models provided for comparison use the same hyper-parameters, but the algorithm (which may be stalled in local minima) is restarted 100 times from different initial solutions, and the best final solution (regarding training error rate) is selected.

5.2 Results

We report mean error rates together with the lower and upper quartiles in table 2. Compared to EM, AdaBoost provides usually good results for low rates of missing labels, but it is always outperformed for the highest rate. The mediocre performances of EM is due to convergence problems and to the criterion used to select a solution which may be inadequate when many examples are unlabeled.

Without missing labels, MixtBoost is equivalent to AdaBoost and thus provides identical error rates. For all settings, the loss l_0 yields equal or worse performances than AdaBoost. Providing the highest margin to unlabeled data results in an exponential forgetting of all unlabeled examples. It is thus puzzling to observe occasional big differences between l_0 and AdaBoost in twonorm and ringnorm. In fact, learning curves (not provided here) show that l_0 performs usually quite as well as AdaBoost in the very first boosting steps, but then its performances degrade heavily while AdaBoost stops rapidly.

For twonorm and ringnorm, the loss l_h never beats significantly AdaBoost, and often provide extremely bad solutions, where nearly the whole space is assigned to a single class, resulting in 50% error rates. The algorithm is extremely unstable; the combination coefficients diverge: at the base classifier level, unlabeled examples are classified with hard margins to opposite classes throughout the boosting process. The smoother version $l_{\overline{h}}$ provides seemingly better results, but the instability remains, and worse results should be expected when increasing the number of boosting steps. For Banana, the good performance of $l_{\overline{h}}$ (and

Table 2. Error rates (in %) obtained with 5 percentages of missing labels for mixture models (EM), AdaBoost (AB) and MixtBoost with the five margins defined in section 3. The first figure is the mean error rate, and the interval displays the interquartile range.

Banana	0%	50%	75%	90%	95%
EM	15.2[14.4,15.3]	18.2[16.7,18.6]	21.8[18.0,25.0]	26.1[20.7,29.8]	31.7[23.8,35.8]
AB	11.7[11.3,12.0]	12.6[11.7,13.1]	15.2[13.0,16.8]	22.1[18.0,24.3]	37.5[32.2,42.2]
l_0	11.7[11.3,12.0]	12.4[11.7,12.8]	15.3[13.0,16.9]	22.0[18.2,24.0]	35.2[23.9,40.4]
l_h	11.7[11.3,12.0]	12.8[12.2,13.3]	16.5[14.6,17.8]	23.9[20.1,24.7]	30.7[25.7,34.3]
$l_{\overline{h}}$	11.7[11.3,12.0]	12.3[11.7,12.9]	14.7[12.9,15.2]	17.5[15.3,19.3]	25.6[21.3,25.7]
l_g	11.7[11.3,12.0]	12.5[11.8,13.1]	15.0[13.3,15.4]	18.4[15.9,19.7]	32.5[22.1,35.6]
$l_{\overline{g}}$	11.7[11.3,12.0]	12.7[12.1,13.1]	15.1[13.6,16.4]	19.7[15.6,21.9]	35.3[26.8,40.7]
Twonorm	**0%**	**50%**	**75%**	**90%**	**95%**
EM	2.7[2.5, 2.9]	3.2[2.7, 3.1]	6.5[3.0, 9.0]	20.6[10.3,22.5]	24.8[18.3,31.9]
AB	3.2[2.9, 3.3]	3.2[2.9, 3.2]	3.2[3.0, 3.5]	11.0[5.2,14.2]	38.9[29.4,50.0]
l_0	3.2[2.9, 3.3]	3.1[2.7, 3.2]	2.9[2.7, 3.0]	28.0[6.0,38.5]	38.5[24.0,48.4]
l_h	3.2[2.9, 3.3]	50.0[50.0,50.0]	50.0[50.0,50.0]	49.6[49.7,50.0]	47.4[49.7,50.1]
$l_{\overline{h}}$	3.2[2.9, 3.3]	3.9[3.2, 3.8]	13.2[6.0,16.4]	16.1[8.3,20.6]	17.3[9.7,22.5]
l_g	3.2[2.9, 3.3]	2.9[2.6, 3.0]	3.0[2.8, 3.2]	4.2[3.4, 4.9]	6.2[3.6, 7.7]
$l_{\overline{g}}$	3.2[2.9, 3.3]	2.8[2.6, 3.0]	3.1[2.9, 3.2]	3.9[2.8, 3.5]	7.2[3.4, 8.4]
Ringnorm	**0%**	**50%**	**75%**	**90%**	**95%**
EM	1.9[1.7, 2.0]	2.1[1.7, 2.1]	4.3[1.9, 5.7]	9.5[2.7,12.0]	23.7[14.5,27.0]
AB	1.8[1.6, 1.9]	1.8[1.6, 2.0]	3.1[1.9, 4.1]	11.5[4.2,12.1]	28.7[11.5,37.6]
l_0	1.8[1.6, 1.9]	1.7[1.6, 1.7]	2.9[1.9, 2.5]	24.0[2.2,44.4]	45.8[42.9,49.7]
l_h	1.8[1.6, 1.9]	50.0[50.0,50.0]	49.9[49.9,50.0]	49.9[49.8,50.0]	50.0[50.0,50.0]
$l_{\overline{h}}$	1.8[1.6, 1.9]	2.0[1.7, 2.3]	8.2[3.1,10.3]	13.2[3.3,17.9]	29.9[11.9,46.8]
l_g	1.8[1.6, 1.9]	1.7[1.6, 1.7]	2.0[1.8, 2.2]	3.7[2.0, 3.8]	5.8[2.3, 6.3]
$l_{\overline{g}}$	1.8[1.6, 1.9]	2.0[1.7, 1.9]	2.2[2.0, 2.5]	3.7[1.9, 4.0]	4.5[2.9, 4.9]

to a lesser extent l_h), are also an artifact due to early stopping, as instability can also be observed on the combination coefficients.

Regarding the versions of MixtBoost based on the estimation of posterior probabilities by g, the results are qualitatively similar for twonorm and ringnorm, with very high improvements on both mixture models and AdaBoost for high rates of missing labels. These versions benefit from the consistent responses of base classifiers with the aggregated classifier g. On the other hand, one would expect $l_{\overline{g}}$ to be consistently more accurate than the hard version l_g, but there is no sign of such (nor opposite) systematic behavior. For banana, MixtBoost is not very efficient: its results are only slightly superior to AdaBoost for high rates of missing labels. We conjecture that the complex class structure prevents the base classifier to extract much information from unlabeled examples.

6 Conclusion

MixtBoost, a new boosting scheme for semi-supervised learning was proposed. It generalizes Adaboost by consistently extending the definitions of loss and margin to unlabeled data. Our experimental results show that boosting can improve the base classifier performances, especially when only a few labeled examples are provided. MixtBoost is thus attractive in applications such as web and text mining.

Large-scale experiments have yet to be performed, but in other experiments not reported here for lack of space, high improvements are still obtained for high Bayes error rates. MixtBoost can thus deal with difficult classification problems, provided data are structured into clusters, *i.e.* when unlabeled data convey relevant information for the classification task.

A semi-supervised task involves examples for which the label is precisely known or completely unknown. When labeling is performed by an expert, the true state of knowledge is better represented by membership to class subsets. For example, in medical diagnosis, a physician is sometimes able to discard some diseases, but not to pinpoint the precise illness of his patient. As mixture models can handle partial class information, MixtBoost should be easily extended to these partially supervised problems.

References

1. C. Ambroise and G. Govaert. EM algorithm for partially known labels. In *IFCS 2000*, july 2000.
2. C. M. Bishop and M. E. Tipping. A hierarchical latent variable model for data vizualization. *IEEE PAMI*, 20:281–293, 1998.
3. L. Breiman. Prediction games and arcing algorithms. Technical Report 504, Statistics Department, University of California at Berkeley, 1997.
4. Y. Freund and R. E. Schapire. A decision-theoretic generalization of on-line learning and an application to boosting. *Journal of Computer and System Sciences*, 55(1):119–139, 1997.
5. J. Friedman, T. Hastie, and R. Tibshirani. Additive logistic regression: a statistical view of boosting. *The Annals of Statistics*, 28(2):337–407, 2000.
6. L. Mason, J. Baxter, P. L. Bartlett, and M. Frean. Functional gradient techniques for combining hypotheses. In *Advances in Large Margin Classifiers*. MIT, 2000.
7. G. J. McLachlan and T. Krishnan. *The EM algorithm and extensions*. Wiley, 1997.
8. K. Nigam, A. McCallum, S. Thrun, and T. Mitchell. Using EM to classify text from labeled and unlabeled documents. *Machine Learning*, to appear.
9. G. Rätsch, T. Onoda, and K.-R. Müller. Soft margins for AdaBoost. *Machine Learning*, 42(3):287–320, 2001.
10. R. E. Schapire, Y. Freund, P. Bartlett, and W. S. Lee. Boosting the margin: A new explanation for the effectiveness of voting methods. *The Annals of Statistics*, 26(5):1651–1686, 1998.

Bagging Can Stabilize without Reducing Variance

Yves Grandvalet

Heudiasyc, UMR CNRS 6599,
Université de Technologie de Compiègne,
BP 20.529, 60205 Compiègne cedex, France
Yves.Grandvalet@hds.utc.fr

Abstract. Bagging is a procedure averaging estimators trained on bootstrap samples. Numerous experiments have shown that bagged estimates almost consistently yield better results than the original predictor. It is thus important to understand the reasons for this success, and also for the occasional failures. Several arguments have been given to explain the effectiveness of bagging, among which the original "bagging reduces variance by averaging" is widely accepted. This paper provides experimental evidence supporting another explanation, based on the stabilization provided by spreading the influence of examples. With this viewpoint, bagging is interpreted as a case-weight perturbation technique, and its behavior can be explained when other arguments fail.

1 Introduction

Bagging, introduced by Breiman in 1994, is a procedure building an estimator by a resample and combine technique [1]. From an original estimator, its bagged version is produced by averaging several replicates trained on bootstrap samples.

Regarding prediction error, it seems that bagging is an almost universal procedure for improving upon a single predictor. Hence, it is very important to understand the reasons for its successes, and also for its occasional failures. In many studies, the method almost systematically compares favorably with the original predictor, on artificial as on real data [1, 5, 10, 11]. Other ensemble methods such as boosting and arcing are often more effective in reducing test errors, but in situations with substantial noise, bagging performs better [5, 10].

Available explanations for bagging success are briefly reviewed in section 2, where the present explanation is also summarized. We then provide new experimental evidence supporting that bagging stabilizes prediction, in the sense that it equalizes the influence of each example in building the predictor. Two illustrative examples are given in regression and classification respectively in sections 3 and 4. Our findings are finally discussed in section 5.

2 How Does Bagging Work?

Breiman [1] presents bagging as a variance reduction procedure *mimicking* averaging over several training sets. The approximation taking place should be

G. Dorffner, H. Bischof, and K. Hornik (Eds.): ICANN 2001, LNCS 2130, pp. 49–56, 2001.

kept in mind: averaging is performed on bootstrap replicates of a single training set, and not on different training sets. Thus, although experimental results often shown the expected variance reduction [1, 11], several other arguments have been given to explain the success of bagging.

Schapire et al. [11] provide bounds for voting algorithms, including bagging, relating the generalization performance of aggregated classifiers to the margin distribution of examples. Unlike boosting, bagging does not explicitly maximizes margins, but experiments show that for complex classifiers, bagging produces rather large margins [11]. However, the obtained bounds are acknowledged to be loose, and even qualitatively misleading according to Breiman [2].

Friedman and Hall [7] provide an asymptotical analysis which concludes that, in the limit of infinite samples, bagging reduces a non-linear component of the predictor variance, while the linear part is unaffected. Bühlmann and Yu [3] present another theoretical study, focused on neighborhood of discontinuities of decision surfaces. From these two studies, we retain that bagging asymptotically performs some smoothing on the estimate.

Smoothing also occurs for finite samples, but it may not be the major effect of bagging. In particular, asymptotic analyses are not suited to the treatment of outliers, which are particularly well handled by bagging [5, 10]. In a previous study, the author [8] showed how bagging affects the potential influence of training examples on the predictor.

The limits of the "reduce variance by averaging" argument were already illustrated in point estimation on a very simple mean estimation [8]. On the one hand, bagging was shown to increase or decrease the estimator variance according to the experimental setup. On the other hand, each example of the training sample was shown to be reweighed, and this reweighing systematically equalized the influence of training examples. This paper presents two "classical" examples, one in regression and one in classification, where the variance argument also breaks down, while the explanation based on stabilization still holds.

3 Regression

The first example showing the limits of the variance argument was given by Breiman himself, in a linear regression experiment [1]. The original aim was to illustrate the benefits of bagging for unstable estimation procedures such as subset selection. A surprising by-product is that bagging is harmful for ordinary least squares (OLS). Breiman explains this failure by stating that, for OLS, averaging predictors trained on several independent datasets is better approximated by averaging over a single dataset drawn from the data distribution (original predictor) than by averaging over several bootstrap samples (bagged predictor). This statement implies that bagging fails when the variance argument reaches its limits. Here, we illustrate that the stabilization effect of bagging explains its successes and failures.

The experimental setup (see [1] for more details) consists in replicating 250 times: 1) draw samples of size 60 from the model $y = \beta^T \mathbf{x} + \varepsilon$, where ε is drawn from a normal distribution $\mathcal{N}(0,1)$, $\mathbf{x} \in \mathbb{R}^{30}$ and β has 27 non-zero coefficients (results are qualitatively equivalent for the two other setups described in [1]); 2) compute estimates of β by forward subset selection. 3) generate 50 bootstrap samples to compute bagged estimates.

The difference in quadratic prediction error is highly in favor of bagging for little subset sizes. With a single variable, the average error over the 250 experiments is 3.26 without bagging and 2.69 with bagging. This advantage decreases as the subset size increases, and there is a cross-over point past which bagging is detrimental. For OLS, the average error is 2.03 without bagging and 2.11 with bagging.

First, we stress that, in the present setting, the failure of bagged OLS can be proved to be due to an *increase* of variance, since both unbagged and bagged estimates are unbiased. For subset selection, there is no analytic expression for bias and variance. Their plug-in estimates show however that bias is not affected by bagging, while variance is reduced. In this experiment, the effect of bagging regarding variance is irregular. A regular effect on the potential influence of examples, which turns out to have different consequences in terms of variance, is exhibited below.

In OLS, potentially highly influential points are known as leverage points, and the smoothing or hat matrix provides the statistics commonly used to flag them. Let $\mathcal{T} = \{(\mathbf{x}_i, y_i)\}_{i=1}^n$ be the training set. The n-dimensional vector $\widehat{\mathbf{f}}$ of fitted values at $\{\mathbf{x}_i\}_{i=1}^n$ ($\widehat{f}_i = \widehat{f}(\mathbf{x}_i)$) can be written as $\widehat{\mathbf{f}} = \mathbf{S}\mathbf{y}$, where $\mathbf{y}$ is the n-dimensional vector of response variables and $\mathbf{S}$ is the $n \times n$ smoothing matrix [9]. The ith row of $\mathbf{S}$ represents the sequence of weights (or equivalent kernel) used to compute $\widehat{f}(\mathbf{x}_i)$. Each element of $\mathbf{S}$ is thus relevant for detecting leverage, but a good summary is provided by the diagonal elements S_{ii}, since $\partial\widehat{f}(\mathbf{x}_i)/\partial y_i = S_{ii}$. As trace($\mathbf{S}$) is a possible definition of the degrees of freedom of the smooth [9], S_{ii} can also be interpreted as the degrees of freedom spent to fit $(\mathbf{x}_i, y_i)$. Leverage points are thus associated with large S_{ii}. These statistics can be computed for bagged OLS [8], but not for subset selection, which is *not* a linear estimate as $\mathbf{y}$ intervenes in the subset choice. For the latter, generalized statistics $\widetilde{S}_{ii}$, based on a data perturbation approach similar to [4] are computed.

Fig. 1 compares the histograms of S_{ii} and S_{ii}^{bag} over all experiments. The distribution for OLS (left) is unimodal, centered on 0.5 (30 free parameters are set by 60 points). Bagging provides a similar distribution, also centered on 0.5, meaning that complexity is not modified. The spread is however halved, rendering the equalization of influence among examples.

The distribution is more complex for subset selection (right). A log-scale is used to highlight that it is bimodal (on a linear scale, the high spread of the minor mode makes it hardly visible). The main mode contains about 90% of points with small $\widetilde{S}_{ii}$ values (mean about 1/60). It gathers points with little influence on the choice of the element entering the subset. The other mode contains highly influential points, with a mean $\widetilde{S}_{ii}$ value of about 0.6. These points play a leading part in choosing the variable entering the subset. The mean number of degrees of freedom is about 4.6 with a single variable: this

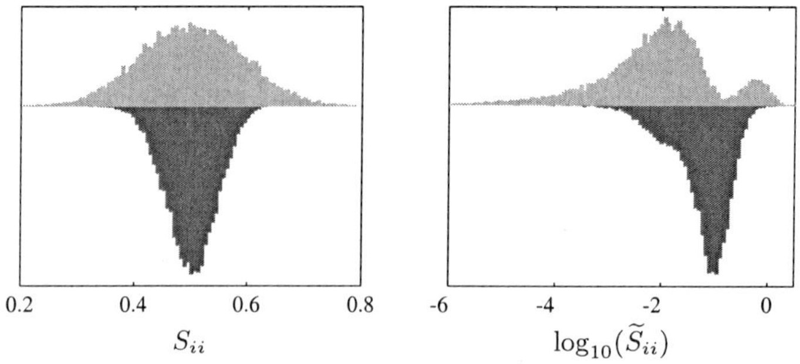

Fig. 1. Left: histograms of S_{ii} (up, light grey) and S_{ii}^{bag} (down, dark grey) for OLS (all variables); right: histograms of $\log_{10}(\widetilde{S}_{ii})$ (up, light grey) and $\log_{10}(\widetilde{S}_{ii}^{\text{bag}})$ (down, dark grey) for subset selection (one variable)

inflation is due to the subset choice, as discussed in [12]. The distribution for the bagged subset selection estimate is unimodal, centered on 0.1, with an average 5.9 degrees of freedom. Influence equalization is summarized by the halving in $\widetilde{S}_{ii}^{\text{bag}}$ spread.

The stabilization effect of bagging being displayed, its relationship with prediction error is investigated. The expected difference in prediction error between the OLS and the bagged estimate is a quadratic function in smoothing matrices $\mathbf{S}$ and $\mathbf{S}^{\text{bag}}$. The difference in normalized spread of leverage statistics,

$$\text{Spread}(\widehat{f}) = \frac{\frac{1}{n}\sum_{i=1}^{n} S_{ii}^2}{\left(\frac{1}{n}\sum_{i=1}^{n} S_{ii}\right)^2} - 1 \ , \tag{1}$$

is used as a (far from exhaustive) summary statistic to depict the trend in the relationship between leverage statistics and prediction error. Fig. 2 displays the difference in prediction error between bagged and original estimates according to this statistic. Each point represents one of the 250 experiments. For OLS (left), differences in prediction error and differences in leverage statistics are positively correlated, while the opposite trend is observed for subset selection (right).

For both regression schemes, performances of bagged predictors are thus shown to be related to the equalization of leverage statistics. The latter is performed blindly, without any assessment of the relevance of such a stabilization. In the considered experimental setting, stabilization happens to be detrimental for OLS (due to the absence of outliers) and beneficial for subset selection. For the latter, using more variables in the bagged estimate is not essential to bagging improvements, as bias is not affected in the process. The crux is that more points intervene in the subset choice, so that variance is reduced.

4 Classification

Failures of bagging in classification are rare. Schapire et al [11] provide however an example in the ringnorm benchmark. They report that stumps (single node trees) achieve a test error rate of 40.6%, compared to 41.4% for bagged stumps.

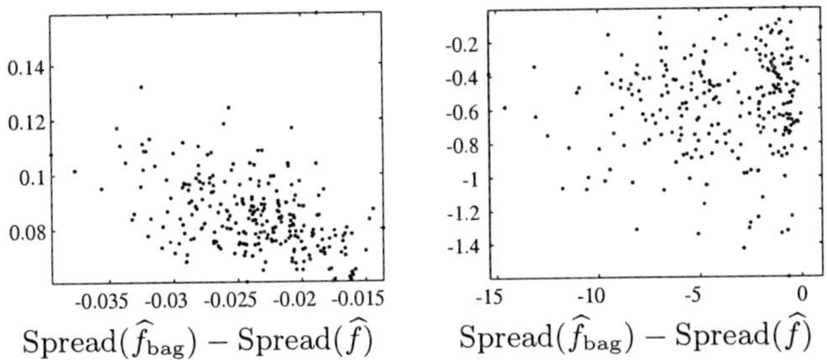

Fig. 2. left: expected difference in prediction error $\mathbb{E}(\text{PE}^{\text{bag}} - \text{PE})$ *vs* (Spread($\widehat{f}_{\text{bag}}$) − Spread($\widehat{f}$)) for OLS (all variables); right: observed difference in prediction error ($\text{PE}^{\text{bag}} - \text{PE}$) *vs* (Spread($\widehat{f}_{\text{bag}}$) − Spread($\widehat{f}$)) for subset selection (one variable).

Ringnorm was proposed by Breiman as a difficult task for decision trees, since the Bayes boundary is spherical. There are 20 features and 2 classes. Class 1 is multivariate normal with mean zero and covariance matrix four times the identity. Class 2 is multivariate normal with mean $(a\ a\ a \ldots a)$, $a = 1/\sqrt{20}$ and covariance matrix identity. The training sets comprise 300 examples, prediction error is estimated on test sets of size 10000, and 250 trials are performed.

In our experiments, the mean prediction error of the bagged estimate is identical to the one of the original estimate (40.4%, with an insignificant difference of 0.05% between the two means). Significant differences between the two estimates are however observed in 35% of experiments (at the 5% level), with a standard error of differences in prediction error of 1.6%. The results are more variable with the bagged predictor, with a standard deviation of prediction error of 1.4% versus 0.7% for unbagged stumps.

There is no general agreement on what the bias/variance decomposition should be for classification problems (see [6] for a review in the binary case), and such a discussion is out of the scope of this paper. However, for all definitions retaining an additive decomposition where variance represents the variable part of prediction error, our results show that bagging increased variance.

As no classical leverage statistic exists in classification, the influence of example i is measured by the difference between the predictor $\widehat{f}$ trained with the whole training sample and the predictor $\widehat{f}^{-i}$ trained with the sample deprived of example i, $\mathcal{T}^{-i} = \{\mathbf{x}_j, y_j\}_{j \neq i}$. The most relevant difference measure in classification is the expected disagreement in decisions, *i.e* the probability $\Pr(\widehat{f}(X) \neq \widehat{f}^{-i}(X))$, which is easily accurately estimated in simulated experiments.

Fig. 3 displays the cumulative distribution of $\Pr(\widehat{f}(X) \neq \widehat{f}^{-i}(X))$ for stumps and bagged stumps. Histograms provide a poor visualization, since the distribution for stumps has a step at zero and is furthermore heavy-tailed.

For stumps, about 94% of examples have no influence on the predictor, in the sense that the classifier is not modified when one of them is deleted. The absence of the other examples causes important changes, with $\widehat{f}^{-i}$ disagreeing

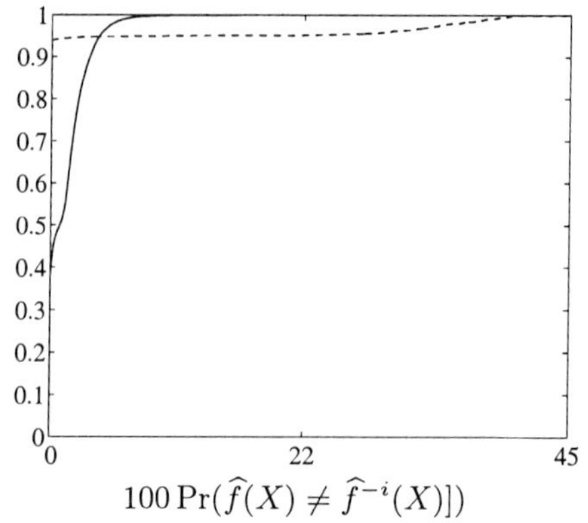

Fig. 3. Cumulative distribution function of $\Pr(\widehat{f}(X) \neq \widehat{f}^{-i}(X))$ (in %) for stumps (dashed) and bagged stumps (solid)

with $\widehat{f}$ in up to 45% of test cases. Overall, the mean disagreement is 1.7%, with a standard deviation of 7.3%[1].

For bagged stumps, only 37% of examples have no influence on the predictor, and $\widehat{f}^{-i}$ disagree with $\widehat{f}$ at most in 16% of test cases. Here, the mean disagreement is reduced to 1.3%, and its standard deviation is only 1.6%. Bagging is thus again shown to equalize the influence of examples.

The plot of differences in prediction error versus the variance of normalized influence does not show any trend. Influential points are sometimes beneficial and sometimes detrimental, resulting in identical prediction errors for stumps and bagged stumps. This plot is not reproduced here for lack of space; instead, we illustrate how influence is distributed on the margin distribution.

In this binary classification problem, coding labels and stump classifier outputs in $\{-1, 1\}$, the margin is defined as $\rho(\widehat{f}_{\text{bag}}(\mathbf{x}), y) = y\widehat{f}_{\text{bag}}(\mathbf{x})$ [11]. It represents the confidence by which the example is correctly classified (negative margins correspond to misclassification). The margin distribution is essential in the analysis of arcing or boosting algorithms [2, 11]. In particular, AdaBoost is shown to shift the margin distribution towards positive values by placing most weight on examples with small margins [11].

Fig. 4 displays the mean influence of points according to their margins. For stumps, most influential examples correspond to intermediate margins. Bagging results in a spread on the margin distribution, increasing the influence of hard and easy examples, while decreasing the one given to intermediate cases. Unlike AdaBoost, it does not concentrate specifically on difficult examples. This behavior may explain why bagging is less effective than AdaBoost when there is little classification noise, and why it is better otherwise [5, 10].

[1] Note that these important differences are not reflected by the prediction error of $\widehat{f}^{-i}$, as many different predictors can provide an error rate of about 40%.

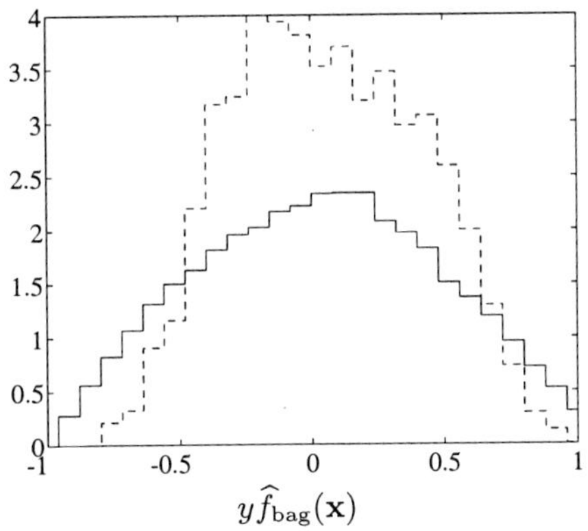

Fig. 4. Mean disagreement $\Pr(\widehat{f}(X) \neq \widehat{f}^{-i}(X))$ (in %) according to the margin of training examples for stumps (dashed) and bagged stumps (solid)

5 Discussion

In bagging, each example appearing the same number of times over all bootstrap samples, it may seem paradoxical that its weight is modified. Our experiments however illustrate that bagging equalizes the influence of examples in setting a predictor. Less point have a small influence, while the highly influential ones are down-weighted.

These observations support one of Breiman's early statements : the vital element for gaining accuracy thanks to bagging is the instability of the prediction method [1]. However, stability is no more related to variance, but to the presence of influential examples in the dataset for a given prediction method.

In many situations, highly influential points are outliers, and their down-weighting reduces the variance of the estimator. If an "unstable predictor" is characterized by severe leverage points, then our analysis is in accordance with Breiman's: bagging stabilizes estimation and reduces variance. But the present explanation, which is not rooted in the averaging argument, also applies when bagging fails: the stabilization effect of bagging is harmful when estimation accuracy benefit from influential points (the so-called good leverage points).

Regarding margins, the classification experiment showed that bagging does not concentrate specifically on difficult examples, but spreads the influence of examples on the margin distribution. This behavior may explain why bagging is much less effective than AdaBoost when difficult examples indicate class boundaries, and why it is more robust to classification noise.

This experimental study also illustrated the limits of the asymptotic analyses stating that a linear estimate is not affected by bagging in the limit of infinite datasets [7, 3]. Bagging was shown to modify the ordinary least squares (linear) predictor for a finite sample size. On the other hand, the up-weighting of points with little influence may be responsible for the smoothing effect which is observed at finite sample sizes and asymptotically remains.

With the influence equalization viewpoint, bagging can be interpreted as a perturbation technique aiming at improving the robustness against outliers. Indeed, averaging predictors trained on perturbated training samples is a means to favor invariance to these perturbations. Bagging applies this heuristic to case-weight perturbations. Bootstrap sampling, which is a central element of bagging according to the variance reduction argument, is only one possibility among others to provide these perturbations. The effectiveness of other resample and combine schemes involving subsampling [3, 7] can be understood within this perspective.

References

1. L. Breiman. Bagging predictors. *Machine learning*, 24(2):123–140, 1996.
2. L. Breiman. Prediction games and arcing algorithms. Technical Report 504, Statistics Department, University of California at Berkeley, 1997.
3. P. Bühlmann and B. Yu. Explaining bagging. Technical Report 92, Seminar für Statistik, ETH, Zürich, 2000.
4. A. N. Burgess. Estimating equivalent kernels for neural networks: A data perturbation approach. In M.C. Mozer, M.I. Jordan, and T. Petsche, editors, *Advances in Neural Information Processing Systems 9*, pages 382–388. MIT Press, 1997.
5. T.G. Dietterich. An experimental comparison of three methods for constructing ensembles of decision trees: Bagging, boosting and randomization. *Machine Learning*, 40(2):1–19, 2000.
6. J. H. Friedman. On bias, variance, 0/1 loss, and the curse of dimensionality. *Data Mining and Knowledge Discovery*, 1(1):55–77, 1997.
7. J. H. Friedman and P. Hall. On bagging and non-linear estimation. Technical report, Stanford University, Stanford, CA., January 2000.
8. Y. Grandvalet. Bagging down-weights leverage points. In S.-I. Amari, C. Lee Giles, M. Gori, and V. Piuri, editors, *IJCNN*, volume IV, pages 505–510. IEEE, 2000.
9. T. J. Hastie and R. J. Tibshirani. *Generalized Additive Models*, volume 43 of *Monographs on Statistics and Applied Probability*. Chapman & Hall, 1990.
10. R. Maclin and D. Opitz. An empirical evaluation of bagging and boosting. In *Proceedings of the Fourteenth National Conference on Artificial Intelligence*, pages 546–551. AAAI Press, 1997.
11. R. Schapire, Y. Freund, P. Bartlett, and W. S. Lee. Boosting the margin: A new explanation for the effectiveness of voting methods. *The Annals of Statistics*, 26(5):1651–1686, 1998.
12. R. J. Tibshirani and K. Knight. The covariance inflation criterion for adaptive model selection. *Journal of the Royal Statistical Society, B*, 61(3):529–546, 1999.

Symbolic Prosody Modeling by Causal Retro-causal NNs with Variable Context Length

Achim F. Müller and Hans Georg Zimmermann

Siemens Corporate Technology, Otto-Hahn-Ring 6, D-81739 Munich, Germany
{achim.mueller,georg.zimmermann}@mchp.siemens.de

Abstract. In this paper the application of causal retro-causal neural networks (NN) to accent label prediction for speech synthesis is presented. Within the proposed NN architecture gating clusters are applied enabeling the dynamic adaptation of a network structure depending on the actual input to the NN. In the proposed causal retro-causal NN, gating clusters are used to adapt the network structure such that the network has a variable context length. This way only available input feature vectors from the actual context window are treated. The proposed NN architecture has been successfully applied for accent label prediction within our text-to-speech (TTS) system. Prediction accuracy ranges at 83%. This result ranges higher than results achieved with tree-based (CART) methods on a corpus with similar complexity.

1 Introduction

In many TTS systems the complex task of f0-generation is split into two parts [3]. In the first part symbolic prosody labels are generated. These labels are used as inputs to the second part that generates f0-contours. Symbolic prosody labels can be separated into two types of labels: phrase break labels and accent labels. In our system phrase break labels are generated for a whole sentence prior to accent labels, using the method presented in [6]. In this paper the data-driven prediction of accent labels will be discussed. As input parameters part-of-speech (POS) tags and phrase break information are used.

For fast and easy adaptation to new languages and/or speakers a data-driven approach is favorable. For symbolic accent label prediction prominent data-driven approaches are based on classification and regression trees (CARTs) [7]. In [8] feed-forward NNs and tree-based learning methods have been used to predict word and syllable prominence. The mentioned approaches are based on previously fixed tree- or NN-structures. In our approach structures can be adapted depending on the actual input to the NN.

Our system works on sentence level, i.e. our goal is to generate a neutral prosody for each sentence as if the sentence was read isolated from surrounding sentences. Further, we determine context windows. The context windows are shifted across each sentence. A problem occurs at sentence boundaries: at sentence beginning no left context is available and at sentence end no right

G. Dorffner, H. Bischof, and K. Hornik (Eds.): ICANN 2001, LNCS 2130, pp. 57–64, 2001.

context is available. By applying gating clusters within the presented NN architecture, a dynamic adaptation of the network structure is proposed to handle the variable context information within the context windows. The architecture is embedded in a causal retro-causal NN as presented in [10]. In this framework time-interdependence of variables can be captured. This way the complex dynamical process of accent label prediction can possibly be modeled more accurate than with tree-based methods or simple feed-forward NNs.

The paper is organized as follows: In section 2 the applied causal retro-causal NN structure is explained and the problem encountered at sentence boundaries when applying this structure is described. In section 3 a solution to this problem is presented by applying gating clusters that can be used for dynamic network adaptation. In section 4 a method is presented to find an optimal context window length data-driven and results are discussed. Finally, in section 5 a conclusion of the presented work is given.

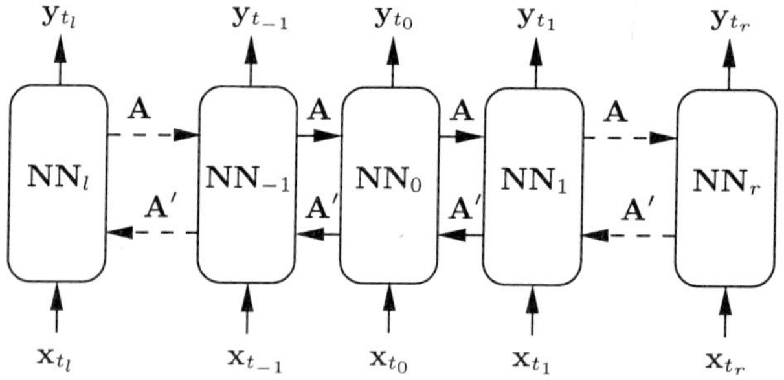

Fig. 1. Block diagram of a causal retro-causal NN.

2 Causal and Retro-causal NN

Figure 1 shows the block diagram of a causal retro-causal NN [10]. As can be seen the system is build up by connecting subnetworks $\mathbf{NN}_i$, $i \in \{l, \ldots, r\}$, $l < 0$, $r > 0$. Each subnetwork is associated with one time step. The connection of these subnetworks is realized by shared weight matrices $\mathbf{A}$ and $\mathbf{A}'$ (shared weights means to use the same set of weights in each connector denoted by the same upper case bold face letter). $\mathbf{A}$ propagates the causal information flow (from past to future) and $\mathbf{A}'$ propagates the retro-causal information flow (from future to past). Within the architecture of figure 1 shared weight matrices allow a symmetric handling of past and future information. In our application past refers to left context and future refers to right context.

As can be seen in figure 1, each $\mathbf{NN}_i$ has a vector $\mathbf{x}_{t_i}$ as input and a vector $\mathbf{y}_{t_i}$ as output. $\mathbf{x}_{t_i} \in \mathcal{R}^m, \forall i$ and $\mathbf{y}_{t_i} \in \mathcal{R}^n, \forall i$, i.e. the input and output vectors for each subnetwork $\mathbf{NN}_i$ are of the same dimension. The input to the whole system is given by a sequence of input vectors $\mathbf{x}_{t_l}, \ldots, \mathbf{x}_{t_0}, \ldots, \mathbf{x}_{t_r}$ and the output is given by a sequence of output vectors $\mathbf{y}_{t_l}, \ldots, \mathbf{y}_{t_0}, \ldots, \mathbf{y}_{t_r}$.

In our application the input vectors $\mathbf{x}_{t_l}, \ldots, \mathbf{x}_{t_0}, \ldots, \mathbf{x}_{t_r}$ of figure 1 represent a sequence of feature vectors. Feature vectors are calculated on word level, i.e. each time step t_i in figure 1 is associated with one word. A feature vector contains the information mentioned in section 1, i.e. POS information and phrase break information. Output vectors are also determined on word level and contain the accent information for the word. Both, input and output information are of nominal structure and are therefore coded with 1-out-of-k codes.

As mentioned in section 1 our system works on sentence level, i.e. no sentence interdependencies are modeled. Therefore, a problem occurs at sentence boundaries: At the beginning (end) of a sentence no left (right) context is available, i.e. no defined input can be determined for the first (last) l (r) feature vectors $\mathbf{x}_{t_i}$ and the first (last) l (r) output vectors $\mathbf{y}_{t_i}$ of the input/output sequence. The broader the left and right context, i.e. the greater l and r, the more feature vectors $\mathbf{x}_{t_i}$ and output vectors $\mathbf{y}_{t_i}$ are not defined.

If the used network structure is static, i.e. the structure cannot be adapted dynamically at runtime, a well defined input signal must be determined for each input neuron. In figure 1 this means an input vector $\mathbf{x}_{t_i}$ must be determined for each subnetwork $\mathbf{NN}_i$. This problem can be overcome by dynamic adaptation of the network structure as presented in the next section.

3 Dynamic Adaptation of Causal Retro-causal NN

In the first part of this section gating clusters are presented that can in principle be applied within any NN architecture for dynamic adaptation. In the second part it is explained how gating clusters are applied within our causal retro-causal NN.

3.1 Gating Clusters

When applying NNs it is generally desirable to design a problem adapted network structure, i.e. to incorporate prior information into the neural network design [4]. The better this adaptation, the better a process can possibly be modeled. Adaptation can e.g. mean optimization of the number of hidden layers or different clustering of neurons. These adaptations are all static, i.e. the adaptations are fixed prior to runtime of the system. Gating clusters enable an adaptation of the network structure at runtime, i.e. a dynamic adaptation.

Figure 2 illustrates how gating clusters can be applied[1]. Gating clusters are applied at the output of a hidden cluster. On the left hand side of figure 2 a cut out hidden cluster of a NN without a gating cluster is displayed. The cluster is connected with the rest of the network (denoted by **NN**) by the weight matrices

[1] Remark on notation of NN in this paper: ellipses denote a cluster of neurons and lower case bold face letters denote the output vector of such a cluster of neurons. The dimension of an output vector is thus equal to the number of hidden neurons in the cluster. Upper case bold face letters denote weight matrices.

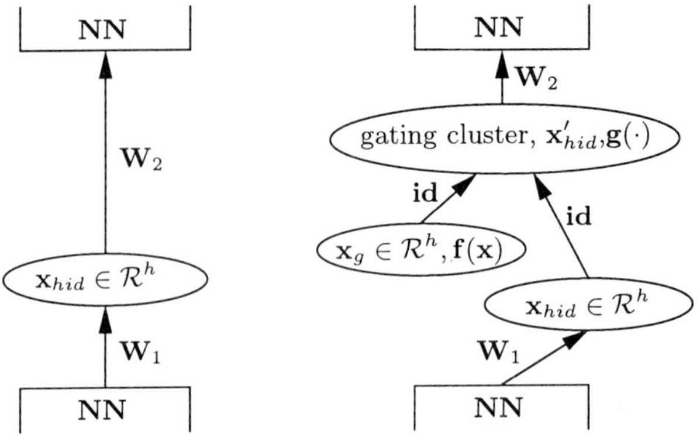

Fig. 2. Left: no gating cluster is used. Right: the gating cluster controls the output signal of hidden cluster $\mathbf{x}_{hid}$.

$\mathbf{W}_1$ and $\mathbf{W}_2$. The cluster contains h neurons and the overall output from all neurons is given by vector $\mathbf{x}_{hid} \in \mathcal{R}^h$.

On the right hand side of figure 2 the same hidden cluster associated with the output signal vector $\mathbf{x}_{hid}$ as on the left hand side of the figure can be found. The input to the hidden cluster did not change. The output signal vector $\mathbf{x}_{hid}$, however, is now propagated to a gating cluster via the weight matrix **id**, identity matrix, $\mathbf{id} \in \mathcal{R}^{h \times h}$. The gating cluster is controlled by a gating vector $\mathbf{x}_g \in \mathcal{R}^h$. The transfer function $\mathbf{g}(\mathbf{x}_g, \mathbf{x}_{hid})$ (denoted by $\mathbf{g}(\cdot)$ in figure 2) of the gating cluster is defined by

$$\mathbf{g}\left(\begin{bmatrix} x_{g,1} \\ \vdots \\ x_{g,h} \end{bmatrix}, \begin{bmatrix} x_{hid,1} \\ \vdots \\ x_{hid,h} \end{bmatrix}\right) = \begin{bmatrix} x_{g,1} \cdot x_{hid,1} \\ \vdots \\ x_{g,h} \cdot x_{hid,h} \end{bmatrix} . \tag{1}$$

The output of the gating cluster given by $\mathbf{x}'_{hid} = \mathbf{g}(\mathbf{x}_g, \mathbf{x}_{hid})$ is propagated to the rest of the network via the same weight matrix $\mathbf{W}_2$ as on the left hand side of figure 2. If $x_{g,i} \in \{0, 1\}$, $i = 1, \ldots, h$, then the gating cluster can be used to gate (fade out) the output of neurons of the hidden cluster. If $\mathbf{x}_g = \mathbf{f}(\mathbf{x})$ (as indicated in figure 2), i.e. $\mathbf{x}_g$ is a function of the input vector $\mathbf{x}$ to the NN, then gating clusters can be used as an elegant way to adapt the network structure dynamically.

If whole clusters of neurons need to be faded out for some $\mathbf{x}$, $\mathbf{f}(\mathbf{x})$ must be chosen such that $\mathbf{x}_g = [0, \ldots, 0]^{\mathrm{T}}$. For $\mathbf{x}_g = [0, \ldots, 0]^{\mathrm{T}}$ no signal of $\mathbf{x}_{hid}$ is propagated to the rest of the network. Further, during learning weight adaptation in $\mathbf{W}_1$ and $\mathbf{W}_2$ is prevented since no error signal can pass the gating cluster. Thus, the network behaves as if the cluster $\mathbf{x}_{hid}$ did not exist.

For $\mathbf{x}_g = [1, \ldots, 1]^{\mathrm{T}}$ the gating cluster has no effect, i.e. the network can be simplified as displayed on the left hand side of figure 2.

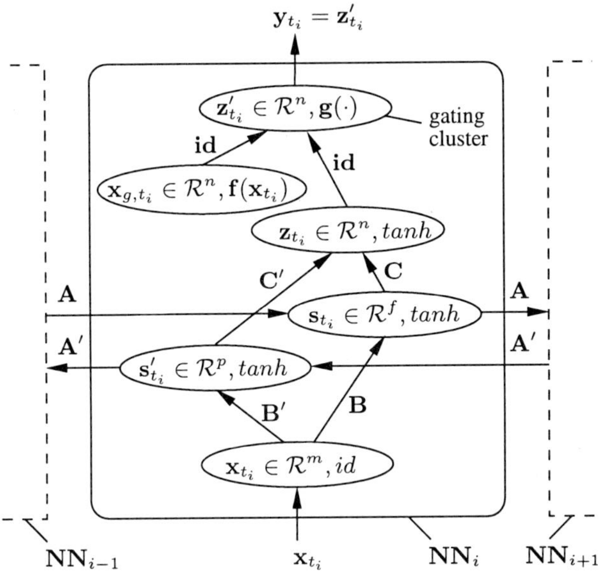

Fig. 3. Subnetwork $\mathbf{NN}_i$ using a gating cluster.

3.2 Embedding in Causal Retro-causal NN

The problem encountered at sentence boundaries (cf. section 2) can be overcome by dynamic adaptation of the network structure depending on the sequence length of defined input vectors $\mathbf{x}_{t_i}$, i.e. to shorten the sequence of $\mathbf{NN}_i$ when necessary. This can be achieved by choosing subnetworks $\mathbf{NN}_i$ using gating clusters as displayed in figure 3.

For a defined $\mathbf{x}_{t_i}$, we choose

$$\mathbf{f}(\mathbf{x}_{t_i}) = \mathbf{x}_{g,t_i} = [1, \ldots, 1]^{\mathrm{T}} \quad . \tag{2}$$

As a consequence $\mathbf{y}_{t_i} = \mathbf{z}'_{t_i} = \mathbf{z}_{t_i}$. This means an output of $\mathbf{NN}_i$ is generated and during training the error signal is back-propagated to $\mathbf{NN}_i$. This in turn leads to weight adaptation within $\mathbf{NN}_i$ and the rest of the network (as can be seen, the error signal is propagated to the rest of the network via $\mathbf{C} \rightarrow \mathbf{A}$ and $\mathbf{C}' \rightarrow \mathbf{A}'$, respectively). The rest of the network means all $\mathbf{NN}_j$ with $j \neq i$.

For an undefined $\mathbf{x}_{t_i}$ we let

$$\mathbf{f}(\mathbf{x}_{t_i}) = \mathbf{x}_{g,t_i} = \mathbf{0} \quad \text{and} \quad \mathbf{x}_{t_i} = \mathbf{0} \tag{3}$$

This way there are no signals from $\mathbf{NN}_i$ to the rest of the network: The gating cluster prevents the back-propagation of an error signal. Therefore, weight adaptation according to an invalid error signal is prevented. Further, by choosing $\mathbf{x}_{t_i} = \mathbf{0}$ no signals are propagated to the rest of the network via $\mathbf{B} \rightarrow \mathbf{s}_{t_i} \rightarrow \mathbf{A}$ and $\mathbf{B}' \rightarrow \mathbf{s}'_{t_i} \rightarrow \mathbf{A}'$ respectively.

The state transition equations for the complete NN architecture (cf. figures 1, 3) are given by:

$$\mathbf{s}_{t_i} = tanh(\mathbf{A}\mathbf{s}_{t_{i-1}} + \mathbf{B}\mathbf{x}_{t_i}) \quad (4)$$

$$\mathbf{s}'_{t_i} = tanh(\mathbf{A}'\mathbf{s}'_{t_{i+1}} + \mathbf{B}'\mathbf{x}_{t_i}) \quad (5)$$

$$\mathbf{y}_{t_i} = \mathbf{C}\mathbf{s}_{t_i} + \mathbf{C}'\mathbf{s}'_{t_i} \quad (6)$$

4 Experiments and Results

The accent labeling scheme used to label our training and testing corpus [2] for the German language is the same as in [1], i.e. perceptual-acoustic labels are assigned to each word by listening only. The corpus contains 1000 sentences (21549 words) taken from the German newspaper *Frankfurter Allgemeine Zeitung.* As in [7], a professional radio news speaker was chosen for the recordings, since news reading is our target domain. For our experiments, we adopt the strategy of [1] and [5] and only distinguish between words perceived accented and words perceived deaccented.

4.1 Optimization of the Number of Time Steps

When applying the presented NN architecture an important choice is the size of the context window, i.e. the size of l and r in figure 1. l determines the maximum number of past time steps (left context) and r determines the maximum number of future time steps (right context). In the following, a method is presented to determine l and r data-driven using the proposed NN architecture (figures 2 and 3).

In [9] the following method is presented to detect how many past time steps contribute information to a causal NN, i.e. no $\mathbf{A}'$-connections in figure 1: For l past time steps the error is observed for each output signal $\mathbf{y}_{t_l}, \mathbf{y}_{t_{l+1}}, \ldots, \mathbf{y}_{t_0}$. If e_{t_i} denotes the error value for $\mathbf{y}_{t_i}$ then one should be able to observe a declining error from past to present time steps, i.e. $e_{t_l} > e_{t_{l+1}} > \ldots$. This has the following reason: $\mathbf{y}_{t_l}$ is only computed with $\mathbf{x}_{t_l}$ as input. For the next output $\mathbf{y}_{t_{l+1}}$ information from $\mathbf{x}_{t_{l+1}}$ and additionally from $\mathbf{x}_{t_l}$ is available (information from $\mathbf{x}_{t_l}$ is transmitted via $\mathbf{A}$, see figure 1). This superposition of more and more information should lead to $e_{t_l} > e_{t_{l+1}} > \ldots$.

If this is not the case for all time steps, i.e. $e_{t_j} \leq e_{t_{j+1}}$ at some time step t_j, $j \in \{l, \ldots, -1\}$, then input $\mathbf{x}_{t_{j+1}}$ did not contribute relevant information to the model and the *maximal inter-temporal connectivity* (MIC) [9] is reached, i.e. inputs longer ago in time do not add information and can be omitted.

In our application the concept of [9] is applied to find MIC for past inputs (left context). Further, the concept is extended symmetrically to analyze MIC for future inputs (right context).

Figure 4 displays the results for MIC analysis for left and right context. The two plots of figure 4 result from independent experiments.

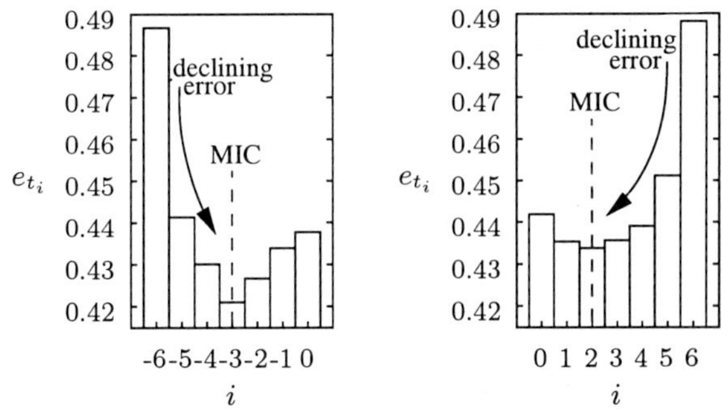

Fig. 4. MIC analysis for left and right context.

In the first experiment (related to the left hand side of figure 4) a causal NN using gating clusters is used to determine MIC for the left context. For the experiment a temporal context of six was used. In figure 1 this means no $\mathbf{A}'$-connections are used and $l = -6$ and $r = 0$. The plot on the left hand side of figure 4 shows the error values e_{t_i} of the output vectors $\mathbf{y}_{t_i}$. As can be seen the error declines for $-6 \leq i \leq -3$ and rises for $i > -3$.

In the second experiment (related to the right hand side of figure 4) a retro-causal NN using gating clusters is used to determine MIC for the right context. For the experiment a temporal context of six time steps was used. In figure 1 this means no $\mathbf{A}$-connections are used and $l = 0$ and $r = 6$. The plot on the right hand side of figure 4 shows the error values e_{t_i} of the outputs $\mathbf{y}_{t_i}$. As can be seen the error declines (now from future to past time steps, i.e. from right to left) for $2 \leq i \leq 6$ and rises for $i < 2$. MIC is thus reached for a right context of four. This demonstrates that MIC analysis can be extended symmetrically for future context analysis as proposed.

The two experiments are used to determine the optimal size of a context window. The context window is chosen such that the two time steps where MIC was determined join at t_0. Thus, the lowest error and highest prediction accuracy is found at t_0. From the experiments related to figure 4 this is achieved by choosing a left context of three time steps and a right context of four time steps. In figure 1 this means $l = -3$ and $r = 4$.

4.2 Prediction Accuracy

For training and testing, the above described corpus (German language) and the corpus used in [7] (English language) were used. Both corpora have been separated into three subsets which contain approximately the following percentage of data: training set: 70%, validation set: 10% and test set (independent testing data): 20%. All results reported below are determined on the test set. The validation set is used to avoid over-fitting to the training data, i.e. training is stopped when a rising error is observed on the validation set (early stopping).

Prediction accuracy for the proposed NN architecture (figures 1, 3) was measured at 84.5% for the English corpus used in [7]. In [7] prediction accuracy for accents on word level obtained with a CART-based approach is reported at 82.5%. The comparison of the results show an improvement for the proposed method. This improvement can be seen as significant because of the large test set.

In a second experiment the above described German corpus was used. Prediction accuracy for this corpus was measured at 83.1%. The German corpus has rougly twice the size of that used in [7]. Thus the test set used has also twice the size of that used in [7] and may contain more complex sentence structures. Therefore, the high prediction accuracy for this large test set demonstrate the good generalization ability of the proposed method.

5 Conclusion

In this paper a new NN architecture for accent label prediction was presented. The architecture applies gating clusters in a causal retro-causal NN. With gating clusters the number of time steps of a causal retro-causal NN can be adapted dynamically.

It was shown how the architecture can be used to determine an optimal size for a context window. With the optimized context window size a prediction accuracy of 84.5% was achieved for an English corpus and a prediction accuracy of 83.1% for a German corpus. This demonstrates the good performance of the proposed method.

References

1. A. Batliner, M. Nutt, V. Warnke, E. Nöth, J. Buckow, R. Huber, and H. Niemann. Automatic annotation and classification of phrase accents in spontaneous speech. In *Eurospeech*, 1999.
2. Institut für Phonetik und sprachliche Kommunikation. Siemens synthese korpus - si1000p. corpus available at http://www.phonetik.uni-muenchen.de/Bas/.
3. Ralf Haury and Martin Holzapfel. Optimization of a neural network for speaker and task dependent f0-generation. In *ICASSP*, 1998.
4. Simon Haykin. *Neural Networks — A Comprehensive Foundation*, chapter 1.7 — Knowledge Representation. Prentice Hall International, 1999.
5. Julia Hirschberg. Pitch accent in context: Predicting prominence from text. *Artificial Intelligence*, 63:305–340, 1993.
6. Achim F. Müller, Hans G. Zimmermann, and R. Neuneier. Robust generation of symbolic prosody by a neural classifier based on autoassociators. In *ICASSP*, 2000.
7. K. Ross and M. Ostendorf. Prediction of abstract prosodic labels for speech synthesis. *Computer Speech and Language*, 10:155–185, 1996.
8. Christina Widera, Thomas Portele, and Maria Wolters. Prediction of word prominence. In *Eurospeech*, 1997.
9. Hans G. Zimmermann, R. Neuneier, and R. Grothmann. *Modeling and Forecasting Financial Data, Techniques of Non-linear Dynamics*, chapter Modeling of Dynamic Systems by Error Correction Neural Networks. Kluwer Academic, 2000.
10. Hans Georg Zimmermann, Achim F. Müller, Çağlayan Erdem, and Rüdiger Hoffmann. Prosody generation by causal retro-causal error correction neural networks. In *Workshop on Multi-Lingual Speech Communication*, Advanced Telecommunications Research Institute International (ATR), 2000.

Discriminative Dimensionality Reduction Based on Generalized LVQ

Atsushi Sato

Multimedia Research Labs., NEC Corporation
1-1, Miyazaki 4-chome, Miyamae-ku,
Kawasaki, Kanagawa 216-8555, Japan
asato@ay.jp.nec.com

Abstract. In this paper, a method for the dimensionality reduction, based on generalized learning vector quantization (GLVQ), is applied to handwritten digit recognition. GLVQ is a general framework for classifier design based on the minimum classification error criterion, and it is easy to apply it to dimensionality reduction in feature extraction. Experimental results reveal that the training of both a feature transformation matrix and reference vectors by GLVQ is superior to that by principal component analysis in terms of dimensionality reduction.

1 Introduction

One of the important roles of feature extraction is to map the data into a lower dimensional space. In general, such a reduction in dimensionality will be accompanied by a loss of some information which discriminates between different classes. However, feature extraction leads to improve performance, because we can reduce noise and the curse of dimensionality as well as incorporating prior knowledge. The goal in dimensionality reduction is therefore to preserve as much of the relevant information to classification as possible.

Principal component analysis (PCA), in which a data set in d dimensional space is assumed to have an intrinsic dimensionality $q < d$, is often used for dimensionality reduction [1]. Fisher discriminant analysis (FDA), in which the ratio of between-class scatter to within-class scatter is maximized, can also be applied to dimensionality reduction [1]. It is noteworthy that we are unable to find more than $K - 1$ bases in FDA; K is the number of classes, because of the rank deficiency of the between-class scatter matrix. FDA is thus not applicable to dimensionality reduction in case of a small number of classes such as digit recognition. In recent years, independent component analysis (ICA), in which the data are projected onto non-orthogonal independent bases, has been investigated [2, 3]. However, finding independent bases in the high dimensional space is still a hard problem, so we have to reduce dimensionality with PCA or FDA before using ICA.

The common problem with the above methods is that the optimality of classification is not guaranteed, because the kinds of classifiers that are used are not

G. Dorffner, H. Bischof, and K. Hornik (Eds.): ICANN 2001, LNCS 2130, pp. 65–72, 2001.

taken into account. Discriminative feature extraction (DFE) is a framework for solving this problem [4]. In DFE, the feature extraction module can be trained, based on the minimum classification error (MCE) criterion [5], for a given classifier. The author has proposed a new form of discriminative training, based on MCE, called generalized learning vector quantization (GLVQ) [6–8]. In recent years, GLVQ's effectiveness has been evaluated in character recognition [9, 10]. GLVQ can be applied to feature extraction as well as to classification, and the effectiveness of combining FDA and GLVQ has been evaluated in Chinese character recognition [11].

In this paper, the combination of PCA and GLVQ is applied to the recognition of handwritten digits, and a feature transformation matrix is trained as well as reference vectors on the basis of GLVQ. Experimental results for digit recognition reveal that, in terms of dimensionality reduction, the training of both the feature transformation matrix and reference vectors by GLVQ is superior to that by PCA. A detailed introduction of MCE and GLVQ is provided in Sec. 2, and dimensionality reduction on the basis on GLVQ is presented in Sec. 3. Experimental results for digit recognition and a discussion are given in Sec. 4. Finally, some conclusions are given in Sec. 5.

2 Minimum Classification Error

MCE is a criterion for discriminative training that minimizes the overall risk in Bayes decision theory by using a gradient descent procedure [12, 5], which can be applied to any other classifier structures. In this section, a brief introduction to Bayes decision theory is provided, and we then explain the formulation for classifier design on the basis of MCE.

2.1 Bayes Decision Theory

According to Bayes decision theory, the classifier should be designed so that it minimizes the overall risk (or the expected loss) [13], which is defined by

$$R = \sum_{k=1}^{K} \int \ell(\alpha(\boldsymbol{x})|\omega_k) P(\omega_k|\boldsymbol{x}) p(\boldsymbol{x}) d\boldsymbol{x}, \tag{1}$$

where K denotes the number of classes, $p(\boldsymbol{x})$ is the probability density of an input vector $\boldsymbol{x}$, $P(\omega_k|\boldsymbol{x})$ is the *a posteriori* probability, and $\ell(\alpha(\boldsymbol{x})|\omega_k)$ is a loss for the decision $\alpha(\boldsymbol{x})$ for a given class ω_k. If $\ell(\alpha(\boldsymbol{x})|\omega_k)$ is a zero-one function such that $\ell(\omega_j|\omega_k) = 1 - \delta_{jk}$, the expected loss can be written as

$$R = 1 - \int P(\omega_j|\boldsymbol{x}) p(\boldsymbol{x}) d\boldsymbol{x}. \tag{2}$$

If we adopt the MAP rule; $\alpha(\boldsymbol{x}) = \omega_j$ if $P(\omega_j|\boldsymbol{x}) = \max_k P(\omega_k|\boldsymbol{x})$, the second term on the right of Eq.(2) has the maximum, and then the expected loss is minimized. Since the *a posteriori* probability can be written as follows, according to Bayes formula:

$$P(\omega_k|\boldsymbol{x}) = \frac{p(\boldsymbol{x}|\omega_k)P(\omega_k)}{p(\boldsymbol{x})}, \tag{3}$$

the estimation of the probability density $p(\boldsymbol{x}|\omega_k)$ is one of the central problems of statistical pattern recognition.

2.2 Generalized Probabilistic Descent

Since the estimation of probability density is difficult because of the curse of dimensionality, one good way of designing the classifier is to minimize the expected loss directly without densities estimating. If we assume that $p(\boldsymbol{x}) = \frac{1}{N}\sum_n \delta(\boldsymbol{x} - \boldsymbol{x}_n)$ and $P(\omega_k|\boldsymbol{x}) = 1(\boldsymbol{x} \in \omega_k)$, the following empirical loss can be obtained:

$$R_e = \frac{1}{N}\sum_{n=1}^{N}\sum_{k=1}^{K} \ell(\alpha(\boldsymbol{x}_n)|\omega_k)1(\boldsymbol{x}_n \in \omega_k), \tag{4}$$

where N is the number of training samples, and $1(\cdot)$ is an indicator function such that $1(true) = 1$ and $1(false) = 0$.

Minimum classification error (MCE) is to directly minimize the empirical loss by using a gradient search procedure. To make the empirical loss differentiable, a continuous loss function is employed instead of the zero-one function; $\ell(\alpha(\boldsymbol{x})|\omega_k) = \ell(\rho_k(\boldsymbol{x};\boldsymbol{\theta}))$, where $\rho_k(\boldsymbol{x};\boldsymbol{\theta})$ is called the misclassification measure, and $\boldsymbol{\theta}$ denotes the classifier's parameter. In the formulation called generalized probabilistic descent (GPD) [5], the loss function $\ell(\cdot)$ is defined by

$$\ell(\rho) = \frac{1}{1 + e^{-\xi\rho}}, \tag{5}$$

where ξ is a positive number. For the misclassification measure, a differentiable one is employed as follows:

$$\rho_k(\boldsymbol{x};\boldsymbol{\theta}) = -g_k(\boldsymbol{x};\boldsymbol{\theta}) + \left[\frac{1}{K-1}\sum_{i\neq k} g_i(\boldsymbol{x};\boldsymbol{\theta})^{\eta}\right]^{1/\eta}, \tag{6}$$

where η is a positive number and $g_k(\boldsymbol{x};\boldsymbol{\theta})$ denotes the discriminant function of a class ω_k. When there are two classes or when $\eta \to \infty$, Eq. (6) becomes simply $\rho_k(\boldsymbol{x};\boldsymbol{\theta}) = -g_k(\boldsymbol{x};\boldsymbol{\theta}) + g_l(\boldsymbol{x};\boldsymbol{\theta})$. Here, $g_l(\boldsymbol{x};\boldsymbol{\theta}) = \max_{i\neq k} g_i(\boldsymbol{x};\boldsymbol{\theta})$. Obviously, $\rho_k(\boldsymbol{x};\boldsymbol{\theta}) > 0$ implies misclassification and $\rho_k(\boldsymbol{x};\boldsymbol{\theta}) < 0$ means a correct decision for a given $\boldsymbol{x} \in \omega_k$.

2.3 Generalized Learning Vector Quantization

If we apply GPD to minimum-distance classifiers, the same update rule for reference vectors as is used in learning vector quantization (LVQ) can be obtained.

However, this approach has a drawback in that the reference vectors diverge in the same way as in LVQ2.1. This problem arises because the solution which minimizes R_e is at infinity. To solve this problem, the following misclassification measure was introduced in generalized learning vector quantization (GLVQ) [6–8]:

$$\rho_k(\boldsymbol{x};\boldsymbol{\theta}) = \frac{-g_k(\boldsymbol{x};\boldsymbol{\theta}) + g_l(\boldsymbol{x};\boldsymbol{\theta})}{|g_k(\boldsymbol{x};\boldsymbol{\theta})| + |g_l(\boldsymbol{x};\boldsymbol{\theta})|}. \tag{7}$$

Moreover, ξ in Eq. (5) increases with time to approximate the zero-one loss function. This is because the minimization of R_e is not equivalent to the minimization of the average probability of error as long as $\xi < \infty$.

The empirical loss can be minimized directly by a gradient descent procedure as follows:

$$\boldsymbol{\theta}(t+1) = \boldsymbol{\theta}(t) - \epsilon(t) \left.\frac{\partial R_e}{\partial \boldsymbol{\theta}}\right|_{\boldsymbol{\theta}=\boldsymbol{\theta}(t)} \tag{8}$$

where $\epsilon(t)$ is some small positive number in the time domain. According to probabilistic descent, the derivative of R_e is given by

$$\frac{\partial R_e}{\partial \boldsymbol{\theta}} = \boldsymbol{U} \sum_{k=1}^{K} \frac{\partial \ell(\rho_k(\boldsymbol{x}_n;\boldsymbol{\theta}))}{\partial \boldsymbol{\theta}} 1(\boldsymbol{x}_n \in \omega_k). \tag{9}$$

If $\boldsymbol{U}$ is a positive definite symmetric matrix, that is $\boldsymbol{x}^T\boldsymbol{U}\boldsymbol{x} > 0$ and $\boldsymbol{U}^T = \boldsymbol{U}$, it is guaranteed that the expectation of R_e decreases [12]. Moreover, if $\epsilon(t)$ satisfies the following conditions, it is ensured by the Robbins-Monro method that $\boldsymbol{\theta}$ converges to the solution which minimizes R_e:

$$\lim_{t\to\infty} \epsilon(t) = 0, \quad \sum_{t=1}^{\infty} \epsilon(t) = \infty, \quad \sum_{t=1}^{\infty} \epsilon(t)^2 < \infty. \tag{10}$$

3 Dimensionality Reduction

3.1 Principal Component Analysis

Principal component analysis (PCA) is commonly used for dimensionality reduction. The input vector $\boldsymbol{x} \in \mathcal{R}^d$ is transformed into $\boldsymbol{y} \in \mathcal{R}^q$ $(q < d)$ in the following way:

$$\boldsymbol{y} = \boldsymbol{A}(\boldsymbol{x} - \boldsymbol{\mu}), \tag{11}$$

where $\boldsymbol{\mu}$ is the mean of all training samples; $\boldsymbol{\mu} = \frac{1}{N}\sum_{n=1}^{N} \boldsymbol{x}_n$. The $q \times d$ transformation matrix $\boldsymbol{A}$ is defined by $\boldsymbol{A} = (\boldsymbol{\phi}_1, \cdots, \boldsymbol{\phi}_q)^T$, in which $\boldsymbol{\phi}_i$ is the eigenvector that corresponds to a given eigenvalue λ_i $(\lambda_i > \lambda_{i+1})$ of the covariance matrix:

$$\boldsymbol{\Sigma} = \frac{1}{N}\sum_{n=1}^{N} (\boldsymbol{x}_n - \boldsymbol{\mu})(\boldsymbol{x}_n - \boldsymbol{\mu})^T. \tag{12}$$

Since there will be just a few large eigenvalues, q is regarded as the intrinsic dimensionality of the subspace that governs the data.

3.2 GLVQ Training

GLVQ can be applied to any classifier structure, but we apply it to the minimum-distance classifier in this paper. The discriminant function is given by

$$g_k(\boldsymbol{x};\boldsymbol{\theta}) = -d_k(\boldsymbol{x};\boldsymbol{\theta}) = ||\boldsymbol{y} - \boldsymbol{m}_{ki}||^2, \tag{13}$$

where $\boldsymbol{m}_{ki}$ is the i-th reference vector in class ω_k, which is the nearest one to the transformed input vector $\boldsymbol{y}$ given by Eq. (11). The misclassification measure is then written as

$$\rho_k(\boldsymbol{x};\boldsymbol{\theta}) = \frac{d_k(\boldsymbol{x};\boldsymbol{\theta}) - d_l(\boldsymbol{x};\boldsymbol{\theta})}{d_k(\boldsymbol{x};\boldsymbol{\theta}) + d_l(\boldsymbol{x};\boldsymbol{\theta})}, \tag{14}$$

where $d_k(\boldsymbol{x};\boldsymbol{\theta}) = ||\boldsymbol{y} - \boldsymbol{m}_{ki}||^2$ and $d_l(\boldsymbol{x};\boldsymbol{\theta}) = ||\boldsymbol{y} - \boldsymbol{m}_{lj}||^2$. If we employ the unit matrix for $\boldsymbol{U}$, Eq. (9) can be written for $\boldsymbol{x} \in \omega_k$ as follows:

$$\frac{\partial R_e}{\partial \boldsymbol{\theta}} = \frac{\partial \ell(\rho_k(\boldsymbol{x};\boldsymbol{\theta}))}{\partial \boldsymbol{\theta}}. \tag{15}$$

Since the parameter $\boldsymbol{\theta}$ is a set of reference vectors and the transformation matrix, the followings can be obtained:

$$\frac{\partial R_e}{\partial \boldsymbol{m}_{ki}} = -4\ell'(\rho_k(\boldsymbol{x};\boldsymbol{\theta}))\frac{d_l(\boldsymbol{x};\boldsymbol{\theta})}{\{d_k(\boldsymbol{x};\boldsymbol{\theta}) + d_l(\boldsymbol{x};\boldsymbol{\theta})\}^2}(\boldsymbol{y} - \boldsymbol{m}_{ki}), \tag{16}$$

$$\frac{\partial R_e}{\partial \boldsymbol{m}_{lj}} = +4\ell'(\rho_k(\boldsymbol{x};\boldsymbol{\theta}))\frac{d_k(\boldsymbol{x};\boldsymbol{\theta})}{\{d_k(\boldsymbol{x};\boldsymbol{\theta}) + d_l(\boldsymbol{x};\boldsymbol{\theta})\}^2}(\boldsymbol{y} - \boldsymbol{m}_{lj}), \tag{17}$$

$$\frac{\partial R_e}{\partial \boldsymbol{m}} = 0 \quad \text{for} \quad \boldsymbol{m} \neq \boldsymbol{m}_{ki} \;\text{ or }\; \boldsymbol{m} \neq \boldsymbol{m}_{lj}\,, \tag{18}$$

$$\frac{\partial R_e}{\partial \boldsymbol{A}} = -\left\{\frac{\partial R_e}{\partial \boldsymbol{m}_{ki}} + \frac{\partial R_e}{\partial \boldsymbol{m}_{lj}}\right\}(\boldsymbol{x} - \boldsymbol{\mu})^T. \tag{19}$$

Note that the denominator $\{d_k(\boldsymbol{x};\boldsymbol{\theta}) + d_l(\boldsymbol{x};\boldsymbol{\theta})\}^2$ in Eqs. (16) (17) should be replaced with $d_k(\boldsymbol{x};\boldsymbol{\theta}) + d_l(\boldsymbol{x};\boldsymbol{\theta})$ to be robust with respect to the initial values of the reference vectors [8]. Moreover, we should employ a constraint such that $\sum_{i=1}^{q} \phi_i^2 = const$ to prevent divergence of $\boldsymbol{A}$. The update of $\boldsymbol{A}$ can be written as

$$\delta \boldsymbol{A} = -\epsilon(t)\frac{\partial R_e}{\partial \boldsymbol{A}} = \{-\alpha_{ki}(\boldsymbol{y} - \boldsymbol{m}_{ki}) + \alpha_{lj}(\boldsymbol{y} - \boldsymbol{m}_{lj})\}(\boldsymbol{x} - \boldsymbol{\mu})^T, \tag{20}$$

where $\alpha_{ki} > 0$ and $\alpha_{lj} > 0$. It is easy to understand the above equation if we apply $(\boldsymbol{x} - \boldsymbol{\mu})$ to it from the right. As shown in Fig. 1, the transformed vector $\boldsymbol{y}$ and the reference vector $\boldsymbol{m}_{ki}$ that belong to the same class approach each other, while $\boldsymbol{y}$ and $\boldsymbol{m}_{lj}$ that belong to different classes diverge from each other.

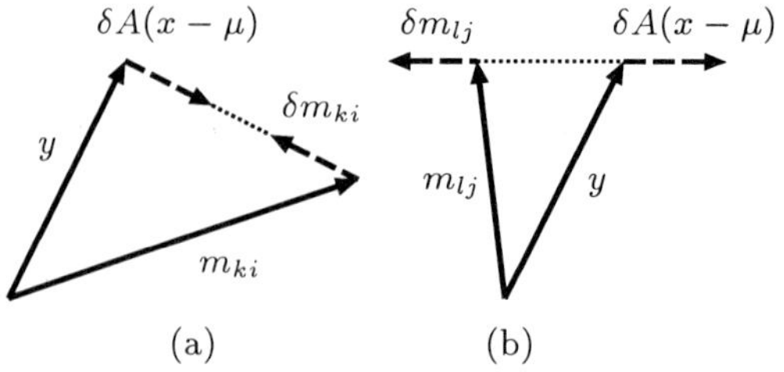

Fig. 1. Update of transformation matrix and reference vectors.

Table 1. Error rates for test samples.

Dimension	Error Rates (%)		
q	Type A	Type B	Type C
2	53.76	53.10	12.06
4	24.02	19.84	1.42
8	10.82	3.78	0.68
16	7.68	1.44	0.56
32	7.42	0.90	0.54
64	7.26	0.54	0.52
128	7.26	0.52	0.50
400	7.26	0.50	0.50

4 Handwritten Digit Recognition Experiments

The dimensionality reduction method based on GLVQ was applied to handwritten digit recognition to evaluate its performance. An isolated character database assembled in our laboratory was used in the experiments. The sample size was 1000 per class and was divided into 500 training samples and 500 test samples per class. 400-dimensional feature vectors were extracted from each image by using the weighted direction code histogram method [14]. One reference vector was assigned to each class, and these reference vectors were used with a minimum-distance method to classify the transformed vectors. The following three combinations of dimensionality reduction and classification were evaluated in the experiments:

- Type A : transformation matrix obtained by PCA, and mean vectors,
- Type B : transformation matrix obtained by PCA, and reference vectors trained by GLVQ, and
- Type C : transformation matrix and reference vectors trained simultaneously by GLVQ.

Type A was used to obtain initial values for training in both Type B and Type C. The dimensionality of reference vectors was reduced from the original 400 to 2, and training and evaluation were carried out for each run.

Table 1 shows the error rates for test samples, and these results imply that

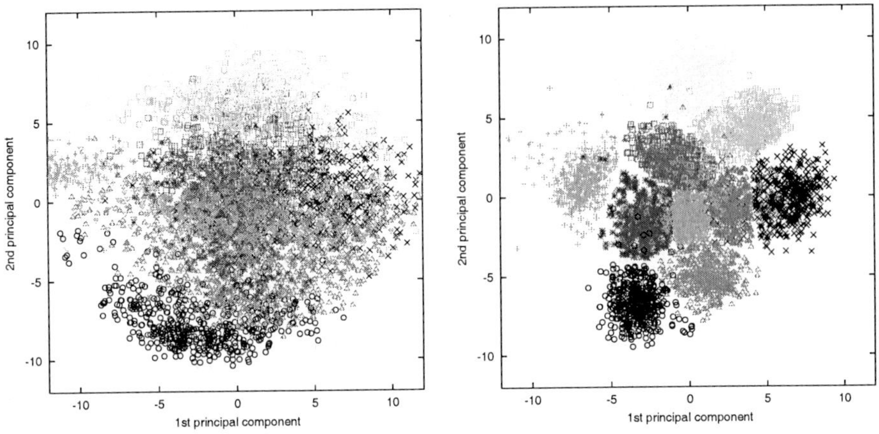

(a) PCA (b) Type C

Fig. 2. Feature transformation into two dimensional space.

– the training of reference vectors by GLVQ is effective in improving recognition accuracy even though the dimensionality was reduced by PCA,
– the training of both the transformation matrix and the reference vectors further improves the recognition accuracy for every number of features, and
– the recognition accuracy declines with the dimensionality.

Figure 2 shows how the feature vectors were mapped to the space of lower dimension. All training samples were mapped to the 2-dimensional space by PCA on the left, and by Type C on the right, respectively. The horizontal axis and the vertical axis denote the values projected to the 1st and 2nd principal components, respectively. In the figure on the right, we can find ten clusters that correspond to each numeral, and the decision border can be approximated by the Voronoi diagram. This implies that the transformation matrix was designed to be appropriate to minimum-distance classifiers. Note that the bases trained by GLVQ are no longer orthogonal to each other.

5 Conclusion

In this paper an application of the discriminative dimensionality reduction based on generalized learning vector quantization (GLVQ) to handwritten digit recognition was described. The experimental results showed that the training of both the feature transformation matrix and the reference vectors by GLVQ is superior to principal component analysis in terms of dimensionality reduction. This implies that the transformation matrix designed by this method was appropriate to a given classifier. However, the recognition accuracy declined with the dimensionality in the experiments. This is a problem for further investigation. Studies on the discriminative training of non-linear feature transformation also remains as an interesting topic for future work.

References

1. K. Fukunaga, "Introduction to Statistical Pattern Recognition," 2nd ed., San Diego: Academic Press, 1990.
2. C. Jutten and J. Herault, "Blind Separation of Sources," *Signal Processing*, Vol. 24, pp. 1–10, 1991.
3. A. Hyvärinen and E. Oja, "A Fast Fixed-Point Algorithm for Independent Component Analysis," *Neural Computation*, Vol. 9, No. 7, pp. 1483–1492, 1997.
4. A. Biem, S. Katagiri and B.-H. Juang, "Pattern Recognition using Discriminative Feature Extraction," *IEEE Trans. on Signal Processing*, Vol. 45, No. 2, pp. 500–504, 1997.
5. B.H. Juang and S.Katagiri, "Discriminative Learning for Minimum Error Classification," *IEEE Trans. on Signal Proc.,* Vol. 40, No. 12, pp. 3043–3054, 1992.
6. A. Sato and K. Yamada, "Generalized Learning Vector Quantization," In *Advances in Neural Information Processing Systems 8*, pp. 423–429, MIT Press, 1996.
7. A. Sato and K. Yamada, "An Analysis of Convergence in Generalized LVQ," In *Proc. of the Int. Conf. on Artificial Neural Networks*, Vol. 1, pp. 171–176, 1998.
8. A. Sato, "An Analysis of Initial State Dependence in Generalized LVQ," In *Proc. of the Int. Conf. on Artificial Neural Networks*, Vol. 2, pp. 928–933, 1999.
9. C.-L. Liu and M. Nakagawa, "Prototype Learning Algorithms for Nearest Neighbor Classifier with Application to Handwritten Character Recognition," In *Proc. of the Int. Conf. on Document Analysis and Recognition*, pp. 378–381, 1999.
10. T. Fukumoto, T. Wakabayashi, F. Kimura and Y. Miyake, "Accuracy Improvement of Handwritten Character Recognition by GLVQ," In *Proc. of the Int. Workshop on Frontiers in Handwriting Recognition*, pp. 271–280, 2000.
11. M.-K. Tsay, K.-H. Shyu and P.-C. Chang, "Feature Transformation with Generalized Learning Vector Quantization for Hand-Written Chinese Character Recognition," *IEEE Trans. on Information and Systems*, Vol. E82-D, No, 3, pp. 687–692, 1999.
12. S. Amari, "A Theory of Adaptive Pattern Classifiers," *IEEE Trans. Elec. Comput.*, Vol. EC–16, No. 3, pp. 299–307, 1967.
13. R.O. Duda and P.E. Hart, "Pattern Classification and Scene Analysis," John Wiley & Sons, 1973.
14. F. Kimura, T. Wakabayashi, S. Tsuruoka and Y. Miyake, "Improvement of Handwritten Japanese Character Recognition Using Weighted Direction Code Histogram," *Pattern Recognition*, Vol. 30, No. 8, pp. 1329–1337, 1997.

A Computational Intelligence Approach to Optimization with Unknown Objective Functions

Hirotaka Nakayama[1], Masao Arakawa[2], and Rie Sasaki[1]

[1] Konan University, Dept. of Information Science and Systems Engineering, Kobe 658-8501, JAPAN
nakayama@konan-u.ac.jp
[2] Kagawa University, Dept. of Reliability-based Information Engineering, Kagawa 761-0396, JAPAN
arakawa@eng.kagawa-u.ac.jp

Abstract. In many practical engineering design problems, the form of objective function is not given explicitly in terms of design variables. Given the value of design variables, under this circumstance, the value of objective function is obtained by some analysis such as structural analysis, fluidmechanic analysis, thermodynamic analysis, and so on. Usually, these analyses are considerably time consuming to obtain a value of objective function. In order to make the number of analyses as few as possible, we suggest a method by which optimization is performed in parallel with predicting the form of objective function. In this paper, radial basis function networks (RBFN) are employed in predicting the form of objective function, and genetic algorithms (GA) in searching the optimal value of the predicted objective function. The effectiveness of the suggested method will be shown through some numerical examples.

1 Introduction

Our aim in this paper is to optimize objective functions whose forms are not explicitly known in terms of design variables. In many engineering design problems under this circumstance, values of objective functions are obtained by some complicated analysis of large scale with multi-peaked characters that are often seen in elastic-plastic analysis and so on. These analyses usually require considerably expensive computational time. Therefore, if these functions are optimized by existing methods, it takes an unrealistic order of time to obtain a solution. For this situation, the number of necessary analyses should be as few as possible. To this end, we suggest a method consisting of two stages. The first stage is to predict the form of objective function by RBFN(Radial Basis Function Networks). The second stage is to optimize the predicted objective function by GA(Genetic Algorithms). A major problem in this method is how to get a good approximation of the objective function based on as few sample data as possible. To this end, the form of objective function is revised by relearning on the basis of additional data step by step. Our discussion will be focused on this additional learning and how to select additional data in the following sections.

G. Dorffner, H. Bischof, and K. Hornik (Eds.): ICANN 2001, LNCS 2130, pp. 73–80, 2001.

2 Additional Learning in RBFN

Since the number of sample points for predicting objective functios should be as few as possible, we adopt some additional learning techniques which predict objective functions by adding learning data step by step. RBFN are effective to this end. The output of RBFN is given by

$$f(\boldsymbol{x}) = \sum_{j=1}^{m} w_j h_j(\boldsymbol{x}).$$

where $h_j (j = 1, \ldots, m)$ are radial basis functions, e.g.,

$$h_j(\boldsymbol{x}) = e^{-\|\boldsymbol{x}-\boldsymbol{c}_j\|^2/r_j}.$$

Given the training data $(\boldsymbol{x}_i, \hat{y}_i)$ $(i = 1, \cdots, p)$, the learning of RBFN is usually made by solving

$$E = \sum_{i=1}^{p} (\hat{y}_i - f(\boldsymbol{x}_i))^2 + \sum_{j=1}^{m} \lambda_j w_j^2 \rightarrow \text{ Min}$$

where the second term is introduced for the purpose of regularization.

Letting $A = (H_p^T H_p + \Lambda)$, we have as a necessary condition for the above minimization

$$A\hat{\boldsymbol{w}} = H_p^T \hat{\boldsymbol{y}}.$$

Here

$$H_p^T = [\boldsymbol{h}_1 \ \cdots \ \boldsymbol{h}_p]$$

where $\boldsymbol{h}_j^T = [h_1(\boldsymbol{x}_j), \ldots, h_m(\boldsymbol{x}_j)]$, and Λ is a diagonal matrix whose diagonal components are $\lambda_1 \ \cdots \ \lambda_m$.

Therefore, the learning in RBFN is reduced to finding

$$A^{-1} = (H_p^T H_p + \Lambda)^{-1}$$

The additional learninig in RBFN can be made by adding new data and/or a basis function, if necesary. Since the learning in RBFN is equivalent to the matrix inversion A^{-1}, the additional learning here is reduced to the incremental calculation of matrix inversion. The following algorithm can be seen in [7]:

(i) Adding a New Training Data

Adding a new data $\boldsymbol{x}_{p+1}$, the additional learning in RBFN can be made by the following simple update formula: Let

$$H_{p+1} = \begin{bmatrix} H_p \\ \boldsymbol{h}_{p+1}^T \end{bmatrix}$$

where $\boldsymbol{h}_{p+1}^T = [h_1(\boldsymbol{x}_{p+1}), \ldots, h_m(\boldsymbol{x}_{p+1})]$.

Then

$$A_{p+1}^{-1} = A_p^{-1} - \frac{A_p^{-1}\boldsymbol{h}_{p+1}\boldsymbol{h}_{p+1}^T A_p^{-1}}{1 + \boldsymbol{h}_{p+1}^T A_p^{-1}\boldsymbol{h}_{p+1}}$$

(ii) Adding a New Basis Function
In cases in which we need a new basis function to improve the learning for a new data, we have the following update formula for the matrix inversion: Let

$$H_{m+1} = \begin{bmatrix} H_m & \boldsymbol{h}_{m+1} \end{bmatrix}$$

where $\boldsymbol{h}_{m+1}^T = [h_{m+1}(\boldsymbol{x}_1), \ldots, h_{m+1}(\boldsymbol{x}_p)]$.

Then

$$A_{m+1}^{-1} = \begin{bmatrix} A_m^{-1} & \mathbf{0} \\ \mathbf{0}^T & 0 \end{bmatrix}$$

$$+ \frac{1}{\lambda_{m+1} + \boldsymbol{h}_{m+1}^T(I_p - H_m A_m^{-1} H_m^T)\boldsymbol{h}_{m+1}} \times \begin{bmatrix} A_m^{-1} H_m^T \boldsymbol{h}_{m+1} \\ -1 \end{bmatrix} \begin{bmatrix} A_m^{-1} H_m^T \boldsymbol{h}_{m+1} \\ -1 \end{bmatrix}^T$$

Remark It is important to decide parameters $\boldsymbol{c}_i$ and r_i of radial basis functions moderately. We assign a basis function at each learning data $(\boldsymbol{x}_i, \hat{y}_i)$ $(i = 1, \cdots, p)$ so that $\boldsymbol{c}_i = \boldsymbol{x}_i$ $(i = 1, \cdots, p)$. Modifying the formula given by [2] slightly, the value of r_j is determined by

$$r = d_{max} / \sqrt[n]{nm}$$

where d_{max} is the maximal distance among the data, n is the dimension of data, m is the number of basis function.

3 How to Select Additional Data

If the current solution is not satisactory, namely if our stop condition is not satisfied, we add some data in order to improve the approximation of objective function. Now, how to select such additional data becomes our issue.

If the current optimal point is taken as such additional data, the estimated optimal point tends to converge to a local maximum (or minimum) point. This is due to lack of global information in predicting the objective function.

On the other hand, if addtional data are taken to be far from the existing data, it is difficult to obtain more detailed information near the optimal point. Therefore, it is hard to obtain a solution with a high precision. This is because of insufficient information near the optimal point.

We suggest a method which gives both global information for predicting the objective function and local information near the optimal point at the same time. To this end, we take two kinds of additional data for relearning the form of objective function. One of them is selected from a neighborhood of the current optimal point in order to add local information near the (estimated) optimal point. The size of this neighborhood is controlled during the convergence process.

The other one is selected to be far from the currenct optimal value in order to give a better prediction of the form of objective function. The former additional data gives more detailed information near the current optimal point. The latter data prevents from converging to local maximum (or minimum) point.

In this paper, the neighborhood of the current optimal point is given by a square S, whose center is the current optimal point, with the length of a side l. Let S_0 be a square, whose center is the current optimal point, with the fixed length of a side l_0. The square S is shrinked according to the number C_x of optimal points appeared continuously in S_0 in the past. Namely,

$$l = l_0 \times \frac{1}{C_x + 1} \tag{1}$$

The first additional data is selected inside the square S at random. The second additional data is selected in a part, in which the existing learning data are sparse, outside the square S. A part with sparse existing data may be found as follows: First, a certain number (N_{rand}) of data are generated randomly outside the square S. Denote d_{ij} the distance between this random data p_i $(i = 1, \dots, N_{rand})$ and the existing learning data q_j $(j = 1, \dots, N)$. Select the shortest k distance $\tilde{d}_{ij}$ $(j = 1, \dots, k)$ for each p_i, and sum up these k distances, i.e., $D_i = \sum_{j=1}^{k} \tilde{d}_{ij}$. Take p_t which maximizes $\{D_i\}_{(i=1,\dots,N_{rand})}$ as an addtional data outside S.

The algorithm is sumarized as follows:

Step 1: Predict the form of objective function by RBFN on the basis of given training data.
Step 2: Estimate an optimal point for the predicted objective function by GA.
Step 3: Count the number of optimal points appeared continuously in the past in S_0. This number is represented by C_x.
Step 4: Terminate the iteration,
 (i) if C_x is larger than or equal to the given C_x^0 a priori,
 or
 (ii) if the best value of objective function obtained so far is identical during the last certain numbrer (C_f^0) of iterations.
 Otherwise calculate l by (1), and go to the next step.
Step 5: Select an additional data near the current optimal value, i.e., inside S.
Step 6: Select another additional data outside S in a place in which the density of training data is low as stated above.
Step 7: Go to Step.1.

4 Numerical Example

As a practical problem, consider the pressure vessel design problem given by Sandgren (1990), Kannan and Kramer (1994), Qian *et al.* (1993), Hsu, Sun *et al.* (1995), Lin Zhang *et al.* (1995), Arakawa *et al.* (1997). A cylindrical pressure vessel is capped at both ends by hemispherical heads. The shell is to be made in two halves, of rolled plates which are joined by two longitudinal welds to form a cylinder. Each head is forged and then welded to the shell. All welds are single welded but joints with a backing strip. The material used in the vessel is carbon steel ASME SA203 grade B. The pressure will be oriented such that the axis of the cylindrical shell is vertical. The pressure vessel is to be a compressed air storage tank with a working pressure of 2.068×10^7 Pa and a minimum volume of 1.229×10^{-2} m^3. The objective is to minimize the total cost of manufacturing the pressure vessel, including the cost of material and cost of forming and welding. The design variables are R and L, the inner radius and length of the cylindrical selection respectively, and Ts and Th are integer multiples of 0.0015875 m, the available thickness of rolled steel plates. The constraints g_1 and g_2 correspond to ASME limits on the geometry while g_3 corresponds to a minimum volume limit.

We can summarize our problem as follows:

Design variables

R : inner radius (continuous)
L : length of the cylindrical section (continuous)
Ts : thickness of rolled steel plate (discrete)
Th : thickness of rolled steel plate (discrete)

Constraints

g_1: minimum shell wall thickness

$$\frac{0.00049022\ R}{Ts} \leq 1.0$$

g_2: minimum head wall thickness

$$\frac{0.000242316\ R}{Th} \leq 1.0$$

g_3: minimum volume of tank

$$\frac{-\frac{4}{3}\ \pi\ R^3 + 21.24}{\pi\ R^2\ L} \leq 1.0$$

Side constraints

$R_{LB} \leq R \leq R_{UB}$
$L_{LB} \leq L \leq L_{UB}$
$Ts_{LB} \leq Ts \leq Ts_{UB}$
$Th_{LB} \leq Th \leq Th_{UB}$

Objective

Minimize

$$f = 37982.2\ Ts\ R\ L + 108506.3\ Th\ R^2 + 193207.3\ Ts^2\ L + 1210711 Ts^2\ R$$

In this example, the objective function is explicitly given in terms of design variables. However, many researchers use this problem as a bench mark, and hence we use the problem in order to see the effectiveness of our method by

Table 1. Numerical result ($C_x^0 = 60,\ C_f^0 = 30$)

—	# of data	R (m)	L (m)	Ts (cm)	Th (cm)	f ($)
1	83	1.1515	3.5643	2.2225	1.1112	6092.1030
2	143	0.9855	5.6505	1.9050	0.9525	5862.2705
3	153	1.1499	3.5803	2.2225	1.1113	6099.2701
4	143	0.9700	5.9066	1.9050	0.9525	5958.4034
5	139	1.1499	3.5803	2.2225	1.1113	6099.2701
6	139	1.0677	4.5620	2.0638	1.1113	6118.7685
7	147	1.1499	3.5803	2.2225	1.1113	6099.2701
8	131	1.0677	4.5620	2.0638	1.1113	6118.7685
9	165	1.0693	4.4873	2.0638	1.1113	6060.3558
10	133	1.1515	3.5643	2.2225	1.1113	6092.1030
AVERAGE	137.6					6060.0580
σ	21.6					83.6750

treating it in such a way that the objective function is not known. The parameters of RBF network are given by $\lambda = 0.01$, and the width r is decided by the conventional form.

For optimizing the predicted objective function, genetic algorithms may be applied due to simplicity of their algorithms and ability of getting an approximate global optimal solution. Although several sophisticated methods have been developed so far (e.g., Arakawa *et al.*[1]), a simple GA is adopted in this example because our main concern in this paper is rather in how well our method can approximate the objective function on the basis of as few data as possible: In our GA, continuous variables are transformed into binary ones of 11 bit. On the other hand, discrete varibles are transformed into binary ones of 4 bit; the population is 80; the generation 100; the rate of mutation is 10%. The constraint functions are penalized by

$$F = f(x) + \sum_{i=1}^{3} p_j \times [P(g_j(x))]^a,$$

where we set $p_j = 100$(penalty coefficient), $a = 2$(penalty exponent) and $P(y) = max\{y, 0\}$. Our problem is to minimize the augmented objective function F.

The data for initial learning is 17 points at the corners of the region given by the side constraints and its center. Finally, each side of the rectangle controlling the convergence and the neighborhood of the point under consideration is half the upper bound - the lower bound of each side constraint. One of results of numerical simulation for various stop conditions are given in Table 1.

Fig. 1 shows the convergence process in the second trial of Table 1. The black line shows the best value obtained so far, while the gray line shows the optimal value of the predicted objective function. Table 2 shows the comparison among existing methods. In this table, the best solution by each method is listed (Sandgren 1990, Kannan and Kramer 1994, Qian *et al.* 1993, Hsu, Sun *et al.*

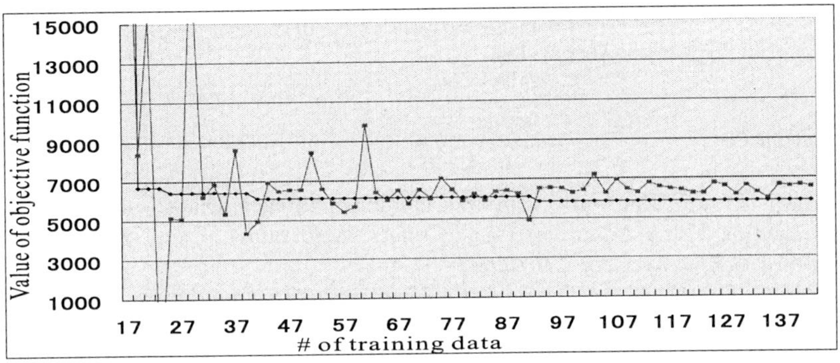

Fig. 1. Convergence process

Table 2. Comparison among existing methods

—	R (m)	L (m)	Ts (cm)	Th (cm)	f ($)
Sandgren	1.212	2.990	2.858	1.588	8129.80
Qian(GAs)	1.481	1.132	2.858	1.588	7238.83
Kannan	1.481	1.198	2.858	1.588	7198.20
Lin	N/A	N/A	N/A	N/A	7197.70
Hsu	1.316	2.587	2.540	1.270	7021.67
Lewis	0.985	5.672	1.905	0.953	5980.95
Arakawa	0.987	5.626	1.905	0.953	5851.78
suggested meth.	0.986	5.651	1.905	0.953	5862.27

1995, Lin Zhang *et al.* 1995, Arakawa *et al.* 1997). Sandgren (1990) and Kannan *et al.* (1994) use gradient type optimization techniques which treat constraints by penalty function or augmented Lagrangean methods. Lin *et al.* (1995) uses simulated annealing method and genetic algorithm, and Qian *et al.* (1993) and Arakawa *et al.* (1997) genetic algorithms.

It can be seen that our suggested method gives a reasonable solution in much fewer analyses than other methods (almost one twenties of genetic algorithms (e.g., Arakawa 1997)).

5 Concluding Remarks

We suggested a method for optimizing objective functions which are not given explicitly in terms of design variables. To this end, RSM(Response Surface Method) is well known (Myers and Montgomery 1995). However, RSM uses polynomial functions (usually, quadratic functions for the simplicity in implementation) for predicting objective functions. This causes a difficulty in getting global information, in particular, in highly nonlinear cases. It seems that the suggested method takes advantage because it considers adding global information and local information simultaneously. Comparative studies among those methods will be subjected to further researches.

References

1. Arakawa, M. and Hagiwara, I. (1997), Nonlinear Integer, Discrete and Continuous Optimization Using Adaptive Range Genetic Algorithms, *Proc. of ASME Design Technical Conferences (in CD-ROM)*
2. Haykin, S. (1994), *Neural Networks: A Comprehensive Foundation*, Macmillan College Publishing Company
3. Hsu, Y.H., Sun, T.L. and Leu, L.H. (1995), A Two-stage Sequential Approximation Method for Nonlinear Discrete Variable Optimization, *ASME Design Engineering Technical Conference*, Boston, MA, 197-202
4. Kannan, B.K. and Kramer, S.N. (1994), An Augmented Lagrange Multiplier Based Method for Mixed Integer Discrete Continuous Optimization and Its Appliations to Mechanical Design, *Transactions of ASME, J. of Mechanical Design*, Vol. 116, 405-411
5. Myers, R.H. and Montgomery,D.C. (1995), *Response Surface Methodology: Process and Product Optimization using Designed Experiments*, Wiley
6. Nakayama, H., Yanagiuchi, S., Furukawa, K., Araki, Y., Suzuki, S. and Nakata, M. (1998), Additional Learning and Forgetting by RBF Networks and its Application to Design of Support Structures in Tunnel Construction *Proc. International ICSC/IFAC Symposium on Neural Computation (NC'98)*, 544-550
7. Orr, M.J.L. (1996), Introduction to Radial Basis Function Networks, http://www.cns.ed.ac.uk/people/mark.html
8. Qian, Z., Yu, J. and Zhou, J. (1993), A Genetic Algorithm for Solving Mixed Discrete optimization Problems, *DE-Vol. 65-1, Advances in Design Automation*, Vol. 1, 499-503
9. Sandgren, E. (1990), Nonlinear Integer and Discrete Programming in Mechanical Engineering Systems, *J. of Mechanical Design*, Vol. 112, 223-229
10. Zhang, C. and Wang, H.P. (1993), Mixed-Discrete nonlinear Optimization with Simulated Annealing, *Engineering Optimization*, Vol. 21, 277-291

Clustering Gene Expression Data by Mutual Information with Gene Function

Samuel Kaski, Janne Sinkkonen, and Janne Nikkilä

Neural Networks Research Centre
Helsinki University of Technology
P.O.Box 5400, FIN-02015 HUT, Finland
{samuel.kaski,janne.sinkkonen,janne.nikkila}@hut.fi

Abstract. We introduce a simple on-line algorithm for clustering paired samples of continuous and discrete data. The clusters are defined in the continuous data space and become local there, while within-cluster differences between the associated, implicitly estimated conditional distributions of the discrete variable are minimized. The discrete variable can be seen as an indicator of relevance or importance guiding the clustering. Minimization of the Kullback-Leibler divergence-based distortion criterion is equivalent to maximization of the mutual information between the generated clusters and the discrete variable. We apply the method to a time series data set, i.e. yeast gene expressions measured with DNA chips, with biological knowledge about the functions of the genes encoded into the discrete variable.

1 Introduction

Clustering methods depend crucially on the criterion of similarity. In this work we present a new algorithm which uses additional information to decide the relevance of a dissimilarity in the data. This information is available as *auxiliary samples* c_k; they form pairs $(\mathbf{x}_k, c_k)$ with the primary samples $\mathbf{x}_k$. In this study the extra information is discrete and consists of labels of functional classes of genes.

It is assumed that differences in the auxiliary data indicate what is important in the primary data space. Intuitively speaking, the auxiliary data guides the clustering to emphasize the (locally) important dimensions of the primary data space, and to disregard the rest. This automatic relevance detection is the main practical motivation for the present work.

The distance between two close-by points $\mathbf{x}$ and $\mathbf{x}'$ is measured by the differences between the *distributions* of c, given $\mathbf{x}$ and $\mathbf{x}'$. Our aim is to minimize the within-cluster dissimilarities in terms of the (implicitly estimated) distributions $p(c|\mathbf{x})$ and still define the clusters by localized basis functions within the primary space.

Put in different words, we supervise the metric of the data space to concentrate on the important or relevant aspects of the data. The main goal is data-driven unsupervised exploration of the data in the new metric to make

G. Dorffner, H. Bischof, and K. Hornik (Eds.): ICANN 2001, LNCS 2130, pp. 81–86, 2001.

new discoveries of its statistical properties. Here the changes in the distribution of the auxiliary data provide the supervised metric, and the unsupervised task is clustering. The method could perhaps be called *semisupervised*, or clustering "guided" by the supervisory signal.

In an earlier work [8] we constructed explicit estimates for the conditional distributions $p(c|\mathbf{x})$ and used them to compute the local distances. In the present work we use an alternative method [11] which does not need an estimate of the conditional distributions $p(c|\mathbf{x})$ as an intermediate step.

Minimizing the within-cluster distortion is equivalent to maximizing the mutual information between the clusters (where cluster membership is interpreted as a value of a multinomial random variable) and the auxiliary data. Maximization of mutual information has been previously used for constructing neural representations [1, 2]. Other related works and paradigms include learning from (discrete) dyadic data [12], and distributional clustering [10] with the information bottleneck [13] principle. These methods start from an estimate of the joint distribution $p(c, x)$, which must be available for all discrete x. Our method places and defines clusters directly in a continuous input space; new data $\mathbf{x}$ can then be assigned to clusters even if the auxiliary data is not available.

In this work the method is applied to gene expression data, which has previously been clustered with conventional clustering algorithms (see, e.g., [5]), and classified with supervised methods (Support Vector Machines, [3]). Our goal is to use the known functional classification of the genes to implicitly define which aspects of the expression data are important in the data, and only utilize the important aspects, local factors or dimensions, in clustering. The difference from ordinary supervised classification methods is that while they cannot surpass the original classes our method is not tied to the classification and may reveal substructures within and relations between the known functional classes.

In this case study we compare our method empirically with alternative methods, and demonstrate the potential usefulness of the results. More detailed biological interpretation of the results will be presented in subsequent papers.

2 Clustering Based on the Kullback-Leibler Divergence

We seek to categorize items $\mathbf{x}$ of the data space by utilizing the information within a set of samples $(\mathbf{x}_k, c_k)$ from a random variable pair (X, C). Here $\mathbf{x} \in \mathbb{X} \subset \mathbb{R}^n$, and the c_k are discrete, multinomial values. We wish to *define* the clusters in terms of $\mathbf{x}$ only, and to keep the clusters local with respect to $\mathbf{x}$. The distortions minimized by the clustering are, however, measured between distributions in the auxiliary space, although it is not necessary to explicitly compute these distributions.

Vector quantization (VQ) or, equivalently, K-means clustering, is one approach to categorization. In ("soft") VQ the goal is to minimize the average distortion E between the data and the prototypes or codebook vectors $\mathbf{m}_j$, defined by

$$E = \sum_j \int y_j(\mathbf{x}) D(\mathbf{x}, \mathbf{m}_j)\, p(\mathbf{x})\, d\mathbf{x} \,. \tag{1}$$

Here $D(\mathbf{x}, \mathbf{m}_j)$ is the distortion between $\mathbf{x}$ and $\mathbf{m}_j$, and $y_j(\mathbf{x})$ is the "soft" cluster membership function that fulfills $0 \leq y_j(\mathbf{x}) \leq 1$ and $\sum_j y_j(\mathbf{x}) = 1$.

We measure distortions by the Kullback–Leibler divergence $D_{KL}(p, \boldsymbol{\psi}) \equiv \sum_i p_i \log(p_i/\psi_i)$, where p_i is the multinomial distribution in the auxiliary space that corresponds to the data $\mathbf{x}$, that is, $p_i \equiv p(c_i|\mathbf{x})$. The second distribution is a prototype for the cluster; let us denote the jth prototype by $\boldsymbol{\psi}_j$.

Although the distortions are measured between distributions of the auxiliary variable, the cluster memberships are still defined in the primary data space, by the membership functions $y_j(\mathbf{x})$. In our application the data lies on a hypersphere, and hence we use membership functions that are defined for spheres. The membership functions are normalized von Mises–Fisher (vMF$(\mathbf{x}; \boldsymbol{\theta}_j)$) kernels [9], spherical analogues of Gaussians, parameterized by their centers $\boldsymbol{\theta}_j$ (and widths κ): $y_j(\mathbf{x}; \boldsymbol{\theta}_j) = \text{vMF}(\mathbf{x}; \boldsymbol{\theta}_j) / \sum_k \text{vMF}(\mathbf{x}; \boldsymbol{\theta}_k)$, where $\text{vMF}(\mathbf{x}; \boldsymbol{\theta}_j) = \text{const.} \times \exp(\kappa \boldsymbol{x}^T \boldsymbol{\theta}_j / \|\boldsymbol{\theta}_j\|)$.

The final cost function to be optimized is (1), with $y_j(\mathbf{x})$ replaced by the parameterized $y_j(\mathbf{x}; \boldsymbol{\theta}_j)$, and $D(\mathbf{x}, \mathbf{m}_j)$ replaced by the divergence $D_{KL}(p(c|\mathbf{x}), \boldsymbol{\psi}_j)$.

Connection to mutual information. Denote by V the random variable that indicates which cluster a data sample belongs to. That is, the value of V is v_i if the sample belongs to cluster i. It can be shown that our cost function is equal to the (negative) mutual information between V and C, $-I(C; V)$, up to an additive constant.

The algorithm. It can be shown [7] that the average distortion can be minimized with stochastic approximation that samples from $y_j(\mathbf{x}) y_l(\mathbf{x}) p(c_i, \mathbf{x}) = p(v_j, v_l, c_i, \mathbf{x})$. This leads to an on-line algorithm in which the following steps are repeated for $t = 0, 1, \ldots$ with $\alpha(t)$ gradually decreasing towards zero:

1. Draw a data sample $(\mathbf{x}(t), c(t))$. The discrete sample $c(t) = c_i$ defines the value of i in the following steps.
2. Draw two basis functions, j and l, from a multinomial distribution with probabilities $\{y_k(\mathbf{x}(t))\}_k$.
3. Adapt the parameters $\boldsymbol{\theta}_l$ and γ_{lm}, $m = 1, \ldots, N_c$, by

$$\boldsymbol{\theta}_l(t+1) = \boldsymbol{\theta}_l(t) - \alpha(t)(\mathbf{x}(t) - \mathbf{x}(t)^T \boldsymbol{\theta}_l(t) \boldsymbol{\theta}_l(t)) \log \frac{\psi_{ji}(t)}{\psi_{li}(t)} \tag{2}$$

$$\gamma_{lm}(t+1) = \gamma_{lm}(t) - \alpha(t)(\psi_{lm}(t) - \delta_{mi}) \,, \tag{3}$$

where N_c is the number of possible values of the random variable C, δ_{mi} is the Kronecker delta, and the γ_l are prototype distributions reparameterized to keep their sum at unity: $\log \psi_{ji} = \gamma_{ji} - \log \sum_m exp(\gamma_{jm})$. Due to the symmetry between j and l, it is possible to adapt the parameters twice for each t by swapping j and l in (2) and (3) for the second adaptation step. Note that $\boldsymbol{\theta}_l(t+1) = \boldsymbol{\theta}_l(t)$ if $j = l$.

3 Case Study: Clustering of Gene Expression Data

We clustered 2,467 genes of the budding yeast *Saccharomyces cerevisiae* by their activity or expression in eight experimental conditions (see [5]), measured as a time series with DNA chips. The total of 79 time points were collected into the feature vector. The data was preprocessed in the same way as in [3], and divided into a training set (two thirds) and a test set (the remaining third).

3.1 Alternative Models

We compared our model with two standard state-of-the-art mixture density models. The goal of these models is slightly different from ours but we use them to provide baseline results. According to our knowledge they are the most closely related existing models.

The first is a totally unsupervised mixture of von Mises–Fisher distributions. The model is analogous to the usual mixture of Gaussians; the Gaussian mixture components are simply replaced by the von Mises–Fisher components.

In the second model, MDA2 [6], the joint distribution between the functional classes c and the expression data $\mathbf{x}$ is modeled by a set of additive components denoted by u_j:

$$p(c_i, \mathbf{x}) = \sum_j p(c_i|u_j)p(\mathbf{x}|u_j)p_j \ , \tag{4}$$

where $p(c_i|u_j)$ and p_j are parameters to be estimated, and $p(\mathbf{x}|u_j) = \mathrm{vMF}(\mathbf{x}; \boldsymbol{\theta}_j)$. Both models are fitted to the data by likelihood maximization with the EM algorithm [4].

3.2 The Experiments

We compared the three models, each with 8 clusters and optimized to near convergence. All models were run three times with different random initializations, and the best of the three results was chosen.

The quality of the resulting clusterings was measured by the average distortion error or, equivalently, the empirical mutual information. The results in Figure 1 **(A)** show that our model clearly outperforms the others for a wide range of widths of the kernels and produces the best overall performance. The clusters of our model therefore convey more information about the functional classification of the genes than the alternative models.

There is a somewhat surprising sideline in the results: The vMF mixture model is slightly better than the MDA2, although it does not utilize the class information at all. The reason is probably in some special property of the data since the likelihood of the paired data (not shown), the measure that the MDA2 optimizes, was larger for MDA2 than for the simple mixture model in which the class distribution in the clusters was estimated afterwards. Moreover, for other data sets the MDA2 has outperformed the plain mixture model.

The distribution of some functional *subclasses* of genes in the clusters is shown in Figure 1 **(B)**. Note that these subclasses were not included in the values of the auxiliary variable C. Instead, they were picked from the second

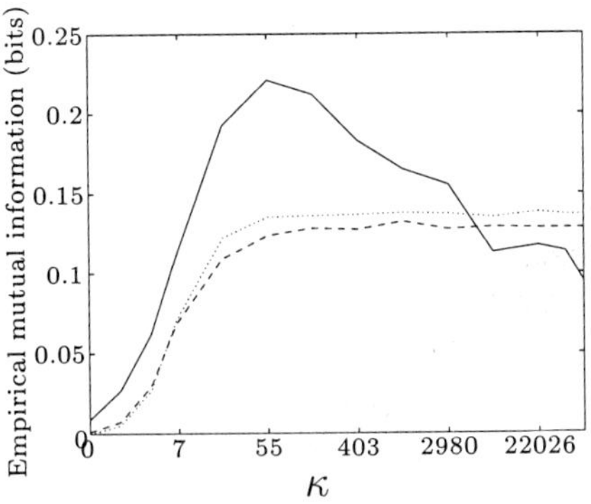

(A)

Class	Cluster number 1	2	3	4	5	6	7	8
a	0	1	6	0	1	0	0	0
b	0	1	16	0	0	0	0	0
c	1	5	39	1	1	4	14	3
d	0	3	8	1	0	0	0	2
e	122	1	0	2	0	2	44	2
f	3	3	1	20	0	46	2	12
g	0	1	0	21	0	0	0	7

(B)

Fig. 1. (A) Empirical mutual information between the generated gene expression clusters and the functional classes of the genes, as a function of parameter κ which governs the width of the basis functions. Solid line: our model, dotted line: mixture of vMFs, dashed line: MDA2. **(B)** Distribution of genes (learning and test set combined) of sample functional subclasses into the 8 clusters. These subclasses were not used in supervising the clustering. a: the pentose-phosphate pathway, b: the tricarboxylic-acid pathway, c: respiration, d: fermentation, e: ribosomal proteins, f: cytoplasmic degradation, g: organization of chromosome structure.

and third level of the functional classification, and they include groups that are known to be regulated in the experimental settings measured by the data [5]. Most of the subclasses are concentrated in one of the clusters. The first four (a-d) belong to the same first-level class and are placed (mostly) in the same cluster, number 3.

Three of the subclasses (c, e, and f) have been clearly divided into more than one cluster, indicating some potentially biologically interesting subdivision of genes inside the subclasses. It remains to be seen in further biological inspection whether such subdivisions provide new biological information; in this paper our goal is to demonstrate that the "semisupervised" clustering approach can be used to explore the data set and provide potential further hypotheses about its structure.

4 Conclusions

We have presented a new clustering method for continuous data which minimizes the within-cluster distortion between distributions of associated, discrete auxiliary data [11]. A simple on-line algorithm was used for optimizing the distortion measure.

We clustered a yeast gene expression data set in which auxiliary information was available as an independent, functional classification of the genes. In our semisupervised task the algorithm performed better than other algorithms available for continuous data, the mixture of vMFs and the MDA2 which models the joint density of the expression data and the classes.

Note that the number of extracted clusters can be arbitrary and it need not depend on the auxiliary data (here the number of classes). Depending on whether the number of clusters is larger or smaller than the number of classes the relationships of the classes or the substructures within the classes can be studied. Such studies are not possible with supervised methods.

It was shown that the obtained clusters mediate information about the function of the genes and, although the results have not yet been biologically analyzed, potentially suggest novel cluster structures for the yeast genes.

Acknowledgments. This work was supported by the Academy of Finland, in part by the grant 50061. We wish to thank Petri Törönen for his help with the data and Jaakko Peltonen with some of the programs.

References

1. S. Becker. Mutual information maximization: models of cortical self-organization. *Network: Computation in Neural Systems*, 7:7–31, 1996.
2. S. Becker and G. E. Hinton. Self-organizing neural network that discovers surfaces in random-dot stereograms. *Nature*, 355:161–163, 1992.
3. M. P. Brown, W. N. Grundy, D. Lin, N. Cristianini, C. W. Sugnet, T. S. Furey, M. Ares, Jr., and D. Haussler. Knowledge-based analysis of microarray gene expression data by using support vector machines. *Proceedings of the National Academy of Sciences, USA*, 97:262–267, 2000.
4. A. P. Dempster, N. M. Laird, and D. B. Rubin. Maximum likelihood from incomplete data via the EM algorithm. *Journal of the Royal Statistical Society, Series B*, 39:1–38, 1977.
5. M. B. Eisen, P. T. Spellman, P. O. Brown, and D. Botstein. Cluster analysis and display of genome-wide expression patterns. *Proceedings of the National Academy of Sciences, USA*, 95:14863–14868, 1998.
6. T. Hastie, R. Tibshirani, and A. Buja. Flexible discriminant and mixture models. In J. Kay and D. Titterington, editors, *Neural Networks and Statistics.* Oxford University Press, 1995.
7. S. Kaski. Convergence of a stochastic semisupervised clustering algorithm. Technical Report A62, Helsinki University of Technology, Publications in Computer and Information Science, Espoo, Finland, 2000.
8. S. Kaski, J. Sinkkonen and J. Peltonen. Bankruptcy Analysis with Self-Organizing Maps in Learning Metrics *IEEE Transactions on Neural Networks*, 2001, in press.
9. K. V. Mardia. Statistics of directional data. *Journal of the Royal Statistical Society. Series B*, 37:349–393, 1975.
10. F. Pereira, N. Tishby, and L. Lee. Distributional clustering of English words. In *Proceedings of the 30th Annual Meeting of the Association for Computational Linguistics*, pages 183–190. 1993.
11. J. Sinkkonen and S. Kaski. Clustering based on conditional distributions in an auxiliary space. *Neural Computation*, 2001, in press.
12. T. Hofmann, J. Puzicha, and M. I. Jordan. Learning from dyadic data. In M. S. Kearns, S. A. Solla, and D. A. Cohn, editors, *Advances in Neural Information Processing Systems 11*, pages 466–472. Morgan Kauffmann Publishers, San Mateo, CA, 1998.
13. N. Tishby, F. C. Pereira, and W. Bialek. The information bottleneck method. In *37th Annual Allerton Conference on Communication, Control, and Computing*, Urbana, Illinois, 1999.

Learning to Learn Using Gradient Descent

Sepp Hochreiter[1], A. Steven Younger[1], and Peter R. Conwell[2]

[1] Department of Computer Science
University of Colorado, Boulder, CO 80309–0430
[2] Physics Department
Westminster College, Salt Lake City, Utah

Abstract. This paper introduces the application of gradient descent methods to meta-learning. The concept of "meta-learning", i.e. of a system that improves or discovers a learning algorithm, has been of interest in machine learning for decades because of its appealing applications. Previous meta-learning approaches have been based on evolutionary methods and, therefore, have been restricted to small models with few free parameters. We make meta-learning in large systems feasible by using recurrent neural networks with their attendant learning routines as meta-learning systems. Our system derived complex well performing learning algorithms from scratch. In this paper we also show that our approach performs non-stationary time series prediction.

1 Introduction

Phrases like "I have experience in ...", "This is similar to ...", or "This is a typical case of ..." imply that the person making such statements learns the task at hand faster or more accurately than an unexperienced human. This learning enhancement results from solution regularities in a problem domain. In a conventional machine learning approach the learning algorithm mostly does not take into account previous learning experiences despite the fact that methods similar to human reasoning are expected to yield better performance. The use of previous learning experiences in inductive reasoning is known as "knowledge transfer" [4, 1, 14] or "inductive bias shifts" [15, 6, 12].

In the research field of "knowledge transfer" we focus on one of the most appealing topics: "meta-learning" or "learning to learn" [4, 14, 13, 11]. A meta-learner searches out and finds appropriate learning algorithms tailored to specific learning tasks. To find such learning methods, a supervisory algorithm that reviews and modifies the training algorithm must be added. In contrast to the subordinate learning scheme, the supervisory routine has a broader scope. It must ignore the details unique to specific problems, and look for symmetries over a long time scale, i.e. it must perform "knowledge transfer". For example consider a human as the supervisor and a kernel density estimator as the subordinate method. The human has previous experiences with overfitting and tries to avoid it by adding a bandwidth adaptation and improving the estimator. We want to automatically obtain such learning method improvements by replacing

G. Dorffner, H. Bischof, and K. Hornik (Eds.): ICANN 2001, LNCS 2130, pp. 87–94, 2001.

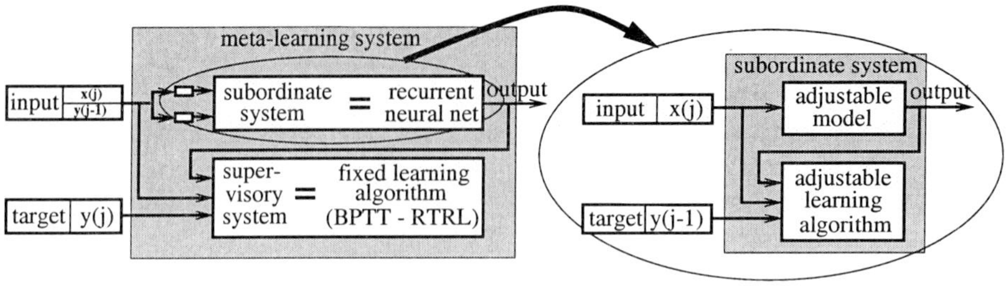

Fig. 1. The meta-learning system consists of the supervisory and the subordinate system (sequence element j is processed). The subordinate system is a recurrent network. Its attendant learning algorithm represents the fixed supervisory system. Target function arguments x are mapped to results y, e.g. $y(j) = f(x(j))$. The previous function result $y(j-1)$ is supplied to the subordinate system so that it can determinate the previous error of the subordinate model. Subordinate and supervisory outputs are identified.

the human part with an appropriate system. This automatic system must include an objective function to judge the performance of the learning algorithm and rules for the adjustment of the algorithm. Meta-learning is known in the reinforcement learning framework [11, 12]. This paper reports on our work on meta-learning in a supervised learning framework where a model is supposed to approximate a function after being trained on examples. Our meta-learning system consists of the supervisory procedure, which is fixed, and of the adjustable subordinate system, which must be run on a certain medium (see left hand side of Figure 1). To exemplify this, for this medium we might have used a Turing machine (i.e. a computer) where the subordinate model and the subordinate training routine is represented by a program (see right hand side of Figure 1). Any changes to the program amount to changes in the subordinate learning algorithm[1]. However, the output of the discrete Turing machine is not differentiable. Thus, only deductive or evolutionary strategies can be used to improve the Turing machine program. Instead of executing the subordinate learning algorithm with a Turing machine, our method executes the algorithm with a recurrent neural network in order to get a differentiable output. This is possible because a (sufficiently large) recurrent neural network can emulate a Turing machine. The differentiable output allows us to apply gradient descent methods to improve the subordinate routine. A recurrent network with random initial weights can be viewed as a learning machine with a very poor subordinate learning algorithm. We hypothesize that gradient based optimization approaches can be used to derive a learning algorithm from a random starting point.

The capability of recurrent networks to execute the subordinate system was proved and demonstrated in [3, 19]. Several researchers have suggested meta-

[1] It should be mentioned that in general, the coded model and the coded learning algorithm cannot be separated. Accordingly, with the term "learning algorithm" we mean both.

learning systems based on neural networks and used genetic algorithms to adjust the subordinate learning algorithm [2, 10, 19]. Our goal is to obtain complex subordinate learning algorithms which need a large recurrent network with many parameters. Genetic algorithms are infeasible due to the large number of computations required. This paper introduces gradient descent for meta-learning to handle such large systems, and, thus, to provide an optimization technique in the space of learning algorithms.

Every recurrent neural network architecture with its attendant learning procedure is a possible meta-learning system. One may choose for example backpropagation through time (BPTT [18, 16]) or real-time recurrent learning (RTRL [9, 17]) as attendant learning algorithms. The meta-learning characteristic of these networks is only determined by the special kind of input-target sequences as described in section 2.1. Both BPTT and RTRL applied to standard recurrent nets do not yield good meta-learning performance as will be seen in section 3. The reason for this poor performance is given in section 2.2. In the same section, the use of the Long Short-Term Memory (LSTM [8]) architecture is suggested to achieve better results. Section 2.3 gives an intuition how the "inductive bias shift" ("knowledge transfer") takes place during meta-learning. The experimental section 3 demonstrates how different learning procedures for different problem domains are automatically derived by our meta-learning systems.

2 Theoretical Considerations

2.1 The Data-Setup for Meta-learning with Recurrent Nets

This section describes the kind of input-target sequences that allow meta-learning in recurrent nets. The training data for the meta-learning system is a set of sequences $\{s_k\}$, where sequence s_k is obtained from a target function f_k. At each time step j during processing the kth sequence, the meta-learning system needs the function result $y_k(j) = f_k(x_k(j))$ as a target. The input to the meta-learning system consists of the current function argument vector $x_k(j)$ and a supplemental input which is the previous function result $y_k(j-1)$. The subordinate learning algorithm needs the previous function result $y_k(j-1)$ so that it can learn the presented mapping, e.g. to compute the subordinate model error for input $x_k(j-1)$. We cannot provide the current target $y_k(j)$ as an input to the recurrent network since we cannot prevent the model from cheating by hard-wiring the current target to its output. Figure 1 illustrates the inputs and targets for the different learning systems.

The meta-learning system is penalized at each time point when it does not generate the correct target value, i.e. when the subordinate procedure was yet not able to learn the current function. This forces the meta-learning system to improve the subordinate algorithm so that it becomes faster and more exact. Figure 2 shows test sequences after successful meta-learning. New sequences start at 513, 770, and 1027 when the subordinate learning method produces large errors because the new function is not yet learned. After a few examples the subordinate system learned the new function.

The characteristics of the derived subordinate algorithms can be influenced by the sequence length (more examples per function give more precise but slower algorithms), the error function, and the architecture.

2.2 Selecting a Recurrent Architecture for Meta-learning

For simplicity we consider one function f giving the sequence $(x_1, y_1), \ldots, (x_J, y_J)$, where $y_j = f(x_j)$. All training examples (x_j, y_j) contain equal information about f. The indices correspond to the time steps of the recurrent net. We want to bias our meta-learning system towards this prior knowledge. The information in the last output O_J (indicated by J) is determined by the entropy $H(O_J \mid X_J)$. Here probability variables are denoted by capital letters, e.g. X for the input, Y for the target, and O for the output. $H(A)$ is the entropy of A and the conditional entropies are denoted by $H(A \mid B)$ The last output is obtained by $o_J = g(y_j; x_J, x_j) + \epsilon$, where g is a bijective function with variable y_j and parameters x_J, x_j. ϵ expresses disturbances during input sequence processing. We assume noisy mappings to avoid infinite entropies. Neglecting ϵ, we get

$$H(O_J \mid X_J, X_j) = H(Y_j \mid X_j) + E_{Y_j, X_j, X_J}\left(\log\left(\left|\frac{\partial g(Y_j; X_j, X_J)}{\partial Y_j}\right|\right)\right),$$

where $p(Y_j \mid x_j, x_J) = p(Y_j \mid x_j)$, $\left|\frac{\partial g(Y_j; X_j, X_J)}{\partial X_j}\right|$ is the absolute value of the g's Jacobian determinant, and $E_{A,B,\ldots}$ is the expectation over variables $A, B, \ldots$.

The hidden state at time j is $s_j = u(s_{j-1}, x_j, y_{j-1})$ and the output is $o_j = v(s_j)$. With $i < j < J$ we get

$$\frac{\partial o_J}{\partial y_j} = \frac{\partial o_J}{\partial s_{j+1}} \frac{\partial s_{j+1}}{\partial y_j}, \quad \frac{\partial o_J}{\partial y_i} = \frac{\partial o_J}{\partial s_{j+1}} \frac{\partial s_{j+1}}{\partial s_{i+1}} \frac{\partial s_{i+1}}{\partial y_i}, \quad \frac{\partial s_{j+1}}{\partial s_{i+1}} = \prod_{l=i+1}^{j} \frac{\partial s_{l+1}}{\partial s_l}.$$

Our prior knowledge says that exchanging example i and j should not affect the output information. That is $H(O_J \mid X_J, X_j) = H(O_J \mid X_J, X_i)$, and also ϵ should not change. In this case $Y_j = Y_i$, $X_j = X_i$, $p(Y_j \mid x_j) = p(Y_i \mid x_i)$ for $x_j = x_i$, and $H(Y_j \mid X_j) = H(Y_i \mid X_i)$. At learn begin with arbitrary weight initialization $E_{Y_j, X_j}\left(\frac{\partial S_{j+1}}{\partial Y_j}\right) = E_{Y_i, X_i}\left(\frac{\partial S_{i+1}}{\partial Y_i}\right)$. Thus, we obtain

$$E_{Y_j, X_j, X_J}\left(\log\left(\left|\frac{\partial g(Y_j; X_j, X_J)}{\partial Y_j}\right|\right)\right) - E_{Y_i, X_i, X_J}\left(\log\left(\left|\frac{\partial g(Y_i; X_i, X_J)}{\partial Y_i}\right|\right)\right) = 0, \text{ or}$$

$$\sum_{l=i+1}^{j} E_{Y_i, X_i}\left(\log\left(\left|\frac{\partial S_{l+1}}{\partial S_l}\right|\right)\right) = 0, \quad \text{e.g. with } \left|\frac{\partial s_{l+1}}{\partial s_l}\right| = 1.$$

u restricted to a mapping from s_l to s_{l+1} should be volume conserving. An architecture which incorporates such a volume conserving substructure should outperform other architectures. An architecture fulfilling this requirement is Long Short-Term Memory (LSTM [8]).

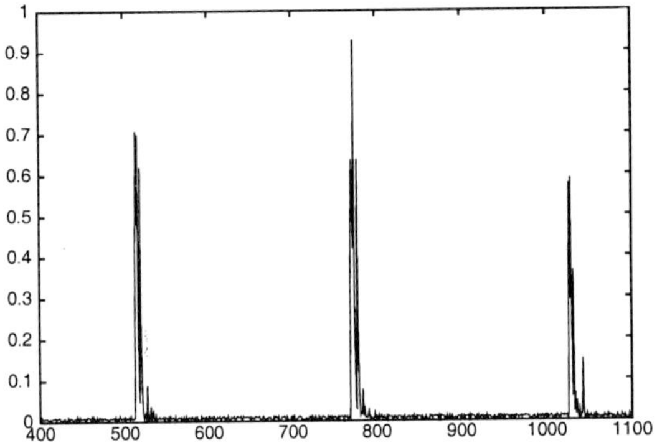

Fig. 2. Error vs. time after meta-learning.

2.3 Bayes View on Meta-learning

Meta-learning can be viewed as constantly adapting and shifting the hyper-parameters and the prior ("inductive bias shift") because the subordinate learning algorithm is adapted to the problem domain during meta-learning. As the experiments confirm, also the prior of subordinate learning algorithms is data dependent. This was suggested in [7], too. Therefore different previously observed examples might lead to different current learning.

3 Experiments

We choose a squared error function for the supervisory learning routine. All networks possess 3 input and 1 non-recurrent output units. All non-input units are biased and have sigmoid activation functions in $[0, 1]$. Weights are randomly initialized from $[-0.1, 0.1]$. All networks are reset after each sequence presentation.

3.1 Boolean Functions

Here we consider the set B_{16} of all Boolean functions with two arguments and one result. The linearly separable Boolean functions set $B_{14} = B_{16} \setminus \{XOR, \neg XOR\}$ is used to evaluate meta-learning architectures.

$\mathbf{B_{14}}$ Experiments
We compared following methods: **(A)** Elman network [5]. **(B)** Recurrent network with fully recurrent hidden layer trained with Back Propagation Through Time (BPTT [18, 16]) truncated after 2 time steps and with Real Time Recurrent Learning (RTRL [9, 17]). **(C)** Long Short-Term Memory (LSTM [8]) with its corresponding learning procedure. The cell input is squashed to $[-2, 2]$ by a sigmoid and the cell output is a sigmoid in $[-1, 1]$. For input gates the bias is set to -1.0. Table 1 gives the results. Only LSTM was successful.

Table 1. The B_{14} experiments for Elman nets ("Elman"), recurrent networks with fully recurrent hidden layer ("Rec."), and LSTM. The columns show: (1) architecture ("arch."), (2) number of hidden units – for LSTM "6/6(1)" means 6 hidden units and 6 memory cells of size 1 –, (3) learning method – for Elman nets and LSTM their learning methods are used, and BPTT is truncated after 2 time steps –, (4) batch ("b") or online ("o") update, (5) learning rate α – 0.001-0.1 means different learning rates in this range –, (6) training epochs, (7) training and (8) test mean squared error per time step ("MSEt"), and (9) successful training ("success").

arch.	hid. units	learning method	up-date	α	train time	train MSEt	test MSEt	suc-cess
Elman	15	Elman	b	0.001-0.1	5000			NO
Rec.	20	BPTT(2)	b & o	0.001-0.01	40000			NO
Rec.	10	RTRL	b & o	0.001-0.1	20000	0.22	0.21	NO
LSTM	6/6(1)	LSTM	o	0.001	1000	0.033	0.038	YES

B_{16} Experiments

The results are shown in Table 2. The mean squared errors per time step (MSEts) are lower than at B_{14} because the large error at the beginning of a new function scales down with more examples. See Figure 2 for absolute error meta-learning. The peaks at 513, 770, and 1027 indicate large error when the function changes.

3.2 Semi-linear Functions

We obtain functions $0.5\,(1.0 + \tanh\,(w_1 x_1 + w_2 x_2 + w_3))$ with input vector $x = (x_1, x_2)$ by choosing each parameter w_l randomly from $[-1, 1]$. Table 2 presents the results. With more examples per function the pressure on error reduction for the first examples is reduced which leads to slower but more exact learning.

3.3 Quadratic Functions

The problem domain are the quadratic functions $a\;x_1^2\;+\;b\;x_2^2\;+\;c\;x_1\;x_2\;+\;d\;x_1\;+\;e\;x_2\;+\;f$ scaled to the interval $[0.2, 0.8]$. The parameters $a, \ldots, f$ are randomly chosen from $[-1, 1]$. We introduced another hidden layer in the LSTM architecture which receives incoming connections from the first standard LSTM hidden layer, and has outgoing connections to the output and the first hidden layer. The first hidden layer has no output connections. The second hidden layer might serve as a model which is seen by the first hidden layer. The standard LSTM learning algorithm is used after the error is propagated back into the first hidden layer.

LSTM has a 6/12(1) architecture in the first hidden layer (notation as in Table 1) and 40 units in the second hidden layer (5373 weights). To speed up learning, we first trained on 100 examples per function and then increased this number to 1000. This corresponds to a bias towards fast learning algorithms. The results are listed in Table 2. The authors are not aware of any iterative learning algorithm with to the derived subordinate method comparable performance.

Table 2. LSTM results for the B_{14}, B_{16}, semilinear ("semil.") and quadratic functions ("quad."). The columns show: (1) experiment name, (2) number of training sequences ("# functions"), (3) length of training sequences (examples per function – "# examples"), (4) training epochs, (5) training MSEt, (6) test MSEt, (7) training time for the derived algorithm ("train time subordinate"), and (8) maximal training mean squared error per example of the subordinate system after training ("train MSE subordinate"). The B_{14} architecture was used except for "quad." (see text for details).

experiment	# functions	# examples	train time	train MSEt	test MSEt	train time subordinate	train MSE subordinate
B_{14}	128	64	1000	0.033	0.038	6	0.003
B_{16}	256	256	800	0.0055	0.0058	6	0.002
semil.	128	64	10000	0.0007	0.0008	10	0.07
semil.	128	1000	5000	0.0020	0.0025	50	0.05
quad.	128	1000	25000	0.00061	0.00068	35	0.02

3.4 Summary of Experiments

The experiments demonstrate that our system automatically generates learning methods from scratch and that the derived online learning algorithms are extremely fast. The test and the training sequence for the meta-learning system contains rapidly changing dynamics, i.e. the changing functions, what can be viewed as a very non-stationary time series. Our system was able to predict well on never seen changing dynamics in the test sequence. The non-stationary time series prediction is based on rapid learning if the dynamic changes.

4 Conclusion

Previous approaches to meta-learning are infeasible for a large number of system parameters. To handle many free parameters this paper presented the application of gradient descent to meta-learning by using recurrent nets. Our theoretical analysis indicated that LSTM is a good meta-learner what was confirmed in the experiments. With an LSTM net our system derived a learning algorithm able to approximate any quadratic function after only 35 examples.

Our approach requires a single training sequence, therefore, it may be relevant for lifelong learning and autonomous robots. The meta-learner proposed in this paper performed non-stationary time series prediction. We demonstrated how a machine can derive novel, very fast algorithms from scratch.

Acknowledgments

The *Deutsche Forschungsgemeinschaft* supported this work (Ho 1749/1-1).

References

1. R. Caruana. Learning many related tasks at the same time with backpropagation. In G. Tesauro, D. Touretzky, and T. Leen, editors, *Advances in Neural Information Processing Systems 7*, pages 657–664. The MIT Press, 1995.
2. D. Chalmers. The evolution of learning: An experiment in genetic connectionism. In D. S. Touretzky, J. L. Elman, T. J. Sejnowski, and G. E. Hinton, editors, *Proc. of the 1990 Con. Models Summer School*, pages 81–90. Morgan Kaufmann, 1990.
3. N. E. Cotter and P. R. Conwell. Fixed-weight networks can learn. In *Int. Joint Conference on Neural Networks*, volume II, pages 553–559. IEEE, NY, 1990.
4. H. Ellis. *Transfer of Learning*. MacMillan, New York, NY, 1965.
5. J. L. Elman. Finding structure in time. Technical Report CRL 8801, Center for Research in Language, University of California, San Diego, 1988.
6. D. Haussler. Quantifying inductive bias: AI learning algorithms and Valiant's learning framework. *Artificial Intelligence*, 36:177–221, 1988.
7. S. Hochreiter and J. Schmidhuber. Flat minima. *Neural Comp.*, 9(1):1–42, 1997.
8. S. Hochreiter and J. Schmidhuber. Long short-term memory. *Neural Computation*, 9(8):1735–1780, 1997.
9. A. J. Robinson and F. Fallside. The utility driven dynamic error propagation network. Technical Report CUED/F-INFENG/TR.1, Camb. Uni. Eng. Dep., 1987.
10. T. P. Runarsson and M. T. Jonsson. Evolution and design of distributed learning rules. In *2000 IEEE Symposium of Combinations of Evolutionary Computing and Neural Networks, San Antonio, Texas, USA*, page 59. 2000.
11. J. Schmidhuber, J. Zhao, and M. Wiering. Simple principles of metalearning. Technical Report IDSIA-69-96, IDSIA, 1996.
12. J. Schmidhuber, J. Zhao, and M. Wiering. Shifting inductive bias with success-story algorithm, adaptive levin search, and incremental self-improvement. *Machine Learning*, 28:105–130, 1997.
13. J. Schmidhuber. Evolutionary principles in self-referential learning, or on learning how to learn: The meta-meta-... hook. Inst. für Inf., Tech. Univ. München, 1987.
14. S. Thrun and L. Pratt, editors. *Learning To Learn*. Kluwer Academic Pub., 1997.
15. P. Utgoff. Shift of bias for inductive concept learning. In R. Michalski, J. Carbonell, and T. Mitchell, editors, *Machine Learning*, volume 2. Morgan Kaufmann, 1986.
16. P. J. Werbos. Generalization of backpropagation with application to a recurrent gas market model. *Neural Networks*, 1, 1988.
17. R. J. Williams and D. Zipser. A learning algorithm for continually running fully recurrent networks. Technical Report ICS 8805, Univ. of Cal., La Jolla, 1988.
18. R. J. Williams and D. Zipser. Gradient-based learning algorithms for recurrent networks and their computational complexity. In Y. Chauvin and D. E. Rumelhart, editors, *Back-propagation: Theory, Architectures and Applications*. Hillsdale, 1992.
19. A. S. Younger, P. R. Conwell, and N. E. Cotter. Fixed-weight on-line learning. *IEEE-Transactions on Neural Networks*, 10(2):272–283, 1999.

A Variational Approach to Robust Regression

Anita C. Faul and Michael E. Tipping*

Microsoft Research, Cambridge, U.K.

Abstract. We consider the problem of regression estimation within a Bayesian framework for models linear in the parameters and where the target variables are contaminated by 'outliers'. We introduce an explicit distribution to explain outlying observations, and utilise a variational approximation to realise a practical inference strategy.

1 Introduction

We study the problem of interpolating through data where the dependent variables are assumed to be noisy ('regression', 'curve fitting' or 'signal estimation'). Our data comprises pairs $\{\mathbf{x}_n, t_n\}$, where $n = 1, \ldots, N$, and we consider linear interpolation models where the predictor $y(\mathbf{x})$ is expressed as a linearly-weighted sum of a set of M fixed basis functions $\phi_m(\mathbf{x})$, $m = 1, \ldots, M$:

$$y(\mathbf{x}) = \sum_{m=1}^{M} w_m \phi_m(\mathbf{x}). \tag{1}$$

We assume that the 'target' variables deviate from this mapping under some additive noise process: *i.e.* $t_n = y(\mathbf{x}_n) + \epsilon_n$. Typically, we assume independent Gaussian noise: $\epsilon_n \sim \mathcal{N}(0, \sigma^2)$. The linear nature of this form of model facilitates a Bayesian treatment of the parameters $\mathbf{w} = (w_1, \ldots, w_M)^{\mathrm{T}}$, since, if we adopt a Gaussian prior thereover, and with a Gaussian noise model, the posterior and marginal likelihood are both similarly Gaussian. For example, if we choose a prior: $p(\mathbf{w}|\mathbf{A}) = \mathcal{N}(\mathbf{0}, \mathbf{A})$, where the weight probabilities depend on some hyperparameterised variance model $\mathbf{A}$, then we can often obtain excellent results utilising the 'type-II maximum likelihood' method: by integrating out the weights and optimising the resulting *marginal likelihood* with respect to $\mathbf{A}$. An excellent article outlining this approach, for the parameterisation $\mathbf{A} = \alpha\mathbf{I}$, is provided by [2]. In this paper we instead utilise a prior of the form $\mathbf{A} = \mathrm{diag}\,(\alpha_1, \ldots \alpha_M)$, recently introduced to realise the *relevance vector machine* [4].

A well-known difficulty with a Gaussian noise model is that it is not *robust*, in that if the target values are contaminated by *outliers*, the accuracy of the predictor $y(\mathbf{x})$ can be significantly compromised. In such circumstances, it is common to utilise a noise (ϵ) distribution with heavier tails, such as a *Student-t*. However, this prevents the analytic marginalisation over $\mathbf{w}$, limiting the application of the Bayesian framework.

* Author to whom correspondence should be directed: `mtipping@microsoft.com`

G. Dorffner, H. Bischof, and K. Hornik (Eds.): ICANN 2001, LNCS 2130, pp. 95–102, 2001.

Here, in Section 2, we describe an alternative approach. We define an *outlier distribution*, $p_{\mathrm{O}}(t)$, which is incorporated in a *mixture* with the standard likelihood model with the express purpose of explaining the outliers in the data. The use of the mixture prevents us evaluating the marginal likelihood analytically, but we utilise a *variational* approximation to obtain a lower bound which we can optimise effectively with respect to the hyperparameters. We demonstrate the utility of our method on some test data in Section 3.

2 Detail of the Method

2.1 Bayesian Interpolation

First, we define our data model, $p_1(t)$. Conventionally, we assume that the training examples have been generated independently by a Gaussian with mean given by (1) and with standard deviation σ_1. So, letting $\mathbf{t} = (t_1, \ldots, t_N)^{\mathrm{T}}$ and, for convenience, $\beta_1 \equiv \sigma_1^{-2}$:

$$p_1(\mathbf{t}|\mathbf{w}, \beta_1) = \prod_{n=1}^{N} p_1(t_n|\mathbf{w}, \beta_1) = \prod_{n=1}^{N} \left(\frac{\beta_1}{2\pi}\right)^{1/2} \exp\left\{-\frac{\beta_1}{2}(t_n - \boldsymbol{\phi}_n^{\mathrm{T}}\mathbf{w})^2\right\}, \quad (2)$$

where $\boldsymbol{\phi}_n = (\phi_1(\mathbf{x}_n), \ldots, \phi_M(\mathbf{x}_n))^T$. In conjunction with this likelihood model, we utilise a Gaussian prior over the parameters:

$$p(\mathbf{w}|\boldsymbol{\alpha}) = \prod_{m=1}^{M} \left(\frac{\alpha_m}{2\pi}\right)^{1/2} \exp\left\{-\frac{\alpha_m}{2} w_m^2\right\} \quad (3)$$

The use of an individual hyperparameter α_m to moderate the contribution of individual basis functions is a form of Bayesian 'automatic relevance determination', as utilised in the 'relevance vector machine' [4]. The reader is referred to the appropriate references for details of the method, but in summary, learning proceeds by integrating out the weights to compute the marginal likelihood:

$$p(\mathbf{t}|\boldsymbol{\alpha}, \beta_1) = \int p_1(\mathbf{t}|\mathbf{w}, \beta_1) p(\mathbf{w}|\boldsymbol{\alpha})\, d\mathbf{w} = \mathcal{N}(\mathbf{0}, \mathbf{C}), \quad (4)$$

where $\mathbf{C} = \beta_1^{-1}\mathbf{I} + \boldsymbol{\Phi}\mathbf{A}^{-1}\boldsymbol{\Phi}^{\mathrm{T}}$ and defining $\boldsymbol{\Phi} = (\boldsymbol{\phi}_1(\mathbf{x}), \ldots, \boldsymbol{\phi}_N(\mathbf{x}))^{\mathrm{T}}$. We then maximise (4) over both $\boldsymbol{\alpha}$ and β_1. The weight posterior can then be computed using these maximising values of the hyperparameters, and we generally utilise the posterior mean values of the weights to obtain the interpolant.

In Figure 1 (left), we show an example interpolant of 100 examples from the univariate function $\operatorname{sinc} x = \sin|x|/|x|$, where *uniform* noise of ± 0.2 has been added to the targets. For $y(\mathbf{x})$, we utilise N 'Gaussian' basis functions, one located at each training example: $\phi_n(x) = \exp\{-(x - x_n)^2/r^2\}$, with $r = 2$ here.

We now consider the case of 'outlying' $\{\mathbf{x}, t\}$ pairs, and in Figure 1 (right), we show the resulting interpolant after 25 contaminated examples with $t = 0.8$ have been added. In common with practically all non-robust approaches, performance deteriorates significantly. The error increases four-fold, and the noise estimate is considerably exaggerated.

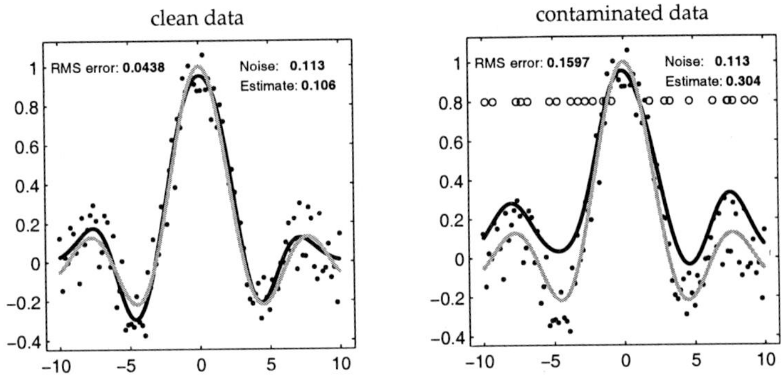

Fig. 1. *Left:* relevance vector regression with 100 examples of 'clean' data. *Right:* with 25 added examples with t fixed at 0.8. The data, generated from the 'sinc' function shown in grey, is plotted as filled circles, while the 'outliers' are denoted by open circles. The RMS deviation from the true function is given, as well as the 'true' and estimated noise standard deviation.

2.2 Explaining Outliers

To alleviate this problem within the presented Bayesian framework, we introduce a second generative model $p_O(t)$ of the data that is intended to explain the outlying observations. We consider two alternatives:

A. Where $p_O(t)$ is Gaussian centred at $y(\mathbf{x})$. *i.e.*

$$p_O(t) = \left(\frac{\beta_O}{2\pi}\right)^{1/2} \exp\left\{-\frac{\beta_O}{2}(t - \boldsymbol{\phi}^{\mathrm{T}}\mathbf{w})^2\right\}, \tag{5}$$

with $\beta_O \equiv \sigma_O^{-2}$ and where $\sigma_O \gg \sigma_1$ and is considered fixed.

B. Where $p_O(t)$ is independent of $y(\mathbf{x})$. *e.g.*

$$p_O(t) \sim \mathrm{Uniform}[\min(\{t_n\}), \max(\{t_n\})]. \tag{6}$$

The observation density for a single example is then defined as a *mixture*:

$$p(t_n) = \theta p_1(t_n) + (1 - \theta) p_O(t_n), \tag{7}$$

where θ is the proportion of 'valid' data, which we may attempt to learn. Note that $p_O(t)$ in A above thus corresponds to a special case of the Gaussian mixture framework recently analysed by [3] for autoregressive models.

Hence, since $\sigma_O \gg \sigma_1$, case A above gives a heavy-tailed generative distribution centred at $y(\mathbf{x})$, and can be considered a very rough approximation to the robust Student-t distribution (which itself is equivalent to a mixture of an *infinite* number of Gaussians). That $p_O(t)$ is Gaussian is necessary for analytic simplicity. Case B is simply a 'general purpose' distribution, and may take any form providing its parameters remain fixed. The present choice is simply intended as a minimally informative (maximally entropic) explanation of the data.

To write down the overall likelihood under this mixture, we introduce hidden variables z_n, $n = 1, \ldots, N$, which take the value 0 if t_n was generated by the outlier distribution, and 1 otherwise. Letting $\mathbf{z} = (z_1, \ldots, z_N)^{\mathrm{T}}$ the likelihood is

$$p(\mathbf{t}|\,\beta_1, \mathbf{w}, \mathbf{z}) = \prod_{n=1}^{N} [p_0(t_n)]^{1-z_n} \left[\left(\frac{\beta_1}{2\pi}\right)^{1/2} \exp\left\{-\frac{\beta_1}{2}(t_n - \boldsymbol{\phi}_n^{\mathrm{T}}\mathbf{w})^2\right\}\right]^{z_n}. \tag{8}$$

The marginal likelihood $p(\mathbf{t}|\boldsymbol{\alpha}, \beta_1, \theta)$ we desire is now given by

$$p(\mathbf{t}|\boldsymbol{\alpha}, \beta_1, \theta) = \int p(\mathbf{t}|\,\beta_1, \mathbf{w}, \mathbf{z})\, p(\mathbf{w}|\boldsymbol{\alpha})\, P(\mathbf{z}|\theta)\, d\mathbf{w} d\mathbf{z}. \tag{9}$$

The probability of the indicator variables $\mathbf{z}$ is simply $P(\mathbf{z}|\theta) = \prod_{n=1}^{N} P(z_n|\theta)$, where $P(z_n = 0) = 1 - \theta$ and $P(z_n = 1) = \theta$, but in conjunction with (8) and our previous prior (3) over the parameters $\mathbf{w}$, we see that the integration (9) is analytically intractable. We thus utilise a *variational* procedure which will give us a *lower bound* on $p(\mathbf{t}|\boldsymbol{\alpha}, \beta_1, \theta)$ which we will maximise to find $\boldsymbol{\alpha}$, β_1 and θ.

2.3 A Variational Approximation for the Robust Case

We note first that $\ln p(\mathbf{t})$ can be expressed as the difference of two terms:

$$\ln p(\mathbf{t}) = \ln p(\mathbf{t}, \mathbf{w}, \mathbf{z}) - \ln p(\mathbf{w}, \mathbf{z}|\,\mathbf{t}), \tag{10}$$

from which we write

$$\ln p(\mathbf{t}) = \ln\left\{\frac{p(\mathbf{t}, \mathbf{w}, \mathbf{z})}{Q(\mathbf{w}, \mathbf{z})}\right\} - \ln\left\{\frac{p(\mathbf{w}, \mathbf{z}|\,\mathbf{t})}{Q(\mathbf{w}, \mathbf{z})}\right\}, \tag{11}$$

where we have introduced an arbitrary 'approximating' distribution $Q(\mathbf{w}, \mathbf{z})$. Integrating both sides of (11) with respect to $Q(\mathbf{w}, \mathbf{z})$ gives

$$\begin{aligned} \ln p(\mathbf{t}) &= \int Q(\mathbf{w}, \mathbf{z}) \ln\left\{\frac{p(\mathbf{t}, \mathbf{w}, \mathbf{z})}{Q(\mathbf{w}, \mathbf{z})}\right\} d\mathbf{w} d\mathbf{z} - \int Q(\mathbf{w}, \mathbf{z}) \ln\left\{\frac{p(\mathbf{w}, \mathbf{z}|\,\mathbf{t})}{Q(\mathbf{w}, \mathbf{z})}\right\} d\mathbf{w} d\mathbf{z} \\ &= \mathcal{L}\left[Q(\mathbf{w}, \mathbf{z})\right] + \mathrm{KL}\left[Q(\mathbf{w}, \mathbf{z}) || \, p(\mathbf{w}, \mathbf{z}|\,\mathbf{t})\right], \end{aligned} \tag{12}$$

since $Q(\mathbf{w}, \mathbf{z})$ is a distribution and integrates to one. The second term is the Kullback-Leibler divergence between the approximating distribution $Q(\mathbf{w}, \mathbf{z})$ and the posterior $p(\mathbf{w}, \mathbf{z}|\,\mathbf{t})$. Since $\mathrm{KL}[Q(\mathbf{w}, \mathbf{z})||\,p(\mathbf{w}, \mathbf{z}|\,\mathbf{t})] \geq 0$, it follows that $\mathcal{L}[Q(\mathbf{w}, \mathbf{z})]$ is a rigorous lower bound on $\ln p(\mathbf{t})$. The approach is to maximize $\mathcal{L}[Q(\mathbf{w}, \mathbf{z})]$ with respect to $Q(\mathbf{w}, \mathbf{z})$. Since $\ln p(\mathbf{t})$ is independent of $Q(\mathbf{w}, \mathbf{z})$, this is equivalent to minimizing the Kullback-Leibler divergence. It is well-known that the Kullback-Leibler divergence is minimized for $Q(\mathbf{w}, \mathbf{z}) = p(\mathbf{w}, \mathbf{z}|\,\mathbf{t})$. We thus see that the optimal $Q(\mathbf{w}, \mathbf{z})$ distribution is given by the true posterior, in which case $\mathrm{KL}(Q(\mathbf{w}, \mathbf{z})\,||\,p(\mathbf{w}, \mathbf{z}|\,\mathbf{t})) = 0$ and the bound becomes exact.

While one can adopt some parameterised form for $Q(\mathbf{w}, \mathbf{z})$, we note that if we simply assume that $\mathbf{w}$ and $\mathbf{z}$ are separable, *i.e.* $Q(\mathbf{w}, \mathbf{z}) = Q_{\mathbf{w}}(\mathbf{w})Q_{\mathbf{z}}(\mathbf{z})$, then the KL divergence is minimized by inspection [5]:

$$Q_{\mathbf{w}}(\mathbf{w}) = \frac{\exp\langle \ln p(\mathbf{t}, \mathbf{w}, \mathbf{z})\rangle_{Q_{\mathbf{z}}(\mathbf{z})}}{\int \exp\langle \ln p(\mathbf{t}, \mathbf{w}, \mathbf{z})\rangle_{Q_{\mathbf{z}}(\mathbf{z})} d\mathbf{w}}, \tag{13}$$

and

$$Q_{\mathbf{z}}(\mathbf{z}) = \frac{\exp\langle \ln p(\mathbf{t}, \mathbf{w}, \mathbf{z})\rangle_{Q_{\mathbf{w}}(\mathbf{w})}}{\int \exp\langle \ln p(\mathbf{t}, \mathbf{w}, \mathbf{z})\rangle_{Q_{\mathbf{w}}(\mathbf{w})} d\mathbf{z}}. \tag{14}$$

Note that these solutions are mutually dependent, so in practice we must iteratively cycle through them, improving (raising) the lower bound with each such iteration. We now give the expressions for the approximating distributions, the lower bound and the update equations.

The Form of the Q-Distributions. We consider case A first where, for our model, it follows that $Q_{\mathbf{w}}(\mathbf{w})$ is Gaussian with covariance matrix:

$$\boldsymbol{\Sigma}_A = \left(\sum_{n=1}^{N} \left[\beta_0 (1 - \langle z_n \rangle_{Q_{\mathbf{z}}(\mathbf{z})}) + \beta_1 \langle z_n \rangle_{Q_{\mathbf{z}}(\mathbf{z})} \right] \boldsymbol{\phi}_n \boldsymbol{\phi}_n^T + \mathbf{A} \right)^{-1}, \tag{15}$$

and mean

$$\boldsymbol{\mu}_A = \boldsymbol{\Sigma}_A \left\{ \sum_{n=1}^{N} \left[\beta_0 (1 - \langle z_n \rangle_{Q_{\mathbf{z}}(\mathbf{z})}) + \beta_1 \langle z_n \rangle_{Q_{\mathbf{z}}(\mathbf{z})} \right] t_n \boldsymbol{\phi}_n \right\}. \tag{16}$$

The distribution $Q_{\mathbf{z}}(\mathbf{z})$ is given by the product $\prod_{n=1}^{N} Q_{z_n}(z_n)$, where

$$Q_{z_n}(z_n) = \frac{\left[\beta_1^{1/2} \theta \exp\left(-\beta_1 a_n / 2\right)\right]^{z_n} \left[\beta_0^{1/2} (1 - \theta) \exp\left(-\beta_0 a_n / 2\right)\right]^{1 - z_n}}{\beta_1^{1/2} \theta \exp\left(-\beta_1 a_n / 2\right) + \beta_0^{1/2} (1 - \theta) \exp\left(-\beta_0 a_n / 2\right)}, \tag{17}$$

and where

$$a_n = (t_n - \boldsymbol{\phi}_n^{\mathrm{T}} \boldsymbol{\mu}_A)^2 + \boldsymbol{\phi}_n^T \boldsymbol{\Sigma}_A \boldsymbol{\phi}_n. \tag{18}$$

For case B the expressions differ since $p_0(t_n)$ is independent of $\mathbf{w}$:

$$\boldsymbol{\Sigma}_B = \left(\beta_1 \sum_{n=1}^{N} \langle z_n \rangle_{Q_{\mathbf{z}}(\mathbf{z})} \boldsymbol{\phi}_n \boldsymbol{\phi}_n^{\mathrm{T}} + \mathbf{A} \right)^{-1}, \tag{19}$$

$$\boldsymbol{\mu}_B = \boldsymbol{\Sigma}_B \left(\beta_1 \sum_{n=1}^{N} \langle z_n \rangle_{Q_{\mathbf{z}}(\mathbf{z})} t_n \boldsymbol{\phi}_n \right), \tag{20}$$

$$Q_{z_n}(z_n) = \frac{\left[\theta (\beta_1 / 2\pi)^{1/2} \exp\left(-\beta_1 a_n / 2\right)\right]^{z_n} \left[(1 - \theta) p_0(t_n)\right]^{1 - z_n}}{\theta (\beta_1 / 2\pi)^{1/2} \exp\left(-\beta_1 a_n / 2\right) + (1 - \theta) p_0(t_n)}. \tag{21}$$

The Variational Lower Bound. The lower bound on $\ln p(\mathbf{t})$ is then given by

$$\begin{aligned}
\mathcal{L}[Q(\mathbf{w},\mathbf{z})] &= \int Q_{\mathbf{w}}(\mathbf{w})Q_{\mathbf{z}}(\mathbf{z}) \ln p(\mathbf{t},\mathbf{w},\mathbf{z}) d\mathbf{w}d\mathbf{z} \\
&\quad - \int Q_{\mathbf{w}}(\mathbf{w}) \ln Q_{\mathbf{w}}(\mathbf{w})\, d\mathbf{w} - \int Q_{\mathbf{z}}(\mathbf{z}) \ln Q_{\mathbf{z}}(\mathbf{z})\, d\mathbf{z}, \qquad (22) \\
&= L_0 + \frac{1}{2}\sum_{n=1}^{N} \langle z_n \rangle_{Q_{\mathbf{z}}(\mathbf{z})} (\ln \beta_1 - \beta_1 a_n) \\
&\quad + \frac{1}{2}\left[\ln |\mathbf{A}\boldsymbol{\Sigma}| - \mathrm{tr}\left\{\mathbf{A}(\boldsymbol{\mu}\boldsymbol{\mu}^{\mathrm{T}} + \boldsymbol{\Sigma})\right\}\right] \\
&\quad + \sum_{n=1}^{N} \langle z_n \rangle_{Q_{\mathbf{z}}(\mathbf{z})} (\ln \langle z_n \rangle_{Q_{\mathbf{z}}(\mathbf{z})} - \ln \theta) \\
&\quad + \sum_{n=1}^{N} \left[1 - \langle z_n \rangle_{Q_{\mathbf{z}}(\mathbf{z})}\right] (\ln[1 - \langle z_n \rangle_{Q_{\mathbf{z}}(\mathbf{z})}] - \ln[1-\theta]), \qquad (23)
\end{aligned}$$

where we have ignored constant terms and where

$$L_0 = \begin{cases} \frac{1}{2}\sum_{n=1}^{N}[1 - \langle z_n \rangle_{Q_{\mathbf{z}}(\mathbf{z})}] (\ln \beta_0 - \beta_0 a_n) & \text{for case A,} \\ \sum_{n=1}^{N} \left[1 - \langle z_n \rangle_{Q_{\mathbf{z}}(\mathbf{z})}\right] \ln p_0(t_n) & \text{for case B.} \end{cases} \qquad (24)$$

(Hyper)parameter Update Formulae. There is no obstacle to continuing with the variational formalism and defining prior distributions over $\boldsymbol{\alpha}$, β_1 (see [1]) and θ. Instead, however, for reasons of computational efficiency and based on our experience from [1], we choose to *optimise* $\mathcal{L}[Q(\mathbf{w},\mathbf{z})]$ over those parameters.

Differentiating (23) with respect to α_m, equating to zero and defining quantities $\gamma_m = 1 - \alpha_m \boldsymbol{\Sigma}_{mm}$ (see [2, 4] for motivation for this) leads to the following update

$$\alpha_m^{\text{new}} = \frac{\gamma_m}{\mu_m^2}, \qquad (25)$$

Differentiating with respect to β_1 and θ and equating to zero gives re-estimates

$$\beta_1^{\text{new}} = \frac{\sum_{n=1}^{N} \langle z_n \rangle_{Q_{\mathbf{z}}(\mathbf{z})}}{\sum_{n=1}^{N} \langle z_n \rangle_{Q_{\mathbf{z}}(\mathbf{z})}\, a_n}, \qquad \theta^{\text{new}} = \frac{1}{N}\sum_{n=1}^{N} \langle z_n \rangle_{Q_{\mathbf{z}}(\mathbf{z})} \qquad (26)$$

Algorithm Sketch. The above computations and updates give an algorithm:

1. Initialise $\boldsymbol{\alpha}$ (*e.g.* $\alpha_m = M/\mathrm{var}[t]$), σ_1 (*e.g.* $\sigma_1^2 = \mathrm{var}[t]$), σ_0 (*e.g.* $\sigma_0^2 = 2\mathrm{var}[t]$) and θ (*e.g.* $\theta = 0.5$).
2. Perform several (*e.g.* ≈ 5) updates of the approximating distributions $Q_{\mathbf{w}}(\mathbf{w})$ and $Q_{\mathbf{z}}(\mathbf{z})$ using (15)-(17) or (19)-(21).
3. Re-estimate $\boldsymbol{\alpha}$, β_1 and θ using (25) and (26).
4. Loop back to step 2 until all updates have stabilised.

Note that in step 2, $\mathcal{L}[Q(\mathbf{w},\mathbf{z})]$ should never decrease after an individual update and so the evaluation of (23) serves as a useful check.

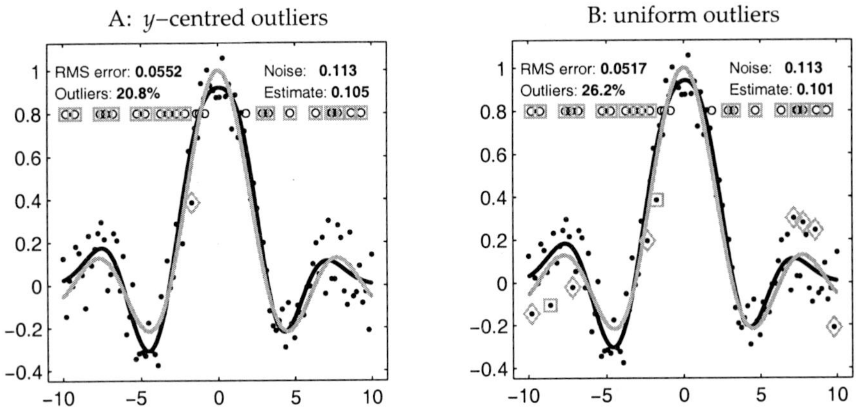

Fig. 2. Interpolation using the y-centred Gaussian model (left) and the uniform-outlier model (right). The data is identical to that of Figure 1. The estimate for θ is given, in terms of the proportion of 'outliers' estimated: *i.e.* $1-\theta$. Note that the 'true' value of $1-\theta$ should be 20%. Data points are overlaid with a grey box if $\langle z_n \rangle < 0.5$, and with a diamond if $0.5 \leq \langle z_n \rangle < 0.75$.

3 Examples

For our first example of application, we return to the data of Figure 1 earlier. Figure 2 shows the results using the two methods presented above. Compared with Figure 1 (right), the case B (uniform) $p_0(t)$ model dramatically reduces the error from 0.1597 to 0.0517, which is close to the 'optimal' clean-data result of 0.0438. Note that no additional information was used by the training algorithm in this latter case, since θ was estimated from the data. The improvement results from the fact that $p_0(t)$ is a more probable explanation of the outlying data, and so the interpolant (specifically, the weights) does not need to 'fit' those points.

We can see that in Figure 2, the y-centred Gaussian $p_0(t)$ is slightly inferior to the uniform case, although still much better than Figure 1 (right). This is not unexpected; the uniform $p_0(t)$ intuitively seems a better model of the contamination. However, we can consider the case where the noise process on the data is heavy-tailed, and in Figure 3 we utilise a Student-t distribution, with 4 degrees of freedom, to generate ϵ_n. Here, although of course only an approximation, the more appropriate y-centred $p_0(t)$ performs marginally better than the uniform case, and both are much superior to the non-robust model.

4 Summary

We have detailed a Bayesian interpolation mechanism which is robust to 'outliers', by which we imply the case where either the data is 'contaminated' (fig. 2) or where the noise process is heavy-tailed (fig. 3). Certainly, the distributions $p_0(t)$ that we utilise can only be approximations to the outlier-generating mechanism, but are nevertheless intended to be sensible, general-purpose, choices,

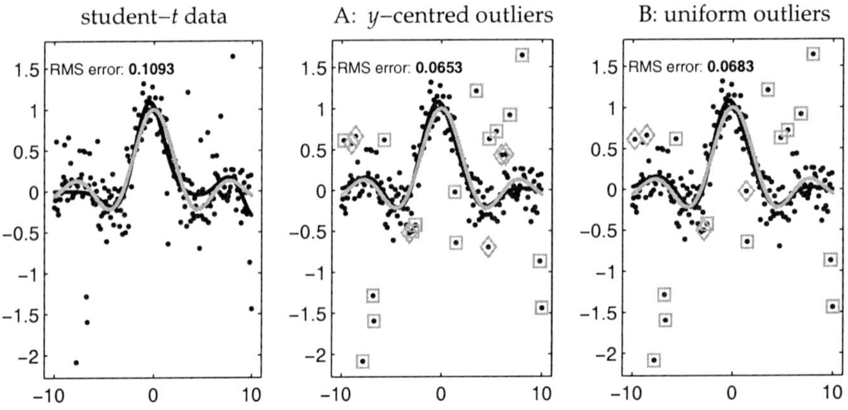

Fig. 3. *Left:* interpolation using a standard Gaussian noise model on data generated from the 'sinc' function with additive Student-t noise. *Centre:* using the y-centred outlier model. *Right:* using the uniform model.

and we were careful to only show examples where the random samples did *not* conform to either our assumed noise or outlier distributions. We are currently obtaining results for other realistic contamination scenarios as well as real-world data, and are working to extend the variational procedure to the *infinite* Gaussian mixture case, corresponding to a Student-t observation model.

References

1. C. M. Bishop and M. E. Tipping. Variational relevance vector machines. In C. Boutilier and M. Goldszmidt, editors, *Proceedings of the 16th Conference on Uncertainty in Artificial Intelligence*, pages 46–53. Morgan Kaufmann, 2000.
2. D. J. C. MacKay. Bayesian interpolation. *Neural Computation*, 4(3):415–447, 1992.
3. W. Penny and S. J. Roberts. Variational Bayes for non-Gaussian autoregressive models. In *Neural Networks for Signal Processing X*, pages 135–144. IEEE, 2000.
4. M. E. Tipping. The Relevance Vector Machine. In S. A. Solla, T. K. Leen, and K.-R. Müller, editors, *Advances in Neural Information Processing Systems 12*, pages 652–658. MIT Press, 2000.
5. S. Waterhouse, D. J. C. MacKay, and T. Robinson. Bayesian methods for mixtures of experts. In M. C. Mozer, D. S. Touretzky, and M. E. Hasselmo, editors, *Advances in Neural Information Processing Systems 8*. MIT Press, 1996.

Minimum-Entropy Data Clustering Using Reversible Jump Markov Chain Monte Carlo

Stephen J. Roberts[1], Christopher Holmes[2], and Dave Denison[2]

[1] Robotics Research, Department of Engineering Science, University of Oxford, UK
[2] Department of Mathematics, Imperial College, London, UK

1 Introduction

Many problems in data analysis, especially in signal and image processing, require the *unsupervised* partitioning of data into a set of 'self-similar' classes or clusters. An ideal partitioning unambiguously assigns each datum to a single class and one thinks of the data as being generated by a number of data generators, one for each class. Many algorithms have been proposed for such analysis and for the estimation of the optimal number of partitions. The majority of popular and computationally feasible techniques rely on assuming that classes are hyper-ellipsoidal in shape. In the case of Gaussian mixture modelling [15, 6] this is explicit; in the case of dendogram linkage methods (which typically rely on the L_2 norm) it is implicit [9]. For some data sets this leads to an over-partitioning. Alternative methods, based on valley seeking [6] or maxima-tracking in scale-space [16, 18, 13] for example, have the advantage that they are free from such assumptions. They can be, however, sensitive to noise and computationally intensive in high-dimensional spaces. In this paper we re-consider the issue of data partitioning from an information-theoretic viewpoint and show that minimisation of partition entropy may be used to evaluate the most probable set of data generators. Rather than formulate the problem as one of a traditional model-order estimation to infer the most probable number of classes we employ a reversible jump mechanism in a Markov-chain Monte Carlo (MCMC) sampler which explores the space of different model sizes.

2 Theory

In a previous publication, [14] we have detailed the use of minimum-entropy criteria to construct candidate partition models for data sets. We briefly describe the relevant theory in this section as well as exploring the issues of Bayesian learning and reversible jump MCMC methods.

2.1 Partitioning Entropy

Consider a partitioning of the data into a set of $k = 1..K$ classes. The probability density function (pdf) of a single datum $\mathbf{x}$, conditioned on this set of classes, is given by:

G. Dorffner, H. Bischof, and K. Hornik (Eds.): ICANN 2001, LNCS 2130, pp. 103–110, 2001.

$$p(\mathbf{x}) = \sum_{k=1}^{K} p(\mathbf{x} \mid k) p(k) \tag{1}$$

We consider the overlap between the contribution of the k-th class to this density function and the unconditional density $p(\mathbf{x})$. This overlap may be measured by the Kullback-Liebler measure between these two distributions. The latter is defined, for distributions $q(x)$ and $p(x)$ as:

$$KL\left(q(x) \parallel p(x)\right) \stackrel{\text{def}}{=} \int q(x) \ln \left(\frac{q(x)}{p(x)} \right) dx \tag{2}$$

Note that this measure reaches a minimum of zero if, and only if, $p(x) = q(x)$. For any other case it is strictly positive and increases as the overlap between the two distributions *decreases*. What we desire, therefore, is that the KL measure be maximised. We hence write our overlap measure for the k-th class as:

$$v_k = -KL\left(p(\mathbf{x} \mid k) \parallel p(\mathbf{x})\right) \tag{3}$$

which will be minimised when the k-th class is well-separated from all others. We define a total overlap, V, as:

$$V \stackrel{\text{def}}{=} \langle v_k \rangle = \sum_{k=1}^{K} p(k) v_k, \tag{4}$$

hence, combining the above equations,

$$V = -\sum_{k=1}^{K} p(k) \int p(\mathbf{x} \mid k) \ln \left(\frac{p(\mathbf{x} \mid k)}{p(\mathbf{x})} \right) d\mathbf{x} \tag{5}$$

We therefore seek the partitioning for which V is minimal. Using Bayes' theorem we may rewrite Equation 5 such that:

$$V = \int \left(-\sum_{k=1}^{K} p(k \mid \mathbf{x}) \ln p(k \mid \mathbf{x}) \right) p(\mathbf{x}) d\mathbf{x} + \sum_{k=1}^{K} p(k) \ln p(k) \int p(\mathbf{x} \mid k) d\mathbf{x} \tag{6}$$

in which we note that the first term is the expected Shannon entropy of the class posteriors and the second the negative entropy of the class priors (the integral component of the second term is unity).

2.2 Classes as Mixture Models

Consider a partitioning of the data into a set of $k = 1...K$ classes. In the case of the well-known Gaussian mixture model, each class density is taken to be a single Gaussian. This does not offer the degree of flexibilty we desire and we hence model the pdf of the data, conditioned on the k-th class, via a semi-parametric mixture model with J basis functions. Each mixture basis may be, for example, a Gaussian and hence *each class* in the candidate partitioning of the data is represented as a mixture of these. The class-conditional posteriors may thence be written as a linear combination of the kernel posteriors, i.e. $\mathbf{p} = \mathbf{W}\Phi$ where $\mathbf{p}$ is the set of *class* posterior probabilities (in vector form), $\mathbf{W}$ is a transform matrix (not assumed to be square) and Φ is the set of *kernel* posteriors (in vector form). Hence the k-th class posterior may be written as:

$$p_k \stackrel{\text{def}}{=} p(k \mid \mathbf{x}) = \sum_{j=1}^{J} W_{k,j}\phi(j \mid \mathbf{x}) \tag{7}$$

If we are to interpret $\mathbf{p}$ as a set of class posterior probabilities we require $p_k \in [0,1]\ \forall k$ and $\sum_k p_k = 1$. As $\phi(j \mid \mathbf{x}) \in [0,1]$ so the first of these conditions is met if each $W_{k,j} \in [0,1]$. The second condition is met when each row of $\mathbf{W}$ sums to unity. Each class in a candidate partitioning of the data is hence formed from a mixture model whose kernels are fixed but whose coupling matrix, $\mathbf{W}$, is adapted so as to minimise the overall entropy of the partitioning.

2.3 Reversible Jump Sampling

In contrast to traditional model-order determination, the reversible-jump approach, put forward in [7, 12], takes an alternative formalism in which the number of model components (classes in this case) is considered as an unknown variable whose posterior density is estimated via a sampling process.

Consider a move from $\mathbf{W} \to \mathbf{W}'$ associated with a change in intrinsic dimension of $K \to K'$ (i.e. a change in the hypothesised number of classes in the partitioning). We utilise the results of [7, 12, 2, 8] to determine the acceptance probability of each proposed move. The acceptance probability for a proposed model change $(\mathbf{W}, K) \to (\mathbf{W}', K')$ is given via a modified Metropolis-Hastings equation as:

$$p(\text{accept}) = \min\left(1, \underbrace{\frac{p(K', \mathbf{W}' \mid \Phi)}{p(K, \mathbf{W} \mid \Phi)}}_{\text{target density}} \underbrace{\text{Jac}(\mathbf{W}, K \to \mathbf{W}', K')}_{\text{Jacobian}} \underbrace{\frac{q(K, \mathbf{W} \mid \Phi)}{q(K', \mathbf{W}' \mid \Phi)}}_{\text{proposal}}\right) \tag{8}$$

Using Bayes' theorem we may expand the terms in the above Equation such that:

$$p(K, \mathbf{W} \mid \Phi) \propto p(\Phi \mid \mathbf{W}, K)p(\mathbf{W} \mid K)p(K)$$

and

$$q(K, \mathbf{W} \mid \Phi) \propto q(\mathbf{W} \mid \Phi, K)q(K \mid \Phi).$$

If we take the prior over the model-order, $p(K)$ to be the flat between $K = 1...J$ and $q(K \mid \Phi) = q(K) = q(K')$ then we obtain an acceptance measure as,

$$\min\left(1, \frac{p(\Phi \mid \mathbf{W}', K')p(\mathbf{W}' \mid K')}{p(\Phi \mid \mathbf{W}, K)p(\mathbf{W} \mid K)} \text{Jac}(\mathbf{W}, K \to \mathbf{W}', K') \frac{q(\mathbf{W} \mid \Phi, K)}{q(\mathbf{W}' \mid \Phi, K')}\right) \tag{9}$$

The likelihood measure, $p(\Phi \mid \mathbf{W}, K)$, represents the evidence that the candidate partitioning (uniquely defined via knowledge of $\mathbf{W}$ and K) conforms to our objective of minimal entropy. We hence define the log-likelihood measure as the change in entropy given the set of fixed kernel responses, Φ, which from Equation (6) gives:

$$\log p(\Phi \mid \mathbf{W}, K) = H(K) - H(K \mid \mathbf{W}, \Phi) \stackrel{\text{def}}{=} \Delta H \quad (10)$$

in which $H(K)$ is the prior entropy of the candidate partitioning and $H(K \mid \mathbf{W}, \Phi)$ is the posterior entropy[1].

Noting that each row of $\mathbf{W}$ corresponds to the mixing fractions which couple Φ to an output class, so a convenient form for the prior and proposal densities is a Dirichlet [3], namely:

$$q(\mathbf{W} \mid \Phi, K) = \prod_{k=1}^{K} \mathcal{D}_J(W_{k,1:J} \mid \boldsymbol{\gamma}),$$

$$p(\mathbf{W} \mid K) = \prod_{k=1}^{K} \mathcal{D}_J(W_{k,1:J} \mid \boldsymbol{\alpha}). \quad (11)$$

in which $W_{k,1:J}$ represents the row of $\mathbf{W}$ coupling the J kernel reponses, Φ, to the k-th output class. The hyperparameter sets are defined as $\boldsymbol{\gamma} = \gamma \mathbf{1}_J$ and $\boldsymbol{\alpha} = \alpha \mathbf{1}_J$ in which $\mathbf{1}_J$ is a vector of J 1s.

Sampler Moves Our sampler has three moves from model $(\mathbf{W}, K)$ to $(\mathbf{W}', K')$:

update: No change in K takes place but the parameters of the existing model are updated.
birth: An increase in model complexity such that the number of classes changes as $K' = K + 1$.
death: A decrease in the number of classes such that $K' = K - 1$.

At each iteration in the Markov chain one of these three moves is selected with probabilities p_u, p_b, p_d. These sum to unity and are fixed, in our implementation, each at 1/3 save at $K = 1$ and $K = J$ (the minimum and maximum number of classes) where we set p_d and p_b to zero respectively.

The proposed moves in the Markov chain are applied to a randomly chosen row of $\mathbf{W}$, whose elements we denote as $\mathbf{w} = W_{k^*,1:J}$. The changes in the parameter set associated with each of the three moves are as follows:

update: The update step is made using a *hybrid* proposal[2]
(a) Within mode 'tweak', $\mathbf{w}' \sim \mathcal{D}_J(\mathbf{w}' \mid \beta \mathbf{w})$ i.e. $\mathbf{w}'$ is re-drawn around a mean given by $\mathbf{w}$. The hyperparameter β is chosen to be large (see later) so that $\mathbf{w}'$ lies close to $\mathbf{w}$.

---

[1] Note that maximization of the log-likelihood in Equation (10) is hence equivalent to the minimization of the overlap measure in Equation (6).

[2] This usage of 'hybrid' is meant in the sense of a multiple component proposal density [17], rather than in the (later) usage of Neal [11] in which he refers to the use of gradient information in MCMC as hybrid-MCMC.

(b) Mode hop: redraw row from prior, $\mathbf{w}' \sim \mathcal{D}_J(\mathbf{w}' \mid \boldsymbol{\alpha})$.

birth: add a row drawn from the proposal, $\mathbf{W}' = \mathbf{W} \overset{\text{row}}{+} \mathbf{w}'$ where $\mathbf{w}' \sim \mathcal{D}_J(\mathbf{w}' \mid \boldsymbol{\gamma})$.

death: remove the row specified by $\mathbf{w}$, i.e. $\mathbf{W}' = \mathbf{W} \overset{\text{row}}{-} \mathbf{w}$.

We note that as the birth and death moves are independent of the current state the Jacobian term in the acceptance probability is unity. Furthermore, as the elements of $\mathbf{W}$ remain unchanged other than for those in $\mathbf{w}$, so the ratios of densities in Equation 9 cancel for all but prior and proposal densities over $\mathbf{w}$.

Writing the resultant acceptance probability as $\min(1, r)$, so combining Equations 9,10 and 11:

update (a)

$$r_u = \exp[\Delta H' - \Delta H] \frac{\mathcal{D}_J(\mathbf{w}' \mid \boldsymbol{\alpha})}{\mathcal{D}_J(\mathbf{w} \mid \boldsymbol{\alpha})} \frac{\mathcal{D}_J(\mathbf{w}' \mid \beta\mathbf{w}')}{\mathcal{D}_J(\mathbf{w} \mid \beta\mathbf{w})}$$

update (b)

$$r_u = \exp[\Delta H' - \Delta H]$$

birth

$$r_b = \exp[\Delta H' - \Delta H] \frac{\mathcal{D}_J(\mathbf{w}' \mid \boldsymbol{\alpha})}{\mathcal{D}_J(\mathbf{w}' \mid \boldsymbol{\gamma})}$$

death

$$r_d = \exp[\Delta H' - \Delta H] \frac{\mathcal{D}_J(\mathbf{w} \mid \boldsymbol{\gamma})}{\mathcal{D}_J(\mathbf{w} \mid \boldsymbol{\alpha})}$$

where ΔH and $\Delta H'$ are the log-likelihood measures (Equation 10) of the models specified by $(\mathbf{W}, K)$ and $(\mathbf{W}', K')$ respectively. The Dirichlet density over $\mathbf{w}$ with hyperparameters $\boldsymbol{\gamma} = \gamma \mathbf{1}_J$ is:

$$\mathcal{D}_J(\mathbf{w} \mid \boldsymbol{\gamma}) = \frac{\Gamma(J\gamma)}{\prod_{j=1}^{J} \Gamma(\gamma_j)} \prod_{j=1}^{J} w_j^{\gamma_j - 1} \tag{12}$$

and hence, for example, r_b is given in full as:

$$r_b = \exp[\Delta H' - \Delta H] \frac{\Gamma(J\gamma)(\Gamma(\alpha))^J}{\Gamma(J\alpha)(\Gamma(\gamma))^J} \prod_{j=1}^{J} w_j^{\gamma - \alpha} \tag{13}$$

In all the examples given in this paper we set $\alpha = 1$ (thus giving a flat reference prior) whilst γ, β are chosen to give acceptance rates of around 30% (we find that $\gamma = 1, \beta = 1000$ worked well and these values are used in all examples in this paper). Convergence of the model is rapid and for the examples given in this paper only run the chain over some 1500 iterations, the first 1000 of which are used as a burn-in period. On the examples presented in this paper this short sample time is sufficient.

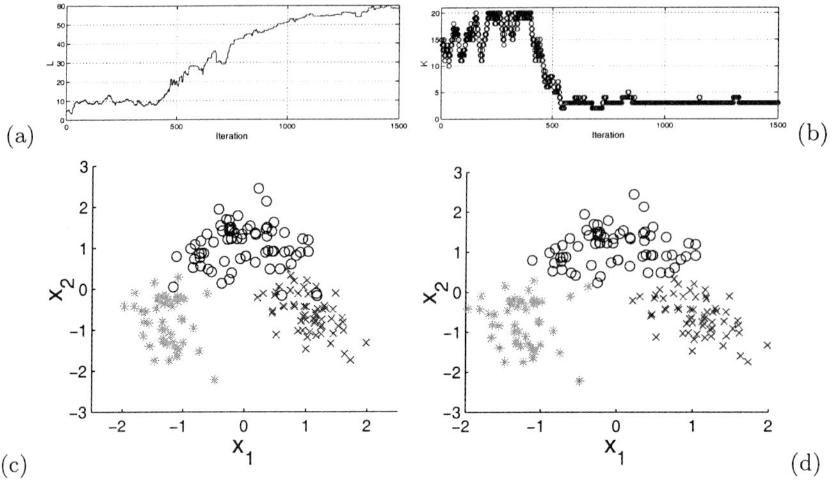

Fig. 1. Wine data set: (a) Likelihood and (b) K during sampling. The first 1000 iterations are burn in. (c) True class labels and (d) final partitioning results. There are 5 errors in this example.

3 Results

3.1 Wine Recognition Data

As a first example we present results from a wine recognition problem. The data set consists of 178 13-dimensional exemplars which are a set of chemical analyses of three types of wine. We utilise the first three principle components from this data set and fit an initial 20-kernel set using the Expectation-Maximization (EM) algorithm [5, 4] and perform a minimum-entropy clustering. Figure 1(a) shows the evolution of the model likelihood in the Markov chain and plot (b) the posterior model-order (number of classes). Plot (c) shows the true class partitions (projected onto the first two dimensions of the data) and (d) the resultant partitioning. We note that the few differences between (c) and (d) are due to 'outlying' points from the class distributions and as such cannot be determined correctly by an unsupervised method. In this example there are 5 errors, corresponding to an equivalent classification performance of 97.2%. Supervised analysis has been reported for this data set and our result is surprisingly good considering that supervised first-nearest neighbour classification achieves only 96.1%, and multivariate linear-discriminant analysis 98.9% [1].

3.2 Infra-red Spectra Data

In the first example we formed a basis set from a Gaussian mixture, so that each class in the resultant partitioning was modelled as a Gaussian mixture. In this example, however, we utilise a highly non-Gaussian basis set formed via a non-negative matrix factorisation of the data [10]. Our resultant class models

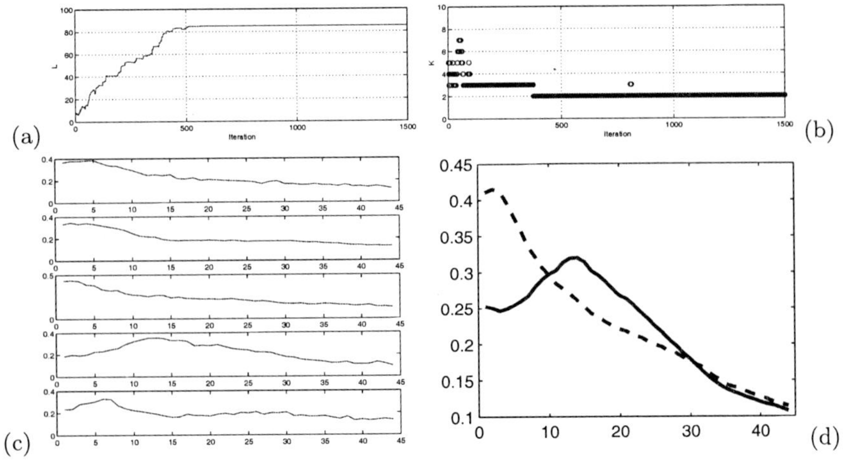

Fig. 2. IR spectra data: (a) Likelihood and (b) K during sampling. The first 1000 samples are burn in. (c) example spectra, (d) resultant two-component spectra. The spectra range linearly from 8-13μm [1-44 on the plot].

are thence mixtures of these considerably more complex bases. The data set we consider comes from the Infra-Red Astronomy Satellite (IRAS) which maps regions of the sky at infra-red wavelengths. The data set consists of 531 examples of slightly overlapping blue-band (8-13μm) and red-band (11-22μm) spectra. We consider here only the blue-band data. Figure 2(a) and (b) show the evolution of the model likelihood and order during sampling. Plot (c) shows randomly selected examples from the data set and (d) the most-probable two-class spectral profiles as determined from the $K = 2$ hypothesis.

4 Conclusions

We have presented a computationally simple technique for data partitioning based on a linear mixing of a set of fixed basis functions. The technique is shown to give good results on a variety of problems. The methodology is general and non-Gaussian basis kernels may be employed in which case the estimated class-conditional densities will be mixture models of the chosen basis functions. The method, furthermore, scales favourably with the dimensionality of the data space and the entropy-minimisation algorithm is efficient even with large numbers of samples.

Acknowledgement

Part of this work was funded by the UK Engineering & Physical Sciences Research Council (grant EESD PRO 264) whose support is gratefully acknowledged.

References

1. S. Aeberhard, D. Coomans, and O. de Vel. Comparative-Analysis of Statistical Pattern-Recognition Methods in High-Dimensional Settings. *Pattern Recognition*, 27(8):1065–1077, 1994.
2. C. Andrieu, N. de Freitas, and A. Doucet. Sequential MCMC for Bayesian Model Selection. IEEE Signal Processing Workshop on Higher Order Statistics. Ceasarea, Israel, June 14-16., 1999.
3. J.M. Bernardo and A.F.M. Smith. *Bayesian Theory.* John Wiley, 1994.
4. C.M. Bishop. *Neural Networks for Pattern Recognition.* Oxford University Press, Oxford, 1995.
5. A.P. Dempster, N.M. Laird, and D.B. Rubin. Maximum Likelihood from Incomplete Data via the EM Algorithm. *J. Roy. Stat. Soc.*, 39(1):1–38, 1977.
6. K. Fukunaga. *An Introduction to Statistical Pattern Recognition.* Academic Press, 1990.
7. P. Green. Reversible jump Markov chain Monte Carlo computation and Bayesian model determination. *Biometrika*, 82:711–732, 1995.
8. C. Holmes and B.K. Mallick. Bayesian Radial Basis Functions of variable dimension. *Neural Computation*, 10:1217–1233, 1998.
9. A.K. Jain and R.C. Dubes. *Algorithms for Clustering Data.* Prentice Hall, 1988.
10. D. D. Lee and H. S. Seung. Learning the parts of objects by non-negative matrix factorisation. *Nature*, 401:788–791, October 1999.
11. R.M. Neal. *Bayesian learning for neural networks.* Lecture notes in statistics. Springer, Berlin, 1996.
12. S. Richardson and P.J. Green. On Bayesian analysis of mixtures with an unknown number of components. *Journal of the Royal Statistical Society (Series B)*, 59(4):731–758, 1997.
13. S.J. Roberts. Parametric and non-parametric unsupervised cluster analysis. *Pattern Recognition*, 30(2):261–272, 1997.
14. S.J. Roberts, R. Everson, and I. Rezek. Maximum Certainty Data Partitioning. *Pattern Recognition*, 33(5):833–839, 2000.
15. S.J. Roberts, D. Husmeier, I. Rezek, and W. Penny. Bayesian Approaches To Mixture Modelling. *IEEE Transaction on Pattern Analysis and Machine Intelligence*, 20(11):1133–1142, 1998.
16. K. Rose, E. Gurewitz, and G.C. Foz. A Deterministic Annealing Approach to Clustering. *Pattern Recognition Letters*, 11(9):589–594, September 1990.
17. L. Tierney. Markov Chains for exploring Posterior Distributions. *Annals of Statistics*, 22:1701–1762, 1994.
18. R. Wilson and M. Spann. A New Approach to Clustering. *Pattern Recognition*, 23(12):1413–1425, 1990.

Behavioral Market Segmentation of Binary Guest Survey Data with Bagged Clustering

Sara Dolničar[1] and Friedrich Leisch[2]

[1] Department of Tourism and Leisure Studies,
Vienna University of Economics and Business Administration,
A-1090 Wien, Austria
Sara.Dolnicar@wu-wien.ac.at
[2] Department of Statistics and Probability Theory,
Vienna University of Technology, A-1040 Wien, Austria
Friedrich.Leisch@ci.tuwien.ac.at

Abstract. Binary survey data from the Austrian National Guest Survey conducted in the summer season of 1997 were used to identify behavioral market segments on the basis of vacation activity information. Bagged clustering overcomes a number of difficulties typically encountered when partitioning large binary data sets: The partitions have greater structural stability over repetitions of the algorithm and the question of the "correct" number of clusters is less important because of the hierarchical step of the cluster analysis. Finally, the bootstrap part of the algorithm provides means for assessing and visualizing segment stability for each input variable.

1 Introduction

The importance of binary data in social sciences is growing due to manifold reasons. Yes-no questions are simpler and faster to answer for respondents. Not only does this fact increase the chances of the respondents finishing a questionnaire and answering it in a concentrated, spontaneous and motivated manner, binary question format also allows the designer of the questionnaire to pose more questions, as the single answer is less tiring. This is especially important for studies, where attitudes towards a multitude of objects are questioned, thus dramatically increasing the number of answers expected from the respondents as it is typically the case with guest surveys within the field of tourism.

These developments lead to an increasing number of medium to large empirical binary data sets available for data analysis. Turning to the field of market segmentation, empirical binary survey data sets exclude a number of clustering techniques viable for analysis due to their size which seems to be too large for hierarchical and too small for parametric approaches. Most parametric approaches require very large amounts of data in relation to the number of variables, growing exponentially. For the use of latent class analysis, Formann [4] recommends a sample size of 5×2^k, a very strict requirement, especially when item batteries of

G. Dorffner, H. Bischof, and K. Hornik (Eds.): ICANN 2001, LNCS 2130, pp. 111–118, 2001.

20 items are not unusual, as it is the case in market segmentation, be it with demographic, socioeconomic, behavioral or psycho-graphic variables. Unless these huge data sets are available, exploratory clustering techniques will broadly be applied to analyze the heterogeneity underlying the population sample.

Among the exploratory approaches, the hierarchical clustering techniques require the data sets to be rather small, as all pairwise distances need to be computed in every single step of the analysis. This leaves us with partitioning approaches like learning vector quantization (LVQ) within the family of cluster analytic techniques. However, partitioning cluster methods typically give less insight into the structure of the data, as the number of clusters has to be specified a-priori and solutions for different number of clusters can often not be easily compared. Myers & Tauber [9] state in their classic book on market structure analysis that hierarchical clustering better shows how individuals combine in terms of similarities and partitioning methods produce more homogeneous groups.

2 The Bagged Clustering Approach

The central idea of bagged clustering [7, 8] is to stabilize partitioning methods like K-means or LVQ by repeatedly running the cluster algorithm and combining the results. Bagging [1], which stands for bootstrap aggregating, has been shown a very succesfull method for enhancing regression and classification algorithms. Bagged clustering applies the main idea of combining several predictors trained on bootstrap sets in the cluster analysis framework. K-means is an unstable method in the sense that in many runs one will not find the global optimum of the error function but a local optimum only. Both initializations and small changes in the training set can have big influence on the actual local minimum where the algorithm converges.

By repeatedly training on new data sets one gets different solutions which should on average be independent from training set influence and random initializations. We can obtain a collection of training sets by sampling from the empirical distribution of the original data, i.e., by bootstrapping. We then run any partitioning cluster algorithm—called the *base cluster method* below—on each of these training sets.

Bagged clustering explores the independent solutions from several runs of the base method using hierarchical clustering. Hence, it can also be seen as an evaluation of the base method by means of the bootstrap. This allows the researcher to identify structurally stable (regions of) centers which are found repeatedly.

The algorithm works as follows:

1. Construct B bootstrap training samples $\mathcal{X}_N^1, \ldots, \mathcal{X}_N^B$ by drawing with replacement from the original sample $\mathcal{X}_N$.
2. Run the base cluster method (K-means, LVQ, ...) on each set, resulting in $B \times K$ centers $c_{11}, c_{12}, \ldots, c_{1K}, c_{21}, \ldots, c_{BK}$ where K is the number of centers used in the base method and c_{ij} is the j-th center found using $\mathcal{X}_N^i$.

3. Combine all centers into a new data set $\mathcal{C}^B = \mathcal{C}^B(K) = \{c_{11}, \ldots, c_{BK}\}$.
4. Run a hierarchical cluster algorithm on $\mathcal{C}^B$ (or $\mathcal{C}^B_{\text{prune}}$), resulting in the usual dendrogram.
5. Let $c(x) \in \mathcal{C}^B$ denote the center closest to x. A partition of the original data can now be obtained by cutting the dendrogram at a certain level, resulting in a partition $\mathcal{C}^B_1, \ldots, \mathcal{C}^B_m$, $1 \leq m \leq BK$, of set $\mathcal{C}^B$. Each point $x \in \mathcal{X}_N$ is now assigned to the cluster containing $c(x)$.

The algorithm has been shown to compare favorably to several standard clustering methods on binary and metric benchmark data sets [7]; please see [8] for a detailed analysis and experiments using artificial data with known structure, as space constraints do not allow us to include the results in this paper. Especially the exploratory nature of the approach is attractive for practitioners [3].

3 Behavioral Segmentation of Tourist Survey

Our application consists of the segmentation of tourist surveys for marketing purposes. A data set including 5365 respondents and 12 variables was used. The respondents were tourists spending their vacation in the rural area of Austria during the summer season of 1997, city tourists were excluded from the study. These visitors were questioned in the course of the Austrian National Guest Survey. The vacation activities used for behavioral segmentation purposes were:

Activity	Agreement (%)
cycling	30.21
swimming	62.65
going to a spa	14.61
hiking	75.62
going for walks	93.25
organized excursions	21.62
excursions	77.04
relaxing	80.17
shopping	71.50
sightseeing	78.02
museums	45.09
using health facilities	13.61

The task is to find market segments having homogeneous preferences in some of the activities. In addition to the variables that were used as segmentation base, a number of demographic, socioeconomic, behavioral and psycho-graphic background variables is available in the extensive guest survey data set: age, daily expenditures per person, monthly disposable income, length of stay, intention to revisit Austria, intention to recommend Austria, number of prior vacations in Austria, etc. These variables were not used as input in the cluster analysis, as only homogeneous groups with respect to vacation activities were of interest. The background variables are only used to describe the market segments in more detail.

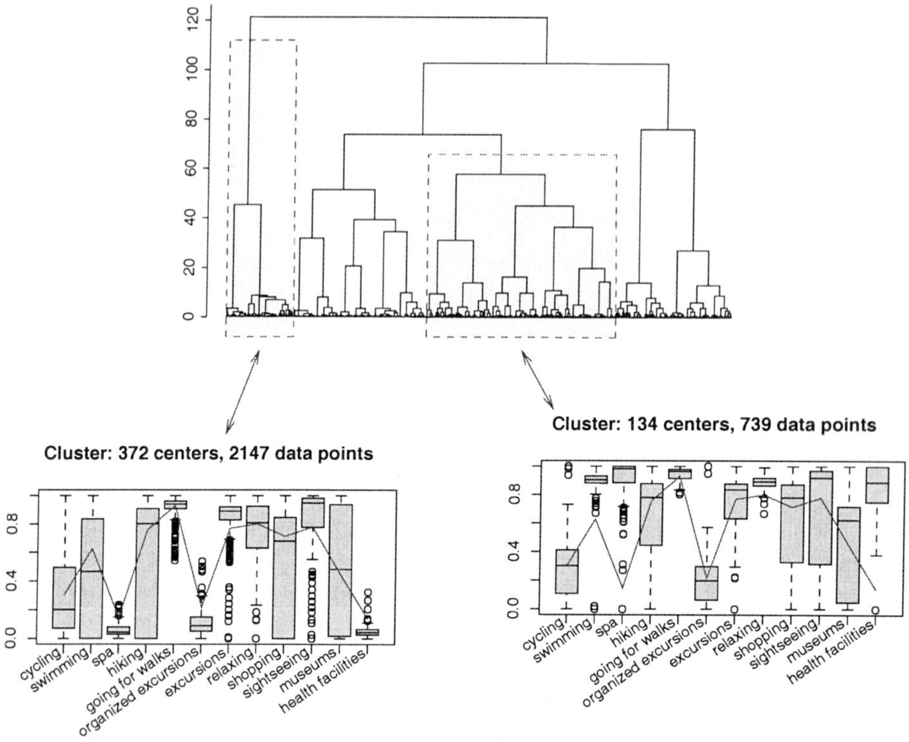

Fig. 1. Bagged clustering dendrogram together with boxplots for two selected clusters

The upper part of Figure 1 depicts the dendrogram resulting from a bagged clustering analysis. Learning vector quantization (e.g., [10]) was used as base method with $K = 20$ centers in each run on $B = 50$ training sets. The resulting 1000 centers were then hierarchically clustered using Euclidean distance and Ward's linkage method (e.g., [6]). We also tried other parameter combinations, but results were very similar, as the algorithm is not very sensitive to B and K once these are large enough.

Bagged clustering has been implemented using the R package for statistical computing (`http://www.R-project.org`), a free implementation of the S language; R functions for bagged clustering can be obtained from the authors upon request. The software allows interactive exploration of the dendrogram: by clicking on a subtree of the dendrogram one gets (in another window) a box-whisker plot of the centers in the corresponding cluster $\mathcal{C}_i^B$. Figure 1 shows as example boxplots corresponding to two such subtrees.

The boxes range from the 25% quantile to the 75% quantile, the line in the middle represents the median, the whiskers and circles depict outliers. See the documentation of any modern statistics package for more details on box-whisker plots. The horizontal polygon depicts the overall sample mean such that one can easily compare which variables are so-called marker variables of the segment, i.e., are different in the segment than in the overall population and can be repeatedly found having similar values such that the corresponding boxes are small.

The market segments corresponding to the two boxplots in Figure 1 can be described as follows:

- **Individual sightseers (left plot):** This large segment (40 percent of the tourists questioned) have a clear focus when visiting Austria: They want to hop from sight to sight. Therefore both the items sightseeing and excursions are strongly and commonly agreed upon in this group. Neither sports nor shopping are of central importance, although some members do spend some of their leisure time undertaking those activities. Well reflecting the individualist character of this group is the heterogeneity of this segment concerning a number of activities, as e.g. swimming, hiking, shopping or visiting museums.
- **Health oriented holiday-makers (right plot):** This niche segment represents a very stable and distinct interest group. Clearly, these tourists spend their vacation swimming and relaxing in spas and health facilities. Also, they all seem to enjoy going for a walk (after the pool is closed?). As far as the remaining activities are concerned, homogeneity decreases as indicated by the large dispersion of mean values.

The information which variables "define" a segment (small boxes) and with respect to which variables a segment is heterogenous (large boxes) is unique to the bagged cluster approach. A small box indicates that the corresponding cluster center was stably found over all repetitions of the base method. Bagged clustering bootstraps the base cluster method, hence the sizes of the boxes visualize the dispersion of the segment mean for each input variable and indicate how "correlated" an input variable is with the segment. This information is not available if the data set is partitioned only once. Note that we have only $\{0,1\}$-valued data, hence data dispersion in a segment can also not be used.

The analysis of the background variables shows that the sightseeing tourists are rather young (median 48 years) very fond of Austria, intend to revisit the country to a high extent and spend an average amount of money per day (52 Euro). The health-oriented tourists are moderately older (median 53 years), have similar intent to revisit the country, however spend significantly more money per day (68 Euro).

4 Stability Analysis

We have also compared the stability of standard K-means and LVQ with bagged versions thereof. K-Means and LVQ were independently repeated 100 times using $K = 3$ to 10 clusters. Runs where the algorithms converged in local minima (SSE more than 10% larger than best solution found) were discarded. Then 100 bagged solutions were computed using $K = 20$ for the base method and $B = 50$ training sets. The resulting dendrograms were cut into 3 to 10 clusters.

All partitions of each method were compared pairwise using one compliance measure from supervised learning (Kappa index, [2]) and one compliance measure from unsupervised learning (corrected Rand index, [5]). Suppose we want to compare two partitions summarized by the contingency table $T = [t_{ij}]$ where $i, j = 1, \ldots, K$ and t_{ij} denotes the number of data points which are in cluster i in the first partition and in cluster j in the second partition. Further let $t_{i\cdot}$ and $t_{\cdot j}$ denote the total number of data points in clusters i and j, respectively:

		Partition 2				
		1	2	...	K	$\sum$
	1	t_{11}	t_{22}	...	t_{1K}	$t_{1\cdot}$
Partition 1	2	t_{21}	t_{22}	...	t_{2K}	$t_{2\cdot}$
	$\vdots$	$\vdots$	$\vdots$	$\ddots$	$\vdots$	$\vdots$
	K	t_{K1}	t_{K2}	...	t_{KK}	$t_{K\cdot}$
	$\sum$	$t_{\cdot 1}$	$t_{\cdot 2}$	...	$t_{\cdot K}$	$t_{\cdot\cdot} = N$

In order to compute the Kappa index for an unsupervised classification problem, we first have to match the clusters from the two partitions such that they have maximal agreement. We do this by permuting the columns (or rows) of matrix T such that the trace $\sum_{i=1}^{K} t_{ii}$ of T gets maximal. In the following we assume that T has maximal trace.

Then the Kappa index is defined as

$$\kappa = \frac{N^{-1}\sum_{i=1}^{K} t_{ii} - N^{-2}\sum_{i=1}^{K} t_{i\cdot}t_{\cdot i}}{1 - N^{-2}\sum_{i=1}^{K} t_{i\cdot}t_{\cdot i}}$$

which is the agreement between the two partitions corrected for agreement by chance given row and column sums.

The Rand index measures agreement for unsupervised classifications and hence is invariant with respect to permutations of the columns or rows of T. Let A denote the number of all pairs of data points which are either put into the same cluster by both partitions or put into different clusters by both partitions. Conversely, let D denote the number of all pairs of data points that are put into one cluster in one partition, but into different clusters by the other partition. Hence, the partitions disagree for all pairs D and agree for all pairs A and $A + D = \binom{N}{2}$. The original Rand index is defined as $A/\binom{N}{2}$, we use a version corrected for agreement by chance [5] which can be computed directly from T as

$$\nu = \frac{\sum_{i,j=1}^{K}\binom{t_{ij}}{2} - \sum_{i=1}^{K}\binom{t_{i\cdot}}{2}\sum_{j=1}^{K}\binom{t_{\cdot j}}{2}/\binom{N}{2}}{\frac{1}{2}\left[\sum_{i=1}^{K}\binom{t_{i\cdot}}{2} + \sum_{j=1}^{K}\binom{t_{\cdot j}}{2}\right] - \sum_{i=1}^{K}\binom{t_{i\cdot}}{2}\sum_{j=1}^{K}\binom{t_{\cdot j}}{2}/\binom{N}{2}}$$

Figure 2 shows the mean and standard deviation of κ and ν for $K = 3, \ldots, 10$ clusters and $100 * 99/2 = 4950$ pairwise comparisons for each number of clusters. Bagging considerably increases the mean agreement of the partitions for all number of clusters while simultaneously having a smaller variance. Hence, the procedure stabilizes the base method. It can also be seen that LVQ is more stable than K-Means on this binary data set.

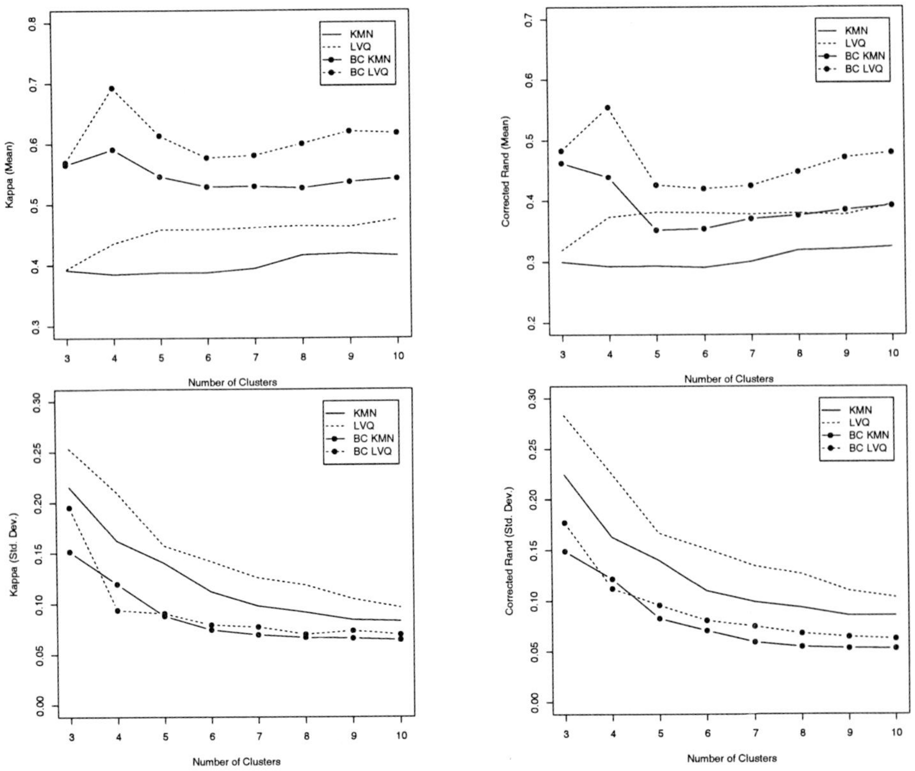

Fig. 2. Stability of clustering algorithms over 100 repetitions for 3 to 10 clusters: Mean kappa (top left), mean corrected Rand (top right), standard deviation of kappa (bottom left) and standard deviation of corrected Rand index (bottom right).

5 Summary

The bagged cluster algorithm has been applied to a binary data set from tourism marketing. Categorical data sets with very few categories are very common in the marketing sciences, yet most cluster algorithms are designed for metric input spaces, especially with Gaussian distributions. Hierarchical cluster methods allow for arbitrary distance measures (and hence arbitrary input spaces) but get quickly infeasible with increasing numbers of observations.

Bagged clustering overcomes these difficulties by combining hierarchical and partitioning methods. This allows for new exploratory data analysis techniques and cluster visualizations. Clusters can be split into sub-segments, each branch of the tree can be explored and the corresponding market segment identified and described.

By bootstrapping partitioning cluster methods we can measure the variance of the cluster centers for each input variable, which is especially important for binary data where usually only cluster centers without any variance information are available. This leads to easy seperation of variables in which a segment is homegenous, and variables where a segment is rather heterogenous.

Finally, building complete ensembles of partitions also has a stabilizing effect on the base cluster method. The average agreement between repetitions of the algorithm is considerably increased, while the variance is reduced. The partitions found in 2 independent runs are more similar to each other, reducing the need for subjective decisions of the practitioner which solution to choose.

Our current work tries to generalize the approach to partitioning methods which are not necessarily represented by centers, e.g., fuzzy clusters. Using distance measures that operate on partitions directly (instead of representatives) these could then also be clustered using hierarchical techniques.

Acknowledgments

This piece of research was supported by the Austrian Science Foundation (FWF) under grant SFB#010 ('Adaptive Information Systems and Modelling in Economics and Management Science').

References

1. Leo Breiman. Bagging predictors. *Machine Learning*, 24:123–140, 1996.
2. J. Cohen. A coefficient of agreement for nominal scales. *Educational and Psychological Measurement*, 1960(20):37–46, 1960.
3. Sara Dolnicar and Friedrich Leisch. Behavioral market segmentation using the bagged clustering approach based on binary guest survey data: Exploring and visualizing unobserved heterogeneity. *Tourism Analysis*, 5(2–4):163–170, 2000.
4. Anton K. Formann. *Die Latent-Class-Analyse: Einfhrung in die Theorie und Anwendung*. Beltz, Weinheim, 1984.
5. Lawrence Hubert and Phipps Arabie. Comparing partitions. *Journal of Classification*, 2:193–218, 1985.
6. Leonard Kaufman and Peter J. Rousseeuw. *Finding Groups in Data*. John Wiley & Sons, Inc., New York, USA, 1990.
7. Friedrich Leisch. *Ensemble methods for neural clustering and classification*. PhD thesis, Institut für Statistik, Wahrscheinlichkeitstheorie und Versicherungsmathematik, Technische Universität Wien, Austria, 1998.
8. Friedrich Leisch. Bagged clustering. Working Paper 51, SFB "Adaptive Information Systems and Modeling in Economics and Management Science", Wirtschaftsuniversitt Wien, Austria, August 1999.
9. James H. Myers and Edward Tauber. *Market Structure Analysis*. American Marketing Association, Chicago, 1977.
10. Brian D. Ripley. *Pattern recognition and neural networks*. Cambridge University Press, Cambridge, UK, 1996.

Direct Estimation of Polynomial Densities in Generalized RBF Networks Using Moments

Evangelos Dermatas

Department of Electrical Engineering and Computer Technology,
dermatas@george.wcl2.ee.upatras.gr

Abstract. We present a direct estimation method of the output layer weights in a polynomial extension of the generalized radial-basis-function networks when used in pattern classification problems. The estimation is based on the L_2-distance minimization of the density and the population moments. Each synaptic weight in the output layer is derived as a non-linear function of the training data moments. The experimental results, using one- and two-dimensional simulated data and different polynomial orders, show that the classification rate of the polynomial densities is very close to the optimum rate.

1 Introduction

Among the single hidden layer feed-forward neural networks, the most widely used in applications is the Radial-basis-function network (RBFN). The RBFN consists of two layers; in the hidden layer, the pattern vector is non-linearly transformed using a family of radial or kernel functions, and the output layer performs a linear transformation. In a popular generalization of the RBFNs, the Generalized RBFNs construct hyperellipses around the centers of the basis functions [8]. It has been shown that the GRBFNs are excellent approximators in curve-fitting problems (regression, classification and density estimation) due to their simple network structure, and the development of fast learning algorithms [5], based on the K-Means algorithm for the estimation of the synaptic weights in the hidden layer. Direct estimation of the weights in the output layer can be easily derived by minimizing the mean square error between the actual and the desired output.

Polynomial feedforward neural networks such as Sigma-pi networks [6] [7], high-order networks [4], functional link architectures [10], the Ridge polynomial networks [9] and the product units [3] are capable to solve complex pattern recognition and regression problems using variances of the backpropagation learning rule. Most of them, even in the case of a single layer network, construct complex polynomial models which quickly lead to combinatorial explosion of the input components. This problem, introduced by Richard Bellman[1], appears frequently in the theory of universal approximation and it is known as the *curse of the dimensionality.* Only few pruning strategies have been proposed [6] to adjust the network structure in the training environment and to decrease the number of synaptic weights.

G. Dorffner, H. Bischof, and K. Hornik (Eds.): ICANN 2001, LNCS 2130, pp. 119–126, 2001.

Recently, a polynomial extension of the GRBFNs (PeGRBFN)has been used in density estimation and pattern classification problems. The synaptic weights are estimated using a direct method based on moments [2]. The only structural difference from the GRBFN, is the inclusion of weighted high-order terms within the stimuli at each output neuron:

$$O(x, A, M, S) = \sum_{j=1}^{J} \sum_{n=0}^{Nj} a_{jn} g^n(\|\mathbf{x} - \mathbf{m}_j\|, \mathbf{s}_j), \tag{1}$$

where $g(\|\mathbf{x} - \mathbf{m}_j\|, \mathbf{s}_j)$ is a radial-basis function, J is the number of hidden neurons, $A = \{a_{jn}\}$ is the weights of the output neuron connecting the nth-order stimuli of the j hidden neuron, and N_j is the number of high-order terms. A computational simple radial-basis function $g : \Re^{3M} \to (0, 1]$ has been used in [2]:

$$y_j = g(\|\mathbf{x} - \mathbf{m}_j\|, \mathbf{s}_j) = \frac{1}{1 + \sum_{i=1}^{M} s_{ji}(x_i - m_{ji})^2}, \tag{2}$$

where $\mathbf{m}_j, \mathbf{s}_j$ are the synaptic weights of the jth neuron, and M is the dimensionality of the feature vector.

One of the network weakness is the limited description capabilities of the network transfer function; high-order dependencies among the hidden layer outputs are excluded (eq. 1). Moreover, in the training process the optimization criterion ensures exact matching between the population and the density moments. Therefore, the uninvolved moments may differ significantly.

This paper presents a PeGRBF network and a direct estimation method of the output layer weights overcoming the previous work disadvantages. More specific, we study the general type of a full connected polynomial network using also a novel optimization criterion based on the mean square error of all moment differences between the network pdf approximation and the statistical estimation. We prove that a direct estimation of the synaptic weights can be achieved by solving a system of linear equations, eliminating also the infinitive power series. In the last section, two examples in one- and two-dimensional classification problems demonstrate the accuracy of the moment based training method.

2 The Full-Connected Polynomial GRBF Network

The full-connected PeGRBFN is an extension of the classic RBF network. In the hidden layer, the feature vector $\mathbf{x}, \mathbf{x} \in \Re^M$ is transformed to a vector of directional weighted distances using a finite set of points $\mathbf{m}_k, k = 1, J$ defined also in the feature space ($\mathbf{m}_k \in \Re^M$). In the output layer, instead of linear neurons, full-connected polynomials are used. The total transfer function is given by:

$$O(\mathbf{y}, A) = \sum_{n_1=0}^{N} \dots \sum_{n_J=0}^{N} a_{n_1 \dots n_J} \prod_{k=1}^{J} g^{n_k}(\|\mathbf{x} - \mathbf{m}_k\|, \mathbf{s}_k) \tag{3}$$

$$= \sum_{n_1=0}^{N} \dots \sum_{n_J=0}^{N} a_{n_1 \dots n_J} \prod_{k=1}^{J} y_k^{n_k}, \tag{4}$$

where $y_k = g(\|\mathbf{x} - \mathbf{m}_k\|, \mathbf{s}_k)$ is the kth neuron output in the hidden layer. the hidden layer outputs y_k is assumed to be bounded in the interval $(0, 1]$, and $A = \{a_{n_1 \dots n_J} : 1 \leq n_1, \dots, n_J \leq N\}$.

Obviously, the PeGRBFN is a generalization of the GRBFN. In case of developing efficient training algorithms for the PeGRBFN we expect better approximation capabilities than the GRBFN in density estimation and pattern classification applications.

In the rest of this paper we discuss the problem of estimating the weights in the output layer of the PeGRBFNs. The unknown parameters in the hidden layer $(\mathbf{m}_k, \mathbf{s}_k, k = 1, J)$ can be estimated independently using self-organized clustering methods. Among them, the K-Means algorithm is an excellent choice due to its convergence properties.

3 Density Estimation Using Moments

Assuming that T examples are available for training $X = \{\mathbf{x}_1, \mathbf{x}_2, \dots, \mathbf{x}_T\}$, and the hidden layer weights have already been estimated. In this case, the training patterns can be non-linearly transformed to the corresponding extended patterns at the output of the hidden layer neurons (eq. 3), $Y = \{\mathbf{y}_i : \mathbf{y}_i = \mathbf{g}(\mathbf{x}_i), \forall \mathbf{x}_i \in X\}$.

Let equation 3 represents a density of a random vector $\mathbf{y}$ which has A unknown parameters. If the kernel output is restricted in the area of $(0, 1]$, the $m_1, m_2, \dots, m_J$ moment of the full-connected polynomial density is

$$M_p(m_1, m_2, \dots, m_J) = \int_0^1 \dots \int_0^1 y_1^{m_1} y_2^{m_2} \dots y_J^{m_J} p_{\mathbf{y}}(\mathbf{y}, A) dy_1 \dots dy_J \tag{5}$$

$$\simeq \int_0^1 \dots \int_0^1 \prod_{k=1}^{J} y_k^{m_k} O(\mathbf{y}, A) dy_1 \dots dy_J$$

$$= \int_0^1 \dots \int_0^1 \prod_{k=1}^{J} y_k^{m_k} \left(\sum_{n_1=0}^{N} \dots \sum_{n_J=0}^{N} a_{n_1 \dots n_J} \prod_{k=1}^{J} y_k^{n_k} \right) dy_1 \dots dy_J$$

$$= \sum_{n_1=0}^{N} \dots \sum_{n_J=0}^{N} a_{n_1 \dots n_J} \int_0^1 \dots \int_0^1 \prod_{k=1}^{J} y_k^{m_k + n_k} dy_1 \dots dy_J$$

$$M_p(m_1, m_2, \dots, m_J) = \sum_{n_1=0}^{N} \dots \sum_{n_J=0}^{N} a_{n_1 \dots n_J} \prod_{k=1}^{J} \frac{1}{m_k + n_k + 1}. \tag{6}$$

The population moments for the set of training examples Y is given by:

$$M_s(m_1, m_2, ..., m_J, Y) = \frac{1}{T}\sum_{t=1}^{T}\prod_{k=1}^{J} y_{tk}^{m_k}.$$

The L_2-distance of the population and the network density moments for all moments is:

$$E(A,Y) = \sum_{m_1=0}^{+\infty} ... \sum_{m_J=0}^{+\infty} (M_p(m_1, m_2, ..., m_J) - M_s(m_1, m_2, ..., m_J, Y))^2 \tag{7}$$
$$= \sum_{m_1=0}^{+\infty} ... \sum_{m_J=0}^{+\infty} (\sum_{n_1=0}^{N} ... \sum_{n_J=0}^{N} a_{n_1 ... n_J} \prod_{k=1}^{J} \frac{1}{m_k + n_k + 1} - \frac{1}{T_i}\sum_{t=1}^{Ti}\prod_{k=1}^{J} y_{tk}^{m_k})^2.$$

The unique minimum of the quadratic error function can be reached by solving the equations:

$$\frac{\partial E(A,Y)}{\partial a_{q_1 ... q_J}} = 0$$

$$\sum_{m_1=0}^{+\infty} ... \sum_{m_J=0}^{+\infty} \prod_{k=1}^{J} \frac{1}{m_k + q_k + 1}$$

$$(\sum_{n_1=0}^{N} ... \sum_{n_J=0}^{N} a_{n_1 ... n_J} \prod_{k=1}^{J} \frac{1}{m_k + n_k + 1} - \frac{1}{T}\sum_{t=1}^{T}\prod_{k=1}^{J} y_{tk}^{m_k}) = 0.$$

Therefore, a total number of $(N+1)^J$ linear equations is derived:

$$\sum_{n_1=0}^{N} ... \sum_{n_J=0}^{N} a_{n_1 ... n_J} \sum_{m_1=0}^{+\infty} ... \sum_{m_J=0}^{+\infty} \prod_{k=1}^{J} \frac{1}{m_k + n_k + 1} \frac{1}{m_k + q_k + 1} \tag{8}$$
$$= \frac{1}{T}\sum_{t=1}^{T}\sum_{m_1=0}^{+\infty} ... \sum_{m_J=0}^{+\infty} \prod_{k=1}^{J} \frac{y_{tk}^{m_k}}{(m_k + q_k + 1)}, \qquad 0 \le q_1 ... q_J \le N.$$

In a more compact form, the direct solution is given by

$$\sum_{n_1=0}^{N} ... \sum_{n_J=0}^{N} a_{n_1 ... n_J} F(P_n, P_q, J) = U(P_q, J, Y) \Leftrightarrow FA = U \Leftrightarrow A = F^{-1}U,$$

where:

$$F(P_n, P_q, J) = \sum_{m_1=0}^{+\infty} ... \sum_{m_J=0}^{+\infty} \prod_{k=1}^{J} \frac{1}{m_k + n_k + 1} \frac{1}{m_k + q_k + 1} \tag{9}$$

$$U(P_q, J, Y) = \frac{1}{T}\sum_{t=1}^{T}\sum_{m_1=0}^{+\infty} ... \sum_{m_J=0}^{+\infty} \prod_{k=1}^{J} \frac{y_{tk}^{m_k}}{(m_k + q_k + 1)}, \tag{10}$$

Table 1. The first 12 values of the function $f(n,n)$

n,	$f(n,n)$	n,	$f(n,n)$	n,	$f(n,n)$
0,	1.644934066	4,	0.221322955	8,	0.117512014
1,	0.644934066	5,	0.181322955	9,	0.105166335
2,	0.394934066	6,	0.153545177	10,	0.0951663356
3,	0.283822955	7,	0.133137014	11,	0.0869018728

and P_n, P_q, is a sequence of positive integer numbers: $P_n = (n_1, n_{2,}...n_J : n_i \in \aleph \cup \{0\}, 0 \le n_i \le N, i = 1, J)$ and $P_q = (q_1, q_2, ...q_J : q_i \in \aleph \cup \{0\}, 0 \le q_i \le N, i = 1, J)$.

The infinitive series in equations 9 and 10 can be eliminated, decreasing also the computational complexity.

Initially, we prove that the function $F(P_n, P_q, J)$ can be factorized,

$$F(P_n, P_q, J) = \sum_{m_1=0}^{+\infty} \cdots \sum_{m_J=0}^{+\infty} \prod_{k=1}^{J} \frac{1}{m_k + n_k + 1} \frac{1}{m_k + q_k + 1} \tag{11}$$

$$= \underbrace{\left(\sum_{m_1=0}^{+\infty} \frac{1}{m_1 + n_1 + 1} \frac{1}{m_1 + q_1 + 1} \right) \cdots \left(\sum_{m_J=0}^{+\infty} \frac{1}{m_J + n_J + 1} \frac{1}{m_J + q_J + 1} \right)}_{J \text{ terms}}$$

$$= \prod_{k=1}^{J} f(n_k, q_k). \tag{12}$$

The function $F(P_n, P_q, J)$ is symmetric:

$$F(P_n, P_q, J) = \prod_{k=1}^{J} f(n_k, q_k) = \prod_{k=1}^{J} f(q_k, n_k) = F(P_q, P_n, J).$$

Further more, the infinitive series in $f(n, q)$ is reduced as follows:

$$f(n,q) = \sum_{m=0}^{+\infty} \frac{1}{m+n+1} \frac{1}{m+q+1} = \begin{cases} \frac{1}{n-q} \sum_{m=1}^{\infty} \left(\frac{1}{m+n} - \frac{1}{m+q} \right), & n \ne q \\ \sum_{i=1}^{+\infty} \left(\frac{1}{i+n} \right)^2, & n = q \end{cases}$$

$$= \begin{cases} \frac{1}{n-q} \left(\sum_{m=1}^{\infty} \frac{1}{m+n} - \sum_{m=1}^{\infty} \frac{1}{m+q} \right), & n \ne q \\ \sum_{i=1}^{+\infty} \left(\frac{1}{i+n} \right)^2, & n = q \end{cases} = \begin{cases} \left| \frac{1}{q-n} \right| \sum_{m=\min(n,q)+1}^{\max(n,q)} \frac{1}{m}, & n \ne q \\ \sum_{i=1}^{+\infty} \left(\frac{1}{i+n} \right)^2, & n = q \end{cases}.$$

The series $\sum_{i=1}^{+\infty} \left(\frac{1}{i+n} \right)^2$ converges for all $n \in \aleph \cup \{0\}$. In table 1 the most useful function values in applications are given.

In a critical action from a practical point of view, the infinitive series involving in $U(q, p, J, Y)$ are transformed to a finite power series and a natural logarithm function of the training dataset Y,

$$U(P_q, J, Y) = \frac{1}{T} \sum_{t=1}^{T} \sum_{m_1=0}^{+\infty} \ldots \sum_{m_J=0}^{+\infty} \prod_{k=1}^{J} \frac{y_{tk}^{m_k}}{m_k + q_k + 1}$$

$$= \frac{1}{T} \sum_{t=1}^{T} \underbrace{\left(\sum_{i=0}^{+\infty} \frac{y_{t1}^{i}}{i + q_1 + 1} \right) .. \left(\sum_{i=0}^{+\infty} \frac{y_{tJ}^{i}}{i + q_J + 1} \right)}_{J \text{ terms}} = \frac{1}{T} \sum_{t=1}^{T} \prod_{k=1}^{J} \left(\sum_{i=0}^{+\infty} \frac{y_{tk}^{i}}{i + q_k + 1} \right)$$

$$= \frac{1}{T} \sum_{t=1}^{T} \prod_{k=1}^{J} \left(y_{tk}^{-q_k-1} \sum_{i=0}^{+\infty} \frac{y_{tk}^{i+q_k+1}}{i + q_k + 1} \right) = \frac{1}{T} \sum_{t=1}^{T} \prod_{k=1}^{J} y_{tk}^{-q_k-1} \left(\sum_{i=1}^{+\infty} \frac{y_{tk}^{i}}{i} - \sum_{i=1}^{q_k} \frac{y_{tk}^{i}}{i} \right) \tag{13}$$

$$U(P_q, J, Y) = (-1)^J \frac{1}{T} \sum_{t=1}^{T} \prod_{k=1}^{J} \left(y_{tk}^{-q_k-1} \ln(1 - y_{tk}) + \sum_{i=1}^{q_k} \frac{y_{tk}^{i-q_k-1}}{i} \right). \tag{14}$$

In case of full-connected PeGRBFN, the matrix F depends on the order of the polynomials and the number of neurons in the hidden layer. It remains to prove that the inverse of the matrix F always exists. Experiments, carried out in computers, showed that the matrix F is inverted in typical network structures. The inversion process is independent to the training data and therefore can be derived off-line. Thus, the computational complexity of the proposed training method is $O(\max((N+1)^{2J}, T\,(N+1)^{J}))$.

4 Examples in Pattern Classification Problems

The proposed density estimation method is evaluateed in a simulation study based on Gaussian distributed one- and two-dimensional examples. The first dataset consists of one-dimensional examples produced by two Gaussian distributions. The generators mean value was set to 0.3 and 0.7 respectively, and the standard deviation was 0.1. In case of one-dimensional data, the constant parameters of the system of linear equations (eq.9 and 10) are rewriting

$$F(n, q, J) = f(n, q), \qquad q = 0, N\,, \tag{15}$$

If $\langle . \rangle_t$ denote the expectation value, then

$$U(q, 1, Y) = -\left\langle y_t^{-q-1} \ln(1 - y_t) \right\rangle_t - \sum_{i=1}^{q} \frac{1}{i} \left\langle y_t^{i-q-1} \right\rangle_t, \qquad q = 0, N. \tag{16}$$

The results of the two classification experiments are shown in table 2. In the first experiment the probabilistic optimum threshold of 0.5 (assuming equal *a priori* probability for each class) was used to classify the data. In the second experiment the examples were classified according to the polynomial approximation. The same order polynomial density was estimated from the training data of each individual class. Four data sizes of 200, 500, 1000 and 2000 examples were tested and the polynomial order varies from 2 to 6.

In figure 1 the Gaussian generators pdf and the density approximation are shown in case of two, three and four order polynomials using the set of 2000

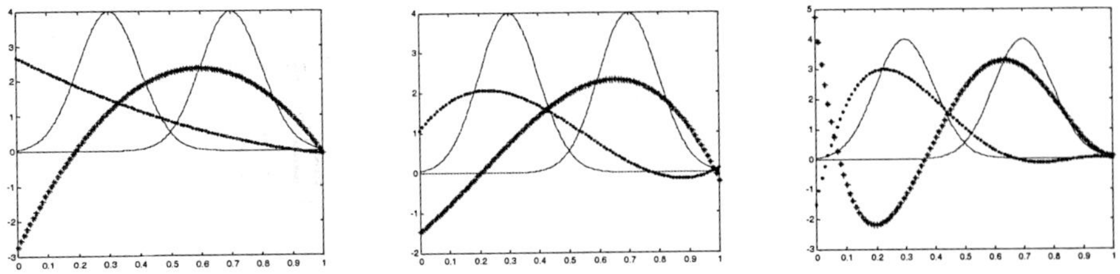

Fig. 1. Gaussian pdf and two, three and four order polynomial densities

Table 2. Percent classification rate in uni-dimensional data using the Gaussian pdfs and the polynomial approximation

Data size	Gaussian	2	3	4	5	6
200	96.0	81.0	94.0	94.0	97.0	97.0
500	98.6	94.0	95.8	95.8	97.0	86.6
1000	98.1	81.4	94.6	91.1	91.1	90.6
2000	97.5	81.1	95.7	96.4	96.8	55.8

Table 3. Percent classification rate in two-dimensional data using the Gaussian pdfs and the polynomial approximation using two up to six order terms

Data size	Gaussian	2	3	4	5	6
200	87.00	87.00	91.50	60.00	48.00	58.00
500	89.80	89.80	87.00	87.60	57.80	48.20
1000	91.20	91.20	83.40	77.00	43.10	52.20
2000	90.80	84.85	89.25	89.50	48.50	43.90

examples. The accuracy of the proposed weights estimation method is very sensitive to the number of training data and the order selection for the polynomial approximation as shown in the experimental results presented in table 2. The appropriate polynomial order gives excellent classification rate which is very close to the optimum rate estimated using the minimum error criterion of the Gaussian classifier. In the second experiment, the polynomial density approximation is evaluated in a two-classes, two-dimensional, pattern classification problem. The same data sizes were tested using Gaussian pdf generators, $(x_1, x_2) \sim (N(0.3, 0.2), N(0.3, 0.2))$, and $(x_1, x_2) \sim (N(0.7, 0.2), N(0.7, 0.2))$. In table 3, the classification rate of the maximum probability criterion using the Gaussian pdf data and the classification rate obtained by the maximum polynomial density rule are shown in four data sets of 200, 500, 1000 and 2000 examples. As in case of one-dimensional experiment, the polynomial order varies from 2 to 6.

The number of unknown weights increases significantly in case of high-order terms and full-connected polynomial terms (the *curse of the dimensionality* problem). Consequently, the size of matrix F increases and computer restriction in the matrix inversion process decreases significantly the accuracy of the density approximation; the maximum classification accuracy in two-dimensional data is 58% when the density is approximated using polynomial order greater than 10, instead of the optimum 87%.

References

1. Bellman, R. Adaptive Control Processes: A Guided Tour. Princeton University Press. (1961)
2. Dermatas, E.: Polynomial extension of the Generalized Radial-Basis Function Networks in Density estimation and Classification Problems. Neural Network World **10(4)** (2000) 553–564
3. Durbin R., and Rumelhart D.: Product units: A computationally powerful and biologically plausible extension to backpropagation networks. Neural Computation **1** (1989) 133–142
4. Giles C., and Maxwell T.: Learning, invariance, and generalization in a high-order neural network. Applied Optics **26(23)** (1987) 4972–4978
5. Haykin S.: Neural Networks: A comprehensive foundation. Prendice Hall International Inc. (1999) 298–305
6. Heywood M., and Noakes P.: A Framework for Improved Training of Sigma-Pi Networks. IEEE Tr. on Neural Networks **6(4)** (1995) 893–903
7. McClelland J., and Rumelhart D.: Parallel Distributed Processing 1. MIT Press (1987)
8. Poggio T., and Girosi F.: Networks for approximation and learning. Proc. IEEE **78(9)** (1990) 1481–1497
9. Shin Y., and Ghosh G.: Ridge Polynomial Networks. IEEE Tr. on Neural Networks **6(3)** (1995) 610–622
10. Pao Y.: Adaptive Pattern Recognition and Neural Networks. Reading MA:Addison-Wesley (1989)

Generalisation Improvement of Radial Basis Function Networks Based on Qualitative Input Conditioning for Financial Credit Risk Prediction

Xavier Parra[1], Núria Agell[2] and Xari Rovira[2]

[1]ESAII, Technical University of Catalonia, Av. Víctor Balaguer, s/n
08800 Vilanova i la Geltrú – Barcelona – Catalonia – Spain
Xavier.Parra@upc.es
[2]ESADE, Ramon Llull University, Av. Pedralbes, 62
08034 Barcelona – Catalonia – Spain
{agell, rovira}@esade.edu

Abstract. The rating is a qualified assessment about the credit risk of bonds issued by a government or a company. There are specialised rating agencies, which classify firms according to their level of risk. These agencies use both quantitative and qualitative information to assign ratings to issues. The final rating is the judgement of the agency's analysts and reflects the probability of issuer default. Since the final rating has a strong dependency on the experts knowledge, it seems reasonable the application of learning based techniques to acquire that knowledge. The learning techniques applied are neural networks and the architecture used corresponds to radial basis function neural networks. A convenient adaptation of the variables involved in the problem is strongly recommended when using learning techniques. The paper aims at conditioning the input information in order to enhance the neural network generalisation by adding qualitative expert information on orders of magnitude. An example of this method applied to some industrial firms is given.

1 Introduction

The present paper aims at applying Neural Network techniques to predict credit of bonds issued by a government or a company. Predicting the rating of a firm therefore requires a thorough knowledge of the ratios and values that indicate the firm's situation and, also, a thorough understanding of the relationships between them and the main factors that can alter these values [1].

The application of learning based techniques to acquire the analyst's knowledge seems reasonable due to the special nature of the problem. The strong dependency on the expert's knowledge is the main reason that brought us to the connectionist approach proposed. However, how to represent the input and output variables of a learning problem in a neural network implementation of the problem is one of the key decisions influencing the quality of the solutions one can obtain. Moreover, this is especially important when qualitative information is available during training.

G. Dorffner, H. Bischof, and K. Hornik (Eds.): ICANN 2001, LNCS 2130, pp. 127–134, 2001.

This application shows how Neural Network and Qualitative Reasoning techniques, and particularly orders of magnitude calculus [2], can be useful in the financial domain [3].

The paper gives a brief introduction to the neural network architecture used. Next, the process to prepare the reference scales of different qualitative variables to be operated is established. In section 5 an application of this neural network to credit risk evaluation of a firm or an issue of bonds is presented. The paper finishes with some conclusions and also with some comments about the implementation of this application and the first results obtained with it.

2 Radial Basis Function Networks Architecture

Radial basis function networks (RBF) are especially interesting for the problem proposed since they are universal classifiers [4]. RBF have been traditionally associated with a simple architecture of three layers [5] (see Fig. 1). Each layer is fully connected to the following one and the hidden layer is composed of a number of nodes with radial activation functions called radial basis functions. Each of the input components feeds forward to the radial functions. The outputs of these functions are linearly combined with weights into the network output. Each radial function has a local response (opposite to the global response of sigmoid function) since their output only depends on the distance of the input from a centre point.

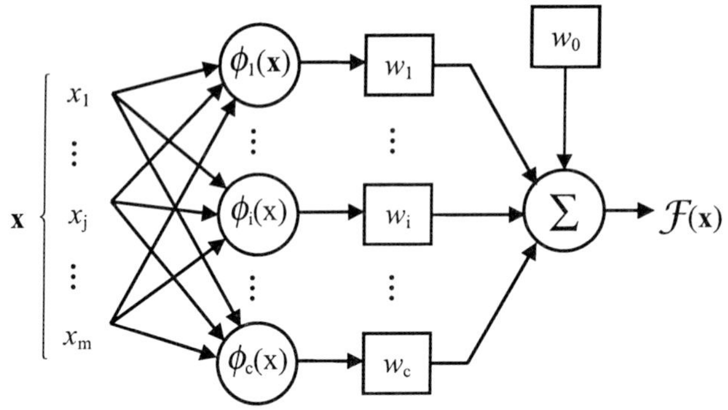

Fig. 1. Radial basis function network architecture

Radial functions in the hidden layer have a structure that can be represented as follows:

$$\phi_i(\mathbf{x}) = \varphi((\mathbf{x}-\mathbf{c}_i)^{\mathsf{T}}\mathbf{R}^{-1}(\mathbf{x}-\mathbf{c}_i)) \tag{1}$$

where φ is the radial function used, $\{\mathbf{c}_i \mid i = 1, 2,\ldots, c\}$ is the set of radial function centres and **R** is a metric. The term $(\mathbf{x} - \mathbf{c})^{\mathsf{T}}\,\mathbf{R}^{-1}\,(\mathbf{x} - \mathbf{c})$ denotes the distance from the input **x** to the centre **c** on the metric defined by **R**. There are several common types of functions used, though the Gaussian function is the most typical choice, combined with the Euclidean metric. In this case, the output of the RBF network is:

$$\mathcal{F}(x) = w_0 + \sum_{i=1}^{c} w_i \,\mathbf{exp}\left(-\frac{\left\| \mathbf{x} - \mathbf{c}_i \right\|^2}{r^2} \right) \tag{2}$$

where c is the number of basis functions, $\{w_i \mid i = 1, 2,..., c\}$ are the synaptic weights, $\| \cdot \|$ denotes the Euclidean norm and r is the radius of the radial function.

The RBF learning algorithm is an incremental and evolutionary process. Its mathematical foundation is called *subset selection* and consists in comparing models made up of different subsets of elements drawn from the same fixed set of candidates. To find the best subset is usually intractable so heuristics must be used to search for a small but hopefully interesting fraction of the space of all subsets. However, the use of these heuristics does not guarantee that the solutions we get include the least number of elements needed to reduce the approximation error to a fixed value.

The heuristic method called *forward selection* is widely used with RBF networks [6]. According to this method, the subset that must be determined is the subset of centres that fix the location of the radial functions in the input space. The method begins with an empty subset to which is added one basis function at a time. The centre of the radial function added is selected among the whole set of input patterns and is the one that most reduces the approximation error. The learning process continues until some chosen criterion stops decreasing (e.g. generalised cross-validation).

3 Qualitative Input Conditioning

The problem of having to extract some information from qualitative values represented in heterogeneous references is not unusual. Many qualitative reasoning techniques are used to manage this kind of references. However, when the values have to be processed with a neural network it is not clear how to prepare the values in order to take into account the experts' knowledge.

In general, the performance obtained when using a neural network depends on the problem representation, i.e., the input and output representations. Moreover, when the neural network used is an RBF network this dependency on the representation is more critical since, as it is shown in 2, the radial function output is a function of a distance defined in the input space. Depending on the kind of problem, there may be several different kinds of variables that can be represented. Unfortunately, there is not a unique method to represent all of these variable kinds. However, it is common to use some of the following hints:

- *Real-valued attributes* are usually rescaled by some function that maps the value into the range 0...1 or –1...+1, in a way that makes roughly even distribution within that range. It is also common to rescale the values to mean 0 and standard deviation 1 by using a linear transformation.
- *Nominal attributes* with m different values are usually either represented using a 1-of-m code or a binary code.

- *Ordinal attributes* with m different values are represented by m-1 variables of which the leftmost k have value 1 to represent the k-th attribute value while all others are 0.

As it is pointed in 1, financial problems involve a strong understanding of ratios and values that, somehow, indicate different financial situations. Often, these ratios and values are represented through a real value though quite frequently the expert extracts information not strictly from that numerical value but from a more qualitative representation. For example, in front of a concrete value the expert is more likely to think in terms of *good, bad, very good, …* than in any other way. Thus, there is the financial information represented with real values and the expert knowledge that treats the same information in a qualitative sense. A way of combining both representations is needed to prepare the variables for a learning process.

Let's suppose that each one of these variables is qualitatively described via a different set of labels, which are intervals of the real line, with an odd number of landmarks given by the experts. This allows having a central point ℓ_i in each set of landmarks. In order to prepare these variables for the neural network training, two steps will first be taken:

Step 1 Transformation of the central landmark ℓ_i to 0, through a translation $t_i: \mathbb{R} \longrightarrow \mathbb{R},\ t_i(x) = x - \ell_i$

Step 2 Transformation of the values through a linear transformation $f: \mathbb{R} \longrightarrow \mathbb{R}$,

$$f(x) = \begin{cases} +s \cdot t_i(x)/\ell_r & \text{if } t_i(x) \geq 0 \\ -s \cdot t_i(x)/\ell_\ell & \text{if } t_i(x) < 0 \end{cases}$$

where s is the sign of the linear transformation determined by the expert and ℓ_r and ℓ_ℓ denotes the right and left landmark, respectively.

After these two steps, all the values of the variables are described in a similar range, but the discretizations of the real line are different since the set of landmark is different for each variable (see Fig. 2).

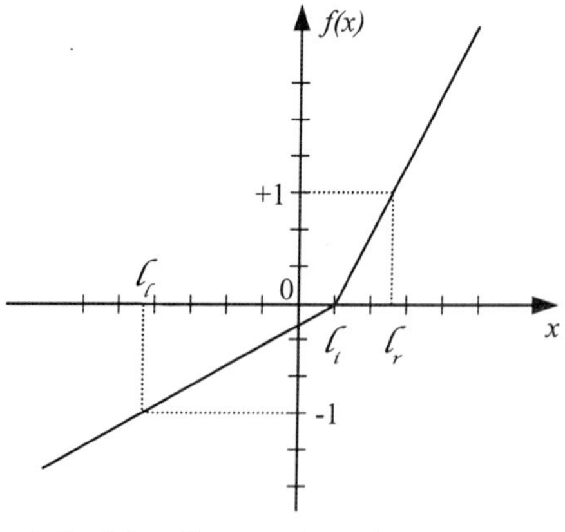

Fig. 2. Real line discretization using expert landmarks

4 A Credit Risk Prediction: Rating Evaluation

Our objective in this application is double, in one hand the goal is to replicate the rating evaluation of the agencies by using radial basis function neural networks, and in the other hand is compare if this evaluation improves when there is available some qualitative information.

If we consider the rating agency Standard & Poor's, it uses the following labels to assign a rating to the firms: {AAA,AA,A,BBB,BB,B,CCC,CC,C}. From left to right these rankings go from high to low credit quality, i.e., the high to low capacity of the firm to return debt, thus they permit to classify firm according to their level of risk.

The processes employed by Standard & Poor's annalists to assign these labels are highly complex. Decision on final rating is based on the information given by the financial data, the industry and the country where the firm operates, forecast on the possibilities of the firm's growth, and its competitive position. As a result the agency provides a global evaluation based on its own expertise to determine the rating.

To this end we use an initial database that include some of the firms on the index Dow Jones 500. For each firm we have financial ratios as quantitative variables and the sector as a qualitative one. A s it can be seen in next table, the firms are considered in seven different sectors.

Table 1. Sectors

Sector	Cyclical consumer	Non-cyclical consumer	Technology	Utilities	Basic Materials	Industrial	Energy
Number	1	2	3	4	6	7	8
Firms	71	80	42	38	33	58	31

The quantitative variables being used are: Interest coverage (IC); Market value over debt (MV/DBT), to see the indebtedness of the firm; Debt over assets (DBT/ATN), as a measure of leverage; Cashflow over debt (CF/DBT), which gives the idea of how many years will need the company to pay the debt; Return on assets (ROA); self financing percent (SELFIN), calculated as the profit over assets; Short term over long term debt (DC/DL), and sales growth (SALES).

The experts agree that some of these variables have a strongly dependence. For that reason and after a statistical study, they were reduced to the following five: V1=IC, V2=MV/DEBT, V3=ROA, V4=SELFIN, V5=DC/DL. Each one of the variables has different landmarks, as it can be seen in table 3, according to expert's knowledge.

5 Experiments and Results

Initially, we started with a database that included a total of 353 patterns. There were 12 input qualitative variables, 1 qualitative variable (sector) for each pattern and 1 output. Since many instances had missing values, all those instances that had one or more missing values were deleted from the database. Following the experts' recommendations and due to the especial peculiarity of a sector of activity (technological sector), the technological companies were also deleted from the set of patterns. Next step was, according to the experts' knowledge, to select those variables that were the most relevant in computing credit risk. The input space was reduced from 12 to 5 variables, and from 495 to 244 instances. All 5 input variables are real-valued while the rating, i.e. the output variable, is a nominal variable with 6 different classes {AAA, AA, A, BBB, BB, B}, and has been represented using a 1-of-6 code. At this point, there were at least two options: to train a single RBF network with 5 inputs and 6 outputs, or to train 6 different RBF networks with 5 inputs each one but only 1 output. The former option is not too appropriate because of the low number of patterns available for training. The latter option is more efficient from the point of view of resource optimisation. Although the final size of the architecture training will be probably smaller for the single RBF network and the training faster, its generalisation will be worse, and to get a good generalisation is more important than the size or the training time. Thus, the experiments had been performed considering that the initial problem of classifying a pattern in 1-of-6 classes has been transformed in 6 different problems of classifying a pattern in a single class. The network will say whether the pattern is or is not of the class for which the network has been trained.

Simulations have been carried out following the PROBEN1 standard rules [7]. The data set available has been sorted by the company name before partitioning it into three subsets: training set, validation set and test set, with a relation of 50%, 25% and 25% respectively. Table 2 shows the pattern distribution in each data subset. Note that for class AAA there are no patterns available in the test subset, and for class B there are no patterns to perform the training or the validation.

Table 2. Pattern distribution over data subsets

Rating	AAA	AA	A	BBB	BB	B	Total
Training	5	18	53	41	5	0	122
Validation	2	10	28	18	3	0	61
Test	0	7	23	27	3	1	61
Total	**7**	**35**	**104**	**86**	**11**	**1**	**244**

To study and analyse the effect that qualitative input conditioning had over RBF generalisation, two different kinds of training have been done. First training (referred to as *blind* training) rescale all the input values to mean 0 and standard deviation 1, but do not take into account the experts' knowledge. Second training (*expert* training) performs the input transformation described in section 3, i.e. it considers the information on orders of magnitude to rescale the values (see Table 3).

Table 3. Expert landmarks and signs

	ℓ_r	ℓ_i	ℓ_v	s
V_1	1	4	10	+1
V_2	1	2	8	+1
V_3	0.02	0.07	0.15	+1
V_4	0.2	1	10	−1
V_5	0.0	0.1	0.3	+1

Initially, networks are trained on the training set while the validation set is used to adjust the radial function width (r). To perform this adjustment of the radial width, a total of 4000 simulations have been done for each class. Widths checked are from 0'0001 to 0'1 with increments of 0'0001, from 0'101 to 1'1 with increments of 0'001, from 1'11 to 11'1 with increments of 0'01 and 11'2 to 111'1 with increments of 0'1.

The final width (see Table 4) is selected among the 4000 widths trained by applying the next criteria:

(a) Choose the width that maximises classification accuracy for the validation set.
(b) Choose the width that, with criterion (a), produces the smallest network.
(c) Choose the width that, with criterion (b), minimises mean squared error for the validation set.
(d) Choose the width that, with criterion (c), minimises mean squared error for the training set
(e) Choose the width that, with criterion (d), maximises classification accuracy for the training set.

Table 4. Radial function width (r) and classification accuracy for validation data set (CA_{va}) and test data set (CA_{te})

	Blind training			*Expert* training		
	r	CA_{va}	CA_{te}	r	CA_{va}	CA_{te}
AAA	1,055	96,7%	100,0%	51,2	98,4%	100,0%
AA	11,2	82,0%	88,5%	68,1	85,2%	90,2%
A	0,831	73,7%	50,8%	4,7	70,5%	59,0%
BBB	80,6	63,9%	57,4%	32,5	75,4%	59,0%
BB	5,11	95,1%	95,1%	19,9	95,1%	95,1%
B	111,1	100,0%	98,4%	111,1	100,0%	98,4%

Once the radial width is determined, networks are trained on training and validation sets while the test set is used to assess the generalisation ability of the final solution. As can be seen in Table 4, classification accuracy for the *expert* training is better, or at least equal, than for the *blind* training. Since the only difference is the use of the expert landmarks during the input conditioning, it seems that the use of this kind of information during training can be useful. However, the initial problem was not to make six classifications independently, but just one. Moreover, each one of the six RBF networks trained say if a pattern is or is not of the class for which the network has been trained. Thus, the networks output can be: *Yes, it is* or *No, it is not.* Unfortunately, if we combine the classification we get from each one of the six networks, the answer is not necessary one of the six classes, but could be more than one or even none of them. This means that each input pattern could be correctly or incorrectly classified or even not classified. Table 5 collects this triple classification for the test set and, as can be seen, the *expert* training is again better than the *blind* training. The number of patterns correctly classified is almost a 40% better for the *expert*. At the same time, *expert* training has a lower indetermination in the classification (41,0% in front of the 47,5% of the *blind* training) and the same could be said for the number of patterns incorrectly classified (29,5% for 31,2%).

Table 5. Final classification for test data set

	Blind training		*Expert* training	
Correctly classified	13	21,3%	18	29,5%
Incorrectly classified	19	31,2%	18	29,5%
Not classified	29	47,5%	25	41,0%

6 Conclusion and Future Work

This paper presents an on-going work, which provides strategies for synthesising qualitative information from variables, each one of which is qualitatively described in a different way. It has been proved that using the expert information we enhance the network generalisation.

The system is applied in the financial domain to evaluate and simulate credit risk. But this approach may also be applicable to problems in other areas where the involved variables are described in terms of orders of magnitude. The limitations of the method presented cannot be evaluated until the implementation is completed and sufficiently tested. The proposed method is currently being implemented to be applied to available data referring to the most important American and European firms, whose Moody's rating is known.

Some of the future tasks consist in using the landmarks given by the experts to codify the input variables to use orders of magnitude labels and using the experts' landmarks to define a qualitative distance in order to build qualitative Gaussian density functions. Discovering alternative methods for building a homogenised reference that takes advantage of expert's knowledge. It is also intended to compare the obtained results with the results furnished by other classifiers used in artificial intelligence.

References

1. Agell, N., Ansotegui, C., Prats F., Rovira, X., and Sánchez, M.: Homogenising References in Orders of Magnitude Spaces: An Application to Credit Risk Prediction. 14th International Workshop on Qualitative Reasoning. Morelia, Mexico (2000)
2. Piera, N. Current Trends in Qualitative Reasoning and Applications, Monografía CIMNE, 33. International Centre for Numerical Methods in Engineering, Barcelona (1995)
3. Goonatilake, S and Treleaven, Ph.: Intelligent Systems for Finance and Business, John Wiley & Sons (1996)
4. Poggio, T. and Girosi, F.: Networks for Approximation and Learning, Proc. IEEE, vol. 78, pp. 1481-1497 (1990)
5. Broomhead, D.S. and Lowe, D.: Multivariable Functional Interpolation and Adaptive Network, Complex Systems, Vol. 2, pp. 321-355 (1988)
6. Chen, S., Cowan, C.F.N. and Grant, P.M.: Orthogonal Least Squares Learning for Radial Basis Function Networks. Transactions on Neural Networks, Vol. 2, pp. 302-309 (1991)
7. Prechelt, L.: PROBEN1: Set of Neural Network Benchmark Problems and Benchmarking Rules. Technical Report 21/94, University of Karlsruhe (1994)

Approximation of Bayesian Discriminant Function by Neural Networks in Terms of Kullback-Leibler Information

Yoshifusa Ito[1] and Cidambi Srinivasan[2]

[1] Department of information and policy studies,
Aich-Gakuin University,
Iwasaki, Nisshin-shi, Aichi-ken 470-0195, Japan
ito@psis.aichi-gakuin.ac.jp

[2] Department of statistics, Patterson Office Tower,
University of Kentucky,
Lexington, Kentucky 40506, USA
srini@ms.uky.edu

Abstract. Following general arguments on approximation Bayesian discriminant functions by neural networks, rigorously proved is that a three layered neural network, having rather a small number of hidden layer units, can approximate the Bayesian discriminant function for the two category classification if the log ratio of the a posteriori probability is a polynomial. The accuracy of approximation is measured by the Kullback-Leibler information. An extension to the multi-category case is also discussed.

1 Introduction

We treat the problem of pairwise approximation of the a posterior probabilities $P(\omega_i|x)$ by the outputs $F_W(\omega_i|x)$ of a three-layered neural network having d linear units on the input layer and c output layer units, where x is the feature, W is the weight vector, d is the dimension of the feature space $\mathbf{R}^d$ and c is the number of categories $\omega_1, ..., \omega_c$. It is known that the a posteriori probabilities can be used as discriminant functions in the Bayesian decision theory, and the L^2-norm and the Kullback-Leibler information have been two main measures to evaluate the accuracy of the approximation [2], [3], [5], [6], [7], [8]. In [8], the square mean $E_a(W)$ of the difference of the network output and the desired output is decomposed: $E_a(W) = \Sigma_i \int [F_W(\omega_i|x) - P(\omega_i|x)]^2 p(x)dx + \Sigma_i \int P(\omega_i|x)(1 - P(\omega_i|x))p(x)dx$, where $p(x)$ is the p.d.f of the feature. Since the second term is independent of W, minimizing $E_a(W)$ implies the respective convergences of the outputs $F_W(\omega_i|x)$ toward the a posteriori probabilities $P(\omega_i|x)$. This decomposition is used in [2] and [5]. However, as pointed out in [3], there is a disadvantage of using the L^2-norm in learning.

In this paper, we use a cross entropy $E(W) = -\int \Sigma_i p(x, \omega_i) \log F_W(\omega_i|x)dx = K(p||f_W) + I(p)$ decomposed similarly to $E_a(W)$, where $p(x, \omega_i)$ is the prob-

G. Dorffner, H. Bischof, and K. Hornik (Eds.): ICANN 2001, LNCS 2130, pp. 135–140, 2001.

ability density on $\mathbf{R}^d \times \Omega$, $\Omega = \{\omega_1, ..., \omega_c\}$, and the first term is the Kullback-Leibler information. The meaning of this information is detailed in [9] and merits of using it as a cost function is discussed in [3]. In Section 3, we treat the two-category case and prove rigorously that a three layered neural network, having a small number of hidden layer units, can approximate the a posteriori distributions in the sense of the Kullback-Leibler information if the log ratio of the a posteriori distributions is a polynomial of low degree. Extension of the result to the multi-category case is discussed in Section 4.

Since there is a restriction $\Sigma_{i=1}^{c} F_W(\omega_i|x) = 1$, one of them is a linear function of others. We use nonlinear units for the output and hidden layers. Hence, the outputs are

$$F_W(\omega_i|x) = \begin{cases} \varphi(\sum_{j=1}^{N} a_{ij}\phi(w_j \cdot x + t_j) + a_{i0}), & i = 1, ..., c-1, \\ 1 - \sum_{i=1}^{c-1} F_W(\omega_i|x), & i = c. \end{cases}$$

Unless $F_W(\omega_i|x)$ as well as $P(\omega_i|x)$ are positive on the support of $p(x)$, the arguments in this paper are all meaningless. We will show later that this assumption cannot be violated while learning.

2 Bayesian Decision Theory and Neural Networks

Note that $f_W(x, \omega) = F_W(\omega|x)p(x)$ is also a probability distribution on $\mathbf{R}^d \times \Omega$. The accuracy of approximation of a distribution $p(x, \omega)$ by $f_W(x, \omega)$ can be measured with the Kullback-Leibler information:

$$K(p||f_W) = \int_{\mathbf{R}^d} \sum_{i=1}^{c} p(x, \omega_i) \log \frac{p(x, \omega_i)}{f_W(x, \omega_i)} dx$$

$$= \int_{\mathbf{R}^d} p(x) \sum_{i=1}^{c} P(\omega_i|x) \log \frac{P(\omega_i|x)}{F_W(\omega_i|x)} dx. \tag{1}$$

To compare category-wise the accuracy of the approximation of $P(\omega_i|x)$ by $F_W(\omega_i|x)$ based on the kullback-Leibler information with that based on the L^2-distance, we define

$$K_m(P(\omega_i|\cdot)||F_W(\omega_i|\cdot))$$

$$= \min \left\{ \int_{\mathbf{R}^d} p(x) \sum_{j=1}^{c} P(\omega_j|x) \log \frac{P(\omega_j|x)}{F_W(\omega_j|x)} dx | F_W(\omega_i|x) \text{ is fixed} \right\}$$

$$= \int_{\mathbf{R}^d} \left\{ P(\omega_i|x) \log \frac{P(\omega_i|x)}{F_W(\omega_i|x)} + (1 - P(\omega_i|x)) \log \frac{1 - P(\omega_i|x)}{1 - F_W(\omega_i|x)} \right\} p(x) dx. \tag{2}$$

In the case of two category classification, this coincides with $K(p||f_W)$. We can prove that K_m bounds the L^2-norm:

$$||P(\omega_i|\cdot) - F_W(\omega_i|\cdot)||_{L^2(\mathbf{R}^d, p)} \leq \frac{1}{2} K_m(P(\omega_i|\cdot)||F_W(\omega_i|\cdot)), \quad i = 1, ...c. \tag{3}$$

Here, we omit the proofs of (2) and (3).

Our cost function is the cross entropy:

$$E(W) = -\int_{\mathbf{R}^d} p(x) \sum_{i=1}^{r} p(\omega_i|x) \log F_W(\omega_i|x) dx, \tag{4}$$

which can be asymptotically approximated by

$$E^{(n)}(W) = -\frac{1}{n} \sum_{j=1}^{n} \log F_W(\omega^{(j)}|x^{(j)}), \tag{5}$$

where $(x^{(j)}, \omega^{(j)})_{j=1}^{\infty}$ is a sequence of independent teacher signals with distribution $p(x, \omega)$. If the p.d.f. $p(x)$ is rapidly decreasing and $\log F_W(\omega_i|x_i)$ are bounded by a polynomial as in the case of most familiar distributions, $E^{(n)}(W)$ converges to $E(W)$ almost everywhere by the strong law of large number. While learning, a new signal $(x^{(j)}, \omega^{(j)})$ must be incorporated and simultaneously the sum (5) must be subdued to the gradient descent. Since $F_W(\omega_i|x) \downarrow 0$ implies $E(W) \uparrow \infty$, the condition $F_W(\omega_i|x) > 0$ cannot be violated while the gradient descent method is applied.

We can decompose $E(W)$:

$$E(W) = K(p||f_W) + I(p), \tag{6}$$

where

$$I(p) = \int_{\mathbf{R}^d} p(x) \sum_{i=1}^{c} P(\omega_i|x) \log P(\omega_i|x) dx. \tag{7}$$

This corresponds to the decomposition of $E_a(W)$ by Ruck et al. [8]. The second term $I(p)$ is nothing but the mean negative entropy of the a posteriori probabilities. Since this term is independent of learning, minimization of $E(W)$ implies that of the first term.

3 Two Category Classification

A theorem below is proved in a more general form in [4]. The proof of this simplified theorem for $n = 2$ is described and used in [5].

Theorem 1. *Let $n \geq 1$, let $1 \leq p \leq \infty$, let μ be a measure on $\mathbf{R}^d$ such that $|x|^n \in L^p(\mathbf{R}^d, \mu)$ and let ϕ be an n times continuously differentiable function such that $\phi^{(i)}$, $i = 0, \cdots, n$, are bounded and $\phi^{(n)} \not\equiv 0$. Then, for any polynomial Q of degree n and any $\varepsilon > 0$, there are constants a_i, $i = 0, \cdots, N$, t_i and vectors V_i in $\mathbf{R}^d$, $i = 1, \cdots, N$, for which*

$$\bar{Q}(x) = \sum_{i=1}^{N} a_i \phi(V_i \cdot x + t_i) + a_0 \tag{8}$$

satisfies $\|\bar{Q} - Q\|_{L^p(\mathbf{R}^d, \mu)} < \varepsilon$.

In particular, if $n = 1$, $N = 1$ and, if $n = 2$, $N = d + 1$.

The N can be rather a small number. Funahashi estimated N to be $2d$ for $n = 2$, consuming two $\phi's$ to approximate t^2 [2]. Actually it can approximated by a linear sum of a single ϕ, t and a constant, and $N = d + 1$ is obtained [5].

In the case of two-category classification, the discriminant function can be $g(x) = \log P(\omega_1|x)$ - $\log P(\omega_2|x)$ and any monotone function of $g(x)$ can also be a decision function [1]. If the logistic function $\sigma(t) = (1 + e^{-t})^{-1}$ is used as the monotone function, $\sigma(g(x)) = P(\omega_1|x)$ [2].

In Section 2, we have stated that the K_m bounds the L^2-distance. Here, we prove that the L^1-distance of the log likelihood ratio bounds the K_m.

Proposition 2. *Let $p(x, \omega)$ and $q(x, \omega)$ be p.d.f's of mutually continuous probability measures defined on $\mathbf{R}^d \times \{\omega_1, ..., \omega_c\}$ such that $P(\omega_i|x) \neq 0$, $Q(\omega_i|x) \neq 0$ and $p(x) = q(x)$. Then, for each i, we have that*

$$K_m(P(\omega_i|\cdot)||Q(\omega_i|\cdot) \leq \| \log \frac{P(\omega_i|\cdot)}{1 - P(\omega_i|\cdot)} - \log \frac{Q(\omega_i|\cdot)}{1 - Q(\omega_i|\cdot)} \|_{L^1(\mathbf{R}^d, p)}. \tag{9}$$

Proof. Set

$$\delta g(x) = \log \frac{P(\omega_i|x)}{1 - P(\omega_i|x)} - \log \frac{Q(\omega_i|x)}{1 - Q(\omega_i|x)}.$$

Then,

$$Q(\omega_i|x) = \frac{P(\omega_i|x)}{P(\omega_i|x) + (1 - P(\omega_i|x))e^{\delta g(x)}}.$$

Accordingly,

$$K_m(P(\omega_i|\cdot)||Q(\omega_i|\cdot)) = \int_{\mathbf{R}^d} p(x)P(\omega_i|x) \log(P(\omega_i|x) + (1 - P(\omega_i|x))e^{\delta g(x)})dx$$

$$+ \int_{\mathbf{R}^d} p(x)(1 - P(\omega_i|x)) \log(1 - P(\omega_i|x) + P(\omega_i|x)e^{-\delta g(x)})dx.$$

We have that, if $\delta g(x) > 0$, $0 < \log(P(\omega_i|x) + (1 - P(\omega_i|x))e^{\delta g(x)}) < \delta g(x)$, and if $\delta g(x) < 0$, $0 > \log(P(\omega_i|x) + (1 - P(\omega_i|x))e^{\delta g(x)}) > \delta g(x)$. With similar inequalities for the second integral, we have that

$$K_m(P(\omega_i|\cdot)||Q(\omega_i|\cdot)) \leq \int_{\mathbf{R}^d} |\delta g(x)|p(x)dx = \|\delta g\|_{L^1(\mathbf{R}^p, p)}. \tag{10}$$

Now we are ready for proving the main theorem below.

Theorem 3. *Let $p(x, \omega)$ be the p.d.f of a probability distribution on $\mathbf{R}^d \times \{\omega_1, \omega_2\}$ such that $\log p(x, \omega_1)/p(x, \omega_2)$ is a polynomial of degree n on the support of $p(x) = p(x, \omega_1) + p(x, \omega_2)$, and let ϕ be an n times continuously differentiable function such that $\phi^{(i)}$, $i = 0, \cdots, n$, are bounded and $\phi^{(n)}(0) \not\equiv 0$. Suppose that $|x|^n \in L^1(\mathbf{R}^d, p)$. Then, for any $\varepsilon > 0$, there are constants a_i, $i = 0, \cdots, N$, t_i and vectors V_i in $\mathbf{R}^d$, $i = 1, \cdots, N$, for which*

$$F_W(\omega_1|x) = \sigma(\sum_{i=1}^{N} a_i\phi(V_i \cdot x + t_i) + a_0), \tag{11}$$

satisfies

$$K_m(P(\omega_1|\cdot)||F_W(\omega_1|\cdot)) < \varepsilon, \tag{12}$$

where σ is the logistic function.

In particular, if $n = 1$, $N = 1$ and, if $n = 2$, $N = d + 1$.

Proof. Set $Q(x) = \log P(\omega_1|x)/P(\omega_2|x)$. Then, by assumption and Theorem 1, there are constants a_i, t_i and vectors V_i in $\mathbf{R}^d$, $i = 1, \cdots, N$, for which

$$\|\bar{Q} - Q\|_{L^1(\mathbf{R}^d,p)} < \varepsilon, \tag{13}$$

where $\bar{Q}(x)$ is defined by (9). Set $F_W(\omega_1|x) = \sigma(\bar{Q}(x))$, then $\bar{Q}(x) = \log F_W(\omega_1|x)/F_W(\omega_2|x)$. By Proposition 2,

$$K_m(P(\omega_1|\cdot)||F_W(\omega_1|\cdot)) \leq \| \log \frac{P(\omega_1|\cdot)}{P(\omega_2|\cdot)} - \log \frac{F_W(\omega_1|\cdot)}{F_W(\omega_2|\cdot)} \|_{L^1(\mathbf{R}^d,p)} \tag{14}$$

holds. Combining (13) and (14), we obtain (12).

This theorem implies that the a posteriori probability can be approximated by the output of a three layered neural network having an output unit with activation function σ and rather a small number of hidden layer units, in the sense of the Kullback-Leibler information.

4 Multi-category Case and Discussions

The log ratio of the a posteriori probabilities is the log likelihood ratio biased by a constant, and the log likelihood ratios of many probability distributions of the exponential family are polynomials of low degrees. In the case of the binomial, polynomial or gamma distribution, the ratio is a linear function in x and, in the case of the normal distribution, it is a quadratic form. Consequently, if the probability distribution is one of them and the neural network has at least $d + 1$ hidden layer units, its output may approximate the a posteriori probability in the case of two category classification. It is unnecessary to teach the network the type of the probability distribution beforehand. Contrarily, we may obtain information about the probability distribution, observing the output after training.

For simplicity, we now restrict discussions to the case $n = 2$. Let $N = \frac{1}{2}d(d+1)$. There are N kinds of quadratic monomials and the same number of linearly independent squares $(V_i \cdot x)^2$. Hence, any number of nonhomogeneous polynomial of degree 2 in x can be expressed as a sum

$$\sum_{i=1}^{N} a_i(V_i \cdot x)^2 + \sum_{i=1}^{d} b_i x_i + c$$

simultaneously by adjusting only a_i, b_i, c, if V_i are appropriately chosen beforehand. A linear sum $\alpha_i\varphi(V_i \cdot x) + \Sigma_{j=1}^d \alpha_{ij}x_j + \alpha_{i0}$ can approximate $(V_i \cdot x)^2$. Hence, a neural network having d linear input units, $N + d = \frac{1}{2}d(d+3)$ hidden layer units and c linear output units can approximate c nonhomogeneous polynomials of degree 2 simultaneously, if the connection weights between the input and hidden layers are fixed appropriately beforehand. Hence, only adjustment of the weights between the hidden and output layers is necessary, which may be advantageous. In the case where the state-conditional p.d.fs $p(x|\omega_i)$ are normal distributions, pairwise decision functions $g_{ij}(x) = g_i(x) - g_j(x)$ are quadratic forms and $\sigma(g_{ij}(x)) = P(\omega_i|x)$ (see Funahashi (1998)). Consequently, if the output units are replaced by those having σ as activation function, the outputs approximate $P(\omega_i|x)$. One way to estmate $P(\omega_i|x)$ may be to take an average of $\sigma(\bar{g}_{ij}(x))$, $j = 1, \cdots, c$, $j \neq i$, where $\bar{g}_{ij}$ are approximations of g_{ij} by the network respectively. Once we can approximate $P(\omega_i|x)$, $i = 1, \cdots, c$, it is not difficult to decide the category ω_i whose a posterior probability is the maximum.

In the Kullback-Leibler information (1), the difference between $F_W(\omega_i|\cdot)$ and $P(\omega_i|\cdot)$ over $\mathbf{R}^d \times \Omega$ is measured with the homogeneous weights with respect to ω_i. However, there may be a case where nonhomogeneous weight is preferable in applications. Then, a weighted average

$$\sum_{i=1}^{c} a_i K_m(P(\omega_i|\cdot)||F_W(\omega_i|\cdot))$$

can be used to estimate the difference.

References

1. Duda, R.O. and Hart, P.E.: Pattern classification and scene analysis. John Wiley & Sons, New York (1973)
2. Funahashi, K.: Multilayer neural networks and Bayes decision theory. Neural Networks **11** (1998) 209-213
3. Hinton, G.E.: Connectionist learning procedures. Artificial intelligence **40** (1989) 185-234
4. Ito, Y.: Simultaneous L^p-approximations of polynomials and derivatives on the whole space. Proceedings of ICANN99 (1999) 587-592
5. Ito, Y., Srinivasan, C.: Bayesian decision theory and three layered neural networks, Proceedings of ESANN2001 (2001) 377-382
6. Richard, M.D. and Lippmann, R.P.: Neural network classifiers estimate Bayes- ian a posteriori probabilities. Neural Compt. **3** (1991) 461-483
7. Ripley, B.D.:Statistical aspect of neural networks. Networks and chaos - Statistical and Probabilistic Aspects. ed. Barndorff-Nielsen, O.E., Jensen, J.L., Kendall, W.S., Chapman & Hall. London (1993) 40-123
8. Ruck,M.D., Rogers, S., Kabrisky, M., Oxley, H., Sutter, B.: The multilayer perceptron as an approximator to a Bayes optimal discriminant function. IEEE Transactions on Neural Networks. **1** (1990) 296-298
9. M.J. Schervish, Theory of statistics. Springer-Verlag, Berlin, New York (1995)

The Bias-Variance Dilemma of the Monte Carlo Method

Zlochin Mark and Yoram Baram

Technion - Israel Institute of Technology, Technion City, Haifa 32000, Israel
{zmark,baram}@cs.technion.ac.il

Abstract. We investigate the setting in which Monte Carlo methods are used and draw a parallel to the formal setting of statistical inference. In particular, we find that Monte Carlo approximation gives rise to a bias-variance dilemma. We show that it is possible to construct a biased approximation scheme with a lower approximation error than a related unbiased algorithm.

1 Introduction

Markov Chain Monte Carlo methods have been gaining popularity in recent years. Their growing spectrum of applications ranges from molecular and quantum physics to optimization and learning. The main idea of this approach is to approximate the desired target distribution by an empirical distribution of a sample generated through the simulation of an ergodic Markov Chain. The common practice is to construct a Markov Chain in such a way as to insure that the target distribution is invariant. Much effort has been devoted to the development of general methods for the construction of such Markov chains. However, the fact that approximation accuracy is often lost when invariance is imposed has been largely overlooked. In this paper we make explicit the formal setting of the approximation problem, as well as the required properties of the approximation algorithm. By analogy to statistical inference, we observe that the desired property of the algorithms is a good rate of convergence, rather than unbiasedness. We demonstrate this point by numerical examples.

2 Monte Carlo Method

2.1 The Basic Idea

In various fields we encounter the problem of finding the expectation:

$$\hat{a} = E[a] = \int a(\theta)Q(\theta)d\theta \tag{1}$$

where $a(\theta)$ is some parameter-dependent quantity of interest and $Q(\theta)$ is the parameter distribution (e.g. in Bayesian learning $a(\theta)$ can be the vector of output values for the test cases and $Q(\theta)$ is the posterior distribution given the data).

G. Dorffner, H. Bischof, and K. Hornik (Eds.): ICANN 2001, LNCS 2130, pp. 141–147, 2001.

Since an exact calculation is often infeasible, an approximation is made. The basic Monte Carlo estimate of $\hat{a}$ is:

$$\bar{a}_n = \frac{1}{n}\sum_{i=1}^{n} a(\theta_i) \qquad (2)$$

where θ_i are independent and distributed according to $Q(\theta_i)$. In the case where sampling from $Q(\theta)$ is impossible, an *importance sampling* [2] can be used:

$$\bar{a}_n = \frac{1}{n}\sum_{i=1}^{n} a(\theta_i)w(\theta_i) \qquad (3)$$

where θ_i are distributed according to some $P_0(\theta_i)$ and $w(\theta_i) = \frac{Q(\theta_i)}{P_0(\theta_i)}$.

In many problems, especially high dimensional ones, independent sampling cannot provide accurate estimates in reasonable time. An alternative is the Markov Chain Monte Carlo (MCMC) approach, where the same estimate as in (2) is used, but the sampled values $\theta^{(i)}$ are not drawn independently. Instead, they are generated by a homogeneous ergodic distribution Q [5].

2.2 Approximation Error and the Bias-Variance Dilemma

Let us consider a yet more general approximation scheme where the invariant distribution of the Markov Chain is not Q but some other distribution Q'. Moreover, we may consider non-homogeneous Markov Chains with the transition probabilities $T_t(\theta^{(t+1)} \mid \theta^{(t)})$ depending on t.

Since $\bar{a}_n$ is a random variable, its *bias, equilibrium bias and variance* are defined as follows:

$$\begin{aligned} Bias_n &= E\{\bar{a}_n\} - \hat{a} \\ Bias^{eq} &= \lim_{n\to\infty} Bias_n \\ Var_n &= E\{\bar{a}_n - E\{\bar{a}_n\}\}^2 \end{aligned}$$

where E denotes expectation with respect to the distribution of $\bar{a}_n$.

In addition, we may define the *initialization bias*, $Bias_n^{init} = Bias_n - Bias^{eq}$, as a measure of how far the Markov Chain is from equilibrium.

The common practice is to try to construct a homogeneous (i.e. $T(\cdot \mid \cdot)$ independent of time) reversible Markov Chain with $Q(\theta)$ as invariant distribution, hence with the equilibrium bias equal to zero.

However the quality of the approximation is measured not by the bias, but by an average squared estimation error:

$$Err_n = E[\bar{a}_n - \hat{a}]^2 = Bias_n^2 + Var_n = (Bias^{eq} + Bias_n^{init})^2 + Var_n \qquad (4)$$

From the last equation it can be seen that the average estimation error has three components. Therefore, there is a possibility (at least potentially) to reduce the total approximation error by balancing those three. Moreover it shows that, since the initialization bias and the variance depend on the number of iterations, the algorithm may depend on the number of iterations as well.

Batch Versus Online Estimation. A similar trade-off between the components of the error appears in statistical inference, where it is known as "the bias-variance dilemma". The analogy to the statistical inference setting can be further extended by making a distinction between two models of MCMC estimation:

Online estimation - the computation time is potentially unlimited and we are interested in as good an approximation as possible at any point in time or, alternatively, as rapid a rate of convergence of Err_n to zero as possible .

Batch estimation - the allowed computation time is limited and known in advance and the objective is to design an algorithm with as low an average estimation error as possible.

Note that the batch estimation paradigm is more realistic than the online estimation, since, in practice, the time complexity is of primary importance. In addition, in many instances of Bayesian learning, the prior distribution is not known but is chosen on some *ad-hoc* basis and the data is usually noisy. Therefore, it makes little sense to look for a time-expensive high accuracy approximation to the posterior. Instead, a cheaper rough approximation, capturing the essential characteristics of the posterior, is needed.

The traditional MCMC method overlooks the distinction between the two models. In both cases, the approximation is obtained using a homogeneous (and, usually, reversible) Markov Chain with the desired invariant distribution. It should be clear, however, that such an approach is far from optimal. In the online estimation model, a more reasonable alternative would be considering a non-homogeneous Markov Chain that rapidly converges to a rough approximation of Q and whose transition probabilities are modified with time so as to insure the asymptotic invariance of Q (hence consistency). A general method for designing such Markov Chains is described in the next subsection.

In the batch estimation model the invariance of Q may be sacrificed (i.e. equilibrium bias may be non-zero) if this facilitates a lower variance and/or initialization bias for the finite computation time.

Mixture Markov Chains. Suppose that we are given two Markov Chains - the first, M_1, is unbiased, while the second, M_2, is biased, but more rapidly mixing. Let their transition probabilities be T_t^1 and T_t^2 respectively. It will usually be the case that for small sample sizes, M_2 will produce better estimates, but as n grows, the influence of the bias becomes dominant, making M_1 preferable.

In order to overcome this difficulty, we may define a *mixture Markov chain* with mixing probabilities p_t, for which the transition is made according to T_t^1 with probability $1 - p_t$ and according to T_t^2 with probability p_t. If the sequence $\{p_t\}$ starts with $p_1 = 1$ and decreases to zero, then for small t the resulting Markov chain behaves similarly to M_2, hence producing better estimates than M_1, but as t grows, its behavior approaches that of M_1 (in particular, the resulting estimate is asymptotically unbiased).

Clearly, if both Markov Chains are degenerate (i.e. produce IID states), it can be easily seen that the bias behaves asymptotically as

$$Bias_n = O(\frac{1}{n}\sum_{i=1}^{n} p_i) \tag{5}$$

and it can be shown that (5) also holds for any *uniformly ergodic* Markov Chains.

In order to balance between the variance, which typically decreases as $O(\frac{1}{n})$, and the bias, the sequence $\{p_i\}$ should be chosen so that $Bias_n = O(\frac{1}{\sqrt{n}})$, i.e. $\sum_{i=1}^{n} p_i = O(\sqrt{n})$.

This mixture Markov Chain approach allows to design efficient sampling algorithms both in online and in batch settings (in the latter case the sequence $\{p_i\}$ may depend on the sample size, n).

3 Examples

In this section we present two approaches for designing approximation algorithms with a small controlled bias. Section 3.1 describes a post-processing scheme based on the smoothing techniques, which reduces the variability of the importance weights in the Annealed Importance sampling algorithm [6]. Section 3.2 compares the well-known Hybrid Monte Carlo algorithm with a simple biased modification and a mixture Markov Chain. In both cases, the empirical comparison shows superiority of the biased algorithms.

3.1 Independent Sampling

Annealed Importance Sampling. In the Annealed Importance sampling algorithm [6], the sample points are generated using some variant of simulated annealing and then the importance weights, $w^{(i)}$, are calculated in a way that insures asymptotic unbiasedness of the weighted average.

The annealed importance sampling (AIS) estimate of $E_Q[a]$ is:

$$\bar{a}_n = \sum_{i=1}^{n} a(x^{(i)}) w^{(i)} \; / \; \sum_{i=1}^{n} w^{(i)} \tag{6}$$

Smoothed Importance Sampling. While implicitly concentrating on the regions of high probability, AIS estimate can still have a rather high variance, since its importance weights are random and their dispersion can be large.

Let us observe that $E_Q[a]$ may be written as

$$E_Q[a] = \frac{E_P[aw]}{E_P[w]} = E_{P(a)} \left\{ a \frac{E_{P(w|a)}[w]}{E_P[w]} \right\} \tag{7}$$

where $a = a(x)$ and P is the distribution defined by the annealing algorithm. The estimate (6) is simply a finite sample approximation to the first part of (7).

Next, the second part of (7) suggests using an estimate:

$$\hat{a}_n = \sum_{i=1}^{n} a(x^{(i)}) \hat{w}(a(x^{(i)})) \tag{8}$$

where $\hat{w}(a)$ is an estimate of $\frac{E_{P(w|a)}[w]}{E_P[w]}$. on the value of a). It can be shown that if $\hat{w}(a)$ was known exactly, than (8) would have a lower variance than (6). In

Table 1. Average estimation error of posterior output prediction for different sample sizes. The results for sample size n are based on $10^4/n$ trials.

n	Annealed IS	Smoothed IS
10	3.04	1.09
100	0.694	0.313
1000	0.0923	0.0594

practice, a good estimate of $\hat{w}(a)$ can be obtained using some data smoothing algorithm such as kernel smoothing [3]. The resulting method henceforth is referred to as *Smoothed Importance sampling.* The usage of smoothing introduces bias (which can be made arbitrary small by decreasing the smoothing kernel width to zero as $n \to \infty$), but in return can lead to a significant reduction of the variance, hence, a smaller estimation error than (6), as demonstrated by the following numerical experiment.

A Bayesian Linear Regression Problem. The comparison was carried out using a simple Bayesian learning problem from [6]. The data for this problem consisted of 100 independent cases, each having 10 real-valued predictor variables, $x_1, \ldots, x_{10}$, and a real-valued response variable, y, which is modeled by $y = \sum_{k=1}^{10} \beta_k x_k + \epsilon$, where ϵ is zero-mean Gaussian noise with unknown variance σ^2. The detailed description of the data-generation model can be found in [6].

The two methods were used to estimate the posterior prediction for a test set of size 100. The "correct" posterior prediction were estimated using Annealed Importance sampling with sample size 10000.

As can be seen in Table 1, the average estimation error of Smoothed Importance sampling was uniformly lower than that of Annealed Importance sampling. It should also be noted that the modeling error, i.e. the squared distance between the Bayesian prediction and the correct test values, was 5.08, which is of the same order magnitude as the estimation error for $n = 10$. This confirms our claim that, in the context of Bayesian learning, high accuracy approximations are not needed.

3.2 MCMC Sampling

HMC Algorithm. The Hybrid Monte Carlo (HMC) algorithm [1] is one of the state-of-the-art asymptotically unbiased MCMC algorithms for sampling from complex distributions. The algorithm is expressed in terms of sampling from a *canonical* distribution defined in terms of the energy function $E(q)$:

$$P(q) \propto \exp(-E(q)) \tag{9}$$

To allow the use of dynamical methods, a "momentum" variable, p, is introduced, with the same dimensionality as q. The canonical distribution over the "phase space" is defined to be:

$$P(q,p) \propto \exp(-H(q,p)) = \exp(-E(q) - \sum_{i=1}^{n} \frac{p_i^2}{2}) \tag{10}$$

where $p_i, i = 1, \ldots, n$ are the momentum components.

Sampling from the canonical distribution can be done using the *stochastic dynamics* method, in which simulation of the *Hamiltonian dynamics* of the system using *leapfrog* discretization is alternated with Gibbs sampling of the momentum.

In the hybrid Monte Carlo method, the bias, introduced by the discretization, is eliminated by applying the Metropolis algorithm to the candidate states, generated by the stochastic dynamic transitions. The candidate state is accepted with probability $\max(1, \exp(\Delta H))$, where ΔH is the difference between the Hamiltonian in the beginning and and the end of the trajectory [5].

A modification of the Gibbs sampling of the momentum, proposed in [4], is to replace p each time by $p \cdot \cos(\theta) + \zeta \cdot \sin(\theta)$, where θ is a small angle and ζ is distributed according to $N(0, I)$. While keeping the canonical distribution invariant, this scheme, called *momentum persistence*, allows the use of shorter (hence cheaper) trajectories. However, in order to insure invariance of the target distribution, the momentum has to be reversed in case the candidate state has been rejected by the Metropolis step. This causes occasional backtracking and slows the sampling down.

Stabilized Stochastic Dynamics. As an alternative to the HMC algorithm we consider stochastic dynamics with momentum persistence, but without the Metropolis step. In order to insure stability of the algorithm, if the momentum size grows beyond a certain percentile of its distribution (say, 99%), it is replaced using Gibbs sampling.

The resulting Stabilized Stochastic Dynamics (SSD) algorithm is biased. However, since it avoids backtracking, it can produce a lower estimation error, as found in experiments described next.

Bayesian Neural Networks. We compared the performances of SSD, HMC and mixture Markov Chain with

$$p_t^n = \begin{cases} 1, & \text{if } t \leq \sqrt{200n} \\ 0, & \text{otherwise} \end{cases} \tag{11}$$

on the following Bayesian learning problem. The data consisted of 30 input-output pairs, generated by the following model:

$$y_i = \frac{\sin(2.5|x_i|)}{2.5|x_i|} + \epsilon_i, i = 1, \ldots, 20 \tag{12}$$

where x_i are independently and uniformly generated from $[-\pi, \pi]$ and ϵ_i are Gaussian zero-mean noise variables with standard deviation 0.1.

We have used multilayer perceptron with one input node, one hidden layer containing 5 tanh nodes and one linear output node. The prior for each weight was taken to be zero-mean Gaussian with inverse variance 0.1.

The comparison was carried out using two test statistics: average log-posterior and average prediction for 100 equidistant points in $[-\pi, \pi]$. Since the correct values of those statistics are not known, they were estimated using a sample of

Table 2. Average estimation error of mean log-posterior and mean output prediction for different sample sizes.

	Log-posterior			Output prediction		
n	SSD	HMC	MIX	SSD	HMC	MIX
1000	3.55	8.52	5.51	0.053	0.102	0.061
3000	0.75	1.34	0.85	0.015	0.031	0.024
10000	0.23	0.43	0.37	0.0049	0.0089	0073

size 10^6 generated by HMC. For each algorithm we performed 100 runs with initial state generated from the prior.

As can be seen from the results in Table 2, both SSD and mixture Markov Chain produced results which are uniformly better than those of HMC. It may be hypothesized that, for very large sample sizes, HMC will become superior to SSD, because of the influence of the bias. However, the mixture Markov Chain for these large sample sizes is expected to produce results which are very similar to HMC. More importantly, the large sample behavior of the algorithms is not very relevant, as the modeling error, i.e. the squared distance between the Bayesian prediction and the correct test values, in this experiment was 0.187, which is larger than the estimation error for $n = 1000$, meaning that, once again, there is no use in making a high accuracy large sample approximation.

4 Conclusion

We have shown that in Monte Carlo estimation there is a "bias-variance dilemma", which has been largely overlooked. By means of numerical examples, we have demonstrated that bias correcting procedures can, in fact, increase the estimation error. As an alternative, approximate (possibly asymptotically biased) estimation algorithms, with lower variance and convergence bias should be considered. Once the unbiasedness requirement is removed, a whole range of new possibilities for designing sampling algorithms opens up, such as automatic discretization step selection, different annealing schedules, alternative discretization schemes etc.

References

1. S. Duane, A. D. Kennedy, B. J. Pendleton, and D. Roweth. Hybrid Monte Carlo. *Physics Letters B*, 195:216–222, 1987.
2. W. R. Gilks, S. Richardson, and D. J. Spiegelhalter, editors. *Markov Chain Monte Carlo in Practice*. Chapman&Hall, 1996.
3. W. Hardle. *Applied nonparametric regression*. Cambridge University Press, 1990.
4. A. M. Horowitz. A generalized guided Monte Carlo algorithm. *Physics Letters B*, 268:247–252, 1991.
5. R. M. Neal. *Bayesian Learning for Neural Networks*. Springer-Verlag, 1996.
6. R. M. Neal. Annealed importance sampling. Technical Report No. 9805, Dept. of Statistics, University of Toronto, 1998.

A Markov Chain Monte Carlo Algorithm for the Quadratic Assignment Problem Based on Replicator Equations

Takehiro Nishiyama, Kazuo Tsuchiya, and Katsuyoshi Tsujita

Dept. of Aeronautics and Astronautics, Graduate School of Engineering,
Kyoto University, Kyoto, Japan
{nisiyama,tsuchiya,tsujita}@kuaero.kyoto-u.ac.jp

Abstract. This paper proposes an optimization algorithm for the Quadratic Assignment Problem (QAP) based on replicator equations. If the growth rate of a replicator equation is composed of the performance index and the constraints of the QAP suitably, by increasing the value of the control parameter in the growth rate, the equilibrium solutions which correspond to the feasible solutions of the QAP become stable in order from the one with smaller value of the performance index. Based on the characteristics of the system, the following optimization algorithm is constructed; the control parameter is set so that the equilibrium solutions corresponding to the feasible solutions with smaller values of the performance index become stable, and then in the solution space of the replicator equations, a Markov chain Monte Carlo algorithm is carried out. The proposed algorithm is applied to many problem instances in the QAPLIB. It is revealed that the algorithm can obtain the solutions equivalent to the best known solutions in short time. Especially, for some large scale instances, the new solutions with the same cost as the best known solutions are obtained.

1 Introduction

The Quadratic Assignment Problem (QAP) [1] is one of the hardest combinatorial optimization problems. The QAP is formulated as a problem to find a N dimensional permutation matrix which minimizes a performance index, where N is the size of the problem. Among approximation methods for the QAP, there are dynamical systems approaches; a dynamical system consisting of the elements of a $N \times N$ matrix is constructed. The mutual interactions between the elements are determined so that equilibrium solutions of the system become permutation matrices with smaller values of the performance index, i.e. approximate solutions for the QAP. In many studies, the dynamical system is constructed as a gradient system [2, 3]; a potential function is composed of the performance index and the constraints, and the dynamical system is constructed as a gradient vector field of the potential function. The system has equilibrium solutions as the minima of the potential function which correspond to the approximate solutions of the QAP. On the other hand, we have constructed the dynamical

G. Dorffner, H. Bischof, and K. Hornik (Eds.): ICANN 2001, LNCS 2130, pp. 148–155, 2001.

system as a replicator equation [4]. The replicator equation is the equation in which derivatives of the variables are proportional to the state of the variables. The proportional coefficients are called growth rates. When the growth rates are determined based on the performance index and the constraints of the QAP suitably, the system has the following characteristics; all feasible solutions of the QAP are the equilibrium solutions of the system. By increasing the value of the parameter (control parameter) in the growth rate, the feasible solutions become stable in order, from the one having the smallest value of the performance index to the largest one. This means that when the control parameter is set suitably, the dynamical system has only solutions with smaller values of the performance index as the stable solutions. In this paper, we propose the following Markov chain Monte Carlo algorithm based on the characteristics of the replicator equations; the growth rates of the replicator equations are designed based on the performance index and the constraints of the QAP. The control parameter in the growth rate is set so that the equilibrium solutions corresponding to the feasible solutions with smaller values of the performance index become stable. The replicator equations are calculated with some initial values to obtain an equilibrium solution. Then, setting the initial values in some neighborhood of the solution, the replicator equations are calculated again to obtain the next equilibrium solution. The obtained solution is accepted according to some probability. This procedure is repeated to give a sequence of solutions. The proposed algorithm is applied to some problem instances in the QAPLIB [5], and in many cases, gives solutions comparable to the best known solutions. Especially, for some large scale instances, the proposed algorithm gives new solutions having the same values of the performance index as the best known solutions. Haken et al. have proposed an optimization algorithm based on the replicator equations [6]. But in their method, the above-mentioned characteristics of the equilibrium solutions of the replicator equations are not used explicitly. On the other hand, Ishii and Niitsuma have proposed a dynamical systems approach in which the search space is restricted [7]. But their method does not utilize the characteristics that the search space is composed of the good approximate solutions of the QAP which our method utilizes to improve the performance.

2 Quadratic Assignment Problem (QAP)

The Quadratic Assignment Problem (QAP) [1] is considered one of the hardest combinatorial optimization problems. Given a set $\mathcal{N} = \{1, 2, \cdots, N\}$ and $N \times N$ matrices $A = (a_{ij}), B = (b_{kl})$, the QAP is defined as follows:

$$\min_{p \in \Pi_{\mathcal{N}}} L(p), \quad L(p) = \sum_{i,j} a_{ij} b_{p(i)p(j)}, \tag{1}$$

where $\Pi_{\mathcal{N}}$ is the set of all permutations of $\mathcal{N}$, and p is an element of it. Letting $\Pi_{N \times N}$ the set of all $N \times N$ permutation matrices and $X = (x_{ij})$ an element of it, the QAP is also represented as the following matrix form:

$$\min_{X \in \Pi_{N \times N}} L(X), \quad L(X) = \mathrm{trace}(A^T X^T B X) = \sum_{i,i',j,j'} a_{jj'} b_{ii'} x_{ij} x_{i'j'}. \tag{2}$$

A typical example of the QAP is the facility location problem; consider assigning N facilities to N locations, where a_{ij} represents the flow of materials from facility i to facility j and b_{kl} is the distance from location k to location l. The cost of assigning facility i to location k and facility j to location l is $a_{ij} b_{kl}$. The objective of the problem is to find an assignment of all facilities to all locations such that the total cost is minimized.

3 Proposed Dynamical System and Its Characteristics [4]

For the QAP, we have proposed the following replicator equation:

$$\dot{u}_{ij} = f_{ij}(u_{i'j'}, \alpha_0, \alpha_1) u_{ij}, \tag{3a}$$

$$\begin{aligned} f_{ij} = (1 - u_{ij}^2) &- \frac{\alpha_0}{2}\Big(\sum_{i' \neq i} u_{i'j}^2 + \sum_{j' \neq j} u_{ij'}^2\Big) \\ &- \frac{\alpha_1}{2} \sum_{i',j'} (a_{jj'} b_{ii'} + a_{j'j} b_{i'i}) u_{i'j'}^2 \\ &\qquad (i, j = 1, \cdots, N), \end{aligned} \tag{3b}$$

where f_{ij} is called the growth rate, and the parameters are $\alpha_0 > 0$, $0 \leq \alpha_1 \ll 1$. The first term of the growth rate f_{ij} leads each u_{ij}^2 to unity. The second term represents the competition between elements having same subscripts i (j), and the parameter α_0 determines the strength of the competition. The third term is derived from the gradient of the performance index:

$$\frac{1}{2} \frac{\partial L(U)}{\partial u_{ij}} = \sum_{i',j'} (a_{jj'} b_{ii'} + a_{j'j} b_{i'i}) u_{i'j'}^2 u_{ij}, \tag{4}$$

where $U = (u_{ij}^2)$, and suppresses the solutions having larger values of the performance index.

The dynamical system (3) has equilibrium solutions $u_{ij}^{(p)}$ ($p \in \Pi_N$):

$$u_{ij}^{(p)2} \begin{cases} \simeq 1 \ (i = p(j)) \\ = 0 \ (i \neq p(j)) \end{cases} (\forall i). \tag{5}$$

The solutions $U^{(p)} = (u_{ij}^{(p)})$ are called the feasible solutions which corresponds to permutation matrices $X^{(p)} = (x_{ij}^{(p)})$:

$$x_{ij}^{(p)} \begin{cases} = 1 \ (i = p(j)) \\ = 0 \ (i \neq p(j)) \end{cases} (\forall i). \tag{6}$$

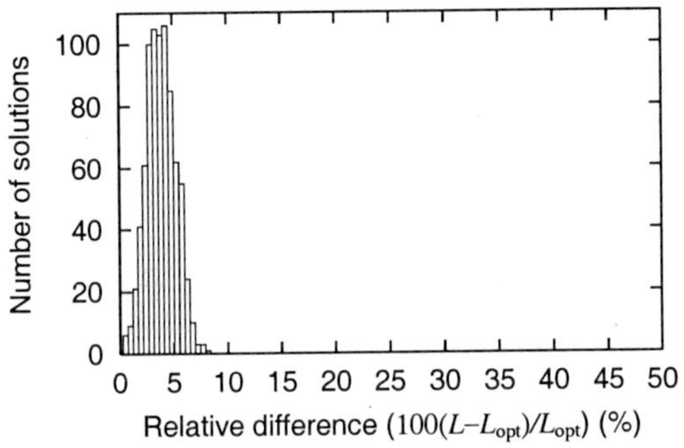

(a) $\alpha_0 = 1.01$

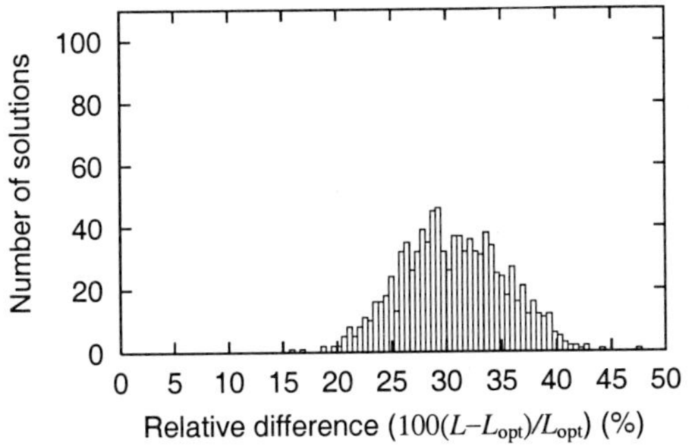

(b) $\alpha_0 = 3.0$

Fig. 1. Distribution of the feasible solutions (L_{opt} is the optimal value of L)

The stability condition for a feasible solution $U^{(p)}$ corresponding to a permutation matrix $X^{(p)}$ is approximately given as follows:

$$\alpha_0 > 1 + \frac{\alpha_1}{N-1}(L(X^{(p)}) - \bar{L}), \quad \bar{L}\text{: constant} \tag{7}$$

The condition (7) indicates that when α_0 is close to 1, only feasible solutions having smaller values of the performance index L, i.e. good approximate solutions of the QAP, are stable.

To verify the stability condition (7), the dynamical system (3) is computed with many sets of random initial values using a problem instance called "Nug20" ($N = 20$) in the QAPLIB [5]. Figures 1 (a), (b) are the results for $\alpha_0 = 1.01$ and $\alpha_0 = 3.0$ respectively. In the case of $\alpha_0 = 1.01$, only good solutions are obtained as compared with the case of $\alpha_0 = 3.0$.

4 Optimization Algorithm

Based on the above analysis, we propose an optimization algorithm for the QAP using the Markov chain Monte Carlo algorithm.

1. $M :=$ size of the neighborhood
2. $T_0 :=$ initial temperature
3. $X^{(p_0)} :=$ initial permutation matrix
4. $n := 0$ (iteration)
5. $\alpha_0 := 1 + \epsilon$ ($\epsilon \ll 1$)
6. **while** $n < n_{\max}$ **do**
   - 6.1. Choose M rows and columns randomly in the matrix U
   - 6.2. Give random initial values to the corresponding $M \times M$ elements u_{ij} where the rest of the elements are fixed to the value of $U^{(p_n)}$
   - 6.3. Calculate (3) to obtain the equilibrium solution $U^{(p_{n+1})}$ and corresponding permutation matrix $X^{(p_{n+1})}$. If the solution is not feasible, go back to Step 6.1
   - 6.4. Apply 2-opt method to $X^{(p_{n+1})}$
   - 6.5. Accept $X^{(p_{n+1})}$ with probability $e^{-[L(X^{(p_{n+1})})-L(X^{(p_n)})]^+/T_n}$, where $[a]^+ = \max\{a, 0\}$. If rejected, $X^{(p_{n+1})} := X^{(p_n)}, U^{(p_{n+1})} := U^{(p_n)}$
   - 6.6. $T_{n+1} = b \cdot T_n$ ($b < 1$)
   - 6.7. $n := n + 1$

Fig. 2. Algorithm

First, the search space is constructed as follows; the replicator equation (3) is derived according to the given QAP. The control parameter α_0 is set close to 1 so that (3) has stable equilibrium solutions $U^{(p)}$ corresponding to permutation matrices $X^{(p)}$ with smaller values of the performance index of the QAP. The search space is composed of these stable equilibrium solutions $U^{(p)}$. Next, given a solution $U^{(p_n)}$ corresponding to the permutation matrix $X^{(p_n)}$, new solution $U^{(p_{n+1})}$ is searched in the 'M-neighborhood' of the current solution as follows; M rows and columns are randomly chosen in the matrix U of (3). Random values are given to the corresponding elements u_{ij} as the initial values and then (3) is calculated, where the rest of the elements are fixed to the values of $U^{(p_n)}$. The new solution $U^{(p_{n+1})}$ and the corresponding permutation matrix $X^{(p_{n+1})}$ are obtained. A local search method called the 2-opt method is applied to the obtained solution. The 2-opt method is a simple heuristic method and adopted to slightly modify the solution to be the local minimum in the 2-neighborhood of the solution. The new solution $X^{(p_{n+1})}$ is accepted or rejected based on the Metropolis method [8]; if the performance index value is decreased by the change of the solution, the new solution is accepted, and if the value is increased, the new solution is accepted with the probability $\exp(-\Delta L/T)$ according to the amount of the increase ΔL. The parameter T called the temperature is decreased every step of the algorithm by multiplying a constant b (< 1). The whole procedure of the proposed algorithm is shown in Fig. 2.

This algorithm is not a true simulated annealing [9] in two respects. First, since the new solution is searched in the M-neighborhood of the current solution among the stable equilibrium solutions of (3), the ergodicity does not always hold. Second, since the obtained solution is modified by the 2-opt method in each step of the algorithm, the detailed balance is not satisfied. Therefore, the

Table 1. Solutions of the proposed algorithm (%)

M	Mean	Standard deviation	Minimum
5	0.014	0.022	0
10	0.0021	0.0019	0
15	0.015	0.027	0

Table 2. Solutions of the proposed algorithm with dynamics and without dynamics (random) (%)

	Mean	Standard deviation	Minimum
Dynamics	0.0021	0.0019	0
Random	0.015	0.017	0

convergence properties of the proposed algorithm are not guaranteed. But, numerical experiments mentioned below reveal that the algorithm can obtain good solutions for many instances in the QAPLIB.

5 Numerical Experiments

Numerical experiments are carried out using some large scale problem instances in the QAPLIB. First, the effect of the size M of the neighborhood was checked using an instance "Wil100" ($N = 100$). The parameters are $\alpha_0 = 1.01, \alpha_1 = 0.003, b = 0.99995, n_{\max} = 50000$, and $T_0 = 300.0$. The resulted means, standard deviations and minima of ten separate trials for each of $M = 5, 10, 15$ are shown in Table 1. These values are the relative differences $100(L - L_{\text{opt}})/L_{\text{opt}}$ (%) from the best known solution L_{opt}. The best result was obtained when $M = 10$. The reason is considered that good parts of the solution may be changed when a new solution is obtained if M is too large. But it is necessary to consider further about the optimum value of M.

Next, we compared the proposed method with the method where the new solution is randomly generated in the neighborhood of the current solution. The computation was carried out using the problem instance "Wil100" with $M = 10$. The results are shown in Table 2. The proposed algorithm gives better performance on average. This indicates that the proposed method searches effectively in the relatively large neighborhood using the dynamical system.

Finally, the performance of the proposed algorithm was verified using some large scale problem instances in the QAPLIB, i.e. "Sko100a", "Sko100b", "Sko100f", "Wil100", "Tai100a" ($N = 100$) and "Tho150" ($N = 150$). The results are shown in Table 3, where L_{opt} is the best known solution given in the QAPLIB and in the parentheses the name of the algorithms which gave the solutions, i.e. Genetic Hybrids (GEN), Reactive Tabu Search (Re-TS), Simulated

Table 3. Performance

Name	N	L_{opt} (algorithm)	L	Difference(%)
Sko100a	100	152002 (GEN)	152002	0
Sko100b	100	153890 (GEN)	153890	0
Sko100f	100	149036 (GEN)	149036	0
Tai100a	100	21125314 (Re-TS)	21146176	0.099
wil100	100	273038 (GEN)	273038	0
Tho150	150	8133484 (SIMJ)	8135474	0.024

Jumping (SIMJ), are shown. L is the solution by the proposed algorithm and in the last column the relative difference from L_{opt} is shown. The solutions having the same values of the performance index as the best known solutions were obtained for the four of the six problem instances. [1]

In the proposed method, since the size of the neighborhood is set to $M(\sim 10)$, the dynamical system with only $M \times M$ elements is calculated even if the size of the problem is $N(\sim 100)$. Therefore the computation time for one step of the algorithm is very short. On a COMPAQ AlphaStation XP900 computer, the total computation time for 50000 steps of the algorithm was only about 1–2 hours.

6 Conclusion

In this paper, we proposed a Markov chain Monte Carlo algorithm based on the replicator equations. In the proposed dynamical system, only good approximate solutions are obtained by appropriately setting a control parameter in the system. Therefore, using the system, good solutions are efficiently searched in relatively large neighborhood in each step of the algorithm. The proposed algorithm is applied to some large scale benchmark problems of the QAP. It was shown that the algorithm can obtain the solutions equivalent to the best known solutions in short time. Especially, the new solutions with the same performance index values as the best known solutions were obtained for some problem instances.

References

1. Pardalos, P. M., Rendl, F. and Wolkowicz, H.: The Quadratic Assignment Problem: A Survey and Recent Developments. In: Pardalos, P. and Wolkowicz, H. (eds.): Quadratic Assignment and Related Problems, Vol. 16 of DIMACS Series in Discrete Mathematics and Theoretical Computer Science. American Mathematical Society (1994) 1–42

[1] The obtained permutation matrices were different from the ones given in the QAPLIB.

2. Hopfield, J. J. and Tank, D. W.: "Neural" Computation of Decisions in Optimization Problems. Biological Cybernetics **52** (1985) 141–152
3. Ishii, S. and Sato, M.: Constrained Neural Approaches to Quadratic Assignment Problems. Neural Networks **11** (1998) 1073–1082
4. Tsuchiya, K., Nishiyama, T. and Tsujita, K.: A Deterministic Annealing Algorithm for a Combinatorial Optimization Problem Using Replicator Equations. Physica D **149** (2001) 161–173
5. Burkard, R. E., Karisch, S. and Rendl, F.: QAPLIB – A Quadratic Assignment Problem Library. Journal of Global Optimization **10** (1997) 391–403
6. Haken, H., Schanz, M. and Starke, J.: Treatment of Combinatorial Optimization Problems using Selection Equations with Cost Terms - Part I: Two-Dimensional Assignment Problems. Physica D **134** (1999) 227–241
7. Ishii, S. and Niitsuma, H.: λ-opt Neural Approaches to Quadratic Assignment Problems. Neural Computation **12** (2000) 2209–2225
8. Metropolis, N., Rosenbluth, A. W., Rosenbluth, M. N., Teller, A. H. and Teller, E.: Equation of state calculation by fast computing machines. Journal of the Chemical Physics **21** (1953) 1087–1092
9. Kirkpatrick, S., Gelatt, C. D., Jr. and Vecchi, M. P.: Optimization by Simulated Annealing. Science **220** (1983) 671–680

Mapping Correlation Matrix Memory Applications onto a Beowulf Cluster

Michael Weeks, Jim Austin, Anthony Moulds, Aaron Turner,
Zygmunt Ulanowski, and Julian Young

Advanced Computer Architecture Group,
Computer Science Department,
University of York,
Heslington, York, UK
{mweeks,austin}@cs.york.ac.uk

Abstract. The aim of the research reported in this paper was to assess the scalability of a binary correlation Matrix Memory (CMM) based on the PRESENCE (PaRallEl StructurEd Neural Computing Engine) architecture. A single PRESENCE card has a finite memory capacity, and this paper describes how multiple PCI-based PRESENCE cards are utilised in order to scale up memory capacity and performance. A Beowulf class cluster, called Cortex-1, provides the scalable I/O capacity needed for multiple cards, and techniques for mapping applications onto the system are described. The main aims of the work are to prove the scalability of the AURA architecture, and to demonstrate the capabilities of the architecture for commercial pattern matching problems.

1 Introduction

This paper investigates methods to distribute pattern-matching tasks over a number of parallel CMMs, implemented as hardware PRESENCE cards within a Beowulf cluster. In theory, the performance of the hardware CMM should scale with the number of PRESENCE cards in use. However, this project will provide results on the effects of software, device driver, and communications overheads on performance. It must be noted that the distributed CMM project is still in progress, with results still forthcoming.

The paper first introduces the AURA architecture and the history behind it. A brief overview of the PRESENCE PCI-card is then given, followed later by techniques for mapping large CMM applications over multiple PRESENCE cards in the Cortex-1 cluster. Next, the specification of Cortex-1 is given, as well as the reasons for choosing a Beowulf class cluster. The AURA library is then discussed followed by the seamless method by which the distributed CMM is implemented over the cluster. The paper closes with conclusions from the work.

2 AURA

AURA (Advanced Uncertain Reasoning Architecture) is a generic family of techniques and implementations intended for high-speed approximate search and

G. Dorffner, H. Bischof, and K. Hornik (Eds.): ICANN 2001, LNCS 2130, pp. 156–163, 2001.

match operations on large unstructured datasets [4]. AURA technology is fast, economical, and offers unique advantages for finding near-matches not available with other methods.

AURA is based on a high-performance binary neural network called Correlation Matrix Memory (CMM). Typically, several CMM elements are used in combination to solve soft or fuzzy pattern-matching problems.

AURA takes large volumes of data and constructs a special type of compressed index. AURA finds exact and near-matches between indexed records and a given query, where the query itself may have omissions and errors. The degree of nearness required during matching can be varied through thresholding techniques. AURA implicitly supports powerful combinatorial queries, which accept a match between, for example, any 5 from 10 fields in the query against the stored records. A degree of index compression is preset, allowing a trade-off between storage efficiency and accuracy of recall. In practice, this means that AURA is guaranteed to find all genuine matches but will typically find additional false matches, depending on the degree of index compression used. The false matches are easily detected in the relatively small result set by conventional (but computationally slow) matching techniques.

The increasing range of applications for AURA includes:

- postal address matching;
- high-speed rule-matching systems [5];
- high-speed classifiers (e.g. novel k-NN implementations) [9];
- structure-matching (e.g. 3D molecular structures) [8];
- trademark-database searching [1].

Other applications under development include data analysis and case-based reasoning systems.

3 PRESENCE Hardware

This section gives a brief overview of the operation of the PRESENCE architecture. A more detailed discussion was presented at MicroNeuro'99 [7]. PRESENCE is the current family of hardware designs which accelerate the core CMM computations needed in AURA applications. PRESENCE incorporates several improvements over the first hardware prototype:

- 5-stage pipelined Sum-and-Threshold (SAT) processor;
- 128 concurrent bit-summations;
- 50 nanosecond summing cycle;
- 128 MByte DRAM memory.

The PRESENCE card utilises the PCI-bus interface, to allow it to be used on standard desktop computer systems, and to give the CPU fast access to the card. The card is a PCI slave device, though future versions will include PCI bus-mastering capability in order to reduce CPU processing.

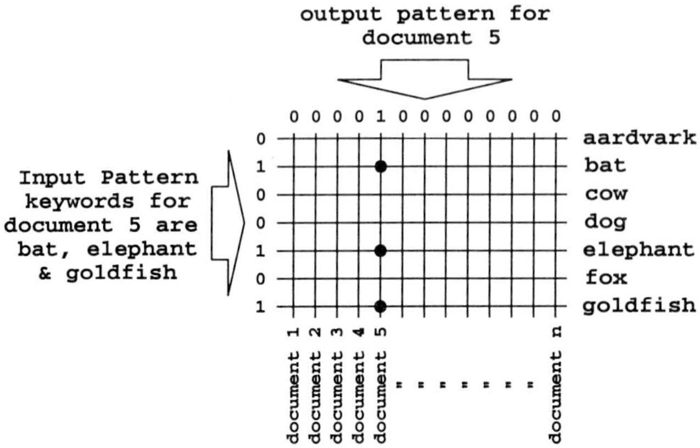

Fig. 1. An example of an inverted index teach

The PRESENCE architecture consists basically of a binary correlation matrix neural network implemented in memory, for the storage and retrieval of vector patterns. Each column of the matrix is seen as a neuron, and each row represents an input and synapses to each neuron. In common with the CMM concept, the PRESENCE card has two main modes of operation; teach and recall. In teach mode, the input and output binary pattern vectors are supplied to the card over the PCI bus. Recall is achieved by issuing only the input binary vector, and on completion, the resulting summed column data of the CMM can be read unprocessed from the card for post-processing, or hardware thresholding can be applied to the data to sort the best matches. Two types of hardware thresholding can be applied (Willshaw or Lmax) dependent upon the application. Willshaw [10] thresholding compares the summed columns with a thresholded level, whilst Lmax [11] retrieves the top L matches from all of the summed columns. A detailed operation of CMM neural networks can be found at [12].

CMM techniques lend themselves for use in such applications as inverted indexes, whereby objects are stored in the CMM categorised by certain attributes. A keyword search of documents, for instance, where the output pattern associates one or more bits with a list of documents, and the input pattern describes keywords in the documents. Figure 1 illustrates how documents and keywords are taught into a CMM. A recall operation applies selected keywords as the input pattern, and the output pattern contains matching documents.

4 Cortex-1 Beowulf Cluster

The number of PRESENCE cards that can be used on a single PC is limited to five slots, due to restrictions imposed by the PCI-bus standard. PCI-to-PCI bridges allow secondary slots to be created (for the loss of one primary slot), but are impractical due to added latencies and the physical dimensions of the resulting system. In order to gain scalable I/O, the PRESENCE cards are distributed across the nodes of a Beowulf cluster (Cortex-1). Communication between the cluster nodes is via sockets.

Cortex-1 is an eight-node Linux Beowulf cluster connected via fast-ethernet, whilst four nodes of the cluster are also connected via SCI (Scalable Coherent Interface) links, though this is not utilised as yet. Each node in the cluster has a 500MHz Pentium-III processor with 384MByte of system memory. Using five PRESENCE cards per node we can utilise 30 cards in six nodes, allowing a maximum possible CMM size of 3.84 GByte.

5 The AURA Library

The object-oriented AURA library, written in C++, is a collection of objects for creating and accessing CMM's, in addition to pre and post-processing algorithms. The library has been written for several machine-OS-compiler combinations, though we are initially concerned with Linux and gcc for this project.

AURA CMM's can be instantiated so that they map onto the PRESENCE hardware, or are simulated in software. Simulated CMM classes were initially designed to investigate the CMM concept, to experiment with new features, and to enable application development whilst the hardware progressed.

The hardware PRESENCE cards in a node are encapsulated inside a NodeCMM class. This is based upon a low-level hardware driver (\dev\presdrv) that has been developed for insertion into the Linux kernel as a module. A static (hw_ops) library enables the NodeCMM class to access the driver via the ioctl() system routine. Currently the low-level driver makes use of polling and as such all operations are blocking. An interrupt version of the driver is currently under development to make the driver non-blocking.

A simulated CMM object's storage capacity scales according to the amount of usable system memory, but the NodeCMM is limited to a maximum of 640MByte with five PRESENCE cards. A *distributedCMM* class has been designed that can seamlessly replace a simulated CMM object. This class encapsulates the underlying cluster infrastructure and each nodes NodeCMM's, so that they are hidden to the application programmer. In this way, converting the existing applications to distributed PRESENCE cards should simply be a matter of re-declaring the CMM objects and re-compiling.

Two methods were considered for the specification of the distributedCMM, depending upon the levels of abstraction handled by the master node. One level of abstraction allows the master to know everything about the slave nodes, basically addressing individual boards within the slave nodes, which are operating in an almost transparent manner. The second option appears to be the most flexible and scalable design for the distributed CMM and is shown in figure 2. Here there are two levels of resource management increasing in abstraction towards the master. The lowest level is the Slave Process (SP) which manages the slave nodes resources and supplies abstract information about its overall "CMM

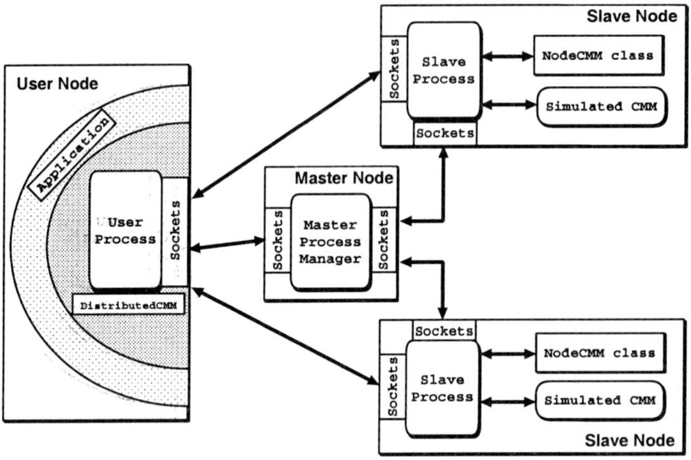

Fig. 2. DistributedCMM object mapped onto Cortex-1 cluster

capacity". In this way a nodes resources are encapsulated, allowing the flexibility for the Slave Process to decide to change where the CMM is, and so on, without the DistributedCMM needing to know this information. Information on the nodes capabilities are then passed to the Master Management Process (MMP), which resides centrally on the master node. The MMP handles all requests for resource information and CMM creation. However, requests for recalls are handled via direct connection from the DistributedCMM to the SP's in order to avoid a bottleneck at the MMP and enhance concurrency. This scheme hides the complexity of the communications inside the DistributedCMM object whilst maximising communication efficiency. The latter is especially important as the number of nodes are scaled up.

With spare processing capacity available on each slave node, we can further enhance the performance of the cluster by implementing a CMM in software. With sufficient memory resources, a software CMM is of comparable performance to a single PRESENCE card. This hybrid approach can only be possible when the driver supports interrupts.

6 NodeCMM Class Implementation

The first step in implementation is to determine how a CMM is mapped onto a single card. The memory on a single PRESENCE card is organised as 8Mwords, where each word is 128-bit wide. Figure 3 illustrates how a CMM with an output pattern greater than 128-bits maps into this memory. Obviously, increasing the output pattern width to $k \cdot 128$, decreases the input pattern width by k (or $\frac{8Mlocations}{k}$).

Whilst it is possible to map a CMM that requires less than 128MByte onto a single card, it is preferable to allocate a CMM over multiple cards by striping its

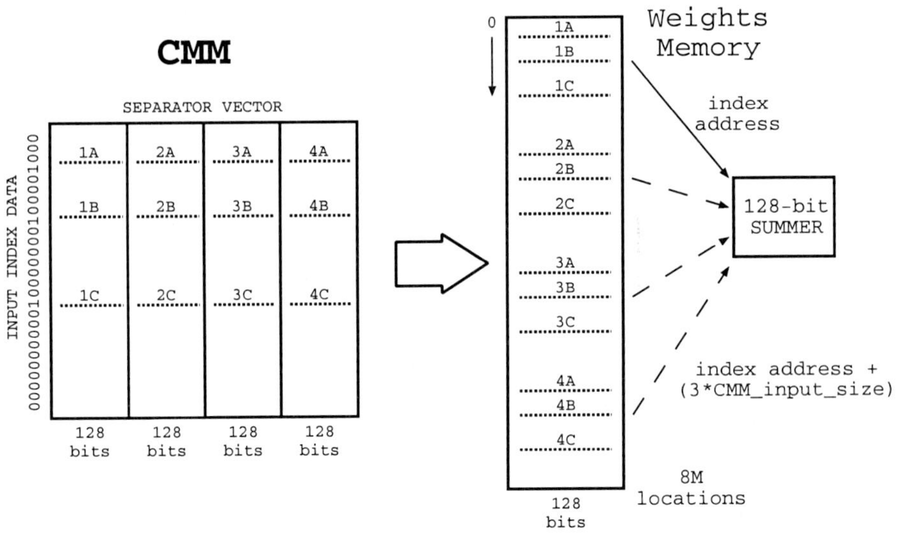

Fig. 3. An example of a CMM mapping to weights memory in a single PRESENCE card

output vector. Firstly, available memory scales with the number of cards utilised, so five PCI PRESENCE cards in a PC crate allows 640MByte. Secondly, the time taken for a recall decreases if its output pattern is allocated across multiple cards. Figure 4 illustrates how the performance of a CMM recall increases when multiple card striping is employed. Note that the speedup does not scale linearly. This is due to a PCI bottleneck when collecting the data from the PRESENCE cards, plus additional preprocessing that is required to reconstitute the partial output patterns into the full CMM output pattern.

Whilst the output pattern is scattered over the PRESENCE cards. The input pattern is broadcast to all cards as it is unlikely that the number of index terms used will exceed the 8 million possible on a PRESENCE card. If an input vector were to be split across multiple cards, upon recall their individual summed columns would be meaningless. To apply thresholding to the result of the CMM recall, the node must gather all raw data results, summing the appropriate column sub-totals, before thresholding. Obviously, the large quantity of data retrieved and the post processing required makes this an inefficient method and is therefore not implemented.

Due to the natural scalability of Willshaw-thresholded recalls, summed column thresholding can remain in hardware. Lmax thresholded recalls, however, require the L highest summed columns on the whole output pattern. These highest matches may be all on one card, or more likely spread across multiple cards in the CMM. Since the thresholding engine on each card does not provide the raw column count values, this can not be scaled across multiple cards. The alter-

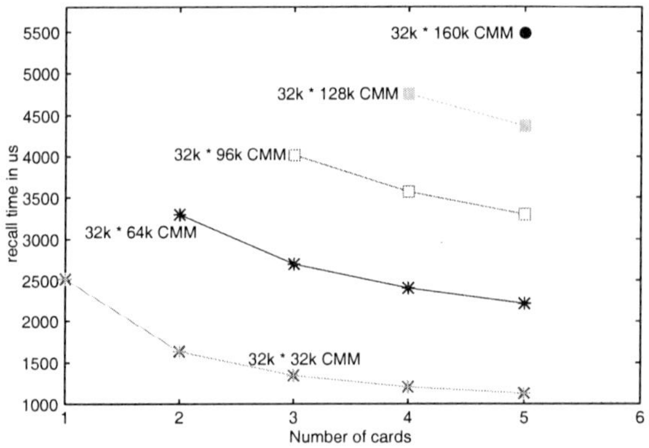

Fig. 4. Time taken for a 30 index-term recall on various size CMMs, when distributed across multiple cards

native approach is to perform multiple Willshaw thresholding at various levels, until there are L columns in the retrieved set.

7 Test Application

The application used for test purposes is a simple inverted index whereby objects are stored in the CMM categorised by certain attributes. The inverted index application used is a keyword search of documents.

Other, more practical applications have been implemented using the AURA library's simulated CMM's, and it is planned that these will be adapted for later use with the cluster. Such applications include address matching, company trademark database matching, and molecular matching.

Performance tests on the NodeCMM class have returned a time of 5.4ms for a 30 index term Willshaw recall on a CMM of input size 32768-bits and output size 163840-bits. The CMM was equally distributed over five cards, using a striped output pattern. When used in the document keyword search application, this relates to an effective search rate of 30.34 million documents per second per node.

Applying the results attained for the NodeCMM class, the distributedCMM performance will theoretically scale with the number of nodes. However, there will be some degradation due to communications and post-processing overheads.

8 Conclusion

The techniques discussed in this paper are intended to scale up CMM performance and storage capacity when used with the PRESENCE architecture. It provides performance data for a node-level CMM, discusses how the cluster is used to provide scalable CMM's, and how the spare processing capacity at each node can be utilised to provide a hybrid hardware/software distributed CMM. This is wrapped in an AURA library class to allow seamless incorporation into current AURA applications.

References

1. Sujeewa Alwis and Jim Austin. A novel architecture for trademark image retrieval systems. In *Electronic Workshops in Computing*. Springer, 1998.
2. Jim Austin. Adam: A distributed associative memory for scene analysis. In *First Int. Conference on Neural Networks*, volume IV, page 285, San Diego, June 1987.
3. Jim Austin and John Kennedy. The design of a dedicated adam processor. In *IEE Conference on Image Processing and its Applications*, 1995.
4. Jim Austin, John Kennedy, and Ken Lees. The advanced uncertain reasoning architecture. In *Weightless Neural Network Workshop*, 1995.
5. Jim Austin, John Kennedy, and Ken Lees. A neural architecture for fast rule matching. In *Artificial Neural Networks and Expert Systems Conference (ANNES '95)*, Dunedin, New Zealand, December 1995.
6. Anthony Moulds. Evaluation of multiple presence hardware in large systems. Technical report, ACA Group, Computer Science, University of York, 2000.
7. Anthony Moulds, Richard Pack, Zygmount Ulanowski, and Jim Austin. A high performance binary neural processor for PCI and VME bus-based systems. In *Weightless Neural Networks Workshop*, 1999.
8. Aaron Turner and Jim Austin. Performance evaluation of a fast chemical structure matching method using distributed neural relaxation. In *Fourth International Conference on Knowledge-Based Intelligent Engineering Systems*, August 2000.
9. P. Zhou and J Austin. A pci bus based corelation matrix memory and its application to k-nn classification. In *MicroNeuro'99*, Granada, Spain, April 1999.
10. D.J.Willshaw, O.P.Buneman, H.C.Longuet-Higgins. Non-holographic associative memory. Nature 222(1969), 960-962.
11. D.P.Casasent and B.A.Telfer. High capacity pattern recognition associative processors. Neural Networks 5(4), 251-261, 1992.
12. Jim Austin. Distributive associative memories for high speed symbolic reasoning. Int. J. Fuzzy Sets Systems, 82, 223-233, 1996.

The research detailed in this paper was funded by EPSRC grant numbers GR/L74651 and GR/K41090.

Accelerating RBF Network Simulation by Using Multimedia Extensions of Modern Microprocessors

Alfred Strey and Martin Bange

Department of Neural Information Processing
University of Ulm, D-89069 Ulm, Germany
strey@neuro.informatik.uni-ulm.de

Abstract. All modern microprocessors offer multimedia extensions that can accelerate many applications based on vector and matrix operations. In this paper the suitability of such units for the fast simulation of RBF networks is analyzed. It is shown that the reduced arithmetic precision is sufficient for recognition and training if rounding is supported. An experimental performance study revealed a high speedup in the range from 2 to 10 compared to sequential implementations.

1 Motivation

Current microprocessors are operating at high clock frequencies of about 1 GHz and achieve a satisfactory performance for most neural network applications. However for the simulation of neural networks that are embedded in real-time systems the power of a single microprocessor is often still insufficient if complex pattern recognition tasks must be executed at high speed. This lack of performance becomes far more evident if also the training must be performed online to adapt the neural network parameters to the current environment in real-time.

Many modern microprocessors contain a special data-parallel execution unit that is called either multimedia, vector or SIMD (*Single Instruction Multiple Data*) unit. It allows the simultaneous execution of arithmetic operations on several short data-elements packed in 64-bit or 128-bit registers. The user must explicitly program the parallel execution in the code by using special SIMD instructions. Although the architecture and the instruction sets of most multimedia units are mainly designed to accelerate popular multimedia algorithms, also many neural network operations can easily be mapped onto the SIMD units. However, no detailed analysis about the performance gain that can be achieved for neural network applications has been published so far. Merely Gaborit et al. have shown, that the calculation of the Mahalanobis distance in a generalized RBF network can be accelerated by a factor of 3 on Intel's MMX unit [2].

In this paper, an RBF network trained by gradient descent is used as typical benchmark application. The following section summarizes the RBF algorithm and explains the main differences of five SIMD units available in current microprocessors. Section 3 discusses if the arithmetic precision offered by multimedia

G. Dorffner, H. Bischof, and K. Hornik (Eds.): ICANN 2001, LNCS 2130, pp. 164–169, 2001.

units is sufficient for the simulation of RBF networks. Section 4 presents the results of an experimental performance study and Section 5 concludes this paper.

2 Approach

The RBF network represents a typical artificial neural network model suitable for many approximation or classification tasks. To achieve good results, it requires a proper initialization of all prototypes c_{ij} and of the widths σ_j of the Gaussian radial basis functions in all RBF neurons. After initialization, all network parameters can be adapted by a gradient-descent training algorithm:

$$x_j = \sum_{i=1}^{m}(u_i - c_{ij})^2 \quad (1)$$
$$y_j = e^{-x_j/2\sigma_j^2} = e^{-x_j s_j}$$
$$z_k = \sum_{j=1}^{h} y_j w_{jk} \quad (2)$$
$$\delta_k = t_k - z_k \quad (3)$$
$$\delta_j = \sum_{k=1}^{n} \delta_k w_{jk} \quad (4)$$
$$s_j = s_j - \eta_s x_j y_j \delta_j \quad (5)$$
$$w_{jk} = w_{jk} + \eta_w y_j \delta_k \quad (6)$$
$$c_{ij} = c_{ij} + \eta_c (u_i - c_{ij}) \delta_j y_j s_j \quad (7)$$

Eq. 1 to 7 represent the basic neural operations that will be analyzed throughout this paper. They are also relevant for many other neural network models. The calculation of the exponential function cannot be accelerated by the multimedia units. In all formulas $s_j = 1/2\sigma_j$ is used instead of the standard deviation σ_j to eliminate divisions that cannot be realized efficiently on most multimedia units. This modification also improves the numeric stability of the algorithm and is essential for the implementation on architectures with limited precision.

Five different multimedia units have been selected for this experimental study (see Table 1): Intel's MMX [4] (also available in actual AMD processors) and Sun's VIS [7] allow only operations on integer data. Intel's SSE [6] and AMD's 3DNow! [3] support only the 32-bit floating point data format, but also provide a few additional integer instructions for improving MMX. Motorola's AltiVec [1] operates on integer and floating-point data. The parallelism degree p varies in the range from 2 to 16 depending on the selected data size and the the available register width. Table 1 lists all SIMD instructions required for the implementation of Eq. 1 to 7. Each instruction operates simultaneously on p corresponding data elements stored in two registers. Especially integer multiplications must be realized differently. A general $16 \times 16 \rightarrow 32$ bit multiplication is available only on AltiVec. All other SIMD units require a sequence of partial multiplications, additions and reorder operations to generate the correct 32-bit product.

All neural operations according to Eq. 1 to 7 were implemented on the five selected multimedia either in assembly language (on Intel and AMD processors) or with a C language interface (on Sun and Motorola processors). For reference, all operations were implemented also in C using the `float` data type. The Gnu C compiler and assembler were used for Intel and AMD processors, the Sun Workshop 5.0 C Compiler for Sun and the Metrowerks CodeWarrior for Motorola's PowerPC. Compiler optimizations have been switched on. As hardware platforms standard PCs with either 500 MHz Pentium III or 700 MHz Athlon, a Sun workstation with 400 MHz Ultra II processor and a Macintosh with a 500 MHz G4 PowerPC processor were used. The execution time of all seven neural

Table 1. Characteristics and selected instructions of several multimedia units (used abbreviations: sat = saturated, m = modulo 2^n, r = rounded, h = only higher result bits, l = only lower result bits, a = arbitrary result bits)

	MMX (Intel)	SSE (Intel)	3DNow! (AMD)	VIS (Sun)	AltiVec (Motorola)
available in	Pentium II	Pentium III	K6-2	Ultra I/II	PowerPC G4
register width	64	128	64	64	128
data types	8,16,32 bit	float	float	16,32 bit	8,16,32 bit, float
parallelism	2-8	4	2	2-4	4-16
no of instr.	57	70	45	85	162
latency cycles	1-3	1-5	2	1-3	1-4
instructions:					
16 ± 16	16[m], 16[sat]			16[m]	16[m], 16[sat]
32 ± 32	32[m], 32[sat]			32[m]	32[m], 32[sat]
float $\pm$ float		+	+		+
8×16				16[h], 24	
16×16	16[h], 16[l]		16[h,r]		16[h,r], 32
float$\times$float		+	+		+
$16\pm16\times16$					16[h,r,sat]
float$\pm$float$\times$float					+
$16\times16+16\times16$	32[m]				32[sat]
reduction by $\sum$			+		32[sat]
pack (32$\rightarrow$16)	+[l,sat]			+[a,sat]	+[l,sat]
merge/unpack	8, 16, 32	+	+	8	16, 32
permutation		16, 32	16, 32		8

operations and the total RBF network simulation time related to the three different network sizes 16-104-16, 64-416-64 and 128-832-128 were measured on the multimedia and the floating-point units of all processors.

3 Analysis of Precision Requirements

To exploit the fast SIMD integer operations on MMX, VIS and AltiVec, all neural network variables must be encoded as fixed-point numbers and mapped onto 8-bit or 16-bit integer data elements. However, Vollmer et al. have demonstrated that for the training of an RBF network applied to a complex pattern recognition task more than 20 bit precision are required [8]. In their experimental study all intermediate results were computed with a high precision and then *truncated* to a lower precision according to the selected data size. To analyze also the implication of *rounding* that is supported by a few SIMD units, an RBF network was trained by gradient descent for a classification task. The precision of all variables was varied and both mean squared error and classification rate were determined experimentally on training set (memorization) and test set (generalization). Figure 1 illustrates that with truncation especially the weight update steps (Eq. 5, 6 and 7) require a high precision of 18 to 20 bit because the quantization error is biased and accumulated during many training epochs. Rounding drastically reduces the precision demands: 11 to 14 bit are sufficient to achieve results comparable to those of the reference implementation.

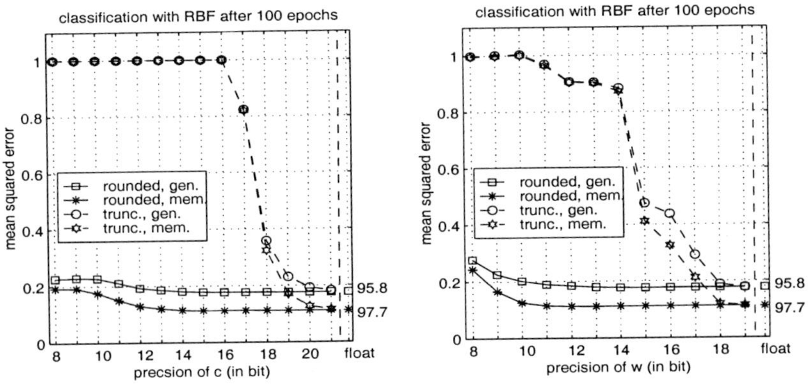

Fig. 1. Precision requirements of the RBF network variables c_{ij} and w_{jk}

Thus, a precision of 16 bit was selected for the SIMD-parallel integer implementation of all neural operations on MMX, VIS and AltiVec. Rounded arithmetic operations were always preferred, although they are not available in all instruction sets (compare Table 1): AltiVec offers a rounding option for the result of all multiplications, MMX supports rounding only in some integer arithmetic instructions added by AMD's 3DNow! extension. Besides of rounding, also *saturation* is required for the correct fixed-point implementation of neural network operations. Fortunately, most multimedia units support at least saturated additions/subtractions (apart from Sun's VIS) and allow the saturated extraction of 16-bit words out of 32-bit results by special *pack* instructions.

Altogether, only the fixed-point implementations on Motorola's AltiVec and on the MMX unit of AMD processors promise acceptable precision for a high-quality RBF implementation including the training phase. Nevertheless, all neural operations were evaluated on all multimedia units because in some cases (e.g. if only the recognition phase is required or if a lower quality is sufficient) also the implementations on the other units may be justified.

4 Analysis of Performance

Fig. 2 shows the total time for calculating the RBF network output and adapting all parameters according to the presentation of a new input vector **u** on all SIMD units. It is evident that the SIMD fixed-point implementations are always faster than their SIMD floating-point counterparts.

To study in more detail the suitability of the five multimedia units for the acceleration of certain neural operations, also the speedup of all single neural operations according to Eq. 1 to 7 compared to the reference implementation was calculated. In the first line of Table 2 the parallelism degree p is listed that represents a kind of theoretical speedup. It can be seen that for certain operations a high speedup can be obtained that exceeds the theoretical speedup by far. On MMX the computation of x_j and z_k required in the recognition phase is 6.3 to 9.8 times faster than the reference float implementation. Also for AMD's

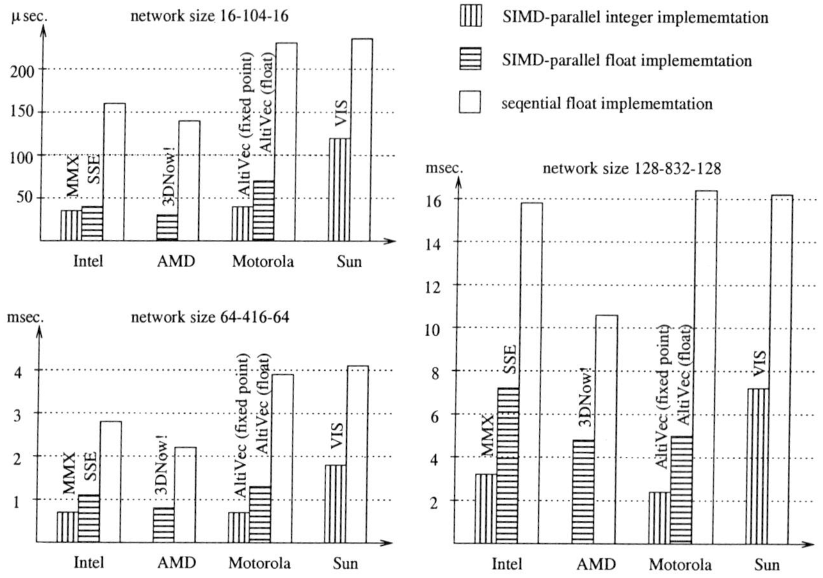

Fig. 2. Total RBF network simulation time for three different network sizes

Table 2. Measured speedup for seven neural operations on five multimedia units (with three network sizes 16-104-16 / 64-416-64 / 128-832-128)

	MMX (Intel)	SSE (Intel)	3DNow! (AMD)	AltiVec (Motorola) (fixed)	AltiVec (Motorola) (float)	VIS (Sun)
p	4	4	2	8	4	2-4
x_j	6.7/6.3/6.9	3.7/3.5/3.2	5.6/3.6/3.2	6.7/8.6/10.0	3.0/3.5/3.9	2.1/2.4/2.4
z_k	6.6/7.4/9.8	4.7/4.2/3.5	5.6/4.6/2.4	7.9/6.6/6.8	3.9/3.0/3.1	1.7/1.7/1.7
δ_k	1.7/2.0/2.3	1.3/1.6/1.8	1.1/1.0/1.0	3.0/5.0/5.5	3.0/2.5/2.7	1.8/2.2/2.4
δ_j	2.6/2.7/4.5	3.7/3.6/3.4	4.7/4.1/3.0	5.3/4.3/4.5	3.5/2.8/2.8	2.0/2.0/2.0
s_j	2.1/2.3/2.4	2.1/2.3/2.3	0.9/1.0/1.0	4.1/4.3/4.5	2.7/2.7/2.7	0.8/0.8/0.8
w_{jk}	5.7/3.7/5.4	3.7/2.0/2.0	3.3/1.9/1.5	7.8/6.9/7.5	3.8/3.4/3.5	1.5/1.7/1.8
c_{ij}	2.9/2.5/3.2	3.3/1.5/1.1	2.3/1.6/1.4	3.6/4.5/5.5	2.7/2.6/3.0	2.5/3.2/3.2

3DNow! which offers only a parallelism degree of $p = 2$ the measured speedup for many operations (e.g. the computation of x_j, z_k or δ_j for small networks) is surprisingly high (up to 5.6 times faster than the float implementation). This anomaly can be explained by special SIMD instructions (such as multiply&add or vector reduction, compare Table 1) that replace more than p sequential float instructions and by shorter latencies provided by SIMD arithmetic instructions (compared to the sequential instructions on the corresponding processor core).

On all units not more than half the theoretical speedup can be achieved for δ_k and s_j calculations. This may be due to the simplicity of Eq. 3 and 5: Only one operation is performed with each element loaded from memory. For the remaining three neural operations (calculation of w_{jk}, δ_j and c_{ij}) the performance is only average. Here the theoretical speedup cannot be reached because a high number of reorder steps (e.g. for the replication of scalar operands) are necessary. In case of w_{jk} and c_{ij} calculations, also the slow 16×16 bit fixed-point multiplications of some SIMD units reduce the performance.

Two effects appear when the network size is varied. On the one hand the speedup is increased for some fixed point implementations (e.g. for the calculation of z_k and δ_j on MMX or for x_j and c_{ij} on AltiVec) if the network is enlarged. On the other hand there are floating point implementations of several operations (e.g. x_j, z_k on SSE or most operations on 3DNow!) that show the reverse effect. This behavior results from cache effects and is analyzed in [5].

5 Results and Conclusions

This case study demonstrates that the multimedia units of modern microprocessors are fairly well suited for the acceleration of neural network simulations. A high speedup in the range from 1.9 (on Sun's VIS with small networks) to 6.6 (for the fixed-point realization on Motorola's AltiVec with large networks) can be achieved for the simulation of a complete RBF training step. Furthermore, applications that require only the fast recognition of presented patterns can be accelerated by a factor of approximately 10 on some multimedia units.

The SIMD instruction sets turned out to be not optimal for neural network simulation. By introducing a few additional instructions (such as more general 16×16 bit multiplications, faster replication of scalar operands in registers or more powerful vector reductions) the speedup could be increased further. Also the missing possibility to round low precision results restricts the applicability of some multimedia units to very simple pattern recognition tasks.

Unfortunately, the neural operations must be encoded on all multimedia units in low-level languages. As long as high-level compiler support will be missing, the high programming effort can be justified only for time-critical applications.

References

1. K. Diefendorff, P.K. Dubey, R. Hochsprung, and H. Scales. AltiVec extension to PowerPC accelerates media processing. *IEEE Micro*, 20(2):85–95, 2000.
2. L. Gaborit, B. Granado, and P. Garda. Evaluating micro-processors' multimedia extensions for the real time simulation of RBF networks. In *Proceedings of MicroNeuro*, pages 217–221. IEEE, 1999.
3. Stuart Oberman, Greg Favor, and Fred Weber. AMD 3DNow! technology: Architecture and implementations. *IEEE Micro*, 19(2):37–48, 1999.
4. A. Peleg, S. Wilkie, and U. Weiser. Intel MMX for multimedia PCs. *Communications of the ACM*, 40(1):24–38, 1997.
5. A. Strey and M. Bange. Performance Analysis of Intel's MMX and SSE: A Case Study. In *Proceedings of Euro-Par 2001*, Manchester, August 28-31, 2001.
6. S.T. Thakkar and T. Huff. Internet streaming SIMD extensions. *Computer*, 32(12):26–34, December 1999.
7. M. Tremblay, J.M. O'Connor, V. Narayanan, and L. He. VIS speeds new media processing. *IEEE Micro*, 16(4):10–20, 1996.
8. U. Vollmer and A. Strey. Experimental study on the precision requirements of RBF, RPROP and BPTT training. In *Proceedings of ICANN 99*, pages 239–244. IEE Conference Publication No. 470, 1999.

A Game-Theoretic Adaptive Categorization Mechanism for ART-Type Networks*

Wai-keung Fung and Yun-hui Liu

Department of Automation and Computer Aided Engineering,
The Chinese University of Hong Kong,
Shatin, N. T., Hong Kong
{wkfung,yhliu}@acae.cuhk.edu.hk

Abstract. A game-theoretic formulation of adaptive categorization mechanism for ART-type networks is proposed in this paper. We have derived the game-theoretic model Γ_{AC} for competitive processes of categorization of ART-type networks and an update rule for vigilance parameters using the concept of learning automata. Numbers of clusters generated by ART adaptive categorization are similar regardless of the initial vigilance parameters ρ assigned to the ART networks as demonstrated in the experiments provided. The proposed ART adaptive categorization mechanism can thus avoid the problem of choosing suitable vigilance parameter a priori for pattern categorization.

1 Introduction

ART-type (Adaptive Resonance Theory) networks [1][2] are a class of well-known categorization neural networks for its incremental categorization capability. The cluster granularity generated by ART-type networks is controlled by a fixed scalar called *vigilance parameter* ρ. This paper incorporates adaptive categorization (variable-sized clustering) into ART-type networks by adjusting vigilance parameter ρ. Few approaches have been proposed for changing the vigilance parameter of ART-type networks and existing methods just blindly increase the vigilance parameter by a fixed amount when all committed F_2 neurons are exhausted [3]. This approach will eventually set vigilance parameter to 1 and as a result any new pattern will form its own cluster. In order to solve the problem, we propose a game-theoretic formulation on the adaptive vigilance parameter strategy in ART-type networks [4] with the help of learning automata theory [5]. The proposed ρ-adaptation scheme can be easily added on the original design of ART-type networks. The game-theoretic vigilance adaptation strategy improves the clustering performance of ART-type networks in the aspect of category number stability, despite of what the pre-specified initial vigilance parameter is. Therefore, it is possible to avoid the problem of choosing suitable vigilance parameter a priori for data categorization by the trial-and-error approach [4].

* This work is supported in part by the Hong Kong Research Grant Council under grant CUHK 4151/97E

G. Dorffner, H. Bischof, and K. Hornik (Eds.): ICANN 2001, LNCS 2130, pp. 170–176, 2001.

This paper is organized as follows. Section 2 describes the mathematical formulation of the adaptive categorization game Γ_{AC} and the derived vigilance parameter update rule. Section 3 presents categorization experiment results with and without the proposed adaptive categorization mechanism. In additions, summary of this paper is given in Section 4.

2 Adaptive Categorization in ART-Type Networks

Original ART-type networks only use single, fixed vigilance parameter for all clusters. Fixed size clusters are difficult to represent thoroughly the data subspace and misclassification often happens in categorization-based classification. On the other hand, variable size clusters, that are generated by adaptive vigilance parameter mechanism, have the capability of approximating data pattern subspace well and even rendering decision boundaries, and thus misclassification can be avoided. Moreover, adaptive categorization helps preventing misclassification with data patterns from disjoint distributions. This paper presents a mathematical formulation of game-theoretic vigilance parameter adaptation mechanism for adaptive categorization in ART-type networks. Each F_2 neuron has its individual vigilance parameter ρ_i and an update rule for adaptively adjusting vigilance parameter during categorization process.

2.1 Game-Theoretic Formulation

The competitive clustering mechanism in ART-type networks is formulated as an *infinite n-person non-cooperative game*, which is characterized by the index set $\mathcal{P} = \{1, 2, \ldots, n\}$ of F_2 neurons (player of the game Γ_{AC}, the *strategy set* $\mathcal{R}^{(i)}$ of the i-th player and the *payoff* function $\pi^{(i)}$ of the i-th player. The vigilance parameter for each F_2 neuron forms its strategy $\rho_i \in \mathcal{R}^{(i)}$ in ART-type networks and ρ_i is usually bounded in $[0, 1]$. Vigilance parameter adaptation strategy will be derived based on the *Nash Equilibrium* of the game-theoretic formulation of ART-type networks.

Each F_2 neuron (player of the game Γ_{AC}) must attend two independent tests for each pattern presentation: the *matching score test* (MST) and the *vigilance test* (VT) [4]. F_2 neurons can then be classified into three groups when a data pattern is presented to an ART network according to three possible states, $\Sigma = \{\mathbf{R}, \mathbf{r}, \mathbf{f}\}$, after every data presentation (see Fig. 1):

RESONANCE (R) : Only one F_2 neuron can be in the *resonance* state as it passed both the MST and VT. The presented data pattern is assigned to the category represented by this particular F_2 neuron.

RESET (r) : F_2 neurons in the reset state have passed the MST but failed the VT. Denote the number of F_2 neurons in *reset* state by k.

FAIL (f) : F_2 neurons in the *fail* state have failed both the MST and VT. There are $(n - k - 1)$ F_2 neurons in this state.

In our algorithm, only F_2 neurons, that are in states **RESONANCE** or **RESET** may have their vigilance parameters updated for next pattern. If a pattern is categorized with *direct access* (no F_2 neuron is in **RESET** state), no update on vigilance parameters will be conducted. The basic derivation steps of adaptive categorization Γ_{AC} follows the Cournot game for oligopoly market model [6]. Each F_2 neuron incurs *costs* when it attends MST and VT and acquires *rewards* if it passes the tests. In the MST, the cost incurred and reward received by a F_2 neuron depend only on the matching score μ_i of that F_2 neuron. On the other hand, the cost incurred and reward received by a F_2 neuron depend only on the vigilance parameter ρ_i of that F_2 neuron in the VT. The incurred costs of the MST $c_{MST}^{(i)}$ and the VT $c_{VT}^{(i)}$ by the i-th F_2 neuron form linear relations with μ_i and ρ_i respectively as follows,

$$c_{MST}^{(i)} = \alpha_{MST} + \beta_{MST}\mu_i \qquad \text{and} \qquad c_{VT}^{(i)} = \alpha_{VT} - \beta_{VT}^{(i)}\rho_i \tag{1}$$

where α_{MST}, β_{MST}, α_{VT} and $\beta_{VT}^{(i)}$ are positive constants. The cost $c_{VT}^{(i)}$ increases with decreasing ρ_i as the F_2 neurons are encouraged to have fine clustering (high ρ) for better approximation to data subspace. The rewards of the MST $r_{MST}^{(i)}$ and the VT $r_{VT}^{(i)}$ obtained by the i-th F_2 neuron are given as,

$$r_{MST}^{(i)} = \mu_i \left((n-1)\,\mu_i - \sum_{j \neq i} \mu_j \right) \qquad \text{and} \qquad r_{VT}^{(i)} = \rho_i \left(\sum_{\substack{j \in \mathcal{I}(t) \\ j \neq i}} \rho_j - k\rho_i \right) \tag{2}$$

where n is the total number of F_2 neurons involved in categorization and k is the number of F_2 neurons in **RESET** state.

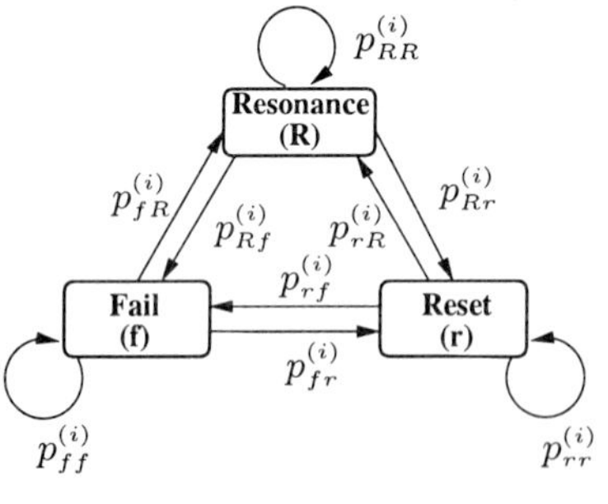

Fig. 1. State Transitions of a F_2 neuron.

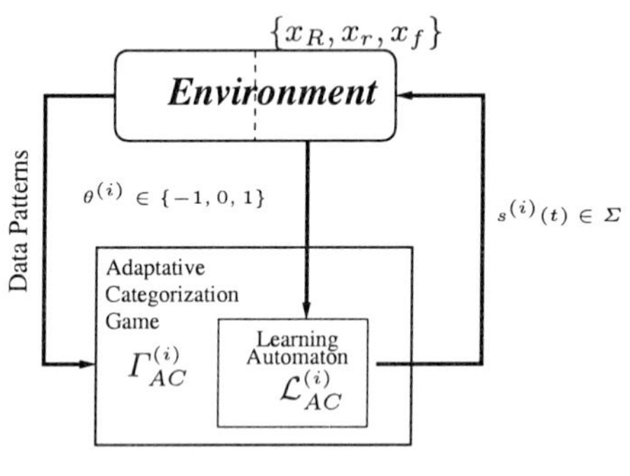

Fig. 2. $\Gamma_{AC}^{(i)}$ and the environment.

Each F_2 neuron will try its best to win the MST and tend to put as small effort (ρ_i) as possible to win the VT. The net gain of F_2 neurons in state $\sigma_i \in \Sigma$ are then given as follows,

$$\pi_R^{(i)} = (r_{VT}^{(i)} - c_{VT}^{(i)}) + (r_{MST}^{(i)} - c_{MST}^{(i)}), \quad \pi_r^{(i)} = -c_{VT}^{(i)} + (r_{MST}^{(i)} - c_{MST}^{(i)}),$$
$$\pi_f^{(i)} = -c_{MST}^{(i)}$$

Denote the state of the i-th F_2 neuron at the t-th pattern presentation by $s^{(i)}(t)$ and let $\mathcal{I}(t) \subset \mathcal{P}$ be the index set of F_2 neurons in states **RESONANCE** or **RESET** after the t-th pattern presentation. The payoff function $\pi^{(i)}$ of the i-th ($i \in \mathcal{I}(t)$) F_2 neuron at the t-th pattern presentation is then defined as the expected gain of that F_2 neuron in the three possible states at the t-th pattern presentation as $\pi^{(i)}(t) \triangleq \sum_{\sigma_u \in \Sigma} \mathbf{Prob}\left(s^{(i)}(t) = \sigma_u\right) \pi_u^{(i)}(t)$, where $\mathbf{Prob}(\cdot)$ denotes the probability of the given outcome to occur.

2.2 State Probability Dynamics of F_2 Neuron

A *learning automaton* $\mathcal{L}_{AC}^{(i)}$, $i = 1, 2, \ldots, n$, is constructed for each F_2 neuron to track the variations of state probabilities with time [5]. Each $\mathcal{L}_{AC}^{(i)}$ in the adaptive categorization game Γ_{AC} for each F_2 neuron consists of a set of internal states $\Sigma = \{\mathbf{R}, \mathbf{r}, \mathbf{f}\}$[1], a set of input reinforcement signals $\theta^{(i)} \in \Theta = \{-1, 0, 1\}$, a state transition probability matrix $\mathbf{P} \in \mathbb{R}^{3\times 3}$ that governs state transitions of $\mathcal{L}_{AC}^{(i)}$ in any two consecutive time instants and a reinforcement scheme $\Re$ for action probability update.

Fig. 2 depicts the interactions between the game Γ_{AC} or learning automaton and the environment. The environment is assumed to generate data patterns that the ART-type network will categorize and reinforcement signals $\theta^{(i)}$ to categorization. The purpose of reinforcement signal, which depends on the current state $s^{(i)}(t) \in \Sigma$ of $\mathcal{L}_{AC}^{(i)}$, is to guide the state probability adjustment according to the tracking capability of the learning automaton to its environment.

Reinforcement scheme $\Re$ of the learning automaton $\mathcal{L}_{AC}^{(i)}$ of the i-th F_2 neuron provides the update rule for state probabilities. The derivation is as follows. The *confirmatory* transition probability $q_{vu}^{(i)} \triangleq \mathbf{Prob}\left(s^{(i)}(t) = \sigma_u \,\middle|\, s^{(i)}(t+1) = \sigma_v\right)$ can be easily deduced from Bayes Theorem. The reinforcement signal $\theta^{(i)}$, reflecting the tracking performance of the learning automaton on its situated environment, is defined based on the state transition probabilities $p_{uv}^{(i)}$ and $q_{vu}^{(i)}$,

$$\theta^{(i)} = \begin{cases} -1 & \text{if } s^{(i)} = \underset{\sigma_u \in \Sigma}{\operatorname{argmin}} \sum\limits_{\sigma_v \in \Sigma} p_{uv}^{(i)} q_{vu}^{(i)} \\ 1 & \text{if } s^{(i)} = \underset{\sigma_u \in \Sigma}{\operatorname{argmax}} \sum\limits_{\sigma_v \in \Sigma} p_{uv}^{(i)} q_{vu}^{(i)} \\ 0 & \text{otherwise} \end{cases} \tag{3}$$

The sum-and-product term $\sum_{\sigma_v \in \Sigma} p_{uv}^{(i)} q_{vu}^{(i)}$ measures the amount of evidence that supports the transition from state σ_u at the t-th pattern presentation to state σ_v at the $(t+1)$-th pattern presentation. The reinforcement scheme $\Re$ for learning automaton $\mathcal{L}_{AC}^{(i)}$ is proposed to have linear reward and penalty functions pair and it is listed as follows,

$$\xi_u^{(i)}(t+1) = \begin{cases} \xi_u^{(i)}(t) - \frac{1}{2}a(1+\theta^{(i)})\xi_u^{(i)}(t) + \frac{1}{2}b(1-\theta^{(i)})\left(\frac{1}{2} - \xi_u^{(i)}(t)\right) & \text{if } s^{(i)}(t) \neq \sigma_u \\ \xi_u^{(i)}(t) + \frac{1}{2}a(1+\theta^{(i)})(1 - \xi_u^{(i)}(t)) - \frac{1}{2}b(1-\theta^{(i)})\xi_u^{(i)}(t) & \text{if } s^{(i)}(t) = \sigma_u \end{cases}$$

[1] The state of a F_2 neuron is reflected in the state of learning automaton associated to it.

2.3 ρ-Adaptation – Nash Equilibrium of Γ_{AC}

Vigilance parameters are adapted at the Nash Equilibrium of the game Γ_{AC}. The Nash equilibrium ρ^* of Γ_{AC} is defined as a strategy that satisfies the best response functions of all players (F_2 neurons), which gives the best reply to strategies $\overline{\rho_i} = \rho \setminus \rho_i$ of other F_2 neurons [6]. The best response function of the i-th F_2 neuron is then given by setting $\frac{\partial \pi^{(i)}}{\partial \rho_i} = 0$. Therefore, the Nash equilibria of Γ_{AC} are given as pairs of $(\rho^*, \beta_{\mathbf{VT}})$ that satisfies the equation $\boldsymbol{\Psi}\rho^* = \boldsymbol{\Omega}\beta_{\mathbf{VT}}$, or

$$\underbrace{\begin{bmatrix} 2k & -1 & \cdots & -1 \\ -1 & 2k & \cdots & -1 \\ \vdots & \vdots & \ddots & \vdots \\ -1 & -1 & \cdots & 2k \end{bmatrix}}_{\boldsymbol{\Psi} \in \mathbb{R}^{(k+1)\times(k+1)}} \underbrace{\begin{bmatrix} \rho^*_{\mathcal{I}_1} \\ \rho^*_{\mathcal{I}_2} \\ \vdots \\ \rho^*_{\mathcal{I}_{k+1}} \end{bmatrix}}_{\rho^* \in \mathbb{R}^{k+1}} = \underbrace{\begin{bmatrix} 1 + \frac{\xi_r^{(\mathcal{I}_1)}}{\xi_R^{(\mathcal{I}_1)}} & & \bigcirc \\ & \ddots & \\ \bigcirc & & 1 + \frac{\xi_r^{(\mathcal{I}_{k+1})}}{\xi_R^{(\mathcal{I}_{k+1})}} \end{bmatrix}}_{\boldsymbol{\Omega} \in \mathbb{R}^{(k+1)\times(k+1)}} \underbrace{\begin{bmatrix} \beta_{VT}^{(\mathcal{I}_1)} \\ \beta_{VT}^{(\mathcal{I}_2)} \\ \vdots \\ \beta_{VT}^{(\mathcal{I}_{k+1})} \end{bmatrix}}_{\beta_{\mathbf{VT}} \in \mathbb{R}^{k+1}} \tag{4}$$

where $\mathcal{I}_j$ is the j-th element in the index set $\mathcal{I}(t)$.

The "kinetic" energy of the i-th F_2 neuron is the same in magnitude of its payoff at Nash equilibrium $\pi_{NE}^{(i)}$, which is defined as $\mathbb{K}_i \triangleq \pi_{NE}^{(i)} = k\xi_R^{(i)}(\rho_i^*)^2 + \mathcal{A}_i$. Every F_2 neuron is eager to gain as much payoff $\pi_{NE}^{(i)}$ as possible in the competition for being in the **RESONANCE** state by tuning its ρ_i^* to 1 during categorization. However, it is not economical because the total energy supplied by all F_2 neurons in categorization process is not minimized so that all F_2 will eventually change their vigilance parameters to the extreme values[2]. Vigilance parameters are adapted so that minimum energy is consumed by the F_2 neurons so as to overcome the potential barrier in becoming the winning F_2 neuron during the categorization of data patterns. The potential barrier $\mathbb{P}_i$ of avoiding the i-th F_2 neuron from becoming a winning neuron (*ie.* in **RESONANCE** state) is $\mathbb{P}_i \triangleq (1 - \xi_R^{(i)})\rho_i^*$. Intuitively, the potential barrier increases with increasing vigilance parameter and the state probability $\xi_R^{(i)}$ indicates the degree of easiness for the F_2 neuron to overcome the potential barrier. $\xi_R^{(i)}$ introduces an inhibitory effect on the F_2 neuron **RESONANCE** state potential barrier. The difference between $\mathbb{K}_i$ and $\mathbb{P}_i$ is minimized subject to vigilance parameters of F_2 neurons, ρ_i^* $i \in \mathcal{I}(t)$, so that the F_2 neurons can consume the minimal energy to overcome the potential barriers in the next pattern presentation. By defining the Lagrangian $\mathbb{L}_i$ for each F_2 neuron, $i \in \mathcal{I}(t)$, $\mathbb{L}_i \triangleq \mathbb{K}_i - \mathbb{P}_i = k\xi_R^{(i)}(\rho_i^*)^2 - (1-\xi_R^{(i)})\rho_i^* + \mathcal{A}_i$. The updated vigilance parameter, ρ_i^* is given by setting $\frac{\partial \mathbb{L}_i}{\partial \rho_i^*} = 0$. Then the vigilance parameter for the i-th F_2 neuron at the t-th pattern presentation is updated by

[2] This argument is analogous to the "principle of least action" hypothesis proposed by Pierre-Louis Moreau de Maupertuis (1698-1759) in the field of analytical dynamics [7].

$$\rho_i^*(t) = \begin{cases} \dfrac{1-\xi_R^{(i)}}{2k\xi_R^{(i)}} & \text{if } \xi_R^{(i)} > \dfrac{1}{2k+1} \\ \rho_i^*(t-1) & \text{otherwise} \end{cases},$$ where $i \in \mathcal{I}(t)$. The condition imposed in the vigilance parameters update law is to restrict each ρ_i^* to lie in its nominal range $(0,1)$.

3 Simulations

Simulations results are presented to compare the performances of Fuzzy ART networks [2] with and without using the proposed game-theoretic ρ-adaptation. Learning rates in the reinforcement scheme are set as $a = 0.75$ and $b = 0.1$. 2000 uniformly distributed random two-dimensional data patterns, which are confined in a pair of *disjoint* distributions, are generated for simulations. Tests on Fuzzy ART networks with and without ρ-adaptation are performed with starting vigilance parameters at 0.4, 0.55, 0.7 and 0.85 with Fuzzy ART learning rate of 0.9. Fig. 3 depicts the categorization results in the tests. The number of categories formed in the tests without ρ-adaptation are 4, 8, 14 and 43 with starting ρ at 0.40, 0.55, 0.70 and 0.85 respectively. On the other hand, the number of categories formed in the tests with ρ-adaptation are 74, 71, 73 and 73 with starting ρ at 0.40, 0.55, 0.70 and 0.85 respectively.

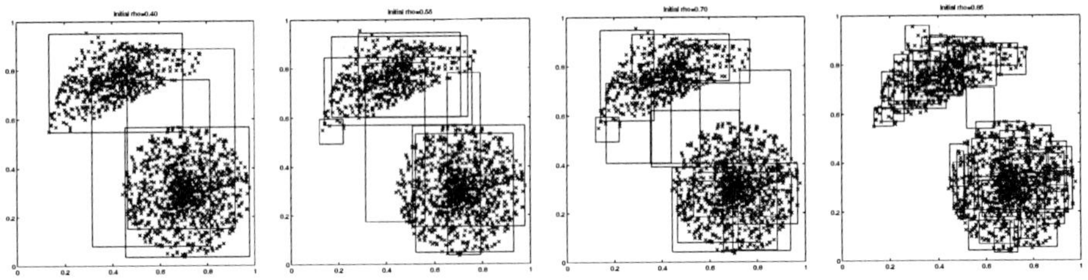

(a) $\rho = 0.40$ (b) $\rho = 0.55$ (c) $\rho = 0.70$ (d) $\rho = 0.85$

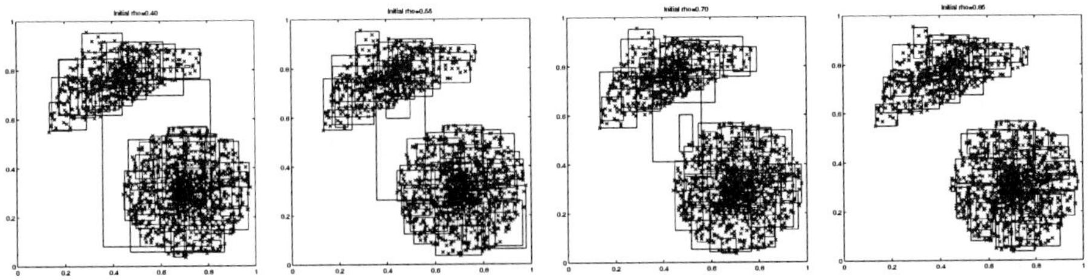

(e) $\rho = 0.40$ (f) $\rho = 0.55$ (g) $\rho = 0.70$ (h) $\rho = 0.85$

Fig. 3. Categorization experiments by Fuzzy ART networks with (lower row) and without (upper row) ρ-adaptation.

As shown in Fig. 3, the Fuzzy ART prototypes [2] generated are displayed as rectangles. The prototype rectangles generated by Fuzzy ART without ρ-

adaptation render the pattern distribution boundary poorly, especially for lower fixed vigilance parameters. On the other hand, the prototype rectangles generated by Fuzzy ART with ρ-adaptation render the pattern distribution boundary well no matter what value the starting vigilance parameter is. Moreover, the number of categories generated by Fuzzy ART without ρ-adaptation grows geometrically with increasing starting vigilance parameters while the number of categories generated by Fuzzy ART with ρ-adaptation is much insensitive to the starting vigilance parameter chosen. Thus ρ-adaptation remedies the difficulties in choosing a prior of vigilance parameters in data clustering using ART-type networks. The categories generated with ρ-adaptation cover far less patterns from either of the disjoint distributions than that generated by conventional Fuzzy ART network, as shown in Fig. 3. The categories generated by ρ-adaptive Fuzzy ART network can even be divided into two distinct groups that contain patterns from one and only one distribution when the starting vigilance parameter is 0.85. This helps to avoid misclassification in classifier systems that are constructed from Fuzzy ART networks.

4 Summary

This paper proposed a mathematical formulation of adaptive categorization of ART-type networks based on the game theory. We have derived the game-theoretic model Γ_{AC} for competitive processes of clustering of ART-type networks and an update rule for vigilance parameters using the concept of learning automata. Categorization experiments demonstrated that the game-theoretic vigilance parameter adaptation can improve the clustering performance of ART networks in the aspect of category number stability and avoid the problem of choosing suitable vigilance parameter a priori for pattern categorization. Moreover, the coverage of clusters generated by ART networks with ρ-adaptation can reflect the shape of pattern distribution and thus prevent misclassification in categorization-based classification.

References

1. G. A. Carpenter and S. Grossberg. A Massively Parallel Architecture for a Self-Organizing Neural Pattern Recognition Machine. *Computer Vision, Graphics and Image Processing,* 37:54–115 1987.
2. G. A. Carpenter, S. Grossberg, and D. B. Rosen. Fuzzy ART: Fast Stable Leaming and Categorization of Analog Patterns by an Adaptive Resonance System. *Neural Networks,* 4:759–71 1991.
3. N. Vlajic and H. C. Card. Categorizing Web pages using Modified ART. In *1998 IEEE Canadian Conference on Electrical and Computer Engineering,* volume 1, pages 313-316, 1998.
4. W. K. Fung and Y. H. Liu. A Game-theoretic Formulation on Adaptive Categorization in ART Networks. In *Proceedings of 1999 International Joint Conference on Neural Networks IJCNN'99,* volume 2, pages 1081–1086, 1999.
5. K. S. Narendra and M. A. L. Thathachar. *Learning Automata: An Introduction.* Prentice Hall, 1989.
6. D. Fudenberg and J. Tirole. *Game Theory.* The MIT Press, 1991.
7. Jr. J. H. Williams. *Fundamentals of Applied Dynamics.* John Wiley and Sons, 1996.

Gaussian Radial Basis Functions and Inner-Product Spaces

Irwin W. Sandberg

Department of Electrical and Computer Engineering
The University of Texas at Austin
Austin, Texas 78712, USA
sandberg@ece.utexas.edu

Abstract. An approximation result is given concerning gaussian radial basis functions in a general inner-product space. Applications are described concerning the classification of the elements of disjoint sets of signals, and also the approximation of continuous real functions defined on all of $\mathbb{R}^n$ using RBF networks. More specifically, it is shown that an important large class of classification problems involving signals can be solved using a structure consisting of only a generalized RBF network followed by a quantizer. It is also shown that gaussian radial basis functions defined on $\mathbb{R}^n$ can uniformly approximate arbitrarily well over *all* of $\mathbb{R}^n$ any continuous real functional f on $\mathbb{R}^n$ that meets the condition that $|f(x)| \to 0$ as $\|x\| \to \infty$.

1 Introduction

Radial basis functions are of interest in connection with a variety of approximation problems in the neural networks area, and in other areas as well. Much is understood about the properties of these functions (see, for instance, [1]–[3]). It is known [2], for example, that arbitrarily good approximation in $L_1(\mathbb{R}^n)$ of a general $f \in L_1(\mathbb{R}^n)$ is possible using uniform smoothing factors and radial basis functions generated in a certain natural way from a single g in $L_1(\mathbb{R}^n)$ if and only if g has a nonzero integral. As another example, in [3] it is proved that gaussian radial basis functions can uniformly approximate arbitrarily well any continuous real functional defined on a compact convex subset of $\mathbb{R}^n$.

Here we give an approximation result concerning gaussian radial basis functions in a general inner product space. This result, Theorem 1 in Section 2, has two applications that are felt to be interesting. In particular, we show that gaussian radial basis functions defined on $\mathbb{R}^n$ can in fact uniformly approximate arbitrarily well over *all* of $\mathbb{R}^n$ any continuous real functional f on $\mathbb{R}^n$ that meets the condition that

$$\lim_{\|x\| \to \infty} |f(x)| = 0.$$

This generalizes the result in [3] because, by the Lebesgue-Urysohn extension theorem [4, p. 63] (sometimes attributed to Tietze), any continuous real func-

G. Dorffner, H. Bischof, and K. Hornik (Eds.): ICANN 2001, LNCS 2130, pp. 177–182, 2001.

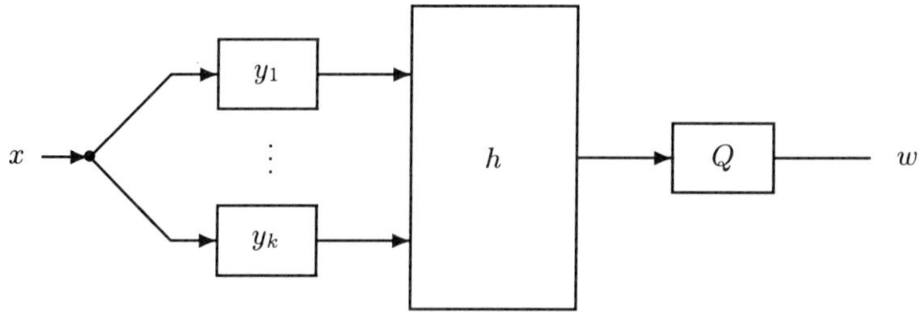

Fig. 1. Classification network.

tional defined on a bounded closed subset of $I\!R^n$ can be extended so that it is defined and continuous on all of $I\!R^n$ and meets the above condition.[1]

Our second application concerns the problem of classifying signals. This problem is of interest in several application areas (e.g., signal detection, computer-assisted medical diagnosis, automatic target identification, etc.). Typically we are given a finite number m of pairwise disjoint sets $C_1, \ldots, C_m$ of signals, and we would like to synthesize a system that maps the elements of each C_j into a real number a_j so that the numbers $a_1, \ldots, a_m$ are distinct. In [5] (see also [6]) it is shown that the structure shown in Fig. 1 can perform this classification (for any prescribed distinct $a_1, \ldots, a_m$) assuming only that the C_j are compact subsets of a real normed linear space. In the figure, x is the signal to be classified, $y_1, \ldots, y_k$ are functionals that can be taken to be linear[2], h denotes a continuous memoryless nonlinear mapping that, for example, can be implemented by a neural network having one hidden nonlinear layer, the block labeled Q is a quantizer that for each j maps numbers in the interval $(a_j - 0.25\rho, a_j + 0.25\rho)$ into a_j, where $\rho = \min_{i \neq j}|a_i - a_j|$, and w is the output of the classifier[3]. Here, in Theorem 3 of Section 2, we show that a similar classification problem can be solved in a very different way using just a generalized radial basis function network and a quantizer of the type in Fig. 1. Theorem 3 establishes in addition that the network structure considered can also classify signals x whose critical property is that the norm of x is sufficiently large. Some additional related results are given in Section 2.

---

1 For example, let a continuous f_0 defined on a bounded closed subset A of $I\!R^n$ be given, and let A be contained in an open ball B centered at the origin of $I\!R^n$. Then, since the complement C of B with respect to $I\!R^n$ is closed (and thus $A \cup C$ is closed), by the Lebesgue-Urysohn extension theorem there is a continuous extension f of f_0 defined on $I\!R^n$ such that $f(x) = 0, x \in C$.

2 Examples are given in [5] and [6].

3 The proof given in [5],[6] expands on a brief remark in [7, p. 274] that any continuous real functional on a compact subset of a real normed linear space can be uniformly approximated arbitrarily well using only a feedforward neural network with a linear-functional input layer and one memoryless hidden nonlinear (e.g., sigmoidal) layer.

2 Approximation and Classification Using Generalized RBF Structures

2.1 Preliminaries

Let S be a real inner-product space (i.e., a real pre-Hilbert space) with inner product $\langle \cdot, \cdot \rangle$ and norm $\|\cdot\|$ derived in the usual way from $\langle \cdot, \cdot \rangle$. Let X be a metric space whose points are a subset of the points of S, and denote the metric in X by d. With V any nonempty convex subset of S such that $\{ \|\cdot + v\| : v \in V\}$ separates the points of X (i.e., such that for x and y in X with $x \neq y$ there is a $v \in V$ for which $\| x + v\| \neq \|y + v\|$), and with P any nonempty subset of $(0, \infty)$ that is closed under addition, let $\mathcal{X}_0$ denote the set of functions g defined on X that have the representation

$$g(x) = \alpha \exp\{-\beta \|x - v\|^2\} \tag{1}$$

in which $\alpha \in I\!R$, $\beta \in P$, and $v \in V$.

Let $\mathcal{X}$ stand for the set of continuous functions f from X to the reals $I\!R$ with the property that for each $\epsilon > 0$ there is a compact subset $X_{f,\epsilon}$ of X such that

$$|f(x) - f(y)| < \epsilon$$

for $x, y \notin X_{f,\epsilon}$. And let $\mathcal{X}_\infty$ denote the family of f's in $\mathcal{X}$ such that for each f and each $\epsilon > 0$ there is a compact subset $X_{f,\epsilon}$ of X such that $|f(x)| < \epsilon$ for $x \notin X_{f,\epsilon}$.

2.2 Approximation

Our main result concerning approximation is the following.

Theorem 1: Assume that X is locally compact but not compact,[4] and that X is such that $\mathcal{X}_0 \subset \mathcal{X}_\infty$. Then for each $f \in \mathcal{X}_\infty$ and each $\epsilon > 0$ there are a positive integer q, real numbers $\alpha_1, \ldots, \alpha_q$, numbers $\beta_1, \ldots, \beta_q$ belonging to P, and elements $v_1, \ldots, v_q$ of V such that

$$|f(x) - \sum_{k=1}^{q} \alpha_k \exp\{-\beta_k \|x - v_k\|^2\}| < \epsilon \tag{2}$$

for all $x \in X$.

Proof:
All proofs are omitted in this version of the paper.

Comments: Since the condition that $\|x - v\| \neq \|y - v\|$ is equivalent to the condition that $2\langle x - y, v\rangle \neq \|x\|^2 - \|y\|^2$, and assuming that $x \neq y$ in X implies that $x \neq y$ in S, we see that V can be taken to be, for instance, any convex

[4] X is *locally compact* if each point of X is interior to some compact subset of X.

subset of S that contains the points of an open ball in S. Of course, P can be taken to be $(0,\infty)$ or $\{1,2,3,\ldots\}$, etc.

Since $\exp\{-\beta\|x-v\|^2\}\to 0$ as $\|x\|\to\infty$ when $\beta>0$, we have $\mathcal{X}_0\subset\mathcal{X}_\infty$ when for each $\gamma>0$ there is a compact subset X_ϵ of X such that $\|x\|\geq\gamma$ for $x\in X$ with $x\notin X_\epsilon$.

In the following theorem (and as suggested earlier), n is an arbitrary positive integer and $\mathbb{R}^n$ stands for the linear space of real n-vectors with the usual Euclidean norm. This norm is also denoted by $\|\cdot\|$.

Theorem 2: Let U be any convex subset of $\mathbb{R}^n$ that contains the points of an open ball in $\mathbb{R}^n$, and let Q be any set of positive numbers that is closed under addition. Let f be a continuous function from $\mathbb{R}^n$ to $\mathbb{R}$ such that

$$\lim_{\|x\|\to\infty}|f(x)|=0. \tag{3}$$

Then for each $\epsilon>0$ there are a positive integer q, real numbers $\alpha_1,\ldots,\alpha_q$, numbers $\beta_1,\ldots,\beta_q$ belonging to Q, and elements $v_1,\ldots,v_q$ of U such that

$$|f(x)-\sum_{k=1}^{q}\alpha_k\exp\{-\beta_k\|x-v_k\|^2\}|<\epsilon \tag{4}$$

for all $x\in\mathbb{R}^n$.

As mentioned earlier, Theorem 2 generalizes the result in [3] concerning approximation on compact convex subsets of $\mathbb{R}^n$.

2.3 Classification

Our main result concerning classification, which follows, is an application of Theorem 1.

Theorem 3: Assume that X is locally compact but not compact, that for each $\gamma>0$ there is a compact subset X_γ of X such that $\|x\|\geq\gamma$ for $x\in X$ with $x\notin X_\gamma$, and that every bounded subset of X is bounded in S.[5] Let $C_1,\ldots,C_m$ be pairwise disjoint compact subsets of X, and let $C_0=\{x\in X:\|x\|\geq\xi\}$ where $\xi>\max_j\sup\{\|x\|:x\in C_j\}$. Let $a_0,a_1,\ldots,a_m$ be distinct real numbers with $a_0=0$ and, with $\rho=\min_{i\neq j}|a_i-a_j|$, let $Q:\cup_j(a_j-0.25\rho,a_j+0.25\rho)\to\mathbb{R}$ be specified by $Q(a)=a_j$ for $a\in(a_j-0.25\rho,a_j+0.25\rho)$ and each j. Then there are a positive integer q, real numbers $\alpha_1,\ldots,\alpha_q$, numbers $\beta_1,\ldots,\beta_q$ belonging to P, and elements $v_1,\ldots,v_q$ of V such that

$$Q\left(\sum_{k=1}^{q}\alpha_k\exp\{-\beta_k\|x-v_k\|^2\}\right)=a_j$$

[5] A subset Y of X is *bounded* if $\sup\{d(x,z):x\in Y\}<\infty$ for some $z\in X$, and Y is *unbounded* if it is not bounded.

for $x \in C_j$ and $j = 0, 1, \ldots, m$.

The interpretation of Theorem 3 is that the family of classification problems addressed by the theorem can be solved using a structure consisting of only a generalized RBF network followed by a quantizer. This family of classification problems is more general than the one studied in [5],[6] in that here we consider also the problem of classifying signals x whose critical property is that the norm of x is sufficiently large. This additional degree of generality is interesting from the viewpoint of understanding the capabilities of the structure considered, but it may not be of much significance from a practical viewpoint. A result similar to Theorem 3, in which the additional class of signals is not considered and whose proof is along the same lines but is more direct, is given in Section 2.4.

Examples:

There are two particularly important examples of applications of Theorem 3:

We may take $S = X = I\!R^n$, in which case the classifier described in the theorem classifies elements of $I\!R^n$.

Alternatively, we may take S to be the set of continuous real-valued functions on the n-dimensional interval $[0,1]^n$, with the inner product given by

$$\langle x, y \rangle = \int_{[0,1]^n} x(w)y(w)\, dw.$$

In this case X can be selected to be any unbounded subset of S consisting of equicontinuous functions with d the uniform (i.e., max) metric, and $C_1, \ldots, C_m$ can be chosen to be any family of pairwise disjoint subsets of X that are closed and bounded. In this case the classifier classifies functions (e.g., images when $n = 2$ or 3). In particular, we can take the set of points of X to be any unbounded subset of S consisting of all Lipschitz continuous functions with a fixed Lipschitz constant.

Other examples can be given in which functions that are not necessarily continuous can be classified.

2.4 Related Results

Theorem 4 below is a version of Theorem 3 whose proof is along the same lines but is more direct. Theorem 4 is included because its proof, which is omitted in this version of the paper, provides additional understanding of the origin if its conclusion – that classification of the elements of pairwise disjoint compact sets can be achieved as indicated.

Theorem 4: Let $C_1, \ldots, C_m$ be pairwise disjoint compact subsets of X. Let $a_1, \ldots, a_m$ be distinct real numbers and, with $\rho = \min_{i \neq j}|a_i - a_j|$, let $Q: \cup_j (a_j - 0.25\rho, a_j + 0.25\rho) \to I\!R$ be specified by $Q(a) = a_j$ for $a \in (a_j - 0.25\rho, a_j + 0.25\rho)$ and each j. Then there are a positive integer q, real numbers $\alpha_1, \ldots, \alpha_q$, numbers

$\beta_1, \ldots, \beta_q$ belonging to P, and elements $v_1, \ldots, v_q$ of V such that

$$Q\left(\sum_{k=1}^{q} \alpha_k \exp\{-\beta_k \|x - v_k\|^2\}\right) = a_j$$

for $x \in C_j$ and $j = 1, \ldots, m$.

Concluding Comment: While natural questions arise concerning the results in this paper and specific practical problems and detailed implementations, such questions are not addressed here. We make no apologies for not having considered these questions. We are interested in this paper in questions concerning what is possible. Answers to such questions are often of considerable value.

References

1. J. Park and I. W. Sandberg, "Universal Approximation Using Radial Basis-Function Networks," *Neural Computation*, vol. 3, no. 2, pp. 246–257, 1991.
2. J. Park and I. W. Sandberg, "Approximation and Radial-Basis Function Networks," *Neural Computation*, vol. 5, no. 2, pp. 305–316, March 1993.
3. E. J. Hartman, J. D. Keeler, and J. M. Kowalski, "Layered Neural Networks with Gaussian Hidden Units as Universal Approximators," *Neural Computation*, vol.2, no.2, pp. 210–215, 1990.
4. M. H. Stone, "A Generalized Weierstrass Approximation Theorem," In *Studies in Modern Analysis*, ed. R. C. Buck, vol. 1 of *MAA Studies in Mathematics*, pp. 30–87, Englewood Cliffs, NJ: Prentice-Hall, March 1962.
5. I. W. Sandberg, "General Structures for Classification," IEEE Transactions on Circuits and Systems I, vol. 41, no. 5, pp. 372–376, May 1994.
6. I. W. Sandberg, J. T. Lo, C. Francourt, J. Principe, S. Katagiri, and S. Haykin *Nonlinear Dynamical Systems: Feedforward Neural Network Perspectives*, New York: John Wiley, 2001.
7. I. W. Sandberg, "Structure theorems for nonlinear systems," *Multidimensional Systems and Signal Processing*, vol. 2, no. 3, pp. 267–286, 1991. (See also the Errata in vol. 3, no. 1, p. 101, 1992.) A conference version of the paper appears in *Integral Methods in Science and Engineering-90* (Proceedings of the International Conference on Integral Methods in Science and Engineering, Arlington, Texas, May 15-18, 1990, ed. A. H. Haji-Sheikh), New York: Hemisphere Publishing, pp. 92–110, 1991.
8. R. A. DeVore and G. G. Lorentz, *Constructive Approximation*, New York: Springer-Verlag, 1993.
9. W. A. Sutherland, *Introduction to Metric and Topological Spaces*, Oxford: Clarendon Press, 1975.

Mixture of Probabilistic Factor Analysis Model and Its Applications

Masahiro Tanaka

Department of Information Science and Systems Engineering
Faculty of Science and Engineering, Konan University
8-9-1 Okamoto, Higashinada, 658-8501, Kobe, JAPAN
m_tanaka@konan-u.ac.jp
http://lotus.mis.konan-u.ac.jp/~tanaka/

Abstract. In this paper, the regression analysis is treated when the output estimate may take more than one value. This is an extension of the usual regression analysis and such cases may happen when the output is affected by some unknown input. The stochastic model used in this paper is the mixture of probabilistic factor analysis model whose identification scheme has been already developed by Tipping and Bishop. We will show the usefulness of our method by a numerical example.

1 Introduction

In many cases in regression analysis, it is necessary for the estimator to have the capability to express the nonlinear relation. Multilayer neural networks [3] are often used for this purpose. However, it is not necessarily useful for some cases. For example, if the output is very noisy, a deterministic output does not mean much. Another case neural network is useless is when the output may take certain separate values stochastically.

Since the probability density function (PDF) describes the underlying distribution of the data, it can be used in various analysis including regression analysis. Moreover, it is often more powerful than multilayer neural networks in the problems mentioned above.

Gaussian mixture model may be used for a wide class of non-Gaussian model. However, observing the data locally, it is often the case the data only exist in a subspace. Thus numerical problems may arise. In such cases a lower dimensional model should be used. Thus probabilistic factor analysis, which is often called "principal component analysis model" [5], is a good candidate.

This paper consists of the following sections. In section 2, the relation between the PDF and the regression analysis is explained. In section 3, the mixture of probabilistic factor analysis (PFA) model is introduced where the data may concentrate on subspaces locally. In section 4, the regression analysis corresponding to the mixture of PFA model is shown. In section 5, a result of the proposed regression analysis will be demonstrated. Section 6 is the conclusion.

G. Dorffner, H. Bischof, and K. Hornik (Eds.): ICANN 2001, LNCS 2130, pp. 183–188, 2001.

2 Probability Density Function and Regression Analysis

Regression analysis is a statistical method to estimate a variable z based on the vector consisting of the observed explanatory variables $\boldsymbol{p}$, i.e. the problem is to give the function

$$\hat{z} = f(\boldsymbol{p}) \tag{1}$$

where the function $f(\cdot)$ is linear or nonlinear.

Fig.1 shows an example of our motivation. If the contour of the joint distribution of the output and input is like this, the estimate of the output should be the bold line, not the dashed line that lies where the density is very low. However, the standard minimum variance estimate (e.g. [1])

$$\hat{z}(\boldsymbol{p}) = E[z|\boldsymbol{p}] = \int zp(z|\boldsymbol{p})d\boldsymbol{p} \tag{2}$$

yields the output expressed by the dashed line.

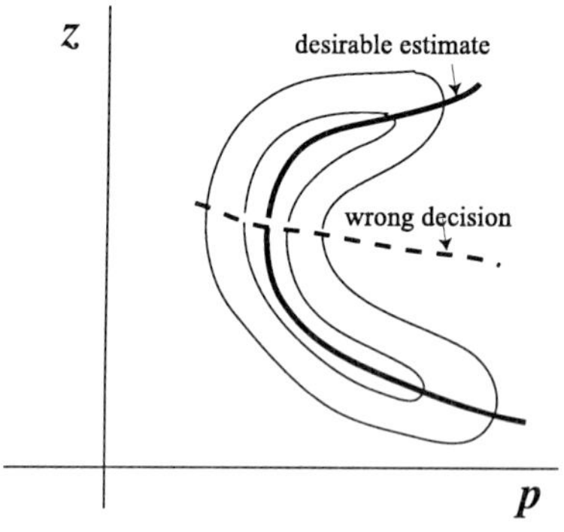

Fig. 1. PDF and regression analysis

So, when such a distribution may arise, it is obviously important to estimate the bold line. Our problem of this paper is to give it by using the conditional PDF $p(z|\boldsymbol{p})$, or equivalently, joint PDF $p(z, \boldsymbol{p})$ based on the training data set $\{(z(k), \boldsymbol{p}(k)); k = 1, \cdots, N\}$ because

$$p(z|\boldsymbol{p}) = \frac{p(z, \boldsymbol{p})}{p(\boldsymbol{p})} \tag{3}$$

3 Mixture of Probabilistic Factor Analysis Model

Suppose the data is expressed by one of the models

$$\boldsymbol{x} = W\boldsymbol{y}^{(i)} + \boldsymbol{\mu}^{(i)} + \boldsymbol{e}^{(i)} \quad \text{with probability } a^{(i)} \tag{4}$$

where $\boldsymbol{x} = [z \mid \boldsymbol{p}']'$, W is the factor loadings, $\boldsymbol{x}$ is the n-dimensional observation data, and $\boldsymbol{y}$ is an m-dimensional $(m < n)$ latent variable. The latent variable is a stochastic variable, and is assumed to obey the normal distribution

$$\boldsymbol{y}^{(i)} \sim N(0, I) \tag{5}$$

and

$$\boldsymbol{e}^{(i)} \sim N(0, (\sigma^{(i)})^2 I) \tag{6}$$

Let us return to the model (4). Using this model, the PDF of $\boldsymbol{x}$ can be written as

$$p(\boldsymbol{x}) = \sum_i a^{(i)} (2\pi)^{-n/2} |C^{(i)}|^{-1/2} \exp\left(-\frac{1}{2} (\boldsymbol{x} - \boldsymbol{\mu}^{(i)})^T (C^{(i)})^{-1} (\boldsymbol{x} - \boldsymbol{\mu}^{(i)}) \right) \tag{7}$$

where

$$C^{(i)} = W^{(i)} (W^{(i)})^T + (\sigma^{(i)})^2 I \tag{8}$$

The model description and the identification algorithm for this mixture of PPCA model has been shown by Tipping and Bishop [5]. However, the dimension of each kernel (submodel) was assumed to be known *a priori*.

First we denote how the identification procedure of PFA model is derived by using the EM algorithm([5]).

The EM algorithm developed by Dempster et al. [2] is an algorithm of ML estimate of the parameters where some missing data exist in the model. The missing data is a stochastic variable. EM algorithm consists of two steps, and iterate them until the estimate converges. E-step is to take expectation of the complete likelihood, based on the observation and the model parameters that have been obtained by now. M-step is to update the parameters so that the expectation of the complete likelihood is to be maximized.

As we mentioned already, it is necessary to determine the dimensions of the latent variables. We propose to use AIC (Akaike Information Criterion), which is given by

$$AIC = -2L + 2P \tag{9}$$

where L is the likelihood function and P is the number of free parameters. The problem is defined as to minimize this criterion.

If the number of the kernels is small and the observation dimension is also small, it is possible to calculate it by an exhaustive search. However, as these numbers grow, the search space expands exponentially, hence some heuristic method may be necessary. Meta-heuristics such as the genetic algorithm also may be useful in some cases. A preliminary experiments have been done by [4].

4 Regression Analysis Based on Mixture of PFA

The estimate of z based on the observation $\boldsymbol{p}$ is given by

$$\begin{aligned} \hat{z} = E[z|\boldsymbol{p}] &= \int z p(z|\boldsymbol{p}) dz \\ &= \sum_i E[z|\boldsymbol{p}, i] P(i|\boldsymbol{p}) \end{aligned} \tag{10}$$

where the conditional expectation of the kernel i given the input is

$$E[z|\boldsymbol{p},i] = W_z^{(i)}(W_p^{(i)})^T\left(W_p^{(i)}(W_p^{(i)})^T + (\sigma^{(i)})^2 I\right)^{-1}(\boldsymbol{p}-\boldsymbol{\mu}_p^{(i)}) + \mu_z^{(i)} \quad (11)$$

the *a posteriori* probability of the event occurrence from the kernel i is

$$P(i|\boldsymbol{p}) = \frac{p^{(i)}(\boldsymbol{p})a^{(i)}}{\sum_i p^{(i)}(\boldsymbol{p})a^{(i)}} \quad (12)$$

further where the stochastic density function of the input $\boldsymbol{p}$ for the kernel i is

$$p^{(i)}(\boldsymbol{p}) = \sum_i (2\pi)^{-n/2}|C^{(i)}|^{-1/2}\exp\left(-\frac{1}{2}(\boldsymbol{p}-\boldsymbol{\mu}_p^{(i)})^T(C_p^{(i)})^{-1}(\boldsymbol{p}-\boldsymbol{\mu}_p^{(i)})\right) \quad (13)$$

$$C_p^{(i)} = W_p^{(i)}(W_p^{(i)})^T + (\sigma^{(i)})^2 I \quad (14)$$

and $P^{(i)}$ is the *a priori* probability of the kernel i.

In certain cases, it is better not to mix the estimate of all the kernels. As was mentioned in the section 1, we may sometimes encounter a case where the output takes separate groups and taking the average of the estimates may severely degrade the output estimate. Such a case can happen when an important attribute is not included in the model. The output looks like taking quite distinct values stochastically. In such cases, it is obviously better to propose the estimate as a set. To do this, it is necessary to group the kernels, and within the group the outputs are mixed. This could be done by checking the sum of the *a posteriori* probabilities. If it comes to extremely low values, we can judge that the output is disconnected there.

Fig. 2 shows an idea to do this. The horizontal axis is the output value, and the vertical axis is the joint PDF of the input and output. Note that the input is fixed to the value of our interest. By scanning the joint PDF along the output value, we may find the point where the joint PDF takes a very low value. Then we take this as a disjoint point, and the mixture of the output is done in the right part and the left part, separately. Although this processing is time consuming, it is not so troublesome because we only have to compute it for a small set of $\boldsymbol{x}$ of our interest.

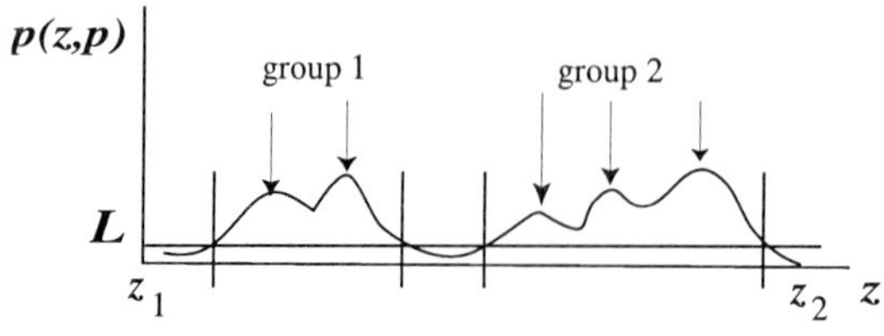

Fig. 2. Boundaries

The following is the algorithm we propose.

Step 1. Fix $\boldsymbol{p}$.
Step 2. Scan the joint PDF $p^{(i)}(z, \boldsymbol{p})$ along z for a certain interval $[z_1, z_2]$ with a pre-determined increment value h.
Step 3. Find the boundary points of y corresponding to the threshould as shown in Fig.2.
Step 4. Partition the kernels into groups by using the information on which interval the ridge of the kernel density falls for the same $\boldsymbol{x}$.
Step 5. Suppose there arose m groups of kernels. Then we estimate the outputs y_i by the equations (10)-(12) where the probability is normalized within each group.

5 Numerical Example

We generated data based on

$$\boldsymbol{x} = W^{(i)}\boldsymbol{y} + \boldsymbol{\mu}^{(i)} + \boldsymbol{e} \tag{15}$$

for $i = 1, \cdots, 5$, and $W^{(i)}$, $\boldsymbol{\mu}^{(i)}$ are all distinct for the different i, where all the variables are 2-dimensional.

In this experiment, we assume that all the parameters are already known, and our problem is to get the estimate of the output z based on the input p.

Fig. 3 shows the estimate of the output z given the input p.

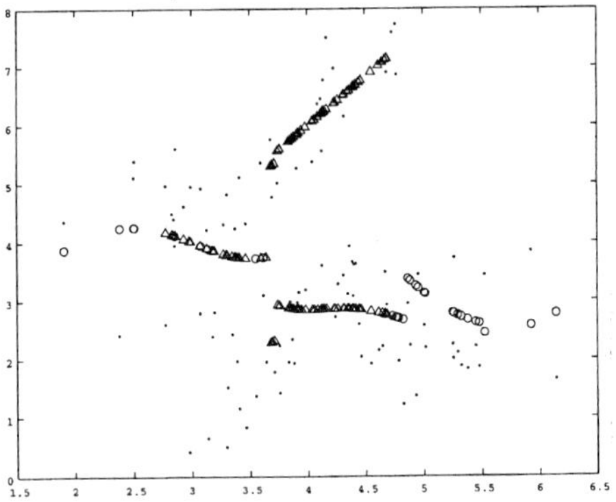

Fig. 3. Observation Data and the Estimates of the Outputs(larger threshold).

The small dot denotes the observation output. The symbol $\circ$ denotes the output estimate when the probability $P(i|\boldsymbol{p})$ is more than 0.8, $\triangle$ denotes the ones when $0.2 \leq P(i|\boldsymbol{p}) < 0.8$ where i denotes the group number of the distribution subset. This shows that our estimation scheme yields an appropriate result for the estimate of multiple outputs with probability. Fig. 4 also shows the estimate of the output for a smaller threshold. For a smaller threshold, the distribution are well-separated, and we have more number of estimates than the case for a larger threshold.

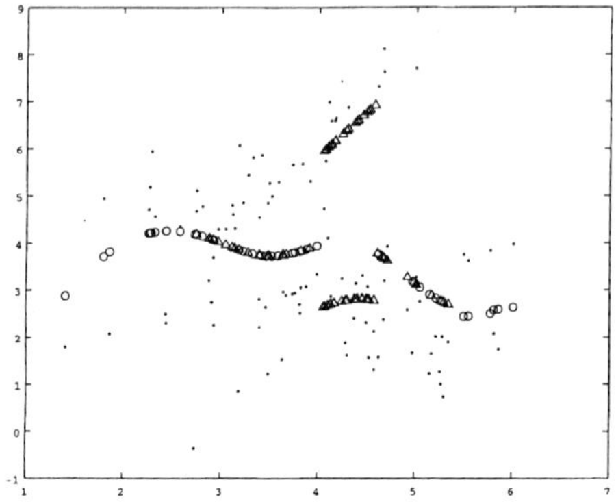

Fig. 4. Observation Data and the Estimates of the Outputs(smaller threshold).

6 Conclusions

In this paper, the regression scheme has been shown for the mixture of PFA model. This model can treat wide class of data distribution where output may take multiple distinct candidates. Such a case can happen when an important explanatory variable is not included in the model. In the numerical example, the model was assumed to be known and it was used in the estimation of the ouptut. In the future research, both the identification and the estimation of the output should be done simultaneously.

References

1. B.D.O. Anderson,B.D.O., Moore,J.B.: Optimal Filtering Prentice-Hall (1979)
2. Dempster,A.P., Laird,N.M., Rubin,D.B.: Maximum Likelihood from Incomplete Data via the EM Algorithm, Journal of the Royal Statistical Society **B39-1** (1977) 1-38
3. Rumelhart,D.E. Hinton,G.E., Williams, R.J.: Learning internal representation by error propagation. in Rumelhart,D.E. et al.: Parallel Distributed Processing: Explorations in the Microstructure of Cognition. **1**, The MIT Press (1986) 321-362
4. Tanaka,M.: Modeling of mixtures of principal component analysis model with genetic algorithm. Proc. 31st ISCIE International Symposium on Stochastic Systems Theory and Its Applications (2000) 157-162
5. Tipping,M.E., Bishop,C.M.: Mixtures of Probabilistic Principal Component Analysers. Technical Report NCRG/97/003/, Aston University, UK (1998) 29 pages

Deferring the Learning for Better Generalization in Radial Basis Neural Networks

José María Valls, Pedro Isasi, and Inés María Galván

Carlos III University of Madrid, Computer Science Department. Avd Universidad, 30, 28911, Leganés, Madrid
jvalls@inf.uc3m.es

Abstract. The level of generalization of neural networks is heavily dependent on the quality of the training data. That is, some of the training patterns can be redundant or irrelevant. It has been shown that with careful dynamic selection of training patterns, better generalization performance may be obtained. Nevertheless, generalization is carried out independently of the novel patterns to be approximated. In this paper, we present a learning method that automatically selects the most appropriate training patterns to the new sample to be predicted. The proposed method has been applied to Radial Basis Neural Networks, whose generalization capability is usually very poor. The learning strategy slows down the response of the network in the generalisation phase. However, this does not introduces a significance limitation in the application of the method because of the fast training of Radial Basis Neural Networks.

1 Introduction

The radial basis neural networks (RBNN) [1,2] are originated from the use of radial basis functions, as the Gaussian functions, in the solution of the real multivariate interpolation problem. RBNNs can be used for a wide range of application primarily because they can approximate any regular function [3]. Generally, the generalisation capability of the RBNN is poor because they are too specialised in the data training.

Some authors have paid attention to the nature and size of the training set in order to improve the generalization ability of the networks. There is no guarantee that the generalization performance is improved by increasing the training set size [4]. It has been shown that with careful dynamic selection of training patterns, better generalisation performance may be obtained [5,6].

The idea of selecting the patterns to train the network from the available data about the domain is close of our approach. However, the aim in this work is to develop learning mechanisms such that the selection of patterns used in the training phase is based on novel samples, instead of based on other training patterns. Thus, the network will use its current knowledge of the new sample to have some deterministic control about what patterns should be used for training.

In this work a selective training strategy has been developed to improve the generalisation capabilities of RBNN inspired on lazy strategies [7,8]. The learning

G. Dorffner, H. Bischof, and K. Hornik (Eds.): ICANN 2001, LNCS 2130, pp.189–195, 2001.

method proposed involve finding relevant data to answer a particular novel pattern and defer the decision of how to generalise beyond the training data until each new sample is encountered. Thus, the decision about how to generalise is carried out when a test pattern needs to be answered constructing local approximations. The main idea is to recognise from the whole training data set, the most similar patterns for each new pattern to be processed.

2 Automatic Selection of Training Data

The method proposed in this work to train RBNN consists of selecting, from the whole training data, an appropriate subset of patterns in order to improve the answer of the network for a novel test pattern. The general idea for the selection of patterns is to include only and several times those patterns close to the novel sample. To develop this idea, let us consider q an arbitrary test pattern described by a n-dimensional vector, $q=(q_1,..., q_n)$, where q_i represents the attributes of the instance q. Let $X = \{(x_i, y_i)/i = 1,..., N\}$ be the whole available training data set, where x_i are the input patterns and y_i their respective target outputs. The steps to select the training set associated to the pattern q, which is named X_q, are the following:

Step 1. A real value, d_k, is associated to each training pattern (x_k, y_k). That value is defined in terms of the standard Euclidean distance.

Step 2. A measure of frequency, $f_k=1/d_k$, where k=1,...N ,is associated to each training pattern (x_k,y_k). In order to obtain a relative frequency, the values f_k are normalised in such way that the sum of the frequencies is equal to the number of training patterns in X. The relative frequencies, named as fn_k, are obtained by:

$$fn_k = \frac{f_k}{S} \quad \text{where } S = \frac{1}{N}\sum_{i=1}^{N} f_k \qquad \text{Thus: } \sum_{i=1}^{N} fn_k = N$$

Step 3. The values fn_k's previously calculated will be used to indicate how many times the training pattern (x_k, y_k) is repeated into the new training subset. Hence, they are transformed to natural numbers as: $n_k=int(fn_k)$.

At this point, each training pattern in X has associated a natural number, n_k, which indicates how many times the pattern (x_k, y_k) has been used to train the RBNN when the new instance **q** is reached.

Step 4. A new training pattern subset associated to the test pattern **q**, X_q, is built up. Once the training patterns are selected, the RBNN is trained with the new subset of patterns, X_q. Training a RBNN involves to determine the centers, the dilations or widths, and the weights. The centers are calculated in an unsupervised way using the K-means algorithm to classify the input space. After that, the dilations coefficients are calculated as the square root of the product of the distances from the respective center to its two nearest neighbours. Finally, the weights of the RBNN are estimated in a supervised way to minimise the mean square error measured in the training subset X_q.

Let's $W_q \subset X_q$ the set resulting of removing the repeated patterns. In this work, the K-means algorithm has been modified in order to avoid the situation where many classes have no patterns at all. Thus, the initial values of the centers are set as:

- M_q, the centroid of the set W_q, is evaluated.
- k centers are randomly generated $(c_{1q}, c_{2q}, \ldots c_{kq})$,

such as $\left| c_{jq} - M_q \right| < \varepsilon$, j=1,...k, and ε is a very small real number.

3 Experimental Results

The deferred learning method proposed in this work has been applied to two different approximation problems. In order to validate the proposed method, RBNNs have also been trained as usual, this is, the network is trained using the whole training data set.

3.1 Hermite Polynomial Approximation

The Hermite polynomial is given by the equation:

$F(x)=1.1(1-x+2x^2)\exp(-1/2x^2)$

A random sampling of the interval [-4 , 4] is used in obtaining 160 input-output points for the training set and 65 input-output data for the test set.

RBNNs with different number of neurones have been trained using the whole training data. The training process is stopped either when 150 cycles are performed or when the derivative of the train error equals zero. The generalization capability of the trained networks has been measured using the test set and the mean square errors achieved by the networks. In figure 1 (a), the mean errors obtained for different architectures are shown. The best results have been achieved using a RBNN with 25 neurones, although no significant differences have been found for networks between 10 and 25 neurones. It is observed that the test error can not be improved even if more learning cycles are performed using the whole training data set.

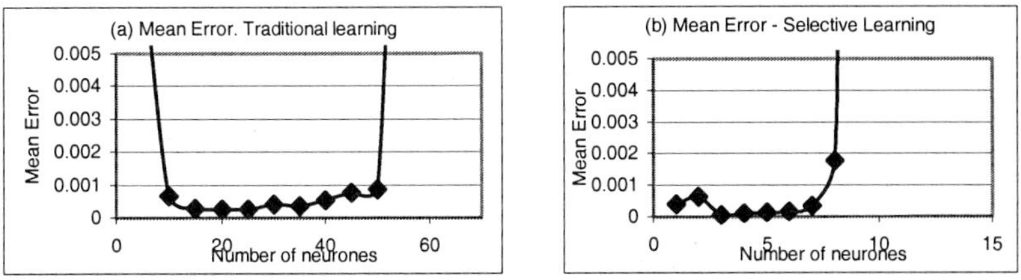

Fig. 1. Mean error on the test set for the Hermite function achieved by different architectures.

The selective learning method proposed in this work has also been used to train RBNNs with different architectures during 150 learning cycles, and their generalization capability has been tested. Mean square errors on the test set achieved by these networks are shown in figure 1 (b). In this case, only 3 neurones are necessary to obtain the best results, and using more than 8 increases the error to the level of the RBNN trained with the classical method.

The value of the mean square test error achieved by the best network, the one with 25 neurones, trained with the whole training set, is shown in Table 1. Using the specific learning method proposed in this work, the best network, the one with 3 neurones, has also been evaluated. In that case, the mean square error over the test set is 2.5 times lower.

Table 1. Performance of different training methods for the Hermite function

	Mean square error	Number of Neurones
Training with the whole data set	2.49×10^{-4}	25
Training with a selection of data	0.649×10^{-4}	3

To show the difference between both learning methods, the errors for each test pattern are represented in figure 2. In this figure it can be observed that the generalization error corresponds initially to some difficult regions of the function. It is specially in these regions where our approach is able to drastically reduce the error and to find a good approximation.

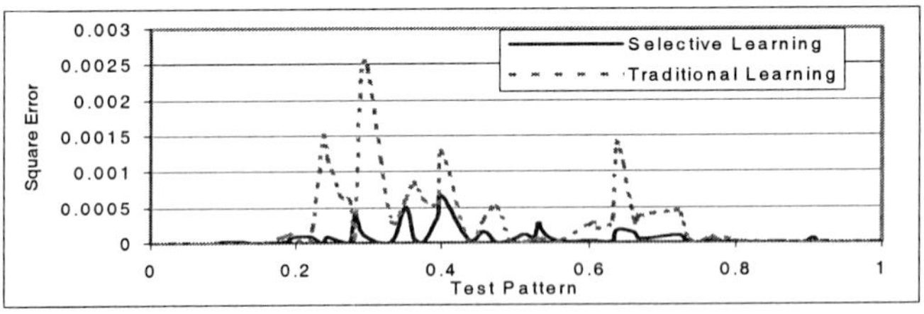

Fig. 2. Hermite function: Square errors for each test patterns.

3.2 Piecewise-Defined Function Approximation

This function has been chosen because of the poor generalisation performance that RBNN presents when approximating it. The function is given by the equation:

$$f(x) = \begin{cases} -2.186x - 12.864 & \text{if } -10 \le x < -2 \\ 4.246x & \text{if } -2 \le x < 0 \\ 10e^{(-0.05x - 0.5)} \sin\left[(0.03x + 0.7)x\right] & \text{if } 0 \le x \le 10 \end{cases}$$

The original training set is composed by 120 input-output points randomly generated by an uniform distribution in the interval [-10,10]. The test set is composed by 80 input-output points generated in the same way as the points in the training set.

As in the previous experiment, RBNNs with different number of neurones have been trained, using the whole training data until the convergence of the network has been reached, that is, either when 150 cycles are performed or when the derivative of the train error equals zero. In figure 3 (a), the mean square errors obtained for different architectures are shown. The best results have been achieved using a RBNN with 20 neurones, although no significant differences have been found for networks between 2 and 30 neurones.

The selective learning method proposed in this work has also been used to train RBNNs with different architectures during 150 learning cycles, and their generalization capability has been tested. Mean square errors on the test set achieved by these networks are shown in figure 3 (b). In this case, only 4 neurones are necessary to obtain the best results, and using more than 8 increases the error to the level of the RBNN trained with the clasical method.

The mean square errors obtained in both cases over the test set, and the learning cycles involved are shown in table 2. As it is possible to observe in table 2, the mean square error over the test set is significantly reduced when an appropriate selection of patterns is made.

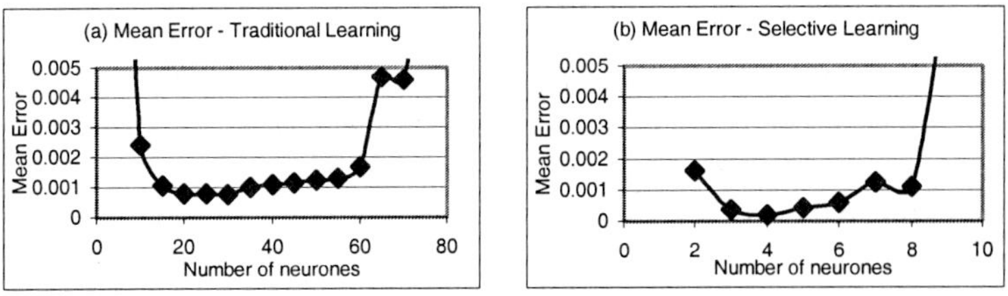

Fig. 3. Mean error on the test set for the Piecewise function achieved by different architectures.

Table 2. Performance of different training methods for the piecewise function.

	Mean square error	Number of Neurones
Training with the whole data set	$7.8x10^{-4}$	20
Training with a selection of data	$2.07x10^{-4}$	4

As in the previous experiments, the computational cost is higher when the deferred training method is used, although, on the other hand, the number of neurones is smaller, and the RBNN is trained in a shorter time. In that case, as in the previous approximation problem, the RBNNs has been trained until they reach the convergence. Thus, the generalisation capability of the network using the whole training data can not be improved if it is trained for more learning cycles.

It has been observed how the network trained with the whole training data has some difficulties to approximate the points in which the function changes their tendency. This can be described as a deficiency in the generalisation capabilities of the network. However, the generalisation is improved when an appropriate selection of patterns is made. The proposed method is able to provide better approximations of points even when the tendency of the function is changed. Figure 4 shows the errors committed by the different learning strategy for each test pattern. Most of the test patterns are better approximated when the specific learning method is used to train RBNN.

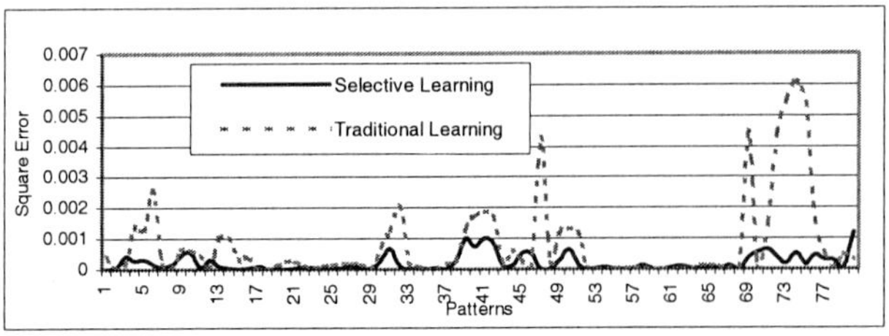

Fig. 4. Piecewise-defined function: Square errors for each set patterns.

4 Conclusions

The results presented in the previous section show that if RBNNs are trained with a selection of training patterns, the generalisation performance of the network is improved. The selection of the most relevant training patterns, helps to obtain RBNN's able to better approximate complex functions. The selective method seems to be more sensitive to the number of hidden neurones than the traditional one. However, when the selective method is used less experiments must be realized because it gets better generalization results with less hidden neurones.

Due to the reduced input space, the election of the initial centeres for the K-means algorithm is extremely important. Previous experiments have shown that if the k-means algorithm is used as usual, a lot of clusters remain empty and many hidden neurones in the RBNN are useless prejudicing the network behaviour. The proposed method to determine the initial centroid of the cluster avoids this problem.

The specific learning methods proposed in this work involves storing the training data in memory, and finding relevant data to answer a particular test pattern. Thus, the decision about how to generalise is carried out when a test pattern needs to be answered constructing local approximations. That implies a large computational cost because the network has to been trained when a new sample test is presented. However, that is not a disadvantage of the method because in many cases that computational effort can be broached and to achieve lower approximation errors is an important advantage. Moreover, the number of neurones of the network trained with the selective method is much lower, thus, the computational effort is not as high as it appears to be.

The proposed method uses the Euclidean distance as similarity measure. However, the method is flexible to incorporate other similarity measures, which will be studied in the future.

References

1. Moody J.E. and Darken C.J.: Fast Learning in Networks of Locally-Tuned Processing Units. Neural Computation 1, (1989), 281-294.

2. The Cluster Detection and Labeling (CDL)

2.1. CDL Structure and Principle

The CDL network is a feedforward network, developed in 1998 by Eltoft and DeFigueiredo [8]. It consists of four layers as shown in figure 1.

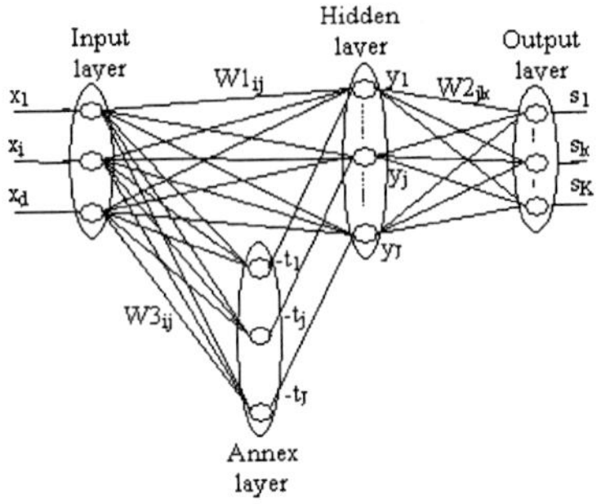

Fig. 1. CDL structure.

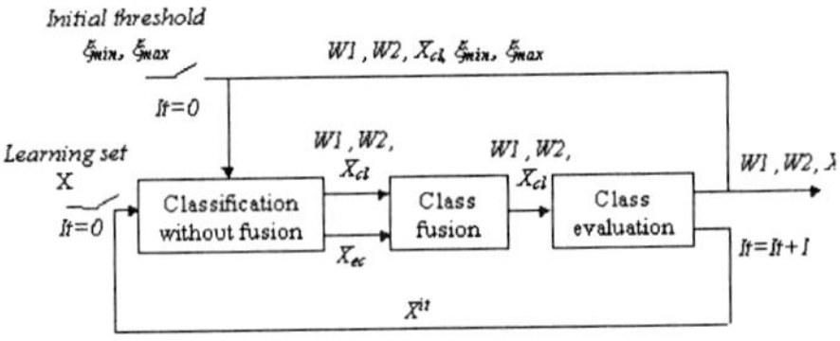

Fig. 2. CDL principle.

The input layer consists of as many neurons as components of the input vectors. The hidden layer is totally connected to the input layer, each neuron represents a prototype of a class, the number of neurons is variable. A prototype regroups a set of close input vectors. The output layer consists of as many neurons as detected classes. The annex layer makes possible the determination of the similarity threshold between an input vector and each prototype defined by neurons on the hidden layer. The connection weights between the hidden layer and the output layer characterize the relation between the prototypes and their class.

The learning and the adaptation of the network are made in three stages, as described in figure 2.

First Stage

The first and main stage is called "classification without fusion". The particularity of this stage is to make possible the creation of a new prototype if the similarity between the presented input vector and the known prototypes is very low. Similarly, a new class is created when a newly created prototype is very different from the already known prototypes. The main criterion for the creation of new prototypes or new classes is based on a distance between a new example X_i and the known prototypes P_j. In the original architecture, the similarity criterion (1) is defined as the inverse squared Euclidean distance.

$$s(P_j, X_i) = \frac{1}{D(P_j, X_i)^2} \tag{1}$$

The similarity criterion between the example and each prototype P_j is compared to two similarity thresholds $\mathbf{t}^j_{min}$ and $\mathbf{t}^j_{max}$, the output of the jth neuron of the annex layer. As a matter of fact, these thresholds are computed by the annex layer by using ξ_{min} and ξ_{max}, which are defined for each iteration, as described in figure 2.

For example, when a new sample is presented, the annex layer calculates for each prototype the two corresponding thresholds $\mathbf{t}^j_{min}$ and $\mathbf{t}^j_{max}$. Then, the hidden layer makes the comparison with the similarity criterion. So, different cases are possible, they are summarized in Table 1.

Table 1. Different cases for the first stage.

	If	Then
1st case	The similarity criterion between an example and any known prototype is smaller that the first threshold $\left(s\left(P_j, X_i\right) < t^j_{\min} \text{ for all } P_j\right)$.	The example is not close to any prototype. It is necessary to create a new prototype (a new neuron on the hidden layer) and a new class (a new neuron on the output layer).
2nd case	The similarity criterion is larger than the first threshold but smaller than the second threshold for prototypes P_j that belong to the same class. $\left(t^j_{\min} < s\left(P_j, X_i\right) < t^j_{\max}\right)$	The example is close to these prototypes, but not enough close to be associated with one of them. It is necessary to create a new prototype. This new prototype and prototypes P_j belong to the same class.
3rd case	The similarity criterion is larger than the two thresholds for some prototypes P_j that belong to the same class $\left(t^j_{\min} < t^j_{\max} < s\left(P_j, X_i\right) \text{ for } P_j\right)$	The example is associated to the prototype P_j and to its class.
4th case	The similarity criterion is larger than the first threshold for multiple prototypes that belong to different classes.	The ambiguity will be analyzed during the next stage of the learning procedure.

Second Stage

The "class fusion" stage regroups prototypes and classes that are close in the space representation. If an ambiguity is detected during the previous stage, as shown in the 4th case, the "class fusion" stage resolves it by merging the different ambiguous prototypes into the same class. So, the output layer is modified by the elimination of the neurons that defined the ambiguous classes in order to let a unique neuron. This neuron is the result of the merged classes.

Third Stage

The "class evaluation" stage makes possible the characterization of the different classes in order to modify the thresholds $\xi_{\min}$ and $\xi_{\max}$, that are used in the computation of the similarity criterion.

For example, a threshold could be used to eliminate classes containing too few assigned examples. These examples are marked as "unclassed", the similarity thresholds are modified in an iterative manner, and the "unclassed" examples are again presented to the neural network. Further information are found in reference [8].

2.2. Advantages and Constraints of the CDL

The principal advantage of the CDL is its auto-adaptive architecture thanks to the two thresholds, used for the creation of prototypes and classes. But on the other hand, the similarity computation with the definition of the annex layer is too much. Furthermore, the choice of the two similarity thresholds is not easy, in the way that the original structure of the CDL is extremely sensitive in the very close neighborhood of the prototypes.

3. Our Adaptation of the CDL

3.1. A New Activation Function for the Neurons of the Hidden Layer

Our first adaptation have consisted in modifying the activation function of the neurons of the hidden layer. We have introduced an Hyper-Elliptical function of activation, so the output of the hidden layer defines a membership degree (2) of the prototypes, by comparison with the similarity criterion of the CDL.

$$\mu(P_j, X_i) = \exp(-\frac{1}{2\alpha_{Pj}}(d(P_j, X_i))^2) \quad (2)$$

where α_{Pj} is a normalization factor defined in section 3.2.

The choice of this new activation function has two advantages. In a first time, it's less sensitive than the inverse squared Euclidean distance for close neighborhood of the prototypes. More over, by introducing an estimation of membership probability for each prototype, we can define a membership degree for each class. Lastly, for the distance, we have chosen the Mahalanobis distance. The Mahalanobis distance is chosen to give the prototypes an Hyper-Elliptical shape thanks to its mean vector M_{Pj} and its covariance matrix Σ_{Pj}.

3.2. An Adaptive and Iterative Learning of the Prototypes

In the proposed adaptation, two thresholds $\mu_{\min}$ et $\mu_{\max}$ for creating new prototypes or new classes are still used, but they are fixed for all iterations and defined for all prototypes, furthermore the annex layer is no longer use and eliminated. More over with our method, we are able to adapt directly the existing prototypes. When a new example X_i is associated to a known prototype P_j, that's to say $\mu(P_j, X_i) > \mu_{\max}$, we adapt only this prototype, in a same way than [7], but with the full covariance matrix Σ_{Pj} [5]. This adaptation is performed in an iterative manner using the following relations.

$$M_{Pj}^{k+1} = \frac{k}{k+1} * M_{Pj}^{k} + \frac{1}{k+1} * X_i \quad (3)$$

$$\Sigma_{Pj}^{k+1} = \frac{k-1}{k}\Sigma_{Pj}^{k} + \frac{1}{(k+1)}(X_i - M_{Pj}^{k})^T(X_i - M_{Pj}^{k}) \quad (4)$$

In addition, we have defined a coefficient α_{Pj} in the relation (2). [5] has introduced an equivalent coefficient as a smoothing parameter. In our situation, α_{Pj} assures a membership degree greater than $\mu_{\max}$ for all the samples that have been associated to the adapted prototype. This coefficient is also modified when we adapt a prototype, according to (5).

$$\alpha_{Pj}^{k+1} = \max_{Xi\in Pj}\left(\frac{(X_i - M_{Pj}^{k+1})^T (X_i - M_{Pj}^{k+1})}{-2\ln(\mu_{max})}\right) \qquad (5)$$

So, our first work has focused on the modification of the first stage, with the definition of the elliptical shape of the prototypes and especially their auto-adaptation.

The use of an Elliptical Basis Function for the definition of the hidden layer, leads to a better definition of the prototypes, as shown in figure 3, in the way that a prototype is not defined by one sample of the learning data and two thresholds, but by several samples of the data set.

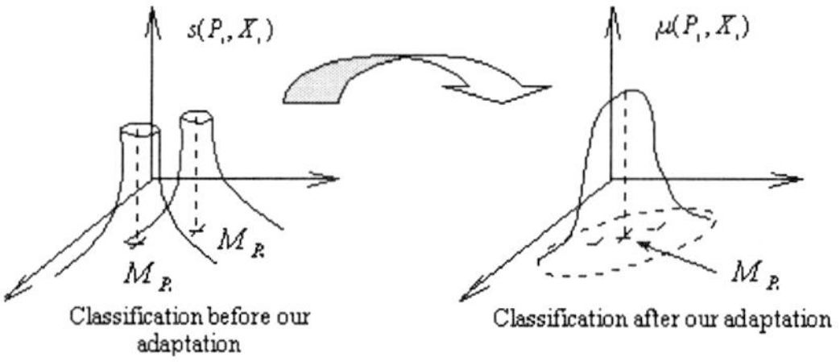

Fig 3. Its figure illustrates by a theoretic way the benefits of our adaptation. The number of prototypes is reducing, a unique prototype is creating.

3.3. Evolution of the Different Stages

We have described, in section 3.1 and 3.2, our adaptation of the CDL in terms of architecture. Here, we're going to present the modifications of the principle of learning by comparison with the initial three stages.

In the first stage "classification without fusion", the different cases defined in Table 1 are the same, except the third. In the case of $\mu(P_j, X_i) > \mu_{max}$ for a prototype P_j, then we associate the example to the prototype and to its class, but we also adapt the prototype P_j, according to section 3.2. However, the use of (3) and (4) imposes to have already a minimum number of examples associated to the prototype. That is the reason why we have defined a threshold N_{min}^P on the number of examples associated with a prototype. In the same way, we have defined a value Σ_{ini} for the initialization of the covariance matrix.

In the second stage "class fusion", no modification has been realized.

In the third stage "class evaluation", we have kept the original criterion and added a stage of "prototype evaluation" in order to verify the threshold N_{min}^P on the number of examples for the definition of a prototype. In our version, no threshold is recomputed for the next iteration.

4. Experimental Results

So as to validate the benefits of our different adaptations of the CDL, we have experimented it on simulated data.

4.1 Irregular and Elongated Classes

The simulated data are composed of three classes which are irregular and elongated, as shown in figure 4a), where the samples are the dots.

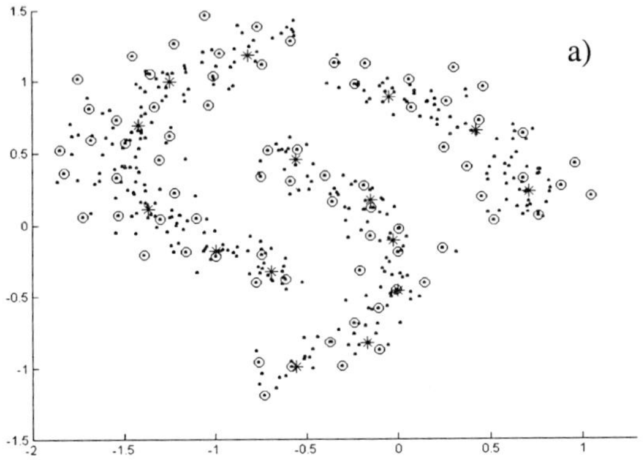

b)

	CDL Classic	CDL Adapted
Number of Prototypes	74	15
Number of Classes	3	3
Number of Iterations	2	5
N° of prototypes defined for less than 12 samples	24	1

Fig 4. Its shows a comparison between the classic and adapted CDL use, on simulated data, with some results summarized in figure 4b). The CDL classic is initialized with $\xi_{\min} = 20$; $\xi_{\max} = 40$; $N^C_{\min}$ =15. Its prototypes are represented by *circles* in fig 4a). Our adaptation of the CDL is used with $\mu_{\min} = 0.3$; $\mu_{\max} = 0.6$; $\Sigma_{ini} = 0.2^2 I$; $N^P_{\min} = 7$; $N^C_{\min}$ =10. It defines only 15 prototypes (*cross dots*)

Our adaptation of the CDL has the advantage to create less prototypes than the original version even if the number of iterations is higher. After computing, our algorithm found three classes whose each is defined by several prototypes which have been adapted in position by M_{Pj}, orientation by Σ_{Pj}, and volume by α_{Pj}. The figure 5a) illustrates the found prototypes, their positioning and their shape for the two thresholds. As a matter of fact, no one is defined at the boundary of classes by opposition with the classic CDL.

The use of elliptical basis function for the definition of the function of activation of the prototypes (hidden layer) permits to define a membership degree Ψ for each class (output layer).

$$\Psi(C_k, X_i) = \max(1, \sum_{P_j \in C_k} \mu(P_j, X_i)) \tag{6}$$

So, in figure 5b), we have represented the boundary of each class by using the definition of membership described in (6). Each boundary is defined with a threshold $\psi_{\min}$ or $\psi_{\max}$ likeness the thresholds $\mu_{\min}$ and $\mu_{\max}$ for the prototypes.

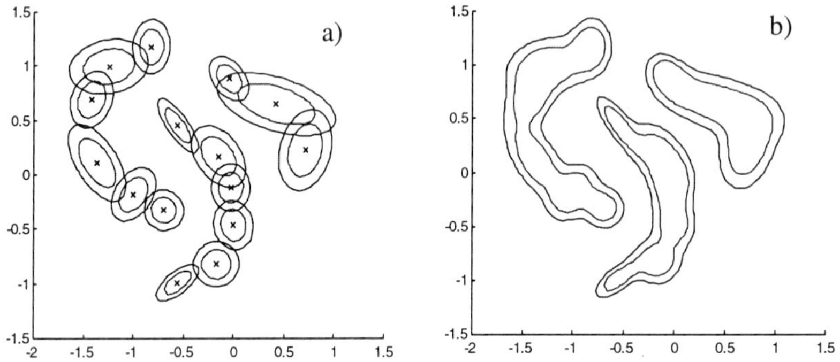

Fig. 5. a) Prototypes and their two membership thresholds, b) Definition of a boundary for each class with membership degree of the classes.

5. Conclusion

As conclusion, the introduction of the Elliptical Basis Function for the definition of the activation function, has several advantages, whose the first is to simplify the architecture of the CDL by eliminating the annex layer. Moreover, we have shown that with our version we can define a boundary for each class, thanks to the shape defined by the prototypes.

The use of our adaptation should be interesting for use in diagnosis, for detection of class evolution, and will be the subject of our future works.

References

1. Looney, C.: Pattern Recognition using Neural Networks. Theory and Algorithms for Engineers and Scientists. Oxford University Press (1997)
2. Bishop, C.: Neural Networks for Pattern Recognition. Oxford University Press (1995)
3. Gupta, L., McAvoy, Phegley, J.: Classification of temporal sequences via prediction using the simple recurrent neural network. Pattern Recognition, Vol. 33. N°10 (2000) 1759-1770
4. Gomm, J.B., Yu, D.L.: Selecting Radial Basis Function Network Centers with Recursive Orthogonal Least Squares Training. IEEE Transactions on Neural Networks, Vol. 11. N°2. (2000) 306-314
5. Mak, M.W., Kung, S.Y.: Estimation of Elliptical Basis Function Parameters by the EM Algorithm with Application to Speaker Verification . IEEE Transactions on Neural Networks, Vol. 11. N°4. (2000) 961-969
6. Mao, J., Jain, A.K.: A Self-Organizing Network for Hyper-Ellipsoïdal Clustering (HEC). IEEE Transactions on Neural Networks. Vol. 11. N°4. (1996) 16-29
7. Zheng, N., Zhang, Z., Shi, G., Qiao, Y.: Self-Creating and Adaptive Learning of RBF Networks : Merging Soft-Competition Clustering Algorithm with Network Growth Technique. International Joint Conference on Neural Networks (1999)
8. Eltoft, T., Rui Defigueiredo, J.P.: A New Neural Network for Cluster-Detection-and-Labeling. IEEE Transactions on Neural Networks. Vol. 9. N°5. (1998) 1021-1035

Independent Variable Group Analysis

Krista Lagus, Esa Alhoniemi, and Harri Valpola

Neural Networks Research Centre, Helsinki University of Technology
P.O. Box 5400, FIN-02015 HUT, Finland
{krista.lagus,esa.alhoniemi,harri.valpola}@hut.fi

Abstract. When modeling large problems with limited representational resources, it is important to be able to construct compact models of the data. Structuring the problem into sub-problems that can be modeled independently is a means for achieving compactness. In this article we introduce Independent Variable Group Analysis (IVGA), a practical, efficient, and general approach for obtaining sparse codes. We apply the IVGA approach for a situation where the dependences within variable groups are modeled using vector quantization. In particular, we derive a cost function needed for model optimization with VQ. Experimental results are presented to show that variables are grouped according to statistical independence, and that a more compact model ensues due to the algorithm.

1 Introduction

The goal of unsupervised learning is to extract an efficient representation of the statistical structure implicit in the observations. A good model is both accurate and simple in terms of model complexity, i.e., it forms a *compact representation* of the input data. Sparse coding, independent components, etc., can all be justified from the point of view of constructing compact representations. Learning amounts to searching in a model space by optimization of some cost function that measures both the accuracy of representation and—ideally at least—the model complexity.

In problems with a large number of diverse observations there are often groups of variables which have strong mutual dependences within the group but which can be considered practically independent of the variables in other groups. It can be expected that the larger the problem domain, the more independent groups there are[1]. Estimating a model for each group separately produces a more compact representation than applying the model to the whole set of variables. Compact representations are computationally beneficial and, moreover, offer better generalization.

1.1 Independent Variable Group Analysis

We suggest an approach, independent variable group analysis (IVGA), where the dependences of variables within a group are modeled, whereas the dependences

[1] Consider, for instance, creating a model of the whole world...

G. Dorffner, H. Bischof, and K. Hornik (Eds.): ICANN 2001, LNCS 2130, pp. 203–210, 2001.

between the groups are neglected. This generates a pressure towards building groups whose variables are mutually dependent but are largely independent of the variables in other groups.

Usually such variable grouping is performed by a domain expert, prior to modeling with automatic, adaptive methods. However, we claim that it is worthwhile and feasible to try to obtain groupings automatically in order to create structured, efficient large-scale models automatically.

In IVGA, each separate group can be modeled using any method, as long as a cost function that measures the quality of the representation, including both compactness and faithfullness, is derived for the model family. In particular, such a cost function can be derived within the statistical framework to generative modeling. In this approach, the best model is the one that gives the highest probability to the observations.

Furthermore, we demonstrate the approach by describing and evaluating an algorithm, $\mathrm{IVGA}_{\mathrm{VQ}}$, that can be used for computing IVGA when the model space used for modeling dependences between variables consists of vector quantizers (VQs). Variational EM-algorithm is used for adapting the VQs and for computing the cost function.

1.2 Structure of the Article

The structure of the rest of the article is as follows: first we describe two related models in section 2. In section 3 we introduce an algorithm for computing IVGA where VQs are used in modeling each variable group. Section 4 describes an experiment performed to validate the feasibility of the approach and to demonstrate the algorithm. Conclusions are presented in section 5.

2 Related Models

In multidimensional independent component analysis (MICA) [1], and its subsequent development independent subspace analysis (ISA) [4], the idea is to find independent linear feature subspaces that can be used to reconstruct the data efficiently. Thus each subspace is able to model the linear dependences in terms of the latent directions defining the subspace. The approach bears resemblance to IVGA which can be seen as a nonlinear version of MICA with the additional requirement which restricts the subspaces to be spanned by subsets of the variable axes.

Factorial vector quantization (FVQ) [3, 7] can be seen as a nonlinear version of MICA. It uses several different VQs that cooperate to reconstruct the observations and is thus very similar to $\mathrm{IVGA}_{\mathrm{VQ}}$. The structural differences of the FVQ model compared to the $\mathrm{IVGA}_{\mathrm{VQ}}$ model are: in FVQ (1) each vector in each pool or group contains *all* the components (variables) of the original input vector, and (2) these are summed to produce the value of a single output. In contrast, in $\mathrm{IVGA}_{\mathrm{VQ}}$ each variable group is modeled by exactly one vector quantizer (VQ). This leads to efficient computation as each VQ can operate independently as opposed to FVQ where the winners are found iteratively.

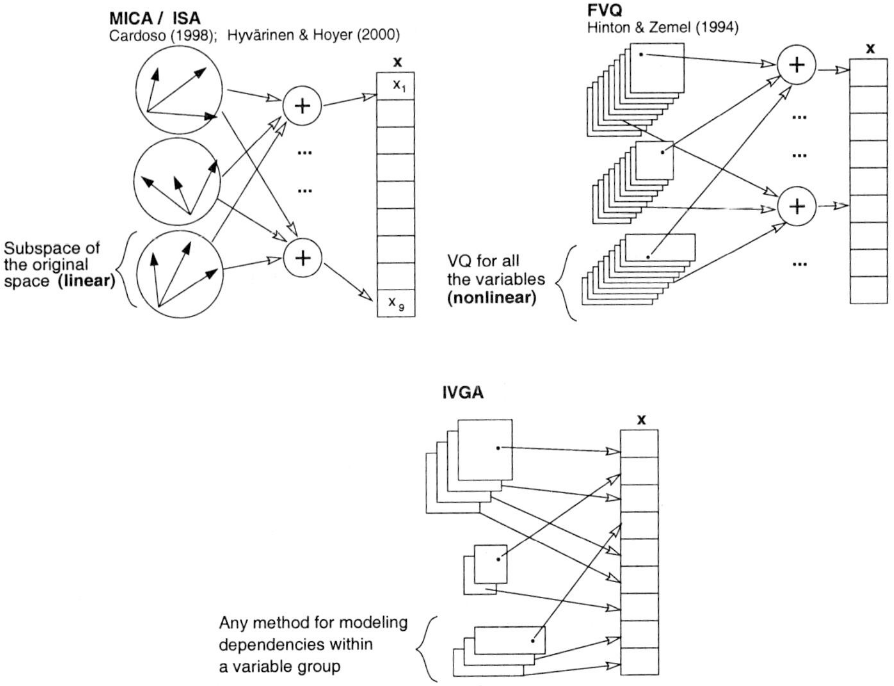

Fig. 1. Schematic illustrations of IVGA and related algorithms, namely MICA/ISA and FVQ that each look for multi-dimensional feature subspaces in effect by maximizing a statistical independency criterion. The input $\mathbf{x}$ is here 9-dimensional. The numbers of squares in FVQ and IVGA denote the numbers of variables modeled in each sub-model, and the numbers of black arrows in MICA the dimensionality of the subspaces.

Figure 1 illustrates the model structures of MICA/ISA, FVQ, and IVGA.

3 IVGA$_{\mathrm{VQ}}$: Algorithm for IVGA with VQs as Data Models

Any IVGA algorithm consists of two parts, (1) grouping variables, and (2) constructing a separate model for each variable group. An independent variable grouping is obtained by comparing models with different groupings using a suitable cost function. In principle any model can be used, if the necessary cost function is derived for the model family.

As a cost function one can use negative log-likelihood of the data given the model, namely $-\ln p(\boldsymbol{x}|H)$. The total model cost L_{tot} needed for comparing variable groupings is the sum of costs of individual variable groups $L_{\mathrm{tot}} = \sum_g L_g = \sum_g -\ln p(\boldsymbol{x}_g|H_g)$, where g is the index of a group of variables, and x_g and H_g are the data and the model related to that variable group, respectively.

In all cases where the IVGA approach is used, the same problem arises: it is computationally infeasible to try all possible different groupings of D variables into G distinct groups, where G varies from 1 to D. Thus, especially when D is large, some heuristic optimization strategy has to be utilized.

In our experiments, all the variables were initially assigned to groups of their own and then moved one by one from group to another if the movement reduced the value of the cost L_{tot}. In addition, every now and then (1) IVGA was recursively run for one group or the union of two groups (the depth of recursion was limited to one) or (2) a merge of two groups was considered.

3.1 The VQ Model and the Cost Function

To demonstrate the approach, we will describe IVGA$_{\text{VQ}}$ where each variable group is modeled using a different VQ. From now on, we will refer to the cost function of a single variable group leaving out the group index g.

The vector quantization model used for a single variable group consists of codebook vectors $\boldsymbol{\mu}(i)$, $i = 1, \dots, C$ (described by their means and a common variance) and indices of the winners $w(t)$ for each data vector $\boldsymbol{x}(t)$, $t = 1, \dots, N$.

For finding $-\ln p(\boldsymbol{x}|H)$ we use variational EM-algorithm with $\theta = \{\boldsymbol{\mu}, w\}$ as missing observations. In the E-phase, an upper bound of the cost is minimized [2, 5]. The rest of the parameters are included in H, i.e. we use ML estimates for the following: c, the hyper parameter governing the prior probability for a codebook vector to be a winner; $\boldsymbol{\sigma}^2_{\boldsymbol{x}}$, the diagonal elements of the common covariance matrix; and $\boldsymbol{\mu}_\mu$ and $\boldsymbol{\sigma}^2_{\boldsymbol{\mu}}$, the hyper parameters governing the mean and variance of the prior distribution of the codebook vectors.

The upper bound for $-\ln p(\boldsymbol{x}|H) = -\ln \int p(\boldsymbol{x}, \theta|H) d\theta$ is obtained from

$$-\ln \int p(\boldsymbol{x}, \theta|H) d\theta = -\ln \int q(\theta) \frac{p(\boldsymbol{x},\theta|H)}{q(\theta)} d\theta \leq -\int q(\theta) \ln \frac{p(\boldsymbol{x},\theta|H)}{q(\theta)} d\theta \,,$$

where the inequality follows from the convexity of $-\ln x$ by Jensen's inequality. H and $q(\theta)$ are alternately optimized but instead of finding the intractable global optimum $q(\theta) = p(\theta|\boldsymbol{x}, H)$, we restrict to finding the optimum among the tractable family of distributions of the form $q(\boldsymbol{\mu}, w) = q(\boldsymbol{\mu})q(w)$. Substituting $p(\boldsymbol{x}, \theta|H) = p(\boldsymbol{x}|w, \boldsymbol{\mu}, \boldsymbol{\sigma}^2_{\boldsymbol{x}})p(w|c)p(\boldsymbol{\mu}|\boldsymbol{\mu}_\mu, \boldsymbol{\sigma}^2_{\boldsymbol{\mu}})$, denoting the integral over $q(\theta)$ by $E\{\}$ and arranging the terms then yields

$$L = \overbrace{E\left\{-\ln p(\boldsymbol{x}|w, \boldsymbol{\mu}, \boldsymbol{\sigma}^2_{\boldsymbol{x}})\right\}}^{\text{errors}} + \overbrace{E\left\{\ln \frac{q(w)}{p(w|c)}\right\}}^{\text{winners}} + \overbrace{E\left\{\ln \frac{q(\boldsymbol{\mu})}{p(\boldsymbol{\mu}|\boldsymbol{\mu}_\mu, \boldsymbol{\sigma}^2_{\boldsymbol{\mu}})}\right\}}^{\text{codebook}} .$$

After substituting with the gaussian model for the codebook vectors the cost function becomes[2]

[2] A similar derivation for the cost function for FVQ is given in [3, 7]. For simplicity, we approximate the posterior probability of winners by the best vector only. Without significant additional computational cost all the winners could be taken into account. Individual variances for model vectors could be included as well, and we plan to do this in the future.

$$L = \sum_{j=1}^{d}\sum_{t=1}^{N} \frac{\tilde{\mu}_j(w(t)) + (x_j(t) - \bar{\mu}_j(w(t)))^2}{2\sigma_{xj}^2} + \frac{N}{2}\sum_{j=1}^{d} \ln 2\pi\sigma_{xj}^2 - \sum_{t=1}^{N} \ln c(w(t))$$
$$-\frac{Cd}{2} + \sum_{j=1}^{d}\sum_{i=1}^{C}\left[\ln\sqrt{\frac{\sigma_{\mu_j}^2(i)}{\tilde{\mu}_j(i)}} + \frac{\tilde{\mu}_j(i) + (\bar{\mu}_j(i) - \mu_{\mu_j}(i))^2}{2\sigma_{\mu_j}^2(i)}\right].$$

where $\bar{\boldsymbol{\mu}}$ and $\tilde{\boldsymbol{\mu}}$ are the mean and variance of the posterior distribution of the codebook vectors, and d is the dimensionality of the subspace, i.e. the size of the variable group.

Minimization of the Cost Function. Minimization of Eq. 1 is carried out so that C is fixed and the cost is minimized with respect to each of the variables $c, w, \boldsymbol{\sigma}_{\boldsymbol{x}}^2, \boldsymbol{\mu}_{\mu}, \boldsymbol{\sigma}_{\boldsymbol{\mu}}^2, \bar{\boldsymbol{\mu}}, \tilde{\boldsymbol{\mu}}$. The following steps are repeated iteratively as long as the value of the cost function decreases or maximum iteration count is reached (in our experiments 100 iterations). This is repeated for various values of C.

1. *Winner selection.*

$$\frac{\partial L}{\partial w(t)} = 0 \Longrightarrow w(t) = \arg\min_i \left\{ -\ln c(i) + \sum_{j=1}^{d} \frac{\tilde{\mu}_j(i) + (x_j(t) - \bar{\mu}_j(i))^2}{2\sigma_{xj}^2} \right\}$$

2. *Update of the winner.*
   (a) *Update of the posterior mean of the codebook vector.*

$$\frac{\partial L}{\partial \bar{\mu}_j(i)} = 0 \Longrightarrow \bar{\mu}_j(i) = \frac{\sigma_{xj}^2 \mu_{\mu_j} + \sigma_{\mu_j}^2 \sum_{w(t)=i} x_j(t)}{\sigma_{xj}^2 + \sigma_{\mu_j}^2 f(i)}$$

   (b) *Update of the posterior variance of the codebook vector.*

$$\frac{\partial L}{\partial \tilde{\mu}_j(i)} = 0 \Longrightarrow \tilde{\mu}_j(i) = \frac{\sigma_{xj}^2 \sigma_{\mu j}^2}{\sigma_{xj}^2 + \sigma_{\mu_j}^2 f(i)}$$

3. *Update of the data variance.*

$$\frac{\partial L}{\partial \sigma_{xj}^2} = 0 \Longrightarrow \sigma_{xj}^2 = \frac{1}{N}\sum_{t=1}^{N} \left[\tilde{\mu}_j(w(t)) + (x_j(t) - \bar{\mu}_j(w(t)))^2\right]$$

4. *Updates of codebook frequency prior and parameters of the prior distribution.*

$$c(i) = \frac{f(i)+1}{N+C}, \quad \mu_{\mu_j} = \frac{1}{C}\sum_{i=1}^{C} \bar{\mu}_j(i), \quad \sigma_{\mu_j}^2 = \frac{1}{C}\sum_{i=1}^{C} [\tilde{\mu}_j(i) + (\bar{\mu}_j(i) - \mu_{\mu_j})^2]$$

Here $f(i)$ is the number of hits of $\boldsymbol{\mu}(i)$, i.e., $f(i) = \#\{t | w(t) = i\}$.

4 Experiments

Four small experiments were run both to (1) verify the general IVGA principle (Experiments 1 vs. 2), i.e, that it is useful to model independent variable groups separately, and (2) to study the performance of the presented IVGA_{VQ} algorithm in optimizing the model for the data and in finding independent variable groups (Experiments 1 vs. 3, Experiments 2 vs. 4).

In each experiment, the best one out of several trials is reported (for VQs this meant about 10000 trials, for IVGA_{VQ}, considerably fewer). With each approach we used a roughly equal amount of computation time. In reporting results we report the number of codebook vector parameters—the number of the rest of the parameters is substantially smaller and constant in all experiments.

Data and Variables. As a data set we used features from 1000 images that have been used for content-based image retrieval in [6][3]. Each image is represented by a vector of 144 features (variables). The features fall naturally into three categories related to their origin: FFT, RGB, and texture. It is reasonable to assume that dependences between variables in different categories are weak[4]. Thus, the data provides an ideal validation for both the IVGA_{VQ} algorithm as well as the general IVGA approach. For the experiments we chose randomly a subset of 50 features: the variable set A consisted of 27 FFT features (numbered 1–27 below), B of 7 RGB features (28–34) and C of 16 texture features (35–50).

Experiment 1: VQ(A+B+C). A single VQ was used to model the combined set A+B+C of 50 variables. The optimized number of codebook vectors was 44 (number of parameters was thus $44 \times 50 = 2200$) and the cost was -138115.5.

Experiment 2: VQ(A)+VQ(B)+VQ(C). The sets A, B and C were each modeled with a separate VQ. For set A, 16 codebook vectors were used (cost -116397.7), for B, 19 codebook vectors (cost -3921.6) and for C, 30 codebook vectors (cost -25476.9). The total number of parameters was 1045 and the total cost was -145796.2. The model cost improved compared to Experiment 1 which shows that the general IVGA approach is feasible.

Experiment 3: IVGA_{VQ}(A+B+C). The feature sets A, B and C were concatenated and IVGA_{VQ} was applied for modeling and grouping. The results are presented in Table 1. Twelve groups were found, and in terms of grouping the result was excellent: each group contained variables from one variable category only. Furthermore, the improved model cost shows that presented IVGA_{VQ} algorithm performs well in the IVGA task.

---

[3] We are grateful to the PicSOM project members for kindly providing us with the image data set.

[4] This assumption has been made in [6] prior to modeling each variable category with a different SOM.

Table 1. Results of the grouping when $IVGA_{VQ}$ was run for the combined data set A+B+C. Negative costs are due to leaving out the constant factor in Eq. 1.

Group	Variables	Belong to data set	Cost	Codebook vectors	Parameters
1	1,6	A	-4760.6	7	14
2	2–5,7–8,11	A	-33213.3	10	70
3	9–10,13,15–16,18–21,24–25	A	-50997.9	12	132
4	12,14,17	A	-14201.3	8	24
5	22,26	A	-8233.6	5	10
6	23,27	A	-4920.4	10	20
7	28–34	B	-3889.7	20	140
8	35–38	C	-6678.2	7	28
9	39–42	C	-7248.3	15	60
10	43–44	C	-3140.6	9	18
11	45–46	C	-2708.1	13	26
12	47–50	C	-7215.0	19	76
Total			**-147206.8**	**135**	**618**

Experiment 4: $IVGA_{VQ}$(A) + $IVGA_{VQ}$(B) + $IVGA_{VQ}$(C) Next, we simulated a situation where the three main groups of variables (A, B and C) are known based on prior information. Now the $IVGA_{VQ}$ algorithm was used to find the best possible sub-groupings for A, B and C separately. The results are shown in Table 2. Note also that the variable groupings obtained are almost, but not exactly, identical in Experiments 3 and 4. The slight improvement in total cost compared to Experiment 3 is as expected, since computational resources needed not be allocated for separating the groups A, B and C.

The results of the four experiments are summarized in Table 3.

5 Conclusions

In this article, the basic idea of IVGA, i.e., grouping variables according to dependences within the data, was presented and motivated, as well as shown to work in practice. An efficient algorithm was suggested for computing IVGA and shown to work rather well on real data. In the experiments vector quantization (VQ) was used to model the dependences within variable groups. In the calculation of the cost function most of the unknown variables are marginalized out and therefore the cost function can reliably be used for model selection.

One should keep in mind that we used VQ just as a simple example: it could be replaced with any method as long as the necessary cost function is derived for the method. Each variable group could even be modeled using a different method.

The presented approach has implications both for constructing better models of large and complex phenomena and for the efficient computation of such models. However, further research is needed to study these issues in depth.

Table 2. Results when IVGA_{VQ} was run separately for each set A, B and C.

Variable set	Group	Variables	Cost	Codebook vectors	Parameters
A	1	1,6	-4773.4	9	18
A	2	2–5, 7–8, 11	-33432.8	12	84
A	3	9–10,13,16,18,20,25	-32122.6	13	91
A	4	12,14–15,17,19,21–24,26–27	-46778.0	14	154
B	1	28–34	-3868.6	17	119
C	1	35–38	-6646.1	8	32
C	2	39–42,45–46	-9934.4	26	156
C	3	43–44	-3190.7	7	14
C	4	47–50	-7187.7	11	44
Total			**-147934.3**	**117**	**712**

Table 3. Summary of results of all experiments.

Experiment	Total cost	#VQs	Parameters
1. VQ(A+B+C)	-138115.5	1	2200
2. VQ(A) + VQ(B) + VQ(C)	-145796.2	3	1045
3. IVGA_{VQ}(A+B+C)	-147206.8	12	618
4. IVGA_{VQ}(A) + IVGA_{VQ}(B) + IVGA_{VQ}(C)	-147934.3	9	712

References

1. J.-F. Cardoso. Multidimensional independent component analysis. In *Proceedings of ICASSP'98*, Seattle, 1998.
2. G. E. Hinton and D. van Camp. Keeping neural networks simple by minimizing the description length of the weights. In *Proceedings of the COLT'93*, pp. 5–13, Santa Cruz, California, USA, July 26–28, 1993.
3. G. E. Hinton and R. S. Zemel. Autoencoders, minimum description length and Helmholtz free energy. In J. et al, ed., *Neural Information Processing Systems 6*, San Mateo, CA, 1994. Morgan Kaufmann.
4. A. Hyvärinen and P. Hoyer. Emergence of phase and shift invariant features by decomposition of natural images into independent feature subspaces. *Neural Computation*, 12(7):1705–1720, 2000.
5. H. Lappalainen and J. W. Miskin. Ensemble learning. In M. Girolami, ed., *Advances in Independent Component Analysis*, pp. 76–92. Springer-Verlag, Berlin, 2000.
6. E. Oja, J. Laaksonen, M. Koskela, and S. Brandt. Self-organizing maps for content-based image database retrieval. In E. Oja and S. Kaski, eds., *Kohonen Maps*, pp. 349–362. Elsevier, 1999.
7. R. S. Zemel. *A Minimum Description Length Framework for Unsupervised Learning*. PhD thesis, University of Toronto, 1993.

Weight Quantization for Multi-layer Perceptrons Using Soft Weight Sharing

Fatih Köksal[1], Ethem Alpaydın[1], and Günhan Dündar[2]

[1] Department of Computer Engineering
[2] Department of Electrical and Electronics Engineering
Boğaziçi University, Istanbul Turkey

Abstract. We propose a novel approach for quantizing the weights of a multi-layer perceptron (MLP) for efficient VLSI implementation. Our approach uses *soft weight sharing*, previously proposed for improved generalization and considers the weights not as constant numbers but as random variables drawn from a Gaussian mixture distribution; which includes as its special cases k-means clustering and uniform quantization. This approach couples the training of weights for reduced error with their quantization. Simulations on synthetic and real regression and classification data sets compare various quantization schemes and demonstrate the advantage of the coupled training of distribution parameters.

1 Introduction

Since VLSI circuits must be produced in large amounts for economy of scale, it is necessary to keep the storage capacity as low as possible to come up with cheaper products. In an artificial neural network, the parameters are the connection weights and if they can be stored using fewer bits, storage need will be reduced and we gain from memory.

In this work, we try to find a good quantization scheme to achieve a reasonable compression ratio for parameters of the network, without significantly degrading accuracy. Our proposed method finds a method to partition the weights of a neural network into a number of *clusters* so that only one value is used for one cluster of weights. Thus, the actual memory which stores the real-numbered values will be small and weights will be pointers to this memory. Then, the weights in a cluster will point to the same location in the real memory. For example, given an MLP with 10,000 weights that can be grouped into 32 clusters, for each weight only five bits are used.

An analogy is the color map. To get 16 million colors, one requires 24 bits for each pixel. Graphics adapters have color maps, e.g., of size 256, where each entry is 24 bits and is one of 16 million colors. Then using eight bits for each pixel, we can index one entry in the color map. So the storage requirement for an image is one third. Although this means that an image can contain only 256 of the 16 million possible colors, if quantization [1] is done well, there will not be a degradation of quality. Our aim is to do a similar quantization of weights of a large MLP.

G. Dorffner, H. Bischof, and K. Hornik (Eds.): ICANN 2001, LNCS 2130, pp. 211–216, 2001.

The benefit is obvious for digital storage in reducing the memory size. For analog storage of weights, capacitors are perhaps the most popular method, and in such circuits, the area of the capacitors generally dominates the total area. Thus, it is very important to reduce the number of weights.

There is a great amount of research on the quantization and efficient hardware implementation of neural networks (specifically MLP). The aim is to get an MLP with weights which are represented by less number of bits, which consumes less memory at the expense of loss of precision; example studies are given in [2–6]. In our approach which is novel, we do not reduce the precision; the operations are still with full precision. We decrease storage by clustering of the weights.

The organization of the paper is as follows: In Section 2, we discuss the possible methods for quantization and divide them into two groups: after-training methods and during-training methods. Section 3 contains experimental design and the results of these methods, and in Section 4, we conclude and discuss future work.

2 Soft Weight Sharing

We can generally classify the applicable methods for weight quantization into two. The simplest method would be training the neural network normally and then directly applying quantization to the weights of the trained network. We call these *after-training* methods. However, the application of quantization without considering the effect of quantization leads to large error and therefore combining quantization and training is a better alternative which we call *during-training* methods [5].

The methodology we use is *soft weight sharing* [7] where it is assumed that the weights of an MLP are not constants but are random variables drawn from a mixture of Gaussians

$$p(w) = \sum_{j=1}^{M} \alpha_j \phi_j(w) \tag{1}$$

where w is a weight, α_j are the mixing coefficients (prior probabilities), and the component densities $\phi_j(w)$ are Gaussians, i.e., of the form $\phi_j(w) \sim \mathcal{N}(\mu_j, \sigma_j^2)$. The main reason for choosing this type of distribution for weights is its generality and analytical simplicity. There are three types of parameters in the mixture distribution, namely, prior probabilities, α_j, means, μ_j, and variances, σ_j. Assuming that the weights are independent, the likelihood of the sample of weights is given by

$$L = \prod_{i=1}^{W} p(w_i) \tag{2}$$

In *after-training*, once the training of the MLP is complete, Expectation-Maximization (EM) method can be used to determine the parameters [8]. It is known that when all priors and variances are equal, the well-known quantization method *k-means clustering* is equivalent to EM. One can even view *uniform quantization* in this framework where additional to all priors and variances being equal, means are equally spaced and fixed.

Table 1. The test set error values (average±standard deviation) on the regression dataset for several quantization levels. The unquantized MLP has an error of 4.10±0.06.

	1 Bit	2 Bits	3 Bits
Uniform	245.33±137.60	104.87±60.33	40.64±23.01
K-means	173.34±87.59	57.60±20.97	16.85±13.71
SWS	18.85±2.27	9.99±4.91	4.61±1.05

In *during-training* methods, to couple the training of weights with their quantization, we take the negative log likelihood converting it to an error function to be minimized

$$\Omega = -\sum_i \ln \left(\sum_{j=1}^{M} \alpha_j \phi_j(w_i) \right) \tag{3}$$

This then is added as a penalty term to the usual error function (mean square error in regression and cross-entropy in classification)

$$\overline{E} = E + \upsilon \Omega \tag{4}$$

to get the augmented error function $\overline{E}$ which is then minimized, e.g., using gradient-descent, to learn both the weights w_i, and also the parameters of their distribution, i.e., μ_j, σ_j, and α_j. We do not give the update equations here due to lack of space but they can be found in [7].

3 Experimental Results

We have used one synthetic regression dataset for visualization and two real datasets for speech phoneme and handwritten digit recognition. All datasets are divided as training and test sets. For all problems, we first train MLPs and find the optimum number of hidden units and the number of parameters, then they are quantized with different number of bits. We have run each model ten times, starting gradient-descent from different random initial weights and report the average and standard deviation of error on the test set.

The MLP of regression problem has one input, one output and four hidden units thus using 13 weights. Thus without any quantization, we need four bits. We try quantizing with two, four, and eight clusters (Gaussians) corresponding to one, two, and three bits. An example graph of regression function as quantized by soft weight sharing is given in Figure 1. In this figure, the effect of quantization on the regression function is easily observed.

The means and the variances of error with the three methods are given in Table 1. k-means is an after-training method and works better than uniform quantization. Still better though is the result with soft weight sharing (SWS), which is a *during-training* method and clearly demonstrates the advantage of coupled training.

The speech phoneme is represented as a 112 dimensional input which contains 200 samples from six distinct classes (six speech phonemes) and is equally divided

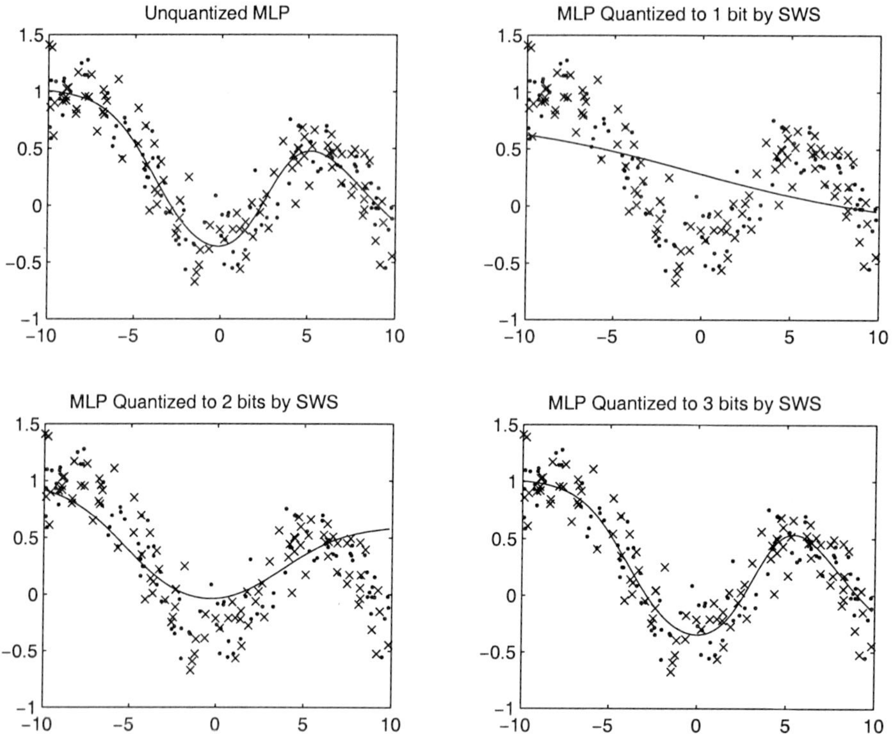

Fig. 1. The graphs of regression function; unquantized and quantized by soft weight sharing (SWS) with one, two, and three bits.

into two as training and test sets. The MLP has ten hidden units with 1196 parameters and thus unquantized, we need 10 bits. We have used one, two, three, four and five bits for our performance evaluation. Table 2 reports the simulation results. In Figure 2, we draw the scatter of weights and the fitted probability distribution using soft weight sharing.

Note that the probability distribution of weights is clearly a Gaussian distribution centered near zero. Although, there are some papers (e.g. [3]) which claim and assume that the weights of an MLP are distributed uniformly, we face a zero-mean normal distribution. This must not be surprising because there are more than 1000 parameters and they are initialized randomly to be close to zero and a large majority are not much updated later on. Note that if we were using a pruning strategy, the connections belonging to a cluster which is centered very close to zero would be the ones to be pruned. The significant ones are those which are distant from zero, since they have large error gradient and are updated much larger than the other parameters.

The handwritten digit dataset contains 1,200 samples each of which is a 16×16 bitmap. The MLP has ten hidden units and we have 2,680 parameters for quantization. The distribution of weight values after the training phase of

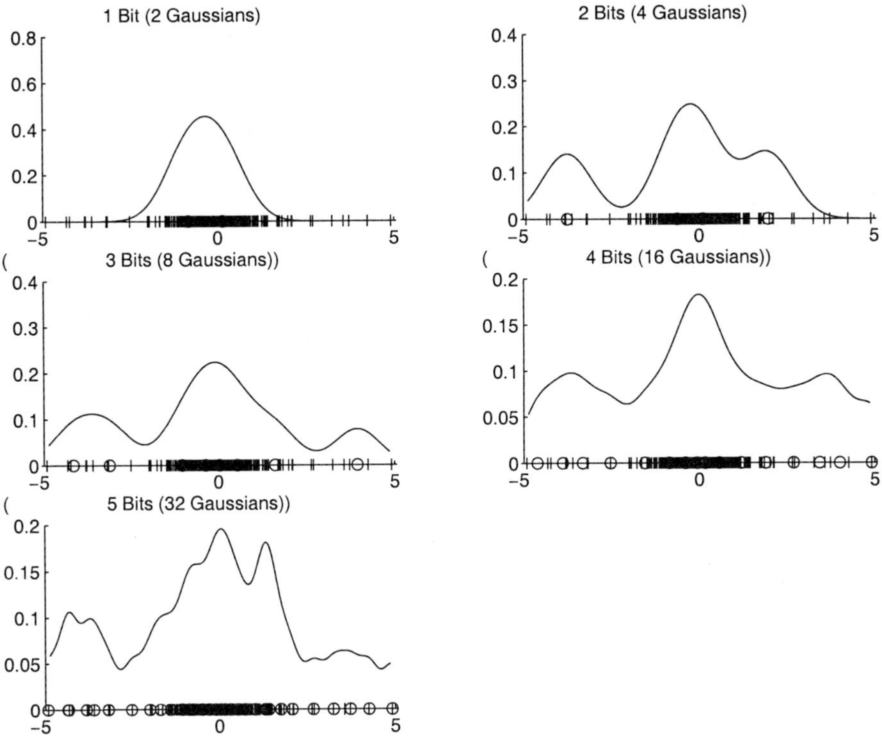

Fig. 2. The distribution of weights of the MLP used for speech phoneme recognition and the fitted Gaussian mixture probability distribution by soft weight sharing. Vertical bars are the weights; circles are the centers of the Gaussians.

MLP is again a normal distribution like with the speech data set. Table 3 contains the quantization results.

We see in both classification datasets that when the number of bits is large, even uniform quantization is good; the advantage of soft weight sharing, i.e., coupled training, becomes apparent with small number of clusters.

4 Conclusions

We propose to use soft weight sharing, previously proposed for improved generalization, for quantizing the weights of an MLP for efficient VLSI implementation and compare it with the previously proposed methods of uniform quantization and k-means. Our results indicate that soft weight sharing, because it couples the training of the MLP with quantization, leads to more accurate networks at the same level of quantization. Once quantization is done, the results also indicate the *saliency* of weights which can be used further to prune the unnecessary connections; we leave this as future work.

Table 2. On the speech phoneme recognition problem, average±standard deviation of number of misclassifications out of 600 on the test set are given. For comparison, with the unquantized MLP using 10 bits, the misclassification error is 40.70±3.79.

	1 Bit	2 Bits	3 Bits	4 Bits	5 Bits
Uniform	409.29±80.36	376.50±77.40	219.89±55.35	104.19±33.69	55.50±8.96
K-means	365.10±82.93	248.00±69.31	123.40±48.74	51.59±8.66	45.50±5.93
SWS	387.50±74.10	209.39±37.55	89.00±45.40	52.00±9.85	44.79±5.92

Table 3. On the handwritten digit recognition problem, average±standard deviation of number of misclassifications out of 600 of the MLP on the test set are given. For comparison, with the unquantized MLP, the misclassification error is 23.50±3.58.

	1 Bit	2 Bits	3 Bits	4 Bits	5 Bits
Uniform	526.70±21.75	512.70±42.37	441.70±70.21	95.19±37.08	29.29±5.67
K-means	448.39±114.14	263.29±75.21	58.79±27.43	27.70±5.38	24.50±2.80
SWS	397.20±139.41	244.19±53.66	64.19±37.24	30.70±7.28	27.90±6.47

Acknowledgment

This work is supported by Grant 00A0101D from Boğaziçi University Research Funds.

References

1. Gersho, A. and R. Gray, *Vector Quantization and Signal Compression* Norwell, MA:Kluwer, 1992.
2. Choi, J. Y. and C. H. Choi, "Sensitivity Analysis of Multilayer Perceptron with Differentiable Activation Functions," *IEEE Transactions on Neural Networks*, Vol. 3, pp. 101–107, 1992.
3. Xie, Y. and M. A. Jabri, "Analysis of the Effects of Quantization in Multi-Layer Neural Networks Using a Statistical Model," *IEEE Transactions on Neural Networks*, Vol. 3, pp. 334–338, 1992.
4. Skaue, S., T. Kohda, H. Yamamato, S. Maruno, and Y. Shimeki, "Reduction of Required Precision Bits for Back Propagation Applied to Pattern Recognition," *IEEE Transactions on Neural Networks*, Vol. 4, pp. 270–275, 1993.
5. Dündar, G. and K. Rose, "The Effects of Quantization on Multi Layer Neural Networks," *IEEE Transactions on Neural Networks*, Vol. 6, pp. 1446–1451, 1995.
6. Anguita, D., S. Ridella and S. Rovetta, "Worst Case Analysis of Weight Inaccuracy Effects in Multilayer Perceptrons," *IEEE Transactions on Neural Networks*, Vol. 10, pp. 415–418, 1999.
7. Nowlan, S. J. and G. E. Hinton, "Simplifying Neural Networks by Soft Weight Sharing," *Neural Computation*, Vol. 4, pp. 473–493, 1992.
8. Alpaydın, E. "Soft Vector Quantization and the EM Algorithm," *Neural Networks*, Vol. 11, pp. 467–477, 1998.

Voting-Merging: An Ensemble Method for Clustering

Evgenia Dimitriadou, Andreas Weingessel, and Kurt Hornik

Institut für Statistik, Wahrscheinlichkeitstheorie
und Versicherungsmathematik
Technische Universität Wien
Wiedner Hauptstraße 8–10/1071
A-1040 Wien, Austria
{dimi,weingessel,hornik}@ci.tuwien.ac.at
http://www.ci.tuwien.ac.at

Abstract. In this paper we propose an unsupervised voting-merging scheme that is capable of clustering data sets, and also of finding the number of clusters existing in them. The voting part of the algorithm allows us to combine several runs of clustering algorithms resulting in a common partition. This helps us to overcome instabilities of the clustering algorithms and to improve the ability to find structures in a data set. Moreover, we develop a strategy to understand, analyze and interpret these results. In the second part of the scheme, a merging procedure starts on the clusters resulting by voting, in order to find the number of clusters in the data set[1].

1 Introduction

Clustering is the partitioning of a set of objects into groups so that objects within a group are "similar" and objects in different groups are "dissimilar". Thus, the purpose of clustering is to identify "natural" structures in a data set. In real life clustering situations usually the following main problems are encountered: First, the true structure, especially the number and shapes of the clusters, is unknown. Second, different cluster algorithms and even multiple replications of the same algorithm result in different solutions due to random initializations and stochastic learning methods. Moreover, there is no clear indication which of the different solutions of the replications of the algorithm is the best one.

Every cluster algorithm tries to optimize some criterion, like minimizing the mean-square error. If the task of clustering is to compress the data by mapping every data point to a prototype, then the minimization of an appropriate error measure is the right thing to do, but if the task is to find structures in the data, these optimization criterion might help to find a good solution, but they are

[1] This piece of research was supported by the Austrian Science Foundation (FWF) under grant SFB#010 ('Adaptive Information Systems and Modeling in Economics and Management Science').

G. Dorffner, H. Bischof, and K. Hornik (Eds.): ICANN 2001, LNCS 2130, pp. 217–224, 2001.

not necessarily the right measures to optimize. Especially, if the (generally non-Gaussian) probability distribution of the given data set is completely unknown, it is not clear which criterion to use.

To tackle these problems, we handle results of various runs by using the existing idea of voting in classification [1, 3]. We develop an algorithm which allows a combination between several results of a cluster algorithm (voting) resulting in a common partition. The idea is that the voting procedure can be applied to any existing algorithm that has instable results. The result of our voting is a "fuzzy" partition of the data. We propose the idea that there are cases where an input point has a structure with a certain degree of confidence and may belong to more than one cluster with a certain degree of "belongingness". With this method the output of several single classifiers can be combined so as to reduce the variance of the error between the different runs and to get an overall decision made by the combined classifiers. Additionally, voting can react to the tendency of every algorithm to create clusters with a specific kind of structure (e.g., k-means is creating round clusters) and may result also to non convex ones.

In the following steps, we take advantage of that "fuzzy" partition of the data to introduce some new measures for handling and understanding these results, and to develop a method for finding the right number of clusters in a data set. This technique is followed by a merging procedure, where clusters resulting from voting are merged according to the highest probability of their data points to belong to another cluster. This procedure continues until every cluster is merged, and then a decision according to some criteria is taken in order to specify the optimal number of clusters in the data set.

Consequently, our Voting-Merging Algorithm (VMA) represents a complete scheme of a clustering algorithm, able to partition data points of a set in clusters and also to find the optimal number of classes existing in the set.

The paper is organized as follows. In Sections 2 we present the Voting-Merging Algorithm (VMA) and its implementation. Section 3 demonstrates our experimental results and some comments on them. Finally a conclusion of the paper is given in Section 4.

2 Description of the VMA

Generally, the VMA is a scheme consisting of 3 procedures, namely the repeated runs of a clustering algorithm, the voting procedure receiving these results, and finally the merging procedure which concludes finding the number of clusters in a data set. The 3 levels of the algorithm are applied sequentially. They do not interfere with each other, but they just receive the results from the previous levels. No feedback process happens, and the algorithm terminates after the completion of all procedures.

2.1 Voting Procedure

In classification, there is a fixed set of labels which are assigned to the data. Therefore, we can compare for every input x the results of the various classifiers, i.e., the labels assigned to x and apply a voting procedure between these results. Things are different in clustering, because different runs of a cluster algorithm can result in different clusters, which might partition the input data in totally different ways. Thus, there is the problem to decide which cluster of one run corresponds to which in another run.

As an example, suppose that we have the results of several runs i of cluster algorithms which partition our data into 3 clusters $C_{ij}, j = 1, 2, 3$. When combining the first two runs we have to decide which of the clusters C_{11}, C_{12}, C_{13} of the first run is similar to which cluster C_{21}, C_{22}, C_{23} of the second run, if there is any similarity at all. Note that as opposed to classification the numbers $1, 2, 3$ are arbitrarily assigned to the clusters and their order can be interchanged. If we combine more than 2 runs, the additional difficulty arises that the similarity relation is not transitive. That is, cluster C_{11} of the first run might be similar to cluster C_{22} of the second run which might again be similar to cluster C_{31} of the third run and so on, but this does not mean that C_{31} is again similar to C_{11}. It might even turn out that C_{31} is more similar to C_{12} for example.

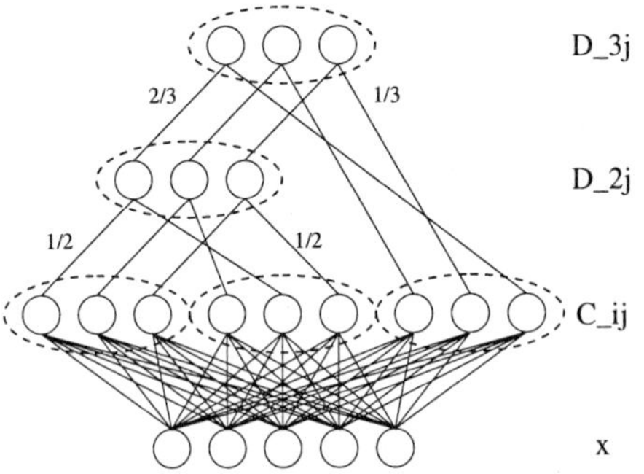

Fig. 1. The Voting Procedure

Since there seems to be no obvious way for combining more than two runs at once, we developed the following procedure, which is depicted by the network in Figure 1. Our input data x is clustered several times, the second layer (C_{ij}) of Figure 1 symbolizes three runs with three clusters each. Let C_{ij} denote the jth cluster in the ith run and D_{ij} denote the jth cluster in the combination of the first i runs. Our procedure works on an iterative basis. The first two runs are combined in the following way. First a mapping between the clusters of the two runs is defined. To do this we compute for each cluster C_{2j} how many percent of its points have been assigned to which cluster C_{1k}. Then, the two clusters with the highest percentage of common points are assigned towards each other. Of the remaining clusters, again the two with the highest similarity are matched and so on. After renumbering the clusters of the second run so that C_{2j} corresponds to $C_{1j}, \forall j$, we assign the points to the common clusters D_{2j} in the following way. If a data point x has been assigned to both C_{1j} and C_{2j} it will be assigned to D_{2j}. If x has been assigned to C_{1j} in the first run and to C_{2k} with $j \neq k$ in the second run then it will be assigned to both, D_{2j} and D_{2k}, with weight 0.5. This step is shown in the connection between the second layer (C_{ij}) and third layer (D_{2j}) in Figure 1. Every cluster of D_{2j} receives input from exactly one cluster of the first run and one of the second run. The weights of these connections are set to $\frac{1}{2}$.

If we have already combined the first n runs in a common clustering D_{nj} we add an additional run $C_{(n+1)j}$ by combining it with D_{nj} in the same way as for two runs, but give weight $n/(n+1)$ to the common clusters of the first n runs and weight $1/(n+1)$ to the new cluster.

Note that this voting procedure gives equal weights to all repetitions of the cluster algorithm and there is no additional gating network [9] which learns to decide between the different results. The reason is that, as opposed to supervised learning, we can not decide which of the various results is the best.

2.2 The Partition Resulting from the Voting Procedure

For the results of the voting procedure the data points are typically not uniquely assigned to one cluster, but there is a "fuzzy" partition. That is, after voting of N runs we get for every data point x and every cluster j a value $D_{Nj}(x)$ which gives the fraction of times x has been assigned to cluster j. For interpreting the final result we can either accept this fuzzy decision or assign every data point x to that cluster $k = \text{argmax}_j D_{Nj}(x)$ where it has been assigned most often. We define the *sureness* of a data point as the percentage of times it has been assigned to its "winning" cluster k, that is $sureness(x) = \max_j D_{Nj}(x)$. Then we can not only see how strong a certain point belongs to a cluster but we can also compute the average sureness of a cluster ($Avesure$) as the average sureness of all the points of a cluster that belong to it, i.e., $avesure(k) = \text{mean}_{x \in k} D_{Nk}(x)$. In this way we notice which clusters have a clear data structure and which not.

2.3 Merging Procedure

As already mentioned in the previous section, after voting every data point x belongs to more than one cluster j with a certain degree of "belongingness" $D_{Nj}(x)$. If we set $n(k,j) := \text{mean}_{x \in k} D_{Nj}(x)$ we get a measure of how strong the points of cluster k belong to cluster j. Thus, $n(k,j)$ defines a (non-symmetric) neighborhood relation between two clusters. So, we can say that a cluster l is the closest cluster to cluster k, if $n(k,l) = \max_{j \neq k} n(k,j)$[2].

We can use this neighborhood relation to develop a merging procedure that starts with many clusters and merges clusters which are closest to each other. More specifically, two clusters k and l are merged together, if k is the closest cluster to l and l is the closest cluster to k. Additionally, we merge "chains" of clusters. That is, a set of clusters $k_1, k_2, \ldots, k_n$ is merged if k_{i+1} is the closest cluster to k_i ($i = 1, \ldots, (n-1)$) and k_1 is the closest cluster to k_n.

[2] Note that cluster k is not necessarily the closest cluster to cluster l.

After the merging step, the *sureness* of the points and *avesure* of the clusters are recomputed by adding up the values of the merged clusters. Then, the merging step repeats until some stopping criterion is met.

The criterion that derives directly from this procedure is that merging will stop when all clusters end up to have an *avesure* of 100%. Since in practice this condition is not always possible to be realized, for example due to number precision problems, we decide to stop the merging procedure when the *avesure* of every cluster is greater than 99%. That is, on average every point is assigned less than once to a "wrong" cluster, if voting has been applied with 100 runs. Thus, we have in that case a significant decision that every point belongs to the "right" cluster. In this case the probability of a cluster to be merged with another is extinguished, which means that the procedure terminates normally. Unfortunately, this happens only in the case where a simple and clear structure, especially a non overlapping one, exists in a data set. In other situations clusters are able to reach a "very sure" structure (of more than 90%) but not the one of 100% *avesure*. Note that clusters with one data point are not considered as clusters but the point is partitioned to the next cluster with the highest probability to "win" it. That is due to the fact that, obviously, one point clusters can become very easily 100% "sure" and stop merging.

Since it is non realistic in real-world data sets to reach a 100% "sure" solution, it is still probable to happen that the "real" clusters existing in a data set will increase all the time their *avesure* during their construction by the merging procedure and the rest will stay quite "unsure" in comparison. The problem logically arising is that after reaching by merging the "real" clusters existing in the data set, every merging step will also continue increasing the mean *avesure* of all the clusters and consequently it is not always clear when the merging should be stopped. However, this increase that takes place is not significant, because the clusters are already to that point quite "sure".

Thus, to tackle these problems we suggest the following criterion (see [2]):

$$\begin{aligned} devsure(i) := & \big[numsure(i) - numsure(i-1)\big] \\ & - \big[numsure(i+1) - numsure(i)\big] \end{aligned}$$

where $numsure(i)$ is the sum of the $sureness(x)$ of all points in all the n clusters in the ith step of the merging. This second order difference shows how much the solution for the n clusters existing in the i step of merging deviates from the general increase (when n decreases) of the *sureness*. The solution with the maximum value for $devsure(i)$ is the one whose cluster structure found by voting and merging is the optimal for the algorithm.

2.4 Implementation

We used different clustering algorithms in our experiments, like k-means (also known as LBG, [4]) as an example of an off-line algorithm, hard competitive learning as an online version of k-means (for example [10]) and neural gas [6].

The experiments presented here consist of 100 runs of one clustering algorithm with a big number of clusters specified in the beginning. There is no formula to determine the value of this number, but experimental results show that it should be bigger proportional to the size of the data set. This is logical, in the way that if a small data set is clustered with a big number of clusters then it will be unavoidable that we will have many clusters of few points (for example 2 points) and also many empty ones. In the case of too many clusters we have the problem of having in the beginning of the algorithm structures being 100% "sure" with only very few points (for example 2), something that does not correspond to reality, when in the case of many empty clusters, it makes simply the decision of clustering with a big number of clusters meaningless. Moreover, in the case of big data sets it is proposed to cluster them with many centers, to avoid points belonging to different clusters but being close neighbors to be clustered together. For the same reason the number of the clusters, with which a clustering algorithm runs, should be also big the more overlapping the structures of the data sets appear to be, for example generally the real-world data sets.

After this step a voting between these runs follows and then a merging according to voting results. Merging stops when we reach the 2 clusters in a data set. Since we are for our data sets aware of the true cluster structure, we can treat the result of a cluster algorithm as a classification problem, [8], although we do not use the class information during clustering. That is, we can compute how many points have been assigned by the cluster algorithm to the right cluster.

All our experiments have been performed in R, a system for statistical computation and graphics, which implements to well-known S-language for statistics. R runs under a variety of Unix platforms (such as Linux) and under Windows95/98/NT. It is available freely via CRAN, the Comprehensive R Archive Network, whose master site is at `http://www.R-project.org`.

3 Experimental Results

An artificial data set as well as a real-world set are used to demonstrate the performance of the VMA. A description of these data sets, some figures, as well as some comments on the results, follow in order to make clear and demonstrate the performance of the algorithm. Each voting run has been performed with the results of 100 new cluster runs.

3.1 A 2-Dimensional Example

This data set is also a 2-dimensional one, and it is a simplification of the one described and used in [5]. It consists of 2000 data points belonging to 2 elongated, curved clusters with opposite curvature (see Figure 2(a)). The two curves with 1000 data points each are generated in the following way; one according to $y = 2(x-1)^2 + \epsilon$ where the x are uniformly distributed in $[0,2]$ and ϵ is a normally distributed noise term with zero mean and standard deviation 0.1; the other is $y = 3 - 2(x-2)^2 + \epsilon$ with x uniformly distributed in $[1,3]$ and ϵ as above. This data set is not linearly separable.

After voting between 100 results of hard competitive learning runs (starting with 30 clusters), merging also results partitioning perfectly the data points to their clusters (100% classification rate). Here, since we reach 100% "avesure" clusters merging stops without the help of the criterion.

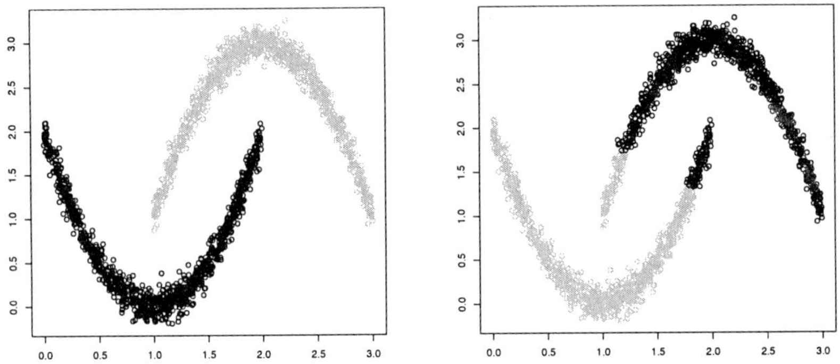

(a) 2-Dimensional Data Set (b) Clustering Result

Fig. 2. Elongated Curves

Also by this data set, hard competitive single runs, with the number of the 2 clusters specified, can never find the two curved clusters by misclassifying 10% of the data points on average (a typical result of hard competitive in Figure 2(b)).

3.2 Real-World Data Set: DNA

We apply a clustering algorithm to the DNA data set (see [7]). Basically, the DNA set is considered as a classification problem because the true classes are known. But because there are no well established benchmark data sets for cluster problems, we treat the problem as a clustering one. It consists of 2000 data points (splice junctions) on a DNA sequence at which "superfluous" DNA is removed during protein creation. The data points are described by 60 indicator binary variables and the problem is to recognize the 3 classes (ei, ie, neither), i.e., the boundaries between exons (the parts of the DNA sequence retained after splicing) and introns (the parts of the DNA sequence that are spliced out).

VMA also here succeeds in detecting the right number of clusters in the data set with a classification rate of 85.56%. The voting procedure has been performed on 100 runs of the hard competitive algorithm (starting with 90 clusters).

4 Conclusion

In this paper we present an unsupervised scheme, able to cluster data points in a given data set and also to find the number of clusters existing. The voting procedure of the algorithm combines the results of several independent cluster runs by voting between their results. This allows us to deal with the problem of local minima of cluster algorithms and to find a partition of the data which is supported by repeated applications of the cluster algorithm and is not influenced by the randomness of initialization or the cluster process itself. Moreover, we develop a strategy of analyzing and interpreting the results of the voting algorithm in such a way that existing clusters as well as data points belonging to a cluster can be identified as "sure" or "clear" concerning their structure and their partition respectively. The next part of the algorithm consists of a merging process of the clusters resulting by voting, while a criterion is computed in order to make the decision in favor of a specific number of clusters. Concluding, the VMA does not suffer from thresholds definitions, since only one parameter need to be defined and it diminishes to the minimum possible the instabilities of other clustering methods and can support non convex solutions.

References

1. Leo Breiman. Bagging predictors. *Machine Learning*, 24:123–140, 1996.
2. Evgenia Dimitriadou, Andreas Weingessel, and Kurt Hornik. Voting in clustering and finding the number of clusters. In H. Bothe, E. Oja, E. Massad, and C. Haefke, editors, *Proceedings of the "International Symposium on Advances in Intelligent Data Analysis (AIDA 99)" ("International ICSC Congress on Computational Intelligence: Methods and Applications (CIMA 99)"*, pages 291–296. ICSC Academic Press, 1999.
3. Yoav Freund and Robert E. Schapire. A decision-theoretic generalization of on-line learning and an application to boosting. *Lecture Notes in Computer Science*, 904, 1995.
4. Yoseph Linde, Andrs Buzo, and Robert M. Gray. An algorithm for vector quantizer design. *IEEE Transactions on Communications*, COM-28(1):84–95, January 1980.
5. Jianchang Mao and Anil K. Jain. Artificial neural networks for feature extraction and multivariate data projection. *IEEE Transactions on Neural Networks*, 6(2):296–317, March 1995.
6. Thomas M. Martinetz, Stanislav G. Berkovich, and Klaus J. Schulten. "Neural-Gas" network for vector quantization and its application to time-series prediction. *IEEE Transactions on Neural Networks*, 4(4):558–569, July 1993.
7. D. Michie, D. J. Spiegelhalter, and C. C. Taylor, editors. *Machine Learning: Neural Statistical Classification.* Ellis Horwood, 1994.
8. Hans-Joachim Mucha. *Clusteranalyse mit Mikrocomputern.* Akademie Verlag, 1992.
9. A. S. Weigend, M. Mangeas, and A. N. Srivastava. Nonlinear gated experts for time series: Discovering regimes and avoiding overfitting. *International Journal of Neural Systems*, 6:373–399, 1995.
10. Lei Xu, Adam Krzyzak, and Erkki Oja. Rival penalized competitive learning for clustering analysis RBF net and curve detection. *IEEE Transactions on Neural Networks*, 4(4):636–649, July 1993.

The Application of Fuzzy ARTMAP in the Detection of Computer Network Attacks

James Cannady[1] and Raymond C. Garcia[2]

[1] School of Computer and Information Sciences
Nova Southeastern University
Fort Lauderdale, FL 33314
cannady@nova.edu

[2] School of Electrical and Computer Engineering
Georgia Institute of Technology
Atlanta, GA 30332
raymond@gcatt.gatech.edu

Abstract. The timely and accurate detection of computer and network system intrusions has always been an elusive goal for system administrators and information security researchers. Existing intrusion detection approaches require either manual coding of new attacks in expert systems or the complete retraining of a neural network to improve analysis or learn new attacks. This paper presents a new approach to applying adaptive neural networks to intrusion detection that is capable of autonomously learning new attacks rapidly using feedback from the protected system.

Introduction

The timely and accurate detection of computer network attacks is one of the most difficult problems facing system administrators. The individual creativity of attackers, the wide range of computer hardware and operating systems, and the ever-changing nature of the overall threat to targeted systems have contributed to the difficulty in effectively identifying intrusions. While intrusion detection systems (IDS) generally rely on rule-based expert systems to identify attacks a limited amount of research has been conducted on the application of neural networks to address the inherent weaknesses in rule-based approaches. Debar (1992) and Fox (1990) proposed neural networks as alternatives to the statistical analysis component of anomaly detection systems. These neural networks identify the typical characteristics of system users and identify statistically significant variations from the user's established behavior. Cannady (1998) demonstrated the use of multi-level perceptron/SOM hybrid neural networks in the identification of computer attacks and Bonifacio (1998) demonstrated the use of a neural network in an integrated detection system. However, each of these approaches required the complete retraining of the neural networks to learn new attacks. In Cannady (2000) a cerebellar model articulation controller (CMAC) neural network (Albus, 1975) was applied in the detection of denial-of-service (DoS) attacks. While the CMAC approach was very effective in detecting the DoS attacks, this category of network intrusion is relatively simplistic. There are numerous other types of network attacks that are much more

G. Dorffner, H. Bischof, and K. Hornik (Eds.): ICANN 2001, LNCS 2130, pp. 225–230, 2001.

difficult to detect. The detection of these complex forms of network attack may require some form of clustering to accurately identify the large variety of potential attacks.

This paper presents the results of a research effort that investigated the application of a Fuzzy ARTMAP (FAM) neural network in the detection of network-based attacks. The objective of this research was to evaluate the ability of a FAM neural network to autonomously learn new attack patterns in a simulated network data stream. The inability of existing systems to autonomously identify new complex attack patterns increases the long-term cost of the systems due to the requirement for dedicated personnel to identify and implement the necessary updates. However, more significant is the fact that the lack of autonomous learning of complex network attacks by existing approaches results in a IDS that is only as current as the most recent update and therefore becomes progressively more ineffective over time.

Fuzzy ARTMAP Approach

The FAM neural network architecture was selected for this application primarily because of the enhanced clustering capabilities of the algorithm. There is also an ease of use with FAMs, they have a small number of parameters and require no problem-specific system crafting or choice of initial weight values. Learning of each pattern occurs as it is received on-line as opposed to performing an off-line optimization of a criterion function. FAMs provide mathematical precision in terms of ART concepts that outperform other soft-computing techniques. The strengths include autonomous learning, recognition, and prediction with fast learning of rare events, stable learning of large nonstationary databases, efficient learning of morphologically variable events, and associative learning of many-to-one and one-to-many maps.

A detailed overview of the FAMs learning process is presented in (Carpenter, 1992). By definition, resonance occurs if the match function meets the vigilance criterion. The match function is defined as $|I \wedge w_J|/|I|$ where I is the input activity vector and w_J is a vector of adaptive weights. The vigilance criterion itself is $|I \wedge w_J|/|I| \geq \rho$ where $\rho \in [0,1]$ is the vigilance parameter. For both definitions, $\wedge$ represents the fuzzy AND minimum operation. Once the vigilance criterion is met, the weight vector is updated in the following way:

$$w_J^{(new)} = \beta\left(I \wedge w_J^{(old)}\right) + (1-\beta)\, w_J^{(old)} \tag{1}$$

The initial series of experiments were conducted to evaluate the ability of the FAM algorithm to distinguish typical event patterns. A Java-based prototype application was created based on the description of the FAM algorithm provided in (Carpenter, 1992). A network-monitoring tool was used to collect a data sample of 100,000 packets from an Ethernet network on the campus of Nova Southeastern University. Thirty-seven attacks sequences were included in the data sample. These attacks included several forms of denial of service attacks, kernel panic attacks, and attacks related to the unauthorized access of system-level resources by an intruder. A sub-group of 10,000 packets was collected from the initial sample, including examples of the 37 different forms of network attacks. The data were arranged in the order that they were received into data vectors consisting of 10 elements each. The input vectors were ordered sets of normalized floating point numbers that represented types of Ethernet network packets, (e.g., ping, telnet, FTP, etc.). These packet sequences were then categorized within a range from 0.0 (normal data) to .99 (attack sequence) based on the effect of the sequence on the state (s) of the protected system. Binary representations of the packet sequence categories were used as inputs to the ART_B module and the packet sequence vectors were used as inputs to the ART_A module. The evaluation of the FAM algorithm was based on the average mean difference between the output of the ART_A module and the desired output for each of the packet sequence vectors. The FAM algorithm that was implemented in the prototype application achieved a average mean difference of 0.0066 in the test of the data sample. The evaluation demonstrated the ability of the FAM algorithm to accurately identify typical event sequences from a network data stream.

The second series of tests were conducted to evaluate the ability of the FAM architecture to autonomously learn and identify attack sequences in a network data stream. While the FAM architecture demonstrated the ability to accurately identify sample network events, a modern computer network data stream is an extremely complex and dynamic environment. The need for an autonomous learning capability in this application is based on the wide variety of potential attacks and the need to recognize new threats quickly. As a result, the FAM architecture that was used in these experiments was modified to incorporate a limited reinforcement learning capability. The ART_B input was replaced with a variable (f) that represented the feedback from the computer system that was being protected by the application.

Each feedback value was based on the combined effect of the input packets on a variety of system state indicators that could be monitored to provide an indication of an attack. These characteristics included the current CPU load, the available network bandwidth, the available memory, and the number of active connections. The feedback variable also included an indication of the relative compliance of the current system state with an established security policy. For example, the policy of the protected system in these experiments included the prohibition against the modification of the system log (syslog) files. An attacker who is able to write to the syslog can overload the buffer with erroneous data and render the syslog daemon useless. As a result, any attempts to modify the syslog by a non-root process is immediately flagged as a violation of the security policy and incorporated into the feedback variable. The system state (s) of the protected host was in the range 0.0 (system stopped) to 0.99 (system optimal). f was computed by using the inverse of the state of the protected host $(1 - s)$. While receiving normal network data the state of the protected host should be nominal, (e.g., 0.75 - 0.99), and the corresponding ART_a output should be small, (e.g., 0.0 – 0.25). However, during an attack the state of the protected host becomes degraded as the system reacts to the intrusion and s is reduced. The use of f provided the FAM with a non-supervised method of learning new patterns. This is a particularly critical capability for a system that must quickly identify new attack patterns before significant damage is done to the protected system.

The revised FAM approach was initially tested by initializing the FAM weights and using a simulated network data stream that contained 100,000 packets. The data sample included single instances of the 37 attack sequences that were included in the initial evaluation. The FAM utilized a map field (F^{AB}) vigilance parameter of 0.1. In this case a low vigilance parameter was chosen to limit the number of ART_a nodes that are created. With a high vigilance parameter the number of ART_a nodes in an actual network implementation could quickly exceed the available memory resources of the system running the FAM. The FAM was evaluated based on the average mean difference between the ART_a output for each of the 37 attack sequences and the f value for each of the input vectors. A radial basis function (RBF) neural network was used to simulate the state (s) of the protected system. The RBF neural network was trained to provide an output for each of the input vectors that were consistent with the expected state of a Linux-based host that was receiving the same network data stream.

After the FAM processed the 100,000 data samples the application achieved an average mean difference of 0.335. This error was significantly higher than the average error rate of 0.15 in commercial intrusion detection systems, (Bonifacio, 1998). While the initial evaluation demonstrated some capabilities for on-line learning of *a priori* attacks, the average mean error would be unacceptable in an actual network implementation.

In an effort to enhance the detection capability of the FAM approach an adaptive F^{AB} vigilance parameter was used in place of the static 0.1 vigilance parameter that was used in the initial evaluation. In the modified algorithm f was used as the F^{AB} vigilance parameter:

$$|I \wedge w_J| / |I \geq f \qquad (2)$$

The variability of the F^{AB} vigilance parameter was considered as a viable method of allowing a system to increase or decrease the rate in which new ART_a clusters are created in response to circumstances in a dynamic environment. By using f as F^{AB} vigilance parameter the FAM implemented in this research was designed to rapidly learn new attack patterns when the state of the protected host was degraded and create nodes representing new patterns at a slower rate when the state was nominal. This results in an adaptive learning process that responds to the current level of potential threat to the protected host. However, to avoid the potential adverse impact of the performance of the application that may be caused by a constantly changing vigilance parameter a limited level of adaptability was utilized. In the revised application the static F^{AB} vigilance parameter is replaced with f when the state of the protected system dropped below 0.25. In this way the FAM retained the ability to limit the creation of new ART_a nodes during normal activity, while it also possesses the flexibility of recognizing new attack patterns quickly when the state of the protected system is degraded.

The revised FAM algorithm was re-tested using the same data set after the weights were initialized. In this test the FAM achieved a mean average difference of 0.043. This was a significant improvement over the use of a static F^{AB} vigilance parameter, (Fig 1), and a level of detection accuracy that is higher than currently available commercial applications.

Conclusions and Future Work

Research and development of IDSs has been ongoing since the early 1980's and the challenges faced by designers increase as the targeted systems because more diverse and complex. The results demonstrate the potential for a powerful new analysis component of a complete IDS that would be capable of identifying *priori* and *a priori* network attack patterns. Based on the results of the tests that were conducted on this approach there were several significant advances in the detection of network attacks:

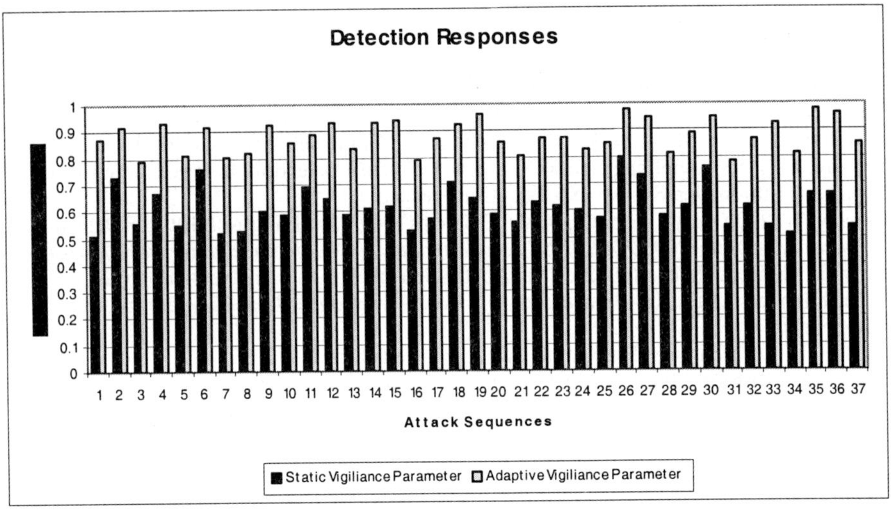

Fig. 1. Attack Detection Test Results

- On-line learning of complex attack patterns – The approach has demonstrated the ability to rapidly learn new complex attack patterns without the complete retraining required in other neural network approaches. This is a significant advantage that could allow the IDS to continually improve its analytical ability without the requirement for external updates.

- Extremely accurate in identifying *priori* attack patterns – The use of the dynamic F^{AB} vigilance parameter resulted in an average error of 0.04, compared with an average error of 0.15 in existing IDSs. Because other information security components rely on the accurate detection of computer attacks the ability to accurately identify network events could greatly enhance the overall security of computer systems.

- Adaptive learning algorithm – The use of an adaptive F^{AB} vigilance parameter, based on the current state of the protected host, provides the ability to rapidly learn new attacks, thereby significantly reducing learning time in periods when rapid attack identification is required.

The results of the tests of this approach shows significant promise, and our future work will involve the development of a full-scale integrated intrusion detection and response system that will incorporate the Fuzzy ARTMAP-based approach as the analytical component.

References

Albus, J.S. (1975, September). A New Approach to Control: The Cerebellar Model Articulation Controller (CMAC). *Transactions of the ASME.*

Bonifacio, J.M, Cansian, A.M., de Carvalho, A., & Moreira, E. (1998). Neural Networks Applied in Intrusion Detection. In *Proceedings of the International Joint Conference on Neural Networks.*

Carpenter, G.A., Grossberg, S., Markuzon, N., Reynolds, J.H., and Rosen, D.B. (1992). Fuzzy ARTMAP: A Neural Network Architecture for Incremental Supervised Learning Of Analog Multidimensional Maps, In IEEE Transactions on Neural Networks, 3, 698-713.

Cannady, J. (1998). Applying Neural Networks to Misuse Detection. In Proceedings of the 21st National Information Systems Security Conference.

Cannady, J. (2000, July). Applying CMAC-based On-line Learning to Intrusion Detection. In *Proceedings of the 2000 IEEE/INNS Joint International Conference on Neural Networks.*

Debar, H. & Dorizzi, B. (1992). An Application of a Recurrent Network to an Intrusion Detection System. In *Proceedings of the International Joint Conference on Neural Networks.*

Fox, Kevin L., Henning, Rhonda R., & Reed, Jonathan H. (1990). A Neural Network Approach Towards Intrusion Detection. In *Proceedings of the 13th National Computer Security*

Transductive Learning: Learning Iris Data with Two Labeled Data

Chun Hung Li and Pong Chi Yuen

Department of Computer Science, Hong Kong Baptist University, Hong Kong
{chli,pcyuen}@comp.hkbu.edu.hk

Abstract. This paper presents two graph-based algorithms for solving the transductive learning problem. Stochastic contraction algorithms with similarity based sampling and normalized similarity based sampling are introduced. The transductive learning on a classical problem of plant iris classification achieves an accuracy of 96% with only 2 labeled data while previous research has often used 100 training samples. The quality of the algorithm is also empirically evaluated on a synthetic clustering problem and on the iris plant data.

1 Introduction

In recent years, important classification tasks have emerged with enormous volume of data. The labeling of a significant portions of the data for training is either infeasible or impossible. A number of approaches have been proposed to combine a set of labeled data with unlabeled data for improving the classification rate. The co-training approach has been proposed to solve the problem of web page classification where the web pages can be represented by two independent representations [1]. Subsequently, a similar co-training method is invented for combining labeled and unlabeled data by co-training with two learning algorithms [2]. Instead of using two representations of the data, this co-training algorithm uses two learning algorithms. The naive Bayes classifier and the EM algorithm have been combined for classifying text using labeled and unlabeled data [3]. A modified support vector machine and non-convex quadratic optimization approaches have been studied for optimizing semi-supervised learning [4]. The transductive learning is suggested to simplify learning problem where there is a restricted amount of data available [5]. One of the essential properties of transductive learning is the estimation of the function of interest at only some given points rather than the estimation of the complete decision surface. The support vector machines have been extended with transductive inference to classify text [6]. The transductive learning is also applied to improve the color tracking in video sequences [7].

Graph based clustering has received a lot of attention recently. A factorization approach has been proposed for clustering[8], the normalized cuts have been proposed as a generalized method for clustering [9] and [10]. In this paper, we investigated the use of stochastic graph-based sampling approach for solving

G. Dorffner, H. Bischof, and K. Hornik (Eds.): ICANN 2001, LNCS 2130, pp. 231–236, 2001.

the transductive learning problem. Graph-based clustering is also shown to be closely related to similarity-based clustering [11].

2 Graph Based Clustering

The data is modeled using an undirected weighted graph $G(V, E)$ consisting of a set of vertices and a set of edges. The data to be clustered are represented by vertices in the graph. The weights of the edge of vertices represents the similarity between the object indexed by the vertices. The similarity matrix $S \in R^{N^2}$ where s_{ij} represents the similarity between the vector x_i and x_j. A popular choice of the similarity matrix is of the form

$$s_{ij} = \exp\left(-\frac{|x_i - x_j|}{\sigma^2}\right) \tag{1}$$

where $|x|$ is an euclidean distance of x and σ is a scale constant. The exponential similarity function have been used for clustering in [8], [9] and [10]. We can also employ the k-nearest neighbour graph where the similiarity matrix includes the edge (i, j) only if j is the one of the k-th nearest neighbour of i. This reduces the number of the edges in the graph from N^2 to kN. The minimum cut algorithm partitions the vertices in the graph G into two disjoint sets (A, B), that minimizes the following objective function,

$$f(A, B) = \sum_{i \in A} \sum_{j \in B} S_{ij}. \tag{2}$$

Karger has proposed a randomized contraction algorithm for finding the minimum cuts [12].

3 Graph Contraction Algorithm

The transductive learning algorithm assumes a set of given labeled samples. Initially, assign an empty label to all nodes,i.e. $L_i = \phi$ for all nodes i. Then assign the labels of the given labeled nodes to their respective labels. After this initialization, the following contraction algorithm is applied:

1. Randomly select an edge (i, j) from G with probability proportional to S_{ij}, depending on the labels of i and j, do one of the following:
   (a) If $L_i = \phi$ and $L_j = \phi$, then $L_{ij} = \phi$
   (b) If $L_i = \phi$ and $L_j \neq \phi$, then assign $L_{ij} = L_j$
   (c) If $L_j = \phi$ and $L_i \neq \phi$, then assign $L_{ij} = L_i$
   (d) If $L_i \neq L_j$ and $L_i \neq \phi$ and $L_j \neq \phi$, then remove the edge (i, j) from G, and return to step 1
2. Contract the edge (i, j) into a meta node ij
3. Connect all edges incident on i and j to the meta-node ij while removing the edge (i, j)

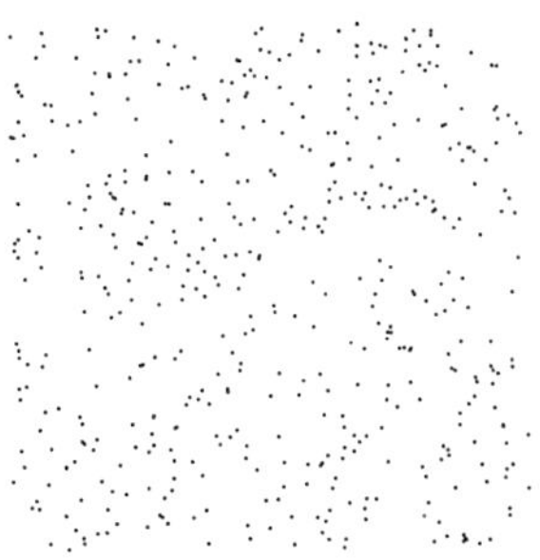

Fig. 1. Two clusters

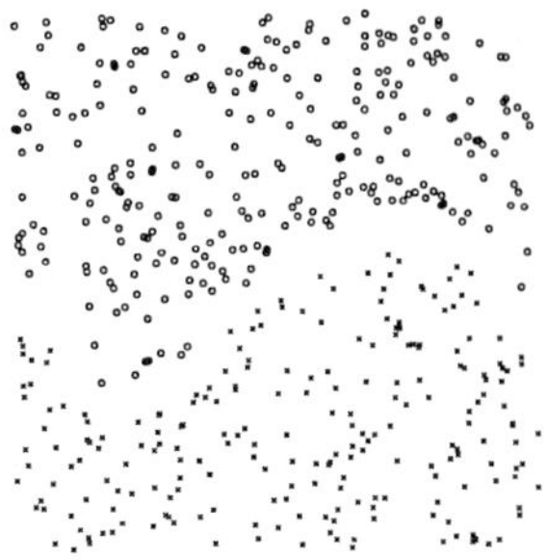

Fig. 2. Solution I

This contraction is repeated until all the nodes are contracted into a specified number of clusters or until all edges are removed. This semi-supervised contraction guarantees the consistency in the labeling outcome in merging the meta-nodes. Furthermore, both of the algorithms above can be repeated as separate randomized trials on the data. The results of each trial can be considered as giving the probability of connectivity between individual nodes and then be combined to give more accurate estimation of the probability of connectivity.

Another version of the transductive learning can be conducted through the use of a weighted cuts sampling method where the random selection of edge to be contracted are weighted by the number of elements in the nodes containing the edge. The normalized similarity between two metanodes is given by s'_{ij} where

$$s'_{ij} = \frac{s_{ij}}{n_i n_j}. \tag{3}$$

where n_i and n_j is the number of vertices in the metanodes i and j. The normalized graph sampling algorithm selects an edge $\{i, j\}$ for contraction with probability proportional to $s'_i j$. This selection process ensures that the nodes with small number of elements are selected first.

4 Results and Discussions

4.1 Testing on Synthetic Data

A synthetic dataset is used for testing the transductive learning algorithm. Figure 1 shows the data for a two cluster clustering problem. The two clusters are seperated with a sinusoidal boundary. To test the performance of the transductive learning algorithm on the synthetic data, the training samples with true labels used for training are Case I: one sample; Case II: two samples; Case III: ten samples; from each classes. With a small number of labeled training samples from each class, one can judge from the figures that inference based/decision surface based classifier will not be able to determine accurately the sinusoidal boundary between the cluster. Four hundred randomized trials are performed

Table 1. Performance of transductive learning on synthetic data: (1) Errors of combined estimation (2) Standard Deviation of estimation

	Sampling		Normalized Sampling	
	Error	SD	Error	SD
Case I	6.6%	0.16	1.8%	0.09
Case II	0.6%	0.09	1.2%	0.05
Case III	0.4%	0.02	0.4%	0.02

for each experiments. The average error percentage of the transductive learning is shown in Table 1. From Table 1, we can see that the combined estimation is very accurate, having errors of less than one percent.

4.2 Iris Dataset

The iris dataset is also used for testing the transductive learning. The iris dataset consists of 150 samples of measurement of the iris plant. There are three species of iris in the dataset and each species has 50 samples. Typical approaches uses 100 samples for training and 50 untrained samples for testing. The iris dataset is well studied and results can be found in numerous literature [13]. In the first experiment in transductive learning for iris dataset, we considered using only 2 labeled data. We pick the first sample in class 2 and first sample in class 3 as the labeled sample and denote this experiment as experiment (a). We perform transductive learning until 3 metanodes are left after successive contractions. All objects that are contracted into the metanodes with the labeled data are assigned to the class of the labeled data. The metanode that does not has any of the labeled data as elements are considered as class one. The above trial are repeated 200 times and we approximate the probability of an object being the three classes by the number of times that the object is being assigned in the three classes divided by 200. Figure 3 shows the estimated probability of different classes after transductive learning with 2 labeled samples. The x-axis in the graph refers to the iris plant in the sequence of iris plant dataset file. The plot of P_1 shows that the first 50 elements are all correctly classified as class one in each of the 200 trials. The plot of P_2 shows that the probability for iris plant 51 to 100 have high probabilities of around 0.8 being class 2. The probability of plant 101 to 150 being class two shows a large amount of confusion. A number of plants shows a high probability of being class two which are incorrect. The plot of P_3 shows the similar situations. In the second experiment, experiment (b), we use the iris plant 52 as a sample for class 2 and plant 102 as sample from class three. As we observed in the previous experiments that there are a large number of class 3 samples that are incorrectly classified as class two, further learning of class 3 data seems to be necessary. In experiment (c), we include another data from class three, namely, the iris plant data 102, in addition to 51 plant data and 101 plant data. There are four errors in this experiment. We also perform the fourth experiment, experiment (d) with similar results. Table 2 shows the

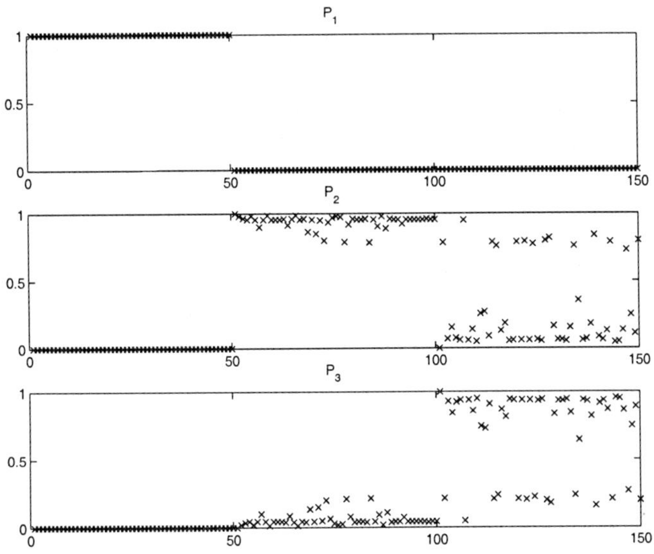

Fig. 3. Probability of class membership after transductive learning: Experiment (a) with normalized sampling ; P_1: probability of class two, (b) P_2: probability of class two, (c) P_3: probability of class three

Table 2. Labeled samples used in transductive learning

	class two	class three
Exp. (a)	51	101
Exp. (b)	52	102
Exp. (c)	51	101,102
Exp. (d)	52	101,102

Table 3. Number of errors after transductive learning on iris dataset

	samp.	norl. samp.
Exp. (a)	16	14
Exp. (b)	38	6
Exp. (c)	8	4
Exp. (d)	4	4

labeled samples used in the four experiments in transductive learning of iris data. Table 3 shows the number of misclassifications of the two experiments in transductive learning of the iris dataset. We can see from the table that the errors of the normalized sampling approach are significantly lower than the unweighted approach. Moreover the accuracy of learning with normalized sampling using two labeled samples can reach 96% while the training accuracy when using 3 labeled samples can reach 97%. Similar results are obtained using other labeled samples as training data. In general, it is found that if the labeled samples represents typical features in the respective class, the learning performances would be acceptable. If a selected sample has features vector that is similar to samples from the other class, the performances of the learning would be significantly lower.

Two transductive learning algorithms are introduced that achieves learning through the stochastic combination of learning with labeled data and the clustering of unlabeled data. The transductive learning algorithm excels in situations where a limited amount of labeled data is available. In comparing the weighted sampling and unweighted sampling approach to graph-based clustering, we find that the normalized sampling approach has a better performance than the unweighted sampling method.

References

1. A. Blum and T Mitchell. Combining labeled and unlabeled data with co-training. In *Proceedings of the Eleventh Annual Conference on Computational Learning Theory*, pages 92–100, 1998.
2. S. Goldman and Y. Zhou. Enhancing supervised learning with unlabeled data. In *Proceedings of the Seventeenth International Conference on Machine Learning*, 2000.
3. K. Nigam, A. McCallum, Sebastian Thrun, and Tom Mitchell. Text classification from labeled and unlabeled documents using em. *Machine Learning*, 34(1), 1999.
4. T. S. Chiang and Y. Chow. Optimization approaches to semi-supervised learning. In M. C. Ferris, O. L. Mangasarian, and J. S. Pang, editors, *Applications and Algorithms of Complementarity*. Kluwer Academic Publishers, 2000.
5. Vladimir N. Vapnik. *The Nature of Statistical Learning Theory*. Springer, New York, second edition, 2000.
6. Thorsten Joachims. Transductive inference for text classification using support vector machines. In *International Conference on Machine Learning (ICML)*, pages 200–209. Morgan-Kaufman, 1999.
7. Ying Wu and Thomas S. Huang. Color tracking by transductive learning. In *Proceedings of the International Conference on Computer Vision and Image Processing*, volume 1, pages 133–138, 2000.
8. P. Perona and W. T. Freeman. A factorization approach to grouping. In *Proceedings of European Conference on Computer Vision*, pages 655–670, 1998.
9. J. Shi and J. Malik. Normalized cuts and image segmentation. *IEEE Trans. Pattern Analysis and Machine Intelligence*, 21(8):888–905, 2000.
10. Y. Gdalyahu, D. Weinshall, and M. Werman. Stochastic image segmentation by typical cuts. In *Proceedings of the IEEE CVPR 1999*, volume 2, pages 596–601, 1999.
11. J. Puzicha, T. Hofmann, and J. M. Buhmann. A theory of proximity based clustering: structure detection by optimization. *Pattern Recognition*, 33:617–634, 2000.
12. D. R. Karger and C. Stein. A new approach to the minimum cut problem. *Journal of the ACM*, 43(4):601–640, 1996.
13. M. Nadler and E. P. Smith. *Pattern Recognition Engineering*. Wiley-interscience, New York, 1993.

Approximation of Time-Varying Functions with Local Regression Models

Achim Lewandowski and Peter Protzel

Chemnitz University of Technology
Dept. of Electrical Engineering and Information Technology
Institute of Automation
09107 Chemnitz, Germany
achim.lewandowski@alewand.de
peter.protzel@e-technik.tu-chemnitz.de

Abstract. Industrial or robot control applications which have to cope with changing environments require adaptive models. The standard procedure of training a neural network off-line with no further learning during the actual operation of the network is not sufficient in those cases. Therefore, we are concerned with developing algorithms for approximating time-varying functions. We assume that the data arrives sequentially and we require an immediate update of the approximating function. The algorithm presented in this paper uses local linear regression models with adaptive kernel functions describing the validity region of a local model. While the method is developed to approximate a time-variant function, naturally it can also be used to improve the fit for a time-invariant function. An example is used to demonstrate the learning capabilities of the algorithm.

1 Introduction

Many industrial applications we are working on are characterized by highly nonlinear systems which also vary with time. Some systems change slowly e.g. due to aging, other systems have very fast timevariances e.g. due to sudden changes in the used tools or materials. Advanced control and prediction of these systems require accurate models, which leads to the problem of approximating nonlinear and time-variant functions. In contrast to time-series prediction, we need to predict not just the next output value at time t+1 but the complete functional relationship between outputs and inputs.

Global models (e.g. multilayer perceptrons with sigmoid activation functions) generally suffer from the Stability-Plasticity-dilemma, i.e. a model update with new training data in a certain region of the input space can change the model also in other regions of the input space which is undesirable. In contrast, local models are only influenced by data points which fall into their validity regions. In the recent years, several algorithms were developed which work with local regression models. The restriction of a model to be only a local learner can be implemented in an easy way by using weighted regression whereby the weights

G. Dorffner, H. Bischof, and K. Hornik (Eds.): ICANN 2001, LNCS 2130, pp. 237–243, 2001.

are given by a kernel function. Schaal and Atkeson [1] introduced an algorithm which can adjust the number and the shape of the kernel functions to yield a better approximation in the case of a time-invariant function. Recently, another algorithm (Vijayakumar and Schaal [2]) was presented, which can deal with some special cases of high-dimensional input data, again for time-invariant functions. While these algorithms are capable to learn online (i.e. pattern-by-pattern update), optimal results will be achieved only if the sample is repeatedly presented. We also developed an algorithm (Lewandowski et al. [3]) for the time-invariant case based on competition between the kernel functions. The algorithm presented in this paper is a modification and extension of our approach with the goal of approximating time-varying functions.

It should be noticed that our and the other mentioned algorithms are not so-called "lazy learners" (Atkeson, Moore and Schaal [4]). For every request to forecast the output of a new input, the already constructed local models are used. In contrast, a lazy learner working with kernel functions would construct the necessary model(s) just at the moment, when a new input pattern arrives.

2 The Algorithm

2.1 Statistical Assumptions

We assume a standard regression model

$$y = f(x,t) + \epsilon \;, \tag{1}$$

with x denoting the n-dimensional input vector (including the constant input 1) and y the output variable. The errors are assumed to be independently distributed with variance σ^2 and expectation zero. The unknown function $f(x,t)$ is allowed to depend on the time t. Each local model i ($i = 1, \ldots, m$) is given by a linear model

$$y = \beta_i^T x \;. \tag{2}$$

The parameters for a given sample are estimated by weighted regression, whereby the weighting kernel for model i is given by

$$w_i(x) = \exp\left(-\frac{1}{2}(x-c_i)^T D_i (x-c_i)\right) \;. \tag{3}$$

The matrix D_i describes the shape of the kernel function and c_i the location. Suppose now, that all input vectors are summarized row-by-row in a matrix X, the outputs y in a vector Y and the weights for model i in a diagonal matrix W_i, then the parameter estimators are given by

$$\hat{\beta}_i = (X^T W_i X)^{-1} X^T W_i Y \;. \tag{4}$$

With $P_i = (X^T W_i X)^{-1}$ a convenient recursion formula exists. After the arrival of a new data point (x, y), according to Ljung and Söderström [5] the following incremental update formula exists:

$$\hat{\beta}_i^{new} = \hat{\beta}_i + e_i w_i(x) P_i^{new} x \;, \tag{5}$$

with

$$P_i^{new} = \frac{1}{\lambda}\left(P_i - \frac{P_i x x^T P_i}{\frac{\lambda}{w_i(x)} + x^T P_i x}\right) \tag{6}$$

and

$$e_i = y - \hat{\beta}_i^T x \ . \tag{7}$$

Please notice, that a forgetting factor λ $(0 < \lambda \leq 1)$ is included. This forgetting factor is essential for our algorithm to allow faster changes during the kernel update, as the influence of older observations must vanish, if the approximating function is expected to follow the true model sufficiently fast.

2.2 Update of Receptive Fields

Assume, that a new query point x arrives. Each model gives a prediction $\hat{y}_i$. These individual predictions are combined into a weighted mean, according to the activations of the belonging kernels:

$$\hat{y} = \frac{\sum w_i(x)\hat{y}_i}{\sum w_i(x)} \ . \tag{8}$$

We will now describe the way how new kernels are integrated and old kernel functions are updated. The basic idea is that models, which give better predictions for a certain area than their neighbors, should get more space while the worse models should retire. This is done by changing the width and shape of the belonging kernel functions.

A new kernel function is now inserted, when no existing kernel function has an activation $w_i(x)$ exceeding a user-defined threshold $\eta > 0$. As a side effect, the user could be warned that the prediction $\hat{y}$ of the existing models must be judged carefully. The center c_i of the new kernel function is set to x and the matrix D_i is set to a predefined matrix D, usually a multiple of the identity matrix $(D = kI)$. High values of k produce narrow kernels.

If no new kernel function is needed, the neighboring kernel functions are updated. We restrict the update to kernel functions which exceed a second threshold $\delta = \eta^2$. After the true output y is known, the absolute errors $|e_i|$ of the predictions of all activated models are computed. If the model with the highest activation belongs to the lowest error, nothing happens. If for example only one kernel function is activated sufficiently, the kernel shapes remain unchanged. Otherwise, if $\bar{e}$ denotes the mean of these absolute errors, the matrix D_i of a model i, which is included in the shape update is slightly changed. Define

$$g_i = \min\left(\frac{|e_i| - \bar{e}}{\bar{e}}, 1\right) , \tag{9}$$

then the shape matrix is updated by the following formula:

$$D_i^{new} = D_i + \frac{\rho^{g_i} - 1}{(x - c_i)^T D_i (x - c_i)} D_i (x - c_i)(x - c_i)^T D_i \ . \tag{10}$$

The parameter $\rho \geq 1$ controls the amount of changes. The essential part of the activation, $a_i = (x - c_i)^T D_i (x - c_i)$, is given after the update by

$$a_i^{new} = \rho^{g_i} a_i \ . \quad (11)$$

The activation of kernel functions belonging to models which perform better than the average will therefore rise for this particular x. The update of D_i is constructed in that way, that activations for other input values z, which fulfil

$$(z - c_i)^T D_i (x - c_i) = 0 \ , \quad (12)$$

are not changed. Inputs z, for which the vector $z - c_i$ is in a certain sense perpendicular to $x - c_i$, receive the same activation as before.

After the kernel update *all* linear models are updated as described in the last section. For a model with very low activation ($w_i(x) \approx 0$) the parameter estimator, computed as in (5), will not change, but according to equation (6) P_i will grow by the factor $1/\lambda$, so that the belonging model gets more and more sensitive if it has not seen data for a long time. We could further use a function $\lambda = \lambda(x)$, expressing our prior believe of variability.

3 Empirical Results

Vijayakumar and Schaal [2] and Schaal and Atkeson [1] used the following time-invariant function to demonstrate the capabilities of their algorithm:

$$z = \max\left(e^{-10x^2}, e^{-50y^2}, 1.25e^{-5(x^2+y^2)}\right) + N(0, 0.01) \ . \quad (13)$$

In their example, a sample of 500 points, uniformly drawn from the unit square was used to fit a model. The whole sample was repeatedly presented pattern-by-pattern whereby from epoch to epoch the sample was randomly shuffled. As a test set, a 41×41 grid over the unit square was used, with the true outputs without noise.

For our purposes we will modify this approach. Each data point will be presented just once. The first 3000 data points will be generated according to (13). From $t = 3001$ to $t = 7000$ the true function will change gradually according to

$$z(t) = \max\left(\frac{7000 - t}{4000} e^{-10x^2}, e^{-50y^2}, 1.25e^{-5(x^2+y^2)}\right) + N(0, 0.01) \ . \quad (14)$$

From $t = 7001$ to $t = 10000$ the function will be time-invariant again:

$$z(t) = \max\left(e^{-50y^2}, 1.25e^{-5(x^2+y^2)}\right) + N(0, 0.01) \ . \quad (15)$$

We set $D = 120I$, $\lambda = 0.999$, $\eta = 0.10$ and $\rho = 1.05$. Figure 1 shows the true function (left) and our approximation (right) for $t = 3000$ (top), $t = 5000$ (middle) and $t = 10000$ (bottom). The approximation quality is good, even for the case $t = 5000$ where the function is in a changing state.

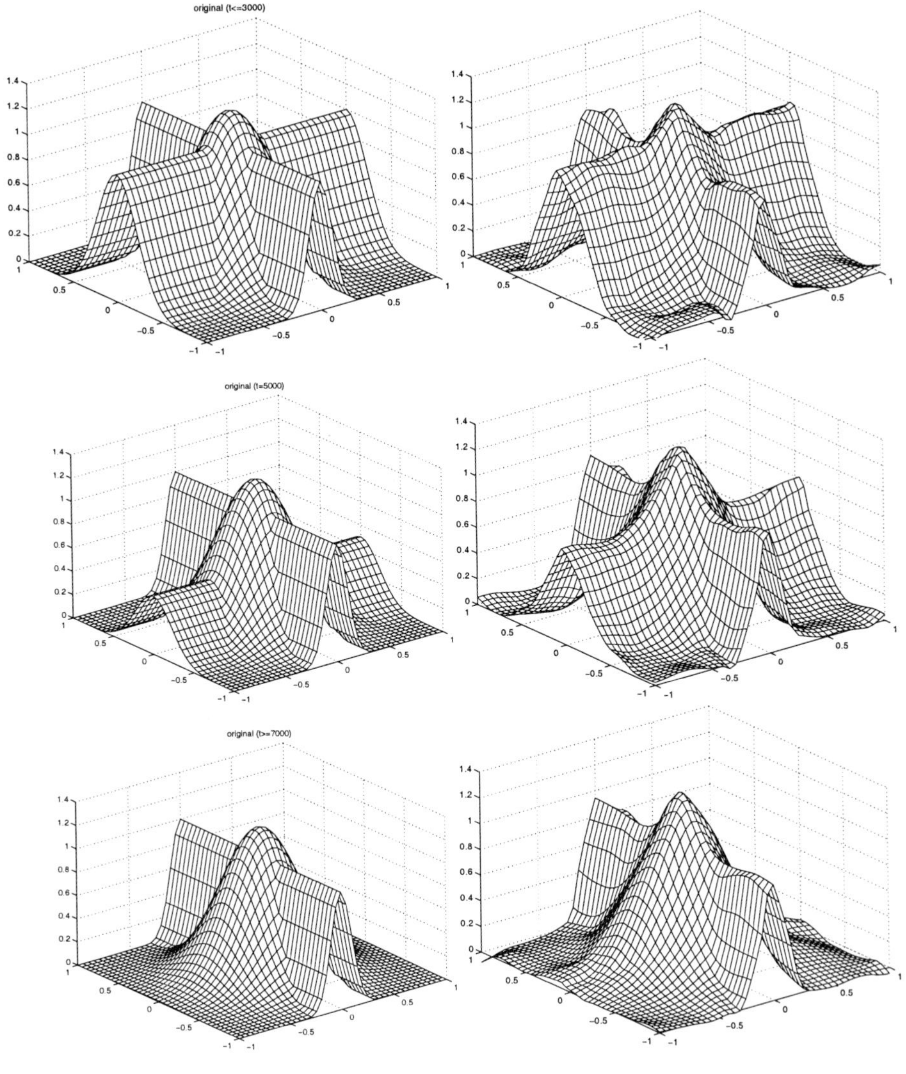

Fig. 1. True function (left) and approximation (right) for $t = 3000$ (top), $t = 5000$ (middle) and $t = 10000$ (bottom)

We started a second run with ρ left to 1, so that no kernel shape update was performed. The same 10000 data points were used. Figure 2 shows the development of the mean squared errors for $\rho = 1.05$ (solid) and $\rho = 1.00$ (dashed). With updating the kernel shapes, the error on the test set was less than half as large, after processing all 10000 data points, compared to the same algorithm with fixed kernel shapes.

Figure 3 shows a contour plot ($w_i(x) \equiv 0.7$) of the activation functions of our local models, after all data points have been presented ($\rho = 1.05$). It is

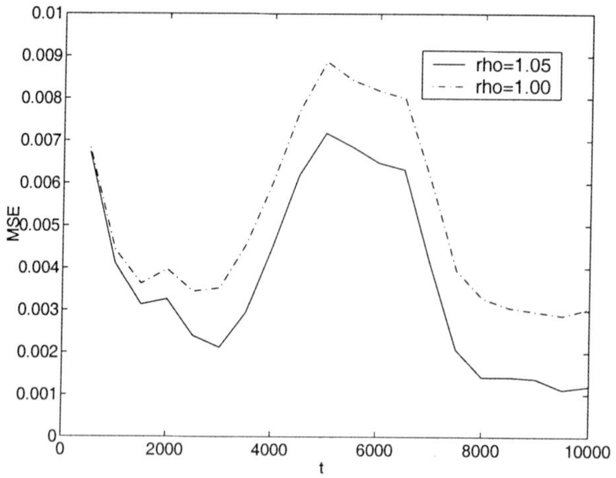

Fig. 2. Mean squared error when updating kernel shapes ($\rho = 1.05$) and without update ($\rho = 1.00$)

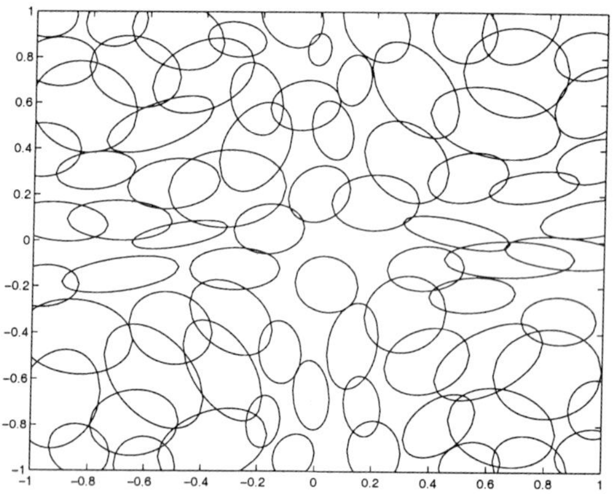

Fig. 3. Contour plot for $t = 10000$ ($w_i(x) \equiv 0.7, \rho = 1.05$)

clearly visible, that the kernel functions have adjusted their shape according to the appearance of our example function.

As the function is time-invariant for the first 3000 points, we can compare our results with the results given in [1] and [2]. After presenting 500 data points sequentially and only once, our mean squared error on the test set was 0.0067 which corresponds to a normalized error, as it's defined by Vijayakumar and Schaal [2], of 0.0473. This is half the value they reached after showing 500 points twice and about half the error we achieved in [3] under the same conditions. This quick reduction of the error while presenting each data point only once is a prerequisite for the time-variant function approximation, which is our actual goal.

4 Conclusions

The increasing need in industrial applications to deal with nonlinear and time-variant systems is the motivation for developing algorithms that approximate nonlinear functions which vary with time. We presented an algorithm that works with local regression models and adaptive kernel functions based on competition. There is no need to store any data points which we consider as an advantage. The results visualized for an example show a quick adaptation and good convergence of the model after presenting each data point from the time-varying function sequentially and only once.

For future work, it will be necessary to develop self-adjusting parameters. In our algorithm, the choice of the initialization matrix D in combination with the threshold value η is crucial for the performance and generalization capabilities. We are currently working on an approach to optimize these parameters online. An even more challenging problem we are working on is to *anticipate* changes of the true function instead of just following it.

References

1. Schaal, S., Atkeson, C.: Constructive incremental learning from only local information. Neural Computation, 10(8) (1998) 2047–2084
2. Vijayakumar, S., Schaal,S.: Locally weighted projection regression. Technical Report (2000).
3. Lewandowski, A., Tagscherer, M., Kindermann, L., Protzel, P.: Improving the fit of locally weighted regression models. Proceedings of the 6th International Conference on Neural Information Processing, Vol. 1 (1999) 371–374
4. Atkeson, C., Moore, A., Schaal, S.: Locally weighted learning. Artificial Intelligence Review, 11(4) (1997) 76–113
5. Ljung, L., Söderström, T.: Theory and practice of recursive identification. MIT Press, Cambridge (1986)

Part III

Theory

Complexity of Learning for Networks of Spiking Neurons with Nonlinear Synaptic Interactions*

Michael Schmitt

Lehrstuhl Mathematik und Informatik, Fakultät für Mathematik
Ruhr-Universität Bochum, D–44780 Bochum, Germany
http://www.ruhr-uni-bochum.de/lmi/mschmitt/
mschmitt@lmi.ruhr-uni-bochum.de

Abstract. We study model networks of spiking neurons where synaptic inputs interact in terms of nonlinear functions. These nonlinearities are used to represent the spatial grouping of synapses on the dendrites and to model the computations performed at local branches. We analyze the complexity of learning in these networks in terms of the VC dimension and the pseudo dimension. Polynomial upper bounds on these dimensions are derived for various types of synaptic nonlinearities.

1 Introduction

In many paradigms of neural computation the interaction of synaptic inputs is modeled in terms of linear operations. This is evident for standard neuron models such as the threshold, the sigmoidal, and the linear unit, but it also holds, in most cases, for the biologically closer leaky integrate-and-fire neurons and the spiking neuron models. Linearity is known to be sufficient for capturing the passive properties of the dendritic membrane where synaptic inputs occur in the form of currents that are combined using a summing operation. In neurobiology, however, it is well established that synaptic inputs can interact nonlinearly, for instance, when the synapses are colocalized on patches of dendritic membrane with specific properties (see, e.g., Koch [3]). In contrast to linear interactions, nonlinearities can reflect the spatial grouping of synapses (the "synaptic clustering") on the dendritic tree and the computations performed at local branches.

In this paper we study models for spiking neurons that use nonlinearities to combine synaptic inputs. We analyze the complexity of learning in networks of these model neurons in terms of the Vapnik-Chervonenkis (VC) dimension and the pseudo dimension. We establish bounds for these dimensions that can be used to derive results on the sample complexity and generalization capability of these networks. (For the relationship of the VC dimension and the pseudo dimension to the complexity of learning see, e.g., Anthony and Bartlett [1].)

* Work supported by the ESPRIT Working Group in Neural and Computational Learning II, NeuroCOLT2, No. 27150. A longer (8-page) version of this paper is available from http://www.ruhr-uni-bochum.de/lmi/mschmitt/.

G. Dorffner, H. Bischof, and K. Hornik (Eds.): ICANN 2001, LNCS 2130, pp. 247–252, 2001.

2 Spiking Neurons with Nonlinear Synaptic Interactions

In a network of *spiking neurons* each node v receives inputs in the form of spikes through its synaptic connections from other nodes. Assume that v has n input connections, each of them being characterized by a synaptic weight $w_i \in \mathbb{R}$ and a transmission delay $d_i \in \mathbb{R}^+$, where $\mathbb{R}^+ = \{x \in \mathbb{R} : x \geq 0\}$. If the neuron presynaptic to connection i emits a spike at time $t_i \in \mathbb{R}^+$, this generates in v a pulse described by the function $t \mapsto h_i(t - t_i)$ where $h_i : \mathbb{R}^+ \to \mathbb{R}$ is defined as

$$h_i(t) = \begin{cases} w_i & \text{if } d_i \leq t < d_i + 1 \ , \\ 0 & \text{otherwise} \ . \end{cases}$$

Hence, the pulse has unit duration and its height corresponds to the strength, or efficacy, of the synapse. We consider d_i and w_i, for $i = 1, \ldots, n$, as adjustable parameters of the neuron.

A *synaptic cluster* of v is a subset $I \subseteq \{1, \ldots, n\}$ of its synapses spatially grouped together on the dendritic tree. They interact nonlinearly where the nonlinearity is specified for each subset $\{i_1, \ldots, i_l\} \subseteq I$ by a rational function $f_{\{i_1,\ldots,i_l\}}$ in l variables. We call $f_{\{i_1,\ldots,i_l\}}$ the *synaptic interaction* of $\{i_1, \ldots, i_l\}$. A synapse may participate in more than one cluster. Let $F \subseteq \{1, \ldots, n\}$ be the set of synapses that receive a spike during the computation of v and define $J_I : \mathbb{R}^+ \to 2^I$ as

$$J_I(t) = \{i \in I \cap F : h_i(t - t_i) \neq 0\} \ ,$$

that is, $J_I(t)$ is the set of synapses in cluster I that are simultaneously active at time t. The interaction of the synapses in I is then modeled by the function $M_I : \mathbb{R}^+ \to \mathbb{R}$ with

$$M_I(t) = f_{\{i_1,\ldots,i_l\}}(w_{i_1}, \ldots, w_{i_l}), \quad \text{where } \{i_1, \ldots, i_l\} = J_I(t) \ .$$

Assume that v has k synaptic clusters $I_1, \ldots, I_k$. They interact linearly to yield the *membrane potential* $P : \mathbb{R}^+ \to \mathbb{R}$ defined by

$$P(t) = \sum_{j=1}^{k} M_{I_j}(t) \ .$$

Neuron v fires when the membrane potential reaches the threshold θ, that is, the *firing time* t_v of v is defined by $t_v = \min\{t : P(t) \geq \theta\}$. If t_v is undefined then v does not fire. The *networks of spiking neurons* that we study here use temporal coding for computations, that is, they encode information in the timing of single firing events. Precisely, if neuron v fires at time t_v this is assumed to encode the real number t_v. We consider feedforward networks with a given number of input nodes and one output node. Input vectors enter the network in terms of firing times of the input nodes. The output value of the network is the firing time of the output node. If the output node does not fire, the network output is defined to be 0. The *depth* of the network is the length of the longest path from an input node to the output node.

The model introduced here generalizes that of [5] by allowing nonlinear interactions of synaptic inputs in terms of clusters that represent the spatial grouping of synapses. Models of spiking neurons with linear interactions are discussed, e.g., in [2, 4]. We have included synaptic nonlinearities to take into account the well established neurobiological finding that the nature of the interaction of synapses colocalized on the dendritic tree is essentially nonlinear (see [3, Chapter 5]). The nonlinearities we employ here are based on arithmetic operations that can easily be implemented in analog VLSI circuits and thus be used for hardware designs of pulsed neural networks such as described in [6].

In this paper we calculate bounds on the VC dimension and the pseudo dimension of the networks introduced above. These dimensions are defined as follows (see also [1]): A class $\mathcal{F}$ of $\{0,1\}$-valued functions in n variables is said to *shatter* a set of vectors $S \subseteq \mathrm{I\!R}^n$ if $\mathcal{F}$ induces all dichotomies on S (that is, if for every partition of S into two disjoint subsets (S_0, S_1) there is some $f \in \mathcal{F}$ satisfying $f(S_0) \subseteq \{0\}$ and $f(S_1) \subseteq \{1\}$). The *Vapnik-Chervonenkis (VC) dimension* of $\mathcal{F}$ is the cardinality of the largest set shattered by $\mathcal{F}$. If $\mathcal{F}$ is a class of real-valued functions in n variables, the *pseudo dimension* of $\mathcal{F}$ is defined as the VC dimension of the class $\{g : \mathrm{I\!R}^{n+1} \to \{0,1\} \mid \exists f \in \mathcal{F} \forall x \in \mathrm{I\!R}^n \forall y \in \mathrm{I\!R} : g(x,y) = \mathrm{sgn}(f(x) - y)\}$ (where $\mathrm{sgn}(z) = 1$ if $z \geq 0$, and 0 otherwise). The VC dimension (pseudo dimension) of a network of spiking neurons is defined to be the VC dimension (pseudo dimension) of the set of functions computed by the network with all possible assignments of values to its adjustable parameters, i.e. its delays, weights, and thresholds. To define the VC dimension for networks with real-valued output we assume that the output values are mapped to $\{0,1\}$ using some (non-trivial) threshold. Clearly, the VC dimension of a network is not larger than its pseudo dimension.

3 Bounded Depth Networks

In the following we provide an upper bound on the pseudo dimension for networks of spiking neurons that is almost linear in the number of network parameters (delays, weights, and thresholds) and the network depth.

Theorem 1. *Suppose $\mathcal{N}$ is a network of spiking neurons with W parameters and depth D where each neuron has at most k synaptic clusters and p is a bound on the degree of the synaptic interactions. Then the pseudo dimension of $\mathcal{N}$ is at most $O(WD\log(WDkp))$.*

The main idea of the proof is first to derive from $\mathcal{N}$ and a given set of input vectors a set of rational functions in the parameter variables of $\mathcal{N}$. Next, the connected components into which the parameter domain is partitioned by the zero-sets of these rational functions are considered. The construction guarantees that the number of dichotomies that $\mathcal{N}$ induces on the set of input vectors is not larger than the number of connected components generated by these functions.

Thus a bound on the number of connected components also limits the number of dichotomies. We require Definitions 7.4 and 7.5 from [1]: A set $\{f_1, \ldots, f_k\}$ of differentiable real-valued functions on $\mathbb{R}^d$ is said to have *regular zero-set intersections* if for every non-empty set $\{i_1, \ldots, i_l\} \subseteq \{1, \ldots, k\}$ the Jacobian (that is, the matrix of the partial derivatives) of $(f_{i_1}, \ldots, f_{i_l}) : \mathbb{R}^d \to \mathbb{R}^l$ has rank l at every point of the set

$$\{a \in \mathbb{R}^d : f_{i_1}(a) = \cdots = f_{i_l}(a) = 0\} \ .$$

A class $\mathcal{G}$ of real-valued functions defined on $\mathbb{R}^d$ has *solution set components bound* B if for every $k \in \{1, \ldots, d\}$ and every $\{f_1, \ldots, f_k\} \subseteq \mathcal{G}$ that has regular zero-set intersections the number of connected components of the set

$$\{a \in \mathbb{R}^d : f_1(a) = \cdots = f_k(a) = 0\}$$

is at most B. Further, a class $\mathcal{F}$ of real-valued functions is said to be *closed under addition of constants* if for every $c \in \mathbb{R}$ and $f \in \mathcal{F}$ the function $z \mapsto f(z) + c$ is also in $\mathcal{F}$. The following Lemma gives a stronger formulation of a bound stated in Theorem 7.6 of [1].

Lemma 2. *Let $\mathcal{F}$ be a class of real-valued functions $(y_1, \ldots, y_d, x_1, \ldots, x_n) \mapsto f(y_1, \ldots, y_d, x_1, \ldots, x_n)$ that is closed under addition of constants and where each function in $\mathcal{F}$ is C^d in the variables $y_1, \ldots, y_d$. If the class $\mathcal{G} = \{(y_1, \ldots, y_d) \mapsto f(y_1, \ldots, y_d, s) : f \in \mathcal{F}, s \in \mathbb{R}^n\}$ has solution set components bound B then for any sets $\{f_1, \ldots, f_k\} \subseteq \mathcal{F}$ and $\{s_1, \ldots, s_m\} \subseteq \mathbb{R}^n$, where $m \geq d/k$, the function $T : \mathbb{R}^d \to \{0,1\}^{mk}$ defined by*

$$T(a) = (\mathrm{sgn}(f_1(a, s_1)), \ldots, \mathrm{sgn}(f_1(a, s_m)), \ldots, \mathrm{sgn}(f_k(a, s_1)), \ldots, \mathrm{sgn}(f_k(a, s_m)))$$

partitions $\mathbb{R}^d$ into at most $B \sum_{i=0}^{d} \binom{mk}{i} \leq B(emk/d)^d$ equivalence classes (where $a_1, a_2 \in \mathbb{R}^d$ are equivalent iff $T(a_1) = T(a_2)$).

Proof (Theorem 1). Due to the length restriction we give only a brief sketch. Assume $\mathcal{N}$ as supposed, let $\{s_1, \ldots, s_m\}$ be a set of input vectors and $u_1, \ldots, u_m$ be real numbers. The main task is to bound the number of dichotomies induced on $S = \{(s_1, u_1), \ldots, (s_m, u_m)\}$ by functions of the form $(x, y) \mapsto \mathrm{sgn}(f(x) - y)$ where f is computed by $\mathcal{N}$. We proceed by induction on the levels of the network nodes where the level of a node v is defined as the length of the longest path from an input node to v. As induction hypothesis we assume that $\mathcal{G}_{\lambda-1}$, $\lambda \geq 1$, is a set of functions in the network parameters for the nodes of level at most $\lambda - 1$ that partitions the parameter domain into equivalence classes such that for parameters from the same class the computations of these nodes do not change on inputs from S in the following sense: The sequence of subsets of simultaneously active synapses remains the same for each node and the firing of a node is triggered by the same set of (starting or ending) synaptic pulses. Performing induction we finally get the bound $2^D(2em)^W(16e^2m^2Wkp)^{WD}$ for the number of equivalence classes. If S is shattered this implies that

$$m \leq D + (W + 2WD)\log m + W\log(2e) + WD\log(16e^2Wkp) \ .$$

We conclude that $m = O(WD\log(WDkp))$. □

If we restrict the type of functions used to model the synaptic interactions to the class of polynomials, we can derive a bound that improves the previous one to the effect that it does not depend on the number of synaptic clusters.

Theorem 3. *Suppose $\mathcal{N}$ is a network of spiking neurons with W parameters and depth D where the synaptic interactions are polynomials of degree at most p. Then the pseudo dimension of $\mathcal{N}$ is at most $O(WD\log(WDp))$.*

Proof. (Sketch) The proof follows the reasoning of the previous one. The only place where the number of clusters was taken into account was in the solution set components bound. For the class of degree p polynomials in d variables the solution set components bound $2(2p)^d$ can be used instead. This leads to the claimed result. □

If the depth and the degree of synaptic interactions are fixed, the previous theorem yields a tight bound. The same holds for rational interactions if the number of synaptic clusters in each node is fixed. We further note that the following almost linear upper bound improves the quadratic bound for networks with linear interactions given in [5].

Corollary 4. *Consider the class of networks of spiking neurons with fixed depth and with polynomial synaptic interactions of fixed degree. Suppose $\mathcal{N}$ is a network from this class with W parameters. Then the pseudo dimension of $\mathcal{N}$ is $\Theta(W\log W)$. This bound is also valid for the class of networks with rational synaptic interactions where the depth, the degree of synaptic interactions, and the number of synaptic clusters in each neuron are fixed.*

Proof. The upper bound is due to Theorem 1, the lower bound is due to Theorem 2.2 in [5] where it was shown that a single spiking neuron with W delay parameters has VC dimension $\Omega(W\log W)$. □

4 Networks with Arbitrary Depth

Clearly, we obtain from Theorem 1 a bound not depending on the depth of the network using that the depth is not larger than the number of parameters. A direct method yields a better bound. Its proof is omitted here.

Theorem 5. *Let $\mathcal{N}$ be a network of spiking neurons with W parameters and rational synaptic interactions of degree at most p where each neuron has at most k synaptic clusters. Then the pseudo dimension of $\mathcal{N}$ is at most $O(W^2\log(kp))$. For networks with polynomial interactions the bound $O(W^2\log p)$ holds.*

This leads to the following result analogous to Corollary 4. The lower bound is due to Theorem 3.1 in [5].

Corollary 6. *Consider the class of networks of spiking neurons with polynomial synaptic interactions of fixed degree. Suppose $\mathcal{N}$ is a network from this class with W parameters. Then the pseudo dimension of $\mathcal{N}$ is $\Theta(W^2)$. This bound is also valid for the class of networks with rational synaptic interactions where the degree of these interactions and the number of synaptic clusters in each neuron are fixed.*

5 Networks with Unrestricted Order

The networks considered thus far require restricted degrees of synaptic interactions. We next establish a polynomial pseudo dimension bound even for unlimited degree. Let the interaction of synpatic cluster $\{i_1, \ldots, i_l\}$ be modeled by the function $f^{(q_1,\ldots,q_l)}_{\{i_1,\ldots,i_l\}}$ in l variables and parameters $q_1, \ldots, q_l \in \mathbb{R}$ with

$$f^{(q_1,\ldots,q_l)}_{\{i_1,\ldots,i_l\}}(w_1, \ldots, w_l) = w_1^{q_1} w_2^{q_2} \cdots w_l^{q_l} \ .$$

(We require that q_i is an integer if $w_i < 0$, to keep the network working in the real domain.) A bound on the pseudo dimension is given by the following result based on new solution set components bounds for function classes defined in terms of polynomials, exponentials, and logarithms, which we omit here.

Theorem 7. *Suppose $\mathcal{N}$ is a network of spiking neurons with W parameters and nonlinear synaptic interactions of the form $f^{(q_1,\ldots,q_l)}_{\{i_1,\ldots,i_l\}}$ where each neuron has at most k synaptic clusters. Then $\mathcal{N}$ has pseudo dimension at most $O((Wk)^4)$.*

6 Conclusions

We have derived upper bounds on the pseudo dimension of networks of spiking neurons allowing for nonlinear synaptic interactions. All bounds are polynomials of low degree. The pseudo dimension is well known to be an upper bound for the so-called fat-shattering dimension. Both dimensions give bounds on the so-called covering numbers. The latter can be employed to obtain estimates for the number of training examples required for the networks to have low generalization error. The results presented here show that even when the computational power of networks of spiking neurons increases considerably due to nonlinear interactions, the sample complexity remains close to that of linear interactions.

Not all bounds were shown to be tight, however, and future work could lead to improvements. It might also be possible to obtain better estimates for the sample complexity by deriving bounds on the fat-shattering dimension directly.

References

1. M. Anthony and P. L. Bartlett. *Neural Network Learning: Theoretical Foundations.* Cambridge University Press, Cambridge, 1999.
2. W. Gerstner. Spiking neurons. In W. Maass and C. M. Bishop, editors, *Pulsed Neural Networks*, chapter 1, pages 3–53. MIT Press, Cambridge, Mass., 1999.
3. C. Koch. *Biophysics of Computation.* Oxford University Press, New York, 1999.
4. W. Maass. Computing with spiking neurons. In W. Maass and C. M. Bishop, editors, *Pulsed Neural Networks*, chapter 2, pages 55–85. MIT Press, Cambridge, Mass., 1999.
5. W. Maass and M. Schmitt. On the complexity of learning for spiking neurons with temporal coding. *Information and Computation*, 153:26–46, 1999.
6. A. F. Murray. Pulse-based computation in VLSI neural networks. In W. Maass and C. M. Bishop, editors, *Pulsed Neural Networks*, chapter 3, pages 87–109. MIT Press, Cambridge, Mass., 1999.

Product Unit Neural Networks with Constant Depth and Superlinear VC Dimension*

Michael Schmitt

Lehrstuhl Mathematik und Informatik, Fakultät für Mathematik
Ruhr-Universität Bochum, D–44780 Bochum, Germany
http://www.ruhr-uni-bochum.de/lmi/mschmitt/
mschmitt@lmi.ruhr-uni-bochum.de

Abstract. It has remained an open question whether there exist product unit networks with constant depth that have superlinear VC dimension. In this paper we give an answer by constructing two-hidden-layer networks with this property. We further show that the pseudo dimension of a single product unit is linear. These results bear witness to the cooperative effects on the computational capabilities of product unit networks as they are used in practice.

1 Introduction

Product units are formal neurons that are different from the most widely used neuron types in that they multiply their inputs instead of summing them. Furthermore, their weights operate as exponents and not as factors. Product units have been introduced by Durbin and Rumelhart [2] to allow neural networks to learn multiplicative interactions of arbitrary degree. Product unit networks have been proven computationally more powerful than sigmoidal networks in many learning applications by showing that they solve problems using less units than networks with summing units. Furthermore, the success of product unit networks has manifested itself in a rich collection of learning algorithms ranging from local ones, like gradient descent, to more global ones, such as simulated annealing and genetic algorithms [3, 5].

In this paper we investigate the Vapnik-Chervonenkis (VC) dimension of product unit networks. The VC dimension, and the related pseudo dimension, is well established as a measure for the computational diversity of analog computing mechanisms. It is further known to yield bounds for the complexity of learning in various well-studied models such as agnostic [1] and on-line learning [7]. For a large class of neuron types a superlinear VC dimension has been established for networks, while being linear for single units [4, 6, 8]. This fact gives evidence that the computational power of these neurons is considerably

* Work supported by the ESPRIT Working Group in Neural and Computational Learning II, NeuroCOLT2, No. 27150. A longer (9-page) version of this paper is available from http://www.ruhr-uni-bochum.de/lmi/mschmitt/. Complete proofs also appear in [9].

G. Dorffner, H. Bischof, and K. Hornik (Eds.): ICANN 2001, LNCS 2130, pp. 253–258, 2001.

enhanced when they cooperate in networks. Using methods due to Koiran and Sontag [4], who considered sigmoidal networks, it is not hard to show that there exist product unit networks with superlinear VC dimension. The drawback with this result is, however, that it requires the networks to have unbounded depth. Such architectures are rarely used in practical applications, where one or two hidden layers are almost always found to be satisfactory.

We show here that constant-depth product unit networks can indeed have superlinear VC dimension. In particular, we construct networks with two hidden layers of product and linear units that have VC dimension $\Omega(W \log k)$, where W is the number of weights and k the number of network nodes. The result is obtained by first establishing a superlinear lower bound for one-hidden-layer networks of a new, sigmoidal-type, summing unit. We contrast these lower bounds by showing that the pseudo dimension, and hence the VC dimension, of a single product unit is no more than linear.

2 Product Unit Networks and VC Dimension

A *product unit* has the form

$$x_1^{w_1} x_2^{w_2} \cdots x_p^{w_p}$$

with input variables $x_1, \ldots, x_p$ and real weights $w_1, \ldots, w_p$. The weights are the adjustable parameters of the product unit. (If $x_i < 0$, we require that w_i is an integer in order for the outputs to be real-valued.) In a monomial they are fixed positive integers. Thus, a product unit is computationally at least as powerful as any monomial. Moreover, divisive operations can be expressed using negative weights. What makes product units furthermore attractive is that the exponents are suitable for automatic adjustment by gradient-based and other learning methods.

We consider *feedforward networks* with a given number of input nodes and one output node. A network consisting solely of product units is equivalent to a single product unit. Therefore, product units are mainly used in networks where they occur together with other types of units. Here each non-input node may be a product or a linear unit (computing an affine combination of its inputs). We require that the networks have constant depth, where the *depth* is the length of the longest path from an input node to the output node.

VC dimension and pseudo dimension are defined as follows (see also [1]): A class $\mathcal{F}$ of $\{0,1\}$-valued functions in n variables is said to *shatter* a set of vectors $S \subseteq \mathbb{R}^n$ if $\mathcal{F}$ induces all dichotomies on S (that is, if for every partition of S into two disjoint subsets (S_0, S_1) there is some $f \in \mathcal{F}$ satisfying $f(S_0) \subseteq \{0\}$ and $f(S_1) \subseteq \{1\}$). The *Vapnik-Chervonenkis (VC) dimension* of $\mathcal{F}$ is the cardinality of the largest set shattered by $\mathcal{F}$. If $\mathcal{F}$ is a class of real-valued functions in n variables, the *pseudo dimension* of $\mathcal{F}$ is defined as the VC dimension of the class $\{g : \mathbb{R}^{n+1} \to \{0,1\} \mid \exists f \in \mathcal{F}\, \forall x \in \mathbb{R}^n \forall y \in \mathbb{R} : g(x,y) = \mathrm{sgn}(f(x) - y)\}$ (where $\mathrm{sgn}(z) = 1$ if $z \geq 0$, and 0 otherwise). The VC dimension (pseudo dimension) of a network is defined to be the VC dimension (pseudo dimension) of the set of functions computed by the network with all possible assignments of values to its adjustable parameters. For networks with real-valued output we assume that the output values are mapped to $\{0,1\}$ using some (non-trivial) threshold. Clearly, the pseudo dimension of a network is not smaller than its VC dimension.

3 Superlinear Lower Bound for Product Unit Networks

We show in this section that product unit networks of constant depth can have a superlinear VC dimension. The following result is the major step in establishing this bound. We note that hidden layers are numbered here from the input nodes toward the output node.

Theorem 1. *Let n, k be natural numbers satisfying $k \leq 2^{n+2}$. There is a network $\mathcal{N}$ with the following properties: It has n input nodes, at most k hidden nodes arranged in two layers with product units in the first hidden layer and linear units in the second, and a product unit as output node; furthermore, $\mathcal{N}$ has $2n\lfloor k/4 \rfloor$ adjustable and $7\lfloor k/4 \rfloor$ fixed weights. The VC dimension of $\mathcal{N}$ is at least $(n - \lfloor \log(k/4) \rfloor) \cdot \lfloor k/8 \rfloor \cdot \lfloor \log(k/8) \rfloor$.*

For the proof we require the following definition: A set of m vectors in $\mathbb{R}^n$ is said to be *in general position* if every subset of at most n vectors is linearly independent. Obviously, sets in general position exist for any m and n. We also introduce a new type of summing unit that computes its output by applying the activation function $\tau(y) = 1 + 1/\cosh(y)$ to the weighted sum of its inputs. (The new unit can be considered as the standard sigmoidal unit with σ replaced by τ). We observe that τ has its maximum at $y = 0$ with $\tau(0) = 2$ and satisfies $\lim \tau(y) = 1$ for $y \to -\infty$ as well as for $y \to \infty$.

Lemma 2. *Let h, m, r be arbitrary natural numbers. Suppose $\mathcal{N}$ is a network with $m + r$ input nodes, one hidden layer of $h + 2^r$ nodes which are summing units with activation function $1 + 1/\cosh$, and a monomial as output node. Then there is a set of cardinality $h \cdot m \cdot r$ that is shattered by $\mathcal{N}$.*

Proof. We choose a set $\{s_1, \ldots, s_{h \cdot m}\} \subseteq \mathbb{R}^m$ in general position and let $e_1, \ldots, e_r$ be the unit vectors in $\mathbb{R}^r$, that is, they have a 1 in exactly one component and 0 elsewhere. Clearly then, the set

$$S = \{s_i : i = 1, \ldots, h \cdot m\} \times \{e_j : j = 1, \ldots, r\}$$

is a subset of $\mathbb{R}^{m+r}$ with cardinality $h \cdot m \cdot r$. We show that it can be shattered by the network $\mathcal{N}$ as claimed.

Assume that (S_0, S_1) is a dichotomy of S. Let $L_1, \ldots, L_{2^r}$ be an enumeration of all subsets of the set $\{1, \ldots, r\}$ and define the function $g : \{1, \ldots, h \cdot m\} \to \{1, \ldots, 2^r\}$ to satisfy

$$L_{g(i)} = \{j : s_i e_j \in S_1\} ,$$

where $s_i e_j$ denotes the concatenated vectors s_i and e_j. For $l = 1, \ldots, 2^r$ let $R_l \subseteq \{s_1, \ldots, s_{h \cdot m}\}$ be the set

$$R_l = \{s_i : g(i) = l\} \ .$$

For each R_l we use $\lceil |R_l|/m \rceil$ hidden nodes for which we define the weights as follows: We partition R_l into $\lceil |R_l|/m \rceil$ subsets $R_{l,p}, p = 1, \ldots, \lceil |R_l|/m \rceil$, each of which has cardinality m, except for possibly one set of cardinality less than m. For each subset $R_{l,p}$ there exist real numbers $w_{l,p,1}, \ldots, w_{l,p,m}, t_{l,p}$ such that every $s_i \in \{s_1, \ldots, s_{h \cdot m}\}$ satisfies

$$(w_{l,p,1}, \ldots, w_{l,p,m}) \cdot s_i - t_{l,p} = 0 \quad \text{if and only if} \quad s_i \in R_{l,p} \ . \tag{1}$$

This follows from the fact that the set $\{s_1, \ldots, s_{h \cdot m}\}$ is in general position. (In other words, $(w_{l,p,1}, \ldots, w_{l,p,m}, t_{l,p})$ represents the hyperplane passing through all points in $R_{l,p}$ and through none of the other points.) With subset $R_{l,p}$ we associate a hidden node with threshold $t_{l,p}$ and with weights $w_{l,p,1}, \ldots, w_{l,p,m}$ for the connections from the first m input nodes. Since among the subsets $R_{l,p}$ at most h have cardinality m and at most 2^r have cardinality less than m, this construction can be done with at most $h + 2^r$ hidden nodes.

Thus far, we have specified the weights for the connections outgoing from the first m input nodes. The connections from the remaining r input nodes are weighted as follows: Let $\varepsilon > 0$ be a real number such that for every $s_i \in \{s_1, \ldots, s_{h \cdot m}\}$ and every weight vector $(w_{l,p,1}, \ldots, w_{l,p,m}, t_{l,p})$:

$$\text{If} \quad s_i \notin R_{l,p} \quad \text{then} \quad |(w_{l,p,1}, \ldots, w_{l,p,m}) \cdot s_i - t_{l,p}| > \varepsilon \ .$$

According to the construction of the weight vectors in (1), such an ε clearly exists. We define the remaining weights $w_{l,p,m+1}, \ldots, w_{l,p,m+r}$ by

$$w_{l,p,m+j} = \begin{cases} 0 & \text{if } j \in L_l \ , \\ \varepsilon & \text{otherwise} \ . \end{cases} \tag{2}$$

We show that the hidden nodes thus defined satisfy the following:

Claim 3. *If $s_i e_j \in S_1$ then there is exactly one hidden node with output value 2; if $s_i e_j \in S_0$ then all hidden nodes yield an output value less than 2.*

According to (1) there is exactly one weight vector $(w_{l,p,1}, \ldots, w_{l,p,m}, t_{l,p})$, where $l = g(i)$, that yields 0 on s_i. If $s_i e_j \in S_1$ then $j \in L_{g(i)}$, which together with (2) implies that the weighted sum $(w_{l,p,m+1}, \ldots, w_{l,p,m+r}) \cdot e_j$ is equal to 0. Hence, this node gets the total weighted sum 0 and, applying $1 + 1/\cosh$, outputs 2. The input vector e_j changes the weighted sums of the other nodes by an amount of at most ε. Thus, the total weighted sums for these nodes remain different from 0 and, hence, the output values are less than 2.

On the other hand, if $s_i e_j \in S_0$ then $j \notin L_{g(i)}$ and the node that yields 0 on s_i receives an additional amount ε through weight $w_{l,p,m+j}$. This gives a total weighted sum different from 0 and an output value less than 2. All other nodes fail to receive 0 by an amount of more than ε and thus have total weighted sum different from 0 and, hence, an output value less than 2. Thus Claim 3 is proven.

Finally to complete the proof, we do one more modification with the weight vectors and define the weights for the ouptut node. Clearly, if we multiply all weights and thresholds defined thus far with any real number $\alpha > 0$, Claim 3 remains true. Since $\lim(1 + 1/\cosh(y)) = 1$ for $y \to -\infty$ and $y \to \infty$, we can find an α such that on every $s_i e_j \in S$ the output values of those hidden nodes that do not output 2 multiplied together yield a value as close to 1 as necessary. Further, since $1 + 1/\cosh(y) \geq 1$ for all y, this value is at least 1. If we employ for the output node a monomial with all exponents equal to 1, it follows from the reasoning above that the output value of the network is at least 2 if and only if $s_i e_j \in S_1$. This shows that S is shattered by $\mathcal{N}$. □

Proof (Theorem 1). Due to the length restriction we omit the proof. The idea is to take a set S' constructed as in Lemma 2 and, as shown there, shattered by a network $\mathcal{N}'$ with a monomial as output node and one hidden layer of summing units that use the activation function $1 + 1/\cosh$. Then S' is transformed into a set S and $\mathcal{N}'$ into a network $\mathcal{N}$ such that for every dichotomy (S'_0, S'_1) induced by $\mathcal{N}'$ on S' the network $\mathcal{N}$ induces the corresponding dichotomy (S_0, S_1) of S. □

From Theorem 1 we obtain the superlinear lower bound for constant depth product unit networks. (We omit its derivation due to lack of space.)

Corollary 4. *Let n, k be natural numbers where $16 \leq k \leq 2^{n/2+2}$. There is a network of product and linear units with n input units, at most k hidden nodes in two layers, and at most $W = nk$ weights that has VC dimension at least $(W/32)\log(k/16)$.*

4 Linear Upper Bound for Single Product Units

We now show that the pseudo dimension, and hence the VC dimension, of a single product unit is indeed at most linear.

Theorem 5. *The VC dimension and the pseudo dimension of a product unit with n input variables are both equal to n.*

Proof. That n is a lower bound easily follows from the fact that the class of monomials with n variables shatters the set of unit vectors from $\{0,1\}^n$. We derive the upper bound by means of the pseudo dimension of a linear unit. This is omitted here. □

Corollary 6. *The VC dimension and the pseudo dimension of the class of monomials with n input variables are both equal to n.*

5 Conclusions

We have established a superlinear lower bound on the VC dimension of constant depth product unit networks. This result has implications in two directions: First, it gives theoretical evidence of the finding that product unit networks employed in practice are indeed powerful analog computing devices. Second, the VC dimension yields lower bounds for the complexity of learning in several models of learnability, e.g., on the sample size required for low generalization error. The result presented here can now directly applied to obtain such estimates. There are, however, models of learning for which better bounds can be obtained in terms of other dimensions, such as, e.g., the fat-shattering dimension and covering numbers. A topic for future research is therefore to determine good bounds for the generalization error of product unit networks in these learning models.

We have also presented here a superlinear lower bound for networks with one hidden layer of a new type of summing unit and have shown that VC and pseudo dimension of single product units are linear. This raises the interesting open problem to determine the VC dimension of product unit networks with one hidden layer. Furthermore, we do not know if the lower bounds given here are tight. Thus far, the best known upper bounds for networks with differentiable activation functions are low order polynomials (degree two for polynomials, four for exponentials). But they even hold for networks of unrestricted depth. The issue of obtaining tight upper bounds for product unit networks seems therefore to be closely related to a general open problem in the theory of neural networks.

References

1. M. Anthony and P. L. Bartlett. *Neural Network Learning: Theoretical Foundations.* Cambridge University Press, Cambridge, 1999.
2. R. Durbin and D. Rumelhart. Product units: A computationally powerful and biologically plausible extension to backpropagation networks. *Neural Computation*, 1:133–142, 1989.
3. A. Ismail and A. P. Engelbrecht. Global optimization algorithms for training product unit neural networks. In *International Joint Conference on Neural Networks IJCNN'2000*, vol. I, pp. 132–137, IEEE Computer Society, Los Alamitos, CA, 2000.
4. P. Koiran and E. D. Sontag. Neural networks with quadratic VC dimension. *Journal of Computer and System Sciences*, 54:190–198, 1997.
5. L. R. Leerink, C. L. Giles, B. G. Horne, and M. A. Jabri. Learning with product units. In G. Tesauro, D. Touretzky, and T. Leen, editors, *Advances in Neural Information Processing Systems 7*, pp. 537–544, MIT Press, Cambridge, MA, 1995.
6. W. Maass. Neural nets with super-linear VC-dimension. *Neural Computation*, 6:877–884, 1994.
7. W. Maass and G. Turán. Lower bound methods and separation results for on-line learning models. *Machine Learning*, 9:107–145, 1992.
8. A. Sakurai. Tighter bounds of the VC-dimension of three layer networks. In *Proceedings of the World Congress on Neural Networks*, vol. 3, pp. 540–543. Erlbaum, Hillsdale, NJ, 1993.
9. M. Schmitt. On the complexity of computing and learning with multiplicative neural networks. *Neural Computation*, to appear.

Generalization Performances of Perceptrons

Gérald Gavin

Laboratoire E.R.I.C – Université Lumière,
5 Av. Pierre Mendès France, 69676 Bron
ggavin@univ-lyon2.fr

Abstract. This paper presents new results about the confidence bounds on the generalization performances of perceptrons. It deals with regression problems. It is shown that the probability to get a generalization error greater than the empirical error plus a precision ε, depends on the number of inputs and on the magnitude of the coefficients of the combination. The result presented does not require to bound *a priori* the magnitude of these coefficients, the size and the number of layers.

1 Introduction

Classical results from statistical learning theory give confidence bounds on the deviation between generalization and empirical errors in terms of combinatorial dimensions related to the class of functions used by the learning system [1]. For multi-layer perceptrons, many studies (see for instance [2]) have shown that these dimensions grow at least as quickly as the number of parameters. These results are unsatisfactory since the number of parameters is often very large and the learning systems perform well with a relatively small training set. In [3], Bartlett has shown that for classification, "the size of the weights is more important than the size of the network". This paper aims at providing the same for regression. As far as we know, it is the first explicit and general result about linear perceptrons.

2 Preliminaries

2.1 Notations and Definitions

In the following, X will denote an arbitrary input space and Y will refer to an output space which will be taken as $[-1, 1]$. Let $z_n = \{(x_1, y_1)...(x_n, y_n)\} \in (X \times Y)^n$ be a finite sample i.i.d. according to an unknown distribution D.

Given these observations, a classical goal of supervised learning is to find a functional relation g, in a set G of functions, such that the quadratic generalisation error $er_D(g)$ is as low as possible. $er_D(g)$ is defined as the expectation of the square of the difference between $g(x)$ and y.

$$er_D(g) = \int (g(x) - y)^2 dD(x, y)$$

G. Dorffner, H. Bischof, and K. Hornik (Eds.): ICANN 2001, LNCS 2130, pp. 259–264, 2001.

However, as D is supposed unknown, $er_D(g)$ can not be computed but only estimated. A classical way to estimate the generalisation error is to consider the empirical one defined as the quadratic error on the training sample z_n

$$er_{z_n}(g) = \frac{1}{n} \sum_{(x_i,y_i)\in z_n} (g(x_i) - y_i)^2$$

Our goal is then to bound the difference between the empirical and the generalization error of a function g. We are interested in bounding in probability :

$$e_{G,z_n} = sup_{g\in G}|er_D(g) - er_{z_n}(g)| \tag{1}$$

The less e_{G,z_n} will be, the more generalization performances will be controlled. Here we will focuse on the bounded linear combination of a set $H \in [-1;1]^X$: $G_\infty = \{\sum_i w_i h_i,\ h_i \in H \quad \sum_i |w_i| < \infty\}$. According to the set H, G_∞ will be for example the set of the multilayer perceptron of linear regression. For the later $G_A = \{\sum_i w_i h_i,\ h_i \in H \quad s.t. \sum_i |w_i| < A\}$.

2.2 Covering Numbers

Covering numbers appear naturally in the study of generalization properties of learning machines. They are central to many bounds concerning both classification and regression. Their definition is related to ε-covers in a pseudo-metric space (E, ρ) : a set $T \subseteq E$ is an $\varepsilon-$cover of a set $A \subset E$ with respect to ρ if for all $x \in A$ there is a $y \in T$ such that $\rho(x, y) \leq \varepsilon$. Given ε, the covering number $N(\varepsilon, A, \rho)$ of A is then the size of the smallest $\varepsilon - cover$ of A. In learning theory, the distance ρ is usually defined from a sequence $X_n = (x_1, ..., x_n)$, $x_i \in X$. The average l_1 semi-norm $l_1^{X_n}(f,g) = \frac{1}{n}\sum_{i=1}^n |f(x_i) - g(x_i)|$ is an example of such a distance. The following theorem adapted from Vidyasagar [8] sheds light on the role of covering numbers.

Theorem 1. *Let D be a distribution over $X \times [-1, 1]$. Let $G \subset [-1, 1]^X$, for $n \geq \frac{2}{\varepsilon^2}$, we have :*

$$P_{D^n}(e_{G,n} \geq \varepsilon) \leq 4E_{D^{2n}}\left(N\left(\frac{\varepsilon}{64}, G, l_1^{X_{2n}}\right)\right) e^{\frac{-n\varepsilon^2}{128}}$$

Bounding the covering numbers is thus a central issue. Many studies aim at bounding its value mostly by using combinatorial dimensions such as Pseudo-dimension or fat-shattering dimension [7]. As far as we know, very few bounds relating the covering number of G_A to the covering number of H exist. Such bounds are however very interesting since they may be used for any linear combination of learning machines. By looking at the proof of Theorem 17 in the technical report related to [3], it is not difficult to derive the following bound (see section 6.2):

$$N(\varepsilon, G_A, l_1^{X_{2n}}) \leq \left(2N(\frac{\varepsilon}{2A}, H, l_1^{X_{2n}}) + 1\right)^{\frac{2A^2}{\varepsilon^2}} \tag{2}$$

which in turn can be used in theorem to give confidence bounds. The resulting bound is however too large to be practical. This drawback is of importance and motivates the search for better bounds. Next section presents a new approach which leads to better results when learning specifically with quadratic loss functions.

3 New Confidence Bounds

3.1 General Confidence Bound Dealing with Linear Combinations

Let us consider the set G_∞. We are then concerned with,

$$e_{G_\infty, z_n} = sup_{g \in G_\infty} \left(\frac{|er_D(g) - er_{z_n}(g)|}{(A_g+1)^2} \right)$$

Unlike previously (1), the deviation between the empirical error and the generalization one is scaled by $1/(A_g+1)^2$. By introducing this factor, we avoid to do assumptions about G_∞ and stay in a general framework. This will be discussed shortly. We prove now the main result of this paper.

Theorem 2. *Let D a distribution over $X \times [-1,1]$, if $n \geq \frac{2}{\varepsilon^2}$, we have :*

$$P_{D^n}(e_{G_\infty, z_n} \geq \varepsilon) \leq 12 E_{D^{2n}} \left(N^2(\frac{\varepsilon}{32}, H, l_1^{X_{2n}}) \right) e^{\frac{-n\varepsilon^2}{128}}$$

The proof is detailed in section 4.

Discussion
The theorem may be rephrased as : with a certain probability, the inequality

$$er_D(g) \leq er_{z_n}(g) + \varepsilon (A_g+1)^2$$

holds for all $g \in G_\infty$. This inequality concerns a wide class of functions which does not contain necessarily the target function.

The value A_g is not bounded a priori. Hence, the confidence concerns the whole space G_∞ instead of a restricted one like G_A defined in previous section. The structure of the functions appears a posteriori in e_{G_∞, z_n}. This implies when applying this theorem, it is not needed to do assumptions about the value of A_g. A structural risk minimization is hence not required.

This theorem suggests to minimize at one and the same time the empirical error and the magnitude of the output weights. It corresponds to a natural intuition : if the output weights are large, then the difference between two functions in G_∞ is significant and hence the control is less precise.

The approach developped in this theorem is direct and does not require to bound the covering number of the set G_∞. From this point of view, it is an original approach. To compare our bounds to the one we obtain in combining Bartlett's bound (2) with the theorem 1 , we consider $G_A \subset G_\infty$ such that

$\forall g \in G_A,\ A_g < A$. That is, we impose an a priori over the set G_∞. Then, according to theorem 2 ,we have :

$$P_{D^n}(e_{G,z_n} \geq \varepsilon) \leq 12 E_{D^{2n}}\left(N^2(\frac{\varepsilon}{32A}, H, l_1^{X_{2n}})\right) e^{\frac{-n\varepsilon^2}{(A+1)^2 128}}$$

whereas Bartlett's bounds combined with theorem 1 leads to :

$$P_{D^n}(e_{G,z_n} \geq \varepsilon) \leq 4 E_{D^{2n}}\left((2N(\frac{\varepsilon}{128A}, H, l_1^{X_{2n}}) + 1)\right)^{\frac{8192A^2}{\varepsilon^2}} e^{\frac{-n\varepsilon^2}{128}}$$

Hence, when minimizing the quadratic error, our bound gives a better confidence than what is derived by using the approach developped in [3]. The leading exponent $\frac{8192A^2}{\varepsilon^2}$ is indeed very large for small ε compared to the exponent 2 in our bound. Our result is hence more accurate.

To obtain practical bounds over the generalization error, one has to bound the covering number of the set H. Many existing results may be plugged in. We do not present them since they strongly depend on the set H. See for instance [5],[1].

3.2 One Hidden-Layer Perceptron

A one hidden layer perceptron can be seen as a linear combination of functions. The previous results could be directly used in order to get confidence bounds.

Let's consider one hidden layer perceptrons defined by :

- p inputs
- One hidden layer (size I not fixed a priori; sigmoid activation function[1])
- One linear output .
  We denote by w_i the output weights and h_i the output of the ith neuron belonging to the hidden layer. The function g computed by this perceptron can be written as a linear combination of function h_i, i.e. $g = \sum_{i=1}^{I} w_i h_i$.

We denote by H the set of all the possible functions h_i (H is the set of the functions computed by a perceptron with p inputs) and by G_∞ the set of linear combination of H. In order to apply theorem 3.1, we need to bound the covering number $N(H, \varepsilon, l_1^X)$.

The Pollard dimension is an extension of the VC-dimension in the regression case. The Pollard dimension of H is equal to $p+1$. The finitude of this dimension allows to bound the covering number. Haussler [6] bounds the covering number $N(F, \varepsilon, l_1^{X_n})$ for a set F. We denote by d the Pollard dimension of F and one gets the following result

$$N(F, \varepsilon, l_1^{X_n}) \leq e(d+1)\left(\frac{2e}{\varepsilon}\right)^d \qquad (3)$$

Inequality 3 and theorem 3.1 lead to the following result

[1] Same results for other classical activation functions

Theorem 3. *Let D a distribution over $\mathbb{R}^p \times [-1,1]$, if $n \geq \frac{2}{\varepsilon^2}$, we have :*

$$P_{D^n}(e_{G_\infty, z_n} \geq \varepsilon) \leq 12\,(d+1)\left(\frac{64e}{\varepsilon}\right)^{2p+2} e^{\frac{-n\varepsilon^2}{128}}$$

This result can be written in terms of sample complexity $n(\varepsilon,\delta)$, i.e the smallest training sample size n needed to ensure $P_{D^n}(e_{G_\infty, z_n} \geq \varepsilon) \leq \delta$. We show that

$$n(\varepsilon,\delta) = O\left(\frac{1}{\varepsilon^2}\left(p\ln\frac{1}{\varepsilon} + \ln\frac{1}{\delta}\right)\right)$$

4 Proof of Theorem 2

For sake of simplicity, we first suppose that y is a function of x and that the distribution D is over X. By definition, we have $er_D(g) = \int_X \left(\sum_{i\in I} w_i h_i(x) - y(x)\right)^2 dD$ where I is a finite or infinite countable index set. We develop the integral and permute sum and integral (bounded convergence theorem can be used here since everything is bounded), we have then, for $n \geq 2/\varepsilon^2$,

$$|er_D(g) - er_{z_n}(g)| \leq \sum_{(i_1,i_2)\in I^2} (|w_{i_1}|.|w_{i_2}|)\,|E_D(h_{i_1}h_{i_2}) - E_{z_n}(h_{i_1}h_{i_2})| + \dots$$
$$\dots + \sum_{i\in I}(|w_i|)\,|E_D(y.h_i) - E_{z_n}(y.h_{i_1})| +$$
$$\dots + \left|E_D(y^2) - E_{z_n}(y^2)\right| \quad (*)$$

We bound then the covering numbers for each set $G_0 = \{h_{i_1}h_{i_2},\ s.t.\ h_{i_j} \in H\}$, $G_1 = \{y.h,\ s.t\ h \in H\}$ and $G_2 = \{y^2\}$ by the covering number of H (see [4] for details),

$$1 = N(\varepsilon, G_2, l_1^{X_{2n}}) \leq N(\varepsilon, G_1, l_1^{X_{2n}}) \leq N(\varepsilon, G_0, l_1^{X_{2n}}) \leq N(\frac{\varepsilon}{2}, H, l_1^{X_{2n}})^2$$

According to theorem 1, we have then, for each set G_k, $k = 0,1,2$

$$P_{D^n}(\sup_{g\in G_k} |E_D(g) - E_{z_n}(g)| \geq \varepsilon) \leq 4E_{D^{2n}}\left(N\left(\frac{\varepsilon}{32}, H, l_1^{X_{2n}}\right)^2\right) e^{\frac{-n\varepsilon^2}{128}}$$

We choose thus a sample z_n such that,

$$\bigwedge_{k=0}^{2} \sup_{h\in G_k} |E_D(h) - E_{z_n}(h)| \leq \varepsilon$$

The probability to have such a sample is at least $1 - 12.E_{D^{2n}}\left(N(\frac{\varepsilon}{32}, H, l_1^{X_{2n}})^2\right)$ $e^{\frac{-n\varepsilon^2}{128}}$. For such a sample z_n, the inequality $(*)$ becomes then

$$|er_D(g) - er_{z_n}(g)| \leq \sum_{(i_1,i_2)\in I^2} (|w_{i_1}|.|w_{i_2}|)\,\varepsilon + \sum_{i\in I} |w_i|\varepsilon + \varepsilon$$

and by refactorizing, we have $\forall g \in G_\infty$

$$|er_D(g) - er_{z_n}(g)| \leq \varepsilon\left(\sum_{i\in I}|w_i| + 1\right)^2$$

We now can extend our result to distributions over $X \times Y = [-1,1]$. Indeed, for any set F of functions $f \in [-1,1]^X$ we consider the set $\widetilde{F}$ of functions $\widetilde{f} \in [-1,1]^{X \times Y}$ such that $\widetilde{f}(x,y) = f(x)$, and the target function $f_t \in [-1,1]^{X \times Y}$ such that $f_t(x,y) = y$. As the covering number of $\widetilde{F}$ are equal to the ones of F and $er_D(f) = E_D\left(\left|\widetilde{f}(x,y) - f_t(x,y)\right|^2\right)$, we can apply the above result.

and this finish the proof. □

5 Conclusion

The quadratic loss function, are very used in neural networks to solve regression problems. From Bartlett's results, we can derive [3] a confidence bound independent of the size of the linear combination corresponding, in the case of neural networks, to the number of hidden units of the hidden layer. Our approach, because devoted to quadratic loss functions, gives much better bounds in a quite simplier way, without using complex combinatorial notions as the fat shattering dimension. As far as we know our result is the first, in regression, concerning G_∞, which gives a pratical confidence bound on the deviation between the generalization and empirical error. Unlike Bartlett, our result does not limit, a priori, the absolute sum of the weights of the linear combination. One can use it to make regularization in reducing the structural risk $er_{z_n}(g) + \varepsilon(A_g + 1)^2$. We get the same result for the multi-layer perceptrons case, by considering an extended computation of A_g.

References

1. N. Alon, S. Ben-David, N. Cesa-Bianchi, and D. Haussler. Scale sensitive dimensions, uniform convergence, and learnability. In *Proceedings of the ACM Symposium on foundations of Computer Science*, 1993.
2. M. Anthony. Probabilistic analysis of learning in artificial neural networks: The pac model and its variants. Technical report, The London School of Economics and Political Sience, 1995.
3. P. Bartlett. The sample complexity of pattern classification with neural networks: the size of the weights is more important than the size of the network. Technical report, Department of Systems Engineering, Australian National University, 1996.
4. G. Gavin and A. Elisseeff. Confidence bounds for the generalization performances of linear combinations of functions. Technical report, ERIC - Universite Lumiere, Bron, 1999. http://eric.univ-lyon2.fr/eric.
5. L. Gurvits and P. Koiran. Approximation and learning convex superpositions. In *Computationnel Learning Theory: EUROCOLT'95*, 1995.
6. D. Haussler. Sphere packing numbers for subsets of the boolean n-cube with boundes vapnik-chervonenkis dimension. *Journal of Combinatorial Theory*, 69:217–232, 1995.
7. M. Kearns and R. Schapire. Efficient distribution-free learning of probabilistic concepts. In *Proceedings of the 31st Symposium on the Foundations of Computer Science*, pages 382–391. IEEE Computer Society Press, Los Alamitos, CA, 1990.
8. M. Vidyasagar. *A Theory of Learning and Generalization*. Springer, 1997.

Bounds on the Generalization Ability of Bayesian Inference and Gibbs Algorithms

Olivier Teytaud and Hélène Paugam-Moisy

Institut des Sciences Cognitives, UMR CNRS 5015
67 boulevard Pinel, F-69675 Bron cedex, France
teytaud,hpaugam@isc.cnrs.fr

Abstract. Recent theoretical works applying the methods of statistical learning theory have put into relief the interest of old well known learning paradigms such as Bayesian inference and Gibbs algorithms. Sample complexity bounds have been given for such paradigms in the zero error case. This paper studies the behavior of these algorithms without this assumption. Results include uniform convergence of Gibbs algorithm towards Bayesian inference, rate of convergence of the empirical loss towards the generalization loss, convergence of the generalization error towards the optimal loss in the underlying class of functions.

1 Introduction

Recent works about the Bayes Point Machine ([5, 6]), reporting good results on a few benchmarks, have put into relief the interest of old well known learning paradigms such as Bayesian inference. Haussler, Kearns and Shapire [4] give results upon Bayesian learning in the zero error case, and Devroye, Gyorfi and Lugosi [3] recall negative results upon similar paradigms.

We consider learning algorithms working on independent identically distributed samples $D = \{(X_1, Y_1), \ldots, (X_m, Y_m)\}$, with $X_i \in \mathcal{X}$ and $Y_i \in \{0, 1\}$. We consider an *a priori* law Pr on the space W of hypothesis (this does *not* mean, in any of the following results, that the underlying dependency is drawn according to this distribution!). We define the following loss function, for $w \in W$: $L(Y, X, w) = \ln((1-Y)w(X) + Y \times (1-w(X)))$. The empirical loss $L_E(w)$ is the average of $L(Y_i, X_i, w)$ on the learning sample. $L(w)$ (the loss in generalization) is the expectation of $L(Y, X, w)$. This means that w is not simply considered as a classifier, but that $w(X) = \frac{2}{3}$ when $Y = 1$ is worst than $w(X) = \frac{9}{10}$ even if thresholding would lead to 1 in both cases. Notice that a good behavior of this loss function implies a good behavior of the loss function of the associated classifier, which chooses class 1 if and only if $f_{D_m}(x) > \frac{1}{2}$. The converse is not necessarily true and an important point is that Bayesian inference can achieve a better error rate than any classifier of the underlying family W of functions. For clarity, we denote $R(Y, X, w) = \ln(1 - \exp L(Y, X, w))$ and $R(w) = E(R(Y, X, w))$ the expectation of $R(., ., w)$ for (X, Y) samples. We call *Bayesian inference* the algorithm which proposes as classifier $f_{D_m}(x) = \int_W V(w, D_m)w(x)dPr(w)$ with

G. Dorffner, H. Bischof, and K. Hornik (Eds.): ICANN 2001, LNCS 2130, pp. 265–270, 2001.

$V(w, D_m) = \frac{\Pi_{i=1}^m (\exp R(Y_i, X_i, w))}{\int_W \Pi_{i=1}^m (\exp R(Y_i, X_i, w)) dPr(w)}$, normalized likelihood of w. Note that this function doesn't necessarily lies in W. We call *first Gibbs algorithm* ([8]) of order k the algorithm which proposes as classifier $g_{D_m}^k(x) = \frac{1}{k}\sum_{i=1}^k w_i(x)$, the k classifiers w_i being drawn randomly according to $V(w, D_m)Pr(w)$. We call *second Gibbs algorithm* of order k the algorithm which proposes as classifier $\tilde{g}_{D_m}^k(x) = \frac{\sum_{i=1}^k w_i V(w_i, D_m)}{\sum_{i=1}^k V(w_i, D_m)}$, w_i being drawn randomly according to $Pr(w)$. As Gibbs algorithms are non deterministic, this inference can be used in two ways, the first one consisting in choosing once the classification function, the second one consisting in doing new drawns of w_i's at each example to be classified. Next we consider one random choice of the w_i's, during the learning phase. The expectation of the loss of this algorithm is the probability of error of the non-deterministic equivalent algorithm before stochastic choice of the w_i's. $N(\epsilon, \mathcal{F}, L_p)$ denotes the ϵ-covering number of a space $\mathcal{F}$ for distance L^p.

The following notions are used: Donsker classes, central limit theorems for Donsker classes, VC-dimension (the VC-dimension of a class of real-valued functions is the VC-dimension of their subgraphs), fat-shattering dimension, entropy integral, covering numbers, weak convergence, weak $O(.)$. An introduction to these notions can be found in [12]. We recall that a class $\mathcal{F}$ of functions is Donsker for a distribution of examples X_i if $\frac{1}{\sqrt{m}}\sum_{i=1}^m X_i$ converges weakly with m to a tight limit in $l^\infty(\mathcal{F})$. This implies uniform weak convergence of empirical means to the expectations, in $O(1/\sqrt{m})$. The main difference with VC-theory is the non-uniform (in the distribution) and asymptotic nature of this convergence. For ϕ lipschitzian, we recall that $\{\phi \circ f / f \in \mathcal{F}\}$ is Donsker provided that f is Donsker. This is usefull for proving that the family of loss functions (in particular, in the case of our loss function, which is Lipschitzian provided that w and $1 - w$ are lower bounded by $c > 0$) is Donsker provided that the family of functions used for prediction is Donsker.

A few easy results can be stated. The ϵ-covering numbers for L_1 of the class of functions in which $g_{D_m}^k$ is chosen are bounded by $N(\epsilon, W, L_1)^k$. If the VC-dimension V of W is finite, then the VC-dimension of this class of functions is bounded by $k \times V$. Lipschitz coefficients in Bayesian inference or Gibbs algorithm are preserved, what can lead to bounds in terms of fat-shattering dimension [1]. Bayesian averaging, in Bayes Point Machine for example, consists in using as classifier the expectation of the $w \in W$ such that all the data points are classified correctly by w - this is not equivalent to Bayesian inference, and some examples [10] can be given, in which Bayesian inference is significantly asymptotically better than Bayesian averaging. Results of [9] explain how to compute Bayesian averaging.

Part 2 gives bounds on the difference between the empirical loss and the loss in generalization. Part 3 gives bounds on the rate of convergence of Bayesian inference towards the optimal error rate in W. Part 4 gives a bound on the convergence of Gibbs algorithm towards Bayesian inference. The conclusion opens a question about the second version of Gibbs algorithm.

---

[1] For examples of use of fat-shattering dimension, the reader is referred to [2] or [1].

2 Bayesian Inference: Generalization Error

Assume that W is included in a space in which integrals of functions of $W \to \mathbb{R}$ or $W \to W$ can be computed as Rieman sums. In particular, $\exists w_{k,i}/i \leq k, k \in \mathbb{N}$

$$f_{D_m}(x) = \frac{\lim_{k\to\infty} \sum_{i=1}^{k} \theta_{k,i} \Pi_{j=1}^{m} \exp R(Y_j, X_j, w_{k,i}) w_{k,i}}{\lim_{k\to\infty} \sum_{i=1}^{k} \theta_{k,i} \Pi_{j=1}^{m} \exp R(Y_j, X_j, w_{k,i})} = \lim_{k\to\infty} \sum_{i=1}^{k} \alpha_{k,i,D_m} w_{k,i}$$

with $\alpha_{k,i,D_m} = \frac{\theta_{k,i} \times \Pi_{j=1}^{m} \exp R(Y_j, X_j, w_{k,i})}{\sum_{l=1}^{k} \theta_{k,l} \Pi_{j=1}^{m} \exp R(Y_j, X_j, w_{k,l})}$, which leads to $\sum_{i=1}^{k} \alpha_{k,i,D_m} = 1$.

This assumption implies that any f_{D_m} lies in the closure W' of the convex hull of W for the pointwise convergence. The following results are derived from this statement.

2.1 Case 1: *W* Has Finite VC-Dimension

If the VC-dimension of W is finite, then the logarithms of covering numbers of $\overline{conv}\ W$ are bounded by $O(1/\epsilon^p)$ with $p < 2$. This condition is verified in the case of a finite VC-dimension $V+1$, but a non-distribution-free bound on covering numbers is sufficient.

Theorem 1 *(see for example [12, p142]) If W verifies $N(\epsilon, W, L_2) = O((\frac{1}{\epsilon})^V)$, then*

$$\log N(\epsilon, \overline{conv}\ W, L_2) = O((\frac{1}{\epsilon})^{\frac{2V}{V+2}})$$

Corollary 1 *If W has finite VC-dimension V, then the logarithm of the ϵ-covering numbers of $\overline{conv}\ W$ for L^2 are $O(\epsilon^{-\frac{2V+2}{V+3}})$.*

A remarkable fact is that this exponent is lower than 2. This implies boundedness of the entropy integral, and so the Donsker nature of $\overline{conv}\ W$. Moreover, results upon covering numbers (in [11]) give the non-asymptotic bound of theorem 2(which is *not* distribution-free), provided that w and $1-w$ are both lower bounded by c (this condition can be removed if we consider $\exp L$ instead of L; indeed, any Lipschitzian cost function can be used, and some others as well). This result is derived from [11, p188] where the author considers empirical risk minimization. Notice that the result can be used thanks to the Lipschitzian nature of the loss function. Covering numbers for $L_1(\mu)$ need prior knowledge about the density of examples. For example, the ϵ-covering numbers for $L_1(\mu)$, for a density μ bounded by K, are bounded by the $\frac{\epsilon}{K}$-covering numbers. Notice that no prior knowledge on the distribution of Y is required; the only assumption is the law of the marginal distribution of X. This leads to the corollary 2. The dependency in ϵ is lower than $1/\epsilon^4$. This is uniform on distributions of examples verifying the hypothesis on the density. The following result, which is not uniform, gives a better non-uniform bound, whenever W has infinite VC-dimension, provided that W is Donsker.

Theorem 2 (Non-asymptotic bounds resulting of covering numbers) *Let the x_i's be drawn according to density μ. Let N be the $(\epsilon \times c)$-covering number for $L_1(\mu)$ of $\overline{conv}\ W$. Then with probability $\geq 1-\delta$, $|L(f_{D_m}) - L_E(f_{D_m})| \leq \epsilon$ if $m \geq \frac{\ln(2N/\delta)}{2c^2\epsilon^2}$, with f_{D_m} chosen among an ϵ-cover (this implies a discretization).*

Corollary 2 *If W has a finite VC-dimension V, there exists a universal constant M such that if the density of examples is bounded by K, then $|L(f_{D_m}) - L_E(f_{D_m})| \leq \epsilon$ with probability $\geq 1-\delta$ if $m \geq \frac{K^2}{\epsilon^2} \times (M(\frac{K}{\epsilon})^{\frac{2V+2}{V+3}} - 8\log(\delta))$.*

2.2 Case 2: *W* Is Donsker

Theorem 3 *(see for example [12, p190]) If W is Donsker, then the convex hull of W is Donsker and the pointwise closure of W is Donsker.*

Corollary 3 *$\overline{conv}\ W$ is Donsker whenever W is Donsker for the underlying distribution.*

conv W is Donsker since W is Donsker. $\overline{conv}\ W$ is Donsker since *conv* W is Donsker. This implies weak convergence of empirical loss to the generalization loss in $O(1/\sqrt{m})$. This is stated in the following corollary.

Corollary 4 *If W is Donsker and if the integrability hypothesis stated above holds, then weak convergence of the difference between empirical loss and generalization loss to 0 in $O(1/\sqrt{m})$ holds for Bayesian inference.*

3 Bayesian Inference: Convergence Rate

Next we assume that $Pr(GC(\epsilon)) \geq K'\epsilon^d$, for ϵ small enough [2], with $GC(\epsilon)$ the set of classifiers w such that $R(w) \geq R^* - \epsilon$, with $R^* = \sup_{w \in W} R(w)$. Moreover, we assume that $\forall (w,x) \in W \times \mathcal{X}\ 0 < c \leq w(x) \leq 1 - c' < 1$. The empirical mean $\frac{1}{m}\sum_{i=1}^m (R(Y_i, X_i, w))$ converges uniformly towards $\log R(w)$, at rate $O(K/\sqrt{m})$ provided that W is Donsker (e.g., W has finite VC-dimension). The expression $\inf_{w \in GC(\epsilon)} \Pi_{i=1}^m R(Y_i, X_i, w)$ is then lower bounded by $\exp(m(R^* - \epsilon)) \times \exp(K\sqrt{m})$. This leads to equation 1. On the other hand, with $\neg GC(\epsilon)$ the complementary set of $GC(\epsilon)$ in W, equation 2 holds. This implies equation 3, which is the quotient between the "weight", in the integral defining f_{D_m}, of "good classifiers" (in $GC(\epsilon)$), and "bad classifiers" (in $\neg GC(\epsilon)$), and expresses the fact that "good classifiers" have much better likelihood. $A(\epsilon, m)$ is bounded by ϵ if $\exp(-\epsilon/2)^m \exp(2K\sqrt{m}) \leq \frac{K'}{x^d}\epsilon^{d+1}$. The ln of this, divided by m, leads to equation 4, hence theorem 4.

$$\int_{GC(\epsilon)} \Pi_{i=1}^m \exp R(Y_i, X_i, w) dPr(w) \geq K'(\exp(R^* - \epsilon/2))^m \exp(-K\sqrt{m})(\frac{\epsilon}{2})^d \quad (1)$$

$$\int_{\neg GC(\epsilon)} \Pi_{i=1}^m \exp R(Y_i, X_i, w) dPr(w) \leq (\exp(R^* - \epsilon))^m \exp(K\sqrt{m}) \quad (2)$$

$$\frac{\int_{\neg GC(\epsilon)} V(w) dPr(w)}{\int_{GC(\epsilon)} V(w) dPr(w)} \leq A(\epsilon, m) = \frac{\exp(\frac{\epsilon}{2})^m \exp(2K\sqrt{m})}{K'(\epsilon/2)^d} \quad (3)$$

$$\epsilon/2 \geq \frac{\ln \frac{2^d}{K'}}{m} + \frac{(d+1)\ln\left(\frac{1}{\epsilon}\right)}{m} + 2K/\sqrt{m} \quad (4)$$

[2] This condition can be relaxed to $Pr(GC(\epsilon)) = \exp(-o(\frac{1}{\epsilon}))$, which does not reduce the asymptotic rate of the convergence $O(1/\sqrt{m})$.

Theorem 4 (Rate of convergence) *The generalization loss converges to the optimal loss in W at rate $O(1/\sqrt{m})$, under the assumptions above. The result is asymptotic if W is Donsker. The result is non-asymptotic and distribution-free if W has finite VC-dimension.*

4 Approximation by Gibbs Algorithms

One can consider Gibb's algorithms as approximations of Bayesian Inference through Monte-Carlo integration. We study the efficiency of this approximation.

Consider the first version of Gibbs algorithm:

$$g^k_{D_m}(x) = \frac{1}{k}\sum_{i=1}^{k} w_i(x)$$

The convergence rate is in $O(1/\sqrt{m})$ if the set $W^\perp = \{w \mapsto w(x)/x \in \mathcal{X}\}$ is Donsker. The convergence is uniform in the distribution of w_i provided that $W^\perp$ has finite VC-dimension. This means that W has finite VC-codimension, which is true in particular if W has finite VC-dimension [3]. This means that under this assumption of finite VC-dimension, choosing $k = \Theta(m)$ leads to a convergence as fast as the convergence of the loss function.

Theorem 5 (Dependency in k of the Gibbs algorithm) *The convergence of $g^k_{D_m}$ towards f_{D_m} is distribution-free at rate $O(1/\sqrt{k})$ whenever W has finite VC-codimension. This is true even if $\overline{conv}\ W$ has infinite VC-dimension.*

A similar conclusion has not been proved in the case of the second version of Gibbs algorithm. If such a result holds, the interest would be a formal justification of an algorithm with the following advantages:

- Easily implementable.
- Computational complexity: results above suggest $k = \Theta(m)$, leading to $\Theta(m^2)$ evaluations of $w(x_i)$, ensuring precision $O(1/\sqrt{m})$ on the loss.
- Asymptotic convergence towards the optimal loss in $O(1/\sqrt{m})$ (that is not the case in most algorithms, in which the theoretical bound to be minimized is modified for the sake of computational cost, as in Support Vector Machines).

5 Conclusion and Further Issues

This paper states positive general results for Bayesian inference: control of the difference between the generalization loss and the empirical loss, good asymptotic properties (convergence to the optimal loss), good non-asymptotic distribution-free properties when the VC-dimension is finite, efficient approximation through Gibbs algorithms provided sufficient conditions on k such that Gibbs algorithm well approximates Bayesian inference. More generally, the structure of $\overline{conv}\ W$,

[3] We recall that the VC-codimension is bounded by 2^V, with V the VC-dimension.

family of functions used by Bayesian inference, is expressed as a Donsker class (that holds independently of the i.i.d assumption), whenever W is Donsker, and has ϵ-entropy bounded in the case of finite VC-dimension. Possible improvements could include non-asymptotic bounds on classes of distribution, in the spirit of [7] or some results of [12].

The most interesting issue would have been a similar result in the case of the second Gibbs algorithm, which is still an open problem. This is strongly challenging (and probably false, as put into relief by an anonymous referee), since the second algorithm presents the main advantage of being much more tractable, as it only requires random drawns of w_i's according to Pr. Such a positive result would lead to an algorithm actually minimizing a loss function, whereas usual learning algorithms, such as Support Vector Machines or Bayes Point Machines contain approximations which lead to undesired behaviors[4].

References

1. N. Alon, S. Ben-David, N. Cesa-Bianci, D. Haussler, *Scale-sensitive dimensions, uniform convergence and learnability. Journal of the ACM, 44(4):615-631,1997.*
2. P.-L. Bartlett, *The sample complexity of pattern classification with neural networks: the size of the weights is more important than the size of the network, IEEE transactions on Information Theory, 44:525-536, 1998*
3. L. Devroye, L. Györfi, G. Lugosi, *A Probabilistic Theory of Pattern Recognition, Springer, 1997*
4. D. Haussler, M. Kearns, R.-E. Schapire, *Bounds on the Sample Complexity of Bayesian Learning Using Information Theory and the VC Dimension, Machine Learning, 14(1):83-113, 1994.*
5. R. Herbrich, T. Graepel, C. Campbell, *Bayes Point Machines: Estimating the Bayes Point in Kernel Space, in Proceedings of IJCAI Workshop on Support Vector Machines, pages 23-27, 1999*
6. R. Herbrich, T. Graepel, C. Campbell, *Robuts Bayes Point Machines, Proceedings of ESANN 2000, pp49-54, 2000*
7. B.-K. Natarajan, *Learning over classes of distributions. In proceedings of the 1988 Workshop on Computational Learning Theory, pp 408-409, San Mateo, CA, Morgan Kaufmann 1988.*
8. M. Opper, D. Haussler, *Calculation of the learning curve of Bayes optimal classification algorithm for learning a perceptron with noise. In Computational Learning Theory: Proceedings of the Fourth Annual Workshop, p75-87. Morgan Kaufmann, 1991.*
9. P. Ruján *Playing Billiard in version space. Neural Computation, 1997*
10. O. Teytaud, *Bayesian learning/Structural Risk Minimization, Research Report RR-2005, Eric,* `http://eric.univ-lyon2.fr`, *2000.*
11. M. Vidyasagar, *A theory of learning and generalization, Springer 1997.*
12. A.-W. van der Vaart, J.-A. Wellner, *Weak Convergence and Empirical Processes, Springer, 1996.*

[4] Support Vector Machines are biased by the distance of points to the separating hyperplane, as the number of errors is replaced, for the sake of computational time, by a sum of distances, and Bayes Point Machines, for similar reasons, work on Bayesian *averaging* instead of Bayesian inference - f_{D_m} is chosen as the centre of the version space through an algorithm developped in [9].

Learning Curves for Gaussian Processes Models: Fluctuations and Universality

Dörthe Malzahn and Manfred Opper

Department of Computer Science and Applied Mathematics
Aston University, Birmingham B4 7ET, United Kingdom
{malzahnd,opperm}@aston.ac.uk

Abstract. Based on a statistical mechanics approach, we develop a method for approximately computing average case learning curves and their sample fluctuations for Gaussian process regression models. We give examples for the Wiener process and show that universal relations (that are independent of the input distribution) between error measures can be derived.

1 Introduction

Gaussian process (GP) models have gained considerable interest in the Neural Computation Community (see e.g.[1–4]) in recent years. However, being non-parametric models by construction their theoretical understanding is less well developed compared to simpler parametric models like neural networks. In this paper we present new results for approximate computation of learning curves by further developing our framework from [5] which was based on a statistical mechanics approach. In contrast to most previous applications of statistical mechanics to learning theory the method is *not* restricted to the so called "thermodynamic" limit which would require a high dimensional input space.

Our approach has the advantage that it is rather general and may be applied to different likelihoods and allows for a systematic computation of corrections.

In this contribution we will rederive our approximation in an new way based on a general variational method. We will show that we can compute other interesting quantities like the sample fluctuations of the generalization error. Nevertheless, one may criticise this and similar approaches of statistical physics as being not relevant for practical situations, because the analysis requires the knowledge of the input distribution which is usually not available. However, we will show (so far for a toy example) that our approximation predicts universal relations (that are *independent* of the input distribution) between different error measures. We expect that similar relations may be obtained for more practical situations.

2 Regression with Gaussian Processes

Regression with Gaussian processes is based on a statistical model [2] where observations $y(x) \in R$ at input points $x \in R^D$ are assumed to be corrupted values

G. Dorffner, H. Bischof, and K. Hornik (Eds.): ICANN 2001, LNCS 2130, pp. 271–276, 2001.

of an unknown function $f(x)$. For independent Gaussian noise with variance σ^2, the likelihood for a set of m example data $D = (y(x_1), \ldots, y(x_m)))$ (conditioned on the function f) is given by

$$P(D|f) = \frac{\exp\left(-\sum_{i=1}^{m} \frac{(y_i - f(x_i))^2}{2\sigma^2}\right)}{(2\pi\sigma^2)^{\frac{m}{2}}} \tag{1}$$

where $y_i \doteq y(x_i)$. To estimate the function $f(x)$, one supplies the *a priori* information that f is a realization of a Gaussian process (random field) with zero mean and covariance $C(x, x') = E[f(x)f(x')]$, where E denotes the expectation over the Gaussian process prior. Predictions $\hat{f}(x)$ for the unknown function f are computed as the posterior expectation of $f(x)$, i.e. by

$$\hat{f}(x|D) = E\{f(x)|D\} = \frac{Ef(x)P(D|f)}{Z_m} \tag{2}$$

where the partition function Z_m normalises the posterior.

In the sequel, we call the true data generating function f^* in order to distinguish it from the functions over which we integrate in the expectations. We will compute approximations for the learning curve, i.e. the generalization (mean square) error averaged over independent draws of example data, i.e. $\varepsilon_g = [(f^*(x) - \hat{f}(x|D))^2]_{(x,D)}$ as a function of m, the sample size. We will use brackets $[\ldots]$ to denote averages over data sets where we assume that the inputs x_i are drawn *independently* at random from a density $p(x)$. The index at the bracket denotes the quantities that are averaged over. For example, $[\ldots]_{(x,D)}$ denotes both an average over example data D and a *test* input drawn from the same density. We will also approximate the sample fluctuations of the generalization error defined by $\Delta\varepsilon_g = \sqrt{[[(f^*(x) - \hat{f}(x|D))^2]_x^2]_D - \varepsilon_g^2}$.

3 The Partition Function

As typical of statistical mechanics approaches, we base our analysis on the averaged "free energy" $[-\ln Z_m]_D$ where the partition function Z_m (see Eq. (2)) is $Z_m = EP(D|f)$. $[\ln Z_m]_D$ serves as a generating function for suitable posterior averages. The computation of $[\ln Z_m]_D$ is based on the replica trick $[\ln Z_m]_D = \lim_{n\to 0} \frac{\partial \ln [Z_m^n]_D}{\partial n}$, where we compute $[Z^n]_D$ for integer n and perform the continuation at the end. We have

$$Z_n(m) \doteq [Z_m^n]_D = E_n \left[\frac{\exp\left(-\sum_{a=1}^{n} \frac{(f_a(x) - y)^2}{2\sigma^2}\right)}{\sqrt{2\pi\sigma^2}^{n}} \right]_{(x,y)}^{m}, \tag{3}$$

where E_n denotes the expectation over the GP measure for the n-times replicated GPs $f_a(x)$, $a = 1, \ldots, n$.

For further analytical treatment, it is convenient to introduce the "grand canonical" free energy

$$\Xi_n(\mu) = \sum_{m=0}^{\infty} \frac{e^{\mu m}}{m!} Z_n(m) = E_n \exp[-H_n] \tag{4}$$

where the energy H_n is a functional of $\{f_a\}$

$$H_n = -e^{\mu} \left[\frac{\exp\left(-\sum_{a=1}^{n} \frac{(f_a(x)-y)^2}{2\sigma^2}\right)}{\sqrt{2\pi\sigma^2}^{n}} \right]_{(x,y)} . \tag{5}$$

This represents a "poissonized" version of our model where the number of examples is fluctuating. For sufficiently large m, the relative fluctuations are small and both models will give the same answer, provided the "chemical potential" μ and the desired m are related by $m = \frac{\partial \ln \Xi_n(\mu)}{\partial \mu}$. Using a Laplace argument for the sum in Eq. (4), we have $\ln Z_n(m) \approx \ln \Xi_n(\mu) + m(\ln m - 1) - m\mu$. Note that as a result of the data average, the model defined by H_n is no longer Gaussian and we cannot compute $\ln \Xi_n(\mu)$ exactly. We will therefore resort to a variational approximation.

4 Variational Approximation

Our goal is to approximate H_n by a simpler quadratic Hamiltonian of the form $H_n^0 = \frac{1}{2}\sum_{a,b=1}^{n} \eta_{ab}[(f_a(x) - y)(f_b(x) - y)]_{(x,y)}$, where η_{ab} are parameters to be optimised. Assuming η_{ab} to be fixed for the moment, we can expand the free energy in a power series in the deviations $H - H_n^0$

$$-\ln \Xi_n(\mu) = -\ln E_n \exp[-H_n^0] + \langle H - H_n^0 \rangle_0 - \frac{1}{2}\left(\langle (H - H_n^0)^2 \rangle_0 - \langle H - H_n^0 \rangle_0^2 \right) \pm \ldots \tag{6}$$

The brackets $\langle \ldots \rangle_0$ denote averages with respect to the effective Gaussian measure induced by the replicated prior and $e^{-H_n^0}$. As is well known [6], the first two terms in Eq. (6) are an upper bound to $-\ln \Xi_n(\mu)$. We will optimise H_n^0, by choosing the matrix η_{ab} such that this upper bound is minimised. Thereafter, a replica symmetric continuation to real n is achieved by restricting the variations to the form $\eta_{ab} = \eta$ for $a \neq b$ and $\eta_{aa} = \eta_0$. Note however, that after this continuation we can no longer establish a bound on $-\ln \Xi_n(\mu)$. To compute the generalization error and other quantities we will use the effective Gaussian measure induced by H_n^0. The variational equations on η_0 and η can be expressed as functionals of the local generalization error

$$\varepsilon_g(x) = \lim_{n \to 0} \langle (f_1(x) - f^*(x))(f_2(x) - f^*(x)) \rangle_0 \tag{7}$$

and the local posterior variance

$$v_p(x) = \lim_{n \to 0} \langle (f_1(x) - f^*(x))^2 \rangle_0 - \varepsilon_g(x). \tag{8}$$

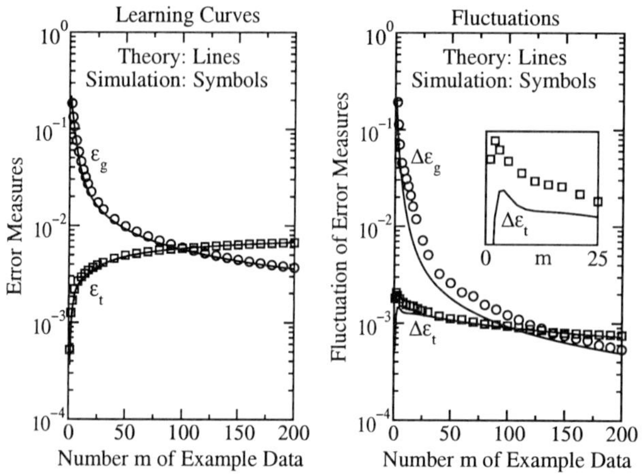

Fig. 1. Learning curves (*left*) and their fluctuations (*right*) for the Wiener Process. Comparison between theory (*lines*) and simulation results (*symbols*)

By neglecting variations of these quantities with x we arrive at the following set of equations

$$[\hat{C}(x,x)]_x + \sigma^2 = \frac{m}{(\eta_0 - \eta)} \tag{9}$$

$$[\tilde{E}^2(f(x) - y)]_{(x,y)} - \eta[\hat{C}^2(x,x')]_{(x,x')} = -\frac{m\eta}{(\eta_0 - \eta)^2} \tag{10}$$

that determine the values of the variational parameters η_0 and η. The mean $\tilde{E}(f(x) - y)$ and the covariance $\hat{C}(x,x') = \hat{E}(f(x)f(x'))$ in Eqs. (9,10) are defined with respect to the Gaussian measures $\tilde{E} \propto E\exp\left(-\frac{(\eta_0-\eta)}{2}[(f(x)-y)^2]_{(x,y)}\right)$ and $\hat{E} \propto E\exp\left(-\frac{(\eta_0-\eta)}{2}[f^2(x)]_x\right)$.

5 Results for Learning Curves and Fluctuations

We compare our analytical results for the generalization error ε_g, the training error $\varepsilon_t = \frac{1}{m}[\sum_{i=1}^m(\hat{f}(x_i|D) - y(x_i))^2]_D$ and for their sample fluctuations $\Delta\varepsilon_g$, $\Delta\varepsilon_t$ with simulations of GP regression. For simplicity, we have chosen the Wiener process $C(x,x') = min(x,x')$ with $x, x' \in [0,1]$ as a toy model. For Fig. 1, the target function f^* is a fixed but random realisation from the prior distribution and the data noise is Gaussian with variance $\sigma^2 = 0.01$. The left panel of Fig. 1 shows learning curves while their fluctuations are displayed in the right panel. Symbols represent simulation results and our theory is given by lines. The training error ε_t converges to the noise level σ^2. As one can see from the pictures our theory is very accurate when m is sufficiently large. It also predicts the initial increase of $\Delta\varepsilon_t$ for small values of m (see *inset* of Fig. 1, right panel).

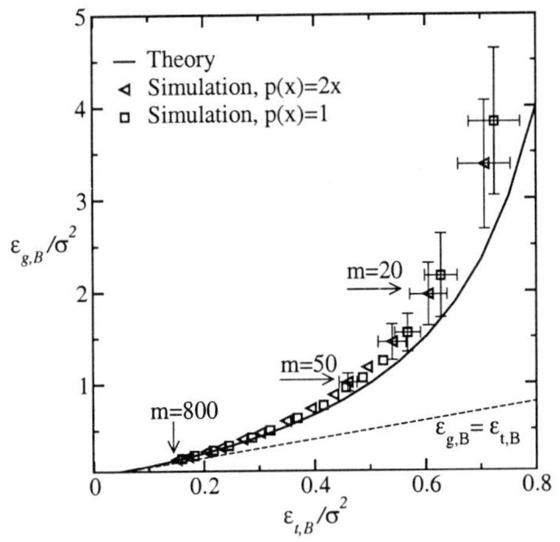

Fig. 2. Bayes errors for the Wiener Process. Theory (*bold line*) versus Simulations (*Symbols*). Simulations have been performed for two input distributions $x \in [0,1]$ and $p(x) = 1$ (*squares*) or $p(x) = 2x$ (*triangles*). Arrows indicate the number m of example data. As m increases, the Bayesian generalization error $\varepsilon_{g,B}$ and its error bars decrease. For $m \to \infty$ holds the trivial limes $\varepsilon_{g,B} \approx \varepsilon_{t,B}$ (*dashed line*)

6 Universal Relations

Although the explicit computations of our results requires the knowledge of the data distribution, we can establish universal relations (valid in the framework of our approximation) which are independent of this density. We restrict ourselves to the full Bayesian scenario where all quantities are averaged over the prior distribution of true functions f^*. The uncertainty of the prediction at a point x is measured by the posterior variance $\varepsilon_B(x) = E(\hat{f}(x|D) - f(x))^2$. Approximations to Bayesian generalization errors defined as $\varepsilon_{g,B} = [\varepsilon_B(x)]_{(x,D)}$ for this scenario were computed previously by Peter Sollich [7]. For the special case of a uniform input distribution, our results turn out to be *identical* to Sollich's result. However, extending our framework to arbitrary input densities, we find that the Bayesian generalization error and its empirical estimate $\varepsilon_{t,B} = \frac{1}{m}\sum_{i=1}^{m}[\varepsilon_B(x_i)]_D$ are expressed by a single variational parameter of our model only. This can be eliminated to give the following surprisingly simple relation

$$\bar{\varepsilon}_{g,B} \approx \frac{\bar{\varepsilon}_{t,B}}{1 - \bar{\varepsilon}_{t,B}} \qquad (11)$$

where $\bar{\varepsilon}_{t,B} = \varepsilon_{t,B}/\sigma^2$ and $\bar{\varepsilon}_{g,B} = \varepsilon_{g,B}/\sigma^2$. Fig. 2 displays simulation results for Wiener process regression with Gaussian noise of variance $\sigma^2 = 0.01$. We used two different input distributions $p(x) = 1$ (*squares*) and $p(x) = 2x$ (*triangles*), $x \in [0,1]$. The number m of example data is indicated by arrows. Eq. (11) is represented by the bold line and holds for sufficiently large m.

7 Future Work

In the future, we will extend our method in the following directions:

- Obviously, our method is not restricted to a regression model but can also be directly generalized to other likelihoods such as the classification case [4, 8]. A further application to Support Vector Machines should be possible.
- We will establish further universal relations between different error measures for the more realistic case of a fixed (unknown) function $f^*(x)$. It will be interesting if such relations may be useful to construct new methods for model selection, i.e. hyper-parameter estimation.
- By computing the influence of the first neglected term in Eq. (6) which is quadratic in $H_n - H_n^0$, we will estimate the region in which our approximation is valid.

Acknowledgement

This work has been supported by EPSRC grant GR/M81601.

References

1. D. J. C. Mackay, Gaussian Processes, A Replacement for Neural Networks, NIPS tutorial 1997, May be obtained from `http://wol.ra.phy.cam.ac.uk/pub/mackay/`.
2. C. K. I. Williams and C. E. Rasmussen, Gaussian Processes for Regression, in *Neural Information Processing Systems 8*, D. S. Touretzky, M. C. Mozer and M. E. Hasselmo eds., 514-520, MIT Press (1996).
3. C. K. I. Williams, Computing with Infinite Networks, in *Neural Information Processing Systems 9*, M. C. Mozer, M. I. Jordan and T. Petsche, eds., 295-301. MIT Press (1997).
4. D. Barber and C. K. I. Williams, Gaussian Processes for Bayesian Classification via Hybrid Monte Carlo, in *Neural Information Processing Systems 9*, M . C. Mozer, M. I. Jordan and T. Petsche, eds., 340-346. MIT Press (1997).
5. D. Malzahn, M. Opper, Learning curves for Gaussian processes regression: A framework for good approximations, in *Neural Information Processing Systems 13*, T. K. Leen, T. G. Dietterich and V. Tresp, eds., MIT Press (2001), to appear.
6. R. P. Feynman and A. R. Hibbs, Quantum mechanics and path integrals, Mc Graw-Hill Inc., 1965.
7. P. Sollich, Learning curves for Gaussian processes, in *Neural Information Processing Systems 11*, M. S. Kearns, S. A. Solla and D. A. Cohn, eds. 344 - 350, MIT Press (1999).
8. L. Csató, E. Fokoué, M. Opper, B. Schottky, and O. Winther. Efficient approaches to Gaussian process classification. In *Neural Information Processing Systems 12*, MIT Press (2000).

Tight Bounds on Rates of Neural-Network Approximation*

Věra Kůrková[1] and Marcello Sanguineti[2]

[1] Institute of Computer Science, Academy of Sciences of the Czech Republic
Pod Vodárenskou věží 2, P.O. Box 5 – 182 07, Prague 8, Czech Republic
vera@cs.cas.cz

[2] Department of Communications, Computer, and System Sciences (DIST)
University of Genoa – Via Opera Pia 13, 16145 Genova, Italy
marcello@dist.unige.it

Abstract. Complexity of neural networks measured by the number of hidden units is studied in terms of rates of approximation. Limitations of improvements of upper bounds of the order of $\mathcal{O}(n^{-\frac{1}{2}})$ on such rates are investigated for perceptron networks with some periodic and sigmoidal activation functions.

1 Introduction

Experience has shown that simple neural network architectures with relatively few computational units can achieve surprisingly good performances in a large variety of high-dimensional problems, ranging from strongly nonlinear controllers to identification of complex dynamic systems and pattern recognition (see, e.g. [13], [14]). The efficiency of such neural-network designs has motivated a theoretical analysis of desirable computational capabilities of neural networks guaranteeing that the number of computational units does not increase too fast with the dimensionality of certain tasks.

The dependence of the accuracy of an approximation by feedforward networks on the number of hidden units can be theoretically studied in the context of approximation theory in terms of *rates of approximation*. Some insight into the reason why many high-dimensional tasks can be performed quite efficiently by neural networks with a moderate number of hidden units has been gained by Jones [4]. The same estimate of rates of approximation had earlier been proven by Maurey using a probabilistic argument (it has been quoted by Pisier [12]; see also Barron [2]). Barron [2] improved Jones's [4] upper bound and applied it to neural networks. Using a weighted Fourier transform, he described sets of multivariable functions approximable by perceptron networks with n hidden units to an accuracy of the order of $\mathcal{O}(\frac{1}{\sqrt{n}})$. Such upper bounds do not depend on

* Both authors were partially supported by NATO Grant PST.CLG.976870. V. K. was partially supported by grant GA ČR 201/00/1489. M.S. was partially supported by the Italian Ministry of University and Research (Project "Identification and Control of Industrial Systems") and by Grant D.R.42 of the University of Genoa.

G. Dorffner, H. Bischof, and K. Hornik (Eds.): ICANN 2001, LNCS 2130, pp. 277–282, 2001.

the number of variables d of the functions to be approximated – in contrast to "curse of dimensionality" rates of the order of $\mathcal{O}(n^{-\frac{1}{d}})$ (d denotes the number of variables) which limit feasibility of linear approximation methods. However, it should be stressed that as the number of variables increases, the sets of multivariable functions to which such estimates apply become more and more constrained (see [3]), and the constants may depend on d (see [9]).

Maurey-Jones-Barron's upper bound is quite general as it applies to *nonlinear approximation of variable-basis type*, i.e., approximation by linear combinations of n-tuples of elements of a given set of basis functions. This approximation scheme has been widely investigated: it includes free-nodes splines, free-poles rational functions and feedforward multilayer neural networks with a single linear output unit. Several authors have further improved or extended these upper bounds and investigated limitations of such improvements. Makovoz [10] improved Maurey's probabilistic argument by combining it with a concept from metric entropy theory, which he also used to show that in the case of Lipschitz sigmoidal perceptron networks, the upper bound cannot be improved to $\mathcal{O}(n^{-\alpha})$ for $\alpha > \frac{1}{2} + \frac{1}{d}$, where d is the number of variables of the functions to be approximated. A similar tightness result for perceptron networks was earlier obtained by Barron [1], who used a more complicated proof technique. For the special case of orthonormal variable-basis, Mhaskar and Micchelli [11] and Kůrková, Savický and Hlaváčková [9] have derived tight improvements of Maurey-Jones-Barron's bound.

In this paper, we extend such tightness results for two types of sets of basis functions: (1) for an orthonormal basis and (2) for a basis satisfying certain conditions on polynomial growth of the number of sets of a given diameter needed to cover such basis and having sufficient "capacity" in the sense that its convex balanced hull has an orthogonal subset that for each integer k contains at least k^d functions with norms greater or equal to $\frac{1}{k}$. The condition (1) is satisfied by perceptrons with sine or cosine activations, and the condition (2) by perceptrons with Lipschitz sigmoidal activations. We show that in the first case, the Maurey-Jones-Barron's upper bound can be at most improved by a factor depending on the ratio between two norms (a norm in which the approximation error is measured and a norm tailored to the set of basis functions), but the term $n^{-\frac{1}{2}}$ remains essentially unchanged (it is only replaced by $\frac{1}{2}(n-1)^{-\frac{1}{2}}$), and in the second case, the upper bound cannot be improved beyond $\mathcal{O}(n^{-(\frac{1}{2}+\frac{1}{d})})$.

2 Preliminaries

Approximation by feedforward neural networks can be studied in a more general context of *approximation by variable-basis functions*. In this approximation scheme, elements of a real normed linear space $(X, \|.\|)$ are approximated by linear combinations of at most n elements of a given subset G. The set of such combinations is denoted by $span_n G = \{\sum_{i=1}^n w_i g_i ; w_i \in \mathcal{R}, g_i \in G\}$; it is equal to the union of n-dimensional subspaces generated by all n-tuples of elements of G. G can represent the set of functions computable by *hidden units* in neural networks. Recall that a *perceptron* with an activation function $\psi : \mathcal{R} \to \mathcal{R}$ computes functions of the form $\psi(\mathbf{v} \cdot \mathbf{x} + b) : \mathcal{R}^{d+1} \times \mathcal{R}^d \to \mathcal{R}$, where $\mathbf{v} \in \mathcal{R}^d$ is an *input weight* vector and $b \in \mathcal{R}$ is a *bias*. By $P_d(\psi) = \{f : [0,1]^d \to \mathcal{R}; f(\mathbf{x}) = \psi(\mathbf{v}\cdot\mathbf{x}+b), \mathbf{v} \in \mathcal{R}^d, b \in \mathcal{R}\}$ we denote the set of functions on $[0,1]^d$ computable by ψ-perceptrons. The most common activation functions are *sigmoidals*, i.e., functions $\sigma : \mathcal{R} \to [0,1]$ such that $\lim_{t\to-\infty} \sigma(t) = 0$ and $\lim_{t\to\infty} \sigma(t) = 1$. A function $\sigma : \mathcal{R} \to \mathcal{R}$ is Lipschitz if there exists $a > 0$ such that $|\sigma(t) - \sigma(t')| \leq a|t - t'|$ for all $t, t' \in \mathcal{R}$.

Rates of approximation of functions from a set Y by functions from a set M can be studied in terms of the *worst-case error* formalized by the concept of *deviation of* Y *from* M defined as $\delta(Y, M) = \delta(Y, M, (X, \|.\|)) = \sup_{f\in Y} \|f - M\| = \sup_{f\in Y} \inf_{g\in M} \|f - g\|$. To formulate estimates of deviation from $span_n G$ we need to introduce a few more concepts and notations. If G is a subset of $(X, \|.\|)$ and $c \in \mathcal{R}$, then we define $cG = \{cg; g \in G\}$ and $G(c) = \{wg; g \in G, w \in \mathcal{R} \,\&\, |w| \leq c\}$. The *closure* of G is denoted by $cl\, G$ and defined as $cl\, G = \{f \in X; (\forall\varepsilon > 0)(\exists g \in G)(\|f - g\| < \varepsilon)\}$. The *convex hull* of G, denoted by $conv\, G$, is the set of all convex combinations of its elements, i.e., $conv\, G = \{\sum_{i=1}^n a_i g_i; a_i \in [0,1], \sum_{i=1}^n a_i = 1, g_i \in G, n \in \mathcal{N}_+\}$. $conv_n\, G$ denotes the set of all convex combinations of n elements of G, i.e., $conv_n\, G = \{\sum_{i=1}^n a_i g_i;\, a_i \in [0,1], \sum_{i=1}^n a_i = 1, g_i \in G\}$. $B_r(\|.\|)$, $S_r(\|.\|)$ denotes the ball, sphere, resp., of radius r with respect to the norm $\|.\|$, i.e., $B_r(\|.\|) = \{x \in X; \|x\| \leq r\}$ and $S_r(\|.\|) = \{x \in X; \|x\| = r\}$.

The following estimate is a version of Jones' result as improved by Barron [2] and also of earlier result of Maurey. Recall that a Hilbert space is a normed linear space with the norm induced by an inner product.

Theorem 1. *Let* $(X, \|.\|)$ *be a Hilbert space,* b *a positive real number,* G *a subset of* X *such that for every* $g \in G$ $\|g\| \leq b$*, and let* $f \in cl\, conv\, G$*. Then, for every positive integer* n*,* $\|f - conv_n G\| \leq \sqrt{\frac{b^2 - \|f\|^2}{n}}$.

As $conv_n G(c) \subset span_n G(c) = span_n G$ for any $c \in \mathcal{R}$, by replacing the set G by $G(c) = \{wg; w \in \mathcal{R}, |w| \leq c, g \in G\}$ we can apply Theorem 1 to all elements of $\cup_{c\in\mathcal{R}_+} cl\, conv\, G(c)$. This approach can be mathematically formulated in terms of a norm tailored to a set G Let $(X, \|.\|)$ be a normed linear space and G its subset, then *G-variation* (variation with respect to G) denoted by $\|.\|_G$ is defined as the Minkowski functional of the set $cl\, conv\, G(1) = cl\, conv(G \cup -G)$, i.e.,

$$\|f\|_G = \inf\{c \in \mathcal{R}_+; f \in cl\, conv\, G(c)\}.$$

G-variation has been introduced by Kůrková [5] as an extension of Barron's [1] concept of variation with respect to half-spaces (more precisely, variation with respect to characteristic functions of half-spaces) corresponding to perceptrons with discontinous threshold (Heaviside) activation function. The following theorem is a corollary of Theorem 1 formulated in terms of G-variation (see [5]). Note that for any G, the unit ball in G-variation is equal to $cl\, conv(G \cup -G)$.

Theorem 2. *Let $(X, \|.\|)$ be a Hilbert space and G be its subset. Then, for every $f \in X$ and every positive integer n, $\|f - span_n G\| \leq \sqrt{\frac{\|f\|_G^2 s_G^2 - \|f\|^2}{n}} = \frac{\|f\|_G s_G}{\sqrt{n}} \sqrt{1 - \frac{\|f\|^2 s_G^2}{\|f\|_G^2 s_G^2}}$ and $\delta(B_1(\|.\|_G), span_n G) \leq \frac{s_G}{\sqrt{n}}$, where $s_G = \sup_{g \in G} \|g\|$.*

3 Tight Bounds for Orthonormal Variable Bases

The results in this section extend to infinite-dimensional spaces tight estimates derived by Kůrková, Savický and Hlaváčková [9] for finite-dimensional spaces as improvements of earlier estimates obtained by Mhaskar and Micchelli [11].

When G is an orthonormal basis of a separable Hilbert space $(X, \|.\|)$, then G-variation is equal to the *l_1-norm with respect to G*, denoted by $\|.\|_{1,G}$ and defined as $\|f\|_{1,G} = \sum_{g \in G} |f \cdot g|$ (see [8]). The following two tight bounds take advantage of this equality (for their proofs see [8]).

Theorem 3. *Let $(X, \|.\|)$ be an infinite-dimensional separable Hilbert space and G be its orthonormal basis. Then, for every positive real number b and every positive integer n, $\delta(B_b(\|.\|_G), span_n G) = \delta(B_b(\|.\|_{1,G}), span_n G) = \frac{b}{2\sqrt{n}}$.*

When G is a countable infinite orthonormal basis, Theorem 3 improves the upper bound from Corollary 2 up to an exact value of the deviation of balls in G-variation from $span_n G$. In contrast to Theorem 2, which expresses an upper bound in terms of both $\|f\|_G$, and $\|f\|$, Theorem 3 does not take $\|f\|$ into account. However, even without using the value of $\|f\|$, it gives a better bound than Theorem 2 when the ratio $\frac{\|f\|}{\|f\|_G}$ is sufficiently large (if $\frac{\sqrt{3}}{2} < \frac{\|f\|}{\|f\|_G}$, then $\sqrt{\frac{\|f\|_G^2 - \|f\|^2}{n}} < \frac{\|f\|_G}{2\sqrt{n}}$). The following theorem gives, for G orthonormal, the maximum possible improvement of Theorem 2, using both $\|.\|_G$ and $\|.\|$.

Theorem 4. *Let $(X, \|.\|)$ be an infinite-dimensional separable Hilbert space, G its orthonormal basis, and b, r real numbers such that $0 \leq r \leq b$. Then, for every positive integer $n \geq 2$:*
(i) if $\frac{r}{b} \geq \frac{1}{\sqrt{2n-1}}$, then $\frac{b}{4\sqrt{n-1}} \left(1 - \frac{r^2}{b^2}\right) \leq \delta((S_b(\|.\|_{1,G}) \cap S_r(\|.\|)), span_n G) \leq \frac{b}{2\sqrt{n-1}} \left(1 - \frac{r^2}{b^2}\right)$;
(ii) if $\sqrt{n} - \sqrt{n-1} \leq \frac{r}{b} < \frac{1}{\sqrt{2n-1}}$, then $\frac{r}{2\sqrt{2}} \leq \delta(S_b(\|.\|_{1,G}) \cap S_r(\|.\|), span_n G) \leq \frac{b}{2\sqrt{n-1}} \left(1 - \frac{r^2}{b^2}\right)$;
(iii) if $\frac{r}{b} < \sqrt{n} - \sqrt{n-1}$, then $\frac{r}{2\sqrt{2}} \leq \delta(S_b(\|.\|_{1,G}) \cap S_r(\|.\|), span_n G) \leq r$.

Thus for an orthonormal set G, Maurey-Jones-Barron's bound can at most be improved by a factor depending on the ratio between the two norms $\|.\|$ and $\|.\|_G$, but the term $\frac{1}{\sqrt{n}}$ remains essentially unchanged: it is only replaced by $\frac{1}{2\sqrt{n-1}}$.

Let $(\mathcal{L}_2([0,1]^d), \|.\|_2)$ denote the Hilbert space of $\mathcal{L}_2$-functions on $[0,1]^d$ with the $\mathcal{L}_2$-norm defined as $\|f\|_2 = \left(\int_{[0,1]^d} f^2(x)\, dx\right)^{1/2}$. Using Theorems 3 and 4, we can derive upper bounds on the deviation from perceptron networks with sine or cosine activations, in terms of the Fourier coefficients of functions in $(\mathcal{L}_2([0,1]^d), \|.\|_2)$ with respect to a suitable orthonormal basis.

Corollary 5 *Let d be a positive integer; then, in $(\mathcal{L}_2([0,1]^d), \|.\|_2)$:*
(i) for every positive real number b and every positive integer n, $\delta(\{f;\ \sum_{k\in\mathcal{N}^d} |\tilde{f}(k)| \le b\}, span_n P_d(\sin)) = \frac{b}{2\sqrt{n}}$;
(ii) for every $0 \le r \le b$ and every positive integer $n \ge 2$, if $\frac{r}{b} \ge \sqrt{n} - \sqrt{n-1}$, then $\delta(\{f;\ \sum_{k\in\mathcal{N}^d} |\tilde{f}(k)| = b\, \&\ \sum_{k\in\mathcal{N}^d} |\tilde{f}(k)|^2 = r\}, span_n P_d(\sin)) \le \frac{b}{2\sqrt{n-1}}\left(1 - \frac{r^2}{b^2}\right)$;
(iii) for every $0 \le r \le b$ and every positive integer $n \ge 2$, if $\frac{r}{b} < \sqrt{n} - \sqrt{n-1}$, then $\delta(\{f;\ \sum_{k\in\mathcal{N}^d} \tilde{f}(k)| = b\, \&\ \sum_{k\in\mathcal{N}^d} |\tilde{f}(k)|^2 = r\}, span_n P_d(\sin)) \le r$.

4 Tightness of the Bound $\mathcal{O}(n^{-\frac{1}{2}})$ for Sigmoidal Perceptrons

The results in this section extend tightness results derived by Barron [1] and Makovoz [10]. We disprove the possibility of an improvement of Maurey-Jones-Barron's upper bound beyond $\mathcal{O}(n^{(\frac{1}{2}+\frac{1}{d})})$ for variable bases satisfying certain conditions defined in terms of (i) polynomial growth of the number of sets of a given diameter needed to cover such basis and (ii) sufficient "capacity" of the basis, in the sense that its convex hull has an orthogonal subset that for each positive integer k, contains at least k^d functions with norms greater or equal to $\frac{1}{k}$. Recall that for $\varepsilon > 0$, the *ε-covering number* of a subset K of a normed linear space $(X, \|.\|)$ is defined as $cov_\varepsilon K = cov_\varepsilon(K, \|.\|) = \min\{n \in \mathcal{N}_+ ; K \subseteq \cup_{i=1}^n B_\varepsilon(x_i, \|.\|), x_i \in K\}$ if the set over which the minimum is taken is nonempty, otherwise $cov_\varepsilon(K) = +\infty$. For a positive integer d (corresponding, in the following, to the number of variables of functions in X), we call a subset A of a normed linear space $(X, \|.\|)$ *not quickly vanishing with respect to d* if $A = \cup_{k\in\mathcal{N}_+} A_k$, where, for each $k \in \mathcal{N}_+$, $card\, A_k \ge k^d$ and for each $h \in A_k$, $\|h\| \ge \frac{1}{k}$.

Theorem 6. *Let $(X, \|.\|)$ be a Hilbert space of functions of d variables and G be its bounded subset satisfying the following conditions:*
(i) there exist a polynomial $p(d)$ and $b \in \mathcal{R}_+$ such that, for every $\varepsilon > 0$, $cov_\varepsilon(G) \le b\left(\frac{1}{\varepsilon}\right)^{p(d)}$;
(ii) there exists $r \in \mathcal{R}_+$ for which $B_r(\|.\|_G)$ contains a set of orthogonal elements which is not quickly vanishing with respect to d.
Then $\delta(B_1(\|.\|_G), conv_n(G \cup -G)) \le \mathcal{O}(n^{-\alpha})$ implies $\alpha \le \frac{1}{2} + \frac{1}{d}$.

For the proof of Theorem 6 and verification that its assumptions are satisfied by perceptrons with Lipschitz sigmoidal activation (see [6]).

Corollary 7 *Let d, n be positive integers and let $\sigma : \mathcal{R} \to \mathcal{R}$ be a Lipschitz sigmoidal function. Then in $(\mathcal{L}^2([0,1]^d), \|.\|_2)$,* $\delta(B_1(\|.\|_{P_d(\sigma)}), conv_n(P_d(\sigma) \cup -P_d(\sigma)) \le \mathcal{O}(n^{-\alpha})$ *implies* $\alpha \le \frac{1}{2} + \frac{1}{d}$.

5 Discussion

We have shown that the upper bounds on worst-case errors following from Maurey-Jones-Barron's theorem cannot be considerably improved for networks with one-hidden-layer of either Lipschitz sigmoidal or cosine activation function. Better rates might be achievable using networks with more than one hidden layer or when sets of functions to be approximated are more restricted than sets defined in terms of the two norms considered in this paper.

References

1. Barron, A.R.: Neural net approximation. *Proc. 7th Yale Workshop on Adaptive and Learning Systems* (K. Narendra, Ed.), pp. 69-72. Yale University Press, 1992.
2. Barron, A.R.: Universal approximation bounds for superpositions of a sigmoidal function. *IEEE Trans. on Information Theory* 39, pp. 930-945, 1993.
3. Girosi, F., Jones, M., and Poggio, T.: Regularization theory and neural networks architectures. *Neural Computation* 7, pp. 219-269, 1995.
4. Jones, L.K.: A simple lemma on greedy approximation in Hilbert space and convergence rates for projection pursuit regression and neural network training. *Annals of Statistics* 20, pp. 608–613, 1992.
5. Kůrková, V.: Dimension-independent rates of approximation by neural networks. In *Computer-Intensive Methods in Control and Signal Processing. The Curse of Dimensionality* (K. Warwick, M. Kárný, Eds.). Birkhauser, Boston, pp. 261-270, 1997.
6. Kůrková, V. and Sanguineti, M.: Covering numbers and rates of neural-network approximation. Research report ICS-00-830, 2000.
7. Kůrková, V. and Sanguineti, M.: Tools for comparing neural network and linear approximation. *IEEE Trans. on Information Theory*, 2001 (to appear).
8. Kůrková, V. and Sanguineti, M.: Bounds on rates of variable-basis and neural-network approximation. *IEEE Trans. on Information Theory*, 2001 (to appear).
9. Kůrková, V., Savický, P., and Hlaváčková, K.: Representations and rates of approximation of real–valued Boolean functions by neural networks. *Neural Networks* 11, pp. 651-659, 1998.
10. Makovoz, Y.: Random approximants and neural networks. *J. of Approximation Theory* 85, pp. 98–1098, 1996.
11. Mhaskar, H. N. and Micchelli, C. A.: Dimension-independent bounds on the degree of approximation by neural networks. *IBM J. of Research and Development* 38, n. 3, pp. 277–283, 1994.
12. Pisier, G.: Remarques sur un resultat non publié de B. Maurey. *Seminaire d'Analyse Fonctionelle*, vol. I, no. 12. École Polytechnique, Centre de Mathématiques, Palaiseau, 1980-81.
13. Sejnowski, T. J. and Rosenberg, C.: Parallel networks that learn to pronounce English text, *Complex Systems* 1, pp. 145-168, 1987.
14. Zoppoli, R., Sanguineti, M., and Parisini, T.: Approximating networks and extended Ritz method for the solution of functional optimization problems. *J. of Optimization Theory and Applications*, 2001 (to appear).

Part IV

Kernel Methods

Scalable Kernel Systems

Volker Tresp[1] and Anton Schwaighofer[1,2]

[1] Siemens AG, Corporate Technology, Otto-Hahn-Ring 6,
81739 München, Germany
{Volker.Tresp,Anton.Schwaighofer.external}@mchp.siemens.de
[2] TU Graz, Institute for Theoretical Computer Science
Inffeldgasse 16b, 8010 Graz, Austria
aschwaig@igi.tu-graz.ac.at
http://www.igi.tu-graz.ac.at/aschwaig/

Abstract. Kernel-based systems are currently very popular approaches to supervised learning. Unfortunately, the computational load for training kernel-based systems increases drastically with the number of training data points. Recently, a number of approximate methods for scaling kernel-based systems to large data sets have been introduced. In this paper we investigate the relationship between three of those approaches and compare their performances experimentally.

1 Introduction

Kernel-based systems such as the support vector machine (SVM) and Gaussian processes (GP) are powerful and currently very popular approaches to supervised learning. Kernel-based systems have demonstrated very competitive performance on several applications and data sets and have great potential for KDD-applications since their degrees of freedom grow with training data size and they are therefore capable of modeling an increasing amount of detail when appropriately many training data points become available. Unfortunately, there are at least three problems when one tries to scale up these systems to large data sets. First, training time increases drastically with the number of training data points, second, memory requirements increase with data set size and third, prediction time is proportional to the number of kernels and the latter is equal to (or at least increases with) the number of training data points. In this presentation, we will concentrate on Gaussian processes which are the basis for Gaussian process regression, generalized Gaussian process regression, and the support vector machine. We analyze and experimentally compare three recently introduced approaches towards scaling Gaussian processes to large data sets using finite-dimensional representations, thus obtaining learning rules which scale linearly in the number of training data points.

The first approach is the *subset of representers method* (SRM) and can be found in the work of Wahba [5], in the work on sparse greedy Gaussian process regression by Smola and Bartlett [2], and in the reduced support vector machine by Lee and Mangasarian [1]. The SRM is based on a factorization of the kernel

G. Dorffner, H. Bischof, and K. Hornik (Eds.): ICANN 2001, LNCS 2130, pp. 285–291, 2001.

functions. The second variant is a reduced rank approximation (RRA) of the Gram matrix introduced in the work of Williams and Seeger [6]. The RRA uses the same decomposition as the SRM but this decomposition is only applied to the Gram matrix. The third variant is the BCM approximation introduced by Tresp [3]. Here the starting point is the optimal projection of the data on a set of base kernels which requires the inversion of a covariance matrix of the size of the number of training data points. The BCM approximation is achieved by a block diagonal approximation of this matrix.

In this paper, we analyze the approaches from a common view point and we will compare the performances of the approximations where we pay particular attention to the issue of the optimal scale parameter.

The paper is organized as follows. In the next section, we will provide a brief introduction into Gaussian processes. In the following sections, we describe the SRM, the RRA and the BCM approximation. Section 6 analyzes the approximations and provides experimental comparisons. Section 7 contains the conclusions.

2 Gaussian Process Regression (GPR)

In GPR one assumes that *a priori* a function $f(x)$ is generated from an infinite-dimensional Gaussian distribution with zero mean and covariance $K(x_i, x_j) = cov(f(x_i), f(x_j))$ defined at input points x_i and x_j. Furthermore, we assume a set of training data $\mathcal{D} = \{(x_i, y_i)\}_{i=1}^N$ where targets are generated according to

$$y_i = f(x_i) + \epsilon_i$$

where ϵ_i is independent additive Gaussian noise with variance σ^2. The optimal regression function $\hat{f}(x)$ takes on the form of a weighted combination of kernel functions

$$\hat{f}(x) = \sum_{i=1}^{N} w_i K(x, x_i). \tag{1}$$

Based on our assumptions, the maximum a posterior (MAP) solution for $w = (w_1, \ldots, w_N)'$ minimizes the cost function

$$\frac{1}{2} w' \Sigma w + \frac{1}{2\sigma^2}(\Sigma w - y)'(\Sigma w - y) \tag{2}$$

where $(\Sigma)_{i,j} = K(x_i, x_j)$ is the $N \times N$ Gram matrix. The optimal weight vector is the solution to a system of linear equation which in matrix form becomes

$$(\Sigma + \sigma^2 I)w = y. \tag{3}$$

Here $y = (y_1, \ldots, y_N)'$ is the vector of targets and I is the $N \times N$-dimensional unit matrix. The experimenter has to specify the positive definite kernel function. A common choice is that

$$K(x_i, x_j) = A \exp(-1/(2\gamma^2)||x_i - x_j||^2) \tag{4}$$

which is a Gaussian with positive amplitude A and scale parameter γ. Other positive definite covariance functions are also used.

3 The Subset of Representers Method (SRM)

Here, one first selects as set of N_b base kernels. These base kernels are typically either defined at a subset of the training data or of the test data. One can now approximate the covariance at x_i and x_j as

$$cov(f(x_i), f(x_j)) \approx (K^b(x_i))'(\Sigma^{b,b})^{-1}K^b(x_j). \tag{5}$$

Here, $\Sigma^{b,b}$ is the covariance matrix for the base kernel points and $K^b(x_i)$ is the vector of covariances between the functional values at x_i and the base kernel points. Since $\Sigma^{b,b}$ contains the covariances at the base kernel points, this approximation is an equality if either x_i or x_j are elements of the base kernel points and is an approximation otherwise. Note that using this approximation, the Gram matrix becomes

$$\Sigma \approx \Sigma^{m,b}(\Sigma^{b,b})^{-1}(\Sigma^{m,b})', \tag{6}$$

where $\Sigma^{m,b}$ contains the covariance terms between all N training data points and the base kernel points. With this approximation, the rank of the Gram matrix Σ cannot be larger than N_b. The regression function is now a superposition

$$\hat{f}(x) = \sum_{i=1}^{N_b} w_i K(x, x_i) \tag{7}$$

of only N_b kernel functions and the optimal weight vector minimizes the cost function

$$\frac{1}{2}(w^b)'\Sigma^{b,b}w^b + \frac{1}{2\sigma^2}(\Sigma^{m,b}w^b - y)'(\Sigma^{m,b}w^b - y), \tag{8}$$

where $w^b = (w_1, \ldots, w_{N_b})'$, and where y is the vector of all training targets. Note that the number of kernels is now N_b instead of N, hence the name subset of representers method (SRM). [1]

Usually, the base kernels are selected from the training data set either randomly or using a clustering algorithm [5]. Smola and Bartlett [2] select an (nearly) optimal subset of base kernel points out of the training data set. Their base kernel point selection procedure does not significantly increase the computationally complexity of the training procedure.

4 A Reduced Rank Approximation (RRA)

In a paper by Williams and Seeger [6], the authors use the decomposition of the Gram matrix of Equation 6 for calculating the kernel weights (Equation 3). Using standard matrix algebra (Woodbury formula), one obtains

$$w_{opt} \approx \frac{1}{\sigma^2}\left(y - \Sigma^{m,b}\left[(\Sigma^{m,b})'\Sigma^{m,b} + \sigma^2\Sigma^{b,b}\right]^{-1}(\Sigma^{m,b})'y\right).$$

[1] Incidentally, the relationship between the full kernel weights and the reduced kernel weights is given by $w^b = (\Sigma^{b,b})^{-1}(\Sigma^{m,b})'w$. Substitution of this identity in the cost function of Equation 8 and using Equation 5 leads to the cost function of Equation 2.

In the SRM method, the decomposition of Equation 5 changes the covariance structures of the kernels, whereas here, the covariance structures defining the kernels are unchanged. The factorization of the Gram matrix is used to obtain an efficient approximation of the optimal kernel weights. As a result, in the RRA approximation the number of kernels with nonzero weights is identical to the number of training data points N (Equation 1), whereas in the SRM method, the number of kernels with nonzero weights is identical to the number of base points N_b (Equation 7).

5 BCM Approximation

The Bayesian committee machine (BCM) was introduced by Tresp [3] and was derived using assumptions about conditional independencies. Here, we will choose a new approach to derive the BCM approximation. Let

$$P(f^b) = G(f^b; 0, \Sigma^{b,b})$$

be the Gaussian prior distribution of the unknown functional values at the base kernel points. Furthermore, let

$$P(y|f^b) = G(y; \Sigma^{m,b} w^b, cov(y|f^b))$$

be the conditional density of the training targets given f^b. Here, w^b is the weights vector defined on the base kernels, and

$$cov(y|f^b) = \sigma^2 I + \Sigma - \Sigma^{m,b} (\Sigma^{b,b})^{-1} (\Sigma^{m,b})' \tag{9}$$

is the covariance of the training data given f^b.

Note that both equations define a joint probability model and allow the calculation of many quantities of interest, e.g. $E(f^b|y)$. To be able to compare the BCM and the SRM, we will use the identity

$$f^b = \Sigma^{b,b} w^b. \tag{10}$$

The optimal w^b then minimizes the cost function

$$\frac{1}{2} (w^b)' \Sigma^{b,b} w^b + \frac{1}{2} (\Sigma^{m,b} w^b - y)' \; cov(y|f^b)^{-1} \; (\Sigma^{m,b} w^b - y). \tag{11}$$

Note that the errors in the likelihood term are correlated.

Equations 10 and 11 can be used to calculate the optimal prediction at the base kernel points but this requires the calculation of the inverse of $cov(y|f^b)$ and the latter has the dimension of $N \times N$. The BCM uses a block diagonal approximation of $cov(y|f^q)$ and the calculation of the weight vector w^b requires the inversion of matrices of only the block size B. The BCM approximation improves if few blocks are used (then a smaller number of elements are set zero) and when N_b is large, since then the last two terms on the right side of Equation (9) tend to cancel and $cov(y|f^b) \approx \sigma^2 I$. Note that the BCM approximation becomes the SRM if we set $cov(y|f^b) = \sigma^2 I$. In the latter, the induced correlations are completely ignored.

With the BCM approximation we obtain

$$w_{opt}^b \approx \left(\Sigma^{b,b} + \sum_{i=1}^{M} (\Sigma_i^{m,b})' cov(y^i|f^b)^{-1} \Sigma_i^{m,b} \right)^{-1} \sum_{i=1}^{M} (\Sigma_i^{m,b})' cov(y^i|f^b)^{-1} y^i$$

which is one particular form of the BCM approximation. Here, M is the number of blocks, y^i is the vector of targets of the i-th module, $cov(y^i|f^b)$ is the i-th diagonal block of $cov(y|f^b)$ and $\Sigma_i^{m,b}$ is the submatrix of $\Sigma^{m,b}$ containing the covariances between the base kernel points and the training data points in the i-th partition. The predictions at the base kernel points can be obtained by substituting w_{opt}^b in Equation 10. The predictions at additional test points can be calculated by substituting w_{opt}^b in Equation 7.

6 Experimental Comparisons

In the first experiment, the base points were selected out of the *test set*. This corresponds to a procedure sometimes referred to as transduction where the user starts training the system only after the inputs to the test points become available. The experimental results presented in Figure 1 (A), (B) show that for the optimal scale parameters, the BCM approximation makes significantly better predictions at the base kernel points if compared to the SRM. For large scale parameters, $cov(y|f^q)$ is approximately diagonal and both approximations give comparable results. For smaller scale parameters, the block diagonal approximation is better than the diagonal approximation and the BCM gives better results than the SRM. The predictions of the RRA at the base points are identical to the predictions of the SRM method. Based on the predictions at the base points, one can calculate prediction at additional test point. The results shown in Figure 1 (C) show that the results of the BCM and the RRA method are comparable, although the latter gives slightly better results.[2] Figure 1 (D) shows the test set error if the standard procedure is used. Here, the base points are randomly chosen out of the *training data set*, weights on the base points in the training data set are calculated, and these are then used to predict at the test points. As expected, the results shown in (D) are comparable to the results in (C). For the experiments in (C) and (D), the results of the RRA (not shown) were considerably worse than the results obtained using the BCM approximation and the SRM method.

7 Conclusions

In this paper, we have compared three approaches for scaling up kernel-based systems. The computational complexity of the presented methods scales as $\mathcal{O}(N \times$

[2] Of course one could calculate the BCM approximation again for the additional test points instead of using Equation 1. This would give better results but would require another $\mathcal{O}(Nm^2)$ operations.

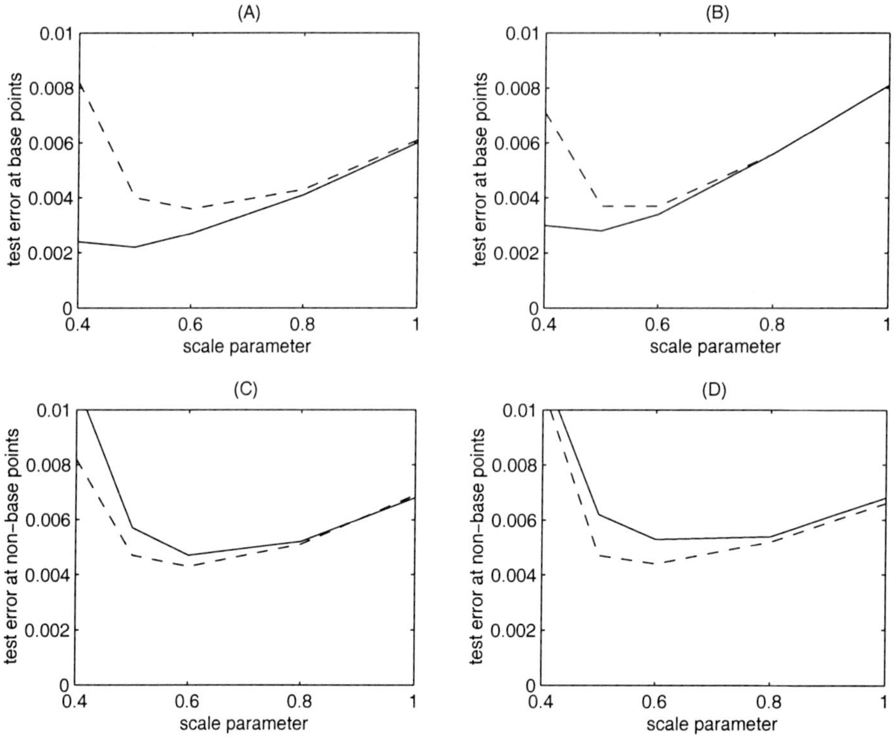

Fig. 1. Test set error is plotted against the scale parameter (width γ) of a Gaussian kernel for the BCM (continuous) and for the SRM (dashed). For (A), (B), and (C), the base points were randomly selected out of the test set. (A) and (B) show the performance at the base points and (C) shows the performance at additional test points. For the experiment in (D), the base points were randomly selected out of the training data set and the error on an independent test set is shown. We used 10000 training data points, 1000 base kernel points and 1000 additional test points. The plots are based on an artificial data set with additive noise with variance $\sigma^2 = 0$ (A, C, D) and $\sigma^2 = 0.001$ (B). The test data are noise free.

N_b^2) where N is the number of training data points and N_b is the number of base kernel points. If training is performed after the test inputs are known (transduction), the BCM outperforms the other approaches. In the more common setting where training is done before the inputs to the test set are available (induction), all three methods perform comparably, although the subset of representers method seems to have a slight advantage in performance.

References

1. Lee, Y.-J. and Mangasarian, O. L.: RSVM: Reduced Support Vector Machines. Data Mining Institute Technical Report 00-07, Computer Sciences Department, University of Wisconsin (2000)

2. Smola, A. J. and Bartlett, P.: Sparse Greedy Gaussian Process Regression. In: T. K. Leen, T. G. Diettrich and V. Tresp, (eds.): Advances in Neural Information Processing Systems 13 (2001)
3. Tresp, V.: The Bayesian Committee Machine. Neural Computation, Vol.12 (2000)
4. Tresp, V.: Scaling Kernel-Based Systems to Large Data Sets. Data Mining and Knowledge Discovery, accepted for publication
5. Wahba, G.: Spline models for observational data. Philadelphia: Society for Industrial and Applied Mathematics (1990)
6. Williams, C. K. I. and Seeger, M.: Using the Nyström Method to Speed up Kernel Machines. In: T. K. Leen, T. G. Diettrich and V. Tresp, (eds.): Advances in Neural Information Processing Systems 13 (2001)

On-Line Learning Methods for Gaussian Processes

Shigeyuki Oba[1], Masa-aki Sato[2], and Shin Ishii[1]

[1] Nara Institute of Science and Technology,
Takayama 8916-5, Ikoma, JAPAN
shige-o@is.aist-nara.ac.jp
http://mimi.aist-nara.ac.jp/~shige-o/
[2] ATR International, Soraku-gun, Kyoto, JAPAN

Abstract. This article proposes two modifications of Gaussian processes, which aim to deal with dynamic environments. One is a weight decay method that gradually forgets the old data, and the other is a time stamp method that regards the time course of data as a Gaussian process. We show experimental results when these modifications are applied to regression problems in dynamic environments.

1 Introduction

Gaussian processes (GPs) [1–3] provide a natural Bayesian framework for regression problems. Especially when data involve large Gaussian noise [4], the generalization performance of GP regressions outweighs those of various other function approximators.

This article considers situations where data are provided to the learning system in an on-line manner.

- At each observation, only one or small number of data are added to $\mathcal{D}_t$. $\mathcal{D}_t$ is a data storage at time t.
- Prediction $p(y_*|\boldsymbol{x}_*, \mathcal{D}_t)$ for a given input $\boldsymbol{x}_*$ is done occasionally based on the current data storage.
- The environment generating data changes over time. Typically, the input-output relationship changes, or the input distribution changes.

In such a situation, the learning system should rapidly adapt to the current environment by putting focus on the recently observed data. This article discusses a couple of GP modifications that realize such on-line adaptation.

2 GP Regression

There is a dataset consisting of N data points: $\mathcal{D} = (\boldsymbol{X}, \boldsymbol{y})$, $\boldsymbol{X} = \{\boldsymbol{x}_n \in \mathbb{R}^d | n = 1, \cdots, N\}$, $\boldsymbol{y} = \{y_n \in \mathbb{R} | n = 1, \cdots, N\}$. A regression problem aims at achieving a predictive function $y = f(\boldsymbol{x})$ according to the dataset. A Bayes inference for this problem is formulated as

G. Dorffner, H. Bischof, and K. Hornik (Eds.): ICANN 2001, LNCS 2130, pp. 292–299, 2001.

$$p(f(\boldsymbol{x})|\mathcal{D}) = \frac{p(\boldsymbol{y}|f(\boldsymbol{x}), \boldsymbol{X})p(f(\boldsymbol{x}))}{p(\boldsymbol{y}|\boldsymbol{X})}. \tag{1}$$

In many parametric models, like MLP or RBF, the predictive function is parameterized by a parameter $\boldsymbol{w}$ and the problem is formulated as a Bayes inference to obtain the posterior distribution of the parameter, $p(\boldsymbol{w}|\mathcal{D})$. On the other hand, a Gaussian process (GP) approach considers a prior distribution of the prediction function, $p(f(\boldsymbol{x}))$.

Instead of directly using the prior distribution of the prediction function, we can equivalently use the prior distribution of output $\boldsymbol{f} = (f(\boldsymbol{x}_1), \cdots, f(\boldsymbol{x}_N))$ for the finite input set $\boldsymbol{X} = \{\boldsymbol{x}_1, \cdots, \boldsymbol{x}_N\}$, which is represented by a zero mean Gaussian distribution with covariance $\boldsymbol{C}$:

$$p(f(\boldsymbol{x}_1), \cdots, f(\boldsymbol{x}_N)|\boldsymbol{C}, \boldsymbol{X}) = \frac{1}{Z_y} \exp\left[-\frac{1}{2}\boldsymbol{f}'\boldsymbol{C}^{-1}\boldsymbol{f}\right], \tag{2}$$

where Z_y is a normalization term. The covariance matrix $\boldsymbol{C}$ is a positive definite function of $\boldsymbol{X}$. Each component is defined by

$$C_{ij} = C(\boldsymbol{x}_i, \boldsymbol{x}_j) + \delta_{ij}\mathcal{N}(\boldsymbol{x}_i), \tag{3}$$

where $C(\boldsymbol{x}_i, \boldsymbol{x}_j)$ is a covariance function which defines the relation between two input points. $\mathcal{N}(\boldsymbol{x}_i)$ is a noise function which defines output noise at individual input $\boldsymbol{x}_i$. We assume $C(\boldsymbol{x}_i, \boldsymbol{x}_j)$ and $\mathcal{N}(\boldsymbol{x}_i)$ are parameterized by a parameter $\boldsymbol{\theta}$.

Under condition (2), the conditional probability distribution of output $y_* = f(\boldsymbol{x}_*)$ for a new input $\boldsymbol{x}_*$ is given by

$$\begin{aligned} p(y_*|\boldsymbol{x}_*, \boldsymbol{X}, \boldsymbol{y}, \boldsymbol{\theta}) &= \frac{p((y_*, \boldsymbol{y})|(\boldsymbol{x}_*, \boldsymbol{X}), \boldsymbol{\theta})}{p(\boldsymbol{y}|\boldsymbol{X}, \boldsymbol{\theta})} \\ &= \frac{1}{Z_*} \exp\left[-\frac{1}{2}\left(\left[\boldsymbol{y}'\ y_*\right] \boldsymbol{C}_*^{-1} \begin{bmatrix} \boldsymbol{y} \\ y_* \end{bmatrix} - \boldsymbol{y}'\boldsymbol{C}^{-1}\boldsymbol{y}\right)\right], \end{aligned} \tag{4}$$

where Z_* is a normalization term and $\boldsymbol{C}_*$ is a covariance matrix for $(\boldsymbol{X}, \boldsymbol{x}_*)$. $\boldsymbol{C}_*$ is given by

$$\boldsymbol{C}_* = \begin{bmatrix} \boldsymbol{C} & \boldsymbol{k} \\ \boldsymbol{k}' & \kappa \end{bmatrix}, \quad \boldsymbol{k} = (C(\boldsymbol{x}_1, \boldsymbol{x}_*), \cdots, C(\boldsymbol{x}_N, \boldsymbol{x}_*)), \quad \kappa = C(\boldsymbol{x}_*, \boldsymbol{x}_*) + \mathcal{N}(\boldsymbol{x}_*). \tag{5}$$

From (4) and (5), we obtain the predictive Gaussian distribution as

$$p(y_*|\boldsymbol{x}_*, (\boldsymbol{X}, \boldsymbol{y}), \boldsymbol{\theta}) = \frac{1}{Z} \exp\left[-\frac{1}{2}\frac{(y_* - E[y_*])^2}{\text{var}\,[y_*]}\right] \tag{6}$$

$$E[y_*] = \boldsymbol{k}'\boldsymbol{C}^{-1}\boldsymbol{y}, \quad \text{var}\,[y_*] = \kappa - \boldsymbol{k}'\boldsymbol{C}^{-1}\boldsymbol{k}. \tag{7}$$

Equation (3) means that the covariance matrix is the summation of the covariance function and the noise function.

As the covariance function $C(\boldsymbol{x}_i, \boldsymbol{x}_j)$, any function form is allowed as long as the covariance matrix is positive definite. When a covariance function:

$$C(\boldsymbol{x}_i, \boldsymbol{x}_j; \boldsymbol{\theta}) = v \exp\left[-\frac{1}{2}\sum_{l=1}^{p} \alpha_l (x_{i,l} - x_{j,l})^2\right], \tag{8}$$

is used, GP can be regarded as an RBF with infinite number of kernel functions [1]. Here, l is an index of the input dimension and $x_{i,l}$ denotes the l-th component of input $\boldsymbol{x}_i$. Parameter v represents the vertical scale of the supposed GP. α_l represents the inverse variance of the l-th component and also indicates the contribution of the l-th component.

The simplest noise function is given by constant $\mathcal{N} = \beta$, corresponding to an input-independent noise. An input-dependent noise can be defined by a noise function like $\mathcal{N} = \exp\left[\sum_{j=1}^{J} (\log \beta_j)\phi_j(\boldsymbol{x})\right]$, where $\phi_j(\boldsymbol{x})$ is a kernel function.

Under the above-mentioned parameterization, the likelihood of dataset $\mathcal{D}$ is given by

$$P(\mathcal{D}|\boldsymbol{\theta}) = \frac{1}{\sqrt{(\log 2\pi)^N |\boldsymbol{C}|}} \exp\left[-\frac{1}{2}\boldsymbol{y}^T \boldsymbol{C}^{-1}\boldsymbol{y}\right]. \tag{9}$$

Using (9), parameter $\boldsymbol{\theta}$ can be obtained by maximum likelihood (ML) estimation or Bayes estimation. A gradient ascent algorithm of the log-likelihood, i.e., an ML estimation, is used in our study. Parameter $\boldsymbol{\theta}$ is updated after observing each datum.

3 Weight Decay Method

We are interested in a situation where the environment changes over time. In order to deal with such a dynamic environment, the learning system needs to put focus on the recent data by gradually forgetting the old data. This section discusses a method that decreases data weight. A weight value indicates the priority of the corresponding datum.

3.1 Covariance Matrix and Overlapping Data

In order to deal with weighted data, we first consider a situation where there are w (> 1) identical data in the dataset. The following theorem states how to deal with such overlapping data.

Theorem 1 *Overlapping w data are equivalent to a single datum whose noise function $\mathcal{N}(\boldsymbol{x})$ is divided by w.*

Proof We assume that there are $N + w$ data points: $(\boldsymbol{x}_1, y_1), \cdots, (\boldsymbol{x}_N, y_N)$, $(\boldsymbol{x}_{N+1}, y_{N+1}), \cdots, (\boldsymbol{x}_{N+w}, y_{N+w})$ and the last w data points are identical; namely, $\boldsymbol{x}_{N+i} = \boldsymbol{x}_0$ and $y_{N+i} = y_0$ for $i = 1, \cdots, w$. Covariance matrix for the $N + w$ data points, $\boldsymbol{C}_+$, is given by

$$\boldsymbol{C}_+ = \begin{bmatrix} \boldsymbol{C} & \boldsymbol{K} \\ \boldsymbol{K}' & \boldsymbol{A} \end{bmatrix}. \tag{10}$$

$\boldsymbol{A} = \mathcal{N}(\boldsymbol{x}_0)\boldsymbol{I}_w + \kappa \boldsymbol{J}_w$ is a w-by-w matrix where $\kappa = C(\boldsymbol{x}_0, \boldsymbol{x}_0)$, $\boldsymbol{I}_w$ is a w-by-w unit matrix and $\boldsymbol{J}_w$ is a w-by-w matrix whose components are all 1. $\boldsymbol{K} =$

$[\boldsymbol{k}, \cdots, \boldsymbol{k}]$ is an N-by-w matrix consisting of N-dimensional columnar vector $\boldsymbol{k} = [C(\boldsymbol{x}_1, \boldsymbol{x}_0), \cdots, C(\boldsymbol{x}_N, \boldsymbol{x}_0)]'$.

Using partial inverse matrix expansion, the inverse of $\boldsymbol{C}_+$ is calculated as

$$\boldsymbol{C}_+^{-1} = \begin{bmatrix} \tilde{\boldsymbol{C}} & \tilde{\boldsymbol{K}} \\ \tilde{\boldsymbol{K}}' & \tilde{\boldsymbol{A}} \end{bmatrix} \tag{11}$$

$$\tilde{\boldsymbol{A}} \stackrel{\text{def}}{=} (\mathcal{N}(\boldsymbol{x}_0)\boldsymbol{I}_w + \kappa \boldsymbol{J}_w - \boldsymbol{K}'\boldsymbol{C}^{-1}\boldsymbol{K})^{-1},$$
$$\tilde{\boldsymbol{K}} \stackrel{\text{def}}{=} -\boldsymbol{C}^{-1}\boldsymbol{K}\tilde{\boldsymbol{A}}, \quad \tilde{\boldsymbol{C}} \stackrel{\text{def}}{=} \boldsymbol{C}^{-1} + \boldsymbol{C}^{-1}\boldsymbol{K}\tilde{\boldsymbol{A}}\boldsymbol{K}'\boldsymbol{C}^{-1}.$$

Using $\boldsymbol{K}'\boldsymbol{C}^{-1}\boldsymbol{K} = (\boldsymbol{k}'\boldsymbol{C}^{-1}\boldsymbol{k})\boldsymbol{J}_w$ and $(a\boldsymbol{I}_w + b\boldsymbol{J}_w)^{-1} = \frac{1}{a} - \frac{b}{a(wb+a)}\boldsymbol{J}_w$, we obtain

$$\tilde{\boldsymbol{A}} = \frac{1}{\mathcal{N}(\boldsymbol{x}_0)}\boldsymbol{I}_w - \frac{\kappa - \boldsymbol{k}'\boldsymbol{C}^{-1}\boldsymbol{k}}{\mathcal{N}(\boldsymbol{x}_0)\{w(\kappa - \boldsymbol{k}'\boldsymbol{C}^{-1}\boldsymbol{k}) + \mathcal{N}(\boldsymbol{x}_0)\}}\boldsymbol{J}_w. \tag{12}$$

Furthermore, using $\boldsymbol{K}\boldsymbol{I}_w = \boldsymbol{K}$, $\boldsymbol{K}\boldsymbol{J}_w = w\boldsymbol{K}$ and $\boldsymbol{K}\boldsymbol{K}' = w\boldsymbol{k}\boldsymbol{k}'$, we obtain

$$\tilde{\boldsymbol{K}} = -\frac{\lambda}{w}\boldsymbol{C}^{-1}\boldsymbol{K}, \qquad \tilde{\boldsymbol{C}} = \boldsymbol{C}^{-1} + \lambda\boldsymbol{C}^{-1}\boldsymbol{k}\boldsymbol{k}'\boldsymbol{C}^{-1}, \tag{13}$$

$$\lambda \stackrel{\text{def}}{=} \frac{1}{\kappa - \boldsymbol{k}'\boldsymbol{C}^{-1}\boldsymbol{k} + \mathcal{N}(\boldsymbol{x}_0)/w}. \tag{14}$$

For a new input $\boldsymbol{x}_*$, the mean output $E[y_*]$ is given as follows using the covariance matrix $\boldsymbol{C}+$:

$$E[y_*] = [\boldsymbol{k}_*' \ \boldsymbol{l}_*'] \, \boldsymbol{C}_+^{-1} \begin{bmatrix} \boldsymbol{y}_N \\ \boldsymbol{y}_w \end{bmatrix}, \tag{15}$$

where $\boldsymbol{k}_* = [C(\boldsymbol{x}_1, \boldsymbol{x}_*) \cdots C(\boldsymbol{x}_N, \boldsymbol{x}_*)]'$, $\boldsymbol{l}_* = [l_0, \cdots, l_0]'$ is a w-dimensional vector consisting of $l_0 = C(\boldsymbol{x}_0, \boldsymbol{x}_*)$, $\boldsymbol{y}_N = [y_1 \cdots y_N]'$ and $\boldsymbol{y}_w = [y_0 \cdots y_0]$ is a w-dimensional vector of y_0. From (11), we obtain

$$\begin{aligned} E[y_*] &= \boldsymbol{k}_*'\tilde{\boldsymbol{C}}\boldsymbol{y}_N + \boldsymbol{k}_*'\tilde{\boldsymbol{K}}'\boldsymbol{y}_w + \boldsymbol{l}_*'\tilde{\boldsymbol{K}}\boldsymbol{y}_N + \boldsymbol{l}_*'\tilde{\boldsymbol{A}}\boldsymbol{y}_w \\ &= \boldsymbol{k}_*'\tilde{\boldsymbol{C}}\boldsymbol{y}_N - \lambda(\boldsymbol{k}_*'\boldsymbol{C}^{-1}\boldsymbol{k}y_0 + l_0\boldsymbol{k}'\boldsymbol{C}^{-1}\boldsymbol{y}_N + l_0 y_0) \\ &= [\boldsymbol{k}_*' \ l_0] \begin{bmatrix} \tilde{\boldsymbol{C}} & -\lambda\boldsymbol{C}^{-1}\boldsymbol{k} \\ -\lambda\boldsymbol{k}'\boldsymbol{C}^{-1} & \lambda \end{bmatrix} \begin{bmatrix} \boldsymbol{y}_N \\ y_0 \end{bmatrix}. \end{aligned} \tag{16}$$

Similarly, $\mathrm{var}[y_*]$ is given by

$$\mathrm{var}[y_*] = \kappa - [\boldsymbol{k}_*' \ l_0] \begin{bmatrix} \tilde{\boldsymbol{C}} & -\lambda\boldsymbol{C}^{-1}\boldsymbol{k} \\ -\lambda\boldsymbol{k}'\boldsymbol{C}^{-1} & \lambda \end{bmatrix} \begin{bmatrix} \boldsymbol{k}_* \\ l_0 \end{bmatrix}. \tag{17}$$

Equations (16) and (17) mean that w identical data can be replaced in the prediction of y_* by a single datum. From equation (14), however, we see that the noise function for the single datum is divided by w. ■

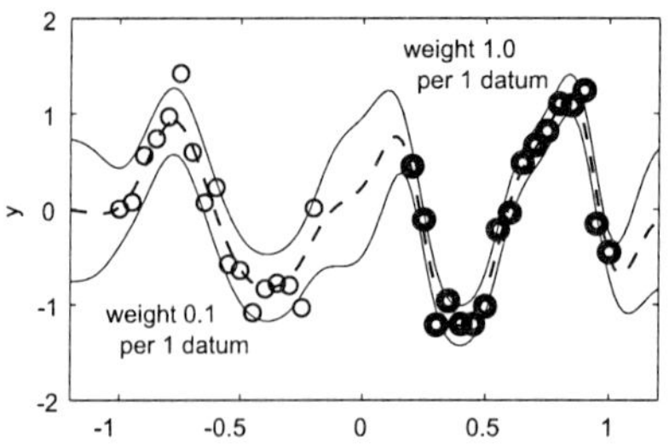

Fig. 1. Thick and thin circles denote data with weight 1.0 and 0.1, respectively. Dash line denotes the expected output, and upper and lower solid lines denote the error bar (expected standard deviation).

The above theorem can be straightforwardly expanded to the case where each datum has a weight value. When there are N weighted data: $(\boldsymbol{x}_1, y_1, w_1), \cdots, (\boldsymbol{x}_N, y_N, w_N)$, its covariance matrix is given by

$$C_{ij} = C(\boldsymbol{x}_i, \boldsymbol{x}_j) + \delta_{ij}\mathcal{N}(\boldsymbol{x}_i)/w_i. \tag{18}$$

Each weight value w_k $(k = 1, \cdots, N)$ is not necessarily an integer any more. Figure 1 shows an experimental result of the GP regression when some data have small weight value (0.1) while the other data have regular weight value (1.0).

3.2 On-Line GP Regression with Weight Decay

In the weight decay method proposed here, the data weight is gradually decreased in order to forget the old data. At time t, a single datum $(\boldsymbol{x}_t, y_t)$ is provided to the learning system with an initial weight value 1.0. After the next time step $t+1$, the weight is multiplied by a decaying factor η; namely, the weight exponentially decays. When a datum is too old and its weight value becomes sufficiently small, the corresponding entry is deleted from the covariance matrix. This can be done by applying the partial matrix inverse calculation reciprocally.

Figure 2 shows an example. The target function for $0 < t \leq 60$ or $60 < t \leq 120$ is $y = \sin(8x)$ or $y = -\sin(8x)$, respectively. At each time step t, an input $\boldsymbol{x}_t$ is randomly generated from a Gaussian distribution whose center is -0.5 (or 0.5) for $0 < t \leq 30$ and $60 < t \leq 90$ (or $30 < t \leq 60$ and $90 < t \leq 120$). For each input $\boldsymbol{x}_t$, target output y_t is generated by the target function disturbed by Gaussian noise. Figure 2(a) shows the whole dataset. Figure 2(b) shows the regression results at $t = 30, 60, 90$ and 120. From the results for $t = 30$ and $t = 60$, the GP regression well approximates the target function for the region where data are observed. At $t = 90$, although the newly observed data in the $x < 0$ region contradict the older data, the regression adapts to the new data. At $t = 120$, the regression comes to well approximate the new target function.

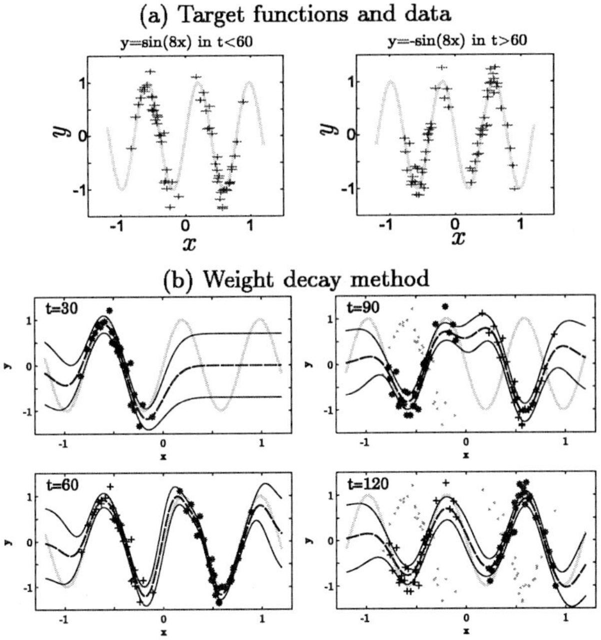

Fig. 2. (a) Target functions and data. (b) The results of the GP regression with weight decay. Data marked by '$*$' denote the recent 30 data, '+' denote the previous 30 data, and '·' denote the older data. Each shadow line shows the target function at that time, dash line is the predicted function and two thin black lines denote the error bar.

4 Time Stamp Method

In a dynamic environment, the prediction based on too old data will be inaccurate in the current prediction. Therefore, the prediction should consider the time interval between the data observation time and the prediction time. One way to do so is that a datum explicitly involves its observation time as its component.

Let a datum at time $t = t_i$ be represented as input, output and observation time, $(\boldsymbol{x}_i, y_i, t_i)$. For the augmented dataset, the covariance function is given by

$$C((\boldsymbol{x}_i, t_i), (\boldsymbol{x}_j, t_j)) = v \exp\left[-\frac{1}{2}\sum_{l=1}^{d} \alpha_l (x_{i,l} - x_{j,l})^2 - \frac{1}{2}\alpha_t (t_i - t_j)^2\right]. \quad (19)$$

The learning of the new parameter α_t is done similarly to that in the ordinary GP regression. This simple method is called the time stamp method.

An experimental result is shown in Figure 3. The target functions and the dataset are the same as those in the previous experiment. Figure 3(b) shows how this method works. Before the target function changes ($t \leq 60$), parameter α_t corresponding to the time variance ('time scale') decreases as the learning proceeds, implying that the GP regression regards the environment as static.

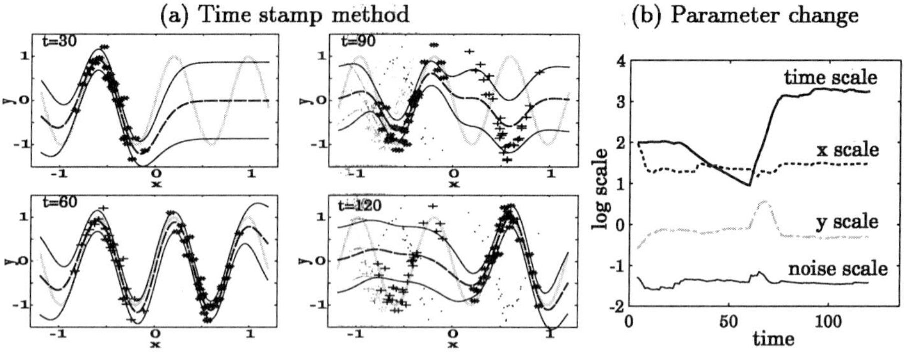

Fig. 3. (a) GP regression results by the time stamp method. (b) Time course of parameters. The ordinate is in log scale. The thick, dash, chain, and thin lines denote parameters α_t (time scale), α_1 (x scale), v (y scale), and β (noise scale), respectively.

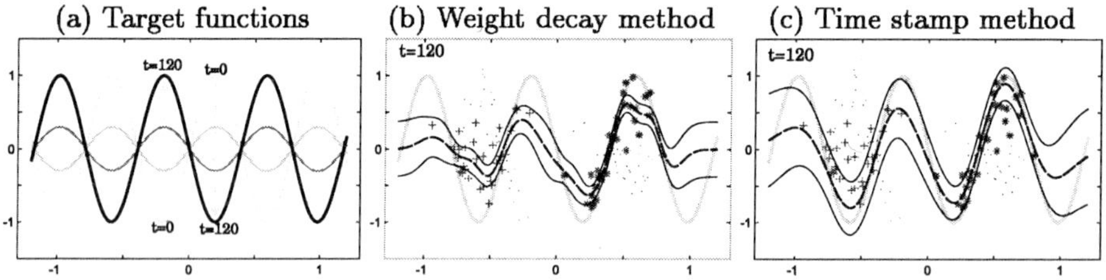

Fig. 4. (a) Target function smoothly changes over time. (b) Result of the weight decay method at $t = 120$. (c) Result of the time stamp method at $t = 120$.

After the target function suddenly changes at $t = 60$, the GP regression becomes to regard the environment as noisy; this can be seen in temporal increase of v and β corresponding to the output variance ('y scale' and 'noise scale'). However, what actually happens is the change of environment instead of the increase of noise. Therefore, the GP regression increases α_t while decreases v and β; namely, the system becomes easier to forget the old data and puts more focus on the recent data. This increase of α_t has side effect. At $t = 90$ or $t = 120$ in Figure 3(a), the error bar for the region where the previous data (mark '+') are provided is fairly large. This is due to the increase of the time scale.

Since the time stamp method handles the observation time as a smooth GP as well as the input, it is advantageous especially when the target function smoothly changes over time. Figure 4 shows an experimental example. Although the weight decay method only follows the recent environment, the time stamp method can predict the future change considering the past environmental change. This improves the prediction along the smooth environmental change. Actually, the final prediction at $t = 120$ is much better in the time stamp method than that in the weight decay method.

5 Concluding Remark

This article proposed two learning methods for GP, i.e., the weight decay method and the time stamp method, aiming at on-line adaptation to environmental changes. Naturally, these two methods can be combined in order to deal with complex environmental changes. On the other hand, on-line learning problems often involve increasingly large number of sample data. Conventional GP algorithms suffer from computational cost with square or cubic order of the sample number. Since our weight decay method employs a heuristic that deletes data whose weight values are sufficiently small, we consider that it can be applied to problems with a fairly large number of data. However, actual applications to real problems are our future work.

Recently GP modifications using sparsely allocated kernels have been proposed in order to deal with a large data set ([6], for example). Since our idea is to manipulate kernel hyper-parameters in an on-line manner, our methods can be combined with such GP modifications.

References

1. MacKay, D. J. C.: Gaussian processes - a replacement for supervised neural networks?. Lecture notes for a tutorial at NIPS 1997.
2. Gibbs, M. N. and MacKay, D. J. C.: Efficient implementation of Gaussian processes. Cavendish Laboratory, Cambridge, UK. Draft manuscript (1997), available from `http://wol.ra.phy.cam.ac.uk/mackay/homepage.html`.
3. Williams, C. K. I.: Prediction with Gaussian processes: from linear regression to linear prediction and beyond. Technical Report NCRG/97/012, Neural Computing Research Group, Department of Computer Science, Aston University (1997).
4. Rasmussen, C. E.: Evaluation of Gaussian processes and other methods for non-Linear regression. PhD Thesis. University of Toronto, Department of Computer Science, Toronto, Canada. `http://www.cs.toronto.edu/pub/carl/thesis.ps.gz`
5. Neal, R. M.: Monte Carlo implementation of Gaussian process models for Bayesian regression and classification.: Technical Report CRG-TR-97-2, Department of Computer Science, University of Toronto (1997).
6. Csatò, L. and Opper, M.: Sparse Representation for Gaussian Process Models. To appear in Advances in Neural Information Processing System 13 (eds. T.K.Leen, T.G. Diettrich, V. Tresp). MIT Press (2001)

Online Approximations for Wind-Field Models

Lehel Csató, Dan Cornford, and Manfred Opper

Neural Computing Research Group, Aston University
B4 7ET Birmingham, United Kingdom
{csatol,cornfosd,opperm}@aston.ac.uk

Abstract. We study online approximations to Gaussian process models for spatially distributed systems. We apply our method to the prediction of wind fields over the ocean surface from scatterometer data. Our approach combines a sequential update of a Gaussian approximation to the posterior with a sparse representation that allows to treat problems with a large number of observations.

1 Introduction

A common scenario of applying online or sequential learning methods [Saad 1998] is when the amount of data is too large to be processed by more efficient offline methods or there is no possibility to store the arriving data. In this article we consider the area of spatial statistics [Cressie 1991], where the data is observed at different spatial locations and the aim is to build a global Bayesian model of the local observations based on a Gaussian Process prior distribution. Specifically, we consider scatterometer data obtained from the ERS-2 satellite [Offiler 1994] where the aim is to obtain an estimate of the wind fields which the scatterometer indirectly measured.

The scatterometer measures the radar backscatter from the ocean surface at a wavelength of approximately 5 cm. The strength of the returned signal gives an indication of the wind speed and direction, relative to scatterometer beam direction. As shown in [Stoffelen and Anderson 1997b] the measured backscatter behaves as a truncated Fourier expansion in relative wind direction. Thus while the wind vector to scatterometer observations map is one-to-one, its inverse is one-to-many [Evans *et al.* 2000]. This makes the retrieval of a wind field a complex problem with multiple solutions. Nabney *et al.* [2000] have recently proposed a Bayesian framework for wind field retrieval combining a vector Gaussian process prior model with local forward (wind field to scatterometer) or inverse models.

One problem with the approach outlined in [Nabney *et al.* 2000] is that the vector Gaussian process requires a matrix inversion which scales as n^3. The backscatter is measured over 50×50 km cells over the ocean and the total number of observations acquired on a given orbit can be several thousand.

In this paper we show that we can produce an efficient approximation to the posterior distribution of the wind field by applying a Bayesian online learning approach [Opper 1998] to Gaussian process models following [Csató and Opper

G. Dorffner, H. Bischof, and K. Hornik (Eds.): ICANN 2001, LNCS 2130, pp. 300–307, 2001.

2001], which computes the approximate posterior by a single sweep through the data. The computational complexity is further reduced by constructing a sparse sequential approximate representation to the posterior process.

2 Processing Scatterometer Data

Scatterometers are commonly used to retrieve wind vectors over ocean surfaces. Current methods of transforming the observed values (scatterometer data, denoted as vector $\mathbf{s}$ or $\mathbf{s}_i$ at a given spatial location) into wind fields can be split into two phases: local wind vector retrieval and ambiguity removal [Stoffelen and Anderson 1997a] where one of the local solutions is selected as the true wind vector. Ambiguity removal often uses external information, such as a Numerical Weather Prediction (NWP) forecast of the expected wind field at the time of the scatterometer observations. We are seeking a method of wind field retrieval which does not require external data.

In this paper we use a mixture density network (MDN) [Bishop 1995] to model the conditional dependence of the local wind vector $\mathbf{z}_i = (u_i, v_i)$ on the local scatterometer observations $\mathbf{s}_i$:

$$p_m(\mathbf{z}_i|\mathbf{s}_i, \boldsymbol{\omega}) = \sum_{j=1}^{4} \beta_{ij}\phi(\mathbf{z}_i|\mathbf{c}_{ij}, \sigma_{ij}) \tag{1}$$

where $\boldsymbol{\omega}$ is used to denote the parameters of the MDN, ϕ is a Gaussian distribution with parameters functions of $\boldsymbol{\omega}$ and $\mathbf{s}_i$. The parameters of the MDN are determined using an independent training set [Evans *et al.* 2000] and are considered known in this application. The MDN which has four Gaussian component densities captures the ambiguity of the inverse problem.

In order to have a global model from the localised wind vectors, we have to combine them. We use a zero-mean vector GP to link the local inverse models [Nabney *et al.* 2000]:

$$q(\underline{\mathbf{z}}) \propto \left(\prod_{i}^{N} \frac{p_m(\mathbf{z}_i|\mathbf{s}_i, \boldsymbol{\omega}) p(\mathbf{s}_i)}{p_G(\mathbf{z}_i|\mathbf{W}_{0i})} \right) p_0(\underline{\mathbf{z}}|\mathbf{W}_0) \tag{2}$$

where $\underline{\mathbf{z}} = [\mathbf{z}_1, \ldots, \mathbf{z}_N]^T$ is the concatenation of the local wind field components, $\mathbf{W}_0 = \{\mathbf{W}_0(x_i, x_j)\}_{ij=1,\ldots,N}$ is the prior covariance matrix for the vector $\underline{\mathbf{z}}$ (dependent on the spatial location of the wind vectors), and p_G is p_0 marginalised at z_i, a zero-mean Gaussian with covariance $\mathbf{W}_{0i}$. The choice of the kernel function $\mathbf{W}_0(x, y)$ fully specifies our prior beliefs about the model. Notice also that for any given location we have a *two-dimensional* wind vector, thus the output of the kernel function *is a* 2×2 *matrix*, details can be found in [Nabney *et al.* 2000]. The link between two different wind field directions is made through the kernel function – the larger the kernel value, the stronger the "coupling" between the two corresponding wind fields is. The prior Gaussian process is tuned carefully to represent features seen in real wind fields.

Since all quantities involved are Gaussians, we could, *in principle*, compute the resulting probabilities analytically, but this computation is *practically* intractable: the number of mixture elements from $q(\underline{\mathbf{z}})$ is 4^N, extremely high even for moderate values of N. Instead, we will apply the online approximation of [Csató and Opper 2001] to have a jointly Gaussian approximation to the posterior at all data points. However, we know that the posterior distribution of the wind field given the scatterometer observations is multi-modal, with in general two dominating and well separated modes. We might thus expect that the online implementation of the Gaussian process will track one of these posterior modes. Results show that this is indeed the case, although the order of the insertion of the local observations appears to be important.

3 Online Learning for the Vector Gaussian Process

Gaussian processes belong to the family of Bayesian [Bernardo and Smith 1994] models. However, contrary to the finite-dimensional case, here the "model parameters" are continuous: the GP priors specify a Gaussian distribution over a function space. Due to the vector GP, the kernel function $\mathbf{W}_0(x,y)$ is a 2×2 matrix, specifying the pairwise cross-correlation between wind field components at different spatial positions.

Simple moments of GP posteriors (which are usually non Gaussian) have a parametrisation in terms of the training data [Opper and Winther 1999] which resembles the popular kernel-representation [Kimeldorf and Wahba 1971]. For all spatial locations x the mean and covariance function of the vectors $\mathbf{z}_x \in \mathbb{R}^2$ are represented as

$$
\begin{aligned}
\langle \mathbf{z}_x \rangle &= \sum_{i=1}^{N} \mathbf{W}_0(x, x_i) \cdot \boldsymbol{\alpha}_{\mathbf{z}}(i) \\
\mathrm{cov}(\mathbf{z}_x, \mathbf{z}_y) &= \mathbf{W}_0(x,y) + \sum_{i,j=1}^{N} \mathbf{W}_0(x, x_i) \cdot \mathbf{C}_{\mathbf{z}}(ij) \cdot \mathbf{W}_0(x_j, y)
\end{aligned}
\tag{3}
$$

where $\boldsymbol{\alpha}_{\mathbf{z}}(1), \boldsymbol{\alpha}_{\mathbf{z}}(2), \ldots, \boldsymbol{\alpha}_{\mathbf{z}}(N)$ and $\{\mathbf{C}_{\mathbf{z}}(ij)\}_{i,j=1,N}$ are parameters which will be updated sequentially by our online algorithm. Before doing so, we will (for numerical convenience) represent the vectorial process by a scalar process with twice the number of observations, i.e. we set

$$
\langle \mathbf{z}_x \rangle = \begin{bmatrix} \langle f_{x^u} \rangle \\ \langle f_{x^v} \rangle \end{bmatrix} \quad \text{and} \quad \mathbf{W}_0(x,y) = \begin{bmatrix} K_0(x^u, y^u) & K_0(x^u, y^v) \\ K_0(x^v, y^u) & K_0(x^v, y^v) \end{bmatrix}
\tag{4}
$$

and write (ignoring the superscripts)

$$
\begin{aligned}
\langle f_x \rangle &= \sum_{i=1}^{2N} K_0(x, x_i) \alpha(i) \\
\mathrm{cov}(f_x, f_y) &= K_0(x,y) + \sum_{i,j=1}^{2N} K_0(x, x_i) C(ij) K_0(x_j, y)
\end{aligned}
\tag{5}
$$

where $\boldsymbol{\alpha} = [\alpha_1, \ldots, \alpha_{2N}]^T$ and $\mathbf{C} = \{C(ij)\}_{i,j=1,\ldots,2N}$ are rearrangements of the parameters from eq. (3).

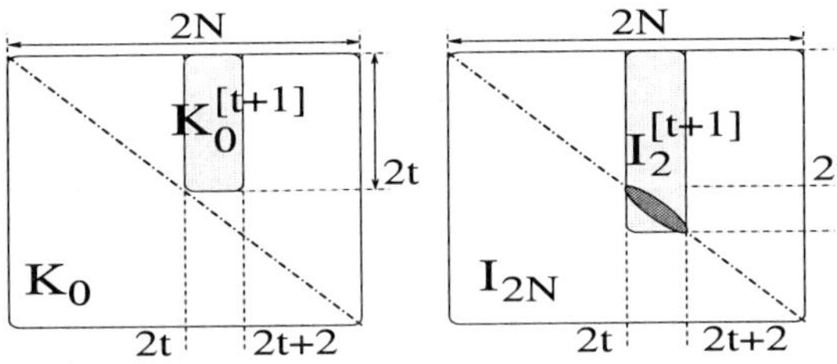

Fig. 1. Illustration of the elements used in the update eq. (6).

The online approximation for GP learning [Csató and Opper 2001] approximates the posterior by a Gaussian at every step. For a new observation $\mathbf{s}_{t+1}$, the previous approximation to the posterior $q_t(\underline{\mathbf{z}})$ together with a local "likelihood" factor (from eq. (2))

$$\frac{p_m(\mathbf{z}_{t+1}|\mathbf{s}_{t+1},\boldsymbol{\omega})p(\mathbf{s}_{t+1})}{p_G(\mathbf{z}_{t+1}|\mathbf{W}_{0,t+1})}$$

are combined into a new posterior using Bayes rule. Computing its mean and covariance enable us to create an updated Gaussian approximation $q_{t+1}(\underline{\mathbf{z}})$ at the next step. $\hat{q}(\underline{\mathbf{z}}) = q_{N+1}(\underline{\mathbf{z}})$ is the final result of the online approximation. This process can be formulated in terms of updates for the parameters $\boldsymbol{\alpha}$ and $\mathbf{C}$ which determine the mean and covariance:

$$\begin{aligned}\boldsymbol{\alpha}_{t+1} &= \boldsymbol{\alpha}_t + \mathbf{v}_{t+1}\frac{\partial \ln g(\langle \mathbf{z}_{t+1}\rangle)}{\partial \langle \mathbf{z}_{t+1}\rangle} \\ \mathbf{C}_{t+1} &= \mathbf{C}_t + \mathbf{v}_{t+1}\frac{\partial^2 \ln g(\langle \mathbf{z}_{t+1}\rangle)}{\partial \langle \mathbf{z}_{t+1}\rangle^2}\mathbf{v}_{t+1}^T\end{aligned} \quad \text{with} \quad \mathbf{v}_{t+1} = \mathbf{C}_t\mathbf{K}_0^{[t+1]} + \mathbf{I}_2^{[t+1]} \qquad (6)$$

with elements $\mathbf{K}_0^{[t+1]}$ and $\mathbf{I}_2^{[t+1]}$ are shown in Fig. 1 and

$$g(\langle \mathbf{z}_{t+1}\rangle) = \left\langle \frac{p_m(\mathbf{z}_{t+1}|\mathbf{s}_{t+1},\boldsymbol{\omega})p(\mathbf{s}_{t+1})}{p_G(\mathbf{z}_{t+1}|\mathbf{W}_{0,t+1})} \right\rangle_{q_t(\mathbf{z}_{t+1})} \qquad (7)$$

and $\langle \mathbf{z}_{t+1}\rangle$ is a vector, implying vector and matrix quantities in (6). Function $g(\langle \mathbf{z}_{t+1}\rangle)$ is easy to compute analytically because it just requires the two dimensional marginal distribution of the process at the observation point $\mathbf{s}_{t+1}$. Fig. 2 shows the results of the online algorithm applied on a sample wind field, details can be found in the figure caption.

3.1 Obtaining Sparsity in Wind Fields

Each time-step the number of nonzero parameters will be increased in the update equation. This forces us to use a further approximation which reduces the number of supporting examples in the representations eq. (5) to a smaller set of basis vectors. Following our approach in [Csató and Opper 2001] we remove the last data element when a certain score (defined by the feature space geometry

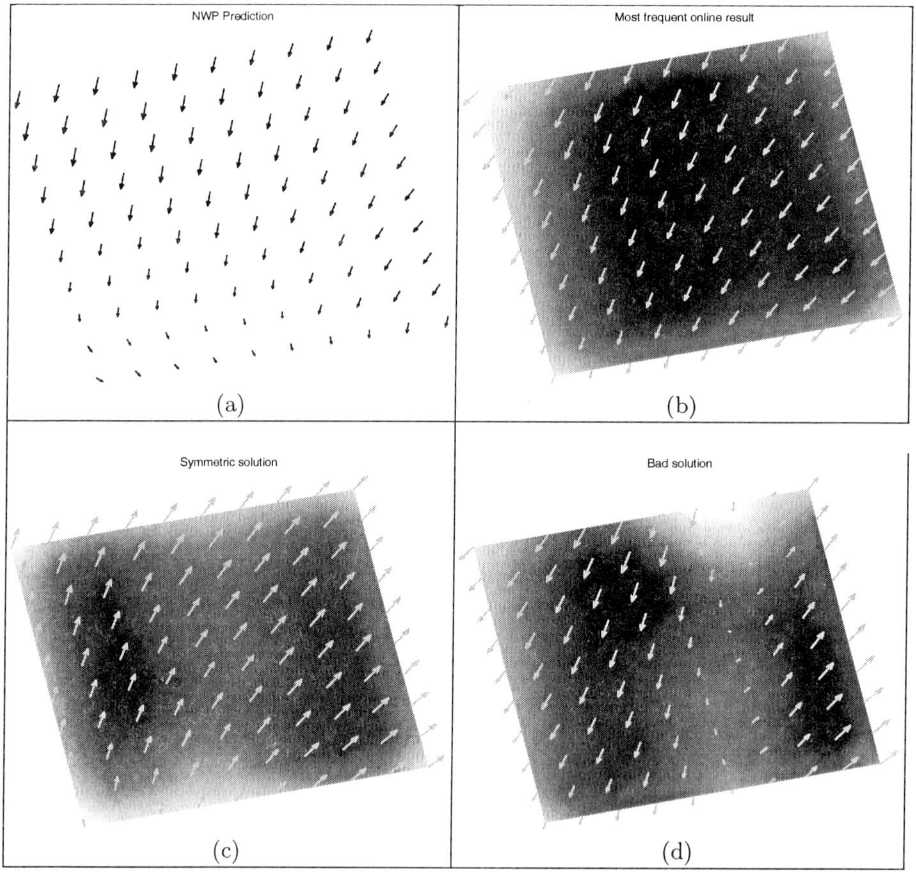

Fig. 2. The NWP wind field estimation (a), the most frequent (b) and the second most frequent (c) online solution together with a bad solution. The assessment of good/bad solution is based on the value of the relative weight from Section 3.2. The gray-scale background indicates the model confidence (Bayesian error-bars) in the prediction, darker shade meaning more confidence.

associated to the kernel $\mathbf{K}_0$) suggests that the approximation error is small. The remaining parameters are readjusted to partly compensate for the removal as:

$$\begin{aligned}\hat{\boldsymbol{\alpha}} &= \boldsymbol{\alpha}^{(t)} - \mathbf{Q}^* \mathbf{q}^{*(-1)} \boldsymbol{\alpha}^* \\ \hat{\mathbf{Q}} &= \mathbf{Q}^{(t)} - \mathbf{Q}^* \mathbf{q}^{*(-1)} \mathbf{Q}^{*T} \\ \hat{\mathbf{C}} &= \mathbf{C}^{(t)} + \mathbf{Q}^* \mathbf{q}^{*(-1)} \mathbf{c}^* \mathbf{q}^{*(-1)} \mathbf{Q}^{*T} - \mathbf{Q}^* \mathbf{q}^{*(-1)} \mathbf{C}^{*T} - \mathbf{C}^* \mathbf{q}^{*(-1)} \mathbf{Q}^{*T}\end{aligned} \tag{8}$$

where $\mathbf{Q}^{-1} = \{\mathsf{K}_0(x_i, x_j)\}_{ij=1,\ldots,2N}$ is the inverse of the Gram matrix, the elements being shown in Fig. 3 ($\boldsymbol{\alpha}^*$, $\mathbf{q}^*$ and $\mathbf{C}^*$ are two-by-two matrices).

The presented update is optimal in the sense that the posterior means of the process at data locations are not affected by the approximation [Csató and

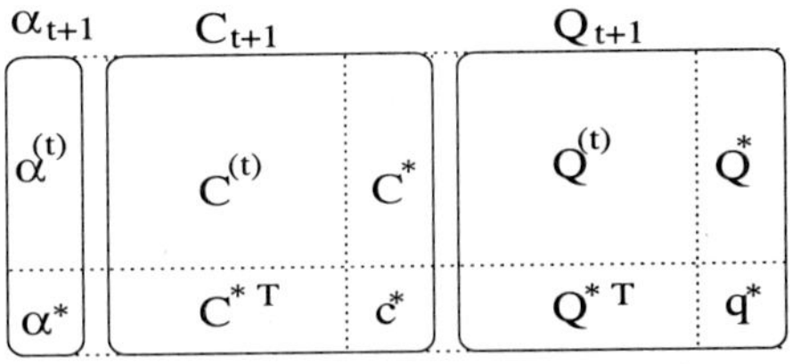

Fig. 3. Decomposition of model parameters for the update equation (8).

Opper 2001]. The change of the mean at the location to be deleted is used as a score which measures the loss. This change is (again, very similar to the results from [Csató and Opper 2001]) measured using the score $\varepsilon = \|(\mathbf{q}^*)^{-1}\boldsymbol{\alpha}^*\|$ (the parameters of the vector GP can have any order, we can compute the score for every spatial location).

Removing the data locations with low score sequentially leaves only a small set of so-called *basis points* upon which all further prediction will depend.

Our preliminary results are promising: Fig. 4 shows the resulting wind field if 85 of the spatial knots are removed from the presentation eq. (5). On the right-hand side the evolution of the KL-divergence and the sum-squared errors in the means between the vector GP and a trimmed GP using eq. (8) are shown. as a function of the number of deleted points. Whilst the approximation of the posterior variance decays fast, the the predictive mean is fairly reliable against deleting.

3.2 Measuring the Relative Weight of the Approximation

An exact computation of the posterior would lead to a multi-modal distribution of wind fields at each data-point. This would correspond to a mixture of GPs as a posterior rather than to a single GP that is used in our approximation. If the individual components of the mixture are well separated, we may expect that our online algorithm will track modes with significant underlying probability mass to give a relevant prediction. However, this will depend on the actual sequence of data-points that are visited by the algorithm. To investigate the variation of our wind field prediction with the data sequence, we have generated many random sequences and compared the outcomes based on a simple approximation for the relative mass of the multivariate Gaussian component.

Assuming an online solution of the marginal distribution $(\hat{\underline{z}}, \hat{\Sigma})$ at a separated mode, we have the posterior at the local maximum expressed:

$$q(\hat{\underline{z}}) \propto \gamma_l \, (2\pi)^{-2N/2} \, |\hat{\Sigma}|^{-1/2} \tag{9}$$

with $q(\hat{\underline{z}})$ from eq. (2), γ_l *the weight of the component* of the mixture to which our online algorithm had converged, and we assume the local curvature is also well approximated by $\hat{\Sigma}$.

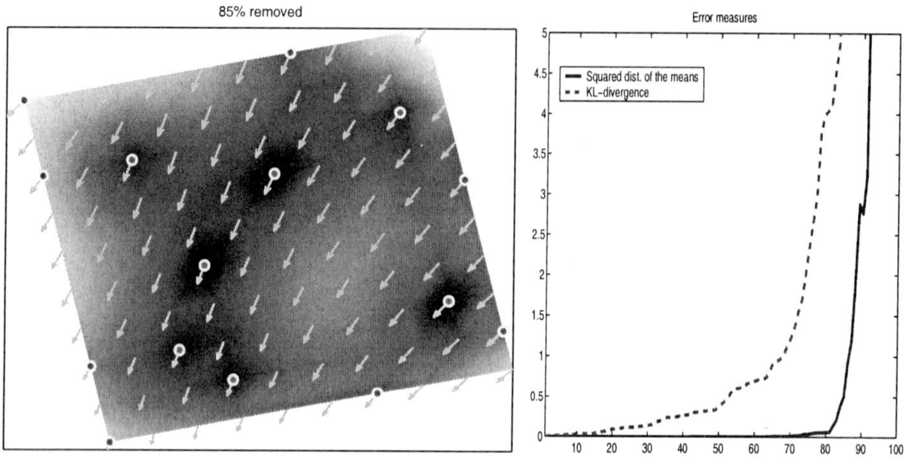

(a) (b)

Fig. 4. (a) The predicted wind fields when 85% of the modes has been removed (from Fig. 2). The prediction is based only on basis vectors (circles). The model confidence is higher at these regions. (b) The difference between the full solution and the approximations using the squared difference of means (continuous line) and the KL-distance (dashed line) respectively.

Having two different online solutions $(\hat{\underline{\mathbf{z}}}_1, \hat{\mathbf{\Sigma}}_1)$ and $(\hat{\underline{\mathbf{z}}}_1, \hat{\mathbf{\Sigma}}_1)$, we find from eq (9) that the proportion of the two weights is given by

$$\frac{\gamma_1}{\gamma_2} = \frac{q(\hat{\underline{\mathbf{z}}}_1)|\hat{\mathbf{\Sigma}}_1|^{1/2}}{q(\hat{\underline{\mathbf{z}}}_2)|\hat{\mathbf{\Sigma}}_2|^{1/2}} \tag{10}$$

This helps us to estimate, up to an additive constant, the "relative weight" of the wind field solutions, helping us to assess the quality of the approximation we arrived at. Results, using multiple runs on a wind field data confirm this expectation, the correct solution (Fig. 2.b) has large value and high frequency if doing multiple runs.

4 Discussion

In the wind field example the online and sparse approximation allows us to tackle much larger wind fields than previously possible. This suggests that we will be able to retrieve wind fields using only scatterometer observations, by utilising all available information in the signal.

Proceeding with the removal of the basis points, it would be desirable to have an improved update for the vector GP parameters that leads to a better estimation of the posterior kernel (thus of the Bayesian error-bars).

At present we obtain different solution for different ordering of the data. Future work might seek to build an adaptive classifier that works on the family of online solutions and utilising the relative weights.

However, a more desirable method would be to extend our online approach to mixtures of GPs in order to incorporate the multi-modality of the posterior process in a principled way.

Acknowledgement

This work was supported by EPSRC grant no. GR/M81608.

References

[Bernardo and Smith 1994] Bernardo, J. M. and A. F. Smith (1994). *Bayesian Theory.* John Wiley & Sons.

[Bishop 1995] Bishop, C. M. (1995). *Neural Networks for Pattern Recognition.* New York, N.Y.: Oxford University Press.

[Cressie 1991] Cressie, N. A. (1991). *Statistics for Spatial Data.* New York: Wiley.

[Csató and Opper 2001] Csató, L. and M. Opper (2001). Sparse representation for Gaussian process models. In T. K. Leen, T. G. Diettrich, and V. Tresp (Eds.), *NIPS*, Volume 13. The MIT Press. http://www.ncrg.aston.ac.uk/Papers.

[Evans *et al.* 2000] Evans, D. J., D. Cornford, and I. T. Nabney (2000). Structured neural network modelling of multi-valued functions for wind retrieval from scatterometer measurements. *Neurocomputing Letters 30*, 23–30.

[Kimeldorf and Wahba 1971] Kimeldorf, G. and G. Wahba (1971). Some results on Tchebycheffian spline functions. *J. Math. Anal. Applic. 33*, 82–95.

[Nabney *et al.* 2000] Nabney, I. T., D. Cornford, and C. K. I. Williams (2000). Bayesian inference for wind field retrieval. *Neurocomputing Letters 30*, 3–11.

[Offiler 1994] Offiler, D. (1994). The calibration of ERS-1 satellite scatterometer winds. *Journal of Atmospheric and Oceanic Technology 11*, 1002–1017.

[Opper 1998] Opper, M. (1998). A Bayesian approach to online learning. See Saad [1998], pp. 363–378.

[Opper and Winther [1999] Opper, M. and O. Winther (1999). Gaussian processes and SVM: Mean field results and leave-one-out estimator. In A. Smola, P. Bartlett, B. Schölkopf, and C. Schuurmans (Eds.), *Advances in Large Margin Classifiers*, pp. 43–65. Cambridge, MA: The MIT Press.

[Saad [1998] Saad, D. (1998). *On-Line Learning in Neural Networks.* Cambridge Univ. Press.

[Stoffelen and Anderson [1997a] Stoffelen, A. and D. Anderson (1997a). Ambiguity removal and assimiliation of scatterometer data. *Quarterly Journal of the Royal Meteorological Society 123*, 491–518.

[Stoffelen and Anderson [1997b] Stoffelen, A. and D. Anderson (1997b). Scatterometer data interpretation: Estimation and validation of the transfer function CMOD4. *Journal of Geophysical Research 102*, 5767–5780.

Fast Training of Support Vector Machines by Extracting Boundary Data

Shigeo Abe and Takuya Inoue

Graduate School of Science and Technology, Kobe University
Rokkodai, Nada, Kobe Japan
{abe,t-inoue}@eedept.kobe-u.ac.jp
http://www.eedept.kobe-u.ac.jp/eedept_english.html

Abstract. Support vector machines have gotten wide acceptance for their high generalization ability for real world applications. But the major drawback is slow training for classification problems with a large number of training data. To overcome this problem, in this paper, we discuss extracting boundary data from the training data and train the support vector machine using only these data. Namely, for each training datum we calculate the Mahalanobis distances and extract those data that are misclassified by the Mahalanobis distances or that have small relative differences of the Mahalanobis distances. We demonstrate the effectiveness of the method for the benchmark data sets.

1 Introduction

Support vector machines are based on the theoretical learning theory developed by Vapnik [1], [2], [3, pp. 47–61]. In support vector machines, an n-class problem is converted into n two-class problems in which one class is separated from the remaining classes. For each two-class problem, the original input space is mapped into the high dimensional dot product space called feature space and in the feature space, the optimal hyperplane that maximizes the generalization ability from the standpoint of the VC dimension is determined.

The high generalization ability compared to other methods has been shown for many applications but the major problem is slow training especially when the number of training data is large. Therefore, many methods for speeding up training have been proposed [2]. If support vectors are known in advance, training of support vector machines can be accelerated using only those data as the training data. Thus, in this paper we calculate the Mahalanobis distances for each data and estimate, as candidates of the boundary data, the training data that are misclassified by the Mahalanobis distances or that have small relative differences of the Mahalanobis distances. Finally, using two benchmark data sets we demonstrate the speedup of training by the proposed method.

2 Architecture of Support Vector Machines

Let m-dimensional inputs $\mathbf{x}_i$ $(i = 1, \ldots, M)$ belong to Class 1 or 2 and the associated labels be $y_i = 1$ for Class 1 and -1 for Class 2. If these data are

G. Dorffner, H. Bischof, and K. Hornik (Eds.): ICANN 2001, LNCS 2130, pp. 308–313, 2001.

linearly separable, we can determine the decision function:

$$D(\mathbf{x}) = \mathbf{w}^t\,\mathbf{x} + b, \tag{1}$$

where $\mathbf{w}$ is an m-dimensional vector, b is a scalar, and

$$y_i\,(\mathbf{w}^t\,\mathbf{x}_i + b) \geq 1 \quad \text{for} \quad i = 1, \ldots, M. \tag{2}$$

The hyperplane $D(\mathbf{x}) = \mathbf{w}^t\,\mathbf{x} + b = c$ for $-1 < c < 1$ forms a separating hyperplane that separates $\mathbf{x}_i$ $(i = 1, \ldots, M)$. The distance between the separating hyperplane and the training datum nearest to the hyperplane is called the margin. The hyperplane $D(\mathbf{x}) = 0$ with the maximum margin for $-1 < c < 1$ is called the optimal separating hyperplane.

Now consider determining the optimal separating hyperplane. The Euclidean distance from a training datum $\mathbf{x}$ to the separating hyperplane is given by $|D(\mathbf{x})|/\|\mathbf{w}\|$. Thus assuming the margin δ, all the training data must satisfy

$$\frac{y_k D(\mathbf{x}_k)}{\|\mathbf{w}\|} \geq \delta \qquad \text{for} \quad k = 1, \ldots, M. \tag{3}$$

If $\mathbf{w}$ is a solution, $a\mathbf{w}$ is also a solution where a is a scalar. Thus we impose the following constraint:

$$\delta\,\|\mathbf{w}\| = 1. \tag{4}$$

From (3) and (4), to find the optimal separating hyperplane, we need to find $\mathbf{w}$ with the minimum Euclidean norm that satisfies (2).

The data that satisfy the equality in (2) are called support vectors.

Now the optimal separating hyperplane can be obtained by minimizing

$$\frac{1}{2}\,\|\mathbf{w}\|^2 \tag{5}$$

with respect to $\mathbf{w}$ and b subject to the constraints:

$$y_i\,(\mathbf{w}^t\,\mathbf{x}_i + b) \geq 1 \qquad \text{for} \qquad i = 1, \ldots, M. \tag{6}$$

The number of variables for the convex optimization problem given by (5) and (6) is the number of features plus 1: $m + 1$. We convert (5) and (6) into the equivalent dual problem whose number of variables is the number of training data.

First we convert the constrained problem given by (5) and (6) into the unconstrained problem:

$$Q(\mathbf{w}, b, \alpha) = \frac{1}{2}\,\mathbf{w}^t\,\mathbf{w} - \sum_{i=1}^{M} \alpha_i\,\{y_i\,(\mathbf{w}^t\,\mathbf{x}_i + b) - 1\}, \tag{7}$$

where $\alpha = (\alpha_1, \ldots, \alpha_M)^t$ is the Lagrange multiplier. The optimal solution of (7) is given by the saddle point where (7) is minimized with respect to $\mathbf{w}$ and b and it is maximized with respect to α_i (≥ 0). Then, we obtain the following dual problem. Namely, maximize

$$Q(\alpha) = \sum_{i=1}^{M} \alpha_i - \frac{1}{2} \sum_{i,j=0}^{M} \alpha_i \alpha_j y_i y_j \mathbf{x}_i^t \mathbf{x}_j \tag{8}$$

with respect to α_i subject to the constraints

$$\sum_{i=1}^{M} y_i \alpha_i = 0, \alpha_i \geq 0 \quad \text{for} \quad i = 1, .., M. \tag{9}$$

Solving (8) and (9) for α_i ($i = 1, \ldots, M$), we can obtain the support vectors for Classes 1 and 2. Then the optimal hyperplane is placed at the equal distances from the support vectors for Classes 1 and 2.

To allow the data that do not have the maximum margin to exist, we introduce the nonnegative slack variables into (2). The resulting optimization problem is similar to the above formulation. The difference is the addition of the upper bound C for α_i.

If the original input $\mathbf{x}$ are not sufficient to guarantee linear separability of the training data, the obtained classifier may not have high generalization ability although the hyperplanes are determined optimally. Thus to enhance linear separability, in the support vector machines, the original input space is mapped into a high-dimensional dot product space called feature space using the kernel function that satisfies Mercer's condition. The kernel functions used in this paper are 1) polynomials with the degree of d: $H(\mathbf{x}, \mathbf{x}') = (\mathbf{x}^t \mathbf{x}' + 1)^d$, and 2) radial basis functions: $H(\mathbf{x}, \mathbf{x}') = \exp(-\gamma \|\mathbf{x} - \mathbf{x}'\|)$.

3 Speeding-up Training by Extracting Boundary Data

According to the architecture of the support vector machine, only the training data that are near the boundaries are necessary. In addition, since the training time becomes longer as the number of training data increases, the training time is shortened if the data that are far from the boundary are deleted. Therefore, if we can delete unnecessary data from the training data efficiently prior to training, we can speed up the training. In the following, we estimate the data that are near the boundaries using the classifier based on the Mahalanobis distance [4] and extracting the misclassified data and the data that are near the boundaries.

3.1 Approximation of Boundary Data

The decision boundaries of the classifier using the Mahalanobis distance are expressed by the polynomials, of the input variables, with the degree of two. Therefore, the boundary data given by the classifier are supposed to well approximate the boundary data for the support vector machine, especially with the polynomials with the degree of two as kernel functions.

For the class i data $\mathbf{x}$, the Mahalanobis distance $d_i(\mathbf{x})$ is given by

$$d_i^2(\mathbf{x}) = (\mathbf{c}_i - \mathbf{x})^t Q_i^{-1} (\mathbf{c}_i - \mathbf{x}), \tag{10}$$

where $\mathbf{c}_i$ and Q_i are the center vector and the covariance matrix for the data belonging to class i, respectively:

$$\mathbf{c}_i = \frac{1}{|X_i|} \sum_{\mathbf{x} \in X_i} \mathbf{x}, \tag{11}$$

$$Q_i = \frac{1}{|X_i|} \sum_{\mathbf{x} \in X_i} (\mathbf{x} - \mathbf{c}_i)\,(\mathbf{x} - \mathbf{c}_i)^t. \tag{12}$$

Here, X_i denotes the set of data belonging to class i and $|X_i|$ is the number of data in the set. The data $\mathbf{x}$ is classified into the class with the minimum Mahalanobis distance. The most important feature of the Mahalanobis distance is that it is invariant for linear transformation of input variables. Therefore, we do not worry about the scaling of each input variable.

For the datum belonging to class i, we check whether

$$r(\mathbf{x}) = \frac{\min\limits_{j \neq i, j=1,\ldots,n} \quad d_j(\mathbf{x}) - d_i(\mathbf{x})}{d_i(\mathbf{x})} \leq \eta \tag{13}$$

is satisfied, where $r(\mathbf{x})$ is the relative difference of distances, $\eta\,(> 0)$ controls the nearness to the boundary. If $r(\mathbf{x})$ is negative, the datum is misclassified. We assume the misclassified data are near the decision boundary. Inequality (13) is satisfied when the second minimum Mahalanobis distance is shorter than or equal to $(1 + \eta)\, d_i(\mathbf{x})$ when the datum is correctly classified.

In extracting boundary data, we set some appropriate value to η and for each class we select the boundary data that are at least equal to or more than the prespecified minimum number N_{min} and that are equal to or smaller than the maximum number N_{max}. Here the minimum number is set so that the number of boundary data is not too small for some classes because the data that satisfy (13) are scarce. The maximum number is set not to allow too many data to be selected. The general procedure for extracting boundary data is as follows.

1. Calculate the centers and covariance matrices for all the classes using (11) and (12).
2. For the training datum $\mathbf{x}$ belonging to class i, we calculate $r(\mathbf{x})$ and we put the data into the stack for class i, S_i, whose elements are sorted in the increasing order of the value of $r(\mathbf{x})$ and whose maximum length is N_{max}. We iterate this for all the training data.
3. If the stack S_i includes more than N_{min} data that satisfy (13), we select these data as the boundary data for class i. Otherwise, we select the first N_{min} data as the boundary data.

3.2 Performance Evaluation

Although the performance varies as kernels vary, the polynomial kernels with the degree of two performed relatively well. Thus in the following, unless otherwise stated, we use the polynomials with the degree of two as the kernel functions in evaluating the iris data [5] and blood cell data [6].

We ran the software developed by Royal Holloway, University of London [7] on a SUN UltraSPARC-IIi (335MHz) workstation. The software used the pairwise classification [8] to resolve unclassified regions that arise by the original two-class formulation.

Iris Data. Since the number of the iris data is small, we checked only the lowest rankings, in the relative difference of the Mahalanobis distances, of support vectors for the pairs of classes. Table 1 lists the results when the boundary data were extracted for each class. The numeral in the ith row and the jth column shows the lowest ranking of the support vector, belonging to class i, that separate class i from class j. The diagonal elements show the number of training data for the associated class. The maximum value among lowest rankings was 8, which was smaller than half the number of class data. Thus, the relative difference of the Mahalanobis distances well reflected the boundary data.

Table 1. The lowest rankings of support vectors for the iris data

Class	1	2	3
1	(25)	1	2
2	8	(25)	3
3	2	3	(25)

Blood Cell Data. We set $N_{\max}$ as the half of the maximum number of class data, namely 200. And we set $N_{\min} = 50$ and evaluated the performance changing η. Table 2 lists the results for the blood cell data. When $\eta \geq 1$, sufficiently good recognition rates were obtained for the test data and training was speeded up two to three times. (The numerals in the brackets in the "Rates" column show the recognition rates of the training data.)

Table 3 lists the speed-up of training for different kernels when $\eta = 2.0$. For each kernel, the upper row shows the results using all the training data and the lower row shows the results using the extracted boundary data. For different kernels, training was speeded up about two times and the recognition rates of the test data were almost the same.

4 Conclusions

We discussed fast training of support vector machines by extracting boundary data that are determined by the relative differences of the Mahalanobis distances.

Table 2. Performance for the blood cell data

η	Data	Rates (%)	Time (s)	Speedup
0.5	1136	90.81 (97.45)	96 (2)	9.4
1.0	1693	92.06 (99.61)	266 (2)	3.4
1.5	1978	92.10 (99.29)	390 (2)	2.4
2.0	2102	92.13 (99.29)	448 (2)	2.1
—	3097	92.13 (99.32)	924	1

Table 3. Performance for the blood cell data for different kernels ($\eta = 2.0$)

Kernel	Parameter	Rates (%)	Time (s)	Speedup
Polynomial	$d = 3$	91.94 (99.94)	937	1
		92.00 (99.81)	461	2.0
	$d = 4$	92.10 (100)	948	1
		92.10 (99.90)	471	2.0
RBF	$\gamma = 1$	92.13 (100)	2736	1
		92.13 (99.97)	1331	2.1
	$\gamma = 0.1$	92.16 (100)	2799	1
		92.13 (99.97)	1387	2.0

The computer simulations using the iris data and blood cell data showed that by this method the boundary data were efficiently extracted and training was speeded up about two times for the blood cell data.

References

1. V. N. Vapnik. *Statistical Learning Theory.* John Wiley & Sons, 1998.
2. B. Schölkopf, C. J. C. Burges, and A. J. Smola, editors. *Advances in Kernel Methods: Support Vector Learning.* The MIT Press, 1999.
3. S. Abe. *Pattern Classification: Neuro-fuzzy Methods and Their Comparison.* Springer-Verlag, 2001.
4. R. O. Duda and P. E. Hart. *Pattern Classification and Scene Analysis.* John Wiley & Sons, 1973.
5. R. A. Fisher. The use of multiple measurements in taxonomic problems. *Annals of Eugenics*, 7:179–188, 1936.
6. A. Hashizume, J. Motoike, and R. Yabe. Fully automated blood cell differential system and its application. In *Proc. IUPAC Third International Congress on Automation and New Technology in the Clinical Laboratory*, pages 297–302, 1988.
7. C. Saunders, M. O. Stitson, J. Weston, L. Bottou, B. Schölkopf, and A. Smola. Support vector machine reference manual. Technical Report CSD-TR-98-03, Royal Holloway, University of London, 1998 (http://svm.cs.rhbnc.ac.uk/).
8. U. H.-G. Kreßel. Pairwise classification and support vector machines. In B. Schölkopf, C. J. C. Burges, and A. J. Smola, editors, *Advances in Kernel Methods: Support Vector Learning*, pages 255–268. The MIT Press, 1999.

Multiclass Classification with Pairwise Coupled Neural Networks or Support Vector Machines

Eddy Nicolas Mayoraz

Human Interface Lab (HIL), Motorola Labs
3145 Porter Drive, Palo Alto, CA 94304, USA
Eddy.Mayoraz@Motorola.com

Abstract. Support Vector Machines (SVMs) are traditionally used for multi-class classification by introducing for each class one SVM trained to distinguish the associated class from all the others. In a recent experiment, we attempted to solve a K-class problem using a similar decomposition with K feedforward binary neural networks. The disappointing results were explained by the fact that neural networks suffer from datasets with a strongly unbalanced class distribution. By opposition to *one-per-class*, *pairwise coupling* introduces one binary classifier for each pair of classes and does not degrade the original class distribution. A few papers report evidences that pairwise coupling gives better results for SVMs than one-per-class. This issue is revisited in this paper where one-per-class class and pairwise coupling decomposition schemes used with both, SVMs and neural networks, are compared on a real life problem. Various methods for aggregating the results of pairwise classifiers are experimented. Beside our online handwriting application, experiments on some databases of the Irvine repository are also reported.

1 Introduction

The context of this work is supervised learning for automatic classification. Formally, the problem consists in finding an approximation $\hat{F}$ of an unknown labelling function F defined from an *input space* Ω onto an unordered set of labels $1, \ldots, K\}$, given a *training set*: $T = \{(\boldsymbol{x}^n, F(\boldsymbol{x}^n))\}_{n=1}^N$. F defines a K-partition of the input space into sets $F^{-1}(k)$ called *classes* and denoted ω_k.

Among the wide variety of methods available to solve such a problem, some of them, *e.g.* perceptron, polynomial classifiers, support vector machine (SVM), are restricted to the discrimination between two classes only. Other algorithms can handle more than two classes but some do not scale up efficiently with the size of the training set or with the number of classes. As many real applications translate into classification problems with a large number of classes and a huge number of data, several techniques have been proposed to decompose a K-class classification problem into a series of smaller 2-class problems. Many studies have demonstrated that even when using an approach which can deal with large scale problems, an adequate decomposition of the classification problem into subproblems can be favorable to the overall computational complexity as well as to the generalization ability of the global classifier [8, 3, 19, 7].

G. Dorffner, H. Bischof, and K. Hornik (Eds.): ICANN 2001, LNCS 2130, pp. 314–321, 2001.

Several *decomposition schemes* have been proposed in the literature to transform a K-class partitioning F of Ω into a series $f_1, \ldots, f_L$ of L bipartitions or *dichotomies* $f_l : \Omega \rightarrow \{-1, +1\}$. A *reconstruction method* is coupled with each decomposition scheme for the selection of one of the K classes, given the answers of all dichotomies on a particular input data.

The simplest decomposition schemes are *one-per-class* (OPC) and *pairwise coupling* (PWC). A K-class classification problem is decomposed by the former method into K dichotomies, each of which separating one class from all the others. The latter will use $K(K-1)/2$ dichotomies, one for each pair of classes, the dichotomy for the pair $(\omega_k, \omega_{k'})$ focuses on the discrimination between classes ω_k and $\omega_{k'}$, ignoring all other classes. More elaborate schemes are proposed in [6, 10], where redundancy is exploited in the decomposition as a way of increasing the error-correcting capability of the reconstruction.

With OPC, the number of dichotomies is small, but each of them involves the whole training data. Moreover, the positive and the negative data are represented in a very unbalanced way in each dichotomy, which can be a problem for some classification methods. In PWC, each dichotomy has a balanced training sample whose size is in average $\frac{K}{2}$ times smaller than for OPC, but the number of dichotomies is $\frac{K-1}{2}$ times larger. In terms of computational time, the latter situation is favorable as most learning methods are super-linear in the training size. Moreover, it has been shown in different studies that if K is large, one can restrict ourselves to a small subset of the $K(K-1)/2$ dichotomies without loosing much in accuracy [9].

This work presents a comparison of OPC and PWC decomposition schemes for both, SVMs and feedforward neural networks in a real application, namely on-line Roman handwriting recognition. Different reconstruction schemes for PWC are evaluated. Beside handwriting recognition, the comparison is carried out on benchmark datasets of very different characteristics.

2 Pairwise Coupling Decomposition and Reconstruction

The application of a decomposition/reconstruction scheme can be formally expressed as decomposing the sought approximation $\hat{F}$ of the unknown function $F : \Omega \rightarrow \{\omega_1, \ldots, \omega_K\}$ as follows:

$$\Omega \xrightarrow{\hat{\boldsymbol{f}}} \mathbb{R}^L \xrightarrow{\sigma} \mathbb{R}^L \xrightarrow{m} \mathbb{R}^K \xrightarrow{\arg\max} \{1, \ldots, K\} \qquad (1)$$

where $\hat{f}_l, l = 1, \ldots, L$ is the hypothesis yielded by the lth classifier learning the lth dichotomy of the decomposition, and m is usually a linear mapping expressed by a matrix $\boldsymbol{M} \in \mathbb{R}^{L \times K}$ transforming the outputs of the L classifiers into K values from which a voting rule will pick the final class label. Some binary classifiers yield a Boolean output, others provide a real output that can express either a confidence, a distance, or a probability measure. Whenever the output is continuous, it can eventually be passed through a non-linear function, either to bound the values into a fix range, or to get a Boolean value. This is the role of

σ that is applied component-wise and which is typically the identical function, a sigmoidal function, or the sign function.

Some papers discuss decomposition/reconstruction methods in a context where the outputs of the classifiers are interpreted as probabilities, while others deal with positive and negative values. The two are equivalent through affine transform. Without loss of generality, it will be assumed hereafter that the outputs of $\hat{f}_l$ are centered around 0, the sign function gives values in $\{-1, 0, +1\}$ and whenever σ is a sigmoidal function, tanh is meant.

The decomposition, specifying the task of each dichotomy, can be conveniently described by a matrix $\boldsymbol{D} \in \{-1, 0, +1\}^{L \times K}$, where $D_{lk} = \pm 1$ means $\omega_k \subset f_l^{-1}(\pm 1)$, and $D_{lk} = 0$ indicates that ω_k is ignored by the dichotomy l. This decomposition matrix also provides a simple and natural choice for the reconstruction matrix $\boldsymbol{M}$.

For OPC, $\boldsymbol{D}$ has +1s on the diagonal and −1s everywhere else and $\boldsymbol{M} = \boldsymbol{D}$ produces the same rule than the identity matrix. When no a priori information on the classes is available, it is the only reasonable choice for $\boldsymbol{M}$.

For PWC and $\sigma = \text{sgn}$, the reconstruction $\boldsymbol{M} = \boldsymbol{D}$ corresponds to the following rule: for each class k, look only at the dichotomies involving ω_k, count how many of these classifiers decide for ω_k, and select the class with the highest score (ties are broken at random). This reconstruction rule is the most widely used in literature on PWC and will be referred to as *hard voting*. By opposition, *soft voting* denotes the case where $\boldsymbol{M} = \boldsymbol{D}$ and $\sigma = \tanh$ (or $\sigma =$ identity). A last version found in [9] is a compromise between the hard and the soft voting and will be refer to as the *fair voting*. The hard voting is applied primarily, and the soft voting is used only to break ties. The author reports that for large K, the chances of ties are rare, thus we expect that the fair rule will give almost the same results as the hard voting rule.

In [14], yet another rule is proposed, which is better expressed in probabilistic terms. If for a given sample $\boldsymbol{x} \in \Omega$, the output $\hat{p}_{kl}$ of the classifier discriminating between classes ω_k and ω_l is in $[0, 1]$ and is interpreted as the probability of $\boldsymbol{x} \in \omega_k$ given that $\boldsymbol{x} \in \omega_k \cup \omega_l$. If $\hat{p}_k$ gives the posterior probability of class ω_k for sample $\boldsymbol{x}$, then we have $\forall k \neq l,\ \hat{p}_{kl} = \frac{\hat{p}_k}{\hat{p}_k + \hat{p}_l}$. Resolving this system of equations for $\hat{p}_k$ one derives the following rule:

$$\hat{F}(\boldsymbol{x}) = \arg\max_k \hat{p}_k = \arg\min_k \sum_{l \neq k} \frac{1}{\hat{p}_{kl}} .$$

This method has been included in the experiments and is referred to as the probabilistic rule.

3 Feedforward Neural Networks and Support Vector Machines

It is a well known theoretical fact that multi-layered feedforward neural networks (MLPs) used for classification are approximating a posteriori probabilities when trained to optimize the mean square error or the cross-entropy functions [18,

2]. However, this is assuming ideal conditions such as: the network has enough parameters, it is trained to converge to a global optimum, the training session has infinitely many data and the a priori probabilities of the test dataset are correctly reflected in the training dataset. In practice, the probabilistic interpretation of neural network outputs has to be made with a lot of care.

Considering the general form for a decomposition/reconstruction scheme presented in equation (1), the analogy with a last layer of an MLP used as K-class classifier is obvious. If in turns the classifiers solving the dichotomies are also MLPs of say depth h, the overall structure is an MLP of depth $h+1$ with a last hidden layer made of L units having an activation function σ. The most important difference is that the units of the last hidden layer are trained to perform some tasks specified ahead of time. So, one could argue that this approach will never perform better than training a single larger MLP. Such argument are invalidated mostly by two considerations. First, the learning in an MLP is sub-optimal, specially in a deep MLP. Second, a large MLP is more likely to overfit the data than a similar architecture with constraints hidden units.

The most famous applications of SVMs for multi-class classification have been using the OPC decomposition scheme [16]. Some works report experiments of SVMs using PWC [15, 9, 13, 17]. A potential problem occurring when using SVMs in any decomposition scheme is due to the unnormalized output ranges of each SVM [11]. This is a potentially serious problem when OPC decomposition is used, as the final voting operates usually on the outputs of the SVMs directly (*i.e.* σ in (1) is the identity function). With PWC, it is essential to use a saturating (non-linear) function as σ such as sgn or tanh, otherwise the approach breaks down completely.

4 Empirical Evaluations and Comparisons

This work was primarily carried out in the context of handwriting recognition, using some of Motorola's databases for on-line Roman handwriting. In order to measure and compare these different approaches in various contexts, we report also some numerical experimentations on smaller, widely used datasets from the Irvine Machine Learning Repository [12].

4.1 Databases

From the Irvine repository, we selected databases of small size (given that the other in-house databases used were large), without missing data and with different characteristics: `glass`, `segmentation`, `ecoli`, `yeast` and `vowel`.

The handwriting databases used in this work represent online handwritten isolated Roman characters, from different writers, from different countries. They result from a preprocessing yielding 19 attributes which were also centered and normalized. Three different benchmarks are considered: `dg` has 91K data equally distributed among 10 classes (10 digits), and `AZ`, respective `az`, contains 200K (resp. 250K) data equally distributed in the 26 classes of upper-case (resp. lower-case) alphabet. Our prior knowledge on these benchmarks is that `dg` is much easier that the other two, and the lower-case problem is slightly more complex than the upper-case one.

4.2 Experiments

The experiments one the Irvine databases were done according to a repetition of five 2-fold cross-validations (denoted 5x2cv in [5]). So, every number presented below is a mean ±standard deviation over 10 runs. The handwriting databases being large, we did only a single 3-fold cross validation. Every class was split once into three equal parts in such a way that the instances of a same writer would all be into the same part (to ensure strict writer independence between training sets and test sets). Thus, all values are averages and standard deviations over 3 runs.

The meta-parameters such as the number of hidden units, the initial learning rate and the annealing process for the MLP, and the standard deviation of the Gaussian (only RBF kernels where experimented here) and the weight C of the regularizing term for the SVM, have been chosen as follows. In the experiments with OPC they were selected, independently for each dataset, through a simple try-and-error process and fixed to the same value for every one of the K binary classifiers. In PWC the characteristics of each of the $\frac{K(K-1)}{2}$ training sets can vary a lot, therefore we tailored these meta-parameters to each binary problem individually.

SVMtorch has been used in all SVMs simulations [4]. The experiments with neural networks have been done with our own software implementing a vanilla backpropagation. In this work, the focus is placed on the accuracy, regardless of the complexity of each system. It is clear that the SVMs constructed are always much more complex that the MLPs, both in terms of memory and in terms of evaulation time. For example, for the handwriting recognition tasks, the best single MLPs found had 200 hidden units, while for the SVMs based on OPC, the total number of support vectors (counting only once the support vectors used in different machines) was typically of several thousands.

4.3 Results

Table 1 summarize the results obtained. The results of a single MLP are reported in the column 'single'. Three reconstruction methods discussed in Section 2 are reported here. Standard deviations are indicated in small fonts. For the soft voting rule, a tanh was used with four different sloaps, and only the best of the three results is reported here. The sloap has an important influence of the efficiency of the method. If it is too high, the soft voting is identical to the fair voting. If it is too low, than the performance drops and in the case of SVM it can get really poor. This is easily explained by the fact that the outputs of the SVM are in a wide range (this is even worse with polynomial kernels) and mostly only their comparison towards -1, 0 and +1 is meaningful..

Except for `segmentation` and `yeast`, SVMs are always outperforming MLPs, and sometimes very significantly (`vowel`, `dg`, `AZ`, and `az`). Replacing a single MLP

Table 1. Means and standard deviations of percentage of error on test sets.

Db	MLP					SVM			
	single	OPC	PWC			OPC	PWC		
			fair	soft	prob		fair	soft	prob
glass	$31.22_{4.3}$	$33.83_{4.0}$	$35.89_{3.3}$	$35.42_{3.9}$	$35.42_{3.6}$	$30.75_{3.1}$	$31.68_{2.8}$	$31.87_{2.7}$	$32.06_{3.0}$
segm.	$3.23_{0.2}$	$3.97_{0.4}$	$5.10_{0.9}$	$5.09_{0.9}$	$5.14_{1.0}$	$4.03_{0.6}$	$3.68_{0.3}$	$3.62_{0.3}$	$3.66_{0.2}$
ecoli	$14.81_{3.1}$	$15.94_{3.1}$	$17.31_{3.0}$	$17.54_{3.0}$	$18.33_{2.0}$	$12.37_{3.1}$	$13.62_{3.3}$	$13.69_{3.3}$	$13.80_{3.2}$
yeast	$39.94_{1.5}$	$47.00_{2.0}$	$43.77_{1.5}$	$43.60_{1.6}$	$44.06_{1.4}$	$40.53_{1.1}$	$41.24_{1.9}$	$41.22_{1.9}$	$41.52_{1.7}$
vowel	$10.42_{2.9}$	$14.16_{2.6}$	$23.13_{2.8}$	$23.19_{2.9}$	$22.97_{2.9}$	$4.42_{1.6}$	$4.04_{1.9}$	$3.98_{1.9}$	$4.08_{1.9}$
dg	$1.37_{0.0}$	$1.32_{0.0}$	$3.08_{0.3}$	$3.06_{0.4}$	$3.10_{0.4}$	$0.95_{0.1}$	$0.98_{0.1}$	$0.95_{0.1}$	$0.96_{0.1}$
AZ	$3.41_{0.3}$	$3.62_{0.3}$	$6.22_{0.4}$	$6.18_{0.4}$	$6.11_{0.4}$	$2.44_{0.1}$	$2.63_{0.1}$	$2.50_{0.1}$	$2.89_{0.1}$
az	$5.21_{0.1}$	$6.18_{0.2}$	$8.15_{0.1}$	$8.12_{0.1}$	$8.04_{0.1}$	$4.23_{0.1}$	$4.29_{0.1}$	$4.27_{0.1}$	$4.33_{0.1}$

by many binary ones, either with OPC or with PWC never helps, except for dg, were a slight improvement was obtained with OPC. This is a surprise however, as the parameters of the single MLP (number of hidden units, initial learning rate, annealing scheme) had been tailored with a great care for this problem in particular.

For both, MLPs and SVMs, OPC often outperforming PWC. This was a suprise to us, specially given that the meta-parameters where adjusted independently for each binary classifier involved in PWC, and that we fine tuned as well the reconstruction method. If it could be justified by too small datasets in the case of Irvine databases, this argument does not hold in the case of handwriting recognition. For dg there is no significant difference, but in the case of AZ and az, the best results obtained with SVM and PWC is worse that OPC, with confidence 0.95, according to the 5x2cv F test [1]. This is in contradiction with previous works showing that PWC always outperforms OPC [9, 17].

After the first submission of the present paper, as an attempt to understand this disagreement between our results and the cited work, we carried out new experiments replacing the RBF kernels by polynomial kernels, which were used in [9, 17]. As a matter of fact, this change degraded the results with one-per-class far below the results with pairwise coupling, which stayed almost unchanged. There is a fundamental difference in nature between these two types of kernels, which makes them appropriate or not for one-per-class or pairwise coupling decomposition schemes. This issue requires further investigation, but a simple explanation of the fact that pairwise coupling does not improve on one-per-class when RBF kernels are used could be that for inputs with no similarities with the ones seen during training (often the case with pairwise coupling), the SVM output is close to zero (does not buy us anything) with RBF kernels, but has a large magnitude for polynomial kernels.

5 Conclusions and Further Research

In this paper, we compare two simple ways of expressing a multi-class classification problem into several 2-class problems. The potentials of such approaches to enhance the classification accuracy of multi-layered neural networks is evaluated.

This is motivated by the excellent results obtained by SVMs which, in order to handle multi-class problems, are used through the same decomposition principles. Among the advantages of training many classifiers on simpler tasks, we can mention the possibility to validate each sub-system individually and fine-tune their training parameters.

The conclusions are three-fold. First, it is not easy to improve the performance of a single neural network by using one binary neural network for each dichotomy of a decomposition scheme. Even though we spent a lot of effort adjusting the training of each individual binary net, retraining the ones whose performances where not completely satisfactory, we succeeded only once out of 8 benchmarks, to achieve significantly better results than a single network.

Second, the common belief that when using SVMs, pairwise coupling always outperforms one-per-class has been revisited. If this seems to be true when polynomial kernels are used, it does not hold at all for RBF kernels, which give scores slightly in favor of one-per-class over pairwise coupling.

Finally, SVMs always outperform neural networks. However, in our experiments the size of the models was deliberately not an issue and each model has been dimensioned to there best recognition performances. The minor improvement on the recognition rates given by SVMs over neural networks costs often one or two order of magnitude in size. It would be interesting to see how the results of SVMs degrade when we progressively suppress the support vectors of lowest coefficients.

Acknowledgements

The author is thankful to Vincent Wan for initiating this research during his visit at Motorola. This work has been made possible by the use of SVM-Torch, which can handle efficiently quite large datasets. The author is also thankful to his colleague Giovanni Seni for providing an efficient code for back-propagation of feedforward neural networks.

References

1. Ethem Alpaydin. Combined 5×2 cv f test for comparing supervised classification learning algorithms. *Neural Computation*, 11(8):1885–1892, 1999.
2. Hervé A. Bourlard and Nelson Morgan. Links between markov models and multilayer perceptrons. In Ed. D. S. Touretzky, editor, *Advances in Neural Information Processing Systems 1 (NIPS*88)*, volume 1, pages 502–510, San Mateo, CA, 1989. Morgan Kaufmann.
3. Pierre J. Castellano, Stefan Slomka, and Sridha Sridharan. Telephone based speaker recognition using multiplt binary classifier and Gaussian mixture models. In *ICASSP*, volume 2, pages 1075–1078. IEEE Computer Society Press, 1997.
4. Ronan Collobert and Samy Bengio. Support vector machines for large-scale regression problems. *Journal of Machine Learning Research*, 1:143–160, 2001. `http://www.idiap.ch/learning/SVMTorch.html`.

5. T. G. Dietterich. Approximate statistical tests for comparing supervised classification learning algorithms. *Neural Computation*, 7, 1998.
6. T. G. Dietterich and G. Bakiri. Error-correcting output codes: A general method for improving multiclass inductive learning programs. In *Proceedings of AAAI-91*, pages 572–577. AAAI Press / MIT Press, 1991.
7. Dominique Genoud, Miguel Moreira, and Eddy Mayoraz. Text dependent speaker verification using binary classifiers. IDIAP-RR 8, IDIAP, 1997. To appear in the Proceedings of the International Conference on Automatic Speech and Signal Processing, ICASSP'98.
8. Stefan Knerr, Leon Personnaz, and Gerard Dreyfus. Handwritten digit recognition by neural networks with single-layer training. *IEEE Trans. on Neural Networks*, 3(6):962–968, November 1992.
9. Ulrich Kreßel. Pairwise classification and support vector machines. In B. Schlkopf, C. J. C. Burges, and A. J. Smola, editors, *Advances in Kernel Methods — Support Vector Learning*. MIT Press, 1999.
10. Eddy Mayoraz and Miguel Moreira. On the decomposition of polychotomies into dichotomies. In Douglas H. Fisher, editor, *The Fourteenth International Conference on Machine Learning*, pages 219–226, 1997.
11. Eddy Mayoraz and Ethem Alpaydin. Support vector machine for multiclass classification. In *Proceedings of the International Workshop on Artificial Neural Networks (IWANN'99)*, 1998. ftp://ftp.idiap.ch/pub/reports/1998/rr98-06.ps.gz.
12. C. J. Merz and P. M. Murphy. UCI repository of machine learning databases. Technical report, Irvine, CA: University of California, Department of Information and Computer Science, 1998. Machine-readable data repository http://www.ics.uci.edu/~mlearn/MLRepository.html.
13. J. Platt, N. Cristianini, and J. Shawe-Taylor. Large margin dags for multiclass classification. In S.A. Solla, T.K. Leen, and K.-R. Muller, editors, *Advances in Neural Information Processing Systems 12 (NIPS*1999)*. MIT Press, 2000.
14. David Price, Stefan Knerr, Leon Personnaz, and Gerard Dreyfus. Pairwise neural network classifiers with probabilistic outputs. In G. Tesauro, D. Touretzky, and T. Leen, editors, *Advances in Neural Information Processing Systems 7 (NIPS*94)*, volume 7, pages 1109–1116. The MIT Press, 1995.
15. M. Schmidt and H. Gish. Speaker identification via support vector classifiers. In *Proceedings of ICASSP'96*, pages 105–108, Atlanta, GA, 1996.
16. B. Schölkopf, C. Burges, and V. Vapnik. Extracting support data for a given task. In U. M. Fayyad and R. Uthurusamy, editors, *Proceedings of the First International Conference on Knowledge Discovery and Data Mining*, pages 252–257. AAAI Press, 1995.
17. Fusi Wang, Louis Vuurpijl, and Lambert Schomaker. Support vector machines for the classification of western handwritten capitals. In *Seventh International Workshop on Frontiers in Handwriting Recognition Proceedings*, pages 167–176, 2000.
18. H. White. Learning in artificial neural networks: A statistical perspective. *Neural Computation*, 1(4):425–464, 1989.
19. Stephen A. Zahorian, Peter Silsbee, and Xihong Wang. Phone classification with segmental features and a binary-pair partitioned neural network classifier. In *ICASSP*, volume 2, pages 1011–1014. IEEE Computer Society Press, 1997.

Incremental Support Vector Machine Learning: A Local Approach

Liva Ralaivola and Florence d'Alché–Buc

Laboratoire d'Informatique de Paris 6,
Université Pierre et Marie Curie,
8, rue du Capitaine Scott, F-75015 Paris, France
{liva.ralaivola,florence.dalche}@lip6.fr

Abstract. In this paper, we propose and study a new on-line algorithm for learning a SVM based on Radial Basis Function Kernel: Local Incremental Learning of SVM or LISVM. Our method exploits the "locality" of RBF kernels to update current machine by only considering a subset of support candidates in the neighbourhood of the input. The determination of this subset is conditioned by the computation of the variation of the error estimate. Implementation is based on the SMO one, introduced and developed by Platt [13]. We study the behaviour of the algorithm during learning when using different generalization error estimates. Experiments on three data sets (batch problems transformed into on-line ones) have been conducted and analyzed.

1 Introduction

The emergence of smart portable systems and the daily growth of databases on the Web has revived the old problem of incremental and on-line learning. Meanwhile, advances in statistical learning have placed Support Vector Machines (SVM) as one of the most powerful family of learners (see [6, 16]). Their specificity lies on three characteristics: SVM maximizes a soft margin criterion, the major parameters of SVM (support vectors) are taken from the training sample and non linear SVM are based on the use of kernels to deal with high dimensional feature space without directly working in it.

However, few works tackle the issue of incremental learning of SVM. One of the main reasons lies on the nature of the optimization problem posed by SVM learning. Although there exist some very recent works that propose ways to update SVM each time new data are available [4, 11, 15], they generally imply to re-learn the whole machine.

The work presented here starts from another motivation: since the principal parameters of SVM are the training points themselves, and as far as a local kernel such as a gaussian kernel is used, it is possible to focus learning only on a neighbourhood of the new data and update the weights of concerned training data. In this paper, we briefly present the key idea of SVM and then introduce incremental learning problem. State of the art is shortly presented and discussed. Then, we present the local incremental algorithm or LISVM and discuss the

G. Dorffner, H. Bischof, and K. Hornik (Eds.): ICANN 2001, LNCS 2130, pp. 322–330, 2001.

model selection method to determine the size of the neighbourhood to be used at each step. Numerical simulations on IDA benchmark datasets [14] are presented and analyzed.

2 Support Vector Machines

Given a training set $\mathcal{S} = \{(\mathbf{x}_i, y_i)\}_{i=1}^{\ell}$, support vector learning [6] tries to find a hyperplane with minimal norm that separates the data mapped into a *feature space* Ψ *via* a nonlinear map $\Phi : \mathbb{R}^n \rightarrow \Psi$, where n denotes the dimension of vectors $\mathbf{x}_i$. To construct such a hyperplane, one must solve the following quadratic problem [3]:

$$\max_{\boldsymbol{\alpha}} \sum_{i=1}^{\ell} \alpha_i - \frac{1}{2} \sum_{i,j=1}^{\ell} \alpha_i \alpha_j y_i y_j k(\mathbf{x_i}, \mathbf{x_j}) \tag{1}$$

$$\text{subject to} \quad \sum_{i=1}^{l} \alpha_i y_i = 0 \quad \text{and} \quad 0 \leq \boldsymbol{\alpha} \leq C \tag{2}$$

with kernel function k defined as $k(\mathbf{x}, \mathbf{x}') = \Phi(\mathbf{x}) \cdot \Phi(\mathbf{x}')$. Solution of this problem provides $\mathbf{w} = \sum_{i=1}^{\ell} \alpha_i y_i \Phi(\mathbf{x}_i)$, a real value b with the help of optimality conditions and the decision rule: $f(\mathbf{x}) = sgn(\mathbf{w} \cdot \Phi(\mathbf{x}) + b)$.

The most efficient practical algorithms to achieve the resolution of the latter problem implement a strategy of iterative subsets selection [9, 12, 13] where only the points in those subsets may see their corresponding lagrange multiplier change. This process of optimizing the global quadratic problem only on a subset of the whole set of variables is a key point of our algorihtm since we use such an optimization scheme on subsets defined as neighbourhoods of new incoming data.

3 Incremental Learning and SVM

In incremental learning, the training dataset is not fully available at the beginning of the learning process as in batch learning. Data can arrive at any time and the hypothesis has to be updated if necessary to capture the class concept. In the following, we suppose that data are drawn from a fixed but unknown distribution. We view this task as a first step towards learning drifting concepts, e.g. learning to classify data that are drawn from a distribution that change over time.

More formally, the problem considered here may be stated as follows. Let H be a hypothesis family (such as the gaussain kernel-based SVM family). Let F be a fixed but unknown distribution over $(\mathbf{x}, y)$ pairs, $\mathbf{x} \in X$ and $y \in \{-1, 1\}$. We suppose that at time t, a training pattern is randomly sampled from F. Let us define S_t the current observed sample at time t. The goal of incremental on-line learning is to find and update an hypothesis $h^t \in H$ using available examples in order to minimize generalization error.

The Kernel-Adatron algorithm [8] is a very fast approach to approximate the solution of the support vector learning and can be seen as a component-wise optimization algorithm. It has been succesfully tested by their authors to dynamically adapt the kernel parameters of the machine, doing model selection in the learning stage. Nevertheless, the only drawback of this work is that it should not be straightforward to extend this work to deal with drifting concepts.

Another approach, proposed in [15], consists in learning new data by discarding all past examples except support vectors. The proposed framework thus relies on the property that support vectors summarize well the data and has been tested against some standard learning machine datasets to evaluate some goodness criteria such as *stability*, *improvement* and *recoverability*.

Finally, a very recent work [4] proposes a way to incrementally solve the global optimization problem in order to find the exact solution. Its reversible aspect allows to do "decremental" unlearning and to efficiently compute leave-one-out estimations.

4 Local Incremental Learning of a Support Vector Machine

We first consider SVM as a voting machine that combines the outputs of experts, each of which is associated with a support vector in the input space. When using RBF kernel or any kernel that is based upon the notion of neighbourhood, the influence of a support vector concerns only a limited area with a high degree. Then, in the framework of on-line learning, when a new example is available it should not be necessary to re-consider all the current experts but only those which are concerned by the localization of the input. In some extent, the proposed algorithm is linked with work of Bottou and Vapnik [1] about local learning algorithm.

4.1 Algorithm

We sketch the updating procedure to build h^t from h^{t-1} when the incoming data $(\mathbf{x_t}, y_t)$ is to be learned, $\mathcal{S}_{t-1}$ being the set of instances learned so far, and $\mathcal{S}_t = \mathcal{S}_{t-1} \cup \{(\mathbf{x_t}, y_t)\}$:

1. *Initialize lagrangian multiplier* α_t *to zero*
2. *If* $y_t f^{t-1}(\mathbf{x}_t) > 1$ *(point is well classified) then terminate (take* h^{t-1} *as new hypothesis* h^t*)*
3. *Build a working subset of size 2 with* x_t *and its nearest example in input space*
4. *Learn a candidate hypothesis* g *by optimizing the quadratic problem on examples in the working subset*
5. *If the generalization error estimation of* g *is above a given threshold* δ*, increase the working subset by adding the next closest point to* x_t *not yet in the current subset and return to step 4*
6. h^t *is set to* g

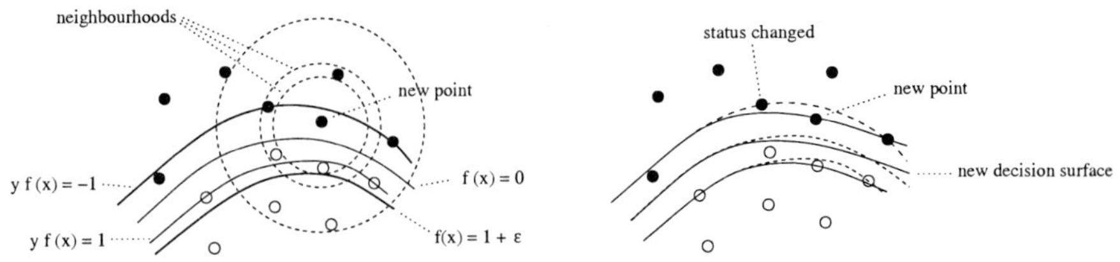

Fig. 1. *Left*: three neighbourhoods around the new data and the interesting small band of points around the decision surface parameterized by ε. *Right*: new decision surface and example of point whose status changed

We stop constructing growing neighbourhoods (see Fig. 1) around the new data as soon as the generalization error estimation falls under a given threshold δ. A compromise is thus performed between complexity in time and the value of the generalization error estimate, as we consider that this estimate is minimal when all data are re-learned. The key point of our algorithm thus lies on finding the size of the neighbourhood (e.g. the number K of neighbours to be considered) and thus on finding a well suited generalization error estimate, what we will focus on in the next section.

To increase computational speed, and implement our idea of locality, we only consider, as shown in Fig. 1, a small band around the decision surface in which points may be insteresting to re-consider. This band is defined upon a real ε: points of class y_t for which $y_t f(\mathbf{x}_t) \leq 1$ and points of class $-y_t$ for which $y_t f(\mathbf{x}_t) \leq 1 + \varepsilon$ are in the band.

4.2 Model Selection and the Neighbourhood Determinaton

The local algorithm requires to choose the value of K or the neighbourhood size. We have several choices to do that. A first simple solution is to fix K *a priori* before the beginning of the learning process. However the best K at time t is obviously not the best one at time $t + t'$. So it may be difficult to choose a single K suitable for all points. A second more interesting solution is therefore to determine it automatically through the minimization of a cost criterion. The idea is to apply some process of model selection upon the different hypothesis h_k^t that can be built.

The way we choose to select models consists in comparing them according to an estimate of their generalization error. One way to do that is to evaluate the estimation error on a test set and thus keeping K as the one for which E_{Test} is the least. In real problems, it is however not realistic to be provided with a test set during the incremental learning process. So this solution cannot be considered as a good answer to our problem.

Elsewhere, there exist some analytical expression of Leave-One-Out estimates of SVMs generalization error such as those recalled in [5]. However, in order to use these estimates, one has to ensure that the margin optimization problem has been solved exactly. The same holds for Joachims'$\xi\alpha-$estimators [10, 11]. This restriction prevents us from using these estimates as we only do a partial local optimization.

To circumvent the problem, we propose to use the bound on generalization provided by a result of Cristianini and Shawe-Taylor [7] for thresholded linear real-valued functions. While the bound it gives is large, it allows to "qualitatively" compare the behaviours of functions of the same family. The theorem states as follows:

Theorem 1. *Consider thresholding real-valued linear functions $\mathcal{L}$ with unit weight vectors on an inner product space X and fix $\gamma \in \mathbb{R}^+$. There is a constant c, such that for any probability distribution $\mathcal{D}$ on $X \times \{-1, 1\}$ with support in a ball of radius R around the origin, with probability $1 - \eta$ over ℓ random examples S, any hypothesis $f \in \mathcal{L}$ has error no more than*

$$\epsilon = err_{\mathcal{D}}(f) \leq B = \frac{c}{\ell}\left(\frac{R^2 + \|\boldsymbol{\xi}\|^2}{\gamma^2}\log^2 \ell + \log\frac{1}{\eta}\right) \tag{3}$$

where $\boldsymbol{\xi} = \boldsymbol{\xi}(\gamma, h, S) = (\xi_1, \xi_2, \ldots, \xi_\ell)$ is the margin slack vector with respect to f and γ defined as $\xi_i = \max(0, \gamma - y_i f(\mathbf{x}_i))$.

We notice that once the kernel parameter σ is fixed, this theorem, directly applied in the feature space Ψ_σ defined by the kernel k_σ, provides an estimate of generalization error for the machines we work on. This estimate is expressed in terms of a margin value γ, the norm of the slack margin vector $\boldsymbol{\xi}$ and the radius of the ball containing the data.

In order to use this theorem, we consider the feature space of dimension $d(K)$ defined by the Gaussian kernel with a fixed value for σ. In this space, we consider $\mathcal{L}$ with unit weight vectors. At step t, different functions h_k^t can be learnt with $k = 1, ..., K_{\max}$. For each k, we get a function f_k^t by normalizing the weight vector of h_k^t. f_k^t belongs to $\mathcal{L}$ and when thresholded provides the same outputs than h_k^t does. The theorem can then be applied to $f = f_k^t$ and data of S_t. It ensures that:

$$\epsilon \leq B(c(\gamma, \mathcal{L}), R_f, \xi_f, \gamma, t). \tag{4}$$

Hence, for each $k = 1...K_{\max}$, we can use this bound as a test error estimate.

However, as R_f is the radius of the ball containing the examples in the feature space [17], it only depends on the chosen kernel and not on k. On the contrary, ξ_f, defined as: $\xi_{i,k}^t = max(0, \gamma - y_i f_k^t(\mathbf{x}_i))$ is the unique quantity which differs among functions f_k. Slack vector ξ_k^t are thus sufficient to compare f_k functions, justifying our choice to use it as a model selection criterion. Looking at the bound, we can see that a value of γ must be chosen: in order to do that, we take a time-varying value defined as $\gamma_t = 1/\|\mathbf{w}_{t-1}\|$.

5 Experimental Results

Experiments were conducted on three different binary classification problems: *Banana* [14], *Ringnorm* [2] and *Diabetes*. Datasets are available at *www.first.gmd.*

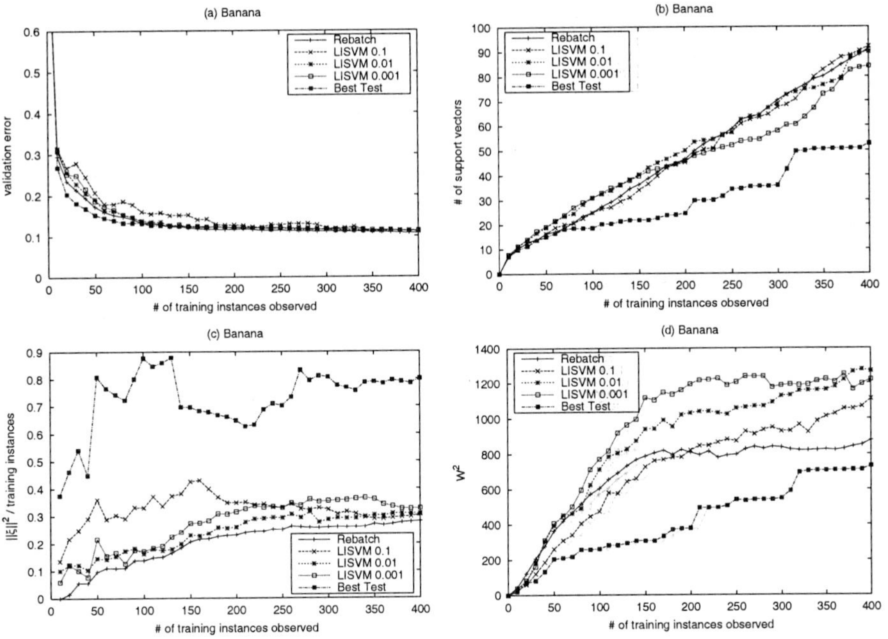

Fig. 2. Evolution of the machine parameters for the *Banana* problem during on-line learning with a potential band (see Fig. 1) defined by $\varepsilon = 0.5$

de/~raetsch/. For each problem, we tested LISVM for different values of the threshold δ. The main points we want to assess are the classification accuracy our algorithm is able to achieve, the appropriateness of the proposed criterion to select the "best" neighbourhood and the relevance of the local approach.

We simulated on-line incremental learning by providing the classifier with one example at a time, taken from a given training set. After each presentation, the current hypothesis is updated and evaluated on a validation set of size two thirds the size of the corresponding testing set. Only the "best test" algorithm uses the remaining third of this latter set to perform neighbourhood selection. An other incremental learning process called "rebatch" is also evaluated: it consists in realizing a classical SVM learning procedure over the whole dataset when a new instance is available. Experiments are run on 10 samples in order to compute means and standard deviations. Quantities of interest are plotted on Fig. 2 and Fig. 3. Table 1 reports the results at the end of the training process.

We led experiments for the same range of δ values in order to show that it is not difficult to fix a threshold that implies correct generalization performance. However, we must be aware that δ reflects our expectation of the error performed by the current hypothesis. Hence, smaller threshold δ should be preferred when the data are assumed to be easily separable (e.g. *Ringnorm*) while bigger values should be fixed when data are supposed to be harder to discriminate. This remark can be confirmed by the observation of *Banana* and *Diabetes* results. While

Table 1. Results on the three datasets. The potential band used (see Fig. 1) is defined by $\varepsilon = 0.5$. Classical SVM parameters are in the top-left cell of each table

Banana $C = 100$, $\sigma = 1$	validation error	Nb of Svs	$\mathbf{w}^2$	$\|\boldsymbol{\xi}(1, h, S)\|^2/\ell$
Batch/Rebatch	0.107 ± 0.004	91.2 ± 9.8	877 ± 166	0.281 ± 0.036
Best test	0.112 ± 0.013	52.7 ± 21.5	732 ± 232	0.801 ± 0.809
LISVM $\delta = 0.001$	0.113 ± 0.008	84.1 ± 18.8	1220 ± 291	0.328 ± 0.073
LISVM $\delta = 0.01$	0.113 ± 0.005	89.9 ± 9.4	1270 ± 244	0.309 ± 0.049
LISVM $\delta = 0.1$	0.111 ± 0.006	92.1 ± 7.9	1110 ± 220	0.305 ± 0.04

Ringnorm $C = 1e9$, $\sigma = 10$	validation error	Nb of Svs	$\mathbf{w}^2$	$\|\boldsymbol{\xi}(1, h, S)\|^2/\ell$
Batch/Rebatch	0.0263 ± 0.004	80.4 ± 8.5	3290 ± 701	$4.5e\text{-}6 \pm 4.5e\text{-}6$
LISVM $\delta = 0.001$	0.0272 ± 0.006	80.4 ± 8.5	3560 ± 809	0.007 ± 0.002
LISVM $\delta = 0.01$	0.0324 ± 0.008	65.9 ± 4	2180 ± 559	0.046 ± 0.012
LISVM $\delta = 0.1$	0.0628 ± 0.013	49.7 ± 5.8	1140 ± 500	0.155 ± 0.035

Diabetes $C = 10$, $\sigma = 20$	validation error	Nb of Svs	$\mathbf{w}^2$	$\|\boldsymbol{\xi}(1, h, S)\|^2/\ell$
Batch/Rebatch	0.228 ± 0.022	275 ± 7.1	382 ± 30.2	0.694 ± 0.017
LISVM $\delta = 0.001$	0.226 ± 0.023	197 ± 27.9	362 ± 70.2	0.646 ± 0.025
LISVM $\delta = 0.01$	0.229 ± 0.021	227 ± 36	382 ± 71.8	0.652 ± 0.031
LISVM $\delta = 0.1$	0.223 ± 0.023	226 ± 39.2	386 ± 66.4	0.652 ± 0.028

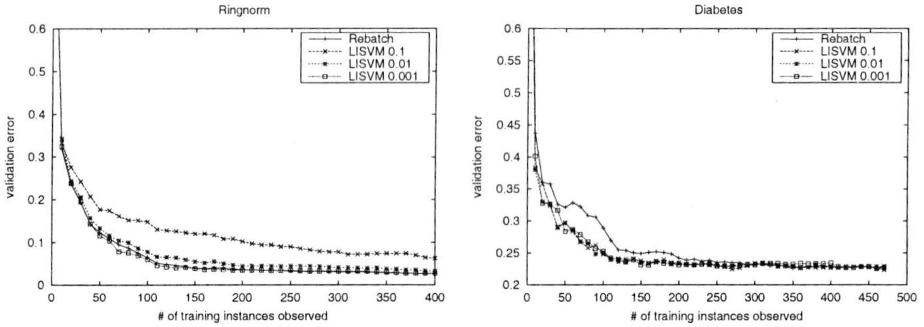

Fig. 3. Evolution of validation error for *Ringnorm* (*left*) and *Diabetes* (*right*)

the chosen values of δ lead to equivalent (or bigger) complexity than the batch algorithm for equivalent validation error, other experiments with $\delta = 0.5$ show that LISVM obtains a lower complexity (55 SVs and $\|\mathbf{w}\|^2 = 677$ for *Banana*, 179 SVs and $\|\mathbf{w}\|^2 = 927$ for *Diabetes*) but with a degraded performance on the validation set (rates of 0.16 and 0.24 respectively). For *Ringnorm*, this observation can also be made in the chosen range of values $[0.001; 0.01; 0.1]$. Relaxing the value of δ leads to a lower complexity at the cost of a higher validation error. These experiments confirm that the relevant range of δ correponds to a balance between a low validation error and a small number of neighbours needed to reach the threshold at each step. CPU time measures provide means to directly evaluate δ for which the local approach sounds attractive. In the *Ringnorm* task for instance, CPU time is of 12.3 s for a large neighbourhood ($\delta = 0.001$) while it is reduced to 4.0 s and 1.9 s for respective smaller δ values of 0.01 and 0.1 and lower complexity.

Several curves reflecting the behaviour of the algorithm during time were drawn for the *Banana* problem. Same curves were measured for the other problems but are omitted for sake of space. The validation errors curves show the convergence of the algorithm. This behaviour is confirmed on the Fig. 2(c) where all the incremental algorithms exhibit a stabilizing value for $\|\boldsymbol{\xi}\|^2/\ell$. For LISVM and "rebatch" algorithm, the number of support vectors linearly increase with the number of new observed instances. This is not an issue, considering that the squared norm of the weight vector increases very much slower, suggesting that if the number of training instances had been bigger, a stabilization should have been observed. At last, let us consider the behaviour of the "best test" algorithm to LISVM on the *Banana* problem. This algorithm performs the SVM selection by choosing the size of the neighbourhood that minimizes the test error and thus is very demanding in terms of CPU time. Nevertheless, it is remarkable to notice that it reaches the same validation performance with twice less support vectors and a restricted norm of the weight vector, illustrating the relevance of the local approach.

6 Conclusion

In this paper, we propose a new incremental learning algorithm designed for RBF kernel-based SVM. It exploits the locality of RBF by re-learning only weights of training data that lie in the neighbourhood of the new data. Our scheme of model selection is based on a criterion derived from a bound an error generalization from [7] and allows to determine a relevant neighbourhood size at each learning step. Experimental results on three data sets show very promising results and open the door to real applications. The reduction in terms of CPU time provided by the local approach should be especially important in case of availability of numerous training instances.

Next further works concern tests on large scale incremental learning tasks like text categorization. The possibility of the δ parameter to be adaptive will also be studied. Moreover, LISVM will be extended to the context of drifting concepts by the use of a temporal window.

References

1. L. Bottou and V. Vapnik. Local learning algorithms. *Neural Computation*, 4(6):888–900, 1992.
2. L. Breiman. Bias, variance and arcing classifiers. Technical Report 460, University of California, Berkeley, CA, USA, 1996.
3. C. Burges. A tutorial on support vector machines for pattern recognition. *Data Mining and Knowledge Discovery*, 2(2):955–974, 1998.
4. G. Cauwenberghs and T. Poggio. Incremental and decremental support vector machine learning. In *Adv. Neural Information Processing*, volume 13. MIT Press, 2001.

5. O. Chapelle, V. Vapnik, O. Bousquet, and S. Mukherjee. Choosing kernel parameters for support vector machines. Technical report, AT&T Labs, March 2000.
6. C. Cortes and V. Vapnik. Support vector networks. *Machine Learning*, 20:1–25, 1995.
7. Nello Cristianini and John Shawe-Taylor. *An Introduction to Support Vector Machines and other kernel-based learning methods*, chapter 4 Generalisation Theory, page 68. Cambridge University Press, 2000.
8. T. Friess, F. Cristianini, and N. Campbell. The kernel-adatron algorithm: a fast and simple learning procedure for support vector machines. In J. Shavlik, editor, *Machine Learning: Proc. of the* 15th *Int. Conf.* Morgan Kaufmann Publishers, 1998.
9. T. Joachims. Making large-scale support vector machine learning practical. In B. Schölkopf, C. Burges, and A. Smola, editors, *Advances in Kernel Methods – Support Vector Learning*, pages 169–184. MIT Press, Cambridge, MA, 1998.
10. T. Joachims. Estimating the generalization performance of a svm efficiently. In *Proc. of the* 17th *Int. Conf. on Machine Learning*. Morgan Kaufmann, 2000.
11. R. Klinkenberg and J. Thorsten. Detecting concept drift with support vector machines. In *Proc. of the* 17th *Int. Conf. on Machine Learning*. Morgan Kaufmann, 2000.
12. E. Osuna, R. Freund, and F. Girosi. Improved training algorithm for support vector machines. In *Proc. IEEE Workshop on Neural Networks for Signal Processing*, pages 276–285, 1997.
13. J.C. Platt. Sequential minimal optimization: A fast algorithm for training support vector machines. Technical Report 98-14, Microsof Research, April 1998.
14. G. Rätsch, T. Onoda, and K.-R. Müller. Soft margins for AdaBoost. Technical Report NC-TR-1998-021, Department of Computer Science, Royal Holloway, University of London, Egham, UK, 1998.
15. N. Syed, H. Liu, and K. Sung. Incremental learning with support vector machines. In *Proc. of the Int. Joint Conf. on Artificial Intelligence (IJCAI)*, 1999.
16. V. Vapnik. *The nature of statistical learning theory*. Springer, New York, 1995.
17. V. Vapnik. *Statistical learning theory*. John Wiley and Sons, inc., 1998.

Learning to Predict the Leave-One-Out Error of Kernel Based Classifiers

Koji Tsuda[1,3], Gunnar Rätsch[1,2], Sebastian Mika[1], and Klaus-Robert Müller[1,2]

[1] GMD FIRST, Kekuléstr. 7, 12489 Berlin, Germany
[2] University of Potsdam, Am Neuen Palais 10, 14469 Potsdam
[3] AIST Computational Biology Research Center, 2-41-6, Aomi, Koutou-ku, Tokyo, 135-0064 , Japan
{tsuda,raetsch,mika,klaus}@first.gmd.de

Abstract. We propose an algorithm to predict the leave-one-out (LOO) error for kernel based classifiers. To achieve this goal with computational efficiency, we cast the LOO error approximation task into a *classification* problem. This means that we need to *learn* a classification of whether or not a given training sample – if left out of the data set – would be misclassified. For this learning task, simple data dependent features are proposed, inspired by geometrical intuition. Our approach allows to reliably select a good model as demonstrated in simulations on Support Vector and Linear Programming Machines. Comparisons to existing learning theoretical bounds, e.g. the span bound, are given for various model selection scenarios.

1 Introduction

Numerous methods have been proposed [7, 8, 3, 5, 9] for model selection of kernel-based classifiers such as Support Vector Machines (SVMs) [7] and Linear Programming Machines (LPMs) [1]. They all try to find a reasonably good estimate of the generalization error to select the proper hyperparameters. The *data dependent* LOO error would in principle be ideal for selecting hyperparameters of learning machines, as it is an (almost) unbiased estimator of the true generalization error [6]. Its computation is, unfortunately, for most practical cases prohibitively slow.

There have been several attempts to approximate the leave-one-out error in closed-form for SVM classifiers [8, 3, 5]. For example, a new type of bound was proposed that relies on the *span* of the SVs and was empirically found to perform best among the learning theoretical bounds [8]. However, such approximations are limited to a special learning machine, i.e. SVM, and it seems difficult to provide a useful approximation that is valid for a more general class of classifiers.

In this work, we introduce a learning approach for approximating the LOO error of general kernel classifers such as SVMs or Linear Programming Machines (LPM). We propose to use geometrical features to cast the leave-one-out error approximation problem for kernel classifiers into a – fast solvable – *classification* problem. Thus, our LOO error approximation problem reduces to *learn a*

G. Dorffner, H. Bischof, and K. Hornik (Eds.): ICANN 2001, LNCS 2130, pp. 331–338, 2001.

classification of whether or not a training sample, if left out the data set, will be misclassified. This task is referred to as a *meta* learning problem because we try to learn about a learning machine. The meta classification task is data rich as a large number of training patterns can be generated from all sorts of different classification problems. Note that we are using features that are meant to reflect the local difficulty or complexity of the meta learning problem across a large set of possible data. In experiments, we show that our approach works well both for SVM and LPM.

After reviewing some popular learning theoretical LOO error bounds we describe the features used as inputs for the meta learning problem and solve it by using a classification approach. Subsequently we perform simulations showing the usefulness of our LOO error approximation scheme in comparison to other bounds and finally conclude with some remarks.

2 Reviewing SVMs, LPMs and Selected LOO Bounds

When learning with SVMs [7] and LPMs [1], one is seeking for the coefficients of a linear combination of kernel functions $K(\boldsymbol{x}_i, \cdot)$, i.e. $f_{\boldsymbol{\alpha},b}(\boldsymbol{x}) = b + \sum_{i=1}^{\ell} \alpha_i K(\boldsymbol{x}_i, \boldsymbol{x})$. This is done by solving the following type of optimization problem [4]:

$$\min_{\boldsymbol{\alpha},b,\boldsymbol{\xi}\geq\mathbf{0}} \|\boldsymbol{\alpha}\|_P + C\textstyle\sum_{i=1}^{\ell}\xi_i \quad \text{with} \quad y_i f_{\boldsymbol{\alpha},b}(\boldsymbol{x}_i) \geq 1 - \xi_i, \quad i = 1, \dots, \ell, \tag{1}$$

where $\|\cdot\|_P$ is the 2-norm of $\boldsymbol{\alpha}$ in feature space for SVMs and the 1-norm of $\boldsymbol{\alpha}$ (in the coefficient space) for LPMs, respectively. The data and labels are denoted by $\boldsymbol{x}_i \in \mathbb{R}^n, y \in \{1, -1\}$ respectively. C is the regularization parameter. When solving (1), one usually introduces Lagrange multipliers λ_i, $i = 1, \dots, \ell$ for the constraints in (1), which are zero if the constraint is not active. For SVMs they turn out to be equal to α_i. For LPM there is no such correspondence.

We will now review some bounds on the LOO error of SVMs that have been proposed. A more complete presentation can be found in e.g. [2, 8]. Let Z be a sample of size ℓ, where each pattern is defined as $z_i = (\boldsymbol{x}_i, y_i)$. Furthermore, define $Z^p = \{z_i \in Z, i \neq p\}$ and $f^p = \mathcal{L}(Z^p)$, i.e. f^p is the decision obtained when learning with the p-th sample left out of the training set. The LOO error is defined as

$$\epsilon^{loo}(Z) = \frac{1}{\ell}\sum_{p=1}^{\ell} \Psi(-y_p f^p(\boldsymbol{x}_p)), \tag{2}$$

where $\Psi(z) = 1$ for $z > 0$ and $\Psi(z) = 0$ otherwise. As $-y_p f^p(\boldsymbol{x}_p)$ is positive only if f^p commits an error on $\boldsymbol{x}_p$, $\epsilon^{loo}(Z)$ is the average number of patterns which are misclassified when they are left out.

Support Vector Count: SVMs have several useful properties that can be exploited for a LOO prediction: The first is that patterns which are not Support

Vectors (SVs) do not change the decision surface and are always correctly classified. Therefore, one has to consider *only the SVs* in the LOO procedure and the LOO error can be easily bounded by

$$\epsilon^{loo}(Z) \leq \frac{\#SV}{\ell}, \tag{3}$$

where $\#SV$ is the number of SVs. However, this is a very rough estimate, because not all SVs will be misclassified when they are removed from the training set. For LPMs, (3) also holds, if one defines the SVs to be the patterns $\boldsymbol{x}_i$ whose expansion coefficients α_i or corresponding Lagrange multipliers λ_i are non-zero.

Jaakkola-Haussler Bound: In [3], Jaakkola and Haussler proposed a tighter bound than Eq.(3)

$$\epsilon^{loo}(Z) \leq \frac{1}{\ell}\sum_{i=1}^{\ell} \Psi(\alpha_p K(\boldsymbol{x}_p, \boldsymbol{x}_p) - y_p f(\boldsymbol{x}_p)), \tag{4}$$

where α_i is the weight coefficient of SVM and K is the kernel function.

Span Predictions: Recently, a sophisticated way of predicting the LOO error using the "span" of support vectors has been proposed [8]. Under the assumption that the set of SVs *does not change* during the LOO procedure, the LOO error can be exactly rewritten as

$$\epsilon^{loo}(Z) = \frac{1}{\ell}\sum_{p \in SV} \Psi(-y_p f(\boldsymbol{x}_p) + \alpha_p S_p^2), \tag{5}$$

where SV denotes the set of support vectors and S_p is a geometric value called "span" of a support vector [8]. Unfortunately, in practice the above assumption made is not always satisfied, however, experimentally it is shown that this approximation works very well [8]. For computing the span one needs to solve an optimization problem which can be very expensive, if the number of SVs is high.

3 Meta Learning: Predicting the Generalization Error from Empirical Data

We will now outline a learning framework to predict the LOO error. The LOO error ϵ^{LOO} depends on the data set Z and the parameters of the learning machine $\boldsymbol{\theta}$. We assume that the LOO errors could be measured for several data sets and various combinations $(Z_1, \boldsymbol{\theta}_1), \cdots, (Z_m, \boldsymbol{\theta}_m)$. Then, a meta-learning machine is trained to predict ϵ^{LOO} on unseen data based on appropriate features extracted from $(Z, \boldsymbol{\theta})$.

Predicting the LOO-Error as a classification problem Recall that the LOO error can be represented by each LOO result $r(\boldsymbol{x}_p)$:

$$\epsilon^{LOO} = \frac{1}{\ell}\sum_{p=1}^{\ell} \Psi(-y_p f^p(\boldsymbol{x}_p)) := \frac{1}{\ell}\sum_{p=1}^{\ell} \Psi(-r(\boldsymbol{z}_p)) \tag{6}$$

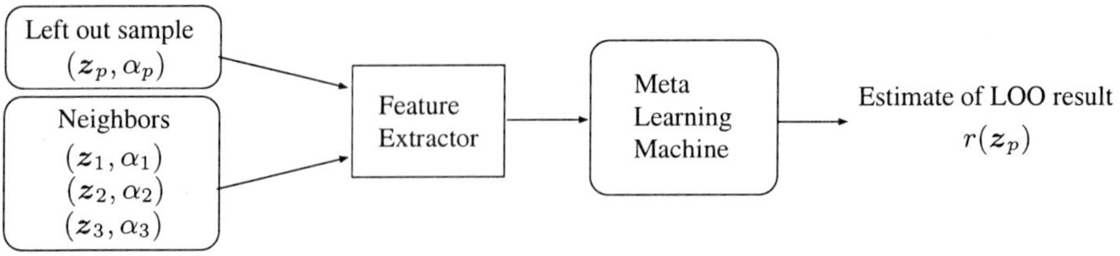

Fig. 1. Learning scheme for predicting the LOO result.

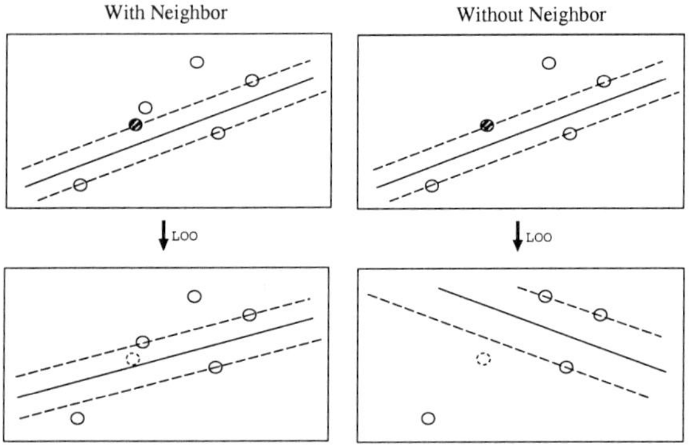

Fig. 2. Consider the case with a non-SV near the SV of interest as in the left panel. In the leave-one-out procedure, the boundary is re-estimated without this SV, but the non SV takes its part and the boundary does not change significantly. The LOO boundary could show a large difference, if there was no close non-SV neighbor as in the right panels.

where $r(\boldsymbol{z}_p) = \mathrm{sgn}(y_p f^p(\boldsymbol{x}_p))$. So, to predict the LOO error, it is sufficient to predict the result of each LOO procedure, i.e. *whether or not* an error will be made, if a certain pattern is left out. This meta learning problem is a binary classification problem.

For kernel classifiers, the learning scheme can be designed as Fig.1: Here, a coefficient α_i is attached to every training sample $\boldsymbol{x}_i$. Features are extracted for the left-out sample $(\boldsymbol{x}_p, \alpha_p)$ as well as for the neighboring samples. The neighbors are included since they are likely to affect the LOO result as shown in Fig. 2. In taking the neighbors, we do not care whether they are support vectors or not. So, the features also include information from non-support vectors whereas the span bound (5) is derived from the support vectors only.

Features for meta classification To include the local geometry around a left-out sample, the features for LOO are extracted as follows: for the left-out sample $\boldsymbol{x}_p$, we use

– Weight α_p, Dual weight λ_p (LPM only), Margin $f(\boldsymbol{x}_p)/\|\boldsymbol{w}\|^2$.

Additionally, we calculate the following quantities from the 3 nearest neighbors $\boldsymbol{x}_k$ of $\boldsymbol{x}_p$:

– Distance in input space $\|\boldsymbol{x}_k - \boldsymbol{x}_p\|/D_i$,
– Distance in feature space $\|\Phi(\boldsymbol{x}_k) - \Phi(\boldsymbol{x}_p)\|/D_f$,

– Weight α_k, Dual weight λ_k (LPM only), Margin $f(\boldsymbol{x}_k)/\|\boldsymbol{w}\|^2$,

where D_i, D_f are the maximum distances between training samples in the input space and the feature space, respectively. All these features of $\boldsymbol{x}_p$ form a vector $\boldsymbol{v}_p$ which is used together with the label $r(\boldsymbol{z}_p)$ to *learn* the LOO error prediction $\{\boldsymbol{v}_p\} \rightarrow \{r(\boldsymbol{z}_p)\}$ for all p patterns. In this work we built the meta-classifier as a linear programming machine with polynomial kernel of degree 2, which employs the 1-norm regularization (in feature space) leading to sparse weight vectors. This turns out to be beneficial as the (meta) LPM automatically selects the relevant features.

4 Experiments

The motivation of the following experiment is to answer the following questions: (1) Does the LOO error predictor learn from a given data set to generalize well on unseen data sets? (2) Is the prediction good enough for reliable model selection?

4.1 Two Class Benchmarks

In our study, we considered three data sets[1]: `twonorm`, `ringnorm` and `heart`. Here, `twonorm` and `ringnorm` are quite similar because the input dimensionality is 20 and the number of training samples is 400 for both data sets. But, `heart` is a quite different data set, where the input dimensionality is 13 and the number of training samples is only 170. For the evaluation of our meta classifier we use the following experimental setup: For each data set we considered ten realizations (train/test-splits, used for averaging and obtaining error bars). On each realization, we trained SVMs and LPMs for wide ranges of the regularization constant C and the (RBF) kernel parameter σ, i.e. $K(\boldsymbol{x}, \boldsymbol{y}) = \exp(-\|\boldsymbol{x} - \boldsymbol{y}\|/\sigma^2)$. For each training sample we extracted the features described in Section 3. These features and the corresponding labels (which have been computed by the actual LOO procedure) are used for training and testing our classifier. We learned from two data sets and tested on the third. To evaluate the performance, the *model selection error* is used: First, the kernel parameter σ and regularization constant C are selected where the predicted LOO error is minimized. The model selection error is then defined as the classification error on the test set at the chosen parameters.

The results for SVM and LPM are shown in Fig. 3 and 4, respectively. Results on the span bound and JH bound are not available for LPMs, as the respective bounds simply do not exist. The experiments show that in most cases our LOO predictor performs almost as well as the actual (highly expensive) LOO calculation. Compared to the bounds we observe that both methods achieve similar performance, note however that our method can also be applied to LPMs. Comparing the three cases in Fig. 4, our method performs slightly worse when `heart`

[1] The data sets incl. training/test splits and the LOO results can be obtained at `http://ida.first.gmd.de/~raetsch`.

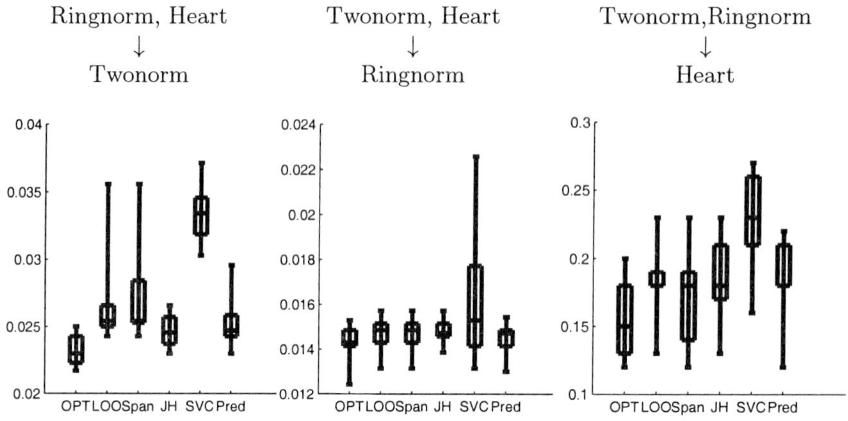

Fig. 3. Model selection errors in SVM. The labels denote: OPT: optimal choice based on the test error, LOO: actual leave-one-out calculation, Span: span bound, JH: Jaakkola-Haussler Bound, SVC: Support Vector Count, Pred: our method. Two data sets are used for training and the test error is assessed on the third one.

is used for testing in comparison with the other two cases. This shows the tendency that LOO error prediction works well if data with a similar characteristics is contained in training set. This indicates also that the statistics of the feature set still varies considerably for different data sets. However, it is still surprising that our simple features can show such a good generalization performance in benchmark problems.

4.2 A Multiclass Example

As one application of our approach, we consider multi-class problems. In solving c-class problems, it is common, e.g. for SVMs, to use c single (two-class) classifiers, each of which is trained to discriminate one class from all the other classes. Here, the hyperparameters of each SVM have to be properly set by some model selection process. Clearly, it takes prohibitively long to perform leave-one-out procedures in all SVMs for all possible hyperparameter settings. To cope with this problem, the leave-one-out procedure is performed with respect to only one of the c classification problems and our meta classifier is then trained based on this result. Then, the hyperparameters of the other SVMs can be efficiently selected according to the LOO error predictions from the meta classifier.
We performed an experiment with the `3dobj` data set[2] with 8 classes and 640 samples (i.e. 80 samples per class). The task is here to choose a proper value of kernel width σ from a prespecified set of 11 values. The sample is randomly divided into 320 training and 320 test samples for obtaining error bars on our results. One class is chosen for the training and the remaining classes are used for testing. For each σ in $\Sigma = \{0.4, 0.6, \cdots, 2.4\}$, we computed the LOO error predictions using the meta classifier.

[2] This dataset will be added to the IDA website.

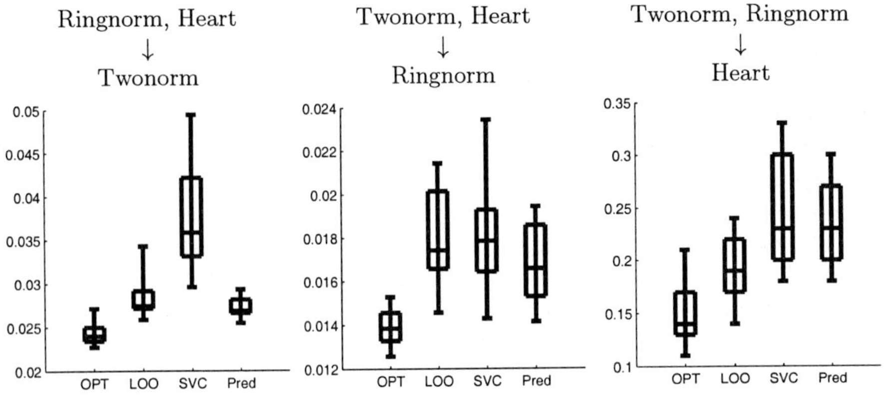

Fig. 4. Model selection errors in LPM.

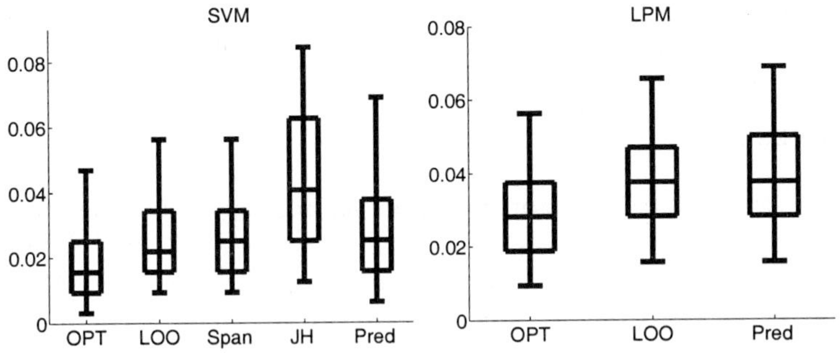

Fig. 5. Model selection errors in the multiclass problem.

Figure 5 shows the model selection errors of SVMs and LPMs. Here, our method performed as well as the actual LOO calculation for both: SVMs and LPMs. The task of model selection in multiclass problems appears very suitable for our method, because the data sets for training and testing are considered to have similar statistical properties.

5 Conclusion

To train a learning machine that learns about the generalization behavior of a set of learning machines seems like an appealing idea and introduces a *meta* level of reasoning. Our goal in this work was to obtain an effective meta algorithm for predicting the LOO error from past experience, i.e. from a variety of data sets. By adding the twist of casting the meta learning problem into a specific classification task allowed to achieve an accurate and fast empirical estimate of the LOO error of SVM and LPM. The crucial point was to use simple geometrical features for classifying whether a given training sample would be misclassified, if left out. Once given a reliable LOO error estimate it can easily be used for model selection.

Careful simulations (using approximately 1.5 CPU years of 800Mhz Pentium III mostly for obtaining the actual LOO error) show that our meta learning framework compares favorably over the conventional bounds. Apparently, our heuristic geometrically motivated features have a good generalization ability for different data sets. We speculate that using these features could provide new ways to improve bounds, in a way, by integrating particularly meaningful features from our meta learning problem into new learning theoretical bounds. Future research will be dedicated to a further exploration of the meta learning idea also for other learning machines and to gaining a better learning theoretical understanding of our findings.

Acknowledgments

We thank for valuable discussions with Jason Weston, Olivier Chapelle and Bernhard Schölkopf. This work was partially funded by DFG under contract JA 379/-71, -91, MU 987/1-1.

References

1. K.P. Bennett and O.L. Mangasarian. Robust linear programming discrimination of two linearly inseparable sets. *Optimization Methods and Software*, 1:23–34, 1992.
2. O. Chapelle and V.N. Vapnik. Choosing kernel parameters for support vector machines. Personal communication, mar 2000.
3. T.S. Jaakkola and D. Haussler. Probabilistic kernel regression models. In *Proceedings of the 1999 Conference on AI and Statistics*, 1999.
4. K.-R. Müller, S. Mika, G. Rätsch, K. Tsuda, and B. Schölkopf. An introduction to kernel-based learning algorithms. *IEEE Transactions on Neural Networks*, 2001. in Press.
5. M. Opper and O. Winther. Gaussian processes and SVM: Mean field and leave-one-out. In A.J. Smola, P.L. Bartlett, B. Schölkopf, and D. Schuurmans, editors, *Advances in Large Margin Classifiers*, pages 311–326, Cambridge, MA, 2000. MIT Press.
6. V.N. Vapnik. *Estimation of Dependences Based on Empirical Data.* Springer-Verlag, Berlin, 1982.
7. V.N. Vapnik. *The nature of statistical learning theory.* Springer Verlag, New York, 1995.
8. V.N. Vapnik and O. Chapelle. Bounds on error expectation for support vector machines. *Neural Computation*, 12(9):2013–2036, September 2000.
9. G. Whaba, Y. Lin, and H. Zhang. Generalized approximate cross-validation for support-vector-machines: another way to look at margin-like quantities. Technical Report TR-1006, Dept. of Statistics, University of Wisconsin, April 1999.

Sparse Kernel Regressors

Volker Roth

University of Bonn, Department of Computer Science III,
Roemerstr. 164, D-53117 Bonn, Germany
roth@cs.uni-bonn.de

Abstract. Sparse kernel regressors have become popular by applying the support vector method to regression problems. Although this approach has been shown to exhibit excellent generalization properties in many experiments, it suffers from several drawbacks: the absence of probabilistic outputs, the restriction to Mercer kernels, and the steep growth of the number of support vectors with increasing size of the training set. In this paper we present a new class of kernel regressors that effectively overcome the above problems. We call this new approach *generalized LASSO* regression. It has a clear probabilistic interpretation, produces extremely sparse solutions, can handle learning sets that are corrupted by outliers, and is capable of dealing with large-scale problems.

1 Introduction

The problem of *regression analysis* is one of the fundamental problems within the field of supervised machine learning. It can be stated as estimating a real valued function, given a sample of noisy observations. The data is obtained as i.i.d. pairs of feature vectors $\{\boldsymbol{x}_i\}_{i=1}^N$ and corresponding targets $\{y_i\}_{i=1}^N$, drawn from an unknown joint distribution $p(\boldsymbol{x}, y)$. Viewed as a function in $\boldsymbol{x}$, the conditional expectation of y given $\boldsymbol{x}$ is called a *regression function* $f_r(\boldsymbol{x}) = E[y|\boldsymbol{x}] = \int_{-\infty}^{+\infty} y\, p(y|\boldsymbol{x})\, dy$.

A very successful approach to this problem is the *support vector machine* (SVM). It models the regression function by way of *kernel functions* $k(\boldsymbol{x}, \boldsymbol{x}_i)$:[1]

$$f_r(\boldsymbol{x}) = \sum_{i=1}^N k(\boldsymbol{x}, \boldsymbol{x}_i)\alpha_i =: K\boldsymbol{\alpha}.$$

However, SV regression bears some disadvantages: (i) The predictions cannot be interpreted in a probabilistic way. (ii) The solutions are usually not very sparse, and the number of support vectors is strongly correlated with the sample size. (iii) The kernel function must satisfy Mercer's condition.

A Bayesian approach to kernel regression that overcomes these drawbacks was presented in [8]. This model is referred to as the *relevance vector machine* (RVM). One of its most outstanding features is the extreme sparsity of the solutions. Once we have

[1] For the sake of simplicity we have dropped the constant term α_0 throughout this paper. If the kernel satisfies Mercer's condition, this can be justified either if the kernel has an implicit intercept, i.e. if $k(\mathbf{0}, \mathbf{0}) > 0$, or if the input vectors are augmented by an additional entry 1 (e.g. for polynomial kernels). If Mercer's condition is violated, which may be possible in the RVM approach, the kernel matrix K itself can be augmented by an additional column of ones.

G. Dorffner, H. Bischof, and K. Hornik (Eds.): ICANN 2001, LNCS 2130, pp. 339–346, 2001.

successfully trained a regression function, this sparsity allows us to make predictions for new observations in a highly efficient way. Concerning the training phase, however, the original RVM algorithm suffers from severe computational problems.

In this paper we present a class of kernel regressors that adapt the conceptual ideas of the RVM and additionally overcome its computational problems. Moreover, our model can easily be extended to *robust loss functions*. This in turn overcomes the sensitivity of the RVM to *outliers* in the data. We propose a highly efficient training algorithm that directly exploits the sparsity of the solutions. Performance studies for both synthetic and real-word benchmark datasets are presented, which effectively demonstrate the advantages of our model.

2 Sparse Bayesian Kernel Regression

Applying a Bayesian method requires us to specify a set of probabilistic models. A member of this set is called a hypothesis $\mathcal{H}_\alpha$ which will have a prior probability $P(\mathcal{H}_\alpha)$. The likelihood of $\mathcal{H}_\alpha$ is $P(\mathcal{D}|\,\mathcal{H}_\alpha)$, where $\mathcal{D}$ represents the data. For regression problems, each $\mathcal{H}_\alpha$ corresponds to a regression function f_α. Under the assumption that the targets y are generated by corrupting the values of f_α by additive Gaussian noise of variance σ, the likelihood of $\mathcal{H}_\alpha$ is

$$\prod_i P(y_i|\,\boldsymbol{x}_i, \mathcal{H}_\alpha) = \prod_i \frac{1}{\sqrt{2\pi\sigma^2}} \exp\left\{-(y_i - f_\alpha(\boldsymbol{x}_i))^2/(2\sigma^2)\right\}. \tag{1}$$

The key concept of the RVM is the use of *automated relevance determination* (ARD) priors over the expansion coefficients of the following form:

$$P(\boldsymbol{\alpha}|\,\boldsymbol{\vartheta}') = N(\mathbf{0}, \Sigma_\alpha) = \prod_{i=1}^N \mathcal{N}(0, {\vartheta'}_i^{-1}). \tag{2}$$

This prior model leads us to a posterior of the form

$$P(\boldsymbol{\alpha}|\boldsymbol{y}, \boldsymbol{\vartheta}', \sigma^2) = (2\pi)^{-\frac{N}{2}}\,|A|^{\frac{1}{2}} \exp\left\{-\tfrac{1}{2}(\boldsymbol{\alpha} - \bar{\boldsymbol{\alpha}})^T A(\boldsymbol{\alpha} - \bar{\boldsymbol{\alpha}})\right\}, \tag{3}$$

with (inverse) covariance matrix $A = \Sigma_\alpha^{-1} + \frac{1}{\sigma^2} K^T K$ and mean $\bar{\boldsymbol{\alpha}} = \frac{1}{\sigma^2} A^{-1} K^T \boldsymbol{y}$. From the form of the (1) and (2) it is clear that the posterior mean vector minimizes the quadratic form

$$M(\boldsymbol{\alpha}) = \|\boldsymbol{y} - K\boldsymbol{\alpha}\|^2 + \sigma^2 \boldsymbol{\alpha}^T \Sigma_\alpha^{-1} \boldsymbol{\alpha} = \|\boldsymbol{y} - K\boldsymbol{\alpha}\|^2 + \sum_{i=1}^N \vartheta_i \alpha_i^2, \tag{4}$$

where we have defined $\boldsymbol{\vartheta} := \sigma^2 \boldsymbol{\vartheta}'$ for the sake of simplicity.
Given the above class of ARD models, there are now different inference strategies:

- In [8] the **Relevance Vector Machine** (RVM) was proposed as a (partially) Bayesian strategy: integrating out the expansion coefficients in the posterior distribution, one obtains an analytical expression for the marginal likelihood $P(\boldsymbol{y}|\,\boldsymbol{\vartheta}', \sigma^2)$, or *evidence*, for the hyperparameters. For ideal Bayesian inference one should define hyperpriors over $\boldsymbol{\vartheta}'$ and σ, and integrate out these parameters. Since there is no closed-form solution for this marginalization, however, it is common to use a Gaussian approximation of the posterior mode. The *most probable* parameters $\boldsymbol{\vartheta}'_{MP}$ are chosen by maximizing $P(\boldsymbol{y}|\,\boldsymbol{\vartheta}', \sigma^2)$. Given a current estimate for the $\boldsymbol{\alpha}$ vector, the parameters ϑ'_k are derived as

$$(\vartheta'_k)^{\text{new}} = (1 - \vartheta'_k(A^{-1})_{kk})/\alpha_k. \tag{5}$$

These values are then substituted into the posterior (3) in order to get a new estimate for the expansion coefficients:

$$(\boldsymbol{\alpha})^{\text{new}} = \left(K^T K + \sigma^2 \text{diag}\left\{\boldsymbol{\vartheta}'\right\}\right)^{-1} K^T \boldsymbol{y}. \tag{6}$$

- The key idea of **Adaptive Ridge** (AdR) regression is to select the parameters ϑ_j by minimizing (4). Direct minimization, however, would obviously shrink all parameters to zero. This is clearly not satisfactory, and can be avoided by applying a constraint of the form

$$\tfrac{1}{N}\textstyle\sum_{i=1}^{N} \tfrac{1}{\vartheta_i} = \tfrac{1}{\lambda}, \quad \vartheta_i > 0, \tag{7}$$

  where λ is a predefined value, (cf. [5]). This constraint connects the individual variances of the ARD prior by requiring that their *mean variance* is proportional to $1/\lambda$. The idea behind (7) is to start with an ridge-type estimate ($\vartheta_i = \lambda\, \forall i$) and then introduce a method of *automatically balancing* the penalization on each variable.

Before going into details, the reader should notice that both approaches are conceptually equivalent in the sense that they share the same idea of using an ARD prior, and they both employ some pragmatic procedure for deriving the optimal prior parameters $\boldsymbol{\vartheta}$.

For a detailed analysis of **AdR regression**, it is useful to introduce a Lagrangian formalism. The Lagrangian for minimizing (4) under the constraint (7) reads

$$\mathcal{L} = \|\boldsymbol{y} - K\boldsymbol{\alpha}\|^2 + \textstyle\sum_{i=1}^{N} \vartheta_i \alpha_i^2 + \mu \left(\sum_{i=1}^{N} \frac{1}{\vartheta_i} - \frac{N}{\lambda}\right). \tag{8}$$

For the optimal solution, the derivatives of $\mathcal{L}$ with respect to both the primal and dual variables must vanish. This means $\partial\mathcal{L}/\partial\alpha_k = 0,\ \partial\mathcal{L}/\partial\vartheta_k = 0,\ \partial\mathcal{L}/\partial\mu = 0$. It follows that for given parameters $\boldsymbol{\vartheta}$ the optimizing coefficients α_i are derived as

$$(\boldsymbol{\alpha})^{\text{new}} = \left(K^T K + \text{diag}\left\{\boldsymbol{\vartheta}\right\}\right)^{-1} K^T \boldsymbol{y}, \tag{9}$$

and we find $(\vartheta_k)^{\text{new}} = \sqrt{\mu}\,/\,|\alpha_k|$. Together with the stationarity conditions, the optimal parameters are derived as

$$(\vartheta_k)^{\text{new}} = (\lambda/N)\left(\textstyle\sum_{i=1}^{N} |\alpha_i|\right) / \,|\alpha_k|. \tag{10}$$

During iterated application of (9) and (10), it turns out that some parameters ϑ_j approach infinity, which means that the variance of the corresponding priors $p(\alpha_j|\vartheta'_j)$ becomes zero, and in turn the posterior $p(\alpha_j|\,\boldsymbol{y}, \boldsymbol{\vartheta}', \sigma^2)$ becomes infinitely peaked at zero. As a consequence, the coefficients α_j are shrinked to zero, and the corresponding variables (the columns of the kernel matrix K) are removed from the model. Note that besides the conceptual equivalence between AdR regression and the RVM, both approaches are also technically very similar in the following sense: they both share the same update equations for the coefficients α_i, (9) and (6), and the hyperparameters ϑ_k in (10) and

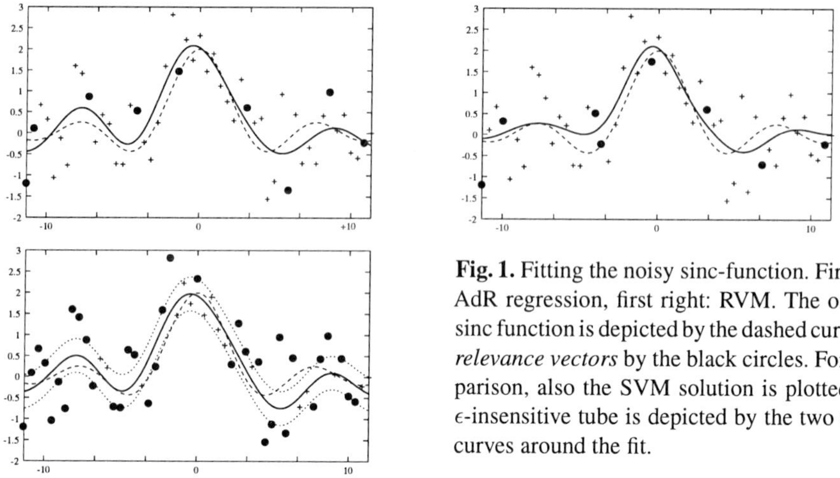

Fig. 1. Fitting the noisy sinc-function. First left: AdR regression, first right: RVM. The original sinc function is depicted by the dashed curve, the *relevance vectors* by the black circles. For comparison, also the SVM solution is plotted. The ϵ-insensitive tube is depicted by the two dotted curves around the fit.

(5) are inverse proportional to the value of the corresponding weights α_k. It is thus not surprising, that both methods produce similar regression fits, see figure 1.

We have chosen the popular example of fitting the noisy sinc-function. For comparison, we have also trained a SVM. This simple yet intuitive experiment nicely demonstrates one of the most outstanding differences between the ARD-models and the SVM: the former produce solutions which are usually much sparser as the SVM solutions, sometimes by several orders of magnitude. This immediately illustrates two advantages of the ARD models: (i) the extreme sparsity may be exploited to develop highly efficient training algorithms, (ii) once we have a trained model, the prediction problem for new samples can be solved extremely fast.

3 An Efficient Algorithm for AdR Regression

Both the above iterative AdR algorithm and the original RVM algorithm share two main drawbacks: (i) convergence is rather slow, thus many iterations are needed, (ii) solving eq. (9) or (6) for the new α_i means solving a system of N linear equations in N variables, which is very time consuming if N becomes large.

In the case of AdR regression, the latter problem can be overcome by applying approximative conjugate gradient methods. Because of (i), however, the over-all procedure still remains rather time consuming. For the original RVM, even this speedup is not suitable, since here the update equations of the hyperparameters require us to *explicitly* invert the matrix $\left(K^T K + \text{diag}\,\{\vartheta\}\right)$ anyway.

However, these computational problems can be overcome by exploiting the equivalence of AdR regression and the so called *Least Absolute Shrinkage and Selection Operator* (LASSO), see [5, 7]. Since space here precludes a detailed derivation, we only notice that it can be derived directly from the Lagrangian formalism introduced in the last section. Minimizing (8) turns out to be equivalent to solving the LASSO problem, which can be interpreted as ℓ_1-penalized least-squares regression:

$$\text{minimize} \quad \sum_{i=1}^{N} (y_i - (K\boldsymbol{\alpha})_i)^2 \qquad \text{subject to} \qquad \|\boldsymbol{\alpha}\|_1 < \kappa. \tag{11}$$

It is worth noticing that the equivalence holds for any differentiable loss function. In particular, this allows us to employ *robust* loss functions which make the estimation process less sensitive to outliers in the data. We will return to this point in section 4. The real payoff of reformulating AdR regression in terms of the LASSO is that for the latter problem there exist highly efficient subset algorithms that directly make use of the sparsity of the solutions. Such an algorithm was originally introduced in [6] for linear least squares problems. In the following it will be generalized to both nonlinear kernel models and to general loss functions. Denoting a differentiable loss function by L, the Lagrangian for the general LASSO problem can be rewritten as

$$\mathcal{L}(\boldsymbol{\alpha}, \lambda) = \sum_{i=1}^{N} L\left(y_i - (K\boldsymbol{\alpha})_i\right) - \mu(\kappa - \sum_{j=1}^{N} |\alpha_j|). \tag{12}$$

According to the Kuhn-Tucker theorem, the partial derivatives of $\mathcal{L}$ with respect to α_i and μ have to vanish. Introducing the function $\omega(t) = \partial L(t)/(t\,\partial t)$ and the diagonal matrix $\Omega(\boldsymbol{\alpha}) = \text{diag}\left\{\omega([K\boldsymbol{\alpha} - \boldsymbol{y}]_i)\right\}$, the derivative of $\mathcal{L}$ w.r.t. $\boldsymbol{\alpha}$ reads (cf. [3, 6])

$$\nabla_{\boldsymbol{\alpha}}\mathcal{L} = K^T\Omega(\boldsymbol{\alpha})\boldsymbol{r} + \mu\boldsymbol{v} = \boldsymbol{0}, \qquad \text{with } v_i = \begin{cases} \text{sign}(\alpha_i) & \text{if } \alpha_i \neq 0 \\ a \in [-1, 1] & \text{if } \alpha_i = 0. \end{cases} \tag{13}$$

In the above equation $\boldsymbol{r} = \boldsymbol{y} - K\boldsymbol{\alpha}$ denotes the vector of residuals. For the derivation it is useful to introduce some notations: from the form of $\boldsymbol{v}$ it follows that $\|v\|_\infty = 1$ which implies $\mu = \|K^T\Omega\hat{\boldsymbol{r}}\|_\infty$. To deal with the sparsity of the solutions, it is useful to introduce the permutation matrix P. It collects the non-zero coefficients of $\boldsymbol{\alpha}$ in the first $|\sigma|$ components, i.e. $\boldsymbol{\alpha} = P^T\binom{\alpha_\sigma}{0}$. Furthermore, we denote by $\boldsymbol{\theta}_\sigma$ a *sign* vector, $\boldsymbol{\theta}_\sigma = \text{sign}(\boldsymbol{\alpha}_\sigma)$. An efficient subset selection algorithm, which heavily draws on [6], can now be outlined as follows: given the current estimate $\boldsymbol{\alpha}$, the key idea is to calculate a new search direction $\boldsymbol{h} = P^T\binom{h_\sigma}{0}$ *locally* around $\boldsymbol{\alpha}$. This local problem reads

$$\underset{\boldsymbol{h}}{\text{minimize}} \quad \sum_{i=1}^{N} L\left([y_\sigma]_i - [K_\sigma(\boldsymbol{\alpha}_\sigma + \boldsymbol{h}_\sigma)]_i\right) \text{ s.t. } \quad \theta_\sigma^T(\boldsymbol{\alpha}_\sigma + \boldsymbol{h}_\sigma) \leq \kappa \tag{14}$$

For quadratic loss function this problem can be solved analytically, otherwise it defines a simple nonlinear optimization problem in $|\sigma|$ variables. The problem is simple, because (i) it is usually a low-dimensional problem, $|\sigma| \ll N$, (ii) for a wide class of robust loss functions it defines a *convex* optimization problem, (iii) either the constraint is inactive, or the solution lies on the constraint boundary. In the latter case, (if the unconstrained solution is not feasible) we have to handle only one simple linear equality constraint, $\theta_\sigma^T\boldsymbol{h}_\sigma = \tilde{\kappa}$. (iv) Our experiments show that the number of nonzero coefficients $|\sigma|$ is usually *not correlated* to the sample size N. It follows that even large-scale problems can be solved efficiently.

The iteration is started from $\boldsymbol{\alpha} = \boldsymbol{0}$ by choosing an initial s to insert into σ and solving the resulting one-variable problem. With the concept of *sign feasibility* (cf. [1]), the algorithm proceeds as follows:

Check if $\boldsymbol{\alpha}^\dagger := \boldsymbol{\alpha} + \boldsymbol{h}$ is sign feasible, i.e. if $\text{sign}(\boldsymbol{\alpha}_\sigma^\dagger) = \boldsymbol{\theta}_\sigma$. Otherwise:

- (A1) Move to the first new zero component in direction $\boldsymbol{h}$, i.e. find the smallest γ, $0 < \gamma < 1$ and corresponding $k \in \sigma$ such that $0 = \alpha_k + \gamma h_k$ and set $\boldsymbol{\alpha} = \boldsymbol{\alpha} + \gamma \boldsymbol{h}$.
- (A2) There are two possibilities:
  1. Set $\theta_k = -\theta_k$ and recompute $\boldsymbol{h}$. If $(\boldsymbol{\alpha} + \boldsymbol{h})$ is sign feasible for the revised θ_σ, set $\boldsymbol{\alpha}^\dagger = \boldsymbol{\alpha} + \boldsymbol{h}$ and proceed to the next stage of the algorithm.
  2. Otherwise update σ by deleting k, resetting α_k and θ_k accordingly, and recompute $\boldsymbol{h}$ for the revised problem.
- (A3) Iterate until a sign feasible $\boldsymbol{\alpha}^\dagger$ is obtained.

Once sign feasibility is obtained, we can test optimality by verifying (13): calculate

$$\boldsymbol{v}^\dagger = \frac{K^T \Omega(\boldsymbol{\alpha}^\dagger)\boldsymbol{r}^\dagger}{\|K_\sigma^T \Omega(\boldsymbol{\alpha}^\dagger)\boldsymbol{r}^\dagger\|_\infty} = P^T \binom{\boldsymbol{v}_1^\dagger}{\boldsymbol{v}_2^\dagger}.$$

By construction $(\boldsymbol{v}_1^\dagger)_i = \theta_i$ for $i \le |\sigma|$, and if $-1 \le (\boldsymbol{v}_2^\dagger)_i \le 1$ for $1 \le i \le N - |\sigma|$, then $\boldsymbol{\alpha}^\dagger$ is the desired solution. Otherwise, one proceeds as follows:

- (B1) Determine the most violated condition, i.e. find the index s such that $(\boldsymbol{v}_2^\dagger)_s$ has maximal absolute value.
- (B2) Update σ by adding s to it and update $\boldsymbol{\alpha}_\sigma^\dagger$ by appending a zero as its last element and θ_σ by appending $\text{sign}(\boldsymbol{v}_2^\dagger)_s$.
- (B3) Set $\boldsymbol{\alpha} = \boldsymbol{\alpha}^\dagger$, compute a new direction $\boldsymbol{h}$ by solving (14) and iterate.

4 Experiments

In a first example, the *prediction performance* of LASSO regression is demonstrated for **Friedman's benchmark functions**, [4]. Since only relatively small learning sets are considered, we postpone a detailed analysis of *computational costs* to the experiments below. The results are summarized in table 1.

Table 1. Results for Friedman's functions. Mean prediction error (100 randomly generated 240/1000 training/test splits) and #(support/relevance vectors). SVM/RVM results are taken from [8].

Dataset	SVM	RVM	LASSO
#1	2.92 / 116.6	2.80 / 59.4	2.84 / 73.5
#2	4140 / 110.3	3505 / 6.9	3808 / 14.2
#3	0.0202 / 106.5	0.0164 / 11.5	0.0192 / 16.4

It should be noticed that all three models attain a very similar level of accuracy. Distinct differences, however, occur in the number of support/relevance vectors: the models employing ARD priors produce much sparser solutions than the SVM, in accordance with the results from figure 1. As real-world examples, we present results for the "house-price-8L" and "bank-32-fh" datasets from the **DELVE benchmark repository**[2]. We compared both the prediction accuracy and the computational costs of RVM, SVM[3] and LASSO for different sample sizes. The results are summarized in table 2. From the table we conclude, that (i) the prediction accuracy of all models is comparable, (ii) the ARD models are sparser than the SVM by 1-2 orders of magnitude, (iii) the RVM has

[2] The datasets are available via `http://www.cs.toronto.edu/~delve/delve.html`

[3] We used the *SVMTorch* V 3.07 implementation, see [2].

severe computational problems for large training sets, (iv) the LASSO combines the advantages of efficiently handling large training sets and producing extremely sparse solutions, see also figure 2. Concerning the training times, the reader should notice that we are comparing the highly tuned *SVMTorch* optimization package, [2], with a rather simple LASSO implementation, which we consider to yet possess ample opportunities for further optimization.

Table 2. Results for the "house-price-8L" and "bank-32-fh" datasets from the DELVE repository. In all experiments RBF kernels are used. The times are measured on a 500 MHz PC. The last 3 columns show the time in seconds for predicting the function value of 4000 test examples.

sample	MSE			# SV/RV			$t_{\text{learn}}[s]$			$t_{\text{test}}[s]$		
	Rvm	Svm	Lasso	Rvm	Svm	Lasso	Rvm	Svm	Lasso	Rvm	Svm	Lasso
	$(\cdot 10^3)$			**house-price-8L**								
1000	1099	1062	1075	33	597	61	$4.2 \cdot 10^3$	33	26	0.1	1.4	0.2
2000	1048	1022	1054	36	1307	63	$3.5 \cdot 10^4$	101	72	0.1	3.5	0.2
4000	-	1012	1024	-	2592	69	-	428	312	-	8	0.2
	$(\cdot 10^{-3})$			**bank-32-fh**								
2000	7.41	7.82	7.39	14	1638	22	$3 \cdot 10^4$	15	24	0.07	6	0.1
4000	-	7.75	7.49	-	3402	23	-	83	102	-	13	0.1

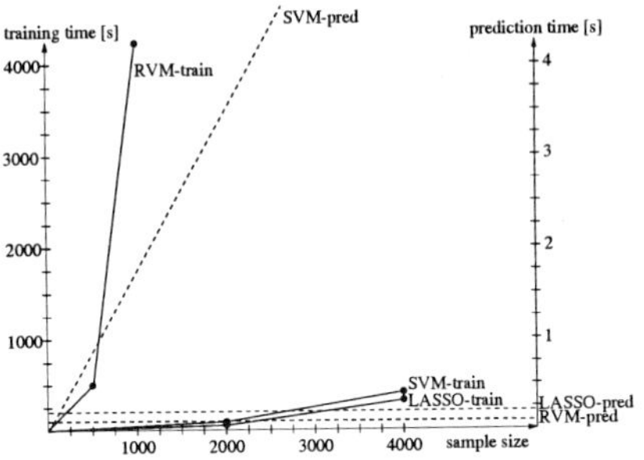

Fig. 2. Computation times for the "house-8L" dataset. Solid lines depict training times, dashed lines depict prediction times for a test set of size 4000. In the training phase, both SVM and LASSO are clearly advantageous over the RVM. For predictions, the ARD models outperform the SVM drastically. Note that the prediction time solely depends on the number of nonzero expansion coefficients, which for ARD models roughly remains constant with increasing size of the training set.

In the experiments presented so far we exclusively used quadratic loss functions. It is worth noticing that within the *whole DELVE archive* we could not find a single problem for which a robust loss function significantly improved the accuracy. This, however, only means that the problems considered are too "well-behaved" in the sense that they obviously contain no or only very few outliers. Here, we rather present an intuitive artificial example with a **high percentage of outliers**. Applications to relevant real-world problems will be subject of future work. We return to the problem of fitting the noisy sinc-function, this time however with additional 20% outliers drawn from a uniform distribution. The situation both for standard- and robust LASSO is depicted in figure 3. The non-robust LASSO approach is very sensitive to the outliers, which results from the quadratic growth of the loss function. The robust version employing a loss function of *Huber's type* (see e.g. [3]) overcomes this drawback by penalizing distant outliers only linearly.

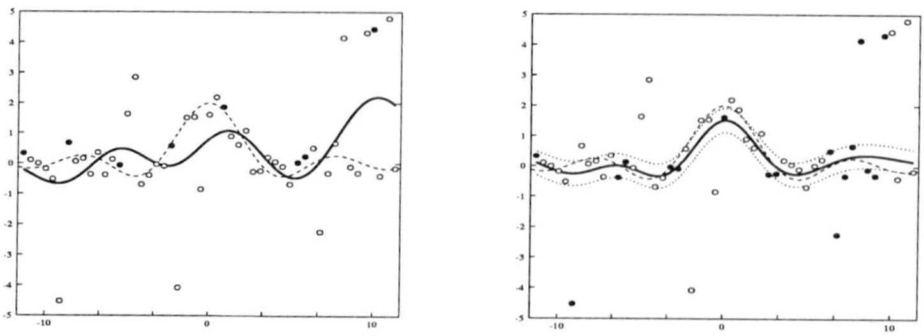

Fig. 3. LASSO results for fitting the noisy sinc-function with 20 % outliers. Left: quadratic loss, right: Huber's robust loss (region of quadratic growth depicted by the two dotted curves).

5 Discussion

Sparsity is an important feature of kernel regression models, since it simultaneously allows us to efficiently learn a regression function and to efficiently predict function values. For the SVM, highly tuned training algorithms have been developed during the last years. However, the SVM approach still suffers from the steep growth of the number of support vectors with increasing training sets. Experiments in this paper demonstrate that ARD models like RVM and LASSO produce solutions that are much sparser that the SVM solutions. Moreover, the number of *relevance vectors* is almost uncorrelated with the size of the training sample. Within the class of ARD models, however, the original RVM algorithm suffers from severe computational problems during the learning phase. We could demonstrate that the "kernelized" LASSO estimator overcomes this drawback while adopting the advantageous properties of the RVM. In addition, we have shown that *robust* LASSO variants employing loss functions of Huber's type are advantageous for situations in which the learning sample is corrupted by outliers.

Acknowledgments. The author would like to thank J. Buhmann and V. Steinhage for helpful discussions. Thanks for financial support go to German Research Council, DFG.

References

1. D.I. Clark and M.R. Osborne. On linear restricted and interval least-squares problems. *IMA Journal of Numerical Analysis*, 8:23–36, 1988.
2. Ronan Collobert and Samy Bengio. Support vector machines for large-scale regression problems. Technical Report IDIAP-RR-00-17, IDIAP, Martigny, Switzerland, 2000.
3. J.A. Fessler. Grouped coordinate descent algorithms for robust edge-preserving image restoration. In *SPIE, Image reconstruction and restoration II*, volume 3170, pages 184–194, 1997.
4. J.H. Friedman. Multivariate adaptive regression splines. *Annals of Stat.*, 19(1):1–82, 1991.
5. Y. Grandvalet. Least absolute shrinkage is equivalent to quadratic penalization. In L. Niklasson, M. Bodén, and T. Ziemske, editors, *ICANN'98*, pages 201–206. Springer, 1998.
6. M.R. Osborne, B. Presnell, and B.A. Turlach. A new approach to variable selection in least squares problems. *IMA Journal of Numerical Analysis*, 20(3):389–404, July 2000.
7. R.J. Tibshirani. Regression shrinkage and selection via the lasso. *Journal of the Royal Statistical Society*, B 58(1):267–288, 1996.
8. M.E. Tipping. The relevance vector machine. In S.A. Solla, T.K. Leen, and K.-R. Müller, editors, *Neural Information Processing Systems*, volume 12, pages 652–658. MIT Press, 1999.

Learning on Graphs in the Game of Go

Thore Graepel, Mike Goutrié, Marco Krüger and Ralf Herbrich

Computer Science Department
Technical University of Berlin
Berlin, Germany
{guru,mikepg,grisu,ralfh}@cs.tu-berlin.de

Abstract We consider the game of Go from the point of view of machine learning and as a well-defined domain for learning on graph representations. We discuss the representation of both board positions and candidate moves and introduce the common fate graph (CFG) as an adequate representation of board positions for learning. Single candidate moves are represented as feature vectors with features given by subgraphs relative to the given move in the CFG. Using this representation we train a support vector machine (SVM) and a kernel perceptron to discriminate good moves from bad moves on a collection of life-and-death problems and on 9×9 game records. We thus obtain kernel machines that solve Go problems and play 9×9 Go.

1 Introduction

Go (Chinese: Wei-Qi, Korean: Baduk) is an ancient oriental game about surrounding territory that originated in China over 4000 years ago. Its complexity by far exceeds that of chess, an observation well supported by the sucess of the computer program `Deep Blue`. In contrast, numerous attempts at reproducing such a result for the game of Go have been unsuccessful. As a consequence, Go appears to be an interesting testing ground for machine learning [2]. In particular, we consider the problem of *representation* because an adequate representation of the board position is an essential prerequisite for the application of machine learning to the game of Go.

A particularly elegant statement of the rules of Go is due to Tromp and Taylor[1] and is only slightly paraphrased for our purpose: (1) Go is played on an $N \times N$ square grid of points, by two players called Black and White. (2) Each point on the grid may be coloured *black*, *white* or *empty*. A point P is said to reach a colour C, if there exists a path of (vertically or horizontally) adjacent points of P's colour from P to a point of colour C. Clearing a colour is the process of emptying all points of that colour that do not reach *empty*. (3) Starting with an empty grid, the players alternate turns, starting with Black. A turn is either a pass; or a move that does not repeat an earlier grid colouring. A move consists of colouring an *empty* point one's own colour; then clearing the

[1] See `http://www.cwi.nl/~tromp/go.html` for a detailed description.

G. Dorffner, H. Bischof, and K. Hornik (Eds.): ICANN 2001, LNCS 2130, pp. 347–352, 2001.

opponent colour, and then clearing one's own colour. (4) The game ends after two consecutive passes. A player's score is the number of points of her colour, plus the number of empty points that reach only her colour. The player with the higher score at the end of the game is the winner. Equal scores result in a tie.

While the majority of computer Go programs work in the fashion of rule-based expert systems [4], several attempts have been made to apply machine learning techniques to Go. Two basic learning tasks can be identified: (1) Learning an evaluation function for board positions. (2) Learning an evaluation function for moves in given positions.

The first task was tackled in the framework of reinforcement learning by Schraudolph and Sejnowski [7] who learned a pointwise evaluation function by the application of $TD(\lambda)$ to a multi-layer perceptron (MLP). The second task found an application to *tsume* Go in [6] who used an MLP to find problem-solving moves.

All the known approaches suffer from a rather naive representation. In this paper we introduce a new representation for both board positions and candidate moves based on what we call a *common fate graph* (CFG). Our CFG representation builds on ideas first presented by Markus Enzensberger in the context of his Go program `Neurogo II`[2]. While our discussion is focused on the game of Go, our considerations about representation should be of interest with regard to all those domains were the standard feature vector representation does not adequately capture the structure of the learning problem. A natural problem domain that shares this characteristics is the classification of organic chemical compounds represented as graphs of atoms (nodes) and bounds (edges) [5].

2 Representation

An adequate representation of the learning problem at hand is an essential prerequisite for successful learning. One could even go as far as saying that an adequate representation should render the actual learning task trivial. If the objects to be classified are represented such that the intra-class distances are zero, while the inter-class distances are strictly greater than zero a simple nearest neighbour classifier would be able to solve the learning problem perfectly. More realistically, we aim at finding a representation that captures the structure of board positions by mapping similar positions (in the sense of the learning problem) to similar representations. Another desirable feature of a representation is a reduction in complexity: only those features relevant to the learning problem at hand should be retained.

Common Fate Graph The value of a given Go position is invariant under rotation and mirroring of the board. Also the rules of the game refer essentially only to the local neighbourhood structure of the game. A board position is thus adequately represented by its *full graph representation* (FGR), a graph with the

[2] "The Integration of A Priori Knowledge into a Go Playing Neural Network" available via `http://www.cgl.ucsf.edu/home/pett/go/Programs/NeuroGo-PS.html`

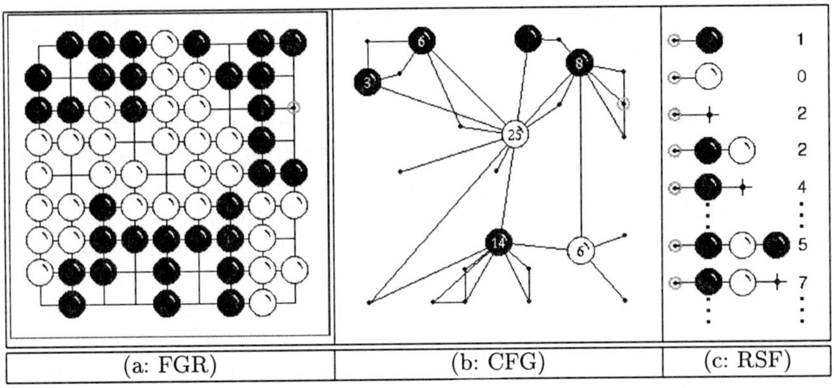

Figure 1. Illustration of the feature extraction process: **(a)** board position (FGR), **(b)** corresponding common fate graph (CFG), and **(c)** a selection of extracted relative subgraph features (RSF) w.r.t. the node marked by a gray circle in (a) and (b).

structure of an $N \times N$ square grid. Let us define the FGR $G_{\mathrm{FGR}} = (P, E)$ as an undirected connected graph $G_{\mathrm{FGR}} \in \mathcal{G}_{\mathrm{uc}}$. The set $P = \{p_1, \ldots, p_{N_P}\}$ of nodes p_i represents the points on the board. Also, each node $p \in P$ has any of three given labels $l : P \to \{black, white, empty\}$. The symmetric binary "edge" relation $E = \{e_1, \ldots, e_{N_E}\}$ with $e_i \in \{\{p,p'\} : p, p' \in P\}$ represents vertical or horizontal neighbourhood between points.

However, we observe that *black* or *white* points that belong to the same chain will always have a *common fate*: either all of them remain on the board or all of them are cleared. In any case we can represent them in a single node. We also reduce the number of edges by requiring that any two nodes may be connected by only a single edge representing their neighbourhood relation. The resulting *reduced graph representation* will be called a *common fate graph* (CFG) and will serve as the basic representation for our experiments. More formally, we define the graph transformation $T : \mathcal{G}_{\mathrm{uc}} \to \mathcal{G}_{\mathrm{uc}}$ by the following rule: given two nodes $p, p' \in P$ that are neighbours $\{p, p'\} \in E$ and that have the same non-empty label $l(p) = l(p') \neq empty$, perform the following transformation (1) $P \mapsto P \setminus \{p'\}$ to melt the node p' into p. (2) $E \mapsto (E \setminus \{\{p', p''\} \in E\}) \cup \{\{p, p''\} : \{p', p''\} \in E\}$ to connect the remaining node p to those nodes p'' formerly connected to p'.

Repeated application of the transformation T to G_{FGR} until no two neighbouring nodes have the same colour leads to the common fate graph G_{CFG}. The result of such a transformation is shown in Figure 1 (b). Clearly, the complexity of the representation has been greatly reduced while retaining essential structural information in the representation.

In how far is the CFG a suitable representation for learning in the game of Go? Go players' intuition is often driven by a certain kind of aesthetics that refers to the local structure of positions and is called *good or bad shape*. As an example consider the two white groups in Figure 1 (a) and (b). Although they look quite distinct to the layman in Figure 1 (a) they share the property of *being alive,* because they both have *two eyes* (i.e. two isolated internal empty points called *liberties*).

Relative Subgraph Features Unfortunately, almost all algorithms that aim at learning on graph representations directly suffer from severe scalability problems (see [5]). Most practically applicable learning algorithms operate on object representations known as feature vectors $\mathbf{x} \in \mathbb{R}^d$. We would thus like to extract feature vectors $\mathbf{x}$ from G_{CFG} for learning. Both learning tasks mentioned in the introduction can be formulated in terms of mappings from single points to real values. In both cases we would like to find a mapping $\boldsymbol{\phi} : \mathcal{G}_{\mathrm{uc}} \times P \to \mathbb{R}^d$ that maps a particular node $p \in P$ to a feature vector $\mathbf{x} \in \mathbb{R}^d$ given the context provided by the graph $G = (P, E) \in \mathcal{G}_{\mathrm{uc}}$, an idea inspired by [5], who apply context dependent classification to the mutagenicity of chemical compounds. We enumerate d possible connected subgraphs $\tilde{G}_i = (\tilde{P}_i, \tilde{E}_i) \in \mathcal{G}_{\mathrm{uc}}$, $i = \{1, \ldots, d\}$ of G such that $p \in \tilde{P}_i$. The relative subgraph feature $x_i = \phi_i(p)$ is then taken proportional to the number of times n_i the subgraph $\tilde{G}_i$ can be found in G and normalised to $\|\mathbf{x}\| = 1$.

Clearly, finding and counting subgraphs $\tilde{G}_i$ of G becomes quickly infeasible with increasing subgraph complexity as measured, e.g. by $|\tilde{P}_i|$and $|\tilde{E}_i|$. We therefore restrict ourselves to connected subgraphs $\tilde{G}_i$ with the shape of chains without branches or loops. In practice, we limit the features to a local context $|\tilde{P}_i| \leq s$, which — given the other two constraints — also limits the number d of distinguishable features.

3 Experimental Results

Learning Algorithms In our experiments we focused on learning the distinction between "good" and "bad" moves. However, we are really interested in the real-valued output of the classifier which enables us to order and select candidate moves according to their quality. We choose the class of binary kernel classifiers for learning. The predictions of these classifiers are given by $\hat{y}(\mathbf{x}) = \mathrm{sign}(f(\mathbf{x})) = \mathrm{sign}\left(\sum_{i=1}^{m} \alpha_i k(\mathbf{x}_i, \mathbf{x})\right)$. The α_i are the adjustable parameters and $k : \mathbb{R}^d \times \mathbb{R}^d \to \mathbb{R}$ is the kernel function. These classifiers have recently gained a lot of popularity due to the success of the support vector machine (SVM) [8]. We decided to use an RBF kernel with diagonal penalty term of the form $k(\mathbf{x}, \mathbf{x}') = \exp\left(-\|\mathbf{x} - \mathbf{x}'\|^2 / \sigma^2\right) + \lambda \mathbf{I}_{\mathbf{x}=\mathbf{x}'}$, where σ^2 controls the width of the kernel, and the second term makes the kernel matrix more well-conditioned. Two methods of "learning" the expansion coefficients α_i were employed: (1) A simple kernel perceptron [1] (2) A soft margin support vector machine[3] [3]

Life and Death An interesting challenge in the game of Go — referred to as *tsume Go* — is to kill opponent groups and to save one's own groups from

[3] Publicly available via `http://www.kernel-machines.org/`

test \ train	white	black	w&b	#probs	#moves	%good
white	65.8/65.3	48.0/48.8	63.8/62.6	1455	10.4	13.4
black	57.7/57.3	66.4/64.5	65.9/65.9	1711	9.8	23.0
w&b	61.5/61.0	58.0/57.3	64.9/64.4	3166	10.0	18.0
# pts	2718	2682	5400			

Table 1. Results of the selection of *tsume* Go moves by an SVM (left) and a kernel perceptron (right). Shown is the percentage of finding a problem-solving move. The last column indicates the average percentage of moves that solve the problem.

being killed. Our *tsume* Go study was based on a database of 40000 computer-generated Go problems with solution by Thomas Wolf [9]. In order to calibrate the parameters of both the representation and the learning schemes, we created a simple task: The training set consisted of $m = 2700$ moves, where we took the best move provided by Wolf's problem solver GoTools [9] to be "good" and the worst one to be "bad". The length s of the subgraph chains was chosen to be $s = 6$ resulting in $d \approx 400$ features. Using a held-out data set of size $m_{\text{heldout}} \approx 3000$ we systematically scanned parameter space. For the SVM we set $\lambda = 0$ (relying on C for regularisation) and found the optimal parameter values to be $\sigma = 1$ and $C = 10$. For the kernel perceptron we used the same value of σ and found $\lambda = 0.05$ to be optimal. Both solutions had a sparsity level $\|\boldsymbol{\alpha}\|_0 / m$ of approximately 50% and lead to a success rate of up to 85% for discriminating between the best and the worst move in a given problem.

In a similar setting, we applied the resulting classifier to all the possible moves in a problem and used the real-valued output $f(\mathbf{x})$ for each legal move $\mathbf{x}$ for ranking. We counted a success when the move ranked highest by the classifier was indeed one of the winner moves. The results are given in Table 1. Focusing on the full training and test sets (*w&b*) the success rate is more than 3 times that of random guessing. Although we were not able to obtain the data-base used in [6] due to copyright problems, our result of a 65% success rate compares favourably with the 50% reported in [6] on similar problems using a naive local context representation.

Game Play We applied the same learning paradigm as used above, but this time to a collection of high quality (Shodan and better) 9×9 Go game records collected from the Internet Go Server (IGS) and pre-processed and archived by Nicol Schraudolph. For each of the $m \approx 2500$ moves played we randomly generated an arbitrary counterpart as a "bad" move. We provide two samples of the SVMs play against Gnu-Go[4] in Figure 2. Connoisseurs of the game will appreciate that the machine plays amazingly coherent considering that it takes into account only local shape and has no concept of territory and urgency.

[4] Publicly available via `http://www.gnu.org/software/gnugo/`

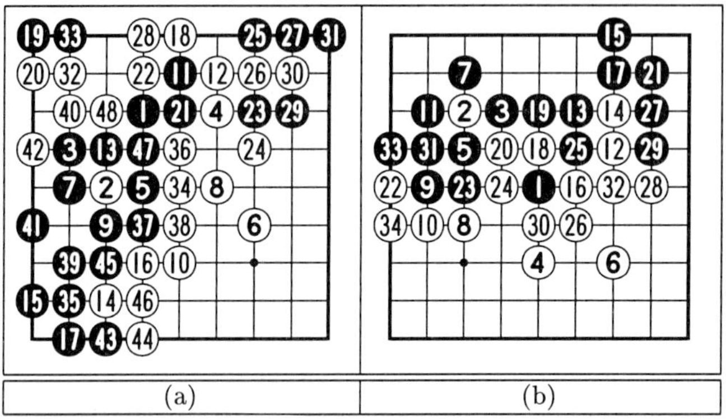

Figure 2. Two examples of games played by an SVM (Black) against Gnu-Go (White). The SVM was trained on 9 × 9 game records from good amateur games.

Conclusion and Acknowledgements We presented Go as a successful application of learning based on feature vectors that are extracted from a graph representation. While the approach appears to be particularly suitable for the game of Go it seems that other domains could benefit from these ideas as well.

We would like to thank Nici Schraudolph and Thomas Wolf for providing datasets. Also we thank the members of the computer-go mailing list for invaluable comments and Klaus Obermayer for hosting our project.

References

1. M. Aizerman, E. Braverman, and L. Rozonoer. Theoretical foundations of the potential function method in pattern recognition learning. *Automation and Remote Control*, 25:821–837, 1964.
2. J. Burmeister and J. Wiles. The challenge of go as a domain for ai research: A comparison between go and chess. In *Proceedings of the 3rd Australian and New Zealand Conference on Intelligent Information Systems*, 1994.
3. C. Cortes and V. Vapnik. Support Vector Networks. *Machine Learning*, 20:273–297, 1995.
4. D. Fotland. Knowledge representation in The Many Faces of Go, 1993.
5. P. Geibel and F. Wysotzki. Learning relational concepts with decision trees. In *Machine Learning: Proceedings of the Thirteenth International Conference*, pages 1141–1144. Morgan Kaufmann Publishers, 1998.
6. N. Sasaki and Y. Sawada. Neural networks for tsume-go problems. In *Proceedings of the Fifth International Conference on Neural Information Processing*, pages 1141–1144, 1998.
7. N. N. Schraudolph, P. Dayan, and T. J. Sejnowski. Temporal difference learning of position evaluation in the game of go. In J. D. Cowan, G. Tesauro, and J. Alspector, editors, *Advances in Neural Information Processing Systems*, volume 6, pages 817–824. Morgan Kaufmann Publishers, Inc., 1994.
8. V. Vapnik. *Statistical Learning Theory.* John Wiley and Sons, New York, 1998.
9. T. Wolf. The program GoTools and its computer-generated Tsume Go database. In *Proceedings of the 1st Game Programming Workshop*, 1994.

Nonlinear Feature Extraction Using Generalized Canonical Correlation Analysis

Thomas Melzer, Michael Reiter, and Horst Bischof*

The authors are with the Pattern Recognition and Image Processing Group, Vienna University of Technology, Vienna, Austria
{melzer,rei,bis}@prip.tuwien.ac.at

Abstract. This paper introduces a new non-linear feature extraction technique based on *Canonical Correlation Analysis* (CCA) with applications in regression and object recognition. The non-linear transformation of the input data is performed using kernel-methods. Although, in this respect, our approach is similar to other *generalized* linear methods like kernel-PCA, our method is especially well suited for relating two sets of measurements. The benefits of our method compared to standard feature extraction methods based on PCA will be illustrated with several experiments from the field of object recognition and pose estimation.

1 Introduction

When dealing with high-dimensional observations, linear mappings are often used to reduce the dimensionality of the data by extracting a small (compared to the superficial dimensionality of the data) number of linear features, thus alleviating subsequent computations. A prominent example of a linear feature extractor is *Principal Component Analysis* (PCA [4]). Among all linear, orthonormal transformations, PCA is optimal in the sense that it minimizes, in the mean square sense, the reconstruction error between the original signal $\mathbf{x}$ and the signal $\hat{\mathbf{x}}$ reconstructed from its low-dimensional representation $\mathbf{f}(\mathbf{x})$. During the recent years, PCA has been especially popular in the object recognition community, where it has successfully been employed in various applications such as face recognition [13], illumination planning [9], visual inspection and even visual servoing [11].

Although this demonstrates the broad applicability of PCA, one has to bear in mind that the goal of PCA is minimization of the reconstruction error; in particular, PCA-features are not well suited for regression tasks. Consider a mapping $\phi : \mathbf{x} \mapsto \mathbf{y}$. There is no reason to believe that the features extracted by PCA on the variable $\mathbf{x}$ will reflect the functional relation between $\mathbf{x}$ and $\mathbf{y}$ in any way; even worse, it is possible that information vital to establishing this relation is discarded when projecting the original data onto the PCA-feature space.

* This work was supported by the Austrian Science Foundation (FWF) under grant no. P13981-INF.

G. Dorffner, H. Bischof, and K. Hornik (Eds.): ICANN 2001, LNCS 2130, pp. 353–360, 2001.

There exist, however, several other linear methods that are better suited for regression tasks, for example *Partial Least Squares* (PLS [3]), *Multivariate Linear Regression* (MLR, also referred to as *Reduced Rank Wiener Filtering*, see for example [2]) and *Canonical Correlation Analysis* (CCA [5]). Among these three, only MLR gives a direct solution to the linear regression problem. PLS and CCA will find pairs of directions that yield maximum covariance resp. maximum correlation between the two random variables $\mathbf{x}, \mathbf{y}$; regression can then be performed on these features. CCA, in particular, has some very attractive properties (for example, it is invariant w.r.t. affine transformations - and thus scaling - of the input variables) and can not only be used for regression purposes, but whenever we need to establish a relation between two sets of measurements (e.g., finding corresponding points in stereo images [1]).

As an example for CCA, consider constructing a parametric manifold for pose estimation [10]. Fig. 1(a) shows two extreme views of an object, which was acquired with two varying pose parameters (pan and tilt). Let $\mathbf{X}$ denote the set of training images and $\mathbf{Y}$ the set of corresponding pose parameters.The visualization of the manifold given in Fig. 1(b) is obtained by plotting the projections of the training set onto the first three eigenvectors obtained by standard PCA, whereby neighboring (w.r.t. the pose parameters) projections are connected.

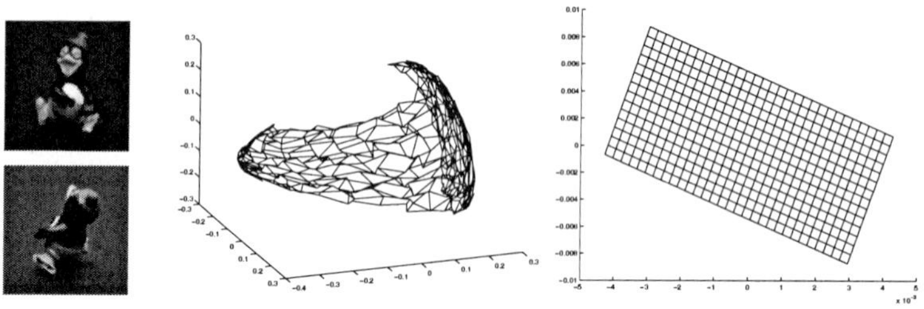

(a) (b) (c)

Fig. 1. Extreme views of training set (a) and the parametric manifold obtained with PCA (b) and CCA (c).

The parametric manifold serves as a starting point for computing pose estimates for new input images. The standard-approach for retrieving these estimates is to resample the manifold using, e.g., bicubic spline interpolation and then to perform a nearest neighbor search for each new image [10].

Fig. 1(c) shows the manifold obtained by projecting the training images onto the first two directions found by computing CCA on $\mathbf{X}$ and $\mathbf{Y}$ (the number of factors obtained by CCA is limited by the dimensionality of the lower-dimensional set). In contrast to the PCA-manifold, the CCA-factors span a perfect grid; one could also say that projections of the training images onto the two linear features found by CCA are topologically ordered w.r.t. their associated pose parameters. It is obvious that pose estimation on the manifold obtained by CCA is much easier than on the PCA-manifold.

Recently, we proposed a non-linear extension of CCA by the use of kernel-methods [7]. Kernel-methods have become increasingly popular during the last few years, and have already been applied to PCA [12] and the *Fisher Discriminant* [8]. In our derivation of kernel-CCA we have used the fact that the solutions (principal directions) of CCA can be obtained as the extremum points of an appropriately chosen *Rayleigh Quotient* (this is also true for the other linear techniques discussed thus far, see [1]).

In this paper we will demonstrate the benefits of kernel-CCA with an application in the field of appearance-based pose estimation. To this end we will compare the performance of features obtained by PCA, standard CCA and kernel-CCA.

The rest of this paper is organized as follows: in section 2, we will give a brief introduction to "classical" CCA and show how kernel-methods can be applied to CCA. In section 3 we will apply kernel-CCA for pose estimation. Conclusions will be given in section 4.

2 Canonical Correlation Analysis (CCA)

2.1 What Is CCA?

Given two zero-mean random variables $\mathbf{x} \in \mathbb{R}^p$ and $\mathbf{y} \in \mathbb{R}^q$, CCA finds pairs of directions $\mathbf{w}_x$ and $\mathbf{w}_y$ that maximize the correlation between the projections $x = \mathbf{w}_x^T\mathbf{x}$ and $y = \mathbf{w}_y^T\mathbf{y}$ (in the context of CCA, the projections x and y are also referred to as *canonical variates*). More formally, CCA maximizes the function:

$$\rho = \frac{E[xy]}{\sqrt{E[x^2]E[y^2]}} = \frac{E[\mathbf{w}_x^T\mathbf{x}\mathbf{y}^T\mathbf{w}_y]}{\sqrt{E[\mathbf{w}_x^T\mathbf{x}\mathbf{x}^T\mathbf{w}_x]E[\mathbf{w}_y^T\mathbf{y}\mathbf{y}^T\mathbf{w}_y]}}, \tag{1}$$

$$\rho = \frac{\mathbf{w}_x^T\mathbf{C}_{xy}\mathbf{w}_y}{\sqrt{\mathbf{w}_x^T\mathbf{C}_{xx}\mathbf{w}_x\mathbf{w}_y^T\mathbf{C}_{xy}\mathbf{w}_y}}. \tag{2}$$

Let

$$\mathbf{A} = \begin{pmatrix} \mathbf{0} & \mathbf{C}_{xy} \\ \mathbf{C}_{yx} & \mathbf{0} \end{pmatrix}, \quad \mathbf{B} = \begin{pmatrix} \mathbf{C}_{xx} & \mathbf{0} \\ \mathbf{0} & \mathbf{C}_{yy} \end{pmatrix}. \tag{3}$$

It can be shown [1] that the stationary points $\mathbf{w}^* = (\mathbf{w}_x^{*T}, \mathbf{w}_y^{*T})^T$ of ρ (i.e., the points satisfying $\nabla\rho(\mathbf{w}^*) = \mathbf{0}$) coincide with the stationary points of the *Rayleigh Quotient*:

$$r = \frac{\mathbf{w}^T\mathbf{A}\mathbf{w}}{\mathbf{w}^T\mathbf{B}\mathbf{w}}, \tag{4}$$

and thus, by virtue of the *Generalized Spectral Theorem* [2], can be obtained as solutions (i.e., eigenvectors) of the corresponding generalized eigenproblem:

$$\mathbf{A}\mathbf{w} = \mu\mathbf{B}\mathbf{w}. \tag{5}$$

The extremum values $\rho(\mathbf{w}^*)$, which are referred to as *canonical correlations*, are equally obtained as the corresponding extremum values of Eq. 4 or the eigenvalues of Eq. 5, respectively, i.e., $\rho(\mathbf{w}^*) = r(\mathbf{w}^*) = \mu(\mathbf{w}^*)$.

2.2 Kernel CCA

In this section we will briefly summarize the formulation of kernel-CCA, which can be used to find non-linear dependencies between two sets of observations. A detailed derivation of the algorithm can be found in [6].

Given n pairs of mean-normalized observations $(\mathbf{x}_i^T, \mathbf{y}_i^T)^T \in \mathbb{R}^{p+q}$, and data matrices $\mathbf{X} = (\mathbf{x}_1..\mathbf{x}_n) \in \mathbb{R}^{p\times n}$, $\mathbf{Y} = (\mathbf{y}_1..\mathbf{y}_n) \in \mathbb{R}^{q\times n}$, we obtain the estimates for the covariance matrices $\mathbf{A}, \mathbf{B}$ in Eq. 3 as

$$\hat{\mathbf{A}} = \frac{1}{n}\begin{pmatrix} \mathbf{0} & \mathbf{X}\mathbf{Y}^T \\ \mathbf{Y}\mathbf{X}^T & \mathbf{0} \end{pmatrix}, \; \hat{\mathbf{B}} = \frac{1}{n}\begin{pmatrix} \mathbf{X}\mathbf{X}^T & \mathbf{0} \\ \mathbf{0} & \mathbf{Y}\mathbf{Y}^T \end{pmatrix} \tag{6}$$

If the mean was estimated from the data, we have to replace n by $n-1$ in both equations.

As we know from section 2.1, computing the CCA between the data sets $\mathbf{X}, \mathbf{Y}$ amounts to determining the extremum points of the *Rayleigh Quotient* (see Eq. 4). It can be shown [6], that for all solutions $\mathbf{w}^* = (\mathbf{w}_x^{*T}, \mathbf{w}_y^{*T})^T$ of Eq. 5, the component vectors $\mathbf{w}_x^*, \mathbf{w}_y^*$ lie in the span of the training data (i.e., $\mathbf{w}_x^* \in span(\mathbf{X})$ and $\mathbf{w}_y^* \in span(\mathbf{Y})$). Under this assumption for each eigenvector $\mathbf{w}^* = (\mathbf{w}_x^{*T}, \mathbf{w}_y^{*T})^T$ solving Eq. 5, there exist vectors $\mathbf{f}, \mathbf{g} \in \mathbb{R}^n$, so that $\mathbf{w}_x^* = \mathbf{X}f$ and $\mathbf{w}_y^* = \mathbf{Y}g$. Thus, CCA can completely be expressed in terms of dot products. This allows us to reformulate the *Rayleigh Quotient* using only inner products:

$$\frac{\begin{pmatrix} \mathbf{f}^T & \mathbf{g}^T \end{pmatrix} \begin{pmatrix} \mathbf{0} & \mathbf{KL} \\ \mathbf{LK} & \mathbf{0} \end{pmatrix} \begin{pmatrix} \mathbf{f} \\ \mathbf{g} \end{pmatrix}}{\begin{pmatrix} \mathbf{f}^T & \mathbf{g}^T \end{pmatrix} \begin{pmatrix} \mathbf{K}^2 & \mathbf{0} \\ \mathbf{0} & \mathbf{L}^2 \end{pmatrix} \begin{pmatrix} \mathbf{f} \\ \mathbf{g} \end{pmatrix}}, \tag{7}$$

where $\mathbf{K}, \mathbf{L}$ are Gram matrices defined by $\mathbf{K}_{ij} = \mathbf{x}_i^T \mathbf{x}_j$ and $\mathbf{L}_{ij} = \mathbf{y}_i^T \mathbf{y}_j$, $\mathbf{K}, \mathbf{L} \in \mathbb{R}^{n\times n}$. The new formulation makes it possible to compute CCA on non-linearly mapped data without actually having to compute the mapping itself. This can be done by substituting the gram matrices $\mathbf{K}, \mathbf{L}$ by kernel matrices ${\mathbf{K}^\phi}_{ij} = \phi(\mathbf{x}_i)^T \phi(\mathbf{x}_j) = k_\phi(\mathbf{x}_i, \mathbf{x}_j)$ and ${\mathbf{L}^\theta}_{ij} = \theta(\mathbf{y}_i)^T \theta(\mathbf{y}_j) = k_\theta(\mathbf{y}_i, \mathbf{y}_j)$. $k_\phi(.,.), k_\theta(.,.)$ are the kernel functions corresponding to the nonlinear mappings $\phi : \mathbb{R}^p \to \mathbb{R}^{m_1}$ resp. $\theta : \mathbb{R}^q \to \mathbb{R}^{m_2}$.

The projections onto $\mathbf{w}_\phi^*$ can be computed using only the kernel function, without having to evaluate ϕ itself:

$$\phi(\mathbf{x})^T \mathbf{w}_\phi^* = \sum_{i=1}^{n} f_{\phi i} \phi(\mathbf{x})^T \phi(\mathbf{x}_i) = \sum_{i=1}^{n} f_{\phi i} k(\mathbf{x}, \mathbf{x}_i). \tag{8}$$

The projections onto $\mathbf{w}_\theta^*$ are obtained analogously.

Note that using kernel-CCA, we can compute more than $min(p, q)$ (p, q being the dimensionality of the variable $\mathbf{x}$ and $\mathbf{y}$, respectively) factor pairs, which is the limit imposed by classical CCA.

3 Experiments

In the following example we apply CCA to a pose estimation problem, where we relate images of objects at varying pose to corresponding pose parameters. Experiments were conducted on three test objects, detailed quantitative figures are given only for object 2(a). However, results for the remaining objects are similar.

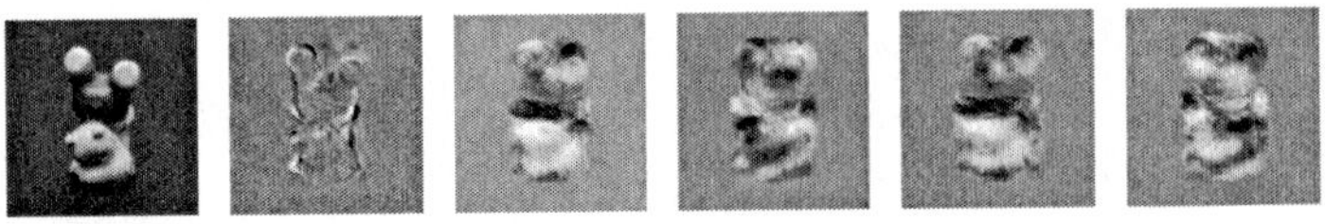

Fig. 2. (a) Image $\mathbf{x}_1$ of one of our test objects. (b) Canonical factor $\mathbf{w}_x^*$ computed on $\mathbf{X}_8^t$. (c),(d) Canonical factors obtained on the same set using a non-linear (trigonometric) representation of the output parameter space. (e),(f) 2 factors obtained by kernel-CCA with canonical correlations $\rho = 1.0$.

Let $\mathbf{X} = \langle \mathbf{x}_i | 1 \leq i \leq 180 \rangle$ denote the set of images and $\mathbf{Y} = \langle y_i | y_i \in [0, 358] \rangle$ corresponding pose parameters (horizontal orientation of the object w.r.t. the camera in degrees). The images are represented as 128^2-dimensional vectors that are obtained by sequentially stacking the image columns.

Each image set $\mathbf{X}$ was subsampled to obtain a subset of images that was used as a training set. The remaining images were assigned to the corresponding test set. Let $\mathbf{X}_k$ denote a training set that was generated by subsampling $\mathbf{X}$ at every kth position[1]. Since y_{i_k} is a scalar, standard CCA yields only one feature vector (canonical factor) $\mathbf{w}_x^*$ (see figure 2(b)). Figure 4 (in the upper 2 plots) gives a quantitative comparison of the pose estimation error ϵ (linear regression model) using the PCA and standard CCA. The pose estimation error ϵ was calculated on the test set as the absolute difference between known and estimated parameter value (orientation in degrees).

Figure 3(a) shows a plot of pose estimates for the complete set of images $\mathbf{X}$ obtained using CCA. The dotted line indicates the true pose parameter values y_i. Pose parameter values of $\mathbf{X}_k$ are marked by filled circles. In figure 3 a linear least-square mapping was used to map the projections of the training set to pose parameters.

Note that the pose estimation error grows rapidly around $i = 180$. This problem is due to the fact that the scalar representation for y_i has a discontinuity at $y_i = 360$. From figure 2(a) it can be seen that the main part of information held in $\mathbf{w}_x^*$ is about the transition of image $\mathbf{x}_{180}$ to image $\mathbf{x}_1$.

For this reason we chose an periodic, trigonometric representation of pose parameters $\mathbf{y}_i = [sin(y_i), cos(y_i)]^T$. The object's pose is now characterized by 2

[1] $\mathbf{X}_k$ contains all images $\mathbf{x}_{i_k}$ with $i_k \in \{1, k, 2k, \ldots, \lfloor 180/k \rfloor\}$. Since the original set shows the object in 2 degrees steps, $\mathbf{X}_k$ shows the object at $2k$ degrees steps.

parameters, and thus CCA yields 2 canonical factors (see figure 2 (c) and (d)) in the image vector space. Estimates obtained for these intermediate output values (again using linear regression) are given in 3(b), while the final pose estimates (obtained by combining these two estimates using atan2) are given in figure 3(c). Thus, by using *a priori* knowledge about the problem domain a significant increase in pose estimation accuracy could be obtained.

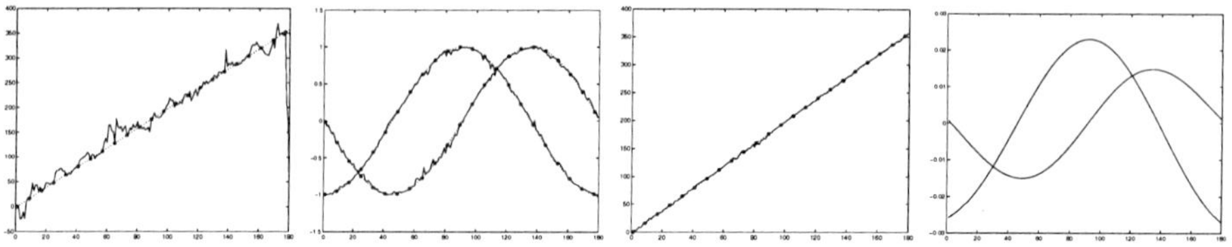

(a) (b) (c) (d)

Fig. 3. Output parameter estimates obtained from feature projections by the linear regression function of the training set. Horizontal axes correspond to the image indices, vertical axes to estimated output parameter values. The dotted line indicates the true pose parameter values. Parameter values of the training set are marked by filled circles. (a) shows estimates using scalar representation of orientation (b) using trigonometric representation. (c) Estimated pose obtained from trigonometric representation using the four-quadrant arc tangens. Note that the accuracy of estimated pose parameters can be improved considerably. (d) Projections onto factors obtained using kernel-CCA: optimal factors can be obtained automatically.

The last experiment shows the relative performance of CCA, kernel-CCA and PCA when using spline interpolation and resampling [10] for pose estimation. For standard CCA we used the hard-coded trigonometric pose representation (yielding 2 factor pairs), while kernel-CCA (using a RBF-kernel with $\sigma = 1.6$) obtained similar factors automatically from the original scalar pose representation (see figures 2 (e)-(f) and 3(d)). Results are given in figure 4 (middle row). The lowermost two plots of figure 4 show results for PCA. The pose estimation error is significantly larger compared to CCA when using the same number of features.

4 Conclusion and Outlook

Although little known in the field of pattern recognition and signal processing, CCA is a very powerful and versatile statistical tool that is especially well suited for relating two sets of measurements. CCA, like PCA, can also be regarded as a linear feature extractor. CCA-features are, however, much better suited for regression tasks than features obtained by PCA; this was demonstrated in section 1 for the task of computing a parametric object manifold for pose estimation.

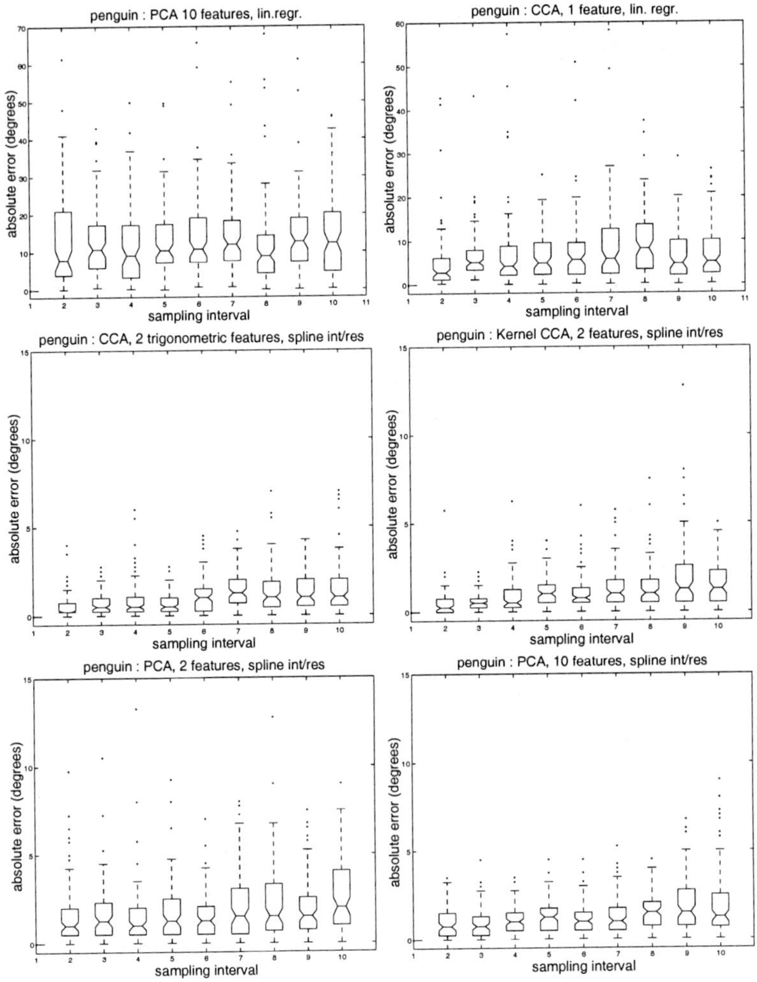

Fig. 4. Uppermost two plots: Comparison of pose estimation error when using a linear regression from factor projections to pose parameters. The left plot shows the errors for a 10 dimensional PCA-feature space. The right plot shows the results when using only one CCA-feature. Image sets have been subsampled at different intervals (horizontal axis) to obtain increasingly smaller training sets. Subsampling was done at at 9 different sampling intervals ($k = \langle 2, ..., 10 \rangle$). Note that the plots have been scaled differently.
Plots in the middle row: Pose estimation error for CCA-features. The left column shows results for "classical" CCA when using a 2 dimensional trigonometric representation of output parameters. The right column shows results when using 2 features obtained by kernel-CCA and a scalar pose representation.
Lower two plots: Pose estimation error using the first 10 eigenvectors (left plot) and only 2 eigenvectors. (right plot). Estimates were obtained from feature projections using spline interpolation and resampling.

In section 2, we discussed how to non-linearly extend CCA by using kernel-functions. **Kernel-CCA** is a efficient non-linear feature extractor, which also overcomes some of the limitations of classical CCA.

Finally, in section 3, we applied kernel-CCA to an object pose estimation problem. There, it was also shown that kernel-CCA will automatically find an optimal, periodic representation for a training set containing object views ranging from 0 to 360 degrees (i.e., for *periodic* data).

Currently, we are investigating methods for obtaining the desired pose parameters (or, in general, output estimates) from the projections onto the transformed output basis vectors $\mathbf{w}_\phi$, i.e., for inverting a kernelized output representation.

References

1. Magnus Borga. *Learning Multidimensional Signal Processing.* Linköping Studies in Science and Technology, Dissertations, No. 531. Department of Electrical Engineering, Linköping University, Linköping, Sweden, 1998.
2. Konstantinos I. Diamantaras and S.Y. Kung. *Principal Component Neural Networks.* John Wiley & Sons, 1996.
3. A. Höskuldsson. PLS regression methods. *Journal of Chemometrics*, 2:211–228, 1988.
4. H. Hotelling. Analysis of a complex of statistical variables into principal components. *Journal of Educational Psychology*, 24:498–520, 1933.
5. H. Hotelling. Relations between two sets of variates. *Biometrika*, 8:321–377, 1936.
6. Thomas Melzer, Michael Reiter, and Horst Bischof. Kernel CCA: A nonlinear extension of canonical correlation analysis. *Submitted to IEEE Trans. Neural Networks*, 2001.
7. Thomas Melzer and Micheal Reiter. Pose estimation using parametric stereo eigenspaces. In Tomas Svoboda, editor, *Czech Pattern Recognition Workshop*, pages 77–80. Czech Pattern Recognition Society Praha, 2000.
8. S. Mika, G. Rätsch, J. Weston, B. Schölkopf, and K.-R. Müller. Fisher discriminant analysis with kernels. In Y.-H. Hu, J. Larsen, E. Wilson, and S. Douglas, editors, *Neural Networks for Signal Processing*, volume 9, pages 41–48. IEEE, 1999.
9. Hiroshi. Murase and Shree K. Nayar. Illumination planning for object recognition using parametric eigenspaces. *IEEE Trans. Pattern Analysis and Machine Intelligence*, 16(12):1219–1227, December 1994.
10. Hiroshi Murase and Shree K. Nayar. Visual learning and recognition of 3-d objects from appearance. *International Journal of Computer Vision*, 14(1):5–24, January 1995.
11. Shree K. Nayar, Sameer A. Nene, and Hiroshi Murase. Subspace methods for robot vision. *IEEE Trans. Robotics and Automation*, 12(5):750–758, October 1996.
12. Bernhard Schölkopf, Alex Smola, and K.-R. Müller. Nonlinear component analysis as a kernel eigenvalue problem. *Neural Computation*, 10:1299–1319, 1998.
13. Matthew Turk and Alexander P. Pentland. Eigenfaces for recognition. *Journal of Cognitive Neuroscience*, 4(1):71–86, 1991.

Gaussian Process Approach to Stochastic Spiking Neurons with Reset

Ken-ichi Amemori[1,2] and Shin Ishii[1,2]

[1] Nara Institute of Science and Technology,
Takayama 8916-5, Ikoma, Nara 630-0101, Japan
[2] CREST, Japan Science and Technology Corporation
keniti-a@is.aist-nara.ac.jp

Abstract. This article theoretically examines the behavior of spiking neurons whose input spikes obey an inhomogeneous Poisson process. Since the probability density of the membrane potential converges to a Gaussian distribution, the stochastic process becomes a Gaussian process. With a frequently-used spike response function, the process becomes a multiple-Markov Gaussian process. We develop a method which can precisely calculate the dynamics of the membrane potential and the firing probability. The effect of reset after firing is also considered. We find that the synaptic time constant of the spike response function, which has often been ignored in existing stochastic process studies, has significant influence on the firing probability.

1 Introduction

Stochastic nature of spikes has been reported in many areas of the cortex. Spike variability that seemingly obeys a Poisson process is observed not only at a low spontaneous firing rate but also at a high firing rate [7]. In spite of this variablity, stimulus-induced spike modulation can be observed in temporal patterns of spikes averaged over trials (e.g., *post-stimulus time histogram*). For instance, MT neurons of a behaving monkey exhibit precisely modulated spikes, and the modulation is almost invariant over trials [3]. In the visual cortex, the neuronal responses vary a lot even when they are evoked by the same stimulus. However, by adding the response averaged over trials to the initial neuron state, the actual single response can be well estimated [2]. Accordingly, spikes are non-stational stochastic events, and the temporal behaviors can be extracted from the trial average of spikes. Here we assume that the spikes are approximately independent of each other, and the spike frequency may temporally change. Such stochastic nature is theoretically modeled by an *inhomogeneous* Poisson process.

In this article, we theoretically examine the behavior of spiking neurons whose input spikes obey an inhomogeneous Poisson process. First, we develop a Gaussian process approach in order to precisely calculate the dynamics of the membrane potential and the firing probability. Our forward type calculation method is able to consider the effect of reset, which has not been dealt with in the existing backward type calculation method [4]. Second, we find that the synaptic time constant of the spike response function, which has often been ignored in existing stochastic process studies [8], has significant influence on the firing probability.

G. Dorffner, H. Bischof, and K. Hornik (Eds.): ICANN 2001, LNCS 2130, pp. 361–368, 2001.

2 Stochastic Spiking Neuron Model

Spiking Neurons Spiking neuron models assume that neuronal activities are described based on the spikes that are input to or emitted by the neuron. The spike response model [5] is one of the spiking neuron models. In that model, membrane potential of a neuron is defined by the weighted summation of post-synaptic potentials, each of which is induced by a single input spike. The membrane potential is given by

$$v(t) = \sum_{j=1}^{N} w_j \sum_{f=1}^{n_j(t)} u(t - t_j^f).$$

Here, N is the number of synapses projecting to the neuron, w_j is the transmission efficacy of synapse j, $n_j(t)$ is the number of spikes that enter through synapse j until time t and t_j^f is the time at which the f-th spike enters through synapse j. $u(s)$ is called the spike response function.

In this article, we assume that the spike response function has the following function form:

$$u(s) = \frac{\tau_m}{\tau_m - \tau_s}\left[\exp\left(-\frac{s}{\tau_m}\right) - \exp\left(-\frac{s}{\tau_s}\right)\right]\mathcal{H}(s) = \frac{\beta}{\beta - \alpha}(e^{-\alpha s} - e^{-\beta s})\mathcal{H}(s), \quad (1)$$

where step function $\mathcal{H}(s)$ is defined by $\mathcal{H}(s) = \begin{cases} 1 \ (s \geq 0) \\ 0 \ (s < 0) \end{cases}$. $\tau_m = 1/\alpha$ is called the membrane time constant, and $\tau_s = 1/\beta$ is called the synaptic time constant. When $\beta \to \alpha$, function (1) approaches

$$u(s) = \frac{s}{\tau_m}\exp\left(-\frac{s}{\tau_m}\right)\mathcal{H}(s) = \alpha s e^{-\alpha s}\mathcal{H}(s), \quad (2)$$

which is called alpha-function. When $\beta \to \infty$, function (1) approaches

$$u(s) = \exp\left(-\frac{s}{\tau_m}\right)\mathcal{H}(s) = e^{-\alpha s}\mathcal{H}(s). \quad (3)$$

The shapes of the response functions are shown in Figure 1 (left).

Poissonian Spikes Based on the spike response model, we examine stochastic properties of a single neuron, when the spikes input to the neuron obey an inhomogeneous Poisson process. The input spike sequence, $\{t_j^f | f = 1, \ldots, n_j(t)\}$, is described by random variables. Subsequently, random variables are denoted by bold-math fonts. Given the spike response function $u(s)$, we examine the property of membrane potential:

$$\boldsymbol{v}(t) = \sum_{j=1}^{N} w_j \sum_{f=1}^{\boldsymbol{n}_j(t)} u(t - \boldsymbol{t}_j^f) \equiv \sum_{j=1}^{N} \boldsymbol{x}_j(t),$$

where $\boldsymbol{x}_j(t)$ is the summation of the *post-synaptic potentials* (PSPs) corresponding to synapse j. When the membrane potential $\boldsymbol{v}(t)$ becomes larger than the firing threshold θ,

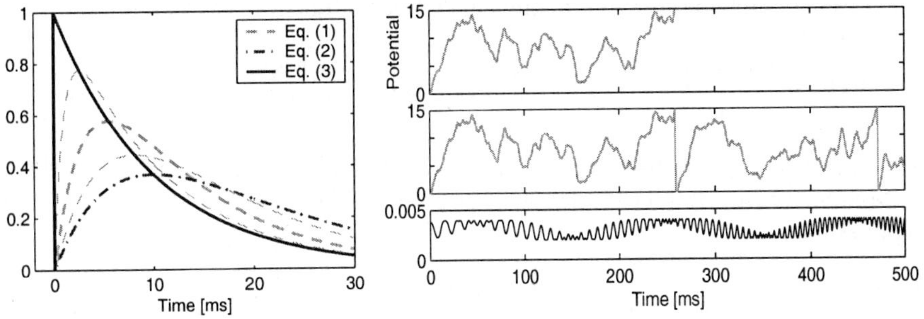

Fig. 1. (left) Shape of response function $u(s)$, where $\alpha = 0.1$ [ms]. The dash lines denote function (1), where $\beta = 1.0, 0.3,$ and 0.15 [ms] from left to right. The dash-dotted line denotes function (2), and the solid line denotes function (3). All of the functions are normalized so that their integrals are $1/\alpha$. (right) Examples of the dynamics of the membrane potential. Upper figure denotes the membrane potential when the absorbing threshold is assumed. Middle figure denotes the membrane potential when the reset after firing is considered. The input spikes obey an inhomogeneous Poisson process whose intensity is described in the lower figure. $\theta = 15$ and β =0.3.

the neuron fires and the membrane potential is reset to zero. We examine the stochastic process of an ensemble of such neurons.

If the number of synapses, N, is large, the density of the membrane potential $\boldsymbol{v}(t)$ can be approximated by a Gaussian distribution [4]. When spikes are assumed to obey an inhomogeneous Poisson process, all the spikes are mutually independent. $\boldsymbol{x}_j(t)$ then becomes the summation of the independent random variables, and the probability density of $\boldsymbol{x}_j(t)$ converges to a Gaussian distribution due to the central limit theorem. When the Poisson intensity is relatively large (e.g., $\lambda(t) \approx 3$ [Hz] like a spontaneous firing rate) in comparison to the decay of the membrane potential (e.g., the membrane time constant $\tau_m \approx 10$[ms]), the Gaussian approximation describes the membrane potential very well [6]. This approximation is valid even when N is not large.

3 Gaussian Process

In order to examine the stochastic process whose density is described by a Gaussian distribution for every time, we first calculate the joint probability density of the membrane potential. In the followings, the neuron behavior is assumed to be observed at every Δt time step, i.e., $t_m = m\Delta t$, where Δt is sufficiently small.

Joint Probability Density For expression simplicity, v^{t_m} is abbreviated as v^m and the sequence of variables, $\{v^m, \ldots, v^1\}$, is denoted by $\{v\}_1^m$. The joint probability density is defined by

$$g(\{v\}_1^m) = \langle \delta(v^m - \boldsymbol{v}(t_m)) \ldots \delta(v^1 - \boldsymbol{v}(t_1)) \rangle =$$
$$\int \frac{d\xi_1}{2\pi} \cdots \frac{d\xi_m}{2\pi} \exp\Big\{\sum_{k=1}^{m} \mathrm{i}\xi_k v^k\Big\} \Big\langle \exp\Big(-\sum_{k=1}^{m} \mathrm{i}\xi_k \boldsymbol{v}(t_k)\Big)\Big\rangle.$$

By expanding the logarithm of the joint characteristic function, we obtain

$$\ln\Big\langle \exp(-\sum_{k=1}^{m} \mathrm{i}\xi_k v(t_k))\Big\rangle = -\sum_{k=1}^{m} \mathrm{i}\xi_k \eta(t_k) - \frac{1}{2}\sum_{k=1}^{m}\sum_{l=1}^{m} \xi_k \xi_l C_{kl} + O(\xi_k^3),$$

where $\eta(t_k) \equiv \langle v(t_k)\rangle$ and $C_{kl} \equiv \langle v(t_k)v(t_l)\rangle - \langle v(t_k)\rangle\langle v(t_l)\rangle$. Under the Gaussian assumption, we can omit the third order terms $O(\xi_k^3)$. Then, the joint probability density is approximated as

$$g(\{v\}_1^m) \approx \int_{-\infty}^{\infty} \frac{d\boldsymbol{\xi}}{(2\pi)^m} \exp\Big(\mathrm{i}\boldsymbol{\xi}'(\mathbf{v}-\boldsymbol{\eta}) - \frac{1}{2}\boldsymbol{\xi}' C \boldsymbol{\xi}\Big),$$

where $\mathbf{v} = (v^{t_1}, \ldots, v^{t_m})'$, $\boldsymbol{\eta} = (\eta(t_1), \ldots, \eta(t_m))'$, $\boldsymbol{\xi} = (\xi_1, \ldots, \xi_m)'$, and C is the m-by-m auto-correlation matrix of $\mathbf{v}$. C is positive definite. Prime ($'$) denotes a transpose. By defining $\boldsymbol{\xi} = \mathbf{x} + \mathrm{i}\mathbf{y}$, we obtain

$$g(\{v\}_1^m) = \int_{-\infty}^{\infty} \frac{d\boldsymbol{\xi}}{(2\pi)^m} \exp\Big(-\frac{1}{2}\mathbf{x}'C\mathbf{x} - (\mathbf{v}-\boldsymbol{\eta})'\mathbf{y} + \frac{1}{2}\mathbf{y}'C\mathbf{y} + \mathrm{i}\mathbf{x}'(\mathbf{v}-\boldsymbol{\eta} - C\mathbf{y})\Big).$$

When $\mathbf{y} = C^{-1}(\mathbf{v}-\boldsymbol{\eta})$, the integral is done only for the real part $\mathbf{x}$. Then,

$$g(\{v\}_1^m) = \frac{1}{\sqrt{(2\pi)^m |C|}} \exp\left(-\frac{1}{2}(\mathbf{v}-\boldsymbol{\eta})'C^{-1}(\mathbf{v}-\boldsymbol{\eta})\right), \tag{4}$$

where $|C|$ is the determinant of C. Subsequently, we analyze a Gaussian process defined by the auto-correlation matrix C.

Time Evolution of Gaussian Process In order to see the characteristics of the auto-correlation matrix C, its incremental (recursive) definition along the time sequence $\{t_1, \ldots, t_m\}$ is derived here. Let $C^n \equiv (c_{ij}^n)$ be the auto-correlation matrix at time t_n $(1 \le n \le m)$. Because synaptic inputs are statistically independent with each other, C^n can be described by the weighted summation of auto-correlation of single synaptic inputs. Namely,

$$c_{ij}^n = w^2 \sum_{k=1}^{\min(i,j)} u(t_i - t_k)u(t_j - t_k)\lambda(t_k)\Delta t \qquad (1 \le i, j \le n),$$

where $w \equiv \sqrt{\sum_{j=1}^N w_j^2}$. Because C^n is symmetric and positive definite, C^n can be decomposed into $(U^n)'U^n$, where $U^n = (u_{ij})$ is an upper triangular matrix. The components of U^n are described as

$$u_{ij} = \begin{cases} wu(t_j - t_i)\sqrt{\lambda(t_i)\Delta t} & i \le j \\ 0 & i > j. \end{cases}$$

If response function (3) is used, C^n is regular. If response function (1) or (2) is used, however, C^n is singular, because the diagonal elements of u^n are zeros due to $u(0) = 0$. In order to avoid the singularity, we redefine the components of U^m by

$$u_{ij} = \begin{cases} wu(t_{j+1} - t_i)\sqrt{\lambda(t_i)\Delta t} & i \le j \\ 0 & i > j. \end{cases} \tag{5}$$

In both cases, the inverse of U^n, Γ^n, can be incrementally defined by

$$\Gamma^n = \begin{pmatrix} \Gamma^{n-1} & -\dfrac{1}{u_{nn}}(\Gamma^{n-1}\boldsymbol{u}^n) \\ \mathbf{0}' & \dfrac{1}{u_{nn}} \end{pmatrix},$$

where $\boldsymbol{u}^n \equiv (u_{1n}, \ldots, u_{n-1\,n})'$. The n-th column of Γ^n is constructed by the Gram-Schmidt orthogonalization. The inverse of C^n is then obtained by

$$(C^n)^{-1} = \Gamma^n(\Gamma^n)' \qquad (n = 1, \ldots, m). \tag{6}$$

Equation (6) introduces an incremental calculation of the inverse of the auto-correlation matrix C along the discretized time sequence.

When response function (3) is used, the inverse auto-correlation matrix becomes a tridiagonal matrix. For response function (1) or (2), the inverse auto-correlation matrix becomes a band diagonal matrix whose bandwidth is five. This is because these response functions are the solutions of the first or second order linear ordinary differential equations. This simplicity also implies that the transition probability density defined by the auto-correlation matrix is strictly determined by a multiple Markov process along the discretized time sequence. Namely, function (3) introduces a single Markov process, and function (1) or (2) introduces a double Markov process.

Transition Probability Density Time evolution is determined by transition probability density $g(v^m|\{v\}_1^{m-1}) = g(\{v\}_1^m)/g(\{v\}_1^{m-1})$. From equation (4), the transition probability density becomes

$$g(v^m|\{v\}_1^{m-1}) = \frac{1}{\sqrt{2\pi\gamma^m}} \exp\left[-\frac{\left(v^m - \eta(t_m) - \sum_{i=1}^{m-1} \kappa_i^m (v^i - \eta(t_i))\right)^2}{2\gamma^m}\right],$$

where conditional variance γ^m and correlation coefficients κ_i^m are obtained from equation (6) as

$$\gamma^m = |C^m|/|C^{m-1}| = (u_{mm})^2, \quad \kappa_i^m = u_{mm}(\Gamma^m)_{im}.$$

By defining $w_0 = \sum_{j=1}^N w_j$, we can prove (proof is omitted) $\sum_{i=1}^{m-1} \kappa_i^m u_{ni} = u_{nm}$ and $\eta(t_n) = \frac{w_0}{w} \sum_{i=1}^n u_{in} \sqrt{\lambda(t_i)\Delta t}$. Then, $\eta(t_m) = \sum_{i=1}^{m-1} \kappa_i^m \eta(t_i) + \frac{w_0}{w} u_{mm} \sqrt{\lambda(t_m)\Delta t}$. If a d-ple Markov process is assumed, $\kappa_i^m = 0$ $(i = 1, \ldots, m-d)$. In this case, the transition probability becomes

$$g(v^m|\{v\}_1^{m-1}) = \frac{1}{\sqrt{2\pi\gamma^m}} \times \exp\left[-\frac{\left(v^m - \frac{w_0}{w} u_{mm}\sqrt{\lambda(t_m)\Delta t} - \sum_{i=m-d}^{m-1} \kappa_i^m v^i\right)^2}{2\gamma^m}\right]. \tag{7}$$

This equation means that the transition probability density is described by $\lambda(t_m)$ and $u_{mm}, \ldots, u_{m-d\,m}$. Namely, the transition probability is determined by the present intensity and the membrane properties for previous d steps. This is actually the d-ple Markov assumption itself.

Now the effect of the reset after firing is considered. If the membrane potential of a sample neuron exceeds the threshold, the potential is reset to zero. Since the threshold does not have any other effect, the transition probability at d steps after the firing can also be given by (7). Therefore, the density of the membrane potential at each time can be described using two terms, the (joint) probability density whose samples have fired and have not fired during the past d steps from the present time.

Firing Probability and Threshold Effect. The joint probability density is decomposed as $g(\{v\}_1^m) = g(v^m|\{v\}_1^{m-1})g(v^{m-1}|\{v\}_1^{m-2})\dots g(v^1)$. The density at time t_m, whose sample has not reached the threshold until time t_{m-1}, is described as

$$g_0(v^m) = \int_{-\infty}^{\theta} dv^{m-1} \dots \int_{-\infty}^{\theta} dv^1 g(v^m|\{v\}_1^{m-1})g(v^{m-1}|\{v\}_1^{m-2})\dots g(v^1).$$

The firing probability at time t_m is given by $\rho(t_m) = \int_\theta^\infty g_0(v^m)dv^m$. If a sample neuron fires at time t_m, its potential is reset to zero. Therefore, the density for potential zero increases by $\rho(t_m)$.

Under the single Markov assumption (i.e., response function (3) is used): $g(v^m|\{v\}_1^{m-1}) = g(v^m|v^{m-1})$, density $g_0(v^m)$ is calculated by very simple equations

$$\rho(t_m) = \int_\theta^\infty g_0(v^m)dv^m$$
$$g_0(v^m) = \int_{-\infty}^{\theta} g(v^m|v^{m-1})g_0(v^{m-1})dv^{m-1} + \rho(t_{m-1})\delta(v^m),$$

where $\delta(\cdot)$ is the Dirac's delta function.

Under the double Markov assumption (i.e., response function (1) or (2) is used): $g(v^m|\{v\}_1^{m-1}) = g(v^m|v^{m-1}, v^{m-2})$, density $g_0(v^m)$ is calculated by

$$\rho(t_m) = \int_\theta^\infty g_0(v^m)dv^m$$
$$g_0(v^m) = \int_{-\infty}^{\theta}\int_{-\infty}^{\theta} g(v^m|v^{m-1}, v^{m-2})g_0(v^{m-1}, v^{m-2})dv^{m-1}dv^{m-2}$$
$$g_0(v^m, v^{m-1}) = \int_{-\infty}^{\theta} g(v^m|v^{m-1}, v^{m-2})g_0(v^{m-1}, v^{m-2})dv^{m-2}$$
$$+\rho(t_{m-1})g(v^m, v^{m-1}).$$

The joint density of reset states, $g(v^m, v^{m-1})$, is obtained by (4). Note that it is difficult to consider the reset effect in the backward type calculation method [4], because the method does not obtain the density whose potential is lower than the threshold.

4 Experiment and Discussion

As a simplified model of the cerebral cortex, we consider a network of excitatory and inhibitory neurons. The number of excitatory inputs and inhibitory inputs are set at

$N_E = 6000$ and $N_I = 1500$, respectively. We also set the synaptic efficacy values to be the same within each of the excitatory and inhibitory input groups, and they are given by $w_E = 0.21$ and $w_I = 0.63$, respectively. These experimental conditions follow [1].

As an example for the inhomogeneous Poisson process, we assume that the inhomogeneous Poisson intensity $\lambda(t)$ [Hz] changes with time in accordance with $\lambda(t) = \lambda_0 + \lambda \sin\left(2\pi\frac{t+\tau}{\tau} + \sin\left((2\pi\frac{t+\tau}{\tau})^2\right)\right)$, where $\lambda_0 = 3.0$ [Hz], $\lambda = 1.0$ [Hz] and $\tau = 0.2$[s]. The numerical calculation is done with the time step $\Delta t = 0.5$ [ms]. We set $\tau_m = 10$ [ms]. Typical time-series (sample paths) of the membrane potential with and without resets are shown in Figure 1 (right). Figure 2 shows the firing probability for various values for the synaptic time constant τ_s and threshold θ. In sub-figures (a), (b) and (c), the reset effect is considered. These theoretical results agree very well with those of Monte-Carlo simulations. It should be noted that only the approximation used in our theory is the Gaussian approximation of the membrane potential, which is a valid assumption.

Figure 2 shows that the firing probability significantly depends on the synaptic time constant τ_s. This effects is prominent especially when threshold θ is large. This change of the firing probability is mainly due to the change of the potential variance $\sigma^2(t)$. Mean $\eta(t)$ does not change very much because the integral of the response function is fixed, while variance $\sigma^2(t)$ decreases as τ_s increases. Conventionally, it has been considered that the change of the synaptic efficacy w_j is important for neuronal information processing (like in the Hebb's prescription). On the other hand, our theoretical result implies another possibility for the neuronal information processing, namely, the modulation of the firing probability by changing the synaptic time constant. When the threshold is relatively low in comparison to the rate of the input spikes, the firing probability is mainly determined by the input rate. In this case, the change of the synaptic time constant is not very important. When the threshold is relatively high, on the other hand, the above-mentioned modulation is possible. In the cortical areas at which the firing rate is low and coincidental spikes often occur, neurons might process their information by temporally changing the shapses of PSPs.

5 Conclusion

The responses of spiking neurons to input spikes obeying an inhomogeneous Poisson process were theoretically examined. We developed Gaussian process approach in order to precisely calculate the dynamics of the membrane potential and the firing probability. Our forward type calculation method can consider the effect of resets. We have found that the synaptic time constant of the spike response function has significant influence on firing probability especially when the threshold is high.

References

1. Amit, D. & Brunel, N. (1997) Dynamics of a recurrent network of spiking neurons before and following learning. *Network*, **8**: 373.
2. Arieli, A., Sterkin, A., Grinvald A., & Aertsen A. (1996) Dynamics of ongoing activity: explanation of the large variability in evoked cortical responses. *Science* **273**: 1868-1871.

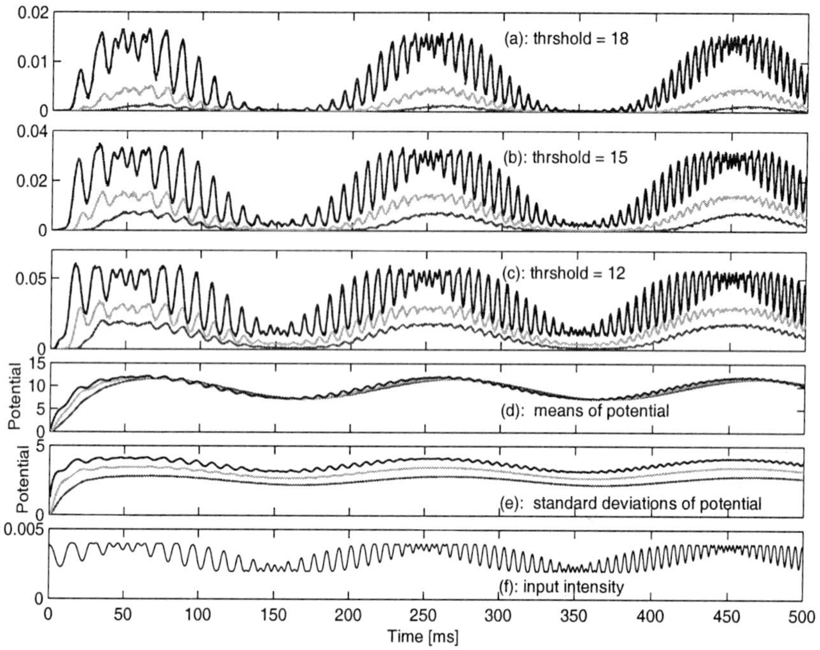

Fig. 2. Variation of the firing probability due to variation of the synaptic time constant $\tau_s \equiv 1/\beta$. (a)-(c): Firing probability density when the firing and reset are considered. The threshold is 18 for (a), 15 for (b), and 12 for (c). In each figure, solid lines are for $\beta = \infty$ (function (3)), 3.0 (function (1)), and α (function (2)) from top to bottom. (d): Means of the membrane potential when the firing is not considered. τ_s is not very important for the mean. (e): Standard deviations (SDs) of the membrane potential when the firing is not considered. SDs decreases as τ_s increases. (a-e): The results of Monte-Carlo simulations over 160000 trials are overwritten by dashed lines in all the sub-figures. Difference between the theoretical results (solid) and the Monte-Carlo results (dashed) is too small to observe in each sub-figure. The *maximum* difference is about 0.002 for the firing probability density, and 0.02 for the mean and SD, which will decrease as the Monte-Calro trials increase. (f) The input Poisson intensity.

3. Bair, W. & Koch, C. (1996) Temporal precision of spike trains in extrastriate cortex of the behaving macaque monkey. *Neural Computation* **8**:1185-1202.
4. Burkitt, A.N. & Clark, G.M. (1999) Analysis of the synchronization of synaptic input and spike output in neural systems. *Neural Computation* **11**:871-901.
5. Gerstner, W. (1998) Spiking Neurons. In *Pulsed Neural Networks*, pp. 3-49. MIT Press.
6. Papoulis, A. (1984) *Probability, Random Variables, and Stochastic Process*. McGraw-Hill.
7. Shadlen, M.N. & Newsome W.T. (1998) The variable discharge of cortical neurons: implications for connectivity, computation, and information coding. *Journal of Neuroscience* **18**: 3870-3896.
8. Tuckwell, H.C. (1988) *Introduction to Theoretical Neurobiology: volume 2, Nonlinear and Stochastic Theories*. Cambridge University Press.

Kernel Based Image Classification

Olivier Teytaud[1] and David Sarrut[2]

[1] Institut des Sciences Cognitives, UMR CNRS 5015
67 boulevard Pinel, F-69675 Bron cedex, France
teytaud@isc.cnrs.fr
[2] ERIC, Université Lumière Lyon 2,
4 av. P. Mendès-France, F-69676 Bron Cedex, France
dsarrut@univ-lyon2.fr

Abstract. In this study, we consider low-level image classification, with several machine learning algorithms adapted to high dimension problems: kernel-based algorithms. The first is Support Vector Machines (SVM), the second is Bayes Point Machines (BPM). We compare these algorithms based on strong mathematical results and nice geometrical arguments in a feature space to the simplest algorithm we could imagine working on the same representation. We use different low-level data, experimenting low-level preprocessing, including spatial information. Our results suggest that the kernel representation is more important than the algorithms used (at least for this task). It is a positive result because it exists much more simpler and faster algorithms than SVM. Our additive low-level preprocessings only improved success rate by few percents.

1 Introduction

In multiclass categorization of images, two families of approaches coexist; the first is based on high level extracted features and the second on low level algorithms. In this work, we study the second approach, made practical, as shown by Chapelle et al [4], by the use of machine learning algorithms able of working in high-dimension datasets: kernel algorithms. Jalam et al. [10] reported that a naive approach on the same kernel representation could be as efficient as Support Vector Machines in the case of low-level text categorization. In this paper, we propose similar experiments in the case of images. Our goal is to test if fast well-known algorithms could be as efficient as other algorithms provided they use the same kernel representation. We also propose other low-level features.

In a first part, we describe classical kernel-based algorithms. Section 3 describes naive kernel based algorithms. We finally present experimental results and discuss our results.

2 Classical Kernel-Based Algorithms

In the following the *training set* is $D = \{(x_1, y_1), ..., (x_m, y_m)\}$, with the x_i's elements of the input space X, and the y_i's elements of the finite output space

G. Dorffner, H. Bischof, and K. Hornik (Eds.): ICANN 2001, LNCS 2130, pp. 369–375, 2001.

Y. The x_i's are images, encoded through features (see section 4), the y_i's can belong to $\{-1, 1\}$ for a two-classes discrimination, or to $\{1, 2, ..., q\}$ for q-classes discrimination. The goal is to find an application from X to Y, minimizing the misclassification probability of new samples. We denote by $K : X \times X \to \mathbb{R}$ a kernel on X.

2.1 Support Vector Machines and Multiclass Support Vector Machines

Support Vector Machines (SVM) are defined in [15] and in many tutorials; they are the most usual algorithm for kernel learning. They are based on *margin maximization*, and choose linear separations in a feature space, using the *kernel trick*. For brevity, this algorithm and its justifications won't be detailed here. The extension of SVM to the multiclass case was done by one-against-all in Vapnik's book [15]; hence $x \mapsto f_k(x) = sign(\sum y_i \alpha_i K(x_i, x) + b)$ was computed for each class k against all the others, and x was classified in class k such that $f_k(x)$ is maximum. Guermeur et al. in [9] proposes another version of multiclass SVM, mathematically justified but more computationally expensive, sometimes leading to better results.

2.2 Bayes Point Machines

Herbrich et al. [6] consider that SVM work mainly as approximators of a Bayesian algorithm defined by Rujan [11]. Their experiments suggest that, in the separable case, the algorithm called Bayes Point Machine, is more efficient than SVM. The principle of the algorithm is the following one: 1) Let K be a kernel. $x \mapsto K(x, x)$ must be constant equal to 1, 2) Consider the set H of kernel classifiers $H = \{x \mapsto \sum_{i=1}^{m} \alpha_i K(x_i, x) \; / \alpha_i \in \mathbb{R}\}$, 3) The version space V is defined as the space of consistent hypothesis with norm 1: $V = \{w \in H \; / \parallel w \parallel = 1 \text{ and } \forall i \; y_i w(x_i) \geq 0\}$, 4) The goal it to select as classifier the average of V, for the metric in $H : \parallel w \parallel = \sum \alpha_i \alpha_j K(x_i, x_j) y_i y_j$. This is done by *playing billiard* in the version space. The algorithm is the following one: 1) Find an initial point $x_0 \in V$, 2) Choose randomly a direction d_0 in H. Let i be equal to 0, 3) Compute d_{i+1}, new direction after bounce on the boundary, 4) Replace i by $i + 1$ and return to 3).

Choose as centre of mass the average point of this billiard (taking into account the metric; see Rujan [11] for details). As the billiard is not necessarily ergodic, a random step is added in reflections in order that the whole version space is uniformly covered (this "artificial" ergodicity hasn't been proved). The algorithm is the following one. Assuming that all vectors have norm 1 (for example $K(x, y) = exp\left(-\frac{\|x-y\|^2}{\sigma^2}\right)$), one can see that hard-margin SVM (i.e. minimization of w^2 under constraint $\xi_i = 0$) choose the w in the version space which is the centre of the maximum inscribed sphere. Herbrich et al. [6] explains that such point is an approximation of the average of the version space, and that when the version space has unregular shape the approximation is not efficient. In these cases, their algorithm gives better results than SVM. The extension to the multiclass case can be done in a one-against-all approach.

We here propose a modified Bayes Point Machine, in which we do not average wx but $sign(wx)$. Such technique has the drawback of not choosing a classifier among a linear family of classifiers, but is theoretically closer to Bayesian inference and was experimentally slightly better on this particular benchmark.

3 Naive Kernel Based Algorithms

Description The Naive Kernel-Based Algorithm (NKBA) used in our experiments is summarized below:

• Let $(x_i)_{i \in I}$ be the family of classified examples (used for training). Let (x'_i) be the family of points to be classified.

• Let O be the matrix such that $O_{i,j} = 1$ if x_i belongs to class j, -1 otherwise.

• Let $K_{i,j} = K(x_i, x_j)$ and let $K'_{i,j} = K(x'_i, x_j)$

• Let W be such that $K1 \times W = O$. $K1$ is the matrix K, plus one column filled with 1's ($K'1$ below defined similarly). W is chosen by Singular Value Decomposition with minimal square norm[1].

• Let $O' = K'1 \times W$. We classify x'_i in class $\hat{k} = \arg\max_k \; O'(i,k)$.

With K a Radial Basis Function, this is a classical RBF algorithm. Here we decide to use other functions as well. Notice that results of Schölkopf et al. [13], emphasizing the superiority of SVM onto classical RBF approaches, do not apply here as we do not use back-propagation for any layer of weights.

One can discuss the mathematical justification of our algorithm, as we cannot justify it by the notion of geometrical margin w^2 (as a regularization term) and VC-dim bounds based upon this. However, Bartlett [1] recalls that the γ-empirical error (number of points in D for which $y_i f(x_i) \leq \gamma$) in classification is bounded by the mean squared error divided by $(1-\gamma)^2$. For fixed γ, the γ-empirical error is so linearly bounded by the mean square error. A bound is easily derived, depending upon the size of weights, by combining results of [1]: the set of functions $H = \{x \mapsto \exp(-\frac{d^2(x,y)}{\sigma^2})\}$ for any d considered here are L-Lipschitz; so their γ-fat-shattering dimension is bounded by the $\frac{\gamma}{2L}$-covering number of X, and so is $O(\frac{1}{\gamma^{dim}})$ with dim the dimension of X. If the sum of weights is bounded by A, then with d the $\frac{\gamma}{32A}$-fat-shattering dimension of H, then the γ-fat-shattering dimension of $F = \left\{\sum_{i=1}^{M} w_i f_i / M \in q\mathbb{N}, w_i \in \mathbb{R}, f_i \in H, \sum_i |w_i| \leq A\right\}$ is bounded by $\frac{cA^2 d}{\gamma^2} \log^2(Ad/\gamma)$, with c a universal constant. So a trade-off between empirical error and norm of weights is justified. One can notice that in our experiments (part 4), allowing high values of M (number of functions) increases the empirical success rate, but never leads to over-fitting. This suggests that the fat-shattering dimension is more adapted to build bounds for this algorithm than arguments depending upon the number of weights. It was also the conclusion of [1] in another framework, with rigorous mathematical arguments.

[1] Gaussian elimination is the fastest, but Singular Value Decomposition and House Holder are much better because of numerical stability reasons, see [5]

Improvements A common possible improvement (both in computational time and memory space) consists in defining $K_{i,j} = K(x_i, x_j)$ and $K'_{i,j} = K(x'_i, x_j)$ only for a subset $S \subset I$ of possible values for j. This subset can be chosen by K-means (see [2]) or randomly. We decide to use a random set in order to preserve the fastness and simplicity of our approach. The size of the subset is chosen as the minimal size ensuring an optimal empirical success rate.

Another improvement inspired by [11] and [6] has been tried: averaging O' over multiple runs (Bayesian Inference). This is possible by introducing random noise on O, preserving the sign of its elements (multiplying each element by a random variable with values between 0 and 1). We notice that this technique combines two learning paradigms: Bayesian Inference (averaging among different possible models) and addition of noisy examples to the training set (solution to over-fitting/resistance to noise). We can also average on another random step: the choice of the subset S.

This Bayesian Inference can be fastly computed by computing $K1^{-1}$. Then computing a new O' is done simply by:

• Computing $K1^{-1}$ (pseudo-inverse of $K1$, computational cost $O(m^3)$ with m the number of examples, by Singular Value Decomposition).

• Generating O (depends on the random generator, not very time-consuming).

• Computing W by a matrix product ($O(mpQ)$ with m the number of examples, p the number of selected examples, the elements x_j for which $K(x_i, x_j)$ is computed, and Q the number of classes)

• Verifying $K1 \times W = O$ (because of the heuristic consisting in choosing the smallest cardinal of S which ensures null empirical error rate)

• computing O' by a matrix product : $O(tpQ)$ with t the number of examples to be tested.

With this improvement, our Bayesian inference becomes as fast as Bayes Point Machines.

4 Experiments: Results in Image Classification

Material and Method We used the image database called Corel14, described in [4, 3], composed of 1400 images classified in 14 classes of 100 images (examples of classes are: "polar bears", "air shows" ...). An image is denoted by $I : \prod_{i=1}^{2}[1..n_i] \mapsto \mathcal{D}$, with n_i the number of sample along the ith axes, $\mathcal{D}$ the domain of pixel value. Here the images are 24 bits color images and $\mathcal{D}$ is composed of three 8 bits channels R,G and B (Red, Green, Blue). These images are real-world images.There are numerous features which can be extracted to describe images. We decided to focus on low-level, high-dimension, translation and rotation invariant features: typically histogram-based features. The features we selected are presented table 1 (with $\boldsymbol{x} = (x, y)$ a coordinate pixel).

In table 1, $\#E$ denotes the size of set E. The two first histograms H_{RGB} and H_{HSV} are simple color histograms based on different color spaces. HSV (Hue Saturation Value) is considered as *user-oriented* (by contrast with RGB considered as *hardware-oriented*), based on the intuitive notion of tint, shade

Table 1. Set of considered low-level image features

H_{RGB}	$\mathcal{D} \mapsto \mathbb{R}$	$H(i) = \#\{\boldsymbol{x} \,/ I(\boldsymbol{x}) = i\}$
H_{HSV}	$\mathcal{D} \mapsto \mathbb{R}$	$H(i) = \#\{\boldsymbol{x} \,/ I(\boldsymbol{x}) = i\}$
H_{∇}	$[0:2\pi] \mapsto \mathbb{R}$	$H(\theta) = \sum \parallel (\nabla_x I, \nabla_y I) \parallel / angle\,(\nabla_x I, \nabla_y I) = \theta$
H_{Δ}	$\mathcal{D} \mapsto \mathbb{R}$	$H(i) = \#\{\boldsymbol{x} \,/ \Delta(\boldsymbol{x}) = i\}$
$H_{\bigcirc,d}$	$\mathcal{D} \times \mathcal{D} \mapsto \mathbb{R}$	$H(i,j) = \#\{(\boldsymbol{x}, \boldsymbol{x}') \,/ I(\boldsymbol{x}) = i\,,\, I(\boldsymbol{x}') = i$ and $dist(\boldsymbol{x}, \boldsymbol{x}') \leq d\}$

and tone. Such color space is generally considered as a better image feature for machine learning than RGB (see [8] for RGB to HSV conversion). $\nabla_x I = \frac{\delta I}{\delta x}$ and $\nabla_y I = \frac{\delta I}{\delta y}$ denotes the two orthogonal gradient values at point $I(\boldsymbol{x})$, $\Delta(\boldsymbol{x}) = \frac{\delta^2 I}{\delta^2 x} + \frac{\delta^2 I}{\delta^2 y}$ the Laplacian. The gradient orientation histogram H_{∇} is forced to be translation and rotation invariant by ordering the angles θ according to $\theta_{max} = \arg\max_{\theta} H(\theta)$ as origin. H_{Δ} is the Laplacian histogram. $H_{\bigcirc,d}(i,j)$ is the number of occurrence of color i and j for points closer than a distance d. Except the two first, each histogram is computed after a previous image smoothing by a Gaussian filter (of size 55×55 and $\sigma = 9$), in order to decrease potential noisy pixels. No other image preprocessing has been applied. We think $H_{\bigcirc,d}$ brings interesting features in the machine learning algorithm, because it takes into account a notion of neighborhood. However, the value of d is difficult to choose a priori (we used $d = 25$). Moreover, in this first study, we did not consider multi-resolution version of histograms. Note that, as images are RGB images, there are different ways to consider pixel value: for example, it would be possible to combine the three channels. However, we decided to take into account the three channels in order to make benefit of the high-dimension capabilities of the kernel-based machine learning. Hence, gradient or Laplacian are computed independently on each channel and each histogram takes his value in $\mathcal{D}$ which has 3 dimensions. In a future work, we will take into account several values of d and multi-resolution features. The images was thus described by the set of presented features and the three presented methods (SVM, BPM, NKBA) have been applied. The learning set is composed of the two third of the images and the validation set is composed of the remaining images. Results are averaged on multiple runs. Tested kernels are the linear kernel ($K(x,y) =< x|y >$) and the Jambu kernel (a symmetrized version of χ^2 dissimilarity) $K(x,y) = exp\left(-\frac{\sum \frac{(x_i - y_i)^2}{x_i + y_i}}{\sigma^2}\right)$ with σ an heuristically chosen parameter (such that values of $K(x_i, x_j)$ for $(i,j) \in I^2$ are nearly homogeneous).

Results Chapelle et al. [4] suggest some possible preprocessings (such as elevating frequencies at power $\frac{1}{8}$). Probably classical improvements in text categorization would be usefull as well (see for example [12], replacing frequency α by $\frac{1}{1+\alpha}$ for example). We do not study this preprocessings here. On the same database, with different approaches (based on feature extraction and decision trees), Carson et al. [3] report nearly 50 % of success (cited in Chapelle et al. [4]). k-nearest neighbour are not better, at least with the similarities used in our kernel algorithms. Table 2 presents the success rates according to different methods and features. Using SVM and H_{RGB}, Chapelle et al. [4] present slightly better results on the same benchmark with the same algorithm. We suppose a slight difference in preprocessing explains this difference. Different approaches (RGB, Gradient, Laplacian, $H_{\bigcirc}$, all with Jambu kernel) are combined through linear combination (using a Singular Value Decomposition algorithm to choose weights of combination); this leads to results of line "Combined". Our Bayes Inference and modified Bayes Point Machines algorithms achieve respectively 82.0% and 82.8% in the case of Jambu Kernel on RBG. Linear algorithms did not achieve as good results as Jambu kernel; their strength lies in recognition speed.

Table 2. Different success rates according to different methods and different features used. The two first lines use the linear kernel, the others the Jambu kernel.

Comparison SVM/NKBA		
Data	SVM	NKBA
RGB (linear)	57.6 %	60.0 %
HSV (linear)	63.4 %	63.4 %
RGB (Jambu)	81.4 %	83.6 %
HSV (Jambu)	81.4 %	83.5 %

Further experiments with NKBA	
Gradient (Jambu)	58.5 %
Laplacian (Jambu)	60.0 %
$H_{\bigcirc}$ (Jambu)	73.4 %
Combined	85.9 %

5 Conclusion

We did not achieve a significant improvement by addition of information about proximity of colors, gradients, laplacian. The question of whether the 95% success rate can be achieved by such low-level tools remains opened. Further work might include local direction evaluation, Spatial Gray Level Dependencies, Gabor filters. The difference with the results given in Chapelle et al. ([4]) could be attributed to the difference of preprocessing or could also be due to a weakness of our implementation of SVM. SVM are a still recent learning algorithm, and different implementations give different results, because different quadratic-programming algorithms have different behaviour (mainly, almost "horizontal" objective function). As already observed in text categorization (Jalam et al. [10]), one can replace complex algorithms like Bayes Point Machines or Support Vector Machines, by old well-known widely available algorithms. For SVM, Mercer's condition doesn't seem to be the reason which makes the difference between a good and a bad kernel. Both from a theoretical (thanks to covering numbers coming from fat-shattering dimension) and practical point of view, simple algorithms can be used in this field. In particular, computational time was much higher for SVM or BPM than for NKBA, even using a fast algorithm as SMO (Smola, [14]) for Support Vector Machines. An interesting point for a further work would be the use of the Lp-machine, a kernel-based algorithm using linear programming instead of quadratic, and which appears theoretically (because of weights-based bounds like the ones used above, section 3) and practically (for its numerical behavior) interesting. In text categorization we did not get any improvement using ECOC (see Dietterich et al. [7]), a classical paradigm for extending two-classes categorization to multi-class categorization. Moreover, the multiclass version of SVM proposed in [9] achieved better results than one-against-all multiclass SVM, whereas [16] doesn't mention any significant improvement of one multi-class SVM onto the other. We suppose the large number of classes and the high-dimensionnality nature of the problem explains the observed gap. In a further work we shall verify whether these results can be stated in histogram-based image classification.

References

1. Bartlett P.L. 1998, *The sample complexity of pattern classification with neural networks: the size of the weights is more important than the size of the network, IEEE transactions on Information Theory, 44:525-536*
2. Bishop C.M. 1995, *Neural Networks for Pattern Recognition, Oxford*
3. Carson C., S. Belongie, H. Greenspan, and J. Malik 1998, *Color and texture-based images segmentation using EM and its application to image querying and classification, submitted to Pattern Anal. Machine Intell.*
4. Chapelle O., P. Haffner, and V.-N. Vapnik 1999, *Support Vector Machines for Histogram Based Image Classification, IEEE transactions on Neural Networks, Vol 10*
5. Golub G., and C. Van Loan 1996, *Matrix computations, third edition, The Johns Hopkins University Press Ltd., London*
6. Herbrich R., T. Graepel, and C. Campbell 1999, *Bayes Point Machines: Estimating the Bayes Point in Kernel Space, in Proceedings of IJCAI Workshop Support Vector Machines, pages 23-27, 1999*
7. Dietterich T.-G., and G. Bakiri 1995, *Solving Multiclass Learning Problems via Error-Correcting Output Codes. Journal of Artificial Intelligence Research 2: 263-286,1995*
8. Foley J., A. van Dam, S.Feiner and J. Hughes 1990, *Computer graphics – Principles and Practice–* Addison-Wesley, 2nd edition
9. Guermeur Y., A. Elisseeff, and H. Paugam-Moisy 2000, *A new multi-class SVM based on a uniform convergence result. Proceedings of IJCNN'00, IV-183*
10. Jalam R., and O. Teytaud 2000, *Text Categorization based on N-grams, research report, Laboratoire ERIC*
11. Rují́n P. 1997, *Playing Billiard in Version Space. Neural Computation 9, 99-122*
12. Sahami M. 1999, *Using Machine Learning to Improve Information Access, Ph.D. in Computer Science, Stanford University*
13. Schölkopf B., K. Sung, C. Burges, F. Girosi, P. Niyogi, T. Poggio, and V.N. Vapnik 1997, *Comparing Support Vector Machines with Gaussian Kernels to Radial Basis Function Classifiers. IEEE Transactions on Signal Processing, 45(11):2758-2765*
14. Smola, A.J. 1998, *Learning with kernels, Ph.D. in Computer Science, Technische Universität Berlin*
15. Vapnik V.N. 1995, *The Nature of Statistical Learning, Springer*
16. Weston J., and C. Watkins 1998, *Multiclass Support Vector Machines, Univ. London, U.-K.,m Tech. Rep. CSD-TR-98-04*

Gaussian Processes for Model Fusion

Mohammed A. El-Beltagy[1] and W. Andy Wright[2]

[1] Southampton University*,
Southampton, UK
m.a.el-beltagy@soton.ac.uk
[2] BAE SYSTEMS (ATC Sowerby), FPC 267, PLC, PO Box 5,
Filton, Bristol, UK
Andy.Wright@bris.ac.uk

Abstract. We present here a novel use of Gaussian processes for the fusion of results of computational simulations with varying degrees of accuracy and computational loads. We demonstrate the efficacy of the approach on one toy problem and two real word examples.

1 Introduction

For many problems in engineering design there are usually a number of different computational models that simulate the problem at hand. However, it is often the case that these models are compromises between accuracy and computational expense. Computationally cheap model evaluations can yield results with systematic errors, whereas computationally expensive models may be much more accurate but can only be used sparingly. In engineering design there is a strong need for cheap and accurate surrogate models of computational expensive simulations, that could be used for purposes of optimisation and visualisation of the problem domain. An example of this is in the design of aerodynamic components. Here accurate aerodynamic CFD calculations using Euler codes may be several orders of magnitude more expensive than the less accurate linearised potential flow approximations with boundary layer calculation. There are also similar parallels in structural and electro-magnetic analysis. The ability to be able to obtain accurate model data at low computational expense is of practical importance in engineering design where it is often necessary to evaluate different designs under the same conditions. Furthermore, the growing use of optimisation methods requires the repeated evaluation of such models. For instance, in aerodynamic design where computationally expensive CFD (e.g. an Euler model can take several hours to perform an evaluation) methods have to be used, this can severely limit the number of optimisation iterations that can practically be computed and so the viability of the optimisation process. Typically less expensive but much more inaccurate *panel methods* have to be used. The model produced by the panel method not only has a greater covariance (which is input dependant) but is also bias since it does not allow correctly for all forms of

* Now at Bios Group Inc., 317 Paseo de Peralta, Santa Fe, NM 87501, USA
mohammed@biosgroup.com

G. Dorffner, H. Bischof, and K. Hornik (Eds.): ICANN 2001, LNCS 2130, pp. 376–383, 2001.

drag. One approach to this problem is to learn a regression model that relates the problem parameters to the desired evaluation. However, to obtain sufficient accuracy requires many expensive evaluations

This paper demonstrates how Gaussian processes (GP) can be used to combine boundary value information with the output from a quicker, but systematically inaccurate model, to obtain better generalisation over data from a slower but more accurate model. The approach is differs from methods, such as "mixtures of experts", where a number of regressors are fused using some weighting. Here it is always assumed that the quicker evaluation method is less accurate than the more expensive method. Furthermore, the quicker method is used directly as in input to the regressor and not fused at the output. A GP approach was chosen for a number of reasons. Firstly, GPs bear very close similarities to *Kriging* techniques that have been used by engineers for some time [9, 4]. GPs, therefore, could be considered a natural extension of an existing method and so would be taken up more easily. Secondly, some of priors in the GP can be interpreted physically and so can be readily set by the engineer, who may have good understanding about the characteristics of the function. Lastly, with the GP, it is possible to obtain the derivative information of the regression function and this is particularly useful to the engineer.

2 Prediction with Gaussian Processes

Given a training data set $\mathcal{D}$ consisting of N pairs of inputs $\mathbf{x}_n$ and scalar outputs t_n for $n = 1, \ldots, N$ a Gaussian process (GP) prediction model is concerned with evaluating the probability $P(t_{N+1}|\mathcal{D}, \mathbf{x}_{N+1})$. Where the input vector for the point to be predicted is denoted by $\mathbf{x}_{N+1}$ and t_{N+1} is the corresponding prediction. Here $P(\mathbf{t}_N|\{\mathbf{x}_N\})$ is assumed to follow a Gaussian distribution given by

$$P(\mathbf{t}_N\,|\mathcal{D}\,,\mathbf{x}) = \frac{1}{\sqrt{(2\pi)^N\,|\mathbf{C}_N|}} \exp\left[-\frac{1}{2}(\mathbf{t}-\mu)^T\mathbf{C}_N^{-1}(\mathbf{t}-\mu)\right]. \tag{1}$$

Where $\mathbf{C}_N$ is the covariance matrix for $P(\mathbf{t}_N|\mathcal{D}, x)$, and μ is the mean. $\{\mathbf{x}_N\}$ and $\mathbf{t}_N$ are the sets of all inputs and outputs in $\mathcal{D}$ respectively. For normalised data, it can be assumed that $\mu = 0$. It can be shown [1] that the predictive distribution is Gaussian with mean and variance

$$\hat{t}_{N+1} = \mathbf{k}^T\mathbf{C}_N^{-1}\mathbf{t}_N; \;\; \sigma^2_{\hat{t}_{N+1}} = \kappa - \mathbf{k}^T\mathbf{C}_N^{-1}\mathbf{t}_N \tag{2}$$

where $\mathbf{k}^T = [C(\mathbf{x}_1, \mathbf{x}_{N+1}), C(\mathbf{x}_2, \mathbf{x}_{N+1}), \cdots, C(\mathbf{x}_N, \mathbf{x}_{N+1})]$ and $\kappa = C(\mathbf{x}_{N+1}, \mathbf{x}_{N+1})$. Elements of the covariance matrix $\mathbf{C}_N$ as well as in the above two expressions are calculated using the covariance function $C(\mathbf{x}_i, \mathbf{x}_j)$. The covariance function used here is quite popular in the literature [3] for the interpretability of its hyperparameters. It is given by

$$C(\mathbf{x}_i, \mathbf{x}_j) = \theta_1 \exp\left[-\frac{1}{2}\sum_{l=1}^{L}\frac{\left(x_i^{(l)} - x_j^{(l)}\right)}{r_l^2}\right] + \theta_2 + \delta_{ij}\theta_3. \tag{3}$$

Here, $x_n^{(l)}$ is the lth component of $\mathbf{x}_n$. The hyperparameters are defined as $\theta = \log(\theta_1, \theta_2, \theta_3, \mathbf{r})$. For irrelevant inputs, the corresponding $\mathbf{r}_l$ will be large and the model will ignore that input. The hyperparameters are defined as the log of the variables in equation 3 to restrict their values to be positive.

Given the covariance function, the log likelihood of the training data is given by

$$L = -\frac{1}{2}\log|\mathbf{C}_N| - \frac{1}{2}\mathbf{t}_N^T\mathbf{C}_N^{-1}\mathbf{t}_N - \frac{N}{2}\log 2\pi + \ln P(\theta) \qquad (4)$$

Here $P(\theta)$ is the prior probability of the hyperparameters. There are a variety of ways by which the likelihood could be maximised. For the result presented here, a conjugate gradient optimiser was used [7]. Note that since the focus here is on deterministic models, the noise level term θ_3 in equation 3 is not considered in covariance calculations.

3 GP Based Model Fusion

Consider two models, $f_e(\mathbf{x})$ representing datum values obtained from an exact and computationally expensive computation, while $f_a(\mathbf{x})$ are the datum values given by an approximate, but computationally cheap computation. The fusion approach constructs a GP, for the accurate model, using input vectors containing both the problem parameters $\mathbf{x}$ and the returned objective function value from the cheaper model $f_a(\mathbf{x})$. This exploits the assumption that $f_e(\mathbf{x})$ and $f_a(\mathbf{x})$ must have a reasonable level of correlation such that the more cheaply obtained $f_a(\mathbf{x})$ value provides valuable information for the GP predictor. Hence, for points used for model fusion, $\widehat{\mathbf{x}}_i = [\mathbf{x}_i, f_a(\mathbf{x}_i)]$ becomes the input vector for which $f_e(\mathbf{x}_i)$ is the output. For many engineering problems considered here this is not be an unreasonable assumption since both methods produce results by modelling the same underlying (non-linear) physical process. Constructing the fused GP involves the following steps:

1 Initially M sample inputs $\mathbf{x}_i$ are evaluated using f_a. Of those M points, N points are evaluated using f_e. Depending on the relative computational cost it could sometimes be the case that $N \ll M$.
2 A GP model is constructed for f_a, using the full M sample points. During likelihood maximisation, the prior term $P(\theta)$ in equation 4 is set to zero.
3 A GP model is constructed for f_e using $\widehat{\mathbf{x}}_i$ as input. For simplicity a Gaussian prior with a diagonal covariance matrix is used in equation 4 with the mean values of length scales r_i for $i = 1, \ldots, L$ set at those obtained from the f_a GP model construction. The corresponding variances on those L variables are fairly small. The mean prior values for r_{L+1}, θ_1, and θ_2 are set to zero and a very large corresponding variance values.

The assumption here is that the relative importance of different dimensions represented in the r_i values are unlikely to be drastically different between the f_a and f_e GP models. This could generally be true if f_a is not a severe distortion of f_e. The fact that N is a relatively small sample that doesn't allow for adequate estimation of $\mathbf{r}$ in the GP training further justifies the approach.

4 Results

4.1 Toy Problem:- Multi-modal Bump Function

A 2-D bump function described by [5] was used for $f_e(\mathbf{x})$. Here:

$$f_e(\mathbf{x}) = \left| \sum_{i=1}^{L} \cos^4(x_i) - 2 \prod_{i=1}^{L} \cos^2(x_i) \right| \left(\sum_{i=1}^{L} i x_i^2 \right)^{-1/2} \quad ; i = 1, \ldots, L, \quad (5)$$

where $0 < x_i < 10$ subject to $\prod_{i=1}^{L} x_i > 0.75$ & $\sum_{i=1}^{L} x_i < 15L/2$. The less accurate $f_a(\mathbf{x})$ was generated by distorting $f_e(\mathbf{x})$ such that:

$$f_a(\mathbf{x}) = f_e(\mathbf{x}) \exp\left[-\frac{1}{2} \sum_{i=1}^{2} \frac{(x_i - 5)}{\sigma_{ap}^2} \right] + h_{ap} \quad (6)$$

Where parameters σ_{ap} and h_{ap} control the distortion.

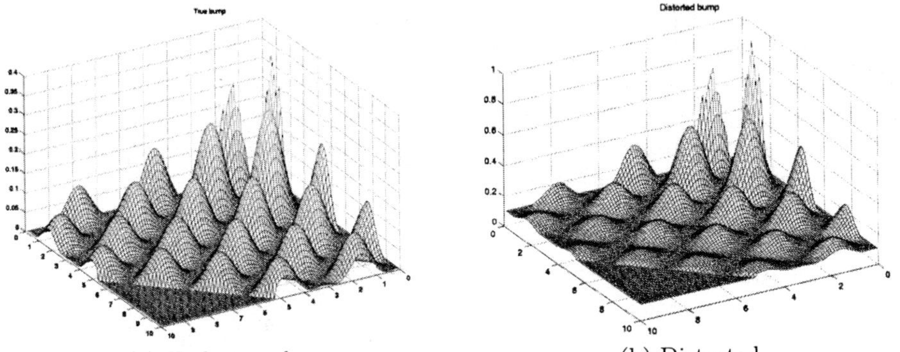

(a) Undistorted (b) Distorted

Fig. 1. A comparison between the distorted and undistorted bump. For the distorted case $\sigma_{ap} = 4$ and $h_{ap} = 0.1$.

A total of 80 points were used for training the GP fused model. For comparison, a non-fused GP was constructed using the same points (i.e. for the non-fused GP, the input vector is simply $\mathbf{x}$ and the output is $f_e(\mathbf{x})$). Considering figure 2, which illustrates the absolute error surfaces between the fused and non-fused results, the advantage of using $f_a(\mathbf{x})$ as an extra input is clear. The point is made even more vividly in figure 3 which shows the average sum squared error (ASSE) for these two approaches plotted as a function of σ_{ap}, which controls the

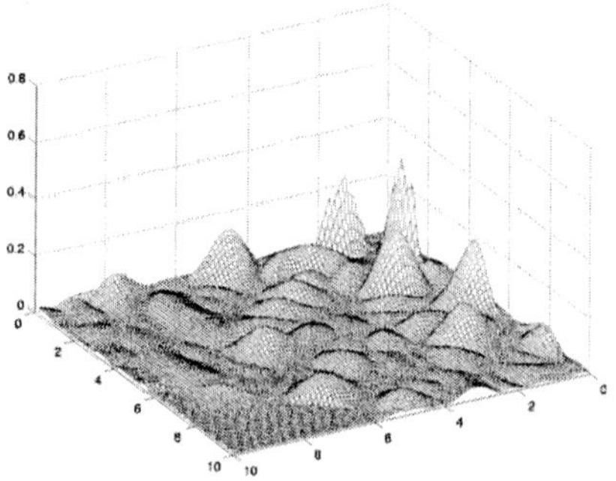

(a) fused

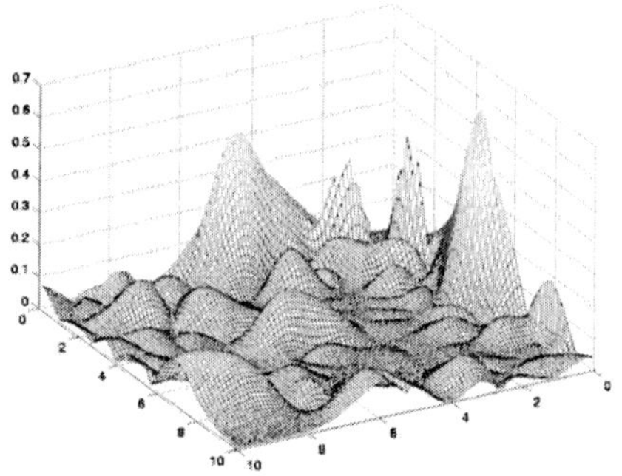

(b) non-fused

Fig. 2. Absolute difference between $f_e(\mathbf{x})$ and (a) fused GP (b) non-fused.

distortion. For this experiment the training set size was set to 160 points, while the test set had 400 points. When $f_a(\mathbf{x})$ was used as a predictor, the resultant ASSE varied exponentially with either σ_{ap} or h_{ap}. The fused model, on the other hand, which makes use of $f_a(\mathbf{x})$, resulted in a relatively well bounded range of ASSE values.

4.2 Satellite Beam Problem

In this problem, the frequency response of a two-dimensional satellite boom structure is considered. This structure (figure 4) consists of 40 individual Euler-Bernoulli beams connected at 20 joints [6]. Each of the 40 beams has the same properties per unit length. By exciting the beam at one end, the propagation of the vibrations through the structure can be considered and the frequency-averaged response at the other end predicted.

The frequency response is a function of the (x, y) co-ordinates of the inner 18 joints in the structure. This gives 36 parameters upon which any regression to predict the vibrational characteristics of the beam must be based. The design requirement for this structure is that the vibration energy between 150 and 250 Hz be reduced. The design optimisation procedure, therefore, requires the integral under the frequency response curve to be calculated for 150-250Hz. The vibration analysis is carried out using matrix receptance methods based on the Green functions of the individual beam elements. By combining these receptance methods with the well-known characteristics of Euler-Bernoulli beams, it is possible to directly solve for the energetic quantities of interest [10]. A typical frequency response is shown in figure 5. In general, accurate predictions can only be obtained with a large number of evaluation points (typically 11).

To demonstrate this fusion method five and eleven integration points of the frequency response curve were used for $f_a(\mathbf{x})$ and $f_e(\mathbf{x})$, respectively. Table 1 shows the average sum of the squared error (ASSE) decreasing monotonically with training set size for both fused and non-fused models. Here, the test set of 800 independent points was used to calculate the ASSE. If $f_a(\mathbf{x})$ was used as

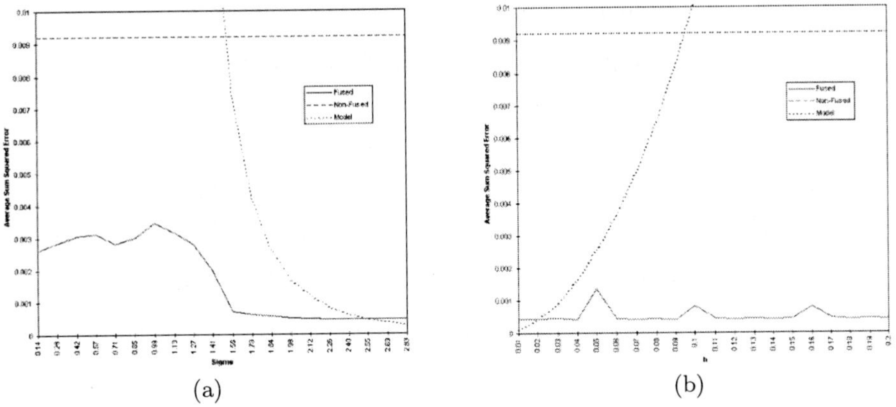

Fig. 3. The average sum squared error as a function of (a) σ_{ap} or (b) h_{ap} for the fused model (solid), the none fused (dashed), and error if $f_a(\mathbf{x})$ was used as the predictor (dotted). In (a), h_{ap} was set to zero, while in (b), σ_{ap} was set to 10^{10}.

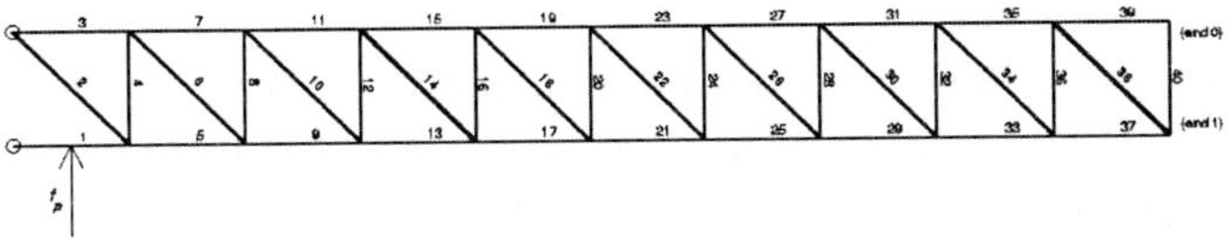

Fig. 4. Schematic of the satellite 2D beam structure.

the predictor, it would result in an ASSE of 0.9707. It is hence clear that the fused model resulted in a significantly better prediction.

4.3 2D Airfoil Wave Drag Prediction

Here the problem was to predict the coefficient of wave drag (the drag induced by the formation of a shock on the wing) using three inputs: 1) the wing camber 2) thickness to cord ratio and 3) desired angle of attack [8]. Two commercial aerodynamic codes were used to evaluate the wave drag values for a set of parameterised NACA aerofoils. Here the quick but inaccurate *linearised potential flow* code, VSAERO was used for f_a ($\sim 5s$ on a Silicon Graphics Challenge). For the computationally more expensive but accurate method an *Euler CFD* code, MGAERO was used for f_e ($\sim$ 2 hours). A test set of 500 points were used to calculate the ASSE. Table 2 shows the average sum of the squared error (ASSE) decreasing with increased training set size, with the fused model out-performing the non-fused one.

5 Conclusions

The approach presented here is based on using information from a less accurate, but much less computationally expensive evaluation method, as input to a re-

Table 1. Model fusion on the beam problem.

Training Set Size	ASSE Fused	Non-fused
250	0.966	1.263
275	0.867	1.156
300	0.852	1.136
325	0.849	1.015

Table 2. Training results for the 2D Airfoil problem.

Training Set Size	ASSE Fused	Non-fused
100	5.36e-4	7.62e-4
200	4.72e-4	5.17e-4
300	3.97e-4	4.75e-4
400	3.06e-4	3.68e-4

gression process that models the output of a more accurate, but more expensive evaluation. From an engineering perspective, this is an important capability to have. As designs become more complex, the optimisation of the designs becomes progressively involved, and some level of automation in the design selection is desirable. However, for automated optimisation procedures to be used effectively, the designer must have access to quick and accurate evaluations of any proposed design. Current accurate evaluation methods are often computationally expensive, thus limiting the utility of this approach. As computers become progressively faster, the scale and complexity of engineering designs attempted by engineers expands. This problem is unlikely to disappear. This is particularly the case in areas of engineering design that rely on increasingly complex numerical models. A potential alternative to this problem is presented here. It is shown that the accuracy of a GP regression can be enhanced by using not only relevant

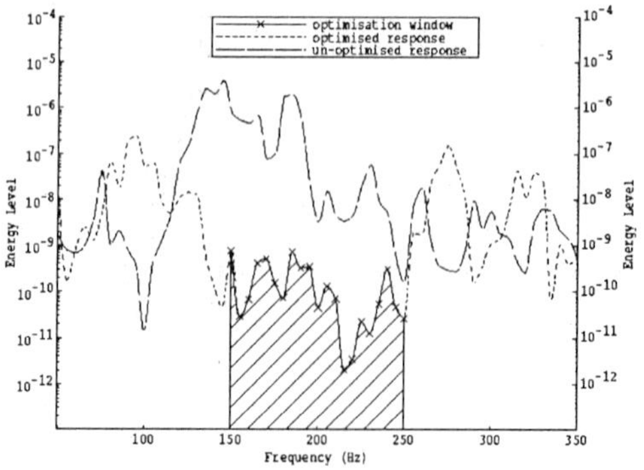

Fig. 5. Frequency response of the 2D-beam structure.

design information, but also evaluation results from a quick, but less accurate evaluation method. This approach is demonstrated on an artificial test problem and two, limited, real-world engineering problems with promising results. The GP model fusion is shown to be capable of maintaining well-bounded prediction errors even as the actual errors between $f_e(\mathbf{x})$ and $f_a(\mathbf{x})$ are varied exponentially. Current research is aimed at incorporating the model fusion approach in smart optimisation strategies that make sparing use of the computational resources, while updating the fused model [2]. Finally, it is stressed that this fusion method does not preclude the use of other regression methods. GPs were chosen here for a number of engineering reason. Consequently, this approach could be adapted for other regression methods such as feed forward neural networks.

Acknowledgments

We would like to thank Chris Williams regarding his original discussions about this work. This work was supported under EPSRC grant no GR/L04733.

References

1. Williams C.K.I and Ramussen CE. Gaussian processes for regression. In Mozer M.C. Touretzky D.S and Hasselmo M.E., editors, *Neural Information Processing Systems 8*, pages 514–520, MIT Press, 1996.
2. Mohammed A. El-Beltagy. *A natural approach to multilevel optimization in engineering design.* PhD thesis, University of Southampton, England, 2000.
3. M.N. Gibbs. *Bayesian Gaussian Processes for Regression and Classification.* PhD thesis, Cambridge University, 1997.
4. M. Handcock and M. Stein. A Bayesian analysis of Kriging. *Technometrics*, 35:403–410, 1993.
5. A.J Keane. Experiences with optimizers in structural design. proceedings of the conference on adaptive computing in engineering design and control. In I. Parmee, editor, *ACEDC 94*, pages 14–27, University of Plymouth, UK, Plymouth, 1994.
6. A.J. Keane and A.P Bright. Passive vibration control via unusual geometries: Experiments on model aerospace structures. *Journal of Sound and Vibration*, 190:713–719, 1996.
7. David J.C. MacKay. *Macopt-a nippy wee optimizer.* University of Cambridge. http://wol.ra.phy.cam.ac.uk/mackay/c/macopt.html, 1999.
8. G. M. Robinson and A. J. Keane. A case for multi-level optimization in aeronautical design. In *Proceedings of the Royal Aeronautical Society Conference on Multidisciplinary Design and Optimization.* 1998.
9. J. Sacks, W. J. Welch, T. J. Mitchell, and H. P. Wynn. Design and analysis of computer experiments. *Statistical Science*, 4(4):409–435, 1989.
10. K. Shankar and A. J. Keane. Energy flow prediction in a structure of rigidly joined beams using receptance theory. *Journal of Sound and Vibratation*, 185(5):867–890, 1995.

Kernel Canonical Correlation Analysis and Least Squares Support Vector Machines

Tony Van Gestel, Johan A.K. Suykens, Jos De Brabanter,
Bart De Moor, and Joos Vandewalle

K.U.Leuven, Dept. of Electrical Engineering, ESAT-SISTA,
Kasteelpark Arenberg 10, B-3001 Leuven, Belgium
{tony.vangestel,johan.suykens}@esat.kuleuven.ac.be

Abstract. A key idea of nonlinear Support Vector Machines (SVMs) is to map the inputs in a nonlinear way to a high dimensional feature space, while Mercer's condition is applied in order to avoid an explicit expression for the nonlinear mapping. In SVMs for nonlinear classification a large margin classifier is constructed in the feature space. For regression a linear regressor is constructed in the feature space. Other kernel extensions of linear algorithms have been proposed like kernel Principal Component Analysis (PCA) and kernel Fisher Discriminant Analysis. In this paper, we discuss the extension of linear Canonical Correlation Analysis (CCA) to a kernel CCA with application of the Mercer condition. We also discuss links with single output Least Squares SVM (LS-SVM) Regression and Classification.

1 Introduction

Support Vector Machines (SVMs) for linear classification aim at constructing a linear large margin classifier in the input space [4, 13]. The idea is extended to nonlinear classification problems by mapping the inputs first in a nonlinear way to a high (possibly infinite) dimensional feature space in which one constructs the large margin classifier. Mercer's condition is then applied to avoid an explicit expression for the nonlinear mapping. The expression for the resulting classifier in the dual space follows from the corresponding finite dimensional QP-problem involving only related kernel function. By taking the 2-norm and using equality instead of inequality constraints, the Least Squares Support Vector Machine (LS-SVM) formulation [11] for classification is obtained, where the solution follows from a linear Karush-Kuhn-Tucker system instead of a QP-problem. In a similar way, the LS-SVM regression formulation boils down to solving a linear system, while typically the SVM regressor follows from a QP-problem.

The key idea of extending a linear technique towards a nonlinear kernel technique is to use this linear technique in the feature space and to apply the Mercer condition in a second step in order to obtain practical expressions. Kernel Principal Component Analysis (kernel PCA) [10] is obtained by applying PCA in the feature space. After applying the Mercer condition, the corresponding principal components are obtained from a generalized eigenvalue problem. Using

G. Dorffner, H. Bischof, and K. Hornik (Eds.): ICANN 2001, LNCS 2130, pp. 384–389, 2001.

Fisher's Linear Discriminant analysis in the feature space one obtains kernel Fisher Discriminant analysis [2, 8]. In the dual space, the solution follows from a generalized eigenvalue problem.

In this paper, we perform Canonical Correlation Analysis (CCA) with regularization term [6], in the feature space to obtain kernel CCA. After applying the Mercer condition also here the solution follows from a generalized eigenvalue problem. We further discuss the relation with kernel Fisher Discriminant analysis and with single output LS-SVM regression and classification.

This paper is organized as follows. Kernel CCA is introduced in Section 2. The LS-SVM formulation for regression is revised and related with kernel CCA in Section 3. An example is given in Section 4.

2 Kernel Canonical Correlation Analysis

In Canonical Correlation Analysis one looks for linear combinations between two variables that maximize the correlation between them. In the case of kernel CCA, the correlation analysis is performed between the features $\varphi(x) \in \mathbb{R}^{n_f}$ and the targets $y = [y^{(1)}, ..., y^{(L)}] \in \mathbb{R}^L$. For regression, the targets correspond to the observed or measured quantities. Classification formulations are obtained by using, e.g., binary class labels $y \in \{-1, +1\}$. The mapping $\varphi : \mathbb{R}^n \to \mathbb{R}^{n_f}$ maps the inputs $x \in \mathbb{R}^n$ to the features $\varphi(x) \in \mathbb{R}^{n_f}$ in a high (possibly infinite) dimensional feature space. This mapping is, however, never explicitly calculated because the Mercer condition $\varphi(x_1)^T\varphi(x_2) = K(x_1, x_2)$ is applied [4, 11, 13]. For the kernel function $K(\cdot,\cdot)$ one has, e.g., the following choices: $K(x, x_i) = x_i^T x$ (linear kernel) and $K(x, x_i) = \exp\{-\|x - x_i\|_2^2/\sigma^2\}$ (RBF kernel), with $\sigma \in \mathbb{R}^+$. Observe that one can also use a nonlinear transformation of the outputs y [7].

We are looking for linear combinations $t_{\text{KCCA}} = w_{\text{KCCA}}^T(\varphi(x) - m_{\varphi(x)})$ of the features and of the class labels $z_{\text{KCCA}} = v_{\text{KCCA}}^T(y - m_y)$ such that the correlation between the canonical variates t_{KCCA} and z_{KCCA} is maximized:

$$\max_{w_{\text{KCCA}}, z_{\text{KCCA}}} \quad \mathcal{E}[t_{\text{KCCA}}^T z_{\text{KCCA}}] = w_{\text{KCCA}}^T \Sigma_{\varphi,y} v_{\text{KCCA}}, \tag{1}$$

subject to the constraints

$$\mathcal{E}[w_{\text{KCCA}}^T(\varphi(x) - m_{\varphi(x)})(\varphi(x) - m_{\varphi(x)})^T w_{\text{KCCA}}] = w_{\text{KCCA}}^T \Sigma_{\varphi,\varphi} w_{\text{KCCA}} = 1 \tag{2}$$

$$\mathcal{E}[v_{\text{KCCA}}^T(y - m_y)(y - m_y)^T v_{\text{KCCA}}] = v_{\text{KCCA}}^T \Sigma_{y,y} v_{\text{KCCA}} = 1. \tag{3}$$

Using the technique of Lagrange multipliers, the following conditions for optimality are obtained [1]:

$$\begin{bmatrix} 0 & \Sigma_{\varphi,y} \\ \Sigma_{\varphi,y}^T & 0 \end{bmatrix} \begin{bmatrix} w_{\text{KCCA}}^{(i)} \\ v_{\text{KCCA}}^{(i)} \end{bmatrix} - \lambda_i \begin{bmatrix} \Sigma_{\varphi,\varphi} & 0 \\ 0 & \Sigma_{y,y} \end{bmatrix} \begin{bmatrix} w_{\text{KCCA}}^{(i)} \\ v_{\text{KCCA}}^{(i)} \end{bmatrix} = 0, \tag{4}$$

with $i = 1, ..., \min(n_f, L)$. We are looking for $w_{\text{KCCA}}^{(i)}$ and $v_{\text{KCCA}}^{(i)}$ corresponding to $\lambda_{max} = \max_i(\lambda_i)$ in order to have maximal correlation. Substituting the expression for $v_{\text{KCCA}}^{(i)}$ from the second equation into the first equation and vice versa, one obtains the following generalized eigenvalue problems: $(\Sigma_{\varphi,y}\Sigma_{y,y}^{-1}\Sigma_{\varphi,y}^T - \lambda^2\Sigma_{\varphi,\varphi})w_{\mathrm{KCCA}}^{(i)} = 0$ and $(\Sigma_{\varphi,y}^T\Sigma_{\varphi,\varphi}^{-1}\Sigma_{\varphi,y} - \lambda^2\Sigma_{y,y})v_{\mathrm{KCCA}}^{(i)} = 0$.

In practice, the covariance matrices $\Sigma_{\varphi,\varphi}$, $\Sigma_{\varphi,y}$ and $\Sigma_{y,y}$ are unknown and are estimated from the data $D = \{(x_i, y_i)\}_{i=1}^N$, using the (regularized) covariance matrices. Defining $\Phi = [\varphi(x_1)^T; ...; \varphi(x_N)^T] \in \mathbb{R}^{N\times n_f}$ and $Y = [y_1; ...; y_N] \in \mathbb{R}^{N\times L}$, we can write

$$\hat{\Sigma}_{\Phi,\Phi,r} = \tfrac{1}{N}\textstyle\sum_{i=1}^N(\varphi(x_i) - \hat{m}_\Phi)(\varphi(x_i) - \hat{m}_\Phi)^T + \tfrac{1}{N\gamma}I_{n_f} = \Phi^T M\Phi + \tfrac{1}{N\gamma}I_{n_f} \quad (5)$$

$$\hat{\Sigma}_{\Phi,Y} = \tfrac{1}{N}\textstyle\sum_{i=1}^N(\varphi(x_i) - \hat{m}_\Phi)(y_i - \hat{m}_y)^T = \Phi^T MY \quad (6)$$

$$\hat{\Sigma}_{Y,Y} = \tfrac{1}{N}\textstyle\sum_{i=1}^N(y_i - \hat{m}_Y)(y_i - \hat{m}_Y)^T = Y^T MY, \quad (7)$$

with $M = \frac{1}{N}I_N - \frac{1}{N^2}1_v 1_v^T \in \mathbb{R}^{N\times N}$ and $1_v = [1; 1; ...; 1] \in \mathbb{R}^N$. The sample means are $\hat{m}_\Phi = 1_v^T\Phi/N$ and $\hat{m}_y = 1_v^T Y/N$, respectively. The parameter $\gamma > 0$ controls the trade-off between error minimization and regularization in (5) [3, 13, 11]. This form of regularization corresponds to an empirical Bayesian approach and has, e.g., been applied in [6] in order to obtain better estimates of the covariance matrices in classification problems.

Substituting (5), (6) and (7) into the generalized eigenvalue problem for w_{KCCA}, we obtain:

$$[\Phi^T MY\hat{\Sigma}_{Y,Y}^{-1}Y^T M\Phi - \lambda_i^2(\Phi^T M\Phi + \tfrac{1}{N\gamma}I_{n_f})]w_{\mathrm{KCCA}}^{(i)} = 0. \quad (8)$$

Since the mapping $\varphi(\cdot)$ is unknown, this generalized eigenvalue problem cannot be solved in its current form. For $\lambda > 0$, we can write $w_{\mathrm{KCCA}}^{(i)} = \Phi^T\alpha_{\mathrm{KCCA}}^{(i)}$ with $\alpha_{\mathrm{KCCA}}^{(i)} \in \mathbb{R}^N$ because $\frac{\lambda^2}{N\gamma}w_{\mathrm{KCCA}}^{(i)} = (\Phi^T MY\hat{\Sigma}_{Y,Y}^{-1}Y^T M\Phi - \lambda_i^2(\Phi^T M\Phi))w_{\mathrm{KCCA}}^{(i)} \propto \Phi^T\alpha_{\mathrm{KCCA}}^{(i)}$. Substituting $w_{\mathrm{KCCA}}^{(i)} = \Phi^T\alpha_{\mathrm{KCCA}}^{(i)}$ into (8) and multiplying to the left with Φ, we obtain

$$\Omega[MY\hat{\Sigma}_{Y,Y}^{-1}Y^T M\Omega - \lambda_i^2(M\Omega + \tfrac{1}{N\gamma}I_N)]\alpha_{\mathrm{KCCA}}^{(i)} = 0, \quad (9)$$

where Mercer's condition is applied within $\Omega = \Phi\Phi^T \in \mathbb{R}^{N\times N}$: $\Omega_{ij} = \varphi(x_i)^T\varphi(x_j) = K(x_i, x_j)$. It can be shown that Ω on the left hand side of (9) can be omitted [10, 12]. The normalization follows from (2)

$$\alpha_{\mathrm{KCCA}}^{(i)T}(\Omega M\Omega + \tfrac{1}{N\gamma})\alpha_{\mathrm{KCCA}}^{(i)} = 1. \quad (10)$$

The corresponding $v_{\mathrm{KCCA}}^{(i)}$ can be obtained following a similar reasoning as for $w_{\mathrm{KCCA}}^{(i)}$. Given $\alpha_{\mathrm{KCCA}}^{(i)}$ and $w_{\mathrm{KCCA}}^{(i)}$, one can also use (4). Applying the Mercer condition in the second equation, one obtains:

$$v_{\mathrm{KCCA}}^{(i)} = \lambda_i^{-1}\hat{\Sigma}_{Y,Y}^{-1}Y^T M\Omega\alpha_{\mathrm{KCCA}}^{(i)}. \quad (11)$$

In the case of a one-dimensional target vector $y \in \mathbb{R}$ (single output regression or binary classification), we have $L = 1$. Since $\Sigma_{y,y}, \hat{\Sigma}_{Y,Y} \in \mathbb{R}$ we can simplify the expressions for w_{KCCA} and α_{KCCA} into

$$w_{\text{KCCA}} = [\hat{\Sigma}_{\Phi,Y}^T \hat{\Sigma}_{\Phi,\Phi,r}^{-1} \hat{\Sigma}_{\Phi,Y}]^{-1/2} \hat{\Sigma}_{\Phi,\Phi,r}^{-1} \hat{\Sigma}_{\Phi,Y} \tag{12}$$

$$\alpha_{\text{KCCA}} = [Y^T M \Omega(\Omega M \Omega + \tfrac{1}{N\gamma}\Omega)^{-1} \Omega M Y]^{-1/2} (\Omega M \Omega + \tfrac{1}{N\gamma}\Omega)^{-1} \Omega M Y, \tag{13}$$

respectively. The normalization follows from (2) as in (10). One obtains v_{KCCA} from (3) corresponding to a scaling to unit variance. These expressions can be used to make predictions taking into account the preprocessing from $\varphi(x) \to t_{\text{KCCA}}$ and $y \to z_{\text{KCCA}}$, as will be discussed in the next Section. One can also show that the linear discriminant w_{KCCA} corresponds to Kernel Fisher Discriminant analysis [2, 8], while the bias term is obtained from regression interpretations [3, 11] or by cross-validation [2, 8].

3 LS-SVM Regression and Kernel CCA

In Least Squares Support Vector Machines for Regression [11, 12], one estimates $y = f(x) = w^T \varphi(x) + b$ by minimizing the following cost function

$$\min_{w_{\text{R}}, b_{\text{R}}} \mathcal{J}(w_{\text{R}}, b_{\text{R}}) = \tfrac{1}{2} w_{\text{R}}^T w_{\text{R}} + \tfrac{\gamma}{2} \textstyle\sum_{i=1}^N e_i^2, \tag{14}$$

$$\text{s.t.} \quad e_i = y_i - (w_{\text{R}}^T \varphi(x) + b_{\text{R}}), \qquad i = 1, ..., N. \tag{15}$$

Observe that the choice of targets $y_i \in \{-1, +1\}$ corresponds to the LS-SVM classifier formulation since $e_i^2 = (y_i e_i)^2 = (1 - y_i(w_{\text{R}}^T \varphi(x_i) + b_{\text{R}}))^2$.

In order to avoid the explicit expression for $\varphi(\cdot)$, one solves the minimization problem (14)-(15), by constructing the Lagrangian and taking the conditions for optimality. This yields the following linear system in terms of the support values $\alpha_{\text{R}} \in \mathbb{R}^N$ and the bias term b_{R}: [11]

$$\left[\begin{array}{c|c} 0 & 1_v^T \\ \hline 1_v & \Omega + \gamma^{-1} I_N \end{array}\right] \left[\begin{array}{c} b_{\text{R}} \\ \hline \alpha_{\text{R}} \end{array}\right] = \left[\begin{array}{c} 0 \\ \hline Y \end{array}\right], \tag{16}$$

with the identity matrix $I_N \in \mathbb{R}^{N \times N}$, $e_{\text{R}} = [e_1; ...; e_N]$ and $\alpha_{\text{R}} = [\alpha_{\text{R}1}; ...; \alpha_{\text{R}N}]$ and where Mercer's condition is applied within the Ω matrix. The LS-SVM regressor is then obtained by applying the Mercer condition:

$$\hat{y} = w_{\text{R}}^T \varphi(x) + b_{\text{R}} = \textstyle\sum_{i=1}^N \alpha_{\text{R}i} K(x, x_i) + b. \tag{17}$$

The ridge regression formulation (14)-(15) in the feature space is equivalent to the regularized form of CCA (12). Expressing (14)-(15) in terms of w_{R} and b_{R}, one obtains the following conditions for optimality in the feature space

$$\begin{bmatrix} (\Phi^T \Phi + \gamma^{-1} I_N) & \Phi^T 1_v \\ 1_v^T \Phi^T & N \end{bmatrix} \begin{bmatrix} w_{\text{R}} \\ b_{\text{R}} \end{bmatrix} = \begin{bmatrix} \Phi^T Y \\ 1_v^T Y \end{bmatrix}. \tag{18}$$

Eliminating b_{R} from (18), we obtain

$$(N \Phi^T M \Phi + \gamma^{-1} I_N) w_{\text{R}} = N \Phi^T M Y, \tag{19}$$

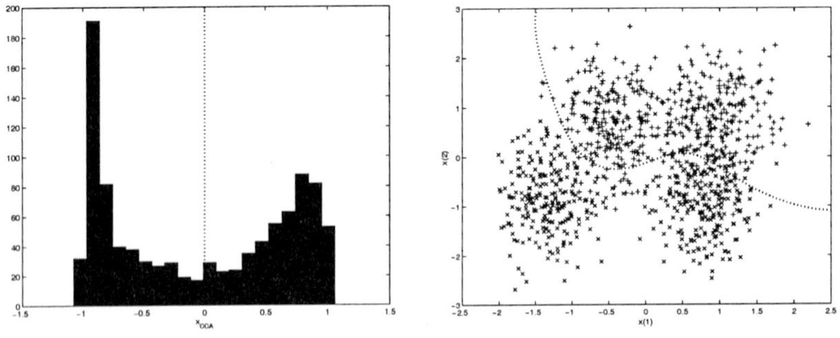

Fig. 1. Representation of the test set for the synthetic binary classification problem [9]. Left: histogram of the $t_{\text{KCCA}} = w_{\text{KCCA}}^T \varphi(x)$ coordinates of the test data with decision line at $t_{\text{KCCA}} = 0$. Right: the corresponding decision boundary in the input space.

or $w_{\text{R}} = \hat{\Sigma}_{\Phi,r}^{-1} \hat{\Sigma}_{\Phi,Y} \propto w_{\text{KCCA}}$. Hence, also in the dual space formulations α_{KCCA} and α_{R} give equivalent regressors (17). A formal prove in the dual space that (16) and (13) yield equivalent regressors involves more complicated matrix algebra.

For the single output case, the use of the regularized covariance matrix $\hat{\Sigma}_{\Phi,r}$ from (5) in kernel CCA corresponds to the LS-SVM formulation (14)-(15) with ridge regression. This form of regularization can also be related to regularization networks [5]. Omitting the bias term in the LS-SVM formulation, as proposed in kernel ridge regression [4], the expressions in the dual space correspond to Gaussian Processes [14].

While CCA is performed on zero mean variables $\varphi(x) - m_{\varphi(x)}$ and $y - m_y$, this is not necessarily the case for the regression formulation which involves $\varphi(x)$ and y. Since there is no regularization applied on the bias term b_{R} in the SVM and LS-SVM formulation [4, 11, 13], the condition for optimality imposes a zero mean training set error [11, 12]. It follows from (16) that $b_{\text{R}} = \hat{m}_Y - w_{\text{R}}^T \hat{m}_\Phi$. Hence (17) can be understood as

$$\hat{y} - \hat{m}_Y = w_{\text{R}}^T (\varphi(x) - \hat{m}_\Phi), \tag{20}$$

using the sample means $\hat{m}_\Phi$ and $\hat{m}_Y$.

4 Example

We illustrate kernel CCA on the synthetic binary classification problem described in [9]. The data points of the test set $D_{\text{test}} = \{([x_i(1)x_i(2)], y_i)\}_{i=1}^{1000}$ corresponding to $\mathcal{C}_-$ $(y = -1)$ and $\mathcal{C}_+$ $(y = -1)$ are denoted by $\times$ and $+$ in Figure 1, respectively. While PCA is often used as a means to perform unsupervised dimensionality reduction, we know apply CCA to perform supervised dimensionality reduction. Within the evidence framework [12] we obtain the hyperparameters $\sigma = 1.311$ and $\gamma = 1.645$, while the corresponding test set t_{KCCA} coordinates are depicted in Figure 1. The classification decision is made by $y = \text{sign}[t_{\text{KCCA}}]$, while the corresponding nonlinear decision boundary is depicted in Figure 1. One achieves a test set accuracy of 90.6%.

5 Conclusions

In this paper, we have proposed a regularized form of Canonical Correlation Analysis (CCA) between the feature space and the targets. As for kernel PCA and kernel Fisher Discriminant Analysis (FDA), the solution follows from a generalized eigenvalue problem in the dual space. However, for scalar targets the solution follows from a linear system and links can be made with LS-SVM Regression or Classification and with kernel FDA.

Acknowledgments

T. Van Gestel and J.A.K. Suykens are a Research Assistant and a Postdoctoral Researcher with the Fund for Scientific Research-Flanders (FWO-Vlaanderen), respectively. This work was partially supported by the Flemish Gov.: (Res. council KULeuven: GOA-Mefisto 666; FWO-Vlaanderen: res. proj. G.0240.99, G.0256.97, and comm. (ICCoS and ANMMM); IWT: STWW Eureka, IMPACT); the Belgian Fed. Gov. (IUAP-IV/02, IV/24; Program Dur. Dev.); the Eur. Comm. (Alapedes, Niconet, ERNSI).

References

1. Anderson, T.W.: An introduction to multivariate analysis. Wiley, New York (1966).
2. Baudat, G., Anouar, F.: Generalized Discriminant Analysis Using a Kernel Approach. Neural Computation **12** (2000) 2385–2404
3. Bishop, C.M.: Neural Networks for Pattern Recognition. Oxford Univ. Press (1995)
4. Cristianini, N., Shawe-Taylor, J.: An Introduction to Support Vector Machines. Cambridge University Press (2000)
5. Evgeniou, T., Pontil, M., Poggio, T.: Regularization Networks and Support Vector Machines. Advances in Computational Mathematics **13** (2001) 1–50
6. Friedman, J.: Regularized Discriminant Analysis. Journal of the American Statistical Association **84** (1989) 165–175
7. Lai, P.L., Fyfe, C.: Kernel and nonlinear canonical correlation analysis. International Journal of Neural Systems (2001), accepted for publication
8. Mika, S., Rätsch, G., Weston, J., Schölkopf, B., & Müller, K.-R.: Fisher discriminant analysis with kernels. In: Hu, Y.-H., Larsen, J., Wilson, E., Douglas, S. (Eds.): Proc. Neural Networks for Signal Processing Workshop IX, NNSP-99 (1999)
9. Ripley, B.D.: Neural Networks and Related Methods for Classification. Journal Royal Statistical Society B **56** (1994) 409–456
10. Schölkopf, B., Smola, A., Müller, K.-M.: Nonlinear component analysis as a kernel eigenvalue problem. Neural Computation **10** (1998) 1299–1319
11. Suykens, J.A.K., Vandewalle, J.: Least squares support vector machine classifiers. Neural Processing Letters **9** (1999) 293–300
12. Van Gestel, T., Suykens, J.A.K., Baestaens, D.-E., Lambrechts, A., Lanckriet, G., Vandaele, B., De Moor, B., Vandewalle, J.: Predicting Financial Time Series using Least Squares Support Vector Machines within the Evidence Framework. IEEE Transactions on Neural Networks (2001), to appear
13. Vapnik, V.: Statistical Learning Theory. Wiley, New-York (1998)
14. Williams, C.K.I.: Prediction with Gaussian Processes: from Linear Regression to Linear Prediction and Beyond. In: Jordan, M.I.: Learning and Inference in Graphical Models. Kluwer Academic Press (1998)

Learning and Prediction of the Nonlinear Dynamics of Biological Neurons with Support Vector Machines

Thomas Frontzek, Thomas Navin Lal, and Rolf Eckmiller

Department of Computer Science VI, University of Bonn
Römerstraße 164, D 53117 Bonn, F. R. Germany
{frontzek,lal,eckmiller}@nero.uni-bonn.de

Abstract. Based on biological data we examine the ability of Support Vector Machines (SVMs) with gaussian kernels to learn and predict the nonlinear dynamics of single biological neurons. We show that SVMs for regression learn the dynamics of the pyloric dilator neuron of the australian crayfish, and we determine the optimal SVM parameters with regard to the test error. Compared to conventional RBF networks, SVMs learned faster and performed a better iterated one-step-ahead prediction with regard to training and test error. From a biological point of view SVMs are especially better in predicting the most important part of the dynamics, where the membranpotential is driven by superimposed synaptic inputs to the threshold for the oscillatory peak.

1 Introduction

Modeling on the basis of biological data extracted via intracellular recording is still a challenging task. A biological neural network well studied by physiologists is the stomatogastric ganglion (STG) of crayfish. This neural network consists of a manageable number of specifiable individual neurons. In the literature the dynamics of single neurons or small parts of the STG is modeled on different levels of detail [2]. On the most detailed level ion channels of the membrane of individual neurons are considered for conductance-based Hodgkin-Huxley-models [1]. So called "neuronal caricatures" are on a more abstract level. Here, the dynamics of a neuron are reduced to a few differential equations. Coupled simple binary neurons build the other end of the modeling hierarchy.

Within our paper we focus on the measured time series of specified individual neurons of the STG to model its dynamics. Former studies show that the Support Vector Machine (SVM) technique is very well suited for learning and predicting nonlinear time series. Applied to different benchmarks SVMs perform better than established regression techniques [3] [4]. In contrast to those approaches we apply SVMs to biological data and compare the results generated by our SVMs with the results of conventional RBF networks.

G. Dorffner, H. Bischof, and K. Hornik (Eds.): ICANN 2001, LNCS 2130, pp. 390–398, 2001.

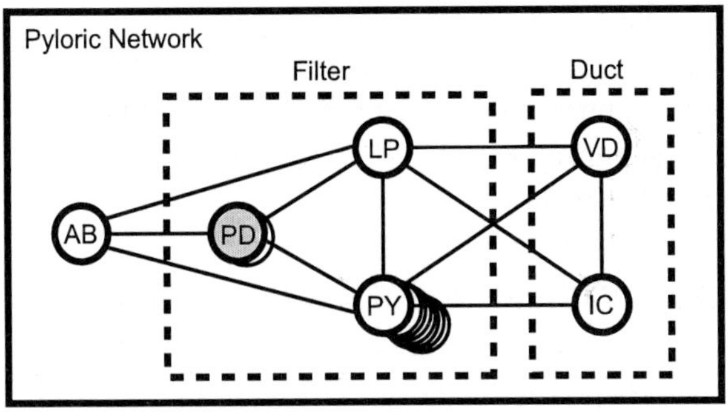

Fig. 1. Pyloric network of crayfish. This network consists of 14 cells: 13 motor neurons of five different cell types (two PD, eight PY, LP, VD, IC) and one interneuron (AB). The synaptic connections are illustrated schematically. Within our paper we focus on the activity of the pyloric dilator (PD) neuron.

2 The Stomatogastric Ganglion of Crayfish

The stomatogastric ganglion (STG) of crayfish is a model system for biological neural networks. A manageable number of specifiable individual neurons generates several rhythmic motor patterns. The STG consists of the gastric network, which controls the movement of the gastric teeth, and the pyloric network (see fig.1), which controls the movement of the pyloric filter and the cardiopyloric duct.

The pyloric rhythm is driven by the anterior burster (AB) interneuron, which is able to oscillate autonomously. Three phases of the pyloric rhythm can be distinguished: pyloric dilator (PD) neuron, lateral pyloric (LP) neuron, and pyloric (PY) neuron. These neurons are connected with electrical and (mostly inhibitory) chemical synapses.

One can measure the activity of the pyloric network neurons in two different ways: either potentials of single neurons via intracellular recording, or the sum potential of PD, LP and PY via extracellular recording at the dorsal ventricular nerve (dvn). Within our paper we focus on the intracellular measured activity of the PD neuron (see fig.2).

3 Learning and Predicting Time Series with SVMs

Let T be some time series that is generated by an unknown mapping $f : \mathbb{N} \to \mathbb{R}^d$:

$$T = (f(1), f(2), ..., f(n)) =: (x_1, x_2, ..., x_n).$$

Often there is no or little information available about f, so x_{n+1} can only be determined by the information included in T. One common approach is:

1. choose $m, h \in \mathbb{N}$ and a set $L \subseteq \{1, ..., n\}$,

2. build training data:
   $D = \{((x_{i-mh}, ..., x_{i-2h}, x_{i-h}), x_i) \in \mathbb{R}^{md} \times \mathbb{R}^d \mid i \in L\}$,
3. choose a class of functions $\mathcal{F} := \{f : \mathbb{R}^{md} \to \mathbb{R}^d\}$, and then
4. try to find a minimizer $\widehat{f}$ of $I : \mathcal{F} \to \mathbb{R}$, $I[f] := \sum_{(x,y)\in D} V(y, f(x))$ with some appropriate loss function V,
5. set $\widetilde{x}_{n+1} := \widehat{f}(x_{n+1-mh}, ..., x_{n+1-2h}, x_{n+1-h})$.

Minimizing I using SVMs with the ϵ-*insensitive* loss function:

$$V(y, f(x)) := \begin{cases} 0, & \text{if } |y - f(x)| \leq \epsilon, \\ |y - f(x)| - \epsilon, & \text{otherwise.} \end{cases}$$

leads to the quadratic minimization problem:

$$\min_{\alpha,\alpha^* \in A_\lambda} \left\{ \frac{1}{2} \sum_{i,j=1}^{|L|} (\alpha_i^* - \alpha_i)(\alpha_j^* - \alpha_j) K(x_i, x_j) - \sum_{i=1}^{|L|} \alpha_i^*(y_i - \epsilon) - \alpha_i(y_i + \epsilon) \right\}$$

with $K : \mathbb{R}^{2d} \to \mathbb{R}$ satisfying the Mercer-condition, $\sum_{i=1}^{|L|}(\alpha_i^* - \alpha_i) = 0$, $A_\lambda := \{\alpha \in \mathbb{R}^{|L|} : 0 \leq \alpha_i \leq \frac{1}{\lambda}, 1 \leq i \leq |L|\}$, ($\lambda$ regularization parameter), which is equivalent to

$$\min_{\beta \in \mathbb{R}^{2|L|}} \left\{ -c^t\beta + \frac{1}{2}\beta^t H \beta \right\}, \quad \text{with } A\beta = b, l \leq \beta \leq u.$$

The matrix H consists of pairwise scalar products of the input data in the feature space:

$$H = \begin{pmatrix} & \vdots & & \vdots & \\ \cdots & +K(x_i, x_j) & \cdots\cdots & -K(x_i, x_j) & \cdots \\ & \vdots & & \vdots & \\ & \vdots & & \vdots & \\ \cdots & -K(x_i, x_j) & \cdots\cdots & +K(x_i, x_j) & \cdots \\ & \vdots & & \vdots & \end{pmatrix}, \quad c = \begin{pmatrix} -y_1 - \epsilon \\ \vdots \\ -y_{|L|} - \epsilon \\ +y_1 - \epsilon \\ \vdots \\ +y_{|L|} - \epsilon \end{pmatrix}.$$

The vector β is defined as:

$$\beta = \begin{pmatrix} \alpha_1 \\ \vdots \\ \alpha_{|L|} \\ \alpha_1^* \\ \vdots \\ \alpha_{|L|}^* \end{pmatrix}, \quad A = \begin{pmatrix} 1 \\ \vdots \\ 1 \\ -1 \\ \vdots \\ -1 \end{pmatrix}, \quad b = 0, \quad l = 0, \quad u = \frac{1}{\lambda}\begin{pmatrix} 1 \\ \vdots \\ 1 \end{pmatrix}, (x_i, y_i) \in T.$$

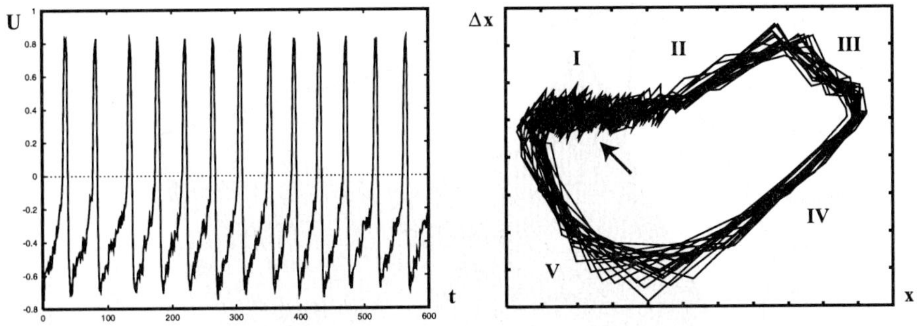

Fig. 2. Time series of the pyloric dilator neuron (left), and phase portrait of its dynamics (right). One can distinguish five sections within the phase portrait, marked I to V. Section I is most important, because here the membranpotential is driven by superimposed synaptic inputs to the threshold for the oscillatory peak (sections II to V).

4 Results

4.1 Biological Data

We used data from the pyloric network of the australian crayfish *cherax destructor albidus*. The intracellular recordings were done using glass-microelectrodes, filled with $3MKCl$ solution and with resistors ranging from $16M\Omega$ to $25M\Omega$. The data was recorded with an A-D-converter and a video-8-recorder[1].

To reduce the amount of data we replaced every 1000 consecutive datapoints by their average. Then we scaled the results to fit into $[-0.9, 0.9]$. Our training set consisted of 200 patterns (patterns $21-220$ of fig.2) and the test set of 100 patterns (patterns $221-320$ of fig.2). For our iterated one-step-ahead prediction we wanted to learn the dynamics of the given time series from the last 20 data points and we therefore set $m = 20$, $h = 1$, and $L = \{21, 22, ..., 220\}$.

To test the generalization ability of our approximator A (SVM or RBF), which was trained to learn the data $(x_{21}, x_{22}, \ldots, x_{220})$, we recursivly set

$$\widetilde{x}_i := x_i, \qquad i = 1..20$$
$$\widetilde{x}_{20+i} := A(\widetilde{x}_{(20+i)-20}, \widetilde{x}_{(20+i)-19}, \ldots \widetilde{x}_{(20+i)-1}), \qquad i = 1..300$$

and then calculated the test error

$$E[A] := \frac{1}{100} \sum_{i=221}^{320} |\widetilde{x}_i - x_i|. \tag{1}$$

4.2 Simulations with Support Vector Machines

We adapted the *LOQO* algorithm by Vanderbei [5] which was implemented by Smola and set the regularization parameter to $\lambda = 0.05$.

[1] Recordings were provided by G. Heinzel, H. Böhm, C. Gutzen, Department of Neurobiology, University of Bonn.

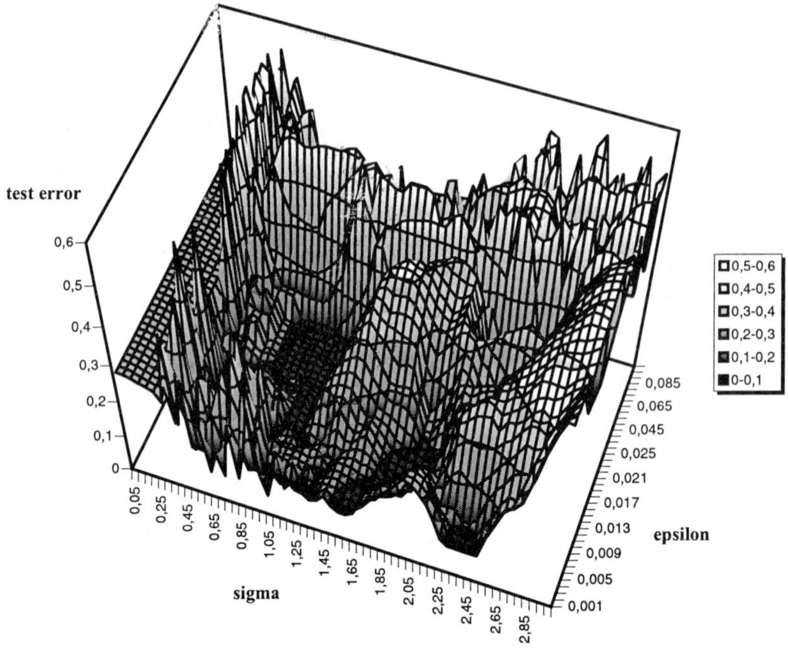

Fig. 3. Test error of SVM with gaussian kernel subject to variance σ and exactness ϵ. One can observe a valley of low error values as well as several peaks of very low error values for certain combinations of σ and ϵ.

Varying the variance σ from 0.05 to 100 and the exactness ϵ from 0.001 to 0.8 we performed a large number of simulations. After the learning process we calculated the training and test errors (see equation 1). Fig.3 shows the test error for the most interesting area, which is $\sigma = [0.05, 3.0]$ and $\epsilon = [0.001, 0.01]$.

One can observe a valley of low error values as well as several peaks of very low error values for certain combinations of σ and ϵ.

Regarding the test error the best SVM we found was SVM_{best} with $\sigma = 2.85$ and $\epsilon = 0.04$. SVM_{best} used 196 support vectors and had a test error of 0.04622. Fig.4 shows the predicted values of SVM_{best} compared to the original values.

We also performed simulations using other than the gaussian kernel. As we expected *tanh*-kernels and polynomial kernels showed worse results. This could be explained by the more global properties of those functions.

4.3 Simulations with RBF Networks

In a first step the centers of the RBF neurons were determined by a *k-means-clustering* algorithm for a given k. The weights of the output layer were set randomly in $[-0.0001, 0.0001]$. Then, in a second step these weights were learned using the delta rule. We stopped the learning process when the change of error was less than 10^4 for the last 20 epochs, compared to the actual epoch.

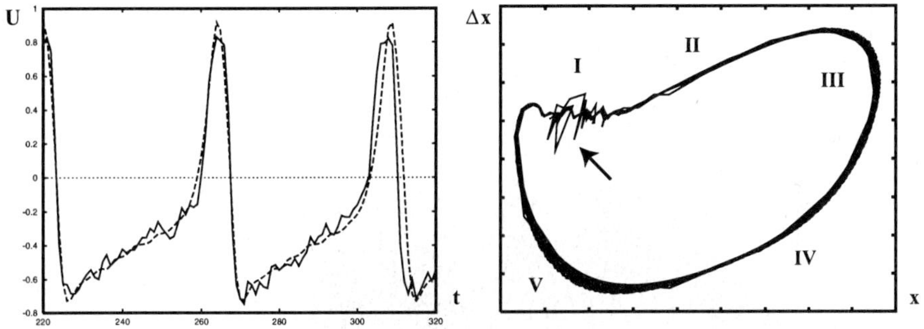

Fig. 4. Iterated one-step-ahead prediction of the PD time series (solid line) with SVM_{best} (dashed line)(left), and phase portrait of the corresponding dynamics (right). Parameters of SVM_{best} are: $\sigma = 2.85$, $\epsilon = 0.04$, number of support vectors = 196. Primarily the increasing sections ($t = [230, 250]$ and $t = [275, 295]$) resp. section I of the phase portrait are predicted better than those predicted by RBF_{best} (see fig.6).

Varying the variance σ from 0.1 to 100 and the number of RBF neurons from 5 to 200 we again performed a large number of simulations. After the learning process we calculated the training and test errors (see equation 1). Fig.5 shows the test error for the most interesting area, which is $\sigma = [0.1, 4.0]$ and number of RBF neurons = $[5, 29]$.

The landscape of test error has two plateaus, one with $\sigma < 0.5$ and one with $\sigma > 3.5$. In between one can observe some mountain ranges of high error values as well as several peaks of very low error values.

Regarding the test error the best RBF network we found was RBF_{best} with 26 RBF neurons and $\sigma = 0.7$. RBF_{best} had a test error of 0.07295. Fig.6 shows the predicted values of RBF_{best} compared to the original values. Primarily the increasing sections ($t = [230, 250]$ and $t = [275, 295]$) are predicted worse than those predicted by SVM_{best} (see fig.4), whereas the oscillatory peak (sections II to V of the phase portrait) is predicted better by RBF_{best} (see fig.6).

4.4 Comparison of the SVM and RBF Results

The results of the previous sections showed that SVMs obtained lower test errors than RBF networks, leading to better predictions. For further comparison we examined the test error subject to the number of support vectors resp. the number of RBF neurons. In fig.7 we plotted for every number of support vectors resp. RBF neurons the lowest test error that was achieved by the corresponding network. Best results for SVMs were obtained using a large number of support vectors (compared to the number of training patterns), whereas best results for RBF networks were obtained using few RBF neurons. Nevertheless, SVMs with about 115 support vectors and even SVMs with about 25 support vectors had test errors which were as low as those of RBF_{best}.

The following table summarizes our results for the lowest training error, test error and the sum of training and test error. It also shows the learning times

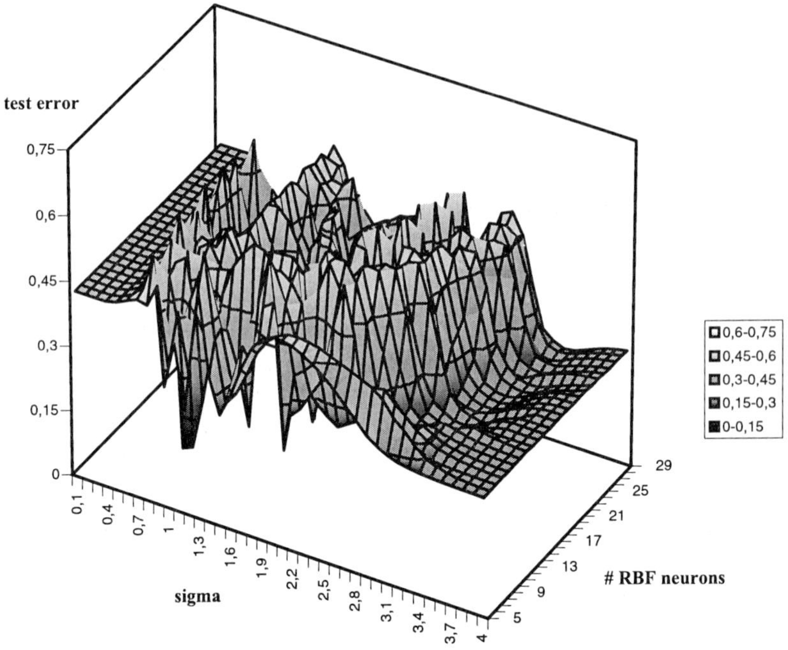

Fig. 5. Test error of RBF network subject to variance σ and the number of RBF neurons. The landscape of test error has two plateaus, one with $\sigma < 0.5$ and one with $\sigma > 3.5$. In between one can observe some mountain ranges of high error values as well as several peaks of very low error values.

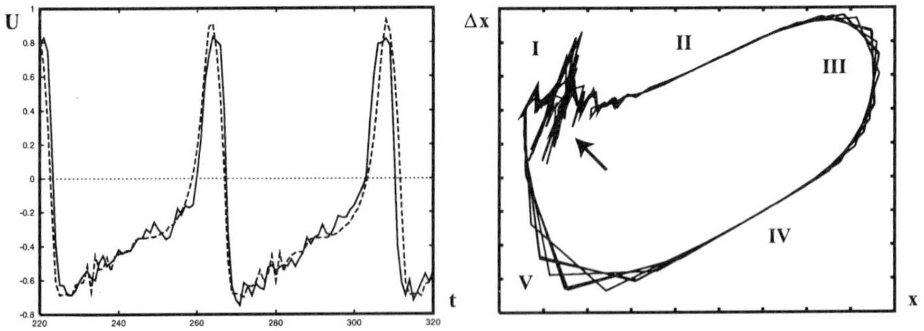

Fig. 6. Iterated one-step-ahead prediction of the PD time series (solid line) with RBF_{best} (dashed line)(left), and phase portrait of the corresponding dynamics (right). Parameters of RBF_{best} are: $\sigma = 0.7$, number of RBF neurons $= 26$. Primarily the increasing sections ($t = [230, 250]$ and $t = [275, 295]$) resp. section I of the phase portrait are predicted worse than those predicted by SVM_{best} (see fig.4), whereas the oscillatory peak (sections II to V of the phase portrait) is predicted better by RBF_{best}.

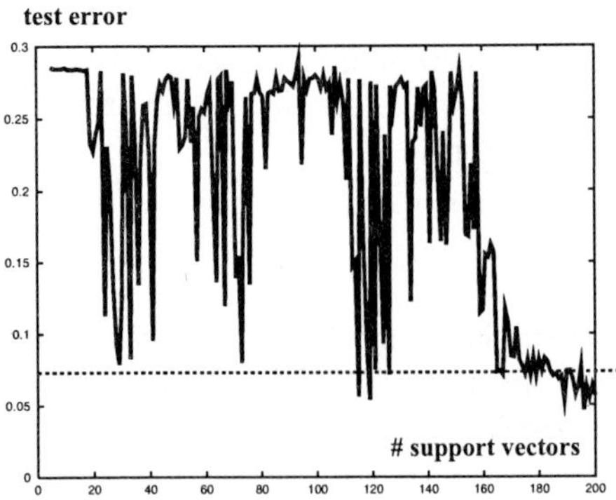

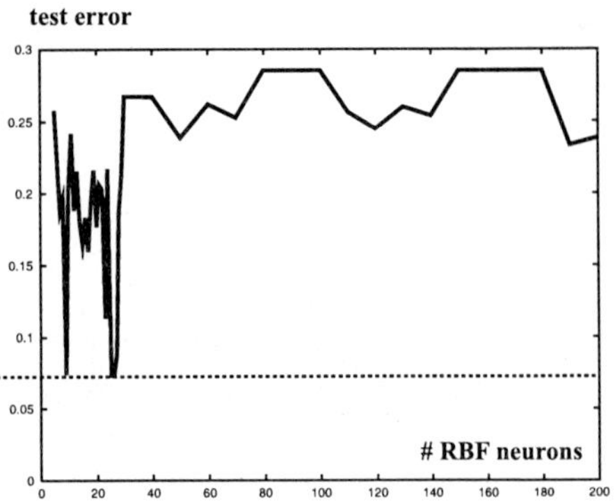

Fig. 7. Test error subject to the number of support vectors (left) and the number of RBF neurons (right). For every number of support vectors resp. RBF neurons we plotted the lowest test error that was achieved by the corresponding network.

(using a Pentium-III computer with 700MHz speed and 128MBytes memory) of the listed networks. All SVMs learned much faster than the corresponding RBF networks.

	SVM (# support vectors)	RBF (# RBF neurons)
best training error	0.02752 (200)	0.17105 (18)
best test error	0.04622 (196)	0.07295 (26)
best error sum	0.19868 (198)	0.33106 (09)
learning times [s]	15/15/15	309/93/27

From a biological point of view SVMs were especially better in predicting the most important part of the dynamics, where the membranpotential is driven by superimposed synaptic inputs to the threshold for the oscillatory peak, whereas RBF networks were better in predicting the different sections of the oscillatory peak.

5 Conclusions

On the basis of the results of our large number of simulations we conclude:

- SVMs for regression were able to learn the nonlinear dynamics of biological data. This was demonstrated with the dynamics of the pyloric dilator neuron of the crayfish *cherax destructor albidus.*
- Compared to conventional RBF networks SVMs with gaussian kernels performed a better iterated one-step-ahead prediction with regard to training and test error. SVMs were especially better in predicting the most important part of the dynamics, where the membranpotential is driven by superimposed synaptic inputs to the threshold for the oscillatory peak.

- SVMs learned faster than RBF networks.
- Best results for SVMs were obtained using a large number of support vectors (compared to the number of training patterns), whereas best results for RBF networks were obtained using few RBF neurons.

Acknowledgments

Parts of this work have been supported by the Federal Ministry for Education, Science, Research, and Technology (BMBF), project LENI. Thanks to Hartmut Böhm and Mikio L. Braun for useful comments.

References

1. Jorge Golowasch, Michael Casey, L. F. Abbott, and Eve Marder. Network stability from activity-dependent regulation of neuronal conductances. *Neural Computation*, (11):1079–1096, 1999.
2. Eve Marder and Allen I. Selverston. Modeling the stomatogastric nervous system. In Ronald M. Harris-Warrick, Eve Marder, Allen I. Selverston, and Maurice Moulins, editors, *Dynamic Biological Networks*, pages 161–196. The MIT Press, 1992.
3. K.-R. Müller, A. J. Smola, G. Rätsch, B. Schölkopf, J. Kohlmorgen, and V. Vapnik. Predicting time series with support vector machines. In *Proceedings of ICANN'97, Lausanne*, pages 999–1004, 1997.
4. Sayan Mukherjee, Edgar Osuna, and Federico Girosi. Nonlinear prediction of chaotic time series using support vector machines. In *Proceedings of IEEE NNSP'97*, pages 511–519, 1997.
5. Robert J. Vanderbei. *LOQO: An Interior Point Code for Quadratic Programming.* Technical Report SOR94-15. Princeton University, 1994.

Close-Class-Set Discrimination Method for Recognition of Stop_Consonant-Vowel Utterances Using Support Vector Machines

Chellu Chandra Sekhar, Kazuya Takeda, and Fumitada Itakura

CIAIR, Itakura Laboratory, Dept. of Information Electronics,
Graduate School of Engineering, Nagoya University, Nagoya-4648603, Japan
chandra@itakura.nuee.nagoya-u.ac.jp

Abstract. 5In conventional approaches for multi-class pattern recognition using Support Vector Machines (SVMs), each class is discriminated against all the other classes to build an SVM for that class. We propose a *close-class-set* discrimination method suitable for large class set pattern recognition problems. The proposed method is demonstrated for recognition of isolated utterances belonging to 80 Stop_Consonant-Vowel (SCV) classes. In this method, an SVM is built for each SCV class by discriminating that class against only 10 classes close to it phonetically.

1 Introduction

Support Vector Machines (SVMs) have been shown to give a good generalization performance in solving pattern recognition problems [1]. The main idea of a support vector machine for pattern classification is to construct a hyperplane as the decision surface in such a way that the margin of separation between the positive and negative examples of a class is maximized [2]. Two approaches to solve multi-class pattern recognition problems using SVMs [3] are: *One-against-the-rest* method and *One-against-one* method. In the first approach one SVM is built for each of the M classes. The SVM for a particular class is constructed by using the training examples of that class as positive examples and the training examples of the rest of $(M-1)$ classes as negative examples. When the number of classes is large, a hyperplane has to be constructed to separate a small number of positive examples from a relatively much larger number of negative examples. In the second approach an SVM is built for every pair of classes to construct a hyperplane for separating the examples of the two classes. Since the number of SVMs required in this approach is $M(M-1)/2$, it is not suitable for large class set pattern recognition.

When the number of classes is large, each class may be close to only a subset of classes. A *close-class-set* for a class includes the classes close to it. The *close-class-sets* may be identified using the description of the classes or using the similarity among patterns of the classes. It is expected that the data points belonging to a particular class and the classes in its *close-class-set* fall in adjacent regions in the feature space. The data points of the other classes are expected to

G. Dorffner, H. Bischof, and K. Hornik (Eds.): ICANN 2001, LNCS 2130, pp. 399–404, 2001.

Table 1. List of SCV classes in Indian languages.

Place of Articulation	Manner of Articulation	Vowel				
		/a/	/i/	/u/	/e/	/o/
Velar	*Unvoiced unaspirated*	ka	ki	ku	ke	ko
	Unvoiced aspirated	kha	khi	khu	khe	kho
	Voiced unaspirated	ga	gi	gu	ge	go
	Voiced aspirated	gha	ghi	ghu	ghe	gho
Alveolar	*Unvoiced unaspirated*	ṭa	ṭi	ṭu	ṭe	ṭo
	Unvoiced aspirated	ṭha	ṭhi	ṭhu	ṭhe	ṭho
	Voiced unaspirated	ḍa	ḍi	ḍu	ḍe	ḍo
	Voiced aspirated	ḍha	ḍhi	ḍhu	ḍhe	ḍho
Dental	*Unvoiced unaspirated*	ta	ti	tu	te	to
	Unvoiced aspirated	tha	thi	thu	the	tho
	Voiced unaspirated	da	di	du	de	do
	Voiced aspirated	dha	dhi	dhu	dhe	dho
Bilabial	*Unvoiced unaspirated*	pa	pi	pu	pe	po
	Unvoiced aspirated	pha	phi	phu	phe	pho
	Voiced unaspirated	ba	bi	bu	be	bo
	Voiced aspirated	bha	bhi	bhu	bhe	bho

be far from the region of the class under consideration and may not be important in construction of a decision surface for that class. In such a case, a decision surface may be constructed by discriminating a class against the classes in its *close-class-set* only.

Support vector machines have been explored for phonetic classification tasks such as vowel recognition, phone identification [4] and stop consonant detection [5]. In this paper, we study the performance of support vector machines for recognition of syllable-like utterances. In our studies we consider combinations of 16 stop consonants with 5 vowels in Indian languages leading to 80 Stop_Consonant-Vowel (SCV) classes. In the next section, we give the phonetic description of SCV classes and explain a method for identification of *close-class-sets* for an SCV class. In Section 3, we present the studies on recognition of SCV utterances.

2 Identification of Close-Class-Sets for SCV Classes

The three components of phonetic description for a Stop_Consonant-Vowel (SCV) class in Indian languages are: Manner of Articulation (MOA) of the stop consonant, Place of Articulation (POA) of the stop consonant, and the vowel in it. The four manners of articulation are: *Unvoiced unaspirated*, *Unvoiced aspirated*, *Voiced unaspirated* and *Voiced aspirated*. The four places of articulation for stop consonants are: *Velar*, *Alveolar*, *Dental* and *Bilabial*. The five vowels considered in our studies are: /a/, /i/, /u/, /e/ and /o/. Combinations of all MOAs, POAs and vowels result in 80 different SCV classes listed in Table 1.

The description based on MOA, POA and the vowel can be used to identify the *close-class-sets* for each SCV class. An SCV class differs with another class in one, two or all the three description components. For example, consider the classes /ka/, /kha/, /ṭha/ and /ṭhi/. The class /ka/ differs with /kha/ in MOA only, with /ṭha/ in both MOA and POA, and with /ṭhi/ in all the three components. The description of SCV classes can be used to identify the classes close to a particular class. Fig.1 shows the classes close to /ka/. The classes shown in each circle have a particular description component as common among them. There are totally 43 classes that have at least one component in common with /ka/. Based on the criterion of at least one common description component for definition of closeness, the *close-class-set* for /ka/ includes all these classes. The remaining 36 classes, not shown in the figure, differ with /ka/ in all the three components. In a similar way, a *close-class-set* of 43 classes can be identified for each SCV class.

Each of the 10 classes shown inside the triangle in Fig.1 has two description components in common with the class /ka/. These classes can be considered to be closer to /ka/ than the classes outside the triangle. Based on the criterion of at least two common components for definition of closeness, a *close-class-set* consisting of 10 classes can be identified for each SCV class. In the next section, we study the performance of the classification models trained using different *close-class-sets*.

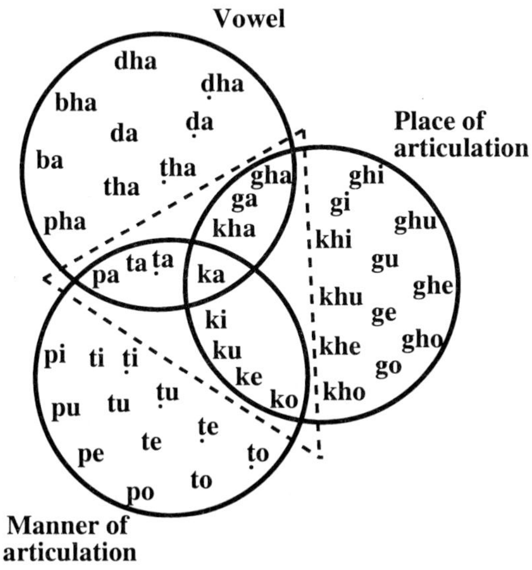

Fig. 1. Subset of SCV classes that differ with /ka/ in only one or two of the three description components (Manner of articulation, Place of articulation and Vowel). The other 36 classes, not shown in the figure, differ with /ka/ in all the three components. A class inside the triangle differs with /ka/ in only one component.

close to a class are adequate to construct the optimal decision surface for that class. This result demonstrates the effectiveness of the proposed *close-class-set* discrimination method in building the support vector machines for large class set pattern recognition tasks. Considering the large number (80) of SCV classes and the similarity among several classes, the 5-best performance of about 94% by the SVMs built using the 10-class *close-class-sets* is significantly good.

4 Summary and Conclusions

In this paper, we proposed a *close-class-set* discrimination method suitable for solving large class set pattern recognition problems using support vector machines. The effectiveness of the method has been demonstrated for recognition of isolated utterances belonging to 80 Stop_Consonant-Vowel classes. Combination of generative models (such as HMMs) and SVMs [10] may be considered for recognition of varying length patterns to avoid the loss of information in extraction of fixed length patterns from SCV utterances. Approaches for detection of acoustic events [5] [7] and the proposed method for classification using SVMs can be combined for recognition of SCV segments in continuous speech.

References

1. C.J.C.Burges, "A tutorial on support vector machines for pattern recognition," *Data Mining and Knowledge Discovery*, vol. 2, no. 2, pp. 121–167, 1998.
2. V.N.Vapnik, *The Nature of Statistical Learning Theory*, Springer, 2000.
3. J.Weston and C.Watkins, "Multi-class support vector machines," Tech. Rep. CSD-TR-98-04, Royal Holloway, University of London, May 1998.
4. P.Clarkson and P.J.Moreno, "On the use of support vector machines for phonetic classification," in *ICASSP*, Mar. 1999, pp. 585–588.
5. P.Niyogi, C.Burges, and P.Ramesh, "Distinctive feature detection using support vector machines," in *ICASSP*, Mar. 1999, pp. 425–428.
6. C. Chandra Sekhar, *Neural Network Models for Recognition of Stop Consonant-Vowel (SCV) Segments in Continuous Speech*, Ph.D. thesis, Indian Institute of Technology, Madras, 1996.
7. J.Y.Siva Rama Krishna Rao, C.Chandra Sekhar, and B.Yegnanarayana, "Neural networks based approach for detection of vowel onset points," in *International Conference on Advances in Pattern Recognition and Digital Techniques*, Calcutta, Dec. 1999, pp. 316–320.
8. L. R. Rabiner and B. H. Juang, *Fundamentals of Speech Recognition*, Prentice-Hall, Englewood Cliffs, 1993.
9. R.Collobert and S.Bengio, "Support vector machines for large-scale regression problems," Tech. Rep. IDIAP-RR-00-17, IDIAP, Switzerland, 2000.
10. T.S.Jaakkola and D.Haussler, "Exploiting generative models in discriminative classifiers," in *Advances in Neural Information Processing Systems*, M.S.Kearns, S.A.Solla, and D.A.Cohn, Eds., vol. 11. MIT Press, 1998.

Linear Dependency between ϵ and the Input Noise in ϵ-Support Vector Regression

James T. Kwok

Department of Computer Science,
Hong Kong University of Science and Technology, Hong Kong
jamesk@cs.ust.hk

Abstract. In using the ϵ-support vector regression (ϵ-SVR) algorithm, one has to decide on a suitable value of the insensitivity parameter ϵ. Smola *et al.* [6] determined its "optimal" choice based on maximizing the statistical efficiency of a location parameter estimator. While they successfully predicted a linear scaling between the optimal ϵ and the noise in the data, the value of the theoretically optimal ϵ does not have a close match with its experimentally observed counterpart. In this paper, we attempt to better explain the experimental results there, by analyzing a toy problem with a closer setting to the ϵ-SVR. Our resultant predicted choice of ϵ is much closer to the experimentally observed value, while still demonstrating a linear trend with the data noise.

1 Introduction

In recent years, uses of the support vector machines (SVMs) on various classification and regression problems have been increasingly popular. SVMs are motivated by results from the statistical learning theory and, unlike other machine learning methods, their generalization performances do not depend on the dimensionality of the feature space. In this paper, we focus on the ϵ-support vector regression (ϵ-SVR) algorithm [7], which has shown promising results in a number of different applications (e.g. [4]).

One issue about ϵ-SVR is on how to set the insensitivity parameter ϵ. Data-resampling techniques such as cross-validation can be used [4], but they are usually very expensive in terms of computation and/or data. A more efficient approach is to use another variant called the ν-SVR [5], which determines ϵ by using another parameter ν. It is claimed that ν may be easier to specify than ϵ. Another approach by Smola *et al.* [6] is to find the "optimal" choice of ϵ based on maximizing the statistical efficiency of a location parameter estimator. They showed that the asymptotically optimal ϵ should scale linearly with the noise in the data, and this was verified experimentally. However, their predicted value of the optimal ϵ does not have a close match with their experimental results.

In this paper, we attempt to better explain the experimental results in [6] by analyzing a toy problem with a closer setting to the ϵ-SVR. The rest of this paper is organized as follows. Brief introduction to the ϵ-SVR is given in Section 2. The specification and analysis of the toy problem are given in Section 3, while the last section gives some concluding remarks.

G. Dorffner, H. Bischof, and K. Hornik (Eds.): ICANN 2001, LNCS 2130, pp. 405–410, 2001.

Setting its partial derivative w.r.t. w to zero, it can be shown that $\hat{w}$ satisfies

$$\alpha\hat{w} + \beta N \left(\int_{-\infty}^{\hat{w}-\epsilon} \tilde{p}(y)dy - \int_{\hat{w}+\epsilon}^{\infty} \tilde{p}(y)dy \right) = 0.$$

To find suitable values of β and ϵ, we use the method of type II maximum likelihood prior[3] (ML-II) in Bayesian statistics. The ML-II solutions of β and ϵ are those that maximize $p(D|\beta,\epsilon)$, which in turn can be obtained by integrating out w, as $p(D|\beta,\epsilon) = \int p(w,D|\beta,\epsilon)dw = \int p(D|w,\beta,\epsilon)p(w)dw$. Assume $p(D|w,\beta,\epsilon)p(w)$ to be sharply peaked at its maximum $\hat{w}$, maximizing $p(D|\beta,\epsilon)$ w.r.t. β and ϵ thus becomes maximizing $p(D|\hat{w},\beta,\epsilon)p(\hat{w})$. This is the same as maximizing $\log p(\hat{w}|D,\beta,\epsilon)$, which equals $M(\hat{w},\beta,\epsilon)$ approximately. Taking the derivative of M w.r.t. ϵ and β to zero, we obtain

$$\frac{1}{1+\epsilon\beta} = \int_{-\infty}^{\hat{w}-\epsilon} \tilde{p}(y)dy + \int_{\hat{w}+\epsilon}^{\infty} \tilde{p}(y)dy = P(y \notin [\hat{w}-\epsilon, \hat{w}+\epsilon]), \tag{5}$$

and

$$\frac{1}{\beta(1+\epsilon\beta)} = E(|y-\hat{w}|_\epsilon). \tag{6}$$

Here, we have assumed that both $\frac{\partial\hat{w}}{\partial\epsilon}$ and $\frac{\partial\hat{w}}{\partial\beta}$ are negligible, which can be shown to be valid as $\tilde{p}(\cdot)$ is symmetrical around $\tilde{w}$. Finally, (5) and (6) can then be used to solve for β and ϵ.

3.3 Linear Dependency between ϵ and the Input Noise

From (5), (6), we obtain

$$\ell \equiv \frac{1}{\beta} = \frac{E(|y-\hat{w}|_\epsilon)}{P(y \notin [\hat{w}-\epsilon, \hat{w}+\epsilon])}, \tag{7}$$

and

$$\frac{\epsilon}{\ell} = \frac{P(y \in [\hat{w}-\epsilon, \hat{w}+\epsilon])}{P(y \notin [\hat{w}-\epsilon, \hat{w}+\epsilon])}. \tag{8}$$

ℓ can be interpreted as the distance from the ϵ-tube averaged over those points that are outside. The denominator acts as a normalizing factor as only that portion of points contribute towards $E(|y-\hat{w}|_\epsilon)$. Moreover, recall our probability model in (4), for $y \in [\hat{w}-\epsilon, \hat{w}+\epsilon]$,

$$p(y|\hat{w},\beta,\epsilon) = \frac{\beta}{2(1+\epsilon\beta)} = \frac{1}{2} \cdot \frac{P(y \notin [\hat{w}-\epsilon, \hat{w}+\epsilon])}{\ell},$$

using (7) and (8). As the total probability mass outside (both sides of) the ϵ-tube is $P(y \notin [\hat{w}-\epsilon, \hat{w}+\epsilon])$, ℓ can be interpreted as the "spread" of the density function outside the tube, by extending the density at the boundary of the ϵ-tube outside (Figure 1). This interpretation can also be seen from (8).

[3] Notice that MacKay's evidence framework [3], which has been popularly used in the neural networks community, is computationally equivalent to the ML-II method.

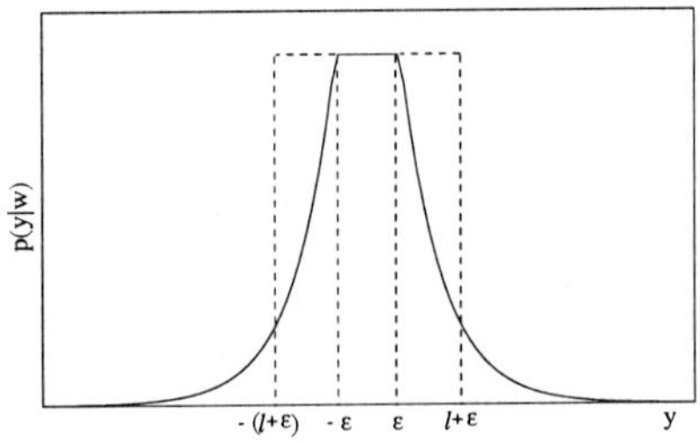

Fig. 1. The role of ℓ as the effective spread of the density function $p(y|\hat{w}, \beta, \epsilon)$ outside the ϵ-tube. Here, for simplicity, $\hat{w}$ is shifted to zero.

It was observed in [6] that there exists a linear dependency between ϵ and the scale parameter of the input noise model. This can be seen as a sufficient condition for ensuring the relationship in (8). When the input noise is changed by a factor λ, $\tilde{p}(\cdot)$ becomes stretched/compressed around $\tilde{w}$ by the same factor. By having a linear relationship between ϵ and the input noise, ϵ becomes $\epsilon' = \lambda\epsilon$. If we assume that $\hat{w}$ is close to $\tilde{w}$ and remains relatively unchanged, then $P(y \in [\hat{w} - \epsilon', \hat{w} + \epsilon'])$ will be unchanged, while $E(|y - \hat{w}|_{\epsilon'}) = \lambda E(|y - \hat{w}|_\epsilon)$. Hence, by (7), $\ell' = \lambda\ell$, and (8) still holds for the new values of ϵ' and ℓ'.

3.4 Applications to the Gaussian and Laplacian Noise Models

For the Gaussian noise model, $\phi(\eta) = \frac{1}{\sqrt{2\pi}\sigma}\exp(-\frac{\eta^2}{2\sigma^2})$. Define $\delta = \hat{w} - \tilde{w}, b_1 = \frac{1}{\sqrt{2}}(\frac{\epsilon}{\sigma} - \frac{\delta}{\sigma}), b_2 = \frac{1}{\sqrt{2}}(\frac{\epsilon}{\sigma} + \frac{\delta}{\sigma})$. Using (5) and (6), it can be shown that

$$\frac{1}{1+\epsilon\beta} = \frac{1}{2}(\mathrm{erfc}(b_1) + \mathrm{erfc}(b_2)), \tag{9}$$

and

$$\frac{1}{\beta(1+\epsilon\beta)} = -\frac{\sigma b_1}{2}\mathrm{erfc}(b_1) - \frac{\sigma b_2}{2}\mathrm{erfc}(b_2) + \frac{\sigma}{\sqrt{2\pi}}\exp(-b_1^2) + \frac{\sigma}{\sqrt{2\pi}}\exp(-b_2^2), \tag{10}$$

where $\mathrm{erfc}(x) = \frac{2}{\sqrt{\pi}}\int_x^\infty \exp(-t^2)dt$ is the complementary error function. Substituting (9) into (10), we see that ϵ always appears as the factor ϵ/σ in the solution, and thus there is always a linear relationship between ϵ and σ. If $|\delta|$ is small compared to ϵ relative to σ, $b_1 \simeq b_2 \simeq \epsilon/\sqrt{2}\sigma$, and we obtain $\mathrm{erfc}(\frac{\epsilon}{\sqrt{2}\sigma})(1 + \sqrt{\frac{\pi}{2}}\frac{\epsilon}{\sigma}\exp(\frac{\epsilon^2}{2\sigma^2})) = 1$. Solving, we have

$$\epsilon = 0.8485\sigma.$$

While the theoretical results in [6] suggest the "optimal" choice of $\epsilon = 0.6166\sigma$, our predicted choice is much closer to the experimentally observed optimal value of $\epsilon = 0.9\sigma$ in [6].

We next consider the Laplacian noise model, $\phi(\eta) = \frac{1}{2\sigma}\exp(-\frac{|\eta|}{\sigma})$. Using (5), (6) and assume that $|\hat{w} - \tilde{w}| \leq \epsilon$, we obtain $\frac{1}{1+\epsilon\beta} = \frac{1}{2}(\exp(-\frac{\epsilon}{\sigma} + \frac{\hat{w}-\tilde{w}}{\sigma}) +$

$\exp(-\frac{\epsilon}{\sigma} - \frac{\hat{w}-\tilde{w}}{\sigma}))$, and $\frac{1}{\beta(1+\epsilon\beta)} = \frac{\sigma}{2}(\exp(-\frac{\epsilon}{\sigma} + \frac{\hat{w}-\tilde{w}}{\sigma}) + \exp(-\frac{\epsilon}{\sigma} - \frac{\hat{w}-\tilde{w}}{\sigma}))$. Solving, then ϵ satisfies the equation

$$\frac{1}{1+\frac{\epsilon}{\sigma}} = \frac{1}{2}\left(\exp(-\frac{\epsilon}{\sigma} + \frac{\hat{w}-\tilde{w}}{\sigma}) + \exp(-\frac{\epsilon}{\sigma} - \frac{\hat{w}-\tilde{w}}{\sigma})\right).$$

Hence, again, there is a linear dependency between ϵ and σ. In the special case when $\hat{w} = \tilde{w}$, the above equations reduce to $\frac{1}{1+\epsilon\beta} = \exp(-\frac{\epsilon}{\sigma})$, and $\frac{1}{\beta(1+\epsilon\beta)} = \sigma\exp(-\frac{\epsilon}{\sigma})$. Solving, we obtain $\epsilon = 0$, which was also obtained in [6].

4 Conclusion

In this paper, we analyze the location estimation problem with a similar setting as for ϵ-SVR, and then derive its optimal choice of ϵ. In comparison to [6], their results are based on maximum likelihood estimation, while our analysis employs the same regularization term as in ϵ-SVR and is based on MAP estimation. Our predicted optimal choice of ϵ under Gaussian noise is also much closer to its experimentally observed optimal value. This prediction will be most accurate when the estimator $\hat{w}$ is close to the true value $\tilde{w}$. Besides, we accord with [6] in that ϵ should scale linearly with the input noise of the training data.

A pitfall of the results here is that their practical use is limited to cases when the noise model is known or at least can be estimated. Nevertheless, they can still be of tremendous help especially in performing simulation experiments. Typically, a researcher/practitioner may be experimenting with new techniques on ϵ-SVR, while not addressing the issue of finding the optimal value of ϵ. The results here can then be used to ascertain that a suitable value/range of ϵ has been chosen in the simulation.

References

1. C. Goutte and L.K. Hansen. Regularization with a pruning prior. *Neural Networks*, 10(6):1053–1059, 1997.
2. L.K. Hansen and C.E. Rasmussen. Pruning from adaptive regularization. *Neural Computation*, 6:1223–1232, 1994.
3. D.J.C. MacKay. Bayesian interpolation. *Neural Computation*, 4(3):415–447, May 1992.
4. K.R. Müller, A.J. Smola, G. Rätsch, B. Schölkopf, J. Kohlmorgen, and V. Vapnik. Predicting time series with support vector machines. In *Proceedings of the International Conference on Artificial Neural Networks*, 1997.
5. B. Schölkopf and A.J. Smola. New support vector algorithms. NeuroCOLT2 Technical Report NC2-TR-1998-031, GMD FIRST, 1998.
6. A.J. Smola, N. Murata, B. Schölkopf, and K.-R. Müller. Asymptotically optimal choice of ϵ-loss for support vector machines. In *Proceedings of the International Conference on Artificial Neural Networks*, 1998.
7. A.J. Smola and B. Schölkopf. A tutorial on support vector regression. NeuroCOLT2 Technical Report NC2-TR-1998-030, Royal Holloway College, 1998.

The Bayesian Committee Support Vector Machine

Anton Schwaighofer[1,2] and Volker Tresp[2]

[1] TU Graz, Institute for Theoretical Computer Science
Inffeldgasse 16b, 8010 Graz, Austria
aschwaig@igi.tu-graz.ac.at
http://www.igi.tu-graz.ac.at/aschwaig/
[2] Siemens AG, Corporate Technology, Otto-Hahn-Ring 6,
81739 München, Germany
{Anton.Schwaighofer.external,Volker.Tresp}@mchp.siemens.de

Abstract. Empirical evidence indicates that the training time for the support vector machine (SVM) scales to the square of the number of training data points. In this paper, we introduce the Bayesian committee support vector machine (BC-SVM) and achieve an algorithm for training the SVM which scales linearly in the number of training data points. We verify the good performance of the BC-SVM using several data sets.

1 Introduction

Kernel-based systems such as the support vector machine (SVM) and Gaussian processes (GP) are powerful and currently very popular approaches to supervised learning [1]. Unfortunately, there are at least three problems when one tries to scale up these systems to large data sets. First, training time increases drastically with the number of training data points, second, memory requirements increase with data set size and third, prediction time is proportional to the number of kernels and the latter is equal to (or at least increases with) the number of training data points.

One approach to scale up kernel systems to large data sets is the Bayesian committee machine (BCM [6]). In the BCM the data set is divided up into M sets of approximately the same size and M models are developed on the individual data sets. The predictions of the individual models are combined using a weighting scheme which is derived from a Bayesian perspective. The computational complexity of the BCM scales linearly in the number of training data points. The BCM was developed in the context of Gaussian process regression (GPR) and was later applied to generalized Gaussian processes regression (GGPR). GGPR uses measurement processes derived from the exponential family of distributions and can be applied to classification and the prediction of counts and lifetimes.

In this paper we apply the BCM approximation to the support vector machine (SVM) in form of the Bayesian committee support vector machine (BC-SVM). The SVM is a kernel-based classifier which is derived from VC learning theory [1]

G. Dorffner, H. Bischof, and K. Hornik (Eds.): ICANN 2001, LNCS 2130, pp. 411–417, 2001.

and minimizes a bound on the generalization error. A particular property of the SVM is that it achieves sparse solutions, i.e. many kernel weights are equal to zero. The basis for the applicability of the BCM to the SVM was made possible by a recent probabilistic interpretation of the SVM by Sollich [4].

The paper is organized as follows. In the next sections we will review Gaussian process regression, the BCM and the SVM. In Section 5 we will derive the BC-SVM. In Section 6 we present experimental results and in Section 7 we present conclusions.

2 Gaussian Process Regression (GPR)

In contrast to the usual parameterized approach to regression, in Gaussian process regression we specify the prior model directly in function space. In particular we assume that *a priori* f is Gaussian distributed (in fact it would be an infinite-dimensional Gaussian) with zero mean and a covariance $K(x_1, x_2) = cov(f(x_1), f(x_2))$, where the experimenter has to specify the covariance function.

We assume that we can only measure a noisy version $y(x) = f(x) + \epsilon(x)$ of the underlying function f, where $\epsilon(x)$ is independent Gaussian distributed noise with zero mean and variance σ_ψ^2.

The goal is now to predict the functional values $f^q = (f_1^q, \dots, f_{N_Q}^q)'$ at a set of N_Q test or query points based on a set of N training data $\mathcal{D} = \{(x_1, y_1), \dots (x_N, y_N)\}$. Let $\Psi^{mm} = \sigma_\psi^2 I$ be the noise variance of the targets of the measurements, where I is the unit matrix. Furthermore, Σ^{mm} is the covariance matrix of the functional values at the training data points, Σ^{qq} is the covariance matrix at the query points and Σ^{qm} is the covariance matrix between training points and the query points.

Under these assumptions the conditional density of the response variables at the query points is Gaussian distributed with mean $E(f^q|\mathcal{D})$ and covariance $cov(f^q|\mathcal{D})$, given by

$$E(f^q|\mathcal{D}) = \Sigma^{qm}(\Psi^{mm} + \Sigma^{mm})^{-1}y^m \tag{1}$$

$$cov(f^q|\mathcal{D}) = \Sigma^{qq} - \Sigma^{qm}\,(\Psi^{mm} + \Sigma^{mm})^{-1}\,(\Sigma^{qm})'. \tag{2}$$

Note, that following Eq. 1, the estimated function at input x becomes

$$\hat{f}(x) = \sum_{i=1}^{N} w_i K(x, x_i). \tag{3}$$

where $w = (\Psi^{mm} + \Sigma^{mm})^{-1}y^m$. That is, the prediction of the GPR system is a weighted combination of kernels on the training data points.

3 The Bayesian Committee Machine Applied to GPR

The calculation of the optimal predictions at the query points (Eq. 1) requires the inversion of an $N \times N$ matrix and therefore scales to the third power of the number of training data points. As a consequence, GPR is limited to problems up to may be 1000 training data points. In the Bayesian Committee Machine (BCM, [6]), the data are partitioned into M data sets $\mathcal{D} = \{D^1, \ldots, D^M\}$ (of typically approximately same size) and M GPR systems are trained on the individual data sets D^i. When applied to a set of query points, each of the M GPR systems outputs a prediction $E(f^q|D^i)$ together with covariance $cov(f^q|D^i)$, calculated employing GPR formulas Equations 1 and 2.

The BCM combines the M estimates and calculates an approximation to the expected values $\hat{E}(f^q|\mathcal{D})$ of the functional values at the query points as

$$\hat{E}(f^q|\mathcal{D}) = {C_{BCM}}^{-1} \sum_{i=1}^{M} cov(f^q|D^i)^{-1} E(f^q|D^i) \tag{4}$$

$$C_{BCM} = \widehat{cov}(f^q|\mathcal{D})^{-1} = -(M-1)(\Sigma^{qq})^{-1} + \sum_{i=1}^{M} cov(f^q|D^i)^{-1}. \tag{5}$$

We recognize that the prediction of each GPR system i is weighted by the inverse covariance of its prediction. In [6] it was shown that the BCM approximation becomes equal to the correct expected value when the number of query points is equal to the effective number of parameters of the kernel system. If we set $N/M = N_Q$ (i.e. the number of data points in each module is equal to the number of query points which is a typical choice) then the computational complexity of the BCM scales as $\mathcal{O}(N \times N_Q^2)$, i.e. linear in the training data set size.

4 The Support Vector Machine (SVM)

The SVM is typically formulated as a linear classifier which is applied to a high-dimensional feature space using the kernel transformation. The SVM takes on the form of a kernel system as in Eq. 3 and a new pattern is classified according to the sign of $\hat{f}(x) + b$ where b is a constant. The optimal weight vector w is found by minimizing the cost function

$$cost = \frac{1}{2} w' \Sigma^{mm} w + \frac{C}{2} \sum_{i=1}^{N} \left[(1 - y_i(f(x_i) + b))_+\right]^{\alpha}.$$

Here, $y \in \{-1, 1\}$ is the class label and, as before, $(\Sigma^{mm})_{ij} = K(x_i, x_j)$ is the kernel matrix. The operation $()_+$ sets all negative values equal to zero and C is a constant that determines to what degree a violation of the margin constraints is penalized. In most applications $\alpha = 1$ has been used (1-norm soft margin) but $\alpha = 2$ (2-norm soft margin) is also a common choice [1].

The state of the art optimization routines (SMO, SVM^{light} [1]) scale approximately $\mathcal{O}(N^2)$, which limits the applicability of the SVM to data sets up to a few 10000 training data points.

Sollich [4] has shown how the SVM can be formulated in the context of Gaussian processes. As in GPR, we assume that the function is derived from an infinite dimensional prior Gaussian distribution. In the SVM, we use a different measurement process[1]. Instead of assuming Gaussian measurement noise, we set

$$P(y|f(x)) \propto \exp\left(-\frac{C}{2}\left[(1 - y(f(x) + b))_+\right]^\alpha\right).$$

The cost function that is minimized during SVM training can then be interpreted as the negative log of the posterior probability $-\log P(f^m|\mathcal{D})$ and becomes

$$-\log P(f^m|\mathcal{D}) = \frac{1}{2}(f^m)'(\Sigma^{mm})^{-1}f^m + \frac{C}{2}\sum_{i=1}^{N}\left[(1 - y_i(f(x_i) + b))_+\right]^\alpha. \quad (6)$$

Here, f^m is the vector of SVM outputs $f(x_i)$ at the training data points x_i.

5 The Bayesian Committee Support Vector Machine (BC-SVM)

For deriving the BC-SVM, we assume that the posterior density $P(f^m|\mathcal{D})$ can be approximated by a Gaussian density (Laplace approximation). The mean of the Gaussian corresponds to the prediction of the SVM at these points and can be calculated using Eq. 3. The covariance matrix is derived from the Hessian matrix of the cost function. We make use of Sollich's re-formulation of the SVM cost function $-\log P(f^m|\mathcal{D})$ (Eq. 6).

Note, that the SVM cost function with $\alpha = 1$ is non-differentiable at the points with $1 - yf(x) = 0$. This is particularly unpleasant since the points with the property $1 - yf(x) = 0$ are the support vectors. Sollich [4] proposes a Gaussian approximation in this case but we experienced instabilities when we used this approximation. For this reason, we used $\alpha = 2$ in our experiments. Here, the second derivative of the likelihood term is equal to C for the support vectors and is 0 otherwise.

Computing the BC-SVM approximation thus consists of the following steps:

1. Train SVM module i on partition $\mathcal{D}^i$ of the training data, using some standard SVM training algorithm.
2. The SVM outputs at the query points, as given by Eq. 3, replace $E(f^q|\mathcal{D}^i)$ in Eq. 4.
3. Compute the Hessian matrix H of the cost function Eq. 6 evaluated at the training points. By analogy to GPR, we calculate $cov(f^q|\mathcal{D}^i)$ (Eq. 2) by identifying $H = (\Sigma^{mm})^{-1} + (\Psi^{mm})^{-1}$.
4. Use the predictions and their covariances $cov(f^q|\mathcal{D}^i)$ of all modules in the BCM (Eq. 4) to obtain the prediction of the BC-SVM.

For the BC-SVM Eq. 2 further simplifies such that the covariances are defined only for the support vectors: Σ^{mm} contains the covariances between only

[1] Properly normalizing the conditional probability density is somewhat tricky and is discussed in detail in [4].

the support vectors, Σ^{qm} contains the covariances between the support vectors and the query points, and $\Psi^{mm} = \frac{1}{C}I$ where I is the unit matrix with the dimensionality of the number of support vectors.

All 2-norm SVMs have been trained using adapted versions of the decomposition method and working set selection in the style of SVM^{light}, as proposed by Joachims [2].

6 Experiments

We demonstrate the performance of the BC-SVM in experiments on four artificial data sets (with different levels of noise) and on four real world data sets.

The ART data sets were generated by randomly drawing $d = 5$ dimensional input points uniformly from $[-1, 1]^d$, using them as inputs to a map defined by 5 normalized Gaussian basis functions, and finally adding independent Gaussian noise with variance σ_ψ^2 to the map. The assigned class corresponds to the sign of the generated outputs. In this way, the degree of overlap between classes can be influenced by varying σ_ψ^2. (This is the same data set as used in [7], see the reference for a detailed description.)

The CHECKER data set is sampled uniformly from a 4×4 checkerboard grid on $[0, 1] \times [0, 1]$. CHECKER is often used to demonstrate highly complex decision surfaces. Note that data sets CHECKER and ART with $\sigma_\psi = 0$ have non-overlapping classes.

Table 1 shows test set errors for different classification methods on those data sets. Shown are results for an SVM trained on the whole of the training data and for BC-SVMs with different module and query set size. It can be seen that the BC-SVM, even with a small module size such as 100, performs comparable to a full SVM on all data sets with noise (ART with $\sigma_\psi > 0$). For the noise free sets CHECKER and ART with $\sigma_\psi = 0$, a larger module size is required to obtain a performance that is comparable with the full SVM.

The evolution of the test set error, as the number of training points (and thus the number of modules in the BC-SVM) increases, is shown in Figure 1 for ART $\sigma_\psi = 0.1$. The graph furthermore shows that the combination scheme employed by the BC-SVM performs significantly better than averaging the predictions of the individual modules.

The performance of the BC-SVM shows up more clearly in Table 2, where the results for four real world data sets are given. For the three smaller data sets PIMA, BUPA and WAVEFORM, the BC-SVM performs comparable to a full SVM and no statistically significant difference in performance can be seen. For the large ADULT data set, the BC-SVM has a slightly higher error rate than the full SVM, yet still performs excellently.

For the ADULT data set we have also logged the training time. Training the BC-SVM takes about 3 minutes, while training the full SVM lasts a few hours. Both SVM and BC-SVM have been implemented in Matlab. (Comparing the BC-SVM runtime directly with methods such as SMO and SVM^{light} would require a C/C++ implementation.)

Table 1. BC-SVM on 4 artificial data sets. Shown is the number of input dimensions d, the number of training examples N, and the average test set error for different classification methods. BC-SVM(i, j) uses a module size i with query set size j. All test set errors have been computed on independent 10000 point test sets.

	ART $\sigma_\psi = 0$	ART $\sigma_\psi = 0.1$	ART $\sigma_\psi = 0.5$	CHECKER
d	5	5	5	2
N	2000	2000	2000	5000
SVM	2.58%	7.25%	19.32%	1.26%
BC-SVM(100,200)	3.30%	7.87%	19.75%	2.67%
BC-SVM(200,200)	3.54%	7.44%	19.54%	2.65%
BC-SVM(500,200)	—	—	—	1.82%

Table 2. BC-SVM on 3 real-world data sets. Shown is the number of input dimensions d, the number of training examples N, and the average test set error for different classification methods. For BUPA, PIMA and WAVEFORM, test set errors have been computed using 7fold cross-validation, and are given together with a 95% confidence interval. The results for ADULT are on an independent test set of 16281 points.

	BUPA	PIMA	WAVEFORM	ADULT
d	6	8	21	124
N	345	700	2000	32561
SVM	$27.7 \pm 2.7\%$	$24.3 \pm 2.7\%$	$8.33 \pm 0.99\%$	14.77%
BC-SVM(50,50)	$27.1 \pm 3.7\%$	$24.4 \pm 2.9\%$	$8.33 \pm 0.89\%$	—
BC-SVM(100,100)	$27.4 \pm 3.9\%$	$25.7 \pm 2.6\%$	$8.94 \pm 1.5\%$	16.88%
BC-SVM(200,200)	—	—	$8.66 \pm 0.98\%$	15.76%

It should be mentioned that predictions of the BC-SVM on test data are more time consuming than predictions of a standard SVM. This results from the matrix operations in the combination scheme (Eq. 4), and from the higher total number of support vectors in the BC-SVM. The modules in a BC-SVM, trained on small portions of the data, can not exploit sparseness as efficiently as an SVM that has been trained on the full data.

7 Conclusions

We have shown a principled way of scaling support vector machines to large data sets by training SVMs on small subsets of the training data and combining their predictions through a Bayesian scheme. The module size in this BC-SVM is small enough that they can be trained with off-the-shelf quadratic optimizers without the necessity for highly optimised SVM training methods (decomposition, SMO, etc). We found the predictions of the BC-SVMs to be comparable with standard SVMs trained on the full data.

An interesting direction for further work would be to apply the BCM to relevance vector machines (RVM [5]). The RVM uses a probabilistic approach

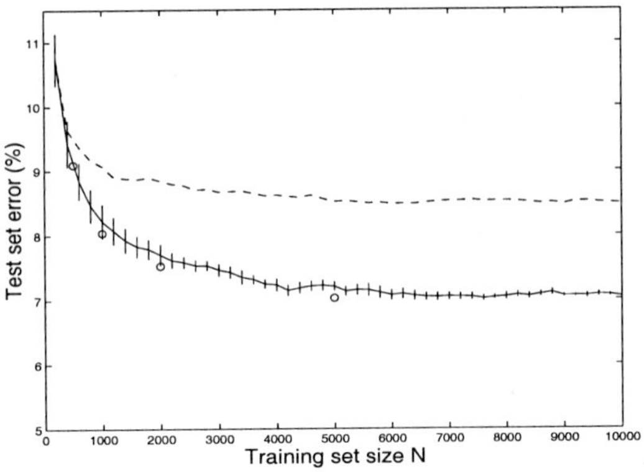

Fig. 1. Artificial data with noise $\sigma_\psi = 0.1$: Test set error of SVM compared to BC-SVM(200,200) as a function of the number of training points N. The test set errors of the SVM's, trained on various training set sizes, are shown as circles. The solid line is the error of the BC-SVM, together with (two-sided) 95% confidence intervals. The BC-SVM does not perform significantly worse than the full SVM. In particular, the BC-SVM clearly outperforms a combination scheme that averages the predictions of the BC-SVM modules (shown dashed).

to obtain a sparse solution of a kernel system. So far, the RVM idea is lacking a method for scaling up to large data sets, since decomposition methods are not applicable.

Acknowledgements

AS gratefully acknowledges support through an Ernst-von-Siemens scholarship.

References

1. Cristianini, N., Shawe-Taylor, J.: Support Vector Machines. Cambridge University Press, 2000
2. Joachims, T.: Making Large-scale Support Vector Machine Learning Practical. In: Schölkopf, B., Burges, C.J., Smola, A.J. (eds.), *Advances in Kernel Methods – Support Vector Learning.* MIT Press, 1998 pages 169–184
3. Solla, S.A., Leen, T.K., Müller, K.R. (eds.): Advances in Neural Information Processing Systems 12. MIT Press, 2000
4. Sollich, P.: Probabilistic Methods for Support Vector Machines. In: Solla et al. [3], pages 349–355, 2000
5. Tipping, M.E.: The Relevance Vector Machine. In: Solla et al. [3], 2000
6. Tresp, V.: A Bayesian Committee Machine. Neural Computation 12 (2000) 2719–2741
7. Tresp, V.: The Generalized Bayesian Committee Machine. In: *Proceedings of the Sixth ACM SIGKDD International Conference on Knowledge Discovery and Data Mining.* Boston, MA USA, 2000

Part V

Topographic Mapping

Using Directional Curvatures to Visualize Folding Patterns of the GTM Projection Manifolds

Peter Tiňo, Ian Nabney, and Yi Sun

Neural Computing Research Group, Aston University,
Aston Triangle, Birmingham B4 7ET
United Kingdom
{tinop,nabneyit,suny}@aston.ac.uk
http://www.ncrg.aston.ac.uk/

Abstract. In data visualization, characterizing local geometric properties of non-linear projection manifolds provides the user with valuable additional information that can influence further steps in the data analysis. We take advantage of the smooth character of GTM projection manifold and analytically calculate its local directional curvatures. Curvature plots are useful for detecting regions where geometry is distorted, for changing the amount of regularization in non-linear projection manifolds, and for choosing regions of interest when constructing detailed lower-level visualization plots.

1 Introduction

Most visualization algorithms project points from a high-dimensional data space onto a two-dimensional projection space. Loosely speaking, algorithms like self-organizing maps (SOM) [7] or the Generalized Topographic Mapping (GTM) [6], a probabilistic reformulation of the SOM, identify the computer screen with a two-dimensional "rubber sheet" that is injected into the high-dimensional data space. The sheet is supposed to "cover" the cloud of data points by locally stretching, contracting and curving. The visualization plot is obtained by first projecting the data points onto the rubber sheet and then letting the rubber sheet relax to its original form of the computer screen. We refer to the injected (possibly curved and stretched) two-dimensional rubber sheet in the data space as the *projection manifold*.

Besides the visualization plot itself, the user is often interested in additional information about the structure of the projection manifold in the high-dimensional data space. For example, local magnification factors describe how small regions on the computer screen are stretched or compressed when mapped to the data space. Magnification factors can be used for detecting (on the visualization plot) separate clusters in the data space (see [4]).

SOM discretizes the "rubber sheet" into a grid of "nodes" and represents the projection manifold only by a grid of nodes mapped to the data space (code-book

G. Dorffner, H. Bischof, and K. Hornik (Eds.): ICANN 2001, LNCS 2130, pp. 421–428, 2001.

vectors). On the other hand, GTM forms a *smooth* two-dimensional projection manifold. This allows us to use techniques of differential geometry to analytically describe (local) geometric properties *anywhere* on the manifold. Local magnification factors of GTM models were *analytically* computed by Bishop, Svensén and Williams in [4] [5]. In SOM, the magnification factors can only be approximated by distances between the code-book vectors.

Magnification factors represent the extent to which the areas are magnified on projection to the data space. However, when injecting a two-dimensional rubber sheet into a high dimensional data space, the projection manifold may form complicated folds that cannot be detected by using magnification factors alone. To provide the user with a tool for monitoring the amount of folding and neighborhood preservation in the projection manifold, we need second-order quantities, such as local curvatures. Neighborhood preservation and folding issues in the context of SOM were studied e.g. in [2] [11]. Such studies present largely heuristic techniques for computing higher-order geometric properties of the SOM projection manifold based on a discrete grid of code-book vectors (representing the manifold) in the data space. In contrast, as shown in this paper, the smooth nature of the GTM projection manifold allows us to analytically compute directional curvatures in any point on the manifold.

2 Generative Topographic Mapping

The Generative Topographic Mapping (GTM) belongs to a family of latent space models that model a probability distribution in the (observable) data space by means of latent, or hidden variables. For the purposes of data visualization, we identify the visualization space (i.e. the "rubber sheet") with the latent space. The latent space is usually a bounded subset of the two-dimensional Euclidean space, such as the (two-dimensional) interval $[-1, 1] \times [-1, 1]$.

Consider an L-dimensional latent space $\mathcal{H} \subset \Re^L$ and represent points in $\mathcal{H}$ as column vectors $\mathbf{x} = (x_1, x_2, ..., x_L)^T$. We allow $\mathcal{H}$ to be covered by an array of K latent space centres $\mathbf{x}_i \in \mathcal{H}$, $i = 1, 2, ..., K$.

Let the data space $\mathcal{D}$ be the D-dimensional Euclidean space $\Re^D$. We define a non-linear transformation $f : \mathcal{H} \to \mathcal{D}$ from the latent space to the data space using a radial basis function network (see e.g. [3]). To this end, we cover the latent space with a set of M fixed non-linear basis functions $\phi_j : \mathcal{H} \to \Re$, $j = 1, 2, ..., M$. As usual in the GTM literature, we choose to work with spherical Gaussian functions of the same width σ, although other choices are possible and require only simple modifications[1]. Usually, the centres of the Gaussian basis functions ϕ_j are positioned in the latent space on a regular grid. Given a point $\mathbf{x} \in \mathcal{H}$ in the latent space, its image under the map f is[2]

$$f(\mathbf{x}) = \mathbf{W}\ \phi(\mathbf{x}), \qquad (1)$$

[1] GTM with other choices of basis functions can be easily constructed using NETLAB available from `http://www.ncrg.aston.ac.uk/netlab/`

[2] We assume that the data set has been normalized to zero mean. Equivalently, we could include a constant basis function $\phi_0(\mathbf{x}) = 1$.

where $\mathbf{W}$ is a $D\times M$ matrix of weight parameters and $\phi(\mathbf{x}) = (\phi_1(\mathbf{x}), ..., \phi_M(\mathbf{x}))^T$.

GTM creates a generative probabilistic model in the data space by placing a radially-symmetric Gaussian with zero mean and inverse variance β around images, under f, of the latent space centres $\mathbf{x}_i \in \mathcal{H}$, $i = 1, 2, ..., K$:

$$P(\mathbf{t}|\ \mathbf{x}_i, \mathbf{W}, \beta) = \left(\frac{\beta}{2\pi}\right)^{D/2} \exp\left\{-\frac{\beta}{2}\|f(\mathbf{x}_i) - \mathbf{t}\|^2\right\}. \quad (2)$$

Defining a uniform prior over $\mathbf{x}_i$, the density model in the data space provided by the GTM is

$$P(\mathbf{t}|\ \mathbf{W}, \beta) = 1/K \sum_{i=1}^{K} P(\mathbf{t}|\ \mathbf{x}_i, \mathbf{W}, \beta). \quad (3)$$

Given a data set $\zeta = \{\mathbf{t}_1, \mathbf{t}_2, ..., \mathbf{t}_N\}$ of independently generated points in the data space, the adjustable parameters $\mathbf{W}$ and β of the model can be fitted to the data by maximum likelihood using the expectation-maximization algorithm [6].

For the purpose of data visualization, we use Bayes' theorem to invert the transformation f from the latent space $\mathcal{H}$ to the data space $\mathcal{D}$. The posterior distribution on $\mathcal{H}$, given a data point $\mathbf{t}_n \in \mathcal{D}$, is a sum of delta functions centered at centres $\mathbf{x}_i$, with coefficients equal to the posterior probability R_{in} that the i-th Gaussian (corresponding to the latent space centre $\mathbf{x}_i$, eq. (2)) generated $\mathbf{t}_n$ [6]. The latent space representation of the point $\mathbf{t}_n$, i.e. the *projection of* $\mathbf{t}_n$, is taken to be the mean $\sum_{i-1}^{K} R_{in}\ \mathbf{x}_i$ of the posterior distribution on $\mathcal{H}$.

The f–image of the latent space $\mathcal{H}$,

$$\Omega = f(\mathcal{H}) = \{f(\mathbf{x}) \in \Re^D |\ \mathbf{x} \in \mathcal{H}\}, \quad (4)$$

forms a *smooth* L-dimensional manifold in the data space. We refer to the manifold Ω as the *projection (regression) manifold* of the GTM.

3 Local Directional Curvatures

The idea of directional curvature is explained in figure 1.

Consider a point $\mathbf{x}_0 \in \mathcal{H}$. Let $\mathbf{x}(b)$, $b \in \Re$, be a straight line passing through $\mathbf{x}_0$ along a unit directional vector $\mathbf{h} = (h_1, h_2, ..., h_L)^T$:

$$\mathbf{x}(b) = \mathbf{x}_0 + b\mathbf{h}, \quad b \in \Re. \quad (5)$$

As the parameter b varies, the image of the line $\mathbf{x}(b)$ generates on Ω the curve

$$\mu(b) = f(\mathbf{x}(b)). \quad (6)$$

The tangent to this curve at $f(\mathbf{x}_0) = \mu(0)$ is

$$\dot{\mu}(0) = \left[\frac{\mathrm{d}\ \mu(b)}{\mathrm{d}\ b}\right]_{b=0} = \left[\sum_{r=1}^{L} \frac{\partial f(\mathbf{x})}{\partial x_r} \frac{\mathrm{d}\ x_r(b)}{\mathrm{d}\ b}\right]_{\mathbf{x}=\mathbf{x}_0, b=0}$$

$$= \sum_{r=1}^{L} \mathbf{\Gamma}_r^{(1)}\ h_r = \mathbf{\Gamma}^{(1)}\ \mathbf{h}, \quad (7)$$

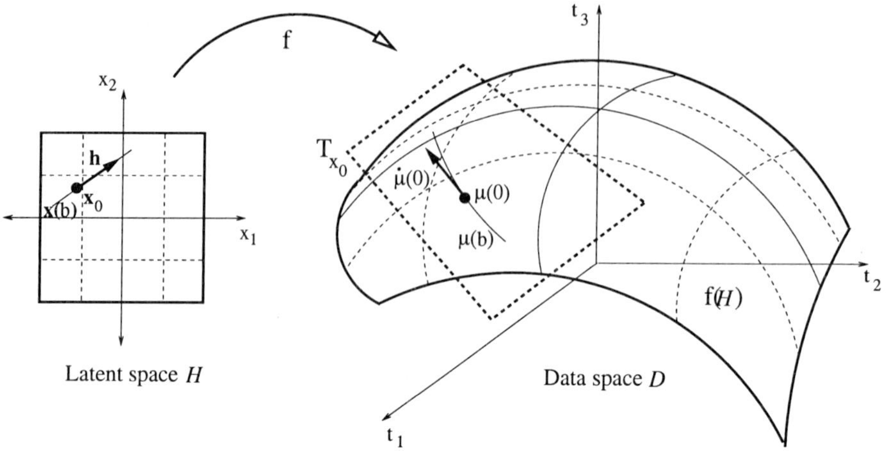

Fig. 1. An explanation of local directional derivative of the visualization manifold. A straight line $\mathbf{x}(b)$ passing through the point $\mathbf{x}_0$ in the latent space $\mathcal{H}$ is mapped via f to the curve $\mu(b) = f(\mathbf{x}(b))$ in the data space $\mathcal{D}$. Curvature of μ at $f(\mathbf{x}_0) = \mu(0)$ is related to the directional curvature of the projection manifold $f(\mathcal{H})$ with respect to the direction $\mathbf{h}$. The tangent vector $\dot{\mu}(0)$ to μ at $\mu(0)$ lies in $\mathbf{T}_{\mathbf{x}_0}$ (dashed rectangle), the tangent plane of the manifold $f(\mathcal{H})$ at $\mu(0)$.

where

$$\mathbf{\Gamma}_r^{(1)} = \mathbf{W} \left(\frac{\partial \phi_1(\mathbf{x}_0)}{\partial x_r}, \frac{\partial \phi_2(\mathbf{x}_0)}{\partial x_r}, ..., \frac{\partial \phi_M(\mathbf{x}_0)}{\partial x_r} \right)^T \tag{8}$$

is a (column) vector of partial derivatives of the GTM map f (at $\mathbf{x}_0 \in \mathcal{H}$) with respect to the r-th latent space variable x_r, and $\mathbf{\Gamma}^{(1)}$ is the $D \times L$ matrix

$$\mathbf{\Gamma}^{(1)} = [\mathbf{\Gamma}_1^{(1)}, \mathbf{\Gamma}_2^{(1)}, ..., \mathbf{\Gamma}_L^{(1)}]. \tag{9}$$

The tangent vector $\dot{\mu}(0)$ to the lifted line $\mu(b)$ is a linear combination of the columns of $\mathbf{\Gamma}^{(1)}$, and so the range of the matrix $\mathbf{\Gamma}^{(1)}$ is the tangent plane $\mathbf{T}_{\mathbf{x}_0}$ of the projection manifold Ω at $f(\mathbf{x}_0) = \mu(0)$.

The second directional derivative [9] of $\mu(b)$ at $\mu(0)$ is

$$\begin{aligned} \ddot{\mu}(0) &= \left[\sum_{s=1}^{L} \frac{\partial}{\partial x_s} \left\{ \sum_{r=1}^{L} \frac{\partial f(\mathbf{x})}{\partial x_r} h_r \right\} \frac{\mathrm{d}\, x_s(b)}{\mathrm{d}\, b} \right]_{\mathbf{x}=\mathbf{x}_0, b=0} \\ &= \left[\sum_{r=1}^{L} \sum_{s=1}^{L} \frac{\partial^2 f(\mathbf{x})}{\partial x_r \partial x_s} h_r h_s \right]_{\mathbf{x}=\mathbf{x}_0} = \sum_{r=1}^{L} \sum_{s=1}^{L} \mathbf{\Gamma}_{r,s}^{(2)} \, h_r \, h_s, \end{aligned} \tag{10}$$

where $\mathbf{\Gamma}_{r,s}^{(2)}$ is a column vector of second-order partial derivatives of f (at $\mathbf{x}_0 \in \mathcal{H}$) with respect to the r-th and s-th latent space variables,

$$\mathbf{\Gamma}_{r,s}^{(2)} = \mathbf{W} \left(\frac{\partial^2 \phi_1(\mathbf{x}_0)}{\partial x_r \partial x_s}, \frac{\partial^2 \phi_2(\mathbf{x}_0)}{\partial x_r \partial x_s}, ..., \frac{\partial^2 \phi_M(\mathbf{x}_0)}{\partial x_r \partial x_s} \right)^T. \tag{11}$$

We decompose $\ddot{\mu}(0)$ into two orthogonal components, one lying in the tangent space $\mathbf{T}_{\mathbf{x}_0}$, the other lying in its orthogonal complement $\mathbf{T}^{\perp}_{\mathbf{x}_0}$,

$$\ddot{\mu}(0) = \ddot{\mu}^{\parallel}(0) + \ddot{\mu}^{\perp}(0), \qquad \ddot{\mu}^{\parallel}(0) \in \mathbf{T}_{\mathbf{x}_0},\ \ddot{\mu}^{\perp}(0) \in \mathbf{T}^{\perp}_{\mathbf{x}_0}. \tag{12}$$

The component $\ddot{\mu}^{\parallel}(0)$ describes changes in the first-order derivatives due to "varying speed of parameterization", while the direction of the first-order derivatives remains unchanged. Changes in the first-order derivatives that are responsible for curving of the projection manifold Ω are described by the component $\ddot{\mu}^{\perp}(0)$.

Orthogonal projection onto $\mathbf{T}_{\mathbf{x}_0}$ is a linear operator described by the projection matrix $\mathbf{\Pi} = \mathbf{\Gamma}^{(1)} \left(\mathbf{\Gamma}^{(1)}\right)^+$, where $\left(\mathbf{\Gamma}^{(1)}\right)^+$ is the Moore-Penrose generalized inverse of $\mathbf{\Gamma}^{(1)}$ (see e.g. [8]). So, $\ddot{\mu}^{\perp}(0) = (\mathbf{I} - \mathbf{\Pi})\ \ddot{\mu}(0)$, where $\mathbf{I}$ is the $D \times D$ identity matrix.

The directional curvature at $\mu(0)$ associated with the latent space direction $\mathbf{h}$ is the (Euclidean) norm of the vector $\ddot{\mu}^{\perp}(0)$. It measures the degree to which the visualization manifold Ω (locally) "curves" in the data space $\mathcal{D}$ [1]. It is the embedding curvature of $\Omega \subset \mathcal{D}$ at $f(\mathbf{x}_0)$, evaluated with respect to the latent space direction $\mathbf{h}$.

4 Experiments

In the experiments reported here, the GTM latent space $\mathcal{H}$ was the square $\mathcal{H} = [-1,1] \times [-1,1]$, the latent space centres $\mathbf{x}_i \in \mathcal{H}$ were positioned on a regular 15×15 square grid and there were 16 basis functions ϕ_j centered on a regular 4×4 square grid. The basis functions were spherical Gaussian functions of the same width $\sigma = 0.44$. Magnification factors and directional curvatures were evaluated at each latent space centre $\mathbf{x}_i$.

In the first experiment we randomly generated 2000 points in $\Re^3$ lying on the two-dimensional manifold shown in figure 2a. As expected, after training, the GTM projection manifold shown in figure 2b closely followed the two-dimensional distribution of the data points. Latent space layouts of local magnification factors and directional curvatures are shown in figures 2c and 2d, respectively. The magnification factor at a centre $\mathbf{x}_i$, which is the Jacobian of the GTM map f at $\mathbf{x}_i$ [4], is represented by the degree of shading of the corresponding patch. In the curvature plot, we show for each latent space centre $\mathbf{x}_i$, the direction $\mathbf{h}$ yielding the maximal norm of $\ddot{\mu}^{\perp}(0)$. The length of the direction line and the degree of shading of the corresponding patch are proportional to the maximal norm of $\ddot{\mu}^{\perp}(0)$.

The curvature and expansion patterns in the projection manifold (figure 2b) are clearly reflected in the magnification factor and curvature plots (figures 2c,d). The form of the projection manifold can be approximately guessed on the basis of its local first- and second-order characterizations.

In the second experiment, we trained GTM on an oil flow data set[3]. This 12-dimensional data set arises from a physics-based simulation of a non-invasive

[3] `http://www.ncrg.aston.ac.uk/GTM/3PhaseData.html`

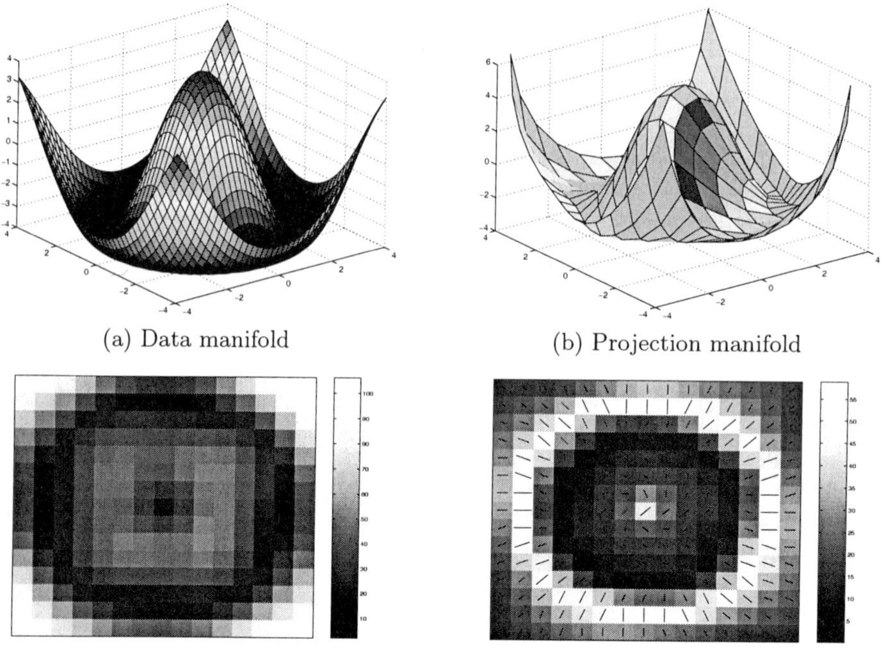

(a) Data manifold (b) Projection manifold

(c) Magnification factors (d) Curvatures

Fig. 2. Toy data experiment.

monitoring system, used to determine the quantity of oil in a multi-phase pipeline containing a mixture of oil, water and gas. The data set consists of 1000 points obtained synthetically by simulating the physical process in the pipe. Points in the data set are classified into three different multi-phase flow configurations, namely *homogeneous*, *annular* and *laminar*.

The top level curvature plot in figure 3b reveals that the two-dimensional projection manifold folded three times in order to "capture" the distribution of points in the 12-dimensional space. Interestingly, the three multi-phase flow configurations seem to be roughly separated by the folds (compare the top level visualization plot in figure 3a with the corresponding curvature plot). We confirmed this hypothesis by constructing three local lower level visualization plots initiated in the regions between the folds. The lower level plots correspond to a mixture of GTMs (see [10]) and are shown in figure 3a. The lower level plots are numbered left-to-right and were initiated in points shown as circles in the top level plot (the number inside each circle indicates the index of the corresponding lower level plot). Curvature plots of the lower level GTMs reveal that, compared to the top level GTM, the lower level projection manifolds are almost flat. We do not show the magnification factor plots here, but there are no significant expansive/contractive tendencies in the lower level projection manifolds. In this example the use of curvature information is crucial to successful modeling.

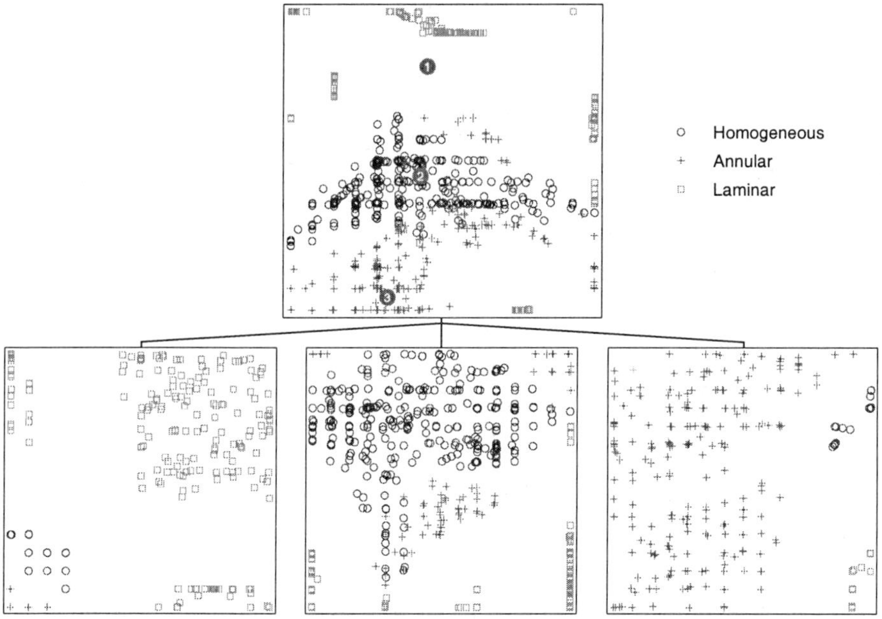

(a) Data projections

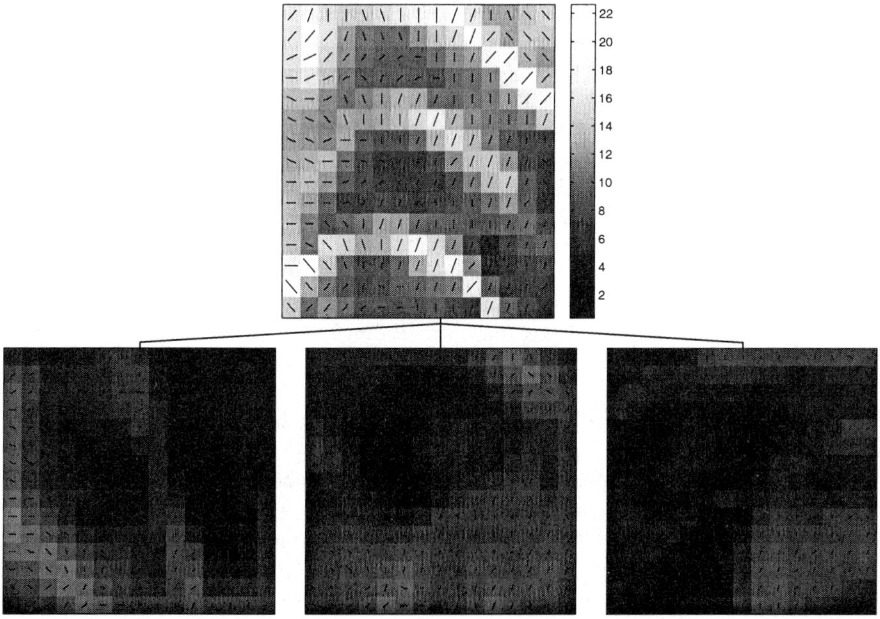

(b) Curvatures

Fig. 3. Oil data experiment.

5 Conclusion

Compared to linear projection methods such as PCA, non-linear visualization techniques are more capable of revealing the nature of data distribution in a high-dimensional space. Characterization of local geometric properties of the non-linear projection manifolds provides user with a valuable additional information that can influence further steps in the data analysis.

In this paper, we extended the work of Bishop, Svensén and Williams [4] [5] on magnification factors in GTM describing local expansion tendencies of the projection manifold. We analytically calculate and graphically represent local directional curvatures of the GTM manifold. Curvature plots are useful for detecting regions where geometry is distorted, for changing the amount of regularization in non-linear projection manifolds and, as illustrated in our experiment, for choosing regions of interest when constructing detailed lower-level visualization plots.

References

1. Bates, D.M., Watts, D.G.: Relative curvature measures of nonlinearity (with Discussion). J. R. Stat. Soc. B **42** (1980) 1–25
2. Bauer, H.U., Pawelzik, K.: Quantifying the neighborhood preservation of self-organizing feature maps. IEEE Transactions on Neural Networks **3** (1992) 570–579
3. Bishop, C.M.: Neural Networks for Pattern Recognition. Oxford University Press, Oxford, UK (1995)
4. Bishop, C.M., Svensén, M., Williams, C.K.I.: Magnification Factors for the SOM and GTM Algorithms. In: Proceedings 1997 Workshop on Self-Organizing Maps, Helsinki, Finland. (1997)
5. Bishop, C.M., Svensén, M., Williams, C.K.I.: Magnification Factors for the GTM Algorithm. In: Proceedings IEE Fifth International Conference on Artificial Neural Networks. IEE, London (1997) 64–69
6. Bishop, C.M., Svensén, M., Williams, C.K.I.: GTM: The Generative Topographic Mapping. Neural Computation **1** (1998) 215–235
7. Kohonen, T.: The self–organizing map. Proceedings of the IEEE **9** (1990) 1464–1479
8. Horn, R.A., Johnson, C.R.: Matrix Analysis. Cambridge University Press, Cambridge, England (1985)
9. Seber, G.A.F., Wild, C.J.: Nonlinear Regression. John Wiley and Sons, New York, NY (1989)
10. Tiňo, P., Nabney, I.: Constructing localized non-linear projection manifolds in a principled way: hierarchical Generative Topographic Mapping. Technical Report NCRG/2000/011, Neural Computation Research Group, Aston University, UK. (2000)
11. Willmann, T., Der, R., Martinez, T.: A new quantitative measure of topology preservation in Kohonen's feature maps. In: ICNN'94 Proceedings. IEEE Service Center (1994) 645–648

Self Organizing Map and Sammon Mapping for Asymmetric Proximities

Manuel Martin-Merino[1] and Alberto Muñoz[2,*]

[1] Universidad Pontificia de Salamanca,
c/Compañia 3, 37002 Salamanca, Spain
manuel@upsa.es
[2] Universidad Carlos III de Madrid
c/Madrid 126, 28903 Getafe, Spain
albmun@est-econ.uc3m.es

Abstract. Self Organizing Maps (SOM) and Sammon Mapping (SM) are two information visualization techniques widely used in the data mining community. These techniques assume that the similarity matrix for the data set under consideration is symmetric. However there are many interesting problems where asymmetric proximities arise, like text mining problems are. In this work we propose modified versions of SOM and SM to deal with data where the proximity matrix is asymmetric. The algorithms are tested using a real document database, and performance is reported using appropriate measures. As a result, the asymmetric algorithms proposed outperform their symmetric counterparts.

1 Introduction

Let X be an $n \times m$ data matrix representing n objects in $\mathbb{R}^m$. Let S be the $n \times n$ matrix made up of object similarities (using a given similarity measure). Sometimes the matrix S is asymmetric, that is, $s_{ij} \neq s_{ji}$. Examples of such matrices are sociometric interaction data, citations among journals and word associations [18]. Most clustering, projection and visualization algorithms assume that $s_{ij} = s_{ji}$. When distances are used instead similarities, $d_{ij} = d_{ji}$ is assumed. Such algorithms can be applied in the asymmetric case if the similarity matrix is first symmetrized (substitute s_{ij} by s^*_{ij}, where $s^*_{ij} = \frac{s_{ij}+s_{ji}}{2}$). However, information provided by asymmetry is lost.

A very interesting example of asymmetric proximities arises when considering word relations [4]. Consider, for instance, the pair 'bayesian statistics'; most people will relate 'bayesian' to 'statistics' more strongly than conversely. Any similarity measure to model the relation between these two words should not obey the constraint $s_{ij} = s_{ji}$.

* Financial support from DGICYT grant BEC2000-0167 (Spain) is gratefully appreciated.

G. Dorffner, H. Bischof, and K. Hornik (Eds.): ICANN 2001, LNCS 2130, pp. 429–435, 2001.

SOM and Sammon mapping (SM) are often used to visualize such word relations [9],[15] producing word maps. Word maps represent words (codified as vectors in a high dimensional space) as points of a two dimensional euclidean space. Several articles have empirically shown that word maps are useful tools to discover vocabulary related to a given topic and are also valuable to model relations among different topics in databases [5],[10].

In this work we propose modified versions of SOM and SM to deal with data where the proximity (or distance) matrix is asymmetric. To this aim appropriate asymmetry coefficients are introduced in the updating equations of the algorithms. Such coefficients convey the asymmetry information of the proximity matrix. In the case of dealing with symmetric matrices, such coefficients equal to one and the asymmetric algorithms reduce to their symmetric counterparts.

The paper is organized as follows. Section 2 introduce two commonly used asymmetric similarity measures and specific asymmetry coefficients for the case of word map generation. Section 3 presents the modified SOM and SM for the asymmetric case. In section 4 we compare the asymmetric algorithms with their symmetric counterparts and finally, section 5 summarizes and points out some directions for future work.

2 Asymmetry

Symmetric measures have been widely used in the context of information retrieval [16]. These measures fail to accurately model similarity between words in the sense that semantically close words often have low similarity coefficients (see [14] for details).

In this section we first introduce two commonly used asymmetric similarity measures that do not suffer from this drawback and then we define asymmetry coefficients that will be used in algorithms presented in section 3.

2.1 Asymmetry Similarity Measures

The first similarity measure, fuzzy logic similarity, is widely used in the context of neural ART models [3],[11] and information retrieval [16]. The second is a dissimilarity measure, the well-known Kullback-Leibler (K-L) divergence [7].

Let x_i denote the ith row of data matrix X.

1. Fuzzy logic similarity is defined as

$$s_{ij} = \frac{|x_i \bigwedge x_j|}{|x_i|} = \frac{\sum_k |min(x_{ik}, x_{jk})|}{\sum_k |x_{ik}|}$$

   where the fuzzy $\bigwedge$ operator is the min operator, and L_1 norm is used. In the context of text mining, s_{ij} can be interpreted as the grade with which topic represented by word x_i is a subtopic of that represented by word x_j [11].

2. K-L divergence is defined as

$$d_{ij} = D(x_i \| x_j) = \sum_{x_k} p(x_k|x_i) log \left(\frac{p(x_k|x_i)}{p(x_k|x_j)} \right)$$

where $p(x_k|x_i)$ denotes the probability that word x_k precedes word x_i. Intuitively, the K-L divergence measures the distance between words using their context in documents: if many words that occur together with x_i also occur with x_j, then d_{ij} will be small. d_{ij} distances can be transformed to similarities, as will detailed in next section.

2.2 Asymmetry Coefficients

Asymmetry coefficients convey the information provided by asymmetry. We define two coefficients that will be used in section 3 to incorporate asymmetry into classic SOM and SM. Performance of different coefficients will be checked in section 4.

Any square non-symmetric matrix M can be decomposed into a symmetric and a skew-symmetric matrix, $M = S^s + S^a$, taking $s^s_{ij} = \frac{m_{ij}+m_{ji}}{2}$ and $s^a_{ij} = \frac{m_{ij}-m_{ji}}{2}$. Obviously, for a symmetric matrix, $S^a = 0$. Let $S = S^s + S^a$ the corresponding decomposition for the word similarity matrix. In experiments with many different databases (including the Reuters benchmark [12]) we have empirically checked that the Spearman correlation coefficient between S when a symmetric similarity measure is used (for example, Jaccard) and S^s when an asymmetric similarity is used (for example, fuzzy logic), is about 0.98. This fact suggest that, at least for text data, symmetrization of an asymmetric similarity does not provide more information than a symmetric similarity. Therefore asymmetry can be explained only by S^a.

On the other hand, it can be easily deduced that $s^a_{ij} \propto |x_i| - |x_j|$ just substituting m_{ij} by the fuzzy logic similarity. This suggest the use of the L_1 norm as a coefficient of asymmetry: $|x_i| = \sum_k |x_{ik}|$. Intuitively, the most asymmetric words will be those with large norms, corresponding to very general topics.

The other coefficient proposed is related to the K-L divergence. First we define similarities $k_{ij} = 1 - \frac{d_{ij}}{\max d_{ij}}$ so that $k_{ij} \in [0, 1]$. Then we define $k_i = \sum_k k_{ki}$. This coefficient informs about the context size of x_i.

3 Algorithms Based on Asymmetric Measures

We present in this section the asymmetric versions of SOM and SM. They use the asymmetric coefficients proposed in section 2 to improve the word map coordinates generated by their symmetric counterparts.

Our approach relies on the experimental evidence that general structure of visual maps is determined by objects with larger L_1 norms. This can be shown by first mapping only words of larger norms and next repeating the experiment without cutting words less frequent. General map structure does not change. Thus words with large asymmetry coefficients will be more influential to determine the structure of maps.

3.1 Asymmetric SOM

Our asymmetric SOM algorithm is a modification of the SOM batch algorithm [13]. We have substituted the non-parametric regression step by a weighted regression algorithm that gives a weight to each observation proportional to the asymmetric coefficient. Specifically, we have used a SOM where each codebook $m(\theta_i)$ is actualized using the following equation:

$$m(\theta_j) = \frac{\sum_{k=1}^{L} \omega_k x_k h(\theta_{i(k)} - \theta_j)}{\sum_{k=1}^{L} \omega_k h(\theta_{i(k)} - \theta_j)}$$

where θ_j is the coordinate along the discrete network associated to the SOM, h is the Gaussian kernel used in the regression step and ω_k is one of the asymmetry coefficients defined in section 2.2. Note that in problems outside the text mining field different asymmetry coefficients could be used, but the proposed model would work in the same way.

3.2 Asymmetric Sammon Mapping

Consider the data matrix X, where each row $x_i \in \mathrm{IR}^m$. Sammon's mapping (SM) [17] was originally proposed as the map $\mathbf{R}^m \mapsto \mathbf{R}^d$ $(d = 2, 3)$ arising from (gradient descent) minimization of

$$E = \frac{\sum\sum_{i<j} \frac{(d_{ij} - d_{ij}^*)^2}{d_{ij}}}{\sum\sum_{i<j} d_{ij}} ,$$

where d_{ij} and d_{ij}^* are the distances between patterns $\mathbf{x}_i$ and $\mathbf{x}_j$ in $\mathbf{R}^m$ and $\mathbf{R}^d$ respectively. In our context, d_{ij} are obtained transforming the available similarities s_{ij} to dissimilarities(eg. $d_{ij} = 1 - s_{ij}$). The asymmetric SM modifies the error function E to incorporate the above defined asymmetry coefficient:

$$E = \frac{1}{\sum_{j \neq i} d_{ij}^s} \sum_{j \neq i} \omega_{ij} \frac{(d_{ij}^s - d_{ij}^*)^2}{d_{ij}^s}$$

where $d_{ij}^s = \frac{d_{ij} + d_{ji}}{2}$ and ω_{ij} are asymmetry coefficients defined from asymmetry coefficients introduced in section 2.2. There are two choices for ω_{ij}: $\omega_{ij} = min(s_i, s_j)$ or $\omega_{ij} = s_i$ where s_i can be the norm $|x_i|$ or the k_i introduced in 2.2. Introducing weights in the error function E has the effect of putting more emphasis on words with large asymmetry coefficients to construct the map, but the first choice gives more influence to word pairs where both words have large asymmetry coefficients. Notice that this method also serves to introduce asymmetry into MDS algorithms that minimizes a similar error function [6].

4 Experimental Results

Assessing the performance of algorithms that produce visual maps may not be an easy task. To be sure of obtaining objective results, we need a thesaurus that allows us to evaluate word relations induced by such word maps, checking if words close each other in the map are also related in the thesaurus.

To this aim we have built a database made up of 2000 documents from three different sources (LISA, INSPEC and Sociological Abstracts), for which thesauri are available. Documents group in three main topics ('Library Science', 'Science and Technology' and 'Sociology') and there are seven well-defined subtopics. After standard text preprocessing [1] we get about 2000 words represented as vectors in $\mathbb{R}^{1774}$.

To assess the quality of word maps produced by our algorithms we have used two well known performance measures. The first is the Spearman rank correlation coefficient [2] that measures the preservation of neighbors order by the mapping algorithm. Before computing this coefficient for SOM , words are first submitted to a quantization algorithm so that the number of vectors equal the number of neurons. The second is the F measure [1], that is widely used by the IR community. F measure is a compromise between "Recall" and "Precision". "Recall" gives the average maximum probability that a word of class i is assigned to a cluster j ($j = 1, 2 \ldots 7$) and "Precision" is the average maximum probability that a word of cluster j is assigned to class i ($i = 1, 2 \ldots 7$). Intuitively this measure evaluates the quality of the word partition into topics induced by the final map. To calculate the F measure we only need to know the predefined classification of words into seven groups and the partition of the word set projected to $\mathbb{R}^2$ (using SOM or SM) induced by a clustering algorithm (k-means in this case). Before using SOM a random projection [8] was first used to represent words in a lower dimensionality space ($\mathbb{R}^{100}$). K-means was run several times and the best result was stored.

Results of the experiments are shown in table 1. The first row shows results measured by Spearman coefficient and F measure using traditional SOM and Sammon Mapping. Rows (2) to (5) show results for asymmetric algorithms, each row corresponding to a different choice of the asymmetry coefficients explained in section 2.2. Higher coefficients are considered better. Notice that the min(modules) and min(Kullback) are only defined for proximity data, and then make sense only for SM algorithm, but not for SOM.

It is apparent from table 1 that asymmetric versions of SOM and SM outperform their symmetric counterparts in all cases but one. Also notice that SOM performs better than SM, both in their symmetric and asymmetric versions. Although it seems a contradiction with the previous work [2] the fact that Spearman coefficient is larger for SOM than for SM. this can be explained by the different definition of the SOM mapping algorithm. Notice also that Spearman measure of topology preservation is stronger than the F measure and therefore asymmetric algorithms improve most specially the first one.

Table 1. Evaluation measures for different word mapping algorithms. Higher coefficients are considered better.

	Sam(Spear)	Sam(F)	SOM (Spear)	SOM (F)
(1) Symmetric	0.19	0.40	0.59	0.41
(2) Asym(modules)	0.25	0.43	0.68	0.40
(3) Asym(min(modules))	0.29	0.45		
(4) Asym(Kullback)	0.25	0.44	0.67	0.54
(5) Asym(min(Kullback))	0.30	0.45		

Finally for the sake of clarity, we show in figure 1 the map generated by the asymmetric Sammon algorithm. To avoid visual confusion only a reduced number of words (those with higher asymmetry coefficients) are drawn. Notice that the three main database groups are easily distinguished.

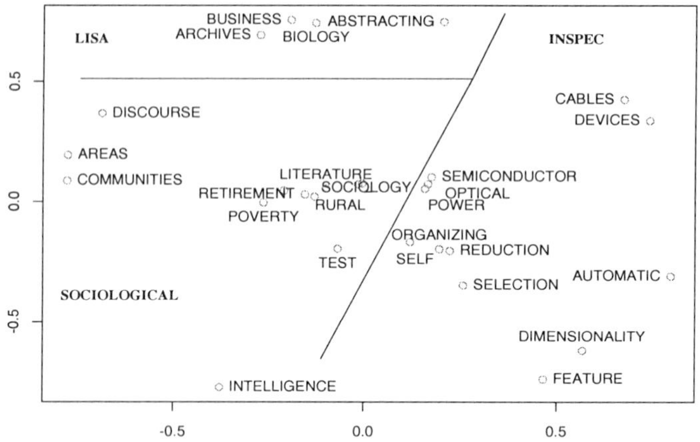

Fig. 1. Visual word map generated by asymmetric Sammon algorithm

5 Summary and Directions for Future Work

In this work we propose modified versions of two well known projection algorithms to deal with data where the proximity (or dissimilarity) matrix is asymmetric. The proposed algorithms have been tested on the interesting text mining problem of building word maps. Performance is assessed using appropriate evaluation measures. Results show that algorithms that make use of asymmetric proximities outperform their symmetric counterparts. Future research will focus on the study of asymmetric hierarchical models for text mining problems.

References

1. R. Baeza-Yates and B. Ribeiro-Neto. Modern Information Retrieval. Addison Wesley, 1999.
2. J. C. Bezdek and N. R. Pal . "An Index of Topological Preservation for Feature Extraction", Pattern Recognition, 28(3):381-391, 1995.
3. G.A. Carpenter and S. Grossberg and N. Markuzon and J.H. Reynolds and D.B. Rosen. Fuzzy ARTMAP: A neural network architecture for incremental supervised learning of analog multidimensional maps. IEEE Transactions on Neural Networks, 3(5):698-713, 1992.
4. H. Chen, B. Schatz, T. Ng, J. Martinez, A. Kirchhoff and C. Lin. A Paralell Computing Approach to Creating Engineering Concept Spaces for Semantic Retrieval: The Illinois Digital Library Initiative Project. IEEE Transactions on Pattern Analysis and Machine Intelligence, 18(8):775-782, August 1996.
5. H. Chen and A. L. Houston. Internet Browsing and Searching: User Evaluations of Category Map and Concept Space Techniques. Journal of the American Society for Information Science, 49(7):582-603, 1998.
6. T. F. Cox and M. A. A. Cox, Multidimensional Scaling, Chapman and Hall/CRC, 2nd ed., USA, 2001.
7. I. Dagan, L. Lee and F. C. N. Pereira. Similarity-Based Models of Word Coocurrence Probabilities. Machine Learning, 34:48-69, 1999.
8. S. Kaski. Dimensionality Reduction by Random Mapping: Fast Similarity Computation for Clustering. 1:413-418, Proceedings of IJCNN'98.
9. T. Kohonen, S. Kaski, K. Lagus, J. Salojarvi, J. Honkela, V. Paatero and A. Saarela. Organization of a Massive Document Collection. IEEE Transactions on Neural Networks, 11(3):574-585, May 2000.
10. A. Kopcsa and E. Schievel. Science and Technology Mapping: A New Iteration Model for Representing Multidimensional Relationships. Journal of the American Society for Information Science, 49(1): 7-17, 1998.
11. B. Kosko. Neural Networks and Fuzzy Systems: A Dynamical Approach to Machine Intelligence. Prentice Hall, Englewood Cliffs, New Jersey, 1991.
12. David D. Lewis. The Reuters-21578, Distribution 1.0 test collection. AT&T Labs. Available at http://www.research.att.com/~lewis
13. F. Mulier and V. Cherkassky. Self-Organization as an Iterative Kernel Smoothing Process. Neural Computation, 7:1165-1177, 1995.
14. A. Muñoz. Extracting term associations from a text database using a hierarchical ART architecture. Proceedings of the ICANN 96, LNCS 866:376-385, Springer-Verlag.
15. A. Muñoz. Compound key word generation from document databases using a hierarchical clustering ART model. Journal of Intelligent Data Analysis, 1(1), 1997.
16. M. Rorvig. Images of Similarity: A Visual Exploration of Optimal Similarity Metrics and Scaling Properties of TREC Topic-Document Sets. Journal of the American Society for Information Science. 50(8):639-651, 1999.
17. J.W. Sammon. A nonlinear mapping for data structure analysis. IEEE Transactions on Computers, 18(5):401-409, 1969.
18. B. Zielman and W.J. Heiser. Models for asymmetric proximities. British Journal of Mathematical and Statistical Psychology, 49:127-146, 1996.

Active Learning with Adaptive Grids

Michele Milano[1], Jürgen Schmidhuber[2], and Petros Koumoutsakos[1,⋆]

[1] Institute for Computational Sciences, ETH Zürich, Switzerland
{milano,petros}@inf.ethz.ch
[2] IDSIA, Galleria 2 Manno, CH-6928, Switzerland
juergen@idsia.ch

Abstract. Given some optimization problem and a series of typically expensive trials of solution candidates taken from a search space, how can we efficiently select the next candidate? We address this fundamental problem using adaptive grids inspired by Kohonen's self-organizing map. Initially the grid divides the search space into equal simplexes. To select a candidate we uniform randomly first select a simplex, then a point within the simplex. Grid nodes are attracted by candidates that lead to improved evaluations. This quickly biases the active data selection process towards promising regions, without loss of ability to deal with "surprising" global optima in other areas. On standard benchmark functions the technique performs more reliably than the widely used covariance matrix adaptation evolution strategy.

1 Introduction

A central problem of optimization is to select promising solution candidates without wasting too much time on others. This problem of efficient active data selection is an essential motivation of a wide variety of optimization techniques including genetic algorithms, evolution strategies, reinforcement learning algorithms, tabu search, etc.

To illustrate the problem we focus on stochastic optimization techniques such as evolution strategies (ES) [2], which are standard tools used for the optimization of multivariate, possibly non-continuous functions.

Given an n-dimensional search space, ES uses a population of points (n-dimensional "parents") to generate offspring at random from gaussian distributions with the parents as mean, and a standard deviation (step size) that is continually adapted based on information about past successes. This so-called *mutation* process provides ES with the ability to escape from local minima.

To improve efficiency in terms of convergence speed, however, adaptation of the step size during the optimization process is of crucial importance [2] [3]. The set of all mutation steps yielding improvements is called the *evolution path* of the ES [4]. Clearly, it makes sense to exploit the information embedded in the evolution path to accelerate convergence.

A widely used technique called *covariance matrix adaptation evolution strategy* (CMA-ES) embeds the information about the evolution path in a covariance

⋆ also with CTR, NASA Ames 202A-1, Moffett Field, California 94035

G. Dorffner, H. Bischof, and K. Hornik (Eds.): ICANN 2001, LNCS 2130, pp. 436–442, 2001.

matrix describing correlations between previous successful mutation steps. Subsequent mutation steps are forced to be uncorrelated with previous ones, thus optimizing step size.

One drawback of this approach is that information is acquired by means of a *local* process which does not improve the global performance properties of the ES [4].

Here we use an adaptive grid or self-organizing map (SOM [1]) to trace the evolution path. We define an SOM-based mutation operator to generate new offspring; the SOM is continuously trained on successful offsprings only, thus reflecting the evolution path in a way that allows for selection of further successful candidates with high probability. We will see that the information collected by the SOM during the optimization process is global rather than local, thus overcoming problems of CMA-ES.

Section 2 will describe the new method in detail; Section 3 will briefly review CMA-ES and present results on benchmark functions for both methods; Section 4 will conclude and provide an outlook.

2 The SOM-ES Algorithm

We start by introducing relevant definitions.

Definition 1. *$f(.)$ is the scalar valued objective function to be optimized*

Definition 2. *$\boldsymbol{x} \in \mathbb{R}^n$ denotes a parameter vector in which $f(.)$ is evaluated*

Definition 3. *$\boldsymbol{x}_{best}$ is a time-varying, variable parameter denoting the $\boldsymbol{x}$ corresponding to the best currently known $f(\boldsymbol{x})$*

Definition 4. *$F_{best} = f(\boldsymbol{x}_{best})$; T is a threshold value for F_{best} used to decide if satisfactory convergence has been attained*

Definition 5. *$\boldsymbol{X}_i \in \mathbb{R}^n$, $i \in \{1, \ldots, m\}$ are the m SOM codebook vectors or nodes of the adaptive grid*

The SOM has n-dimensional connectivity. Its nodes or codebook vectors partition the search space into simplexes; nodes on the search space boundary are fixed. In Fig.1 a subdivision for $n = 2, m = 36$ is shown.

We are now ready to define a mutation operator $\mathbf{M}$ as follows:

1. Select a simplex at random in the SOM
2. generate a random point $\boldsymbol{x}$ uniformly distributed within the simplex.

Given an optimization problem, the SOM-ES algorithm can be outlined as follows:

1. Initialization:
   (a) distribute the codebook vectors such that they are roughly equally spaced in the search volume

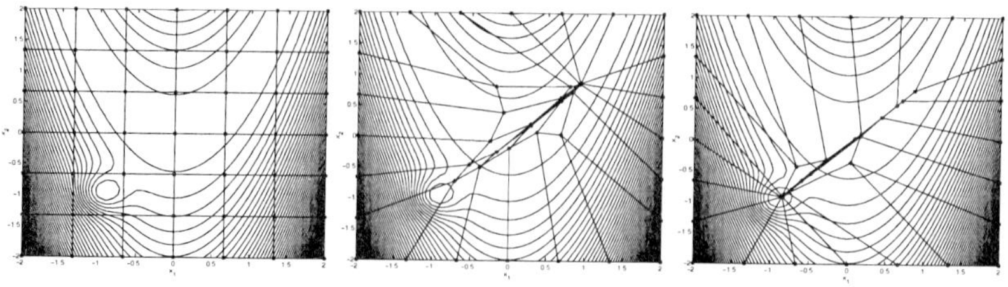

Fig. 2. Evolving SOM superimposed on contour plots of function 1. From left to right: situation after initialization, after 60 iterations, and after 100 iterations. Only neighbourhood connections are shown.

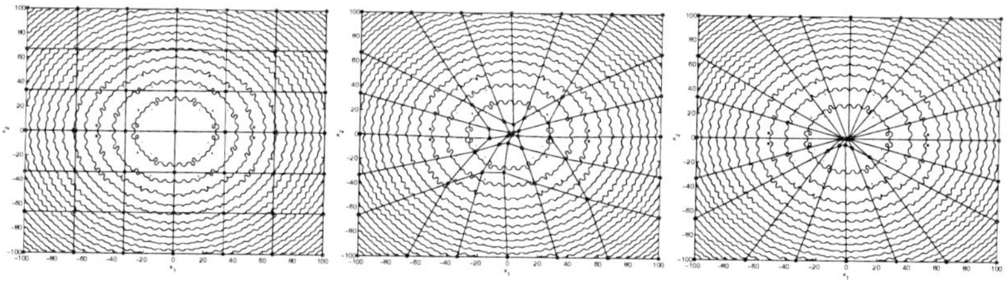

Fig. 3. Evolving SOM superimposed on contour plots of function 2. From left to right: situation after initialization, after 300 iterations and after 700 iterations.

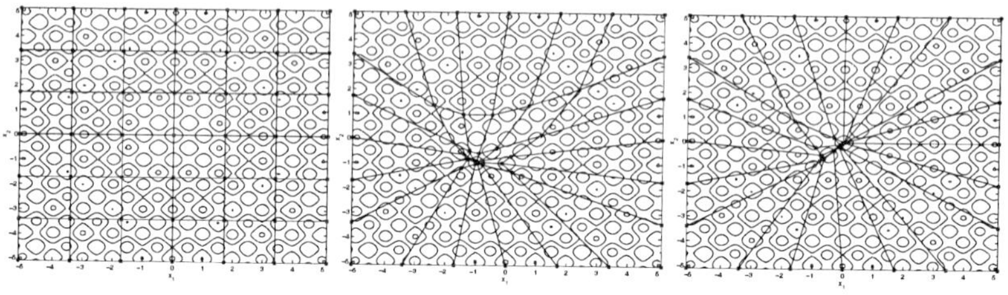

Fig. 4. Evolving SOM superimposed on contour plots of function 3. From left to right: situation after initialization, after 100 iterations and after 200 iterations.

In our initial study the SOM parameters were kept constant during the optimization process, in contrast to what is usually done by the more sophisticated standard training algorithm [1]. The possibility of implementing an adaptive parameter adjustment process will be subject of further studies.

Figures 2-4 trace the evolution of the grid used by 7×7 SOM-ES. We observe that the SOM codebook vectors quickly focus the search on areas that contain the most promising points.

To perform a fair comparison with CMA-ES we used the same convergence criterion for both methods. The thresholds for satisfactory convergence were fixed to 40 for function 1, and to 10^{-3} for functions 2 and 3. There was a maximum of 5000 iterations per run.

Table 1. Performance comparison: SOM-ES vs CMA-ES on 3 test functions

	Test 1		Test 2		Test 3	
Method	Av. Iter.	% Succ.	Av. Iter.	% Succ.	Av. Iter.	% Succ.
5×5 SOM-ES	130	93%	1600	79%	180	94%
7×7 SOM-ES	100	97%	750	94%	200	97%
CMA-ES	50	24%	340	28%	100	61%
CMA-ES with restarts	80	41%	580	67%	140	89%

To obtain statistically significant results we performed a total of 10000 optimization runs per test function and method, each run with a different initial point for CMA-ES (unnecessary for SOM-ES). To alleviate initialization dependence and convergence problems of CMA-ES, we also evaluated a CMA-ES variant that reinitializes and restarts CMA-ES whenever it fails to converge within 500 iterations. The number 500 actually was chosen with hindsight ("cheating" in favor of CMA-ES), as it is roughly 1.5 times the maximum average number of iterations required for CMA-ES to converge on the test functions (compare Table 1).

The results in table 1 demonstrate that SOM-ES consistently outperforms CMA-ES in terms of convergence frequency. On the other hand, the average number of iterations required to achieve convergence is smaller for CMA-ES. This is not surprising as CMA-ES is a local method devised for optimal exploitation of local information. CMA-ES's local focus also is evident from the fact that the restart variant of CMA-ES exhibits improved convergence frequency. Its average number of required iterations tends to become comparable with the one of SOM-ES; SOM-ES's convergence frequency is still better though.

The advantages of SOM-ES are most evident in case of the most difficult function 1 whose difficulty stems from the additional gaussian bump containing the global minimum: the local minimum basin is much larger than the global minimum basin, and the global minimum is not centered in the search domain (a centered global optimum with large basin facilitates the task of many optimization algorithms [5]).

We expect that the harder the optimization problem the more pronounced the advantages of SOM-ES — therefore we have started to apply the method to much more complex optimization problems from the field of computational fluid dynamics, to be reported elsewhere. But even on the easier test functions SOM-ES achieves convergence on global minima much more frequently than both CMA-ES and the restart variant of CMA-ES.

4 Conclusions and Outlook

We proposed a conceptually simple yet novel approach to active data selection. It is based on adaptive grids or self-organizing maps trained to reflect relevant structure of an error or fitness landscape reveiled by successful candidates encountered during an optimization process.

Although the method tends to focus computational resources on the neighborhood of previously observed successful candidates, it never loses the ability to discover global optima. Unlike widely used CMA-ES, the novel algorithm does not require random initialization. On standard test functions it consistently outperforms CMA-ES in terms of reliability, without necessity for parameter tuning.

We believe the principles put forward here are quite general and suggest a number of closely related promising algorithms. One of the most general approaches that we are currently evaluating experimentally is this:

Given an n-dimensional confined search space S, define m initially equally spaced, variable grid points $\in S$, *without* any prewired topology. The set of grid points P includes a subset $B \subset P$ of points that may move only on the boundary of S. Like in the present paper, points in P are attracted by solution candidates yielding improved performance on some optimization problem, where the movements of points in B are restricted to the boundary of S. Each time we select a new solution candidate, we simply perform an n-dimensional Voronoi tesselation of the entire search space based on the current positions of all points in P, then uniform randomly select one of the simplexes in the search space partition, then a point within the selected simplex. Preliminary experiments with this topology-free approach already led to excellent results; a detailed analysis will be published elsewhere.

References

1. T. Kohonen, *Self-Organizing maps*, Springer Verlag 1995
2. H. P. Schwefel, *Evolution and Optimum Seeking*,Wiley 1995
3. T. Bäck, U. Hammel, H. P. Schwefel, "Evolutionary Computation. Comments on the History and Current State", IEEE Trans. on Evolutionary Computation, vol. 1, n. 1, 1997, pp. 3-17
4. N. Hansen, A. Ostermeier, "Adapting Arbitrary Normal Mutation Distributions in Evolution Strategies: The Covariance Matrix Adaptation", IEEE Intern. Conf. on Evolutionary Computation (ICEC) Proceedings, 1996, pp. 312-317
5. D. Whitley, K. Mathias, S. Rana, J. Dzubera, "Building Better Test Functions", Proc. of the 6th Int. Conf. on GAs, Morgan Kaufmann, 1995, pp. 239-246

Complex Process Visualization through Continuous Feature Maps Using Radial Basis Functions

Ignacio Díaz, Alberto B. Diez, and Abel A. Cuadrado Vega

Área de Ingeniería de Sistemas y Automática
Universidad de Oviedo
Campus de Viesques 33204 Gijón, España

Abstract. In this paper we propose a method for complex process visualization using a continuous mapping from the space of measurements or features of the process onto a continuous visualization space. To construct this mapping we suggest a continuous extension of the self organizing map using a kernel regression approach. We also describe a method for continuous condition monitoring based on the proposed continous mapping. We finally illustrate the proposed method with experimental data from an induction motor working in different fault conditions.

1 Introduction

Complex industrial processes usually yield large amounts of data coming from different sensors from where the actual state of the process has to be determined. Visualization and dimension reduction techniques have received considerable attention in the literature in recent years for the analysis of large sets of multidimensional data [1],[2],[3],[4],[5]. The idea of this approach is to obtain a mapping from the input space on a low dimensional visualization space where complex relationships among input variables can be easily represented preserving information significant to a given problem.

The Self Organizing Map (SOM) can be used to project the measurements or features of the process on a low dimensional space (usually 2-D) for visualization [6],[3],[4]. The SOM projection algorithm can be described as a nonlinear, ordered mapping from a high-dimensional continuous input space $\mathbb{R}^n$ onto a low-dimensional, discrete output space $\mathbb{Z}^p$. The SOM is defined by a set of K neurons. Each neuron i has associated to it a *codebook vector* in the input space, $\mathbf{m}_i \in \mathbb{R}^n$, and a *grid position* in a discrete output space, $\mathbf{g}_i \in \mathbb{Z}^p$. The SOM mapping can be defined as

$$S(\mathbf{x}) = \mathbf{g}_c, \quad \text{with} \quad c = \arg\min_i \{d(\mathbf{x}, \mathbf{m}_i)\} \tag{1}$$

where $d(\cdot, \cdot)$ is some distance function defined in the input space. This mapping transforms a trajectory $\mathbf{f}(t)$ defined by the measurement or feature vectors in the input (feature) space F, into a sequence of positions $\mathbf{f}_g(t)$ in the grid defined in

G. Dorffner, H. Bischof, and K. Hornik (Eds.): ICANN 2001, LNCS 2130, pp. 443–449, 2001.

the output (visualization) space V. Supposed that the SOM has been previously trained with data corresponding to different situations, the condition of the machine or process can be assessed in terms of the regions of the grid which the trajectory reaches [6],[3],[4].

Unfortunately, mapping (1) leads to a quantization process as the target point in the output space is the same for all points in the input space for which neuron c wins. When (1) is applied to the trajectory described by a feature vector $\mathbf{f}(t)$ of the process, it produces a discontinuous projected trajectory $\mathbf{f}_g(t) = S(\mathbf{f}(t))$ which seldom reflects slow or little changes on the process's condition such as trends, drifts or limit cycles near a stable point, whose detection and/or visualization might be critical to avoid an imminent fault or catastrophic condition. This problem can be alleviated by increasing the number of neurons in the map; however, this results in a larger computational burden while the quantization process, though in a lesser extent, still exists.

2 Kernel Regression Approach for the Construction of a Continuous Mapping

2.1 Kernel Regression Estimate

The projection algorithm (1) can be viewed as an approximation of a nonlinear projection of each point $\mathbf{x}$ in the input space onto a 2-D manifold formed by the positions of the codebook vectors $\mathbf{m}_i$. A continuous extension of the SOM projection algorithm (1) can be expressed using the Specht's [7] *general regression neural network* (GRNN). The GRNN is a mapping $\mathcal{S}_{\mathbf{x}_i \to \mathbf{y}_i} : X \to Y$ from an input space $X \in \mathbb{R}^n$ onto an output space $Y \in \mathbb{R}^p$ defined as follows

$$\hat{\mathbf{y}}(\mathbf{x}) = \mathcal{S}_{\mathbf{x}_i \to \mathbf{y}_i}(\mathbf{x}) = \frac{\sum_{i=1}^{n} \exp\left(-\frac{(\mathbf{x}-\mathbf{x}_i)^T(\mathbf{x}-\mathbf{x}_i)}{2\sigma^2}\right)\mathbf{y}_i}{\sum_{i=1}^{n} \exp\left(-\frac{(\mathbf{x}-\mathbf{x}_i)^T(\mathbf{x}-\mathbf{x}_i)}{2\sigma^2}\right)} = \frac{\sum_i \Phi(\mathbf{x}-\mathbf{x}_i)\mathbf{y}_i}{\sum_i \Phi(\mathbf{x}-\mathbf{x}_i)} \quad (2)$$

where σ is the *kernel bandwidth* and can be tuned according to some heuristic (typically on a K-nearest neighobours basis) or in a principled approach such as the EM algorithm. Expression (2) can be shown [7],[8] to be a consistent estimator of the regression function $f(\mathbf{x})$ given a set of sample points $\mathbf{x}_i$ from the input space $\mathbb{R}^n$ and their corresponding points $\mathbf{y}_i$ in the output space $\mathbb{R}^p$

$$f(\mathbf{x}) = E[\mathbf{y}|\mathbf{x}] = \int_{\mathbb{R}^p} \mathbf{y}p(\mathbf{y}|\mathbf{x})d\mathbf{y} \quad (3)$$

where $p(\mathbf{y}|\mathbf{x})$ is the conditional pdf of vector $\mathbf{y}$ in the output space given a vector $\mathbf{x}$ in the input space. The conditional expectation in Eq. (3) can be obtained from the knowledge of the joint pdf of $\mathbf{x}$ and $\mathbf{y}$. This joint pdf can be estimated from the samples using nonparametric estimators such as the Parzen kernel used by Specht in [7] to obtain Eq. (2). The SOM training algorithm yields a set $\{\mathbf{m}_i, \mathbf{g}_i\}$ which shows several desirable statistical properties [6], two of which are of essential importance:

1. *Input pdf approximation.* Codebook vectors $\mathbf{m}_i$ are jointly distributed in the input space in a similar manner as the examples in the input space.
2. *Topology preservation.* Codebook vectors $\mathbf{m}_i, \mathbf{m}_j$ which are close in the input space, are mapped into nodes $\mathbf{g}_i, \mathbf{g}_j$ which are also close in the output space according to the distance measure defined on it. This property resembles some mild form of *uniform continuity*, and provides essential information to define a smooth manifold from the codebook vectors.

Both properties are completely described by the joint pdf of the codebook vectors and their grid positions $p_{\mathbf{m}_i,\mathbf{g}_i}(\mathbf{x}, \mathbf{y})$. Kernel regression techniques, such as (2) derive a continuous mapping on the basis of a consistent estimation of the joint pdf of the example points and therefore it is expected to preserve the statistical properties of the SOM.

Using this approach it is possible to obtain a continuous *forward mapping*, $\mathcal{S}_{\mathbf{m}_i \to \mathbf{g}_i} : F \to V$ from the input or feature space F onto the visualization space V using

$$\hat{\mathbf{y}}(\mathbf{x}) = \mathcal{S}_{\mathbf{m}_i \to \mathbf{g}_i}(\mathbf{x}) = \frac{\sum_i \Phi(\mathbf{x} - \mathbf{m}_i)\mathbf{g}_i}{\sum_i \Phi(\mathbf{x} - \mathbf{m}_i)} \tag{4}$$

where $\hat{\mathbf{y}}$ is a point in the visualization space V and $\mathbf{x}$ is a point in the feature space F. In a similar way, we can obtain a *backward mapping*, $\mathcal{S}_{\mathbf{g}_i \to \mathbf{m}_i} : V \to F$ which relates the visualization space to the feature space, as

$$\hat{\mathbf{x}}(\mathbf{y}) = \mathcal{S}_{\mathbf{g}_i \to \mathbf{m}_i}(\mathbf{y}) = \frac{\sum_i \Phi(\mathbf{y} - \mathbf{g}_i)\mathbf{m}_i}{\sum_i \Phi(\mathbf{y} - \mathbf{g}_i)} \tag{5}$$

Expressions (4) and (5) provide a way to relate in a continuous fashion the feature and visualization spaces and lay down the basis for a continous condition monitoring technique which will be described in section 3.

2.2 Relationship to Other Methods

The proposed method is an alternative to other approaches such as the GTM [9] and the PSOM [10], providing a more straightforward generalization of the SOM. The GTM yields a continuous mapping from a latent space onto the input space which maximizes the likelihood of the sample data using the EM algorithm. However, it does not allow to consider alternative geometries of the approximating manifold as in topology variant or growing SOM's. The PSOM makes use of polinomial interpolation to relate the output space to the input space. However, this approach is not well behaved for SOMs with a large number of neurons. While solutions based on local models have been proposed by the authors to overcome this problem, these make the algorithm more complicated. Besides this, the *forward mapping* in the PSOM requires an iteration process which can make the algorithm computationally heavier. The kernel regression approach, in turn, does not need any iteration and, as the PSOM, allows *auto-associative completion* in a natural way just by using $\hat{\mathbf{x}}^u = \mathcal{S}_{\mathbf{x}_i^a \to \mathbf{x}_i^u}(\mathbf{x}^a)$, where the superindex a denotes the *available* dimensions and superindex u is used for *unavailable* dimensions.

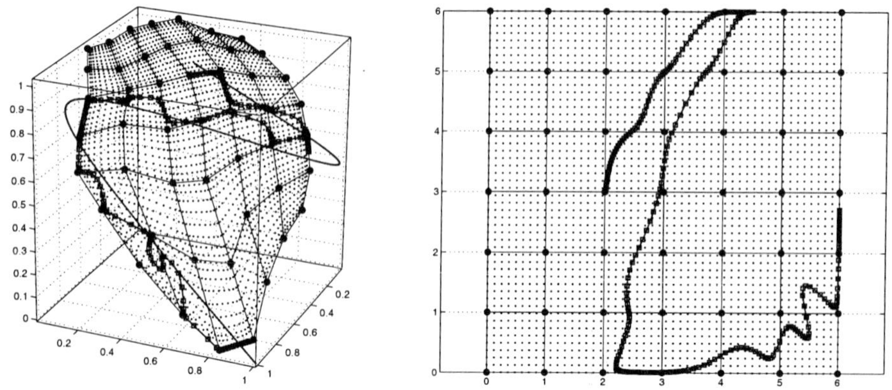

(a) Input (feature) space (b) Output (visualization) space

Fig. 1. Smooth mapping from the visualization space onto the input space obtained with GRNN *projections* (5), (4). A trajectory $\mathbf{f}(t)$ of process's features in the input space and its projection in the visualization space.

3 Continuous Process Condition Monitoring

In this section a methodology for continuous visualization of machine or process states and condition monitoring is described. This methodology is rooted in the method proposed by Kohonen and in most aspects both share similar concepts. However, due to the continuous nature of the mapping, it overcomes some of the problems previously addressed.

Projection Algorithm. The projection algorithm is of essential importance in condition monitoring as the projected trajectory is the element which describes the condition of the machine. Given a feature vector $\mathbf{f}(t)$ from the process, according to (4) the projection algorithm is

$$\mathbf{p}(t) = \mathcal{S}_{\mathbf{m}_i \to \mathbf{g}_i}(\mathbf{f}(t)) = \frac{\sum_i \Phi(\mathbf{f}(t) - \mathbf{m}_i)\mathbf{g}_i}{\sum_i \Phi(\mathbf{f}(t) - \mathbf{m}_i)} \tag{6}$$

where the projected point $\mathbf{p}(t)$ describes a continuous trajectory in the visualization space.

Component Planes. In the original SOM projection algorithm defined in [3], component planes can be constructed in a 2-D space using gray or pseudocolor levels to represent any feature of the process, in pixels or cells defined by each position in the SOM grid. Continuous component planes can be obtained using the backward mapping defined in (5). Thus, for each point $\mathbf{u} = (u_1, u_2)^T \in \mathbb{R}^2$ from the visualization space, one can obtain the corresponding point in the input (feature) space

$$\mathcal{F}(\mathbf{u}) = \mathcal{S}_{\mathbf{g}_i \to \mathbf{m}_i}(\mathbf{u}) \tag{7}$$

The so generated vector $\mathcal{F}(\mathbf{u})$ has n components $f_1(\mathbf{u}), f_2(\mathbf{u}), \cdots, f_n(\mathbf{u})$ which define the n continous component planes.

Interneuron Distance Map. Interneuron distance represents a good indication of the density of input data and helps to recognize clusters of data in the input space and regions with low probability of occurrence. A representative value of the mean interneuron distance for each neuron can be obtained weigthing each neuron distance by the neighborhood function.

$$D_i = \frac{\sum_j h_{ij} \cdot d(\mathbf{m}_i, \mathbf{m}_j)}{\sum_j h_{ij}} \tag{8}$$

A continous interneuron distance map can be obtained using the same approach as above,

$$\mathcal{D}(\mathbf{u}) = \mathcal{S}_{\mathbf{g}_i \to D_i}(\mathbf{u}) \tag{9}$$

Estimation Errors. In [6] and [3], Kohonen defines the *quantization error* as the difference between the actual point $\mathbf{x}$ and its best approximation given by the model, $\mathbf{m}_c$

$$q(\mathbf{x}) = \|\mathbf{x} - \mathbf{m}_c\| \tag{10}$$

The most plausible approximation to the actual data $\mathbf{x}$ in the continuous approach can be obtained using the GRNN to build an autoassociative map in the input space

$$\hat{\mathbf{x}} = \mathcal{S}_{\mathbf{m}_i \to \mathbf{m}_i}(\mathbf{x}) \tag{11}$$

Indeed, vector $\hat{\mathbf{x}}$ represents the estimation of $\mathbf{x}$ obtained by the model, from the information provided by vector $\mathbf{x}$ itself. We can therefore define the *estimation error* for vector $\mathbf{x}$ as

$$\varepsilon(\mathbf{x}) = \|\mathbf{x} - \hat{\mathbf{x}}\| = \|\mathbf{x} - \mathcal{S}_{\mathbf{m}_i \to \mathbf{m}_i}(\mathbf{x})\| \tag{12}$$

This quantity, as Kohonen's quantization error, gives an indication of the inadequacy of the model to predict $\mathbf{x}$.

4 Experimental Results

To illustrate the proposed methodology we will briefly describe an experiment carried out on a 4 kW, 2 pole-pair asyncronous motor. In order to visualize the condition of the motor two kind of faults were induced to the motor: a) an *electrical fault* consisting of a power supply unbalance, provoked by the inclusion of a variable resistance R on a phase line and b) a *mechanical fault* provoked by the presence of an asymmetric mass on the axis. Both types of faults give rise to different vibration and current patterns. As R varies from 0 to ∞, the

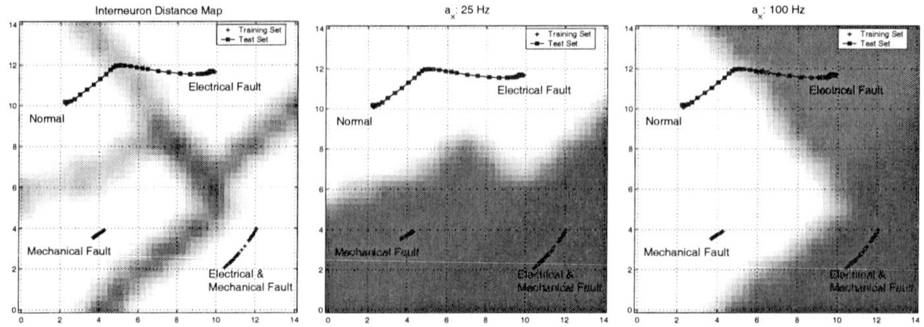

(a) Distance map (b) 25 Hz, X-Vibration (c) 100 Hz, X-Vibration

Fig. 2. Distance map and the a_x 25 Hz and 100 Hz continuous component planes. A smooth trajectory from the normal region to the electrical fault region is generated when the power supply unbalance appears gradually.

process's state (and hence, these patterns) change in a smooth manner from a normal state to a faulty state. Five sensors were installed in the motor: three vibration accelerometers and two current sensors

$$a_{bear}(t),\ a_x(t),\ a_y(t),\ i_r(t),\ i_s(t) \tag{13}$$

Feature extraction was performed obtaining the harmonic content of these variables at 25Hz and 100Hz in the vibration accelerometers and at 50Hz in the currents. The training set was composed of data from the four possible situations combining the electrical fault ($R = 0\Omega$, $R = \infty\Omega$) and the mechanical fault (with and without asymmetric mass). To show the advantages of the continuous mapping the test set consisted of data recorded during a smooth transition of R from 0Ω to 25Ω without asymmetric mass.

A 15×15 SOM was trained using the batch algorithm, with a gaussian neighborhood σ_n monotonically decreasing from 7 to 1. Input data were normalized to a $[-1, +1]$ range. Width factors of $\sigma = 0.35$ and $\sigma = 0.40$ for the forward and backward mapping respectively. Fig. 2(a) shows the continuous interneuron distance map obtained after SOM training. Four zones can be distinguished which reveal the four conditions considered in the training set. Fig. 2(b) shows one of the component planes defined by the continuous mapping. As we see from figures 2(a) to 2(c), the test set yields a continuous smooth trajectory from the original motor state (good condition) towards the final fault state (power supply unbalance) allowing to detect incipient transition to a faulty state.

5 Conclusions

This paper suggests a generalization of SOM to the continuous case for complex process visualization purposes by means of kernel regression which is an alternative to other approaches such as the GTM and the PSOM. The proposed approach has its main strength in its simplicity and robustness and provides a more straightforward generalization of the SOM than the GTM and the PSOM approaches. On the basis of the proposed algorithm we suggest an improvement to the process monitoring methodology based on the SOM described in [3] and [4], which allows to visualize smooth transitions between different process states. To illustrate the proposed method it is applied to real data from a 4 kW motor where a progressive power unbalance fault has been induced.

References

1. Zhuo Meng and Yoh-Han Pao. Visualization and self-organization of multidimensional data thorough equalized ortogonal mapping. *IEEE Transactions on Neural Networks*, 11(4):1031–1038, July 2000.
2. David J. H. Wilson and George W. Irwin. RBF principal manifolds for process monitoring. *IEEE Transactions on Neural Networks*, 10(6):1424–1434, November 1999.
3. Teuvo Kohonen, Erkki Oja, Olli Simula, Ari Visa, and Jari Kangas. Engineering applications of the self-organizing map. *Proceedings of the IEEE*, 84(10):1358–1384, october 1996.
4. Esa Alhoniemi, Jaakko Hollmén, Olli Simula, and Juha Vesanto. Process monitoring and modeling using the self-organizing map. *Integrated Computer Aided Engineering*, 6(1):3–14, 1999.
5. Jianchang Mao and Anil K. Jain. Artificial neural networks for feature extraction and multivariate data projection. *IEEE Transactions on Neural Networks*, 6(2):296–316, March 1995.
6. Teuvo Kohonen. *Self-Organizing Maps*. Springer-Verlag, 1995.
7. Donald F. Specht. A general regression neural network. *IEEE Transactions on Neural Networks*, 2(6):568–576, November 1991.
8. Simon Haykin. *Neural Networks, a Comprehensive Foundation*. Prentice-Hall, Inc., 1999.
9. C. M. Bishop, M. Svensen, and C. K. I. Williams. GTM: a principled alternative to the self-organizing map. In C. von der Malsburg, W. von Seelen, J. C. Vorbruggen, and B. Sendhoff, editors, *Artificial Neural Networks—ICANN 96. 1996 International Conference Proceedings*, pages 165–70. Springer-Verlag, Berlin, Germany, 1996.
10. Helge Ritter. Parametrized self-organizing maps. In S. Gielen and B. Kappen, editors, *Proc. ICANN'93, International Conference on Artificial Neural Networks, Amsterdam*, pages 568–577, Berlin, 1993. Springer Verlag.

A Soft k-Segments Algorithm for Principal Curves

Jakob J. Verbeek, Nikos Vlassis, and Ben Kröse

Computer Science Institute, University of Amsterdam
Kruislaan 403, 1098 SJ Amsterdam, The Netherlands

Abstract. We propose a new method to find principal curves for data sets. The method repeats three steps until a stopping criterion is met. In the first step, k (unconnected) line segments are fitted on the data. The second step connects the segments to form a polygonal line, and evaluates the quality of the resulting polygonal line. The third step inserts a new line segment. We compare the performance of our new method with other existing methods to find principal curves.

1 Introduction

Principal curves form the natural generalization of Principal Component Analysis (PCA) [4]. The first principal component can be thought of as the 'optimal' linear 1-d summarization of the data. A principal *curve* generalizes this idea to 1-d *non-linear* summarizations of the data. Since for finite data-sets there are always curves passing exactly through all the data, some regularization of the complexity of the curve is inevitable. Possible applications of principal curves include dimension reduction for feature extraction or visualization of data.

Several definitions of principal curves have been proposed [3, 5]. In [5] the Polygonal Line Algorithm (PLA) is provided as a method to find principal curves. In a probabilistic setting [10] principal curves are defined as curves minimizing a penalized log-likelihood measure. Let $\mathbf{x} \in \mathbb{R}^D$ and assume a mixture density

$$p(\mathbf{x}) = \int_0^l p(\mathbf{x}|t)p(t)\mathrm{d}t \tag{1}$$

where t is a latent variable distributed on an arc-length parameterized curve of length l. $p(\mathbf{x}|t)$ is, for example, a spherical Gaussian modeling the noise located on point t of the curve.

Several other methods from the field of unsupervised learning can also be employed to find principal curves: Generative Topographic Mapping (GTM) [1], Self Organizing Maps (SOM) [6], Growing Cell Structures (GCS) [2] and Isomap [9]. Situations in which the data-set *is* concentrated along a non-linear curve but the aforementioned methods perform poorly or fail completely typically consist of data-sets that are concentrated along a highly curved or self-intersecting curves,

G. Dorffner, H. Bischof, and K. Hornik (Eds.): ICANN 2001, LNCS 2130, pp. 450–456, 2001.

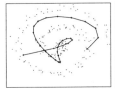

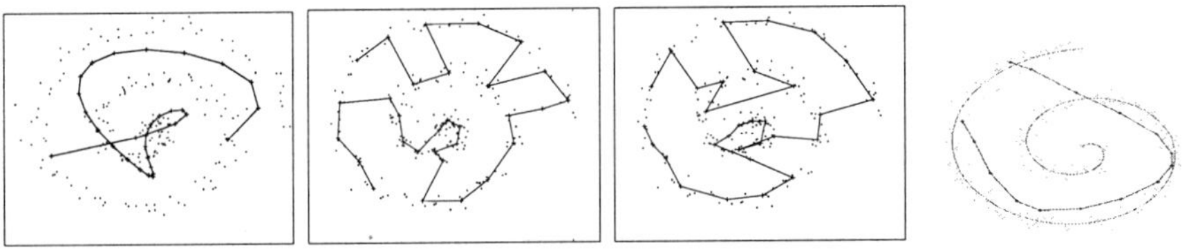

Fig. 1. Results on synthetic data for (left to right) GTM, SOM, GCS and PLA.

see Figure 1. The poor performance in such situations is mainly due to the local-search strategies employed by these methods.[1] By local-search we mean that the methods start with some initial curve and gradually improve upon it. The result is that the methods end up in rather poor local optima in the space of curves.

We propose an alternative method to find principal curves that is able to overcome some of these local optima. The key feature is that our method can make *jumps* in the space of curves. This is achieved by splitting the search for curves in two parts. First, k unconnected line segments $s_1, \ldots, s_k$ are fitted on the data by means of local-search. Note that we can insert $(k-1)$ line segments to join $s_1, \ldots, s_k$ to form a polygonal line, we can do so in $2^{k-1}k!$ ways. The second search consists of a local search among the $2^{k-1}k!$ candidate polygonal lines. In the next section we discuss exactly what curves we are looking for. The subsequent sections discuss the algorithm in detail. Section 6 ends the paper with a discussion and conclusions.

2 Principal Curves

In this paper, we use a probabilistic setting to find principal curves by means of maximizing log-likelihood, resembling the method by Tibshirani [10]. We use the model (1), where $p(t)$ is uniform along the curve and $p(x|t)$ is a spherical Gaussian located on point t of the curve, with constant variance σ_*^2 over all t. The variance σ_*^2 is a smoothing parameter, to be set by the user. If σ_*^2 tends to zero, the preference of the algorithm is drawn towards curves passing exactly through all data. We restrict the considered curves to Polygonal Lines (PLs).

Suppose our data-set lives in $\mathbb{R}^D$. Let s be a line segment, defined as: $s = \{\mathbf{f}(t) | t \in [-a, a]\}$, with $\mathbf{f}(t) = \mathbf{c} + \mathbf{u}t$, $\mathbf{u}, \mathbf{c} \in \mathbb{R}^D$ and $\|\mathbf{u}\| = 1$. Let $\mathbf{x}_\parallel = \mathbf{c} + ((\mathbf{x} - \mathbf{c})^\top \mathbf{u})\mathbf{u}$ and $\mathbf{x}_\perp = \mathbf{x} - \mathbf{x}_\parallel$. If the PL is a line segment and $p(t)$ and $p(\mathbf{x}|t)$ are as defined above, then (1) can be written as the product: $p_\parallel(\mathbf{x}_\parallel) p_\perp(\mathbf{x}_\perp)$, where $p_\perp$ is a (D-1) dimensional spherical Gaussian with zero mean and variance σ_*^2. We approximate $p_\parallel$ with a function that is constant $1/(2a)$ on the interval $[-a+\sigma_*, a-\sigma_*]$ and drops to zero outside this interval like a Gaussian with variance σ_*^2,

[1] This does not hold for Isomap. However, Isomap is not able to find self-crossing curves. This is not surprising since Isomap is based on multi dimensional scaling (MDS).

hence $-\log p_{\|}(\mathbf{x}_{\|}) \approx \log 2a + \max\left(\|\mathbf{x}_{\|}\| - a + \sigma_*, 0\right)^2/(2\sigma_*^2) + \text{const.}$[2] Combining this approximation with $p_{\perp}$ and letting $d^2(\mathbf{x}, s') = \max\left(\|\mathbf{x}_{\|}\| - a + \sigma_*, 0\right)^2 + \|\mathbf{x}_{\perp}\|^2$ denote the squared distance between $\mathbf{x}$ and the line segment $s' = \{\mathbf{f}(t) | t \in [-a + \sigma_*, a - \sigma_*]\}$ we have:

$$-\log p(\mathbf{x}) \approx d^2(\mathbf{x}, s')/(2\sigma_*^2) + \log 2a + \text{const} \tag{2}$$

We use (2) to measure the negative log-likelihood of the data given a principal curve. We neglect effects of higher density occurring in bends of the curve.

Note that, using our density model, computing the negative log-likelihood involves measuring the *Euclidean distance between a data point and a line segment*, whereas using a multivariate Gaussian, like in [1, 10, 11], we measure *Mahalanobis distance to the centroid.*

3 Local Search I: Good Unconnected Segments

The task of the first part of the algorithm is to fit k line segments to a data-set. In an accompanying paper [12], we discussed how we can extend the well-known k-means or Generalized Lloyd algorithm to a k-lines or k-segments algorithm. Here, we take another approach where we replace the hard partitioning of the data-set with a 'soft' weighting scheme that determines how much a data point contributes to the configuration of a line segment.

We interpret the fitting of the line segments as a mixture density estimation problem which we solve in an EM style. Each segment defines a mixture component of the form (1). Let π_i be the mixing weight of component i and let D denote the dimensionality of the data space. It is easy to derive the following update equations for σ^2 (recall that σ is assumed constant along the curve) and π_i [12]:

$$\sigma^2_{new} = \frac{\sum_{j=1}^{k} \sum_{\mathbf{x}} p_{old}(j|\mathbf{x}) d(\mathbf{x}, s_j)^2}{nD} \tag{3}$$

$$\pi_{i,new} = \frac{\sum_{\mathbf{x}} p_{old}(i|\mathbf{x})}{n} \tag{4}$$

We do not have such update equations for the parameters that determine the length, location and orientation of the segments. However, given the posterior distribution over the components given a data point, $p(j|x)$, we *can* compute the optimal center and covariance matrix if we were fitting a multivariate Gaussian. We do so and derive, heuristically, parameter updates for the line segments. The center of the segment is taken equal to the center of the Gaussian. The direction is taken along the eigenvector with maximal eigenvalue λ_{max} of the covariance matrix of the Gaussian. The length of the segment is taken as $3\sqrt{\lambda_{max}}$, this

[2] One can approximate $p_{\|}$ closer with a sigmoid function of $\|x_{\|}\|$. However, the Gaussian approximation allows for simplifications in the computation of the log-likelihood that are not obtained when using the sigmoid.

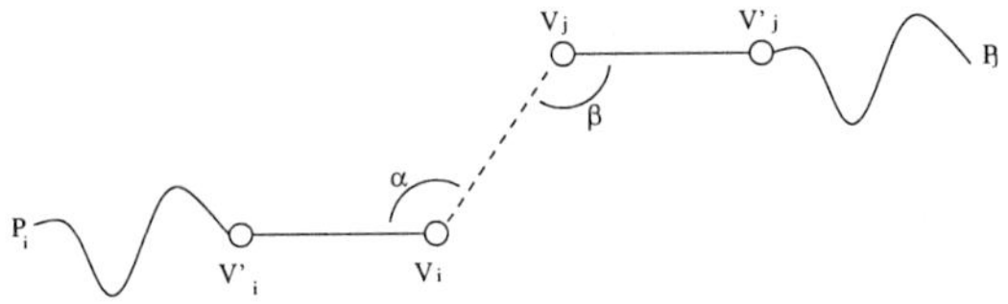

Fig. 2. Connecting two sub-PLs: the angle penalty is the sum of the angles between an edge and the adjacent edges, i.e. $\alpha + \beta$.

value was experimentally. Using longer segments may yield segments that cover subsequent parts of the curve to cross. Shorter segments describe the data less well. In both cases the construction of the curve (see next section) is made more difficult.

This method is not guaranteed to provide a sequence of pdf's with increasing log-likelihood on the data due to its heuristic nature. Therefore, we only update the parameters of the mixture if this results in an increase in log-likelihood. We stop updating if the increase in log-likelihood drops below a certain threshold.

4 Local Search II: Good Combinations of Segments

To achieve a fast algorithm to find the PL from a set of segments, we do not consider the data in the construction of the PL. The only requirements on the PL are that it is reasonably short and does not contain too sharp angles between adjacent segments. To each PL P we assign a cost $l(P) + \lambda a(P)$, where the first term is the length of the PL and the second is the sum of all angles between adjacent segments of the PL. The value of λ determines the importance of a 'smooth' curve. We construct a first PL in a greedy fashion as follows: First, we start with all original segments as sub-PLs. Then at each step we connect those two sub-PLs P_i and P_j that can be connected with minimal cost. The cost is given by the length of the connection and λ times the total incurred angle penalty, see Figure 2. We repeat this procedure until just one complete PL is left. The angle penalty is a heuristic used to prevent the construction algorithm from constructing very long curves. Suppose the algorithm can connect a segment s to several other segments at approximately the same distance, then the heuristic articulates a preference for the segment that is most alligned with s. In our experiments we found this heuristic to improve performance significantly, especially when modeling data originating from crossing curves.

Due to the greedy nature of the construction discussed above, it might well happen that we can obtain a PL with lower cost by some simple modifications of the initial PL. To find such improvements we use a simple result from graph theory: finding the optimal (in terms of cost) Hamiltonian Path for k vertices, can be expressed as finding the optimal solution for a $k+1$ cities traveling salesman problem (TSP). We use the 2_{opt} TSP optimization scheme [7] to improve our initial PL. This procedure is discussed in more detail in [12].

5 Inserting New Segments

When fitting mixture models we often do not know the 'optimal' number of components. Furthermore, the more components we use the greater the risk becomes of getting stuck at a local minimum. Li and Barron [8] recently provided evidence that an approach that interleaves greedy search and component insertion may provide a promising alternative to local search. These reasons motivate why we start our algorithm with one segment and then insert a new segment after both aforementioned local searches (fitting the segments and combining them) have finished. To apply this strategy, we need an efficient method to determine the parameters of a new mixture component given a mixture and a data-set.

Since we cannot compute directly the optimal parameters for the new segment, we use a heuristic to find reasonable initial values for the new segment. We limit the possible locations of the new segment to points in the data-set and choose the location $\mathbf{x}_{i^*}$ that minimizes an upper bound on the total squared distance from data points to the closest segment. We initialize the new component using the data points for which $\mathbf{x}_{i^*}$ is closer than any of the k segments, we denote this subset by X^*. The center of the new segment s_{k+1} is taken as the centroid of X^*. The direction of s_{k+1} is taken along the direction with maximal variance in X^*, denote this variance by λ_{max}. The length of s_{k+1} is taken again $3\sqrt{\lambda_{max}}$. The mixing weight π_{k+1} of the new component is set to $1/(k+1)$, all other mixing weights are rescaled, so that $\sum \pi_i = 1$. The noise variance σ is updated according to $\sigma^2_{new} = (k\sigma^2_{old} + \sigma^2_{k+1})/(k+1)$, where σ^2_{k+1} is the mean squared distance from points in X^* to the new segment.

Alternatively, we could initialize the new component as a spherical Gaussian with variance equal to the variance of the other segments. We could then compute $p(k+1|x)$ and compute new parameters for the new component accordingly.

In addition to computing/guessing initial values for the new component once, we could in an EM style optimize the parameters of the new component and the mixing weight. In our experiments we found that applying this extra optimization does not very much improve and sometimes even worsens performance.

Once the new segment is initialized, we can again start the iterative update scheme discussed in Section 3. The insertion of the new component is not guaranteed to increase the likelihood of the data. Hence, we monitor the change in log-likelihood due to the insertion and only accept the insertion if it leads to increased log-likelihood.

6 Demonstration and Discussion

Experiments. To test our algorithm, we conducted several experiments. We compared the performance of our algorithm with the performance of GTM, SOM, GCS and PLA, see Figure 1. We also compared results with Isomap [9], as expected Isomap is not able to find the crossing structures due to its MDS nature. Data-sets distributed along a number of different curves embedded in both $\mathbb{R}^2$ and $\mathbb{R}^3$ were used. Most of the data-sets used were data-sets on which the aforementioned methods performed poorly. Figure 3 provides results obtained for our

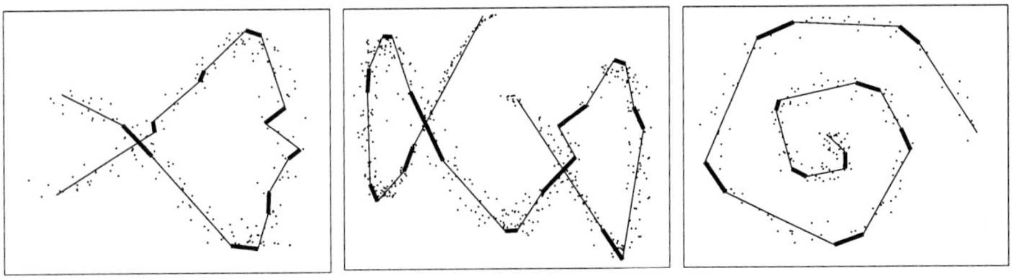

Fig. 3. Some results of our method. In the middle and right-hand picture the thin line segments are the unconnected fitted segments.

method, both on the data used for the other methods (right) and for two other data sets. All other methods failed on the other data sets, we did not include illustrations of these results.

Conclusions. We observed that our method can avoid poor local minima in many of the situations where the other methods gave poor results. This is due to the combination of local search for the unconnected segments and the local search for good polygonal lines built from those segments. The combination of the two local search steps allows for a non-local search in the space of curves.

The soft weighting scheme used to fit the components as compared to the hard Voronoi partitions used in [12] provides more robustness to the method by making it less sensitive to the details of the configuration of the data-set.

The first local search, the fitting of the line segments, resembles a procedure by Tipping et al. [11]. By using the density model of the form (1) we obtain a mixture density that resembles the final density generated by the curve more closely than when using a simple mixture of Gaussians. We also found

Experimentally we found that if the noise estimation was off up to 1/4 of the real noise level the algorithm still found correct curves. The weighting factor of the angle penalty is typically set to $\lambda = 1$, although values in the range $(1/2, 2)$ still gave good performance. The 'correct' setting of λ depends on the actual data set.

Note that in cases where one has a preference for smooth curves over the polygonal lines, the algorithm is still of interest. One can simply use the projection indices of the data on the polygonal line as an new variable, and then use standard regression techniques to find a smooth curve for the data.

Future Research. An objective function that does not require a smoothing parameter σ_* to be set by the user is a subject of further study. We hope to achieve this by implementing an objective function based on Rissanen's Minimum Description Length principle. A different approach to find the 'right' number of segments is to base the objective function on preservation of pairwise distances between (a subset of) the datapoints obtained in a neighbourhood graph.

Finally, we hope to replace some of the heuristic update rules by closed form update equations. Especially for the center and direction of the segments we speculate closed form update equations exist.

Acknowledgment

This research is supported by the Technology Foundation STW (project nr. AIF4997) applied science division of NWO and the technology programme of the Dutch Ministry of Economic Affairs.

References

1. C. M. Bishop, M. Svensén, and C. K. I Williams. GTM: The generative topographic mapping. *Neural Computation*, 10:215–234, 1998.
2. B. Fritzke. Growing cell structures - a self-organizing network for unsupervised and supervised learning. *Neural Networks*, 7(9):1441–1460, 1994.
3. T. Hastie and W. Stuetzle. Principal curves. *Journal of the American Statistical Association*, 84(406):502–516, 1989.
4. I.T. Jolliffe. *Principal Component Analysis*. Springer-Verlag, New York, 1986.
5. B. Kégl, A. Krzyzak, T. Linder, and K. Zeger. Learning and design of principal curves. *IEEE Transactions on Pattern Analysis and Machine Intelligence*, 22(3):281–297, 2000.
6. T. Kohonen. *Self-Organizing Maps*. Springer-Verlag, 1995.
7. E.L. Lawler, J.K. Lenstra, A.H.G. Rinnooy Kan, and D.B. Shmoys, editors. *The Traveling Salesman Problem*. Series in Discrete Mathematics and Optimization. John Wiley & Sons, 1985.
8. J. Q. Li and A. R. Barron. Mixture density estimation. In *Advances in Neural Information Processing Systems 12*. The MIT Press, 2000.
9. J.B. Tenenbaum, V. de Silva, and J.C. Langford. A global geometric framework for nonlinear dimensionality reduction. *Science*, 290(5500):2319–2323, December 2000.
10. R. Tibshirani. Principal curves revisited. *Statistics and Computing*, 2:183–190, 1992.
11. M. E. Tipping and C. M. Bishop. Mixtures of probabilistic principal component analysers. *Neural Computation*, 11(2):443–482, 1999.
12. J.J. Verbeek, N. Vlassis, and B. Kröse. A k-segments algorithm to find principal curves. Technical Report IAS-UVA-00-11, Computer Science Institute, University of Amsterdam, 2000.

Product Positioning Using Principles from the Self-Organizing Map

Chris Charalambous[1], George C. Hadjinicola[1], and Eitan Muller[2]

[1] University of Cyprus, Department Of Public and Business Administration, Kallipoleos 75 Str., P.O. Box 20537, CY 1678 Nicosia, Cyprus
[2] Tel Aviv University, Faculty of Management, Tel Aviv 69978, Israel

Abstract. This paper presents a methodology that identifies the position of a new product in the attribute space. The methodology uses principles from Kohonen's self-organizing feature map. The algorithm presented is robust and can be used for a number of objective functions commonly used in the product positioning problem. The method can also be used in competitive environments where other competing products are already present in the market. Furthermore, the algorithm can accommodate single-choice models (the consumer purchases the product "closest" to his/her preferences) and probabilistic-choice models (the consumer assigns to each product a probability for purchasing it).

1 Introduction: The Product Positioning Problem

New product design is considered by the contemporary business world as the activity that sustains a firm's competitive advantage. As such, designing a new product, implies that the firm must select the product's features that best "meet" the needs of the consumers in the targeted market segment. The problem of selecting the level of the various features of a product is referred to as the product positioning problem. Due to its importance, product positioning is an area in the marketing literature that has received a great deal of attention.

The origins of the analytical research on product positioning are traced back to Shocker and Srinivasan [9] who first represented products and consumer preferences as points in a joint attribute space. In this framework, the dimensions of the attribute space represent the various product features and each consumer is associated with a vector in the attribute space which represents his/her most preferred product features. This vector is referred to as the consumer ideal point. The objective is to position a product in the attribute space so that it is "closer" to the preferences (ideal points) of a larger number of consumers. Shocker and Srinivasan [9] modeled the distance between the product and the consumer ideal points as a weighted Euclidean distance. Their framework and all subsequent that were based on their work, assumed that as the distance between the product and the consumer ideal point decreases, the propensity that the consumer buys the product increases. The work by Shocker and Srinivasan was the inspiration for further research whose basis was multidimensional scaling techniques. For reviews on the topic see [10, 4, 6, 5].

G. Dorffner, H. Bischof, and K. Hornik (Eds.): ICANN 2001, LNCS 2130, pp. 457–463, 2001.

In this paper we propose a method for solving the product positioning problem by using artificial neural networks. The neural architecture we employ is known as the Kohonen's Self-Organizing Map (SOM) [7]. There are three major advantages for employing the adapted SOM in product positioning. First, the method can easily be applied by managers to obtain new product positions in a time efficient manner. Second, the method is robust since it can handle various utility-related objective functions. Third, the method can be used for the positioning of any number of new products both in single-choice and multi-choice consumer behavior environments. As such, the method is a decision-making tool, providing managers the capability to easily identify regions where new products should be positioned.

The neural network architecture of the SOM has not been applied in business-related problems, but it has been successfully applied in numerous engineering applications [8]. However, other neural architectures have been adopted in business-related problems such in accounting and finance. In marketing, multi-layer neural networks have been used to classify consumers according to their decision-making behavior [11].

2 Formulation of the Product Positioning Problem

2.1 Single-Choice Product Positioning and the SOM

Consumer choice models can be classified into two broad categories, namely, single-choice models and multi-choice or probabilistic models [3]. Under the single-choice model, consumers buy only the product which is "closest" to their most preferred product also referred to as the consumer's *ideal point* [9].

The SOM can be adapted to position new products in an attribute space, given the consumers' ideal points. To apply the SOM in the product positioning problem we define as $\mathbf{x_i(t)} = [x_{i1}(t), x_{i2}(t), \ldots, x_{iZ}(t)]^T \in R^Z, \quad i = 1, 2, \ldots, K$, to be the ideal point of consumer i at time t. Note that there are K consumers in the market and that the products have Z product attributes. Furthermore, t is the time coordinate with $t = 1, 2, \ldots, t_{max}$. Since the time argument in the above definition refers to time periods of the SOM algorithm, we may assume that the consumer ideal points during the application of the algorithm remain fixed. Henceforth, we will denote by $\mathbf{x_i}$ to be the ideal point of consumer i. In addition, we define as $\mathbf{m_j}(t) = [m_{j1}(t), m_{j2}(t), \ldots, m_{jZ}(t)]^T \in R^Z$, where $j = 1, 2, \ldots, M$, to be the coordinates of the new product j in the attribute space. This implies that M new products will be introduced in the market.

Let $U_{ij}(\| \mathbf{x_i} - \mathbf{m_j}(t) \|)$ represent the utility derived by consumer i from product j. The utility is a function of some distance norm between the consumer's ideal point $\mathbf{x_i}$ and the product $\mathbf{m_j(t)}$. Following the principles of classic economic theory, it is implied that U_{ij} should increase when $\| \mathbf{x_i} - \mathbf{m_j}(t) \|$ decreases. The general objective is to position the new products (neurons) so as to maximize the total utility derived by all consumers. Equivalently, the problem is to find the coordinates of the new products in the attribute space so as to minimize the total disutility derived by all consumers [1]. The mathematical formulation of the above problem can be written as follows:

$$\min_{\mathbf{m_{wi}}} \sum_{i=1}^{K} \min_{j}\{-U_{ij}[\| \mathbf{x_i} - \mathbf{m_j}(t) \|]\}, \qquad j = 1, 2, \ldots, M. \tag{1}$$

The algorithm below solves the product positioning using the SOM.

Step 1: Do steps 3.1-3.3.
- 3.1 Set the initial values of $\mathbf{m_j}(t)$, $j = 1, 2, \ldots, M$, i.e., set $\mathbf{m_j}(0)$.
- 3.2 Set the initial learning rate $\alpha(t)$, i.e. set $\alpha(0)$.
- 3.3 Set the initial radius of the neighborhood $N_j(t)$ of neuron (product) j, i.e set $N_j(0)$.

Step 2: While the stopping condition is false, perform Steps 3 and 4, otherwise stop.

Step 3: For each consumer ideal point $\mathbf{x_i}$, $i = 1, 2, \ldots, K$, perform steps 2.1-2.3.
- 3.1 Determine the winner neuron $\mathbf{m_{wi}}(t)$ for the consumer ideal point $\mathbf{x_i}$ such that $-U[\| \mathbf{x_i} - \mathbf{m_{wi}}(t) \|] = \min_j\{-U[\| \mathbf{x_i} - \mathbf{m_j}(t) \|]\}$.
- 3.2 Identify all neurons that belong in the neighborhood $N_{wi}(t)$.
- 3.3 Update the neuron reference vectors as follows:

$$\mathbf{m_j}(t+1) = \begin{cases} \mathbf{m_j}(t) + \alpha(t)[\mathbf{x_i} - \mathbf{m_j}(t)] & if\ j \in N_{wi}(t), \\ \mathbf{m_j}(t) & if\ j \notin N_{wi}(t), \end{cases}$$

Step 4: Do steps 4.1-4.3.
- 4.1 Compute the new value of the learning rate for $t+1$, i.e. compute $\alpha(t+1)$.
- 4.2 Reduce the value of the neighborhood radius.
- 4.3 Go to Step 2.

At the end of the algorithm, consumer i will purchase the product which is closest to his/her ideal point. This product is represented by his/her winner neuron $\mathbf{m_{wi}}$ which satisfies $-U[\| \mathbf{x_i} - \mathbf{m_{wi}}(t) \|] = \min_j\{-U[\| \mathbf{x_i} - \mathbf{m_j}(t) \|]\}$.

2.2 Probabilistic-Choice Product Positioning and the SOM

In the probabilistic choice framework, each product carries some probability of being purchased by a consumer. We define $P_{ij} = f[\| \mathbf{x_i} - \mathbf{m_j}(t) \|]$ to be the probability that consumer i purchases product j. The objective now becomes the minimization of the total expected disutility, which is given by:

$$\min_{\mathbf{m_{wi}}} \sum_{i=1}^{K} \min_{j}\{-P_{ij}U_{ij}[\| \mathbf{x_i} - \mathbf{m_j}(t) \|]\}, \qquad j = 1, 2, \ldots, M. \tag{2}$$

Depending on the nature of the utility function, the program given in equation (2) can solve problems with different objective functions. For example if $U_{ij} = p_j$ where p_j represent the revenues derived from the sale of unit of product j, the above program becomes the maximization of the expected revenues derived from the sale of the new products. Similarly, if $U_{ij} = 1$, equation (2) becomes a market share maximization program. The SOM algorithm presented for the product positioning problem in the single-choice model must be adapted to accommodate the multi-choice scenario. More specifically, Step 3, must be adjusted as follows:

Step 3: For each consumer ideal point $\mathbf{x_i}$, $i = 1, 2, \ldots, K$, perform steps 3.1-3.3:
3.1 Determine the winner neuron $\mathbf{m_{wi}}(t)$ for the consumer ideal point $\mathbf{x_i}$ such that $-P_{ij}U[\| \mathbf{x_i} - \mathbf{m_{wi}}(t) \|] = \min_j\{-P_{ij}U[\| \mathbf{x_i} - \mathbf{m_j}(t) \|]\}$.
3.2 Identify all neurons that belong in the neighborhood $N_{wi}(t)$.
3.3 Update the neuron reference vectors as follows:

$$\mathbf{m_j}(t+1) = \begin{cases} \mathbf{m_j}(t) + \alpha(t)P_{ij}[\mathbf{x_i} - \mathbf{m_j}(t)] & if\ j \in N_{wi}(t), \\ \mathbf{m_j}(t) & if\ j \notin N_{wi}(t), \end{cases} \qquad (3)$$

The other steps of the algorithm are the same with those of the algorithm in the single-choice case. Note that in the above step of the multi-choice scenario, the winner neuron for each consumer ideal point is the one that has the minimum expected disutility. Furthermore, as evident in equation (3), the updating of the neurons belonging in the neighborhood of the winner neuron is also "weighted" by the probability that consumer i purchases the products in the neighborhood. The reason for adjusting the updating process by the probability of purchase, is to assist the SOM algorithm move the neurons (products) that have a high probability of being purchased by a consumer closer to that consumer and at a faster rate than those neurons that have a lower probability of being purchased by the same consumer.

2.3 Product Positioning and Competition

In the case where competitors have already positioned their products, the SOM product positioning algorithm can easily accommodate competition by adjusting Step 3 of the algorithm. To do so, let C^* be the set of neurons representing the products of the competition and M^* the set of neurons representing the new products to be positioned. We also define a constant β such that $0 < \beta < 1$. The new Step 3 that can accommodate the presence of competition becomes:

Step 3: For each consumer ideal point $\mathbf{x_i}$, $i = 1, 2, \ldots, K$, perform the steps below:
3.1 Determine the winner neuron $\mathbf{m_{wi}}(t)$ for the consumer ideal point $\mathbf{x_i}$ such that $-P_{ij}U[\| \mathbf{x_i} - \mathbf{m_{wi}}(t) \|] = \min_j\{-P_{ij}U[\| \mathbf{x_i} - \mathbf{m_j}(t) \|]\}$.
3.2 If the winning neuron belongs to M^*, go to step 3.4, otherwise go to 3.3.
3.3 If the winning neuron belongs to C^*, i.e the product belongs to a competitor, do 3.3.1-3.3.3.
3.3.1 If $P_{ij} > \gamma$, go to Step 3 of the algorithm.
3.3.2 If $\delta < P_{ij} \leq \gamma$, do steps 3.3.2.1-3.3.2.3
3.3.2.1 Identify all neurons that belong in the neighborhood $N_{wi}(t)$.

3.3.2.2 Update the neuron reference vectors as follows:

$$\mathbf{m_j}(t+1) = \begin{cases} \mathbf{m_j}(t) + \alpha(t) P_{ij} \beta [\mathbf{x_i} - \mathbf{m_j}(t)] \\ \qquad if\ j \in N_{wi}(t)\ \ and\ \ j \notin C^*, \\ \mathbf{m_j}(t) \\ \qquad if\ j \notin N_{wi}(t), \end{cases}$$

3.3.2.3 Go to Step 3 of the algorithm.

3.3.3 If $P_{ij} \leq \delta$, go to step 3.4.

3.4 Do steps 3.4.1-3.4.2.

3.4.1 Identify all neurons that belong in the neighborhood $N_{wi}(t)$.

3.4.2 Update the neuron reference vectors as follows:

$$\mathbf{m_j}(t+1) = \begin{cases} \mathbf{m_j}(t) + \alpha(t) P_{ij} [\mathbf{x_i} - \mathbf{m_j}(t)] \\ \qquad if\ j \in N_{wi}(t)\ \ and\ \ j \notin C^*, \\ \mathbf{m_j}(t) \\ \qquad if\ j \notin N_{wi}(t), \end{cases}$$

3.4.3 Go to Step 3 of the algorithm.

The important point in the above step lies in the way the SOM algorithm updates the neuron (product) positions in the attribute space. More specifically, if the winning neuron of a consumer ideal point represents a new product, then the algorithm updates all neurons in the neighborhood, except the neurons that represent competing products.

On the other hand, if the winning neuron represents a competitor's product, the algorithm evaluates the probability that the consumer purchases the product represented by the winning neuron. Three possible cases are examined. During the first case, if this probability is greater than a threshold value, then no updating of the neurons in the neighborhood of the winning neuron is performed (step 3.3.1). The logic behind this step is that the algorithm does not move the new products closer to consumers that are in a way "locked' by a competing product. In other word, no effort is exerted to capture consumers that are more likely to buy an existing product offered by a competitor.

During the second case, if the probability that the consumer purchases the product represented by the competitor's winning neuron is between two threshold values, the updating of the neurons in the neighborhood of the winning neuron is performed using a discounting factor β where $0 < \beta < 1$. As such, some effort is made to move the new products closer to this consumer, however, given than this consumer has a high probability of purchasing a competing product, the effort exerted is discounted by some factor (step 3.3.2).

Finally, in the third case where the winning neuron still represents a competitor' product, but the probability that the consumer purchasing this product is less than a threshold value, then all new products in the neighborhood of the winning neuron are moved closer to the consumer ideal point without any discount factor (step 3.3.3).

The algorithm presented above can also accommodate the case where a consumer considers only the L closest products ([4]) or considers only products within some predefined radius. In this case, we can easily adapt Step 3 of the SOM product positioning algorithm, both in the single-choice and multi-choice cases, to consider only the L closest products or those within the predefined radius.

2.4 Common Functions of Utility

The adapted SOM algorithm to produce meaningful results must receive as inputs utility functions that could further result the probability that consumer will purchase the product. A frequently used utility function (see for example [1]) in product positioning studies is as follows:

$$U_{ij} = -\sum_{n=1}^{Z} b_{in}[x_{in} - m_{jn}]^2, \tag{4}$$

where b_{in} represents the weight consumer i places on product attribute n. Note that there are Z attribute dimensions. If in the utility function in equation (4) we set $b_{in} = 1$, $\forall n$, the single-choice product positioning program given in equation (1) becomes a simple distance minimizing program.

The multinomial logit model [3] is a common modeling framework relating the utility derived by the consumer and the probability that the consumer purchases the product. It has been widely used in estimating market shares [2] and also in the product positioning literature [1]. Using the multinomial logit model, the probability that consumer i purchases product p is given by:

$$P_{ip} = \frac{e^{U_{ip}}}{\sum_{j=1}^{M} e^{U_{ij}}} \tag{5}$$

References

1. Choi, S.C., Desabro, W.S., Harker, P.T.: Product positioning under price competition. Management Science **36** (1990) 175-199.
2. Cooper, L.G., Nakanishi, M.: Market Share Analysis. Kluwer Academic Press, Boston (1988).
3. Corstjens, M.L., Gautschi, D.A.: Formal choice models in marketing. Marketing Science **2** (1983) 19-56.
4. Green, P.E., Krieger, A.M.: Recent contributions to optimal product positioning and buyer segmentation. European Journal of Operational Research. **41** (1989) 127-141.
5. Hadjinicola, G.C.: Product positioning under production cost considerations. Decision Sciences **30** (1999) 849-864.
6. Kaul, A., Rao, V.R.: Research for product positioning and design decisions: An integrative review. International Journal of research in Marketing **12** (1995) 293-320.

7. Kohonen, T.: The self-organizing map. Proceedings of the IEEE **78** (1990) 1464-1480.
8. Kohonen, T.: Self-Organizing Maps. Springer-Verlag, Berlin (1995).
9. Shocker, A.D., Shrinivasan, V.: A consumer-based methodology for the identification of new product ideas. Management Science **20** (1974) 921-937.
10. Sudharshan, D., May, J.H., Shocker, A.D.: A simulation comparison of methods for new product location. Marketing Science **6** (1987) 182-201.
11. West, P.M., Brockett, P.L., Golden, L.L.: A comparative analysis of neural networks and statistical methods for predicting consumer choice. Marketing Science **16** (1997) 370-391.

Combining the Self-Organizing Map and K-Means Clustering for On-Line Classification of Sensor Data

Kristof Van Laerhoven

Starlab Research Laboratories
Englandstraat 555
B-1180 Brussels, Belgium
kristof@starlab.org

Abstract. Many devices, like mobile phones, use contextual profiles like "in the car" or "in a meeting" to quickly switch between behaviors. Achieving automatic context detection, usually by analysis of small hardware sensors, is a fundamental problem in human-computer interaction. However, mapping the sensor data to a context is a difficult problem involving near real-time classification and training of patterns out of noisy sensor signals. This paper proposes an adaptive approach that uses a Kohonen Self-Organizing Map, augmented with on-line k-means clustering for classification of the incoming sensor data. Overwriting of prototypes on the map, especially during the untangling phase of the Self-Organizing Map, is avoided by a refined k-means clustering of labeled input vectors.

1 Introduction

Human-computer interaction research suggests that applications should make obvious decisions autonomously, so that the user does not have to be bothered. One crucial step towards this paradigm is to enable the system to capture its context (location, activity of the user, state of the device itself, etc.) autonomously, which is usually established by adding a module that does pattern matching on incoming data from hardware sensors. Context awareness (see [1] for an overview) is a particularly hot topic in research fields where user interfacing is less obvious, like wearable computing and ubiquitous computing.

Well-known examples where context awareness would improve the application are mobile phones that have different notification behaviors (e.g. ringing or vibrating) for different contexts (e.g. "having a conversation" or "at home"), or handheld computers that start up applications depending on the context (proactive scheduling). Recognizing more (useful) contexts requires more information about the context, and thus more sensors and better recognition algorithms. As hardware sensors and their supporting processing components become cheaper, more reliable, efficient and smaller, they can be effortlessly embedded in clothes, small devices, furniture, etc.

The bottleneck in context awareness is not the hardware, but the algorithm that has to interpret the sensor data. An on-line adaptive algorithm to recognize contexts (as proposed in [2]) would result in a very efficient way to deal with these problems. Not the designer of the application, but the user decides what contexts the system should

G. Dorffner, H. Bischof, and K. Hornik (Eds.): ICANN 2001, LNCS 2130, pp. 464–469, 2001.

distinguish, by labeling the incoming sensor data with a context description. The source of information often contains noise, and processing should be computationally inexpensive, so artificial neural networks are well-suited algorithms to tackle this problem.

2 An Adaptive Algorithm for Context Awareness

As already mentioned in the previous section, sensors should be as effective as possible. It is usually very easy to enhance raw sensor data by preprocessing simple features like standard deviation or mean to replace the raw values. The advantages of this approach are that the algorithm does not have to deal with time series, and throughput of data is considerably slowed down. The drawback is that some information can be lost due to a small interval in which the features are calculated or because of a lack of quality of features themselves. Features solve the throughput problem of input vectors, but also increase their dimension, which leads to a slower learning algorithm due to the *curse of dimensionality* [3].

Once the fast stream of raw sensor data has been filtered by calculating the features of each sensor, context recognition becomes more achievable. The objective is then to map the high-dimensional, possibly noisy, features to the most likely context description. Central to our approach is a Kohonen Self-Organizing Map (KSOM) [4] that builds up a 2D topological map using the input data.

2.1 The Kohonen Self-Organizing Map

The Kohonen Self-Organizing Map is based upon earlier work by von der Malsburg [5] and can be classified as a Winner-Take-All unsupervised learning neural network. It stores prototypes w_{ij} (also known as codebook vectors) of the input vectors at time t, x(t), in the neurons (or cells) of a 2D grid. At each iteration, the neuron that stores the closest prototype to the new input vector (according to the Euclidean metric for instance) is chosen as the winner o_i:

$$o_i = \arg\min_j \left(\sqrt{\sum_{i=1}^{n} \left| x_i - w_{ij} \right|^2} \right) \tag{1}$$

The prototype vectors are usually initialized with small random values.

The winner neuron updates its prototype vector, making it more sensitive for latter presentation of that type of input. This allows different cells to be trained for different types of data. To achieve a topological mapping, the neighbors of the winner neuron can adjust their prototype vector towards the input vector as well, but in a lesser degree, depending on how far away they are from the winner. Usually a radial symmetric Gaussian neighborhood function η is used for this purpose:

$$\forall j : w_{ij}(t) = w_{ij}(t-1) + \alpha(t) \cdot \eta(t') \cdot \left(x_i(t) - w_{ij}(t-1) \right) \tag{2}$$

The learning rate α and neighborhood function η decrease as the value of t', the time that was spend in the current context, increases. The decrease rate is crucial in on-line clustering since it can lead to overwriting of prototype vectors on the map (known as the *plasticity-stability dilemma* [6]). Different neurons on the map will become more sensitive to different types of input as more input vectors are presented. Neurons that are closer in the grid tend to respond to inputs that are closer in the input space. In short, the KSOM learns a smooth mapping of the input space to neurons in the grid space. The map that gets created would be especially useful as visual feedback to the user of the network state, which is for us the main reason why the KSOM was used. Although more traditional methods yield more accurate results for clustering and multidimensional scaling [7], they were not found to be as suitable for on-line, real-time processing of sensor data as KSOM.

2.2 On-line K-Means Clustering and Classification

The k-means clustering algorithm is mainly used for minimizing the sum of squared distances between the input and the prototype vectors. However, it does not perform topological mapping like KSOM does. Most variants of the k-means clustering algorithm involve an iterative scheme that operates on a fixed number of k clusters to minimize the sum of Euclidean distances between the input and prototype vectors.

The original, on-line version (by MacQueen [8]) first finds the cluster with the closest prototype vector to the input, and then updates it just like KSOM, without further updating of neighboring clusters.

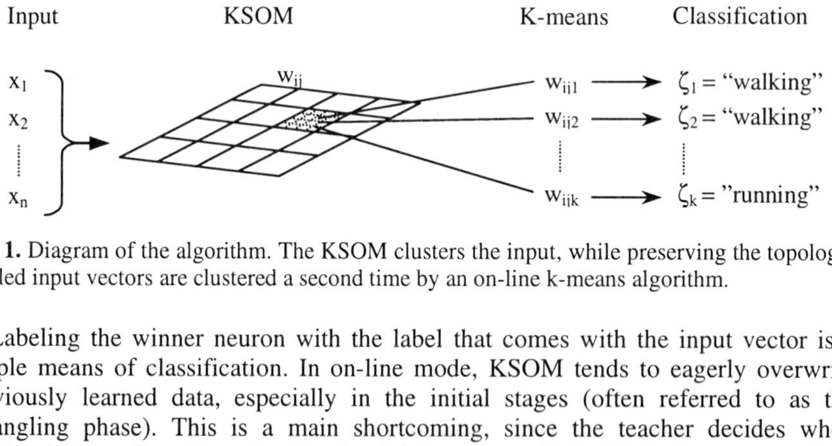

Fig. 1. Diagram of the algorithm. The KSOM clusters the input, while preserving the topology; labeled input vectors are clustered a second time by an on-line k-means algorithm.

Labeling the winner neuron with the label that comes with the input vector is a simple means of classification. In on-line mode, KSOM tends to eagerly overwrite previously learned data, especially in the initial stages (often referred to as the untangling phase). This is a main shortcoming, since the teacher decides when training is done (i.e. when classification labels are supplied with the input vectors). A KSOM neuron that was initially trained for one class, could after a while have its prototype vector overwritten and should belong to another class, or it could be labeled for two or more classes since its prototype vector has not converged yet to one class or another. These errors cause the KSOM to be highly unstable in the initial phase, and make it even possible to stay unstable afterwards.

To avoid overwriting, a second layer with on-line k-means clusters of labeled input vectors (see Figure 2) is introduced for every winning KSOM neuron. The label can thus be found by first searching the winner and then retrieving the label of the closest k-means sub-cluster for that neuron. Only labeled input vectors are stored in this second layer and a small number is usually chosen for k, so the second layer does not cause a large computational overhead.

If the incoming data is labeled, the input vector x will be clustered to the k-means of the winning KSOM neuron:

$$w_{ijl}(t) = w_{ijl}(t-1) + \lambda \cdot (x_{ti} - w_{ijl}(t-1)), \text{ with } l \in [1,k] \quad \textbf{(3)}$$

The algorithm initializes the clusters by simply filling them with the first k inputs. Unlabeled data is classified by searching the *k* closest labeled cells on the KSOM. A vote among the k resulting labels determines the final classification.

3 Experiments and Results

It is hard to evaluate the exact performance of the algorithm, since many requirements have to be fulfilled. One way is to track the percentage of correct classifications as the experiment progresses. This allows verifying the stability of the algorithm and its convergence towards an acceptable recognition.

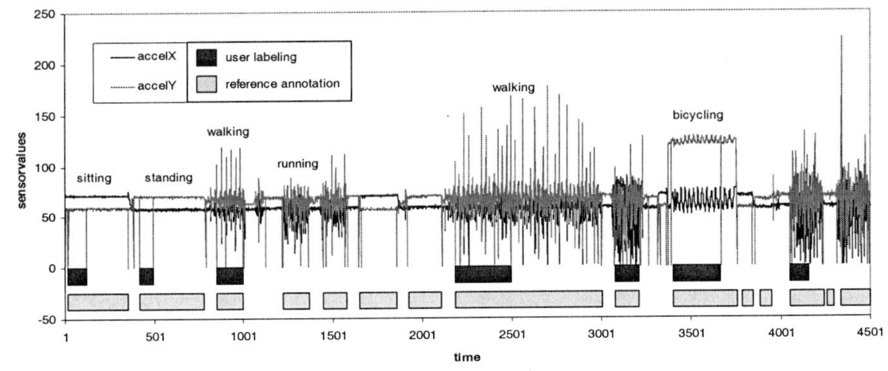

Fig. 2. Part of the raw sensor data stream, generated by two accelerometers. The black and gray boxes indicate the labeling by the user and by the experimenter after the recording.

The data from only two sensors was used in the experiments since the stability of the algorithm in the initial phase is the focus of this paper. Two accelerometers (sensors that measure motion and position) placed just above the knee of a person were used to recognize simple, everyday activities: sitting, standing, walking, running and bicycling. The data was labeled sporadically by the user during capturing. Afterwards most of the recorded data was annotated to enable an evaluation of the algorithm (see Figure 2). The minimum, the maximum, the mean and the standard deviation were recalculated over a sliding window of 3 seconds (or 100 data points) at one tenth of the speed the raw data is streaming in.

To illustrate the detrimental effect of overwriting in the KSOM, a 10 × 10 KSOM was used. Initial learning rate and neighborhood radius were set to α=0.05 and η=2, decreasing linearly in function of the time that was spent in a particular context. Performance is quite good immediately after the KSOM is trained for a context (as can be seen in Figure 3), but it becomes overwritten as other contexts gets trained. Convergence occurs only after the winning KSOM neurons for a certain context are far enough apart on the map, and is slightly less qualitative (<5%) on the long run.

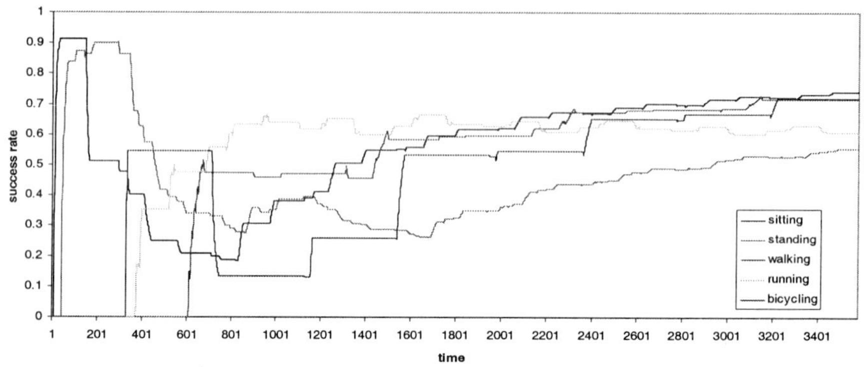

Fig. 3. Performance of a 10 × 10 Kohonen Self-Organizing Map (KSOM) with initial learning rate α=0.05 and neighborhood radius η=2, decaying linearly per class (or context). Due to overwriting, initially learned concepts deteriorate which leads to slow convergence.

The same data was also used with the k-means layer added to the KSOM, as illustrated in Figure 1. The parameter k was set at a low 5, theoretically limiting the maximum number of clusters to be 500. This worst-case number was not met at all, however (up to 210 clusters were formed during the experiments). Figure 4 confirms that the performance is substantially better than the KSOM's, showing no traces of overwriting. The main reason for this is the hierarchical structure of the algorithm: the k-means sub-clusters refine the early KSOM clustering and secure data preservation, when the KSOM's topological mapping interferes with the clustering.

5 Conclusions

The typically large dimension of the input space, noisy signals, and the irregular training make context awareness a challenging problem to learning algorithms. This paper proposed an algorithm to cope with real-time processing of sensor data and on-line training, while the user can decide when the training should occur.

The Kohonen Self-Organizing Map can be very unstable, especially in the initial untangling stage when the parameters that control the flexibility are high. To avoid overwriting, KSOM was extended with a second layer of labeled on-line k-means sub-clusters. The k-means clustering assures rapid vector quantization and a better safekeeping of memorized prototypes, while the KSOM gradually creates its topological map. Experiments show that unlearning is indeed kept to a minimum, resulting in a stable topological mapping that is executed in real-time on sensor data.

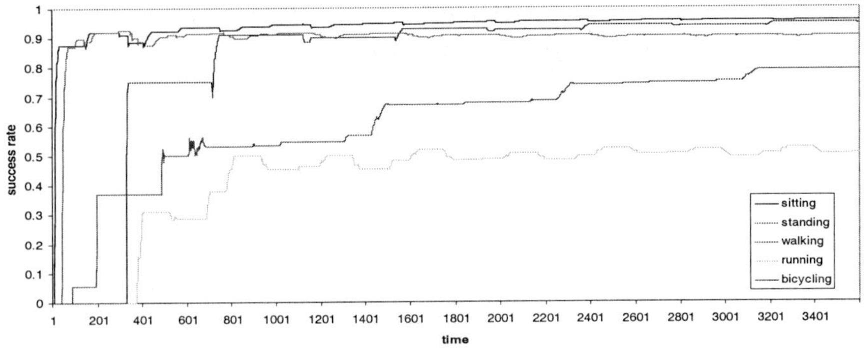

Fig. 4. Average performance over three experiments of a 10×10 KSOM, with identical parameters as in Figure 3, and a second k-means clustering layer with k = 5 and $\alpha = 0.2$.

Acknowledgements

Thanks to all people from Starlab, especially Erol Sahin and Ronald Schrooten for reviewing the drafts of this paper, and Steven Lowette for prototyping the software that was used during the experiments. The research described in this paper is partly funded by Starlab's i-Wear project, thanks go out to all partners for their support.

References

1. Abowd, G.D., Dey, A.K., Brotherton, J. Context Awareness in Wearable and Ubiquitous Computing. Proceedings of the First International Symposium on Wearable Computing (ISWC), Boston, MA: IEEE Press. (1997) 179-180
2. Van Laerhoven, K. & Cakmakci, O. What shall we teach our pants? Proceedings of the Fourth International Symposium on Wearable Computing (ISWC), Atlanta, USA: IEEE Press (2000) 77-84
3. Bellman, R. Adaptive Control Processes. Princeton University Press (1961)
4. Kohonen, T. Self-Organizing Maps. Springer-Verlag Berlin-Heidelberg (1994)
5. von der Malsburg. Self-organization of orientation sensitive cells in the striate cortex. Kybernetic. Vol. 14, Springer-Verlag (1973) 85-100
6. Grossberg, S. Adaptive pattern classification and universal recoding, I: Parallel development and coding of neural feature detectors. Biological Cybernetics 23. (1976) 121-134
7. Flexer, A. Limitations of Self-organizing Maps for Vector Quantization and Multidimensional Scaling, in Mozer M.C., et al.(eds.), Advances in Neural Information Processing Systems 9, MIT Press/Bradford Books, (1997) 445-451
8. MacQueen, J. On convergence of k-means and partitions with minimum average variance, Ann. Math. Statist. 36. (1965) 1084

Histogram Based Color Reduction through Self-Organized Neural Networks

Antonios Atsalakis, Ioannis Andreadis, and Nikos Papamarkos

Department of Electrical and Computer Engineering
Democritus University of Thrace
67100 Xanthi, Greece

Abstract. A new technique suitable for reduction of the number of colors in an image is presented in this paper. It is based on histogram processing and the use of Kohonen Self Organizing Feature Map (SOFM) neural networks. Initially, the dominant colors of each primary image are extracted through a simple linear piece-wise histogram approximation process. Then, using a SOFM the dominant color components of each primary color band are obtained and a look up table is constructed containing all possible color triplets. The final dominant colors are extracted from the look-up table entries using a SOFM by specifying the number of output neurons equal to the number of the dominant colors. Thus, the final image has all the dominant color classes. Experimental and comparative results demonstrate the applicability of the proposed technique.

1. Introduction

Reduction of the image colors is a very useful tool for segmentation, compression, presentation and transmission of images. In many cases, it is easier to process and understand an image with a limited number of colors. However, the main reason to use color reduction, is to reduce the image storage requirements.

Color reduction can be achieved by using a color quantization technique. Such a technique consists of two main steps: palette design step, in which a reduced number of palette colors is specified, and pixel mapping step in which each color pixel is assigned to one of the colors in the palette. The problem can be considered as an unsupervised classification of the 3-D color space, each class being represented by one palette color. Since an RGB image can contain millions of distinct colors, the classification problem involves a large number of data points in a low dimensional space. Several techniques have been proposed for color quantization. First, there is the class of splitting algorithms that divide the color space into disjoint regions, by consecutive splitting up the color space. From each region a color is chosen to represent the region in the color palette. In this case belong the median-cut algorithm [8], octree quantization [1,7] and the variance-based algorithm [14]. Other splitting algorithms, which incorporate techniques to take into account the human visual system, were introduced [2-3]. A good comparison of color quantization techniques is given in [12].

Another class of quantization techniques consider the problem as a clustering approach and performs clustering of the color space to construct the color palettes.

G. Dorffner, H. Bischof, and K. Hornik (Eds.): ICANN 2001, LNCS 2130, pp. 470–476, 2001.

Usually the centres of the color class obtained are taken as the dominant image colors. In this category the classifiers used can be based on statistical or neural networks approaches. In the first category belongs the quantization technique that is based on the C-means clustering algorithms [4]. In the second category belong methods such as those based SOFM neural networks [5,9,10,11].

In general, the color quantization techniques are suitable for eliminating the uncommon colors in an image but they are ineffective for image analysis and segmentation. The quality of the resultant image varies, depending on the number of the final colors and the algorithm intelligence. Color quantization techniques are usually used for the reduction of the colors of an image to a large number of colors. These techniques are not good enough to produce images with very limited number of colors. In addition, none of the color quantization techniques offers a simple procedure for estimation of the optimal number of image dominant colors. This is important especially for segmentation tasks, where the image color tones are well separated. For example, in complex color documents the colors of text regions may be not solid but are well separated from background.

The color reduction technique proposed in this paper uses in the first stage a linear Piece-Wise Histogram Approximation (PWHA) procedure, which estimates the optimal number of image colors. In the next stage, the image dominant colors or each primary color component are estimated by clustering the image histograms with SOFM neural networks. The number of the output neurons that correspond to the number of the dominant colors for each color component is defined by using the PWHA algorithm. In order to reduce both the storage requirements and computation time, the training set can be a representative sample of the image pixels. In our approach, the sub-sampling is performed via a fractal scanning technique, based on the Hilbert's space filling curve [13]. The proposed method was tested using a variety of images. The experimental and comparative results confirm the effectiveness of the proposed method.

2. Estimation of the Number of Dominant Colors

According to the PWHA algorithm, the shape of a histogram can be approximated through a set of straight-line segments that are determined by means of a simple iterative process. This technique is similar to the curve approximation technique introduced in [6]. This approach is relatively immune to small histogram bin variation (due to noise) by suitably choosing through suitable selection of the fitting parameters. At the end of the fitting process the dominant peaks of the histogram are easily obtained and their number specifies the optimal number of the image dominant colors. It is well-known that the Hue color component includes the principal of the image color information. Therefore, the dominant color estimation process is applied first to the Hue color component of the image. After this, the PWHA technique is used for the estimation of the dominant colors in the three RGB color histograms. Fig. 1 illustrates the entire procedure for the peps image.

The PWHA fitting process is analyzed as follows. If $h(j)$ with $j = j_0 \cdots j_1$, $j_0 = 0$ then the histogram can be approximated as follows. Initially we define line ε_1 which passes through points $(j_0, h(j_0))$ and $(j_1, h(j_1))$. Then, the histogram point

$(j_2,h(j_2))$ with $j_0 < j_2 < j_1$ having the maximum distance from ε_1 is specified. The histogram curve is now approximated by the lines ε_2 and ε_3 defined by the points $(j_0,h(j_0))$-$(j_2,h(j_2))$ and $(j_2,h(j_2))$-$(j_1,h(j_1))$, respectively. This fitting procedure is repeated until some convergence conditions are satisfied. At the end of the PWHA procedure, the histogram shape is approximated by a set of line segments.

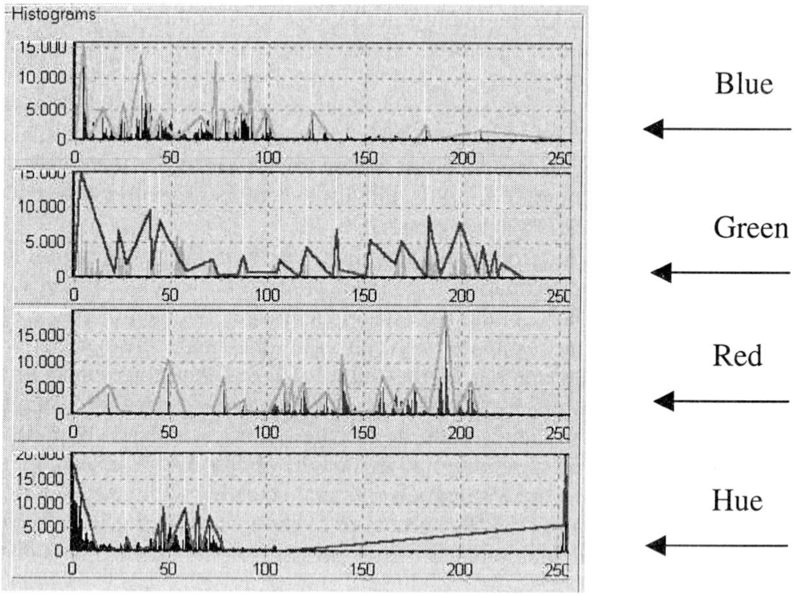

Fig. 1. Illustration of the PWHA technique.

The convergence conditions for the PWHA process are:

(i) A line is not divided if the maximum distance is smaller that a predefined threshold d_0.

(ii) A line segment $(k_1,h(k_1))$-$(k_2,h(k_2))$ is not divided more if $k_3 - k_1 < d_1$ and $|h(k_3) - h(k_1)| < d_2$ or $k_2 - k_3 < d_1$ and $|h(k_2) - h(k_3)| < d_2$, where $(k_3,h(k_3))$ the point with the maximum distance. Threshold d_2 is defined as a percentage of the maximum value of the histogram.

3. Description of the Method

A color image could be considered as a set of *n*x*m* pixels, where the color of each pixel is a point in the color space. There are several color spaces used to represent color images. The proposed method can be applied to any type of color space. A color space can be considered as a 3-D vector space, where each pixel (*i,j*) is associated with an ordered triple of color components $(c_1(i,j), c_2(i,j), c_3(i,j))$. Therefore, a general color image function can be defined by the relation:

$$I_k(i,j) = \begin{cases} c_1(i,j), & if\ k=1 \\ c_2(i,j), & if\ k=2 \\ c_3(i,j), & if\ k=3 \end{cases} \tag{1}$$

Each primary color component represents an intensity, which varies from zero to a maximum value. For example, in the RGB space, color allocation is defined over a domain consisting of a cube, with opposite vertices at (0,0,0) and (N_R, N_G, N_B). Usually, $N_R=N_G=N_B=255$.

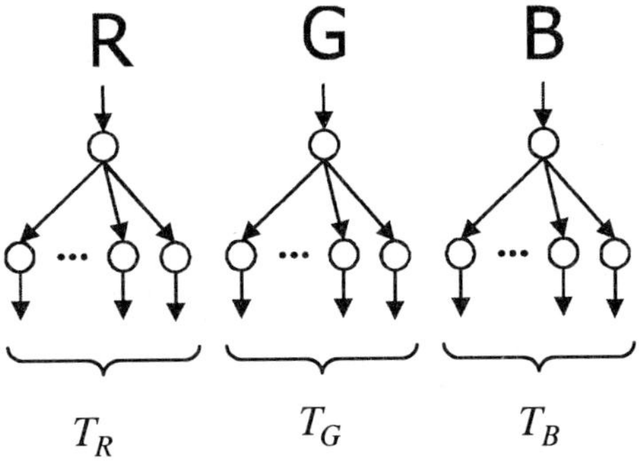

Fig. 2. Palette construction through SOFM.

In order to estimate the optimal number of the image colors, the PWHA algorithm is applied to the Hue color component histogram. The Hue color image component is obtained through the standard RGB→IHS color space transform. The PWHA procedure is simultaneously applied to each of the RGB color component histograms independently. Through this, the number of the dominant colors for each component are determined. Taking into account the previously estimated number of different colors, the SOFM neural network is used to obtain the dominant color components for each primary band. It must be noticed that the number of the output neurons of the SOFM specifies the number of the final dominant colors for each primary band. In this way we obtain three matrices of dominant colors $T_R(k_R)$, $T_G(k_G)$ and $T_B(k_B)$, with $k_R \in [0 \ldots N_R - 1]$, $k_G \in [0 \ldots N_G - 1]$, $k_B \in [0 \ldots N_B - 1]$. All the possible combinations of the dominant colors form a color palette $\vec{P}$ with $N_R \cdot N_G \cdot N_B$ colors, where $\vec{P} = [T_R(k_R) \quad T_G(k_G) \quad T_B(k_B)]$. Fig. 2 illustrates the use of a SOFM for the palette construction.

For the selection of the N_i most appropriate combinations of the palette colors we use again a SOFM neural network fed by the palette colors. Each color is a 3-D vector and thus the SOFM has the form shown in Fig. 3.

The training of the SOFM is according to the following procedure. For each pixel of the image we find the entry of the palette *P(j)* that exhibits the minimum Euclidean distance from the palette colors and this value is introduced in the neural network. After training the weights of the synapses Wj will provide the centers of the color clusters which will be finally specified. Then the entries of the initial palette are classified in each one of the final dominant color classes according to their Euclidean distance from the centers of the classes. In this way N_i unique colors are specified for the final image. Finally, in order to extract the image with the reduced number of colors we convert the colors of all pixels of the original image to the dominant colors obtained using the Euclidean distance criterion.

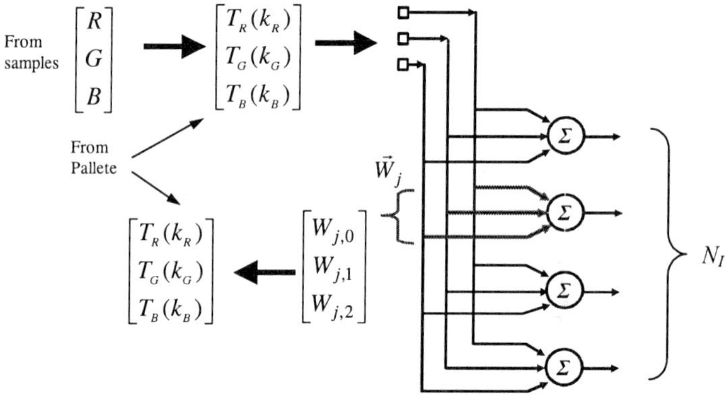

Fig. 3. Structure of the SOFM for the final estimation of the number of dominant colors.

4. Experimental Results

An illustrative example of reduction of the number of colours in an 8-bit per primary channel colour image is shown in Fig. 4(a). The optimal number of colors obtained by PWHA is 8 and the final resulting image is shown in Fig, 4(b). For comparative reasons, we perform color reduction using other well-known techniques. Specifically, the images obtained (for only 8 colors) using the method of median-cut, NeuQuant [5], and Wu [15] are shown in Fig. 4. As it can be observed, the proposed method gives satisfactory results. This is confirmed by means of Table 1, where using the mean squared error distance criterion it is proven that the proposed method provides superior results in most cases.

5. Conclusions

This paper proposes a new color quantization technique which is based on histogram processing and the use of Kohonen SOFM neural networks. An advantage of the proposed technique is the estimation of the image dominant colors. This is achieved by using a histogram PWHA technique which is relatively immune to small histogram bin variation. The entire color quantization process is unsupervised and is suitable for the reduction of the image colors to a small number. The proposed method was successfully tested with a variety of images. The experimental results presented in this paper show the efficiency of the method, which can be useful for grey-level and colour image segmentation, compression and interpretation.

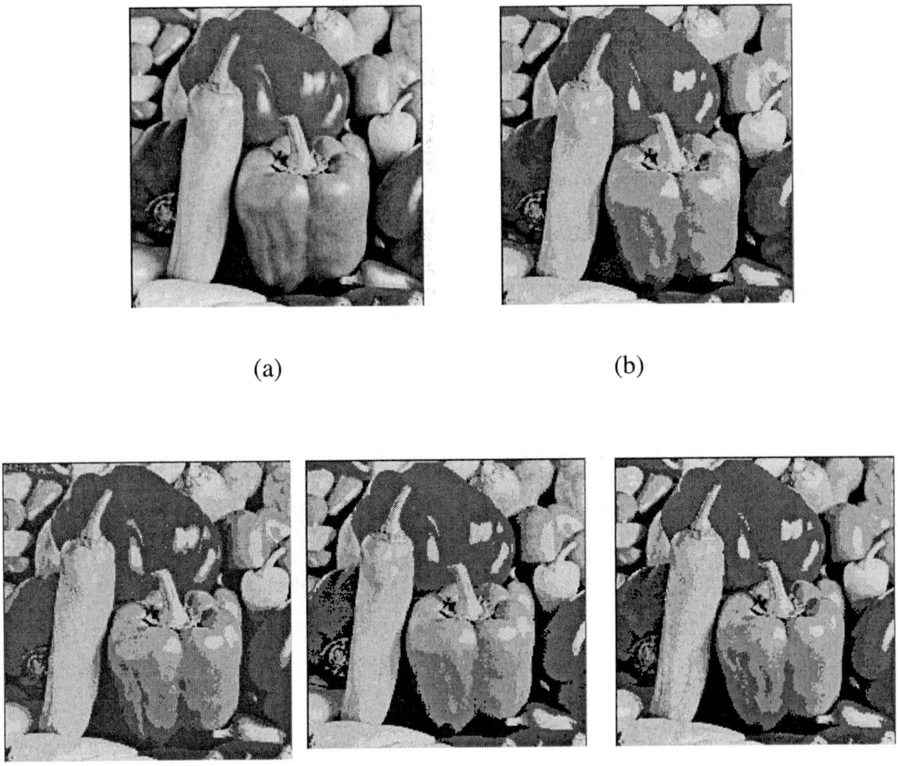

(a) (b) (c) (d) (e)

Fig. 4. Color reduction results: (a) original image, (b) proposed method, (c) median-cut, (d) NeuQuant and (e) Wu.

Table 1. MSE comparative results

Proposed	Median-Cut	NeuQuant	Wu
28.40	33.30	31.65	28.32

References

1. Ashdown I., "*Octree color quantization", from the book: Radiosity-A Programmer's Perspective*, Wiley, New York, 1994.
2. R. Balasubramanian and J.P. Allebach,"A new approach to palette selection for color images", *J. Imag. Technol.* 17, 284–290, 1990.
3. R. Balasubramanian, J.P. Allebach and C. Bouman, "Color-image quantization with use of a fast binary splitting technique", *J. Opt. Soc. Amer.* A 11 11 , 2777–2786, 1994.
4. M. Celenk, "A color clustering technique for image segmentation", *Comput. Vision Graphics Image Process.* 52, pp. 145–170, 1990.

5. A.H. Dekker, "Kohonen neural networks for optimal colour quantization", *Network: Computation in Neural Systems*, Vol. 5, pp. 351-367, 1994.
6. R. O. Duda and P. E. Hart, *Pattern Classification and Scene Analysis, John* Wiley, New York, 1973.
7. M. Gervauz and W. Purgathofer, "A simple method for color quantization: Octree quantization," *in Graphic Gems*, pp. 287-293, Academic Press, New York, 1990.
8. P. Heckbert, "Color image quantization for frame buffer display", *Comput. Graphics* 16 3 , pp. 297–307, 1982.
9. N. Papamarkos, "Color reduction using local features and a SOFM neural network", *Int. Journal of Imaging Systems and Technology*, Vol. 10, No 5, pp. 404-409, 1999.
10. N. Papamarkos, C. Strouthopoulos and I. Andreadis, "Multithresholding of color and color images through a neural network technique", *Image and Vision Computing*, vol. 18, pp. 213-222, 2000.
11. N. Papamarkos and A. Atsalakis, "Gray-level reduction using local spatial features", *Computer Vision and Image Understanding*, CVIU-78, pp. 336-350, 2000.
12. P. Scheunders, "A comparison of clustering algorithms applied to color image quantization", *Pattern Recognition Letters* 18, pp. 1379–1384, 1997.
13. H. Sagan, *Space-Filling Curves*, Springer-Verlag, New York, 1994.
14. S.J. Wan, P. Prusinkiewicz and S.K.M. Wong, "Variance-based color image quantization for frame buffer display", *Color. Res. Appl.* 15, 52–58, 1990.
15. X. Wu, "Color Quantization by Dynamic Programming and Principal Analysis", *ACM Trans. on Graphics*, vol. 11, no. 4 (TOG special issue on color), p. 348-372, 1992.

Sequential Learning for SOM Associative Memory with Map Reconstruction*

Motonobu Hattori, Hiroya Arisumi, and Hiroshi Ito

Yamanashi University, 4-3-11, Takeda, Kofu, 400-8511, Japan
hattori@media.yamanashi.ac.jp

Abstract. In this paper, we propose a sequential learning algorithm for an associative memory based on Self-Organizing Map (SOM). In order to store new information without retraining weights on previously learned information, weights fixed neurons and weights semi-fixed neurons are used in the proposed algorithm. In addition, when a new input is applied to the associative memory, a part of map is reconstructed by using a small buffer. Owing to this remapping, a topology preserving map is constructed and the associative memory becomes structurally robust. Moreover, it has much better noise reduction effect than the conventional associative memory.

1 Introduction

It is well known that when a neural network is trained on one set of patterns and then attempts to add new patterns to its repertoire, catastrophic interference, or the complete loss of all of its previously learned information may result. This type of radical forgetting is unacceptable both for a model of human brain and for practical applications. French has pointed out that catastrophic forgetting is a direct consequence of the overlap of distributed representations and can be reduced by reducing this overlap [7],[8]. Since most of conventional associative memories employ mutually connected structures, information is stored in weights in completely distributed fashion [1]-[4]. Therefore, catastrophic forgetting is inevitable and a serious problem for them.

Yamada et al. have proposed a learning algorithm for an associative memory based on Self-Organizing Map (SOM) [6]. In order to prevent catastrophic forgetting and enable sequential learning with only new information, weights fixed neurons and weights semi-fixed neurons are used in the learning algorithm [12]. That is, weights of well-trained neurons on the map are fixed and those of neurons surrounding the weights fixed neurons are semi-fixed. Information is stored by the weights fixed neurons and their weights are not changed if once fixed. Hence, in the SOM associative memory, it is not necessary for the old information to be continually relearned by the network. In the SOM associative memory, however, it becomes very difficult to preserve topology from the space

* This work was supported by the Telecommunications Advancement Organization of Japan. The authors would like to thank Masayuki Morisawa for valuable discussions.

G. Dorffner, H. Bischof, and K. Hornik (Eds.): ICANN 2001, LNCS 2130, pp. 477–484, 2001.

of inputs to the plane of the output units as the number of additional data to be learned increases.

In this paper, we propose a novel sequential learning algorithm for the SOM associative memory. In the proposed learning algorithm, when a new training data is applied to the SOM associative memory, it is stored in a small buffer with some previously learned data which are extracted from weights of fixed neurons. Then the data stored in the buffer are learned by the SOM algorithm [6] within a small circular area of the map. Hence, the topological relationships of the data stored in the buffer are preserved within this area. Since this remapping within a small area of the map is occurred every time a new data is applied to the SOM associative memory, it can gradually construct a topology preserving map for all inputs. The ability in topology preserving contributes the structural robustness of the SOM associative memory. Moreover, we can expect that the SOM associative memory can be applied to various intelligent tasks such as categorization, concept formation, knowledge discovery and so on.

A number of computer simulation results show the effectiveness of the proposed learning algorithm.

2 SOM Associative Memory

2.1 Structure of SOM Associative Memory

Ichiki et al. have proposed a supervised learning algorithm for Self-Organizing Map (SOM) and they have shown that the SOM can behave as an associative memory [10].

The SOM associative memory regards the input vector consisting of different several components. Figure 1 shows the structure of a SOM associative memory when the input layer is divided into two parts. The learning algorithm for the SOM associative memory is based on the conventional SOM algorithm [6].

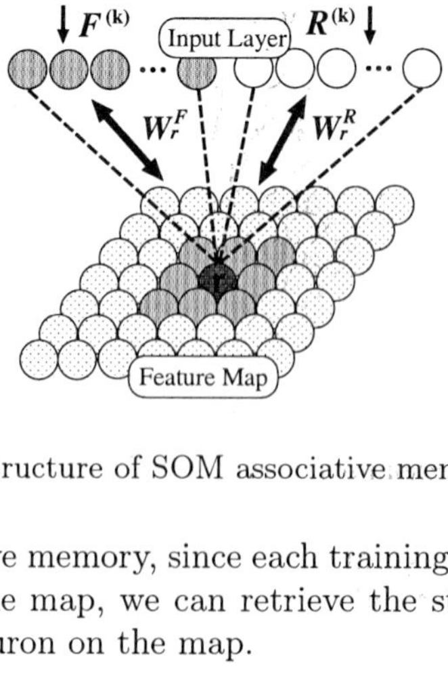

Fig. 1. Structure of SOM associative memory [10].

In the SOM associative memory, since each training data is stored by a small number of neurons on the map, we can retrieve the stored data by examining weights of the winner neuron on the map.

2.2 Conventional Sequential Learning for SOM Associative Memory

Yamada et al. have proposed a sequential learning algorithm for the SOM associative memory and they have shown that new information can be stored without retraining previously learned information [12]. In the learning algorithm, weights of well-trained neurons of the SOM associative memory are fixed [9] and weights of neurons surrounding a weights fixed neuron become hard to be learned in inverse proportion to the distance from the nearest fixed neuron. Here, we briefly review the conventional sequential learning algorithm [12].

Let us consider that the learning vector $\boldsymbol{X}^{(k)}$ consists of two components:

$$\boldsymbol{X}^{(k)} = \begin{bmatrix} \boldsymbol{F}^{(k)} \\ \mathbf{0} \end{bmatrix} + \begin{bmatrix} \mathbf{0} \\ \boldsymbol{R}^{(k)} \end{bmatrix} \tag{1}$$

where $\boldsymbol{X}^{(k)} \in \{-1,1\}^{M+N}$, $\boldsymbol{F}^{(k)} \in \{-1,1\}^{M}$ and $\boldsymbol{R}^{(k)} \in \{-1,1\}^{N}$. For efficiency of the process and mathematical convenience, $\boldsymbol{X}^{(k)}$ is normalized to u:

$$\boldsymbol{X}^{(k)*} = \frac{u}{\sqrt{M+N}}\boldsymbol{X}^{(k)} = \frac{u}{\sqrt{M+N}}\begin{pmatrix} \boldsymbol{F}^{(k)} \\ \mathbf{0} \end{pmatrix} + \frac{u}{\sqrt{M+N}}\begin{pmatrix} \mathbf{0} \\ \boldsymbol{R}^{(k)} \end{pmatrix} . \tag{2}$$

Let $\boldsymbol{W_i}$ be the connection weights between the input layer and the map:

$$\boldsymbol{W_i} = \begin{bmatrix} \boldsymbol{W_i}^F \\ \mathbf{0} \end{bmatrix} + \begin{bmatrix} \mathbf{0} \\ \boldsymbol{W_i}^R \end{bmatrix} \tag{3}$$

where $\boldsymbol{i}$ is a position vector indicating the position of a neuron on the map. Then, the connection weights $\boldsymbol{W_i}$ are learned by the following procedure:

1. Choose the initial values of weights randomly.
2. Calculate the Euclidean distance between $\boldsymbol{X}^{(k)*}$ and $\boldsymbol{W_i}$, $d(\boldsymbol{X}^{(k)*}, \boldsymbol{W_i})$.
3. Find a neuron which satisfies $\min[d(\boldsymbol{X}^{(k)*}, \boldsymbol{W_i})]$ and let $\boldsymbol{r}$ be the position vector of the neuron.
4. If $d(\boldsymbol{X}^{(k)*}, \boldsymbol{W_r}) < d_f$, the weights of the neuron $\boldsymbol{r}$ are fixed. Weights except those of fixed neurons are learned by the following rule:

$$\boldsymbol{W_i}(t+1) = \boldsymbol{W_i}(t) + H(d) \times \alpha(t) \cdot h_{\boldsymbol{ri}} \cdot (\boldsymbol{X}^{(k)*} - \boldsymbol{W_i}(t)) \tag{4}$$

where t shows discrete time and $h_{\boldsymbol{ri}}$ is the neighborhood function:

$$h_{\boldsymbol{ri}} = \exp\frac{-\|\boldsymbol{r}-\boldsymbol{i}\|^2}{2\sigma(t)^2} \tag{5}$$

$\alpha(t)$ and $\sigma(t)$ shows the following decreasing functions:

$$\alpha(t) = \frac{-\alpha_0(t-T)}{T} \tag{6}$$

$$\sigma(t) = \frac{t(\sigma_f - \sigma_i)}{T} + \sigma_i \tag{7}$$

where T shows the upper limit of learning iterations, α_0 is the initial value of $\alpha(t)$ and $\sigma(t)$ varies from σ_i to σ_f $(\sigma_i > \sigma_f)$. $H(d)$ shows effect from the nearest fixed neuron:

$$H(d) = \frac{1 - \exp(-d \cdot k)}{1 + \exp(-d \cdot k)} \tag{8}$$

where d shows the Euclidean distance between neuron $\boldsymbol{i}$ and the nearest weights fixed neuron on the map and k defines a slope of $H(d)$. Owing to $H(d)$, weights of neurons close to the fixed neurons are semi-fixed, that is, they become hard to be learned.

5. 2–4 is iterated until $d(\boldsymbol{X}^{(k)*}, \boldsymbol{W_r}) < d_f$ is satisfied.
6. 2–5 is iterated for all k.

Information is stored by the weights fixed neurons and their weights are not changed if once fixed. Hence, the SOM associative memory can store new information without retraining previously learned information.

In the recall process, an output can be obtained by the following rule:

$$\boldsymbol{O} = \begin{cases} \boldsymbol{W}_{\boldsymbol{r}}^{R} \cdot \frac{\sqrt{\|\boldsymbol{F}\|^2 + \|\boldsymbol{R}\|^2}}{u} & \text{if } \boldsymbol{F} \text{ is inputed} \\ \boldsymbol{W}_{\boldsymbol{r}}^{F} \cdot \frac{\sqrt{\|\boldsymbol{F}\|^2 + \|\boldsymbol{R}\|^2}}{u} & \text{if } \boldsymbol{R} \text{ is inputed} \end{cases} . \tag{9}$$

Owing to the process above, the SOM associative memory can behave as a bidirectional associative memory (BAM) [3].

Figure 2 shows the structure of the SOM associative memory using weights fixed and semi-fixed neurons.

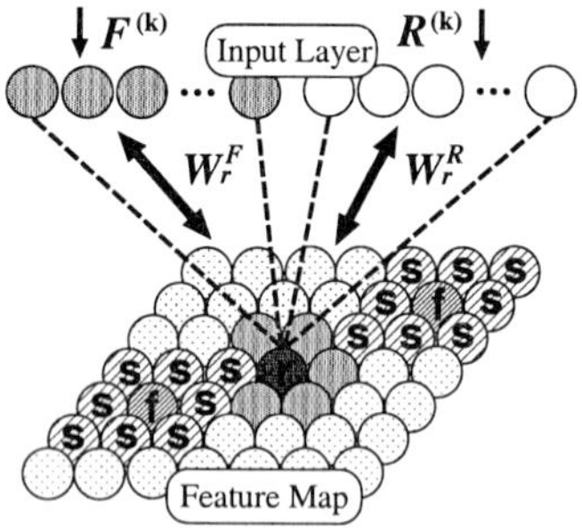

Fig. 2. Structure of SOM associative memory using weights fixed neurons (**f**) and weights semi-fixed neurons (**s**) [12].

3 Sequential Learning with Map Reconstruction

In the conventional sequential learning algorithm [12], it becomes very difficult to preserve topology from the space of inputs to the plane of the output units as the number of additional data to be learned increases because each data is learned by the SOM associative memory one by one. If the SOM associative memory can construct a topology preserving map, its structural robustness can be much improved. Moreover, the SOM associative memory can be applied to categorization of memory items, concept formation, knowledge discovery [11] and so on. Namely, owing to the inherent features of the SOM, we can expect that the SOM associative memory can handle various intelligent tasks which are difficult for the conventional associative memories [1]-[4].

In order to preserve the topological relationships of input data, we use a buffer of small capacity in the proposed sequential learning algorithm. When a new data to be learned is applied to the SOM associative memory, some previously learned data are extracted from the weights of fixed neurons which exist near the winner neuron to the new data and they are stored in the buffer with the new data. Then, all of the data stored in the buffer are learned by the SOM algorithm [6] within a radius of the farthest fixed neuron from the winner neuron on the map. Since this remapping within small area of the map is occurred whenever a new training data is applied to the SOM associative memory, it can gradually construct a topology preserving map for all inputs.

In the proposed learning algorithm, weights are learned as follows:

1. Choose the initial value of weights randomly.
2. Calculate $d(\boldsymbol{X}^{(k)*}, \boldsymbol{W}_i)$.
3. Find the winner neuron which satisfies $\min[d(\boldsymbol{X}^{(k)*}, \boldsymbol{W}_i)]$ and let $\boldsymbol{r}$ be the position vector of the neuron.
4. $\boldsymbol{X}^{(k)*}$ is stored in the buffer which has a capacity of B.
5. Find $(B-1)$ fixed neurons near $\boldsymbol{r}$, store their weights in the buffer in order of nearness to $\boldsymbol{r}$ and release their fixations of weights. Let E be the Euclidean distance from $\boldsymbol{r}$ to the farthest fixed neuron.
6. Define a circular area C on the map, the center and radius of which are $\boldsymbol{r}$ and E, respectively.
7. Calculate $d(\boldsymbol{X}^{(k_B)*}, \boldsymbol{W}_{i_C})$, where $\boldsymbol{X}^{(k_B)*}$ is a data stored in the buffer and $\boldsymbol{W}_{i_C}$ shows weights of neuron $\boldsymbol{i}$ in C.
8. Find the winner neuron in C which satisfies $\min[d(\boldsymbol{X}^{(k_B)*}, \boldsymbol{W}_{i_C})]$ and let $\boldsymbol{r}_C$ be the position vector of the neuron.
9. Weights of all neurons in C are changed by
$$\boldsymbol{W}_{i_C}(t+1) = \boldsymbol{W}_{i_C}(t) + H(d) \times \alpha(t) \cdot h_{\boldsymbol{r}_C \boldsymbol{i}_C} \cdot (\boldsymbol{X}^{(k_B)*} - \boldsymbol{W}_{i_C}(t)) \quad (10)$$
where d shows the Euclidean distance from $\boldsymbol{i}_C$ to the nearest fixed neuron in the outside of C.
10. 7–9 is iterated until $d(\boldsymbol{X}^{(k_B)*}, \boldsymbol{W}_{\boldsymbol{r}_C}) < d_f$ is satisfied for all $k_B = 1, \cdots, B$.
11. Fix the weights of each winner neuron for $\boldsymbol{X}^{(k_B)*}$.
12. 2–11 is iterated when a new input to be learned is applied to the SOM associative memory.

4 Computer Simulation Results

In the computer simulation, we used the following parameters: $\alpha_0 = 0.1, \sigma_i = 3.0, \sigma_f = 0.5, T = 3000, d_f = 10^{-3}, k = 1.0$. The size of map were set to 9×9.

4.1 Comparison of Topological Mapping

In this experiment, we examined the ability in topology preserving of the SOM associative memory learned by the proposed learning algorithm by using 25 two dimensional training data shown in Fig.3. The capacity of the buffer was set to $B = 9$. Figures 4 and 5 show the obtained maps by the conventional and proposed learning algorithm, respectively: numbers on each map show the winner neurons for inputs.

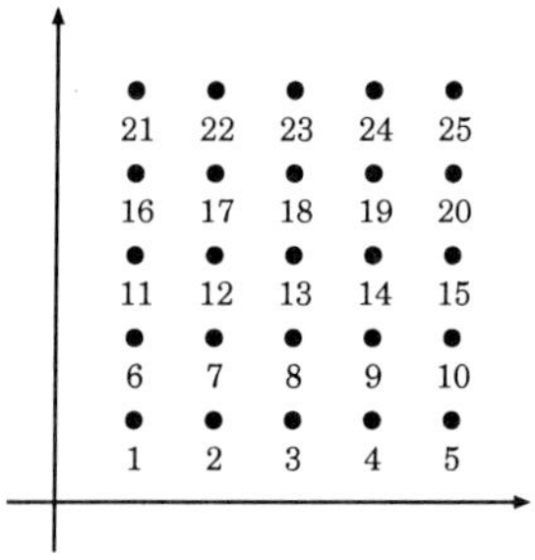

Fig. 3. Training data.

20	24		13	15		14		4
19	18				5		3	
25								
			2			9		10
23								
					7			8
22		17						
			16				1	
12		21	11	6				

Fig. 4. A mapping by conventional algorithm.

25		20		15		10		5
								4
24		19		14		9		
			13					3
23		18				8		
				12				2
22		17			7			
21		16		11		6		1

Fig. 5. A mapping by proposed algorithm.

As shown in these figures, the ability in topology preserving of the SOM associative memory becomes much better by using the proposed algorithm. The average of correlation coefficients between input data and the position vectors of winner neurons was 0.310 for the conventional algorithm and 0.945 for the proposed algorithm based on 10 trials. Even when $B = 5$, the proposed algorithm showed much better performance than the conventional algorithm: the average of correlation coefficients was 0.675.

4.2 Structural Robustness

In this experiment, we compared the structural robustness between the SOM associative memory learned by the proposed algorithm and that learned by the conventional algorithm. The capacity of the buffer was set to $B = 5$. 20 pairs of letters shown in Fig.6 were stored by both sequential learning algorithms and then some neurons on the maps were made to fail randomly.

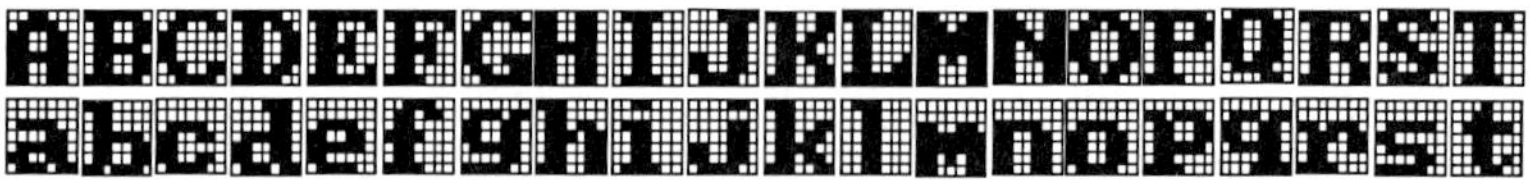

Fig. 6. Alphabetical patterns.

Figure 7 shows the result of this experiment based on 50 trials. In this figure, the failure rate means the rate of neurons on the map which lose their functions, that is, the rate of dead neurons on the map. As shown in this figure, the SOM associative memory learned by the proposed algorithm can greatly improve the robustness because neurons surrounding each winner neuron store similar information to the winner neuron by the ability to preserve topology. Therefore, even if a winner neuron loses its function, its neighbor can play a role of the winner neuron.

4.3 Noise Reduction Effect

In this experiment, we examined sensitivity to noise of the SOM associative memories. $B = 5$ was used in the proposed algorithm. Figure 8 shows the result when 20 training pairs shown in Fig.6 were stored in each network. In the learning of the BAM [3], the pseudo-relaxation learning algorithm [5] was used. We can see that the SOM associative memories can greatly improve the noise reduction effect in comparison with the BAM. In the SOM associative memory learned by the conventional algorithm, almost only winner neurons store training data. In contrast, training data are stored by not only winner neurons but also neurons surrounding them in the proposed algorithm. However, some neurons surrounding winner neurons are likely to store slightly different information from training data because of the ability to preserve topology and such neurons can respond to noisy inputs. Therefore, the SOM associative memory learned by the proposed algorithm shows a little worse performance than that learned by the conventional algorithm. That is, a tradeoff exists between the structural robustness and the noise reduction effect in the SOM associative memory.

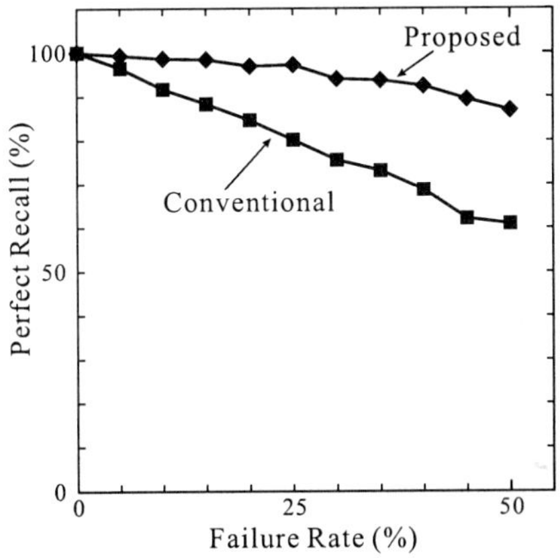

Fig. 7. Structural robustness of SOM associative memory.

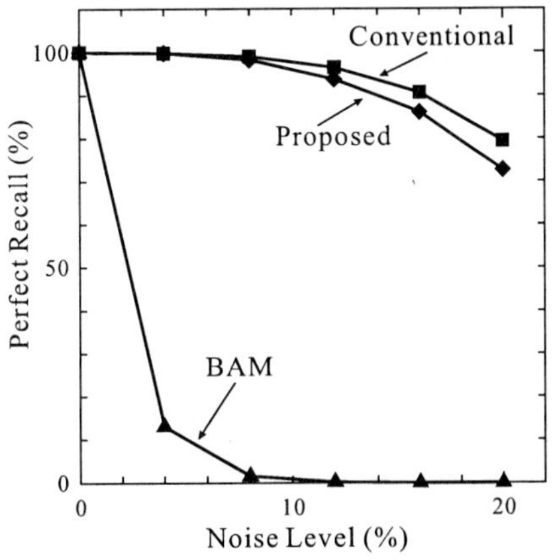

Fig. 8. Sensitivity to noise.

5 Conclusions

We have proposed a sequential learning algorithm for an associative memory based on Self-Organizing Map (SOM). In the proposed algorithm, when a new data to be learned is applied, it is learned within a small area of map with some previously stored data which are extracted from weights of fixed neurons. Repeating this partial remapping every time a new training data is applied, the SOM associative memory can preserve the topological relationships of input data. This contributes the structural robustness. Moreover computer simulation results showed that the noise reduction effect of the SOM associative memory is much better than that of the conventional BAM. In addition, we have confirmed that the SOM associative memory can easily improve the storage capacity by increasing the size of map and can deal with one-to-many associations [12].

Since information once stored in the SOM associative memory learned by the proposed algorithm is not destroyed by newly learned information, we can regard this model as a long term memory. In the future research, we apply our method to much more intelligent tasks such as concept formation, categorization of memory items and so on.

References

1. Nakano, K.: Associatron - a model of associative memory. IEEE Trans. Syst. Man. Cybern. **SMC-2** 1 (1972) 380–388
2. Hopfield, J.J.: Neural Networks and Physical Systems with Emergent Collective Computational Abilities. Proc. of National Academy Sciences USA. **79** (1982) 2554–2558
3. Kosko, B.: Bidirectional Associative Memories. IEEE Trans. Neural Networks. **18** 1 (1988) 49–60
4. Hagiwara, M.: Multidirectional Associative Memory. Proc. of IEEE and INNS International Joint Conference on Neural Networks. **1** (1990) 3–6
5. Oh, H. and Kothari, S.C.: Adaptation of the Relaxation Method for Learning in Bidirectional Associative Memory. IEEE Trans. Neural Networks. **5** 4 (1994) 576–583
6. Kohonen, T.: Self-Organized Formation of Topologically Correct Feature Map. Biological Cybernetics. **43** (1982) 59–69
7. French, R.M.: Using Semi-Distributed Representation to Overcome Catastrophic Forgetting in Connectionist Networks. Proc. of the 13th Annual Cognitive Science Society Conference. (1991) 173–178
8. French, R.M.: Dynamically constraining connectionist networks to produce distributed, orthogonal representations to reduce catastrophic interference. Proc. of the 16th Annual Cognitive Science Society Conference. (1994) 335–340
9. Kondo, S., Futami, R. and Hoshimiya, N.: Pattern recognition with feature map and the improvement on sequential learning ability. IEICE Japan Technical Report. **NC95-161** (1996) 55–60
10. Ichiki, H., Hagiwara, M. and Nakagawa, M.: Kohonen feature maps as a supervised learning machine. Proc. of IEEE International Conference on Neural Networks. (1993) 1944–1948
11. Simula, O., Vesanto, J. and Vasara, P.: Analysis of Industrial Systems Using the Self-Organizing Map. Proc. of Second International Conference on Knowledge-Based Intelligent Electronic Systems. **1** (1998) 61–68
12. Yamada, T., Hattori, M., Morisawa, M. and Ito, H.: Sequential Learning for Associative Memory using Kohonen Feature Map. Proc. of IEEE and INNS International Joint Conference on Neural Networks. (1999) paper no.555

Neighborhood Preservation in Nonlinear Projection Methods: An Experimental Study

Jarkko Venna and Samuel Kaski

Helsinki University of Technology
Neural Networks Research Centre
P.O. Box 5400, FIN-02015 HUT, Finland
{jarkko.venna,samuel.kaski}@hut.fi

Abstract. Several measures have been proposed for comparing nonlinear projection methods but so far no comparisons have taken into account one of their most important properties, the trustworthiness of the resulting neighborhood or proximity relationships. One of the main uses of nonlinear mapping methods is to visualize multivariate data, and in such visualizations it is crucial that the visualized proximities can be trusted upon: If two data samples are close to each other on the display they should be close-by in the original space as well. A local measure of trustworthiness is proposed and it is shown for three data sets that neighborhood relationships visualized by the Self-Organizing Map and its variant, the Generative Topographic Mapping, are more trustworthy than visualizations produced by traditional multidimensional scaling-based nonlinear projection methods.

1 Introduction

Nonlinear projection methods map a set of multivariate data samples into a lower-dimensional space, usually two-dimensional, in which the samples can be visualized. Such visualizations are useful especially in exploratory analyses: They provide overviews of the similarity relationships in high-dimensional data sets that would be hard to acquire without the visualizations.

The methods differ in what properties of the data set they try to preserve. The simplest methods, such as the principal component analysis (PCA) [3], are based on linear projection. A more complex set of traditional methods, that are based on multidimensional scaling (MDS) [10], try to preserve the pairwise distances of the data samples as well as possible. That is, the pairwise distances after the projection approximate the original distances. In a variant of nonlinear MDS, nonmetric MDS [8], only the rank order of the distances is to be preserved. Another variant, Sammon mapping [9], emphasizes the preservation of local (short) distances relative to the larger ones.

The Self-Organizing Map (SOM)[6, 7] is a different kind of a method that fits a discretized nonlinear surface to the distribution of the input data. The data

G. Dorffner, H. Bischof, and K. Hornik (Eds.): ICANN 2001, LNCS 2130, pp. 485–491, 2001.

samples are projected to the map display by finding the closest location on the discretized surface. In the end of the fitting procedure each discrete grid point on the surface is located in the centroid of all data projected onto it and its neighbors on the surface. The procedure implicitly defines a mapping from the input space onto the map grid that aims at keeping close-by grid points close-by in the original data space.

There exist several related algorithms and variants. In this study we will only consider the generative topographic mapping (GTM) [1], a probabilistic variant of the SOM. The GTM is a generative model of the probability distribution of the data, defined by postulating a latent space, a mapping from the latent space to the input space, and a noise model. The latent space corresponds to the SOM grid, and samples can be projected to it by finding their maximum a posterior locations.

In literature the methods have been compared using several measures. Since the methods are obviously good for different purposes it is not sensible to search for the overall best method but to use the "goodness measures" to characterize in which kinds of tasks each method is good at.

Most of the methods try to preserve the pairwise distances of the data samples [2], even though the emphasis of the methods varies. The SOM does not belong to this group, however. Several measures have been used to compare the SOM to other nonlinear projection methods and to measure the properties of the SOM, but according to our knowledge one key property has not been measured thus far: Whether the data points mapped close-by on the displays are generally close-by in the input space as well. The goal of this study is to test empirically which methods are best with respect to this key property of trustworthiness of the neighborhood relationships in the resulting displays.

2 Neighborhood Preservation

Preservation of neighborhoods or proximity relationships has often been mentioned as the key property of SOMs, and several different kinds of more or less heuristic measures of neighborhood preservation have been proposed. In this paper the main purpose is not to compare these methods or to develop yet another method. Our goal is to point out that, according to our knowledge, a key aspect related to neighborhood preservation in nonlinear projection methods has not been measured so far, and to compare the methods with respect to that aspect that we call the trustworthiness of the neighborhoods in the visualization.

For discrete data we define *neighborhoods* of data vectors as sets consisting of the k closest data vectors, for relatively small k. Topological concepts have been defined in terms of "arbitrarily small" neighborhoods but for discrete data we have to resort to finite neighborhoods. When the data are projected, the neighborhood is preserved if the set of the k nearest neighbors does not change.

Two kinds of errors are possible. Either new data may enter the neighborhood of a data vector in the projection, or some of the data vectors originally within the neighborhood may be projected further away on the 2D graphical display.

The latter kinds of errors result from *discontinuities* in the mapping, and they have been measured extensively when quantifying the neighborhood preservation of SOMs (see e.g. [5]). As a result of discontinuities not all of the proximities in the original data are visible after the projections.

We argue that the former kinds of errors are even more harmful since they reduce the *trustworthiness* of the proximities or neighborhood relationships that are visible on the display after the projection: Some data points that seem to be close to each other may actually be quite dissimilar. According to our knowledge this property of the projection methods has not been measured previously; for SOMs the accuracy has been measured as "quantization errors" in the input space, but not in the output space.

If the data manifold is higher-dimensional than the display, then both kinds of errors cannot be avoided and all projection methods must make a tradeoff. In this paper we will concentrate on measuring the new property of trustworthiness. The discontinuities in the mapping will be measured by the preservation of neighborhoods to show the tradeoffs.

In principle, the errors could be measured simply as the average number of data items that enter or leave the neighborhoods in the projection. We have used slightly more informative measures: Trustworthiness of the neighborhoods is quantified by measuring how far from the original neighborhood the new data points entering a neighborhood come. The distances are measured as rank orders; similar results have been obtained with Euclidean distances as well (unpublished). Using the notation in Table 1 the measure for trustworthiness of the projected result is defined as

$$M_1(k) = 1 - \frac{2}{Nk(2N-3k-1)} \sum_{i=1}^{N} \sum_{\mathbf{x}_j \in U_k(\mathbf{x}_i)} (r(\mathbf{x}_i, \mathbf{x}_j) - k) , \tag{1}$$

where the term before the summation scales the values of the measure between zero and one[1].

Preservation of the original neighborhoods is measured by

$$M_2(k) = 1 - \frac{2}{Nk(2N-3k-1)} \sum_{i=1}^{N} \sum_{\mathbf{x}_j \in V_k(\mathbf{x}_i)} (\hat{r}(\mathbf{x}_i, \mathbf{x}_j) - k) . \tag{2}$$

3 Test Setting

Three datasets were used: UCI[2] 9-dimensional Glass identification database (Glass) having 214 data vectors; UCI 34-dimensional ionosphere database (Ionosphere) having 350 data vectors; and 20-dimensional Phonetic[3] dataset having 1962 data vectors.

[1] For clarity we have only included the scaling for neighborhoods of size $k < N/2$

[2] `http://www.ics.uci.edu/~mlearn/MLRepository.html`

[3] Included in LVQ_PAK, available at http://www.cis.hut.fi/research/software.shtml

Table 1. Symbols used in defining the error measures

$\mathbf{x}_i \in \mathbb{R}^n, i = 1, \ldots, N$	data vector
$C_k(\mathbf{x}_i)$	the set of those k data vectors that are closest to $\mathbf{x}_i$ in the original data space
$\hat{C}_k(\mathbf{x}_i)$	the set of those k data vectors that are closest to $\mathbf{x}_i$ after projection
$U_k(\mathbf{x}_i)$	the set of data vectors $\mathbf{x}_j$ for which $\mathbf{x}_j \in \hat{C}_k(\mathbf{x}_i) \wedge \mathbf{x}_j \notin C_k(\mathbf{x}_i)$ holds
$V_k(\mathbf{x}_i)$	the set of data vectors $\mathbf{x}_j$ for which $\mathbf{x}_j \notin \hat{C}_k(\mathbf{x}_i) \wedge \mathbf{x}_j \in C_k(\mathbf{x}_i)$ holds
$r(\mathbf{x}_i, \mathbf{x}_j), i \neq j$	the rank of $\mathbf{x}_j$ when the data vectors are ordered based on their Euclidean distance from the data vector $\mathbf{x}_i$ in the original data space
$\hat{r}(\mathbf{x}_i, \mathbf{x}_j), i \neq j$	the rank of $\mathbf{x}_j$ when the data vectors are ordered based on their distance from the data vector $\mathbf{x}_i$ after projection

The resolution and flexibility of SOM and GTM can be freely selected, and need be fixed for the comparison. The number of grid points on the map or latent space, which governs the resolution of the projection, was set about equal to the number of data points. The flexibility or stiffness of the SOM is determined by the final radius σ of the neighborhood function that governs adaptation during the learning process. The radius was selected according to the goodness measure given in [4].[4] We used the neighborhood function $h(d) = 1 - d^2/\sigma^2$, for $d <= \sigma$ and $h(d) = 0$ otherwise. The argument d refers to the distance on the map grid, the unit being the distance between neighboring map nodes.

The flexibility of the GTM is governed by the number and width of the basis functions that map the latent grid to the input space. For each data set the values that gave the best log-likelihood on a validation set were chosen.

Note that since the map grids of SOM and GTM are discrete, several data points may be projected onto the same points and hence there will be ties in the rank ordering of the distances between the projected data points. To make these methods comparable to the rest, we assume that in the case of ties all rank orders are equally likely, and compute averages of the error measures.

4 Results

Trustworthiness of the Visualized Neighborhood Relationships. When interpreting the visualizations it is particularly important that the small neighborhoods, having a small value of k, are trustworthy. They are the most salient relationships between the data items. In Fig. 1a the trustworthiness of the projected proximities is measured for the three data sets. The SOM and GTM preserve proximities (at small values of k up to 5-10% of the size of the datasets tested) better than the other methods, whereas the MDS-based methods are mostly better at preserving the global ordering (large values of k). NMDS is the best of

[4] Because the measure works better if there is more data per grid point, the number of grid points on the map was set to about 1/10 of the number data points (but to at least 50) for the selection of the stiffness. The results were scaled back to the larger map.

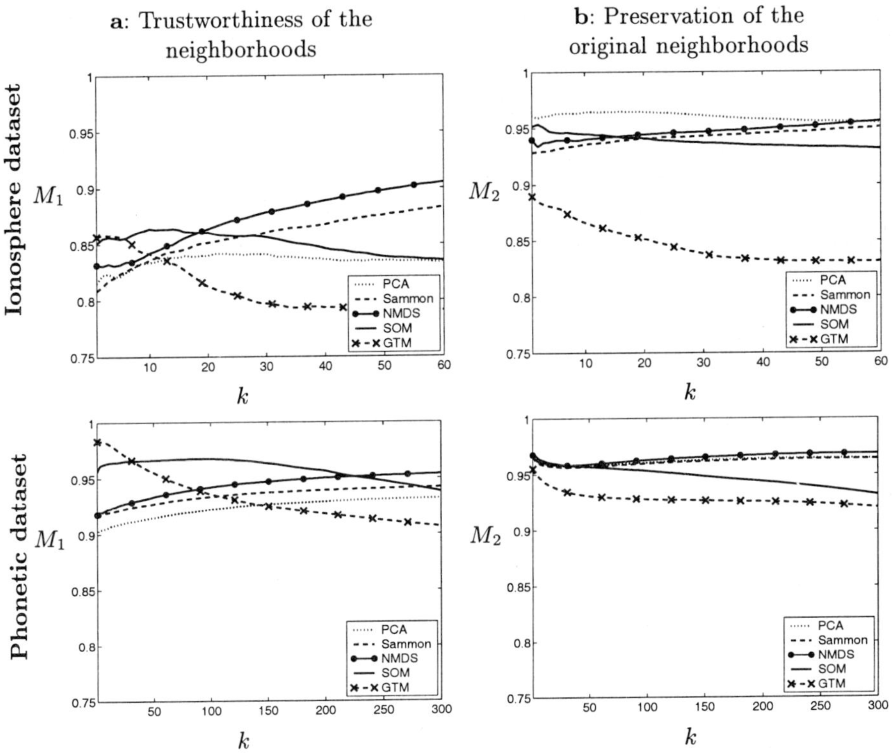

Fig. 1. Trustworthiness of the neighborhoods after projection (**a**) and preservation of the original neighborhoods (**b**) for two data sets as a function of the neighborhood size k. The results on the third data set (Glass) were similar.

the traditional methods, being consistently better than Sammon mapping and PCA. The relative order of the SOM and the GTM differs on different datasets; the explanation is probably related to the different methods used in selecting their stiffness (cf. the discussion about stiffness and trustworthiness below).

Preservation of the Original Neighborhoods. While preservation of the original neighborhoods, measured in Fig. 1**b**, is not as important as the trustworthiness of the projection, it gives insight to the tradeoffs done in the projection. PCA and NMDS are the overall best methods: They perform similarly on all datasets and are consistently among the best methods. For small neighborhoods (small k) the SOM is among the best methods as well.

Flexibility of Mapping Increases Trustworthiness of Small Neighborhoods. It seems plausible that there is a connection between the stiffness of the mappings and the different kinds of errors. If the mapping is stiff (the linear PCA is the extreme example), then close-by points will be projected close to each other. Unless the data lies within a low-dimensional (linear) subspace of the original space, however, data points originally far away may "collapse" into the same location. We next investigated how the stiffness of the SOM affects the kinds of errors made.

The more flexible the SOM is (the smaller the σ) the more trustworthy the small neighborhoods are (Fig. 2**a**), as expected. There is a tradeoff in that the trustworthiness of the larger neighborhoods decreases.

The relation between the stiffness and the preservation of the original neighborhoods is not as clear, however, and requires more investigation. For larger neighborhoods it holds that the stiffer the map the better the neighborhoods are preserved (Fig. 2**b**), but for very small neighborhoods the stiffest map ($\sigma = 20$) makes most errors.

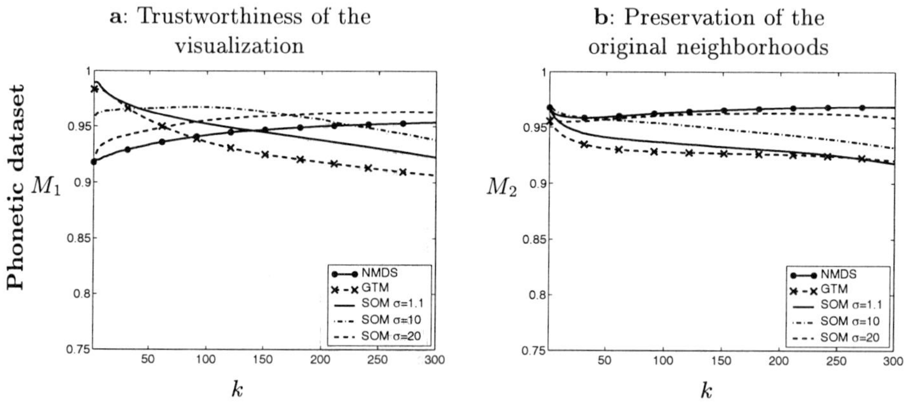

Fig. 2. The effect of the stiffness of the SOM on trustworthiness of the visualized neighborhoods (**a**) and preservation of the original neighborhoods (**b**), as a function of the neighborhood size k. The stiffness was varied by varying the final neighborhood radius σ on a SOM of a fixed size. The stiffness of the GTM shown for comparison was selected by maximum likelihood on a validation set. Data set: Phonetic.

5 Conclusions

The experimental results showed that the neighborhood relationships on the SOM and GTM displays are more trustworthy than on MDS-based displays. That is, if two data samples are close to each other on the visualizations, they are more likely to be close to each other in the original high-dimensional space as well. Moreover, the capacity of the SOM in preserving the original neighborhoods, the other aspect of neighborhood preservation, seems to be comparable to the other methods.

The trustworthiness of the local neighborhoods can still be improved by increasing the flexibility of the mapping, by reducing the radius of the neighborhood function of the SOM. The cost is that larger neighborhoods (longer distances) are not preserved as well.

References

[1] C. M. Bishop, M. Svensén, and C. K. I. Williams. GTM: The generative topographic mapping. *Neural Computation*, 10:215–234, 1998.
[2] G. J. Goodhill and T. J. Sejnowski. A unifying objective function for topographic mappings. *Neural Computation*, 9:1291–1303, 1997.
[3] H. Hotelling. Analysis of a complex of statistical variables into principal components. *Journal of Educational Psychology*, 24:417–441,498–520, 1933.
[4] S. Kaski and K. Lagus. Comparing self-organizing maps. In C. von der Malsburg, W. von Seelen, J. C. Vorbrüggen, and B. Sendhoff, editors, *Proceedings of ICANN'96, International Conference on Neural Networks*, pages 809–814, Berlin, 1997. Springer.
[5] K. Kiviluoto. Topology preservation in self-organizing maps. *Proceedings of IEEE International Conference on Neural Networks.*, volume 1, pages 294–299, 1996.
[6] T. Kohonen. Self-organized formation of topologically correct feature maps. *Biological Cybernetics*, 43:59–69, 1982.
[7] T. Kohonen. *Self-Organizing Maps.* Springer-Verlag, Berlin, 1995 (third, extended edition 2001).
[8] J. B. Kruskal. Multidimensional scaling by optimizing goodness of fit to a nonmetric hypothesis. *Psychometrica*, 29(1):1–26, Mar 1964.
[9] J. W. Sammon, Jr. A nonlinear mapping for data structure analysis. *IEEE Transactions on Computers*, C-18(5):401–409, May 1969.
[10] W. S. Torgerson. Multidimensional scaling I—theory and methods. *Psychometrica*, 17:401–419, 1952.

A Topological Hierarchical Clustering: Application to Ocean Color Classification

Méziane Yacoub[1], Fouad Badran[2,1], and Sylvie Thiria[1]

[1] Laboratoire d'Océanographie Dynamique et de Climatologie (LODYC) case 100, Université Paris 6, 4 place Jussieu 75252 Paris cedex 05 France
{yacoub,badran,thiria}@lodyc.jussieu.fr
[2] CEDRIC, Conservatoire National des Arts et Métiers, 292 rue Saint Martin 75003 Paris France
{badran}@cnam.fr

Abstract. We propose a new criteria to cluster the referent vectors of the self-organizing map. This criteria contains two terms which take into account two different errors simultaneously: the square error of the entire clustering and the topological structure given by the Self Organizing Map. A parameter T allows to control the corresponding influence of these two terms. The efficiency of this criteria is illustrated on the problem of top of the atmosphere spectra of satellite ocean color measurements.

1 Introduction

The aim of Self-Organizing Map (SOM) is to provide a "refined" partition of the data space using a huge number of neurons and to induce a topological order between them. The main goal of this partition is to reduce the information provided by the data using a vector quantization method. For practical application, one often looks for a limited number of significant clusters on the data space. Thus the problem is to reduce the number of clusters and to define a new partition of clusters from the initial SOM partition. This can be done by clustering the referent vectors of SOM using a hierarchical clustering algorithm [7, 8, 2].

In the present paper, we look for a new dissimilarity measure which allows us to take into account the two informations provided by SOM: the square error for the entire clustering and the existing topological order on the map. An adequate decomposition of the cost function which determines SOM suggests that some new criteria will be able to do it.

2 A New Hierarchical Clustering Criteria

2.1 SOM Quantization

The standard Self Organizing Map (SOM) [4] consists of a discrete set $\mathcal{C}$ of neurons called the map. This map has a discrete topology defined by an undirected

G. Dorffner, H. Bischof, and K. Hornik (Eds.): ICANN 2001, LNCS 2130, pp. 492–499, 2001.

graph; usually it is a regular grid in one or two dimensions. We denote N_{neuron} the number of neurons in $\mathcal{C}$. For each pair of neurons (c,r) on the map, the distance $\delta(c,r)$ is defined as the shortest path between c and r on the graph. For each neuron c, this distance allows us to define a neighborhood of order d: $V_c(d) = \{r \in \mathcal{C}/\delta(c,r) \leq d\}$. In the following, in order to control the neighborhood order, we introduce a Kernel positive function K ($\lim_{|x|\to\infty} K(x) = 0$) and its associated family K_T parametrized by T:

$$K_T(\delta) = [1/T]K(\delta/T)$$

Let $\mathcal{D}$ be the data space ($\mathcal{D} \subset \mathcal{R}^n$) and $\mathcal{A} = \{\mathbf{z}_i; i = 1, \ldots, N\}$ the training data set ($\mathcal{A} \subset \mathcal{D}$). The standard SOM algorithm defines a mapping from $\mathcal{C}$ to $\mathcal{D}$ where each neuron c is associated to its referent vector $\mathbf{w}_c$ in $\mathcal{D}$. The set of parameters $\mathcal{W} = \{\mathbf{w}_c; c \in \mathcal{C}\}$, which fully determines the SOM, have to be estimated from $\mathcal{A}$. This is done iteratively by minimizing a cost function:

$$J^T_{som}(\chi, \mathcal{W}) = \sum_{c\in C}\sum_{\mathbf{z}_i\in\mathcal{A}} K_T(\delta(c, \chi^T(\mathbf{z}_i)))\|\mathbf{z}_i - \mathbf{w}_c\|^2 \tag{1}$$

Where $\chi^T(\mathbf{z}_i)$ represents a particular neuron of $\mathcal{C}$ assigned to $\mathbf{z}_i$. This minimization can be done using a "batch" version of the standard SOM algorithm ([5, 6, 4, 1]).

For a given value of T, the batch algorithm minimizes (1) and leads to a local minima of this cost function with respect to both χ^T and $\mathcal{W}$. Using the batch version iteratively, with decreasing values of T, provides the standard SOM model. The nature of the SOM model reached at the end of the algorithm, the quality of the clustering (or quantization) and those of the topological order induced by the graph depend on the first value of T (T^{max}), its final value (T^{min}) and the number of iterations (N_{iter}) of the batch algorithm.
SOM uses the neighborhood system, whose size is controlled by T, in order to introduce the topological order[4, 9]. When the value of T is large, an observation $\mathbf{z}_i$ will modify a large number of referent vectors $\mathbf{w}_c$, in opposite to small values of T allowing few changes. At the end of the learning algorithm (when T^{min} is reached), two neighbor neurons on the map have close referent vectors in the euclidian space ($\mathcal{R}^n$). In that sense, the map provides a topological order; the clustering associated to this topological order is defined by taking $T = T^{min}$ [9]. If T^{min} is such that the neighborhood of a neuron is reduced to itself for any distance d ($V_c(d) = \{c\}$) the cost function $J^{T^{min}}_{som}$ minimized at the end of the learning phase is the k-means distortion function. So, the successive iterations allow to reach a k-means solution which takes into account the topological constraint. In this case one shows [9] that, for each neuron c, the referent vectors $\mathbf{w}_c$ is just the mean vector $\mathbf{g}_c$ of $P_c = \{\mathbf{z}_i \in \mathcal{A}/\chi^{T^{min}}(\mathbf{z}_i) = c\}$, in the following we denote by n_c the cardinality of P_c.

2.2 A Topological Hierarchical Clustering

Rewriting J^T_{som} gives: $J^T_{som} = \sum_c \sum_r \sum_{\mathbf{z}_i \in P_r} K_T(\delta(c,r)) \|\mathbf{z}_i - \mathbf{w}_c\|^2$

$$= \left[\sum_c \sum_{r \neq c} \sum_{\mathbf{z}_i \in P_c} K_T(\delta(c,r)) \|\mathbf{z}_i - \mathbf{w}_r\|^2 \right] + \left[K_T(\delta(c,c)) \sum_c \sum_{\mathbf{z}_i \in P_c} \|\mathbf{z}_i - \mathbf{w}_c\|^2 \right] \quad (2)$$

Since usually at the end of the learning phase, $\mathbf{w}_c$ is no more that the mean vector of P_c (see section 2.1), we can decompose J^T_{som} using the square error of each individual cluster (or neuron): $I_c = \sum_{\mathbf{z}_i \in P_c} \|\mathbf{z}_i - \mathbf{w}_c\|^2$ ($I_c = 0$ for $P_c = \emptyset$), and (2) gives [9]

$$J^T_{som} = \left[\frac{1}{2} \sum_c \sum_{r \neq c} K_T(\delta(c,r))(n_c + n_r) * \|\mathbf{w}_r - \mathbf{w}_c\|^2 \right] + \left[\sum_c \left(\sum_r K_T(\delta(c,r)) \right) I_c \right] \quad (3)$$

The first term of the decomposition of J^T_{som} takes into account the topological order, the second term corresponds to a weighted square error for the entire clustering and is similar to Ward criteria, which minimizes the intra-class inertia.

The hierarchical clustering, presented in this paper and denoted HC_{som}, proceeds by successive aggregations of neurons reducing by one, at each time, the cardinality of the preceding partition. At each iteration a new partition is defined. We denote by $\mathcal{P}_K$, such a partition made of K clusters, each cluster being denoted by an index c. The partition $\mathcal{P}_K = \{P_c / c \in \mathcal{C}_K\}$ is such that the set of index $\mathcal{C}_K$ has a graph structure which induce a discrete topology between the different clusters. For every c in $\mathcal{C}_K$, the cluster P_c is represented by its mean vector $\mathbf{g}_c$, its cardinality n_c and its square error I_c. We use J^T_{som} as a measure of the "quality" of the partition $\mathcal{P}_K$. Using $\mathcal{C}_K$, the dedicated measure becomes a sum of two terms:

$$J^T_{hc} = \left[\frac{1}{2} \sum_c \sum_{r \neq c} K_T(\delta(c,r))(n_c + n_r) * \|\mathbf{g}_r - \mathbf{g}_c\|^2 \right] + \left[\sum_c \left(\sum_r K_T(\delta(c,r)) \right) I_c \right] \quad (4)$$

Where c and r belong to $\mathcal{C}_K$ and $\delta(c,r)$ represents the distance on the graph $\mathcal{C}_K$ which will be defined below, as in (3) the first term of (4) (a) involves the topological order of the graph $\mathcal{C}_K$ and the second term (b) is similar to Ward criteria.

The initial partition $\mathcal{P}_{K_0}$ is given by the SOM map at the end of the learning algorithm. The graph $\mathcal{C}_{K_0}$ is the sub-graph of the map, where all the neurons such that $n_c = 0$ are removed. The initial distance $\delta(c,r)$ on $\mathcal{C}_{K_0}$ is defined as in section 2.1 by the length of the shortest path on the map. In general, the hierarchical clustering reduce $\mathcal{P}_K$ to $\mathcal{P}_{K-1}$ aggregating two vertices of $\mathcal{C}_K$ which allows us to determine the graph $\mathcal{C}_{K-1}$ of $\mathcal{P}_{K-1}$. If we denoted by $\{c_1, c_2\}$ the new index which aggregate c_1 and c_2 and $P_{\{c_1,c_2\}}$ its related cluster, $P_{\{c_1,c_2\}}$ is represented by its mean and its cardinality on the map:

$$\mathbf{g}_{\{c_1,c_2\}} = \frac{(n_{c_1}*g_{c_1})+(n_{c_2}*g_{c_2})}{n_{c_1}+n_{c_2}},$$

$$n_{\{c_1,c_2\}} = n_{c_1} + n_{c_2}$$

and its individual square error

$$I_{\{c_1,c_2\}} = n_{c_1} * \|g_{c_1} - g_{\{c_1,c_2\}}\|^2 + n_{c_2} * \|g_{c_2} - g_{\{c_1,c_2\}}\|^2 + I_{c_1} + I_{c_2}.$$

The new distances δ on the graph $\mathcal{C}_{K-1}$ is defined by:

$$\delta(c, \{c_1, c_2\}) = min\{\delta(c, c_1), \delta(c, c_2)\}.$$

HC_{som} is looking for the best aggregation; as we compute the criteria J_{hc}^T, among all the possible pairs of $\mathcal{C}_K$ and the possible resulting partitions, we select the pair for which the value of J_{hc}^T is minimal. This pair gives rise to the new partition $\mathcal{P}_{K-1} = \{P_c/c \in \mathcal{C}_{K-1}\}$. Doing so, the parameter T defines a family of criteria whose characteristics are related to its value. Taking T small (as $T = T_{min}$), cancels the first term (a) of (4); in this case HC_{som} is the Ward criteria. Using a large value of T (as $T = T_{max}$), cancels the term (b); the method classify using only the topological order given by SOM. In this later case, HC_{som} becomes similar to the single link criteria. The intermediate values of T represent a compromise between these two alternatives. The 'best' value of T has to be specified, as any hyper-parameter.

In the following, we show the behavior of HC_{som} when applied to simulated and a real applications.

3 Application

3.1 Simulated Data

HC_{som} described in the previous section was applied on a simulated data set of $\mathcal{R}^2$ shown in figure 1(a). Clearly this data set presents two different clusters which are irregularly shaped. First, we trained a two-dimensional map of size 10×10 with SOM algorithm, the set of referent vectors and the graph induced by $(\mathcal{C})$ is presented in figure 1(b). We used HC_{som} to cluster the set $\mathcal{W}$ of referent vectors given by SOM and select the partition $\mathcal{P}_2$ with 2 clusters. The experiment was repeated, varying the value of the parameter T. For a given value of T, the behavior of the hierarchy can be seen, during the aggregation process, looking simultaneously at the evolution of J_{hc}^T and at those of its first and second terms (a) and (b). The results we obtain for $T = 0.001$, $T = 0.2$ and $T = 5$, are shown in figure 1(c,d,e). It can be seen that for $T = 0.001$, J_{hc}^T is approximatively equal to the second term (b). In this case $HC_{som}^{0.001}$ is similar to the Ward hierarchy. For $T = 5$, J_{hc}^T is approximatively equal to its first term (a), and the topology provided by SOM predominate. For $T = 0.2$, J_{hc}^T can be seen as a Ward criteria regularized by the topological order of the graph, the two terms of the sum have a particular impact. In figure 1 (f,g) we show the two clusters obtained using HC_{som}^T for $T = 0.001$, and for $T = 0.2$. Clearly the intermediate value $T = 0.2$ allows us to find the underlying structure of the data set; the use of the Ward criteria gives more spherical classes and is unable to extract the two circles.

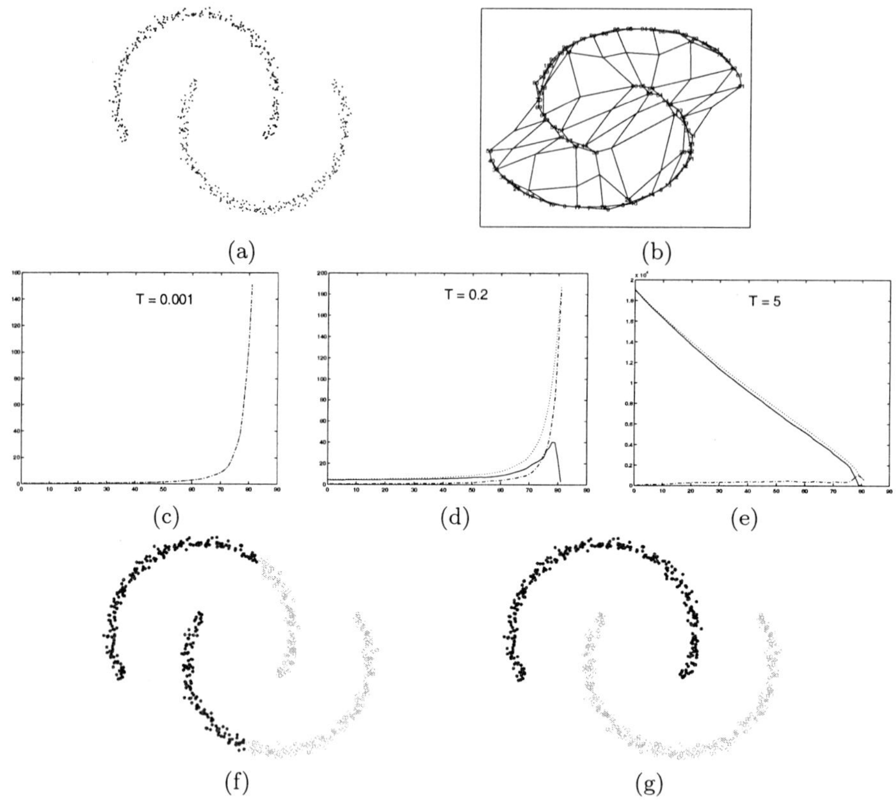

Fig. 1. (a): data set (b): the set of referent vectors with the induced topological order. (c, d, e): the evolution of J_{hc}^{T} (dotted ligne), and its two terms a (solid ligne) and b (dashdot ligne) during the aggregation of neurons, for three diffrent temperatures. (f, g) The two clusters obtained for $T = 0.001$ and $T = 0.2$.

3.2 Classification of Ocean Color Remote Sensing Measurements

Satellite ocean color sensors which are now operational or under preparation, measure ocean reflectance, allowing us to a quantitative assessment of geophysical parameters (e.g., chlorophyll concentration).

The major problem for ocean color remote sensing processing is that interactions of the solar light with aerosols in the atmosphere and water constituents are responsible for the observed spectrum of Top Of the Atmosphere (TOA) reflectances for a cloud-free pixel. The critical issue is thus to remove the aerosol contribution to the measured signals in order to retrieve the actual spectrum of marine reflectance which is directly related to the chlorophyl and water constituent content. Ocean color is determined by the interactions of the solar light with substances or particles present in the water, and provides useful indications on the composition of water.

In the following we used the SeaWiFS data products. The SeaWiFS[1] on board the SeaStar satellite is a color-sensitive optical sensor used to observe color variations of the ocean. It contains 8 spectral bands in the visible and near-infrared wavelengths [2]. SeaWiFS ocean color data is available in several different types. We used level 1 GAC data: it consists of raw radiances measured at the top of the atmosphere.

We studied the region shown in figure 2(a). It represents the Atlantic sea in the west north African coast. The image ($536 \star 199 = 106664$ pixels) was taken on January 1999. We removed the land pixels and some other erroneous pixels detected by the SeaWiFS product (the black region in figure 2(b)), and used our method to classify the remaining pixels (the white region in figure 2(b)).

First, we trained a two-dimensional map of size 10×10 with SOM algorithm. Then, we used HC_{som} to cluster the 100 referent vectors given by SOM and select the partition with 3 clusters. The experiment was repeated, varying the value of the parameter T. In figure 3, we show the three areas obtained using HC^T_{som} for $T = 0.0001$ (which correspond to the Ward criteria), $T = 0.1$, $T = 0.3$, $T = 0.5$, and $T = 2$. For technical reasons, we used three colors to show these partitions (white color, grey color, and black color). Thus one of the three partitions will have the same color than the removed region. Hereafter, by black region we means the black region without the removed pixels. First it is clear that the 5

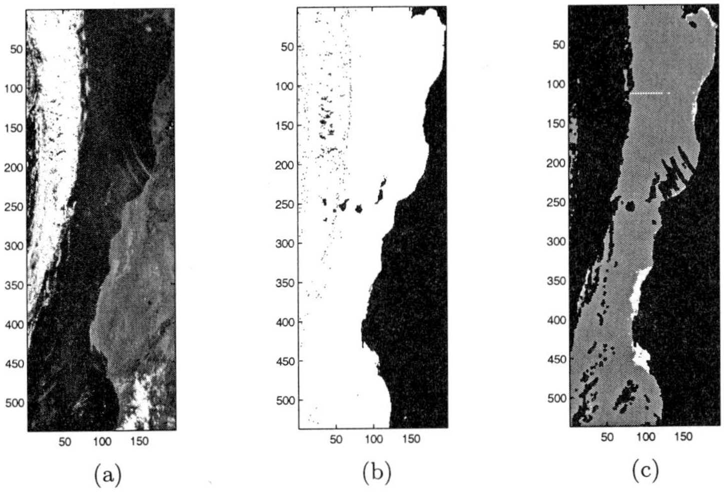

Fig. 2. (a) the studied region. (b) removed pixels according to the classification proposed by SeaWiFS (the black region) and the region to be classified using HC_{som} (the white region) (c) The three classes given by an expert on the studied region without removed pixels.

[1] SeaWiFS Project Home Page http://seawifs.gsfc.nasa.gov/SEAWIFS.html

[2] 412nm, 443nm, 490nm, 510nm, 555nm, 670nm, 765nm and 865nm

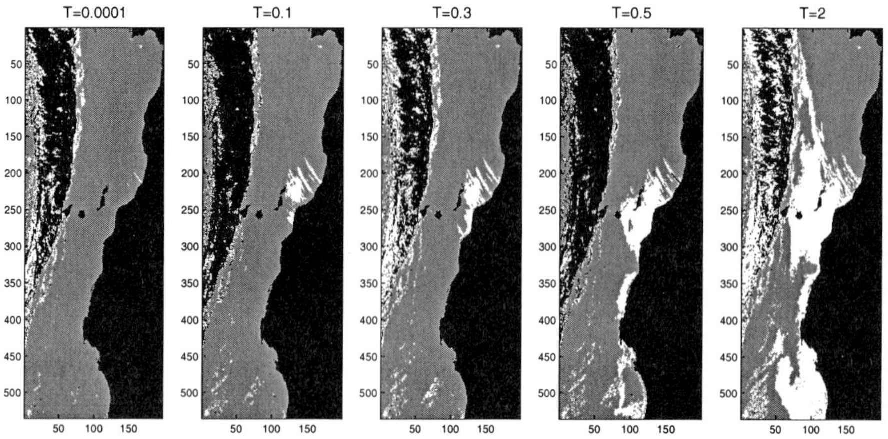

Fig. 3. The three classes obtained using HC_{som} for different value of T

different classifications proposed when using 5 different values of T correspond to different partitions which give rise to different possible interpretations. The expert gives a physical meaning to each classification. For $T = 0.0001$, the thick clouds became visible bringing together the black and white aereas. For $T = 0.1$ and $T = 0.3$ the black and white aereas represent thick clouds and desertic aerosols. For $T = 0.5$ the black and white areas represent thick clouds, desertic aerosols, and case 2 waters. The expert noticed that desertic aerosols have similar spectrum than case 2 waters. He provides a labeled map where he labeled pixel by pixel the image using physical models of aerosols. The labeled image is shown in figure 2(c). The three proposed classes are: the grey area representing the sea under a clear sky, the case 2 waters (white areas), and the cloud pixels (black areas). Clearly T=0.5 provides a classification similar to the one given by the expert. As case 2 waters and desertic aerosols have similar signature, HC_{som} put them into the same class. So the expert choose the case where $T = 0.5$ as being the most significant classification.

4 Conclusion

In this paper, we introduce a family of new criteria to perform hierarchical clustering. This family presents the new properties to mix two different criteria: the square error of the entire clustering and a graph approach which allows us to take into account the structure of the data set. This approach greatly takes advantage of the neural approach, the Self organizing Map provided an ordered codebook of the initial data and suggest a particular criteria in order to cluster this codebook. Some experiments on the problem of satellite ocean color classification shows that this hierarchical clustering can be useful for identifying different coherent regions.

References

1. Anouar F., Badran F. and Thiria S., (1997): Self Organized Map, A Probabilistic Approach. proceedings of the Workshop on Self-Organized Maps. Helsinki University of Technology, Espoo, Finland, June 4–6.
2. Ambroise C., Seze G., Badran F. and Thiria S. (2000): Hierarchical clustering of self-organizing maps for cloud classification. Neurocomputing, vol 30, number 1-4, January 2000. 47–52.
3. Diday E. and Simon J.C. (1976): Clustering Analysis. In Digittal Pattern Recognition, Edited by K.S.FU. Springer-Verlag
4. Kohonen T. (1995): Self-organizing maps. Springer Series in Information Sciences, Springer.
5. Luttrel S.P. (1994): A bayesian analysis of self-organizing maps. Neural comput. 6.
6. Ritter, Martinez and Schulten (1992) Neural computation and self organizing maps. Addison Wesley.
7. Thiria S., Lechevallier Y., Gascuel O. et Canu S., (1997): Statistique et méthodes neuronales. Dunod
8. Yacoub M., D. Frayssinet, F. Badran and S. Thiria, (2000): Clustering and Classification Based on Expert Knowledge Propagation Using a Probablistic Self-Organizing Map: Application to Geophysics. Data Analysis, edyted by W.GAUL, O.OPITZ and M.SCHADER, Springer, 67-78.
9. Yacoub M., N. Niang, F. Badran and S. Thiria, (2001): A New Hierarchical Clustering Method using Topological Map. ASMDA (Applied Stochastic Models and Data Analysis)

Hierarchical Clustering of Document Archives with the Growing Hierarchical Self-Organizing Map

Michael Dittenbach[1], Dieter Merkl[2], and Andreas Rauber[2]

[1] E-Commerce Competence Center – EC3,
Siebensterngasse 21/3, A–1070 Wien, Austria
[2] Institut für Softwaretechnik, Technische Universität Wien,
Favoritenstraße 9–11/188, A-1040 Wien, Austria
www.ifs.tuwien.ac.at/{~mbach, ~dieter, ~andi}

Abstract. With the increasing amount of information available in electronic document collections, methods for organizing these collections to allow topic-oriented browsing and orientation gain increasing importance. In this paper, we present the *Growing Hierarchical Self-Organizing Map*, which allows an automatic hierarchical decomposition and organization of documents. We present a case study based on a 3-month article collection from an Austrian daily newspaper.

1 Introduction

With the increasing amount of textual information available in digital document archives, methods for accessing and providing orientation in these vast information repositories have gained significant importance. While research in information retrieval provides us with sophisticated methods for selecting relevant subsets of collections, only few systems address the need for interactive navigation in topically organized libraries. Still, this way of organization is the most common and thus most natural one to users of conventional libraries.

With the SOMLib Digital Library [11] we developed a system focusing on this aspect of information access by automatically organizing documents by topic on a two-dimensional map using a popular neural network model, namely the *Self-Organizing Map* (*SOM*) [5, 6]. Documents covering similar topics are found in neighboring regions on the trained *SOM*, which very much resembles the conventional bookshelf metaphor of document organization. While this approach allows easy orientation in an unknown text collection and facilitates the integration of distributed repositories [12], the resulting map-based metaphor poses some constraints on the size of the document collection that can feasibly be displayed on a single two-dimensional map. The more topics are present, the larger the according map has to be. Furthermore, the size of the map has to be determined in advance, which turns out to be impractical for unknown document collections. Instead, we are interested in a hierarchical representation of the various topics covered in the document archive.

G. Dorffner, H. Bischof, and K. Hornik (Eds.): ICANN 2001, LNCS 2130, pp. 500–505, 2001.

Approaches like the *Hierarchical Feature Map* [10] provide such a hierarchical organization. Albeit, the architecture of the resulting system has to be determined in advance, i.e. the number of layers as well as the size of the maps on each layer has to be specified prior to training. This turns out to be difficult for document collections, where the precise nature and topic distribution of a collection is unknown. Furthermore, some topics are present more prominently than others, thus requiring more map space for representation.

Another hierarchical modification of the *SOM* is constituted by the *TS-SOM* [7, 8]. Yet, the focus of this model lies primarily with speeding up the training process by providing faster winner selection using a tree-based organization of units. However, it does not focus on providing a hierarchical organization of data as all data are organized on one single flat map.

Various approaches address the problem of having to define the size of a SOM in advance, such as *Dynamic Self-Organizing Maps* [1], *Incremental Grid Growing* [2] or *Growing Grid* [4], where new units are added to map areas where the data, i.e. the topics, are not represented at a satisfying degree of granularity. However, since all of these methods rely on the creation of a single flat map, they tend to produce huge maps limiting their usability for the display of and navigation in document archives.

To overcome these limitations we developed the *Growing Hierarchical Self-Organizing Map* (*GHSOM*) [3], which automatically creates a hierarchical organization of a set of documents. This allows the network architecture to determine the topical structure of the given document repository during the training process, creating a hierarchy of *Self-Organizing Maps*, each of which provides a topologically sorted representation of a topical subset. Starting from a rather small high-level *SOM*, which provides a coarse overview of the various topics present in the collection, subsequent layers are added where necessary to display a finer subdivision of topics. Each map in turn grows in size until it represents its topic to a sufficient degree of granularity. Since usually not all topics are present equally strong in a collection, this leads to an unbalanced hierarchy, assigning more representation space to topics that are more prominently present in a given collection. This allows the user to approach and intuitively browse a document collection in a way similar to conventional libraries. The benefits of the *GHSOM* is shown by organizing a real-world document collection according to semantic similarities.

The remainder of this paper is structured as follows. Section 2 presents describes architecture and training process of the *Growing Hierarchical Self-Organizing Map*. Next, we present the hierarchical organization of the newspaper articles and the various topical sections discovered in Section 3. Finally, some conclusions are drawn in Section 4.

2 Growing Hierarchical Self-Organizing Map

The key idea of the *GHSOM* is to use a hierarchical structure of multiple layers where each layer consists of a number of independent *SOMs*. One *SOM* is used at the first layer of the hierarchy. For every unit in this map a *SOM* might be added to the next layer of the hierarchy. This principle is repeated with the third and any further layers of the *GHSOM*.

Since one of the shortcomings of *SOM* usage is its fixed network architecture we rather use an incrementally growing version of the *SOM*. This relieves us from the burden of predefining the network's size which is rather determined during the unsupervised training process. We start with a layer 0, which consists of only one single unit. The weight vector of this unit is initialized as the average of all input data. The training process basically starts with a small map of, say, 2×2 units in layer 1, which is self-organized according to the standard *SOM* training algorithm.

This training process is repeated for a fixed number λ of training iterations. Ever after λ training iterations the unit with the largest deviation between its weight vector and the input vectors represented by this very unit is selected as the error unit. In between the error unit and its most dissimilar neighbor in terms of the input space either a new row or a new column of units is inserted. The weight vectors of these new units are initialized as the average of their neighbors.

An obvious criterion to guide the training process is the quantization error q_i, calculated as the sum of the distances between the weight vector of a unit i and the input vectors mapped onto this unit. It is used to evaluate the mapping quality of a *SOM* based on the mean quantization error (*MQE*) of all units in the map. A map grows until its *MQE* is reduced to a certain fraction τ_1 of the q_i of the unit i in the preceding layer of the hierarchy. Thus, the map now represents the data mapped onto the higher layer unit i in more detail.

As outlined above the initial architecture of the *GHSOM* consists of one *SOM*. This architecture is expanded by another layer in case of dissimilar input data being mapped on a particular unit. These units are identified by a rather high quantization error q_i which is above a threshold τ_2. This threshold basically indicates the desired granularity level of data representation as a fraction of the initial quantization error at layer 0. In such a case, a new map will be added to the hierarchy and the input data mapped on the respective higher layer unit are self-organized in this new map, which again grows until its *MQE* is reduced to a fraction τ_1 of the respective higher layer unit's quantization error q_i. Note that this may not necessarily lead to a balanced hierarchy. The depth of the hierarchy will rather reflect the diversity in input data distribution which should be expected in real-world data collections.

Depending on the desired fraction τ_1 of *MQE* reduction we may end up with either a very deep hierarchy with small maps, a flat structure with large maps, or – in the most extreme case – only one large map, which is similar to the *Growing Grid*. The growth of the hierarchy is terminated when no further units are available for expansion.

A graphical representation of a *GHSOM* is given in Figure 1. The map in layer 1 consists of 3×2 units and provides a rough organization of the main clusters in the input data. The six independent maps in the second layer offer a more detailed view on the data. Two units from one of the second layer maps

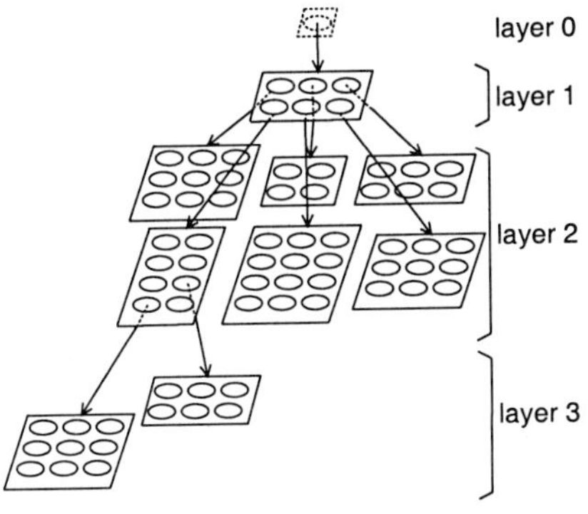

Fig. 1. *GHSOM* reflecting the hierarchical structure of the input data.

have further been expanded into third-layer maps to provide sufficiently granular input data representation.

3 A Hierarchy of Newspaper Articles

In the experiments presented hereafter we use an article collection of the Austrian daily newspaper *Der Standard* covering the second quarter of 1999. The 11.627 articles are represented by automatically extracted 1104-dimensional feature vectors of word histograms weighted by a $tf \times idf$ weighting scheme.

Since it is impossible to present the complete topical hierarchy of three months of news articles, we will concentrate on a few sample topical sections. The top-level map of the *GHSOM*, which evolved to the size of 3×4 units during training by adding two rows and one column, is depicted in Figure 2. Here, we find an overview of the main subject areas such as the war on the Balkan on unit (1/1)[1], labeled[2] with *Kosovo* and *NATO*. Austrian internal politics (2/1), economy (3/1), and the European Union (3/2) are other examples of main subjects. Cultural topics are located on three units in the lower left corner, representing sports, fashion and theater, respectively.

For the following discussion, we restrict ourselves to the topical cluster of the Balkan War. The second-layer map, depicted in Figure 3, shows a more detailed topical structuring of the subject Balkan War. The top-row sections concentrate on the general political and social situation, while unit (1/3) deals with the situation of refugees. Slobodan Milosevic is the main topic of the sections mapped onto the units (2/3), (3/2) and (3/3) in the bottom right corner of the map, whereas unit (2/2) covers political negotiations to resolve the crisis and Russia's role in the Balkan War.

[1] We refer to a unit located in column x and row y as unit (x/y), starting with (1/1) in the upper left corner.

[2] The labeling is performed with the *LabelSOM* method as described in [9].

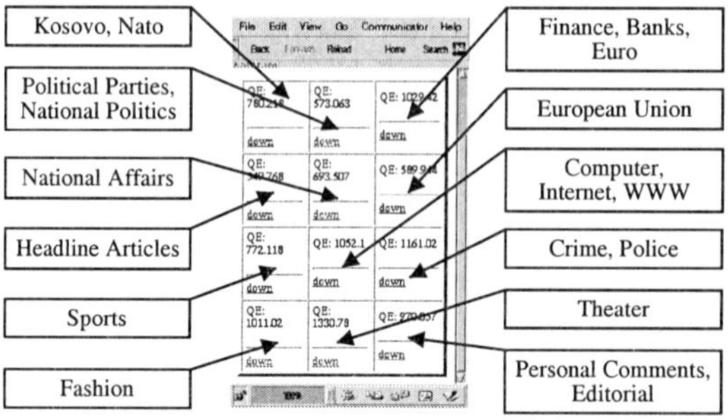

Fig. 2. First layer map

Fig. 3. Balkan War: Second-layer map

In the third-layer map shown in Figure 4 we find the actual articles on the negotiations concerning the role of Russia. In particular, these articles cover issues such as reports on the G-8 meetings following the Russian approach towards the airport of Pristina and reactions of European politicians on that topic.

Similar topic organization can be found in the other areas of the *GHSOM* hierarchy with, e.g., the section on the European Union being composed of subjects such as the relationship between the EU and the USA on imports of US beef or penal taxes, or Austria-specific EU topics such as transit regulations or the agriculture subsidies within the Agenda 2000 pact.

4 Conclusions

With the *Growing Hierarchical Self-Organizing Map*, a dynamically growing hierarchical extension to the *Self-Organizing Map*, we are able to provide a hierarchical organization of topics, where each topic is assigned the required map space according to its prominence in a document archive. The resulting organization allows for an even more intuitive navigation and orientation within an unknown document collection. It allows users to find out, which topics are present, and where to find specific documents, putting the emphasis on browsing as the means for interactive exploration.

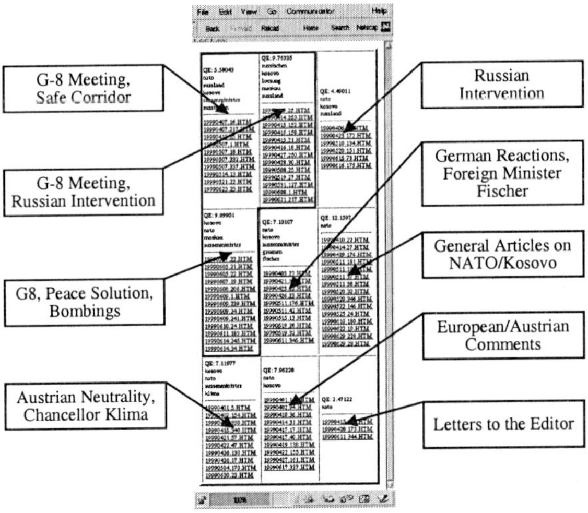

Fig. 4. Balkan War: Third-layer map

References

1. D. Alahakoon, S. K. Halgamuge, and B. Srinivasan. Dynamic self-organizing maps with controlled growth for knowledge discovery. *IEEE Trans Neural Networks*, 11(3), 2000.
2. J. Blackmore and R. Miikkulainen. Incremental grid growing: Encoding high-dimensional structure into a two-dimensional feature map. In *Proc Int'l Conf Neural Networks (ICANN'93)*, San Francisco, CA, 1993.
3. M. Dittenbach, D. Merkl, and A. Rauber. The Growing Hierarchical Self-Organizing Map. In *Proc Int'l Joint Conf Neural Networks (IJCNN00)*, Como, Italy, 2000.
4. B. Fritzke. Growing grid – a self-organizing network with constant neighborhood range and adaption strength. *Neural Processing Letters*, 2, No. 5:1 – 5, 1995.
5. T. Kohonen. Self-organized formation of topologically correct feature maps *Biological Cybernetics* (43), 1982.
6. T. Kohonen. *Self-Organizing Maps.* Springer Verlag, Berlin, Germany, 1995.
7. P. Koikkalainen and E. Oja. Self-organizing hierarchical feature maps. In *Proc Int'l Joint Conf Neural Networks*, San Diego, CA 1990.
8. P. Koikkalainen. Fast deterministic self-organizing maps. In *Proc Int'l Conf Neural Networks*, Paris, France, 1995.
9. D. Merkl and A. Rauber. Automatic Labeling of Self-Organizing Maps for Information Retrieval. In *Proc Int'l Conf Neural Information Processing (ICONIP'99)*, Perth, Australia, 1999.
10. R. Miikkulainen. Script recognition with hierarchical feature maps. *Connection Science*, 2:83 – 101, 1990.
11. A. Rauber and D. Merkl. The SOMLib Digital Library System. In *Proc. Europ. Conf. on Research and Advanced Technology for Digital Libraries (ECDL99)*, Paris, France, 1999. LNCS, Springer Verlag.
12. A. Rauber and D. Merkl. Creating an Order in Distributed Digital Libraries by Integrating Independent Self-Orgaizing Maps. In *Proc Int'l Conf Artificial Neural Networks (ICANN'99)*, Skövde, Sweden, 1999.

Part VI

Independent Component Analysis

Blind Source Separation of Single Components from Linear Mixtures

Roland Vollgraf, Ingo Schießl, and Klaus Obermayer

Department of Computer Science
Technical University of Berlin, Germany
{vro,ingos,oby}@cs.tu-berlin.de

Abstract. We present a new method, that is able to separate one or a few particular sources from a linear mixture, performing source separation and dimensionality reduction simultaneously. This is in particular useful in situations in which the number of observations is much larger than the number of underlaying sources, as it allows to drastically reduce the number of the parameters to estimate. It is well applicable for the long time series recorded in optical imaging experiments. Here one is basically interested in only one source containing the stimulus response. The algorithm is based on the technique of joint diagonalization of cross correlation matrices. To focus the convergence to the desired source, prior knowledge is incorporated. It can be derived, for instance, from the expected time course of the metabolic response in an optical imaging experiment. We demonstrate the capabilities of this algorithm on the basis of toy data coming from prototype signals of former optical recording experiments and with time courses that are similar to those obtained in optical recording experiments.

1 Introduction

Blind source separation (BSS) algorithms usually assume, that that a number of unknown sources $\mathbf{s}(\mathbf{r}) = (s_1(\mathbf{r}), \ldots, s_M(\mathbf{r}))^T$ is linearly mixed by an unknown mixing matrix $\mathbf{A}$. $\mathbf{r}$ is an index in the domain of the signals, i.e. a time index for temporal signals or, as in our examples, a spatial index for image signals. Additionally, sensor noise $\boldsymbol{\eta}$ may be included into the model.

$$\mathbf{x}(\mathbf{r}) = \mathbf{A}\mathbf{s}(\mathbf{r}) + \boldsymbol{\eta} \tag{1}$$

The task of the BSS is to find an appropriate demixing matrix $\mathbf{W}$ that reverts the mixture as good as possible under consideration of the noise, so that

$$\mathbf{s}(\mathbf{r}) \approx \hat{\mathbf{s}}(\mathbf{r}) = \mathbf{W}\mathbf{x}(\mathbf{r}) = \mathbf{W}\mathbf{A}\mathbf{s}(\mathbf{r}) + \mathbf{W}\boldsymbol{\eta} \ . \tag{2}$$

Many different separation algorithms have been developed and successfully tested on mixtures of sources fulfilling the assumption of independently distributed source amplitudes and moderate noise level [1, 2]. There are, however, situations in which these assumptions about the sources don't hold.

G. Dorffner, H. Bischof, and K. Hornik (Eds.): ICANN 2001, LNCS 2130, pp. 509–514, 2001.

In optical imaging, a cortical area of interest is illuminated with monochromatic light of wavelengths usually between 500 - 800 nm. This area is then recorded with a sensitive CCD- or video camera. Changes in reflectance of this light from the cortex are mainly due to variations in the light scattering properties of the tissue and to variations in the local concentrations of deoxigenated and oxygenated hemoglobin. Typically these changes are very small and do not exceed 0.1% of the reflected light [3]. Because of these small signal intensities the signal to noise ratios are of the order of (0 dB). A further problem of real world measurements is the basic assumption of the independence, because several source signals may be evoked by the same event.

We showed, that for this situation BSS algorithms, that are based on zero cross correlation functions, are superior to those that rely on factorizing source distributions. In [4] we applied a method called *Multi Shift Extended Spatial Decorrelation (ESD)* on optical imaging data. It is based on joint diagonalization of cross correlation matrices and relates to the *SOBI*-algorithm introduced in [5]. The assumption is that the sources have zeros cross correlation functions at the considered lags, i.e. the *shifted cross correlation matrices*

$$\mathbf{C_s}(\Delta\mathbf{r}) = \left\langle \mathbf{s}(\mathbf{r})\mathbf{s}^T(\mathbf{r}+\Delta\mathbf{r})\right\rangle_\mathbf{r} = \frac{1}{R}\sum_{\Delta\mathbf{r}} \mathbf{s}(\mathbf{r})\mathbf{s}^T(\mathbf{r}+\Delta\mathbf{r}) \tag{3}$$

are diagonal. $\langle\cdot\rangle_\mathbf{r}$ denotes the expectation or average over the signals domain. So the task is to find an appropriate demixing matrix $\mathbf{W}$ that diagonalizes the shifted cross correlation matrices $\mathbf{C}_{\hat{\mathbf{s}}}(\Delta\mathbf{r})$ for the recovered sources as good as possible. Note that these matrices transform in the following way

$$\mathbf{C}_{\hat{\mathbf{s}}}(\Delta\mathbf{r}) = \mathbf{W}\mathbf{C_x}(\Delta\mathbf{r})\mathbf{W}^T = \mathbf{W}\mathbf{A}\mathbf{C_s}(\Delta\mathbf{r})\mathbf{A}^T\mathbf{W}^T + \mathbf{W}\mathbf{C}_\eta(\Delta\mathbf{r})\mathbf{W}^T \ . \tag{4}$$

However, $\mathbf{C}_\eta(\Delta\mathbf{r})$ vanishes for white Gaussian noise and lags $\Delta\mathbf{r} \neq 0$, as it has zero cross correlation functions and a delta peaked auto correlation function. Thus, from now on we omit this term. More general reflections about correlation based source separation and the influence of noise we gave in [7].

Here we develop a method, that allows to separate one or a few components from the mixture only, which enables us to deal with situations, where the number of observations is much higher than the number of sources and where the estimation of the complete square demixing matrix with reasonable precision and effort is no longer possible. The recording of much more mixtures than sources one is interested in allows to decrease the influence of the sensor noise as it will be shown in the experiment.

We introduce regularization terms that allow us, to take an influence on which source is separated and improve the stability of the algorithm. For the evaluation of the separation quality we used a dataset of three prototype patterns from optical imaging recordings (Fig. 1), [8].

2 Cost Function for Extracting Single Components

In order to separate one component we have to find a N-dimensional vector $\mathbf{w}$, that yields zero for all components except one, when it is multiplied with the

Fig. 1. The toy datasets used for simulation. These are reconstructed sources from a former optical recording experiment.

unknown and rectangular $N \times M$ mixing matrix $\mathbf{A}$,

$$\mathbf{w}^T \mathbf{A} = (c, 0, \ldots, 0) \ . \tag{5}$$

Without loss of generality we constrain $\mathbf{w}$ to unit length and assume that the first component of $\mathbf{w}^T\mathbf{A}$ is different from zero, thus $\mathbf{w}$ separates the first source $\mathbf{w}^T\mathbf{x}(\mathbf{r}) = cs_1(\mathbf{r}) + \mathbf{w}^T\boldsymbol{\eta}$ from the mixture, while $\mathbf{A}$ is not known. We not only attempt to fulfill (5) but also want a large value c to get the variance of the recovered source domineering the noise.

Let us consider a matrix $\mathbf{M}$,

$$\mathbf{M} = \mathbf{C_s}(\Delta\mathbf{r})\mathbf{A}^T(\mathbf{a}, \mathbf{b}_1, \ldots, \mathbf{b}_{N-1}) \ , \tag{6}$$

$\mathbf{C_s}(\Delta\mathbf{r})$ is the shifted source cross correlation matrix, $\mathbf{a}$ is the normalized first[1] column vector of $\mathbf{A}$ and the $\mathbf{b}_i$ are $N-1$ orthogonal vectors of unit length that span the subspace orthogonal to $\mathbf{a}$. Then a multiplication from the right of (5) with $\mathbf{M}$ keeps its form unchanged, i.e. $\mathbf{v}^T = \mathbf{w}^T\mathbf{AM} = (\bar{c}, 0, \ldots, 0)$. Using $\mathbf{C_x}(\Delta\mathbf{r}) = \mathbf{AC_s}(\Delta\mathbf{r})\mathbf{A}^T$, we obtain the cost function, which is the squared sum of all elements of $\mathbf{v}$ except the first.

$$\begin{aligned}
E &= \sum_{\Delta\mathbf{r}} \sum_{i=2}^{N} \left(\mathbf{w}^T\mathbf{C_x}(\Delta\mathbf{r})\,(\mathbf{a}, \mathbf{b}_1, \ldots, \mathbf{b}_{N-1})\right)_i^2 \\
&= \sum_{\Delta\mathbf{r}} \Big(\left(\mathbf{w}^T\mathbf{C_x}(\Delta\mathbf{r})\,(\mathbf{a}, \mathbf{b}_1, \ldots, \mathbf{b}_{N-1})\right) \left(\mathbf{w}^T\mathbf{C_x}(\Delta\mathbf{r})\,(\mathbf{a}, \mathbf{b}_1, \ldots, \mathbf{b}_{N-1})\right)^T - \\
&\quad -\mathbf{w}^T\mathbf{C_x}(\Delta\mathbf{r})\mathbf{a}\mathbf{a}^T\mathbf{C_x}(\Delta\mathbf{r})^T\mathbf{w}\Big) \\
&= \sum_{\Delta\mathbf{r}} \mathbf{w}^T\mathbf{C_x}(\Delta\mathbf{r})\left(\mathbb{1} - \mathbf{a}\mathbf{a}^T\right)\mathbf{C_x}(\Delta\mathbf{r})^T\mathbf{w}
\end{aligned} \tag{7}$$

E has to be minimized with respect to $\mathbf{w}$ and $\mathbf{a}$ under the constraint $\|\mathbf{w}\| = 1$. In estimating a single source we so reduced the number of parameters from N^2 for a full demixing matrix $\mathbf{W}$ to $2N$ for the vectors $\mathbf{w}$ and $\mathbf{a}$.

3 Prior Knowledge for a and w

Successful minimization of (7) will lead to the separation of one source from the mixture, but we don't have control yet about which source is being extracted.

[1] without loss of generality

Therefore, some prior knowledge is needed, that allows to add regularization terms to the cost function (7),

$$E = E_{(7)} + E_{\mathbf{w}}(\mathbf{w}) + E_{\mathbf{a}}(\mathbf{a}) \; . \tag{8}$$

Although it is in general possible to use any functional form for the regularization term, that reflects the knowledge about $\mathbf{w}$ and $\mathbf{a}$ respectively, one is interested in using such, that are computationally convenient. Following the concept of Gaussian priors we therefore suggest to use a quadratic function

$$E_{\mathbf{a}}(\mathbf{a}) = \mathbf{a}^T \Sigma_{\mathbf{a}}^{-1} \mathbf{a} \; . \tag{9}$$

$\Sigma_{\mathbf{a}}$ is chosen to have its largest eigenvalue in direction of $\mu_{\mathbf{a}}$, the most probable value of $\mathbf{a}$. The other eigenvalues have to be small compared to the largest one. The smallest value of $E_{\mathbf{a}}$, given $\mathbf{a}$ is constrained to unit length, is obtained if $\mathbf{a}$ is in direction of $\mu_{\mathbf{a}}$.

$E_{\mathbf{w}}$ must be constructed in a different way, because we usually only have prior information about the mixing process and not its inverse. We know from (5), that $\mathbf{w}$ has to be orthogonal to all those column vectors of $\mathbf{A}$, that represent the mixing process of the unwanted components. If prior knowledge is available for any of them, a quadratic function $E_{\bar{\mathbf{a}}_i}(\mathbf{a}) = \mathbf{a}^T \Sigma_{\bar{\mathbf{a}}_i}^{-1} \mathbf{a}$ could be set up as for equation (9). Then, $E_{\mathbf{w}}$ must yield large values if $E_{\bar{\mathbf{a}}_i}$ is small and vice versa. Hence the quadratic form must be $E_{\mathbf{w}} = \mathbf{w}^T \Sigma_{\bar{\mathbf{a}}_i} \mathbf{w}$. For a couple of $\bar{\mathbf{a}}_i$ then the regularization term for $\mathbf{w}$ becomes

$$E_{\mathbf{w}}(\mathbf{w}) = \mathbf{w}^T \Sigma_{\mathbf{w}} \mathbf{w}, \qquad \text{with} \quad \Sigma_{\mathbf{w}} = \left(\sum_i \Sigma_{\bar{\mathbf{a}}_i} \right) \; . \tag{10}$$

This results in the regularized error function

$$E = \sum_{\Delta \mathbf{r}} \mathbf{w}^T \mathbf{C}_{\mathbf{x}}(\Delta \mathbf{r})(\mathbb{1} - \mathbf{a}\mathbf{a}^T)\mathbf{C}_{\mathbf{x}}(\Delta \mathbf{r})^T \mathbf{w} \; + \; \mathbf{a}^T \Sigma_{\mathbf{a}}^{-1} \mathbf{a} \; + \; \mathbf{w}^T \Sigma_{\mathbf{w}} \mathbf{w} \; . \tag{11}$$

This loss function has to be minimized with respect to $\mathbf{w}$ and $\mathbf{a}$. The additive terms perform a regularization according to the prior knowledge. Their strength directly depends on the determinant of $\Sigma_{\mathbf{a}}$ and $\Sigma_{\bar{\mathbf{a}}_i}$, which have to be set to reasonable values.

Note, that the cost function (11) is quadratic in $\mathbf{w}$ and in $\mathbf{a}$ (but not jointly in $\mathbf{w}$ and $\mathbf{a}$). Minimization can be robustly performed by conjugate gradient methods under the constraint $\|\mathbf{w}\| = \|\mathbf{a}\| = 1$.

4 Extraction of Additional Sources

The algorithm can be applied iteratively to separate additional sources. After one source is successfully extracted using $\hat{s}_i = \mathbf{w}^T \mathbf{x}$, it is also possible to remove that source from the remaining mixtures using the parameter $\mathbf{a}$. To do that, we construct a $(N-1 \times N)$-dimensional Matrix $\mathbf{W}' = (\mathbf{w}'_1, \ldots, \mathbf{w}'_{N-1})^T$. The $\mathbf{w}'_i$ are chosen to be orthogonal and span the subspace orthogonal to $\mathbf{a}$. For the separated source, $\mathbf{a}$ is the corresponding column vector of the mixing matrix $\mathbf{A}$. Thus $\mathbf{W}'\mathbf{A}$ has that column completely zero and hence $\mathbf{x}' = \mathbf{W}'\mathbf{A}\mathbf{s}$ is a mixture of $\mathbf{s}$, that does not contain the separated source. Then $\mathbf{x}'$ is used as input to separate the next source. It is important to note, that with every step the priors have to be projected into the subspace of $\mathbf{x}'$ using $\mathbf{W}'$.

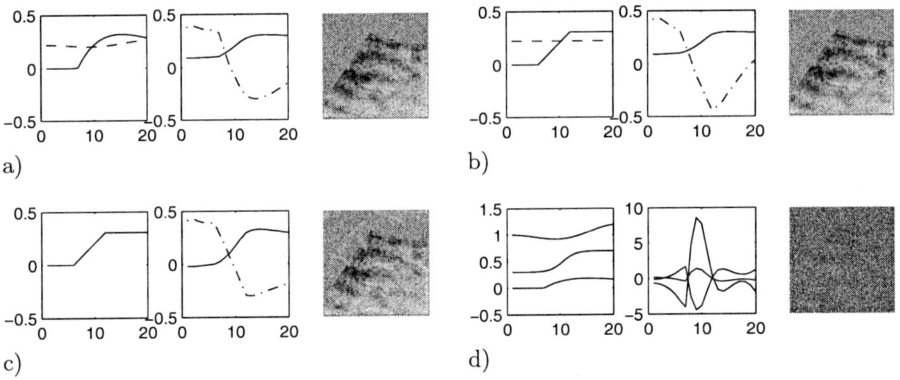

Fig. 2. Results on a 20×3 mixture of the sources of Fig. 1. Figures a-c) one separated source using different priors. left: prior for $\mathbf{a}$ (solid), to $\mathbf{w}$ orthogonal prior (dashed), middle: estimated course of $\mathbf{a}$ (solid) and $\mathbf{w}$ (dash-dotted), right: the separated source. Figure d) left: course of the column vectors of the mixing matrix $\mathbf{A}$, middle: course of the row vectors of the pseudo inverse of $\mathbf{A}$, right: the separated source using the first row of the pseudo inverse

5 Results and Discussion

To evaluate the properties of the algorithm in separating one source, we created 20 mixtures of the sources of Fig. 1, using a 20×3-mixing matrix $\mathbf{A}$. Each column represents the time course of a single source (Fig. 2d, left) and was motivated by results of optical imaging experiments [9]. Finally, white noise with approximately 3dB SNR was added to the mixture. In three experiments $\mathbf{w}$ and $\mathbf{a}$ where estimated using a set of 8 cross correlation matrices with shifts $\Delta\mathbf{r}$ arranged in a star like pattern. In Fig. 2a, the first column of $\mathbf{A}$ was used as the prior for $\mathbf{a}$ and the third column was used as the orthogonal prior for $\mathbf{w}$. The eigenvalues of the matrices $\Sigma_{\mathbf{a}}$ and $\Sigma_{\bar{\mathbf{a}}_i}$ where chosen as 1 for the largest one and 0.1 for the remaining. The separated source is displayed next to the estimated values of $\mathbf{w}$ and $\mathbf{a}$. The algorithm performs well, even if the shape of the priors only roughly corresponds to columns of $\mathbf{A}$ (Fig. 2b). If the orthogonal prior was omitted, the result is only slightly worse (Fig. 2c). This shows that the priors just select on which source the algorithm focuses. The particular shape of $\mathbf{w}$ is not much influenced by the priors.

It is important to note, that even exact knowledge of $\mathbf{a}$ is not sufficient to recover the source, since no $\mathbf{w}$ can be derived from $\mathbf{a}$ alone. Only the knowledge about mixing process of all sources, i.e. the whole matrix $\mathbf{A}$ would allow to compute it's pseudo inverse. However, source reconstruction using the first row of the pseudo inverse of $\mathbf{A}$ fails completely (Fig. 2d). Although the pseudo inverse would perform a perfect separation, it dramatically amplifies the noise and yields a poor reconstruction. The single source separating algorithm, however, is able to exploit information from the higher number of mixtures to achieve a much better SNR in the separated source.

Acknowledgments

This work was supported by DFG OB102/3-1 and Welcome Trust 050080/Z/97

References

[1] Anthony J. Bell and Terrence J. Sejnowski, "An information-maximization approach to blind separation and blind deconvolution," *Neural Computation*, vol. 7, no. 6, pp. 1129–1159, 1995.

[2] A. Hyvärinen and E. Oja, "A fast fixed point algorithm for independent component analysis.," *Neural Comput.*, vol. 9, pp. 1483–1492, 1997.

[3] G. G. Blasdel and G. Salama, "Voltage-sensitive dyes reveal a modular organization in monkey striate cortex.," *Nature*, vol. 321, pp. 579–585, 1986.

[4] H. Schöner, M. Stetter, I. Schießl, J. Mayhew, J. Lund, N. McLoughlin, and K. Obermayer, "Application of blind separation of sources to optical recording of brain activity," in *Advances in Neural Information Processing Systems NIPS 12*, S.A. Solla, T.K. Leen, and K.-R. Müller, Eds. 2000, pp. 949–955, MIT Press, In press.

[5] A. Belouchrani, K. Abed-Meraim, J. Cardoso, and E. Moulines, "A blind source separation technique using second-order statistics," *IEEE transactions on Signal Processing*, , no. 45, pp. 434–444, 1997.

[6] L. Molgedey and H. G. Schuster, "Separation of a mixture of independent signals using time delayed correlations," *Phys. Rev. Lett.*, vol. 72, pp. 3634–3637, 1994.

[7] R. Vollgraf, M. Stetter, and K. Obermayer, "Convolutive decorrelation procedures for blind source separation," in *Proceedings of the 2. International Workshop on ICA, Helsinki*, P. Pajunen and J. Karhunen, Eds., 2000, pp. 515–520.

[8] I. Schießl, H. Schöner, M. Stetter, and K. Obermayer, "Regularized second order source separation," in *Proceedings of the 2. International Workshop on ICA, Helsinki*, P. Pajunen and J. Karhunen, Eds., 2000, pp. 111–116.

[9] D. Malonek and A. Grinvald, "Interactions between electrical activity and cortical microcirculation revealed by imaging spectroscopy: implications for functional brain mapping.," *Science*, vol. 272, pp. 551–554, 1996.

Blind Source Separation Using Principal Component Neural Networks

Konstantinos I. Diamantaras

Depatrment of Informatics
Technological Education Institute of Thessaloniki
54101, Sindos, Thessaloniki, Greece
kdiamant@it.teithe.gr
http://www.it.teithe.gr/~kdiamant

Abstract. Blind source separation (BSS) is approached from the second order statistics point of view. In particular, it is shown that temporal filtering by an arbitrary filter combined with PCA leads to the solution of the problem provided that the sources are colored and have different spectra. This result is demonstrated by applying a neural PCA model such as APEX to BSS problems with artificially created, randomly colored data.

Keywords: Blind source separation, principal component analysis, independent component analysis

1 Introduction

Blind Source Separation (BSS) is the task of recovering n hidden source signals from their linear mixtures observed at m sensors without knowledge of the underlying mixing operator. When the linear mixing operator is memoryless we talk about the instantaneous BSS problem which relates to many important applications including digital communications, signal processing, and medical science. In fact BSS is a part of a large family of Blind Problems including blind deconvolution, blind system identification, blind channel equalization, etc.

Starting from the early paper of Jutten and Herault [10] where two signals were separated using a nonlinear feedback neural model, there has been a wide range of neural and non-neural approaches for BSS. Most existing algorithms for BSS are based on higher-order statistical properties of the sources. These algorithms are classified into three major categories (a) nonlinear principal component analysis (b) independent component analysis (ICA) and (c) entropy maximization. In the first approach [11, 9], nonlinear PCA is seen as an extension of PCA capable of analyzing the observable mixtures not in orthogonal components (as in PCA) but in independent components. Nonlinear PCA is typically implemented using nonlinear neural models and learning algorithms which are often extensions of their linear PCA counterparts. Nonlinear PCA is directly related with the second approach: independent component analysis [5]. ICA defines criteria for the statistical independence between the sources. Provided that the sources are independent a suitable maximization algorithm based

G. Dorffner, H. Bischof, and K. Hornik (Eds.): ICANN 2001, LNCS 2130, pp. 515–520, 2001.

on these criteria leads to the desired separation. Various criteria, including the equivariance [4] have been proposed. Finally, in the third approach, the entropy (a measure of the information content of the output) is maximized [2, 16]. The statistical efficiency and dynamic stability of these algorithms is studied in [1].

The use of second order statistics (SOS) for blind problems was first noted by Tong, e.a. [14] in the early '90's. The problem studied in this classic paper was blind equalization. Although blind equalization is related to BSS it was not until 1997 that second order methods were proposed for BSS by Belouchrani e.a. [3]. We must note that second order methods do not actually replace higher order ones since each approach is based on different assumptions. For example, second order methods assume that the sources be temporally colored whereas higher order methods assume white sources. Another difference is that higher order methods do not apply on Gaussian signals but second order methods do not have any such constraint.

An adaptive second order Hebbian-like approach for BSS has been proposed in [7, 8]. This approach is based on neural SVD analyzers known as Asymmetric PCA models. One such model, the Cross-Coupled Hebbian rule proposed in [6], has been applied for the blind extraction of sources such as images or speech signals using a pre-selected time lag l.

In this paper we demonstrate that, in fact, standard (symmetric) PCA can be used instead of Asymmetric PCA provided that the observed data are temporally prefiltered. Almost any filter is sufficient except for the trivial impulse or delayed impulse function. Since PCA is a well studied problem there are a number of fast adaptive/neural algorithms including GHA [13], APEX [12], the PASTd algorithm [15], etc.

2 PCA Solution

Consider n signals x_1, ..., x_n, resulting from the linear combination of an equal number of source signals s_1, ..., s_n,

$$\begin{bmatrix} x_1(k) \\ \vdots \\ x_n(k) \end{bmatrix} = \begin{bmatrix} h_{11} & \cdots & h_{1n} \\ \vdots & & \vdots \\ h_{n1} & \cdots & h_{nn} \end{bmatrix} \begin{bmatrix} s_1(k) \\ \vdots \\ s_n(k) \end{bmatrix} \tag{1}$$

$$\mathbf{x}(k) = \mathbf{H}\mathbf{s}(k) \tag{2}$$

The sources s_i contain information that needs to be extracted. The only available data are the signals x_i which are observed at the outputs of some receiving devices, e.g. microphones, antennas, etc. Since the mixing parameters h_{ij} are unknown the solution to the BSS problem posed above is not unique. This is so because we can multiply source s_i by any non-zero scaling factor α and at the same time divide the i-th column of $\mathbf{H}$ by α without affecting the observation sequence $\mathbf{x}$. Furthermore, we can change the order of the sources and similarly permute the columns of $\mathbf{H}$ and still obtain the same observation sequence $\mathbf{x}$. Thus the scale and the order of the sources is unobservable.

Our assumptions are as follows:

A1. The observation vector sequence $\mathbf{x}(k)$ is spatially white (sphered)

$$\mathbf{R}_x(0) = E\{\mathbf{x}(k)\mathbf{x}(k)^T\} = \mathbf{I}. \tag{3}$$

Any vector sequence $\mathbf{x}(k)$ can be sphered by a transformation of the form $\mathbf{x}(k) \leftarrow \mathbf{R}_x(0)^{-1/2}\mathbf{x}(k)$.

A2. The sources are unit variance, pairwise uncorrelated, and each one is temporally colored

$$E\{s_i(k)s_j(k)\} = \begin{cases} 0 & \text{if } i \neq j \\ r_i(0) = 1 & \text{if } i = j \end{cases} \tag{4}$$

$$E\{s_i(k)s_j(k-1)\} = \begin{cases} 0 & \text{if } i \neq j \\ r_i(1) \neq 0 & \text{if } i = j \end{cases} \tag{5}$$

$$\mathbf{R}_s(0) = E\{\mathbf{s}(k)\mathbf{s}(k)^T\} = \mathbf{I} \tag{6}$$

$$\mathbf{R}_s(1) \triangleq E\{\mathbf{s}(k)\mathbf{s}(k-1)^T\} = \text{diag}[r_1(1), \cdots r_n(1)] \neq 0 \tag{7}$$

The assumption that the sources are unit variance removes the part of the scaling ambiguity related with the magnitude of the scaling. There remains the sign ambiguity: if we change both the sign of s_i and the sign of the i-th column of $\mathbf{H}$ then neither the observed sequence $\mathbf{x}$ nor the variance of s_i will change. It follows that if $\hat{\mathbf{H}}$ is a solution to the BSS problem, then so is $\hat{\mathbf{H}}\mathbf{P}$, for any a signed permutation matrix $\mathbf{P}$.

Consider now the PCA of the signal

$$\mathbf{y}(k) = \mathbf{x}(k) + \mathbf{x}(k-1)$$

The covariance matrix of $\mathbf{y}$ is easily computed to be

$$\begin{aligned} \mathbf{R}_y(0) &= E\{[\mathbf{x}(k) + \mathbf{x}(k-1)][\mathbf{x}(k) + \mathbf{x}(k-1)]^T\} \\ &= 2\mathbf{R}_x(0) + \mathbf{R}_x(1) + \mathbf{R}_x(1)^T \\ &= 2\mathbf{I} + \mathbf{R}_x(1) + \mathbf{R}_x(1)^T \end{aligned}$$

where we define the lagged covariance

$$\mathbf{R}_x(1) = E\{\mathbf{x}(k)\mathbf{x}(k-1)^T\} = \mathbf{H}\,\mathbf{R}_s(1)\,\mathbf{H}^T \tag{8}$$

So

$$\mathbf{R}_y(0) = \mathbf{H}\mathbf{D}\mathbf{H}^T \tag{9}$$

Using assumption A2 the matrix

$$\mathbf{D} = 2\mathbf{I} + \mathbf{R}_s(1) + \mathbf{R}_s(1)^T \tag{10}$$

is diagonal. In addition, $\mathbf{H}$ is orthogonal since $\mathbf{H}\mathbf{H}^T = \mathbf{H}\mathbf{R}_s(0)\mathbf{H}^T = \mathbf{R}_x(0) = \mathbf{I}$. Hence, Equation (9) represents the eigenvalue decomposition of $\mathbf{R}_y(0)$. Moreover, the eigenvalues are given by the formula

$$d_i = 2 + 2E\{s_i(k)s_i(k-1)\} = 2(1 + r_i(1))$$

If d_1, ..., d_n, are distinct then the eigen-decomposition is unique upto a permutation and sign of the eigenvectors.

The consequences of this observation are twofold. First, it is obvious that the Principal Component Analysis of $\mathbf{y}$ leads to an estimate of $\mathbf{H}$. If the columns of the matrix $\mathbf{U}$ are the n principal eigenvectors of $\mathbf{R}_y(0)$ then

$$\mathbf{U} = \mathbf{HP}$$

where $\mathbf{P}$ is a signed permutation matrix. We can take this further and extract the source signals with arbitrary order and sign

$$\hat{\mathbf{s}}(k) = \mathbf{U}^T\mathbf{x}(k) = \mathbf{P}^T\mathbf{s}(k)$$

$$\hat{s}_i(k) = \pm s_{\pi(i)}(k)$$

Since the order and sign of the source signals are unobservable this is the best estimate we can hope for.

Secondly, our analysis holds for any temporally filtered version of the observed signal $\mathbf{y}(k) = \sum_{l=0}^{L} \alpha_l \mathbf{x}(k-l)$. In that case we would have $\mathbf{R}_y(0) = \mathbf{H}\sum_{p=-L}^{L} \beta_p \mathbf{R}_s(p)\mathbf{H}^T$ where β_p is the convolution $a_l \star a_l$. However, the key equation (9) would still be valid with $d_i = \sum_{p=-L}^{L} \beta_p r_i(p)$ while the subsequent analysis would remain the same. Figure 1 shows the block diagram of the overall BSS system using PCA.

3 PCA Neural Networks for BSS

An adaptive solution of the BSS problem can be implemented using neural PCA models that can extract a complete set of principal eigenvectors. Such models include GHA, APEX, etc. Figure 2 shows the results of a blind separation experiment using the APEX model. The data were artificially created in two steps. First, we randomly generated $N = 2000$ samples of four Pulse Amplitude Modulated (PAM) signals with 5 amplitude levels. Then the signals were colored using the following, randomly generated, FIR filters

$$g_1 = [0.8559,\ -0.8510,\ 0.8119,\ 0.7002,\ 0.7599,\ -1.7129,\ 1.5370,\ -1.6098,\ 1.1095,\ -1.1097]$$
$$g_2 = [0.3855,\ 0.9652,\ 0.8183,\ 0.0370,\ -0.9260,\ -0.1119,\ -0.8030,\ -1.6650,\ -0.9014,\ 0.5883]$$
$$g_3 = [0.5542,\ -0.4152,\ 0.0618,\ 0.4574,\ 0.1990,\ 0.2576,\ 2.0807,\ -2.2772,\ 0.3390,\ 0.2899]$$
$$g_4 = [0.6623,\ -0.5809,\ 0.8878,\ 0.1719,\ 0.8488,\ 0.9638,\ 1.3219,\ -0.0643,\ 1.3171,\ 0.2280]$$

thus creating the sources s_1, s_2, s_3, s_4. The mixing matrix

$$\mathbf{A} = \begin{bmatrix} -0.7283 & -0.5760 & 0.8701 & 0.8784 \\ -0.8719 & -0.6425 & 0.7825 & 0.9195 \\ -0.6235 & 0.8974 & -0.5981 & 0.5525 \\ -0.7057 & 0.6310 & -0.5692 & -0.6003 \end{bmatrix}$$

was used to derive the four observed signals x_1, x_2, x_3, x_4 The vector sequence $\mathbf{x}(k)$ was spatially whitened $\mathbf{x}'(k) = \mathbf{R}_x(0)^{-1/2}\mathbf{x}(k) = \mathbf{A}'\mathbf{s}(k)$ and then the

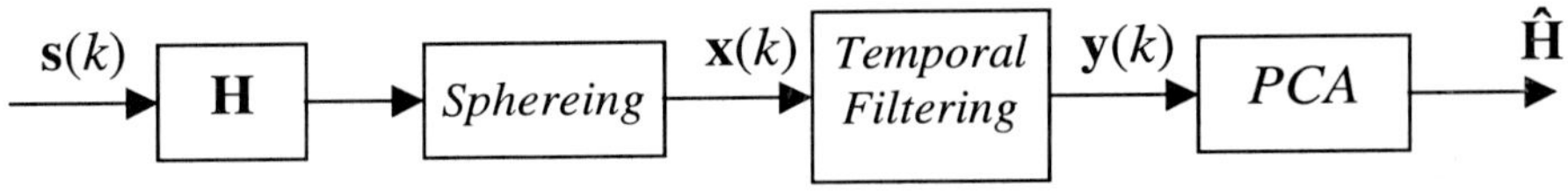

Fig. 1. A model for Blind Source Separation using PCA.

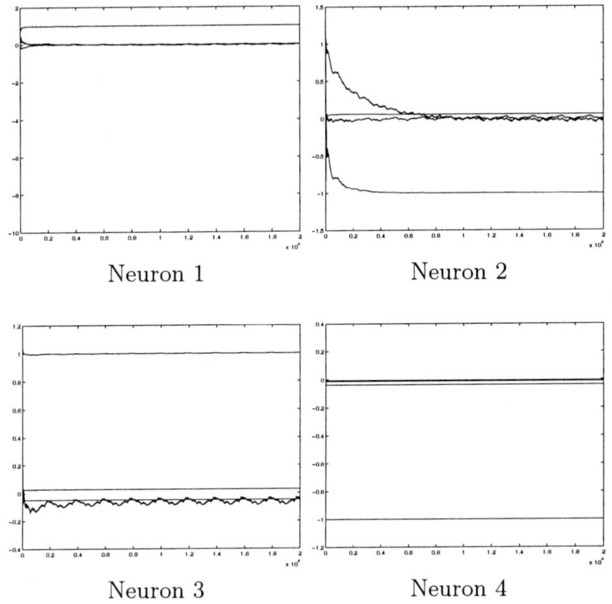

Fig. 2. Blind separation of four sources using 10 sweeps of the APEX neural PCA model. The data length is $N = 2000$) samples per source. The plots depict the inner product between the weights of each neuron with the whitened mixing matrix.

sources were blindly separated using a neural PCA model. In particular, we ran 10 sweeps of the sequential APEX model [12] with input $\mathbf{y}(k) = \mathbf{x}'(k) + \mathbf{x}'(k-1)$. The steady-state forward weight matrix $\mathbf{W}$ attains the following unmixing performance

$$\mathbf{WA}' = \begin{bmatrix} 0.0206 & 0.0169 & 1.0004 & 0.0046 \\ -1.0001 & -0.0400 & 0.0530 & -0.0152 \\ -0.0481 & 0.9997 & 0.0257 & -0.0466 \\ -0.0126 & -0.0378 & -0.0054 & -0.9996 \end{bmatrix}$$

Figure 2 shows the convergence behavior of each neuron in this particular example.

References

1. Amari S., Chen T.P., and Cichocki A.: Stability analysis of adaptive blind source separation. Neural Networks, **10** (1997) 1345–1351.

2. Bell, A.J. and Sejnowski, T.J.: An information maximization approach to blind separation and blind deconvolution. Neural Computation, **7** (1995) 1129–1159.
3. Belouchrani, A., Abed-Meraim, K., Cardoso, J.-F., and Moulines, E.: A Blind Source Separation Technique Using Second-Order Statistics. IEEE Trans. Signal Processing, **45**(2) (1997) 434–444.
4. Cardoso, J.F., Laheld, B.: Equivariant adaptive source separation. IEEE Trans. Signal Processing, **44** (1996) 3017–3030.
5. Comon, P.: Independent component analysis, a new concept?. Signal Processing, **36** (1994) 287–314.
6. Diamantaras, K.I. and Kung, S.Y.: Cross-correlation Neural Network Models. IEEE Transactions on Signal Processing, **42**(11) (1994) 3218-3223.
7. Diamantaras, K.I.: Asymmetric PCA Neural Networks for Adaptive Blind Source Separation. In Proc. IEEE Workshop on Neural Networks for Signal Processing (NNSP'98), Cambridge, UK, (1998) 103–112.
8. Diamantaras, K.I.: Second Order Hebbian Neural Networks and Blind Source Separation. In Proc. EUSIPCO'98 (European Signal Processing Conference), Rhodes, Greece (1998) 1317–1320.
9. Hyvarinen, A. and Oja, E.: Independent component analysis by general nonlinear Hebbian-like learning rules. Signal Processing, **64** (1998) 301–313.
10. Jutten, C. and Herault, J.: Blind Separation of Sources, Part I: An Adaptive Algorithm Based on Neuromimetic Architecture. Signal Processing, **24**(1) (1991) 1–10.
11. Karhunen, J. and Joutsensalo, J.: Representation and separation of signals using nonlinear PCA type learning. Neural Networks, **7** (1994) 113–127.
12. Kung, S.Y, Diamantaras, K.I., and Taur, J.S.: Adaptive Principal Component Extraction (APEX) and Applications IEEE Transactions on Signal Processing, **42**(5) (1994) 1202–1217.
13. Sanger, T.D.: Optimal Unsupervised Learning in a Single-Layer Linear Feedforward Neural Network. Neural Networks, **2**(6) (1989) 459–473.
14. Tong, L., Xu, G., and Kailath, T.: Blind Identification and Equalization Based on Second-Order Statistics: A Time Domain Approach. IEEE Trans. Information Theory, **40**(2) (1994) 340–349.
15. Yang, B.: Projection Approximation Subspace Tracking. IEEE Trans. Signal Processing, **43**(1) (1995) 95–107.
16. Yang, H.H., and Amari, S.: Adaptive on-line learning algorithms for blind separation–Maximum entropy and minimum mutual information. Neural Computation, **9** (1997) 1457–1482.

Blind Separation of Sources by Differentiating the Output Cumulants and Using Newton's Method

Rubén Martín-Clemente[1], José I. Acha[1], and Carlos G. Puntonet[2]

[1] Área de Teoría de la Señal y Comunicaciones, Universidad de Sevilla, Avda. de los Descubrimientos s/n., 41092-Sevilla, Spain
ruben@cica.es, acha@viento.us.es

[2] Departamento de Arquitectura y Tecnología de Computadores, Universidad de Granada, E-18071 Granada, Spain
carlos@atc.ugr.es

Abstract. This paper addresses the Blind Separation of non-Gaussian Sources for instantaneous mixtures. It is proven that the estimation of the separating system can be based on the cancellation of some second partial derivatives of the output cross-cumulants. These equations can be seen as a subset of those solved by JADE. Finally, we present an algorithm which is based on Newton's method and exhibits a fast rate of convergence.

1 Introduction

Blind Source Separation (BSS) is the problem of recovering a set of unknown source signals from the observation of another set of linear mixtures of the sources. The 'blind' qualification refers to the coefficients of the mixture: no information is assumed to be available about them. Since the inverse system has to be learned with the sole knowledge of the observed mixtures, this problem can be termed "unsupervised" or "self-learning". It has many applications in diverse fields, like communications, speech separation (the 'cocktail party problem'), processing of arrays of radar or sonar signals, in biomedical signal processing (e.g., EEG and MEG analysis), etc.

The key property that makes separation possible is mutual independence of the sources. In the past ten years, numerous solutions to this problem have been proposed, starting from the seminal work of Jutten and Herault [8]. In the Independent Component Analysis (ICA) approach, which is closely related to BSS, the goal is to linearly transform the data such that the transformed variables are as statistically independent from each other as possible [1, 2, 7]. Other approaches are concerned with batch (off-line) algorithms based on higher-order statistics [4]. The properties of the vectorial spaces of sources and mixtures also lead to geometrical methods for BSS [11, 12]. For further reading, see the review paper [3] and the references therein.

In this paper, we present a set of necessary and sufficient conditions for BSS which lead to a new dependence measure. Then, an algorithm which is based on Newton's method is proposed to minimize it.

G. Dorffner, H. Bischof, and K. Hornik (Eds.): ICANN 2001, LNCS 2130, pp. 521–526, 2001.

2 Problem Statement

The simplest BSS model is represented by the equation:

$$\mathbf{x}(t) = A\mathbf{s}(t) \tag{1}$$

where $\mathbf{s}(t) = [s_1(t),...,s_n(t)]^T$ is a vector of n source signals, $A = (a_{ij})$ is an unknown $n \times n$ invertible *mixing matrix* and vector $\mathbf{x}(t)$ collects the observed signals, being the only data available. The aim of BSS is to determine a $n \times n$ *separating matrix* $B = (b_{ij})$ such that:

$$\mathbf{y}(t) = B\mathbf{x}(t) = G\mathbf{s}(t) \tag{2}$$

is an estimate of the source signals, *i.e.*, the *global matrix* $G = (g_{ij})$ has one and only one non-zero coefficient per row and column, where $G = A \cdot B$. In this case, the term *'generalized permutation matrix'* is used to denote matrix G. The only assumptions are that the sources must be statistically independent, stationary, unit-variance and zero-mean. In addition, *at most* one source is gaussian-distributed.

3 Cumulant-Based criterion

By using the multi-linearity property of the cumulants, the output cross-cumulant $cum_{31}(y_i(t), y_j(t)) = cum(y_i(t), y_i(t), y_i(t), y_j(t))$ can be expanded as follows:

$$cum_{31}(y_i(t), y_j(t)) = \sum_{l=1}^{n} g_{il}^3\, g_{jl}\, \kappa_l \tag{3}$$

where κ_l stands for the kurtosis (fourth-order cumulant) of the l-th source. By setting (3) to zero for all $i \neq j$, we obtain a set of equations with respect to the coefficients of the matrix B. These equations are satisfied when G is a *generalized permutation matrix*. However, spurious roots are also possible depending on the source statistics.

Nevertheless, that (3) should vanish when G is a *generalized permutation matrix* is due to the fact that all the terms '$g_{il}^3\, g_{jl}\, \kappa_l$' that appear in (3) take the value of zero but not to the fact that the coefficients 'g_{il}' are raised to the power of three. On the other hand, cumulant (3) can be simplified by differentiation, without losing any fundamental property. Thus, let us introduce the following second-order derivatives of (3):

$$\Delta_{ij} \stackrel{\Delta}{=} \frac{1}{6}\frac{\partial^2 c_{31}}{\partial b_{ij}^2} = \sum_{l=1}^{n} g_{il}\, g_{jl}\, a_{jl}^2\, \kappa_l, \quad (i \neq j) \tag{4}$$

Then, by using $cum(x_l,x_m,x_j,x_j) = \sum_p a_{lp}a_{mp}a_{jp}^2\kappa_p$ (which is a consequence of the multi-linearity property), it is possible to express (4) as:

$$\Delta_{ij} = \sum_{l=1}^{n}\sum_{m=1}^{n} b_{il}b_{jm}cum(x_l, x_m, x_j, x_j), \tag{5}$$

We maintain that the set of equations $\Delta_{ij} = 0$ for all $i \neq j$ provide us with a set of *necessary* and *sufficient* conditions to assure the source separation, i.e., spurious roots are not possible, provided that at most one source is gaussian (e.g. for mixtures of two sources, a direct solution can be obtained in agreement with previous results [9]).

Proof. Trivial non-separating solutions, in which any of the output signals vanishes, can occur when the separating matrix is not full row rank. These solutions can be avoided by assuming that the signals in $\mathbf{x}(t)$ are uncorrelated and have unit variance since, in this case, A, B and G must necesarily be orthogonal matrices. This assumption can be easily accomplished by means of the so-called *whitening* or *sphering* of the original mixtures, in which uncorrelated linear combinations of the observed signals are obtained [4].

Then, observe that (4) can be written compactly in a matrix form:

$$\Delta_{ij} = \mathbf{g}_i^T \Gamma_j \mathbf{g}_j \tag{6}$$

where $\mathbf{g}_k^T$ is the k-th row of G and Γ_j is the diagonal matrix whose (l,l)-entry is equal to $a_{jl}^2 \kappa_l$. If at most one source is gaussian (i.e., its kurtosis equals zero), we can also assume without loss of generality that each diagonal element of Γ_j is different (i.e. Γ_j has no repeated eigenvalues) since A (and thus Γ_j) can be properly adjusted by multipliying the observations $\mathbf{x}(t)$ by any orthogonal matrix.

Since both A and B are orthogonal matrices, it follows that $\mathbf{g}_i^T \mathbf{g}_j = \delta_{ij}$, where δ_{ij} stands for the Kronecker delta. Now, we prove our main result:

(*Necessity*). If B is a separating matrix, then each vector $\mathbf{g}_i$ is a different canonical vector. Therefore, it follows from (6) that $\Delta_{ij} = 0$ for all i, j $(i \neq j)$.

(*Sufficiency*). Let us assume that $\Delta_{ij} = 0$ for all $i \neq j$. The vector $\Gamma_j \mathbf{g}_j$ can be expressed as a linear combination of the vectors $\mathbf{g}_i$ since the set $\{\mathbf{g}_1, \mathbf{g}_2, \ldots, \mathbf{g}_n\}$ forms an orthonormal basis of the space. Therefore:

$$\Gamma_j \mathbf{g}_j = \Delta_{jj} \mathbf{g}_j + \sum_{i \neq j} \Delta_{ij} \mathbf{g}_i \tag{7}$$

where $\Delta_{ij} = \mathbf{g}_i^T \Gamma_j \mathbf{g}_j$ as was defined before and $\Delta_{jj} = \mathbf{g}_j^T \Gamma_j \mathbf{g}_j$. Since, $\Delta_{ij} = 0$ for all $i \neq j$, we obtain that $\Gamma_j \mathbf{g}_j = \Delta_{jj} \mathbf{g}_j$, which implies that each $\mathbf{g}_i$ is an eigenvector of a diagonal matrix, i.e., a canonical vector. As a consequence, B is a separating matrix.

4 Solving the Equations

From above, we propose to minimize the following cost function:

$$f = \sum_{i \neq j} \Delta_{ij}^2, \tag{8}$$

under constraint

$$B^T B = I \tag{9}$$

where $\mathbf{x}(t)$ must satisfy $E\{\mathbf{x}^T(t)\mathbf{x}(t)\} = I$ (see above). Different algorithms can be chosen to minimize f. Specifically, consider finding a minimum by using Newton's method to search for a zero of the gradient of the function [6]. For the sake of simplicity, let $\mathbf{z} = [\Delta_{12}, \Delta_{13}, \ldots, \Delta_{N,N-1}]^T$ be a vector which collects all the functions Δ_{ij} $(i \neq j)$ given in (5) and $\mathbf{b} = [b_{11}, b_{12}, \ldots, b_{NN}]^T$ be a 'stacking' of the rows of B. It is straightforward to check that the gradient of (8) is precisely given by

$$\nabla_{\mathbf{b}} f = 2\, J^T\, \mathbf{z} \tag{10}$$

where J is the Jacobian matrix, *i.e.* $J_{ij} \equiv \frac{\partial z_i}{\partial b_j}$, where z_i and b_j are the i-th and j-th components of $\mathbf{z}$ and $\mathbf{b}$, respectively. In Newton's method, the Hessian matrix of f is required to compute the iteration points. Nevertheless, we do not use the actual Hessian matrix, but instead use a current approximation of it. Specifically, in [5], it is shown that near the solution the Hessian matrix H can be approximated by $H \approx 2\, J^T J$. Then, it can be shown that Newton's method leads to the following update rule:

$$\tilde{\mathbf{b}}_i = \mathbf{b}_i + \mu\; \Delta\tilde{\mathbf{b}}_i, \quad \mu \in (0, 2) \tag{11}$$

where

$$(2 J_i^T\, J_i)\, \Delta\tilde{\mathbf{b}}_i = -2 J_i^T\, \mathbf{z}_i \tag{12}$$

and the subscript i recalls that both J and $\mathbf{z}$ must also be updated. Observe that

$$\nabla_{\mathbf{b}}^T f \Delta\tilde{\mathbf{b}}_i = -2\mathbf{z}_i^T\, J_i\, (J_i^T J_i)^{-1} J_i^T \mathbf{z}_i < 0 \tag{13}$$

since $J_i\, (J_i^T J_i)^{-1} J_i^T$ is a positive definite matrix. Hence, $\Delta\tilde{\mathbf{b}}_i$ is a 'truly' descent direction of f. Finally, the new point is projected onto the constraint set of orthogonal matrices [10] to get:

$$B_i = \tilde{B}_i (\tilde{B}_i^T \tilde{B}_i)^{-1/2}, \tag{14}$$

where $\tilde{B}_i$ is the matrix obtained by 'unstacking' vector $\tilde{\mathbf{b}}_i$ and $(\tilde{B}_i^T \tilde{B}_i)^{-1/2}$ is obtained from the eigenvalue decomposition of $\tilde{B}_i^T \tilde{B}_i = U\, V\, U^T$ as $(\tilde{B}_i^T \tilde{B}_i)^{-1/2} = U\, V^{-1/2}\, U^T$ [7].

5 Links

First, note that (5) can be written in a matrix form as:

$$\Delta_{ij} = \mathbf{b}_i^T\, C_j\, \mathbf{b}_j \tag{15}$$

where $\mathbf{b}_k = [b_{k1}, \ldots, b_{kN}]^T$ and C_j is the $N \times N$ matrix whose (n, m)-entry equals $cum(x_n, x_m, x_j, x_j)$. Hence, the same argument that led to eqn. (7) now gives the identity:

$$C_j\, \mathbf{b}_j = \Delta_{jj} \mathbf{b}_j + \sum_{i \neq j} \Delta_{ij} \mathbf{b}_i = \Delta_{jj} \mathbf{b}_j \tag{16}$$

where $\Delta_{jj} = \mathbf{b}_j^T C_j \mathbf{b}_j$, which means that $\mathbf{b}_j$ is also an *eigenvector* of C_j. Interestingly, following a different approach, Cardoso and Souloumiac proposed the minimization of the sum of squares of non-diagonal entries of matrices $B^T C_j B$ in JADE [4]. Clearly, (8) comes from a *sufficient* subset of those equations which are solved by JADE. Consequently: *i.)* a significant computational cost saving is expected in our approach and *ii.)* since JADE satisfies more statistical equations, it may also be numerically more robust. The experiments provide support for both intuitions.

6 Simulations

The validity of the procedure has been corroborated by many simulations. Figure 1 shows the mean *Signal to Noise Ratio* of the estimated sources averaged over 20 independent experiments for a mixture of 6 sources (three had a uniform distribution -negative kurtosis- and the other three were generated as the cube of gaussian variables -positive kurtosis-), where the cumulants were estimated over 10000 samples. The Figure illustrates the rate of convergence of the algorithm. It is noteworthy that a small number of iterations seems to be enough to separate the sources.

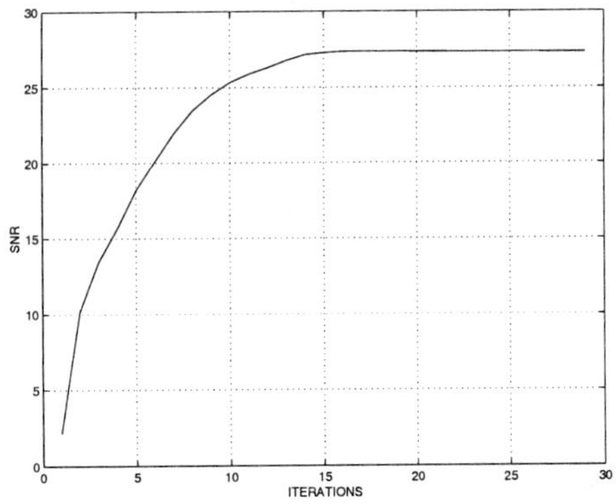

Fig. 1. Signal to Noise Ratio of the estimated sources vs. iterations.

7 Conclusions

In this Communication, we present a set of conditions for BSS which exhibits two main properties: (1) absence of spurious solutions and (2) Simplicity: Δ_{ij} is a quadratic function of the coefficients of B whereas $cum_{31}(y_i, y_j)$ depends on fourth-order powers of these coefficients. It was shown that they are a sufficient subset of those conditions presented in JADE. The proposed algorithm exhibits a fast convergence, since it is a quasi-Newton method. Finally, notice that the simplification of cumulant-based criterions for BSS by differentiating the cumulants seems to be a general procedure. We are currently studying its application to the separation of convolutive mixtures.

References

1. S.Amari, A. Cichocki and H.H. Yang "A new Learning Algorithm for Blind Source Separation", in *Advances in Neural Information Processing 8*, Cambridge, MA:MIT Press, 757-763, 1996.
2. A.J.Bell and T. Sejnowski "An Information-Maximization Approach to blind separation and blind deconvolution", *Neural Computation*, vol. 7, pp.1129-1159, 1995.
3. J.-F. Cardoso "Blind Signal Separation: Statistical Principles", *Proc. of the IEEE*, vol. 86, No. 10, pp.2009-2025, Oct.1998.
4. J.-F. Cardoso and A. Souloumiac "Blind Beamforming for Non Gaussian Signals", *IEE Proc.-F*, vol. 140, No. 6, pp.362-270, Dec. 1993.
5. S. Cruces and L. Castedo, "A Gauss-Newton method for Blind Source Separation of Convolutive Mixtures", in *Proceedings of the Int. Conf. On Acoustic, Speech and Signal Processing (ICASSP98)*, Seattle, USA, Vol IV, pp. 2093-2096, 1998.
6. R. Fletcher, "Practical methods of optimization", *John Willey and Sons*, second edition, 1987.
7. A. Hyvarinen and E.Oja "A fast fixed-point algorithm for independent component analysis", *Neural Computation*, vol. 9, pp.1483-1492, 1997.
8. C. Jutten and J. Herault, "Blind Separation of Sources, Part I: an adaptive algorithm based on neuromimetic architecture", *Signal Processing*, vol. 24, pp.1-10, 1991.
9. R. Martín-Clemente and J.I.Acha, "Blind Separation of Sources using a New Polynomial Equation", *Electronics Letters*, Vol.33, No.1, pp.176-177,1997.
10. E. Oja "Nonlinear PCA Criterion and Maximum Likelihood in Independent Component Analysis", in *First Intern. Workshop on Independent Component Analysis and Signal Separation*, Aussois, France, pp. 143-148, 1999.
11. C. Puntonet, A. Prieto, C.Jutten and J. Rodriguez-Alvarez, "Separation of Sources: A Geometry-Based Procedure for reconstruction of N-Valued Signals", *Signal Processing*, vol. 46, No. 3, pp.267-284, 1995.
12. A. Prieto, C.Puntonet and B. Prieto, "A Neural Learning Algorithm for Blind Separation of Sources based on Geometrical Properties", *Signal Processing*, vol. 64, No. 3, pp.315-331, 1998.

Mixtures of Independent Component Analysers

Stephen J. Roberts[1] and William D. Penny[2]

[1] Robotics Research, Department of Engineering Science, University of Oxford, UK
[2] Wellcome Department of Cognitive Neurology, University College London, UK

1 Introduction

Gaussian Mixture Models (GMMs) are widely used throughout the fields of machine learning and statistics. Despite their popularity, however, GMMs suffer from two serious drawbacks. The first is that, as the dimensionality d of the problem space increases, the size of each covariance matrix, d^2, becomes prohibitively large. This can be dealt with by assuming isotropic Gaussians (ie. ignoring the covariance structure) but this greatly reduces the flexibility of the model class. This problem has been solved by Tipping and Bishop [12] who replaced each Gaussian with a probabilistic Principal Component Analysis (PCA) model. This allowed the dimensionality of each covariance to be effectively reduced whilst maintaining the richness of the model class. The second problem with GMMs is that each component is a Gaussian, an assumption which is often violated in many natural clustering problems [5]. It is this second problem which we address in this paper. A solution is reached by extending the mixtures of probabilistic PCA model to a mixtures of Independent Component Analysis (ICA) model.

2 Mixture Models

The probability of generating a data point $p(\mathbf{x}_n)$ from a K-component mixture model is:

$$p(\mathbf{x}_n) = \sum_{k=1}^{K} p(k)p(\mathbf{x}_n|k) \tag{1}$$

Furthermore, having observed datum $\mathbf{x}_n$ we can compute the probability that it was generated from component k as

$$p(k|\mathbf{x}_n) = \frac{p(k)p(\mathbf{x}_n|k)}{p(\mathbf{x}_n)} \tag{2}$$

Learning in mixture models takes place using an Expectation-Maximisation (EM) strategy where the objective is to maximise the joint log-likelihood of the observed and hidden variables, averaged over $p(k|\mathbf{x}_n) \stackrel{\text{def}}{=} \gamma_k(n)$. Using Jensen's inequality this maximisation can be shown to maximise the likelihood of the data (see eg. [2]). We therefore maximise

G. Dorffner, H. Bischof, and K. Hornik (Eds.): ICANN 2001, LNCS 2130, pp. 527–534, 2001.

$$L = \sum_{n=1}^{N} \sum_{k=1}^{K} \gamma_k(n) \log \left[p(k) p(\mathbf{x}_n|k) \right] \tag{3}$$

which splits up into two terms

$$L = \sum_{n=1}^{N} \sum_{k=1}^{K} \gamma_k(n) \log p(k) + \sum_{n=1}^{N} \sum_{k=1}^{K} \gamma_k(n) \log p(\mathbf{x}_n|k) \tag{4}$$

Maximising the first term yields an update rule for the mixing coefficients $p(k) \stackrel{\text{def}}{=} \pi_k$, namely:

$$\pi_k = \gamma_k / N \tag{5}$$

where we define

$$\gamma_k \stackrel{\text{def}}{=} \sum_{n=1}^{N} \gamma_k(n) \tag{6}$$

Maximising the second term gives update formulae for the parameters of the component densities. In this paper, the component densities derive from ICA models which are described in the next section.

3 ICA

In an ICA mixture model each component has a density derived from an ICA model. For the kth component the data are generated according to

$$\mathbf{x} = \mathbf{A}_k \mathbf{s}_k + \boldsymbol{\mu}_k + \mathbf{e}_k \tag{7}$$

where $\mathbf{e}_k$ is zero-mean Gaussian observation noise having an isotropic covariance matrix with variance σ_k^2, $\mathbf{A}_k$ is the mixing matrix and $\boldsymbol{\mu}_k$ is the mean of component k. Essentially, $\boldsymbol{\mu}_k$ determines the location of the cluster and $\mathbf{A}_k \mathbf{s}_k$ its shape. Given particular source values the *conditional* likelihood of a data point is

$$p(\mathbf{x}|k, \mathbf{s}_k) = \mathcal{N}(\mathbf{x}; \mathbf{A}_k \mathbf{s}_k - \boldsymbol{\mu}_k, \sigma_k^2 I) \tag{8}$$

where $\mathcal{N}(\mathbf{x}; \boldsymbol{\mu}, \boldsymbol{\Sigma})$ is a multivariate normal distribution with mean $\boldsymbol{\mu}$ and covariance $\boldsymbol{\Sigma}$. The likelihood of a data point is hence

$$p(\mathbf{x}|k) = \int p(\mathbf{x}|k, \mathbf{s}_k) p(\mathbf{s}_k) d\mathbf{s}_k \tag{9}$$

in which, as the underlying sources $\mathbf{s}_k$ are taken as statistically independent,

$$p(\mathbf{s}_k) = \prod_{i=1}^{M} p(s_{k,i}) \tag{10}$$

where M is the hypothesised number of sources.

To compute the likelihood of an observation we must be able to calculate the integral in equation 9. If we assume that the distribution over sources is dominated by a single peak, $\hat{\mathbf{s}}_k$, then the integral can be performed using Laplace's method. This gives [8, 10]

$$\log p(\mathbf{x}|k) = \frac{-(d-M)}{2}\log\left(2\pi\sigma_k^2\right) - \frac{1}{2\sigma_k^2}(\mathbf{z}_k - \mathbf{A}_k\hat{\mathbf{s}}_k)^{\mathsf{T}}(\mathbf{z}_k - \mathbf{A}_k\hat{\mathbf{s}}_k) + \log p(\hat{\mathbf{s}}_k) - \frac{1}{2}\log|\det(\mathbf{A}_k^{\mathsf{T}}\mathbf{A}_k)| \tag{11}$$

where $|\cdot|$ denotes the absolute value and $\mathbf{z}_k = (\mathbf{x} - \boldsymbol{\mu}_k)$. The likelihood for the kth ICA component in a mixture model is given by the second term in equation 4,

$$L(k) = \sum_{n=1}^{N} \gamma_k(n)\log p(\mathbf{x}_n|k) \tag{12}$$

which may be combined with equation 11 to give:

$$L(k) = \frac{-\gamma_k(d-M)}{2}\log\left(2\pi\sigma_k^2\right) - \frac{1}{2\sigma_k^2}E_k + \sum_{n=1}^{N}\gamma_k(n)\log p(\hat{\mathbf{s}}_k(n)) - \frac{\gamma_k}{2}\log|\det(\mathbf{A}_k^{\mathsf{T}}\mathbf{A}_k)| \tag{13}$$

in which we have summed over N data points and we define

$$E_k \stackrel{\text{def}}{=} \sum_{n=1}^{N}\gamma_k(n)(\mathbf{z}_k(n) - \mathbf{A}_k\hat{\mathbf{s}}_k(n))^{\mathsf{T}}(\mathbf{z}_k(n) - \mathbf{A}_k\hat{\mathbf{s}}_k(n)) \tag{14}$$

as the total reconstruction error. Maximising the above likelihood term results in an update equation for the mean of the k-th component as

$$\boldsymbol{\mu}_k = \frac{\sum_{n=1}^{N}\gamma_k(n)\mathbf{x}_n}{\gamma_k} \tag{15}$$

3.1 The Decorrelating Manifold

In previous work we have constrained the unmixing matrix to be a decorrelating matrix [3]. The motivation for this is that, for sources to be statistically independent, they must be at least linearly decorrelating. This has the further advantage that the unmixing matrix lies on a lower-dimensional manifold which makes the approach considerably faster. The corresponding mixing matrix is defined as follows.

Consider $\mathbf{Z}_k$ as an $N \times T$ matrix of zero-mean data vectors, with each entry weighted by $\sqrt{\gamma_k(n)/\gamma_k}$. The Singular Value Decomposition (SVD) of $\mathbf{Z}_k$ is given by [3]

$$\mathbf{Z}_k = \mathbf{U}_k\boldsymbol{\Lambda}_k\mathbf{V}_k \tag{16}$$

hence $\mathbf{U}$ contains the principal components of the weighted observation covariance matrix and $\boldsymbol{\Lambda}_k = [\lambda_1(k), \lambda_2(k), ...\lambda_d(k)]$ contains the standard deviations of the corresponding principal components. The mixing matrix is then given by

$$\mathbf{A}_k = \mathbf{U}_k^M \boldsymbol{\Lambda}_k^M \mathbf{Q}_k^\mathsf{T} \tag{17}$$

where $\mathbf{U}_k^M$ and $\boldsymbol{\Lambda}_k^M$ are the first M columns of $\mathbf{U}_k$ and $\boldsymbol{\Lambda}_k$. The transform is also parameterised by an orthonormal matrix $\mathbf{Q}_k$ of dimension $M \times M$ which constitutes the ICA transform proper[1].

3.2 Source and Noise Estimation

We estimate the sources at their Maximum Likelihood (ML) values. These are recovered via an unmixing matrix given by the pseudo-inverse of the mixing matrix

$$\mathbf{W}_k = (\mathbf{A}_k^\mathsf{T} \mathbf{A}_k)^{-1} \mathbf{A}_k^\mathsf{T} = \mathbf{Q}_k (\boldsymbol{\Lambda}_k^M)^{-1} (\mathbf{U}_k^M)^\mathsf{T} \tag{18}$$

and the sources are then estimated as

$$\mathbf{s}_k = \mathbf{W}_k \mathbf{z}_k \tag{19}$$

By defining

$$\hat{\mathbf{z}}_k \stackrel{\text{def}}{=} (\boldsymbol{\Lambda}_k^M)^{-1} (\mathbf{U}_k^M)^\mathsf{T} \mathbf{z}_k \tag{20}$$

we can then write $\mathbf{s}_k = \mathbf{Q}_k \hat{\mathbf{z}}_k$. The reconstructed observations are hence given by:

$$\mathbf{A}_k \mathbf{s}_k = (\mathbf{U}_k^M \mathbf{U}_k^M)^\mathsf{T} \mathbf{z}_k \tag{21}$$

As the average reconstruction error (under an L_2 norm) is equal to the variance of the data not explained by the first M components [8]

$$E_k = \sum_{i=M+1}^{d} \lambda_i(k)^2 \tag{22}$$

so the ML estimate of the noise variance (found from the first derivative of equation 13) is [10]

$$\hat{\sigma}_k^2 = \frac{E_k}{d - M} \tag{23}$$

hence,

$$\hat{\sigma}_k^2 = \frac{1}{d - M} \sum_{i=M+1}^{d} \lambda_i(k)^2 \tag{24}$$

This is the same as the noise estimate in the probabilistic PCA model [11]. Therefore, if ICA is constrained to the decorrelating manifold and ML source estimates are used, the observation noise level is not dependent on $\mathbf{Q}_k$ ie. the ICA transform proper. It can therefore be calculated ahead of optimising $\mathbf{Q}_k$ and fixed to its calculated value.

[1] In this paper, because we have a flexible source model - see later - the rows of our mixing matrix can be constrained to unit length, so it is not necessary to also include a diagonal scaling matrix [3].

3.3 Mixing Matrix Estimation

The matrix $\mathbf{Q}_k$ is constrained to be orthonormal by writing it as

$$\mathbf{Q}_k = \exp(\mathbf{S}_k) \tag{25}$$

where $\mathbf{S}_k$ is a skew-symmetric matrix ($\mathbf{S}_k^\top = -\mathbf{S}_k$) whose upper off-diagonal entries are known as Cayley coordinates [6].

Substituting the maximum likelihood estimates for the noise variance from equation 23 and the unmixing matrix from equation 18 into equation 13 and noting that since $\mathbf{W}$ is the psuedo-inverse of $\mathbf{A}$ we have $\det(\mathbf{W}^\top\mathbf{W}) = 1/\det(\mathbf{A}^\top\mathbf{A})$, so the log-likelihood term becomes:

$$L(k) = \frac{-\gamma_k(d-M)}{2}\log\left(2\pi\hat{\sigma}_k^2\right) - \frac{\gamma_k(d-M)}{2} + \sum_{n=1}^{N}\gamma_k(n)\log p(\mathbf{s}_k(n)) + \frac{\gamma_k}{2}\log|\det(\mathbf{W}_k^\top\mathbf{W}_k)| \tag{26}$$

The second term has been simplified due to the introduction of the ML noise variance estimate. The last term has the effect of preventing the unmixing matrix from becoming co-linear (ie. it prevents the components from pointing in the same direction).

The only term in the likelihood that depends on $\mathbf{Q}_k$ is the log source density where the dependence is introduced via equations 18 and 19. In previous work [3] we show how to compute the derivative of this term and combine it with an expression for $d\mathbf{Q}_k/d\mathbf{S}_k$. In this paper we use the same expression save for weighting the data with $\gamma_k(n)$. This then gives the gradient of the likelihood with respect to the Cayley coordinates. Fitting the ICA model therefore corresponds to following this gradient using, for example, a quasi-Newton optimisation.

3.4 Generalised Exponential Sources

In Bell and Sejnowski's original ICA formulation [1] the source densities were (implicitly) assumed to be reciprocal-cosh densities. More recent work (see, for example, [3]) however, shows that the inverse-cosh form cannot adequately model sub-Gaussian densities[2]. A more general parametric form which can model super-Gaussian, Gaussian *and* sub-Gaussian forms is the 'Exponential Power Distribution' or 'Generalised Exponential' (GE) density

$$G(s; R, \beta) = \frac{R\beta^{1/R}}{2\Gamma(1/R)}\exp(-\beta|s|^R) \tag{27}$$

where $\Gamma(\cdot)$ is the gamma function [9]. This density has zero mean, a kurtosis determined by the parameter R and a variance which is then determined by $1/\beta$. This source model has been used in an ICA model developed in [3].

[2] Sub/super densities have lighter/heavier tails than Gaussians.

The log-likelihood of the ith source model is given by the third term in equation 26. By dividing by γ_k we obtain the normalised log-likelihood

$$L_i(k) = \sum_{n=1}^{N} \frac{\gamma_k(n)}{\gamma_k} \log p(s_{k,i}(n)) \tag{28}$$

Substituting in the GE density gives

$$L_i(k) = \frac{1}{\gamma_k} \sum_{n=1}^{N} \gamma_k(n) \log \left[\frac{R_i \beta_i^{1/R_i}}{2\Gamma(1/R_i)} \right] - \frac{1}{\gamma_k} \sum_{n=1}^{N} \gamma_k(n) \beta_i |s_{k,i}(n)|^{R_i} \tag{29}$$

The source density parameters for each state can now be estimated component-wise by evaluating the derivatives $\frac{\partial L_i(k)}{\partial \beta_i}$ and $\frac{\partial L_i(k)}{\partial R_i}$ and setting them to zero. This gives

$$\frac{1}{\beta_i} = \frac{R_i}{\gamma_k} \sum_{n=1}^{N} \gamma_k(n) |s_{k,i}(n)|^{R_i} \tag{30}$$

and R_i can be found by solving the one-dimensional problem

$$\frac{\partial L_i(k)}{\partial R_i} = \frac{1}{R_i} + \frac{1}{R_i^2} \log \beta_i + \frac{1}{R_i^2} \varphi(1/R_i) - \frac{\beta_i}{\gamma_k} \sum_{n=1}^{N} \gamma_k(n) |s_{k,i}(n)|^{R_i} \log |s_{k,i}(n)| = 0 \tag{31}$$

where $\varphi(x) = \Gamma'(x)/\Gamma(x)$ is the digamma function [9]. Equations 30 and 31 are generalised versions of the equations in Appendix A of [3] and are applied iteratively until a consistent solution is reached (this whole process is identical to the optimisation of GE sources in a Hidden Markov ICA context [7]).

4 Practical Issues

The ICA-mixture model is initialised using a K-means clustering routine [2]. If at the end of this process any cluster is assigned less than 2 points, the K-cluster model is deemed unviable.

Otherwise, the K-means process produces K data sets. A PCA model order selection criterion [6] is then applied to each to determine the optimal latent dimension, M_k, and an ICA model is then fitted to each group. The responsibilities are then calculated (E-step) and the full EM optimisation of the ICA-mixture model begins. This consists of repeated application of equations 5, 14, 30, 31 and setting the mixing matrix so as to maximise equation. 26.

It is also possible to perform model order selection on the number of clusters. This is easily achieved by substituting the ICA-mixture likelihood into a Bayesian Information Criterion (BIC) as described in [8].

5 Results

We applied our algorithm to two data sets[3]. The first consists of a set of spectra from quasar emissions. The data is 4-D and the algorithm selects a two-

[3] Both from the UCI archives.

component ICA model (Fig. 1(a)) with each component having a latent dimension of 2. The resultant contours are shown in Fig. 1(b). The two components correspond to background and line emission spectra. The second dataset we

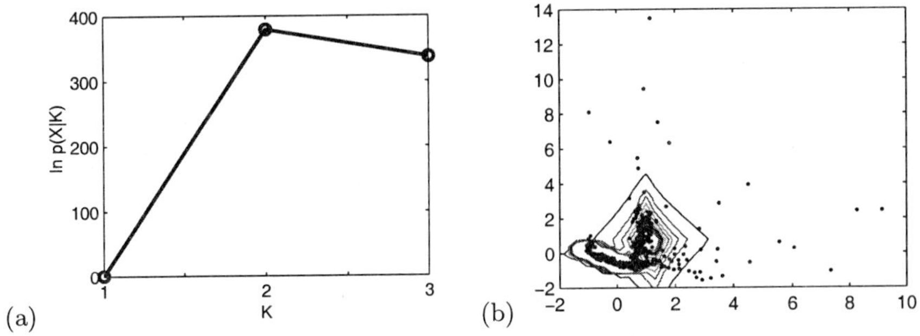

Fig. 1. Quasar data: *(a)* $\ln p(X|K)$ *against number of components,* K *and (b) resultant* $K = 2$ *density plot. Note the non-Gaussianity of the sources.*

consider is a 13-D set of wine composition data. Fig. 2(a) shows that a 3 component model is (correctly) favoured. Each component has a latent dimension of two. The resultant unsupervised partitioning is shown in Fig. 2(b) and is 97% accurate (in this case we know the true labels).

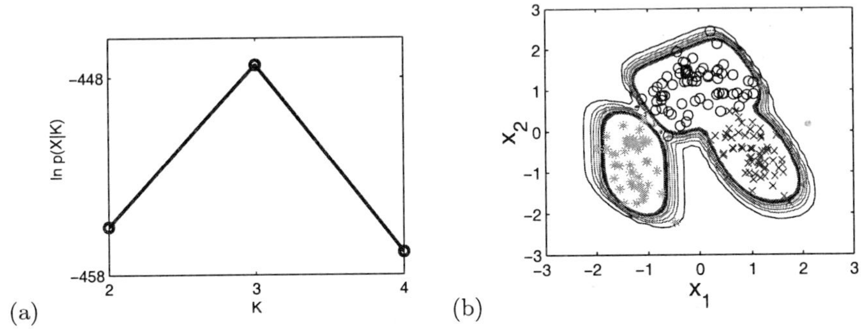

Fig. 2. Wine data: *(a)* $\ln p(X|K)$ *against number of components,* K *and (b) resultant* $K = 3$ *data partition and density.*

6 Discussion

We have proposed a clustering algorithm in which each cluster may be non-Gaussian. This has been achieved by extending the mixtures of probabilistic PCA model to a mixtures of ICA model. The model is a special case of Hidden

Markov ICA [7], which itself may thence be viewed as MoICA for time-series data. An ICA mixture model has also been proposed in [5]. This paper extends their approach by allowing the dimensionality of the latent space to be smaller than the observation space and for the ICA components to be evaluated on the decorrelating manifold. Furthermore, the appropriate latent dimension is set using a model order criterion.

References

1. A. J. Bell and T. J. Sejnowski. An information-maximization approach to blind separation and blind deconvolution. *Neural Computation*, 7(6):1129–1159, 1995.
2. C. M. Bishop. *Neural Networks for Pattern Recognition.* Oxford University Press, Oxford, 1995.
3. R. Everson and S.J. Roberts. Independent Component Analysis: A flexible non-linearity and decorrelating manifold approach. *Neural Computation*, 11(8), 1957–1984, 1999.
4. R.E. Kass and A.E. Raftery. Bayes factors and model uncertainty. Technical Report 254, University of Washington, 1993. http://www.stat.washington.edu/tech.reports/tr254.ps.
5. T. Lee and M.S. Lewicki. Image Processing methods using ICA mixture models. In S. Roberts and R. Everson, editors, *Independent Component Analysis: Principles and Practice.* Cambridge University Press, 2001.
6. T.P. Minka. Automatic choice of dimensionality for PCA. Technical Report 514, MIT Media Laboratory, Perceptual Computing Section, 2000.
7. W.D. Penny, R.M. Everson, and S.J. Roberts. Hidden Markov Independent Components Analysis. In M. Girolami, editor, *Advances in Independent Component Analysis.* Springer, 2000.
8. W.D. Penny, S.J. Roberts, and R. Everson. ICA: model order selection and dynamic source models. In S.J. Roberts and R. Everson, editors, *Independent Component Analysis: Principles and Practice.* Cambridge University Press, 2001.
9. W. H. Press, S.A. Teukolsky, W.T. Vetterling, and B.V.P. Flannery. *Numerical Recipes in C.* Cambridge, 1992.
10. S.J. Roberts. Independent Component Analysis: Source assessment and Separation, a Bayesian Approach. *IEE Proceedings on Vision, Image and Signal Processing*, 145(3):149–154, 1998.
11. M. E. Tipping and C. M. Bishop. Probabilistic Principal Component Analysis. Technical report, Neural Computing Research Group, Aston University, 1997.
12. M. E. Tipping and C. M. Bishop. Mixtures of Probabilistic Principal Component Analyzers. *Neural Computation*, 11(2):443–482, 1999.

Conditionally Independent Component Extraction for Naive Bayes Inference

Shotaro Akaho

Neuroscience Research Institute of AIST,
Central 2, 1–1–1 Umezono, Tsukuba, Ibaraki 305-8568, Japan
s.akaho@aist.go.jp,
http://staff.aist.go.jp/s.akaho/index.html

Abstract. This paper extends the framework of independent component analysis (ICA) to supervised learning. The key idea is to find a conditionally independent representation of input variables for given output. The representation is useful for the naive Bayes learning which has been reported to perform as well as more sophisticated methods. The learning algorithm is derived in a similar criterion to ICA. Two dimensional entropy takes an important role, while one dimensional entropy does in ICA.

1 Introduction

Independent component analysis (ICA) has been successfully applied to data analysis, signal processing and modeling of brain functions, especially early vision[3]. ICA has been restricted to unsupervised learning and we extend the framework of ICA to supervised learning. As an application of this extension, we focus on naive Bayes learning in this paper. Naive Bayes learning scheme performs well especially on most classification tasks, and is often significantly more accurate than more sophisticated methods. However, for real valued prediction, the assumption of conditional independence is sensitive to the performance. This paper proposes a method to find conditionally independent components.

2 Naive Bayes Learning

We consider the problem of predicting an output value y for an input vector $\boldsymbol{x}$ with d elements $x_1, \ldots, x_d$. If the probability density function $p(y \mid \boldsymbol{x})$ is known, we can choose y to minimize the expected prediction error. However, $p(y \mid \boldsymbol{x})$ is not usually known and it must be estimated from a set of training samples. If the dimensionality of $\boldsymbol{x}$ is large, the learning is considered to be difficult, which is often called "curse of dimensionality". Naive Bayes learning solves this problem by assuming the conditional independence of x_i's for given y. Under this assumption, $p(y \mid \boldsymbol{x})$ can be factorized as

$$p(y \mid \boldsymbol{x}) = \frac{p(x_1 \mid y) \cdots p(x_d \mid y)\, p(y)}{\int p(x_1 \mid y) \cdots p(x_d \mid y)\, p(y)\, dy}, \tag{1}$$

G. Dorffner, H. Bischof, and K. Hornik (Eds.): ICANN 2001, LNCS 2130, pp. 535–540, 2001.

thus the estimation of $p(y \mid \boldsymbol{x})$ is reduced to the estimation of individual functions $p(x_i \mid y)$ and $p(y)$, which makes the problem much easier.

Despite its simplicity, the naive Bayes scheme performs well on most classification tasks, even if the assumption of independence is seriously violated[5]. Although this scheme has been tried to real value estimation as well, it is reported that the performance is rather sensitive to the validity of the assumption of independence than the discrete case[7, 6]. In order to reduce this sensitivity, we propose a method to find a linear transformation of $\boldsymbol{x}$ which provides a representation satisfying the independence assumption.

3 Conditionally Independent Component Extraction

3.1 Problem

We assume that there is an unknown vector of source signals $\boldsymbol{s} = (s_1, \ldots, s_{d^*})$ whose elements are mutually independent conditionally for given y,

$$p(\boldsymbol{s} \mid y) = p(s_1 \mid y) \cdots p(s_{d^*} \mid y). \tag{2}$$

Observed $\boldsymbol{x}$ is considered to be a linear mixture of $\boldsymbol{s}$. For the sake of simplicity, we assume $d^* = d$. The case $d^* < d$ is dealt with as a dimension reduction problem in section 5. If $d^* = d$, the observation is generated by

$$\boldsymbol{x} = A\boldsymbol{s}, \tag{3}$$

where $A \in \mathcal{R}^{d \times d}$ is an unknown nonsingular matrix. Without knowing $\boldsymbol{s}$ and A, we want to recover $\boldsymbol{s}$ from $\boldsymbol{x}$ by a linear transformation

$$\boldsymbol{z} = W\boldsymbol{x}, \tag{4}$$

where $W \in \mathcal{R}^{d \times d}$ is a matrix. If $W = A^{-1}$, we obtain $\boldsymbol{z} = \boldsymbol{s}$. However, by the same discussion as ICA, W can not be uniquely determined because of the ambiguity of the order and the amplitude of signals. Our goal is to find W to recover $\boldsymbol{s}$ except for this ambiguity. In addition, it is necessary to assume that at most one conditional distribution $p(s_i \mid y)$ is Gaussian. Otherwise, there remains continuous freedom of rotation transformation.

3.2 Cost Function

In order to evaluate the degree of the conditional independence, we define the cost function to be minimized by the Kullback-Leibler divergence between the left and the right hand side of (2), which is averaged over the whole y,

$$L(W) = \mathrm{E}_y \left[K[p(\boldsymbol{z} \mid y) \| \prod_{i=1}^{d} p(z_i \mid y)] \right] = K[p(\boldsymbol{z}, y) \| p(y) \prod_{i=1}^{d} p(z_i \mid y)], \tag{5}$$

where $K[p\|q)] = \int p \log(p/q)$.

In terms of entropy, the cost function can be written in the form,

$$L(W) = -H(\boldsymbol{z}, y) + \sum_{i=1}^{d} H(z_i, y) + (1-d)H(y). \qquad (6)$$

Since $H(\boldsymbol{z}, y) = H(\boldsymbol{x}, y) + \log |W|$, we have

$$L(W) = -\log |W| + \sum_{i=1}^{d} H(z_i, y) + \text{const}. \qquad (7)$$

where $|W|$ denotes the determinant of W,

3.3 Learning Algorithm

The cost function is similar to that of ICA based on the mutual information minimization[9], except the entropy term of one variable in ICA is replaced by the joint entropy of z_i and y. We can derive the gradient descent learning algorithm for the cost function, CICA (conditionally independent component analysis) algorithm.

CICA algorithm (general form)

$$\Delta W \propto -\frac{\partial L(W)}{\partial W} W^T W = \left(\mathrm{I}_d - \mathrm{E}_{y,\boldsymbol{z}}[\boldsymbol{\varphi}(\boldsymbol{z}, y)\boldsymbol{z}^T]\right) W, \qquad (8)$$

where I_d is a unit matrix of order d, and $\boldsymbol{\varphi}$ is a vector with elements

$$\varphi_i(\boldsymbol{z}, y) = -\frac{\partial \log p(z_i, y)}{\partial z_i} = \frac{-\partial p(z_i, y)/\partial z_i}{p(z_i, y)}, \qquad (9)$$

In the above algorithm, the gradient is calculated in the sense of natural gradient which makes the algorithm simpler and faster.

There are various possibilities how the probability distribution $p(z_i, y)$ is estimated from training samples. We use kernel density estimation described in the next section both for CICA and for the naive Bayes learning. In the above algorithm, the average $\mathrm{E}_{y,\boldsymbol{z}}[u]$ can be considered in three ways: the first one is that u is numerically integrated by the estimated $p(z_i, y)$, whose cost is the most expensive and might not be practical. The second one, which we use, is batch learning where u is averaged over training samples. The third one is on-line learning where $\mathrm{E}_{y,\boldsymbol{z}}$ is eliminated, which still ensures the convergence in the sense of 'stochastic approximation'.

4 Kernel Density Estimation

Suppose we have n training samples $(\boldsymbol{x}^{(1)}, y^{(1)}), \ldots, (\boldsymbol{x}^{(n)}, y^{(n)})$. Let $\hat{W}$ denote an estimation of the demixing matrix W, and $\boldsymbol{z}^{(j)}$ denote the transformed value of $\boldsymbol{x}^{(j)}$ by $\hat{W}$, $\boldsymbol{z}^{(j)} = \hat{W}\boldsymbol{x}^{(j)}$.

Kernel density estimation is a nonparametric method[8], which is written for the joint distribution of z_i and y as

$$\hat{p}(z_i, y) = \frac{1}{nh_ih_y} \sum_{j=1}^{n} K\left(\frac{z_i - z_i^{(j)}}{h_i}\right) K\left(\frac{y - y^{(j)}}{h_y}\right), \tag{10}$$

where K is a kernel function, and h_i and h_y are parameters defining kernel widths for z_i and y.

A typical choice for K is the Gaussian kernel $K(u) = \exp(-u^2)/\sqrt{2\pi}$, and we use it in our experiments. Various methods to determine the kernel widths have been proposed. However, since performances of those methods depend on applications, we choose simple one, $h_i = \sigma_i n^{-1/6}, h_y = \sigma_y n^{-1/6}$, where σ_i^2 and σ_y^2 are sample variances of z_i and y. Although this kernel widths are derived under the assumption that z_i and y are jointly Gaussian, it does not affect seriously even if the assumption is not correct.

In the case of Gaussian kernel, φ_i in (9) can be written as

$$\varphi_i(\boldsymbol{z}, y) = \frac{\sum_{j=1}^{n}(z_i - z_i^{(j)})K_i(z_i - z_i^{(j)})K_y(y - y^{(j)})}{h_i^2 \sum_{j=1}^{n} K_i(z_i - z_i^{(j)})K_y(y - y^{(j)})}, \tag{11}$$

where $K_i(u) = K(u/h_i)$, $K_y(u) = K(u/h_y)$.

It is necessary to estimate $p(y)$ and $p(z_i \mid y)$ for the naive Bayes, they can be estimated by $\hat{p}(y) = K_y(y - y^{(j)})/(nh_y)$ and $\hat{p}(z_i \mid y) = \hat{p}(z_i, y)/\hat{p}(y)$. Although a kernel width of order $n^{-1/5}$ achieves better performance for one dimensional density estimation, we use the same kernel width for the consistency that the marginal distribution of $\hat{p}(z_i, y)$ is equal to $\hat{p}(y)$.

5 Dimension Reduction

In this section, we consider the case that the dimensionality of observations is larger than that of sources, i.e. $d > d^*$. In order to apply the CICA algorithm, we reduce the dimensionality of $\boldsymbol{x}$ by linear transformation, $\boldsymbol{u} = B\boldsymbol{x}$, where B is a $d^* \times d$ matrix and $\boldsymbol{u}$ is a transformed value. How should we choose B? There are two requirements for $\boldsymbol{u}$: one is that $\boldsymbol{u}$ is preferred to preserve as much information on y as possible. The other is that applying the CICA algorithm to $\boldsymbol{u}$ can achieve the conditional independence as well as possible. In this paper, we deal with mainly the former one, since the latter may need some change to the CICA algorithm which is left as a future work.

Although it is difficult to find the map that preserves the most information on y in a general setting, the canonical correlation analysis (CCA) provides the map that we want if a pair of multivariates are jointly Gaussian. Since we have only one dimensional output variable y, CCA gives a map onto just one dimensional space. Therefore, we use CCA which is nonlinearly extended by basis functions: let $c_1(y), \ldots, c_k(y)$ be $k(\geq d^*)$ functions of y which are linearly independent. We apply CCA between $\boldsymbol{x}$ and $\boldsymbol{c}(y) = (c_1(y), \ldots, c_k(y))^T$. We obtain $\boldsymbol{u}$ by choosing d dimensional canonical subspace.

In order to design a good set of functions $\boldsymbol{c}(y)$, we have to check how transformed variable $\boldsymbol{u}$ fulfills the two requirements described above.

6 Simple Experiments

We show a simple experiment result for artificial data to ascertain the algorithm to work. In this experiment, we does not investigate the case of dimension reduction.

Training data and test data are generated as follows: first output value y is uniformly distributed from $u[-3,3]$ and two dimensional source signals are generated by $s_1 = y + \varepsilon_1$, $s_2 = y^2 + \varepsilon_2$, where ε_1 and ε_2 are independently generated from $u[-1,1]$. Observation signals are generated by mixing source signals, $x_1 = s_1 + 0.5s_2$ and $x_2 = 0.5s_1 + s_2$.

We applied the CICA algorithm to 20 training samples. After each step of the CICA, we evaluated the prediction error by using 100 test samples. Prediction is taken by the posterior average of y, $\hat{y} = \int y\hat{p}(y \mid \boldsymbol{x}; \hat{W})\, dy$, which achieves least mean square error, where $\hat{p}$ is an estimated predictive distribution by the CICA and the naive Bayes learning.

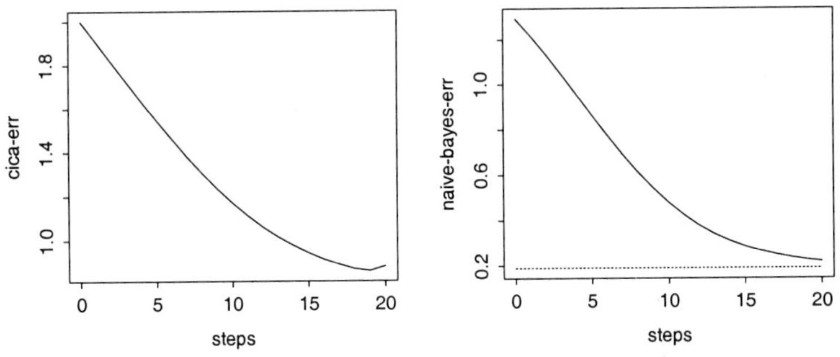

Fig. 1. Error curves for the CICA and the naive Bayes

Figure 1 shows the convergence of the algorithm. The error for the CICA (the left figure) is evaluated by

$$\sum_{i=1}^{d}\left(\sum_{j=1}^{d} \frac{|P_{ij}|}{\max_k |P_{ik}|} - 1\right) + \sum_{j=1}^{d}\left(\sum_{i=1}^{d} \frac{|P_{ij}|}{\max_k |P_{kj}|} - 1\right), \tag{12}$$

which represents the recovery rate of source signals, where $P = \hat{W}A$. The right figure shows the mean square error of the naive Bayes learning in each step of the CICA. The lower line is the minimum error when the source signals are known.

This experiment shows that although the naive Bayes is rather sensitive to the assumption of conditional independence, the CICA algorithm can successfully recover the conditional independence.

7 Concluding Remarks

In this paper, we have proposed an extended framework of ICA to supervised learning scheme and applied to the naive Bayes learning. Applying the algorithm to real world data remains as a future work.

The method is useful not only for the naive Bayes learning, but also for appropriate representation of graphical models[4]. Conditionally independent representation might simplify the structure of the graphical models and the table of conditional probabilities.

The CICA is similar to the multimodal ICA (MICA) proposed by the author[1], which extends the canonical correlation analysis. The assumption of MICA is that the pairs of source signals are jointly independent which is different from the assumption in this paper. However, it is technically similar because the cost functions of both algorithms include two dimensional entropy.

ICA can be considered as a model of lower level systems in the brain such as early vision. Becker et al[2] has proposed the higher level model where the system receives signals from several kinds of lower level systems. the CICA and the MICA can be a candidate of models for such a higher level systems.

References

1. Akaho, S., Kiuchi, Y., Umeyama, S.: MICA: Multimodal independent component analysis. In *Proc. of IJCNN* (1999) 927–932
2. Becker, S.: Mutual Information Maximization: Models of Cortical Self-Organization. *Network: Computation in Neural Systems*, **7** (1996)
3. Bell, A. J., Sejnowski, T.J.: The 'independent components' of natural scenes are edge filters. *Vision Research*, **37** (1997) 3327–3338
4. Cowell, R.G., Dawid, A.P., Lauritzen, S.L., Spiegelhalter, D.J.: *Probabilistic Networks and Expert Systems.* Springer (1999)
5. Domingos, P. and Pazzani, M.: On the optimality of the simple Bayesian classifier under zero-one loss. *Machine Learning*, **29** (1997) 103–130
6. Frank, E., Leonard, T., Holmes, G., Witten, I.H.: Naive Bayes for regression. *Machine Learning*, **41** (2000) 5–25
7. Friedman, J.: On bias, variance, 0/1-loss, and the curse-of-dimensionality. *Data Mining and Knowledge Discovery*, **1** (1997) 55–77
8. Simonoff, J.S.: *Smoothing Methods in Statistics.* Springer-Verlag (1998)
9. Yang H., Amari, S.: Adaptive online learning algorithms for blind separation: Maximum entropy and minimum mutual information. *Neural Computation*, **9** (1997) 1457–1482

Fast Score Function Estimation with Application in ICA*

Nikos Vlassis

Computer Science Institute
University of Amsterdam
Kruislaan 403, 1098 SJ Amsterdam
The Netherlands
vlassis@science.uva.nl

Abstract. We describe an efficient method for accurate estimation of the score function of a random variable, which can be regarded as an extension of the FFT-based fast density estimation method of Silverman (1982), and which scales no more than linearly with the sample size. We demonstrate the utility of our approach in a real-life ICA problem involving the separation of eight sound signals, where better results are observed than using state-of-the-art ICA methods.

Keywords: Score function estimation, Independent component analysis, Blind signal separation.

1 Introduction

It has been recognized that the asymptotic behavior of an algorithm for Independent Component Analysis (ICA) should depend on the 'true' probability density function of each source, but in practice these densities are unknown. However, for large sample sizes it is possible to estimate directly from the sample the distributions of the sources and obtain the same asymptotic performance as if these distributions were known [1].

Although an efficient algorithm exists [12] for fast estimation of an unknown probability density function, in actual ICA implementations we need to estimate the *score* functions of the unknown sources [5]. In most existing ICA algorithms, approximate models of the score functions are employed based on moment expansions which are often prone to outliers, simple parametric models [4, 7], or mixtures of simple models [10, 9, 15, 3] which are in general considered expensive to compute.

In this paper we describe a method for fast estimation of score functions. We model the unknown densities of the sources with mixtures of Gaussian models and then show how we can efficiently estimate the score functions associated with each density. The approach can be regarded as an extension of the fast FFT-based kernel density estimation method of Silverman [12, 14]. The estimated

* This research is supported by the Dutch Technology Foundation STW, applied science division of NWO, and the technology programme of the Ministry of Economic Affairs, project AIF 4997.

G. Dorffner, H. Bischof, and K. Hornik (Eds.): ICANN 2001, LNCS 2130, pp. 541–546, 2001.

score functions are then plugged in a natural gradient ICA algorithm to search for the separating matrix. A real-life experiment shows that in some cases our method can outperform methods based on approximations. A more detailed analysis can be found in [13].

2 Score Function Estimation

A general class of semi-parametric density models is the Gaussian mixture model [8]. This models the unknown density with a sum of m Gaussian kernels in the form

$$f(y) = \sum_{k=1}^{m} \pi_k g(y; \mu_k, \sigma_k^2) \tag{1}$$

where $g(y; \mu_k, \sigma_k^2)$ is the Gaussian kernel

$$g(y; \mu_k, \sigma_k^2) = \frac{1}{\sqrt{2\pi}\sigma_k} \exp\left[-\frac{(y-\mu_k)^2}{2\sigma_k^2}\right] \tag{2}$$

with center μ_k and variance σ_k^2. Under the Gaussian mixture model, the score function of y reads

$$\phi(y) = -\frac{f'(y)}{f(y)} = \frac{\sum_{k=1}^{m} \pi_k g(y; \mu_k, \sigma_k) \frac{(y-\mu_k)}{\sigma_k^2}}{f(y)} = \sum_{k=1}^{m} P(k|y) \frac{(y-\mu_k)}{\sigma_k^2} \tag{3}$$

where $P(k|y)$ is the posterior probability that y was sampled from the kernel k

$$P(k|y) = \frac{\pi_k g(y; \mu_k, \sigma_k^2)}{f(y)} \tag{4}$$

and for any y holds $\sum_{k=1}^{m} P(k|y) = 1$.

For fast score function estimation we use a large number of mixing kernels, e.g., m=400, with their centers equidistantly positioned over the range of y and with constant variance. If $\mu_1 \leq \min(y)$ is the center of the first kernel then the center of the k-th kernel is $\mu_k = \mu_1 + (k-1)\delta$ where $\delta = (\mu_m - \mu_1)/(m-1)$ is the distance between centers. This is in effect a limited Gaussian mixture model where the only degree of freedom is on the mixing weights π_k, and it precisely corresponds to the approximate kernel smoothing density estimate in [12]. In this case the score function (3) becomes

$$\phi(y) = \frac{1}{\sigma^2}[y - \sum_{k=1}^{m} P(k|y)\mu_k] \tag{5}$$

with σ^2 the common variance to all kernels.

The choice of σ is crucial in such nonparametric applications. From [6, p. 583] we know that in projection pursuit applications that involve nonparametric density estimation, the error of the direction estimates is minimized by choosing a kernel bandwidth σ between $O(n^{-1/3})$ and $O(n^{-1/4})$. Our choice for the ICA problem was $\sigma = n^{-2/7}\text{std}(y)$ where $\text{std}(y)$ is the sample standard deviation of each source y. This value satisfies the above bounds and it was found to give good results in practice.

To estimate the score function $\phi(y)$ in (5) on all points y_i of a sample, we need to compute for all points the quantity $\sum_{k=1}^{m} P(k|y)\mu_k \equiv h(y)/f(y)$ where

$$h(y) = \sum_{k=1}^{m} \pi_k \mu_k \exp[-0.5(y - \mu_k)^2/\sigma^2] \tag{6}$$

$$f(y) = \sum_{k=1}^{m} \pi_k \exp[-0.5(y - \mu_k)^2/\sigma^2]. \tag{7}$$

However, instead of computing directly by substituting the y_i in the formulas above, we can evaluate both $h(y)$ and $f(y)$ only at the points μ_k and then interpolate for computing the values of the functions at the points y_i. This offers considerable speedup since the first evaluation can be carried out with the FFT very quickly. Using the μ_k from above, the value of $h(y)$ and $f(y)$ on the kernel centers becomes

$$h(\mu_l) = \sum_{k=1}^{m} \pi_k \mu_k \exp[-0.5(l - k)^2\delta^2/\sigma^2] \tag{8}$$

$$f(\mu_l) = \sum_{k=1}^{m} \pi_k \exp[-0.5(l - k)^2\delta^2/\sigma^2]. \tag{9}$$

Both quantities are in the form $b_l = \sum_{k=1}^{m} a_k g_{l-k}$ which is a discrete convolution and can be efficiently carried out with the FFT algorithm [11]. Moreover, the mixing weights π_k can be approximated by the histogram of y which has linear complexity.

Having the values $h(y_k)$ and $f(y_k)$ on all kernel centers, we can interpolate to find $h(y_i)$ and $f(y_i)$ for all y_i, and efficient interpolation methods exist [11]. If m' is the next power of 2 larger than m, and n is the sample size, then the above procedure has cost $O(\max\{n, m' \log_2 m'\})$ which, for $m = 400$ and relatively large data sets $n \geq 5000$, is no more than linear in n. Some examples of density and score function approximation with $n = 5000$ and $m = 400$ are shown in Fig. 1 for a sub-Gaussian, a two-component mixture, and a super-Gaussian distribution. The approximation is good in areas of high probability.

3 Plugging into an ICA Algorithm

In an ICA context we assume the existence of a $d \times 1$ zero-mean random vector $\mathbf{s}$ that has pairwise statistically independent components (corresponding to the sources), and which are linearly mixed to form the $d \times 1$ vector $\mathbf{x}$ by

$$\mathbf{x} = \mathbf{A}\mathbf{s} \tag{10}$$

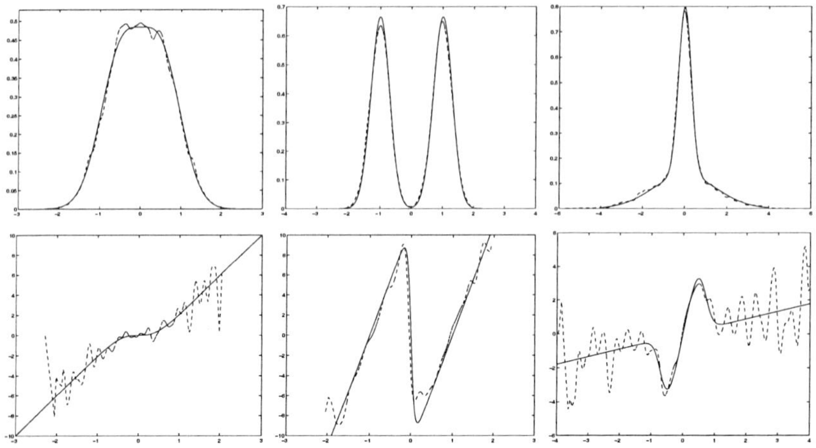

Fig. 1. Three density and score functions, and their approximations (dashed lines).

where $\mathbf{A}$ is a nonsingular $d \times d$ 'mixing' matrix. Our task is to compute from a sample of i.i.d. observations $\mathbf{x}_i$, $1 \leq i \leq n$, of the random vector $\mathbf{x}$ the 'separating' matrix $\mathbf{W} = \mathbf{A}^{-1}$ that successfully recovers the original sources.

If $p(\mathbf{x})$ denotes the density of $\mathbf{x}$ and $f(\mathbf{y})$ is a model for the joint density of

$$\mathbf{y} = \mathbf{W}\mathbf{x} \tag{11}$$

then the identity $p(\mathbf{x}) = |\det \mathbf{W}| f(\mathbf{W}\mathbf{x})$ allows the computation of the log-likelihood of the sample as a function of the unknown matrix $\mathbf{W}$. Maximization of this log-likelihood leads to the natural (relative) gradient iterative scheme in the batch version

$$\mathbf{W} := \mathbf{W} - \eta(E[\boldsymbol{\phi}(\mathbf{y})\mathbf{y}^T] - \mathbf{I})\mathbf{W} \tag{12}$$

where η is a learning rate, E is the expectation operator (in practice substituted by an average over the sample), $\mathbf{I}$ is the $d \times d$ identity matrix, and

$$\boldsymbol{\phi}(\mathbf{y}) = [\phi_1(y^{(1)}), \ldots, \phi_d(y^{(d)})]^T \tag{13}$$

is the $d \times 1$ vector of the score functions associated with each source (for derivations and details we refer to [5]). In our case, we apply the gradient iteration scheme (12) using the estimated score functions from (5).

4 Demonstration and Discussion

To show the validity of our method in separating mixed signals we carried out an experiment involving eight real-life signals corresponding to different sounds of duration about one second and 10000 sample points. The distributions of these signals are shown in Fig. 2. Note that this is a difficult problem that involves many distributions that are close to Gaussian.

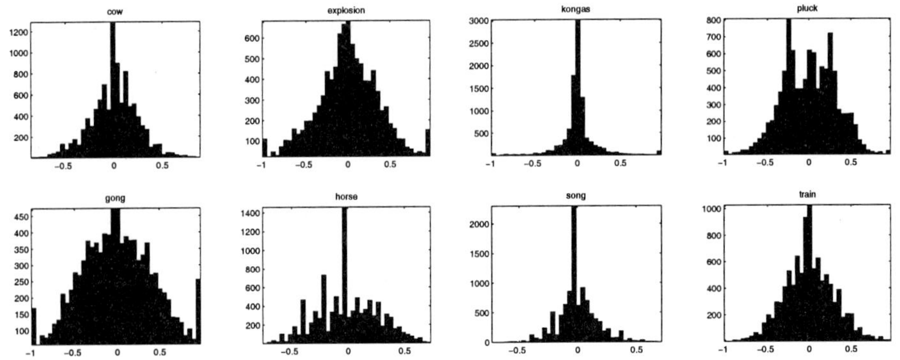

Fig. 2. The histograms of the eight sound signals.

These signals were mixed with a random matrix $\mathbf{A}$ with two-norm condition number 7.5 (ratio of the largest singular value to the smallest). The performance of the algorithm during learning was monitored by the error measure [2]

$$E = \sum_{i=1}^{d} \left(\sum_{j=1}^{d} \frac{|p_{ij}|}{\max_k |p_{ik}|} - 1 \right) + \sum_{j=1}^{d} \left(\sum_{i=1}^{d} \frac{|p_{ij}|}{\max_k |p_{kj}|} - 1 \right) \qquad (14)$$

where p_{ij} are the elements of the matrix $\mathbf{P} = \mathbf{WA}$ which measures how close $\mathbf{W}$ is to the true mixing matrix $\mathbf{A}^{-1}$.

To avoid solutions corresponding to local minima of the contrast function we first applied the extended infomax algorithm [7] for approximately 40 steps, starting with the solution $\mathbf{W} = 0.1\ \mathbf{I}$ and using no sphering, a block size of 300, and a learning rate of $\eta = 0.0005$. As we see in Fig. 3, the extended infomax algorithm converges fast to a solution with error approximately $E = 10$.

After convergence we applied (i) the extended infomax in batch mode (the score function was computed using all data points) with the same learning parameters, and (ii) the gradient descent scheme (12) using our score estimation method and constant learning rate $\eta = 0.05$. In Fig. 3 we plot the error measure when using the extended infomax score function model (crosses) and our score function estimation (circles). We see that our method successfully manages to reduce the error to a small number, corresponding to a solution with almost all signals perfectly separated, with a slight noise in the separation of the cow and train sounds which have distributions very close to Gaussian.

From this experiment it became clear that our approach can often outperform methods that use approximate models of the score functions of the sources. In general, it is advisable to use an ICA algorithm like the extended infomax for roughly locating the neighborhood of the correct solution, and then switching to our score function estimation procedure for 'fine tuning' of the separating matrix. For more details about the proposed method we refer to [13]. Implementation in Matlab of the score function estimation method can be found at `http://www.science.uva.nl/~vlassis`

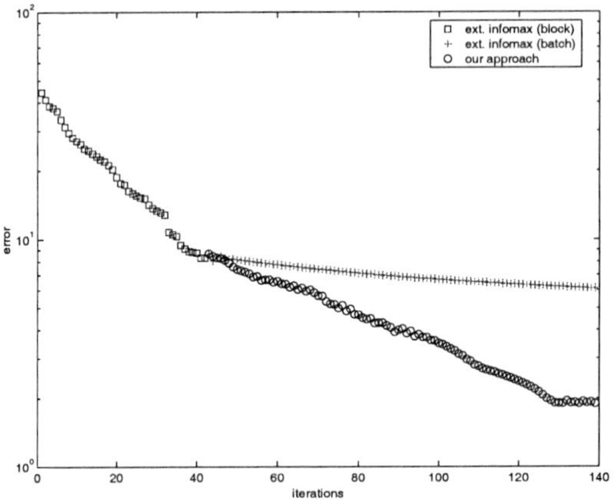

Fig. 3. The error measure vs. the number of iterations in the eight sounds problem.

References

1. S. Amari and J.-F. Cardoso. Blind source Separation-semiparametric statistical approach. *IEEE Trans. Signal Processing*, 45(11):2692–2700, 1997.
2. S. Amari, A. Cichocki, and H. H. Yang. A new learning algorithm for blind signal Separation. In D. S. Touretzky, M. C. Mozer, and M. E. Hasselmo, editors, *Advances in Neural Information Processing Systems*, volume 8, pages 757–763. The MIT Press, 1996.
3. H. Attias. Independent factor analysis. *Neural Computation*, 11:803–851, 1999.
4. A. J. Bell and T. J. Sejnowski. An information-maximization approach to blind separation and blind deconvolution. *Neural Computation*, 7:1129–1159, 1995.
5. J.-F. Cardoso. Blind signal Separation: statistical principles. *Proc. IEEE*, 9(10):2009–2025, Oct. 1998.
6. P. Hall. On projection pursuit regression. *Ann. Statist.*, 17(2):573–588, 1989.
7. T.-W. Lee, M. Girolami, and T. Sejnowski. Independent component analysis using an extended infomax algorithm for mixed sub-Gaussian and super-Gaussian sources. *Neural Computation*, 11(2):409–433, 1999.
8. G. J. McLachlan and D. Peel. *Finite Mixture Models.* Wiley, New York, 2000.
9. É. Moulines, J.-F. Cardoso, and E. Gassiat. Maximum likelihood for blind separation and deconvolution of noisy signals using mixture models. In *Proc. ICASSP*, pages 3617–3620, Munich, Germany, 1997.
10. B. A. Pearlmutter and L. C. Parra. A context-sensitive generalization of ICA. In *Proc. Int. Conf. on Neural Information Processing*, Hong Kong, 1996.
11. W. H. Press, S. A. Teukolsky, B. P. Flannery, and W. T. Vetterling. *Numerical Recipes in C.* Cambridge University Press, 2nd edition, 1992.
12. B. W. Silverman. Kernel density estimation using the Fast Fourier Transform. *Appl. Statist.*, 31:93–99, 1982.
13. N. Vlassis and Y. Motomura. Efficient source adaptivity in independent component analysis. *IEEE Trans.* Neural Networks, 12(3), May 2001.
14. M. P. Wand and M. C. Jones. *Kernel Smoothing.* Chapman & Hall, London, 1995.
15. L. Xu, C. C. Cheung, and S. Amari. Learned parametric mixture based ICA algorithm. *Neurocomputing*, 22:69–80, 1998.

Health Monitoring with Learning Methods

Alexander Ypma*, Co Melissant, Ole Baunbæk-Jensen**, and Robert P.W. Duin

Pattern Recognition Group, TU Delft, Lorentzweg 1, 2628 CJ Delft, The Netherlands
ypma@mbfys.nl
www.ph.tn.tudelft.nl/~ypma

Abstract. In this paper we provide a framework for the design of a practical monitoring method with learning methods. We demonstrate that three medical and industrial monitoring problems involve subproblems that can be tackled with our approach. Application of interference removal, novelty detection and learning of a signature leads to a feasible monitoring method in these cases.

1 A Framework for Learning Health Monitoring

Automatic health monitoring of eal-world systems is becoming increasingly important. Examples are machines, plants, cars and engines. Human health monitoring may also be included in this list, e.g. monitoring of the depth of the narcosis during surgery or on-line detection of a disease using measurements of biomedical quantities. We take a *learning* approach, where the particularities of the monitored subject are used: the behaviour of the system is learned from available examples instead of formulating (nonflexible) rules on the basis of human experience beforehand. A practical method for on-line health monitoring comprises the stages shown in figure 1. After measuring indicators of system behaviour (usually with an ensemble of transducers), we aim at improving signal-to-noise ratios and suppressing interfering sources by using the spatial diversity in a multichannel measurement (*interference removal*). The measurement time series are then represented by feature vectors in order to use them for health indication (*feature extraction*). Taking several measurements of the system operating normally under varying (operating) conditions and finding a suitable characterization of the measurements (using *prior knowledge* about the system and monitoring heuristics) leads to a description of the *domain* that represents normal behaviour of the system (*state space signature* or *fingerprint*). This description can be used to detect deviating measurement patterns (*novelty detection*). Knowledge about the parts of the space occupied by anomalies can

* This work is supported by Technology Foundation STW, project DTN-44.3584. We thank TechnoFysica b.v. and Drs. Groeneveld, Stam and Frietman for cooperation.

** A. Ypma and O. Baunbæk-Jensen are with the Dutch Foundation for Neural Networks SNN Nijmegen and the TU Denmark, respectively.

G. Dorffner, H. Bischof, and K. Hornik (Eds.): ICANN 2001, LNCS 2130, pp. 547–553, 2001.

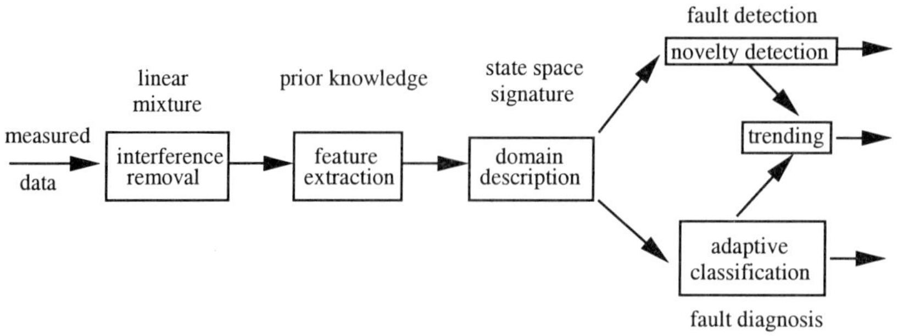

Fig. 1. Stages in automatic health monitoring

then be used for diagnosis of system faults or illnesses. This knowledge can either be present beforehand, e.g. in the form of heuristics and knowledge from previous 'case studies', or can be acquired on-the-fly (i.e. during the course of system operation), involving an *adaptive classification.* Ultimately, the transition of the system state during wear (machines) or aging (humans) and fault/ illness development can be used as an indicator of the health *trend.*

2 Eyeblink Detection in EOG Measurements

An important indicator of the severity of Tourette's syndrom[1] is the eyeblink activity of the subjects under examination. Eyeblinks and -movements can be registered automatically using ElectroOculoGraphy (EOG). An eyeblink detection system should be able to discriminate between eyeblink-events and all other events. Since the eyeblinks of a subject are assumed to be fairly repetitive, a *novelty detection* approach is taken. In earlier work [6] we have used the Self-Organizing Map (SOM) for this purpose. An eyeblink in the EOG signal consists of a negative peak, followed by a positive peak. The negative peak is fairly consistent in shape and duration. The positive peak shows more variability and can be characterized by its peak size. A fixed-length window is slided over the univariate time series in order to obtain segments that are processed with the SOMs. We used two different SOMs in the detection method: one SOM (NPSOM) detects the negative peak using examples of negative peaks only; the other SOM (FV-SOM) uses information about negative and positive peak size (i.e. it is trained on feature vectors describing the signal segments) to take also the positive peak into account. The SOM describes the domain of the set of eyeblinks in the feature space. If it is used for novelty detection, a rejection threshold has to be set. We do this by cross-validation, using a training and a testset. The raw data is segmented manually by the medical expert. When a suffient amount of eyeblinks has been segmented (say 200), they are devided into a training set and a test set, and furthermore a set of events that are not eyeblinks is also segmented from

[1] Tourette patients suffer from involuntary verbal activity and muscle contractions

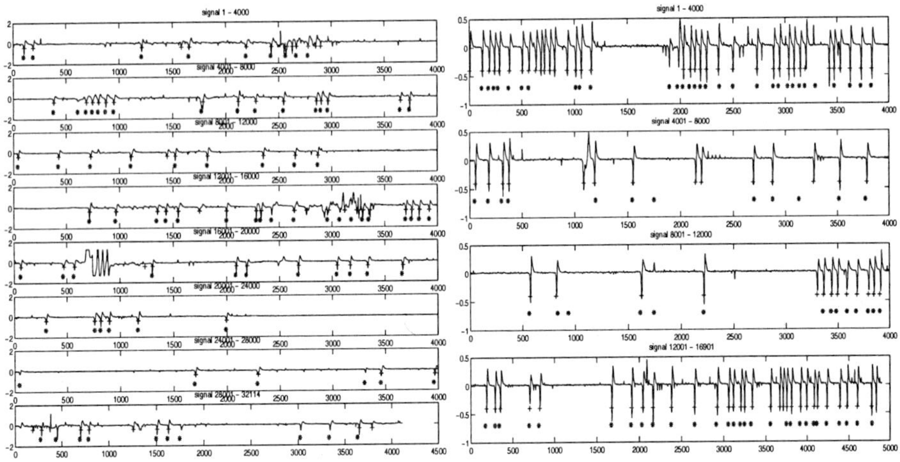

Fig. 2. Automatic detection of eyeblinks in subject 5 (left) and 6 (right); manually labeled eyeblinks labeled are denoted with *, detected eyeblinks with +

the signal. The negative peaks are segmented out and the features are extracted for each blink for in the "true-eyeblink class". The NPSOM is then trained with negative peak data, and the FVSOM is trained with the feature data. After training, our detection system is then applied to the training set and a variety of threshold combinations for both SOMs are evaluated on the test set, in order to find the optimal combination. Around 200 blinks were segmented out of the protocol measurements[2] from *subject 5*. 150 blinks are used for training and 50 blinks are used for threshold setting; 45 non-blink events were segmented as well. The detection system has several adjustable parameters: NPSOM map size, FVSOM map size, size of negative peak window and thresholds for both maps. Experiments suggested that the size of the negative peak used in the NPSOM does not have great influence on the detection results. The detection result for map sizes NPSOM = 8x8, FVSOM = 3x3 led to a detection rate of 94% and a false detection rate[3] of 7%. The detection result on an unseen segment of non-protocol measurements from subject 5 using previously mentioned parameters is shown in figure 2a. For the protocol-measurements of a second subject (*subject 6*), we used a negative peak window size of 20, NPSOM size of 8x8 and FVSOM size 6x6. We varied the thresholds for NPSOM and FVSOM and inspected the results. Setting the thresholds to 1.0 and 0.09 respectively, allowed for 90% detection rate at the expense of 11% false alarms. The detection result of the system on an unseen set of non-protocol measurements from subject 6 using the above set of parameters is shown in figure 2b. Comparing the labeling of events

[2] Protocol: the subject was asked to perform tasks like eye blinking, -closing, and -moving; non-protocol: involuntary activity during conversation and watching a movie

[3] By false detection rate we mean the ratio of the number of detected events (samples) from the non-blink set to the total number of samples in the non-blink set

by a medical expert with our automatic detection we achieved a performance of 97% correct detection at the expense of 3% false alarms. In the measurements at hand, there are almost exclusively blink-events present, which can explain these good results. The overall detection results for both subjects are comparable to the detection results reported in [4].

3 Early Detection of Alzheimer's Disease

Development of Alzheimer's disease can be detected from ElectroEncephalo-Graphic (EEG) measurements of a subject. Automatic analysis of EEG measurements is a highly nontrivial matter. The activity of the underlying sources is often distorted by noise and interferences; the number of sources is usually unknown. Moreover, diagnostic results may depend on the choice of the sensor(s) that are incorporated in the analysis. Inclusion of multiple sensors will introduce redundant information or lead to high-dimensional feature spaces, which may have a negative effect on the generalization performance of a diagnostical system. In several papers [2, 3, 5] it was demonstrated that Independent Component Analysis (ICA) is an effective method to remove artifacts from an ensemble of EEG measurements. Application of this method assumes that the electric dipoles in the cortex can be modelled as independent sources and that the combined activities of all dipoles are measured on the scalp as a linear instantaneous mixture. *Saccadic movements* of the eyes introduce electric activity that distorts the EEG measurements. For quantification of Alzheimer's disease this is an unwanted distortion, since it is not related to the electrical fields in the cortex that are indicative of the disease. In figure 3a two measurements are shown from sensors F7 and F8 (the 10-20 measurement configuration was used). The saccadic activity is present in both channels, and ICA analysis of the two sensors leads to the saccadic movement component that is shown in the middle subfigure. The contribution of the component to the 21 measurement channels is shown in the figure of the head, left of the component plot. It can be observed that the saccade causes an electric field over the scalp that is a dipole. This is due to the change of the polarity caused by the movement of the eyeball. The lower subfigure shows the measurements, where the contribution of the saccadic movements is removed. Two datasets were used during these experiments. Set #1 consists of a control group and a patient group. The control group contains 21 healthy volunteers (12 female, 9 male) of mean age 63 and the patient group consists of 15 severe demented patients (6 female, 9 male), also of mean age 63. All healthy volunteers were in self-reported good health and no relevant abnormalities were found during neurological examination. Set #2 consists of a control group containing 10 subjects (5 male, 5 female) with age associated memory impairment and a patient group consisting of 28 subjects diagnosed as having early Alzheimer's disease. Both groups have mean age 73. The acquired EEG signals were low-pass filtered with cut-off frequency 70 Hz and time constant 1 second at a sampling frequency of 500 Hz. The EEGs were recorded during a "no-task" condition with eyes closed. To test the influence of *correcting* an EEG recording with ICA, a classification experiment is done using features extracted from the original EEGs (from now on referred to as "raw EEGs") and ICA-corrected EEGs (referred to as "ICA EEGs"). The removal of artifacts using ICA-correction is performed in three steps: **1.** computing the ICA components, using the wavelet ICA algorithm by Koehler [3]; **2.** manually selecting the component(s) that represent an artifact; **3.** removing the selected components from the original recording. The selection of components that represent an artifact is not trivial: whether a deduced ICA-component represents an artifact, an EEG source (or both) is deduced from the morphology. The kurtosis of the deduced ICA component can be used for detecting eye-blinks. The spectrum can give an indication whether an ICA component is noise, interference or an EEG source. Classification experiments are performed with set #1 and set #2. For set #1 the leave-one-out method is used[4]. Set #2 has been tested on classifiers that are trained on all measurements in set #1. We trained classifiers to distinguish severe Alzheimer patients from healthy people (i.e. on set #1) and tested the classifier to distinghuish early Alzheimer patients from health subjects (i.e. set #2). We trained a linear discriminant classifier (LDC), a nearest neighbour classifier (NNC) and a feedforward neural network classifier (ANNC) with 5 hidden neurons. Results presented from ANNC are averaged results over several runs, to diminish the influence of initial settings of the neural network. Each preprocessed segment is then represented by a spectrum-based feature[5]. Several features were tested, and the relative spectral power in the θ frequency band (4-8 Hz) proved to be very suitable. The results in Table 1. indicate that ICA-correction improves the classification performance in case of beginning Alzheimer's disease; the choice of the classifier seems to have no significant effect. On average our results are slightly worse compared to [1], which can be explained by the *different approach for marking a patient* as being in an early stage of Alzheimer's disease. Preprocessing with ICA, however, generally improves the detection of early Alzheimer's disease. In the case of severe Alzheimer's disease (set #1) less improvement was obtained with ICA preprocessing, which can be explained from the larger influence of interfering sources in the early stages of the disease.

4 Machine Condition Monitoring in a Pumping Station

chine under investigation is quite large and measurement directions are not always the same, this assumption may not hold. Incorporation of multiple channels in the above manner might improve robustness (less dependence on particular sensor to be selected), but on the other hand might introduce class overlap, because uninformative channels are treated equally important as informative channels. In a pumping station at Urk, The Netherlands, a gasmotor drives either a pump or a generator. At two different moments, vibration

[4] A classifier is trained with all samples but one and the left-out sample is used as test set and thus classified. This is repeated until all samples are classified

[5] It is known that Alzheimer's disease leads to a 'slowing' of the EEG signal

Table 1. Detection of early-stage Alzheimer's disease with and without interference removal; the format of the entries in the table is: *total classification performance (sensitivity / specificity)* (%); relative theta power is chosen as feature

sensor combination	combination method	LDC	NNC	ANNC
Fz, Pz and Cz:	raw	66(54/100)	63(50/100)	65(53/100)
	ICA	71(61/100)	71(61/100)	71(60/99)
17 channels:	raw	55(39/100)	63(50/100)	65(55/93)
	ICA	55(43/90)	71(61/100)	70(60/99)
21 channels:	raw	55(43/90)	66(54/100)	71(62/97)
	ICA	61(54/80)	74(64/100)	69(59/100)

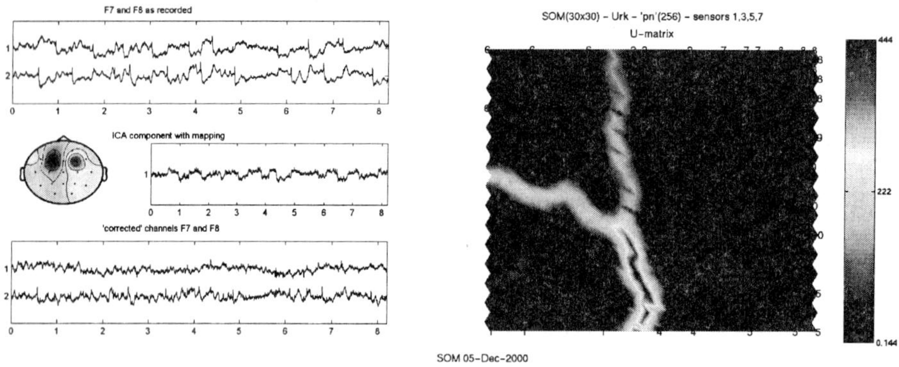

Fig. 3. a: Removal of saccadic activity from ensemble of EEG measurements; **b:** SOM of spectra indicating a developing fault at Urk pumping station; in generator mode, no significant differences are observed in the measured patterns

measurements on several positions of the machinery were performed at different operating modes (pumping; generator mode at four different power generation levels). Power spectra with 256 bins were estimated on segments of 4096 samples. The dataset was reduced in dimensionality with PCA to 100 features. Then a 30x30 SOM was trained using the SOM-toolbox from TU Helsinki, `http://www.cis.hut.fi/projects/somtoolbox`. The resulting map (visualized with the U-matrix method) is shown in figure 3b. Labels 1 - 5 denote measurements in March 2000 (1 = pumping mode, 2 - 5 = generator modes). Labels 6 - 10 denote measurements in May 2000 (6 = pumping mode, 7 - 10 = generator modes). Patterns with label 1 are mapped in the lower-left part of the map, patterns with label 6 are mapped in the upper-left part and all generator mode patterns are mapped close to each other in the right part of the map. During pumping, a gearbox failure has developed in the second measurement series; it is absent in the first measurement series. However, in generator mode no significant differences in condition between both measurements are observed by the human expert. This is visible in the map as small distances between the 'generator mode neurons' (blue colours), whereas the 'pumping mode neurons' are separated with large distance (the light ridge between the '1' and '6' areas). gekleurde scheidingslijn tussen beide gebieden). De gebieden '2' tot '5' en '7' tot '10' liggen veel dichter bij elkaar. Tenslotte wijken de metingen tijdens pompbedrijf sterk af van de metingen tijdens generatorbedrijf.

5 Conclusion

We presented a framework for practical health monitoring with learning methods. A particular monitoring problem will often incorporate several subproblems that are present in our framework. We demonstrated the feasibility of the learning approach to system monitoring in three real-world monitoring problems.

References

1. P. Anderer. Discrimination between demented patients and normals based on topographic eeg slow wave activity. *EEG and Clin. Neurophys.*, (91):108 – 117, 1994.
2. T. P. Jung et al. Removing EEG artifacts by BSS. *Psychophys.*, (37):163–178, 2000.
3. B.-U. Koehler and R. Orglmeister. Ica of electroencephalographic data using wavelet decomposition. In *Proc. of VIII Mediterranian Conference on MBEC*, Cyprus, 1998.
4. X. Kong and G. Wilson. A new eog-based eyeblink detection algorithm. *Behavior Research Methods, Instruments & Computers*, 30(4):713 – 719, 1998.
5. R. N. Vigario. Extraction of ocular artifacts from eeg using independent component analysis. *Electroencephalography and Clinical Neurophysiology*, 103:395 – 404, 1997.
6. A. Ypma and R. P. W. Duin. Novelty detection using self-organizing maps. In *Proc. of ICONIP'97*, pages 1322–1325. Springer-Verlag Singapore, 1997.

Breast Tissue Classification in Mammograms Using ICA Mixture Models

Ioanna Christoyianni, Athanasios Koutras, Evangelos Dermatas, and George Kokkinakis

WCL, Electrical & Computer Engineering Dept., University of Patras
26100 Patras, Hellas
ioanna@ios.wcl2.ee.upatras.gr

Abstract. In this paper we present a novel method for recognizing all kinds of abnormalities in digital mammograms using Independent Component Analysis mixture models and two sets of statistical features based on texture analysis. Our approach is concentrated on finding the ICA mixture model parameters that describe in an exclusive and effective way the abnormal and the normal tissue, and with the aid of a supervised probabilistic classifier we are able to successfully recognize suspicious regions in mammograms. Extensive experiments using the MIAS database have shown great accuracy of 98.33% in classifying an unknown regions of suspicion as abnormal and 62.71% as healthy tissue.

1 Introduction

Breast cancer is a leading cause of fatality among all cancers for women. Therefore, it is crucial to be detected in its early stages of development because early detection allows treatment before it is spread to other parts of the body. Mammography is currently the single most effective and low-cost technique for reliable detection of early, non-palpable breast cancer [1]. The mammograms are searched for signs of abnormality by expert radiologists but complex structures in appearance and signs of early disease are often small or subtle. That's the main cause of many missed diagnoses that can be mainly attributed to human factors [1,2]. Since the consequences of errors are costly, there has been a tremendous interest in developing methods for automatically classifying suspicious areas of breast tissue, as a means of aiding radiologists by improving the efficacy of screening programs and avoiding unnecessary biopsies.

Among the various types of breast abnormalities, clustered microcalcifications and masses are the most important ones that characterize early breast cancer [3] and are detectable in mammograms before a woman or the physician can palp them. Masses appear as dense regions of varying sizes and properties and can be characterized as circumscribed, spiculated, or ill defined. On the other hand, microcalcifications, appear as small bright arbitrarily shaped regions on the large variety of breast texture background. Finally, asymmetry and architectural distortion is very a important, yet difficult abnormality to detect. The great variability in their appearance in digital mammograms is the main obstacle for building a unified detection method.

G. Dorffner, H. Bischof, and K. Hornik (Eds.): ICANN 2001, LNCS 2130, pp. 554–560, 2001.

Neural network computer-aided diagnosis (CAD) for detecting masses or microcalcifications in mammograms, has been widely used [4-6]. However, the exploration of other types of classifiers for detecting all kinds of abnormalities has been very limited. In this study, we propose a novel method to classify any region of suspicion (ROS) in mammograms using Independent Component Analysis (ICA) mixture models.

ICA is a signal processing technique that has shown promise in various signal processing applications [7], and its purpose is to decompose a set of multivariate data into a set of Independent Components. The task of detecting all kinds of abnormalities in digital mammograms can be assisted by ICA, by estimating the appropriate ICA mixture models that describe effectively the abnormal and the normal tissue. Using such models [8] we have implemented a novel probabilistic classifier capable of successfully categorizing the observed data into two mutually exclusive classes; the abnormal and the healthy tissue.

To estimate the ICA mixture models first a feature extraction technique is implemented and then using ICA techniques, we estimate the model parameters that describe effectively the two classes. The classification procedure is carried out by processing each unknown ROS with both models parameters and computing the probability of each class. The final diagnosis is made according to the highest-class probability. Extensive experiments using the MIAS database, showed that our proposed method has an accuracy of 98.33% in classifying an unknown ROS as abnormal and 62.71% as healthy tissue.

The structure of this paper is as follows: In the next section, a brief description of the overall method is presented while in section 3 a detailed description of the features extracted from mammograms is given. In section 4, the implementation of the ICA mixture model classifier is given. In section 5 we present the experimental results and finally, in section 6, some conclusions are drawn.

2 Overall Proposed Method

In Figure 1 we present the basic scheme of the proposed ICA mixture model classifier. In the training procedure a subset of abnormal, together with an equal number of normal regions are fed into the feature extraction module the output of which is a set of N-dimensional feature vectors for each category. Each set of the normal and the abnormal feature vectors is next processed by an ICA neural network that estimates the mixture model parameters for the abnormal $\{\mathbf{A}_{ab},\mathbf{b}_{ab}\}$ and the normal $\{\mathbf{A}_{n},\mathbf{b}_{n}\}$ tissue. For the testing procedure, each unknown ROS is fed into the feature extraction module, and the features are post-processed by the ICA based classifier using the learned parameters $\{\mathbf{A}_{ab},\mathbf{b}_{ab}\}$ and $\{\mathbf{A}_{n},\mathbf{b}_{n}\}$. The probability of each class $p(\mathrm{NORMAL}|\mathrm{ROS}_i,\{\mathbf{A}_{n},\mathbf{b}_{n}\})$ and $p(\mathrm{ABNORMAL}|\mathrm{ROS}_i,\{\mathbf{A}_{ab},\mathbf{b}_{ab}\})$ is computed in the ICA classifier stage and the final diagnosis is made, based on the higher class probability. In the next section each stage of the proposed method is analyzed in detail.

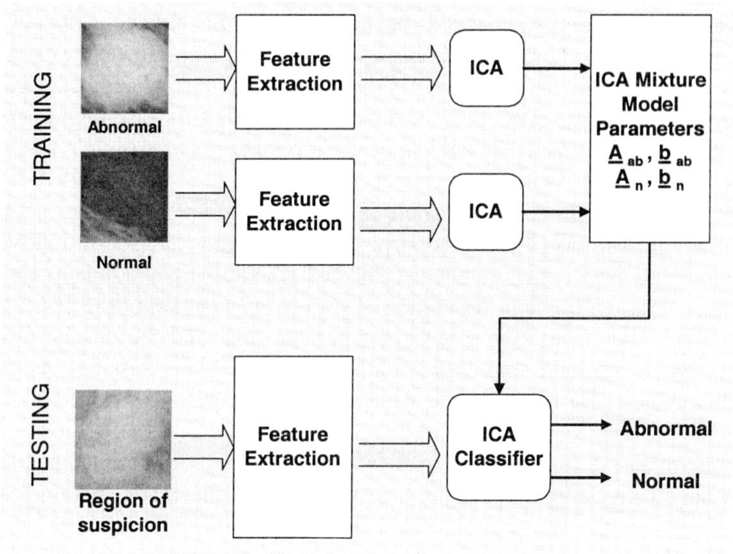

Fig. 1. Basic scheme of the proposed method

3 Feature Extraction

The implemented feature extraction procedure relies on the texture, which is the main descriptor for all kinds of mammograms. In this paper, we concentrate on statistical descriptors that include averages, standard deviations, and higher-order statistics of intensity values and on the Spatial Gray Level Dependence Matrix (SGLD) for texture description, which are widely used [5,6].

3.1 Gray Level Histogram Moments

Four gray level sensitive histogram moments are extracted from the pixel value histogram of each region and are defined as follows:

$$\text{Mean: } \mu = \sum_{k=1}^{N} f_k p_f(f_k) \quad , \text{ Variance: } \sigma^2 = \sum_{k=1}^{N} (f_k - \mu)^2 p_f(f_k) \tag{1}$$

$$\text{Skewness: } \mu_3 = \frac{1}{\sigma^3} \sum_{k=1}^{N} (f_k - \mu)^3 p_f(f_k) \text{, Kurtosis: } \mu_4 = \frac{1}{4} \sum_{k=1}^{N} (f_k - \mu)^4 p_f(f_k) \text{,} \tag{2}$$

where: N denotes the number of gray levels in the mammogram, f_k is the k-th gray-level and $p_f(f_k) = n_k / n$, where n_k is the number of pixels with f_k gray-level and n is the total number of pixels in the region.

3.2 Spatial Gray Level Dependence Matrix

The Spatial Gray Level Dependence Matrix is constructed by counting the number of occurrences of pixel pairs at a given displacement. In particular, to compute the SGLD matrix for an image *H(i,j)*, a displacement vector *d=(x,y)* is defined. The *(i,j)-th* element of the SGLD matrix *P* is defined as the joint probability that gray levels *i* and *j* occur at a distance *d* in *H(i,j)*:

$$P(i,j,d,\theta^o) = Count\{((u,v),(s,t)) \in WxW \mid ((f(u,v)=i \wedge f(s,t)=\mathrm{j}) \vee (f(u,v)=i \wedge f(s,t)=j)) \wedge |\mathrm{s}-\mathrm{u}|+|\mathrm{t}-\mathrm{v}| = d \wedge \theta^o = \arctan((t-v)/(s-u))\}, \quad (3)$$

where, *Count{x}* is the number of elements of the set *x*, $W=\{0,1, \ldots, n_{gl}-1\} \times \{0,1, \ldots, n_{gl}-1\}$, and n_{gl} is the size of the rectangular window *W*. The matrix is then normalized so that it can be treated as a probability density function (pdf).

$$p(i,j,d,\theta^o) = \frac{P(i,j,d,\theta^o)}{\sum_{i,j} P(i,j,d,\theta^o)}. \quad (4)$$

In this paper for a fixed displacement *d* and angle θ^o, ten texture features are extracted from the density function *p*, the definitions of which can be found in [9].

4 ICA Mixture Model Classifier

When a mixture model is used in classification problems [10-11], the observed data are categorized into several mutually exclusive classes. The utilized ICA mixture model, which is a generalization of the Gaussian mixture model (GMM), models each class with independent variables. The main superiority that the ICA mixture model presents, is that it allows modeling of classes with non-Gaussian structure (i.e. leptokyrtic or platykyrtic), which finds exact implementation in the case of breast tissue characterization.

Let assume that the observed data $\mathbf{X}=\{\mathbf{x}_1,\mathbf{x}_2,\ldots,\mathbf{x}_T\}$ (with $\mathbf{x}_i$ an N-dimensional vector) are drawn independently and are generated by a mixture density model. The likelihood of the data is given by:

$$p(\mathbf{X}|\Theta) = \prod_{t=1}^{T} p(\mathbf{x}_t|\Theta) \quad (5)$$

and the model's mixture density is

$$p(\mathbf{x}_t|\Theta) = \sum_{k=1}^{K} p(\mathbf{x}_t|C_k,\theta_k)p(C_k) , \quad (6)$$

where $\Theta = \{\theta_1, \theta_2, \ldots, \theta_K\}$ are the unknown parameters for each component density $p(\mathbf{x}|C_k,\theta_k)$. By C_k we denote the class k, and it is assumed throughout the paper that the number of classes is known beforehand. The data $\mathbf{x}_t$ within each class are considered to be generated by the source vectors $\mathbf{s}_k$ by

$$\mathbf{x}_t = \mathbf{A}_k \mathbf{s}_k + \mathbf{b}_k$$

where $\mathbf{A}_k$ is a NxN scalar matrix and $\mathbf{b}_k$ is a vector containing the biases for each class.

For the estimation of the mixture model parameters $\{\mathbf{A_k}, \mathbf{b_k}\}$ for each class, we process the training data of each class separately in the following way: Let us assume we want to find the model parameters for class C_k. The observations $\mathbf{X_k}$ are fed into the ICA network shown in Figure 1, which results to the independent source data $\mathbf{s_k}$ for the class C_k:

$$\mathbf{s}_k = \mathbf{W}_k \cdot \mathbf{X}_k .$$

The matrix $\mathbf{W_k}$ is a scalar NxN matrix, called separating matrix. The separating matrix is learnt adaptively by [12]:

$$\Delta\mathbf{W}_k = n \cdot \left(\mathbf{I} - f(\mathbf{s}_k) \cdot \mathbf{s}_k^{\ T}\right) \cdot \mathbf{W}_k \ , \tag{7}$$

where the nonlinear function $f(.)$ is directly associated with the probability density function of the source data $\mathbf{s_k}$ in each class and plays an important role to the efficacy of the ICA network's performance. For the case of normal and abnormal regions of mammograms, we have concluded experimentally that the best approximation for their probability density function is a Laplacian density of the form $p(s_k)=exp(-|s_k|)$. For this case, the choice of $f(\mathbf{s}_k) = sign(\mathbf{s}_k)$ proved to be the best for the adaptation rule in (7). The model parameter $\mathbf{A_k}$ is the inverse of the separating matrix $\mathbf{W_k}$ and can be calculated by $\mathbf{A_k}=\mathbf{W_k}^{-1}$. The bias vector $\mathbf{b_k}$ is estimated as the mean value of the observations $\mathbf{X}_k$ and it's role is to allow non-zero-meaned data to be classified as well.

At the end of the learning procedure, we estimate the K sets of parameters for our ICA mixture model $\theta_k=\{\mathbf{A_k}, \mathbf{b_k}\}$. These are used in the classification procedure, which consists of the following steps:

- First we compute the log-likelihood of the testing data for each one of the classes using:

$$\log p(\mathbf{x}_t \mid C_k, \theta_k) = \log p(\mathbf{s}_k) - \log(\det|\mathbf{A}_k|) \ . \tag{8}$$

  Note that in the above equation $\mathbf{s_k}$ is calculated by $\mathbf{s}_k = \mathbf{A}_k^{-1} \cdot (\mathbf{x}_t - \mathbf{b}_k)$.
- Next the probability for each class is computed given the testing vector $\mathbf{x}_t$ as:

$$p(C_k \mid \mathbf{x}_t, \theta_k) = \frac{p(\mathbf{x}_t \mid Ck, \theta_k) p(C_k)}{\sum_k p(\mathbf{x}_t \mid Ck, \theta_k) p(C_k)} \ . \tag{9}$$

- Finally, the classification of the unknown data $\mathbf{x}_t$ is performed using the following decision rule based on the maximum class probability:

$$\mathbf{x}_t \in \arg\max_{C_k} (p(C_k \mid \mathbf{x}_t, \theta_k)) \ . \tag{10}$$

5 Experimental Results

In our experiments the MIAS MiniMammographic Database [13], provided by the Mammographic Image Analysis Society (MIAS), was used. The mammograms have been digitized at 200- micron pixel edge, resulting to a 1024x1024-pixel resolution.

In the MIAS Database there is a total of 119 ROS containing all kinds of existing abnormal tissue from masses to clustered microcalcifications. These 119 ROS along with another 119 randomly selected sub-images from entirely normal mammograms were used throughout our experiments. From a total number of 238 ROS we used 119 regions for the training procedure (*jackknife* method): 59 groundtruthed abnormal regions along with 60 randomly selected normal ones. For the evaluation we used the remaining 119 regions that contain 60 groundtruthed abnormal regions together with 59 entirely normal regions. Our experiments were run 10 times with random initial conditions and the averaged classification results were computed.

Two sets of features based on texture analysis were tested. In the first set of experiments we examined the effectiveness of the gray level histogram moments (GLHM). The ICA mixture model based classifier recognized correctly 42 out of 60 abnormal and 50 out of 59 normal ROS, giving a total number of 92 out of 119 correct classifications. The average classification results are shown in Table 1.

Table 1. Recognition Rate for GLHM fetures

	Gray Level Moments
ABNORMAL	70 %
NORMAL	84.75 %
TOTAL	77.31 %

In the second set of experiments we examined the effect of the SGLD features on the classification accuracy at different phases of the same classifier. Since the image matrix is descrete, the displacement vector used for feature extraction was chosen to have the following phase and displacement values: (0°,1), (45°,1), (-45°,1). The best results were obtained using the SGLD matrix features for a displacement of 45° achieving a total recognition score of 80.67% (96 ROS out of 119 were correctly classified). Analytically, 59 ROS out of 60 were accurately classified as abnormal and 37 out of 59 ROS as normal, as can also be seen in the results in Table 2.

Table 2. Recognition Rate for SGLD Matrix fetures

	ABNORMAL	**NORMAL**	**TOTAL**
-45°	93.33 %	62.71 %	78.15 %
0°	90 %	67.79 %	78.99 %
45°	**98.33 %**	62.71 %	**80.67 %**

6 Conclusion

In this paper we investigated the performance of a probabilistic classifier based on ICA mixture models and two sets of statistical features, in the problem of recognizing breast cancer in digital mammograms. It is well known that the disease diagnosis on mammograms is a very difficult task even for experienced radiologists due to the great variability of the mass appearance. The novel approach used in this work, is concentrated on finding the ICA mixture model parameters that describe effectively and in an ample way the abnormal and the normal tissue, and with the aid of a probabilistic classifier we are able to successfully detect cancerous regions in mammograms. Extensive experiments showed great performance for both tested feature types, reaching in some cases a recognition accuracy of 98.33%.

References

1. Martin, J., Moskowitz, M. and Milbrath, J.: Breast cancer missed by mammography. AJR, Vol. 132. (1979) 737
2. Kalisher, L.: Factors influencing false negative rates in xero-mammography. Radiology, Vol.133. (1979) 297
3. Tabar, L. and Dean, B.P.: Teaching Atlas of Mammography.2nd edition, Thieme, NY (1985)
4. Christoyianni, I., Dermatas, E., and Kokkinakis, G.: Fast Detection of Masses in Computer-Aided Mammography. IEEE Signal Processing Magazine, vol. 17, no 1. (2000) 54-64
5. Sonka, M., Fitzpatrick, M.,: Handbook of Medical Imaging, Vol 2, SPIE Press (2000).
6. Doi, K., Giger, M., Nishikawa, R., and Schmidt, R. (eds.): Digital Mammography 96. Elsevier Amsterdam (1996)
7. Lee, Te-Won: Independent Component Analysis: Theory and Applications. Kluwer Academic Publishers (1998)
8. Lee, T.-W., Lewicki, M: The Generalized Gaussian Mixture Model Using ICA. International Workshop on Independent Component Analysis. Helsinki (2000) 239-244
9. Haralick, R.: Statistical and Structural Approaches to Texture. Proc. IEEE, Vol. 67, No. 4. (1979) 786-804
10. Lee, T.-W., Lewicki, M and Sejnowski, T.: ICA Mixture Models for Unsupervised Classification of Non-Gaussian Sources and Automatic Context Switching in Blind Signal Separation. IEEE Trans. on Pattern Recognition and Machine Intelligence 22(10) (2000) 1-12
11. Duda, R and Hart, P.: Pattern Classification and Scene Analysis. New York Wiley (1973)
12. Cardoso, J.: Blind Signal Separation: statistical principles. IEEE Proceedings 9 (10) (1998) 2009-2025
13. http://s20c.smb.man.ac.uk/services/MIAS/MIASmini.html

Neural Network Based Blind Source Separation of Non-linear Mixtures

Athanasios Koutras, Evangelos Dermatas, and George Kokkinakis

WCL, Electrical and Computer Engineering, University of Patras, Hellas
koutras@giapi.wcl2.ee.upatras.gr

Abstract. In this paper we present a novel neural network topology capable of separating simultaneous signals transferred through a memoryless non-linear path. The employed neural network is a two-layer perceptron that uses parametric non-linearities in the hidden neurons. The non-linearities are formed using a mixture of sigmoidal non-linear functions and present greater adaptation towards separating complex non-linear mixed signals. Simulation results using complex forms of non-linear mixing functions prove the efficacy of the proposed algorithm when compared to similar networks that use standard non-linearities, achieving excellent separation performance and faster convergence behavior.

1. Introduction

Blind Source Separation (BSS) consists in recovering a set of statistically independent signals given only a number of observed mixture data. The term "blind" is justified by the fact that the only *a-priori* knowledge that we have for the signals is their statistical independence and no information about the mixing model parameters is available beforehand. The BSS problem has been intensively studied for the linear instantaneous (memoryless) mixing case and more recently for the linear convolutive case [1,2] showing very good and promising results. However, up to now, presentation of algorithms that deal with more realistic mixing models, especially non-linear ones, has been very limited.

In its full generality, the non-linear BSS (NLBSS) problem is intractable since the indeterminacies in the separating solutions are much more severe than those existing in the linear mixing case [3]. Basically, this can be mainly attributed to the fact that any non-linear transformations $f(x)$ and $g(y)$ of two independent random variables x and y are still independent. As a result, without any prior knowledge about the mixing function of the non-linear mixing model, criteria using only statistical independence cannot be considered efficient for recovering the source signals without distortions. Under these facts, to limit the function class of the de-mixing functions is equivalent to assume some prior knowledge about the mixing function.

In the last years a few algorithms that deal with the NLBSS problem have been presented, some of them based on neural networks [4-6]. Lee *et al.* [7] have addressed the problem of non-linear separation based on information maximization criteria, using a

G. Dorffner, H. Bischof, and K. Hornik (Eds.): ICANN 2001, LNCS 2130, pp. 561–567, 2001.

parametric sigmoidal non-linearity and higher order polynomials. Taleb and Jutten [8] have dealt with a special case of post non-linear mixtures (PNL) using entropy maximization criteria. Finally, Yang *et al.* [9] have proposed a two-layer perceptron with parametric sigmoidal non-linearities to separate non-linear mixtures. A recent survey that describes the NLBSS problem can be also found in [3].

In this paper, we present a neural network based approach towards solving the non-linear signal separation problem of unknown sources. The employed neural network is a two-layer perceptron that uses a generalized form of a mixture of parametric sigmoidal non-linearities in the hidden neurons. This non-linearity is a generalization of previous solutions [9] and presents a number of advantages compared to the standard sigmoid function used in neural networks. In particular, it shows a greater flexibility towards fitting more complex non-linear mixing functions, thus it contributes to a better demixing accuracy and performance by the separation network as our experimental results have shown. The separation network's learning rules have been derived using the Maximum Likelihood Estimation (MLE) criterion and show great efficacy when used to separate non-linearly mixed signals.

The structure of this paper is as follows: In section 2, we present the general NLBSS model and derive the learning rules for the neural network parameters using the MLE criterion. Additionally, the network's performance index is introduced as a mean to measure the efficacy of the separation performance. In section 3 our experiments are presented and finally in the last section some conclusions and remarks are given.

2. Non-linear Blind Source Separation

Let us have N statistically independent sources $\{s_1, s_2,..,s_N\}$ that are mixed by a non-linear memoryless model. The mixed signals $\{x_1,x_2,..,x_N\}$ are formulated by:

$$\mathbf{x}(t)=f(\mathbf{s}(t)), \tag{1}$$

where f is an unknown non-linear mixing function.

In this paper, we assume that f is invertible and it's inverse f^{-1} can be approximated using the neural network shown in Fig. 1. The output $\mathbf{y}(t)$ of the neural network is given by

$$\mathbf{y}(t)=\mathbf{W}_1 g(\mathbf{W}_2\mathbf{x}(t)) \, , \tag{2}$$

where $\mathbf{g}$ denotes the hidden neurons non linear transfer function and $\mathbf{W}_1$, $\mathbf{W}_2$ are non-singular square matrices containing the neural network's weights. The non-linear function $\mathbf{g}$ was chosen to be a mixture of parametric sigmoidal functions in the form of:

$$g(x) = \sum_{i=1}^{K} \tanh(a_i \cdot x + b_i) \, . \tag{3}$$

The parameters a_i and b_i of the non-linear function $g(x)$ control the slope and the position of each component in the mixture of the sigmoids and are learnt adaptively together with the separating matrices $\mathbf{W}_1$, $\mathbf{W}_2$ by the proposed adaptation rules which are introduced in the next section.

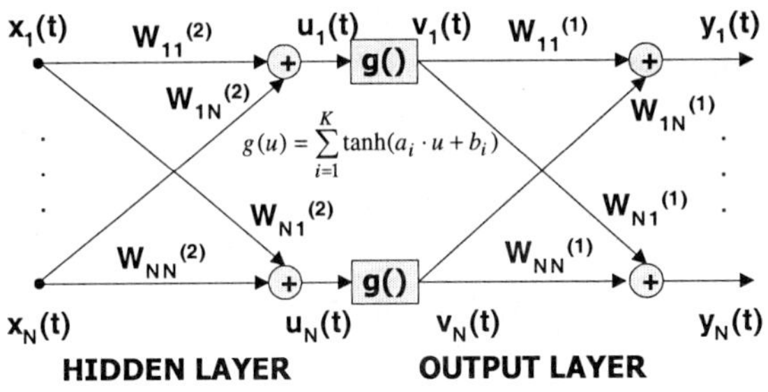

Fig. 1. The non-linear blind signal separation neural network

The parameters a_i and b_i of the non-linear function $g(x)$ control the slope and the position of each component in the mixture of the sigmoids and are learnt adaptively together with the separating matrices $\mathbf{W}_1$, $\mathbf{W}_2$ by the proposed adaptation rules which are introduced in the next section.

2.1 The Maximum Likelihood Estimator

We apply the MLE criterion to estimate the unknown networks parameters, namely the matrices $\mathbf{W}_1$, $\mathbf{W}_2$, $[\boldsymbol{a}]_{ij}$ and $[\boldsymbol{b}]_{ij}$ $\{i=1..N, j=1..K\}$ in an unsupervised manner. From (2), the log-likelihood of the observed data is given by:

$$L = \log(p(\mathbf{x})) = \log(|\mathbf{W}_2|) + \log(|\mathbf{W}_1|) + \sum_{i=1}^{N} \log\left(\frac{\partial g_i(u_i)}{\partial u_i}\right) + \log(p(\mathbf{y}; \mathbf{W}_1, \mathbf{W}_2, \boldsymbol{\alpha}, \mathbf{b})) \quad (4)$$

The weights of the hidden and the output layer of the network and the unknown parameters of the mixtures of sigmoids can be estimated using the stochastic gradient of L with respect to the $\mathbf{W}_1$, $\mathbf{W}_2$, $\boldsymbol{a}$ and $\boldsymbol{b}$ as follows.

2.1.1 Output Weights

The stochastic gradient of the log-likelihood with respect to the output layer weights $\mathbf{W}_1$ is given in a matrix form:

$$\Delta\mathbf{W}_1 \propto \frac{\partial L}{\partial \mathbf{W}_1} = \left[\mathbf{W}_1^T\right]^{-1} - \mathbf{\Phi}_1(\mathbf{y}) \cdot \mathbf{v}^T, \quad (5)$$

where the matrix $\mathbf{\Phi}_1(\mathbf{y})$ is given by:

$$\mathbf{\Phi}_1(\mathbf{y}) = -\left[\frac{p'_1(y_1; \mathbf{W}_1, \mathbf{W}_2, \boldsymbol{\alpha}, \mathbf{b})}{p_1(y_1; \mathbf{W}_1, \mathbf{W}_2, \boldsymbol{\alpha}, \mathbf{b})}, \ldots, \frac{p'_N(y_N; \mathbf{W}_1, \mathbf{W}_2, \boldsymbol{\alpha}, \mathbf{b})}{p_N(y_N; \mathbf{W}_1, \mathbf{W}_2, \boldsymbol{\alpha}, \mathbf{b})}\right]^T. \quad (6)$$

If we approximate the marginal pdf of the output signals $\mathbf{y}$ using the Gram-Charlier expansion as in [9], the *i-th* element of $\mathbf{\Phi}_1$ becomes:

$$\mathbf{\Phi}_1(y_i) = \left(\frac{-k_3}{2} + \frac{9k_3k_4}{4}\right)y_i^2 + \left(\frac{-k_4}{6} + \frac{3k_3^2}{2} + \frac{3k_4^2}{4}\right)y_i^3, \quad (7)$$

where $k_3=m_3$ and $k_4=m_4-3$ and $m_k=E[(y_i^k)]$. The cumulants k_3 and k_4 are estimated adaptively using the equations:

$$\Delta k_3^{(i)} = -n(k_3^{(i)} - (y_i)^3), \Delta k_4^{(i)} = -n(k_4^{(i)} - (y_i)^4 + 3). \tag{8}$$

To avoid the computationally expensive inverse matrix operations in (5), we use the natural gradient approach and derive the following weight adaptation rule for $\mathbf{W}_1$:

$$\Delta \mathbf{W}_1 \propto \frac{\partial L}{\partial \mathbf{W}_1} \mathbf{W}_1{}^T \mathbf{W}_1 = \left[\mathbf{I} - \mathbf{\Phi}_1(\mathbf{y})\mathbf{y}^T\right] \cdot \mathbf{W}_1. \tag{9}$$

2.1.2 Hidden Layer Weights

Using similar calculations as before, we derive the stochastic gradient of the log-likelihood L with respect to the weights $\mathbf{W}_2$:

$$\Delta \mathbf{W}_2 \propto \frac{\partial L}{\partial \mathbf{W}_2} = \left[\mathbf{W}_2^T\right]^{-1} - \mathbf{\Phi}_2(\mathbf{u})\mathbf{x}^T - \mathbf{D}(\mathbf{u})\mathbf{W}_1^T \mathbf{\Phi}_1(\mathbf{y})\mathbf{x}^T. \tag{10}$$

The elements of $\mathbf{\Phi}_2$ and the diagonal matrix $\mathbf{D}$ are given by:

$$\mathbf{\Phi}_2(u_i) = -\frac{\sum_{j=1}^{K} \frac{\partial^2}{\partial^2 u_i}\left(\tanh(a_{ij}u_i + b_{ij})\right)}{\sum_{j=1}^{K} \frac{\partial}{\partial u_i}\left(\tanh(a_{ij}u_i + b_{ij})\right)}, \quad \mathbf{D}(u_i) = \sum_{j=1}^{K} \frac{\partial}{\partial u_i}\left(\tanh(a_{ij}u_i + b_{ij})\right). \tag{11}$$

Using the natural gradient approach, the following adaptation rule is derived for the estimation of the hidden layer's network weights:

$$\Delta \mathbf{W}_2 \propto \frac{\partial L}{\partial \mathbf{W}_2} \mathbf{W}_2{}^T \mathbf{W}_2 = \left[\mathbf{I} - \mathbf{\Phi}_2(\mathbf{u})\mathbf{u}^T - \mathbf{D}(\mathbf{u})\mathbf{W}_1^T \mathbf{\Phi}_1(\mathbf{y})\mathbf{u}^T\right] \cdot \mathbf{W}_2 \tag{12}$$

2.1.3 Sum of Parametric Sigmoidal Non-linearities

Let $\boldsymbol{a}_j = [a_{1j},..,a_{Nj}]^T$ denote the vector with the slope parameters of the *j-th* component of the network's non-linearities. By taking the stochastic gradient of L with respect to the vector $\boldsymbol{a}_j$ we find:

$$\Delta \boldsymbol{\alpha}_j \propto \frac{\partial L}{\partial \boldsymbol{\alpha}_j} = \left[\mathbf{\Phi}_a^{(j)}(\mathbf{u}) + \mathbf{D}_a^{(j)} \mathbf{W}_1^T \mathbf{\Phi}_1(\mathbf{y})\right], \tag{13}$$

where

$$\mathbf{\Phi}_a^{(j)}(u_i) = -\frac{\frac{\partial^2}{\partial a_{ij}\partial u_i}\left(\tanh(a_{ij}u_i + b_{ij})\right)}{\sum_{k=1}^{K} \frac{\partial}{\partial u_i}\left(\tanh(a_{ik}u_i + b_{ik})\right)}, \quad \mathbf{D}_a^{(j)}(u) = \frac{\partial}{\partial a_{ij}}\left(\tanh(a_{ij}u_i + b_{ij})\right). \tag{14}$$

Similarly, the adaptation equation for the biases of the non-linearities $\boldsymbol{b}_j$ is:

$$\Delta \mathbf{b}_j \propto \frac{\partial L}{\partial \mathbf{b}_j} = \left[\mathbf{\Phi}_b^{(j)}(\mathbf{u}) + \mathbf{D}_b^{(j)} \mathbf{W}_1^T \mathbf{\Phi}_1(\mathbf{y})\right], \tag{15}$$

where,

$$\mathbf{\Phi}_b^{(j)}(u_i) = -\frac{\frac{\partial^2}{\partial b_{ij}\partial u_i}\left(\tanh(a_{ij}u_i + b_{ij})\right)}{\sum_{k=1}^{K}\frac{\partial}{\partial u_i}\left(\tanh(a_{ik}u_i + b_{ik})\right)}, \quad \mathbf{D}_b^{(j)}(u_i) = \frac{\partial}{\partial b_{ij}}\left(\tanh(a_{ij}u_i + b_{ij})\right). \qquad \mathbf{(16)}$$

2.2 Performance Index

The performance of the proposed NLBSS method can be measured by the mutual information of the output **y,** $I(\mathbf{y};\mathbf{W}_1,\mathbf{W}_2,\boldsymbol{\alpha},\mathbf{b}) \approx -H(\mathbf{x}) + PI$, where

$$PI = -\log\left(\left|\mathbf{W}_1\mathbf{W}_2\right|\right) - \sum_{i=1}^{N}\left(E\left[\log\left(\frac{\partial}{\partial u_i}\sum_{j=1}^{K}\tanh(a_{ij}u_i + b_{ij})\right)\right]\right) + \sum_{i=1}^{N}F(k_3^{(i)}, k_4^{(i)}), \qquad \mathbf{(17)}$$

$$F(k_3^{(i)}, k_4^{(i)}) = \frac{1}{2}\log(2\pi e) - \frac{\left(k_3^{(i)}\right)^2}{2\cdot 3!} - \frac{\left(k_4^{(i)}\right)^2}{2\cdot 4!} + \frac{3}{8}\left(k_3^{(i)}\right)^2 k_4^{(i)} + \frac{1}{16}\left(k_4^{(i)}\right)^3. \qquad \mathbf{(18)}$$

Since $H(\mathbf{x})$ does not change in the learning process, we may use PI as the performance index which may take negative values. The difference between $I(\mathbf{y};\mathbf{W}_1,\mathbf{W}_2,\boldsymbol{\alpha},\mathbf{b})$ and PI is $-H(\mathbf{x})$.

3. Experiments

To evaluate the proposed non-linear signal separation network and the convergence behavior of the derived learning rules, we performed several experiments in various kinds of non-linear mixing scenarios. These included testing of our proposed method with post non-linear mixtures (PNL) as in [7], with a simple sigmoid non-linear mixing function as in [9] and finally in a more difficult situation where the non-linear mixing function was formed by a sum of three sigmoid non-linearities. Indicatively for the last case, three independent source signals $s_1(t)=sin(2\ 10t)$, $s_2(t) = sin(2\ 30t+6cos(2\ 6t))$ and $s_3(t)$: a uniformally distributed white noise, were chosen and mixed using a non-linear function $\boldsymbol{f}$ that was simulated by a two layer perceptron of 3 input nodes, 3 nodes in the hidden layer and 3 output neurons, resulting in the mixed signals $\boldsymbol{x(t)=A_2f(A_1s(t))}$, where $\boldsymbol{A_1}$, $\boldsymbol{A_2}$ are two 3x3 matrices with their elements randomly selected:

$$\mathbf{A}_1 = \begin{bmatrix} 0.15 & .30 & -0.41 \\ -0.44 & -0.21 & -0.33 \\ 0.35 & 0.37 & 0.05 \end{bmatrix}, \ \mathbf{A}_2 = \begin{bmatrix} 0.45 & -0.89 & -0.42 \\ 0.66 & 0.97 & -0.91 \\ 0.99 & -0.31 & -0.01 \end{bmatrix} \text{ and } f(x) = \sum_{i=1}^{3}\tanh(a_i \cdot x + b_i),$$

with $\boldsymbol{a_i}$=[1 2 3] and $\boldsymbol{b_i}$=[-.5 1 –1.1] randomly chosen as well.

The learning rules in (9,12,13,15) were applied using successively one to ten components in the mixture of the non-linear sigmoid functions. However, experiments utilizing more than 5 components showed no further improvement of the separation performance. The experimental results indicated that when using more components to form the non-linear function **g**, a better fitting of the non-linear mixing function is achieved, resulting to better separation performance. In Fig. 2 we present the experimental results. In Fig. 2a and Fig. 2b we show the original independent signals and their non-linear mixture. In figure Fig. 2d the separating signals are shown when 5 components are used in the mixture of the non-linearities. It is evident that the proposed algorithm works well and manages to separate the 3 independent signals satisfactorily. In figure Fig. 2c we show the separating signals that were obtained using the standard BSS algorithm for linear mixtures [2] and is clearly evident that this algorithm cannot separate the non-linear mixtures at all.

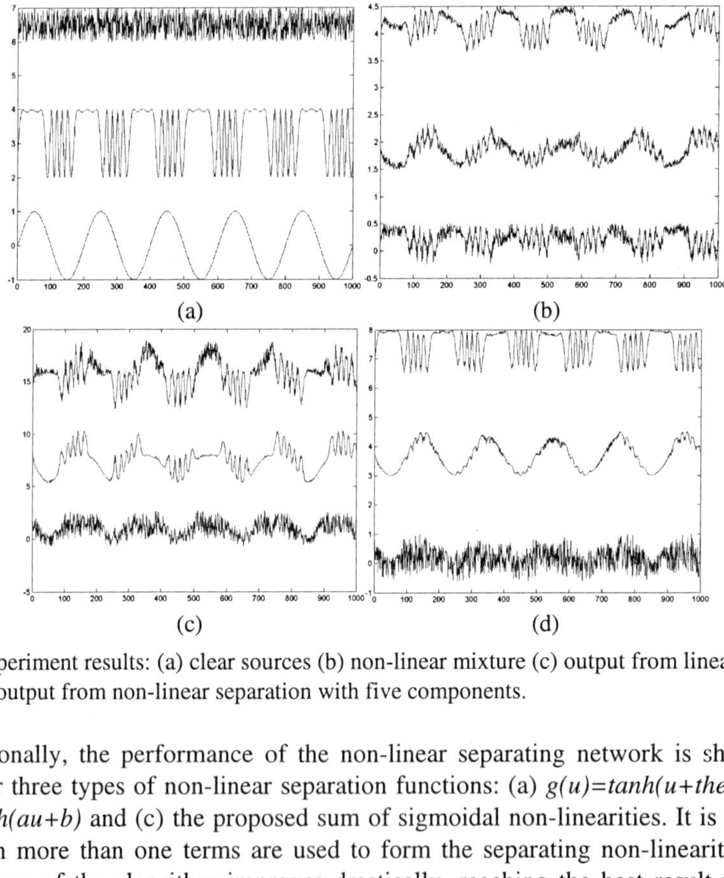

Fig. 2. Experiment results: (a) clear sources (b) non-linear mixture (c) output from linear separation (d) output from non-linear separation with five components.

Additionally, the performance of the non-linear separating network is shown in Fig. 3 for three types of non-linear separation functions: (a) *g(u)=tanh(u+theta)*, (b) *g(u)=tanh(au+b)* and (c) the proposed sum of sigmoidal non-linearities. It is evident that when more than one terms are used to form the separating non-linearities, the performance of the algorithm improves drastically, reaching the best result when 5 terms are used. In the same figure we show the performance index when the non-linearity used has the form *g(u)=tanh(u+theta)* as in [9]. It is clear that the proposed generalized non-linear function outperforms the one used in [9], presenting faster and smoother convergence. Similar results were obtained for the experiments that employed simple non-linear mixing functions, confirming that by increasing the number of terms in the mixture of the sigmoid non-linearity, better separation results can be achieved.

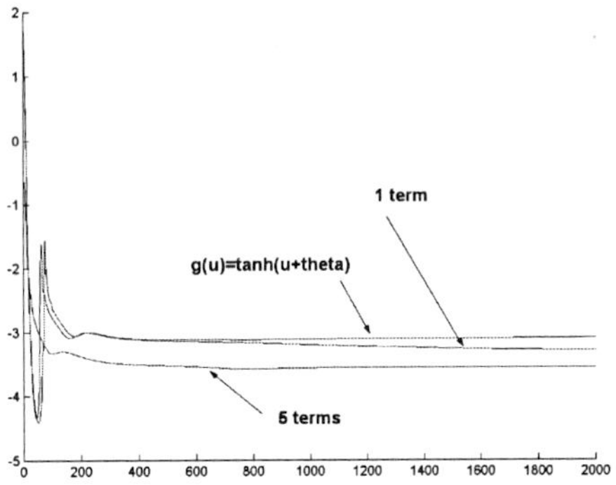

Fig. 3. The performance index of the NLBSS network for various types of non-linear separation functions.

4. Conclusion

In this paper we presented a novel solution to the non-linear blind separation problem, which is based on a two-layer neural network using a mixture of parametric non-linearities in the hidden nodes. Assuming that the non-linear mixing function can be approximated by a two-layer perceptron, our network is able to extract successfully source signals from non-linear as well as post non-linear mixtures, showing a better separation performance in complex non-linear mixtures than when using simple sigmoid non-linearities.

References

1. Haykin, S., Unsupervised Adaptive Filtering, Vol. I Blind Source Separation, J. Wiley & Sons (2000).
2. Lee, T. W., Independent Component Analysis, Kluwer Academic Publishers (1998).
3. Karhunen, J., Nonlinear Independent Component Analysis, in Roberts & Everson (eds.), Independent Component Analysis: Principles and Practice, Cambridge University Press (2001).
4. Burel, G.: Blind separation of sources: A nonlinear neural algorithm. Neural Networks, vol 5. (1992) 937-947.
5. Pajunen, P., Hyvarinen, A., Karhunen, J.: Nonlinear blind source separation by self-organizing maps. ICONIP Vol 2 (1996) 1207-1210.
6. Deco, G., Brauer, W.: Nonlinear higher-order statistical decorrelation by volume-conserving architectures. Neural Networks, Vol 8 (1995) 525-535.
7. Lee, T.W., Koehler, B, Orglmeister, R.: Blind source separation of nonlinear mixing models. Neural Network for signal processing. (1997) 406-415.
8. Taleb, A., Jutten, C.: Nonlinear source separation: the post-nonlinear mixtures. ESANN (1997) 279-284.
9. Yang, H., Amari, S., Cichocki, A.: Information theoretic approach to blind separation of sources in non-linear mixture. Signal Processing, 64 (1998) 291-300.

Feature Extraction Using ICA

Nojun Kwak, Chong-Ho Choi, and Jin Young Choi

School of Electrical Eng., ASRI, Seoul National University
San 56-1, Shinlim-dong, Kwanak-ku, Seoul 151-742, Korea
{triplea,chchoi}@csl.snu.ac.kr, jychoi@neuro.snu.ac.kr

Abstract. In manipulating data such as in supervised learning, we often extract new features from original features for the purpose of reducing the dimensions of feature space and achieving better performances. In this paper, we propose a new feature extraction algorithm using independent component analysis (ICA) for classification problems. By using ICA in solving supervised classification problems, we can get new features which are made as independent from each other as possible and also convey the output information faithfully. Using the new features along with the conventional feature selection algorithms, we can greatly reduce the dimension of feature space without degrading the performance of classifying systems.

1 Introduction

In supervised learning, one is given an array of attributes to predict the target value or output class. These attributes are called features, and there may exist irrelevant or redundant features to complicate the learning process, thus leading to wrong prediction. Even in the case when the features presented contain enough information about the output class, they may not predict the output correctly because the dimension of feature space is so large that it requires numerous instances to figure out the relationship. This problem is commonly referred to as the 'curse of dimensionality' [1], and can be avoided by selecting only the relevant features or extracting new features containing the maximal information about the output from the original ones. The former methodology is called feature selection or subset selection, while the latter is named feature transformation whose variants are feature extraction and feature construction [2].

This paper considers the feature extraction problem since it often results in improved performance, especially when small dimensions are required, thus can be directly applied to make oblique decision borders as optimal classifiers like support vector machines do.

Recently, in neural networks and signal processing societies, independent component analysis (ICA), which was devised for blind source separation problems, has received a great deal of attention because of its potential applications in various areas. Bell and Sejnowski have developed an unsupervised learning algorithm performing ICA based on entropy maximization in a single-layer feed-forward neural network [3]. Though this ICA method can be directly applied

G. Dorffner, H. Bischof, and K. Hornik (Eds.): ICANN 2001, LNCS 2130, pp. 568–573, 2001.

to transform the original feature set into a new set of features as in [4]-[5], the performance cannot be guaranteed, since it does not utilize output class information like the PCA. These ICA-based feature extraction methods just focus on the visual representation of the data [6].

In this paper, we propose a new ICA-based feature extraction algorithm which utilizes the output class information in addition to having the advantages of the original ICA method. This method is well-suited for classification problems in the aspect of constructing new features that are strongly related to output class. By combining the proposed algorithm with existing feature selection methods, we can greatly reduce the dimension of feature space while improving classification performance.

This paper is organized as follows. In Section 2, we propose a new feature extraction algorithm. In Section 3, we give some simulation results showing the advantages of the proposed algorithm. Conclusions follow in Section 4.

2 Feature Extraction Based on ICA

The main idea of the proposed feature selection algorithm is very simple. In applying ICA to feature extraction, we include output class information in addition to input features.

ICA is classified as unsupervised learning because it outputs a set of maximal independent component vectors. This unsupervised method by nature is related to the input distribution, but cannot guarantee good performance in classification problems though the resulting independent vectors may find some application in such areas as visualization [4] and source separation [3]. Instead of using only the input vectors, we treat the output class information as one of the input features and append it to the inputs for ICA. To illustrate the advantage of this method, consider the following simple problem.

Suppose we have two input features x_1 and x_2 uniformly distributed on [-1,1] for a binary classification, and the output class y is determined as follows:

$$y = \begin{cases} 0 & \text{if } x_1 + x_2 < 0 \\ 1 & \text{if } x_1 + x_2 \geq 0 \end{cases}$$

Plotting this problem on a three dimensional space of (x_1, x_2, y) leads to Fig. 1 where the class information as well as the input features correspond to each axis respectively. The data points are located in the shaded areas in this problem. As can be seen in the figure, this problem is linearly separable and we can easily pick out $x_1 + x_2$ as an important feature.

Performing ICA on a set of data provides us with N-dimensional vectors indicating the directions of independent components on feature space. If we take the output class as an additional input feature, it forms an *(N+1)*-dimensional space, and applying ICA to this dataset gives us *(N+1)*-dimensional vectors. These vectors are made to well describe the data distribution over the extended *(N+1)*-dimensional feature space and can be useful for classification problems if we project the vectors onto the original N-dimensional feature space. In the problem above, the vector shown as a thick arrow in Fig. 1 will surely indicate the apparent discontinuity in the data distribution over the extended feature space. This vector can provide us with a great deal of information about the problem and projecting it onto the (x_1, x_2) feature space gives us a new feature relevant to the output class.

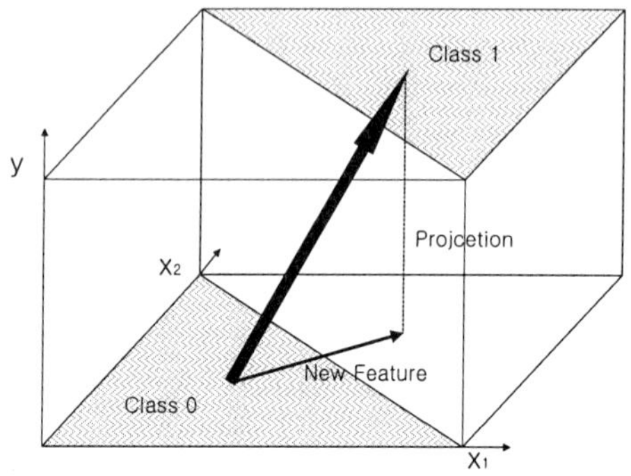

Fig. 1. Concept of ICA based feature extraction

Based on this idea, a feature extraction algorithm for classification problem is proposed as follows:

1. **Preparing data**
   (a) With N input features $\mathbf{x}= [x_1, \cdots, x_N]^T$ and one output class c, compose a new *(N+1)*-dimensional input dataset $\{(\mathbf{x}^T, c)^T\}$.
   (b) Normalize each feature f_i by $(f_i - m_i)/2\sigma_i$ where m_i and σ_i are the mean and the standard deviation of f_i, respectively.
2. **Performing ICA**
   Apply ICA to the new dataset, and store the resulting weight matrix $\mathbf{W}$ of dimension $(N+1) \times (N+1)$.
3. **Shrinking small weights**
   (a) For each $N+1$ independent row vector W_i of $\mathbf{W}$, compute the absolute mean $a_i = \frac{1}{N+1}\sum_{j=1}^{N+1} |w_{ij}|$.
   (b) For all w_{ij} in $\mathbf{W}$, if $|w_{ij}| < \alpha \cdot a_i$, then shrink $|w_{ij}|$ to zero. Here, α is a small positive number [1].
4. **Extracting feature candidates**
   (a) For each weight vector W_i, project it onto the original input feature space, i.e., delete weights $w_{ic}(= w_{i,N+1})$ corresponding to the output class, resulting in new N-dimensional row weight vector W_i'.
   (b) By mutiplying new weight matrix $\mathbf{W}'$ of dimension $(N+1) \times N$ to the original input data $\mathbf{x}$, construct a *(N+1)*-dimensional vector whose components f_i's are new feature candidates.

[1] In the simulations, we set $\alpha = 0.2$ or 0.5.

5. **Deleting unappropriate features**
   (a) Form a set of feature candidates $F = \{f_i = W_i'\mathbf{x}, \quad i \in 1 \cdots N{+}1\}$. Then set $F_S = F$.
   (b) For each feature candidate f_i, if the corresponding weight for class w_{ic} is zero, then exclude it from F_S.
   (c) For each feature candidate f_i, if corresponding weights $w_{ij} = 0$ for all $j \in 1 \cdots N$, then exclude f_i from F_S.
   (d) Resulting F_S contains final N' extracted features.

Step 3 is useful for excluding the effect of features which do not contribute significantly to the new features. The effect of the exclusion is controlled by a suitable choice of α.

When we applied the algorithm to the problem illustrated in Fig. 1, with 1,000 samples, we obtained 3 raw weight vectors $W = \{(w_1, w_2, w_c)\} = \{(7.55, 1.74, -2.24), (2.10, 7.88, -2.48), (-0.75, -0.78, 3.53)\}$. Note that the last one leads to a new feature $-0.75x_1 - 0.78x_2$ which is nearly equivalent to the optimal feature $x_1 + x_2$.

Though we consider only two-class problems in the algorithm, the algorithm can easily be extended to multi-class problems which can be set up as a series of two-class problems, where in each two-class problem, one class is distinguished from the others.

3 Experimental Results

In this section we apply the proposed algorithm on UCI dataset [8] and will present some experimental results which show the characteristics of the proposed algorithm. For the classification problems in this section, to show the effectiveness of the proposed algorithm, we selected appropriate numbers of original features and extracted features by MIFS-U [9] and compared the classification performance. We tested the proposed algorithm for several problems, but present the results of only two cases for space limitation. The other cases show the similar enhancement in performance.

Wiscosin Breast Cancer Diagonosis. The dataset consists of nine numerical attributes and two classes which are benign and malignant. It contains 699 instances with 458 benign and 241 malignant. There are 16 missing values and in our experiment we replaced these with average values of corresponding attributes.

The proposed algorithm extracted seven new features. We selected appropriate numbers of original features and extracted features by MIFS-U [9] and trained the data with C4.5 [10] and multilayer perceptron with standard back-propagation(BP). MLP was trained for 500 iterations with three hidden nodes, zero momentum and learning rate of 0.2. For verification, 10-fold cross validation is used. The classification results are shown in Table 1 and the numbers in parentheses are the tree sizes of C4.5.

Table 1. Classification performance for Breast Cancer data (%)

no. of features selected	Original Features (C4.5/MLP)	Extracted Features (C4.5/MLP)
1	91.13(3)/92.42	95.42(3)/95.42
2	94.71(13)/96.28	95.57(9)/95.42
3	95.85(23)/96.14	95.71(11)/96.57
9	94.56(31)/96.42	–

Table 2. Classification performance for Chess data (%)

no. of features selected	Original Features (MLP)	Extracted Features (MLP)
1	67.43	96.23
2	74.52	97.98
3	91.08	98.80
36	98.54	–

It shows that with only one extracted feature, we can get near the maximum classification performance that can be achieved with at least two or three original features. Also, note that the tree size does not expand much for the case of extracted features, which is desirable for generating simple rule.

Chess End-Game Data. There are 36 attributes with two classes, *white-can-win ('won')/white-cannot-win ('nowin')*. There are 3,196 instances of which 1,669 are *'won'* and the other 1,527 are *'nowin'* with no missing values.

In conducting feature extraction algorithm, we set $\alpha = 0.5$ and obtained six additional features. After feature selection process with MIFS-U, we trained the data for 500 iterations by the standard BP with 0.2 learning rate and zero momentum. We divided the data into two sets, one half for training and the other half for testing. Table 2 shows the classification performances on the test data. The proposed method shows significant improvement in performance over the original features.

4 Conclusions

In this paper, we have proposed an algorithm for feature extraction. The proposed algorithm is based on ICA and have the ability to generate appropriate features for classification problems.

As the name indicates, ICA is characterized by its ability of identifying statistically independent components based on input distribution. Although ICA can be directly used for feature extraction, it does not generate useful information because of its unsupervised learning nature. In the proposed algorithm, we added class information in training ICA. The added class information plays a critical role in the extraction of useful features for classification. For problems with original N features, we get $N+1$ weight vectors in $(N+1)$-dimensional extended feature space. Projecting them onto original feature space leads to new feature candidates and by deleting unappraite ones which have little correlation with the class information, we can extract better features.

Since it uses the standard feed-forward structure and learning algorithm of ICA, it is easy to implement and train. If we use the proposed algorithm with the conventional feature selection algorithm, we can get simpler classifier with better performance. Experimental results for several datasets show that the proposed algorithm generates good features which outperforms the original features for classification problems. Though the algorithm is proposed for two-class problems, it is also useful for multi-class problems with simple modification.

Acknowledgments

This project is partially supported by the Brain Science and Engineering Program of the Korea Ministry of Science and Technology and the Brain Korea 21 Program of the Korea Ministry of Education.

References

1. V. Cherkassky and F. Mulier, *Learning from Data : Concepts, Theory, and Methods*, pp. 60-65, John Wiley & Sons, 1998.
2. H. Liu and H. Motoda, "Less is more," *Feature Extraction Construction and Selection*, pp. 3-11, Boston: Kluwer Academic Publishers, 1998.
3. A.J. Bell and T.J. Sejnowski, "An information-maximization approach to blind separation and blind deconvolution," *Neural Computation*, vol 7, pp. 1129-1159, 1995.
4. A. Hyvarinen, E. Oja, P. Hoyer and J. Hurri, "Image feature extraction by sparse coding and independent component analysis," *Fourteenth International Conference on Pattern Recognition*, vol. 2, pp. 1268 -1273, 1998.
5. A.D. Back and T.P. Trappenberg, "Input variable selection using independent component analysis," *The 1999 Int'l Joint Conf. on Neural Networks*, July 1999.
6. H.H. Yang and J. Moody, "Data visualization and feature selection: new algorithms for nongaussian data," *Advances in Neural Information Processing Systems*, vol 12, 2000.
7. T.M. Cover, and J.A. Thomas, *Elements of Information Theory*, John Wiley & Sons, 1991.
8. P.M. Murphy and D.W. Aha, UCI repository of machine learning databases, 1994. For information contact ml-repository@ics.uci.edu or http://www.cs.toronto.edu/~delve/.
9. N. Kwak and C.-H. Choi, "Improved mutual information feature selector for neural networks in supervised learning," *The 1999 Int'l Joint Conf. on Neural Networks*, July 1999.
10. R. Quinlan, *C4.5: Programs for Machine Learning*, Morgan Kaufmann Publishers, San Mateo, CA., 1993.

Part VII

Signal Processing

Continuous Speech Recognition with a Robust Connectionist/Markovian Hybrid Model

Edmondo Trentin and Marco Gori

Dipartimento di Ingegneria dell'Informazione
Università di Siena, V. Roma, 56 - Siena, Italy

Abstract. This paper introduces a novel combination of Artificial Neural Networks (ANNs) and Hidden Markov Models (HMMs) for Automatic Speech Recognition (ASR), relying on ANN non-parametric estimation of the *emission* probabilities of an underlying HMM. A gradient-ascent *global training* technique aimed at maximizing the likelihood (ML) of acoustic observations given the model is presented. A *maximum a-posteriori* variant of the algorithm is also proposed as a viable solution to the "divergence problem" that may arise in the ML setup. A 46.34% relative word error rate reduction with respect to standard HMMs was obtained in a speaker-independent, continuous ASR task with a small vocabulary.

1 Introduction

The problem of acoustic modeling in Automatic Speech Recognition (ASR) is mostly approached relying on Hidden Markov Models (HMMs) [6]. Yet effective to a significant extent, HMMs are inherently limited [8]. In particular, the assumption of a given parametric form for the probability density functions (*pdfs*) that represent the *emission* probabilities associated with HMM states is arbitrary and constraining. In this respect, the use of artificial neural networks (ANNs) appears promising. Unfortunately ANNs historically failed as a general framework for ASR [8], due to their lack of ability to model long-term time dependencies, even when recurrent architectures are considered [1]. The failure led to the idea of combining HMMs and ANNs within *hybrid* ANN/HMM systems. A variety of hybrid paradigms for ASR were proposed in the literature [8]. This paper introduces a novel approach to the combination of HMM and ANN, related to some extent to Bourlard and Morgan's architecture [3], as well as to Bengio's optimization scheme [2].

In [3], for each state q_ℓ in an HMM $\mathcal{M}$, an output unit of a Multilayer Perceptron (MLP) is trained to estimate the corresponding *conditional transition probability*, $Pr(q_\ell \mid \mathbf{y}_{\ell-k}, \cdots, \mathbf{y}_{\ell+k}, q_{\ell-1})$, given a *window* of acoustic vectors $\mathbf{y}_{\ell-k}, \ldots, \mathbf{y}_{\ell+k}$ centered in the current observation $\mathbf{y}_\ell$), and the previous state $q_{\ell-1}$. One problem with this approach is that the optimization of the system parameters is not accomplished according to a globally-defined criterion, but relies on a *heuristic* procedure: (i) initial segmentation; (ii) iterative Backpropagation (BP)/Viterbi [6] scheme.

G. Dorffner, H. Bischof, and K. Hornik (Eds.): ICANN 2001, LNCS 2130, pp. 577–582, 2001.

In Bengio's hybrid model [2] the ANN extracts acoustic features $\mathbf{x}_1^L$ for a standard HMM. The HMM parameters and the ANN weights are jointly estimated over the same criterion L, i.e. the likelihood of the observations given the model, realizing a global-training algorithm. Gradient-ascent is applied to learn the weights w of the ANN, relying on the computation of $\frac{\partial L}{\partial x_t}$ (here x_t denotes a generic component of t-th observation $\mathbf{x}_t$). Unfortunately, using the ANN as a feature extractor for a standard HMM does not allow for any solutions to the intrinsic problems of the HMM itself.

This paper proposes an ANN/HMM paradigm relying on connectionist non-parametric estimation of the *emission* probabilities of an HMM, with a gradient-ascent *global training* algorithm aimed at maximizing the likelihood L. The ANN has an output unit for each of the HMM states, which provides an estimate of the corresponding emission probability given the current acoustic observation in the feature space. Training of the other probabilistic quantities in the underlying HMM, i.e. *initial* and *transition* probabilities, still relies on the *Baum-Welch* algorithm [6]. The Viterbi algorithm is eventually applied to the recognition step.

2 The Global Training Algorithm

The criterion function to be maximized by the model during training is the *likelihood* L of the acoustic observations given the model, where $L = \sum_{i \in \mathcal{F}} \alpha_{i,T}$. The sum is extended to the set $\mathcal{F}$ of all possible *final* states within the HMM [2] corresponding to the current phonetic transcription, which is supposed to involve Q states, and T is the length of the current observation sequence $Y = \mathbf{y}_1, \ldots, \mathbf{y}_T$. *Forward* and *backward* terms [6] for state i at time t are defined as $\alpha_{i,t} = b_{i,t} \sum_j a_{ji} \alpha_{j,t-1}$ and $\beta_{i,t} = \sum_j b_{j,t+1} a_{ij} \beta_{j,t+1}$, respectively, where a_{ij} denotes the transition probability between i-th and j-th states, $b_{i,t}$ denotes emission probability associated with i-th state over t-th acoustic observation $\mathbf{y}_t$, and the sums are extended to all possible states within the HMM. The initialization of the forward and backward terms is accomplished as in [2].

Assuming that the ANN may represent the emission *pdfs* in a proper manner, and given a generic weight w of the ANN, hill-climbing gradient-ascent over L prescribes a learning rule of the kind:

$$\Delta w = \eta \frac{\partial L}{\partial w} \tag{1}$$

where η is the *learning rate*. Borrowing the scheme proposed by Bridle [4] and Bengio, the following theorem can be proved to hold true [2]: $\frac{\partial L}{\partial \alpha_{i,t}} = \beta_{i,t}$, for each $i = 1, \ldots, Q$ and for each $t = 1, \ldots, T$. In addition, taking the partial derivatives of $\alpha_{i,t}$ w.r.t. $b_{i,t}$ we obtain $\frac{\partial \alpha_{i,t}}{\partial b_{i,t}} = \frac{\alpha_{i,t}}{b_{i,t}}$. Given this property and the theorem above, repeatedly applying the chain rule we can expand $\frac{\partial L}{\partial w}$ by writing:

$$\frac{\partial L}{\partial w} = \sum_i \sum_t \beta_{i,t} \frac{\alpha_{i,t}}{b_{i,t}} \frac{\partial b_{i,t}}{\partial w} \tag{2}$$

where the sums are extended over all states $i = 1, \ldots, Q$ of the HMM corresponding to the correct transcription of the training utterance under consideration, and to all $t = 1, \ldots, T$, respectively. Let us consider a multilayer feed-forward ANN, the j-th output of which, computed over t-th input observation $\mathbf{y}_t$, is expected to be a non-parametric estimate of the emission probability $b_{j,t}$ associated with j-th state of the HMM at time t. An activation function $f_j(x_j(t))$ is attached to each unit j of the ANN, where $x_j(t)$ denotes input to the unit itself at time t. The corresponding output $o_j(t)$ is given by $o_j(t) = f_j(x_j(t))$. The net is assumed to have n layers $\mathcal{L}_0, \mathcal{L}_1, \ldots, \mathcal{L}_n$, where $\mathcal{L}_0$ is the input layer and is not counted, and $\mathcal{L}_n$ is the output layer. For notational convenience we write $i \in \mathcal{L}_k$ to denote the index of i-th unit in layer $\mathcal{L}_k$.

Consider first the case of a generic weight w_{jk} between k-th unit in layer $\mathcal{L}_{n-1}$ and j-th unit in the output layer. We can write:

$$\frac{\partial b_{j,t}}{\partial w_{jk}} = f_j'(x_j(t))o_k(t). \tag{3}$$

By defining the quantity[1]

$$\delta_i(j,t) = \begin{cases} f_j'(x_j(t)) & \text{if } i = j \\ 0 & \text{otherwise} \end{cases} \tag{4}$$

for each $i \in \mathcal{L}_n$, we can rewrite Equation (3) in the form:

$$\frac{\partial b_{j,t}}{\partial w_{jk}} = \delta_j(j,t)o_k(t). \tag{5}$$

Consider now the case of connection weights in the first hidden layer. Let w_{kl} be a generic weight between l-th unit in layer $\mathcal{L}_{n-2}$ and k-th unit in layer $\mathcal{L}_{n-1}$.

$$\begin{aligned} \frac{\partial b_{j,t}}{\partial w_{kl}} &= f_j'(x_j(t)) \sum_{i \in \mathcal{L}_{n-1}} \frac{\partial w_{ji} o_i(t)}{\partial w_{kl}} \\ &= f_j'(x_j(t)) w_{jk} f_k'(x_k(t)) o_l(t). \end{aligned} \tag{6}$$

Similarly to definition (4), we introduce the quantity:

$$\delta_k(j,t) = f_k'(x_k(t)) \sum_{i \in \mathcal{L}_n} w_{ik}\delta_i(j,t) \tag{7}$$

for each $k \in \mathcal{L}_{n-1}$, which allows us to rewrite Equation (6) in the form:

$$\frac{\partial b_{j,t}}{\partial w_{kl}} = \delta_k(j,t)o_l(t) \tag{8}$$

which is formally identical to Equation (5). In a similar manner, by carrying out the calculations for a generic weight w_{lm} holding between m-th unit in layer

[1] Dependency on the specific HMM state j under consideration and on current time t is written explicitly, since this will turn out to be useful in the final formulation of the learning rule.

$\mathcal{L}_{n-3}$ and l-th unit in layer $\mathcal{L}_{n-2}$, but the same results apply also to subsequent layers ($\mathcal{L}_{n-4}, \mathcal{L}_{n-5}, \ldots, \mathcal{L}_0$), it can be seen that:

$$\frac{\partial b_{j,t}}{\partial w_{lm}} = f'_j(x_j(t)) \sum_{i \in \mathcal{L}_{n-1}} w_{ji} f'_i(x_i(t)) w_{il} f'_l(x_l(t)) o_m(t). \tag{9}$$

Again, as in (4) and (7), for each $l \in \mathcal{L}_{n-2}$ we define

$$\delta_l(j,t) = f'_l(x_l(t)) \sum_{i \in \mathcal{L}_{n-1}} w_{il} \delta_i(j,t) \tag{10}$$

and we rewrite Equation (9) in the familiar, compact form:

$$\frac{\partial b_{j,t}}{\partial w_{lm}} = \delta_l(j,t) o_m(t) \tag{11}$$

that is formally identical to Equations (5) and (8). Using the above developments to expand Equation (2) and substituting it into Equation (1), the latter can now be restated in the form of a general *learning rule* for a generic weight w_{jk} of the ANN, by writing:

$$\Delta w_{jk} = \eta \sum_{i=1}^{Q} \sum_{t=1}^{T} \beta_{i,t} \frac{\alpha_{i,t}}{b_{i,t}} \delta_j(i,t) o_k(t) \tag{12}$$

where the term $\delta_j(i,t)$ is computed via Equation (4), Equation (7) or Equation (10), according to the fact that the layer under consideration is the output layer, the first hidden layer, or one of the other hidden layers, respectively.

The learning rule is expected to be effective under the assumption that the ANN actually realizes a model of a *pdf*. Whereas ANNs *may* reasonably approximate any continuous *pdf* as close as desired, in general they completely lack of any constraints ensuring that, at any given time during training, the function they compute actually *is* a *pdf*. In fact, the training criterion (ML) simply encourages the ANN to yield high output values over all input patterns, and the (pseudo)likelihood of the training sequences may diverge toward infinity. We refer to such a phenomenon as the "divergence problem".

In practice, the ANN is not necessarily required to realize a *pdf*, but *divergence* of its output values should rather be avoided. The adoption of a "discriminative" training criterion, namely the *Maximum A Posteriori* (MAP) criterion [2] appears promising in this respect. We formally define the MAP criterion $\mathcal{C}_{MAP}$ as follows:

$$\mathcal{C}_{MAP} = P(\mathcal{M} \mid Y) = \frac{P(Y \mid \mathcal{M}) P(\mathcal{M})}{P(Y)} \tag{13}$$

where Y is the observation sequence and $\mathcal{M}$ is the specific model under consideration . Bayes' theorem was applied in order to obtain the right-end side of Equation (13): $P(\mathcal{M})$ is the *prior* probability of the model, and $P(Y)$ is the overall likelihood of the acoustic observation sequence Y, that does not depend on the specific choice for the model $\mathcal{M}$. $P(\mathcal{M})$ is independent of the ANN outputs, since it depends on the language model only, and it can be computed separately. A gradient-based learning rule for a generic connection weight w can be obtained by taking the partial derivative of $\mathcal{C}_{MAP}$ w.r.t. w. The quantity $P(Y \mid \mathcal{M})$ and its derivative w.r.t. w are computed according to the calculations outlined above, since $P(Y \mid \mathcal{M})$ is the usual likelihood of the observations given the "correct" model $\mathcal{M}$, i.e. the model *constrained* by the (known) transcription of acoustic observation sequence Y. The quantity $P(Y)$ and its derivative w.r.t. w are computed by applying the same calculations to an *unconstrained* model, as explained in [7]. The latter is the same model used for *recognition* with the Viterbi algorithm, where no prior knowledge on the phonetic transcription of current sequence is given.

3 Experimental Results

Clean speech signals from the `cdigits` part of the *SPK* database[2], collected at ITC-irst (Trento, Italy), were considered. It is a continuous speech task, namely 1000 utterances of connected Italian digit strings having length 8 (for a total of 8000 words), acquired over 40 different speakers (21 male and 19 female). The whole dataset was divided into two equally-sized subsets, to be used for training (10 male and 10 female) and test (the other 11 male and 9 female), respectively. A close-talk microphone was used for the recordings, under quiet laboratory conditions. Spectral analysis of the speech signals (acquired at 16kHz) was accomplished over 20ms Hamming windows having an overlap of 10ms, extracting 8 *Mel Frequency Scaled Cepstral Coefficients* (MFSCCs) [5] and the signal log-energy as acoustic features. The latter were normalized in order to have input values distributed in a uniform manner over the $[0, 1]$ interval, before feeding them into the connectionist models.

Words in the dictionary were modeled using individual, left-to-right HMMs having 3 to 6 states, according to the phonetic transcription of each Italian digit, plus a 1-state model for the "silence", for a total of 40 states. Mixtures of 8 Gaussian components were used to model emission probabilities for each state of the standard HMM, that was initialized via *Segmental k-Means* [6] and trained by applying the Baum-Welch algorithm. After a preliminary cross-validation step, the topology of the MLP included 93 sigmoids in the hidden layer, and 40 output sigmoids (one per each state of the underlying HMM).

Experimental results are reported in Table 1 in terms of *Word Recognition Rate* (WRR) and *String Recognition Rate* (SRR), i.e. the fraction of test digit-strings that are recognized without any errors. A comparison with Bourlard and Morgan's hybrid architecture is provided, carried out over the same ANN topology, and applying the variant "Case 4" described in [3]. Normalization of ANN outputs, i.e. division by the corresponding state-priors (estimated from the training set) in order to reduce to "likelihoods", was accomplished at recognition time, as recommended in [3], Chapter 7. It is seen that the proposed ANN/HMM

[2] SPK is available from the European Language Resources Association (ELRA).

Table 1. String recognition rate (SRR) and word recognition rate (WRR) on test set.

Architecture/algorithm	*SRR (%)*	*WRR (%)*
Standard HMM	46.60	**90.03**
Bourlard and Morgan's hybrid	47.40	**90.20**
Proposed ANN/HMM hybrid	68.60	**94.65**

recognizer trained with the MAP version of the algorithm yields a 46.34% relative WER reduction over the standard HMM.

4 Conclusion

This paper introduced a novel ANN/HMM hybrid system where a feed-forward ANN carries out estimation of emission probabilities associated with the states of an underlying HMM. Initial and transition probabilities are estimated via Baum-Welch, while gradient-ascent techniques aimed at the extremization of the ML or MAP criteria were developed for ANN training. The MAP allowed us to tackle the "divergence problem". The usual Viterbi decoding strategy is then applied at recognition time. ASR experiments were accomplished on speaker-independent, continuous speech from the *SPK* database. The proposed model turned out to overcome significantly the limitations of standard HMMs. In particular, the MAP algorithm applied to an MLP with sigmoidal hidden and output units compared favorably with standard HMMs, as well as with Bourlard and Morgan's approach.

References

1. Y. Bengio, P. Simard, and P. Frasconi. Learning long-term dependencies with gradient descent is difficult. *IEEE Trans. on Neural Networks*, 5(2):157–166, 1994.
2. Y. Bengio. *Neural Networks for Speech and Sequence Recognition.* International Thomson Computer Press, London, UK, 1996.
3. H. Bourlard and N. Morgan. *Connectionist Speech Recognition. A Hybrid Approach.* Kluwer Academic Publishers, Boston, 1994.
4. J.S. Bridle. Alphanets: a recurrent 'neural' network architecture with a hidden Markov model interpretation. *Speech Communication*, 9(1):83–92, 1990.
5. S. B. Davis and P. Mermelstein. Comparison of parametric representations of monosyllabic word recognition in continuously spoken sentences. *IEEE Trans. on Acoustics, Speech and Signal Processing*, 28(4):357–366, 1980.
6. X. D. Huang, Y. Ariki, and M. Jack. *Hidden Markov Models for Speech Recognition.* Edinburgh University Press, Edinburgh, 1990.
7. E. Trentin, Y. Bengio, C. Furlanello, and R. De Mori. Neural networks for speech recognition. In R. De Mori, editor, *Spoken Dialogues with Computers*, pages 311–361, London, UK, 1998. Academic Press.
8. E. Trentin and M. Gori. A survey of hybrid ANN/HMM models for automatic speech recognition. *Neurocomputing*, 37(1–4):91–126, 2001.

Faster Convergence and Improved Performance in Least-Squares Training of Neural Networks for Active Sound Cancellation

Martin Bouchard

School of Information Technology and Engineering, University of Ottawa
161 Louis-Pasteur, P.O. Box 450, Station A, Ottawa (Ontario)
K1N 6N5, Canada
bouchard@site.uottawa.ca

Abstract. This paper introduces new recursive least-squares algorithms with faster convergence and improved steady-state performance for the training of multilayer feedforward neural networks, used in a two neural networks structure for multichannel non-linear active sound cancellation. Non-linearity in active sound cancellation systems is mostly found in actuators. The paper introduces the main concepts required for the development of the algorithms, discusses why it is expected that they will outperform previously published steepest descent and recursive least-squares algorithms, and shows the improved convergence produced by the new algorithms with simulations of non-linear active sound cancellation.

1. Introduction

Adaptive linear filtering techniques have been extensively used for the active cancellation of sound and vibration, and many of today's implementations of active cancellation use those techniques. However, those linear digital controllers may not perform well in cases where non-linearity is found in an active cancellation system. The most common source of non-linearity in active cancellation systems is the actuator. The use of neural networks for the active cancellation of sound and vibration with non-linear plants has been reported in [1],[2], using a control structure with two multilayer feedforward neural networks: one as a non-linear controller and one as a non-linear plant model, as shown in Fig. 1 for a multichannel active cancellation system. The controller neural network basically generates control signals from reference signals, and these control signals are sent to the plant through actuators. The plant model neural network is required to perform the backpropagation of the error signals.

The training of the plant model neural network of Fig. 1 can be done with classical neural networks algorithms, and will not be further discussed here. Some conditions for the plant model to be valid over a wide range of control signals are discussed in [2]. The training of the controller neural network in Fig.1 can not be done

G. Dorffner, H. Bischof, and K. Hornik (Eds.): ICANN 2001, LNCS 2130, pp. 583–592, 2001.

with those classical algorithms, because of the delay lines between the two neural networks. In [1],[2], steepest descent algorithms based on two distinct gradient approaches were introduced for the training of the controller network. Some recursive least-squares algorithms were also introduced in [2] for the training of the controller neural network. In this paper, new recursive least-squares algorithms are introduced for the training of the controller neural network, in order to improve the convergence performance.

In Section 2 of this paper, the development of the new recursive least-squares algorithms is described. In Section 3, simulation results compare the convergence performance of the new algorithms with previously published learning algorithms for the controller neural network. A conclusion will follow.

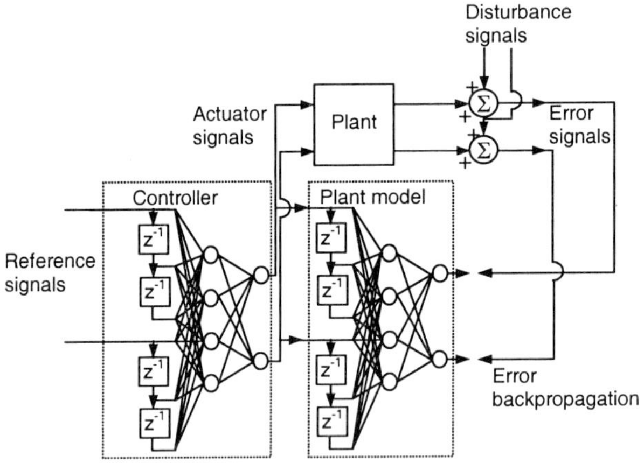

Fig. 1. A two neural networks structure for on-line multichannel non-linear active cancellation of sound or vibration.

2. Development of the New Recursive-Least-Squares Algorithms

In this paper, bold variables are used to represent vectors, and capital bold variables are used to represent matrices. Let us first define the cost function to be minimized as:

$$J(n) = \frac{1}{2} \sum_{k=1}^{K} e_k(n)^2 \tag{1}$$

where:

$J(n)$ is the sum of the squared instantaneous values of the error signals $e_k(n)$

$e_k(n)$ is the kth error signal at time n in the active cancellation system (see Fig. 1)

K is the total number of error signals.

Two different approaches can be used to compute a gradient of the cost function $J(n)$ as a function of the controller neural network coefficients:

$$gradient\ \Delta(n) = \sum_{l=0}^{L_2} \frac{\partial J(n)}{\partial\, w(n-l)} \tag{2}$$

$$gradient\ \Delta(n) = \sum_{l=0}^{L_2} \frac{\partial\, J(n-L_2+l)}{\partial\, w(n-L_2)} \tag{3}$$

where

$w(n-l)$ is the value of one particular weight, or a vector of several weights, or a vector of all the weights in the controller neural network at time *n-l*

L_2 is the number of delays in each delay line between the two neural networks in Fig. 1.

The gradient from equation (2) will be called the filtered-x gradient approach, because the simplification of the control system to the linear case (with no hidden layers and with linear activation functions) produces the well-known filtered-x LMS algorithm [3], when (2) is combined with a steepest descent algorithm. On the other hand, the gradient from equation (3) will be called the adjoint gradient approach, because the simplification of the control system to the linear case (again with no hidden layers and with linear activation functions) produces the adjoint-LMS algorithm [4], when (3) is combined with a steepest descent algorithm. Intuitively, the filtered-x gradient approach of (2) computes a gradient based on the fact that the error signals at time *n* are affected by the values of the controller network weights from the last L_2+1 samples (weight values at times *n*, *n*-1, ..., *n*- L_2). The adjoint approach of (3) computes a gradient based on the fact that the controller network weights at time *n* affect the error signals for the L_2+1 next samples (at times *n*, *n*+1, ..., *n*+ L_2). This statement can be delayed to produce a causal relation as in (3): the control network weights at time *n*- L_2 affect the error signals at times *n*- L_2, *n*- L_2+1, ..., *n*-1, *n*.

In [1], the filtered-x gradient approach was used and combined with a steepest descent algorithm to produce a training algorithm for the controller neural network of Fig. 1. In [2], the adjoint gradient approach was combined with both steepest descent algorithms and recursive least-squares algorithms, to develop several training algorithms for the controller network. In this paper, recursive least-squares algorithms are developed with the filtered-x gradient approach, and it will be explained and showed through simulations that these new algorithms can outperform the previously published algorithms. To introduce the new algorithms, the following notation needs to be defined:

L_1 : number of delays in each delay line at the input of the controller network (Fig. 1)

I : number of reference signals in the active cancellation system

J : number of actuators in the active cancellation system

M : number of hidden layers in the controller neural network (Fig. 1)

P : number of hidden layers in the plant model neural network (Fig. 1)

$x_i(n-l)$: value of the i th reference signal at time *n-l*

$y_j^m(n)$: value of the j th output in the m th layer of the controller neural network ($y_j^0(n)$ is the j th input of the input layer and $y_j^{M+1}(n)$ is the value of the j th actuator signal in the active cancellation system). The value of $y_j^m(n)$ before the activation function of a neuron is $s_j^m(n)$.

$w_{i,j}^m(n)$: value at time n of the (adaptive) weight linking the i th neuron of layer m-1 to the j th neuron of layer m in the controller neural network ($1 \le m \le M+1$)

$z_j^p(n)$: value of the j th output in the p th layer of the plant model neural network ($z_j^1(n)$ is the j th output of the first hidden layer and $z_j^{P+1}(n)$ is the j th output of the output layer). The value of $z_j^p(n)$ before the activation function of a neuron is $t_j^p(n)$.

$h_{i,j}^p$: value of the (fixed) weight linking the i th neuron of layer p-1 to the j th neuron of layer p in the plant model neural network ($2 \le p \le P+1$)

$h_{i,j,l}^1$: value of the (fixed) weight linking at time n the actuator signal $y_i^{M+1}(n-l)$ to the j th neuron of the first hidden layer in the plant model neural network

$d_k(n)$: value at time n of the k th disturbance signal (i.e primary field, controlled by the active cancellation system)

$f()$ is the activation function of a neuron

$f'()$ is the derivative of the activation function of a neuron

Using either (2) or (3), it is straightforward to have a steepest descent algorithm to update the weights in the controller neural network of Fig. 1:

$$\mathbf{w}(n+1) = \mathbf{w}(n) - \mu\, \Delta(n) \tag{4}$$

where μ is a scalar gain.

The gradient in the filtered-x gradient approach of equation (2) can be expressed as:

$$\Delta(n) = \sum_{l=0}^{L_2} \frac{\partial J(n)}{\partial z^{P+1}(n)} \frac{\partial z^{P+1}(n)}{\partial w(n-l)} \tag{5}$$

where

$w(n-l)$: value at time n-l of one particular weight that affects the value of the intermediate variable z^{P+1} in (5), or a vector of a few weights that affect this variable, or a vector of all the weights that affect this variable

$z^{P+1}(n)$: vector containing all the values of $z_j^{P+1}(n)$

The choice of $z^{P+1}(n)$ as the intermediate variable in (5) is not the only possibility, but it has some advantages as we will see. First of all, the index l is found only in one term inside of the summation. Thus one term can be taken out of the summation, and (5) can be re-written as:

$$\Delta(n)=\left(\sum_{l=0}^{L2}\frac{\partial z^{P+1}(n)}{\partial w(n-l)}\right)\frac{\partial J(n)}{\partial z^{P+1}(n)} \tag{6}$$

$$=\left(\sum_{l=0}^{L2}\frac{\partial z^{P+1}(n)}{\partial w(n-l)}\right)\begin{bmatrix} e_1(n) \\ \vdots \\ e_K(n)\end{bmatrix}=\boldsymbol{INPUT}(n)\,\boldsymbol{error}^T(n)$$

The form $\Delta(n)=\boldsymbol{INPUT}(n)\boldsymbol{error}^T(n)$ is the familiar form of the stochastic gradient in the LMS algorithm [3], that uses a steepest descent approach with a stochastic gradient for linear FIR adaptive filtering, as described by equations (7),(8):

$$\boldsymbol{error}(n)=\boldsymbol{desired}(n)+\boldsymbol{w}(n)^T\,\boldsymbol{INPUT}(n) \tag{7}$$

$$\boldsymbol{w}(n+1)=\boldsymbol{w}(n)-\mu\,\Delta(n)=\boldsymbol{w}(n)-\mu\,\boldsymbol{INPUT}(n)\;\boldsymbol{error}^T(n)\,. \tag{8}$$

The sign in equations (7)-(8) is the opposite of what is typically found in the literature. This is simply to reflect the fact that for the active cancellation of sound and vibration, it is a summation as in (7) that occurs (it is the physical superposition of waves). For a linear equation with the form of (7), it is well known that recursive least-squares algorithms can achieve a faster convergence speed than stochastic descent algorithms, at least for stationary or quasi-stationary signals and systems, because the convergence speed of the recursive least-squares algorithms does not depend on the eigenvalue spread of the input signals correlation matrix [5]. For the training of non-linear neural networks, recursive least-squares algorithms may not only provide an increase of convergence speed, it was also reported in [6],[7],[2] that they can also improve the steady-state performance achieved by the trained neural network. Combining the $\boldsymbol{INPUT}(n)$ and $\boldsymbol{error}(n)$ terms computed from (6) with recursive least-squares equations produces recursive least-squares algorithms which will minimize the weighted sum over time of the signal $\frac{1}{2}\boldsymbol{error}(n)^T\boldsymbol{error}(n)$, through the well known Riccati equations [5]. This signal is exactly equivalent to the cost function $J(n)$ defined in (1), and this partly explains why the new recursive least-squares algorithms outperform the recursive least-squares algorithms previously published in [2] (where $J(n)$ was only indirectly minimized through the minimization of other cost functions). The other reason for the improved convergence of the new recursive least-squares algorithms is that a weight change in the controller neural network has a shorter effect on the gradient with the filtered-x gradient approach of equation (2) than with the adjoint gradient approach of equation (3) [8] (the adjoint gradient approach was used in the previously published recursive least-squares algorithms [2]). This leads to faster convergence because a larger adaptive gain μ can then be used in the weights update equation [8].

Using weight vectors $w(n-l)$ containing all the weights directly connected to a specific neuron in the controller network of Fig. 1, a training algorithm called the filtered-x NEKA (or FX-NEKA) can be developed (from the NEKA algorithm [6]). The equations for this algorithm are found in Table 1 and Table 2. It is straightforward to modify the FX-NEKA algorithm to use a single global vector $w(n-l)$ containing all the weights in the controller network, to produce the global FX-EKF algorithm (from the EKF algorithm [7]). This algorithm is described by Table 1, by Table 2 and by replacing the last four equations of Table 2 by Table 3. Note that it is possible to reduce the memory requirements of the algorithms by assuming $w_{j,k}^{m+1}(n-l) \approx w_{j,k}^{m+1}(n)$ in the equations. This assumption is valid if the adaptation rate is slow compared to the value of L_2, which is true for most systems. The computational load of the new algorithms and the previously published steepest descent and recursive least-squares algorithms can be found in [8].

Table 1. Forward equations for the two feedforward neural networks of Fig. 1.

common forward equations	
$y_{i(L_1+1)+l}^{0}(n) = x_i(n-l)$	$0 \le l \le L_1$, $0 \le i \le I-1$
$s_j^m(n) = \sum_i w_{i,j}^m(n) y_i^{m-1}(n)$	(includes offset input)
$y_j^m(n) = f(s_j^m(n))$	$1 \le m \le M+1$
$t_j^1(n) = \sum_i \sum_{l=0}^{L_2} h_{i,j,l}^1 y_i(n-l)$	(includes offset input)
$z_j^1(n) = f(t_j^1(n))$	
$t_j^p(n) = \sum_i h_{i,j}^p z_i^{p-1}(n)$	(includes offset input)
$z_j^p(n) = f(t_j^p(n))$	$2 \le p \le P+1$
$e_k(n) = d_k(n) + z_k^{P+1}(n)$ (computed in off-line/simulated systems only)	

Table 2. Equations for the FX-NEKA algorithm.

FX-NEKA algorithm
$\delta^{P+1}_{k,k}(n) = f'(t^{P+1}_{k}(n)) \qquad (= \partial z^{P+1}_{k}(n) / \partial t^{P+1}_{k}(n))$ $\delta^{P+1}_{i,j}(n) = 0 \quad i \neq j \qquad 1 \leq (i,j,k) \leq K$
$\delta^{p}_{j,h}(n) = \left(\sum_{k} \delta^{p+1}_{k,h}(n) h^{p+1}_{j,k} \right) f'(t^{p}_{j}(n))$ $(= \partial z^{P+1}_{h}(n) / \partial t^{p}_{j}(n)) \qquad 1 \leq p \leq P$
$\phi^{M+1}_{j,h,l}(n) = \left(\sum_{k} \delta^{1}_{k,h}(n) h^{1}_{j,k,l} \right) f'(s^{M+1}_{j}(n-l))$ $(= \partial z^{P+1}_{h}(n) / \partial s^{M+1}_{j}(n-l))$
$\phi^{m}_{j,h,l}(n) = \left(\sum_{k} \phi^{m+1}_{k,h,l}(n) w^{m+1}_{j,k}(n-l) \right) f'(s^{m}_{j}(n-l))$ $(= \partial z^{P+1}_{h}(n) / \partial s^{m}_{j}(n-l)) \qquad 1 \leq m \leq M$ $\phi^{m}_{j,h,l}(n) \approx \left(\sum_{k} \phi^{m+1}_{k,h,l}(n) w^{m+1}_{j,k}(n) \right) f'(s^{m}_{j}(n-l))$
$\boldsymbol{H}^{m}_{j,l}(n) = \left[\phi^{m}_{j,1,l}(n) \boldsymbol{y}^{m-1}(n-l) \quad \cdots \quad \phi^{m}_{j,K,l}(n) \boldsymbol{y}^{m-1}(n-l) \right]$ $(= \partial z^{P+1}(n) / \partial \boldsymbol{w}^{m}_{j}(n-l))$ (includes offset inputs)
$\boldsymbol{K}^{m}_{j}(n) = \lambda^{-1} \boldsymbol{P}^{m}_{j}(n) \left(\sum_{l=0}^{L_2} \boldsymbol{H}^{m}_{j,l}(n) \right) \times$ $\left(\boldsymbol{I} + \lambda^{-1} \left(\sum_{l=0}^{L_2} \boldsymbol{H}^{m}_{j,l}(n) \right)^{T} \boldsymbol{P}^{m}_{j}(n) \left(\sum_{l=0}^{L_2} \boldsymbol{H}^{m}_{j,l}(n) \right) \right)^{-1}$
$\boldsymbol{w}^{m}_{j}(n+1) = \boldsymbol{w}^{m}_{j}(n) - \mu \boldsymbol{K}^{m}_{j}(n) [e_1(n) \cdots e_K(n)]^{T}$
$\boldsymbol{P}^{m}_{j}(n+1) = \lambda^{-1} \left(\boldsymbol{P}^{m}_{j}(n) - \boldsymbol{K}^{m}_{j}(n) \left(\sum_{l=0}^{L_2} \boldsymbol{H}^{m}_{j,l}(n) \right)^{T} \boldsymbol{P}^{m}_{j}(n) \right)$

3. Convergence Performance

Simulations of active cancellation of sound in a duct using a non-linear actuator were performed, with a tonal single frequency disturbance to cancel. The tonal frequency was 50 Hz, while the sampling rate used was 450 Hz. The same non-linear plant model as the one experimentally found in [2] was used: a neural network with a configuration (number of neurons) of 8-4-1. The simulated system was a monochannel system, with 1 reference signal, 1 actuator signal and 1 error signal. In the controller network, the configuration used was 2-6-1. The activation functions used in the simulations were:

$$f(x) = x \quad \text{for neurons of an output layer} \qquad \textbf{(9)}$$

$$f(x) = \tanh(x) \quad \text{for all other neurons} \qquad \textbf{(10)}.$$

Table 3. Equations for the FX-EKF algorithm

FX-EKF algorithm
$H_l(n) = \begin{bmatrix} \phi^m_{j,1,l}(n) y^{m-1}(n-l) & \cdots & \phi^m_{j,K,l}(n) y^{m-1}(n-l) \\ \vdots & \vdots & \vdots \\ \phi^m_{j,1,l}(n) y^{m-1}(n-l) & \cdots & \phi^m_{j,K,l}(n) y^{m-1}(n-l) \end{bmatrix}$ $\left(= \partial z^{P+1}(n) \big/ \partial w(n-l)\right)$ for all values of j,m (includes offset inputs)
$K(n) = \lambda^{-1} P(n) \left(\sum_{l=0}^{L_2} H_l(n) \right) \times \left(I + \lambda^{-1} \left(\sum_{l=0}^{L_2} H_l(n) \right)^T P(n) \left(\sum_{l=0}^{L_2} H_l(n) \right) \right)^{-1}$
$w(n+1) = w(n) - \mu K(n) \begin{bmatrix} e_1(n) \\ \vdots \\ e_K(n) \end{bmatrix}$
$P(n+1) = \lambda^{-1} \left(P(n) - K(n) \left(\sum_{l=0}^{L_2} H_l(n) \right)^T P(n) \right)$

The convergence of the different algorithms was evaluated by computing the ratio of the (moving) average power of the error signal divided by the average power of the disturbance signal. The weight update gains μ, forgetting factors λ and inverse correlation $\boldsymbol{P}$ matrices initial values that produced the best performance for each algorithm were found by trial and error. Table 4 shows a summary of the results obtained with the previously published steepest descent algorithms (FX-BP, ADJ-BP [1],[2]), recursive least-squares algorithms (ADJ-NEKA, ADJ-EKF [2]) and the new FX-NEKA and FX-EKF algorithms. For more details, curves showing the complete learning dynamics of all the algorithms can be found in [8]. From Table 4, it can be seen that:

- recursive least-squares algorithms produce better steady-state performance than steepest descent algorithms for the training of the controller neural network
- the global recursive least-squares algorithms (FX-EKF, ADJ-EKF) outperform the local recursive least-squares algorithms (FX-NEKA, ADJ-NEKA), by producing better steady-state performance. However, the computational load of the global algorithms is much higher [8].
- the new FX-EKF outperforms the previously published global recursive least-squares algorithm (ADJ-EKF) in convergence speed. It is expected that for the cancellation of broadband disturbances the performance gain would be even higher (including steady-state performance) because the two advantages of the new recursive least-squares algorithms over the previously published recursive least-squares algorithms would be even more important (see Section 2).
- the new FX-NEKA greatly outperforms the previously published local recursive least-squares algorithm (ADJ-NEKA) in steady-state performance. Since local recursive least-squares algorithms have a much lower computational complexity that global algorithms [8], the FX-NEKA may be the best overall choice among the different algorithms.

4. Conclusion

In this paper, two new recursive least-squares algorithms for the training of the controller neural network in Fig. 1 were introduced. It was discussed and then shown with simulations that these algorithms outperform previously published steepest descent and recursive least-squares algorithms for the training of the controller neural network. Moreover, it is expected that for systems with broadband signals, the convergence gain of the introduced algorithms over the previously published algorithms would be even greater.

Table 4. Simulation results: convergence performance.

Algorithm	Steady-state convergence
FX-BP	-9 dB
ADJ-BP	-9 dB
FX-NEKA	-39 dB
ADJ-NEKA	-21 dB
FX-EKF	-43 dB (in 3000 iterations)
ADJ-EKF	-43 dB (in 6000 iterations)

References

1. Snyder, S.D., Tanaka N.: Active Control of Vibration using a Neural Network, IEEE Trans. on Neural Networks **6** (1995) 819-828

2. Bouchard M., Paillard, B., Le Dinh, C.T.: Improved Training of Neural Networks for Non-Linear Active Control of Noise and Vibration. IEEE Trans. on Neural Networks. **10** (1999) 391-401
3. Widrow, B., Stearns, S.D.: Adaptive signal processing. Prentice-Hall, Upper Saddle River (1985)
4. Wan, E.A.: Adjoint-LMS: An Efficient Alternative to the Filtered-X LMS and Multiple Error LMS algorithms. Proc. Int. Conf. Acoust. Speech and Signal Process. 1996, Atlanta. **3** 1842-1845
5. Haykin, S.: Adaptive filter theory. 3rd edn. Prentice Hall, Englewood Cliffs (NJ) (1996)
6. Shah, S., Palmieri, F., Datum, M.: Optimal Filtering Algorithms for Fast Learning in Feedforward Neural Networks. Neural Networks. **5** (1992) 779-787
7. Singhal S., Wu, L.: Training Feed-Forward Networks with the Extended Kalman Filter., Proc. Int. Conf. Acoust. Speech and Signal Process. 1989. **2** 1187-1190
8. Bouchard, M.: New Recursive-Least-Squares Algorithms for Nonlinear Active Control of Sound and Vibration using Neural Networks. IEEE Trans. on Neural Networks. **12** (2001) 135-147

Bayesian Independent Component Analysis as Applied to One-Channel Speech Enhancement

Ilyas Potamitis, Nikos Fakotakis, George Kokkinakis,

Wire Communications Lab., Electrical & Computer Engineering Dept., University of Patras, Rion-26 500, Patras, Greece
potamitis@wcl.ee.upatras.gr

Abstract. Our work applies a unifying Bayesian-Independent Component Analysis (BICA) framework in the context of speech enhancement and robust Automatic Speech Recognition (ASR). The corrupted speech waveform is reshaped in overlapping speech frames, and is assumed to be composed as a linear sum of the underlying clean speech and noise. Subsequently, a linear sum of latent independent functions is proposed to span each clean frame. Two different techniques are applied following a Bayesian formulation: In the first case the posterior probability of a clean speech frame is formed conditioned on the noisy one on which a maximum *a posteriori* (MAP) approach is applied, leading to Sparse Code Shrinkage (SCS) - a fairly new statistical technique originally presented to applied mathematics and image denoising, but its much promising potential for speech enhancement has not yet been exploited. In the second case, viewed within the Variational Bayes framework, the model for noisy speech generation is stated in a block-based fashion as a noisy, blind source separation problem from which we infer the independent basis functions that span the space of a speech frame and their mixing matrix, thus reconstructing directly the corresponding clean frames.

1 Introduction

The primary objective of noise compensation methods as applied in the context of speech processing is to reduce the effect of any signal which is alien to and disruptive of the message conveyed among participants in a communicative event (whether humans or ASR machines). Depending on the application, speech enhancement methods aim at improving the quality of perception and/or robust speech recognition.

The subtractive and the attenuating type of filters are well-established, one-channel noise compensation methods that predominate in speech enhancement literature [1]. Namely: a) spectral subtraction b) least mean square, adaptive filtering c) filter-based parametric approaches d) model-based, short-time spectral magnitude estimation, e) Hidden Markov Model (HMM)-based speech enhancement techniques. In what follows we foreground the main presuppositions of these techniques from which the BICA framework departs:

Most speech enhancement algorithms are based on a transformation, which facilitates the estimation of the clean speech model parameters such as Discrete

G. Dorffner, H. Bischof, and K. Hornik (Eds.): ICANN 2001, LNCS 2130, pp. 593–600, 2001.

Fourier Transform, Discrete Cosine Transform, Karhunen-Loève Transform. The transformation itself is more or less determined on an ad hoc basis. BICA is based on a data driven transformation kernel adapted to the structure of speech data.

Most methods focus on the distorted short-time amplitude of the speech signal leaving the phase unprocessed. In the BICA framework the speech signal is processed uniformly, meaning that there is no inherent need that compels us to concentrate on amplitude and ignore phase processing.

The noise reduction process in the subtractive types of algorithms introduces a trade-off between distortion of spectral balance of the processed speech signal and noise suppression factor. The BICA framework is not a subtractive technique and musical noise is not perceivable.

Spectral Subtraction and some versions of Wiener/Kalman and HMM-based algorithms require an accurate estimate of corrupting noise statistics. This is acquired during speech pauses through the use of a Voice Activity Detector (VAD). The construction of a robust VAD at low SNRs is a task still open to research. The framework of BICA has no need of a VAD to estimate speech-presence.

A linear modelling approach allows the corrupted speech frame to be decomposed in a linear sum of the underlying speech waveform and noise. Subsequently, the speech waveform is assumed to be composed of a linear sum of independent functions. This generation process can be viewed within a Bayesian Independent Component Analysis framework in two ways.

In the first case we apply ICA to a large ensemble of frames derived from clean, phonetically balanced recordings, revealing their underlying independent component structure based on Bells' seminal work on the higher order structure of images and sounds [2]. ICA is a statistical technique that determines a linear coordinate system, whose axes are defined by all higher moments of the data to which it is applied. Projecting the overlapping frames of a noisy recording on this basis, the time-domain observations are linearly transformed in order to reduce their dependency. Based on Hyvärinen's MAP formulation on the independent bases of images and Maximum Likelihood denoising of non-Gaussian data [3], we apply MAP inference on the posterior pdf of the ICA transformed clean frame conditioned on the noisy one, which leads to a threshold function optimally derived from the statistics of each independent component that effectively reduces white and coloured Gaussian noise. Finally, an inverse transformation from the ICA transformed domain back to time domain, taking into account the frame overlap, reconstructs the enhanced signal.

SCS requires the derivation of the unmixing matrix from clean data prior to applying the method. In the second technique there is no need to preprocess clean speech recordings, as the mixing matrix and the independent functions are directly inferred from the degraded observations. The key idea is to state the one-channel speech enhancement problem as a noisy, blind source separation case. As in SCS, the underlying clean speech frame is proposed to be constructed by the mixing of independent functions. The posterior distribution of the model for the generation of observed noisy frames defines a probability distribution over all sets of parameters. The posterior pdf is expressed as a computationally feasible approximation. The variables of interest, (i.e the mixing matrix and independent sources), are inferred in the process of fitting an approximating posterior to the true posterior, where the accuracy of the fit is assessed by the Kullback-Leibler (KL) divergence measure between the two distributions. The clean frame is then reconstructed by multiplying the mixing matrix with the estimated independent components.

We support theoretical derivations by extensive experimentation using recorded speech signals and real noise sources from the NOISEX-92 database[1]. Assessment criteria are based on objective and subjective measurements. The former category includes the segmental signal to noise ratio (SNR) measure, Itakura-Saito distortion measurements (allegedly correlated with subjective perception of speech quality), as well as word recognition accuracy of a speech recognition system. The latter category includes visual comparison of speech signals and informal listening tests.

2 Problem Formulation

Consider a clean, time-domain speech signal $\mathbf{x_t}$ corrupted by additive Gaussian noise $\mathbf{n_t}$, where (t) denotes the sample index. We process $\mathbf{x_t}$ in overlapping blocks of N=80 samples with 79 samples overlap (N corresponds to 10 ms frame at 8 kHz sampling rate). The resulting matrices are:

$$\mathbf{Y}=\mathbf{X}+\mathbf{N} \tag{1}$$

where $\mathbf{Y}=[\mathbf{y}_1,..,\mathbf{y}_F]$, $\mathbf{X}=[\mathbf{x}_1,..,\mathbf{x}_F]$, $\mathbf{N}=[\mathbf{n}_1,..,\mathbf{n}_F]$. Each $\mathbf{x}_i$, $\mathbf{y}_i$, $\mathbf{z}_i$, i=1,..,F, corresponds to a frame and F denotes the number of frames. N has been selected in such a way that the speech stationarity assumption is accurate enough and the window is small enough to capture rapid varying phenomena as transients and stop bursts.

We assume the existence of an underlying process that is responsible for the composition of every frame $\mathbf{x}_i \in \mathbf{X}$. This generative process is summarized as:

$$\mathbf{X}=\mathbf{A}\mathbf{S} \tag{2}$$

where $\mathbf{A} \in \Re^{NxN}$ is an unknown constant matrix called the mixing matrix and $\mathbf{S} \in \Re^{NxF}$. Eq. 2 suggests that any frame $\mathbf{x}$ is a linear combination of 'templates' (columns of $\mathbf{A}$), weighted by the corresponding coefficients of $\mathbf{s}$, or equivalently; the columns of $\mathbf{A}$ span the space of 10 ms frames and the independent coefficients $\mathbf{s} \in \mathbf{S}$ are responsible for the linear combination of the columns of $\mathbf{A}$ that reconstruct every phonetic segment $\mathbf{x}$. In SCS, in order to derive the unmixing matrix $\mathbf{W}$, which, when applied on any speech frame provides its variables with maximum independency, we make use of 1000 clean, phonetically balanced recordings uttered by an equal number of speakers of both genders at 8 kHz sampling rate. One can observe that a time varying mixing/unmixing matrix is more proper for the time varying case of speech. However, the unmixing matrix aims at reducing the dependency of consecutive speech samples and forming sparser distributions of the resulting pdf's of the ICA transformed speech frames. Sparsity can be achieved with a constant unmixing matrix as well. In the case of Variational Approximation the speech waveform is segmented in small blocks for which a different mixing matrix is derived and the enhanced signal is reconstructed in a block by block basis. Finally, heavy overlapping in both techniques is necessary, in order to result in a smooth, enhanced speech waveform.

[1] Varga A., "NOISEX-92 study," Technical Report, DRA Speech Research Unit, 1992.

3 Speech Enhancement Using the SCS Technique

There is a broad class of algorithmic versions implementing ICA for the instantaneous noiseless mixing case. We made use of a batch version of the Fastica algorithm [4] that derives **W** iteratively, through successive presentations of the frames of each recording separately until convergence in order to avoid processing at once the huge amount of total observational data. After estimating **W** we proceeded in an orthogonalizing transformation:

$$\mathbf{W}_{orth}=\mathbf{W}(\mathbf{W}^T\mathbf{W})\ . \tag{3}$$

The orthogonalization procedure normalizes the eigenvectors of **W** resulting in a more numerically-stable matrix. **W**orth was found superior to **W** on the grounds of better performance as assessed using the objective and subjective criteria of section 5. By applying $\mathbf{W}_{orth}$ to Eq. 1 we map the time domain matrices to the ICA domain:

$$\mathbf{Z}=\mathbf{W}_{orth}\mathbf{Y}=\mathbf{W}_{orth}\mathbf{X}+\mathbf{W}_{orth}\mathbf{N}=\mathbf{U}+\mathbf{N}'\ . \tag{4}$$

Gausssian distributions are invariant to linear transformations, therefore, $\mathbf{N}'=\mathbf{W}_{orth}\mathbf{N}$ is also zero mean Gaussian. Each $\mathbf{u}\in U$:$\mathbf{U}=\mathbf{W}_{orth}\mathbf{X}$, where $\mathbf{U}=[\mathbf{u}_1,..,\mathbf{u}_F]$, corresponds to a clean speech frame in the ICA transform domain. The components composing each **u** can be considered almost independent and can be factorized in terms of pdf marginals. The Bayesian form of the posterior pdf of **u** is (index (i) is dropped):

$$f_{\mathbf{u}/\mathbf{z}}(\mathbf{u}/\mathbf{z})=\frac{f_{\mathbf{z}/\mathbf{u}}(\mathbf{z}/\mathbf{u})f_{\mathbf{u}}(\mathbf{u})}{f_{\mathbf{z}}(\mathbf{z})}=\frac{1}{f_{\mathbf{z}}(\mathbf{z})}f_{\mathbf{n}'}(\mathbf{z}-\mathbf{u})f_{\mathbf{u}}(\mathbf{u})\ . \tag{5}$$

Assuming that there is no correlation of noise between components (the assumption holds approximately for non white Gaussian noise) and based on the factorization of $f_{\mathbf{u}}(\mathbf{u})$, we obtain the MAP estimate of **u** by minimizing the uniform cost function:

$$\hat{\mathbf{u}}=\underset{u}{\operatorname{argmax}}\left\{\prod_{i=1}^{N} f_{n'}(z_i-u_i)\prod_{i=1}^{N} f_{ui}(u_i)\right\}\ . \tag{6}$$

As $\hat{\mathbf{u}}$ is now factorized, it can be decomposed and the denoising method can be applied to each individual component as soon as a closed-form of the densities $f_{ui}(u_i)$ is derived. Evaluating Eq. 6 for i=1,..,N allows $\hat{\mathbf{u}}_i=[\hat{u}_1,\ldots,\hat{u}_N]^T$ to be constructed and subsequently $\bar{\mathbf{U}}=[\hat{\mathbf{u}}_1,\ldots,\hat{\mathbf{u}}_F]$. By making use of the relation $\bar{\mathbf{U}}=\mathbf{W}_{orth}\mathbf{X}$ and the orthogonality of $\mathbf{W}_{orth}$, we return back to time domain by: $\mathbf{X}=\mathbf{W}^T_{orth}\bar{\mathbf{U}}$.

We reconstruct the enhanced waveform by reshaping matrix **X** to vector form, taking into account the overlapping of vectors. The appropriate density representing each independent component $u_i(t)$ i=1,..,N in Eq. 6, can be selected according to the smallest KL divergence between the non-parametric density estimate of each component (histogram method) and a suitable, fitted parametric density (e.g Gaussian mixtures, double Laplacian). However, this selection procedue was found unnecessary. We estimated the normalized kurtosis of each $u_i(t)$ for a large number of recordings and we found with no exception that all components were very sparse. (Let $\mathbf{K}_{rec}$ denote the normalized kurtosis over all recordings, then mean$\mathbf{K}_{rec}$ =15, min$\mathbf{K}_{rec}$=8, max$\mathbf{K}_{rec}$=48 and it is obvious that they diverge significantly from the zero value of normalized kurtosis of a Gaussian pdf). Therefore, the representative of a family of very sparse super-Gaussian pdfs' is selected in advance as described in [3].

$$f_{ui}(u_i) = \frac{1}{2d} \frac{(a+2)[a(a+1)/2]^{a/2+1}}{[\sqrt{a(a+1)/2} + |u_i/d|]^{a+3}} \tag{7}$$

d denotes standard deviation and $a=(2-k+(k(k+4))^{1/2})/(2k-1)$ where $k=d^2 f_{ui}(0)$ controls the sparsity of the distribution. Substituting Eq. 7 in Eq. 5 and setting the derivative of the log-likelihood to zero results to a non-linearity applied to z_i.

$$\bar{u}_i = \mathrm{sign}(z_i)\max(0, \frac{|z_i| - bd}{2} + \frac{1}{2}\sqrt{(|z_i| + bd)^2 - 4\sigma^2(a+3)}) \tag{8}$$

where i=1,..,N and $b = \sqrt{a(a+1)/2}$. As can be observed from the coefficient $|z_i|$-bd , the non-linearity has a threshold, 'shrinking' effect by setting small values to zero.

4 Speech Enhancement Using Variational Bayes Approximation

Allowing Eq. 1 and Eq. 2 to be combined, the problem of speech enhancement can be re-stated as a problem of noisy, blind signal separation. We segment the noisy speech signal in blocks in order to infer a different mixing matrix **A** and independent sources for each segment, allowing clean speech to be constructed by a simple multiplication of the mixing matrix and the independent components in a block by block basis. For each segment we assign a mixture of Gaussian priors to the sources as in [6], and [5].

$$f(s_m) = \sum_{c=1}^{Nc} p_{mc} N(s_m \mid 0, b_{mc}) \tag{9}$$

We adopt the same zero-mean Gaussian noise model as in Eq. 4. The prior pdfs of the source and noise models and their parameters are stated in conjugate forms (see Appendix). A simple separable formulation of the posterior distribution is selected:

$$Q(\mathbf{s}, p, b, \mathbf{A}, a, \mathbf{L}) = Q(\mathbf{s})Q(p)Q(b)Q(\mathbf{A})Q(a)Q(\mathbf{L}) \tag{10}$$

Applying Jensen's inequality to the log evidence one can obtain an upper bound of the Maximum Likelihood for a model with visible variables the frame observations and hidden variables and their parameters as expressed in Eq. 1, 2, 4, 9. Jensen's inequality leads to a simplified cost function C_{KL} (Eq. 11), which can be minimized with respect to each of the Q distributions (see also [7]). The model parameters are inferred by a free-form functional minimization of the Kullback-Leibler divergence between the approximate distribution of Eq. 10 and the true posterior f(**s**, p, b, **A**, a, **x**, **L**), (refer to [5] for a detailed derivation and implementation).

$$C_{KL} = \left\langle \log \frac{Q(\mathbf{s})Q(p)Q(b)}{f(\mathbf{s}, p, b)} \right\rangle_q + \left\langle \log \frac{Q(\mathbf{A})Q(a)}{f(\mathbf{A}, a)} \right\rangle_q + \left\langle \log \frac{Q(\mathbf{L})}{f(\mathbf{z}, \mathbf{L})} \right\rangle_q \tag{11}$$

The minimization process of the cost function results in set of coupled equations that are updated iteratively until the parameters of interest (i.e **A** and **S**), are inferred.

5 Experimental results

The BICA framework makes the assumption that the distribution of noise is Gaussian. Therefore, we expect some impact on the effectiveness of BICA for noise cases exhibiting divergence from the normality assumption. Nevertheless, from Eq. 4, it becomes obvious that every $\mathbf{z} \in \mathbf{Z}$: $\mathbf{Z=WX+N'}$ accumulates a number of noise components. According to the Central Limit Theorem for each $\mathbf{z}$ component the density of noise will be closer to Gaussian than the distribution of each of the noise components. Therefore, we expect small impact on the effectiveness of BICA framework for noise cases exhibiting moderate divergence from the normality assumption. As regards the SNR of the recordings: each noise type is added to 34 clean speech files of 5 sec. mean duration so that the corrupted waveform ranges from –10 to 20 SNR_{dB}. The objective criteria are based on the mean value over all recordings. Let s(t) and ŝ(t) be the original undegraded and enhanced speech signals:

Segmental SNR provides simple error estimation over time and frequency and is calculated for (M), non-overlapping frames of 15 ms duration while the result is averaged over all waveform segments. Fig. 1 and Fig. 2 illustrate that best performance is achieved for the Gaussian type of noise. We attribute the extensive denoising capability to the fact that Gaussian noise complies with the assumptions of the algorithm, more than any other type of noise. Good performance is observed for the 'Operations Room' and 'STITEL' noise cases which are slowly varying, though, the algorithm cannot suppress the impulsively occurring components.

$$SNR_{dB} = \frac{1}{M}\sum_{j=0}^{M-1} 10\log_{10}\frac{\sum s_M{}^2(t)}{\sum (s_M(t)-\hat{s}_M(t))^2} \tag{12}$$

Itakura-Saito (IS) distortion measure is based on the spectral distance between the AR coefficient (a_c, a_p) of the clean and enhanced speech waveforms over synchronous frames of 15ms duration (g_c, g_p: all-pole gains). (IS) measure is closely associated with speech quality assessment and it is very sensitive to spectrum variations. It confirms our observation based on total SNR measure. (IS) measure reveals that although noise reduction is high at low SNRs the mismatch between the true and the enhanced version of the signal is augmented as the noise variance increases.

$$d_{IS}(a_p, a_c) = \left[\frac{g^2{}_c}{g^2{}_p}\right]\left[\frac{a_p R_c a^T{}_p}{a_c R_c a^T{}_c}\right] + \log\left[\frac{g^2{}_p}{g^2{}_c}\right] - 1 \tag{13}$$

Word Recognition accuracy was assessed by using a speech recognition module built with HTK Hidden Markov Models toolkit (Young S., "The HTK Book", Cambridge University, 1995). The basic recognition units are tied state context dependent triphones of five states each. Given this set of HMMs, and the corresponding dictionary, the HTK recognition unit produces the best path of the word network using the Frame Synchronous Viterbi Beam Search algorithm. The testing set consists of 100 files, part of the identity card corpus of the SpeechDat database (one speaker for each recording). The results of the tests conducted are depicted in Fig 1 and 2 and demonstrate that BICA enhancement method cooperates well with the HMMs framework, achieving consistently good results in low SNRs.

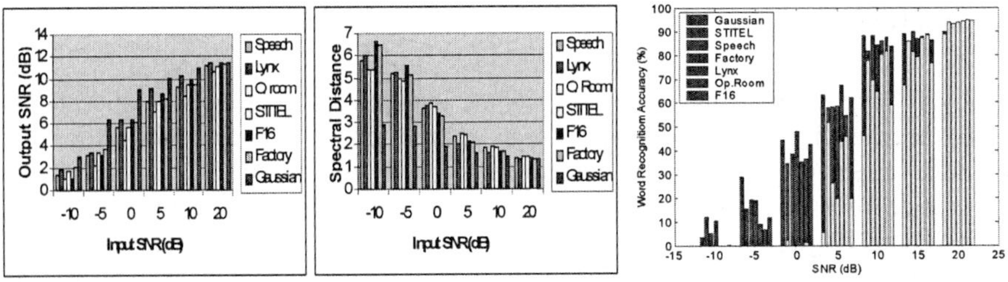

Fig. 1. SCS technique. (*From left to right*): Seg. SNR, Itakura-Saito, Word Recognition Accuracy (Black: SCS enhancement applied, White: no enhancement applied)

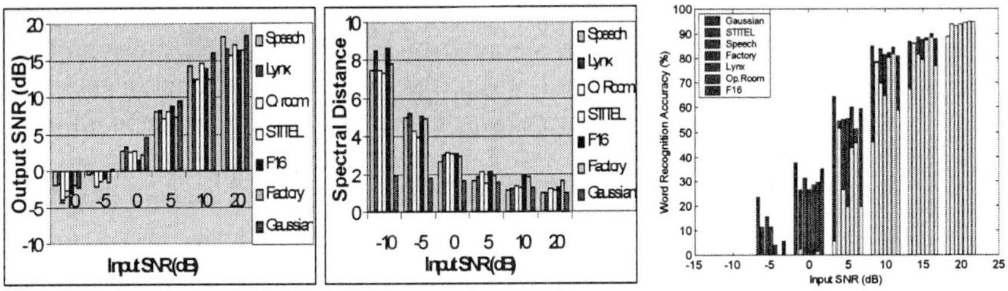

Fig. 2. Variational Bayes. (*Left to right*): Seg. SNR, Itakura-Saito, Word Recognition Accuracy (Black: SCS enhancement applied, White: no enhancement applied)

Listening tests. What we find most interesting is that at low SNRs as –10 dB, musical noise is hardly perceivable, though, the enhanced signal suffers due to distortions induced on the speech waveform. In such adverse conditions the algorithm shows its best performance in the case of Gaussian noise, since it is the noise that complies the most with the normality assumption (refer also to the time domain plots of noisy and enhanced signals of Fig. 3). At –5 dB the enhanced signal is quite perceptible though speech distortion is still present. The quality of the enhanced speech is quite high at higher SNRs, though, the enhanced coloured-noise speech is less clear. The non-stationary, impulsively occurring components of some noises cannot be suppressed, (e.g. slums of door in factory noise, voices in Operation room noise-case), a fact that we attribute to the violation of the stationarity assumption.

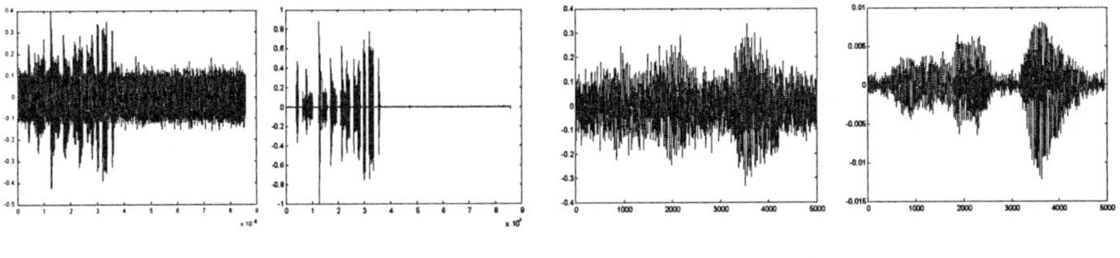

Fig. 3. Time domain plots for noisy and enhanced signals degraded with Gaussian noise at 0 dB input SNR. From Left to Right (a),(b): SCS enhancement, (c), (d): Variational Bayes method.

6 Conclusions

A novel view for the enhancement of signals is applied successfully to speech. It is based on the idea of decomposing clean speech frames corresponding to 10 ms phonetic segments into their independent basis functions. We proposed the application of a unifying BICA framework of two different techniques. In the first case an unmixing projection matrix is derived from a large ensemble of clean frames. This ensemble facilitates the application of MAP formulation that leads to a shrinkage function optimally derived from the statistics of each component. In the second case we demonstrated how a speech-enhancement problem can be stated as a noisy blind source separation problem faced with Bayesian Variational Approximation. Extensive experimentation with these techniques gave excellent results in the case of white Gaussian noise. They proved very effective in coloured types of Gaussian noise that do not diverge significantly from the stationarity assumption, and, most important, they preserve natural sound. As regards SCS, the most time-consuming task of the algorithm is the computation of the projection matrix **W**, which is computed off-line and only once for all subsequent restorations, whereas the Variational Bayes approach, though more computationally demanding, infers the clean frames directly from the noisy observations without the need of a preprocessing stage.

Appendix: Conjugate Priors and Optimal Posterior Distributions

The conjugate priors (left side) of the optimal posterior distributions (right side) are:

$f(b_{mc}) \sim G(b_{mc} \mid b^{(b)}, c^{(b)})$	$Q(b_{mc}) \sim G(b_{mc} \mid b_{mc}^{(b)}, c_{mc}^{(b)})$
$f(A_{nm}) \sim N(A_{nm} \mid 0, a_m)$	$Q(\mathbf{s}) \sim G(\mathbf{s} \mid \hat{\mathbf{s}}, \check{\mathbf{s}})$
$f(\{\pi_{mc}\}_{c=1,..,N}) \sim D(\{\pi_{mc}\}_{c=1,..,N} \mid c_{(p)})$	$Q(\{p_{mc}\}_{c=1,..,N}) \sim D(\{p_{mc}\}_{c=1,..,N} \mid c^{(p)}{}_{c=1,..N})$
$f(a_m) \sim G(a_m \mid b^{(a)}, c^{(a)})$	$Q(a_{mc}) \sim G(a_{mc} \mid b_{mc}{}^{(a)}, c_{mc}{}^{(a)})$
$f(L_{nn}) \sim G(L_{nn} \mid b^{(L)}, c^{(L)})$	$Q(L_{nm}) \sim G(L_{nn} \mid b^{(L)}, c^{(L)})$

where N, G, D denote Normal, Gamma and Dirichlet pdfs respectively.

References

1. Deller, J., Proakis, G., Hansen, J.: Discrete Time Processing of Speech Signals. New York, Macmillan Publishing Company (1993)
2. Bell, A.: The Independent Components of Natural Scenes are Edge Filters. Vision Research (1997) 3327-3338
3. Hyvärinen, A., Hoyer, O., Oja, E.: Image Denoising by Sparse Code Shrinkage. S. Haykin and B. Kosko (eds), Intelligent Signal Processing, IEEE Press, (2000)
4. Hyvärinen, A., Oja, E.: A Fast Fixed-Point Algorithm for Independent Component Analysis. Neural Computation, 9(7) (1997) 1483-1492
5. Miskin, J., MacKay, D.: Ensemble Learning for Blind Source Separation. Principles and Practice, Cambridge University Press
6. Attias, H.: Independent Factor Analysis. Neural Computat., (11) (1999) 803-851
7. Ghahramani, Z.: On Structured Variational Approx., Techn. Report CRG-TR-97-1

Massively Parallel Classification of EEG Signals Using Min-Max Modular Neural Networks

Bao-Liang Lu, Jonghan Shin, and Michinori Ichikawa

Lab. for Brain-Operative Device, RIKEN Brain Science Institute
2-1 Hirosawa, Wako-shi, Saitama 351-0198, Japan
{lu,shin,ichikawa}@brainway.riken.go.jp

Abstract. This paper presents a massively parallel method for classifying electroencephalogram (EEG) signals based on min-max modular neural networks. The method has several attractive features. a) A large-scale, complex EEG classification problem can be easily broken down into a number of independent subproblems as small as the user needs. b) All of the subproblems can be easily learned by individual smaller network modules in parallel. c) The classification system acts quickly and facilitates hardware implementation. To demonstrate the effectiveness of the proposed method, we perform simulations on a set of 2,127 non-averaged single-trial hippocampal EEG data. Compared with a traditional approach based on multilayer perceptrons, our method converges very much faster and recognizes with high accuracy.

1 Introduction

Neurophysiologists generate large amounts of time-series data such as EEG data as they record the electrical activity in the brain. Artificial neural networks have been applied to neurophysilogical data analysis such as classification of EEG signals [1, 2, 10]. However, training large networks on large sets of high-dimensional EEG data is a hard problem because no efficient algorithm is available for training large networks and a very long training time is required to achieve satisfactory learning accuracy [10]. To avoid this problem, existing methods usually use a few number of features extracted from EEG data as inputs. For example, only eight features were used in [10]. However, extremely reducing the number of features will cause the loss of useful information of the original EEG signals and the decrease in correct classification rate.

In this paper, we present a massively parallel method for classifying EEG signals based on the min-max modular (M^3) neural network, an alternative committee machine proposed in our previous work [4]. The method has the following several attractive features. a) A large-scale, complex EEG classification problem can be easily broken down into a number of independent subproblems as small as the user needs. b) All of the subproblems can be simply learned by individual smaller network modules in parallel, and therefore, a large set of high-dimensional EEG data can be learned efficiently. c) The classification system acts quickly and facilitates hardware implementation; consequently, it can

G. Dorffner, H. Bischof, and K. Hornik (Eds.): ICANN 2001, LNCS 2130, pp. 601–608, 2001.

be applied to implementing a hybrid brain-machine interface [5], which relies on the real-time sampling and processing of large-scale brain activity to control artificial devices.

2 EEG Data

It has been shown that hippocampal EEG signals are associated to cognitive process and behaviors, such as attention, learning, and voluntary movement [7]. The hippocampal EEG signals used in this study were recorded from eight adult male hooded rats between 300 and 400 g. These rats were housed in individual cages with food and water provided until the behavioral training. One week after surgery for implanting hippocampal electrodes, the rats were water-deprived and trained in a chamber by means of an oddball paradigm [7], in which occasional 'target' stimuli have to be detected in a train of frequent 'non-target' stimuli. We used a low frequency tone (so called odd tone) as 'target' stimuli and a high frequency tone (so called frequent tone) as 'non-target' stimuli. The animals were rewarded by water whenever they discriminate 'target' tone and cross the light beam in the water tube.

A total of 2,127 non-averaged single-trial hippocampal EEG signals were recorded from the rats. Each of the EEG signals is 6 sec in duration and belongs to one of the four classes, namely FR, FW, OR, and OW, where 'FR' means frequent tone and right behavior (no go), 'FW' frequent tone and wrong behavior (go), 'OR' odd tone and right behavior (go), and 'OW' odd tone and wrong behavior (no go). Figure 1 illustrates four non-averaged single-trial EEG signals belonging to FR, FW, OR, and OW, respectively. In the simulations below, we use 1,491 EEG signals for training and the rest of 636 EEG signals for testing. Table 1 shows the distributions of the training and test data sets.

Table 1. Distributions of the training and test data

	FR	FW	OR	OW
Training	1027	136	307	21
Test	430	68	128	10

3 Method

3.1 Feature Extraction with Wavelet Transform

In order to quantify changes in single-trial hippocampal EEG signals in both frequency and amplitude, we use wavelet transform techniques [9] to extract the features of EEG signals. The original EEG signals were convolved by the Morlet wavelet $w(t, w_o)$ with Gaussian shape both in the time domain and in the frequency domain around its central frequency w_o:

$$W(t, w_o) = \exp\left(jw_0 t - \frac{t^2}{2}\right) \tag{1}$$

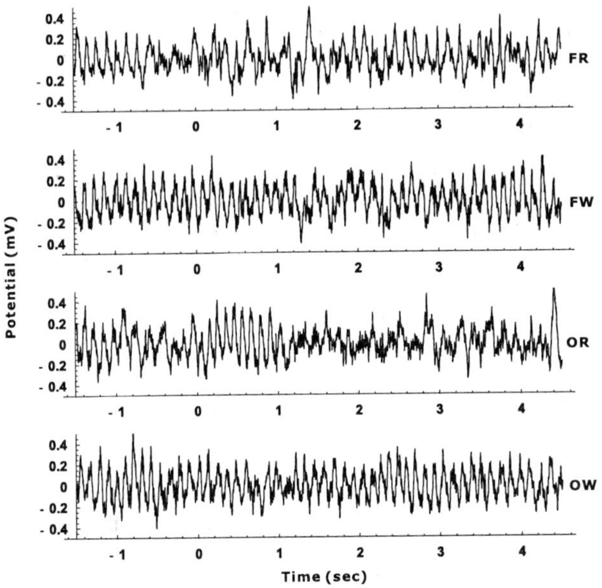

Fig. 1. Four EEG signals belonging to FW, FR, OR, and OW, respectively.

These wavelets can be compressed by a scale factor a and a shifted in time by a parameter b. Convolving the signal and the shifted and dilated wavelet leads to a new signal,

$$S_a(b) = \frac{1}{\sqrt{a}} \int W\left(\frac{t-b}{a}\right) x(t)dt \tag{2}$$

where W is the conjugate of the complex wavelet and $x(t)$ is the hippocampal EEG signal.

The new signals $S_a(b)$ are computed for different scaling factors a. In order to generate maps of hippocampal theta activity, the features of EEG signals were extracted between 5 Hz and 12 Hz out of the time-frequency maps. By selecting different numbers of samples in the time domain and using the same five wavelet coefficients in the theta frequency bandwidth, two sets of data were created, which have 200 and 2,000 features, respectively. Figure 2 shows the contour plots of the time-frequency representations of the four EEG signals illustrated in Fig. 1 with 2,000 features.

3.2 Task Decomposition

By using the task decomposition method proposed in our previous work [4], a K-class classification problem can be divided into $\binom{K}{2}$ two-class subproblems as follows:

$$\mathcal{T}_{ij} = \{(X_l^{(i)},\, 1-\epsilon)\}_{l=1}^{L_i} \cup \{(X_l^{(j)},\, \epsilon)\}_{l=1}^{L_j} \tag{3}$$

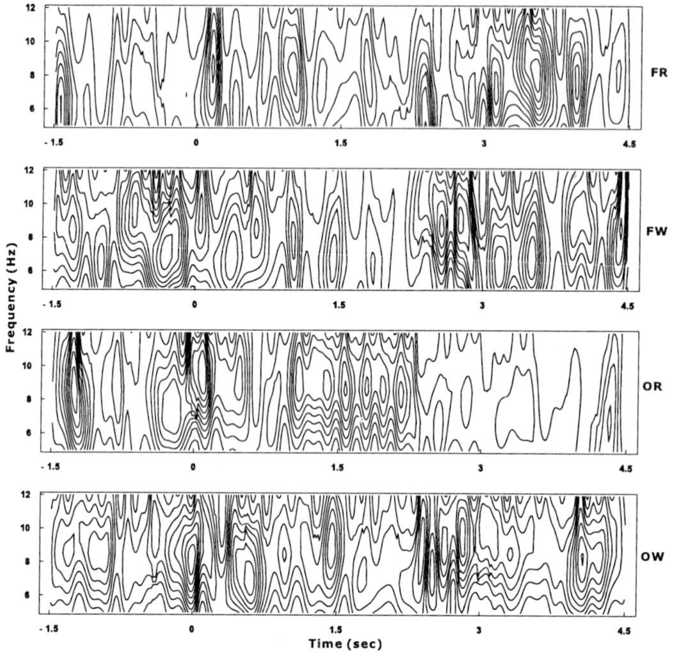

Fig. 2. The contour plots of the time-frequency representations of the four EEG signals illustrated in Fig. 1. Here, the number of features is 2,000.

where $i = 1, \cdots, K$, $j = i+1, \cdots, K$, ϵ is a small real positive number, $X_l^{(i)} \in \mathcal{X}_i$ and $X_l^{(j)} \in \mathcal{X}_j$ are the training inputs belonging to class $\mathcal{C}_i$ and class $\mathcal{C}_j$, respectively, $\mathcal{X}_i$ is the set of training inputs belonging to class $\mathcal{C}_i$, L_i denotes the number of data in $\mathcal{X}_i$, $\sum_{i=1}^{K} L_i = L$, and L is the total number of training data.

If some of the two-class problems defined by (3) are still large and hard to be learned, each of these subproblems can be further decomposed into a number of two-class problems as small as the user needs. Assume that $\mathcal{X}_i$ is partitioned into N_i $(1 \leq N_i \leq L_i)$ subsets in the form

$$\mathcal{X}_{ij} = \{X_l^{(ij)}\}_{l=1}^{L_i^{(j)}} \tag{4}$$

where $j = 1, \cdots, N_i$, $i = 1, \cdots, K$, and $\cup_{j=1}^{N_i} \mathcal{X}_{ij} = \mathcal{X}_i$. According to the above partition of $\mathcal{X}_i$, the two-class problem $\mathcal{T}_{ij}$ defined by (3) can be further divided into $N_i \times N_j$ much smaller and simpler two-class subproblems as follows:

$$\mathcal{T}_{ij}^{(u,v)} = \{(X_l^{(iu)}, 1-\epsilon)\}_{l=1}^{L_i^{(u)}} \cup \{(X_l^{(jv)}, \epsilon)\}_{l=1}^{L_j^{(v)}} \tag{5}$$

where $u = 1, \cdots, N_i$, $v = 1, \cdots, N_j$, $i = 1, \cdots, K$, $j = i+1, \cdots, K$, $X_l^{(iu)} \in \mathcal{X}_{iu}$ and $X_l^{(jv)} \in \mathcal{X}_{jv}$ are the training inputs belonging to class $\mathcal{C}_i$ and class $\mathcal{C}_j$, respectively. From (3) and (5), we see that a K-class problem can be decomposed into $\sum_{i=1}^{K} \sum_{j=i+1}^{K} N_i \times N_j$ two-class subproblems in a top-down approach.

According to (3), the four-class EEG classification problem is divided into $\binom{4}{2} = 6$ two-class subproblems, namely $\mathcal{T}_{1,2}$, $\mathcal{T}_{1,3}$, $\mathcal{T}_{1,4}$, $\mathcal{T}_{2,3}$, $\mathcal{T}_{2,4}$, and $\mathcal{T}_{3,4}$. From Table 1, we see that the number of training data for the smallest two-class subproblem $\mathcal{T}_{2,4}$ is 157, while the number of training data for the largest two-class subproblem $\mathcal{T}_{1,3}$ is 1,334. Although these two-class subproblems are smaller than the original problem, they are not adequate for massively parallel computation and efficient learning due to the following reasons. a) The subproblems are rather 'load imbalanced'. Since the speed of parallel learning is limited by the speed of the slowest subproblem, the unduly burdening of even a single subproblem can dramatically degrade the overall performance of the learning. b) Some of the subproblems are still too big for training. c) Some of the subproblems are very imbalanced, i.e., the training set contains many more data of the 'dominant' class than the other 'subordinate' class.

To speed-up learning, the bigger subproblems are further decomposed into a number of relatively smaller and simpler subproblems. According to (4), three larger training input data sets belonging to FR, FW, and OR are randomly broken down into 49, 6, and 15 subsets, respectively. As a result, the original four-class EEG classification problem is divided into $\sum_{i=1}^{4}\sum_{j=i+1}^{4} N_i \times N_j = 1,189$ balanced, two-class subproblems, where $N_1 = 49$, $N_2 = 6$, $N_3 = 15$, and $N_4 = 1$. The number of training data for each of the subproblems is about 40.

3.3 Massively Parallel Learning

An important feature of the above task decomposition method is that each of the two-class subproblems can be treated as a completely independent, non-communicating subproblem in the learning phase. Consequently, all of the subproblems can be learned in parallel.

In comparison with existing parallel implementations of neural network paradigms [8], the advantage of our massively parallel learning scheme is that it can be easily implemented not only on general-purpose parallel computers, but also on large numbers of individual serial machines and distributed Internet applications [3].

3.4 Module Combination

After learning, all of the trained individual network modules are integrated into an M^3 network by using three integrating units, namely MIN, MAX, and INV units, according to two simple module combination laws [4].

In contrast to the task decomposition mentioned above, the module combination process is bottom-up and involves the following two main steps.

a) The $N_i \times N_j$ network modules corresponding to the subproblems $\mathcal{T}_{ij}^{(u,v)}$ defined by (5) are integrated by using N_i MIN and one MAX units as follows:

$$\mathrm{MIN}_{ij}^{(u)} = \min(\mathrm{M}_{ij}^{(u,1)}, \mathrm{M}_{ij}^{(u,2)}, \cdots, \mathrm{M}_{ij}^{(u,N_j)}) \text{ for } u = 1, \cdots, N_i \quad (6)$$

and

$$\mathrm{M}_{ij} = \max(\mathrm{MIN}_{ij}^{(1)}, \mathrm{MIN}_{ij}^{(2)}, \cdots, \mathrm{MIN}_{ij}^{(N_i)}) \quad (7)$$

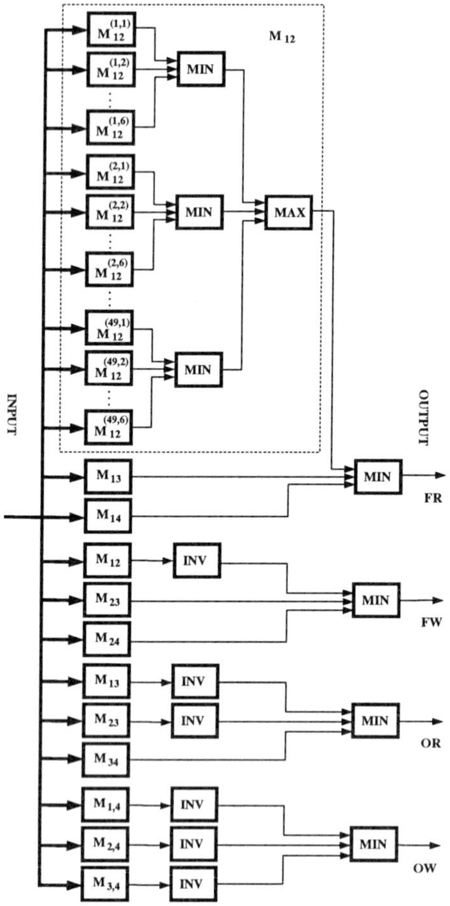

Fig. 3. The min-max modular network used for solving the four-class EEG classification problem, where thin arrowed-lines represent scalar inputs or outputs, and thick arrowed-lines represent vector inputs. Note that only module $M_{1,2}$ is plotted in detail, and the other modules are roughly illustrated due to space requirements.

where $M_{ij}^{(u,\,v)}$ denotes both the name of trained module corresponding to the subproblem $\mathcal{T}_{ij}^{(u,\,v)}$ and its actual output, and $MIN_{ij}^{(u)}$ denotes the output of the combination of N_j network modules. For example, according to (6) and (7), the 294 modules $M_{12}^{(u,v)}$ for $u = 1, \cdots, 49$ and $v = 1, \cdots, 6$ are integrated into the module M_{12} illustrated by the open dashed box in Fig. 3.

b) The $\binom{K}{2}$ network modules corresponding to $\mathcal{T}_{ij}$ and their $\binom{K}{2}$ inversions are integrated as

$$\mathrm{MIN}_i = \min(\mathrm{M}_{i1}, \cdots, \mathrm{M}_{ij}, \cdots, \mathrm{M}_{ik}) \text{ for } i = 1, \cdots, K \text{ and } i \neq j \qquad (8)$$

where MIN_i denotes the output of the combination of the $K-1$ modules by using the MIN unit. The M^3 network is guaranteed to produce solutions to the

Table 2. Performance of M^3 networks and single MLPs

No. of features	Method	No. of modules	No. of hidden units	CPU time (s.)		Correct rate (%)	
				Max	Total	Training	Test
200	MLP	1	60	120,011	120,011	100.0	78.9
	M^3	1189	2	7	2,209	100.0	79.7
2000	MLP	1	80	727,355	727,355	100.0	80.5
	M^3	1189	5	103	11,919	100.0	81.8

original problem as follows:

$$\mathcal{C} = \arg\max_i \{\text{MIN}_i\} \text{ for } i = 1, \cdots, K \tag{9}$$

where $\mathcal{C}$ is the class that the M^3 network has assigned to the input.

4 Computer Simulations

In the simulations, 1,189 three-layer MLPs were selected as network modules to learn the corresponding 1,189 subproblems. To compare M^3 networks with traditional MLPs, the original problems were also learned by three-layer MLPs. All of the network modules and the three-layer MLPs were trained by standard back-propagation algorithm [6]. All of the simulations were performed on an HP 9000 C240 workstation.

The simulation results are shown in Table 2, where 'Max' means the maximum CPU time required for training any networks. Figures 4(a) and 4(b) show the distribution of the 1,189 trained network modules measured by the number of epochs and CPU time (sec), respectively. From these figures we see that over 98% of the network modules converged within 300 epochs. From Table 2, we can see that our method is about 17,144 and 7,061 times faster than the method based on traditional three-layer perceptrons for learning the two EEG data sets that have 200 and 2,000 features, respectively, providing that learning in our method was performed in parallel. Even though all of the network modules were trained in serial, the results show that our method is still very much faster than the conventional method.

After the learning, the 1,189 individual trained network modules were integrated into the M^3 network shown in Fig. 3. Looking at Fig. 3, we see that our method has two useful features. a) The response time of the classification system is almost independent of the number of modules, that is, the system response time is almost independent of the problem size. b) The system might be easily implemented in hardware because of its modular structure.

5 Conclusions

In this paper we have presented a massively parallel method for classifying high-dimensional EEG data based on min-max modular neural networks. The advantages of the method over existing approaches are its high modularity and

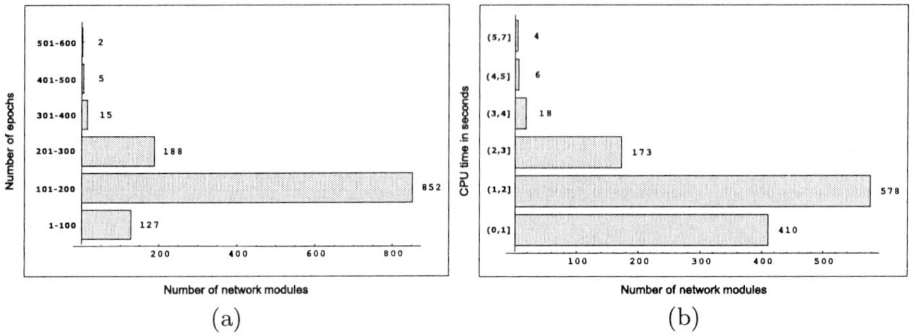

Fig. 4. Summary of the 1,189 network modules used for classifying the EEG data with 200 features. a) Distributions of trained modules measured by the number of epochs, and b) distributions of trained modules measured by CPU time in seconds.

parallelism, good scalability, and very fast learning speed. We have also demonstrated that the method is superior to the traditional approach based on multilayer perceptrons in generalization performance. By using the proposed method, we have begun doing computer experiments on a big data set containing 10,635 single-trial hippocampal EEG signals.

References

1. Bankman, I. N., Sigillito, V. G., Wise, R. A., Smith, P. L.: "Feature-based detection of the K-complex wave in the human electroencephalogram using neural networks", *IEEE Trans. Biomedical Engineering*, vol. 39, no. 12, pp. 1305-1310, 1992.
2. Jansen, B. H., Desai, P. R.:"K-complex detection using multi-layer perceptrons and recurrent networks", *Bio-Medical Computing*, vol. 37, pp. 249-257, 1994.
3. Lawton, W: "Distributed net applications create virtual supercomputers", *Computer*, vol. 33, no. 6, pp. 16-21, 2000.
4. Lu, B. L., Ito, M.: "Task decomposition and module combination based on class relations: a modular neural network for pattern classification", *IEEE Trans. Neural Networks*, vol. 10, no. 5, pp. 1244-1256, 1999.
5. Nicolelis, M. A. L.: "Actions from thoughts", *Nature* (insight feature), vol. 409, pp. 403-407, 2001.
6. Rumelhart, D. E. Hinton, G. E., Williams, R. J.: "Learning internal representations by error propagation", in 'Parallel Distributed Processing: Exploration in the Microstructure of Cognition', Vol. 1, (MIT Press, Cambridge, Mass, 1986)
7. Shin, J., Lu, B. L., Talnov, A., Matsumoto, G., Brankack, J.: "Reading auditory discrimination behavior of freely moving rats from hippocampal EEG", *Neurocomputing*, vol. 38-40, pp. 1557-1566, 2001.
8. Sundararajan, N., Saratchandran, P.: *Parallel Architectures for Artificial Neural Networks*, Los Alamitos, California, IEEE Computer Society Press, 1998.
9. Torrence, C., Compoo, G. P. : "A practical guide to wavelet analysis", *Bulletin of the American Meteorological Society*, 1998, Vol. 79, pp. 61-78
10. Tsoi, A. C., So, D., Sergejew, A. A.: "Classification of electroencephalogram using artificial neural networks", in 'Advances in Neural Information Processing Systems', Vol. 6, (Morgan Kaufmann, San Mateo, CA 1994), pp. 1151-1158

Single Trial Estimation of Evoked Potentials Using Gaussian Mixture Models with Integrated Noise Component

Arthur Flexer[1], Herbert Bauer[2], Claus Lamm[2], and Georg Dorffner[1]

[1] The Austrian Research Institute for Artificial Intelligence
Schottengasse 3, A-1010 Vienna, Austria
arthur@ai.univie.ac.at
[2] Department of Psychology, University of Vienna
Liebiggasse 5, A-1010 Vienna, Austria

Abstract. Gaussian Mixture Models with integrated noise component are a method developed for speech analysis to estimate signals hidden in background noise. We apply this technique to estimate single trial evoked potentials which are buried in noise up to five times stronger than the signal. An empirical study using artificial data is presented and results are compared to the standard technique of averaging.

1 Introduction

An evoked potential (EP) is the electro cortical potential measurable in the human electro encephalogram (EEG) before, during and after sensoric, motoric or cognitive events. An EP is defined as the combination of the brain electric activity that occurs in association with the eliciting event and 'noise', which is brain activity not related to the event together with inference from non-neural sources. Since the noise contained in single trial EPs is up to five times stronger than the signal, the common approach is to compute an average across several EPs recorded under equal conditions to improve the signal-to-noise ratio (SNR). For averaging of a set of N EPs, the assumed model for the ith EP is the following (see e.g. [Glaser & Ruchkin 1976]):

$$\vec{z}_i(t) = \vec{x}(t) + \vec{y}_i(t); \quad i = 1, 2, \ldots N; \quad 0 \leq t < T \tag{1}$$

where $\vec{z}_i(t)$ is the ith recorded EP, $\vec{x}(t)$ the underlying signal, $\vec{y}_i(t)$ the noise associated with the ith EP, and T the duration over which each EP is recorded. The average $\hat{\vec{x}}(t)$ over the sample of N EPs is used to estimate the underlying signal $\vec{x}(t)$:

$$\hat{\vec{x}}(t) = \frac{1}{N} \sum_{i=1}^{N} \vec{z}_i(t) = \vec{x}(t) + \frac{1}{N} \sum_{i=1}^{N} \vec{y}_i(t) \tag{2}$$

Averaging will attenuate the noise $\vec{y}_i(t)$ and not the signal $\vec{x}(t)$ given that signal and noise linearly sum together to produce the recorded EPs $\vec{z}_i(t)$ and the

G. Dorffner, H. Bischof, and K. Hornik (Eds.): ICANN 2001, LNCS 2130, pp. 609–616, 2001.

evoked signal $\vec{x}(t)$ is the same for each recorded EP $\vec{z}_i(t)$ and the noise contributions $\vec{y}_i(t)$ can be considered to constitute statistically independent samples of a random process. If the above assumptions do not hold, the averaging will result in a biased, distorted estimate of the signal. This distorted estimate also prohibits analysis of unique trial to trial information, like different latencies or amplitudes of components in single trials. What is needed is a method which is able to improve the SNR based on single trial EPs.

The problem is closely related to the estimation of speech model parameters in noisy acoustic environments. [Rose et al. 1994] present a general framework of estimating integrated models of signal and background noise. The authors distinguish between 3 processes: a process **Z** containing both the signal and the noise background, this is what usually can be directly observed (in our case this would be the single trial EP recordings); the signal **X** alone, in speech analysis this could be recordings with very low or zero background activity; the noise background **Y** alone, in speech analysis this could be segments of the recordings without any word utterances (at the beginning or in between).

The authors develop a general model around the assumption that all three processes can be modelled via mixtures of Gaussians. If **Z** can be observed it is necessary to either observe **X** or **Y** to be able to estimate the respective other "single" process. The authors present solutions for the case of additive Gaussian mixture models for **X** and **Y**. In this work we adopt the [Rose et al. 1994] approach to the problem of enhancement of single trial evoked potentials. One Gaussian kernel will be reserved for modeling of the noise process **Y** using parameters of the so-called residual noise obtained via averaging. This will allow us to estimate the signal process **X**.

Our work is structured as follows: first we describe a testbed of artificial EPs which allows for systematic variation of noise characteristics; then we present all technical details of the Gaussian Mixture Model with integrated noise component; in the results section we give quantitative measures for the achieved improvement in SNR; finally we discuss our approach by comparing it to other methods dealing with improvement of SNR in single trial EP analysis.

2 Data

The goal of our work is to devise a method that estimates EP signals hidden in considerable amounts of background noise. Since for real EP data the hidden signal is actually not known, we resort to simulated artificial EP data. Only the use of artificial data with known signal and noise proportions allows us to really quantify the gain in signal-to-noise ratio.

We simulated $500msec$ of a visually evoked EP recorded via $d = 22$ electrodes mounted according to the international 10-20 system. One artificial EP is of length $T = 125$ since we simulated a $250Hz$ recording. First we defined eight topographical patterns (each consisting of $d = 22$ values) corresponding to well known EP components like the $N100$ or $P300$. These topographical patterns have distinct amplitudesignal values and temporal latencies (see Fig. 1 (a)).

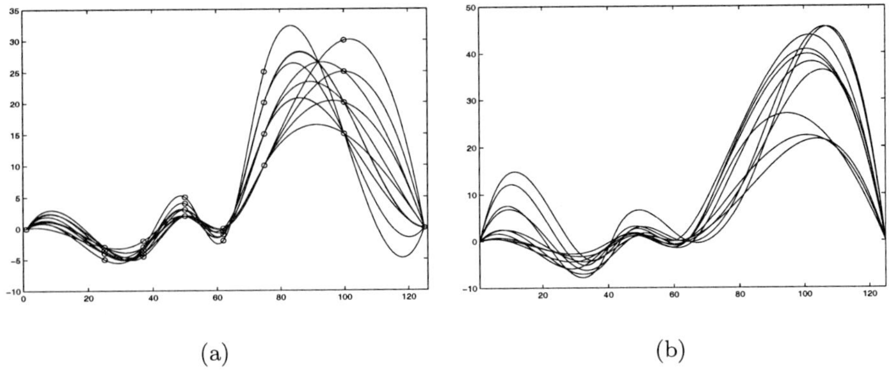

Fig. 1. (a) All $d = 22$ channels of an artificial EP plotted in one graph (x-axis time, y-axis amplitude). The underlying topographical patterns are shown as circles at $t = 1, 25, 37, 50, 62, 75, 100, 125$. Note the earlier onset of the $P300$ at more frontal electrodes relative to more posterior electrodes (i.e. amplitude peaks of $\sim 30\mu V$ at $t \approx 80$ and 100). (b) Ten artificial EP signals at electrode O2. Notice the between-trial variation in terms of latency and amplitude.

By computing a cubic spline interpolation for each signal dimension d we obtain values at all 125 sample points. To simulate between-trial variation we randomly shifted the topographical patterns both in time and amplitude prior to spline interpolation (change of temporal position: $t_{new} = t_{old} + 3 * \mathcal{N}(0,1)$, change of amplitude: $x_{new} = x_{old} + 0.33 * \mathcal{N}(0,1) * \mid x_{old} \mid$, $\mathcal{N}$ is the Gaussian distribution). The between-trial variation of ten EP signals is depicted in Fig. 1 (b).

We computed 5×100 of these artificial EPs with latency and amplitude variation. We added random Gaussian noise to each of these 5 EP sets in five different signal-to-noise ratios (SNR): 1:1, 1:2, 1:3, 1:4, 1:5. Gaussian noise was produced by limiting a $4000Hz$ Gaussian random signal to a bandwidth between 0 and $30Hz$ via FIR filtering and then down-sampling it to $250Hz$. A specific SNR was achieved by scaling the mean powers of signal and noise to yield the desired SNR. This yielded empirical mean *Input* SNR ratios of $3.34, -10.36, -18.45, -24.18$ and $-28.86dB$.

3 Methods

In the case of an GMM with integrated noise component by [Rose et al. 1994], both the signal $\vec{x}$ and the background noise $\vec{y}$ are modelled as mixtures of Gaussians, independent in time:

$$p(\vec{x}|\lambda_s) = \sum_{i=1}^{M} p_i b_i(\vec{x}) \qquad p(\vec{y}|\lambda_b) = \sum_{j=1}^{N} q_j a_j(\vec{y}) \tag{3}$$

where the signal model $\lambda_s = \{p_i, \mu_i[d], \sigma_i[d]\}$ is a mixture of M Gaussian densities b_i and the background noise model $\lambda_b = \{q_j, \mu_j[d], \sigma_j[d]\}$ is a mixture of

N Gaussian densities a_j, both with diagonal covariance matrices. The p_i and q_j are the mixing coefficients with $\sum_{i=1}^{M} p_i = 1$ and $\sum_{j=1}^{N} q_j = 1$.

If the observable noisy signal $\vec{z}$ is given via a general function of signal and background noise $\vec{z} = f(\vec{x}, \vec{y})$, than the density of the noisy signal for state i and j at time step t is given by:

$$p(\vec{z}_t | i_t, j_t, \lambda_s, \lambda_b) = \iint_{C_t} b_{i_t}(\vec{x}_t) a_{j_t}(\vec{y}_t) d\vec{x}_t d\vec{y}_t \tag{4}$$

where C_t denotes the contour defined by the general function $\vec{z} = f(\vec{x}, \vec{y})$ introduced above. The likelihood of a noise corrupted observation $\vec{z}_t$ is given by:

$$p(\vec{z}_t | \lambda_s, \lambda_b) = \sum_{i=1}^{M} \sum_{j=1}^{N} p_i q_j p(\vec{z}_t | i, j, \lambda_s, \lambda_b) \tag{5}$$

In the following, reference to λ_b will be dropped as a notational shorthand. Therefore the signal model is simply $\lambda = \{p_i, \vec{\mu}_i, \vec{\sigma}_i\}$. Our goal is now to obtain estimates of the underlying signal model by maximizing the likelihood of a set of noisy observations Z. Estimation in the maximum likelihood sense involves finding the model λ which maximizes $p(Z|\lambda)$ with

$$P(Z|\lambda) = \sum_I \sum_J \iint_C P(X, Y, I, J|\lambda) dX dY \tag{6}$$

and

$$P(X, Y, I, J|\lambda) = \prod_{t=1}^{T} b_{i_t}(\vec{x}_t) p_{i_t} a_{j_t}(\vec{y}_t) q_{i_t}. \tag{7}$$

Application of the Expectation-Maximization (EM) technique [Dempster et al. 1977] yields the following general expressions for the estimation of the model parameters. Since all densities are assumed to have diagonal covariance matrices, each vector component can be estimated independently. The notational shorthands $\bar{p}_i$, $\bar{\mu}_i$ and $\bar{\sigma}_i$ therefore refer to arbitrary vector components.

$$\bar{p}_i = \frac{1}{T} \sum_{t=1}^{T} \sum_{j=1}^{N} p(i_t = i, j_t = j | z_t, \lambda) \tag{8}$$

$$\bar{\mu}_i = \frac{\sum_{t=1}^{T} \sum_{j=1}^{N} p(i_t = i, j_t = j | z_t, \lambda) E\{x_t | z_t, i_t = i, j_t = j, \lambda\}}{\sum_{t=1}^{T} \sum_{j=1}^{N} p(i_t = i, j_t = j | z_t, \lambda)} \tag{9}$$

$$\bar{\sigma}_i^2 = \frac{\sum_{t=1}^{T} \sum_{j=1}^{N} p(i_t = i, j_t = j | z_t, \lambda) E\{x_t^2 | z_t, i_t = i, j_t = j, \lambda\}}{\sum_{t=1}^{T} \sum_{j=1}^{N} p(i_t = i, j_t = j | z_t, \lambda)} - \bar{\mu}_i^2 \tag{10}$$

where $p(i_t = i, j_t = j | z_t, \lambda)$ is given by

$$p(i_t = i, j_t = j | z_t, \lambda) = \frac{p(z_t | i_t = i, j_t = j, \lambda) p_i q_j}{\sum_{i=1}^{M} \sum_{j=1}^{N} p(z_t | i_t = i, j_t = j, \lambda) p_i q_j}. \tag{11}$$

If we assume an additive noise model with $f(x_t, y_t) = x_t + y_t$ and a background process which can be modelled by a single Gaussian (i.e. $N = 1$ in all above formulas), we get:

$$E\{x_t|z_t, i_t = i, \lambda\} = \frac{\sigma_i^2}{\sigma_i^2 + \sigma_b^2}\left[z_t + \left(\frac{\sigma_b^2}{\sigma_i^2}\mu_i - \mu_b\right)\right] \tag{12}$$

$$E\{x_t^2|z_t, i_t = i, \lambda\} = \frac{\sigma_i^2\sigma_b^2}{\sigma_i^2 + \sigma_b^2} + E\{x_t|z_t, i_t = i, \lambda\}^2 \tag{13}$$

$$p(z_t|i_t = i, \lambda) = \mathcal{N}[z_t, \mu_i + \mu_b, \sigma_i^2 + \sigma_b^2] \tag{14}$$

where $\mathcal{N}[z_t, \mu_i + \mu_b, \sigma_i^2 + \sigma_b^2]$ is the Gaussian distribution corresponding to the observed noisy signal. Since both the signal x_t and the background y_t are random variables with Gaussian distribution their sum is also a Gaussian distribution with the above parameters.

4 Results

We applied Gaussian Mixture Models with integrated noise component (GMMnoise) to all 5 data sets described in Sec. 2. We used mixtures of $M = 10$ Gaussians for the signal $\vec{x}$ and a single Gaussian ($N = 1$) for the background noise $\vec{y}$. Using mixtures of $M = 20$ Gaussians yielded only minor improvements. As an estimate of the noise mean μ_b and variance σ_b^2 we used the respective parameters of the residual noise obtained via averaging [1]. For each set of 100 artificial EPs we computed an average according to Equ. 2 and subtracted it from all 100 artificial EPs. The remainder is the residual noise.

Once a GMMnoise model is trained we can use the component densities given in Equ. 14 to weigh the means of the Gaussians $\vec{\mu}_i$ at each time step t and obtain a denoised estimate $\hat{\vec{x}}'_t$ of the signal:

$$\hat{\vec{x}}'_t = \sum_{i=1}^{M} p(\vec{z}_t|i_t = i, \lambda)\vec{\mu}_i \tag{15}$$

Since our data are artificial EPs we know exactly which signals $\vec{x}$ are hidden in the noisy signals $\vec{z}$. We can compare the denoised estimates obtained via GMMnoise as well as the common average given in Equ. 2 to these signals $\vec{x}$. We obtain noise residuals by subtracting the real signals $\vec{x}$ from the denoised signal estimates $\hat{\vec{x}}'_t$ and $\hat{\vec{x}}$. The ratio in dB between the real signals $\vec{x}$ and these noise residuals are called *Output* SNRs and are given in Tab. 1 for both GMMnoise and averaging. The same information is depicted in Fig. 2(a).

[1] Since averaging does not fully remove the noise from the signal the variance of the noise residual is always underestimating the true noise variance. One could try to think about accounting for this to get more accurate estimates.

Table 1. Results for GMMnoise and averaging applied to the five EP data sets given as SNR in dB.

data set	1:1	1:2	1:3	1:4	1:5
Input SNR	3.34	– 10.36	– 18.45	– 24.18	– 28.86
Output SNR GMMnoise	16.43	9.92	4.38	– 0.02	– 3.17
Output SNR average	16.92	15.20	14.11	12.38	10.47

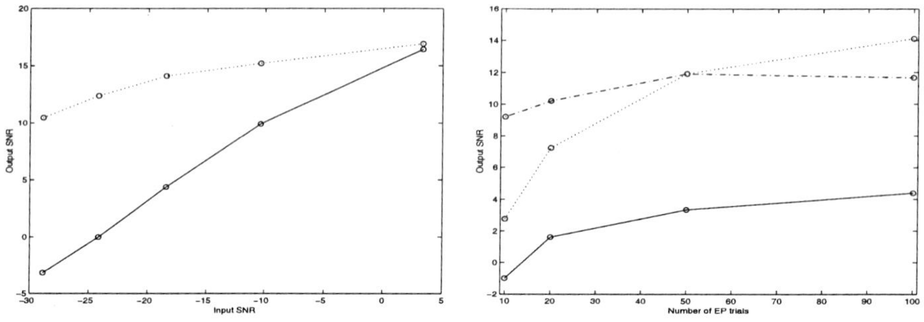

(a) (b)

Fig. 2. (a) Input SNR (x-axis) vs. Output SNR (y-axis) in dB for GMMnoise (solid) and averaging (dotted). (b) Number of EP trials (x-axis) vs. Output SNR (y-axis) for GMMnoise (solid), averaging (dotted) and GMMnoise+ave (dash-dot).

If the power of the underlying signal is approximately equal to the power of the background noise (data set "1:1"), GMMnoise is able to recover the information based on single EP trials as good as averaging which uses the full 100 trials of the data set. At lower SNRs, the performance of GMMnoise decreases relative to averaging. But even for data set "1:5", the improvement in SNR is around $25dB$. Our results are comparable to those achieved with other more involved approaches. [Cerutti et al. 1988] employ an ARX (autoregressive with exogenous input) model for single trial estimation using separate EEG and EOG (electro-oculogram) noise sources. The authors report an improvement in SNR of around $20dB$ for an Input SNR of $-30dB$ which is very similar to what GMMnoise achieves for data set "1:5". [Westerkamp & Aunon 1987] employ an a posteriori Wiener filter which is time-varying and uses information from multiple electrodes to estimate single trial EPs. Contrary to our GMMnoise this is a non-stationary technique taking into account temporal variations of noise and signal statistics. In their study on simulated EP data, the best improvement in SNR achieved for their lowest Input SNR of $-9dB$ is around $12dB$. GMMnoise achieved an improvement of almost $20dB$ for an Input SNR of $-10.36dB$ (see Tab. 1, data set "1:2").

Another application of GMMnoise besides estimation of single trial EPs is to use it as a preprocessing technique prior to averaging. The improvement in SNR might make it possible to compute averages of high quality using less EP trials.

This would greatly benefit experimental work since recording of EPs is time and cost consuming.

We compared GMMnoise ($M = 10, N = 1$ just as before), averaging and averaging after application of the GMMnoise method (GMMnoise+ave) on subsets of the "1:3" data set containing 10, 20, 50 and 100 EP trials. Estimates of the noise means μ_b and variances σ_b^2 were based on the respective parameters of the residual noise obtained via averaging of 10, 20, 50 or 100 trials. Results are depicted in Fig. 2(b). Results achieved by averaging are worse for numbers of trials smaller than 50 when compared to results obtained by GMMnoise+ave. GMMnoise+ave shows considerable improvements of SNR already using only 10 or 20 EP trials.

5 Conclusion

We presented an adaption of the speech analysis method of Gaussian mixture models with integrated noise component (GMMnoise) to the improvement of SNR in single trial evoked potentials (EP). GMMnoise is basically a mixture of Gaussian approach where one Gaussian kernel is reserved for noise modeling and its parameters are fixed prior to estimation of the whole model. Using a testbed of simulated visually evoked potentials we showed that GMMnoise performs at the level of other state of the art techniques. We also showed that GMMnoise could also be used as a preprocessing method prior to averaging to enable computation of high quality averages using less EP trials.

Up till now we used GMMs with diagonal covariance matrices only. Employing full covariances might allow more accurate modeling of signals. Another possibility for improving GMMnoise would be to account for the non-stationarity of EP signals by switching from atemporal GMMs to Hidden Markov Models (HMM). [Logan 1998] presents an extension of the GMMnoise framework to the non-stationary HMM case where auto-regressive observation densities are used to model signal and noise. Although it was necessary to use artificial data to accurately measure the gain in SNR obtained via GMMnoise the ultimate test would of course be its application to real EP data. An EP data set where considerable trial to trial variance is expected would be an ideal testbed.

Acknowledgements

The Austrian Research Institute for Artificial Intelligence is supported by the Austrian Federal Ministry of Education, Science and Culture. This research is supported by the Austrian Scientific Research Fund (FWF), project P14100-MED "Improving analysis of brain electric potentials to enable monitoring of temporally extended cognition".

References

[Cerutti et al. 1988] Cerutti S., Chiarenza G., Liberati D., Mascellani P., Pavesi G.: A parametric method for identification of single-trial event related potentials in the brain, IEEE Trans. on Biomedical Engineering, vol. BME-35, pp. 701-711, 1988.

[Dempster et al. 1977] Dempster A.P., Laird N.M., Rubin D.B.: Maximum likelihood from incomplete data via the EM algorithm, Journal of the Royal Statistical Society, B, 39:1-38, 1977.

[Glaser & Ruchkin 1976] Glaser E.M., Ruchkin D.S.: Principles of Neurobiological Signal Processing, Academic Press, London/New York/San Francisco, 1976.

[Logan 1998] Logan B.T.: Adaptive Model-Based Speech Enhancement, University of Cambridge, PhD Thesis, 1998.

[Rose et al. 1994] Rose R.C., Hofstetter E.M., Reynolds D.A.: Integrated Models of Signal and Background with Application to Speaker Identification in Noise, IEEE Transactions on Speech and Audio Processing, Vol. 2, No.2, pp.245-257, 1994.

[Westerkamp & Aunon 1987] Westerkamp J., Aunon J.I.: Optimum multielectrode a posteriori estimates of single-response evoked potentials, IEEE Transactions on Biomedical Engineering, 34, 13-22, 1987.

A Probabilistic Approach to High-Resolution Sleep Analysis

Peter Sykacek[1], Stephen Roberts[1], Iead Rezek[1], Arthur Flexer[2], and Georg Dorffner[2]

[1] Robotics Research Group, Dept. Eng. Sci., University of Oxford, Parks Road, Oxford OX1 6PJ, UK
psyk@robots.ox.ac.uk

[2] Austrian Research Institute for Artificial Intelligence (ÖFAI), Schottengasse 3, A-1010 Vienna, Austria

Abstract. We propose in this paper an entirely probabilistic approach to sleep analysis. The analyser uses features extracted from 6 EEG channels as inputs and predicts the probabilities that the sleeping subject is either awake, in deep sleep or in rapid eye movement (REM) sleep. These probability estimates are provided for different temporal resolutions down to 1 second. The architecture uses a "divide and conquer" strategy, where the decisions of simple experts are fused by what is usually refered to as "naïve Bayes" classification. In order to show that the proposed method provides viable means for sleep analysis, we present some results obtained from recordings of good and bad sleep and the corresponding manual scorings.

1 Introduction

This paper proposes a sleep analyser that maps all night EEG recordings to 3 probability plots. We model the probabilities of being awake, in deep sleep and in REM sleep with temporal resolutions down to one second. Allowing for probabilities and for high temporal resolutions is motivated by prior experiences with a similar method which were reported in [PRTJ97]. The hope is that using such an approach allows us to improve upon classical Rechtschaffen and Kales (R&K) rules [RK68]. Although useful for many purposes, this standard has raised dissatisfaction (e.g. [PSKH91]).

- The rules are based on rather short events in the EEG. In the worst case up to 98% of the EEG signal in a scoring window could be ignored.
- R&K rules use amplitude based criteria and rather large scoring windows, which is valid for healthy young subjects. The rules do however not account for aging effects and they fail to work for some important sleep-disturbances (e.g. sleep apnoea).

The analyser proposed in this paper differs from that in [PRTJ97] by allowing for an entirely probabilistic approach. This has the advantage that all uncertainties (e.g. from noise contamination of the EEG recordings) will lead to less certain decisions about the state of the sleeping subject. The analyser consists of three major building blocks:

G. Dorffner, H. Bischof, and K. Hornik (Eds.): ICANN 2001, LNCS 2130, pp. 617–624, 2001.

- a preprocessing stage,
- a classification stage implemented as a generative model
- and a sensor fusion stage.

The paper also presents some first results obtained from four recordings. Two of the recordings represent "good sleep"[1] (a young male and an elderly female) and two recordings represent "bad sleep" (two middle aged females, one control and a patient with an anxiety disorder). We conclude from this evaluation that the proposed method captures these overall aspects. Contrary to the R&K scoring, which does not find any stage 4 in the recording of the elderly female, the proposed method shows a clear sleep cycles.

2 Methods

The first step in the analyser design was to decide upon an optimal preprocessing technique. We performed an extensive investigation with different feature subset selection (FSS) techniques. Details of this analysis were reported in [SRR+99] and [Syk00b]. The investigation revealed that we should prefer the use of complexity measures and autoregressive (AR) model parameters (without preferring either of these techniques) as opposed to classical FFT based features. We decided for theoretical reasons[2] to use AR models in a lattice filter representation. The second result of this evaluation is that the optimal number of AR coefficients for *classifying* EEG is as small as 3. These lattice filter coefficients are used as inputs in a modular design, which is motivated by several interesting properties.

- We build several simple classifiers (experts) on top of each electrode and hence avoid the curse of dimensionality of fitting one large model.
- The low input dimension allows us to use a fully generative model which is trained using labelled and unlabelled data[3].
- Fusing multiple experts will increase the reliability of the overall system because information from such segments where the experts disagree (e.g. caused by electrode failure) will be downweighted.
- The proposed architecture can be extended very efficiently.

2.1 A Bayesian Lattice Filter

The lattice filter is a representation of an auto regressive (AR) model. The parameters of the lattice filter are the so called refection coefficients, below denoted as ρ_m. Equation (1) shows an AR model, where $y[t]$ is a sample of a time series at time t, a_m is the m-th AR coefficient and $e[t]$ a sample of an independently

[1] That was judged by human experts.
[2] A probabilistic formulation of complexity measures has so far not been successful.
[3] We use labels only for such segments that were unanimously classified by three human experts. Hence we have plenty of unlabelled data.

identically distributed (i.i.d.) Gaussian innovation sequence with zero mean and precision (inverse variance) β.

$$y[t] = \sum_{m=1}^{M} y[t-m]a_m + e[t] \tag{1}$$

We refer to [Lju99] for details of how to reparameterise Equation (1) in terms of reflection coefficients. Using a lattice filter representation increases the computational efficiency and allows us to use a "stability prior" [4]. The reflection coefficients of stable AR models must lie in the intervall $[-1, 1]$. This stability requirement may be coded by a flat prior $p(\rho_m) = 0.5$. For the precision β we use the prior $p(\beta) = 1/\beta$ as is suggested in [Jef61]. The posterior in Equation (2) is obtained by integrating the AR coefficients and β out of the joint distribution over all coefficients.

$$p(\rho_m|\mathcal{Y}, I_m) \propto p(\mathcal{Y}|I_m)\frac{1}{\sqrt{2\pi}s}\exp\left(-\frac{1}{2s^2}(\rho_m - \hat{\rho}_m)^2\right), \tag{2}$$

Equation (2) denotes the time series used for inference as $\mathcal{Y}$. The model indicator, I_m, represents the hypothesis that an m-th order AR model has generated $\mathcal{Y}$. The multiplicative constant $p(\mathcal{Y}|I_m)$ is the so called *model evidence.* Further aspects of how to calculate and to use the evidence for model selection can be found in [Syk00a].

Equation (3) shows the most probable value of ρ_m and its variance. The equations use N as number of samples in $\mathcal{Y}$. We define $\underline{\epsilon}_m$ as the vector of forward prediction errors of the $(m-1)$-th order AR model in $\mathcal{Y}$ and $\underline{r}_m$ as the vector of the corresponding backward prediction errors[5]

$$\hat{\rho}_m = -\frac{\underline{r}_m^{\mathsf{T}}\underline{\epsilon}_m}{\underline{r}_m^{\mathsf{T}}\underline{r}_m} \text{ and } s^2 = \frac{1-(\hat{\rho}_m)^2}{(N-1)} \tag{3}$$

Inference of reflection coefficients was performed with the following settings:

- Recordings with different sampling rates were resampled to 200 Hz.
- Calculations use a two seconds window that is shifted by one second offsets.
- The data in each window are linearly detrended and normalised to unit standard deviation.
- The time delay operation which is performed in the order recursion is coupled with pre and post windowing (see [Lju99]).

2.2 Classification

As already mentioned, the classifier models the joint density over inputs and labels. We use a fully generative model with K classes, where each of the class

[4] Sleep EEG is necessarily generated by stable AR models.

[5] The backward prediction errors are the errors when precicting one time step into the past. That is with model order $m-1$, we predict $y[t-m]$.

conditional densities are modelled by a mixture model with D Gaussian kernels. Equation (4) abreviates the set of reflection coefficients which acts as input into the classifier as $\underline{x} = \{\rho_1, \rho_2, ..., \rho_m\}$. Hence the classifier is expressed as

$$p(\underline{x}) = \sum_{k=1}^{K} P_k p(\underline{x}|k), \text{ where } p(\underline{x}|k) = \sum_{d=1}^{D} w_{d,k} p(\underline{x}|\underline{\mu}_d, \underline{\Sigma}_d). \tag{4}$$

In Equation (4) P_k are K class priors and $p(\underline{x}|k)$ are K class conditional densities. Hence we may express the posterior probabilities for classes by Bayes' theorem to give $P(k|\underline{x}) = P_k p(\underline{x}|k)/p(\underline{x})$. The D component densities $p(\underline{x}|\underline{\mu}_d, \underline{\Sigma}_d)$ are normal densities with mean, $\underline{\mu}_d$, and diagonal covariance matrix, $\underline{\Sigma}_d$. The $w_{d,k}$ denote the D class conditional kernel priors. We use this model for two reasons.

- Generative models give us the possibility to solve missing data problems. In particular we exploit their ability to cope with missing target labels.
- Generative models allow for probabilistic cluster assessment and hence provide deeper insight into the problem structure.

The key problem in training this classifier is to choose an appropriate number of Gaussian components, which is implicit when learning is performed in a Bayesian setting ([Bis95]). We use here a variational Bayesian framework ([JGJS99]). Recently [Att99] applied variational learning to Gaussian mixture models, which are similar to the classifier we use here.

Bayesian inference requires that we specify a likelihood function. Using t_n to denote known class labels and $\underline{\varphi}$ for the model coefficients, we obtain the joint likelihood of all labelled, $\underline{\mathcal{T}}$, and unlabelled data, $\mathcal{X}$.

$$p(\mathcal{T}, \mathcal{X}|\underline{\varphi}) = \prod_{n=1}^{N} P(t_n) \sum_{d=1}^{D} w_{d,t_n} p(\underline{x}_n|\underline{\mu}_d, \underline{\Sigma}_d) \prod_{m=1}^{M} \sum_{k=1}^{K} P(k) \sum_{d=1}^{D} w_{d,k} p(\underline{x}_m|\underline{\mu}_d, \underline{\Sigma}_d) \tag{5}$$

The next step in any Bayesian analysis is to specify priors over all model coefficients. We adopt here a setting that was proposed in [RG97]:

- Each component mean, $\mu_{d,i}$, is given a Gaussian prior: $\mu_{d,i} \sim \mathcal{N}_1(\xi_i, \kappa_{ii}^{-1})$,
- The inverse variances are given a Gamma prior: $\sigma_{d,i}^{-2} \sim \Gamma(\alpha, \beta_i)$,
- The hyper-parameter, β_i, is given a Gamma prior: $\beta_i \sim \Gamma(g, h_i)$,
- The class conditional kernel prior, $\underline{W}_k$, is given a Dirichlet prior: $\underline{W}_k \sim \mathcal{D}(\delta_W, ..., \delta_W)$,
- The class prior, $\underline{P}$, is given a Dirichlet prior: $\underline{P} \sim \mathcal{D}(\delta_P, ..., \delta_P)$,

in which i denotes a particular input dimensions. The hyper-parameters have to be set a-priori. Values for α are between 1 and 2, g is usually between 0.2 and 1 and h_i is typically between $1/R_i^2$ and $10/R_i^2$, with R_i denoting the range of the i-th input. The mean, μ_i, is given a Gaussian prior centred at the range midpoint, ξ_i, with inverse variance $\kappa_{ii} = 1/R_i^2$. Both the prior counts δ_P and δ_W are set to 1 to give the corresponding probabilities the most uninformative proper Dirichlet prior. We note however that our inference results did never depend critically on the setting of the hyper-parameters. In fact we used the same setting successfully for a range of different problems.

Having specified a likelihood and priors, we are ready to derive the variational approximation of the posterior. The key idea is to obtain an approximation of the Bayesian posterior by maximising a lower bound of the logarithmic *model evidence*, $\int_{\underline{\varphi}} \log(p(\underline{\varphi})p(\mathcal{T},\mathcal{X}|\underline{\varphi}))d\underline{\varphi}$. We approximate the posterior by a mean field expansion, $p(\underline{\varphi}|\mathcal{T},\mathcal{X}) = Q(\underline{\varphi}) \approx \prod_{\forall \underline{\varphi}_l} Q(\underline{\varphi}_l)$, and bound the logarithmic evidence by $\mathcal{F}(Q(\underline{\varphi})) = \int_{\underline{\varphi}} \log(\frac{p(\underline{\varphi})p(\mathcal{T},\mathcal{X}|\underline{\varphi})}{Q(\underline{\varphi})})Q(\underline{\varphi})d\underline{\varphi}$. For our classifier, the natural factorisation is $Q(\underline{\varphi}) = \prod_d (Q(\underline{\mu}_d) \times \prod_i Q(\sigma_{d,i})) \times \prod_k Q(W_k) \times Q(P)$. This finally leads to the functional shown in Equation (6), where we use n to denote labelled samples, m to denote unlabelled samples and I as the number of inputs.

$$
\begin{aligned}
\mathcal{F}(Q) = &\int_{\mu,\sigma,P,W,\beta} Q(\mu)Q(\sigma)Q(P)Q(W)Q(\beta) \qquad (6)\\
&\left(\log\left[\frac{P(\mu)P(\sigma)P(P)P(W)P(\beta)}{Q(\mu)Q(\sigma)Q(P)Q(W)Q(\beta)}\right]\right.\\
&+ \sum_n \left(\log(P(t_n)) + \sum_{d_n}\left(Q(d_n)\left[\log(W_{t_n,d_n}) - \frac{I}{2}\log(2\pi)\right.\right.\right.\\
&\left.\left.\left. - 0.5\sum_i \log(\sigma^2_{d_n,i}) - 0.5(\underline{x}_n - \underline{\mu}_{d_n})^T \Sigma^{-1}_{d_n}(\underline{x}_n - \underline{\mu}_{d_n}) - \log(Q(d_n))\right]\right)\right)\\
&+ \sum_m \sum_{k_m} Q(k_m)\left[\log(P_{k_m}) + \sum_{d_m}\left(Q(d_m)\left(\log(W_{k_m,d_m}) - \frac{I}{2}\log(2\pi)\right.\right.\right.\\
&- 0.5\sum_i \log(\sigma^2_{d_m,i}) - 0.5(\underline{x}_m - \underline{\mu}_{d_m})^T \Sigma^{-1}_{d_m}(\underline{x}_m - \underline{\mu}_{d_m})\\
&\left.\left.\left.\left. - \log(Q(d_m))\right)\right) - \log(Q(k_m))\right]\right) d\mu d\sigma d\beta dP dW
\end{aligned}
$$

The mean field assumption (i.e. the independence assumption among the Q-functions) allows the maximisation of Equation (6) to be done separately w.r.t. every Q-function. Maximising w.r.t. one Q-function involves an E-step, where we take the expectation of the functional $\mathcal{F}(Q)$ w.r.t. all other Q-functions, and an M-step which involves minimising a Kullback-Leibler (KL) divergence. After each iteration we evaluate $\mathcal{F}(Q)$ and test for convergence. Usually we observe convergence after at most 100 iterations. Although $\mathcal{F}(Q)$ is only a lower bound of the logarithm of the model evidence, [Syk00a] contains several examples where model selection gave the correct result. Similar iterations are done for predicting probabilities of test samples. The main difference is that the Q-functions over model coefficients are fixed. The maximum of $\mathcal{F}(Q)$ is then found in between two and five iterations.

2.3 Sensor Fusion

Sensor fusion is used to combine predictions of multiple experts and across time to obtain a desired resolution. We use a sliding window and assume that the observations in the window are class conditionally independent. In this case we obtain the resulting probability by a "naïve Bayes" expression, which is identical to one of the techniques advocated in [PHR99].

$$P(t|x_{w1}, x_{w2}, ..., x_{wn}) = \frac{\prod_{wi} P(t|x_{wi})}{\sum_k \prod_{wi} P(k|x_{wi})} \quad (7)$$

The optimal window size depends on the resolution we aim at. In this paper we use 30 seconds, which is equivalent to the R&K window length that was used by the human experts.

3 Experiments

The experiments have been peformed using recordings of 14 healthy subjects otained at at different sleep centres. As already mentioned we use only such labels of R&K stages wake, REM sleep and stage 4 that were unanimously classified by three human scorers, one of them providing a consensus scoring. These labels were further subsampled to get equal class priors[6]. The inputs were obtained by extracting three reflection coefficients from each of the EEG electrodes (Fp1, C3, O1, Fp2, C4 and O2).

Using these data, we trained one classifier for each electrode. Across electrodes we find that the optimal number of Gaussian components is between 10 and 19. The plots shown in Figures 1 and 2 illustrate the information extracted by the proposed approach. The left plots in Figure 1 were obtained from a recording of a young male. The right plots in Figure 1 were obtained from a female aged 65 years. It is evident that the probabilities obtained with our sleep analyser show less aging effects than do the corresponding R&K scorings, where there is no sign that the elderly person reaches stage 4 at all. The plots in figure 2 show examples of bad sleep - mostly because of rather long wake phases during the night. Also for these examples we see that the structure of sleep is preserved in our plots. We would like to emphasise that we can easily distinguish between REM and non-REM, which is known to be difficult from EEG alone.

4 Conclusion

We proposed in this paper an entirely probabilistic approach to sleep analysis. The proposed classifier allows the use of unlabelled data for inference. Model selection is part of the training procedure.

Applying the classifier to some sleep recordings we can certainly assess that the proposed method preserves the structure of sleep that is also found in manually labeled R&K scorings. We see a clear indication of deep sleep also for elderly subjects and with that respect have overcome a known problem of R&K scoring rules.

[6] As inference is based both on labelled and unlabelled data this does *not* imply that the optimal priors are equal.

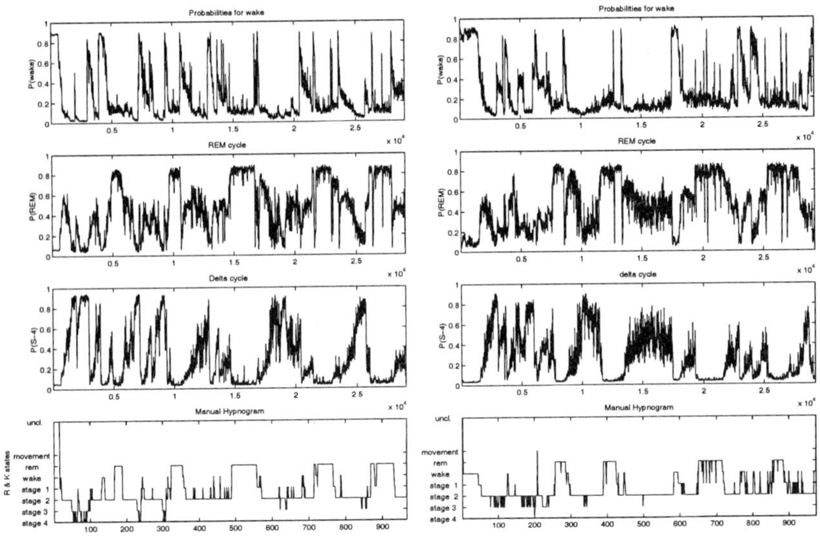

Fig. 1. Probability traces showing good sleep. The probabilities were obtained by combining all EEG channels with a 30 seconds sliding window. An R&K scoring is added to allow for a comparison with the classical scoring technique. Both plots shows a good perfect correspondence between our plots and the R&K labels. The right hand plots were obtained from an elderly subject. They correctly show phases of deep sleep which are missing in the R&K scoring.

Acknowledgements

This work has been done within the project SIESTA, funded by the EC Dg. XII grant BMH4-CT97-2040. Peter Sykacek is currently funded by grant Nr. F46/399 kindly provided by the the BUPA foundation. The authors want to express gratitude to all partners who were involved in this project.

References

[Att99] H. Attias. Inferring parameters and structure of latent variable models by variational Bayes. In *Proc. 15th Conf. on Uncertainty in AI, 1999*, 1999.

[Bis95] C. M. Bishop. *Neural Networks for Pattern Recognition.* Clarendon Press, Oxford, 1995.

[Jef61] H. Jeffreys. *Theory of Probability.* Clarendon Press, Oxford, third edition, 1961.

[JGJS99] M. I. Jordan, Z. Ghahramani, T. S. Jaakkola, and L. K. Saul. An introduction to variational methods for graphical models. In M. I. Jordan, editor, *Learning in Graphical Models.* MIT Press, Cambridge, MA, 1999.

[Lju99] L. Ljung. *System Identification, Theory for the User.* Prentice-Hall, Englewood Cliffs, New Jersey, 1999.

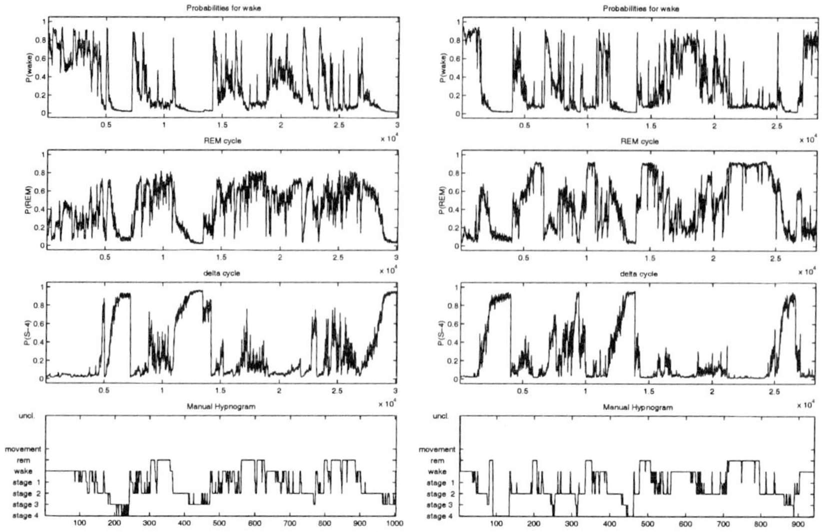

Fig. 2. The traces shown in this figure represent bad sleep. We see that our plots preserve the structure also found in the manual scoring.

[PHR99] W. Penny, D. Husmeier, and S. Roberts. Covariance based weighting for optimal combination of model predictions. In *Proceedings of the 8th Conference on Artificial Neural Networks*, pages 826–831, London, 1999. IEE.

[PRTJ97] J. Pardey, S. J. Roberts, L. Tarassenko, and J.Stradling. A new approach to the analysis of the human sleep/wakefulness continuum. *J. of Sleep. Res.*, 5:201–210, 1997.

[PSKH91] T. Penzel, K. Stephan, S. Kubicki, and W.M. Herrmann. Integrated sleep analysis with emphasis on automati methods. *Epilepsy Research Supplement*, 2:177–204, 1991.

[RG97] S. Richardson and P. J. Green. On Bayesian analysis of mixtures with an unknown number of components. *Journal Royal Stat. Soc. B*, 59:731–792, 1997.

[RK68] A. Rechtschaffen and A. Kales. *A manual of standardized terminology, techniques and scoring system for sleep stages of human subjects.* NIH Publication No. 204, US Government Printing Office, Washington, DC., 1968.

[SRR$^+$99] P. Sykacek, S. J. Roberts, I. Rezek, A. Flexer, and G. Dorffner. Bayesian wrappers versus conventional filters: Feature subset selection in the Siesta project. In *Proceedings of the European Medical & Biomedical Engineerinmg Conference*, pages 1652–1653, 1999.

[Syk00a] P. Sykacek. *Bayesian inference for reliable biomedical signal processing.* PhD thesis, Technical University, Vienna, 2000.

[Syk00b] P. Sykacek. On input selection with reversible jump Markov chain Monte Carlo sampling. In S.A. Solla, T.K. Leen, and K.-R. Müller, editors, *Advances in Neural Information Processing Systems 12*, pages 638–644, Boston, MA, 2000. MIT Press.

Comparison of Wavelet Thresholding Methods for Denoising ECG Signals

Vladimir Cherkassky[1] and Steven Kilts[2]

[1] Department of Electrical and Computer Engineering
University of Minnesota
Minneapolis, MN 55455
cherkass@ece.umn.edu
[2] Medtronic, Inc
Minneapolis, MN 55432
steven.kilts@medtronic.com

Abstract. We present empirical comparisons of several wavelet-denoising methods applied to the problem of removing (denoising) myopotential noise from the observed noisy ECG signal. Namely, we compare the denoising accuracy of several wavelet thresholding methods (VISU, SURE and soft thresholding) and a new thresholding approach based on Vapnik-Chervonenkis (VC) learning theory. Our findings indicate that the VC-based wavelet approach is superior to the standard thresholding methods in that it achieves higher denoising accuracy (in terms of both MSE measure and visual quality) as well as a more robust and compact representation of the denoised signal (i.e., it uses fewer wavelets).

1 Introduction

A good electrocardiogram (ECG) signal is shown in the stylized representation in Figure 1. The ECG is mainly used for cardiac arrhythmia detection, where the P, R, and T waves constitute the most critical signal components representing atrial and ventricular contractions respectively. In practice, wideband myopotentials from pectoral muscle contractions will cause a noisy overlay with this ECG signal where:

$$\textit{Observed signal} = \textit{ECG} + \textit{myopotential} \quad (1)$$

In the above expression (1), the myopotential component of a signal corresponds to additive noise, so obtaining the true ECG signal from noisy observations can be formulated as the problem of signal estimation or signal denoising. Fig.2 shows the ECG recording with myopotential noise. In this example, the noise occurs between sample #8000 and sample #14000. Note that the sampling rate is 1kHz and the total number of samples in the ECG under consideration is 16,384.

G. Dorffner, H. Bischof, and K. Hornik (Eds.): ICANN 2001, LNCS 2130, pp. 625–630, 2001.

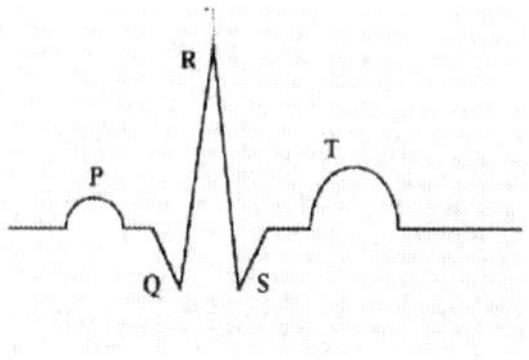

Fig. 1. A Stylized ECG signal

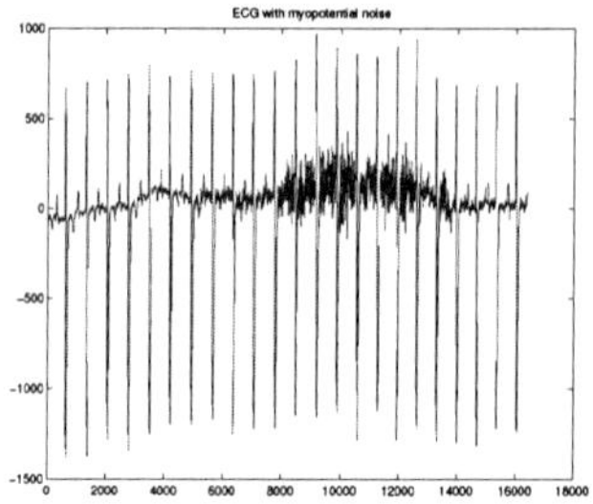

Fig. 2. ECG with myopotential noise

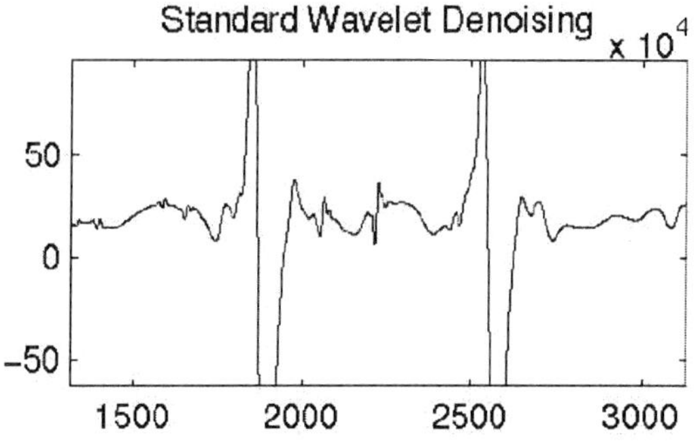

Fig. 3. Denoised ECG signal using Soft Thresholding. DOF = 325. MSE = 1.79e5

This paper investigates applicability of several wavelet-based denoising methods to the problem of myopotential denoising. It can be readily seen from the data in Fig 2 that myopotential denoising of ECG signals is a challenging problem because:

- The useful signal (ECG) itself is non-stationary
- The myopotential noise is applied only to localized sections of a signal

Consequently, we can expect that standard (linear) filtering methods for signal denoising are not appropriate for this application. On the other hand, wavelet methods are more suitable for denoising non-stationary signals.

2 Signal Denoising and Learning from Samples

Under most signal denoising approaches, a given function (signal) with certain smoothness/ frequency characteristics is represented (approximated) as a linear combination of orthogonal basis functions (i.e., Fourier, wavelets etc.). The recently proposed *wavelet thresholding* methods for signal denoising [Donoho and Johnstone, 1994; Donoho et al, 1998; Donoho, 1995] use special basis functions (wavelets) and adaptively selecting 'relevant' wavelet coefficients in a data-dependent fashion. In many cases, such (nonlinear) methods have been shown to outperform traditional linear methods based on projections onto fixed subspaces.

It is well known in statistics that accurate signal estimation requires an optimal trade-off between the model complexity and the accuracy of fitting the training data. Vapnik-Chervonenkis (VC) theory [Vapnik, 1995] has recently emerged as a general theory for function estimation from finite samples. VC-theory provides a new theoretical and conceptual framework for model complexity control. Cherkassky and Shao [1998, 2001] applied the framework of VC-theory to signal denoising and have proposed the following *VC-based signal denoising procedure*:

1. Apply Discrete Wavelet Transform (DWT) to n noisy samples yielding n wavelet coefficients

2. Order the wavelet coefficients according to their magnitude penalized by frequency:

$$\frac{|w_{k1}|}{freq_{k1}} \geq \frac{|w_{k2}|}{freq_{k2}} \geq \ldots \frac{|w_{km}|}{freq_{km}} \geq \ldots \tag{2}$$

The rationale for such ordering is that non-stationary signals contain many high-frequency components. In order to distinguish these 'true' high-frequency terms from the noise, we need to select only the components with sufficiently large coefficient values.

3. Select m most important wavelets from ordering (2) using VC bound on prediction risk

$$R(w) < Remp(w) * r(p, n) \tag{3}$$

Where the penalization factor is computed as Vapnik's measure:

$$r(p,n) = \left(1 - \sqrt{p - p\ln p + \frac{\ln n}{2n}}\right)_+^{-1} \tag{4}$$

and $p = (m+1)/n$. For a given noisy signal, we try exhaustively all m-values and select the optimal number of wavelets m^* providing minimal prediction risk according to bound (3). Note that the empirical risk in (3) corresponds to the MSE fitting error of the denoised signal with m most important coefficients.

4. Generate the denoised signal via inverse DWT from m coefficients selected in step (3).

Notice that many existing wavelet thresholding methods follow the same general procedure as outlined above, except that they use:

- An ordering of coefficients according to their magnitude: $|w_{k1}| \geq |w_{k2}| \geq \ldots |w_{km}| \geq \ldots$
- Other thresholding rules motivated by various statistical modeling assumptions.

3 Empirical Comparisons

This section describes application of several denoising methods to ECG signals.

Data sets used. All of the experiments presented in this paper are done with an ECG sampled at 1 kHz and a resulting digitized signal with 16,384 (2^14) samples. For sections where „clean“ and „noisy“ regions are analyzed, a slice of this larger sample is taken with 4096 (2^12) samples. These sample regions correspond to samples 261 through 4356 (clean region) and 8000 through 12095 (noisy region) respectively.

The experimental procedure was consistent among all learning methods and across all regions. The procedure consisted of applying each method to each (clean or noisy) region, and comparing methods' estimation (denoising) accuracy. In all cases, mean squared error relative to the clean region was used as a measure of denoising accuracy followed by verification via visual inspection.

Wavelet-thresholding methods. Standard wavelet thresholding techniques are implemented in the publicly available software from Stanford University (located at http://playfair.stanford.edu/~wavelab). The wavelet family used throughout the study was the symmlet due to its visual similarity to the ECG and its standard implementation. The methods utilized from this package were SURE, VISU, and Soft Thresholding (with meta-parameters set to default values for each method). The table of MSE values obtained from denoising the noisy section only is shown in Table 1, that also shows the number of wavelet coefficients (degrees of freedom) selected by each method. Comparisons in Table 1 indicate that Soft Thresholding works better than SURE and VISU for our application. Hence, we only show the visual result obtained by soft thresholding in Figure 3. Even though the P, R, and T waves can be identified with soft thresholding in many cases, there are still regions in which waves (P and T waves in particular) are ambiguous to the noise.

Table 1. Performance of standard wavelet thresholding methods for noisy part of ECG

Thresholding method	MSE (x10^5)	Degrees of Freedom
VISU	1.93	4096
SURE	1.95	2072
Soft Threshold	1.79	325

VC-based wavelet denoising. VC-based denoising performs automatic thresholding from an ordering (2) using VC bound (3). Empirical results shown in Fig. 4 indicate that the optimal number of wavelet coefficients chosen by VC method with respect to the noisy region equals DOF = 76. This is much smaller than the DOF=325 obtained by soft thresholding. Also, VC denoising approach results in a much better quality of the denoised signal, based on evaluation performed by medical experts from Medtronic. Since VC-based estimates result in models of much lower complexity (i.e.,

DOF) in comparison with standard wavelet denoising approaches, we conclude that VC denoising is much better in terms of robustness and generalization.

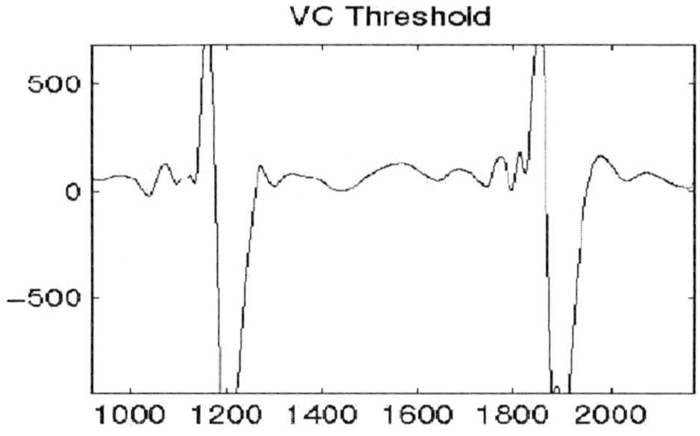

Fig. 3. Denoised ECG signal using VC-based method. DOF = 76. MSE = 1.76e5

4 Summary

Currently, everyone in the industry simply lives with myopotential noise, as there has been no workable method to filter it. We have shown that VC-based denoising produces better results than standard wavelet thresholding methods, in terms of MSE error measure and visual identification of P, R, and T signals. Hence, wavelet denoising, and especially VC-based denoising, appears to have much potential for industrial applications in biomedical engineering.

References

V. Cherkassky and X. Shao (1998), Model selection for wavelet-based signal estimation, in Proc. IJCNN-98, Anchoradge, Alaska, 1998

V. Cherkassky and X. Shao (2001), Signal estimation and denoising using VC-theory, *Neural Networks*, 14 (1), 37-52

Donoho, D.L., M. Vetterli, R.A, DeVore, and I Daubechies (1998), Data Compression and Harmonic Analysis, *IEEE Trans. Information Theory*, 2435-2476

Donoho, D. L. (1995), Denoising by soft thresholding, *IEEE Trans. Info Theory*,41,3, 613-627

Donoho, D. L., and I. M. Johnstone(1994), Ideal denoising in an orthonormal basis chosen from a library of bases, Technical Report 461, Dept. of Statistics, Stanford University

Vapnik , V. (1995), The Nature of Statistical Learning Theory, Springer

Evoked Potential Signal Estimation Using Gaussian Radial Basis Function Network

G. Sita and A.G. Ramakrishnan

Dept. of Electrical Engg., Indian Institute of Science, Bangalore, India
{sita,ramkiag}@ee.iisc.ernet.in

Abstract. Gaussian radial basis function neural networks are used to capture the functional mapping of the evoked potential (EP) signal buried in the additive electroencephalographic noise. The kernel parameters are obtained from the signal edges detected using direct spatial correlation at several adjacent scales of its undecimated non-orthogonal wavelet transform (WT). The segment of the data, where the WT is highly correlated across Scales, is considered a component region and its width is employed as the variance of the Gaussian kernel placed at the center of the region. The weights of the kernels are computed using gradient descent algorithm. Results obtained for both simulated and real brainstem auditory EPs show the superior performance of the technique. Because the technique incorporates signal knowledge into the network design, the number of hidden nodes reduces, and a more accurate estimation of signal components ensues.

1 Introduction

Evoked potentials (EP) are the responses of the brain to specific external sensory stimuli. EP's are used for the diagnosis of a variety of neurological disorders, and also in psychophysics research. The times of occurrence (latencies) and the amplitudes of individual components are the information of interest in EPs. These responses are buried in the ongoing electroencephalogram (EEG) with a signal-to-noise ratio (SNR) less than 0 dB. Hence, an ensemble of hundreds of responses to identical stimuli must be averaged to yield a reliable signal estimate. The focus of EP research has been on improving the ensemble average (EA) of fewer number of EP sweeps resulting in reduction in experimental time and improvement of the SNR of EA.

Most of the earlier research in EP estimation have used linear parametric models, both time [1] and frequency domain [2] ones. A relatively new approach uses neural network filters, which possess built-in nonlinear processing elements. Fung et al [3] used a Gaussian radial basis function network (GRBFN) as a model for the EP signal to account for the nonlinear nature of the signal. They assume that EP responses can be modelled by a finite number of Gaussian RBFs with their centers evenly distributed in time. Their weights are adaptively determined using LMS algorithm.

G. Dorffner, H. Bischof, and K. Hornik (Eds.): ICANN 2001, LNCS 2130, pp. 630–635, 2001.

However, EP's are usually complex, and consist of both low and high frequency components of longer and shorter durations, respectively. Hence, an uniform parameter RBF network (U-RBFN), with its centers uniformly placed over the length of signal and with equal variances, cannot model the signal components efficiently. We propose a method, termed as W-RBFN, where the RBF parameters are deduced from the wavelet transform (WT) of the noisy signal to enable the network to learn the functional form of the underlying EP signal effectively. Once the number of hidden nodes and the kernel parameters are fixed, their weights are computed using gradient descent algorithm.

2 GRBFN for EP Estimation

It is assumed that EP responses can be modelled by a finite number of Gaussian RBFs. The recorded j^{th} evoked response, $\mathbf{X}_j$, may be vectorially denoted as,

$$\mathbf{X}_j = \mathbf{S}_j + \mathbf{V}_j \tag{1}$$

where $\mathbf{S}_j = [s_j(1) \ \cdots \ s_j(M)]$ and $\mathbf{V}_j = [v_j(1) \ \cdots \ v_j(M)]$ denote the signal and noise content in the jth sweep with the time index ranging from 1 to M. The signal and the noise are assumed to be stationary random processes with Gaussian distribution and uncorrelated to each other. We choose *Gaussian function* as the RBF to capture the underlying dynamics of $\mathbf{S}_j$ in (1) resulting in

$$\mathbf{Y}_j = y(\mathbf{X}_j) = \sum_{k=1}^{N} w_k exp\left(-\frac{\| \mathbf{X}_j - C_k \|^2}{{\sigma_k}^2}\right) \tag{2}$$

where C_k and σ_k are the Gaussian kernel parameters and N is the number of hidden nodes. The output of the GRBFN may be written as,

$$\mathbf{Y}_j = \sum_{k=1}^{N} w_k \mathbf{H}_{kj} \tag{3}$$

In the above equation, $\mathbf{H}^T = [\mathbf{H}_1 \ \mathbf{H}_2 \ \cdots \ \mathbf{H}_N]$ is the hidden layer output matrix, and $\mathbf{H}_k = [h_j(1) \ h_j(2) \cdots \ h_j(M)]$ represents the output of the kth hidden node. The weights $\mathbf{W} = [w(1), \cdots, w(N)]$ can be determined using optimization methods such as gradient search algorithm by providing the input $\mathbf{X}_j$ and the corresponding desired output $\mathbf{Y}_j$ for few input-output measurements. A schematic of the GRBF network with N hidden nodes is shown in Fig. 1. The key problem is the placement of the centers C_k and determination of the radial dilation factors (usually called *widths*) to achieve best performance. This problem is often approached by clustering the data points and then using the cluster centers as the RBF centers C_k. In the proposed method, the number of hidden nodes and the Gaussian kernel parameters are obtained from the WT of the noisy signal.

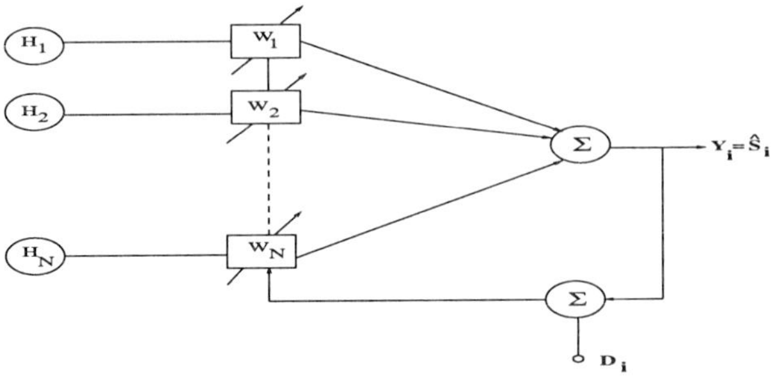

Fig. 1. Schematic of the GRBF network

3 Network Parameters from the Wavelet Transform

The discrete wavelet transform $W(m,n)$ of a 1-D function $f(x)$ is defined as the projection of the function onto the wavelet set $\psi_{m,n}(x)$.

$$W(m,n) = \int_{-\infty}^{+\infty} \psi_{m,n}(x) f(x) dx \tag{4}$$

Since the set of $\psi_{m,n}(x)$ spans the space containing $f(x)$, the reconstruction of $f(x)$ from $W(m,n)$ is possible as,

$$\hat{f}(x) = \sum_m \sum_n \psi'_{m,n}(x) W(m,n) \tag{5}$$

where $\psi'_{m,n}$ is the normalized dual basis of $\psi_{m,n}(x)$. In our case, $\psi' \approx \psi$.

The WT gives a scale-space decomposition of signals into different resolution scales, with m indexing the scale and n indexing position in the original signal space. In wavelet domain, features are well localized in space and sharp transitions in signals are preserved faithfully as the WT modulus maxima, and the evolution of the latter across the scales characterizes the local regulariy of the signal. Hence, we use the correlation between adjacent scales to identify the signal transition regions.

The idea of using the strength of scale space correlation of the subband decompositions of a signal to distinguish significant edges from noise was introduced by Witkin [4]. A similar concept was used by Xu et al. [5] to filter noise from images. Direct spatial correlation $Corr_l(m,n)$ of WT is defined as

$$Corr_l(m,n) = \prod_{i=0}^{l-1} W(m+i,n), \qquad n = 1, \cdots, N \tag{6}$$

where N is the length of the signal and $m = 1, \cdots, M$ is the scale. For a signal, significant features are strongly correlated across scales in the wavelet domain, whereas noise is poorly correlated [4]. We use this fact in detecting the signal related coefficients in each scale m at any position $n = 1, \cdots, N$ for which

$$|Corr_2(m,n)| > |W(m,n)| \tag{7}$$

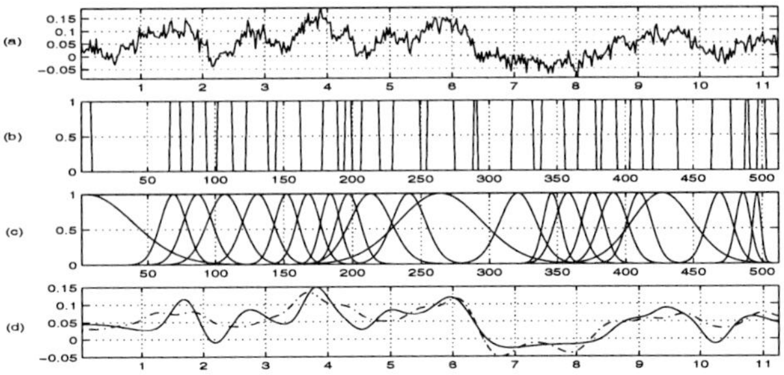

Fig. 2. Signal estimation using GRBF network (a)Simulated BSAEP signal, EA of 100 sweeps. (b) Binary mask vector in wavelet domain (c) Gaussian kernels (d) Estimated signals (x-axis: time in msec. y-axis: amplitude —:WRBFN - - -:URBFN)

A binary mask vector of length N is created, in which only those elements meeting the above condition across all the scales except the finest one, are set to 1. The smallest scale is not considered because noise dominates it in low SNR signals, resulting in too many edges being extracted [7]. The mask indicates the important signal regions and their widths. The technique to detect edges in the wavelet domain is well explained in [5] and [6]. We used the number of edges detected in the scale-space domain as the required number of GRBF kernels; their centers are determined as the mid point of the estimated signal regions. The spacing between adjacent edges determines the variances of the kernels.

The individual sweeps of brainstem auditory evoked potentials (BSAEP) have a low SNR of -10 dB or less. Since the technique is based on edges detected in the wavelet domain from the input signal, we start with the EA of the first 100 sweeps. This improves the SNR to a reasonable degree and facilitates good results. This avoids detection of noise edges, which may dominate at very low SNR. Subsequently, we use the current EA as the desired signal in the GRBF network weight adaptation so that the model parameters converge fast.

4 Results

4.1 Results for Simulated Data

The proposed technique has been tested both on simulated and actual EP data. Results thus obtained are compared with those of the U-RBFN estimator. EEG is simulated as a 4-th order autoregressive (AR) process [3], with a spectral shape comparable to that of true EEG. In order to meet the assumption of uncorrelatedness, each trial of the simulated EEG is obtained from the steady-state response of an independently initialized AR process. The SNR of the simulated sweeps is 0 dB.

Fig. 2 illustrates an instantaneous average of 100 BSAEP sweeps and the signal estimated using the proposed technique. Significant regions, detected in the wavelet domain using the inter-scale correlation, are shown as a binary vector

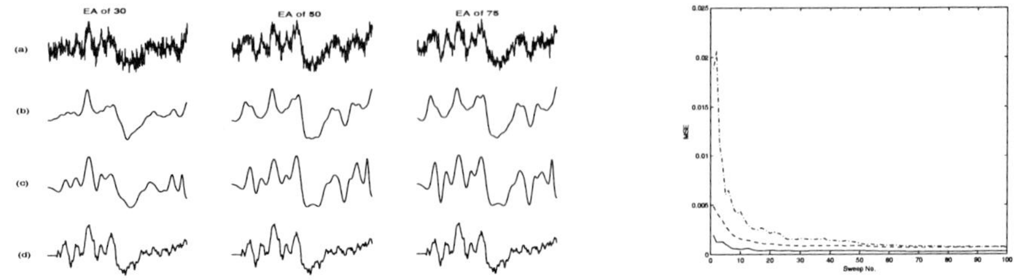

(1) (2)

Fig. 3. (1) BSAEP signals estimated from different instantaneous averages. Row(a) Noisy BSAEP instantaneous EA signals (b) signals estimated using U-RBFN (c) signals estimated using W-RBFN (d)Original signal (2) MSE comparison for simulated data. (-.-.) EA (- -) U-GRBF (—)W-GRBF

Table 1. Latencies of BSAEP components, as detected by various techniques

Component peaks and their latencies in msec	I	II	III	IV	V
True	1.7	2.8	3.9	5.0	5.9
W-RBF	1.65	2.75	3.81	4.98	5.95
U-RBF	1.92	-	3.7	-	6.12

in Fig. 2(b). The number of significant regions detected is 22 and hence those many hidden nodes are used in the GRBF network. The corresponding Gaussian kernels used in time domain are shown in Fig.(c). It can be seen from Fig. 2 that the component latencies are well preserved in the W-RBFN method, whereas the U-RBFN method fails to estimate some of the components. A comparison of the component latencies estimated using both the methods is presented in Table 1, along with the true values. Fig. 3(1) presents both the signals, estimated using WRBFN and URBFN, from different instantaneous averages. The comparison, in terms of MSE, as shown in Fig. 3(2), shows almost equal performance by both methods as it reflects only the mean error over the entire length of the signal. The plots show that GRBF entails a good approximation of the underlying signal from noisy observations, for any length of the ensemble. Further, wavelet preprocessing results in an accurate approximation of the component peaks of the underlying signal with fewer number of hidden nodes.

4.2 Results for Human Data

The proposed technique has been tried on clinically recorded BSAEP's. BSAEP occurs within 12 msec of presentation of auditory clicks of 0.1 ms duration at 60 dB above auditory threshold, at the rate of 20 Hz. The subject keeps his eyes closed throughout the session to minimise ocular artifacts. Post stimulus data, sampled at 40 kHz, is collected in each sweep. Clinically, BSAEP is estimated from more than 1000 sweeps. Fig. 4(a) presents the EA of 100 sweeps of a BSAEP and the signal estimated using our method is shown in Fig. 4(b). It can be seen

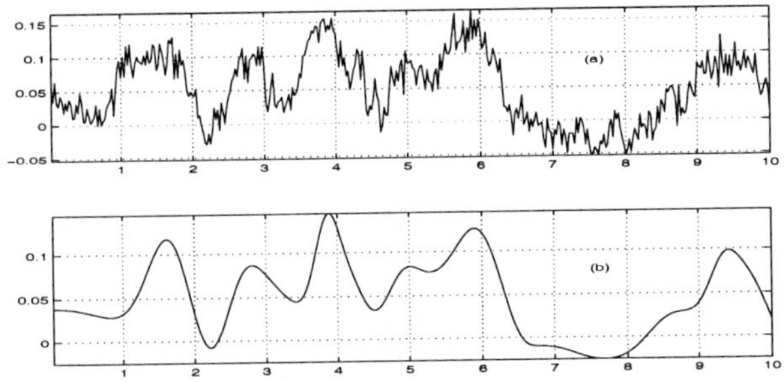

Fig. 4. Results using real BSAEP data (a) EA of 100 sweeps (b) Estimated signal

that the latencies and amplitudes of the components can be easily measured from the estimated signal, which is not possible using the EA shown in Fig. 4(a).

5 Conclusion

The powerful capability of RBF networks is used to model the non-stationary characteristics of EP's. Results show that a network, whose parameters are derived from the signal itself, performs better than a network with fixed parameters. Results obtained from simulated and real EP data demonstrate the estimator's ability to suppress the noise even at low SNR. The advantages of using signal dependent kernel parameters for the GRBFNN are in terms of model accuracy, and fewer nodes in hidden layer, and thus, in the speed of convergence of the network too. The proposed technique entails enhanced signal components.

References

1. K.B. Yu, C.D. McGillem, "Optimum filters for estimating evoked potential waveforms," IEEE Trans. Biomed. Engg., **30**, pp. 730-736, Nov. 1983.
2. J.P.C. De Weerd, and W.L.J. Martens, "Theory and practice of a posteriori Wiener filtering of average evoked potentials," Biol. Cybernet., **30**, pp. 81-94, Jan. 1978.
3. K.S.M. Fung, F.H.Y. Chan, F.K. Lam and P.W.F. Poon, "A tracing evoked potential estimator," Med. Biol. Engg. Comput., **37**, pp. 218-227, Feb. 1999.
4. A. Witkin, "Scale space filtering," Proc. 8th Internat. Joint Conf. on Artificial Intell., Karlsruhe, Germany, pp. 1019-1022, 1983.
5. Y. Xu, J.B. Weaver, D. M. Healey,Jr., and J. Lu, "Wavelet domain filters: A spatially selective noise filtration," IEEE Trans. Img. Proc. **3**, pp. 747-757, 1994.
6. G. Sita and A. G. Ramakrishnan "Wavelet domain nonlinear filtering for evoked potential signal enhancement," Comp. and Biomed. Res. **33**, pp. 431-446, 2000.
7. D. L. Donoho and I. M. Johnstone, "Ideal spatial adaptation via wavelet shrinkage," *Biometrika*, Vol. 81, 1994, pp. 425-455.

'Virtual Keyboard' Controlled by Spontaneous EEG Activity

Bernhard Obermaier[2], Gernot Müller[2], and Gert Pfurtscheller[1]

[1] Institute for Biomedical Engng., Department of Medical Informatics, University of Technology Graz, Inffeldgasse 16a, A-8010 Graz, Austria
[2] Ludwig Boltzmann Institute of Medical Informatics and Neuroinformatics, University of Technology Graz, Inffeldgasse 16a, A-8010 Graz, Austria
obermai@dpmi.tu-graz.ac.at

Abstract. A 'Virtual Keyboard' (VK) is a letter spelling device operated by spontaneous EEG, whereby the EEG is modulated by mental hand and leg motor imagery. We report on a tetraplegic patient initially trained on an EEG-based orthosis control system operating the VK. He achieved an outstanding ability in the use of the VK and can operate it with 0.95 letters per minute.

Keywords: Virtual Keyborad, Brain-Computer Interface, EEG

1 Introduction

Recently the group of Birbaumer presented a spelling device [1] (Thought Translation Device (TTD)) for a patient suffering from late stage Amyotrophic Lateral Sclerosis (ALS) causing a locked-in syndrome (LIS). Patients suffering under LIS have the ability to perceive their environment via optical, acoustic and somatosenoric stimuli but they are not able to have any interaction with it. LIS can also be caused by severe cerebral palsy, head trauma and spinal injuries. Note, however, that the intellectual functions stay unaffected [2] and therefore the need of communication is essential. The TTD is currently used by 3 ALS patients, trained over a period of some months in more than 100 sessions. The patients have to produce slow cortical potential (SCP) shifts either in a positive or a negative direction to operate the TTD. Two of the three patients achieved a classification accuracy of 70-80% allowing them to write with a speed of approximately 1 letter every 2 minutes. The spelling rate σ is defined as the number of correctly selected letters per minute and is therewith 0.5 for these patients.

Even though the TTD returns a certain level of communication abilities to the ALS patients, there is the problem of the low σ caused by the relatively high classification error of these patients. The TTD is based on a binary decision, and therefore a classification error of 30% corresponds to 1 wrong decision out of 3. Wrong decisions have to be corrected and therefore decrease σ.

To increase σ the classification error has to be decreased. The Graz Brain-Computer Interface (Graz-BCI) [3–5] based on the classification of motor imagery demonstrated that after a training period of some months a low classification error can be achieved [6]. This makes the Graz-BCI predestinated as

G. Dorffner, H. Bischof, and K. Hornik (Eds.): ICANN 2001, LNCS 2130, pp. 636–641, 2001.

the base of a spelling device. In this paper we present an alternative spelling device called "Virtual Keyboard" (VK) based on the Graz-BCI [7] and the first outstanding results achieved by a tetraplegic patient using it.

2 Methods

2.1 The Graz-BCI

The Graz-BCI [3–5] was developed from the group of Pfurtscheller at the Department of Medical Informatics at the University of Technology Graz. It is a small and portable system optimized for daily use of patients suffering under any kind of severe motor disabilities. The system gets its control information from the on-line classification of motor-imagery-related EEG patterns. Mainly the motor imagery corresponds to left and right hand movement but the combination of one hand and the feet is also feasible. The classification of two EEG patterns (classes) corresponds to a binary decision, and therefore the control signals are of binary type too.

The Graz-BCI Paradigm. The motor imagery is performed in a fixed repetitive scheme called trial. Each trial starts with the presentation of a fixation cross at the center of the monitor, followed by a short warning tone at second 2. At second 3 an arrow is displayed at the center of the monitor for 1.25 seconds. Depending on the direction of the arrow (left or right) the subject is instructed to imagine a movement of either the left or the right hand. In the period from second 4.25 to second 8.0 a bar (feedback) either extending to the left or right is displayed. During feedback the subject's task is to extend the bar via mental activity to the side that was indicated by the cue stimulus. The trial ends after 8 seconds and a blank screen is shown until the beginning of the next trial. The inter-trial period (ranging from 0.5 up to 2.5 seconds) as well as the sequence of left and right trials is randomized to avoid adaptation.

Electrode Setup and Feature Estimation. The bipolar EEG signals were recorded from 4 Ag/AgCl electrodes placed 2.5 cm anterior and posterior to electrode positions C3 and Cz, respectively. The ground electrode was located on the forehead. The EEG signals were band-pass filtered between 0.5 Hz and 30 Hz and sampled at 128 Hz. The used feature set comprises AAR parameters of 6th order estimated for every EEG channel at every sample time instance using the Recursive Least Square algorithm [8].

Classification of the EEG Patterns. The classification of the EEG patterns is done using the linear discriminant (LD) [9, 10]. The LD projects the data from the v dimensional feature space into a one dimensional feature space. The feature vector $\mathbf{y}$ is projected into a value d by

$$d = \mathbf{w}^T \mathbf{y} + w_0 \tag{1}$$

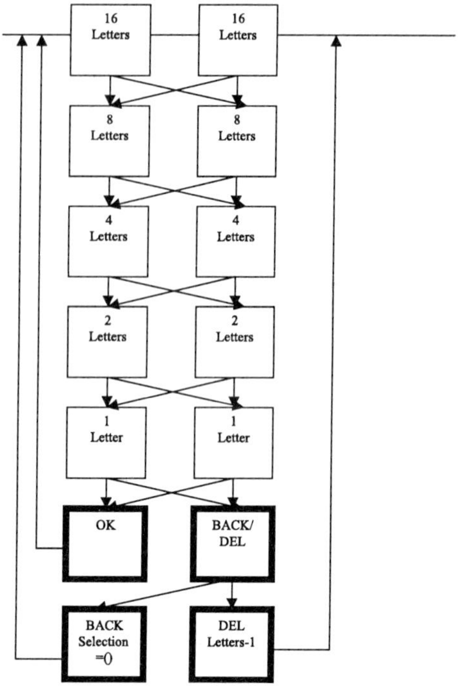

Fig. 1. VK realized by dichotomous search procedure followed by two levels for confirmation and correction.

where $\mathbf{w}^T$ is a transposed vector consisting of adjustable weights and w_0 is called bias or threshold. Classification of an unknown feature vector is simply a calculation of the inner product and one addition. This calculation is again performed sample by sample and is the base for a continuous feedback during on-line sessions.

2.2 Virtual Keyboard (VK)

The Virtual Keyboard is an application of the Graz-BCI. The spelling consists of the selection of a letter using successive steps of isolation. Starting with the complete set of letters, successive subsets are selected via mental activity until the selection of the desired letter.

Structure of the VK. The structure of the VK (see Fig. 1) contains 5 consecutive levels of selection and two further levels of confirmation (OK) and correction "BACK/DEL". In a dichotomous structure with 5 levels 32 letters can be selected. So the alphabet was extended by 6 special characters. The correction has to be performed in a further level due to two different actions needed: Firstly "BACK" allowing a cancellation of the actual selection with a subsequently return to the first selection level. Secondly "DEL" which deletes the last written letter, the selection done during the five preceding steps is not considered. If the user makes a wrong decision during the selection it is mandatory to reach the correction level. Further intermediate correction possibilities are not considered in this presented VK structure.

The structure of a trial during spelling was identical to the one described in Section 2.1 except that no arrow indicates the motor imagery to be performed. The patient himself has to decide which motor imagery has to be performed to select the subset containing the desired letter. At the end of the trial the selection region is updated according to the classification result. The break between two trials was omitted in order to increase the number of trials resulting in a maximum σ of 1.25.

Screen content of the VK. As a final step the Graz-BCI and the structure of the VK was combined. The screen content of the Graz-BCI was extended by 2 areas.

1. Text region: The text region at the top of the screen displays the text already written.
2. Selection region: The middle region of the screen displays the contents of the subsets in a selection level or the commands of the function level.
3. Feedback region: The bottom region displays a feedback to help the user.

2.3 The Patient

The tetraplegic patient T.S. is a 25 year old man, paralyzed due to an accident in the year 1998. The lesion was at C4/C5 and the patient is now just able to move his left arm solely with the biceps. It is of importance that, the VK was driven solely based on the motor imagery of the totally paralyzed limbs (right hand vs. both feet).

3 Results

Two different types of results are presented, firstly the propagation of the classification accuracy in Graz-BCI sessions for patient T.S.(see Fig. 2). Patient T.S. performed these sessions (a session comprises 160 trials of motor imagery) during another study concerning the development of a hand orthosis [6] for him. T.S. changed the strategy of motor imagery two times during training. The error rates are calculated based on a 10 times 10 cross-validation method. In session 62 he could achieve zero percent classification error. The error rates of the succeeding sessions stabilized close to zero percent.

The VK was introduced to the patient after session 62 and he wrote the following text taken from Goethes Faust: Der Tragödie erster Teil:Nacht (HAB-NUN-ACH! JURISTEREI-UND-MEDIZIN-PHILOSOPHIE-UND-LEIDER-AUCH-THEOLOGIE) in his first 4 consecutive VK sessions (performed on two different days). T.S. could achieve an average spelling rate of 0.95 letters per

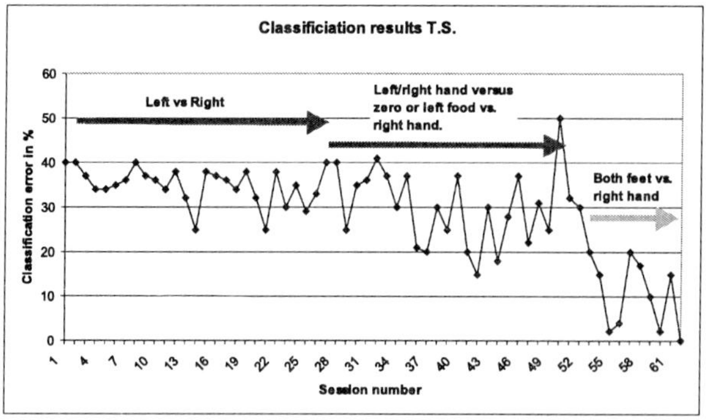

Fig. 2. Propagation of the classification error of 62 performed Graz-BCI sessions of patient T.S. He altered his strategy of motor imagery 2 times. In Session 62 he achieved 0% classification error.

minute. The BCI was trained once prior the VK sessions and was not changed during the presented VK study. The training itself comprised 160 trials (80 feet/80 right hand) of motor imagery.

4 Discussion

The propagation of the classification error (see Fig. 2), show a decreasing tendency, until 0% classification error was achieved in session 62. It is interesting to note that T.S. changed the types of motor imagery two times (see Fig. 2) to determine which strategy results in the most discriminateable EEG patterns. The fluctuations are mainly due to differences in the daily constitution. Differences due to bad electrode setup can be excluded, because they were done with high accuracy. It can be seen that T.S. was not able to control the Graz-BCI in the beginning, could achieve 0% classification error within 62 sessions. This is one of the major advantages of the Graz-BCI compared to the one based on SCP achieving a classification error of 30-20%. Based on the low classification error a σ of 0.95 could be achieved by T.S. in his first VK sessions. This is almost doubling the σ of the TTD.

Furthermore it has to be emphasized, that the additional task of selecting a letter, does not confuse the patient. T.S. explained this fact by the circumstance, that the break between two trials was sufficiently long to search the desired letter and then to concentrate to perform the corresponding motor imagery. As a second reason the feedback style is kept the same and therefore the patient is familiar to it.

Finally we want to point out that the VK has a potential for further increasing the number of letters spelled per minute. This could be either done by decreasing the break between two imaginations, or by the shortening of the motor imagery phase. Expanding the number of classes to more than 2 seems feasible based on a classification of the EEG using Hidden Markov Models [11]. It could also be considered that the letters have a probability of occurrence (e.g. the letter "e" has the highest probability of occurrence in the German language) and a consideration of these probabilities will also improve the spelling rate. These aspects will have a great impact on the spelling rate of a future version of the VK. To conclude we can state that the T.S. is the fastest person to our knowledge solely writing letters by changing his spontaneous EEG activity.

Acknowledgements

The project was partly supported by the Federal Ministry of Transport, Innovation and Technology. We like to thank Dr. Guger for technical and Dr. Neuper for psychological support.

References

1. N. Birbaumer, N. Ghanayim, T. Hinterberger, I. Iversen, B. Kotchoubey, A. Kübler, J. Perelmouter, E. Taub, and H. Flo, "A spelling device for the paralysed," *Nature*, vol. 398, pp. 297–298, 2000.
2. C.M.C. Allen, "Conscioius but paralized:releasing the locked-in," *The Lancet*, vol. 342, pp. 130–131, 1993.
3. G. Pfurtscheller, C. Neuper, C. Guger, W. Harkam, H. Ramoser, A. Schlögl, B. Obermaier, and M. Pregenzer, "Current trends in graz brain-computer-interface (BCI) research," *IEEE Transactions On Rehabilitation Engineering*, vol. 8, no. 2, pp. 216–219, 2000.
4. G. Pfurtscheller, D. Flotzinger, M. Pregenzer, J.R. Wolpaw, and D.J. McFarland, "EEG-based brain computer interface (BCI)," *Medical Progress trough Technology*, vol. 21, pp. 111–121, 1996.
5. C. Guger, A. Schlögl, Ch. Neuper, D. Walterspacher, T. Strein, and G. Pfurtscheller, "Rapid prototyping of an EEG-based brain-computer interface (BCI)," *IEEE Transactions On Rehabilitation Engineering*, 2000 in press.
6. G. Pfurtscheller, C. Guger, G. Mueller, G. Krausz, and C. Neuper, "Brain oscillations control hand orthosis in a teraplegic," *Neurosience Letters*, vol. 292, pp. 211–214, 2000.
7. B. Obermaier, *Design and Implementation of an EEG based Virtual Keyboard using Hidden Markov Models*, Ph.D. thesis, TU-Graz, 2001.
8. S. Haykin, *Adaptive Filter Theory*, Prentice Hall, 1996.
9. C. M. Bishop, *Neural Networks for Pattern Recognition*, Clarendon Press, 1995.
10. K. Fukunaga, *Introduction to statistical pattern recognition*, Academic Press, 1990.
11. B. Obermaier, C. Neuper, and G. Pfurtscheller, "Information transfer rate in a 5 classes brain-computer interface," *submitted to IEEE Transactions On Rehabilitation Engineering*, 2000.

Clustering of EEG-Segments Using Hierarchical Agglomerative Methods and Self-Organizing Maps

David Sommer, Martin Golz

University of Applied Sciences, Department of Computer Science, Postfach 182
D-98574 Schmalkalden, Germany
{sommer, golz}@informatik.fh-schmalkalden.de

Abstract. EEG segments recorded during microsleep events were transformed to the frequency domain and were subsequently clustered without the common summation of power densities in spectral bands. Any knowledge about the number of clusters didn't exist. The hierarchical agglomerative clustering procedures were terminated with several standard measures of intracluster and intercluster variances. The results were inconsistent. The winner histogram of Self-organizing maps showed also no evidence. The analysis of the U-matrix together with the watershed transform, a method from image processing, resulted in separable clusters. As in many other procedures the number of clusters was determined with one threshold parameter. The proposed method is working fully automatically.

1 Introduction

Eleven subjects (3 females, 8 males), aged 19 to 36 years, participated in an overnight driving simulator study. Their task was intentionally monotonous, simply to avoid major lane deviations. Electrooculogram (EOG, oblique) and electroencephalogram (EEG) were recorded in two unipolar and two bipolar recordings.

Dangerous short attention-loss phases during driving are characterized by a transition from struggling to remain awake to an involuntary short sleep episode. These phases, sometimes called microsleeps [1], are associated in many cases with an increased activity of slow eye movements (SEM) [2, 3]. The EEG during SEM, as in the whole transition phase from drowsiness to sleep, was found to have much more complex and variable patterns than the wakeful EEG [2]. Our investigations were focused on clustering the spectral features of the EEG during the SEM, to show if there are differences in EEG indicative of different functional states of the brain.

2 Clustering with agglomerative hierarchical methods

The input vectors for the following analysis consisted of 47 spectral components (2 to 25 Hz; 0.5 Hz steps). A principal component analysis (PCA) was routinely computed,

G. Dorffner, H. Bischof, and K. Hornik (Eds.): ICANN 2001, LNCS 2130, pp. 642–649, 2001.

but there was no reason to assume input vectors in a linear subspace, because the last ten principal components had a residual variance of approximately 8%. A Scree test and the Kaiser criteria for the covariance matrix (number of eigenvalues greater than one) indicated 13 principal components, but they explained only 30% of the total variance.

Therefore, all 47 variables were included in a cluster analysis, performed with the SAS package [4]. Five different agglomerative procedures were applied (Tab. 1) and were terminated with six different measures. The measures, except the elbow measure, are based on estimates of the intracluster and intercluster variances and are all implemented in SAS. In [5] 24 different measures were applied to data sets with known numbers of clusters. The measure with the best reliability was 'Pseudo-F'.

Measure	Method				
	Single Linkage	Complete Linkage	Average Linkage	Centroid	Ward
Elbow	3 (7, 11)	3 (9)	12 (6, 9)	6 (5)	4 (7, 9)
R^2	---	8, 4	(10)	7	3 (4)
R^2_{semi}	9, 11 (6)	8, 4	---	7	3, 4
Pseudo F	---	8, 4	4 (10)	7	(3)
Pseudo t^2	11 (9, 6, 3)	9, 5	8, 10	(11, 9, 5)	(10, 8)
RMSSTD	---	8	---	---	7, 12

Measure	Method				
	Single Linkage	Complete Linkage	Average Linkage	Centroid	Ward
Elbow	3	5, 9	4, 8	3, 6	6 (3)
R^2	(10)	7	6, 9, 7	7	(6)
R^2_{semi}	(5, 7, 10)	7	9, 6, 7	7	6
Pseudo F	5 (7, 10)	7	9, 11	7	(6)
Pseudo t^2	12 (3, 5)	9 (7)	7, 11	7 (5)	12, 4 (7)
RMSSTD	---	12, 8	(9)	---	10, 3, 6

Table 1. Estimated numbers of clusters obtained from five different agglomerative hierarchical methods with six different measures. Left: for the standardized data set. Right: for the non-standardized data set.

The estimated numbers of clusters were inconsistent (Tab. 1) depending on the method used on the termination measure and on the data set. We used two data sets, one without and one with standardization of the data, recommended by several authors [6, 7]. A number of clusters greater than 12 was excluded, because it is difficult for an EEG-expert to interpret such a large number of clusters.

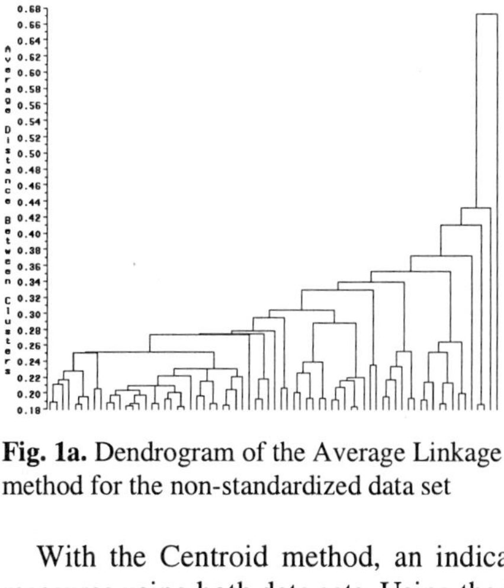

Fig. 1a. Dendrogram of the Average Linkage method for the non-standardized data set

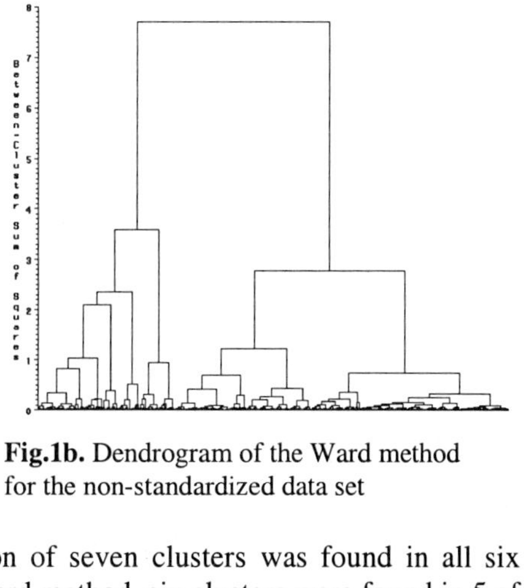

Fig.1b. Dendrogram of the Ward method for the non-standardized data set

With the Centroid method, an indication of seven clusters was found in all six measures using both data sets. Using the Ward method, six clusters were found in 5 of the 6 measures in the non-standardized data set only. In the standardized data set, three clusters were found in only three of the measures.

The dendrograms obtained for three linkage methods showed a lot of very small clusters (Fig. 1a). Dendrograms without such miniclusters were computed for Centroid and Ward methods (Fig. 1b).

A factor analysis using Varimax rotation prior to clustering led to a reduction to 13 variables. But all subsequently applied agglomerative methods achieved higher inconsistencies in the number of clusters compared to the analysis of original input vectors.

The simple model underlying all measures for terminating the agglomerative process is disadvantageous. Specifically, the linkage methods are not well characterized with variance measures.

3 Clustering with SOM

The self-organizing feature map (SOM) [8] was applied to the non-standardized data set for clustering. Using the principle of competitive learning, the weight vectors can be adapted to the probability density function of the input vectors [9]. The similarity between the input vector $\mathbf{x}$, and the weight vector $\mathbf{w}$, was calculated by the Euclidian distance. During training an arbitrary weight vector $\mathbf{w}_j$ was updated at iteration index t by:

$$\Delta \mathbf{w}_j(t) = \eta(t)\, h_{cj}(t)\, [\mathbf{x}(t) - \mathbf{w}_j(t)] \qquad (1)$$

Where $\eta(t)$ is a learning rate factor decreasing during training, and $h_{cj}(t)$ is a neighborhood function between $\mathbf{w}_c$, the weight vector winning the competition, and the weight vector $\mathbf{w}_j$. $h_{cj}(t)$ is also decreasing during training. The neighborhood relationships are defined by a topological structure and are fixed during training. We used a two-dimensional tetragonal relationship. In the final phase of training, the fine-adjustment [9], the neighborhood radius is very small, leading to updates of the winning weight vectors $\mathbf{w}_c$ and of their nearest neighbors. For a one-dimensional topological structure it can be shown [9] that the training rule Eq. (1) leads to an approximation of a monotonous function of the probability density function of the input vectors. The two-dimensional topology results in a compromise between a density approximation and a minimal mean squared error of vector quantization [10].

In the case of existing compact regions of input vectors and of existing density centers, as for Gaussian mixtures, the evaluation of the relative winner frequency of the neurons leads to a visualization of clusters. Fig. 2a shows such a gray-level-coded winner histogram. Five areas with increased winner frequency are evident. The Gaussian mixture data were generated by fixing five cluster centers and five covariance matrices, and adding normal distributed noise in a 47-dimensional space, as in our experimental data set. The size of the generated data set and of the experimental data set was equal, and their estimated total covariance matrices were approximately equal. The black colored units in Fig. 2a are dead neurons, which make it easy to separate clusters. Fig. 2b shows the relative winner frequency for the experimental data set. A separation of regions with increased winner frequency is not possible.

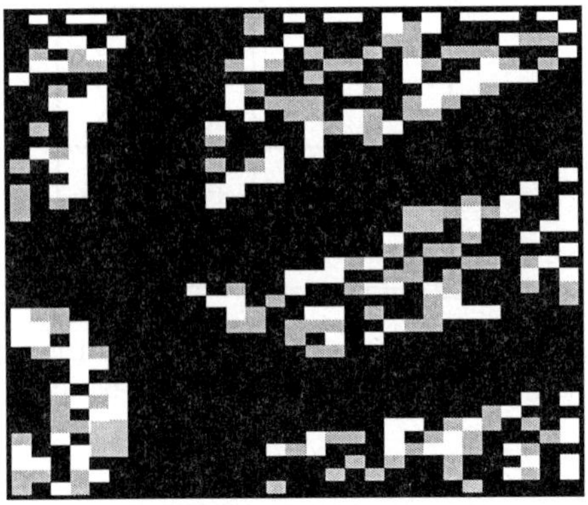

Fig. 2a. Relative winner frequency for a SOM with 30x40 neurons for Gaussian mixture data

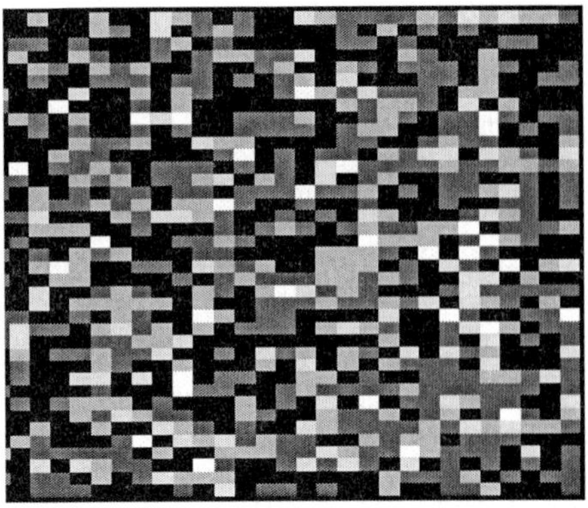

Fig. 2b. Relative winner frequency for a SOM with 30x40 neurons for SEM-EEG data

For a topology-preserving mapping neighbored, weight vectors in input space are also neighbored in output space (the two-dimensional map) [11]. The mapping from input to output space is quasi-continuous, but must not be continuous in the reversed direction. Therefore, two topologically neighbored weight vectors must not represent one cluster. If their distance is small, then they probably represent one cluster, otherwise they probably represent different clusters. The visualization of the distances between neighbored weight vectors was introduced as the unified distance matrix (U-matrix) [12]. In the following, only two-dimensional tetragonal topologies are considered. For every weight vector $\mathbf{w}_{x,y}$, where x and y are the topological indices, the Euclidian distances dx and dy between two neighbors and the distance dxy to the next but one neighbor is calculated:

$$dx(x,y) = \left\| \mathbf{w}_{x,y} - \mathbf{w}_{x+1,y} \right\|$$

$$dy(x,y) = \left\| \mathbf{w}_{x,y} - \mathbf{w}_{x,y+1} \right\| \tag{2}$$

$$dxy(x,y) = \frac{1}{2}\left(\frac{\left\| \mathbf{w}_{x,y} - \mathbf{w}_{x+1,y+1} \right\|}{\sqrt{2}} + \frac{\left\| \mathbf{w}_{x,y+1} - \mathbf{w}_{x+1,y} \right\|}{\sqrt{2}} \right)$$

The distance du was calculated as the mean over eight surrounding distances. With the four distances for each neuron dx, dy, dxy and du, the U-matrix is well defined and has the size $(2n_X - 1) \times (2n_Y - 1)$ (Fig. 3).

In Fig. 4 the U-matrix elements were mapped on a gray scale. Light-gray levels indicate low values, and dark-gray levels indicate high values.

Visual scoring of five clusters in the U-matrix of Gaussian mixture data (Fig. 4a) is evident. As expected, the cluster regions on the map are regions of small distances between the weight vectors, which are separated by small regions of large distances.

The U-matrix of the SEM-EEG data (Fig. 4b) has much more complexity and it is difficult to define borders.

$$\mathbf{U} = \begin{bmatrix} du(1,1) & dx(1,1) & du(2,1) & \cdots & du(n_x,1) \\ dy(1,1) & dxy(1,1) & dy(2,1) & \cdots & dy(n_x,1) \\ du(1,2) & dx(1,2) & du(2,2) & \cdots & du(n_x,2) \\ dy(1,2) & dxy(1,2) & dy(2,2) & \cdots & dy(n_x,2) \\ \vdots & \vdots & \vdots & \ddots & \vdots \\ du(1,n_y) & dx(1,n_y) & du(2,n_y) & \cdots & du(n_x,n_y) \end{bmatrix}$$

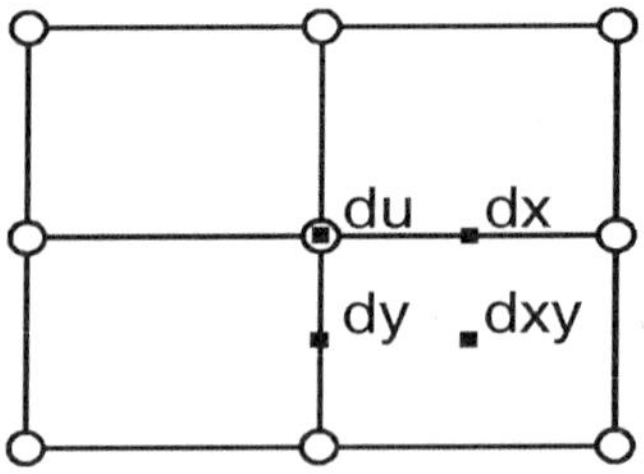

Fig. 3. Definition of the U-matrix and localization on the tetragonal topological structure, shown for the neuron in the center only. Circles: positions of neurons; black squares: positions of U-matrix elements

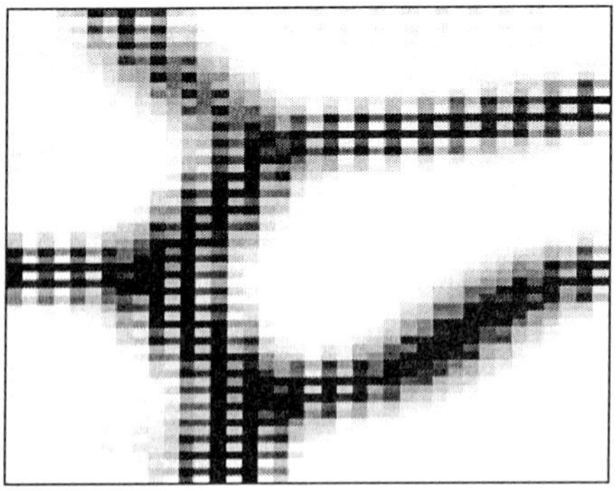

Fig. 4a. U-matrix for SOM from Fig. 2a

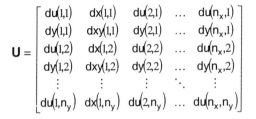

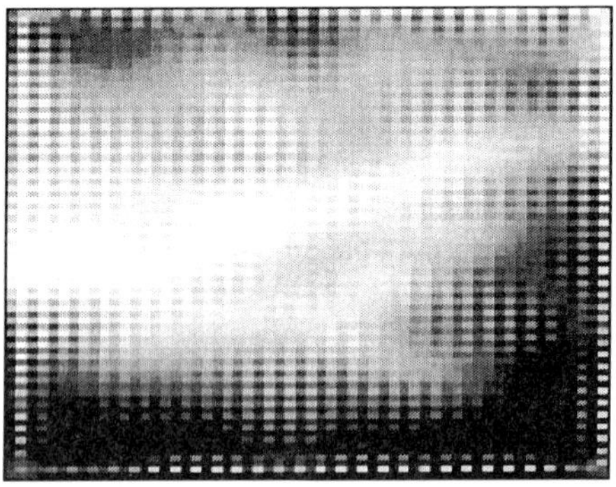

Fig. 4b. U-matrix for SOM from Fig. 2b

Costa et al. [13] propose an automatic segmentation of the U-matrix using the watershed algorithm of gray scale image processing [14]. Regarding high values as mountains and low values as valleys, the algorithm can be illustrated by flooding the valleys with water; watersheds will be built up where the water converges (Fig. 5a). This algorithm leads to closed borders. All weight vectors in one segmented region represent one cluster, and the fusion of their Voronoi sets leads to all items of a cluster.

The results of segmentation are dependent on the size of the SOM. With a relatively large number of weight vectors, many clusters are obtained. Smoothing the gray level function with a two-dimensional filter reduces the risk of over-segmentation; here a 3x3-gaussian filter was applied. The size of the SOM was considerably restricted when the topographic product [15] was taken into account. The topographic product is a measure for the correspondence of the input and output space dimensions. We obtained an optimal value for a SOM size of 4 x 6, but in the case of such small maps the segmentation of the U-matrix failed. Therefore, we extended the size but retained the ratio of approximately 70%. On the other hand, it was shown, that the topographic product was a good measure for approximately linear manifolds only [11].

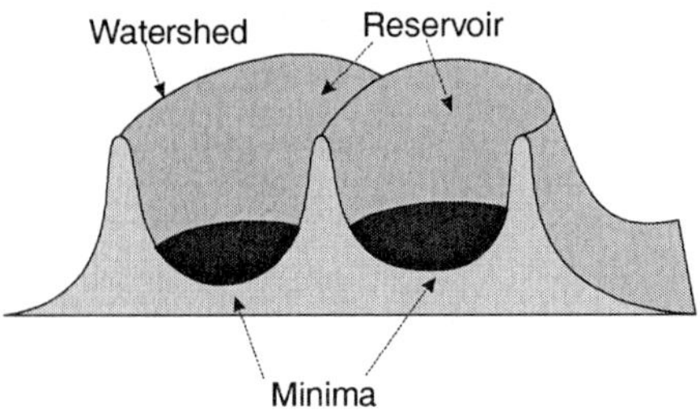

Fig. 5a. Cross section of gray scale mountains

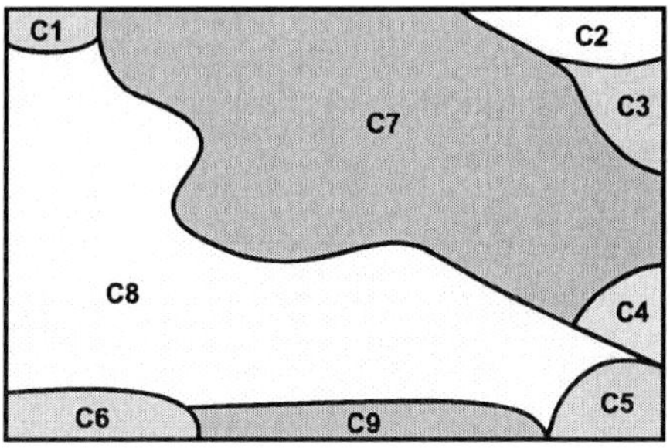

Fig. 5n. U-matrix from Fig. 4b after watershed transformation

In several regions of the U-matrix (Fig. 5a, 5b) black-and-white textures are observable. They describe relatively large differences between dx and dy, and are connected with local stretchings of the SOM along one topological coordinate. If, on the other hand, the dx-elements of the U-matrix, for example, are visualized only, some segment borders disappear. Therefore we restricted our investigations to the function du(x,y).

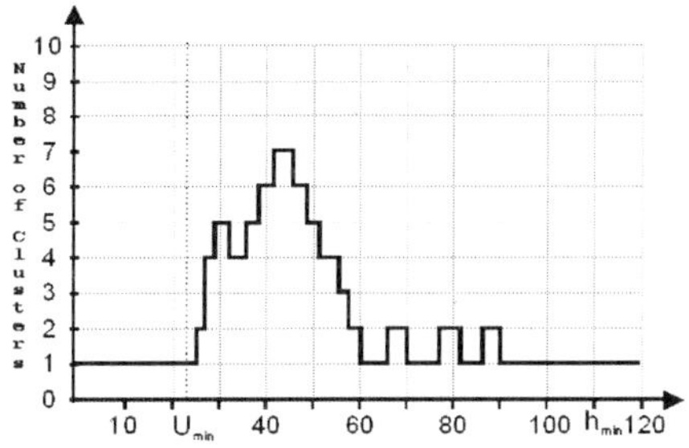

Fig. 6a. Number of clusters vs. h_{min} for the U-matrix of Fig. 5b with generation of new regions [13].

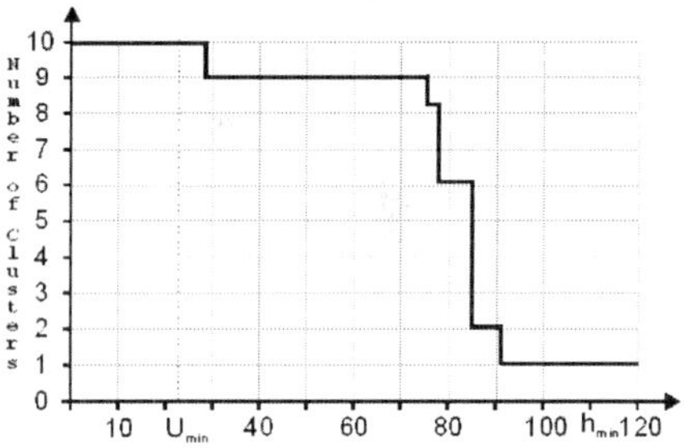

Fig. 6b. The same as in Fig. 6a but without generation of new regions

Flooding the reservoirs has to be started at an initial level h_{min} [14]. It was not straightforward to find suitable values for h_{min}. Starting with $h_{min} = U_{min} = \min(du(x,y))$, the global minimum, as well as with to high values for h_{min} led to one cluster only. With a given value h_{min} the number of clusters is fixed and is preserved during flooding.

The number of clusters versus h_{min} showed no plateau (Fig. 6a), however, distinctive plateaus are required to get a reliable number of clusters [13]. Therefore, we propose a modification allowing the generation of new minima regions during flooding. All local minima regions of the function du(x,y) with values greater than h_{min} are considered as autonomous clusters. To avoid over-segmentation only significantly extended regions were taken into account. Such regions could correspond to clusters with lower probability densities.

Under this modification an extended plateau was calculated at 9 clusters (Fig. 6b). The assumption of 10 clusters is misleading because h_{min} can not be below the global minimum U_{min}.

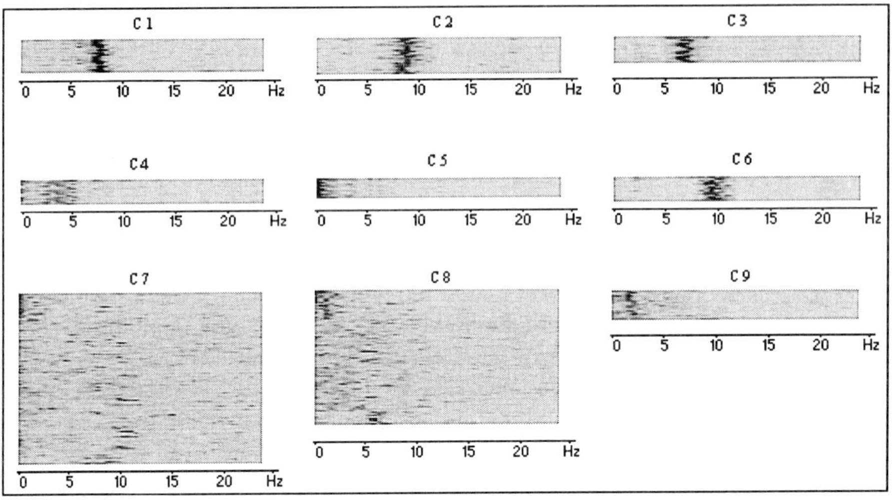

Fig. 7. All 1652 input vectors grouped in 9 different clusters; horizontal: frequency; vertical: item; gray scale: spectral power density.

4 Results

The 9-cluster solution of the described automatic clustering and segmentation procedure is shown in Fig. 5b. The segments contain different numbers of weight vectors. The fusion of their Voronoi sets, mentioned above, leads to the clusters (Fig. 7). Cluster 1 and 2 contain input vectors with large magnitudes in the alpha1 band (7.5-10.5 Hz), cluster 6 in the alpha2 band (10.5-12.5 Hz), cluster 3 in the theta band (3.5-7.5 Hz) and cluster 9 in the delta band (1.0-3.5 Hz). In contrast to the usual summation in frequency bands, greater detail can be seen. The input vectors of cluster 1 and 2, for example, are large in the same spectral band, but they differ in the magnitude range and differ in other spectral bands.

From Fig. 7 one may get the visual impression of homogeneous clusters, with the exception of cluster 7 and 8. The described method is working fully automatically and has a low number of free parameters. Beside the common parameters for the SOM, there is only one parameter h_{min} for segmentation. Clustering is possible with a high reproducibility. Many reruns with 80% of all input vectors, arbitrary selected, lead to always nine clusters. With the proposed segmentation method, a visualization of the SOM is not necessary and the limitation to two-dimensional maps can be left behind. It is possible to extend the whole clustering method to higher dimensional topologies.

References

[1] Thorpy, MJ; Yager, J; The Encyclopedia of Sleep and Sleep Disorders; NewYork: Facts on File, 1991.

[2] Santamaria, J; Chiappa, KH; The EEG of Drowsiness in Normal Adults; J Clin Neurol; 4 (4), 1987, 327-382.

[3] Liberson, WT; Liberson, CT; EEG recordings, reaction times, eye movements, respiration and mental content during drowsiness; Proc Soc Biol Psychiat, 19, 1966, 295-302.

[4] SAS Institute Inc., SAS/STAT User's Guide, Version 6, Fourth Edition, Volume 1, Cary, NC: SAS Institute Inc., 1989, 943 pp.

[5] Milligan, GW; Cooper, MC; An examination of procedures for determining the number of clusters in a data set; Psychometrika, 50 (2), 1985, 159-179.

[6] Deichsel, G; Trampisch, H; Clusteranalyse und Diskriminanzanalyse; Gustav Fisher Verlag, Stuttgart, 1985.

[7] Backhaus, K; Erichson, B; Plinke, W; Weiber, R; Multivariate Analyseverfahren; (6. Aufl.), Berlin, Heidelberg, New York.Springer, 1996.

[8] Kohonen, T; Self-organized formation of topologically correct feature maps; Biol Cybern, 43, 1982, 59-69.

[9] Kohonen, T; Self-Organizing Maps; 3rd edition, Springer, Berlin, 2000.

[10] Fritzke, B; Wachsende Zellstrukturen – ein selbstorganisierendes neuronales Netzwerkmodell; PhD thesis; University of Erlangen, 1992 (in german).

[11] Villmann, T; Topologieerhaltung in selbstorganisierenden neuronalen Merkmalskarten; PhD thesis, University of Leipzig, 1996 (in german).

[12] Ultsch, A; Siemon, HP; Exploratory Data Analysis: Using Kohonen Networks on Transputers; Univ. of Dortmund, Technical Report 329, Dortmund, Dec. 1989.

[13] Costa, JAF; Netto, MLA; Estimating the Number of Clusters in Multivariate Data by Self-Organizing Maps; Int. J Neural Systems, 9 (3), 1999, 195-202.

[14] Vincent, L; Soille, P; Watersheds in Digital Spaces: An Efficient Algorithm Based on Immersion Simulation; IEEE Transaction on Pattern Analysis and Machine Intelligence, 1991.

[15] Bauer, HU; Pawelzik, KR; Quantifying the neighborhood preservation of Self-Organizing Feature Maps. IEEE Trans. Neural Networks, 3 (4), 1992, 570-579.

Nonlinear Signal Processing for Noise Reduction of Unaveraged Single Channel MEG Data

Wei Lee Woon and David Lowe

Neural Computing Research Group,
Aston University,
Aston Triangle,
Birmingham B4 7ET,
United Kingdom
{woonw,d.lowe}@aston.ac.uk

Abstract. A method for isolating and removing noise sources in single channel, unaveraged MEG recordings is presented. By combining tools from the fields of information theory and dynamical systems, we seek to decompose MEG time series into their underlying generating components. A simple technique is introduced for extracting components of interest and results from the application of these methods are presented. Traditional low pass filtering is also performed as a comparison to demonstrate the advantage of our methodology.

1 Introduction

In this paper we propose a new approach to MEG signal processing. By focussing on the underlying brain processes that generate the observed signals, rather than these signals themselves, we show that it is possible to perform useful analysis using just single channel, unaveraged MEG data.

The essence of our approach is that the complexity in the MEG recordings is the result of the interaction between a relatively small number of underlying nonlinear generators, rather than the combination of a large number of linear sources.

The starting point is the assumption that the brain can be modelled as a nonlinear dynamical system. In accordance with our main assumption, we postulate the existence of a relatively small number of underlying nonlinear source generators, $\mathbf{s}_t$, where each member of the vector $\mathbf{s}$ corresponds to some independent cortical-process. Further, the evolution of $\mathbf{s}$ can be described in discretised time by the equation:

$$\mathbf{s}_{t+1} = f(\mathbf{s}_t) + \eta_t \tag{1}$$

f is the nonlinear transition function and η is the time dependent dynamical noise term. We envisage the existence of an unobservable phase space within which these dynamics are embedded. Generally, the dynamics of such systems are unlikely to fill the entire phase space but rather exist on a *manifold* within the phase space, as depicted in figure 1.

G. Dorffner, H. Bischof, and K. Hornik (Eds.): ICANN 2001, LNCS 2130, pp. 650–657, 2001.

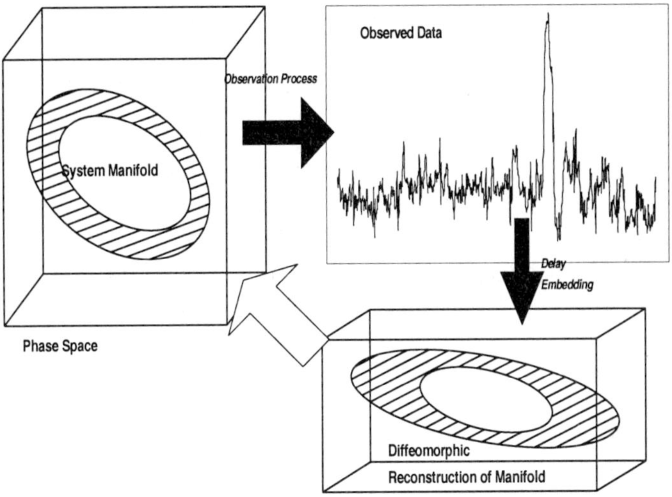

Fig. 1. Graphic depicting the relationship between the true underlying brain processes (the manifold), the observed data and the intended reconstruction

This leads us to the crucial shortcoming with traditional methods of studying MEG - the heavy dependence on frequency analysis. Firstly though, it is important to note that there is nothing wrong with this *per se* - in fact, frequency based methods have led to many useful applications and insights into the workings of the brain. The trouble begins when any activity that cannot be classified in this fashion is automatically dismissed as external noise. Such an approach implicitly assumes that the content of the MEG time series, and hence the evolution rules of the underlying brain dynamics, is linear. In view of existing knowledge regarding brain function on both the microscopic and macroscopic scales [5], this is clearly an unrealistic assumption.

2 Objectives

Due to the extremely high levels of noise that are invariably present, the bulk of existing research into MEG employs signal averaging as a method of noise reduction. This involves repeating experiments a large number of times before averaging over the trials. The number of measurement channels employed has also increased dramatically, resulting in bulkier, more expensive, and increasingly inaccessible machines.

In [8], we demonstrated how the MEG time series can be decomposed into a set of constituent signals. Using nonlinear visualisation tools, we were able to group the extracted components into clusters which exhibited similar morphology, indicating the presence of a small number of unobservable generators. In this paper, we apply similar analysis but to the more practical problem of noise reduction. More specifically, we show that by focussing on the underlying dynamical system, rather than on the observed time-domain signals, we are able to overcome many of the crippling technical problems associated with noise and artifacts, and extract useful information from single channel, unaveraged data.

Starting with the scalar time series, the first step is to reconstruct the manifold. Due to the proof by Takens [6], it is known that time delay embedding provides a simple way of doing this. While it is not possible to completely reproduce the manifold, delay embedding is proven to generate a topologically consistent representation, as illustrated in figure 1.

By itself, the embedding matrix is not particularly useful. The reconstructed manifold will contain contributions from a wide variety of sources including environmental noise sources, artifacts, non-cortical biological signals, as well as brain activity associated with different stimuli. In addition, all of these components will be embedded in a high-dimensional data matrix.

This problem can be readily solved if we assume a locally linear mixing model. Provided we analyse short enough segments of MEG at a time, this assumption should be reasonable. For a set of components $\mathbf{x}_t$, this model can readily be represented by the equation:

$$\mathbf{y}_t = \mathbf{A}\mathbf{x}_t + \nu_t \tag{2}$$

Boldface lowercase characters represent vector quantities, indexed by time t. $\mathbf{y}_t$ are the component axes of the reconstructed phase space whereas $\mathbf{x}_t$ are the constituent sources. $\mathbf{A}$ is the mixing matrix, containing the mixing proportions for each member of $\mathbf{x}$, and ν_t is an external noise source. Also, $\mathbf{x}$ should be distinguished from $\mathbf{s}$: the former refers to any independent signal that has been mixed into the final reconstruction and will include both the actual brain-related sources $\mathbf{s}$, as well as noise sources and artifacts.

Clearly, the aim should be to recover the original sources $\mathbf{s}$ from the reconstructed phase space, which would allow us to study the actual dynamics of the brain. While we are generically interested in an independent factor analysis of this data, for the purpose of demonstration in this paper, it is adequate to consider ICA. ICA attempts to find an optimal rotation for a multivariate dataset by minimising the statistical dependence between the separated components, and has been used in EEG [1, 2] and MEG [7] analysis previously in the literature. This has been proven to result in a maximum likelihood fit for a generative model of the form in equation 2. For the results in this paper, the FastICA algorithm was used. For details about this as well as other ICA algorithms, please refer to [3, 4].

3 Data

The data used was recorded using a CTF Omega-151 system. The recordings were done in one minute trials with a sampling frequency of 1250Hz. The data was then downsampled by a factor of five, resulting in an effective sampling rate of 250Hz. Low pass filtering was applied prior to downsampling with a cut-off frequency of 100hz to avoid aliasing effects. Because it is unlikely that signals

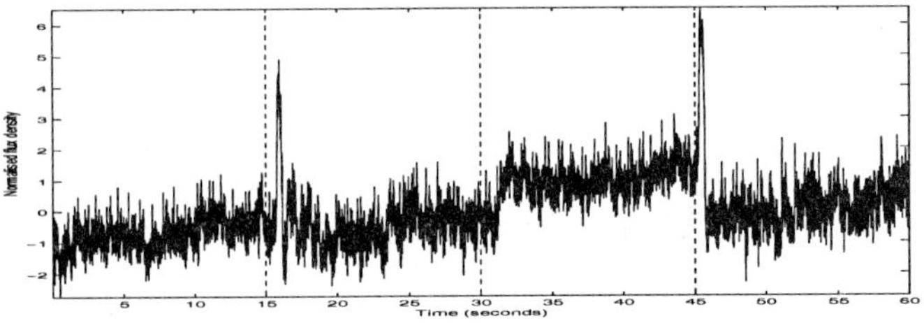

Fig. 2. MEG time series - Right temporal region

of biological origin would have frequency components higher than 100 Hz, this was deemed to be acceptable. However, note that we are still allowing for the existence of higher frequency brain activity than is traditionally assumed.

The data recorded was from a simple steady state experiment in which the subject was required to repeatedly open and close her eyes: She was first instructed to sit quietly and to fixate on a point displayed on a screen. After fifteen seconds, she was instructed to close her eyes, then to open them again, then to close them again.

4 Extraction of Independent Components and Noise Reduction

As a demonstration of the usefulness of the framework described, a simple noise reduction application will now be discussed, though noise reduction is just one of the techniques we are studying in this framework.

First, a suitable channel was selected (in theory, the choice of channel is not important). For this paper, we used a channel located in the right temporal region, shown in figure 2.

To ensure the stationarity of the data set, it was necessary to divide the data into smaller chunks to be processed separately. From figure 2, it can be seen that the data divides naturally into four separate chunks of 15 seconds each, delimited by the points when the subject was instructed to either open or close her eyes. The next step was to construct a delay embedding matrix. A practical issue that needs to be addressed is the number of embedding dimensions that should be used. Unfortunately, there is no real satisfactory solution to this problem but in general we should aim to fix the size of the embedding window to be at least as large as the longest observable feature in the time-series to enable the dynamics to be properly studied. In the channel that we are studying, the largest discernible features are heartbeats, which span a period of about 25 samples (0.1 seconds). Hence, we choose 30 to be a suitable choice of embedding dimension. The embedding matrix was constructed and the FastICA algorithm applied to each of these segments separately.

For each 15-second segment of the time series we now have a thirty dimensional data matrix of ICA components. The challenge is now to find a way

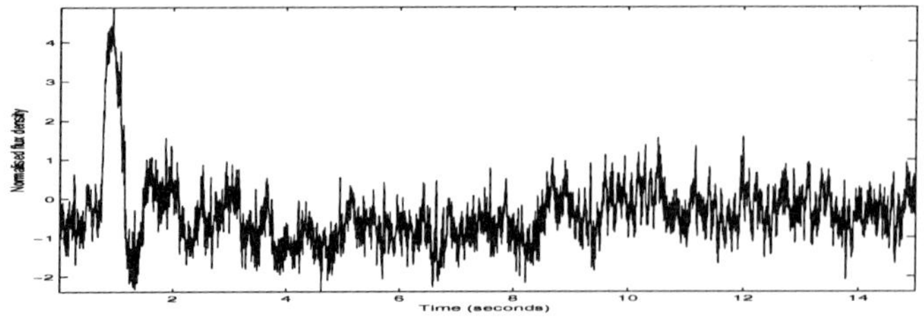

Fig. 3. Chunk 2: Time Step 3751–7500.

of separating the noise components from those actually containing useful information. As it turned out however, this was straightforward since the components tended to fall into a very small number of relatively easily identified categories. These were heartbeat associated signals, 50Hz mains contamination, Alpha Waves, white noise as well two signals that clearly possessed structure and were possibly neuronal in origin.

As it is not possible to view the entire time series in any detail given the available space, only the components extracted from the second chunk of data will be considered henceforth. This segment of the data is shown in figure 3. For this segment, two extracted independent components of particular interest are shown in figure 4. The rest of the components are very clearly alpha waves, 50Hz noise and white noise components. One example from each group is shown in figure 5

Finally, the components were recombined by setting mixing matrix elements corresponding to heartbeat, mains contamination and white noise to zero. As a comparison, a filtered time series was generated using low pass filtering with a cut-off frequency of 40 Hz. The two time series are shown in figure 6. A closer view (at 4 to 8 seconds) of the two signals is given in figure 7.

5 Discussion and Further Work

While the main purpose of this paper has been as a proof-of-concept, the preliminary results already clearly demonstrate that interesting information may be extracted from *single channel* unaveraged MEG. The two time series shown in figure 6 highlight the great difference between the traditional approach (low pass filtering) and the proposed approach. In particular, low pass filtering was incapable of filtering out the heartbeat activity since this was a broadband component with low-frequency content. Consequently, this is still very evident in the filtered signal and can be clearly observed in figure 7. In addition, the ICA approach produces a visibly cleaner signal than low pass filtering, while still retaining fine detail and structure in the time series.

A natural next step will be to fuse the method presented here with the work described in [8]. By clustering the components into recognisable categories, we should be able to greatly simplify and practically automate the choice of components to retain when performing the reconstruction.

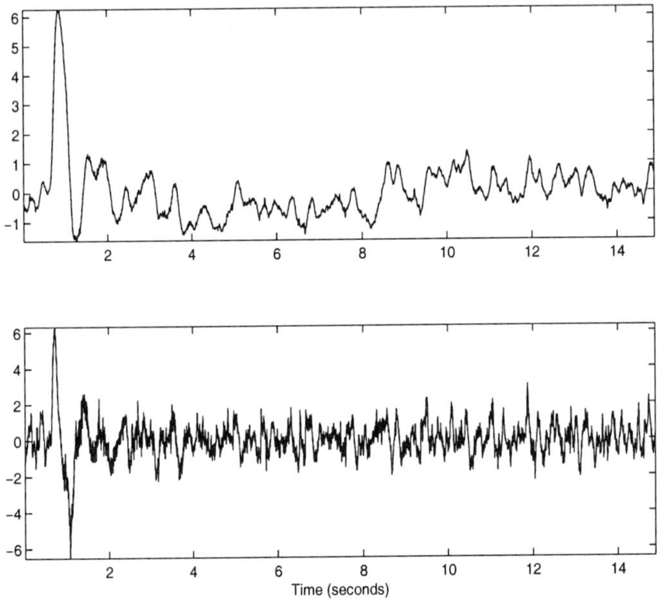

Fig. 4. Chunk 2: Time Step 3751–7500. Independent Components No.1 and 5

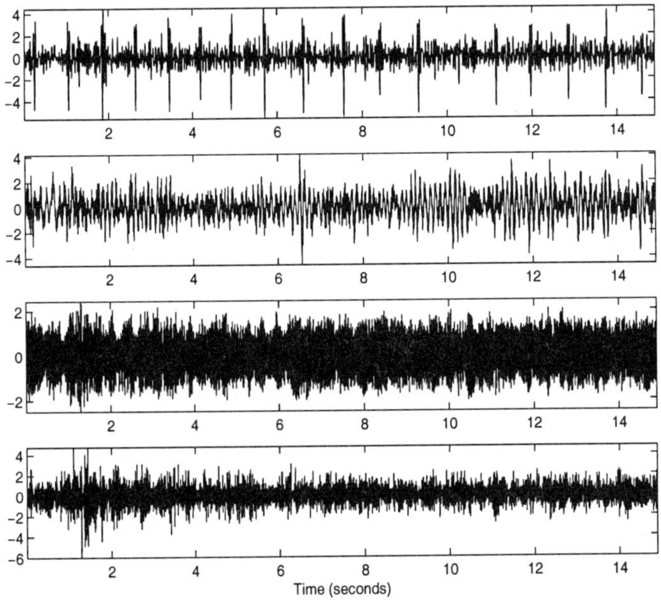

Fig. 5. Chunk 2: Time Step 3751–7500. Independent Components No.2, 26, 21 and 22, corresponding to heartbeat, alpha waves, mains hum and white noise respectively (from top to bottom)

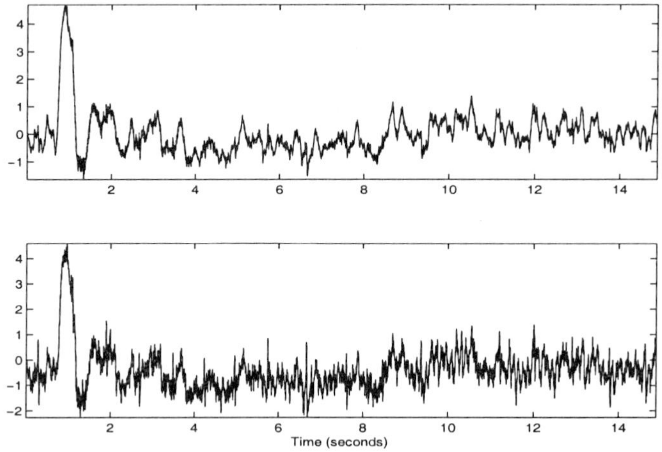

Fig. 6. Top: Result of ICA decomposition and subsequent reconstruction. Bottom: Result using low pass filtering

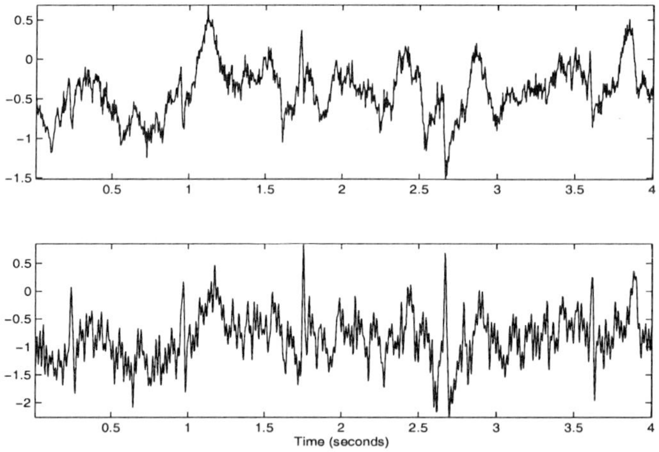

Fig. 7. Enlarged view of ICA decomposition-reconstruction (top) and low pass filtering (bottom)

Furthermore, while this is already a useful achievement, the noise reduction demonstrated here is only one of a large number of possible applications of this methodology. Being able to decompose the MEG time-series in this manner without the need for multiple evoked potential response experiments, will be hugely useful for future research. We believe that this approach holds potential for much more rewarding and complex research activity beyond simple preprocessing.

Acknowledgements

The research detailed in this paper has been sponsored by the UK EPSRC grant GR/L94673. We thank the EPSRC for supporting this research. We also wish to thank the staff of the Clinical Neurophysiology Unit at Aston University. In particular, we thank Dr.Ian Holliday, Dr.Gareth Barnes and Dr.Arjan Hillebrand for much-needed assistance and advice in the MEG data collection process, as well as regarding various MEG theoretical issues in general.

References

1. C.James and D.Lowe. Using dynamical embedding to isolate seizure compoenents in the ictal EEG. *IEE Proc. Sci. Meas. Technol.*, 146(6), November 2000.
2. C.James, J.Kobayashi, and D.Lowe. Isolating epileptiform discharges in the unaveraged EEG using independent component analysis. In *Medical Applications of Signal Processing colloquium (IEE)*, 1999.
3. Aapo Hyvarinen and Erkki Oja. A fast fixed-point algorithm for independent component analysis. *Neural Computation*, 9:1483–1492, 1997.
4. Aapo Hyvrinen and Erkki Oja. Independent component analysis: algorithms and applications. *Neural Networks*, 13:411–430, 2000.
5. P.A. Robinson, C.J. Rennie, and J.J.Wright. Propagation and stability of waves of electrical activity in the cerebral cortex. *Physical Review E*, 56(1):826, 1997.
6. F. Takens. Detecting strange attractors in turbulence. *Lecture notes in mathematics (dynamical systems and turbulence, warwick 1980)*, 898:366–381, 1981.
7. Ricardo Vigario, Veikko Jousmaki, Matti Hamalainen, Riitta Hari, and Erkki Oja. Independent component analysis for identification of artifacts in magnetoencephalographic recordings. *Proc. NIPS*, pages 229–235, 1997.
8. Wei Lee Woon and David Lowe. A nonlinear dynamical systems apporach to MEG signal processing: an example of brain source elicitation. *4th International Conference 'Neural Networks and Expert Systems in Medicine and Healthcare' (NNESMED)*, June 2001.

Part VIII

Time Series Processing

A Discrete Probabilistic Memory Model for Discovering Dependencies in Time

Sepp Hochreiter and Michael C. Mozer

Department of Computer Science
University of Colorado
Boulder, CO 80309–0430
{hochreit,mozer}@cs.colorado.edu

Abstract. Many domains of machine learning involve discovering dependencies and structure over time. In the most complex of domains, *long-term temporal dependencies* are present. Neural network models such as LSTM have been developed to deal with long-term dependencies, but the continuous nature of neural networks is not well suited to discrete symbol processing tasks. Further, the mathematical underpinnings of neural networks are unclear, and gradient descent learning of recurrent neural networks seems particularly susceptible to local optima. We introduce a novel architecture for discovering dependencies in time. The architecture is formed by combining two variants of a hidden Markov model (HMM) – the factorial HMM and the input-output HMM – and adding a further strong constraint that requires the model to behave as a latch-and-store memory (the same constraint exploited in LSTM). This model, called an MIOFHMM, can learn structure that other variants of the HMM cannot, and can generalize better than LSTM on test sequences that have different statistical properties (different lengths, different types of noise) than training sequences. However, the MIOFHMM is slower to train and is more susceptible to local optima than LSTM.

1 Introduction

Many domains of machine learning involve discovering dependencies and structure over time. Example domains include speech recognition, process control, and time series prediction. In the most complex of domains, *long-term temporal dependencies* are present. A long-term dependency is one in which the observation at time t_u, $o(t_u)$, and the observation at some time in the future, $o(t_v)$, are dependent, where $t_v \gg t_u$, and there is no time t_w, $t_u < t_w < t_v$, such that the dependency between the $o(t_u)$ and $o(t_v)$ can be described in terms of the dependency between $o(t_u)$ and $o(t_w)$ plus the dependency between $o(t_w)$ and $o(t_v)$. To capture the structure present in the temporal sequence, it is therefore necessary to construct a memory holding information about $o(t_u)$ during the time intervening between the observations.

Hidden Markov Models (HMMs) and Recurrent Neural Networks (RNNs) are natural candidates to encode long-term dependencies. However, theoretical and

G. Dorffner, H. Bischof, and K. Hornik (Eds.): ICANN 2001, LNCS 2130, pp. 661–668, 2001.

empirical work argues that learning these dependencies is difficult; for RNNs, see [8, 4, 6], and for HMMs, see [2]. A constrained form of the RNN architecture, called LSTM, has been proposed to learn long-term dependencies using standard learning procedures such as gradient descent [7]. LSTM succeeds because it imposes an inductive bias via hidden units with fixed linear self-recurrent connections of strength 1.0. These units behave as memory cells, responding to learned inputs, and then remaining active indefinitely.

LSTM has three weaknesses. First, LSTM was designed to address many tasks that are intrinsically discrete – they involve classifying sequences of input symbols. A neural network with continuous activation levels does not seem well suited to a discrete domain. Second, gradient-descent learning is slow and in the case of RNNs is particularly prone to encountering local optima. Third, the mathematical underpinnings of neural networks are shaky; for example, the semantics of "activation levels" are ill defined. None of these weaknesses are found in HMMs: HMMs are well suited for discrete inputs and outputs, they use EM procedures for training instead of gradient descent, the HMM has a probabilistic interpretation.

In this paper, we take the inductive bias provided by the LSTM model and incorporate it into a HMM, with the goal of obtaining the benefits of each. Rather than abandoning neural networks for the increasingly popular graphical models, we believe it valuable to exploit the inductive biases discovered by the RNN community in the design of constrained variations of HMMs. The constraint suggested by LSTM involves a fixed state transition probability matrix that implements a latch-and-hold memory.

2 A Discrete Probabilistic Memory Model

A standard HMM generates output sequences. To handle temporally-varying input as well as temporally-varying output, we use an extension known as an input-output HMM [3], in which the state at t, $s(t)$ is conditionally dependent on the previous state, $s(t-1)$, as well as the current input, $x(t)$, and the output, $y(t)$, is conditionally dependent on $s(t)$ and $x(t)$. Further, we allow for a state with compositional structure using a factorial HMM [5]. The particular sort of compositional state we explore in our model is one consisting multiple non-resettable flip-flops – memory elements which can be triggered by particular inputs and will remain unchanged in time thereafter; this same sort of *latch* is the heart of LSTM. Thus, our model is an *memory-based input-output factorial HMM*, which we shorten to MIOFHMM.

Consider factorizing the state into H components, denoted $s_1 \dots s_H$, each of which we wish to behave as a latch-and-hold memory. Each component is a multinomial random variable with N values. Initially, all components have value $\mathbf{U}$ for "uncommitted"; various inputs can trigger the component to take on values $\mathbf{2}, \dots, \mathbf{N}$. The constraint on the MIOFHMM is to fix the state transition function, $p(s_i(t) = a | s_1(t-1) = b_1, \dots, s_H(t-1) = b_H, x(t) = c)$, to δ_{a,b_i} if $b_i \neq \mathbf{U}$, where δ is the Kronecker delta. Once component s_i takes on values

2...**N**, the component can not change its value – it behaves as a memory for the occurrence of an input event. Thus, it has $N-1$ *memory states.*

The restriction on the state transition function that allows each component to store its value indefinitely should have significant benefits in learning: the fixed transition probabilities prevent the transition matrix from becoming irreducible, and hence the limitations on learning temporal dependencies discussed in [2] are not applicable. Each component is further restricted in that it cannot be reset to **U** or any other value, and therefore cannot be re-used. However, we skirt this limitation by allowing multiple components that can be used to store different facets of the input sequence.

To avoid the possibility that all hidden variables become committed at a certain time point and the MIOFHMM becomes unable to track dynamics, we could add conventional hidden variables, i.e., hidden variables without constraints. Another possibility is to soften the constraints by adding a penalty for transitions that were neither 0 or 1, allowing learning to produce non-binary transition probabilities if it was warranted by improved performance.

2.1 Training the MIOFHMM

Training data for the MIOFHMM consists of a set of input and output sequence pairs. The goal of training is to determine model parameters – discrete conditional probability distributions – that maximize the likelihood of the training output sequences given the corresponding training input sequences.

We train the MIOFHMM using the Baum-Welch algorithm [1]. The complexity of the MIOFHMM training procedure is exponential in the number of memory components. Ignoring the memory constraint, the complexity of the Baum-Welsh algorithm for the MIOFHMM is $O\left(T\ N^{2H}\right)$, where T is the sequence length. However, exploiting the memory constraint reduces the complexity to $O(T\ [2N-1]^H)$ which is a savings of a factor $(N/2)^H$. Approximations to Baum-Welsh updating [5] might be used to further accelerate training, although we did not explore such approximations in the present work.

3 Experiments

We perform two sets of experiments. First, we compare our MIOFHMM to conventional IOHMMs and IOFHMMs on the detection of long-term dependencies. The tasks involve a nondeterministic mapping from input sequences to output sequences. Second, we compare the generalization performance of our MIOFHMM to the LSTM recurrent neural network. We use a classification task in which the model must produce an output indicating class membership following the entire input sequence. Each result we present is the average of twenty replications of a model, excluding replications that yielded local optima (as determined by a validation set). In all experiments, the HMM conditional distributions are initialized randomly.

3.1 Comparing the MIOFHMM, IOFHMM, and IOHMM

We begin with a study of learning long-term dependencies using a simple *latch task* that has been used to test various approaches to this problem, e.g., [4, 6]. The essence of the task is that a sequence of inputs are presented, beginning with one of two symbols, **A** or **B**, and after a variable number of time steps, the model must output a corresponding symbol – **U** if the original input was **A**, or **V** if the original input was **B**. Thus, the task requires memorizing the original input. The end of the input sequence is marked by the symbol **E**, and in the intervening time steps, symbols are chosen at random from $\{\mathbf{C}, \mathbf{D}\}$. Except for the final output symbol, any output from $\{\mathbf{X}, \mathbf{Y}, \mathbf{Z}\}$ is allowed. We vary T, the number of time steps intervening between the first input and the final output. A sample input sequence for $T = 6$ is **A–C–C–D–D–C–E**, and an allowed output sequence for this input is **Y–Z–X–Y–Z–Y–U**. Five hundred sequences were generated for training and for validation.

We compared the MIOFHMM against the IOHMM and the IOFHMM. The IOHMM is given 5 hidden states, and the IOFHMM and MIOFHMM are given 2 components with 5 hidden states each. (We also tested a version of the MIOFHMM with a single component – essentially an MIOHMM– and the performance was comparable to that of the MIOFHMM.) For each simulation, we record the number of updates required for the model to produce the correct output on the final time step for all examples in the the validation set. If a model does not process the validation set correctly within reasonable number of updates (multiple standard deviations above the mean), we treat the run as having become stuck in a local optimum. We report the mean number of update required for learning and the frequency of becoming stuck in local optima.

Figure 1 shows the number of updates required to train the three models, as a function of the sequence length T. Experiments with the IOHMM and IOFHMM with $T > 5$ were terminated due to lack of CPU cycles. The training time for the IOHMM and IOFHMM appears to scale exponentially with the sequence length for the IOHMM and IOFHMM, consistent with the theoretical results in [2], but training time for the MIOFHMM is flat. The IOHMM and IOFHMM also yielded many local optima: For $T = 5$, the IOHMM and IOFHMM discovered local optima on 35% and 85% of training runs, respectively, whereas the MIOFHMM yielded no local optima. The MIOFHMM clearly outperforms conventional HMM models on tasks involving long-term temporal dependencies. The key feature necessary for the success of the MIOFHMM is the constraint that the state components behave as memory, which is absent from the otherwise identical IOFHMM.

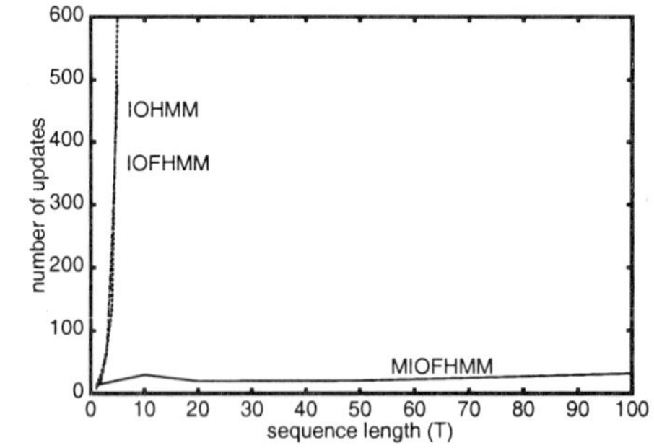

Fig. 1: Number of updates required to learn the latch task for three architectures, as a function of the sequence length (the number of time steps over which an input element must be remembered).

In a second study, we used another task that has previously been explored in the neural net community [6]. The task involves discovering a classification rule for input sequences that depends on the temporal order of events. Each sequence begins with the start symbol **S** and terminates with the end symbol **E**. Embedded in each sequence are two *critical* symbols chosen with replacement from $\{\mathbf{A}, \mathbf{B}\}$. All other input symbols are random, chosen with replacement from $\{\mathbf{C}, \mathbf{D}, \mathbf{F}, \mathbf{G}\}$. The input alphabet thus consists of eight symbols – two start symbols, two critical symbols, and four random symbols. A sample input sequence is **S–C–A–G–B–F–E**. The classification of the sequence depends on the identity and order of the two critical symbols: sequences containing an **A** followed by a **B** are assigned to class 1, a **B** followed by an **A** to class 2, an **A** followed by an **A** to class 3, and a **B** followed by a **B** to class 4. On receiving the final input, the task involved outputting the class label; prior to this input, the task required outputting a special "no class" label.

We compared the IOFHMM to the MIOFHMM. The models had two state components, each having two memory states. We trained the models with 500 examples, and used a validation set of 500 further examples to determine when the model had learned the task. The IOFHMM was never able to learn the task to a criterion of 0 classification errors. Although the MIOFHMM ran into local optima on 65% of trials, it needed only 162 model-parameter updates on average to learn on the remaining 35%. (The local optima obtained by the MIOFHMM in this and other experiments is actually a form of overfitting: the model performs very well on the training set, but not on the validation set. But we call this a local optimum nonetheless because there is a solution for which the model would perform better on training and validation set.) Regardless, the MIOFHMM can succeed on a difficult sequence-ordering task where the IOFHMM fails, due to the constraint imposed on the MIOFHMM that it implement a latch-and-hold memory.

To summarize our two experiments, the latch-and-hold memory constraint imposes a strong inductive bias on the MIOFHMM, which allows it to learn more efficiently and reliably than models such as the IOFHMM and IOHMM which do not exploit this constraint. Of course, the benefit extends only to tasks for which this bias is appropriate – tasks involving storing and remembering sequence elements and their ordering.

3.2 Comparison of MIOFHMMs and LSTM

The experiments in this section explore the generalization capabilities of the MIOFHMM as compared to those of LSTM [6], the neural network model with a latch-and-hold memory constraint. We consider a generalization task that is particularly difficult for machine learning systems, and for which no guarantees of good generalization are possible – where the distributions from which the training and test examples are drawn differ from one another.

In these experiments, we study a variation of the latch task. The input at each time consists of two real-valued elements, a *value* and a *marker*, both in [0,1]. The marker having value 1.0 indicates that the current value is to be stored, and the marker having value 0.0 indicates that the previously stored value should be retrieved and outputted; a marker value of 0.5 indicates "no action".

Because the MIOFHMM is intrinsically discrete, input values were quantized into one of E equal width intervals in [0,1]. For example, with $E = 10$, the intervals were [0,.1], [.1,.2], etc. Each interval corresponded to a unique input value, which was crossed with the three distinct markers for a total of $3E$ input symbols. The output consisted of E symbols. The LSTM, in contrast, required only two continuous inputs and one continuous output. Its output was judged to be correct if it lay in the correct interval. Although the two architectures are quite different, it is not clear whether one has an advantage over the other on this task. The MIOFHMM benefits from the fact that it receives inputs that are quantized in a task-appropriate manner, whereas the LSTM benefits from the fact that its input has a compositional structure which is task appropriate.

In a first experiment, we trained the models on sequences with lengths between 2 and 10, sampled uniformly, with $E = 10$ intervals, and with the value to be stored always the first element of the sequence. Both models were supplied with 1000 training examples and 1000 validation examples. The models were tested on 1000 generalization examples for various lengths between 10 and 1000. Thus, the challenge was to extrapolate to longer sequences, and hence, to form a memory that could persist over long time intervals.

For this experiment, LSTM was provided with two memory cells. Weights in the LSTM were initialized randomly from $[-.1, .1]$, with an initial bias of -1.0 on each input gate. A learning rate of 0.1 was used. Following each sequence, the weights were updated and the network was reset. The MIOFHMM utilized one state component with 10 memory states. For both models, training continued until all examples in the validation set were classified correctly (i.e., in the correct interval); if this did not occur within a reasonable amount of time, then the training run was considered to have become stuck in a local optimum.

LSTM learned the task efficiently and reliably: training took was 36 seconds of CPU time on a 400 MHz PC (corresponding to 130 training epochs), and never encountered local optima. In contrast, the MIOFHMM required 77 minutes of CPU time (15 updates), and became stuck in local maxima on 47% of runs. In testing, however, the MIOFHMM outshone the LSTM. Figure 2 shows generalization error on test sequences with lengths ranging from 10 to 1000 elements. The test sequence length can be extended to 1000 without any affect on performance for the MIOFHMM, whereas the error rate increases rapidly for the LSTM for sequences having lengths greater than 30.

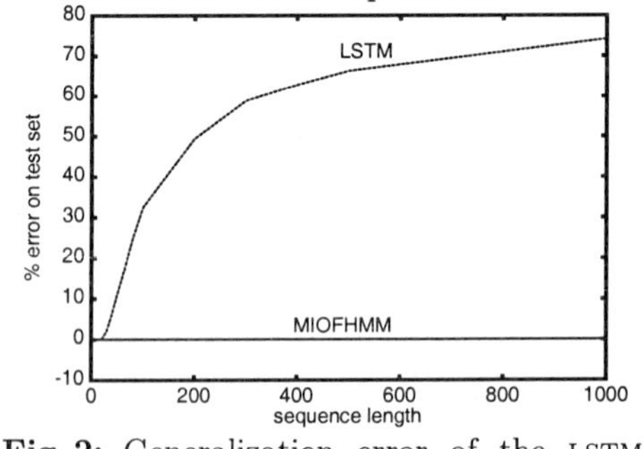

Fig. 2: Generalization error of the LSTM and the MIOFHMM on a latch task with test sequences longer than the longest training sequence.

In a second experiment, we simplified the latch task by presenting sequences whose element to be stored had only $E = 2$ discriminable values, but made the task more difficult in that the value to be stored could occur on any of the first 5 sequence elements. Sequences ranged in length from 5 to 20. As in the first experiment, a marker input of 1.0 was a signal to store an input, and a marker input of 0.0 was a signal to retrieve the stored value. However, we modified the task by replacing the neutral marker value of 0.5 with values ranging from 0.025 to 0.975. In the training set, the neutral marker value was randomly chosen from a uniform distribution over 39 discrete values evenly spaced in [0.025, 0.975]. In the test set, the neutral marker value was randomly chosen from a rectified, discretized Gaussian distribution over the 39 discrete values. The variance of the Gaussian was chosen based on a parameter a, such that with 99% probability the a largest values will be chosen. Consequently, as a is decreased, more marker values in the sequence will become confused with the store (1.0) markers.

As in the first experiment, LSTM training was faster and more reliable: LSTM required 7.75 minutes on average to train (976 epochs), whereas the MIOFHMM required 19 hours (50 updates). LSTM never encountered local optima, whereas MIOFHMM did on 15% of trials. However, in terms of generalization performance, MIOFHMM once again beat out LSTM. Figure 3 shows percentage error on a test set as a function of the noise parameter a. Even for large values of a, MIOFHMM produces fewer errors than LSTM, but as a is decreased, LSTM errors increase dramatically. For small a, the neutral marker was more likely to be a value close to that of the store marker, and consequently, acted as a lure to confuse LSTM. MIOFHMM benefits from the fact that the marker 1.0 and the marker 0.975 are two different symbols, and the similarity structure of the numerical values is therefore irrelevant to performance.

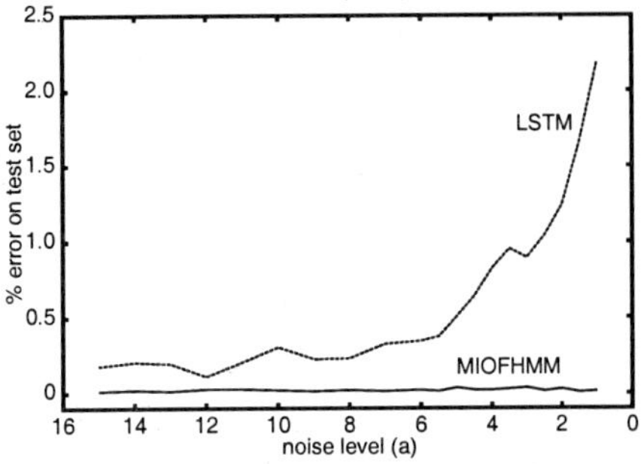

Fig. 3: Generalization error of the LSTM and the MIOFHMM on a latch task with increased noise.

To summarize these two experiments, the discrete nature of the MIOFHMM allows it to reliably hold information for longer periods of time than the continuous LSTM, and also prevents the MIOFHMM from becoming confused by noise, even noise whose statistics in the training and test sets are quite different. Although the two experimental tasks we presented are somewhat contrived, they emphasize that the discrete nature of the MIOHMM can be a virtue that distinguishes from any continuous recurrent neural network model.

4 Conclusions

In this paper, we have introduced a novel architecture for classifying input sequences and for mapping input sequences to output sequences, the MIOFHMM.

The MIOFHMM combines two of the virtues of hidden Markov models – the explicit probabilistic framework and the powerful Baum-Welsh training procedure – with a form of inductive bias found to be valuable in recurrent neural network models such as LSTM. The bias forces the MIOFHMM to behave as a latch-and-store memory. We explored four tasks that involved discovering key elements in an input sequence whose detection or temporal order was critical to performance. The MIOFHMM performed well on all these tasks. In contrast, variants of the architecture without the memory constraint – the IOHMM and the IOFHMM – scaled poorly as a function of the temporal span of dependencies. Further, the discrete nature of the MIOFHMM makes it particularly well suited to reaching stable, robust fixed points in state space, and consequently, it appears less susceptible than a continuous model like LSTM to disruption by noise, either in the form of interspersed irrelevant sequence elements or variability in the input at a particular point in time. The one weakness of the MIOFHMM is that – in contrast to our original intuitions – it takes significantly longer than LSTM to train and is more susceptible to local optima.

Acknowledgments

The work was supported by the *Deutsche Forschungsgemeinschaft* (Ho 1749/1-1), McDonnell-Pew award 97-18, and NSF award IBN-9873492.

References

1. L. E. Baum. An inequality and associated maximization technique in statistical estimation for probabilistic functions of a Markov process. *Inequalities*, 3:1–8, 1972.
2. Y. Bengio and P. Frasconi. Diffusion of context and credit information in markovian models. *Journal of Artificial Intelligence Research*, 3:249–270, 1995.
3. Y. Bengio and P. Frasconi. An input output HMM architecture. In G. Tesauro, D. S. Touretzky, and T. K. Leen, editors, *Advances in Neural Information Processing Systems 7*, pages 427–434. MIT Press, Cambridge MA, 1995.
4. Y. Bengio, P. Simard, and P. Frasconi. Learning long-term dependencies with gradient descent is difficult. *IEEE Trans. on Neural Networks*, 5(2):157–166, 1994.
5. Z. Ghahramani and M. I. Jordan. Factorial hidden markov models. *Machine Learning*, 29:245, 1997.
6. S. Hochreiter and J. Schmidhuber. Long short-term memory. *Neural Computation*, 9(8):1735–1780, 1997.
7. S. Hochreiter and J. Schmidhuber. LSTM can solve hard long time lag problems. In M. C. Mozer, M. I. Jordan, and T. Petsche, editors, *Advances in Neural Information Processing Systems 9*, pages 473–479. MIT Press, Cambridge MA, 1997.
8. M. C. Mozer. Induction of multiscale temporal structure. In J. E. Moody, S. J. Hanson, and R. P. Lippman, editors, *Advances in Neural Information Processing Systems 4*, pages 275–282. San Mateo, CA: Morgan Kaufmann, 1992.

Applying LSTM to Time Series Predictable through Time-Window Approaches

Felix A. Gers — felix@idsia.ch
Douglas Eck — doug@idsia.ch
Jürgen Schmidhuber — juergen@idsia.ch

IDSIA, Galleria 2, 6928 Manno, Switzerland, www.idsia.ch

Abstract. Long Short-Term Memory (LSTM) is able to solve many time series tasks unsolvable by feed-forward networks using fixed size time windows. Here we find that LSTM's superiority does *not* carry over to certain simpler time series prediction tasks solvable by time window approaches: the Mackey-Glass series and the Santa Fe FIR laser emission series (Set A). This suggests to use LSTM only when simpler traditional approaches fail.

1 Overview

Long Short-Term Memory (LSTM) is a recurrent neural network (RNN) architecture that had been shown to outperform traditional RNNs on numerous temporal processing tasks [1–4].

Time series benchmark problems found in the literature, however, often are conceptually simpler than many tasks already solved by LSTM. They often do not require RNNs at all, because all relevant information about the next event is conveyed by a few recent events contained within a small time window. Here we apply LSTM to such relatively simple tasks, to establish a limit to the capabilities of the LSTM-algorithm in its current form. We focus on two intensively studied tasks, namely, prediction of the Mackey-Glass series [5] and chaotic laser data (Set A) from a contest at the Santa Fe Institute (1992) [6].

LSTM is run as a "pure" autoregressive (AR) model that can only access input from the current time-step, reading one input at a time, while its competitors — e.g., multi-layer perceptrons (MLPs) trained by back-propagation (BP) — simultaneously see several successive inputs in a suitably chosen time window. Note that Time-Delay Neural Networks (TDNNs) [7] are not purely AR, because they allow for direct access to past events. Neither are NARX networks [8] which allow for several distinct input time windows (possibly of size one) with different temporal offsets.

We also evaluate stepwise versus iterated training (IT) [9, 10].

2 Experimental Setup

The task is to use currently available points in the time series to predict the future point, $t + T$. The target for the network t_k is the difference between

G. Dorffner, H. Bischof, and K. Hornik (Eds.): ICANN 2001, LNCS 2130, pp. 669–676, 2001.

the values $x(t+p)$ of the time series p steps ahead and the current value $x(t)$ multiplied by a scaling factor f_s: $t_k(t) = f_s \cdot (x(t+p) - x(t)) = f_s \cdot \Delta x(t)$. The value for f_s scales $\Delta x(t)$ between –1 and 1 for the training set; the same value for f_s is used during testing. The predicted value is the network output divided by f_s plus $x(t)$. During iterated prediction with $T = n * p$ the output is clamped to the input (self-iteration) and the predicted values are fed back n times. For direct prediction $p = T$ and $n = 1$; for single-step prediction $p = 1$ and $n = T$.

The error measure is the normalized root mean squared error: NRMSE = $\langle (y_k - t_k)^2 \rangle^{\frac{1}{2}} / \langle (t_k - \langle t_k \rangle)^2 \rangle^{\frac{1}{2}}$, where y_k is the network output and t_k the target. The reported performance is the best result of 10 independent trials.

LSTM Network Topology. The input units are fully connected to a hidden layer consisting of memory blocks with 1 cell each. The cell outputs are fully connected to the cell inputs, to all gates, and to the output units. All gates, the cell itself and the output unit are biased. Bias weights to input and output gates are initialized block-wise: -0.5 for the first block, -1.0 for the second, -1.5 for the third, and so forth. Forget gates are initialized with symmetric positive values: $+0.5$ for the first block, $+1$ for the second block, etc. These are standard values that we use for all experiments. All other weights are initialized randomly in the range $[-0.1, 0.1]$. The cell's input squashing function g is a sigmoid function with the range $[-1, 1]$. The squashing function of the output unit is the identity function.

To have statistically independent weight updates, we execute weight changes every 50+rand(50) steps (where rand(max) stands for a random positive integer smaller than max which changes after every update). We use a constant learning rate $\alpha = 10^{-4}$.

MLP. The MLPs we use for comparison have one hidden layer and are trained with BP. As with LSTM, the one output unit is linear and Δx is the target. The input differs for each task but in general uses a time window with a time-space embedding. All units are biased and the learning rate is $\alpha = 10^{-3}$.

3 Mackey-Glass Chaotic Time Series

The Mackey-Glass (MG) chaotic time series can be generated from the MG delay-differential equation [5]: $\dot{x}(t) = \frac{\alpha x(t-\tau)}{1+x^c(t-\tau)} - \beta x(t)$. We generate benchmark sets using a four-point Runge-Kutta method with step size 0.1 and initial condition $x(t) = 0.8$ for $t < 0$. Equation parameters were set at $a = 0.2$, $b = 0.1$, $c = 10$ and $\tau = 17$. The equation is integrated up to $t = 5500$, with the points from $t = 200$ to $t = 3200$ used for training and the points from $t = 5000$ to $t = 5500$ used for testing.

MG Previous Work. In the following sections we list attempts to predict the MG time series. To allow comparison among approaches, we did not consider works where noise was added to the task or where training conditions were very different from ours. When not specifically mentioned, an input time window with time delays t, $t-6$, $t-12$ and $t-18$ or larger was used. Summary of previous approaches:

BPNN [11]: A BP continuous-time feed forward NNs with two hidden layers and with fixed time delays. **ATNN** [11]: A BP continuous-time feed forward NNs with two hidden layers and with adaptable time delays. **DCS-LMM** [12]: Dynamic Cell Structures combined with Local Linear Models. **EBPTTRNN** [13]: RNNs with 10 adaptive delayed connections trained with BPTT combined with a constructive algorithm. **BGALR** [14]: A genetic algorithm with adaptable input time window size (Breeder Genetic Algorithm with Line Recombination). **EPNet** [15]: Evolved neural nets (Evolvable Programming Net). **SOM** [16]: A Self-organizing map. **Neural Gas** [17] : The Neural Gas algorithm for a Vector Quantization approach. **AMB** [18]: An improved memory-based regression (MB) method [19] that uses an adaptive approach to automatically select the number of regressors (AMB). The results from these approaches are found in Table 1.

Table 1. Results for the MG task, showing (from left to right) the number of units, the number of parameters (weights for NNs), the number of training sequence presentations, and the NRMSE for prediction offsets $T \in \{1, 6, 84\}$.

Reference	Units	Para.	Seq.	NMSE		
				$T=1$	$T=6$	$T=84$
Linear Predictor	-	-	-	0.0327	0.7173	1.5035
6th-order Polynom.	-	-	-	-	0.04	0.85
BPNN	-	-	-	-	0.02	0.06
FTNN	20	120	$7 \cdot 10^7$	-	0.012	-
ATNN	20	120	$7 \cdot 10^7$	-	0.005	-
Cascade-Correlation	20	≈ 250	-		0.04	0.17
DCS-LLM	200	200^2	$\approx 1 \cdot 10^5$	-	0.0055	0.03
EBPTTRNN	6	65	-	-	0.0115	-
BGALR	16	≈ 150	-	-	0.2373	0.267
EPNet	≈ 10	≈ 100	$\approx 1 \cdot 10^4$	-	0.02	0.06
SOM	-	10x10	$\approx 1.5 \cdot 10^4$	-	0.013	0.06
	-	35x35	$\approx 1.5 \cdot 10^4$	-	0.0048	0.022
Neural Gas	400	3600	$2 \cdot 10^4$	-	-	0.05
AMB	-		-	-	-	0.054
MLP, $p=T$	16	97	$1 \cdot 10^4$	0.0113	0.0502	0.4612
MLP, $p=1$	16	97	$1 \cdot 10^4$	$p=T=1$	0.0252	0.4734
MLP, $p=1$, IT	16	97	$1 \cdot 10^4$	0.0094	0.0205	0.3929
MLP, $p=6$	16	97	$1 \cdot 10^4$	-	$p=T=6$	0.1466
MLP, $p=6$, IT	16	97	$1 \cdot 10^4$	-	0.0945	0.2820
LSTM, $p=T$	4	113	$5 \cdot 10^4$	0.0214	0.1184	0.4700
LSTM, $p=1$	4	113	$5 \cdot 10^4$	$p=T=1$	0.1981	0.5927
LSTM, $p=1$, IT	4	113	$1 \cdot 10^4$	s. text	0.1970	0.8157
LSTM, $p=6$	4	113	$5 \cdot 10^4$	-	$p=T=6$	0.2910
LSTM, $p=6$, IT	4	113	$1 \cdot 10^4$	-	0.1903	0.3595

MG Results. The LSTM results are listed at the bottom of Table 1. After six single-steps of iterated training ($p = 1$, $T = 6$, $n = 6$) the LSTM NRMSE for single step prediction ($p = T = 1$, $n = 1$) is 0.0452. After 84 single-steps of iterated training ($p=1$, $T=84$, $n=84$) the LSTM NRMSE single step prediction ($p = T = 1$, $n = 1$) is 0.0809. Increasing the number of memory blocks did not significantly improve the results.

The results for AR-LSTM approach are clearly worse than the results for time window approaches, for example with MLPs. Why did LSTM perform worse than the MLP? The AR-LSTM network does not have access to the past as part of its input and therefore has to learn to extract and represent a Markov state. In tasks we considered so far this required remembering one or two events from the past, then using this information before over-writing the same memory cells. The MG equation, contains the input from $t-17$, hence its implementation requires the storage of all inputs from $t-17$ to t (time window approaches consider selected inputs back to at least $t-18$). Assuming that any dynamic model needs the event from time $t-\tau$ with $\tau \approx 17$, we note that the AR-RNN has to store all inputs from $t-\tau$ to t and to overwrite them at the adequate time. This requires the implementation of a circular buffer, a structure quite difficult for an RNN to simulate.

MG Analysis It is interesting that for MLPs ($T=6$) it was more effective to transform the task into a one-step-ahead prediction task and iterate than it was to predict directly (compare the results for $p = 1$ and $p = T$). It is in general easier to predict fewer steps ahead, the disadvantage being that during iteration input values have to be replaced by predictions. For $T=6$ with $p=1$ this affects only the latest value. This advantage is lost for $T=84$ and the results with $p=1$ are worse than with $p=6$, where fewer iterations are necessary. For MLPs, iterated training did not in general produce better results: it improved performance when the step-size p was 1, and worsened performance for $p=6$. For LSTM iterated training decreased the performance. But surprisingly, the relative performance decrease for one-step prediction was much larger than for iterated prediction. This indicates that the iteration capabilities were improved (taking in consideration the over-proportionally worsened one-step prediction performance).

The single-step predictions for LSTM are not accurate enough to follow the series for as much as 84 steps. Instead the LSTM network starts oscillating, having adapted to the strongest eigen-frequency in the task. During self-iterations, the memory cells tune into this eigen-oscillation, with time constants determined by the interaction of cell state and forget gate.

4 Laser Data

This data is set A from the Santa Fe time series prediction competition [6][1]. It consists of one-dimensional data recorded from a Far-Infrared (FIR) laser in a

[1] The data is available from http://www.stern.nyu.edu/~aweigend/Time-Series/SantaFe.html.

chaotic state [20]. The training set consists of 1,000 points from the laser, with the task being to predict the next 100 points. We run tests for stepwise prediction and fully iterated prediction, where the output is clamped to the input for 100 steps.

For the experiments with MLPs the setup was as described for the MG data but with an input embedding of the last 9 time steps as in Koskela, Varsta and Heikkonen [21].

FIR-laser Previous Work Results are listed in Table 2. Linear prediction is no better than predicting the data-mean. Wan [22] achieved the best results submitted to the original Santa Fe contest. He used a Finite Input Response Network (FIRN) (25 inputs and 12 hidden units), a method similar to a TDNN. Wan improved performance by replacing the last 25 predicted points by smoothed values (sFIRN). Koskela, Varsta and Heikkonen [21] compared recurrent SOMs (RSOMs) and MLPs (trained with the Levenberg-Marquardt algorithm) with an input embedding of dimension 9 (an input window with the last 9 values). Bakker et. al. [10] used a mixture of predictions and true values as input (Error Propagation, EP). Then Principal Component Analysis (PCA) was applied to reduce the dimensionality of the time embedding. Kohlmorgen and Müller [23] pointed out that the prediction problem could be solved by pattern matching, if it can be guaranteed that the best match from the past is always the right one. To resolve ambiguities they propose to up-sample the data using linear extrapolation (as done by Sauer [24]). The best result to date, according to our knowledge, was achieved by Weigend and Nix [25]. They used a nonlinear regression approach in a maximum likelihood framework, realized with feed-forward NN (25 inputs and 12 hidden units) using an additional output to estimate the prediction error. McNames [26] proposed a statistical method that used cross-validation error to estimate the model parameters for local models, but the testing conditions were too different to include the results in the comparison. Bontempi et. al. [27] used a similar approach called " Predicted Sum of Squares (PRESS)" (here, the dimension of the time embedding was 16).

FIR-laser Results The results for MLP and LSTM are listed in Table 2. The results for these methods are not as good as the other results listed in Table 2. This is true in part because we did not replace predicted values by hand with a mean value where we suspected the system to be lead astray.

Iterated training yielded improved results for iterated prediction, even when single-step prediction became worse, as in the case of MLP.

FIR-laser Analysis The LSTM network could not predict the collapse of emission in the test set (Figure 1). Instead, the network tracks the oscillation in the original series for only about 40 steps before desynchronizing. This indicates performance similar to that in the MG task: the LSTM network was able to track the strongest eigen-frequency in the task but was unable to account for high-frequency variance. Though the MLP performed better, it generated inaccurate amplitudes and also desynchronized after about 40 steps. The MLP did however manage to predict the collapse of emission.

Table 2. Results for the FIR-laser task, showing (from left to right): The number of units, the number of parameters (weights for NNs), the number of training sequence presentations, and the NRMSE.

Reference	Units	Para.	Seq.	NMSE	
				stepwise	iterated
Linear Predictor	-	-	-	1.25056	-
FIRN [22]	26	≈ 170	-	0.0230	0.0551
sFIRN [22]	26	≈ 170	-	-	0.0273
MLP [21]	70	≈ 30	-	0.01777	-
RSOM [21]	13	-	-	0.0833	-
EP-MLP Bakker et. al. (2000) [10]	73	>1300	-	-	0.2159
Sauer (1994) [24]	-	32	-	-	0.077
Weigend and Nix (1994) [25]	27	≈ 180	-	0.0198	0.016
Bontempi (1999) [27]	-	-	-	-	0.029
MLP	32	353	$1 \cdot 10^4$	0.0996017	0.856932
MLP IT	32	353	$1 \cdot 10^4$	0.158298	0.621936
LSTM	4	113	$1 \cdot 10^5$	0.395959	1.02102
LSTM IT	4	113	$1 \cdot 10^5$	0.36422	0.96834

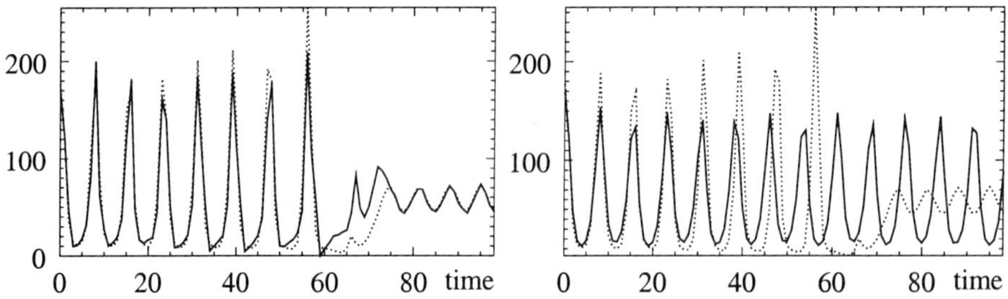

Fig. 1. Test run with LSTM network solution after iterated training for the FIR-laser task. Left: Single-Step prediction. Right: Iteration of 100 steps.

5 Conclusion

A time window based MLP outperformed the LSTM pure-AR approach on certain time series prediction benchmarks solvable by looking at a few recent inputs only. Thus LSTM's special strength, namely, to learn to remember single events for very long, unknown time periods, was not necessary here.

LSTM learned to tune into the fundamental oscillation of each series but was unable to accurately follow the signal. The MLP, on the other hand, was able to capture some aspects of the chaotic behavior. For example the system could predict the collapse of emission in the FIR-laser task.

Iterated training has advantages over single-step training for iterated testing only for MLPs and when the prediction step-size is one. The advantage is evident when the number of necessary iterations is large.

Our results suggest to use LSTM only on tasks where traditional time window based approaches must fail. One reasonable *hybrid* approach to prediction of unknown time series may be this: start by training a time window-based MLP, then freeze its weights and use LSTM only to reduce the residual error if there is any, employing LSTM's ability to cope with long time lags between significant events.

LSTM's ability to track slow oscillations in the chaotic signal may be applicable to cognitive domains such as rhythm detection in speech and music.

Acknowledgment. This work was supported by SNF grant 2100-49'144.96.

References

1. S. Hochreiter and J. Schmidhuber, "Long short-term memory," *Neural Computation*, vol. 9, no. 8, pp. 1735–1780, 1997.
2. F. A. Gers, J. Schmidhuber, and F. Cummins, "Learning to forget: Continual prediction with LSTM," *Neural Computation*, vol. 12, no. 10, pp. 2451–2471, 2000.
3. F. A. Gers and J. Schmidhuber, "Recurrent nets that time and count," in *Proc. IJCNN'2000, Int. Joint Conf. on Neural Networks*, (Como, Italy), 2000.
4. F. A. Gers and J. Schmidhuber, "LSTM recurrent networks learn simple context free and context sensitive languages," *IEEE Transactions on Neural Networks*, 2001. accepted.
5. M. Mackey and L. Glass, "Oscillation and chaos in a physiological control system," *Science*, vol. 197, no. 287, 1977.
6. A. Weigend and N. Gershenfeld, *Time Series Prediction: Forecasting the Future and Understanding the Past.* Addison-Wesley, 1993".
7. P. Haffner and A. Waibel, "Multi-state time delay networks for continuous speech recognition," in *Advances in Neural Information Processing Systems* (J. E. Moody, S. J. Hanson, and R. P. Lippmann, eds.), vol. 4, pp. 135–142, Morgan Kaufmann Publishers, Inc., 1992.
8. T. Lin, B. G. Horne, P. Tiño, and C. L. Giles, "Learning long-term dependencies in NARX recurrent neural networks," *IEEE Transactions on Neural Networks*, vol. 7, pp. 1329–1338, Nov. 1996.
9. J. C. Principe and J.-M. Kuo, "Dynamic modelling of chaotic time series with neural networks," in *Advances in Neural Information Processing Systems* (G. Tesauro, D. Touretzky, and T. Leen, eds.), vol. 7, pp. 311–318, The MIT Press, 1995.
10. R. Bakker, J. C. Schouten, C. L. Giles, F. Takens, and C. M. van den Bleek, "Learning chaotic attractors by neural networks," *Neural Computation*, vol. 12, no. 10, 2000.
11. S. P. Day and M. R. Davenport, "Continuous-time temporal back-progagation with adaptive time delays," *IEEE Transactions on Neural Networks*, vol. 4, pp. 348–354, 1993.
12. L. Chudy and I. Farkas, "Prediction of chaotic time-series using dynamic cell structuresand local linear models," *Neural Network World*, vol. 8, no. 5, pp. 481–489, 1998.
13. R. Bone, M. Crucianu, G. Verley, and J.-P. Asselin de Beauville, "A bounded exploration approach to constructive algorithms for recurrent neural networks," in *Proceedings of IJCNN 2000*, (Como, Italy), 2000.

14. I. de Falco, A. Iazzetta, P. Natale, and E. Tarantino, "Evolutionary neural networks for nonlinear dynamics modeling," in *Parallel Problem Solving from Nature 98*, vol. 1498 of *Lectures Notes in Computer Science*, pp. 593–602, Springer, 1998".
15. X. Yao and Y. Liu, "A new evolutionary system for evolving artificial neural networks," *IEEE Transactions on Neural Networks*, vol. 8, pp. 694–713, May 1997.
16. J. Vesanto, "Using the SOM and local models in time-series prediction," in *Proceedings of WSOM'97, Workshop on Self-Organizing Maps, Espoo, Finland, June 4–6*, pp. 209–214, Espoo, Finland: Helsinki University of Technology, Neural Networks Research Centre, 1997.
17. T. M. Martinez, S. G. Berkovich, and K. J. Schulten, "Neural-gas network for vector quantization and its application to time-series prediction," *IEEE Transactions on Neural Networks*, vol. 4, pp. 558–569, July 1993.
18. H. Bersini, M. Birattari, and G. Bontempi, "Adaptive memory-based regression methods," in *In Proceedings of the 1998 IEEE International Joint Conference on Neural Networks*, pp. 2102–2106, 1998.
19. J. Platt, "A resource-allocating network for function interpolation," *Neural Computation*, vol. 3, pp. 213–225, 1991.
20. U. Huebner, N. B. Abraham, and C. O. Weiss, "Dimensions and entropies of chaotic intensity pulsations in a single-mode far-infrared nh3 laser," *Phys. Rev. A*, vol. 40, p. 6354, 1989.
21. T. Koskela, M. Varsta, J. Heikkonen, and K. Kaski, "Recurrent SOM with local linear models in time series prediction," in *6th European Symposium on Artificial Neural Networks. ESANN'98. Proceedings. D-Facto, Brussels, Belgium*, pp. 167–72, 1998.
22. E. A. Wan, "Time series prediction by using a connectionist network with internal time delays," in *Time Series Prediction: Forecasting the Future and Understanding the Past* (W. A. S. and G. N. A., eds.), pp. 195–217, Addison-Wesley, 1994.
23. J. Kohlmorgen and K.-R. Müller, "Data set a is a pattern matching problem," *Neural Processing Letters*, vol. 7, no. 1, pp. 43–47, 1998.
24. T. Sauer, "Time series prediction using delay coordinate embedding," in *Time Series Prediction: Forecasting the Future and Understanding the Past* (A. S. Weigend and N. A. Gershenfeld, eds.), Addison-Wesley, 1994.
25. A. S. Weigend and D. A. Nix, "Predictions with confidence intervals (local error bars)," in *Proceedings of the International Conference on Neural Information Processing (ICONIP'94)*, (Seoul, Korea), pp. 847–852, 1994.
26. J. McNames, "Local modeling optimization for time series prediction," in *In Proceedings of the 8th European Symposium on Artificial Neural Networks*, pp. 305–310, 2000.
27. B. H. Bontempi G., Birattari M., "Local learning for iterated time-series prediction," in *Machine Learning: Proceedings of the Sixteenth International Conference* (B. I. and D. S., eds.), (San Francisco, USA), pp. 32–38, Morgan Kaufmann, 1999.

Generalized Relevance LVQ for Time Series

Marc Strickert, Thorsten Bojer, and Barbara Hammer

University of Osnabrück, Department of Mathematics/Computer Science,
Albrechtstraße 28, D-49069 Osnabrück, Germany
marc@informatik.uni-osnabrueck.de

Abstract. An application of the recently proposed generalized relevance learning vector quantization (GRLVQ) to the analysis and modeling of time series data is presented. We use GRLVQ for two tasks: first, for obtaining a phase space embedding of a scalar time series, and second, for short term and long term data prediction. The proposed embedding method is tested with a signal from the well-known Lorenz system. Afterwards, it is applied to daily lysimeter observations of water runoff. A one-step prediction of the runoff dynamic is obtained from the classification of high dimensional subseries data vectors, from which a promising technique for long term forecasts is derived.[1]

1 Introduction

The learning vector quantization (LVQ) algorithm introduced by Kohonen has been successfully applied to various supervised classification tasks such as radar data extraction, or the mapping of EEG signals to body movement [9]. The general application of LVQ is the assignment of labels to high dimensional input vectors, which is technically done by representing the labels as regions of the data space, associated with adjustable prototype vectors. Several versions of the prototype update rule are known as LVQ1, LVQ2, LVQ3, or OLVQ, which differ in robustness and the rates of convergence. We will focus on further aspects of vector quantization: the relevance weighting of the components of the input data applied to the analysis of scalar time series.

The task of time series prediction occurs in a wide range of applications and various methods have been proposed for adequate processing: standard feedforward and recurrent networks, ARIMA modeling, or unsupervised algorithms, to name just a few [3, 6, 7, 9]. Modeling temporal structure is essentially connected with the question about the contribution of the past to the current system state. Common measures of temporal dependencies are the autocorrelation function or the mutual information, which both compare the sequence of observations to a time-shifted version; these methods, operating on data pairs, may therefore be inadequate for describing higher order temporal correlations. Speaking in terms of LVQ, a mapping of real-valued past observations to the index of the target partition bin of the current state, the class label, has to be trained; information about the influence of a system's history to its future can be gained from an extended LVQ algorithm that additionally provides component weighting.

We will use generalized relevance LVQ (GRLVQ) for this purpose, a new algorithm that has been suggested in [5], and which has successfully been used for the determination

[1] We gratefully acknowledge the MWK, Lower Saxony, for financial support, and Michael Brandenburg, Dept. of Ecology, Münster, for supplying the runoff data.

G. Dorffner, H. Bischof, and K. Hornik (Eds.): ICANN 2001, LNCS 2130, pp. 677–683, 2001.

of relevant sensors for satellite data classification. Here, attention is paid to those time steps in the past with much or disappearing significance concerning the next system state. Thereby, a nonlinear measure of temporal correlations can be established. Also, an embedding of the scalar time series into a canonic phase space can be obtained after pruning irrelevant dimensions from the time window on the past. We will demonstrate the merits of GRLVQ and its applications on real-life data.

2 Experimental Environment

We start our analysis with a time series obtained from the well-known Lorenz System, which is a differential equation based description of atmospheric turbulences [10]. From a chaotic trajectory in the according three dimensional phase space, we pick out the dynamic of the z-component which refers to the derivation of the vertical temperature profile from its equilibrium; this sequence serves as a test series, here. Our investigation concentrates on extracting temporal correlations, aiming at finding an embedding alternative to the commonly used delay embedding technique. By means of recurrence analysis [2], we study the signal of water runoff observed for a large grass lysimeter at St. Arnold, Germany. The series reflects observations of rainfall runoff through an isolated soil cube of 3.5m depth with a grass covered surface of 20m x 20m. Recordings have been taken daily with an accuracy of 0.01mm/d, and a subset of 24 years from 11/01/1968 to 10/31/1992 without missing data has been used for analysis.

3 The GRLVQ Algorithm

Generalized relevance LVQ is an extension of the standard LVQ algorithm which introduces weighting [5]. Assume that a finite training set is given by $X = \{(x^i, y^i) \subset \mathbb{R}^n \times \{1, \ldots, C\} \,|\, i = 1, \ldots, m\}$, which describes the mapping of the vectors x^i to their class y^i. Each class y^i is associated with a number of prototype vectors, the codebooks $c^l \subset \mathbb{R}^n, l \in \{1, \ldots, C\}$ to class $l = y^i$. Hence, codebooks belonging to the same class define regions in the data space by the union of their receptive fields. A data vector x^j will therefore fall into the category $l = y^j$ if x^j is closer to one of the c^l rather than to any other from the set $c^m, m \neq l$. Training adapts both the closest correct prototype W_1 and the closest wrong prototype W_2 according to:

$$W_1 := W_1 + \epsilon f' \cdot \frac{d_2}{(d_1 + d_2)^2}(x - W_1),\; W_2 := W_2 - \epsilon f' \cdot \frac{d_1}{(d_1 + d_2)^2}(x - W_2)\,. \quad (1)$$

The learning rate ϵ is taken from the interval $(0, 1)$. The squared *weighted* Euclidean distances $d_1 = d^2(x, W_1, \lambda)$ and $d_2 = d^2(x, W_2, \lambda)$ are calculated between the input pattern and the prototypes; $f' = \text{sgd}'((d_1 - d_2)/(d_1 + d_2))$ denotes the derivative of the sigmoid. Distance weighting refers to an extension of the Euclidean metric, generally given by

$$d^2(x, y, \lambda) := \sum_{i=1}^{n} \lambda_i (x_i - y_i)^2\,, \quad \lambda_i \geq 0\,. \quad (2)$$

For $\lambda_i \equiv 1, i = 1 \ldots n$ the update rule (1) is an instance of the generalized LVQ suggested in [11]. The authors apply a stochastic gradient descent in order to minimize the misclassification costs induced by $\sum_x f((d_1 - d_2)/(d_1 + d_2))$ with respect to the codebooks. A canonic extension to the constant choice of the λ_i is weight adaptation based on a gradient descent of the metric (2), now with respect to the λ_i:

$$\lambda_l := \lambda_l - \epsilon_1 f' \cdot \left(\frac{d_2}{(d_1 + d_2)^2} (x_l^i - W_l^1)^2 - \frac{d_1}{(d_1 + d_2)^2} (x_l^i - W_l^2)^2 \right) . \quad (3)$$

The weight adaptation rate is $\epsilon_1 \in (0, 1)$. Numerical instabilities of the algorithm can be avoided by normalizing the weights to $\lambda_l := \lambda_l / |\lambda|$ after the update procedure. Training is done until the classification error has reached a desired level, since the overall convergence cannot be granted. Especially with noisy data, fluctuations occur due to the fact that GRLVQ degenerates to a special case of perceptron learning which may theoretically lead to cyclic states. In practice, though, only well behaved classification results have been found. An important addition to the above GRLVQ training is the implementation of a pruning scheme: set the smallest weight $\lambda_{\min} \neq 0$ from a sufficiently trained classification task to constant zero, which leads to masking the related input component, retrain the codebooks, and repeat both steps pruning and retraining until a major degradation in the classification error curve is displayed.

4 Applications of GRLVQ

4.1 Data from the Lorenz System

Embedding by Relevance: For nonlinear time series analysis, the usual assumption is that a scalar sequence of observations only reflects one of several differentiable components of an unknown high dimensional attractor. Referring to the embedding theorem of Whitney [13], it is known that we can use an observed series for creating a manifold that is similar to the unknown attractor: the obtained reconstruction will exhibit the same topological properties and it will also maintain the structure of temporal changes, i.e. periodicities and transient states. For real-life data, the delay embedding technique suggested by Takens [12] constitutes an approach for reconstruction: equidistant observations $\mathbf{x} = (x_i)_{i=1\ldots N}$ are grouped into vectors of dimension m according to $\Phi_j(\mathbf{x}) = \{x_j, x_{j+\tau}, x_{j+2\tau}, \ldots, x_{j+(m-1)\tau}\}$, $j = 1 \ldots N - (m-1)\tau$. This construction of delay coordinates depends on two parameters, the dimension m and the time delay τ, which have to be chosen properly in order to achieve a good attractor reconstruction. The choice of the lag τ is motivated by the aim to maximize the independence of the delay vectors' components, but still keeping the value of τ small. Some heuristics to determine τ are: the value where the autocorrelation function decays to $1/e$, where it decays to zero, where it falls into the 95% intervall of confidence, and where the lag of the mutual information takes its first minimum [4]. Usually, different τ are suggested, but in principle, none of them can avoid the side effect of aliasing, because τ, set to a fixed value, may still interfere with periodicities of the observations.

GRLVQ provides a data-oriented alternative: the appropriately discretized components of the embedding vectors are taken according to their relevance for explaining the next observation. Thus, we can use the λ_i from GRLVQ, which, after training, reflect the global significance of observations in the past for a one-step prediction. Actually, we get a canonic decorrelation length where the most recent λ_i exhibit their first minimum.

For the Lorenz system, top panel of Fig. 1, the resulting λ_i after GRLVQ training of a 25 time steps history to a 17 bins data partition is shown in the lower left plot of

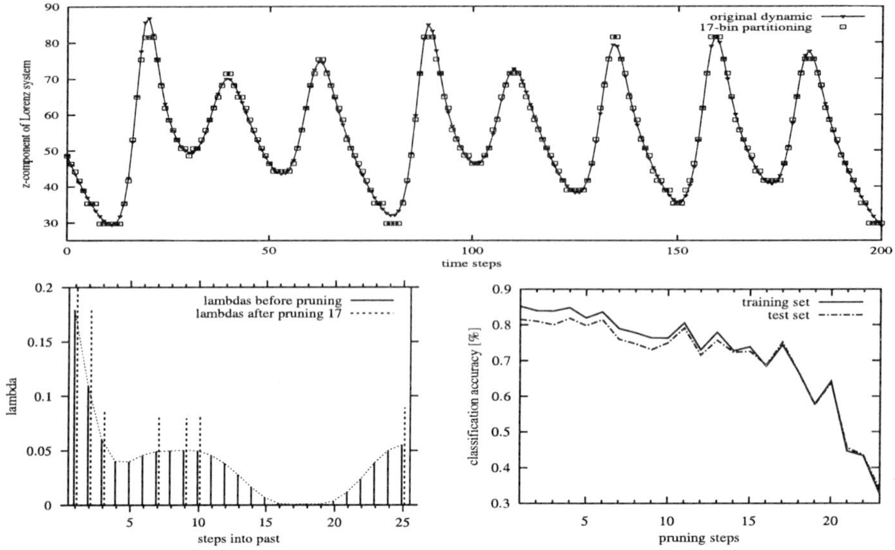

Fig. 1. Top: Time series of Lorenz z-component and partitioning. Lower left: Relevance of the ith step from the past. Lower right: Classification accuracy for successive pruning.

Fig. 1. Connected line segments indicate a relevance profile for the contribution of the past in the case without explicit pruning the λ_i. An expected observation is the very high influence of latest system states on the current process realization, emphasizing the strong temporal correlations of successive observations. The minimum at step 5 is associated with a characteristic decorrelation time that is in accordance with a decay of the autocorrelation function to zero at lag 6 and a minimum of the mutual information at lag 5, both calculated separately. Beyond the absolute minimum at about 18, the significance increases again, because of the periodic dynamic that appears in the time series plot above. As seen in the lower right panel, we still obtain an accuracy of 75% at pruning step 17 for both training set and validation set. Taking away more λ_i induces a rapid decay of accuracy; the remaining seven most prominent components, shown in the left plot, confirm those being already most significant before pruning.

4.2 Runoff Data

Embedding by Relevance: For the runoff data we follow the above steps: performing a k-median cut such that the number of data in $k+1$ bins or classes is equally distributed, choosing a window length of l days, training, and finally performing an s-step pruning, s being the critical location in the accuracy graph. The number of partitions has been set to $k+1=15$, a value large enough to catch minor changes in the runoff process, but small enough yield a statistically significant support of $(24-1)*366.25/15 \approx 561$ samples per class (23 years not 24, because we consider one year of history). With the respective results from Fig. 2 we detect the most important points of time within an average year past. For all three levels of pruning: no pruning, and after 311 and 354 steps, the recent past of 10 days is similarly rated. The last day is very important, then

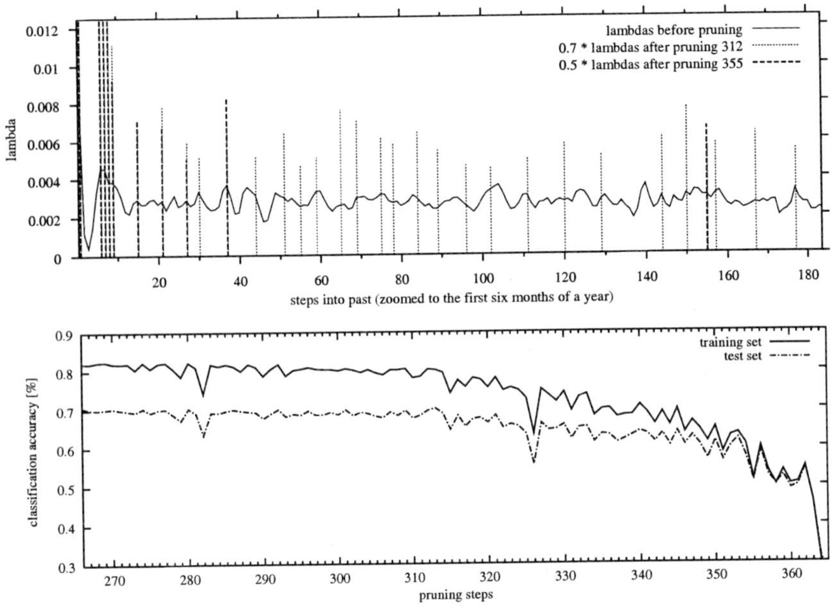

Fig. 2. Top: Relevance of the ith step from the past. Bottom: Accuracy for successive pruning.

four days of minor significance, and days 6 to 9 in the past are again of high influence on the presence. The poor structure of the none-pruned case for the further past is turned into a richer profile by pruning at change points in the accuracy graph. Pruning 311 components, hence maintaining 54, provides a high percentage of correct classification, and a high density of the remaining λ_i is located in the past of about 100 days. For comparison, additional computations showed a decay of the autocorrelation function to zero at a lag of 96, while the first minimum of the mutual information was found at about 70. Interestingly, ranges of high values for the λ_i led to merging and focusing on a single $\lambda_{\hat{i}}$ under pruning stress, thus, reducing the problem that potential initial noise of the λ_i is amplified during pruning. Leaving only 11 components puts much emphasis on the recent past, but also a 155-day long term component remains.

Recurrence analysis with vectors of dimension 366 but weighted by the remaining 11 λ_i and therefore projected to dimension 11 is contrasted to a delay embedding with $m = 11$, $\tau = 366/(m-1) \approx 37$ in Fig. 3; thus, both types of embedding vectors represent a year. The choice of the dimension m for delay embeddings is not critical as long as it is large enough; previous studies with the method of the false nearest neighbors detection [8] led to a dimension requirement of at least $m = 8$. Setting the lag to $\tau = 37$ is only about half of the usually taken value from the autocorrelation decay in order to maintain temporal compatibility, yielding the same time spanned by the embedding vectors. The plots in Fig. 3 show a 12-year period, with pixels at time indices i, j where the Euclidean distance of the embedding vector pair $(\Phi_i(\mathbf{x}), \Phi_j(\mathbf{x}))$ is smaller than a critical threshold r_C. This distance has been set such that 20% of all vector pairs fulfill the recurrence condition, i.e. that they are close enough in the embedding space. Both recurrence plots look quite different: the delay embedding displays a rather determin-

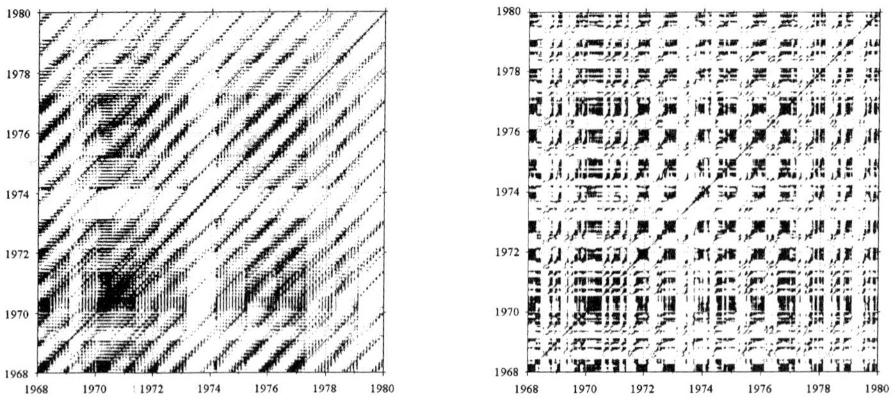

Fig. 3. Left: Recurrence plot with delay embedding. Right: Relevance embedding.

istic pattern, dominated by diagonal lines, whereas the relevance embedding exhibits a checkerboard structure. Since line segments parallel to the main diagonal in a distance of d indicate a d-periodic dynamic, statistics of their lengths and their frequencies are the typical task of recurrence analysis. As said before, aliasing effects cannot be avoided with *delay* embedding, and also deterministic temporal interlocking of the delay vectors is inevitable. This leads to an attractor reconstruction with embedding artefacts, which is reflected by blurred diagonal lines in the left recurrence plot; setting $\tau = 70$ does slightly reduce these artefacts on the annual scale, but induces new false recurrences on larger scales. The right plot shows more details about presence and absence of periodic dynamic (squares with long corners in the lower left) and transient states (bands of white). The attractor reconstruction reveals more structure, and it is therefore promising to review techniques that base on delay embeddings, such as nonlinear filtering or forecasting, with the proposed relevance embedding.

Long Term Prediction: The temporal structure reflected in the weight vector λ after pruning, can be exploited in another promising way: since generalization in the pruning phase is displayed in the classification accuracy plot of Fig. 2, we expect an appropriate representation of the main dynamic with respect to the trained one-step prediction. We use the well-trained codebooks and the pruned weight vector to the difficult task of long term prediction [1] by iterating the one-step classification: starting with an initial real-data vector, we obtain a predicted bin of which the associated median value is reinserted as a most recent system state, after the data vector has been shifted one step back into the past. In case of the runoff data, the repeated application of the classification-reinsertion rule, a simulated dynamic shown in Fig. 4 has been obtained. The initial data vector represents one year past of the day 7150 beyond which the simulation follows quite well the decay phase of the runoff intensity, and, with a delay and with smaller amplitude, the increase is modeled as well. Damping effects arise from the successive reinsertions of median values, which leads to the replacement of the originally large range of observations by the smaller range of medians. Additional simulations have shown, though, that the simulations produce reasonable annual cycles, if stochastic noise is added to the medians in order to better reflect the data distribution.

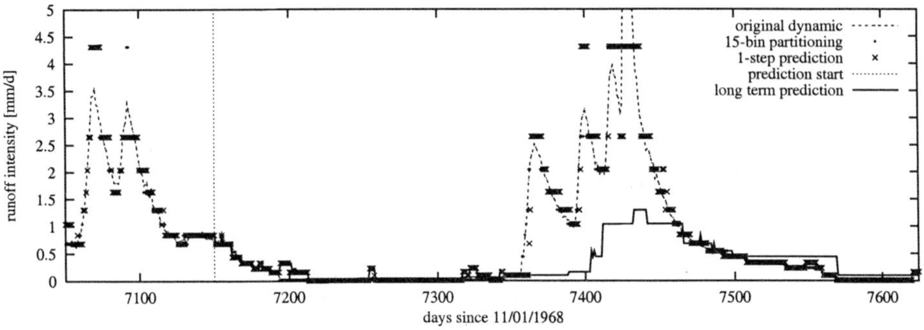

Fig. 4. Runoff dynamic, data partitioning, one-step and long term prediction.

5 Conclusions

The GRLVQ algorithm yields a weighting of input data, which for time series modeling has been used for two applications, the embedding of time series, and the long term prediction. Many traditional methods crucially depend on the autocorrelation function, as in the case of the lag determination for the delay embeddings, or the long term modeling by autoregressive processes, which quantifies linear correlations only. In contrast, the weight vector from GRLVQ reflects the temporal structure in a data-oriented way in terms of the contribution of the past to a one-step prediction. Automatic relevance determination leads to promising results for long-term prediction as well.

References

1. G. Bontempi, M. Birattari: A Multi-Step-Ahead Prediction Method Based on Local Dynamic Properties, in M. Verleysen ed., ESANN'2000, pp. 311-316, D-Facto, 2000.
2. M.C. Casdagli: Recurrence Plots Revisited, Physica D 108, pp. 12-44, 1997.
3. G. Dorffner: Neural Networks for Time Series Processing, Neural Network World, 6(4), pp. 447-468, 1996.
4. A.M. Fraser, H.L. Swinney: Independent Coordinates for Strange Attractors from Mutual Information, Physical Review A 33 (1986), pp. 1134-1140, 1986.
5. B. Hammer, T. Villmann: Estimating relevant input dimensions for self-organizing algorithms, accepted at WSOM 2001.
6. S. Hochreiter, J. Schmidhuber: Long short-term memory, Neural Computation, 9(8), pp. 1735-1780, 1997.
7. H. Kantz, T. Schreiber: Nonlinear time series analysis, Cambridge University Press, 1997.
8. M. Kennel, R. Brown, H. Abarbanel: Determining embedding dimension for phase-space reconstruction using a geometrical construction, Phys. Rev. A 45, 3403-3411, 1992.
9. T. Kohonen, Self-Organizing Maps, Springer, 1997.
10. E.N. Lorenz: Deterministic Nonperiodic Flow, Journ. of Atmospheric Sc. 20, p. 130f., 1963.
11. A. S. Sato, K. Yamada, Generalized learning vector quantization, in G. Tesauro, D. Touretzky, and T. Leen, eds., Advances in Neural Information Processing Systems, Vol. 7, pp. 423-429, MIT Press, 1995.
12. F. Takens: Detecting Strange Attractors in Turbulence, Lecture Notes in Mathematics Vol. 898, Springer: New York, 1981.
13. H. Whitney: Differentiable Manifolds. Annals of Mathematics 37, p. 645, 1936.

Unsupervised Learning in LSTM Recurrent Neural Networks

Magdalena Klapper-Rybicka[1], Nicol N. Schraudolph[2], and Jürgen Schmidhuber[3]

[1] Institute of Computer Science, University of Mining and Metallurgy, al. Mickiewicza 30, 30-059 Kraków, Poland
mklapper@uci.agh.edu.pl
[2] Institute of Computational Sciences, Eidgenössische Technische Hochschule (ETH), CH-8092 Zürich, Switzerland
nic@inf.ethz.ch
[3] Istituto Dalle Molle di Studi sull'Intelligenza Artificiale (IDSIA), Galleria 2, CH-6928 Manno, Switzerland
juergen@idsia.ch

Abstract. While much work has been done on unsupervised learning in feedforward neural network architectures, its potential with (theoretically more powerful) recurrent networks and time-varying inputs has rarely been explored. Here we train Long Short-Term Memory (LSTM) recurrent networks to maximize two information-theoretic objectives for unsupervised learning: Binary Information Gain Optimization (BINGO) and Nonparametric Entropy Optimization (NEO). LSTM learns to discriminate different types of temporal sequences and group them according to a variety of features.

1 Introduction

Unsupervised detection of input regularities is a major topic of research on feedforward neural networks (FFNs). Most of these methods derive from information-theoretic objectives, such as maximizing the amount of preserved information about the input data at the network's output. Typical real-world inputs, however, are not static but sequential, full of temporally extended features, statistical regularities, and redundancies. FFNs therefore necessarily ignore a large potential for compactly encoding data. This motivates our research on unsupervised feature detection with recurrent neural networks (RNNs), whose computational abilities in theory exceed those of FFNs by far.

Hitherto little work has been done, however, on this topic [1, 2]. One of the reasons for the conventional focus on FFNs may be the relative maturity of this architecture, and the algorithms used to train it. Compared to FFNs, traditional RNNs [3–5] are notoriously difficult to train, especially when the interval between relevant events in the input sequence exceeds about 10 time steps [6–8]. Recent progress in RNN research, however, has overcome some of these problems [8, 9], and may pave the way for a fresh look at unsupervised sequence learning.

G. Dorffner, H. Bischof, and K. Hornik (Eds.): ICANN 2001, LNCS 2130, pp. 684–691, 2001.

Here, for the first time, we will plug certain information-theoretic objectives into a recent RNN architecture called Long Short-Term Memory (LSTM), which dramatically outperforms other RNNs on a wide variety of *supervised* sequence learning tasks. We will begin with a brief description of LSTM, then describe two unsupervised learning methods: Binary Information Gain Optimization [10, 11] and Nonparametric Entropy Optimization [11, 12]. We will then present examples of LSTM performance with each unsupervised objective function applied to both artificial and real-world data sequences.

2 Unsupervised Training of LSTM

Long Short-Term Memory (LSTM) is a novel efficient type of recurrent neural network architecture [8] whose advantages over other RNN models have been demonstrated in various areas: learning regular and context-free languages [13], predicting continual time series [9], motor control and rhythm detection [14].

While previous work trained LSTM networks on explicitly defined targets, here we will show that LSTM can also handle unsupervised problems. We used two unsupervised learning algorithms, Binary Information Gain Optimization (BINGO) and Nonparametric Entropy Optimization (NEO), to train the LSTM network to discriminate between sets of temporal sequences, and cluster them into groups. The network is expected to autonomously detect some relevant aspect of the input sequences, according to the particular learning rule used. The learning algorithm strongly determines which aspects of the input sequences the network considers to be important; this can be formalized as a maximization of *relevant* information about the input data at the network's output [11].

2.1 LSTM Architecture

The basic unit of an LSTM network is the *memory block* containing one or more *memory cells* and three adaptive, multiplicative gating units shared by all cells in the block (Fig. 1). Each memory cell has at its core a recurrently self-connected linear unit (CEC) whose activation is called the cell *state*. The CECs enforce *constant* error flow and overcome a fundamental problem plaguing previous RNNs: they prevent error signals from decaying quickly as they propagate "back in time". The adaptive gates control input and output to the cells and learn to reset the cell's state once its contents are out of date. Peephole connections connect the CEC to the gates. All errors are cut off once they leak out of a memory cell or gate, although they do serve to change the incoming weights. The effect is that the CECs are the only part of the system through which errors can flow back forever, while the surrounding machinery learns the nonlinear aspects of sequence processing. This permits LSTM to bridge huge time lags (1000 steps and more) between relevant events, while traditional RNNs already fail to learn in the presence of 10-step time lags, despite requiring more complex update algorithms. See [14] for a detailed description.

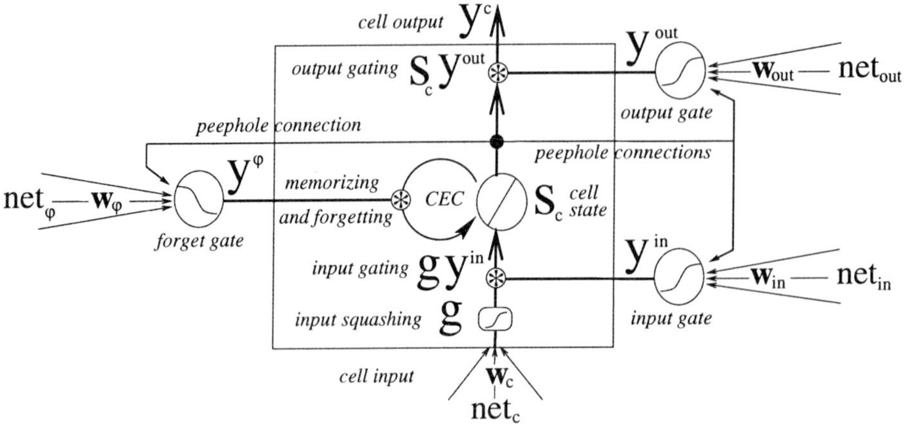

Fig. 1. LSTM memory block with one cell. From [14].

2.2 Binary Information Gain Optimization

The BINGO algorithm clusters unlabeled data with linear adaptive discriminants stressing the gaps between clusters [10, 11]. The method maximizes the information gained from observing the output of a single-layer network of logistic nodes, interpreting their activity as stochastic binary variables. The resulting weight update formula (see [10, 11] for complete derivation) for node i is given by

$$\Delta \boldsymbol{w}_i \propto f'(y_i)\, \boldsymbol{x}\, (y_i - \hat{y}_i)\,, \tag{1}$$

where $f'(y_i)$ is the derivative of the logistic squashing function f, $\boldsymbol{x}$ the presynaptic input, and $(y_i - \hat{y}_i)$ the difference between actual and estimated network output. We used a linear second-order estimator for multiple outputs:

$$\hat{\boldsymbol{y}} = \boldsymbol{y} + (Q_{\boldsymbol{y}} - 2I)(\boldsymbol{y} - \bar{\boldsymbol{y}})\,, \tag{2}$$

where $\bar{\boldsymbol{y}}$ denotes the average of $\boldsymbol{y}$ over the data, and $Q_{\boldsymbol{y}}$ its autocorrelation. The binary information gain is maximal (namely, 1 bit/node) when the network's outputs are uncorrelated, and approach '1' (resp. '0') for half the data points. Thus BINGO seeks to identify independent dichotomies in the data.

2.3 Nonparametric Entropy Optimization

NEO, in contrast to parametric unsupervised learning techniques, is a differential learning rule that optimizes signal entropy by way of kernel density estimation, so no additional assumption about a density's smoothness or functional form is necessary [11, 12]. The nonparametric *Parzen window* density estimate is

$$\hat{p}(\boldsymbol{y}) = \frac{1}{|T|} \sum_{\boldsymbol{y}_j \in T} K_\sigma(\boldsymbol{y} - \boldsymbol{y}_j)\,, \tag{3}$$

where T is a sample of points $\boldsymbol{y}_j$ and K_σ is a kernel function, in our case an isotropic Gaussian with variance σ^2. The kernel width σ that best regularizes the density estimate can be found by maximizing the log-likelihood

$$\hat{L} = \sum_{\boldsymbol{y}_i \in S} \log \sum_{\boldsymbol{y}_j \in T} K_\sigma(\boldsymbol{y}_i - \boldsymbol{y}_j) - |S| \log |T| , \tag{4}$$

whose derivative with respect to the kernel width is given by

$$\frac{\partial}{\partial \sigma} \hat{L} = \sum_{\boldsymbol{y}_i \in S} \frac{\sum_{\boldsymbol{y}_j \in T} \frac{\partial}{\partial \sigma} K_\sigma(\boldsymbol{y}_i - \boldsymbol{y}_j)}{\sum_{\boldsymbol{y}_j \in T} K_\sigma(\boldsymbol{y}_i - \boldsymbol{y}_j)} . \tag{5}$$

The maximum likelihood kernel makes a second sample S derived from $p(\boldsymbol{y})$ most likely under the estimated density $\hat{p}(\boldsymbol{y})$ computed from the sample T.[1] A nonparametric estimate of the empirical entropy (given optimal kernel shape) of a neural network's output $\boldsymbol{y}$ can then be calculated as:

$$\hat{H}(\boldsymbol{y}) = -\frac{1}{|S|} \sum_{\boldsymbol{y}_i \in S} \log \sum_{\boldsymbol{y}_j \in T} K_\sigma(\boldsymbol{y}_i - \boldsymbol{y}_j) + \log |T| , \tag{6}$$

and minimized by gradient descent in the neural network's weights $\boldsymbol{w}$:

$$\frac{\partial}{\partial \boldsymbol{w}} \hat{H}(\boldsymbol{y}) = -\frac{1}{|S|} \sum_{\boldsymbol{y}_i \in S} \frac{\sum_{\boldsymbol{y}_j \in T} \frac{\partial}{\partial \boldsymbol{w}} K_\sigma(\boldsymbol{y}_i - \boldsymbol{y}_j)}{\sum_{\boldsymbol{y}_j \in T} K_\sigma(\boldsymbol{y}_i - \boldsymbol{y}_j)} . \tag{7}$$

Low $\hat{H}(\boldsymbol{y})$ is achieved by clustering the data. See [11, 12] for further details.

3 Experimental Setup

The LSTM networks were applied to unsupervised discrimination of groups of temporal sequences. Two types of data were used: artificial (random sequences) and real (fragments of clarinet sounds). For each data type two experiments were run, using the BINGO and NEO objective functions, respectively.

3.1 Training Sets

Artificial data sequences were generated by independently drawing random values from a Gaussian distribution with mean and variance as shown in Table 1. For each of four distributions a group of five 100-pattern long sequences was generated, producing a training set of 20 sequences. The mean and variance values were chosen such a way that the ranges of all four groups of sequences overlapped one another. The means of two groups (symbols $\circ$ and $+$ in Table 1) were set to the same value in order to force the network to distinguish between them based only on variance information.

In the real data case we used sequences constructed of 100-sample fragments of clarinet sounds at four different pitches, sampled at 44.1kHz (Table 2). As

[1] To make efficient use of avaliable training data, we let $\boldsymbol{y}_j$ in the inner sums of equations (5) and (7) range over $T = S \setminus \{\boldsymbol{y}_i\}$ for each $\boldsymbol{y}_i \in S$ in the outer sum [11].

Sym.	*Mean*	*Var.*	*Output range*
○	0.4	0.3	12.93 – 13.68
+	0.4	0.1	11.43 – 12.49
∗	0.2	0.2	7.50 – 7.69
△	0.7	0.1	29.16 – 30.18

Table 1. The artificial data (from left to right): symbol used in diagrams, mean and variance of the random sequences, and range of outputs produced by the NEO network.

Sym.	*Pitch*		*# of*	*Output range*
	Note	*[Hz]*	*cycles*	*(×10^6)*
○	A3	440	1	-3.57 – -3.54
+	A4	880	2	-3.40 – -3.35
∗	E5	1320	3	-3.47 – -3.43
△	F♯5	1480	3.4	-3.66 – -3.62

Table 2. The real data (from left to right): symbol used in diagrams, base frequency of the clarinet sounds in musical notation and in Hertz, number of cycles in a sequence, and range of outputs produced by the NEO network.

with the artificial data, the fragments of five different sounds were selected for each of four pitches, for a training set of 20 sequences. In this case the network should discover an entirely different feature of the training set: groups of sounds are no longer distinguished by mean and variance, but by the number of cycles in the sequence, corresponding to the sound's pitch. Two groups (symbols $*$ and $\triangle$ in Table 2) have very similar pitch and are thus hard to distinguish from the small number of samples we supplied.

3.2 Network Architecture and Training

The LSTM networks trained with the BINGO algorithm consisted of two single-cell memory blocks and two output units with sigmoidal activation function. For each sequence the values of the two outputs at the end of the sequence can be interpreted as a point in the unit square. The networks were expected to group these points (corresponding to the groups of sequences), and spread them towards the four corners of the square, equivalent to assigning a unique two-bit binary code to each group.

For the NEO algorithm LSTM networks with just one memory cell appeared sufficient to learn both tasks. The single linear output unit produced just one real value for each sequence and the networks were supposed to cluster the output at the end of each sequence, according to the group it belonged to.

After training with the artificial data the network was expected to discriminate between groups of the sequences by detecting their mean and (especially in case of groups with equal mean) variance. For the clarinet sounds, where mean and variance were no longer informative, the networks had to discover the quite different feature of varying pitch (number of cycles) between groups of sequences.

In both experiments the most recent LSTM network model with forget gates and peephole-connections was used (Fig. 1) [8, 9, 14]. The network was trained until the error dropped below an empirically chosen threshold, or until further training did not noticeably improve the result. In order to make variance information accessible to the network, a second input was added, whose value at any given time was set to the square of the first (main) input [15].

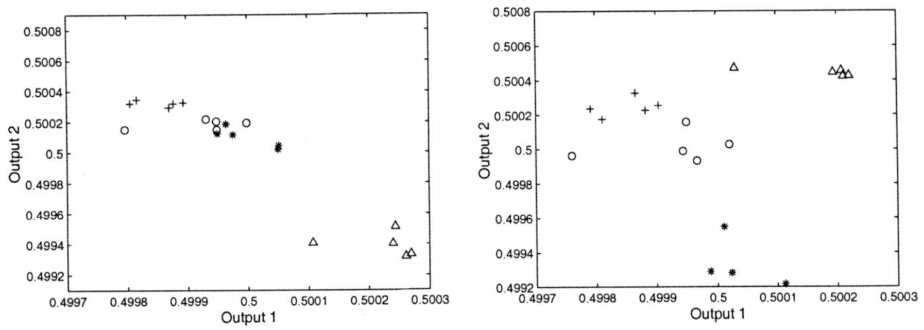

Fig. 2. Network output for artificial data and BINGO objective function. Left: beginning of training, right: after 860 epochs.

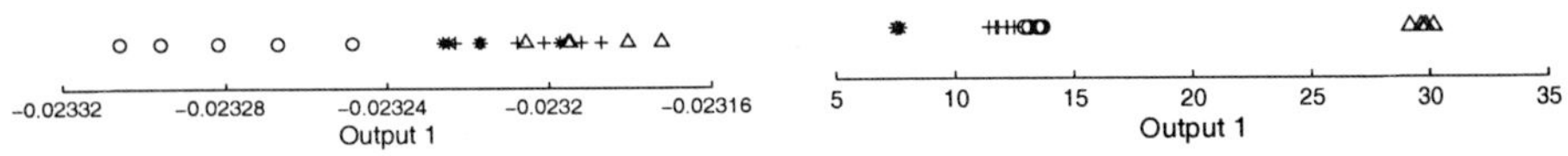

Fig. 3. Network output for artificial data and NEO objective function. Left: beginning of training, right: after 14 epochs.

4 Results

For both real and artificial data, with application of either of the two previously described unsupervised training algorithms, the LSTM networks were able to distinguish between four groups of sequences. Figures 2, 3, 4, and 5 present the results of BINGO and NEO training with artificial data and real clarinet sounds, respectively. For each experiment two diagrams are presented, showing the network output before and after training.

4.1 Training with the BINGO Algorithm

Artificial Data (Fig. 2). From the very beginning of training, points corresponding to the group marked △ are well separated. The other groups overlap each other, and their correlation coefficient is high. After 860 training epochs all four groups are clearly separated, and spread out across the unit square. The two groups with identical mean remain close to each other, with one occupying a more central location.

Clarinet Sounds (Fig. 4). At the beginning of training all points are distributed along a line, with a correlation coefficient near one, and do not form distinct clusters. After 690 epochs, however, all four groups of points are well separated.

4.2 Training with the NEO Algorithm

Artificial Data (Fig. 3). Before training the outputs for the group marked ∘ are separated, though spread over a comparatively large range. The other three

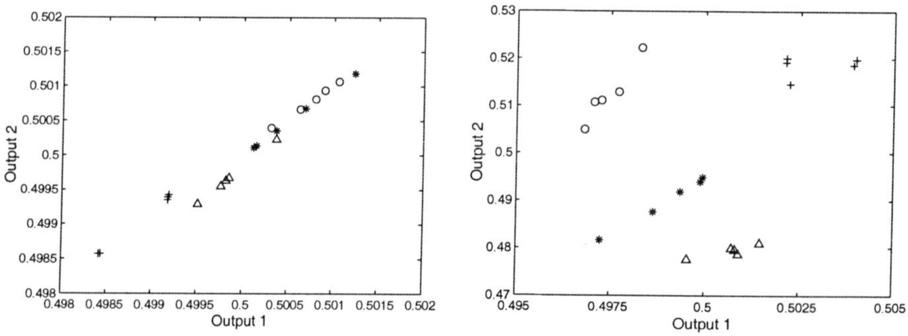

Fig. 4. Network output for clarinet sounds and BINGO objective function. Left: beginning of training, right: after 690 epochs.

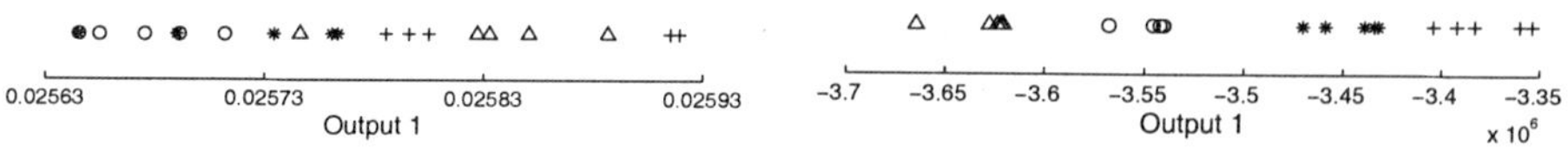

Fig. 5. Network output for clarinet sounds and NEO objective function. Left: beginning of training, right: after 6910 epochs.

groups overlap each other. After just 14 epochs of training all groups are well clustered, with very small range compared to the gaps between them (Table 1). Again the two groups with identical mean remain close to each other.

Clarinet Sounds (Fig. 5). The initial distributions of outputs for all groups overlap heavily. After training, however, the network's outputs form four clearly separated groups, though they are not as dense as in case of the artificial data (Table 1). The result was obtained only after almost 7000 epochs, suggesting that in this experiment the correct discriminants were harder to find than for the artificial data set.

5 Conclusion

Previous work presented many successful applications of Long Short-Term Memory (LSTM) networks to various supervised tasks, where LSTM definitely outperforms other recurrent neural network architectures. In this paper we showed that in conjunction with appropriate objective functions, LSTM can also solve unsupervised problems. We combined LSTM with two unsupervised learning methods, Binary Information Gain Optimization and Nonparametric Entropy Optimization, and showed that the resulting system performs unsupervised discrimination and classification of temporal sequences. Our experiments with both artificial and real data showed a remarkable ability of unsupervised LSTM to cluster temporal sequences according to a variety of reasonable features.

Combining the advantages of the LSTM recurrent network model with various objective functions for unsupervised learning should result in promising techniques for numerous real-world tasks involving unsupervised detection of input sequence features spanning extended time periods.

Acknowledgement

This research was supported by the Stefan Batory Foundation, and by the Swiss National Science Foundation under grants 2100-49'144.96 and 2000-052'678.97/1.

References

1. S. Lindstädt, "Comparison of unsupervised neural networks for redundancy reduction," Master's thesis, University of Colorado at Boulder, 1993.
2. P. Tiňo, M. Stancik, and L. Beňušková, "Building predictive models on complex symbolic sequences with a second-order recurrent BCM network with lateral inhibition," in *Proc. Int. Joint Conf. Neural Networks*, vol. 2, pp. 265–270, 2000.
3. A. J. Robinson and F. Fallside, "The utility driven dynamic error propagation network," Tech. Rep. CUED/F-INFENG/TR.1, Cambridge University Engineering Department, 1987.
4. P. J. Werbos, "Generalization of backpropagation with application to a recurrent gas market model," *Neural Networks*, vol. 1, 1988.
5. R. J. Williams and D. Zipser, "A learning algorithm for continually running fully recurrent networks," Tech. Rep. ICS 8805, Univ. of California, San Diego, 1988.
6. S. Hochreiter, " Untersuchungen zu dynamischen neuronalen Netzen. Diploma thesis, Institut für Informatik, Lehrstuhl Prof. Brauer, Technische Universität München," 1991. See www7.informatik.tu-muenchen.de/~hochreit.
7. Y. Bengio, P. Simard, and P. Frasconi, "Learning long-term dependencies with gradient descent is difficult," *IEEE Transactions on Neural Networks*, vol. 5, no. 2, pp. 157–166, 1994.
8. S. Hochreiter and J. Schmidhuber, "Long short-term memory," *Neural Computation*, vol. 9, no. 8, pp. 1735–1780, 1997.
9. F. A. Gers, J. Schmidhuber, and F. Cummins, "Learning to forget: Continual prediction with LSTM," *Neural Computation*, vol. 12, no. 10, pp. 2451–2471, 2000.
10. N. N. Schraudolph and T. J. Sejnowski, "Unsupervised discrimination of clustered data via optimization of binary information gain," in *Advances in Neural Information Processing Systems* (S. J. Hanson, J. D. Cowan, and C. L. Giles, eds.), vol. 5, pp. 499–506, Morgan Kaufmann, San Mateo, CA, 1993.
11. N. N. Schraudolph, *Optimization of Entropy with Neural Networks.* PhD thesis, University of California, San Diego, 1995.
12. P. A. Viola, N. N. Schraudolph, and T. J. Sejnowski, "Empirical entropy manipulation for real-world problems," in *Advances in Neural Information Processing Systems* (D. S. Touretzky, M. C. Mozer, and M. E. Hasselmo, eds.), vol. 8, pp. 851–857, The MIT Press, Cambridge, MA, 1996.
13. F. A. Gers and J. Schmidhuber, "Long Short-Term Memory learns simple context free and context sensitive languages," *IEEE Transactions on Neural Networks*, 2001 (forthcoming).
14. F. A. Gers and J. Schmidhuber, "Recurrent nets that time and count," in *Proc. IJCNN'2000, Int. Joint Conf. on Neural Networks*, IEEE Computer Society, 2000.
15. G. W. Flake, "Square unit augmented, radially extended, multilayer perceptrons," in *Neural Networks: Tricks of the Trade* (G. B. Orr & K.-R. Müller, eds.), vol. 1524 of *Lecture Notes in Computer Science*, pp. 145–163, Berlin: Springer Verlag, 1998.

Applying Kernel Based Subspace Classification to a Non-intrusive Monitoring for Household Electric Appliances

Hiroshi Murata and Takashi Onoda

Communication & Information Research Lab. CRIEPI,
2-11-1 Iwado kita, Komae-shi, Tokyo 201-8511, Japan
{murata,onoda}@criepi.denken.or.jp

Abstract. A non-intrusive load monitoring system that estimates the behavior of individual electrical appliances from the measurement of the total household load demand curve is useful for the forecast of electric energy demand and better customer services. Furthermore, this system will become important for power companies to control peak electric energy demand in the near future. We have already reported the system using Support Vector Machines (SVM) and SVM could establish sufficient accuracy for the non-intrusive load monitoring system. However, SVM needs too much computational cost for training to establish sufficient accuracy. This paper shows Kernel based Subspace Classification can solve this problem with an equal accuracy of classification to SVM.

1 Introduction

Japanese household electric appliance manufacturers have begun to enter the market with products of much higher energy-efficiency, following up on the first product of this kind, which was an air conditioner with inverter circuits[1] in 1982. Now all air conditioners for households on the market have inverter circuits to reach a higher energy efficiency. Recently, also electric appliances with inverter circuits, such as refrigerators, vacuum cleaners and microwave ovens, have emerged on the market, aiming to save energy. However, overall demand increases too greatly, to be able to compensate this additional demand with inverter circuits.

So, power companies need means to control the electric energy demand in order to avoid (1) a big power cut in the near future and (2) the construction of in principle unnecessary additional power plants. A prerequisite for controlling the electric energy demand, however, is the ability to correctly and non-intrusively detect and classify the operating state of electric appliances of individual households. Our goal is therefore to develop exactly such a non-intrusive measuring system for assessing the state of electric appliances with and without inverters (cf. left side image of figure 2). Particularly for inverter type electric appliances

[1] An inverter circuit controls e.g. the rotation speed of a motor (as in air-conditioner) by changing the frequency of the voltage.

G. Dorffner, H. Bischof, and K. Hornik (Eds.): ICANN 2001, LNCS 2130, pp. 692–698, 2001.

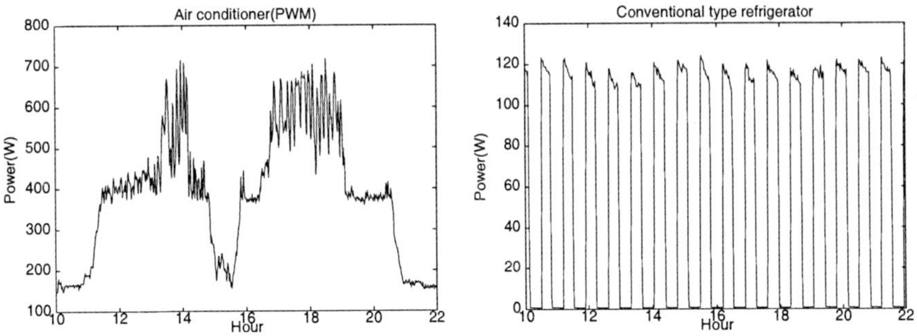

Fig. 1. Load Curves of Household Electric Appliances

this is a hard problem, whereas non-intrusive measuring systems have already been developed for conventional on-off (non-inverter) operating electric appliances [1, 2]. Figure 1 illustrates the electric load curves of an air conditioner with inverter circuit and a refrigerator without inverter circuit. The load curve of the air conditioner is more complex than that of the refrigerator. The operating behavior of the refrigerator is characterized by repeated on-off operation, which is a comparatively simple pattern, whereas the operating behavior of the air conditioner depends on the modulation of the inverter circuit and has a more complicated electric load curve. We have already proposed the non-intrusive monitoring system using Support Vector Machine (SVM) and shown this system established a sufficient accuracy for the estimation of on-off operation [4]. However, this system based on SVM has one big problem about too much computational cost for training.

This paper presents the evaluation of a Kernel based Subspace Classification (KSC) technique to reduce the training time and keep a sufficient accuracy for estimation of the operating state of electric appliances (with and without inverter). Section 2 describes a summary of the system and the problem of the system based on SVM, when this system is applied in a real world. The KSC algorithms are shown in section 3 and section 4 gives an experimental study evaluating the capability of KSC algorithm in this task.

2 Non-intrusive Monitoring System Based on SVM

The left side image of figure 2 shows an outline image of a non-intrusive monitoring system. This system estimates the operating state of all appliances from the electric current and voltage data at the feeder line. These data are analyzed into frequency features that are affected by the inverter type appliances. The features which are used as input data for the learning algorithms are odd-side harmonic currents and respective the phase angels.

The non-intrusive monitoring system is trained to estimate the operating state of appliances for each household. For practical purposes, the training is performed to obtain the best parameters of classification model for the estima-

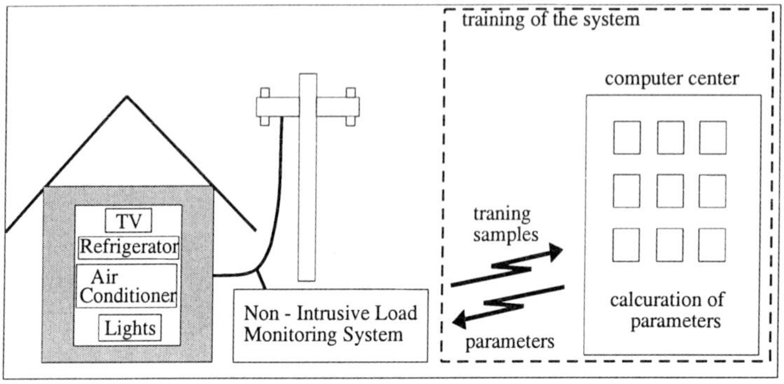

Fig. 2. Outline of training of a non-intrusive monitoring system

tion. Before the system works, these parameters are calculated for respective households. This system, moreover, needs the updating of parameters when a new appliance which affects a electric load curve is applied to a household. The procedures of training are as follows: (1) training samples which are consists of total electric load data and operating state of appliances at the point of time are sent computer center, (2) parameters of the system are calculated using training samples, (3) calculated parameters are sent the system. For that reason, if updating of parameters is delayed, the accuracy of the system are decreased significantly in the updating time. So the updating period is expected at most a day.

We already reported the system applying SVM [4]. SVM has sufficient accuracy of estimation for the system. SVM, however, need a high computational cost for training. By using wireless communications technologies, it will be possible to collect a lot of training samples easily. The training using many samples are very important to construct a good classification model. Therefore, a classification method which have sufficient accuracy in less computational time dealing with a lot of training samples is expected for non-intrusive monitoring system. KSC is more rapidly trained classification method than SVM for the difference of solving method. We applied KSC to a non-intrusive monitoring system and studied accuracy of classification.

3 Kernel Based Subspace Classification

In this section, KSC is explained briefly. First, we explain the conventional subspace classification method. Next, subspace classification using kernels is explained.

3.1 Subspace Classification Method

The subspace classification method is multi-class classification method. In a training phase, subspaces which are characterized for each class by training data are determined using Principal Component Analysis (PCA). In a classifying phase, an unlabeled data is classified a class of which the subspace has a maximum projection distance of the data vector.

A training data matrix $\mathbf{X}$ which belongs to a class is defined as $\mathbf{X} = (\mathbf{x}_1, \ldots, \mathbf{x}_n)$, where $\mathbf{x}_i$ is a training data vector. Eigenvalues of $\mathbf{X}^t\mathbf{X}$ is denoted as $\lambda_i(\lambda_1 \geq \cdots \geq \lambda_n)$. The normalized orthogonal eigenvectors of $\mathbf{X}^t\mathbf{X}$ and $\mathbf{X}\mathbf{X}^t$ that correspond to an eigenvalue λ_i are denoted as $\mathbf{u}_i$ and $\mathbf{v}_i$ respectively, and matrices $\mathbf{U}, \mathbf{V}$ and Λ are defined as

$$\begin{aligned} \mathbf{U} &= (\mathbf{u}_1, \ldots, \mathbf{u}_{d'}), \\ \mathbf{V} &= (\mathbf{v}_1, \ldots, \mathbf{v}_{d'}), \\ \Lambda &= \mathrm{diag}(\sqrt{\lambda_1}, \ldots, \sqrt{\lambda_{d'}}), \end{aligned} \tag{1}$$

where $d' \leq \mathrm{rank}(\mathbf{X})$. These matrices have the relation as follows:

$$\mathbf{X} = \mathbf{V}\Lambda\mathbf{U}. \tag{2}$$

The projection of unlabeled data vector $\mathbf{y}$ to the $\mathbf{v}_i$ is written as

$$\mathbf{v}_i^t\mathbf{y} = \frac{1}{\sqrt{\lambda_i}}\mathbf{u}_i^t\mathbf{X}^t\mathbf{y}. \tag{3}$$

The projection distance $P(\mathbf{y})$ of a data $\mathbf{y}$ to a subspace spanned by $\mathbf{V}$ is written as follows;

$$P^2(\mathbf{y}) = \sum_{i=1}^{d'}(\mathbf{v}_i^t\mathbf{y})^2 = \|\mathbf{V}^t\mathbf{y}\|^2 = \|\Lambda^{-1}\mathbf{U}^t\mathbf{X}^t\mathbf{y}\|^2. \tag{4}$$

Therefore Λ^{-1} and $\mathbf{U}^t$ are calculated in a training phase, and $\mathbf{y}$ is classified into a class which has maximum $P(\mathbf{y})$ in a classifying phase.

3.2 Kernel Based Subspace Classification [3, 5]

The subspace classification method has two problems. The first is that the subspace which is obtained from the distribution along a nonlinear axis is meaningless. The second is that the performance of classification is reduced when the ratio of feature dimensions to class dimensions is small.

The Kernel based Subspace Classification is effective in these problems because the kernel function is the nonlinear transformation to the high dimensions.

The kernel function $k(\mathbf{x}, \mathbf{y})$ is defined as

$$k(\mathbf{x}, \mathbf{y}) = \phi(\mathbf{x})^t\phi(\mathbf{y}) = \sum_{i=1}^{d_\phi} \phi_i(\mathbf{x})\phi_i(\mathbf{y}), \tag{5}$$

where $\phi : \mathcal{R}^d \mapsto \mathcal{R}^{d_\phi}$.

Using a set of d dimensional patterns $\mathbf{x}_1, \ldots, \mathbf{x}_n$ and a function ϕ, a pattern matrix $\mathbf{X}_\phi$ is defined as

$$\mathbf{X}_\phi = (\phi(\mathbf{x}_1), \ldots, \phi(\mathbf{x}_n)). \tag{6}$$

Table 1. Considered Electric Appliances

Appliances	Standard Watt
air conditioner with inverter	500W
refrigerator	150W
incandescent lamps	50W×5
fluorescence lamps with inverter	50W×5
television set	80W

The kernel matrix $K(\mathbf{X},\mathbf{Y})$ of which the (i,j) element is $k(\mathbf{x}_i,\mathbf{y}_j)$ is written as

$$K(\mathbf{X},\mathbf{Y}) = \mathbf{X}_\phi^t \mathbf{Y}_\phi. \tag{7}$$

Equation 4 can be rewritten as

$$P^2(\mathbf{y}) = \|\Lambda'^{-1}\mathbf{U}'^t K(\mathbf{X},\mathbf{y})\|^2, \tag{8}$$

where Λ' and $\mathbf{U}'$ are calculated from $K(\mathbf{X},\mathbf{X})$.

4 Experimental Result and Discussion

The household electric appliances which are used for the measurement are shown in Table 1. The data of harmonic waves are obtained from the combined condition of operation states of these appliances. The number of lamps are changed at the "on" state.

We compare KSC with k nearest neighbor (kNN) and SVM in the accuracy of classification and computational cost. The training time of kNN does not exist practically because the classification phase of kNN is the calculation of distances between an unlabeled data and training data. Therefore, kNN is compared in the classification accuracy only.

In this experiment we use a RBF kernel

$$k(\mathbf{x},\mathbf{y}) = \exp\left(\frac{\|\mathbf{x}-\mathbf{y}\|^2}{\sigma^2}\right) \tag{9}$$

for KSC and SVM.

KSC has two parameters that are the kernel parameter σ and the dimension parameter d' (cf. equation 1). SVM also has two parameters that are σ and regularization parameter C. We use a cross-validation method to find the good model parameters. The cross-validation method is rather reliable, but very costly for computational cost. We scan the following range in the model parameter space:

KSC	$\sigma = 5, 10, 15, 30, 50, 100$	$d' = 10, 15, 20, 30, 50, 100$
SVM	$\sigma = 5, 10, 30, 50, 100, 150, 200$	$C = 10^0, 10^1, 10^2, 10^3$

Table 2. Classification Errors for KSC, SVM, and kNN

Method	KSC	SVM	kNN
air conditioner	0.68%	**0.25%**	0.60%
refrigerator	2.05%	1.90%	**1.78%**
incandescent	**0.75%**	1.00%	1.83%
fluorescence	0.68%	**0.65%**	1.95%
television set	**0.15%**	0.25%	0.33%
all appliances	3.70%	**3.63%**	5.18%

In this experiment we generate 10 data sets of 400 training and 400 test data. A data set is made into 800 patterns that are chosen from whole pattern (1112 patterns) randomly and split into training and test data.

KSC and kNN are multi-category classification methods. SVM, however, is a binary classification method. Therefore we make a binary classifier for each appliance to compare these methods.

All experiments are performed at a Linux machine of PentiumIII–1GHz.

Table 2 shows classification errors that are averaged over 10 data sets.

We find that KSC has equal accuracy of classification to SVM from Table 2. In the total error of classification, SVM has the best accuracy. KSC, however, has better accuracy than SVM in classifying incandescent lamps and television set.

Next, we observe the calculation time of training for KSC and SVM.

Figure 3 shows relations between calculation time and number of training data for KSC and SVM. This results are averaged over 10 types data which are trained on fixed parameters (KSC: σ=15, d'=100; SVM: σ=30, C=100).

The training time of KSC is about three times as short as SVM for 100 training data. However, the difference of training time between the two methods multiply as increase of training data. For 800 training data, KSC is about thirty times as fast as SVM. In actual parameter tuning for non-intrusive monitoring system, the cross-validation method is very useful for the model selection of SVM and KSC. Suppose observed training samples to be 600. Practically, these training samples are split into 20 subsets and free parameters are changed 20 times to make cross-validation for classification models of non-intrusive monitoring system. In this case, SVM takes about 8 days to construct good classification model and KSC takes only about 7 hours. Therefore, KSC has great advantage of parameter tuning of a non-intrusive monitoring system using a cross-validation method which has to repeat a training phase.

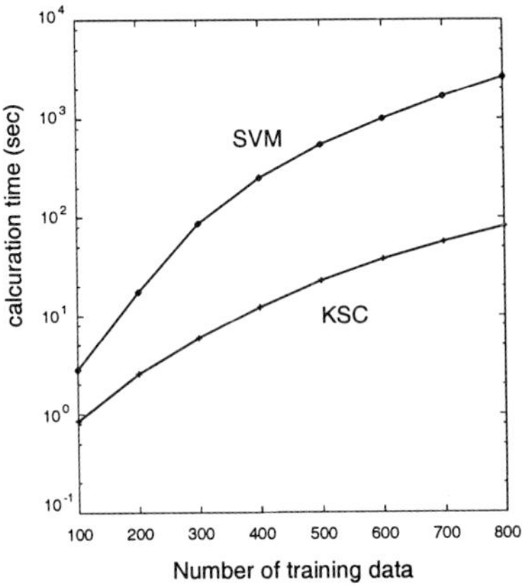

Fig. 3: Calculation time for number of training data

5 Conclusion

In order to combine sufficient accuracy and rapidly training for the non-intrusive monitoring system, we compared with KSC, kNN, and SVM. In consequence of experiments, we found that the KSC has a rapidly computational time and a equal accuracy of classification to the SVM. Therefore, KSC is useful in training of this system particularly using cross-validation method which is reliable but computationally expensive.

References

1. J. Carmichael.: Non-intrusive appliance load monitoring system. *EPRI Journal, Electric Power Research Institute*, 1990.
2. W. Hart.: Non-intrusive appliance load monitoring. *Proceedings of the IEEE*, vol. 80, no. 12, 1992.
3. E.Maeda et al.: Multi-category Classification by Kernel based Nonlinear Subspace Method. *ICASSP'99*, 2, 1025-1028, 1999.
4. T.Onoda et al.: Applying Support Vector Machines and Boosting to a Non-Intrusive Monitoring System for Household Electric Appliances with Inverters. *NC'2000*, 2000.
5. K.Tsuda.: Subspace Classifier in Reproducing Kernel Hilbert Space. *IJCNN'99*, 1999.
6. V. N. Vapnik.: *The Nature of Statistical Learning Theory.* Springer, 1995.

Neural Networks in Circuit Simulators

Alessio Plebe[1], A. Marcello Anile[1], and Salvatore Rinaudo[2]

[1] University of Catania Department of Mathematics and Informatics V.le Andrea Doria, 8 I-95125 Catania, Italy
[2] ST Microelectronics Str. Primosole, 50 I-95121 Catania, Italy

Abstract. Artificial Neural Networks (ANN) are gaining attention in the semiconductor modeling area, as alternative to physical modeling of high speed devices. A fundamental issue when including ANN's in a circuit simulator is how to manage the time dependency. One elegant solution recently proposed is the *Dynamic Neural Network* concept, where neurons are instances of differential equations. In this work the dynamic approach and further variations has been compared with classical static ANN, applied to the modeling of high performance bipolar junction transistor.

1 Introduction

The increasing difficulties in deriving models for circuit simulators of devices directly from the underlying physical principles have called for researches in alternative directions [11, 12, 5]. ANN is one of the possible alternatives, widespread in many other areas, recently gaining space also in in the semiconductor modeling field [1, 5, 6, 8].

It is necessary for this field to be able to model dynamic phenomena, therefore to represent time, which is not a new feature for ANN's, since the seminal work of Elman [3] introducing recurrent networks at the beginning of '90. Five years later a survey [4] listed ten different ANN architectures able to deal with time, using various schemes of recursion. Unfortunately all those networks can only work with a fixed time discretization, which is not acceptable in circuit simulators.

Recently Meijer [5, 6] introduced a new neural architecture where all neurons are instances of second order differential equations, and ideally can behave as a continuous-time dynamic system. Plebe and coworkers [8] have extended this approach using a mixed neural architecture with various kind of static and dynamic neurons. There is actually an other simpler route that may be followed when only DC and small signal responses of the device are required, which is very common for high frequency BJT (Bipolar Junction Transistor). A pure static network can provide numerical values of characteristic AC parameters of the device, such as S or Y parameters, as a function of the DC bias point and the current frequency only. Dynamic ANN are an attractive paradigm, where the modeling problem is expressed in a form close to the classical representation of dynamic systems, however their training is not an easy job. It has to be carefully verified whether recasting the problem in neural pattern-recognition form, as in

G. Dorffner, H. Bischof, and K. Hornik (Eds.): ICANN 2001, LNCS 2130, pp. 699–705, 2001.

practice when using static ANN to output parameter values, can at the end be more effective.

This is the purpose of the work here presented. In the next section the candidate neural architectures will be briefly described, then the modeling results will be presented.

2 Dynamic Neural Networks

The dynamic neural network originally introduced by Meijer, which will be hereafter referred as **DNN**, is composed by identical neurons, each with associated the following equation:

$$\tau_2 \frac{\partial^2 y_i}{\partial t^2} + \tau_1 \frac{\partial y_i}{\partial t} + y_i = \mathcal{F}(a_i, \delta_i) \tag{1}$$

where y_i is the neuron output, τ_1, τ_2 are constant parameters, a_i is the neural activation produced by the N neurons x_j in the underlying layer, $a_i = \sum_{j=1}^{N} w_{ij} x_j + \sum_{j=1}^{N} v_{ij} \frac{\partial x_j}{\partial t} - \theta_i$, with θ_i the threshold level, and $\mathcal{F}$ is a non linear function of type:

$$\mathcal{F}(a, \delta) = \frac{1}{\delta^2} \log \frac{e^{\frac{\delta^2 (a+1)}{2}} + e^{\frac{-\delta^2 (a+1)}{2}}}{e^{\frac{\delta^2 (a-1)}{2}} + e^{\frac{-\delta^2 (a-1)}{2}}} \tag{2}$$

This is a special neural function, since for proper choices of the parameter δ can approximate the basic physical equation of the DC current across a generic diode junction, and δ is typically included in the parameters to be learned for each neuron.

In the extension proposed by Plebe et al., referred as **HNN** (Higher-level dynamic Neural Network), a mixture of several neuron types is allowed, including the ones following equation (1), and in particular including neurons with varying dynamic, described by the equation:

$$\mathcal{T}_2(o_i, \omega_i) \frac{\partial^2 y_i}{\partial t^2} + \mathcal{T}_1(o_i, z_i, \omega_i, \zeta_i) \frac{\partial y_i}{\partial t} + y_i = \mathcal{F}(a_i, \delta_i) \tag{3}$$

which requires the additional activation $o_i = \sum_{j=1}^{N} u_{ij} x_j - v_i$ and $z_i = \sum_{j=1}^{N} s_{ij} x_j - \sigma_i$, and uses two more non linear functions:

$$\mathcal{T}_1(o, z, \omega, \zeta) = \frac{2\zeta(1 + e^{-o})}{\omega(1 + e^{-z})} , \tag{4}$$

$$\mathcal{T}_2(o, \omega) = \frac{1 + 2e^{-o} + e^{-2o}}{\omega^2} . \tag{5}$$

In the HNN architecture there are four different types of connections: w is the usual *proportional* neural connection, v is a *derivative* connection, u and s connect *natural frequency* and *damping ratio* contributions. The non linear functions

$\mathcal{T}_{\{1,2\}}$ are shaped by the parameters ω and ζ, adapted individually for each neuron. Both DNN and HNN typically include more parameters per neuron than ordinary feed-forward multilayer networks, and in general are not easy to train against device data. The HNN should offer more flexibility, and has been developed for high performance BJT, but is more difficult to train than the DNN.

A particular instance of HNN, here called MNN (Multiple-level dynamic Neural Network) has been introduced by Plebe and coworkers [9] in order to make the training problem tractable. It is an association of a higher-level architecture $\mathcal{A}$, with less flexibility then an ordinary HNN, with a couple of derived lower-level architectures $\mathcal{A}_\mathcal{D}$ and $\mathcal{A}_\mathcal{S}$ used for training u and s weights. $\mathcal{A}$ is a restriction of the HNN, with 3 layers only, higher-level neurons (those following equation (3)) allowed in the second layer only, and a constant number K of connections of type u and s to each dynamic neuron. An example is visible in Fig. 1, with static neuron neurons represented with the symbol , dynamic neuron neurons represented with the symbol , and higher-level neurons with 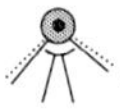.

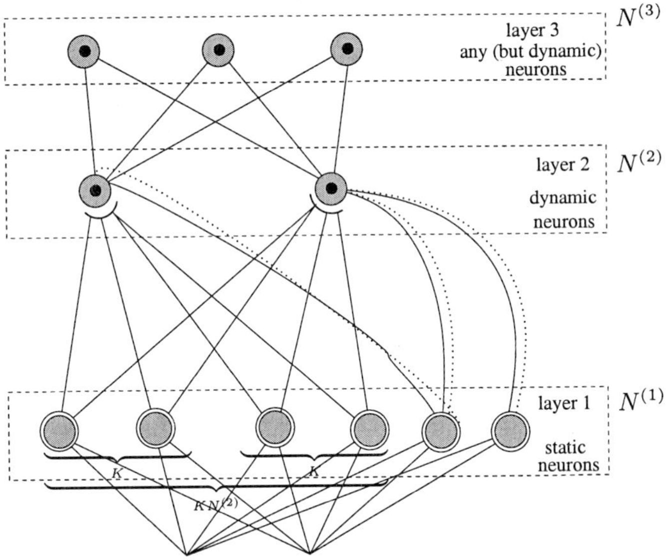

Fig. 1. Detail of a sample $\mathcal{A}$ architecture, $N^{(k)}$ is the number of neurons in layer k

The auxiliary architecture $\mathcal{A}_\mathcal{D}$ is made substituting all higher-level nodes of $\mathcal{A}$ with simple dynamic ones, this architecture is replicated for all the available DC bias points, and at each points the fixed τ_1 and τ_2 of equation (1) are learned. With the other auxiliary architecture $\mathcal{A}_\mathcal{S}$, the static layer of $\mathcal{A}$ is trained in such way to produce the set of τ_1 and τ_2 generated from $\mathcal{A}_\mathcal{D}$ as function of the DC bias point. The remaining parameters of $\mathcal{A}$ shall be trained normally using the full data set of the device, but this strategy relieves significantly the difficulty. Architectures $\mathcal{A}_\mathcal{D}$ and $\mathcal{A}_\mathcal{S}$ for the example of Fig 1 are shown in Fig. 2.

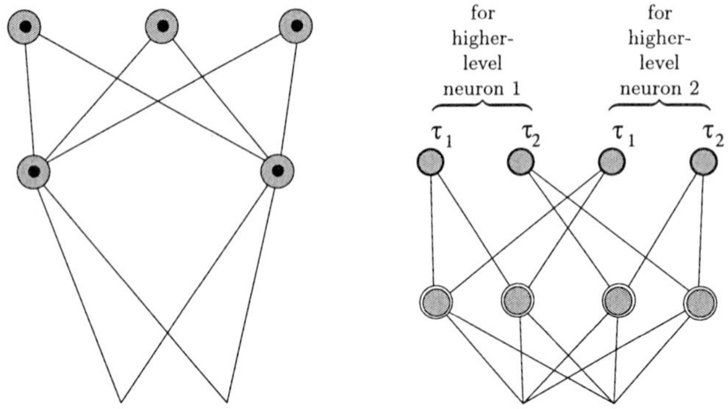

Fig. 2. Auxiliary architectures $\mathcal{A}_\mathcal{D}$ on the left and $\mathcal{A}_\mathcal{S}$ on the right to be used for training the sample architecture of Fig. 1.

Eventually, for simulations limited to the small-signals mode, a simple static network can be considered, for computing the complex characteristic parameters of the device, for example the components of the $\boldsymbol{Y}$ matrix, as a function of DC bias conditions and frequency. Those values are enough for modeling AC current components inside the circuit simulator at the device pins. An architecture has been taken as candidate, called SNN (Static Neural Network) with one hidden layer, and neurons with sigmoidal non-linear function ($\frac{1}{1+e^{-a}}$) mixed with neurons which nonlinearity follows equation (2). In this case the network will not try to learn a functional, which should express the transfer function of the device, rather to recognize "patterns" of DC point + frequency, producing as output the values to be used as parameters of the device.

3 Experimental Results

The four neural approaches yet described, DNN, HNN, MNN and SNN, have been tested in the modeling of a real device from ST-Microelectronics, a high frequency BJT, using a set of measurements with 273 frequencies in range 1-40GHz, 3 values of V_{CE} and 11 values of V_{BE}. For this device there is no standard model yet available, and all results here reported are direct comparison with the measured data of the BJT, without external components. For all networks, the training is performed using the average error over the training set as an objective function, whose parameters are optimized in two stages. First with a global method, the *Controlled Random Search* [10], to generate an approximate parameter point, which is used as first guess in the second method, a quasi-Newton local optimization [7].

Table 1. Description of the architectures used in table 2. Layers are separated by "–", and in each layer the neuron type is coded with a single letter. In the last column is listed the total number of parameters of the architectures to be trained.

architecture name	architecture description s = static with sigmoidal non-linearity S = static with non-linearity of eqn (2) d = dynamic following eqn (1) D = dynamic following eqn (3)	# of parameters
DNN_1	dddd--dd	58
DNN_2	dddd--ddd--dd	91
HNN_1	SSSS--DD--dd	72
HNN_2	SSSSSS--DD--dd	96
HNN_3	SSSSSSSS--DDDD--dd	212
MNN_1	SSSSS-DD-dd	92
MNN_2	SSSSSSSS--DDDD--dd	212
SNN_1	SSSSSSSSSSSSSSSS--SSSSSSSS	128
SNN_2	sSsSsSsSsSsSsSsS--ssssssss	112

The architectures used in the comparison here presented are a selection of the best over a larger number which has been experimented, and are described in table 1. Note that the difference between HNN and MNN is not just in the type of neurons, but also the training strategy, which for MNN requires the additional architectures $\mathcal{A}_\mathcal{D}$ and $\mathcal{A}_\mathcal{S}$, table 1 refers to the main $\mathcal{A}$ only. All dynamic architectures have input/output matching the BJT device pins: two outputs for V_{CE}, V_{BE} and two outputs for I_B, I_C. For the SNN it is different, there are three inputs: $V_{BE}|_{DC}$, $V_{BE}|_{DC}$, and the frequency; and eight outputs: the two complex parts of the four elements of $\boldsymbol{Y}$. It has been found more convenient to train against module and phase, instead of imaginary and real part.

In table 2 the results of the selected architectures are compared, with respect to the errors on the $\boldsymbol{Y}$ components, and the computing time spent for the training (in thousands of seconds on a 850MHz Pentium machine running Linux). The error is computed on a number of points equal to the training set, but at frequencies and DC bias points between those used for the training, and is expressed as normalized root of mean square:

$$\mathcal{E} = \frac{1}{\sigma}\sqrt{\frac{1}{N}\sum_b\sum_f\left|\widehat{Y} - Y\right|^2}$$

where Y is a generic component of $\boldsymbol{Y}$, and σ its standard deviation, summations are over frequencies and biases.

For several component of the $\boldsymbol{Y}$ matrix, the lowest errors are found for the static architectures SNN_1 and SNN_2. Despite the large number of neurons in the hidden layer, those architectures have less parameters to train compared with dynamic architectures like MNN_2. Since the non-linear function (2) belongs to

Table 2. Comparison of final error and training time of various architectures

architecture	Y_{11}	Y_{12}	Y_{21}	Y_{22}	CPU time [s $\times 10^3$]
DNN_1	0.274	0.186	0.157	0.389	130.3
DNN_2	0.0308	0.0730	0.0887	0.129	530.2
HNN_1	0.131	0.200	0.195	0.210	79.5
HNN_2	0.124	0.153	0.174	0.162	118.1
HNN_3	0.0791	0.225	0.0530	0.332	480.4
MNN_1	0.0432	0.0920	0.0529	0.0812	55.3
MNN_2	0.00721	0.0653	0.0264	0.0428	212.7
SNN_1	0.0804	0.00220	0.0319	0.0332	280.2
SNN_2	0.0438	0.00211	0.0405	0.0213	276.7

the set of generalized sigmoids [13] for all SNN networks there is no advantage in using more then one hidden layer, as known for conventional feedforward networks since 1989 [2]. This is not true for dynamic networks, on the contrary, it is often necessary to use at least two hidden layers to achieve reasonable performances. There is no significant difference between DNN and HNN, the latter can use less dynamic nodes, but the overall number of parameters is larger, for both the best results are achieved using a considerable amount of CPU time. The real difference is the training strategy of MNN, allowing reduced computational time and lower errors, close to that of static networks. The results clearly reflect that the dynamic and the static networks are actually learning different things: the errors are distributed in different ways among $\boldsymbol{Y}$ components. For the static network the error on Y_{11} is the worst, and larger then for MNN, while Y_{21} is the easier component for the static networks and not for the others.

4 Conclusions

To cope with the current scenario of high speed sub-micron bipolar devices, and their use in CAD applications, new and more efficient modeling techniques are needed. ANN's can be valid candidates. In this work it has been addressed the advantage of recent ANN architectures with embedded time representation, which include functions familiar to the classical linear dynamic system theory, as well as non linear functions close to the semiconductor junction laws. Such advantages can be jeopardized by the complexity of the architectures and the difficulty in training. When only AC simulation is requested, static ANN's used as "pattern-matchers" over frequency and DC bias input, seems to produce better results.

This is not a conclusive word, dynamic neural networks are a very young tool, and there is still a wide range of variations and further development to explore. The training method used in MNN has proven that auxiliary strategies can bring significant improvements, with results close and in some case better then the static ANN. An other aspect to be investigated is the computational performance of the neural approaches inside a circuit simulator, when dealing with complex circuits. Modules of the MNN and SNN architectures for the ELDO circuit simulator are currently under development.

References

1. G. Creech, J. Zurada, and P. Aronhime. Feedforward neural networks for estimating IC parametric yield and device characterisation. In *Proc. of the IEEE Int. Symposium on Cyrcuits and Systems*, Seattle, WA, 1995.
2. G. Cybenko. Approximation by superpositions of a sigmoidal function. *Mathematics of Control, Signals and Systems*, 2(4), 1989.
3. J. L. Elman. Finding structure in time. *Cognitive Science*, 14:179–221, 1990.
4. B. G. Horne and C. L. Giles. An experimental comparison of recurrent neural networks. In G. Tesauro, D. Touretzky, and T. Leen, editors, *Neural Information Processing Systems, 7.* MIT Press, Cambridge (MA), 1995.
5. P. Meijer. *Neural Network Applications in Device and Subcircuit Modelling for Circuit Simulation.* PhD thesis, Technical University of Eindhoven, 1996.
6. P. Meijer. Neural networks for device and circuit modelling. In *Proc. of the Third Int. Workshop on Scientific Computing in Electrical Engineering*, Warnemünde, Germany, 2000.
7. J. Nocedal and S. J. Wright. *Numerical Optimization.* Springer-Verlag, New York, 1999.
8. A. Plebe, A. M. Anile, and S. Rinaudo. Sub-micrometer bipolar transistor modeling using neural networks. In *Proc. of the Third Int. Workshop on Scientific Computing in Electrical Engineering*, Warnemünde, Germany, 2000.
9. A. Plebe, A. M. Anile, and S. Rinaudo. Neural networks with higher level architecture for bipolar device modeling. In *Proc. of the Eleventh ECMI Conference*, Palermo, Italy, 2001.
10. W. L. Price. Global optimization algorithms for a cad workstation. *Journal of Optimization Theory and Applications*, 55:133–146, 1987.
11. K.-G. Rauh. A table model for circuit simulation. In *Proc. of the Twelfth European Solid-State Circuit Conference*, pages 211–213, Delft, The Netherlands, 1986.
12. T. Shima. Device and circuit simulator integration techniques. In W. L. Engl, editor, *Process and devics modeling*, pages 433–459, Amsterdam, The Netherlands, 1986. North Holland.
13. M. Stinchombe. Neural network approximation of continuous functionals and continuous functions on compatifications. *Neural Networks*, 12:467–477, 1999.

Neural Networks Ensemble for Cyclosporine Concentration Monitoring

Gustavo Camps[1], Emilio Soria[1], José D. Martín[1], Antonio J. Serrano[1], Juan J. Ruixo[2], and N. Víctor Jiménez[2]

[1] Grupo de Procesado Digital de Señales, Universitat de València, Spain
gustavo.camps@uv.es, http://gpds.uv.es/
[2] Departamento de Farmacia y Tecnología Farmacéutica, Universitat de València (Spain) and with the Pharmacy Service of the Hospital Universitario Dr. Peset, València (Spain)
victor.jimenez@uv.es*

Abstract. This paper proposes the use of neural networks ensemble for predicting the cyclosporine A (CyA) concentration in kidney transplant patients. In order to optimize clinical outcomes and to reduce the cost associated with patient care, accurate prediction of CyA concentrations is the main objective of therapeutic drug monitoring.
Thirty-two renal allograft patients and different factors (age, weight, gender, creatinine and post-transplantation days, together with past dosages and concentrations) were studied to obtain the best models. Three kinds of networks (multilayer perceptron, FIR network, Elman recurrent network) and the formation of neural-network ensembles were used. The FIR network, yielding root-mean-squared errors (RMSE) of 41.61 ng/mL in training (22 patients) and 52.34 ng/mL in validation (10 patients) showed the best results. A committee of trained networks improved accuracy (RMSE = 44.77 ng/mL in validation).

1 Introduction

At present, cyclosporine A (CyA) is still the cornerstone of immunosuppression in renal transplant recipients and its use continues to expand globally. Despite significant advances in dose formulation and therapeutic drug monitoring and guidelines for the last years, CyA is still considered a drug with a narrow therapeutical index. For this reason, the clinical management of transplant patients may result in underdosing situations with a graft loss or overdosing situations that causes kidney damage, increases opportunistic infections, systolic and diastolic pressure, and cholesterol [1].

The pharmacokinetic (PK) behavior of CyA presents a substantial inter- and intra-individual variability, several important clinical drug interactions and different patient compliance rates [2, 3]. This variability is particularly evident in

* This paper has been partially supported by the European FEDER Project IFD1997-0935 entitled "Desarrollo de Sistemas Neuronales Aplicados en Atención Farmacéutica".

G. Dorffner, H. Bischof, and K. Hornik (Eds.): ICANN 2001, LNCS 2130, pp. 706–711, 2001.

the earlier post-transplantation period, when the risk and clinical consequences of acute rejection are higher. In order to optimize clinical outcomes, patient's quality of life and health system cost, accurate prediction of CyA concentrations is the main objective of therapeutic drug monitoring. Prediction of CyA concentration by mathematical models allows the patient dose individualization. Several studies with pharmacokinetic assumptions have been done in this setting (see [4–6]), but none, to our knowledge, using neural networks.

In this paper, we propose artificial neural networks models (ANN) ensembles to predict CyA blood concentration in kidney transplant patients. Data collection and the time series prediction techniques used, are introduced in the next section. The results are presented in Section 3 and Section 4 contains concluding remarks and a proposal for further work.

2 Material and Methods

2.1 Patients and Data Collection

Thirty-two renal allograft recipients treated in Hospital Universitario Dr. Peset (Spain) were included in this study. Immunosuppressive treatment was based in lipidic microemulsion of CyA (Neoral), mycophenolate mofetil (2 g/d) and prednisone (0.5-1 mg/kg/d)*. Initial oral dose of CyA (5 mg/kg b.i.d) was reduced according to the measured CyA blood concentration and the desired target range (150-300 ng/mL) [7].

Table 1. Statistics of the predictors used in the models.

Predictor	Range	Mean	STD
Post-transplantation days, PTD (d)	170-197	183	8.2
CyA concentration, C (ng/mL)	45.8-803.3	270.4	83.9
Body Weight, WE (kg)	37.5-106.0	69.9	14.7
Creatinine, CR (mg/dL)	0.7-10.2	1.7	1.0
Age, AG (yr)	28.8-68.9	49.2	9.9
Gender, GE (0, male; 1, female)	0-1	0.3	0.5
Dosage, DD (mg/kg)	2.1-10.8	5.4	2.0

The collected data along with some basic population statistics are shown in Table 1. The follow up was made during the first six month after trasplant and we obtained from 13 to 22 samples per patients. Each *pattern* or example was constituted by the present and past values of the variables described. In our problem it is only necessary to make a one-step ahead prediction or estimation.

* Patients who received metabolic inducers or inhibitors were excluded because they modify the pharmacokinetic profile of CyA.

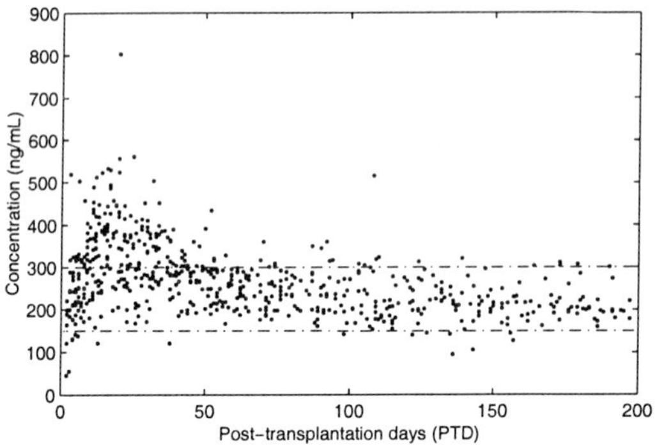

Fig. 1. Distribution of the blood concentration (ng/mL) of CyA. Dashed lines indicate the desired target range.

Blood concentration within the population is shown in Fig. 1, where the high inter-subjects variability observed in the early post-transplantation days makes it necessary to raise or lower dosage while closely monitoring the patient's blood concentration.

Multilayer Perceptron. The traditional model of the multilayer neural network is composed of a layered arrangement of artificial neurons in which each neuron of a given layer feeds all the neurons of the next layer. The first layer contains the input nodes, which are usually fully-connected to hidden neurons and these, in turn, to the output layer. Training of the network can be accomplished using the familiar *backpropagation* algorithm [8]. The multilayer perceptron can be used for prediction tasks if patterns are introduced in a delayed form, i.e. an input vector $\mathbf{x}$ must be defined in terms of the past samples $x(n-1), x(n-2), ...x(n-t_d)$. We refer to t_d as the *prediction order* or the *time delay*. But this mapping is nevertheless, a *static* mapping; there are no internal dynamics.

FIR Networks. The FIR network models each weight as the FIR basic linear filter which can be modeled with a tapped delay line (see Fig. 2). The simple analogies between the FIR network and the multilayer perceptron carry through when comparing standard backpropagation for static networks with *temporal backpropagation* for FIR networks [9].

Elman Recurrent Network. In recurrent networks connections are added from higher layers in a network to lower ones. The feedback connections introduce iteration into the network or, in other words, a kind of memory of the past. The Elman recurrent network has a simple architecture and can be trained using the standard backpropagation learning algorithm [10]. In this architecture there are also context units that are used only to memorize the previous activations of the hidden units.

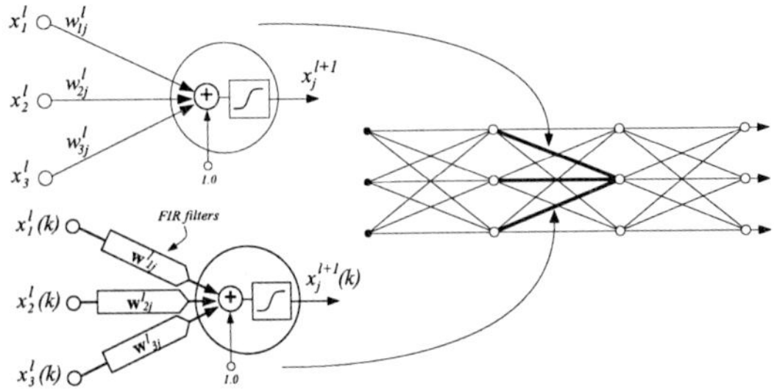

Fig. 2. Static and FIR neuron along with the feedforward network. In a static neuron each line represents a synaptic connection. In the temporal model, input signals are passed through synaptic filters, summed up and then passed through a sigmoid function to form the output of the neuron.

Combined Forecasts. Previous work has shown that an effective ensemble should consist of networks that are not only highly correct, but ones that make their errors on different parts of the input space as well. Therefore uncorrelated predictions are desired and it is assumed that using different kinds of prediction techniques, in principle, can be more appropiate to this purpose. We will focus on simple linear combinations such as the equal weights (EW) method, the minimum-variance (MV) method and the Optimal Linear Combination (OLC). All of them are extensively described in [11].

3 Results and Discussion

A comparison of the performance of all the models is shown in Table 2. We have used the mean prediction error (ME) as a measure of bias and the root mean squared error (RMSE) and the correlation coefficient (r) as a measure of precision. The cross-validation method together with variation of the topology parameters, the weight initialization range and the learning rate were used to determine the best network. Two thirds of the population was used for training the models and the rest for their validation.

With neural-network ensembles, outcomes are clearly improved when simple combined forecasts are considered. The best mixture was formed by the OLC, which reduces the ME by 8.24% and the RMSE of the best single model by 14.46%. More complex models using different number of each kind of experts have also been attempted but results were similar.

A useful way to show results graphically is to make predicted versus observed representations [12]. A plot of the best model is shown in Fig. 3a. The solid line represents the line of identity, and the dotted one is the regression line (RL). We also show the predicted values against the residuals (Fig. 3b). The slope of the linear regression of prediction errors versus the predicted value can also be used as a measure of divergence. Squared correlation coefficient (R^2) of the linear regression is also shown. The FIR network was slightly less biased (the

Table 2. Models comparison for concentration prediction in the validation set. Weights are indicated in brackets (MLP,FIR,ELMAN).

Performance	r	ME (ng/mL) (95% CI)	RMSE (ng/mL) (95% CI)
MLP	0.746	1.286	58.53
(t_d=2)		(0.696,1.876)	(-12.19,129.25)
FIR[†]	0.759	4.687	52.34
(7x2x10x1,1:1:1)		(4.177,5.196)	(-71.10,175.78)
ELMAN	0.750	2.528	55.701
(RN=3)		(1.173,3.883)	(-147.85,259.255)
EW	0.799	1.910	49.83
		(1.601,2.220)	(-15.41,115.07)
MV	0.794	0.980	48.71
(0.3,0.4,0.3)		(0.300,1.660)	(-10.25,107.67)
OLC	0.814	1.180	44.77
(4.5,12.85,7.31)		(-0.510,1.700)	(-10.01,99.55)

† FIR Topology: (Number of Inputs×Hidden Neurons×Outputs, Order of FIR synapsis in each layer)

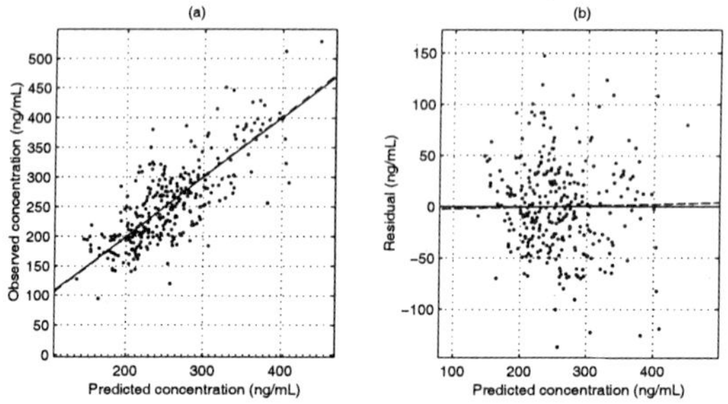

Fig. 3. Best models representations. (a) Predicted versus observed. RL: $y = 1.007x - 0.829$, $R^2(T) = 0.654$, $R^2(V) = 0.577$ (b) predicted versus residuals.

difference is 0.51 ng/mL) than the MLP (0.59 ng/mL) and the Elman net (1.36 ng/mL) but showed better precision in validation.

4 Conclusions

In this paper we have presented several neural approaches to a pharmacokinetic prediction problem: MLP with delayed inputs, the FIR network and the Elman recurrent network. Neural models have proven to be well suited to this task although the results, even being acceptable, must be improved and considered as a preliminary work before their clinical application. The best results are obtained with the FIR network. We have also shown how combined forecasts can be used to easily improve the accuracy and bias of predictions. Our future work is tied to the use of more sophisticated and dynamic models (IIR and gamma structures and more complex recurrent networks) and will focus on the construction of committees of trained networks.

References

1. P. Belitsky. Neoral used in the renal transplant recipient. *Transplantation Proceedings*, 32(3A Suppl. Review.):S10–S19, May 2000.
2. A. Lindholm. Factors influencing the pharmacokinetics of cyclosporine in man. *Therapeutic Drug Monitoring*, 13(6):465–477, Nov 1991.
3. C. Campana, M. B. Regazzi, and Buggia I. Clinically significant drug interactions with ciclosporin. *Clinical Pharmacokinetics*, 30(2):141–179, Feb 1996.
4. S. Vozeh, P. O. Maitre, and D. R. Stanski. Evaluation of population (NONMEM) pharmacokinetic parameter estimates. *Journal of Pharmacokinetics and Biopharmaceutics*, 18(2):161–173, April 1990.
5. J. Parke and B. G. Charles. NONMEM population pharmacokinetic modeling of orally administered cyclosporine from routine drug monitoring data after heart transplantation. *Therapeutic Drug Monitoring*, 20(3):284–293, Jun 1998.
6. B. Charpiat, I. Falconi, V. Bréant, R. W. Jellife, J. M. Sab, C. Ducerf, N. Fourcade, A. Thomasson, and J. Baulieux. A population pharmacokinetic model of cyclosporine in the early postoperative phase in patients with liver transplants, and its predictive performance with bayesian fitting. *Therapeutic Drug Monitoring*, 20:158–164, 1998.
7. M. Oellerich, V. W. Armstrong, B. Kahan, L. Shaw, D. W. Holt, R. Yatscoff, A. Lindholm, P. Halloran, K. Gallicano, and K. Wonigeit. Lake Louise consensus conference on cyclosporin monitoring in organ transplantation: report of the consensus panel. *Therapeutic Drug Monitoring*, 17:642–654, Dec 1995.
8. S. Haykin. *Neural Networks: A Comprehensive Foundation.* Prentice Hall, 1999.
9. E. A. Wan. *Finite Impulse Response Neural Networks with Applications in Time Series Prediction.* PhD thesis, Department of Electrical Engineering. Stanford University, November 1993.
10. J. L. Elman. Finding structure in time. *Cognitive Science*, 14:179–211, 1988.
11. S. Hashem. Optimal linear combinations of neural networks. *Neural Networks*, 10(4):599–614, 1997.
12. M. E. Brier, J. M. Zurada, and G. R. Aronoff. Neural network predicted peak and trough gentamicin concentrations. *Pharmaceutical Research*, 12(3), 1995.

Efficient Hybrid Neural Network for Chaotic Time Series Prediction

Hirotaka Inoue[1], Yoshinobu Fukunaga[1], and Hiroyuki Narihisa[2]

[1] Graduate School of Engineering,
Okayama University of Science,
1-1 Ridai-cho, Okayama-shi, Okayama, 700-0005, Japan
{inoue,fukunaga}@ice.ous.ac.jp

[2] Department of Information & Computer Engineering, Faculty of Engineering,
Okayama University of Science,
1-1 Ridai-cho, Okayama-shi, Okayama, 700-0005, Japan
narihisa@ice.ous.ac.jp

Abstract. We propose an efficient hybrid neural network for chaotic time series prediction. The hybrid neural network is constructed by a traditional feed-forward network, which is learned by using the backpropagation and a local model, which is implemented as a time delay embedding. The feed-forward network performs as the global approximation and the local model works as the local approximation. Experimental results using Mackey-Glass data and K.U. Leuven competition data show that the proposed method can predict the more long term than each of predictors.

1 Introduction

The time series prediction problems have been solved by using the statistical procedures such as ARIMA model and/or others [1]. However, their prediction power is limited due to their inability to model the evolutionary dynamics of the process. Recently, neural networks and local models have been directly applied for chaotic time series prediction and comparatively satisfied results are reported [2–4].

Neural networks are useful for nonlinear adaptive signal processing. They have many important characteristics such as adaptability, flexibility, and universal nonlinear input-output mapping capability [5]. Therefore, they can perform as the global approximation for time series prediction.

Local models are good for a short term chaotic time series prediction. They reconstruct the m dimensional attractor from the one dimensional time series. This technique is called as the time delay embedding [6]. After the time delay embedding, the future values are predicted by using k nearest neighbor points in the reconstructed space. Therefore, they can perform as the local approximation for time series prediction. Hence, it is impossible to predict the long term of the chaotic time series because the chaos has the feature of the sensitivity dependent of initial conditions.

G. Dorffner, H. Bischof, and K. Hornik (Eds.): ICANN 2001, LNCS 2130, pp. 712–718, 2001.

In this paper, we propose a hybrid neural network of the traditional feed-forward network and the simple time delayed embedding model. The advantage of this method is a robust prediction for more long term than the each of predictors. We use Mackey-Glass data and K.U. Leuven competition data which are well used on this field as benchmark problems [7, 8].

2 Hybrid Neural Network

2.1 Definitions

The hybrid neural network which is described here as a designed to predict time series generated by nonlinear dynamic systems that can be described by a set of nonlinear ordinary differential equations as follows:

$$\boldsymbol{x}_t = f(\boldsymbol{x}_{t-1}), \tag{1}$$

$$y_t = g(\boldsymbol{x}_t), \tag{2}$$

where $\boldsymbol{x}_t \in \mathbb{R}^{n_s}$ is the state of the system, n_s is the order of the system, and $y_t \in \mathbb{R}^1$ is the time series to be predicted. Here, we assumed that the sampling period is one for ease of presentation.

Many equivalent mathematical representations exist for any given set of ordinary differential equations. Takens has shown that an equivalent representation exists where the equivalent state is defined as a finite window of past values of the time series [9]. This window is called a time delay embedding and is defined as

$$\boldsymbol{a}_t = (x_t, x_{t-\tau}, \ldots, x_{t-(m-1)\tau}), \tag{3}$$

where m is the embedding dimension and τ is the delay, both of them are natural numbers. Takens' work was later generalized and shown to apply to a broader class of systems by Casdagli and others [6]. Takens' theorem is important because it implies that an exact one-step ahead prediction is possible given a finite window of previous values:

$$\begin{aligned} y_{t+1} &= g(\boldsymbol{a}_t) \\ &= g(x_t, x_{t-\tau}, \ldots, x_{t-(m-1)\tau}) \end{aligned} \tag{4}$$

for some unknown function $g(\boldsymbol{a}_t)$. Multi-step ahead prediction can be generated by making further iterations. For example, if the embedding delay is 1 then the following equation illustrate how iterative prediction can be used to predict more one-step ahead.

$$\begin{aligned} y_{t+1} &= \hat{g}(x_t, x_{t-1}, x_{t-2}, \cdots, x_{t+1-m}) \\ y_{t+2} &= \hat{g}(y_{t+1}, x_t, x_{t-1}, \cdots, x_{t+2-m}) \\ y_{t+3} &= \hat{g}(y_{t+2}, y_{t+1}, x_t, \cdots, x_{t+3-m}) \\ &\vdots \end{aligned} \tag{5}$$

The generalization to other values of τ is straightforward.

Any of the standard nonlinear models can be used to create an approximation for $g(\boldsymbol{a}_t)$. We use the hybrid neural network of the feed-forward network and the local model to acquire the high approximation for $g(\boldsymbol{a}_t)$ efficiently.

2.2 Structure of the Hybrid Neural Network

Figure 1 shows the structure of the hybrid neural network. In this figure, we assumed that the embedding delay is one for ease of presentation. Each of predictor's outputs is integrated as the predicted value which is the network output. We compute the output of the hybrid neural network, y_t for the desired output (x_t) as follows:

$$y_t = w_{c_1} y_{c_1 t} + w_{c_2} y_{c_2 t}, \tag{6}$$

where w_{c_1} is the contribution weight of the feed-forward network (c_1) and w_{c_2} is the contribution weight of the local model (c_2). We compute these contribution weights as follows:

$$w_{c_1} = \frac{e_{c_2}}{e_{c_1} + e_{c_2}}, \quad w_{c_2} = \frac{e_{c_1}}{e_{c_1} + e_{c_2}}, \tag{7}$$

e_{c_1} and e_{c_2} are the evaluation variables for c_1 and c_2, respectively. Although many variables can use as the evaluation variable, the following NMSE (normalized mean squared error) is used in this paper.

$$e_{c_1} = \frac{\sum_{t \in T} (y_{c_1 t} - x_t)^2}{\sum_{t \in T} (x_t - \mathrm{E}[x])^2}, \quad e_{c_2} = \frac{\sum_{t \in T} (y_{c_2 t} - x_t)^2}{\sum_{t \in T} (x_t - \mathrm{E}[x])^2}, \tag{8}$$

t is the discrete time variable which is belong to the set T.

In order to evaluate how the predictors are well constructed, we select five random points in the training interval $0.5N < t \leq 0.95N$ and compute e_{c_1} and e_{c_2} for the following $0.05N$ samples. Here, N denotes the number of total training samples. We use all N samples for the feed-forward network training, and $0.5N$ samples for estimation of e_{c_2} in the local model. Note that the selected samples for feed-forward network are added Gaussian noise to reduce the effect of the over-training. The average of five estimate variables is used as the estimate valuable in our hybrid neural network.

Feed-Forward Network. In order to compute $y_{c_1 t}$, we use the feed-forward network. The output of the feed-forward network, $y_{c_1 t}$ can be shown by

$$y_{c_1 t} = f\left(\sum_{j=1}^{H} \left(w_{1j}^{(2)} f\left(\sum_{i=1}^{m} \left(w_{ji}^{(1)} x_{t-i} + \theta_i\right)\right) + \theta_j\right)\right), \tag{9}$$

where $w_{ji}^{(1)}$ is the connection weight from the input i to the hidden unit j, $w_{1j}^{(2)}$ is the connection weight from the hidden unit j to the output unit 1, $\theta_{i,j}$ is

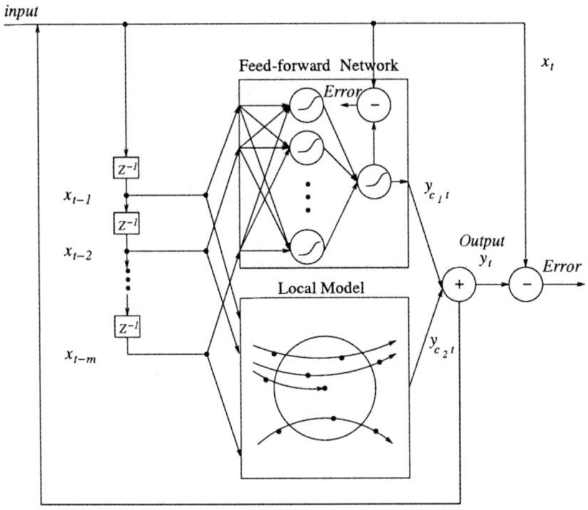

Fig. 1. Structure of the hybrid neural network

the bias of the corresponding unit, and H is the number of hidden units. The input-output function, f, used the sigmoid function as follows:

$$f(x) = \frac{1}{1 + \exp(-\varepsilon x)} , \tag{10}$$

where ε denotes the gain coefficient.

In order to minimize MSE (mean squared error), the interconnection weights of the epoch p ($1 \leq p \leq P$) are updated using backpropagation [10]. In this paper, we use the modified gradient descent which is added a momentum term.

Local Model. In order to compute y_{c_2t}, we use the local model. Generating a prediction with the local model can be divided into two steps.

First, the k nearest neighbor in the data set of a given input query vector $\boldsymbol{a}_t$ are found. Here $\boldsymbol{a}_t$ is the time delay embedding given as Eq.(3).

Second, a simple model is constructed using only the k neighboring points to generate the prediction y_{c_2t}. Although the two most common types of local models are local averaging models and local linear models, we use the local averaging model because of its simplicity. We compute y_{c_2t} for x_t as follows:

$$y_{c_2t} = \frac{1}{k} \sum_{i=1}^{k} x_{a_i} , \tag{11}$$

where a_i is the data set index of the i th nearest neighbor.

3 Experimental Results

In order to evaluate the prediction capability of the hybrid neural network, we use Mackey-Glass ([7]) data[1] and K.U. Leuven time series competition([11]) data.[2] In both problems, We use 2000 samples for training and predict following 200 samples iteratively. We select five validation sets from the training set at random.

Table 1 shows the parameters on our experimental results. All computations of the algorithms written in C are performed on an IBM PC-AT machine (CPU: Intel Pentium III 600MHz, Memory: 128MB). We compute 20 trials for each of problems.

Table 1. Values of parameters

parameter	**value**
Embedding dimension (m)	20
Time interval (τ)	1
Hidden neurons (H)	20
Learning rate (η)	0.05
Gain coefficient (ε)	0.5
Momentum parameter (μ)	0.5
Learning iterations (P)	500000
Nearest neighbors (k)	3

3.1 Mackey-Glass Data

Figure 2 shows the prediction of the best of the hybrid neural network and predictions of the feed-forward network and the local model which are used in this hybrid neural network. In this case, w_{c_1} is 0.104 and w_{c_2} is 0.896. Although, the result of the feed-forward network saturates gradually (MSE=0.0358, NMSE=0.6874), the result of the hybrid neural network catches the trajectory (MSE=0.0028, NMSE=0.053) better than the result of the local model (MSE=0.0103, NMSE=0.1975). The accumulated squared errors of these models are shown in Figure 3.

3.2 K.U. Leuven Competition Data

Figure 4 shows the prediction of the best of the hybrid neural network and predictions of the feed-forward network and the local model which are used in this hybrid neural network. In this case, w_{c_1} is 0.426 and w_{c_2} is 0.574. The result of the feed-forward network saturates gradually (MSE=0.0826, NMSE=1.442). The

[1] The data set was available at http://www.ece.pdx.edu/~mcnames/DataSets/ .

[2] The data set was available at http://www.esat.kuleuven.ac.be/sista/workshop/.

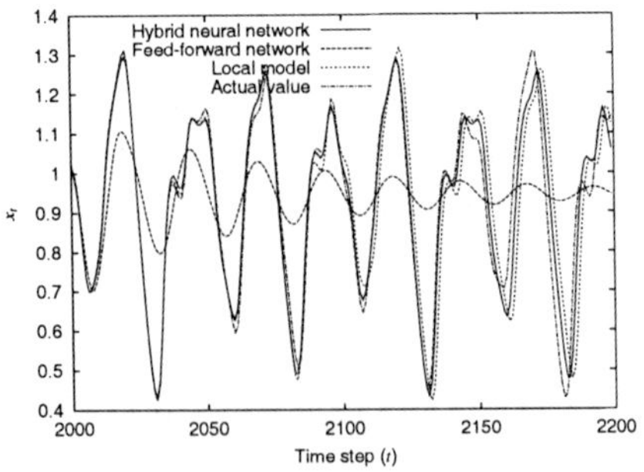

Fig. 2. Time series predictions of Mackey-Glass data

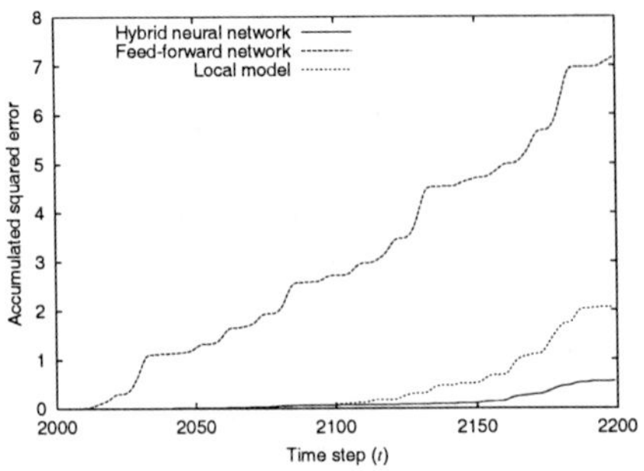

Fig. 3. Evolution of the accumulated squared errors of Mackey-Glass data

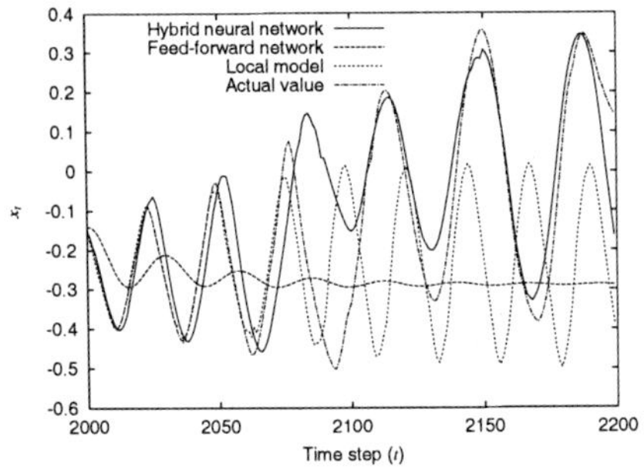

Fig. 4. Time series predictions of K.U.Leuven competition data

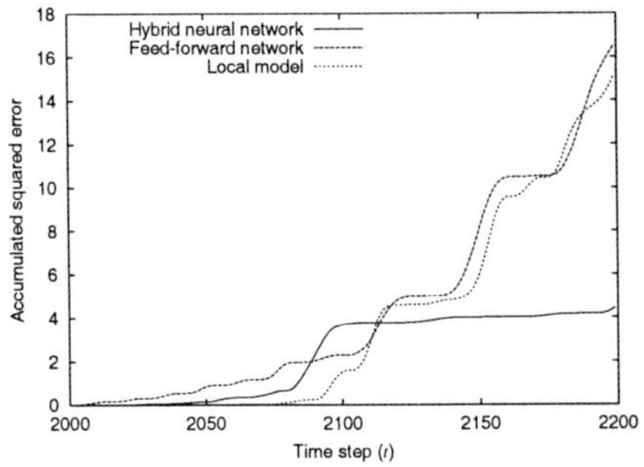

Fig. 5. Evolution of accumulated squared errors of K.U. Leuven competition data

result of the local model catches the trajectory until 2080, but the following predicted wave shifts the periodic trajectory (MSE=0.0756, NMSE=1.32). On the other hand, the result of the hybrid neural network catches the trajectory with the exception of the interval from 2080 to 2100 (MSE=0.0222, NMSE=0.3884). The accumulated squared errors of these models are shown in Figure 5. The result of this hybrid neural network is also superior to the second prize winner in the contest (MSE=0.0475) [12].

4 Conclusions

In this paper, we proposed a hybrid neural network for chaotic time series prediction. The experimental results show that the hybrid neural network can predict the middle term by fusion of the output of the global model, which is implemented as the standard feed-forward network and the local model, which is implemented as the simple nearest neighbor method successively. In the future work, we will consider the optimization of the model parameters using the heuristic algorithm and the cross-validation method.

References

1. Brockwell, P. J. and Davis, R. A.: Introduction to Time Series and Forecasting. Springer-Verlag, New York (1996)
2. Lapedes, A. and Farber, R.: Nonlinear signal processing using neural networks: Prediction and system modelling. Technical Report LA-UR–87-2662, Los Alamos National Laboratory, Los Alamos (1987)
3. Casdagli, M.: Nonlinear Prediction of Chaotic Time Series. Physica D **35** (1989) 335–356
4. Weigend, A. S. and Gershenfeld, N. A. (eds.): Time Series Prediction: Forecasting the Future and Understanding the Past. Santa Fe Institute Studies in the Sciences of Complexity, Vol. 15. Perseus Books Publiching, Reading (1994)
5. Haykin, S.: Neural Networks: A comprehensive foundation. 2nd edn. Prentice-Hall, Upper Saddle River (1999)
6. Casdagli, M., Sauer, T. and Yorke, J.: Embedology. Journal of Statistical Physics **65** (1991) 579–616
7. Mackey, M. C. and Glass, L.: Oscillation and Chaos in Physiologist Control Systems. Science **197** (1977) 287–289
8. Suykens, J. and Vandewalle, J. (eds.): Nonlinear Modeling: Advanced Black-box Techniques. Kluwer Academic Publishers, Boston (1998)
9. Takens, F.: Detecting strange attractors in turbulence. In: Rand, D. and Young, L. (eds.): Dynamical Systems and Turbulence, Warwick 1980. Lecture Notes in Mathematics, Vol. 898. Springer-Verlag, Berlin (1981) 366–381
10. Rumelhart, D. E., Hinton, G. E. and Williams, R. J.: Learning Internal Representations by Error Propagation. In: Rumelhart, D. E., McClelland, J. L. and the PDP Research Group (eds.): Parallel Distributed Processing: Explorations in the Microstructure of Cognition. The MIT Press, Cambridge (1986) 318–362
11. McNames, J., Suykens, J. and Vandewalle, J.: Winning Entry of the K. U. Leuven Time Series Prediction Competition. International Journal of Bifurcation and Chaos **9** (1999) 1485–1500
12. Suykens, J. and Vandewalle, J.: The K.U. Leuven Competition Data: a Challenge for Advanced Neural Network Techniques. In: European Symposium on Artificial Neural Networks, ENNS, D-Facto public, Bruges (2000) 299–304

Online Symbolic-Sequence Prediction with Discrete-Time Recurrent Neural Networks*

Juan Antonio Pérez-Ortiz, Jorge Calera-Rubio, and Mikel L. Forcada

Departament de Llenguatges i Sistemes Informàtics,
Universitat d'Alacant, E-03071 Alacant, Spain

Abstract. This paper studies the use of discrete-time recurrent neural networks for predicting the next symbol in a sequence. The focus is on *online* prediction, a task much harder than the classical offline grammatical inference with neural networks. The results obtained show that the performance of recurrent networks working online is acceptable when sequences come from finite-state machines or even from some chaotic sources. When predicting texts in human language, however, dynamics seem to be too complex to be correctly learned in real-time by the net. Two algorithms are considered for network training: real-time recurrent learning and the decoupled extended Kalman filter.

1 Introduction

Discrete-time recurrent neural networks (DTRNN) [5] are generally accepted as a good alternative to feedforward networks for temporal-sequence processing. Feedback enables DTRNN to develop state representations of this kind of sequences. This work analyses the use of DTRNN for predicting online the next component in a symbolic sequence.

Arithmetic compression [7] is used to evaluate the quality of the predictor. Different symbolic-sequence sources ranging from finite-state machines to texts in human language are considered in the experiments. Unlike previous works [3, 11, 13] which performed neural *offline* prediction on the kind of sequences studied here, in this paper we concentrate on *online* prediction.

2 Prediction with DTRNN

We have chosen two classical DTRNN: Elman's *simple recurrent network* (SRN) [4], and the *recurrent error propagation network* (REPN) [10]. Both architectures are applied to online prediction of symbolic sequences. The networks work in real-time trying to produce an output as correct as possible to every component of the sequence supplied at each time step; this output is considered as a prediction of the probabilities of the next symbol in the sequence.

* Work supported by the Generalitat Valenciana through grant FPI-99-14-268 and the Spanish Comisión Interministerial de Ciencia y Tecnología through grant TIC97-0941. This paper completes the results in [8].

G. Dorffner, H. Bischof, and K. Hornik (Eds.): ICANN 2001, LNCS 2130, pp. 719–724, 2001.

Two online supervised training algorithms are considered: the real-time recurrent learning (RTRL) [14], and the decoupled extended Kalman filter (DEKF) [9]. Both of them update weights according to an error measure $E(t)$ whenever a new target or desired output is supplied. The RTRL algorithm performs gradient descent along the instantaneous error hypersurface. On the other hand, the DEKF computes recursively and efficiently a solution to the *least-squares method*: finding the curve of best fit for a given set of data in terms of minimizing the average distance between the data and the curve. The DEKF also needs the derivatives of $E(t)$, which may be calculated the same way as in the RTRL algorithm.

Grammatical Inference with DTRNN. Grammatical inference (GI) is a task where DTRNN have been widely used [3] and that is related to our problem, although it also has strong differences.

Even though one possible approach to GI consists of training the network to predict the next symbol in the sequence, the modus operandi is far different. In the case of GI, it can be easily shown (by writing the total quadratic prediction error as a sum over all the prefixes of all the sequences in the sample) that the ideal neural probability model obtained for a finite set of finite sequences through exhaustive *offline* training, *global* quadratic error function, and exclusive coding of outputs (see later), gives an output that approximates as much as possible the next-symbol relative frequencies observed in the finite sample, which could be used as approximate probabilities when treating unseen sequences.

Our problem is different in some respects: the processing is done *online*, a relatively-long *single* sequence is used (whereas in GI the length of the sequences is usually *small*), and as a result of online processing, the error function is *local*. This work is based on the conjecture, similar to Elman's [4], that even under these different conditions the output of the model may still be considered as an approximation to next-symbol probabilities.

Neural Online Prediction. This section shows briefly how DTRNN can be used to predict online the next symbol in a sequence. Consider we have an alphabet $\Sigma = \sigma_1, \ldots, \sigma_{|\Sigma|}$ and a temporal sequence to process $s[1], \ldots, s[t], \ldots, s[L]$. The number of neurons in the input and output layers is set equal to the size of the alphabet, $|\Sigma|$. Inputs and targets are coded by means of *exclusive coding*, that is, the symbol σ_i is coded with a unary vector with all the components but the i-th set to zero.

At time t the symbol $s[t]$ is coded in the input vector $\boldsymbol{u}[t]$ using exclusive coding. Introducing $\boldsymbol{u}[t]$ into the network, the output vector $\boldsymbol{y}[t]$ is obtained and normalized so as all its components add to one. Now, $\boldsymbol{y}_i[t]$ can be interpreted as the probability of next symbol being σ_i. After that, the next observed symbol $s[t+1]$ is exclusively coded in $\boldsymbol{d}[t]$, which is used as the target for weight updating by the online training algorithm.

We will study empirically whether the probabilities obtained this way are a good approximation to the real ones. Online learning makes this difficult because the net is simultaneously learning how much history to keep and how to predict from that history, and proving convergence to the real probabilities is not trivial; this is worth further theoretical study.

3 Measure of Prediction Quality

Mean squared error may be a way of measuring the quality of a predictor. But in the common case where we do not know the real next-symbol probabilities, the error may only be based on the difference between the predicted symbol (usually the one with the highest probability) and the symbol observed at next time step. An alternative error measure which considers the whole vector of predicted probabilities is necessary: arithmetic compression is a possible solution.

An *arithmetic compressor* [7] derives its performance from a correct model of next-symbol probabilities. The important property that makes it appropiate for evaluating the quality of a predictor (neural or not) is that the better this probability model, the larger the *compression ratio* (CR) obtained. Besides that, this model can be adaptive, which is the case when the processing is online and probabilities change dynamically at every time step.

With the object of comparising, in addition to DTRNN, *n-grams* are used as an alternative probability model for arithmetic compression. In an n-gram model, next-symbol probability depends only on the $n-1$ symbols preceding it. Very good results are obtained when various models of different orders n are maintained simultaneously and the probability estimation is computed by mixing them [7]. The predictor considered in our experiments uses $n \in [0,4]$.

4 Results

This section presents the results of the experiments developed with some symbolic sequences. The graphs illustrate the CR versus the number of state neurons, denoted with n_X.[1] Results are shown for both the RTRL and the DEKF training algorithms (using a quadratic prediction error), and SRN and REPN architectures. Note that $n_X = 0$ means that there is no recurrence at all (in the case of SRN, only output biases are adjusted).

RTRL was used with learning rate 0.9 and momentum 0.4 (chosen after preliminary experiments). In the case of the DEKF, some experiments were carried out in order to determine correct values for the tunable parameters of the algorithm; the values proposed by Haykin [5, p. 771] proved to be correct.

The results are the average of 7 experiments (the variance was very small in all cases so it is omitted). The initial values for the weights were taken randomly from a uniform distribution in $[-0.2, 0.2]$. We consider three different kinds of symbolic sequences: generated by finite-state machines, chaotic, and texts in human language. A discussion of one particular sequence of each group and the results obtained follows.

[1] The code used for the arithmetic coder was written by Nelson [7] and is freely available at `http://dogma.net/markn/articles/arith`.

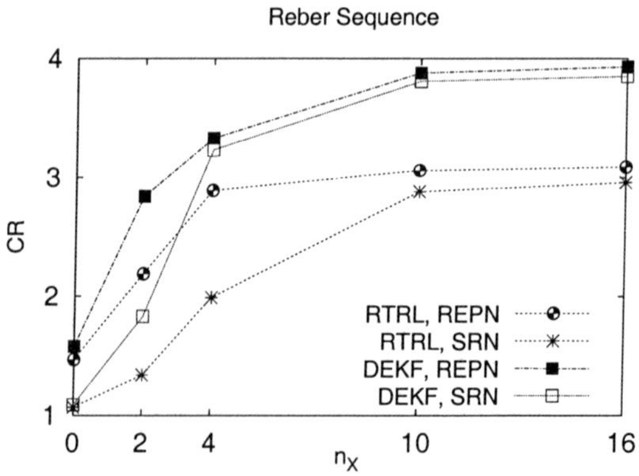

Fig. 1. Compression ratios for the continual embedded Reber sequence. The [0, 4]-gram model gives a CR of 4.23.

Finite-State Sequences. The experiments consider the continual embedded Reber sequence [12]. From the outset, good results are expected since DTRNN are capable of learning *offline* regular languages [3] and even emulating finite-state machines [2]. On the other hand, languages derived from this automaton have been proved [12] to be hard to learn with common recurrent architectures due to the existence of long-term dependencies [1] (see also [6]).

The graph in Fig. 1 illustrates the CR for this sequence (20000 symbols long, $|\Sigma| = 7$). The CR with the $[0, 4]$-gram model is 4.23. As can be seen, the number of state neurons n_X affects the CR attained, although for values of $n_X \geq 10$ the influence is not very significant. Both architectures, REPN and SRN, give comparable results. The DEKF training algorithm gives CR near those of the $[0, 4]$-gram model, whereas those of RTRL are lower.

Chaotic Sequences. We show the results for a symbolic sequence composed of the activation measures of a chaotic laser [13]. Figure 2 shows the CR for this sequence (length 10000, $|\Sigma| = 4$). The $[0, 4]$-gram model gives a CR of 2.73.

When using RTRL, differences between REPN and SRN are bigger than before. Again, the DEKF surpasses the results of RTRL. Anyway, RTRL and the REPN give CR (for $n_X > 4$) similar to those of the $[0, 4]$-gram model, and the DEKF (independently of the architecture) attains much higher ones (near 4).

Human Language Texts. Finally, we considere an essay in English about Polish film director Krzysztof Kieslowski. Figure 3 illustrates the results for this sequence in human language (length 62648, $|\Sigma| = 27$). The CR with the $[0, 4]$-gram model is 1.85.

As can be seen, the networks are not capable of developing a useful state representation of sequence dynamics. The influence of n_X on the prediction with the REPN and both training algorithms is not noticeable, although DEKF

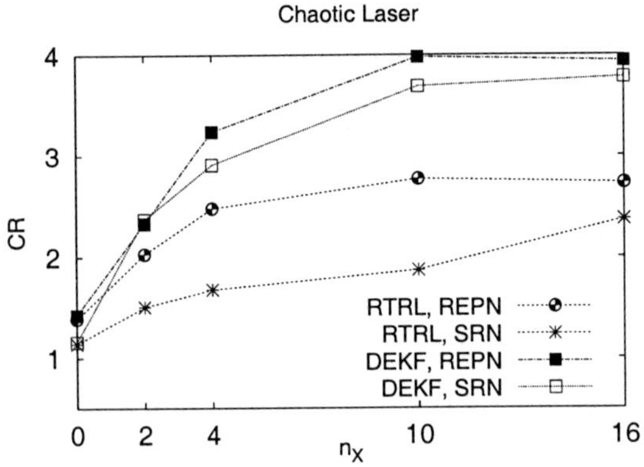

Fig. 2. Compression ratios for the chaotic laser sequence. The [0, 4]-gram model gives a CR of 2.73

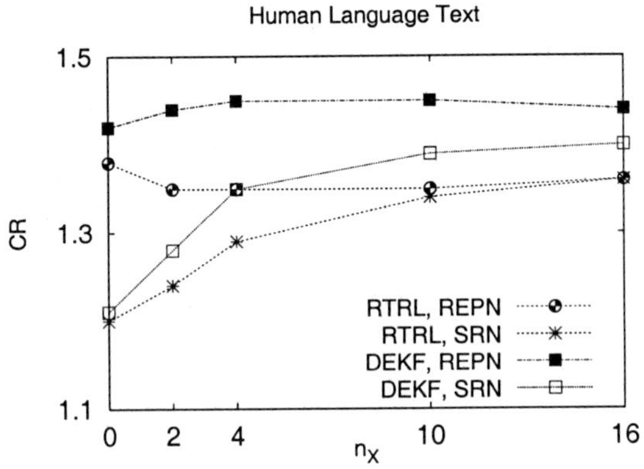

Fig. 3. Compression ratios for the natural language sequence. The [0, 4]-gram model gives a CR of 1.85

consistently gives better results. The SRN partially overcomes this limitation, although the CR are lower than those of the REPN. The results of the DEKF are once more slightly better than the results of RTRL, but they are still far lower than those achieved with the [0, 4]-gram model.

5 Concluding Remarks

The DEKF training algorithm outperforms RTRL when predicting online the symbols in a sequence. It has to be pointed out, however, that the DEKF has greater complexity because it has to compute the same number of derivatives plus some matrix operations (including an inversion) at every time step.

In the case of the regular sequence the results of the DEKF are comparable to those obtained with n-grams. When processing the chaotic sequence, even the results obtained with RTRL are similar to those of an [0, 4]-gram probability model, and the DEKF clearly outperforms it.

When dealing with sequences in human language, however, DEKF results are much worse than [0, 4]-gram results. Prediction over texts in human language seems to be a difficult task for RTRL and the DEKF. It has to be noted, anyway, that the number of free parameters used in a [0, 4]-gram model is far greater than the number of parameters in the recurrent neural networks used in this paper.

A more detailed and theoretical study has to be carried out in order to analyse how a DTRNN approximates symbol probabilities when working online. We have found that acceptable approximations are possible for regular or simple chaotic sequences and that the problem is far more arduous with more complex sequences. Besides that, it would be very interesting to analyse the evolution of the network state while processing the different kind of sequences used in this paper.

References

1. Bengio, Y., P. Simard and P. Frasconi (1994), "Learning long-term dependencies with gradient descent is difficult", *IEEE Transactions on Neural Networks*, **5**(2).
2. Carrasco, R. C. et al. (2000), "Stable-encoding of finite-state machines in discrete-time recurrent neural nets with sigmoid units", *Neural Computation*, **12**(9).
3. Cleeremans, A., D. Servan-Schreiber and J. L. McClelland (1989), "Finite state automata and simple recurrent networks", *Neural Computation*, **1**(3), 372–381.
4. Elman, J. L. (1990), "Finding structure in time", *Cognitive Science*, **14**, 179–211.
5. Haykin, S. (1999), *Neural networks: a comprehensive foundation*, Chapter 15, Prentice-Hall, New Jersey, 2nd edition.
6. Hochreiter, J. (1991), *Untersuchungen zu dynamischen neuronalen Netzen*, Diploma thesis, Institut für Informatik, Technische Universität München.
7. Nelson, M. (1991), "Arithmetic coding + statistical modeling = data compression", *Dr. Dobb's Journal*, Feb. 1991.
8. Pérez-Ortiz, J. A., J. Calera-Rubio, M. L. Forcada (2001), "Online text prediction with recurrent neural networks", *Neural Processing Letters*, in press.
9. Puskorius, G. V. and L. A. Feldkamp (1991), "Decoupled extended Kalman filter training of feedforward layered networks", in *International Joint Conference on Neural Networks*, volume 1, pp. 771–777.
10. Robinson, A. J. and F. Fallside (1991), "A recurrent error propagation speech recognition system", *Computer Speech and Language*, **5**, 259–274.
11. Schmidhuber, J. and S. Heil (1996), "Sequential neural text compression", *IEEE Transactions on Neural Networks*, **7**(1), pp. 142-146.
12. Smith, A. W. and D. Zipser (1989), "Learning sequential structures with the real-time recurrent learning algorithm", *International Journal of Neural Systems*, **1**(2).
13. Tiňo, P., M. Köteles (1999), "Extracting finite-state representations from recurrent neural networks trained on chaotic symbolic sequences", *IEEE Transactions on Neural Networks*, **10**(2), pp. 284–302.
14. Williams, R. J. and R. A. Zipser (1989), "A learning algorithm for continually training recurrent neural networks", *Neural Computation*, **1**, 270–280.

Prediction Systems Based on FIR BP Neural Networks*

Stanislav Kaleta, Daniel Novotný, and Peter Sinčák

Center for Intelligent Technologies, Computational Intelligence Group
Department of Cybernetics and AI
Faculty of Electrical Engineering & Informatics
TU Košice, Letná 9, 040 00 Košice
SLOVAK REPUBLIC
http://neuron-ai.tuke.sk/cig
cig@neuron-ai.tuke.sk

Abstract. Paper deals with experience of application of FIR filter (Finite Impulse Response) in BP neural networks used for prediction system. Prediction can be defined as extrapolation of unknown function determine by representing data sets. One of the most difficult problems in prediction systems is determination of the size of the time series variable input. It is always very difficult to determine the right size of the "history" of input variable necessary for prediction – extrapolation of the modeled function. So in general prediction problem can have input e.g. ($\mathbf{x}(t)$, $\mathbf{x}(t-1)$, …,$\mathbf{x}(t-\tau)$) , where τ - determine the width of the input window size and output ($\mathbf{y}(t+1)$, $\mathbf{y}(t+2)$, …, $\mathbf{y}(t+\lambda)$), where λ - is size of the length of the time series of the predicted variable and $\mathbf{x}$ and $\mathbf{y}$ are vectors. Usually both τ and λ depend on the problem and in case of output the size of the λ is small then there is bigger chance that prediction will be more reliable. The avoid a problem of τ determination a FIR neural networks offer solution to gather "**history**" of the unknown function inside of neural network topology , namely on neural networks synapses in the form of FIR filters.

1 Introduction

Intelligent technologies represent a very interesting direction in current and future technology. These technologies are based on AI techniques including neural networks and other intelligent approaches. Prediction problems are actual in any part of everyday life including technological processes in industrial as well as non-industrial domain. Prediction is in fact an extrapolation of the unknown function, which is being approximated by neural network or any statistical tools. Usually we have in the input

$$(\mathbf{x}(t), \mathbf{x}(t-1), \ldots, \mathbf{x}(t-\tau))$$

where τ - determine the width of the as named input window size and output ($\mathbf{y}(t+1)$, $\mathbf{y}(t+2)$, …, $\mathbf{y}(t+\lambda)$), where λ - is size of the length of the time series of the predicted variable and $\mathbf{x}$ and $\mathbf{y}$ are vectors. Always the $\tau \gg \lambda$ and the easiest situation is

* Research is supported via project # 9433 awarded by Slovak National Agency for Science "Computational Intelligence in Decision Procedures" for 1999-2001

G. Dorffner, H. Bischof, and K. Hornik (Eds.): ICANN 2001, LNCS 2130, pp. 725–730, 2001.

when **λ=1** or only one value is predicted. Certainly problems if **λ>>1** are more difficult and prediction task is more complex and difficult.

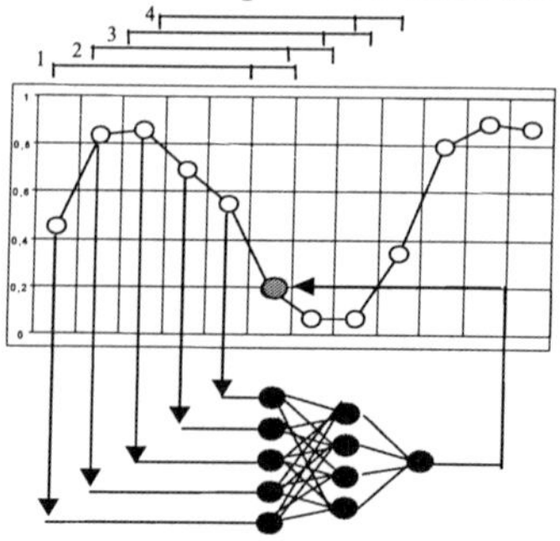

Fig. 1. Example of the time series window (t-1), (t-2), (t-τ) as the input vector. The output is a single value related to time (t+1). the input window is shifted during iterations and is related to the parts of training curve as it is shown on above figure (iterations 1-4). Many very useful sources about prediction and neural networks can be found in [1,2,3].

2 Motivation of the Project

The main objective of the project was to obtain practical experience with prediction systems, which do not require having a window (time sequence) of input variable to provide "historical" information for prediction (extrapolation) purposes. Whereas simple feed-forward network can be trained to accomplish pattern classification tasks with complex nonlinear boundaries, they are limited to processing static patterns – patterns that are fixed rather than temporal in nature. We test (with some modifications) here a network that overcomes this limitation and that is capable of processing temporally modulated signals. Networks with this capability can play an important role in applications domains that have naturally time-varying properties.

3 Error Backpropagation for FIR Neural Networks

Application of Error back propagation includes temporal error BP approach to the FIR neural training . If we assume that the normal BP neural networks was based on regular Error function as follows:

$$J^{P} = \frac{1}{2}\sum_{i=1}^{N_O} (ev_i^{P} - x_i^{P})^2 \quad \textbf{(1)}$$

where ev_i^p is expected value on the neural network for i-th output neuron concerning input pattern "p". Also x_i^p represents the real output of on i-th output neuron concerning input pattern "p" while N_0 is number of neuron in output layer .

Then a usual BP- like adaptation rule is used as follows:

$$\Delta\overline{w}_{ij}^{l}(n) = -\gamma\overline{\delta}_{j}^{l+1}(n)\overline{x}_{i}^{l}(n) \tag{2}$$

where i - pre-synaptic neuron; j - post-synaptic neuron; n - iteration step; w - synaptic weight; γ - is learning parameter of learning; δ_j^{l+1} - is an error signal associated to neuron "j" which is in layer "$l+1$"; x_i^l - is value of activation of pre-synaptic neuron "i" in layer "l"

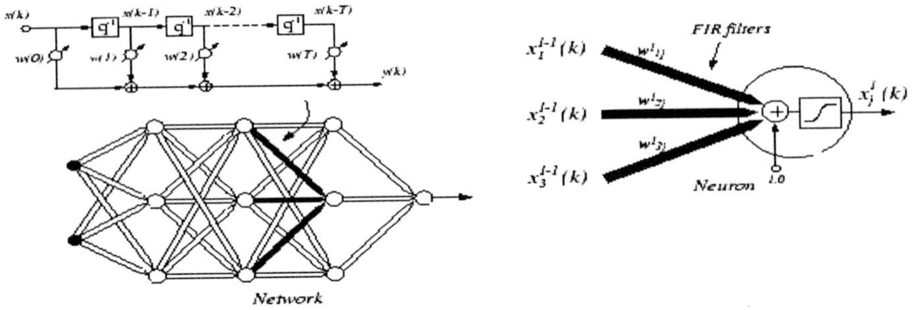

Fig. 2. Practical implementation of FIR filters into synapses of ANN.

There is basic difference between regular BP approach and this applied for FIR neural network is as follows:

- It is associated a vector synaptic weight to each synapse.
- The propagation of Error is being effective after number of inputs, which will propagate through the overall neural network up to the end. Only after then an Error will occur and can be propagated back to the neural networks.

Concerning the FIR filters on neural network which have length of the FIR filter "M" and there are L layer of the neural networks then we can express the adaptation as follows:

$$\Delta\overline{w}_{ij}^{L-1}(n) = -\gamma\delta_{j}^{L-n}(n-(L-1)M)x_{i}^{L-1-n}(n-(L-1)M) \tag{3}$$

and adaptation starts for n > $(L-1)M$ as pattern index.

4 AFIR - Adaptive Time-Delays Neural Network

Both time delays τ_{jik} and connection weights w_{jik} are adapted on-line according to a gradient descent approach. AFIR as adaptive FIR network in meaning of adaptation of time-delays in weight connections is equivalent to ATNN (Adaptive time-delays neural network), presented in [2]. This network adapts its time delay values as well its weights during training, to better adapt to changing temporal patterns, and provide more flexibility for optimal tasks. Furthermore, time-windows are not used as in previous (FIR) example. After adaptation of ATNN (AFIR) is structure of network transformable to FIR one, only some of connections according some of time-delays are cut or added. Because of way of implementation and similarity of the weight structure, name AFIR was chosen (Adaptive FIR neural network). In AFIR are retro operators from FIR $\mathbf{q}^{-1}$ replaced by τ_{jik}, where $\mathbf{q}^{-1}$ corresponds to memory cells for one iteration and τ_{jik} as variable, holds number of iteration for memory cell for which to hold signal value. Simply, one τ_{jik} means as many $\mathbf{q}^{-1}$ time-delays to past.

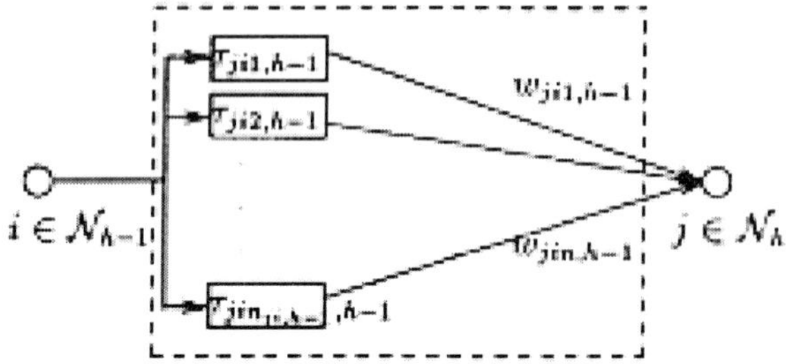

Fig. 3. Schematic description of <u>one</u> synapse connection as delay block in ATNN resp. AFIR. The entire network is constructed by the delay blocks, which connect neurons layer by layer.

There exists transformation of vector of weights, from FIR filter weights vector (Fig.2) to AFIR (resp. ATNN) based vector of weights (Fig.3). The adaptation of weight "w" and time-delay variables "τ" are related to Error function of the neural network. So the main aim is to define all proper weights and lengths of the adaptive FIR filters, which produce the best Error function results. So according to the careful analysis we consider the following equations. The adaptation of $\tau_{jik}^{h-1}(t_n)$ is as follows:

$$\tau_{jik}^{h-1}(t_n+1) = \tau_{jik}^{h-1}(t_n) + \Delta\tau_{jik}^{h-1}(t_n) \tag{4}$$

then the change of the $\tau_{jik}^{h-1}(t_n)$ is related to the Error function in typical way :

$$\Delta\tau_{jik}^{h-1}(t_n) = -\eta\frac{\partial J(t_n)}{\partial\tau_{jik}^{h-1}} \tag{5}$$

and the final formula for time-delay adaptation is as follows:

$$\Delta\tau_{jik}^{h-1} = \eta\rho_j^h(t_n)w_{jik}^{h-1}x_j^{h-1}(t_n - \tau_{jik}^{h-1}) \tag{6}$$

where ρ_j^h is error signal related to adaptation of "τ" and η is the learning parameter for the same adaptation. The adaptation of the synaptic weight is as follows

$$\Delta w_{jik}^{h-1} = \gamma\delta_j^h(t_n)x_j^{h-1}(t_n - \tau_{jik}^{h-1}) \tag{7}$$

More details can be found in [5].

5 Experiments and Results

Both FIR and AFIR network with one hidden layer were trained on the same data, with more different topologies. These networks were trained on the same data sets. The "A" set from Santa-Fe data for prediction problems and also real-world data related to power engineering domain - electricity load forecast were chosen. The first set consists of 1000 points and indicate chaotic intensity pulsations of an NH_3 laser in chaotic state. The second set of 2000 points as hour samples of electricity load measured in one junction of energetic system.
Topologies of networks had one input and one output neuron with different numbers of neurons in hidden layer and there were used more different initial parameters for trained networks:

number of hidden neurons: 5; 25
(max) length of the filters: 10; 30
sigmoid activation function: 0.1; 1
momentum: 0.02;0.2
function values of sigmoid: <0;1>; <-1;1>
adaptation coefficient for w: 0.001; 0.01; 0.1
adaptation coefficient for τ: 0.001; 0.01; 0.1

There are parameters used for experiments and all combinations of them were investigated. The topology of neural network was designed with a single input and single output and with one hidden layer with various numbers of neurons. All networks were trained on the data for some small number of iterations. When error lowered, network was retrained for next 1000 iterations. For comparison purposes was used normalised mean error:

$$ME = \frac{1}{N}\sum_{i=1}^{N} | x_i - d_i | \qquad (8)$$

where x_i is pattern value, d_i is predicted value and N is number of patterns. Best results for recursive prediction on test data are in following tables and graphs.

	ME- FIR MLP	ME-AFIR MLP
Laser	0.043	0.0076
Energy	0.062	0.0376

Table 1. Comparing of errors (ME on testing set average for 3 steps) on laser and electricity load data

6 Conclusion

The obtained experience confirms that idea of avoiding a problem of the size of input window is possible to solve and FIR and AFIR approach were tested on real-world data. Some implementation modifications were done and we obtained the prediction results from both approaches. The results show that AFIR is useful for more "complex" problems when prediction is more difficult and results on simple FIR are not promising. Adaptation of FIR filters ends up with different lengths of FIR filters in each synapse and therefore we assume that this approach is useful and effective to obtain the best prediction results. Generally it is believed that this type of research should be progressing and some more experiments on various data are underway in close future.

References

1. A. Robel :The dynamic pattern selection algorithm: Effective training and controlled generalization of backpropagation neural networks. Technical report, Technische Universität Berlin, 1994
2. J.K. Lee, W. Ch. Jhee: Performance of neural networks in managerial forecasting, In E.~Turban~E. Trippi~R.R., editor, Neural Networks in Finance a Investing. IRWIN, 1996
3. S. Andreas : Time Series Prediction: Forecasting the Future and Understanding the Past, Addison-Wesley, 1993. [[3]
4. E.A. Wan : Time series prediction by using a connectionist network with internal delay lines, time series prediction. forecasting the future and understanding the past.In Gershenfeld~N. Weigend~A., editor, SFI Studies in the Science of Complexity, 1994.
5. Daw-Tung Lin; Judith E. Dayhoff; Panos A. Ligomenides, *A Learning Algorithm for Adaptive Time-Delays in Temporal Neural Network*, System Research Center, Department of Electrical Engineering, Univesity of Maryland, 1992
6. D. Novotny, S. Kaleta: Prediction systems based on FIR Filters–TechnicalReport – in Slovak, Center for Intelligent technologies, TU Kosice December 2000.

Appendix

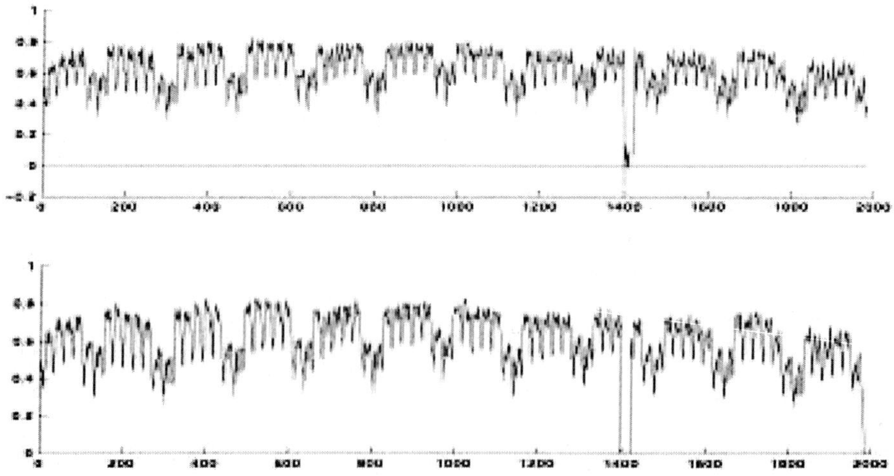

Graph. 1. Prediction results on real-world data from electricity load forecast in Slovak Republic. The lower graph is the original data and the upper graph is a predicted curve. From 1-1600 was used for training 1601-1800 was a validation set and 1801-2000 was a testing set.

On the Generalization Ability of Recurrent Networks

Barbara Hammer

Department of Mathematics/Computer Science,
University of Osnabrück
D-49069 Osnabrück, Germany
hammer@informatik.uni-osnabrueck.de

Abstract. The generalization ability of discrete time partially recurrent networks is examined. It is well known that the VC dimension of recurrent networks is infinite in most interesting cases and hence the standard VC analysis cannot be applied directly. We find guarantees for specific situations where the transition function forms a contraction or the probability of long inputs is restricted. For the general case, we derive posterior bounds which take the input data into account. They are obtained via a generalization of the luckiness framework to the agnostic setting. The general formalism allows to focus on repppresentative parts of the data as well as more general situations such as long term prediction.

1 Introduction

Recurrent networks (RNNs) constitute an elegant way of enlarging the capacity of feedforward networks (FNNs) to deal with complex data, sequences of real vectors, as opposed to simple real vectors. They process data where time is involved in a natural way – such as language or time series [5, 7]. Moreover, sequences constitute a simple form of symbolic data. Indeed, RNNs can easily be generalized such that general symbolic data, terms and formulas, can be processed as well [4, 6, 8]. One hopes that the advantages of FNNs transfer to RNNs directly; in particular, the theoretical guarantees for the success of FNNs should hold for RNNs, i.e. their universal approximation property, the possibility of efficient training, and their generalization ability [2, 11, 18]. There exist well known results showing the universal approximation capability; remarkable algorithms for efficient training have been developed, though training is more complex for RNNs than for FNNs [15]; some work has been done on the generalization ability [9]. However, the latter question is not yet answered satisfactory and we make a step further into the direction of establishing the generalization ability of RNNs in this paper.

We first examine restricted situations where the standard VC arguments can be applied directly: restricted transition functions or restricted input distributions, respectively. Afterwards, we derive posterior bounds on the generalization capability which depend on the concrete training set. For this purpose we generalize the luckiness framework to the general agnostic setting [10, 12, 16]. Unlike in [9], we obtain results which cover long-term prediction and allow the restriction to representative parts of data.

2 The Learning Scenario

A single-layer FNN computes a function $f : \mathbb{R}^m \to \mathbb{R}^n, \mathbf{x} \mapsto \sigma(\mathbf{A} \cdot \mathbf{x} - \mathbf{\Theta})$ where $\mathbf{A} \in \mathbb{R}^{n\times m}$, $\mathbf{\Theta} \in \mathbb{R}^n$, and σ denotes the componentwise application of some *activation function* $\sigma : \mathbb{R} \to \mathbb{R}$. $\mathbf{A}$ and $\mathbf{\Theta}$ are the *weights* of the network. Popular choices for σ are the sigmoidal function $\text{sgd}(x) = (1 + \text{e}^{-x})^{-1}$, or piecewise polynomial functions.

G. Dorffner, H. Bischof, and K. Hornik (Eds.): ICANN 2001, LNCS 2130, pp. 731–736, 2001.

For $m > n$, recursive application of f induces a *recursive network* which computes $\tilde{f} : (\mathbb{R}^{m-n})^* \to \mathbb{R}^n$, ($(\mathbb{R}^{m-n})^*$ denoting sequences of arbitrary length with elements in $\mathbb{R}^{m-n}$) defined by $[] \mapsto 0$, $[x_1, \ldots, x_t] \mapsto f(x_1, \tilde{f}([x_2, \ldots, x_n]))$, $[]$ denoting the empty sequence and $[x_1, \ldots, x_t]$ denoting a sequence of length t. Hence starting with 0 we recursively apply the transition function f to the already computed value, the so called *context*, and the next entry in the sequence. n is the number of *context neurons*. Often we implicitly assume that $\tilde{f}$ is combined with a projection to a lower dimensional vector space. It is easy to see that various more general definitions of RNNs can be simulated in the above formalism such that the argumentation presented in this paper holds for various realistic architectures [8]. A neural *architecture* only specifies the structure but not the weights $\mathbf{A}$ and $\mathbf{\Theta}$. Sometimes, RNNs are used for long term predictions. I.e., given an input sequence $\mathbf{x}$ we are interested in the output sequence with entries $\tilde{f}([0^t, \mathbf{x}])$, $t \in \mathbb{N}$ obtained via further application of f.

The so-called pseudodimension plays a key role in learning theory. It measures the richness of a function class. Formally, the *pseudodimension* $\mathcal{VC}(\mathcal{F})$ of a class $\mathcal{F} : X \to [0,1]$ is the largest cardinality (possibly infinite) of $X' \subset X$ which can be shattered, which means that for each $x \in X'$ a reference point $r_x \in \mathbb{R}$ exists such that for every $d : X' \to \{0,1\}$ some $f \in \mathcal{F}$ exists with $f(x) \geq r_x$ iff $d(x) = 1$ for all $x \in X'$. Upper and lower bounds on $\mathcal{VC}(\mathcal{F})$ where $\mathcal{F}$ is computed by a RNN can be found in the literature or easily derived from the arguments given therein [3, 8, 13]. For a sigmoidal or piecewise polynomial RNN, W weights, and inputs of length at most t the bounds are polynomial in W and t. Since all lower bounds depend on t we obtain an *infinite* pseudodimension for arbitrary inputs in contrast to standard FNNs.

We consider the following so called agnostic setting [17]. All sets are assumed to be measurable. Denote by X the input set, by U the output set, by $\mathcal{F}$ a function class mapping from X to U used for learning which may be given by a neural architecture. A set of probabilities $\bar{\mathcal{P}}$ on $X \times Y$ is to be learned. Some *loss function* $l : U \times Y \to [0,1]$ measures the distance of the real and the prognosed output. This may be the euclidian distance, if U and Y are subsets of $[0,1]$. Associated to each $f \in \mathcal{F}$ is the function $l_f(x,y) = l(f(x), y)$ and associated to $\mathcal{F}$ is the class $l_{\mathcal{F}} = \{l_f \mid f \in \mathcal{F}\}$. Given a finite set of examples $(x_i, y_i)_{i=1}^m$ chosen according to some unknown $\bar{P} \in \bar{\mathcal{P}}$ we want to find a function $f \in \mathcal{F}$ which models the regularity best. I.e. the error $E_{\bar{P}}(l_f) = \int_{X \times Y} l(f(x), y) d\bar{P}(x,y)$ should be as small as possible. Since this quantity is not known, often simply the empirical error $\hat{E}_m(l_f, (\mathbf{x}, \mathbf{y})) = \sum_i l(f(x_i), y_i)/m$ is minimized. In order to guarantee valid generalization, the so called uniform convergence of empirical means (UCEM) property for $\mathcal{F}$ w.r.t. l should hold:

$$\sup_{\bar{P} \in \bar{\mathcal{P}}} \bar{P}^m((\mathbf{x}, \mathbf{y}) \mid \sup_{f \in \mathcal{F}} |E_{\bar{P}}(l_f) - \hat{E}_m(l_f, (\mathbf{x}, \mathbf{y}))| > \epsilon) \to 0 \quad (m \to \infty).$$

For standard FNNs the UCEM property is guaranteed for the distribution independent case, i.e. $\bar{\mathcal{P}}$ may be arbitrary, since FNNs have finite pseudodimension. The situation is more difficult for recursive networks since the pseudodimensions is infinite for unrestricted inputs. In general, distribution independent learnability cannot be guaranteed [8]. However, one can equivalently characterize the UCEM property w.r.t. $\bar{\mathcal{P}}$ which may be a proper subset of all probabilities, i.e. for the so called distribution dependent setting, via

$$\lim_{m \to \infty} \sup_{\bar{P} \in \bar{\mathcal{P}}} \frac{E_{\bar{P}^m}(\lg L(\epsilon, l_{\mathcal{F}}|(\mathbf{x}, \mathbf{y}), |\cdot|_\infty))}{m} \to 0 \quad (m \to \infty).$$

Here L denotes the *external covering number* of $l_{\mathcal{F}}$ on inputs $(\mathbf{x}, \mathbf{y})$ measured with the pseudometric $|\cdot|_\infty$ mapping two functions to the maximum absolute pointwise distance. Explicit bounds on the generalization error arise. The key observation for the distribution independent case is that for a function class with finite pseudodimension d the external covering number on m points can be limited by $2\,(2em/\epsilon\ln(2em/\epsilon))^d$. Hence the distribution independent UCEM property is guaranteed. Additionally, the covering numbers of $l_{\mathcal{F}}$ and $\mathcal{F}$ can be related under mild conditions on l [1, 17].

3 Restricted Architectures or Distributions

Recurrent architectures with step activation function and binary input sequences have finite pseudodimension [13]. Moreover, noise may reduce the capacity of general architectures to a finite pseudodimension [14]. An analogous result holds for contractions:

Theorem 1. *Assume the class $\mathcal{F} : X \times U \to U$ induces the recursive class $\tilde{\mathcal{F}}$. Assume $C < 1$ exists such that for each $f \in \mathcal{F}$ and x, u_1, u_2 $|f(x,u_1) - f(x,u_2)| \le C|u_1 - u_2|$ holds. Then*

$$L\left(\frac{\epsilon}{1-C}, \mathcal{R}, |\cdot|_\infty\right) \le L(\epsilon, \mathcal{F}, |\cdot|_\infty).$$

Proof. Assume $\{f_1, \ldots, f_n\}$ is an external ϵ-covering of $\mathcal{F}$. Choose $f \in \mathcal{F}$. Assume f_i is a function from the cover with distance smaller than ϵ to f. For contexts u_1 and u_2 with $|u_1 - u_2| \le \epsilon(1 + C + C^2 + \ldots + C^t)$ we find for any x

$$\begin{aligned}|f(x,u_1) - f_i(x,u_2)| &\le |f(x,u_1) - f(x,u_2)| + |f(x,u_2) - f_i(x,u_2)| \\ &\le \epsilon(C + \ldots + C^{t+1}) + \epsilon\end{aligned}$$

and hence by induction $|(\tilde{f} - \tilde{f}_i)([x_1, \ldots, x_t])| \le \epsilon(1 + \ldots + C^t) \le \epsilon/(1-C)$. □

Hence RNNs induced by a contraction provide valid generalization with respect to every loss l, such that $l : U \times Y \to [0,1]$ is equicontinuous with respect to U, i.e., $\forall\epsilon > 0 \exists\delta(\epsilon)$ such that $\forall y \in Y, u_1, u_2 \in U$ if $|u_1 - u_2| \le \delta(\epsilon)$, $|l(u_1,y) - l(u_2,y)| \le \epsilon$. We call such a loss function *admissible*. This holds due to the inequality $L(\epsilon, l_{\mathcal{F}}|(\mathbf{x},\mathbf{y}), |\cdot|_\infty) \le L(\delta(\epsilon), \mathcal{F}|\mathbf{x}, |\cdot|_\infty)$ as proved in [17]. Admissible loss functions are the popular choices $|u - y|^s$, for $s \ge 1$ and $U = Y = [0,1]$, as an example. Sufficient guarantees for the contraction property of the transition function can be stated easily for a large number of activation functions including the sigmoidal function:

Lemma 1. *Any recurrent architecture with Lipschitz continuous activation function with constant K, n context neurons, and weights bounded by $C/(Kn)$ for $C < 1$, fulfills the distribution independent UCEM property with respect to admissible loss functions.*

A second guarantee for valid generalization can be obtained for restricted input distributions. Denote by X_t the input sequences restricted to length at most t and choose values p_t such that $0 \le p_t \uparrow 1$. Assume that we restrict to those probabilities $\bar{P}$ on $X \times Y$ such that the marginal distribution on X assigns a probability of at least p_t to X_t, hence prior information on the probability of long input sequences is available. It is shown in [9] that the UCEM property with respect to the linear loss holds under these conditions if the $\mathcal{VC}(\mathcal{F}|X_t) < \infty$. One can generalize the result as follows:

Theorem 2. *Assume $\mathcal{F}$ is a function class with inputs in $X = \cup X_t$ with $X_t \subset X_{t+1}$ and outputs in $[0,1]$. Assume $0 \le p_t \uparrow 1$. Assume $\bar{\mathcal{P}}$ are distributions on $X \times Y$ such that the marginal distributions assign a value $\ge p_t$ to X_t. Then $\mathcal{F}$ fulfills the UCEM property w.r.t. any admissible loss function and $\bar{\mathcal{P}}$ if the restrictions $\mathcal{F}|X_t$ fulfill the UCEM property w.r.t. the probabilities induced by $\bar{\mathcal{P}}$ on X_t and the linear loss.*

Proof. Assume the parameter $\delta(\epsilon)$ describes the admissible loss. As proved in [17], we can equivalently substitute the metric $|\cdot|_\infty$ in the UCEM characterization by the euclidian metric $|\cdot|$. Denote the restrictions of the marginal distribution of $\bar{P}$ to X_t by P_t and the restriction of $\mathcal{F}$ to inputs X_t by $\mathcal{F}_t$. One can estimate

$$\begin{aligned}
&E_{\bar{P}^m}(\lg L(\epsilon, l_\mathcal{F}|(\mathbf{x},\mathbf{y}), |\cdot|))/m \leq E_{P^m}(\lg L(\delta(\epsilon)/2, \mathcal{F}|\mathbf{x}, |\cdot|))/m \\
&\leq P(\leq m(1-\delta(\epsilon)/2) \text{ points in } X_t) \lg(2/\delta(\epsilon))^m + E_{P_t^m}(\lg L(\delta(\epsilon), \mathcal{F}_t|\mathbf{x}, |\cdot|))/m \\
&\leq \lg(2/\delta(\epsilon)) \cdot 16 p_t(1-p_t)/(m\delta(\epsilon)^2) + E_{P_t^m}(\lg L(\delta(\epsilon)/2, \mathcal{F}_t|\mathbf{x}, |\cdot|))/m
\end{aligned}$$

for $p_t \geq 1 - \delta(\epsilon)/4$ due to the Tschebychev inequality. The second term tends to zero for $m \to \infty$ if $\mathcal{F}_t$ fulfills the UCEM property with respect to P_t. □

Hence RNNs with finite pseudodimension for *restricted* input length t fulfill the UCEM property for *arbitrary* inputs with respect to admissible loss and input distributions where sequences of length $\leq t$ have a probability $\geq p_t \uparrow 1$.

4 Posterior Bounds

However, it may happen that neither prior information about the input distribution is available nor the weights fulfill appropriate restrictions. A different approach allows to derive *posterior, data dependent* bounds on the generalization ability. For this purpose we use a modification of the luckiness framework proposed in [16]. The key idea is very simple: Generalization does not hold for the general setting. However, in lucky situations, restriucted maximum input length of sequences, for example, very good generalization bounds can be obtained. Hence the situation is stratified according to the concrete setting. Some luckiness function $\mathcal{L} : X^m \times \mathcal{F} \to \mathbb{R}^+$which measures the concrete luckiness of the output of a learning algorithm on the training data. In lucky situations, i.e. for large t, the covering number $L_{\mathcal{L}\geq t}(\epsilon, l_\mathcal{F}|(\mathbf{x},\mathbf{y}), |\cdot|_\infty)$ of functions in $\mathcal{F}$, such that $\mathcal{L}$ yields at least t on $\mathbf{x}$ should be small. Additionally, the luckiness function should be smooth in the following sense: Denote by $\mathcal{P}$ the marginal distributions of $\bar{\mathcal{P}}$ on X. Denote by $B : \mathbb{R}^+ \to \mathbb{R}^+$ some monotonic, increasing function. Then

$$\sup_{P\in\mathcal{P}} P^{2m}(\mathbf{x}\mathbf{x}' \mid \exists f \in \mathcal{F} \forall \mathbf{X} \subset_\eta \mathbf{x}, \mathbf{X}' \subset_\eta \mathbf{x}' \; \mathcal{L}(\mathbf{X}\mathbf{X}', f) < B(\mathcal{L}(\mathbf{x}, f))) \leq \delta$$

should hold for some $\eta = \eta(m, \mathcal{L}(\mathbf{x}, f), \delta) : \mathbb{N}^2 \times \mathbb{R}^+ \to \mathbb{R}^+$. $\mathbf{X} \subset_\eta \mathbf{x}$ indicates that a fraction η of the points of $\mathbf{x}$ is canceled in $\mathbf{X}$. This tells us that we can bound the luckiness of a double sample with regard to the luckiness on the first half of the sample.

Theorem 3. *Assume p_t are positive values which add up to 1. Assume $\alpha > 0$. If $\mathcal{L}$ is a smooth luckiness function then one can estimate under the above conditions*

$$\sup_{\bar{P}\in\bar{\mathcal{P}}} \bar{P}^m((\mathbf{x},\mathbf{y}) \mid \exists f \exists t (\mathcal{L}(\mathbf{x}, f) \geq t \wedge |\hat{E}_m(l_f, (\mathbf{x},\mathbf{y})) - E_{\bar{P}}(l_f)| > \epsilon)) \leq \delta$$

for $\eta = \eta(m, \mathcal{L}(x, f), p_t\delta/4)$ and $\epsilon = 6\eta + 4\alpha +$

$$2\sqrt{4\eta \lg m + \frac{1}{m}\left(2\lg\frac{16}{p_t\delta} + 2\lg \sup_{\bar{P}\in\bar{\mathcal{P}}} E_{\bar{P}^{2m}}(L_{\mathcal{L}\geq B(t)}(\alpha, l_\mathcal{F}|(\mathbf{X}\mathbf{X}', \mathbf{Y}\mathbf{Y}'), |\cdot|_\infty)\right)}.$$

(The proof is similar to [1, 17] and is omitted due to space limitations.)

We want to apply this result to recurrent networks. For this purpose, we consider the dual formalism, an unluckiness function. Let $\tilde{\mathcal{F}}$ be given by a recurrent architecture. We define the unluckiness function $\mathcal{L}_\beta(\mathbf{x}, f)$ = maximum length of all but a fraction β of the sequences in $\mathbf{x}$ and outputs in $[0, 1]$. We consider the absolute loss function. The expected external α covering number of $\tilde{\mathcal{F}}$ on m points restricted to situations with unluckiness at most t can be limited by (X_t are the sequences of length $\leq t$)

$$\alpha^{-\beta m} \cdot 2\left(\frac{2\mathrm{e}(1-\beta)m}{\alpha} \ln \frac{2\mathrm{e}(1-\beta)m}{\alpha}\right)^{\mathcal{VC}(\tilde{\mathcal{F}}|X_t)}$$

Corollary 1. *The inequality* $\sup_{\bar{P}\in\bar{\mathcal{P}}} \bar{P}^{2m}(\mathbf{z} \mid |\hat{E}_m(l_{h(\mathbf{z})}, \mathbf{z}) - E_{\bar{P}}(l_{h(\mathbf{z})})| > \epsilon(\mathbf{x})) \leq \delta$ *is valid in the above situation for any output* $h(\mathbf{z})$ *of a learning algorithm* h *on* $\mathbf{z}$ *and for*

$$\epsilon(\mathbf{x}) = \mathrm{O}\left(\beta + \sqrt{\beta \ln \frac{1}{\beta} + \frac{t_x}{m} \ln \frac{1}{\delta} + \ln m \sqrt{\frac{1}{m} \ln \frac{1}{\delta}} + \frac{\mathcal{VC}(\tilde{\mathcal{F}}|X_t)}{m} \ln\left(\frac{m}{\beta} \ln \frac{m}{\beta}\right)}\right)$$

t_x *being the maximum length of all but a fraction* β *in the sample* $\mathbf{x}$.

Proof. The above unluckiness function is smooth with the identity B and $\eta(m, L, \delta) = \sqrt{1/(2m) \ln(2/\delta)}$ which follows from Hoeffding's inequality. Hence the result follows directly from Theorem 3 and the estimation of the covering number setting $p_t = 1/2^t$ and $\alpha = \beta$. □

Hence we obtain posterior bounds which depend on the concrete training sample. Moreover, we may drop a fraction if measuring the maximum length compared to [8, 16]. Note that this possibility is essential in order to obtain good bounds since the fraction of long input sequences will approximate their respective probability in the limit.

A second advantage lies in the possibility to deal with general loss functions. One interesting topic are long term predictions where the outputs consist in sequences. One possible loss function computes $d(x, y) = \sum_t p_t |x_t - y_t|$ where p_t are positive values adding up to 1. x_t and y_t are the t's component of sequences x or y, respectively, if existing, or 0 otherwise. Theorem 3 yields the following result:

Theorem 4. *Assume a recurrent sigmoidal architecture with* W *weights is given. We consider long term prediction, i.e. inputs and outputs are contained in* $[0, 1]^*$. *Choose the above loss function and choose* T *such that* $\sum_{t>T} p_t < \beta$. *Assume* $\mathbf{x}$ *is the training sample and* t_x *is the maximum length of all but a fraction* β *in* $\mathbf{x}$. *Then the empirical and real error deviate by at most* ϵ *with confidence* δ *for* $\epsilon(\mathbf{x}) =$

$$\mathrm{O}\left(\beta + \sqrt{\beta \ln \frac{1}{\beta} + \frac{t_x}{m} \ln \frac{1}{\delta} + \ln m \sqrt{\frac{1}{m} \ln \frac{1}{\delta}} + \frac{W^4(T^3 + T t_x^2)}{m} \ln\left(\frac{m}{\beta} \ln \frac{m}{\beta}\right)}\right).$$

Proof. As above we choose the smooth unluckiness which measures the length of all but a fraction β of the sample. The covering number $L_{\mathcal{L}\geq B(t)}(\beta, l_{\mathcal{F}}|(\mathbf{XX}', \mathbf{YY}'), |\cdot|_\infty)$ is to be estimated. T is chosen such that sequences of length $> T$ do not contribute to the minimum number of covering functions. For outputs of a varying length we can upper bound the covering number of the entire class via the product of the covering numbers restricted to outputs of a fixed length [1]. Hence we obtain the bound $\beta^{-\beta m} \cdot 2\prod_{t\leq T}(2em/\beta \ln 2em/\beta)^{d_{t+t_x}}$ where d_{t+t_x} is the pseudodimension of the sigmoidal recurrent architecture with inputs of length at most $t+t_x$, t_x being the maximum length of all but a fraction β in x which is limited by $O(W^4(t+t_x)^2)$. □

5 Conclusions

We have considered the in principle ability of learning neural mappings dealing with sequences, more precisely, the generalization ability of RNNs. We obtained prior bounds for restricted weights or restricted probabilities of long input sequences. Alternatively, posterior bounds depending on the concrete sample could be derived. Since the general agnostic scenario is considered, the results apply to general loss functions as demonstrated for the example of long term prediction.

References

1. M. Anthony and P. Bartlett *Neural Network Learning: Theoretical Foundations.* Cambridge University Press, 1999.
2. E. B. Baum and D. Haussler. What size net gives valid generalization? *Neural Computation*, 1, 1989.
3. B. Dasgupta and E. D. Sontag. Sample complexity for learning recurrent perceptron mappings. *IEEE Transactions on Information Theory*, 42, 1996.
4. P. Frasconi, M. Gori, and A. Sperduti. A general framework for adaptive processing of data sequences. *IEEE Transactions on Neural Networks*, 9(5), 1997.
5. C. L. Giles, G. M. Kuhn, and R. J. Williams. Special issue on dynamic recurrent neural networks. *IEEE Transactions on Neural Networks*, 5(2), 1994.
6. M. Gori (chair). Special session: Adaptive computation of data structures. *ESANN'99*, 1999.
7. M. Gori, M. Mozer, A. C. Tsoi, and R. L. Watrous. Special issue on recurrent neural networks for sequence processing. *Neurocomputing*, 15(3-4), 1997.
8. B. Hammer. *Learning with Recurrent Neural Networks.* Springer Lecture Notes in Control and Information Sciences 254, 2000.
9. B. Hammer. *Generalization ability of folding networks.* To appear in IEEE Transactions on Knowledge and Data Engineering.
10. D. Haussler. Decision theoretic generalizations of the PAC model for neural net and other learning applications. *Information and Computation*, 100, 1992.
11. K. Hornik, M. Stinchcombe, and H. White. Multilayer feedforward networks are universal approximators. *Neural Networks*, 2, 1989.
12. M. J. Kearns, R. E. Schapire, and L. M. Sellie. Toward Efficient Agnostic Learning. *Machine Learning*, 17, 1994.
13. P. Koiran and E. D. Sontag. Vapnik-Chervonenkis dimension of recurrent neural networks. *Discrete Applied Mathematics*, 86(1), 1998.
14. W. Maass and P. Orponen. On the effect of analog noise in discrete-time analog computation. *Neural Computation*, 10(5), 1998.
15. S. Hochreiter and J. Schmidhuber. Long short-term memory. *Neural Computation*, 9(8), 1997.
16. J. Shawe-Taylor, P. L. Bartlett, R. Williamson, and M. Anthony. Structural risk minimization over data dependent hierarchies. *IEEE Transactions on Information Theory*, 44(5), 1998.
17. M. Vidyasagar. *A Theory of Learning and Generalization.* Springer, 1997.
18. P. Werbos. *Beyond Regression: New Tools for Prediction and Analysis in the Behavioral Science.* PhD thesis, Harvard University, 1974.

Finite-State Reber Automaton and the Recurrent Neural Networks Trained in Supervised and Unsupervised Manner

Michal Čerňanský and Lubica Beňušková

Department of Computer Science and Engineering, FEI Slovak Technical University, Ilkovičova 3, 812 19 Bratislava 1, Slovakia
{cernansky,benus}@dcs.elf.stuba.sk, http://www.dcs.elf.stuba.sk/~benus

Abstract. We investigate the evolution of performance of finite-context predictive models built upon the recurrent activations of the two types of recurrent neural networks (RNNs), which are trained on strings generated according to the Reber grammar. The first type is a 2nd-order version of the Elman simple RNN trained to perform the next-symbol prediction in a supervised manner. The second RNN is an interesting unsupervised alternative, e.g. the 2nd-order RNN trained by the Bienenstock, Cooper and Munro (BCM) rule [3]. The BCM learning rule seems to fail to organize the RNN state space so as to represent the states of the Reber automaton. However, both RNNs behave as nonlinear iteration function systems (IFSs) and for a large enough number of quantization centers, they give an optimal prediction performance.

1 Introduction

State-vector clustering serves as the standard mechanism for the extraction of finite-state automata from RNNs [4] as well as for the extraction of predictive models in case of complex symbolic sequences [8]. It was shown, that when trained via the real time recurrent learning (RTRL), RNNs organize their state space so that close recurrent activation vectors correspond to histories of symbols yielding similar next-symbol distributions. Incidentally, by means of the theory of IFSs, Kolen [6] demonstrated that randomly initialized RNNs display a surprising amount of clustering before training .

We investigate the 2nd-order Elman RNN trained to perform the next-symbol prediction and the 2nd-order RNN trained by an unsupervised BCM rule (BCM RNN) [1]. Recently, we found that the latter type of RNN gives comparable predictive performance on the chaotic time series as the former one [9]. This time, both RNNs are trained on strings generated according to the Reber grammar (Fig. 1a). It was already shown that the Elman SRN can learn the Reber automaton perfectly [4]. In this work, we observe the evolution of performance of predictive models built upon the activation of the recurrent layer in both types of RNNs during training as a function of the amount of training and the number of quantization centers in the recurrent state space.

G. Dorffner, H. Bischof, and K. Hornik (Eds.): ICANN 2001, LNCS 2130, pp. 737–742, 2001.

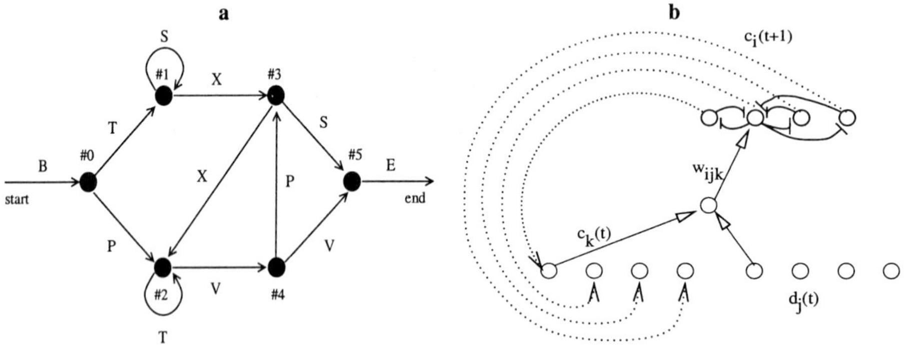

Fig. 1. (a) The Reber automaton. We start with B, and move from one node to the next. If there are two paths we can take, e.g. after B we can take either T or P, we choose one with equal probability, and so on for every node, except for the end node #5. All the Reber states are uniquely determined by each pair of symbols in the Reber string. **(b)** The 2nd-order recurrent BCM neural network with lateral inhibition

2 The 2nd-Order BCM RNN with Lateral Inhibition

First, let us consider one isolated BCM neuron, i, with the output activity $c_i(t) = \sigma(\sum_j w_{ij}(t)d_j(t))$, where σ denotes the sigmoid activation function and $d_j(t)$ is the jth input component. Then, according to the BCM theory [3], the jth synaptic weight changes as

$$\frac{dw_{ij}(t)}{dt} = -\eta\frac{\partial L_i(t)}{\partial w_{ij}(t)} = \eta\phi_i(t)d_j(t) \ , \tag{1}$$

where η is the learning speed and $\phi_i(t) = c_i(t)[c_i(t) - \theta^i(t)]$ is the synaptic modification function [2]. When the neuronal activity $c_i(t) > \theta^i(t)$, all active synapses potentiate (provided $d_j > 0$). On the other hand, when $0 < c_i(t) < \theta^i(t)$, all active synapses weaken. The variable $\theta^i(t)$ is the moving synaptic modification threshold defined as in neuro-computational models [2]

$$\theta^i(t) = E[c_i^2(t)] = \frac{1}{\tau}\int_{-\infty}^{t} c_i^2(t')e^{-\frac{t-t'}{\tau}}dt' \ , \tag{2}$$

where τ is the averaging period. Sliding of θ^i guarantees the upper boundedness of synaptic weights without placing artificial constraints on them. The quantity $L_i(t)$ is the loss function to be minimized [1], i.e.

$$L_i(t) = -\left\{\frac{1}{3}c_i^3(t) - \frac{1}{4}E[c_i^2(t)]\ c_i^2(t)\right\} . \tag{3}$$

However, the output of the ith neuron in the 2nd-order BCM RNN with the so-called feedforward lateral inhibition of constant strength μ between each pair of neurons (Fig. 1b) is [1]

$$c_i(t+1) = \sigma\left\{\sum_{j,k}\left[w_{ijk}(t) - \mu\sum_{\beta\neq i} w_{\beta jk}(t)\right]d_j(t)c_k(t)\right\} . \tag{4}$$

Synaptic weights w_{ijk} are updated at each time step t, i.e. after each presentation of a consecutive symbol in the temporal sequence. Weight changes are derived by means of the gradient descent minimization of the total loss function $L(t) = \sum L_i(t)$ [1] yielding a more complicated rule than (1), namely

$$\frac{dw_{ijk}(t+1)}{dt} = -\eta \frac{\partial L(t+1)}{\partial w_{ijk}(t)} = \eta \Big[\sum_{\alpha} \phi_\alpha(t+1) \frac{\partial c_\alpha(t+1)}{\partial w_{ijk}(t)} \Big] . \qquad (5)$$

3 Normalized Negative Log-Likelihood

We evaluate the performance or quality of predictive models $\mathcal{M}$s extracted from the RNNs by means of normalized negative log-likelihood [7]. The predictive models based on internal state vectors are constructed as follows. After each training epoch, all synaptic weights are fixed and stored. Next, sliding through the training sequence, for each symbol in S_{train}, using these weights, we record the recurrent activations $\mathbf{c}$(t+1) $= \{c_i(t+1)\}$. This creates an activation sequence $S_{\text{train}}^{\text{act}}$ for each network. Then we perform a vector quantization on $S_{\text{train}}^{\text{act}}$ for N quantization centers. In our predictive models, the quantization centers are identified with predictive states. To calculate the state-conditional probabilities, we associate with each center A counters, one for each symbol from the input alphabet $\mathcal{A}$. Sliding through the recurrent activations' sequence, $S_{\text{train}}^{\text{act}}$, the closest center C is found for the current activation vector. Next, we identify the next symbol s_{next} in S_{train}. For the center C, the counter associated with s_{next} is raised by one. After seeing the whole training sequence S_{train}, for each quantization center (prediction state) C, we normalize the counters to calculate the conditional next-symbol probabilities $P_{\mathcal{M}}(\cdot|C)$.

On the test sequence $S_{\text{test}} = s_1 s_2 ... s_t ... s_m$, the next-symbol probabilities are determined as follows: for each $t = 1, 2, ..., m-1$, given the network state $\mathbf{c}(t) = \{c_i(t)\}$, the symbol $s_t \in S_{\text{test}}$ drives the network to a new recurrent activation vector $\mathbf{c}$(t+1) $= \{c_i(t+1)\}$. We find the quantization center C_{t+1} closest to $\mathbf{c}$(t+1). The next-symbol probabilities are $P_{\mathcal{M}}(s_{t+1}|C_{t+1})$.

For each model $\mathcal{M}$, we evaluated its performance by means of the normalized negative log-likelihood NNL [7] on the test sequence S_{test}:

$$\text{NNL}_{\mathcal{M}}(S_{\text{test}}) = \frac{-\sum_{t=1}^{m-1} \log_A P_{\mathcal{M}}(s_{t+1}|C_t)}{m-1} , \qquad (5)$$

where the base of the logarithm is the number of symbols A in the alphabet $\mathcal{A}$. The higher are the next correct-symbol probabilities the smaller is NNL, with NNL $= 0$ corresponding to the 100% correct next-symbol prediction.

The metric in vector quantization and nearest-center detection is Euclidean. We use the hierarchical vector quantization inspired by [5]. The first activation vector becomes the first quantization center. Given a certain number of centers, we iteratively find the corresponding maximal cluster radius γ. For every internal state vector in turn, we find the closest center. If their distance is less than γ, the vector belongs to the center. Otherwise, this vector becomes a new center.

4 Experiments and Results

We trained the networks on randomly generated strings, starting with the "B". The activations of recurrent neurons were reset to the same small random values at the beginning of each string. Before training and after each 10×10^3 symbols we evaluated the quality of the next-symbol prediction by means of NNL. For the NNL calculation we used the test set of the length 20×10^3 symbols consisting of newly generated strings. The input of both networks and output of the Elman RNN had the dimension of 6 with the one-hot encoding of symbols. "E" is equally coded as "B". The values of parameters of the 2nd-order BCM RNN were set to: the number of (recurrent) neurons $n = 8$ or 12, $\tau = 100$ iterations, $\eta = 0.001$, $\mu = 1/n$, $\lambda = 0.4$ for the unipolar '0-1' sigmoid. The values of parameters of the 2nd-order RNN were set to: $n = 6$ or 8, $\eta = 0.03$, momentum $= 0.03$, and $\lambda = 1$ for the unipolar '0-1' sigmoid. The initial (reset) activations of recurrent neurons and initial weights were randomly generated from a uniform distribution over $[-0.5, 0.5]$, for both RNNs.

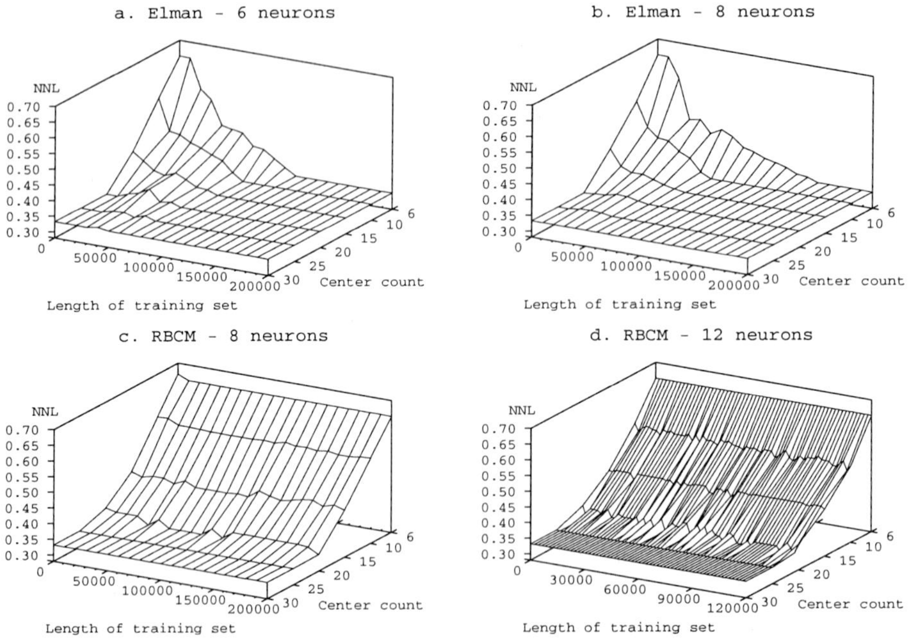

Fig. 2. NNL results for the next-symbol prediction of the Elman RNN for **(a)** 6 and **(b)** 8 recurrent neurons, and of the BCM RNN for **(c)** 8 and **(d)** 12 recurrent neurons. NNL at time 0 expresses the prediction performance of an untrained "naïve" network. Length of training set means the total number of symbols over all training strings. Center count means the number of quantization centers in the hierarchical vector quantization. The lowest NNL matches the theoretically calculated NNL for the occurence of symbols in Reber strings, e.g. 0.331.

In Fig. 2, we present NNLs of the next-symbol prediction on the test Reber strings for both types of RNNs, as they evolve during training for different numbers of quantization centers. All presented values are averages over 5 different runs. The standard deviations (not shown) may reach up to 20% of the corresponding means.

5 Discussion

One can show that the Elman RNN behaves as a kind of nonlinear IFS consisting of a sequence of transformations that map the state space represented by activations of recurrent neurons into separate subspaces of the state space [6]. Each next state of an RNN, represented by the recurrent activations, is mostly determined by the last performed transformation, e.g. by the last presented symbol (input). Within the subspace belonging to the last transformation, the network state is determined by the last previous transformation, and thus by the last previous symbol (input). In this way, the RNN "theoretically" codes an infinite time window to the past, e.g. its current state uniquelly represents the history of inputs. The more distant the input is in the past the less it determines the current RNN state. It turns out, that two subsequences with the common suffix will correspond to the two states that are close to each other in the state space. The longer the common suffix the smaller the distance between the corresponding states in the state space.

All the Reber states are uniquely determined by each pair of symbols in the Reber string. There are 20 of these pairs including the "SE" and "VE" pairs. Thus, even in the "naïve" untrained RNN, the 20 quantization centers obtained by means of the hierarchical clusterization described above, correspond to the RNN states determined by the last 2 symbols. It means that there is a unique mapping between the automaton states and the RNN states and the prediction model gives the minimal possible NNL, in this case 0.331 (Fig. 2). In an untrained case, the individual states (recurrent activities) evoked by the last two symbols are well separated in the state space. Given more than 20 (e.g. 25 or 30) quantization centers, these recurrent activities can become split, due to the third last input. Therefore, the quality of the next-symbol prediction does not get worse.

The Elman RNN is trained by RTRL, thus the weights are updated after each symbol in turn. The errors that backpropagate from the output layer cause the weights to modify in such a way that the recurrent activities which ought to belong to the same automaton state get closer to each other step by step. They get closer, thus reflecting the probability that the continuation of the sequence will be the same. This movement may cause a temporary disturbance of the coding based on the last pair of symbols as we can observe for the large number of quantization centers (Fig. 2a, b).

After the training, the network state reflects not only the history of inputs but, which is perhaps more important, also the probability of a certain continuation. Each network state belongs to a certain automaton state. It can be shown that the optimal prediction model can be created with only 6 clusters in the state space (see also [4]).

The BCM RNN behaves also like a kind of IFS, albeit different from the previous one because of the lateral inhibition. In this case, we can also obtain the optimal prediction with a large number of quantization centers (e.g. ≥25) for the predictive model built upon the "naïve" untrained network (Fig. 2c, d). Then the weights are modified after each presentation of input according to the BCM rules for the recurrent network [1]. The recurent neurons become sensitive to particular statistical characteristics of the input set (see e.g. eq. 2), and they become selective only to individual last symbols and thus we are not able to observe the improvement of prediction for the number of quantization centers smaller than 20.

Acknowledgments

Supported by the VEGA grants 2/6018/99 and 1/7611/20. We thank Peter Tiňo and anonymous reviewers for their valuable comments.

References

1. Bachman, C.M., Musman, S.A., Luong, D., Shultz, A.: Unsupervised BCM projection pursuit algorithms for classification of simulated radar presentations. Neural Networks **7** (1994) 709–728
2. Beňušková, L., Diamond, M.E., Ebner F.F.: Dynamic synaptic modification threshold: computational model of experience-dependent plasticity in adult rat barrel cortex. Proc. Natl. Acad. Sci. USA **91** (1994) 4791–4795
3. Bienenstock, E.L., Cooper, L.N, Munro, P.W.: Theory for the development of neuron selectivity: orientation specificity and binocular interaction in visual cortex. J. Neurosci. **2** (1982) 32–48
4. Cleeremans, A., Servan-Schreiber, D., McClelland, J.L.: Finite state automata and simple recurrent networks. Neural Comp. **1** (1989) 372–381
5. Das, S., Das, R.: Induction of discrete-state machine by stabilizing a simple recurrent network using clustering. Comp. Sci. Inf. **21** (1991) 35–40
6. Kolen, J.F.: The origin of clusters in recurrent neural network space. In: The Proceedings of the 16th Annual Conference of the Cognitive Science Society, Atlanta, GA, August 13-16, 1994
7. Ron, D., Singer, E., Tishby, N.: The power of amnesia: learning probabilistic automata with variable memory length. Machine Learning **25** (1996) 117–149.
8. Tiňo, P., Köteleš, M.: Extracting finite state representations from recurrent neural networks trained on chaotic symbolic sequences. IEEE Trans. Neural Net. **10** (1999) 284–302
9. Tiňo, P., Stančík, M., Beňušková, L.: Building predictive models on complex symbolic sequences with a second-order recurrent BCM network with lateral inhibition. Proc. Intl. Join Conf. Neural Net. vol. 2 (2000) 265–270

Estimation of Computational Complexity of Sensor Accuracy Improvement Algorithm Based on Neural Networks

Volodymyr Turchenko, Volodymyr Kochan, and Anatoly Sachenko

Institute of Computer Information Technologies
Ternopil Academy of National Economy
3 Peremoga Square, 46004, Ternopil, UKRAINE
vtu@tanet.edu.te.ua

Abstract. The estimation method of computational complexity of sensor data acquisition and processing algorithm based on neural networks is considered in this paper. An application of this method allows to propose a three-level structure of distributed sensor network with improved accuracy.

1 Introduction

The problem of accuracy improvement of sensor data acquisition and processing [1] is relevant in distributed sensor networks (DSN) [2]. Thus, sensor drift is the predominant compound of common error [3]. Using structure-algorithmic methods of accuracy improvement [4], for example testing and calibration, requires the interruption of data acquisition process that limits the usage of these methods. A prediction method [3] does not have this lack and allows correcting sensors instability during the system's operation. However, the specific features of sensor drift don't provide necessary prediction accuracy in the usage non-adaptive mathematical models. Therefore, it is expedient to combine testing or calibration methods with a prediction method [5]. The authors propose to use artificial neural networks (NN) for sensor drift prediction and accuracy improvement of sensor data acquisition and processing in DSN [6]. However, the use of neural networks requires the corresponding computational resources. Therefore, it is necessary to evaluate computational complexity of sensor data acquisition and processing algorithm based on neural networks in order to develop DSN hardware structure.

2 Sensor Data Acquisition and Processing Algorithm Based on Neural Networks

Sensor data acquisition and processing algorithm based on neural networks uses three kinds of neural networks: (i) integrating "historical" data NN, (ii) approximating NN and (iii) predicting NN. These neural networks were mainly considered in other previous publications of the same team of authors [7-9].

The model of heterogeneous neural network (Fig. 1) is used as predicting NN (PNN), which consists of N input neurons, M neurons of the hidden layer with the logistic activation function and one output neuron with the linear activation function [7]. The output neuron value

G. Dorffner, H. Bischof, and K. Hornik (Eds.): ICANN 2001, LNCS 2130, pp. 743–748, 2001.

$$y = \sum_{j=1}^{M} w_{j3} h_j - T_o , \tag{1}$$

where w_{j3} is the synapse from j-neuron of hidden layer to output neuron; T_o is the threshold of the output neuron; $h_j = F(\sum_{i=1}^{N} w_{ij} x_i + \sum_{k=1}^{M} w_{kj} h_k(t-1) + w_{3j} y(t-1) - T_j)$ is the output value of j-neuron of hidden layer in t moment of time; w_{ij} is the synapse from i-input neuron to j-neuron of the hidden layer; x_i is the i-element of input vector p; w_{kj} is the synapse from k-neuron of the hidden layer to j-neuron of the same layer; $h_k(t-1)$ is the output value of k-neuron of hidden layer in the previous moment of time $t-1$; w_{3j} is the synapse from output neuron to j-neuron of the hidden layer; $y(t-1)$ is the value of output neuron in the previous moment of time $t-1$; T_j is the threshold of j-neuron of the hidden layer.

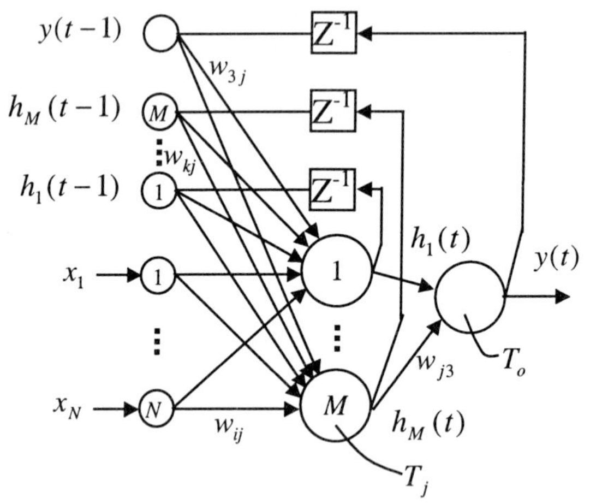

Fig. 1. Structure of Predicting Neural Network

The back propagation error algorithm [7, 9] is used for PNN training. The sum-squared error $E^p(t)$ for each training vector p and common training error $E(t)$ is calculated as

$$E^p(t) = 0.5(y^p(t) - d^p)^2 , \; E(t) = \sum_{p=1}^{P} E^p(t) , \tag{2}$$

where $y^p(t)$ is current and d^p is desired PNN output value for training iteration t for each training vector p, the error of output neuron is $b_3^p(t) = y^p(t) - d^p$.

The adaptive learning rate [7, 9] for output neuron with linear activation function

$$\alpha_3^p(t) = 1/(1 + \sum_{j=1}^{M} h_j^p(t)) \tag{3}$$

and for neurons of the hidden layer with the logistic activation function

$$\alpha_2^p(t) = 4/(R^2 \sum_{j=1}^{M} (w_{j3}^p(t))^2 h_j^p(t)(1 - h_j^p(t)) , \tag{4}$$

where $R = 1 + \sum_{i=1}^{N} (x_i^p(t))^2 + (y(t-1))^2 + \sum_{j=1}^{M} (h_j^p(t-1))^2$.

The synapses and neuron's thresholds are changed during the training process taking into account (3) and (4)

$$w_{ij}^{p}(t+1)=w_{ij}^{p}(t)-\alpha_{2}^{p}(t)b_{3}^{p}(t)w_{j3}^{p}(t)h_{j}^{p}(t)(1-h_{j}^{p}(t))x_{i}, \quad (5)$$

$$w_{j3}^{p}(t+1)=w_{j3}^{p}(t)-\alpha_{3}^{p}(t)b_{3}^{p}(t)h_{j}^{p}(t), \quad (6)$$

$$w_{kj}^{p}(t+1)=w_{kj}^{p}(t)-\alpha_{2}^{p}(t)b_{3}^{p}(t)w_{j3}^{p}(t)h_{j}^{p}(t)(1-h_{j}^{p}(t))h_{k}^{p}(t-1), \quad (7)$$

$$w_{3j}^{p}(t+1)=w_{3j}^{p}(t)-\alpha_{2}^{p}(t)b_{3}^{p}(t)w_{j3}^{p}(t)h_{j}^{p}(t)(1-h_{j}^{p}(t))y^{p}(t-1), \quad (8)$$

$$T_{o}^{p}(t+1)=T_{o}^{p}(t)+\alpha_{3}^{p}(t)b_{3}^{p}(t), \quad (9)$$

$$T_{j}^{p}(t+1)=T_{j}^{p}(t-1)+\alpha_{2}^{p}(t)b_{3}^{p}(t)w_{j3}^{p}(t)h_{j}^{p}(t)(1-h_{j}^{p}(t)). \quad (10)$$

The training process of PNN according to the expressions (1-10) is executed while the common training error $E(t)$ does not become less than the required value.

The experimental researches (by simulation modeling) of the proposed algorithms have been done using generalized mathematical models of sensors drift "with acceleration" (Fig. 2). The data from 0-30 calibrations with step 5 on each curve are used for NN training. The prediction interval is 30-60 calibrations with step 1. The average and the maximum percentage prediction errors do not exceed 14% and 27% (Fig. 3) for sensors drift "with acceleration" using integrating "historical" data NN, approximating NN and predicting NN. It allows the improvement of sensor data acquisition and processing accuracy in 3-5 times with a simultaneous increase of interesting interval in 6-12 times [6, 9].

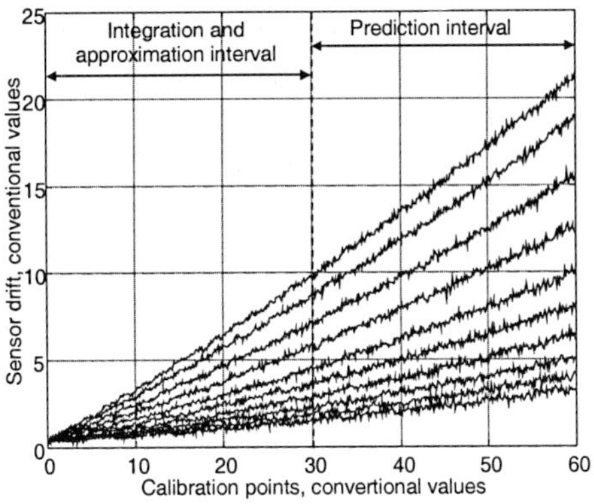

Fig. 2. Sensor drift "with acceleration"

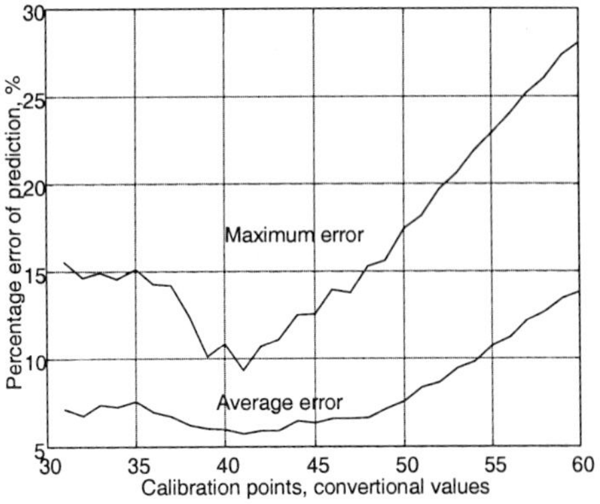

Fig. 3. Prediction results of sensor drift

3 An Approach to Estimation of Computational Complexity of Algorithm

Using the proposed NN-based algorithms of sensor data acquisition and processing defines the corresponding requirements to DSN structure and the necessary computing components. The analysis of time parameters of sensors and known DSN structures [1, 10] allows defining three scales of real time, which is necessary to consider during the analysis of computational productivity of the DSN components:

- The required scanning period for the majority of sensors is not less than 50 μs. Therefore, the data acquisition frequency is $2 \cdot 10^4$ Hz;

- The necessary period of sensor drift correction factor calculation is not less than 20 min. Therefore the frequency is $8.3 \cdot 10^{-4}$ Hz;
- The necessary period of replacement of the sensor drift mathematical model is not less than 20 hours. Therefore, the frequency is $1.4 \cdot 10^{-5}$ Hz.

All the above procedures should run in the correspondent real time scale.

Let us consider a technique of computational complexity (and the necessary productivity of computing components) estimation for the algorithm of sensor drift mathematical model replacement using the PNN structure.

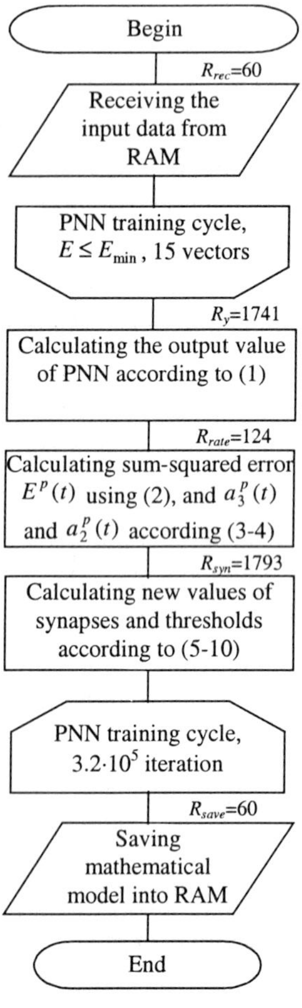

Fig. 4. General algorithm of sensor drift mathematical model replacement

In a general case, productivity P of the computing device [11]

$$P = K \cdot F \cdot R / N , \tag{11}$$

where K is the number of input data channels, F is the frequency of the data entrance on the any of K input channels, $R = R(N)$ is the computational complexity of the algorithm (number of addition/multiplication operations (float-point operations) per second) and N is the number of input data elements. Furthermore, the analysis of computational complexity will be provided for the situation, when only one value from input arrays (one value from each appropriate data array) goes to the computing device on one input channel in order to simplify the calculation of operations.

The productivity of the necessary computing components for the algorithm of sensor drift mathematical model replacement according to Fig. 4 and (11)

$$P_{DRFMODEL} = K \cdot F \cdot R_{DRFMODEL} / N = 2.5 \cdot 10^5 \text{ op/sec},$$

where K =1; F =$1.4 \cdot 10^{-5}$ Hz is the frequency of the mathematical model of correction factor replacement;

$$R_{DRFMODEL} = R_{rec} + 3.2 \cdot 10^5 \cdot 15 \cdot (R_y + R_{rate} + R_{syn}) + R_{save} = 1.8 \cdot 10^{10} \tag{12}$$

operations - is the computational complexity of algorithm ($3.2 \cdot 10^5$ is the number of required iterations; 15 is the number of training vectors; $R_y = 1741$ operations is the computational complexity of PNN output value calculation according to (1); $R_{rate} = 124$ operations is the computational complexity of sum-squared error calculation (2) and the adaptive learning rate for neurons of output (3) and hidden (4) layers; $R_{syn} = 1793$ operations is the computational complexity of the synapses and the thresholds modification for all layers according to (5-10)); N =1.

The estimation of computational complexity of algorithms is necessary in order to choose the optimum DSN structure, which executes NN-based algorithms in order to improve the accuracy of sensor data acquisition and processing. The preliminary analysis has shown that the data acquisition algorithm (•), the algorithm of correction factor of sensor drift mathematical model replacement (*B*), the algorithms of PNN training set formation using additional approximating NN (*C*) [7, 9] and the use of a set of integrating "historical" data neural networks (*D*) [6, 8] are required the largest computational recourses. The results of productivity P_s estimation for the above algorithms for single data acquisition channel of DSN are presented in Table 1.

4 Experimental Researches for DSN Structure Design

A personal computer (PC) is the main computing DSN component [10, 12]. Therefore, it is necessary to evaluate productivity of modern PCs on the basis of different processors using the proposed approach. The fragment of PNN calculation routine according to (1-10) is used as the test routine for the estimation of computational complexity and productivity of the algorithm. The computational complexity of the test routine

$$R_{TEST} = 1\cdot10^6\cdot(R_{y1}+R_{rate}+R_{syn}) = 3.7\cdot10^9 \text{ operations,}$$

where $1\cdot10^6$ is the iteration number of the test routine, R_y, R_{rate}, R_{syn} are operations according to (12). The productivity of the tested PCs $P_{TEST} = R_{TEST}/t_{TEST}$, where t_{TEST} is the execution time of the test routine in seconds (Table 2).

Table 1. Computational complexity for single data acquisition channel

Algorithm	P_s, op	R_s, op/s
A	27	$5.4\cdot10^5$
B	$1.8\cdot10^{10}$	$2.5\cdot10^5$
C	$8.6\cdot10^8$	$1.2\cdot10^4$
D	$1.3\cdot10^{10}$	$1.8\cdot10^5$

Table 2. Productivity estimation of modern PCs

PC	t_{TEST}, s	P_{TEST}, op/s
Intel Pentium	412	$8.9\cdot10^6$
AMD Pro	282	$1.3\cdot10^7$
Intel Celeron	69	$5.4\cdot10^7$
Intel P-III	52	$7.1\cdot10^7$

The productivity analysis of modern PCs (see Table 2) has shown that a single PC doesn't provide the fulfillment of all the considered above NN-based algorithms in real time scale for a multi-channel DSN, especially when it is together with user interaction. The known data acquisition modules [12, 13] fulfill sensor signal processing using only permanent algorithms stored into ROM. Therefore, it is expedient to add the third (middle) level of sensor signal processing [1] into a two-level structure of known DSN. This level should fulfill the algorithms of the current error correction. The elements of the middle level should execute the algorithms, which change during the DSN operation. For this purpose, they should provide a remote reprogramming mode that allows them to be recognized as intelligent nodes [14].

5 Conclusions

The proposed approach for a computational complexity estimation of neural network based algorithms and an appropriate productivity of necessary computing components allows to propose three-level DSN structure with the improved accuracy and effectiveness of sensor data processing in 3-5 times.

Acknowledgments

The authors wish to acknowledge Prof. A. Melnyk from National University "Lvivska Politechnika", Ukraine for the fruitful discussions during the preparation of this paper. The authors also would like to thank the European organization INTAS for its financial support, grant reference number INTAS-OPEN-97-0606.

References

1. S.Iyengar: Distributed Sensor Network - Introduction to the Special Section, Transaction on Systems, Man, and Cybernetics, Vol. 21, No.5 (1991) 1027-1031
2. A.Sachenko, V.Kochan, V.Turchenko: Intelligent Distributed Sensor Network, Proceedings of the 15th IEEE Instrument. and Measurement Tech. Conf. St. Paul, USA (1998) 60-66
3. A.Sachenko, J.Tverdyj: Perfection of Temperature Measurement Methods. Technika, Kiev Ukraine (1983)
4. Tuz Y: Structural Methods of Accuracy Improvement of Measuring Devices. Vyshcha Shkola, Kiev Ukraine (1976)
5. A. Sachenko: Development of Measurement Precision Algorithm for Intelligent Sensor System, 41 Internationalen Wissenschaft Kolloquium, Ilmenau, Germany (1996) 645-650
6. V.Turchenko, R.Kochan, A.Karachka: Prediction of Sensor Drift Conversion Characteristics Using Neural Networks, Journal of Ternopil State Technical University, Vol 5, No 4. Ternopil, Ukraine (2000) 102-108
7. V.Golovko, J.Savitsky, A.Sachenko, V.Kochan, V.Turchenko, T.Laopoulos, L.Grandinetti: Intelligent System for Prediction of Sensor Drift, Proceedings of the International Conference on Neural Networks and Artificial Intelligence. Brest, Belarus (1999) 126-135
8. A.Sachenko, V.Kochan, V.Turchenko: Sensor Drift Prediction Using Neural Networks, Proceedings of the International Workshop on Virtual and Intelligent Measurement Systems. Annapolis, USA (2000) 88-92
9. A.Sachenko, V.Kochan, V.Turchenko, V.Golovko, J.Savitsky, A.Dunets, T. Laopoulos: Sensor Errors Prediction Using Neural Networks, Proceedings of the IEEE-INNS-ENNS International Joint Conference on Neural Networks, Vol. IV. Como, Italy (2000) 441-446
10. V.Golovko, L.Grandinetti, V.Kochan, T.Laopoulos, A.Sachenko, V.Turchenko: Sensor Signal Processing Using Neural Networks, Proceedings of the IEEE Region 8 International Conference Africon'99. Cape Town, South Africa (1999) 339-344
11. A. Melnyk: Real-Time Specialized Computer Systems: Lectures notes, State University "Lvivska Politechnika" (1996)
12. http://content.honeywell.com/sensing/prodinfo/sds/prodsheets/micro.pdf
13. http://www.fluke.com/products/home.asp?SID=7&AGID=6&PID=5308
14. A.Sachenko, V.Kochan, V.Turchenko, V.Tymchyshyn, N.Vasylkiv: Intelligent Nodes for Distributed Sensor Network, Proceedings of the 16th IEEE Instrumentation and Measurement Technology Conference. Venice, Italy (1999) 1479-1484

Fusion Architectures for the Classification of Time Series

Christian Dietrich, Friedhelm Schwenker, and Günther Palm

Department of Neural Information Processing - University of Ulm,
Oberer Eselsberg, D-89069 Ulm
{dietrich,schwenker,palm}@neuro.informatik.uni-ulm.de

Abstract. The classification of time series based on local features is discussed in this paper. In this context we discuss the topics data fusion, decision fusion, and temporal fusion. Three different classifier architectures for these fusion tasks are proposed. Some local features are automatically derived form the time and frequency domain and categorized through a fuzzy-k-nearest-neighbour rule. Soft decisions are combined to a crisp decision of the whole time series. Numerical results for all architectures are given for a data set (songs of crickets, recorded in Thailand and Ecuador) containing 35 different categories.

1 Introduction

In real world applications information or features extracted from time series is used for the categorization. For example, in medical diagnosis a patient may be classified into one of two or more classes using an electrocardiogram (ECG) recording. A pattern recognition problem may the identification of an individual based on its speech recording. One difficulty with the classification of time series is that the number of measurements is typically large in comparison to the number of objects and varies from recording to recording. To overcome these problems some kind of preprocessing and feature extraction on these time series has to be performed. In principle there are two approaches of feature extraction for time series:

1. **Global features**. These features based on the information in the whole time series, e.g. the mean frequency, mean energy, etc.
2. **Local features**. These are derived from subsets of the whole time series, which are usually defined through a local time windows W^t. A set of features is calculated within this window W^t. The window is then moved by a time step Δt into $W^{t+\Delta t}$ and the next set of features is calculated.

In this paper we focus on the classification of time series based on local features, i.e. on sequences of locally derived feature vectors.

2 Classification Fusion for Local Features

In this section we propose three different types of classifier architectures for the fusion of local features and decisions based on these local features. The situation

G. Dorffner, H. Bischof, and K. Hornik (Eds.): ICANN 2001, LNCS 2130, pp. 749–755, 2001.

is illustrated in Figure 1. Here a window W^t is covering a small part of the time series. For each window W^t, $t = 1, ..., T$ a set of p features $F_i^t \in \mathbb{R}^{d_i}$, $i = 1, ..., p$ and $d_i \in \mathbb{N}$, is extracted from the time series. Typically T, the number of time windows varies from time series to time series.

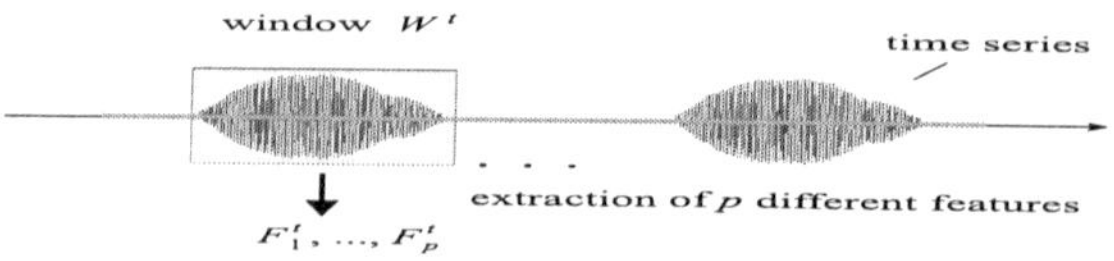

Fig. 1. A set of p features $F_1^t, ..., F_p^t$ is extracted from a local time window W^t.

A. Architecture **CDT** (see Figure 2a)
The classification of the time series is performed in the following three steps:
1.) Classification of single feature vectors (C-step)
For each feature F_i^t, $i = 1, ..., p$ derived from the time window W^t a classifier c_i is given through a mapping $c_i : \mathbb{R}^{d_i} \rightarrow \Delta$ (see Section 3) where the set Δ is defined through

$$\Delta := \{(q_1, ..., q_l) \in [0, 1]^l | \sum_{i=1}^{l} q_i = 1\}. \quad (1)$$

Here l is the number of classes. Thus for time window W^t, the p classification results are $c_1(F_1^t), ..., c_p(F_p^t)$.
2.) Decision fusion of the local decisions (D-step)
The classification results are combined into a decision $C_A^t \in \Delta$ for the time window W^t through a fusion mapping $\mathcal{F} : \Delta^p \rightarrow \Delta$

$$C_A^t := \mathcal{F}(c_1(F_1^t), ..., c_p(F_p^t)), \quad t = 1, ..., T \quad (2)$$

which calculates the fused classification result of the p different decisions.
3.) Temporal fusion of decisions over the whole time series (T-step)
The classification for the whole set of time windows W^t, $t = 1, ..., T$ is given through

$$C_A^o = \mathcal{F}(C_A^1, ..., C_A^T) \quad (3)$$

again $\mathcal{F} : \Delta^T \rightarrow \Delta$ is a fusion mapping.

B. Architecture **DCT** (see Figure 2b)
The classification of the whole time series is determined through the classification of the fused features and decision fusion over the whole time series:
1.) Data fusion of feature vectors (D-step)
Here the features $F_1^t, ..., F_p^t$ inside the window W^t are simply concatenated into a feature vector $F^t = (F_1^t, ..., F_p^t) \in \mathbb{R}^P$, with $P = \sum_{i=1}^{p} d_i$.

2.) Classification (C-step)
The combined feature vector F^t is classified into $C_B^t \in \Delta$ using a classifier mapping $c^t : \mathbb{R}^P \rightarrow \Delta$.

$$C_B^t := c^t(F^t) \tag{4}$$

3.) Temporal fusion of decisions over the whole time series (T-step)
Here

$$C_B^o = \mathcal{F}(C_B^1, ..., C_B^T) \tag{5}$$

is the classification result for the set of time windows W^t, $t = 1, ..., T$.

C. Architecture **CTD** (see Figure 2c)
Here, the final classification result is determined through temporal fusion followed by decision fusion.

1.) Each of the p feature vectors $F_1^t, ..., F_p^t$ within W^t is classified (C-step)

$$F_i^t \rightarrow c_i(F_i^t) \in \Delta, i = 1, ..., p \tag{6}$$

2.) Temporal fusion based on each feature $i = 1, ..., p$ (T-step)

$$C_C^i = \mathcal{F}(c_i(F_i^1), ..., c_i(F_i^T)) \in \Delta \tag{7}$$

again $\mathcal{F}$ is a fusion mapping $\mathcal{F} : \Delta^T \rightarrow \Delta$.

3.) Decision fusion over all p features C_C^i (D-step)

$$C_C^o = \mathcal{F}(C_C^1, ..., C_C^p) \tag{8}$$

here $\mathcal{F}$ is a fusion mapping $\mathcal{F} : \Delta^p \rightarrow \Delta$.

For the integration of of n classifier decisions $C^1, ..., C^n \in \Delta$ for an input x different fusion mappings may be considered. Temporal fusion (see Eq. 3, 5, 7) and decision fusion (see Eq. 2, 8) are performed by *average fusion* (AF) [4] or *symmetrical probabilistic fusion* (SPF). The average of the classification results is given by

$$\mathcal{F}(C^1, ..., C^n)_j := \frac{1}{n} \sum_{i=1}^{n} C_j^i. \tag{9}$$

A probabilistic approach to combine classification results is to apply Bayes' rule under the assumption that the classification results are independent [6]. The symmetrical probabilistic fusion is given by

$$\mathcal{F}(C^1, ..., C^n)_j := 1 - \left(1 + \alpha \left(\frac{1 - P_j}{P_j}\right)^{n-1} \prod_{i=1}^{n} \frac{C_j^i}{1 - C_j^i}\right)^{-1}. \tag{10}$$

Here P_j denotes the prior probability for class ω_j and α is a normalizing constant.

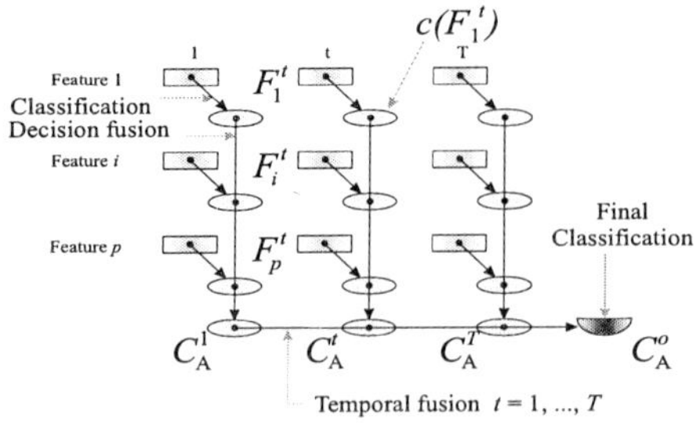

(a) CDT - Classification, Decision fusion, Temporal fusion.

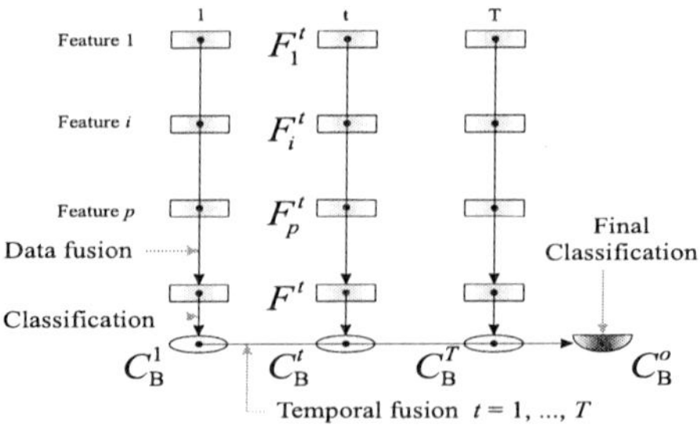

(b) DCT - Data fusion, Classification, Temporal fusion.

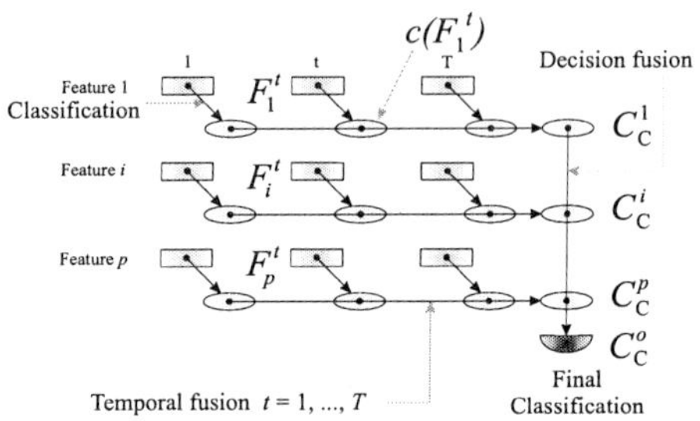

(c) CTD - Classification, Temporal fusion, Decision fusion.

Fig. 2. Classifier architectures for the classification of time series

3 Classification with the k-Nearest-Neighbour Rule

One of the most elegant and simplest classification techniques is the *k-nearest-neighbour rule* [2]. The classifier searches for the k nearest neighbours among a set of m prototypes $\{x^1, ..., x^m\} \in \mathbb{R}^d$ and an input vector $x \in \mathbb{R}^d$. The k nearest neighbours are calculated utilizing the L_p-norm.

$$d_j^p(x, x^j) = ||x - x^j||_p = (\sum_{i=1}^{d} |x_i - x_i^j|^p)^{\frac{1}{p}} \tag{11}$$

between x and x^j. The class which occurs most often among the k nearest neighbours is the classification result.

Let l be the number of classes. To determine the membership of an input vector x to each class, a *fuzzy-k-nearest-neighbour classifier* is applied. A fuzzy classifier $\mathcal{N}$ is a mapping $\mathcal{N} : \mathbb{R}^d \rightarrow [0,1]^l$, with output $\mathcal{N}(x) = (\delta_1(x), ..., \delta_l(x)) \in \Delta$.

For each class $\omega = 1, ..., l$ let m_ω be the number of prototypes with class-label ω. Then for input x and each class ω there is a sequence $(\tau_i^\omega)_{i=1}^{m_\omega}$ with

$||x^{\tau_1^\omega} - x||_p \leq, ..., \leq ||x^{\tau_{m_\omega}^\omega} - x||_p$. Here m_i denotes the number of prototypes of class ω. The k nearest neighbours ($k \leq m_\omega$) of x to class ω are then given by

$$\mathcal{N}_k^\omega(x) := \{x^{\tau_1^\omega}, ..., x^{\tau_k^\omega}\}. \tag{12}$$

Let

$$\tilde{\delta}_\omega(x) = \frac{1}{\sum_{x_i \in \mathcal{N}_k^\omega(x)} ||x - x_i||_p + \alpha}, \quad \alpha > 0 \tag{13}$$

be the support for the hypothesis that the true label of x is ω. The parameter α is used to determine how to grade low values of $||x - x_i||_p$.

After normalisation by $\delta_j(x) := \tilde{\delta}_j / \sum_{i=1}^{l} \tilde{\delta}_i$ we call the classifier outputs *soft labels* [4]. For input x the classification result $\mathcal{N}(x) := (\delta_1(x), ..., \delta_l(x)) \in \Delta$ is containing the membership of x for each class.

4 Application

We present results achieved by testing the algorithms on a dataset which contains sound patterns from 35 different cricket species. The dataset contains recordings of 3 or 4 different individuals per species. The recordings are from Thailand (used by Ingrisch [3]) and from Ecuador (used in the doctoral thesis by Nischk [5]). The cricket songs consist of sequences of pulses. Based on these pulses the crickets may be classified [1, 5]. Therefore we analyse the structure of single pulses in the cricket songs, and so the first processing step is the pulse detection.

The start position and end position of the pulses are located using an modified algorithm of Rabiner and Sambur [1, 7]. These pulses are used to calculate the local features. The following three local features are extracted from the single pulses and then used as inputs for the classifiers:

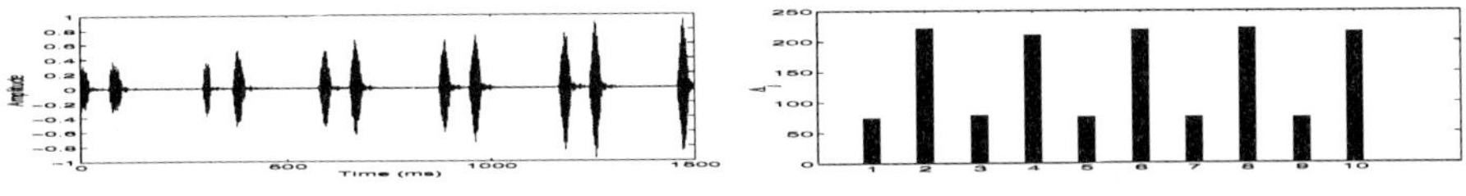

(a) Filtered cricket waveform (b) Distances between pulses

Fig. 3. Waveform and energy of the species *Noctitrella glabra* (time window 1500 ms).

1. **Pulse length**: The length L_i is given by the start position and the end position of the pulse.
2. **Pulse distances**: The pulse distances are extracted using a d−tuple encoding scheme producing datapoints $T_i \in \mathbb{R}^d$
3. **Pulse frequency** : The pulse frequency F_i is extracted from the spectra of the single pulses.

Table 1. Classification results of single features: Local features after temporal fusion with the average function.

feature	error in %
Pulse length (L_i)	85.265 ± 1.893
Pulse frequency (F_i)	75.049 ± 1.435
Time structure (T_i)	22.003 ± 2.274

Table 2. Classification results of locally fused features for the DCT architecture dependent on the fusion function (AF_T temporal fusion with the average function and SPF_T temporal fusion with the symmetrical probabilistic function).

architecture	AF_T	SPF_T
DCT	9.332 ± 0.973	8.644 ± 0.703

5 Results and Discussion

Because of the limited data sets we utilize the cross validation method to evaluate classifiers. The training set has been used to design the classifiers, and the test set for testing the performance of the classification task. The test is done in a *k-fold cross validation test* with $k = 4$ cycles using 2-3 records of each species for training and 1 record of each species for the classification test. In the numerical evaluation 25 different cross validation tests have been made. The training and test records are randomly splitted and are always recordings from different individuals. Table 1 shows the cross-validation results for the single features.

The best feature for the classification is the distance between pulses T_i. Table 2 depicts the classification rates of the previously introduced DCT architecture. For this architecture the fusion functions average fusion and symmetrical probabilistic fusion (see Eq. 9 and 10) are only be used for the temporal fusion (see Eq. 5), because data fusion is the first step.

The best classification rates for the described application are achieved with the DCT architecture. It seems that in this application the symmetrical probabilistic fusion function is better suited for decision fusion than average fusion (see Table 3).

For decision fusion with average fusion and temporal fusion with the symmetrical probabilistic function (AF_D/SPF_T) the CDT architecture outperforms the CTD architecture. The reason for that is that the symmetrical probabilistic function is sensitive to errors in the probability estimations C_j^i (see Eq. 10). These probability estimations may be better if the classifier results are combined through decision fusion (CDT) before temporal fusion is applied (see Table 3). The same effect is observed for the AF_D/SPF_T fusion function combination.

We observed that in our numerical experiments the CDT architecture outperforms the CTD architecture. For temporal fusion it seems that average fusion is better suited than symmetrical probabilistic fusion.

Table 3. Classification results of locally fused features for the CDT and CTD architecture dependent on the fusion function (AF average fusion (see Eq. 9) and SPF symmetrical probabilistic fusion (see Eq. 10)). The indices D and T indicate if the fusion function is used for D decision fusion or T temporal fusion.

architecture	AF_D/AF_T	AF_D/SPF_T	SPF_D/AF_T	SPF_D/SPF_T
CDT	12.180 ± 1.720	17.741 ± 1.273	11.493 ± 1.740	11.591 ± 1.563
CTD	12.180 ± 1.720	22.003 ± 2.346	11.493 ± 1.740	22.593 ± 2.509

Acknowledgment

DORSA (`www.dorsa.de`) forms part of the Entomological Data Information System (EDIS) and is funded by the German Ministry of Science and Education (BMBF). We are greatful to Klaus Riede (ZFMK Bonn, Germany), Sigfrid Ingrisch (ZFMK Bonn, Germany), Frank Nischk and Klaus Heller for providing their sound recordings, suggestions and discussions.

References

1. C. Dietrich, F. Schwenker, K. Riede, and G. Palm. Automated classification of cricket songs utilizing pulse detection, time and frequency features and data fusion. *submited to Journal of the Acoustical Society of America, 2001.*
2. K. Fukunaga. *Introduction to Statistical Pattern Recognition.* Academic Press, New York, 1990.
3. S. Ingrisch. Taxonomy, stridulation and development of podoscirtinae from thailand. *Senckenbergiana biologica,* 77:47–75, 1997.
4. L. I. Kuncheva and J. C. Bezdek. An integrated framework for generalized nearest prototype classifier design. *Journal of Uncertainty, Fuzziness and Knowledge-Based Systems,* 6(5):437–457, 1998.
5. F. Nischk. *Die Grillengesellschaften zweier neotropischer Waldökosysteme in Ecuador.* PhD thesis, University of Köln in Germany, 1999. Memorandum UCB/ERL-M89/29.
6. J. Pearl. *Probabilistic Reasoning in Intelligent Systems.* Morgan Kaufmann Publishers, San Francisco, 1988.
7. L. R. Rabiner and S. E. Sambur. An algorithm for determining the endpoints of isolated utterances. *Bell Syst. Technical Journal* 54(2):297–315, 1997.

Part IX

Special Session: Agent-Based Economic Modeling

The Importance of *Representing* Cognitive Processes in Multi-agent Models

Bruce Edmonds and Scott Moss

Centre for Policy Modelling,
Manchester Metropolitan University,
Aytoun Building, Aytoun Street, Manchester, M1 3GH, UK.
http://www.cpm.mmu.ac.uk/

Abstract. We distinguish between two main types of model: predictive and explanatory. It is argued (in the absence of models that predict on unseen data) that in order for a model to increase our understanding of the target system the model must credibly *represent* the structure of that system, including the relevant aspects of agent cognition. Merely "plugging in" an existing algorithm for the agent cognition will not help in such understanding. In order to demonstrate that the cognitive model matters, we compare two multi-agent stock market models that differ only in the type of algorithm used by the agents to learn. We also present a positive example where a neural net is used to model an aspect of agent behaviour in a more descriptive manner.

1. Introduction – Types of Modeling and Their Goals

There are two ways in which our models are constrained by reality: *verification* and *validation*. Verification is where the *design* of a model is constrained by *prior* knowledge, for example when a chemist's model of interacting atoms is constructed so that it is consistent with atomic physics. Validation occurs *after* the model has been constructed – it is a check that the output of the model is consistent with known behaviour, for example by testing to see whether the error of a model's predictions with respect to unseen data can be explained by noise (A hold-out sample is insufficient for a predictive model as (Mayer 1975) graphically illustrates.).

Different types of models relate to different kinds of goals and involve verification and validation in different ways. For example in a predictive model, where the goal is to predict unseen data, the validation is all important and comes first. Later the construction of the model (i.e. the verification) may be examined to try and find out *why* the model successfully predicts the phenomena.

Another important type is the explanatory model, here the goal is to explain the outcomes in terms of the design – the validation does not need to be done with unseen data, but the design must be a credible *representation* of the structure of the process if the explanation is to be of any use. The methodology of modelling is discussed in greater detail in (Edmonds, 2000).

This approach differs significantly from the standard modelling procedures in the social sciences – especially conventional economics and its derivatives in political science and sociology. For example, in conventional economic theory, competition is

G. Dorffner, H. Bischof, and K. Hornik (Eds.): ICANN 2001, LNCS 2130, pp. 759–766, 2001.

an unmodelled process that is claimed to drive all economic actors to behave *as if* they were constrained optimisers. What is actually modelled in conventional economic theory is a competitive equilibrium that is said to capture the *result* of the unmodelled competitive process. A survey by Moss (forthcoming) of the 14 papers mentioning "game theory in title, abstract or keywords, published in 1999 in the *Journal of Economic Theory* showed that seven of the papers proved the existence of a Nash or similar equilibrium for an n-person game, six papers reported models and results for two-person games (sometimes in round robin tournaments) and one paper reported results for a three-person game. There were no papers reporting the *process* of any game with more than three players.

A social process involves a process of interaction among individuals. By ignoring the process, or considering only very limited numbers of agents (e.g. 2 agents), economic modellers do not even seek to capture social interaction. In consequence, it is not possible to validate their models against observed social behaviour. Instead, with extremely strong assumptions they hope to take a "short cut" to capture the outcomes only – in other words to make a predictive model. The trouble is that they do not predict. because they fail on unseen data. There seems to be a sort of "double-think" going on – the use of hold-out sets rather than unseen data would be acceptable for an explanatory model and the lack of representational content would be acceptable for a predictive model, but many economic models fail to be either.

The nature of a social process depends upon the behaviour of the individuals in the course of their social interactions. In multi agent systems, that behaviour is represented by *some* specification of individual cognition. That specification will not capture all of the behaviour of the individual, of course, but only what is relevant to the modelling purpose – i.e. what elements of the modelling target is going to explain the outcomes. For an explanation to be effective these elements need to be present in the target systems and they have to be necessary for the outcomes.

The use of multi-agent modelling techniques is a step towards descriptive modelling, because the separate entities (agents) in the model correspond to the separate entities in the target domain (people or institutions). In other words the *structure* of the target system is *represented* in the model (at least to some extent) as an essential part of the model design. In practice, because the agents in the system are identifiable as entities in the target domain, the interactions in the model are also programmed so that they also correspond to those between the entities modelled.

It is still common in multi-agent modelling to take an existing algorithm (designed with another purpose in mind) and use it for the cognition of the agent, regardless of whether this can be justified in terms of what is known about the behaviour of the modelled entities. For, example there have been many economic models which have utilised standard Genetic Algorithm techniques without regard for whether this can be justified by reference to the target domain (Chattoe 1998).

In this paper we will argue that neural network techniques can be a valuable part of the multi-agent modellers palette, but that simply "plugging in" existing techniques will not help in *understanding* what we model. In the absence of models that predict unseen data this seems to be simply an anachronism. Rather, if we construct models whose structure *including that of the cognition* corresponds to a description of the target domain, then the behaviour of the whole system can provide an explanation of this behaviour in terms of the elements of that description.

Thus the algorithm that is to drive an agent's behaviour must be, at least, *chosen* and *adapted* to correspond to what is known concerning the modelled entity's behaviour and cognition. This is as true for neural network techniques as for other algorithms – the fact that at a micro level there might be a correspondence between brain structure and NN techniques does not give any guarantee that this is true at the macro level of agent behaviour or in social phenomena arising from agent interaction.

2. A Negative Example – A Model of Stock Market Traders

To emphasis that the *type* of the cognitive algorithm is crucial to the behaviour of a whole system (and not merely its adjustment or parameterisation), I will describe two versions of a model of stock market traders (roughly following Palmer et al. 1994) where the only difference is that different types of algorithm were used for their decision making mechanism were used: a GP based algorithm and a NN algorithm. The different cognitive models lead to qualitatively different outcomes – differences that I suggest will not be eliminated by any simple "tuning" of the algorithms. The corollary of this is that it matters that we use *appropriate representations* of cognition, and do not simply "plug in" algorithms "off the shelf".

What is common to both versions of this model is what is exterior to the trading agents - their environment and the structure and type of their interactions.

There are seven traders, three stocks and one market maker. Each time cycle the market maker posts prices for the stocks and traders can buy (if they have cash or are also selling stocks) or sell (if they have that stock) at the price posted in each of the three stocks. They pay a fixed cost for each trade they make. They do not have to trade. There is a dividend for each stock which varies upwards and downwards in a slow random walk, so that different stocks have the best dividend rate at different times. This dividend is paid each cycle and is the only fundamental underlying the value of each stock. Cash and levels of the stocks are adjusted at the cycle's end.

The market maker always accepts trades, unless it has run out of stock to sell, in which case it sells what it has left. The prices it sets are initially random but from then on are moderated according to demand – that is, if there traders are buying a stock it raises its price and if there are net sales it lowers the price.

The traders are initialised with 100 units of cash at the start and a small but random level of each stock. During the first five time cycles the traders can learn about the market but make small random purchases or sales. There after the assets of each trader depends on its trading decisions (to buy, sell or keep stocks). The traders' learning algorithms attempt to find actions that will increase its assets or, equivalently, to correctly predict stock price movements. The internal algorithm decides on its intentions, which is then modified by what is possible given its resources – this moderation is the same for both versions. The maximum purchase or sale of each stock is 5 units.

The inputs that the learning algorithm has are as follows:

- The prices of the three stocks during the previous two cycles and 5 cycles ago;
- The value of the stock index (average of the prices) during the previous two cycles and 5 cycles ago;
- The last actions of its contacts in terms of buying and selling the three stocks;
- And one constant: the Boolean value of "true".

At the beginning of the simulation a randomised "contact structure" is imposed upon the traders, with each trader having three contacts, that is a specified and fixed set of three of the other traders. Each trader can observe its contacts, that is each cycle the actions of its contacts in terms of buying or selling each of the three stocks are employed as additional inputs for its decision making algorithm. Thus each trader could learn to imitate another trader's actions.

2.1 The GP Cognitive Algorithm Version

The GP learning algorithm is a strongly typed GP algorithm following (Montana 1995). Each trader has 20 "models", where each model is made up of a GP tree for each stock. The nodes are as follows:

```
IDidLastTime indexLastTime indexNow maxHistoricalPrice
AND divide doneByLast greaterThan lastBoolean
lastNumeric lessThan minus NOT OR plus priceDownOf
priceLastWeek priceNow priceUpOf times
```

The terminals are consist of the names of its contacts, the names of the stocks, *True* and *False*. The evaluation of the models is done by assessing what the trader's total assets would be if it had employed the model over the last 5 time periods. The operations are: 85% crossover, 5% new random models and 10% propagation.

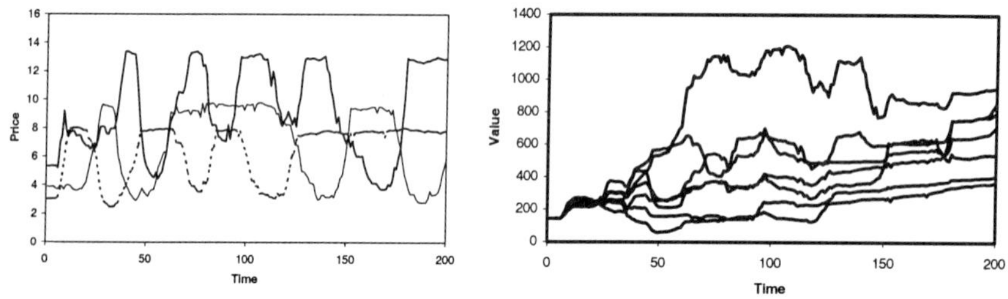

Fig. 1. & **Fig. 2.** GP-Version: Price of stocks *and* Total value of trader's assets

Figures 1 and 2 give a flavour of the resulting interactions of the agents. Figure 1 below shows the prices of the three stocks and figure 2 the total assets (at current stock prices) of the agents. There is quite a lot of market activity – traders continue to drop in out of stocks resulting in speculative "cycles". As a result the fortunes of the different traders vary over time in comparison with each other.

2.2 The NN Cognitive Algorithm Version

The neural net version of the model broadly follows (Chialvo and Bak 1999). This has a standard fixed and layered topology with adjustable weights on the arcs. This is a winner-take-all network where only the strongest node in each layer (after the input) fires. Feedback is entirely on the basis of error, the arcs that fired leading to an erroneous prediction are slight depressed. The NN used is a variation of this to allow more than one set of arcs to fire in parallel – a "critical level" is set and all those nodes whose inputs sum to over this level fire, the feedback is as before but with the addition that the critical level is adjusted if the network is generally under firing.

In this network there were 47 inputs two intermediate layers of 35 nodes each and 11 output nodes. The inputs were as follows:

- For each of the three stocks, for the last and previous time whether it has gone significantly up, down or not changed;
- For the index, for the last and previous time whether it has gone significantly up, down or not changed;
- Whether the index is historically high or low;
- For each contact and each stock whether they bought, sold or neither.
- The predictive outputs were:
- For each of the 3 stocks whether it went significantly up, down or neither;

The arcs were initialised with weights around 100 (the initial critical level) and decremented by a factor of 0.99 when leading up to an erroneous prediction. Here we have a very different picture – traders invest early, predicting that prices in a couple of stocks will go up, which turns out to be a self-fulfilling prophesy. Those traders that bought into those stocks first did best. The assets of traders increase largely as a result of the dividends they gain on those stocks.

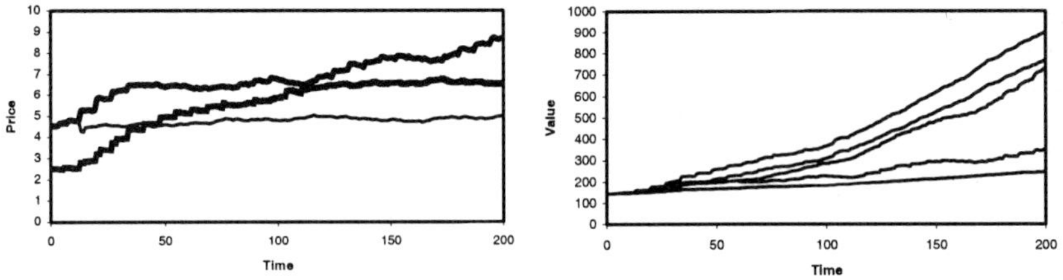

Fig. 3. & Fig. 4. NN-Version: Price of stocks *and* Total Value of trader's assets

2.3 Some Speculations about the Causes of the Different Outcomes

It would be almost impossible to prove that the two algorithms could not be "tuned" so they produced qualitatively similar results, but I would argue that this unlikely. It is unlikely because the qualitative difference in outcomes can be explained in terms of the different characteristics of the two algorithms.

The GP algorithm is far more "brittle" and noisy – there is a continual stream of new models being created as part of its operation, and small changes in a model's content can mean a very large change in its impact. The NN algorithm is also a critical system – paths that are just below the critical level are not further depressed and have the potential for "resurfacing" when the current paths fail, but whilst there *is* a small probability that a successful model in the GP algorithm can be forgotten, this is not the case in the NN, where a successful path will remain as long as it is successful. Finally, while the GP algorithm employs a retrospective evaluation of models, the NN is incremental – it can build up the feedback a little bit each time period. The incrementally means that the NN will tend to be more consistent in its behaviour whilst the GP can be more context-sensitive (for example if the market entered a new phase exactly 5 periods ago).

The point is *not* that one algorithm is *better* than the other, but which accurately reflects the happenstance of real traders' behaviour. In this case the one algorithm might be a better representation of one kind of trader and the NN of another.

3. A Positive Example – A Model of Multilateral Negotiation

The model of multilateral negotiation reported in this section is under development as part of a project on using multi agent systems to inform and support the management of water resources across Europe under anticipated conditions of climate change. A characteristic of water issues is that they involve conflicting interests. Extraction of water for irrigation conflicts with leisure and environmental interests; industrial water use conflicts with domestic water quality etc.. The purpose of the model is to identify, both in general terms and in relation to particular cases, how to structure negotiations over a large number of conflict-laden issues among interested stakeholders in order to achieve a commonly acceptable set of outcomes. The implementation was designed specifically to investigate the importance of different attitudes among negotiations towards the behaviour of others.

It is possible – and certainly should not be excluded by assumption – that over the course of any complicated negotiation process, there will be changing criteria for determining with which other agents one should seek agreement. Consequently, any representation of cognition that supports the selection of negotiating partners must allow for flexibility and the evolution of selection criteria. The Chialvo-Bak (Chialvo and Bak 1999) learning algorithm is a clear candidate in this regard. Because synaptic strengths are only altered by small amounts when an expected or required result is *not* realised, agents incorporating this representation of cognition remain flexible and adaptive, but remain fairly faithful as long as they don't encounter negative experiences. Moreover, the winner-takes-all strategy, whereby the strongest of alternative synapses is always chosen, ensures fast execution.

Negotiators' cognition is represented by a problem space architecture within which the Chialvo-Bak algorithm is used to implement learning. There is a middle layer network for each agent composed of a random network of 200-300 neurons.

An abstract, canonical model has been implemented that can represent specific local sets of issues. This was achieved by ascribing to each agent a digit string to represent its negotiating position. Each digit in the string could take any integer value within a user-specified range. A second digit string of the same length indicated the importance to the agent of the values of the digits at corresponding positions on the string representing the negotiating position. So if the negotiating position string were "1 5 3 7 6 9 2" and the string indicating the importance of the elements of the negotiating position were "3 1 2 5 8 3 4" then the value of the first string at position 5 (with a value of 8 in the second string) would be the most important element in the agent's position. The least important would be the issue represented by the digit at position 2 corresponding to the value 1 in the importance string.

The inputs to the neural network are collections of endorsements of other agents. Another agent will be endorsed as "trustworthy" if it reaches an agreement and is subsequently found to have observed the agreement. An agent will be endorsed as:

"reliable" if it responds positively to an offer made by the endorsing agent; "helpful" if it brings together two other agents that reach some sort of agreement; "similar" if it shares more of the same negotiating position than any other agent known to the endorsing agent; and "known" if it is known to the endorsing agent. There are also three negative endorsements: "untrustworthy", "unreliable" and "unhelpful".

Each combination of endorsements is associated with an input neuron with synaptic connections to seven neurons of the middle layer network. The initial order of goodness is determined by randomly ordering the individual endorsements from best (value 5) to least good (value 1). The pejorative endorsements are given the negative of the value of their positive counterparts. Then all combinations of endorsements are valued at the sum of the values of their components and, for each value, an output neuron is created.

When all of the neurons are created, each neuron that is not an output neuron is randomly linked by synapses to seven other neurons within that network. Each synapse is given a strength of 1 ± a small random perturbation in the interval [0, 0.02). On the first occasion when an agent encountered a set of endorsements, the network was trained to associate that set with its predetermined initial value. The training took the form of finding a synaptic path from the neuron associated with the endorsement set to an output neuron. The synapse with the largest weight from each neuron was selected to form the path. Following the Chialvo-Bak algorithm, if the output neuron did not correspond to the correct value of the endorsement set, weight of every synapse in that path was reduced by 0.02 and the sum of the weight reductions was distributed among all of the other synapses in the network. This procedure was repeated until the input neuron associated with the endorsement set was linked by the strongest synapses to the correct output neuron. The choice of which agent's negotiation offer to accept is determined by which agent is associated with the highest ranked set of endorsements. Having chosen an agent's offer in this way, if an agreement on any of the negotiating issues results, the synapse values are left unchanged. If, however, no agreement is reached, then the weights of the synapses connecting the input neuron to the rank-determining output neuron are reduced and the weights redistributed.

The simulation experiments were run with either 30 or 40 negotiating positions which could take any of five values. There were nine negotiating agents.

In line with the conflict resolution literature, each agent would seek agreement with other agents on the positions which it found least important and then on the next least important, and so on until full agreement were reached. Agents will agree to less important positions held by another agent when, by so doing, the other agent agrees to a position that the first agent finds more important.

It is not too surprising that we have found the *sine qu non* of successful negotiation to be a common commitment to achieve some agreement. We have also found that the standard prescription for conflict resolution – agree on the least important elements first – does not result in overall agreement in a setting with more than two negotiators. This is because, once differences have been resolved with one agent, resolving differences with another agent might involve reopening or opening up new differences with the previous negotiating partner. It seems likely that a necessary characteristic of the negotiation process will be the emergence and evolution of *coalitions* among negotiators. Further simulations and development of the model will be required to investigate appropriate means of creating such.

4. Conclusion

Neural network algorithms can be a valuable addition to the palette of techniques available to the multi-agent modeller. However their use needs to be justified by reference to the behaviour of individuals in the target domain if the model is to have any value in terms of understanding, just as with any other approach. It is theoretically possible that a model might be predictively successful on unseen data but, at present this seems unlikely without greater understanding being gained first.

There are some problem with using algorithms that are produced by fields such as neural networks or evolutionary computing, and that is that there is very little understanding about their domain of applicability. This makes it very difficult to know when a particular technique is appropriate. The tendency in many papers is to imply that a certain technique is *generally* better, rather than it has advantages for certain types of problem. This specificity of advantage seems inevitable given the "No Free Lunch" theorems of (Wolpert and Macready 1995), which show that no technique is better than the others over *all* possible problem domains.

References

Edmonds, B. (2000) The Use of Models - making MABS actually work. *Lecture Notes in Artificial Intelligence*, **1979**:15-32.

Chattoe, E. (1998) Just How (Un)realistic are Evolutionary Algorithms as Representations of Social Processes?, *Journal of Artificial Societies and Social Simulation*, **1**(3), <http://www.soc.surrey.ac.uk/JASSS/1/3/2.html>

Chialvo, D. R., Bak, P. (1999) Learning from mistakes, *Neuroscience* 90:1137-1148.

Mayer, T. (1975). Selecting Economic Hypotheses by Goodness of Fit. *Econ. J.*, **85**:877-883.

Montana, D. J. (1995). Strongly Typed Genetic Programming, *Evol. Comput.*, **3**:199-230.

Moss, S. (2001). Game Theory: Limitations and an Alternative, *Journal of Artificial Societies and Social Simulation.* **4**(2), <http://www.soc.surrey.ac.uk/JASSS/4/2/2.html>

Palmer, R.G. et. al. (1994). Artificial Economic Life - A Simple Model of a Stockmarket. *Physica D*, 75:264-274.

Wolpert, D. H. and Macready, W. G. (1995) No Free Lunch Theorems for Search, Technical Report, Santa Fe Institute, Number SFI-TR-95-02-010, 1995.

Multi-agent FX-Market Modeling Based on Cognitive Systems

Georg Zimmermann, Ralph Neuneier, and Ralph Grothmann

Siemens AG, Otto-Hahn-Ring 6, 81730 Munich, Germany
Georg.Zimmermann@mchp.siemens.de

Abstract. We present an approach of multi-agent market modeling on the basis of cognitive systems with three functionality features. These features are perception, internal processing and acting. A cognitive system is structurally represented by an error correction neural network. On the mirco-level we describe agents decisions behavior by combining cognitive systems with a framework of multi-agent market modeling. By aggregating agents decisions we are able to capture the underlying market dynamics on the macro-level. As an application, we apply our approach to the DEM / USD FX-Market. Fitting real-world data, our approach is superior to more conventional forecasting techniques.

1 Introduction

The microeconomic structure of a market consists of a plenty of interacting agents. (Dis-)Equilibrium conditions for the market as well as the price formation can be described on basis of agents activities. Hence, the natural way to explain market prices is to give a detailed causal analysis of agents decisions and their interaction. An overview of multi-agent based approaches is given in [1, 7, 14].

We model agents decision making by cognitive systems. The proposed cognitive system has three functionalities (perception, internal processing and acting), which are structurally mapped by an error correction neural network (ECNN).

This paper consists of *five* parts: First, we introduce the theory of ECNN. Secondly, we propose an uncommon view on cognitive systems. The functionalities of such a system are represented by an ECNN. The third part concerns multi-agent modeling. The fourth part combines multi-agent modeling and cognitive systems. Finally, we apply our model to the DEM / USD FX-market.

2 Error Correction Neural Networks (ECNN)

Most dynamical systems are driven by a superposition of autonomous development and external influences [15, 16]. For discrete time grids, such a dynamic system can be described by a recurrent state transition and an output equation:

$$\begin{aligned} s_t &= f(s_{t-1}, u_t, y_{t-1} - y^d_{t-1}) && \text{state transition eq.} \\ y_t &= g(s_t) && \text{output eq.} \end{aligned} \tag{1}$$

If the last model error $(y_{t-1} - y^d_{t-1})$ is zero, we have a perfect description of the dynamics. In this case, s_t is a mapping from the prior state s_{t-1} and externals u_t. The output equation gives rise to the model output y_t.

G. Dorffner, H. Bischof, and K. Hornik (Eds.): ICANN 2001, LNCS 2130, pp. 767–774, 2001.

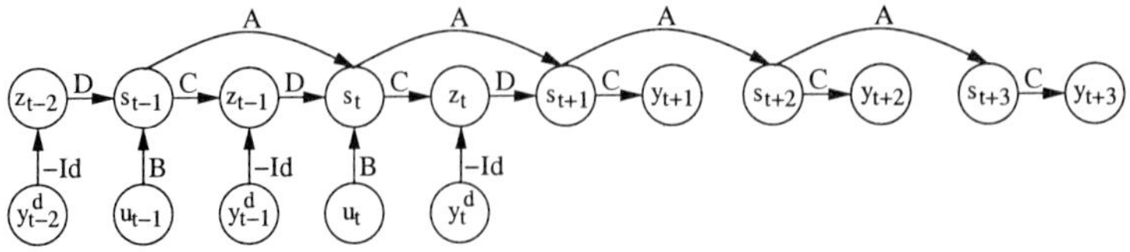

Fig. 1. ECNN using *unfolding in time* and *overshooting.*

Else, if we are unable to identify the dynamics due to missing externals u_t or noise, the last model error $(y_{t-1}-y_{t-1}^d)$ serves as an indicator of the models misfit. Using such a measure of unexpected shocks, the learning of false causalities might be decreased and models generalization ability should be improved [16].

A non-ambiguous neural network approach of Eq. 1 can be formulated as

$$\begin{aligned} s_t &= \tanh(As_{t-1} + Bu_t + D\tanh(Cs_{t-1} - y_{t-1}^d)) \\ y_t &= Cs_t \end{aligned} \tag{2}$$

$$\frac{1}{T} \cdot \sum_{t=1}^{T} \left(y_t - y_t^d\right)^2 \rightarrow \min_{A,B,C,D} \tag{3}$$

In Eq. 2, the last output y_{t-1} is computed by Cs_{t-1} and compared to the observation y_{t-1}^d. Matrix D adjusts different dimensions in the state transition. The system identification (Eq. 3) is a parameter optimization task of A, B, C, D [16]. For an overview of algorithmic solution techniques see e. g. [12].

A spatial solution of Eq. 3 is shown in Fig. 1. Here, Eq. 2 is translated into an *unfolding in time* neural network architecture using *shared weights* [6].

The ECNN (Fig. 1) is best to comprehend by analyzing the dependencies of s_{t-1}, u_t, $z_{t-1} = Cs_{t-1}-y_{t-1}^d$ and s_t. There are two different inputs to the ECNN: (*i.*) the externals u_τ, which have a direct impact on the state transition s_τ, and (*ii.*) the targets $y_{t-\tau}^d$. Only the difference between $y_{t-\tau}$ and $y_{t-\tau}^d$ influences the state s_t [16]. Note, that $-Id$ is the fixed negative of an identity matrix. The system identification is approximately solved by *finite* unfolding which truncates the unfolding after some time steps (e. g. $t-2$ in Fig. 1). For details see [15, 16].

In future time, we have no compensation of the internal expectation. Thus the system offers forecasts $y_{t+\tau} = Cs_{t+\tau}$ [16]. In order to optimize the compensation mechanism, $z_{t-\tau}$ are target clusters with values of zero. The autonomous part of the ECNN is extended into future direction by *overshooting*, which gives extra information about the system and regularizes the learning. Due to shared weights, we have the same number of parameters [15, 16].

3 Necessary Conditions for Cognitive Systems

Let us proceed with an uncommon view on cognitive systems, which might help to model the decision making of economic agents. We propose a cognitive system with three interdependent functionality features. These are (*i.*) perception, (*ii.*) internal processing and (*iii.*) action. These features are necessary conditions for cognitive systems [5, 8, 13]. We believe, that an ECNN is a native structural representation of the proposed cognitive system and its functionality features.

Approaching the functionality features, *Perception* is either done consciously or unconsciously. In case of conscious perception, selective observations are compared to an internal expectation. Attention solely focuses on this deviation. In contrast, unconscious perceptions are not within systems attention, because there is no comparison to an internally expected value. These perceptions have a direct impact on the system. This complies to a stimulus response mechanism. Note, that perception does not discriminate between input and target information. Compared to traditional modeling, this is an uncommon property.

The *internal processing* mainly consists of an internal model of the external world, which combines perception and memory. It can be seen as a reflection of the surrounding environment. In particular, the internal model sets up certain expectations of the external world, which are used in conscious perception.

The internal model accesses the systems memory, which is based on observation based learning. The memory is used to evaluate / understand perceptions, e. g. it supports pattern recognition and stores a chronology of perceptions. Without a certain chronology, many events are incomprehensible. For example, the internal model refers to the memory in order to identify an object along different time periods. Otherwise, an object changing its location from t to $t+1$ would be recognized as a completely new object. Accessing the systems memory, the internal model also is able to complete imperfect conscious perceptions.

As a result, the internal processing delivers decision support and preparation, which is the basis for the functionality feature of acting. Note, that the internal model might also stay in a dynamical stable state, i. e. evaluating perceptions without giving any kind of decision support. This is referred to as *homeostasis.*

Actions are initiated by evaluating the decision support of the internal model with regards to a particular objective function depending on the preferences of the individual. Thus, the internal processing and action overlap. Similar, action and perception overlap, since the system immediately perceives its own actions.

Due to the objective function, actions are always goal-oriented. Actions are either performed consciously or mechanically, i. e. if we get familiar with an action, there is a shift form conscious to unconscious acting. The objective function may also interpret the decision support such that an action is not helpful.

Now, we transfer the proposed functionalities into a structural framework, which is given by the ECNN of Fig. 1: *Perception* is structurally mapped by the different inputs of ECNN. Conscious perceptions refer to the error correction mechanism, while unconscious perceptions are modeled by the external inputs. Since all perceptions are given to the ECNN as inputs, there is no distinction between input and target information. Incomplete or missing conscious perceptions are automatically corrected by the ECNN. At all future time steps, the ECNN offers forecasts, because there is no compensation of the internal expectation.

The ECNN can be seen as a model for *all* conscious perceptions. Such 'world models' can be avoided (i.) by focusing the learning on major conscious perceptions or (ii.) by omitting the error correction part for most perceptions.

The *internal processing* is structurally mapped by the autonomous part of the ECNN (coded in matrix A). Internal expectations of the cognitive system are mapped by the outputs $y_\tau = Cs_\tau$ of the ECNN. Due to its recurrent formulation, the ECNN also includes a memory constructed by observation based learning.

Acting is not included in the ECNN (Fig. 1), since there is neither an action variable nor an objective function. However, acting can be integrated directly into the ECNN by a separate error correction branch (Fig. 2). In doing so, the internal model reflects not only the environment, but is also conscious of itself.

4 Modeling a FX-Market by a Multi-agent Approach

Intending to model the DEM / USD FX-market by a multi-agent approach, let us assume, that each agent $a = 1, \dots, N$ has two accounts x_t^a(in USD) and y_t^a(in DEM), which reflect his dispositions in USD and DEM [17]. For simplicity, let us assume, that there are no transaction costs. Recognizing that p_t is the DEM / USD FX-rate and Δ_t^a is the amount of money an agent desires to exchange, the upcoming cash balances of an agent a can be written as

$$x_{t+1}^a = x_t^a + \Delta_t^a \qquad y_{t+1}^a = y_t^a - p_t \Delta_t^a \tag{4}$$

Periodically an agent chooses among three alternatives. These are (1.) purchase USD ($\Delta_t^a > 0$), (2.) sell USD ($\Delta_t^a < 0$), and (3.) retain investments unchanged ($\Delta_t^a = 0$). To solve this problem, the agent refers to his objective function. Assuming, that each agent has an expectation of the upcoming price shift π_{t+1}, their actions are driven by the profit maximization task of Eq. 5 [17].

$$\frac{1}{T} \sum_t \pi_{t+1} \Delta_t^a \to \max \quad \text{where} \quad \pi_{t+1} = \ln\left(p_{t+1}/p_t\right) \tag{5}$$

Each agent tries to to expand his net capital balance by maximizing the average profit gained from trades Δ_t^a. For example, an agent expecting a positive DEM / USD FX-rate shift ($\pi_{t+1} > 0$) would buy USD ($\Delta_t^a > 0$).

Up to now, the agents are unable to interact, i. e. the activities of agent a have no effect on the buying / selling decisions of other agents. The interaction of the agents can by established by the *market clearance condition* of Eq. 6 [17].

$$\sum_a \Delta_t^a = 0 \quad \text{for} \quad t = 1, \dots, T \tag{6}$$

While agents decisions concern the *micro-level*, Eq. 6 maps the *macro structure*: An agent can only buy USD if there are other agents who are selling a congruent amount. Hence, agents plans are indirectly associated with each other.

In a market equilibrium no price adjustments are needed, because agents activities match (Eq. 7). Otherwise, a price shift will balance demand and supply. Noticing, that $c > 0$ is a constant, the price shift is estimated by Eq. 8 [17].

$$\sum_a^A \Delta_t^a = 0 \quad \Rightarrow \quad \pi_{t+1} = 0 \tag{7}$$

$$\sum_a^A \Delta_t^a \neq 0 \quad \Rightarrow \quad \pi_{t+1} = c \cdot \sum_a^A \Delta_t^a \tag{8}$$

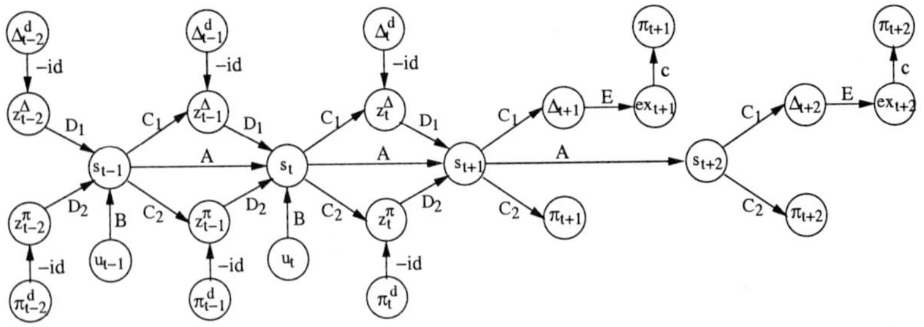

Fig. 2. FX-market modeling by a multi-agent approach based on cognitive systems

Eq. 8 models the price formation mechanism of the FX-market. For alternative approaches see [1, 7]. The price shift π_{t+1} is estimated by scaling the excess demand / supply $\sum \Delta^a_t$ with a constant $c > 0$. For example, an *excess supply* of USD ($\sum \Delta^a_t < 0$) leads to a devaluation of USD ($\pi_{t+1} < 0$). In this case, the negative price shift $\pi_{t+1} < 0$ is computed by scaling $\sum \Delta^a_t < 0$ with $c > 0$ [17].

Note, that the proposed multi-agent framework is not only a descriptive model of the underlying market dynamics, but also an explanatory approach.

5 Multi-agent Modeling Based on Cognitive Systems

The ECNN of Fig. 2 serves our purpose of modeling the DEM / USD FX-market by a cognitive system based multi-agent approach.

The ECNN (Fig. 2) includes *all* agents and the market price formation.

First, let us focus on the modeling of a single agent. Each agent is represented by a cognitive system resp. its structural mapping ECNN. As an extension, the ECNN (Fig. 2) has *two* error correction branches concerning (*i.*) the actions Δ_τ of the agent and (*ii.*) his expectations π^e_τ about FX-rate shifts. At each unfolding time step, the internal expectations of the agent concerning an action Δ_τ and the price shift π^e_τ are compared to observations of Δ^d_τ and π^d_τ. Δ^d_τ is derived from π^d_τ, e. g. buy USD if $\pi^d_\tau > 0$. In future time, the agent estimates an action Δ_τ and a price shift π^e_τ. For the action variable, we use the *prof-max* cost function [11], while we use squared errors for the price shift.

The internal processing of the agent is mapped by the autonomous part of the ECNN. Agents memory depends on the dimension s of the state transition, i. e. higher-dimensional states accumulate more memory than smaller ones. The interior processing is supported by *overshooting*. Thus, the agent predicts the FX-rate stepwise several time periods ahead and plans trading activities as well.

The agents are separated by block matrices A, C_1, C_2, D_1 and D_2. A block of s states represents one agent, e. g. states $1, \ldots, s$ constitute the first agent, while the Nth agent is determined by states $sN - s + 1, \ldots, 2s$. Due to the block matrices, there is no interaction among the cognitive systems of the agents.

In particular, matrix A is a block diagonal matrix with N separate blocks of $s \times s$ weights located on the main-diagonal. The dimension of A is therefore $sN \times sN$. A block maps the internal model of an agent. Matrices C_1 and C_2

calculating the expected values $\Delta_\tau = C_1 s_\tau$ and $\pi^e_\tau = C_2 s_\tau$ consist of a sequence of s weights in each column, e. g. weights $c_{1,1}, \ldots, c_{s,1}$ are associated with the first agent. The dimension of $C_{1,2}$ is $sN \times N$. The structure of matrix D_1 resp. D_2 is the transpose of C_1 resp. C_2. Note, that the compensation mechanisms z^Δ_τ and z^π_τ consist of N targets according to the number of agents.

Unconscious perceptions enter the internal models of the agents through matrix B, whereby each input is available for every agent. In order to strew the information among the agents, which is likely to cause heterogeneous decision behavior, we apply the concept of information filtering [17]. This means, that we designed B as a randomly initialized sparse connector (see Fig. 4).

Finally, let us focus on the market price formation. Due to the error correction of the action branch, the price formation is only applied to all future time steps, where the trading activities Δ_{t+1}, Δ_{t+2}, ... are computed. For a single time step, the mechanism is as follows: First, the excess demand ex_τ is computed out of the agents decisions Δ_τ using fixed matrix $E = [1, ..., 1]$. Thereafter the FX-rate shift π_τ is estimated by scaling the excess demand ex_τ with a constant $c > 0$ (see Eq. 8). Note, that the actual FX-rate shift π_τ might differ form the price expectation π^e_τ of some agents. This is due to a wrong market evaluation, which is e. g. caused by agents insufficient knowledge about external influences.

6 Empirical Study

Now, we apply our multi-agent approach (Fig. 2) to the DEM / USD FX-market. The empirical study is as follows: We work on basis of weekly data from Jan. 91 to Oct. 97 to forecast the DEM / USD FX-rate. The data is divided into two sets (training set from Jan. 91 to Sep. 96; test set from Oct. 96 to Oct. 97).

The selection of the inputs u_t is due to the idea, that a major currency is driven by at least *four* influences. These are the development of (*i.*) other currencies, (*ii.*) bond, (*iii.*) stock and (*iv.*) commodity markets [10]. The data set consists of 11 technical and fundamental inputs (e. g. Dow Jones, US-Bonds).

Due to the recurrent modeling, no complicated input preprocessing is needed. We only calculated the momentum of each input. The transformed inputs are scaled such that they have a mean of zero and a variance of one. The ECNN was trained until convergence using *vario-eta* learning [11].

We compare our model with a naive strategy and an 3-layer MLP. Referring to the naive strategy, one assumes that an observed market trend will last for more than one time period. For example, if the price rises from time period $t-1$ to t, one might as well expect another positive price shift from t to $t+1$. The 3-layer MLP consists of an input layer, a hidden layer (20 neurons with tanh(.) activation function, see [11]) and an output layer (ln cosh(.) error function, see [11]). Inputs to MLP are the (time-lagged) external influences u_t, u_{t-1} and u_{t-2}. The network was trained until convergence with pattern-by-pattern learning [11].

The performance is measured on the basis of hit rates and accumulated model return curves. The 'hit rate' counts how often the sign of the forecast is correct. The accumulated model return is calculated across the generalization set by cumulating the actual FX-rate shifts which are multiplied by the sign of the corresponding model forecasts. We included transaction costs of 0.01% per trade.

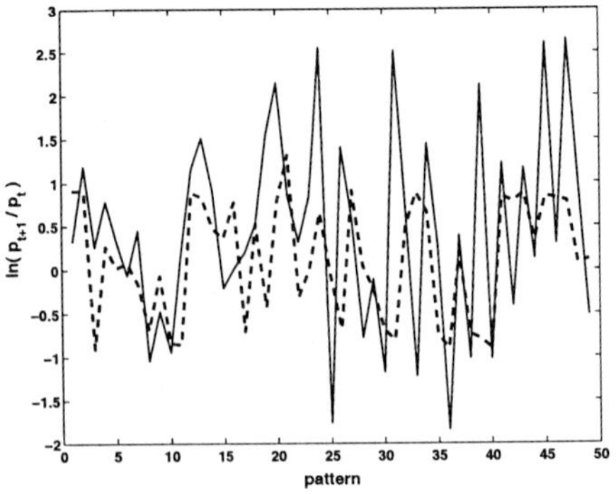

Fig. 3.

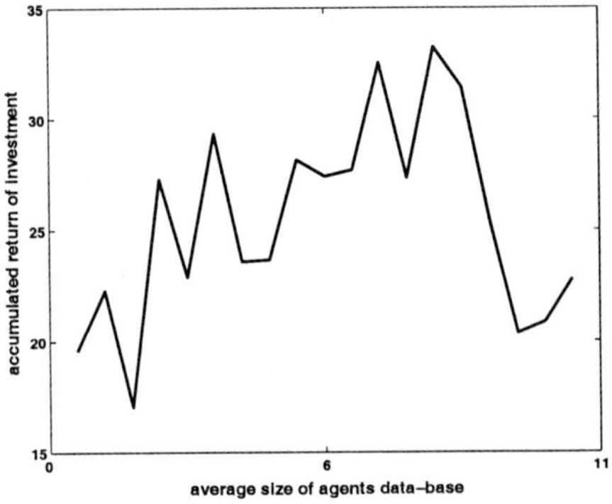

Fig. 4.

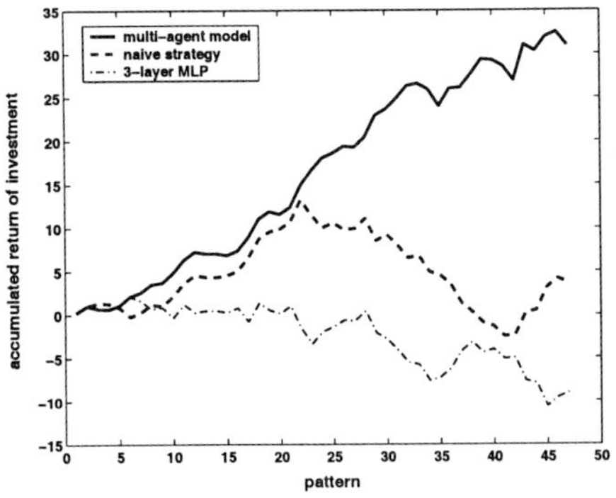

Fig. 5.

Fig.3. Comparison of model forecasts and actual DEM / USD price shifts (generalization set)

Fig.4. Accumulated model return subjected to the average size of agents data subset (generalization set)

Fig.5. Comparison of profit / loss curves (generalization set)

In Fig. 3 we compare the output of our multi-agent model to the actual price movement of the DEM / USD FX-rate. As it can be seen, our model allows a fairly good fitting of the DEM / USD FX-rate on the generalization set.

Fig. 4 depicts the dependency of the accumulated model return form the average size of agents data subset (sparseness of matrix B). If the size of agents data subset is too large (B is fully occupied), their decision behavior is too homogeneous. The market tends to be very volatile. Otherwise, if the average data set size is too small (B is too sparse), agents decisions are too heterogeneous and one can only observe small market movements. Both situations do not reflect real world behavior. Our experiments indicate, that 8 inputs on the average are recommendable, while a number of 200 agents is a suitable FX-market size.

Concerning the performance, the hit rate of the multi-agent model is 81.1%, while the naive strategy achieves 56.5% and the MLP obtains 50.0%. Examining Fig. 5, it turns out, that our approach is superior to the benchmarks. Among the benchmarks, the naive strategy is superior. This is caused by an up-warding trend market in the first part of the generalization set. The MLP loses its generalization abilities due to learning by heart and thus operates at a deficit.

7 Conclusion

We combined cognitive systems and explanatory multi-agent modeling. The decision making of an agent is mapped by a cognitive system with three functionalities (perception, internal processing and action). By the application of ECNN, we merge multi-agent models, cognitive systems and econometrics. Our model concerns semantic specifications instead of ad-hoc functional relationships.

The proposed ECNN has a mechanism to handle external shocks. *Overshooting* enforces the autonomous part of network, which allows long-term forecasting. The structural representation of a cognitive system by an ECNN is natural.

Our approach allows the fitting of real world data (DEM / USD FX-market). Compared to classical benchmarks, it turns out, that our approach is superior.

There are a lot of extensions and further analysis, e. g. the inclusion of interest rates, utility functions or transaction costs. The architecture of the ECNN can be extended by *unfolding in space and time* [16].

References

[1] Beltratti, A., Margarita, S. and Terna, P.: *Neural Networks for economic and financial modeling*, International Thomson Computer Press, London, UK. 1996.
[2] Chalmers, D.: *Facing up the problem of consciousness, in: Explaining Consciousness: The Hard Problem*, ed. J. Shear, MIT Press, 1997.
[3] De Grauwe, P., Dewachter, H. and Embrechts, M.: *Exchange Rate Theory: Chaotic Models of Foreign Exchange Markets*, Blackwell, Oxford. 1993.
[4] Epstein, J. M. and Axtell, R.: *Growing Artificial Societies*, MIT Press, 1997.
[5] Grossberg, S.: *Neural Networks and Natural Intelligence.* MIT Press, 1988.
[6] Haykin S.: *Neural Networks. A Comprehensive Foundation.*, Macmillan College Publishing, New York, 1994, 2nd edition 1998.
[7] LeBaron, B.: *Agent-based computational finance: Suggested readings and early research*, Journal of Economic Dynamics and Control, Vol. 24, 2000, pp. 679 - 702.
[8] MacPhail, E.: *The Evolution of Consciousness*, Oxford Uni. Pr., New York, 1998.
[9] Metzinger, Th.: *Neural Correlates of Consciousness*, Empirical and Conceptual Questions, MIT Press, Cambridge, Massachusetts, London 2000.
[10] Murphy, J.J.: *Intermarket Technical Analysis*, New York.
[11] Neuneier, R. and Zimmermann, H.G.: *How to Train Neural Networks, in: Tricks of the Trade: How to make algorithms really to work*, Springer, Berlin 1998.
[12] Pearlmatter, B.:*Gradient Calculations for Dynamic Recurrent Neural Networks: A survey*, in: IEEE Transactions on Neural Networks, Vol. 6, 1995.
[13] Perlovsky, L.I.: *Neural Networks and intellect: using model-based concepts.* Oxford Uni. Pr., New York, 2001.
[14] Tesfatsion, L.:*Agent-Based Computational Economics: A Brief Guide to the Literature*, Reader's Guide to the Social Sciences, Fitzroy-Dearborn, London, 2000.
[15] Zimmermann, H.G. and Neuneier, R.:*Neural Network Architectures for the Modeling of Dynamical Systems*, in: A Field Guide to Dynamical Recurrent Networks, Eds. Kolen, J.F.; Kremer, St.; IEEE Press 2001.
[16] Zimmermann, H.G., Neuneier, R., Grothmann, R.: *Modeling of Dynamical Systems by Error Correction Neural Networks*, in: Modeling and Forecasting Financial Data, Techniques of Nonlinear Dynamics, Eds. Soofi, A. and Cao, L., Kluwer 2001.
[17] Zimmermann, H.G., Neuneier, R., Grothmann, R.: *Multi-Agent Market Modeling of Multiple FX-Markets by Neural Networks*, IEEE Trans. on Neural Networks, special issue, 2001 forthcoming.

Speculative Dynamics in a Heterogeneous-Agent Model

Taisei Kaizoji[1,*]

Department of Economics, University of Kiel,
Olshausenstr. 40, 24118 Kiel, Germany
kaizoji@icu.ac.jp
http://www.bwl.uni-kiel.de/vwlinstitute/gwrp/german/team/kaizoji.html

Abstract. This paper proposes a model of a stock market which is composed of three different types of traders: fundamentalists, chartists, and noise traders. We investigate the speculative price dynamics though the simulation, and show that a *non-stationary chaos* that is considered as speculative bubble is caused by the heterogeneity of traders' strategies, and furthermore that the distributions of stock returns generated from our heterogeneous agent model have the fat tails which is a stylized fact observed in almost all the financial markets.

1 Introduction

A number of studies have found that the empirical distributions of relative price changes (returns) in almost of financial data such as stock prices and exchange rates posse higher peaks around the mean as well as fatter tails than the Normal distribution. This phenomenon is called as *fat tail phenomenon* [1]. Recently several authors have attempted to explain the stylized facts including the fat tails by demonstrating that their interacting heterogeneous agent models can generate the artificial time series with the feature [c.f. Brock and Hommes (1999), Gaunersdorfer and Hommes (2000), Lux (1998), Lux and Marchesi (1999)]. In the models cited above the dynamical systems that describes financial markets have chaotic attractors, and the asset prices fluctuate irregularly around the fundamental values. They reported that the mechanisms by which the chaotic attractors are created, more generally, the bifurcation phenomena the dynamical systems undergo may help account for the stylized facts such as the fat tails and clustering volatility.

In this paper we concerned with speculative bubbles that is an interesting phenomenon that is often observed in the financial markets. The speculative bubbles has often been touched in the rational expectations literature but little attention has been given to this in the recent heterogeneous agent literature. To this end we propose a simple heterogeneous agent model in which speculative

* On research leave from Division of Social Science, International Christian University, Osawa, Mitaka, Tokyo, 181-8585, Japan.

[1] Pagan (1996) provides a excellent survey of the stylized facts on financial markets.

G. Dorffner, H. Bischof, and K. Hornik (Eds.): ICANN 2001, LNCS 2130, pp. 775–781, 2001.

bubbles can be caused by the heterogeneity of traders' trading strategies, and show the distributions of the relative price changes generated from the model have fat tails.

2 The Model

We think of a stock market in which a large number of agents trade a stock. The stock market is composed of three groups of traders having the different trading strategies; *fundamentalists*, *chartists* and *noise traders*. Traders can either invest in money or in the stock. We consider the trade in a very short term. Thus, we omit the dividend and the interest rate.

2.1 Fundamentalists

We assume that fundamentalist maximizes the utility function, $U(x_t^f, y_t^f) = a(y_t^f + p_{t+1}^f x_t^f) + b(x_t^f - (1+x_t^f)\log(1+x_t^f))$, subject to the budget constraint: $y_t^f + p_t x_t^f = 0$, where x_t^f and y_t^f represent the fundamentalist excess demands for the stock, and for money at period t. p_t denotes the stock price at period t. p_{t+1}^f denotes the expected price by fundamentalists for the next period $t+1$. The fundamentalists are assumed to anticipate the price for the next period $t+1$ using $p_{t+1}^f = p_t + \nu(p^* - p_t)$, that is, the fundamentalists believe that the price will move in the direction of the fundamental price p^*, so that they correct their expectation by a factor ν. The problem are solved for the fundamentalist excess demand function for the stock by

$$x_t^f = \exp(\alpha(p_{t+1}^f - p_t)) - 1, \quad \alpha = \frac{a}{b} > 0. \tag{1}$$

This excess demand function means the following: if the stock price p_t is below the expected price p_{t+1}^f, then fundamentalists try to buy the stock until the stock price p_t is equal to the expected price, because he thinks that the stock is undervalued, and vice versa.

2.2 Chartists

The behavior of a chartist is formalized as maximizing the utility function, $V(x_t^c, y_t^c) = a(y_t + p_{t+1}^c x_t^c) + b(x_t^c - (1+x_t^c)\log(1+x_t^c))$ subject to the budget constraint $y_t^c + p_t x_t^c = 0$ where x_t^c and y_t^c represent the chartist excess demand for the stock and for money at period t, and p_{t+1}^c denotes the stock price for the next period expected by chartist. The chartist utility function is identical to the fundamentalist one utility function except for the term on the expected price. The chartist excess demand function for the stock is

$$x_t^c = \exp(\alpha(p_{t+1}^c - p_t)) - 1, \quad \alpha = \frac{a}{b} > 0. \tag{2}$$

It follows from this excess demand function that chartists try to buy the stock when they anticipate that the stock price will rise over the next period, and to sell the stocks when they expect the stock price will fall over the next period. Chartists forecast the price for the next period p_{t+1}^c according to the so-called *adaptive expectation* given by $p_{t+1}^c = p_t^c + \mu(p_t - p_t^c)$ where the parameter μ is called the error correction coefficient.

2.3 Noise Traders

The noise traders decide their behavior on base of noisy information ϵ_t. The behavior of noise trader is formalized as maximizing the following utility function given by $W(x_t^n, y_t^n) = g(y_t^n + (p_t + \epsilon_t)x_t^n) + kx_t^{n2}$ subject to the budget constraint $y_t^n + p_t x_t^n = 0$, where x_t^n and y_t^n represent the noise trader desired excess demand for the stock and for money at period t. The noisy information at period t ϵ_t follows IID. The noise trader excess demand function for the stock is

$$x_t^n = \gamma\epsilon_t, \quad \gamma = \frac{g}{k} > 0 \tag{3}$$

2.4 The Adjustment Process of the Stock Price

Let us now consider the adjustment process of the stock price in the stock market. We assume that the existence of a market-maker who mediates the trading. If the excess demand in period t is positive (negative), the market maker raises (reduces) the price for the next period $t+1$. The process of adjustment of the stock price can be rewritten as

$$p_{t+1} - p_t = \theta n[(1 - \kappa - \xi)x_t^f + \kappa x_t^c + \xi x_t^n], \tag{4}$$

where θ denotes the speed of the adjustment of the price, and n, the total number of traders, and κ and ξ, the proportions of chartists and the noise traders in the total number of traders n.

2.5 Strategy Switching

The question which we must consider next is how traders choose between the fundamentalist and the chartist strategy. To put it in the model we assume that the proportion of chartist κ depends on difference between the squared prediction errors each strategy made. Formally we write the dynamics of the proportion of chartists as follows:

$$\kappa_t = \frac{(1-\xi)}{1 + \exp(\psi(R^c - R^f))} \qquad R^c = (p_t - p_t^c)^2, \quad R^f = (p_t^f - p_t) \tag{5}$$

The equation expresses that if the squared prediction error by chartist R^c is greater than the squared prediction error by fundamentalist R^f, then some of chartists become fundamentalists, and visa versa.

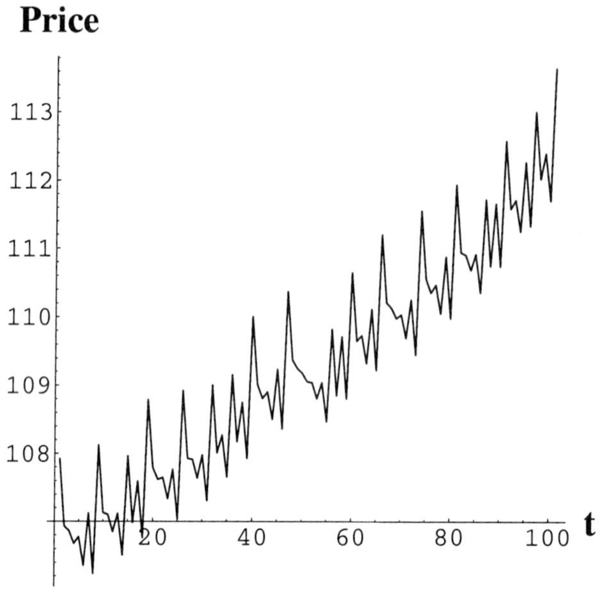

Fig. 1. Non-stationary chaos of the stock price

3 Dynamics of Speculative Price

It follows that the dynamical system we have seen has the unique equilibrium price which is equal to the fundamental price p^*. The global dynamics of the model are quite complex according to the results of the computer simulation to the various values of the parameters. In order to illustrate a typical result from the model put the parameters below as follows: $\alpha = 6$, $\theta = 0.001$, $n = 1000$, $\mu = 0.5$, $\nu = 0.5$, $\gamma = 1$, and $\psi = 0.01$. For a while we also assume that there is no noise trader, that is, $\xi = 0$, and the fundamental price is constant, i.e. $p^* = 10$, so that the price dynamics become deterministic. Figure 1 and Figure 2 indicate typical time series of the price and of the price change obtained from the model with the above set of the parameters. As we see in Figure 1, the dynamic behavior of the stock price is *non-stationary chaos*. Namely, the time series have a upward trend, and fluctuates irregularly around the trend. We may consider the non-stationary chaos as *speculative bubble* that have often been observed in the actual financial markets. On the other hand the price change fluctuates irregularly within a certain range. Namely, the movement of the price change is stationary. Results of the simulation experiments show that continuous increases of α lead the dynamics of the price changes to chaos through the well-known period-doubling bifurcation[2].

[2] For a detailed analysis of the non-stationary chaos see Kaizoji (2001).

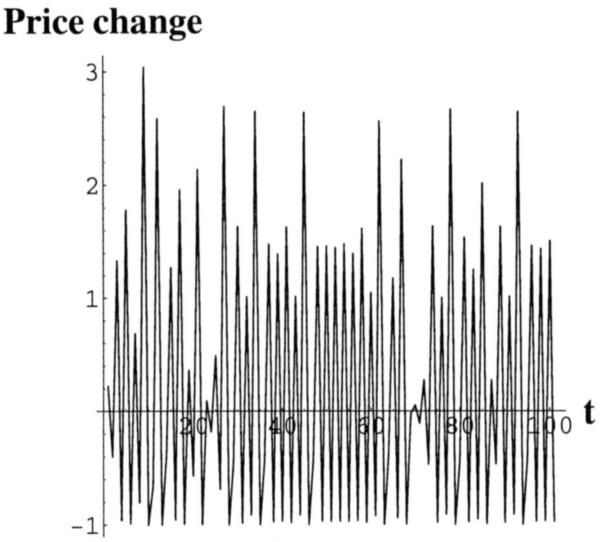

Fig. 2. Stationary chaos of the stock price change

Table 1. Statistics of return from artificial data

α	*Mean*	*S.D.*	*Skewness*	*Kurtosis*	*Price*	*Return*	ρ
1	0	0.014	0.023	0.061	stable	stable	4.83
2	0	0.016	0.210	0.323	stable	stable	4.14
3	0	0.02	0.703	1.552	periodic	periodic	3.2
4	0	0.026	1.411	4.277	periodic	periodic	2.32
5	0	0.004	2.963	17.605	bubble	chaotic	4.58

3.1 Fat Tail Phenomenon

To bring our model close to the actual stock markets let us here assume the existence of noise traders. We put $\xi = 0.3$ below. A number of empirical studies on financial markets have found that the empirical distributions of relative price changes (the so called returns) in almost all financial data show uni-modal bell shapes, and posse excessive fourth moments, so called *leptokurtosis*, (cf. Pagan (1996)). Figure 3 indicates the distribution of the stock returns[3] generated from the model with the above setting of the parameters. The distribution has an uni-modal bell-shape. Table 1 presents sample statistics of stock return. The mean value is very close to zero. The standard deviation is small. Skewness is positive, and kurtosis is significantly positive. The important point to note is that the return distribution exhibits leptokurtosis apparently.

The recent literature finds that the distribution of returns is not only leptokurtotic, but also belongs to the class of *fat tailed* distributions. This finding is quite universally accepted as an universal characteristic of practically all fi-

[3] The stock return is defined as $r_t = \log p_t - \log p_{t-1}$.

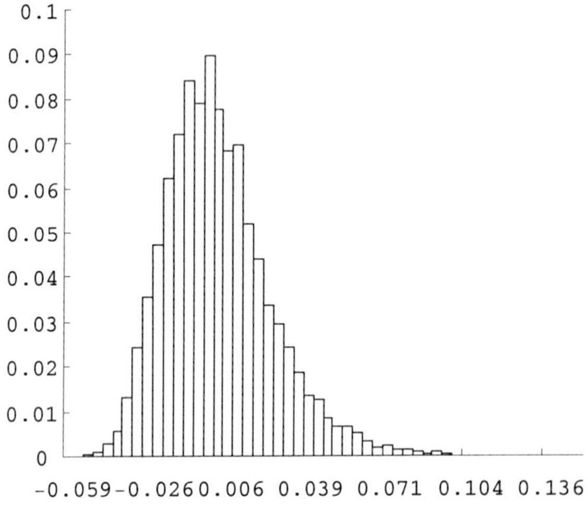

Fig. 3. The distribution of the stock returns

nancial returns. More formally it has been shown that the tails of the return distribution follow approximately a power law: $F(|z_t| > x) \sim cx^{-\rho}$, where $F()$ denotes the cumulative probability distribution of the normalized stock returns. Most of empirical studies show estimates of the *tail index* ρ falling in the range 2 to 4 (Pagan (1996)). The last column of the table 1 shows the values of ρ of the return distributions generated by our model under the various values of α[4]. The values of ρ for tail sizes up to 5 per cent of the size of the data fall in the range 2 to 4 in the cases that periodic solutions are observed in the artificial stock market.

4 Concluding Remarks

In this paper we propose a heterogeneous agent model of a stock market. We showed that the non-linearity of the excess demand functions derived from the traders' optimization behavior might generate non-stationary chaos that is regarded as speculative bubbles. Furthermore we showed that the return distributions derived from the model share important characteristics of the empirical financial data: they exhibit leptokurtosis as well as fat tails.

Acknowledgements

I would like to thank T. Aoki, S-H. Chen, J. Holyst, A. Matsumoto, T. Misawa, T. Onozaki, P. Flaschel R. Frank, W. Zemmler, and T. Asada for helpful sug-

[4] The values of ρ are computed by the Hill tail index estimate that is standard work tool for estimation of the Pareto exponent of the tails.

gestions and comments, and the Alexander Humboldt Foundation for financila support.

References

1. Brock, W. A., and Hommes, C. H.: Heterogeneous beliefs and routes to chaos in a simple asset pricing model, Journal of Economic Dynamics and Control, 22 (1998) pp.1235-1274.
2. Gaunersdorfer, A., and Hommes, C. H.: A Nonlinear Structural Model fro Volatility Clustering, Working Paper No. 63, Adaptive Information System and Modelling in Economics and Manegement Science (2000).
3. Kaizoji, T.: Speculative Bubbles and Fat Tail Phenomena in A Heterogeneous Model, forthcoming into International Symposia in Economic Theory and Econometrics, Cambridge University Press (2001).
4. Lux, T.: The socio-economic dynamics of speculative markets: interacting agents, chaos and the fat tails of return distributions, Journal of Economic Behaviour and Organization, 33 (1998) pp. 143-165.
5. Lux, T., and M. Marchesi, M.: Volatility Clustering in Financial Markes: A Micro-Simulation of Interacting Agents, International Journal of Theoretical and Applied Finance 3 (2000) pp. 675 - 702.
6. Pagan, A.: The Econometrics of Financial Markets, Journal of Empirical Finance, 3 (1996) 15-102.

Nonlinear Adaptive Beliefs and the Dynamics of Financial Markets: The Role of the Evolutionary Fitness Measure

Andrea Gaunersdorfer[1] and Cars H. Hommes[2]

[1] Department of Business Studies, University of Vienna
Brünner Straße 72, A-1210 Wien, Austria
Andrea.Gaunersdorfer@univie.ac.at
http://finance2.bwl.univie.ac.at
[2] CeNDEF, University of Amsterdam
Roetersstraat 11, NL-1018 WB Amsterdam, The Netherlands
hommes@fee.uva.nl
http://www.fee.uva.nl/cendef/

Abstract. We introduce a simple asset pricing model with two types of adaptively learning traders, fundamentalists and technical traders. Traders update their beliefs according to past performance and to market conditions. The model generates endogenous price fluctuations and captures some stylized facts observed in real returns data, such as excess volatility, fat tails of returns distributions, volatility clustering, and long memory. We show that the results are quite robust w.r.t. to different choices for the performance measure.

1 Introduction

Traditional models in finance rest on the assumptions of informationally efficient markets (efficient market hypothesis, EMH) and rational agents that have symmetric information, identical (homogenous) preferences and build their expectation on all available information (rational expectations hypothesis). In a world with rational expectations and efficient markets everything is driven by unanticipated uncorrelated shocks (technological shocks, changing preferences, etc.). However, these models fail to explain many so-called stylized facts observed in financial time series, for example, excess volatility, fat tailed distributions of stock returns, and volatility clustering (that is, high/small price changes are mostly followed by high/small price changes). Though asset returns contain little serial correlation (which is consistent with the weak form of the EMH) there is substantially more correlation between absolute or squared returns. These facts indicate that prices in financial markets are not solely determined by economic fundamentals. Markets seem to have internal dynamics of their own and are influenced by market psychology and investors 'animal spirits'.

In the last decade an increasing number of models with heterogeneous, boundedly rational agents has been introduced. Such models typically distinguish be-

G. Dorffner, H. Bischof, and K. Hornik (Eds.): ICANN 2001, LNCS 2130, pp. 782–789, 2001.

tween two types of traders, fundamentalists, who believe that prices are determined by economic fundamentals, and chartists, who believe that prices can be predicted by simple trading rules based upon past price patterns in the recent past.

Brock and Hommes [BH1] have introduced the concept of *Adaptive Belief Systems*. Agents adapt their predictions by choosing among a finite number of predictors (beliefs, expectations), which are functions of past information. Based on a performance measure attached to every predictor, which is publically available, agents make a boundedly rational choice between the predictors. This leads to the so-called *Adaptive Rational Equilibrium Dynamics*, a nonlinear, evolutionary dynamics across competing beliefs or trading strategies, coupled to the dynamics of the endogenous variables. In a series of papers Brock and Hommes [BH2], [BH3], [BH4], Gaunersdorfer [G1], and Gaunersdorfer and Hommes [GH] have applied this concept to a simple asset pricing model. For a comprehensive survey on these models see [H].

In this paper we review the model of [GH] and focus on the influence of different performance measures on the dynamics. In section 2 we recall the model and discuss a second evolutionary fitness measure. Section 3 concludes.

2 The Model

Our adaptive belief system is based on a standard asset pricing model extended to *heterogeneous* beliefs. Agents can either invest in a riskless asset, which is perfectly elastically supplied with gross rate of return R, or in a risky asset with price p_t per share (ex-dividend), which pays a stochastic dividend y_t at time t. Agents are assumed to be mean-variance maximizers, so that the demand z_{ht} of trader type h for the risky asset solves

$$\max_{z_{ht}}\{E_{ht}(W_{h,t+1}) - \frac{a}{2}V_{ht}(W_{h,t+1})\}, \tag{1}$$

where $W_{h,t+1} = RW_{ht} + (p_{t+1} + y_{t+1} - Rp_t)z_{ht} =: RW_{ht} + R_{t+1}z_{ht}$ describes wealth dynamics and E_{ht} and V_{ht} denote the *beliefs* or forecasts of trader type h about conditional expectation E_t and conditional variance V_t, based on a publically available information set such as past prices and dividends. $a > 0$ is the risk aversion parameter, $R_t = p_t + y_t - Rp_{t-1}$ is the excess return per share. Assuming that beliefs about conditional variances are constant,[1] $V_{ht} \equiv \sigma^2$, demand for the risky asset is given by

$$z_{ht} = \frac{E_{ht}R_{t+1}}{a\sigma^2}. \tag{2}$$

Let z_{st} and n_{ht} denote supply of outstanding shares of the risky asset per investor and fraction of trader type h, respectively. Equilibrium of demand and supply yields

[1] [G1] studies a model version with time varying beliefs about conditional variances and shows that the dynamics are quite similar as in the case with constant beliefs.

$$\sum_h n_{ht} z_{ht} = z_{st} \,. \tag{3}$$

We focus on the special case of zero suppy of outside shares, $z_{st} \equiv 0$. Further assume that dividends follow an iid process with constant mean $E_t(y_{t+1}) \equiv E(y_t) \equiv \bar{y}$ and traders have homogeneous beliefs about future dividends $E_{ht}(y_{t+1}) = E_t(y_{t+1}) = \bar{y}$. Thus, the market equilibrium equation (3) can be rewritten as

$$Rp_t = \sum_h n_{ht} E_{ht}(p_{t+1}) + \bar{y} \,. \tag{4}$$

Market equilibrium thus states that the price of the risky asset equals the discounted sum of tomorrow's expected price and tomorrow's expected dividends, averaged over all trader types.

Let us first consider the special case where traders have homogeneous rational expectations, that is, $E_{ht}(p_{t+1}) = E_t(p_{t+1})$ $\forall h$. This allows us to define a benchmark 'fundamental solution' (EMH fundamental rational expectations price)

$$p^* = \sum_{k=1}^{\infty} \frac{\bar{y}}{R^k} = \frac{\bar{y}}{R-1} \,. \tag{5}$$

Note that in case of homogeneous rational expectations (5) is the only solution of (4) which fulfills the transversality condition $\lim_{t\to\infty} E_t(p_{t+k})/R^k = 0$ ('no bubbles condition').

We consider a simple example with two types of traders choosing simple linear prediction rules. The first type are so-called fundamentalists, believing that prices will move towards the fundamental rational expectations value p^*,

$$E_{1t}(p_{t+1}) = p^e_{1,t+1} = p^* + v(p_{t-1} - p^*), \qquad 0 \le v \le 1. \tag{6}$$

In the special case $v = 1$, traders use the latest observed price as predictor, $E_{1t}(p_{t+1}) = p^e_{1,t+1} = p_{t-1}$. We call this type of traders *EMH believers*, since the naive forecast is consistent with an efficient market, where prices follow a random walk. The second type of traders are technical traders extrapolating the latest observed price change,

$$E_{2t}(p_{t+1}) = p^e_{2,t+1} = p_{t-1} + g(p_{t-1} - p_{t-2}) \,, \qquad g \ge 0 \,. \tag{7}$$

Market equilibrium equation (4) now becomes

$$Rp_t = n_{1t}(p^* + v(p_{t-1} - p^*)) + n_{2t}(p_{t-1} + g(p_{t-1} - p_{t-2})) + \bar{y} + \varepsilon_t \,, \tag{8}$$

where a dynamic iid noise term ε_t has been added to equation (4), representing model approximation error or noise coming from noise traders, traders whose behavior is not explained by the model but considered as exogenously given.

The second, conditionally evolutionary part of the model describes how beliefs, i.e. fractions, change over time. The basic idea is that fractions are updated based upon a performance measure of forecasting rules, conditioned on the deviation of the actual price from the fundamental price p^*. In the first step (evolutionary part) fractions are determined as discrete choice probabilities according to past performance or fitness,

$$\tilde{n}_{ht} = \exp(\beta U_{h,t-1})/Z_{t-1}\,, \qquad Z_{t-1} = \sum_h \exp(\beta U_{h,t-1}), \tag{9}$$

where Z_{t-1} is a normalization factor such that fractions add up to one. U_{ht} ist the 'fitness function' or 'performance measure'. $\beta \geq 0$ is called 'intensity of choice', it measures how fast the mass of traders will switch to the optimal strategy. [BH1], [BH2], and [G1] show that this parameter plays a crucial role in the route to complicated dynamical behavior.

A natural candidate for evolutionary fitness is (accumulated) realized (excess) profits $R_t z_{ht}$, given by

$$U_{ht}^{(1)} = \frac{1}{a\sigma^2}(p_t + y_t - Rp_{t-1})(p_{ht}^e + \bar{y} - Rp_{t-1}) + \eta U_{h,t-1}\,, \tag{10}$$

where $0 \leq \eta \leq 1$ is a memory parameter measuring how fast past realized fitness is discounted for strategy selection and z_{ht} is given by (2).

However, this performance measure is inconsistent with investors being risk averse mean-variance maximizers, since it ignores the variance term in (1). Thus, another natural candidate for the fitness function is utility derived from realized profits, that is, risk adjusted realized profits. Utilities of realized profits in period t of investortype h are given by

$$\pi_{ht} := \pi(R_{t+1}, E_{ht}[R_{t+1}]) = R_{t+1} z(E_{ht}[R_{t+1}]) - \frac{a}{2}\sigma^2 z^2(E_{ht}[R_{t+1}]),$$

where

$$z(E_{ht}[R_{t+1}]) = \arg\max \pi_{ht} = \frac{E_{ht}[R_{t+1}]}{a\sigma^2} = z_{ht}.$$

Note that discrete choice probabilities are independent of utility level, i.e. they do not change if the same term π_t is subtracted off all exponents in (9). Thus, we may consider

$$\begin{aligned}\pi_{ht} - \pi_t &= \pi(R_{t+1}, E_{ht}[R_{t+1}]) - \pi(R_{t+1}, R_{t+1}) \\ &= -\frac{1}{2a\sigma^2}(E_{ht}[R_{t+1}] - R_{t+1})^2 \\ &= -\frac{1}{2a\sigma^2}(p_{t+1} - p_{h,t+1}^e + \delta_{t+1})^2,\end{aligned}$$

where $\pi_t = \pi(R_{t+1}, R_{t+1})$ is the utility of profits of perfect foresight investors and δ_t is the noise term of the stochastic dividend process, i.e. $y_t = \bar{y} + \delta_t$. Note that in the absence of random shocks, $\delta_t \equiv 0$, these differences simply reduce to a negative constant times the squared prediction errors. Thus, we define the second fitness measure as

$$U_{ht}^{(2)} = -\frac{1}{2a\sigma^2}(p_t - p_{ht}^e + \delta_{t+1})^2 + \eta U_{h,t-1}. \tag{11}$$

In the second step of updating of fractions, technical traders condition their charts upon information about fundamentals, that is, on price deviations from the fundamental value p^*,

$$n_{2t} = \tilde{n}_{2t} \exp[-(p_{t-1} - p^*)^2/\alpha], \quad \alpha > 0, \qquad n_{1t} = 1 - n_{2t}. \tag{12}$$

As long as prices are close to the fundamental value, fractions are almost completely determined by evolutionary fitness. However, when price deviations from the fundamental become large the correction term $\exp[-(p_{t-1}-p^*)^2/\alpha]$ becomes small, representing the fact that more and more traders believe that a correction towards the fundamental is about to occur.

The noisy conditional evolutionary asset pricing model with fundamentalists versus technical traders is now given by (8), (9), (12) and one of the evolutionary performance measures (10) or (11). This defines a six-deminsional (in the special case of $\eta = 0$ a four-dimensional) dynamical system. A detailed bifurcation analysis of the *deterministic skeleton* (where $\varepsilon_t \equiv 0$ and $\delta_t \equiv 0$) for performance measure (11) is given in [GHW]. Figure 1 gives an overview of the dynamics (our simulations show that the model with performance measure (10) has similar properties, in particular, local behavior around the fundamental steady state is the same). The fundamental value p^* is a unique steady state which is stable for $g < 2R$. As g is increased it is destabilized by a Hopf bifurcation at $g = 2R$ and a stable invariant circle with periodic or quasi-peridic dynamics emerges. This circle may undergo bifurcations as well, turning into a strange attractor.

For some value of $v =: v_{\text{Ch}1}$ on the Hopf bifurcation line $g = 2R$ the Hopf bifurcation changes from super- to subcritical, a so-called *Chenciner bifurcation* occurs. That is, when the Hopf bifurcation line is crossed from right to left (decreasing g beyond $2R$) for $v_{\text{Ch}1} < v < v_{\text{Ch}2}$ an unstable invariant circle emerges out of the (unstable) equilibrium and the equilibrium becomes stable, wheras the stable invariant circle created in the supercritical Hopf bifurcation still exists. Decreasing g further, the stable und unstable circles collide and disappear in a saddle-node bifurcation of invariant circles at curve S. Thus, there is a parameter region where a stable equilibrium and a stable invariant circle coexist, bounded by the Hopf bifurcation line and curve S. When the system is buffeted with dynamic noise irregular switching between close to fundamental steady state fluctuations (with small volatility) and (almost) periodic price fluctuations (with high volatility) occurs. We call this region *'volatility clustering region'*.

A second phenomenon in our model is *intermittency*, that is, small (almost regularly) price fluctuations are suddenly interrupted by large price fluctuations triggered by technical trading. On the right hand side of figure 1 a strange attractor and the corresponding price series are plotted, showing intermittent chaos. Both phenomena, coexistence of attractors and intermittency, appear to be extremly well suited as description of the phenomenon of volatility clustering.

Figure 2 compares returns series of daily S&P 500 data over 40 years with those generated by our model. In the chosen examples parameter v is set equal to 1, that is, EMH-believers and technical traders interact. In these examples prices are highly persistent and the system is *close* to having an *unit root*. In fact,

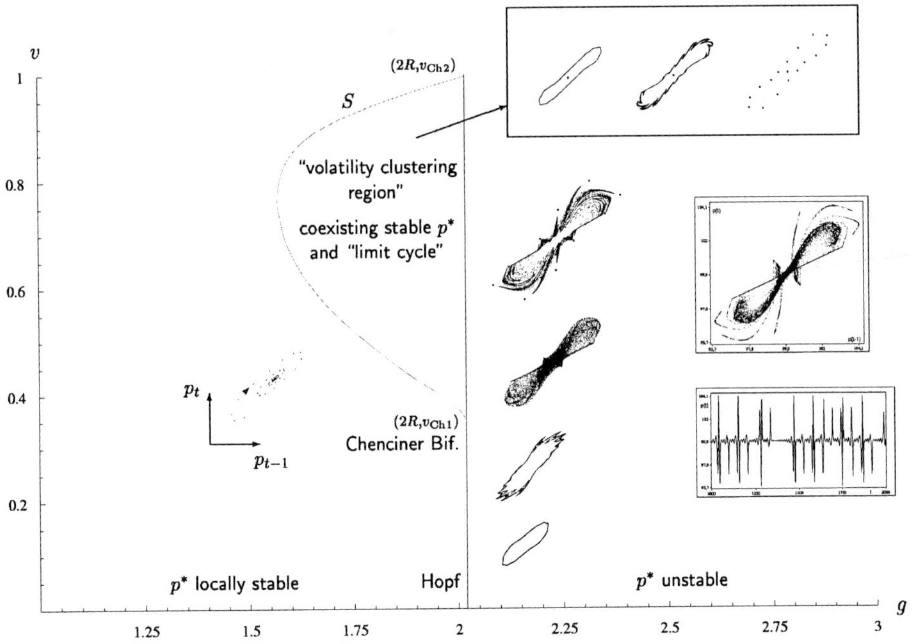

Fig. 1. Bifurcation diagram in the g-v-plane. Projections of attractors to the p_{t-1}-p_t-plane in the corresponding parameter regions are plotted. At points $(2R, v_{\mathrm{Ch}i})$, $i = 1, 2$, *Chenciner bifurcations* occur. These points lie on the Hopf bifurcation line $g = 2R$ and are connected by a curve S corresponding to a saddle-node bifurcation curve of invariant circles. Parameter values: $\beta = 3$, $R = 1.01$, $\bar{y} = 1$, $\alpha = 10$, $a\sigma^2 = 1$, parameter values for the attractor and corresponding price series on the right hand side, showing intermittent chaos: $\beta = 20$, $g = 2.1$, $v = 0.5$.

in case $v = R = 1$ the characteristic polynomial of the Jacobian at the steady state p^* has an eigenvalue equal to 1. Thus, price dynamics are characterized by interaction between unit root behavior far from the the fundamental (when EMH-believers dominate), with relatively small price changes driven only by exogenous news, and larger price changes due to amplification by trend following rules in some neighborhood of the fundamental price.[2]

[2] The deterministic skeletons of these examples have a stable steady state and a co-existing limit cycle. The dynamics along the cycle is characterized as follows. Let us start with a price slightly below p^*. Prices slowly increase and trend followers perform slightly better than EMH-believers. Thus their fraction increases, which reinforces the increase of prices and in turn the increase of the fraction of trend followers. During this phase returns increase and volatility jumps to a high value. However as prices deviate too much from the fundamental value the fraction of trend followers declines and EMH-believers dominate the market. This causes prices to decline and to move slowly in the direction of the fundamental price again, etc. (see [GH]). Adding a dynamic noise term destroys the regularity, but clustered volatility remains.

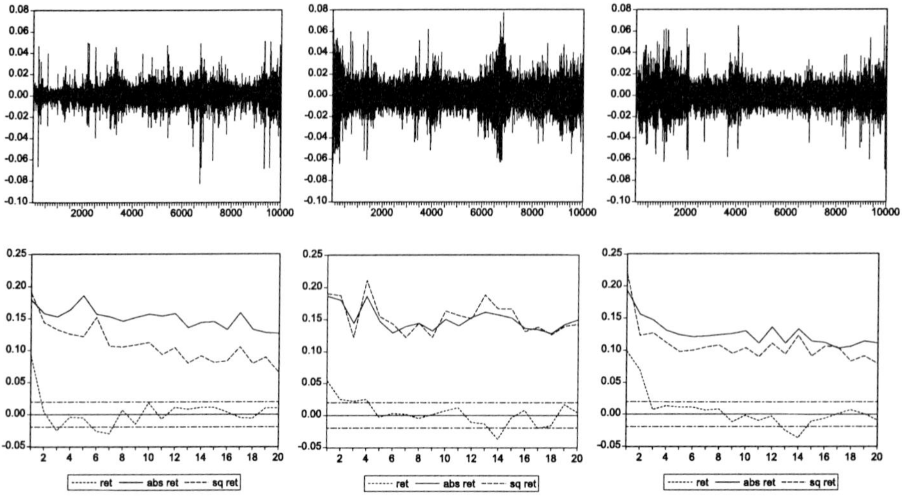

Fig. 2. Returns series (top panel) and autocorrelations of returns, absolute returns and squared returns (bottom panel) for daily S&P 500 data (08/17/1961–05/10/2000, the October 1987 crash and the two days thereafter have been excluded; left panel) and data generated by our model. Middle: performance measure minus squared prediction errors (11), right: performance measure accumulated profits (10); parameter values: $\beta = 2$, $R = 1.001$, $v = 1$, $g = 1.89$ resp. 1.9, $\bar{y} = 1$, $p^* = 1000$, $\alpha = 2000$ resp. 1800, $\eta = 0$ resp. 0.99, $a\sigma^2 = 1$, $\varepsilon_t \sim N(0, 10)$, $\delta_t \equiv 0$.

The return series of both simulations exhibit fat tails with kurtosis coefficient 5.39 resp. 5.58 (versus 8.51 for the S&P 500). Autocorrelations for our simulations are very similar to those of the S&P 500, with almost no significant autocorrelations in returns, but highly significant positive, slowly decaying autocorrelations in absolute and squared returns. That is, our model exhibits long memory and captures the phenomenon of volatility clustering (see [GH] for further details).[3] Our examples also demonstrate that the characteristics of the model is very robust w.r.t the choice of the performance measure in (9).

3 Conclusions

We have modeled financial markets as nonlinear adaptive belief systems. Even in the simple case with only two types of traders, fundamentalists and technical traders, evolutionary competition between different trading strategies may generate endogenous fluctuations causing excess volatility. Our model is able to generate some of the stylized facts observed in many financial time series, such as unpredictability of returns, fat tails, and volatility clustering. These phenom-

[3] We get the strongest form of volatility clustering for $v = 1$. However, our results are fairly robust w.r.t. other model parameters. [G2] presents examples for $v < 1$ and with positive information costs for the fundamental predictor.

ena are created or reinforced by the trading process itself, which seems to be in line with common financial practice. Deviations from fundamentals triggered by *small* shocks may be *reinforced* by boundedly rational beliefs or strategies. Random news plays a role as a trigger event for volatility clustering, which may be amplified by evolutionary forces. This demonstrates that small fundamental causes may occasionally have big consequences and trigger large changes in asset prices. This observation is important for risk management and regulatory policies, but much more research in this area is needed for a better understanding of these forces.

Acknowlegdements

This research was supported by the Austrian Science Foundation (FWF) under grant SFB#010 ('Adaptive Information Systems and Modelling in Economics and Management Science') and by the Netherlands Organization for Scientific Research (NWO) under a NWO-MaG Pionier grant.

References

[BH1] Brock, W.A., Hommes, C.H.: A rational route to randomness. Econometrica **65** (1997a) 1059–1095

[BH2] Brock, W.A., Hommes, C.H.: Models of complexity in economics and finance. In: Hey, C., Schumacher, J.M., Hanzon, B., Praagman, C. (eds.): System Dynamics in Economic and Financial Models. Wiley Publ., New York (1997b) 3–41

[BH3] Brock, W.A., Hommes, C.H.: Heterogeneous beliefs and routes to chaos in a simple asset pricing model. Journal of Economic Dynamics and Control **22** (1998) 1235–74

[BH4] Brock, W.A., Hommes, C.H.: Rational Animal Spirits. In: Herings, P.J.J., Laan, van der G., Talman, A.J.J. (eds.): The Theory of Markets. North-Holland, Amsterdam (1999) 109–137

[G1] Gaunersdorfer, A.: Endogenous fluctuations in a simple asset pricing model with heterogeneous beliefs. Journal of Economic Dynamics and Control **24** (2000) 799–831

[G2] Gaunersdorfer, A.: Adaptive Beliefs and the Volatility of Asset Prices. Central European Journal of Operations Research (2001) (to appear)

[GH] Gaunersdorfer A., Hommes C.H.: A Nonlinear Structural Model for Volatility Clustering. SFB Working Paper No. 63, University of Vienna and Vienna University of Economics and Business Administration, and CeNDEF Working Paper WP 00-02 (2000)

[GHW] Gaunersdorfer A., Hommes C.H., Wagener F.O.J.: Bifurcation routes to volatility clustering. SFB Working Paper No. 73, University of Vienna and Vienna University of Economics and Business Administration, and CeNDEF Working Paper WP 00-04 (2000)

[H] Hommes C.H.: Financial Markets as Nonlinear Adaptive Evolutionary Systems. Quantitative Finance **1** (2001) 149–167

Analyzing Purchase Data by a Neural Net Extension of the Multinomial Logit Model

Harald Hruschka[1], Werner Fettes[2], and Markus Probst[1]

[1] University of Regensburg, D-93053 Regensburg, Germany
harald.hruschka@wiwi.uni-regensburg.de
http://www.wiwi.uni-regensburg.de/hruschka
[2] debis Systemhaus GEI, D-80992 Munich, Germany

Abstract. Brand choice models used to analyze households purchase data as a rule have a linear (deterministic) utility function, i.e. they conceive utility as linear combination of predictors like price, sales promotion variables, brand name and other product attributes. To discover nonlinear effects on brands' utilities in a flexible way we specify deterministic utility by means of a certain type of neural net. This feedforward multilayer perceptron is able to approximate any continuous multivariate function and its derivatives with the desired level of precision. In an empirical study the neural net based choice model leads to better results with regard to both estimation and test data, and implies different choice elasticities.

Choice models as introduced by McFadden [9, 10] are the state-of-the-art approach used in marketing research to analyze the effect of marketing instruments like price, advertising etc. on households' purchases. These models are based on the assumption that households maximize utility when deciding which brand out of several alternatives to buy.

Applications of brand choice models as a rule specify a linear deterministic utility function, i.e. utility as linear combination of predictors [6, 3]. This approach seems to be restrictive as it precludes modeling of nonlinear effects of metric predictors on utility, e.g. threshold effects (utility changes only if a certain value of a predictor is exceeded) or saturation effects (utility does not change if a certain value of a predictor is exceeded).

Artificial neural networks (ANN) of the feedforward multilayer perceptron type approximate any continuous multivariate function and its derivatives with the desired level of precision given a sufficient number of hidden units with S-shaped activation functions [5, 8]. That is why we use feedforward multilayer perceptrons to identify nonlinear effects on utility. By combining such a neural net with the most widespread choice model, the multinomial logit, we remain within the predominant utility-maximizing framework[1].

[1] This research is supported by the DFG.

G. Dorffner, H. Bischof, and K. Hornik (Eds.): ICANN 2001, LNCS 2130, pp. 790–795, 2001.

1 Model Specification

From the assumption that each household chooses the brand perceived to have the largest utility (formed by adding an iid type I extreme value distributed error term to a deterministic component) follows the multinomial logit (MNL) model [9]. According to this model the conditional choice (purchase) probability $p(y_{h,m,t}|B)$ of brand m at occasion t by household h is:

$$p(y_{h,m,t}|B) = \frac{\exp(V_{h,m,t})}{\sum_{m^* \in \mathcal{M}_{h,t}} \exp(V_{h,m^*,t})} \tag{1}$$

$y_{h,m,t}$ denotes a binary purchase indicator (one if brand m is purchased, else zero), B the parameter vector, $V_{h,m,t}$ the deterministic utility of brand m at occasion t for household h, $\{\boldsymbol{c}_{x,h,t}\}$ the set of predictors of all brands, $\mathcal{M}_{h,t}$ the set of brands available at the outlet visited by household h at occasion t.

For the linear specification to which we refer as (linear utility) MNL model the deterministic utility may be written as:

$$V_{h,m,t} = \boldsymbol{\beta}_1 \cdot \boldsymbol{c}_{x,h,m,t} + \boldsymbol{E}_m \cdot \boldsymbol{\beta}_3 \tag{2}$$

$\boldsymbol{E}_m$ is a $m-1$ dimensional vector of zero-one dummy variables which only for brand $m^{'} > 1$ attains the value one, i.e. $\boldsymbol{E}_m = (0, \ldots, 0, 1, 0, \ldots 0)$. $\beta_{3m'}$ gives the difference of deterministic utilities between brand $m^{'}$ and brand 1 holding predictors constant.

Instead of restricting the MNL to a linear utility form, we want to discover nonlinear effects of predictors on utility in a flexible manner. To this end we approximate deterministic utility by means of a feedforward multilayer perceptron with Q hidden units and still remain within the MNL framework. We specify deterministic utility as follows:

$$V_{h,m,t} = \sum_{j=1}^{Q} (\beta_{2,j} \cdot g(\boldsymbol{\beta}_{1,j} \cdot \boldsymbol{c}_{x,h,m,t})) + \boldsymbol{E}_m \cdot \boldsymbol{\beta}_3 \tag{3}$$

Putting deterministic utilities formed according to equation 3 into the basic MNL equation 1 finally gives the combined model which we call artificial neural net–multinomial logit (ANN-MNL) model.

2 Estimation and Evaluation

Both linear utility MNL and ANN-MNL choice models are estimated by maximizing the following log likelihood function:

$$L = \sum_{h=1}^{H} \sum_{t=1}^{T_h} \sum_{m \in \mathcal{M}_{h,t}} y_{h,m,t} \cdot \ln p(y_{h,m,t}|B) \tag{4}$$

H denotes the number of households, T_h the number of purchases of household h, p the conditional choice probability, $y_{h,m,t}$ the purchase indicator of household h of brand m at occasion t and $\mathcal{M}_{h,t}$ the set of available brands at occasion t.

Estimation of ANN-MNL choice models consists of two steps. It starts with stochastic gradient descent to avoid local maxima [7, 11]. We set the number of iterations equal to five times the number of parameters. One iteration selects each of the observations (purchases) according to a random sequence. For each observation parameters are changed by $\eta = 0.02$ times its gradient (i.e. vector of first derivatives). To approach the exact (local) maximum faster stochastic gradient descent is followed by a quasi-Newton optimization method (BFGS) in the second step which only needs information on parameters' first derivatives [4]. To this end an implementation of BFGS developed for multilayer perceptrons is used [12]. Estimation of linear utility MNL models only requires the BFGS step.

Like other more complex models ANN models may suffer from overfitting problems. That is why we randomly generate ten different assignment to estimation (training) and test data [4]. Results presented in the next section (log likelihood values, AIC values, elasticities) are in fact averages over assigments. To evaluate models we look at log likelihood values according to equation 4 for test data and AIC values for estimation data [2].

3 Empirical Study

3.1 Data

We analyze purchase data of the six largest brands in terms of market share for two product groups (ketchup and peanut butter). In each product group we consider only one package size. For a time span of 123 weeks the data refer to households making at least ten purchases. This way 811 and 960 households remain for ketchup and peanut butter, respectively. Predictors of the choice models are point-of-sale advertising (binary), newspaper advertising (binary), loyalty, (observed) price, reference price, price gain (price is less than the reference price) and price loss (price is greater than the reference price). Gains (losses) increase (decrease) any household's choice probability.

Following Guadagni and Little [6] we measure loyalty values as exponentially smoothed past purchases $y_{h,m,t}$ for each household h of brand m at occasion t using smoothing constant α_l.

Reference prices are defined as expected brand-specific prices to which households compare observed prices [13]. This expectation is equivalent to a forecast of prices for the next purchase. Reference price models are estimated by minimizing the error sum of squares between expected and observed prices with lagged prices and a time index as predictors. Linear models are estimated by ordinary least squares, nonlinear ANN reference price models by the same two-step procedure as ANN-MNL models.

Table 1. *Model Performance*

Choice Model	Reference Price Model	Ketchup AIC	Ketchup L	Peanut Butter AIC	Peanut Butter L
MNL	no	-6053.80	-1012.06	-11744.6	-1959.12
	linear	-5710.31	-954.23	-11510.0	-1920.19
	ANN	-5712.53	-954.54	-11528.0	-1923.55
ANN-MNL					
$Q = 3$	no	-5975.37	1000.26	-11589.3	-1937.23
	linear	-5476.88	-920.80	-11302.1	-1890.50
	ANN	-5463.73	-915.62	-11314.4	-1897.97
$Q = 7$	no	-5895.00	-990.59	-11526.3	-1933.25
	linear	-5184.18	-881.47	-11076.5	-1863.23
	ANN	-5182.54	-874.21	-11074.0	-1867.34
$Q = 10$	no	-5932.83	-995.86	-11507.1	-1934.11
	linear	-5166.74	-889.62	-11026.2	-1862.64
	ANN	-5117.47	-879.76	-11029.3	-1871.30

3.2 Estimation Results

Table 1 contains AIC values of the estimation data and log likelihood values of the test data for the various models. Inclusion of reference prices as predictors of choice models increases log likelihood values. This effect is more pronounced for ANN-MNL than MNL models. The most complex ANN-MNL models (with ten hidden units) attain the best AIC values. We evaluate models by their log likelihood values for the test data. That is why we select ANN-MNL models with seven and ten hidden for ketchup and peanut butter brands, respectively. These ANN-MNL models are superior to all MNL models considered. The best model possesses an ANN reference price model for ketchup, a linear reference price model for peanut butter.

To aid interpretation we compute arithmetic means of choice elasticities for metric predictors (a closed form expression for elasticities is easily derived) and probability differences for binary predictors over purchase occasions.

The elasticity of predictor $c_{x_i,h,m,t}$ w.r.t. choice of brand m is defined by:

$$\frac{\partial p(y_{h,m,t}|B,\{\mathbf{c}_{x,h,m,t}\})}{\partial c_{x_i,h,m,t}} \frac{c_{x_i,h,m,t}}{p(y_{h,m,t}|B,\{\mathbf{c}_{x,h,m,t}\})} \tag{5}$$

Elasticities are ratios of relative changes (e.g. an elasticity of -2.5 (0.5) means that a 1% increase of the predictor's value leads to a 2.5 (0.5)% decrease (increase) of choice probability).

Table 2. *Average Elasticities/Probability Differences*

	Ketchup Brands 1	2	3	4	5	6	Peanutbutter Brands 1	2	3	4	5	6
	Reference Price											
MNL	-1.33	-1.87	-1.87	-1.74	-1.99	-1.84	-5.79	-6.76	-6.84	-6.73	-6.97	-7.52
ANN-MNL	-1.24	-1.92	-2.26	-2.90	-2.97	-3.09	-5.35	-6.59	-6.16	-6.42	-6.91	-7.03
	Gain											
MNL	-0.01	-0.02	-0.04	-0.08	-0.08	-0.11	0.04	0.03	0.02	0.05	0.04	0.02
ANN-MNL	0.04	0.07	0.13	0.27	0.22	0.21	0.13	0.12	0.06	0.16	0.13	0.07
	Loss											
MNL	-0.46	-0.64	-0.58	-0.16	-0.20	-0.09	-0.37	-0.44	-0.57	-0.44	-0.47	-0.61
ANN-MNL	-0.25	-0.36	-0.37	-0.15	-0.18	-0.12	-0.20	-0.25	-0.29	-0.23	-0.26	-0.30
	Loyalty											
MNL	0.37	0.38	0.26	0.18	0.18	0.17	0.52	0.39	0.38	0.42	0.34	0.28
ANN-MNL	0.38	0.39	0.28	0.21	0.23	0.26	0.48	0.44	0.44	0.45	0.39	0.38
	Feature											
MNL	0.24	0.24	0.23	0.22	0.20	0.18	0.23	0.21	0.20	0.21	0.19	0.15
ANN-MNL	0.25	0.24	0.20	0.15	0.13	0.10	0.22	0.21	0.21	0.20	0.19	0.18
	Display											
MNL	0.11	0.10	0.10	0.10	0.08	0.08	0.09	0.07	0.07	0.07	0.07	0.05
ANN-MNL	0.07	0.10	0.10	0.11	0.09	0.07	0.09	0.09	0.09	0.09	0.08	0.07

For binary predictors feature and display we consider the increase of choice probabilities achieved by using the respective sales promotion instrument:

$$p(y_{h,m,t}|B, \{\mathbf{c}_{x,h,t}\}, c_{x_i,h,m,t} = 1) - p(y_{h,m,t}|B, \{\mathbf{c}_{x,h,m,t}\}, c_{x_i,h,m,t} = 0) \quad (6)$$

Average elasticities and probability differences obtained on the basis of the MNL model and the ANN-MNL models (with seven and ten hidden units for ketchup and beanut butter, respectiveley) can be looked up in table 2. Reference price elasticities are negative and differ between the two models. In absolute terms the ANN-MNL models as a rule imply higher elasticities for ketchup and somewhat lower elasticities for peanut butter.

We get positive gain elasticities as postulated when using the ANN-MNL models, but implausible negative elasticities for the MNL model in the case of ketchup brands. Moreover elasticities of the ANN-MNL models are much greater compared to those of the MNL model. Loss elasticities are negative as postulated in section 1, their absolute values are smaller for the ANN-MNL models. Note that for the ANN-MNL models gain and loss elasticities have the expected signs and are in agreement with prospect theory, as absolute values of loss elasticities are greater than gain elasticities.

For loyalty positive elasticities are computed. ANN-MNL models provide higher elasticities for ketchup brands.

MNL and ANN-MNL models show that choice probabilities increase if a brand is featured or displayed with features having more effect. The ANN-MNL model indicates smaller effects for ketchup brands with lower market share compared to the MNL model.

4 Conclusions

For empirical choice data of two product groups the ANN-MNL model is superior to a convential linear utility MNL model. This result not only applies to AIC values computed for estimation data, it remains valid for different test data sets not used for estimation.

The estimated ANN-MNL models imply elasticities different from those of the linear utility MNL model for most of the predictors. Contrary to the MNL model average elasticities conform to prospect theory.

This contribution used the linear utility MNL model as comparison standard. Work on the performance of other nonlinear utility extensions of the MNL model (e.g. semiparametric models of the GAM type [1]) is currently under way. Moreover, we also study if mixtures of ANN-MNL models are necessary to deal with the heterogeneity of households.

References

1. Abe, M.: A Generalized Additive Model for Discrete-Choice Data. Journal of Business and Economic Statistics. **17** (1999) 271–284
2. Akaike, H.: A New Look at the Statistical Model Identification. IEEE Transactions on Automatic Control. **19** (1974) 716–723
3. Allenby G.M., Lenk P.: Modeling Household Purchase Behavior with Logistic Normal Regression. Journal of the American Statistical Association. **89** (1994) 1218–1231
4. Bishop, C.M.: Neural Networks for Pattern Recognition. Clarendon Press, Oxford, UK (1995)
5. Cybenko, G.: Continuous Value Neural Networks with Two Hidden Layers are Sufficient. Mathematics of Control, Signals, and Systems. **2** (1989) 303–314
6. Guadagni, P.M., Little, J.D.C.: A Logit Model of Brand Choice Calibrated on Scanner Data. Marketing Science. **2** (1983) 203–238
7. Hertz, J., Krogh, A., Palmer, R.G.: Introduction to the Theory of Neural Computation. Addison-Wesley, Redwood City, CA (1991)
8. Hornik, K., Stinchcombe, M., White, H.: Multilayer Feedforward Networks are Universal Approximators. Neural Networks. **3** (1989) 359–366
9. McFadden, D.: Conditional Logit Analysis of Qualitative Choice Behavior. In: Zarembka, P. (ed.): Frontiers in Econometrics. Academic Press, New York (1973) 105–142
10. McFadden, D.: Econometric Models for Probabilistic Choice among Products. Journal of Business **53** (1980) S13–S34
11. Ripley, B.D: Pattern Recognition and Neural Networks. Cambridge University Press, Cambridge, UK (1996)
12. Saito, K., Nakano, R.: Partial BFGS Update and Efficient Step-Length Calculation for Three-Layer Neural Networks. Neural Computation. **9** (1997) 123–141
13. Winer, R.S.: Behavioral Perspective on Pricing. In: Devinney, T.M. (ed.): Issues in Pricing. Lexington Books, Lexington, MA (1988) 35–57

Part X

Selforganization and Dynamical Systems

Using Maximal Recurrence in Linear Threshold Competitive Layer Networks

Heiko Wersing[1] and Helge Ritter[2]

[1] HONDA R&D Europe (Germany) GmbH, Future Technology Research, Carl-Legien-Str.30, 63073 Offenbach/Main, Germany
[2] University of Bielefeld, Faculty of Technology, P.O.-Box 10 01 31, D-33501 Bielefeld, Germany

Abstract. We demonstrate the application of recent theoretical results on the stability of linear threshold (LT) networks to the competitive layer model architecture (CLM). LT networks can be efficiently built in silicon and exhibit the interesting behavior of coexistence of digital selection and analogue amplification in a single circuit. The CLM provides a large-scale network based on LT units, which was successfully applied to complex perceptual grouping tasks. We show that recent results on LT networks can be employed to operate the CLM with maximal recurrence close to its stability limits, causing strong contextual integration and improved grouping quality.

1 Introduction

A prominent model of the coupled dynamics of networks of neurons is the additive recurrent model

$$\dot{x}_i = -x_i + \sigma\Big(\sum_j w_{ij} x_j + h_i\Big), \tag{1}$$

with neuron activities $x_i \geq 0$, weights w_{ij}, external inputs h_i and transfer function σ. In recent years there has been a growing amount of literature on biologically-based models [2, 6, 1], where the transfer function is piecewise linear and non-saturating, given by $\sigma(x) = \max\big(0, k(x-\theta)\big)$. Threshold θ and gain $k > 0$ characterize this nonlinearity, denoted as semilinear or linear threshold (LT) function. It has been argued [3] that this transfer function is more appropriate than saturating nonlinearities, because cortical neurons rarely operate close to saturation, despite strong recurrent excitation. This indicates that the saturation may not be involved in the actual computations of the recurrent network, since the activation is dynamically bounded due to effective net interactions.

Recently Hahnloser et al.[4] demonstrated an efficient silicon design for LT networks and discussed the coexistence of analogue amplification and digital selection in their circuit. They showed that the stability analysis for symmetric LT networks can be reduced to the discussion of sets which contain the active neurons, where an active superset of an unstable attractor is also unstable and a

G. Dorffner, H. Bischof, and K. Hornik (Eds.): ICANN 2001, LNCS 2130, pp. 799–805, 2001.

subset of a stable attractor is also stable. They also stated a condition for non-divergence of symmetric systems, which, however, depends on eigenvalues of all possible subsystems, and is therefore difficult to evaluate for complex networks.

The coexistence of analog context-dependent activation and digital selection in an LT-circuit is achieved by a combination of avoiding runaway excitation and multistability. In the following we state two new conditions, proved in [7], which allow to estimate parameter bounds on the regime of maximal recurrent modulation in multistable LT networks. We then show how this can be employed for improved contextual integration in the CLM perceptual grouping model.

2 General LT Stability Conditions

The essential property that must be achieved to avoid runaway excitation is boundedness of the dynamics. Our first condition applies to arbitrary symmetric or non-symmetric LT networks described by (1).

Theorem 1. *If the self-coupling weights satisfy* $w_{ii} < 1 - \sum_{j \neq i} \max(0, w_{ij})$ *for all* i*, then the LT dynamics is bounded.*

The condition states that local inhibition with a strength corresponding to the sum of incoming excitatory weights is sufficient to avoid runaway excitation.

If the weights are symmetric with $w_{ij} = w_{ji}$ for all i, j, an additional criterion can be given that allows to improve the stability margins by taking into account also the non-local inhibitory contribution. The result is based on the energy function of the system which can be constructed for symmetric weights.

Theorem 2. *The LT dynamics is bounded if there exists a symmetric matrix* $\hat{W}$ *with* $\hat{w}_{ij} \geq w_{ij}$ *for all* i, j*, for which* $\lambda_{\max}\{\hat{W}\} < 1$*, where* $\lambda_{\max}\{\hat{W}\}$ *is the maximum eigenvalue of* $\hat{W}$.

This condition is different from the one given by Hahnloser et al. [4], which is based on considering submatrices for subsystems of the LT circuit. Note also the difference to the condition $\lambda_{\max}\{W\} < 1$ which would impose global asymptotic stability on the system, not allowing any multistability [7].

3 The CLM Architecture

The CLM [5, 8] consists of a set of L layers of feature-selective neurons (see Fig. 1). The activity of a neuron at position r in layer α is denoted by $x_{r\alpha}$, and a *column* r denotes the set of the neuron activities $x_{r\alpha}$, $\alpha = 1, \ldots, L$ that share a common position r. With each column a particular "feature" is associated, which is described by a set of parameters like e.g. local edge elements characterized by position and orientation (x_r, y_r, θ_r). A binding between two features, represented by columns r and r', is expressed by simultaneous activities $x_{r\hat{\alpha}} > 0$ and $x_{r'\hat{\alpha}} > 0$ that share a common layer $\hat{\alpha}$. All neurons in a column r are equally driven by an external input h_r which represents the significance of the detection of feature

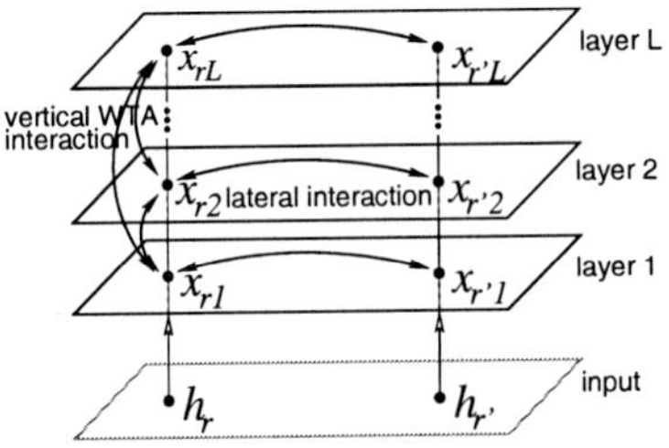

Fig. 1. The competitive layer model architecture (see text for description).

r by a preprocessing step. The afferent input h_r is fed to the activities $x_{r\alpha}$ with a connection weight $J > 0$.

Within each layer α the activities are coupled via the lateral connections $\mathsf{f}^{\alpha}_{rr'}$ which corresponds to the degree of compatibility between features r and r' and which is a symmetric function of the feature parameters, thus $\mathsf{f}^{\alpha}_{rr'} = \mathsf{f}^{\alpha}_{r'r}$. The purpose of the layered arrangement in the CLM is to enforce an assignment of the input features to the layers by the dynamics, using the contextual information stored in the lateral interactions. The unique assignment to a single layer is realized by a columnar Winner-Take-All (WTA) circuit, which uses mutual symmetric inhibitory interactions with absolute strength $J > 0$ between neural activities $x_{r\alpha}$ and $x_{r\beta}$ that share a common column r. Due to the WTA coupling, for a stable equilibrium state of the CLM only a neuron from one layer can be active within each column [8]. The number of layers does not predetermine the number of active groups for a particular run, since for sufficiently many layers only those are active that carry a salient group.

The combination of afferent inputs and lateral and vertical interactions is combined into the standard linear threshold additive activity dynamics

$$\dot{x_{r\alpha}} = -x_{r\alpha} + \sigma\Big(J(h_r - \sum_{\beta} x_{r\beta}) + \sum_{r'} \mathsf{f}^{\alpha}_{rr'} x_{r'\alpha} + x_{r\alpha}\Big) \qquad (2)$$

The CLM dynamics (2) is of the LT form as in (1), if a simple indexing correspondence is established with $i \leftrightarrow (r, \alpha)$. If we then apply Theorem 1, we obtain the following condition for the CLM:

Theorem 3. *The CLM dynamics is bounded and convergent if the condition* $J > \sum_{r' \neq r} \max(0, \mathsf{f}^{\alpha}_{rr'}) + \mathsf{f}^{\alpha}_{rr}$ *is satisfied.*

This can be derived by constructing the corresponding matrix of diagonal and nonpositive off-diagonal weights as in Theorem 1 for the whole CLM architecture. We note that this condition for the CLM was also derived in [8] using a different approach. The parameter J controls the influence of lateral contextual effects on the resulting activity pattern [8]. For J large compared to the lateral weights $\mathsf{f}^{\alpha}_{rr'}$, the single active neuron in a column reproduces its afferent input, $x_{r\alpha} \approx h_r$. For J small, the activity is modulated by the context of coactivated neurons in its layer with $x_{r\alpha} = h_r + J^{-1} \sum_{r'} \mathsf{f}^{\alpha}_{rr'} x_{r'\alpha}$. Therefore, if we choose J close to the stability bound of Theorem 3, the activity is strongly influenced by lateral effects. Theorem 2 gives a new, alternative condition for the CLM:

Theorem 4. *The CLM dynamics is bounded and convergent if* $J > \lambda_{\max}\{f^{\alpha}_{rr'}\}$ *for all* α*, where* $\lambda_{\max}\{f^{\alpha}_{rr'}\}$ *is the largest eigenvalue of the lateral interaction matrix within layer* α*.*

Proof. Construct a matrix $\hat{W}$ (see Theorem 2) by omitting all off-diagonal vertical interactions with $\hat{W}^{\alpha\beta}_{rr'} = -\delta_{rr'}\delta_{\alpha\beta}(J-1) + \delta_{\alpha\beta}f^{\alpha}_{rr'}$. The Kronecker delta is defined by $\delta_{ij} = 1$ if $i = j$ and $\delta_{ij} = 0$ if $i \neq j$. The matrix $\hat{W}$ satisfies $\hat{W}^{\alpha\beta}_{rr'} \geq W^{\alpha\beta}_{rr'}$ for all r, r', α, β. Since $\hat{W}$ contains no cross-layer coupling it has a layer-wise block structure, where all entries corresponding to a coupling of neurons which are in different layers are zero. The eigenvalues of $\hat{W}$ are then given by the eigenvalues of the lateral interaction within layers, given by $\lambda_i\{f^{\alpha}_{rr'}\} - J + 1$. Therefore, if $J > \lambda_{\max}\{f^{\alpha}_{rr'}\}$, then $\lambda_{\max}\{\hat{W}\} < 1$ and the dynamics is bounded according to Theorem 2.

4 Results

We now apply the above stability conditions and investigate with quantitative measures the corresponding enhancement of grouping quality through strong contextual effects. We introduce two measuring functions to assess the quality of the final assignment state that is obtained from the CLM. Suppose, that for a synthetically generated feature set, each feature can be assigned in advance to one of M groups indexed by $l = 1, \ldots, M$. If l_r denotes the group index of feature r then the *label grouping measure* of an assignment state $\hat{\alpha}$ is defined by

$$Q^l = \frac{1}{N^2}\sum_{rr'} q_{rr'}, \qquad \text{where } q_{rr'} = \begin{cases} 1 & \text{if } l_r = l_{r'} \text{ and } \hat{\alpha}(r) = \hat{\alpha}(r'), \\ 1 & \text{if } l_r \neq l_{r'} \text{ and } \hat{\alpha}(r) \neq \hat{\alpha}(r'), \\ 0 & \text{else }. \end{cases} \tag{3}$$

The quantity Q^l measures the proportion of correctly grouped pairs of features, if $Q^l = 1$ then all features are grouped in accordance with the a priori known object relations. If grouping on real images is performed, the object labels are generally not available. In that case the lateral contribution of the CLM energy function provides an indirect measure for the grouping quality. The *energy grouping measure* is defined by

$$Q^E = \sum_{\alpha}\sum_{rr'} f^{\alpha}_{rr'} g(x_{r\alpha}) g(x_{r'\beta}), \qquad \text{where } g(x) = \begin{cases} 1 & \text{if } x > 0, \\ 0 & \text{if } x = 0. \end{cases} \tag{4}$$

The activities are passed through the squashing function $g(.)$ to make the value of Q^E independent of the contextual activity modulation, since here we are only interested in the binary group relations.

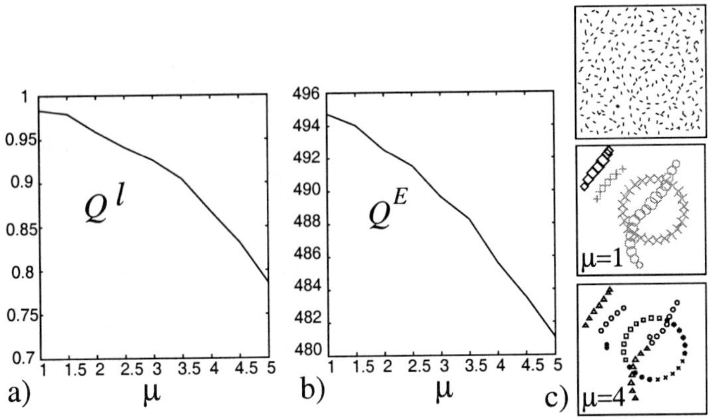

Fig. 2. Grouping performance dependent on contextual modulation for a synthetic image averaged over 100 simulations. a) shows the dependence of the labeling measure Q^l on μ. The energy grouping measure in b) shows an analog behavior with best results for $\mu \to 1$. c) Artificially generated feature set for contour grouping, for which the results in a) and b) were computed. Two grouping results for $\mu = 1$ and $\mu = 4$ are shown with symbols coding for different layers/groups. Symbol size represents activation, the background layer is omitted.

Theorems 3 and 4 give critical values for J, that characterizes the influence of recurrent modulation and lateral effects. For the CLM contour grouping setup, which we consider in the following, the eigenvalue bound of Theorem 4 results in a lower bound on J. Therefore we define $J_c = \max_\alpha \lambda_{\max}\{\mathsf{f}^\alpha_{rr}\}$ as the critical lower value for J and vary $\mu = J/J_c \geq 1$. The detailed lateral coupling scheme f^α_{rr} for all r, r', α for contour grouping with figure-ground segmentation is described in [8].

Figure 2 shows the dependency of Q^l and Q^E for the contour grouping of a synthetic pattern which consists of four objects and a noisy background. The energy measure Q^E varies in accordance with Q^l, which justifies its application in cases where Q^l is not available. The grouping quality increases by letting J approach the critical value J^c. This effect is mainly caused by the contextual destabilization of small fragments. Since larger fragments can develop stronger activity, they tend to attract small fragments into their group for $\mu \to 1$. The numerical simulation reveals that values $\mu < 1$ without instability are possible, since the conditions are sufficient and not necessary. We observed, however, that usually values less than $\mu < 0.95$ cause runaway instability, which indicates that the conditions are close to the maximal possible value for J. Figure 3 shows the corresponding results for a large feature set derived from a real image.

5 Discussion

Our results emphasize the flexibility of LT networks, if an application requires both digital selection for symbolic operations like assignment to a group/layer

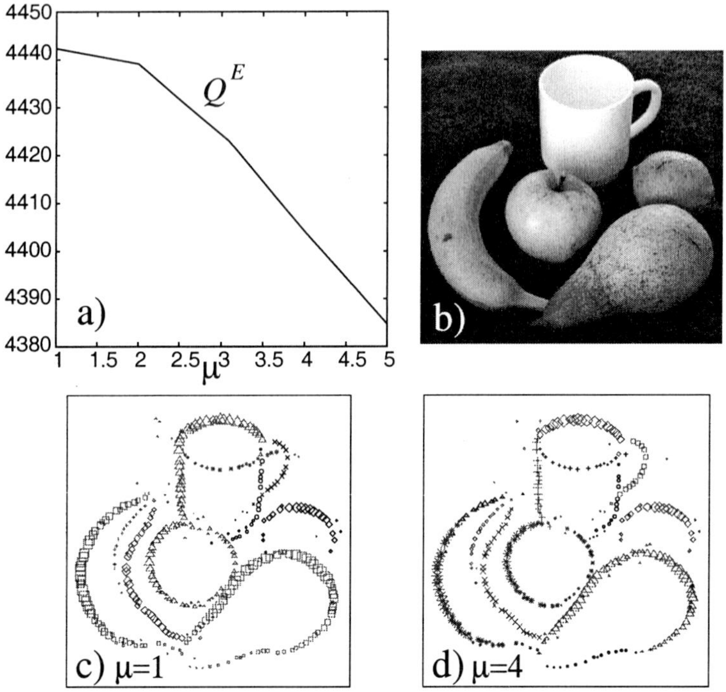

Fig. 3. a) Grouping quality Q^E dependent on contextual modulation for a real image shown in b) averaged over 50 simulations. The quality is enhanced by operating the CLM at $\mu \to 1$. c) and d) show the resulting groupings (background layer omitted).

and analogue modulation for context integration. The new LT stability conditions allow to operate LT networks close to their stability limits, which causes strong dominance of recurrent activation over afferent external inputs. In the context of generation of orientation selectivity this has also been discussed as a symmetry-breaking process[2], which requires multistable systems. As the perceptual grouping applications of the CLM demonstrate, large-scale LT networks can provide powerful sensory processing architectures. They are also interesting for hardware implementations, if new approaches to their VLSI design [4] are applied to large-scale networks.

Acknowledgment

This work was supported by DFG grant GK-231.

References

1. U. Bauer, M. Scholz, J. B. Levitt, K. Obermayer, and J. S. Lund. A biologically-based neural network model for geniculocortical information transfer in the primate visual system. *Vision Research*, 39:613–629, 1999.

2. R. Ben-Yishai, R. Lev Bar-Or, and H. Sompolinsky. Theory of orientation tuning in visual cortex. *Proc. Nat. Acad. Sci. USA*, 92:3844–3848, 1995.
3. R. Douglas, C. Koch, M. Mahowald, K. Martin, and H. Suarez. Recurrent excitation in neocortical circuits. *Science*, 269:981–985, 1995.
4. R. Hahnloser, R. Sarpeshkar, M. A. Mahowald, R. J. Douglas, and H. S. Seung. Digital selection and analogue amplification coexist in a cortex-inspired silicon circuit. *Nature*, 405:947–951, 2000.
5. H. Ritter. A spatial approach to feature linking. In *Proc. Int. Neur. Netw. Conf. Paris Vol.2*, pages 898–901, 1990.
6. A. Salinas and L. F. Abbott. A model of multiplicative responses in parietal cortex. *Proc. Nat. Acad. Sci. USA*, 93:11956–11961, 1996.
7. H. Wersing, W.-J. Beyn, and H. Ritter. Dynamical stability conditions for recurrent neural networks with unsaturating piecewise linear transfer functions. *Neural Computation 13(8)*, 2001.
8. H. Wersing, J. J. Steil, and H. Ritter. A competitive layer model for feature binding and sensory segmentation. *Neural Computation*, 13(2):357–387, 2001.

Exponential Transients in Continuous-Time Symmetric Hopfield Nets

Jiří Šíma[1,*] and Pekka Orponen[2]

[1] Institute of Computer Science, Academy of Sciences of the Czech Republic, P.O. Box 5, 182 07 Prague 8, Czech Republic
sima@cs.cas.cz

[2] Department of Mathematics, University of Jyväskylä, P.O. Box 35, FIN-40351 Jyväskylä, Finland
orponen@math.jyu.fi

Abstract. We establish a fundamental result in the theory of continuous-time neural computation, by showing that so called continuous-time symmetric Hopfield nets, whose asymptotic convergence is always guaranteed by the existence of a Liapunov function may, in the worst case, possess a transient period that is exponential in the network size. The result stands in contrast to e.g. the use of such network models in combinatorial optimization applications.

1 Introduction

Continuous-time recurrent neural networks are an attractive class of computational models with applications in, e.g., control, optimization, and signal processing (cf. [1, 5]). Recently there has also been increasing theoretical interest towards achieving a general understanding of the capabilities and limitations of these and other continuous-time computation models. (For overviews of this work, see e.g. [6, 7].)

Probably the best-known, and most widely-used continuous-time recurrent network model is that popularized by John Hopfield in 1984 [4], and known as the "continuous-time Hopfield model".[1] A fundamental property of this model is that if a given network has a symmetric coupling weight matrix, then its dynamics is governed by a *Liapunov*, or *energy function* [2, 4]. In particular, such a symmetric network always converges from any initial state towards some stable equilibrium state. This is a very useful property for obtaining guaranteed behavior in practical applications, but would at first sight seem to severely limit the networks' general dynamical capabilities. For instance, nondamping oscillations of the network state obviously cannot be created under this constraint, whereas such oscillations are easily obtained in networks with *asymmetric* coupling weights.

* Research supported by grants GA AS CR B2030007, GA ČR No. 201/01/1192, and by a personal grant from the University of Jyväskylä.

[1] Although in fact the dynamics of this model were already analyzed earlier by Cohen and Grossberg in a more general setting [2].

G. Dorffner, H. Bischof, and K. Hornik (Eds.): ICANN 2001, LNCS 2130, pp. 806–813, 2001.

Because of the apparent simplicity of symmetric Hopfield network dynamics, one might also assume that they always converge rapidly—an assumption that seems to often be implicitly made in e.g. discussing the potential of such networks as "fast analog solvers" for optimization problems. Contrary to this expectation, we shall in this paper construct for every n a Hopfield network $\mathcal{C}_n$ of $6n+1$ units with a symmetric coupling weight matrix and a saturated-linear "activation function" that simulates an $(n+1)$-bit binary counter and thus produces a sequence of $2^n - 1$ well-controlled oscillations before it converges. Besides suggesting some caution in applying neural networks to optimization problems, this result provides to our knowledge the first known example of a continuous-time, Liapunov-function controlled dynamical system with an exponential transient period. Such an exponential-transient oscillator can also be used to support a general Turing machine simulation by symmetric Hopfield networks [9].

In terms of bit representations, our convergence time lower bound can be compared to a general upper bound for discrete Hopfield networks [10]. It turns out that the continuous-time system $\mathcal{C}_n$ converges later than any discrete symmetric Hopfield network of the same description length, assuming that the time interval between two subsequent discrete updates corresponds to a continuous time unit. This suggests that continuous-time analog models of computation may be worth investigating more for their gains in representational efficiency than for their (theoretical) capability for arbitrary-precision real number computation [8].

2 A Simulated Binary Counter

A *(symmetric) Hopfield network*[2] consists of m computational *units* or "neurons" $p = 1, \ldots, m$, whose *states* are represented by real variables $y_1, \ldots, y_m \in [0, 1]$. The dynamics of such a network is given by a system of m symmetrically coupled ordinary differential equations:

$$\frac{dy_p}{dt}(t) = -y_p(t) + \sigma(\xi_p(t)), \quad p = 1, \ldots, m, \tag{1}$$

where $\xi_p(t) = \sum_{q=0}^{m} v(p,q) y_q(t)$ is the real-valued *excitation* for unit $p = 1, \ldots, m$. Here, the real coupling coefficient $v(p,q) = v(q,p)$ corresponds to the *weight* on an edge connecting unit p to unit q whereas $v(0,p)$ is a local *bias* $v(0,p)$, associated with a formal constant variable $y_0(t) \equiv 1$. Further, σ is some nonlinear *activation function*, which we fix to be the *saturated linear* map: $\sigma(\xi) = 1$ for $\xi \geq 1$, $\sigma(\xi) = \xi$ for $0 < \xi < 1$, and $\sigma(\xi) = 0$ for $\xi \leq 0$. The *initial network state* $\mathbf{y}(0) \in [0,1]^m$ determines the boundary condition for the system (1).

A Hopfield network $\mathcal{C} = \mathcal{C}_n$ with $m = 6n+1$ neurons will now be constructed which simulates an $(n+1)$-bit binary counter, and thus has a transient period that is exponential in the parameter m. The original idea for a corresponding discrete-time counter network stems from [3]. In our simulation, the binary states of the counter will be represented by excitations of the corresponding real-valued

[2] We shall henceforth discuss only symmetric networks.

units in $\mathcal{C}$ that are either above the upper saturation threshold of 1 or below the lower saturation threshold of 0 for the activation function σ. For brevity, we shall simply say that a unit p is *saturated* at 0 or 1 at time t if its excitation satisfies $\xi_p(t) \leq 0$ or $\xi_p(t) \geq 1$, respectively. We also say that p is *unsaturated* when $0 < \xi_p(t) < 1$. (Note that we use the *excitations*, not the actual *states* of the units to represent binary values.) The following theorem summarizes the result:

Theorem 1. *For every integer $n \geq 0$ there exists a continuous-time symmetric Hopfield net $\mathcal{C}$ with $m = 6n + 1$ neurons whose global state transition from saturation at 0 to saturation at 1 requires continuous time $\Omega(2^{m/6}/\varepsilon)$, for any $0 < \varepsilon < 0.05$ such that $2^{m/2} < \varepsilon 2^{1/\varepsilon}$. This convergence bound translates to $2^{\Omega(g(M))}$ time units, where M represents the number of bits that are sufficient for encoding the weights in $\mathcal{C}$ and $g(M)$ is an arbitrary continuous function such that $g(M) = o(M)$, $g(M) = \Omega(M^{2/3})$, and $M/g(M)$ is increasing.*

Proof. (Sketch.) The construction of symmetric Hopfield net $\mathcal{C} = \mathcal{C}_n$ with $m = 6n + 1$ units and zero initial state $\mathbf{y}(0) = 0^m$ simulating an $(n+1)$-bit binary counter will be described by induction on n. The operation of the network will first be discussed intuitively, and its correctness will then be formally verified. The induction starts with a network $\mathcal{C}_0$ containing only a single unit c_0, with bias $v(0, c_0) = \varepsilon$ and feedback coupling $v(c_0, c_0) = 1 + \varepsilon$. This represents the first counter bit of "order 0". Because of its positive feedback the state of c_0 gradually grows from initial 0 towards 1. Eventually c_0 saturates at 1, at which point we say that the unit c_0 becomes *active* or *fires*. This trick of gradual transition from 0 to 1 (see Lemma 3 below) is used repeatedly throughout our construction of $\mathcal{C}$.

For the induction step depicted in Figure 1 (the edges in this graph drawn without an originating unit correspond to the biases), assume that an "order $(k-1)$" counter network $\mathcal{C}_{k-1}$ $(1 \leq k \leq n)$ has been constructed, containing the first k counter units $c_0, \ldots, c_{k-1}$, together with auxiliary units $a_\ell, x_\ell, b_\ell, d_\ell, z_\ell$ $(\ell = 1, \ldots, k-1)$, for a total of $m_k = 6k - 5$ units. Then the next counter unit c_k is connected to all the m_k units $p \in \mathcal{C}_{k-1}$ via unit weights which, together with c_k's bias, make c_k to fire shortly after all these units are active, i.e. when the simulated counting from 0 to $2^k - 1$ has been accomplished. In addition, unit c_k is connected to a sequence of five auxiliary units a_k, x_k, b_k, d_k, z_k, which are being, one by one, activated after c_k fires (Lemma 3). The purpose of the auxiliary units a_k, b_k, d_k is only to slow down the continuous-time state flow. The unit x_k is used to reset all the lower-order units in $\mathcal{C}_{k-1}$ back to values near 0 after c_k fires (Lemma 2.2b). To achieve this effect, x_k is linked with each $p \in \mathcal{C}_{k-1}$ via a large negative weight $v(x_k, p) = -\lceil v(c_k, p) + \sum_{q \in \mathcal{C}_{k-1};\, v(q,p)>0} v(q, p) \rceil$ that exceeds the mutual positive influence of units in $\mathcal{C}_{k-1} \cup \{c_k\}$. The value of parameter $V_k = 1 - \sum_{p \in \mathcal{C}_{k-1}} v(x_k, p)$ is determined so that the state of x_k is independent of the states of $p \in \mathcal{C}_{k-1}$. Finally, unit z_k balances the negative influence of x_k on $\mathcal{C}_{k-1}$ so that the first k counter bits can again count from 0 to $2^k - 1$ but now with c_k being active. This is achieved by exact weights $v(z_k, p) = -v(x_k, p) - 1$ for $p \in \mathcal{C}_{k-1}$ in which the -1 compensates for $v(c_k, p) = 1$. Clearly, units $p \in \mathcal{C}_{k-1}$ cannot reversely affect z_k since their maximal contribution $\sum_{p \in \mathcal{C}_{k-1}} v(p, z_k) =$

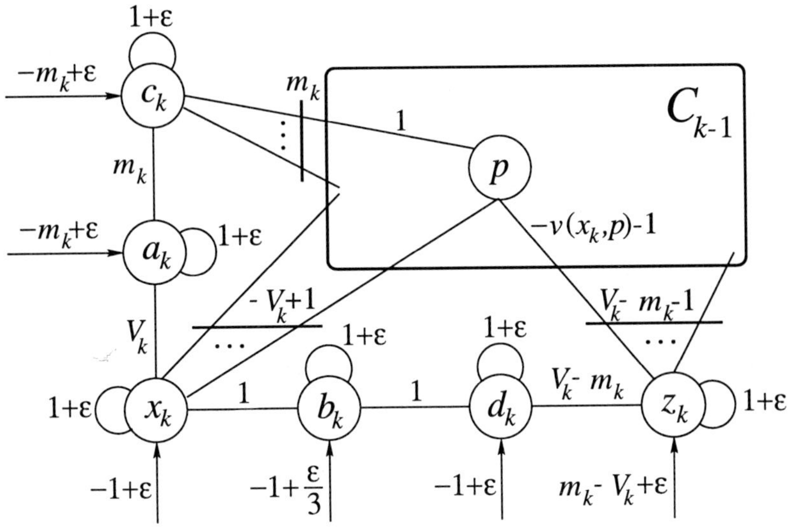

Fig. 1. Inductive construction of $\mathcal{C}_k$

$-m_k - \sum_{p \in \mathcal{C}_{k-1}} v(x_k, p) = V_k - m_k - 1$ to the excitation of z_k cannot overcome its bias. This completes the inductive step of the counter network construction.

Now the correct state evolution of the Hopfield network $\mathcal{C}$ described above needs to be verified. Thus, a sequence of lemmas analyzing the corresponding system (1) is presented. Due to lack of space, the proofs are only sketched here. Lemma 1 first upper bounds the maximum sum of absolute values of weights incident on any unit in $\mathcal{C}$. Lemma 2 then describes explicitly the continuous-time state evolution for saturated units. An analysis of how the decreasing *defects*, i.e. distances from limit values in the states of saturated units, affect the excitation of any other unit reveals that the units in $\mathcal{C}$ actually approximate the discrete update rule of corresponding threshold gates after a certain transient time. The proof of Lemma 2 follows from the dynamics equations (1) and Lemma 1. Furthermore, the transfer of the activity in $\mathcal{C}$ from a unit to a subsequent one, when all the incident units are saturated, will be analyzed explicitly and its duration time will be calculated in Lemma 3. (But note that the analysis for c_0 at $t = 0$ slightly differs.) The result is also generalized to the case when some of the incident units may become unsaturated.

Lemma 1. *For any unit $p \in \mathcal{C}$ in the Hopfield network constructed above, the sum of absolute values of its incident weights (excluding its local bias) is upper bounded by $\Xi_p = \sum_{q=1}^{m} |v(q,p)| < \varepsilon 2^{1/\varepsilon}$.*

Proof. (Sketch.) The maximum value of Ξ_p among $p \in \mathcal{C}$ is reached by unit x_n of the highest order n, that is $\Xi_{x_n} = 2V_n + 1 + \varepsilon$. Parameter $V_n = 2(11 \cdot 7^{n-1} - 5)/3$ is computed by induction on n in which recursive formula $v(x_k, p) = 2v(x_{k-1}, p)$ for $p \in \mathcal{C}_{k-2}$ $(k > 1)$, and Figure 1 are employed. Hence, $\Xi_p \leq 4(11 \cdot 7^{n-1} - 5)/3 + 1 + \varepsilon < \varepsilon 2^{1/\varepsilon}$ by assumptions on ε in Theorem 1. □

Lemma 2.
1. *Let $p \in C$ be a unit saturated at $b \in \{0,1\}$ with a* defect $\delta_p(t) = |y_p(t) - b|$, *for the duration of a continuous time interval $\tau = [t_0, t_f]$ for some $t_0 \geq 0$. Then the state dynamics of p converging towards value b can be explicitly solved as $y_p(t) = |b - \delta_p e^{-(t-t_0)}|$ for $t \in \tau$, where $\delta_p = \delta_p(t_0)$ is p's initial defect.*
2a. *Let $Q \subseteq C$ be a subset of units saturated for the duration of time interval $\tau = [t_0, t_f]$. Then the dynamics of $\xi_p(t)$ for any unit $p \in C$ can be described as*

$$\xi_p(t) = v(0,p) + \sum_{q\in Q;\, \xi_q(t)\geq 1} v(q,p) + \sum_{q\notin Q} v(q,p)y_q(t) + \Delta_{pQ}e^{-(t-t_0)} \qquad (2)$$

for $t \in \tau$, where $\Delta_{pQ} = \sum_{q\in Q;\, \xi_q(t_0)\leq 0} v(q,p)\delta_p - \sum_{q\in Q;\, \xi_q(t_0)\geq 1} v(q,p)\delta_p$ is the initial total weighted defect *of Q affecting $\xi_p(t_0)$.*
2b. *In addition, let $t_f > t_0 + t_1$ where $t_1 = (\ln 2)/\varepsilon$, and assume the respective weights in C satisfy either $v(0,p)+\sum_{q\in Q;\, \xi_q(t_0)\geq 1} v(q,p)+\sum_{q\notin Q;\, v(q,p)>0} v(q,p) < -\varepsilon$ or $v(0,p)+\sum_{q\in Q;\, \xi_q(t_0)\geq 1} v(q,p)+\sum_{q\notin Q;\, v(q,p)<0} v(q,p) > 1+\varepsilon$. Then p is saturated at either 0 or 1, respectively, for the duration of time interval $[t_0+t_1, t_f]$.*

Lemma 3.
1. *Consider a situation where a unit $p \in C$ (e.g. $c_0, c_k, a_k, b_k, d_k, z_k$ for $1 \leq k \leq n$) with fractional part of bias $\varepsilon' \in \{\varepsilon, \varepsilon/3\}$ and feedback weight $v(p,p) = 1+\varepsilon$ is supposed to activate and transfer a signal to the subsequent unit r (i.e. $c_k, a_k, x_k, d_k, z_k, c_0$, respectively) with bias fraction ε and $v(r,r) = 1+\varepsilon$ via weight $v(p,r) \geq 1$. Let all the units incident on p, r excluding p, r be saturated for the duration of some sufficiently large time interval $\tau = [t_0, t_f]$ (e.g. $t_f > t_0 + t_2$ where t_2 is defined below), starting at a time $t_0 \geq 0$ when $\xi_p(t_0) = 0$. Assume that the initial defects $\delta_p + \Delta_{rQ} < \varepsilon$ for $Q = C \setminus \{p\}$ are bounded. Further assume that the respective weights satisfy $v(0,p) + \sum_{q\in Q;\, \xi_q(t_0)\geq 1} v(q,p) = \varepsilon'$ and $v(0,r) + \sum_{q\in Q;\, \xi_q(t_0)\geq 1} v(q,r) = \varepsilon - v(p,r)$. Then p is unsaturated with the state dynamics*

$$y_p(t) = \frac{\varepsilon'\left(e^{\varepsilon(t-t_0)} - 1\right)}{\varepsilon(1+\varepsilon)} - \frac{\varepsilon' + \Delta_{pQ}e^{-(t-t_0)}}{1+\varepsilon} \qquad (3)$$

exactly for the duration of time interval $(t_0, t_0 + t_1')$, where $t_1' = (\ln(1+\varepsilon/\varepsilon'))/\varepsilon$ (note $t_1' = t_1$ for $\varepsilon' = \varepsilon$ and $t_1' = 2t_1$ for $\varepsilon' = \varepsilon/3$), while r is saturated at 0. In addition, p is saturated at 1 for the duration of $[t_0 + t_1', t_f]$, while r unsaturates from 0 at time t_0+t_2 where $t_2 = \ln((v(p,r)\delta_p(t_0+t_1')(1+\varepsilon/\varepsilon')^{1/\varepsilon} - \Delta_{rQ})/\varepsilon) \geq t_1'$.
2. *Consider a situation in C where unit x_k ($1 \leq k \leq n$) is supposed to receive a signal from preceding unit a_k, activate itself, and further transfer the signal to subsequent unit b_k while units in C_{k-1} incident on x_k may unsaturate from 1 after x_k unsaturates from 0. Let all the other units incident on x_k, b_k excluding x_k, b_k and C_{k-1} be saturated for the duration of a sufficiently large time interval $\tau = [t_0, t_f]$ (e.g. at least until b_k unsaturates from 0) starting at a time $t_0 \geq 0$ when $\xi_{x_k}(t_0) = 0$. Assume that the initial defects meet $\delta_{b_k}, \Delta_{b_kQ'} < \varepsilon 2^{-1/\varepsilon}$ for $Q' = C\setminus(C_{k-1}\cup\{x_k\})$, and also $(1+\varepsilon)\delta_{x_k} - \sum_{p\in C_{k-1}} v(p,x_k)\delta_p \leq \varepsilon 2^{-1/\varepsilon}$ outside*

Q', are bounded. Further, assume that the respective weights satisfy $v(0,x_k) + \sum_{q\in Q';\,\xi_q(t_0)\geq 1} v(q,x_k) + \sum_{p\in \mathcal{C}_{k-1}} v(p,x_k) = \varepsilon$ and $\sum_{q\in Q';\,\xi_q(t_0)\geq 1} v(q,b_k) = 0$. Then x_k saturates at 1 in time at most $t_0 + 2t_1$, remaining then saturated until time at least t_f, and b_k unsaturates from 0 only after x_k is saturated at 1.

Proof. (Sketch.)
1. Excitation $\xi_p(t) = \varepsilon' + (1+\varepsilon)y_p(t) + \Delta_{pQ}e^{-(t-t_0)}$ of p for $t \in [t_0, t_0+t_2]$ is obtained from (2) which determines p's state dynamics (1) by differential equation $(dy_p/dt)(t) = -y_p(t) + \varepsilon' + (1+\varepsilon)y_p(t) + \Delta_{pQ}e^{-(t-t_0)}$ when p is unsaturated. The corresponding initial condition $y_p(t_0) = (-\varepsilon' - \Delta_{pQ})/(1+\varepsilon) = \delta_p$ comes from $\xi_p(t_0) = 0$ which also bounds the initial defect as $-1-\varepsilon-\varepsilon' \leq \Delta_{pQ} \leq -\varepsilon' < 0$, due to $1 \geq \delta_p \geq 0$. Hence solution (3) follows, which provides dynamics $\xi_p(t) = \varepsilon'(e^{\varepsilon(t-t_0)}-1)/\varepsilon > 0$, ensuring that p is unsaturated exactly for the duration of (t_0, t_0+t_1'), even though its state $y_p(t)$ is initially decreasing for $t \in (t_0, t_0+t_g)$ where $t_g = (\ln(-\Delta_{pQ}/\varepsilon'))/(1+\varepsilon) < t_1'$. Excitation $\xi_r(t) = \varepsilon - v(p,r) + v(p,r)y_p(t) + \Delta_{rQ}e^{-(t-t_0)}$ should prove to be nonpositive for all $t \in (t_0, t_0+t_1')$. By using $v(p,r) \geq 1$, $\delta_p + \Delta_{rQ} < \varepsilon$, and dynamics (3) in which $-\Delta_{pQ} = \varepsilon' + (1+\varepsilon)\delta_p$, this reduces to $\varepsilon(\varepsilon'+\varepsilon(1+\varepsilon))e^{-(t-t_0)} + \varepsilon'(e^{\varepsilon(t-t_0)}-1) - \varepsilon \leq \varepsilon(\varepsilon'-\varepsilon^2)$. For $t \in [t_0, t_0+t_\varepsilon]$ where $t_\varepsilon = \ln((\varepsilon'+\varepsilon(1+\varepsilon))/(\varepsilon'-\varepsilon^2))$, term $e^{\varepsilon(t-t_0)}$ reaches its maximum at t_0+t_ε which implies the underlying inequality. For $t \in [t_0+t_\varepsilon, t_0+t_1']$, term $\varepsilon(\varepsilon'+\varepsilon(1+\varepsilon))e^{-(t-t_0)}$ achieves its maximum $\varepsilon(\varepsilon'-\varepsilon^2)$ at t_0+t_ε while $\varepsilon'(e^{\varepsilon(t-t_0)}-1) - \varepsilon \leq 0$. Hence, r is saturated for the duration of (t_0, t_0+t_1'). Furthermore, $\xi_p(t) = 1 + (\varepsilon+\varepsilon')(1-e^{-(t-t_0-t_1')}) \geq 1$ of saturated p derived from Lemma 2.1 ensures that p stays saturated at 1 at least for the duration of $[t_0+t_1', t_0+t_2]$, where t_2 comes from $\xi_r(t_0+t_2) = 0$. It must also be checked that $\xi_p(t) = \varepsilon' + 1 + \varepsilon + v(p,r)y_r(t) - (1+\varepsilon)\delta_p(t_0+t_2)e^{-(t-t_0-t_2)} + (\Delta_{pQ} - v(p,r)\delta_r)\,e^{-(t-t_0)} \geq 1$ for all $t \in [t_0+t_2, t_f]$. Here, $v(p,r)y_r(t) > 0$ whereas the respective defect terms having the least value at t_0+t_2 can be lower bounded by $-\varepsilon'-\varepsilon$ when the explicit formulas are substituted for $\delta_p(t_0+t_2)$, t_2, Δ_{pQ}, and inequalities $\delta_p + \Delta_{rQ} < \varepsilon$, $\delta_r \leq 1$, $v(p,r) \geq 1$, $\varepsilon' \geq \varepsilon/3$ are applied.
2. Notice that unit a_k saturates at 1 before x_k is unsaturated from 0 according to Lemma 3.1. Excitation $\xi_{x_k}(t) \geq \varepsilon + (1+\varepsilon)y_{x_k}(t) + \Delta_{x_kQ'}\,e^{-(t-t_0)}$ of x_k for $t \in \tau$ is lower bounded from formula (2) and $v(p,x_k) < 0$ for all $p \in \mathcal{C}_{k-1}$, which gives $(dy_{x_k}/dt)(t) \geq \varepsilon y_{x_k}(t) + \varepsilon + \Delta_{x_kQ'}\,e^{-(t-t_0)}$ for x_k unsaturated, according to (1). In the beginning of interval τ, state $y_{x_k}(t)$ is determined by (3) before the first $p \in \mathcal{C}_{k-1}$ unsaturates, since the assumption of Lemma 3.1 concerning the weights incident on x_k coincides with that of Lemma 3.2 due to $\varepsilon' = \varepsilon$ and $\xi_p(t_0) \geq 1$ for all $p \in \mathcal{C}_{k-1}$. Hence, $\Delta_{x_kQ'}$ can appropriately be expressed in terms of $\Delta_{x_kQ} = -\varepsilon - (1+\varepsilon)\delta_{x_k}$ for $Q = \mathcal{C}\setminus\{x_k\}$ from Lemma 3.1 so that the bound assumed on the initial defect outside Q' can be used to lower bound $\Delta_{x_kQ'} \geq -\varepsilon(1+2^{-1/\varepsilon})$ which gives $(dy_{x_k}/dt)(t) \geq \varepsilon y_{x_k}(t) + \varepsilon - \varepsilon(1+2^{-1/\varepsilon})e^{-(t-t_0)}$. It follows that $(dy_{x_k}/dt)(t) \geq \varepsilon - \varepsilon^2 > 0$ for $t \geq t_0 + t_d$ where $t_d = \ln((1+2^{-1/\varepsilon})/\varepsilon)$, provided that x_k is still unsaturated. This implies that $y_{x_k}(t)$ grows at least as fast as the straight line with equation $(\varepsilon-\varepsilon^2)(t-t_0-t_d) - y = 0$ until x_k saturates at 1. Thus, x_k saturates at 1 certainly before $t_0+t_d+t_s < t_0+2t_1$ where $t_s = 1/(\varepsilon-\varepsilon^2)$ because $\xi_{x_k}(t) > y_{x_k}(t)$ from (1) due to its state derivative is positive for $t \geq t_0+t_d$.

Similarly, $\xi_{b_k}(t) = -1 + \varepsilon/3 + y_{x_k}(t) + \Delta_{b_kQ'}\, e^{-(t-t_0)}$ for b_k saturated at 0. Let $t_y > 0$ be the least local time instant at which $y_{x_k}(t_0+t_y) = 1-\varepsilon/3-\Delta_{b_kQ'}\, e^{-t_y}$ when b_k is still saturated at 0 since $\xi_{b_k}(t_0+t_y) = 0$. Excitation $\xi_{x_k}(t_0+t_y) \geq \varepsilon + (1+\varepsilon)(1-\varepsilon/3-\Delta_{b_kQ'}\, e^{-t_y}) + \Delta_{x_kQ'}\, e^{-t_y}$ of x_k at t_0+t_y can be lower bounded by 1 from $\Delta_{x_kQ'} \geq -\varepsilon(1+2^{-1/\varepsilon})$ and $\Delta_{b_kQ'} < \varepsilon 2^{-1/\varepsilon}$, ensuring x_k is already saturated at 1 at t_0+t_y. Finally, it must be checked that $\xi_{x_k}(t) \geq \varepsilon+(1+\varepsilon)(1-(\varepsilon/3+\Delta_{b_kQ'}\, e^{-t_y})e^{-(t-t_0-t_y)})+y_{b_k}(t)+(\Delta_{x_kQ'}-\delta_{b_k})e^{-(t-t_0)} \geq 1$ for all $t \in [t_0+t_y, t_f]$ when b_k may unsaturate, which follows from $y_{b_k}(t) \geq 0$ and the respective defect bounds for δ_{b_k}, $\Delta_{b_kQ'}$, and $\Delta_{x_kQ'}$. □

The correct timing of the counter simulation must ensure a sufficiently fast decrease of the defects as assumed in Lemma 3. According to Lemma 2.2b, the absolute value of the total weighted defect affecting any unit in C is bounded by ε after time t_1, decreasing further to $\varepsilon 2^{-1/\varepsilon}$ by time $2t_1$. On the other hand, t_1 lower bounds the time necessary for activating unit p in Lemma 3.1. Hence, the subsequent unit r has always time at least t_1 for decreasing the defect induced by its incident saturated units below ε even before unit p starts its activation. Similarly, the stronger defect bounds in Lemma 3.2 are met since time $2t_1$ is guaranteed before unit x_k unsaturates. The lower bound $\Omega(2^n/\varepsilon) = \Omega(2^{m/6}/\varepsilon)$ on the total simulation time follows immediately from the previous time analysis.

From the proof of Lemma 1, the maximum integer weight parameter in C is of order $2^{O(m)}$. This corresponds to $O(m)$ bits per weight that is repeated $O(m^2)$ times, and thus yields at most $O(m^3)$ bits in the representation. In addition, the biases and feedbacks of the m units include fraction ε (or $\varepsilon/3$), and taking this into account requires $\Theta(m\log(1/\varepsilon))$ additional bits, say at least $\kappa m\log(1/\varepsilon)$ bits for some constant $\kappa > 0$. By choosing $\varepsilon = 2^{-f(m)/(\kappa m)}$ in which f is a continuous increasing function whose inverse is defined as $f^{-1}(\mu) = \mu/g(\mu)$, where g satisfies $g(\mu) = \Omega(\mu^{2/3})$ (implying $f(m) = \Omega(m^3)$) and $g(\mu) = o(\mu)$, it follows that $M = \Theta(f(m))$, especially $M \geq f(m)$ from $M \geq \kappa m\log(1/\varepsilon)$. The convergence time $\Omega(2^{m/6}/\varepsilon)$ can be translated to $\Omega(2^{f(m)/(\kappa m)+m/6}) = 2^{\Omega(f(m)/m)}$ which can be rewritten as $2^{\Omega(M/f^{-1}(M))} = 2^{\Omega(g(M))}$ since $f(m) = \Omega(M)$ from $M = \Theta(f(m))$ and $f^{-1}(M) \geq m$ from $M \geq f(m)$. This completes the proof of the theorem. □

3 A Simulation Example

A computer program HCOUNT has been created to automate the construction from Theorem 1. For input $n \geq 0$, the program generates system (1) describing the Hopfield net dynamics in the form of a FORTRAN subroutine corresponding to the $(n+1)$-bit binary counter to be simulated. This FORTRAN procedure is then presented to a solver from the NAG library that provides a numerical solution for the system. For example, implementing a 4-bit counter on the HCOUNT generator results in a continuous-time symmetric Hopfield net C_3 with 19 variables. Figure 2 shows the state evolution of counter units c_0, c_1, c_2, c_3 for a period of $2^3 - 1 = 7$ simulated discrete steps confirming the correctness of the construction. A parameter value of $\varepsilon = 0.1$ was used in this numerical simulation, showing that the theoretical estimate of ε in Theorem 1 is actually quite conservative.

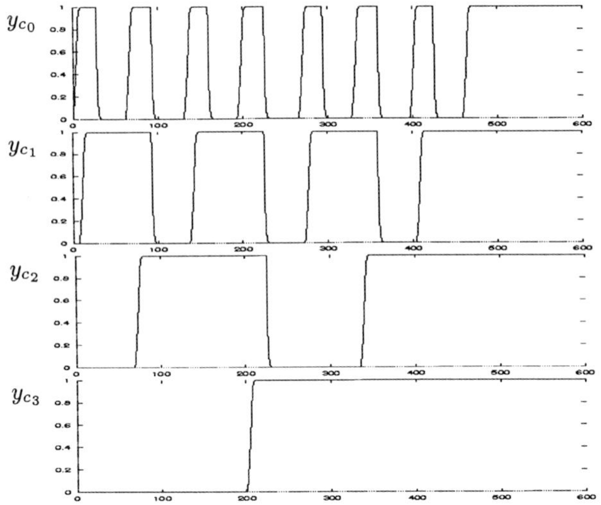

Fig. 2. Continuous-time simulation of 4-bit binary counter for $\varepsilon = 0.1$

References

1. A. Cichocki and R. Unbehauen: *Neural Networks for Optimization and Signal Processing.* John Wiley & Sons, Chichester, 1993.
2. M. A. Cohen and S. Grossberg: Absolute stability of global pattern formation and parallel memory storage by competitive neural networks. *IEEE Transactions on Systems, Man, and Cybernetics* **13**:815–826, 1983.
3. E. Goles and S. Martínez: Exponential transient classes of symmetric neural networks for synchronous and sequential updating. *Complex Systems* **3**:589–597, 1989.
4. J. J. Hopfield: Neurons with graded response have collective computational properties like those of two-state neurons. In *Proceedings of the National Academy of Sciences*, volume **81**, pages 3088–3092, 1984.
5. L. R. Medsker and L. C. Jain (editors): *Recurrent Neural Networks: Design and Applications.* The CRC Press, Boca Raton, FL, 2000.
6. C. Moore: Finite-dimensional analog computers: flows, maps, and recurrent neural networks. In *Proceedings of the 1st International Conference on Unconventional Models of Computation (Auckland, January)*, Springer-Verlag, Berlin, 1998.
7. P. Orponen: A survey of continuous-time computation theory. In D.-Z. Du and K.-I. Ko, editors, *Advances in Algorithms, Languages, and Complexity*, pages 209–224, Kluwer Academic Publishers, Dordrecht, 1997.
8. H. T. Siegelmann: *Neural Networks and Analog Computation: Beyond the Turing Limit.* Birkhäuser, Boston, MA, 1999.
9. J. Šíma and P. Orponen: Computing with continuous-time Liapunov systems. In *Proceedings of the 33rd Annual ACM Symposium on Theory of Computing (Crete, July 2001).* Association for Computing Machinery, New York, NY, 10 pp, to appear.
10. J. Šíma, P. Orponen, and T. Antti-Poika: On the computational complexity of binary and analog symmetric Hopfield nets. *Neural Computation* **12**:2965-2989, 2000.

Initial Evolution Results on CAM-Brain Machines (CBMs)

Hugo de Garis[1], Andrzej Buller[1], Leo de Penning[1], Tomasz Chodakowski[2], and Derek Decesare[3]

[1] Brain Builder Group, STARLAB,
Rue Engeland 555, B-1180, Brussels, Belgium,
{degaris,buller,ldp}@starlab.net,
http://foobar.starlab.net/~degaris, ~buller, ~ldp
[2] Information Systems Dept., ATR,
2-2 Hikari-dai, Seika-cho, Soraku-gun, Kyoto, 619, Japan,
tomasz@isd.atr.co.jp
[3] Computer Science Dept, Utah State University,
Logan, UT, USA
derek.decesare@usu.edu

Abstract. This paper presents results of some of the first evolution experiments undertaken on actual CAM-Brain Machines (CBM), using the hardware itself and not software simulations. A CBM is a specialised piece of programmable (evolvable) hardware that uses Xilinx XC6264 programmable FPGA chips to grow and evolve, at electronic speeds, 3D cellular automata (CA) based neural network circuit modules of some 1000 neurons each. A complete run of a genetic algorithm (e.g. with 100 generations and a population size of 100) is executed in a few seconds. 64000 of these modules can be evolved separately according to the fitness definitions of human "EEs" (evolutionary engineers) and downloaded one by one into a gigabyte of RAM. Human "BAs" (brain architects) then interconnect these modules "by hand" according to their artificial brain architectures. The CBM then updates the binary neural signaling of the artificial brain (with 64000 "hand" interconnected modules, i.e. 75 million neurons) at a rate of 130 billion CA cell updates a second, which is fast enough for real time control of robots. Before such multi-moduled artificial brains can be constructed, it is essential that the quality of the evolution (the "evolvability") of individual modules be adequate. This paper reports on the first evolution results obtained on CBM hardware.

1 Introduction

It has been a decade long dream of the first author that it might be possible to evolve neural network circuits at electronic speeds and in such large numbers that building artificial brains would become practical. In 1993, a research project was conceived [1] which aimed to evolve neural networks embedded in a medium of cellular automata. By storing the state of each CA cell with 1 or 2 bytes, todays technology, with its gigabytes of RAM in workstations, would mean a potential

G. Dorffner, H. Bischof, and K. Hornik (Eds.): ICANN 2001, LNCS 2130, pp. 814–819, 2001.

space of billions of CA cells in which to grow and evolve neural nets - more than enough in which to place an artificial brain. Software simulation studies showed that such a thing was possible. The initial CA based neural net model was too complex to be implemented in electronics, so a simplified model was conceived in mid 1996 [6]. Soon after, a contract was signed between Genobyte, a hardware engineering company in Boulder, Colorado, USA, and the first author's previous lab ATR in Kyoto, Japan, to design and construct a specialised piece of programmable (evolvable) hardware (called a CAM-Brain Machine (CBM)) that uses Xilinx XC6264 chips to grow and evolve 3D CA based neural network circuit modules in about a second or so each (see Fig 5). With limited man power and budget, the first CBM was delivered in early 1999, and work on debugging and improving the CBM's performance continued throughout that year and 2000. There are now (February 2001) 4 CBMs in the world, 2 in Belgium, 1 in Japan, and 1 in the US. After many technical and commercial problems (e.g. Xilinx took the XC6200 family of chips off the market, so that Genobyte had to work with untested chips, which created all kinds of delays), the CBMs have finally come on line, and are beginning to function as designed, i.e. they implement the algorithms as originally planned, bug free. Of course, since this is an ongoing research project, the fitness definitions used to evolve the modules, and perhaps even the neural net model itself [6] used in the CBM, may need to be changed in the future, depending on the quality of the experimental results. The CBM is a programmable hardware machine, so changing its functionality is not too difficult.

The content of this paper is summarized as follows. Section 2 presents some initial results on the "evolvable capacity (EC)" of a module (i.e. the module's "MEC" = modular evolvable capacity [7]), in other words, for how long (how many ticks of the clock) can an evolved output signal of a module follow closely some randomly specified target waveform. If the CBM cannot evolve such an output with reasonable quality (i.e. have a reasonable "evolvability"), then the whole CBM approach is not worth much. Section 3 presents a rather simple 1D (1 dimensional) dynamic input example of a frequency halver which generalizes. This example gives a feel for what can be evolved by the CBM. Section 4 gives an example of a 2D dynamic input pattern detector (in this case detecting the direction of motion, up or down, of a line moving across a 2D input grid). Section 5 discusses future plans and hopes.

2 Modular Evolvable Capacities (MECs)

This section presents an attempt to see just how well the CBM, in its current state (i.e. with its current global module fitness definition [4, 5] and its current neural net model [6], both of which can be hardware reprogrammed if necessary), is able to evolve modules with "acceptable" evolvability levels. By "acceptable" we mean, for the example illustrated in this section, that the evolved output signal of a module follows closely some arbitrary target waveform. There have already been a string of papers published on the basic principles of how the CBM works [2–7]. A very brief summary will be provided here. CA based neural nets grow and evolve in a 3D CA space. Once the circuits are grown, the binary neural signaling spreads over the grown network. The output signal(s) are convoluted with a digitised convolution function (which looks a bit like a gaussian or hill curve). The result of this convolution is a ditigal to analog conversion. The resulting analog output curve is then matched with a user specified target curve. The closer the match, the higher the fitness.

Figs. 1,2,3 show examples of target wave forms (the spikier ones) and their corresponding evolved waveforms (the flatter ones) which are supposed to be as close as possible and for as long as possible, to the former. Common sense says that there is a limit to how long such an evolved waveform can follow closely some target waveform - a finite number of bits used to evolve the module cannot generate a waveform of infinite length that remains accurate to an infinitely long target curve. This limit we call the module's MEC (modular evolvable capacity) [7]. The smaller the number of clock cycles (the duration) of the target waveform, the easier it is for the evolved waveform to follow its target waveform closely, e.g. Figs. 1 and 2 (50 and 70 clocks (i.e. clockticks) respectively. As the number of clocks increases, the evolvable capacity of the module approaches its MEC (defined as the maximum number of clock ticks over which the evolved curve follows the target curve closely). Fig. 3 (100 clocks) shows a form of averaging effect, where the evolved curve takes the middle path along a fluctuating target curve.

We are not satisified with these results. We feel we can improve upon them by changing our global fitness definition (used to measure how closely the evolved and target waveforms match). At present, the earlier and later spikes in the output evolved spiketrain are equally weighted. This may be a mistake, resulting in the above averaging effect. Perhaps with a more appropriate weighting, a better accuracy, and hence a better evolvability may be achieved.

3 Frequency Halver

One of the attractions of a CBM module is that it can accept static or dynamic input (in the sense of a waveform (before it is deconvoluted into a spike train [4, 5]) which does not change or does change its value in time). Modules can be evolved which detect dynamic input patterns (in 1D or 2D). In this subsection we present a case of a 1D dynamic input signal, whose spike train frequency (a spike every N clocks) is halved at the output. Interestingly, the resulting evolved module generalizes, in the sense that if one inputs to this evolved module, a spike train with a spike frequency of 1 spike every M clocks, it will output (approximately) a spike train with a spike frequency of 1 spike every 2M clocks.

Figure 4 shows a very good evolution result. The lower line shows the response of the evolved module and the upper line shows the target. The two lines are depicted with a slight vertical offset for comparative purposes. Also, a slight horizontal offset is seen in Fig. 4. This is due to the delay of the binary signals flowing through the module. (The 1 bit signals move one CA cell per clock tick).

After this module was evolved we applied signals of various frequencies to its input and the response was indeed half of the input frequency.

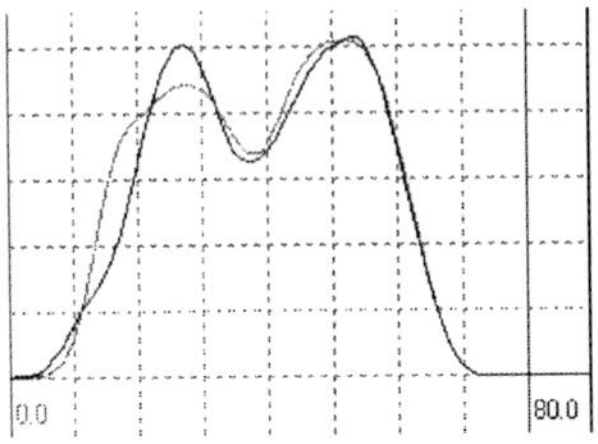

Fig. 1. Target (dark line) and response (grey line) of evolved module over 50 clocks

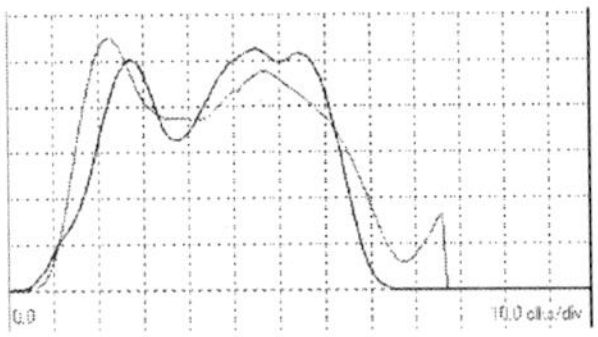

Fig. 2. Target (dark line) and response (grey line) of evolved module over 70 clocks

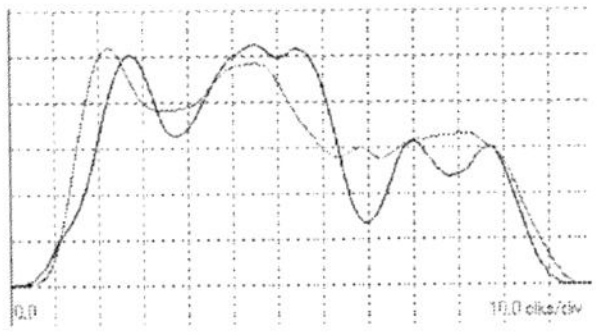

Fig. 3. Target (dark line) and response (grey line) of evolved module over 100 clocks

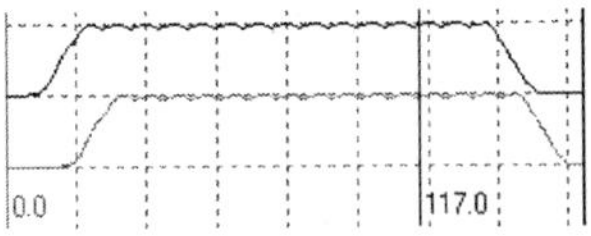

Fig. 4. Evolved and target signals of the best module after evolution of a frequency halver (lower line is the evolved signal, upper line is the target signal).

4 Moving Line Direction Detector

The experiment discussed in this section is an example of how the CBM can evolve a 2D dynamic pattern detector. An 8-by-8 input point array on one face of a CBM module cube is presented with a series of images representing a moving line. The images are binary coded with a '1' presenting a part of the line. The module was evolved to detect whether the line moves up or down across the array. If the line moves up, the module should generate a high output signal frequency (= high convoluted analog output value) and a low value if the line moves down.

The stimulus, a binary image of a line moving up and down several timens, was applied to the front face of the module (a space of 24*24*24 CA cells) as can be seen in figure 5.

The CBM evolved a module capable of distinguishing between a line moving up and a line moving down, although the target frequencies were not produced

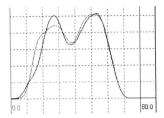

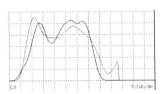

exactly (see figure 6). Nevertheless, the response frequencies are much higher when the bar moves up than when it moves down. So if one were to use a threshold function (or module) we could build a multi-module circuit that distinguishes movement (i.e. a dynamic 2-D input pattern detector).

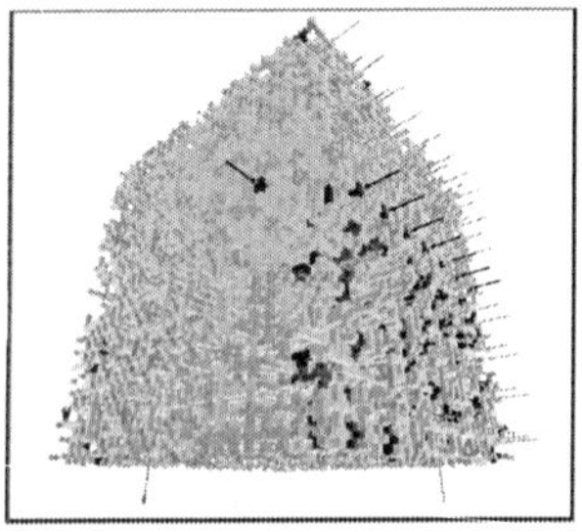

Fig. 5. Stimulus signal of one time step applied to the front face of the moving line direction detector

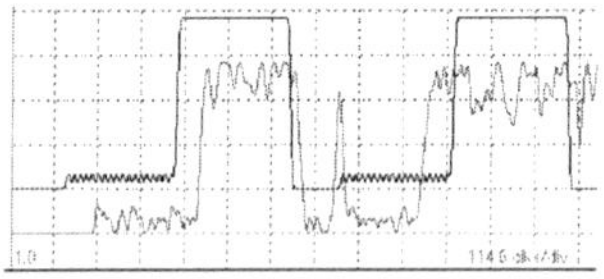

Fig. 6. Evolved and target signal of the best module after evolution of a moving line direction detector. (The lower line is the evolved signal. The upper line is the target signal).

From several tests performed on the evolved module we could conclude that the module "fired" more vigorously when the bar was moving up than when the bar was moving down. This is encouraging. Further work should show to what extent the CBM is able to evolve dyanamic 2D pattern detectors (and in a few seconds).

5 Future Plans and Hopes

Although the initial results look promising, a lot more experimentation needs to be done. For example, the convolution technique we use in measuring the fitness value of an evolved curve may be inappropriate for generating higher levels of evolvability. Figs 1,2,3 show that the evolved curves are approximately accurate but not impressively so. Future work will be needed to choose better ways to evolve more accurate curves (with higher quality and longer MECs). Perhaps we may have to change the global fitness definition (used by the CBM) so that correctly evolved earlier spikes are given a heavier weighting, thus forcing the initial part of the evolved curve to follow the target curve more accurately, and then use step wise (incremental) evolution (where the evolved population in a GA run "N" becomes the starting population for a GA run "N+1" with a different fitness definition). Since the CBMs have only very recently come on line at the time of writing (Feb 2001), we still dont really know what they are capable of, so a thorough testing of their capacities and a study of their limitations are needed. In parallel with this, software development packages need to be built to facilitate the evolution, running and testing process of modules and artificial brains. The "automation" of brain building will be very important, because it will allow the full exploitation of the considerable capacities of the CBM.

The unrelenting pressure of Moore's law now forces us to think hard about the principles and architecture of the second generation brain building machine, that we call "BM2" (Brain Machine 2nd Generation). We are thinking of using a self-configuring electronics approach, that according to initial estimates, ought to provide a performance level of about 1000 times that of the CBM, by about the year 2004, and for a development cost of perhaps several million dollars. BM2 should be a lot more biological brain based, due to the collaboration of some senior people in the neuro science and neural network fields. In the meantime, we feel we have not yet scraped the surface of what the CBM has to offer. With 64000 modules to evolve, we plan to give brain builder teams around the world remote access to the Starlab CBM(s), so that the machine(s) can be used while the Starlab team sleeps.

References

1. Hugo de Garis: Evolvable Hardware: The Genetic Programming of Darwin Machines. Int. Conf. on Artificial Neural Nets and Genetic Algorithms, ICANNGA (1993) Innsbruck, Austria
2. Michael Korkin, Hugo de Garis, Felix Gers, Hitoshi Hemmi: CAM-Brain Machine (CBM): A Hardware Tool which Evolves a Neural Net Module in a Fraction of a Second and Runs a Million Neuron Artificial Brain in Real Time. Genetic Programming Conf. (July 1997), Koza John R., Deb Kalyanmoy, Dorigo Marco, Fogel David B., Garzon Max, Iba Hitoshi, Riolo Rick L. (eds.), Stanford University, San Francisco, CA, USA
3. Hugo de Garis, Felix Gers, Michael Korkin, Arvin Agah, Norberto Eiji Nawa: CAM-Brain, ATR's billion neuron artificial brain project : A three year progress report. Artificial Life and Robotics Journal, Vol.2, (1998) 56–61
4. Hugo de Garis, Michael Korkin, Felix Gers, Eiji Nawa, Michael Hough: Building an Artificial Brain Using an FPGA Based CAM-Brain Machine. Applied Mathematics and Computation Journal, Special Issue on Artificial Life and Robotics, Artificial Brain, Brain Computing and Brainware
5. Hugo de Garis, Michael Korkin: The CAM-Brain Machine (CBM): Real Time Evolution and Update of a 75 Million Neuron FPGA-Based Artificial Brain. Journal of VLSI Signal Processing Systems (JVSPS), Special Issue on Custom Computing Technology
6. Felix Gers, Hugo de Garis, Michael Korkin: CoDi-1Bit: A Simplified Cellular Automata Based Neuron Model. Proceedings of AE97, Artificial Evolution Conference (October 1997) Nimes, France
7. Hugo de Garis, Andrzej Buller, Thierry Dob, Jean Honlet, Padma Guttikonda, Derek Decesare: Building Multimodule Systems with Unlimited Evolvable Capacities from Modules with Limited Evolvable Capacities (MECs). 2nd DoD/NASA Workshop on Evolvable Hardware EH2000, Silicon Valley, California, USA (July 2000)

Self-Organizing Topology Evolution of Turing Neural Networks

Christof Teuscher and Eduardo Sanchez

Logic Systems Laboratory, Swiss Federal Institute of Technology,
CH-1015 Lausanne, Switzerland
{name.surname}@epfl.ch
http://lslwww.epfl.ch

Abstract Turing's neural-network-like structures (unorganized machines) are presented and compared to Kauffman's random boolean networks (RBN). A self-organizing topology evolving algorithm is applied to Turing's networks and it is shown that the network evolves towards an average connectivity of $K_C = 2$ for large systems ($N \to \infty$).

1 Introduction

In a little-known paper entitled "Intelligent Machinery" [7, 13], Turing had already investigated connectionist networks at the end of the forties, almost simultaneously with McCulloch and Pitts [9]. Copeland and Proudfoot write: "Turing was probably the first person to consider building computer machines out of simple, neuron-like elements connected together in to networks in a largely random manner" [5].

Since Turing, many a researcher has been interested in randomly connected networks of simple elements. Allanson [1] investigated the properties of randomly connected networks of simple, biologically plausible neurons. Rozonoér [10] analyzed the properties of networks consisting of elements whose properties depend on parameters chosen at random. The connections among the elements were also established at random. He suggested that "[...] objects of this type may present a certain interest in connection with physiological models and, possibly, will have direct technical applications in the future" [10]. In 1971, Amari [2] published a paper on the characteristics of randomly connected threshold-element networks with the intention of understanding some aspects of information processing in nervous systems. He showed that two statistical parameters are sufficient to determine the characteristics of such networks. Random boolean networks (RBN) have been seriously investigated many years after Turing by Weisbuch [14] and Kauffman [8]. Asynchronous RBN have been investigated in [6].

Above all, Turing's neural-network-like machines are interesting because they are build up from all and the same very simple element (neuron) with the same logical function and because the interconnections might be easily configured (opened or closed binary switch). The goal of this paper is to present a self-organizing topology evolving algorithm that evolves the average connectivity of the network towards a critical value.

G. Dorffner, H. Bischof, and K. Hornik (Eds.): ICANN 2001, LNCS 2130, pp. 820–826, 2001.

In section 2, we present Turing's connectionist machines, an extension made by the authors, and some basic properties. Section 3 presents Kauffman's RBN and their properties. They are compared to Turing's unorganized machines. Section 4 presents a self-organizing topology evolving algorithm for Turing networks that organizes the connectivity of a network towards a critical value. Section 5 concludes the paper.

2 Unorganized Machines

In this section, we shall present the neuron-like machines Turing described in his 1948 paper [13]. For a more detailed description, see also [5,11,12].

The term *unorganized machine* has been defined by Turing in a rather inaccurate and informal way. An *unorganized machine* is a machine that is built up "[...] in a comparatively unsystematic and random way from some kind of standard components" [13, p. 9]. Modern neural networks are often organized in layers. For unorganized machines, the layered structure does not really exist since the machines are normally constructed at random and recurrent connections are allowed with no constraints.

In terms of a digital system, the basic unit which Turing described can be straightforwardly defined as an edge-triggered D flip-flop preceded by a two-input NAND gate. A positive-edge-triggered D flip-flop samples its D input and changes its Q output only at the rising edge of the controlling clock (CLK) signal. A global clock generator is used to synchronize the units.

An A-*type unorganized machine* is a machine built up from N units (neurons). Each unit receives inputs from exactly two other units. A second type of unorganized machine is called B-*type unorganized machine.* A B-type machine is an A-type machine where each connection between two nodes has been replaced by a small A-type machine as shown in Figure 1(a). This small A-type machine functions as a sort of switch (modifier) that is either *opened* or *closed* (b). Turing simply wanted to create a switch based on the same basic element as the rest of the machine. The B-type link can be in three different states of operation: (1) it may invert the incoming signal (closed/enabled connection), (2) it may interrupt the incoming signal and put a constant 1 on its output (opened/disabled connection), or (3) it may act as in (1) and (2) alternately.

So far, the two unorganized machines we have described are, once initialized, no longer modifiable. It would clearly be interesting to modify the machine's interconnection switches (synapses) at runtime since such a possibility would allow the use of an online learning algorithm to train the net. Turing proposed to replace a B-type link by a connection as shown in Figure 1(c). Each interconnection has two additional *interfering inputs* I_A and I_B, i.e., inputs that affect the internal state of the link. By supplying appropriate signals at I_A and I_B we can set the interconnection into state (1), (2), or (3). We will call this type of link a BI-type link (the I stands for "interference"). By means of this type of link, an external or internal agent can organize an initially random BI-type machine by disabling (opening) and enabling (closing) connections.

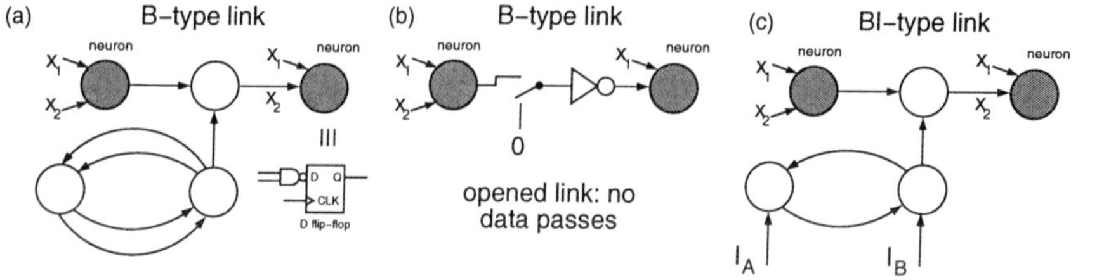

Figure 1. Turing's B-type links. Each B-type link is a small 3-node A-type machine. Part (b) shows a functional diagram of the switch incorporated in each node-to-node link (the example shows a disabled link). Part (c) shows a link with two interfering inputs that affect the internal state of the link.

Unfortunately, Turing made a silly mistake in the definition of his B-type nets: they are not universal [5] and it its thus not possible to realize every logical function. However, it is generally desirable to work with networks that offer universal computational capabilities (such as A-type nets). We therefore propose the TB-type link as shown in Figure 2(a)—which is functionally equivalent to Copeland's and Proudfoot's proposal [5], but simpler. The additional node inverts the input signal. Figure 2(b) shows how two interfering inputs may be added to the introverted pair of the TB-type link. The resulting links is called TBI-*type link*.

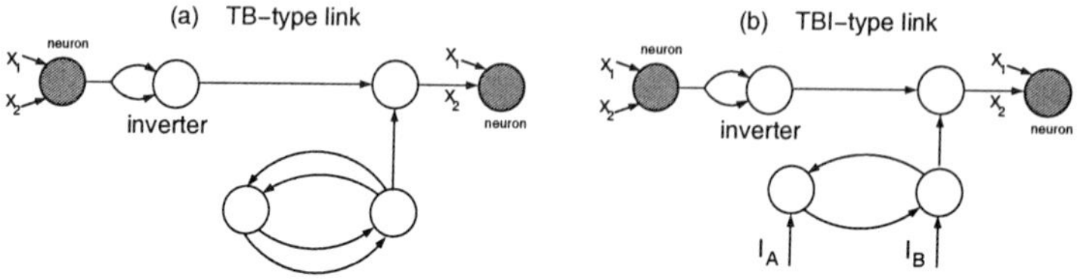

Figure 2. The universal TB-type (a) and TBI-type (b) link proposed by Teuscher.

3 Random Boolean Networks

Random boolean networks form a class of networks in which the links between nodes and the boolean functions are specified at random. They are often specified by two parameters: N, the number of nodes and K, the number of incoming links per node (sometimes, K indicates the average number of links). Synchronous RBN have been seriously investigated by Weisbuch [14] and Kauffman [8] and have been used as models for biological phenomena such as genetic regulatory networks. Turing's unorganized machines might in fact be considered as a very particular subset of synchronous RBN ($K = 2$, NAND functions only, special and fixed interconnections for B-, BI-, and TBI-types). A Turing unorganized machine can always be regarded as a RBN. The inverse statement is not true since it is very unlikely that one gets a Turing unorganized machine when randomly constructing a boolean network. Furthermore, Kauffman's RBN do not allow an external agent to configure the network directly.

Networks built of many interacting units (nodes, neurons) are used to study complex dynamical systems in may areas (e.g., gene regulation networks, neural networks, economics, etc.). Kauffman's studies have revealed surprisingly ordered structures in randomly constructed networks. In particular, the most highly organized behavior appeared to occur in networks where each node—like in Turing's unorganized networks—receives inputs from two other nodes ($K = 2$). It turned out that the networks exhibit three major regimes of behavior: *ordered* (solid), *complex* (liquid), and *chaotic* (gas). The most complex and interesting dynamics correspond to the liquid interface, the boundary between order and chaos. In the ordered regime, little computation can occur. In the chaotic phase, dynamics are too disordered to be useful. The most important and dominant results of Kauffman's numerical simulations can be summarized as follows [8]: (1) The expected median state cycle length is about $\sqrt{N}$ (where N=number of network nodes). (2) Most networks have short state cycles, while a few have very long ones. (3) The number of state cycle attractors is about $\sqrt{N}$. (4) The most interesting dynamics appear with an average connectivity of $K = 2$ (the boundary between order and chaos).

Very few work has been done around asynchronous random boolean networks (ARBN). Harvey and Bossomaier [6] have shown that they behave radically different from the deterministic synchronous version. For many physical and biological phenomena, the assumption of asynchrony seems more plausible.

4 Topological Evolution of TBI-Type Networks

Topological evolution of dynamical networks has been studied in a recent paper by Bornholdt and Rohlf [4]. They evolved network topologies of an asymmetrically connected threshold network by a simple local rewiring rule: quiet nodes grow links, active nodes lose links. They have shown that this leads to an average connectivity of the networks towards the *critical* value $K_C = 2$ for large systems.

Kauffman postulated that gene regulatory networks may exhibit properties of complex dynamical networks near criticality [8], however, without providing an algorithm able to generate a topology near the critical point. Bornholdt and Rohlf asked "[...] whether connectivity may be driven towards a critical point by some dynamical mechanism" [4] and presented an approach that evolves the connectivity towards the critical point. In the remainder of this section we present a self-organizing topology evolving rule adapted for Turing's boolean networks that might also be applied to RBN.

So far, all TBI-type networks had an connectivity of exactly $K = 2$ connections per node. Algorithm 1 presents one of several possible iterating rules that evolves the average connectivity of TBI-type networks towards the critical point. Instead of creating and removing interconnections, the algorithm only operates the switches incorporated in each link and thus *configures* the network topology. The connectivity $K(i)$ of a given node i is equivalent to the number of incoming links that have closed switches. At the end of a successful topology evolution, links with opened switches might simply be removed.

Algorithm 1 Topology evolution of TBI-type networks

```
Choose a fully connected network (K = N)
Choose a random switch configuration with an average connectivity K_ini
for t = 1 to T time steps do
  Choose a random initial state vector s(0)
  Find a dynamical attractor and determine its length L
  for l = 1 to L do
    Run the network and record the activity A(i) of each node
  end for
  for all nodes i do
    if A(i) = 0 then
      A switch on link l_ji (where node j is chosen at random) is enabled/closed.
    else
      A switch on link l_ji (where node j is chosen at random) is disabled/opened.
    end if
  end for
end for
```

The activity $A(i)$ of a node i is defined as the number of changes it makes in the time interval $T_2 - T_1$. When algorithm ?? is applied to a TBI-type network, a typical picture as shown in figure 3 (to the left) arises. A 30-node network self-organizes towards an average connectivity of about $K = 2.5$ (after 10^3 time steps), independent of the initial connectivity (switch configuration). For the simulations we ran, the average connectivity approximatively obeys the scaling law presented by Bornholdt and Rohlf [4]: $K_{ev}(N) - 2 = cN^{-\delta}$ where $c = 12.4$ and $\delta = 0.47$. Thus, when $N \to \infty$, the network evolves towards the critical connectivity $K_C = 2$. Figure 3 (to the right, $N = 20$ nodes, 10^3 time steps) shows the evolution of the average attractor length and the average number of attractors when algorithm ?? is applied to a network. It can be seen that both values essentially remain constant.

The above topology evolution algorithm presents a new and different type of mechanism compared to the phenomenon of self-organized criticality [3]. The networks self-organize towards criticality and "[...] exhibit considerable robustness against noise in the system" [4]. Bornholdt and Rohlf have shown that the mechanism is based on a topological phase transition in the networks.

5 Conclusion

The application of a topology evolving algorithm to universal TBI-type networks has shown that the network self-organizes towards an average connectivity of

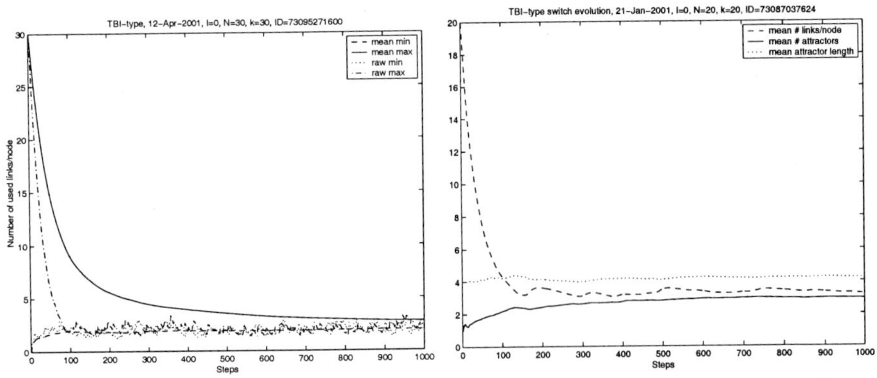

Figure 3. Evolution of the average connectivity of a TBI-type network (on the left, $N = 30$ nodes). Typical average attractor length and average number of attractors during TBI-type topology evolution (on the right, $N = 20$ nodes).

about $K = 2$. For large systems ($N \to \infty$), the connectivity evolves towards the critical value $K_C = 2$. It is of a particular interest that, although Turing TBI-type networks are quite different compared to the asymmetrically connected threshold networks presented in [4], both models evolve towards the same critical connectivity $K_C = 2$ for large systems. Already Kauffman experimentally showed that the most interesting dynamics of RBN appear with an average connectivity of $K = 2$ (the boundary between order and chaos!) [8], however, he did not self-organize the topology of his networks.

The crucial question is "[...] whether a comparable mechanism may occur in natural complex systems" [4]. Bornholdt and Rohlf further state "[...] that this form of global evolution of a network structure towards criticality might be found in natural complex systems". So far, there are still a lot of open questions. Kauffman's gene regulation networks are a good example of a biologically plausible model. It is still one of the greatest mysteries how the genetic code influences the development and growth of organisms. The synaptic changes in the brain are another example where the connectivity of a neural systems is regulated. Turing himself rather naively suggested that an A-type unorganized machine might be the simplest model of a nervous system with a random arrangement of neurons [13, p. 10]. However, today it is clear that his neural networks were by far too simple to model biological neurons, nevertheless, they offer many interesting properties of dynamical complex systems (see also [11]).

Future work will concentrate on the development of a completely local self-organizing rule that not only evolves the network towards an average critical connectivity but also optimizes the network to perform a given task.

References

1. J. T. Allanson. Some properties of randomly connected neural nets. In C. Cherry, editor, *Proceedings of the 3rd London Symposium on Information Theory*, pages 303–313, Butterworths, London, 1956.

2. S.-I. Amari. Characteristics of Randomly Connected Threshold-Element Networks and Network Systems. *Proceedings of the IEEE*, 59(1):35–47, January 1971.
3. P. Bak and L. Chen. Self-organized criticality. *Scientific American*, 265:26–33, January 1991.
4. S. Bornholdt and T. Rohlf. Topological evolution of dynamical networks: Global criticality from local dynamics. *Physical Review Letters*, 84(26):6114–6117, June 2000.
5. B. J. Copeland and D. Proudfoot. On Alan Turing's anticipation of connectionism. *Synthese: An International Journal for Epistemology, Methodology and Philosophy of Science*, 108:361–377, 1996. Kluwer Academic Publishers.
6. I. Harvey and T. Bossomaier. Time out of joint: Attractors in asynchronous random boolean networks. In P. Husbands and I. Harvey, editors, *Proceedings of the Fourth European Conference on Artificial Life*, pages 67–75. MIT Press, Cambridge, MA, 1997.
7. D. C. Ince, editor. *Mechanical Intelligence: Collected Works of A. M. Turing*, chapter Intelligent Machinery, pages 107–128. North-Holland, 1992.
8. S. A. Kauffman. *The Origins of Order: Self–Organization and Selection in Evolution.* Oxford University Press, New York, Oxford, 1993.
9. W. S. McCulloch and W. H. Pitts. A logical calculus of the ideas immanent in neural nets. *Bulletin of Mathematical Biophysics*, 5:115–133, 1943.
10. L. I. Rozonoér. Random logical nets I. *Automation and Remote Control*, 5:773–781, 1969. Translation of Avtomatika i Telemekhanika.
11. C. Teuscher. Study, Implementation, and Evolution of the Artificial Neural Networks Proposed by Alan M. Turing. A Revival of his "Schoolboy" Ideas. Master's thesis, Swiss Federal Institute of Technology Lausanne, Logic Systems Laboratory, EPFL-DI-LSL, CH-1015 Lausanne, February 2000.
12. C. Teuscher and E. Sanchez. A Revival of Turing's Forgotten Connectionist Ideas: Exploring Unorganized Machines. In *Proceedings of the 6th Neural Computation and Psychology Workshop, NCPW6*, University of Liège, Belgium, September 16–18 2000. Springer-Verlag. (To appear).
13. A. M. Turing. Intelligent Machinery. In B. Meltzer and D. Michie, editors, *Machine Intelligence*, volume 5 of *National Physical Laboratory Report*, pages 3–23. Edinburgh University Press, Edinburgh, 1969.
14. G. Weisbuch. *Complex Systems Dynamics: An Introduction to Automata Networks*, volume 2 of *Lecture Notes, Santa Fe Institute, Studies in the Sciences of Complexity.* Addison-Wesley, Redwood City, CA, 1991.

Efficient Pattern Discrimination with Inhibitory WTA Nets

Brijnesh J. Jain and Fritz Wysotzki

Dept. of Computer Science, Technical University Berlin, Germany

Abstract. A mathematical analysis of a special class of winner-take-all networks is given. Starting with a solution in closed form describing the dynamics of the network, we show that an inhibitory winner-take-all net efficiently discriminates input patterns with respect to a canonical measure. This result in combination with further properties suggests an upgrading of the system by incorporating fault tolerance and efficiently generated evidential response into the system.

1 Introduction

The classical way to select the maximum or minimum from a set of inputs within a connectionist framework are *winner-take-all* (WTA) networks ([8]). The operation of these networks is a mode of contrast enhancement and pattern normalization where only the unit with the highest activation fires and all other units in the network are inhibited after some setting time.

Such architectures are designed to signal the particular unit to select, but are not capable to provide information about the evidence in the decision made. This information may be used to improve the classification performance of the system in case of ambiguous decisions. Furthermore WTA networks are not robust to failure of single units. If a unit or its connecting links are damaged, perfomance is impaired or faulty and the network behaves unreliable. Finally, a direct access to completely analyze the general tendency of the network's dynamics is missing, although there is a growing body of mathematical results on competitive neural systems ([1], [2], [3], [5], [9], [10], [11]).

To remove the drawbacks mentioned, we investigate the dynamical behavior of a special class of inhibitory WTA networks for which the underlying system of differential equations is solved exactly (Sect. 2). From the solution we derive and prove four properties of the networks behavior: (P_1) Pattern normalization; (P_2) contrast enhancement; (P_3) evidential response; and (P_4) efficient pattern discrimination. Counterparts of (P_1) and (P_2) are well known ([7]) and if an exact solution is given, restating the proof is straightforward. Property (P_3) was first established in [6]. For this reason we briefly restate (P_1)–(P_3) in Sect. 3 omitting the proofs. The novel and main part of this paper comprises property (P_4). It states that the system *efficiently* generates information about how evident the choice of a particular pattern is (Sect. 4). The properties (P_1)–(P_4) initiate an upgrading of the system to a fault tolerant network with evidential responses (Sect. 5).

G. Dorffner, H. Bischof, and K. Hornik (Eds.): ICANN 2001, LNCS 2130, pp. 827–834, 2001.

Notation: For any vector $\boldsymbol{x} = (x_1, \ldots, x_n) \in \mathbb{R}^n$ we set $\bar{x} := \frac{1}{n}\sum_i x_i$ and $\bar{\boldsymbol{x}} := (\bar{x}, \ldots, \bar{x})$. For $\boldsymbol{x}$, $\boldsymbol{y} \in \mathbb{R}^n$ the scalar product is denoted by $\langle \boldsymbol{x}, \boldsymbol{y} \rangle$ and the Euclidean distance by $\|\boldsymbol{x} - \boldsymbol{y}\|$.

2 The WTA Model

The dynamical system we will work with is a linear WTA model with laterally inhibitory connections of the form

$$\dot{x}_i(t) = -dx_i(t) + \sum_{j=1,\, j\neq i}^{n} w_{ij} f(x_j(t)) + I_i, \qquad x_{0i} := x_i(0) \tag{1}$$

where $x_i(t)$ is the activation of unit i, $w_{ij} = -w < 0$ denotes the inhibitory strength of the synapse connecting unit i and unit j, $d < 0$ is the self-decay rate and I_i an external input. Here we assume $w > d \geq 0$. The function f is a piecewise linear transfer function of the form

$$f(x) = \begin{cases} \phi & : \text{ for } x \geq \phi \\ x & : \text{ for } x \in \,]\psi, \phi[\\ \psi & : \text{ for } x \leq \psi \end{cases}$$

where $\psi < \phi$ are arbitrary constants. This network is a special case of an *additive short-term-memory* model ([4]). The aim of this network is to discriminate input patterns, i.e. the components of an input vector. The input patterns may be feeded in either by an external input layer into the decision net or by initializing the units of the decision net accordingly.

In the following we focus on the dynamics of the net where the activations $x_i(t)$ are in the range of the open interval $]\psi, \phi[$. To investigate the general tendency of the network's behavior we will assume ϕ and ψ to be sufficiently large or shifted towards $+\infty$ and $-\infty$. So we are concerned with a linear transfer function $f(x) = x$. Then a solution of (1) can be given in closed form. The vectors $\boldsymbol{v_1} = (-1, 1, 0, \ldots, 0)^T$, …, $\boldsymbol{v_{n-1}} = (-1, 0, \ldots, 0, 1)^T$, $\boldsymbol{v_n} = (1, \ldots, 1)^T$ form an eigenbasis of $\boldsymbol{W}$ with distinct eigenvalues $\lambda_1 = w - d$ and $\lambda_2 = -(n-1)w - d$ where $\boldsymbol{v_1}, \ldots, \boldsymbol{v_{n-1}} \in \mathrm{Eig}(\boldsymbol{W}, \lambda_1)$ and $\boldsymbol{v_n} \in \mathrm{Eig}(\boldsymbol{W}, \lambda_2)$ (see [6]). Knowledge of the eigenvectors and their corresponding eigenvalues enables us to derive a solution in closed form.

Proposition 1. ([6], Proposition 1) *The solution of the differential equation* $\dot{\boldsymbol{x}} = \boldsymbol{W}\boldsymbol{x} + \boldsymbol{I}$ *in* $\mathbb{R}^n$ *with initial condition* $\boldsymbol{x}(0) = \boldsymbol{x_0}$ *is given by the formula*

$$\boldsymbol{x}(t) = \left(\boldsymbol{x_0} - \bar{\boldsymbol{x}}_{\boldsymbol{0}} + \frac{\boldsymbol{I} - \bar{\boldsymbol{I}}}{\lambda_1}\right) e^{\lambda_1 t} + \left(\bar{\boldsymbol{x}}_{\boldsymbol{0}} + \frac{\bar{\boldsymbol{I}}}{\lambda_2}\right) e^{\lambda_2 t} - \frac{\boldsymbol{I} - \bar{\boldsymbol{I}}}{\lambda_1} - \frac{\bar{\boldsymbol{I}}}{\lambda_2}.$$

To keep the illustrations of our ideas clear we restrict our analysis to networks without self-decay ($d = 0$), without external input ($\boldsymbol{I} = \boldsymbol{0}$), and with initial activation $1 > x_{01} > x_{02} > \cdots > x_{0n} > 0$. Then Proposition 1 simplifies to

Corollary 1. *The solution of the associated homogeneous differential equation* $\dot{\boldsymbol{x}} = \boldsymbol{W}\boldsymbol{x}$ *with initial condition* $\boldsymbol{x}(0) = \boldsymbol{x_0}$ *is of the form*

$$\boldsymbol{x}(t) = (\boldsymbol{x_0} - \bar{\boldsymbol{x}}_0)e^{\lambda_1 t} + \bar{\boldsymbol{x}}_0 e^{\lambda_2 t} \ .$$

A global stabilty analysis shows that the origin $\mathbf{0}$ is a saddle point. Trajectories with $\boldsymbol{x_0} = \bar{\boldsymbol{x}}_0$ go to $\mathbf{0}$ on the line in direction of positive and negative multiples of the eigenvector $\boldsymbol{v_n}$, and those with $\bar{\boldsymbol{x}}_0 = \mathbf{0}$ go to ∞ on the hyperplane defined by $E_1 = \mathrm{Eig}(\boldsymbol{W}, \lambda_1)$. All other trajectories are superpositions of these motions.

3 Properties (P_1)–(P_3)

In this section we briefly replicate some properties of the activations $x_i(t)$ from [6]. The system stores the information provided in form of the initial activation. This information is processed by normalizing patterns (P_1) and enhancing contrasts (P_2), i.e. the intensity $\sum_i x_i(t)$ converges to 0 and the absolute differences $|x_i(t) - x_j(t)|$ increases with time t, respectively. At a specific point of time the system signals the evidence of favorite units i using an approximation of the canonical discrimination measure $\delta_i := x_{0i} - \bar{x}_0$ (P_3). The rest of this section specifies property (P_3).

Let us call units with initial activation $x_{0i} > \bar{x}_0$ *dominating units.* The activations of dominating units have a global minimum while the activations of all other units are monotonously decreasing. If $x_{0i} > \bar{x}_0$, i.e. if unit i is dominating, then the global minimum of $x_i(t)$ is located at

$$t_i = -\frac{1}{nw} \ln\left(\frac{x_{0i} - \bar{x}_0}{(n-1)\bar{x}_0}\right) \quad \text{with} \quad x_i(t_i) = n{\cdot}\bar{x}_0^{1/n}{\cdot}\left(\frac{x_{0i} - \bar{x}_0}{n-1}\right)^{(n-1)/n} . \tag{2}$$

Since $x_i(t) > x_{0i} - \bar{x}_0 = \delta_i$ for $t \geq 0$, the activation $x_i(t)$ of a dominating unit is closest to δ_i when $x_i(t)$ arrives at its minimum. Thus the global minimum $x_i(t_i)$ is the best approximation of δ_i. This result is of no use if this minimum $x_i(t_i)$ is still far away from δ_i. But for an increasing number n of units we find that

$$\lim_{n\to\infty} x_i(t_i) = x_{0i} - \bar{x} \ . \tag{3}$$

Hence **at specific points of time the system signals how evident the corresponding choices are in comparison to all possible choices.**

As we will see later, (3) suggests two extensions of our model. To justify the extensions we need two order preserving properties of the activations:

Lemma 1. ([6], Lemma 1, 2) *Let* $x_{0i} > x_{0j} > \bar{x}_0$. *Then* $x_i(t) > x_j(t)$ *for* $t \geq 0$ *and* $t_i < t_j$.

4 (P_4) Efficient Discrimination

In the last section we recaptured the properties (P_1)–(P_3). In this section we show that the network *efficiently* approximates the canonical measure δ_i. We proceed as follows: First we compute the velocity of the solution curve and subsequently we analyze the direction of evolution. From the results we find a critical point of time t_+ marking the boundary between an efficient and inefficient operation of the system in order to discriminate input patterns. With knowing t_+ we are able to prove property (P_4), i.e. the capability of the net to discriminate input patterns efficiently.

The following definition turns out to be useful for our considerations. Let

$$\sigma_{\boldsymbol{x}}^2 := \frac{\sum_{i=1}^n (x_i - \bar{x})^2}{n} \tag{4}$$

be the *variance* of $\boldsymbol{x} \in \mathbb{R}^n$. In the sequel we assume $\sigma_{\boldsymbol{x}}^2 \neq 0$ and $\bar{x} \neq 0$.

In order to talk about efficient and inefficient states of the network we will simultaneously observe the velocity $\nu(t)$ with which a trajectory approachs the hyperplane $E_1 := \text{Eig}(\boldsymbol{W}, \lambda_1)$ and the velocity $\nu_{\boldsymbol{x}}(t)$ of a trajectory $\boldsymbol{x}(t)$ moving through the state space $\mathbb{R}^n$.

Let us first consider the velocity $\nu(t)$. Asymptotic convergence of a trajectory to E_1 corresponds to the endeavour of the net to normalize patterns. This process is achieved by gradually shifting the intensity to 0. Hence the velocity $\nu(t)$ with which a trajectory converges to E_1 can be determined by using the transformation $T(\boldsymbol{x}(t)) = \langle \boldsymbol{x}(t), \boldsymbol{v_n} \rangle / \|\boldsymbol{v_n}\|$. With increasing time the change of the intensity and therefore the velocity $\nu(t)$ decreases. This is caused by an increasing influence of the eigenvectors $\boldsymbol{v_i} \in E_1$ on the evolving system pulling the curve parallel to E_1. This consideration is summarized in Proposition 2.

Proposition 2. *The velocity $\nu(t)$ of $T(\boldsymbol{x}(t))$ is monotoneously decreasing.*

Proof. The velocity at t of $(T \circ \boldsymbol{x}) : \mathbb{R}_+ \to \mathbb{R}^n$, $t \mapsto T(\boldsymbol{x}(t))$ is given by $|\dot{T}(\boldsymbol{x}(t))|$. Substituting the solution $x_i(t) = (x_{0i} - \bar{x}_0)e^{\lambda_1 t} + \bar{x}_0 e^{\lambda_2 t}$ into $T(\boldsymbol{x}(t)) = \frac{1}{\sqrt{n}} \langle \boldsymbol{x}(t), \boldsymbol{v_n} \rangle = \frac{1}{\sqrt{n}} \sum_{i=1}^n x_i(t)$ leads to

$$T(\boldsymbol{x}(t)) = \frac{1}{\sqrt{n}} \left(\sum_{i=1}^n (x_{0i} - \bar{x}_0)e^{\lambda_1 t} + n\bar{x}_0 e^{\lambda_2 t} \right) = \sqrt{n}\bar{x}_0 e^{\lambda_2 t} \ .$$

Thus from $|\dot{T}(\boldsymbol{x}(t))| = \sqrt{n}|\lambda_2|\bar{x}_0 e^{\lambda_2 t}$ follows the assertion. □

Now let us turn to the velocity $\nu_{\boldsymbol{x}}(t)$ of a trajectory $\boldsymbol{x}(t)$ moving through the state space $\mathbb{R}^n$. For a given network the velocity of $\boldsymbol{x}(t)$ differ subject to the variance $\sigma_{\boldsymbol{x_0}}^2$ of the initial activation with constant average $\bar{x}_0$.

Proposition 3. *The velocity $\nu_{\boldsymbol{x}}(t)$ of $\boldsymbol{x}(t)$ at time t is given by*

$$\nu_{\boldsymbol{x}}(t) = ||\dot{\boldsymbol{x}}(t)|| = \sqrt{n \left(\{\lambda_1 \sigma_{\boldsymbol{x_0}} e^{\lambda_1 t}\}^2 + \{\lambda_2 \bar{x}_0 e^{\lambda_2 t}\}^2 \right)} \ .$$

Proof. Differentiating the solution $\boldsymbol{x}(t)$ yields $\dot{\boldsymbol{x}}(t) = \lambda_1(\boldsymbol{x_0} - \bar{\boldsymbol{x}}_0)e^{\lambda_1 t} + \lambda_2 \bar{\boldsymbol{x}}_0 e^{\lambda_2 t}$. Then the square of the length of the velocity vector $\dot{\boldsymbol{x}}(t)$ at t is $||\dot{\boldsymbol{x}}(t)||^2 = \sum_{i=1}^n \left\{ \lambda_1 (x_{0i} - \bar{x}_0)e^{\lambda_1 t} + \lambda_2 \bar{x}_0 e^{\lambda_2 t} \right\}^2$. After some equivalence transformations using $\sum_{i=1}^n (x_{0i} - \bar{x}_0) = 0$ we find $||\dot{\boldsymbol{x}}(t)||^2 = n\lambda_1^2 \sigma_{\boldsymbol{x_0}}^2 e^{2\lambda_1 t} + n\lambda_2^2 \bar{x}_0^2 e^{2\lambda_2 t}$. □

In contrast to the trajectory in the transformed state space the velocity $\nu_{\boldsymbol{x}}(t)$ of $\boldsymbol{x}(t)$ slows down until it is moving slowest and then accelerates. For later purposes the point of time t_{V} of minimal velocity of a trajectory is determined.

Proposition 4. *The velocity* $\nu_{\boldsymbol{x}}(t)$ *has a global minimum at*

$$t_{\mathrm{V}} = -\frac{1}{2nw}\left\{2\ln\left|\frac{\sigma_{\boldsymbol{x}0}}{\bar{x}_0}\right| + 3\ln\left|\frac{1}{n-1}\right|\right\} . \tag{5}$$

Proof. It is sufficient to show the assertion for $\nu_{\boldsymbol{x}}^2(t) = \|\dot{\boldsymbol{x}}(t)\|^2$. Differentiating $\|\dot{\boldsymbol{x}}(t)\|^2$ and equating with 0 leads to

$$\left(\lambda_1\left\{\lambda_1\sigma_{\boldsymbol{x}_0}e^{\lambda_1 t}\right\}^2 + \lambda_2\left\{\lambda_2\bar{x}_0 e^{\lambda_2 t}\right\}^2\right) = 0 .$$

Solving the equation to t yields

$$t_{\mathrm{V}} = \frac{1}{2(\lambda_2-\lambda_1)}\ln\left(-\left\{\frac{\lambda_1}{\lambda_2}\right\}^3\left\{\frac{\sigma_{\boldsymbol{x}0}}{\bar{x}_0}\right\}^2\right) = -\frac{1}{2nw}\ln\left(\left\{\frac{1}{n-1}\right\}^3\left\{\frac{\sigma_{\boldsymbol{x}0}}{\bar{x}_0}\right\}^2\right) .$$

Rearraniging the terms gives us the equation to prove. It is left to show, that t_{V} is indeed a minimum. The second derivative of $\|\dot{\boldsymbol{x}}(t)\|^2$ is of the form

$$4n\left(\left\{\lambda_1^2\sigma_{\boldsymbol{x}_0}e^{\lambda_1 t}\right\}^2 + \left\{\lambda_2^2\bar{x}_0 e^{\lambda_2 t}\right\}^2\right) > 0 .$$

Clearly, t_{V} is a global minimum, since $\|\dot{\boldsymbol{x}}(t)\|$ is continuously differentiable and t_{V} is the unique zero of the first derivative. □

The movement of a trajectory through the state space is controlled by the eigenvectors of the connectivity matrix. When the net starts evolving at initial point $\boldsymbol{x}_0$ the trajectory is guided into direction to the hyperplane $E_1 = \mathrm{Eig}(\boldsymbol{W}, \lambda_1)$ by the vector $\bar{\boldsymbol{x}}_0$ with small superposition in direction parallel to E_1 by the vector $\boldsymbol{x}_0 - \bar{\boldsymbol{x}}_0$. The influence of $\boldsymbol{x}_0 - \bar{\boldsymbol{x}}_0$ is increasing with the time whereas at the same time the influence of $\bar{\boldsymbol{x}}_0$ is decreasing. There is a critical point of time t_+ where the influence of both eigenvectors $\bar{\boldsymbol{x}}_0$ and $\boldsymbol{x}_0 - \bar{\boldsymbol{x}}_0$ is equal. This point of time marks the transition from an efficient to an inefficient system as we will see at the end of this subsection. Thus we define:

Definition 1. *Consider an inhibitory WTA network described by the differential equation (1). The network* operates efficiently *if* $t \leq t_+$. *Otherwise we call the network* inefficient.

The influence of an eigenvector $\boldsymbol{v}$ to direct the run of the trajectory $\boldsymbol{x}(t)$ at time t can be measured by the angle between the tangential vector $\dot{\boldsymbol{x}}(t)$ and $\boldsymbol{v}$. The smaller the angle is, the stronger is the line of trace guided by the corresponding eigenvector.

Proposition 5. *The critical point of time is given by*

$$t_+ = -\frac{1}{nw}\left\{\ln\left|\frac{\sigma_{\boldsymbol{x}0}}{\bar{\boldsymbol{x}}_0}\right| + \ln\left|\frac{1}{n-1}\right|\right\} . \tag{6}$$

Proof. Let $\alpha(t)$ and $\beta(t)$ be the angles between the tangential vector $\dot{\boldsymbol{x}}(t)$ and the vector $\boldsymbol{x_0}-\bar{\boldsymbol{x}}_0$ and $\bar{\boldsymbol{x}}_0$ at time t. Clearly, if $\alpha(t) = \beta(t)$ then $\cos(\alpha(t)) = \cos(\beta(t))$. This leads to the equation

$$\frac{\langle \dot{\boldsymbol{x}}(t), \boldsymbol{x_0} - \bar{\boldsymbol{x}}_0 \rangle}{||\dot{\boldsymbol{x}}(t)|| \cdot ||\boldsymbol{x_0} - \bar{\boldsymbol{x}}_0||} = \frac{\langle \dot{\boldsymbol{x}}(t), \bar{\boldsymbol{x}}_0 \rangle}{||\dot{\boldsymbol{x}}(t)|| \cdot ||\bar{\boldsymbol{x}}_0||} \tag{7}$$

where $||\dot{\boldsymbol{x}}(t)||$ can be cancelled. Now observe that with $\dot{\boldsymbol{x}}(t) = \lambda_1(\boldsymbol{x_0} - \bar{\boldsymbol{x}}_0)e^{\lambda_1 t} + \lambda_2 \bar{\boldsymbol{x}}_0 e^{\lambda_2 t}$ and $(\boldsymbol{x_0} - \bar{\boldsymbol{x}}_0) \perp \bar{\boldsymbol{x}}_0$ the following relations hold:

$$\langle \dot{\boldsymbol{x}}(t), \boldsymbol{x_0} - \bar{\boldsymbol{x}}_0 \rangle = n\lambda_1 \sigma^2_{\boldsymbol{x_0}} e^{\lambda_1 t} \tag{8}$$
$$\langle \dot{\boldsymbol{x}}(t), \bar{\boldsymbol{x}}_0 \rangle = n\lambda_2 \bar{x}_0^2 e^{\lambda_2 t} \tag{9}$$
$$||\boldsymbol{x_0} - \bar{\boldsymbol{x}}_0|| = \sqrt{n}\, \sigma_{\boldsymbol{x_0}} \tag{10}$$
$$||\bar{\boldsymbol{x}}_0|| = \sqrt{n}\, \bar{x}_0 \ . \tag{11}$$

Plugging eqns. (8-11) into eqn. (7) leads to $n\lambda_1 \sigma_{\boldsymbol{x_0}} e^{\lambda_1 t} = n\lambda_2 \bar{x}_0 e^{\lambda_2 t}$. Finally, solving this equation to t gives us the desired equation for t_+. □

Now we are able to turn to the key result of this paper as stated in Corollary 2. The statement is of interest for the follwing reasons: Firstly, it justifies our definition of efficient and inefficient states of a network and therefore establishes a basis for property (P_4). Secondly, it proves property (P_4).

Corollary 2. *If $n > 2$ and $1 > x_{01} > \cdots > x_n \geq 0$ then $t_V > t_+ > t_1$ holds.*

Proof. The first inequality $t_V > t_+$ follows from (5) and (6), the other inequality $t_+ > t_1$ from (2) and (6). For the latter inequality first show $\sigma_{\boldsymbol{x_0}} < x_{01} - \bar{x}_0$. □

Let us point out the significant statements of Corollary 2 in detail. From our discussion in Sect. 2 we know that the net strives to achieve a normalized state in the long-term (P_1). Now our question is whether the net achieves this objective efficiently. Intuitively, a system is considered to be inefficient, if it spends increasing energy for less avail. Conversely if a system progresses succesfully with low energy, it can be considered to be efficient. In analogy to physics we may interpret that increasing velocity of a trajectory results in increasing energy consumption. Now note, that by Proposition 2 the velocity of approaching a normalized state decelerates continuously. The same holds for the speed of a trajectory moving through the state space until the velocity arrives at its minimum at time t_V (Proposition 4). Since $t_+ < t_V$, progresses to achieve a normalized state slow down for two reasons. Firstly, energy consumption is reduced in terms of decelerating velocity of the trajectory in the state space and secondly, the influence of the power pulling the trajectory to a normalized pattern is decreasing. Thus, beyond t_+ most energy is used to keep the current average activation $\bar{x}(t)$, i.e. to strive for moving parallel to the hyperplane E_1. Beyond t_V the trajectory accelerates and with it energy consumption increases with little effect on gaining the networks long-term objective, namely $\bar{x}_0 = 0$ (P_1). Hence it is suggestive to mark t_+ as the critical point of time at which a networks changes from an efficient to an inefficient state with respect to pattern normalization.

As our main result in this paper, Corollary 2 also shows property (P_4): Discrimination is pursued by normalizing the input patterns and enhancing the contrasts until at time t_1 the system signals the evidence of the selected unit in terms of an approximation of the canonical measure δ_1. Since $t_+ > t_1$, **the system efficiently generates an evidential response of the future winner.** Furthermore the difference $t_+ - t_1 = \frac{1}{nw} \ln \left| \frac{x_1 - \bar{x}_0}{\sigma_{\bar{x}_0}} \right|$ reveals that the operating efficiency of the system is exhausted increasingly the more similar the initial activations are and vice versa.

5 Conclusion

So far we have shown that the network efficiently processes the input patterns by normalizing patterns and enhancing contrast until it signals the evidence of potential choices at a specific point of time. Our goal is to design a model which enables us to read out the evidential response and with which we can control the focus on the given problem. The sharpness of the focus is measured in terms of the accuracy with which the evidential response $x_i(t_i)$ approximates the canonical measure δ_i.

In order to control the focus, that is to obtain a better approximation of δ, (3) suggests to increase the number of units by means of distributed redundancy. As a readout device for evidential response we employ differentiating units which are able to detect local minima. Besides fault tolerance and evidential response the extended system keeps the activations bounded and guarantees stability within finite time. In the following, we will discuss both extensions separately.

(1) Robust, Fault Tolerant Networks: The network can be enlarged without changing the behavior of the activations by the following procedure. Increase the number n of units by adding artifical units without changing the average activation $\bar{x}_0$. This can be achieved by connecting a fixed number k of copies of each unit to the net where the activations of the copies are slighty perturbed by a random noise. This enlarges the system to kn mutually inhibited units consisting of n groups each with k units of nearly equal initial activation. If the noise is Gaussian distributed with expectation 0 and small variance σ^2 compared to the initial activation of the original units, then the expectation of $\bar{x}_0$ of the enlarged system corresponds to the average of the original system. Now let x_{0i} be the initial activation of a representative of group i. Then at t_i the activations of group i retrieve a better averaged approximation of $x_{0i} - \bar{x}$ than unit i of the original system (see (3)). Furthermore from (2) we can conclude $\lim_{n \to \infty} t_1 = 0$. Thus, **distributed redundancy of the input patterns does not only provide a system which is robust against failure of single units but also efficiently sharpens the focus on the given problem in terms of a faster generated and more accurate evidential response.**

(2) Differentiating Units: We call units which are able to detect local minima of their activations *differentiating units*. With differentiating units we can construct a network keeping the activations bounded and leading to a stable state within finite time. In mathematical terms, we may describe a differentiating unit i by the following pair of iterative equations:

$$x_i(t+1) = (1-d)x_i(t) + \sum_{j \neq i} w_{ij} y_j(t) + I_i$$
$$y_i(t+1) = x_i(t+1) \cdot f[x_i(t) - x_i(t+1)]$$

where $x_i(t+1)$ is the activation of unit i at time $t+1$, $f(x)$ the threshold function with $f(x) = 0$, if $x \leq 0$ and $f(x) = 1$ otherwise. Finally $y_i(t+1)$ is the output signal of unit i.

Let us assume w to be sufficiently small, such that the iterative system is a good approximation of the corresponding continuous system in the interval $]0, t_i]$. From Lemma 1 we know, that the activation of unit i with maximal initial activation arrives first at its minimum. This leads to an ouptut signal $y_i(t_i) = 0$. In the next two iteration steps the net will arrive at a stable state. The winning unit i is the one first entering a stable state with activation approximately $x_i(t_i)$. Hence **a network using differentiating units is capable to signal the winner and provides efficiently generated information about the evidence of the selected unit.**

References

1. M.A. Cohen and S. Grossberg. Absolute stability of global pattern formation and parallel memory storage by competitive neural networks. *IEEE Transactions on Systems, Man, and Cybernetics*, 13(5):815–826, 1983.
2. B. Ermentrout. Complex dynamics in winner-take-all neural nets with slow inhibition. *Neural Networks*, 5:415–431, 1992.
3. Y. Fang, M.A. Cohen, and T.G. Kincaid. Dynamics of a winner-take-all neural network. *Neural Networks*, 9(7):1141–1154, 1996.
4. S. Grossberg. Nonlinear neural networks: Principles, mechanisms, and architectures. *Neural Networks*, 1:17–61, 1988.
5. R. Hahnloser. On the piecewise analysis of networks of linear threshold neurons. *Neural Networks*, 11:691–697, 1998.
6. B.J. Jain and F. Wysotzki. On the short-term-memory of wta nets. In M. Verleysen, editor, *Proc. ESANN'01*. D-Facto, Brussels, 2001. To appear.
7. D.S. Levine. *Introduction to neural and cognitive modelling.* Lawrence Erlbaum Associates, Inc., Hillsday, New Jersey, 1991.
8. R.P. Lippman. An introduction to computing with neural nets. *IEEE ASSP Magazine*, pages 4–22, April 1987.
9. W. Maass. On the computational power of winner-take-all. *Neural Computation*, 12(11):2519–2536, 2000.
10. H. Wersing, J.J. Steil, and H. Ritter. A competitive layer model for feature binding and sensory segmentation. *Neural Computation*, 13(2):357–387, 2001.
11. X. Xie, R. Hahnloser, and S. Seung. Learning winner-take-all competition between groups of neurons in lateral inhibitory networks. In Todd K. Leen, Thomas G. Dietterich, and Volker Tresp, editors, *Advances in Neural Information Processing Systems 13*, pages 350–356. MIT Press, 2001.

Cooperative Information Control to Coordinate Competition and Cooperation

Ryotaro Kamimura[1] and Taeko Kamimura[2]

[1] Information Science Laboratory, Tokai University, 1117 Kitakaname, Hiratsuka, Kanagawa, 259-1292, Japan
ryo@cc.u-tokai.ac.jp
[2] Department of English, Senshu University, 2-1-1 Higashimita, Tama-ku, Kawasaki, Kanagawa, 214-8580, Japan
taekok@isc.senshu-u.ac.jp

Abstract. This paper proposes a novel information theoretic approach to self-organization called *cooperative information control.* The method aims to mediate between competition and cooperation among neurons by controlling the information content in the neurons. Competition is realized by maximizing information content in neurons. In the process of information maximization, only a small number of neurons win the competition, while all the others are inactive. Cooperation is implemented by having neighboring neurons behave similarly. These two processes are unified and controlled in the framework of cooperative information control. We applied the new method to political data analyses. Experimental results confirmed that competition and cooperation are flexibly controlled. In addition, controlled processes can yield a number of different neuron firing patterns, which can be used to detect macro as well as micro features in input patterns.

1 Introduction

In this paper, we propose a new type of self-organization method called *cooperative information control.* In this method, competition among neurons is realized by maximizing the information content in neurons. In addition, this information maximization process synchronizes with the process of cooperation in which neighboring neurons behave similarly. In contrast to conventional self-organization methods, cooperative information control can flexibly mediate between competition and cooperation. Self-organization maps developed by Kohenen [1], [2] have been used in a variety of fields because of their simplicity and powerful visualization. However, in the self-organization maps, because the firing rates of neurons are automatically given in the process of approximating input patterns, and because they have no methods to control cooperation and competition, final neuron firing patterns and accompanying internal representations are sometimes ambiguous, uncertain and dependent upon detailed parts of input patterns. For this reason, a great number of theoretic and heuristic techniques have been proposed to enhance internal representations obtained by the

G. Dorffner, H. Bischof, and K. Hornik (Eds.): ICANN 2001, LNCS 2130, pp. 835–842, 2001.

conventional self-organization maps [3], [4], [5], [6], to cite a few. If it is possible to control and mediate between competition and cooperation with flexibility, final neuron firing patterns and obtained internal representations can be more easily interpreted.

2 Cooperative Information Control

2.1 Competitive Information Control

We realize competition by controlling information content in neurons. In this paper, we focus upon information stored in neural systems [7], [8]. Information is defined as the decrease in uncertainty from an initial state to another state after receiving input patterns. This uncertainty decrease or the information gained is defined by

$$I = -\sum_j p(j) \log p(j) + \sum_s \sum_j p(s) p(j \mid s) \log p(j \mid s), \tag{1}$$

where $p(j)$, $p(s)$, $p(j \mid s)$ denote the probability of the jth unit being turned on, the probability of the input signal s and the conditional probability of the jth unit after receiving the sth input signal.

As information is increased, a neuron tends to fire in response to some input patterns. In this way, information maximization is used to simulate competition among neurons.

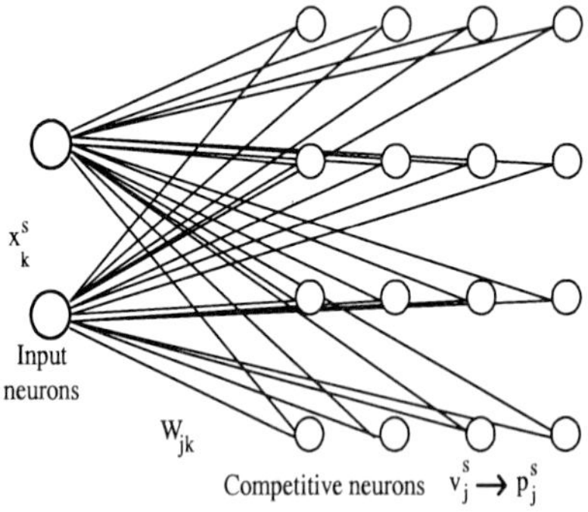

Fig. 1. A network architecture for self-organization.

Next, we apply information gain to actual neural networks. As shown in Figure 1, a network architecture is composed of input neurons x_k^s and competitive neuron v_j^s. Let us define information for competitive neurons and try to control the neuron firing patterns. The jth competitive neuron receives a net input from input neurons, and an output from the jth competitive neuron can be computed by

$$v_j^s = f(\sum_k w_{jk} x_k^s), \tag{2}$$

where w_{jk} denotes a connection from the kth input neuron to the jth competitive neuron, and $f(t) = 1/(1+\exp(-t))$. One of the easiest ways to model competition among neurons is to normalize the outputs from the competitive neurons as follows:

$$p_j^s = \frac{v_j^s}{\sum_m v_m^s}. \tag{3}$$

The conditional probability $p(j \mid s)$ is approximated by this normalized competitive neuron output, that is,

$$p(j \mid s) \approx p_j^s. \tag{4}$$

Since input patterns are supposed to be uniformly given to networks, the probability of the jth competitive neuron is approximated by

$$\begin{aligned} p(j) &= \sum_s p(s)p(j \mid s) \\ &\approx \frac{1}{S}\sum_s p_j^s \\ &= p_j. \end{aligned} \tag{5}$$

By using these probabilites, the information is approximated by

$$I \approx -\sum_j p_j \log p_j + \frac{1}{S}\sum_s \sum_j p_j^s \log p_j^s. \tag{6}$$

In the process of information maximization, neurons compete with each other, and finally one neuron wins the competition. At the initial stage of learning, no information is given to a network, meaning that all competitive neurons are uniformly distributed. When information is maximized, only one competitive neuron is turned on, while all the others are off. By maximizing information, we can simulate competitive learning.

2.2 Cooperated Information Control

The cooperation process is realized by having neurons behave similarly to neighboring neurons. In living systems, it is well known that a group of neurons fire to a specific input pattern [9], [10]. This phenomenon is a good example of cooperation among neurons in living systems. For example, Figure 2(a) and (b) show two examples of this cooperation process. In (a), the surrounding neurons' firing rates around the target neuron at the center are relatively high. The target neuron must increase its firing rate up to the level of the neighboring neurons. In (b), the firing rates of the neighboring neurons are relatively small, while the rate of the target neuron in the center is high. As such, the target neuron must decrease its firing rate down to the level of the neighboring neurons.

Now, let us consider update rules to realize this cooperation. Following conventional self-organization maps, we define a topological neighborhood function

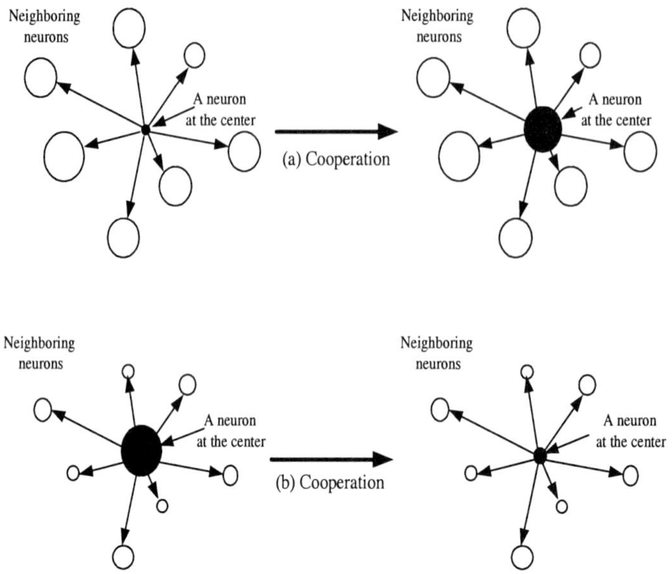

Fig. 2. A schematic representation of cooperative processes in cooperative information maximization. In (a), the target neuron at the center should increase its firing rate, while in (b), the target neuron must decrease its rate down to the level of the surrounding neurons.

ϕ_{jm} between the jth and mth neuron. For this purpose, we introduce the lateral distance d_{jm} between the jth and mth neurons as follows:

$$d_{jm}^2 = \| \mathbf{r}_j - \mathbf{r}_m \|^2, \tag{7}$$

where the discrete vector $\mathbf{r}_j$ denotes the jth neuron position in the two dimensional lattice. By using this distance function, we have the topological neighborhood function:

$$h_{jm} = \exp\left(-\frac{d_{jm}^2}{2\sigma^2} \right), \tag{8}$$

where σ denotes a parameter, representing the effective width of the topological neighborhood. For realizing cooperation, we consider the following modified topological neighborhood function:

$$\phi_{jm} = p_m h_{jm}, \tag{9}$$

where p_m denotes the mth neuron's firing rate. This neighborhood function becomes stronger, as the distance becomes smaller and at the same time the neighboring neurons' firing rate becomes higher. By using this topological function, we can simulate cooperation by making connections into the jth neuron similar to those into the mth neuron. Differentiating information function I with respect to connections w_{jk} and adding the cooperation term, we have the final update rules:

$$\Delta w_{jk} = -\beta \sum_{s} \left(\log p_j - \sum_{m} p_m^s \log p_m \right) Q_{jk}$$
$$+\beta \sum_{s} \left(\log p_j^s - \sum_{m} p_m^s \log p_m^s \right) Q_{jk}$$
$$+\theta \sum_{m} \phi_{jm} \left(w_{mk} - w_{jk} \right), \quad (10)$$

where

$$Q_{jk} = \frac{1}{S} p_j^s (1 - v_j^s) x_k^s. \quad (11)$$

3 Application to Political Analysis

We attempted to classify congressmen in terms of their voting attitudes. The data were represented as a qualitative matrix of the voting attitudes of U.S. congressmen toward 19 environmental bills [11]. The first 8 congressmen are Democrats, while the latter 7 are Republicans. Competitive neurons are arranged in a two-dimensional square, meaning that the numbers of rows and columns are the same. The number of input unit was 19, and the number of competitive neurons was set to 25 (5 × 5).

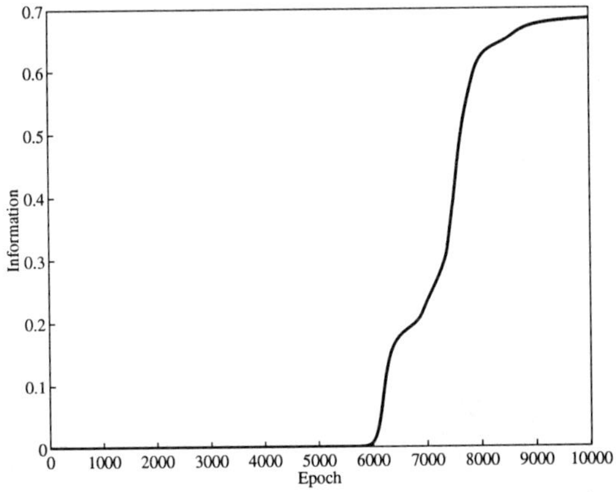

Fig. 3. Information as a function of the number of epochs for the political analysis. Information was normalized for its values to range between 1 and 0.

Figure 3 shows normalized information ranging between 0 and 1 as a function of the number of epochs. Information is almost zero before the number of epochs reaches 6,000. After this point, information rapidly increases and reaches about 70 percent of maximum information. Figure 4 shows neuron firing patterns p_j^s obtained in 5,000 epochs (a) and 10,000 epochs (b). The magnitude of squares represents the strength of firing. The largest squares show the highest firing rate

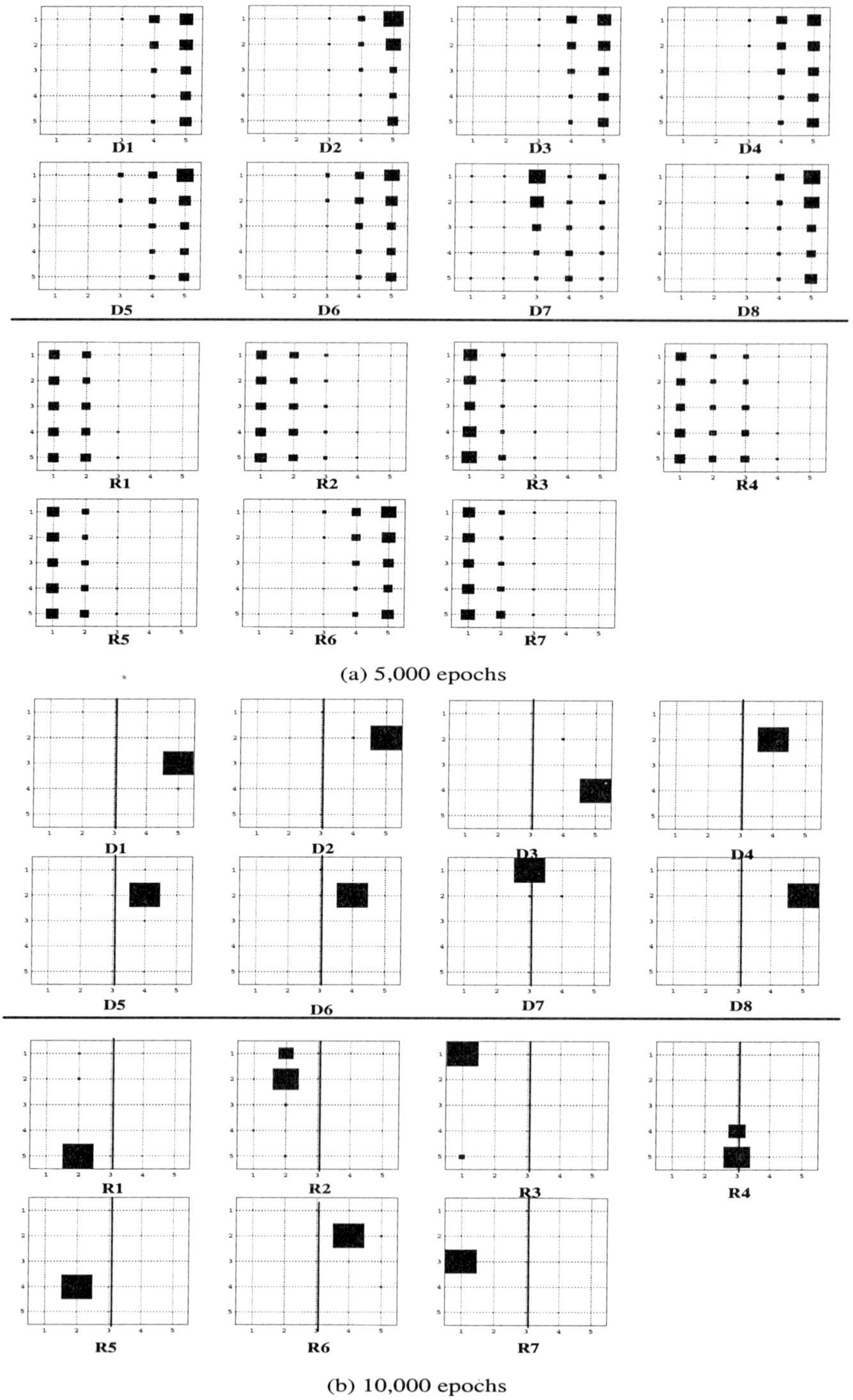

Fig. 4. Competitive neuron firing patterns p_j^s for 15 senators. The number of epochs was 5,000 (a) and 10,000 (b). Black squares represent the magnitude of firing of neurons.

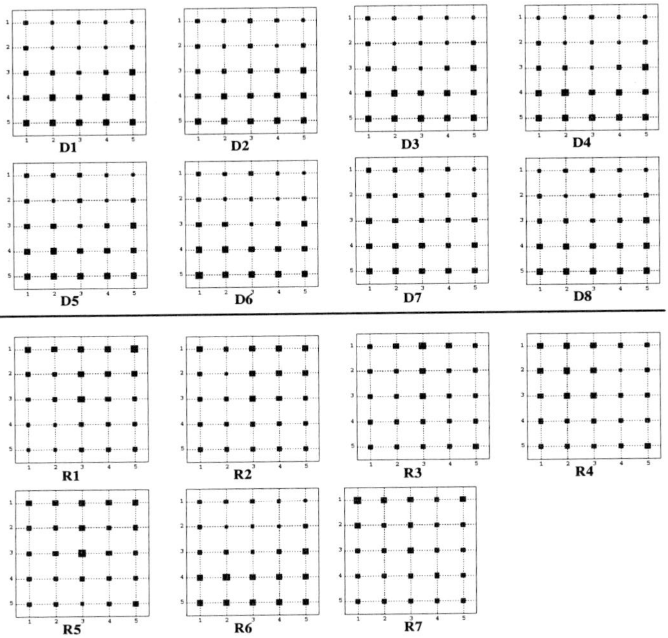

Fig. 5. Competitive neuron firing patterns p_j^s for 15 senators by the onventional self-organization map. The number of epochs was 100,000.

of 1. In 5,000 epochs, we can see a clear tendency that the Democrats (D1-D7) are represented by firing some neurons on the right hand side, while the Republicans (R1-R7) are represented by firing some neurons on the left hand side. However, we can find that Republican No.6 is classified as a Democrat because he or she shows the same firing pattern as that for the Democrats. In 10,000 epochs (Figure (b)), each neuron tends to respond to a different input pattern. At this stage, for D7 and R4, neurons in the central position fire strongly. This means that these two congressmen take a slightly different voting attitude from that of the other senators. In addition, we gave the network the data of the league of conservation voters, which was not included in the training data. In 5,000 epochs, some neurons located on the right hand side fire strongly, corresponding to the region of the Democrats. In 10,000 epochs, only two neurons on the upper right hand side respond strongly to the league of the conservation. This indicates that the voting attitude of the league is similar to that of the Democrats.

Lastly, we plotted neuron firing patterns by the conventional self-organization map, as shown in Figure 5. All firing rates are normalized between 0 and 1. If we look closely at the firing patterns in the figure, we can find that some neurons on the lower side fire for the Democrats, while neurons on the upper side tend to fire for the Republicans. However, this difference between the two is not so clear as that shown by our cooperative information control. In addition, it is impossible to enhance neuron firing patterns by the conventional method. Even if the number of learning cycles is increased, and learning parameters are changed appropriately, it is not possible to obtain clearer firing patterns by conventional methods. This is the reason why the conventional self-organization methods need to be developed with a great number of heuristic techniques which will allow for a much easier interpretation, as mentioned in the Introduction.

References

1. T. Kohonen, *Self-Organization Maps.* Springer-Verlag, 1995.
2. T. Kohonen, "The self-organization map," *Proceedings of the IEEE*, vol. 78, no. 9, pp. 1464–1480, 1990.
3. J. Versanto, "Som-based data visualization methods," *Intelligent Data Analysis*, vol. 3, pp. 111–126, 1999.
4. S. Kaski, J. Nikkila, and T. Kohonen, "Methods for interpreting a self-organized map in data analysis," in *Proceedings of European Symposium on Artificial Neural Networks*, (Bruges, Belgium), 1998.
5. I. Mao and A. K. Jain, "Artificial neural networks for feature extraction and multivariate data projection," *IEEE Transactions on Neural Networks*, vol. 6, no. 2, pp. 296–317, 1995.
6. A. Ultsch and H. P. Siemon, "Kohonen self-organization feature maps for exploratory data analysis," in *Proceedings of International Neural Network Conference*, (Dordrecht), pp. 305–308, Kulwere Academic Publisher, 1990.
7. L. L. Gatlin, *Information Theory and Living Systems.* Columbia University Press, 1972.
8. R. Kamimura and S. Nakanishi, "Hidden information maximization for feature detection and rule discovery," *Network*, vol. 6, pp. 577–622, 1995.
9. E. L. Bienenstock, L. N. Cooper, and P. W. Munro, "Theory for the development of neuron selectivity," *Journal of Neuroscience*, vol. 2, pp. 32–48, 1982.
10. D. H. Hubel and T. N. Wisel, "Receptive fields, binocular interaction and functional architecture in cat's visual cortex," *Journal of Physiology*, vol. 160, pp. 106–154, 1962.
11. H. C. Romesburg, *Cluster Analysis for Researchrers.* Florida: Krieger Publishing Company, 1984.

Qualitative Analysis of Continuous Complex-Valued Associative Memories

Yasuaki Kuroe, Naoki Hashimoto, and Takehiro Mori

Department of Electronics and Information Science
Kyoto Institute of Technology
Matsugasaki, Sakyo-ku, Kyoto 606-8585, Japan
kuroe@dj.kit.ac.jp

Abstract. This paper presents a model of self-correlation type associative memories using complex-valued continuous neural networks and studies its qualitative behaviors theoretically. The proposed model is an extension of the conventional real-valued associative memories of self-correlation type. We investigate the structures and asymptotic behaviors of solution orbits near each memory pattern. We also discuss a recalling condition of each memory pattern, that is, a condition which assures that each memory pattern is correctly recalled.

1 Introduction

In recent years, there have been increasing research interests of artificial neural networks and many efforts have been made on applications of neural networks to various fields. One of the most useful and most investigated areas of applications of neural networks addresses implementations of associative memories. As applications of the neural networks spread more widely, developing neural network models which can deal with complex numbers is desired in various fields. Several models of complex-valued neural networks have been proposed and their abilities of information processing have been investigated [1–6].

The purpose of this paper is to present a model of self-correlation type associative memories using complex-valued continuous neural networks and to investigate its qualitative behaviors theoretically. Some models of complex-valued associative memories of self-correlation type have already been proposed [3–5]. The model in [3] is a partially complex-valued model of self-correlation type associative memories. It is shown that by using the arguments of complex numbers as supplementary information carriers the model becomes capable of possessing twice or more the memory capacity of the real-valued one. The literature [4] proposes three types of complex-valued neural networks and discusses their applicability to constructing self-correlation associative memories. In [5] a complex-valued extension of the self-correlation associative memories is proposed and experimental analysis on its behavior has been done.

Our proposed model is the continuous-time associative memories of self-correlation type. We treat the proposed model as a nonlinear dynamical system

G. Dorffner, H. Bischof, and K. Hornik (Eds.): ICANN 2001, LNCS 2130, pp. 843–850, 2001.

and study its qualitative behavior theoretically. In particular, we investigate the structures and asymptotic behaviors of solution orbits near each memory pattern. We also discuss a recalling condition, that is, a condition that assures that each memory pattern is correctly recalled.

In the following, the imaginary unit is denoted by i ($i^2 = -1$). The n-dimensional complex (real) space is denoted by $\boldsymbol{C}^n(\boldsymbol{R}^n)$ and the set of $n \times m$ complex (real) matrices is denoted by $\boldsymbol{C}^{n\times m}(\boldsymbol{R}^{n\times m})$. For $A \in \boldsymbol{C}^{n\times m}$ ($\boldsymbol{a} \in \boldsymbol{C}^n$), its real and imaginary parts are denoted by A^R ($\boldsymbol{a}^R$) and A^I ($\boldsymbol{a}^I$), respectively.

2 Continuous Complex-Valued Associative Memory

2.1 Model

Let m be the number of memory patterns to be stored, and each memory pattern is an N dimensional complex vector, denoted by $\boldsymbol{s}^{(\gamma)} \in \boldsymbol{C}^N$, $\gamma = 1, 2, \cdots, m$. Suppose that the set of memory patterns satisfies the following orthogonal relations.

$$\boldsymbol{s}^{(\gamma)*}\boldsymbol{s}^{(l)} = \begin{cases} N, & \gamma = l \\ 0, & \gamma \neq l \end{cases}, \qquad l = 1, 2, \cdots, m \tag{1}$$

$$|s_j^{(\gamma)}| = 1, \qquad j = 1, 2, \cdots, N \tag{2}$$

where $\boldsymbol{s}^*$ is the conjugate transpose of $\boldsymbol{s}$ and $s_j^{(\gamma)}$ denotes the jth element of vector $\boldsymbol{s}^{(\gamma)}$. We consider a complex-valued continuous neural network described by differential equations of the form:

$$\begin{cases} \tau \dfrac{du_j(t)}{dt} = -u_j(t) + \displaystyle\sum_{k=1}^{N} w_{jk} x_k(t) \\ x_j(t) = f(u_j(t)) \end{cases}, \qquad j = 1, \cdots, N \tag{3}$$

where $u_j(t) \in \boldsymbol{C}$ and $x_j(t) \in \boldsymbol{C}$ are the state and the output of the jth neuron at time t, respectively, τ is the time constant (positive real number), $w_{jk} \in \boldsymbol{C}$ is a connection weight from the kth neuron to the jth neuron and $f(\cdot)$ is an activation function which is a nonlinear complex function ($f : \boldsymbol{C} \to \boldsymbol{C}$).

Let us determine the weight matrix $W = \{w_{jk}\} \in \boldsymbol{C}^{N\times N}$ and the activation function $f(\cdot)$ so that the neural network (3) can store the set of memory vectors. Taking account of the orthogonal structure of the set of memory vectors $\{\boldsymbol{s}^{(\gamma)}\}$ satisfying (1) and (2), we determine the weight matrix W as the sum of the autocorrelation matrix of each memory vector $\boldsymbol{s}^{(\gamma)}$:

$$W = \frac{1}{N} \sum_{\gamma=1}^{m} \boldsymbol{s}^{(\gamma)} \boldsymbol{s}^{(\gamma)*}. \tag{4}$$

As a candidate of the activation function $f(\cdot)$ we will choose a complex function which satisfies the conditions: (i) $f(\cdot)$ is a smooth and bounded function, by analogy with the sigmoidal function of real-valued neural networks, and (ii) each memory vector $s^{(\gamma)}$ becomes an equilibrium point of (3). The condition (ii) is accomplished by choosing a function $f(\cdot)$ which satisfies

$$f(s_j^{(\gamma)}) = s_j^{(\gamma)}, \qquad j = 1, 2, \cdots, N, \quad \gamma = 1, 2, \cdots, m. \tag{5}$$

In regard to the condition (i), we recall the Liouville's theorem, which says that 'if $f(z)$ is analytic at all $z \in \boldsymbol{C}$ and bounded, then $f(z)$ is a constant function'. Since a suitable $f(z)$ should be bounded, it follows from the theorem that if we choose an analytic function for $f(z)$, it is constant, which is clearly not suitable [1]. We choose

$$f(z) = \frac{\eta z}{\eta - 1 + |z|}, \quad \eta - 1 > 0, \ z \in \boldsymbol{C} \tag{6}$$

as the activation function in which η is a tunable parameter satisfying $\eta > 1$. The function (6) is not analytic, but has the continuous partial derivatives $\partial f^R/\partial z^R$, $\partial f^R/\partial z^I$, $\partial f^I/\partial z^R$ and $\partial f^I/\partial z^I$ and is bounded:($|f(z)| < \eta, \forall z \in \boldsymbol{C}$ ($|z| < \infty$)). Note also that, the function (6) satisfies (5).

2.2 Memory Patterns as Equilibrium Sets

In the proposed model of complex-valued associative memories (3), (4) and (6), each memory vector $s^{(\gamma)}$ to be stored becomes an equilibrium point of the network. However $s^{(\gamma)}$ is not an isolated equilibrium point as shown below. For each memory vector $s^{(\gamma)}$, we consider the vector $e^{i\alpha}s^{(\gamma)}$ where α is a real number ($\alpha \in \boldsymbol{R}$). By using the function (6) we can check that $e^{i\alpha}s^{(\gamma)}$ satisfies

$$e^{i\alpha}s_j^{(\gamma)} = f(\sum_{k=1}^{N} w_{jk}e^{i\alpha}s_k^{(\gamma)}), \qquad j = 1, 2, \cdots, N \tag{7}$$

for any real number α. This implies that $e^{i\alpha}s^{(\gamma)}$ is also an equilibrium point of the network. Hence each memory vector $s^{(\gamma)}$ to be stored is not an isolated equilibrium point but a point in the equilibrium set defined by

$$\Phi^{(\gamma)} = \{e^{i\alpha}s^{(\gamma)} : \forall\alpha \in \boldsymbol{R}\} \subset \boldsymbol{C}^N \tag{8}$$

which is a closed curve in $\boldsymbol{C}^N$. Considering this fact, we treat the proposed model as associative memories as follows. For each memory vector $s^{(\gamma)}$, we identify all the points in the equilibrium set $\Phi^{(\gamma)}$ and regard $\Phi^{(\gamma)}$ as a memory pattern. Hence the proposed model is an associative memory which has equilibrium sets $\Phi^{(\gamma)}$, $\gamma = 1, 2, \cdots, m$ as memory patterns.

3 Analysis of Qualitative Behaviors Near Memory Patterns

3.1 Linearized Model Near Memory Patterns

The qualitative behaviors of a nonlinear dynamical system near an equilibrium point can be studied via linearization with respect to that point. We will derive the linearized model of the complex-valued neural network (3), (4) and (6) at a point $e^{i\alpha}\boldsymbol{s}^{(\gamma)}$ in each memory pattern $\Phi^{(\gamma)}$. Note that the activation function $f(z)$ given by (6) is not differentiable with respect to z, but its real and imaginary parts, f^R and f^I, are continuously partially differentiable with respect to z^R and z^I. It is, therefore, possible to derive the linearized model by separating the model (3) into its real and imaginary parts.

Choose a real number α arbitrarily and let $\boldsymbol{q}^{(\gamma)} = e^{i\alpha}\boldsymbol{s}^{(\gamma)}$. We separate the model (3) into its real and imaginary parts and linearize them around each equilibrium point $\boldsymbol{q}^{(\gamma)}$. Let $\Delta x_i(t) := x_i(t) - q_i^{(\gamma)}$ and define $\boldsymbol{y}(t) \in \boldsymbol{R}^{2N}$ by

$$\boldsymbol{y}(t) = (\Delta x_1^R(t), \Delta x_2^R(t), \cdots, \Delta x_N^R(t), \Delta x_1^I(t), \Delta x_2^I(t), \cdots, \Delta x_N^I(t))^T \quad (9)$$

where $(\cdot)^T$ denotes the transpose of $(\cdot)$. The linearized model is given by

$$\frac{d\boldsymbol{y}(t)}{dt} = D(\boldsymbol{q}^{(\gamma)})\boldsymbol{y}(t), \qquad \gamma = 1, 2, \cdots, m \quad (10)$$

where

$$D(\boldsymbol{q}^{(\gamma)}) = \frac{1}{\tau}\left[F(\boldsymbol{q}^{(\gamma)})Y - I_{2N}\right] \in \boldsymbol{R}^{2N\times 2N} \quad (11)$$

$$F(\boldsymbol{q}^{(\gamma)}) = \begin{bmatrix} F_{RR} & F_{RI} \\ F_{IR} & F_{II} \end{bmatrix} \in \boldsymbol{R}^{2N\times 2N}, \ Y = \begin{bmatrix} W^R & -W^I \\ W^I & W^R \end{bmatrix} \in \boldsymbol{R}^{2N\times 2N}$$

$$F_{RR}=\mathrm{diag}(\left.\frac{\partial f^R}{\partial z^R}\right|_{z=q_1^{(\gamma)}}, \cdots, \left.\frac{\partial f^R}{\partial z^R}\right|_{z=q_N^{(\gamma)}}),\ F_{RI}=\mathrm{diag}(\left.\frac{\partial f^R}{\partial z^I}\right|_{z=q_1^{(\gamma)}}, \cdots, \left.\frac{\partial f^R}{\partial z^I}\right|_{z=q_N^{(\gamma)}})$$

$$F_{IR}=\mathrm{diag}(\left.\frac{\partial f^I}{\partial z^R}\right|_{z=q_1^{(\gamma)}}, \cdots, \left.\frac{\partial f^I}{\partial z^R}\right|_{z=q_N^{(\gamma)}}),\ F_{II}=\mathrm{diag}(\left.\frac{\partial f^I}{\partial z^I}\right|_{z=q_1^{(\gamma)}}, \cdots, \left.\frac{\partial f^I}{\partial z^I}\right|_{z=q_N^{(\gamma)}})$$

$$\frac{\partial f^R}{\partial z^R}=\eta\frac{(\eta-1)|z|+{z^I}^2}{|z|(\eta-1+|z|)^2},\ \frac{\partial f^R}{\partial z^I}=\frac{\partial f^I}{\partial z^R}=\eta\frac{-z^R z^I}{|z|(\eta-1+|z|)^2},\ \frac{\partial f^I}{\partial z^I}=\eta\frac{(\eta-1)|z|+{z^R}^2}{|z|(\eta-1+|z|)^2}$$

and I_{2N} is the $2N$ dimensional identity matrix.

3.2 Structure of Solution Orbits Near Memory Patterns

We now analyze the qualitative behavior and structure of the solution orbits near each memory pattern. This can be done by investigating the eigenvalues and eigenvectors of the coefficient matrix $D(\boldsymbol{q}^{(\gamma)})$ of the linearized model (10).

Theorem 1. *For an arbitrary real number β, the matrices $D(\boldsymbol{q}^{(\gamma)})$ and $D(e^{i\beta}\boldsymbol{q}^{(\gamma)})$ are orthogonally similar, that is, there exists an orthogonal matrix T such that*

$$D(e^{i\beta}\boldsymbol{q}^{(\gamma)}) = TD(\boldsymbol{q}^{(\gamma)})T^{-1}. \quad (12)$$

Proof. Choose the orthogonal matrix T as

$$T = T(\beta) = \begin{bmatrix} I_N\cos(\beta) & -I_N\sin(\beta) \\ I_N\sin(\beta) & I_N\cos(\beta) \end{bmatrix} \quad (13)$$

where I_N is the N dimensional unit matrix. It is easy to check that (12) holds for the T.

This theorem implies that all the eigenvalues of the matrix $D(\boldsymbol{q}^{(\gamma)})$ are the same for all point $\boldsymbol{q}^{(\gamma)} = e^{i\alpha}\boldsymbol{s}^{(\gamma)} \in \Phi^{(\gamma)}$ and the corresponding eigenvectors relate with each other by the orthogonal matrix. Furthermore each matrix $D(\boldsymbol{q}^{(\gamma)})$ has the following characteristic properties with respect to its eigenstructure.

Theorem 2. *Let $\lambda_i(D(\boldsymbol{q}^{(\gamma)}))$ be the ith eigenvalue of $D(\boldsymbol{q}^{(\gamma)})$. $\lambda_i(D(\boldsymbol{q}^{(\gamma)}))$, $i = 1, 2, \cdots, 2N$ are all real and their real parts are all less than or equal to zero:*

$$Im\{\lambda_i(D(\boldsymbol{q}^{(\gamma)}))\} = 0 \quad and \quad Re\{\lambda_i(D(\boldsymbol{q}^{(\gamma)}))\} \leq 0, \quad i = 1, 2, \cdots, 2N \tag{14}$$

for all $\gamma = 1, 2, \cdots, m$.

Theorem 3. *The matrix $D(\boldsymbol{q}^{(\gamma)})$ has at least one eigenvalue 0 and at least one eigenvalue $-1/(\tau\eta)$. It also has $2(N-m)$ eigenvalues $-1/\tau$. The corresponding eigenvector to the eigenvalue 0 is*

$$\boldsymbol{p} = ((\{i\boldsymbol{q}^{(\gamma)}\}^R)^T, (\{i\boldsymbol{q}^{(\gamma)}\}^I)^T)^T \tag{15}$$

and that to the eigenvalue $-1/(\tau\eta)$ is

$$\boldsymbol{r} = ((\boldsymbol{q}^{(\gamma)R})^T, (\boldsymbol{q}^{(\gamma)I})^T)^T. \tag{16}$$

The proofs of Theorems 2 and 3 are omitted. Note that $-1/(\tau\eta) < 0$ and $-1/\tau < 0$ because $\tau > 0$ and $\eta > 1$. It is important to note that Theorems 1, 2 and 3 hold for all $\Phi^{(\gamma)}$, $\gamma = 1, 2, \cdots, m$. The proposed associative memory is homogeneous in this sense.

It is known that qualitative behavior of solutions near an equilibrium point of a nonlinear dynamical system is determined by its linearized model at the point if it is hyperbolic (the coefficient matrix of the linearized model has no eigenvalues with zero real part) [7]. From Theorem 3, however, all the equilibrium points $\boldsymbol{q}^{(\gamma)}$, $\gamma = 1, 2, \cdots, m$ are not hyperbolic. We introduce the center manifold theory in order to investigate qualitative behavior of solutions near the equilibrium points $\boldsymbol{q}^{(\gamma)}$, $\gamma = 1, 2, \cdots, m$.

Center Manifold Theory [7] *Let $\boldsymbol{g}$ is a $C^r (r \geq 2)$ vector field on $\boldsymbol{R}^n$. Let $\bar{\boldsymbol{x}}$ be an equilibrium point; $g(\bar{\boldsymbol{x}}) = 0$ and define $A := D\boldsymbol{g}(\bar{\boldsymbol{x}}) = \partial \boldsymbol{g}(\bar{\boldsymbol{x}})/\partial \boldsymbol{x}$. Divide the spectrum of A into three parts, σ^s, σ^u and σ^c with $Re\lambda < 0$ if $\lambda \in \sigma^s$, $Re\lambda > 0$ if $\lambda \in \sigma^u$ and $Re\lambda = 0$ if $\lambda \in \sigma^c$. Let the generalized eigenspaces of σ^s, σ^u and σ^c be E^s, E^u and E^c, respectively. Then there exist C^r stable and unstable invariant manifolds W^s, W^u tangent to E^s, E^u at $\bar{\boldsymbol{x}}$ and a C^{r-1} center manifold W^c tangent to E^c at $\bar{\boldsymbol{x}}$. W^s, W^u and W^c are invariant for the flow of $\boldsymbol{g}$.*

It follows from the center manifold theory that, from Theorems 2, each point $\boldsymbol{q}^{(\gamma)}$ in each memory pattern $\Phi^{(\gamma)}$ has no unstable invariant manifold W^u, and from Theorem 3, it has stable and center manifolds; W^s and W^c. Let c be the number of eigenvalues 0 which $D(\boldsymbol{q}^{(\gamma)})$ has. Then there exist $2N - c$ dimensional stable invariant manifold W^s and c dimensional center manifold W^c near the equilibrium point $\boldsymbol{q}^{(\gamma)}$. If $c = 1$, that is, $D(\boldsymbol{q}^{(\gamma)})$ has only one eigenvalue 0, the following theorems hold.

Theorem 4. *Consider the linearized model (10). If $D(\boldsymbol{q}^{(\gamma)})$ has only one eigenvalue 0, the corresponding eigenspace denoted by $E^{c(\gamma)}$ is given by*

$$E^{c(\gamma)} = \{a\boldsymbol{p} : \forall a (\in \boldsymbol{R})\} \subset \boldsymbol{R}^{2N} \tag{17}$$

and $E^{c(\gamma)}$ is stable, that is, every solution $\boldsymbol{y}(t)$ of the linearized model (10) approaches $E^{c(\gamma)}$ as $t \to \infty$.

Proof. From Theorem 3, $\boldsymbol{p}$ is the eigenvector associated with the eigenvalue 0. The stability of $E^{c(\gamma)}$ comes from the fact that the real parts of all the other eigenvalues are less than 0 (from Theorem 2).

Consider the neural network (3), (4) and (6) to be a $2N$ dimensional real dynamical system by separating it into its real and imaginary parts. Denote the expressions of the equilibrium point $\boldsymbol{q}^{(\gamma)}$ and the memory pattern $\Phi^{(\gamma)}$ in the $2N$ dimensional space by $\boldsymbol{q}^{(\gamma)_{2N}}$ and $\Phi^{(\gamma)_{2N}}$, respectively. They are given by $\boldsymbol{q}^{(\gamma)_{2N}} = ((\boldsymbol{q}^{(\gamma)R})^T, (\boldsymbol{q}^{(\gamma)I})^T)^T$ and

$$\Phi^{(\gamma)_{2N}} = \{T(\beta) \begin{pmatrix} \boldsymbol{q}^{(\gamma)R} \\ \boldsymbol{q}^{(\gamma)I} \end{pmatrix} : \forall \beta \in \boldsymbol{R}\} \subset \boldsymbol{R}^{2N}. \tag{18}$$

Theorem 5. *If $D(\boldsymbol{q}^{(\gamma)})$ has only one eigenvalue 0, the equilibrium point $\boldsymbol{q}^{(\gamma)_{2N}}$ has one dimensional center manifold, denoted by $W^{c(\gamma)}$, and $2N-1$ dimensional stable invariant manifold, denoted by $W^{s(\gamma)}$, and $W^{c(\gamma)} = \Phi^{(\gamma)_{2N}}$.*

Proof. The first half of the theorem is obvious. We will show $W^{c(\gamma)} = \Phi^{(\gamma)_{2N}}$. Note that $\Phi^{(\gamma)_{2N}}$ is invariant for the dynamical system (3), (4) and (6) because it is its equilibrium set. Calculating the gradient of $\Phi^{(\gamma)_{2N}}$ with respect to β at $\beta = 0$, we obtain

$$\left. \frac{d\Phi^{(\gamma)_{2N}}}{d\beta} \right|_{\beta=0} = \begin{pmatrix} -\boldsymbol{q}^{(\gamma)I} \\ \boldsymbol{q}^{(\gamma)R} \end{pmatrix} = \boldsymbol{p}. \tag{19}$$

Then $\Phi^{(\gamma)_{2N}}$ is tangent to $E^{c(\gamma)}$ in (17) at $\beta = 0$. This completes the proof.

Theorem 5 implies that if $D(\boldsymbol{q}^{(\gamma)})$ has only one eigenvalue 0, the memory pattern $\Phi^{(\gamma)_{2N}}$ itself is the center manifold of the equilibrium point $\boldsymbol{q}^{(\gamma)_{2N}} \in \Phi^{(\gamma)_{2N}}$. Note that this theorem holds for all points in the memory pattern $\Phi^{(\gamma)_{2N}}$ if it holds for any one point $\boldsymbol{q}^{(\gamma)_{2N}}$ in $\Phi^{(\gamma)_{2N}}$.

4 Discussion

We now discuss the recalling condition of each memory pattern of the proposed model (3), (4) and (6). We have shown that, if $D(\boldsymbol{q}^{(\gamma)})$ has only one eigenvalue 0 for a point $\boldsymbol{q}^{(\gamma)}$ in a memory pattern $\Phi^{(\gamma)}$, the structure of solution orbits near each point in $\Phi^{(\gamma)_{2N}}$ consists of $2N-1$ dimensional stable invariant manifold and one dimensional center manifold. Furthermore the memory pattern $\Phi^{(\gamma)_{2N}}$ itself is the center manifold which is tangent to $E^{c(\gamma)}$ defined in Theorem 4. It can be concluded, therefore, that '$D(\boldsymbol{q}^{(\gamma)})$ has only one eigenvalue 0' is a condition for each memory pattern $\Phi^{(\gamma)}$ to be correctly recalled, that is, every solution starting in the neighborhood of $\Phi^{(\gamma)}$ approaches $\Phi^{(\gamma)}$ as $t \to \infty$. Then we can determine if each memory pattern is correctly recalled or not by evaluating the eigenvalues of $D(\boldsymbol{q}^{(\gamma)})$. In order to verify this, we have carried out numerical experiments. Note that for each memory pattern $\Phi^{(\gamma)}$ it is enough to evaluate the eigenvalues of $D(\boldsymbol{q}^{(\gamma)})$ at any one point in $\Phi^{(\gamma)}$.

Table 1. Orthogonal vectors $\boldsymbol{s}^{(\gamma)}$, $\gamma = 1, 2, \cdots, 6$.

	$\boldsymbol{s}^{(1)}$	$\boldsymbol{s}^{(2)}$	$\boldsymbol{s}^{(3)}$	$\boldsymbol{s}^{(4)}$	$\boldsymbol{s}^{(5)}$	$\boldsymbol{s}^{(6)}$
$s_1^{(\gamma)}$	$1.000 + i0.000$	$1.000 + i0.000$	$1.000 + i0.000$	$1.000 + i0.000$	$1.000 + i0.000$	$1.000 + i0.000$
$s_2^{(\gamma)}$	$-0.021 - i0.999$	$-0.959 - i0.282$	$-0.294 + i0.955$	$0.997 - i0.071$	$0.086 + i0.996$	$0.961 - i0.274$
$s_3^{(\gamma)}$	$-0.659 - i0.752$	$0.937 + i0.347$	$0.600 - i0.799$	$-0.513 + i0.858$	$-0.886 + i0.461$	$0.059 - i0.998$
$s_4^{(\gamma)}$	$0.127 - i0.991$	$-0.497 + i0.867$	$-0.164 + i0.986$	$-0.014 + i0.999$	$-0.731 - i0.682$	$-0.796 - i0.605$
$s_5^{(\gamma)}$	$-0.844 + i0.535$	$-0.766 - i0.642$	$0.525 + i0.850$	$-0.453 + i0.891$	$0.807 + i0.590$	$0.425 - i0.904$
$s_6^{(\gamma)}$	$-0.541 - i0.840$	$0.822 + i0.569$	$-0.800 - i0.599$	$0.549 + i0.835$	$0.941 - i0.335$	$0.903 - i0.428$
$s_7^{(\gamma)}$	$0.440 + i0.897$	$0.440 + i0.897$	$-0.944 - i0.328$	$-0.983 - i0.182$	$0.347 + i0.937$	$-0.026 - i0.999$
$s_8^{(\gamma)}$	$0.820 - i0.570$	$-0.993 + i0.110$	$-0.313 - i0.949$	$0.998 - i0.050$	$-0.128 + i0.991$	$-0.785 - i0.619$
$s_9^{(\gamma)}$	$0.444 - i0.895$	$0.444 - i0.895$	$0.971 + i0.238$	$-0.825 - i0.564$	$0.500 + i0.865$	$-0.986 - i0.164$

Let the dimension of the neural network be $N = 9$ and let $\tau = 0.1$ and $\eta = 2$ in the nonlinear function (6). Six orthogonal vectors $\boldsymbol{s}^{(1)}, \boldsymbol{s}^{(2)}, \cdots, \boldsymbol{s}^{(6)}$ shown in Table 1 are specified. First we let $m = 5$ and choose the vectors $\boldsymbol{s}^{(1)}, \boldsymbol{s}^{(2)}, \cdots, \boldsymbol{s}^{(5)}$ and construct the network (3), (4) and (6). We evaluate the eigenvalues of $D(\boldsymbol{q}^{(\gamma)})$ for all $\gamma = 1, 2, \cdots, 5$. It turns out that each $D(\boldsymbol{q}^{(\gamma)})$ has one eigenvalue 0, one eigenvalue $-1/(\tau\eta) = -5.0$, eight eigenvalues $-1/\tau = -10.0$ and eight eigenvalues with negative real parts. Since each $D(\boldsymbol{q}^{(\gamma)})$ has only one eigenvalue 0, we can determine that all the 5 memory patterns are correctly recalled. To verify this, the recalling test has been performed. We ran the network (3) starting from the various initial conditions $\boldsymbol{x}(0) := [x_1(0), x_2(0), \cdots, x_N(0)]^T$ in the neighborhood of each memory pattern. Figure 1 shows examples of the obtained recalling process. In Fig. 1, the directional cosine $a(t)$ between $\Phi^{(\gamma)}$ and the solution $\boldsymbol{x}(t) := [x_1(t), x_2(t), \cdots, x_N(t)]^T$ of the network (3) versus time t is plotted for $\gamma = 1$, where the directional cosine $a(t)$ is defined by $a(t) := ||\Phi^{(\gamma)*}\boldsymbol{x}(t)||/(||\Phi^{(\gamma)}||\ ||\boldsymbol{x}(t)||)$. Note that $a(t) \to 1$ implies that the memory pattern is correctly recalled. It is seen from the figure that, if we choose the initial condition $\boldsymbol{x}(0)$ close enough to the memory pattern $\Phi^{(1)}$, the solution $\boldsymbol{x}(t)$ approaches $\Phi^{(1)}$, that is, the memory pattern $\Phi^{(1)}$ is correctly recalled. Similarly, we have checked that all the other memory patterns are correctly recalled.

Next, we increase the number of the vectors to be stored to $m = 6$ and choose $\boldsymbol{s}^{(1)}, \boldsymbol{s}^{(2)}, \cdots, \boldsymbol{s}^{(6)}$ in Table 1, and construct the network (3), (4) and (6). We evaluate eigenvalues of $D(\boldsymbol{q}^{(\gamma)})$ for all $\gamma = 1, 2, \cdots, 6$. It turns out that each $D(\boldsymbol{q}^{(\gamma)})$ has three eigenvalues 0, three eigenvalues $-1/(\tau\eta) = -5.0$, six eigenvalues $-1/\tau = -10.0$ and six eigenvalues with negative real parts. Hence in this case, we cannot determine if each memory pattern is correctly recalled. We have performed the recalling test. Examples of the results for the memory pattern

$\Phi^{(1)}$ are shown in Fig. 2. It is seen from the figure that, no matter how close we choose the initial condition $x(0)$ to the memory pattern $\Phi^{(1)}$, the solution $x(t)$ does not approach the memory pattern $\Phi^{(1)}$, that is, the memory pattern is not recalled. These experimental results support our recalling condition.

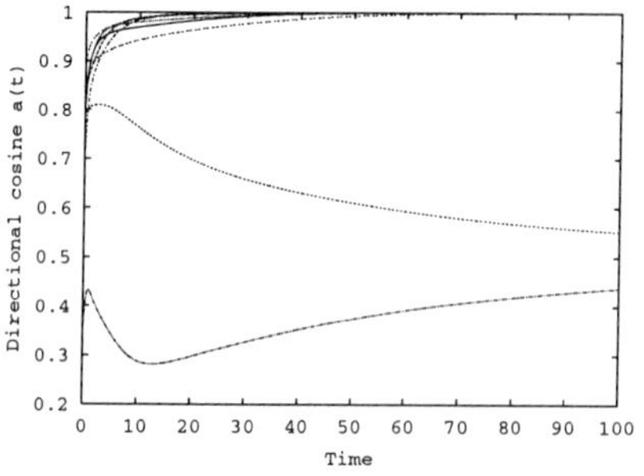

Fig. 1. Time evolution of directional cosine $a(t)$ between $\Phi^{(\gamma)}$ and $x(t) (N = 9, m = 5)$.

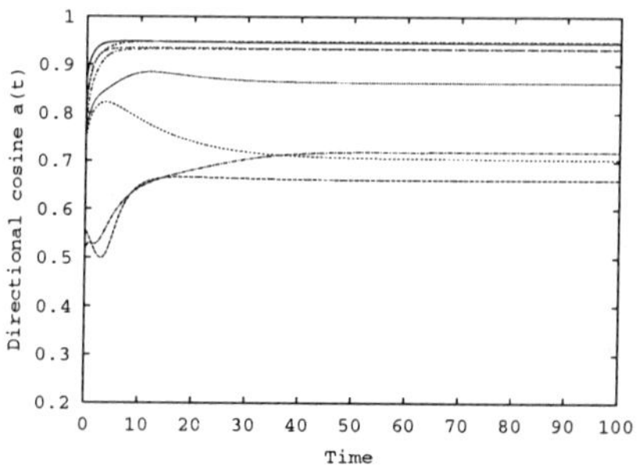

Fig. 2. Time evolution of directional cosine $a(t)$ between $\Phi^{(\gamma)}$ and $x(t) (N = 9, m = 6)$.

5 Conclusion

In this paper we proposed a model of associative memories using complex-valued continuous neural networks and studied its qualitative behaviors theoretically. The proposed model is an extension of the conventional real-valued associative memories of self-correlation type. In the proposed model, each memory pattern is not an isolated equilibrium point of the network but an equilibrium set, which is a closed curve in the complex state space. We investigated the eigenstructures and asymptotic behaviors of solution orbits near each memory pattern by linearizing the proposed model at a point of each memory pattern. A recalling condition of each memory pattern was also discussed.

References

1. G. M. Georgiou and C. Koutsougeras, "Complex domain backpropagation," IEEE Trans. on Circuits and Systems, vol. 39, no. 5, pp. 330–334, May 1992.
2. N. Benvenuto and F. Piazza, "On the complex backpropagation algorithm" IEEE Trans. on Signal Processing, vol. 40, no. 4, pp. 967–969, Apr. 1992.
3. I. Nemoto and T. Kono, "Complex-valued neural network," The Trans. of IEICE, vol. J74-D-II, no. 9, pp. 1282-1288, 1991 (in Japanese).
4. A.J.Noest, "Phaser neural network," Neural Informaion Processing Systems, D.Z.Anderson,ed., pp. 584–591, AIP, New York, 1988.
5. A. Hirose, "Dynamics of fully complex-valued neural networks," Electronics Letters, vol.28, no.16, pp.1492-1494, 1992.
6. S.Jankowski, A.Lozowski, and J.M.Zurada, "Complex-valued multistate neural associative memory," IEEE Trans. Neural Networks, vol. 7, no. 6, pp. 1491–1496, Nov. 1996.
7. J. Guckenheimer and P. Holmes, Nonlinear Oscillations, Dynamical Systems, and Bifurcations of Vector Fields, pp. 16–22, Springer-Verlag, New York, 1986.

Self Organized Partitioning of Chaotic Attractors for Control

Nils Goerke, Florian Kintzler, and Rolf Eckmiller

Department of Computer Science VI, University of Bonn
Römerstraße 164, D-53117 Bonn, F.R. Germany
{goerke,kintzler,eckmiller}@nero.uni-bonn.de

Abstract. We propose a method to use self organizing neural networks to extract information out of nonlinear dynamic systems for control. Nonlinear strange attractors are educed by these systems or the attractors can be reconstructed. These attractors are partitioned by a newly developed self organizing neural network. Thus the stream of system states is transformed into a stream of symbols, which can now serve as basis for further investigation or control. We are convinced, that controlling and understanding such nonlinear or chaotic systems is easier, when using the information within the stream of extracted symbols.

1 Introduction

Nonlinear systems can be found in a wide variety of applications. Nonlinearity seems to be a favorite of "mother nature"; therefore it is desirable to get control over nonlinear dynamical systems, although modeling isn't available or appropriate. Controlling these systems mostly requires profound knowledge about the governing dynamics [1, 2, 6]. Since nonlinear dynamic systems tend to educe chaotic attractors that determine the systems development most linear methods fail directly while attempting to gain control over the system. As presented in this paper, these attractors are the key-components in a new control-model, which utilizes almost the entire complexity of the system and is more convenient than linear methods.

The complete dynamics of the attractor is represented within the temporal evolution of the state variables [1, 7, 8]. A time series of these system variables contains the characteristics of the nonlinear, sometimes even chaotic system. [1, 8]. It has been shown, that the reconstruction of an attractor by temporal delayed sampling of the state variables is possible [7]. Such attractors are a reconstruction of the original attractor, following the same dynamics. Thus it is an alternative representation of the dynamics of the system, suitable for control.

To take further advantage of the information carried within the attractor, respectively the time series, we transform this information into a stream of symbols.

2 Method

Our purpose is to control nonlinear dynamical systems. Transforming information from a nonlinear, sometimes even chaotic attractor, into a stream of symbols is the method we propose.

G. Dorffner, H. Bischof, and K. Hornik (Eds.): ICANN 2001, LNCS 2130, pp. 851–856, 2001.

In traditional control theory often a priori knowledge about the application or the process is used, e.g. a threshold value. We want our model of control to be self organized and purely data supported. So we decided to use self organizing neural networks to do the job of knowledge extraction.

We tested several different self organizing neural network paradigms to discriminate regions of chaos of attractors. We exerted the Self Organizing Map (SOM) as proposed by T. Kohonen [4], the Neural Gas [5] and the Growing SOM (G-SOM) as proposed by B. Fritzke [3]. In some cases, clustering the Roessler or the Lorenz attractor, the SOM twisted and the results weren't usable. It has been often observed that twisting depends on the SOM size [4], and small SOMs ($\leq 5x5$) normally don't twist at all. The Neural Gas model on the other hand, has the disadvantage that it does not realize a topology preserving mapping into a low-dimensional space [3], which is desirable for easy handling of a system of higher dimensionality. To overcome these described disadvantages, we developed an extention of the SOM, a Multiple Self Organizing Map (M-SOM).

The Multi-SOM (M-SOM) is a flock of individual Partner-SOMs having identical or different topology, size and dimension. To avoid the unwanted effect of twisting and to increase adaptation speed, each of these partner SOMs should be kept small in size. Learning with a Multi-SOM is done using the same methods as for classical SOMs, to benefit from the complete variety of published modifications and enhancements to the basic SOM algorithm.

Two basic M-SOM learning principles arise directly from the Multi-SOM architecture: **a)** pure un-supervised M-SOM and **b)** self-organized-supervised M-SOM.

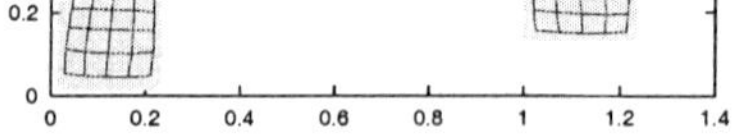

Fig. 1. M-SOM containing six 5×5 partner-SOMs learning a segmented distribution in 2D.

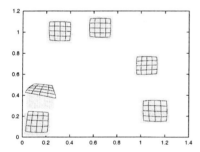

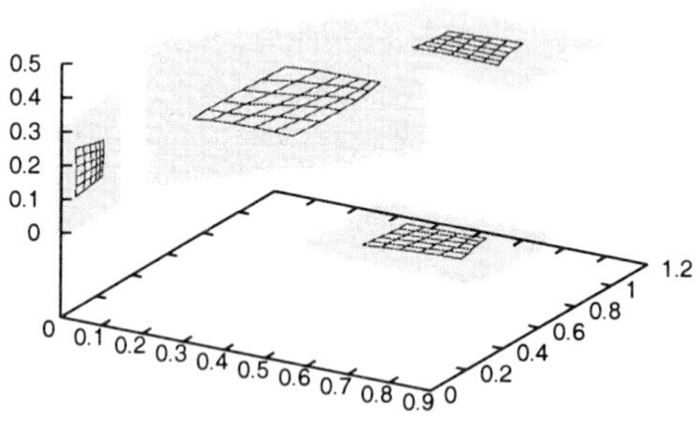

Fig. 2. M-SOM containing four $6 \times 6 \times 1$ partner-SOMs learning a segmented distribution in 3D.

For pure un-supervised M-SOM adaptation, only the partner SOM that contains the winning neuron (winning SOM) is changed, following the approved methods of SOM-adaptation. Thus only the reference vectors of the winning SOM are modified to form a topographic mapping of the input space in the direct vicinity of the actual input vector. The other partner SOMs within the M-SOM remain unchanged, and can thus "concentrate" their topological mapping to other regions of the input space. Since only a small fraction of the M-SOM has to be modified during each training step, the adaptation speed is high in terms of elapsed time, but not necessarily in terms of number of iterations.

In the case of self-organized-supervised M-SOM, regarding each individual partner SOMs as a representative of an individual class of data, the M-SOM adaptation can be expanded to self-organized-supervised learning. Once the winning SOM has been determined and adapted, the other partner SOMs are trained following the idea of Learning Vector Quantization. LVQ is designed to be a supervised method for classification which makes benefit of the class-membership of a given input value. Within the self-organized-supervised M-SOM learning the self organized information of being a member of the winning SOM or not is used instead. Of course the complete variety of LVQ methods and enhancements can be applied to the M-SOM.

The M-SOM topology has two extraordinary cases. If the number of SOMs shrinks to M=1, the M-SOM equals a classical SOM. On the other hand, if the number of neurons in every Partner-SOM shrinks to 1, the M-SOM becomes a Neural Gas.

In our task of classifying regions of chaos in the selected attractors, we regarded every net as one class of it's own. Classifying a stream of derived values then only means noticing which net contains the winner-neuron. This leads to a fast computation and good classification-results, as demonstrated below.

Progress work on the derived stream of symbols can lead to further information extraction. Principles from signal processing can be used to characterize the development within the stream of symbols and thus the dynamic system. Since the underlying dynamics is deterministic and concentrated to reside on the attractor, it is likely to find a grammar on this stream of extracted symbols.

3 Simulation

To evaluate and demonstrate the capabilities of our approach, we choose two well investigated nonlinear chaotic systems, the Roessler and the Lorenz-attractor [8]. These attractors do have the advantage of being embedded into the 3-dimensional space, so that visualization is possible. The developed method of information-extraction is not limited to three dimensions. It can of course also be adopted for control of nonlinear dynamic systems of a higher dimensionality.

So we calculated data from the chaotic attractors, which - faced as a time-series - represent the evolution of state variables of a fictive nonlinear dynamic system. Starting the calculation with arbitrary parameters thus equals starting this fictive system at an arbitrary starting-configuration. These values were used to train self organizing neural networks, generating a partitioning scheme. In the case of the M-SOM (described above), we tested variuos sizes and quantities of SOMs.

The time series, we calculated from the differential equation system, was then translated into a stream of symbols, using the developed partitioning scheme.

Further work on the M-SOM model will lead to more sophisticated results and finer shaped self organizing neural representations.

Additionally we applied the developed method on an attractor, reconstructed by time-delayed sampling. Therefore we reconstructed the Roessler attractor out of a time series (≈ 10000 values), which contained the development of one component of the systems state, by time delayed sampling [8].

4 Results

Figure 3 and 5 show the chosen attractors clustered by M-SOM. These M-SOMs generate the streams of symbols as visualized in Figure 4 and 6. Regular structures in this visualizations are obvious. This validated our assumption, that transforming of information into a stream of symbols via self organizing neural networks represent the underlying dynamic behavior of a nonlinear and even chaotic system.

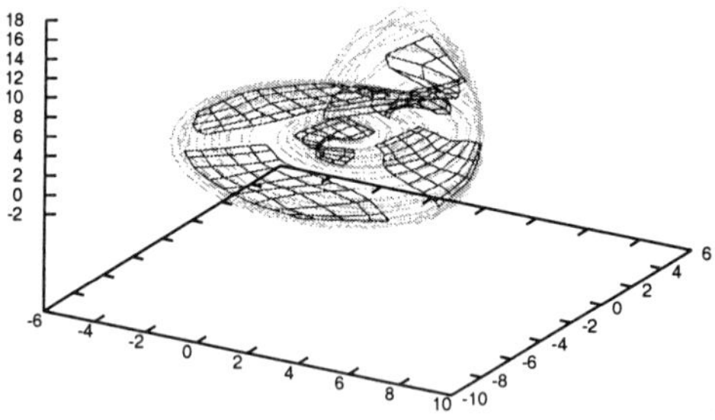

Fig. 3. The Roessler-attractor clustered by a 5-5×10 M-SOM.

...BBBBBBBBBAADDDDDDDDDDDDCCCCCCCC
BBAAAAAADDDDDDDDDDCCCCCCCCCBBBBBBB
AAAADDDDDDDDDDDDCCCCCCCCCBBAAAAAEE
EEEEBBBBBBBBAAADDDDDDDDDDDDCCCCCCC
CCBBBBAAAAAEEEEEEEEEEEEEEEEEEEEEEE
EEEEEEEEEEEEEEEEEEEEEEEEEBBBBBBBA
AAADDDDDDDDDDCCCCCCCCCBBBBBBAAAAAD
DDDDDDDDDDDCCCCCCCCCBBAAAAAEEEEEEE
BBBBBBBBAAAADDDDDDDDDDDCCCCCCCCCBB
BBAAAAAADDDDDDDDDDDCCCCCCCCCBBBBAA
AAAAEEDDDDDDDDCCCCCCCCCBBBBBBBAAAA
DDDDDDDDDDDDCCCCCCCCCBBBAAAAAEE...

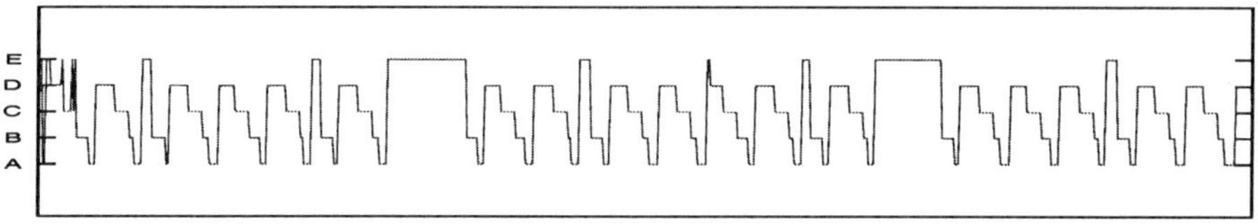

Fig. 4. Stream of symbols derived by clustering the Roessler-attractor of figure 3.

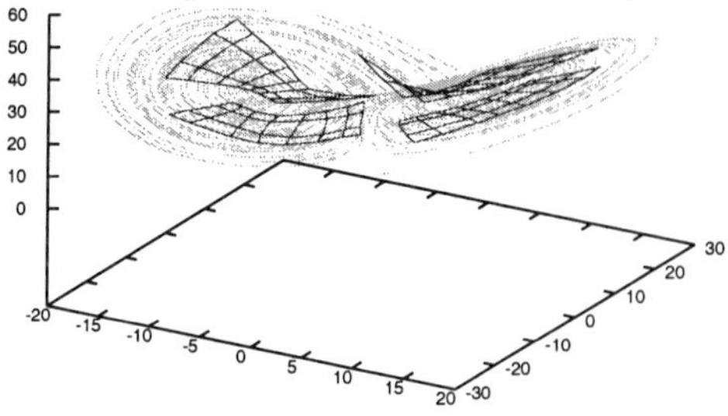

Fig. 5. The Lorenz-attractor clustered by a 4-4×8 M-SOM.

...CCCCCCCCCCCCCCCCCCCCCCCCCCCCCCC
CCCDDDDDDDDDDDDBBBBBBBBBBBBBBBBBBB
BBBBBBBBBBBBBBBBBBBBAAAAAAAAAAAAAAA
AAAAAAAAAAAAAAAAAAAAAAAABBBBBBBBBB
BBBBBBBBBBBBBDDDDAAAAAAAAAAAAAAAAA
ACCCCAAACCCCCCCCCCCCCCCCCCCCCCCCCC
CCCCCCCCCCCCCCDDDDDDDDDDDDDBBBBBBBB
BBBBBBBBBBBBBBBBBBAAAAAAAAAAAAAAAAB
BBBBBBBBBBBBBBBBBBBBBBBBBBBBBBBBBB
BBBBBBBBBAAAAAAAAAAAAAAAAAAAAAABBBBB
BBBBBBBBBBBBBBBBBBBBBBBBBBBBBBBB...

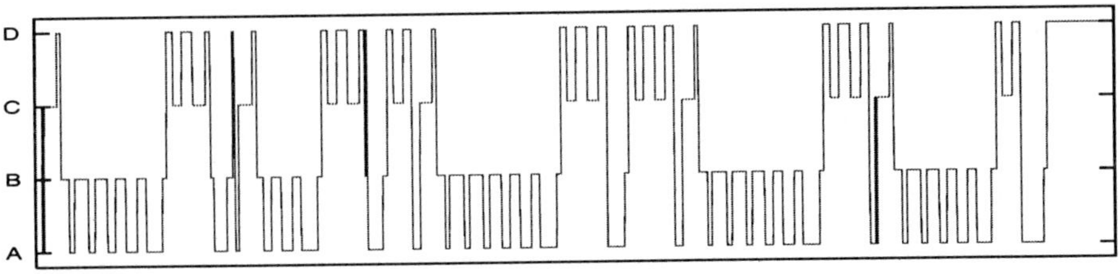

Fig. 6. Stream of symbols derived by clustering the Lorenz-attractor of figure 5

For testing our approach we used the x-component of the Roessler-system and reconstructed attractor by time delayed sampling. Figure 7 shows the result of the reconstruction together with the trained 5-5×10 M-SOM.

In figure 7 we show the reconstructed Roessler-Attractor. The coordinates of the reconstructed attractor are derived as following: $a_1 = x(t-0\cdot\tau)$, $a_2 = x(t-1\cdot\tau)$, $a_3 = x(t-2\cdot\tau)$. We choose the x-trajectory as the base time-series $x(t)$ and time-steps $\tau = 10$. The visualization of the derived stream in figure 8 is very similar to the one of the original attractor shown above.

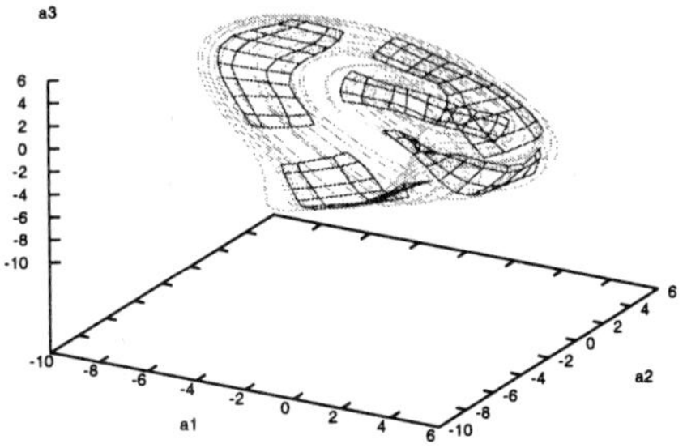

Fig. 7: Reconstructed Roessler-attractor clustered by a 5-5×10 M-SOM.

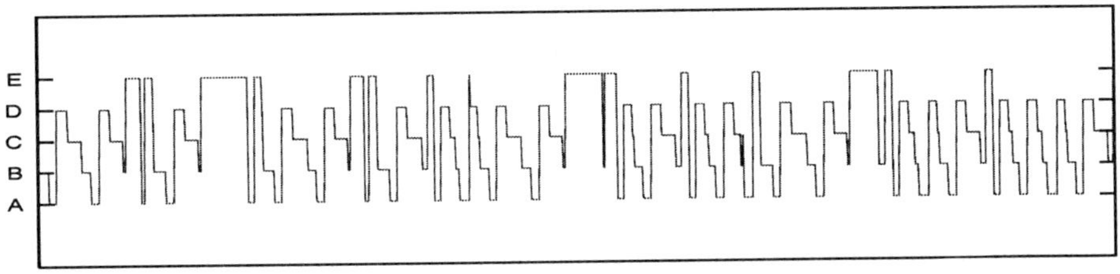

Fig. 8. Stream of symbols derived by clustering the reconstructed Roessler-attractor of figure 7. The stream of symbols reflect the dynamics of the attractor.

5 Conclusion

Within this paper we have presented a method of partitioning chaotic attractors of nonlinear dynamical systems into a set of classes. Each class is thereby represented by an individual symbol, accessible for further classification and as a basis for controlling the dynamical system. The dynamic evolution of the system along the attractor is reflected within the stream of symbols generated by the partitioning network. The symbols can now be used for further analysis and control of the dynamical system.

The process of partitioning was performed using different neural self organizing paradigms (SOMs, Neural-Gas, Growing SOMs). Applying Multi-SOMs for assigning classes to different regions of the attractor, showed to be a powerful and efficient method to guarantee a pre-definable number of classes.

Several nonlinear dynamical attractors that are known to be chaotic (Lorenz- and Roessler attractor) have been partitioned and transformed into streams of symbols. The stream of extracted symbols can now serve for various further tasks of analysis and control:

- classification of different forms or quality of chaos - qualitative characterization of the attractor into different types,
- quantitative calculation of dynamic parameters (fractal dimension, ljapunov exponents, ...),
- input basis for a controller to obtain control over the dynamical system.

The different partitions (classes) the attractor has been subdivided into, are solely formed by the probability distribution of the un-disturbed attractor itself. It is unlikely that such type of segmentation will be optimal suited to control everytype of the dynamical system. Therefore the self organization mechanism of the M-SOM will be extended for taking the controlability of the system into account. Currently we are investigating several approaches of incorporating the effectiveness of applied control into the adaptation process of the neural self organizing maps.

References

1. Bergé, P.; Pomeau, Y.; Vidal, Ch.: Order within Chaos, Towards a deterministic approach to turbulence, John Wiley & Sons, New York 1984.
2. Chen, G.; Dong, X.: From Chaos to Order, Methodologies, Perspectives and Applications, World Scientific Series on Nonlinear Science, Series A, Vol.24., World Scientific Publ., 1998.
3. Fritzke, Bernd: Wachsende Zellstrukturen - ein selbstorganisierendes neuronales Netzwerkmodell, PhD Thesis, Erlangen (1992).
4. Kohonen, T.: Self organizing maps, Springer, Berlin Heidelberg 1995.
5. Martinez, Thomas M.; Berkovich, Stanislav G.; Schulten, Klaus J.: "Neural Gas" Network for Vector Quantization and its Application to Time-Series Prediction, IEEE 1993.
6. Nicolis, G.; Prigogine, I.: Die Erforschung des Komplexen, Pieper, München 1987.
7. Takens, F.: Detecting strange attractors in turbulence. In: Rand,D.; Young, L.S. (Eds.): Neural Information Processing Systems 7, volume 898 of Lecture Notes in Mathematics, pp. 366-381. Springer Verlag, Berlin, 1981.
8. Thomson, J.; Stewart, H.: Nonlinear Dynamics and Chaos, Geometrical Methods for Engeneers and Scientists. John Wiley & Sons, Chinchester, 1986.

A Generalisable Measure of Self-Organisation and Emergence

W. Andy Wright, Robert E. Smith, Martin Danek, and Pillip Greenway

[1] British Aerospace Systems ATC Sowerby, Bristol, UK
{andy.wright,phil.greenway}@src.bae.co.uk
[2] The Intelligent Computing Systems Centre, Bristol, UK
{robert.smith,martin.danek}@uwe.ac.uk

Abstract. In adaptive systems that involve large numbers of entities, emergent, global behaviours that arise from localised interactions are a critical concept. Understanding and shaping emergence may be essential to such systems' success. To aid in this understanding, this paper introduces a measure gleaned from non-linear systems theory. The paper discusses how this measure can be used in reinforcing self organising behaviours in adaptive systems. Further, it is shown that the measure can be successfully employed as feedback to a system employing evolutionary computation (EC) and using this to design in desired self organising behaviours in an approximation to a biological plausible collective system.

1 Introduction

As the complexity of computing systems has increased, the presence of undesirable behaviours that are a result of unforseen non-linear interactions with the different components of these systems has grown. These behaviours can have catastrophic consequences, as in the sudden collapse of the eastern seaboard telephone system in the USA [4]. It has also been argued that similar behaviour is also apparent in other complex systems, such as: crowds of people, traffic systems (i.e., the stop start behaviour in motorway traffic jams). Biological systems also appear to demonstrate such behaviour, such as the sudden dispersion a schools of fish in the presence of a predator [3]. In this biological case, what may be termed as an emergent property of the system appears to provide a system-level behaviour that is advantageous to the group as a whole. Interest in the behaviour of social animals, coupled with the possibility of utilising these behaviours to design complex systems (including communications, robotic [6], and artificial intelligence systems), has prompted the study of the self-organisation in collective biological systems, such as flocks and herds of animals.

Although a substantial element of this research has successfully concentrated on building mathematical [12, 9] and rule-based [7] models that simulate the dynamics of these systems, the control of these models is difficult. To utilise the inherent non-linear interactions that are present in many complex systems, one would ideally like to be able to produce a desired behaviour by simply choosing the appropriate parameters values for a system. This has not been possible. An

G. Dorffner, H. Bischof, and K. Hornik (Eds.): ICANN 2001, LNCS 2130, pp. 857–864, 2001.

alternative is to adjust the parameters (via some optimisation routine), such that the system emulates a desired behaviour. However, to do this requires some qualitative measure of the behaviours of the system (as has recently been pointed out by Zaera et al [14]). In the past, Lyapunov exponents have been used to measure the behaviour of flocking systems [11]. Unfortunately, this approach is limited, since the Lyapunov exponent only measures the stability of a system's dynamics, and does not consider the level of coherence (or structure) in the behaviours exhibited by a system. This paper presents an alternative measure based on Takens' method of delays [10] and the embedding approach of Broomhead and King [1], to derive an entropic measure that describes the complexity of a system. This measure is then used as direct feedback to evolutionary computation (EC) systems that evolve swarm-like behaviours in a simple, Hamiltonian-based flocking model. The implications of this work for other complex systems are discussed in future work.

2 Continuous Time Model of Flocking

This section briefly describes a simple continuous time model of a flock of "particles" that is used to illustrate the utility of complexity measure derived in this paper. In this model it is assumed the a self organising set of N particles can be described by a number of local state-space variables. For the simple flocking model these are the set of positions and velocities for all the flock members. The dynamics of these agents are controlled by local interactions between each agent and its neighbours. There interactions, in turn, indirectly control the spatial and velocity coherence of the system.

The spatial coherence between the particles is modelled using the Hamiltonian

$$\mathcal{H} = 1/2m \sum_i^N |\dot{\boldsymbol{x}}_i|^2 - V_0 \sum_i^N \sum_{j \neq i}^N V(r_{ij}), \tag{1}$$

where m is the particle mass, $\boldsymbol{x}_i$ is the positional vector for the particle i, r_{ij} is the Euclidian distance between the ith and j^{th} particle, and $V(r_{ij})$ is a hydrogen potential with depth V_0. The velocity coherence is modelled using a Stokes viscous term

$$\mathcal{R} = 1/2\mu \sum_i^N \sum_{j:r_{ij} \leq r_1}^N |\dot{\boldsymbol{x}}_i - \dot{\boldsymbol{x}}_j|^2, \tag{2}$$

where μ is the viscosity coefficient and r_1 is a effective range of this term. Solving for the equations of motion gives:

$$\ddot{x}_i = \gamma \sum_{j \neq i} \frac{\partial V(r_{ij})}{\partial x_i} + \delta \sum_i^N \sum_{j:r_{ij} \leq r_1}^N (\dot{x}_i - \dot{x}_j), \tag{3}$$

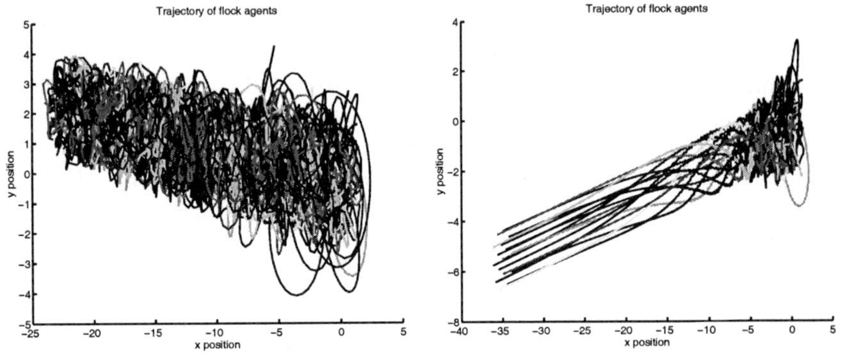

(a) Swarm (b) Crystal

Fig. 1. Trajectories of 17 particles illustrates two different behaviours. Note that the particles start at the right of the plot, and proceed to the left.

where the couplings between the spatial and damping terms that control the dynamics have been rewritten as:

$$\gamma = \frac{V_0}{r_0^2 m}, \tag{4}$$

and

$$\delta = \frac{\mu}{m}. \tag{5}$$

The parameter r_0 is derived from the hydrogen potential and controls the optimal spacing of the particles. Solving this continuous time model gives rise to a number of distinctly different behaviours as the two control parameters γ and δ are varied. Two extreme behaviours (an uncoordinated "swarm" like behaviour and a coordinated "crystal" like behaviour) are shown in Figure 1

3 Entropic Measure

The measure derived in this paper is based on the idea that, although a complex system may have many degrees of freedom, the self-organised or emergent behaviour of that system can be modelled using noticeably fewer degrees of freedom. This is illustrated in the "crystalline" behaviour of the flock model shown in Figure 1. Here the particles (which have in total 17×4 degrees of freedom) eventually move as a single body with 4 degrees of freedom. In non-linear dynamical systems this self-organising behaviour is likened to the system moving on some attractor that has a much smaller dimension m than that of the whole system.

Takens [10] has shown that it is possible to determine the dimension of the attractor that generates the time-series from successive observations of the time-series. One straight forward way to estimate the attractor dimension is the approach of Broomhead and King [1], where a bound on the attractor dimension is estimated from the singular values σ of the embedding matrix. Unlike some other bounding methods, this approach is robust to additive noise. It has been shown that an additive noise process significantly influences only the smaller singular values, allowing noise and signal to be separated. For the 2D flocking system described above, the embedding matrix is constructed from M time delay samples of the time-series, generated by tracking each particle over time, to give:

$$Z = \begin{pmatrix} x_1^1, \; y_1^1, \; \dot{x}_1^1, \; \dot{y}_1^1, \; \ldots, \; x_1^N, \; y_1^N, \; \dot{x}_1^N, \; \dot{y}_1^N \\ \vdots \\ x_M^1, \; y_M^1, \; \dot{x}_M^1, \; \dot{y}_M^1, \; \ldots, \; x_M^N, \; y_M^N, \; \dot{x}_M^N, \; \dot{y}_M^N . \end{pmatrix} \tag{6}$$

Here x_j^i is the jth time sample of the x coordinate of the ith agent, and N is the total number of agents in the flock. To ensure that any correlations in the data are removed, the mean position and velocity are subtracted from the positional and velocity data during the analysis, giving a total of $4 \times (N-1)$ degrees of freedom. These values can then be used to construct an entropy

$$S = \sum_i^M \sigma_i' \log_2 \sigma_i', \tag{7}$$

where σ' is the normalised singular value [8]. It is stressed that this is not the only measure of complexity that can be used. A specific advantage of the entropy measure is that it can be seen as a log count of the number of effective states (i.e., the number of degrees of freedom) in the system. The complexity of the system may therefore be expressed as

$$\Omega = 2^{\sigma'} . \tag{8}$$

An alternative may be to use the *free energy* (TS). However, in this model there is no clear analogy to the system's temperature (T), and thus, this measure would be difficult to interpret. Other recent work has suggested determining the *Fisher information* [5] from the from the singular spectra. The utility of this approach is left for further study.

Using the Ω measure, it is possible to show that the flocking system described above radically changes behaviours for smooth changes in parameter values. Figure 2 shows a plot of Ω versus two dimensionless parameters: $\alpha = \frac{\mu^2 r_0^2}{m V_0}$ (which measures the relative strength of the spatial potential and viscous term), and $\beta = r_1/r_0$, (which is the ratio of the effective range of these terms). From this it can be seen that there is an abrupt change (possibly a phase transition) between the "swarm" (the upper surface with large Ω) and the "crystal" behaviours (the lower surface with small Ω).

To summarise and clarify, we associate a system's ability to exhibit emergent behaviour with sudden transitions in the Ω measure, relative to smooth changes in system parameters. Emergent behaviours are those whose parameter values

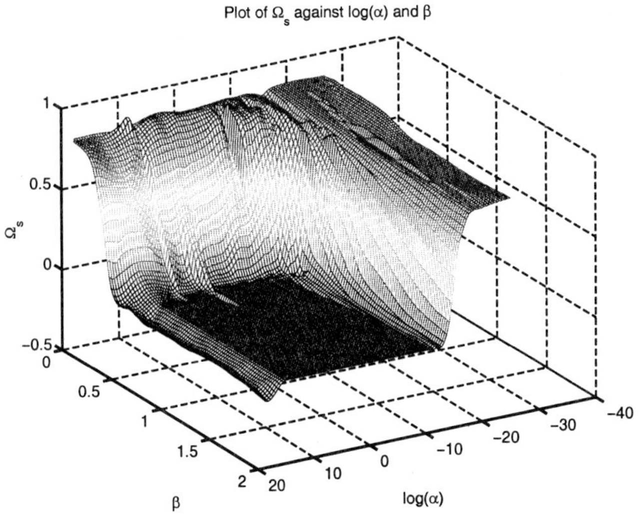

Fig. 2. Behaviour of our simple, illustrative system versus system parameters. This is a plot of Ω scaled by the total number of degrees of freedom versus two dimensionless parameters.

are associated with these abrupt transitions in Ω. Further, note that this measure is generalisable to any complex system comprised of many elements, in that it only relies on the selection of a set of local state variables that can be to associated with each element. Its use of the method of delays allows consideration of second or higher order effects, without the need to explicitly include second or higher-order derivatives as state variables for each element. Therefore, the measure is likely to have broad utility in the design and analysis of many types of complex systems, including communications, robotic, and artificial intelligence systems.

4 Using the Measure as Feedback for Adaptation

General-purpose adaptive and collective systems (e.g. collective mobile robotics, adaptive rule-based systems, etc.) each involve the ability to exploit the complex (non-linear) interactions that exist between different system elements. In a stable environment, direct programming (explicitly allowing for the different interactions between all the agents) may be possible, albeit complex. However, direct programming is much more difficult if the system adapts to a changing environment. In an adaptive setting, it is necessary to include some form of feedback, or use some type of online learning to promote desirable behaviour without direct programming.

In this section, we show how the Ω measure can be used as feedback in an adaptive system to encourage self-organisation. We do this by using the measure as feedback to a genetic algorithm (GA) [2]. Initial experiments demonstrate that it is possible to achieve desired behaviours in the simple flocking model, and that these can be seen as mixtures of the two main behaviours (crystal and swarm). Figure 3 shows such a hybrid behaviour[1] that was generated using a an intermediate value of the scaled omega ($\Omega = 0.4$). Here the particles oscillate around each other. Other, more complex behaviours have been obtained using heterogenous systems of particles [13].

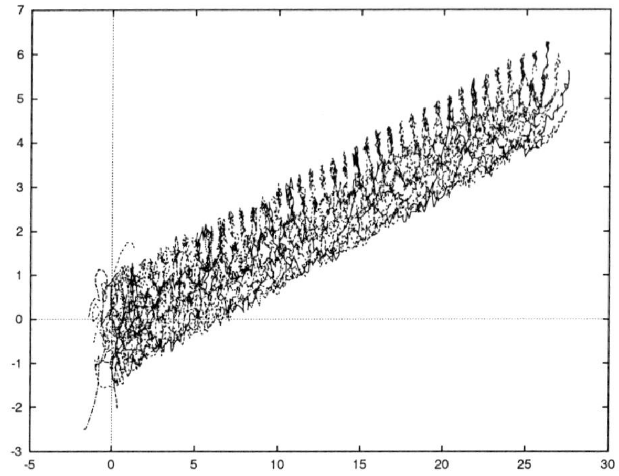

Fig. 3. Trajectories of 17 particles illustrating a hybrid behaviour.

More recent investigations have looked at the use of this measure for the online adaptation of a collections of interacting particles in a changing environment. In these experiments, the system of particles is placed in an environment with periodic boundary conditions (a toroid). This environment contains a small region where the particles are able to pick up some "food". Once an particle has this food, it local attraction to other particles is increased by a proportion v. The amount of food that an individual particle holds decays exponentially with time, with a half-life λ. When one particle interacts with another by coming within radius r_0 of it, a proportion p of the food available to the particle is passed to the other. The aim of this experiment is to determine what type of behaviour maximises the propagation of food through the population of particles. Examination of this system has shown that the maximum propagation of the food throughout the population corresponds to $\Omega \sim 2.5$ for $\lambda = 0.5$ (see Figure 4a). Then, when this "target" Ω is used as a basis for fitness, a GA optimisation procedure finds other results for differing λ, v and p. Generally these results show a behaviour where the particles swarm initially to find the food, and then orbit the food source, with sufficient oscillatory behaviour to ensure the propagation of the

[1] It is difficult to see the full dynamic behaviour from this still figure. However, animations of the dynamics will be presented with the full paper.

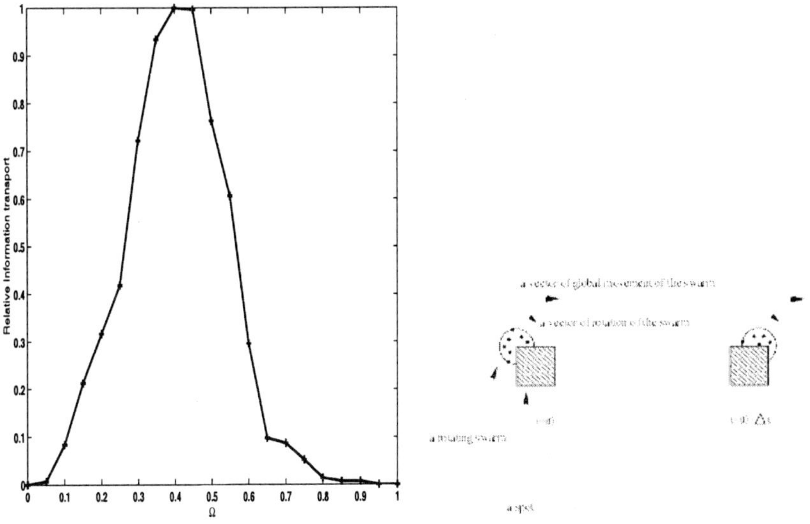

(a) (b)

Fig. 4. (a) graph showing the Ω plotted against the transport of food in the population. (b) a schematic of the swarm behaviour around the food source.

food to particles outside the food region[2] (see Figure 4b). Furthermore, as the proportion of food in the population drops, the strategy adopted by the particles is to increase the size of the swarm, thus increasing the swarm's cross section relative to the food source. Current work is using this measure to adjust a heterogenous group of particles, and looking at how adapting the behaviour of the individual particles can maximise the transport of food through the population.

5 Final Comments and Future Directions

We have introduced a means of quantifying the notion of self-organisation in multi-element systems, and illustrated how that measure can be used to obtain a desired emergent behaviour, via feedback to a GA. The measure, which is based on ideas borrowed from dynamical systems analysis, can be calculated directly from data pertaining to the individual elements in a system. Consequently, although the utility of the measure has only been demonstrated on a model with very limited biological plausibility, its hope that it can be applied in a much more general fashion. We believe this measure (and ones like it) could be used

[2] Again it is difficult to show the dynamic behaviour of this system. However, an animation of this result can be obtained at http://dual.felk.cvut.cz/ danek/research/mas.html

as general-purpose feedback to evaluate and encourage desirable behaviours in a variety of complex systems, including artificial neural systems, evolutionary computation systems, and reinforcement learning systems. These applications are a focus of the authors' current investigations.

References

1. D.S. Broomhead and G.P. King. Extracting quantitative dynamics from experimental data. *Physica D*, 20:217–236, 1998.
2. D. E. Goldberg. *Genetic Algorithms in Search, Optimization, and Machine Learning*. Addison-Wesley, 1989.
3. S.A. Gueron, S. Levin, and D.I. Rubenstein. The dynamics of herds: from individuals to aggregations. *Journal of Theoretical Biology*, 182:85–98, 1996.
4. D. R. Kuhn. Sources of failure in the public switched telephone network. *Computer*, 3(4):31–36, 1997.
5. D. Lowe. Private communication, 2001.
6. J. M. Maja. Behavior-based robotics as a tool for synthesis of artificial behavior and analysis of natural behavior. *Trends in Cognitive Science*, 2(3):82–87, 1998.
7. C. W. Reynolds. Flocks, herds, and schools: A distributed behavioral model. In *ACM SIGGRAPH '87 Conference Proceedings*, volume 21, pages 25–34. ACM Press, 1987.
8. S.J. Roberts, W. Penny, and I. Rezek. Temporal and spatial complexity measures for eeg-based brain-computer interfacing. *Medical & Biological Engineering & Computing.*, 37(1):93–99, 1998.
9. N. Shimoyama, K. Sugawara, T. Mizuguchi, Y. Hayakawa, and M. Sano. Collective motion in systems of mobile elements. *Physical Review Leters*, 1996.
10. F. Takens. Detecting strange attractors in turbulence. In D.A. Rand and L.S. Yong, editors, *Lecture notes in mathematics*, page 366. Springer, Berlin, 1981.
11. J. Toner and Y. Tu. Flocks, herds, and schools: A quantitative theory of flocking. *Phys Rev E*, 58(4), 1998.
12. T. Vicsek, A. Czirol, E. Ben-Jacob, I. Cohen, and O. Shochet. Novel type of phase transition in system of self-driven particles. *Phy Rev E*, 75(6):1226–1229, 1995.
13. W. A. Wright, R. E. Smith, M. A. Danek, and P. Greenway. A measure of emergence in an adapting, multi-agent context. In *Supplmental Proceedings of the Sixth Intenational Conference on the Simulation of Adaptive Behaviour*, pages 20–27, Honolulu, 2000. ISAB.
14. N. Zaera, D. Cliff, and J. Bruten. Not evolving collective behaviours in synthetic fish. In *Proceedings of the Fourth International Conference on the Simulation of Adaptive Behaviour*, pages 635–644. MIT Press, 1996.

Market-Based Reinforcement Learning in Partially Observable Worlds

Ivo Kwee, Marcus Hutter, and Jürgen Schmidhuber

IDSIA, Manno CH-6928, Switzerland
{ivo,marcus,juergen}@idsia.ch*

Abstract. Unlike traditional reinforcement learning (RL), market-based RL is in principle applicable to worlds described by partially observable Markov Decision Processes (POMDPs), where an agent needs to learn short-term memories of relevant previous events in order to execute optimal actions. Most previous work, however, has focused on reactive settings (MDPs) instead of POMDPs. Here we reimplement a recent approach to market-based RL and for the first time evaluate it in a toy POMDP setting.

1 Introduction

One major reason for the importance of methods that learn sequential, event-memorizing algorithms is this: sensory information is usually insufficient to infer the environment's state (*perceptual aliasing*, Whitehead 1992). This complicates goal-directed behavior. For instance, suppose your instructions for the way to the station are: "Follow this road to the traffic light, turn left, follow that road to the next traffic light, turn right, there you are." Suppose you reach one of the traffic lights. To do the right thing you need to know whether it is the first or the second. This requires at least one bit of memory — your current environmental input by itself is not sufficient.

Any learning algorithm supposed to *learn* algorithms that store relevant bits of information from training examples that only tell whether the current attempt has been successful or not — this is the typical reinforcement learning (RL) situation — will face a major temporal credit assignment problem: which of the many past inputs are relevant (and should be represented in some sort of internal memory), which are not? Most traditional work on RL requires Markovian interfaces to the environment: the current input must provide all information about probabilities of possible inputs to be observed after the next action, e.g., [15, 16]. These approaches, however, essentially learn how to react in response to a given input, but cannot learn to identify and memorize important past events in complex, partially observable settings, such as in the introductory example above.

Several recent, non-standard RL approaches in principle are able to deal with partially observable environments and can learn to memorize certain types of

* This work was supported by SNF grants 21-55409.98 and 2000-61847.00

G. Dorffner, H. Bischof, and K. Hornik (Eds.): ICANN 2001, LNCS 2130, pp. 865–873, 2001.

relevant events [7, 8, 6, 3, 9, 10, 13, 18, 17, 14]. None of them, however, represents a satisfactory solution to the general problem of learning in worlds described by *partially observable Markov Decision Processes* (POMDPs). The approaches are either ad hoc, or work in restricted domains only, suffer from problems concerning state space exploration versus exploitation, have non-optimal learning rate or have limited generalization capabilities. Most of these problems are overcome in [5] but the model is yet computationally intractable. Here we consider a novel approach to POMDP-RL called *market-based RL* which in principle does not suffer from the limitations of traditional RL. We first give an overview of this approach and its history, then evaluate it in a POMDP setting, and discuss its potential and limitations.

2 Market-Based RL: History & State of the Art

Classifier Systems. Probably the first market-based approach to RL is embodied by Holland's Classifier Systems and the Bucket Brigade algorithm [4]. Messages in form of bitstrings of size n can be placed on a global message list either by the environment or by entities called classifiers. Each classifier consists of a condition part and an action part defining a message it might send to the message list. Both parts are strings out of $\{0, 1, _\}^n$ where the '_' serves as a 'don't care' if it appears in the condition part. A non-negative real number is associated with each classifier indicating its 'strength'. During one cycle all messages on the message list are compared with the condition parts of all classifiers of the system. Each matching classifier computes a 'bid' based on its strength. The highest bidding classifiers may place their message on the message list of the next cycle, but they have to pay with their bid which is distributed among the classifiers active during the last time step which set up the triggering conditions (this explains the name bucket brigade). Certain messages result in an action within the environment (like moving a robot one step). Because some of these actions may be regarded as 'useful' by an external critic who can give payoff by increasing the strengths of the currently active classifiers, learning may take place. The central idea is that classifiers which are not active when the environment gives payoff but which had an important role for setting the stage for directly rewarded classifiers can earn credit by participating in 'bucket brigade chains'. The success of some active classifier recursively depends on the success of classifiers that are active at the following time ticks. Bankrupt classifiers are removed and replaced by freshly generated ones endowed with an initial amount of money.

PSALMs. Holland's approach suffers from certain drawbacks though. For instance, there is no credit conservation law — money can be generated out of nothing. Pages 23-51 of [11] are devoted to a more general approach called *Prototypical Self-referential Associating Learning Mechanisms* (PSALMs). Competing/cooperating agents bid for executing actions. Winners may receive external reward for achieving goals. Agents are supposed to learn the credit assignment process itself (meta-learning). For this purpose they can execute actions for collectively constructing and connecting and modifying agents and for transferring credit (reward) to agents. PSALMs are the first systems that enforce the important constraint of total credit conservation (except for consumption and external reward) - this constraint is not enforced in Holland's classifier economy, which may cause money inflation and other problems. PSALMs also inspired the money-conserving *Neural bucket brigade*, where the money is "weight substance" of a reinforcement learning neural net [12].

Hayek Machines. [1] are designed to avoid further loopholes in Holland's credit assignment process that may allow some agent to profit from actions that are not beneficial for the system as a whole. Property rights of agents are strictly enforced. The current owner of the right to act in the world computes the minimal price for this right, and sells it to the highest bidder. Agents can create children and invest part of their money into them, and profit from their children's success. Wealth and bid of an agent are separate quantities. All of this makes Hayek more stable than the original bucket brigade. There have been impressive applications — Hayek learned to solve complex blocks world problems [1]. Hayek is reminiscent of PSALM3 [11], but a crucial difference between PSALM3 and Hayek may be that PSALM3 does not strictly enforce individual property rights. For instance, agents may steal money from other agents and temporally use it in a way that does not contribute to the system's overall progress.

3 The Hayek4 System

We briefly describe our reimplementation of Hayek4 [2] — the most recent Hayek machine variant. Unlike earlier Hayek versions, Hayek4 uses a new representation language.

Economic Model. Hayek4 consists of a collection of agents that each have a rule, a possible action, a wealth and a numerical bid. The agents are required to solve an instance of a complex task. But because each agent can only perform a single action, they need to cooperate. In a single instance, computation proceeds in a series of *auctions.* The highest bidding agent wins and "owns" the world, pays an amount equal to its bid to the previous owner and may then perform its action on the world. The owner of the world collects any reward from it, plus the payment of the next owner. Only if the actual revenue is larger than its bid, an agent can increase its *wealth.*

Birth, Death and Taxes. Agents are allowed to create *children* if their wealth is larger than some predetermined number (here 1000). Parents endorse their offspring with a minimum amount of money which is subtracted from their wealth, but in return parents receive a share c of their children's future profit. Children's rules are randomly created, copied or mutated from their parents with probabilities p_r, p_c and p_m, respectively. See Table 1 for their numerical values.

By the definition of wealth, birth processes automatically focus on rather successful agents. To improve efficiency, after each instance all agents pay a

Table 1. Parameter settings for our Hayek4 reimplementation. For a detailed explanation see [2].

symbol	value	description
W	10.0	Baum and Durdanovic's W parameter
R	100.0	reward value
ϵ	0.01	children always bid this amount higher than highest bid
u	0.25	increase level if (1-u) instances are solved; decrease if less than u.
t	1e-3	execution tax
c	0.25	copyright share
m	0.5	mutation ratio
p_r	0.3	probability of random rule for new child
p_c	0.3	probability of copying rule for new child
p_m	0.3	probability of mutation of rule for new child

small amount of *tax* proportional to the amount of computation time they have used. To remove unuseful agents, earlier Hayek versions required agents to pay a small fixed amount of tax; agents without money were removed from the system. Hayek4 removes them once they have been inactive for the last, say, 100 instances.

Post Production System. The agents in Hayek4 are essentially a *Post* production system. Baum and Durdanovic argue that this provides a richer representation than the S-expressions in earlier Hayek versions because Post systems are Turing complete (like the representation used by PSALMs [11]). Each agent forms a Post rule of the form $L \rightarrow R$ where L is the antecedent (or rule) and R is the consequent (the action). The state is encoded in a string S; the agent is only allowed to perform R when L matches string S. This procedure is iterated until no rules match, and computation halts.

4 Implementation

We found Hayek4 rather difficult to reproduce. Several parameters require tuning — making Hayek stable still seems to remain an art. Parameter settings of our Hayek system are summarized in Table 1; for a detailed explanation see [2]. For small task sizes our program tends to find reliable solutions, but larger sizes cause stability problems, or prevent Hayek4 from finding a solution within several days.

We tried to reimplement Hayek4 as closely as possible, except for certain deviations justified by improved performance or implementation issues:

- *Single rule:* instead of agents having multiple rules, we allowed only one rule per agent.
- *Fixed reward:* instead of paying off reward proportional to the size of the task, we used a fixed reward of 100.
- *TD backup:* we additionally applied time differential (TD) [15] backup of the agent's bid, as traditionally done in reinforcement learning systems, to accelerate convergence.

Blockworld. Baum and Durdanovic reported a set of Hayek agents that collectively form a universal solver for arbitrary blockworld problems [2] involving arbitrarily high stacks of blocks. The set consists of five rules listed in the table below.

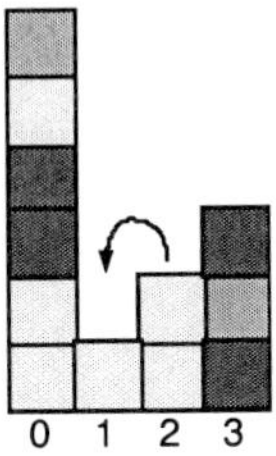

Fig. 1. Example of Blockworld instance with stack height=6. The objective is to replicate in stack 1 the color sequence represented by the blocks in stack 0, by moving blocks between stacks 1-3. Here the optimal next move would be to move the top block of stack 2 to stack 1.

	agent		
world	rule	action	bid
babc:cbb::a	*:*:*:*	1→3	7.78
babc:cb::ab	*:*:*:*	1→3	7.78
babc:c::abb	*:*:*:*	1→3	7.78
babc:::abbc	*.*:1:*:*2*	3→2	8.07
babc::c:abb	*.*:1:*:*2*	3→2	8.07
babc::cb:ab	*.*:1:*2:*	2→1	8.07
babc:b:c:ab	*.*:1:*:*2*	3→2	8.07
babc:b:cb:a	*.*:1:*:*2*	3→2	8.07
babc:b:cba:	*.*:1:*2:*	2→1	8.07
babc:ba:cb:	*.*:1:*2:*	2→1	8.07
babc:bab:c:	*.:1:*2:*	2→1	35.80
babc:babc::			

Fig. 2. A typical Blockworld solution for stack height=4. Each colon-separated field of the world string represents a stack of blocks of colors {a,b,c}. The agent's rule matches the world string; "*" represents a "don't care" symbol; numbers in the rule encode replacements of matched substrings of stack 0. Actions move blocks between stacks 1, 2 and 3.

We tested Hayek4 by first hardwiring a system consisting of those five so-called universal agents only. Indeed, it reliably generated stacks of increasing size, reaching stack level 20 within 800 instances, and level 30 within 1000 instances. Figure 2 shows a typical run resulting in a stack of height=4. Then we switched on the creation of children but kept mutation turned off. This caused Hayek to progress at a much slower pace, probably because new children disrupt the universal solver as they always win the auctions using the ϵ-bid scheme.

Problems. Unfortunately, our system was not able to find the universal solver by itself. It typically reached stack size 5, but then failed to progress any further. We believe that with increasing task size the system is encountering stability problems. Much remains to be done here.

5 Adding Memory to Hayek

Unlike traditional RL, market-based RL is in principle applicable to POMDPs. Previous work, however, has usually focused on reactive settings (MDPs) instead of POMDPs (see [19] for a notable exception though). Here we add a memory register to the Hayek machine and apply it to POMDPs. The memory register

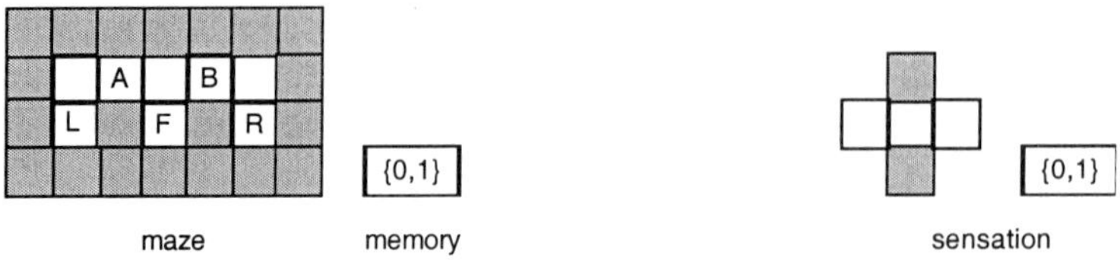

Fig. 3. Woods101 is a partially observable environment; grey positions represent walls. Left: The agent starts at one of the positions L or R and has to reach the food at F. The two aliased positions are A and B. A 1-bit memory register is available for writing/reading. Right: the agent senses the four neigbouring positions and the value of the memory register; here, for example, the agent might be either at A or at B.

acts as a additional environmental variable that can be manipulated by the agent using *write* actions.

Test Problem. To study Hayek4's performance on POMDPs, we focus on one of the simplest possible POMDP problems. Figure 3 shows our toy example which is an adapted version of Woods101 [3], a standard test case for non-Markov search problems. Agents start in either of the two positions, L or R, and have to reach the food at F. Agents have only limited perception of their four immediate neighbouring positions *north*, *east*, *south* and *west* plus the value of the memory register. Without memory, there is no Markov solution to this problem because there are two aliased positions A and B with identical environmental input where the optimal agent needs to execute different actions: in A the optimal action is *east*; in B the optimal action is *west*. Note that the starting positions L and R are also aliased. However, in both L and R the optimal action is coincidentally the same (i.e. going *north*).

Results. We ran Hayek4 on our test problem. After about 1000 instances it learned a policy that solved the POMDP problem using a 1-bit memory. In early instances the number of agents grew up to 1000 or more but finally only 8 agents remained. While the bids of all agents converged to the actual reward value of 100, their wealth values varied considerably. Sample histories of solved instances starting at L are shown in Table 2; histories of solved instances starting at R are shown in Table 3.

The results can be summarized as follows. When an instance starts from L, Hayek assures that the memory is set to 0 before it reaches aliased position A. If the memory was initially set to m=1, a write agent sets the memory to 0; if the memory was already m=0, nothing is done. When the agent starts from the right, the opposite occurs: before Hayek reaches aliased position B, it checks if the memory bit is properly set to m=1. Now, two different agents — with identical perception of the environment but different perception of the memory — act at aliased positions A and B: an *east* agent at A if the memory bit is 0, and a *west* agent if the memory is set to m=1.

Table 2. Two policies from starting position L=(1,1); auctions proceed from the first line downwards. Left: the memory is set to m=1 before going *north* but again reset to m=0 before the aliased position (2,2) is reached. Right: the memory is initially set to m=0 by a memory agent at (1,2). The first four fields of the rule represent the bit values of *north*, *east*, *south* and *west*, respectively; the fifth field matches the value of the memory register. Star ('*') variables represent "don't care" symbols.

state		agent			
x y	m	rule	action	bid	wealth
1 1	0	1:0:0:0:0	m1	100.00	119.51
1 1	1	1:*:*:0:1	n	100.00	184.04
1 2	1	*:*:1:0:1	m0	100.00	301.70
1 2	0	0:1:1:0:0	e	100.00	116.76
2 2	0	0:1:0:1:0	e	100.00	951.08
3 2	0	0:1:1:1:*	s	100.00	103.52
3 1	0				

state		agent			
x y	m	rule	action	bid	wealth
1 1	1	1:*:*:0:1	n	100.00	185.01
1 2	1	*:*:1:0:1	m0	100.00	302.15
1 2	0	0:1:1:0:0	e	100.00	117.22
2 2	0	0:1:0:1:0	e	100.00	951.40
3 2	0	0:1:1:1:*	s	100.00	103.51
3 1	0				

Table 3. Two policies from starting position R=(5,1); auctions proceed from the first line downwards. Left: the memory is initially set to m=1 and the agent proceeds directly *north* to F=(5,2). Right: the memory was initially set to m=0 and a memory agent sets m=1 before proceeding *north*.

state		agent			
x y	m	rule	action	bid	wealth
5 1	1	1:*:*:0:1	n	100.00	183.80
5 2	1	0:0:1:1:1	w	100.00	143.04
4 2	1	0:1:0:*:1	w	100.00	378.54
3 2	1	0:1:1:1:*	s	100.00	103.52
3 1	1				

state		agent			
x y	m	rule	action	bid	wealth
5 1	0	1:0:0:0:0	m1	100.00	120.21
5 2	1	1:*:*:0:1	n	100.00	185.33
5 2	1	0:0:1:1:1	w	100.00	143.75
4 2	1	0:1:0:*:1	w	100.00	379.05
3 2	1	0:1:1:1:*	s	100.00	103.50
3 1	1				

Hayek also evolves one agent executing action *south* at sequence ends at (3,2). It disregards the value of the memory by using the '*'-symbol ("don't care"). "Don't care" symbols are useful to prevent unnecessary over-specialization, i.e., we do not need *two* separate agents for each memory value. In fact, such over-specialization occurred with the starting agent going *north*: Hayek did not use a "don't care" symbol for the memory register and therefore needed a extra memory agent ($m1$) if the bit was initially unset.

Discussion. This simple test problem shows that Hayek, in principle, is able to solve POMDPs. More precisely, using a memory register, Hayek "de-aliases" aliased positions in the POMDP by tagging these situations *before* they actually occur. We note that the solution found by Hayek is not unique: it does not matter whether it chooses to write a "0" or a "1" for the left path, as long as the right path uses the complement. Once a policy seems to perform well, Hayek converges toward this local optimum and sticks to it.

Note that the found solution is nearly but not exactly optimal; an extra $m1$-memory agent was necessary for the left policy in Table 2 because Hayek was not able to generalize optimally in case of the agent going *north*, failing to use a “don't care” symbol for the memory value.

6 Conclusion

We have started to evaluate market-based RL in POMDP settings, focusing on the Hayek machine as a vehicle for learning to memorize relevant events in short-term memory. Using a memory bit register, Hayek was able to distinguish aliased positions in a toy POMDP and evolve a stable solution. The approach is promising, yet much remains to be done to make it scalable.

References

1. E. B. Baum and I. Durdanovic. Toward a model of mind as an economy of agents. *Machine Learning*, 35(2):155–185, 1999.
2. E. B. Baum and I. Durdanovic. An evolutionary Post production system. Technical report, NEC Research Institute, Princeton, NJ, January 2000.
3. D. Cliff and S. Ross. Adding temporary memory to ZCS. *Adaptive Behavior*, 3:101–150, 1994.
4. J. H. Holland. Properties of the bucket brigade. In *Proceedings of an International Conference on Genetic Algorithms.* Hillsdale, NJ, 1985.
5. M. Hutter. Towards a universal theory of artificial intelligence based on algorithmic probability and sequential decision theory. *Submitted to the* 17th *International Joint Conference on Artificial Intelligence (IJCAI-2001)*, (IDSIA-14-00), December 2000.
6. T. Jaakkola, S. P. Singh, and M. I. Jordan. Reinforcement learning algorithm for partially observable Markov decision problems. In G. Tesauro, D. S. Touretzky, and T. K. Leen, editors, *Advances in Neural Information Processing Systems 7*, pages 345–352. MIT Press, Cambridge MA, 1995.
7. L.P. Kaelbling, M.L. Littman, and A.R. Cassandra. Planning and acting in partially observable stochastic domains. Technical report, Brown University, Providence RI, 1995.
8. M.L. Littman, A.R. Cassandra, and L.P. Kaelbling. Learning policies for partially observable environments: Scaling up. In A. Prieditis and S. Russell, editors, *Machine Learning: Proceedings of the Twelfth International Conference*, pages 362–370. Morgan Kaufmann Publishers, San Francisco, CA, 1995.
9. R. A. McCallum. Overcoming incomplete perception with utile distinction memory. In *Machine Learning: Proceedings of the Tenth International Conference.* Morgan Kaufmann, Amherst, MA, 1993.
10. M. B. Ring. *Continual Learning in Reinforcement Environments.* PhD thesis, University of Texas at Austin, Austin, Texas 78712, August 1994.
11. J. Schmidhuber. Evolutionary principles in self-referential learning, or on learning how to learn: the meta-meta-... hook. Institut für Informatik, Technische Universität München, 1987.
12. J. Schmidhuber. A local learning algorithm for dynamic feedforward and recurrent networks. *Connection Science*, 1(4):403–412, 1989.

13. J. Schmidhuber. Reinforcement learning in Markovian and non-Markovian environments. In D. S. Lippman, J. E. Moody, and D. S. Touretzky, editors, *Advances in Neural Information Processing Systems 3*, pages 500–506. San Mateo, CA: Morgan Kaufmann, 1991.
14. J. Schmidhuber, J. Zhao, and M. Wiering. Shifting inductive bias with success-story algorithm, adaptive Levin search, and incremental self-improvement. *Machine Learning*, 28:105–130, 1997.
15. R. S. Sutton. Learning to predict by the methods of temporal differences. *Machine Learning*, 3:9–44, 1988.
16. C. J. C. H. Watkins and P. Dayan. Q-learning. *Machine Learning*, 8:279–292, 1992.
17. M. Wiering and J. Schmidhuber. HQ-learning. *Adaptive Behavior*, 6(2):219–246, 1998.
18. M.A. Wiering and J. Schmidhuber. Solving POMDPs with Levin search and EIRA. In L. Saitta, editor, *Machine Learning: Proceedings of the Thirteenth International Conference*, pages 534–542. Morgan Kaufmann Publishers, San Francisco, CA, 1996.
19. S.W. Wilson. ZCS: A zeroth level classifier system. *Evolutionary Computation*, 2:1–18, 1994.

Sequential Strategy for Learning Multi-stage Multi-agent Collaborative Games

W. Andy Wright

BAE SYSTEMS (ATC Sowerby)* FPC 267, PLC, PO Box 5,
Filton, Bristol, UK
Andy.Wright@bris.ac.uk

Abstract. An alternative approach to learning decision strategies in multi-state multiple agent systems is presented here. The method, which uses a game theoretic construction of "best response with error" does not rely on direct communication between the agents in the system. Limited experiments show that the method can find Nash equilibrium points at least for a 2 player multi-stage coordination game and converges more quickly than a comparable co-evolution method.

1 Introduction

The use of machine learning to determine appropriate decision strategies for multi-agent systems is becoming increasingly important within a number of fields. These include:

- *mobile robotics* where it has been shown that the coordinated use of multiple platforms provides both computational and performance advantages for navigation and map building [4], and in the performance of manipulative tasks [10].
- *software systems* where the use of coordinated software agents provide an alternative approach to the construction of programs for large involved problems [8].

These brief examples provide an illustration of the problems that are common in many multi-agent systems. Often the environment that these agents will be placed in is unstructured and uncertain. Furthermore, the system may be complex involving many states and possible actions, this is particularly the case in robotics. The design of a control strategy for such systems, therefore, is expensive since it is often the case that hand crafted b-spoke solutions have to be produced.

An alternative approach is to learn appropriate strategies for the individual agents such that they learn to function as a coherent whole. For sequential decision processes reinforcement learning has been shown to provide both a robust and general approach to this problem for single decision agents [13].

* Currently visiting Fellow in Department of Mathematics University of Bristol.

G. Dorffner, H. Bischof, and K. Hornik (Eds.): ICANN 2001, LNCS 2130, pp. 874–881, 2001.

For multiple agent systems this is not the case. Here by definition the reward given to any agent will not only be a function of its own actions but also the actions of other agents. However, these actions, either because they are taken simultaneously or because of limited communication between the agents may not be known. Consequently the decision system is partially observed. Because of these difficulties there has been very little research addressing the problem of learning collaborative strategies for multi-agent systems.

Possibly the most active area of research concerning this problem is in game theory. Here there are a number of approaches [5] that have been put forward to learn strategies for individual players in a multi-agent system or game. Possibly the most common learning framework is fictitious play [2]. Here an agent, monitors previous plays by other agents, given this information calculates the most probable action another agent may take and then plays its *best response* to this. Unfortunately, these game theoretic frameworks are not model free and require all the agent to be are aware of the actions taken by the other agents together with the rewards these agents receive.

To overcome some of these restrictions the use of reinforcement learning (RL) to learn behaviour in 2 player games has attracted a significant level of attention [12, 3, 6, 8] and a number of alternative strategies have been put forward. Initial work [12] allowed policies learnt by an agent to be shared with other agents. However, this approach is limited since it only able to consider collaborative strategies where both (or all) agents share the same strategy. Collaborative strategies where the agents adopt different but complimentary strategies are not included. An alternative RL approach, that overcomes this, is to allow each agent to share its history of actions with the other agents. An agent then *co-evolves* its Q value with the other agents allowing for the strategies of the other agents by also calculating their Q value [6]. Another co-evolution strategy where agents to share their past actions treats the decision process as partially observed and writes the Q value for an agent as a function of its actions and a "belief" about the potential actions of the other agents [3]. This is termed a joint action learner (JAL) and can be seen as a combination of RL and fictitious play. The convergence of the method relies on each agent calculating sufficient statistics regarding the potential actions of the other agents given the past histories. However, even for a system with a *single* state this method scales as $O(||\mathcal{A}||^N)$, where $||\mathcal{A}||$ is the size of an individuals action space and N is the number of collaborating agents. For large systems involving large numbers of agents, and more importantly systems with multiple states, generating sufficient statistics rapidly becomes impracticable. This reflected in the currently literature which has concentrated on the application of these methods to two player *single* stage games. An added disadvantage of these approaches is that they require substantial communication between the agents, which may not be practically realisable.

An alternative co-evolution approach, which converges almost as well as the JAL approach, treats the agents as weakly coupled or "independent learners" (IL). Here there is no sharing of action spaces. Coupling between the agents is through the common reward structure of the game since this will depend on the actions of all the agents in the system. The approach does not explicitly suffer from the combinatorial explosion inherent in the JAL approach nor does it require direct communication between the agents [3].

This paper presented an alternative sequential approach to learning strategies in a multi agent system which, like the IL approach, does not require communication between agents. The method, is based on the game theoretic ideas of best response with error [9]. This paper presents a limited empirical study that shows this approach can converge to a game equilibrium point and in some cases is several orders of magnitude quicker to converge than the IL co-evolution approach.

2 Multi-player Games and Equilibrium

The focus of this paper is to consider the utility of a sequential multi-agents RL framework. In a multi-agent system the outcome of the game is not controlled by any single agent but is a function of strategies adopted by all the agents. Any learning system, therefore, that attempts to learn an appropriate strategy for a single agent must take into consideration (directly or indirectly) the strategy adopted by the other agents. The difficulty in doing this is ensuring that the process adopted converges. Convergence in such system is a much more difficult proposition than in a single agent decision system because here the system self-couples. That is an agent will adjusts its strategy to allow for the strategy of others who in turn are likewise adjusting theirs. Fortunately, game theory provides a structure that allows us to view this problem more formally.

Consider a collection $\mathcal{C}$ of n (heterogenous) agents, where each agent $i \in \mathcal{C}$ has a finite set of possible actions $a \in \mathcal{A}^i(s)$ in each state $s \in \mathcal{S}$ of the system. The vector of actions from all the agents forms a joint action $\mathbf{a} = \{a^1, \ldots, a^n\} \in \mathcal{A}$ where

$$\mathcal{A} = \times_{i \in \mathcal{C}} \mathcal{A}_i$$

is the set of all joint actions. For each state s an agent i can have a mixed strategy $\sigma(a^i, s) = \sigma^i(s) \in \mathbf{\Sigma^i}$. Simply, given some randomised selection process, this prescribes a probability of adopting action a^i in that state s. Combining these strategies gives a *mixed strategy profile* which is denoted $\sigma \in \mathbf{\Sigma} = \times_{i \in \mathcal{C}} \mathbf{\Sigma^i}$. Likewise for any agent i there will be a mixed strategy profile σ^{-i} which is the combination of the strategies adopted by the *other* agents.

If it assumed that all the agents are rational then for any state s an agent i will attempt to adopt a mixed strategy σ^i that maximises its expected the reward $R(s, \mathbf{a})$ which is a function of the state of the system and the joint action. The mixed strategy that obtains a maximal expected reward is termed the *best response* and here is denoted as $\hat{\sigma}^i \in BR^i(\sigma^{-i})$. If all agents are playing their best response then the strategy profile $\hat{\sigma}$ where

$$\hat{\sigma}^i \in BR^i(\hat{\sigma}^{-i}), \forall i$$

prescribes a Nash equilibrium for the system. Thus a system is in Nash equilibrium if all the agents adopt a set of strategies where any deviation by any agent from their strategy results in an decrease in the expected reward to that agent. It has been proven that any game will have at least one Nash equilibrium point [11]. Where a game has more than one equilibrium point the profile that gives the highest reward is termed Pareto optimal. The focus of any multi agent learning method, therefore, is to find a profile that is a Nash equilibrium of the system and preferably one that is Pareto optimal.

3 Sequential Reinforcement Learning for Multiple Agent

Typically reinforcement learning is applied to *single* agent sequential decision problems. Here the decision system is modelled as a series of Markov transitions, with transition probability $P^{a^i}_{s,s'} \in P_t : \mathcal{S} \times \mathcal{A}^i \times \mathcal{S} \rightarrow [0,1]$, where an agent i moves the system from one state $s \in \mathcal{S}$ to another $s' \in \mathcal{S}$ under the influence of some policy, $\sigma(s, a^i)$

The value of a particular action in a particular state is given by the expected reward

$$Q^i(a^i, s) = \sum_{s'} P^a_{s,s'} \left[r_{s,s'} + \gamma V^i(s') \right], \tag{1}$$

which is termed the Q value. Here $\gamma \in [0,1]$ is a discount function, $r_{s,s'} \in \mathcal{R} : \mathcal{S} \rightarrow \Re$ is the reward given to the agent when the system moves from state s to s' and

$$V^i(s) = \max_{a^i \in \mathcal{A}^i} Q(a^i, s). \tag{2}$$

Where the N tuple $\langle \mathcal{S}, \mathcal{A}^i, P_t, \mathcal{R} \rangle$ is known then dynamic programming can be used to find the best response. However, when this is not the case, either because P_t or $\mathcal{R}$ are uncertain or not known, then RL can be used. Here an agent iteratively estimates $Q(s, a^i)$ by moving through the decision process and obtaining samples $\{a^i, s, r\}$. In a *multi-agent* system further difficulty arises because if there is no communication between agents then an agent cannot observe the state of the system and further is denied knowledge of the actions taken by the other agents. Under these circumstances it follows that the decision process is partially observed. However, this can be overcome if the agent learning process is treated as a process with a non-stationary reward structure and adopt a form of Q learning that can track these rewards.

Here an on policy temporal difference learning TD(0) (Sarsa) approach was used. This method has the added advantage that, unlike a Monte Carlo approach, it can be applied to repeated games which do not have a finite horizon. Here the Q values are updates using

$$Q_{t+1}(s, a^i) = Q_t(s, a^i) + \alpha_t \left[r + \lambda Q_t(s', a^{i'}) - Q_t(s, a^i) \right] \tag{3}$$

where r is the observed reward, s' is the state observed by the agent by taking action a^i at state s, actions a^i and $a^{i'}$ are obtained by sampling from the Boltzmann distribution

$$\pi(s', a^{i'}) = \frac{\exp(Q_t(s', a^{i'})/T)}{\sum_{b^i \in \mathcal{A}^i} \exp(Q_t(s', b^i)/T)}, \tag{4}$$

$\alpha_t = 1/t$ is a time varying *learning parameter*[1]. To ensure convergence the temperature T in equation 4 was reduced exponentially per iteration t of the Sarsa algorithm [7].

The multi-agent component of this reinforcement learning approach is constructed as follows. The algorithm iteratively selects an agent i and updates Q^i for N iterations to give $\hat{Q}^i$ while the other agents' Q values, Q^{-i}, are fixed. To overcome any cycles in the learning process the agents play what is termed " best response with error" [9] by adopting a strategy based on the Boltzmann distribution (equation 4). The Q^i is then updated such that

$$Q^i_{\tau+1}(s, a^i) = \delta \hat{Q}^i(s, a^i) + (1 - \delta) Q^i_\tau(s, a^i), \tag{5}$$

where $0 < \delta < 1$ is a constant. This process is then repeated for all $i \in \mathcal{C}$ this gives one epoch enumerated by τ. The whole procedure then repeated until the value functions for all the agents has converged.

In the results presented two versions of the sequential learning framework are used.

1. *Complete convergence.* The Sarsa algorithm is run to convergence with the other agents Q values fixed.
2. *Early stopping.* The Sarsa algorithm is only run for a short number of iterations N. This methods relies on the general observation that the Sarsa algorithm will find the appropriate action for each state long before obtaining a fully accurate value function.

4 Results

This sequential learning approach was compared empirically against the IL approach for the repeated two player multi-state team game [1] shown schematically in figure 1. The game may be described as having six states $s_1, \ldots, s_6$ and each agent has two actions a^1, b^1 for *agent 1* and a^2, b^2 for *agent 2*. The transition $s_1 \longrightarrow s_2$ and $s_1 \longrightarrow s_3$ is governed only by the actions of *agent I.* Action a^1 takes the system to s_2 whereas action b^1 takes the system for s_3. Transition $s_3 \longrightarrow s_6$ then occurs with probability 1 irrespective of the actions taken by either agent in s_3. On reaching s_6 a reward of $+5$ is given to both agent. However, the transition $s_2 \longrightarrow s_4$ depends on the cooperation between *agent 1* and *agent 2.* State s_4 is obtained under action combinations of (a^1, a^2) and (b^1, b^2) and a reward of $+10$ received by both agents. Conversely s_5 is obtained and a reward of -10 give to the agents when they do not cooperate. That is when action combination (a^1, b^2) or (b^1, a^2) are selected. From states s_4, s_5, and s_6 the system returns to state s_1 and repeats infinitely.

If this game is viewed only from *agent 1's* perspective then the expected reward in s_2 ($V^1(s_2)$), since it does not know what action *agent 2* will take,

[1] Note that $\sum_t^\infty \alpha_t = \infty$ and $\sum_t^\infty (\alpha_t)^2 < \infty$ ensuring convergence.

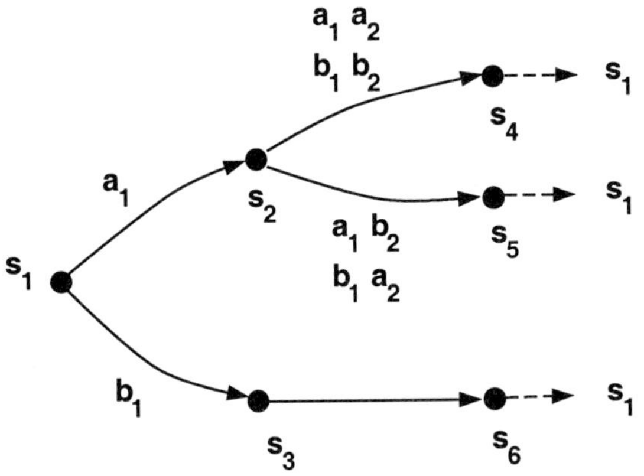

Fig. 1. Schematic of a simple multi-state team game. Note this is not an extensive form.

is 0. Consequently without regard to the action of the other agent this simple approach suggests that the optimal sequence of states is

$$s_1 \longrightarrow s_3 \longrightarrow s_6, \tag{6}$$

since $V^1(s_2) < V^1(s_3)$ (i.e. $0 < 5$). This is the game's *security strategy* and it is *not* a stable Nash equilibrium profile of the game. The Nash equilibrium strategic profiles are given by the profiles $((a_1,(a_1,a_2))$ and $((a_1,(b_1,b_2))$. Since the expected reward for these profiles are the same they are said to be equivalent. A successful multi-agent learning method should converge to either of these collaborative profiles avoiding the security strategy which although Nash is sub-optimal. The convergence of the IL and sequential learner on this game is illustrated in figure 2. This shows the proportion of times the learning methods find the Nash profiles as a function of iteration number for $\gamma = 0.7$. Figure 3 shows the difference in the Q values for s_1 for three sequential learners using early stopping of 1 and 5 Sarsa iterations and complete convergence at the RL stage together with the IL method. It is noticeable that the 1 and 5 iteration versions, allowing for the extra computation involved, converge orders of magnitude faster that the IL method. Typically IL method takes ~ 1500 iterations to converge (and not always to the correct value $Q(s_1,a^1) - Q(s_1,b^1) \sim 3.2$). This difference in convergence is even more apparent for smaller γ. Here the IL method fail to converge after 2000 iterations for some values of γ while experiments with the sequential learning approach show full convergence (see figure 2 (b).

5 Conclusions

This paper presents a reinforcement learning approach for multiple agents based on sequential learning using a game theoretic best response with errors approach.

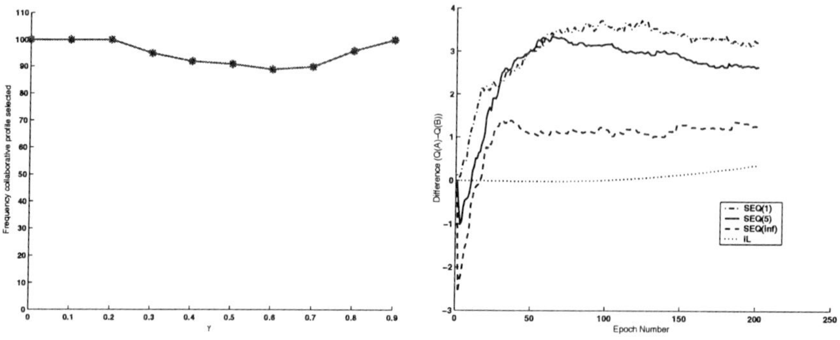

(a) IL Convergence (b) Convergence of Methods

Fig. 2. Convergence results for showing the selection of the cooperative strategy for (a) For the IL method as a function of γ (b) the IL and sequential methods as a function of epoch number.

The convergence of the method is compared empirically against the co-evolution independent learner (IL) method where all agents are updated simultaneously. This was undertaken on a 2 player multi-stage repeated coordination game. The results suggest that the sequential approach provides a learning method that can converge significantly faster than IL. From a practical point of view the neither method relies on direct communication between the agents in the system. However, the IL method does have one advantage, it can be used online. Since the sequential approach requires the agents not being updated to freeze their Q value the method has to be used in batch mode. However, for circumstances where a batch training can be used the results presented here suggest that the sequential approach will converge more quickly. One area that is unresolved for both these learning methods is the theoretical proof of their convergence. This remains the focus for future work.

Acknowledgments

This work was supported by Royal Society Industrial Fellowship.

References

1. C. Boutiler. Sequential optimality and coordination in multiagent systems. In *IJCAI-99*, volume 5, pages 393–400, 1999.
2. G.W. Brown. Iterative solution of games by fictitious play. In *Activity Analysis of Production and Allocation.* Wiley, New York, 1951.
3. C. Claus and C. Boutiler. The dynamics of reinforcement learning in cooperative multiagent systems. In *AAAI-98*, volume 1, 1998.
4. R.H. Deaves, D. Nicholson, D. Gough, and L. Binns. Multiple robot systems for decentralised SLAM investigations. In *SPIE*, volume 4196, page 360, 2000.

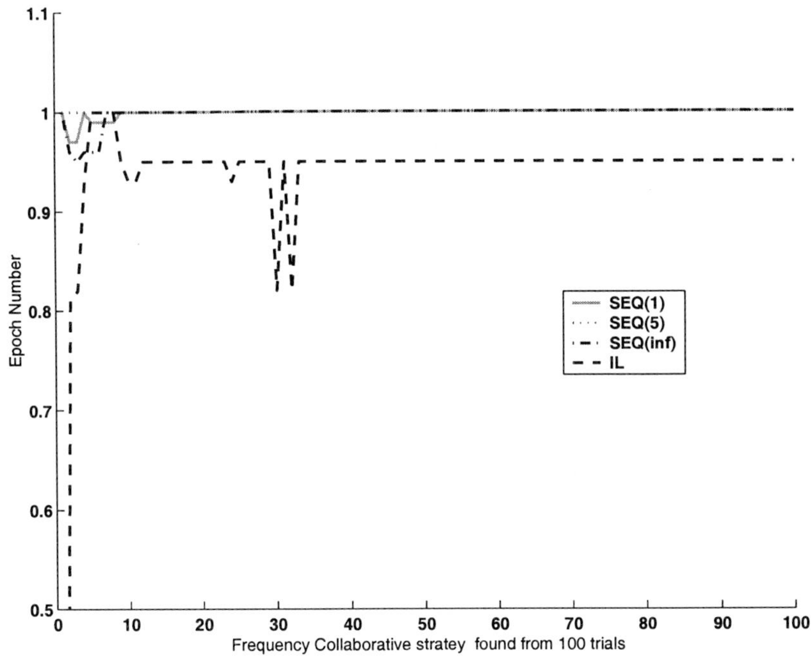

Fig. 3. Plot of the difference of the Q values in s_1 for the sequential learners with 1, 5 RL iterations and complete convergence, together with the result for the IL method.

5. D. Fudenberg and D.K. Levine. *The Theory of Learning in Games.* Economic Learning and Social Evolution. MIT Press, 1998.
6. J. Hu and M.P. Wellman. Multiagent reinforcement learning: Theoretical frameworkand an algorithm. In *Proceedings of fifteenth inbternational conference on machine learning*, pages 242–250, 1998.
7. T. Jaakkola, M.I. Jordan, and S.P. Singh. On the convergence of stochastic iterative dynamic programming algorithms. *Neural Computation*, 6:1185–1201, 1994.
8. M. Kearns, Y. Mansour, and S. In UAI 2000 Singh. Fast planning in stochastic games. In *UAI*, 2000.
9. J.M. MacNamara, J.N. Webb, T. Collins, E.J. Slzekely, and A. Houstton. A general technique for computing evolutionary stable strategies based on errors in decision making. *Theoretical Biology*, 189:211–225, 1997.
10. J. M. Maja. Behavior-based robotics as a tool for synthesis of artificial behavior and analysis of natural behavior. *Trends in Cognitive Science*, 2(3):82–87, 1998.
11. J.F. Nash. Non-cooperative games. *Annals of Mathematics*, 54:286–295, 1951.
12. M. Tan. Multi-agent reinforcement learning: Independant vs cooperative agents. In *Proceedings of the tenth international conference on machine learning*, pages 330–337, 1993.
13. C.J.C.H. Watkins and P. Dyan. Q-learning. *Machine Learning*, 8:279–292, 1992.

Part XI

Robotics and Control

Neural Architecture for Mental Imaging of Sequences Based on Optical Flow Predictions

Volker Stephan and Horst-Michael Gross

Ilmenau Technical University, Department of Neuroscience, P.O.B. 100565,
D-98684 Ilmenau, Germany
{vstephan,homi}@informatik.tu-ilmenau.de

Abstract. In this paper we present a neural architecture for a mental imaging like generation of image sequences. Mental imaging plays a central role for various perception processes. Thereto, we investigated mechanisms to model this ability of biological systems at a functional level for sequences of images. Because it is impossible to memorize many experienced sequences, we developed an universal, general and very powerful approach based on the ability to predict optic flow fields as consequences of the systems own actions and tested the resulting architecture on a real mobile system.

1 Introduction

Mental imaging is understood as a process, which generates conscious images in mind without any sensory stimulation[1]. This is possible, since experienced sensory (visual) impressions are stored by some neural memory structure and can afterwards be reconstructed.

In literature there is broad consens about the central aspect of these imaging processes. Many authors [1–4] propose, that perception and mental imaging share common mechanisms and support each other. In [5–7] there is described, that mentally generated images are used as an expectation for what the system sees. This so called top-down expectation is then used to enhance the sensory bottom-up data in order to become more robust against noise or sensory dropouts. Other authors postulated, that mental imagery is also used for internal simulation of the consequences of hypothetically executed actions in order to find an optimal action sequence [5, 8, 9].

In general, we and almost any perceiving biological system do not operate on sensory snapshots. Instead, we operate on a continuous flow of data, which requires a continuous flow of corresponding expectations. The generation of single mental images already requires a large memory to store the experienced images. The extension to *sequences* of mental images seems to be unrealizable through memorization, because of the exponential growth of data to be stored. In consequence, a more general approach is required.

By learning universally valid sensorimotor relationships between subsequent images and the executed motor actions of the mobile system such an approach

G. Dorffner, H. Bischof, and K. Hornik (Eds.): ICANN 2001, LNCS 2130, pp. 885–890, 2001.

could be developed. The resulting neural architecture is able to anticipate sequences of images, which are expected after the hypothetical execution of a given action trajectory. This ability can be utilized to generate the desired continuous flow of top-down expectations or to realize an internal simulation.

2 Scenario

We investigated our approach within a scenario with the miniature robot KHEPERA equipped with an omnidirectional camera. Using a real robot in favor to stored image sequences from any video stream is very important, since the executed action *causes* the changes within the image and thus defines the direction of the development of the whole image sequence. This emphasizes the fact, that perception does not end in itself, instead it is a sensorimotor process integrating the generation of behavior [10]. For our experiments we used the scenario depicted in figure 1. The omnidirectional camera images are polar-transformed and then used to estimate optic flow using a correlation based method [11].

Fig. 1. Scenario used for our experiments with the mobile robot KHEPERA equipped with an omnidirectional camera.

3 Architecture

Our approach to mental imaging of sequences avoids the need of memorization of all experienced image sequences by using an iterative process of subsequent image transformations. Therefore, the required image transformation represents the changes within the image that are caused by egomotion of the observing system. If this transformation can be found, the system can generate sequences of subsequent sensory visual impressions starting from a real or an initially memorized image.

3.1 How to Find this Transformation?

The required transformation has to comply the following criteria. It must be invariant to color, shape, texture, and other object or scene specific features. In contrast, motion parameters of the mobile system and the spatial configuration of the objects within the scene are very important. With respect to these requirements, optic flow seems to be well suited for this task, since it exclusively represents spatial relations and movements of objects in the scene caused by egomotion of the system.

To generate a hypothetically image sequence for a given action sequence, an anticipation of the action consequences is crucially required. Instead of realizing a scene depending mapping from the current image $I(t)$ to the following one $I(t+1)$, our system learns the scene-independent mapping from past optic flow fields and actions to the following one by a time-delayed neural network (TDNN):

$$OF(t-n)\ldots OF(t) \quad \times \quad a(t-n)\ldots a(t) \quad \mapsto \quad OF(t+1)$$

Thus, the resulting image transformation depicted in figure 2 is based on universally valid sensorimotor relationships, represented by optic flow.

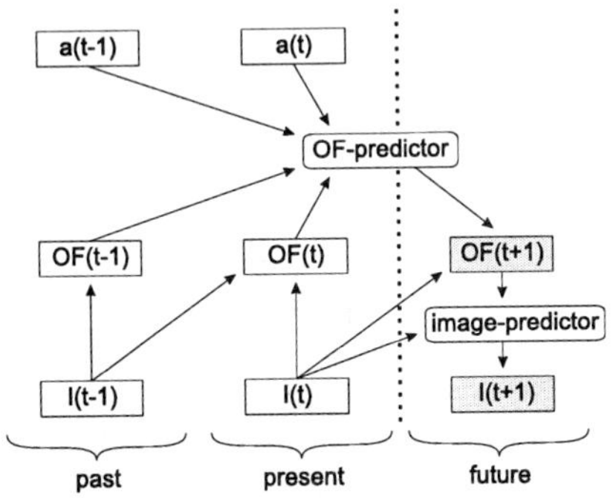

Fig. 2. Architecture for mental imaging. Based on universally valid sensorimotor relationships the next optic flow $OF(t+1)$ is predicted by a neural network (TDNN). The image-prediction uses the predicted optic flow $OF(t+1)$ to transform the current image $I(t)$ to the mentally generated image $I(t+1)$.

3.2 Image Generation

Based on the predicted optic flow field $OF(t+1)$ and the current image $I(t)$ it is possible to generate the predicted image $I(t+1)$. This is realized by shifting the pixels along their corresponding predicted flow field vectors. First, for each pixel of the new image $I(t+1)$ a source region in $I(t)$ is computed. For this purpose those flow vectors $\underline{f}^{pq}$ are searched, which point most closely to the new pixel position (i,k) (figure 3).

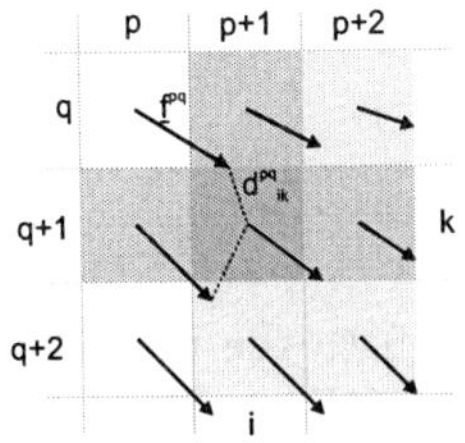

Fig. 3. Computation of distance d_{ik}^{pq} (dashed line) between the end-point of the neighbored optic flow vector $\underline{f}^{pq}$ originating from the pixel at position (p,q) and the pixel-coordinates at position (i,k).

To compute the source region for the new pixel at position (i,k) all distances d_{ik}^{pq} of the flow vectors $\underline{f}^{pq}$ out of the neighborhood N to the new position (i,k) are computed (equation 1). Then, the color of the new pixel $\underline{c}_{i,k} = (r,g,b)^T$ results from a superposition of all pixels out of neighborhood N weighted by a distance-dependent factor (equation 2).

$$d_{ik}^{pq} = \sqrt{(i - f_x^{pq} + p)^2 + (k - f_y^{pq} + q)^2} \tag{1}$$

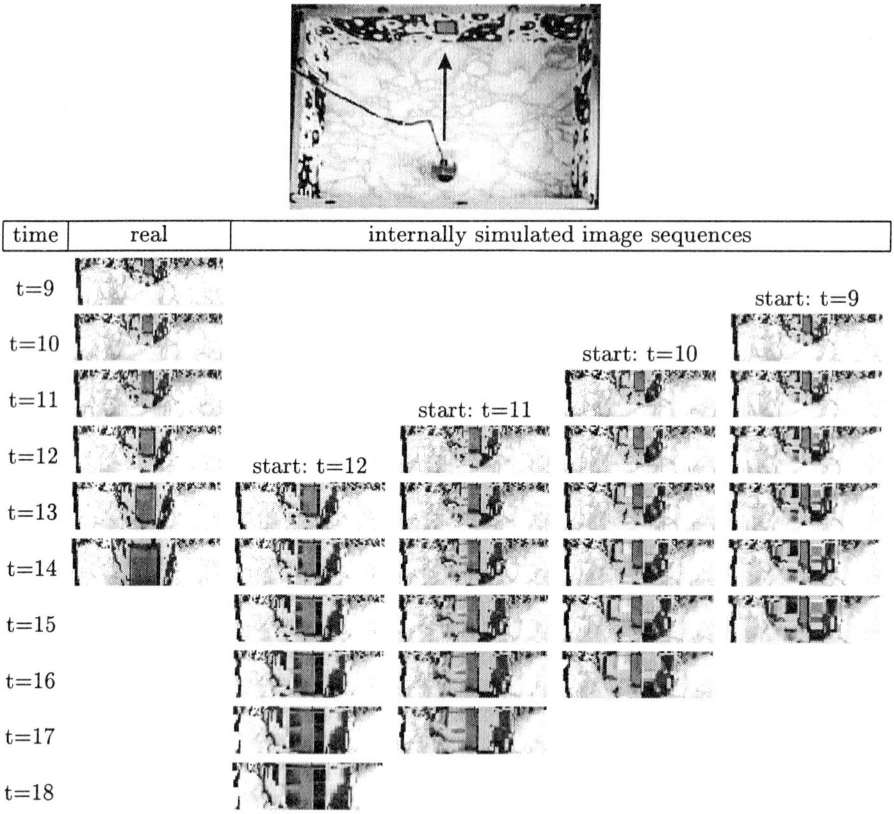

Fig. 4. Internal simulation of images realized by our anticipatory system during straight ahead movement of the mobile robot towards the blue rectangular mark (top). The really experienced image sequence (leftmost column) plotted since time $t = 9$ and four image sequences with different starting points, generated by means of our mental imaging approach, are plotted on the right.

$$\underline{c}_{ik} = \sum_{p,q \in N} \frac{1}{(d_{ik}^{pq})^8} \cdot \underline{c}_{pq} \tag{2}$$

Thus, an interpolated transformation of the image $I(t)$ to the predicted image $I(t+1)$ is defined, which uses the predicted optic flow $OF(t+1)$.

4 Results

We tested the above described image transformation approach for several image sequences. First, we put the robot into the situation illustrated in figure 4 top. At each time step the process of mental imaging generates a sequence of images (vertical columns), that reflect the expected image changes caused by the locomotion of the mobile system.

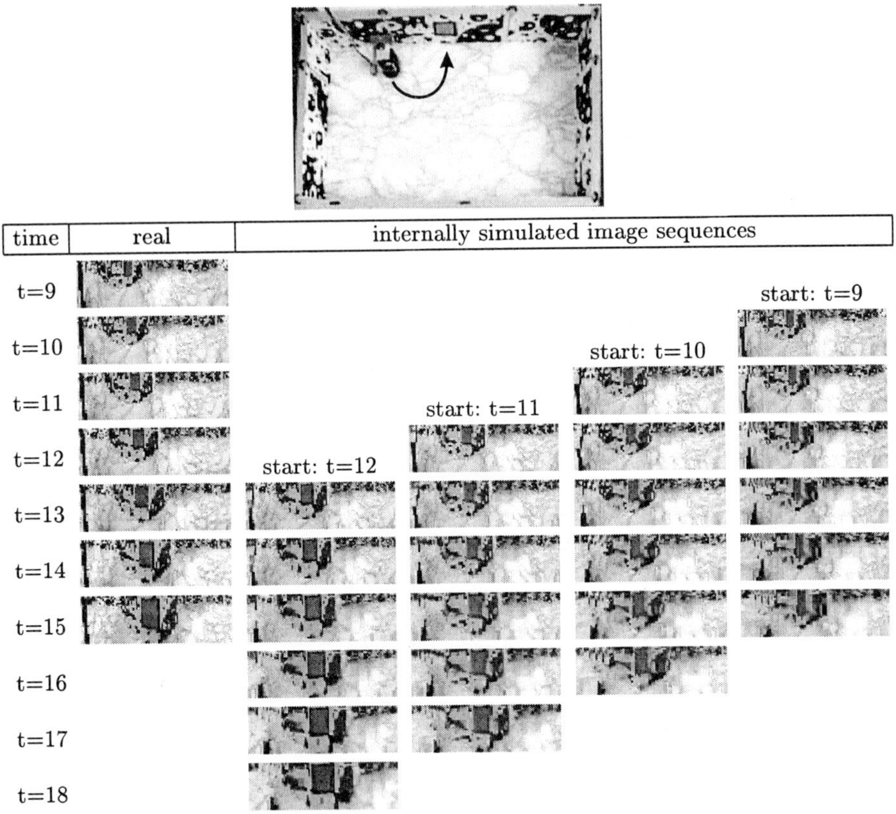

Fig. 5. Internal simulation of images realized by our anticipatory system during a left-turn movement of the mobile robot towards the blue rectangular mark (top). Image arrangement as in figure 4.

The leftmost column of really experienced images shows the growing rectangular blue mark in the image caused by the oncoming opposite marked wall. The rightmost mentally generated sequence based on the real image at time step $t = 9$ shows in contrast to the real sequence only a vertical growth of the central blue mark. The later the starting point for mental imagination, the better the similarity to the last really experienced image at time $t = 14$ can be seen. Likewise, the image quality decreases over simulation time, since both optic flow prediction errors and image interpolation errors are accumulated over simulation time.

Nevertheless, the principal ability of our approach to internally generate image sequences representing the course of the changing reality caused by the systems actions can be demonstrated.

Similar results are depicted in figure 5, where the robot drove a left turn towards the blue mark. In contrast to the previous experiment, the driven curve causes a horizontal shift of the polar-transformed image over time, which also is predicted by our mental imaging architecture.

5 Outlook

The presented neural architecture for optic flow-based mental sequence-imaging internally generates image sequences representing the consequences of a mobile systems own hypothetically executed actions. These image sequences can be used to generate the desired continuous flow of top-down expectations or to realize an internal simulation process for action planning on image sequences.

In future work we intend to investigate these challenging anticipative mechanisms for mobile systems at the behavioral level in more detail.

References

1. M.J. Farah. The neurobiological basis of mental imagery: A componential analysis. *Frontiers in Cognitive Neuroscience*, pages 559–573, 1992.
2. S.M. Kosslyn. *Image and Brain: The Resolution of the Imagery Debate.* MIT Press, 1996.
3. J. Demiris and G. Hayes. Active imitation. In *AISB'99 Symposium on Imitation in Animals and Artifacts*, 1999.
4. L. Deecke. Planning, preparation, execution and imagery of volitional action. *Cognitive Brain Research*, 3(2):59–64, 1996.
5. S.M. Kosslyn and A.L. Sussman. Roles of imagery in perception: Or, there is no such thing as immaculate perception. In M.S. Gazzangia, editor, *The Cognitive Neuroscience*, pages 1035–1042. MIT Press, 1995.
6. Y. Miyashita, M. Ohbayashi, M. Kameyama, Y. Naya, and M. Yoshida. Origin of visual imagery: role of temporal cortex and top-down signal. In M. Ito and Y. Miyashita, editors, *Integrative and Molecular Approach to Brain Function*, pages 133–145. Elsevier, 1996.
7. V. Stephan and H.-M. Gross. Improvement of optical flow estimates by visuomotor anticipation. In *Proceedings of Workshop Dynamische Perzeption, Ulm, Germany, 2000.*, pages 191–194. akademische Velagsgesellschaft, infix, 2000.
8. A.R. Damasio and H. Damasio. Making images and creating subjectivity. In R. Llinas and P. Churchland, editors, *The Mind-Brain Continuum*, pages 19–27. MIT Press, 1996.
9. H.-M. Gross, A. Heinze, T. Seiler, and V. Stephan. Generative character of perception: A neural architecture for sensorimotor anticipation. *Neural Networks*, 12:1101–1129, 1999.
10. R. Pfeifer and C. Scheier. From perception to action: The right direction? In *Proc. PerAc'94*, pages 1–11. IEEE Computer Society Press, 1994.
11. J.L. Barron, D.J. Fleet, and S.S. Beauchemin. Performance of optical flow techniques. *International Journal of Computer Vision, 12:1*, pages 43–77, 1994.

Visual Checking of Grasping Positions of a Three-Fingered Robot Hand

Gunther Heidemann and Helge Ritter

Universität Bielefeld, AG Neuroinformatik, Germany
gheidema@techfak.uni-bielefeld.de

Abstract. We present a computer vision system for judgement on the success or failure of a grasping action carried out by a three-fingered robot hand. After an object has been grasped from a table, an image is captured by a hand camera that can see both the object and the fingertips. The difficulty in the evaluation is that not only identity and position of the objects have to be recognized but also a qualitative judgement on the stability of the grasp has to be made. This is achieved by defining sets of prototypic "grasping situations" individually for the objects.

1 Introduction

Grasping objects with robot grippers or even anthropomorphic artificial hands is one of the most challenging subjects in current robotics. Up to now, the abilities of robotic systems in grasping objects are surpassed by human skills by far: We can grasp objects of almost arbitrary shape, many different sizes and we are able to adapt easily the applied forces to light or heavy weight. This ability relies on a complex interaction of the controlling neural system and the "sensors", which are mainly haptic and visual. Especially force sensing in the fingertips plays an important role. Unfortunately, there is no technical equivalent to human fingertip force sensing by now, though progress has been made e.g. using piezo-resistive pressure sensitive foils [6] for tactile imaging [5, 9] or piezo-electrical pressure sensitive foils [1] for dynamic sensing [2, 14].

Though it is no problem to provide high quality sensors in general, sensors the size and shape of a human fingertip are extremely difficult to produce (and apply). In our experience such fingertip sensors are still unreliable and coarse in measurement. Consequently, other sensory sources have to be exploited. A suitable means is visual control as miniature cameras and image processing hardware are easily available. Our approach is therefore to supplement fingertip sensing by a visual check of the points where the fingers touch the object.

In this contribution we deal with the evaluation of images captured by a hand camera which can see the grasped object and the fingertips. I.e., the object has been grasped already – this can verified by sensors –, but it is still unknown how stable the object is held. Evaluating this "grasping situation" is a challenge to computer vision since it is not a classical classification or pose estimation task but instead a complex situation has to be judged. Moreover, the relative

G. Dorffner, H. Bischof, and K. Hornik (Eds.): ICANN 2001, LNCS 2130, pp. 891–898, 2001.

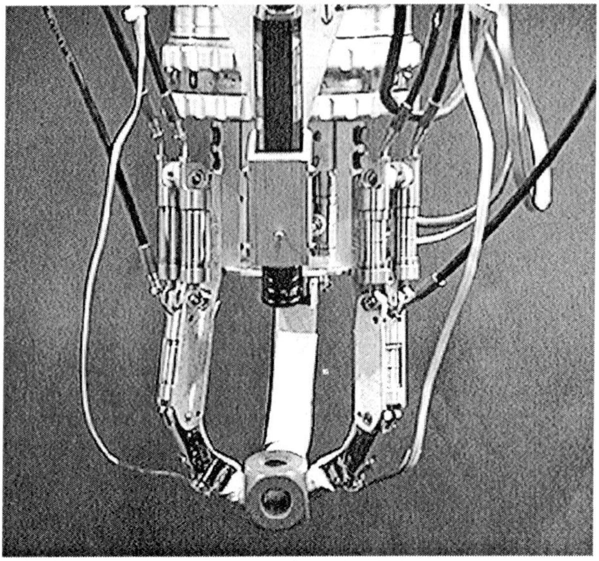

Fig. 1. The three-fingered hydraulic hand holding an object. A camera in mounted in front, looking at the object and the fingertips.

fingertip positions on the object are sometimes even difficult to see for humans (Fig. 2). In our approach, these problems are solved defining typical success- or failure-situations known from robotic experience as classes for a neural classifier, thus achieving a qualitative judgement on the situation.

2 The Robot Setup

Our system is based on a standard industrial robot arm (PUMA 500) with six degrees of freedom. The end-effector is an oil-hydraulic robot hand developed by Pfeiffer et al. [10]. It has three equal fingers mounted in an equilateral triangle, pointing in parallel directions (Fig. 1). Each finger has three degrees of freedom: bending the first joint sideways and inward at the wrist, and bending the coupled second and third joint. The oil pressure in the hydraulic system serves as a sensory feedback. A detailed description of the setup can be found in [7]. Fig. 1 shows also a force/torque wrist sensor which will not be used in these experiments.

Currently the robot system is used within a man-machine interaction scenario where a human instructor tells the robot to grasp an object on a table. An active stereo camera head apart from the robot detects the hand movements of the instructor as well as the objects. The world coordinates of the indicated object are given to the robot system as soon as a pointing gesture could be detected. The robot arm then carries out an approach movement until the hand camera detects the object to control the final approach. The object is then grasped by the three-fingered hand. This system is described in [11].

Up to recently, the only feedback on the success of the grasping movement was the sensed hydraulic oil pressure. By this data, however, it can only be judged if the object is still in the grasp or if it was lost. Though there are additional position sensors on the hydraulic motor pistons, these position measurements cannot be used to estimate the resulting finger positions within the required accuracy due to mechanical hysteresis.

Consequently, a supplementary system has been developed which checks the grasping position visually using the hand camera. We will first characterize the task of this system by some examples. Fig. 2 shows a cube shaped object with a hole at each side in the grasp of the hand. The row above shows as an overview the hand from the side, below are the corresponding views of the hand camera. Besides checking if the object is in the hand at all, the questions relevant to robotics which the hand camera should answer are the following:

1. In which position or pose is the object?
2. Which of the fingertips have supporting contact?

It should be noted that both these questions can only in part be answered in general, like that e.g. three contact points give usually more stability than two. However, whether a certain fingertip position does really support the object or not depends on the type of the object. Similarly, the absolute position of the object held in the hand is not necessarily relevant for all tasks. Whether a position is useful depends on the further proceeding, e.g. putting the object down or attaching it to other objects. Therefore, object and task specific knowledge must be introduced by defining appropriate judgement-classes to be trained by the recognition system. This will be outlined in section 4.

3 Visual Recognition System

The recognition system for the hand camera is composed of two modules (left and right in Fig. 3): (a) a simple location module to find the fingertip regions and (b) a classification module (VPL-classifier, section 3.3) of which two instances (VPL1 and VPL2) are used: one for the classification of the grasped object and one for the evaluation of the three fingertip regions.

Image processing is carried out on two resolutions (images I_1 and I_2, not shown in Fig. 3): The camera input is a grey level image sub-sampled to 192×144 pixels (image I_1), which is used as input for the classifier. I_2 is an even further sub-sampled version of the camera image of 96×72 pixels, it is used for ROI selection.

3.1 Locating the Robot Fingertips

In order to keep ROI selection as simple and fast as possible, we use black background and the robot fingers are covered with white tape. The objects have

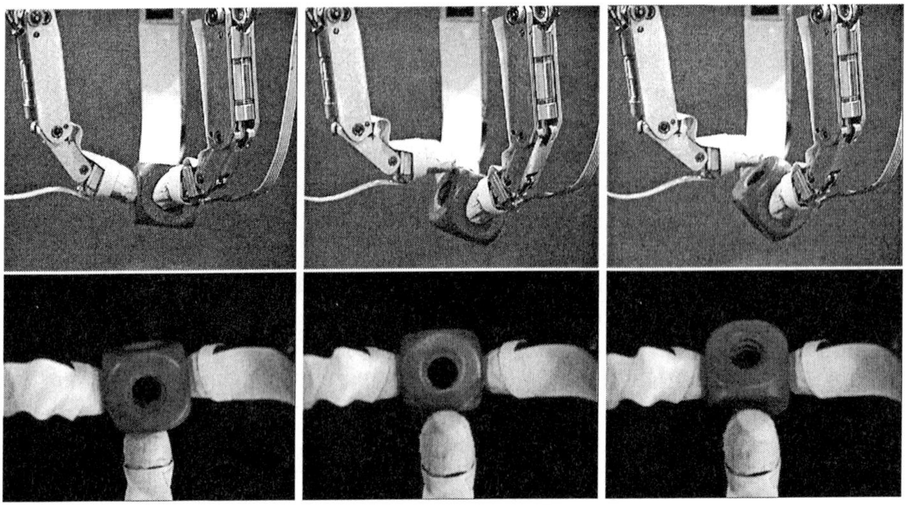

Fig. 2. Three grasping positions, above view from the side, below view of the hand camera. Left: Object held correctly by all three fingers. Middle: Object held only by two fingers (left finger has lost its grip) but still stable and in similar position like (Left). Right: Like (Middle), but object in a different position. Note how small the differences appear from the position of the hand camera.

intensity values between those of the background and the fingers, thus the finger-regions can be found by a simple intensity based binarization: Starting with a high threshold $\vartheta = \vartheta_{start}$ ($\approx$ finger intensity) on image I_2, ϑ is decreased until three blobs of a minimum size and a priori defined form features can be found by a connectivity analysis of the binarized image. Since the fingers are brighter than both objects and background, it is made sure the three blobs correspond to the fingers.

To locate the finger*tips*, the centers of gravity $\boldsymbol{C}_i, i = 1, 2, 3$ and the second moments of the blobs are calculated. Starting from the $\boldsymbol{C}_i$, pixels are tested along the major axis towards the image center until the first non-blob pixel is encountered, thus fingertip i is reached. Of course, this procedure is highly special purpose but justified by the computational speed. It could be easily replaced by other localization methods as e.g. Eigenspace projection ("Eigenfingertips"). However, the localization is not the main topic of this paper.

3.2 Windows for Classification

Four classifications have to be carried out to get the full information on the scene under the robot hand: VPL1 classifies a 65×65 window of I_1 centered between the fingertips at $\boldsymbol{C} = 1/3 \sum_i \boldsymbol{C}_i$. The result is the identity and position of the object.

VPL2 classifies three windows of size 45×45 from I_1 independently of each other, which are centered at the fingertips. The size is chosen such that a part of the finger is included as well as a part of the object. Thus the window provides information on the fingertip position, the contacted part of the object, the object position and the relative fingertip-object position. This is the basis for a judgement on the type of "grasping situation" by the classifier. The choice of a set of such prototypical grasping situations is crucial for the performance (section 4).

3.3 VPL-Classifier

The classifier is a trainable system based on neural networks which performs a mapping $\boldsymbol{x} \rightarrow \boldsymbol{y}, \boldsymbol{x} \in \mathbb{R}^M, \boldsymbol{y} \in \mathbb{R}^N$. In this case, the input dimension M is the number of pixels of the windows. The window vector $\boldsymbol{x}$ has to be mapped to a discrete valued output k that denotes the class. There is one separate output channel for each of the N output classes. Training is performed with hand-labeled sample windows $\boldsymbol{x}_i^{Tr}$ and binary output vectors $y_i^{Tr} = \delta_{ij}, i = 1 \ldots N$, coding the class $1 \leq j \leq N$. Classification of unknown windows $\boldsymbol{x}$ is carried out by taking the class k of the channel with maximal output: $k = \arg\max_i(\boldsymbol{y}_i(\boldsymbol{x}))$.

The VPL-classifier combines visual feature extraction and classification. It consists of three processing stages which perform a local principal component analysis (PCA) as feature extraction followed by a classification by neural expert networks. Local PCA can be viewed as a nonlinear extension of simple, global PCA [13]. "VPL" stands for the three stages: **V**ector quantization, **P**CA and **L**LM-network. The vector quantization is carried out on the raw image windows to provide a first data partitioning with N_V reference vectors $\boldsymbol{r}_i \in \mathbb{R}^M, i = 1 \ldots N_V$. For vector quantization the Activity Equalization Algorithm is used, which is an extension of the well-known "winner takes all" method that prevents node under-utilization by monitoring the average node activities. [3].

To each reference vector $\boldsymbol{r}_i$ a single layer feed forward network for the successive calculation of the principal components (PCs) as proposed by Sanger [12] is attached which projects the input $\boldsymbol{x}$ to the $N_P < M$ PCs with the largest eigenvalues: $\boldsymbol{x} \rightarrow \boldsymbol{p}_l(\boldsymbol{x}) \in \mathbb{R}^{N_P}, l = 1 \ldots N_V$. To each of the N_V different PCA-nets one expert neural classifier is attached which is of the Local Linear Map – type (LLM-network), see e.g. [4] for details. It does the final mapping $\boldsymbol{p}_l(\boldsymbol{x}) \rightarrow \boldsymbol{y} \in \mathbb{R}^N$. The LLM-network is related to the self-organizing map [8]. It can be trained to approximate a nonlinear function by a set of locally valid linear mappings.

The three processing stages are trained successively, first vector quantization and PCA-nets (unsupervised), finally the LLM-nets (supervised). Classification of input $\boldsymbol{x}$ is carried out by finding the best match reference vector $\boldsymbol{r}_{n(\boldsymbol{x})}$, then mapping $\boldsymbol{x} \rightarrow \boldsymbol{p}_{n(\boldsymbol{x})}(\boldsymbol{x})$ by the attached PCA-net and finally mapping $\boldsymbol{p}_{n(\boldsymbol{x})}(\boldsymbol{x}) \rightarrow \boldsymbol{y}$.

The major advantage of the VPL-classifier is its ability to form many highly specific feature detectors (the $N_V \cdot N_P$ local PCs), but needing to apply only $N_V + N_P - 1$ filter operations per classification, thus large object domains can be represented and detected without the need of many (and computationally expensive) filter operations. The VPL-classifier has been successfully applied

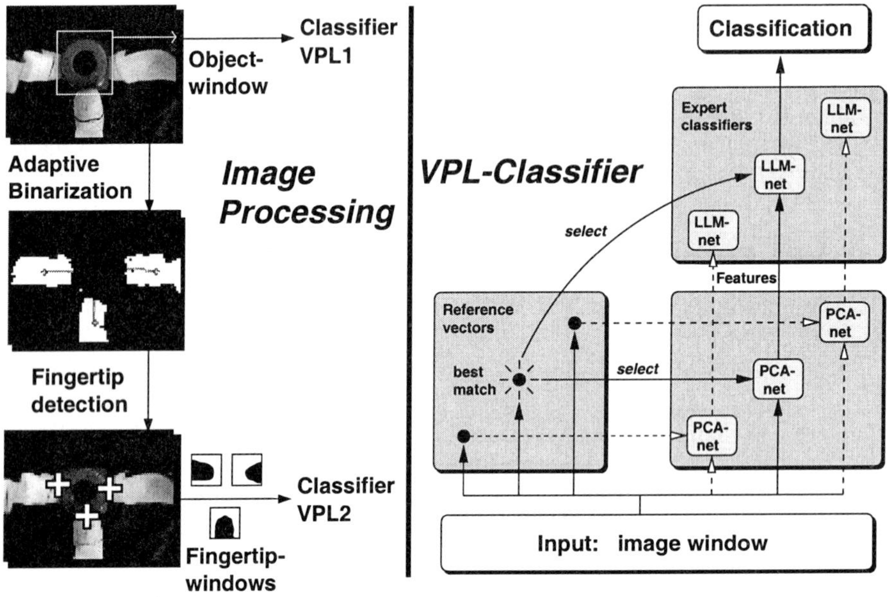

Fig. 3. System architecture. Left: From the input image the regions of the fingers can be easily segmented and the fingertips located (section 3.1). Around each fingertip a window is cut out of the input image (section 3.2). The three windows are fed to the VPL-classifier which classifies the "grasping situation" for each fingertip separately (section 3.3).

to several vision tasks, for details see [4]. Especially it could be shown that classification performance and generalization properties are well-behaved when the main parameters, i.e. N_V, N_P and the number of nodes in the expert LLM-nets N_L are changed.

4 Results

VPL1 classifies the N_O different objects and additionally up to N_Q qualitatively different classes for the position, so the VPL output dimension is $N = N_O + N_Q$. Fig. 4 shows typical grasping situations in which the object position matters most: The first two situations (from left to right) are stable (classes 1,2). The third is unstable though all fingers touch the object because the object can fall out of the grasp to the right (class 3). The fourth is an unstable version of the first (object tilted, class 4). So there are four position classes for this object. VPL2 needs not to be evaluated.

By contrast, Fig. 2 shows an object where the stability of the grasp cannot be easily derived from the object position. Hence, after recognition of the object type by VPL1, the "grasping situation" on the fingertips has to be judged by VPL2. The left and right fingers are in a correct position in all three situations,

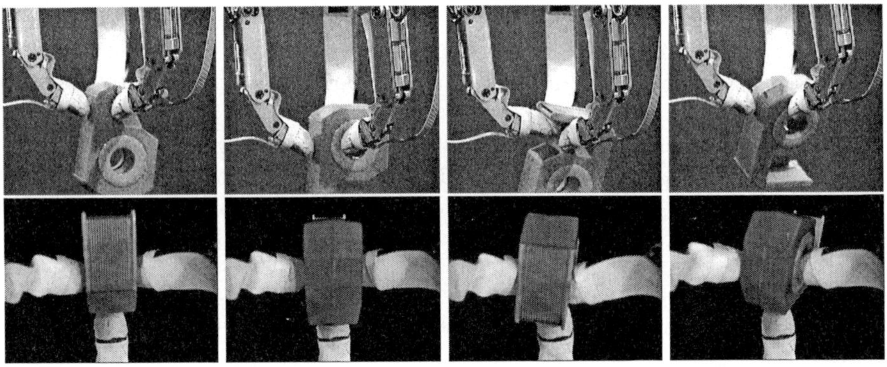

Fig. 4. Typical grasping situations in which the object position is decisive for grasp stability. Above view from the side, below the same situations as seen by the hand camera. From left to right: (1,2) stable position, (3) unstable though all fingers touch the object because the object can fall out of the grasp to the left, (4) unstable version of the (1), since object is tilted.

but the middle finger is lose in the second and third scene due to too strong bending. Note this causes a noticeable tilt of the object only in the third situation, not in the second (in the view of the hand camera!). So this object needs two classes for VPL2: finger correct / above.

We used in total $N_O^{VPL1} = 5$ objects with up to $N_P^{VPL1} = 4$ position classes for VPL1. Note the definitions of the position classes differ from object to object, also there may be fewer classes than N_P^{VPL1}. The architecture could be refined by using one VPL to classify the object type and then select another VPL specialized to the position classes of this one object. But the current scenario is simple enough to be covered by a single VPL1. For VPL2, $N^{VPL2} = 5$ classes were used to judge the position of the fingertips relative to the objects: "correct", "below", "above", "right" and "left".

The system was tested on a series of 60 images using the "leaving one out" method. For VPL1, $N_V = 3, N_P = 6, N_L = 9$ proved to be sufficient to reach a rate of 96% correct classifications. For VPL2, with $N_V = 4, N_P = 8, N_L = 20$ 88% correct classifications could be reached. The slightly larger nets (as compared to VPL1) and the lower classification rate reflect the greater number of degrees of freedom since the views of the fingertips are from different directions. Moreover, there are partial occlusions and overshadowing.

5 Discussion and Acknowledgement

A system for the qualitative visual judgement on the success or failure of a grasping action of a robot hand was presented. As the judgement has to provide information that can be used for further robotic actions, a set of "grasping situations" was defined for each object. Still, the number of objects and grasping situations is small. As we will get more images and more complex scenes, the architecture will have to be enlarged by a single object classifier which activates specific object-position– and fingertip-position– classifiers individual for each object. Moreover, in the future sensor fusion between visual and oil pressure data will have to be developed.

This work was supported by the Deutsche Forschungsgemeinschaft (DFG) within research Project SFB 360 "Situated Artificial Communicators".

References

1. AMP Incorporated, Valley Forge, PA 19482. *Piezo Film Sensors Technical Manual*, 1993.
2. G. Canepa, M. Campanella, and D. de Rossi. Slip detection by a tactile neural network. In *Proc. ICIROS 94*, volume 1, pages 224–231, 1994.
3. G. Heidemann and H.J. Ritter. Efficient Vector Quantization using the WTA-rule with Activity Equalization. *Neural Processing Letters*, 13(1):17–30, 2001.
4. G. Heidemann. *Ein flexibel einsetzbares Objekterkennungssystem auf der Basis neuronaler Netze.* PhD thesis, Univ. Bielefeld, Techn. Fak., 1998. Infix, DISKI 190.
5. R.D. Howe and M.R. Cutkosky. Dynamic tactile sensing: Perception of fine surface features with stress rate sensing. *IEEE Transactions on Robotics and Automation*, 9(2):140–150, 1993.
6. Interlink Electronics, Europe, Echternach, G.D. de Luxemburg. *The force sensing resistor*, 1990.
7. Ján Jockusch. *Exploration based on Neural Networks with Applications in Manipulator Control.* PhD thesis, Universität Bielefeld, Technische Fakultät, 2000.
8. T. Kohonen. Self-organization and associative memory. In *Springer Series in Information Sciences 8*. Springer-Verlag Heidelberg, 1984.
9. H. Liu, P. Meusel, and G. Hirzinger. A tactile sensing system for the DLR three-finger robot hand. In *Proc. ISMCR 95*, pages 91–96, 1995.
10. R. Menzel, K. Woelfl, and F. Pfeiffer. The development of a hydraulic hand. In *2nd Conf. on Mechatronics and Robotics*, pages 225–238, 1993.
11. Robert Rae. *Gestikbasierte Mensch-Maschine-Kommunikation auf der Grundlage visueller Aufmerksamkeit und Adaptivität.* PhD thesis, Universität Bielefeld, Technische Fakultät, 2000.
12. T.D. Sanger. Optimal unsupervised learning in a single-layer linear feedforward neural network. *Neural Networks*, 2:459–473, 1989.
13. Michael E. Tipping and Christopher M. Bishop. Mixtures of probabilistic principal component analyzers. *Neural Computation*, 11(2):443–482, February 15 1999.
14. M.E. Tremblay and M.R. Cutkosky. Estimating friction using incipient slip sensing during a manipulation task. In *Proc. ICRA 93*, volume 1, pages 429–434, 1993.

Anticipation-Based Control Architecture for a Mobile Robot*

Andrea Heinze and Horst-Michael Gross

Ilmenau Technical University, Department of Neuroinformatics, P.O.B. 100565, D-98684 Ilmenau, Germany
{heinze,homi}@informatik.tu-ilmenau.de

Abstract. We present a biologically motivated computational model that is able to anticipate and evaluate multiple hypothetical sensorimotor sequences. Our Model for Anticipation based on Cortical Representations (MACOR) allows a completely parallel search at the neocortical level using assemblies of rate coded neurons for grouping, separation, and selection of sensorimotor sequences. For a vision-controlled local navigation of a mobile robot Khepera, we can demonstrate that our anticipative approach outperforms a reactive one. We also compare our explicitely planning approach with the implicitely planning Q-learning.

1 Introduction

Based on findings for the sensorimotor character of perception [1, 2], we developed an alternative approach to perception that avoids the common separation of perception and generation of behavior and fuses both aspects into one consistent neural process. In this approach, perception of space and shape in the environment is regarded to be an active process which anticipates the sensory consequences of alternative hypothetical interactions with the environment, that could be performed by a sensorimotor system, starting from the current sensory situation. This approach is supported by biological findings. For example, it was shown that such planning of motor actions takes place in the secondary motor areas [3]. Thach (1996) found that the premotor parts of the brain are active both in planning movements to be executed as well as in thinking about movements that shall not be executed.

Based on these findings, we developed our Model for Anticipation based on Cortical Representations MACOR, presented in sec.2. It is intended as a general scheme for sensorimotor anticipation in a neural architecture. The model does not attempt to provide a detailed description of a specific cortical or subcortical structure, but we try to capture some general properties that are relevant to our "perception as anticipation"-approach in brain-like systems (for details see [4]). The objective of this paper is to demonstrate the efficiency of our anticipatory approach for a real-world sensorimotor control problem, the local navigation and obstacle avoidance of a vision-controlled mobile robot showing non-holonom movement characteristics. We compare the achieved navigation behavior with

* This work was partially supported by DFG Graduate college GRK 164.

G. Dorffner, H. Bischof, and K. Hornik (Eds.): ICANN 2001, LNCS 2130, pp. 899–905, 2001.

other non-anticipative approaches like reactive control and Q-learning-based control. It is important to note that the MACOR-concept is also suitable for other sensorimotor or cognitive tasks that must consider alternative action sequences and their hypothetical results.

2 Model for Anticipation

In the framework of a visually-guided navigation task (sec.2.3), our architecture (see fig.1) processes as sensorimotor information a visual input yielded from an omnicamera, supplemented by the last motor command executed. Within the architecture, those inputs are represented by a Fuzzy ART architecture [5]. The visual part is represented by the F2-nodes of a Fuzzy ART network. Each of these nodes contains several neurons, which represent the motor commands available to bring the system into the considered visual situation (visuomotor column). Between the resulting visuomotor representation of the current situation and that of the preceeding one, connections incorporating the certainty of that transition and its evaluation are adapted (sec.2.1).

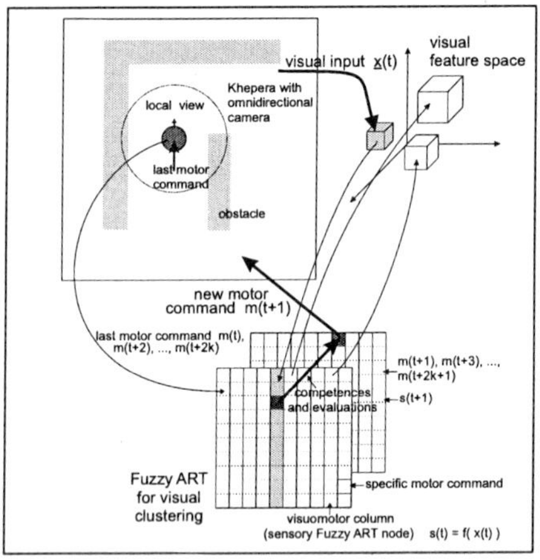

Fig. 1. Overview of the architecture. To represent visuomotor inputs, each Fuzzy ART node representing a specific visual input contains several neurons to represent motor commands available to bring the system in this situation. Connections between the present and the preceeding visuomotor situations are adapted to reflect the reinforcement incurred during that particular transition and the competence gained. By activation of a neuron and the subsequent spread of activity through its weights to activate other neurons, alternative sensorimotor sequences may be generated.

This architecture embeds the functionality required for a *parallel* generation of sequences of sensorimotor hypotheses (sec.2.2). In the context of the biological foundations of the architecture, this parallel generation is realized by a spread of activity through connections between the visuomotor columns. In the following investigations, we use a restricted set of discrete motor commands to investigate the effect of anticipative behavior (see sec.2.3, too).

2.1 Learning within the Map

The learning of sensorimotor connections between the present and the preceding visuomotor situations takes place after each executed motor command. To evaluate the certainty of the existance of sensorimotor connections, we investigated several approaches, for example actual transition probabilities. Because of the clustering of the input space, high competence weights between neurons in the same sensorimotor assembly are established. The correspondingly lower competence weights between different situations yield lower sequence evaluations. These sequences are subsequentely not selected, but important for the consideration of movements resulting in a collision and thus for achieving of system goals.

Using the simple learning equ.1 realized as a sigmoidal function for **competence weights**, a specific weight can quickly reach a maximum value $w^{\mathrm{c}}_{\mathrm{max}}$, determined by the parameter μ and σ. Thus all weights hold the same value after a sufficiently high number of adaptation steps, such that the action selection (sec.2.2) depends only on the values of the evaluation weights. Therefore, the competence weights are only meaningful for unknown transitions, which are devalued by the competence. In equ.1, z_{ij} is the number of adaptations of the respective weight.

$$w^{\mathrm{c}}_{ij}(t) = \frac{1}{1+e^{\frac{z_{ij}-\mu}{\sigma}}} \qquad \text{where} \quad \mu = 2.8 \quad \text{and} \quad \sigma = 0.4 \tag{1}$$

For learning of **evaluation weights**, a simple form of reinforcement learning is used. Thus, the evaluation weights hold only the expected evaluation for the next transition, yielding a purely reactive behavior. Only by means of internal simulation, an anticipative behavior can be generated. The reinforcements r yield the evaluation of the generated behavior. In our experiments, they were chosen to reward a collision free, straight navigation behavior and to punish turns and collisions. For the adaptation of evaluation connections w^{r}_{ij}, the number z of adaptations of the respective weight w_{ij} is used to determine an adaptive learning rate $\frac{1}{z^r_{ij}}$.

$$w^{\mathrm{r}}_{ij}(\mathrm{t{+}1}) = w^{\mathrm{r}}_{ij}(\mathrm{t}) + \frac{1}{z^r_{ij}} \cdot (\mathrm{r} - w^{\mathrm{r}}_{ij}(\mathrm{t})) \tag{2}$$

For comparison, a special form of reinforcement learning, Q-learning [6, 7] was also investigated, because it is a model free, but implicitely planning approach.

$$w^{\mathrm{r}}_{ij}(\mathrm{t{+}1}) = w^{\mathrm{r}}_{ij}(\mathrm{t}) + \frac{1}{z^r_{ij}} \cdot (\mathrm{r} + \gamma \cdot w^{\mathrm{r}}_{ki}(t) - w^{\mathrm{r}}_{ij}(\mathrm{t})) \tag{3}$$

$$w^{\mathrm{r}}_{ki}(t) = \max_l w^{\mathrm{r}}_{li}(t) \tag{4}$$

By choosing the parameter γ to control the planning horizon $\gamma > 0$, the prediction of evaluations of subsequent transitions can be stored in one connection (strong Q-learning, equ.3 and 4). Q-learning becomes a planning free approach by choosing $\gamma = 0$ (weak Q-learning, equ.2).

2.2 Generation and Evaluation of Sensorimotor Sequences

For generation of sensorimotor sequences, a specific neuron in a Fuzzy ART assembly is activated by the current sensory situation and the last executed motor command. This neuron propagates its activity y_j^α to all other interconnected neurons $i \in S = [0, n \cdot m - 1]$ using its competence connections w_{ij}^{c}, where n is the number of sensory assemblies and m the number of motor neurons within each assembly. The activated neurons may in turn activate further neurons, resulting in a mechanism of internal simulation and thus the generation of whole sequences of sensorimotor hypothesis (equ.5).

$$y_i^\alpha(t+1) = \max_{j \in S} w_{ij}^{\mathrm{c}}(t) \cdot y_j^\alpha(t) \tag{5}$$

Since the maximal value of the competence connection is less than 1.0, a subsequent neuron will always be less activated than its predecessor. Also, a neuron can only be activated by its maximum input activity. This supplies a simple stopping criterion for the propagation of interconnecting sequences. Simultaneous to the parallel generation of sequences of sensorimotor hypotheses, the model realizes the selection of the best evaluated sequence by a backpropagation of local sequence evaluations, as shown in equ.6. This backpropagation starts as the start neuron activates further neurons and each evaluation is backpropagated to the respective sequence predeccesor.

$$y_i^\beta(t+1) = w_{ij}^{\mathrm{r}}(t) \cdot y_i^\alpha(t+1) + \max_{r \in S} y_r^\beta(t) \tag{6}$$

The activity backpropagated to the start neuron represents the highest sequence evaluation in each time step. Thus in each time step, an action selection is possible, which improves as internal simulation goes on. This mechanism yields high cumulated sequence evaluations for well known and highly evaluated transitions. To realize a reactive behavior, the process of internal simulation runs only for one time step. This means the starting neuron activates further neurons which then propagate their evaluations directly back onto the starting neuron.

2.3 Experimental Scenario

Because of their embodiment and situatedness, robots are ideal systems to demonstrate the advantages of an anticipation based sensorimotor control compared to a reactive one. To navigate successfully, for example, to avoid obstacles or to go through narrow passages, they have to consider their physical and mechanical properties and constraints (e.g. inertia, holonom or nonholonom kinematics). A mobile system, a robot or an animal, that is not able to learn and consider its constraints and their sensorimotor consequences, will not be able to evolve successful navigation behaviors.

To demonstrate the advantages of an anticipation based sensorimotor control, we used the mobile robot Khepera as a non-holonom system. Because of the used restricted action space, our system has to consider its constraints and is forced to start an early obstacle avoidance by internal simulation to realize a successful navigation behavior.

In our investigations, the visual sensory inputs were provided by an omnicamera. After a transformation into a physiological color space [8] specially tuned receptive fields extract the color distribution around the robot, which gives an implicit description of the obstacle arrangement.

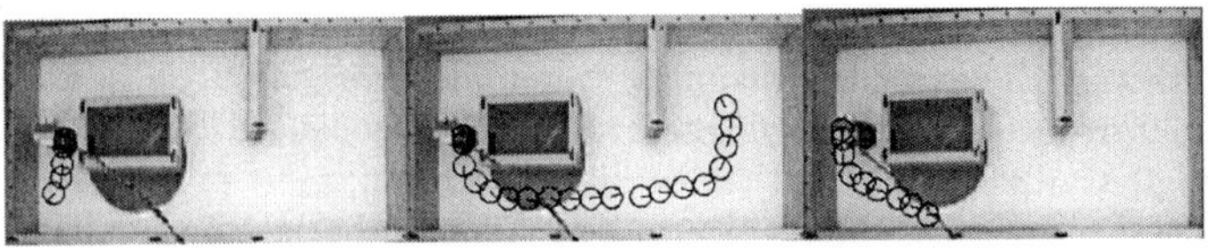

Fig. 2. In the experiments, a trial ended either after a collision or a maximum of 40 steps. The generated behavior shows that only the explicitly planning system (**middle**) was able to realize a successful obstacle avoidance considering the physical limitations of the non-holonom robot with a very limited action space. After the very first detection of a curve or obstacle, this system began to move to the left (as second selected action) and pursued that movement until the turn was negotiated. In contrast, the reactive system (**left**) chose to drive straight forward until almost hitting the wall. Only directly in front of the wall it chose to turn right, thus unable to avoid a collision. The Q-learning system (**right**) initiated also an early turn to the left and continued until the turn was negotiated. Subsequently, the action 'go straight' was repeatedly chosen, because the visual situation in the middle of the floor is represented by the same Fuzzy ART node as the floor situation near the wall. Only in the situation directly in front of the wall, another Fuzzy ART node was activated and a turn initiated. Like Q-learning, our reactive approach also learned the action 'go straight' for the floor situation, but by means of internal simulation, it chose mostly turns.

3 Results

To demonstrate the advantages of anticipative systems, i.e. systems featuring explicit planning, we provide a comparison between the following approaches:

- the presented system which explicitely plans by internal simulation,
- a purely reactive system which operates exclusively on the current sensorimotor situation, and
- an implicitly planning approach, Q-learning.

For the investigation of anticipation-based behavior, we used Fuzzy-ART modules with 50 nodes (vigilance $\rho = 0.9$) for sensory clustering and three executable actions per sensory assembly. A sequential learning process was employed to structure the sensory clusters and to adapt the competence and evaluation weights involving more than 10.000 typical sensorimotor situations. In Q-learning we used a planning horizon of $\gamma = 0.9$. Systems except Q-learning, which does not use competence weights, utilized equ.1 to adapt their competence weights. The behavior generated by the different systems was recorded over 10 trials each and is exemplarily depicted in fig.2.

Further, in each trial the achievable total reinforcement was determined as the sum of the individual reinforcements received after each motor command. The respective mean values and variances together with typical trial lengths and the mean reinforcements for a single motor command are shown in table 1.

These results give a strong indication that only the explicitly planning system is able to realize an obstacle avoidance for the inert, non-holonom roboter system. In comparison, the reactive system did not produce any avoiding actions at all.

Table 1. Mean reinforcement and standard deviation (std) for 10 trials of the different systems. Although the explicitly planning system achieves the highest mean reinforcement per trial, it also yielded the highest std. This was caused by 4 of those 10 trials (maximal trial length was 40) that ended due to collision, most probably as a matter of the hardware fixation of the omnicamera. The trials of the reactive and the Q-learning system all ended by collision and therefore in lower trial lengths. Further the mean reinforcements per step (not considering the step that resulted in a collision) show that the explicitely planning system got the lowest reinforcement per step as a result of more turns executed, which yield lower reinforcements than the movement 'go straight'.

criteria		reactive system	explicitly anticip. system	Q-learning
reinforcement	mean	2.8	15.6	4.2
per trial	variance	0.1	8.6	2.4
trial length	mean	7.2	29.2	11.7
	variance	0.4	14.2	4.2
reinforcement per step	mean	0.61	0.55	0.57

Though the Q-Learning system chose mostly proper motor commands, it could not overcome the perhaps non-optimal sensorimotor representation.

4 Conclusions and Outlook

We presented MACOR that is able to anticipate and evaluate multiple hypothetical sensorimotor sequences. Using a framework of local navigation of a nonholonome robot, the advantage of anticipation based control was demonstrated empirically. The investigations performed demonstrate the advantage of anticipative systems compared to reactive systems and especially, the advantage of explicitly planning systems compared to implicitly planning ones.

This comparison is the basis of further investigations targeting a more complicated scenario, dealing with the aspect of changing system goals. These investigations are intended to show the advantages of reactively stored evaluation weights compared to weights whose values represent an implicit planning horizont, like Q-learning. These advantages become obvious for system goals that change over time, and which thus require readaptations of the sensorimotor transitions. With reactive evaluation weights in combination with internal simulation, only the actually different sensorimotor transitions must be relearned, which are only a few, to reflect the new system goal. So we want to show that an explicitly planning system may realize a flexible behavior according to a new task much faster than an implicitly planning system like Q-learning.

Further we want to investigate obvious similarities to probabilistic modelling techniques involving Hidden Markow Models [9] in detail. Besides some similarities in structure and purpose, there are a number of conceptual differences concerning, e.g., the learning strategies, the biological plausibility, or a prospective parallel feasibility.

References

1. M.A. Arbib, P. Erdi, and J. Szentagothai. *Neural Organization: Structure, Function and Dynamics.* MIT Press, 1998.
2. R. Pfeifer and C. Scheier. From Percept. to Action:The Right Direction? In *Proc.of PerAc'94*, pages 1–11. IEEE Comp.Soc.Press, 1994.
3. W. T. Thach. On the specific role of the cerebellum in motor learning and cognition: Clues from PET activation and lesion studies in man. *BBS*, 19(3):411–431, 1996.
4. H.-M. Gross, A. Heinze, T. Seiler, and V. Stephan. Generative Character of Percept.: A Neural Architecture for Sensorimotor Anticipation. *NN*, 12:1101–1129, 1999.
5. G. Carpenter et.al. Fuzzy ARTMAP: A Neural Netw. Architure for Incr. Supervised Learning of Analog Multidimensional Maps. In *NN*, pages 698–719, 1992.
6. C. Watkins. *Learn. from delayed rewards.* PhD thesis, King's Coll., Cambridge, UK, 1989.
7. R. S. Sutton and A. G. Barto. *Reinforcement Learning: An introduction.* MIT Press: A Bradford Book, Cambridge, MA, USA, 1998.
8. T. Pomierski and H.-M. Gross. Biological Neural Architecture for Chromatic Adaptation Resulting in Constant Color Sensations. In *Proc.IEEE Int. Conf. on NN*, pages 734–739, 1996.
9. L. R. Rabiner. A Tutorial on Hidden Markov Models and Selected Applications in Speech Recognition. *Proc. of IEEE*, 77(2):257–286, 1989.

Neural Adaptive Force Control for Compliant Robots

N. SAADIA(*), Y. AMIRAT(**), J. PONTNAUT(**) and A. RAMDANE-CHERIF(***)
(*) Laboratoire de robotique, Institute d'éléctronique. Université, USTHB, BP 32 Babs Ezzouar ,Alger, Algeria.
(**) LIIA-IUT de Creteil, 122 Rue Paul Armangot 94400 Vitry Sur Seine Cedex France
(***) Lab. PRiSM, Université de Versailles St-Quentin en Yvelines, 45 av. des Etats Unis 78035 Versailles, France
Email: saadia_nadia@hotmail.com
Email: amar.ramdane-cherif@prism.uvsq.fr

Abstract. *This paper deals with the methodological approach for the development and implementation of force-position control for robotized systems. We propose a first approach based on neural network to treat globally the problem of the adaptation of robot behavior to various classes of tasks and to actual conditions of the task where its parameters may vary. The obtained results are presented and analyzed in order to prove the efficiency of the proposed approach*

1 Introduction

Recently, the robot control problem during the phase of contact with the environment has received considerable attention in the literature [8] due to its complexity. Assembly tasks constitute a typical example of such an operation. These tasks are especially complex to achieve considering the stern constraints that they imply as the intrinsic parameters type of the task: mechanical impedance of the environment is non-linear and submitted to temporal and spatial changes, variable contact geometry according to the shape of parts, variable nature of surfaces. These difficulties increase when the system has to manipulate a class of different objects. In this paper, we focalize on the problematic of the force control of assembly tasks which constitutes a class of compliance tasks.

To obtain a robot's behavior, satisfying different situations of utilization, it necessary to design non linear adaptive control laws. The elaboration of these controls allows to overcome the difficulty to synthesize suitable control lows that are satisfactory over a wide range of operating points. Thus, several approaches are possible such as conventional adaptive controls [10], alternate methods or modern adaptive control such as intelligent control which uses experimental knowledge. The different approach analysis like H_∞ [6] [5] are not suited to our context because most of these strategies are complex to implement in real time context due to the dynamics of the system which requires short time sampling (less than 10 milliseconds) [7].

Learning adaptive control based on the utilization of connectionist techniques allows to solve directly the problem of non linearity of unknown systems by functional approximation. The objective of this paper is to propose tools which allow us to increase the control performances of contact interactions established between a robot and its environment, in complex situations by means of adaptability.

G. Dorffner, H. Bischof, and K. Hornik (Eds.): ICANN 2001, LNCS 2130, pp. 906–913, 2001.

In the following of this article, we first present in section 2 a structure of type external forces control, which use the technique of reinforcement learning. In section 3, we describe the implementation of the proposed approach for the realization of various piece insertion according to different tolerances and speeds. In the last section, the experimental results obtained are presented and analyzed.

2 Neural Adaptive Control Structure

A control structure based on an external force control scheme is proposed (Figure 1). Indeed, this structure of control consists in hierarchical force control loop with regard to the position control loop. The law of force control that intervenes in the diagram works out, from the mistake by force (ΔF) between the desired contact force torque F_d and the measured one F_{cm}, an increment of position ΔX that modifies the task vector Xd according to the following relationship:

$$\Delta X = C_f(\Delta F) \tag{1}$$

where C_f represents the forces control law.

Considering the existing coupling between the different types of forces to which the robot is submitted (contact forces, gravitational forces end forces owed to the system's dynamics), for extract the contact force values from the measured ones, a FFNNM is used to estimate free motion forces.

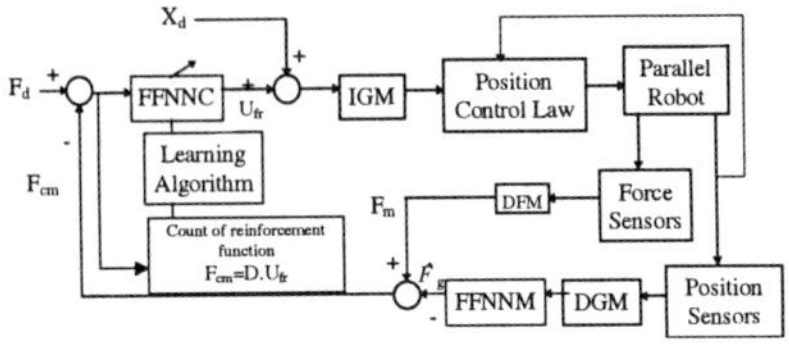

Fig. 1. Adaptive controller using associative reinforcement paradigm: DGM: the direct geometric model of parallel robot; IGM: the inverse geometric model of parallel robot; DFM: direct force model; X_d: the desired position trajectory; F_d: the desired forces trajectory; $\hat{F}_g$: the estimate free motion force torque; F_m: the measured forces torque; F_{cm} the estimate contact force torque.

2.1 Estimation of Contact Forces

In order to estimate the contact forces from the measured ones in a compliant motion, we propose an identification model based on a multi-layer neural network. In our approach, the task is released with constant velocity. This FFNNM is trained off-line using the data obtained from the robot's displacements in the free space. In such a situation, gravity and contact forces are de-coupled as contact forces are null. To build a suitable identification model (Fig 2), we proceed as follows: when the input $X_{m,k}$ is applied to the FFNNM, the corresponding output is $\hat{F}_{g,k}$ which approximates F_g such as:

$$\left\| \hat{F}_g - F_g \right\| = \left\| \hat{F}_g(X_{m,k}) - F_g(X_{m,k}) \right\| \leq \varepsilon \tag{2}$$
$$X_{m,k} \in P$$

where $\boldsymbol{\varepsilon}$ is the mistake of optimization, F_g the free motion force tensor in the operational space, $\hat{F}_g$ the estimated one and $X_{m,k}$ the position and orientation end-effector vector.

The input vector I_m of the neural network identifier model of free motion force is the position vector $X_m(k)$ of the end effector. The out put of the FFNNM can be computed using the input and the neural network's parameters W_m as:

$$\hat{F}_g = N_m(W_m, I_m) \tag{3}$$

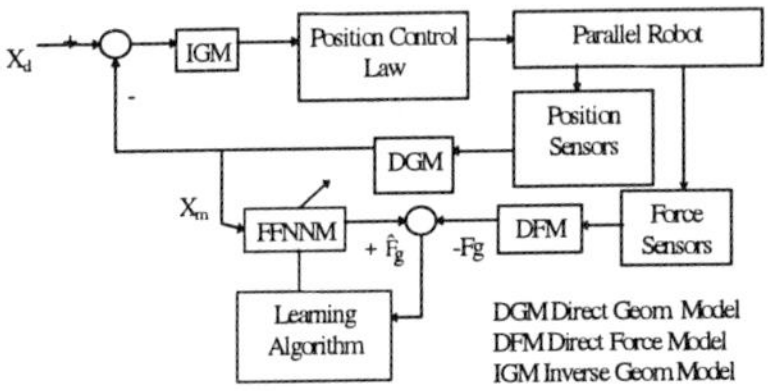

Fig. 2. FFNNM learning a free motion forces

2.2 Initialization of Neural Controller

The initialization of neural networks controller is done off line. The learning structure is given figure 3. While achieving a control of the system in the task space with a conventional controller of PID type, we constructed a basis of training (figure 3). This one is constituted of measured forces torque Fcm, and wanted ones F_d and control vector U_{fd} generated by the conventional controller. Values of mistakes to the previous sampling instants are taken in amount in the learning. Parameters of the controller PID have been adjusted experimentally for a speed of insertion of 5 mm/s of way to get a satisfactory behavior of the point of view of the minimization of contact forces. The period of sampling of the control is 5 ms. The input vector Ic of neural controller is constituted by force error vector $e = F_d - F_{cm}$. In this step the FFNNC is trained off-line using the data obtained from the robot's displacements in the constrained space under classical external force control. The learning structure is presented figure 3.

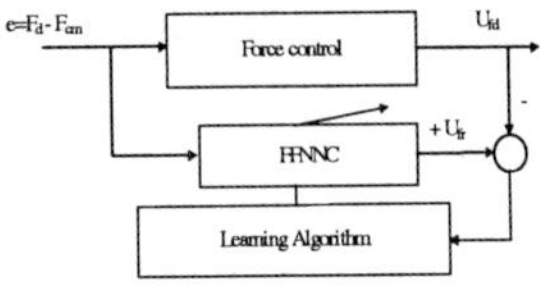

Fig. 3. Initialization of neural networks controller (FFNNC) scheme

The learning process involves the determination of a vector W_c* which optimizes a performance function J_1:

$$J_1 = \frac{1}{2}(U_{fd} - U_{fr})^T R_c (U_{fd} - U_{fr}) \tag{4}$$

where U_{fd} is the desired output, U_{fr} is the corresponding actual output and R_c is the weighting diagonal matrix. This last is function of the input of the neural network controller « Ic » and of its weight matrix W_c:

$$U_{fr} = N_c(W_c, I_c) \tag{5}$$

2.3 Adaptive Neural Network Controller

A dynamic learning process may be formulated as follows:

$$W_{c,k+1} = W_{c,k} - \mu \frac{\partial J_{k+1}}{\partial W_{c,k}} \tag{6}$$

where $W_{c,k}$ is the estimation of the weight vector et time t_k et μ is the learning parameter.

The output of the neural networks at time t_k can be obtained only by using the state and input of the network at past time. The learning process imply the determination of an $W_{c,k+1}$ vector which optimize the cost function J. This function is based on state criterion for process output and is as follow:

$$J = \frac{e^T * R * e}{2} \tag{7}$$

where: $e = F_d - F_{cm}$; e^T is the transpose vector of e and R is the weighting diagonal matrix.

The learning algorithm consists in minimizing J [Fcm, k+1] with regard to weights of the network, its application permits to write :

$$\frac{\partial J}{\partial W_c} = -e^T * R * \frac{\partial F_{cm,k+1}}{\partial U_{fr}} * \frac{\partial U_{fr}}{\partial W_c} \tag{8}$$

In the proposed structure (Figure 1), the back propagation method is based on the computation of the gradient of the criterion J using the learning paradigm called associative reinforcement learning [3].
In the proposed control approach, we chose a D matrix as reinforcement function. This choice permits to assimilate the system to control to a three-dimensional spring. The D matrix represents the model of behavior in contact with the environment. The dynamics of the interaction robot environment is described then below by the matrix relation:

$$F_{cm,\,k} = D * U_{fr,\,k} \tag{9}$$

where:

$$\frac{\partial F_{cm,\,k}}{\partial U_{fr,\,k}} = D \tag{10}$$

However, in the case of assembly tasks it appears discriminating to choose a linear type reinforcement function assimilating so the system to adaptable three-dimensional spring.
The learning algorithm consists in minimizing J with regard to weights of the network:

$$\frac{\partial J_k}{\partial W_{c,k}} = -e^T * R * D * \frac{\partial U_{f,k}}{\partial W_{c,k}} = -e^T * R * D * \frac{\partial N_{c,k}}{\partial W_{c,k}} \tag{11}$$

This last equation shows that the implementation on-line of an adaptation algorithm of neural controller parameters makes that this last doesn't depend any more on the robot model but rather of the interaction robot-environment model. Thus, the proposed control structure present the advantage of not being subject to mistakes of identification of the system.

3 Implementation and Experimentation

In order to validate the proposed structure and to evaluate its performances, we have implemented it for the control of the parallel robot.
This first step consist to compute an optimal neural networks and to use it to control the system. The neural networks controller initialized off line using the structure presented figure 3 is implemented in order to control the parallel robot of assembling cell. This experimentation consist of inserting cylindrical part in a bore for a velocity of 5 mm/s, and a tolerance of 3/10 mm. The temporal evolution of the measured and desired contact force tensor are show figure 4a. The force component according to the axis of insertion (F_x) exhibits a peak because of the appearance of a contact between the part and the bore. Then, the components of contact force tensor decrease by oscillating around a corresponding desired value. Components F_y et M_z, exhibits a

peak because of the appearance of a contact between the part and the bore. Then, they decrease by oscillation around a residual value.

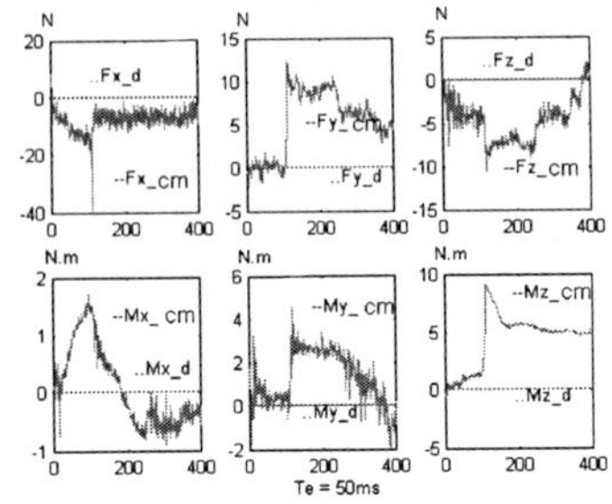

Fig. 4. a. Desired force torque (_d) and measured ones (_cm)

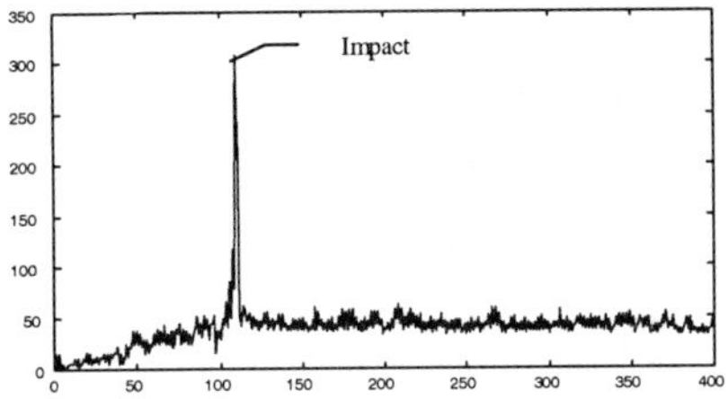

Fig. 4 b. Temporal evolution of criterion (off line learning)

The figure 4b shows the temporal evolution of criterion which toward to an asymptote of 45 when as its value at the impact time is 300. This asymptotic phase corresponds to a mistake on force torque components according to X of 5 NS, according to Y of 5 NS, according to Z of 2 NS, according to Mx of –0.5 Nm, according to My of –1 Nm and according to Mz of 5Nm. We can also notice that the decrease of the criteria is slow, since the final value is only reached after 300 samples. We have presented in this first step the off line learning of en external force control. The thus gotten neural controller has been tested on line to achieve the insertion of a cylindrical piece in a bore. Tests of insertion achieved show that the training out line of the controller permitted its generalization. The minimization of the criteria is not however not quite optimal since it exists a weak mistake in end of insertion.

To optimize the neural network controller (FFNNC), in real situation of utilization, an on line adjustment algorithm of FFNNC parameters is implemented. The on line training permits to refine the learning of the networks controller worked out off line. The adaptive controller performances have been evaluated on insertion tasks for different tolerances and velocities values. The comparison of temporal evolution of criterion obtained in the cases of an off-line and on-line learning (Figure 5) shows that the on-line adaptation algorithm leads to a better minimization of the criterion.

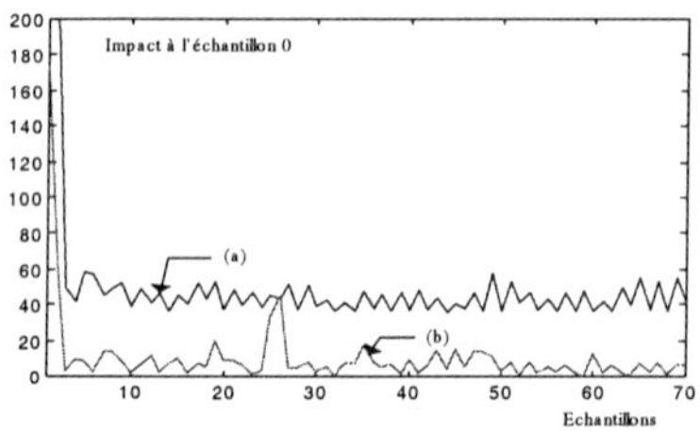

Fig. 5. Comparison of temporal evolution of criteria
(a) off line learning (V=5mm/s, T=3/10mm, cylindrical section part)
(b) on line learning (V=10mm/s, T=1/10mm, cylindrical section part)

In the case of an on-line adaptation, the criterion presents at the end of insertion an error weaker than in the case of an off-line learning. Indeed, in the case of the on line adaptation the criterion toward to an asymptote of 5, what corresponds to a mistake on force components according to X of 0.5 NS, according to Y of 2 NS, according to Z of 2 NS, according to Mx of 0.2 Nm, according to My of 2 Nm, according to Mz of 0.5Nm. Moreover, if velocity insertion increases, figure 6 shows that the criterion trace varies little.

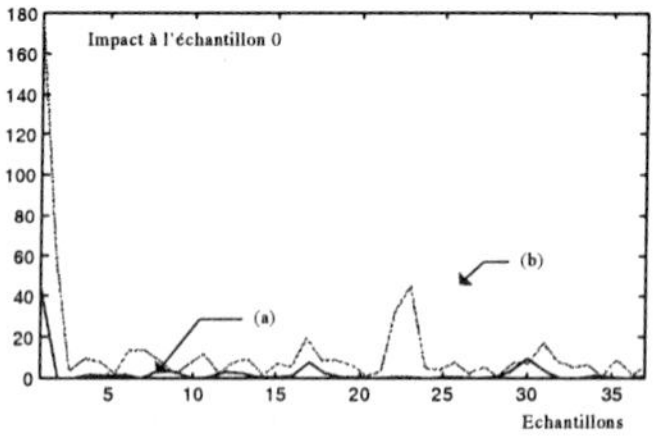

Fig. 6. Comparison of temporal evolution of criterion
(a) on line learning (V=15mm/s, T=1/10mm, cylindrical section part)
(b) on line learning (V=10mm/s, T=1/10mm, cylindrical section part)

Therefore, the on-line adaptation makes it possible to minimize the criterion with a residual error due to the sensors noise and perturbations that undergoes the system (for example frictions due to surface states). For the insertion of a parallelepiped part, figure 7 shows two phases: a phase where the criterion presents oscillations and a phase where it presents a closely identical behavior to the one corresponding to the cylindrical part insertion. Indeed, the learning base which has been constructed during the insertion of a cylindrical part, is not consistent in information on momentum. Thus, during the insertion of a parallelepiped part, oscillations of short duration and weak amplitude can appear and be considered as a normal phenomenon. In the case of parallelepiped section part the temporal evolution of the criteria toward an asymptote of 5, what corresponds to a mistake on force tensor components according to X of 2 NS, according to Y of -9N, according to Z of 0 NS, according to Mx of 0.002 Nm, according to My of 0.8 Nm, according to Mz of –0.5Nm,.

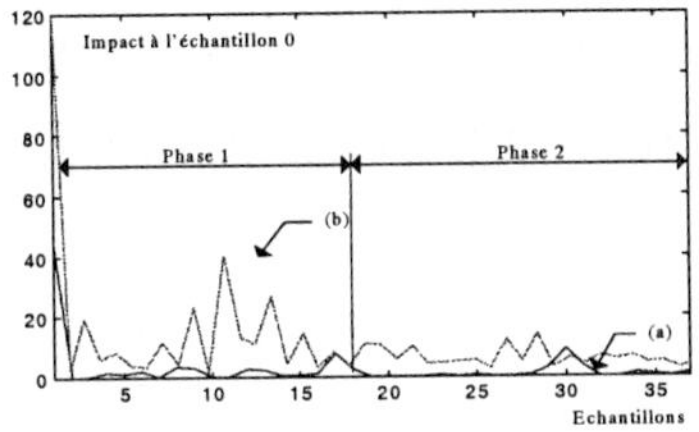

Fig. 7. Comparison of temporal evolution of criteria
(a) on line learning (V=15mm/s, T=1/10mm, cylindrical section part)
(b) on line learning (V=10mm/s, T=1/10mm, parallelepiped section part)

4 Conclusion

We presented in this paper a new approach of controller to treat the problem of the force control of active compliance robots witch are in interaction with their environment. The connectionist approach that is proposed rests on a methodology of control while considering the treatment of the task according to aspects off line and on line. Thus we proposed an external force control structure that uses the reinforcement paradigm for the on-line adaptation of the controller's parameters to variations of parameters of the task.

References

1. M.Y. Amirat. Contribution à la commande de haut niveau de processus robotisés et à l'utilisation des concepts de l'IA dans l'interaction robot-environnement. *PHD thesis* University Paris XII, Janvier 1996.
2. A. G. Barto and P. Anandan. Pattern recognizing stochastic learning automata. In *IEEE Transaction on Systems, Man, and Cybernetics*, 360-375, 1985.
3. N. Chatenet and H. Bersini. Economical reinforcement learning for non stationary problems. In Lecture notes in Computer Science 1327, Artificial Neural Network ICANN'97, 7th International Conference Lausanne, Switzerland Proceeding, pp. 283-288, October 1997.
4. G. Cybenco. Approximations by superposition of sigmoidal function. In *Advanced Robotics, Intelligent Automation and Active Systems*, pp 373-378, Bremen, September 15-17, 1997.
5 E. Dafaoui and Y. Amirat and J. Pontnau and C. François. Analysis and Design of a six DOF Parallel Manipulator. Modelling, Singular Configurations and Workspace. In *IEEE Transactions on Robotics And Automation*, vol. 14, pp. 78-92, Février 1998.
6. J.C. Doyle and K. Glover and P.P. Khargonecker and B.A. Francis. State space solutions to the standard H^2 and $H\infty$. In *IEEE Trans. On Automat. Contr.*, vol. 34, pp. 831-847, 1989.
7. B. Karan. Robust position/force control of robot manipulator in contact with linear dynamic environment, In *Proc. of the Third ECPD International Conference on Advanced Robotics, Intelligent Automation and Active Systems,* pp 373-378, Bremen, September 15-17, 1997.
8. O. Khatib. A unified approach to motion and force control of robot manipulators. In *IEEE J. Robot Automation,* 43—53, 1987.
9. S. Komada and al.. Robust force control based on estimation of environment. In *IEEE Int. Conf. on Robotics and Automation,* pp. 1362-1367, Nice, France, May,1992.
10. L. Laval and N.K. M'sirdi. Modeling, identification and robust force control of hydraulic actuator using $H\infty$ approach. In *Proceeding of IMACS,* Berlin, Germany, 1995.

A Design of Neural-Net Based Self-Tuning PID Controllers

Michiyo Suzuki[1], Toru Yamamoto[1], Kazuo Kawada[2], and Hiroyuki Sogo[2]

[1] Department of Technology & Information Education,
Graduate School of Education, Hiroshima University
1-1-1 Kagamiyama, Higashi-Hiroshima, Hiroshima, 739-8524, Japan
{yama,suzuki}@hiroshima-u.ac.jp
[2] Department of Electro-Mechanical Systems Engineering,
Takamatsu National College of Technology
355 Chokushi, Takamatsu, 761-8058, Japan
{kawada,sogo}@takamatsu-nct.ac.jp

Abstract. Recently, neural network techniques have widely used in designing adaptive and learning controllers for nonlinear systems. However, it costs a lot of time for learning of the neural network included in the control system. Furthermore, the physical meaning of neural networks constructed as a result, is not obvious. In this paper, a design scheme of self-tuning PID controllers is proposed, which has a structure of fusing self-tuning and neural network techniques. The newly proposed scheme enables us to adjust PID gains quickly.

1 Introduction

In recent years, the structure and the mechanism of the human brains have been partially clarified, and their knowledge is formulated as artificial neural networks [1]-[3]. Furthermore, the neural networks have been applied for pattern matching problems, pattern recognition and learning control. Especially, they play an important role in the field of nonlinear control systems, that is, the neural network technique enables us to deal with nonlinear control systems [4]-[10]. There are many works on design schemes of neural-net based controllers. However, to the best of our knowledge, almost of these works are based on the way to calculate control inputs directly. Since the control input is only generated in the output layer, it is difficult to understand a physical meaning of neural networks constructed as a result. Furthermore, generally it costs a lot of time for learning considerably, in the case where the only neural network technique is applied to control systems.

This paper deals with a design scheme of self-tuning PID controllers, which has a structure of fusing self-tuning and neural network schemes. According to the proposed control scheme, since PID gains are tuned in an adaptive manner, these gains are adjusted quickly. Furthermore, this method enables us to understand a physical meaning of the control parameters. This paper is organized as follows. The structure of the newly proposed self-tuning PID control system

G. Dorffner, H. Bischof, and K. Hornik (Eds.): ICANN 2001, LNCS 2130, pp. 914–921, 2001.

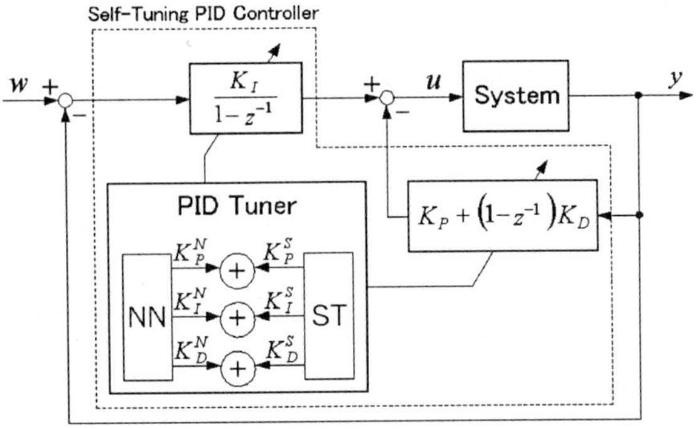

Fig. 1. Schematic diagram of the proposed self-tuning PID control system.

is briefly explained. The controller is designed by using the self-tuning scheme together with the neural network, and the controller design scheme is discussed in detail. Finally, in order to show the effectiveness of the newly proposed self-tuning PID control scheme, a numerical simulation example is illustrated.

2 Self-Tuning PID Controller

2.1 Outline

The discrete-time PID controller can be expressed in the following form:

$$u(k) = \frac{K_I}{1 - z^{-1}}\{w(t) - y(t)\} - F(z^{-1})y(t) \tag{1}$$

with

$$F(z^{-1}) := K_P + (1 - z^{-1})K_D, \tag{2}$$

where $u(t)$, $y(t)$ and $w(t)$ denote the input, the output and the reference signals at time t, respectively. Furthermore, K_P is the proportional gain; K_I is the integral gain and K_D is the derivative gain. The control law (1) is so-called a velocity-type form. In this paper, a self-tuning PID controller is designed, whose PID gains are tuned adaptively by fusing both the self-tuning control technique and the neural network. Thus, PID gains are composed as follows:

$$\left.\begin{aligned} K_P &:= K_P^S + K_P^N \\ K_I &:= K_I^S + K_I^N \\ K_D &:= K_D^S + K_D^N, \end{aligned}\right\} \tag{3}$$

where superscripts S and N denote PID gains calculated in the self-tuning(ST) and the neural network(NN) parts, respectively. Figure 1 shows the schematic diagram of the proposed self-tuning PID control system.

2.2 System Description

Consider the mathematical model of the form:

$$A(z^{-1})y(t) = z^{-1}B(z^{-1})u(t) + g(u, y) \tag{4}$$

$$\left.\begin{aligned} A(z^{-1}) &= 1 + a_1 z^{-1} + a_2 z^{-2} \\ B(z^{-1}) &= b_0 + b_1 z^{-1} + \cdots + b_m z^{-m} \end{aligned}\right\} \tag{5}$$

where $g(u, y)$ denotes the unmodeled dynamics, that can not be expressed by the second-order linear model.

Suppose the following assumptions for the system (4):

Assumptions

A.1 The degree of $B(z^{-1})$, m, is known.
A.2 Parameters a_i and b_i are unknown.
A.3 The time delay, k, may be unknown, but the following relation is satisfied:

$$0 \leq k \leq m. \tag{6}$$

A.4 $g(u, y)$ is bounded, that is, the following relation is satisfied:

$$\|g(u, y)\| \leq M \leq \infty. \tag{7}$$

A.5 The reference signal $w(t)$ is given by piecewise constantsD

In tuning PID gains, it is expected that the self-tuning control mechanism works on the linear part, and the neural network on the unmodeled part, respectively. These tuning schemes are discussed below.

2.3 Self-Tuning Control Part

The self-tuning control algorithm is derived based on the relationship between the PID control law and the pole-assignment control law.

First, separate $u(t)$ in (4) temporarily as follows:

$$u(k) := u_S(t) + u_N(t), \tag{8}$$

where $u_S(t)$ and $u_N(t)$ denote inputs calculated by the self-tuning and the neural network parts, respectively. Next, introduce the following feedback control law:

$$P(z^{-1})y(t) + Q(z^{-1})\Delta u_S(t) - P(1)w(t) = 0 \tag{9}$$

$$\left.\begin{aligned} P(z^{-1}) &= p_0 + p_1 z^{-1} + p_2 z^{-2} \\ Q(z^{-1}) &= 1 + q_1 z^{-1} + q_2 z^{-2} + \cdots + q_m z^{-m}, \end{aligned}\right\} \tag{10}$$

where $\Delta := 1 - z^{-1}$ and $P(1) = 1 + p_1 + p_2$. Then, substituting (8) and (9) into (4) yields

$$y(t) = \frac{z^{-1}B(z^{-1})P(1)}{T(z^{-1})}w(t) + \frac{\Delta Q(z^{-1})}{T(z^{-1})}\{z^{-1}B(z^{-1})u_N(t) + g(u, y) + \xi(t)\}, \tag{11}$$

where $T(z^{-1})$ is defined as

$$T(z^{-1}) := \Delta A(z^{-1})Q(z^{-1}) + z^{-1}P(z^{-1})B(z^{-1}). \tag{12}$$

It is clear that $T(z^{-1})$ in (11) becomes the characteristic polynomial of the closed-loop systems. Then, the pole-assignment can be realized by choosing desired $T(z^{-1})$ and calculating $P(z^{-1})$ and $Q(z^{-1})$ so as to satisfy (12).

On the other hand, extract the following equation assigned by the self-tuning control part:

$$u_S(t) = \frac{K_I^S}{1 - z^{-1}}\{w(t) - y(t)\} - F_S(z^{-1})y(t) \tag{13}$$

with

$$F_S(z^{-1}) := K_P^S + (1 - z^{-1})K_D^S. \tag{14}$$

Then, (13) can be rewritten in the following form:

$$L_S(z^{-1})y(t) + \Delta u_S(t) - L_S(1)w(t) = 0 \tag{15}$$

where

$$L_S(z^{-1}) = K_I^S + (1 - z^{-1})F_S(z^{-1}). \tag{16}$$

Now, since $Q(z^{-1})$ in (9) becomes $Q(1)$ in the steady state, the following equation which approximately corresponds to (9) is derived:

$$P(z^{-1})y(t) + Q(1)\Delta u_S(t) - P(1)w(t) = 0. \tag{17}$$

Then, $L_S(z^{-1})$ defined as

$$L_S(z^{-1}) := P(z^{-1})/Q(1), \tag{18}$$

can be adjusted based on the pole-assignment scheme, although it is not strictly. Thus, from (14), (16) and (18), PID gains in the self-tuning control part can be obtained. For the parameter estimation of unknown parameters included in (4), the following least squares method with a dead-zone is employed [10][11]:

$$\hat{\theta}(t) = \hat{\theta}(t-1) + \frac{\Gamma(t-1)\psi(t-1)}{1 + \psi^T(t-1)\Gamma(t-1)\psi(t-1)}\eta(t) \tag{19}$$

$$\Gamma(t) = \Gamma(t-1) - \frac{\Gamma(t-1)\psi(t-1)\psi^T(t-1)\Gamma(t-1)}{1 + \psi^T(t-1)\Gamma(t-1)\psi(t-1)} \tag{20}$$

$$\varepsilon(t) = y(t) - \hat{\theta}^T(t-1)\psi(t-1) \tag{21}$$

$$\eta(t) = \begin{cases} \varepsilon(t) > \delta & \rightarrow \eta(t) = \varepsilon(t) - \delta \\ |\varepsilon(t)| \le \delta & \rightarrow \eta(t) = 0 \\ \varepsilon(t) < -\delta & \rightarrow \eta(t) = \varepsilon(t) + \delta \end{cases} \tag{22}$$

$$\hat{\theta}(t) = [\hat{a}_1(t), \hat{a}_2(t), \hat{b}_0(t), \hat{b}_1(t), \cdots, \hat{b}_m(t)]^T \tag{23}$$

$$\psi(t-1) = [-y(t-1), -y(t-2), u(t-1), u(t-2), \cdots, u(t-m-1)]^T \tag{24}$$

where $\varepsilon(t)$ and 2δ denote the prediction error and the width of the dead-zone.

2.4 Neural Network Part

In order to tune PID gains for the unmodeled dynamics, a three-layered neural network is introduced, whose connection weights are adjusted based on the back propagation method without a direct teacher. Then, it is important to consider how to evaluate the construction of the neural network. Therefore, this problem is mainly discussed below.

First, (1) is rewritten as

$$L(z^{-1})y(t) + (1 - z^{-1})u(t) - K_I w(t) = 0, \tag{25}$$

where $L(z^{-1})$ is defined as

$$L(z^{-1}) := K_I + (1 - z^{-1})F(z^{-1}). \tag{26}$$

Here, introduce a new signal $\phi(t+1)$ given by

$$\phi(t+1) := L(z^{-1})y(t) + (1 - z^{-1})u(t), \tag{27}$$

and define the learning error signal $e(t)$ as

$$e(t) := y_m(t) - \phi(t). \tag{28}$$

Note that $e(t)$ differs from $\varepsilon(t)$ in (21). Furthermore, $y_m(k)$ in (28) is calculated by

$$y_m(t) = \frac{K_I S(z^{-1})}{r} y(t) \tag{29}$$

where

$$r := S(1). \tag{30}$$

If $e(t) \to 0$, that is, $\phi(t) \to y_m(t)$, the following equation can be obtained from (25), (27), (28) and (29):

$$y(t) = \frac{z^{-1} r}{S(z^{-1})} w(t). \tag{31}$$

Here, (31) means that the model following control system is realized via the neural network. Based on the previous procedure, the three-layered neural network is designed. The input layer has $p+q+1$ neurons connected with signals $y(t-1)$, $y(t-2)$, $\cdots$, $y(t-p)$, $u(t-1)$, $u(t-2)$, $\cdots$, $u(t-q)$, $w(t)$, and the threshold unit. The hidden layer has $r+1$ neurons which include a threshold unit, and the output layer has three neurons corresponding to PID gains. p, q and r can be selected by trial and error according to the complexity of the system. First, define the following variables:

- I_i : The output signal of the i-th unit in the input layer
- H_j : The output signal of the j-th unit in the hidden layer
- O_k : The output signal of the k-th unit in the output layer
- $V_{i,j}$: The weight between the i-th unit in the input layer and the j-th unit in the hidden layer

$W_{j,k}$: The weight between the j-th unit in the hidden layer and the k-th unit in the output layer
I_0 : The threshold unit included in the input layer
H_0 : The threshold unit included in the hidden layer

Here, O_1, O_2 and O_3 correspond to K_P^N, K_I^N and K_D^N, respectively. Furthermore, the sigmoidal functions are employed in the hidden and the output layers, that is,

$$H_j = f(\sum_{i=0}^{p+q+1} V_{i,j} \cdot I_i) \tag{32}$$

$$O_k = g(\sum_{j=0}^{r} W_{j,k} \cdot H_j), \tag{33}$$

where $f(\cdot)$ and $g(\cdot)$ denote the following sigmoidal functions:

$$f(x) := 2\left(\frac{1}{1+e^{-a\cdot x}} - \frac{1}{2}\right) \quad (a > 0) \tag{34}$$

$$g(x) := c\left(\frac{1}{1+e^{-b\cdot x}} - \frac{1}{2}\right) \quad (b > 0). \tag{35}$$

Next, introduce the cost function as follows:

$$E(t) := \frac{1}{2}e^2(t). \tag{36}$$

Based on the back propagation method, the weights $V_{i,j}$ and $W_{j,k}$ are adjusted so as to minimize the above cost function. The detailed tuning rules of $V_{i,j}$ and $W_{j,k}$ are omitted due to the page limit.

From the above consideration, the self-tuning PID control system can be constructed, whose PID gains are tuned by both the self-tuning and the neural network schemes.

3 Numerical Simulation Example

The following example is discussed to investigate the behavior of the newly proposed self-tuning PID controller.

Example

The system to be considered in this paper is as follows:

$$y(t) = 0.9y(t-1) - 0.01y^2(t-2) + u(t-1) + \sin(u(t-2)). \tag{37}$$

It is clear that this system includes nonlinear components.

The newly proposed self-tuning PID control scheme was employed for the system (37), and then control result as shown in Fig.2 could be obtained where $T(z^{-1})$ and $S(z^{-1})$ were designed as

$$T(z^{-1}) = 1 - 1.638z^{-1} + 0.676z^{-2} \tag{38}$$

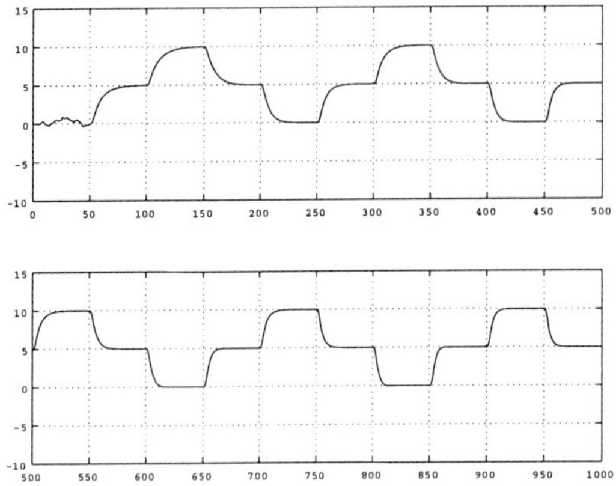

Fig. 2. The control result by using the newly proposed self-tuning PID control scheme.

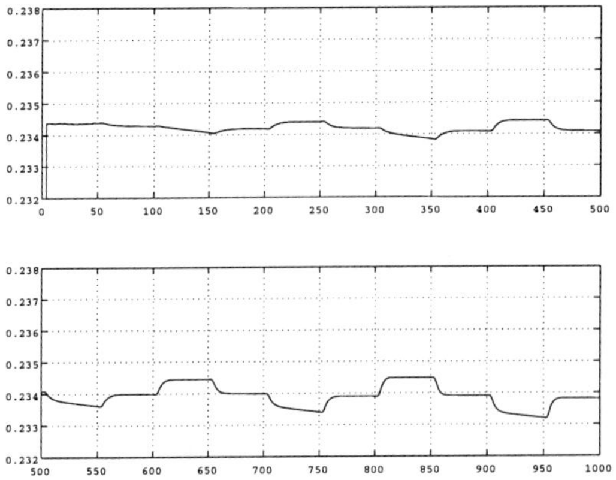

Fig. 3. Trajectory of the proportional gain.

$$S(z^{-1}) = 1 - 1.027z^{-1} + 0.264z^{-2}. \tag{39}$$

Furthermore, the user-specified parameters were set as follows:

$$\delta = 0.5,\ a = 1.0,\ b = 1.0,\ c = 2.0,\ p = 2,\ q = 2,\ r = 30. \tag{40}$$

During first 50 steps, the control input was given by the random sequence for the purpose of improving reliability of the parameter estimation in the self-tuning control part. Fig.3 shows the trajectory of the proportional gain. These figures illustrate that controller parameters are adjusted adequately corresponding the reference signals, and the output signal can track the reference model output quickly.It is clear that the newly proposed self-tuning PID control scheme works well. Due to the page limit, comparative studies with other control schemes are omitted.

4 Conclusions

In this paper, a self-tuning PID controller has been proposed, which is the result of fusing a self-tuning control and a neural network technique. The features of the newly proposed control scheme are summarized as follows:

- PID gains are adequately adjusted corresponding to nonlinearities.
- The model following control system is realized.
- The physical meaning of the controller is obvious.

Furthermore, from the numerical simulation example, the effectiveness of the proposed control scheme has been shown. However, some open problems are remained. Especially, since the control performance depends on some user-specified parameters in this control scheme, for example, δ, a, b, c, and unit numbers (p, q and r), it must be considered how to determine these parameters as our future work.

References

[1] Rumelhart,D.E., McClelland,J.L. and the PDP Research Group : Parallel Distributed Processing. MIT Press, Cambridge (1986)

[2] Dayhoff,J.E. : Neural Network Architectures; An Introduction. Van Nostrand Reinhold, New York (1990)

[3] Haykin,S : Neural Networks. Macmillan College Publishing Company Inc. New York (1994)

[4] Narendra,K.S. and Parthasarathy,K. : Identification and Control of Dynamical Systems Using Neural Networks. IEEE Trans. on Neural Networks, **1**-1 (1990) 1–27

[5] Miller III,W.T., Sutton,R.S. and Werbors,P.J. : Neural Networks for Control. The MIT Press, Cambridge (1990)

[6] Hunt,K.J., Sbarbaro,D., Zbikowski,R. and Gawthrop,P.J. : Neural Networks for Control Systems - A Survey. Automatica, **28**-6 (1992) 1083–1112

[7] Gupta,M.M. and Rao,D.H. : Neuro-Control Systems Theory and Applications. IEEE Press, New York (1993)

[8] Marzuki,K., Omatu,S. and Rubiyah,Y. : MIMO Furnace Control with Neural Networks. IEEE Trans. on Control Systems Technology, **1**-4 (1993) 238–245

[9] Omatu,S. and Akhyar,S. : Self-Tuning PID by Neural Networks. Proc. of IJCNN, Nagoya (1993) 2749–2752

[10] Omatu,S. : Adaptive and Learning Control Neural Networks. Proc. of the Eight Yale Workshop on Adaptive and Learning Systems, New Haven (1996) 184–189

[11] Peterson,B.B. and Narendra,K.S. : Bounded Error Adaptive Control. IEEE Trans. on Automatic Control, **27**-6 (1982) 1161–1168

[12] Goodwin,G.C. and Sin,K.S. : Adaptive Filtering Prediction and Control. Prentice-Hall, New York (1984)

However, the value of JB in (1) could be greater than zero but less than the link length l_i, negative or greater than the link length corresponding to the obstacle located in three different regions relative to the link i as shown in Figs. 1(b), 1(c) and 1(d), respectively. Consequently, r_{oi}, L_{oi} and D_i are determined in different ways according to the three cases as described below. *Case (1)*: $0 \leq JB < l_i$ In this case, as shown in Fig. 1(b), the point r_{oi} is at B. Hence, $L_{oi} = JB$ and $D_i = OB$. *Case (2)*: $JB < 0$ For this case, as shown in Fig. 1(c), the point J should be assigned as r_{oi}. Thus, $L_{oi} = 0$ and $D_i = \sqrt{JB^2 + OB^2}$. *Case (3)*: $JB \geq l_i$ This case is the situation shown in Fig. 1(d), and the point C should be selected as r_{oi}. Therefore, $L_{oi} = l_i$ and $D_i = \sqrt{(JB - l_i)^2 + OB^2}$.

Having determined D_i for each link, define

$$R = \min\{D_i\}, \quad \text{for} \quad i = 1, 2, \cdots, n, \tag{3}$$

which gives the shortest distance between the obstacle and the manipulator, and we could then specify the most dangerous link, which is the link having the smallest distance to the obstacle. If link j is the most dangerous link, then r_{oj} is the obstacle avoidance point and its desired velocity vector $\dot{r}_o$ can be assigned according to the associated obstacle location as follows. If Case (1) is encountered, then $\dot{r}_o$ may be defined as

$$\dot{r}_o = \text{sgn}(z) \begin{bmatrix} -\sin\theta_{1,j} \\ \cos\theta_{1,j} \end{bmatrix},$$

where

$$\text{sgn}(z) = \begin{cases} 1 & \text{for } z > 0, \\ -1 & \text{for } z < 0, \end{cases}$$

and $z = (x_{j-1} - x_{ob})(y_j - y_{ob}) - (x_j - x_{ob})(y_{j-1} - y_{ob})$. If Case (2) and Case (3) are encountered, $\dot{r}_o$ may be defined, respectively, as

$$\dot{r}_o = \begin{bmatrix} (x_{j-1} - x_{ob})/\sqrt{(x_{j-1} - x_{ob})^2 + (y_{j-1} - y_{ob})^2} \\ (y_{j-1} - y_{ob})/\sqrt{(x_{j-1} - x_{ob})^2 + (y_{j-1} - y_{ob})^2} \end{bmatrix}$$

$$\dot{r}_o = \begin{bmatrix} (x_j - x_{ob})/\sqrt{(x_j - x_{ob})^2 + (y_j - y_{ob})^2} \\ (y_j - y_{ob})/\sqrt{(x_j - x_{ob})^2 + (y_j - y_{ob})^2} \end{bmatrix}.$$

2.2 General Solution for Obstacle Avoidance

By knowing the obstacle avoidance point r_o and its velocity vector $\dot{r}_o$, the redundancy resolution for obstacle avoidance is obtained by [2]

$$\dot{\theta} = J_e^+ \dot{r}_e + [J_o(I - J_e^+ J_e)]^+ (\dot{r}_o - J_o J_e^+ \dot{r}_e), \tag{4}$$

where $J_e \in \Re^{m \times n}$, $J_e^+ \in \Re^{n \times m}$ and $\dot{r}_e \in \Re^m$ are the Jacobian, the pseudoinverse of the Jacobian and the velocity of the end-effector respectively; and $J_o \in \Re^{m \times n}$ is the Jacobian for the obstacle avoidance point.

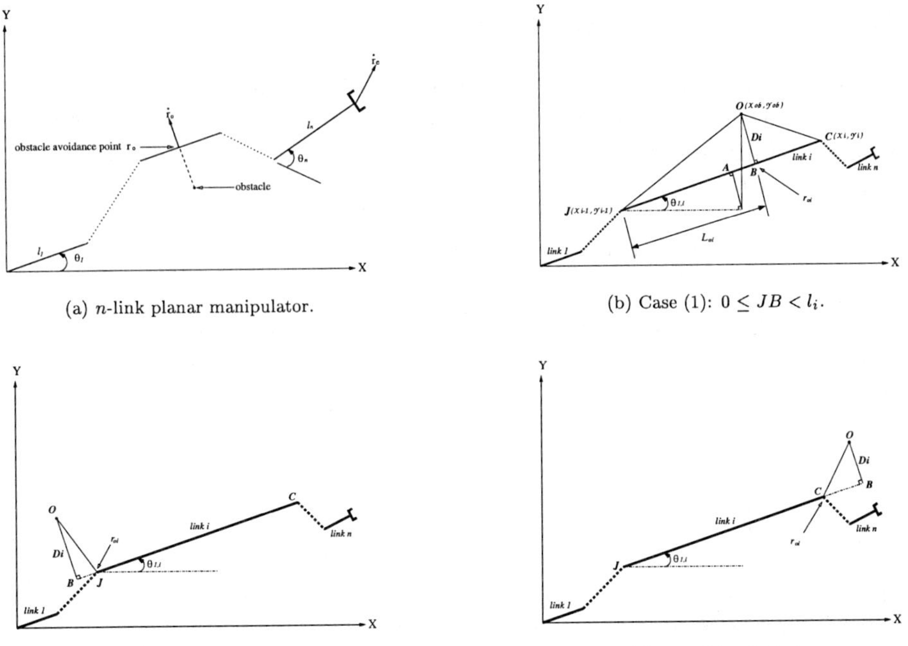

(a) n-link planar manipulator.

(b) Case (1): $0 \leq JB < l_i$.

(c) Case (2): $JB < 0$.

(d) Case (3): $JB \geq l_i$.

Fig. 1. (a) An n-link planar redundant manipulator with the presence of obstacle in its work space. (b), (c) and (d) the location of the obstacle avoidance points corresponding to the three possible obstacle positions relative to the manipulator link.

3 The Lagrangian Network for Obstacle Avoidance

3.1 Time-Varying Quadratic Program

Another method to obtain the redundancy resolution for obstacle avoidance without computing the pseudoinverse is to solve the following quadratic program subject to equality constraints:

$$\begin{array}{l} \text{minimize } \frac{1}{2}\dot{\theta}^T \dot{\theta}, \\ \text{subject to } J_e \dot{\theta} = \dot{r}_e, \; J_o \dot{\theta} = \dot{r}_o, \end{array} \tag{5}$$

where the first constraint relates the desired end-effector velocities to the joint velocities, and the second constraint relates the obstacle avoidance point desired velocities to the joint velocities, which is taken into account only if $R < R_s$. For the existence of more than one obstacle, other relations that associate the corresponding obstacle avoidance point desired velocities to the joint velocities similar to the second constraint should be included as extra constraints.

3.2 Network Dynamics

By the Lagrange optimization method, the Lagrangian of the quadratic program (5) is defined as

$$L(\dot{\theta}, \lambda) = \frac{1}{2}\dot{\theta}^T\dot{\theta} + \lambda^T(J\dot{\theta} - \dot{r}), \tag{6}$$

where $J = [J_e^T, J_o^T]^T$, $\dot{r} = [\dot{r}_e^T, \dot{r}_o^T]^T$ and λ is the Lagrange multiplier vector. The necessary condition for the minimal solution for (5) is found by taking the partial derivatives of $L(\dot{\theta}, \lambda)$ with respect to $\dot{\theta}$ and λ, respectively, and setting them to zero:

$$\frac{\partial L}{\partial \dot{\theta}} = \dot{\theta} + J^T\lambda = 0 \quad \text{and} \quad \frac{\partial L}{\partial \lambda} = J\dot{\theta} - \dot{r} = 0. \tag{7}$$

In view of the Lagrangian optimality conditions (7), and denote u and v as estimated joint velocity vector $\dot{\theta}$ and Lagrange multiplier vector λ respectively, the dynamical equations for the Lagrangian network may be defined as

$$\mu\frac{du}{dt} = -u - J^T v, \tag{8}$$

$$\mu\frac{dv}{dt} = Ju - \dot{r}, \tag{9}$$

where μ is a positive scaling constant to scale the convergence rate of the neural network, and it should be selected as small as possible as long as the hardware is allowed. Clearly, the equilibrium points of system (8) and (9) satisfy the Lagrangian optimality conditions (7) and contain the optimal solution to (5). Furthermore, the system described by (8) and (9) is asymptotically stable [8]. In this context, the desired end-effector velocity $\dot{r}_e$ is combined with $\dot{r}_o$ to form $\dot{r}$ which is fed into the neural network as its input. The positions of the obstacles are used to determine J_o, which is merged with J_e to form the essential element J for the neural network. The joint rate $\dot{\theta}$ that could make the manipulator avoid obstacles is generated as the neural network output. By further taking integration on the joint rate with known initial values yields the joint motions that the manipulator should be followed.

4 Simulation Results

Computer simulations based on a five-link planar manipulator are performed to demonstrate the effectiveness of the approach to obstacle avoidance of redundant manipulators. The manipulator has the lengths of $l_1 = 1$ m, $l_2 = 0.7$ m, $l_3 = l_4 = 0.5$ m and $l_5 = 0.3$ m and inital configuration of $\theta(0) = [\pi/4, -\pi/3, \pi/4, \pi/4, -\pi/3]^T$ rad. The end-effector is to track a straight line with the velocities $\dot{r}_e = [-0.25, 0.5]^T$ m/s for one second. The saftey range R_s is set to be 5 cm. The differential equations (8) and (9) are solved by the fourth-order Runge-Kutta method with the positive scaling constant μ being 2×10^{-5}. Figure 2 shows the joint motion trajectories computed by the recurrent neural network for the cases of having different number of obstacles. In Fig. 2(a), there is no obstacle in the manipulator work space while there are respectively one, two and three obstacles denoted by $*$ in Figs. 2(b), 2(c) and 2(d). The motions

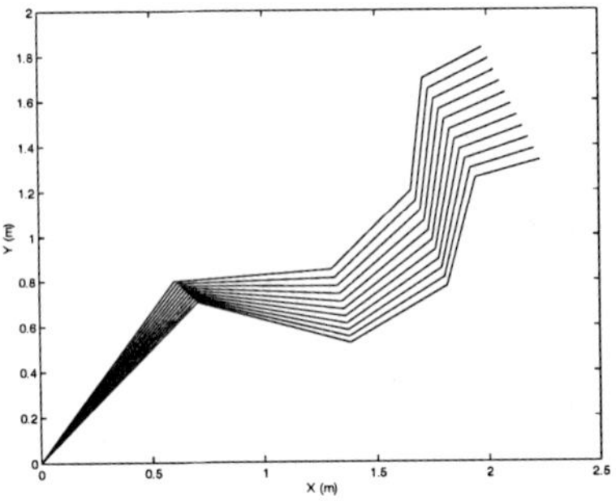

(a) No obstcle

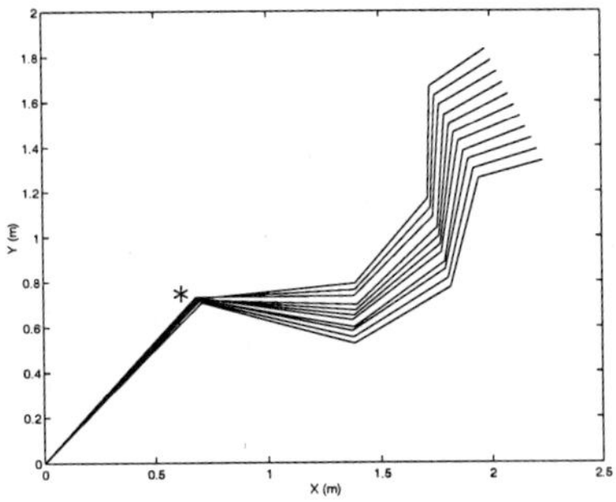

(b) One obstacle

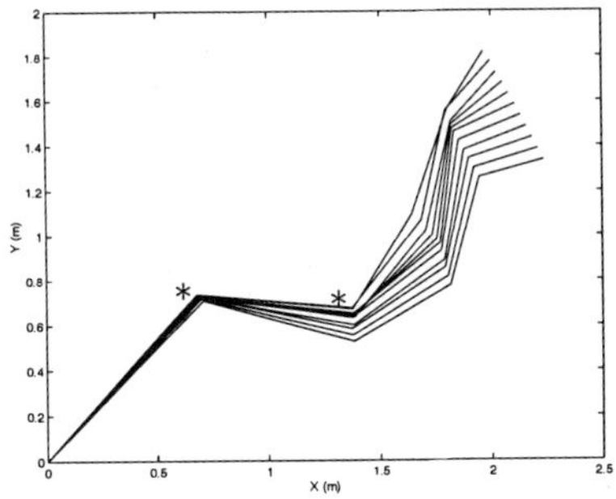

(c) Two obstacles

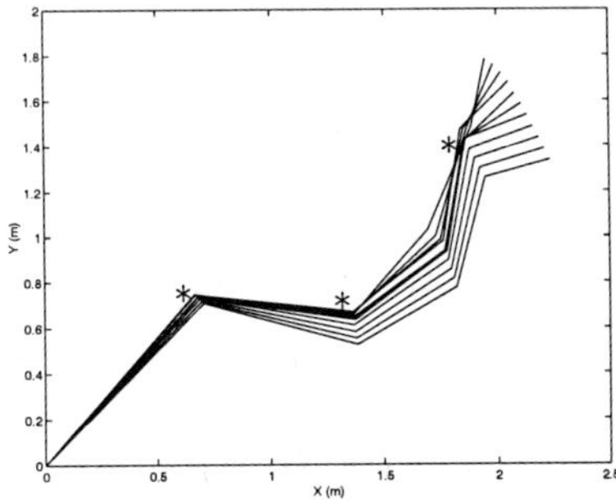

(d) Three obstacles

Fig. 2. Joint motion trjectories computed by the Lagrangian network with or without obstalces.

computed by the Lagrangian network are shown to enable the manipulator to avoid collision with the obstacles while achieving the desired end-effector task.

Next, we consider a 7-DOF redundant manipulator (PA-10) made by Mitsubishi. The mechanical configuration and coordinate system are given in [7]. The objective is to simulate the PA-10 manipulator for a circular motion with the radius of 0.2m in the 3D workspace to avoid an obstacle at $(0.4, 0.1, 0.1)$. In the simulation, $\mu = 5 \times 10^{-7}$, the run time of the manipulator is 10 sec., $R_s = 10cm$, and the initial angular vector is $\theta(0) = [0, -\pi/4, 0, \pi/2, 0, -\pi/4, 0]^T$. Figure 3 shows the 3D view of the motion of the PA-10 manipulator. Illustrated in Figure 5 is the transient behaviors of the PA-10 manipulator and the Lagrange network. Specifically, the eight subplots show respectively $\dot{\theta}(t)$, $\|\dot{\theta}(t)\|_2$, $\theta(t)$, $\|\theta(t)\|_2$, $r(t) - r_d(t)$, and $\dot{r}(t) - \dot{r}_d(t)$, R in eqn. (3), and cond(J) which is the condition number of the Jacobian to measure the singularity. The orientation of the end-effector is required to keep unchanged while the manipulator moves throughout the manipulation. The two subplots in Fig. 4 on position and velocity errors of the end-effector show respectively that the deviations between the actual position and the desired one and between the actual velocity and the desired one are rather small at any time. The subplot in Figure 5 for R shows

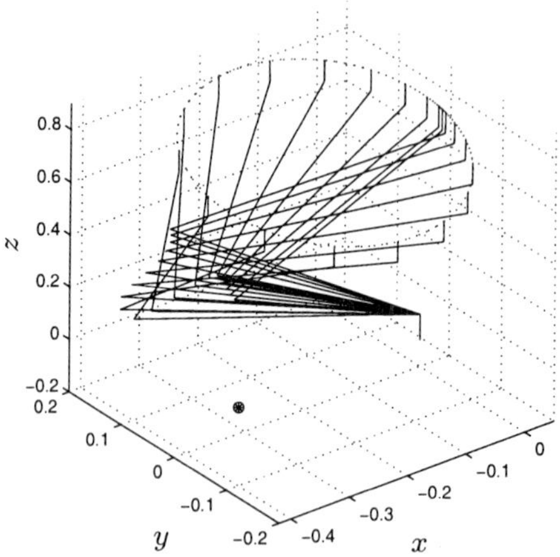

Fig. 3. Simulated circular motion of the PA-10 manipulator.

the obstacle avoidance capability of the neural control system for the PA-10 manipulator.

5 Concluding Remarks

The Lagrangian network is applied for kinematic control and obstacle avoidance of redundant manipulators in this paper. Compared with supervised learning feedforward neural networks, the proposed Lagrangian network eliminates the need for training. Unlike the penalty parameter based recurrent neural networks, the Lagrangian network converges to the exact solution without using any penalty parameter. The presented approach implemented in dedicate hardware such as VLSI circuitry is suitable for real-time kinematic control and obstacle avoidance of redundant manipulators.

References

1. Nenchev, D.N.: Redundancy resolution through local optimization: A review. J. Robot. Syst. **6** (1989) 769-798
2. Maciejewski, A.A, Klein, C.A.: Obstacle avoidance for kinematically redundant manipulators in dynamically varying environments. Int. J. Robot. Res. **4** (1985) 109-117
3. Cheng, F.T., Chen, T.H., Wang, Y.S., Sun, Y.Y.: Obstacle avoidance for redundant manipulators using the compact QP method. Proc. IEEE Int. Conf. Robo. Automat. (1993) 262-269
4. Sciavicco, L, Siciliano, B.: A solution algorithm to the inverse kinematic problem for redundant manipulators. IEEE Trans. Robot. Automat. **4** (1998) 403-410
5. Khatib, O.: Real-time obstacle avoidance for manipulators and mobile robots. Int. J. Robot. Res. **5** (1986) 90-99

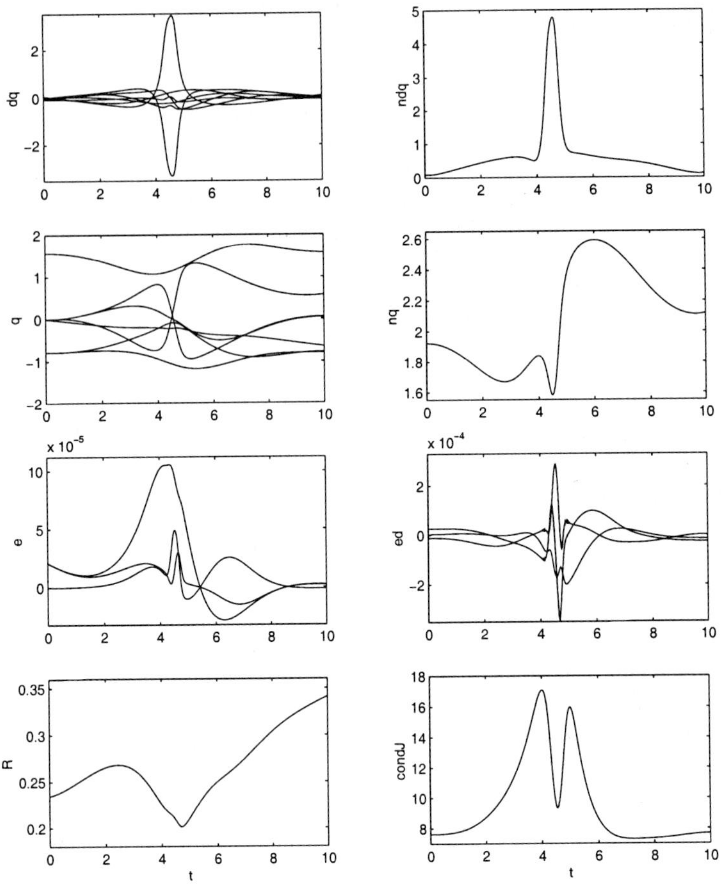

Fig. 4. Transient behaviors of the recurrent neural networks and the PA-10 redundant manipulator.

6. Guo, J, Hsia, T.C.: Joint trajectory generation for redundant robots in an environment with obstacles. J. Robot. Syst. **10** (1993) 199-215
7. Wang, J., Hu, Q, Jiang, D.: A Lagrangian network for kinematic control of redundant robot manipulators. IEEE Trans. Neural Networks. **10** (1999) 1123-1132
8. Tang, W.S., Wang, J.: Two recurrent neural networks for local joint torque optimization of kinematically redundant manipulators. IEEE Trans. Syst. Man, Cybern. B. **30** (2000) 120-128

Adaptive Neural Control of Nonlinear Systems

Ieroham Baruch[1], Jose Martin Flores[1,*],
Federico Thomas[2], and Ruben Garrido[1]

[1] CINVESTAV-IPN, Ave. IPN No. 2508,
A. P. 14470, Mexico D. F., C. P. 07360, Mexico
baruch@ctrl.cinvestav.mx
[2] IRI-UPC, Barcelona, Spain
fthomas@iri.upc.es

Abstract. The aim of the present paper is to integrate a recurrent neural network in two schemes of real-time soft computing neural control. There are applied the following control schemes: an indirect and a direct trajectory tracking control, using the state and parameter information, given by an identification recurrent neural network.

1 Introduction

The Neural Network (NN) modeling and application to system identification, prediction and control is discussed by many authors [1], [2], [3], [4], [5]. Mainly, two types of NN models are used: Feedforward (FFNN) and Recurrent (RNN). The main problem here is the use of different NN mathematical descriptions and control schemes, according to the structure of the plant model. For example, Narendra and Parthasarathy, [1], [2], applied FFNN for system identification and direct model reference adaptive control of various nonlinear plants. They considered four plant models with a given structure and supposed that the order of the plant dynamics is known. Yip and Pao, [3], solved control and prediction problems by means of a flat-type functional FFNN, used for an inverse model learning control. Pham, [4], applied Jordan RNN for robot control. Sastry et all, [5], introduced two types of neurons — Network Neurons and Memory Neurons to solve identification and adaptive control problems, considering that the plant model is also autoregressive one. In [7], some schemes of NN and RNN applications to control, especially of direct model reference adaptive control, are surveyed. All drawbacks of the described in the literature NN models could be summarized as follows: 1) there exists a great variety of NN models and universality is missing, [1], [2], [3], [4], [5]; 2) all NN models are sequential in nature as implemented for systems identification. (The FFNN model uses one or more tap-delays in the input, [1], [2], and the most of the RNN models usually are based on the autoregressive model, [5], which is one-layer sequential one); 3) some of the applied RNN models are not trainable, others are not trainable in the feedback part, [4]. Most of them are dedicated to a SISO and not to a MIMO

* Supported by CONACYT-Mexico

G. Dorffner, H. Bischof, and K. Hornik (Eds.): ICANN 2001, LNCS 2130, pp. 930–936, 2001.

applications, [3]; 4) in most of the cases, the stability of the RNN is not considered, [4], especially during the learning; 5) in the case of FFNN application for systems identification, the plant is given in one of the four described in [1] plant models. The linear part of the plant model, especially the system order, has to be known and the FFNN approximates only the non-linear part of the plant model, [1]; 6) all these NN models are non-parametric ones, in the sense that its parameters are not applicable for an indirect adaptive control systems design; 7) all this NN models does not perform state and parameter estimation in the same time, [6].

The major disadvantage of all this approaches is that the identification NN model applied, is a non parametric one. This means that the NN model parameters are not useful for indirect control systems design objetives. Baruch et all, [7], [8], in their previous paper, applied the state-space approach to describe RNN in an universal way, defining a Jordan canonical two-layer RNN model, named Recurrent Trainable Neural Network (RTNN). This NN model is a parametric one, permitting to use the obtained parameters during the learning, for control systems design. Furthermore, the RTNN model is a system state predictor/estimator, which permits to use the obtained system states directly for state-space control. The aim of this paper is to use the RTNN as an identification and state estimation tool in a direct and an indirect adaptive control of nonlinear plants. The application of RTNN in a direct and an indirect control schemes will be studied by appropriate simulation example.

2 Recurrent Neural Model for Non-linear Systems Identification

In [7] and [8], a discrete-time model of Recurrent Trainable Neural Network (RTNN), and the dynamic Backpropagation (BP) weight updating rule, are given. The RTNN model is described by the following equations:

$$X(k+1) = JX(k) + BU(k) \tag{1}$$

$$Z(k) = S[X(k)] \tag{2}$$

$$Y(k) = S[CZ(k)] \tag{3}$$

$$J \doteq block\ \ diag\ (J_i);\ \ |J_i| < 1 \tag{4}$$

Where $X(\cdot)$ is a n- state vector of the RTNN; $U(\cdot)$ is a m- input vector; $Y(\cdot)$ is a l- output vector; $Z(\cdot)$ is an auxiliary vector variable with dimension l; $S(\cdot)$ is a vector-valued smooth activation function (sigmoid, $tanh$, saturation) with appropriate dimension; J is a weight-state block-diagonal matrix with (1×1) and (2×2) blocks; J_i is an $i-th$ block of J, and (4) is a stability condition. It imply that all eigenvalues of J, which are $\lambda_i = |J_i|$ ought to be placed inside the unit circle. B and C are weight input and output matrices with appropriate dimensions and block structure, corresponding to the block structure of J. As it can be seen, the given RTNN model is a completely parallel parametric one, so it is useful for identification and control purposes. Parameters of that model are the weight matrices J, B and C. The general BP learning algorithm is given in the form:

$$W_{ij}(k+1) = W_{ij}(k) + \eta \Delta W_{ij}(k) + \alpha \Delta W_{ij}(k-1) \tag{5}$$

Where: W_{ij} $(C,\ J,\ B)$ is the $ij-th$ weight element of each weight matrix (given in parenthesis) of the RTNN model to be updated; ΔW_{ij} is the weight correction of W_{ij}; η, α are learning rate parameters. The initial values of the weigths are choosen to be uniformly distributed random values inside the open interval (-1, 1). The updates ΔC_{ij} , ΔJ_{ij} and ΔB_{ij} of model weights C_{ij} , J_{ij} and B_{ij} are given by:

$$\Delta C_{ij}(k) = [T_j(k) - Y_j(k)]S_j'[Z_j(k)]Z_i(k) \tag{6}$$

$$\Delta J_{ij}(k) = R_1 X_i(k-1) \tag{7}$$

$$R_1 = C_i(k)[T(k) - Y(k)]S_j'[X_j(k)] \tag{8}$$

$$\Delta B_{ij}(k) = R_1 U_i(k) \tag{9}$$

Where: T is a target vector with dimension l and $[T - Y]$ is an output error vector also with the same dimension; R_1 is an auxiliary variable; $S_j'(\cdot)$ is the first derivative of the activation function S_j.

The use of the RTNN-model proposed, could be very helpful for a nonlinear plants identification. As it could be seen from equations (1)-(4), the proposed canonical RTNN architecture is a two-layer hybrid one, with one feedforward output layer and one recurrent hidden layer. The linearization of the activation function allows us to study the dynamic properties of this RTNN architecture, such as stability, observability and controllability. To preserve the RTNN stability during the learning, the model poles must remain inside the unity circle. To do this, some restrictions of the weights in the feedback loop J of the hidden layer, are imposed (see equation (4)). The next part of the paper integrates this RTNN in different schemes of adaptive control, illustrated by simulation results.

3 An Indirect Adaptive Trajectory Tracking Neural Control

The state and output variables of the RTNN model, given by the equations (1), (2) and (3), are restricted in the interval, defined by the limits of the activation functions . So, the linearizided equations of this model are given in the form:

$$X(k+1) = JX(k) + BU(k) \tag{10}$$

$$Y(k) = CX(k) \tag{11}$$

The linearized RTNN model is given in state-space Jordan canonical form, so it is a parametric one. This means that its parameters J, B and C, obtained during the learning, could be used for an indirect adaptive trajectory tracking control system design. The output model equation (11) could be rewritten for the $(k+1)-th$ step and the state variable $X(k+1)$ could be substituted in it, which yields:

$$Y(k+1) = CX(k+1) \quad (12)$$

$$Y(k+1) = CJX(k) + CBU(k) \quad (13)$$

Denoting the reference set-point signal as $R(k)$ and following the design method of Isidori, [9], the next control law could be derived:

$$U(k) = [CB]^{-1}\{-CJX(k) + R(k+1) + \gamma[R(k) - Y(k)]\} \quad (14)$$

Where: J, B, C and $X(k)$ are generated from the RTNN during the learning. The substitutions of (14) in (13) give the following difference equation for the system error:

$$E(k) = R(K) - Y(k) \quad (15)$$

$$E(k+1) + \gamma E(k) = 0 \quad (16)$$

So, the stability of the equation (16) depends on the value of the parameter γ and the absolute value of this parameter ought to be less than 1. The block-diagram of this control is given on Fig.1.

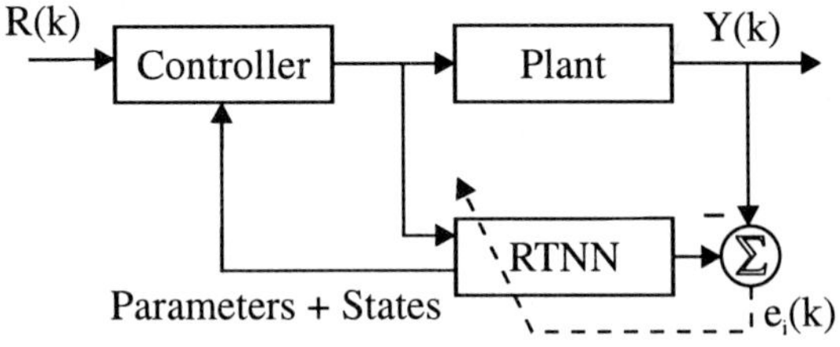

Fig. 1. Block-diagram of an indirect adaptive trajectory tracking control system.

The block diagram of the indirect adaptive trajectory tracking control system, shown on Fig.1, contains one identification and state estimation RTNN, which generates the estimated states and parameters.

4 A Direct Adaptive Trajectory Tracking Neural Control

The block diagram of the direct adaptive trajectory tracking control system, is shown on Fig.2. It contains three RTNN. The first RTNN is a plant model identificator and state estimator, which generates a vector of states, representing an input to the second RTNN. So, the second RTNN is a feedback stabilizing systems controller.

The third RTNN is an inverse system neural model (a feedforward precompensator), that assure the tracking ability of the system output to track the desired set-point trajectory. The control law is given by the following equation:

$$U(k) = -N_1(X(k)) + N_2(R(k)) \quad (17)$$

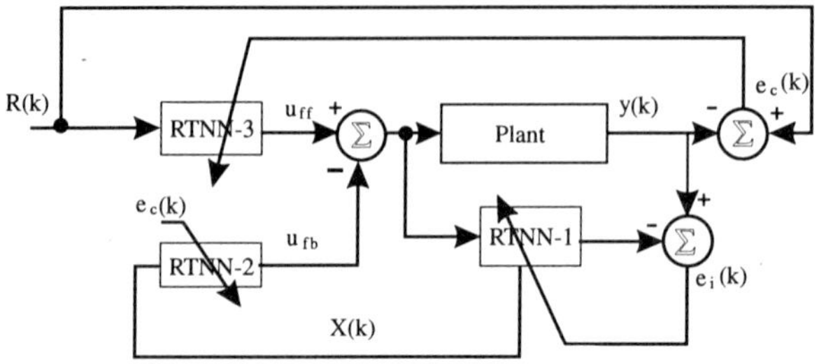

Fig. 2. Block-diagram of a direct adaptive trajectory tracking control system.

Where $N_1(\cdot)$ is the feedback control generated by the RTNN2, which depends on the system states, and $N_2(\cdot)$ is the feedforward control, generated by the RTNN3, which depends on the reference set-point trayectory.

5 Simulation Results

As an example of this control, the following discrete-time nonlinear plant model, taken form [5], is used:

$$Y(k+1) = \frac{Y(k)Y(k-1)Y(k-2)U(k-1)\left[Y(k-2)-1\right]+U(k)}{1+Y(k-1)^2+Y(k-2)^2} \tag{18}$$

The results, obtained from the indirect adaptive neural control scheme are given on Fig.3. The results show a good convergence of the system output to the desired trajectory after 5 seconds of working, due to the good performance of the identification RTNN. The results, obtained using the direct adaptive neural control scheme, are given on Fig.4, and show also a good convergence. This is due to the good performance of the identification RTNN which generates a proper set of states for the feedback systems control, realized by the second RTNN. The learning parameters and the period of discretization of all RTNN's, for both control schemes, are equal and chosen as it is: $\alpha = 0.01$, $\eta = 0.03$, $Ts = 0.001sec.$ The identification RTNN's, for both control schemes, are trained by the error between the output of the plant and the output of this RTNN. The two control RTNN's, for the direct control scheme, are trained by the error between the reference set-point and the output of the plant. The identification RTNN for the indirect control system has the topology (1, 8, 1), and for the direct adaptive control system —(1, 2, 1). The two control RTNN's for the direct adaptive control system have the topology: (2, 2, 1) —for the feedback part and (1, 2, 1) for the feedforward part. As it could be seen on the graphics of the Fig.3b, 4b, the plant model outputs and the outputs of the identification RTNN's are very close, which exhibits a good convergence of the identification RTNN. The good tracking performance could be seen also from the Fig. 3a, 4a, and the small final MSE of tracking (Fig.3d, 4d). The initial performance of the direct control scheme is better than the indirect control scheme, which is more sensitive to the results of identification, and so, needs more neurons in the hidden layer. This could be seen also from the graphics of the MSE (Fig.3d, 4d).

6 Conclusions

The present paper integrates a recurrent trainable neural network in two schemes of real-time soft computing neural control. The following control schemes has been applied: a) an indirect adaptive trajectory tracking control; b) a direct adaptive trajectory tracking control. These control schemes use the state and parameter information, given by an identification recurrent neural network. The given simulation results show a good performance of both control schemes, but the indirect control scheme seems to be more sensitive to the obtained identification results, than the direct one.

References

1. K. S. Narendra, and K. Parthasarathy, "Identification and Control of Dynamic Systems using Neural Networks", IEEE Transactions on NNs, 1(1), 1990, 4-27.
2. K. S. Narendra, and K. Parthasarathy, "Gradient Methods for the optimisation of Dynamical Systems Containing Neural Networks", IEEE Transactions on NNs, 2(2), 1991, 252-262.
3. P. P. C. Yip, and Y. H. Pao,. "A Recurrent Neural Net Approach to One-Step Ahead Control Problems". IEEE Transactions on SMC, 24(4), 1994, 678-683.
4. D. T. Pham,. and S. Yildirim, "Robot Control using Jordan Neural Networks", Proc. of the Internat. Conf. on Recent Advances in Mechatronics, Istanbul, Turkey, Aug. 14-16, 1995, (O. Kaynak, M. Ozkan, N. Bekiroglu, I. Tunay, Eds., Bogaziçi University Printhouse, Istanbul, Turkey, 1995), Vol. II, 888-893.
5. P. S. Sastry, G. Santharam, and K. P. Unnikrishnan, "Memory Networks for Identification and Control of Dynamical Systems". IEEE Transactions on NNs, 5(2), 1994, 306-320.
6. K. J. Hunt, D. Sbarbaro, R. Zbikowski and P. J. Gawthrop. "Neural Network for Control Systems - A Survey". Automatica, vol. 28, No. 6, 1992, pp. 1083-1112.
7. I. Baruch, I. Stoyanov, and E. Gortcheva, "Topology and Learning of a Class RNN". ELEKTRIK, Vol. 4 Supplement, 1996, 35-42.
8. Baruch I., Stoyanov I. and T. Arsenov. "An Improved RTNN Model for Dynamic Systems Identification and Time-Series Prediction". Proc. of the 4-th International Symposium on Methods and Models in Automation and Robotics, 26-29 August., 1997, Miedzyzdroje, Poland, ISBN 83-86359-98-6, vol. 2, pp. 711-717.
9. A. Isidori, "Nonlinear, Control systems", third edition, (London, Springer-Verlag, 1995).

Appendix

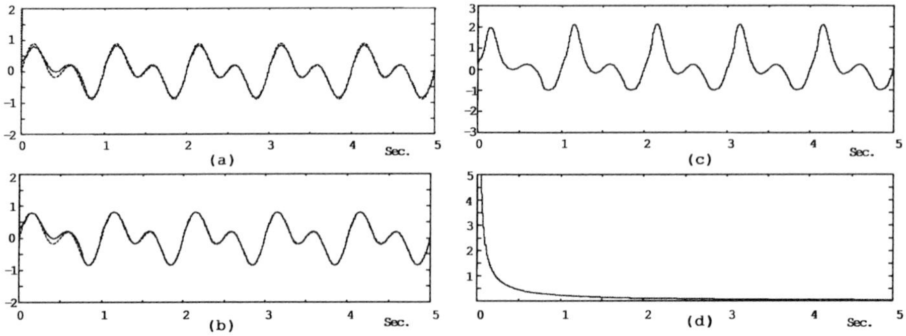

Fig. 3. Indirect adaptive neural control. RTNN topology: (1, 8, 1); parameters: $\alpha = 0.01$, $\eta = 0.03$; $\gamma = 0.001$; reference signal: $u(kTs) = 0.5sin(2\pi kTs) + 0.5sin(4\pi kTs)$; sampling period: $Ts = 0.001Sec.$
a) Control behaviour: — plant output; - - - reference Signal
b) Identification: — plant output; - - - RTNN output
c) Control signal
d) Mean square control error

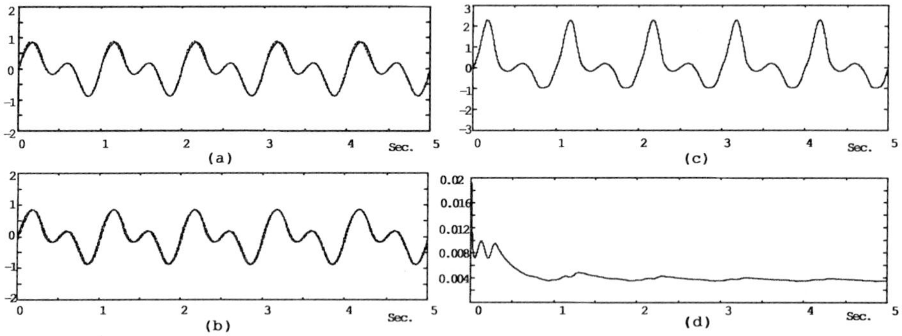

Fig. 4. Direct adaptive neural control. RTNN1, 3 topologies: (1, 2, 1), RTNN2 topology: (2, 2, 1); parameters: $\alpha = 0.01$, $\eta = 0.03$; reference signal: $u(kTs) = 0.5sin(2\pi kTs) + 0.5sin(4\pi kTs)$; sampling period: $Ts = 0.001Sec.$
a) Control behaviour: — plant Output; - - - reference Signal
b) Identification: — plant output; - - - RTNN output
c) Control signal
d) Mean square control error

A Hierarchical Method for Training Embedded Sigmoidal Neural Networks

Jinglu Hu and Kotaro Hirasawa

Department of Electrical and Electronic Systems Engineering
Kyushu University, Hakozaki 6-10-1, Higashi-ku, Fukuoka 812-8581, Japan
{jinglu,hirasawa}@ees.kyushu-u.ac.jp

Abstract. This paper discusses the problem of applying sigmoidal neural networks to identification of nonlinear dynamical systems. When using sigmoidal neural networks directly as nonlinear models, one often meets problems such as model parameters lack of physical meaning, sensitivity to noise in model training. In this paper, we introduce an embedded sigmoidal neural network model, in which the neural network is not used directly as a model, but is embedded in a shield such that part of the model parameters become meaningful. Corresponding to the meaningful part and the meaningless part of model parameters, a hierarchical training algorithm consisting of two learning loops is then introduced to train the model. Simulation results show that such a dual loop learning algorithm can solve the noise sensitivity and local minimum problems to some extent.

1 Introduction

Neural networks have recently attracted much interest in system control community because they learn any nonlinear mapping. Many approaches have been proposed to apply neural networks to prediction and control of general nonlinear systems [5, 6, 8]. In most of these approaches, neural networks are used directly as nonlinear models. However, one of the major criticisms on using neural network models is that they do not have useful interpretations in their parameters, especially for sigmoidal neural network models. It is highly motivated to introduce certain schemes to make part of model parameters meaningful. In this paper, we introduce an embedded sigmoidal neural network model for nonlinear systems. The neural network is not used directly as model, but is embedded in a shield constructed based on application specific knowledge. In this way, it is possible to have a neural network based model that has not only useful interpretations in part of its parameters, but also structures favorable to certain applications.

On the other hand, in practice a model is usually identified using limited, noisy data sets. Since a neural network based model is usually over parameterized, and highly nonlinear one, it is crucial to have an algorithm that prevents overfitting. Various regularization based methods have been introduced for this purpose [7, 3, 9]. However, these methods are found to be not efficient for our embedded sigmoidal neural network model. To solve this problem, we introduce

G. Dorffner, H. Bischof, and K. Hornik (Eds.): ICANN 2001, LNCS 2130, pp. 937–942, 2001.

a hierarchical method consisting of two learning loops, corresponding to the meaningful part and the meaningless part of model parameters.

A simulated numerical problem is designed to demonstrate the effectiveness of the proposed method. Simulation results show that the dual loop learning algorithm can solve the overfitting and local minimum problems to some extent.

2 Embedded Neural Network Model

Let us consider a single input single output (SISO) nonlinear time-invariant systems whose input-output relation described by

$$y(t) = g(\varphi(t)) + e(t) \tag{1}$$

$$\varphi(t) = [y(t-1) ... y(t-n_y)\, u(t-1) ... u(t-n_u)]^T \tag{2}$$

where $y(t)$ is the output at time t $(t = 1, 2, ...)$, $u(t)$ the input, $\varphi(t)$ the regression vector, $e(t)$ the disturbance, and $g(\cdot)$ the unknown nonlinear function. We further introduce the following assumptions

- Assumption 1, $g(0) = 0$ [1];
- Assumption 2, the elements of $\varphi(t)$ are bounded;
- Assumption 3, $g(\cdot)$ is a continuous function, but at a small region around $\varphi(t) = 0$, it is C^∞ continuous.

Performing Taylor expansion to $g(\cdot)$ around the region $\varphi(t) = 0$

$$y(t) = g(0) + g'(0)\varphi(t) + \frac{1}{2}\varphi^T(t)g''(0)\varphi(t) + ... + e(t)$$

where the prime denotes differentiation with respect to $\varphi(t)$, and introducing a vector $\theta(t)$ defined by

$$\begin{aligned}\theta(t) &= g'(0) + \frac{1}{2}\varphi^T(t)g''(0) + ... \\ &= [\theta_1(t)\, \theta_2(t) \ldots \theta_n(t)]^T,\end{aligned}$$

we have

$$y(t) = \varphi^T(t)\theta(t) + e(t) \tag{3}$$

where $n = n_y + n_u$. Then we obtain an embedded neural network (EBNN) model by parameterizing (3) with a multi-input multi-output (MIMO) sigmoidal neural network

$$\mathcal{M}: \quad y(t) = \varphi^T(t)\mathcal{N}(\varphi(t), \Omega) + e(t). \tag{4}$$

where $\mathcal{N}(\cdot,\cdot)$ is a 3-layer sigmoidal neural network with n input nodes, M sigmoidal hidden nodes and n *linear* output nodes. Figure 1 shows the structure of embedded sigmoidal neural network.

[1] Without loss of generality, this assumption is only introduced for simplicity.

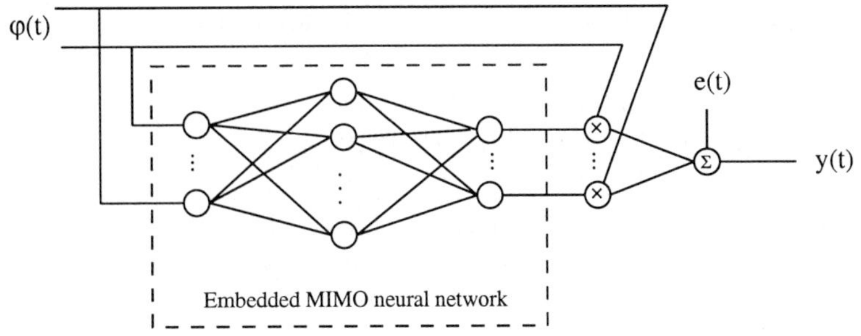

Fig. 1. Structure of embedded sigmoidal neural network model

Let us express the 3-layer sigmoidal neural network by

$$\mathcal{N}(\varphi(t), \Omega) = W_2 \Gamma(W_1 \varphi(t) + B) + \theta \tag{5}$$

where $\Omega = \{W_1, W_2, B, \theta\}$, $W_1 \in \mathcal{R}^{M \times n}$, $W_2 \in \mathcal{R}^{n \times M}$ are the weight matrices of the first and second layers, $B \in \mathcal{R}^{M \times 1}$ is the bias vector of hidden nodes, $\theta \in \mathcal{R}^{n \times 1}$ is the bias vector of output nodes, and Γ is the diagonal nonlinear operator with identical sigmoidal elements σ (i.e. $\sigma(x) = \frac{1-e^{-x}}{1+e^{-x}}$). Then we can rewrite the EBNN model (4) in a form of

$$\mathcal{M}: \quad y(t) = \varphi^T(t) \cdot W_2 \Gamma(W_1 \varphi(t) + B) + \varphi^T(t)\theta + e(t). \tag{6}$$

We notice that the model can be seen to consist of two parts: the one that contains the second and the third terms of the right side of (6) is a linear ARX model part, while the other one including the first term is a nonlinear part. Therefore, in an EBNN model the bias of output nodes θ describes a linear approximation of object system. This makes θ distinctive to other parameters in an EBNN model. In the next section, we develop a dual loop leaning algorithm by using the feature.

From Fig.1, we can see that EBNN model consists of a shield described by (3) and a kernel (5) which is embedded in the shield. As an example, in this paper we use an linear ARX-like shield. The EBNN model obtained in this way contains various linearity properties that are useful in some applications [2]. If application specific knowledge is available, it can be used to construct a shield so that the obtained EBNN model is favorable to the application [1, 2].

3 A Hierarchical Training Algorithm

Since θ describes a linear approximation, it is natural to estimate θ in a different way to other parameters. Let

$$z_L(t) = y(t) - \varphi^T(t) \cdot W_2 \Gamma(W_1 \varphi(t) + B) \tag{7}$$

$$z_N(t) = y(t) - \varphi^T(t)\theta \tag{8}$$

then the model (6) may be decomposed into the following two submodels

$$\mathcal{SM}1: \; z_L(t) = \varphi^T(t)\theta + e(t) \tag{9}$$
$$\mathcal{SM}2: \; z_N(t) = \varphi^T(t) \cdot W_2\Gamma(W_1\varphi(t) + B) + e(t) \tag{10}$$

where $z_L(t)$ can be regarded as the output of linear ARX submodel (9), and $z_N(t)$ the output of nonlinear submodel (10). In this way, the hierarchical training algorithm is described as follows.

- *Step 1*: set $\theta = 0$, and *small* initial values to W_1, W_2, and B.
- *Step 2*: calculate $z_L(t)$ from (7), then estimate θ for submodel $\mathcal{SM}1$ by using least squares (LS) method with a batch or recursive algorithm, as described in Ref.[4].
- *Step 3*: calculate $z_N(t)$ from (8), then estimate W_1, W_2 and B for submodel $\mathcal{SM}2$. This is realized by using the well-known backpropagation (BP) algorithm, but the BP is only performed for a few epochs L.
- *Step 4*: stop if pre-specified condition is met, otherwise go to Step 2 and repeat the estimation of θ, and then W_1, W_2, B.

It is found that adjusting the BP learning step L in each iteration may solve overfitting problem to some extent. For a small noisy data set, the step L should be small. In the numerical simulations discussed in the next section, $L = 40$.

4 Numerical Simulations

To show the effectiveness of the proposed approach, we apply the approach to identification of a simulated system described by

$$\begin{aligned} y(t) = & \frac{\exp(-y^2(t-2) * y(t-1))}{1 + u^2(t-3) + y^2(t-2)} \\ & + \frac{\exp(0.5 * (u^2(t-2) + y^2(t-3))) * y(t-2)}{1 + u^2(t-2) + y^2(t-1)} \\ & + \frac{\sin(u(t-1) * y(t-3)) * y(t-3)}{1 + u^2(t-1) + y^2(t-3)} \\ & + \frac{\sin(u(t-1) * y(t-2)) * y(t-4)}{1 + u^2(t-2) + y^2(t-2)} + u(t-1) + e(t) \end{aligned} \tag{11}$$

where $e(t) \in (0, 0.1)$ is white Gaussian noise. We record 600 input-output data sets by exciting the system with a random input sequence. The first 300 data set is used as training data, while the second 300 data set is using as testing data. The training data set is depicted in Fig.2.

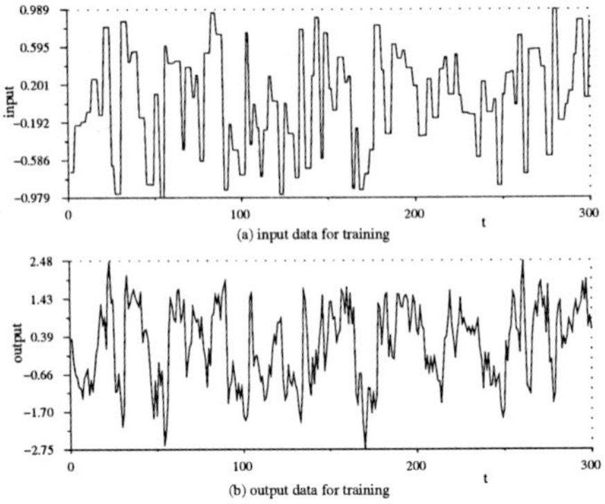

Figure 2. Input-output data set for training.

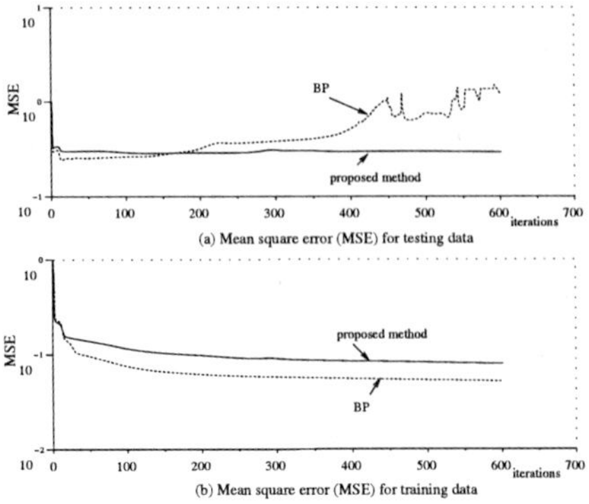

Figure 3. Mean squares error.

To identify the system, we use an EBNN with $n_y = 4$, $n_u = 3$, $M = 8$ ($n = 7$) as model. The model contains 127 parameters to be adjusted. First, we trained the model by using a fast backpropagation (BP) algorithm modified from trainbpx.m for MATLAB Neural Network Toolbox. We found that a BP algorithm for EBNN has much higher probability to be stuck at a local minimum than the one for conventional NN; a mechanism for randomly resetting learning rate is introduced to overcome the difficulty. However, when using the proposed hierarchical algorithm described in Section 3, no local minimum problem appeared in our simulations. This means that the proposed hierarchical training algorithm can more and less solve the local minimum problem.

Figure 3 shows the mean square errors of the BP (dashed lines), the proposed algorithm (solid lines) for (a) testing data, and (b) training data. We can see that using the BP has overfitting problem although it has smaller training error, while this overfitting problem is solved by using the proposed hierarchical algorithm.

To show how well the system has been identified, we excite the system and the identified model using an input sequence described by $u(t) = \sin(2\pi t/250)$ for $t \leq 500$ and $u(t) = 0.8\sin(2\pi t/250) + 0.2\sin(2\pi t/25)$ for $t > 500$ in noise free case. Figure 4 shows the simulation of the model trained using the proposed hierarchical algorithm (dashed line). The solid line shows the system output. We can see that the model represents the system quite well, where RMS (root mean square) error is 0.1473202.

For comparison, Fig.5 shows the results when the model is trained using the BP. We can see that the result is not so good because of overfitting, where RMS error is 0.2098678.

5 Conclusions

In this paper, we discussed the problems of applying sigmoidal neural networks to identification of nonlinear dynamical systems. An embedded neural network model has been introduced, in which the sigmoidal neural network is embedded in a shield constructed using application specific knowledge. The model has useful physical meaning in part of model parameters. This feature has been used to develop a hierarchical training algorithm. Numerical simulations show that the proposed hierarchical algorithm can solve overfitting and local minimum problems to certain extent.

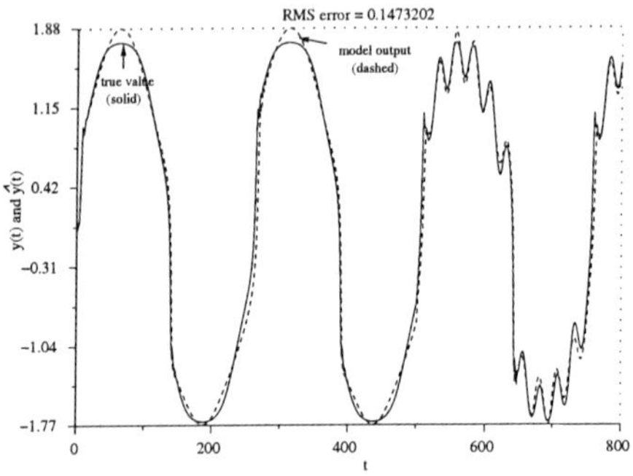

Figure 4. Simulation of the model on testing data in the case where the model is training using the proposed algorithm.

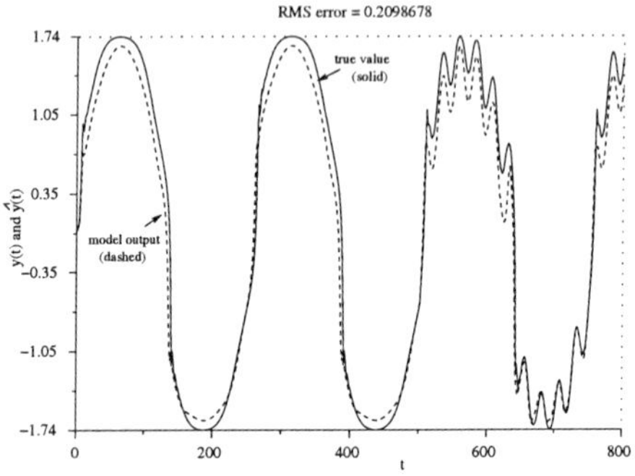

Figure 5. Simulation of the model on testing data in the case where the model is trained using conventional BP.

References

1. J. Hu, K. Hirasawa, and K. Kumamaru. Adaptive predictor for control of nonlinear systems based on neurofuzzy models. In *Proc. of European Control Conference (ECC'99) (Karlsruhe)*, 8 1999.
2. J. Hu, K. Kumamaru, and K. Hirasawa. A quasi-ARMAX approach to modeling of nonlinear systems. *International Journal of Control*, 2001. (to appear in the special issue on nonlinear estimation).
3. M. Ishikawa. Structure learning with forgetting. *Neural Networks*, 9(3):509–521, 1996.
4. L. Ljung and T. Söderström. *Theory and Practice of Recursive Identification.* MIT press, Cambridge, Mass, 1983.
5. W.T. MillerIII, R.S. Sutton, and P.J. Werbos, editors. *Neural Networks for Control.* The MIT Press, Massachusetts, 1990.
6. K.S. Narendra and K. Parthasathy. Identification and control of dynamical systems using neural networks. *IEEE Trans. on Neural Networks*, 1(1):4–27, 1990.
7. J. Sjöberg and L. Ljung. Overtraining, regularization and searching for a minimum, with application to neural networks. *INT. J. Control*, 62(6):1391–1407, 1995.
8. J. Sjöberg, Q. Zhang, L. Ljung, A. Benveniste, B. Deglon, P.Y. Glorennec, H. Hjalmarsson, and A. Juditsky. Nonlinear black-box modeling in system identification: a unified overview. *Automatica*, 31(12):1691–1724, 1995.
9. A.S. Weigend, D.E. Rumelhart, and B.A. Huberman. Generalization by weight-elimination with application to forgetting. *Advances in Neural Information Processing Systems*, 3:875–882, 1991.

Towards Learning Path Planning for Solving Complex Robot Tasks

Thomas Frontzek, Thomas Navin Lal, and Rolf Eckmiller

Department of Computer Science VI, University of Bonn
Römerstraße 164, D-53117 Bonn, F. R. Germany
{frontzek,lal,eckmiller}@nero.uni-bonn.de

Abstract. For solving complex robot tasks it is necessary to incorporate path planning methods that are able to operate within different high-dimensional configuration spaces containing an unknown number of obstacles. Based on Advanced A*-algorithm (AA*) using expansion matrices instead of a simple expansion logic we propose a further improvement of AA* enabling the capability to learn directly from sample planning tasks. This is done by inserting weights into the expansion matrix which are modified according to a special learning rule. For an examplary planning task we show that Adaptive AA* learns movement vectors which allow larger movements than the initial ones into well-defined directions of the configuration space. Compared to standard approaches planning times are clearly reduced.

1 Introduction

The A*-algorithm is an established method for planning tasks. In the literature one can find theoretical studies, mainly dated from the late 60s to the early 80s ([6][5][8]), as well as lots of variations of A*, dated from the 80s until now ([7][10][3][1][4][9]). Most of the applications focus on computing a (slightly suboptimal) path as quickly as possible for limited types of configuration spaces.

In contrast to these approaches we conserve the optimality of computed paths. Furthermore by introducing expansion matrices we make A* more flexible for planning within different types of configuration spaces, and by adding weights into an expansion matrix we enable the capability of learning from sample planning tasks. These capabilities are absolutely needed to implement a flexible and robust tool for complex planning applications, e.g. for path searching in robot control, and for avoidance of self collision and collisions with obstacles in manipulator control.

2 Learning Path Planning with Adaptive AA*-Algorithm

Based on standard A* we explain our concept of Adaptive Advanced A*-algorithm for learning path planning. First, we modify the collision check by abstracting from concrete cell extents. Then we introduce expansion matrices replacing

G. Dorffner, H. Bischof, and K. Hornik (Eds.): ICANN 2001, LNCS 2130, pp. 943–950, 2001.

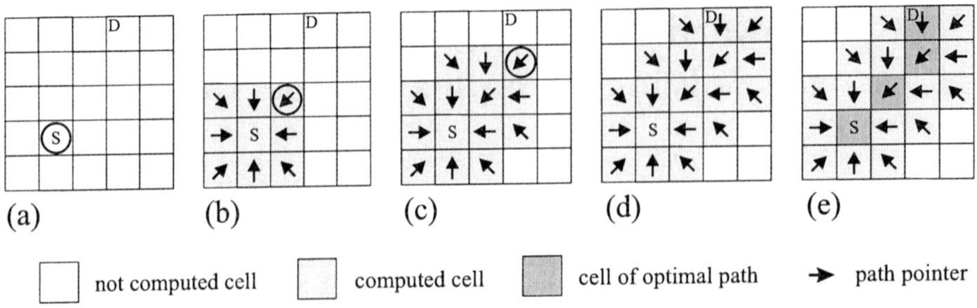

Fig. 1. Proceeding of standard A* from start cell S to goal cell D. First, S is expanded to its adjacent cells (a) and path pointers from these cells to S are set (b). Until reaching D the cell of current minimum costs is expanded (marked with a circle). Finally, an optimal path P_{opt} linked from D to S is computed (e).

the simple expansion logic of A*. At last we further improve the capabilities of expansion matrices by making them adaptive using additional weights and a special learning rule.

2.1 Standard A*-Algorithm

Applying standard A*-algorithm the configuration space representing the actual world model is subdivided into entities called *cells* which usually have the same size. A* searches for an optimal path P_{opt} through the free space from a given start cell S (start) to a given goal cell D (destination). For every considered cell C the costs $f(C,S,D) = g(S,C) + h(C,D)$ are computed, with $g(S,C)$ as costs for the movement from start cell S to the actually considered cell C and $h(C,D)$ as heuristic estimation of the costs from cell C to the goal cell D. Additionally, for every considered cell a check is performed whether the movement to this cell is permitted or not. In real-world scenarios a collision check between the object to be moved and some obstacles has to be performed. Fig.1 visualizes the expansion logic and construction of path pointers of A*.

2.2 Collision Check

Abstracting from concrete cell extents we consider a reference point r_C for each cell C, forming a grid within the configuration space. In addition, a reference point r_O for the object O to be moved is determined. For the collision check first r_O is mapped to the actually considered r_C; then the collision check with respect to the actually used configuration space is computed (fig.2). The underlying world model is only necessary for computing the collision check. Hence, the expansion logic of A* is completely detached from the world model.

2.3 Expansion Matrices

We replaced the expansion logic of standard A* by our novel concept of expansion matrices, forming the Advanced A*-algorithm (AA*) ([4]). This was designed to

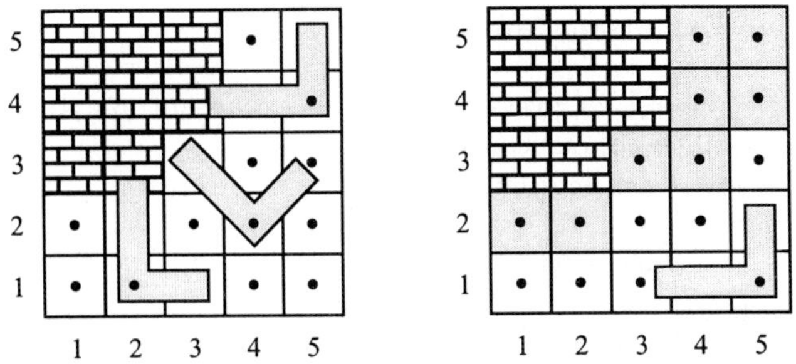

Fig. 2. Collision check for geometric objects considering reference points. The left figure shows an "**L**"-shaped object having orientations of 0°, 45° and 90°. For example cell $(4, 2, 45°)$ is not occupied, cells $(2, 1, 0°)$ and $(5, 4, 90°)$ are defined to be occupied by an obstacle, because at these positions the object intersects an obstacle. The right figure visualizes all cells defined to be occupied by an obstacle (stonewalled and grey cells) given an object orientation of 90°.

overcome the limitations of equally sized cells resp. equidistant reference points and the restriction of symmetric path costs.

An *expansion matrix*:

$$\mathcal{E} := \begin{pmatrix} \Delta x_{11} & \cdots & \Delta x_{1k} & c_1 \\ \vdots & & \vdots & \vdots \\ \Delta x_{m1} & \cdots & \Delta x_{mk} & c_m \end{pmatrix}$$

is an $m \times n$-matrix where m denotes the number of expansion steps and $n = k+1$ with k denoting the number of dimensions of the configuration space $\mathcal{C}$. $c_i \in \mathbb{R}^+$ indicates the costs of expansion step i, and $\Delta x_{ij} \in \mathbb{R}$ indicates the movement difference within dimension j during the expansion step i.

Four parameters of an expansion matrix $\mathcal{E}$ have to be determined:

- *the number of rows*
- *the expansion sequence by permuting rows of* $\mathcal{E}$
- *the length and direction of each movement vector* $(\Delta x_{i1}, \cdots, \Delta x_{ik})$ *for every row* i *of* $\mathcal{E}$
- *the costs* c_i *for every expansion step to model symmetric as well as asymmetric costs*

Starting a new AA* planning process an expansion matrix $\mathcal{E}_{init}$ is initialized appropriating knowledge about the world model. During a single planning process this expansion matrix remains unchanged. With this concept a variety of complex environments and difficult planning tasks can easily be modeled, like different cell extensions resp. distances of reference points or the effect of flow (for example movements with the flow cost nothing, whereas movements against the flow are very expensive).

2.4 Adaptive Expansion Matrices

By adding a weight w_{ij} for every entry of an expansion matrix $\mathcal{E}$ we further improve the capabilities of our concept.

An *adaptive expansion matrix*:

$$\mathcal{A} := \begin{pmatrix} w_{11} \cdot \Delta x_{11} & \cdots & w_{1k} \cdot \Delta x_{1k} & w_{1n} \cdot c_1 \\ \vdots & & \vdots & \vdots \\ w_{m1} \cdot \Delta x_{m1} & \cdots & w_{mk} \cdot \Delta x_{mk} & w_{mn} \cdot c_m \end{pmatrix}$$

is an $m \times n$-matrix using the same notations as $\mathcal{E}$ (see section 2.3). $w_{ij} \in \mathbb{R}^+$ indicates the weight for an entry of $\mathcal{A}$.

Separating a row i of $\mathcal{A}$ we get the *actual movement vector* $\boldsymbol{v}_{act}$ of an expansion matrix:

$$\boldsymbol{v}_{act} := \mathcal{A}^T \cdot \boldsymbol{e}_i \quad \textit{with } \boldsymbol{e}_i \textit{ unit vector } i$$

$\boldsymbol{v}_{act}$ contains the weights $w_{i1}, ..., w_{in} =: \boldsymbol{w}_{act}$.

Inserting $\mathcal{A}$ instead of $\mathcal{E}$ into the expansion logic of AA* the capability of learning from sample planning tasks is enabled by modifying w_{ij}.

At the beginning of a planning process all weights w_{ij} of the expansion matrix $\mathcal{A}$ are initialized with 1.0. Throughout the planning process for every expansion step AA* gets row by row $\boldsymbol{v}_{act}$ of $\mathcal{A}$ describing the movement from the actual considered reference point r_C to a next, possibly new one. After the planning process the computed optimal path P_{opt} is analyzed: Depending on the contribution of the respective $\boldsymbol{v}_{act}$ forming P_{opt} the corresponding weight vector $\boldsymbol{w}_{act}$ is increased by a constant vector. This is done with the aim to strengthen successful movements that is to say movements used forming P_{opt}.

During a single planning process $\mathcal{A}$ stays fixed. After terminating the actual planning process $\mathcal{A}$ is modified taking into account all changes for every weight vector, forming $\mathcal{A}'$. Then $\mathcal{A}'$ is used for subsequent planning processes.

Another variant of our concept, inspired by the paradigm of evolutionary algorithms, is pruning within the adaptive expansion matrix: if a weight vector falls below a certain threshold Φ, the corresponding row of the expansion matrix is substituted by a new, randomly initialized one.

3 Results

We embedded our novel Adaptive Advanced A*-algorithm into a simulation environment which allows us to plan within n-dimensional configuration spaces. First simulations, available for $3D$-configuration spaces ($2D$-position and $1D$-orientation) as well as for $6D$-configuration spaces (6 joint angles of an industrial robot), show encouraging results.

For an exemplary $3D$-planning task, moving an "**L**"-shaped object through a bottleneck, several iterations of our learning procedure were computed (fig. 3).

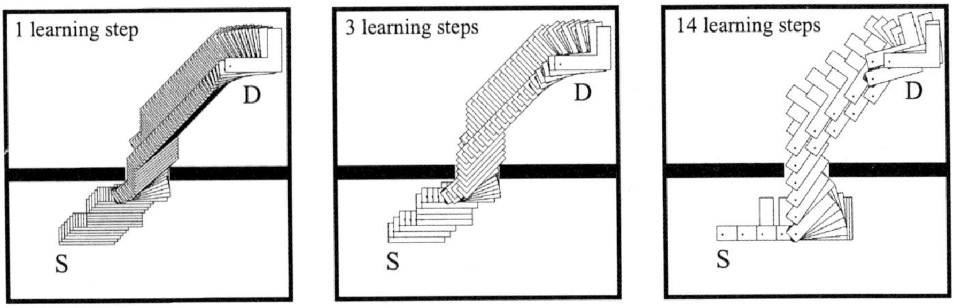

Fig. 3. Visualization of optimal paths computed by Adaptive AA* after one (left), three (middle), and 14 (right) learning steps. The planning task was to move an "**L**"-shaped object from S to D through a bottleneck. The length of the computed optimal path is 133 (left), 77 (middle), and 33 (right) reference points.

Initialized with a standard expansion matrix

$$\mathcal{A}_0 = \begin{pmatrix} -1.0 & 0.0 & 0.0 & 1.0 \\ +1.0 & 0.0 & 0.0 & 1.0 \\ 0.0 & -1.0 & 0.0 & 1.0 \\ 0.0 & +1.0 & 0.0 & 1.0 \\ 0.0 & 0.0 & -1.0 & 1.0 \\ 0.0 & 0.0 & +1.0 & 1.0 \end{pmatrix}$$

and all weights w_{ij} set to 1.0 Adaptive AA* gets the same training planning task for every iteration together with the last updated expansion matrix. Output of the algorithm is an optimal path if one exists, and a modified expansion matrix. Column 1 of $\mathcal{A}_0$ describes the movement along dimension 1 (positions *left* and *right*), column 2 along dimension 2 (positions *down* and *up*), column 3 along dimension 3 (in 5°-steps; orientations *counterclockwise* and *clockwise*); every movement has the same costs 1.0 (column 4 of $\mathcal{A}_0$).

Figure 4 shows the learning progress for our example. The length of the encountered optimal path decreases from 133 reference points using $\mathcal{A}_0$ to 33 reference points after 14 learning steps; the following steps > 14 generate unsteady results. After 14 learning steps the expansion matrix

$$\mathcal{A}_{14} = \begin{pmatrix} -1.0 & 0.0 & 0.0 & 1.0 \\ +7.0 & 0.0 & 0.0 & 7.0 \\ 0.0 & -1.0 & 0.0 & 1.0 \\ 0.0 & +6.0 & 0.0 & 6.0 \\ 0.0 & 0.0 & -2.0 & 2.0 \\ 0.0 & 0.0 & +2.0 & 2.0 \end{pmatrix}$$

was learned by Adaptive AA*. $\mathcal{A}_{14}$ shows that a better partition of the configuration space than the initial one exists: one can solve the planning task moving clearly larger steps *right* and *up*, and moving slightly larger steps *counterclockwise* and *clockwise*. Figure 5 shows the computed optimal paths within the $3D$-configuration space in detail.

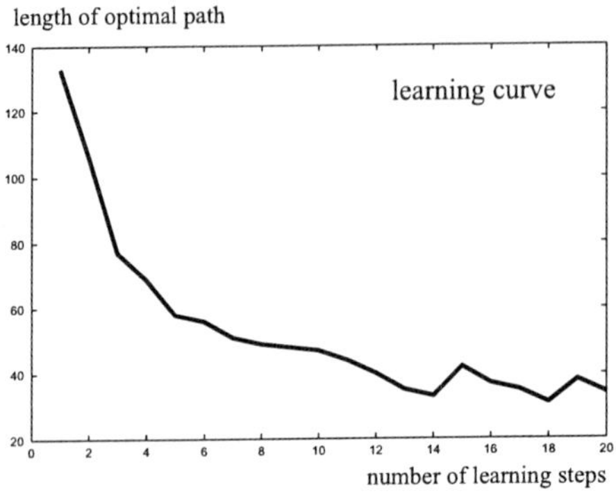

Fig. 4. Learning progress for our sample planning task (see fig. 3). Measure for the proceeding of learning is the length of P_{opt} resp. the number of reference points forming P_{opt}, which decreases from 133 to 33 reference points after 14 learning steps. Steps 15 to 20 generate unsteady results, so in this example the best result is achieved using the expansion matrix learned after step 14.

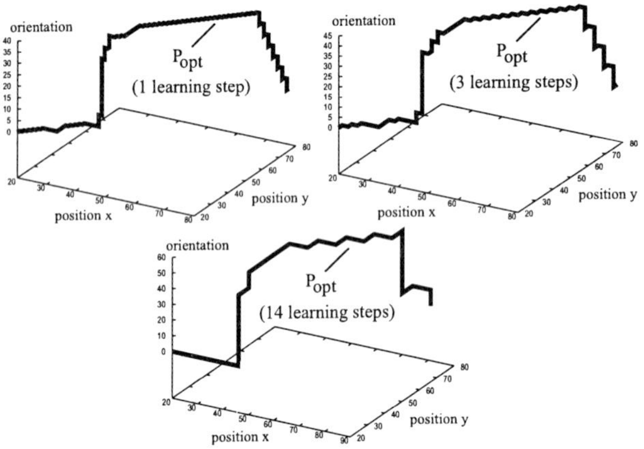

Fig. 5. Visualization of P_{opt} within the 3D-configuration space (dimensions: position x, position y, orientation) after one (left), three (middle), and 14 (right) learning steps.

Comparison with breadth-first-search and standard A* shows the improvements of learning path planning performed by Adaptive Advanced A*-algorithm (see table 1). The increase of $\#r_C$ after 14 learning steps is due to the difficulty of moving through the bottleneck using larger steps *right* and *up* (see figure 3); nevertheless $\#P_{opt}$ decreases.

Performing Adaptive AA* we also computed optimal paths within 6D-joint angle-space for our industrial 6DOF-manipulator *manutec r2*. Here the collision check had to be expanded by coding limitations of the joints as obstacles and by a self-collision check. The results for Adaptive AA* after learning were similar to those specified for our exemplary planning task. In addition to software tests

Table 1. Comparison of breadth-first search, standard A*, and Adaptive Advanced A* after 1, 3, 14 learning steps; $\#P_{opt}$ denotes the length of the encountered optimal path, $\#r_C$ denotes the number of computed reference points during the whole planning process. The $3D$-configuration space consists of 1000000 reference points.

Algorithm	$\#P_{opt}$	$\#r_C$	time [s]
breadth-first search	137	455989	4422
standard A*	137	153811	2190
AA* (1 learning step)	133	3651	2
AA* (3 learning steps)	77	3480	1
AA* (14 learning steps)	33	8723	7

we executed optimal paths generated by our algorithm with the manipulator *manutec r2*.

Altogether our simulation results show that Adaptive AA* can learn from sample planning tasks using adaptive expansion matrices. For our exemplary planning task, moving an "**L**"-shaped object through a bottleneck, a better partition of the configuration space than the initial one was learned. Current investigations focus on a learning rule that substitutes rarely used movement vectors by new, randomly initialized ones.

4 Conclusions

Based on Advanced A*-algorithm we proposed a novel concept of adaptive expansion matrices for learning path planning. Substituting the standard expansion logic of A* we enabled the capability to overcome the limitations of equidistant reference points as well as to learn from sample planning tasks. Adaptive AA* keeps the feature of A* to compute an optimal path, with respect to the defined costs and the actual expansion matrix.

Results obtained by solving exemplary planning tasks underlined the power of AA* using adaptive expansion matrices: movement vectors, which allow larger movements than the initial ones into well-defined directions of the configuration space, were learned. Learning efficient movements within high-dimensional configuration spaces leads to the ability of Adaptive AA* to plan complex robot tasks. This can be done either by using Adaptive AA* as global planner (with some restrictions regarding the "curse of dimensions") or as very flexible local planner within a probabilistic roadmap frame.

Acknowledgments

Parts of this work have been supported by the Federal Ministry for Education, Science, Research, and Technology (BMBF), project LENI. Thanks to Jürgen A. Donath, Nils Goerke, and Stefan Hoffrichter for useful comments.

References

1. N. M. Amato, O. B. Bayazit, L. K. Dale, C. Jones, and D. Vallejo. Choosing good distance metrics and local planners for probabilistic roadmap methods. In *Proc. of the IEEE Int. Conf. on Robotics and Automation (ICRA'98) Leuven, Vol.1*, pages 630–637, 1998.
2. S. E. Dreyfus. An appraisal of some shortest-path algorithms. *Operations Research*, 17:395–412, 1969.
3. T. Frontzek, N. Goerke, and R. Eckmiller. A hybrid path planning system combining the A*-method and RBF-networks. In *Proc. of the Int. Conf. on Artificial Neural Networks (ICANN'97) Lausanne*, pages 793–798, 1997.
4. T. Frontzek, N. Goerke, and R. Eckmiller. Flexible path planning for real-time applications using A*-method and neural RBF-networks. In *Proc. of the IEEE Int. Conf. on Robotics and Automation (ICRA'98) Leuven, Vol.2*, pages 1417–1422, 1998.
5. D. Gelperin. On the optimality of A*. *Artificial Intelligence*, 8:69–76, 1977.
6. P. E. Hart, N. J. Nilsson, and B. Raphael. A formal basis for the heuristic determination of minimum cost paths. In *IEEE Transactions on Systems, Man, and Cybernetics*, volume 2, pages 100–107, 1968.
7. R. E. Korf. Real-time heuristic search. *Art. Int.*, 42:189–211, 1990.
8. J. Pearl. Knowledge versus search: A quantitative analysis using A*. *Artificial Intelligence*, 20:1–13, 1983.
9. L. Shmoulian and E. Rimon. $A_\epsilon^* - DFS$: an algorithm for minimizing search effort in sensor based mobile robot navigation. In *Proc. of the IEEE Int. Conf. on Robotics and Automation (ICRA'98) Leuven, Vol.1*, pages 356–362, 1998.
10. C. W. Warren. Fast path planning using modified A* method. In *Proc. of the IEEE Int. Conf. on Robotics and Automation*, pages 662–667, 1993.

Hammerstein Model Identification Using Radial Basis Functions Neural Networks

Hussain N. Al-Duwaish and Syed Saad Azhar Ali

King Fahd University of Petroleum and Minerals,
Department of Electrical Engineering,
Dhahran 31261, Saudi Arabia
{hduwaish,saadali}@kfupm.edu.sa
http://www.kfupm.edu.sa

Abstract. A new method for the identification of the nonlinear Hammerstein Model consisting a static nonlinearity in cascade with a linear dynamic part, is introduced. The static nonlinearity is modeled by radial basis function neural networks (RBFNN) and the linear part is modeled by an autoregressive moving average (ARMA) model. A recursive algorithm is developed to update the weights of the RBFNN and the parameters of the ARMA model.

1 Introduction

The behavior of many systems can be approximated by a static nonlinearity in cascade with some linear dynamics. This type of model is known as Hammerstein Model, shown in Fig 1, where $u(t)$ is the input to the system, $y(t)$ is the output and $x(t)$ is the intermediate nonmeasureable variable. The examples where such a system was applied are nonlinear filters, nonlinear networks, detection of signals in non-Gaussian noise, nonlinear prediction, nonlinear data transmission channels, control system, identification of nonlinear systems [1], biological systems and many others.

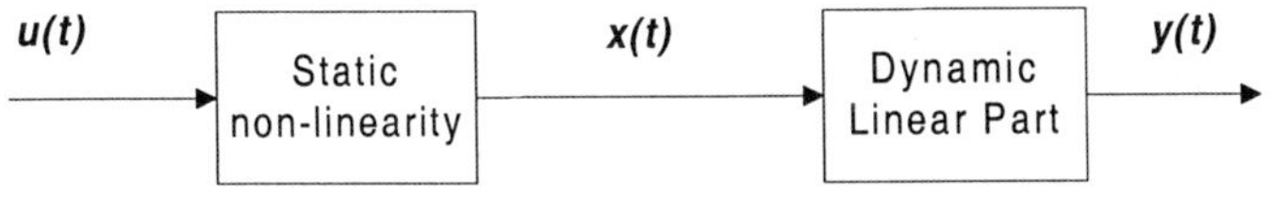

Fig. 1. Structure of Hammerstein Model

Many identification methods have been developed to identify the Hammerstein model [2], [3], [4], [5]. These methods can be classified into parametric and non-parametric techniques. Generally Hammerstein model is represented by the following [2], [3]:

$$y(t) = \frac{B(q^{-1})}{A(q^{-1})} q^{-d} x(t), \tag{1}$$

G. Dorffner, H. Bischof, and K. Hornik (Eds.): ICANN 2001, LNCS 2130, pp. 951–956, 2001.

with

$$B(q^{-1}) = b_o + b_1 q^{-1} + \cdots + b_m q^{-m},$$
$$A(q^{-1}) = 1 + a_1 q^{-1} + \cdots + a_n q^{-n}$$

and the non-measured intermediate variable $x(t)$ is the output of the static nonlinearity given by

$$x(t) = f(u(t), \rho) \tag{2}$$

where q^{-1} is the unit delay operator, $u(t)$ is the input, $y(t)$ is the output, (m, n) represent the order of the linear dynamics, d is the system delay and ρ corresponds to the parameters of the nonlinearity *e.g.* weights of the RBFNN. Thus, the problem of the Hammerstein model identification is to estimate the parameters $a_1, \ldots, a_n$, $b_o, \ldots, b_m$ and ρ from input-output data.

Most of the non-parametric methods use kernels regression estimates to identify the nonlinearity. In these methods, the identification of the nonlinearity is performed separately from the linear part.
A combined model consisting of Multi-layer feed forward network (MFFN) and ARMA model has been implemented previously [6] for identification of the Hammerstein model. MFFN is updated using the famous Back Propagation (BP) algorithm. However, the BP tends to be slower as the error has to be back propagated through all the network layers and used in weight update. For a complex layered network the BP will be even slower. This suggests the replacement of MFFN with RBFNN, which contains only one hidden layer. The convergence of RBFNN is faster than the MFFN as all the computations are done in only one layer.

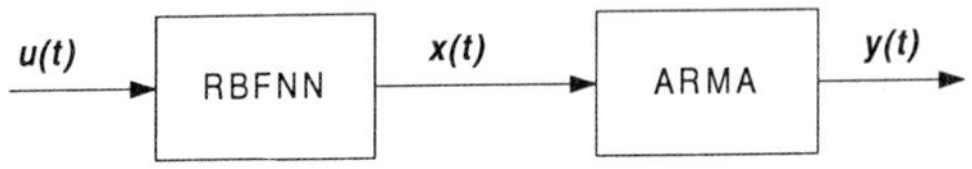

Fig. 2. RBFNN/ARMA identification method Structure

2 RBFNN/ARMA Identification Method Structure

The RBFNN/ARMA identification method consists of an RBFNN in series with an ARMA model, as shown in Fig 2.The use of RBFNN is motivated by its ability to model any nonlinear function. They are used in a wide variety of contexts such as function approximation, pattern recognition and time series prediction.

A single-input single-output RBFNN is shown in Fig 3. It consists of an input node $u(t)$, a hidden layer with n_o nodes and an output node $x(t)$. The input node is connected to all the nodes in the hidden layer through unity weights. While each of the hidden layer nodes is connected to the output node through some weights. Each node finds the distance, using Euclidean norm, between its centre and the input and passes the resulting scalar through a non-linearity. So the

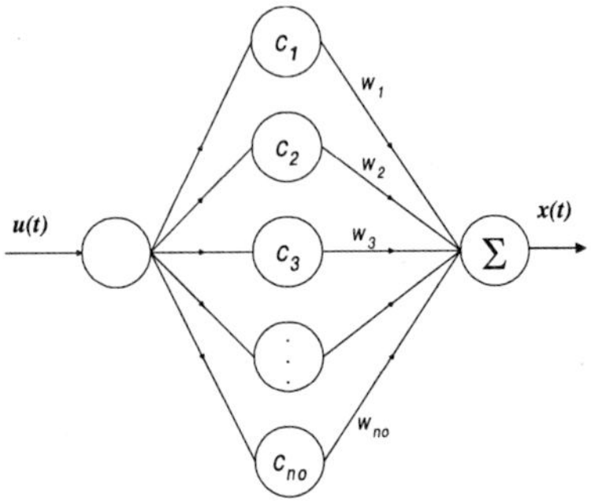

Fig. 3. A general RBF network

output of each hidden node is given by $\phi(\|u(t) - c_i\|)$ where $u(t)$ is the input, c_i is the centre of ith hidden layer node, $i = 1, 2, \dots, n_o$, and $\phi(\|\cdot\|)$ is the nonlinear basis function. Normally this function is taken as a Gaussian function. The output $x(t)$ is a weighted sum of the outputs of the hidden layer, given by

$$x(t) = \sum_{i=1}^{n_o} w_i \phi(\|u(t) - c_i\|) \tag{3}$$

$$x(t) = W\Phi(t) \tag{4}$$

$$\text{where } W = [w_1\, w_2\, \dots\, w_{n_o}] \tag{5}$$

$$\Phi(t) = [\phi\|u(t) - c_1\|\ \phi\|u(t) - c_2\|\ \dots\ \phi\|u(t) - c_{n_o}\|]^T \tag{6}$$

$x(t)$ is the output, w_i is the weight corresponding to the ith hidden layer node. The linear part of the system is modeled by an ARMA, given by 1.

3 Proposed Training Algorithm

Considering Fig 2, the objective is to develop a recursive algorithm that adjusts the weights of the RBFNN and parameters of the ARMA model in such a way that the set of inputs produces a desired set of outputs. This problem is solved by developing a new parameter estimation algorithm based on the least mean squares (LMS) algorithm. The parameters (weights of RBFNN and the coefficients of the ARMA) are updated by minimizing the performance index I given by

$$I = \frac{1}{2}e^2(t),$$

where $e(t) = y(t) - \hat{y}(t)$ and $\hat{y}(t)$ is the estimated output of the Hammerstein model. The coefficients of the ARMA model and the weights of the RBFNN should be updated in the negative direction of the gradient. Keeping the coefficients in a parameter vector θ as $\theta(t) = [a_1\ \dots\ a_n\ b_o\ \dots\ b_m]$ and the regressions in regressor vector ψ as $\psi(t) = [y(t-1)\ \dots\ y(t-n)\ x(t-d)\ \dots\ x(t-d-m)]$ and finding partial derivatives.

$$\begin{aligned}
\frac{\partial I}{\partial \theta(t)} &= \frac{1}{2}\frac{\partial e^2(t)}{\partial \theta(t)} \\
&= e(t)\frac{\partial}{\partial \theta(t)}\left(y(t)-\hat{y}(t)\right) \\
&= e(t)\frac{\partial}{\partial \theta(t)}\left(y(t)-\left(a_1q^{-1}+\cdots+a_nq^{-n}\right)\hat{y}(t)\right. \\
&\qquad\qquad\qquad \left. -\left(b_o+b_1q^{-1}+\cdots+b_mq^{-m}\right)q^{-d}\hat{x}(t)\right) \\
&= -e(t)\left(\hat{y}(t-1)+\hat{y}(t-2)+\cdots+\hat{y}(t-n)\right. \\
&\qquad\qquad\qquad \left. +\hat{x}(t-d)+\hat{x}(t-d-1)+\cdots+\hat{x}(t-d-m)\right) \\
\frac{\partial I}{\partial \theta(t)} &= -e(t)\hat{\psi}(t)
\end{aligned}$$

$\therefore$ the final LMS parameter update equation for ARMA will be,

$$\boxed{\theta_{k+1} = \theta_k + \alpha\, e(t)\psi(t)} \tag{7}$$

α is the learning rate. Now the partial derivative for the weights is derived as

$$\begin{aligned}
\frac{\partial I}{\partial W} &= \frac{1}{2}\frac{\partial e^2(t)}{\partial W} \\
&= e(t)\frac{\partial}{\partial W}\left(y(t)-\hat{y}(t)\right) \\
&= e(t)\frac{\partial}{\partial W}\left(y(t)-\left(a_1q^{-1}+\cdots+a_nq^{-n}\right)\hat{y}(t)\right. \\
&\qquad\qquad\qquad \left. -\left(b_o+b_1q^{-1}+\cdots+b_mq^{-m}\right)q^{-d}W\Phi(t)\right) \\
&= -e(t)\left(b_o+b_1q^{-1}+\cdots+b_mq^{-m}\right)q^{-d}\Phi(t) \\
\frac{\partial I}{\partial W} &= -e(t)B(q^{-1})q^{-d}\Phi(t)
\end{aligned}$$

This gradient can be used in finding the new weights.

$$\therefore \boxed{W_{k+1} = W_k + \alpha\, e(t)B(q^{-1})q^{-d}\Phi(t)} \tag{8}$$

4 Simulation Results

Example 1: The Process used in this example is a second-order system with a saturation nonlinearity at the input.

$$y(t) = 0.4y(t-1) + 0.35y(t-2) + s(u(t)), \text{ where}$$

$$s(u(t)) = \begin{cases} 0.5, & \text{for } u(t) > 0.5 \\ u(t), & \text{for } -0.5 \le u(t) \le 0.5 \\ -0.5, & \text{for } u(t) < -0.5. \end{cases}$$

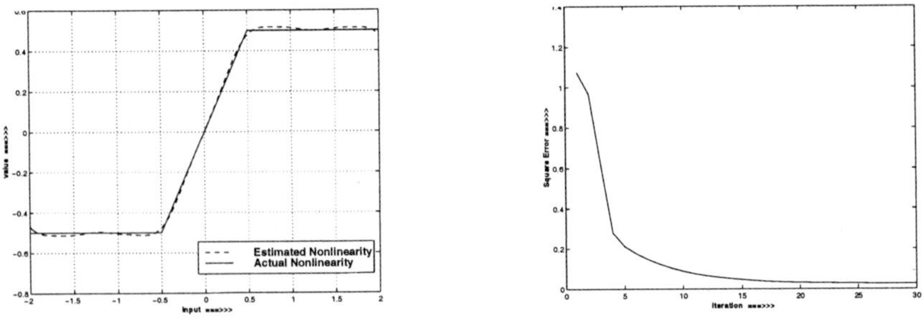

(a) (b)

Fig. 4. (a) actual and identified nonlinearities (b) square error plot

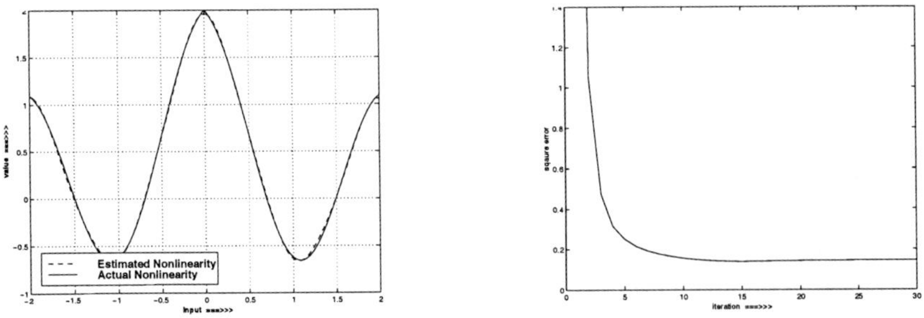

(a) (b)

Fig. 5. (a) actual and identified nonlinearities (b) square error plot

The structure of the identification model consists of an RBFNN in series with an ARMA. The RBFNN has 9 hidden nodes with Gaussian functions and centers evenly distributed in input range. The ARMA model is of second order with $\theta = [\hat{a}_1\ \hat{a}_2\ \hat{b}_o]^T$. Using random inputs uniformly distributed in the interval $[-2, 2]$, the desired outputs are generated using the process model. The proposed learning algorithm is used to update the weights of the RBFNN and the parameters for the ARMA model. The parameters $\hat{a}_1$, $\hat{a}_2$ and $\hat{b}_o$ converged to values 0.3987, 0.3508 and 0.7879 respectively. Fig 4(a) shows the actual and identified nonlinearity and Fig 4(b) shows the square error plot.

Example 2: Consider the same process as in Example 1 with the input nonlinearity replaced by

$$s(u(t)) = \cos(3u(t)) + \exp(-|u(t)|). \tag{9}$$

The parameters $\hat{a}_1$, $\hat{a}_2$ and $\hat{b}_o$ converged to values 0.3997, .3512 and 0.7716 respectively. Fig 5(a) shows the actual and identified nonlinearity of the identified model after training and Fig 5(b) shows the square error plot.

5 Conclusions

A new recursive algorithm based on LMS has been derived to identify the Hammerstein systems. The proposed algorithm is much faster than the MFFN approach, as the simulation time and number of iterations were considerably reduced in addition to lessening of the number of nodes. For comparisons see [6]. The proposed identification scheme provided satisfactory tolerance for different SNRs. Simulation studies suggest robust and satisfactory identification behaviour of the developed method and algorithm. The algorithm is derived for the single-input single-output systems. However, it can be generalized to multi-input multi-output systems.

Acknowledgement

The authors would like to acknowledge the support of King Fahd University of Petroleum & Minerals, Dhahran, Saudi Arabia.

References

1. S. H. Eskinat, E. Johnson and W. L. Luyben. Use of hammerstein models in identification of nonlinear systems. *American Institute of Chemical Engineering Journal*, 37:255–268, 1991.
2. K. Narendra and P. Gallman. An iterative method for the identification of nonlinearsystems using hammerstein model. *IEEE Transactions on Automatic Control*, 11:546–550, 1996.
3. F Chang and R Luus. A noniterative method for identification using hammerstein model. *IEEE Transactions on Automatic Control*, 16:464–468, 1971.
4. H. Al-Duwaish and M. Nazmul Karim. A new method for the identifcation of hammerstein model. *Automatica*, 33:1871–1875, 1997.
5. W. Greblicki and M. Pawlak. Nonparametric identification of hammerstein systems. *IEEE Transactions on Information Theory*, 35:409–418, 1989.
6. M. Nazmul Karim H. Al-Duwaish and V. Chandrasekar. Hammerstein model identification by multilayer feedforward neural networks. *International Journal of Systems Science*, 28(1):49–54, 1997.

Evolving Neural Behaviour Control for Autonomous Robots

Martin Hülse, Bruno Lara, Frank Pasemann, and Ulrich Steinmetz

[1] TheorieLabor, Friedrich Schiller University Jena
07740 Jena, Germany
[2] Max Planck Institute for Mathematics in the Sciences
04103 Leipzig, Germany

Abstract. An evolutionary algorithm for the creation of recurrent network structures is presented. The aim is to develop neural networks controlling the behaviour of miniature robots. Two different tasks are solved with this approach. For the first, the agents are required to move within an environment without colliding with obstacles. In the second task, the agents are required to move towards a light source. The evolution process is carried out in a simulated environment and individuals with high performance are also tested on a physical environment with the use of Khepera robots.

1 Introduction

Within the frame of Synthetic Modeling and Embodied Cognitive Science [8] an evolutionary algorithm is presented. This algorithm, called ENS^3 (for Evolution of Neural Systems by Stochastic Synthesis [10]), evolves the structure and size of artificial neural networks and optimizes the corresponding parameters at the same time. It is designed especially to generate networks with recurrent connectivity. In acquiring both network topology and parameter optimization simultaneously it is similar to the GNARL algorithm [1].

The starting hypothesis for the presented experiments is that internal dynamical properties of recurrent networks are the basis for cognitive capabilities and therefore should be also used for the control of robot behaviour. It is well known that even small recurrent networks present a variety of dynamical characteristics such as bifurcating attractors, hysteresis effects, and chaotic attractors co-existing with periodic or quasi-periodic attractors [9]. So the idea is to use the ENS^3 algorithm to generate recurrent network structures for agents having to exist and survive in a complex environment.

For the experiments described in this paper the agents are Khepera robots [5]. Evolution is carried out using a Khepera simulator [3]. Finally, the recurrent networks with highest performance are tested on the real robot. Section 2 of this paper introduces the ENS^3 evolutionary algorithm. Section 3 describes the experimental set-up and the results are outlined in Section 4 which also concludes that paper with a discussion of these results.

G. Dorffner, H. Bischof, and K. Hornik (Eds.): ICANN 2001, LNCS 2130, pp. 957–962, 2001.

2 ENS^3 Description

The algorithm is inspired by a biological theory of co-evolution. Originally it was designed to study the appearance of complex dynamics and the corresponding structure-function relationship in artificial sensorimotor systems. The basic assumption is that in "intelligent" systems the behaviour relevant features of neural subsystems, called neuro-modules, are due to their internal dynamical properties which are provided by their recurrent connectivity structure. But the structure and interaction of such nonlinear subsystems with recurrences in general can not be designed to produce a reasonably complex dynamical behaviour. Thus, the main focus of the ENS^3-algorithm is to evolve an appropriate structure of artifical neural networks.

To start the algorithm one first has to decide which type of neurons to use for the network. We prefer to have standard additive neurons with sigmoidal transfer functions for output and internal units, that is their dynamics is given by $a_i(t+1) = \theta_i + \sum_{j=1}^{j} w_{ij} \cdot \sigma(a_j(t))$. The input units are used as buffers.

The number of input and output units is chosen according to the definition of the problem; that is, it depends on the pre-processing of input and post-processing of output signals. Nothing else is determined, neither the number of internal units nor their connectivity, i.e. self-connections and every kind of recurrences are allowed, as well as excitatory and inhibitory connections. Because input units are only buffering data, no backward connections to these are allowed.

To evolve the desired neuro-module we consider a population $p(t)$ of $n(t)$ neuro-modules undergoing a variation-evaluation-selection loop, i.e. $p(t+1) = S \cdot E \cdot V \cdot p(t)$. The *variation operator* V is defined as a stochastic operator, and allows for the insertion and deletion of neurons and connections as well as for alterations of bias and weight terms. Its action is determined by fixed per-neuron and per-connection probabilities. The *evaluation operator* E is defined problem-specific, and it is usually given in terms of a fitness function. After evaluating the performance of each individual network in the population the number of network copies passed from the old to the new population depends on the *selection operator* S. This operator performs the differential survival of the varied members of the population according to evaluation results. In consequence of this selection process the average performance of the population will tend to increase. Thus, after repeated passes through the variation-evaluation-selection loop populations with networks solving the problem can be expected to emerge.

3 Experiments

The aim of the experiments is to find recurrent neural networks which are able to control specific behaviour of the Khepera robots. To achieve this the networks are evolved using the ENS^3 algorithm described in Section 2. As aforementioned the only necessary initial conditions, on terms of structure and size, are the input and output neurons. These constrains are defined by the structure of the Khepera robot.

Two basic behaviour were studied using an average population size of 50 individuals, obstacle avoidance and light seeking. The time interval for evaluating these individuals was set to 5000 simulator time steps. Furthermore, we can influence the size of resulting networks by changing the probabilities of increase and/or decrease of neurons and/or synapsis and adding cost terms for neurons and connections to the given fitness functions, as described in Section 2.

For evaluating the performance of the robots the two dimensional Khepera simulator [3] was used. A suitable interface for the communication between the Kephera simulator and the evolutionary program was implemented. In this way every neuro-module of a generation is sent to the simulator and will be directly used for controlling the simulated Khepera. The start position of the robots is randomly set for every generation but fixed for the evaluation of all neuro-modules in one generation. The fitness function for the calculation of the performance is implemented in the simulator. Performance of every neuro-module (controlling the simulated Khepera) is calculated for the life span of each individual in each generation and then sent back to the ENS^3.

We want to evolve neuro-modules which allow the Khepera robot to move in a given environment as long as possible without colliding with any obstacle. A typical environment used for this evolution task is shown in the Fig. 1. Environments are build on the Khepera simulator. It is worth noting that the evolved network structures are expected to behave at least with the same performance in the physical robot as they do in the simulated environments.

For evolving neuro-modules which have an obstacle avoidance behaviour the 8 infra-red proximity sensors of the Khepera are used. The number of sensors determines the number of input neurons of the neuro-controllers. To control the motors governing the robot wheels positive and negative signals are required. A *tanh* transfer function satisfies this requirement and is therefore used for the output neurons as well as for hidden neurons. The bias terms θ_i are set to zero for all of the neurons throughout the whole evolution process. The initial structure of the individual neuro-modules have only 8 linear input neurons, as buffers and two nonlinear output neurons.

To generate an obstacle avoidance behaviour we introduce a fitness function F_1 which states: For a given time t go straight ahead as long and as fast as possible. This is coded in terms of the two network output signals M_0 and M_1 as follows:

$$F_1 := \sum_{t=1}^{5000} M_0(t) + M_1(t), \qquad (1)$$

where $-1 < M_0, M_1 < 1$. A second condition for calculation of the performance is used. If the robot has a collision before 5000 time steps the evaluation of the neuro-module stops and the value of the current performance is taken as the maximum performance of the individual.

At the beginning of the evolutionary process robots do not move or are spinning around having only one wheel active or both with opposite signals. After a few generations displacement of the robots in straight trajectories can be observed. After approx. 30 generations the robots move fast, turning away from the walls and exploring nearly all of the accessible space at the simulated environment. Fig. 1 shows a typical path of such simulated robot controlled by an evolved neuro-module.

During this phase the networks generally grow in size. To make or keep the network size small we increase the probabilities of deleting neurons and synapses within the evolutionary algorithm. This is done around the 50th generation, where it can be seen that the performance remains constant. After the probability change, the network size (number of synapsis and hidden neurons) decreases, but the performance and the observable behaviour remains constant.

For the second experiment in addition to the 8 proximity sensors we use the 8 light sensors of the Khepera; i.e., we now have 16 sensor inputs. The goal in this experiment is to find a light source as fast as possible and move towards it. In the case of a moving light source the robot should follow it. Aiming to find a network structure only for this type of behaviour the individuals should not be concerned with obstacles. The fitness function is coded as:

$$F_2 := \sum_{t=1}^{5000} \alpha \cdot M(t) + \beta \cdot (1 - \Delta_m(t)) + \gamma \cdot (1 - P_{ls}(t)) \; , \qquad (2)$$

where $-2 < M < 2$, $0 \leq \Delta_m < 2$, $0 \leq P_{ls} \leq 1$ and α, β and γ are appropriate constant parameters. It states: move forward, as M is the sum of the output to the motors, minimize turning, as Δ_m is the absolute difference between the output signals, and move towards light, as P_{ls} is a value proportional to the activity of the light sensitive sensor with the highest activity. Note: in this case evaluation does not stop, if a collision occurs before 5000 time steps.

4 Results and Discussion

A network, which generates good behaviour in the simulated environment as well as on the physical robot is shown in figure 1. Input neurons are represented by black circles with the position they have in the robot. This network was randomly selected and represents the best individual of the 150th generation. It uses two hidden neurons, which are self-excitatory, one of this is larger than 1. It is known that for a single neuron with positive self-connection larger than 1 there is an input interval over which hysteresis phenomena (flipping between two stable states) can be observed [9]. It seems that this is used for instance to get out of situations which usually generate deadlocks. The output neuron driving the left motor is self-inhibitory, and has also an inhibitory connection from the other output neuron, which drives the right motor.

The behaviour in the real robot generated by this network is similar to the plotted paths in Fig. 1. The robot presents obstacle avoidance as well as exploration abilities. Letting the robot move in its physical environment after a while it will have visited almost all areas within reach. This can of course not be achieved by pure wall following behaviour, as it is usually learned by robots.

After only a few generations the evolved structures presented light seeking behaviour. The individuals advance towards the light when this is within the

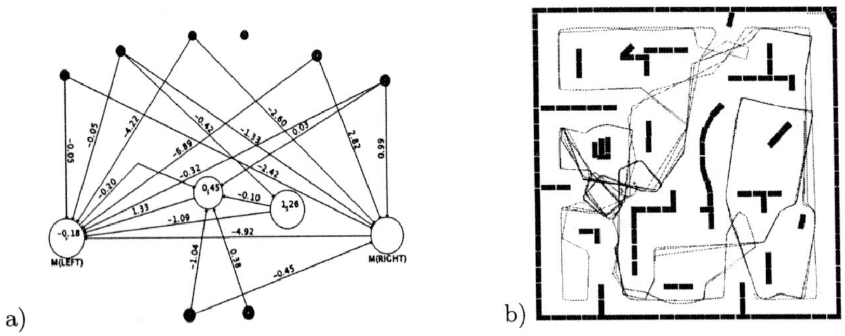

Fig. 1. a) Network with highest performance of the 150th generation. b) Typical paths followed by the simulated robots with evolved neuro-controllers for obstacle avoidance.

reach of the light sensors. The individual with the highest performance of the 50th generation is shown in Fig. 1. Again, black circles show input neurons, the first group of 8 neurons represents the proximity sensors, the second group represents the light sensors. It can be observed that there exist connections from the proximity sensors, even though there was no specific data requirement from their input in the definition of the fitness function. These connections are related to the straight trajectory of the robot in the absence of light.

When the physical robot was used, the individuals go in straight trajectories towards the light. If the light is moved they follow it. When there is no light present in the environment the individuals move in a wide semi-circular path. If the connections coming from the first 8 inputs neurons were manually deleted the robot presented a different behaviour. In the absence of light, it will turn in its own central axis, but when presented with the light source it would advance towards it and keep following it if the light source was in movement. Trajectories, with different starting positions, followed by the best individual of generation 50 are shown in Fig.2.

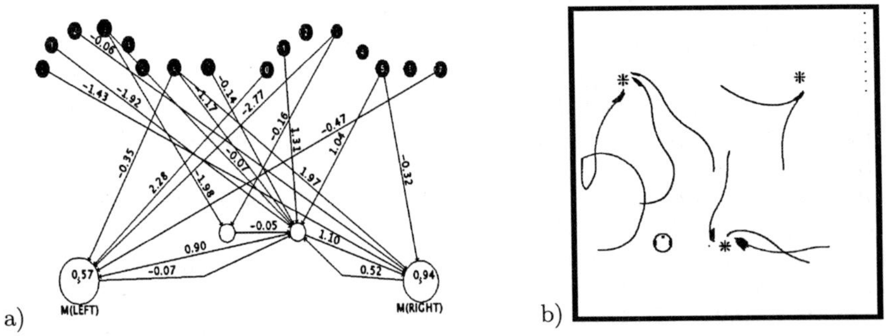

Fig. 2. a) The network with the best performance of the 50th generation. b) Typical paths followed by the simulated robots with evolved neuro-controllers for light seeking.

For the first simulated environment there is only one light source in the center. It is seen that disregarding the starting position the robot will always move towards the light and stay around it. On the second environment the robot moves towards the light source which finds closest at the front.

The two different tasks investigated, were already solved using various approaches e.g. [4],[2], [11] and especially also feed-forward network solutions are well known [7]. But basically the presented results indicate that the ENS^3 algorithm is able to generate working controllers for simulated as well as real autonomous robots. Finally, it is shown that dynamical effects of recurrent connectivity structure provide surprising possibilities to solve control tasks. In particular the case, where a hysteresis element enabled the robot to come out of deadlock situations, like for instance in sharp corners and blind alleys. These effects demonstrate the existence of yet unknown mechanisms which allow more sophisticated solutions for behaviour control of autonomous systems.

References

1. Angeline, P.J., Saunders, G.B., Pollack, J.B. (1994), A Evolutionary Algorithm that Evolves Recurrent Neural Networks, *EEE Transactions on Neural Networks*, **5**, 54-65.
2. Gaussier, P., and Zrehen, S. (1994) A topological neural map for on-line learning: Emergence of obstacle avoidance in a mobile robot, in: Cliff, D., Husbands, P., Meyer, J.-A. and Wilson, S. (eds.), *From Animals to Animats III: Proceedings of the 3. International Conference on Simulation of Adaptive Behavior, Cambridge*, MA: MIT Press-Bradford Books, pp. 282–290.
3. Michel, O., *Khepera Simulator* Package version 2.0: Freeware mobile robot simulator written at the University of Nice Sophia-Antipolis by Oliver Michel. Downloadable from the World Wide Web at `http://wwwi3s.unice.fr/~om/khep-sim.html`
4. Mondada, F. and Floreano, D. (1995), Evolution of neural control structures: Some experiments on mobile robots, *Robotics and Autonomous Systems*, **16**, 183–195.
5. F. Mondala, E. Franzi, and P. Ienne. Mobile robots miniaturization: a tool for investigation in Control Algorithms. In *Proceedings of ISER' 93*, Kyoto, October 1993.
6. Nolfi, S., and Floreano, D. (2000) *Evolutionary Robotics: The Biology, Intelligence, and Technology of Self-Organizing Machines* MIT Press, Cambridge.
7. Pfeifer, R., and Verschure, F.M.J. (1992), Design efficiently navigating non-goal-directed robots, in: Meyer, J.-A., Roitblat, H., and Wilson, S. (eds.), *From Animals to Animats II: Proceedings of the Second International Conference on Simulation of Adaptive Behavior, Cambridge*, MA: MIT Press-Bradford Books, pp. 282–290.
8. Pfeifer, R., and Scheier, C. (2000), *Understanding Intelligence*, MIT Press, Cambridge.
9. Pasemann, F. (1993), Dynamics of a single model neuron, *International Journal of Bifurcation and Chaos*, **2**, 271-278.
10. Pasemann, F. (1998), Evolving neurocontrollers for balancing an inverted pendulum, *Network: Computation in Neural Systems*, **9**, 495–511.
11. Scutt, T.(1994), The fife neuron trick: Using classical conditioning to learn how to seek light, in: Cliff, D., Husbands, P., Meyer, J.-A. and Wilson, S. (eds.), *From Animals to Animats III: Proceedings of the 3. International Conference on Simulation of Adaptive Behavior, Cambridge*, MA: MIT Press-Bradford Books, pp. 210–217.

Construction by Autonomous Agents in a Simulated Environment

Anand Panangadan and Michael G. Dyer

Computer Science Department, University of California at Los Angeles,
Los Angeles CA 90095, USA
{anand,dyer}@cs.ucla.edu

Abstract. A connectionist behavior-based architecture enables a society of agents to navigate efficiently and construct specified structures within a simulated environment. The architecture incorporates both reactive and planning behaviors so that agents can manage multiple goals. The agents use cognitive maps to internally represent the environment for navigational planning. Simulations within a continuous 2-D environment demonstrate the abilities of the group of agents to construct structures. The effect of various placement strategies on the time to complete the construction task are studied.

1 Introduction

In this paper, a group of autonomous agents that can construct 2-D structures in their simulated 2-D environment is presented. The environment contains only disks of the same radius and construction involves arranging these disks to form a given pattern. Construction is chosen as the high-level goal not only because of its potential applications but also for the issues it raises. Any multi-agent system that rearranges the objects in its environment has to address these issues: 1) interference between agents when attempting to place objects at the same location, 2) deciding on the order in which objects are to be placed, 3) the variation of the time to complete the task with respect to number of agents, and 4) the internal representation of the world and the structures to be built.

In this work, a behavior-based connectionist architecture coupled with an egocentric grid-based spatial representation is described that enables agents to perform this construction task. Agents do not have the ability to communicate with each other nor do they explicitly take into account the actions of the other agents. Instead, they decide on the goal location where discs are to be placed by comparing their internal space representation with the desired configuration. Different goal selection strategies lead to different orders of placing the discs (and time to complete the task). The architecture allows easy implementation of these different goal selection strategies and their performance is presented.

2 The Agent and Its Environment

The simulation environment is similar to the one presented in [6]. The agents are situated in a two dimensional and continuous world, with all objects being discs

G. Dorffner, H. Bischof, and K. Hornik (Eds.): ICANN 2001, LNCS 2130, pp. 963–970, 2001.

of uniform radius. The discs represent entities that are relevant to the agent, such as food, water and building blocks, and are distinguished by their color: green discs are food, blue discs are water, and red discs are building blocks.

Each agent perceives the world around it through 36 distance sensors for each color evenly distributed all around it. The sensors have a limited sensing range of 20 units, covering only a small portion of the world. The activation of each sensor is inversely proportional to the distance of the nearest disc in its field of vision. Each agent also has a compass and this is used to align all sensor readings in one global direction. An agent in this world can continuously move forward or turn within a finite range, through motor commands that consist of the speed and the angle of a turn. In order to mimic real world characteristics, the motors do not respond instantaneously to motor commands: it takes time to accelerate and decelerate, and an agent can only turn through a small angle per time step and random noise (up to 1% of the sensed distance) is added to the sensor readings.

Motivations are indicators of the health of the agent and consist of *hunger* and *thirst*. When the internal food or water levels go below a threshold, these motivations are activated. To remain alive, the agent must "eat" and "drink" by touching the appropriate disc. In addition, there is an *avoid* motivation that is always active so that the agent does not collide with objects. The agents also have the ability to "grab" a disc close to them and to "drop" it off later. Construction in this world involves moving scattered building blocks into designated configurations.

3 Architecture

The architecture of the agent is composed of three modules: 1) Sensory/Motor, 2) Navigational Planning, and 3) Action Selection. These modules and their interconnections are shown in figure 1(a). The Sensory/Motor module contains purely reactive behaviors that are responsible for maintaining the viability of the agent by responding to objects within the sensor range. The Navigational Planning module builds and maintains cognitive maps of the environment, learned from the history of sensory inputs. These maps are then used for both construction and self-preservation goals, by planning trajectories to goal locations (such as the source of building blocks). The behaviors within the Sensory/Motor and Navigational Planning modules generate motor actions as outputs, which are fed into the Action Selection module. The Action Selection module then chooses the appropriate action by prioritizing actions responsible for maintaining the agent's health over those required for construction. The basic architecture is described in greater detail in [5].

Sensory/Motor: The architecture of the reactive module is shown in figure 1(b). *Behaviors* are motor primitives of the agent and each one takes the sensor data concurrently and outputs a motor activation. The motor outputs are computed using simple neural networks. Each agent has reactive modules that allow it approach discs and to avoid discs and other agents (obstacle avoidance). The

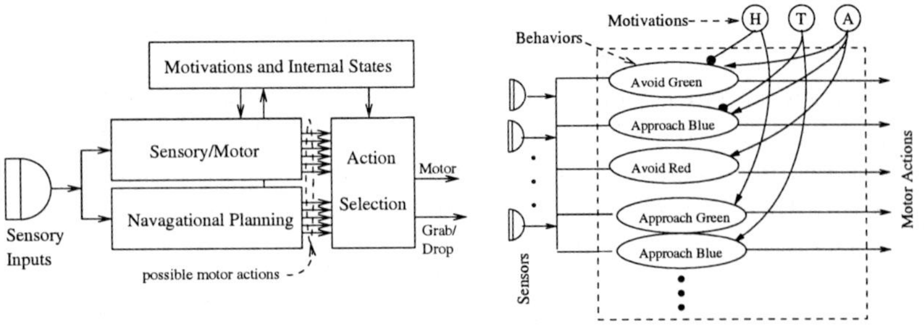

Fig. 1. (a) Block diagram of the architecture. (b) Architecture of the Sensory/Motor module.

connections between motivations and behaviors are used to excite or inhibit behaviors. For instance, *Hunger* motivation excites the "approach green" behavior and inhibits the "avoid green" behavior. This results in the agent approaching food only when it is hungry.

The design of the Sensory/Motor module adheres to the principles of behavior-based robotics [3][1]. Each module has direct access to the sensory data and is able to produce an appropriate motor action. This ensures robustness and fast response times which are important since the Sensory/Motor module is responsible for the viability of agents.

Navigational Planning: To plan a path to objects and locations that are outside the sensor range of an agent, an internal representation of the world is needed. In our model, Egocentric Spatial Maps (ESMs) [4] are used to represent the spatial relationship between the agent and objects in the environment. An ESM contains neurons arranged in an uniform grid (100×100 in this work) to maintain an egocentric view (the center node of the map always represents the current location of the agent) of the world. As the agent moves, the neuron activations are passed to neighboring neurons to maintain egocentricity. The amount by which the activations have to be shifted is obtained from dead-reckoning inputs. Agents keep track of changes in the positions of discs by integrating new sensory inputs to the center portion of the map. To plan a trajectory, neurons that correspond to goal locations initiate a spreading activation. Neurons that represent obstacles inhibit the activation. The gradient created by the spreading activation becomes the planned path to the nearest goal.

Each agent has a separate ESM for each kind of disc in the environment: the *Food ESM*, *Water ESM*, and the *Building Block ESM*. The structure to be built is also represented in a ESM called the *Configuration ESM*. Like the other ESMs, the activations on this map are also shifted as the agent moves, but the activations on the Configuration ESM (that encode the structures to be built) are set a priori and are not updated by the sensors.

Each ESM is associated with a Navigation Map that computes spreading activation to plan paths to goal locations. The Configuration Navigation Map is used to compute a path to desired locations of where building-block discs should be placed. Hence, every node is activated by the corresponding node from the Configuration ESM and inhibited by the activations from the nodes on the Food and Water ESMs (since food and water discs are obstacles). The Configuration Navigation Map is also inhibited by the Building Block ESM because if a building block is already present at a desired location, then the agent should not place another disc there. Similarly, the building block navigation map is excited by the building block ESM and inhibited by the others.

Additionally, simple computations of the activations on the Navigation Maps provide useful information. For example, the sum of all node activations of the Construction Navigation map indicates if any part of the structure is missing discs. This information is encoded into an *Internal State* node, *Needs Repair.* Other internal state nodes are *At Construction Site*, *Agent at Building Block* (active when the central node on the Configuration and the Building Block navigation maps, respectively, are excited) and *Holding Disc.* These nodes are used by the Action Selection module to sequence the different steps in the construction task. Figure 2(b) is a simplified schematic of the Navigation Planning module.

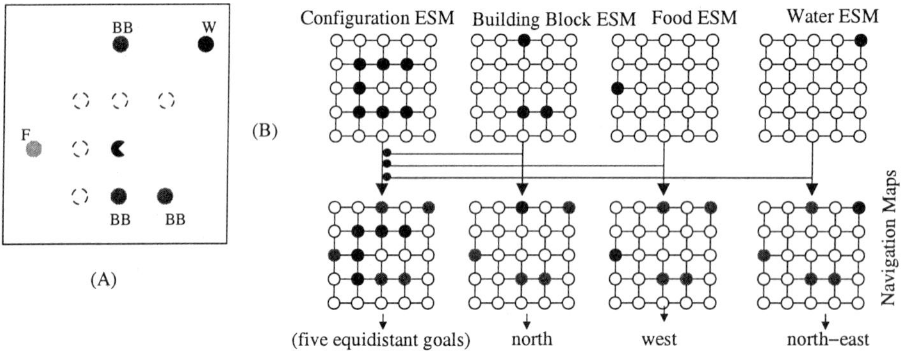

Fig. 2. (A) The agent surrounded by a food, a water and three building block discs. The dashed circles indicate the structure to be built. (B) The four ESMs and their corresponding Navigational Maps (dark circles indicate goal nodes while shaded circles are obstacles). The direction to the nearest goal is listed under the maps. Note that since the two building blocks south of the agent are already part of the structure, the Building Block Navigation map plans a path to the disc in the north. Inhibitory connections from the Food, Water and Building Block ESMs Map to the Configuration navigation map exist between all corresponding nodes.

Action Selection: The Action Selection module chooses one of the motor actions from the Sensory/Motor and Navigational Planning modules to be sent to the motors, based on the actions' priority. Actions that are crucial to the survival of the agent (reactive behaviors) have higher priority. The prioritizing of these actions is implemented by lateral-inhibition links from a high priority action to all lower priority actions.

Construction for an agent is a repeated sequence of moving to the location of a building block disc that is not part of the structure, picking it up, moving to a location that requires a disc and dropping it there. The Action Selection module selects those Sensory/Motor and Navigational Planning actions that execute a particular step through excitatory and inhibitory connections between the motivations, internal state nodes and the outputs from the Sensory/Motor and Navigational Planning modules. These weights are fixed a priori. For instance, if *Holding Disc*, and *Needs Repair* internal state nodes are active, then the motor output of the Configuration navigation map is chosen.

4 Results and Discussion

A series of simulation runs were carried out to show how various goal selection strategies can be implemented on the architecture. The performance of the different goal placement strategies are then compared. The parameters that determine the time to complete construction are:

1. The number of agents: 1 to 4 agents
2. The shape of the structure to be built: The agents have a priori information (as activations in each agent's Configuration ESM) about where the structure (shown in figure 3(b)) has to be built. The particular shape was chosen since building the top row first forces agents to move around it to place subsequent discs.
3. Initial arrangement of building blocks: The discs are initially clustered together in one group, two groups or are scattered all around (figure 3 shows two clusters).
4. Initial positions of agents: Agents are introduced at a random location in the environment and initially explore their world until a building block disc becomes visible. From that point, construction begins.
5. Positions of Food and Water discs: The agents must eat/drink periodically (every 2000 time-steps).

The reported data for each case is obtained from one simulation run. Results from different simulation runs vary slightly depending on the initial positions of the agents (and hence the time taken to explore), but the trends across parameters are similar. The three goal placement strategies that were studied are:

- **Greedy Placement:** The agents pick the closest available building block and place it at the closest location requiring a disc. To implement this, the agent activates *all* the goal nodes in the Configuration Navigation Map. Thus, the Navigation map always plans the path to the nearest goal location.
- **Random Placement:** The agents pick the closest available building block and place it at a random location requiring a building block. The agent randomly activates *one* of the goal nodes in the Configuration Navigation Map. Thus, the Navigation map plans the path to that particular goal, ignoring closer goal locations.

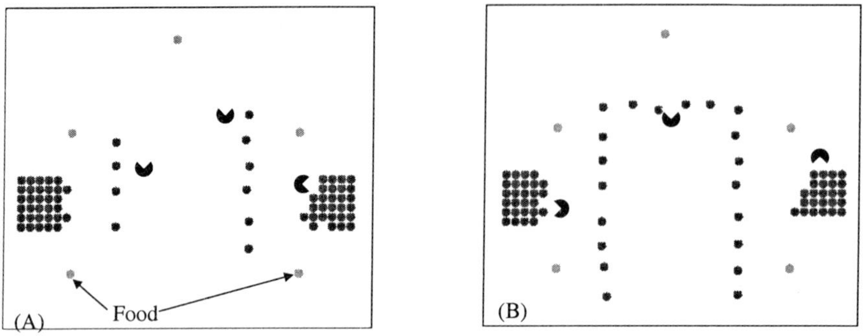

Fig. 3. Greedy placement: (a) Environment at T=1000. Initially all building blocks are clustered in the two sides. (b) The three agents have completed construction at T=2471. The building blocks are in the desired configuration.

- **Localized Placement:** Each agent places building blocks only in a small randomly chosen area of the environment. An agent activates only those nodes that are *neighbors of a randomly chosen goal node* in the Configuration Navigation map. Thus the agent plans paths only to construction locations within the area represented by the chosen nodes. Each agent counts the number of times it comes into contact with other agents in that area and if the count exceeds a threshold it chooses a new area in which to build. In this way, there is no overcrowding of agents in a given area.

Figure 4 shows the number of time-steps taken to complete the construction task using the three placement strategies for each of the three initial environments: building blocks in one cluster, two clusters or randomly distributed. When the building blocks are in one cluster (figure 4(a)), the greedy placement blocks off direct paths to construction locations by constructing the top "wall" (figure 3(a)). Thus, randomly choosing goal locations works better since gaps exist for the agents to move directly to their goal locations. When the number of agents is small, the localized placement strategy is similar to random placement. However, as the number of agents increases to 4, localized placement works significantly better as it tries to keep the goal locations of different agents separate.

As the building block sources get more widely distributed (figure 4(b,c)), the greedy placement strategy does not suffer from the drawback described above because building blocks are always available close to goal locations. Thus, the difference in performance between the strategies is less marked in this case. Distributed construction is a complex task and simple local strategies will not work for all cases. Thus it is important for the architecture to allow for easy implementation of different placement strategies. Figure 4(d) shows the number of discs placed over time by four agents when the building blocks are in one cluster. The number of discs placed per time-step (the slope of the curves) decreases with time for greedy placement, but is relatively constant for the other two strategies.

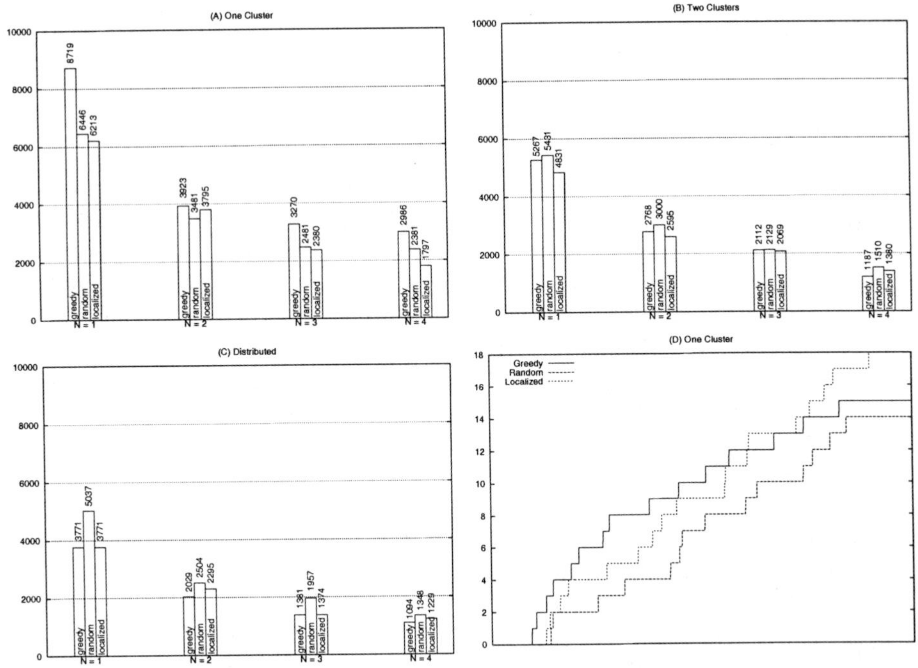

Fig. 4. (a) The time to complete construction task for different initial distributions of building blocks: (a) one cluster, (b) two clusters, (c) widely distributed. N = number of agents. (d) Number of discs placed over time when N=4, and building blocks are initially in one cluster.

5 Conclusions and Future Work

The construction task described in this report brings up several issues that must be addressed when building autonomous multi-agent systems. The connectionist architecture presented here has several features that address some of these issues. For instance, the reactive behaviors and the motivations that gate these behaviors enable the agent to handle more than one goal simultaneously. The Configuration ESM can represent any 2-D structure of arbitrary shape. Since the active nodes on the Configuration Navigation map determine the goal locations, it is easy to implement different placement strategies by just exciting or inhibiting the corresponding nodes. The three placement strategies show how a society of autonomous agents can coordinate to accomplish the construction task without communication or explicit modeling of other agents.

There are other issues of the construction task that are yet to be studied. Some structures like circles can trap the building agent inside. Other structures like concentric circles can only be built in a particular order (the inner circle has to be built first). Mechanisms that address these issues have to be added. Adding communication between agents can increase the range of their spatial maps and reduce the time to explore. Moreover, communication should also lead to more efficient strategies for construction.

6 Related Work

Systems that study embodied agents are often inspired by biological phenomena like ant foraging behavior [11]. [10] provides an overview of such biologically based artificial navigation systems. The construction task can serve as another useful test-bed for multi-agent systems (existing test-beds include the robotic soccer environment [2] and the Predator/Prey pursuit domain [9]). The ESMs used in our system are similar in principle to "evidence grids" developed in [8], which were tested on physical robots. [7] shows how spatial representation can be integrated into a behavior-based architecture. Since the internal representation only stores the information necessary for basic navigation, the system in [7] cannot be extended to the construction task. The construction task has also been studied in [6] without the aid of an internal spatial representation. This is possible by allowing the sensors to have an infinite range and coloring discs to mark crucial locations (like ends of walls).

Acknowledgements

This work is supported in part by an Intel University Research Program grant to the second author.

References

1. R.C. Arkin. *Behavior-based Robotics.* MIT Press, 1998.
2. Tucker Balch. Hierarchic social entropy: An information theoretic measure of robot group diversity. *Autonomous Robots*, 8(3), June 2000.
3. Rodney A. Brooks. A robust layered control system for a mobile robot. *IEEE Journal of Robotics and Automation*, 2:14–23, 1986.
4. G. Chao and M. G. Dyer. Concentric spatial maps for neural network based navigation. In *Proc. of the Ninth Intl. Conf. on Artificial Neural Networks*, 1999.
5. G. Chao, A. Panangadan, and M. G. Dyer. Learning to integrate reactive and planning behaviors for construction. In *From Animals to Animats 6*, 2000.
6. F. L. Crabbe and M. G. Dyer. Second-order networks for wall-building agents. In *Proc. of the Ninth Intl. Conf. on Artificial Neural Networks*, 1999.
7. Maja J. Mataric. Integration of representation into goal-driven behavior-based robots. *IEEE Trans. on Robotics and Automation*, 8:333–344, 1992.
8. Hans Moravec and Martin C. Martin. Robot evidence grids. Technical Report CMU-RI-TR-96-06, Robotics Institute, Carnegie Mellon University, March 1996.
9. Peter Stone and Manuela Veloso. Multiagent systems: A survey from a machine learning perspective. *Autonomous Robots*, 8(3), June 2000.
10. O. Trullier, S. Wiener, A. Berthoz, and J. Meyer. Biologically based artificial navigation systems: review and prospects. *Progress in Neurobiology*, 51, 1997.
11. Barry Werger and Maja J Mataric. Exploiting embodiment in multi-robot teams. Technical Report IRIS-99-378, Institute for Robotics and Intelligent Systems, 1999.

A Neural Control Model Using Predictive Adjustment Mechanism of Viscoelastic Property of the Human Arm

Masazumi Katayama[1,2]

[1] Department of Information and Computer Sciences
Toyohashi University of Technology
1-1 Hibarigaoka, Tempaku, Toyohashi 441-8580, Japan
katayama@ics.tut.ac.jp
[2] Bio-mimetic Control Research Center,
The Institute of Physical and Chemical Research (RIKEN), Japan

Abstract. In this paper, we investigate a strategy for motor learning and propose a learning control model that integrates internal model learning and the viscoelastic adjustment of the human arm. In this model, the value of the viscoelasticity is modulated by the values of a control error and a predictive error. Consequently, the adjustment mechanism eliminates the weak point that for conventional learning control models desired movements cannot be realized at the beginning of learning, by gradually shifting from relatively higher viscoelasticity to lower viscoelasticity through learning.

1 Introduction

We skillfully perform various movement tasks. In this motor control system, feedback control alone cannot execute skillful movements because too large feedback gains cause instability. This is because feedback control is quite limited by the long delay associated with neural information transmission, neural computation and so on. Moreover, feedback control alone cannot explain how deafferented monkeys move their arms to an illuminated target light without concurrent visual and somatosensory information [2]. Thus, it is noted that in the human motor control system the mechanisms of feedforward control, prediction and motor planning is necessary for skillful movement execution at least.

From the above points of view, many types of learning control models based on internal model learning have been proposed in the past decade (e.g., [3, 5]). For those learning control models, desired movement trajectories can be accurately realized after learning. However, those learning control models have a disadvantage that desired movements cannot be realized at the beginning of learning because the arm is mainly controlled by feedback control before learning.

While, our movements become more skillful through learning: movements can be executed accurately, movement speed becomes faster and/or energy consumption becomes low, by adjusting variable parameters such as movement trajectories, arm postures, the viscoelasticity of the human arm and so on. Moreover, by adjusting these variable parameters, we will make possible efforts to achieve

G. Dorffner, H. Bischof, and K. Hornik (Eds.): ICANN 2001, LNCS 2130, pp. 971–976, 2001.

the objective of a movement task even if the internal models of the arm are not acquired at the beginning of learning. However, because conventional studies for learning control have focused on how to learn the internal models, conventional learning control models lack the above adjustment mechanisms.

From this point of view, this paper focuses on an adjustment mechanism for the viscoelasticity of the human arm and we present a strategy for motor learning. Moreover, we propose a learning control model that integrates internal model learning and the viscoelastic adjustment.

2 A Motor Learning Strategy

Feedback control alone cannot execute accurate movements because too large feedback gains cause instability, as mentioned above. While, the viscoelastic modulation in the human motor system is observed in behavioral measurement results for arm movements (e.g., [6, 8]). An advantage of the viscoelastic modulation is that a movement error can be decreased by increasing the value of the viscoelasticity. Even if the value of the viscoelasticity is relatively high, stable arm movements can be realized because the viscoelasticity is mechanically realized without delay. Therefore, at the beginning of learning, a movement error can be decreased by increasing the value of the viscoelasticity without increasing the values of feedback control gains. However, the amount of energy consumption increases as the value of the viscoelasticity increases. Too large value of the viscoelasticity should not be used from the view point of energy consumption. Therefore, it is desirable that after learning our movements are accurately executed with the lower value of the viscoelasticity. There is a possible method that the learning starts from the higher value of the viscoelasticity and then the value decreases gradually through learning (see [4, 7]).

In this study, we propose a motor learning strategy that learns the internal models by adjusting the value of the viscoelasticity by a movement error. The adjustment can be expressed as the following equation.

$$c = f(e) + c_{min}. \tag{1}$$

Here, c expresses the value of the viscoelasticity (or the coactivation level of the agonist and antagonist muscles), c_{min} is the minimal value of c, f is a monotonically increasing function and e is a movement error. In this equation, the value of c decreases as the value of the error decreases through learning. Therefore, relatively accurate movements can be executed at the beginning of learning because the value of the viscoelasticity is relatively high. Consequently, the learning starts from higher viscoelasticity and adapts to correct movements with lower viscoelasticity. This is a motor learning strategy that integrates internal model learning and the viscoelastic modulation.

3 A Neural Control Model

Fig.1 shows a learning control model that learns an inverse dynamics model (IDM) and a forward dynamics model (FDM), by using adjustment mechanisms

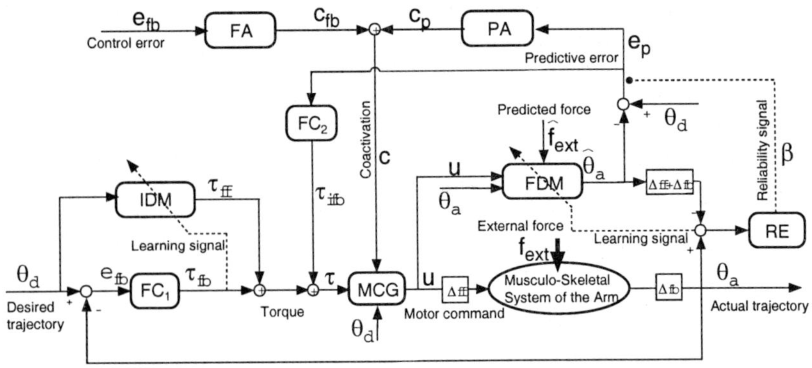

Fig. 1. A neural control model using a predictive adjustment mechanism of the viscoelasticity of the human arm.

of the viscoelasticity. FC_1 and FC_2 are feedback controllers such as a PID controller. Δ_{fb} represents the afferent delay caused by proprioceptors, somatosensory afferent pathway, etc. Δ_{ff} is the efferent delay associated with neural transmission of efferent signals. IDM and FDM are structured by a neural network model (three layered perceptron). The output τ_{ff} of IDM is a joint torque vector, θ_d is a vector that expresses desired joint angle, desired angular velocity and desired angular acceleration. θ_a is a vector of actual joint angle and actual angular velocity. $\hat{\theta}_a$ is the output of FDM and expresses a predicted trajectory. u is a motor command vector and $\hat{f}_{ext}$ expresses a predicted external force. IDM can be trained by a feedback-error learning method proposed by Kawato [5] and FDM can be trained by using the difference between θ_a and $\hat{\theta}_a$. An internal feedback control can be executed by using a trained FDM and FC_2. A total joint torque τ is expressed as $\tau = \tau_{fb} + \tau_{ff} + \tau_{ifb}$. MCG is a motor command generator that computes motor commands u given to the muscles from τ, θ_d and c and MCG expresses the inverse function of the human muscle system.

RE is a reliability estimator that evaluates the reliability of the output of FDM. In this study, the evaluation value β is defined as the following equation.

$$\beta = \exp\left(-k \sum_{t=0}^{t=t_f} \left(\theta_a(t) - \hat{\theta}_a(t)\right)^2\right) \qquad (0 < \beta \leq 1) \tag{2}$$

Here, k is a constant coefficient. The value of β is almost zero before FDM learning.

FA is an adjustment mechanism based on a control error e_{fb} that is the difference between θ_d and θ_a. However, the adjustment based on sensory feedback signals is very limited by the long delay. Therefore, by using a predictive adjustment mechanism (PA) based on a predicted error e_p that is the difference between θ_d and $\hat{\theta}_a$, effective adjustment can be executed. FA and PA calculate the values of c_{fb} and c_p, respectively.

$$c_{fb} = \alpha(1-\beta)|e_{fb}|, \qquad c_p = \alpha\beta|e_p| \tag{3}$$

where α is a constant coefficient and e_{fb} and e_p is weighted by β. The value of the viscoelasticity (or the coactivation level) is expressed as $c = c_{fb} + c_p + c_{min}$. These equations correspond to (1). The value of c is mainly modulated by FA before FDM learning because the value of β is almost zero. Thus, the control error becomes small at the beginning of learning because the value of c_{fb} increases. However, as mentioned above, the adjustment of FA lags behind an actual control error or an external force for the period of the delay. While, after FDM learning, effective adjustment can be achieved by PA. Therefore, the viscoelastic modulation gradually shifts from FA to PA through FDM learning.

4 Simulation Results

We ascertain the effectiveness of the proposed learning control model by computer simulations using a human arm model. In this study, the human arm is modeled as a 2-link manipulator with four single-joint muscles and two double-joint muscles (see [3]). In the human arm model, we assumed that the values of the muscle viscoelasticity linearly depend on the motor command u and a larger value of u results in higher viscoelasticity. The values of the physical parameters in the human arm model were determined from measurement data (see, [3]). And the delays of Δ_{fb} and Δ_{ff} are 20 [msec] and 10 [msec], respectively. The sampling time for control is 1 [msec]. Movement distance is about 0.4 [m] and movement time is 0.75 [sec].

IDM and FDM were trained without an external force at first. In the next step, after IDM and FDM learning, two types of external forces were given to the arm during movements: a viscous force and an instantaneous force.

A viscous force f_{ext} was given to the hand as an external force: $f_{ext} = \begin{pmatrix} 0 & 10 \\ -10 & 0 \end{pmatrix} \begin{pmatrix} \dot{x} \\ \dot{y} \end{pmatrix}$. $(\dot{x}, \dot{y})^T$ expresses a vector of hand velocity. IDM and FDM were trained under the condition of the viscous force. Fig.2 shows the control results for two desired trajectories that transform the joint coordinates (joint angle) into the Cartesian coordinates (x,y). These results show that before learning the differences between the desired and actual trajectories are so large when the adjustment mechanisms (FA and PA) are not used. In this case, the joint stiffness values are fixed to lower values: the values of the shoulder and elbow joint stiffness (R_{ss} and R_{ee}) are 4.6 [Nm/rad] and 3.4 [Nm/rad], respectively. While, by using FA, the differences become smaller before learning. After learning, the differences are quite small and the desired trajectories are accurately realized. The average value of R_{ss} changes from about 40 [Nm/rad] to 4.6 [Nm/rad] through learning and the average value of R_{ee} changes from about 20 [Nm/rad] to 3.4 [Nm/rad].

In addition to the above simulation, in order to ascertain the efficiency of PA, the arm is perturbed by an instantaneous force f_{ext} at time 0.21 [sec]. Fig.3 shows that the amplitude of hand oscillation is so large when the adjustment mechanisms (FA and PA) are not used. While, the amplitude becomes smaller by using FA. Moreover, by the effective adjustment of PA, the movement error becomes small drastically. In this condition, actual trajectories are accurately predicted by inputting the external force into FDM. Fig.4 shows the changes of joint stiffness when the arm is perturbed. In Fig.4(a), the start time of the changes of joint stiffness lags behind the timing of the instantaneous force. While,

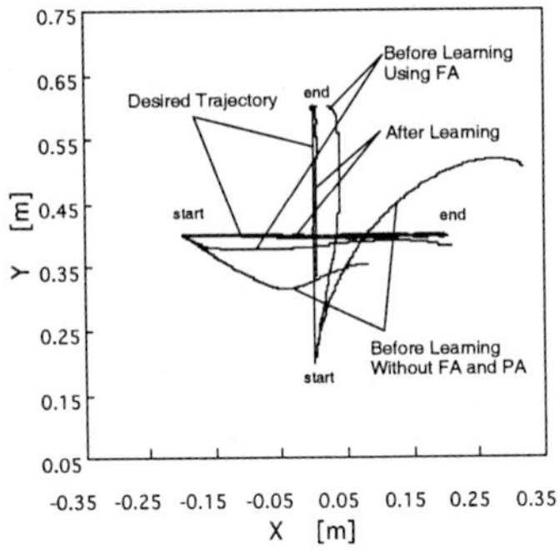

Fig. 2. Control results.

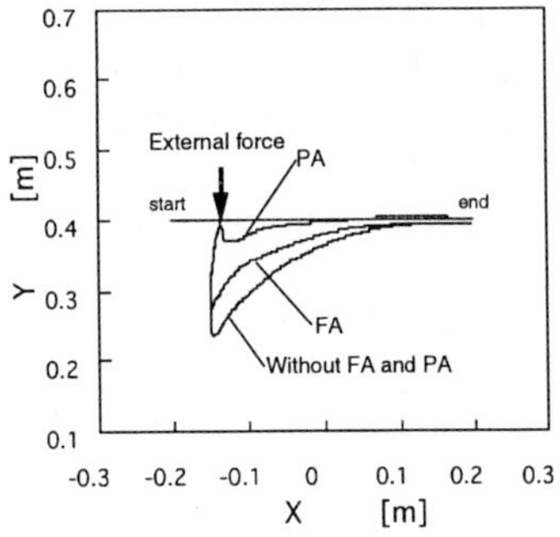

Fig. 3. Control results.

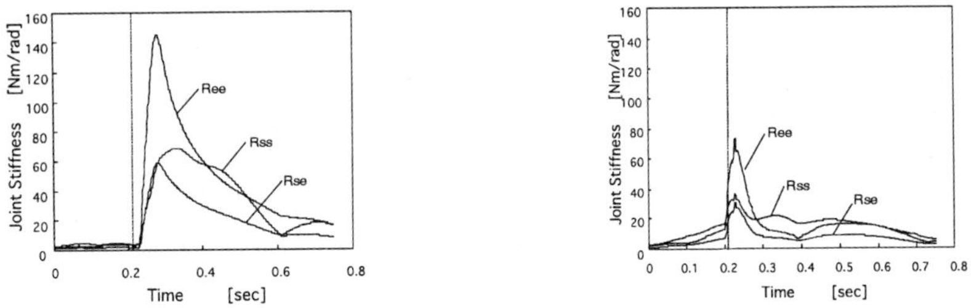

(a) Joint stiffness (FA).

(b) Joint stiffness (PA).

Fig. 4. Joint stiffness in the case that the arm was perturbed at time 0.21 [sec].

in Fig.4(b), the anticipatory increasing of joint stiffness is observed before the timing of the external force and moreover the amplitude of the joint stiffness value is so smaller than the case of Fig.4(a). These results show the effectiveness of the learning control model using the adjustment mechanisms (FA and PA).

5 Discussion and Conclusion

We investigated a motor learning strategy and presented the learning control model that integrates internal model learning and the viscoelastic adjustment. The integrated model eliminates the weak point of conventional learning control models because relatively accurate movements can be realized at the beginning of learning. This model have also the other advantages that a control error becomes smaller when an unpredictable perturbation is given to the arm, when executing an untrained movement and when grasping an unknown object.

While, Bhushan et al. [1] presented that control results predicted by a control model using a pair of IDM and FDM reproduce measurement results of arm movements. Moreover, it is observed that the value of the viscoelasticity becomes gradually small through learning, from measurement experiments of electromyographic signals (e.g., [6, 8]). These findings support the structure of the proposed model and a proposed motor learning strategy. And the proposed adjustment mechanisms can apply to any types of learning control models that trains the internal models. Moreover, the proposed model can be extended to a learning control model with multiple pairs of IDM and FDM that are trained by a modular learning method proposed by Wolpert and Kawato [9].

We discuss a motor learning strategy based on the viscoelastic modulation. A beginner for motor learning makes possible efforts to achieve the objective of a movement task by increasing the value of the viscoelasticity even if the internal models are not trained. This is an appropriate strategy that considers the learning level of a learner although the energy consumption increases. Thus, when we perform various movement tasks, it is very important that the objective of movement can be achieved by the strategy that changes continuously by the learning level.

When the value of the viscoelasticity is small, static and dynamic forces must be compensated. While, it is not necessary to compensate dynamic forces for arm movements when the value of the viscoelasticity is large. In this case, only an internal model that compensates static forces is needed. The learning of the internal model is easier than the learning of an internal model for dynamic forces. Therefore, the learning gradually shifts from relatively easier learning control to more complex learning control. Moreover, by the gradually shifting mechanism, the learning speed may be faster and the learning may be easier because the learning starts from the neighborhood of the optimal solution.

References

1. Bhushan, N., Shadmehr, R. (1999) Computational nature of human adaptive control during learning of reaching movements in force fields.*Biological Cybernetics***81**, pp.39-60.
2. Bizzi, E., Accornero, N., Chapple, W., and Hogan, N. (1984) Posture control and trajectory formation during arm movement. *The Journal of Neuroscience* **4**, 11, 2738-2744.
3. Katayama, M., and Kawato, M. (1991) Learning trajectory and force control of an artificial muscle arm by parallel-hierarchical neural network model. *Advances in Neural Information Processing Systems* **3**, Lippmann, R.P., Moody, J.E., and Touretzky, D.S. (eds.), 436-442, San Mateo: Morgan Kaufmann.
4. Katayama, M., Inoue, S., Kawato, M. (1998) A strategy of motor learning using adjustable parameters for arm movement. *Proceedings of the 20th Annual International Conference on the IEEE Engineering in Medicine and Biology Society*, pp.2370-2373.
5. Kawato, M., Furukawa, K., Suzuki, R. (1987) A hierarchical neural-network model for control and learning of voluntary movement. *Biological Cybernetics* **57**, 169-185.
6. Osu, E., Domen, K., Gomi, H., Yoshioka, T., Imamizu, H. and Kawato, M. (1997) A change of muscle activation in learning propcess. *Technical Report of IEICE Japan*, NC96-139, 201-208, in Japanese.
7. Sanger, T.D. (1994) Neural network learning control of robot manipulators using gradually increasing task difficulty. *IEEE Trans. Robot. Auto.* **10**, 3, 323-333.
8. Thoroughman, K.A., Shadmehr, R. (1999) Electromyographic correlates of learning an internal model of reaching movements. *The Journal of Neuroscience* **19**, 19, pp.8573-8588.
9. Wolpert, D.M., Kawato, M. (1998) Control of Arm and Other Body Movements - Multiple paired forward and inverse models for motor control. *Neural Networks* **11**, 7, pp.1317-1330. Multijoint

Multi-joint Arm Trajectory Formation Based on the Minimization Principle Using the Euler-Poisson Equation

Yasuhiro Wada[1], Yuichi Kaneko[1], Eri Nakano[2], Rieko Osu[3], and Mitsuo Kawato[2,3]

[1] Nagaoka University of Technology, Nagaoka, 940-2188, Japan
ywada@nagaokaut.ac.jp
[2] Advanced Telecommunications Research Institute International, Kyoto, 619-0288, Japan
[3] ERATO Dynamic Brain Project, Kyoto, 619-0288, Japan

Abstract. In previous research, criteria based on optimal theories were examined to explain trajectory features in time and space in multi joint arm movements. Four criteria have been proposed. They were the minimum hand jerk criterion, the minimum angle jerk criterion, the minimum torque change criterion, and the minimum commanded torque change criterion. Optimal trajectories based on the two former criteria can be calculated analytically. In contrast, optimal trajectories based on the minimum commanded torque change criterion are difficult to be calculated even with numerical methods. In some cases, they can be computed by a Newton-like method or a steepest descent method combined with a penalty method. However, for a realistic physical parameter range, a former becomes unstable quite often, and the latter is unreliable about the optimality of the obtained solution. In this paper, we propose a new method to stably calculate optimal trajectories based on the minimum commanded torque change criterion. The method can obtain trajectories satisfying Euler-Poisson equations with a sufficiently high accuracy. In the method, a joint angle trajectory, which satisfies the boundary conditions strictly, is expressed by using orthogonal polynomials. The coefficients of the orthogonal polynomials are estimated by using a linear iterative calculation so as to satisfy the Euler-Poisson equations with a sufficiently high accuracy. In numerical experiments, we show that the optimal solution can be computed in a wide work space and can also be obtained in a short time compared with the previous methods.

Keyword: computational neuroscience, motor control, minimization principle, Euler-Poisson equation, trajectory formation

1 Introduction

The following two characteristics have been well demonstrated about the feature of a point-to-point human arm movement on a plane [1]. (1) The path is a roughly straight line, but is slightly curved. (2) The velocity profile is bell shaped with a single peak. Several models have been proposed to explain these features

G. Dorffner, H. Bischof, and K. Hornik (Eds.): ICANN 2001, LNCS 2130, pp. 977–984, 2001.

[2,3,7,5,4]. In particular, the criteria for trajectory planning based on optimal principles have been proposed.

In general, it is quite difficult to obtain an optimal solution. Nakano et al. [4] reported that the trajectory predicted by the minimum commanded torque change criterion is the trajectory closest to a human trajectory among trajectories predicted by the above criteria. However, it has been very difficult to establish a robust method for obtaining an optimal trajectory based on the minimum commanded torque change criterion at various starting points, final points, and dynamic parameters. The optimal trajectory can be obtained by the Newton-like method [7], and the steepest descent method [4]. The Newton-like method is an algorithm to guarantee the local optimality of the converged trajectory mathematically; however, the algorithm often becomes unstable according to the parameter value of the dynamics such as the viscosity, and then, often diverges. The algorithm also occasionally diverges according to the positions of the starting and final points. On the other hand, the optimal solution can be stable and computed comparatively more easily using the steepest descent method. However, the optimality of the solution cannot be guaranteed. Because it was difficult for Nakano et al. [4] to calculate an optimal trajectory of the minimum commanded torque change, comparisons were made on quasi-optimal trajectories based on the minimum commanded torque change obtained using the steepest descent method and measured trajectories.

Calculating the optimal trajectory is necessary and indispensable for explaining human arm movements by a mathematical model. In this paper, we propose a trajectory calculation method based on the minimum commanded torque change criterion, which satisfies Euler-Poisson equations with a sufficiently high accuracy. These Euler-Poisson equations, which are derived from the functional optimal problem, provide the necessary condition for obtaining the optimal trajectory. We show the numerical experiment results of the minimum commanded torque change trajectory generation for a two-joint arm on two planes (the horizontal and sagittal planes) and also point out that the optimal trajectory, which satisfies the Euler-Poisson equations with a sufficiently high accuracy, can be obtained in a short time, regardless of the positions of the starting and final points or the value of the dynamic parameter.

2 Minimum Commanded Torque Change Criterion

With the minimum commanded torque change criterion, the link dynamics and the muscles are controlled, and the trajectory is planned so as to minimize the time integral of the square of the commanded torque change rate. The object function is given by the following equation.

$$\begin{aligned} C_{CTC} &= \frac{1}{2}\int_0^{t_f} \sum_{i=1}^{N} \left(\frac{d\tau_i^c}{dt}\right)^2 dt \\ &= \frac{1}{2}\int_0^{t_f} F(t,\theta_1,\theta_2,\dot{\theta}_1,\dot{\theta}_2,\ddot{\theta}_1,\ddot{\theta}_2,{\theta_1}^{(3)},{\theta_2}^{(3)})dt \rightarrow Min \end{aligned} \quad (1)$$

Here, t_f indicates the movement time, and N shows the number of joints. τ^c shows the commanded torque. Incidentally, the commanded torque does not correspond to the mechanical output torque. It is an approximated torque for the motor commands. A point-to-point movement based on the criterion is planned in consideration of the dynamics parameters of the mass of the arm and the inertia moment, etc. The feature of the trajectory generated by the criterion changes according to the start posture and final posture. The commanded torque is calculated from an equation of two link manipulators [4]. In the minimum torque change model proposed by Uno et al. [7], only the link dynamics is regarded as the controlled object. The equation of the commanded torque is a high nonlinear equation which contains a trigonometrical function of the joint angle and the second power term, etc. It was extremely difficult to predict a trajectory satisfying the criterion as well as the minimum torque change criterion. In particular, because the value of the viscosity was not set to 0 for the minimum commanded torque change criterion, it became more difficult to obtain the solution by the Newton-like method.

3 A Prediction Algorithm for Trajectories Based on the Minimum Commanded Torque Change Criterion Using the Euler-Poisson Equation

3.1 Euler-Poisson Equation

In the section, a calculation algorithm for trajectories based on the minimum commanded torque change criterion using Euler-Poisson equations is described. Basically, the algorithm calculates a trajectory that satisfies Euler-Poisson equations with a high accuracy and the boundary conditions strictly. By the boundary conditions of the starting point and the final point (position, velocity, and acceleration), the following Euler-Poisson equations for the above optimal problem can be derived as a necessary condition for an extreme value.

$$E_i = \frac{\partial}{\partial \theta_i} F - \frac{d}{dt}\frac{\partial}{\partial \dot{\theta}_i} F + \frac{d^2}{dt^2}\frac{\partial}{\partial \ddot{\theta}_i} F - \frac{d^3}{dt^3}\frac{\partial}{\partial {\theta_i}^{(3)}} F = 0 \quad (i = 1, 2) \tag{2}$$

Here, $F = \sum_{i=1}^{N} (d\tau_i^c/dt)^2$.

Therefore, θ_1 and θ_2, which satisfy these two Euler-Poisson equations for every movement time, are the minimum commanded torque change trajectory.

3.2 A System of Orthogonal Polynomials for Expression of the Minimum Commanded Torque Change Trajectory

The minimum angle jerk trajectory is one of the good approximations of the minimum commanded torque change trajectory. Here, let us assume that the minimum commanded torque change trajectory $\theta_i(t)$ $(i = 1, 2)$ can be expressed by the following equation.

$$\theta_i(t) = \theta_i^{AJ}(t) + \Delta\theta_i(t) \tag{3}$$

Here, $\theta_i^{AJ}(t)$ corresponds to the minimum angle jerk trajectory, and $\Delta\theta_i(t)$ is denoted as the difference between the minimum commanded torque change trajectory and the minimum angle jerk trajectory. The trajectory of equation (3) must satisfy the following boundary conditions of the starting position and final position.

$$\theta_i(0) = \theta_i^s, \dot{\theta}_i(0) = \ddot{\theta}_i(0) = 0, \theta_i(t_f) = \theta_i^f, \dot{\theta}_i(t_f) = \ddot{\theta}_i(t_f) = 0 \quad (i = 1, 2) \quad (4)$$

3.3 A Difference Trajectory $\Delta\theta_i(t)$

$\theta_i^{AJ}(t)$ in Eq. (3) is the minimum angle jerk trajectory, which satisfies the boundary conditions (Eq. (4)). Therefore, $\Delta\theta_i(t)$ should be a function that makes the boundary conditions at the starting point and final point all become 0. We use the Jaccobi polynomial [6] as the set of orthogonal polynomials to express $\Delta\theta_i(t)$. The Jaccobi polynomial can be represented as $\tilde{Q}_k(t) = 64\ t^3\ (1-t)^3 \tilde{P}_k^{(6,6)}(t)$. We define the difference trajectory $\Delta\theta_i(t)$ as the following equations, which are normalized to $0 \le t \le 1$.

$$\Delta\theta_i(t) = \sum_k 64\ a_{ik}\ t^3\ (1-t)^3 \tilde{P}_k^{(6,6)}(t) \quad (5)$$

$\Delta\theta_i(t)$ is composed of the summation of Jaccobi orthogonal polynomials; the boundary conditions of the position, velocity, and acceleration at the starting point and final point all satisfy 0. From the above, the minimum commanded torque change trajectory $\theta_i(t)$ is given by Eq. (6).

$$\theta_i(t) = \theta_i^s + (\theta_i^s - \theta_i^f)(-10t^3 + 15t^4 - 6t^5) + 64\ t^3\ (1-t)^3 \sum_{k=0}^{K} a_{ik}\ \tilde{P}_k^{(6,6)}(t) \quad (6)$$

3.4 An Estimation Algorithm of Parameter a_{ik}

To estimate a_{ik} satisfying the Euler-Poisson equations with a high accuracy, we propose the following two innovations. (1) The left side of the Euler-Poisson equation is expressed as a linear summation concerning a_{ik}, and (2) the coefficient a_{ik} is estimated using an iterative calculation by the discretization of the movement time.

(1) A linear summation concerning parameter a_{ik} : Eq. (6) is substituted for Eq. (2), and this is arranged to become a linear expression of a_{ik} such as Eq. (7).

$$E_j(t, \theta_1, \theta_2, \dot{\theta}_1, \dot{\theta}_2, \cdots, \theta_1^{(3)}, \theta_2^{(3)}) = \sum_{i=1}^{2}\sum_{k=0}^{K} {}_j h_{ik}(t, \theta_1, \theta_2, \dot{\theta}_1, \dot{\theta}_2, \ddot{\theta}_1, \ddot{\theta}_2)\ a_{ik}$$
$$+ J_j(t, \theta_1, \theta_2, \dot{\theta}_1, \dot{\theta}_2, \cdots, \theta_1^{(3)}, \theta_2^{(3)})\ (j = 1, 2) \quad (7)$$

Eq. (7) is obtained according to a standard [8].

(2) A discretization of the movement time : Time t is dispersed to M divisions with equal intervals for the normalized movement time $[t_0, t_1, \cdots, t_m, \cdots,$

$t_M]$ $(t_0 = 0, t_M = 1)$. Here, if a parameter a_{ik} expressing an optimal trajectory exists, the trajectory of Eq. (7) satisfies Eq. (2). That is, Eq. (7) satisfies the following equation. Here, the above equation is rearranged using ${}_l h_{ikm} = {}_l h_{ik}(t_m, \cdots)$ and $J_{lm} = J_l(t_m, \cdots)$ as follows.

$$\sum_{i=1}^{2}\sum_{k=0}^{K} {}_l h_{ikm}\, a_{ik} + J_{lm} = 0 \tag{8}$$

Eq. (8) holds for all discretization time values. Then, $2(M+1)$ linear equations of a_{ik} are derived. These equations are expressed as follows, in matrix form.

$$\begin{bmatrix} {}_1h_{100} & \cdots & {}_1h_{2K0} \\ \vdots & \ddots & \vdots \\ {}_1h_{10M} & \cdots & {}_1h_{2KM} \\ {}_2h_{100} & \cdots & {}_2h_{2K0} \\ \vdots & \ddots & \vdots \\ {}_2h_{10M} & \cdots & {}_2h_{2KM} \end{bmatrix} \begin{bmatrix} a_{10} \\ \vdots \\ a_{1K} \\ a_{20} \\ \vdots \\ a_{2K} \end{bmatrix} + \begin{bmatrix} J_{10} \\ \vdots \\ J_{1M} \\ J_{20} \\ \vdots \\ J_{2M} \end{bmatrix} = \mathbf{0}$$

$$\boldsymbol{Ha} + \boldsymbol{J} = \mathbf{0}$$

a_{ik} is estimated by an iterative calculation of the following method.

(1) The initial value of parameter a_{ik} is set to 0, and joint angle trajectory θ_i is calculated from Eq. (6).
(2) ${}_1h_{ikm}$, ${}_2h_{ikm}$, J_{1m}, and J_{2m} are calculated at every time t_m $(m = 0, 1, \cdots, M)$.
(3) a_{ik} is obtained from Eq. (9).
(4) A new joint angle trajectory θ_i is calculated from Eq. (6) using a_{ik} computed in (3).

Steps (2) - (4) are repeated until a_{ik} converges. In other words, in Eq. (9), let d $(d = 0, 1, 2, \cdots)$ denote an index of the iterative calculation; the iterative calculation can be shown by the following equation. Here, $\boldsymbol{H}^{\#}$ indicates a pseudo inverse matrix of $\boldsymbol{H}$.

$$\boldsymbol{a}^{d+1} = -\boldsymbol{H}(\boldsymbol{a}^d)^{\#}\boldsymbol{J}(\boldsymbol{a}^d) \tag{9}$$

4 Numerical Experiments of Point-to-Point Movement on the Horizontal and Sagittal Plane

The formation of minimum commanded torque change trajectories with the same starting point, final point, and motion duration as measured trajectories by Nakano et al. [4] was carried out. The error of the Euler-Poisson equation was defined by the maximum value of the absolute value of the error, such as Eq. (10), to evaluate each predicted trajectory.

$$\max_{0 \le t_m \le 1} \sum_{i=1}^{2} |E_i^*(t_m) - E_i(t_m)| = \max_{0 \le t_m \le 1} \sum_{i=1}^{2} |0 - E_i(t_m)| = |\boldsymbol{E}\,|_{\max} \tag{10}$$

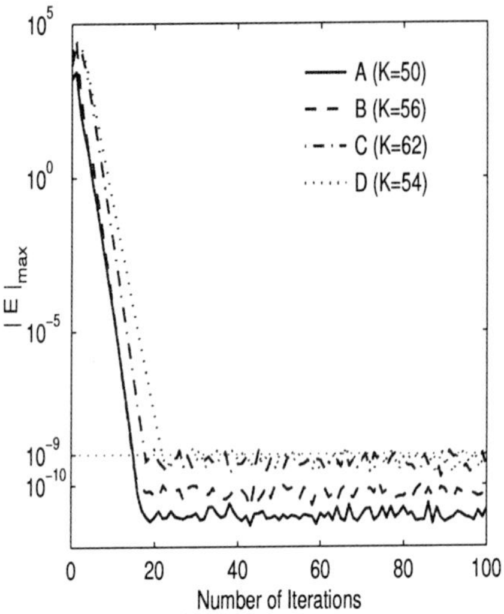

Fig. 1. Convergence of $|E|_{max}$. A, B, C, D: trajectories in Fig. 3.

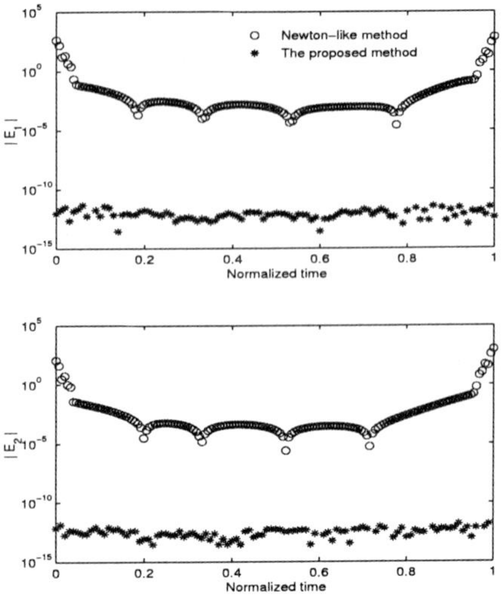

Fig. 2. Absolute values of E_1 and E_2 (Trajectory A in Fig. 3. : Newton-like method, *: the proposed method.

Here, $E_i^*(t_m)(=0)$ shows the optimal value of $E_i(t_m)$. Whether the trajectory satisfies the Euler-Poisson equation can be judged according to the error definition. 1200 minimum commanded torque change trajectories were generated. First, the relationship between the number of parameters K and the error $|\boldsymbol{E}|_{\max}$ was examined in the proposed method. The value of K was changed (i.e, 30, 32, 34,c, 70). All of the initial values of a_{ik} were set to 0. Fig. 1 indicates the convergence of error $|\mathrm{E}|_{\max}$ in 100 times iterative calculations. The horizontal axis shows the iterative number of estimations of parameter a_{ik} and the vertical axis shows $|\boldsymbol{E}|_{\max}$ calculated by Eq. (10). $|\boldsymbol{E}|_{\max}$ for all trajectories converges to a small value below about 10^{-9}. It can therefore be said that the Euler-Poisson equations are satisfied with a high accuracy at every movement time of the trajectory. Moreover, the error clearly converges in a small number of iterative calculations, that is, about 20 times.

To confirm how the proposed method and the Newton-like method satisfy the Euler-Poisson equations, E_1 and E_2 at every movement time were calculated. Fig. 2 shows the absolute values of E_1 and E_2 at every time point for the proposed method and the Newton-like method. The horizontal axis indicates the normalized time, and the vertical axis is the absolute values of E_1 and E_2. It can be understood that the proposed method satisfies the Euler-Poisson equations better than the Newton-like method during the whole movement time.

Table 1 indicates values of the object function (integration of the second power of the commanded torque change) for two trajectories solved by each method on the horizontal plane and the sagittal plane. It can be understood that the object function of the proposed method is smaller than or equal to the

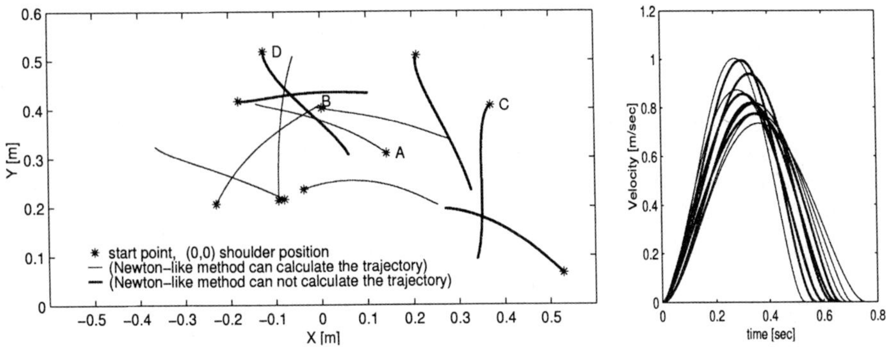

(a) Hand path (b) Hand velocity

Fig. 3. Minimum commanded torque change trajectories calculated using the proposed method in the horizontal plane.

Table 1. Values of the performance index (the time integral of the square of the commanded torque change rate). — shows that the method can not calculate the optimal trajectory

Horizontal plane	1	2	Sagittal plane	1	2
The proposed method	33.4432	92.1030	The proposed method	93.7844	40.3807
Newton-like method	33.4432		Newton-like method	93.7845	
Steepest descent method	33.4619	92.1660	Steepest descent method	93.8458	40.3932

other methods for any case. Moreover, it can be confirmed that the proposed method can also obtain the converged trajectory solution, which is unable to be calculated by the Newton-like method.

The minimum commanded torque change trajectories on the horizontal plane obtained by the proposed method are shown in Fig. 3. The Y-axis of (a) is forward of the body, and the origin is the position of a shoulder. The optimal trajectories shown by the thick line can not be obtained by the Newton-like method. Consequently, optimal trajectories can be obtained by the proposed method regardless of whether the trajectories can be solved by the Newton-like method or not. Moreover, it can be understood that the features of human arm movements, that is, hand trajectories are not straight lines but gradually curved, and the velocity profiles each have a single peak and are reproduced well.

5 Conclusion

It was shown that the proposed method has the following advantages by numerical experiments. The trajectory obtained by the proposed method is similar to the trajectory of the mathematically guaranteed Newton-like method. An optimal trajectory, which can not be obtained by the Newton-like method, is able to be calculated. Even if the value of the viscosity, which is one cause why a converged solution is unable to be obtained by the Newton-like method, is greatly changed, it can be confirmed that the proposed method is able to obtain an optimal solution. This was shown to obtain the converged solution in a very short period of time compared with the previous method. Though we searched for the most appropriate number of parameters K in the paper, a value of around $K = 60$ is sufficient. An optimal solution can be obtained by an iterative calculation of several ten times. The proposed method is a general method for a nonlinear, optimal problem. Therefore, the method can be applied to various optimization problems. However, an important remaining problem is to clarify the convergence conditions of the method theoretically.

Finally, we performed supplementary examinations of the experiments by Nakano et al. [4]. We recalculated the minimum commanded torque change trajectory using the proposed method; the same examinations as Nakano et al. [4] were performed. As a result, it was reconfirmed that the measured trajectory was closest to the minimum commanded torque change trajectory, like the result by Nakano et al. [4].

Acknowledgements

This study was performed through Special Coordination Funds for the promotion of Science and Technology of the Science and Technology Agency, Japan

References

1. W. Abend, E. Bizzi, and P. Morasso. Human arm trajectory formation. *Brain*, 105:331–348, (1982).
2. E. Bizzi, N. Accornero, W. Chapple, and N. Hogan. Posture control and trajectory formation during arm movement. *Journal of Neuroscience*, 4:2738–2744, (1984).
3. T. Flash and N. Hogan. The coordination of arm movements: An experimentally confirmed mathematical model. *Journal of Neuroscience*, 5:1688–1703, (1985).
4. E. Nakano, H. Imamizu, R. Osu, Y. Uno, H. Gomi, T. Yoshioka, and M. Kawato. Quantitative examinations of internal representations for arm trajectory planning: minimum commanded torque change model. *Journal of Neurophysiology*, 81(5):2140–2155, (1999).
5. D. A. Rosenbaum, L. D. Loukopoulos, R. G. Meulenbriek, J. Vaughan, and S. E. Engelbrecht. Planning resaches by evaluating stored posture. *Psychol. Rev.*, 102:28–67, (1995).
6. G. Szego. *Orthogonal polynomials.* American mathematical society, New York, 4th edition, (1975).
7. Y. Uno, M. Kawato, and R. Suzuki. Formation and control of optimal trajectory in human multijoint arm movement – minimum torque change model. *Biological Cybernetics*, 61:89–101, (1989).
8. Y. Wada, Y. Kaneko, E. Nakano, R. Osu, and M. Kawato. Quantitative examinations for multi joint arm trajectory planning by a strict and robust calculation algorithm of minimum commanded torque change trajectory. *Neural Networks*, 14(4):381–393, (2001).

Part XII

Vision and Image Processing

Neocognitron of a New Version: Handwritten Digit Recognition

Kunihiko Fukushima*

Tokyo University of Technology, Hachioji, Tokyo 192-0982, Japan
fukushima@karl.teu.ac.jp

Abstract. The author previously proposed a neural network model *neocognitron* for robust visual pattern recognition. This paper proposes an improved version of the neocognitron and demonstrates its ability using a large database of handwritten digits (ETL-1).
To improve the recognition rate of the neocognitron, several modifications have been applied: such as, the inhibitory surround in the connections from S-cells to C-cells, contrast-extracting layer between input and edge-extracting layers, self-organization of line-extracting cells, supervised competitive learning at the highest stage, and so on. These modifications allowed the removal of accessory circuits that were appended to the previous versions, resulting in an improvement of recognition rate as well as simplification of the network architecture.
The recognition rate varies depending on the number of learning patterns. When we used 3000 patterns (300 patterns for each digit) for the learning, for example, the recognition rate was 98.5% for a blind test set (3000 patterns), and 100% for the learning set.

1 Introduction

The author previously proposed a neural network model *neocognitron* for robust visual pattern recognition [1]. This paper proposes an improved version of the neocognitron and demonstrates its ability using a large database of handwritten digits.

The neocognitron was initially proposed as a neural network model of the visual system that has a hierarchical multilayered architecture similar to the classical hypothesis of Hubel and Wiesel [2],[3]. It acquires the ability to recognize robustly visual patterns through learning. It consists of layers of S-cells, which resemble simple cells in the primary visual cortex, and layers of C-cells, which resemble complex cells. These layers of S-cells and C-cells are arranged alternately in a hierarchical manner.

S-cells are feature-extracting cells, whose input connections are variable and are modified through learning. C-cells, whose input connections are fixed and unmodified, exhibit an approximate invariance to the position of the stimuli presented within their receptive fields.

* The author is also with Nippon Institute of Technology

G. Dorffner, H. Bischof, and K. Hornik (Eds.): ICANN 2001, LNCS 2130, pp. 987–992, 2001.

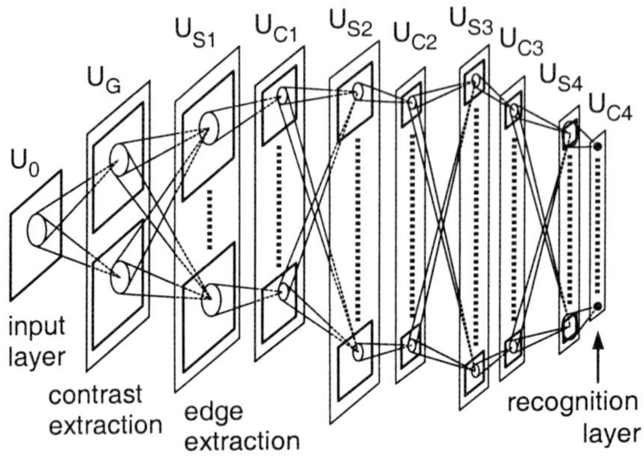

Fig. 1. The architecture of the proposed neocognitron.

The C-cells in the highest stage work as recognition cells and indicate the result of the pattern recognition. After learning, the neocognitron can recognize input patterns robustly, with little effect from deformation, change in size, or shift in position.

The neocognitron of a new version, which this paper proposes, also has a network architecture similar to that of the conventional neocognitron, but several new ideas have been introduced: such as, an inhibitory surround in the connections from S-cells to C-cells, a contrast-extracting layer followed by an edge-extracting layer, self-organization of line-extracting cells, supervised competitive learning at the highest stage, and so on.

This paper shows that these modifications allow the removal of accessory circuits appended to the neocognitron of recent versions [4], resulting in an improvement of recognition rate as well as simplification of the network architecture.

2 Architecture of the Network

2.1 Outline of the Network

Figure 1 shows the architecture of the proposed model, which has 4 stages of S- and C-cell layers.

The stimulus pattern is presented to the input layer U_0. A layer of contrast-extracting cells (U_G), which correspond to retinal ganglion cells or lateral geniculate nucleus cells, follows U_0. The contrast-extracting layer U_G consists of two cell-planes: one cell-plane consisting of cells with concentric on-center receptive fields, and one cell-plane consisting of cells with off-center receptive fields. The former cells extract positive contrast in brightness, whereas the latter extract negative contrast from the images presented to the input layer.

The output of layer U_G is sent to the S-cell layer of the first stage (U_{S1}). The S-cells of layer U_{S1} correspond to simple cells in the primary visual cortex. They have been trained using supervised learning [1] to extract edge components of various orientations from the input image. To be more specific, layer U_{S1} consists of 16 cell-planes, each of which consists of edge-extracting cells of a particular preferred orientation. The learning pattern for each cell-plane is a straight edge of a particular orientation. Preferred orientations of the cell-planes, namely, the orientations of the learning patterns, are chosen at an interval of 22.5°. The threshold of the S-cells, which determines the selectivity in extracting features, is set low enough to accept edges of slightly different orientations.

The output of layer U_{Sl} (S-cell layer of the lth stage) is fed to a C-cell layer U_{Cl}, where a blurred version of the response of U_{Sl} is generated. The connections from S- to C-cell layers will be discussed in detail in Section2.2. The density of the cells in each cell-plane reduces between layers U_{Sl} and U_{Cl}.

The S-cells of intermediate stages (U_{S2} and U_{S3}) are self-organized using unsupervised competitive learning similar to the method used in the conventional neocognitron [1]. Every time a training pattern is presented to the input layer, seed cells are determined by a kind of winner-take-all process. Each input connection to a seed cell is increased by an amount proportional to the output of the cell from which the connection leads. Because of the shared connections within each cell-plane, all cells in the cell-plane come to have the same set of input connections as the seed cell.

Line-extracting S-cells, which were created by supervised learning in the previous version [4], are now automatically generated (or, self-organized) together with cells extracting other features in the second stage (U_{S2}).

Training of the network is performed from the lower stages to the higher stages: after the training of a lower stage has been completely finished, the training of the succeeding stage begins. The same set of training patterns is used for the training of all stages except layer U_{S1}.

The neural networks' ability to recognize patterns robustly is influenced by the selectivity of feature-extracting cells, which is controlled by the threshold of the cells. The author previously proposed the use of higher threshold values for feature-extracting cells in the learning phase than in the recognition phase, when unsupervised learning with a winner-take-all process is used to train neural networks [5]. This method of dual threshold is used for the learning of intermediate layers, U_{S2} and U_{S3}.

Layer U_{C4} at the highest stage is the recognition layer, whose response shows the final result of the pattern recognition. S-cell layer U_{S4} of this stage is trained through supervised competitive learning, which will be discussed in Section 2.3.

2.2 Inhibitory Surround in the Connections to C-cells

The response of an S-cell layer U_{Sl} is spatially blurred in the succeeding C-cell layer U_{Cl}. In the conventional neocognitron, the input connections of a C-cell, which converge from a group of S-cells, consisted of only excitatory components of a circular spatial distribution.

An inhibitory surround is newly introduced around the excitatory connections. The concentric inhibitory surround endows the C-cells with the charac-

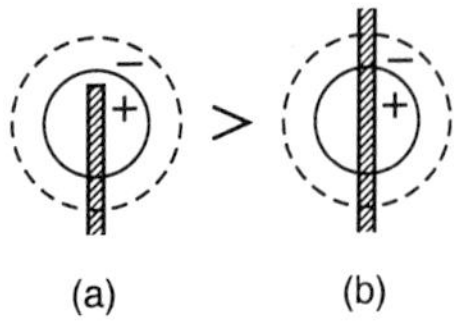

(A) Response like an end-stopped cell. Stimulus (a) produces a larger response than (b).

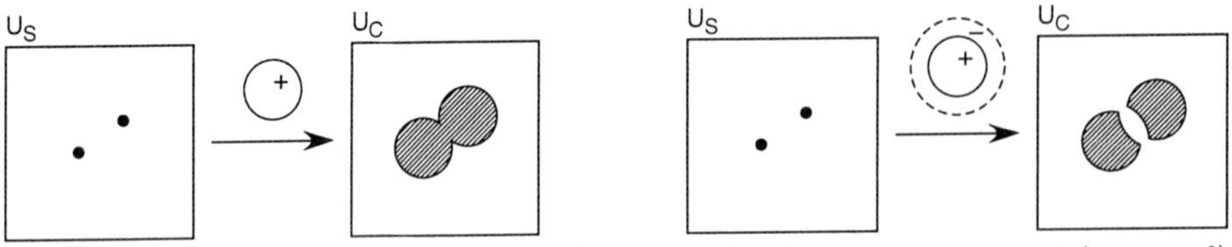

(a) No inhibitory surround (conventional) (b) Inhibitory surround (proposed)
(B) Separation of the blurred responses produced by two independent features.

Fig. 2. The effect of inhibitory surround in the input connections to a C-cell.

teristics of end-stopped cells, and C-cells behave like hypercomplex cells in the visual cortex (Fig. 2(A)).

Bend points and end points of lines are important features for pattern recognition. In the network of previous versions (e.g., [4]), an extra S-cell layer, which was called a bend-extracting layer, was placed after the line-extracting stage to extract these feature points. In the proposed network, C-cells, whose input connections have inhibitory surrounds, participate in extraction of bend points and end points of lines while they are making a blurring operation. This allows the removal of accessory layer of bend-extracting cells, resulting in a simplification of the network architecture and an increased recognition rate as well.

The inhibitory surrounds in the connections also have another benefit. The blurring operation by C-cells, which usually is effective for improving robustness against deformation of input patterns, sometimes makes it difficult to detect whether a lump of blurred response is generated by a single feature or by two independent features of the same kind. For example, a single line and a pair of parallel lines of a very narrow separation generate a similar response when they are blurred. The inhibitory surround in the connections to C-cells creates a non-responding zone between the two lumps of blurred responses (Fig. 2(B)). This silent zone makes the S-cells of the next stage easily detect the number of original features even after blurring.

2.3 Learning of the Highest Stage

S-cells of the highest stage (U_{S4}) are trained using a supervised competitive learning. The learning rule resembles the competitive learning used to train U_{S2} and U_{S3}, but the category names of the training patterns are also utilized for the learning. When the network learns varieties of deformed training patterns through competitive learning, more than one cell-plane for one category is usually generated in U_{S4}. Therefore, when each cell-plane first learns a training pattern, the category name of the training pattern is assigned to the cell-plane. Thus, each cell-plane of U_{S4} is assigned to one of the 10 digits.

Every time a training pattern is presented, competition occurs among all S-cells in the layer. (In other words, the competition area for layer U_{S4} is large enough to cover all cells of the layer.) If the winner of the competition has the same category name as the training pattern, the winner becomes the seed cell and learns the training pattern in the same way as the seed cells of the lower stages. If the winner has a different category name (or if all S-cells are silent), however, a new cell-plane is created and is assigned the category name of the training pattern.

During the recognition phase, the category name of the maximum-output S-cell of U_{S4} determines the final result of recognition. We can also express this process of recognition as follows. Recognition layer U_{C4} has 10 C-cells corresponding to the 10 digits to be recognized. Every time a new cell-plane is created in layer U_{S4} in the learning phase, excitatory connections are created from all S-cells of the cell-plane to the C-cell of that category name. Competition among S-cells occur also in the recognition phase, and only one maximum output S-cell within the whole layer U_{S4} can transmit its output to U_{C4}.

3 Computer Simulation

We tested the behavior of the proposed model by computer simulation using handwritten digits (free writing) randomly sampled from the ETL-1 database. Incidentally, the ETL-1 is a database of segmented handwritten characters and is distributed by Electrotechnical Laboratory, Tsukuba, Japan.

The scale of the simulated network is as follows. The total number of cells (not counting inhibitory V-cells) in each layer is: U_0: 65×65, U_G: $71\times 71\times 2$, U_{S1}: $71\times 71\times 16$, U_{C1}: $39\times 39\times 16$, U_{S2}: $39\times 39\times K_{S2}$, U_{C2}: $23\times 23\times K_{C2}$, U_{S3}: $23\times 23\times K_{S3}$, U_{C3}: $13\times 13\times K_{C3}$, U_{S4}: $7\times 7\times K_{S4}$, U_{C4}: $1\times 1\times 10$. The number of cell-planes in an S-cell layer (K_{Sl}) is determined automatically in the learning phase depending on the learning set. In each stage except the highest one, the number of cell-planes of the C-cell layer (K_{Cl}) is the same as K_{Sl}.

The recognition rate varies depending on the number of learning patterns. When we used 3000 patterns (300 patterns for each digit) for the learning, for example, the recognition rate was 98.5% for a blind test sample (3000 patterns), and 100% for the learning set.

Fig. 3 shows a typical response of the network that has finished the learning using 3000 learning patterns. The responses of layers U_0, U_G and layers of C-cells of all stages are displayed in series from left to right. The rightmost layer, U_{C4}, is the recognition layer, whose response shows the final result of recognition. The responses of S-cell layers are omitted from the figure but can be estimated from the responses of C-cell layers: a blurred version of the response of an S-cell layer appears in the corresponding C-cell layer, although it is slightly modified by the inhibitory surround in the connections.

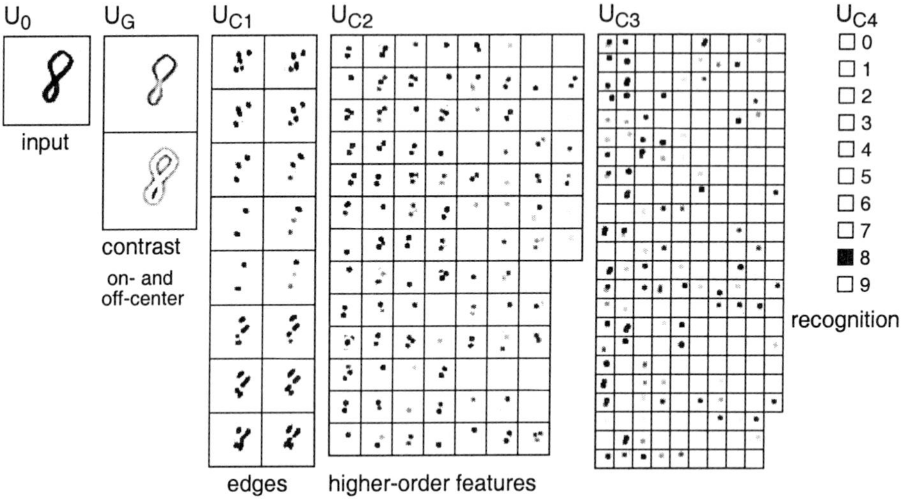

Fig. 3. An example of the response of the neocognitron. The input pattern is recognized correctly as '8'.

4 Conclusions

To improve the recognition rate of the neocognitron, several modifications have been applied: such as, the inhibitory surround in the input connections of C-cells, contrast-extracting layer followed by edge-extracting cells, self-organization of line-extracting cells, supervised competitive learning at the highest stage, and so on. These modifications allowed the removal of accessory circuits appended in the previous versions, resulting in an improvement of recognition rate as well as simplification of the network architecture.

References

1. K. Fukushima: "Neocognitron: a hierarchical neural network capable of visual pattern recognition", *Neural Networks*, **1**[2], 119–130 (1988).
2. D. H. Hubel, T. N. Wiesel: "Receptive fields, binocular interaction and functional architecture in the cat's visual cortex", *J. Physiology (Lond.)*, **106**[1], 106–154 (1962).
3. D. H. Hubel, T. N. Wiesel: "Receptive fields and functional architecture in two nonstriate visual areas (18 and 19) of the cat", *J. Neurophysiol.*, **28**, 229–289 (1965).
4. K. Fukushima, N. Wake: "Improved neocognitron with bend-detecting cells", *IJCNN'92*, Baltimore, MD, USA, Vol. IV, pp. 190–195 (1992).
5. K. Fukushima, M. Tanigawa: "Use of different thresholds in learning and recognition", *Neurocomputing*, **11**[1], 1–17 (1996).

A Comparison of Classifiers for Real-Time Eye Detection

Alex Cozzi, Myron Flickner, Jainchang Mao, and Shivakumar Vaithyanathan

IBM Almaden Research Center, 650 Harry Rd., San Jose, CA, USA
{cozzi,mao,flick,shiv}@almaden.ibm.com

Abstract. We describe a system for detecting and tracking human eyes using a digital camera. The system uses the combination of an active illumination scheme to detect eyes and an appearance-based object classifier to weed out spurious detections. We briefly describe the eye-detection mechanism and then we compare the performances of subspace Gaussian classifiers and support vector machines applied to this task.

1 The BlueEyes System

The eye is the center of attention in human-human interaction: by observing the eyes of the person with whom we are interacting we can learn about his or her state of attention, level of interest and comprehension. It is our belief that enabling computer to understand this non-verbal form of communication would greatly enhance their ability to react appropriately, especially in the case of perceptual user interface, where the human uses gestures and speech to interact with the computer.

The first step towards this goal is to devise a method to detect eyes in images in real time. In this way a camera-equipped computer is able to find the people that are initiating an interaction.

Most previous approaches concentrate on finding faces [8], others use images gradient to search for good candidates of eyes [3]. We use a more active approach.

We have built an eye detection system that exploits the well known red-eye effect, caused when an image is taken with the flash too close to the camera's optical axis. Since very few other objects exhibit this strong directional reflection, its uniqueness can be exploited to detect and localize eyes [4]. Fig. 1 shows a picture of our pupil-detection device. The system is composed by a conventional CCD camera with a visible light blocking filter and two sets of 875nm IR LEDs, which are invisible to the human eye. The filter blocks most of indoor ambient illumination, since most artificial light sources emit weakly in the infrared part of the spectrum, consequently the dominant sources of illumination are the infrared LEDs. The first set of LEDs is mounted very close to the camera's optical axis, thus causing the red-eyes effect. The second set is far enough from the camera's optical axis so as not to produce the red-eye effect. In our system the two rings of LEDs are switched in synchrony with the fields that compose an interlaced video signal, thus generating a frame composed by an odd field with bright pupils and

G. Dorffner, H. Bischof, and K. Hornik (Eds.): ICANN 2001, LNCS 2130, pp. 993–999, 2001.

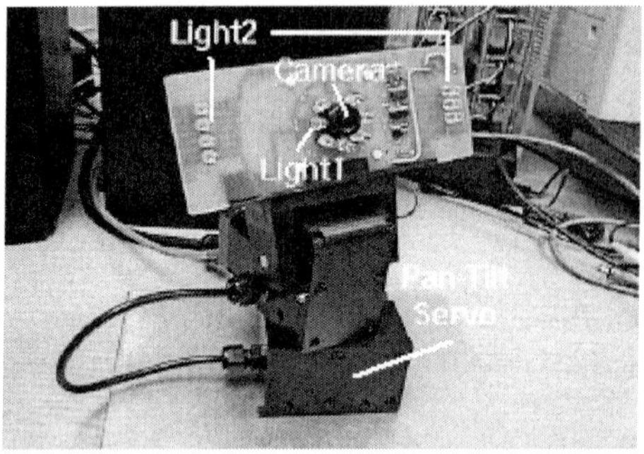

Fig. 1. Camera and IR illumination setting.

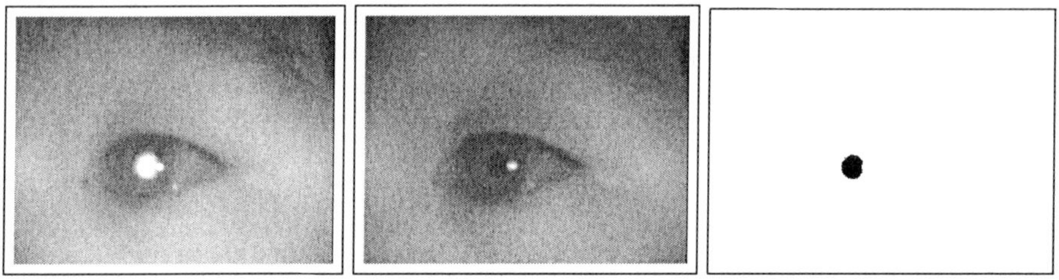

Fig. 2. The light, dark, and difference pupil image.

an even field with normal dark pupils. To identify the eyes, the dark-pupil image is subtracted from the bright-pupil image and thresholded. The resulting binary image consists of the areas that have changed significantly in the 33ms between the video fields[1]. These ares include the pupils and other "noise", such as motion disparities (moving objects) or bright light spots. To separate unwanted noise from the eye candidates we use a two-level approach.

First the binary image is fed into a connected component labelling algorithm. Geometrical constraints based on the shape and size of each connected component eliminate most of the errors (large motions, isolated points, asymmetrical regions). This geometrical constraints ensure that remaining regions are size-compatible with the image of a pupil. However, depending on the amount of motion and other factors, about 5–50% of these regions are not eyes. In order to eliminate the remaining errors, we explored the use of two appearance-based image classification schemes: Gaussian subspace classifier and support vector machine (SVM).

2 Gaussian Subspace Classifier

Gaussian subspace classifier, implemented via principal component analysis (PCA), became popular in computer vision for a wide variety of image-recog-

[1] A NTSC camera works at a frequency of 60 fields per second

nition tasks [8]. One of the reasons of its success is due to the fact that it can generate a small set of decorrelates features.

We use the classical implementation of the technique: using SVD we compute the principal eigenvectors of the covariance matrix of the training set, then we select the first q eigenvectors to project all the samples of the training set to a lower dimensionality space, i.e., from $32 \times 16 = 512$ to q, usually a number between 5 and 20. In this smaller space we build a naive Bayes classifier using a Gaussian model. The first Gaussian distribution is used to model the projections of the eye samples and the second Gaussian distribution is used to model the projections of non-eye samples.

The decision to classify an example as eye or non-eye is taken according to its conditional likelihood given the eye distribution or the non-eyes distribution.

3 Support Vector Machine

One of the most interesting recent developments in classifier design is the introduction of the support vector machine (SVM) by Vapnik [10], which has also been studied by other researchers [1].

It is primarily a two-class classifier. The optimization criterion is the width of the margin between the two classes. A classifier with a large margin separating two classes has a small Vapnik-Chervonenkis (VC) dimension, which yields a good generalization performance. Many successful applications of SVMs have demonstrated the superiority of this objective function over others [2].

Let $\{(\mathbf{x}_i, y_i)|i = 1, 2, \cdots, n\}$ denote a set of training patterns, where $\mathbf{x}_i \in \mathbf{R}^d$ and $y_i \in \{+1, -1\}$ are the d-dimensional feature vector and class label for the ith pattern, respectively. SVM is formulated as the following quadratic programming (QP) problem.

$$\min_{\mathbf{w}}\{\|\mathbf{w}\|^2 + C\sum_{i=1}^{n} \varepsilon_i\} \tag{1}$$

subject to

$$y_i(\mathbf{w}^T\mathbf{x}_i - b) \geq 1 - \varepsilon_i, \quad \forall i. \tag{2}$$

This *primal* QP problem can be formulated as the following *dual* problem using Wolfe duality theory.

$$\max_{\alpha}\{\sum_{i=1}^{n} \alpha_i - \frac{1}{2}\sum_{i=1}^{n}\sum_{j=1}^{n} \alpha_i\alpha_j y_i y_j \mathbf{x}_i^T\mathbf{x}_j\} \tag{3}$$

subject to

$$0 \leq \alpha_i \leq C, \quad \text{and} \quad \sum_{i=1}^{n} \alpha_i y_i = 0, \tag{4}$$

where α_i is the Lagrange multiplier for the ith constraint (corresponding to the ith pattern). The weight vector $\mathbf{w}$ and bias b can be derived from the Lagrange multipliers $\alpha_i's$.

$$\mathbf{w} = \sum_{i=1}^{n} \alpha_i y_i \mathbf{x}_i, \tag{5}$$

$$b = \mathbf{w}^T \mathbf{x}_k - y_k, \text{ for some } k \text{ and } \alpha_k > 0. \tag{6}$$

A nice property of SVM is that the values of all α_i become 0, except for those boundary patterns called support vectors. Eq.(5) indicates that the decision boundary is fully specified by the support vectors.

Although the dual problem is independent of the feature dimensionality, the size of this QP problem can be huge since the number of Lagrange multipliers is equal to the number of training patterns. Existing QP solvers quickly become impractical when the number of training patterns becomes large due to their large memory requirement and slow iterative process. Various training algorithms have been proposed in the literature [1], including chunking [9], Osuna's decomposition method [5], and Sequential Minimal Optimization (SMO) [7].

SMO solves the QP problem in SVM by decomposing it into the minimal QP problem involving only two Lagrange multipliers (i.e., two patterns), which can be solved analytically. SMO selects a pair of patterns at each step and optimizes them. This procedure repeats until all the patterns satisfy the optimality. In this paper, we use an improved version of SMO. The improved version [6] is significantly faster than the standard SMO.

4 The Sample Set

We collected a set of samples, composed of roughly equal number of eyes and non-eyes, for a total of 1621 images, each 32×16 pixels, and used it to train and test the performances of the two classifiers. Some of the samples are shown in Fig. 3. These images were automatically collected from the pupil candidates that satisfy the geometrical constraints and then manually classified as eyes or non-eyes. The training set is composed of eye images from eight different male individuals. Before being fed into the training algorithms some pre-processing is required. In the case of PCA it is necessary to compute the mean of the training samples, and create a new training set with the mean subtracted. In the case of SVM it is necessary to normalize all the training samples to unitary norm.

The training set is composed only of left eyes. Due to the symmetry of the human face, it is easy, once we obtain a classifier for left eyes, to flip it and generate a classifier for right eyes. It is possible to avoid to train two different classifiers by a simple transformation that maps two flipped images to the same transformed vector. The transformation builds a new image composed by taking alternatively the mean and the geometrical mean of the pixels a symmetrical fashion respect to the mid-line of the image:

$$\begin{aligned} p'_{i,j} &= \frac{p_{i,j} + p_{c-i,j}}{2} \\ p'_{i+1,j} &= \sqrt{p_{i,j} p_{c-i,j}} \\ i = 0, \ldots, c;\ j &= 0, \ldots M-1 \end{aligned}$$

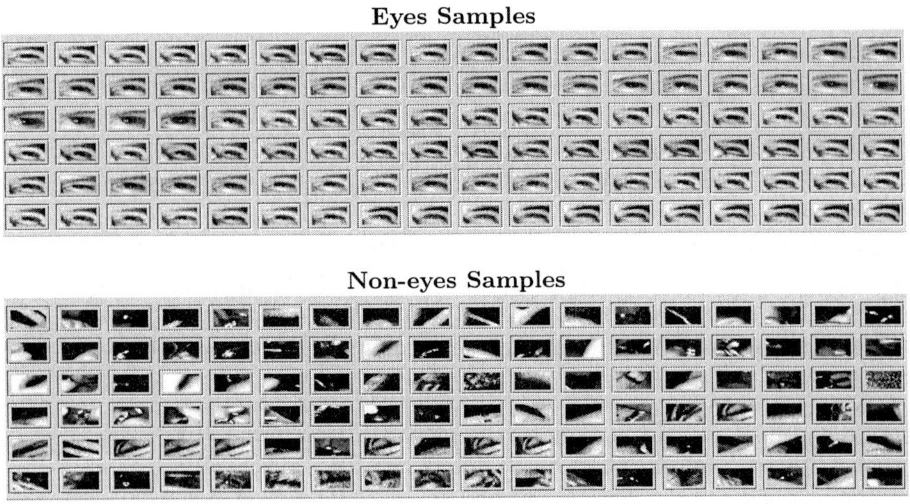

Fig. 3. the training set.

where c is the horizontal position of the center of the image, M the height of the image and $p_{i,j}$ the pixel value of the image at position (i, j). When this transformation is applied to the training set, the performances of the SVM were very similar to the case of no pre-processing. In the case of PCA the results were significantly worse, requiring more eigenvectors in order to obtain the same level of reconstruction error, the reason being that while the transformation does not significantly affects the separability of the samples, it alters their distributions making it more complex, thus requiring more eigenvectors to be reconstructed with the same accuracy.

5 Experiments and Discussion

We performed the experiments using 3-fold cross-validation and observing the influence of the tuning parameters of the two classifiers on our classification task. The tunable parameter of PCA is the number of eigenvectors that define the lower dimensionality space. The number of eigenvectors is also proportional to the computational cost required to classify a new sample. Using the number of multiply-and-add operations as a measure, to classify a new sample using ten eigenvectors costs $10 \times 512 = 5120$ multiply-and-add operations (ten inner products of vectors of length 512). As can be seen from Fig. 4, the optimal value is probably between 20 and 40 eigenvectors, depending how much importance we give to precision, i.e. true positive divided by the sum of true positive and false positives; against recall, i.e., true positives divided by the sum of true positives and true negatives. In the case of SVM, the tunable parameter is the factor C in Eq.(1). In this case the computational cost to classify a new sample does not depend on the parameter C but it is fixed and equivalent to 512 multiply-and-

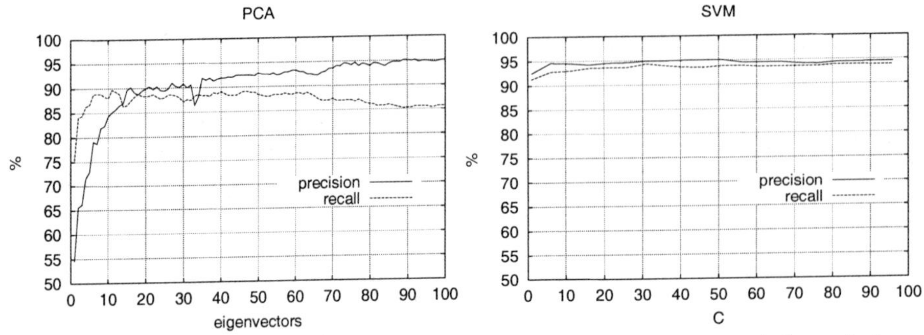

Fig. 4. The influence of the tuning parameters on the performances of the techniques

add operations. As can be seen in the plot, SVM performs pretty consistently across all tested values of C.

Interestingly the techniques proved to be quite robust against changes in illumination and small differences in scale, rotation and different subjects. We also experimented with another scenario where the differences in scale were much larger. In this different conditions the performance of both techniques rapidly degraded to the point of rendering them useless. A way around this problem is to train several classifiers. Another solution is to try to estimate the scale of the object to be recognized and perform an image scaling operation to normalize the scale of the objects to be recognized.

Since SVM has an advantage in classifier performance as well as requiring about 1/10 of the CPU time required by PCA we conclude that SVM classifier has a clear advantage for our problem.

As a final recommendation, if the detection task does not require to cope with significant changes in scale, the 95% classification rate of SVM is more than adequate for most applications. Otherwise the practitioner will have to decide whether to trade speed for accuracy by taking decisions bases on several successive frames.

References

1. C. J. C. Burges. A tutorial on support vector machines for pattern recognition. *Data Mining and Knowledge Discovery*, 2(2):121–167, 1998.
2. M. A. Hearst. Support vector machines. *IEEE Intelligent Systems*, pages 18–28, July/August 1998.
3. R. Kothari and J. Mitchell. Detection of eye locations in unconstrained visual images. In *ICIP 96*, 1996.
4. C.H. Marimoto, D. Koons, A. Amir, and M. Flickner. Frame-rate pupil detector and gaze tracker. In *ICCV 99 Frame-rate workshop*, Corfu, Greece, September 1999.
5. E. Osuna, R. Freund, and F. Girosi. An improved training algorithm for support vector machines. In J. Principe, L. Gile, N. Morgan, and E. Wilson, editors, *Proc. of IEEE Workshop on Neural Networks for Signal Processing*, pages 276–285. IEEE Press, New York, 1997.

6. D. Pavlov, J. Mao, and B. Dom. Scaling-up support vector machines using boosting algorithm. In *Proceedings of ICPR2000*, volume 2, pages 219–222, Barcelona, Catalonia, Spain, September 2000.
7. J. Platt. Fast training of support vector machines using sequential minimal optimization. In B. Scholkopf, C.J.C. Burges, and A.J. Smola, editors, *Advances in Kernel Methods – Support Vector Learning*, pages 185–208. MIT Press, Cambridge, MA, 1999.
8. M. Turk and A. Pentland. Eigenfaces for recognition. *J. Cogn. Neurosci.*, 3:71–86, 1991.
9. V. N. Vapnik. *Estimation of Dependences Based on Empirical Data.* Springer Verlag, Berlin, 1982.
10. V. N. Vapnik. *Statistical Learning Theory.* John Wiley & Sons, New York, 1998.

Neural Network Analysis of Dynamic Contrast-Enhanced MRI Mammography

Axel Wismüller[1], Oliver Lange[1],
Dominik R. Dersch[2], Klaus Hahn[1], and
Gerda L. Leinsinger[1]

[1] Institut für Radiologische Diagnostik,
Ludwig-Maximilians-Universität München,
Klinikum Innenstadt, Ziemssenstr. 1, 80336 München, Germany
Axel.Wismueller@physik.uni-muenchen.de
[2] Hypovereinsbank AG, München, Germany

Abstract. We present a neural network clustering approach to the analysis of dynamic contrast-enhanced magnetic resonance imaging (MRI) mammography time-series. In contrast to conventional extraction of a few voxel-based perfusion parameters, neural network clustering does not discard information contained in the complete signal dynamics time-series data. We performed exploratory data analysis in patients with breast lesions classified as indeterminate from clinical findings and conventional X-ray mammography. Minimal free energy vector quantization provided a self-organized segmentation of voxels w.r.t. fine-grained differences of signal amplitude and dynamics, thus identifying the lesions from surrounding tissue and enabling a subclassification within the lesions with regard to regions characterized by different MRI signal time-courses. We conclude that neural network clustering can provide a useful extension to the conventional visual inspection of interactively defined regions-of-interest. Thus, it can contribute to the diagnosis of indeterminate breast lesions by non-invasive imaging.

1 Introduction

Breast cancer is among the leading causes of death in women in the industrial countries. The analysis of indeterminate breast lesions, therefore, is an issue of enormous clinical importance.

Dynamic susceptibility contrast material-enhanced MRI mammography provides a non-invasive method for the diagnosis of questionable mammographic lesions. It enables high spatial and temporal resolution and avoids the disadvantage of patient exposure to ionizing radiation. After intravenous bolus administration of a nondiffusible paramagnetic substance such as gadolinium, contrast-agent uptake in breast tissue can be studied by observing, as a function of time, the agent-induced changes of proton magnetic properties. Recent publications discuss the potential usefulness of signal intensity time course data analysis for the differential diagnosis of enhancing breast lesions, see e.g. [3] for review.

G. Dorffner, H. Bischof, and K. Hornik (Eds.): ICANN 2001, LNCS 2130, pp. 1000–1005, 2001.

In this paper, we present an approach to the analysis of dynamic contrast-enhanced MRI mammography data that does not imply speculative presumptive knowledge on contrast agent dilution models, but strictly focusses on the observed complete MRI signal time-series. Neural network clustering enables a self-organized data-driven segmentation of dynamic contrast-enhanced MRI mammography time-series w.r.t. fine-grained differences of signal amplitude and dynamics, such as focal enhancement in patients with indeterminate breast lesions, thus providing a useful extension to the conventional visual inspection of interactively defined regions-of-interest. As a result, we obtain both a set of prototypical time-series and a corresponding set of cluster assignment maps. The inspection of these clustering results provides a practical tool for the radiologist to quickly scan the data set for regional differences or abnormalities of contrast-agent uptake.

2 Methods

2.1 Neural Network Clustering

Let n denote the number of subsequent scans in a dynamic contrast-enhanced MRI mammography study. Let K denote the number of pixels in each scan. The dynamics of each pixel $\nu \in \{1, \dots, K\}$, i.e. the sequence of signal values $x^\nu(\tau)$ over all scan acquisition time frames τ can be interpreted as a vector $\boldsymbol{x}^\nu \in \mathbb{R}^n$ in the n-dimensional feature space of possible signal time-series at each pixel (Pixel Time Course, PTC). Clustering identifies N groups j of pixels with similar PTCs denoted by the index $j \in \{1, \dots, N\}$. These groups or clusters are represented by prototypical time-series called Codebook Vectors (CVs) $\boldsymbol{w}_j$ located at the center of the corresponding clusters. VQ procedures determine these cluster centers by an iterative adaptive update according to

$$\boldsymbol{w}_j(t+1) = \boldsymbol{w}_j(t) + \epsilon(t)\, a_j(\boldsymbol{x}(t), W(t), \kappa)\, (\boldsymbol{x}(t) - \boldsymbol{w}_j(t)), \tag{1}$$

where $\epsilon(t)$ denotes a learning parameter, a_j a so-called cooperativity function which, in general, depends on the codebook $W(t)$, a cooperativity parameter κ, and a, in general, randomly chosen feature vector $\boldsymbol{x} \in \{\boldsymbol{x}^\nu | \nu \in \{1, \dots, K\}\}$. In the context of dynamic MRI mammography data analysis, one should clearly note the difference between the time τ of the time-series representing a specific component of the feature vector $\boldsymbol{x}^\nu$ and the time t which denotes the iteration step of the VQ procedure. In the fuzzy clustering scheme proposed by Rose, Gurewitz, and Fox [5], the cooperativity function a_j reads

$$a_j(\boldsymbol{x}(t), W(t), \kappa \equiv \rho(t)) = \frac{\exp(-E_j(\boldsymbol{x}(t))/2\rho^2)}{\mathcal{Z}}. \tag{2}$$

Here, the 'energy' $E_j(\boldsymbol{x}(t)) = \|\boldsymbol{x}(t) - \boldsymbol{w}_j(t)\|^2$ measures the distance between the codebook vector $\boldsymbol{w}_j$ and the data vector $\boldsymbol{x}$. $\mathcal{Z}$ denotes a partition function given by $\mathcal{Z} = \sum_j \exp(-E_j(\boldsymbol{x})/2\rho^2)$ and ρ is the cooperativity parameter of the model. This so-called 'fuzzy range' ρ defines a length scale in data space and is annealed to repeatedly smaller values in the VQ procedure. Noting the analogy to statistical mechanics, ρ can be interpreted as the temperature T in a multiparticle system by $T = 2\rho^2$ [5]. The learning rule (1) with a_j given by (2) describes a stochastic gradient descent on the error function

$$F_\rho(W) = -\frac{1}{2\rho^2} \int P(\boldsymbol{x}) \ln \mathcal{Z} d^n x, \tag{3}$$

which is a free energy in a mean-field approximation [1]. Here, $P(\boldsymbol{x})$ denotes the probability density of feature vectors $\boldsymbol{x}$. For the minimal free energy VQ procedure, the codebook vectors mark local centers of this multidimensional probability distribution.

Thus, for the application to perfusion MRI signal analysis, the codebook vector $\boldsymbol{w}_j$ is the weighted average perfusion MRI signal of all the PTCs $\boldsymbol{x}$ belonging to group j with respect to a fuzzy tessellation of the feature space. In the beginning of the VQ process, there is only one cluster representing the center of the whole data set. As the deterministic annealing procedure continues, phase transitions occur and large clusters split up into smaller ones marking increasingly smaller regions of the feature space. Tracing this repetitive cluster splitting through the whole VQ procedure leads to a 'genealogy' of cluster centers, i.e. a resemblance tree of codebook vectors. Thus, the scope of data analysis resolution can be adapted according to the observer's needs. The procedure can be monitored by various control parameters like the free energy, entropy, reconstruction error etc. which allow an easy detection of cluster splitting [2].

For mathematical details as well as a thorough discussion of application issues with regard to medical image analysis using neural networks, we refer to [8]. Functional MRI data analysis employing minimal free energy clustering has been presented previously [6].

2.2 Imaging Protocol

A total of 31 women with focal breast lesions classified as indeterminate from clinical findings and conventional X-ray mammography were included in the study. MRI was performed with a 1.5 T system (Magnetom Vision, Siemens, Erlangen, Germany) equipped with a dedicated surface coil to enable simultaneous imaging of both breasts. The patients were placed in a prone position. First, transversal images were acquired with a STIR (short TI inversion recovery) sequence (TR = 5600 ms, TE = 60 ms, FA = 90°, TI = 150 ms, matrix size 256 × 256 pixels, slice thickness 4 mm). Then a dynamic T1 weighted gradient echo sequence (3D fast low angle shot sequence) was performed (TR = 12 ms, TE = 5 ms, FA = 25°) in transversal slice orientation with a matrix size of 256 × 256 pixels and an effective slice thickness of 4 mm.

The dynamic study consisted of 6 measurements with an interval of 110 s. The first frame was acquired before injection of paramagnetic contrast agent (gadopentetate dimeglumine, 0.1 mmol/kg body weight, Magnevist™, Schering, Berlin, Germany), immediately followed by the other 5 measurements. The initial localization of suspicuous breast lesions was performed by computing difference images, i.e. subtracting the image data of the first from the fourth acquisition. As a preprocessing step to clustering, each raw gray level time-series $S(\tau)$, $\tau \in \{1, \dots, 6\}$ was transformed into a PTC of relative signal reduction $x(\tau)$ for each voxel, the pre-contrast scan at $\tau = 1$ serving as reference.

3 Results

Clustering results for a 6 scan dynamic contrast-enhanced MRI mammography study in a patient with breast cancer in the upper medial quadrant of the left breast are presented in figs. 1 and 2 for VQ employing 9 CVs. They show cluster assignment maps and CVs for minimal free energy VQ of MRI mammography data covering a supramammilar transversal slice of the left breast containing a suspicuous lesion that has been proven to be malignant by subsequent histological examination.

The procedure is able to segment the lesion from the surrounding breast tissue as can be seen from clusters #0 and #1 in fig. 1. Note the rapid and strong contrast agent uptake followed by subsequent plateau and washout phases in the round central region of the lesion as indicated by the corresponding CV of cluster #0 in fig. 2. Furthermore, clustering results enable a subclassification within this lesion with regard to regions characterized by different MRI signal time-courses: The central cluster #0 is surrounded by the peripheral circular cluster #1 which primarily can be separated from both the central region and the surrounding tissue by the amplitude of its contrast agent uptake ranging between CV #0 and all the other CVs. Furthermore, an additional, even more peripheral ringlike structure can be detected by careful observation of cluster #2.

The clustering results of figs. 1 and 2 represent typical findings: in all 31 patients with indeterminate mammographic lesions investigated with our method, we could differentiate the lesions from the surrounding tissue. Subclassification within the lesions could be achieved in up to 4 clusters with respect to differences in local signal dynamics. In 11/17 malignant lesions, central clusters showed early enhancement with a washout phase. In 8/17 malignant lesions, peripheral ringlike clusters with early enhancement and a plateau phase were found. Two or more concentric peripheral clusters were seen in 6 malignant lesions. In 10/14 benign lesions, central clusters showed early enhancement with a plateau phase, and in 9/14 lesions, peripheral clusters showed delayed enhancement. A detailed presentation and discussion of these findings including clinical implications of our method will be published elsewhere [4], [7].

4 Discussion

Our study shows that neural network clustering can provide a self-organized segmentation of dynamic contrast-enhanced MRI mammography data identifying groups of pixels sharing common properties of signal dynamics. In contrast

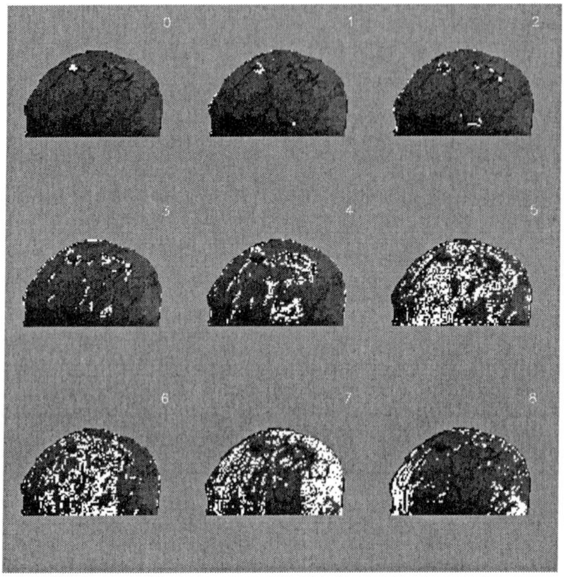

Fig. 1. Cluster assignment maps for minimal free energy vector quantization of a dynamic MRI mammography study in a patient with histologically confirmed breast cancer in the upper medial quadrant of the left breast.

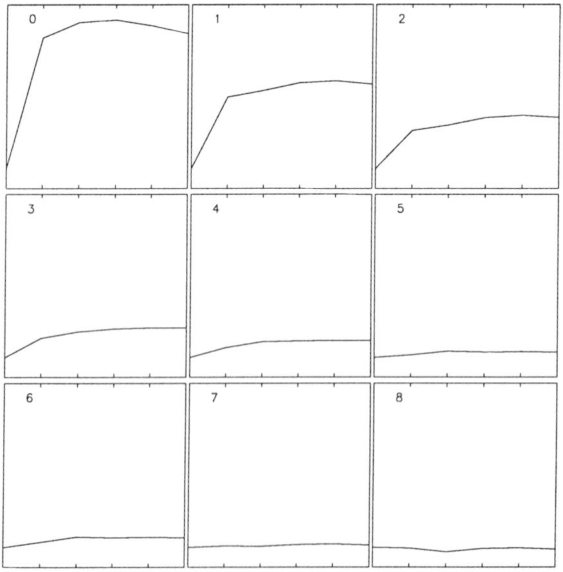

Fig. 2. Codebook vectors for minimal free energy vector quantization of a dynamic MRI mammography study in a patient with histologically confirmed breast cancer in the upper medial quadrant of the left breast. Cluster numbers correspond to fig. 1.

to the conventional extraction of a few perfusion parameters alone, neural network clustering does not discard information contained in the complete signal dynamics of MRI time-series.

In our study, minimal free energy VQ was able to unveil regional properties of contrast-agent uptake characterized by subtle differences of signal amplitude and dynamics. It could provide a segmentation with regard to identification and regional subclassification of pathological breast tissue lesions.

Although we conjecture from our results that neural network clustering can provide a useful extension of conventional data processing in dynamic contrast-enhanced MRI mammography studies, large controlled clinical trials are needed to establish the validity of the method and to finally evaluate its contribution to clinical decision making in the field of breast cancer diagnosis. Nevertheless, we hope that neural network vector quantization can serve as a useful strategy for exploratory data analysis of dynamic MRI mammography studies with applications ranging from biomedical basic research to clinical assessment of patient data.

References

1. D.R. Dersch. *Eigenschaften neuronaler Vektorquantisierer und ihre Anwendung in der Sprachverarbeitung.* Verlag Harri Deutsch, Reihe Physik, Bd. 54, Thun, Frankfurt am Main, 1996. ISBN 3-8171-1492-3.
2. D.R. Dersch and P. Tavan. Load balanced vector quantization. In *Proceedings of the International Conference on Artificial Neural Networks ICANN*, pages 1067–1070. Springer, 1994.
3. C.K. Kuhl, P. Mielcareck, S. Klaschik, C. Leutner, E. Wardelmann, J. Gieseke, and H.H. Schild. Dynamic breast MR imaging: Are signal intensity time course data useful for differential diagnosis of enhancing lesions? *Radiology*, 211:101–110, 1999.
4. G.L. Leinsinger, M. Scherr, A. Wismüller, O. Lange, D.R. Dersch, and K. Hahn. Unsupervised cluster analysis of dynamic contrast-enhanced MR mammography by neural networks – characterization of benign and malignant breast lesions. In preparation, 2000.
5. K. Rose, E. Gurewitz, and G.C. Fox. Vector quantization by deterministic annealing. *IEEE Transactions on Information Theory*, 38(4):1249–1257, 1992.
6. A. Wismüller, D.R. Dersch, B. Lipinski, K. Hahn, and D. Auer. Hierarchical clustering of fMRI time-series by deterministic annealing. *Neuroimage*, 7(4):593, 1998.
7. A. Wismüller, G.L. Leinsinger, D.R. Dersch, O. Lange, M. Scherr, and K. Hahn. Neural network image processing in contrast-enhanced MRI mammography – unsupervised cluster analysis of signal dynamics time-series by deterministic annealing. *Radiology Suppl.*, 217 (P):208–209, 2000.
8. A. Wismüller, F. Vietze, and D.R. Dersch. Segmentation with neural networks. In Isaac Bankman, Raj Rangayyan, Alan Evans, Roger Woods, Elliot Fishman, and Henry Huang, editors, *Handbook of Medical Imaging*, Johns Hopkins University, Baltimore, 2000. Academic Press. ISBN 0120777908.

A New Adaptive Color Quantization Technique

Antonios Atsalakis, Nikos Papamarkos, and Charalambos Strouthopoulos

Electric Circuits Analysis Laboratory, Department of Electrical & Computer Engineering
Democritus University of Thrace, 67100 Xanthi, Greece
papamark@ee.duth.gr

Abstract. This paper proposes a new algorithm for color quantization. The proposed approach achieves color quantization using an adaptive tree clustering procedure. In each node of the tree a self-organized Neural Network Classifier (NNC) is used which is fed by image color values and additional local spatial features. The NNC consists of a Principal Component Analyzer (PCA) and a Kohonen self-organized feature map (SOFM) neural network. The output neurons of the NNC define the color classes for each node. The final image not only has the dominant image colors, but also its texture approaches the image local characteristics used. For better classification, split and merging conditions are used in order to define if color classes must be split or merged. To speed-up the entire algorithm and reduce memory requirements, a fractal scanning subsampling technique is used.

1. Introduction

True type color images can have more than 16 millions colors in a 24 bits full RGB color space. There are many reasons for the reduction of the number of colors in a digital image. Reduction of the number of the image colors is an important task for compression, presentation, transmission and segmentation of color images. In most cases, it is easier to process and understand an image with a limited number of colors.

The most usually used techniques for color reduction are the color quantization approaches. The color quantization techniques group similar colors and replace them with only a single "quantized" color [**1**],[**2**],[**3**],[**5**]. These techniques are suitable for eliminating the uncommon colors in an image but they are ineffective for image analysis and segmentation. The quality of the resultant image varies, depending on the number of the final colors and the algorithm intelligence. A color quantization algorithm consists of two main stages, the color palette generation stage and the pixel mapping stage [**6**]. Color quantization techniques are usually used for the reduction of the colors of an image to a large number of colors. These techniques are not good enough to produce images with very limited number of colors. In addition, they only use the color values and not any other information related to the structure and the texture of the images.

This paper proposes a new adaptive color reduction technique, which not only exploits the colors of the images, but also their local spatial characteristics. The pro-

G. Dorffner, H. Bischof, and K. Hornik (Eds.): ICANN 2001, LNCS 2130, pp. 1006–1012, 2001.

posed approach improves significantly previous techniques [8],[9],[10] by introducing an adaptive clustering scheme, which is based on a tree decomposition procedure suitable for color images. The proposed technique can be used with any color scheme and is suitable for reduction of the colors of the images to a small number. In addition, class split and merging conditions are introduced in order to define if a feature class must be further split or merged. The color values of each pixel are related to local spatial features extracted from its neighboring region. Thus, the 3-D histogram clustering approach of the color multithresholding techniques is extended to a multi-dimensional feature clustering technique.

The RGB (or any other color scheme) color components of each pixel are considered as the first three features. The entire feature set is completed by additional spatial features, which are extracted from neighboring pixels. These features are associated with spatial image characteristics, such as min, max, entropy values, etc. The structure of the new color reduction technique allows the use of a different feature set in each level of the tree. The adaptive clustering scheme has the form of a tree with levels and nodes. According to this scheme, the color reduction technique is applied level-by-level, starting from the root node, until all the tree nodes are examined. In each node, the features (colors and spatial features) of the node are classified in m classes. Then, the pixels of each one of the m classes are further classified by the color reduction technique, using a different spatial feature set. The split and merging conditions define if the classes can be split or merged. Adjusting these conditions, the color reduction algorithm can be used as an estimator of the number of the dominant colors in the images.

In each node of the tree, the feature sets feed a self-organized neural network classifier, which consists of a PCA and a Kohonen SOFM [4]. After training, the output neurons of the SOFM define the proper feature classes. Next, each pixel is classified into one of these classes and resumes the color of the class. In order to reduce storage requirements and cut off computation time, the training set must be a representative sample of the image pixels. In our approach, the sub-sampling is performed via a fractal scanning technique, based on the Hilbert's space filling curve [11].

The proposed method was tested in a variety of images and the results are compared with other color reduction techniques. The entire technique has been implemented in a visual environment using C++ and the program can be downloaded from the site: http://ipml.ee.duth.gr/~papamark/Index.html.

2. Description of the Method

A color image could be considered as a set of $n \times m$ pixels, where the color of each pixel is a point in the color space. The proposed method can be applied to any type of color space. A color space can be considered as a 3-D vector space where each pixel (i,j) is associated with an ordered triple of color components $(c_1(i,j), c_2(i,j), c_3(i,j))$. Each primary color component represents an intensity, which varies from zero to a maximum value. Therefore, a general color image function can be defined by the relation:

$$I_k(i,j) = \begin{cases} c_1(i,j), & if \ k=1 \\ c_2(i,j), & if \ k=2 \\ c_3(i,j), & if \ k=3 \end{cases} \quad (1)$$

Let $N(i,j)$ denotes the local neighboring region of pixel (i,j). We assume that every pixel (i,j) belongs to its neighborhood $N(i,j)$. It is obvious that in most cases, the color of each pixel is associated with the colors of the neighboring pixels and the local texture of the image. Therefore, the color of each pixel (i,j) can be associated with local image characteristics extracted from the region $N(i,j)$. These characteristics can be considered as local spatial features of the image and can be helpful for the color reduction process. That is, using the values of the colors of $N(i,j)$, f_k, $k=4,5,\ldots,K+3$ local features can be defined, which are next considered as image spatial features. As it can be observed in Fig. 1, each pixel (i,j) is related to its color values, which are considered to be the first spatial features, and moreover with K additional features. No restrictions are implied for the type of the local features. However, the features must represent simple spatial characteristics, such as min, max, entropy, median values of the neighboring masks, etc.

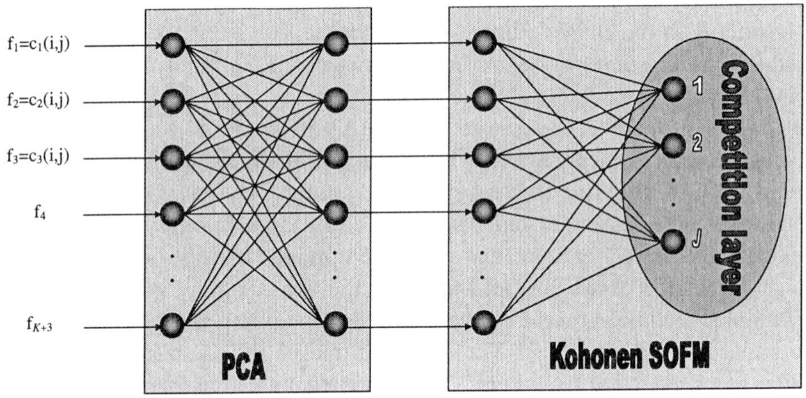

Fig. 1. The structure of the neural network classifier

According to the above analysis, the color quantization problem can be considered as the problem of best transforming the original color image to a new one, with only J colors, so that the final image to approximate not only the principal color values but the local characteristics used as well. To solve this problem, the color quantization procedure must be applied in an adaptive mode. Specifically, this procedure follows a tree structure with levels and nodes. In each level, an additional and proper set of features can be used so that new classes become visible. In each tree node, the color reduction technique is performed only on the pixels of the initial image that correspond to the color class produced in the previous level. The entire procedure is terminated when all the nodes of the tree have been examined. In the final stage, the extracted color image components are merged by a simple ADD procedure.

The classifier used in each tree node is a powerful self-organized neural network classifier with the structure shown in Fig. 1. As it can be observed, it consists of a PCA and a SOFM neural network. Through the PCA, the maximum variance in the input feature space is achieved and hence, the discrimination ability of the SOFM is increased. It is well known that the main goal of a SOFM neural network is the representation of a large set of input vectors with a smaller set of "prototype" vectors, so that a "good" approximation of the original input space can be succeeded. The resultant feature space can be viewed as a representative of the original feature space and therefore it approximates statistical characteristics of the input space.

The PCA has $K+3$ input and N output neurons. Usually, N is taken equal to $K+3$. It is used to increase the discrimination of the feature space. The SOFM has N input and J output neurons. The entire neural network is fed by the extracted features and after training, the neurons in the output competition layer of the SOFM define the J classes. Using the neural network, each image pixel is classified into one of the J classes and it converts its color to the color defined by the class.

2.1 Split and Merging Conditions

To avoid the split of compact classes, split conditions can be used during the adaptive clustering process. The split conditions define when a class must be split more. These conditions are applied locally to each tree node or to each level of the tree. Analytically, the split conditions used are

(a) The variance of each class. A class is not split further if its variance is less than a threshold value.
(b) The variance of the class centers. The classes produced in each node of the tree are not split further if the variance of their centers is less than a threshold value.
(c) The increasing of the variance. In each level of the tree, each class is accepted if its appearances increase the variance of the centers of the classes. Otherwise, the specific class will not be split further.
(d) The minimum number of pixels in each class. The minimum number of pixels classified in each class cannot be less than a threshold value.
(e) The minimum number of samples. In each node of the tree, the number of the training points, i.e. the number of pixels belonging to the node class must exceed an upper bound.

Additionally, in order to increase the clustering capabilities, merging conditions can be used in each level of the tree defining when classes are close enough and must be merged. Specifically, in each level of the tree, the Euclidean distances between the classes are determined and then the classes that have distances less than a threshold value, expressed as a percentage of the mean distance, are merged. This procedure is important because it leads to good color segmentation results by merging color classes that come from different paths of the tree.

3. Experimental Results

The proposed method has been implemented in a visual environment and available from http://ipml.ee.duth.gr/~papamark/Index.html. Due to the space limitation, only one characteristic experimental result is presented. This example demonstrates the application of the proposed method for the color reduction of the mixed-type color document shown in Fig. 2(a). The size of the original image is 1104×616 pixels with 150 dpi resolution and the number of unique colors in the image is 179473. As spatial features, the horizontal edge extraction mask of Prewitt is used:

$$f_{4+k}(i,j) = \frac{1}{3}(I_k(i-1,j+1) + I_k(i,j+1) + I_k(i+1,j+1) \\ - I_k(i-1,j-1) + I_k(i,j-1) + I_k(i+1,j-1)), \quad k = 1,2,3 \tag{2}$$

(a)

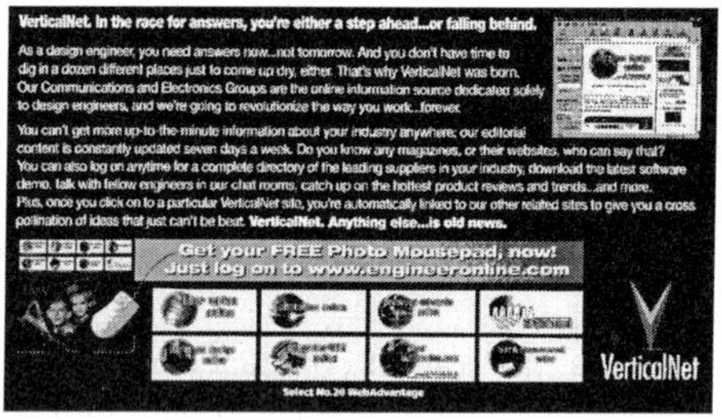

(b)

(c)

Fig. 2. (a) The original image with 179473 unique colors. (b) The document with only 4 colors. (c) The document with 4 unique color obtained using split and merging conditions.

The form of the tree is binary with two levels each one of which has two nodes. No split and merging conditions have been applied and the RGB color scheme is used. The application of the proposed technique results to the image shown in Fig. 2(b) with only the (7,21,101), (114,65,84), (232,230,237), (186,133,105) RGB colors.

The proposed adaptive color reduction technique can be used to determine the proper number of the dominant colors in an image. This can be done if we define suitable split and merging conditions. For example let as a tree of 4 levels each one with 4 nodes and the following split and merging conditions:

The split threshold value for the variance of each class is taken equal to 7.

The threshold value for the variance of class centers is equal to 12.

The minimum number of samples is equal to 20.

The minimum number of pixels in each class must be greater than 100.

The increasing of the variance condition.

The threshold value for the merging condition is taken equal to 70% of the mean distance. Using these parameters, the adaptive color reduction technique results only to the (7,21,102), (178,108,81), (228,225,232) and (28,39,102) RGB colors and the image obtained is depicted in Fig. 2(c). Comparing Fig. 2(b)&(c) we can observe that the two images have similar colors.

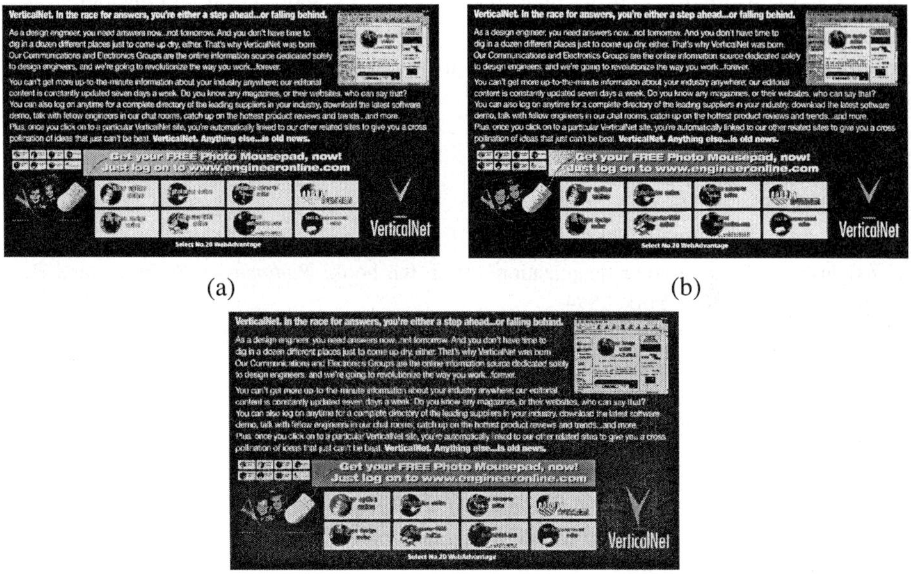

(a) (b)

(c)

Fig. 3. Color reduction using (a) median cut, (b) the Dekker's method and the (c) Paint Shop Pro 5 image processing software.

For comparison reasons the colors of the same image are reduced to four using the method of median cut [**5**], Dekker [**2**], and the Paint Shop Pro image processing software. The images obtained are shown in Fig. 3 (a), (b) and (c) respectively. As it can be optically observed, the proposed technique has better color segmentation results. The median cut algorithm provides good color results but unfortunately, the color area produced is not solid and many regions with dithering appear.

4. Conclusions

This paper proposes a general color quantization technique, which is applicable to any color image. The proposed technique is based on a neural network structure, which consists of a PCA and a Kohonen SOFM. The training set of the neural network consists of the image color values and additional spatial features extracted in the neighborhood of each pixel. The final image has not only the proper colors, but its structure approximates the local characteristics used. The proposed technique can be used with any color scheme and is suitable for drastic reduction of the colors contained in a color image.

The adaptive tree color reduction procedure is accompanied with split and merging conditions that define if the classes must be split or merged. This procedure is important because results to small number of colors. In this way, it can be considered as a technique for the detection of the number of dominant colors in an image.

In order to speed up the entire algorithm, a fractal sub-sampling procedure based on the Hilbert's space filling curve is applied.

References

1. Ashdown I., "Octree color quantization", from the book*: Radiosity-A Programmer's Perspective*, Wiley, New York, 1994.
2. A.H. Dekker, "Kohonen neural networks for optimal colour quantization", *Network: Computation in Neural Systems*, Vol. 5, pp. 351-367, 1994.
3. M. Gervautz and W. Purgathofer, "A simple method for color quantization: Octree Quantization", *in Glassner, A.S. (Ed.), Graphics Gems, Academic Press, San Diego*, pp. 287-293, 1990.
4. S. Haykin, Neural Networks: A comprehensive foundation, MacMillan College Publishing Company, N. York 1994.
5. P. Heckbert, "Color image quantization for frame buffer display", *Computer & Graphics*, Vol. 16, pp. 297-307, 1982.
6. Ing-Sheen Hsieh and Kuo-Chin Fan, "An adaptive clustering algorithm for color quantization", *Pattern Recognition letters*, Vol. 21, pp. 337-346, 2000.
7. T. Kohonen, Self-Organizing Maps, Springer-Verlag, New York, 1997.
8. N. Papamarkos, "Color reduction using local features and a SOFM neural network", *Int. Journal of Imaging Systems and Technology*, Vol. 10, No 5, pp. 404-409, 1999.
9. N. Papamarkos, C. Strouthopoulos and I. Andreadis, "Multithresholding of color and color images through a neural network technique", *Image and Vision Computing*, vol. 18, pp. 213-222, 2000.
10. N. Papamarkos and A. Atsalakis, "Gray-level reduction using local spatial features", *Computer Vision and Image Understanding*, CVIU-78, pp. 336-350, 2000.
11. H. Sagan, Space-Filling Curves, Springer-Verlag, New York, 1994.

Tunable Oscillatory Network for Visual Image Segmentation

Margarita G. Kuzmina[1], Eduard A. Manykin[2], and Irina I. Surina[2]

[1] Keldysh Inst. of Appl. Math., Miusskaya Sq. 4, 125047 Moscow, Russia
kuzmina@spp.keldysh.ru

[2] RRC Kurchatov Institute, Kurchatov sq. 1, 123182 Moscow, Russia
{edmany,surina}@isssph.kiae.ru

Abstract. Recurrent oscillatory network with tunable oscillator dynamics and nonlocal dynamical interaction has been designed. Two versions of the network model have been suggested: 3D oscillatory network of columnar architecture that reflects some image processing features inherent in the brain visual cortex, and 2D version of the model, obtained from the 3D network by proper reduction.

The developed image segmentation algorithm is based on cluster synchronization of the reduced network that is controlled by means of interaction adaptation method. Our approach provides successive separation of synchronized clusters and final decomposition of the network into a set of mutually desynchronized clusters corresponding to image fragments with different levels of brightness. The algorithm demonstrates the ability of automatic gray-level image segmentation with accurate edge detection. It also demonstrates noise reduction ability.

1 Introduction

Since the experimental discovery of synchronous oscillations in the brain visual cortex (VC) of cat and monkey [1,2] the viewpoint was expressed that synchronization-based performance is inherent in VC. Series of attempts was enterprised to elucidate the role of cortical oscillations and synchronization in visual image processing. The hypothesis on the role of dynamical link architecture was initiated since 80th. (C. von der Malsburg, 1983 ; W.Singer et al., 1988). A series of models with various types of oscillators as processing units were designed in the 90th and studied in the context of visual image segmentation problems [3-9] and contour integration task [10].

Bearing in mind a key idea of a version of image processing method that would exploit dynamical (instantly attunable) connections and synchronization similarly to VC, we designed 3D oscillatory network of columnar architecture [13,14]. Processing units of the network are oscillators formed by pairs of neurons. Internal dynamics of single oscillator is parametrically tunable by two visual image characteristics - local brightness and elementary bar orientation. Besides, there is an important internal parameter of oscillator dynamics – receptive field orientation. Resembling stimulus-dependent response of simple cells of VC, the

G. Dorffner, H. Bischof, and K. Hornik (Eds.): ICANN 2001, LNCS 2130, pp. 1013–1019, 2001.

oscillator can demonstrate either stable oscillatory activity (undamped auto-oscillations) or silence (quickly damped oscillations). Dynamical connections in the network are defined in terms of oscillator interaction. They are nonlinearly dependent both on oscillator activities and on receptive field (RF) orientations. Network performance is based on self-controlled synchronization of oscillator ensembles (clusters) encoded by visual image fragments.

3D columnar network has been further reduced to 2D network of idealized oscillator-columns, and image-segmentation algorithm based on clusterized synchronization of the reduced network has been developed. The results of computer simulations demonstrate workability of the algorithm.

2 Statement of the Problem. 3D Columnar Network

We try to imitate one step of "top-down" performance of VC: given a visual image in the retina VC provides image processing via network performance. In our case the performance consists in transfer of recurrent oscillatory network into stable state of clusterized synchronization encoded by visual image fragments.

2.1 Network Architecture

Imitating columnar structure of VC and following [10], we design oscillatory network of columnar architecture consisting of N^2 columns of K oscillators each so that $N^2 \cdot K$ is the total number of oscillators. The bases of the columns are located at the nodes of 2D square lattice G_{N^2}, whereas oscillators of each column are located at the nodes of 1D lattice L^K oriented normally with respect to the plane of G_{N^2}. Thus, the oscillators of the whole network are located at the nodes of 3D lattice $G_{N^2} \times L^K$. The state of the 3D network is defined by $(N \times N \times K)$-array of oscillator states $[u^k_{jm}]$. For each oscillator RF orientation is specified by 2D unit vector $\mathbf{n}^k_{jm}$, located in a plane normal to column direction. The vectors $\mathbf{n}^k_{jm}$ are supposed to be deterministically uniformly distributed over the column. The retina is modelled by 2D square lattice similar to G_{N^2} so that each retina node corresponds just to one oscillator column.
It is assumed that using continuous distribution of brightness $I(x, y)$ corresponding to retinal image, some discretization of $I(x, y)$ is defined in the retina lattice. For our model the matrix of pairs $(I_{jm}, \mathbf{s}_{jm}), j, m = 1, \ldots, N$, should be extracted, where I_{jm} is local value of brightness and $\mathbf{s}_{jm}$ – local orientation of image elementary bar. The $N \times N$ matrix of pairs $\mathcal{M} = [(I_{jm}, \mathbf{s}_{jm})]$ serves further as a set of tuning parameters for oscillatory network.

2.2 Single Oscillator

Following [10], we suppose that single oscillator is formed by a pair of VC neurons interconnected through excitatory and inhibitory connections. Bearing in mind biologically motivated model of such oscillator, previously proposed by W.Freeman [11] and Z.Li & J.Hopfield [12], we designed limit cycle oscillator with qualitatively similar dynamics and appropriate response to local image characteristics $(I, \mathbf{s})$. If oscillator state is defined by a pair of real-valued variables (u_1, u_2), the system of ODE for u_1, u_2 can be written in the form of single equation for complex-valued function $u = u_1 + i \cdot u_2$:

$$\dot{u} = (\rho_0^2 + i\omega - |u - c|^2)(u - c) + g(I) + q(\mathbf{s}, \mathbf{n}). \qquad (1)$$

Here ρ_0, c, ω are constants defining asymptotic parameters of the limit cycle of dynamical equation (1). The size and location of the limit cycle in the plane (u_1, u_2) are controlled by $(I, \mathbf{s})$ via properly constructed functions g and q. The following functions g and q are used in eq.(1):

$$g(I) = 1 - \mathcal{H}(I - h_0), \quad \mathcal{H}(I - h_0) = 1/(1 + e^{-2\nu(I-h_0)}), \qquad (2)$$

$$q(\mathbf{s}, \mathbf{n}) = 1 - \Gamma(|\phi|), \quad \Gamma(|\phi|) = 2e^{-\sigma|\phi|}/(1 + e^{-2\sigma|\phi|}). \qquad (3)$$

Here $\mathcal{H}(x)$ is a continuous step function depending on threshold h_0, and $\Gamma(|\phi|)$ is a narrow symmetrical peak-shaped function depending on the angle $\phi = \beta - \psi$, where β and ψ are the angles defining orientations of vectors $\mathbf{s}$ and $\mathbf{n}$ respectively: $\mathbf{s} = (\cos\beta, \sin\beta), \quad \mathbf{n} = (\cos\psi, \sin\psi)$.

The limit cycle size (radius ρ) is large enough if two conditions are satisfied simultaneously: a) I essentially exceeds the threshold value h_0; b) the angle between $\mathbf{s}$ and $\mathbf{n}$ is sufficiently small. Otherwise, either the limit cycle size is very small, or it degenerates into stable focus.

2.3 Dynamical Interaction

Dynamical system governing the dynamics of the network can be written as:

$$\dot{u}_{jm}^k = f(u_{jm}^k, \mu_{jm}^k) + S_{jm}^k; \quad j, m = 1, \ldots, N, \quad k = 1, \ldots, K. \qquad (4)$$

Here $f(u, \mu) = (\rho^2 + i\omega - |u - c|^2)(u - c) + \mu, \quad \mu_{jm}^k = g(I_{jm}) + q(|\psi_{jm}^k - \beta_{jm}|)$ are the controlling parameters for oscillator dynamics and the term S_{jm}^k specifies oscillator interaction. It has been designed in the form:

$$S_{jm}^k = \sum_{j', m', k'} W_{jj'mm'}^{kk'}(u_{jm}^k, u_{j'm'}^{k'})(u_{j'm'}^{k'} - u_{jm}^k), \qquad (5)$$

where the elements of the matrix of connections W are:

$$W_{jj'mm'}^{kk'}(u, u') = P_{jj'mm'}^{kk'}(u, u') Q_{jj'mm'}^{kk'}(\mathbf{n}, \mathbf{n}') D_{jj'mm'}^{kk'}(|\mathbf{r} - \mathbf{r}'|), \qquad (6)$$

where $\mathbf{n}$ and $\mathbf{n}'$ are RF orientations, $\mathbf{r}$ and $\mathbf{r}'$ are radius-vectors, defining spatial locations of oscillators (j, m, k) and (j', m', k') in the network. Three cofactors introduced in the expression for $W_{jj'mm'}^{kk'}$ provide the following features of dynamical interaction: a) dependence on oscillatory activity; b) dependence on RF orientations; c) dependence on spatial distance between the oscillators in the network.

The cofactors $P^{kk'}_{jj'mm'}$ providing the dependence of oscillator interaction on oscillatory activity are chosen in the form:

$$P^{kk'}_{jj'mm'} = w_0 \cdot \mathcal{H}(\rho^k_{jm}\rho^{k'}_{j'm'} - h), \tag{7}$$

where ρ^k_{jm} is limit cycle radius of the oscillator defined by indices (j, m, k), $\mathcal{H}(x)$ is a continuous step function similar to that one in eq.(2), h is a threshold, w_0 is a constant of interaction. The cofactors $Q^{kk'}_{jj'mm'}$ providing the dependence on RF orientations are defined as $Q^{kk'}_{jj'mm'} = \Gamma(|\psi^k_{jm} - \psi^{k'}_{j'm'}|)$, where $\Gamma(|\phi|)$ is introduced in eq.(3). At last, the cofactors $D^{kk'}_{jj'mm'}$, providing spatially nonlocal character of network interaction, can be defined by any function vanishing at some finite spatial distance.

3 Reduced Oscillatory Network

The 3D columnar oscillatory network can be naturally reduced to its limit version - 2D network defined in the lattice G_{N^2} and consisting of idealized oscillator-columns as processing units. The reduced network can be obtained via inter-column averaging over oscillator responses and special limit analogous to well-known thermodynamical limit in statistical physics. As a result, RF orientations disappear from consideration at all, and internal dynamics of oscillator-column, governed by eq.(1) with $q(\mathbf{s}, \mathbf{n}) \equiv 0$, depends on single image characteristics – local brightness I. The state of the network is defined by $N \times N$ matrix $\hat{u} = [u_{jm}]$, and network dynamical equations are:

$$\dot{u}_{jm} = (\rho^2 + i\omega_{jm} - |u_{jm} - c|^2)(u_{jm} - c) + g(I_{jm}) = \sum_{j',m'=1}^{N} W_{jmj'm'}(u_{j'm'} - u_{jm}), \tag{8}$$

where

$$W_{jj'mm'} = P_{jj'mm'}(u, u')D_{jj'mm'}(|\mathbf{r} - \mathbf{r}'|), \tag{9}$$

The cofactors $P_{jj'mm'}$ and $D_{jj'mm'}$ in eq.(9) are calculated in the same manner as in the case of 3D network, but the cofactor Q is now reduced to 1. We are able to consider further the reduced network as independent 2D model and develop an image segmentation algorithm based on clusterized synchronization of the network.

4 Interaction Adaptation Method. Simulation Results

We introduce new additional controlling parameter γ which will allow to improve algorithm performance due to sequential excluding of synchronized clusters from network interaction. Namely, we introduce modified dynamical connections in 2D network

$$\tilde{W}_{jj'mm'} = W_{jj'mm'}\Gamma(\gamma_{jm} - \gamma_{j'm'}), \tag{10}$$

where $\Gamma(\gamma - \gamma')$ is the function defined in eq. (3), and use the following interaction adaptation method. Let an image to be segmented be defined by piecewise constant matrix $[I_{jm}]$, where I_{jm} belong to some quantized scale of brightness $\mathcal{I} = \{I^{(l)}\}, \quad I^{(1)} > I^{(2)} > \ldots > I^{(L)}$. Put at initial step $\gamma_{jm} \equiv 0$ in the matrix of controlling parameters $\tilde{\mathcal{M}} = [(I_{jm}, \gamma_{jm})]$. We realize a process of interaction strengthening via gradual decreasing of threshold h in eq.(7), starting from sufficiently high initial value at which the network is totally desynchronized. The network demonstrates the following sequence of states at decreasing h. At some value of $h = h_1$ the first synchronized cluster, corresponding to image fragment of brightness $I^{(1)}$, arises whereas the rest of the network remains desynchronized. Such network state is conserved at $h \in [h_1', h_1], \quad h_1' < h_1$. It provides the possibility of separation of the first cluster via excluding it from network interaction. In view of eq. (10) the excluding can be achieved simply by prescribing some nonzero value $\gamma = \gamma_1, \quad \gamma_1 \geq b > 0$, to those components of the matrix $\tilde{\mathcal{M}}$ that correspond to spatial locations of oscillators belonging to the synchronized cluster. Further the process can be repeated until the second cluster will be synchronized and excluded. Finally, all the clusters corresponding to the scale $\mathcal{I}$ will be mutually desynchronized. Accurate separation of clusters corresponds to exact detection of boundaries between image fragments. This is the direct consequence of the essential attribute of oscillator dynamics – monotonic dependence of limit cycle size on image brightness. Several synthetic images were used for testing the algorithm. Three stages of algorithm performance are shown in Fig.1.

5 Summary

Recurrent oscillatory network of columnar architecture for visual image processing has been designed. Internal dynamics of single oscillator is parametrically tunable by visual image characteristics – local brightness and bar orientation. Nonlocal dynamical connections are dependent on oscillatory activities and receptive field orientations. 2D reduced oscillatory network has been extracted. Image segmentation algorithm has been developed based on interaction adaptation for 2D network. The algorithm provides successive separation of synchronized network clusters and demonstrates promising advantages: clearly visible image segmentation, accurate edge detection and noise reduction.

References

[1] Eckhorn R. Bayer R., Jordan W., Brosch M., Kruse W., Munk M., Reitboek H.J.: Biol. Cybern., **60**, (1988), 121-130.
[2] Gray C.M., Singer W.: Proc. Natl. Acad. Sci. USA, **86**, (1989), 1698-1702.
[3] Schuster H.C., Wagner P.: Biol. Cyb., **64**, (1990), 77-85.
[4] Wang D., Buhman J., von der Malsburg C.: Neur. Comp., **2**, (1990), 94-106.
[5] König P., Schillen T.B.: Neur. Comp., **3**, (1991), 155-166.
[6] Sompolinsky H., Golomb D., Kleinfeld D.: Phys. Rev. A, **43**, (1991), 6990-7011.
[7] Wang D.L., Terman D.: IEEE Trans. NN, **6**, (1995), 283-286.

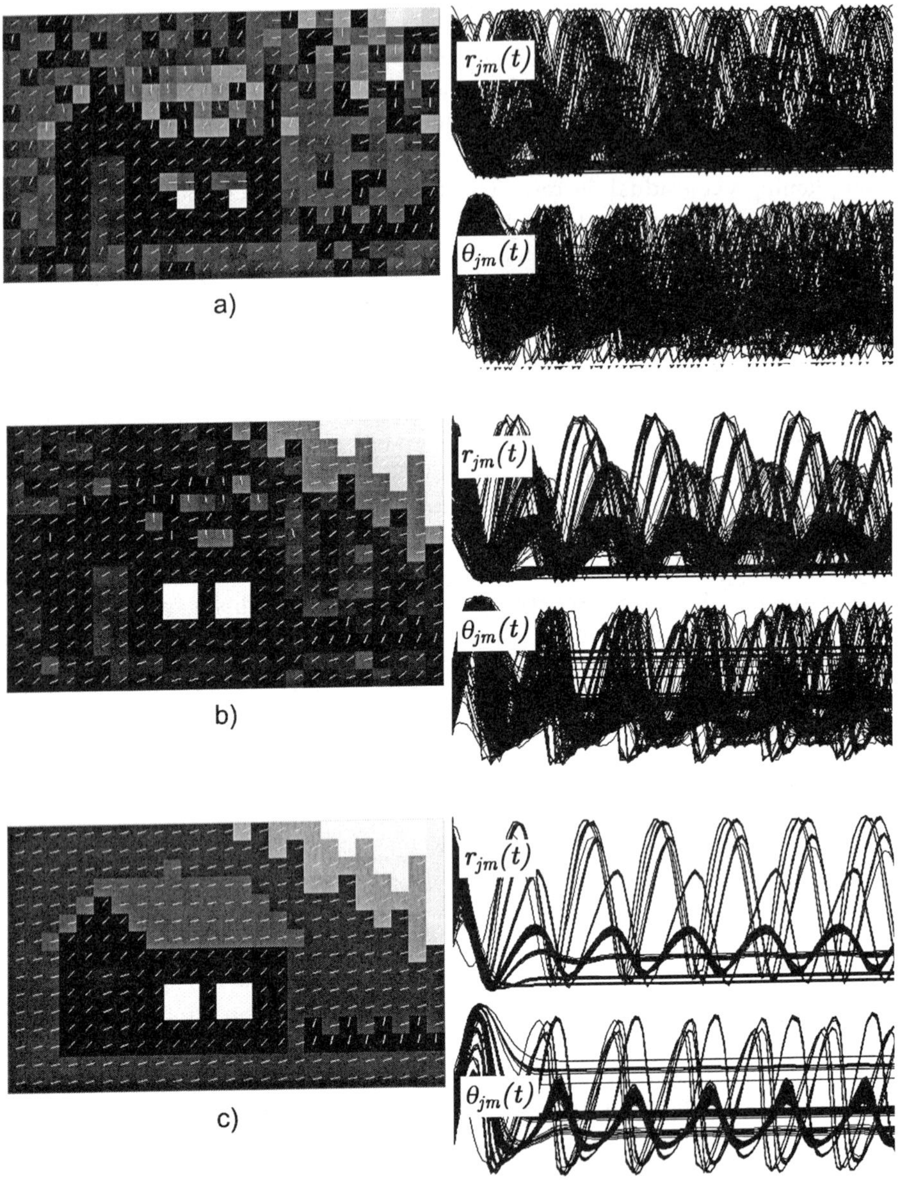

Fig. 1. The results of segmentation of synthetic image consisting of 25×15 pixels. Three states of the network arising in the process of the algorithm running are shown:
a) total desynchronization of the network ($h = 3.0$);
b) partial synchronization (three clusters, corresponding to three brightest image segments, are synchronized ($h = 0.6$));
c) complete synchronization (the clusters, corresponding to all image segments are synchronized ($h = 0.1$)).
Instantaneous network states (left) and the curves $r_{jm}(t) \equiv |u|_{jm}(t)$, $\theta_{jm}(t) \equiv \arg(u_{jm}(t))$ (right) for all oscillators are depicted.

[8] Wang D.L., Terman D.: Neur. Comp., **9**, (1997), 805-836.
[9] Chen K., Wang D.L., Liu X.: IEEE Trans. NN, **11**, (2000), 1106-1123.
[10] Li Z.: Neur. Comp., **10**, (1998), 903-940.
[11] Freeman W.J.: Electroenceph. Clin. Neurophys., **44**, (1978), 586-605.
[12] Li Z., Hopfield J.J.: Biol. Cybern., **61(5)**, (1989), 379-392.
[13] Kuzmina M.G., Surina I.I.: Proc. of AICS'99, Cork, (1999), 37-43.
[14] Kuzmina V.G., Surina I.I.: Proc. of NOLTA'2000, Dresden, Vol.**1**, (2000), 335-338.

Detecting Shot Transitions for Video Indexing with FAM

Seok-Woo Jang, Gye-Young Kim, and Hyung-Il Choi

Soongsil University, 1-1, Sangdo-5 Dong, Dong-Jak Ku, Seoul, Korea
swjang@vision.soongsil.ac.kr, gykim,hic@computing.soongsil.ac.kr

Abstract. We describe a fuzzy inference approach for detecting and classifying shot transitions in video sequences. Our approach basically extends FAM(Fuzzy Associative Memory) to detect and classify shot transitions, including cuts, fades and dissolves. We consider a set of feature values that characterize differences between two consecutive frames as input fuzzy sets, and the types of shot transitions as output fuzzy sets.

1 Introduction

The amount of digital video that is available has increased dramatically in the last few years. For automatic video content analysis and efficient browsing, it is necessary to split the sequences of video into more manageable segments. The transition of camera shot may be a good candidate for such a segmentation. Shot transitions may be classified into three major types [1]. A cut is an instantaneous transition from one scene to the next. A fade is a gradual transition between a scene and a constant image (fade out) or between a constant image and a scene (fade in). A dissolve is a gradual transition from one scene to another, in which the first scene fades out and the second scene fades in.

Many researchers, especially in multimedia community, are working on shot transition detection, and different detection techniques have been proposed and continue to appear [2-4]. But, most of the existing techniques seem to work in restricted situations and lack robustness, since they use simple decision rules like thresholding and they heavily rely on intensity based features like histograms. We believe that the detection of a shot transition intrinsically involves the nature of fuzzyness, as many transitions occur quite gradually and decision about the transitions should consider rather compound aspects of image appearance.

This paper presents a fuzzy inference approach for detecting shot transitions, which basically extends FAM(Fuzzy Associative Memory). FAM provides a framework which maps one family of fuzzy sets to another family of fuzzy sets [5]. This mapping can be viewed as a set of fuzzy rules which associate input fuzzy sets with output fuzzy sets. We consider a set of feature values that characterize differences between two consecutive frames as input fuzzy sets, and the types of shot transitions as output sets.

G. Dorffner, H. Bischof, and K. Hornik (Eds.): ICANN 2001, LNCS 2130, pp. 1020–1025, 2001.

2 Feature Set

We use HSI color model to represent a color in terms of hue, saturation and intensity. So the first step of feature extraction is to convert RGB components of a frame into HSI color representation. Our feature set contains three different types of measures on frame differencing and changes in color attributes. The first one is the correlation measure of intensities and hues between two consecutive frames. If two consecutive frames are similar in terms of the distribution of intensities and hues, it is very likely that they belong to a same scene. This feature is very easy to compute. But it works reasonably well, especially for detecting a cut. We compute area-wise correlation rather than simple differences between consecutive frames. The area-wise operation is to reduce the influence of noises, and the correlation operation is to reflect the distribution of values.

$$F_{Corr} = \alpha \times Corr(BIM_{t-1}, BIM_t) + \beta \times Corr(BHM_{t-1}, BHM_t) \quad (1)$$
$$where \quad 0 \leq F_{Corr} \leq 1, \quad 0 \prec \alpha \leq \beta \prec 1, \quad \alpha + \beta = 1$$

In (1), BIM(Block Intensity Mean) denotes the average of block intensities and BHM(Block Hue Mean) denotes the average of block hue values. The α and β are weighting factors that control the importance of related terms. We assign a higher weight to β, as hues are less sensitive to illumination than intensities are.

The second feature is to evaluate how intensities of successive frames vary in the course of time. This feature is especially useful for detecting fades. During a fade, frames have their intensities multiplied by some value of α. A fade in forces α to increase from 0 to 1, while a fade out forces α to decrease from 1 to 0. In other words, overall intensities of frames transit toward a constant. To detect such a variation, we define a ratio of overall intensity variations as in (2).

$$F_{Diff} = \frac{D_{Idiff}}{D_{IAdiff}} \quad (2)$$

$$D_{Idiff} = \frac{\sum_{i=1}^{M}\sum_{j=1}^{N}\left(I(i,j,t) - I(i,j,t-1)\right)}{K \cdot M \cdot N}$$

$$D_{IAdiff} = \frac{\sum_{i=1}^{M}\sum_{j=1}^{N}\left|I(i,j,t) - I(i,j,t-1)\right|}{K \cdot M \cdot N}$$

$$where\ k = intensity\ level,$$
$$M,\ N = frame\ height,\ width$$

The above ratio ranges from -1 to 1, revealing whether frames become bright or dark. It has negative values during a fade out and positive values during a fade in, while the magnitude of the values approaches to 1. On the other hand, the ratio remains unvaried during a normal situation.

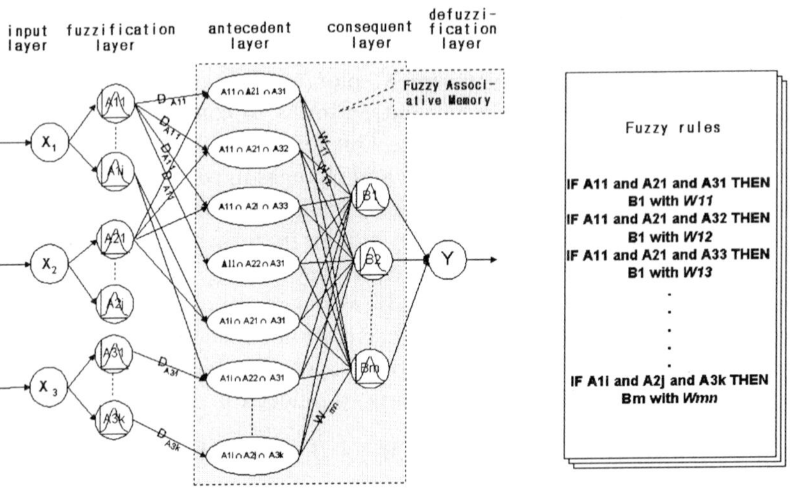

Fig. 1. Model of FAM based fuzzy inference system

The third feature evaluates the difference of differences of saturations along a sequence of frames. This feature is especially useful for detecting a dissolve, since its behavior resembles that of a laplacian(a second derivative). A laplacian usually reveals a change of the direction of variations. A dissolve occurs when one scene fades out and another scene fades in. In other words, the direction of fades switches at the time of a dissolve. Furthermore, in order to make a smooth transition at a dissolve instant, the overall saturation of frames tends to become low. Therefore, the values of (3) crosses a zero at an instant of a dissolve.

$$F_{Laplacian} = S_t - S_{t-1} \tag{3}$$

$$S_t = \frac{\sum_{i=1}^{M}\sum_{j=1}^{N}\left(S(i,j,t) - S(i,j,t-1)\right)}{K \cdot M \cdot N}$$

$$S_{t-1} = \frac{\sum_{i=1}^{M}\sum_{j=1}^{N}\left(S(i,j,t-1) - S(i,j,t-2)\right)}{K \cdot M \cdot N}$$

3 Inferring Model

The extracted feature values are fed into the inference mechanism for detecting and classifying shot transitions. We suggest a fuzzy inference system which employs FAM for implementing fuzzy rules. Figure 1 shows the structure of our inferring model which consists of five layers.

The input layer of Figure 1 just accepts input feature values. Thus, the number of nodes in the input layer becomes 3. The fuzzification layer contains membership functions of input features. The output of this layer then becomes the fit values of input to associated membership functions. The antecedent layer contains antecedent parts of fuzzy rules, which have the form of logical AND of individual fuzzy terms. We allow every possible combinations of fuzzy sets drawn one from each group of p_i fuzzy sets. The consequent layer contains consequent parts of fuzzy rules. This layer contains 5 membership functions(stay, cut, fade in, fade out, dissolve) of an output variable. We allow full connections between the antecedent layer and the consequent layer. But, each connection may have a different value of weight, which represents the degree of credibility of each connection. We basically follow the max-min compositional rule of inference [6]. Thus, when N antecedent nodes $A_1, ..., A_N$ are connected to the j-th consequent node B_j with weight w_{ij}'s, the output of the j-th consequent node becomes a fuzzy set whose membership function is defined as in (4).

$$\mu'_{B_j}(y) = min\left[\max_{1\leq i\leq N}\left\{min\big(w_{ij}, output(A_i)\big)\right\}, \mu_{B_j}(y)\right] \tag{4}$$

where $\mu_{B_j}(y)$ is a membership function contained in the j-th consequent node, and $output(A_i)$ is an output of the i-th antecedent node. The output of each consequent node has the form of fuzzy set. The defuzzification layer then combines incoming results which are in the form of fuzzy sets, and produces a final crisp conclusion. Here we use a centroidal defuzzification technique which computes the center of mass of incoming fuzzy sets [6]. That is, the final output y^* is computed as in (5).

$$y^* = \frac{\sum_{y_i} y_i \cdot \left\{\max_k\left[\mu'_{B_k}(y_i)\right]\right\}}{\sum_{y_i}\left\{\max_k\left[\mu'_{B_k}(y_i)\right]\right\}} \tag{5}$$

4 Learning Model

Our inferring model can work properly only when membership functions as well as synaptic weights are determined in advance. In this section, we propose a learning method which derives the necessary information from given input-output learning data. The first problem associated with fuzzy inference is how to divide the range of each input and output variable into how many subranges. We solve such problems by analyzing histograms. We first define a basic structure of a membership function which is a mixture of trapezoidal and sigmod functions. A membership function will then be refined later by tuning the structure to a constructed histogram. The basic structure G has five parameters as in (6), and these parameters form three basic functions; left and right sigmoid functions, and central base function with a value of 1.

$$G(x, C, L, R, M) = \begin{cases} S(x, C-L, C-L/2, C) & \text{if } x \prec C \\ 1 & \text{if } C \leq x \prec C+M \\ 1 - S(x, C+M, C+M+R/2, C+M+R) & \text{if } x \geq C+M \end{cases} \tag{6}$$

$$s(x, \alpha, \beta, \gamma) = \begin{cases} 0 & \text{if } x \prec \alpha \\ 2((x-\alpha)(\gamma-\alpha))^2 & \text{if } \alpha \leq x \prec \beta \\ 1 - 2((x-\gamma)/(\gamma-\alpha))^2 & \text{if } \beta \leq x \prec \gamma \\ 1 & \text{if } x \geq \gamma \end{cases} \tag{7}$$

In (6), x is a variable on which a membership function is to be defined, M denotes the length of a central base which has a value of 1. L and R denotes the left and right range on which a left and right sigmoid function is to be defined, respectively. One important characteristic of the structure G is that the left and right sigmoid functions as well as the central base function can be adjusted independently. We have formed prototypical membership functions for each input and output variables. We now define a measure which represents the degree of usefulness of input fuzzy sets. We use the degree of homogeneity of the output values as an indicator to usefulness of a relevant input fuzzy set.

As another important factor for determining the usefulness of fuzzy sets, we may consider the amount of separateness between adjacent fuzzy sets. When a fuzzy set is well separated from its neighboring fuzzy sets defined on the same input variable, we may say the fuzzy set is meaningful and also useful in deriving a desired conclusion. Based on the above conjectures, we define $D_{i,j}$, the degree of usefulness of the j-th fuzzy set of the i-th input feature, as in (8).

$$D_{i,j} = 1 - \left[\frac{1}{N_i - 1} \sum_{k, k \neq j} \frac{area(G_{i,j} \cap G_{i,k})}{area(G_{i,j})} \right] \times H_{i,j} \tag{8}$$

$$H_{i,j} = 1 - \frac{number\ of\ O(G_{i,j})\ in\ major\ class}{total\ number\ of\ O(G_{i,j})}$$

In (8), N_i is the number of fuzzy sets defined on the i-th input feature and $O(G_{i,j})$ denotes outputs which are associated with inputs belong to $G_{i,j}$. $D_{i,j}$ has two major components. The first component considers the amount of overlaps between $G_{i,j}$ and its neighboring membership functions. As this overlap becomes smaller, $D_{i,j}$ gets closer to a value of 1. But if this overlap becomes larger, $D_{i,j}$ gets closer to 0. The second component evaluates the homogeneity of $O(G_{i,j})$. In fact, we count the number of $O(G_{i,j})$ whose class index is a major one and divide the counted number by the total number of $O(G_{i,j})$.

5 Experimental Results and Conclusions

The proposed approach that detects shot transitions by the output of the fuzzy inference mechanism is applied to mpeg files. The total number of frames is 7814.

Table 1. Accuracy of shot transition detection

Method	cut			fade in(out)			dissolve		
	N_c	N_m	N_f	N_c	N_m	N_f	N_c	N_m	N_f
intensity histogram	55	10	7	5(4)	3(4)	3(2)	4	4	2
edge	57	8	6	7(7)	1(1)	2(3)	5	3	2
motion vector	61	4	3	5(6)	3(2)	4(3)	4	4	3
proposed method	65	0	0	8(8)	0(0)	0(0)	6	2	1

They include 65 cuts, 8 fades and 8 dissolves. We compared the performance of our approach against those of such approaches as the intensity histogram difference [2], the edge counting [3], and the motion vector detection [4]. Table 1 summarizes the comparison in terms of accuracy, where N_c denotes the number of shot transitions that are correctly detected, N_m denotes the number of misses, and N_f denotes the number of false positives. Our approach was able to detect all of cuts and fades with no false positives or misses. For the dissolve transitions, we had 2 misses and 1 false positive.

To sum up, our fuzzy inference approach seems to work as a promising solution for detecting shot transitions, even though results obviously depend on the involved features.

Acknowledgement

This work was partially supported by the KOSEF through the AITrc and BK21 program (E-0075)

References

1. Hong Heather Yu and Wyne Wolf: Multi-resolution video segmentation using wavelet transformation. Storage and Retrieval for Image and Video Databases, SPIE Vol. 3312, pp. 176-187, 1998.
2. H. J. Zhang, A. Kankanhalli, and S. W. Smoliar: Automatic partitioning of full-motion video. Proc. of ACM Multimedia Systems, Vol. 1, pp. 10-28, 1993.
3. Yi Wu and David Suter: A Comparison of Methods for Scene Change Detection in Noisy Image Sequence. First International Conference on Visual Information Systems, pp. 459-468, 1996.
4. Ramin Zabih, Justin Miller, and Kevin Mai: Feature-based algorithms for detecting and classifying scene breaks. Technical Report CS-TR-95-1530, Cornell University Computer Science Department, July, 1995.
5. Kosko B: Neural Network and Fuzzy Systems. Prentice-Hall International, 1994.
6. Zimmermann HJ: Fuzzy Set Theory and Its Applications. KALA, 1987.

Finding Faces in Cluttered Still Images with Few Examples

Jan Wieghardt and Hartmut S. Loos

Institut für Neuroinformatik,
Ruhr-Universität Bochum,
D-44780 Bochum,
Germany
{Jan.Wieghardt,Hartmut.Loos}@neuroinformatik.ruhr-uni-bochum.de
http://www.neuroinformatik.ruhr-uni-bochum.de/ini/PEOPLE/

Abstract. *Elastic graph matching* and its extension *bunch graph matching* have proven to be among the best methods for face recognition and the interpretation of facial expressions. We here demonstrate for the first time that, in combination with a simple color template, it is also an excellent means for the localization of faces in cluttered still images. The system does not need extensive learning, all information is extracted from a handful of example face images.

1 Introduction

Humans are able to detect faces even under the most difficult visual conditions. Apart from other things faces serve as the primary source of information about persons' identities and their emotional state. *Elastic graph matching* [4] and its extension *bunch graph matching* [11] have proven to be among the best methods for face recognition [8] and the interpretation of facial expressions [1, 3].

But to process a face at all, it must be located first. Most proposed solutions in computer vision either rely on motion detection to separate a person from the background (see, e.g., [5, 10]), require extensive training of specially designed neural network classifiers (see, e.g., [9]), or need a large database of faces to derive Eigenfaces for the local feature analysis (see, e.g., [7]).

We here propose a face finder, which achieves high detection rates on typical snapshots with only a few examples. To solve this complex problem we relied on two distinct visual cues – texture and color. Texture information is processed by a simplified version of *elastic bunch graph matching*. To this end the similarity between a bunch graph based on 14 example images of faces and the given image was calculated at every pixel position for four different scales.

The color cue was processed by first calculating the distance between the color value at each pixel in the image and a reference skin color value in HSI space. The correlation between the resulting skin color distance image and a primitive head template (resembling an upside down 'U') served as color similarity.

The two cues are integrated by taking the weighted average at any given scale. The resulting cue-fused images for each scale are subsequently integrated

G. Dorffner, H. Bischof, and K. Hornik (Eds.): ICANN 2001, LNCS 2130, pp. 1026–1033, 2001.

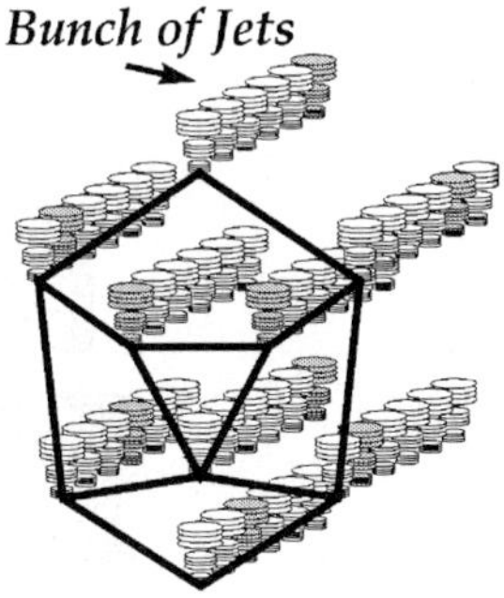

Fig. 1. Schematic view of a bunch graph: the displayed graph consists of 9 nodes (13 edges), each node labeled with 6 jets. During comparison to an image the best matching jet at each node is selected independently, here indicated by gray shading.

into one resulting similarity image by taking the maximum similarity over all scales. The local maxima of the resulting similarity image are then labeled as face hypothesis. The system performed very well on a gallery of very complex scenes described in [6].

2 Method

The proposed method for the localization of faces in complex scenes is based upon two distinct cues – texture and color. Both cues are evaluated separately and are subsequently fused into one coherent interpretation of the current visual scene.

2.1 Texture Cue

The texture cue is based upon the *bunch graph matching* algorithm, first proposed in [11], which was originally developed to solve the correspondence problem for typical mugshot pictures of faces as used in the FERET face recognition test [2].

A face is represented in terms of a labeled graph. The nodes are labeled with feature vectors, so called *jets*, containing local texture information, which is derived from example images. The edges code the topology between the individual nodes. A jet $\boldsymbol{J}$, describing a point $\boldsymbol{x}_0$ in a gray-value image $I(\boldsymbol{x})$, is given by the complex valued responses of a set of Gabor filters of the image at $\boldsymbol{x}_0$.

$$(\boldsymbol{J}(\boldsymbol{x}_0))_{\boldsymbol{k}} = \int I(\boldsymbol{x}') \frac{k^2}{\sigma^2} e^{-\frac{k^2(\boldsymbol{x}_0-\boldsymbol{x}')^2}{2\sigma^2}} \left[e^{i\boldsymbol{k}^T(\boldsymbol{x}_0-\boldsymbol{x}')} - e^{-\frac{\sigma^2}{2}} \right] d^2x' \qquad (1)$$

Two *jets* are compared via the similarity function s_{abs}, which is basically the normalized scalar product between the absolute values $a_{\boldsymbol{k}} = \|(\boldsymbol{J})_{\boldsymbol{k}}\|$ of the complex components of the two jets, i.e.

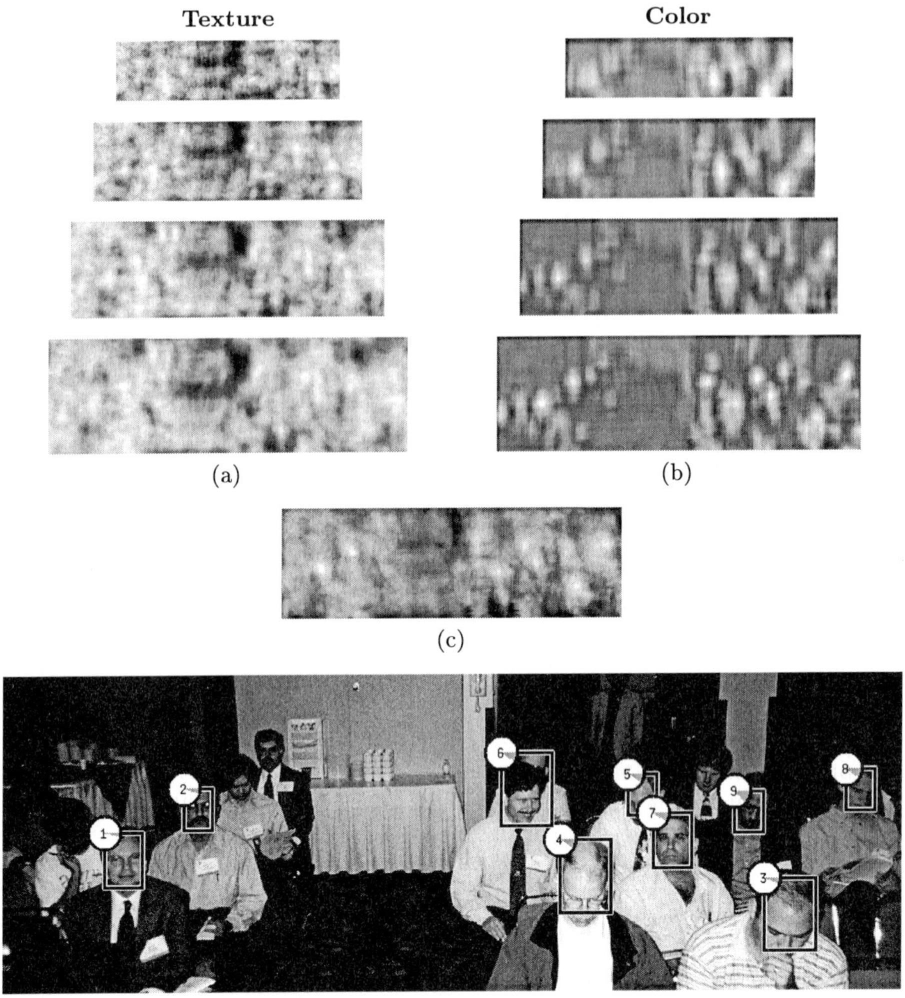

Fig. 2. **(a)** shows the similarity images of the texture cue and **(b)** of the color cue for all four scales. **(c)** represents the final similarity of the example image displayed below. In the upper left corner of each facial hypothesis in the example image the corresponding similarity value (as pie chart) and its relative ranking is displayed. The resolution of the image is 852 x 292 pixels.

$$s = \frac{\sum_k a_k a'_k}{\sqrt{\sum_k {a_k}^2 \sum_k {a'_k}^2}} \tag{2}$$

For the localization of faces four *bunch graphs* are employed. Each containing 16 nodes (33 edges) and each node being labeled with 14 jets, taken from 14 different example images (see figure 1). The relative position of the nodes for each bunch graph are given by the mean positions over all examples. The four bunch graphs differ only in scale (by factors of $\sqrt{2}$) and are derived from scaled versions of the same example images.

Given an input image the bunch graphs are separately scanned across the image and the similarity between the bunch graph and the image is calculated at each sample point. The bunch graph similarity is given by the mean similarity over all nodes, whereas the similarity of a node is given by the maximum of all similarities between the jet derived from the image at a given position, and all jets the node is labeled with.

This way four different similarity images are created, one for each bunch graph (see figure 2a). At last the similarities are normalized such that the final similarity $s_{\boldsymbol{x},s}^{texture}$ at a pixel position $\boldsymbol{x}$ at a scale s is given by

$$s_{\boldsymbol{x},s}^{texture} = \frac{s_{\boldsymbol{x},s} - \min_{\boldsymbol{x},s}\{s_{\boldsymbol{x},s}\}}{\max_{\boldsymbol{x},s}\{s_{\boldsymbol{x},s}\} - \min_{\boldsymbol{x},s}\{s_{\boldsymbol{x},s}\}} \tag{3}$$

2.2 Color Cue

The color cue is processed in a very similar fashion . First a primitive skin color distance is calculated in *HSI* color space as follows:

$$d_{color} = \sqrt{\Delta h^2 + \frac{1}{w_{sh}} \Delta s^2} \tag{4}$$

Δh and Δs are differences between the skin color prototype and the current pixel in the hue and saturation component, respectively. The factor $\frac{1}{w_{sh}}$ allows to assign less weight to the saturation, as it is in general more susceptible to changes in illumination and different pigmentations of different skin types.

Finally, the correlation between the resulting skin color distance image and an approximately head shaped template filter (an upside down 'U') is calculated. This correlation was measured at the four scales separately (see figure 2b). The template is given by the sum of three rectangular filters

$$C(\boldsymbol{x}) = -2r(\boldsymbol{x}, x_{bg}, y_{bg}) + r\left(\boldsymbol{x} - \begin{pmatrix} 0 \\ \frac{9}{8} y_{bg} \end{pmatrix}, x_{bg}, \frac{1}{4} y_{bg}\right) + r\left(\boldsymbol{x}, \frac{3}{2} x_{bg}, y_{bg}\right) \tag{5}$$

$$r(\boldsymbol{x}, w, h) = \begin{cases} 1 & \text{if } |x_1| \leq w \wedge |x_2| \leq h \\ 0 & \text{else} \end{cases} \tag{6}$$

x_{bg} and y_{bg} are the width and the height of the enclosing rectangle of the bunch graph associated with a given scale. The resulting similarity images are transformed in the same way as for the texture cue (see (3)).

2.3 Cue Fusion

The cues are fused by taking the weighted average similarity between the texture and the color cue at each pixel of the similarity image for each scale separately. The resulting average similarity images for each scale are integrated into the

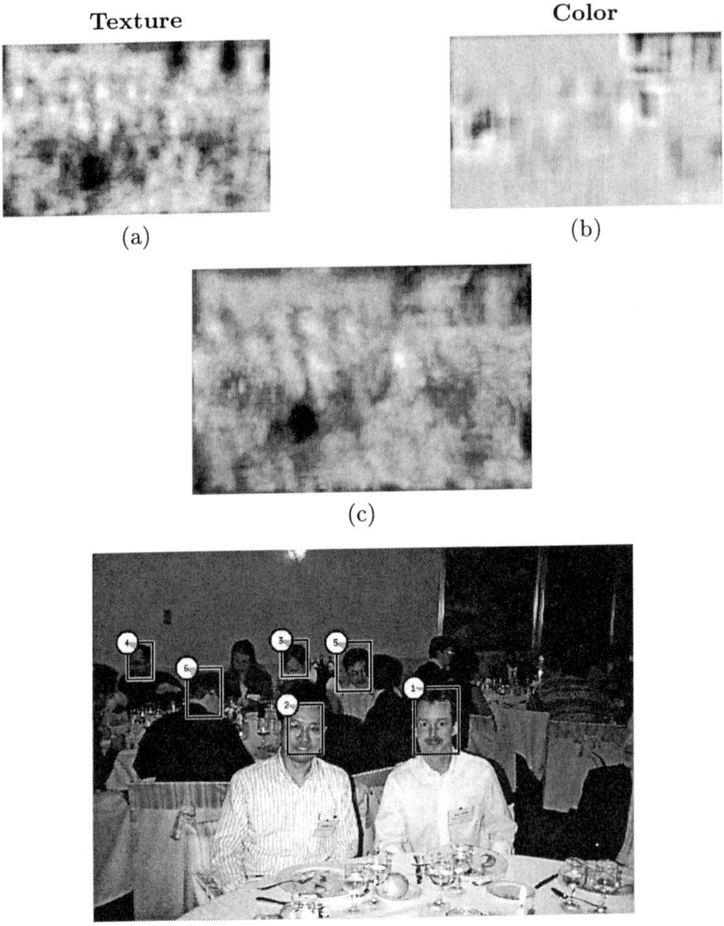

Fig. 3. **(a)** shows the similarity image of the texture cue and **(b)** of the color cue for one scale. **(c)** represents the final similarity of the example image displayed below. In the upper left corner of each facial hypothesis in the example image the corresponding similarity value (as pie chart) and its relative ranking is displayed. The resolution of the image is 756 x 504 pixels.

final similarity image by taking the maximum similarity at each pixel over all scales (see figure 2c), i.e.

$$s_{\boldsymbol{x}}^{fused} = \max_{s} \left\{ 0.3\, s_{\boldsymbol{x},s}^{color} + 0.7\, s_{\boldsymbol{x},s}^{texture} \right\} . \tag{7}$$

It must be noted that due to the normalization of similarities within each cue the resulting averaged similarities are not an absolute measure of similarity in relation to some template but rather a measure of consistency between the cues.

2.4 Hypothesis Selection

From the resulting similarity image $s_{\boldsymbol{x}}^{fused}$ face *hypotheses* are extracted, i.e. those regions, which are believed to contain faces, by calculating the local maxima. From the set of local maxima retrieved from the similarity image those are discarded, whose corresponding image region (the graphs enclosing rectangle) overlaps by more than 40% of its surface with an other face hypothesis of higher similarity. This way multiple detection of a single face is avoided.

3 Results

The algorithm was tested on an image database compiled by *Kodak* [6] showing person engaged in all kinds of activities. It must be noted that the database did not contain simple portraits as for example the often used FERET database [2, 8]. Furthermore, the 14 example images of faces used to create the four scaled bunch graphs are completely unrelated to the database. These greyscale images show only frontal faces of different persons of the same ethnic group without any illumination variation.

To evaluate the quality of the resulting hypothesis regions all faces were manually labeled with a binarized amount of rotation in depth (more or less than 50°) and whether or not they were partially occluded. The table below shows the average detection rate in percent for three categories of decreasing difficulty. Moreover it shows the percentage of faces which were detected and were ranked higher in terms of overall similarity than any of the flawed hypotheses in the same picture.

Face Category	Detected Faces	Faces Ranked Before Non-Faces
All faces	83 %	72 %
Faces rotated up to 50° in depth	87 %	83 %
Unoccluded faces rotated up to 50° in depth	89 %	80 %

Although the employed similarity measure was initially only developed in order to rank the hypothesis regions according to their likelihood of containing a face, it was also tested whether a global threshold could be found, which allows to separate correct classifications from false positives. The table below shows the *equal-error rate*, i.e. the rate of correct classification for the similarity threshold value, which yields as many false positives as undetected faces. It also shows the detection rate achievable without false positives.

Face Category	Equal-Error Rate	No-False-Detection Rate
All faces	68 %	22 %
Faces rotated up to 50° in depth	68 %	28 %
Unoccluded faces rotated up to 50° in depth	70 %	28 %

Fig. 4. A few examples from the face finding process. In all images the best facial hypotheses with their relative ranking are displayed. The resolution of the images in the upper and lower part is 756 x 504 and 400 x 600 pixels respectively.

4 Discussion

Taking the extreme difficulty of the completely unconstrained image database into account, the results are satisfying and comparable to other state-of-the-art face detection algorithms. However the number of labeled example images required to bootstrap the system is at least two orders of magnitude lower. The system performed especially well in ranking image regions according to their likelihood of containing a face. The rejection of flawed hypotheses proved somewhat more challenging. In order to achieve even better results regarding the rejection of false positives further experiments were carried out where the hypothesis regions were subjected to a more detailed bunch graph matching as described in [11]. Using the more detailed bunch graph (48 nodes, 38 example frontal faces completely unrelated to the database) and a more elaborate matching the *equal-error rate* and *no-false-detection rate* were increased by 10 to 15 percentage points.

Acknowledgments

This work has been partially funded by grants from KODAK, the German Federal Ministry for Education and Research (01 IB 001 D) and the European Union (RTN MUHCI). Our special thanks go to Alexander Loui, KODAK for providing the challenging image database and Dr. Rolf Würtz for prove reading.

References

1. Gianluca Donato, Marian Stewart Bartlett, Joseph C. Hager, Paul Ekman, and Terrence J. Sejnowski. Classifying facial actions. *IEEE Transactions on Pattern Analysis and Machine Intelligence*, 21(10):974–989, 1999.
2. J. Phillips et al. The FERET evaluation methodology for face-recognition algorithms. *National Institute of Standards and Technology (NIST)*, 1998.
3. Hai Hong, H. Neven, and C. v. d. Malsburg. Online facial expression recognition based on personalized galleries. In *Intl. Conference on Automatic Face and Gesture Recognition*, pages 354–359, Nara, Japan, April 14–16 1998. IEEE Comp. Soc.
4. M. Lades, J. C. Vorbrüggen, J. Buhmann, J. Lange, C. von der Malsburg, R. P. Würtz, and W. Konen. Distortion invariant object recognition in the dynamic link architecture. *IEEE Transactions on Computers*, 42:300–311, 1993.
5. Hartmut S. Loos. Suchbilder: Computer erkennt Personen in Echtzeit. *c't: magazin für computer technik*, 15:128–131, 2000.
6. Alexander C. Loui, Charles N. Judice, and Sheng Liu. An image database for benchmarking of automatic face detection and recognition algorithms. In *Proceedings of the IEEE International Conference on Image Processing*, Chicago, Illinois, USA, October 4–7 1998.
7. Penio S. Penev and J. J. Atick. Local feature analysis: A general statistical theory for object representation. *Network: Computation in Neural Systems*, 7(3):477–500, 1996.
8. P. Jonathon Phillips, Hyeonjoon Moon, Syed A. Rizvi, and Patrick J. Rauss. The feret evaluation methodology for face-recognition algorithms. *IEEE Transactions on Pattern Analysis and Machine Intelligence*, 22(10):1090–1103, 2000.
9. H.A. Rowley, S. Baluja, and T. Kanade. Neural network-based face detection. *IEEE Transactions on Pattern Analysis and Machine Intelligence*, 20(1):23–38, 1998.
10. J. Steffens, E. Elagin, and H. Neven. Personspotter – A Fast and Robust System for Human Detection, Tracking and Recognition. In *Intl. Conference on Face and Gesture Recognition*, pages 516–521, 1998.
11. Laurenz Wiskott, Jean-Marc Fellous, Norbert Krüger, and Christoph von der Malsburg. Face recognition by elastic bunch graph matching. *IEEE Transactions on Pattern Analysis and Machine Intelligence*, 19(7):775–779, 1997.

Description of Dynamic Structured Scenes by a SOM/ARSOM Hierarchy

Antonio Chella[1,2], Maria Donatella Guarino[2], and Roberto Pirrone[1,2]

[1] DIAI, University of Palermo, Viale delle Scienze,
I-90128 Palermo, Italy
{chella,pirrone}@unipa.it
[2] CERE-CNR, National Research Council,
Viale delle Scienze,
I-90128 Palermo, Italy
guarino@cere.pa.cnr.it

Abstract. A neural architecture is presented, aimed to describe the dynamic evolution of complex structures inside a video sequence. The proposed system is arranged as a tree of self-organizing maps. Leaf nodes are implemented by ARSOM networks as a way to code dynamic inputs, while classical SOM's are used to implement the upper levels of the hierarchy. Depending on the application domain, inputs are made by suitable low level features extracted frame by frame of the sequence. Theoretical foundations of the architecture are reported along with a detailed outline of its structure, and encouraging experimental results.

1 Introduction

Symbolic description of the objects inside a scene, and of their dynamic behavior is a crucial task in whatever vision system. Such a goal can be accomplished, in general, only using some a priori knowledge about the scene context along with various assumptions on the object models, and on their geometrical or physical constraints. We present a novel multi-layer neural architecture for dynamic scenes description where structural relations between the various elements are coded into the network topology. The proposed system has been developed as a part of the architecture prototype for the control of the ASI SPIDER robotic arm that will operate on the International Space Station [11]. The network is arranged to provide a symbolic description of the robot motion from a live video sequence taken by a fixed camera looking at the robot workspace. This description is in form of short assertions regarding the motion direction of the whole arm and of its components. Phrases are obtained after a labelling step performed over the units of the trained network. Inside the general framework of the robotic control architecture developed by some of the authors [1] the proposed system can be regarded as a way to implement the conceptual level: our architecture allows us to fill the gap between low level perceptual data, and their symbolic representation. This in turn can be useful to enrich the knowledge base defined

G. Dorffner, H. Bischof, and K. Hornik (Eds.): ICANN 2001, LNCS 2130, pp. 1034–1041, 2001.

for the application domain, and to guide planning activity and new sensing actions. The architecture consists in a tree of self-organizing maps in order to encode the arm topology. Leaf nodes provide a description of the motion for single joints. This layer is made by ARSOM networks [8] to process dynamic image features extracted from the video input to describe joints motion. At the intermediate level, three SOM networks are used to provide motion description for different groups of joints belonging to the shoulder, the elbow, and the wrist/hand components respectively. The top most SOM is used to encode the motion description of the whole arm starting from the descriptions obtained for its components. Several SOM-based multi-layer networks have been developed by various researchers, but our model is the only one dealing with dynamic inputs, and using a feed-forward scheme to propagate information along the maps hierarchy. In our model a vector representation of the internal state of each map is computed at each training step and is propagated from the child maps to the parent one, starting from the leaves of the tree structure. The most known multi-layer SOM-like networks are the WEBSOM and the PicSOM models [5], [7]. In the WEBSOM case the first SOM (the word category map) is used to obtain an off-line coding for the input pattern of the second one (the document map) so the two SOM are completely independent from each other. In the PicSOM system the T S-SOM multi-layer architecture [6] is used to speed up the search in the image database. In this model the lower layers maps can be regarded as a multi-resolution representation of the clusters that are present in the top most map. In this way is possible to face the problem of obtaining clusters with very complex shapes in the feature space: each map at the lower level is feeded with a subset of the global input patterns that correspond all the inputs mapped into a cluster of the top most map. Many other architectures have been developed that use multi-layer SOM [12], [10], [3], but all of them are very similar to the TS-SOM approach. In the HOSOM network [14] successive layers are dynamically grown starting from overlapping neighbourhoods of the root map, encoding data with some redundancy and increasing classification performance. The HSOM network [9]is a multi-layer SOM where the index value of the best match unit of the lower level map is propagated in a feed-forward scheme to the upper level one. This training algorithm allows to obtain arbitrary cluster shapes. The rest of the paper is arranged as follows. In section 2 theoretical motivations for the choice of the architecture topology are explained. Section 3 is devoted to the detailed system description. In section 4 experimental results are reported, while in section 5 conclusions are drawn.

2 Theoretical Remarks

The main goal of the proposed system is to provide a symbolic interpretation for the dynamic evolution of a pattern of image features, making use of the a priori knowledge about the structure underlying the features themselves in the application domain. Due to the nature of the problem a hybrid architecture has been selected where the subsymbolic part provides a suitable coding for the input data, while the symbolic one is merely a labelling performed on the network activations pattern accordingly to our knowledge about the specific domain. The particular choice of the network topology is mainly inspired to the work that Gori and his associates [2][13] proposed as a general framework to build adaptive models for the symbolic processing of data structures. In this works, the authors focus their attention on a particular type of structure namely the directed ordered acyclic graph (DOAG). A DOAG D is a directed acyclic graph (DAG) with a vertex set $\mathrm{vert}(D)$ and an edge set $\mathrm{edg}(D)$ where a total order relation is defined on the edges leaving from a vertex $v \in \mathrm{vert}(D)$. The class of all DOAG's having at most i edges incoming in a vertex, and at most o edges leaving from it, is denoted by $\#^{(i,o)}$. Structured information can be regarded as a labelled DOAG. We'll call $\mathbf{Y}_v$ the label attached to a vertex v. We'll denote with $\mathcal{Y}$ the general space of the labels, and a structured space with structure in $\#^{(i,o)}$ and labels in $\mathcal{Y}$ is denoted as $\mathcal{Y}^{\#^{(i,o)}}$. We'll call structural transduction $\tau(\cdot)$ a relation that is defined on $\mathcal{U}^{\#} \times \mathcal{Y}^{\#}$ where $\mathcal{U}^{\#}$ and $\mathcal{Y}^{\#}$ are two generic structured spaces. A transduction $\tau(\cdot)$ is *IO-isomorph* if the skeleton of the input DOAG is preserved that is:

$$\mathrm{skel}(\tau(\mathbf{U})) = \mathrm{skel}(\mathbf{U}) \ \forall \mathbf{U} \in \mathcal{U}^{\#}$$

A transduction admits a recursive state representation if there exist a structured space $\mathbf{X} \in \mathcal{X}^{\#}$ with $\mathrm{skel}(\mathbf{X}) = \mathrm{skel}(\mathbf{U}) = \mathrm{skel}(\mathbf{Y})$ and two functions defined as follows for each $v \in \mathrm{skel}(\mathbf{U})$ (here the notation $\mathrm{ch}[v]$ refers to the set of the children of vertex v).

$$\mathbf{X}_v = f(\mathbf{X}_{\mathrm{ch}[v]}, \mathbf{U}_v, v) \tag{1}$$

$$\mathbf{Y}_v = g(\mathbf{X}_v, \mathbf{U}_v, v) \tag{2}$$

3 Description of the System

The previous considerations about processing of structured data led us to develop our architecture as a tree of SOM/ARSOM networks. The tree structure fits to the topological structure of the robotic arm: at the leaves layer the motion of each joint will be described, while at the upper levels a description will be provided for the motion of groups of joints, and for the whole arm. The main difference with the approach proposed in [2] is the use of an unsupervised learning scheme by means of self-organizing maps that have been adopted as a natural way to classify different motions. On the other hand, smoothing between the clusters of trained units is in turn useful to help labelling: in fact complex and more descriptive labels are generated from primitive ones in correspondence of all those units that are not so close to cluster centers. The system performs as explained in the following. At the first level, the system processes the sequences obtained computing, for each frame, 2D projections of the arm joints onto the image plane $\mathbf{q}_i(t) = [x_i(t), y_i(t)]^T$ with $i = 0, \ldots, 4$. We consider only the last five joints even if the robot arm has 6 d.o.f. because the first one is projected always in a fixed position, and we'll use it as a reference point to track the others.

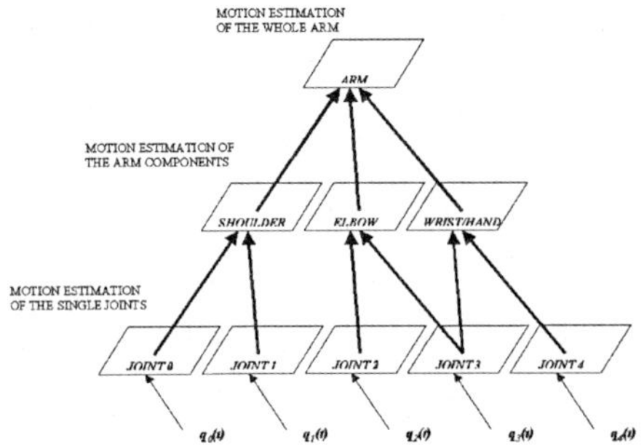

Fig. 1. An outline of the proposed system: Temporal sequences of the joints projections onto the image plane are feeded to the leaf networks. Status is propagated from each set of children to their parent, and every layer is responsible for the motion description of the arm at different resolution.

The leaf networks provide a direct classification of the 2D motion in terms of the clustering they perform on the input data. Information is propagated between the layers using a vector representation of the error distribution over the units of a map: this is the prediction error in the case of the ARSOM layer, and the matching error in case of the SOM layers. This kind of data do not depend directly on time, so we can use simple SOM networks to classify them. In figure 1 is depicted a simple scheme of the architecture. In what follows, we'll proof that is possible to determine an IO-isomorph, causal structural transduction with a recursive state representation between the input feature space and the space of the symbolic output labels. In particular, we'll show that the error distribution over the units can be regarded as the internal state of the system. The input and output spaces have a $\#^{(1,2)}$ structure. Labels for input leaf nodes are given by $\mathbf{U}_i = \{\mathbf{q}_i(t), \mathbf{q}_i(t-1), \ldots, \mathbf{q}_i(t-M)\}$ at the generic time instant t, while the generic SOM node label $\mathbf{U}_s$ consists in the name of the corresponding robotic arm component. The labels $\mathbf{U}_s$ are mapped directly onto the output space, so we don't consider any extern input to the intermediate levels apart from the connections transmitting the state: this is equivalent to set $\mathbf{U}_s = \mathbf{0}\ \forall s$. The previous assertion imply that the transduction is IO-isomorph: the output space has the same tree structure with textual labels $\mathbf{Y}_v$ explaining the motion of each arm part. We'll show that the transduction admits a recursive state representation, and therefore it's causal. If we consider a generic ARSOM node, from [8] we can write for each unit k:

$$\mathbf{q}_i(t) = w_{i1}^{(k)}\mathbf{q}_i(t-1) + \ldots + w_{iM}^{(k)}\mathbf{q}_i(t-M) + \mathbf{e}_i^{(k)}$$

where $\mathbf{e}_i^{(k)}$ is the approximation error that can be rewritten as:

$$\mathbf{e}_i^{(k)} = \mathbf{q}_i(t) - \mathbf{Q}_i^T \tilde{\mathbf{w}}_i^{(k)}$$

here $\tilde{\mathbf{w}}_i^{(k)}$ is the weights vector, while

$$\mathbf{Q}_i = \begin{bmatrix} x_{i-1}(t) & y_{i-1}(t) \\ \vdots & \vdots \\ x_i(t-M) & y_i(t-M) \end{bmatrix}$$

We can rewrite the label attached to the node:

$$\mathbf{U}_i = \begin{bmatrix} x_i(t) & y_i(t) \\ \vdots & \vdots \\ x_i(t-M) & y_i(t-M) \end{bmatrix}$$

then it's possible to express the error as:

$$\mathbf{e}_i^{(k)} = \mathbf{U}_i^T \mathbf{w}_i^{(k)}$$

where $\mathbf{w}_i^{(k)} = [1| - \tilde{\mathbf{w}}_i^{(k)}]^T$. So the expression of the overall error can be written like in (1):

$$\mathbf{E}_i = f(\mathbf{0}, \mathbf{U}_i, \mathbf{W}_i) \tag{3}$$

In the previous equation $\mathbf{E}_i = [\mathbf{e}_i^{(k)}]_{k=1}^L$ is the overall error, while $\mathbf{W}_i = [\mathbf{w}_i^{(k)}]_{k=1}^L$ is the weight matrix, and is the node parametric representation as stated in [2]. It must be noted that we've set the initial state to **0** because it makes no sense to provide initial error values. We provide null initial error that corresponds to the casual distribution of the weights when there is no applied input. For the generic SOM node s, we can write from [4] for each unit k at the training step n:

$$\mathbf{w}_s^{(k)}(n) = \mathbf{w}_s^{(k)}(n-1) + \gamma h_{bk}(\mathbf{e}_{\mathrm{ch}[s]} - \mathbf{w}_s^{(k)}(n-1)) \tag{4}$$

$$b \, : \, \|\mathbf{e}_{\mathrm{ch}[s]} - \mathbf{w}_s^{(b)}\| = \min_k \|\mathbf{e}_{\mathrm{ch}[s]} - \mathbf{w}_s^{(k)}\| \tag{5}$$

here γ is the gain factor, h_{bk} is the neighbourhood coefficient, while b is the best matching unit index, and $\mathbf{e}_{\mathrm{ch}[s]}$ is the input vector that correspond to error computed for all the children of node s. The matching error for the unit k is defined by:

$$\mathbf{e}_s^{(k)} = \mathbf{e}_{\mathrm{ch}[s]} - \mathbf{w}_s^{(k)}(n-1) = \frac{1}{\gamma h_{bk}}(\mathbf{w}_s^{(k)}(n) - \mathbf{w}_s^{(k)}(n-1)) \tag{6}$$

If we rewrite the (6) expanding $\mathbf{w}_s^{(k)}(n)$ according to (4) we obtain:

$$\mathbf{e}_s^{(k)} = \mathbf{e}_{\mathrm{ch}[s]} + \frac{1 - \gamma h_{bk}}{\gamma h_{bk}} \mathbf{w}_s^{(k)}(n-1)$$

Then we can write the expression of the overall error in the same way of (1):

$$\mathbf{E}_s = f(\mathbf{e}_{\mathrm{ch}[s]}, \mathbf{U}_s, \mathbf{W}_s) \tag{7}$$

In (7) $\mathbf{E}_s$ and $\mathbf{W}_s$ have the same meaning of the corresponding quantities in (3), while $\mathbf{U}_s$ has been set to the null vector. The state of the system, we've just defined has the same structure as the input and output spaces: every map in the hierarchy owns an error matrix **E**. The output function g defined in (2) is implemented by means of the labelling process. Using the training episodes, labels are attached to the winner units at each layer of the network. Each label corresponds to the arm component motion description for the particular training episode. The training episodes span a set of possible basic arm motions like rolling, tilt toward the observer, turn left and right, and so on. Not all units receive a label from this process: complex labels are attached to unlabelled units during the test. These labels are formed as a composition of primitive ones that is weighted with the distance of the selected unit from the closer labelled ones in a certain neighbourhood. It can be noted that labelling depends from the units activation that, in turn, is related to the state as in (2).

4 Experimental Results

The proposed architecture has been developed as a monitoring system for the operations of the ASI SPIDER robotic arm. The environment is the outside surface of a module in the International Space Station where the robot arm will be installed to verify and/or replace those portions of the hull that could be damaged by external agents. In our lab we couldn't acquire video sequences of the true arm because of its dimensions, so we have set up our experiments using a model of the real arm made with the ROBIX system (see figure 2). The model has been built with the same joint arrangement as the original one: so it can span a similar workspace. Feature points are extracted as the intersections between the lines corresponding to the projection of the links symmetry axes, when they are visible. In some undecidable cases, like the one in which the arm projects itself as a unique line, we used our knowledge about the arm kinematics, and the projection geometry of the camera in order to estimate the joints positions. In the general case, the input image is blurred with a gaussian filter to reduce noise, then a binary image is obtained subtracting the fixed background image and applying a threshold. The resulting blob suffers from contour imperfections, and presence of holes in the inner region. So a morphological sequence of opening and closure operations is applied to obtain a more regular and connected shape. In the opening step we used a squared structural element 5 pixels wide, while its size was 7 pixels for the closure one. The next step consists in the skeleton determination: in this way we obtain a first outline of the arm, but this is not yet a collection of line segments. To perform line approximation of the skeleton we used a modified version of the Hough transform. At first, the classical Hough transform is applied to the skeleton, then we search the Hough space in order to find clusters of cells with the same slope parameter, but slightly differing in the intercept value. All the cells in the cluster are collapsed into the one with the highest number of accumulated skeleton points: their accumulation values are collected into the winner cell, while the other ones are erased. The algorithm is repeated until a number of segments equal to the arm links is detected, or it's impossible to find other lines. In figure 2 the various preprocessing steps are depicted. Starting from

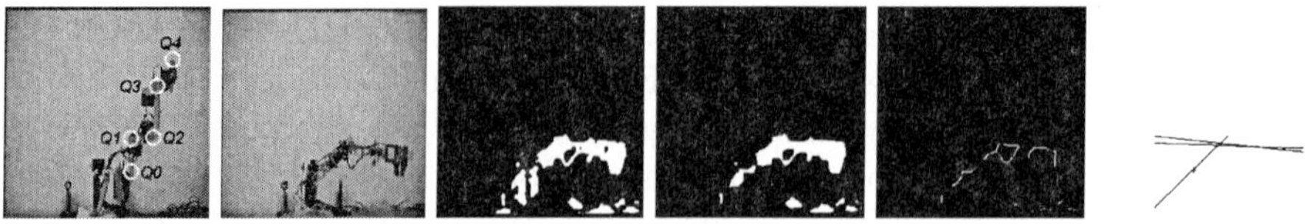

Fig. 2. The ROBIX arm, and the feature points extraction process. From left to right: ROBIX with the feature points outlined, a generic frame to start feature extraction, the difference image, regularized difference image after morphological operation, skeletonisation, line extraction.

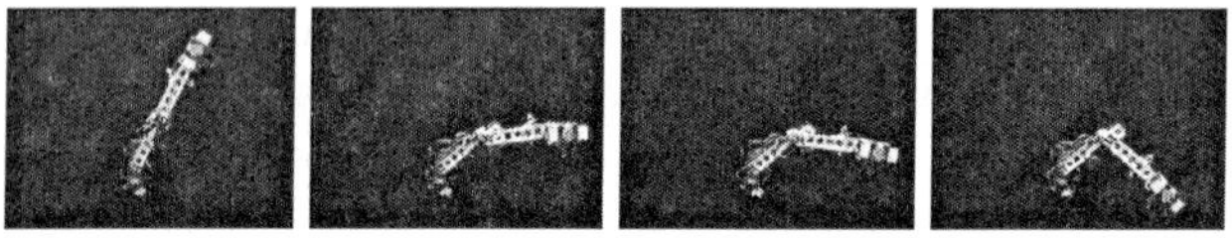

```
frame 1
q0     rotate right      | shoulder   go_down_and_rotate right
q1     rotate right      | elbow      go_down right
q2     go_down right     | hand       go_down right
q3     go_down right     | arm        go_down right
q4     go_down right     |
```

Fig. 3. A few samples from a test episode along with the textual output of the system.

the lines intersections a simple decision scheme is implemented to determine the joints projections that makes use of our assumptions about the arm kinematics, and the projection geometry. The training episodes have been set up from 30 frames sequences regarding some basic movements of the robot arm: turning left, turning right, turn left or right and tilt toward the observer, go down to the left or right with all the arm extended. All the inverse movements have been also used in the training phase. After various trials we obtained a hierarchy with maps 6×6 units wide, a learning rate of 0.1 for the ARSOM's and 0.6 for the SOM's. The neighbourhood initial size was 6 units for all the networks. We performed training in 600 epochs, and tested the system with 10 episodes obtained by partial composition of the training movements. In figure 3 some samples from a test episode are depicted together with the textual output of the system.

5 Conclusions

A novel hybrid approach for the symbolic description of dynamic structured scenes have been presented that can be used in several application domains where some a priori knowledge has been provided about the structure and the relations between the parts of the scene itself. Knowledge about the domain is used to guide the process of building the system structure. We stated that the proposed architecture is able to learn in an unsupervised way complex mappings between structured data spaces. We motivated our theoretical choices and, in particular, it has been proved that the feed-forward propagation scheme of the error pattern can be regarded as way to represent the internal state of the system so it's able to learn structural transductions. The proposed system has to be studied more in detail, and an extensive experimental phase is needed to fully evaluate its performance. In particular, we want to investigate the case of complex transductions between generic DOAG structures, and to experiment the system domains with a large amount of low level data like content based video indexing for multimedia databases.

References

1. A. Chella, M. Frixione, and S. Gaglio. A Cognitive Architecture for Artificial Vision. *Artificial Intelligence*, 89:73–111, 1997.
2. P. Frasconi, M. Gori, and A. Sperduti. A General Framework for Adaptive Processing of Data Structures. *IEEE Trans. on Neural Networks*, 9(5):768–786, September 1998.
3. M. Hermann, R. Der, and G. Balzuweit. Hierarchical Feature Maps and Nonlinear Component Analysis. In *Proc. of ICNN'96*, volume 2, pages 1390–1394, Texas, 1996.
4. T. Kohonen. The Self–Organizing Map. *Proceedings of the IEEE*, 78(9):1464–1480, September 1990.
5. T. Kohonen, S. Kaski, K. Lagus, and T. Honkela. Very Large Two-Level SOM for the Browsing of Newsgroups. In *Proc. of International Conference on Artificial Neural Networks*, pages 269–274, Bochum, Germany, 1996.
6. P. Koikkalainen and E. Oja. Self–Organizing Hierarchical Feature Maps. In *Proc. of International Joint Conference on Neural Networks*, pages 279–284, San Diego, CA, 1990.
7. J. Laaksonen, M. Koskela, and E. Oja. PicSOM: Self–Organizing Maps for Content–Based Image Retrieval. In *Proc. of International Joint Conference on Neural Networks*, Washington D.C., USA, July 1999.
8. J. Lampinen and E. Oja. Self–Organizing Maps for Spatial and Temporal AR Models. In *Proc. of the sixth SCIA Scandinavian Conference on Image Analysis*, pages 120–127, Helsinky, Finland, 1990.
9. J. Lampinen and E. Oja. Clustering Properties of Hierarchical Self–Organizing Maps. *J. Math. Imaging Vision*, 2(3):261–271, 1992.
10. T.C. Lee and M. Peterson. Adaptive Vector Quantization Using a Self–Development Neural Network. *IEEE J. Select Areas Commun.*, 8:1458–1471, 1990.
11. R. Mugnuolo, P. Magnani, E. Re, and S. Di Pippo. The SPIDER Manipulation System (SMS). The Italian Approach to Space Automation. *Robotics and Autonomous Systems*, 23(1–2), 1998.
12. G. Ongun and U. Halici. Fingerprint Classification through Self–Organizing Feature Map Modified to Treat Uncertainty. *Proceedings of the IEEE*, 84(10):1497–1512, October 1996.
13. A. Sperduti and A. Starita. Supervised Neural networks for the Classification of Structures. *IEEE Trans. on Neural Networks*, 8(3):714–735, May 1997.
14. P. N. Suganthan. Hierarchical Overlapped SOM's for Pattern Classification. *IEEE Trans. on Neural Networks*, 10(1):193–196, January 1999.

Evaluation of Distance Measures for Partial Image Retrieval Using Self-Organizing Map

Yin Huang, Ponnuthurai N. Suganthan, Shankar M. Krishnan, and Xiang Cao

School of Electrical and Electronic Engineering,
Nanyang Technological University
Singapore 639798
{pg00130975,epnsugan}@ntu.edu.sg

Abstract. Digital image libraries are becoming more common and widely used as more visual information is produced at a rapidly growing rate. With this immense growth, there is a need to organize and index these databases so that we can efficiently retrieve the desired images. In this paper, we evaluate the performance of the self-organising maps (SOMs) with different distance measures in retrieving similar images when a full or a partial query image is presented to the SOM. Our method makes use of RGB colour histograms. As the RGB colour space is very large, another SOM is employed to adaptively quantise the colour space prior to generating the histograms. Some promising results are reported.

1 Introduction

Due to the rapid advances in data storage technologies, image compression and internet technologies, image and multimedia databases are fast becoming as prevalent as the text document databases. Hence it is becoming important to be able to search these huge databases efficiently in order to retrieve desired objects from them [1][2][3][4][5][6].

In this paper, we evaluate the performances of SOM [7] trained using different distance measures in image database organization, indexing and retrieval. In order to train the SOM, we make use of RGB colour histograms [8]. As the resolution of the images is 256 in R, G and B spaces, the histograms generated using this high resolution will result in a 256x3 dimensional input vector to the SOM. In order to reduce the dimensions of the histogram, we make use of another SOM to perform the quantisation of the RGB colour space. The number of output neurons in this SOM determines the number of bins in the colour histograms. Then we make use of these histograms to train the SOM in order to organize the database images. In fact, a number of SOMs are trained using these histograms extracted from whole images only but with different distance measures. We then performed some experiments in order to evaluate the performances of various distance measures in retrieving partial images based on their histograms. Our results are promising and indicate that the standard Euclidean distance may not yield the best results. Due to long time needed to compute color histograms in case of large number of images, a 4-level tree-structured

G. Dorffner, H. Bischof, and K. Hornik (Eds.): ICANN 2001, LNCS 2130, pp. 1042–1047, 2001.

SOM [6] algorithm is implemented to improve the speed of generation of image color histograms. The result shows the improvement on computation speed is remarkable.

2 The Quantization of RGB Color Space

The quantisation of the RGB color space in an image is 256x256x256. As this is too large to be used to train the SOM, we employed an SOM to quantise the colour space into 64 levels. At the completion of quantisation, the RGB color space is mapped into a 2-dimensional SOM composed of 8x8 neurons. Euclidean distance is used to find the winner neuron:

$$\|x - m_c\| = \min_i \{\|x - m_i\|\} \tag{1}$$

where $x \in \Re^n$, m is reference vector, and c denotes the best-matching node. The weight update in this case is:

$$m_i(t+1) = m_i(t) + h_{ci}(t)[x(t) - m_i(t)] \tag{2}$$

The neighbourhood and learning rate parameters are updated as follows:

$$h_{ci} = \alpha(t) \cdot \exp\left(- \frac{\|r_c - r_i\|^2}{2\sigma^2(t)} \right) \tag{3}$$

$$\alpha(t) = \alpha_0 (1 - t / N) \tag{4}$$

where $t = 0,1,2,\ldots$ is the iteration number, $r_c \in \Re^2, r_i \in \Re^2$ are the location vectors of nodes c and i respectively, α_0 is a constant, N is the total number of training iterations.

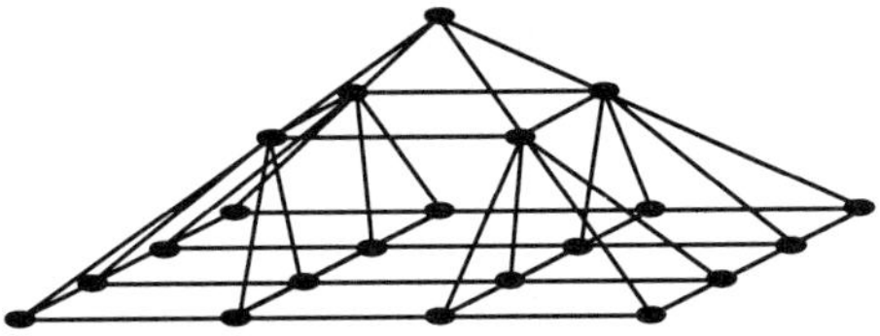

Fig. 1. The structure of TS-SOM

In the first step of retrieval process, the histogram of the query image needs to be constructed. In this step, the 8x8 SOM used for quantisation should be searched for N times, where N is the number of pixels in the query image. In order to speed up the generation of histograms, self-organizing hierarchical feature maps [10] is applied. The figure 1 above shows the structure of TS-SOM [6]. The search space in TS-SOM for the best-matching vector on the underlying SOM layer is restricted to a predefined portion just below the best-matching unit on the above SOM. Therefore, the complexity of the searches in TS-SOM is remarkably lower. So our 8x8 SOM is mapped to 4 layers, with every parent neuron denoting 4 child neurons in the layer below. Thus, the top layer has only one neuron. And searching on the lower level is performed actually in all units that are linked to the "winner" in the upper layer and the neighbors of the winner in the four units in the lower level.

3 Database Organisation Using SOM

After the quantization of RGB color space, a 64-bin normalized-histogram for every image can be obtained. These histograms are used as training vectors for the second-phase SOM, which consists of 6x6 neurons in our experiments. In this paper, we considered the following two distance measures in addition to the standard Euclidean distance.

- **L_1 Norm**

$$L_1(P_D, P_M) = \sum_{i=1}^{n} |P_D(i) - P_M(i)| \tag{5}$$

- **Bhattacharyya Distance**

$$B(P_D, P_M) = -\ln \sum_{i=1}^{n} \sqrt{P_D(i) \times P_M(i)} \tag{6}$$

where the normalized histogram contents for the bin indexed i in model and data histograms are respectively denoted by $p_M(i)$ and $p_D(i)$. $\sum_i P_M(i) = \sum_i P_D(i) = 1$, $P_M(i)\ \&\ P_D(i) \geq 0 \quad \forall i$. Each histogram contains 64 bins. The Bhatacharyya distance is effectively a correlation measure. The update equation for the L_1 norm and Bhattacharyya Distance are as follows [7]:

L_1 Norm update equation:

$$P_D(i,t+1) = P_D(i,t) + \alpha(t) h_{CD}(t) \operatorname{sgn}[P_M(i,t) - P_D(i,t)] \tag{7}$$

Bhattacharyya Distance update equation:

$$P_D(t+1) = \frac{P_D(t) + \alpha(t) h_{CD}(t) \sqrt{P_M(t) \times P_D(t)}}{\sum_{i=1}^{N} \left| P_D(i,t) + \alpha(t) h_{CD}(t) \sqrt{P_M(i,t) \times P_D(i,t)} \right|} \tag{8}$$

Once the training process is completed, the database images are associated with the winning neurons.

4 Experimental Results

We conducted experiments using about 300 colour images with 85x128 pixels per image. These colour images are topographically mapped to 36 neurons. As our objective is to evaluate various distance measures, we trained 3 different two-dimensional SOMs of 36 neurons. [1]Figures 2 and 3 shows some images represented by neurons (1,3) and (6,6). The Euclidean distance is used in training this SOM. Next we conducted image retrieval experiments using partial images (25%, 50%, 75% of full size image) of the database images. In Figure 4, we present an example of image retrieval using a partial query image. The images associated with the winning neuron are also displayed. The retrieval rates are given in Table 1 due to the three distance measures [9]. From the table, it is clear that all three distance-measures do not

---

[1] The visual similarity between images represented by the same neuron and the dissimilarity between images represented by these two neurons are clear in the colour images.

perform equally well and the Bhattacharyya distance consistently outperforms other metrics. The time needed to compute images color histograms for 8x8 single layer SOM and 4-level TS-SOM with standard Euclidean distance is given in Table 2. It shows remarkable time is saved in histogram computation using 4-level TS-SOM compared to 8x8 single layer SOM.

Fig. 2. The images associated with neuron (1,3) in the 6x6 SOM.

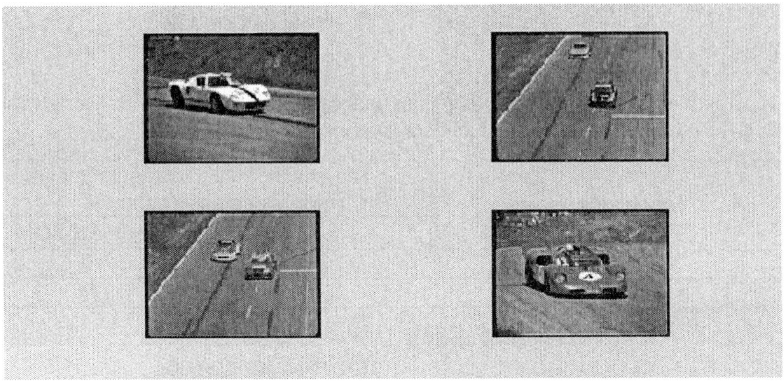

Fig. 3. The images associated with neuron (6,6) in the 6x6 SOM.

Table 1. Comparison of retrieval rates for 3 different distance measures

Fraction	Retrieval Rates		
	Standard Euclidean Distance	L1 Norm Distance	Bhattachar yya Distance
25%	62%	63%	72%
50%	65%	73%	77%
75%	73%	80%	80%
90%	95%	96%	98%

A query image with 85x64 pixels

Retrieved images associated with neuron (6,4) for the above partial query image.

Fig. 4. Retrieval results for a partial query image.

Table 2. Comparison of the speed to compute image color histogram for 8x8 single layer SOM and 4-level TS-SOM with Standard Euclidean Distance

Number of images	The time needed to compute images color histograms (minutes)	
	8x8 SOM	4-level TS-SOM
1	1.5	0.5
300	510	174

Another experiment is conducted for a larger size database organization neural network. In this case, the network is composed of 8x8 neurons. Using the same images in the database to test half size partial image retrieval with Euclidean distance, the retrieval rate is 54%. With larger number of neurons in network, the database organization is actually more accurate, which means less images are associated with winner neuron after training with the same image database. So the possibility of retrieving correct image from images associated with winner neuron is decreased. Thus, the retrieval rate of 8x8 size neural network is lower than the 6x6 size neural network. A better approach is to perform comparison between the query image and images retrieved to rank the results. We also intend to consider performance measures such as recall and precision in our studies to compare against other established methods.

5 Conclusion

An SOM based image database organization and retrieval system is presented in this paper. The objective of this work is to compare the performances of different distance measures in comparing partial histograms. The results show that the Bhattacharyya distance is superior to other distance measures considered. In case of large image database, the searching speed becomes more important. Due to the remarkably lower complexity of the searching in TS-SOM, we implemented hierarchical SOMs for quantisation of the RGB color space.

References

1. S-F. Chang, A. Eletheriadis, and R McClintock. Next-generation content representation, creation and searching for new media applications in education. IEEE Proceedings, 86(5): 602-615, Dept (1998).
2. H. Zhang and D. Zhang: A scheme for visual feature based image indexing, Proc. SPIE Conf. on Storage and Retrieval for Image and Video Databases, San Jose, Feb. (1995).
3. W. Niblack, et al: The QBIC project: Querying images by using color, texture and shape. Image and Visual Storage and Retrieval. (1993) 173-187.
4. T. Gevers and A.W.M.Smeulders: PicToSeek: Combining color and shape invariant features for image retrieval. IEEE Trans. Image Processing, vol. 9, No.1, 102-119.
5. A. K. Jain and V.Vailaya: Image retrieval using color and shape, Pattern recognition, vol.29, No.8, 1233-1244.
6. J Laaksonen, M Koskela and E Oja: PicSOM: Self-Organizing Maps for Content-Based Image Retrieval, Neural Networks, IJCNN '99, Volume 4, 2470 –2473.
7. T.Kohonen: Self-Organizing Maps, volume 30 of Springer Series in Information Sciences. Springer-Verlag, (1997). Second Extended Edition.
8. M. Swain and D. Ballard: Indexing with color histograms. ICCV'98, 390-393.
9. B Huet, E R Hancock: Cartographic Indexing into a Database of Remotely Sensed Images, Proc. 3rd IEEE Workshop on Applications of Comp. Vision. 1996, pp. 8 –14.
10. Koikkalainen, P.; Oja, E: Self-organizing hierarchical feature maps Neural Networks, IJCNN International Joint Conference, vol.2, (1990) 279 –284.

Video Sequence Boundary Detection Using Neural Gas Networks

Xiang Cao and Ponnuthurai N. Suganthan

School of EEE., Nanyang Technological University,
Nanyang Avenue, Singapore 639798
{p148482455,epnsugan}@ntu.edu.sg

Abstract. Video sequence boundary detection is an important first step in the construction of efficient and user-friendly video archives. In this paper, we propose to employ growing neural gas (GNG) networks [7] to detect the shot boundaries, as the neural networks are capable of learning the characteristics of various shots and clustering them accordingly. We represent the image frames by 6-bit color-coded histograms. We make use of the chi-square distances between histograms of neighboring frames as the primary features to train the GNG and to detect the shot boundaries. Experimental results presented in this paper demonstrate the reliable performance of our proposed approach on real video sequences.

1 Introduction

Due to the recent advances in the storage, communication and compression technologies, image and video databases have been created at a rapid pace around the world. To efficiently organize and manage video databases, video sequences are first segmented into shots. This process is known as the video shot boundary detection, temporal video segmentation, or scene change detection. It is considered to be an important first step that is followed by the identification of *key frames* to represent every shot [1]. The visual summary generated by this approach would be a better abstraction of the video sequence for rapid browsing. A *shot* is defined as a sequence of frames captured from a unique and continuous record from a camera. The most common class of shot boundary is known as the abrupt cut boundary. This is the result of splicing two distinct scenes consecutively. However, there exist more subtle shot boundaries collectively known as the gradual shot boundaries where the transition from one shot to the other lasts several frames and usually takes place gradually. They are the products of video effects such as dissolves, fade-ins, fade-outs and wipes [2]. The detection of a gradual shot boundary is much harder than detecting an abrupt cut boundary.

In recent years, some shot boundary detection procedures have been proposed [3], [4], [5]. These proposed schemes compare features such as the frame histograms (grayscale or color, global or local histograms), motion compensated images, edge images, MPEG DCT coefficients, pixel intensities and statistical properties of two neighboring frames. Frame pairs with a large content difference

G. Dorffner, H. Bischof, and K. Hornik (Eds.): ICANN 2001, LNCS 2130, pp. 1048–1053, 2001.

are termed as shot boundaries. However, the video shot boundary detection remains a challenging problem and the performances of the segmentation schemes are not considered satisfactory [5]. Hence, in this paper, we further investigate this problem and propose a GNG-based procedure.

This paper is organized as follows. In the next section, we explain the feature extraction and labeling of feature vectors. The GNG is presented in Section 3. Section 4 describes a procedure to locate and classify shot boundaries by interpreting the output labels generated by the GNG for the test feature vectors. In Section 5, we present some promising experimental results and perform a basic comparison with some recently published results. The paper is concluded in Section 6.

2 Feature Extraction

Histogram based algorithms are the most common video shot boundary detection techniques because of their simplicity and the acceptable quality results that they yield. In the literature several researchers such as Nagasaka and Tanake [6] indicate that the chi-square test of significance is the best histogram metric compared with other measures such as the Yakimovsy likelihood ratio test and the Kolmogorov-Smirnov test. For this reason, we use the normalized χ^2 test to compare the color histograms of two frames j and k, as follows:

$$\chi^2 = \frac{1}{n}\sum_{i=1}^{M}\frac{\{h_j(i) - h_k(i)\}^2}{\{h_j(i) + h_k(i)\}} \tag{1}$$

where $h_j(i)$ is the bin i value of the histogram of frame j, M is the overall number of histogram bins, and n is the number of pixels in one frame. A 6-bit color-coded (M=64) frame image is employed using the two most significant bits from each of the three color bands.

Though the histogram difference between two consecutive frames may show good performance in case of abrupt scene changes, it is intuitively clear that using only this difference cannot detect the gradual shot boundary effectively. To account for this, a sliding window is introduced. If $(2n+1)$ is the window size, the central frame in the window compares its histogram with the n preceding frames and n succeeding frames using (1). Then we get $2n$ features for each frame. In addition to the histogram differences, it is also feasible to incorporate other useful features into the feature vector to investigate various kinds of shot boundaries and improve detection performance. We are currently exploring the possibility of incorporating some motion compensated frame differences. The window size can be varied, to increase the number of features for a frame. A large size window may perform better for detecting gradual transition but needs more computation time. For neural network labeling, we classify each feature vector into one of the six classes ("1", "12", "21", "13", "3", "31") as follows:

- k successive frames just before the abrupt boundary are labeled as "12".
- k successive frames just after the abrupt boundary are labeled as "21".

- m successive frames just before the gradual transition are labeled as "13".
- the frames in the gradual transition are labeled as "3".
- m successive frames just after the gradual transition are labeled as "31".
- other non-boundary frames are labeled as "1".

k and m are preset parameters which are not larger than n.

At the completion of data processing as outlined above, the feature vectors are available to train the neural network. We partition the dataset into a training set and a test set.

3 Growing Neural Gas (GNG)

With the aim to identify the clusters presented in the dataset described in Section 2, a GNG [7] network has been introduced. The incremental self-organizing GNG has its origin in the neural gas (NG) [8] algorithm. It combines the growth mechanism inherited from the growing cell structures (GCS) [9], with the topology generation of competitive Hebbian learning (CHL) [10]. It is capable of generating and removing both lateral connections and neurons. Due to this insertion and removal strategy, it is robust against noise, and this avoids overfitting.

In our model, the learning processing of the GNG network starts with only two units whose weight w_i is initialised randomly. Then the following steps are iterated until a stopping criterion (e.g., net size or some performance measure) is fulfilled:

1. Present an input vector v from the training dataset.
2. Determine the winner s_1 and the second-nearest unit s_2, apply CHL to the unit pair s_1 and s_2. Set the age of the connection between s_1 and s_2 to zero.
3. Increase the age of all edges emanating from s_1 by one.
4. Update the local error of s_1 according to:
$$\Delta E_{s_1} = \|w_{s_1} - v\|^2 \tag{2}$$
5. Adapt weight vectors of s_1 and its direct topological neighbors according to:
$$\Delta w_{s_1} = \varepsilon_b(v - w_{s_1}) \tag{3}$$
$$\Delta w_i = \varepsilon_n(v - w_i) \qquad (\forall i \in N_{s_1}) \tag{4}$$
where ε_b and ε_n are adaptation constants with $\varepsilon_b > \varepsilon_n$, N_{s_1} denotes the set of direct topological neighbors of s_1.
6. Remove edges whose age is larger than a preset number a_{max}. If this results in units having no emanating edges, remove such units.
7. Insert a new unit r every λ adaptation steps halfway between unit q featuring the maximum accumulated local error and its neighbor f with the largest error variable in the direct topological neighborhood of q. The local error of r is set to $(E_q + E_f)/2$; then E_q and E_f are decreased by a fraction α.
8. Decrease all local error variables by multiplying them with a fraction β.

Having completed the unsupervised GNG learning, the neurons in the network are labeled using the labels of the training samples and a simple voting mechanism.

4 Decision Mechanism

Having trained and labeled the GNG, now we can test its performance on shot boundary detection using the test samples. The GNG will associate each frame with the label of corresponding winner in GNG. But the abrupt shot boundary or the gradual shot boundary or non-boundary cannot be declared directly from the output. A decision mechanism has to be applied on the output data to find the shot boundary. Similar to the method we use in feature extraction, for each frame, we adopt a window consisting of some number of past and future output labels. The total number of labels except the label of non-boundary ("1" in our experiments) in the window is calculated. The window moves through the output labels. When the number exceeds a threshold T_N and it lasts more than a predefined number of consecutive frames, a shot boundary will be declared. As the feature vectors corresponding to "12" & "13" as well as feature vectors "21" & "31" are very similar, we have decided to treat them collectively. Furthermore, by counting the number of consecutive windows whose number exceeding T_N, we can classify the abrupt shot boundary and gradual shot boundary approximately. A gradual shot boundary is expected to have several frames exceeding T_N whereas an abrupt cut boundary is expected to have a few frames exceeding T_N and it usually occurs at the transition from "12" to "21". For the digital video organization and browsing purposes, it is sufficient to approximately localize and classify the shot boundaries into abrupt or gradual boundary. A verification procedure can also be used to localize the boundary more accurately after making the decision as explained above.

5 Experimental Results

About 23170 training samples, containing the data from video sequences with various effects like abrupt changes, gradual changes, no changes were used to train the GNG network. All video clips were decompressed before processing. The frames were digitized at 25 frames per second (fps) with a 352 × 288 pixels per frame. Although the number of training samples of each class label was not equal, it was observed that training was good. After training, we tested the system with a test dataset containing 9 video sequences extracted from commercial VCDs. All of the experiments were conducted on a PIII450 PC running Windows-NT system. Two parameters, *recall* and *precision* [4], are chosen to evaluate the performance of the detection. The equation for recall and precision are the following:

$$Recall = \frac{N_c}{N_c + N_m} \times 100\% \tag{5}$$

$$Precision = \frac{N_c}{N_c + N_f} \times 100\% \tag{6}$$

where, N_c is the number of correct detections, N_m is the number of missed detections, N_f is the number of false detections. The detection results for the training dataset and test dataset are shown in the first two rows of Table 1.

Table 1. The overall performance

Video clip	Detected	False	Miss	Precision(%)	Recall(%)
Training Samples	229	6	14	97.4	94.2
Test Samples	220	10	15	95.7	93.6
Action: *Airwolf*	1	0	0	100	100
Action: *Blade Runner*	18	3	2	85.7	90
Action: *StarWars movie trailer*	23	4	2	85.2	92
Animation: *Starwars animation*	7	1	0	87.5	100
Comedy: *Friends sitcom*	10	0	0	100	100
Drama: *A Few Good Men*	11	0	1	100	91.7
Drama: *Alaska movie trailer*	61	9	5	87.1	92.4
Drama: *American President*	56	2	2	96.6	96.6
News: *Endeavor astronauts*	2	0	0	100	100
Sports: *Skateboarding*	4	0	0	100	100
Sports: *Basketball*	8	1	0	88.9	100
Total	650	36	41	94.8	94.1

As noticed in Table 1, very promising results were obtained. Ford et al. [5] have reported their results using different metrics. Though they chose some different parameters to measure the performance, the precision and the recall can be derived from their illustrations approximately. They indicated that the metric with the lowest recorded false-alarm rate (3.2%) at a detection rate of 90% corresponds to 1161 false alarms for 1428 abrupt shot boundary detected out of 1581 actual abrupt boundaries. This means the precision is 55.2% and the recall is 90.3%. Our results compare very favourably. But it must be pointed out that the results were obtained using different datasets. We have obtained a part of the video sequences used in [5]. They have varying but, lower resolutions and sampling rates. We did not use these sequences for training. Some of the results for these sequences using our method are also provided in Table 1 from the third row. Since [5] only lists their overall results for each video category, we intend to use the whole dataset to obtain a faithful comparison.

By analyzing the results, we find many false and missed detections are due to the special features of the video sequence, especially in the movie trailers, such as:

- The movement of people or camera is too fast.
- There is a huge moving object in the field of view. And the appearance or disappearance of the huge object is also sudden. Our procedure detected one false boundary in the sequence shown in Fig. 1.

6 Conclusion

In this paper, we propose a new scheme based on neural network for video sequence shot boundary detection. The GNG plus a decision mechanism has been

Fig. 1. A big object moves across generating a false detection

shown to be efficient for this task. One advantage of our method is the detection robustness because not only does it consider the neighbors of each frame by using a window, (instead of only two consecutive frames), but also a decision mechanism is introduced to the system for reducing the false detections. In our preliminary system, we only use the image histograms differences as the primary features. Due to the flexibility of neural networks, various features can be introduced to enhance the detection process. Currently, we are investigating the merits of incorporating other useful frame features [3], [4], [5] and enhancing the proposed neural network procedure in order to improve the overall system performance and to implement subsequent operations such as key frame extraction and video summarization.

References

1. Rui, Y., Huang, T.S., Mehrotra, S.: Browsing and Retrieving Video Content in a Unified Framework. Proc. IEEE 2nd Workshop on Multimedia Signal Processing (1998) 9–14
2. Bordwell, D., Thompson, K.: Film Art: An Introduction. 4th edn. McGraw Hill, New York, NY (1993)
3. Zhang, H.J., Kankanhalli, A., Smoliar, S.W.: Automatic Partitioning of Full-motion Video. ACM Multimedia Systems, Vol. 1, No. 1 (1993) 10–28
4. Lupatini, G., Saraceno, C., Leonardi, R.: Scene Break Detection: A Comparison. Proc. Eighth Int. Workshop on Reasearch Issues in Data Engineering (1998) 34–41
5. Ford, R.M., Robson, C., Temple, D., Gerlach, M.: Metrics for Shot Boundary Detection in Digital Video Sequences. Multimedia Systems, Vol. 8 (2000) 37–46
6. Nagasaka, A., Tanaka, Y.: Automatic Video Indexing and Full Video Search for Object Appearances. Visual Database Systems II (1992) 113–127
7. Fritzke, B.: A Growing Neural Gas Network Learns Topologies. In: Tesauro, G., Touretzky, D.S., Keen, T.K.(eds.): Advances in Neural Information Processing Systems. MIT Press, Cambridge, MA (1995) 625–632
8. Martinetz, T.M., Schulten, K.J.: A "Neural-Gas" Network Learns Topologies. In: Kohonen, T., Mäkisara, K., Simula, O., Kangas, J. (eds.): Artificial Neural Networks. North-Holland, Amsterdam (1991) 397–402
9. Fritzke, B.: Growing Cell Structures—A Self-organizing Network for Unsupervised and Supervised Learning. Neural Networks, Vol. 7, No. 9 (1994) 1441–1460
10. Martinetz, T.M.: Competitive Hebbian Learning Rule Forms Perfectly Topology Preserving Maps. ICANN'93: International Conference on Artificial Neural Networks, Amsterdam (1993) 427–434

A Neural-Network-Based Approach to Adaptive Human Computer Interaction

George Votsis, Nikolaos D. Doulamis, Anastasios D. Doulamis,
Nicolas Tsapatsoulis, and Stefanos D. Kollias

Image, Video and Multimedia Systems Laboratory
Department of Electrical and Computer Engineering
National Technical University of Athens
Zografou, 15773, Greece
stefanos@cs.ntua.gr

Abstract. A neural-network-based approach is proposed in this paper providing multimedia systems with the ability to adapt their performance to the specific needs and characteristics of their users. An emotion recognition system based on facial cues is used to illustrate the performance of this approach. Although some archetypal emotions are considered to be universal, there is no doubt that the way they are expressed diversifies from one individual to the other. The initial generic system is adapted in order to meet the specific user's traits.

1. Introduction

The widespread use of the Internet, mobile telephony and inexpensive computer equipment has made human computer interaction (HCI), and in general, man machine communication, a usual activity in everyday life. Making this interaction as efficient and friendly as possible, by including intelligence to the machines that interact with people, is a crucial aspect which is currently attracting large R&D efforts worldwide.

Neural networks have not played a significant role in the development of audiovisual coding standards, such as MPEG-1 and MPEG-2. Nevertheless the MPEG-4 and MPEG-7 standards [1], which refer to content-based audio-visual coding, analysis and retrieval, are based on physical object extraction from scenes, focusing on handling multimedia information with an increased level of intelligence. Neural networks, with their superior non-linear classification abilities, can play a major role in these standards, especially due to their ability to learn and adapt their behavior based on real actual data [2, 3]. Probably the most important issue when designing and training artificial neural networks in real life applications is network generalization. Despite the achievements obtained during the last years, most real life applications do not obey some specific probability distribution and may significantly differ from one case to another, mainly due to changes of their environment. In such cases, the training set is not able to represent all possible variations or states of the operational environment to which the network is to be applied. Instead, it would be desirable to have a mechanism, which would provide the network with the capability to be on-line retrained, when its performance is not acceptable. The retraining

G. Dorffner, H. Bischof, and K. Hornik (Eds.): ICANN 2001, LNCS 2130, pp. 1054–1059, 2001.

algorithm should update the network weights taking into account both the former network knowledge and the knowledge extracted from the current input data [4].

In this paper we show that neural networks can form a crucial component of HCI systems responsible for recognition of users' emotional state based on their facial expressions. The major issue that is explored is the ability of the system to adapt its performance to its specific user expressions and emotional states [5]. A well trained expression/emotion recognition system, which is, for example, sold by service providers to their customers, is on-line retrained, adapting its former knowledge to the specific characteristics or expressions of its owner.

2. Retraining Strategy of the Neural System

Let us first consider that a neural-network-based system has been created by a service provider and is being supplied to customers, so as to be used, e.g., for facial expression recognition; it is able to classify each image, or video shot of a human to one, or more specific categories, such as happy, angry or neutral. The neural classifier has obtained the necessary knowledge to perform the task, having been trained with a carefully selected training set, say, S_b. Let us then consider that a specific customer includes the system in his/her own PC and starts to use it. The system will now be facing its true owner, and thus should adapt to its owner's specific characteristics and behavior, while keeping up with its former knowledge.

Let us consider network adaptation through retraining. Let vector $\underline{w}_b$ include all weights of the network before retraining, and $\underline{w}_a$ the new weight vector which is obtained through retraining. A retraining set S_c is assumed to be extracted from the current operational situation composed of, say, m_c feature vectors. The retraining algorithm should compute the new network weights $\underline{w}_a$, by minimizing the following error criterion with respect to the weights,

$$E_a = E_{c,a} + \eta E_{f,a} \qquad \textbf{(1)}$$

where $E_{c,a}$ is the error performed over training set S_c ("current" knowledge), $E_{f,a}$ the corresponding error over training set S_b ("former" knowledge). Parameter η is a weighting factor accounting for the significance of the current training set compared to the former one.

In most real life applications, training set S_c is initially unknown; consequently selection of S_c, as well as detection of the need for training should be provided to the system, either through user interaction, or automatically, when this is possible.

The goal of the training procedure is to minimize (1) and estimate the new network weights $\underline{w}_a$. Let us first assume that a small perturbation of the network weights (before retraining) $\underline{w}_b$ is enough to achieve good classification performance. This assumption leads to an analytical and tractable solution for estimating $\underline{w}_a$, since it permits linearization of the non-linear activation function characterizing each neuron, using a first order Taylor series expansion.

It can be shown [6] that, through linearization, solution of this problem with respect to the weight increments is equivalent solving a set of linear equations

$$\underline{c} = \mathbf{A} \cdot \Delta \underline{w} \tag{2}$$

where vector $\underline{c}$ and matrix **A** are appropriately expressed in terms of the previous network weights $\underline{w}_b$. In particular, $\underline{c}$, indicates the difference between network outputs after and before retraining for all input vectors in S_c. Among all possible solutions that satisfy (2), the one which causes a minimal degradation of the previous network knowledge is selected as the most appropriate. The network weights before retraining, i.e., $\underline{w}_b$, have, however, been estimated as an optimal solution over data of set S_b. Furthermore, the weights after retraining provide a minimal error over all data of the current set S_c. Thus, minimization of the second term of (1), which expresses the effect of the new network weights over data set S_b, is equivalent to minimization of the absolute difference of the error over data in S_b with respect to the previous and the current network weights. This means that the weight increments are minimally modified, resulting in minimization of the following error criterion

$$E_S = \left\| E_{f,a} - E_{f,b} \right\|_2 = \frac{1}{2} (\Delta \underline{w})^T \cdot \mathbf{K}^T \cdot \mathbf{K} \cdot \Delta \underline{w} \tag{3}$$

where $E_{f,b}$ is defined similarly to $E_{f,a}$ and the elements of matrix **K** are expressed in terms of the previous network weights $\underline{w}_b$ and the training data in S_b. Thus, the problem results in minimization of (3) subject to constraints of (2).

The error function defined by (3) is convex since it is of squared type, while the constraints are linear equalities. Thus, the solution should lie on the hyper-surface defined by (2) and simultaneously minimize the error function given in (3). The gradient projection method has been used to solve this problem. The gradient projection method starts from a feasible point and moves in a direction, which decreases E_s and simultaneously satisfies the constraints; a point is called feasible, if it satisfies all constraints. The computational complexity of the retraining algorithm is very small. The total cost of retraining is of order of a few seconds, allowing the efficient use of the proposed scheme to real-life interactive multimedia systems.

3. Recognition of User's Emotional State

Analysis of the emotional expression of a human face requires a number of pre-processing steps which attempt to detect or track the face, to locate characteristic facial regions such as eyes, mouth and nose on it, to extract and follow the movement of facial features, such as characteristic points in these regions, or model facial gestures using anatomic information about the face. Most of the above techniques are based on a well known system for describing "all visually distinguishable facial movements", called the Facial Action Coding System (FACS). The FACS model has inspired the derivation of facial animation and definition parameters in the framework of the ISO MPEG-4 standard. In particular, the Facial Definition Parameter (FDP) set and the Facial Animation Parameter (FAP) set were designed in the MPEG-4 framework to allow the definition of a facial shape and texture, as well as the animation of faces reproducing expressions, emotions and speech pronunciation [1]. By monitoring facial gestures corresponding to FDP and/or FAP movements over time, it is possible to derive cues about user's expressions/emotions. Facial anatomy as well as social issues make the emotional analysis of faces to be more or less user dependent.

Let us assume that prominent MPEG-4 FDP points have been extracted and used to define corresponding FAPs. The latter are normalised according to some fixed face distances and provide a user independent description of the movement of particular facial areas. In the following we use a subset of the FDP feature points that we have identified as relevant to emotion [5]. Based on that, we have created a relation between FAPs that can be used for the description of the archetypal facial expressions and the selected FDP subset [7].

In the next section, we use the above-mentioned techniques and a database generated in [5] with presence of realistic, non-extreme emotional states to extensively train a neural network classifier so as to be able to discern among different human expressions. We will, however, show that, when providing a new user with the trained classifier (also providing the training feature data set, which can serve as the initial set for retraining), the obtained results will greatly improve, if we permit retraining of the system, as described in section 2 of this paper. Retraining will be performed, at the first time that the customer uses the system and, possibly, when a significant change of the environment takes place, that results in deterioration of the system's performance; an interactive framework has been created, where the system captures its user's expressions and interacts with him to get his/her own categorisation of them, so as to create the 'current' (re)training data set.

4. Experimental Results

An experimental study has been conducted in the emotion recognition case using well-known databases [5] such as the MediaLab database, the database of the UCSF Human Interaction Lab, and the database with natural video sequences developed in [5]. The initial training set (S_b) includes faces of various people expressing six archetypal – type emotions, i.e. anger, sadness, happiness, disgust, fear and surprise, as well as their neutral state. More than one thousand images have been incorporated in this set. The emotional content of each image is described through the feature vector described in section 3. We then assumed that these data have been used to train a commercial neural network product that needs to be adapted to its final end user. A second data set has been created with 20 images, portraying a specific end user, in each of the above seven emotional states. This data pool serves as retraining set (S_c) for the weight adaptation procedure. Following retraining, the product will be ready for use by its end user; a third data set composed of 170 other images with expressions of the same end user constitutes the test material (S_t), on which the performance of the retrained neural network is to be examined.

Using the above data sets, two kinds of experiments were carried out. In the first, the scenario of the retraining strategy was tested. In the second, for comparison purposes, the retraining data set was used together with the initial training set to train the neural network de novo. Let us denote by Z_b the initially trained network (on S_b), by Z_c the retrained network (on S_c according to the presented procedure) and by Z_o the overall trained network (on the union of S_b and S_c).

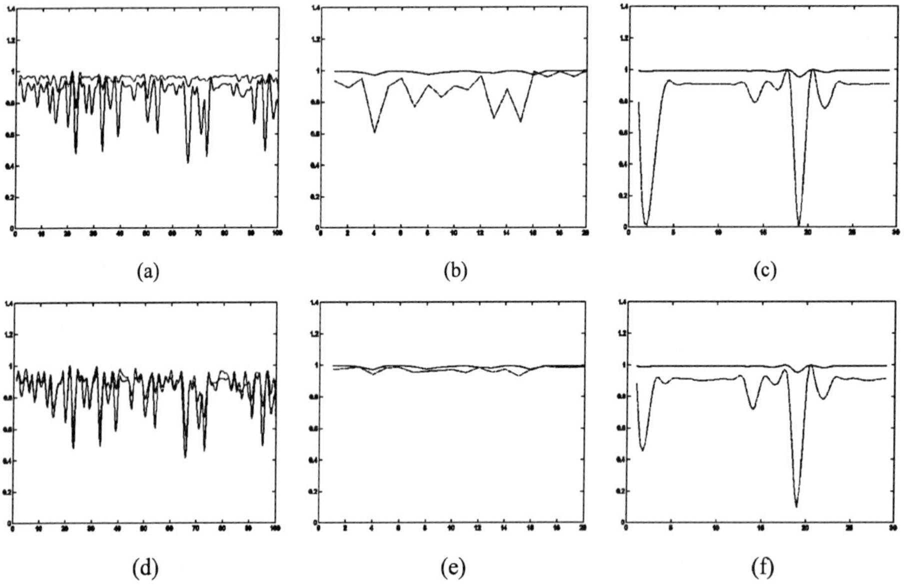

Fig. 1. "Happy" expression. (a) Performance of networks Z_b (dashed) and Z_c (solid) over set S_b. (b) Performance of Z_b (dashed) and Z_c (solid) over set S_c. (c) Performance of networks Z_b (dashed) and Z_c (solid) over set S_t. (d) Performance of networks Z_o (dashed) and Z_c (solid) over set S_b. (e) Performance of networks Z_o (dashed) and Z_c (solid) over set S_c. (f) Performance of networks Z_o (dashed) and Z_c (solid) over set S_t. Horizontal axis in (a)-(f) shows the sample numbering, while vertical axis shows the relevance to the specific class.

The results for the "happy" expression are presented in Figure 1: (a) compares the performance of the network that was trained with S_b (Z_b) with that of the retrained network (Z_c) over a small subset of the initial set (S_b); (b) compares the performance of the two networks when applied to the retraining set, illustrating the better performance of the retrained network. Let us note that (a) and (b) are in one sense complementary as expected; personalization of the network leads Z_c to a lower performance on S_b. However, (a) shows that the retrained network still retains at some degree the original network's integrity, while in (b) it is evident that the reverse statement may not be supported, i.e. the Z_b network does not achieve acceptable performance rates; this verifies that the potential product needs retraining in its new environment (end user). In (c) both networks are applied to the test data set (S_t) and their performance is plotted throughout the data samples. The effectiveness of the retraining procedure is clear: the output of Z_c approaches unit values, which is not the case with network Z_b. Subfigures (d) through (f) compare the performance of the network that has been trained from scratch with both initial and retraining data (Z_o) with that of the retrained network (Z_c). Although the performances of Z_o and Z_b on both S_b and S_c do not significantly differ, subfigure (f) points out that the Z_o has not adapted itself well enough to the new environment (S_t). Such an observation seems quite reasonable, since the number of samples of the retraining set is much smaller that the one of the initial set. Moreover, training from scratch might urge the Z_o network to converge to a point in the weight parameter space, which is far from the one of the Z_b network. In fact, the percentage of change of the weight parameters' norm between Z_b and Z_o has been calculated to rise up to 90%, while at the same time the respective percentage between Z_b and Z_c is as low as 5%.

5. Conclusions

Emotion recognition through facial expression analysis has been examined in this paper, using a novel adaptive framework based on artificial neural networks. In particular, an efficient scheme for on-line retraining of neural network classifiers has been proposed, which provides the above system with the ability to adapt its behavior to the specific needs or characteristics of their users, while retaining its former knowledge.

References

1. Chiariglione, L.: MPEG and Multimedia Communications. IEEE Trans. on Circuits and Systems for Video Techn., Vol. 7 (1997) 5-18.
2. Haykin, S.: Neural Networks: A Comprehensive Foundation. New York: Prentice Hall, 2nd edition (1999).
3. Kung, S.: Neural Networks for Intelligent Multimedia Processing. IEEE Signal Processing Magazine, Vol. 14 , No. 4 (1997).
4. Park, D. C., EL-Sharkawi, M. A. and Marks R. J.: An Adaptively Trained Neural Network. IEEE Trans. on Neural Networks, Vol. 2 (1991) 334-345.
5. Cowie, R., Douglas-Cowie, E., Tsapatsoulis, N., Votsis, G., Kollias, S., Fellenz, W. and Taylor, J.: Emotion Recognition and Human Computer Interaction. IEEE Signal Processing Magazine, No.1 (2001).
6. Doulamis, N., Doulamis, A. and Kollias, S.: On-Line Retrainable Neural Nets: Improving Performance of Neural Networks in Image Analysis Problems. IEEE Trans. on Neural Networks. Vol. 11, No 1 (2000) 137-155.
7. Tsapatsoulis, N., Karpouzis, K., Stamou, G., Piat, F. and Kollias, S.: A Fuzzy System for Emotion Classification based on the MPEG-4 Facial Definition Parameter Set. In Proceedings of the EUSIPCO-2000, Tampere, Finland (2000).

Adaptable Neural Networks for Unsupervised Video Object Segmentation of Stereoscopic Sequences

Anastasios D. Doulamis, Klimis S. Ntalianis,
Nikolaos D. Doulamis, and Stefanos D. Kollias

National Technical University of Athens
Electrical and Computer Engineering Department
9, Heroon Polytechniou str. Zografou 15773, Athens, Greece
{adoulam,kntal,ndoulam}@image.ntua.gr

Abstract. An adaptive neural network architecture is proposed in this paper, for efficient video object segmentation and tracking in stereoscopic sequences. The scheme includes: (A) A retraining algorithm that optimally adapts the network weights to the current conditions and simultaneously minimally degrades the previous network knowledge. (B) A semantically meaningful object extraction module for constructing the retraining set of the current conditions and (C) a decision mechanism, which detects the time instances when network retraining is required. The retraining algorithm results in the minimization of a convex function subject to linear constraints. Furthermore description of the current conditions is achieved by appropriate combination of color and depth information. Experimental results on real life video sequences indicate the promising performance of the proposed adaptive neural network-based video object segmentation scheme.

1 Introduction

Emerging multimedia applications, such as content-based image retrieval and object-based video transmission depend on the development of sophisticated algorithms for efficient segmentation of the visual content [1]. To address these needs, the MPEG-4 standard has introduced Video Objects (VOs) for content-oriented coding of video sequences [2]. A VO is an arbitrarily shaped region, consisting of multiple color, texture or motion characteristics, such as a human, a chair, or an airplane. Nevertheless, although VOs and their functionalities are analytically described by the MPEG-4 standard, techniques to generate them are left to content developers. Generally, identifying semantic entities remains a difficult problem and for this reason most content description schemes use low-level features such as color, texture or motion and do not consider VOPs. On the other hand neural networks, which can be taught to interpret several variations of the same video object, can help in VOP segmentation and play a significant role in the development of the new multimedia oriented standards, such as MPEG-4/7 [3].

Color segmentation is a first attempt towards video content representation. Some typical works include the morphological watershed [4], or the RSST algorithm [5]. However, every color-oriented segmentation scheme oversegments a video object into

G. Dorffner, H. Bischof, and K. Hornik (Eds.): ICANN 2001, LNCS 2130, pp. 1060–1066, 2001.

multiple color regions. Other schemes use motion homogeneity criteria, like the algorithms proposed in [6] and [7]. Nevertheless, in these schemes object boundaries cannot be identified with high accuracy, due to erroneous estimation of the motion vectors. For this reason, some advanced schemes have been recently proposed. In particular, in [8] color segments with coherent motion are merged to form the video object, while in [9], a model of the region of interest is derived using the Canny operator. The aforementioned techniques produce sufficient results when 1) adjacent video objects have different motion characteristics, and 2) color segments of each video object have coherent motion. However in real life sequences, there are several cases when no motion exists or when parts of the object do not have coherent motion.

In this paper, an adaptive neural network architecture is proposed for unsupervised segmentation and tracking of stereoscopic video objects. Initially a decision mechanism is employed which decides whether retraining of the neural network is required or not. When a retraining phase is activated, extraction of the retraining set is performed by a segmentation fusion scheme, which merges color segments according to depth similarity criteria, so that video objects with reliable contours are estimated. Based on the segmentation fusion results, neural network retraining is performed, to adapt the network response to current conditions [10]. After weight updating the network is applied to the following video frames and tracks video objects. Experimental results on real life stereoscopic video sequences indicate the prominent performance of the proposed scheme.

2 Problem Formulation and Retraining Algorithm

2.1 Problem Formulation

Let us consider that video object tracking is a classification problem. More specifically, each video object corresponds only to one class ω_i out of, say p available. Then, for each image region the neural network will produce a p-dimensional output vector $\mathbf{z}(\mathbf{x}_i)$:

$$\mathbf{z}(\mathbf{x}_i) = \left[\, p^i_{\omega_1}\; p^i_{\omega_2} \cdots p^i_{\omega_p} \,\right]^T \tag{1}$$

where $\mathbf{x}_i$ is a feature vector, describing the content of the ith block of the image and $p^i_{\omega_j}$ denotes the likelihood of the ith block to belong to the jth class ω_j (jth VO).

Let us consider that the neural network has been initially trained using the training set, $S_b = \{(\mathbf{x}'_1, \mathbf{d}'_1), \cdots, (\mathbf{x}'_{m_b}, \mathbf{d}'_{m_b})\}$, where vector $\mathbf{x}'_i$, $i = 1,2,\cdots,m_b$ denotes the ith input training vector, while $\mathbf{d}'_i$ is the respective desired output vector.

Let us now suppose that a retraining phase is activated. During this phase new weights $\mathbf{w}_a$ are estimated by taking into consideration both the former knowledge (of set S_b) and the current knowledge of set $S_c = \{(\mathbf{x}_1, \mathbf{d}_1), \cdots (\mathbf{x}_{m_c}, \mathbf{d}_{m_c})\}$ composed of m_c

elements. In particular, the new network weights $\mathbf{w}_a$ are estimated by minimizing the following error criteria,

$$E_a = E_{c,a} + \eta E_{f,a} \tag{2}$$

with

$$E_{c,a} = \frac{1}{2}\sum_{i=1}^{m_c} \left\| \mathbf{z}_a(\mathbf{x}_i) - \mathbf{d}_i \right\|_2 \tag{2a}$$

and

$$E_{f,a} = \frac{1}{2}\sum_{i=1}^{m_b} \left\| \mathbf{z}_a(\mathbf{x}'_i) - \mathbf{d}'_i \right\|_2 \tag{2b}$$

In the previous equations, $\mathbf{z}_a(\mathbf{x}_i)$ and $\mathbf{z}_a(\mathbf{x}'_i)$ are the network outputs when the new weights $\mathbf{w}_a$ are used and feature vectors $\mathbf{x}_i$ (of set S_c) and $\mathbf{x}'_i$ (of set S_b) are fed as network inputs. Therefore, $E_{c,a}$ is the error between the network output and the desired output over all data of set S_c ("current knowledge"), while $E_{f,a}$ the corresponding error over training set S_b ("former" knowledge); Parameter η regulates the significance between the current training set and the former one.

2.2 The Retraining Algorithm

In order to estimate the weights $\mathbf{w}_a$, we assume that a small perturbation of the network weights $\mathbf{w}_b$ before retraining is enough to achieve good classification performance. Then,

$$\mathbf{W}_a^0 = \mathbf{W}_b^0 + \Delta\mathbf{W}^0 \text{ and } \mathbf{w}_a^1 = \mathbf{w}_b^1 + \Delta\mathbf{w}^1 \tag{3}$$

where $\Delta\mathbf{W}^0$ and $\Delta\mathbf{w}^1$ are small weight increments, and $\mathbf{W}_a^0$, $\mathbf{W}_b^0$ are matrices of the form $\mathbf{W}_{\{a,b\}}^0 = \left[\mathbf{w}_{1,\{a,b\}}^0 \ldots \mathbf{w}_{q,\{a,b\}}^0 \right]$ and $\mathbf{w}_a^1$, $\mathbf{w}_b^1$ are vectors containing the weights between the output and the hidden layer neurons after and before retraining respectively. This assumption leads to an analytical and tractable solution for estimating $\mathbf{w}_a$ [10]. Furthermore, in order to stress the importance of current training data, equation (2a) is replaced by the constraint that the actual network outputs are equal to the desired ones:

$$z_a(\mathbf{x}_i) = d_i \quad i = 1,\ldots,m_c, \quad \text{for all data in } S_c \tag{4}$$

It can be shown that solution of (4) with respect to the weight increments is equivalent to a set of linear equations [10]:

$$\mathbf{c} = \mathbf{A} \cdot \Delta\mathbf{w} \tag{5}$$

where $\Delta\mathbf{w} = \left[(\Delta\mathbf{w}^0)^T (\Delta\mathbf{w}^1)^T \right]^T$, $\Delta\mathbf{w}^0 = \text{vec}\left\{ \Delta\mathbf{W}^0 \right\}$, with $\text{vec}\left\{ \Delta\mathbf{W}^0 \right\}$ denoting a vector formed by stacking up all columns of $\Delta\mathbf{W}^0$.

Uniqueness is imposed by an additional requirement due to $E_{f,a}$ in (2b). In particular, equation (2b) is solved with the requirement of minimum degradation of the previous network behavior, i.e., minimizing the following error criterion.

$$E_S = \left\| E_{f,a} - E_{f,b} \right\|_2 \tag{6}$$

It can be shown [10] that (6) takes the form:

$$E_S = \frac{1}{2} (\Delta \underline{w})^T \cdot \mathbf{K}^T \cdot \mathbf{K} \cdot \Delta\mathbf{w} \tag{7}$$

where the elements of matrix **K** are expressed in terms of the previous network weights $\mathbf{w}_b$ and the training data in S_b. The problem results in minimization of (7), which is convex, subject to the linear equalities of (5). Thus, only one global minimum exists and in this paper it is obtained by the gradient projection method.

3 Retraining Set Extraction and Decision Mechanism

The aforementioned retraining algorithm requires the estimation of the retraining set S_c, which provides information about the current conditions. In our approach, the set S_c is provided using a combined color-depth segmentation fusion algorithm. In particular, let us assume that a depth segments and a color segments map are available by performing the technique in [11]. Then color segments are projected onto depth segments so that video objects provided by depth segmentation are retained and, at the same time, object boundaries given by color segmentation are accurately extracted. For this reason, each color segment is associated with a depth segment, so that the area of intersection between the two segments is maximized. This method leads to construction of *K* segment classes, each corresponding to a specific video object. More details of the proposed technique can be found in [11].

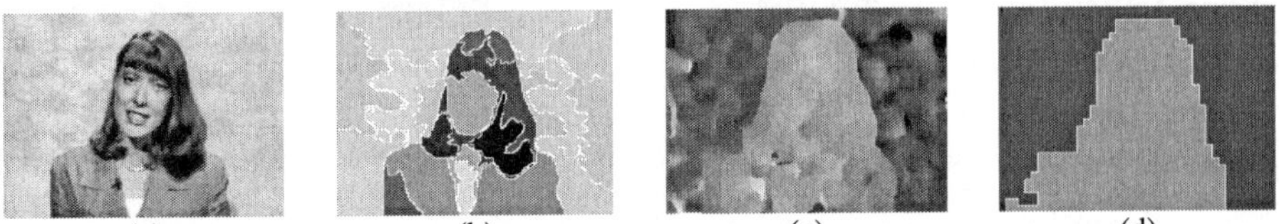

(a) (b) (c) (d)

Fig. 1. a) Original image. (b) Color segmentation. (c) Depth map and (d) Depth segmentation.

On the other hand the purpose of the decision mechanism is to detect when a change of the environment occurs and consequently to activate the retraining algorithm. In this paper retraining is performed every time the beginning of a new shot is detected, since frames within a scene are usually of similar content. For this reason a shot cut detection algorithm is applied. In our approach the algorithm proposed in [12] has been adopted, due to its efficiency and low computational complexity. This algorithm is based on the dc coefficients of the DCT transform of each frame, which are directly available in the case of I frames of MPEG compressed video sequences, while for the P and B frames, they can be estimated with a minimal decoding effort.

4 Experimental Results

In this section the performance of the proposed scheme is investigated. In particular, a stereoscopic sequence consisting of 10 shots was formed, obtained from the stereoscopic television program "Eye to Eye". Each stereo frame was in QCIF format, having 144x176 pixels per color image component. In the following each frame is partitioned into blocks of size 8x8 pixels. The DC coefficient and the first 8 AC coefficients of the zig-zag scanned DCT for each color component of a block, (i.e., 27 elements in total) are used as feature vector $\mathbf{x}_i$ of the respective block. Initially, a set of approximately 1000 image blocks is used to train the neural network.

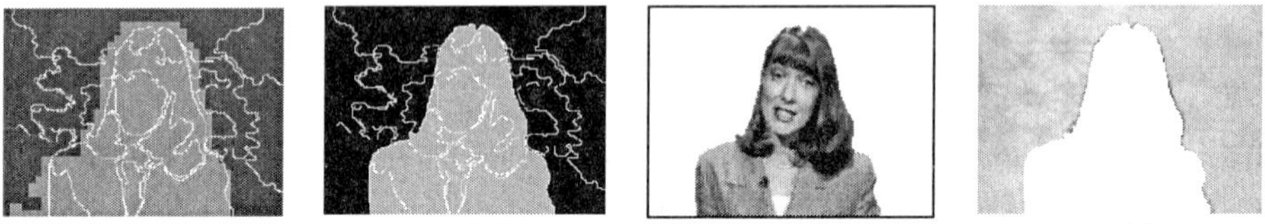

(a) (b) (c) (d)

Fig. 2. (a) Segmentation fusion results for the image of Figure 1(a). (a) Depth segmentation overlaid with color segment contours (in white). (b) Fused segments overlaid with color segment contours. (c) Foreground object. (d) Background object.

(a) (b) (c) (d)

Fig. 3. The tracking performance of the proposed adaptive neural network architecture for several frames at different time instances within the scene of Figure 1. (a,c) The original frames. (b,d) The tracking results of the foreground object.

Let us assume that the decision mechanism activates a new retraining process at the stereo frame, depicted in Figure 1(a). Then, this stereo frame is analyzed and the color and depth segmentation masks are constructed, depicted in Figures 1(b,d). After creation of the color and depth segmentation maps the information fusion scheme is incorporated to extract the retraining set. More specifically, color segments are projected onto the depth map and fused according to depth similarity. Figure 2 describes the proposed segmentation fusion method in case of the image in Figure 1(a). Depth segmentation, shown with two different gray levels as in Figure 1(d), is overlaid in Figure 2(a) with the white contours of the color segments, as obtained from Figure 1(b). In Figure 2(b) all color segments, projected on the foreground object, are indicated with gray, while the "background" color segments are black. Thus, fusion of the two segmentation maps provides accurate extraction of the foreground and background objects, as illustrated in Figures 2(c,d). Based on the color-depth segmentation fusion the retraining set S_c is constructed. Afterwards, the weight adaptation algorithm is activated to update network weights to current conditions. Then video object tracking is performed, until a new retraining process is activated. In order to evaluate the tracking efficiency of the network, in Figures 3(b,d) the tracking results of two different frames are presented, belonging to the same scene with the frame of Figure 1(a). The original left-channel stereo frames are depicted in Figures 3(a,c). It should be mentioned that the video object is extracted in block resolution (8x8 pixels) since each block is classified only to one class.

Results for another video shot are depicted in Figure 4. The scene consists of two actors within a complicated background. The tracking results are presented in Figures 4(b,c) for two different video frames of this scene. As is observed, the neural network performs accurate tracking, though the complicated visual content.

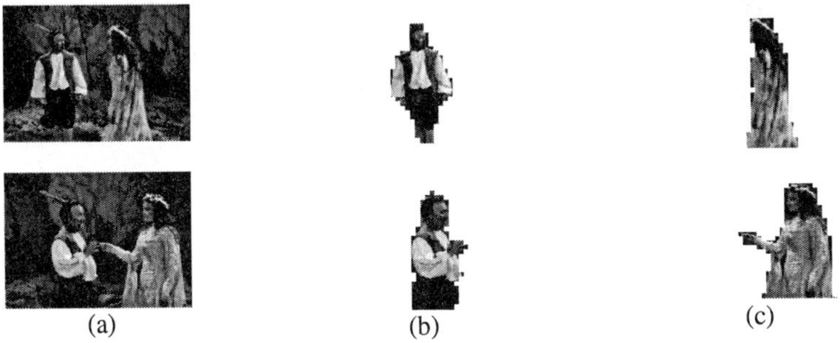

Fig. 4. The tracking performance of the proposed adaptive neural network architecture for two frames at different time instances within a scene. (a) The original frames. (b,c) Tracking results of the two foreground objects.

Acknowledgments

The authors wish to thank Mr. Chas Girdwood for providing the stereoscopic sequence “Eye to Eye” (ACTS MIRAGE project) and Dr. Siegmund Pastoor of the HHI (Berlin), for providing the video sequences of the DISTIMA project. This research is funded by the Institute of Governmental Scholarships of Greece.

References

1. K. N. Ngan, S. Panchanathan, T. Sikora and M.-T. Sun, "Guest Editorial: Special Issue on Representation and Coding of Images and Video," IEEE Trans. CSVT, Vol. 8, No. 7, pp. 797-801, November 1998.
2. ISO/IEC JTC1/SC29/WG11 N3156, "MPEG-4 Overview," Doc. N3156, Maui, Hawaii, December 1999.
3. S. Y. Kung, "Neural Networks for Intelligent Multimedia Processing," IEEE Signal Processing Magazine, vol. 14, no. 4, pp. 44-45, July 1997.
4. Meyer and S. Beucher, "Morphological Segmentation," Journal of Visual Communication on Image Representation, Vol. 1, No. 1, pp.21-46, Sept. 1990.
5. O. J. Morris, M. J. Lee and A. G. Constantinides, "Graph Theory for Image Analysis: an Approach based on the Shortest Spanning Tree," IEE Proceedings, Vol. 133, pp.146-152, April 1986.
6. W. B. Thompson and T. G. Pong, "Detecting Moving Objects," Int. Journal Computer Vision, Vol. 4, pp. 39-57, 1990.
7. J. Wang and E. Adelson, "Representing Moving Images with Layers," IEEE Trans. Image Processing, Vol. 3, pp. 625-638, Sept. 1994.
8. D. Wang, "Unsupervised Video Segmentation Based on Watersheds and Temporal Tracking," IEEE Trans. CSVT, Vol. 8, No. 5, pp. 539-546, 1998.
9. T. Meier, and K. Ngan, "Video Segmentation for Content-Based Coding," IEEE Trans. CSVT, Vol. 9, No. 8, pp. 1190-1203, 1999.
10. A. Doulamis, N. Doulamis, S. Kollias, "On Line Retrainable Neural Networks: Improving the Performance of Neural Networks in Image Analysis Problems," IEEE Trans. on Neural Networks, Vol. 11, No. 1, January 2000.
11. A. Doulamis, N. Doulamis, K. Ntalianis, and S. Kollias, "Efficient Unsupervised Content-Based Segmentation in Stereoscopic Video Sequences," Journ. Artificial Intellig. Tools, World Scientific Press, vol. 9, no. 2, pp. 277-303, June 2000.
12. B. L. Yeo and B. Liu, "Rapid Scene Analysis on Compressed Videos," IEEE Trans. CSVT, Vol. 5, pp. 533- 544, Dec. 1995.

Part XIII

Computational Neuroscience

A Model of Border-Ownership Coding in Early Vision

Masayuki Kikuchi[1] and Youhei Akashi[2]

[1] Institute of Information Sciences and Electronics, University of Tsukuba, Tennoudai 1-1-1, Tsukuba, Ibaraki, 305-8573, Japan
kikuchi@viplab.is.tsukuba.ac.jp
[2] Department of Systems and Human Science, Graduate School of Engineering Science, Osaka University, Machikaneyama 1-3, Toyonaka, Osaka, 560-8531, Japan

Abstract. Zhou et al. had found through physiological experiments that in early visual system there are cells encoding the side of the object to which local contour-elements belong, and they called this way of representation "border-ownership coding" (Zhou et al., 2000). This study shows that a simple neural network model supposing early visual system can encode border-ownership. We confirmed that the cells in our model exhibited responses similar to the cells coding border-ownership found by Zhou et al. by computer simulation

1 Introduction

One of the important characteristics of contours in visual images is that contours divide images into two types of perceptually different regions: regions belonging to figure, and regions belonging to ground. When some parts are recognized as figural regions, the rest are autonomously perceived as ground. This perception is called figure-ground segregation.

Rubin had demonstrated an ambiguous figure that causes two different and alternating interpretations: the one is two faces, and the other is an vase[6]. When two faces are recognized, the rest region is perceived as ground, on the other hand when a vase is recognized, the flanking two regions are perceived as not faces but grounds. Both of the vase and the two faces cannot be recognized simultaneously, though borders between those regions provide form information for both interpretations. Rubin had shown that above face-vase phenomenon is not due to the competition between interpretations based on knowledge processed in higher areas, by a demonstration that when one sees an image containing two regions whose border forms an unfamiliar shape, similar figure-ground reversal effect occurs. This fact suggests that figure-ground segregation is achieved in relatively early stages in the visual system.

Nakayama et al. had discussed from theoretical standpoint that when two regions corresponding to different surfaces meet, only one region can own the border between them[5]. According to this notion, Rubin's figure cannot be interpreted as faces and a vase simultaneously because border-ownership is assigned to only one side of the adjacent regions.

G. Dorffner, H. Bischof, and K. Hornik (Eds.): ICANN 2001, LNCS 2130, pp. 1069–1074, 2001.

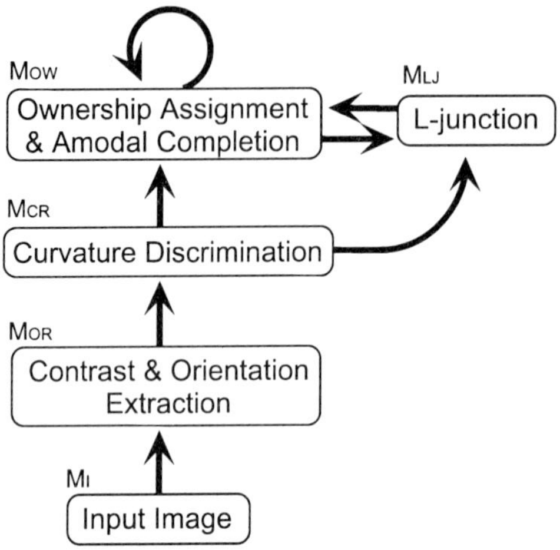

Fig. 1. Model Architecture

Zhou et al. [11] had found through physiological experiments that many cells in visual areas V2 and V4 encode the side of the object to which local contour-elements belong. This physiological finding corresponds to above psychological arguments. Based on some experimental evidences, Zhou et al. suggested that the property of border-ownership coding emerge within the early stages of the visual system, independent of higher visual areas.

Both psychological and physiological findings provide evidences of border-ownership coding in early visual areas. However, the mechanism realizing this coding has not been revealed yet. We propose a neural network model of border-ownership coding in early vision. Our model relies on only local geometrical information such as curvatures of contours, and outer angles at L-junctions. Propagation and competition mechanisms working along contour assign border-ownership. We tested the ability of the model by computer simulation, and confirmed that the model worked correctly.

2 The Model

Figure 1 depicts the architecture of the model. The model has several modules. Each module consists of specific neural networks.

At first, in module M_{OR}, contrast is extracted from input image in M_I by mexican-hat filters, and each orientation components is detected by filters supposing simple cells. These mechanisms are the same as the one in the model proposed previously[4][3].

Next module M_{CR} discriminates curvatures from orientation components M_{OR}. For each position and each orientation, there are some curvature cells tuned to different curvatures. This curvature discrimination is realized by detecting slight differences in orientation and in position from former module M_{OR}. The principle of this process is the same as the one in the previous model[3].

M_{CR} cells send signals for next module M_{OW}, ownership assignment stage. The strength of those signals depends on detected curvature: cells responding to small curvature send weak signals, on the contrary cells responding to large curvature send strong signals. At the same time, in module M_{LJ} L-junctions are detected at the positions where orientation of contour change discontinuously. Cells in M_{LJ} send also signals to cells in ownership assignment module M_{OW}. The strength of those signals is in proportion to their corresponding outer angles at L-junctions.

In ownership assignment module M_{OW}, there are two cells for each position, orientation, and curvature. Each cell corresponds to a particular side of a contour with particular position, orientation and curvature. Cells in this module receive signals from both cells in M_{CR} and cells in M_{LJ} as following manner: on the parts of contour with smooth curve, only cells corresponding to the side to which curvature vectors are pointing receive those signals, and cells corresponding to the other side receive no signals. At each L-junction, cells corresponding to inner side of the angle receive signals in proportion to the value of outer angle. Cells corresponding to the other side receive no signals from M_{LJ}.

Signals from M_{CR} and M_{LJ} are used to determine initial values of the cells in ownership assignment module M_{OW}. After initial values are set, signals start to propagate along contour on each side separately. The local connections for this propagation are set so that as iteration of response calculation goes on, responses of cells approach to the average of initial values along the contour. Those connections for propagation construct reaction-diffusion system. Another important mechanism is mutual inhibition between two sides of the cells along a contour. When an input pattern has concave parts, cells corresponding to outer side of the contour at those parts will respond at the initial state. However, owing to this mutual inhibition between two sides and propagation along each side, signals corresponding to the inner side of the contour at those parts will be eliminated.

The mechanism of this border-ownership assignment is based on a mathematical nature: for arbitrary closed curve, summed outer angles and integral curvatures equal 2π when tracing the curve counterclockwise[8]. At the convex parts, curvatures or outer angles at L-junctions are positive values, and at the concave parts, they are negative values. However, the total value summed along contour is positive. The propagation and competition mechanism in our model can be regarded as summation and integral operation along the contour. The side at which signals remain after propagation and competition process represents inner side of the contour.

At L-junctions, connections for signal propagation along the contour described above cannot convey signals properly to the connecting curve by itself, because orientation of the contour changes discontinuously. At these positions, cells in M_{OW} send signals to the cells in M_{LJ}, and cells in M_{LJ} send the received signals to the cells of the connecting curve in M_{OW} module.

The connections for signal propagation along a contour in M_{OW} also have a function of amodal completion: when a part of a input pattern is occluded by other patterns, these connections convey signals from visible part to occluded part with the curvature of the visible part of the contour. As a result, the occluded part will be completed. In addition, for both visible and occluded part of contour, ownership will be assigned.

3 Computer Simulation

We confirmed the ability of the model by computer simulation. Figure 2 shows simulation results. In each case of (a)–(d), left figure represent the input image presented to the module M_I of the network, and right figure shows the network response in module M_{OW}. Short lines drawn from contour elements indicate the assigned ownership side. In case of (a), the square-region owned the border. Case (b) shows the result when an input pattern with the same border-shape

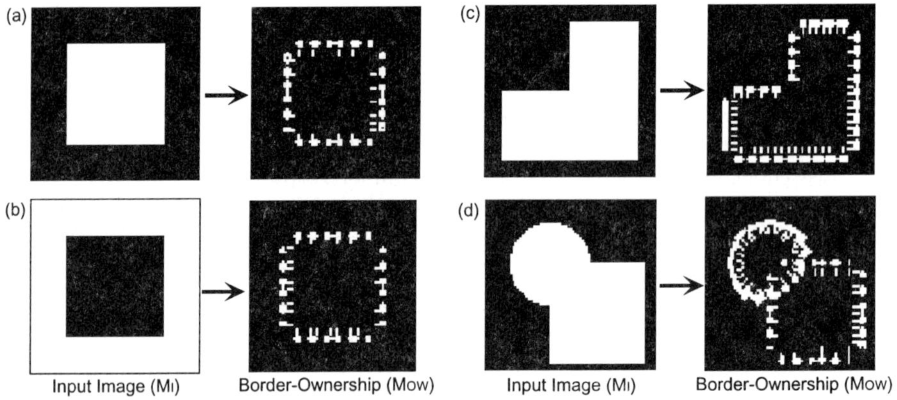

Fig. 2. Simulation results for four cases. In each case, left figure shows input image, and right figure shows responses of the network. Short lines in right figure represent the direction of figure to which contour belongs. (a), (b): Simulations when input pattern had the same shape, but reversed contrast polarity. (c): Simulation when the contour had both convex and concave parts. (d): Simulation when two overlapping patterns are presented.

as the case (a) but reversed contrast polarity was presented. In this case, the network responded as the same manner as the case (a). This simulation explains the behavior of contrast independent border-ownership coding cells. Case (c) shows a result when input pattern contained a concave part. In this case, initial assignments of border-ownership were different at concave part from other parts. However, after signal propagation and competition, all parts along the border indicate identical ownership region. Case (d) shows a result when an image containing two overlapping patterns was presented to the network. Zhou et al. reported that there were cells that could encode border-ownership when occlusion occurred. Our model was also able to assign border-ownership in such a condition. The model exhibited the function of amodal completion in this case. The occluded contour was completed and border-ownership was also assigned for such a part, in addition to visible part.

We had confirmed that the larger the size of an input pattern was, the smaller the responses of the cells encoding border-ownership became. This result is consistent with the property of actual cells coding border-ownership[11]. Such a tendency in our model comes from a reason that average value of responses along a contour decrease as the length of the contour increases, because the value of summed signal along contour is constant regardless of the length of the contour.

Our model requires relatively small number of iteration times till response of the network converges, because of the reaction-diffusion system that works speedy. The convergence time depends on (1) with and strength of connections for propagation along the same side of the contour, (2) strength of mutual inhibition between two sides, and (3) size of input patterns.

4 Discussion

We proposed a neural network model of border ownership coding in early visual system. Our model utilizes local information such as curvatures of border or angles at corners to assign ownership. Local propagation and competition mechanisms along contour determine globally attributed border-ownership. We confirmed that our model worked correctly by computer simulation

The border-ownership assignment problem is closely related to inside/outside problem[1][7][9], because a region inside the closed contour always corresponds to the figure, and a region outside the closed contour corresponds to ground. Ullman[9] had addressed this problem and proposed some mechanisms solving the problem. However, his ideas include sequential operation that does not agree with early visual system. Finkel and Sajda had proposed more biological model. Their model has a mechanism determining "direction of figure", for each contour. That mechanism receives signals from some modules, including closure detection module. However, the mechanism of closure detection is considered to equip very long-range connections, which seems unlikely to exist in early visual system. If it is true that border-ownership is processed in early visual system, we must suppose local mechanisms as a solution for this problem. Our model is constructed by local mechanisms that agree with early visual system.

Kanizsa and Gerbino reported psychological finding that convexity play an important role for grouping process [2]. This fact supports our standpoint that inner side of the corner or curve is the source of the signal assigning border-ownership.

There are cases when a region inside closed contour is perceived as a hole[10]. In such cases, the region outside the contour has ownership. Our current model cannot work correctly in such situations. It is future problem to consider the interaction between the factor that tends to interpret the region inside the closed contour as a figure, and the factor for hole perception.

Acknowledgement

We thank Kunihiko Fukushima, Ko Sakai and Yuzo Hirai for their helpful comments. This research was supported in part by Grant-in Aid #11780272 for Scientific Research from Japan Society for the Promotion of Science.

References

1. Finkel, L.H., Sajda, P.: Object discrimination based on depth-from-occlusion. Neural Computation **4** (1992) 901–921
2. Kanizsa, G., Gerbino, W.: Convexity and symmetry in figure-ground organization. Vision and Artifact, M. Henle ed., New York, Springer (1976) 25–32
3. Kikuchi, M., Anada, K., Fukushima, K.: Neural network model completing occluded contour. ICONIP'98 (Internatinal Conference on Neural Information Processing) (1998) 315–318
4. Kikuchi, M., Fukushima, K.: Neural network model of the visual system: binding form and motion. Neural Networks **9** (1996) 1417–1427
5. Nakayama, K., He, Z.J., Shimojyo, S.: Visual surface representation: a critical link between low-level and higher level vision. Visual Cognition, S.M.Kosslyn, D.N.Osherson eds. MIT Press (1995) 1–70
6. Rubin, E.: Visuall wahrgenommene figuren. Copenhagen (1921)
7. Sajda, P., Finkel, L.H.: Intermediate-level visual representations and the construction of surface perception. Journal of Cognitive Neuroscience **7** (1995) 267–291
8. Shimaya, A.: Perception of complex line drawings. Journal of Experimental Psychology: Human Perception and Performance **23** (1997) 25–50
9. Ullman, S.: High-level vision. MIT Press (1995)
10. Yin, C., Kellman, P.J., Shipley, T.F.: Surface integration influences depth discrimination. Vision Research **40** (2000) 1969–1978
11. Zhou, H., Friedman, H.S., von der Heydt, R.: Coding of border ownership in monkey visual cortex. The Journal of Neuroscience **20** (2000) 6594–6611

Extracting Slow Subspaces from Natural Videos Leads to Complex Cells

Christoph Kayser, Wolfgang Einhäuser, Olaf Dümmer, Peter König, and Konrad Körding

Institute of Neuroinformatics, ETH / University Zürich
Winterthurerstr. 190, 8057 Zürich, Switzerland
{kayser,weinhaeu,olaf,peterk,koerding}@ini.phys.ethz.ch

Abstract. Natural videos obtained from a camera mounted on a cat's head are used as stimuli for a network of subspace energy detectors. The network is trained by gradient ascent on an objective function defined by the squared temporal derivatives of the cells' outputs. The resulting receptive fields are invariant to both contrast polarity and translation and thus resemble complex type receptive fields.

Keywords: Computational Neuroscience, Learning, Temporal Smothness

1 Introduction

A large body of research addresses the problem of obtaining selective responses to a class of stimuli (e.g. Hebb 1949, Grossberg 1976, Oja 1982) but surprisingly few results exist on learning representations invariant to given transformations. But real world problems like recognition tasks do not only require the network to be specific to the relevant stimulus dimensions but also to be insensitive to the irrelevant dimensions (e.g. Fukushima 1988). In this paper we address the problem of learning translation invariance from natural video sequences, pursuing an objective function approach. We implement the temporal smoothness criterion as proposed by Hinton (1989) and used by Földiak (1991). A generative model containing slowly changing hidden variables is assumed. The effect of these hidden variables on linear subspaces can be described by a mixing matrix. This mixing matrix is inverted by the search for slowly varying subspace energy detectors. Instead of mathematically deriving the objective function for these subspaces from an explicit generative model we here explore the effect of a given function on learning of nonlinear detectors. We analyze the obtained slow components and compare them with properties of complex type receptive fields of cortical cells.

2 Methods

The stimuli used to train our network consist of randomly chosen 10 by 10 patches sampled from a natural video recorded by a camera mounted on a cat's

G. Dorffner, H. Bischof, and K. Hornik (Eds.): ICANN 2001, LNCS 2130, pp. 1075–1080, 2001.

head (Betsch et al. submitted). Patches from the same spatial location within the image are taken from 2 subsequent images yielding a pair of intensity vectors I_{t-1} and I_t (images are sampled at 25 Hz). Each vector is normalized to mean zero. The complete stimulus set, consisting of 11000 such pairs, is reduced in dimensionality by PCA and whitened using the procedure described in Hyvärinen and Hoyer (2000). If not stated otherwise, the number of used principal components is 30 (in the following termed PCA dimension).

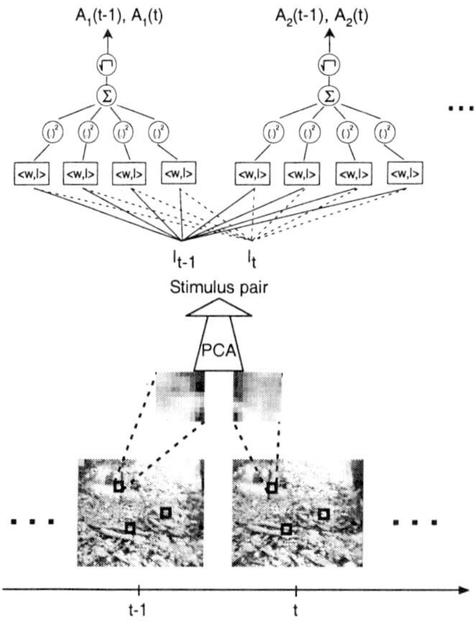

Fig. 1. Network layout. Two cells of the network together with the four sub-units are shown (top) Two images of the natural movie are shown together with patches used as stimuli (bottom).

For the reported results the network consists of 5 neurons each of which summes the input of 4 sub-units (Fig. 1). Each sub-unit has a weight-vector associated and the activity of sub-unit j of neuron i is calculated as the product $A_{ij} = W_{ij} \cdot I$. The neurons are modelled as subspace energy detectors (Kohonen 1996) and their activity is calculated as $A_i = \sqrt{\sum_j A_{ij}^2}$. The analyzed objective function is

$$O_{time} := - \sum_{\text{cells } i} \frac{\left\langle \left(\frac{\mathrm{d}}{\mathrm{d}t} A_i\right)^2 \right\rangle_t}{var_t(A_i)} \qquad (1)$$

where the mean ($<>_t$) and the variance are taken over time. In order to implement this in discrete time, the derivative is approximated by the difference of the activities for two consecutive patches, $A_i(t) - A_i(t-1)$. The variance is furthermore replaced by the product of the standard deviation taken over all the activities for the patches I_{t-1} times the standard deviation for the patches I_t. The network learns by changing the sub-unit weights W_{ij} following the (analytically calculated) gradient of (1) to a local maximum. The gradient ascent is controlled using adaptive stepsizes as described in Hyvärinen and Hoyer (2000) till a stationary state is reached. All sub-units are forced to be orthonormal in whitened space. The weights are randomly initialized with values between 0 and 1. The network layout together with two typical stimuli is shown Figure 1.

In order to quantify the properties of the learned cells their orientation and position specificity is calculated and displayed in θ-r diagrams: The cells are probed with Gaussian bars of defined orientation θ and position r as stimuli and the resulting activities displayed. From these diagrams two parameters are extracted: The *orientation specificity index* (σ_θ) is computed as the mean width of the orientation tuning over all positions. The *position specificity index* (σ_r) is computed by first taking the standard deviation of the activity over all orientations at a fixed position and then averaging over all positions.

3 Results

In order to explore learning of invariant detectors a nonlinear network is implemented (see methods). We use neurons that compute the 2 norm of the corresponding sub-unit activities (Fig. 1). On the activities of these neurons an objective function characterizing their temporal smoothness, O_{time}, is defined and the network is trained till a stationary state is reached (Fig. 2A).

The resulting receptive fields of the sub-units largely resemble those of simple cells (Fig. 2B). After training every neuron receives input from a set of sub-units which all share the same orientation preference but differ in spatial localization. This is shown by the θ - r diagrams for the sub-units (Fig. 2C). Thus the resulting neurons are insensitive to the position of the stimuli and are therefore translation invariant (Fig. 2D). The system is also invariant with respect to the contrast polarity of the stimuli: The response for a bright bar on dark background is the same as for a dark bar on bright background. Note that this contrast polarity invariance is not learned by the network but instead is a built in feature of the transfer function of the neurons (since an even norm is used).

As an important control it is necessary to check that translation invariance is indeed a consequence of the temporal smoothness of the stimuli and not also an inherent network property. The stimulus vectors are randomly shuffled to destroy the temporal coherence of the pairs $\{I_{t-1}, I_t\}$. Figure 3A shows the resulting receptive fields of the sub-units, which no longer exhibit the systematic properties of those obtained with the stimuli in natural order. This shows that the correlations in the time domain of the video sequences are necessary for the learning of the complex like receptive fields.

Since the temporal correlation between patches in natural videos decays gradually over time (Betsch et al. Submitted) we pair frames of larger temporal distances ($\{I_{t-\Delta_n}, I_t\}$ instead of $\{I_{t-1}, I_t\}$). As expected, with growing time shift Δ_n orientation specificity decreases and the cells become more specific to position (Fig. 3B). In the limit of no correlation (large temporal distances or randomly paired frames) position and orientation specificity index become identical within error range.

In the current implementation the stimuli are whitened and all principal components up to the given PCA dimension are amplified to amplitude one whereas the other amplitudes are set to zero. One reason for this preprocessing is the large decrease in computation time when using fewer dimensions. To assess the effect of the choice of the PCA dimension the position and orientation specificity is computed for different dimensions (Fig. 3C). None of these quantities changes significantly. Inspection of the resulting sub-unit receptive fields and θ-r diagrams reveals that still complex like receptive fields are obtained (data not shown). But since the dimension of the stimulus space is now much larger than the number of feature detectors, the coverage of the stimulus space is coarse and most complex cells have similar preferences.

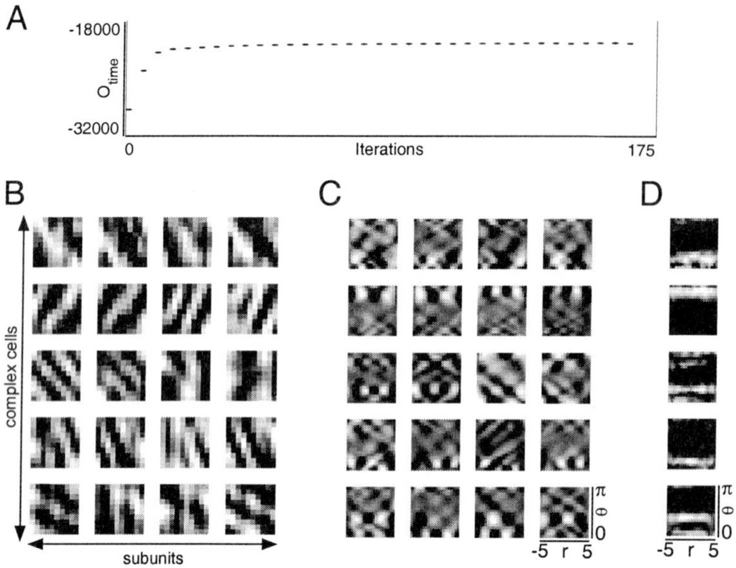

Fig. 2. Results. A) The objective function is optimized till a stationary state is reached. B) Receptive fields of the sub-units after 175 iterations. C) θ -r diagrams for these sub-units. The diagram shows the response strength of the unit for bars of different position (x-axis) and different orientation (y-axis). D) θ -r diagrams for the complex cells.

4 Discussion

The presented results show that complex like receptive fields can be learned by extracting the slowly varying subspaces of natural stimuli. The obtained receptive fields are comparable to those of Hyvärinen and Hoyer (2000) who use a different approach, independent subspace analysis (ISA). ISA uses the same network layout but implements a different objective, independence of the cells' responses, which is comparable to sparse coding. Whereas they use natural photographs taken from PhotoCDs we exploit the temporal domain of natural image sequences.

Another network for learning transformation invariant filters is the adaptive-subspace self-organizing map (ASSOM) proposed by Kohonen (1996). There also the neurons are modelled as sub-space energy detectors but the network learns a two dimensional map such that the activity maximum moves slowly over the network. The cells are implicitely forced to extract slowly varying features resulting in an approach comparable to the work of Foldiak and to the one presented here. Opposed to the ASSOM, the objective function approach incorporates the temporal smoothness in an explicit way and the results shown here were obtained from more natural stimuli.

The fact that quite different objectives lead to similar receptive fields poses the question to which degree the objectives of temporal smoothness and independence are equivalent.

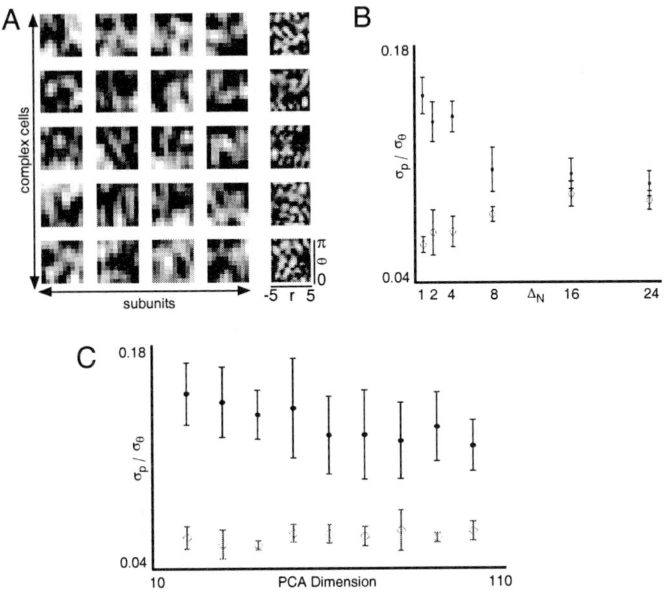

Fig. 3. Controls. A) (Left) Receptive fields of the sub-units for a network trained with randomly paired stimuli (no temporal coherence). (Rightmost column) θ -r diagrams for the (no longer complex) cells. B) Increasing the time lag Δ_N between two subsequent stimuli decreases the orientation specificity σ_θ (circles) and increases the position specificity σ_r (diamonds). Errorbars denote the standard deviation over all cells in the network. C) σ_θ (circles) and σ_r (diamonds) are shown as a function of the PCA Dimension.

It is interesting to note that the temporal smoothness function is very well compatible with a number of physiological mechanisms found in the mammalian cortex (Körding and König 2000). In this respect it is of importance that optimizing the objective function only needs information locally available to the cell.

A number of issues remain for further research: Different PCA dimensions require different subspace sizes and different numbers of neurons for optimal stimulus space coverage. Incorporating a dynamic subspace size in the objective function approach might recruit the optimal number of sub-units needed.

The presented results are obtained by using the 2-Norm of the subspace as transfer function for the cells. In this way the network becomes very similar to the classical energy models for complex cells which are supported by electrophysiological evidence. Some research on the other hand advocates stronger nonlinearities. Riesenhuber and Poggio (2000) for example propose the max function, which corresponds to the infinity norm. It seems likely that this network property can also be learned using the same objective function. Learning the norm of the sub-spaces might be worthwhile since it incorporates learning the nonlinearity of the network. Furthermore this could also lead to an explicitly learned contrast polarity invariance which so far is built in.

Concluding, temporal coherence is a method for learning complex type receptive fields from natural videos, and seems very well suited for learning different network properties of biological systems in which temporal information is ubiquitous.

Acknowledgments

This work was supported by the SNF (CK, PK) and the Boehringer Ingelheim Fonds (KPK). Furthermore we thank A. Hyvärinen and P. Hoyer for making their code available to the public. We are grateful to B. Betsch and C. Arielle for help with the acquisition of the stimulation videos.

References

Betsch, B.Y., Körding, K.P., Einhäuser, W., König, P. What cats see - statistics of natural images. Submited

Földiak, P. (1991). Learning Invariance from Transformation Sequences. Neural Computation, 3(2):194-200.

Fukushima, K. (1988). Neocognitron: A hierarchical neural network capable of visual pattern recognition. Neural Networks, 1, 119-130.

Grossberg, S. (1976) A neuronal model of attention, reinforcement and discrimination learning. International Review of Neurobiology, 18:263-327.

Hebb, D.O. (1949) The Organization of Behaviour: A Neurophysiological Theory. Wiley

Hinton, G.E. (1989) Connectionist Learning Prcedures. Artificial Intelligence, 40:185-234.

Hyvarinen, A., and Hoyer, P.O. (2000). Emergence of phase and shift invariant features by decomposition of natural images into independent feature subspaces. Neural Comput., 12, 1705-20.

Kohonen, T. (1996). Emergence of Invariant-Feature Detectors in the Adaptive-Subspace SOM. Biological Cybernetics, 75(4):281-291.

Körding, K.P., König, P. Learning with two sites of synaptic integration. Network: Computation in neural systems, 11:1-15.

Oja, E. (1982) A simplified neuron model as a principal component analyzer. J. of Mathematical Biology, 15:267-273.

Riesenhuber, M., Poggio, T. (1999) Hierachical model of object recognition in cortex. Nature Neuroscience, 2(11):1019-1025

Neural Coding of Dynamic Stimuli

Stefan D. Wilke

Institut für Theoretische Physik, Universität Bremen,
Postfach 330 440, D-28334 Bremen, Germany
swilke@physik.uni-bremen.de

Abstract. An estimation-theoretical approach is used to characterize the accuracy by which a pair of neurons encodes a time-varying stimulus. Whether high firing rates yield a representational advantage depends on the type of noise involved in spike generation: High rates are favorable for Poissonian noise, but fail to yield improvements in the case of multiplicative Gaussian noise. Moreover, we find that the performance of single-phase stimulus filters increases monotonically with decreasing temporal width, while biphasic filters have a unique optimal time scale that roughly corresponds to the stimulus autocorrelation time. This study demonstrates that estimation-theoretic methods, which have previously been used mainly for static stimuli, can also yield quantitative results on neural codes for dynamic stimuli.

1 Introduction

The search for the neural code by which nervous systems transmit and represent information is still one of the most fascinating aspects of neuroscience. Recent progress in this field has been achieved by applying stimulus reconstruction methods and information theory to neural activity [3]. On the theoretical side, classical estimation theory has proven to be a powerful tool in the analysis of neural codes, since it yields quantitative statements on the achievable reconstruction quality given an encoding scheme [6, 9, 11, 5]. The abovementioned studies share the idea that neural codes optimize reconstruction accuracy, so that aspects of theoretical models that imply an especially good reconstruction (e. g., large receptive fields [11, 5] or uniform correlations [1]) are expected to have an equivalent in real biological systems.

In view of the great computational complexity, theoretical approaches in this direction have often made rather strong assumptions so far. One of them is that the stimulus to be encoded is static in time. In real life, this situation is rather exceptional: animals frequently encounter the problem of tracking continuously (and quickly) varying stimuli, e. g. object positions during navigation tasks.

In this contribution, the encoding of *dynamic* stimuli will be studied using the methods of estimation theory. We have constructed a biologically plausible encoding model that translates a stimulus waveform into a spike train, and a reconstruction algorithm to obtain an estimate of the original stimulus from this spike train. An analysis of the reconstruction quality as a function of the

G. Dorffner, H. Bischof, and K. Hornik (Eds.): ICANN 2001, LNCS 2130, pp. 1081–1086, 2001.

parameters of the encoding model yields that there is an optimal, stimulus-dependent time scale for biphasic filters, and that the exact type of noise involved in the spike train generation qualitatively modifies the reconstruction quality.

2 Stimulus Statistics and Encoding Model

We assume a real-valued random stimulus $s(t)$ with zero temporal mean and Gaussian autocorrelation function

$$R_{ss}(t) := \lim_{T\to\infty} \frac{1}{2T} \int_{-T}^{T} \mathrm{d}\tau\, s(\tau)s(t+\tau) = \frac{\sigma_s^2}{\sqrt{2\pi}} \exp\left(-\frac{t^2}{2\sigma_t^2}\right), \qquad (1)$$

where σ_s and σ_t determine stimulus amplitude and correlation time, respectively.

The stimulus $s(t)$ is encoded by the spike trains (i. e., sums of delta functions) $n_1(t)$ and $n_2(t)$ of two neurons (see Fig. 1a). Their time-dependent mean firing rates are given by a baseline level r_B plus a rectified stimulus-dependent contribution,

$$\langle n_j(t)\rangle =: r_j(t) = r_\mathrm{B} + \left[(-1)^{j+1} r(t)\right]_+, \qquad j = 1, 2, \qquad (2)$$

where $\langle \ldots \rangle$ denotes the average over many repetitions of the same stimulus, and

$$r(t) := \int_{-\infty}^{\infty} \mathrm{d}\tau\, f(\tau)s(t-\tau). \qquad (3)$$

Eq. (2) implies that neuron 1 uses the filter $f(t)$, while neuron 2 uses $-f(t)$ prior to rectification. The function $f(t)$ is assumed to have the biphasic form [2]

$$f(t) := r_0 \Theta(t)\, \alpha \exp(-\alpha t) \left[\frac{(\alpha t)^5}{5!} - \beta \frac{(\alpha t)^7}{7!}\right], \qquad (4)$$

with $\Theta(t)$ denoting the Heaviside step function (see schematic plot in Fig. 1a). The parameter β determines the relative contribution of the negative part, and $1/\alpha$ is the characteristic time scale. Note that the temporal integral over $f(t)$ is independent of α, $\int_{-\infty}^{\infty} \mathrm{d}t\, f(t) = r_0(1-\beta)$.

The model system described above is motivated by the wide-field motion-sensitive H1 neurons in the blowfly's visual system [3]. However, filters of the type (4) are found in reverse correlation experiments in visual systems of various species [8, 3, 7], and therefore appear to originate in some common principle of neural processing.

3 Stimulus Reconstruction from Noisy Spike Trains

An organism faces the task of estimating the signal $s(t)$ from the noisy spike trains $n_1(t)$ and $n_2(t)$. The variable $n(t) := n_1(t) - n_2(t)$ yields $r(t)$ on ensemble averaging, i. e., $\langle n(t)\rangle = r(t)$, and is therefore suited as a starting point of the estimation, since it avoids problems with the non-linearities in eq. (2). A linear estimate $s_{\text{est}}(t)$ of the original stimulus is thus obtained via

$$s_{\text{est}}(t) = \int_{-\infty}^{\infty} \mathrm{d}\tau \, K(\tau) n(t-\tau) \,. \tag{5}$$

The kernel $K(t)$ is chosen to minimize the (time-averaged) mean square error between stimulus and reconstruction, eq. (8), yielding the Kolmogorov-Wiener kernel

$$K(\omega) = \frac{f(-\omega) R_{ss}(\omega)}{|f(\omega)|^2 R_{ss}(\omega) + v_n^2} \,, \tag{6}$$

where the Fourier transform is defined as $g(\omega) := \int_{-\infty}^{\infty} \mathrm{d}t \, \mathrm{e}^{\mathrm{i}\omega t} g(t)$, and v_n^2 is the time-averaged spike-count variance per unit time,

$$v_n^2 := \lim_{T\to\infty} \frac{1}{2T} \int_{-T}^{T} \mathrm{d}\tau \, v_n^2(\tau) \,, \quad v_n^2(t) := \lim_{\Delta t \to 0} \frac{1}{\Delta t} \left\langle [n(t)\Delta t - \langle n(t)\Delta t \rangle]^2 \right\rangle . \tag{7}$$

Here, $n(t)\Delta t$ abbreviates the spike count in $[t; t+\Delta t]$. Note that for low variance v_n^2, $K(\omega) \approx 1/f(\omega)$ essentially inverses the convolution (3), whereas the estimate approaches zero in the limit of large variance. The value of v_n^2 is determined by the stochastic process that generates the spikes. For example, an inhomogeneous Poisson process yields $v_n^2(t) = 2r_{\text{B}} + |r(t)|$, which will be used in the following unless otherwise stated.

The reconstruction error (RE) $\epsilon^2(t)$ achieved by (5) is composed of a bias and a variance component,

$$\epsilon^2(t) := \left\langle [s(t) - s_{\text{est}}(t)]^2 \right\rangle = [\langle s(t) - s_{\text{est}}(t) \rangle]^2 + \left\langle [s_{\text{est}}(t) - \langle s_{\text{est}}(t) \rangle]^2 \right\rangle . \tag{8}$$

The first term (bias, $b(t)^2$) measures deviations of the mean estimate from the true stimulus, while the second term (variance, $v^2(t)$) is sensitive to the scatter of the estimate around its mean. The two components are given by

$$b(t)^2 = \left[s(t) - \int_{-\infty}^{\infty} \mathrm{d}\tau \, K(\tau) \, r(t-\tau) \right]^2 \quad v^2(t) = \int_{-\infty}^{\infty} \mathrm{d}\tau K(\tau)^2 \, v_n^2(t-\tau) \,. \tag{9}$$

Eqs. (9) enable the calculation of the mean RE without "rolling the dice", i. e., without averaging over many realizations of the random variable $n(t)$. In the following, we will write ϵ^2 for the time-average of $\epsilon^2(t)$.

4 Evaluation of Encoding Filters for Dynamic Stimuli

The RE ϵ^2 depends on the rescaled parameters of the encoding model, $r_0 \sigma_t \sigma_s$, $r_{\text{B}} \sigma_t$, $\alpha \sigma_t$, and β. By analyzing these dependencies, it is possible to evaluate encoding strategies within the model class described by (2)–(4), and to find especially favorable filters for a given stimulus statistic. Under the assumption that biological systems also optimize encoding accuracy, this could help understanding features of their neural code.

Baseline Firing Rate. Fig. 1b shows the RE as a function of the stimulus-independent baseline firing rate r_B. For $r_B = 0$, the error is solely due to noise in the spike generation. As r_B increases, more and more "baseline" spikes are added to the spike trains, which gradually corrupts the stimulus estimate. In the limit $r_B \to \infty$, the estimation fails completely and ϵ^2 approaches its upper bound $R_{ss}(t=0)$ given by the error of the "worst" estimate $s_{\text{est}}(t) \equiv 0$. This behavior does not depend on the values of the other parameters.

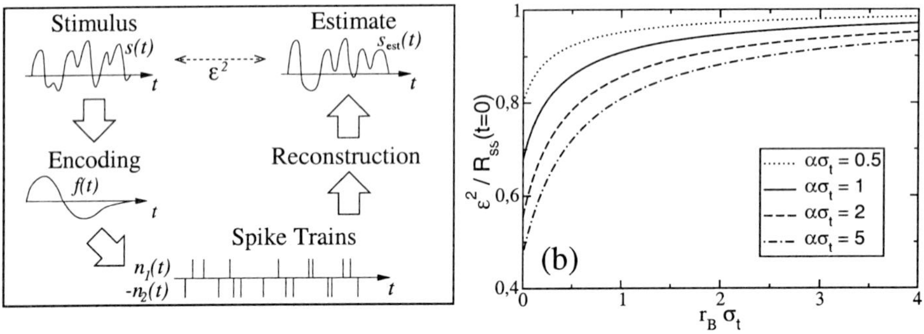

Fig. 1. (a) Sketch of the encoding-decoding scheme. The stimulus waveform $s(t)$ is translated into two spike trains using the filter $f(t)$. An estimate $s_{\text{est}}(t)$ of the stimulus is then constructed from these spike trains. The mean square deviation between $s(t)$ and $s_{\text{est}}(t)$, the RE ϵ^2, is a measure of the encoding quality of the model. (b) Normalized mean squared RE ϵ^2 as a function of baseline firing rate r_B for different values of α. Other filter parameters were $r_0\sigma_t\sigma_s = 1$ and $\beta = 0$.

Firing Rate Scale. A more interesting dependence is expected for the parameter r_0 governing the magnitude of the stimulus-related response $r(t)$. As the firing rate scale becomes small, $r_0\sigma_t\sigma_s \ll 1$, the reconstruction is expected to fail since most stimulus information is erased in the convolution (3) unless $r_B = 0$. In the opposite limit, as the firing rate becomes very large, the behavior of ϵ^2 depends on the spike train variance v_n^2. This becomes obvious if one considers that $f(\omega) \propto r_0$, so that (6) can be written as

$$K(\omega) = \frac{C_1 r_0}{C_2 r_0^2 + v_n^2}, \tag{10}$$

with C_1 and C_2 independent of r_0. If $v_n^2 \propto r_0 + \text{const}$ as for Poissonian noise, v_n^2 can be neglected for large firing rates, and the reconstruction becomes perfect. If, however, the variance grows quadratically with r_0, i. e., $v_n^2 \propto r_0^2 + \text{const}$ as for multiplicative noise, the kernel will approach $K(\omega) \propto 1/r_0$, thus yielding an r_0-independent RE in the limit $r_0\sigma_s \gg r_B$. The dependence of ϵ^2 on the firing rate scale r_0 is demonstrated in Fig. 2a for different noise models.

Time Scale and Filter Form. Fig. 2b shows a plot of the RE as a function of the inverse time scale α for different shapes of the filter function, i. e., for different

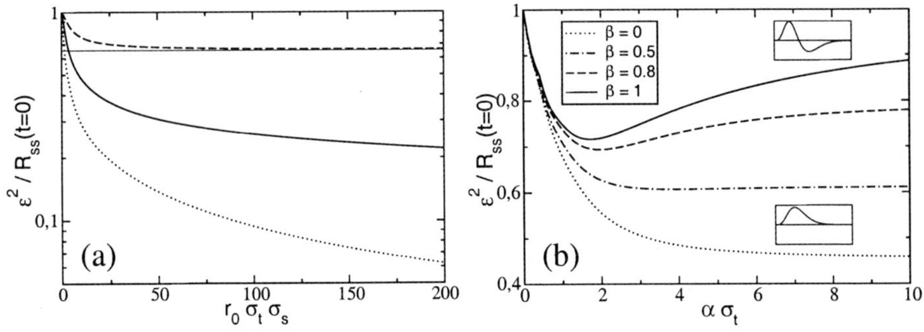

Fig. 2. (a) Normalized mean squared RE as a function of firing rate scaling parameter r_0 for different noise models (additive noise: $v_n^2(t) = 1$ (dotted), Poissonian noise: $v_n^2(t) = 2r_B + |r(t)|$ (solid), multiplicative noise: $v_n^2(t) = [2r_B + |r(t)|]^2$ (dashed)). Other parameters were $\alpha\sigma_t = 1$, $\beta = 0$, $r_B = 1/\sigma_t$. Thin horizontal line: multiplicative noise for zero baseline firing ($r_B = 0$). (b) Mean squared RE error as a function of characteristic frequency α for different values of shape parameter β from $\beta = 0$ (single-phase) to $\beta = 1$ (biphasic). Other parameters were $r_B = 0$, $r_0 = 1/(\sigma_t\sigma_s)$. Small insets indicate shape of filter $f(t)$ for $\beta = 1$ (top) and $\beta = 0$ (bottom).

values of β. The curves for any β converge at $\alpha \ll \sigma_t^{-1}$, where no reconstruction is possible since the filter varies on a much slower time scale than the stimulus. For $\beta = 0$, the RE decreases monotonically with increasing α, which indicates that in this case, sharply peaked filters yield the best representation. This can be understood by considering that $\beta = 0$ implies $f(t) \to r_0\delta(t)$ in the large-α limit, so that the convolution (3) does not involve any smoothing. For larger values of β, ϵ^2 exhibits a qualitatively different α-dependence. In this case, there is a unique minimum around $\alpha \approx 1.5\sigma_t^{-1}$ at which the RE is minimal. Thus, for biphasic filter functions, the wavelength of the filter should match the stimulus autocorrelation time to achieve optimal encoding performance.

At any fixed α, ϵ^2 increases with increasing β, and this effect is strongest at large α. An intuitive argument for this increase of ϵ^2 is that, if the filter is so narrow that the stimulus is essentially constant over its duration (i. e., $\alpha \gg \sigma_t^{-1}$), the biphasic form yields very small values. This effect is most destructive for $\beta = 1$, where the integral over the filter vanishes.

These results must be seen in the context of the task to be performed by the system. For waveform encoding, single-phase filters appear to be generally favorable since ϵ^2 increases monotonically with β for any α. Biphasic filter functions could serve other purposes, such as change detection via signal differentiation. The analysis of a given biological system should take these factors into account by appropriately modifying the error measure (here: eq. (8)) to be optimized.

5 Summary and Conclusions

We have analyzed a class of neural coding models for dynamic stimuli within the framework of estimation theory. This was done by calculating the reconstruction error achievable with the linear minimum mean square error estimator (5) as a function of the encoding filter.

The first result was that, for Poissonian spike generation, a high firing rate yields better performance despite the associated increase of variance. However, this statement does not hold for multiplicative noise in the spike generation mechanism, where the encoding accuracy approaches a finite value at high firing rates. The exact type of noise (Poissonian, additive, or multiplicative) is also important in the context of population coding of static stimuli [10]. Second, it was shown that biphasic filters allow the best reconstruction if their time scale corresponds to the stimulus autocorrelation time, while the performance of single-phase filters always improves with decreasing time scale. This result is in agreement with the observation that empirically measured filter time scales fall into the range of behaviorally relevant decision times [4]: The reconstruction of important features of the stimulus is optimized by these filters, while irrelevant signal components on other time scales are discarded.

We have demonstrated that estimation theory provides a unified framework for the study of neural codes for both static and time-varying stimuli. The method described here can be extended in a straightforward manner to other encoding models, and to populations of neurons. An interesting question in this direction could be, for example, whether diversity of filter waveforms across a population yields a better encoding than identical waveforms. Progress in this direction is hoped to lead to further insights into the mechanisms of neural codes.

Acknowledgment

The author would like to thank C. W. Eurich for helpful discussions. This work was supported by Deutsche Forschungsgemeinschaft, SFB 517.

References

1. L. F. Abbott and P. Dayan. *Neural Comput.*, 11:91–101, 1999.
2. E. H. Adelson and J. R. Bergen. *J. Opt. Soc. Am. A*, 2:284–299, 1985.
3. W. Bialek, F. Rieke, R. R. de Ruyter van Steveninck, and D. Warland. *Science*, 252:1854–1857, 1988.
4. R. de Ruyter van Steveninck and W. Bialek. *Proc. Roy. Soc. Lond. B*, 234:379–414, 1988.
5. C. W. Eurich and S. D. Wilke. *Neural Comput.*, 12(7):1519–1529, 2000.
6. M. A. Paradiso. *Biol. Cybern.*, 58:35–49, 1988.
7. D. L. Ringach, M. J. Hawken, and R. Shapley. *Nature*, 387:281–284, 1997.
8. R. W. Rodieck. *Vision Res.*, 5:583–601, 1965.
9. H. S. Seung and H. Sompolinsky. *Proc. Natl. Acad. Sci. USA*, 90:10749–10753, 1993.
10. S. D. Wilke and C. W. Eurich. *Neural Comput.*, in press, 2001.
11. K. Zhang and T. J. Sejnowski. *Neural Comput.*, 11:75–84, 1999.

Resonance of a Stochastic Spiking Neuron Mimicking the Hodgkin-Huxley Model

Ken-ichi Amemori[1,2] and Shin Ishii[1,2]

[1] Nara Institute of Science and Technology,
Takayama 8916-5, Ikoma, Nara 630-0101, Japan
[2] CREST, Japan Science and Technology Corporation
keniti-a@is.aist-nara.ac.jp

Abstract. This article theoretically examines the stochastic process of a spiking neuron whose spike response is similar to that of the Hodgkin-Huxley (H-H) model. Our simplified model based on a linear differential equation can successfully reproduce the subthreshold membrane properties of the H-H model. Analyzing the stochastic process of this model, we find that the neuron resonates with Poisson input spikes whose intensity oscillates with a certain frequency. This frequency corresponds to the proper frequency of the H-H model.

1 Introduction

Nonlinear dynamics can often be approximated by linear differential equations when weak stimuli are added to the system. Although firing activities of a neuron are highly nonlinear, subthreshold potentials caused by weak inputs to the neuron can be reproduced by linearization [9, 4, 5]. Moreover, the Hodgkin-Huxley model can be reduced into a spiking neuron model [3]. Spiking neuron models assume that neuronal activities are described based on spikes, and the membrane potential of a neuron is given by the weighted summation of post-synaptic potentials, each of which is induced by a single input spike [2, 3]. In this article, we discuss the stochastic process of a spiking neuron model whose response function imitates the spike response of the Hodgkin-Huxley model.

Stochastic Spiking Neuron Model. We first describe a stochastic spiking neuron model. When stochastic spikes are provided, the membrane potential of the i-th stochastic spiking neuron is given by

$$\boldsymbol{v}_i(t) = \sum_{j=1}^{N} w_{ij} \sum_{f=1}^{\boldsymbol{n}_j(t)} u(t - \boldsymbol{t}_j^f) + \sum_{f=1}^{\boldsymbol{n}_i(t)} \eta(t - \boldsymbol{t}_i^f),$$

where random variables are described by bold math fonts. Here, N is the number of synapses projecting to neuron i, w_{ij} is the transmission efficacy of synapse j, $\boldsymbol{n}_j(t)$ is the number of spikes that enter through synapse j until time t, and $\boldsymbol{t}_j^f$ is the time at which the f-th spike enters through synapse j. $\boldsymbol{n}_i(t)$ is the number of spikes emitted by neuron

G. Dorffner, H. Bischof, and K. Hornik (Eds.): ICANN 2001, LNCS 2130, pp. 1087–1094, 2001.

i until time t, and t_i^f is the f-th emission time of neuron i. In this article, we assume that $\eta(t)$, which represents the potential reduction caused by spike emission, takes a constant value η for simplicity. $u(s)$ is called the spike response function, and we assume that this function has the following function form:

$$u(s) = qe^{-\alpha s} \sin(\beta s)\mathcal{H}(s), \tag{1}$$

where $\mathcal{H}(s)$ is a Heaviside (step) function. α, β and q are constant parameters.

The Hodgkin-Huxley Neuron Model. According to the Hodgkin-Huxley (H-H) neuron model [6], the membrane potential of an axon, $v(t)$, is given by

$$C_m \frac{dv}{dt} = \bar{G}_{\rm Na} m^3 h(E_{\rm Na} - v) + \bar{G}_{\rm K} n^4 (E_K - v) + G_m(V_{\rm rest} - v) + I_{\rm input},$$

where the reversal potentials are $E_{\rm Na} = 115$ [mV], $E_{\rm K} = -12$ [mV] and $V_{\rm rest} = 10.6$ [mV], and the maximal conductances are $\bar{G}_{\rm Na} = 120$ [mS/cm^2], $\bar{G}_{\rm K} = 36$ [mS/cm^2] and $G_m = 0.3$ [mS/cm^2]. The channel dynamics are voltage dependent. The dynamics of components corresponding to the potassium and sodium channels are given by

$$\frac{dx}{dt} = \alpha_x(v)(1-x) - \beta_x(v)x,$$

where x is n, m or h. The activating and inactivating probabilities, $\alpha_x(v)$ and $\beta_x(v)$, are given by

$$\begin{aligned} \alpha_n(v) &= (0.1 - 0.01v)/(e^{(1-0.1v)} - 1), & \beta_n(v) &= 0.125e^{-v/80}. \\ \alpha_m(v) &= (2.5 - 0.1v)/(e^{(2.5-0.1v)} - 1), & \beta_m(v) &= 4e^{-v/18} \\ \alpha_n(v) &= 0.07e^{0.05v}, & \beta_n(v) &= 1/(e^{(3-0.1v)} + 1). \end{aligned}$$

The input current is written by $I_{\rm input}(t) = \sum_{j=1}^{N} \sum_{f=1}^{n_j(t)} g_E(t - t_j^f)(v(t) - E_E)$, where $E_E = 65$ [mV] is the reversal potential for excitatory inputs. N is the number of inputs. $g_E(s)$ is the synaptic conductance defined by $g_E(s) = G_{\rm syn} e^{-\gamma s}\mathcal{H}(s)$, where $G_{\rm syn} = 1.0$ [mS] is the conductance gain and $\gamma = 0.5$ [1/ms] is the decay parameter.

Simple neuron models like the leaky integrate-and-fire model intend to describe the membrane behaviors based on the input current $I_{\rm input}(t)$. On the other hand, the membrane potential of the H-H model is significantly dependent on the sodium and potassium currents. In order to mimic the H-H model, a spiking neuron then model needs to employ a spike response function that includes those ionic effects.

Mimicking the Hodgkin-Huxley Model. Here, we present a spiking neuron model which imitates the dynamics of the subthreshold potential of the H-H model. When the input currents do not produce an output spike, the membrane potential is said to keep subthreshold potential. In the subthreshold potential, membrane responses to a single input spike are almost the same. This single spike response can be mimicked by function (1), whose parameters are $\alpha = 0.1$ [ms], $\beta = 0.5818$ [rad/ms], $q = 0.055$ [mV] and $\eta = -1.8$ [mV]. Figure 1 shows that the spike response function successfully mimics

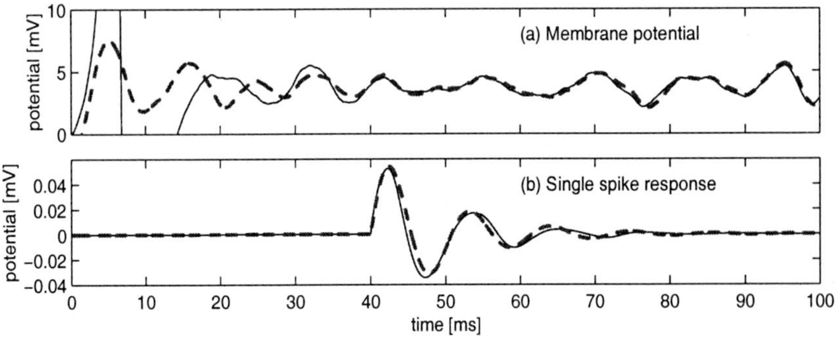

Fig. 1. (a) Dynamics of the membrane potential of the H-H model (solid line) and the spiking neuron model (dash line). The same inputs, generated by Poisson process with fixed intensity, are provided to the both models. (b) Single spike response of the H-H model in the middle of its dynamics (solid line) and the response function (1) we adopted (dash line).

the subthreshold potential of the H-H model. Because the dynamics of the subthreshold potential are composed of responses to weak stimuli, a linear differential equation can well approximate the dynamics. The spike response function (1) is one of the solutions of linear differential equations.

2 Analysis of Subthreshold Potential

We theoretically examine the stochastic process of the membrane potential of the stochastic spiking neuron model, $\boldsymbol{v}(t) = \sum_{j=1}^{N} w_{ij} \sum_{f=1}^{\boldsymbol{n}(t)} u(t - \boldsymbol{t}_j^f) \equiv \sum_{j=1}^{N} w_{ij} \boldsymbol{x}_j(t)$, where input spikes obey an inhomogeneous Poisson process with intensity $\lambda(t)$.

2.1 Dynamics of the Membrane Potential Density

First, we examine the dynamics of the membrane potential density when the firing threshold is ignored.

Single Synaptic Response. For simplicity, synapse index j is omitted in this paragraph. The probability density of the membrane response for a single synapse, $\boldsymbol{x}(t)$, is described as

$$g(x^t) \equiv \langle \delta(x^t - \boldsymbol{x}(t)) \rangle = \int_{-\infty}^{\infty} \frac{d\xi}{2\pi} \exp(\mathrm{i}\xi x^t) \Big\langle \exp\Big(-\mathrm{i}\xi \boldsymbol{x}(t)\Big) \Big\rangle, \tag{2}$$

where $\mathrm{i} = \sqrt{-1}$ and $\langle \cdot \rangle$ denotes an ensemble average over trials. Time $[0, t]$ is partitioned into intervals $[s_l, s_{l+1}]$ ($l = 0, \ldots, T-1$) whose period is Δs; namely, $s_l = l\Delta s$, $s_0 = 0$ and $s_T = T\Delta s = t$. If the time step Δs is small enough, the change of the membrane response at time t, which is caused by the spikes input within $[s_l, s_{l+1}]$, can be approximated as $\Delta \boldsymbol{x}^l(t) \approx u(t - s_l)\Delta \boldsymbol{n}^l$. Here, $\Delta \boldsymbol{n}^l$ denotes the number of spikes input within $[s_l, s_{l+1}]$. Then, the membrane response $\boldsymbol{x}(t)$ can be described by the summation of statistically independent random variables $\{\Delta\boldsymbol{x}^l(t)|l = 0, \ldots, T-1\}$, and the characteristic function of $\boldsymbol{x}(t)$ becomes

$$\left\langle \exp\Big(-\mathrm{i}\xi\boldsymbol{x}(t)\Big)\right\rangle \approx \exp\Big(\sum_{l=0}^{T-1} \ln\Big\langle \exp\Big(-\mathrm{i}\xi u(t-s_l)\Delta\boldsymbol{n}^l\Big)\Big\rangle\Big)$$
$$= \exp\Big(-\mathrm{i}\xi \sum_{l=0}^{T-1} u(t-s_l)\langle\Delta\boldsymbol{n}^l\rangle - \frac{\xi^2}{2}\sum_{l=0}^{T-1} u^2(t-s_l)(\langle(\Delta\boldsymbol{n}^l)^2\rangle - \langle\Delta\boldsymbol{n}^l\rangle^2) + O(\xi^3)\Big).$$

Using the Poisson intensity $\lambda(s)$, we obtain $\langle\Delta\boldsymbol{n}^l\rangle \approx \lambda(s_l)\Delta s$ and $\langle(\Delta\boldsymbol{n}^l)^2\rangle - \langle\Delta\boldsymbol{n}^l\rangle^2 \approx \lambda(s_l)\Delta s$. Therefore, the mean and the variance of the membrane response $\boldsymbol{x}(t)$ are given by

$$\mu(t) = \int_0^t u(t-s)\lambda(s)ds, \quad \sigma^2(t) = \int_0^t u^2(t-s)\lambda(s)ds. \tag{3}$$

If the Poisson intensity $\lambda(t)$ is large enough in comparison to the decay of the membrane potential, α, the membrane response becomes the summation of large number of statistical independent random variables. Therefore, the membrane response density can be represented by a Gaussian distribution [1] due to the central limit theorem.

Total Density. When the input spikes are statistically independent with respect to synapse indices j, the density of the integrated membrane potential becomes

$$g(v^t) = \frac{1}{\sqrt{2\pi\sigma_{\mathrm{tot}}^2(t)}} \exp\Big[-\frac{(v^t - \mu_{\mathrm{tot}}(t))^2}{2\sigma_{\mathrm{tot}}^2(t)}\Big],$$
$$\mu_{\mathrm{tot}}(t) = \sum_{j=1}^{N} w_j\mu_j(t), \quad \sigma_{\mathrm{tot}}^2(t) = \sum_{j=1}^{N} w_j^2\sigma_j^2(t). \tag{4}$$

2.2 Dynamics of the Density Mean and Variance

If the Poisson intensity of inputs can be expanded into finite Fourier series:

$$\lambda(t) = \sum_{n=1}^{M} \{\lambda_n^c \cos(\omega_n t) + \lambda_n^s \sin(\omega_n t)\},$$

the mean and the variance of the density can be analytically obtained by calculating integral (3) with the response function (1).

Mean. Trigonometric functions are analytically integrated, and the mean is obtained as

$$\mu(t) = q\sum_{n=1}^{M} \{A_1(\omega_n)e^{-\alpha t}\cos(\beta t - \theta_1(\omega_n)) + A_2(\omega_n)\cos(\omega_n t - \theta_2(\omega_n))\},$$

where

$$A_1(\omega_n) = \sqrt{\frac{(\lambda_n^c \alpha - \lambda_n^s \omega_n)^2 + (\lambda_n^c)^2 \beta^2}{N_1(\omega_n)}}, \quad A_2(\omega_n) = \beta \sqrt{\frac{(\lambda_n^c)^2 + (\lambda_n^s)^2}{N_1(\omega_n)}},$$
$$\theta_1(\omega_n) = \arctan\left(\frac{\alpha \lambda_n^c(\alpha^2 + \beta^2 + \omega_n^2) + \omega_n \lambda_n^s(\beta^2 - \alpha^2 - \omega_n^2)}{\beta \lambda_n^c(\alpha^2 + \beta^2 - \omega_n^2) - 2\alpha\beta\omega_n \lambda_n^s}\right) + m\pi,$$
$$\theta_2(\omega_n) = \arctan\left(\frac{2\alpha\omega_n \lambda_n^c + \lambda_n^s(\alpha^2 + \beta^2 - \omega_n^2)}{\lambda_n^c(\alpha^2 + \beta^2 - \omega_n^2) - 2\alpha\omega_n \lambda_s}\right) + m\pi,$$
$$N_1(\omega) = (\alpha^2 + (\beta - \omega)^2)(\alpha^2 + (\beta + \omega)^2). \tag{5}$$

Here, m is an integer which is determined by the continuous conditions of $\theta_1(\omega)$ and $\theta_2(\omega)$ with respect to ω. Ignoring the decay term, we find

$$\mu(t) \approx \sum_{n=1}^{M} \beta \sqrt{\frac{(\lambda_n^c)^2 + (\lambda_n^s)^2}{N_1(\omega_n)}} \cos(\omega_n t - \theta_2(\omega_n)).$$

Variance. The variance is also analytically obtained as

$$\sigma^2(t) = q^2 \sum_n \{ e^{-2\alpha t}\{B_1(\omega_n)\cos(2\beta t - \phi_1(\omega_n)) + B_3(\omega_n)\} + B_2(\omega_n)\cos(\omega_n t - \phi_2(\omega_n))\},$$

where

$$B_1(\omega_n) = \frac{1}{2}\sqrt{\frac{(2\lambda_n^c \alpha - \lambda_n^s \omega_n)^2 + 4(\lambda_n^c)^2\beta^2}{(4\alpha^2 + (\omega_n + 2\beta)^2)(4\alpha^2 + (\omega_n - 2\beta)^2)}},$$
$$B_2(\omega_n) = 2\beta^2 \sqrt{\frac{(\lambda_n^s)^2 + (\lambda_n^c)^2}{N_2(\omega_n)}}, \quad B_3(\omega_n) = -\frac{1}{2}\frac{(2\alpha\lambda_n^c - \lambda_n^s \omega_n)}{(4\alpha^2 + \omega_n^2)},$$
$$\phi_1(\omega_n) = \arctan\left(\frac{-2\beta\lambda_n^c(4\alpha^2 + 4\beta^2 - \omega_n^2) + 8\alpha\beta\omega_n\lambda_n^s)}{2\alpha\lambda_n^c(4\alpha^2 + 4\beta^2 + \omega_n^2) - \lambda_n^s \omega(\omega^2 + 4\alpha^2 - 4\beta^2)}\right) + m\pi,$$
$$\phi_2(\omega_n) = \arctan\left(\frac{\lambda_n^c \omega(12\alpha^2 + 4\beta^2 - \omega_n^2) + \lambda_n^s \alpha(-6\omega_n^2 + 8\alpha^2 + 8\beta^2)}{\lambda_n^c \alpha(-6\omega_n^2 + 8\alpha^2 + 8\beta^2) + \lambda_n^s \omega_n(12\alpha^2 + 4\beta^2 - \omega_n^2)}\right) + m\pi,$$
$$N_2(\omega) = (4\alpha^2 + \omega^2)(4\alpha^2 + (\omega + 2\beta)^2)(4\alpha^2 + (\omega - 2\beta)^2).$$

Here, m is an integer which is determined by the continuous conditions of $\phi_1(\omega)$ and $\phi_2(\omega)$ with respect to ω. Ignoring the decay term, we find

$$\sigma^2(t) \approx \sum_{n=1}^{M} 2\beta^2 \sqrt{\frac{(\lambda_n^c)^2 + (\lambda_n^s)^2}{N_2(\omega_n)}} \cos(\omega_n t - \phi_2(\omega_n)).$$

3 Resonance of the Mean and Firing Probability

In this section, we assume that the Poisson intensity $\lambda(t)$ oscillates,

$$\lambda(t) = \lambda(1 - \cos(\omega t)). \tag{6}$$

We examine how the dynamics of the membrane potential depend on frequency ω.

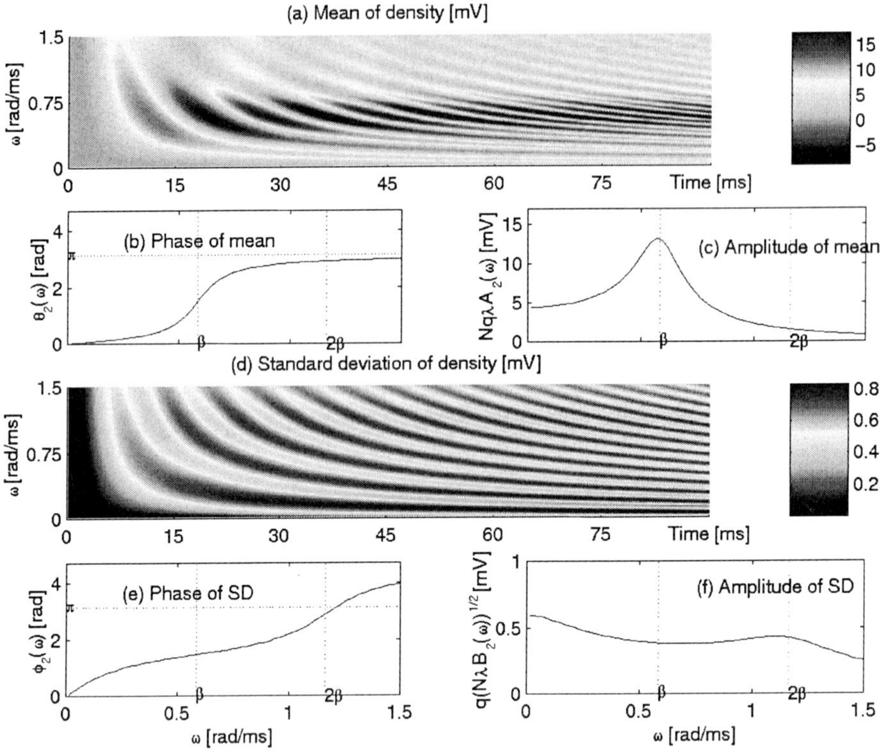

Fig. 2. Mean and variance of the membrane potential density when the Poisson intensity is given by $\lambda(t) = \lambda(1 - \cos(\omega t))$. The number of inputs, N, is 8000, $\lambda = 6$ [Hz] and $w_j = 1$ for all j. (a) Variation of the density mean. (b,c) The phase and amplitude of the leading terms of equation (7), i.e., $\theta_2(\omega)$ in (b) and $Nq\lambda A_2(\omega)$ in (c). The amplitude is maximum around $\omega = \beta \sim 0.5818$ [rad/ms]. (d) Variation of the standard deviation of the density. (e,f) The phase and amplitude of the standard deviation, i.e., $\phi_2(\omega)$ in (e) and $Nq^2\lambda B_2(\omega)$ in (f). The amplitude has two peaks. One is around $\omega = 0$ and the other is around $\omega = 2\beta$.

3.1 Resonance to Oscillatory Inputs

First, we examine how the statistics of the membrane potential density depends on ω. The mean and variance for the Poisson intensity (6) can be estimated as a special case of the solutions obtained in the previous section:

$$\mu_{\text{tot}}(t) \approx \frac{Nq\lambda\beta}{\alpha^2 + \beta^2} - Nq\lambda A_2(\omega)\cos(\omega t - \theta_2(\omega)),$$
$$\sigma^2_{\text{tot}}(t) \approx \frac{Nq^2\lambda\beta^2}{4\alpha(\alpha^2 + \beta^2)} - Nq^2\lambda B_2(\omega)\cos(\omega t - \phi_2(\omega)), \tag{7}$$

where $w_j = 1$ for all j. From equation (5), the amplitude of the density mean, $A_2(\omega)$, has a peak at $\omega = \beta$ (see Fig. 2(c)). Around $\omega = \sqrt{\alpha^2 + \beta^2}$, the phase of the density mean, $\theta_2(\omega)$, drastically changes (see Fig. 2(b)). When ω is close to β, the spiking neuron hence resonates with the oscillatory inputs. This resonance can also be seen as a thick band in Fig. 2(a); namely, the amplitude of the mean is large around $\omega = \beta$. On the other hand, the variance has two peaks, around $\omega = 0$ and 2β. The standard deviation of the density does not significantly depend on ω (see Figs. 2(e) and 2(f)).

3.2 Estimation of the Firing Probability

Although the inducement of action potentials of the H-H model cannot be represented by a fixed firing threshold, rough estimation can be done by a threshold fire model [3]. Here, we approximate the firing probability of our spiking neuron model by using the error function of the density. It should be noted that we consider here firings from the subthreshold potential. Namely, we consider a situation where the firing rate is small.

Firing Probability. If the firing probability is small, it can be simply estimated by the complimentary error function:

$$\rho(t) = \frac{1}{2}\mathrm{erfc}\left(\frac{\vartheta - \eta - \mu_{\mathrm{tot}}(t)}{\sqrt{2\sigma^2_{\mathrm{tot}}(t)}}\right), \tag{8}$$

where $\eta = -1.8$ [mV], $\mathrm{erfc}(x) = 1 - \frac{2}{\sqrt{\pi}}\int_0^x e^{-y^2}dy$ and ϑ is the firing threshold. When the input intensity is given by $\lambda(t) = \lambda(1 - \cos(\omega t))$, which is shown in Fig. 3(b), the firing probability estimated by (8) is shown in Fig. 3(a). White points in Fig. 3(a) signifies high firing probability. This figure shows that the firing probability is high when the input frequency is around $\omega = \beta$. The firing probability is mainly dependent on the variation of the mean rather than that of the variance. The variance is small because the subthreshold potential means weak input stimuli.

Accordingly, the resonance of the density mean remarkably affects the firing probability. From this analysis using spiking neuron model that mimics the H-H model, we can see that the H-H neuron works as a bandpass filter around 92.6 ($\approx 1000\beta/(2\pi)$) [Hz]. This frequency corresponds to the proper frequency of the H-H model, which can be seen in Fig. 1(b). These resonance properties are observed at hair cells in the cochlea of amphibians, which prefer to respond to an acoustic stimulus of a particular frequency [7, 8]. Our results imply that a neuron detects the input frequency rather than the input amplitude.

4 Conclusion and Remark

This article theoretically examines the stochastic process of a spiking neuron model whose spike response imitates that of the H-H model. Our simplified model based on a linear differential equation can mimic the complex nonlinear H-H equation for the subthreshold potential. The stochastic process of the linear model can be theoretically analyzed. The analysis reveals that the neuron resonates with Poisson input spikes whose intensity is oscillatory with a certain frequency. The firing of this neuron is significantly dependent on the input frequency, and the neuron works as a bandpass filter around

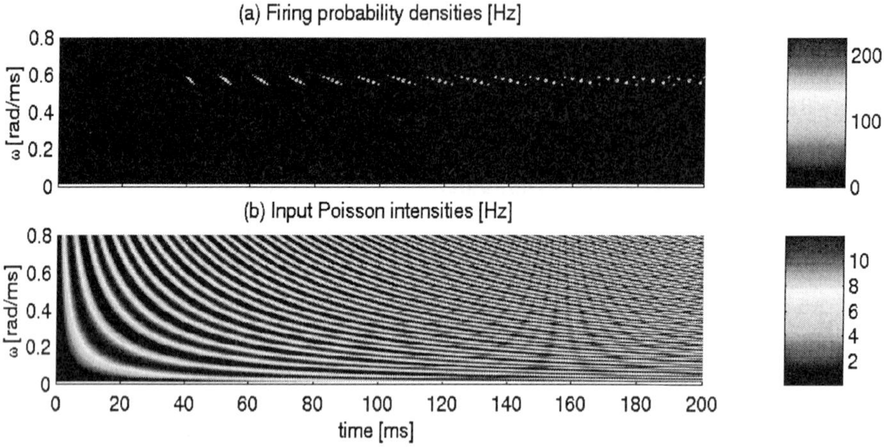

Fig. 3. Estimated firing probability of the spiking neuron model. The number of inputs, N, is 8000. $\lambda = 6$ [Hz]. $\vartheta = 17$ [mV] and $w_j = 1$ for all j. (a) Estimated firing probability density. (b) Input Poisson intensity, $\lambda(t) = \lambda(1 - \cos(\omega t))$.

its proper frequency. Although we approximated the firing probability by using the error function, strict calculation is possible by using a double-Markov Gaussian process analysis [1]. Comparison between them will be discussed elsewhere.

Acknowledgment

We appreciate helpful comments of Yuichi Sakumura and Dai Keyakidani.

References

1. Amemori, K., Ishii, S.: Gaussian process approach to stochastic spiking neurons with reset. In *Proceedings of ICANN 2001*, in press.
2. Gerstner, W.: Spiking neurons. In *Pulsed Neural Networks*, (1998) 3-49. MIT Press.
3. Kistler, W. M., Gerstner W., van Hemmen, J. L: Reduction of the Hodgkin-Huxley equation to a single-value threshold model. *Neural Comp.* **9** (1997) 1015–1045.
4. Koch, C.: Cable theory in neurons with active, linearized membranes. *Biol. Cybern.* **50** (1984) 15–33.
5. Koch, C.: Linearizing voltage-dependent currents. In *Biophysics of Computation.* (1999) 232–247. New York: Oxford Univ. Press.
6. Hodgkin, A. L., Huxley, A. F.: A quantitative description of ion currents and its application to conduction and excitation in nerve membranes. *J. Physiol.* **117** (1952) 500–544.
7. Hudspeth, A. J.: The cellular basis of hearing: The biophysics of hair cells. *Science* **230** (1985) 745–752.
8. Hudspeth, A. J., Lewis, R. S.: A model for electrical resonance and frequency tuning in saccular hair cells of the bullfrog *Rana catesbeiana. J. Physiol.* **400** (1988) 275–297.
9. Mauro, A., Conti, F., Dodge, F., and Schor, R.: Subthreshold behavior and phenomenological impedance of the squid giant axon. *J. Gen. Physiol.* **55** (1970) 497–523.
10. Papoulis, A.: *Probability, Random Variables, and Stochastic Process.* (1984) McGraw-Hill.

Spike and Burst Synchronization in a Detailed Cortical Network Model with I-F Neurons

Baran Çürüklü[1] and Anders Lansner[2]

[1] Department of Computer Engineering, Mälardalen University, S-721 23 Västerås, Sweden
baran.curuklu@mdh.se
[2] Studies of Artificial Neural Systems, Department of Numerical Analysis and Computing Science, Royal Institute of Technology, S-100 44 Stockholm, Sweden
ala@nada.kth.se

Abstract. Previous studies have suggested that synchronized firing is a prominent feature of cortical processing. Simplified network models have replicated such phenomena. Here we study to what extent these results are robust when more biological detail is introduced. A biologically plausible network model of layer II/III of tree shrew primary visual cortex with a columnar architecture and realistic values on unit adaptation, connectivity patterns, axonal delays and synaptic strengths was investigated. A drifting grating stimulus provided afferent noisy input. It is demonstrated that under certain conditions, spike and burst synchronized activity between neurons, situated in different minicolumns, may occur.

1 Introduction

The synchronized activity of the neurons in visual cortex is believed to contribute to perceptual grouping [14,20,21,22,24,28], see also [23]. It is also assumed that neurons in primary visual cortex have small and overlapping receptive field [8]. We assume also that local cortical connectivity (<300 μm) is dense [7]. Dense local connectivity should help neurons to synchronize their activities. This should mean that neurons with same or contiguous receptive fields are active in presence of stimulus.

Studies have shown evidence for long-range horizontal connections in primary visual cortex [10,13]. Recently Bosking et al. also showed evidence for modular and axial specificity of long-range horizontal connections in layer II/III, and suggested that these connections could help neurons respond to a stimulus, in part because they receive input from other layer II/III neurons [11].

We suggest that long-range horizontal connections that exist in layer II/III together with local connections can be used by the neurons for synchronization of their activities over distances of several millimeters on cortex surface.

This follows the same ideas as earlier network model simulations where horizontal connections [16,17,19,25, 27] and synchronization [15,18,26] play an important role.

2 Network Model

We have built a biologically plausible, but sub-sampled, network model. The network consists of neurons situated in six cortical minicolumns (orientation columns), having

G. Dorffner, H. Bischof, and K. Hornik (Eds.): ICANN 2001, LNCS 2130, pp. 1095–1102, 2001.

the same orientation preference. The minicolumns were lined up with a distance of 0.5 mm between two successive ones. We assume that minicolumns are co-linearly positioned in adjacent hyper-columns [8]. The cylinder shaped minicolumns had a height of 300 μm [7] and diameter of 50 μm [8].

Each of the six minicolumns was composed of 12 layer II/III pyramid cells. The neurons were positioned stochastically to fill up the volume of a minicolumn. Connection probability between two neurons was a function of the distance between them [7]. This resulted in a very spread connection probability of 15-80% for neuron pairs, and led to a connectivity of 50-60% between neurons inside a minicolumn [8].

Long-range horizontal connections were defined as connection between two neurons situated in different minicolumns. We computed the connection probability between such pairs of neurons by extrapolating the reported connection probabilities (<500 μm) by Hellwig [7] so that they fitted the findings by Bosking et al. [11]. This resulted in a smooth transition between local and long-range connection probabilities. In average a neuron received 6.2 intra-columnar inputs and 2.1 inter-columnar inputs.

We implemented a leaky Integrate-and-Fire model neuron with noise [1,2]. It was modified to allow adaptation of the membrane time constant [3]. Adaptation is very crucial for the dynamics because of the fact that our network, as many others, does not have inhibitory interneurons. The neuron population was heterogeneous with all values sampled from a uniform distribution with a deviation of 10%.

An axonal diameter of 0.3 μm [8] resulted in a spike propagation velocity of 0.85 m/s [5]. This value together with distances between neurons, and the synaptic delay, resulted in maximum delay inside a minicolumn of approximately 1.36 ms. Maximum delay between neurons situated in two minicolumns was approximately 3.96 ms. This delay corresponds to a distance of approximately 2.52 mm. Maximum values of EPSP:s were in the range of 0.5-2.2 mV for intra-columnar connections and three times those values for inter-columnar connections [4]. The simulation time step was 0.5 ms.

3 Simulation Results

We did three simulations series to demonstrate the influence of long-range horizontal connections on spike and burst synchronization. The initial simulation was done to show that, for our network, synchronization between neurons situated in different minicolumns is not possible, in the absent of the long-range horizontal connections. The network was tested on two different types of inputs. A drifting grating stimulus, modeled as a constant current [6], and a moving bar stimulus, modeled as a Poisson spike train with maximum frequency defined in a Gaussian manner [6]. We saw that in the absence of long-range horizontal connections, the neurons in different minicolumns were not synchronized with each other (Fig. 1). Although the response of the network to the two inputs was different, spike synchronization between neurons inside a minicolumn was present in both cases, especially for constant current input.

The following two simulations were done with the same two input types as described above, and in the presence of long-range horizontal connections. A drifting grating stimulus (constant current) resulted in both spike and burst synchronization for different values on input currents. In the first part of this simulation all the neurons received constant current. Some of the cells displayed repetitive bursting, for higher values of input current, synaptic weights or more dense connections. Spike synchronization as well as burst synchronization could be seen (Fig. 2). It is assume that the repetitive bursting behaviour contributes to synchronous oscillation of the population [12]. But this behaviour destroyed high precision spike synchronization (Fig. 2). In the second part of this simulation we fed only neurons in minicolumns one, two, five and six with constant current to see if neurons in minicolumns one and six still were correlated with each other. Observe that there were no direct connections between minicolumns one and six during this particular simulation, and neurons in the two middle minicolumns were not spiking (not shown here). Spike synchronization between neurons in minicolumn one and six was however present (Fig. 3). There was a tendency for a lag of a couple of millisecons if synaptic weights were low or connections were sparse (not shown here). Observe that the shortest possible delay between two neurons situated in minicolumn one and six was 2.75 ms. Gamma-band oscillation (20-70Hz) could be generated with different values on current input, and we assumed that oscillation was a property of the network.

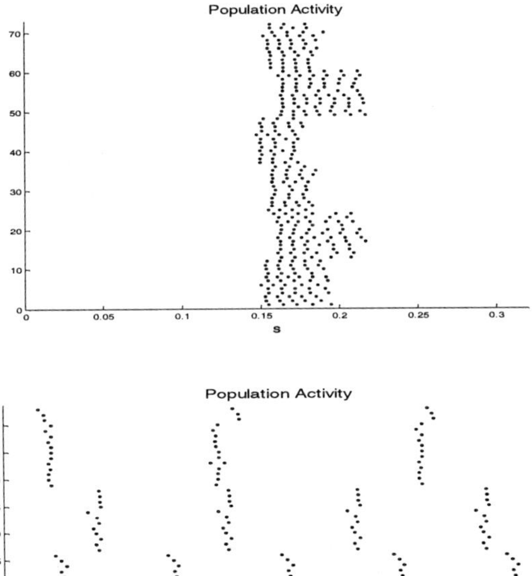

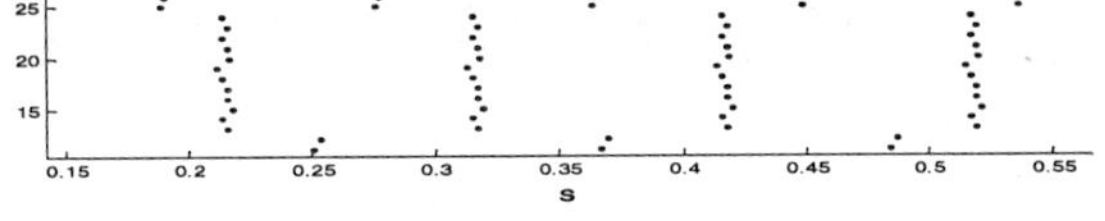

Fig. 1. Population activity in the absent of long-range horizontal connections. Poisson spike train input (*above*), Constant current input (*below*).

The following two simulations were done with the same two input types as described above, and in the presence of long-range horizontal connections. A drifting grating stimulus (constant current) resulted in both spike and burst synchronization for different values on input currents. In the first part of this simulation all the neurons received constant current. Some of the cells displayed repetitive bursting, for higher values of input current, synaptic weights or more dense connections. Spike synchronization as well as burst synchronization could be seen (Fig. 2). It is assume that the repetitive bursting behaviour contributes to synchronous oscillation of the population [12]. But this behaviour destroyed high precision spike synchronization (Fig. 2). In the second part of this simulation we fed only neurons in minicolumns one, two, five and six with constant current to see if neurons in minicolumns one and six still were correlated with each other. Observe that there were no direct connections between minicolumns one and six during this particular simulation, and neurons in the two middle minicolumns were not spiking (not shown here). Spike synchronization between neurons in minicolumn one and six was however present (Fig. 3). There was a tendency for a lag of a couple of millisecons if synaptic weights were low or connections were sparse (not shown here). Observe that the shortest possible delay between two neurons situated in minicolumn one and six was 2.75 ms. Gamma-band oscillation (20-70Hz) could be generated with different values on current input, and we assumed that oscillation was a property of the network.

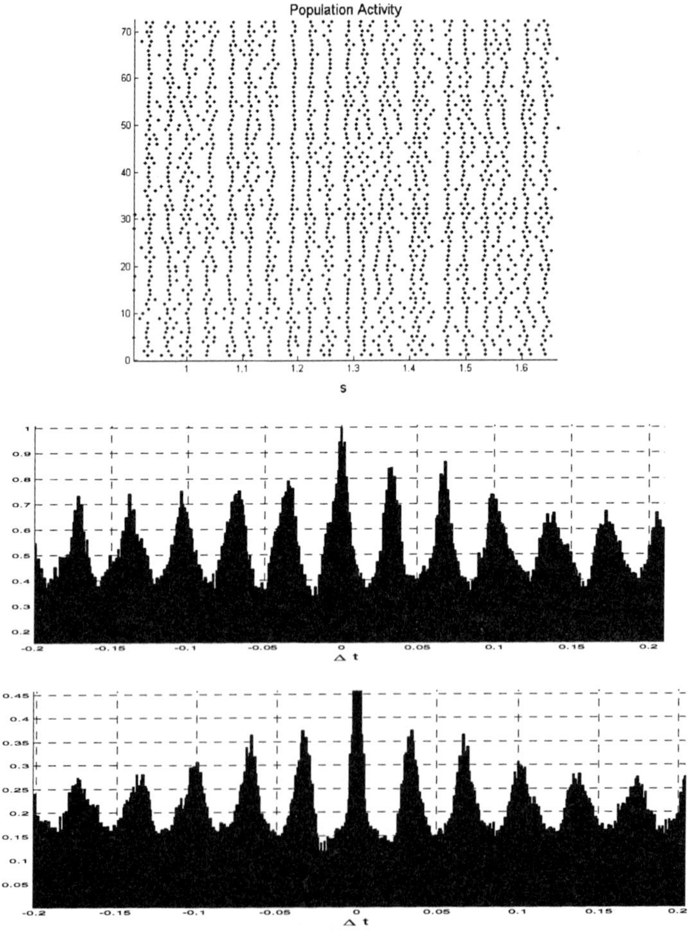

Fig. 2. Population activity (*top*), and cross-correlation (*middle*) between minicolumn one (neurons 1-12) and six (neurons 61-72). Auto-correlation on minicolumn six (*bottom*).

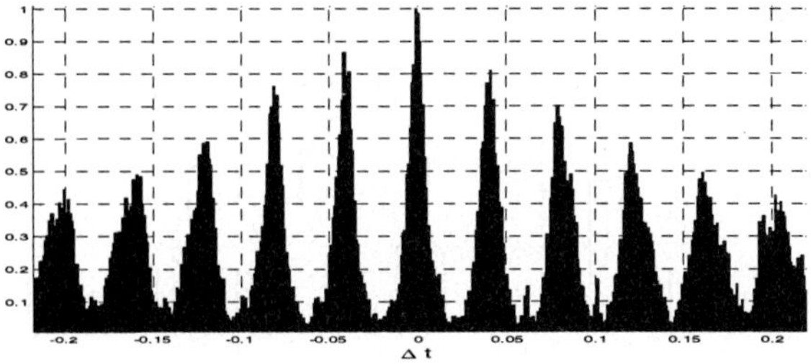

Fig. 3. Cross-correlation between minicolumn one and six. Neurons in the two middle minicolumns were not receiving current input, and were not spiking throughout the simulation. Oscillation frequency is 25 Hz.

The intention of the moving *one* bar-stimulus simulation was to show that, it is possible for neurons to respond to a stimulus that they are not receiving directly. In this simulation all neurons received constant current input. However this input was not sufficient for generating a spike, and corresponded to lowering of the thresholds. Neurons in minicolumns one, two, five and six received uncorrelated Poisson spike trains for a short period of time (80-100 ms) as described above. Poisson spike EPSP:s were in the range of 0.05mV. On average neurons received 65 of these spikes. We saw that neurons in minicolumns three and four were activated by other neurons, with a lag of approximately 4 ms (Fig. 4 and 5).

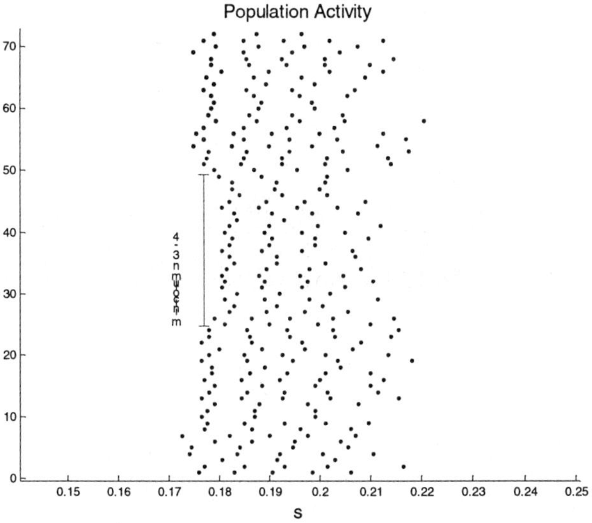

Fig. 4. Population activity. Notice how minicolumns three and four lagged behind.

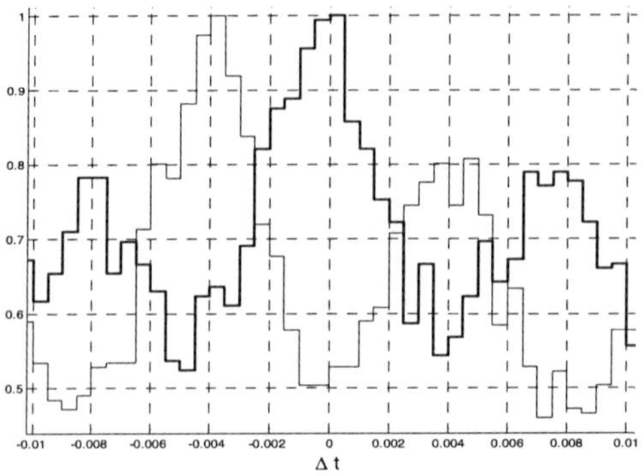

Fig. 5. Cross-correlation between neurons 1-24 and 49-72 (*thick line*). Cross-correlation between neurons 1-24 and 25-48 (*thin line*) for the trial shown in Fig. 3.

High precision spike synchronization was present only during the first two spikes (Fig. 4). With lower values on Poisson spike EPSP:s or larger deviation for the Gaussian distribution high precision spike synchronization could be achieved for longer durations (not shown here). However we could see that burst synchronization and oscillation was present (Fig. 6).

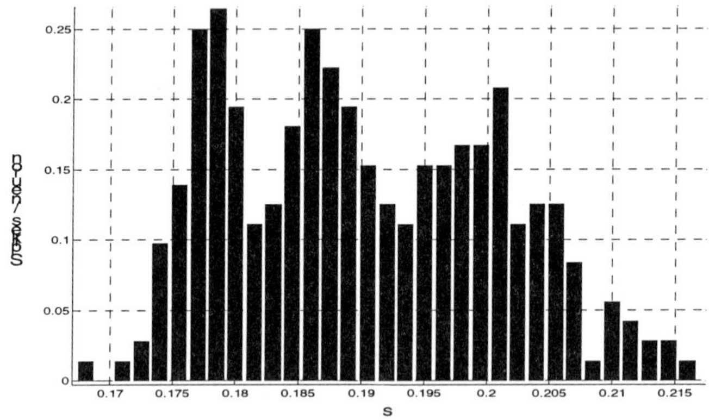

Fig. 6. Average PSTH of six trials.

4 Conclusion

We have shown that phenomena like spike and burst synchronization is possible to simulate with a biologically detailed network of I-F neurons. High precision spike synchronization (<10 ms) was possible to achieve with a constant current input. With increased input some of the neurons displayed repetitive bursting, which helped population oscillation but destroyed high precision spike synchronization.

We saw also that the network behaviour was rather independent of the input type. We assume that more pronounced spike synchronization could be achieved for the one bar stimulus simulation, if the stimulus configuration was different, as stated before.

The long-range horizontal connections played an important role for synchronization. Even with very few connections it was possible to spike synchronize neurons situated in minicolumns 2.5 mm from each other. We would like to stress the fact that, on average, there were at maximum two connections between neurons in minicolumn one and six during the simulations. Spike synchronization was tighter between neurons spatially closer to each other and decreased with distance.

Direction of our future research is to make the network model even more realistic. We are currently testing the network model with Poisson neurons. Cortical neurons are known for their irregularity of the interspike interval [3,9]. Preliminary results have shown that oscillation is possible to achieve with our network using Poisson neurons. Our intention is to expand the model with inhibitory neurons as well. We believe that inhibition will contribute to synchronized activity [4].

References

1. W. Gerstner, Spiking Neurons, in: W. Maass, C. M. Bishop, Pulsed Neural Networks, The MIT Press, 1998.
2. W. Kistler, W. Gerstner, J.L. van Hemmen, Reduction of Hodgkin-Huxley equations to a threshold model, Neural Comp. 9 (1997) 1069-1100.
3. C. Koch, Biophysics of Computation: Information Processing in Single Neurons, Oxford University Press, 1999.
4. E. Fransén, A. Lansner, A model of cortical associative memory based on a horizontal network of connected columns, Network: Comput. Neural Syst. 9 (1998) 235-264.
5. C. Koch, Ö. Bernander, Axonal Modeling, in: M.A. Arbib (Ed.), The Handbook of Brain Theory and Neural Networks, The MIT Press, 1998.
6. R.L. De Valois, N.P. Cottaris, L.E. Mahon, S.D. Elfar, J.A. Wilson, Spatial and temporal fields of geniculate and cortical cells and directional selectivity, Vision Research, 40 (2000) 3685-3702.
7. B. Hellwig, A quantitative analysis of the local connectivity between pyramidal neurons in layer 2/3 of the rat visual cortex, Biol. Cybern. 82 (2000) 111-121.
8. V. Braitenberg, A. Schüz, CORTEX: Statistics and Geometry of Neuronal Connectivity, Springer, 1998.
9. M.N. Shadlen, W.T. Newsome, The Variable Discharge of Cortical Neurons: Implications for Connectivity, Computation, and Information Coding, J. of Neuroscience, 18(10):3870–3896, 1998.
10. C. Lyon, N. Jain, J.H. Kaas, Cortical Connections of Striate and Extrastriate Visual Areas in Tree Shrews, J of Comparative Neurology 401 (1998) 109-128.
11. H. Bosking, Y. Zhang, B. Schofield, D.Fitzpatrick, Orientation Selectivity and the Arrangement of Horizontal Connections in Tree Shrew Striate Cortex, J of Neuroscience, 17(6):2112-2127, 1997.
12. C.M. Gray D.A. McCormick, Chattering cells: superficial pyramidal neurons contributing to the generation of synchronous oscillations in the visual cortex. Science 274 (1996) 109-113.
13. M. Stetter, K. Obermayer, Biology and theory of early vision in mammals, in: H. H. Szu (Ed.), Brains and Biological Networks, INNS Press, 2000.

14. W. Singer, C.M. Gray, Visual feature integration and the temporal correlation hypothesis, Annual Review of Neuroscience, 18 (1995) 555-586.
15. K.E. Martin, J.A. Marshall, Unsmearing Visual Motion: Development of Long-Range Horizontal Intristic Connections, in: S.J. Hanson, J.D. Cowan, C.L. Giles (Eds.) Adv. in Neural Inf. Pro. Sys. 5 (1993) 417-424.
16. S.C. Yen, L.H. Finkel, Extraction of Perceptually Salient Contours by Striate Cortical Networks, Vision Research 38(5):719-741, 1998.
17. S.C. Yen, E.D. Menschik, L.H. Finkel, Perceptual grouping in striate cortical networks mediated by synchronization and desynchronization, Neurocomp. 26-27 (1999) 609-616.
18. J.J. Wright, P.D. Bourke, C.L. Chapman, Synchronous oscillation in the cerebral cortex and object coherence: simulation of basic electrophysiological findings. Bio. Cyber. 83 (2000) 341-353.
19. S.C.Yen, E.D. Menschik, L.H. Finkel, Cortical Synchronization and Perceptual Salience, Neurocomp. 125-130, 1998.
20. S. Friedman-Hill, P.E. Maldonado, C.M. Gray, Dynamics of Striate Cortical Activity in the Alert Macaque: I. Incidence and Stimulus-dependence of Gamma-band Neuronal Oscillations, Cerebral Cortex, 10 (2000) 1105-1116.
21. P.E. Maldonado, S. Friedman-Hill, C.M. Gray, Dynamics of Striate Cortical Activity in the Alert Macaque: II. Fast Time Scale Synchronization, Cerebral Cortex, 10 (2000) 1117-1131.
22. A.R. Haig, E. Gordon, J.J. Wright, R.A. Meares, H Bahramali, Synchronous cortical gamma-band activity in task-relevant cognition, 11 (2000) 669-675.
23. M.N. Shandlen, J.A. Movshon, Synchrony Unbound: A Critical Evaluation of the Temporal Binding Hypothesis, Neuron, 24 (1999) 67-77.
24. W.M. Usrey, R.C. Reid, Synchronous Activity in the Visual System, Annu. Rev. Physiol. 61 (1999) 435–56.
25. S. Grossberg, R. Williamson, A Neural Model of how Horizontal and Interlaminal Connections of Visual Cortex Develop into Adult Circuits that Carry Out Perceptual Grouping and Learning, Cerebral Cortex, 11 (2001) 37-58.
26. T. Wennekers, G. Palm. How imprecise is neuronal synchorization?, Neurocomp. 26-27 (1999) 579-585.
27. T. Hansen, H. Neumann, A model of V1 visual contrast processing utilizing long-range connections and recurrent interactions, ICANN, 61-66, 1999
28. A.K. Engel, P.R. Roelfsema, P. Fries, M. Brecht, W. Singer, Role of the Temporal Domain for Response Selection and Perceptual Binding, Cerebral Cortex, 7 (1997) 571-582.

Using Depressing Synapses for Phase Locked Auditory Onset Detection

Leslie S. Smith

Centre for Cognitive and Computational Neuroscience,
Department of Computing Science and Mathematics, University of Stirling, Stirling
FK9 4LA, Scotland
lss@cs.stir.ac.uk

Abstract. Auditory onsets are robust features of sounds: because the direct path from sound source to ear is the shortest path, the onset is unaffected by reverberation. Many cells in the cochlear nucleus (in the auditory brainstem) are very sensitive to onsets. We propose a neurobiologically inspired spiking onset detector which spikes in phase with the incoming signal, and which can, as a result, be used to measure relatively small inter-aural time differences, permitting sound source direction estimation.

1 Introduction

The auditory system is particularly sensitive to onsets, and this is clear right from the signal on the auditory nerve all the way to cognitive perceptual experience. Hewitt and Meddis [6] suggested a role for what amount to depressing synapses causing onset sensitivity of the auditory nerve. Very strong onset responses have been detected in the bushy and octopus cells of the cochlear nucleus (reviewed in [7]). In auditory signal processing, it has been suggested that onsets may be useful for sound segmentation [8], lip-synchronisation [10], for monaural streaming of signals [3, 4] and for providing interaural temporal differences to enable sound source direction finding which is robust against reverberation [9].

We present results from a novel spiking phase locked onset detector which has a low latency, and whose spikes occur in phase with the incoming signal. This allows it to be used to estimate interaural time differences (ITDs) of the order of 10's or 100's of μs even although actual onset durations are much larger. We present first a brief discussion of why phase locking onset detectors are useful, and of the depressing synapse which underlies this particular implementation, followed by results showing the estimation of shortest path ITDs.

1.1 Phase Locked Onset Detection

Detecting onsets means responding when the input signal increases in intensity. Increases in intensity are characterised by their duration, rate and amount of increase. A useful onset detector must not be too sensitive (or noise will be classified as onsets), nor too insensitive (or onsets will be missed). Further, the onsets

G. Dorffner, H. Bischof, and K. Hornik (Eds.): ICANN 2001, LNCS 2130, pp. 1103–1108, 2001.

it should respond to will have a particular range of increase rate. For example, auditory onset detectors for use with speech will ignore sounds that increase too slowly. Too rapid increases will not normally be a problem in this application because of the way the sound is produced, and also because preprocessing includes bandpass filtering.

Initial onsets are of interest because they are unaltered by reverberation. For incoming sound, the shortest path is always the direct path, so that these onsets come from the direct path. This work is intended as a step towards using onsets for sound streaming [9] where (i) the sound is filtered in many bandpassed channels, (ii) initial onset times and onset based ITDs are computed in each channel and then (iii) those channels in which onsets occur within some small period of time are grouped together. Precise measurement of onset timing is crucial to this approach. Actual onset duration is upwards of 5ms, but ITDs are at maximum about 0.9ms.

To achieve this accuracy needs more than low latency onset detection (where the onset spike is emitted just after the onset). In [5] phase locked spikes are used for ITD estimation: the technique here provides for phase locking of the onset spike. Thus, even although the actual onset takes more than 10ms, the onset spike will always be emitted near a particular point in the phase of the input signal. As a result, we can achieve much better accuracy in computing onset time differences, and hence use onsets to compute interaural time differences relatively accurately. Thus we do not actually use onset time differences, but the phase difference between the signals at onset. The advantage over [5] is that the ITD is computed only for the direct path.

1.2 Modelling Depressing Synapses

Bertram [1] discusses two mechanisms of depressing synapse: vesicle depletion and G-protein inhibition. The mechanism we model is neurotransmitter (vesicle) depletion. The model we are using is a 3 reservoir model used in [6] in the context of inner hair cell to auditory nerve fibers, and later in [11] to model rat neocortex synapses. The model has three interconnected populations of neurotransmitter: M, the presynaptic neurotransmitter reservoir, C, the amount of neurotransmitter in the synaptic cleft, and R, the amount of neurotransmitter in the process of reuptake. These neurotransmitter reservoir levels are interconnected by first order differential equations as follows:

$$\frac{dM}{dt} = \beta R - gM \quad \frac{dC}{dt} = gM - \alpha C \quad \frac{dR}{dt} = \alpha C - \beta R \qquad (1)$$

where α and β are rate constants, and g is positive during a spike, and zero otherwise. We do not model loss or manufacture of neurotransmitter. We take the amount of post-synaptic depolarisation to be directly proportional to C.

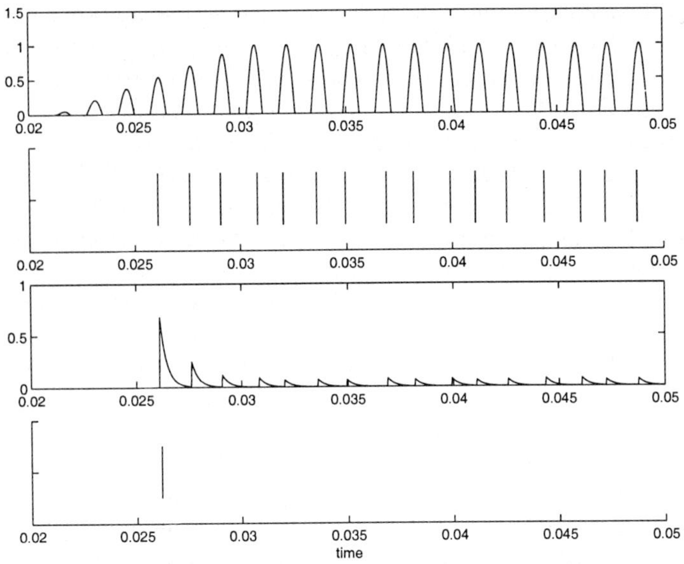

Fig. 1. Top shows 660Hz signal after cochlear filtering($F_c = 660Hz$) and rectification. Next line is the probabilistically generated spikes, followed by the depressing synapse EPSPs. Last line is the onset cell spike output. With above parameters, the onset cell fires on first input spike, but this is not crucial in this application.

2 Experimental Techniques and Results

We first show that a simple depressing synapse allied with an integrate and fire neuron can act as a phase locked onset detector. The audio signal was cochlear filtered, half-wave rectified, then turned into a spike generation probability at each sampled time instant using a logistic function so that when spikes occur they do so at some point during the positive-going part of the rectified signal. This is a much simplified model of the behaviour of the auditory nerve fibres. We are not interested in spike length effects, so we set the spike length to be one sample interval. These spikes were then applied to a depressing synapse, described by equation 1. g was set to 30,000, (which, at a sampling rate of 44100 samples/second is equivalent to 680 (30,000/44.1) for a 1ms pulse), α was set to 3,000, and β to 50. Thus, the first input pulse results in a large EPSP, which rapidly decreases. The EPSP from this synapse was applied to the onset cell, an integrate and fire unit. This unit has a high dissipation (1500) (the cell is very leaky), and has a sufficiently high synaptic weight (20,000) so that a single isolated input pulse causes it to fire. The threshold is set to 1, and the unit has a refractory period of 2ms. The result is illustrated in figure 1 where the input signal is a simple sinusoid with a frequency of 660Hz, and a rise time of 10ms.

From figure 1 it is clear that the output spike occurs at onset and is in phase with the signal because the spike generation maintains phase, and both the depressing synapse and the integrate and fire neuron react quickly. Phase locking will be maintained up to a high frequency, as the auditory nerve signals will remain phase locked, and the there is no delay in either the depressing synapse EPSP or the onset cell itself. These last two factors are not biologically plausible.

To illustrate the use of the technique for ITD estimation, signals were generated digitally (16 bit, 44100 samples/second) using the CoolEdit 2000 package, and played using small amplifier and loudspeaker to the binaural recording system ("Gloria"). This consisted of two matched omni-directional microphones (AKG type C417), placed one in each auditory canal of a model head. This head consisted of a realistic model skull with a latex covering modelling real flesh and skin, and complete with latex rubber pinnae. The head was mounted on a simple model torso. Signals from the microphones were digitised at 44100 samples/second, and stored for later analysis. Signals were played to the binaural recording system from the loudspeaker, at the same height as the model head pinnae, at a distance of 1.5m. They were played at 10° intervals from +30° (right) to −70° (left). The signals used were short tone pulses at 220, 380, 660, 1000, 1500, 2250 and 3250Hz, with (linear) rise times of 10ms. We do not currently have equipment to measure SPL at the pinnae, but the sound level was about that of normal speech.

The stochastic nature of spike generation from the signal means that with a single onset cell for each side, the time difference between onsets for the left and right ears will consist of the estimate of the ITD (EITD) $\pm\frac{n}{f}$ where n is some integer, and f is the frequency of the signal. In addition, the stochastic nature of spike generation means that each onset can occur within one half-period, making the each EITD approximate. To improve the estimate, we model a number of auditory nerve signals and associated onset cells in each channel so that we have a number of onset spikes for each side. To achieve this, we run the simulation a number of times (N_{iter}). We then compute the mean onset firing times for each channel by averaging the near-coincident spike times (i.e. those which fall within $\frac{1}{2f}$ of each other). The ITDs are then calculated from these means. Results with $N_{\text{iter}} = 20$ are shown in figure 2.

3 Discussion and Conclusions

The figure shows that the ITD can be estimated, and that best accuracy is obtained at small ITDs (low angles) using higher frequencies. Although we have not quantified it precisely, it is clear that an accuracy of better than ±5 deg can be obtained at 2250Hz and above for source angles between −10 deg and +10 deg. At higher angles, lower frequencies need to be used. The ITD error becomes larger, and accuracy decreases. We note that 0 deg does not result in an ITD of 0, and this is due to both slight asymmetry in the model head, and possibly differential delays in the audio equipment. We note that the ITD appears to increase as the signal frequency decreases, particularly for signals from the left, but we do not have any explanation for this.

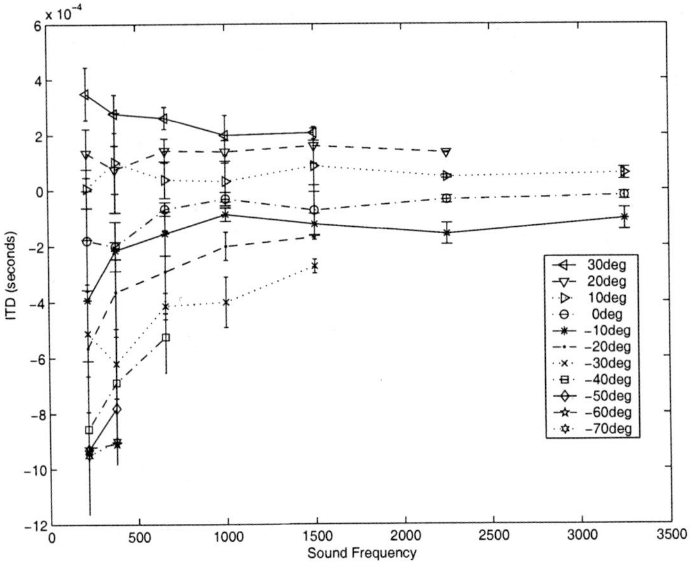

Fig. 2. Graph of ITD found for source angle between −70 deg (i.e. left) and +30 deg (right). Each line shows the ITDs found for stimuli of a particular frequency. Error bars are STDs of between 3 and 5 measurements. Results for low ITD (low angle) at lower frequencies have been omitted as they have a very high STD.

In common with many spike-based algorithms, accuracy requires many parallel measurements: using a larger value for N_{iter} improves estimates. This onset detector system is able to overcome the problem in [8] of latency being affected by signal intensity. The time of occurrence of the onset spikes remains affected by the actual size of the onset, but because of the probabilistic nature of the system, they occur throughout the duration of the onset. Since the onset spikes are phase locked, signal intensity has little effect on the ITD estimate. This also helps when the left and right signals differ in intensity due to head shadow: ITDs can still be computed.

We have concentrated on the ITD at the onsets of these sounds because in an echoing environment, (such as the one the experiments were carried out in), the relative signal phase varies. We note that the relative phase difference between the left and right signals can vary considerably during the sound. This is due to the additive effect of multiple reflections.

The onset detector described has been made as simple as possible. It lacks dynamic range so that if a sound starts, then gets louder, this system will not detect the second onset. Where the second onset occurs soon after the first, this emulates human performance (the precedence effect [2]): however, here, the effect continues for longer. In the auditory system, transduction of the pressure wave to a neural signal is more than rectification: there is onset enhancement as well. The depressing synapse we simulate is logically situated at the onset cell:

in the auditory system, there appear to be two depressing synapses in series, one at the inner hair cell/auditory nerve synapse, and (putatively) one at the auditory nerve/onset cell synapse.

The results here illustrate the technique: we note considerable improvement in human performance with broadband sounds [12], and this suggests we should use wideband sounds, thus permitting multiple concurrent ITD estimations. We note that the spikes from the onset detector code two different pieces of information, namely the signal onset and the signal phase. This suggests that single spikes can be quite rich sources of information, and this may be the source of some novel engineering solutions which emulate the neurobiological system's algorithms.

Acknowledgements

The author thanks David Sterratt, Bruce Graham and Amir Hussain for useful discussions, Frank Kelly, Val Shatwell and Caroline Wilkinson for helping construct Gloria, Nhamo Mtetwa and Frank Kelly for assistance in performing experiments, and the anonymous referees for suggesting improvements.

References

1. R. Bertram. Differential filtering of two presynaptic depression mechanisms. *Neural Computation*, 13:69–85, 2000.
2. J. Blauert. *Spatial Hearing.* MIT Press, revised edition, 1996.
3. A.S. Bregman. *Auditory scene analysis.* MIT Press, 1990.
4. M. Cooke. *Modelling Auditory Processing and Organisation.* Distinguished Dissertations in Computer Science. Cambridge University Press, 1993.
5. W. Gerstner, R. Kempter, J. Leo van Hemmen, and H. Wagner. A neuronal learning rule for sub-millisecond temporal coding. *Nature*, 383(6595):76–78, 1996.
6. M.J. Hewitt and R. Meddis. An evaluation of eight computer models of mammalian inner hair-cell function. *Journal of the Acoustical Society of America*, 90(2):904–917, 1991.
7. R. Romand and P. Avon. Anatomical and functional aspects of the cochlear nucleus. In G. Ehret and R. Romand, editors, *The Central Auditory System.* Oxford, 1997.
8. L.S. Smith. Onset-based sound segmentation. In D.S. Touretzky, M.C. Mozer, and M.E. Hasselmo, editors, *Advances in Neural Information Processing Systems 8*, pages 729–735. MIT Press, 1996.
9. L.S. Smith. Method and apparatus for processing sound. World Patent application number WO 00/01200, published 6 Jan 2000, January 2000.
10. C. Tait. *Wavelet analysis for onset detection.* PhD thesis, Department of Computing Science, University of Glasgow, 1997.
11. M.V. Tsodyks and H. Markram. The neural code between neocortical pyramidal neurons depends on neurotransmitter release probability. *Proc. Nat Acad Sciences*, 94:719–723, 1997.
12. A. van Schaik, C. Jin, and S. Carlile. Human localisation of band-pass filtered noise. *International Journal of Neural Systems*, 5:441–446, 1999.

Controlling Oscillatory Behaviour of a Two Neuron Recurrent Neural Network Using Inputs

Robert Haschke, Jochen J. Steil, and Helge Ritter

University of Bielefeld, Department of Computer Science, Neuroinformatics Group,
P.O.-Box 10 01 31, D-33501 Bielefeld, Germany

Abstract. We derive analytical expressions of codim-1-bifurcations for a fully connected, additive two-neuron network with sigmoidal activations, where the two external inputs are regarded as bifurcation parameters. The obtained Neimark-Sacker bifurcation curve encloses a region in input space with stable oscillatory behaviour, in which it is possible to control the oscillation frequency by adjusting the inputs.

1 Introduction

Since physiological experiments revealed that the brain exhibits complex oscillatory or even chaotic dynamics, research interest concerning the dynamical properties of artificial neural networks has increased. The experiments suggested that the brain intensively uses complex dynamics to solve its computational problems. And because convergent feed-forward networks are relatively well understood, it is natural to investigate oscillatory dynamics of recurrent networks as the next level of complexity.

Because generally oscillatory dynamics of recurrent networks of reasonable size is difficult to describe analytically, several papers numerically proved the existence of a wide range of dynamical behaviour – from a fixed point over a quasiperiodic to a chaotic regime [1]. However, tools from bifurcation theory allow a mathematical analysis of fixed point bifurcations. Since two-neuron networks already exhibit quasiperiodic and chaotic dynamics [6] we restrict our analysis to such simple networks. Because of their low dimension these networks are treatable by analytical methods and thus can be used as well understood, basic building blocks of more complex networks.

Dynamical regimes of continuous-time and discrete-time networks with two varying parameters have been studied for various network architectures and nonlinearities. However, these works consider special connectivities [2] or are restricted to a numerical description [5]. In the present work we consider the external inputs as bifurcation parameters and derive analytical expressions for the bifurcation curves of saddle node, period doubling and Neimark-Sacker bifurcations in dependence of the weights of a fully connected additive network. Our results confirm and extend a numerical investigation of a two-neuron network made for the special case of no selfcoupling in [5]. Izhikevich performed a similar analysis for a continuous-time network, whose Hopf bifurcation curve differs from the Neimark-Sacker bifurcation curve [3].

The Neimark-Sacker curve encloses a region in input space where the network exhibits oscillatory behaviour. Thus crossing this curve allows to switch on and off oscillations. Furthermore input variations within this region allow frequency control.

G. Dorffner, H. Bischof, and K. Hornik (Eds.): ICANN 2001, LNCS 2130, pp. 1109–1114, 2001.

After introducing the investigated network model in section 2 we derive the bifurcation curves in section 3. The obtained expressions are discussed in section 4 with respect to different weight parameters and their corresponding dynamical regimes. Finally in section 5 we investigate the range, in which the oscillation frequency can vary in oscillatory regime. Section 6 concludes with a short discussion.

2 Network Model

We investigate the following class of discrete-time additive recurrent neural networks consisting of two neurons:

$$\begin{aligned} s_1 &\mapsto F_1(\boldsymbol{s}) := \sigma(w_{11}s_1 + w_{12}s_2 + u_1) \\ s_2 &\mapsto F_2(\boldsymbol{s}) := \sigma(w_{21}s_1 + w_{22}s_2 + u_2) \end{aligned} \tag{1}$$

where s_1, s_2 are the state variables of the system describing the activities of each neuron. The parameters w_{ij} and u_i represent the synaptic weight connecting neuron j to neuron i and the external input to neuron i respectively. The activation function σ is assumed to be sigmoid. Especially we consider tanh and Fermi function:

$$\sigma_{\tanh}(x) = \tanh(x) \qquad \sigma_{\text{fermi}}(x) = (1 + \mathrm{e}^{-x})^{-1} = \tfrac{1}{2}(\tanh(x/2) + 1) \tag{2}$$

Note that the tanh and the Fermi system are topologically equivalent. Using the transformation $\hat{\boldsymbol{s}} = \frac{1}{2}(\boldsymbol{s} + \mathbf{1})$, where $\mathbf{1} = [1,1]^t$, one obtains for a tanh system: $\hat{\boldsymbol{s}} \mapsto \sigma_{\text{fermi}}(4W\hat{\boldsymbol{s}} - 2W\mathbf{1} + 2\boldsymbol{u})$. With the inverse transformation the Fermi system can be written as: $\hat{\boldsymbol{s}} \mapsto \tanh(\frac{1}{4}W\hat{\boldsymbol{s}} + \frac{1}{2}W\mathbf{1} + \frac{1}{2}\boldsymbol{u})$.

3 Bifurcation Curves

In the following we assume the weight parameters w_{ij} to be fixed and the external, time-constant inputs u_i as the bifurcation parameters. An input-dependent fixed point $\overline{\boldsymbol{s}} := \overline{\boldsymbol{s}}(\boldsymbol{u})$ of the dynamics (1), i.e. $\boldsymbol{F}(\overline{\boldsymbol{s}}) = \overline{\boldsymbol{s}}$, is asymptotically stable as long as all eigenvalues λ of the Jacobian

$$J := J(\overline{\boldsymbol{s}}) = (D_{\boldsymbol{s}}F)\,(\overline{\boldsymbol{s}}) = D(\overline{\boldsymbol{s}})\,W \tag{3}$$

satisfy $|\lambda| < 1$. $D_{ij} := \delta_{ij}\sigma'(\overline{h}_i)$ is the diagonal matrix of derivatives $\sigma'(\overline{h}_i)$ where $\overline{\boldsymbol{h}} = W\overline{\boldsymbol{s}} + \boldsymbol{u}$. The fixed point becomes unstable and the system undergoes a qualitative change of dynamics, if an eigenvalue λ leaves the unit circle in the complex plane. Depending on this eigenvalue the following basic bifurcations are discriminated [4]:[1]

eigenvalue	bifurcation type	necessary condition	change of dynamics
$\lambda = 1$	saddle node	$\det(J - \mathbf{1}) = 0$	birth of two new fixed points
$\lambda = -1$	period doubling	$\det(J + \mathbf{1}) = 0$	birth of a 2-cycle
$\lambda = \mathrm{e}^{\pm i\theta}$	Neimark-Sacker	$\det(J) = 1$	birth of a limit cycle

[1] As already mentioned continuous-time networks have the same bifurcation conditions only in case of saddle node bifurcations. Period doubling bifurcations do not occur and the necessary condition $\operatorname{tr} J = 0$ for a Hopf bifurcation differs from its discrete-time pendant.

In order to separate the input space into regions of qualitative different dynamics we have to calculate the bifurcation curves, which are separating manifolds of these regions. Thus along a bifurcation curve in the (u_1, u_2)-plane the fixed point equations

$$\overline{s}_i = \sigma\Big(\sum w_{ij}\overline{s}_j + u_i\Big) \quad \Leftrightarrow \quad u_i = \sigma^{-1}(\overline{s}_i) - \sum w_{ij}\overline{s}_j \qquad i = 1, 2 \tag{4}$$

together with the appropriate bifurcation condition hold. To find the bifurcation curves the corresponding system of three nonlinear equations – the bifurcation condition and the two fixed point equations (4) – can be solved due to a special property of the activation functions. More precisely we use the identities

$$\sigma'_{\text{tanh}}(\overline{h_i}) = 1 - \sigma^2_{\text{tanh}}(\overline{h_i}) = 1 - \overline{s_i}^2 \tag{5}$$

$$\sigma'_{\text{fermi}}(\overline{h_i}) = \sigma_{\text{fermi}}(\overline{h_i})\,(1 - \sigma_{\text{fermi}}(\overline{h_i})) = \overline{s_i}\,(1 - \overline{s_i}) \tag{6}$$

to solve each of the bifurcation conditions for $\overline{s}_2$ in dependence of $\overline{s}_1$. This is possible because the equations (5, 6) express the derivatives σ' in terms of the activation function itself and thus ensure together with the fixed point equations (4) that the bifurcation condition does not depend explicitly on the input parameters. Thus we get for the different bifurcation types:

	tanh function	Fermi function
saddle node	$\overline{s}_2 = \pm\sqrt{1 + \dfrac{1 - w_{11}x}{\det W\, x - w_{22}}}$	$\overline{s}_2 = \dfrac{1}{2} \pm \sqrt{\dfrac{1}{4} + \dfrac{1 - w_{11}x}{\det W\, x - w_{22}}}$
period doubling	$\overline{s}_2 = \pm\sqrt{1 + \dfrac{1 + w_{11}x}{\det W\, x + w_{22}}}$	$\overline{s}_2 = \dfrac{1}{2} \pm \sqrt{\dfrac{1}{4} + \dfrac{1 + w_{11}x}{\det W\, x + w_{22}}}$
Neimark-Sacker	$\overline{s}_2 = \pm\sqrt{1 - \dfrac{1}{\det W\, x}}$	$\overline{s}_2 = \dfrac{1}{2} \pm \sqrt{\dfrac{1}{4} - \dfrac{1}{\det W\, x}}$

where we used the abbreviations $x = 1 - \overline{s}_1^2$ in case of tanh function and $x = \overline{s}_1\,(1 - \overline{s}_1)$ in case of Fermi function. Inserting this into the solved fixed point equations (4) we get the bifurcation curves in the (u_1, u_2)-plane parameterized with respect to $\overline{s}_1$ (see Fig. 1).

4 Dynamical Regimes in Dependence of Weights

First we discuss which parameter ranges can exhibit interesting dynamical behaviour. As long as $\sup_{\overline{s}} \|J(\overline{s})\| \leq \sup_{\overline{s}} \|D(\overline{s})\|\,\|W\| < 1$ the system's mapping F is a contraction, so that a unique globally asymptotic stable fixed point exists. This condition especially holds for a small norm $\|W\| \ll 1$ and for large amplitudes of the inputs u_i. In the latter case the activations $\overline{\boldsymbol{h}} = W\overline{\boldsymbol{s}} + \boldsymbol{u}$ are in the saturation region of the sigmoid, such that the diagonal matrix D of derivatives $\sigma'(\overline{h}_i)$ becomes small in magnitude. Thus interesting dynamical behaviour with more than one stable fixed point can be expected for relatively small inputs and weight matrices which satisfy $\|W\| \geq 1$.

Oscillations of arbitrary frequency ω arise, if a Neimark-Sacker bifurcation occurs, whereby the angular frequency $\omega \approx \theta$ is in first order given by the argument of the

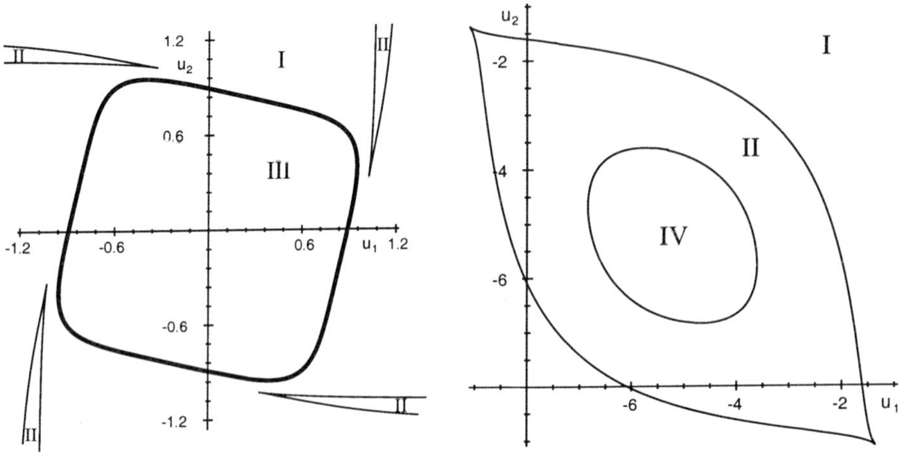

(a) tanh, $w_{ii} = 1.23$, $w_{21} = -w_{12} = 1.15$ (b) Fermi, $w_{ii} = 3$, $w_{12} = w_{21} = 7.45$

Fig. 1. Computed bifurcation curves for tanh function (a) and Fermi function (b) for different weight parameters w_{ij}. Region I corresponds to a globally asymptotic stable fixed point. In region II there exist three fixed points, two of them stable and one unstable. Region III corresponds to oscillatory behaviour evoked by a Neimark-Sacker bifurcation (bold curve) contrasting to region IV, which corresponds to a 2-cycle, evoked by a period doubling bifurcation. Note that due to the topological equivalence of the tanh and Fermi system all bifurcations curves can be observed in both systems, of course with appropriate changes of weights and inputs.

complex Jacobian eigenvalues $\lambda(J) = e^{\pm i\theta}$. If we consider the tanh function with zero inputs, the system's Jacobian at the origin reduces to $J(0) = W$. This allows to choose an appropriate weight matrix with respect to a desired oscillation frequency ω for $\boldsymbol{u} = 0$. This result still holds for inputs belonging to a small neighbourhood of the origin, because eigenvalues continuously depend on matrix entries.

Figure 1 shows the computed bifurcation curves for tanh and Fermi systems at different weight parameters, which are chosen according to the previous considerations. In all subfigures region I corresponds to a globally asymptotic regime of a unique fixed point. As expected, this region always includes parameter ranges, where both inputs u_1 and u_2 are large in magnitude. Because there exist no other bifurcation curves in this region and our expressions cover all possible fixed point bifurcations, the uniqueness of the fixed point is guaranteed. Its existence follows from Brouwer's Fixed Point Theorem.

Region II marks a regime of three fixed points, two of them stable and one unstable. If the input parameters are varied quasi-statically – i.e. slowly enough that the system always approaches a limit set – along a curve crossing region II, a hysteresis effect occurs, as is discussed in [5]. The singular point, where both saddle node bifurcations curves meet, corresponds to a cusp bifurcation, at which all three fixed points coalesce.

In region III an unstable fixed point and a stable limit cycle exist. This cycle is evoked by a Neimark-Sacker bifurcation contrasting to region IV, where a stable 2-cycle – evoked by a period doubling bifurcation – exists beside the unstable fixed point. As is shown in

[1] region III exhibits a wide range of complex dynamical behaviour. There exist subregions where frequency locking occurs, i.e. the quasiperiodic limit cycle reduces to a stable k-cycle. Furthermore special singular weight matrices produce chaotic dynamics in region III, as proven by Wang [6].

Note that the bifurcation curves of saddle node and period-doubling become identical if no self-coupling is allowed. Thus crossing the bifurcation curve besides the existing stable fixed point two new fixed points (a stable and an unstable one) occur together with a stable period-2 orbit. Because $x \geq 0$ and $\det W = w_{12}w_{21}$ in this case, saddle node and period-doubling bifurcations occur only for even, Neimark-Sacker bifurcations only for odd 2-modules. (The bifurcation curves exist only if the derived expressions for $\overline{s}_2$ are defined and assume values within $[0..1]$ or $[-1..1]$ for fermi resp. $\tanh$ function. This result confirms numerically obtained results of Pasemann [5].

5 Frequency Control

As discussed in previous section, crossing the Neimark-Sacker bifurcation curve into region III produces a stable limit cycle, whose frequency equals up to first order the argument of the Jacobian eigenvalue $\lambda(J(\overline{s}))$ on the bifurcation curve. Generally the eigenvalues change along the curve and so does the frequency. Furthermore the oscillation frequency changes, if the input parameters move deeper into region III, because of nonlinear effects.

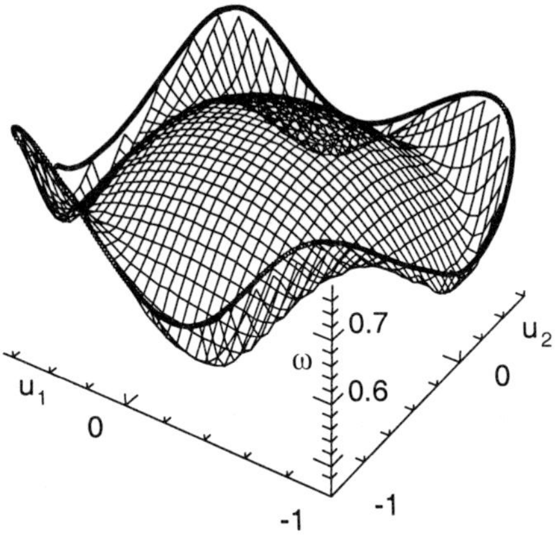

Fig. 2. Frequency drift within region III of neural network of figure 1a.

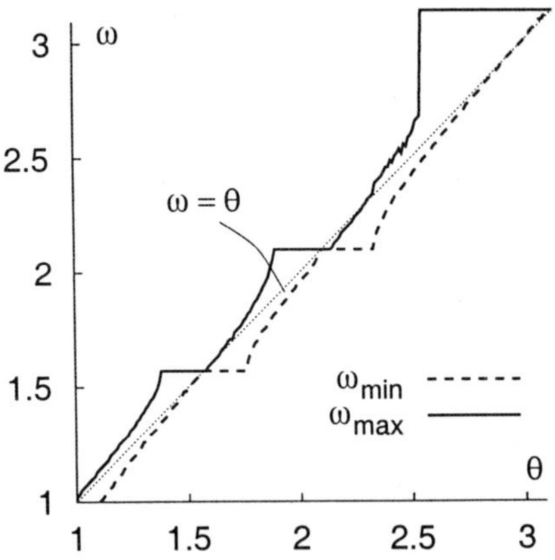

Fig. 3. Frequency range in dependence of rotation angle θ of weight matrix.

Figure 2 shows this frequency drift for a typical configuration with weight parameters equal to those of Figure 1a. The frequencies along the border of the grid are computed using the Jacobian eigenvalues. The frequencies inside the oscillatory region are obtained numerically as the strongest frequency component of a discrete Fourier transformation. As can be seen, they vary within a small range, such that it is possible to adjust the frequency choosing appropriate inputs. Choosing inputs outside the oscillatory region one can switch off oscillations completely.

To investigate the dependence of the frequency range on the weight parameters, we use matrices $W(g,\theta)$ possessing complex eigenvalues $\lambda = g\,\mathrm{e}^{\pm i\theta}$. Numerical studies reveal that the overall width of the possible frequency range increases with the gain g and depends crucial on θ, as shown in Figure 3. The minimally and maximally observable frequencies ω vary around the medium frequency θ. The steps of these curves arise due to frequency locking to $\omega = \frac{1}{2}\pi, \frac{2}{3}\pi$ and π, which corresponds to the 1:4, 1:3 and 1:2 strong resonance respectively.

6 Discussion

We presented analytical expressions for bifurcation curves in the input space of discrete-time two-neuron networks. These expressions enable us to switch on and off oscillatory behaviour by choosing appropriate inputs. Moving quasi-statically within the oscillatory region of the input space we can also continuously modify the frequency of the ongoing oscillation around a medium value defined by the weights. Such networks could be used as pattern generators or basic components of more complex networks, which would have higher-order information processing capabilities using resonance and synchronization effects between its components or with respect to time-dependent inputs. The far goal is to control the dynamics of large networks consisting of small modules using a small number of (input) parameters.

However, even in the simple case of a two-neuron network the whole range of dynamical behaviour cannot be described through the investigation of these basic bifurcations. Especially an investigation of the chaotic regime within region III of oscillatory behaviour requires the development of further mathematical tools.

Acknowledgments

This work was supported by the DFG grants GK-231.

References

1. E. Dauce, M. Quoy, B. Cessac, B. Doyon, and M. Samuelides. Self-organization and dynamics reduction in recurrent networks: stimulus presentation and learning. *Neural Networks*, 11:521–533, 1998.
2. M. di Marco, A. Tesi, and M. Forti. Bifurcations and oscillatory behaviour in a class of competitive cellular neural networks. *International Journal of Bifurcation and Chaos*, 10(6):1267–1293, 2000.
3. Frank C. Hoppensteadt and Eugene M. Izhikevich. *Weakly Connected Neural Networks.* Springer, New York, applied mathematical sciences edition, 1997.
4. Yuri A. Kuznetsov. *Elements of Applied Bifurcation Theory*, volume 112 of *Applied Mathematical Sciences*. Springer-Verlag, New York, Berlin, Heidelberg, 1995.
5. Frank Pasemann. Discrete dynamics of two neuron networks. *Open Systems & Information Dynamics*, 2(1):49–66, 1993.
6. Xin Wang. Period-doublings to chaos in a simple neural network: An analytic proof. *Complex Systems*, 5:425 – 441, 1991.

Temporal Hebbian Learning in Rate-Coded Neural Networks: A Theoretical Approach towards Classical Conditioning

Bernd Porr and Florentin Wörgötter

University of Stirling, Department of Psychology, Stirling FK9 4LA, Scotland
http://www.cn.stir.ac.uk

Abstract. A novel approach for learning of temporally extended, continuous signals is developed within the framework of rate coded neurons. A new temporal Hebb like learning rule is devised which utilizes the predictive capabilities of bandpass filtered signals by using the derivative of the output to modify the weights. The initial development of the weights is calculated analytically applying signal theory and simulation results are shown to demonstrate the performance of this approach. In addition we show that only few units suffice to process multiple inputs with long temporal delays.

1 Introduction

The learning rule by Donald Hebb has always been one of the basic concepts of computational neuroscience [1]. Learning is achieved through changes of the synaptic weights between neurons. The synaptic weight is increased when the pre- and postsynaptic spiking activity is correlated and decreased vice versa. This learning rule is symmetric in time within a certain time window. In contrast to this traditional approach the temporal Hebb rule is asymmetric in time: A weight will be strengthened only if the input precedes the output by a short interval. If the order of input and output is reversed the weight will be decreased [2–4].

We emphasize here the most ubiquitous temporal learning known as "classical conditioning". It takes place on rather long time scales and normally requires two input stimuli, the unconditioned and the conditioned stimulus, which often arise from different sensorial modalities. The unconditioned stimulus (food) is followed by an output event (salivation). After learning the conditioned stimulus (bell), which always precedes the unconditioned stimulus will elicit the same event and, thus, the conditioned stimulus can be interpreted as a predictor of the unconditioned stimulus (the sound of a bell predicts the feeding).

So far temporal sequence learning has in general been achieved by the "temporal difference (TD)-algorithm" [7, 8]. This algorithm is discrete in time and amplitude. Instead, in our approach we have chosen a rate code description as the level of abstraction, which allows to treat continuous signals and this has the advantage that we can develop the theory in an analytically solvable way.

G. Dorffner, H. Bischof, and K. Hornik (Eds.): ICANN 2001, LNCS 2130, pp. 1115–1120, 2001.

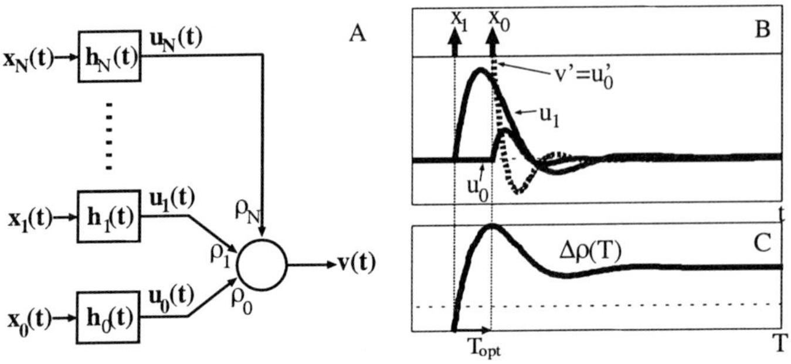

Fig. 1. A) The basic circuit in the time domain. B) shows the inputs x, the impulse responses u for a choice of two different resonators h and the derivative of the output v'. C) shows the initial weight change $\rho_1(T)_{t=0}$. T_{opt} is the optimal time difference between x_1 and x_0 for a maximum change of the weight ρ_1. The time difference of $x_0, x1$ in B is adjusted to T_{opt}, as well. Parameters: $f_0 = 0.1, f_1 = 0.05, Q_0 = Q_1 = Q = 1$.

In order to implement our algorithm we will introduce neuronal filter circuits which will act as damped oscillators and devise a new learning rule. This rule directly uses the temporal change (the derivative) of the neuronal output activity to modify the weights in the network. In addition, our system has the feature, that very few components are required to cover large time intervals (in contrast to the TD-algorithm) and it allows to use multiple inputs, which are both the critical prerequisites for "classical conditioning".

2 The Neuronal Circuit

We consider a system of $N+1$ units h receiving a continuous input signal x and producing a continuous output u. The input units connect with weights ρ to one output unit v (Fig. 1A). All input units are in principle equivalent but we will use h_0 to denote the one unit which transmits the unconditioned stimulus. The output v is then given as:

$$v = \rho_0 u_0 + \sum_{i=1}^{N} \rho_i u_i \qquad \text{with} \qquad u_i(t) = x_i(t) * h_i(t) \tag{1}$$

The transfer function h shall be that of a *resonator* with Frequency f and Quality Q which transforms a δ-pulse input into a damped oscillation (see Fig. 1B). For later use we will define the transfer function of the resonator in the *Laplace* domain as: $H(s) = [(s+p)(s+p^*)]^{-1}$ with $p = a + ib$, $a = \pi f/Q$ and $b = \sqrt{(2\pi f)^2 - a^2}$.

Learning (viz. weight change) takes place according to a new learning rule changing the weights in each time-step by $\Delta\rho_i$, with:

$$\Delta\rho_i = \mu u_i v' \qquad \mu \ll 1 \tag{2}$$

where the weight change depends on the correlation between u_i and the *derivative* of the output v (see Fig. 1B).

3 Performance

3.1 Goal

We state our goal as: After learning, the output unit shall produce a well discernible signal v (e.g., of high amplitude and steeply rising) in response to the earliest occurring conditioned stimulus $x_j, j \geq 1$.

3.2 Analytical Solution of the Initial Weight Change

Similar to other approaches [9] we compute the initial development of the weights as soon as learning starts, because this is indicative of the continuation of the learning process. As usual we require that all weight changes occur on a much longer time scale (i.e., very slowly) as compared to the decay of the oscillatory responses u. This allows us to treat the system in a steady state condition, thus, $\Delta\rho \to 0$ for $\Delta t \to 0$.

In order to calculate this weight change we will now introduce several restrictions: A) The weight of the unconditioned stimulus is set to $\rho_0 = 1$ and kept constant throughout learning. B) We will consider only one unit that can learn, thus, $N = 1$. C) Accordingly we have to deal with only two input functions x_0, x_1 and we define them as (delayed) δ-pulses: $x_0(t) = \delta(t + T), T \geq 0$ and $x_1(t) = \delta(t)$. These restrictions will allow us to develop the theory but can be waived in the end without affecting our basic findings.

Because we assume steady-state, we can rewrite the product in the learning rule (Eq. 2) as a correlation integral between input and output:

$$\Delta\rho_1(T) = \mu \int_0^\infty u_1(T + \tau)v'(\tau)d\tau \tag{3}$$

The value of T represents the delay between conditional and unconditional stimulus. In order to assure optimal learning progress we must, therefore, determine that particular value T_{opt} for which a maximal weight change is observed. Because we are interested in the initial weight change, we can assume $\rho_1(t) = 0$ for $t = 0$. and Eq. 3 turns into:

$$\Delta\rho_1(T)_{t=0} = \mu \int_0^\infty u_1(T + \tau)u_0'(\tau)d\tau \tag{4}$$

In order to solve Eq. 4 we have to switch to the LAPLACE domain where $H(s)$ is the transfer function of the resonator as defined earlier. We get:

$$\Delta\rho_1(T)_{t=0} = \mu\frac{1}{2\pi} \int_{-\infty}^{+\infty} H_1(i\omega)\left[-i\omega e^{Ti\omega} H_0(-i\omega)\right] d\omega \tag{5}$$

We set $P^+ = |p_1|^2 + |p_0|^2$ and $P^- = |p_1|^2 - |p_0|^2$ and get after integration with the method of residues:

$$\Delta\rho_1(T)_{t=0} = \mu \frac{b_1 P^- \cos(b_1 T) + (a_1 P^+ + 2a_0|p_1|^2)\sin(b_1 T)}{b_1(P^+ + 2a_1 a_0 + 2b_1 b_0)(P^+ + 2a_1 a_0 - 2b_1 b_0)} e^{-Ta_1} \quad (6)$$

This result is plotted in Fig. 1. If we assume identical resonators h_0, h_1, we have $a_0 = a_1 = a$ and $b_0 = b_1 = b$ and Eq. 6 simplifies to:

$$\Delta\rho_1(T)_{t=0} = \mu \frac{1}{4ab} \sin(bT) e^{-aT} \quad (7)$$

This is simply the impulse response of the resonator itself. Thus, in this special case the learning curve can be tuned by the choice of a certain impulse response of the resonator. In other words it is possible to implement an *analog temporal Hebb* learning behavior and treat this analytically. Before showing simulation results we note that the above obtained analytical results can be extended to cover the most general system structure as represented in Fig. 1A. With Eq. 1 transformed in the LAPLACE domain the learning rule turns to:

$$V(s) = \sum_{j=0}^{N} \rho_j X_j(s) H_j(s) \quad (8)$$

$$\Delta\rho_i = -\mu \frac{i\omega}{2\pi} \int_{-\infty}^{+\infty} V(-i\omega) U_i(i\omega) d\omega, \quad (9)$$

where Eq. 9 is the general form of Eq. 5. It should be noted that for all $\Delta\rho_i$ this integral can still be evaluated analytically in the same way as in the special case with two resonators discussed above.

3.3 Simulations

In order to validate the approach several simulations of increasing complexity were performed. As the analytical solution treats only the initial learning step we determined the learning behavior for $t >> 0$ with a sequence of two δ pulses x_1, x_0 as inputs. Fig. 2A shows that at the initial learning step the output function v still coincides with the unconditioned stimulus response u_0. After 10 repetitions of the pulse-sequence (Fig. 2 B), the output function has shifted forward to u_1. This forward shift can be interpreted in the sense of classical conditioning: after learning the output signal v predicts the occurrence of u_0 having been conditioned by u_1.

A sufficiently strong correlation occurs in this setup only for a small temporal difference between the input pulses. To improve on this one can use several resonators $h_1, ..., h_n$ with different frequencies which will all receive input from the conditioning stimulus x_1, defining: $T_0 = T; x_j = x_1, T_j = 0$ for $j \geq 1$ (Fig. 1A and Eq. 8). In Fig. 2 C-E we use five resonators to represent x_1 and show the results after 10 learning steps for increasingly larger intervals T which separate the input pulses. We note that in all cases the output is a superposition of the

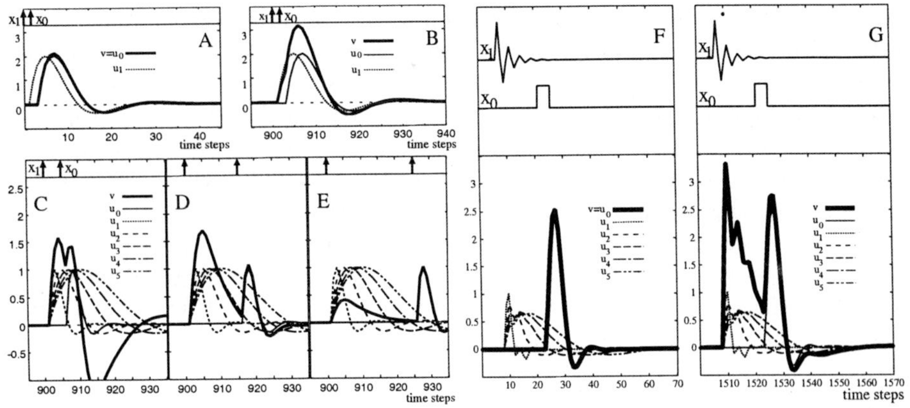

Fig. 2. Simulation results. Arbitrary time steps were used in the simulations. Pulse sequences were repeated every 100 time-steps, the first starting at zero. Resonators were implemented digitally as IIR-filters, which leads to a small onset-delay of 2 time-steps. They had always a value of $a = 0.1\pi$. A,B) are obtained with two identical resonators h_0, h_1 with: $b_{0,1} = 1.09$. The other parameters were $\mu = 0.01$ and $T = 2$. A) Result for $t = 0$, B) for $t = 900$. C-E) are calculated with six resonators, five (curves of $u_1, ..., u_5$) receiving input from x_1 and one (curve of u_0) from x_0. Resonator parameters are: $f_0 = 1.09$, $f_j = \frac{f_0}{j}, j \geq 1$, μ was set to 0.11. Results are shown for $t = 900$. C) $T = 5$. D) $T = 15$, E) $T = 25$. F,G) have the same conditions as D) x_0 was also a square wave but x_1 was sine burst.

different resonator outputs and that a well discernible early signal component exists in response to the conditioning stimulus x_1. This is due to the fact that in this setup the learning process can be seen as a sequential improvement of the rising edge at the onset of the conditioned stimulus.

In the next simulation we copied all conditions from the last simulation but replaced the δ-pulses at x_0 and x_1 by temporally extended signals (Fig 2 F/G). This was done to show that the system is able to deal with complex input signals. In an application the signal x_0 (the unconditioned stimulus) could be a light stimulus and the signal x_1 (the conditioned stimulus) a sound event. After learning the system is again able to detect the sine burst as the predictor of the square pulse x_0.

4 Discussion

From the above results it can be deduced that the system generalizes without problems to more generic input combinations (e.g. more than two inputs). An important feature of other algorithms for sequence learning is that after convergence the output function will approximate an "expectation potential", which is essentially a rectangular function, starting at x_1 and ending at x_0 (see [10], chapter 10). This is the reason, why our theory was developed using resonators (band-passes), because they allow to compose arbitrary shapes of v, such as expectation potentials. We realize, however, that a pure low-pass characteristic would also suffice to induce temporal learning, which relates our approach to the theory of adaptive filters [11]. This helps to understand the predictive properties of our system, because from this context it is known that the derivative of a low-pass filtered signal acts as a predictor of the signal [12]. However, adaptive filters rely on the (gradually shifting) self-similarity of a single signal - hence on its auto-correlation properties. Whereas our approach gains its predictive power from the cross-correlation properties between two (or more) signals such as in biological classical conditioning. Therefore, we expect that this approach will also prove useful in a more technologically oriented context.

References

1. Donald O. Hebb. *The organization of behavior.* Science Ed., New York, 1967.
2. A. H. Klopf. A neuronal model of classical conditioning. *Psychobiology*, 16(2):85–123, 1988.
3. Richard Kempter, Wulfram Gerstner, and J. Leo van Hemmen. Hebbian learning and spiking neurons. *Physical Review E*, 59:4498–4514, 1999.
4. S. Song, K. D. Miller, and L. F. Abbott. Competitive hebbian learning through spike-timing-dependent synaptic plasticity. *Nature Neuroscience*, 3:919–926, 2000.
5. Gerstner, Kempter, van Hemmen, and Wagner. A neuronal learning rule for sub-millisecond temporal coding. *Nature*, 383:76–78, 1996.
6. Ken D. Miller. Receptive fields and maps in the visual cortex: Models of ocular dominance and orientation columns. In E Donnay, J.L. van Hemmen, and K. Schulten, editors, *Models of Neural Networks III*, pages 55–78. Springer-Verlag, 1996.
7. Wolfram Schultz, Peter Dayan, and P. Read Montague. A neural substrate of prediction and reward. *Science*, 275:1593–1599, 1997.
8. R.S. Sutton and A.G. Barto. Toward a modern theory of adaptive networks: expectation and prediction. *Psychol Rev*, 88(2):135–170, 1981.
9. E. Oja. A simplified neuron model as a principal component analyzer. *J Math Biol*, 15(3):267–273, 1982.
10. P Dayan and L. F. Abbott. *Theoretical Neuroscience.* MIT Press, Cambridge MA, 2001.
11. S. S. Haykin. *Adaptive filter theory.* Prentice Hall, Englewood Cliffs, 1991.
12. S. M. Bozic. *Digital and Kalman filtering: an introduction to discrete-time filtering and optimum linear estimation.* E. Arnold, London, 1994.

A Mathematical Analysis of a Correlation Based Model for the Orientation Map Formation

Tadashi Yamazaki

Department of Mathematical and Computing Sciences, Tokyo Institute of Technology, 2-12-1 Meguro Tokyo, 152-8552, JAPAN
tyam@is.titech.ac.jp

Abstract. We consider a correlation based model for the orientation map formation proposed by Miller[3] and study the formation mathematically. We perform the Fourier transform of the model and we observe that the model has the following properties. (1) It develops oriented receptive fields; (2) These oriented receptive fields are arranged in order, and ordered orientation columns are developed; (3) The periodicity of orientations on the cortical surface lacks; (4) The periodicity of phases appears; (5) Singular points appear irregularly on the cortical surface. Our analytical results are justified by computer simulations.

1 Introduction

Since von der Malsburg developed a mathematical model for the orientation map formation[2], a huge number of models have been proposed, and some of them are widely reviewed and discussed in depth[1, 9]. These discussions, however, are based on numerical simulations. There are few attempts to consider the formation mathematically. In this paper, we consider a correlation based model proposed by Miller[3] and discuss mathematically how the receptive fields become oriented and how the oriented receptive fields are arranged in order, i.e., how orientation columns are developed. Relations of the simulated orientation map to the actual cortical data has been intensively investigated by Miller[3].

Although several mathematical analyses of a correlation based model have been made, yet the formation has not been investigated sufficiently. Shouval and Cooper[7] studied a similar correlation based model. They defined an energy function of the model and discussed the structures of receptive fields and orientation columns corresponding to the lowest energy. However, the retinotopy has not been considered in their analysis, that is, in their model, all cells in the visual cortex have the same receptive field. Wimbauer et al.[10] computed the shapes of receptive fields possibly developed by the principal component analysis; however, they have not mentioned how these receptive fields are arranged in two-dimensional cortical surface. That is, the development of orientation columns has not been analyzed. To make clear these points, we take another approach and give some explanations for the above problems.

We consider some simplifications in a way similar to Wimbauer et al.'s paper[10], and then we perform the Fourier transform of the simplified model.

G. Dorffner, H. Bischof, and K. Hornik (Eds.): ICANN 2001, LNCS 2130, pp. 1121–1128, 2001.

By our analysis, we have the following observations about the formation. (1) A Mexican-hat type correlation function determines the spatial frequencies of receptive fields; (2) The combination of a Gaussian arbor function and an Gaussian intracortical function develops oriented receptive fields; (3) The intracortical function develops stripes of ON and OFF like ocular dominance columns on a layer on which each cell has its receptive field if the range is wide enough; (4) Preferred orientations and phases of receptive fields for each cell are determined by the stripes within the receptive fields. From these, we observe that this model obtains the following properties. (1) It develops oriented receptive fields; (2) These oriented receptive fields are arranged in order, and ordered orientation columns are developed; (3) The periodicity of orientations on the cortical surface does not appear; (4) The periodicity of phases appears; (5) Singular points appear irregularly on the cortical surface. Finally, we confirm our results by computer simulations.

This paper is organized as follows. In section 2, we will review Miller's model and consider some simplifications. Section 3 contains our main results. Section 4 is devoted to computer simulations.

2 Miller's Model and Its Simplifications

The model consists of two neuronal layers. The pre- and postsynaptic layers correspond to the LGN and the visual cortex respectively, and cells in each layer are arranged on a two-dimensional grid. Two types of synaptic connections, 'ON' type and 'OFF' type, are assumed, and each corresponds to the synaptic connection from a ON-center LGN cell and a OFF-center LGN cell. Both layers conserve *the retinotopy*, that is, neighboring cells in the cortex and in the LGN correspond to neighboring positions on the retina. Thus, the neighborhood relations of the visual input are conserved. The initial synaptic weights are assigned uniformly at random. The modification rule of synaptic weights is

$$\begin{aligned} \overset{\bullet}{w}{}^{\mathrm{ON}}(\mathbf{x},\mathbf{a}) &= \lambda A(\mathbf{x}-\mathbf{a}) \sum_{\mathbf{y},\mathbf{b}} I(\mathbf{x}-\mathbf{y}) \\ &\quad \left[C^{\text{ON-ON}}(\mathbf{a}-\mathbf{b}) w^{\mathrm{ON}}(\mathbf{y},\mathbf{b}) + C^{\text{OFF-OFF}}(\mathbf{a}-\mathbf{b}) w^{\mathrm{OFF}}(\mathbf{y},\mathbf{b})\right] . \end{aligned} \tag{1}$$

A, I, C, and w are respectively the arbor function, the intracortical interaction function, the correlation functions, and the synaptic weights. The vector $\mathbf{x}, \mathbf{y}$ denote two-dimensional postsynaptic positions while $\mathbf{a}, \mathbf{b}$ label two-dimensional presynaptic positions. The equation for OFF synaptic weights is identical after exchange of 'ON' and 'OFF' everywhere. A, I, and C are the following equations.

$$\begin{aligned} &A(\mathbf{x}) \stackrel{\text{def}}{=} \exp\left(-|\mathbf{x}|^2/2\sigma_A^2\right), \quad I(\mathbf{x}) \stackrel{\text{def}}{=} \exp\left(-|\mathbf{x}|^2/2\sigma_I^2\right), \\ &C(\mathbf{a}) \stackrel{\text{def}}{=} C_1 \exp\left(-|\mathbf{a}|^2/2\sigma_{C1}^2\right) - C_2 \exp\left(-|\mathbf{a}|^2/2\sigma_{C2}^2\right), \end{aligned} \tag{2}$$

$$\begin{aligned} &C^{\text{ON-ON}}(\mathbf{a}) \stackrel{\text{def}}{=} C(\mathbf{a}) \quad (\text{resp., } C^{\text{OFF-OFF}}), \\ &C^{\text{ON-OFF}}(\mathbf{a}) \stackrel{\text{def}}{=} -\varepsilon C(\mathbf{a}) \quad (\text{resp., } C^{\text{OFF-ON}}). \end{aligned} \tag{3}$$

In addition to these, this model includes an upper and a lower bounds for each synaptic weight and the normalization to keep the sum of synaptic weights for each postsynaptic cell a constant.

We consider some simplifications. First we consider the continuous version of the modification rule by going to the continuum limit. Second we omit the lower and upper bounds by assuming that the principal features of the dynamics are established before these bounds are reached. We also omit the normalization as well by assuming that the principal features grows exponentially faster than the other features. The dynamics of the normalization is studied in [4]. Third we introduce $w(\mathbf{x}, \mathbf{a})$ as the difference between $w^{\mathrm{ON}}(\mathbf{x}, \mathbf{a})$ and $w^{\mathrm{OFF}}(\mathbf{x}, \mathbf{a})$ as $w(\mathbf{x}, \mathbf{a}) \stackrel{\mathrm{def}}{=} w^{\mathrm{ON}}(\mathbf{x}, \mathbf{a}) - w^{\mathrm{OFF}}(\mathbf{x}, \mathbf{a})$. Fourth for each $w(\mathbf{x}, \mathbf{a})$, we consider the position $\mathbf{a}$ as the relative position α from the position $\mathbf{x}$. We define $w(\mathbf{x}, \alpha)$ as $w(\mathbf{x}, \alpha) \stackrel{\mathrm{def}}{=} w(\mathbf{x}, \mathbf{a} - \mathbf{x})$. From the above simplifications, it follows that

$$\overset{\bullet}{w}(\mathbf{x}, \alpha) = A(\alpha) \int_{-\infty}^{\infty} \int_{-\infty}^{\infty} I(\mathbf{x} - \mathbf{y}) C(\mathbf{x} - \mathbf{y} + \alpha - \beta) w(\mathbf{y}, \beta) d\beta d\mathbf{y}. \quad (4)$$

Further we consider this equation.

3 Analytical Results

We perform the Fourier transform of Eqn. (4).

$$\overset{\bullet}{\widehat{w}}(\mathbf{u}, \mathbf{n}) = \widehat{I}(\mathbf{u} - \mathbf{n}) \widehat{C}(\mathbf{n}) \int_{-\infty}^{\infty} \widehat{A}(\mathbf{n} - \mathbf{m}) \widehat{w}(\mathbf{u}, \mathbf{m}) d\mathbf{m}, \quad (5)$$

where,

$$\widehat{A}(\mathbf{l}) = \sigma_A^2 e^{-\sigma_A^2 |\mathbf{l}|^2/2}, \quad \widehat{I}(\mathbf{l}) = \sigma_I^2 e^{-\sigma_I^2 |\mathbf{l}|^2/2},$$
$$\widehat{C}(\mathbf{l}) = C_1 \sigma_{C1}^2 e^{-\sigma_{C1}^2 |\mathbf{l}|^2/2} - C_2 \sigma_{C2}^2 e^{-\sigma_{C2}^2 |\mathbf{l}|^2/2}.$$

Once we obtain the solution of Eqn. (5) we have also the solution of Eqn. (4) by the following inverse Fourier transform.

$$\begin{aligned} w(\mathbf{x}, \alpha) &= \int_{-\infty}^{\infty} \int_{-\infty}^{\infty} \widehat{w}(\mathbf{u}, \mathbf{n}) e^{i\mathbf{u}\mathbf{x}} e^{i\mathbf{n}\alpha} d\mathbf{n} d\mathbf{u} \\ &= \int_{-\infty}^{\infty} \int_{-\infty}^{\infty} \widehat{w}_{\cos}(\mathbf{u}, \mathbf{n}) \cos \mathbf{u}\mathbf{x} \cos \mathbf{n}\alpha d\mathbf{n} d\mathbf{u} \\ &+ \int_{-\infty}^{\infty} \int_{-\infty}^{\infty} \widehat{w}_{\sin}(\mathbf{u}, \mathbf{n}) \sin \mathbf{u}\mathbf{x} \sin \mathbf{n}\alpha d\mathbf{n} d\mathbf{u}, \end{aligned} \quad (6)$$

where $\widehat{w}_{\cos}(\mathbf{u}, \mathbf{n})$ (resp., sin) is the Fourier cosine (resp., sine) part of $\widehat{w}(\mathbf{u}, \mathbf{n})$.

We assume that $\widehat{C}(\mathbf{l})$ is a sharp bi-modal function that shows the peaks at $\sigma \stackrel{\mathrm{def}}{=} |\mathbf{l}|$. Then, the integral domain can be restricted to the circle of radius σ because only $w(\mathbf{u}, \mathbf{n})$ such that $|\mathbf{n}| = \sigma$ are effective. Thus we have

$$\overset{\bullet}{\widehat{w}}(\mathbf{u}, \mathbf{n}) = \widehat{I}(\mathbf{u} - \mathbf{n}) \int_{|\mathbf{m}|=\sigma} \widehat{A}(\mathbf{n} - \mathbf{m}) \widehat{w}(\mathbf{u}, \mathbf{m}) d\mathbf{m}. \quad (7)$$

This can not be solved as it is, alternatively we consider an approximation. The t times expansion of Eqn. (7) in the discrete time is the following.

$$\begin{aligned}\Delta w^{(t)}(\mathbf{u},\mathbf{n}) = \widehat{I}(\mathbf{u}-\mathbf{n}) \int_{|\mathbf{n}_0|=\sigma} \cdots \int_{|\mathbf{n}_{t-1}|=\sigma} w^{(0)}(\mathbf{u},\mathbf{n}_0)\widehat{I}(\mathbf{u}-\mathbf{n}_0)\cdots \\ \widehat{I}(\mathbf{u}-\mathbf{n}_{t-1})\widehat{A}(\mathbf{n}_0-\mathbf{n}_1)\cdots\widehat{A}(\mathbf{n}_{t-1}-\mathbf{n})d\mathbf{n}_0\cdots d\mathbf{n}_{t-1}d\mathbf{u}.\end{aligned} \tag{8}$$

$\widehat{I}(\mathbf{u}-\mathbf{n}_0)\cdots\widehat{I}(\mathbf{u}-\mathbf{n}_{t-1})$ strengthens the relation between $\mathbf{u}$ and $\mathbf{n}$, however, $\widehat{I}(\mathbf{u}-\mathbf{n})$ may be enough if we ignore the strength and focus only on the relation. By omitting that, we obtain the following system which has been studied in [11].

$$\begin{aligned}\dot{\widehat{w}}_*(\mathbf{u},\mathbf{n}) &= \int_{|\mathbf{m}|=\sigma} \widehat{A}(\mathbf{n}-\mathbf{m})\widehat{w}_*(\mathbf{u},\mathbf{m})d\mathbf{m}, \\ \widehat{w}(\mathbf{u},\mathbf{n}) &= \widehat{I}(\mathbf{u}-\mathbf{n})\widehat{w}_*(\mathbf{u},\mathbf{n}).\end{aligned} \tag{9}$$

By considering the Fourier transform of $\widehat{w}_*(\mathbf{u},\mathbf{n})$, it follows

$$\widehat{\widehat{w}}_*(\mathbf{u},\xi) = \widehat{\widehat{w}}_*^{(0)}(\mathbf{u},\xi)\exp(\widehat{\widehat{A}}(\xi)t). \tag{10}$$

Since $\widehat{\widehat{A}}(\xi)$ is a Gaussian function, $\xi = \mathbf{0}$ will be mainly extracted after a number of steps. Then we have

$$\begin{aligned}\widehat{w}_*(\mathbf{u},\mathbf{n}) &= \int_{-\infty}^{\infty} \widehat{\widehat{w}}_*^{(0)}(\mathbf{u},\xi)\exp(\widehat{\widehat{A}}(\xi)t)\exp(i\xi\mathbf{n})d\xi \\ &\propto \widehat{\widehat{w}}_*^{(0)}(\mathbf{u},\mathbf{0}) = \int_{-\infty}^{\infty} \widehat{w}_*^{(0)}(\mathbf{u},\mathbf{n})d\mathbf{n}.\end{aligned} \tag{11}$$

Thus, $\widehat{w}_*(\mathbf{u},\mathbf{n})$ will be in proportional to some constant independent of $\mathbf{n}$, which we denote by $\widehat{w}^{(0)}(\mathbf{u})$. Therefore $\widehat{w}(\mathbf{u},\mathbf{n}) \propto \widehat{I}(\mathbf{u}-\mathbf{n})\widehat{w}^{(0)}(\mathbf{u})$.

From Eqn. (6) it follows

$$\begin{aligned}w(\mathbf{x},\alpha) &= \int_{-\infty}^{\infty}\int_{|\mathbf{n}|=\sigma} \widehat{w}^{(0)}(\mathbf{u})\widehat{I}(\mathbf{u}-\mathbf{n})e^{i\mathbf{u}\mathbf{x}}e^{i\mathbf{n}\alpha}d\mathbf{n}d\mathbf{u} \\ &= \int_{-\infty}^{\infty}\int_{|\mathbf{n}|=\sigma} \widehat{w}^{(0)}_{\cos}(\mathbf{u})\widehat{I}_{\cos}(\mathbf{u},\mathbf{n})\cos\mathbf{u}\mathbf{x}\cos\mathbf{n}\alpha d\mathbf{n}d\mathbf{u} \\ &+ \int_{-\infty}^{\infty}\int_{|\mathbf{n}|=\sigma} \widehat{w}^{(0)}_{\sin}(\mathbf{u})\widehat{I}_{\sin}(\mathbf{u},\mathbf{n})\sin\mathbf{u}\mathbf{x}\sin\mathbf{n}\alpha d\mathbf{n}d\mathbf{u},\end{aligned} \tag{12}$$

where

$$\begin{aligned}\widehat{I}_{\cos}(\mathbf{u},\mathbf{n}) &\stackrel{\text{def}}{=} \int_{-\infty}^{\infty}\int_{-\infty}^{\infty} I(\mathbf{x})\cos\mathbf{u}\mathbf{x}\cos\mathbf{n}\mathbf{x}d\mathbf{x}, \\ \widehat{I}_{\sin}(\mathbf{u},\mathbf{n}) &\stackrel{\text{def}}{=} \int_{-\infty}^{\infty}\int_{-\infty}^{\infty} I(\mathbf{x})\sin\mathbf{u}\mathbf{x}\sin\mathbf{n}\mathbf{x}d\mathbf{x}.\end{aligned} \tag{13}$$

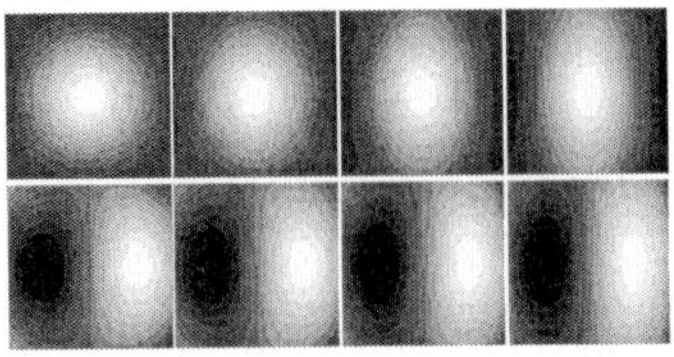

Fig. 1. Shapes of (a) and (b) in Eqn. (14). The upper and lower four figures are these of (a) and (b) respectively, and from left to right, σ_I^2 varies 0, 10.0, 20.0, 30.0. $\mathbf{u}$ is taken as a vertical vector, so these figures show the orientation to the vertical direction.

Here for a fixed $\mathbf{u}$, we consider respectively the shapes of

$$(a): \int_{-\infty}^{\infty} \widehat{I}_{\cos}(\mathbf{u},\mathbf{n}) \cos \mathbf{n}\alpha d\mathbf{n}, \quad \text{and} \quad (b): \int_{-\infty}^{\infty} \widehat{I}_{\sin}(\mathbf{u},\mathbf{n}) \sin \mathbf{n}\alpha d\mathbf{n}. \tag{14}$$

(a) is a weighted sum of cosine functions with respect to $\mathbf{n}$. These weights take the maximum when $\mathbf{n} = (\sigma/|\mathbf{u}|)\mathbf{u}$ and decrease as $\mathbf{n}$ leaves from $(\sigma/|\mathbf{u}|)\mathbf{u}$. Thus the shape of (a) is tri-lobed with the orientation $\theta = \tan^{-1}(\mathbf{u})$ (Fig. 1) while the shape of (b) is bi-lobed with the same orientation (Fig. 1). Sharpness of the orientation tuning is determined by the range of I. Wider I makes the tuning sharper because $\widehat{I}$ becomes narrower.

Next, we consider which $\mathbf{u}$ are the dominant. Since $\mathbf{n}$ are on the circle of the radius σ, by the effect of $\widehat{I}(\mathbf{u}-\mathbf{n})$, only some $\mathbf{u}$ are effective. It is enough to find $\mathbf{u}$ that maximizes

$$f(\mathbf{u}) \stackrel{\text{def}}{=} \int_{|\mathbf{n}|=\sigma} \widehat{I}(\mathbf{u}-\mathbf{n}) d\mathbf{n}. \tag{15}$$

For a fixed unit vector $\mathbf{u}_*$, we consider $\mathbf{u} = l\mathbf{u}_*$. The distance between $\mathbf{u}$ and $\mathbf{n}$ is $\sigma^2 + l^2 - 2l\sigma\cos\theta$, where θ is the angle between $\mathbf{u}$ and $\mathbf{n}$. Thus we consider l that maximizes

$$f(l) = \int_0^{2\pi} \exp\left(-\frac{\sigma_I^2}{2}(\sigma^2 + l^2 - 2l\sigma\cos\theta)\right) d\theta = \pi e^{-\sigma_I^2(\sigma^2+l^2)/2} I_0(\sigma_I^2 \sigma l), \tag{16}$$

where I_0 is the modified Bessel function of the 0th order. Numerically we can see that when σ_I is small $l = 0$ maximizes $f(l)$ while as σ_I increases such l converges to σ. Therefore, when σ_I is small, $\mathbf{u} \approx \mathbf{0}$ is the dominant. Then, as σ_I increases, $\mathbf{u}$ such that $|\mathbf{u}| = \sigma$ (i.e. $l = \sigma$) would be the dominant.

Further we assume that $|\mathbf{u}| = \sigma$ is the dominant. Then, Eqn. (12) can be written as follows.

$$\begin{aligned} w(\mathbf{x},\alpha) = &\int_{|\mathbf{u}|=\sigma} \widehat{w}_{\cos}^{(0)}(\mathbf{u}) \cos \mathbf{ux} \int_{|\mathbf{n}|=\sigma} \widehat{I}_{\cos}(\mathbf{u},\mathbf{n}) \cos \mathbf{n}\alpha d\mathbf{n} d\mathbf{u} \\ &+ \int_{|\mathbf{u}|=\sigma} \widehat{w}_{\sin}^{(0)}(\mathbf{u}) \sin \mathbf{ux} \int_{|\mathbf{n}|=\sigma} \widehat{I}_{\sin}(\mathbf{u},\mathbf{n}) \sin \mathbf{n}\alpha d\mathbf{n} d\mathbf{u}, \end{aligned} \tag{17}$$

Then we consider the shape of portions of Eqn. (17) with respect to $\mathbf{u}$ and $\mathbf{x}$.

$$w(\mathbf{x}) = \int_{|\mathbf{u}|=\sigma} \widehat{w}_{\cos}^{(0)}(\mathbf{u}) \cos \mathbf{ux} + \widehat{w}_{\sin}^{(0)}(\mathbf{u}) \sin \mathbf{ux} d\mathbf{u}. \tag{18}$$

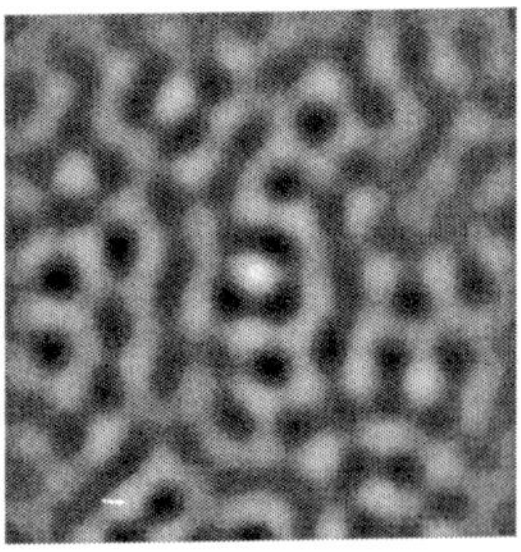

Fig. 2. Signs of $w(\mathbf{x})$ in Eqn. (18). Black and white regions represent negative and positive respectively, which corresponds of the dominance between ON and OFF.

 $\times \cos \mathbf{ux} +$  $\times \sin \mathbf{ux} =$ (19)

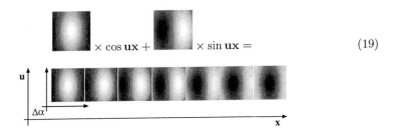

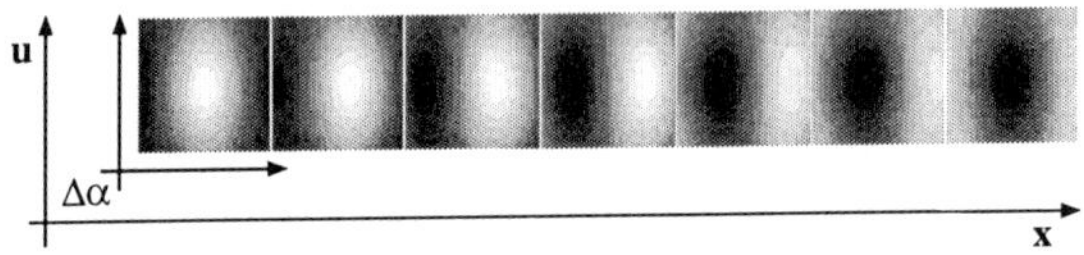

Fig. 3. The change of the shapes of receptive fields. For a vertical unit vector $\mathbf{u}$ (i.e., stripes runs vertically), the shapes of receptive fields arranged in parallel to x-axis change, which depends on the position on the x-axis. Phases of these receptive fields also depends on the position.

This equation is identical to the modification rule of some spectral models of ocular dominance columns [8, 6]. Thus this equation develops stripes or blobs of signs of $w(\mathbf{x})$, corresponding to the dominance between ON and OFF (Fig. 2). Since the initial weights are assigned uniformly at random, the stripes run irregularly.

Now we can summarize the development of orientation map formation as follows. In spite of which each cortical cell has its own set of synaptic weights, we could consider that all cells share one layer of $\mathbf{u}$ and each cell has its "receptive field", i.e., a small region, on the layer. Stripes running irregularly like ocular dominance columns are developed on the layer. The preferred orientation of a cell is determined by the orientation of stripes running within its receptive field, and the phase of a cell is also determined in a similar way (Fig. 3).

Since the preferred orientations are determined by the stripes of $\mathbf{u}$, and since the stripes run irregularly, the periodicity of orientations on the cortical surface lacks, on the other hand, the periodicity of phases appears. Moreover, singular points appear irregularly on the cortical surface.

Finally, it should be noted that this development seems similar to that by Miyashita and Tanaka's thermodynamic model[5]. In their model, all cells on the visual cortex shares a layer called "synaptic terminal point", and synaptic terminals on the layer are arranged like ocular dominance columns or blobs. Cortical cells are assumed existing over the layer with keeping the topography

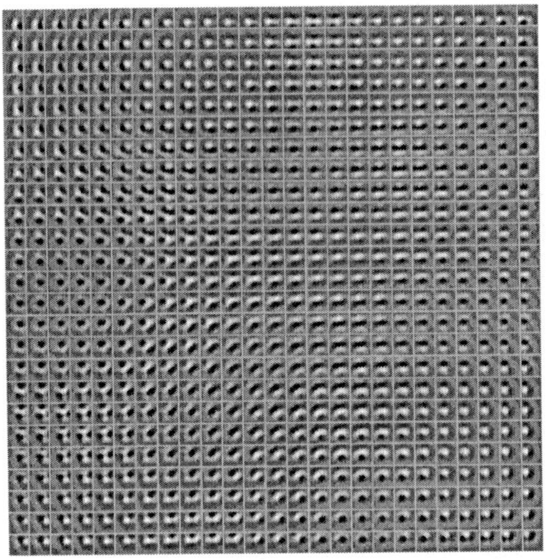

Fig. 4. Simulated orientation map by Eqn. (4).

and each cell has its "receptive field" on the layer defined by clipping a small region by an isotropic function, e.g., a Gaussian function.

4 Computer Simulations

We justify our mathematical analysis by some computer simulations.

We consider 31x31 presynaptic cells and 25x25 postsynaptic cells. Each postsynaptic cell receives synaptic connections from 13x13 presynaptic cells with keeping the retinotopy. We make use of Eqn. (4) as the modification rule, where parameters are the following: $\sigma_A^2 = 10.0$, $\sigma_I^2 = 10.0$, $C_1 = 2.0$, $C_2 = 1.0$, $\sigma_{C1}^2 = 2.0$, and $\sigma_{C1}^2 = 4.0$. Initial synaptic weights are assigned uniformly at random from the range $[-0.4, 0.4]$, and the modification rule is applied 80 times.

Figure 4 shows the result. As can be seen, oriented receptive fields are developed and the change of shapes of receptive fields is similar to Fig. 3. Moreover, we can see that the periodicity of orientations on the cortical surface lacks, on the other hand, the periodicity of phases appears, and singular points appear irregularly on the cortical surface.

5 Conclusion

We considered a correlation based model for the orientation map formation proposed by Miller[3]. We performed the Fourier transform of the model and studied how this model develops the oriented receptive fields and orientation columns mathematically. By our analysis, we observed that (1) A Mexican-hat type correlation function determines the spatial frequencies of receptive fields; (2) The combination of a Gaussian arbor function and an Gaussian intracortical function develops oriented receptive fields; (3) The intracortical function develops stripes like ocular dominance columns on a layer on which each cell has its receptive field if the range is wide enough; (4) The preferred orientations and the phases of receptive fields for each cell are determined by the stripes within the receptive fields. From these results, we saw that the model obtains the following properties. (1) It develops oriented receptive fields; (2) These oriented receptive fields are arranged in order, and ordered orientation columns are developed; (3) The periodicity of orientations on the cortical surface lacks; (4) The periodicity of phases appears; (5) Singular points appear irregularly on the cortical surface.

References

1. Erwin, E., Obermayer, K., Shulten, K.: Models of orientation and ocular dominance columns in the visual cortex: a critical comparison, Neural Comput. **7** (1995) 425–468
2. Malsburg, C. von der.: Self-organization of orientation sensitivity cells in the striate cortex, Kybernetik **14** (1973) 85–100
3. Miller, K.D.: A model for the development of simple cell receptive fields and the ordered arrangement of orientation columns through activity-dependent competition between on- and off-center inputs, J. Neurosci. **14** (1994) 409–441
4. Miller, K.D., MacKay, D.J.C.: The role of constraints in Hebbian learning. Neural Computat. **6**, 100–126
5. Miyashita, M., Tanaka, S.: A mathematical model for the self-organization of orientation columns in visual cortex, NeuroReport **3** (1992) 69–72
6. Rojer, A.S., Schwartz, E.L.: Cat and monkey cortical columnar patters modeled by bandpass-filtered 2d white noise, Biol. Cybern. **62** (1990) 381–391
7. Shouval, H., Cooper, L.N.: Organization of receptive fields in networks with Hebbian learning: the connection between synaptic and phenomenological models, Biol. Cybern. **74** (1996) 439–447
8. Swindale, N.V.: A model for the formation of ocular dominance stripes, Proc. R. Soc. Lond. **B. 208** (1980) 243–264
9. Swindale, N.V.: A development of topography in the visual cortex: a review of models, Network **7** (1996) 161–247
10. Wimbauer, S., Gerstner, W., van Hemmen, J.L.: Analysis of a correlation-based model for the development of orientation-selective receptive fields in the visual cortex, Network **9** (1998) 449-466
11. Yamazaki, T.: A mathematical analysis of development of oriented receptive fields in Linsker's model, Dept. Math. Comput. Sci, Tokyo Inst. Tech. **C-147** (2001)

Learning from Chaos: A Model of Dynamical Perception

Emmanuel Daucé

ENSAE, 10, avenue Edouard Belin,
F-31055 Toulouse cedex, France
dauce@cert.fr
http://www.cert.fr/anglais/deri/dauce

Abstract. We present a process that allows to learn complex spatio-temporal signals, in a large random neural system with a self-generated chaotic dynamics. Our system modelizes a generic sensory structure. We study the interplay between a periodic spatio-temporal stimulus, i.e a sequence of spatial patterns (which are not necessarily orthogonal), and the inner dynamics, while a Hebbian learning rule slowly modifies the weights. Learning progressively stabilizes the initial chaotic dynamical behavior, and tends to produce a reliable resonance between the inner dynamics and the outer signal. Later on, when a learned stimulus is presented, the inner dynamics becomes regular, so that the system is able to simulate and predict in real time the evolution of its input signal. On the contrary, when an unknown stimulus is presented, the dynamics remains chaotic and non-specific.

1 Introduction

An artificial neural network with inner recurrent links can be seen as a dynamical system as it can generate an inner signal (that propagates through inner interactions). In order to perform computation, people often try to avoid interferences between inner signal and input (command) signal. For instance, in classical applications of recurrent networks [9, 4], inner recurrent states work as buffers that memorize a context, but one tries to avoid active inner dynamics for the response to be specified by the input signal (i.e. the same input sequence always produces the same response). On the other side, classical Hopfield model [6] and derivatives [5] are autonomous dynamical systems (i.e there is no interaction in real time with an input signal), and the final attractor only depends on initial conditions. The final attractor is thus strictly determined by the initial conditions.

In this article, we will present an alternative approach for the use of dynamical systems for neural computation. We study the *dynamical coupling* between the input signal and the system response. We propose a model where the system response relies on the interplay between an inner (autonomous) dynamics and outer (command) signal. The inner dynamics is chaotic, so that the response of the system is not fully specified by the input sequence.

G. Dorffner, H. Bischof, and K. Hornik (Eds.): ICANN 2001, LNCS 2130, pp. 1129–1134, 2001.

2 Random Recurrent Models

2.1 Activation Dynamics

We consider a system with $N > 1$ state variables (we often suppose that N is large), where $\mathbf{x}(t)$ is a $N \times 1$ state vector at time t (also called activation vector). We also consider an Input signal $\mathbf{I} = \{\mathbf{I}(t)\}_{t=1,...,+\infty}$, where $\mathbf{I}(t)$ is a $N \times 1$ input vector at time t. According to $\mathbf{I}$ and initial conditions $\mathbf{x}(0)$ (random variable, uniform in $[0,1]^N$), the update of the dynamics is set according to:

$$\begin{aligned} &\forall t \geq 1, \\ &\mathbf{x}(t) = f(\mathbf{J}\mathbf{x}(t-1) + \mathbf{I}(t) - \theta) \end{aligned} \tag{1}$$

where f is a non-linear sigmoid transfer function. In our simulations, we use as transfer function $f_g(h) = (1 + \tanh(gh))/2$, whose gain is equal to $g/2$. Note that our system is not autonomous. It is deterministic if we suppose that $\mathbf{I}(t)$ is set according to a deterministic process. The values of activation threshold vector θ are supposed positive and homogeneous, in order to lower the activity.

In classical RRNN (Random Recurrent Neural Networks) models [3, 1], the connectivity is full, and the weights are homogeneous *in law*, i.e. the weight are set according to a normal law $\mathcal{N}(0, \sigma_J^2/N)$, where the weights standard deviation is such that $E(\text{var}(J_i)) = \sigma_J^2$, where $J_i = \sum_{j=1}^{N} J_{ij}$. The autonomous model (i.e. $\mathbf{I} = 0$) is known to display a generic quasi-periodicity route to chaos while tuning gain parameter g [3].

In a structured RRNN model, one can distinguish P populations of neurons. We have $P \geq 1$ layers, of sizes $N^{(1)}, ..., N^{(P)}$ (so that $N = N^{(1)} + ... + N^{(P)}$). The weights, input signal and thresholds are thus bloc matrices:

$$\forall t \geq 1,$$
$$\mathbf{J}(t) = \begin{pmatrix} \mathbf{J}^{(11)}(t) & \ldots & \mathbf{J}^{(1P)}(t) \\ & \ldots & \\ \mathbf{J}^{(P1)}(t) & \ldots & \mathbf{J}^{(PP)}(t) \end{pmatrix} \quad \mathbf{I}(t) = \begin{pmatrix} \mathbf{I}^{(1)}(t) \\ \ldots \\ \mathbf{I}^{(P)}(t) \end{pmatrix} \quad \theta = \begin{pmatrix} \theta^{(1)} \\ \ldots \\ \theta^{(P)} \end{pmatrix}$$

$\forall p, q \in \{1, ..., P\}^2$, $\mathbf{J}^{(pq)}$ is a $N^{(p)} \times N^{(q)}$ random matrix whose values are set according to $\mathcal{N}(0, (\sigma_J^{(pq)})^2/N^{(q)})$.

2.2 Learning Dynamics

A learning dynamics simulates a slow process of adaptation that takes place on the weights, so that the weights belong to the state variables of the system. Our learning rule is a generalization of Hebbian rule which stands on instantaneous estimates of the covariance between source and target neuron [7]. We will suppose hereafter that we have at each time step local estimates of mean activation, which are stored in vector $\hat{\mathbf{x}}(t)$, and updated according to $\hat{\mathbf{x}}(t) = (1-\beta)\hat{\mathbf{x}}(t-1) + \beta\mathbf{x}(t)$, and we take $\beta = 0.1$ in our simulations. The dynamics of the system is thus defined by a coupled set of equations:

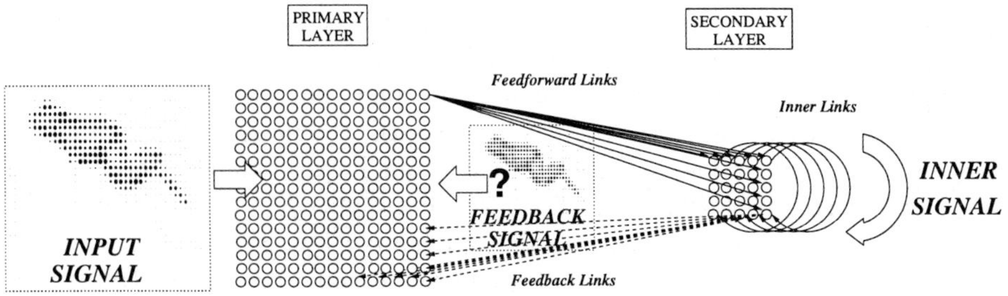

Fig. 1. Sensory 2-populations model. Only few links are represented. The activity on primary layer depends both on the input signal and feedback signal from the secondary layer.

$$\forall t \geq 1,\quad \begin{cases} \mathbf{x}(t) = f_g(\mathbf{J}(t-1)\mathbf{x}(t-1) + \mathbf{I}(t) - \theta) \\ \mathbf{J}(t) = \mathbf{J}(t-1) + \mathcal{E} \otimes \left((\mathbf{x}(t) - \hat{\mathbf{x}}(t)) \, (\mathbf{x}(t-1) - \hat{\mathbf{x}}(t-1))^T \right) \end{cases} \tag{2}$$

where $\mathcal{E}$ is a $N \times N$ normalization matrix and '$\otimes$' is a term to term matrix product. In the P-populations model,

$$\mathcal{E} = \begin{pmatrix} \mathcal{E}^{(11)} & \ldots & \mathcal{E}^{(1P)} \\ & \ldots & \\ \mathcal{E}^{(P1)} & \ldots & \mathcal{E}^{(PP)} \end{pmatrix}$$

and $\forall p, q \in \{1, ..., P\}^2, \forall i \in \{1, .., N^{(p)}\}, \forall j \in \{1, ..., N^{(q)}\}, \mathcal{E}_{ij}^{(pq)} = \varepsilon^{(pq)}/N^{(q)}$. We suppose that $\varepsilon^{(pq)}$'s are small so that weight evolution is slow.

3 Dynamical Adaptation and Learning in a Generic Sensory Structure

According to the framework stated above, we designed a generic "sensory structure", made of two layers (so that $P = 2$). The parameters are thus chosen in order to have a "sensitive" layer (primary layer), which is under the direct influence of the input signal $\mathbf{I}^{(1)}$, and a deeper layer (secondary layer), which is not stimulated by the input (so that $\forall t, \mathbf{I}^{(2)}(t) = 0$), and whose inner links generate autonomous chaotic dynamics (see Fig. 1). The input signal $\mathbf{I}^{(1)}$ is composed of distributed spatial patterns, that activate at each time step about 5% of primary layer neurons. Note however that those spatial patterns are not necessarily orthogonal. We can define three important classes of links: "feedforward links" $\mathbf{J}^{(21)}$ propagate the input signal towards secondary layer, "inner links" $\mathbf{J}^{(22)}$ generate inner signal, and "feedback links" $\mathbf{J}^{(12)}$ send back the activity of secondary layer towards primary layer.

Few parameters are necessary to define the system. For the simulations, the size of primary and secondary layers are set to $N^{(1)} = 1600$ and $N^{(2)} = 200$. The activation thresholds are set to $\theta^{(1)} = 0.5$ and $\theta^{(2)} = 0.4$. The inner links are randomly set according to $\sigma_J^{(22)} = 1$, and gain is chosen in order to have an active chaotic dynamics on secondary layer ($g = 8$), so that about 15% of secondary layer neurons are active at a given time. The feedforward links are also randomly set according to $\sigma_J^{(21)} = 1$. However, the inner influence remains stronger than input influence, mostly because one have a smaller proportion of active neurons on primary layer ($\simeq 5\%$). Feedback links $\mathbf{J}^{(12)}$ and primary layer lateral links $\mathbf{J}^{(11)}$ are initially equal to zero (i.e $\sigma_J^{(12)} = \sigma_J^{(11)} = 0$). Learning takes place on inner and feedback links, and the learning rate is set stronger on feedback links, so that $\varepsilon^{(22)} = 0.02$ and $\varepsilon^{(12)} = 0.1$ ($\varepsilon^{(11)} = \varepsilon^{(21)} = 0$).

Let us now consider that primary layer is continuously stimulated by a signal $\mathbf{I}^{(1)}(t)$, and the dynamics of the system is (2). The weight are thus slowly modified while the inner dynamics interacts with the input dynamics. A significant change in the behavior of the system can be observed if the signal has some statistical regularities, i.e is predictable. This regularity may moreover be repeatidly presented in a session, for the system to memorize it. In our simulations, we use spatio-temporal periodic signals (which repeat every τ time steps, $\tau \in \{3, ..., 10\}$). Two sorts of dynamical changes can thus be observed in the system (Fig. 2). First, the secondary layer activity, which is initially chaotic, gets closer to a periodic behavior (Fig. 2– a –). This increase in regularity can moreover be measured. Second, learning slowly reinforces the feedback links, and produces a feedback signal which directly stimulates the primary layer (Fig. 2– b –).

This feedback signal can be dynamically suppressed if a change occurs on the input signal. Fig. 2– a – and 2– c – present the evolution of inner and outer dynamics while a change occurs on the input at time $t = 601$ (We change the input periodic sequence). During some time steps that follow input change, the first sequence (frog) remains fed back to the primary layer. However, primary layer signal, which superimposes input and feedback signals, misfits inner signal prediction. This misfit tends to modify the dynamical organization in secondary layer, and drives the system towards a new attractor which does not fit with the specific coincidences of activation that have been previously learned. The "frog illusion" thus disappears. This transitory perseverance shows the stability of the learned response: only the repeating of feedback error (or another strong discrepancy) on several time steps allows the inner dynamics to change.

Finally, a new signal is sent on primary layer at time $t = 701$, which is a truncated version of the learned signal (Fig. 2– d –). Even different from the learned signal, this signal is "interpreted" as coherent, so that the system reaches an attractor which is similar to the attractor corresponding to the full sequence. The feedback signal, which is thus activated, helps to fill the missing information on the primary layer, so that the system behaves "as if" the full sequence was present on primary layer. So, the system can interpret its inputs, and the stable and specific response it has learned is extended to a class of input signals which are coherent with the learned signal.

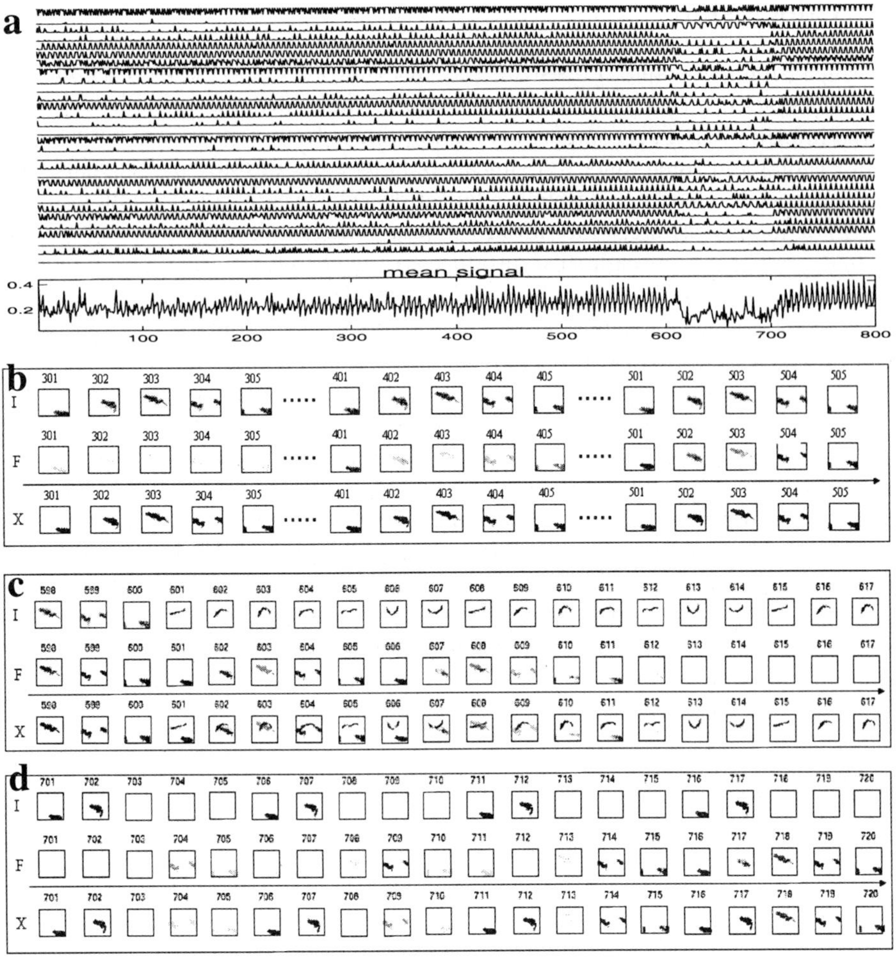

Fig. 2. Learning dynamics, between $t = 1$ and $t = 800$. Input changes takes place at $t = 601$ (new input sequence) and $t = 701$ (presentation of an truncated version of the learned "frog" sequence). – a – Neuronal activity on secondary layer. 30 individual signals (out of 200) have been represented, and their mean activity is represented below. – b – Time evolution of input signal $\mathbf{I}^{(1)}(t)$, feedback signal $\mathbf{F}^{(12)}(t) = f_g(\mathbf{J}^{(12)}(t-1)\mathbf{x}^{(2)}(t-1) - \theta^{(1)})$ and primary layer activation $\mathbf{x}^{(1)}(t)$, between $t = 301$ and $t = 505$ (For readability, most of the time steps have been discarded). At each time step, the 1600 values have been represented as 40×40 images, where white corresponds to 0 and black to 1 (in-between values are gray tones). – c – Same representation, between $t = 598$ and $t = 617$. – d – Same representation, between $t = 701$ and $t = 720$.

4 Conclusion

In conclusion, we want to insist on the fact that very few parameters allow to design a system with a high number of neurons, that can learn complex and structured information. This complex processing comes from a resonance between the input signal and some inner dynamical configuration, initially non-specific, that become specific thanks to a weight adaptation. This weight adaptation, however, does not modify strongly the initial random configuration of inner links. Light changes on the inner weights can produce strong changes in the inner dynamical behavior. Feedback links then help to "materialize" this inner specific response on the primary layer. We have also seen that system's response is not immediate. Due to convergence constraints, it takes some time steps to reach the attractor corresponding to the response of the system. The counterpart of this constraint is a gain in response stability, and the possibility to deal with non-orthogonal series of patterns. Finally, we can say that our system closely modelizes the conditions of real-world perception, where the vital necessity is to discriminate between few relevant stimuli and a bunch of non-relevant stimuli [8].

This dynamical approach of learning may be very suitable for the design of systems that have to adapt in real time to their environment. Some preliminary experiments have yet been carried out (see [2]), that allow to translate inner dynamics in a motor behavior on a mobile robot dealing with real-world data. One can thus observe distinct motor behaviors when the system faces predictable (learned) or unpredictable environment configurations.

References

[1] Daucé, E., Quoy, M., Cessac, B., Doyon, B., and Samuelides, M.: Self-organization and dynamics reduction in recurrent networks: stimulus presentation and learning. Neural Networks, Vol. 11 (1998) 521–533
[2] Daucé, E. and Quoy, M.: Random Recurrent Neural Networks for autonomous system design. In: Meyer, J.-A. et al. (ed.): SAB2000 proceedings supplement. ISAB, 2430 Campus Road, Honolulu, USA (2000)
[3] Doyon, B., Cessac, B., Quoy, M., and Samuelides, M.: Control of the transition to chaos in neural networks with random connectivity. Int. J. of Bif. and Chaos, Vol. 3 **2** (1993) 279–291
[4] Elman, J. L.: Finding structure in time. Cognitive Science, Vol. 14 (1990) 179–211
[5] Herz, A., Sulzer, B., Kuhn, R., and van Hemmen, J. L: Hebbian learning reconsidered: Representation of static and dynamic objects in associative neural nets. Biol. Cybern., Vol. 60 (1989) 457–467
[6] Hopfield, J.: Neural networks and physical systems with emergent collective computational abilities. Proc. Nat. Acad. Sci., Vol. 79 (1982) 2554–2558
[7] Sejnowski, T. J.: Storing covariance with nonlinearly interacting neurons. Journal of Mathematical biology, Vol. 4 (1977) 303–321
[8] Skarda, C. and Freeman, W.: How brains make chaos in order to make sense of the world. Behav. Brain Sci., 10:(1987) 161–195
[9] Williams, R. J. and Zipser, D: A learning algorithm for continually running recurrent neural networks. Neural Computation Vol. 1 **2** (1989) 270–280.

Episodic Memory and Cognitive Map in a Rate Model Network of the Rat Hippocampus

Fanni Misják, Máté Lengyel, and Péter Érdi

Dept. Biophysics, KFKI Res. Inst. for Particle and Nuclear Physics,
Hungarian Academy of Sciences
29-33 Konkoly-Thege M. út, Budapest H-1121, Hungary
lmate@rmki.kfki.hu

Abstract. The hippocampus of rodents seems to play a central role in navigation: place cells are prone to serve as elements of a cognitive map. The hippocampus of primates is apparently involved in episodic memory processes. Here we show how the hippocampi of different species (with their anatomy and physiology being considerably similar) can fulfill both purposes. A network was built from simple rate neurons, synaptic weights were modified by a batch version of the coincidence-based Bienenstock–Cooper–Munro learning rule. The network was trained with temporally contiguous sensory stimuli, and then tested in two different paradigms: (1) for navigational performance, place field profiles were evaluated to allow comparison with experimental data; (2) for memory performance, associative capability was assessed. The same neuron network was found to be able to show place cell properties and function as episodic memory, even when trained with non-topographical stimuli.

1 Introduction

Only few investigators have yet addressed directly the question, how the hippocampi of different species, showing remarkable anatomical and physiological similarity, (or even the hippocampus of the same species) meet demands made by two different functions: navigation and episodic memory. Some of these investigations give only a qualitative description on how this two-fold function could be achived [10, 12]. Computational models are ideal for making quantitative predictions, but those already dealing with this problem did not attempt to test their network explicitly on both tasks [14, 13, 2]. Quantitative measurements on place cells (firing rate profiles, time necessary to establish place fields, etc.) provide a good basis for assessing navigational performance (although intact place cells were found under conditions when navigation was heavily impaired, see [8]). Episodic memory, however, is more loosely defined [3]. From the various definitions, two criteria can be postulated on a network to function as an episodic memory buffer: it has to be able to perform pattern completion (episodes can be recalled from their fragments) and to make cross-associations between memories (recall of an episode leads to the recall of preceding or successive episodes

G. Dorffner, H. Bischof, and K. Hornik (Eds.): ICANN 2001, LNCS 2130, pp. 1135–1140, 2001.

– sequence learning is a prominent demonstration of the latter [5, 7, 17], but is usually restricted to associating from earlier to later memory traces).

Another key problem of hippocampal models is the choice of input. Most models use topographical inputs where they presume the similarity of stimuli perceived by the animal at adjacent locations, i.e. the closer two locations are the more similar input patterns belong to them [16, 6, 4]. This approach is ideal for place cell modeling but does not seem to be realistic when addressing episodic memory: it is hard to see why input patterns belonging to consecutive episodes would be more similar than those belonging to temporally more distant events. Thus, as a "worst case" presumption, non-topographical inputs should rather be picked, i.e. the correlation of subsequent stimuli should be minimal.

Based on these considerations we built a neural network that was fed by non-topographical input during training, when a rat was running around on a circular track and thus input sequences were repeated several times. After training, the network was tested both for navigational performance (indicated by properly formed place fields), and for episodic memory performance (bidirectional cross-association between memory traces). This way we intended to directly show under strict conditions (same network, same non-topographical input, two different performance measures) how the apparently two different functions of the hippocampus can be reconciled.

2 Methods

2.1 Network Dynamics

The model network of CA3 pyramidal neurons consisted of $N = 40$ units. Cells that had a similar afferentation pattern were grouped into a single unit and characterized by a single scalar, the mean firing rate in the group. Activity values of units were arranged into an activity vector $\mathbf{a}$. Activity dynamics was described by the following equation:

$$\mathbf{a}(t+1) = c\mathbf{W}\mathbf{a}(t) + c\mathbf{i}(pos(t)) \ , \tag{1}$$

where $\mathbf{W}$ is the weight matrix of recurrent connections, $\mathbf{i}$ is the input vector of external stimulation and $c = 1$ is a scale factor.

Connection weight between two units was considered as the mean of individual connection strenghts between neurons of the units. A linear version of the Bienenstock–Cooper–Munro learning rule [1] applied on recurrent networks was used to train connections:

$$\frac{d\mathbf{W}(t)}{dt} = -\mu\mathbf{W}(t) + \nu(\mathbf{a}(t) - \vartheta\langle\mathbf{a}\rangle)\mathbf{a}^T(t) \ , \tag{2}$$

where ϑ is an unlearning factor and $\langle\mathbf{a}\rangle$ is the time average of activities. Here we used a batch version of this learning rule, meaning that connection weights were updated after $E = 100$ long epochs of iterations without weight changes (for a derivation of this formula, see [15]):

$$\mathbf{W}(t+E) = \mathbf{W}(t) - \mu\mathbf{W}(t) + \frac{\nu}{E}\sum_{t'=t+1}^{t+E}\left(\mathbf{a}(t') - \frac{\vartheta}{E}\sum_{t''=t+1}^{t+E}\mathbf{a}(t'')\right)\mathbf{a}^T(t')\ , \quad (3)$$

where μ is a decay time constant, and ν is learning speed. Elements of $\mathbf{W}$ were always kept non-negative, diagonal elements were kept at a fixed value of $f = 0.8$, and off-diagonal elements were initialized to 0.

2.2 Stimulus

External stimulus to the network depended on the position of the modeled animal. We examined a rat exploring a one-dimensional circular track in discrete time and space. The circle was divided into $P = 40$ locations. Position on the track was described by $pos(t)$. During exploration the rat moved in one direction along the track, proceeding one step forward with p probability and staying at its position with $1 - p$ probability in each time step.

To describe preprocessed information about the environment we assigned an arbitrary input vector to each position: $\mathbf{i}(post(t))$. Input vectors were uncorrelated, that was achieved by stimulating only one unit per position and by that each unit was only stimulated in one position (the "driving" position of the unit, note that units were just defined on the basis of common input, see previous section), and for the sake of clarity units were ordered based on their driving positions:

$$i_k(pos(t)) = \begin{cases} 1 & \text{if } k = pos(t), \\ 0 & \text{otherwise,} \end{cases} \qquad k = 1,\ 2,\ \ldots N \quad (4)$$

where $i_k(pos(t))$ is the kth element of $\mathbf{i}(pos(t))$.

2.3 Evaluation of Performance

During training the rat explored the track for $T_{\text{training}} = 1000$ time step. After training, connections were no longer modified and performance of the network was assessed on two different tests. For measuring navigational performance, the rat moved around the track for an other $T_{\text{test1}} = 8000$ time steps, and activity profiles of units were calculated corresponding to experimental characterization of place cell firing profiles [11]. To test memory performance, the network was presented the same input vector for $T_{\text{test2}} = 1000$ simulation steps and activities of cells at the end of this period were recorded. This procedure was repeated for every input vector that was encountered by the network during the training session.

A four dimensional parameter exploration was performed on parameters ν, μ, p and ϑ. At $\nu = 0.3$, $\mu = 0.5$, $p = 0.1$ and $\vartheta = 3.2$ parameter values $\mathbf{W}$ was found to be convergent, and the network sufficiently reproduced the experimental place field data [11] and its memory performance was also satisfactory. This parameter set was used in further simulations.

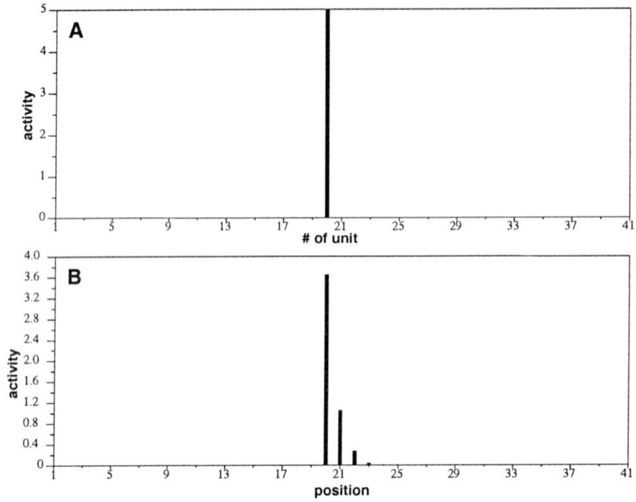

Fig. 1. Performance of the network before training. A: Memory performance. Developing unit activities after the input pattern belonging to position 20 was presented: only unit no. 20 showed activity. (Similar results were obtained with all inputs.) B: Position dependent activity (place field) of unit no. 20. (Similar results were obtained for all units.)

3 Results

When memory performance of the initial network was examined, each unit responded only to one input vector and was not affected by other inputs (Fig. 1A). Navigational performance was similarly poor, but due to high self excitation and unidirectional movement of the rat, each unit had a small tail of activation on positions following its driving position (Fig. 1B).

During exploration, spreading of activities could be observed (data not shown): activity of units began gradually earlier and lasted longer at later phases of exploration than at the beginning. This was caused by potentiation of recurrent connections between units, as units began to recieve excitation through their intra-network connections well before and after they received external stimulation at their driving position, and in turn, due to the coincidence-based learning rule, this spreading lead to further potentiation of recurrent synapses.

Memory and navigational performance of the trained network was different from the initial network. Stimulating the network with an input vector, not only directly driven units showed activity but also units that were driven by stimuli experienced consequently either earlier or later during exploration (Fig. 2A). Similarly, when navigation performance was tested, unit activity did not only occur in the driving position but also at adjacent locations (Fig. 2B), most importantly at locations ahead of the driving position, thus producing realistic firing profiles [11], and reproducing the experience-dependent expansion and

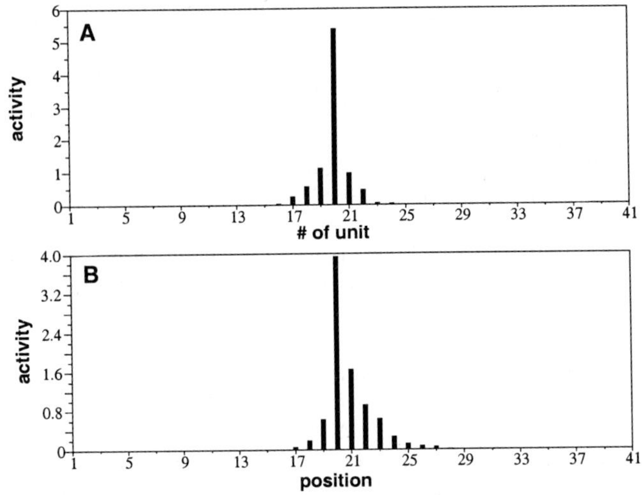

Fig. 2. Performance of the network after training. A: Memory performance. Developing unit activities after the input pattern belonging to position 20 was presented: units driven by preceding and successive inputs were also activated (Similar results were obtained with all inputs.) B: Position dependent activity (place field) of unit no. 20. Note, that its place field expanded asymmetrically backwards compared to its initial size and position of center of mass. (Similar results were obtained for all cells.)

backward-shifting of place fields [9]. Note, that the effects of self-excitations (seen on Fig. 1B) added to the activity at successive positions.

4 Conclusions

We have shown that consistent reocurrence of sensory input patterns may convey enough information to establish place sensitive firing patterns. We found a good match between our model results and experimental place field data. Using non-topographical input vectors we did not assume similarity of consecutive stimuli, thus examining episodic memory properties became available in the same network. Presenting an input vector belonging to an episode to the network resulted in partial recall of earlier and later episodes. Whith this results the same network was shown to have place cell properties and function as episodic memory. This might allow an interpretation that the hippocampus is a more general memory system and is not confined to solve only spatial problems.

References

[1] E L Bienenstock, L N Cooper, and P W Munro. Theory for the development of neuron selectivity: orientation specificity and binocular interaction in visual cortex. *Journal of Neuroscience*, 2:32–48, 1982.

[2] A Bose, V Booth, and M Recce. A temporal mechanism for generating the phase precession of hippocampal place cells. *Journal of Computational Neuroscience*, 9:5–30, 2000.
[3] D Griffiths, A Dickinson, and N Clayton. Episodic memory: what can animals remember about their past? *Trends in Cognitive Sciences*, 3(2):74–80, 1999.
[4] T Hartley, N Burgess, C Lever, F Casucci, and J O'Keefe. Modeling place fields in terms of the cortical inputs to the hippocampus. *Hippocampus*, 10(4):369–379, 2000.
[5] O Jensen and J E Lisman. Theta/gamma networks with slow NMDA channels learn sequences and encode episodic memory: role of NMDA channels in recall. *Learning and Memory*, 3:264–278, 1996.
[6] A Kamondi, L Acsády, X-J Wang, and Gy Buzsáki. Theta oscillation in somata and dendrites of hippocampal pyramidal cells in vivo: activity-dependent phase-precession of action potentials. *Hippocampus*, 8:244–261, 1998.
[7] W B Levy. A sequence predicting CA3 is a flexible associator that learns and uses context to solve hippocampal-like tasks. *Hippocampus*, 6:579–590, 1996.
[8] B L McNaughton, C A Barnes, J Meltzer, and R J Sutherland. Hippocampal granule cells are necessary for normal spatial learning but not for spatially-selective pyramidal cell discharge. *Experimental Brain Research*, 76:485–496, 1989.
[9] M R Mehta, C A Barnes, and B L McNaughton. Experience-dependent, asymmetric expansion of hippocampal place fields. *Proceedings of the National Academy of Sciences of the U.S.A.*, 94(16):8918–21, 1997.
[10] J O'Keefe, N Burgess, J G Donnett, K J Jeffery, and E A Maguire. Place cells, navigational accuracy, and the human hippocampus. *Philosophical transactions of the Royal Society of London. Series B: Biological sciences*, 353:1333–40, 1998.
[11] J O'Keefe and M L Recce. Phase relationship between hippocampal place units and the EEG theta rhythm. *Hippocampus*, 3:317–330, 1993.
[12] M L Shapiro and H Eichenbaum. Hippocampus as a memory map: synaptic plasticity and memory encoding by hippocampal neurons. *Hippocampus*, 9:365–84, 1999.
[13] P E Sharp. Complimentary roles for hippocampal versus subicular/entorhinal place cells in coding place, context, and events. *Hippocampus*, 9:432–43, 1999.
[14] B Shen and B McNaughton. Modeling the spontaneous reactivation of experience-specific hippocampal cell assembles during sleep. *Hippocampus*, 6:685–92, 1996.
[15] Z Szatmáry, M Lengyel, P Érdi, and K Obermayer. Using temporal associatons to model the development of place fields in a novel environment. *European Journal of Neuroscience*, 10(Suppl. 10):39, 1998.
[16] M V Tsodyks, W E Skaggs, T J Sejnowski, and B L McNaughton. Population dynamics and theta phase precession of hippocampal place cell firing: a spiking neuron model. *Hippocampus*, 6:271–280, 1996.
[17] G V Wallenstein and M E Hasselmo. GABAergic modulation of hippocampal activity: sequence learning, place field development, and the phase precession effect. *Journal of Neurophysiology*, 78:393–408, 1997.

A Model of Horizontal 360° Object Localization Based on Binaural Hearing and Monocular Vision

Carsten Schauer and Horst-Michael Gross

Dept. of Neuroinformatics, Ilmenau Technical University, D-98684 Ilmenau, Germany

Abstract. We introduce a biologically inspired localization system, based on a "two-microphone and one camera" configuration. Our aim is to perform a robust, multimodal 360° detection of objects, in particular humans, in the horizontal plane. In our approach, we consider neurophysiological findings to discuss the biological plausibility of the coding and extraction of spatial features, but also meet the demands and constraints of a practical application in the field of human-robot interaction. Presently we are able to demonstrate a model of binaural sound localization using Interaural Time Differences for the left/right detection and spectrum-based features to discriminate between in front and behind. The objective of the current work is the multimodal integration of different types of vision systems. In this paper, we summmarize the experiences with the design and use of the auditory model and suggest a new model concept for the audio-visual integration of spatial localization hypotheses.

1 Introduction

In recent years a lot of promising work on the problem of spatial hearing has been published – many investigations and models of auditory perception exist from neurobiology to psychoacoustics [2, 1]. However, although numerous applications in robotics and human-machine interaction are imaginable, only a few working examples are known. There could be different reasons for that: on the one hand, the models normally can include only a few details of the complex neural coding and processing mechanisms in the real auditory system. On the other hand, when aiming at localization systems working in everyday environments, many acoustic effects arising from very different acoustic characteristics must be faced.

Moreover it is surprising, that there are still hardly any multimodal approaches, even though artificial vision systems provide processing of motion, color or any object specific feature and the mechanisms of spatial hearing and vision complement one another quite obviously (see table). A multimodal approach is both self-evident following the auditory modeling, and promising advantages in the orientation behavior of mobile robots [3], used to demonstrate the models. Furthermore, some remarkable publications on the neurophysiological background of multisensory integration [8], [9] inspire new solutions for our computational models.

G. Dorffner, H. Bischof, and K. Hornik (Eds.): ICANN 2001, LNCS 2130, pp. 1141–1146, 2001.

spatial vision	spatial hearing
topologically organized receptor fields	due to purely temporal receptor, spatial information need further processing (computational map)
activity in topological maps not necessarily corresponds to objects	activity in computational maps mostly corresponds to objects
time–continuous representation	time–discontinuous representation
limited topological range of receptor	complete spatial mapping

2 Modeling Binaural Sound Localization

In contrast to visual perception, hearing starts with one–dimensional, temporal signals, whose phasing and spectrum are essential for the localization. To evaluate spatial information of a sonic field, the auditory system utilizes acoustic effects caused by a varying distance between the sound source and the two ears and the shape of the head and body. We can categorize these effects in intensity differences and time delays. In [1] a comprehensive study of sound localization based on different types of binaural and monaural information is presented, including findings about the localization blur: The achieved precision in the horizontal plane corresponds conspicuously to the relation of azimuth angle variation and interaural intensity differences (ITDs) – a hint for the importance of ITD processing. The assumption, that many localization tasks could be solved just by calculating ITDs and the detailed functional and structural description of ITD processing neural circuits has been the starting point of our modeling.

2.1 Binaural Model Concept

Our work on real–world–capable ITD processing is similar to Lazzaro's neuromorphic auditory localization system [5], but follows a more pragmatic approach. In our simulations, we use digital algorithms for the preprocessing and coincidence detection within spike patterns, as well as an uniform spike–response neuron in the other parts of the model [6]. The motivation to use spikes is, that ITD detection requires a timing that is more precise than the description of neural activities by firing-rates. Moreover, spike patterns can be considered a consistent way of signal coding, which enables a merging of features from different modalities. Figure 1 sketches the architecture, including an extension for a simple in front/behind discrimination. The stages of the model:

1. Microphone signals are filtered by a cochlear model (all–pole–gammatone filter) and coded into spikes (hair-cell model).
2. For every frequency channel, the spike patterns from left and right are cross–correlated (Jeffress coincidence detection model for the medial superior olive (MSO) [4]) - the time–code of binaural delay is transformed into a place code representing interaural phase differences.
3. The resulting pattern is projected onto a non–tonotopic representation of ITDs and thus of azimuthal locations of sound sources (Model of the Inferior

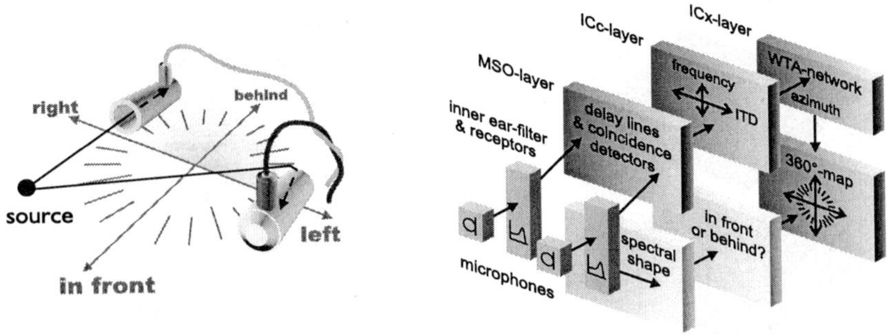

Fig. 1. Left: Microphone configuration: The opposite aligned tubes produce binaural spectral differences, whereby ITDs are not significantly affected. Right: System architecture, composed of the biologically inspired model of ITD detection (dark blocks), a spectrum–based in front/behind discrimination (light blocks). A more detailed description of the ITD-based system is published in [6].

Colliculus, IC). As the result of a winner-take-all (WTA) process, only one direction will be dominant at a time.

4. With the help of a special microphone configuration (see figure 1 left), a simple estimation of interaural spectral differences determines the in front or behind orientation. A 360°-map of horizontal directions is formed.

2.2 Performance of the Sound Localization

Performance tests included all sorts of common sounds (clicks, hand claps, voices, pink noise) and were performed outdoors (without echos) and in an empty, acoustically disadvantageous (echoic) lecture hall. In quiet situations (background noise $<$ -30dB), 100% of the test signals were localized correctly within the accuracy of the discrete system. In additional tests in a shoping center (less echoic, signal-noise ratio 3-5dB) command–words and hand–claps of a person were detected with a probability of 81% and a precission of +/- 10° (90% within +/- 20°). To demonstrate the ability of detecting even moving natural stimuli, the processing of a 12 word long sentence is shown (figures 2), where the tracked speaker position is traveling once around the microphones (performed in a lecture hall without background noise). The experiment demonstrates the superiority of the proposed model over conventional correlation methods that easily fail to process voiced sounds in reverberant conditions. The special properties from which our model benefits can be formulated as follows:

- The spike-pattern mainly codes temporal information like phasing - amplitude is coded indirectly by the spike rate - the effect is a sharp peak as correlation result instead of a smooth sine–response.
- The broadband response of the cochlea filter together with the tonotopically distributed processing and the recombination of frequency bands effectively prevents ambiguous correlation responses.

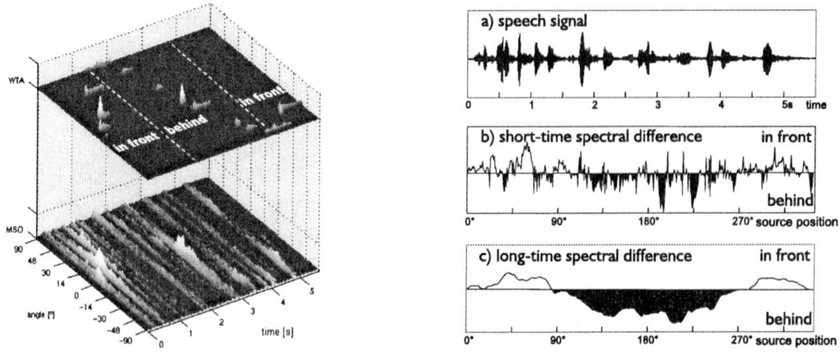

Fig. 2. Left: 360-degree localization of a speaker moving around as described in the text. Visualization of the IC output (bottom) and the resulting activity of the WTA neurons (top). Right: a) Speech–signal, b)short–term (12.5 ms) and c) long–term (0.5 s) spectral difference.

– Because competing sound sources, noise or the disturbance by interferences need energy and time to shift the focus in the WTA–layer, the system shows a hysteresis property and thus prefers onsets.

Based on these qualities the system integrates mechanisms of onset, transient, and ongoing sound processing, and realizes the localization beyond the first wavefront. Due to the hysteresis of the WTA-layer, we become adaptive to the overall loudness and sensitive to onsets or changes.

3 Concept for a Sub-Cortical Auditory-Visual Integration

Usually, the attentive perception, especially the localization of objects, has multi-modal character. Spatial hearing and vision seem to be complementary strategies in localization. Other than the auditory system, vision is based on a receptor, that is already providing topologically organized information and the question becomes in which way objects of interest manifest themselves in the continuous visual representation? In the context of human-robot interaction, we have named feature candidates like pure intensity, motion, color or contour. In contrast to the low-level auditory-space processing in the midbrain we must now distinguish between cortical object recognition and low-level multi-sensor fusion. Firstly we clarify the term "low-level": Since visual features have no interrelations with characteristic frequencies, the first stage for a visual-auditory integration can be found, following the projections from the non-tonotopical spatial maps in the extern IC. Investigations on the mainly visually, but also auditory (via ICx) innervated, Superior Colliculus (SC), provide evidence for a merging of the sensor modalities and the forming of multisensory spatial maps [8]. Visually sensitive neurons found here, are not or less specialized for color or orientation of contours but respond to broadband intensity and certain velocities of moving stimuli (changes in intensity). We use these findings as a basis for our multi-modal model, although we also consider to integrate higher-level feature as an option in applications. According to [9], at least the following properties of the representation and integration of multiple sensory inputs in the SC are to be considered in the model architecture:

Superficial SC (SCs) is responsive only to visual stimuli. Counterpart of a retinotopically ordered map in SCs is a one-dimensional map of horizontally arranged intensity or intensity differences, provided by a wide-angle vision system.

(i) Deep layers of the SC (SCd) respond to visual, somatosensory, auditory and multi-modal stimuli. (ii) visual receptive fields (RF) are significantly larger than in SCs. In the model we propose convergent visual projections from SCs to SCd, where also the auditory input from ICx is received. According to the field of vision, the RFs of the visual projections cover just a part of the resulting multisensory map.

Most SCd multisensory neurons display response enhancement when receiving spatially and temporally coincident stimuli but show response depression if simultaneous stimuli are spatially separated. This actual property of multisensory integration can be realized by a WTA-type network with both auditory and visual afferents and global inhibition. Competing features inhibit each other, aligned stimuli excite one another.

In SCd overlapping multisensory and motor-maps initiate overt behavior. We are going to use the multisensory map to code turn reflexes of a robotic head toward the acoustic or visual stimulus, small moves if the stimuli originate almost from the center, and stronger ones if "something" is to be seen on or heard from the side.

Different modality-specific RFs have to be aligned to allow response enhancement, even if eyes and/or ears can be moved separately. If so, there has to be also an exclusively visual map in SCd, controlling eye movement. This is consistent with the known models of saccade generation [7]. To achieve map alignment every eye-specific motor-command must cause a shift in the auditory map (in the model the weights in a variable ICx–SCd projection change).

4 First Results and Outlook

The implementation of the model as seen in figure 3a) is currently in progress and hardly depends on a realtime capable simulation and the integration of an experimental robotic platform. This will be the supposition for interactive map shifting for RF alignment or tests in real man-machine communication, but we can already demonstrate the behavior of the WTA-implementation of the multisensory feature map based on offline recorded data, Figure 3 b).

The practical aspect of the application of the model is an expected significant advantage in the detection and tracking of users by the multimodal robot. Besides, we will have to discuss in detail the role of sub-cortical multisensory

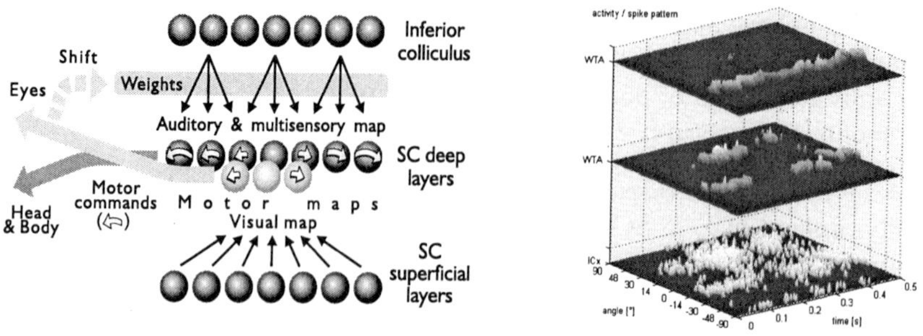

Fig. 3. Left: A simplified and universal model of the superior colliculus, which satisfies all properties, mentioned in the text. Further simplifications can be made, if no separate camera turns are possible (no separate SCd visual map and static ICx-SCd-projections) or if an omnidirectional camera is used (modified SCs-SCd projection). Right: Localization of a narrow band sound (whistle) and a human call setting in about 100 ms later. Response of the auditory model (middle), and the multisensory map (above). The dynamic range of the (spike-based) multisensory map enables local response enhancement and focusing on the multimodal event (an artificial visual stimulus in the form of a lamp-position was applied).

integration in attentive perception or in control of overt behavior. The interaction in real man-machine communication surely will help to gain insights that are hidden to purely theoretically investigations.

References

1. Jens Blauert. *Spatial Hearing : The Psychophysics of Human Sound Localization.* MIT Press, 1996.
2. G. Ehret and R.Romand The Central Auditory System. Oxford University Press, New York 1997.
3. Gross, H.-M., Boehme, H.-J. PERSES - a Vision-based Interactive Mobile Shopping Assistant. in: Proc. IEEE SMC 2000, pp. 80-85.
4. L.A. Jeffress. A place theory of sound localization. *J. Comp. Physiol. Psychol.*, 41:35–39, 1948.
5. John Lazzaro and Carver Mead. A silicon model of auditory localization. Neural Computation, 1(1):41–70, 1989.
6. Schauer, C., Zahn, Th., Paschke, P., Gross, H.-M. Binaural Sound Localization in an Artificial Neural Network. in: Proc. IEEE ICASSP 2000, pp. II 865-868.
7. P.H. Schiller A model for the generation of visually guided saccadic eye movements. in: Models of the visual cortex, D. Rose, V.G. Dobson (Eds). Wiley, 1985, pp 62-70
8. Stein, B.E. and Meredith, M.A. The Merging of the Senses. The MIT Press, Cambridge, Massachusetts
9. M. T. Wallace, L. K. Wilkinson, B. E. Stein. Representation and Integration of multisensory Inputs in Primate Superior Colliculus. Journal of Neurophysiology. Vol. 76, No.2: 1246-1266, 1996

Self-Organization of Orientation Maps, Lateral Connections, and Dynamic Receptive Fields in the Primary Visual Cortex

Cornelius Weber

Dept. of Brain and Cognitive Science, University of Rochester, NY, USA

Abstract. We set up a combined model of sparse coding bottom-up feature detectors and a subsequent attractor with horizontal weights. It is trained with filtered grey-scale natural images. We find the following results on the connectivity: *(i)* the bottom-up connections establish a topographic map where orientation and frequency are represented in an ordered fashion, but phase randomly. *(ii)* the lateral connections display local excitation and surround inhibition in the feature spaces of position, orientation and frequency. The results on the attractor activations after an interrupted relaxation of the attractor cells as a response to a stimulus are: *(i)* Contrast-response curves measured as responses to sine gratings increase sharply at low contrasts, but decrease at higher contrasts (as reported for cells which are adapted to low contrasts [1]). *(ii)* Orientation tuning curves of the attractor cells are more peaked than those of the feature cells. They have reasonable contrast invariant tuning widths, however, the regime of gain (along the contrast axis) is small before saturation is reached. *(iii)* The optimal response is roughly phase invariant, if the attractor is trained to predict its input when images move slightly.

Introduction

It is common for hierarchical maximum likelihood models to assume a factorial prior probability distribution of the hidden variables on each layer. However, statistical dependencies remain to be captured by subsequent hierarchical levels. Lateral connections can help to find this structure, may they serve to statistically de-correlate [10] or to find correlation structure [3] within activities of nearby cells. Lateral connections which form an attractor were shown [5] to recover noisy input with maximum likelihood, given certain noise models. Interpreting shifts in images as noise, may supply an approximate theoretical framework for shift invariant recognition as well as for understanding complex cells in V1.

In our contribution, bottom-up weights W^{bu} of hidden layer neurons ("feature cells") and the top-down feed-back projections V^{td} (Fig. 1) are trained with a sparse coding paradigm to become topographically arranged edge detectors. We train lateral connections V^{lat} to form attractors which capture the correlation structure on the activations of hidden layer neurons. The "attractor cells" receive initial input from the feature cells and then relax freely for a fixed period of

G. Dorffner, H. Bischof, and K. Hornik (Eds.): ICANN 2001, LNCS 2130, pp. 1147–1152, 2001.

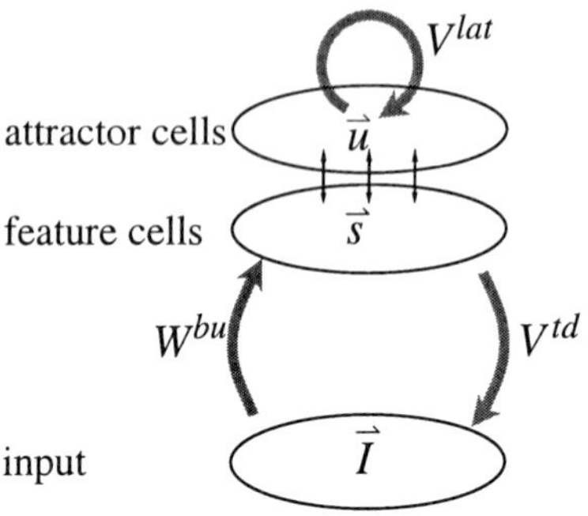

Fig. 1. Model architecture with weights W^{bu}, V^{td}, V^{lat} and activations $\boldsymbol{I}$, $\boldsymbol{s}$, $\boldsymbol{u}$.

time using the lateral connections. The final activation state is defined as their response to the input and compared to biological cell response properties in V1. We find that this simple setup can explain a variety of biological observations.

Theory and Methods

The Helmholtz machine framework [2] is originally intended as a purely hierarchical model: generative, top-down weights are trained to generate the data while bottom-up, recognition weights are trained to statistically invert the generative model. Hereby, the data can be considered as part of the recognition model and supply the target values for training the generative model. Vice versa, the prior on the hidden layer activations is considered as part of the generative model and supplies the target training values for the recognition model.

However, the recognition model should be optimized using activation statistics as delivered by the training data instead of the prior on the hidden layer activations [2]. This can be done by lateral weights [3]: they should learn the statistics of the hidden unit activations during presentation of the data in order to recall these data during the *sleep phase* to be used as target values for training the recognition model. Lateral weights are thus part of the generative model.

Bottom-Up and Top-Down Weights. However, to train the lateral weights, a coarsely functioning recognition model has to exist first. For clarity, we do not use the prior which is learned by the attractor cells for training of the recognition weights at all. Instead, we assume a topographic Gaussian function as a prior on "higher order neurons" [7] to establish a topographically ordered orientation map.

Recently, the square of feature cell responses has been proposed as the relevant feature to arrange neurons topographically on V1 [8][7], because this will produce a random phase distribution as observed experimentally [4]. To mimic this effect, we do not distinguish excited from inhibited cells: in the *sleep phase*, we draw, from the topographic Gaussian prior, values (0, 1, 1′) on the feature cells as the target training values $\boldsymbol{s}^-$ for the recognition model. Hereby, a small number of mostly adjacent units is switched ON. The activity $1'$ is interpreted as an activation of an "inhibited" cell, and is thus used as a target value while the bottom-up input is inverted (by setting β, Eq. 5, to $-\beta$). The bottom-up and top-down connection matrices W^{bu}, V^{td} are used to transform the input $\boldsymbol{I}$ into a hidden representation $\boldsymbol{s}$, and vice versa:

$$I_j = \sum_i v_{ji}^{td} s_i \,, \qquad s_i = f(\sum_j w_{ij}^{bu} I_j) \tag{1}$$

They are trained to functionally invert each other mutually, by maximizing the log-likelihood to reconstruct a stimulus sent through the other weight matrix:

$$\Delta w_{ij}^{bu} \approx (s_i^- - \tilde{s}_i^-)\, I_j^- \,, \qquad \Delta v_{ji}^{td} \approx (I_j^+ - \tilde{I}_j^+)\, s_i^+ \tag{2}$$

where the upper index "−" means that a random hidden vector $\boldsymbol{s}^-$ initiated the activities (*sleep phase*) and "+" denotes initiation by data $\boldsymbol{I}^+$ (*wake phase*).

Lateral Weights. In the *wake phase*, attractor cell activations $\tilde{u}^+(t=0)$ are initialized with the maximum of the (positive) values $\boldsymbol{s}^+$ of an excitatory and its associated inhibitory neuron (this could as well be the sum instead of the maximum, because the smaller of the two values is almost vanishing). Then, the lateral weights V^{lat} define their activation dynamics through time:

$$\tilde{u}_i(t+1) = f(\sum_l v_{il}^{lat} \tilde{u}_l(t)) \tag{3}$$

They are trained in the *wake phase* only, to generate (predict) incoming activities. Learning maximizes the log-likelihood to generate this distribution via Eq. 3:

$$\Delta v_{il}^{lat} \approx \sum_t (u_i^+(t) - \tilde{u}_i^+(t))\, \tilde{u}_l^+(t-1)\,. \tag{4}$$

During each relaxation, input images were shifted by one pixel into a random direction at each time step. Hereby, "target" data $u_i^+(t)$ moved slightly away from $u_i^+(t=0) = \tilde{u}^+(t=0)$ which also initiated the relaxation process.

Parameters and Sparseness. Model neurons are binary with activations $\{0,1\}$ and probabilistic. We use the probability for neuron i to be active as a continuous transfer function:

$$f(h_i) = p_i(1) = \frac{e^{\beta h_i}}{e^{\beta h_i} + n} \tag{5}$$

Parameters $\beta = 2$ scales the slope of the function and n is the degeneracy of the 0-state. Large $n = 64$ for the feature cells reduces the probability of the 1-state favoring sparseness. $n = 8$ for the attractor cells. A weight decay of $-0.03 \cdot w_{ij}$ on W^{bu} and V^{td} adds to Eqs. 2 to hinder weights to compensate for sparseness. W^{bu} and V^{td} were initialized randomly under topographic Gaussian envelopes, V^{lat} randomly, and self-connections were constantly set $v_{ii}^{lat} = 0$. Images were filtered as described in [9]. Input layer size: 16×16, hidden layer size: 24×24.

Results

Anatomy. The hidden neurons have developed localized, Gabor function shaped receptive fields to the input. Fig. 2 shows essential observable parameters which

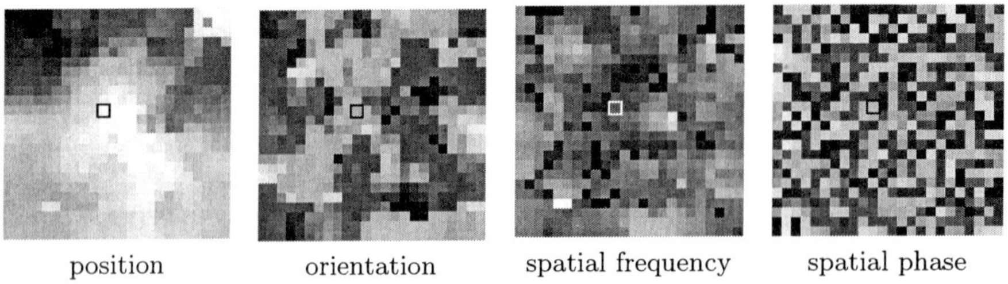

Fig. 2. Individual parameter maps for the input receptive fields of the simple hidden layer cells, from W_{10}. Parameters, except for spatial phase, vary continuously across large regions of the neuronal layer. A box marks the cell at position α (see Figs. 4,5).

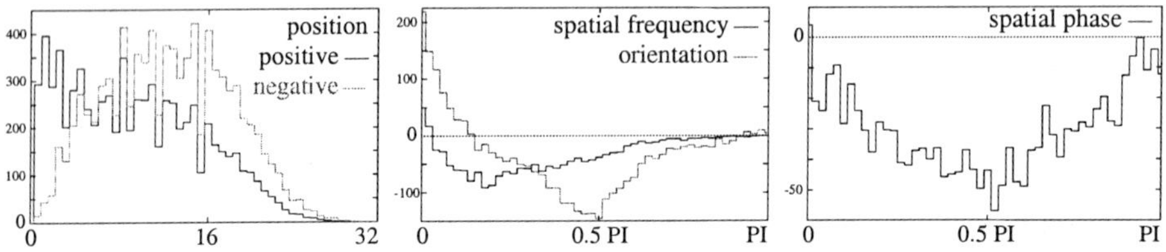

Fig. 3. Densities of lateral connections V^{lat}. **Left:** excitatory and inhibitory connections over distance across the hidden layer. **Middle, right:** functional connections (sum of excitatory and inhibitory) over feature differences. On the x-axes, differences between the features for two feature cell pairs (obtained from the Gabor fits to W_{10}) are displayed, binned into 50 intervals. The y-axis shows the sum of lateral connections which link attractor cell pairs whose feature difference is in the corresponding interval.

were derived from Gabor function fits to the bottom-up connections W_{10}. Position, orientation and spatial frequency of the receptive fields in the input show continuous changes across the hidden layer, spatial phase w.r.t the receptive field center is distributed randomly.

Functional lateral connection densities along individual parameter axes are displayed in Fig. 3. The concept of local excitation and surround inhibition is well attributed to position-, orientation- and spatial frequency difference. Inhibition is for example strongest at cross-orientation, $\pi/2$. Lateral connections integrate over larger parts of phase space.

Physiology. Fig. 4 shows the relaxation of the network activities after initialization with a sinusoidal grating. For learning and for obtaining the response curves below, relaxation was performed until time step 10. The response remains stable for limited further time.

Fig. 5 a) shows contrast response curves of *(i)* a feature cell and *(ii)* the corresponding attractor cell after relaxation ($t = 10$). With increasing contrast (simulated as stimulus amplitude), the feature cell's response follows its transfer function. In contrary, the attractor cell's response within the lateral dynamics saturates, however, unlike biologically observed, near its "physically" maximal

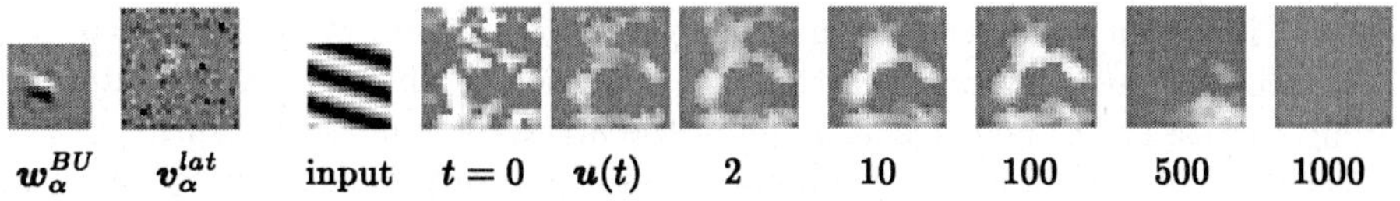

Fig. 4. From left to right: the receptive field w_α^{BU} of feature cell indexed α. The lateral weights v_α^{lat} of the corresponding attractor cell. Bright positive, dark negative values. A sinusoidal vector $\boldsymbol{I}$ shown at the **input**. Starting from $t = 0$, attractor activations $\boldsymbol{u}(t)$ develop after initialization from that stimulus through times given later on.

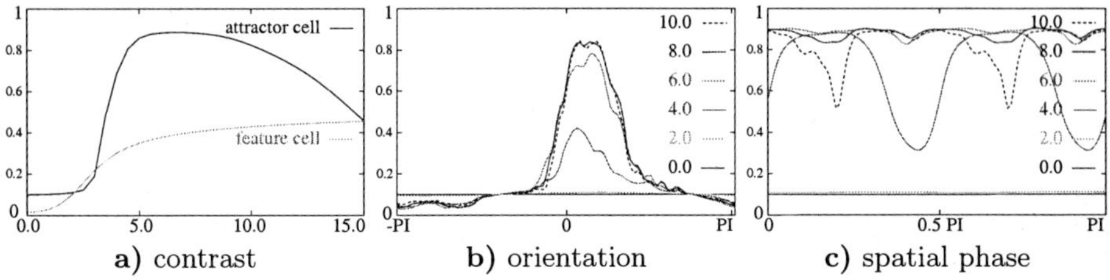

Fig. 5. a) Contrast response curves, for the attractor cell at position α and the corresponding feature cell. The average over phase is taken, which downsizes the output of the feature cell. **b)** Orientation tuning and **c)** phase tuning curves of the attractor cell at different contrasts. Input was a static sine wave pattern with the optimal parameters for the corresponding feature cell (which in **b)** defines the 0-orientation as optimal).

activation. It decreases at higher contrasts, correctly [1]. Note that a sine grating activates more neurons than a typical image as used for training.

Orientation tuning curves of the attractor cell, Fig. 5 b), display a more or less constant and sharp tuning width. At cross-orientations and high contrasts, there is suppression w.r.t. zero-input activity.

Fig. 5 **c)** shows that the attractor neuron responds across a large phase range, with phase dependence only in a small contrast range. To obtain phase invariance, the "target" data $u_i^+(t)$ for the attractor must not be constant in time. Biologically, micro-saccades may deliver slightly shifting input.

Discussion

Single new elements in this work are only a Helmholtz machine which develops edge detectors from images or forms a topography. However, to our knowledge, a network fully trained on natural images and (complex) cells with contrast invariant orientation tuning are for the first time integrated.

The present work investigates the learning mechanisms which produce biologically observed results in this system, assuming that the architectural setting is suited. We show that learning the correlation structure in the activations which are initiated by the data (Eq. 4) produces weights with a structure of local excitation and surround inhibition. This is sufficient to generate sharp, contrast invariant orientation tuning curves. In particular, we do **not** need: *(i)* a sparseness constraint on the attractor activities to sharpen the orientation tuning curves, or *(ii)* de-correlation [6] of the attractor unit activations.

Fortunately, the attractor cannot generate its input data correctly (even if images were not shifted and the identity was learned). Otherwise, broadly tuned, feature cell response properties would remain. A (line-) attractor is instead the only patterns that the attractor cells can maintain. Each attractor is a hill of activity on similar neurons, but more irregular than if constructed analytically [11].

The fact that we observe different orientation tuning curves at different contrasts reflects a dynamic effect of the relaxation procedure. At low contrasts the hill of activity builds up from smaller activations, primarily involving specific excitatory connections to the few active cells. At high contrasts, suppression via less specific but more numerous inhibitory connections dominates.

In our model set-up, the input to the attractor is not controlled by parameters and does not constantly influence the attractor. The question remains, how the attractor cells can *learn* to select their input from the feature cells.

References

1. M. Carandini and D. Ferster. Membrane potential and firing rate in cat primary visual cortex. *J. Neurosci.*, 20(1):470–84, 2000.
2. P. Dayan, G. E. Hinton, R. Neal, and R. S. Zemel. The Helmholtz machine. *Neur. Comp.*, 7:1022–1037, 1995.
3. P. Dayan. Recurrent sampling models for the Helmholtz machine. *Neur. Comp.*, 11:653–77, 2000.
4. G.C. DeAngelis, G.M. Ghose, I. Ohzawa, and R.D. Freeman. Functional micro-organization of primary visual cortex: Receptive field analysis of nearby neurons. *J. Neurosci.*, 19(9):4046–64, 2000.
5. S. Deneve, P.E. Latham, and A. Pouget. Reading population codes: a neural implementation of ideal observers. *Nature Neurosci.*, 2(8):740–5, 1999.
6. D.W. Dong. Associative decorrelation dynamics: A theory of self-organization and optimization in feedback networks. In *Proceedings of NIPS 7*, pages 925–32, 1994.
7. A. Hyvärinen and P.O. Hoyer. Emergence of phase- and shift-invariant features by decomposition of natural images into independent feature subspaces. *Neur. Comp.*, 12:1705–20, 2000.
8. N. Mayer, J.M. Herrmann, H.U. Bauer, and T. Geisel. A cortical interpretation of ASSOMs. In *Proceedings ICANN*, pages 961–966, 1998.
9. B.A. Olshausen and D.J. Field. Sparse coding with an overcomplete basis set: A strategy employed by V1? *Vision Research*, 37:3311–3325, 1997.
10. J. Sirosh and R. Miikkulainen. Topographic receptive fields and patterned lateral interaction in a self-organizing model of the primary visual cortex. *Neur. Comp.*, 9:577–94, 1997.
11. K. Zhang. Representation of spatial orientation by the intrinsic dynamics of the head-direction cell ensemble: A theory. *J. Neurosci.*, 16:2112–26, 1996.

Markov Chain Model Approximating the Hodgkin-Huxley Neuron

Yuichi Sakumura[1], Norio Konno[2], and Kazuyuki Aihara[3,4]

[1] Nara Institute of Science and Technology, Ikoma-shi Nara 630–0101, Japan
[2] Yokohama National University, Hodogaya-ku Yokohama-shi 240–8501, Japan
[3] The University of Tokyo, Bunkyo-ku Tokyo 113–8656, Japan
[4] CREST, Japan Science and Technology Corporation (JST), Kawaguchi-shi, Saitama 332–0012, Japan

Abstract. The Hodgkin-Huxley (HH) model well reconstructs a neuronal membrane potential. However, the HH equations, which are nonlinear ordinary differential equations with four variables, are not appropriate to simulate a large network because of the high complexity. In addition to the dynamical complexity, stochastic factors such as synaptic transmissions should be considered in a model neuron. In this paper, we propose the Markov chain model that statistically approximates firing patterns of the HH model.

1 Introduction

Brain works as the large network of a huge number of neurons. The property of a single neuron has been physiologically studied by many researchers. It is well known that the neuronal response to stimulus is so complex [1]. For example, the spike threshold has a close relation to the slope of a membrane potential [2]. Such complex neurons integrate a large number of synaptic inputs and generate electrical spikes according to the dynamics of sodium ion channels. These spikes, which play a primary role for information transmission, propagate through many neurons when brain processes the information. The problem is that it is difficult to numerically examine the large network composed of complex neurons.

Information transmission between neurons at a synaptic connection site is a stochastic phenomenon [3–5]. Then, the process from presynaptic spiking to postsynaptic spiking can be regarded as stochastic. Therefore, it is reasonable that the simplified neuron model is supposed to generate spikes stochastically. If the stochastic simple neuron model can be proposed, which has a similar property to a complex realistic neuron in the statistical viewpoint, the large network as brain can be easily examined.

The purpose of this study is to construct the Markov chain model which is consistent with the Hodgkin-Huxley model (HH model) in statistics of interspike intervals (ISIs). Although previous studies attempt to simplify the HH model [6, 7], there is no stochastic Markov chain models with only two states. Specific temporal patterns of spikes is not important here.

G. Dorffner, H. Bischof, and K. Hornik (Eds.): ICANN 2001, LNCS 2130, pp. 1153–1160, 2001.

2 Neuron Models

2.1 The Hodgkin-Huxley Equations

We compute the following HH equations [1],

$$C_m \frac{dV}{dt} = -I_{Na} - I_K - I_l - I_{syn}, \tag{1}$$

where V (mV) is the membrane potential, and $C_m (= 1.0\mu\text{F/cm}^2)$ is the membrane capacitance per unit area. The ionic currents (μA) I_{Na} (sodium), I_K (potassium), and I_l (passive leak) are defined as $I_{Na} = 120m^3h(V - V_{Na})$, $I_K = 36n^4(V - V_K)$ and $I_l = 0.3(V - V_l)$, where the reversal potentials are $V_{Na} = 50.0 - E_{rest}$, $V_K = -77.0 - E_{rest}$ and $V_l = -54.4 - E_{rest}$, respectively (the resting potential is $E_{rest} = -65$ mV). The gating variables m, h, and n obey a kinetic equation $dx/dt = \{\alpha_x \cdot (1 - x) - \beta_x \cdot x\}\, 3^{(T-6.3)/10}$, where the temperature $T = 15$, and functions α_x and β_x $(x = m, h, n)$ are $\alpha_m = 0.1(25-V)/\{e^{(25-V)/10} - 1\}$, $\beta_m = 4e^{-V/18}$, $\alpha_n = 0.01(10-V)/\{e^{(10-V)/10} - 1\}$, $\beta_n = 0.125e^{-V/80}$, $\alpha_h = 0.07e^{-V/20}$ and $\beta_h = 1/\{e^{(30-V)/10} + 1\}$. The neuron expressed by equation (1) has $V = 0.0$ (mV) under the resting condition. The term I_{syn} represents the summation of all voltage-dependent synaptic currents $I_{syn} = (\sum_i g_{syn}(t - t_i)) \cdot (V - V_{syn})$ and t_i is the instant of the i-th synaptic input. Here, t_is obey the Poisson process because cortical neurons likely fire irregularly [8]. We suppose $g_{syn}(t)$ as the fast excitatory AMPA ($V_{syn} = -10 - E_{rest}$ mV) types of synaptic conductances per unit area defined by the alpha function [9, 10] $g_{syn}(t) = Ate^{1-t/t_{peak}} (t \geq 0)$ where $A = g_{peak}/4\pi r^2 t_{peak}$ and g_{peak} (nS) is the peak value of the synaptic conductance at the peak time t_{peak} (msec). Here we assume $g_{peak} = 0.5$ and $t_{peak} = 2.0$. For the sake of simplicity, we suppose that the neuron is composed of a single compartment of a sphere shape with radius $r = 20$ μm. A sample time course of the membrane potential is shown in Fig. 1.

2.2 Characteristics of the HH Model

As shown in Fig. 1, the HH model generates spikes under the condition of random synaptic inputs (homogeneous Poisson process). The reverse correlations in the figure imply two types of processes for the spike generation. The first is that a large input after a small input evokes a spike when the membrane potential was at the subthreshold level (**a** and **b**). Such a fluctuation is shown in experimental data of cortical neurons [11]. The other is that succeeding tonic spikes appear just after preceding spikes (**c** and **d**). Tonic spikes are mainly caused by the sodium current (I_{Na} in eq. (1)).

2.3 Two State Markov Model

To construct a stochastic model, time is supposed to be discrete. The width of a time window is so small that there is at most one spike. We can introduce two

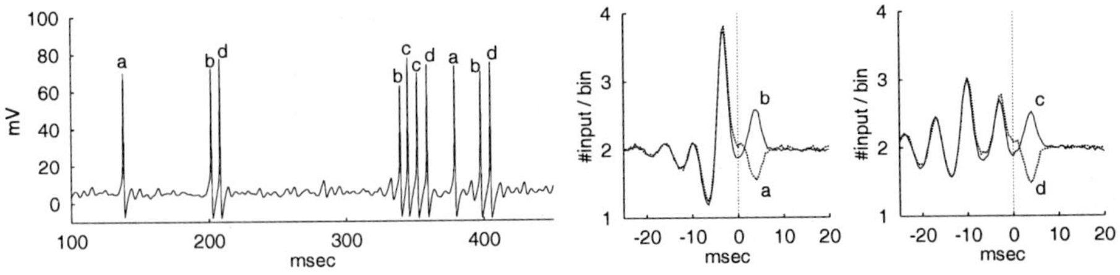

Fig. 1. Sample time course of the membrane potential obtained from eq. (1) with 4 kHz excitatory Poisson synapse (left) and reverse correlation histograms (middle and right) of spikes labeled **a** – **d**, which mean the single spike, initial tonic spike, succeeding tonic spike, and last tonic spike, respectively.

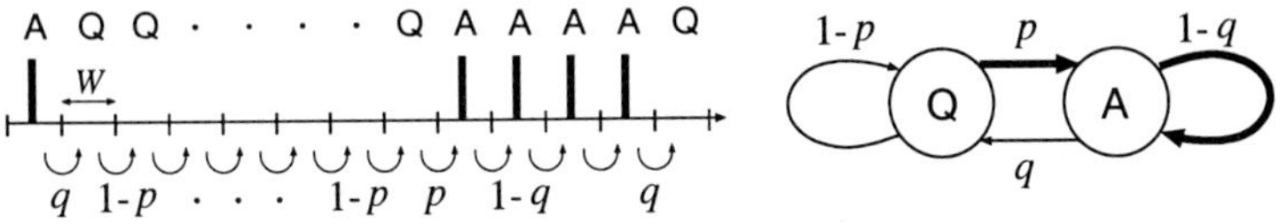

Fig. 2. *left*: Schematic spike train of the HH neuron. The horizontal axis as time is divided into small windows (W msec), in which there is at most one spike. The windows A (active) and Q (quiescent) represent a spike existing window and a window with no spike, respectively. The transition probabilities are $p = \Pr\{\mathsf{Q} \to \mathsf{A}|\mathsf{Q}\}$ and $q = \Pr\{\mathsf{A} \to \mathsf{Q}|\mathsf{A}\}$. *right*: Two state Markov chain model substituting for **a**. It is supposed that the neuron model generates spikes when the state changes from Q to A or from A to A (thick arrow). It takes W msec to transfer to any states.

probabilities of the spike generation in a window when the HH neuron subjected to random synaptic input. One is the probability that a spike occurs in the window after no-spike window (p). The other is the probability that a spike occurs in the window after a spike existing window $1 - q$ (Fig. 2). This equals to the two state Markov chain model shown in the right figure of Fig 2.

2.4 Coefficient of Variation

We prepare the numerical analysis for demonstrating the adequacy by the coefficient of variation C_V which is one of barometers for the irregularity. C_V is defined as $C_V = \sigma/\mu$, where σ and μ are s.d. and mean of ISI, respectively. C_V is equal to zero for periodic spikes and to 1 for the pure Poisson process. It is reported that cortical neurons fire irregularly ($0.5 \leq C_V$) [8].

We can easily obtain C_V values of ISI produced by the stochastic model shown in Fig. 2. The random variable X, number of windows between two spikes, obeys the following probability distribution (Fig. 3),

$$X = \begin{cases} 1 & : \quad 1-q \\ k\ (\geq 2) & : \quad q(1-p)^{k-2}p \end{cases} . \tag{2}$$

Therefore,

$$E(X) = 1 + \frac{q}{p} \tag{3}$$

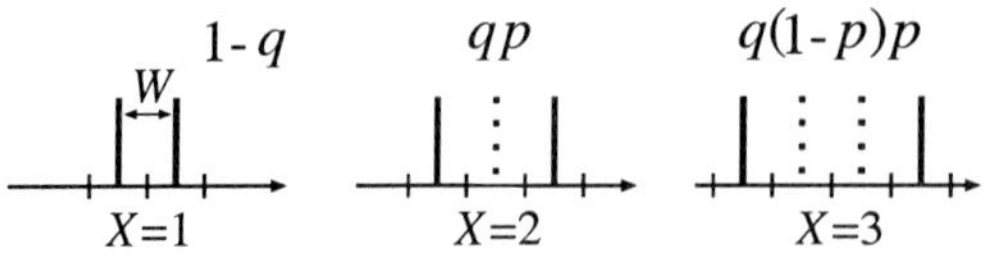

Fig. 3. Probabilities of the random variable X which means the number of windows between two spikes ($X = 1, 2, 3$).

$$E(X^2) = 1 + \frac{q(p+2)}{p^2} \tag{4}$$

$$Var(X) = \frac{q}{p^2}(2 - p - q). \tag{5}$$

Then, C_V of the stochastic model is,

$$C_V = \frac{\sqrt{Var(X)}}{E(X)} = \frac{\sqrt{q(2-p-q)}}{(p+q)}. \tag{6}$$

This value can be calculated if the probabilities p and q are estimated.

3 Simulation

3.1 Estimation of the Transition Probabilities

For estimation of the transition probabilities p and q of the HH model, we shall examine a pair of a quiescent period and a tonic spike assembly (an active period) as shown in Fig. 4**a**. Of all statistics about spikes observed from this pair, we can estimate p and q from:

- mean ISI within tonic spike assemblies. This is supposed to be the time width W of a window.
- mean (μ_1) and s.d. (σ_1) of the number of windows in a quiescent period (the quotient when the period divided by W).
- mean (μ_2) and s.d. (σ_2) of the spike number in a tonic spike assembly.

Since μ_1 and σ_1^2 are the mean and the variance of all ISI except for the tonic case ($X = 1$),

$$\mu_1 = \left(1 + \frac{q}{p}\right) - 1 \cdot (1 - q) = \frac{q(1+p)}{p} \tag{7}$$

$$\sigma_1^2 + \mu_1^2 = \left\{1 + \frac{q(p+2)}{p^2}\right\} - 1^2 \cdot (1 - q) = \frac{q(p+2)}{p^2} + q. \tag{8}$$

By canceling q from these equations,

$$p = \frac{\sqrt{1+4r} - 1}{2}, \tag{9}$$

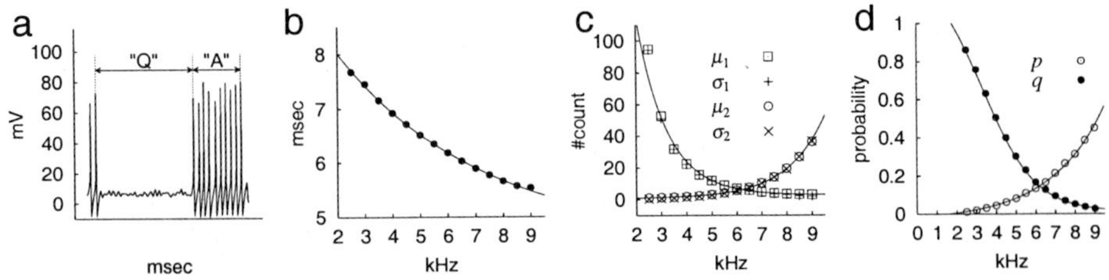

Fig. 4. **a**: Sample pair of a quiescent period and a tonic assembly of the membrane potential obtained from the HH neuron with homogeneous Poisson synaptic inputs. **b**: Mean of ISI in the tonic spike assembly for input frequencies 2.5-9.0 kHz. This mean is supposed to be the time width W of a window for each input frequency. **c**: Mean and s.d. of the number of windows in a quiescent period (μ_1, σ_1) and a tonic spike count (μ_2, σ_2). **d**: Estimated transition probabilities p and q (see text). All the solid lines in **b-d** are fitting curves.

where $r = 2\mu_1/(\mu_1^2 + \sigma_1^2 - \mu_1)$.

On the other hand, the probability that there are k tonic spikes in an assembly is

$$\Pr\{\mathsf{Q} \to \underbrace{\mathsf{A} \to \cdots \to \mathsf{A}}_{k} \to \mathsf{Q}|\mathsf{Q}\} = p(1-q)^{k-1}q. \tag{10}$$

Therefore,

$$\mu_2 = E(k) = \frac{p}{q} \tag{11}$$

$$\sigma_2^2 + \mu_2^2 = \frac{p(2-q)}{q^2}. \tag{12}$$

Thus we obtain q by canceling p from these equations,

$$q = \frac{2}{s+1}, \tag{13}$$

where $s = (\mu_2^2 + \sigma_2^2)/\mu_2$.

By using the observed statistics W, μ_1, σ_1, μ_2 and σ_2 from the HH model (Fig. 4**b** and **c**), the transition probabilities p and q are estimated with eqs. (9) and (13) as shown in Fig. 4**d**. Note that all the variables depend on input frequencies. It is also interesting to note that the s.d. is almost same as the mean both of a quiescent period and of a tonic spike number (Fig. 4**c**).

3.2 Comparison of C_V

Figure 5 shows C_V values of output ISI of the HH model and the Markov model by the numerical calculation. As the figure indicates, C_V values of both model are similar to each other. It follows that two state Markov chain model can approximate the response of the complex HH model as long as evaluating from

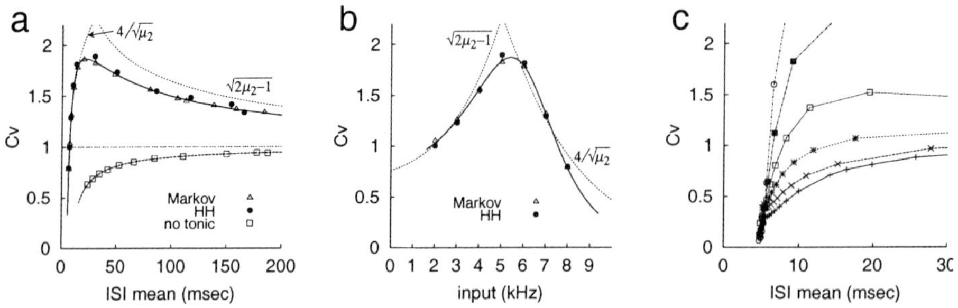

Fig. 5. **a**: Comparison of C_V values obtained from the HH model, the Markov model, and the Markov model without tonic spike ($q = 1$). The thick curve is drawn by eq. (6). The horizontal axis is the mean ISI (eq. (3) for the Markov model). The thin curves $4/\sqrt{\mu_2}$ and $\sqrt{2\mu_2 - 1}$ are approximation curves for large and small μ_2, respectively (see text). **b**: Same as **a** but some points are omitted and the horizontal axis is the input frequency. **c**: C_V values of the HH model for $t_{peak} = 0.5, 1.0, 1.5, 2.0, 2.5, 3.0$ from the bottom. The stimulus current is the sum of the voltage-*independent* synaptic current (I_{in}) and the input frequency (f Hz) dependent shunting current (I_{shunt}) [12]; $I_{in} = \sum_i g_{in}(t - t_i) \cdot (0 - V_{syn})$ where $g_{in}(t) = t \cdot exp(1 - t/t_{peak})/4\pi r^2 t_{peak}^2$. The alpha function $g_{in}(t)$ satisfies $\int_0^\infty g_{in}(t)dt = 0.054$ (mS/cm^2) for any t_{peak}, so that the total input current is independent of t_{peak}. $I_{shunt} = 5.4 \times 10^{-5} f$.

the quadratic statistics, C_V. C_V values are all less than 1.0 for the Markov model with no tonic spike ($q = 1$) as shown in Fig. 5**a**. Since some cortical neurons have C_V more than 1.0 [8], it is considered that cortical neurons often show tonic spikes.

In real neurons, the dendritic attenuation and peak latency of the synaptic current can be considered. The effects of these factors are shown in Fig. 5**c**: C_V is larger for larger t_{peak} (large attenuation and latency, distal synaptic input). Appropriate Markov models for each t_{peak} can be considered by calculating p, q and W (not shown). It may be that the relative distance from an input synapse site to the soma should be estimated from C_V values.

4 Discussion

To discuss the effect of tonic spikes, we shall obtain C_V of random variable X as ISI by a different way. We divided a spike train into m pairs of a quiescent period and a tonic spike assembly as shown in Fig. 6**a** and **b**. Each of these m spike trains can be regarded as the state transitions of the Markov chain model (Fig. 6**c**).

In each spike train of Fig. 6**b**, one ISI with k time windows (a quiescent period) and $n - 1$ ISIs with one time window (period between two tonic spikes) are shown. By averaging all ISIs,

$$E(X) = \frac{m \cdot E(k) \cdot 1 + m \cdot 1 \cdot (E(n) - 1)}{m \cdot 1 + m(E(n) - 1)} = \frac{E(k) + E(n) - 1}{E(n)} \quad (14)$$

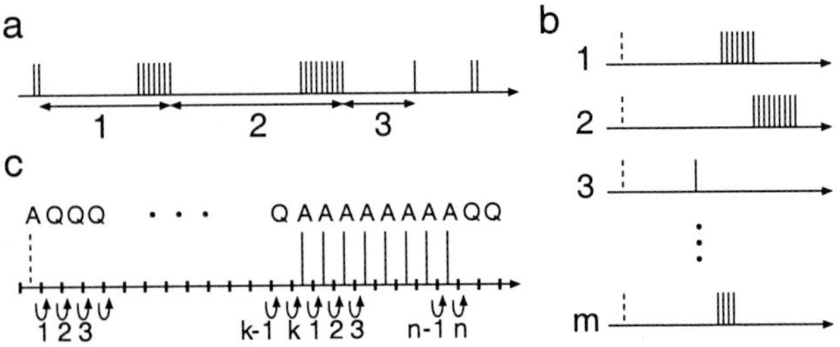

Fig. 6. **a**: Spike train in which pairs of a quiescent period and a tonic spike assembly are labeled numbers (1, 2, 3, ⋯, m). **b**: Spike trains divided into pairs in **a**. **c**: One of pairs of **b**. Time is supposed to be discrete for the Markov model. State transitions (Q→A) and (Q→A) occur at the k-th and n-th window, respectively.

$$E(X^2) = \frac{m \cdot E(k^2) \cdot 1 + m \cdot 1^2 \cdot (E(n) - 1)}{m \cdot 1 + m(E(n) - 1)} = \frac{E(k^2) + E(n) - 1}{E(n)}. \quad (15)$$

Since $E(k) = \mu_1$, $E(n) = \mu_2$ and $Var(k) = \sigma_1^2$,

$$C_V = \frac{\sqrt{\sigma_1^2 + (\mu_2 - 1)\{\sigma_1^2 + (\mu_1 - 1)^2\}}}{\mu_1 + (\mu_2 - 1)}. \quad (16)$$

When there is no tonic spike ($\mu_2 = 1$), eq. (16) is reduced as $C_V = \sigma_1/\mu_1$.

As mentioned above, we can suppose $\sigma_1 = \mu_1$ (Fig. 4**c**). Since $\mu_1 \gg \mu_2 - 1$ and $\mu_1 \gg 1$ at low input frequency, we can approximate C_V as $\sqrt{2\mu_2 - 1}$. This suggests $C_V \to 1$ if the mean number of tonic spikes approaches 1 (Fig. 5**a** and **b**). At high input frequency, since $\mu_1 \ll \mu_2 - 1$ and $\mu_2 \gg 1$, C_V is approximately proportional to $1/\sqrt{\mu_2}$. This suggests that C_V decreases if the mean number of tonic spikes increases at high input frequency (Fig. 5**a** and **b**).

While C_V monotonically increases (at low frequency) and decreases (at high frequency) against μ_2, it has the peak around middle frequency (5-6 kHz, see Fig. 5**b**). Since μ_1 is close to μ_2 around 5-6 kHz (Fig. 4**c**), it is suggested that C_V has the peak when the mean of quiescent periods is close to that of the time width of a tonic spike assembly.

5 Conclusion

We proposed the two state Markov chain model that is similar to the HH model in the temporal statistics of output spikes. We also discussed that tonic spikes increase C_V and that the relative distance from an input synaptic site to the soma may be estimated from C_V. By this simple Markov model, the behavior of a large network as brain can be easily simulated.

Acknowledgments

This research is partially supported by the grant on Research for the Future Program from JSPS (the Japan Society for the Promotion of Science) (No.96I00104).

References

1. A. L. Hodgkin and A. F. Huxley. A quantitative description of membrane current and its application to conduction and excitation in nerve. *J. Physiol.*, 117:500–544, 1952.
2. R. Azouz and C. M. Gray. Dynamic spike threshold reveals a mechanism for synaptic coincidence detection in cortical neurons *in vivo*. *Proc. Natl. Acad. Sci. USA*, 97:8110–8115, 2000.
3. C. F. Stevens and Y. Wang. Facilitation and depression at single central synapses. *Neuron*, 14:795–802, 1995.
4. H. Markram, J. Lübke, M. Frotscher, and B. Sakmann. Regulation of synaptic efficacy by coincidence of postsynaptic APs and EPSPs. *Science*, 275:213–215, 1997.
5. T. Schikorski and C. F. Stevens. Quantitative ultrastructural analysis of hippocampal excitatory synapse. *J. Neuroscience*, 17:5858–5867, 1997.
6. W. Gerstner and J. L. van Hemmen. Coding and information processing in neural networks. In E. Domany, J. L. van Hemmen, and K. Schulten, editors, *Models of Neural Networks* II, chapter 1, pages 1–93. Springer-Verlag, 1994.
7. W. M. Kistler, W. Gerstner, and J. L. van Hemmen. Reduction of the hodgkin-huxley equations to a single-variable threshold model. *Neural Computation*, 9:1015–1045., 1997.
8. W. R. Softky and C. Koch. The highly irregular firing of cortical cells is inconsistent with temporal integration of random EPSPs. *J. Neuroscience*, 13(1):334–350, 1993.
9. W. Rall. Distinguishing theoretical synaptic potentials computed for different soma-dendritic distributions of synaptic inputs. *J. Neurophysiol.*, 30:1138–1168, 1967.
10. J. J. B. Jack, D. Noble, and R. W. Tsein. *Electrical Current Flow in Excitable Cells.* Claredon Press, Oxford, second edition, 1975.
11. Z. F. Mainen and T. J. Sejnowski. Reliability of spike timing in neocortical neurons. *Science*, 268:1503–1506, 1995.
12. Ö. Bernander, R. J. Douglas, A. C. Martin, and C. Koch. Synaptic background activity influences spatiotemporal integration in single pyramidal cells. *Proc. Natl. Acad. Sci. USA*, 88:11569–11573, 1991.

Part XIV

Connectionist Cognitive Science

A Neural Oscillator Model of Auditory Attention

Stuart N. Wrigley and Guy J. Brown

Speech and Hearing Research Group, Department of Computer Science,
University of Sheffield, Regent Court, 211 Portobello Street, Sheffield S1 4DP, UK
{s.wrigley,g.brown}@dcs.shef.ac.uk
http://www.dcs.shef.ac.uk/~stu
Tel: +44(0)114 222 1879 Fax: +44 (0) 114 222 1810

Abstract. A model of auditory scene analysis is proposed, which incorporates an attentional mechanism and is implemented using a network of neural oscillators. The core of the model is a two-layer neural oscillator network which performs stream segregation and selection on the basis of oscillatory correlation. A stream is represented by a sychronised oscillator population, whereas different streams are represented by desynchronised oscillator populations. The output of the model is an 'attentional stream' describing which parts of the auditory scene are in the attentional foreground. The model simulates a number of perceptual phenomena including two tone streaming and the capture of a tone from a harmonic complex.

1 Introduction

In typical listening situations, a mixture of sounds reaches our ears: for example, a party with multiple concurrent conversations, a musical recording or a busy urban environment. Despite this, human listeners can selectively attend to a particular acoustic source, suggesting that they can separate the complex mixture into its constituent components. It has been convincingly argued that the acoustic signal is subjected to a similar form of scene analysis as occurs in vision [2]. Such *auditory scene analysis* (ASA) takes place in two conceptual stages. Firstly, the signal is decomposed into discrete sensory elements. These are then recombined into perceptual *streams* on the basis of the likelihood of them having arisen from the same physical source.

In common usage, the term *attention* usually refers to both selectivity and capacity limitation. It is widely accepted that conscious perception is selective; in other words, we perceive only a small fraction of the information impinging upon the senses. The second phenomenon - that of capacity limitation - can be illustrated by the fact that two tasks when performed individually pose no problem; however, when they are attempted simultaneously, they become very difficult. It would appear, therefore, that attention is a finite resource [5].

In this study we propose a conceptual model [14] of how auditory attention influences auditory scene analysis. Following [7], we propose that attention can be split

G. Dorffner, H. Bischof, and K. Hornik (Eds.): ICANN 2001, LNCS 2130, pp. 1163–1170, 2001.

into two mechanisms: unconscious and conscious allocation (*exogenous* and *endogenous*, respectively). One or more exogenous processes are responsible for grouping of stimuli reaching the ears. The perceptual organisations of these processes are directed to the endogenous selection mechanism. It is at this stage that both conscious decisions and salient information about the incoming organisations are combined, and a selection made which allows for the occasional over-ruling of endogenous mechanisms by exogenous attention (e.g. a sudden loud bang or other 'survival situations'). Only a single perceptual stream can be selected by the mechanism, which simulates the capacity limitation of attention.

The model is based upon the oscillatory correlation theory ([12]; see also [10]), in which neural oscillators representing a single stream are synchronised, and are desynchronised from oscillators representing other streams. The model consists of two stages: peripheral processing and a network of neural oscillators.

2 Peripheral Auditory Processing

Peripheral auditory processing is simulated by a bank of bandpass filters, followed by a model of inner hair cell transduction. Specifically, the frequency selectivity of the basilar membrane is modelled by a gammatone filterbank, in which the output of each filter represents the frequency response of the membrane at a specific position [6]. The gammatone filter is based on an analytical approximation to physiological measurements of auditory nerve impulse responses. The gammatone filter of order n and centre frequency f_0 Hz is given by

$$gt(t) = t^{n-1} e^{-2\pi bt} \cos(2\pi f_0 t + \phi) H(t) \ . \tag{1}$$

where ϕ represents the phase and $H(t)$ is the unit step (Heaviside) function for which $H(t) = 1$ if $t \geq 0$ and $H(t) = 0$ otherwise. The bandwidth b is set to the equivalent rectangular bandwidth (ERB), a psychophysical measurement of critical bandwidth in human subjects [4]. A bank of 32 filters were used with centre frequencies equally distributed on the ERB scale between 50 Hz and 5 kHz. The gain of each filter is adjusted to simulate the pressure gains of the outer and middle ears.

Our oscillator network requires an estimate of firing rate in each auditory filter channel as its input. Accordingly, the envelope of each gammatone filter response is computed, and this is half-wave rectified and compressed to give an approximation of auditory nerve firing rate.

3 Oscillator Network

The three conceptual stages of attentionally modulated computational auditory scene analysis (segmentation, grouping, attentional stream selection) occur within an oscillatory correlation framework (Fig. 1A).

The building block of the network is a single relaxation oscillator [11], which is defined as a reciprocally connected excitatory variable x and inhibitory variable y:

$$\dot{x} = 3x - x^3 + 2 - y + I + S + \eta \ . \tag{2}$$

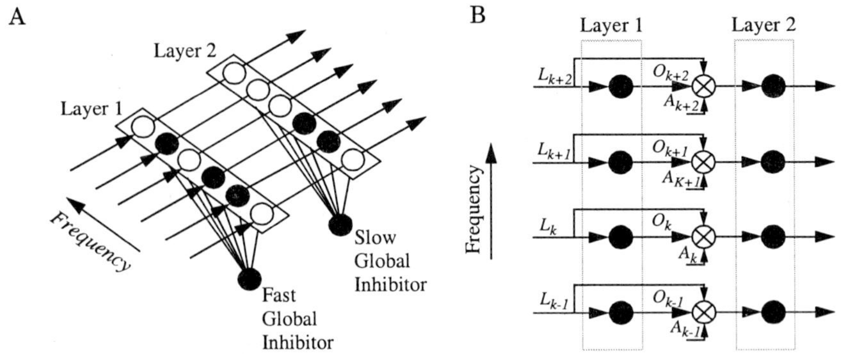

Fig. 1. A. A one-dimensional, two-layer oscillator network performs segmentation and grouping (layer one) followed by stream selection (layer two) to produce an 'attentional stream'. **B.** Section of the network showing how the output of the layer one activity, stimulus intensity and attention highlight are combined. For clarity, excitatory and inhibitory connections within layers are not shown.

$$\dot{y} = \varepsilon\left[\gamma\left(1 + \tanh\frac{x}{\beta}\right) - y\right] . \tag{3}$$

Here, ε, γ and β are parameters, I represents the oscillator's external input, S represents the coupling from other oscillators in the network, and η is a noise term. Ignoring S and η, constant I defines a relaxation oscillator with two time scales. The x-nullcline is a cubic function and the y-nullcline is a sigmoid function. When I>0, the two nullclines intersect only at the middle branch of the cubic in which case the oscillator exhibits periodic activity (a stable limit cycle). When I<0, the nullclines intersect at a stable fixed point and no activity occurs. The oscillator activity is therefore stimulus dependent.

3.1 Layer One: Segmentation and Grouping

The first layer of the network is based upon a one-dimensional locally excitatory globally inhibitory oscillator network (LEGION [8]) composed of relaxation oscillators. Synchronised blocks of oscillators (*segments*) form in this layer in which each segment corresponds to a contiguous region of acoustic energy. The input I to layer one is the estimate of auditory nerve firing rate in each channel. To restrict the spread of activation across frequency, the firing rate is adjusted such that rates below a threshold value are set to zero.

Layer one is a one-dimensional network of neural oscillators with a global inhibitor. The coupling term S in (2) is given by:

$$S = \sum_{k \neq i} W_{ik} H(x_k - \theta_x) - W_z H(z - \theta_z) . \tag{4}$$

Here, W_{ik} is the connection weight between oscillators i and k, and H is the Heaviside function. The parameter θ_x is a threshold above which an oscillator can affect others in the network and W_z is the weight of inhibition from the global oscillator z whose activity is defined as:

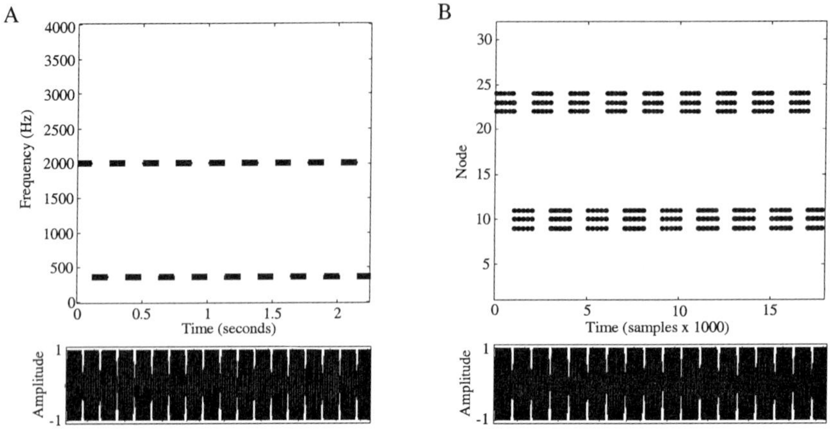

Fig. 2. A. Stimulus spectrogram showing tones at 400Hz and 2kHz. **B.** Layer 1 output showing segments corresponding to tones centred on channels 10 and 23. The time-domain representation of the stimulus is shown below each plot for reference. The parameter values are: ε=0.1, γ=6.0, β=0.1.

$$\dot{z} = \sigma_{\infty} - z . \quad (5)$$

Here, the input to the global inhibitor $\sigma_{\infty} = 1$ if $x_k \geq \theta_z$ for at least one oscillator k, otherwise $\sigma_{\infty} = 0$. Similar to θ_x in (4), θ_z acts a threshold above which an oscillator can affect the global inhibitor. If the global inhibitor receives supra-threshold stimulation from at least one oscillator, $z \rightarrow 1$.

Initially, all oscillators have the same phase (all frequencies belong to a single stream) - an assumption supported by psychophysical evidence which suggests that perceptual fusion is the default perceptual organisation [2].

An excitatory connection weight is placed between adjacent oscillators whose corresponding auditory nerve firing rates exceed a threshold value to generate segments. This allows contiguous regions of acoustic energy to be represented by sychronised groups of oscillators.

Figure 2B shows the response over time of the first layer to a repeating sequence of alternating high and low frequency tones (Fig. 2A). Each dot indicates that the oscillator for that channel is active at that time step. From this, it can be seen that each tone generates a synchronised block of oscillators.

3.2 Layer Two: Attentional Stream Selection

The oscillators in the first layer are connected to oscillators in the second layer by excitatory links. The strength of these connections is modulated by endogenous attention, thus allowing multiple frequency regions to dominate the network. The second layer is also a LEGION, but in this case the global inhibitor operates on a much slower timescale. This causes layer two to act as a winner take all (WTA) network (see [12]). The activity of the layer two global inhibitor z_s is defined as:

$$\dot{z}_s = [\alpha - z_s]^+ - cz_s \ . \quad (6)$$

Here, $[n]^+ = n$ if $n \geq 0$ and $[n]^+ = 0$ otherwise; α is the total activity of the layer two network. This combined with the small value of c result in a quick rise and slow decay for the inhibitor. It is the slow decay that allows this network to behave as a winner take all network. When a group of oscillators are in the active phase, they feed input to the slow inhibitor. The slow inhibitor through its quick rise and slow decay maintains a level of inhibition that must be overcome by a group if the group is to oscillate. The central idea for object selection is that one group sets the level of slow inhibition, which can then be overcome by that group only. The key is to choose an appropriate value for c in (6).

Input to oscillator k in the layer is a weighted version of the corresponding layer one output:

$$I_k = [L_k - T_k]^+ O_k \ . \quad (7)$$

Here, T_k is the attentional threshold which is related to the endogenous interest at frequency k; O_k is the activity of oscillator k in layer one and L_k is the thresholded firing rate (see section 3.1) of the stimulus at frequency k (Fig. 1B).

An important property highlighted by [1] is the build up over time of the two tone streaming effect. It is proposed that the level of endogenous interest builds over time in relation to the input to the network. Attention can be reoriented consciously to other segregated streams (without further buildup); when input ceases, endogenous interest decays such that a second presentation of a stimulus will require a second buildup of streaming. Such behaviour is achieved by use of a leaky integrator with slow rise and decay time constants to weight the endogenous interest:

$$T_k = (1 - A_k)S \ . \quad (8)$$

Here, A_k is the endogenous interest at frequency k and S is the leaky integrator defined as:

$$\dot{S} = d([\sigma_\infty - S]^+ - [1 - H(\sigma_\infty - S)]aS) \ . \quad (9)$$

Here, small values of a and d result in a slow rise and slow decay for the integrator. The time-varying activity of the layer two network can be thought of as an *attentional stream*: a trace through time highlighting which stream is in the attentional foreground at any one time.

L_k is included in (7) to model the over-riding of endogenous attention by loud acoustic events. In normal circumstances, the endogenously chosen group (using the A_k) is the only group that can overcome the level of inhibition present in the network. However, should a sufficiently loud stimulus appear outside the area of interest defined by A_k, this will overcome the inhibition and so re-orientate attention exogenously.

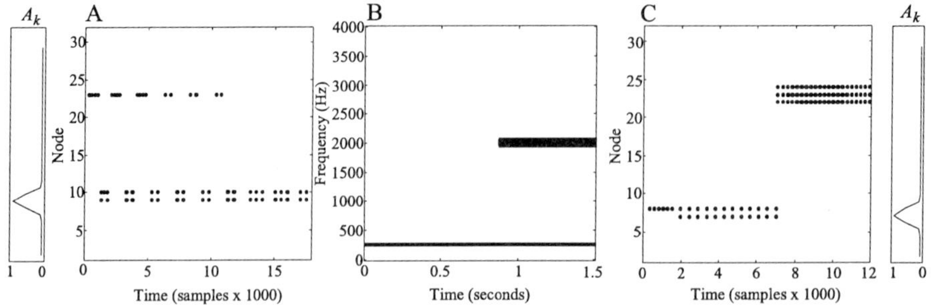

Fig. 3. A. The attentional stream of layer two shows the build-up of streaming [1] with the low frequency stream dominating by time step 11000. **B.** Stimulus spectrogram showing a tone at 300Hz and a much louder tone at 2kHz. **C.** The attentional stream of layer two shows the exogenously redirection of attention by the onset of the louder tone. Attention is directed toward the low frequency tone. Attentional interest A_k is shown beside each layer two diagram.

Figure 3A shows the response over time of the second layer to the repeating alternating tone sequence used earlier (Fig. 2). Again, each dot indicates that the oscillator for that channel is active at that time step. The relationship between frequency proximity and temporal proximity has been studied extensively using the two tone streaming phenomenon [9]. The more distant in frequency two tones are, the more likely it is that they will segregate into two streams with one dominating the other. Similarly, as presentation rate increases, tones of similar frequency group together. From the diagram, it can be observed that activity for the low frequency tones is sustained whereas activity for the high frequency tones gradually decreases until all high frequency oscillator activity ceases. A build-up of streaming has occurred culminating in the low frequency tones being in the attentional foreground. In this case, attention was oriented via the A_k values in (8) to the low frequency tones.

Figure 3C shows how the attentional stream can be subconsciously re-directed by the onset of a new, loud stimulus (Fig. 3B). Attention is directed toward the low frequency tone but when the loud tone begins, this overrides the conscious decision and the new tone becomes the attention foreground. This is achieved by having a very low level of attentional interest across all frequencies which can only be overcome by loud stimuli.

Figure 4 shows the layer two responses to the stimulus used in [3] in which captor tones are used to affect the percept of a two tone complex. Listeners reported that the higher frequency tone of the complex was captured when the frequency separation between the captor tones and the complex was small. This result can been seen in the responses of Figure 4 in which the captor tones capture the high frequency complex tone into a separate stream only for small frequency separations.

4 Discussion

A model of attentionally modulated auditory scene analysis has been described, in which stream segregation and selection occur within an oscillatory framework. The

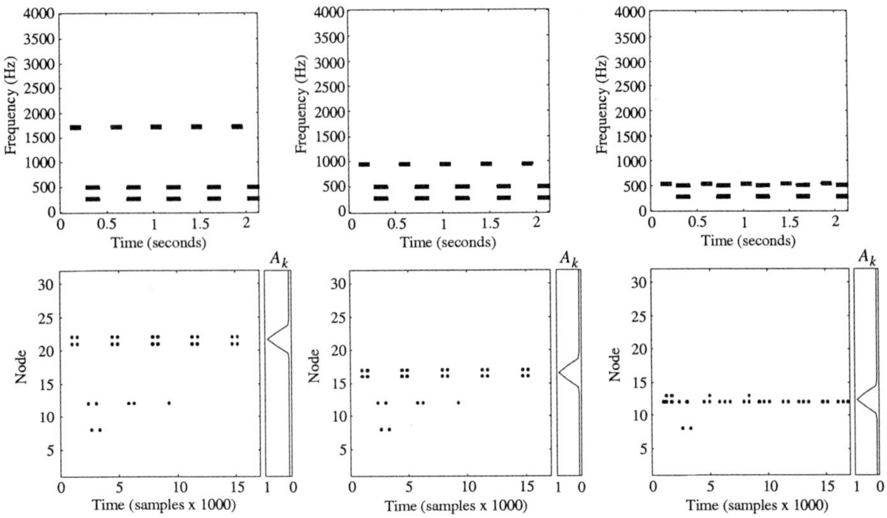

Fig. 4. The attentional stream of layer two (bottom) in response to the stimuli used in [3] (top). Attentional interest A_k is shown to the right of each layer two diagram.

model is based on the previous neural oscillator work of Wang and colleagues [11][13] but incorporates a number of novel factors. First, we use a one-dimensional layer of oscillators, in contrast to the two-dimensional time-frequency grid used in Wang's model. The latter suffers from conceptual difficulties. The input to Wang's two-dimensional network is sampled at regular intervals, such that the network produces a finite duration 'snapshot' of the auditory scene at each epoch. It is unclear how these snapshots should be integrated to produce a time-varying estimate of stream content. The model proposed here produces an attentional stream in continuous time; thus, there is no need to deal with the difficult problem of combining overlapping temporal snapshots. Furthermore, sequential grouping in Wang's model is only possible via lateral connections on the time axis, which do not have a known physiological correlate. Sequential grouping is an emergent property of the model described here; segments occurring over time which appear in the attentional stream are implicitly grouped.

The main contribution of this study is the incorporation of attentional mechanisms within the oscillatory framework. Little work has been conducted on auditory models with an attentional component. Wang's 'shifting synchronisation' theory [11] assumes that '*attention is paid to a stream when its constituent oscillators reach their active phases*'. Such stream multiplexing contradicts experimental findings [2], which show that listeners perceive one dominant stream. Furthermore, his theory cannot explain how attention can be redirected by a sudden stimulus. Such an event would be encoded by Wang's network as an individual stream which would be multiplexed as normal - with no attentional emphasis. As the shifting synchronisation theory allows attention to alternate quickly between each stream in the entire temporal snapshot, it is possible for attention to switch from a stream at the beginning of the snapshot to one at the end (later in time) and back again. This suggests a confusion in Wang's model between 'oscillator time' and 'real time'. Attention exists in real time - but in Wang's system it shifts in oscillator time.

The model proposed here accounts for a number of psychophysical phenomena including the streaming of alternating tone sequences [9] and its associated build-up over time [1]. An attentional stream can also be re-directed by the onset of a new, loud stimulus, and the model demonstrates other perceptual phenomena such as the capture of a tone from a complex [3]. Current work is concentrating on using a pitch analysis technique to allow segments to be created by cross-channel correlation and groups to be formed by common periodicity. The model will then be expanded to account for binaural streaming phenomena.

We would like to thank DeLiang Wang and the three anonymous reviewers for comments on a previous version of this paper. The work of S. N. Wrigley is supported by the University of Sheffield Hossein Farmy Scholarship.

References

[1] Anstis, S., Saida, S.: Adaptation to auditory streaming of frequency-modulated tones. Journal of Experimental Psychology: Human Perception Performance **11** (1985) 257-271

[2] Bregman, A.S.: Auditory Scene Analysis. The Perceptual Organization of Sound. MIT Press (1990)

[3] Bregman, A.S., Pinker, S.: Auditory streaming and the building of timbre. Canadian Journal of Psychology **32**(1) (1978) 19-31

[4] Glasberg, B.R., Moore, B.C.J.: Derivation of auditory filter shapes from notched-noise data. Hearing Research **47** (1990) 103-138

[5] Pashler, H.E.: The Psychology of Attention. MIT Press (1998)

[6] Patterson, R. D., Nimmo-Smith, I., Holdsworth, J., Rice, P.: APU Report 2341: An Efficient Auditory Filterbank Based on the Gammatone Function, Cambridge: Applied Psychology Unit (1988)

[7] Spence, C.J., Driver, J.: Covert spatial orienting in audition: exogenous and endogenous mechanisms. Journal of Experimental Psychology: Human Perception and Performance **20**(3) (1994) 555-574

[8] Terman, D., Wang, D.L.: Global competition and local cooperation in a network of neural oscillators. Physica D **81** (1995) 148-176

[9] van Noorden, L.P.A.S.: Temporal coherence in the perception of tone sequences. Doctoral thesis, Institute for Perceptual Research, Eindhoven, NL (1975)

[10] von der Malsburg, C., Schneider, W.: A neural cocktail-party processor. Biological Cybernetics **54** (1986) 29-40

[11] Wang, D.L.: Primitive auditory segregation based on oscillatory correlation. Cognitive Science **20** (1996) 409-456

[12] Wang, D.L.: Object selection based on oscillatory correlation. Neural Networks **12** (1999) 579–592

[13] Wang, D.L., Brown, G.J.: Separation of speech from interfering sounds based on oscillatory correlation. IEEE Transactions on Neural Networks **10** (1999) 684-697

[14] Wrigley, S.N.: A Model of Auditory Attention. Technical Report CS-00-07, Department of Computer Science, University of Sheffield, UK (2000). Available from http://www.dcs.shef.ac.uk/~stu/pubs.html

Coupled Neural Maps for the Origins of Vowel Systems

Pierre-yves Oudeyer

Sony Computer Science Lab, Paris
py@csl.sony.fr

Abstract. A unified connectionist model of the perceptual magnet effect (the perceptual warping of vowels) is proposed, and relies on the concept of population coding in neural maps. Unlike what has been often stated, we claim that the imprecision of the classical "sum of vectors" coding/decoding scheme is not a drawback and can account for psychological observations. Furthermore, we show that coupling these neural maps allows the formation of vowel systems, which are shared symbolic systems, from initially continuous and uniform perception and production. This has important consequences for existing theories of phonetics.

1 Introduction

Our perception of vowel sounds is biased by our knowledge of our own language sound system. More precisely, Kuhl (1992) showed that when people are asked to evaluate the similarity between couples of vowel sounds, they tend to perceive two vowels closer than they are in a physical space when they are both close to a vowel prototype of their language, and further than they are in the same physical space when they are far from their vowel systems prototypes. In short, there is a perceptual warping that collapses percepts around the vowel prototypes of a given language. As a side effect, the perceptual difference between vowels belonging to different vowel categories is enhanced. This is refered in the literature as the "magnet effect". This is a particular instantiation of the well known and widely spread across modalities phenomenon of categorical perception, which Harnad (1987) defines as: "Categorical perception (CP) occurs when equal-sized physical differences in the signals arriving at our sensory receptors are perceived as smaller within categories and larger between categories".

Several connectionnists models of categorical perception, and in particular for the "magnet effect", have been developed. Among them, there are 2 broad categories: on the one hand, models like the Brain-State-in-a-Box of Anderson et al. (1977) or the backpropagation based model of Harnad (Harnad and Damper, 1997) which consist in training a feedforward neural network to reproduce its input point as outputs, with standard supervised learning training, in particular back-propagation ; on the other hand, models like the one of Guenther and Gadja (1996) which are based on the unsupervised training and self-organization of neural maps whose activity is interpreted with population codes as used in the pionneer work of Georgopoulos and colleagues (1988). Both kind of models have advantages and drawbacks.

G. Dorffner, H. Bischof, and K. Hornik (Eds.): ICANN 2001, LNCS 2130, pp. 1171–1176, 2001.

The first group of models provides more than the simple ability to simulate "magnet effect" like phenomena: it provides also the ability to categorize and identify stimuli, which allows to model on the one hand the ability to categorize, and also other aspects of categorical perception such as those concerning decision boundaries (Harnad et and Damper, 1997). The drawbacks of these models are that they are not very biologically plausible and need supervised training mechanisms (the BSB model is certainly more interesting from a cognitive modeling point of view than back-propagation training, especially for its modeling of categories as attractors, but it is brittle as Harnad and Damper 1997 explain). The model of Guenther does not have these drawbacks since there are many neurological evidences in favor of it, but looses the ability to categorize and identify stimuli. Moreover, no real comparison has been done so far to our knowledge.

In this paper, we propose to reformulate the model of Guenther in such a way that it can be interpreted in the framework of auto-associators, which allows a unified model of the "perceptual magnet effect", and extend it so as to provide the ability to categorize in a way similar to the BSB model. Moreover, we present a new experimental setup in which these neural maps are not used any more only to make models of the environmental vowels, but also for the production of vowels. Additionnally, instead of having one neural map that learns an already existing vowel system, we have many agents, each endowed with the neural system, that interact by producing sounds and listening to each other. They behave exactly as if they were actually learning an existing sound system, except that there is no such system (intially, each agent produces vowel that are spread uniformly across the space). We show that a phenomenon of self-organization arises in which all agents crystallise in a state where the distribution of vowels they produce is not uniform anymore, but comprises a number of sharply defined peaks: they did create effectively a sound system from scratch. We will explain why this has important implications for the current theories of the origins of communication systems, and in particular human language and sound system.

The next section describes the new formulation of Guenther's model, as well as its extensions. The following section describes the crystallization of neural maps when they are coupled. Then the result is replaced in a larger scientific framework in the conclusion.

2 A Neural Map for the Perception of Sounds

This model is based on topological neural maps, as Guenther. This type of neural network has been widely used for many models of cortical maps, which are the neural devices that humans have to represent aspects of the outside world (acoustic, visual, touch etc...). There are 2 neuroscientific findings on which our model relies, and that were initially made popular with the experiments of Georgopoulos (1988): on the one hand, for each neuron/receptive field in the map there exist a stimulus vector to which it responds maximally (and the response decreases when stimuli get further from this vector) ; on the other hand, from the set of activities of all neurons at a given moment one can predict the perceived stimulus or the motor output, by computing what is termed the population vector (see Georgopoulos 1988): it is the sum of all prefered vectors of the neurons ponderated by their activity (normalized like here since we are intereseted in both direction and amplitude of the stimulus vector). When there are many neurons and the preferred vectors are uniformly spread across the space, the population vector corresponds accurately to the stimulus that gave rise to the activities of neurons, while when the distribution is inhomogeneous, some imprecisions appear. This imprecision has been the subjects of rich research, and many people proposed more precise variants (see Abbot and Salinas, 1996) to the formula of Georgopoulos because they assumed the sensory system coded exactly stimuli (and hence the formula of Georgopoulos must be somewhat false). On the contrary here we will show that this imprecision allows the interpretation of "magnet effect" like psychological phenomena, i.e. sensory illusions, and so may be a fundamental characteristic of neural maps.

Hence here a neural map consists of a set of neurons n_i whose "preferred" stimulus vector is noted v_i. The activity of neuron n_i when presented stimulus v is computed with a gaussian function: $act(n_i) = e^{-dist(v_i,v)^2/\sigma^2}$ (1) with sigma being a parameter of the simulation (to which it is very robust). The population vector is then: $pop(v) = \frac{\sum_i act(n_i)*v_i}{\sum_i act(n_i)}$ (2) The normalizing term is necessary here since we are not only interested in the direction of vectors. There are arguments for this being biologically acceptable (see Reggia 199?). Stimuli are here 2 dimensional, corresponding to the first two formants of vowel sounds. The difference with Guenther's model is that we use distances and gaussians while he uses scalar products and normalized vectors with agonist-antagonistic coding. Our formulation allows to unify this neural map model with auto-associators: indeed, (2) corresponds exactly to the Nadaraya-Watson estimator/regression formula (1964): $f(v) = \frac{\sum_i K(v,v_i)*f(v_i)}{\sum_i K(v,v_i)}$ (3) that is used to approximate some function f given a set of data points $(v_i, f(v_i)$, and where K is a kernel function. (2) can be mapped to (3) by taking a gaussian as the kernel function and $f(v_i) = v_i$, i.e. $f = Id$, which means that neural maps behave exactly like an auto-associator.

Initially, a neural map is formed by initializing the preferred vectors of neurons (i.e. their weights in a biological implementations) to random vectors. This is actually done through a babbling phase in a refined version of the model shortly described in next section. It means that the v_i are uniformly spread across the space. This can be visualized by plotting all the v_i as in one of the squares of figure 1. To visualize how agents initially perceive the vowel sound world in a way comparable to Kuhl's experiments, we can plot all the $pop(v)$ corresponding to a set of stimulus whose vectors values are the intersections of a regular grid covering the whole space. Figure 2 is an example of such an initial perceptual initial state: we see that the grid is nearly not deformed, which means as predicted by theoretical results, that the population vector is a rather accurate coding of the input space.

Then the learning mechanism used to updated these weights when presented a vowel sound stimulus v consists in shifting slightly these vectors towards the stimulus vectors when the v_i is close to it (activation is very high), shifting it away when it is in mid-range, and no shifting when it is far. In brief, this is a

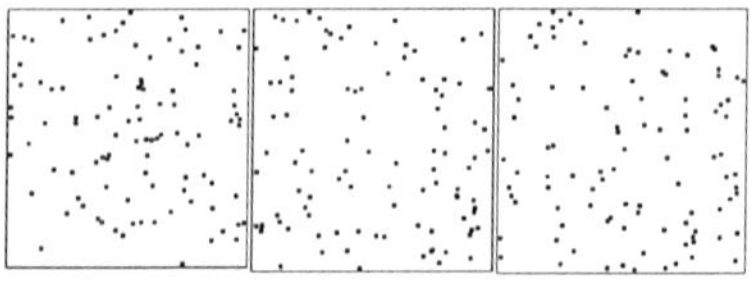

Fig. 1. Neural map at the beginning

sort of mexican hat based competitive learning mechanism. The actual formula is $\delta v_i = mexicanHatFunction(dist(v, v_i)) * (v - v_i)$. Furthermore, here a neural map represents an underlying vowel sound distribution, that one can compute by adding gaussians centered on all v_i's and then normalizing. Learning consists in adjusting neurons so that the coded distribution corresponds to the distribution of heard vowels.

When learning an existing sound system, agents are presented with vowel sounds coming from an existing vowel system (basically consisting of points grouped in as many clusters as there are vowels). As in experiments performed by Guenther, the kind of results we obtain are similar to what one can see on figures 4 (repartition of neurons) and 5 (image of a regular input grid), and fits perfectly well to experimental data found by Kuhl: the perceptual space has warped, i.e. input vowels close to the center of regions corresponding to a vowel of the language are perceive even closer and vice versa. This is due to the uneven repartition of neurons coupled with the imprecision of the population vector. A more precise way to visualize this is to plot the warping function with arrows giving the exact shift for a number of example points. Figure 7 is an example of warping function. What is interesting is that this looks like the plotting of attractor basins of dynamical systems. This leads to a possible refinement of the model that confers it the status of dynamical system and the ability to categorize/identify stimuli: as in the BSB model, the output of the neural map, $pop(v)$ can be re-directed to the input, thus making the map recurrent. The fact that this recurrent network has at the end effectively point attractors corresponding to prototypes for categories of vowels, which is suggested by figure 7, can easily be shown by casting formally the map onto a continuous hopfield network. This will be detailed in a longer paper.

3 Coupling Neural Maps

In the last section, neural maps were used only to perceive the sound world. One can imagine that they are also used in the process of vowel production (which is very sensible according to psychological evidence): more precisely, that the vowels produced by an agent at a given moment of its life follow the distribution coded by its perception map. This can be implemented in a biological plausile way by coupling the acoustic map with an articulatory/motor map, as will be detailed in a longer paper. Still, here we are only interested in the consequence of using perception maps to produce sounds.

The experiment presented consists in having a population of agents (typically 20 agents), each endowed with the neural system described previously. They interact by pairs of two (following the evolutionary cultural scheme devised in many models of the origins of language, see Steels 1997, Steels and

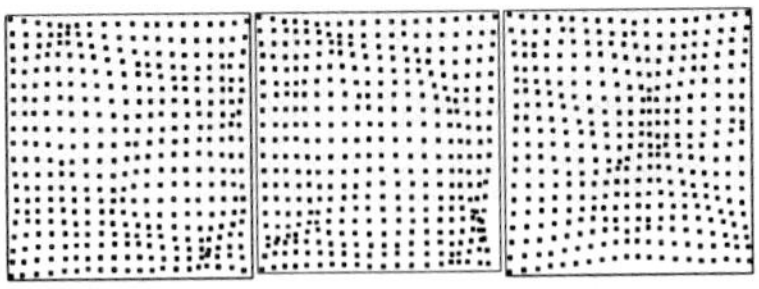

Fig. 2. Perceptual (non-)warping at the beginning (image of a regular grid of stimuli)

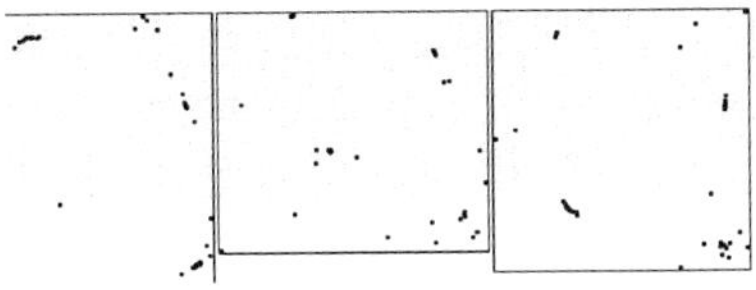

Fig. 3. Neural maps after 1000 interactions

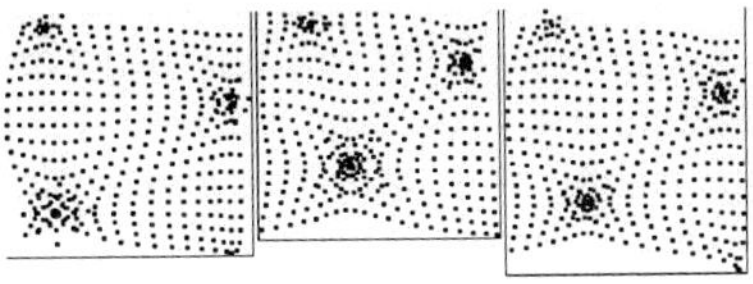

Fig. 4. Perceptual warping after 1000 interactions

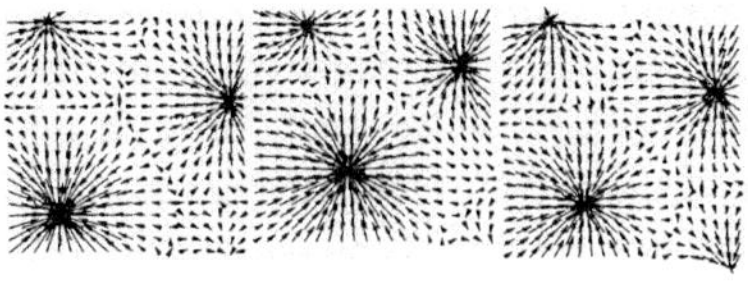

Fig. 5. Representation of the values of the warping at the edges of a regular gfrid

Oudeyer, 2000): at each round, one agent is choosen randomly and produces a sound according to the distribution coded by its map, and another agent is choosen randomly, hears the vowels and updates its map in the same way as above. Their maps are initially built by setting the v_i to random values, which means that the vowel they produce initially follow a quasi-uniform distribution. Figure 1 shows the initial state of neural maps for 3 agents in a simulation, figure 2 shows their initial perceptual warping (quasi non-existant). What is very interesting, is that this situation is not stable: rapidly, agents get in the a situation like on figures 3 and 4 which are the correspondances of figures 2 and 3 after 1000 interactions. Moreover, this final situation is experimentally stable, this is why we say that agents "crystalize". In fact, symmetry has broken and a positive feedback loop made that a distribution composed of a number of very well defined peaks appears, being the same for all agents (but different between 2 experiments). This final state is characteristic of the distribution of vowels in human languages. Moreover, because we added to the network the possibility to relax through recurrent connections, the peaks in vowel distributions are also attractors of the neural maps, and so good models of a categorizing behavior. In brief, this shows how a full fledged vowel system, which means not only a peaked distribution but also a shared system of discrete/symbolic units, can emerge through self-organization out of the coupling of unsupervised learning system.

4 Conclusion

The paper presented a connectionist model of 3 important phenomena: 1) the perceptual warping of vowels, due to the imprecision of population coding when receptive fields are not uniformly distributed ; 2) the emergence of a categorizing

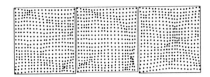

behavior/discrete perception through the combined effect of imprecision and recurrent connections ; 3) the emergence of shared vowel systems with the coupling of unsupervised learning system, through symmetry breaking due to stochasticity and positive feedback loops. The last point is of particular importance to the field of linguistics. Indeed, many researchers (Archangeli and Langendoen 1997), defending nativist theories of language, think humans need to have in their genome many pre-specifications of phonemes, in particular vowel systems, and believe that biological evolution is responsible for their origins. For instance, (Kuhl 2000) proposed that we are innately given a number of vowel prototypes (all those that are "possible" in human languages), specified genetically, and that learning consists in pruning those that are not used in the environment. Our model shows that there is no need for linguistically specific devices. More importantly, nativist theories are faced with the problem of how these specifications got into the genes, to which they never provided some possible operational accounts. Our model shows that very generic neural devices can answer the questions, and need not biological evolution, but only cultural evolution (of course these neural devices were built through biological evolution, but their genericity indicates that language played certainly very little role for their selection).

References

Abbot L., Salinas E. (1994) Vector reconstruction from firing rates, Journal of computational Neuroscience, 1, 89-116.

Anderson J., Silverstein, Ritz, Jons (1977) Distinctive features, categorical perception and probability learning: some applications of a neural model, Psychologycal Review, 84, 413-451.

Watson G.S. (1964) Smooth regression analysis, Sankhya: The Indian Journal of Statistics. Series A, 26 359-372.

Reggia J.A., D'Autrechy C.L., Sutton G.G., Weinrich M. (1992) A competitive distribution theory of neocortical dynamics, Neural Computation, 4, 287-317.

R.I. Damper and S.R. Harnad (2000) Neural network modeling of categorical perception. Perception and Psychophysics, 62 p.843-867.

Georgopoulos, Kettner, Schwartz (1988), Primate motor cortex and free arm movement to visual targets in three-dimensional space. II. Coding of the direction of movement by a neuronal population. Journal of Neurosciences, 8, pp. 2928-2937.

Guenther and Gjaja (1996) Magnet effect and neural maps, Journal of the Acoustical Society of America, vol. 100, pp. 1111-1121.

Kuhl, Williams, Lacerda, Stevens, Lindblom (1992), Linguistic experience alters phonetic perception in infants by 6 months of age. Science, 255, pp. 606-608.

Kuhl (2000) Language, mind and brain: experience alters perception, The New Cognitive Neurosciences, M. Gazzaniga (ed.), The MIT Press.

Steels, L. (1997a) The synthetic modeling of language origins. *Evolution of Communication*, 1(1):1-35.

Steels L., Oudeyer P-y. (2000) The cultural evolution of phonological constraints in phonology, in Bedau, McCaskill, Packard and Rasmussen (eds.), Proceedings of the 7th International Conference on Artificial Life, pp. 382-391, MIT Press.

Archangeli and Langendoen (1997), Optimality theory, an overview, Blackwell Publishers.

Learning for Text Summarization Using Labeled and Unlabeled Sentences

Massih-Reza Amini and Patrick Gallinari

LIP6, Université Paris 6, 4 Place Jussieu, F-75252 Paris cedex 5.
{Massih-Reza.Amini,Patrick.Gallinari}@lip6.fr

Abstract. We describe an original machine learning approach for automatic text summarization; it works by extracting the most relevant sentences from a document. Since labeled corpora are difficult to collect for this task, we propose a semi-supervised method, which makes use of a small set of labeled sentences together with a large set of unlabeled documents, for improving the performances of summary systems. We show that this method is an instance of the Classification EM algorithm in the case of gaussian densities, and that it can also be used in a non-parametric setting. We finally provide an empirical evaluation on the Reuters news-wire corpus.

1 Introduction

With the continuing growth of online text resources, it is becoming more and more important to help users to access this information by developing easy to use information research tools. Text summarization is such a technique, which can be used together with conventional information research engines, and help the user to quickly evaluate the relevance of documents or to navigate through a corpus.

Automated Summarization dates back to the fifties [6]. The different attempts in this field have shown that human-quality text summarization was very complex since it encompasses discourse understanding, abstraction, and language generation [14]. Simpler approaches were then explored which consist in extracting representative text-spans, using statistical techniques and/or techniques based on superficial domain-independent linguistic analysis [15]. In this context, summarization can be defined as the selection of a subset of the document sentences which is representative of its content. Ranking the document sentences and selecting those with higher score typically does this. Query based summarization aims to produce a synthesis of a document with regard to a given query, whereas the goal of generic summarization is to abstract the main ideas of the document.

Since there is an increase need for fully automatic techniques in order to process huge amounts of textual information, machine learning is becoming very popular in this field. [5] consider the problem of sentence extraction as a classification task and propose a generic model based on a Naïve-Bayes classifier. [8] compare different

G. Dorffner, H. Bischof, and K. Hornik (Eds.): ICANN 2001, LNCS 2130, pp. 1177–1184, 2001.

machine learning techniques for both generic and query based summarization. [2] present an algorithm that extracts phrases instead of sentences, in order to increase the summary concision. They also compare three supervised learning algorithms: C4.5, Naïve-Bayes and neural networks.

All these approaches rely on supervised learning. A major difficulty here is that labeling large amounts of documents is very costly since this takes place at the sentence level. From a machine learning perspective, summarization is typically a task for which there is a lot of unlabeled data and very few labeled texts so that semi-supervised learning seems well suited for the task. Early work for semi supervised learning dates back to the 70s. A review of the work done prior to 88 in the context of discriminant analysis may be found in [7]. Most approaches propose to adapt the EM algorithm for handling both labeled and unlabeled and to perform maximum likelihood estimation. Theoretical work mostly focuses on gaussian mixtures, but practical algorithms may be used for more general settings, as soon as the different statistics needed for EM may be estimated. More recently this idea has motivated the interest of the machine learning community and many papers now deal with this subject. For example [11] propose an algorithm which is a particular case of the general semi-supervised EM described in [7], and present an empirical evaluation for text classification. [9] adapt EM to the mixture of experts, [13] propose a Kernel Discriminant Analysis which can be used for semi-supervised classification. The co-training paradigm [10] is also related to semi supervised training. All these approaches rely on density estimation and generative models.

Our approach bears similarities with the well-established decision directed technique, which has been used for many different applications in the field of adaptive signal processing [3]. We present an automatic summarization system that is based on the *extraction* of relevant sentences from a document. This system can be used both for generic and query based summaries, it makes use of a small set of labeled texts together with a large body of unlabeled documents. This method can be interpreted as an instance of the Classification EM algorithm (CEM) [1, 7] under the hypothesis of gaussian conditional class densities. The originality of our work is that instead of estimating conditional densities, it is based on a discriminative approach for estimating directly posterior class probabilities and as such it can be used in a non-parametric context.

We present the formal framework of our model and its interpretation as a CEM instance in section 2, we then describe our approach to text summarization based on sentence segment extraction in section 3. Finally we present a series of experiments in section 4.

2 Model

2.1 Framework

We denote the sentence terms and sentence class labels (relevant or irrelevant for the summary) respectively as $\boldsymbol{x}$ and $c \in \{1,2\}$, the problem we consider is to classify document sentences, using classified data in conjunction with unclassified ones with respect to these 2 classes. Let $D_l = \{(x_i, t_i) \mid i = 1,\dots,n\}$ be the labeled data set, where $t_i=(t_{1i}, t_{2i})$ is the indicator vector for x_i. It is supposed that in addition to D_l, there is available a set of m vectors x_i of unknown origin, $D_u=\{x_i \mid i= n+1,\dots,n+m\}$. The latter are assumed to have been drawn from a mixture with two components C_1, C_2 in some unknown proportions π_1, π_2. Since we are interested in classification, we will consider that unlabeled data have an associated missing parameter t_i ($i=n+1, \dots, n+m$) which is a class indicator vector.

2.2 Classification Maximum Likelihood Approach

Let us first briefly introduce the classification maximum likelihood (CML) approach to clustering. In this unsupervised approach there are N samples generated via a mixture density:

$$f(x,\Theta) = \sum_{k=1}^{c} \pi_k f_k(x,\theta_k) \tag{1}$$

Where the f_k are parametric densities with unknown parameters $\{\theta_k\}_{k=1,\dots,c}$, c is the number of mixture components. The goal here is to cluster the samples into c components. Under the mixture-sampling scheme, the CML criterion is [1]:

$$\log L_{CML}(P,\pi,\theta) = \sum_{i=1}^{N}\sum_{k=1}^{c} t_{ki} \log\{\pi_k . f_k(x_i,\theta_k)\} \tag{2}$$

Where $P = (P_1,\dots,P_c)$ denotes a partition of data into c components P_i. Note that this is different from the mixture maximum likelihood (MML) approach[1], where the goal is to fit the mixture model (1), for MML t_{ki} is an unknown variable whose expectation gives the posterior probability for x to be in the kth component. For CML the mixture indicator t_{ki} for a given data x_i is treated as an unknown parameter and corresponds to a hard decision on the mixture identity. The classification EM algorithm (CEM) [1, 7] has been proposed for maximizing (2), it is similar to the classical EM except for an additional ***C***-step where each x_i is assigned to one and only one component. The algorithm is briefly described below.

CEM

Initialization: start from an initial partition $P^{(0)}$

j^{th} *iteration*, $j \geq 0$:

E –step. Estimate the posterior probability that x_i belongs to $P_k^{(j)}$ ($i=1,\dots, N$; $k=1,\dots,c$):

---

[1] In the mixture maximum likelihood approach the criterion is

$$\log L_M(P,\pi,\theta) = \sum_{i=1}^{N} \log(\sum_{k=1}^{c} \pi_k . f_k(x_i,\theta_k))$$

$$E[t_{ki}^{(j)} / x_i; P^{(j)}, \pi^{(j)}, \theta^{(j)}] = \frac{\pi_k^{(j)} . f_k(x_i; \theta_k^{(j)})}{\sum_{k=1}^{c} \pi_k^{(j)} . f_k(x_i; \theta_k^{(j)})} \quad (3)$$

C – step. Assign each x_i to the cluster $P_k^{(j+1)}$ with maximal a posteriori probability according to (3)

M–step. Estimate the new parameters ($\pi^{(j+1)}$, $\theta^{(j+1)}$) which maximize log $L_{CML}(P^{(j+1)}, \pi^{(j)}, \theta^{(j)})$.

CML can be easily modified to handle both *labeled* and *unlabeled* data, the only difference is that in (2) the t_{ki} for labeled data are known. In this case, (2) becomes:

$$\log L_C(P, \pi, \theta) = \sum_{k=1}^{c} \sum_{x_i \in P_k} \log\{\pi_k . f_k(x_i, \theta_k)\} + \sum_{k=1}^{c} \sum_{i=n+1}^{n+m} t_{ki} \log\{\pi_k . f_k(x_i, \theta_k)\} \quad (4)$$

CEM can also be used for maximizing (4), in this case vector indicators for the labeled data are kept fixed and they are estimated for unlabeled data as in classical CEM (***E*** and ***C***-steps).

2.3 Linking CEM and Mean Squared Error Classification

Instead of estimating the posterior probability (3) as a ratio of densities, one could use a multiple regression approach. Suppose we are at the jth step of the CEM algorithm. Let X denote the matrix whose i^{th} column is x_i and $Y^{(j)}$ the corresponding target vector whose i^{th} element is a, if $x_i \in P_1^{(j)}$ and b, if $x_i \in P_2^{(j)}$. For well chosen values of a and b, e.g. $a = 1/\pi_1^{(j)}$ and $b = -1/\pi_2^{(j)}$ the solution to the minimization of the mean squared error $\left\| Y^{(j)} - W'^{(j)} X \right\|^2$ is

$$W^{(j)} = \alpha . \hat{\Sigma}^{(j)^{-1}} . (m_1^{(j)} - m_2^{(j)}) \quad (5)$$

Where $m_k^{(j)}$ and $\Sigma^{(j)}$ respectively denote the mean of $P_k^{(j)}$ and the variance of the data for the current partition $P^{(j)}$ and α is a constant (see e.g. [3]). With an appropriate threshold $w_0^{(j)}$, the decision is then to choose C_1 if $W'^{(j)} .x + w_0^{(j)} > 0$ and C_2 otherwise. It is also known that for two normal populations with equal covariance matrices $N(\mu_1, \Sigma)$, $N(\mu_2, \Sigma)$, the optimal discriminant function is:

$$g(x) = (\mu_1 - \mu_2) \Sigma^{-1} x + x_0 \quad (6)$$

for some given threshold x_0. If we replace the mean and variance in (6) by their plug in estimate used in (5), the two decision rules are similar up to a threshold. The latter may be easily estimated from the data [3]. Therefore the ***E*** and ***C***-steps in the CEM algorithm may be replaced by a minimum mean squared classification.

Because we are interested only in classification, the ***M***-step is no more necessary, the EM algorithm starts from an initial partition, the jth step consists in classifying unlabeled data according to the mean squared error decision function. It can easily be proved that this EM algorithm converges to a local maximum of the likelihood function (2) for unsupervised training and (4) for semi-supervised training. The algorithm is summarized below:

Regression-CEM

Initialisation: start from an initial partition $P^{(0)}$

j^{th} iteration, $j \geq 0$:

E – step: compute $W^{(j)} = \alpha.\hat{\Sigma}^{(j)^{-1}}.(m_1^{(j)} - m_2^{(j)})$

C – step: classify the x_is according to the sign($W^{'(j)}.x_i + w_0^{(j)}$) into $P_1^{(j+1)}$ or $P_2^{(j+1)}$.

In the above algorithm, posterior probabilities are directly estimated via a discriminant approach. In practice this is more efficient and more robust than the density estimation approach and allows to reach the same performance by making use of fewer labeled data. A more attractive reason for applying this technique is that for non-linearly separable data, more sophisticated regression based classifiers such as non linear NNs or non linear SVMs may be used instead of the linear classifier proposed above.

3 The Text Summary System

A classical measure of similarity between a document sentence s_k and a query q is the *tf-idf* term weighting scheme [4]:

$$Sim(q, s_k) = \sum_{w_t \in s_k \cap q} tf(w_t, q).tf(w_t, s_k).\left(1 - \frac{\log(df(w_t) + 1)}{\log(n+1)}\right)^2$$

Where $tf(w,x)$ is the frequency of term w in x (q or s_k), $df(w)$ is the document frequency of term w and n is the total number of documents in the collection. Representative sentences are then selected by comparing the sentence score for a given document to a preset threshold.

We consider a sentence as a sequence of terms, each term being represented by a set of features. The sentence representation will then be the corresponding sequence of these features. We used four values for characterizing each term w of a sentence s: $tf(w,s)$, $tf(w,q)$, $(1\text{-}(\log(df(w)+1)/\log(n+1))$ and $Sim(q, s)$.

In order to initialize an initial partition $P^{(0)}$, we have labeled 10% of the sentences in the training set using the news-wire summaries as the correct set of sentences. We then train a linear classifier with a sigmoid output function to label all the sentences from the training set, and iterate according to algorithm ***regression-CEM*** for semi-supervised training.

4 Experiments

A corpus of documents with associated summaries is required for the evaluation. We have used the Reuters data set consisting of news-wire summaries [12]: this corpus is composed of 1000 documents and their extracted sentence summaries. The data was split into training and test sets. We generate a query by collecting the most frequent words in the training set. Our method provides a set of relevant document sentences. Taking all the selected sentences, we can build an *extract* for the document. For the evaluation, we compared these extracts with the news-wire summaries and used Precision and Recall measures, defined as follows:

$$\text{Precision} = \frac{\#\text{ of sentences extracted by the system which are in the news - wire summaries}}{\text{total }\#\text{ of sentences extracted by the system}}$$

$$\text{Recall} = \frac{\#\text{ of sentences extracted by the system which are in the news - wire summaries}}{\text{total }\#\text{ of sentences in the news - wire summaries}}$$

We give below the average precision (table 1) for different systems and the precision/recall curves (figure 1). The baseline system gives bottom line performance that allows evaluating the contribution of our training strategy. In order to provide an upper bound of the expected performances, we have also trained a classifier in a fully supervised way: using the news-wire summaries for labeling all the sentences in the training set.

Table 1. Comparison between the baseline system and different learning schemes, using linear sigmoid classifier. Performances are on the test set.

	Precision (%)	Total Average (%)
Baseline system	54,94	56,33
Supervised learning	72,68	74,06
Semi-Supervised learning	63,94	65,32

Semi-supervised learning provides a clear increase of performances (up to 9 %). If we compare these results to fully supervised learning, which is also 9% better, we can infer that with 10% of labeled data, we have been able to extract from the unlabeled data half of the information needed for this "optimal" classification.

11-point precision recall curves allow a more precise evaluation of the system behavior. For the test set, let M be the total number of sentences extracted by the system as relevant (correctly or incorrectly), N_s the total number of sentences extracted by the system which are in the newswire summaries, N_g the total number of sentences in newswire summaries and N_t the total number of sentences in the test set.

Precision and recall are computed respectively as N_s/M and N_s/N_g. For a given document, each sentence *s* is ranked according to $P(R_Q/ s)$, the posterior probability of relevance to a query in regard of *s*. Precision and recall are computed for $M = 1,..,N_t$ and plotted here one against the other as an 11 point curve. The curves illustrate the same behavior as table 1; the semi-supervised system lies just in between the fully supervised and the baseline systems. Semi supervised learning appears as a very promising technique for this application.

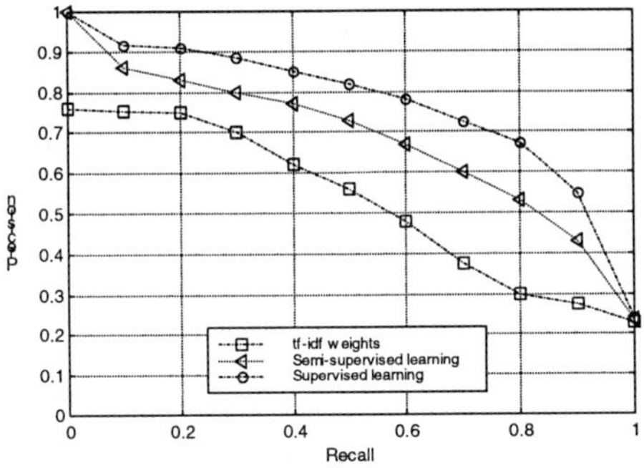

Fig. 1. Precision-Recall curves for base line system (square), semi-supervised learning (triangle) and the supervised learning (circle).

5 Conclusion

We have described a text summarization system in the context of sentence based extraction summaries. The main idea proposed here is the development of a fully automatic summarization system using a semi-supervised learning paradigm. A simple linear sigmoid classifier has implemented this. Experiments on Reuters news-wire have shown a clear performance increase. Semi-supervised learning allows reaching half of the performance increase allowed by a fully supervised system, and is much more realistic for applications. It can also be used, exactly the same way, for query based summaries.

References

1. Celeux G., Govaert G.: A classification EM algorithm for clustering and two stochastic versions. Computational Statistics & Data Analysis. 14 (1992) 315-332.
2. Chuang W.T., Yang J.: Extracting sentence segments for text summarization: a machine learning approach. Proceedings of the 23rd annual international ACM SIGIR. (2000) 152-159.
3. Duda R. O., Hart P. T.: Pattern Recognition and Scene Analysis. Edn. Wiley (1973).
4. Knaus D., Mittendorf E., Schauble P., Sheridan P.: Highlighting Relevant Passages for Users of the Interactive SPIDER Retrieval System. in TREC-4 proceedings. (1994).

5. Kupiec J., Pedersen J., Chen F. A.: Trainable Document Summarizer. Proceedings of the 18th ACM SIGIR conference on research and development in information retrieval. (1995) 68-73.
6. Luhn P.H.: Automatic creation of literature abstracts. IBM Journal (1958) 159-165.
7. McLachlan G.J.: Discriminant Analysis and Statistical Pattern Recognition. Edn. John Wiley & Sons, New-York (1992).
8. Mani I., Bloedorn E.: Machine Learning of Generic and User-Focused Summarization. Proceedings of the Fifteenth National Conference on AI. (1998) 821-826.
9. Miller D., Uyar H.: A Mixture of Experts classifier with learning based on both labeled and unlabeled data. Advances in Neural Information Processing Systems. 9 (1996) 571-577.
10. Blum A., Mitchell T.: Combining Labeled and Unlabeled Data with Co-Training. Proceedings of the 1998 Conference on Computational Learning Theory. (1998).
11. Nigam K., McCallum A., Thrun A., Mitchell T.: Text Classification from labeled and unlabeled documents using EM. In proceedings of National Conference on Artificial Intelligence. (1998).
12. http://boardwatch.internet .com/mag/95/oct/bwm9.html
13. Roth V., Steinhage V.: Nonlinear Discriminant Analysis using Kernel Functions. Advances in Neural Information Processing Systems. 12 (1999).
14. Sparck Jones K.: Discourse modeling for automatic summarizing. Technical Report 29D, Computer laboratory, university of Cambridge. (1993).
15. Zechner K.: Fast Generation of Abstracts from General Domain Text Corpora by Extracting Relevant Sentences. COLING. (1996) 986-989.

On-Line Error Detection of Annotated Corpus Using Modular Neural Networks

Qing Ma[1], Bao-Liang Lu[2], Masaki Murata[1],
Michnori Ichikawa[2], and Hitoshi Isahara[1]

[1] Communications Research Laboratory, Kyoto 619-0289, Japan
[2] RIKEN Brain Science Institute, Wako 351-0198, Japan

Abstract. This paper proposes an on-line error detecting method for a manually annotated corpus using min-max modular (M^3) neural networks. The basic idea of the method is to use guaranteed convergence of the M^3 network to detect errors in learning data. To confirm the effectiveness of the method, a preliminary computer experiment was performed on a small Japanese corpus containing 217 sentences. The results show that the method can not only detect errors within a corpus, but may also discover some kinds of knowledge or rules useful for natural language processing.

1 Introduction

To enable machines to deal with considerably varied natural language texts, it is next to impossible to pre-code all knowledge necessary for them. One solution to this problem is to compile the knowledge needed by the system directly from corpora: very large databases for natural language texts to which several kinds of tags, such as part of speech (POS) and syntactic dependency, have been added instead of one that consists only of plain texts. Corpora have been used successfully in constructing various fundamental natural language processing systems including a morphological analyzer and parser, which can be widely applied in many areas of information processing including preprocessing for speech synthesis, post-processing for OCR and speech recognition, machine translation, and information retrieval and text summarization. Manually annotating very large corpora (the Penn Treebank, e.g., consists of over 4.5 million words of American English with 135 POSs.), however, is a very complex and costly endeavor.

For this purpose, many automatic POS tagging systems using various machine learning techniques have been proposed(e.g., [1, 2]). In our previous work, we have developed a neuro and rule-based hybrid tagger, which reached a practical level in terms of tagging accuracy that requires less training data compared to the other methods [3]. To further improve our systems' tagging accuracy, we can use two approaches; one is to increase the amount of training data and the other is to improve the quality of the corpus to be used for training. However, since in the first approach multilayer perceptrons were used in our tagger, it will suffer from an unconvergent problem. To overcome this inherent drawback, we have adopted a min-max modular (M^3) neural network [4] that can solve large and complex problems by decomposing them into many small and simple

G. Dorffner, H. Bischof, and K. Hornik (Eds.): ICANN 2001, LNCS 2130, pp. 1185–1192, 2001.

subproblems [5]. For the second approach, a POS-error-detecting technique is needed, which is the main issue of this paper.

Words are often ambiguous in terms of their POSs, which have to be disambiguated (tagged) using context of the sentence. POS tagging, however, no matter using manual or automatic methods, usually involves errors. The POSs in manually annotated corpora can basically have three kinds of errors: simple-mistake type (e.g., POS "Verb" is inputed as "Varb"), incorrect-knowledge type (e.g., word *fly* is always tagged as "Verb"), and inconsistent-type (e.g., word *like* in sentence "Time flies like an arrow" is correctly tagged as "Preposition", but in the sentence "The one like him is welcome" it is tagged as "Verb"). The simple-mistake type can be easily detected by only referring to an electronic dictionary. The incorrect-knowledge type, however, is hardly possible to detect using automatic methods. If we consider tagging words with correct POSs as classification or input-output mapping problems of mapping words under the context of POSs, then the inconsistent-type errors can be considered as sets of data with the same input but different outputs (classes), which can be dealt with using the statistical methods proposed so far or the neural-network method proposed in this paper. The work that has been done so far to develop a detection technique with statistical approaches [6, 7] was for off-line use, that is, the detection must be performed ahead of the learning. Off-line detection, however, is expensive for very large corpora because detection must be performed word by word through the whole corpus with no preprocessing to first focus on a few blocks of words or sentences that are certain to include errors, as can be done by the proposed method.

Since the M^3 network consists of modules dealing with very simple and small subproblems, the modules can be constructed with very simple multilayer perceptrons by only either using very few or no hidden units. This means that such modules will basically not suffer from unconvergent problems. In other words, if a module cannot converge, we may basically consider that the module is trying to learn the data, including the inconsistent-type errors. This type of error within an annotated corpus may therefore be detected on-line, in the sense that the detection is performed while learning, that is, by picking out the uncoverged modules and then determining the inconsistent data from the data sets being learned. Since the unconverged modules compared to the converged ones will be very limited when using a corpus of high quality, and the data set each module learns is very small, this on-line error-detecting method will be extremely cost-effective for very large corpora. By using such an on-line error detection method, the quality of the corpus while it is being learned can be improved immediately by a light manual intervention, and the new data can be immediately used to re-train the unconverged modules.

2 The M^3 Network

This section describes in brief the key question for M^3 Network solving classification problems: i.e., how a large and complex K-class problem is decomposed into many smaller and simpler two-class problems, which are individually solved by using modules independent of each other, and how they are integrated to obtain the final solution (for details see [4]).

2.1 Problem Decomposition

Suppose T is a set of learning data for a K-class problem; i.e.,

$$T = \{(X_l, Y_l)\}_{l=1}^{L}, \tag{1}$$

where $X_l \in R^n$ is an input vector, $Y_l \in R^K$ is a desired output, and L is the number of learning data. In general, any K-class problem can be decomposed into $\binom{K}{2}$ two-class problems:

$$T_{ij} = \{(X_l^{(i)}, 1-\epsilon)\}_{l=1}^{L_i} \cup \{(X_l^{(j)}, \epsilon)\}_{l=1}^{L_j}, \quad i = 1, \cdots, K, \ j = i+1, \cdots, K \tag{2}$$

where ϵ is a small positive number, and $X_l^{(i)}$ and $X_l^{(j)}$ are input vectors belonging to C_i and C_j.

In the $\binom{K}{2}$ two-class problems, those that are still too complex can be further decomposed. First, each larger set of input vectors, e.g., $X_l^{(i)}$ [see Eq. (2)], belonging to each class is divided by a random method into N_i $(1 \leq N_i \leq L_i)$ subsets χ_{ij}; i.e.,

$$\chi_{ij} = \{X_l^{(ij)}\}_{l=1}^{L_i^{(j)}}, \quad j = 1, \cdots, N_i, \tag{3}$$

where $L_i^{(j)}$ is the number of input vectors in subset χ_{ij}. If by using such subsets, the two-class problem defined by Eq. (2) can be decomposed into $N_i \times N_j$ smaller and simpler problems as follows.

$$T_{ij}^{(u,v)} = \{(X_l^{(iu)}, 1-\epsilon)\}_{l=1}^{L_i^{(u)}} \cup \{(X_l^{(jv)}, \epsilon)\}_{l=1}^{L_j^{(v)}}, u = 1, \cdots, N_i, \quad v = 1, \cdots, N_j, \tag{4}$$

where $X_l^{(iu)} \in \chi_{iu}$ and $X_l^{(jv)} \in \chi_{jv}$ belong to C_i and C_j, respectively.

Thus, if all the two-class problems defined by Eq. (2) are further decomposed into those defined by Eq. (4), the original K-class problem is decomposed into $\sum_{i=1}^{K} \sum_{j=i+1}^{K} N_i \times N_j$ two-class problems. If the learning data set only includes two different elements, i.e., $L_i^{(u)} = 1$ and $L_j^{(v)} = 1$, then the two-class problem defined by Eq. (4) is obviously a linearly separable problem.

2.2 Module Integration

After learning the decomposed subproblems using the individual modules, they next have to be integrated to give a final solution for the original problem. This section focuses on how the modules are integrated. To read why such an integration can solve problems, see Ref. [4].

For integration, three units called MIN, MAX, and INV are used. Here, denotations M_{ij} and $\mathrm{M}_{ij}^{(u,v)}$ are used to indicate the modules for learning subproblems T_{ij} [Eq. (2)] and $T_{ij}^{(u,v)}$ [Eq. (4)], respectively. In the case that K-class problem

T [Eq. (1)] is solved by decomposing into $\binom{K}{2}$ two-class problems T_{ij} [Eq. (2)], first the following combination is performed with the MIN unit that selects the minimum value from the multiple inputs:

$$\mathrm{MIN}_i = \min(\mathrm{M}_{i1}, \cdots, \mathrm{M}_{ij}, \cdots, \mathrm{M}_{iK}), \quad i = 1, \cdots, K \ (i \neq j) \tag{5}$$

where for convenience, the denotation for the MIN unit is used to express its output and the denotations for the modules are used to express their outputs. Thus, the final solution is obtained from these K outputs of the MIN units:

$$C = \arg\max_i \{\mathrm{MIN}_i\}, \quad i = 1, \cdots, K, \tag{6}$$

where C is the class that the input data belongs to. In the case that two-class problem T_{ij} is further decomposed into $T_{ij}^{(u,v)}$ [Eq. (4)], the module $\mathrm{M}_{ij}^{(u,v)}$ learning $T_{ij}^{(u,v)}$ is combined first with the MIN unit:

$$\mathrm{MIN}_{ij}^{(u)} = \min(\mathrm{M}_{ij}^{(u1)}, \cdots, \mathrm{M}_{ij}^{(uN_j)}), \quad u = 1, \cdots, N_i, \tag{7}$$

and then module M_{ij} is formed by using the MAX unit which selects the maximum value from the multiple inputs:

$$\mathrm{M}_{ij} = \max(\mathrm{MIN}_{ij}^{(1)}, \mathrm{MIN}_{ij}^{(2)}, \cdots, \mathrm{MIN}_{ij}^{(N_i)}). \tag{8}$$

The module M_{ij} formed in such a way is then integrated into Eq. (5). Since the two-class problem T_{ij} is the same as T_{ji}, M_{ji} is constructed by M_{ij} and an INV unit which reverses the input value.

3 Error Detection Using M^3 Network

Since the error detection is performed on-line during the learning of a POS tagging problem, to describe how error detection can be performed we have to first describe what is the POS tagging problem, and how the POS tagging problem is decomposed and learned by M^3 networks.

3.1 POS Tagging Problem

Suppose there is a lexicon $V = \{w^1, w^2, \cdots, w^v\}$, where the POSs that can be served by each word are listed, and there is a set of POSs, $\Gamma = \{\tau^1, \tau^2, \cdots, \tau^\gamma\}$. The POS tagging problem is thus to find a string of POSs $T = \tau_1\tau_2\cdots\tau_s$ ($\tau_i \in \Gamma$, $i = 1, \cdots, s$) by following procedure φ when sentence $W = w_1 w_2 \cdots w_s$ ($w_i \in V$, $i = 1, \cdots, s$) is given.

$$\varphi : W^p \rightarrow \tau_p, \tag{9}$$

where p is the position of the word to be tagged in the corpus, and W^p is a word sequence centered on the target word w_p with (l, r) left and right words:

$$W^p = w_{p-l} \cdots w_p \cdots w_{p+r}, \tag{10}$$

where $p - l \geq s_s$, $p + r \leq s_s + s$, s_s is the position of the first word of the sentence. Tagging can thus be regarded as a classification or mapping problem by replacing the POS with class and can therefore be handled by using supervised neural networks trained with an annotated corpus.

3.2 Decomposition of POS Tagging Problem

The Kyoto University Text Corpus [8] is composed of 19,956 Japanese sentences that include a total of 487,691 words with 30,674 distinct ones. More than half the total words are ambiguous in terms of the 175 kinds of POSs used in the corpus. As a preliminary study to see whether the M^3 Network can detect errors in the on-line mode while learning POS tagging problem, this paper selected only 217 Japanese sentences, each of which had at least one error. These sentences include a total of 6,816 words with 2,410 distinct ones and 97 kinds of POS tags. By regarding POS as a class, the POS tagging problem in this case is therefore a 97-class classification problem.

According to Sec. 2.1, this 97-class problem is first uniquely decomposed into $\binom{K}{2} = 4,656$ two-class problems. Since some of these problems are still too large, they are further decomposed in the random method described in Sec. 2.1. As a result, two-class problem $T_{1,2}$, for example, is decomposed into eight subproblems, but problem $T_{5,10}$ is not further decomposed. In such a way, the original 97-class problem has been decomposed into total of 23,231 smaller two-class problems.

3.3 M^3 Network for Learning POS Tagging Problem

The M^3 Network that learns the POS tagging problem described in the previous section is constructed by integrating modules, as shown in Fig. 1(a). The individual module M_{ij} is further constructed as shown in Fig. 1(b) if the corresponding problem T_{ij} is further decomposed. In the example shown in Fig. 1(b), since problem $T_{7,26}$ is further decomposed into $N_7 \times N_{26} = 25 \times 10 = 250$ subproblems, $M_{7,26}$ is constructed by 250 modules $M_{7,26}^{(u,v)}$ $(u = 1, \cdots, 25,\ v = 1, \cdots, 10)$. And, M_{ji} $(j > i)$ is constructed by M_{ij} and the INV unit.

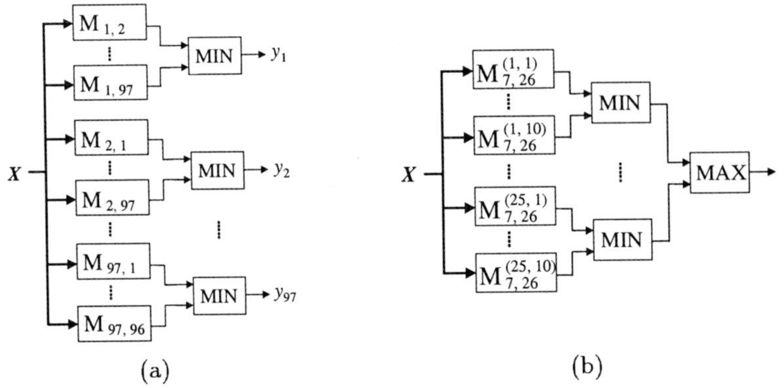

Fig. 1. The M^3 Network: (a) the entire construction and (b) a close-up module $M_{7,26}$.

Input vector X (X_l [Eq. (1)] etc.) in the learning phase is constructed from word sequence W^p [Eq. (10)]:

$$X = (\boldsymbol{x}_{p-l}, \cdots, \boldsymbol{x}_p, \cdots, \boldsymbol{x}_{p+r}). \tag{11}$$

Element $\boldsymbol{x}_p$ is a binary-coded vector with ω dimensions

$$\boldsymbol{x}_p = (e_{w1}, \cdots, e_{w\omega}) \tag{12}$$

for encoding the target word. Element $\boldsymbol{x}_t$ ($t \neq p$) for each contextual word is a binary-coded vector with τ dimensions

$$\boldsymbol{x}_t = (e_{\tau 1}, \cdots, e_{\tau\tau}) \tag{13}$$

for encoding the POS that has been tagged to the word.[1] The desired output is a binary-coded vector with τ dimensions

$$Y = (y_1, y_2, \cdots, y_\tau) \tag{14}$$

for encoding the POS that the target word should be tagged.

3.4 Detection of POS Errors within Corpus

Since each individual module in the M^3 network only needs to learn a very small and simple two-class problem, it can be constructed, for example, with a very simple multilayer perceptron by only using either none or a few hidden units. Thus, individual module basically does not suffer from an unconvergent problem as long as the learning data is correct. In other words, if a module does not converge, it may be considered that it is learning a data set, $T_M = (X_l, Y_l)_{l=1}^{L_M}$, that includes some inconsistent data: i.e., there is at least one pair of data, (X_i, Y_i) and (X_j, Y_j) in the data set, such that

$$X_i = X_j, Y_i \neq Y_j \quad (i \neq j). \tag{15}$$

Here, T_M is used for denoting T_{ij} [Eq. (2)] or $T_{ij}^{(u,v)}$ [Eq. (4)].

Thus, this type of error within an annotated corpus that is being learned may be detected "on-line" by only picking out the unconverged modules and then determining if the data are in conflict with each other; i.e., a set of pairs, (X_i, Y_i) and (X_j, Y_j), that satisfy Eq. (15) from the data set the module is learning with a simple program. Since the unconverged modules compared to the converged ones will be very limited when an high-quality annotated corpus is used and the data set each module learns is very small, this on-line error-detective method has an extremely good cost performance, which will increase with the size of a corpus. By using such an effective on-line error detection method, the quality of the corpus while it is being learned can be improved by a light manual intervention and the new data can be immediately used to re-train the unconverged modules.

[1] This coding is a simplified version of the original coding method actually used in POS tagging system [5]. However, the detective mechanism that determines the patterns that are in conflict with each other will be unchanged.

4 Experimental Results

The data described in Sec. 3.2 was used in our experiment. Because there are 30,674 distinct words and 175 kinds of POSs in the whole corpus, the dimensions of the binary-coded vectors for word and POS, ω and τ, were set at 16 and 8, respectively. The length (l, r) of a word sequence given to M^3 network was set at (2,2). The number of units in the input layers of all modules was therefore $[(l + r) \times \tau] + [1 \times \omega] = 48$ and all modules were basically constructed using three-layer perceptrons whose input-hidden-output layers had 48-2-1 units, respectively. Modules will stop learning as one round when the average-squared error reaches an objective value, 0.05, or the iterations reached 5,000 epochs. For the modules that could not reach the objective error, relearning of up to five rounds was performed by adding hidden layers of two units each in each new round until reaching the objective error.

Experimental results show that 82 modules within the total of 23,231 modules did not finally converge. Of these 82 modules, 81 had exactly 97 pairs of contradictory learning data, which at first reinforced the hypothesis that the M^3 network has basically no unconvergent problems. These 97 pairs of learning data were checked by one of the authors, an NLP expert who knows both Japanese grammar and the Kyoto University Text Corpus well. As a result, we saw that of these 97 pairs of learning data, 94 pairs included true POS errors and the precision rate reached nearly 97%. Table 1 shows a pair of learning data detected from unconverged module $M_{7,26}^{(1,6)}$, a sub-module of $M_{7,26}$ shown in Fig. 1(b). The left column shows in order of sentence and word number the positions of the words that are being checked. Each word sequence shown in the right column is constructed by morphemes that are separated by the denotation ",". Each morpheme was formed by "Japanese word (English equivalent): POS". The underlined Japanese word is the target word being checked. The word sequence tagged with the symbol "*" in the head shows that the target word in this word sequence is tagged incorrectly.

We examined the remaining three pairs that were in exact conflict with each other, but were all are correct and found that they were all tagging word cases "*de* (in , at, on, ...)" functioning as either postpositional particle or copula in various contexts. The word "*de*" in Japanese, however, belongs to a very special case in which it is not sufficient to only use n-gram word and POS information to determine its POS, the grammar of the whole sentence has to be considered

Table 1. An example of detected pair of learning data

No.	Word sequences tagged by POSs
2-18	* *seijika* (politician): common-noun, *gawa* (side): nominal-suffix, <u>*no*</u> (of): case-postpositional-particle, *odate* (flattery): common-noun, *ni* (by): case-postpositional-particle
124-19	*shinario* (scenario): common-noun, *dukuri* (making) : nominal-suffix, <u>*no*</u> (of): conjunctive-postpositional-particle, *houkou* (direction): common-noun, *o* (object): case-postpositional-particle

instead. This result indicates that this method not only detected POS errors with a substantially 100% precision rate, but also discovered some kind of knowledge that is difficult to be found by us in general but is useful for natural language processing. Since each of the 217 sentences had at least one error, respectively, the recall rate was low. The main reason, however, was in the too small corpus, from which too few clues could be extracted and then used to find further conflicting data. In our future full-scale experiment using the entire corpus, the recall rate of error detection can therefore be expected to rise dramatically with an almost unchanged precision rate.

5 Conclusion

We proposed a cost-effective on-line error-detecting method for manually annotated corpus by conversely utilizing an unconvergent problem, which usually causes us distress when using neural networks, and reinforced the hypothesis that M^3 network basically has no unconvergent problems at the same time. Although the scale of the preliminary experiment was small, since the M^3 network can deal with a very large and complex classification problem, it appears that this technique may be very useful not only in further improving our POS tagging system, but also in improving the quality of various manually annotated corpora, and may discover some kinds of knowledge or rules useful for natural language processing.

References

1. Merialdo, B.: Tagging English text with a probabilistic model, *Computational Linguistics*, Vol. 20, No. 2, pp. 155-171, 1994.
2. Brill, E.: Transformation-based error-driven learning and natural language processing: a case study in part-of-speech tagging, *Computational Linguistics*, Vol. 21, No. 4, pp. 543-565, 1994.
3. Ma, Q., Uchimoto, K., Murata, M., and Isahara, H: Hybrid neuro and rule-based part of speech taggers, *Proc. COLING'2000*, Saarbrücken, pp. 509-515, 2000.
4. Lu, B. L. and Ito, M.: Task decomposition and module combination based on class relations: a modular neural network for pattern classification, *IEEE Trans. Neural Networks*, Vol. 10, No. 5, pp. 1244-1256, 1999.
5. Lu, B. L., Ma, Q., Isahara, H., and Ichikawa M.: Efficient part-of-speech tagging with a min-max module neural network model, to appear in *Applied Intelligence*, 2001.
6. Eskin, E.: Detecting errors within a corpus using anomaly detection, *Proc. NAACL'2000*, Seattle, pp. 148-153, 2000.
7. Murata, M, Utiyama, M., Uchimoto, K., and Ma, Q., and Isahara, H.: Corpus error detection and correction using the decision-list and example-based methods, *Proc. Information Processing Society of Japan, WGNL 136-7*, pp. 49- 56, 2000 (in Japanese).
8. Kurohashi, S. and Nagao, M: Kyoto University text corpus project, *Proc. 3rd Annual Meeting of the Association for Natural Language Processing*, pp. 115-118, 1997 (in Japanese).

Instance-Based Method to Extract Rules from Neural Networks

DaeEun Kim[1] and Jaeho Lee[2]

[1] Division of Informatics,
University of Edinburgh, Edinburgh, EH1 2QL
United Kingdom
daeeun@dai.ed.ac.uk

[2] Department of Electrical Engineering
University of Seoul, Tongdaemoon-ku,
Seoul, 151-011, Korea
jaeho@uofs.ac.kr

Abstract. It has been shown that a neural network is better than induction tree applications in modeling complex relations of input attributes in sample data. Those relations as a set of linear classifiers can be obtained from neural network modeling based on back-propagation. A linear classifier is derived from a linear combination of input attributes and neuron weights in the first hidden layer of neural networks. Training data are projected onto the set of linear classifier hyperplanes and then information gain measure is applied to the data. We propose that this can reduce computational complexity to extract rules from neural networks. As a result, concise rules can be extracted from neural networks to support data with input variable relations over continuous-valued attributes.

1 Introduction

The highly nonlinear property of neural networks makes it difficult to describe how they reach predictions. Although their predictive accuracy is satisfactory for many applications, they have long been considered as a complex model in terms of analysis. By using expert rules derived from neural networks, the neural network representation can be more understandable.

It has been shown that neural networks are better than direct application of induction trees in modeling nonlinear characteristics of sample data [3, 10]. Neural networks have the advantage of being able to deal with noisy, inconsistent and incomplete data. A method to extract symbolic rules from neural networks has been proposed to increase the performance of the decision process [4, 11, 12, 10]. The KT algorithm developed by Fu [4] extracts rules from subsets of connected weights with high activation in a trained network. The M-of-N algorithm [12] clusters weights of the trained network and removes insignificant clusters with low active weights. Then the rules are extracted from the weights.

A simple rule extraction algorithm that uses discrete activations over continuous hidden units is presented in by Setiono and Taha [10, 11]. They used in

G. Dorffner, H. Bischof, and K. Hornik (Eds.): ICANN 2001, LNCS 2130, pp. 1193–1198, 2001.

sequence a weight-decay back-propagation over a three-layer feed-forward network, a pruning process to remove irrelevant connection weights, a clustering of hidden unit activations, and extraction of rules from discrete unit activations. They derived symbolic rules from neural networks that include oblique decision hyperplanes instead of general input attribute relations [10]. Normally the direct conversion from neural networks to rules has an exponential complexity when using search-based algorithm over incoming weights for each unit [12]. Similar to the rule extraction from neural networks, the production of oblique decision rules with other methods has been tested [7], but the problem of inducing oblique decision trees has very high computational complexity.

Most of the rule extraction algorithms are used to derive rules from neuron weights and neuron activations in the hidden layer as a search-based method. The search-based algorithms find subsets of incoming weights that exceed the bias on a unit, then rewrite such sets as a pair of preconditions and classes. This search algorithm is one of NP-complete problems, trying to transform threshold logic units to Boolean logic gates.

We propose an instance-based rule extraction method that reduces computation time by escaping search-based methods. After training one or two hidden layer neural networks, the first hidden layer weight parameters are treated as linear classifiers. Many sample data can be projected onto the set of linear classifiers with neural network response surfaces and then we use information gain measure [8] to choose proper linear classifiers to determine decision boundaries for classification.

2 Method

The goal for our approach is to generate rules following the shape and characteristics of neural network response surfaces. Many researchers suggested neural networks should be approximated with oblique decision rules [10, 1, 6]. The oblique decision rules overcome the disadvantage of axis-parallel decision rules by allowing hyperplanes with any orientation in parameter space. Thus, we allow input variable relations for multi-attributes in a set of rules in order to improve classification.

2.1 Neural Network and Linear Classifiers

We use a two-phase method for rule extraction from neural networks. Given a large training set of data points, the first phase, as a feature extraction phase, is to train feed-forward neural networks with back-propagation and collect the weight set over input variables in the first hidden layer. A feature useful in inferring multi-attribute relations of data is found between the input layer and the first hidden layer of neural networks. In the second phase, as a feature combination phase, each extracted feature for a linear classification boundary is combined together using Boolean logic gates. In this paper, we use Quinlan's information gain measure [8, 9] to select and combine each linear classifier.

After training data patterns with a neural network by back-propagation, we can have linear classifiers in the first hidden layer. If we take a high value of α, the activation of each neuron is close to the property of digital logic gates, which has a binary value of 0 or 1. Input to each neuron in the first hidden layer is represented as a linear combination of input attributes and weights, $\sum_i^N a_i W_{ij}$. This forms linear classifiers for data classification as a feature extraction over data distribution. If we have n input attributes $\{a_1, a_2, ..., a_n\}$ and m nodes in the first hidden layer, then we have m classifiers as follows:

$$
\begin{aligned}
L_1 &= W_{01} + a_1 W_{11} + a_2 W_{21} + ... + a_n W_{n1} \\
&\cdots \\
L_m &= W_{0m} + a_1 W_{1m} + a_2 W_{2m} + ... + a_n W_{nm}
\end{aligned}
$$

2.2 Information Gain Measure

After obtaining linear classifiers, all data are projected into a set of linear classifier hyperplanes and form m classifier vectors. Information gain measure selects one of classifier vectors and its thresholds.

Let Y be the set of examples and let C be the set of k classes. Let $p_Y(C_i)$ be the probability of the examples in Y that belong to class C_i. The split probability $p(Y_j^w)$ in continuous-valued attributes among w partitions is given as the probability of the examples that belong to the partition Y_j^w when splitting the attribute. Then information gain of a linear classifier L over a collection of instances Y is defined as

$$
Gain(L) = \frac{E(Y) - \sum_{i=1}^{v} \frac{|Y_i|}{|Y|} E(Y_i)}{\sum_{j=1}^{w} p(Y_j^w) \log p(Y_j^w)}
$$

where $E(Y) = \sum_{i=1}^{k} p_Y(C_i) \log p_Y(C_i)$ is an entropy function, L has v distinct values, and Y_i is the set of instances whose value of linear classifier L is l_i, and Y_j^w is one of w partitions when splitting the linear classifier L .

Each linear classifier L_i has its own threshold τ_i that produces the greatest information gain. This threshold is automatically selected by information gain measurement, and the threshold τ_i is one of the best cut for the linear classifier L_i to make good decision boundaries; it plays a role of bias value for linear hyperplane. The hyperplane divides the parameter space into two non-overlapping subsets, depending on τ_i; it has a dichotomy $L_i > \tau_i$ and $L_i \leq \tau_i$.

3 Experiments

We first tested our method for a set of artificial data with some noise as shown in Fig.1. We used slope $\alpha = 2$ for sigmoid functions. Fig.1 displays a set of 150 sample data with two classes. The artificial data include much noisy data and show complex relations of input attributes for each class distribution.

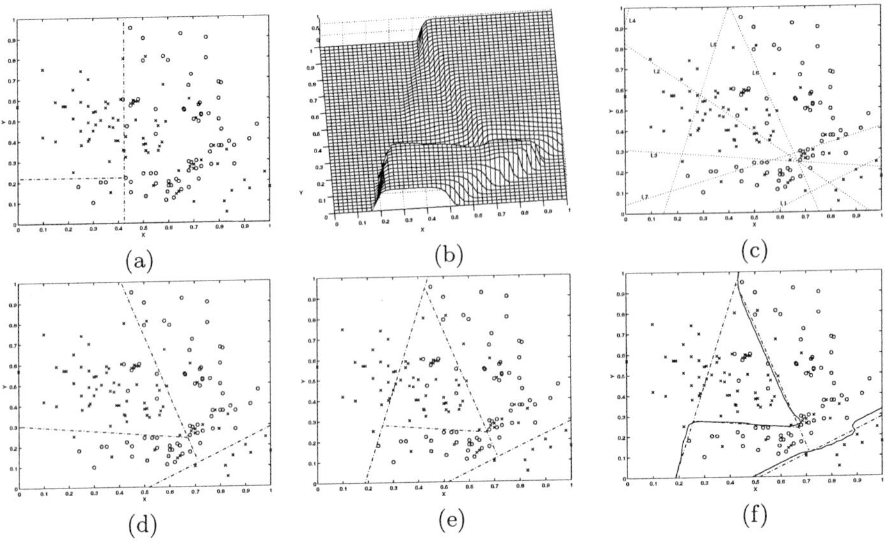

Fig. 1. Example (a) data set and C4.5 decision boundaries (O : class 1, X : class 0) (b) neural network fitting (c) a set of linear classifiers $\{L_1, L_2, ..., L_7\}$ extracted from neural network (d) linear classifier rules with training data size 150 (e) linear classifier rules with training data size 441 (f) neural network decision borders and linear classifier boundaries

When we trained a neural network with one hidden layer and 7 nodes in it over the artificial data, we obtained the response surface shown in Fig.1(b) and 22 examples were misclassified (14.7% errors) with the neural network. A set of the dotted boundary lines in Fig.1(c) is a set of linear classifiers extracted from the neural network. When 150 data points are projected into each linear classifier hyperplane L_i, $i = 1, .., 7$, the information gain measure selects L_1 first, and then L_6 and L_3 in order. They are more dominant for classification in terms of entropy measurement than the other classifiers. With those classifiers, 4 rules represent the decision boundaries in Fig.1(d). We generated 441 example data in uniform distribution with neural network response surface; each axis has 21 scales for the distribution. Then we applied again the information gain measure to select proper linear classifiers among the set $\{L_1, L_2, ..., L_7\}$. At this moment, L_5, L_1, L_6, L_3 were selected in sequence. As a result, we obtained 5 rules and each rule corresponds to one area inside classification boundaries given in Fig.1(e). Fig.1(f) shows the linear classifiers follow very closely neural network decision boundaries. The linear classifier rules obtain high fidelity to approximate the trained neural network. When we compare our linear classifier rules with C4.5 decision tree method, the performance was worse than our linear classifier rules; error rate is 29.3 %. The decision boundary is given in Fig.1(a).

For general case, it has been tested on several sets of data in the UCI depository [2]. These data sets are mostly for machine learning experiments. Table 1 shows error rates to compare neural networks, linear classifier rules, and C4.5

Table 1. Data classification errors in C4.5 and linear classifier method

	neural network		linear classifier rules		*C*4.5	
data	train (%)	test (%)	train (%)	test (%)	train (%)	test (%)
wine	0.0 ± 0.0	2.0 ± 0.4	0.0 ± 0.0	3.6 ± 0.9	1.2 ± 0.1	7.9 ± 1.3
iris	0.6 ± 0.1	4.1 ± 1.3	0.7 ± 0.2	4.7 ± 1.5	1.9 ± 0.1	5.4 ± 0.7
breast-w	0.4 ± 0.1	4.4 ± 0.5	0.8 ± 0.1	4.1 ± 0.4	1.1 ± 0.1	4.7 ± 0.6
ionosphere	1.3 ± 0.2	8.8 ± 1.3	1.2 ± 0.2	8.8 ± 1.5	1.6 ± 0.2	10.4 ± 1.1
pima	12.7 ± 0.4	27.1 ± 0.9	15.8 ± 0.4	27.0 ± 0.9	15.1 ± 0.8	26.4 ± 0.9
glass	4.3 ± 0.9	31.6 ± 1.2	6.6 ± 0.8	34.9 ± 1.6	6.7 ± 0.4	32.0 ± 1.5
bupa	10.2 ± 1.1	32.8 ± 2.1	14.8 ± 1.4	33.7 ± 2.1	12.9 ± 1.5	34.5 ± 2.0

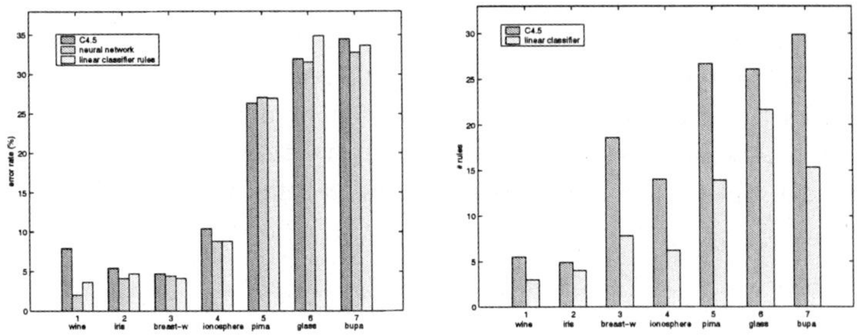

(a) (b)

Fig. 2. Comparison between neural network, linear classifier method, and C4.5 (a) average error rate in test data with C4.5, neural network, and linear classifier (b) the number of rules with C4.5 and linear classifier method

method. The error rates were estimated by running the complete 10-fold cross-validation ten times, and the average and the standard deviation for ten runs were given in the table. The average classification error rate results and the number of rules are shown in Fig.2. Several neural network topologies were tested for each data set. Two hidden layer neural networks are not significantly more effective in terms of both error rates and the number of rules than one hidden layer. The neural network can improve training performance by increasing the number of nodes in the hidden layers. However, it does not guarantee to improve test set performance. In many cases, reducing errors in a training set tends to increase the error rate in a test set by overfitting.

The result of linear classifier rules supports the fact that the methods greatly depend on neural network training. If neural network fitting is not correct, then the fitting errors may mislead the result of linear classifier methods. The error rate difference between a neural network and the linear classifier method explains that some data points are located on marginal boundaries of classifiers. It is due to the fact that our neural network model uses sigmoid functions with high slopes instead of step functions. The activation may be anywhere between 0 and 1, and the weighted sum of activations may lead to different output classes.

Most of the data sets in the UCI depository have a small number of data examples relative to the number of attributes. The significant difference, between C4.5 application rules and linear classifier rules extracted from neural networks, is not seen distinctively in UCI data in terms of error rate, unlike the artificial data in Fig.1. However, the number of rules using our method is significantly smaller than that using conventional C4.5 in all the data sets, as shown in Fig.2(b).

4 Conclusions

This paper presents an efficient method for constructing symbolic rules from neural networks, compared to other search-based algorithms. Information gain measure can be seen as a heuristic method to reduce computational complexity. This method extracts symbolic rules which have neural network features, and thus it has advantages over a simple decision tree method. First, we can obtain good features for a classification boundary from neural networks by training input patterns. Second, because of feature extractions about input variable relations, we can obtain a compact set of rules to reflect input patterns.

References

1. F. Alexandre and J.F. Remm. Knowledge extraction from neural networks for signal interpretation. In *European Symposium on Artificial Neural Networks*, 1997.
2. C. Blake, E. Keogh, and C.J. Merz. *UCI* repository of machine learning databases. In *Preceedings of the Fifth International Conference on Machine Learning*, http://www.ics.uci.edu/ mlearn, 1998.
3. T.G. Dietterich, H. Hild, and G. Bakiri. A comparative study of *ID3* and backpropagation for english text-to-speech mapping. In *Proceedings of the 1990 Machine Learning Conference*, pages 24–31. Austin, TX, 1990.
4. L. Fu. Rule learning by searching on adaptive nets. In *Preceedings of the 9th National Conference on Artificial Intelligence*, pages 590–595, 1991.
5. K. Hornik, M. Stinchrombe, and H. White. Multilayer feedforward networks are universal approximators. *Neural Networks*, 2(5):359–366, 1989.
6. D. Kim and J. Lee. Handling continuous-valued attributes in decision tree using neural network modeling. In *European Conference on Machine Learning*, pages 211–219. Springer Verlag, 2000.
7. P.M. Murthy, S. Kasif, and S. Salzberg. A system for induction of oblique decision trees. *Journal of Artificial Intelligence Research*, 2:1–33, 1994.
8. J.R. Quinlan. Induction of decision trees. *Machine Learning*, 1(1):81–106, 1986.
9. J.R. Quinlan. *C4.5 Programs for Machine Learning.* Morgan Kaufmann, San Mateo, CA, 1993.
10. R. Setiono and H. Liu. Neurolinear: A system for extracting oblique decision rules from neural networks. In *European Conference on Machine Learning*, pages 221–233. Springer Verlag, 1997.
11. I. A. Taha and J. Ghosh. Symbolic interpretation of artificial neural networks. *IEEE Transactions on Knowledge and Data Engineering*, 11(3):448–463, 1999.
12. G.G. Towell and J.W. Shavlik. Extracting refined rules from knowledge-based neural networks. *Machine Learning*, 13(1):71–101, Oct. 1993.

A Novel Binary Spell Checker

Victoria J. Hodge[1] and Jim Austin[1]

University of York, UK
{vicky,austin}@cs.york.ac.uk

Abstract. In this paper we propose a simple, flexible and efficient hybrid spell checking methodology based upon phonetic matching, supervised learning and associative matching in the AURA neural system. We evaluate our approach against several benchmark spell-checking algorithms for recall accuracy. Our proposed hybrid methodology has the joint highest top 10 recall rate of the techniques evaluated. The method has a high *recall* rate and low computational cost.

1 Introduction

Errors, particularly spelling and typing errors are abundant in human generated electronic text. For example, Internet search engines are criticised for their inability to spell check the user's query. This would prevent wasted computational processing, prevent wasted user time and make any system more robust as spelling and typing errors can prevent the system identifying the required information.

We describe an interactive spell checker that identifies spelling errors and recommends alternative spellings. The spell checker uses a hybrid approach to overcome phonetic spelling errors and the four main forms of typing errors: insertion, deletion, substitution and transposition. We use a Soundex-type coding approach (see Kukich [4]) coupled with transformation rules to overcome phonetic spelling errors. We use an n-gram approach [5] to overcome the first two forms of typing error and integrate a Hamming Distance approach to overcome substitution and transposition errors. Our spell checker aims to high recall accuracy possibly at the expense of precision. However, the scoring allows us to rank the retrieved matches so we can limit the number of possibilities suggested to the user to the top 10 matches, giving both high recall and precision.

Some alternative spelling approaches include: Levenshtein Edit Distance (see [4]) which scores similarity by the number of transformations required to turn the misspelt word into each lexicon word but runs slowly; Agrep [7] [6] is based on Edit Distance but uses an approach optimised for speed, however Agrep has the lowest recall in our evaluation; Aspell [1] which integrates transformation rules and phonetic code generation; and, the two benchmark approaches MS Word 97 and MS Word 2000. We evaluate our approach against these alternatives for quality of retrieval both recall - the percentage of correct words retrieved from 600 misspelt words and precision - the number of false matches returned. The reader is referred to Kukich [4] for a thorough treatise of spell checking techniques.

G. Dorffner, H. Bischof, and K. Hornik (Eds.): ICANN 2001, LNCS 2130, pp. 1199–1204, 2001.

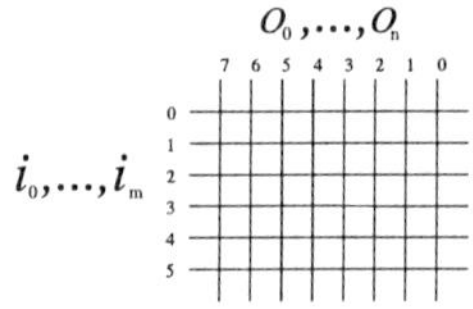

Fig. 1. The input vector i addresses the rows of the CMM and the output vector o addresses the columns.

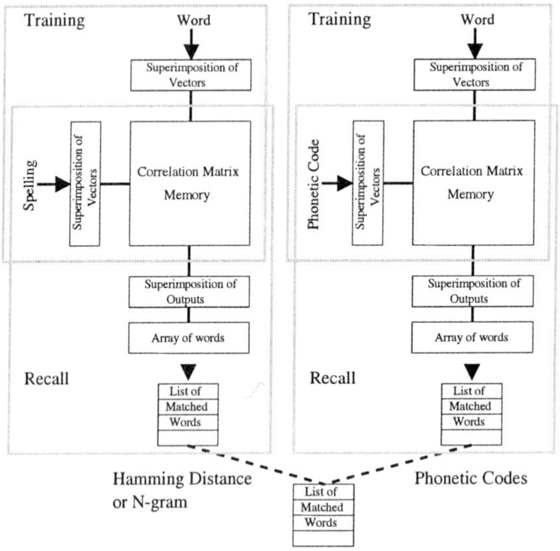

Fig. 2. Diagram of the hybrid spell checker in the AURA modular system.

1.1 Our Integrated Hybrid Modular Approach

AURA [2] is a collection of binary associative neural networks that may be implemented in a modular fashion. AURA utilises Correlation Matrix Memories (CMMs) to map inputs to outputs through a supervised learning rule, similar to a hash function see Fig. 1. In our system shown in Fig. 2, we use two CMMs [2]: one CMM stores the words for n-gram and Hamming Distance matching and the second CMM stores phonetic codes for homophone matching. The CMMs are used independently but the results are combined during the scoring phase of the spell checker to produce an overall score for each word.

Hamming Distance and N-Gram. For Hamming Distance and n-gram, the word spellings form the inputs. We divide a binary vector of into a series of 30-bit chunks to allow 30 characters to be represented. Each word is divided into its constituent characters. The appropriate bit is set in the chunk to represent each character, in order of occurrence. The chunks are concatenated to produce a binary vector to represent the spelling of the word and form the input to the CMM. Any unused chunks are set to all zero bits. Each word in the lexicon has a unique binary vector to represent it with a single bit set corresponding to the word's position in the lexicon. The binary vector forms the output from the CMM for that word so we can identify when the word has been retrieved as a match.

Recalling from the Network – Hamming Distance. Only the spelling input vector is applied to the network. The columns are summed, see (1).

$$output_j = \sum_{\text{all } i} input_i \wedge w_{ji} \tag{1}$$

We apply the Willshaw threshold which sets a bit in the output vector for each column summing to the threshold value or greater. We threshold at the highest summed column value to retrieve the best matches, i.e., words that match as many characters in the input as possible. The single binary vector output after thresholding is a superimposition of the best matching words' vectors.

Recalling from the Network – Shifting N-Grams. We use exactly the same CMM for the n-gram method as we use for the Hamming Distance retrievals. However, we use 1-grams for spellings with less than 4 characters, 2-grams for 4 to 6 characters and 3-grams for spellings with more than 6 characters. Misspelt words with less than four characters are unlikely to have any 2-grams or 3-grams found in the correct spelling. Spellings with 4 to 6 characters may have no common 3-grams but should have common 2-grams and words with more than 6 characters should match 3-grams. For a 3-gram, we take the first three characters of the spelling, input these left aligned to the spelling CMM and threshold the output activation at the value three. We then slide the 3-gram one place to the right, input to the CMM and threshold at three. We continue sliding the 3-gram to the right until the first letter of the 3-gram is in the position of the last character of the spelling. We logically OR the output vector from each 3-gram position to produce an output vector denoting any word that has matched any of the 3-gram positions. We then move onto the second 3-gram, left align, input to the CMM, threshold and slide to the right producing a second 3-gram vector. When we have matched all n 3-grams from the spelling, we will have n output vectors representing the words that have matched each 3-gram respectively. We sum these output vectors to produce an integer vector representing a count of the number of 3-grams matched for each word. We then threshold at the maximum value to produce a thresholded binary vector with bits set corresponding to the best matching words.

Phonetic Spell Checking. Our methodology combines Soundex-type codes with the phonetic transformation rules listed in Table 1 to produce a four-character code for each word.

Table 1. Table of the phonetic transformation rules in our system. Italicised letters are Soundex codes - all other letters are standard alphabetical letters. ^ indicates 'the beginning of a word', $ indicates 'the end of the word' and + indicates '1 or more letters'. Rule 1 is applied before rule 2.

<table>
<tr><td>^hough → h5
^cough → k3
^chough → s3
^laugh → l3</td><td>^ps → s
^pt → t
^pn → n
^mn → n</td><td>1. c(e|i|y|h) → s 2. c → k
gn$ → n
gns$ → ns
1. (i|u)gh(¬a) → _ 2. gh → g
mb$ → m</td></tr>
<tr><td>^rough → r3
^tough → t3
^enough → e83
^trough → tA3</td><td>^wr → r
^kn → n
^gn → n
^x → z</td><td>ph → f
q → k
sc(e|i|y) → s
+ti(a|o) → s
+x → ks</td></tr>
</table>

Table 2. Table giving our codes for each letter.

`01-2034004567809-ABC0D0-0B0000`
`abcdefghijklmnopqrstuvwxyz-'&/`

Any applicable transformation rules are first applied to the word. The phonetic code for the word is then generated according to the algorithm in Fig. 3 using the codes given in Table 2.

For the phonetic vectors, we divide the vector into an initial alphabetical character representation (23 characters as c, q and x are mapped to other letters by transformation rules) and three 13-bit chunks. Each of the three 13-bit chunks

```
First letter of code is set to first letter of word
For all remaining word letters {
   if letter maps to 0 then skip
   if letter maps to same code as previous letter then skip
   Set next code character to value of letter mapping in table 2
}
If the code has less than 4 characters then pad with 0s
Truncate the code at 4 characters
```

Fig. 3. Figure listing our code generation algorithm in pseudocode. Skip jumps to the next loop iteration.

represents a phonetic code from table 2 where the position of the bit set is the hexadecimal value of the code. Each word's output vector is identical to the Hamming Distance and n-gram CMM output vector for uniformity

Recalling from the Network – Phonetic. Recall from the phonetic CMM is essentially similar to the Hamming Distance recall. We input the 4-character

Table 3. The table indicates the recall accuracy of the methodologies evaluated. The present column indicates the number of misspellings present in the dictionary of each method, e.g, 'imbed' included in the dictionary along with 'embed'.

Method	Found (Top 10)	Found (Position 1)	Present	Not Found	% Recall (Top 10)
Hybrid	558	368	6	36	93.9
Aspell	558	429	6	36	93.9
Word 2k	510	432	17	73	87.5
Word 97	504	415	15	81	86.1
Edit	510	367	6	84	85.6
Agrep	481	303	6	113	80.1

phonetic code for the search word into the CMM and recall a vector representing the superimposed outputs of the matching words. The Willshaw threshold is set to the maximum output activation to retrieve all words that phonetically best match the input word.

Integrating the Modules. We produce three separate scores for the Hamming Distance, shifting n-gram and phonetic modules. We separate the Hamming Distance and n-gram scores so the system can utilise the best match and overcome the four forms of typing-error. We add the Soundex score to each producing two word scores. The overall word score is the maximum of these two values. We normalise all scores to ensure that none of the three modules biases the overall score. The score for the word is then given by (2).

$$\begin{aligned}\text{Score} = max(&(\text{ScoreHamming} + \text{ScorePhonetic}),\\ &(\text{ScoreN-gram} + \text{ScorePhonetic}))\end{aligned} \tag{2}$$

2 Evaluation – Quality of Retrieval

We extracted 583 spelling errors from the Aspell [1] and Damerau [3] word lists and combined 17 spelling errors we extracted from the MS word 2000 auto-correct list to give 600 spelling errors. We counted the number of times each algorithm suggested the exact correct spelling among the top 10 matches and also the number of times the exact correct spelling was placed first. We include the score for MS Word 97, MS Word 2000 and Aspell spell-checkers for a benchmark comparison. We used the standard supplied dictionaries for both MS Word and Aspell.

For all other evaluated methodologies we used the standard UNIX dictionary augmented with the correct spelling for each of our misspellings giving 29,178 words in total. We note that the dictionary and spelling errors contain many inflections (singular/plural nouns, verb tenses etc) and we only counted the exact match. The spelling errors range from 1 to 5 error combinations.

3 Analysis

If we consider the final column of table 3, we can see that our hybrid implementation has the joint highest recall with Aspell. Our stated aim in the Introduction was high recall accuracy. Both Aspell and MS Word 2000 have more first place matches than our method. However, MS Word is optimised for first place retrieval and Aspell relies on a much larger rule base than we use. We have minimised our rule base for speed and yet achieved equivalent overall recall. The user will see the correct spelling in the top 10 matches an equal number of times for both Aspell and our method. We note that both Aspell and MS Word were using their standard dictionaries but we assume a valid comparison.

4 Conclusion

Spell checkers are somewhat dependent on the words in the lexicon. Some words have very few words spelt similarly, so even multiple mistakes will retrieve the correct word. Other words will have many similarly spelled words so one error may make correction difficult or impossible. Of the techniques evaluated, our hybrid approach had the joint highest recall rate at 93.9%. Humans averaged 74% for isolated word-spelling correction [4]. Kukich [4] posits that ideal isolated word-error correctors should exceed 90% when multiple matches are returned.

References

1. Aspell. Web page http://aspell.sourceforge.net/.
2. J. Austin. Distributed associative memories for high speed symbolic reasoning. In R. Sun and F. Alexandre, editors, *IJCAI '95 Working Notes of Workshop on Connectionist-Symbolic Integration: From Unified to Hybrid Approaches*, pages 87–93, Montreal, Quebec, Aug. 1995.
3. F. Damerau. A technique for Computer Detection and Correction of Spelling Errors. *Communications of the ACM*, 7(3):171–176, 1964.
4. K. Kukich. Techniques for Automatically Correcting Words in Text. *ACM Computing Surveys*, 24(4):377–439, 1992.
5. J. R. Ullman. A Binary N-Gram Technique for Automatic Correction of Substitution, Deletion, Insertion and Reversal Errors in Words. *Computer Journal*, 20(2):141–147, May 1977.
6. S. Wu and U. Manber. AGREP - A Fast Approximate Pattern Matching Tool. In *Usenix Winter 1992 Technical Conference*, pages 153–162, San Francisco, CA, Jan. 1992.
7. S. Wu and U. Manber. Fast Text Searching With Errors. *Communications of the ACM*, 35, Oct. 1992.

Neural Nets for Short Movements in Natural Language Processing

Neill Taylor and John Taylor

Department of Mathematics, King's College,Strand, London WC2R2LS, UK
ntaylor@mth.kcl.ac.uk; john.g.taylor@kcl.ac.uk

Abstract. A neural model is constructed able to generate short sentences and their moved (question) forms, based on modern linguistics (X-bar theory) and a cartoon version of frontal lobes. Future extensions are mentioned in conclusion.

1 Introduction

Natural language understanding has been much developed after the introduction by Chomsky of universal grammar forty years ago. As language is being processed the dynamics of syntactic and semantic analysis needed to make sense is complex, composed of primed analysers initially activated but then later removed by reduction of ambiguity. Attempts have been made to tackle the problem in modern linguistics of how such dynamics evolves [1, 2, 3, 4]. There is also data from brain imaging that begins to determine the sites and time course of language processing [5], indicating that syntactic processing commences in frontal cortex in the first 200 msecs for a given input word, continues for the next 200 msecs by further semantic processing posteriorly, then involves parietal/frontal interaction for ambiguity resolution.

We have begun to develop [6] neural network systems able to replicate the main features observed, based on models of sequence learning and generation capabilities of frontal lobes [7, 8]. We turn here to the task of generating (or understanding) a question, which we take of simple form 'Can I have...?', which arises from the deep structure 'I can have...' by the movement of the auxiliary verb 'can' to the front of the question sentence. This movement leads to determining how to develop a suitable trace 't' as in the linguist's form of the sentence as 'Can$_t$ I t have...?' The trace 't' denotes the relic of the moved auxiliary, preserving information of the deep structure for understanding. We must develop a suitable syntactic and semantic system able to generate both the deep as well as the surface (moved) question forms. We use the recurrent ACTION neural network [7, 8] as a syntactic generator. It is composed of a cortical module, with coupled striatal, basal ganglia and thalamic modules. This network is repeated for the various lexical and non-lexical phrase structure analysers needed to represent the various components in the analysis [9]; there are also action representations of verbs and nouns, coupled to the ACTION net representations of the associated actions and non-ACTION net image representations for the objects associated with the nouns [6]. We discuss this briefly in section 2. In the following section we develop the phrase structure analysers needed for movement analysis. Traces on weights are also defined so as to hold targeted activity. The overall systems

G. Dorffner, H. Bischof, and K. Hornik (Eds.): ICANN 2001, LNCS 2130, pp. 1205–1210, 2001.

for generating the deep and surface forms of the movement are then more carefully analysed in section 4; dynamical patterns of neural activity are demonstrated. In this way a possible neural system able to implement some aspects of modern linguistic structures are presented. The paper concludes with future extensions to handle longer movements.

2 The ACTION Network Approach to Syntax and Semantics

We consider first the nature of the frontal lobes themselves. There are no comparable subcortical structures accompanying the posterior cortex to those for frontal regions. The basic feature of the sub-cortical components of these frontal areas are the basal ganglia, which is a two layer system activated unidirectionally from the cortex, the upper layer being the striatum (from its appearance), being decomposable into caudate and putamen. Beneath these, and close to the relevant part of thalamus (the mediodorsal nucleus) is the globus pallidus, comprising two components, termed internal and external. The unidirectional flow of information is: cortex $\rightarrow$ striatum (STR)$\rightarrow$ globus pallidus (internal) (GPi) $\rightarrow$ thalamus (TH) $\rightarrow$ cortex so there is a feedback of activity round the above 'long' loop. There may also be similar feedback more rapidly round the 'short' cortex $\leftrightarrow$ TH loop, and although there is known topography in these connections the presence of any reverberation of such feedback activity is unknown. There is also the presence of the indirect loop, involving the globus pallidus external (GPe) as well as the sub-thalamic nucleus (STN), as shown in figure 1, a schematised version of the frontal lobe architecture we have termed elsewhere the ACTION network, since it is so important for the generation of actions.

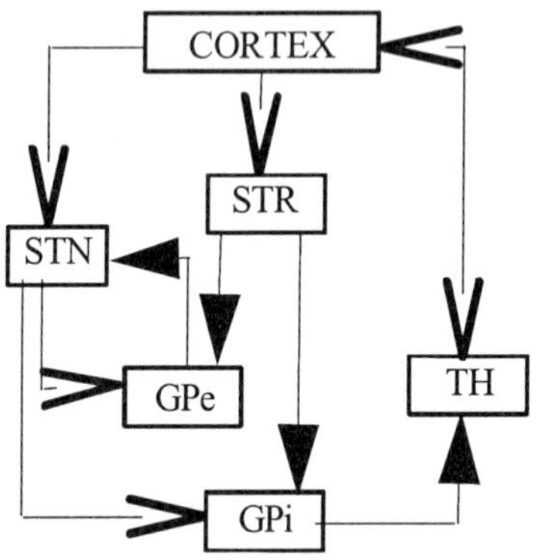

Fig. 1. The ACTION network. Excitatory connections indicated by open arrowheads and inhibitories by closed.

The ACTION network has been used to model temporal sequence storage and generation [7, 8]. In these papers it has been shown how, either in suitably hard-wired architectures or in trained ones, there appear a set of neurones which achieve chunking of the temporal sequences, and which agree with the activity patterns of neurones observed by researchers in single cell recordings in monkeys. These neurones have continued activity during delay periods. We classified them as various sorts of initiator, transition and complex memory neurones. These types of neurones are sequence-specific; in addition pre-movement and movement neurones are recorded. It is these chunking nodes that we use as the crucial components of syntactic phrase structure analysers below, modified to allow for infinite generativity.

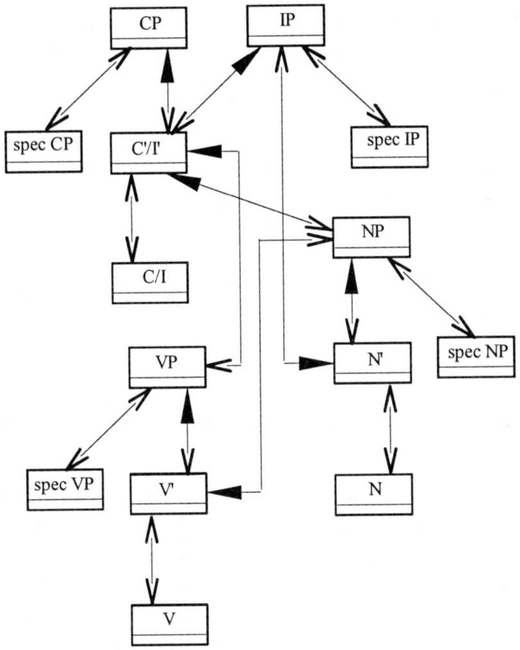

Fig. 2. The wiring diagram of the phrase analysers.

3 The Phrase Structure Analysers for Short Movement

The phrase structure analysers used here are based on X' theory [9]. Only four phrase analysers are used: verb phrase analyser (VP), noun phrase analyser (NP), infinitival phrase analyser (IP) and complementiser phrase analyser (CP); for our sample sentences adjectives and prepositions are not required so their phrase analysers are not included (but could easily be so [6]). We limit the phrase analysers to one level of expansion, so an XP projects to an X' which then has to project to an X, rather than being able to project to another level of X'. All XPs and X's are composed of initiators and transitions, whilst all specifier (spec) XP and X regions are stores of words; all modules are ACTION networks. Specifiers can be determiners for NPs, words such as 'when', 'who', 'how' for spec CP. The general wiring diagram is indicated in figure 2. More specifically (where all excitatory weights are cortico-cortical) the XP initiator is connected via excitatory weights to all words in spec XP, all of which in turn are excitatory back up to the XP transition neurone. All transition neurons act in an inhibitory manner via lateral striatal connections on their associated initiator circuit. The XP transition excites the X' initiator neurone which has an inhibitory connection back. The X' initiator excites all words in the X module, these all having excitatory connections back up to the X' transition circuit; as in the case of XP the transition neurone inhibits the initiator. The X' transition neurone can then excite the initiator of some other phrase analyser YP. The YP initiator inhibits the X' transition circuit. The YP wiring is then expanded in exactly the same manner as XP.

This repeated formation for all phrase analysers only allows the initiator of any XP or X' module to excite a word module, while the transition either projects to an X' or YP module. There is mutual inhibition between words within a module and between word modules. Words can only be generated when there is priming. These modules also have output neurones connected to the cortical nodes, so that when the output nodes exceed a threshold then that word is produced, at which point there is inhibition lasting for 80 time-steps. Not only are the word representations inhibited but also object representations for nouns, action representations for verbs, etc..

In order to preserve knowledge of input activation weight traces are introduced:

$$w(t+1) = \lambda\, w(t) + (1-\lambda)\, \text{Input} \qquad (1)$$

as is usual, this acts as a weight modulation. These give important contributions to the dynamics.

4 The Dynamics of Generation of the Moved Sentence

The short movement we are concerned with is that of an auxilliary verb, which occurs when a statement is changed into a question. This occurs for instance in the change from the statement: "I can have a Z", to the question: "Can I have a Z?".

The rule for this type of movement in X' theory is that 'can' which is an infinitival word is excited in its new position via CP, but the structures that excite it in the statement sentence, IP, are still active in the moved phrase, but do not lead to a second activation of 'can'. This could be caused either by lasting inhibition or by lack of priming. The activation of the IP as the theory states, even though the auxilliary has previously been generated, may explain why young children when learning to form questions by auxilliary movement may say sentences such as:

"Can I can have...?".

The flow of neurones in the model to generate the original sentence is:

NP Initiator (IN) -> spec NP (determiners none needed for 'I') -> NP transition (TR) -> N' IN -> N (generates 'I') -> N' TR->IP IN -> spec IP (none needed) -> IP TR -> I' IN -> I (auxilliary 'can' generated) ->I' TR -> VP IN -> spec VP (none needed) -> VP TR -> V' IN -> V ('have' generated) -> V' TR -> NP IN -> spec NP (determiner 'a' generated) -> NP TR -> N' IN -> N (noun 'Z' generated) -> N' TR.

Once the noun 'Z' is generated the sentence is completed leading to global inhibition, from an assumed higher order 'theme' module.

For the moved sentence the flow is:

CP IN -> spec CP (none needed) -> CP TR -> I' IN -> I (auxilliary 'can' generated) -> I' TR -> NP IN -> spec NP (determiners, none needed) -> NP TR -> N' IN -> N ('I' generated) -> N' TR ->IP IN -> spec IP (not needed) -> IP TR -> I' IN -> I (none generated, as 'can' already produced)-> I' TR -> VP TR -> V' IN -> V ('have' generated) -> V' TR -> NP IN -> spec NP (determiner 'a' generated) -> NP TR -> N' IN -> N (noun 'Z' generated) -> N' TR.

Again when 'Z' is generated there is global inhibition of the model. Thus the temporal dynamics predicted by the model would be for a longer generation of the moved, to the original, sentence.

Figure 3 shows the activation patterns of some of the model's cortical neurones involved in the generation of the original sentence. These show the correct temporal order of activation of the words 'I', 'can' and 'have', as well as a single strong

Using Document Features to Optimize Web Cache

Timo Koskela, Jukka Heikkonen, and Kimmo Kaski

Helsinki University of Technology,
Laboratory of Computational Engineering,
P.O.Box 9400, FIN-02015 HUT, Finland
timo.koskela,jukka.heikkonen,kimmo.kaski@hut.fi
http://www.lce.hut.fi/

Abstract. In this paper Web cache optimization using document features is proposed. The problem in Web cache optimization is to decide which strategy to use in replacement of cache objects. While commonly used policies use heuristic rules, proposed model predicts the value of each Web object by using features collected from the HTTP responses and from the HTML structure of the document. In a case study, generalized linear model and multilayer perceptron committee model are used to classify about 50000 Web documents according to their popularity. Results show that linear model does not find any correlation between the features and document popularity. MLP model gives better results, yielding mean classification percentages of 64 and 74 for the documents to be left or to be removed from the Web cache, respectively.

1 Introduction

Web caching is a technique which reduces WWW network traffic and improves response time for the end users. Modern CPU's, operating systems and distributed filesystems have caches for data accesses between clients and servers. In all of these systems, caches work well because it is very likely that some piece of data will be repeatedly accessed. Access delays are minimized by keeping popular data close to the entity which needs it.

The same holds true for the Web as well. Especially proxy caches have some very attractive advantages when implemented correctly. Proxy caches can be simultaneously accessed and shared by many users, and are often located near network gateways to reduce the bandwidth required over expensive dedicated Internet connections. As a disadvantage, caching can lead to increased latency or to providing outdated documents to users. Also the rate of change in today's Web is very fast, which makes caching less attractive [1].

2 Web Cache Optimization

Objectives of Web caching include reducing network traffic, alleviating server bottleneck, minimizing user access latency, and maintaining transparency to the

G. Dorffner, H. Bischof, and K. Hornik (Eds.): ICANN 2001, LNCS 2130, pp. 1211–1216, 2001.

end user. Different metrics can be used to estimate how the cache performs. Common performance metrics include request hit rate, relative request hit rate, byte hit rate, relative byte hit rate and latency ratio. Let m to be the total number of requests (documents), and let $\delta_i = 1$ if request i is in the cache and $\delta_i = 0$ otherwise. Request hit rate can then be defined as the percentage of documents that are in the cache [2],

$$W^r = \frac{\sum_{i=1}^{M} \delta_i}{m} \tag{1}$$

The relative request hit rate is defined as W^r/W^r_{max}, where W^r_{max} is the request hit rate obtained for a cache with an infinite size. Similarly, the byte hit rate is the percentage of bytes transferred from the cache. Latency ratio can be defined as the ratio of the sum of download time of missing documents over the sum of all downloading latencies.

Optimizing the Web cache involves choosing the appropriate performance metrics to be maximized according to the objectives set. E.g. maximizing the request byte hit rate will usually minimize outbound network traffic, while minimizing the latency ratio will optimize the end user experience. Key issue in maximizing selected performance metrics is using a suitable caching policy.

2.1 Caching Policies

Caching policy is the strategy which is used in replacement of objects in the cache. When a new document must be brought into the cache which is full (or heavily loaded), which document(s) should be removed from the cache? The difficulty is that of finding the value of a document, and whether it should be cached, without dramatically increasing computation and communication overheads.

Most Web caches use traditional memory paging policies like the Least Recently Used (LRU). LRU is simple to implement and efficient in the case of uniform objects like in memory cache, but since it does not consider size or download latency of documents, it does not perform well in Web caching [2]. Several other heuristic policies have been proposed and proved to be optimal using simulation with actual cache access logs [3]. However, most of the policies are suboptimal since they only optimize a particular performance metric. Some policies also include statistics from request rates and server response times, which allows adapting to changing conditions in network or server resources [4].

2.2 Estimating the Value of Cache Objects

More advanced caching policies frequently use statistical methods to estimate the value of each Web object. If the value of each object can be predicted, cache performance can be optimized with respect to any given performance metric or cost criteria. LRV [5] adaptively estimates the lowest relative value for each object by using features collected from the cache server logs. The object, which has the lowest relative value, is replaced first from the cache. In [6] logistic regression models are used in a similar approach, where the value of the document is predicted by using a model. The inputs of the model include document size and type, number of requests and time since last request.

2.3 Using Document Features

In addition to the cache server logs, features can also be extracted from the structure of the HTML document. These features can then be used in predicting the value of the document. This approach was first presented in [7]. The value of the document can be determined by its popularity, e.g. number of requests to the cache. The modelling task can be simplified by transforming it to a classification problem. Each document is assigned either a high or a low value according to its popularity during a chosen time period. A classifier is then used to predict the value of each document. The inputs of the classifier are the features collected from each document, and its output gives the class of the document. If the classification results can be generalized to documents which are not included in the training data, the classifier can be used in Web cache optimization.

To study the feasibility of this approach, a case study was conducted which is described in the next section. The models used include linear and nonlinear classifiers. Generalized linear model (see e.g. [8]) was used as the linear classifier [1]. As a nonlinear classifier multilayer perceptron network was used.

2.4 Multilayer Perceptron (MLP) Network

MLPs are nonlinear models consisting of number of units organized to multiple layers. The complexity of the MLP network can be changed from a simple parametric model to a highly nonlinear model by varying the number of layers and the number of units in each layer. For instance, in a typical one hidden layer MLP network with K outputs the model $f_k(x, w)$, $k = 1, \ldots, K$, with weights w might be

$$f_k(x, w) = w_{k0} + \sum_{j=1}^{m} w_{kj} \tanh \left(w_{j0} + \sum_{i=1}^{d} w_{ji} x_i \right) , \tag{2}$$

where indices j and i correspond to output and hidden units, respectively.

A practical problem with neural networks is selecting the correct complexity of the model, i.e., the correct number of hidden units or correct regularization parameters. Plain maximum likelihood optimization of MLP may lead to severe overfitting. Usually regularization is needed in order to provide a good generalization ability for samples not seen during the training. Traditionally complexity of MLP has been controlled with early stopping or weight decay [8].

MLP is often considered as a generic semiparametric model meaning that the effective number of parameters may be less than the number of available parameters. Effective number of parameters determines the complexity of the

[1] Netlab library implementation of the generalized linear models for Matlab by C.M. Bishop et al. was used.

model. For small weights the network mapping is almost linear and has low effective complexity, since the central region of sigmoidal activation function can be approximated by linear transformation.

In early stopping weights are initialized to very small values. Part of the training data is used to train the MLP and the other part is used to monitor the validation error. Iterative optimization algorithms used for minimizing the training error gradually take parameters in use. Training is stopped when the validation error begins to increase. Since training is stopped before a minimum of the training error, the effective number of parameters remains less than the number of available parameters.

Early stopping in its basic form is rather inefficient, as it is very sensitive to the initial conditions of the network and only part of the available data is used to train the model. These limitations can easily be alleviated by using a committee of early stopping networks, with different partitioning of the data to training and stopping sets for each network. The output of the committee model is calculated as the mean of the all committee member's outputs. When used with caution MLP early stopping committee is good baseline method for neural networks.

A committee of 10 MLP networks was used which were trained with Rprop algorithm [9]. For each MLP network a unique portion of the data was used for stopping the training. Each MLP had 20 hidden units and one logistic output unit.

3 Case Study

To test the proposed approach and the models discussed, a case study was carried out. An access log containing one day requests was obtained from FUNET (Finnish University and Research Network) proxy cache. A Web robot was used to retrieve each HTML document in the log and a HTML parser to parse its structure and store the features in a data file. The data was cleaned from empty HTML documents and from features that had the same value throughout the data (usually either missing or zero). Final data consisted of 48765 HTML documents with 56 features from each document. Table 1. shows the features used in more detail.

HTML documents with two or more requests in the access log were labeled to belong to the class 1 (high value), and all the others to the class 0 (low value). Separate portions of the data were used to estimate the parameters of the model, and to test the ability of the model to classify documents correctly. 5-fold cross-validation was used, where 4/5 of the data were used for training, and the remaining 1/5 for testing.

Table 2. shows the confusion matrices of each cross-validation training case for both linear and MLP models. The upper left corner value represents the classification percentage of the class 0 and lower right corner value represents the classification percentage for the class 1. In the case of the linear model the matrices are almost identical, supposedly because the model cannot learn

Table 1. Features 01-49 contain the frequency of certain HTML tag (shown in the table) existing in the document. Features 50-56 include information extracted mostly from the Web cache access log. Features 51 Last modified and 52 Expires were assigned to 1 if the feature existed in the document headers and 0 otherwise. Feature 55 Path depth states how many "/" characters the document URL contains.

01	02	03	04	05	06	07
<a>	<address>	<area>	<b>	<base>	<big>	<blink>
08	09	10	11	12	13	14
<blockquote>	<body>	 	<center>	<div>	<dl>	<dt>
15	16	17	18	19	20	21
<dd>	<em>	<font>	<frame>	<frameset>	<h1>	<h2>
22	23	24	25	26	27	28
<h3>	<h4>	<h5>	<h6>	<head>	<hr>	<html>
29	30	31	32	33	34	35
<i>	<iframe>	<img>	<li>	<link>	<map>	<meta>
36	37	38	39	40	41	42
<p>	<param>	<pre>	<script>	<select>	<small>	<strong>
43	44	45	46	47	48	49
<td>	<textarea>	<th>	<title>	<tr>	<u>	<ul>
50	51	52	53	54	55	56
Doc length	Last modified	Expires	Length of content	Content type	Path depth	Domain

Table 2. Confusion matrices for each cross-validation training case.

	1	2	3	4	5
Linear	0.67 0.33	0.67 0.33	0.67 0.33	0.66 0.34	0.66 0.34
model	0.44 0.55	0.44 0.55	0.44 0.56	0.45 0.56	0.43 0.57
MLP	0.74 0.26	0.75 0.25	0.75 0.25	0.76 0.24	0.74 0.26
model	0.36 0.64	0.35 0.65	0.37 0.63	0.38 0.62	0.36 0.64

any useful mapping from the training data. For the MLP model, there are some differences. However, differences are quite small regarding the amount of training samples used.

Table 3 shows the classification results. Linear model classifies 67 percent of class 0 and 56 percent of class 1 correctly. Respectively, MLP model classifies 74 percent of class 0 and 64 percent of class 1 correctly. In this case, nonlinear model can distinguish documents that have a high value from those that a have low value much more accurately than linear a model.

4 Conclusions

A novel approach to Web cache optimization was presented. The results of a case study suggest that nonlinear classifier can find some regularities from the collected data, which can be used to predict the value (class) of the document. This value can be used to optimize Web cache. Suggested method also effectively replaces heuristic algorithms with a statistical model which is based on real measurement data. However, further experiments should be carried out to fully evaluate the usefulness of the method. Building the model requires a large

Table 3. Classification percentages of classifiers in the test data. Class 0 refers to a low value of the document (do not cache) and class 1 refers to a high value of the document (store in cache).

Method	Class 0	Class 1
Linear model	0.67	0.56
MLP model	0.74	0.64

amount of data to be collected. Also finding a set of suitable features, the model architecture and estimating the model parameters can be a computationally intensive task.

To be applicable to implementation in Web caches, the prediction of the document's value should be carried out within the caching policy algorithm. Initial parameters of the model can be estimated beforehand, which greatly reduces the computational burden. Since the statistics of requested documents change continuously, model parameters should also be tuned frequently as a part of the caching algorithm.

Acknowledgments

This work is supported by the Academy of Finland, Research Centre for Computational Science and Engineering, project. no. 44897 (Finnish Centre of Excellence Programme 2000-2005).

References

1. Douglis, F., Feldmann, A., Krishnamurthy, B., Mogul, J.: Rate of Change and Other Metrics: a Live Study of the World Wide Web. USENIX Symposium on Internet Technologies and Systems (1997)
2. Williams, S., Abrams, M., Standridge, R.G., Abdulla, G., and Fox, E.A.: Removal Policies in Network Caches for World-Wide Web Documents. ACM SIGCOMM'96 Conference (1996)
3. Davison, B.D.: A Survey of Proxy Cache Evaluation Techniques. 4th International Web Caching Workshop (1999)
4. Cao, P., Irani, S.: Cost-Aware WWW Proxy Caching Algorithms. USENIX Symposium on Internet Technologies and Systems (1997)
5. Rizzo, L., Vicisano, L.. Replacement Policies for a Proxy Cache. IEEE/ACM Transactions on Networking, **8** (2) (2000) 158–170
6. Foong, A.P., Hu, Y-H., Heisey, D.M.: Logistic Regression in an Adaptive Web Cache. IEEE Internet Computing **3**(5) (1999) 27–36
7. Koskela, T., Heikkonen, J., Kaski, K.: Modeling the Cacheability of HTML Documents. 9th International World Wide Web Conference (2000)
8. Bishop, C.M.: Neural Networks for Pattern Recognition. Oxford University Press (1995)
9. Riedmiller, M., Braun, H.: A Direct Adaptive Method for Faster Backpropagation Learning: The RPROP Algorithm. IEEE International Conference on Neural Networks (1993) 586–591

Generation of Diversiform Characters Using a Computational Handwriting Model and a Genetic Algorithm

Yasuhiro Wada[1], Kei Ohkawa[1], and Keiichi Sumita[1]

Nagaoka University of Technology, Nagaoka, 940-2188, Japan
ywada@nagaokaut.ac.jp

Abstract. In pattern recognition, a diversification of characters is necessary for learning, which is like neural network learning. The artificial diversification of characters has been suggested as one means of collecting a variety of characters. Accordingly, we show that a computational handwriting model can be applied to the diversification of characters. It is thought that characters diversified by the model can be used as a database of character images for learning. Wada & Kawato's handwriting model [11] is based on an optimal principle and the feature space of the characters includes sets of via-points extracted from actual handwritten characters. Therefore, if the via-point information is changed, diversiform characters can be generated using the handwriting model. In this paper, we propose a method for generating a diversification of characters by changing via-point information based on a genetic algorithm.

Keyword: data analysis and pattern recognition, diversiform characters, computational handwriting model, genetic algorithm, via-point

1 Introduction

Several handwriting models, which can reproduce human handwritten trajectories, have been proposed in a computational model of the human brain [4] [6] [1] [11] [8]. The handwriting model proposed by Wada & Kawato [11] involves Marr's three levels in the computational theory [5] shown in Fig. 1. The computational level uses the minimum commanded torque change criterion [7][9], the representation level uses sets of via-points to generate characters, and the hardware and algorithm level uses a neural network model called the Forward-Inverse Relaxation Model (FIRM) [10]. FIRM produces a trajectory based on the minimum commanded torque change criterion, which passes through several via-points for a character. The handwriting model can reproduce human handwritten trajectories well.

On the other hand, in handwritten character recognition research, many diversiform handwritten patterns are required to make a highly accurate recognition system, and the character deformation model is important.

Incidentally, we confirm that Wada & Kawato's handwriting model [11] can change the shapes of character trajectories through a perturbation of the positions and times of via-points. In this paper, we propose a method for generating

G. Dorffner, H. Bischof, and K. Hornik (Eds.): ICANN 2001, LNCS 2130, pp. 1217–1224, 2001.

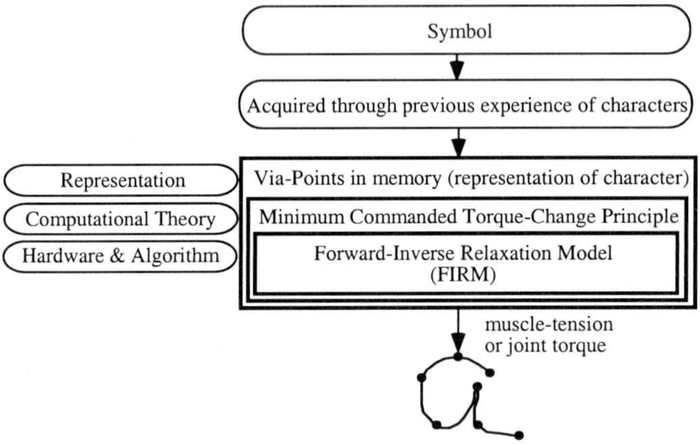

Fig. 1. A handwriting model

a diversification of characters by perturbing via-point information based on a genetic algorithm. We show experimental results obtained for the generation of diversiform characters. Then, we point out that the diversification of the generated characters is more than the diversification of a Japanese character database ETL and also that the generated characters can be used as the training data set for a pattern recognition system.

2 A Computational Handwriting Model Based on an Optimization Principle

Fig. 1 shows the proposed handwriting model, which involves a hierarchical handwriting planning structure. The computational theory of our model is based on a dynamic optimization principle; i.e., the minimum commanded torque change model, shown by the following equation.

$$C_{CTC} = \frac{1}{2}\int_0^{t_f} \sum_{k=1}^{K} \left(\frac{d\tau^k}{dt}\right)^2 dt \rightarrow \min \tag{1}$$

τ^k shows the torque of joint k, K shows the number of joints, and t_f shows the motion duration.

The representation is a set of via-points that expresses a character. In planning a cursive handwritten character, many via-point positions (x coordinates, y coordinates, and z coordinates) function as boundary conditions in the optimization problem. The elapsed time of the via-points is specified in the model. We earlier proposed the algorithm for extracting the via-points of handwritten characters [11]. The via-point extraction problem is formulated as finding the minimum number of via-points able to reproduce the given trajectory within a given bound error. The hardware and algorithm are based on FIRM, which is a neural network model for optimal trajectory formation [10]. FIRM can generate a trajectory with many via-points within quite a small number of iterations such that an approximate optimal trajectory based on the minimum commanded torque change is obtained.

3 A Method for Generating a Diversification of Characters Using a Genetic Algorithm

First of all, we propose a performance index for evaluating character patterns. This is because, to use a genetic algorithm, we need to measure the amount of diversification in the generated characters, that is, we need to define a fitness function in the genetic algorithm. Then, we propose a method for generating a variety of characters using the genetic algorithm and our proposed handwriting model.

3.1 A Fitness Function for Character Shapes

We have not yet proposed a hand trajectory formation model capable of generating a trajectory based on the minimum commanded torque change criterion in three-dimensional space. In this paper, therefore, we use a model based on the minimum jerk criterion [2] for the sake of simplicity, which is an approximation of the minimum commanded torque change criterion. An object function in the handwriting model can be expressed as Eq. (2). The object function corresponds to a performance index that generates a trajectory passing through a set of via-points according to the minimum jerk criterion.

$$\int_0^T \left\{ \left(\frac{d^3x}{dt^3}\right)^2 + \left(\frac{d^3y}{dt^3}\right)^2 + \left(\frac{d^3z}{dt^3}\right)^2 \right\} dt$$
$$+\lambda\Sigma\{(x(t_{via}) - x^*_{via})^2 + (y(t_{via}) - y^*_{via})^2 + (z(t_{via}) - z^*_{via})^2\} \qquad (2)$$

Here x, y, and z indicate the trajectory coordinates of the pattern to be evaluated. x^*_{via}, y^*_{via}, and z^*_{via} correspond to the coordinates of the via-point of a reference pattern. t_{via} indicates the via-point time.

The first term in Eq. (2) is equivalent to the minimum jerk criterion. The second term shows the boundary conditions for via-points to be satisfied in the character formation problem. Conversely, we think that the amount of diversification from the reference character can be evaluated by computing Eq. (2) for the given character, whose representation is expressed as via-point $x(t_{via})$, $y(t_{via})$, and $z(t_{via})$.

Here, let consider the difference between the index values (Eq. (2)) for the reference pattern and the pattern to be evaluated. The second term of Eq. (2) for the reference pattern becomes zero. In this paper, we use Eq. (3) as an exponential evaluation function.

$$W = exp\left\{ -\frac{\{C_J(x,y,z) - C_J(x^*,y^*,z^*)\}^2}{\alpha^2} \right\} \cdot$$
$$\exp\left\{ -\sum_{via=1}^{N_{via}} \frac{(x(t_{via}) - x^*_{via})^2 + (y(t_{via}) - y^*_{via})^2 + (z(t_{via}) - z^*_{via})^2}{\beta^2} \right\} \qquad (3)$$

$C_J(x,y,z)$ denotes the first term in Eq. (2). (x^*,y^*,z^*) and (x,y,z) show the coordinates of the reference pattern and the coordinates of the pattern to be evaluated, respectively. via is the index for the via-points and $via = 1$ corresponds to the starting point. T is the motion duration normalized as $T = 1$.

However, the number and qualitative positions of via-points estimated by Wada & Kawato's algorithm [11] differ according to the trajectory. Therefore, to calculate the second term of the right-hand side in Eq. (3), we need to correspond the reference pattern's via-points and the via-points of the pattern to be evaluated. The following is the via-point correspondence method we adopt. The correspondence values between the via-points via of the reference pattern and the via-points via' of the pattern to be evaluated are computed using Eq. (4). We estimate the via-point correspondence by selecting the via-point via' with the maximum correspondence value (4).

$$p_{via,via'} = \exp\left\{-\frac{(x_{via'}-x^*_{via})^2}{\sigma_x^2}-\frac{(y_{via'}-y^*_{via})^2}{\sigma_y^2}-\frac{(z_{via'}-z^*_{via})^2}{\sigma_z^2}-\frac{(t_{via'}-t^*_{via})^2}{\sigma_t^2}\right\}$$

$$\hat{p}_{via,via'} \leftarrow \max_{via'}\{p_{via,via'}\} \quad (via = 1,2,\cdots,N; via' = 1,2,\cdots,M; N \neq M) \quad (4)$$

Here $x^*_{via}, y^*_{via}, z^*_{via}$ and $x_{via'}, y_{via'}, z_{via'}$ show the coordinates of the reference pattern for the via-point via and the coordinates of the evaluation pattern for the via-point via', respectively.

Basically, Eq. (3) takes the maximum value of '1' when the evaluation pattern is equal to the reference pattern. Eq. (3) becomes '0' when the evaluation pattern is different from the reference pattern.

We examined the validity of the evaluation for various characters using Eq. (3) in a psychological experiment. Twenty-four patterns were used for each character in the experiment performed by 11 subjects. Fig. 2 shows the correlations between the model evaluation and human evaluation for a Japanese kana character. The horizontal axis and the vertical axis indicate the rank according to Eq. (3) and the rank according to subjects, respectively. In the figure, the correlation coefficient is high, that is, approximately 0.80. We have already confirmed the same results for several other Japanese kana characters and English letters.

Accordingly, we propose a method for generating a set of various characters by applying a genetic algorithm with a fitness function that corresponds to Eq. (3). As a result, the evaluation equation gives a relative good quantitation of a human preference for a character shape.

3.2 A Method of Forming Various Characters Based on a Genetic Algorithm

(1) Locus: Information on the locus corresponds to the space coordinates x_{via}, y_{via}, z_{via} and via-point time t_{via}. Here, the space information is normalized by the sum of the distances between via-points and the time information is normalized by the motion duration. Moreover, the origin on the via-point coordinates is located in the center, which is calculated by taking the average of the via-point coordinates. In this paper, the via-point coordinates x, y are given by a relative

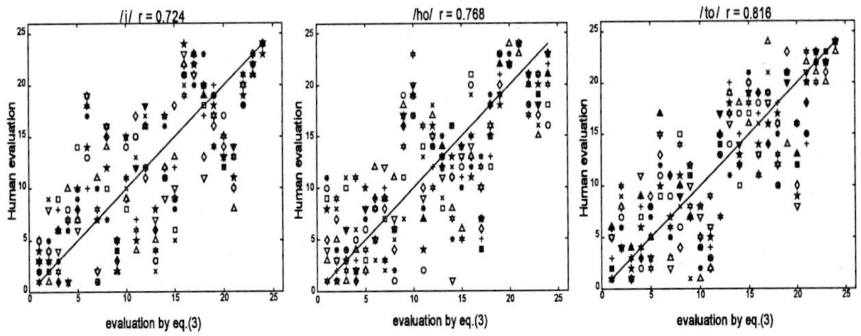

Fig. 2. Results of a psychological experiment

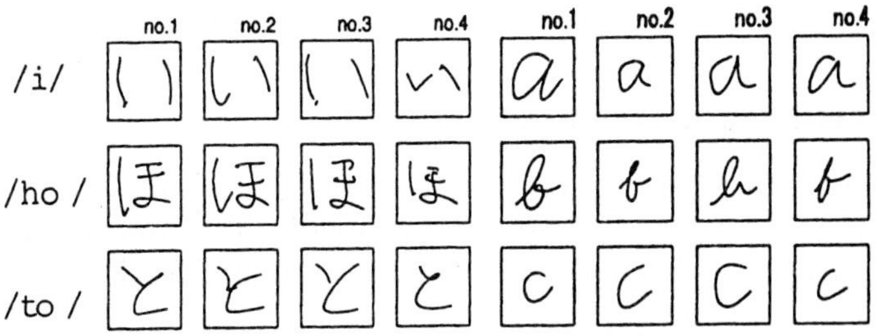

Fig. 3. Patterns in first generation

position coordinate between via-points and z is given by an absolute position coordinate. The time information t_{via} is provided with a relative movement time between via-points.

2) Cross-over: The search for corresponding via-points between two genes is performed using Eqs. (4) since the number and qualitative positions of via-points differ in each gene information. Two new genes are generated by one point cross-over at the position selected by a random number from among all via-points. Here, we do not generate the same gene.

3) Mutation: In mutation, an individual is chosen with probability p_m and via-point information in the chosen individual is chosen with probability p_v. Then, the via-point information is multiplied by coefficient ε.

4) Character generation: New characters are produced by the new gene information generated from the cross-over and mutation using Wada & Kawato's handwriting model.

5) Selection: Selection is performed using the evaluation function (3). We suppose that a pattern with an evaluation value W that is less than a threshold cannot be recognized by humans, so a new character is weeded out when a generated pattern cannot be recognized as a character. The method is based on the genetic algorithm in order to generate as many diversiform characters as possible. It is not our purpose to generate an optimal trajectory based on the criterion. Therefore, we do not apply a selection strategy such as an elite strategy.

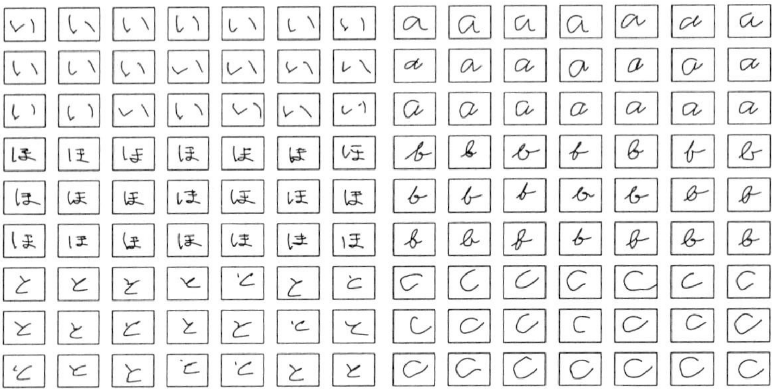

Fig. 4. Experimental results of character generation

4 Experimental Results

The following two points were examined. (1) Is there variety in the character patterns generated by the method compared with the character patterns in an existing handwritten database (Japanese character database ETL) ? (2) Is there availability in composing of a character recognition system that uses handwritten characters generated by the method as training data ?

4.1 Variety

We show several deformation results for the Japanese kana characters /i/, /ho/ and /to/. Four patterns obtained from four subjects were given in the first generation (Fig. 3). In the experiment, 50 cross-overs were performed in each generation and 11 generations were executed. Moreover, a reference pattern used for the selection was chosen among the four patterns in the first generation (the left endpoint in Fig. 3). The parameter values were as follows: the threshold for selection, the mutation probability p_m, and p_v were 5×10^{-4}, $p_m = 0.05$, and $p_v = 0.2$, respectively. Coefficient ε $(0.4 \leq \varepsilon \leq 1.6)$ was given by random numbers.
Fig. 4 shows a part of the generated patterns. Characters with various shapes such as a loop were generated. On the other hand, Fig. 5 shows some selected characters. It is clear that the characters shown in Fig. 5 cannot be recognized by humans.

We also compared the diversification of the generated characters with a Japanese character database ETL. Fig. 6 shows the results of the comparison between the ETL database and the generated trajectories. We measured the diversification level by the variation entropy [3]. We used about 180 patterns for each character from the ETL database, which contains actual human handwritten characters. The same number of generated characters was chosen from a thousand generated characters by random numbers. The variety of the generated characters by the proposed method was found to be greater than the variety

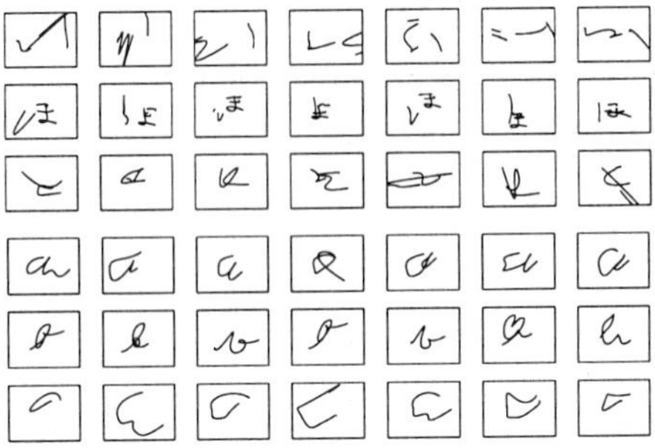

Fig. 5. Examples of selections

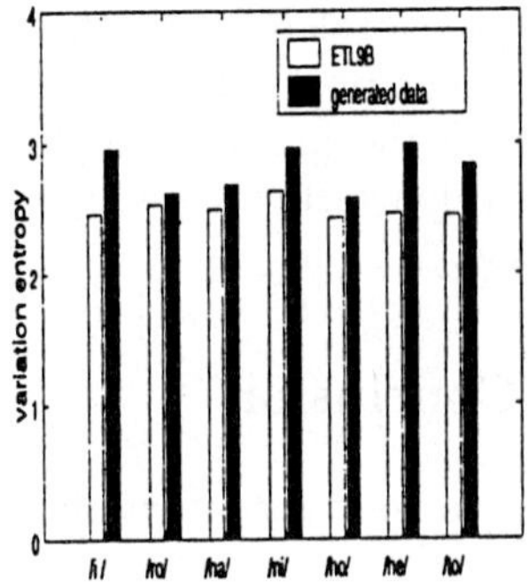

Fig. 6. Comparison between the ETL database and generated trajectories

of the human handwritten characters in the ETL database. We could therefore confirm that the proposed method can produce a variety of character shapes.

4.2 Availability

The effectiveness of the pattern generation by the method was evaluated by comparing the recognition accuracy rates of the following two systems; a neural network learned by using the characters generated by the proposed method and a neural network learned by using the initial characters shown in Fig. 3. The structure of the neural network was a three layered network shown in Fig. 7. The character trajectories obtained from the proposed method were converted into binary images of 64 63 pixels and the image data was inputted to the network. The learning algorithm was the error backpropagation algorithm. The test data for the both networks was selected from the ETL database at random. In learning by using only the initial characters, the recognition rate was not high, that is, the average for five Japanese characters was 72.6%. On the other hand, in learning by using the generated characters, the recognition ability was able to be improved greatly compared with the case only of the initial characters. The

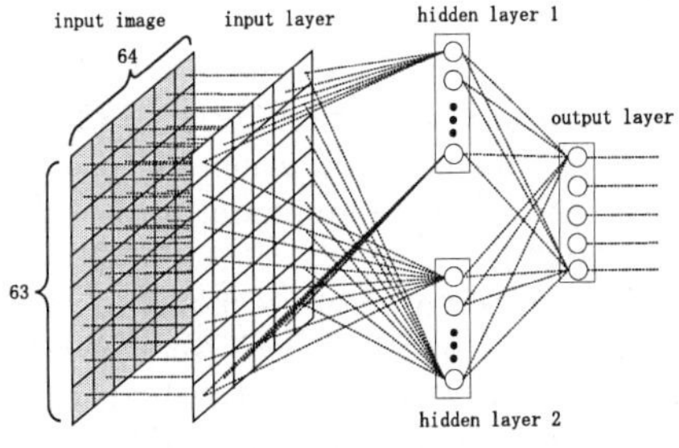

Fig. 7. A neural network for handwritten character recognition

average was 89.6%. In other words, it can be said that the initial characters, insufficient in number to learn, came to be sufficient as data for learning by applying the proposed method. Therefore, the improvement of the character recognition technique and the dictionary, etc., can be reinforced by using the method.

5 Conclusion

We have confirmed that the proposed method can generate characters with a variety of shapes. Moreover, the evaluation function (3) can be applied to an evaluation of the generated characters. It is important that we verify the construction of a robust pattern recognition system using the proposed method.

Acknowledgements

This study was performed through Grant-in -Aid for Scientific Research from Ministry of Education, Culture, Sports, and Science and Technology, Japan and the Kayamori Foundation of Information Science Advancement.

References

1. S. Edelman and T. Flash. A model of handwriting. *Biological Cybernetics*, 57:25–36, (1987).
2. T. Flash and N. Hogan. The coordination of arm movements: An experimentally confirmed mathematical model. *Journal of Neuroscience*, 5:1688–1703, (1985).
3. H. Hase, M. Yoneda, M. Sakai, and J. Yoshida. Evaluation of handprinting variation of characters using variation entropy. *IEICE*, J-71-D(6):1048–1056, (1988). (in Japanese).
4. J. M. Hollerbach. An oscillation theory of handwriting. *Biological Cybernetics*, 39:139–156, (1981).
5. D. Marr. *Vision.* Freeman, W.H. and Company, New York, (1982).
6. P. Morasso and F. A. Mussa Ivaldi. Trajectory formation and handwriting: A computational model. *Biological Cybernetics*, 45:131–142, (1982).
7. E. Nakano, H. Imamizu, R. Osu, Y. Uno, H. Gomi, T. Yoshioka, and M. Kawato. Quantitative examinations of internal representations for arm trajectory planning: minimum commanded torque change model. *Journal of Neurophysiology*, 81(5):2140–2155, (1999).
8. R. Plamondon and W. Guerfali. The generation of handwriting with delta-lognormal synergies. *Biological Cybernetics*, 78:119–132, (1998).
9. Y. Uno, M. Kawato, and R. Suzuki. Formation and control of optimal trajectory in human multijoint arm movement – minimum torque change model. *Biological Cybernetics*, 61:89–101, (1989).
10. Y. Wada and M. Kawato. A neural network model for arm trajectory formation using forward inverse dynamics model. *Neural Networks*, 6:919–932, (1993).
11. Y. Wada and M. Kawato. Theory for cursive handwriting based on the minimization principle. *Biological Cybernetics*, 73:3–13, (1995).

Information Maximization and Language Acquisition

Ryotaro Kamimura[1] and Taeko Kamimura[2]

[1] Information Science Laboratory and Future Science and Technology Joint Research Center, Tokai University, 1117 Kitakaname, Hiratsuka, Kanagawa, 259-1292, Japan
ryo@cc.u-tokai.ac.jp

[2] Department of English, Senshu University, 2-1-1 Higashimita, Tama-ku, Kawasaki, Kanagawa, 214-8580, Japan
taekok@isc.senshu-u.ac.jp

Abstract. In this paper, we propose a new information maximization method for feature discovery and demonstrate that it can discover linguistic rules in unsupervised ways. The new method can directly control competitive unit activation patterns to which input-competitive connections are adjusted. This direct control of the activation patterns permits considerable flexibility for connections and shows the ability to discover salient features not captured by traditional methods. We applied the new method to a linguistic rule acquisition problem. In this problem, unsupervised methods are needed because children acquire rules even without any explicit instruction. Our results confirmed that only by maximizing information content in competitive units linguistic rules can be extracted. These results suggest that linguistic rule acquisition is induced by the processes of information maximization in living systems.

1 Introduction

In this paper, we propose a new information theoretic method for competitive learning and apply it to a problem of linguistic rule acquisition. The new method can contribute to neural computing and linguistic rule acquisition from two perspectives: (1) competition is realized by maximizing information in competitive units; and (2) the method can provide a guiding principle that language acquisition results from the processes of information maximization in living systems. Let us discuss these points in more detail.

First, competition in the new method is realized by maximizing mutual information between input patterns and competitive units. Information theoretic methods applied to neural computing have given promising results in various aspects of neural computing and pattern recognition [1], [2], [3]. However, the complexity of these approaches has severely constrained their range of application. The information theoretic methods have not reached a state in which they can replace traditional unsupervised learning methods such as competitive learning. In our method, information is computed by using competitive unit activation patterns, and the final connections are easily interpreted, because the new method attempts to compress information in compact ways.

G. Dorffner, H. Bischof, and K. Hornik (Eds.): ICANN 2001, LNCS 2130, pp. 1225–1232, 2001.

verbs are used in Japanese is necessary. Any action of transfer can be looked at from two different points of view. When viewed from the giver's perspective, an action of transfer is *a giving event*; on the other hand, the same action becomes *a receiving event* if it is viewed from the receiver's perspective.

In describing a giving event in English, the verb *give* is always used. On the other hand, Japanese has a set of different verbs, and a speaker of Japanese selects an appropriate verb, depending on contextual factors that characterize the relationship between the giver and receiver. The *in-group/out-group* relationship and the difference in social status between the giver and the receiver are the two contextual factors that determine the selection of appropriate donatory verbs in Japanese.

In this paper, we consider a giving event where the second contextual factor, i.e., social status, is kept constant. In this case, either *ageru* or *kureru* is used in Japanese, in accordance with the notion of the in-group/out-group relationship. A giver and a receiver in a giving event can be located on the *in-group/out-group continuum*. The speaker is the core of the in-group, and the in-group consists of those to whom the speaker feels psychological proximity. As a basic rule, when the speaker or an in-group member is a giver and the receiver is from the out-group, *ageru* is chosen; on the other hand, when the giver belongs to the speaker's out-group and the receiver is from the in-group, *kureru* is used.

Next, consider a receiving action. In describing a receiving event in English the verb *receive* is used, while the verb *morau* is used in Japanese. Here again, the notion of in-group/out-group affects the use of the verb *morau*. As a basic principle, when using the verb *morau*, the grammatical subject of a sentence (which corresponds to the receiver in a receiving action) has to be the speaker or an in-group member. It is extremely rare to place an out-group member in the grammatical subject [12].

3.2 Coding System

When dealing with a symbolic system such as language, neural networks cannot immediately detect relations between individual symbols and their meanings. Symbols are arbitrary in nature [13]; therefore, in the process of language acquisition, children need to establish a connection between a given symbol and its meaning by utilizing the context in which the symbol is used. Once children establish this connection, they can communicate by using symbols alone without relying on contextual information. This contextual learning was advocated by Rumelhart and Zipser [14] and discussed fully by Ritter and Kohonen [15].

Following Ritter and Kohonen, we conducted a series of experiments to associate discrete symbols with their meanings by adopting two codes: *symbol code* and *attribute code*. The former corresponds to a label arbitrarily given to a symbol, while the latter provides contextual information in which the former is used. In our experiment, we first trained networks by employing both the symbol code and attribute code. Subsequently, by giving the symbol code alone, we examined whether the network could behave according to the linguistic rules it had acquired just as children do.

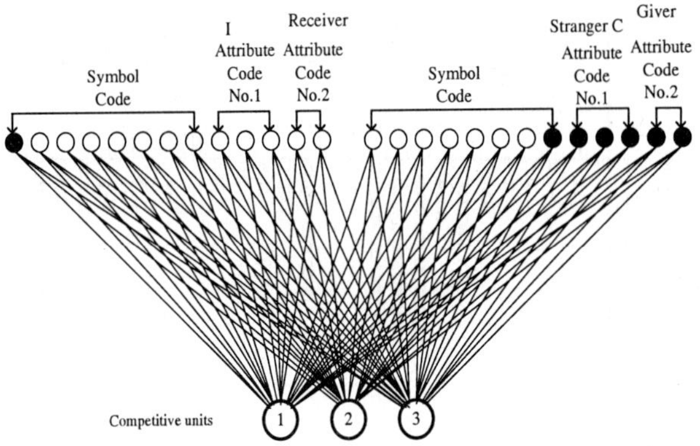

Fig. 3. A network architecture for inferring the three verbs.

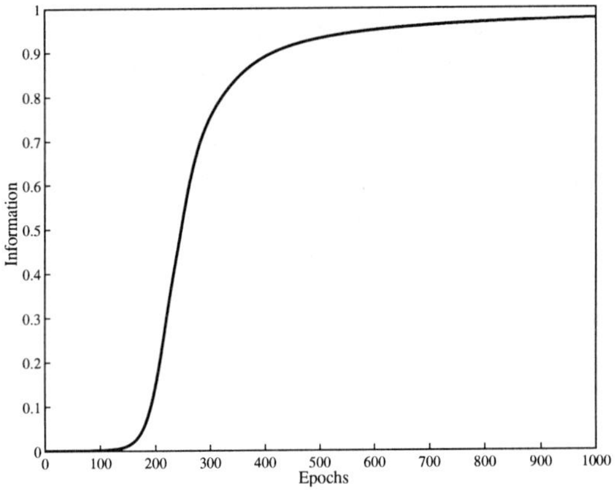

Fig. 4. Information as a function of the learning epochs for the three verbs.

3.3 Experiment

In our experiments, networks were trained to detect the rules inherent in the Japanese donatory verb system. To do so, we used two kinds of codes as mentioned, that is, the symbol code and attribute code. In our experiment, we considered four groups of participants of the same social status, including the speaker, his family members, his acquaintances, and strangers, as shown in Figure 2. We further considered several instances for the respective categories, and this resulted in the total of eight participants. In our coding system, the symbol code differentiated these eight participants from each other, bearing a function of a name given to a person in natural language. The attribute codes were composed of two kinds: *attribute code No. 1* and *attribute code No. 2*. The attribute code No. 1 provides information about the relational distances among the four categories of participants who were located at the different points on the in-group/out-group continuum (see Figure 2). The attribute code No. 2 had

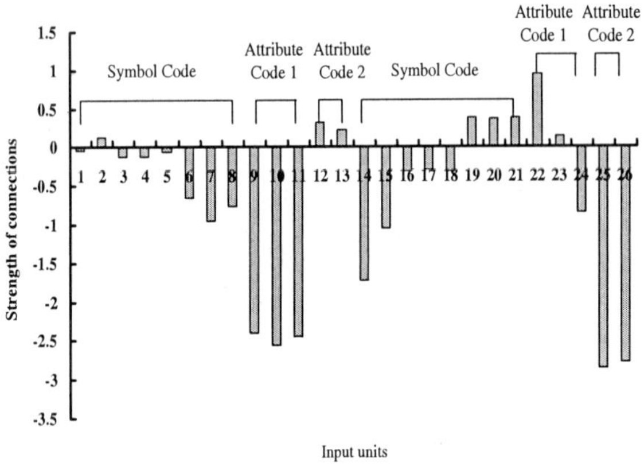

(a) First competitive unit

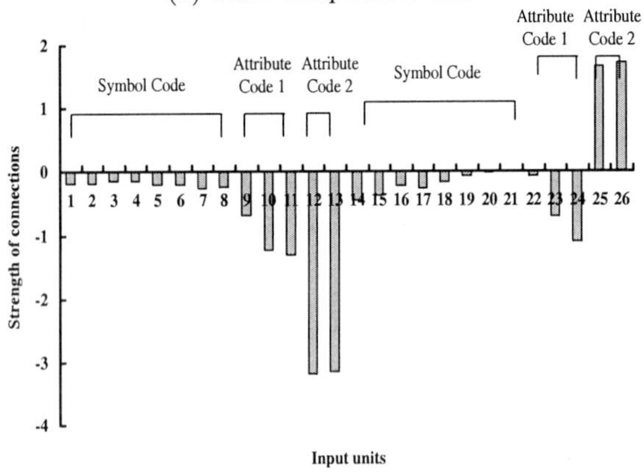

(b) Third competitive unit

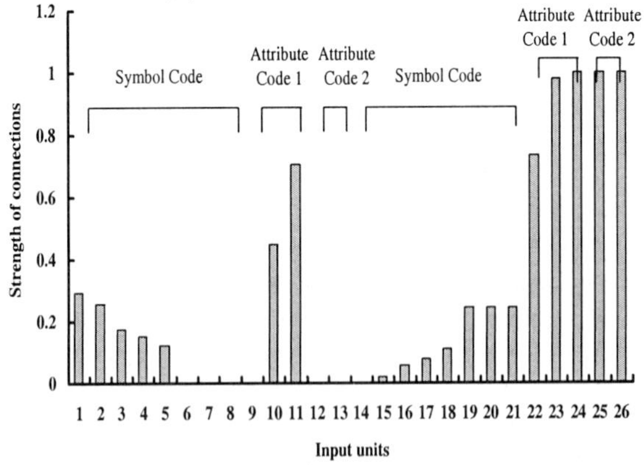

(c) Conventional method

Fig. 5. Connections into the first (a) and the third (b) competitive unit by the information maximization method; and connections by the conventional competitive method (c).

a function to determine from whose perspective a given action of transfer was viewed, i.e., either the giver's or the receiver's perspective. Thus, the attribute code No.2 differentiated giving events from receiving events. The symbol code consisted of eight bits. The attribute code No. 1 consisted of 3 bits, whereas the attribute code No.2 was composed of 2 bits.

As shown in Figure 4, information approaches the maximum value in 1,000 epochs. Figures 5(a) and (b) show the connections into the first (a) and the third competitive unit by the information theoretic method. Figure (c) shows the connections into the third competitive unit by the conventional competitive method. In Figure (a), we can clearly see that the networks respond strongly both to the attribute code No.1 that contains information about the relational distances among the participants and to attribute code No.2 that contains information about the giving and receiving action. Though a detailed explanation is necessary for the full understanding of this unit, we can say that the first competitive unit is on when an in-group member gives something to an out-group member. Thus, the unit corresponds to the verb *ageru.* Figure (b) shows connections into the third competitive unit. This unit is on when the action concerned is a receive event, because the second attribute code No.2 is strongly positive. Thus, this case corresponds to the verb *morau.* Lastly, we removed the attributed codes No1 and No.2, and examined whether the networks could capture the linguistic rules. It was found that without the attribute codes, the networks could infer the correct use of the three donatory verbs.

We find a striking difference between the information theoretic method and the traditional competitive method in terms of the connections into the third competitive unit. Figure 5(c) shows the connections into the third competitive unit by the traditional competitive method. We can see that networks respond almost equally to the attribute codes No.1 and No.2. On the other hand, by using the information theoretic method, connections into the third competitive unit respond much more strongly to the attribute code No.2, as shown in 5(b). These results show that the new method can focus upon certain specific features, and this will allow us to simulate the process of human information processing such as *attention* [16].

References

1. R. Linsker, "Self-organization in a perceptual network," *Computer*, vol. 21, pp. 105–117, 1988.
2. S. Becker, "Mutual information maximization: models of cortical self-organization," *Network: Computation in Neural Systems*, vol. 7, pp. 7–31, 1996.
3. S. Becker and G. E. Hinton, "Learning mixture models of spatial coherence," *Neural Computation*, vol. 5, pp. 267–277, 1993.
4. R. Kamimura, T. Kamimura, and T. R. Shultz, "Information theoretic competitive learning and linguistic rule acquistion," *Transactions on the Japanese Society for Artificial Intelligence*, vol. 16, no. 2, pp. 287–298, 2001.
5. K. Plunkett and V. Marchman, "U-shaped learning and frequency effects in a multilayered perceptron: Implication for child language acquisition," *Cognition*, vol. 38, pp. 43–102, 1991.

6. V. A. Marchman, "Constraints on plasticity in a connectionist model of the english past tense," *Journal of Cognitive Neuroscience*, vol. 5, no. 2, pp. 215–234, 1993.
7. D. E. Rumelhart and J. L. McClelland, "On learning the past tenses of english verbs," in *Parallel Distributed Processing* (D. E. Rumelhart, G. E. Hinton, and R. J. Williams, eds.), vol. 2, pp. 216–271, Cambrige: MIT Press, 1986.
8. R. Brown and C. Hanlon, "Derivational complexity and order of acquisition," in *Cognition and Development of Language* (R. Hayes, ed.), New York: Wiley, 1970.
9. M. J. Demetras, K. N. Post, and C. E. Snow, "Feedback to first language learners: The role of repetitions and clarification questions," *Journal of Child Language*, vol. 13, pp. 275–292, 1986.
10. L. L. Gatlin, *Information Theory and Living Systems.* Columbia University Press, 1972.
11. R. Kamimura and S. Nakanishi, "Hidden information maximization for feature detection and rule discovery," *Network*, vol. 6, pp. 577–622, 1995.
12. S. Makino, *UchiCo to Soto no Gengogaku (A Linguistic-cultural Study of Uchi and Soto.* Tokyo: ALC, 1996.
13. Saussure, *Cours de Linguistique Generale.* Paris: Payot, 1969.
14. D. E. Rumelhart and D. Zipser, "Feature discovery by competitive learning," in *Parallel Distributed Processing* (D. E. Rumelhart, G. E. Hinton, and R. J. Williams, eds.), vol. 1, pp. 151–193, Cambrige: MIT Press, 1986.
15. H. Ritter and T. Kohonen, "Self-organizing semantic map," *Biological Cybernetics*, vol. 61, pp. 241–254, 1989.
16. J. R. Anderson, *Cognitive Psychology and its Implication.* New York: Worth Publishers, 1980.

A Mirror Neuron System for Syntax Acquisition

Steve Womble and Stefan Wermter

University of Sunderland, Centre of Informatics,
St. Peter's Way, Sunderland, SR6 0DD. United Kingdom
www.his.sunderland.ac.uk

Abstract. We investigate the use of a connectionist model of a mirror neuron cortical network for a context free syntax acquisition task. A finite state representation of the context free grammar is learned by an implicit knowledge system (IKS) modelled by a connectionist network. A mirror neuron system (MNS) whose evolutionary pedigree suggests adaptation for goal-directed sequential processing is used to track embedded recursions in a learned finite state model of the grammar. The mirror system modifies the output of the IKS depending on the depth of embedding. Reciprocally the IKS updates the MNS as natural 'goals' occur within a sequence during sentence production. This solves the computationally hard problem of inferring contexts from sequential input.

1 Introduction

Rizzolatti, Gallese and their colleagues have found a class of neurons in the rostral part of the ventral premotor cortex (area F5) in macaque monkeys that are active both when a monkey handles an object and when it observes an experimenter performing similar actions [6, 7, 2]. These mirror neurons are highly selective, firing only when a very specific goal-directed action is witnessed or performed. The homologue to area F5 in humans is usually taken to be Broca's area, an area traditionally associated with language processing and production [4]. Recent brain imaging experiments seem to have confirmed the existence of a similar 'human schematic matching system' [3] located in and around these areas. Consequently it has been suggested that mirror systems originally involved in goal-directed gestural actions may be involved in language processing [5].

Inspired by this work we investigate the properties of a mirror system relevant to a context free grammar acquisition task that would be difficult to learn for a conventional artificial neural network (ANN). We highlight some advantages of applying a mirror system to the acquisition task. Furthermore, we discuss the limitations of the system in a real biological environment where it would be limited by both the difficulty of learning deeper embeddings during acquisition, and by real time demands on working memory during both comprehension and production tasks.

G. Dorffner, H. Bischof, and K. Hornik (Eds.): ICANN 2001, LNCS 2130, pp. 1233–1238, 2001.

2 The Grammar Acquisition Task

We chose for our investigation a stochastic context free noun phrase grammar. It was developed to represent the most frequent syntactic constructions in the NPL corpus, and has been investigated previously using hybrid symbolic-connectionist architectures [9].

It consists of six rule sets which themselves form two distinct sets. The noun group (NG) set consists of a noun group (NG) rule set, an adjective group (ADJG) rule set and a compound noun (CN) rule set. The noun phrase set consists of noun phrases (NP), verb phrases (VP) and prepositional phrases (PP). The construction rules for the NG set are:

NG → CN	CN → N	ADJG → ADJ
NG → DET CN	CN → N CN	ADJG → ADV ADJ
NG → ADJG CN	CN → CN ADJ CN	ADJG → ADJ ADJG
NG → DET ADJG CN		ADJG → ADJG CONJ ADJG

Grouping together these three sets yields a single self-consistent group for which a simple feed-forward ANN can accurately predict the probability distribution for the next term in a noun group by implicitly learning the conditional probability distributions and has been discussed in the context of mirror neuron systems elsewhere [10].

The rules for the noun phrase set of clauses are:

NP → NG	VP → V NP	PP → P NP
NP → NG PP	VP → V PP	PP → P VP
NP → NG VP	VP → VP CONJ VP	PP → PP CONJ PP
NP → NG CONJ NG	VP → V CONJ V NP	PP → P CONJ P NP

What makes this problem interesting is the requirement for phrase agreement either side of a conjugate in the third rule of each of the VP and PP rule sets. If both verb and prepositional phrases are active a construction which locally looks like VP CONJ PP or PP CONJ VP is possible. However if only one of the phrases is embedded (possibly multiply embedded) in a complex phrase either being produced or parsed then the grammar requires strict agreement between the conjoined phrases. The recursive nature of these noun phrases means that such agreement must be achievable over an arbitrary number of intervening elementary symbols. This makes learning with an ANN difficult due to the combinatorial complexity of the search space.

While this task may be tackled using a symbolic system, explicit statistical analysis or an ANN supported by such mechanisms [9] here we investigate how a connectionist system using only local learning rules might solve the problem in an 'online' manner given only information that is cognitively plausible.

3 The Mirror System

For the purposes of this paper our mirror system consists of two interacting parts. These are an implicit knowledge system (IKS) and a mirror neuron system

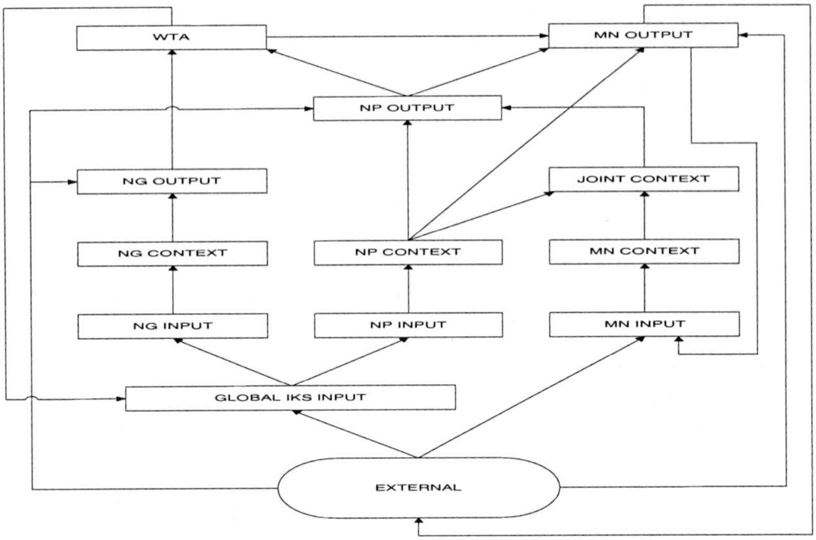

Fig. 1. High level representation of model mirror system.

(MNS). A schematic diagram of the entire system is given in Fig. 1. The IKS system is depicted on the left hand side of the diagram, and consists of the global IKS input, and the layers above it. The mirror system is depicted on the right hand side of the diagram, and consists of the MN input, and the layers above it. Connections within the system are in all cases all-to-all except for the recurrent connections from the WTA to the global IKS input and connections from the MN output to the MN input. These are 1-1 projections and are only active during production. The input layers consist of 'moving windows', containing a sparse binary encoded representation of the current symbol and a number of preceding symbols depending on the actual window width. All context layers contain nodes consisting of normalised exponentials. These learn explicit contexts found in an unsupervised manner using a model of long term heterosynaptic depression [8]. The MN context layer is used to 'clean up' the MN input representation. For the joint context layer the input is the simultaneously occurring contexts produced on each of the IKS and MNS context layers. The outputs of the IKS consist of linear nodes performing standard δ-rule error correction [1]. These learn probability distributions for the next symbol based on the context layer activity. The MN output in our model contains the neurons in our model that possess the characteristic mirror neuron properties identified by Gallese and Rizzolatti. The WTA layer performs a stochastic winner takes all and is only active when the system is used to produce sequences.

If any sequence either being produced or presented is viewed as a set of goals then the mirror neuron paradigm can easily be adapted for use. Success occurs when the resulting sequence is grammatically valid. The IKS will only allow for certain next terms to be selected or accepted as valid. However the system is now in a particular state defined by the MN input pattern, and the IKS probability distributions can then be learned to be modified to rule out incorrect sequences.

The MN input patterns can be fairly arbitrary, requiring only that they form a separable set which is isomorphic to the set of 'goals' i.e. to the relevant higher order symbols ('VP' and 'PP' are sufficient for the NPL grammar)

The mirror output layer is mapped to from the IKS output and IKS context. This layer consists of the actual mirror neurons in our model.

This characteristic output representation which after training will be active either during presentation or production contains the information that a goal, such as completing a particular phrase, has been completed. The MN output patterns are mapped onto the input patterns again using basic δ-rule error correction learning.

4 Results

As a precursor to the acquisition of the full context free grammar the system was trained using just the IKS. This allowed the system to learn the local statistics of the grammar. Once the IKS had asymptoted during training the mirror system was then activated and training continued until the system had again asymptoted. The training set consisted of 1000 strings generated stochastically from the rule sets given above. A further 10000 strings were also stochastically generated and were used as a test set for the full system.

For the mirror system there were four different input patterns; one for 'no goal', VP, PP and VP&PP combined. We used a four bit sparse binary encoding for each of these cases. For this initial investigation we dealt with deeper embeddings using a simple stack to keep track of the depth of multiple embeddings.

This is computationally convenient for the model but it is not trivial to interpret such a computational shortcut as a direct model of a biologically plausible mechanism. We discuss this issue later.

The quantitative results are given in the following table:

	IKS Only	Training Set	Test Set
Negative Log Likelihood Error	0.0644	0.0573	0.0589
One-Norm Error	0.0757	0.0690	0.0698

All figures quoted are for the average (for the given measure) over all strings tested, and are per prediction per output node. Thus the 1-norm measure gives the exact mean distance that each output node is from the correct prediction for each possible next symbol. The residual 'errors' thus include the distance from the probability distributions intrinsic to the stochastic grammar used, which for the negative log likelihood case is proportional to the entropy of the stochastic grammar.

5 Discussion and Conclusion

The reduction in error from the results for the basic IKS to the full mirror system is accounted for by the reduction in the spread of predictions for the next symbol at each point in a given sequence for which the finite state representation was unable to capture the full properties of the context free grammar. In particular with just the IKS running the system learned a mixture of probabilities for the next term in the sequence following a CONJ symbol in the constructions of conjoined VPs and PPs. The mirror system successfully disambiguated the cases for which the CONJ should be followed by a VP, a PP or either. This resulted in a sharpening of the predictions for these occurrences.

The small increase in the error between the training set and test sets highlights the advantage of our particular moving window feed-forward approach over other connectionist networks using more powerful global optimisation techniques such as stochastic gradient descent back propagation with simple recurrent networks, for which the generalisation properties of the system are dependent on the particular local optimum found during the search procedure.

Our model was successful in solving the problem of learning a complex noun phrase context free grammar from example sequences. The system can generalise probability distributions for sequence prediction over multiple complex recursions even if they exist with very low frequency in the set of training examples or are completely novel. Thus it captures the primary strength of rule based computational linguistic approaches to language acquisition using a neural system. However it also captures the cognitively not valid power of such approaches. That is, it can actually deal with arbitrarily deep layers of recursion. However the reason for this is due to our use of a stack to keep count of the depth of VP and PP recursions. This is the exact part of this system that is not yet directly related to either neuroscience evidence concerning language processing in the brain, or to psychological or cognitive models of language cognition and is the focus of our current research.

In this work we have interpreted the generation of phrase constructions as goals. The mirror system processes this goal state information and integrates it with the IKS learned local statistics of the grammar. Further the mirror system learns to identify corresponding goal states in the sequences produced by the system. At which point the activity patterns in our model mirror neurons are the same for goal states recognised either from produced or presented strings. Hence they possess the characteristic responses seen in mirror neurons in the brain. Consequently we conclude that our model mirror system provides computational evidence to support the conjecture that a similar mechanism may be involved in grammar acquisition and processing in the brain.

References

1. Christopher M. Bishop. *Neural Networks for Pattern Recognition.* Oxford University Press, Great Clarendon Street, Oxford, OX2 6DP, United Kigdom, 1995.

2. Vittorio Gallese, Luciano Fadiga, Leonardo Fogassi, and Giacomo Rizzolatti. Action recognition in the premotor cortex. *Brain*, 119:593–609, 1996.
3. Matthew A. Howard, F. McGlone, K. D. Singh, and N. Roberts. Brain activations to passive observation of relevant actions - an fMRI study. In Maxim Stamenov and Vittorio Gallese, editors, *Mirror Neurons and the Evolution of Brain and Language*, Advances In Consciousness Research. John Benjamins, John Benjamins, Netherlands, 2001 (In Press).
4. Ralph-Axel Muller. Innateness, autonomy, universality? neurobiological approaches to language. *Behavioural and brain Sciences*, 19:611–675, 1996.
5. Giacomo Rizzolatti and Michael A. Arbib. Language within our grasp. *Trends in the Neurosciences*, 21:188–194, 1998.
6. Giacomo Rizzolatti, R. Camarda, L. Fogassi, M. Gentilucci, G. Luppino, and M. Matelli. Functional organization of inferior area 6 in the macaque monkey: II. Area F5 and the control of distal movements. *Experimental Brain Research*, 71:491–507, 1988.
7. Giacomo Rizzolatti, Luciano Fadiga, Leonardo Fogassi, and Vittorio Gallese. Premotor cortex and the recognition of motor actions. *Cognitive Brain Research*, 3:131–141, 1996.
8. Edmund T. Rolls and Alessandro Treves. *Neural Networks and Brain Function*, chapter 6, pages 95–135. Oxford University Press, Great Clarendon Street, Oxford, OX2 6DP, United Kigdom, 1998.
9. Stefan Wermter. *Hybrid Connectionist Natural Language Processing.* Chapman & Hall, 2-6 Boundary Row, London., 1995.
10. Steve Womble and Stefan Wermter. Mirror neurons and feedback learning. In Maxim Stamenov and Vittorio Gallese, editors, *Mirror Neurons and the Evolution of Brain and Language*, Advances In Consciousness Research. John Benjamins, Netherlands, 2001 (In Press).

A Network of Relaxation Oscillators that Finds Downbeats in Rhythms

Douglas Eck

Istituto Dalle Molle di Studi sull'Intelligenza Artificiale (IDSIA)
Galleria 2, CH 6928 Manno, Switzerland
doug@idsia.ch, http://www.idsia.ch/~doug

Abstract. A network of relaxation oscillators is used to find downbeats in rhythmical patterns. In this study, a novel model is described in detail. Its behavior is tested by exposing it to patterns having various levels of rhythmic complexity. We analyze the performance of the model and relate its success to previous work dealing with fast synchrony in coupled oscillators.

1 Downbeat Induction

The term *beats* refers to sounds that are perceived as being equally spaced in time. *Downbeats* are particularly salient beats that usually occur at a comfortable tapping rate. When you tap your feet to the radio you are finding downbeats, a skill called *beat induction*. Downbeats act as a unifying force, lending music the feeling of movement by allowing the listener to predict the onset of important musical events. The process of beat induction is influenced by many aspects of music including harmony, melody and rhythm [2, 11]. Because interactions are not always simple, it can be difficult to predict the locations of downbeats. For example, in rock and roll music, even though chord changes (harmonic components) usually occur on the first note of a musical bar, downbeats are usually aligned with the second note due to syncopated drumming style. Furthermore, although beats are *perceived* as being equally-spaced in time, perfect timing is rarely if ever found on the radio. Both motor noise and deviations due to expressive timing are found in performed music.

The task of beat induction can be simplified by considering only patterns of equal-amplitude beeps or clicks. This removes the influence of melody and harmony (and in addition amplitude and timbre variations). However, even in this "rhythm-only" domain, two important influences compete to decide downbeat location. First, the *grouping* of events in the pattern is important. For example, when three or more events are presented in rapid succession, the first and last events are more perceptually salient than the middle one. Second, the *meter* of the pattern is important. Meter is the sense of strong and weak beats that arises from interactions among hierarchical levels in a pattern having nested periodic components. Such a hierarchy is implied in Western music notation, where different levels are indicated by kinds of notes (whole notes, half notes, quarter

G. Dorffner, H. Bischof, and K. Hornik (Eds.): ICANN 2001, LNCS 2130, pp. 1239–1247, 2001.

notes, etc.) and where bars establish measures of an equal number of beats. For an overview see [7].

For these reasons, downbeat induction can be seen as a seemingly simple cognitive skill (tapping along with music) that is in fact challenging to model. We address this challenge by presenting a network of coupled oscillators that performs downbeat induction. The network consists of Fitzhugh-Nagumo oscillators [5, 15] designed to model the dynamics of neural action potential. We show in this paper that they can also "tap along" when forced by rhythmical pulses of voltage. Computational simulations using a popular set of rhythmical patterns reveal that our model succeeds at finding the most likely downbeat in almost all cases. In the paper we analyze model performance and provide a possible explanation of why a small network of simulated neurons would do so well at the task.

In the literature, many have addressed the question of whether a simple model can find downbeats in patterns. Longuet-Higgins & Lee [12] and Povel & Essens [17] offer rules based symbolic models that find downbeats in symbolic patterns. Parncutt [16] considers the importance of presentation rate in beat induction. Large & Kolen [10] and McAuley [14] provide oscillator models that work online to find downbeats in temporal signals.

2 Model Details

The Fitzhugh-Nagumo oscillator is a two-variable caricature of the four-variable Hodgkin-Huxley model of neural action potential [8]. It exhibits the integrate-and-fire spiking dynamics of a real neuron. When fed a constant amount of low voltage, the oscillator gradually accrues its own voltage until it reaches a threshold; upon reaching that threshold it fires and quickly releases the energy. If the energy source is sufficient and is constantly applied, this results in stable limit-cycle oscillation. That is, even after a major perturbation the oscillator will quickly return to a particular path in phase space. A Fitzhugh-Nagumo oscillator is described by the following differential equation:

$$\begin{aligned} \frac{dv}{dt} &= -v(v-\theta)(v-1) - w + \Omega \\ \frac{dw}{dt} &= \epsilon(v - \gamma w) \end{aligned} \tag{1}$$

where v is voltage, w is recovery of voltage, and Omega Ω is external driving voltage. In the simple case Ω is just a small amount of constant external voltage v_c necessary for sustained oscillation. However, when a rhythmical input pattern v_{in} is used, the driving term is $\Omega = v_c + v_{in}$. The frequency of oscillation is controlled by varying epsilon ϵ: a high value yields high-frequency oscillation with a sinus-like waveform while a low value yields low-frequency oscillation with a highly nonlinear waveform. This second, nonlinear oscillation is called *relaxation-mode* oscillation because the limit cycle of the oscillator closely follows or "relaxes onto" the cubic voltage v nullcline of the oscillator. The two

Flow diagram, time series, and coupling function

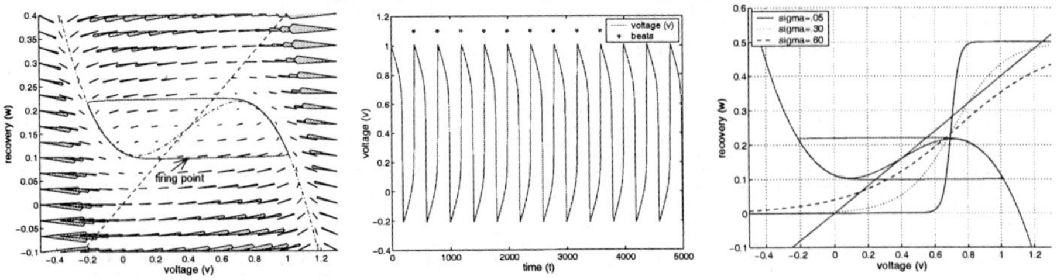

Fig. 1. The flow diagram (left) shows the the two distinct time-scales found in all relaxation oscillators: slow movement along the y axis and fast movement along the x axis. The time series (center) shows the oscillator waveform. In these plots, the firing point for output beats can be seen. The coupling function (right) is the Heaviside function with three values of τ and σ.

remaining parameters are fixed for all simulations. They include a firing threshold parameter θ set to 0.2, and a shunting parameter γ set to 1.2.

To simplify analysis, we modified the oscillator so that it generates a discrete output with every cycle. These outputs are generated each time the model neuron fires and are visible in Figure 1 in the flow diagram (left) and in the time series (center) of the oscillator.

We created a network by coupling the oscillators using a Heaviside coupling function. We chose this function based upon evidence [18] that Heaviside-coupled relaxation oscillators are exceptionally good at rapid and robust synchronization. As our goal is to synchronize quickly with important periodic components in the input, this was a desirable trait for the network to exhibit. The Heaviside function for a Fitzhugh-Nagumo oscillator $H(x) = 0.25(1 + tanh((x - \tau)/\sigma)(v - 1.5))$ is in Figure 1, right. Though several values of τ and σ are shown in the diagram, these parameters were fixed for all simulations at $\tau = .7$ and $\sigma = .05$.

To add Heaviside oscillator-to-oscillator coupling, the driving term Ω in Equation 1 Equation 1 is modified to include voltage v_{net} from other oscillators: $\Omega = v_c + v_{in} - \alpha H(v_{net})$. Oscillator-to-oscillator coupling strength α is a variable that may be optimized. In fact, optimization through gradient descent learning is a current research goal. For these simulations, however, a weak positive connection $\alpha = .1$ is used for all oscillators.

3 Experiment 1: Finding Downbeats in a Set of Rhythmical Patterns

To test the model we used a set of rhythms from Experiment 1 of Povel & Essens [17]. These rhythms are generated by permuting the interval sequence **1 1 1 1 1 2 2 3** terminated by the interval **4**. In this notation an integer indicates how much time follows a note. For example, an interval of 3 consists of 1 note followed by 2 rests. Thus all patterns are of length 16 and consist of nine notes and seven rests. These patterns were used by Povel & Essens in a learning

Table 1. 35 patterns ordered by row. In these patterns, an "x" indicates a note and a "." indicates a rest. The difficulty ratings in the left-hand column are from a model by Povel & Essens and range from easy (1) to hard (7).

1	xxxxx..xx.x.x...	xxx.x.xxx..xx...	x.xxx.xxx..xx...	x.x.xxxxx..xx...	x..xx.x.xxxxx...
2	xxx.xxx.xx..x...	x.xxxx.xx..xx...	xx..xxxxx.x.x...	xx..x.xxx.xxx...	x.xxx.xxxx..x...
3	xxx.xx..xx.xx...	xx.xxxx.x..xx...	xx.xx.xxxx..x...	xx..xx.xx.xxx...	x..xxx.xxx.xx...
4	xx.xxxx.xx..x...	xx.xxx.xxx..x...	xx.xxx..xx.xx...	xx..xx.xxxx.x...	xx..xx.xxx.xx...
5	xxxxx.xx.x..x...	xxxx.x..xxx.x...	xxx..xx.xxx.x...	x.xxx..x.xxxx...	x.x..xxxx.xxx...
6	xxxx.x.x..xxx...	xx.xxx.x..xxx...	xx.x..xxx.xxx...	x.xxxx.x..xxx...	x..xxxxx.xx.x...
7	xxxx.xxx..x.x...	xxxx..xx.xx.x...	xx.xxxx..xx.x...	xx.x..xxxxx.x...	x.x..xxx.xxxx...

experiment that revealed a connection between ease of downbeat induction and ease of pattern recall: though all patterns have the same number of notes and rests, patterns that are easy to tap along with were also easier for listeners to reproduce later. The authors propose a model of beat induction which predicts downbeat induction strength for temporal patterns. The patterns are assigned difficulty rankings from 1 to 7 based upon ease of downbeat induction (Table 1). In the figure, rests shown as . and notes as x.

3.1 Method

Input Patterns: The patterns in Table 1 were translated for the oscillator network by encoding notes as voltage spikes of width 1. So that the network of oscillators would run in low-frequency relaxation mode, the base interval between notes was 125 timesteps. Rests were not encoded in any way. The amplitude of the spikes started at .065 and increased evenly throughout the duration of the simulation to .08 (mean input strength .0725). This strategy of steadily increasing input amplitude kept the network from always preferring early notes in the input pattern. Early-note preference may seem like a weakness in our model, but in fact listeners exhibit the same preference [6].

Network Simulations: 20 Fitzhugh-Nagumo oscillators were instantiated at 4 times the base interval of the patterns, or period 500 ($\epsilon = 0.0015$). This choice of period simplified the task by giving the network the correct tapping rate. But it did not trivialize the problem because the network must still oscillate at the right phase with respect to the input. To achieve this, the oscillators must be stable enough to respond to some notes in the input (those indicating the downbeat) while ignoring all others. Oscillators were fully coupled to one another with a connection strength α of .01. This value was found by trial and error, and is similar to the value used in single-oscillator simulations [4]. At the start of simulation the 20 oscillators were distributed high on the left-hand side of the cubic voltage v nullcline. Through simulation it was verified that the distribution was broad enough to ensure that the oscillators did not have an initial bias towards any particular downbeat. For each input pattern, the system was run for 8 pattern repetitions. All simulations were run using the ODE-solver suite in Matlab 5.3. Several solvers were tested to ensure that computational error did not affect results.

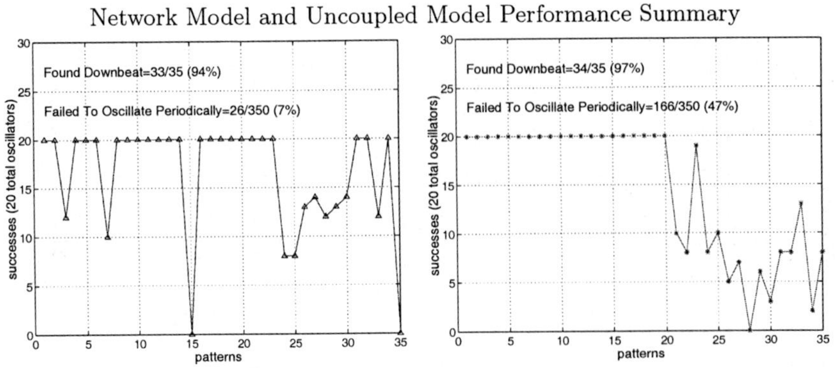

Fig. 2. Performance summary for the network model (left) and the uncoupled model (right).

Measurement: Downbeat induction was measured using a binning method that matched oscillator outputs to locations in the input pattern. After allowing the system to run for 6 pattern repetitions, oscillator output for 2 following pattern repetitions were placed into one of four bins that represented the four possible downbeat phases. These pattern phases are shown at the bottom of Figure 4.

If the bins did not match for both pattern repetitions, an oscillator was declared "failed". Thus, successful oscillators are those which always fired at the same phase and so revealed stable period oscillation. Through a large number of simulations, this method was shown to be an effective measure of beat induction. Namely, using more than 6 repetitions did not yield a significant decrease in failures. Also, using more than 2 binning repetitions did not yield in a significant increase in failures.

3.2 Results

Since downbeat induction is a perceptual phenomenon that changes with musical experience and presentation rate [13], there is no easy way to know in advance desired network behavior. That is, even with close examination it is not always clear where the downbeat *should* fall in a given pattern, especially for difficult patterns. Thus, it would be preferable to compare performance to a large group of listeners performing beat induction on these same patterns. Unfortunately to our knowledge those experiments have not been done; they are the topic of future research. For this analysis we compared the network downbeat inductions to the predictions made by the Povel & Essens model mentioned above. This model was chosen due to its success at predicting ease of pattern learning in the same pattern set. A discussion of the model is not in the scope of this paper and readers are referred to [4] and [3] for our views on the model.

In general the performance of the network was very good[1]. Multiple oscillators locked onto to the most strongly induced downbeat in 33 of 35 patterns

[1] The full set of predictions is available at www.idsia.ch/~doug/publications.html

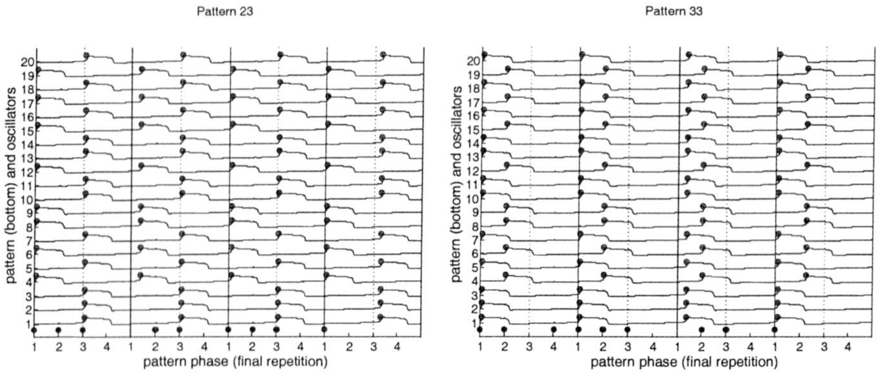

Fig. 3. In patterns 23 (left) and 33 (right), two downbeats are equally likely. The network finds them both.

(94%). In patterns 23 and 33 the Povel & Essens model predicts that two competing downbeats are equally likely. For both of these patterns, 12 oscillators discovered one downbeat while 8 oscillators discovered the other one. In Figure 2 (left) the number of successful oscillators is plotted for the patterns.

4 Discussion

In in previous research we ran the same simulations with coupling turned off to test the usefulness of oscillator-to-oscillator coupling. These results are reported in detail in [4] and are summarized very briefly here.

As can be seen in Figure 2 (right), the uncoupled oscillator model did succeed at finding downbeats in many of the patterns. In fact, the uncoupled model failed completely at only one pattern while the coupled model failed at two of them. However, a closer look at the results shows clear advantages for the network model. In terms of failed oscillators—that is, oscillators that did not settle into a periodic limit cycle—the uncoupled model had 166 of 350 failures (47%) while the network model only had 26 of 350 failures (7%). This reveals a general lack of stability in the uncoupled model. Compare the performance summary of the coupled network model (Figure 2) (left) to that of the uncoupled model (right). Here it is easy to see that the uncoupled model suffers a drastic degradation in performance for patterns 21 through 35.

In patterns 22 and 23 the Povel & Essens model predicts two equally-plausible downbeat phases. In these cases, the network found both downbeats. This can be seen in Figure 3. For pattern 23 (left) 8 oscillators are aligned with phase 1 and 12 with phase 3. For pattern 33 (right) 12 oscillators are aligned with phase 1 and 8 with phase 3. Furthermore, for patterns 17, 18 and 20 the Povel & Essens model predicts that phase 2 is a good downbeat assignment (though for reasonable parameter settings it is not as good as phase 1). For all three of these patterns, most oscillators are aligned with the preferred phase 1 while a small but significant number of oscillators are the second good phase 2. The uncoupled

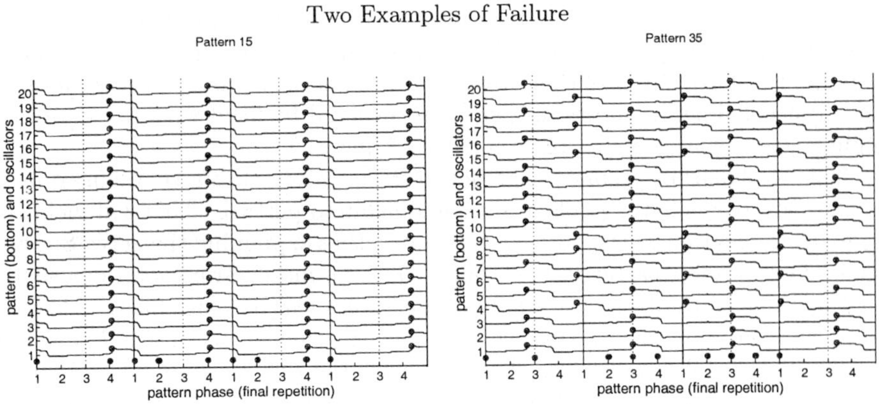

Fig. 4. On the left,(pattern 15) the network fails by finding downbeat at phase 4 instead of phase 15. On the right, (pattern 35) the network fails by oscillating aperiodically.

model does not exhibit this behavior. Though some oscillators are aligned with both solutions for patterns 23 and 33, in the case of patterns 17, 18 and 20 the uncoupled model only finds the primary solution of phase 1.

We believe that the behavior of the network is desirable. In complicated rhythmical patterns there is rarely only one important periodic component. The network model is superior to the uncoupled model at finding and synchronizing with multiple periodic components in parallel. This behavior, we believe, can be useful when more difficult tasks like pattern learning are considered.

The network failed most drastically on pattern 35. For this pattern, the network found two aperiodic solutions that show a sensitivity to grouping influences rather than metrical influences. Specifically, the two groups of oscillators lie close to periodic solutions at phases 1 and 3 but are pulled away due to influences from the In [17] it is suggested that for very difficult patterns like this one, a clock may not be induced at all and that listeners may uses such non-metrical grouping structure to help memorize the pattern. [1] presents evidence that even young children are sensitive to both metrical and phrasal information in rhythms like these.

Two main observations have been made about the network of oscillators. First, it was shown that the network is less prone to drastic loss of efficiency for certain patterns than is the uncoupled model. Second, it was shown that the network finds multiple correct solutions more often than the uncoupled model.

These two phenomena can be attributed to *fractured synchrony* in the network model. Fractured synchrony [18] occurs when a group of coupled oscillators splits into small synchronized pools having stable phase relationships to one another. In networks with no external driving rhythmical pattern, the phase relationships between the pools are arbitrary. Here, with the presence of an input pattern, multiple pools also emerge, but their phase relationships are no longer arbitrary but rather fixed by regularly recurring notes in the temporal pattern.

A fractured synchrony explanation would predict that networks started with random initial phases should find multiple simultaneous solutions by pooling together (and staying together) due to the influence of positive oscillator-to-oscillator coupling; in addition, their phases should be fixed with respect to the input pattern due to oscillator-to-input coupling.

This explanation predicts that networks are more stable than single oscillators because oscillator pools are more difficult to perturb. This will yield less aperiodic oscillation in response to stray notes, thus explaining the network's better performance on patterns 21 through 35. Also, more varied solutions should be found due to the fact that a pool of oscillators will be more likely to *remain* phase-aligned with some recurring energy in the signal once it has found that energy. That is, all pools will not collapse onto the *earliest* predicted solution.

In short, the positive oscillator-to-oscillator coupling acts as a substitute for signal energy when the signal does not provide a reliable note every time the oscillators reach phase zero (i.e. patterns 21 through 35). Uncoupled oscillators have no such mechanism. They either succeed at finding the earliest predicted solution when one is present or they oscillate aperiodically, causing them to be treated as failures by the binning mechanism.

5 Conclusions

These simulations successfully show that networks of Fitzhugh-Nagumo relaxation oscillators can perform beat induction. Given that these oscillators were designed to model the dynamics of neural action potential—not rhythm cognition—this is an interesting finding. This study is only preliminary, and clearly more work needs to be done. A comparison should be performed between this model and similar models (e.g. [10],[9],[14]). Experimental data needs to be collected from listeners performing a similar task. Finally, a pattern set which does a better job of finding the breaking points of this model and others like it would be of great value.

Though only 20 oscillators were used in these simulations, the dynamics of Fast Threshold Modulation predict that much larger networks should work without catastrophic desynchronization. Thus, the research in this chapter can be seen as a base from which to explore network models where a large number of oscillators can work in parallel to find important periodic components in temporal patterns. One obvious application of such a model is in learning long temporal patterns having rhythmical structure.

References

1. Jeanne Bamberger. *The Mind Behind the Musical Ear: How Children Develop Musical Intelligence.* Harvard University Press, 1991.
2. Grosvenor Cooper and Leonard B. Meyer. *The Rhythmic Structure of Music.* The University of Chicago Press, 1960.
3. Douglas Eck. A positive-evidence model for classifying rhythmical patterns. *Journal of New Music Research*, 2001. In press.

4. Douglas Eck. Tracking rhythms with a relaxation oscillator. *Psychological Review*, 2001. In press.
5. R. Fitzhugh. Impulses and physiological states in theoretical models of nerve membrane. *Biophysical Journal*, 1:455–466, 1961.
6. W.R. Garner and R.L. Gottwald. The perception and learning of temporal patterns. *Quarterly Journal of Experimental Psychology*, 20:97–109, 1968.
7. Stephen Handel. *Listening: An introduction to the perception of auditory events*. MIT Press, Cambridge, MA, 1989.
8. A.L. Hodgkin and A.F. Huxley. A quantitative description of membrane current and its application to conduction and excitation in nerve. *Journal of Physiology*, 117:500–544, 1952.
9. Edward W. Large and Mari Reiss Jones. The dynamics of attending: How people track time-varying events. *Psychological Review*, 106(1):119–159, 1999.
10. Edward W. Large and J. F. Kolen. Resonance and the perception of musical meter. *Connection Science*, 6:177–208, 1994.
11. F. Lerdahl and R. Jackendoff. *A Generative Theory of Tonal Music*. M.I.T. Press, Cambridge, Mass., 1983.
12. H.C. Longuet-Higgins and C.S Lee. The perception of musical rhythms. *Perception*, 11:115–128, 1982.
13. J. D. McAuley and Peter Semple. The effect of tempo and musical experience on perceived beat. *Australian Journal of Psychology*, 51(3):176–187, 1999.
14. J.D. McAuley. *On the Perception of Time as Phase: Toward an Adaptive-Oscillator Model of Rhythm*. PhD thesis, Indiana University, Bloomington, IN, 1995.
15. J. Nagumo, S. Arimoto, and S. Yoshizawa. An active pulse transmission line simulating nerve axon. *Proceeding IRE*, 50:2061–2070, 1962.
16. R. Parncutt. A perceptual model of pulse salience and metrical accent in musical rhythms. *Music Perception*, 11:409–464, 1994.
17. D.J Povel and Peter Essens. Perception of temporal patterns. *Music Perception*, 2:411–440, 1985.
18. David Somers and Nancy Kopell. Waves and synchrony in networks of oscillators of relaxation and non-relaxation type. *Physica D*, 89:169–183, 1995.

Knowledge Incorporation and Rule Extraction in Neural Networks

Minoru Fukumi, Yasue Mitsukura, and Norio Akamatsu

Dept. of Information Science and Intelligent Systems,
University of Tokushima
2-1, Minami-Josanjima, Tokushima 770-8506 Japan

Abstract. In this paper a new knowledge incorporation and rule extraction method in neural networks is presented. The rule form of an if–then type can be inserted into a neural network (NN) as knowledge of a problem. NN is then trained by using a set of training samples. In this case the structure learning algorithm with forgetting is used to generate a small-sized NN system. After the NN training, rules are extracted from it.

The results of computer simulations show that this approach can generate obvious network architectures and as a result simple rules compared with conventional rule extraction methods.

1 Introduction

Artificial neural networks (NNs) have been a focus of active research fields. However, NNs possess a drawback without explainability to reformulate as a black-box [1]. In this situation, much work has been devoted to the development of techniques for extracting symbolic rules from trained NNs [1]–[8].

In this paper, a method to incorporate and extract rules in NN is presented. Rules of the if–then type for a kind of pattern classification problem are manually generated considering properties of training data and then are inserted as weight values into NN. This NN is trained to refine the inserted rules by using the training data. In this training the structure learning algorithm with forgetting [5] is used to generate a small-sized NN system. After training, rules are extracted from it.

Furthermore, in examining recognition accuracy of NN hidden units with a signum function to produce binary outputs are utilized. In training, sigmoid functions are used in the hidden layer [4]. It enables us to extract simple rules from trained NN.

Fu [7] presented the rule extraction method for continuous valued inputs. This method, however, is usually limited to rules of a single attribute. Furthermore it is complicated and needs a threshold to generate rules. Ueda et al.[6] presented the rule extraction method using a structure learning. This method can produce rules by combination of some attributes, but needs a relatively large network to retain 100 % classification accuracy. We presented the rule extraction method from NNs evolved by using a genetic algorithm with deterministic mutation [3].

G. Dorffner, H. Bischof, and K. Hornik (Eds.): ICANN 2001, LNCS 2130, pp. 1248–1253, 2001.

It is effective to extract logical functions. However it is time-consuming. On the one hand, there are a few method to perform rule insertion and extraction in NNs [9]. However, an effective and easy rule insertion and extraction method has never been presented as long as the authors know.

From the results of computer simulations, it is shown that the present method can make a simple NN structure based on manually generated rules and, as a result of training simple rules effectively.

2 Rule Insertion and Extraction

2.1 Rule Insertion into NN

Knowledge incorporated in NNs is represented as rule forms, which are inserted into NNs. A rule insertion technique is presented for three-layered NNs, in this paper. The rule considered here is that of an if–then type. This rule is inserted into NN as knowledge of a given classification problem. The rules are produced by considering the properties of training data. The better rules can produce a high recognition accuracy.

The rules generated by the training data are then incorporated into NN weights. A NN structure in this case is composed of signum hidden units. In training, however, the signum units are replaced with sigmoid units to use a popular back-propagation (BP) algorithm [4].

Rules used for rule insertion possess a form with a single attribute and a threshold. The rule, for example, is given as

$$\text{If } x_1 > 0.5, \text{ then class} = 2. \tag{1}$$

where x_1 is an attribute. In this case, this rule shows that the input sample is included in class 2 if $x_1 > 0.5$. This rule is converted into weight values of NN connections as follows. A single hidden unit with the signum output function is used to represent this rule. Its hidden unit can have input connections from input units. The connection corresponding to x_1 in the hidden unit has a weight value of 1.0 and the value of its bias weight is -0.5, as shown in Fig. 1. The other connection weights are assigned small random values. The connection to the output unit of class 2 is set the value of +1 and the connections to the other output ones are -1.

Rules with combination of many attributes can be constructed by using many hidden units. One rule represented as ">" or "<" needs one hidden unit. These relations can be unified as complex rules at output units.

2.2 NN Training

BP with forgetting of link weight (BPWF, but SLF in [5]) is utilized in NN training. BPWF uses as a criterion the following:

$$J_f = J + \varepsilon' \sum_{i,j} |w_{ij}| = \sum_i (d_i - o_i)^2 + \varepsilon' \sum_{i,j} |w_{ij}| \tag{2}$$

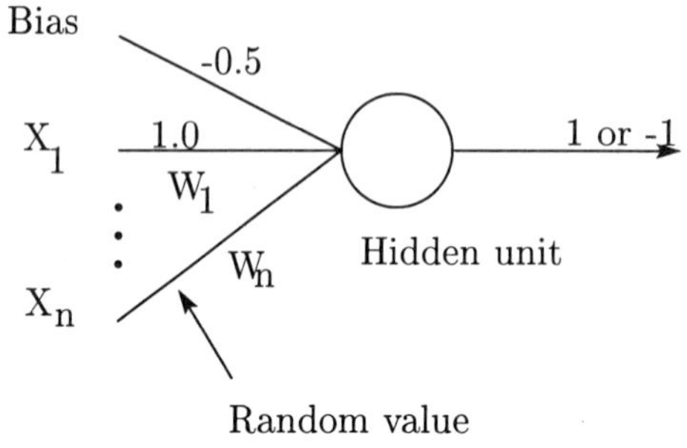

Fig. 1. Rule insertion.

where d_i is the desired signal and o_i is the output value of the i-th unit. The first term in equation (2) is the criterion function of the BP training. The quantity of weight change at time t is

$$\Delta w_{ij}(t) = -\eta' \frac{\partial J_f}{\partial w_{ij}(t)} = \eta\ \Delta w'_{ij}(t) - \varepsilon\ \mathsf{sgn}(w_{ij}(t)) \tag{3}$$

where $\Delta w'_{ij}(t)$ is the quantity of weight change by BP and ε is the forgetting quantity in each iteration cycle.

2.3 Rule Extraction from Trained NN

In this paper, we try to extract simple rules from NN trained using the iris data [5]. To perform this, it is essential to extract rules from data with continuous valued inputs and discrete valued outputs. The network is then constrained to achieve 100 % classification accuracy. In this paper, the resubstitution method is used. Some of the data are very similar as shown in Table 1. If hidden units with a sigmoid function are used in NN, a distributed representation on the hidden layer often arises. Therefore it is difficult to perform the interpretation of hidden units or the discovery of regularities in training data.

To overcome this difficulty, we used the d-CONE models [2] with the signum output function as hidden units. Using this neuron model, the hidden units can produce binary output values. It enables us to extract rules from trained NN. In this paper, the technique to replace sigmoid functions by signum ones in examining NN accuracy is introduced in a hidden layer.

Suppose that the first hidden unit H_0 in NN is connected to the second output unit O_1 after training. The unit O_1 has only a connection from the unit H_0. In this case, a rule is generated by examining the output value of the unit H_0 when the output value of the unit O_1 is 1. Notice that a bias input to an output unit does not affect extracted rules although it makes the role of the unit clear.

Table 1. Iris data

Iris data				
Category	sepal length x_1	sepal width x_2	petal length x_3	petal width x_4
Setosa	5.1	3.5	1.4	0.2
Setosa	5.5	4.2	1.4	0.2
Versicolor	5.9	3.2	4.8	1.8
Versicolor	6.0	2.7	5.1	1.6
Virginica	6.3	2.8	5.1	1.5
Virginica	6.0	3.0	4.8	1.8

3 Computer Simulations

The rules inserted into NN are given as

$$\text{If } x_3 < 2.0, \text{ then class} = 1. \tag{4}$$

$$\text{If } x_3 > 2.0 \text{ and } x_3 < 4.6, \text{ then class} = 2. \tag{5}$$

$$\text{If } x_3 > 4.6, \text{ then class} = 3. \tag{6}$$

There are three rules generated by considering the iris data properties, as a maximum and a minimum values of each class. The iris data set is composed of three classes, each of which includes 50 samples. Note that the above rules can be generated very easily. In this paper, three rules (4)–(6) are used and inserted into NN. Then four hidden units H_1 to H_4 are necessary to represent the rules. Note that another rule could be generated by considering the attribute x_4.

Before BPWF training, NN has ten hidden units with a bias input and the number of connections is 83. If the initial NN architecture with the rules without training is used, its recognition accuracy is 92 %. The rules with 100 % accuracy are remained except changing those weight values through BPWF learning. The number of connections after BPWF training is reduced to 26, as shown in Fig. 2. The result by Ishikawa [5] includes 30 connection weights. This means that the conventional method [6] can generate only a NN structure with 30 connections at a best case. Our results are better in NN size.

The hidden units produce ± 1 outputs. Therefore, it is easy to extract rules. As a result of clustering of hidden outputs in the trained network, the following rules are extracted.

- Rule 1 $H_1 = 1 \Rightarrow$ Setosa
- Rule 2 $H_1 = -1$ and $H_2 = 1$ and $H_3 = -1$ and $H_5 = 1 \Rightarrow$ Versicolor
- Rule 3 $H_1 = -1$ and $H_2 = -1$ and $H_3 = -1$ and $H_5 = -1 \Rightarrow$ Versicolor
- Rule 4 $(H_2 = 1$ or $H_3 = 1)$ and $H_5 = -1 \Rightarrow$ Virginica

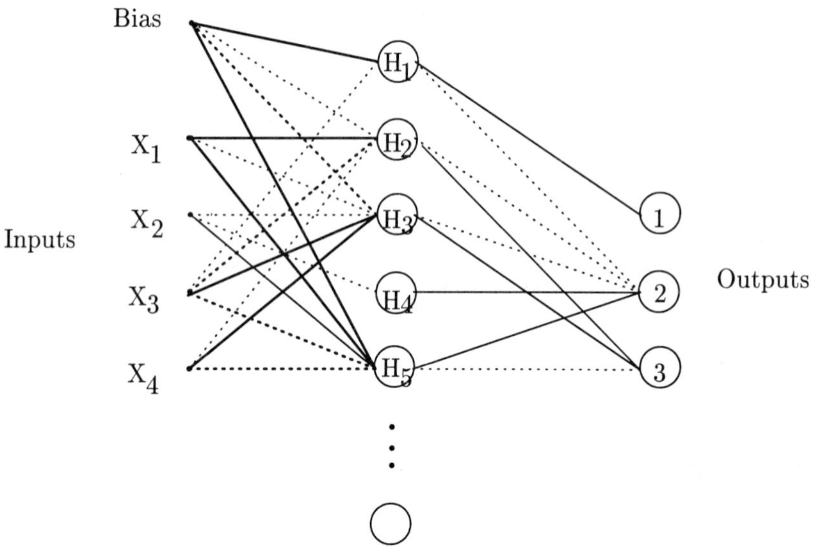

Fig. 2. A neural network architecture generated after BPWF. The dotted lines mean negative value weights. The line width is proportional to the absolute value of its weight.

where "Setosa" corresponds to class 1. In the extracted rules after the clustering, the unit H_4 has only the role of a bias unit. Therefore this unit is not included in the above rules. "Rule 4" is obtained by using the Karnaugh map.

The firing conditions in the hidden units are given as:

- H_1: if $x_3 < 2.39$, then +1 else -1
- H_2: if $(7.32\ x_1 - 6.68\ x_3 - 4.96\ x_4 - 2.90) \geq 0$, then +1 else -1
- H_3: if $(-3.30\ x_1 - 4.14\ x_2 + 5.38\ x_3 + 9.93\ x_4 - 12.39) \geq 0$, then +1 else -1
- H_4: if $x_2 < 0$, then +1 else -1
- H_5: if $(7.94\ x_1 + 1.95\ x_2 - 11.25\ x_3 - 10.24\ x_4 + 16.04) \geq 0$, then +1 else -1

where $x_1 \sim x_4$ show the input attributes left after the BPWF learning.

The incorporated rule (4) is almost the same as Rule 1 extracted from NN and is slightly changed in threshold value based on the NN weight values.

4 Conclusions

This paper presents a new rule insertion and extraction method in neural networks. This method can refine rules (knowledge) and generate better rules for a pattern classification.

This research is partially supported by Suzuki Foundation and The Telecommunications Advancement Foundation in Japan.

References

1. A.B.Tickle et al. : "The Truth Will Come to Light: Directions and Challenges in Extracting the Knowledge Embedded within Trained Artificial Neural Networks", *IEEE Trans. NN*, Vol.9, No.6, pp.1057-1067 (1998)
2. M.Fukumi and N.Akamatsu : "A Method to Design a Neural Pattern Recognition System by Using a Genetic Algorithm with Partial Fitness and a Deterministic Mutation", *Proc. of IEEE Int. Conf. on SMC*, Vol.3, pp.1989-1993 (1996)
3. M.Fukumi and N.Akamatsu : "Rule Extraction from Neural Network Trained Using Evolutionary Algorithms with Deterministic Mutation", *Proc. of IJCNN*, pp.686-689 (1998)
4. M.Fukumi and N.Akamatsu : "A New Rule Extraction Method from Neural Networks", *Proc. of IJCNN*, # 713 (1999)
5. M. Ishikawa : "Structural Learning with Forgetting", *Neural Networks*, Vol. 9, No. 3, pp. 509-521 (1996)
6. H.Ueda and M.Ishikawa :"Rule Extraction from Data with Continuous Valued Inputs and Discrete Valued Outputs Using Neural Networks", Technical Report IEICE Japan, NC96-121, pp.63-70 (1997), in Japanese
7. L.Fu: "Rule Generation from Neural Networks", *IEEE Trans. SMC*, Vol. 24, No. 8, pp.1114-1124 (1994)
8. C. Sano : *Learning and Evolution of Intelligent Artificial Life*, Morikita-shuppan (1996), in Japanese
9. A-H. Tan : "Cascade ARTMAP: Integrating Neural Computation and Symbolic Knowledge Processing", *IEEE Trans. on Neural Networks*, Vol.8, No.2, pp.237-250 (1997)

Author Index

Lecture Notes in Computer Science

For information about Vols. 1–2048
please contact your bookseller or Springer-Verlag

Vol. 1982: S. Näher, D. Wagner (Eds.), Algorithm Engineering. Proceedings, 2000. VIII, 243 pages. 2001.

Vol. 1996: P.L. Lanzi, W. Stolzmann, S.W. Wilson (Eds.), Advances in Learning Classifier Systems. Proceedings, 2000. VIII, 273 pages. 2001. (Subseries LNAI).

Vol. 2049: G. Paliouras, V. Karkaletsis, C.D. Spyropoulos (Eds.), Machine Learning and Its Applications. VIII, 325 pages. 2001. (Subseries LNAI).

Vol. 2060: T. Böhme, H. Unger (Eds.), Innovative Internet Computing Systems. Proceedings, 2001. VIII, 183 pages. 2001.

Vol. 2062: A. Nareyek, Constraint-Based Agents. XIV, 178 pages. 2001. (Subseries LNAI).

Vol. 2064: J. Blanck, V. Brattka, P. Hertling (Eds.), Computability and Complexity in Analysis. Proceedings, 2000. VIII, 395 pages. 2001.

Vol. 2065: H. Balster, B. de Brock, S. Conrad (Eds.), Database Schema Evolution and Meta-Modeling. Proceedings, 2000. X, 245 pages. 2001.

Vol. 2066: O. Gascuel, M.-F. Sagot (Eds.), Computational Biology. Proceedings, 2000. X, 165 pages. 2001.

Vol. 2068: K.R. Dittrich, A. Geppert, M.C. Norrie (Eds.), Advanced Information Systems Engineering. Proceedings, 2001. XII, 484 pages. 2001.

Vol. 2069: C. Peters (Ed.), Cross-Language Information Retrieval and Evaluation. Proceedings, 2000. IX, 389 pages. 2001.

Vol. 2070: L. Monostori, J. Váncza, M. Ali (Eds.), Engineering of Intelligent Systems. Proceedings, 2001. XVIII, 951 pages. 2001. (Subseries LNAI).

Vol. 2071: R. Harper (Ed.), Types in Compilation. Proceedings, 2000. IX, 207 pages. 2001.

Vol. 2072: J. Lindskov Knudsen (Ed.), ECOOP 2001 – Object-Oriented Programming. Proceedings, 2001. XIII, 429 pages. 2001.

Vol. 2073: V.N. Alexandrov, J.J. Dongarra, B.A. Juliano, R.S. Renner, C.J.K. Tan (Eds.), Computational Science – ICCS 2001. Part I. Proceedings, 2001. XXVIII, 1306 pages. 2001.

Vol. 2074: V.N. Alexandrov, J.J. Dongarra, B.A. Juliano, R.S. Renner, C.J.K. Tan (Eds.), Computational Science – ICCS 2001. Part II. Proceedings, 2001. XXVIII, 1076 pages. 2001.

Vol. 2075: J.-M. Colom, M. Koutny (Eds.), Applications and Theory of Petri Nets 2001. Proceedings, 2001. XII, 403 pages. 2001.

Vol. 2076: F. Orejas, P.G. Spirakis, J. van Leeuwen (Eds.), Automata, Languages and Programming. Proceedings, 2001. XIV, 1083 pages. 2001.

Vol. 2077: V. Ambriola (Ed.), Software Process Technology. Proceedings, 2001. VIII, 247 pages. 2001.

Vol. 2078: R. Reed, J. Reed (Eds.), SDL 2001: Meeting UML. Proceedings, 2001. XI, 439 pages. 2001.

Vol. 2079: E. Burke, W. Erben (Eds.), Practice and Theory of Automated Timetabling III. Proceedings, 2001. XII, 359 pages. 2001.

Vol. 2080: D.W. Aha, I. Watson (Eds.), Case-Based Reasoning Research and Development. Proceedings, 2001. XII, 758 pages. 2001. (Subseries LNAI).

Vol. 2081: K. Aardal, B. Gerards (Eds.), Integer Programming and Combinatorial Optimization. Proceedings, 2001. XI, 423 pages. 2001.

Vol. 2082: M.F. Insana, R.M. Leahy (Eds.), Information Processing in Medical Imaging. Proceedings, 2001. XVI, 537 pages. 2001.

Vol. 2083: R. Goré, A. Leitsch, T. Nipkow (Eds.), Automated Reasoning. Proceedings, 2001. XV, 708 pages. 2001. (Subseries LNAI).

Vol. 2084: J. Mira, A. Prieto (Eds.), Connectionist Models of Neurons, Learning Processes, and Artificial Intelligence. Proceedings, 2001. Part I. XXVII, 836 pages. 2001.

Vol. 2085: J. Mira, A. Prieto (Eds.), Bio-Inspired Applications of Connectionism. Proceedings, 2001. Part II. XXVII, 848 pages. 2001.

Vol. 2086: M. Luck, V. Mařík, O. Štěpánková, R. Trappl (Eds.), Multi-Agent Systems and Applications. Proceedings, 2001. X, 437 pages. 2001. (Subseries LNAI).

Vol. 2087: G. Kern-Isberner, Conditionals in Nonmonotonic Reasoning and Belief Revision. X, 190 pages. 2001. (Subseries LNAI).

Vol. 2088: S. Yu, A. Păun (Eds.), Implementation and Application of Automata. Proceedings, 2000. XI, 343 pages. 2001.

Vol. 2089: A. Amir, G.M. Landau (Eds.), Combinatorial Pattern Matching. Proceedings, 2001. VIII, 273 pages. 2001.

Vol. 2090: E. Brinksma, H. Hermanns, J.-P. Katoen (Eds.), Lectures on Formal Methods and Performance Analysis. Proceedings, 2000. VII, 431 pages. 2001.

Vol. 2091: J. Bigun, F. Smeraldi (Eds.), Audio- and Video-Based Biometric Person Authentication. Proceedings, 2001. XIII, 374 pages. 2001.

Vol. 2092: L. Wolf, D. Hutchison, R. Steinmetz (Eds.), Quality of Service – IWQoS 2001. Proceedings, 2001. XII, 435 pages. 2001.

Vol. 2093: P. Lorenz (Ed.), Networking – ICN 2001. Proceedings, 2001. Part I. XXV, 843 pages. 2001.

Vol. 2094: P. Lorenz (Ed.), Networking – ICN 2001. Proceedings, 2001. Part II. XXV, 899 pages. 2001.

Vol. 2095: B. Schiele, G. Sagerer (Eds.), Computer Vision Systems. Proceedings, 2001. X, 313 pages. 2001.

Vol. 2096: J. Kittler, F. Roli (Eds.), Multiple Classifier Systems. Proceedings, 2001. XII, 456 pages. 2001.

Vol. 2097: B. Read (Ed.), Advances in Databases. Proceedings, 2001. X, 219 pages. 2001.

Vol. 2098: J. Akiyama, M. Kano, M. Urabe (Eds.), Discrete and Computational Geometry. Proceedings, 2000. XI, 381 pages. 2001.

Vol. 2099: P. de Groote, G. Morrill, C. Retoré (Eds.), Logical Aspects of Computational Linguistics. Proceedings, 2001. VIII, 311 pages. 2001. (Subseries LNAI).

Vol. 2100: R. Küsters, Non-Standard Inferences in Description Logocs. X, 250 pages. 2001. (Subseries LNAI).

Vol. 2101: S. Quaglini, P. Barahona, S. Andreassen (Eds.), Artificial Intelligence in Medicine. Proceedings, 2001. XIV, 469 pages. 2001. (Subseries LNAI).

Vol. 2102: G. Berry, H. Comon, A. Finkel (Eds.), Computer-Aided Verification. Proceedings, 2001. XIII, 520 pages. 2001.

Vol. 2103: M. Hannebauer, J. Wendler, E. Pagello (Eds.), Balancing Reactivity and Social Deliberation in Multi-Agent Systems. VIII, 237 pages. 2001. (Subseries LNAI).

Vol. 2104: R. Eigenmann, M.J. Voss (Eds.), OpenMP Shared Memory Parallel Programming. Proceedings, 2001. X, 185 pages. 2001.

Vol. 2105: W. Kim, T.-W. Ling, Y-J. Lee, S.-S. Park (Eds.), The Human Society and the Internet. Proceedings, 2001. XVI, 470 pages. 2001.

Vol. 2106: M. Kerckhove (Ed.), Scale-Space and Morphology in Computer Vision. Proceedings, 2001. XI, 435 pages. 2001.

Vol. 2107: F.T. Chong, C. Kozyrakis, M. Oskin (Eds.), Intelligent Memory Systems. Proceedings, 2000. VIII, 193 pages. 2001.

Vol. 2108: J. Wang (Ed.), Computing and Combinatorics. Proceedings, 2001. XIII, 602 pages. 2001.

Vol. 2109: M. Bauer, P.J. Gymtrasiewicz, J. Vassileva (Eds.), User Modelind 2001. Proceedings, 2001. XIII, 318 pages. 2001. (Subseries LNAI).

Vol. 2110: B. Hertzberger, A. Hoekstra, R. Williams (Eds.), High-Performance Computing and Networking. Proceedings, 2001. XVII, 733 pages. 2001.

Vol. 2111: D. Helmbold, B. Williamson (Eds.), Computational Learning Theory. Proceedings, 2001. IX, 631 pages. 2001. (Subseries LNAI).

Vol. 2116: V. Akman, P. Bouquet, R. Thomason, R.A. Young (Eds.), Modeling and Using Context. Proceedings, 2001. XII, 472 pages. 2001. (Subseries LNAI).

Vol. 2117: M. Beynon, C.L. Nehaniv, K. Dautenhahn (Eds.), Cognitive Technology: Instruments of Mind. Proceedings, 2001. XV, 522 pages. 2001. (Subseries LNAI).

Vol. 2118: X.S. Wang, G. Yu, H. Lu (Eds.), Advances in Web-Age Information Management. Proceedings, 2001. XV, 418 pages. 2001.

Vol. 2119: V. Varadharajan, Y. Mu (Eds.), Information Security and Privacy. Proceedings, 2001. XI, 522 pages. 2001.

Vol. 2120: H.S. Delugach, G. Stumme (Eds.), Conceptual Structures: Broadening the Base. Proceedings, 2001. X, 377 pages. 2001. (Subseries LNAI).

Vol. 2121: C.S. Jensen, M. Schneider, B. Seeger, V.J. Tsotras (Eds.), Advances in Spatial and Temporal Databases. Proceedings, 2001. XI, 543 pages. 2001.

Vol. 2123: P. Perner (Ed.), Machine Learning and Data Mining in Pattern Recognition. Proceedings, 2001. XI, 363 pages. 2001. (Subseries LNAI).

Vol. 2124: W. Skarbek (Ed.), Computer Analysis of Images and Patterns. Proceedings, 2001. XV, 743 pages. 2001.

Vol. 2125: F. Dehne, J.-R. Sack, R. Tamassia (Eds.), Algorithms and Data Structures. Proceedings, 2001. XII, 484 pages. 2001.

Vol. 2126: P. Cousot (Ed.), Static Analysis. Proceedings, 2001. XI, 439 pages. 2001.

Vol. 2129: M. Goemans, K. Jansen, J.D.P. Rolim, L. Trevisan (Eds.), Approximation, Randomization, and Combinatorial Optimization. Proceedings, 2001. IX, 297 pages. 2001.

Vol. 2130: G. Dorffner, H. Bischof, K. Hornik (Eds.), Artificial Neural Networks – ICANN 2001. Proceedings, 2001. XXII, 1259 pages. 2001.

Vol. 2132: S.-T. Yuan, M. Yokoo (Eds.), Intelligent Agents. Specification. Modeling, and Application. Proceedings, 2001. X, 237 pages. 2001. (Subseries LNAI).

Vol. 2136: J. Sgall, A. Pultr, P. Kolman (Eds.), Mathematical Foundations of Computer Science 2001. Proceedings, 2001. XII, 716 pages. 2001.

Vol. 2138: R. Freivalds (Ed.), Fundamentals of Computation Theory. Proceedings, 2001. XIII, 542 pages. 2001.

Vol. 2139: J. Kilian (Ed.), Advances in Cryptology – CRYPTO 2001. Proceedings, 2001. XI, 599 pages. 2001.

Vol. 2141: G.S. Brodal, D. Frigioni, A. Marchetti-Spaccamela (Eds.), Algorithm Engineering. Proceedings, 2001. X, 199 pages. 2001.

Vol. 2143: S. Benferhat, P. Besnard (Eds.), Symbolic and Quantitative Approaches to Reasoning with Uncertainty. Proceedings, 2001. XIV, 818 pages. 2001. (Subseries LNAI).

Vol. 2146: J.H. Silverman (Eds.), Cryptography and Lattices. Proceedings, 2001. VII, 219 pages. 2001.

Vol. 2147: G. Brebner, R. Woods (Eds.), Field-Programmable Logic and Applications. Proceedings, 2001. XV, 665 pages. 2001.

Vol. 2149: O. Gascuel, B.M.E. Moret (Eds.), Algorithms in Bioinformatics. Proceedings, 2001. X, 307 pages. 2001.

Vol. 2150: R. Sakellariou, J. Keane, J. Gurd, L. Freeman (Eds.), Euro-Par 2001 Parallel Processing. Proceedings, 2001. XXX, 943 pages. 2001.

Vol. 2154: K.G. Larsen, M. Nielsen (Eds.), CONCUR 2001 – Concurrency Theory. Proceedings, 2001. XI, 583 pages. 2001.

Vol. 2161: F. Meyer auf der Heide (Ed.), Algorithms – ESA 2001. Proceedings, 2001. XII, 538 pages. 2001.

Vol. 2164: S. Pierre, R. Glitho (Eds.), Mobile Agents for Telecommunication Applications. Proceedings, 2001. XI, 292 pages. 2001.